BIG IDEA 3: Living systems store, retrieve, transmit and respond to information essential to life processes.

ESSENTIAL KNOWLEDGE	CHAPTERS/SECTIONS	ILLUSTRATIVE EXAMPLES COVERED
3.A.1 DNA, and in some cases RNA, is the primary source of heritable information.	3.1, 3.2, 3.3, 3.4, 9.1, 9.2, 9.3, 10.1, 10.2, 10.3, 10.4, 10.5, 13.1, 13.2, 13.3, 13.4	• Poly A tail • GTP cap • Excision of introns • Enzymes • Transport by proteins • Synthesis • Degradation • GM foods • Transgenic animals • Cloned animals • Pharmaceuticals • Electrophoresis • Plasmid-based transformation • Polymerase chain reaction
3.A.2 In eukaryotes, heritable information is passed to the next generation via processes that include the cell cycle and mitosis, or meiosis plus fertilization.	7.1, 7.2, 7.3, 7.4, 32.1, 32.2, 32.3	• Mitosis-promoting factor • Cancer and cell cycle control
3.A.3 The chromosomal basis of inheritance provides an understanding of the pattern of passage (transmission) of genes from parent to offspring.	8.1, 8.2, 8.3, 12.4, 41.2	• Down syndrome • X-linked color blindness • Sickle cell anemia • Civic issues
3.A.4 The inheritance pattern of many traits cannot be explained by simple Mendelian genetics.	8.3, 9.3, 30.4	• Sex-linked genes • The Y chromosome carries few genes • In mammals and flies, females are XX and males are XY
3.B.1 Gene regulation results in differential gene expression, leading to cell specialization.	11.1, 11.2, 11.3, 11.4	• Promoter • Terminator • Enhancers
3.B.2 A variety of intercellular and intracellular signal transmissions mediate gene expression.	5.5, 14.3, 26.1, 26.2, 26.3, 30.4, 33.3	• Morphogens stimulate development • Cytokines regulate gene expression • HOX genes and development • Seed germination and gibberellin
3.C.1 Changes in genotype can result in changes in phenotype.	7.4, 9.3	• Antibiotic resistance mutations • Sickle cell disorder and heterozygote advantage
3.C.2 Biological systems have multiple processes that increase genetic variation.	7.4, 8.4, 9.2	• *No illustrative examples listed in Curriculum Framework.*
3.C.3 Viral replication results in genetic variation, and viral infection can introduce genetic variation into the hosts.	8.4, 9.1, 12.3	• Transposons
3.D.1 Cell communication processes share common features that reflect a shared evolutionary history.	5.5, 5.6, 9.2	• Epinephrine stimulation of glycogen breakdown • DNA repair mechanisms
3.D.2 Cells communicate with each other through direct contact with other cells or from a distance via chemical signaling.	4.5, 14.3, 28.1, 30.1, 31.4, 31.5, 34.3, 41.4	• Immune cells interact • Plasmodesmata between plant cells • Plant immune response • Morphogens and embryonic development • Neurotransmitters • Insulin • Quorum sensing in bacteria • Thyroid hormone • Testosterone• Estrogen
3.D.3 Signal transduction pathways link signal reception with cellular response.	5.5, 5.6	• G-protein linked receptors • Ligand gated ion channels • Receptor tyrosine kinases • Second messengers
3.D.4. Changes in signal transduction pathways can alter cellular response.	5.6, 30, 32.1-2	• Diabetes • Effects of neurotoxins • Drugs
3.E.1 Individuals can act on information and communicate it to others.	28.2, 30.2, 41.1-4, 41.6	• Fight or flight response• Predator warnings • Colony behavior • Herbivory responses • Coloration • Parent-offspring interactions • Territorial marking • Plant-plant interactions in herbivory • Courtship and mating behaviors • Bee dances • Bird songs

ESSENTIAL KNOWLEDGE	CHAPTERS/SECTIONS	ILLUSTRATIVE EXAMPLES COVERED
3.E.2 Animals have nervous systems that detect external and internal signals, transmit and integrate information, and produce responses.	34.1-4, 35.1-4, 36.1-3	• Acetylcholine • Epinephrine • Dopamine • Serotonin • GABA • Hearing • Muscle movement • Abstract thought • Neurohormone production • Forebrain, midbrain and hindbrain • Right and left cerebral hemispheres

BIG IDEA 4: Biological systems interact, and these systems and their interactions possess complex properties.

ESSENTIAL KNOWLEDGE	CHAPTERS/SECTIONS	ILLUSTRATIVE EXAMPLES COVERED
4.A.1 The subcomponents of biological molecules and their sequence determine the properties of that molecule.	3.1, 3.2, 3.3, 3.4, 9.1	• *No illustrative examples listed in Curriculum Framework.*
4.A.2 The structure and function of sub-cellular components, and their interactions, provide essential cellular processes.	4.3, 4.4, 6.2, 6.5	• *No illustrative examples listed in Curriculum Framework.*
4.A.3 Interactions between external stimuli and regulated gene expression result in specialization of cells, tissues and organs.	14.2, 14.3, 33.1	• *No illustrative examples listed in Curriculum Framework.*
4.A.4 Organisms exhibit complex properties due to interactions between their constituent parts.	24.1-3, 36.1, 36.3, 39.3	• Plant vascular and leaf • Root, stem and leaf • Kidney and bladder • Respiratory and circulatory • Nervous and muscular • Stomach and small intestines
4.A.5 Communities are composed of populations of organisms that interact in complex ways.	43.1-4, 44.1-4, 45.1-6	• Predator-prey relationship • Symbiotic relationship • Graphical representation of field data • Introduction of species • Global climate change models
4.A.6 Interactions among living systems and with their environment result in the movement of matter and energy.	43.2, 45.3, 46.2	• *No illustrative examples listed in Curriculum Framework.*
4.B.1 Interactions between molecules affect their structure and function.	3.3, 3.4	• *No illustrative examples listed in Curriculum Framework.*
4.B.2 Cooperative interactions within organisms promote efficiency in the use of energy and matter.	29.1, 34.4, 36.2, 37.2, 39.4, 42.1	• Exchange of gases • Circulation of fluids • Digestion of food • Excretion of wastes • Bacterial community in the rumen • Bacterial community in the gut
4.B.3 Interactions between and within populations influence patterns of species distribution and abundance.	43.1, 43.4, 43.5, 44.1-4, 45.1-6	• Loss of keystone species
4.B.4 Distribution of local and global ecosystems changes over time.	42.1-45, 45.2, 46.5	• Continental drift • Impacts of human land use • Effects of introduced species • Volcanic eruption • Impacts of climate change
4.C.1 Variation in molecular units provides cells with a wider range of functions.	5.1, 6.5, 31.4	• Phospholipids in membranes • MHC proteins • Chlorophylls • Molecular diversity in antibodies
4.C.2 Environmental factors influence the expression of the genotype in an organism.	8.2, 11.2, 39.1	• Height and weight in humans • Effect if adding lactose to a Lac$^+$ bacterial culture • Darker fur in cooler regions of the body
4.C.3 The level of variation in a population affects population dynamics.	15.2-4, 28.1, 28.3	• Wheat rust
4.C.4 The diversity of species within an ecosystem may influence the stability of the ecosystem.	42.5, 45.4, 46.5	*No illustrative examples listed in Curriculum Framework.*

PRINCIPLES OF LIFE

HIGH SCHOOL EDITION

David M. Hillis
University of Texas at Austin

David Sadava
Emeritus, The Claremont Colleges

H. Craig Heller
Stanford University

Mary V. Price
Emerita, University of California, Riverside

 Sinauer Associates, Inc.   W. H. Freeman and Company

About the Cover

Harlequin ghost pipefish (*Solenostomus paradoxus*) live among feather stars (a type of echinoderm), where the fish's skin flaps provide excellent camouflage from potential predators. Pipefish (which are related to the seahorses) position themselves upside down and nearly motionless as they seek their prey—tiny crustaceans living near the seafloor. Such intricate interactions among different types of organisms exemplify some of life's overarching principles. Photograph © Fred Bavendam/Minden Pictures.

Principles of Life

Copyright © 2012 by Sinauer Associates, Inc. All rights reserved.
This book may not be reproduced in whole or in part without permission.

Address editorial correspondence to:
Sinauer Associates, Inc., 23 Plumtree Road, Sunderland, MA 01375 U.S.A.

www.sinauer.com

publish@sinauer.com

Address orders to:
MPS / W. H. Freeman & Co., Order Dept., 16365 James Madison Highway, U.S. Route 15, Gordonsville, VA 22942 U.S.A.

Examination copy information: 1-800-446-8923

Orders: 1-888-330-8477

Planet Friendly Publishing
✓ Made in the United States
✓ Printed on Recycled Paper
Text: 10%   Cover: 10%
GREEN EDITION   Learn more: www.greenedition.org

SUSTAINABLE FORESTRY INITIATIVE
Certified Sourcing
www.sfiprogram.org
SFI-00712

This SFI label applies to the text paper.

Library of Congress Cataloging-in-Publication Data

Principles of life High School Edition/ David M. Hillis ... [et al.].
     p. cm.
   Includes index.
   ISBN 1-4292-9117-6
   ISBN 978-1-4292-9117-0
   1. Biology. I. Hillis, David M., 1958–
   QH308.2.P75 2012
   570--dc22   2010047780

Printed in U.S.A.
Fourth Printing June 2012
The Courier Companies, Inc.

The Authors

DAVID M. HILLIS is the Alfred W. Roark Centennial Professor in Integrative Biology and the Director of the Center for Computational Biology and Bioinformatics at the University of Texas at Austin, where he also has directed the School of Biological Sciences. Dr. Hillis has taught courses in introductory biology, genetics, evolution, systematics, and biodiversity. He has been elected to the National Academy of Sciences and the American Academy of Arts and Sciences, awarded a John D. and Catherine T. MacArthur Fellowship, and has served as President of the Society for the Study of Evolution and of the Society of Systematic Biologists. He served on the National Research Council committee that wrote the report *BIO 2010: Transforming Undergraduate Biology Education for Research Biologists*.
His research interests span much of evolutionary biology, including experimental studies of evolving viruses, empirical studies of natural molecular evolution, applications of phylogenetics, analyses of biodiversity, and evolutionary modeling. He is particularly interested in teaching and research about the practical applications of evolutionary biology.

DAVID SADAVA is the Pritzker Family Foundation Professor of Biology, Emeritus, at the Keck Science Center of Claremont McKenna, Pitzer, and Scripps, three of The Claremont Colleges. In addition, he is Adjunct Professor of Cancer Cell Biology at the City of Hope Medical Center. Twice winner of the Huntoon Award for superior
teaching, Dr. Sadava has taught courses on introductory biology, biotechnology, biochemistry, cell biology, molecular biology, plant biology, and cancer biology. In addition to *Life: The Science of Biology*, he is the author or coauthor of books on cell biology and on plants, genes, and crop biotechnology. His research has resulted in many papers coauthored with his students, on topics ranging from plant biochemistry to pharmacology of narcotic analgesics to human genetic diseases. For the past 15 years, he has investigated multi-drug resistance in human small-cell lung carcinoma cells with a view to understanding and overcoming this clinical challenge. At the City of Hope, his current work focuses on new anti-cancer agents from plants.

H. CRAIG HELLER is the Lorry I. Lokey/Business Wire Professor in Biological Sciences and Human Biology at Stanford University. He has taught in the core biology courses at Stanford since 1972 and served as Director of the Program in Human Biology, Chairman of the Biological Sciences Department, and Associate Dean of Research. Dr. Heller is a fellow of the American Association for the Advancement of Science and a recipient of the Walter J. Gores Award for excellence in teaching. His research is on the neurobiology of sleep and circadian rhythms, mammalian hibernation, the regulation of body temperature, the physiology of human performance, and the neurobiology of learning. He has done research on a huge variety of animals and physiological problems ranging from sleeping kangaroo rats, diving seals, hibernating bears, photoperiodic hamsters, and exercising athletes.
Dr. Heller has extended his enthusiasm for promoting active learning through the development of a two-year curriculum in human biology for the middle grades, and at the college level he directed the production of Virtual labs—interactive computer-based modules to teach physiology.

MARY V. PRICE is Professor of Biology, Emerita, at the University of California Riverside and Adjunct Professor in the School of Natural Resources and the Environment at the University of Arizona. In
"retirement," she continues to teach and study, having learned the joy and art of scientific discovery as an undergraduate student at Vassar College and doctoral student at the University of Arizona. Dr. Price has mentored and published with independent-research students and has developed and taught general biology and ecology courses from introductory (majors and nonmajors) to graduate levels. She has particularly enjoyed leading field classes in the arid regions of North America and Australia, and the tropical forests of Central America, Africa, and Madagascar. Dr. Price's research focuses on understanding the ecology of North American deserts and mountains. She has asked why so many desert rodents can coexist, how best to conserve endangered kangaroo rat species, how pollinators and herbivores influence floral evolution and plant population dynamics, and how climate change affects ecological systems.

Brief Table of Contents

Preface

We wrote *Principles of Life* to serve as a textbook for teachers who are changing the way they teach biology to AP biology students and college undergraduates. Many instructors see the value of emphasizing an understanding of major concepts over memorization of details. When students understand how biological concepts have been developed through observation and experimentation, and when they gain experience in applying those concepts to real biological problems, they are more likely to retain an understanding of the big picture. This emphasis on *understanding* rather than *memorizing* biology is especially effective when coupled with active learning approaches, including problem solving, analyzing real data, discussing and synthesizing ideas, and using interactive simulations—all based on a foundational understanding of a few overarching concepts. *Principles of Life* is intended for such an active learning approach to understanding the fundamental concepts of biology.

Existing introductory textbooks for AP students and biology majors are impressive encyclopedic summaries of biological knowledge, and many instructors like having a wealth of examples for their course. Such tomes can be effective for presenting an overview of biology; indeed, most of the authors of *Principles of Life* are also among the authors of one of these comprehensive texts—*Life: The Science of Biology*, now in its ninth edition. We are proud of that textbook and believe it has set a standard for the field for many years. We recognize, however, that comprehensive textbooks contain far more information than is needed for an introductory course in biology. Many students are overwhelmed by the detail, and may have a hard time focusing on the most important concepts.

In *Life: The Science of Biology*, we introduced active problem solving as an important component of learning the major concepts of biology. In *Principles of Life*, we have expanded these efforts at promoting active learning, and to do so we have greatly abbreviated the volume of detail found in all introductory texts. Our intention is not to leave out important concepts, but to give students more opportunities to learn those concepts through applying them. We believe that students who spend their time diligently committing to memory the myriad details and vast terminology of a broad array of biological topics actually learn and retain fewer of the concepts that are the foundation for further study in advanced biology courses.

Voices for Change in AP Biology and Undergraduate Biology Education

Our motivation to write this book comes from the numerous reports published by education agencies and national study groups over the past 10 years that have called for reforms in undergraduate biology education. For example, the National

Research Council published a report called *BIO2010: Transforming Undergraduate Education for Future Research Biologists*, sponsored by the National Institutes of Health and the Howard Hughes Medical Institute. Similar reports, containing many of the same recommendations, have been published by committees of the American Association for the Advancement of Science, the National Academy of Sciences, the National Evolutionary Synthesis Center, the College Board, and many other organizations that are invested in science education. The College Board's revised AP Biology course is consistent with these recommendations and will be among the first to implement them during the 2012–2013 academic year.

These recent reports note that even though science is constantly changing, science education has changed relatively little over several decades. They emphasize a consensus that instructors should move away from teaching biology as a set of facts to be memorized and focus instead on major concepts. They also advise teachers to showcase the logical structure of scientific investigation, including lab, field, and computer modeling approaches. Students should be able to apply the concepts they learn by analyzing original data, and they need to understand the growing relevance of quantitative science in addressing life-science questions. There should be an integration of analyses of experimental and observational data into the course. And finally, biology courses should incorporate inquiry-based approaches that encourage active learning.

We have been participants in some of these study groups, and agree with their conclusions and prescription for change in biology education. Our goal in this new book, *Principles of Life*, is to use our experience as authors and educators to incorporate recommendations from these reports into a new approach to introductory biology. To help us create this new breed of biology textbook, our publishers Sinauer Associates and W. H. Freeman enlisted the help of an Advisory Board made up of leading biology educators and instructors in introductory biology from throughout North America. This Advisory Board met with the authors and publishers to help develop new ideas and features for *Principles of Life*. Board members reviewed the potential contents of the book, identifying material that was less than essential for teaching introductory biology. They then reviewed the emerging chapters and have provided considerable feedback at every stage of the book's development.

How Is *Principles of Life* Different?

Each chapter of *Principles of Life* is organized into a series of *Concepts* that are important for mastering introductory biology. We have carefully chosen these concepts in light of feedback from our colleagues, from the Advisory Board, and from the numerous reports examining introductory biology. Concepts are elaborated upon, but not with the extensive detail found in current introductory texts. *Principles of Life* is focused; it is not meant to be encyclopedic.

Students learn concepts best when they apply the concepts to practical problems. **Each chapter of *Principles of Life* contains exercises, called *Apply the Concept*, that present data for students to analyze.** Each of these exercises uses problem solving to reinforce a concept that is central to the respective chapter. Science students need to understand basic methods for data presentation and analysis, so many of these problems ask students about statistical significance of the results. **To help students understand issues in data presentation and interpretation, we have provided a short introduction to biological statistics in Appendix B.** Although this Appendix is not meant to replace a more formal introduction to statistics, we believe that statistical thinking is an important skill that should be developed in all introductory science courses. We have kept the problems and examples straightforward to emphasize the concepts of statistical analysis rather than the details of any particular statistical test. Some of the *Apply the Concept* exercises are simple enough that they can be presented, analyzed, and discussed in class; others are better suited for homework assignments.

Our *Investigation* figures let students see *how* we know *what* we know. These figures present a Hypothesis, Method, Results, and Conclusion. Most of these *Investigation* figures now include a section titled *Analyze the Data*, in which we have extracted a subset of data from the published experiment. Students are asked to analyze these data and to make connections between observations, analyses, hypotheses, and conclusions. As with *Apply the Concept* problems, students are asked to apply basic statistical approaches to understand the results and draw conclusions. We have also provided extensive online resources for each *Investigation* figure, available on the book's free web site (*thelifewire.com*) and integrated with the online version of the book on the BioPortal (*yourBioPortal.com*). These resources include expanded discussions of the original research, links to the original publications, and discussion and links for any follow-up investigations that have been published. **For many *Investigations*, we have expanded *Working with Data* problems**, which provide students with an opportunity to explore and analyze the original published data in greater depth.

Each chapter begins with an application of a major concept—a true story that illustrates and provides a motivation for understanding the chapter's content, and provides a social, medical, scientific, or historical context for the material. Each of these vignettes ends with an open-ended question that students can keep in mind as they read and study the rest of the chapter. We return to this opening question at the close of the chapter to show how information presented throughout the chapter illuminates the question and helps provide an answer. By pondering these questions as they read and study, students can begin to think like scientists.

At the end of each conceptual discussion we pose a series of questions (*Do You Understand Concept X?*) designed to help students self-evaluate their understanding of the material. These questions span the incremental levels of Bloom's Taxonomy of Cognitive Domains: factual knowledge, comprehension, application, analysis, synthesis, and evaluation.

Another important element for student success is **reinforcement and application of concepts through online resources such as *Animated Tutorials, Activities, and Interactive Tutorials*.** These are provided for each chapter, both in BioPortal and on the free Companion Website. For many concepts, students can conduct their own simulations, explore a concept in greater depth, and understand concepts through active discovery.

The interdisciplinary connections among the different subfields of biology are an emphasis of modern biology education. **To help students build bridges between different portions of the course, and make connections between different areas of knowledge, we have provided *Links* throughout the book.** Using these *Links*, students can see that information they learn about molecular or cell biology is connected and relates to topics in evolution, diversity, physiology, and ecology, for example.

Biologists make new discoveries and apply biological concepts to new applications every day. These applications are often in the news, and students are often excited to see the connections between current developments and the fundamental concepts they are learning. We highlight many such developments in *Frontiers* in each chapter. **These *Frontiers* highlight exciting applications and provide a connection between introductory biology and current technological applications.**

Students need to learn about some of the major *Research Tools* that are used in biology, including major laboratory, computational, and field methods. Our *Research Tools* figures explain these tools and provide a context for how they are used by biologists.

Our art program for *Principles of Life* builds on our success with *Life: The Science of Biology*. We pioneered the use of balloon captions to help students understand and interpret the biological processes illustrated in figures without repeatedly going back and forth between the figure, its legend, and the text. Without these guides, students often skip or miss the important points of figures.

Alternative Electronic Choice

The *Principles of Life* eBook is a complete interactive version of the textbook. Available as a stand-alone and within BioPortal, the eBook offers a substantial discount from the price of the printed textbook. This online version of *Principles of Life* gives students an efficient and rich learning experience by integrating all of the resources from the student website directly into the eBook platform. It also incorporates additional features that allow students and instructors to customize the text.

Media and Supplements

To support both students and instructors, we offer a wide range of media and supplements to accompany *Principles of Life*. Bio-Portal, the online platform that integrates all of the student and instructor resources with the interactive eBook, powerful assessment tools, and Prep-U adaptive quizzing is the most expansive offering. Throughout the textbook, references to *Animated Tutorials*, *Web Activities*, and *Working with Data* exercises refer students to the rich collection of resources available within BioPortal. BioPortal also includes a robust set of assessment and course management features for instructors, making it easy to quickly gauge class and individual progress, build assignments, customize the content of the eBook, and more.

The rich collection of visual resources in the *Principles of Life* Instructor's Media Library and on the Teacher's Resource DVD provides instructors with a wide range of options for enhancing lectures, course websites, and assignments. Highlights include multiple versions of all textbook figures, a wealth of PowerPoint® resources including layered art presentations, a large collection of videos, and in-class active learning exercises that help instructors bring the active learning approach of the textbook into the classroom.

For a complete list of all the media and supplements available for *Principles of Life*, please refer to "Media and Supplements to accompany *Principles of Life*" on page *xii*. Also, refer to the inside of the back cover for a complete list of all the student media resources referenced in the text.

Many People to Thank

Our editor and publisher, Andy Sinauer, has been a source of inspiration and support at every stage of this book's development. He embraced the need for change in introductory biology textbooks and has helped make our vision into a reality. Bill Purves and Gordon Orians, our co-authors on earlier editions of *Life: The Science of Biology*, were instrumental in articulating the concepts developed in *Principles of Life*, and many aspects of this book can be traced back to their critical contributions. In the development of *Principles of Life*, Nickolas M. Waser, Frank E. Price, Kathleen Hunt, and the members of our Advisory Board (see the next page) provided detailed suggestions for organizing our chapters around the fundamental concepts of biology.

For this new book, Sinauer Associates assembled a talented and dedicated team of professionals. Foremost were the contributions of editors Carol Wigg and Laura Green, who participated in the planning and shared the developmental editing. Their meticulous work included many refinements of the new pedagogical features. Norma Roche added her editorial expertise to the unit on biological diversity and provided organizational input for the brand-new ecology unit. Carol and Laura worked closely with two top-notch copyeditors, Jane Murfett and Liz Pierson. Elizabeth Morales, "our" artist, worked with each of us to develop effective and beautiful line art. David McIntyre cheerfully took on the challenge of finding superior images for the photographs in the book. Designer Joanne Delphia brought a fresh look to the book that we find both functional and handsome. Jeff Johnson did a masterful job of assembling all the book's elements into clear and attractive pages. Chris Small coordinated production and imposed his exacting standards on keeping the myriad components consistent. Susan McGlew and Johannah Walkowicz organized and commissioned the many expert academic reviews. Jason Dirks coordinated the ever-growing team that created the vast array of media and supplements that compose BioPortal. Dean Scudder, Director of Sales and Marketing, participated in every stage of the book's development.

At W. H. Freeman, we were fortunate to have the long-term input of Executive Editor Susan Winslow, whose advice on all sorts of issues has been invaluable. Associate Director of Marketing Debbie Clare, in collaboration with the Regional Specialists, Regional Sales Managers, and the Market Development team, did an impressive job coordinating all the stages of informing Freeman's skilled sales force of our book's story. We also wish to thank the Freeman media group for their expertise in producing the e-Book and BioPortal.

DAVID HILLIS

DAVID SADAVA

CRAIG HELLER

MARY PRICE

Advisors and Reviewers

Advisory Board Members

Teri Balser, University of Wisconsin, Madison

Jessica Bolker, University of New Hampshire

Judith L. Bronstein, University of Arizona

Nancy Burley, University of California, Irvine

Patti Christie, Massachusetts Institute of Technology

Laura DiCaprio, Ohio University

Cole Gilbert, Cornell University

Rick Grosberg, University of California, Davis

Matthew B. Hamilton, Georgetown University

Mark Johnston, Dalhousie University

Daniel J. Klionsky, University of Michigan

Richard Londraville, University of Akron

Sharon Long, Stanford University

Jennifer Nauen, University of Delaware

Benjamin B. Normark, University of Massachusetts, Amherst

Brian Olsen, University of Maine, Orono

Ann E. Rushing, Baylor University

Dee Silverthorn, University of Texas, Austin

Kenneth Wilson, University of Saskatchewan

William B. Wood, University of Colorado

Reviewers

Stephen Aley, Univeristy of Texas, El Paso

Peter Alpert, University of Massachusetts, Amherst

Erol Altug, The Stony Brook School, Stony Brook, NY

William J. Anderson, Harvard University

Juan Anguita, University of Massachusetts, Amherst

Joel Bader, Johns Hopkins University

Brian Bagatto, University of Akron

Lisa Baird, San Diego University

Mark Barsoum, Davidson College

Mike Barton, Centre College

Donna Becker, Northern Michigan University

Michael D. Beecher, University of Washington

Douglas G. Bielenberg, Clemson University

Meredith Blackwell, Louisiana State University

Elizabeth Blinstrup Good, University of Illinois

Arnold Bloom, University of California, Davis

Nicole Bournias, California State University, San Bernadino

Jeffrey D. Brown, University of Portland

Pamela Brunsford, University of Idaho

Stephen Burton, Grand Valley State University

Ruth Buskirk, University of Texas, Austin

Mickael Cariveau, Mount Olive College

Ethan Carver, University of Tennessee, Chattanooga

David Champlin, University of Southern Maine

Stylianos Chatzimanolis, University of Tennessee, Chattanooga

Alice Cheung, University of Massachusetts, Amherst

Timothy Clark, University of British Columbia

Reggie Cobb, Nash Community College

William Cohen, University of Kentucky

Elizabeth Connor, University of Massachusetts, Amherst

Helene Cousin, University of Massachusetts, Amherst

Joel Cracraft, American Museum of Natural History

Karen Crow, San Francisco State University

Jill Crowder, Milwaukee Area Technical College

Christopher Cullis, Case Western Reserve University

Stacey Darling-Novak, University of LaVerne

Christine Davis, Durham Technical Community College

Nora Demers, Florida Gulf Coast University

Jean DeSaix, University of North Carolina, Chapel Hill

Leif Deyrup, University of the Cumberlands

Raj Dhindsa, McGill University

Robert Donaldson, George Washington University

Robert Dorit, Smith College

Meredith Dorner, MiraCosta College

Robert Drewell, Harvey Mudd College

Devin Drown, Indiana University, Bloomington

Amy Dykstra, Crown College and University of Minnesota

Ryan Earley, University of Alabama

Gretchen Edwalds-Gilbert, Claremont McKenna, Pitzer, and Scripps Colleges

Debby Filler, Anoka Ramsey Community College

Sylvia Fromherz, Southern Illinois University School of Medicine

Glenn Galau, University of Georgia

Stephen Gehnrich, Salisbury University

Wayne Goodey, University of British Columbia

Carole Gutterman, Molloy College

John Hageman, Thomas More College

Dale Harrington, Caldwell Community College, Watuga

Bernard Hauser, University of Florida

Virginia Hayssen, Smith College

Rick Heineman, University of Texas, Austin

Albert Herrera, University of Southern California

Richard Hill, Michigan State University

Peggy S. M. Hill, University of Tulsa

Jonathan Hillis, Carleton College

Colleen Hitchcock, Boston College

Lawrence Hobbie, Adelphi University

William Hoese, California State University, Fullerton

MaryKate Holden, Greensboro College

Ava Howard, Western Oregon University

Joel Jacobs, Northeastern University

Jeremiah Jarrett, Central Connecticut State University

Bruce Johnson, Cornell University

Steve Jordan, Bucknell University

Walter Judd, University of Florida

Susan Keen, University of California, Davis

Thomas Keller, University of Texas, Austin

Stephen Kilpatrick, Univeristy of Pittsburgh, Johnstown

Nancy Kirkpatrick, Lake Superior State University

Maggie Koopman, Eastern Michigan University

Jennifer Kowalski, Butler University

Thomas Lambert, Frostburg State University

Pamela Lanford, University of Maryland, College Park

Sarah Lang, Indiana University-Purdue University Indianapolis

Paul LeBlanc, University of Alabama

John Lepri, University North Carolina, Greensboro

Ben Liebeskind, University of Texas, Austin

Mark Liles, Auburn University

Chris Little, Kansas State University

David Longstreth, Louisiana State University

Christopher Lortie, York University

Michael Manson, Texas A&M University

Brett McMillan, McDaniel College

Karin Melkonian, Long Island University, CW Post

Jill Miller, Amherst College

Randall Mitchell, University of Akron

Pamela Monaco, Molloy College

Brian Morton, Barnard College

Patricia Mote, Georgia Perimeter College

Kathryn Nette, Cuyamaca College

Dana Newton, College of the Albemarle

Margaret Oliver, Carthage College

Laura Palmer, Pennsylvania State University, Altoona

Susan Parrish, McDaniel College

Michael Peek, William Paterson University

Nancy Pelaez, Purdue University

Patrick Pfaffle, Carthage College

Jennifer Pfannerstill, Tomahawk High School, Tomahawk, WI

Deb Pires, University of California, Los Angeles

Joe Poston, Catawba College

David Puthoff, Frostburg State University

Jennifer Randall, New Mexico State University

Melissa Reedy, University of Illinois, Urbana-Champaign

Julie Reynolds, Duke University

Laurel Roberts, University of Pittsburgh

Kenneth R. Robinson, Purdue University

Deborah Ross, Indiana University-Purdue University, Fort Wayne

Jodie Rummer, University of British Columbia

Amy Russell, Grand Valley State University

Christina T. Russin, Northwestern University

Shereen Sabat, La Sierra University

Susan Safford, Lincoln University

Milton Saier, University of California, San Diego

Nathan Sanders, University of Tennessee, Knoxville

Scott Santagata, Long Island University, CW Post

Daniel Sasson, University of Florida

Andrew Schnabel, Indiana University, South Bend

Marcia Schofner, University of Maryland, College Park

Christopher Schroeder, Milwaukee Area Technical College

Roxann Schroeder, Humboldt State University

Nancy Shontz, Grand Valley State University

Diviya Sinha, Massachusetts Institute of Technology

James Smith, Michigan State University

William Stein, Binghamton University

Asha Stephens, College of the Mainland

Janet Steven, Sweet Briar College

Bethany Stone, University of Missouri

Rema Suniga, Ohio Northern University

Cynthia Surmacz, Bloomsburg University of Pennsylvania

Yun Tao, Emory University

Mark Thogerson, Grand Valley State University

James W. Thomas, Emory University

Kathy S. Thompson, Louisiana State University

Anthony Tolvo, Molloy College

Lars Tomanek, California Polytechnic State University

Paul Trombley, Florida State University

Lowell Urbatsch, Louisiana State University

Randall Walikonis, University of Connecticut

Richard Walker, Virginia Polytechnic Institute and State University

Andrea Ward, Adelphi University

Audra Ward, Marist School, Atlanta, GA

Lisa Webb, Christopher Newport University

Kelly Wentz-Hunter, Roosevelt University

Brad Wetherbee, University of Rhode Island

Carolyn Wetzel, Smith College

Morgan Wilson, Hollins University

Gregory Wray, Duke University

Eve Wurtele, Iowa State University

Aimee Wyrick, Pacific Union College

Tim Xing, Carleton University

Robert Yost, Indiana University-Purdue University Indianapolis

Kathryn Yurkonis, University of North Dakota

Kathy Zarilla, Durham Technical Community College

Jean Claude Zenklusen, George Washington University

Media and Supplements
to accompany *Principles of Life*

For ordering information and various package options, visit:
www.bfwpub.com/highschool

⊕bioportal featuring Prep-U

yourBioPortal.com

BioPortal brings together the extensive teaching and learning resources for *Principles of Life*, including Prep-U Personal Adaptive Quizzing, a fully integrated eBook, dynamic features such as In-Class Active Exercises and BioNews, and additional assessment resources—all in a convenient, fully customizable online course environment that makes organizing and administering your course easy. Access may be purchased for one year or at a discounted three-year price. BioPortal includes:

Principles of Life eBook

- Integration of all activities, animated tutorials, and other media resources
- Quick, intuitive navigation to any section or subsection, as well as any printed book page number
- In-text links to all glossary entries
- Easy text highlighting
- A bookmarking feature that allows for quick reference to any page
- A powerful notes feature that allows students to add notes to any page
- A full glossary and index
- Full-text search, including an additional option to search the glossary and index
- Automatic saving of all notes, highlighting, and bookmarks

Additional eBook features for instructors:

- *Content Customization:* Instructors can easily add pages of their own content and/or hide chapters or sections that they do not cover in their course.
- *Instructor Notes:* Instructors can choose to create an annotated version of the eBook with their own notes on any page. When students in the course log in, they see the instructor's personalized version of the eBook. Instructor notes can include text, Web links, images, links to all BioPortal content, and more.

 Smarter *than the average quiz*

Built by educators, Prep-U focuses student study time exactly where it should be, through the use of personalized, adaptive quizzes that move students toward a better grasp of the material—and better grades. Prep-U is fully integrated into BioPortal, making it easy for instructors to take advantage of this powerful quizzing engine in their course. Features include:

- Personalized adaptive quizzing
- Automatic results reporting into the BioPortal gradebook
- Instant "How's My Class Doing?" reports that include strengths and weaknesses, common misconceptions, and comparisons to national data

Student Resources

DIAGNOSTIC QUIZZING. The diagnostic quiz for each chapter of *Principles of Life* assesses student understanding of that chapter, and generates a Personalized Study Plan to effectively focus student study time. The plan includes links to specific textbook sections, animated tutorials, and activities.

INTERACTIVE SUMMARIES. For each chapter, these dynamic summaries combine a review of important concepts with links to all of the key figures from the chapter as well as all of the relevant animated tutorials, activities, and key terms.

ANIMATED TUTORIALS. A comprehensive set of in-depth animated tutorials present complex topics in a clear, easy-to-follow format that combines a detailed animation with an introduction, conclusion, and quiz.

ACTIVITIES. A variety of activities help students learn important facts and concepts through a wide range of exercises, such as labeling steps in processes or parts of structures, building diagrams, and identifying different types of organisms.

INTERACTIVE TUTORIALS. These tutorial modules help students understand key concepts through the use of problem scenarios, experimental techniques, and interactive models.

INTERACTIVE QUIZZES. Each question includes an image from the textbook, thorough feedback on both correct and incorrect answer choices, references to textbook pages, and links to eBook pages, for quick review.

LECTURE NOTEBOOK. This invaluable resource provides all the artwork and tables from the textbook, with ample space for note-taking. Students can download chapters of the Lecture Notebook as PDF files and then either enter notes electronically into the PDFs, or print out the chapters/sections they need.

BIONEWS FROM SCIENTIFIC AMERICAN. BioNews makes it easy for instructors to bring the dynamic nature of the biological sciences and up-to-the minute currency into their course. Accessible from within BioPortal, BioNews is a continuously updated feed of current news, podcasts, magazine articles, science blog entries, "strange but true" stories, and more.

BIONAVIGATOR. This unique visual resource is an innovative way to access the wide variety of *Principles of Life* animations and activities. A visual interface begins with a whole-Earth view and allows the user to zoom to any level of biological inquiry, encountering links to a wealth of animations, activities, and tutorials on the full range of topics along the way.

WORKING WITH DATA. A companion to the in-text Analyze the Data problems, these exercises are built around some of the original experiments depicted in the Investigation figures. They help students build their quantitative skills and encourage student interest in how scientists do research, by looking at real experimental data and answering questions based on those data.

FLASHCARDS & KEY TERMS. For each chapter of the book, there is a set of flashcards that allows the student to review all the key terminology from the chapter. Students can review the terms in study mode and then quiz themselves on a list of terms.

INVESTIGATION LINKS. For each Investigation figure in the textbook, BioPortal includes an overview of the experiment featured in the figure and related research or applications that followed, a link to the original paper, and links to additional information related to the experiment.

GLOSSARY. The language of biology is often difficult for students taking introductory biology to master, so BioPortal includes a full glossary that features audio pronunciations of all terms.

TREE OF LIFE. An interactive version of the Tree of Life in Appendix A. Includes links to the extensive Discover Life online biodiversity database.

MATH FOR LIFE. A collection of mathematical shortcuts and references to help students with the quantitative skills they need in the laboratory.

SURVIVAL SKILLS. A guide to more effective study habits. Topics include time management, note-taking, effective highlighting, and exam preparation.

Instructor Resources

Assessment

- Diagnostic Quizzing provides instant class comprehension feedback to instructors, along with targeted lecture resources for those areas requiring the most attention.

- Multiple question banks (test bank, diagnostic quiz, interactive quiz, and study guide) include thousands of questions, all referenced to specific textbook sections and ranked according to Bloom's taxonomy.
- Question filtering allows instructors to select questions based on Bloom's category and/or textbook section.
- Easy-to-use customized assessment tools allow instructors to quickly create quizzes and many other types of assignments using any combination of the questions and resources provided in BioPortal, along with their own materials.
- In-class active learning exercises help instructors engage students in the classroom.

Media Resources

(see Instructor's Media Library below for details)

- Videos
- PowerPoint Presentations (Textbook Figures, Lectures, Layered Art, Editable Labels)
- Supplemental Photos
- Active Learning Exercises
- Instructor's Manual
- Answers to the Apply the Concept, Analyze the Data, and Working with Data questions

Course Management

- Complete course customization capabilities
- Custom resources/document posting
- Robust gradebook
- Communication Tools: Announcements, Calendar, Course Email, Discussion Boards

Note: The printed textbook, the eBook, BioPortal, and Prep-U can all be purchased individually as stand-alone items, in addition to being available in a package.

Student Supplements

Companion Website
www.bfwpub.com/highschool/POL1e

For those students who do not have access to Bio-Portal, the *Principles of Life* Companion Website is available free of charge (no access code required). The site provides access **to all of the media referenced in the textbook links,** as well as a variety of additional resources, including flashcards, study ideas, and more.

Strive for a 5: Preparing for the AP Biology Exam (ISBN 1-4292-9849-9)

Prepared by leading authorities in AP Biology, this guide serves both as a study and reading guide for students as they work through the course and as an AP test preparation resource as they prepare for the AP exam. The guide reinforces the topics and key concepts covered in the text and on the AP exam.

The **study guide component** of *Strive for a 5* provides chapter-by-chapter guidance for students taking the AP biology course. The coverage for each chapter includes:

- Guided reading questions to help students focus on the most important concepts in each chapter.

- A running "Connections" feature that reinforces the Four Big Ideas: evolution, biological systems/cell processes, living systems, and interactions and shows how they apply in different contexts and settings.

- Worked problems that ask students to practice using math appropriately.

- Vocabulary review and practice.

- A review of key illustrations to help students gain a better understanding of the processes and concepts being illustrated. Students are asked to interact with the figure by adding labels and explaining steps in the process.

- Additional study questions and sample Free-Response questions to help review the material in the chapter. An Answer Key allows students to check their work.

The **AP preparation section** of *Strive for a 5* is a comprehensive test review resource. It begins with a diagnostic pre-test and instructions to help students determine where to focus their test preparation efforts. Test preparation tips, suggestions for setting a test preparation schedule, and advice on how to study effectively and efficiently in preparation for the AP exam are also featured. Finally, sample practice tests that simulate the AP exam with solutions and sample grading rubrics are provided.

Printed Image and Lecture Notebook
(ISBN 1-4292-9614-3)

Available digitally on the Companion Website, in the eBook, and on yourBioPortal.com, the printed Image and Lecture Notebook may be purchased for use as an in-class, note-taking tool. This invaluable resource provides all the artwork and tables from the textbook, with ample space for notes.

Teacher's Media & Supplements

Teacher's Resource Binder
(ISBN 1-4292-9118-4)

Written by a distinguished panel of AP Biology educators, including former AP Biology Development Committee members, Question Leaders, Table leaders, Readers, and teachers, this comprehensive resource is designed to support every AP Biology teacher using *Principles of Life* to teach a successful course and prepare their students for the redesigned Exam. Whether teachers are changing to *Principles of Life* from another text or teaching AP biology for the first time, the TRB provides extensive support including:

- **An introduction to the College Board's redesigned AP Biology course and suggestions for how to prepare students for the new 2013 exam.**

- A sample syllabus and pacing guide

- A detailed outline of each chapter with notes on common student errors and suggestions on how to cover the concepts efficiently and effectively.

- Suggested resources

- AP tips

- Teaching strategies and suggestions for which of the many media offerings may be most appropriate

- A guide to incorporating labs into the redesigned AP Biology course

Printed Test Bank
(ISBN 1-4292-9116-8)

The Test Bank offers thousands of questions, covering the full range of topics presented in the textbook. All questions are referenced to textbook sections and page numbers, and are ranked according to Bloom's taxonomy. Each chapter includes a wide range of multiple choice and fill-in-the-blank questions. In addition, each chapter features a set of diagram questions that involve the student in working with illustrations of structures, graphs, steps in processes, and more. The electronic versions of the Test Bank (within BioPortal, on the Instructor's Media Library, and in ExamView®) also include all of the BioPortal Diagnostic Quiz questions, Interactive Quiz questions, and additional multiple choice questions.

ExamView® Assessment Suite
(ISBN 1-4292-9115-X)

ExamView Test Generator® guides teachers through the process of creating online or paper tests and quizzes quickly and easily. Users may select from an extensive bank of test questions or use the step-by-step tutorial to write their own questions. Questions may be scrambled to create different versions of tests. Tests may be printed in many different types of formats to provide maximum flexibility or may be administered online using the **ExamView Player**®. When administering online testing through the school's LAN or selected course management systems, student results flow to the **ExamView Test Manager**® to provide a comprehensive assessment management system for the teacher.

Teacher's Resource DVD and Media Library
(ISBN 1-4292-9267-9)

The *Principles of Life* Instructor's Media Library, available both online via BioPortal and on the TRDVD, includes a wide range of electronic resources to help instructors plan their course, present engaging lectures, and effectively assess student comprehension. The following resources are included:

TEXTBOOK FIGURES AND TABLES. Every image and table from the textbook is provided in both JPEG (high- and low-resolution) and PDF formats. Each figure is provided both with and without balloon captions, and large, complex figures are provided in both a whole and split version.

UNLABELED FIGURES. Every figure is provided in an unlabeled format, useful for quizzing and custom presentation development.

SUPPLEMENTAL PHOTOS. The supplemental photograph collection contains over 1,000 photographs (in addition to those in the textbook), giving instructors a wealth of additional imagery to draw upon.

ANIMATIONS. A wide range of detailed animations, all created from the textbook's art program, and viewable in either narrated or step-through mode.

VIDEOS. A collection of video segments that covers topics across the entire textbook and helps demonstrate the complexity and beauty of life.

POWERPOINT RESOURCES. For each chapter of the textbook, several different PowerPoint presentations are available. These give instructors the flexibility to build presentations in the manner that best suits their needs. Included are:

- Textbook Figures and Tables
- Lecture Presentation
- Figures with Editable Labels
- Layered Art Figures
- Supplemental Photos
- Videos
- Animations

ACTIVE LEARNING EXERCISES. These exercises help instructors engage students in the classroom through a variety of questions and problems that include discussion questions, data analysis exercises, and more, in a format designed to be used with clicker systems.

COLLEGE INSTRUCTOR'S MANUAL. (Available in BioPortal only) Includes a wealth of information to help instructors in the planning and teaching of their course. The Instructor's Manual includes the following sections for each chapter of the textbook:

- Chapter Overview
- Key Concepts/Chapter Outline
- Lecture Outline
- Key Terms

MEDIA GUIDE. A visual guide to the extensive media resources available with *Principles of Life*. The guide includes thumbnails and descriptions of every video, animation, activity, and supplemental photo in the Media Library, all organized by chapter.

INTUITIVE BROWSER INTERFACE provides a quick and easy way to preview and access all of the content on the Instructor's Media Library.

Faculty Lounge for Majors Biology is the first publisher-provided website for the majors biology community that lets instructors freely communicate and share peer-reviewed lecture and teaching resources. It is continually updated and vetted by majors biology instructors—there is always something new to see. The Faculty Lounge offers convenient access to peer-recommended and vetted resources, including the following categories: Images, News, Videos, Labs, Lecture Resources, and Educational Research. In addition, the site includes special areas for resources for lab coordinators, resources and updates from the Scientific Teaching series of books, and information on biology teaching workshops.
http://majorsbio.facultylounge.whfreeman.com

Figure Correlation Tool

An invaluable resource for instructors switching to *Principles of Life* from another textbook, this online tool provides easy correlations between the figures in *Principles of Life* and figures in other majors biology textbooks.

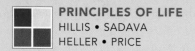

HOW **PRINCIPLES O**

Each chapter of *Principles of Life* focuses on a series of essential biological concepts and the research that led to our understanding of them. Chapters are designed to help you focus on what's important and offer a number of ways to analyze and review what you've read, as you prepare for classtime or exams.

OPENING STORY

How do kangaroo rats, rattlesnakes, owls, bats, and moths "see in the dark"? Each chapter of *Principles of Life* begins with a **Question** like this, giving you a real-world starting point for what comes next.

Question & Answer at the end of the chapter revisits the opening question, explaining the answer by drawing on the material you've just read.

KEY CONCEPTS

This list outlines the key ideas covered in the chapter ahead.

DO YOU UNDERSTAND THE CONCEPT?

After each section in the chapter, a series of questions helps you check your understanding of the section's central ideas.

APPLY THE CONCEPT

These exercises give you the opportunity to work with a concept in a natural world setting by interpreting actual research data and drawing your own conclusions.

SUMMARY

The **Summary** reviews chapter content, including key figures, and directs you to supporting online resources, including:
- Interactive Tutorials
- Animated Tutorials
- Web Activities
- Working with Data problems

Sensors **35**

On moonless, pitch-black nights in the desert, kangaroo rats move silently about to forage and to seek mates. When two kangaroo rats meet, they immediately identify each other's gender and behave accordingly—males fight with other males and try to court females. We may not be able to observe their behavior in total darkness, but they can easily identify each other, interact, and also gather food. However, other denizens of the desert can observe the kangaroo rats and attempt to capture them for food.

If a kangaroo rat chances to hop near a coiled rattlesnake, the snake can detect the rat, accurately fix its location, and with a lightning fast strike attempt to nail the rat. But even as the strike of the rattlesnake is in process, the kangaroo rat senses it

QA QUESTION How do kangaroo rats, rattlesnakes, owls, bats, and moths "see in the dark"?

ANSWER Kangaroo rats have large eyes that give them maximal visual sensitivity even in very dim light, but that is not the only sense that enables them to function on moonless nights. Their long whiskers enable them to use tactile information to explore their environments. Their olfactory sensitivity enables them to locate and identify other kangaroo rats, both as to the identity of familiar individuals and the gender of unfamiliar ones. The acute auditory sensitivity of the kangaroo rat enables it to hear the air disturbance caused by the strike of the rattlesnake and to jump reflexively.

The rattlesnake's "sense" of the kangaroo rat and other small

---

KEY CONCEPTS

16.1 All of Life Is Connected through Its Evolutionary History

16.2 Phylogeny Can Be Reconstructed from Traits of Organisms

16.3 Phylogeny Makes Biology Comparative and Predictive

Do You Understand Concept 16.1?

- What biological processes are represented in a phylogenetic tree?
- Why is it important to consider only homologous characters in reconstructing phylogenetic trees?
- What are some reasons that similar traits might arise independently in species that are only distantly related? Can you think of examples among familiar organisms?

APPLY THE CONCEPT

Phylogeny is the basis of biological classification

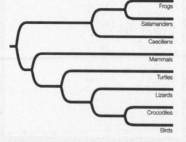

Frogs
Salamanders
Caecilians
Mammals
Turtles
Lizards
Crocodiles
Birds

Classification One:

Named group	Included taxa
Amphibia	Frogs, salamanders,
Mammalia	Mammals
Reptilia	Turtles, lizards, and cr
Aves	Birds

Classification Two:

Named group	Included taxa
Amphibia	Frogs, salamanders,
Mammalia	Mammals
Reptilia	Turtles, lizards, crococ

Classification Three:

Named group	Included taxa
Amphibia	Frogs, salamanders,
Homothermia	Mammals and birds
Reptilia	Turtles, lizards, and cr

Consider the above phylogeny and three possible classifications of the taxa.

1. Which of these classifications contains a paraphyletic group?

2. Which of these classifications contains

3. Which of these classifications is consiste of including only monophyletic groups classification?

4. Starting with the classification you nam how many additional group names wo

---

5 SUMMARY

concept 5.1 Biological Membranes Have a Common Structure and Are Fluid

- Biological membranes consist of lipids, proteins, and carbohydrates. The **fluid mosaic model** of membrane structure describes a phospholipid bilayer in which proteins can move about within the plane of the membrane.

- The two layers of a membrane may have different properties because of their different phospholipid compositions, exposed domains of **integral membrane proteins**, and **peripheral membrane proteins**. Transmembrane proteins span the membrane. Review Figure 5.1, WEB ACTIVITY 5.1, and INTERACTIVE TUTORIAL 5.1

concept 5.2 Some Substances Can Cross the Membrane by Diffusion

- Membranes exhibit **selective permeability**, regulating which substances pass through them.

- A substance can diffuse passively across a membrane by one of two processes: **simple diffusion** through the phospholipid bilayer or **facilitated diffusion**, either through a channel created by a **channel protein** or by means of a **carrier protein**. In both cases, molecules diffuse down their concentration gradients.

- In **osmosis**, water diffuses from a region of higher water concentration to a region of lower water concentration through membrane channels called aquaporins. Ions diffuse across membranes through ion channels. Review Figures 5.3 and 5.4, ANIMATED TUTORIAL 5.1, and WORKING WITH DATA 5.1

- **Carrier proteins** bind to polar molecules such as sugars and amino acids and transport them across the membrane. Review Figure 5.6

LIFE WORKS FOR YOU

INVESTIGATION

FIGURE 5.5 Aquaporin Increases Membrane Permeability to Water A protein was isolated from the membranes of cells in which water diffuses rapidly across the membranes. When the protein was inserted into oocytes, which do not normally have it, the water permeability of the oocytes was greatly increased.

HYPOTHESIS
Aquaporin increases membrane permeability to water.

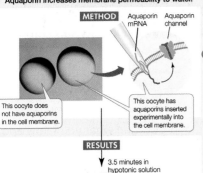

METHOD — Aquaporin mRNA — Aquaporin channel

This oocyte does not have aquaporins in the cell membrane.

This oocyte has aquaporins inserted experimentally into the cell membrane.

RESULTS
3.5 minutes in hypotonic solution

Water does not diffuse into the cell, so it does not swell.

Water diffuses into the cell through the aquaporin channels, and it swells.

CONCLUSION
Aquaporin **increases** the rate of water diffusion across the cell membrane.

ANALYZE THE DATA
Oocytes were injected with aquaporin mRNA (red circles) or a solution without mRNA (blue circles). Water permeability was tested by incubating the oocytes in hypotonic solution and measuring cell volume. After time X in the upper curve, intact oocytes were not visible:

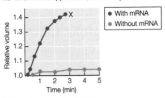

- With mRNA
- Without mRNA

A. Why did the cells increase in volume?
B. What happened at time X?
C. Calculate the relative rates (volume increase per minute) of swelling in the control and experimental curves. What does this show about the effectiveness of mRNA injection?

For more, go to Working with Data 5.1 at **yourBioPortal.com**.

Go to **yourBioPortal.com** for original citations, discussions, and relevant links for all INVESTIGATION figures.

RESEARCH TOOLS

FIGURE 4.4 Centrifugation Structures within cells can be separated from one another on the basis of size and density, and the isolated structures can then be analyzed chemically.

1 A piece of tissue is homogenized by grinding it.

2 The cell homogenate contains large and small cell structures.

3 A centrifuge is used to separate the cell structures based on size and density.

4 The heaviest cell structures can be removed and the remaining suspension re-centrifuged until the next heaviest cell structures reach the bottom of the tube.

- Golgi
- Mitochondria
- Nuclei

LINK Review the discussion of phagocytosis in Concept 4.3

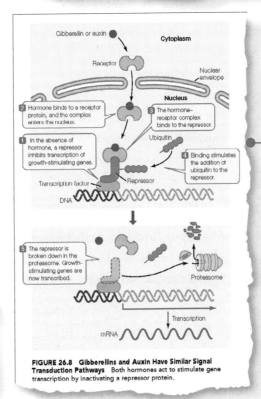

Gibberellin or auxin

Cytoplasm

Receptor

Nuclear envelope

Nucleus

2 Hormone binds to a receptor protein, and the complex enters the nucleus.

3 The hormone–receptor complex binds to the repressor.

Ubiquitin

1 In the absence of hormone, a repressor inhibits transcription of growth-stimulating genes.

4 Binding stimulates the addition of ubiquitin to the repressor.

Transcription factor — Repressor

DNA

5 The repressor is broken down in the proteasome. Growth-stimulating genes are now transcribed.

Proteasome

Transcription

mRNA

FIGURE 26.8 Gibberellins and Auxin Have Similar Signal Transduction Pathways Both hormones act to stimulate gene transcription by inactivating a repressor protein.

FRONTIERS As an animal embryo develops, cells detach from one region, move, and then reattach at another region. This requires that cell membrane glycoproteins with binding specificities for adjacent cells are replaced by new ones with different specificities, so the cells can bind to their new neighbors. To investigate embryonic development, scientists are making genetically modified organisms whose embryonic cell membranes lack one or another surface binding component.

RESEARCH TOOLS

Throughout *Principles of Life*, this feature shows you how professional biologists go about their work, focusing on techniques and quantitative methods used to investigate biological systems.

INVESTIGATION

- **Investigation** figures describe a key experiment's hypothesis, methods, results, and conclusion.

- **Analyze the Data** exercises ask you to analyze a subset of data from the published experiment so you can see the connections between observations, analyses, hypotheses, and conclusions.

- For many Investigations, there are also online **Working with Data** problems, which give you an opportunity to work with the original published data in greater depth.

LINKS

These cross-references connect you to a more thorough discussion of a key term or concept elsewhere in the book.

HELPFUL ART

The numbered **Balloon Captions** allow you to easily follow the important steps as key processes unfold in the figures.

FRONTIERS

In each chapter, **Frontiers** introduces you to ongoing or possible future avenues of research and their importance to society.

BIO**P**ORTAL

yourBioPortal.com

featuring  ✓ *PrepU* *Smarter than the average quiz* and the **Interactive eBook**

BioPortal is your gateway to an extensive set of easy-to-use resources that can have a direct and immediate positive impact on your grade for the course.

Note: The activation card inserted into this textbook shows you how to get started with **BioPorta**. If your copy of the textbook does not include access to **BioPortal**, you can purchase it at **yourBioPortal.com**. The stand-alone eBook can be purchased at **ebooks.bfwpub.com/hillis1**

BIOPORTAL features include:

Prep-U—A powerful quizzing engine that bridges the gap between what you know and what you need to learn, with **personalized**, **adaptive quizzes** that are proven to lead to better grades. Here's how it works: You take a quiz. Prep-U grades it and then creates follow-up quizzes based on how you did. The system adapts to provide questions at the right level and on the right topics for you, so that you improve your mastery of the material as efficiently as possible.

Diagnostic Quizzing and Personalized Study Plans—The Diagnostic Quiz checks your understanding of the full range of concepts presented in each chapter, and then provides you with a **Personalized Study Plan** that helps you focus your study time on the areas that need your attention the most.

A Powerful Interactive eBook—A complete online version of the textbook and all of its media resources, fully integrated with links to animations, activities, and more, right where you need them. You can even personalize the eBook just like you would a printed textbook, with highlighting, bookmarks, and a flexible notes features.

Extraordinary media resources and study tools including **Animated Tutorials**, **Activities**, **Flashcards & Key Terms**, the **Lecture Notebook**, **BioNews** from *Scientific American*, and a lot more.

Available to all students

Free Companion Website at www.whfreeman.com/hillis1e
All students, including those NOT using BioPortal, have open access to the free *Principles of Life* Companion Website. Chapter by chapter, you'll find animations, flashcards, activities, study ideas, help with math, and more.

For more about BioPortal and its features, and to find out about additional student resources to accompany Principles of Life, *see the Preface.*

Table of Contents

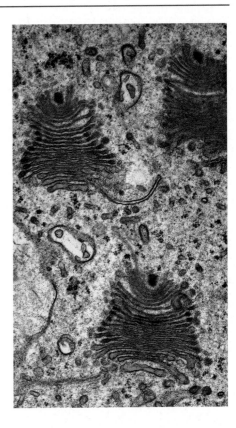

Genetics

PART 3

Evolution

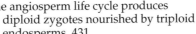

22 The Evolution and Diversity of Fungi 437

23 Animal Origins and Diversity 456

PART 5

Plant Form and Function

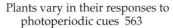

27 Reproduction of Flowering Plants 556

28 Plants in the Environment 572

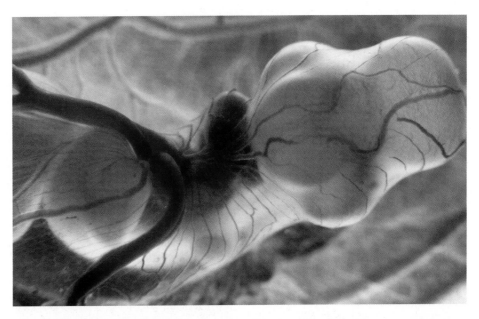

35 Sensors 695

36 Musculoskeletal Systems 712

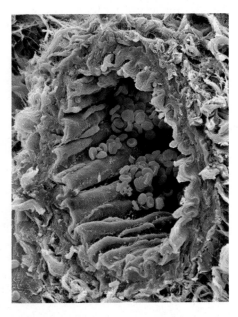

PART 7

Ecology

42 Organisms in Their Environment 822

43 Populations 842

Principles of Life

When you take a walk through the woods and fields or a park near your home, what do you see? If you are like most people, you probably notice the trees, colorful flowers, and some animals. You probably spend little time, however, thinking about how these living things function, reproduce, interact with one another, or affect their environment. An introduction to biology should inspire you to ask questions about what life is, how living systems work, and how the living world came to be as we observe it today.

Biologists have amassed a huge amount of information about the living world, and some introductory biology classes focus on memorizing these details. This book takes a different approach, focusing on the major principles of life that underlie everything in biology.

What do we mean by *principles of life*? Consider the photograph. Why is the view so overwhelmingly green? The color is explained by a fundamental principle of life, namely that all living organisms require energy in order to grow, move, reproduce, and maintain their bodies. Ultimately, most of that energy comes from the sun. The green leaves of plants contain chlorophyll, a pigment that captures energy from the sun and uses it to transform water and carbon dioxide into sugar and oxygen (a process called photosynthesis). That sugar can then be broken down again by the plant, or by other organisms that eat the plant, to provide energy. The frog in the photograph is using energy to grasp the trunk of the tree. That energy came from molecules in the bodies of insects eaten by the frog. The insects, in turn, built up their bodies by ingesting tissues of plant leaves, which grew by capturing the sun's energy through photosynthesis. The frog, like the plants, is ultimately solar-powered.

The photograph illustrates other principles of biology. You probably noticed

the frog and the trees in the photograph above, but did you notice the patches of growth on the trunk of the tree? Most of those are lichens, a complex interaction between a fungus and a photosynthetic organism (in this case, a species of algae). Living organisms often survive and thrive by interacting with one another in complex ways. In lichen, the fungus and the alga live in an obligate symbiosis, meaning that they depend on each other for survival. Many other organisms in this scene are too small to be seen, but they are critical components for keeping this living system functioning over time.

After reading this book, you should understand the main principles of life. You'll be able to describe how organisms capture and transform energy; pass genetic information to their offspring in reproduction; grow, develop, and behave; and interact with other organisms and with their physical environment. You'll also learn how this system of life on Earth evolved, and how it continues to change. May a walk in the park never be the same for you again!

What principles of life are illustrated in this scene?

KEY CONCEPTS

1.1 Living Organisms Share Common Aspects of Structure, Function, and Energy Flow

1.2 Genetic Systems Control the Flow, Exchange, Storage, and Use of Information

1.3 Organisms Interact with and Affect Their Environments

1.4 Evolution Explains Both the Unity and Diversity of Life

1.5 Science Is Based on Quantifiable Observations and Experiments

<table>
<tr><td>

concept

1.1
</td><td>

Living Organisms Share Common Aspects of Structure, Function, and Energy Flow
</td></tr>
</table>

Biology is the scientific study of living things. Most biologists define "living things" as all the diverse organisms descended from a single-celled ancestor that evolved on Earth almost 4 billion years ago. We can image other origins, perhaps on other planets, of self-replicating systems that have properties similar to life as we know it. But the evidence suggests that all of life on Earth today has a single origin—a single *common ancestor*—and we consider all the organisms that descended from that common ancestor to be a part of life.

Life as we know it had a single origin

The overwhelming evidence for the common ancestry of life lies in the many distinctive characteristics that are shared by all living organisms. All organisms:

- *are composed of a common set of chemical components such as nucleic acids and amino acids, and similar structures such as cells enclosed within plasma membranes*
- *contain genetic information that uses a nearly universal code to specify the assembly of proteins*
- *convert molecules obtained from their environment into new biological molecules*
- *extract energy from the environment and use it to do biological work*
- *regulate their internal environment*
- *replicate their genetic information in the same manner when reproducing themselves*
- *share sequence similarities among a fundamental set of genes*
- *evolve through gradual changes in their genetic information*

If life had multiple origins, there would be little reason to expect similarities across gene sequences, or a nearly universal genetic code, or a common set of amino acids. If we were to discover an independent origin of a similar self-replicating system (i.e., life) on another planet, we would expect it to be fundamentally different in these aspects. Organisms from another origin of life might be similar in some ways to life on Earth, such as using genetic information to reproduce. But we would not expect the details of their genetic code or the fundamental sequences of their genomes to be like ours.

The simple list of characteristics above, however, is an inadequate description of the incredible complexity and diversity of life. Some forms of life may not even display all of these characteristics all of the time. For example, the seed of a desert plant may go for many years without extracting energy from the environment, converting molecules, regulating its internal environment, or reproducing; yet the seed is alive.

And what about viruses? Viruses do not consist of cells, and they cannot carry out the functions of life enumerated in the list above on their own; they must parasitize host cells to do those jobs for them. Yet viruses contain genetic information and use the same basic genetic code and amino acids as do other living things, and they certainly mutate and evolve. The existence of viruses depends on cells, and there is strong evidence that viruses evolved from cellular life forms. So, although viruses are not independent cellular organisms, they are a part of life and are studied by biologists.

This book explores the characteristics of life, how these characteristics evolved and how they vary among organisms, and how they work together to enable organisms to survive and reproduce.

Life arose from non-life via chemical evolution

Geologists estimate that Earth formed between 4.6 and 4.5 billion years ago. At first, the planet was not a very hospitable place. It was some 600 million years or more before the earliest life evolved. If we picture the history of Earth as a 30-day month, life first appeared somewhere toward the end of the first week (**FIGURE 1.1**).

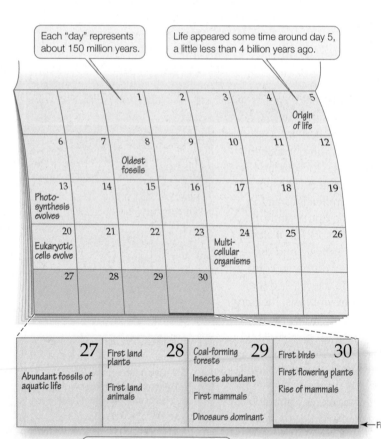

FIGURE 1.1 Life's Calendar Depicting Earth's history on the scale of a 30-day month provides a sense of the immensity of evolutionary time.

When we consider how life might have arisen from nonliving matter, we must take into account the properties of the young Earth's atmosphere, oceans, and climate, all of which were very different than they are today. Biologists postulate that complex biological molecules first arose through the random physical association of chemicals in that environment. Experiments simulating the conditions on early Earth have confirmed that the generation of complex molecules under such conditions is possible, even probable. The critical step for the evolution of life, however, was the appearance of *nucleic acids*—molecules that could reproduce themselves and also serve as templates for the synthesis of large molecules with complex but stable shapes. The variation in the shapes of these large, stable molecules—*proteins*—enabled them to participate in increasing numbers and kinds of chemical reactions with other molecules.

Cellular structure evolved in the common ancestor of life

The next step in the origin of life was the enclosure of complex proteins and other biological molecules by *membranes* that contained them in a compact internal environment separate from the surrounding external environment. Molecules called *fatty acids* played a critical role because these molecules do not dissolve in water; rather, they form membranous films. When agitated, these films can form spherical *vesicles*, which could have enveloped assemblages of biological molecules. The creation of an internal environment that concentrated the reactants and products of chemical reactions opened up the possibility that those reactions could be integrated and controlled within a tiny *cell*. Scientists postulate that this natural process of membrane formation resulted in the first cells with the ability to reproduce—the evolution of the first cellular organisms.

For more than 2 billion years after cells originated, every organism consisted of only one cell. These first unicellular organisms were (and are, as multitudes of their descendants exist in similar form today) **prokaryotes**. Prokaryotic cells consist of genetic material and other biochemicals enclosed in a membrane (**FIGURE 1.2**). Early prokaryotes were confined to the oceans, where there was an abundance of complex molecules they could use as raw materials and sources of energy. The ocean shielded them from the damaging effects of ultraviolet (UV) light, which was intense at that time because there was little or no oxygen (O_2) in the atmosphere, and hence no protective ozone (O_3) layer in the upper atmosphere.

Photosynthesis allowed living organisms to capture energy from the sun

To fuel their cellular metabolism, the earliest prokaryotes took in molecules directly from their environment and broke these small molecules down to release and use the energy contained in their chemical bonds. Many modern species of prokaryotes still function this way, and very successfully.

About 2.7 billion years ago, the emergence of **photosynthesis** changed the nature of life on Earth. The chemical reactions of photosynthesis transform the energy of sunlight into a form of biological energy that can power the synthesis of large molecules. These large molecules are the building blocks of cells, and they can be broken down to provide metabolic energy. Photosynthesis is the basis of much of life on Earth today because its energy-capturing processes provide food for other organisms. Early photosynthetic cells were probably similar to present-day prokaryotes called *cyanobacteria* (**FIGURE 1.3**). Over time, photosynthetic prokaryotes became so abundant that vast quantities of O_2, which is a by-product of photosynthesis, slowly began to accumulate in the atmosphere.

During the early eons of life on Earth, there was no O_2 in the atmosphere. In fact, O_2 was poisonous to many of the prokaryotes that lived at that time. Those organisms that did tolerate O_2, however, were able to proliferate, and the presence of O_2 opened up vast new avenues of evolution. *Aerobic metabolism* (energy production using O_2) is more efficient than *anaerobic* (non-O_2-using) *metabolism*, and it allowed organisms to grow larger. Aerobic metabolism is used by the majority of living organisms today.

Oxygen in the atmosphere also made it possible for life to move onto land. For most of life's history, UV radiation falling on Earth's surface was so intense that it destroyed any living cell that was not well shielded by water. But the accumulation of photosynthetically generated O_2 in the atmosphere for more than 2 billion years gradually produced a layer of ozone in the upper atmosphere. By about 500 million years ago, the ozone layer was sufficiently dense and absorbed enough of the sun's UV radiation to make it possible for organisms to leave the protection of the water and live on land.

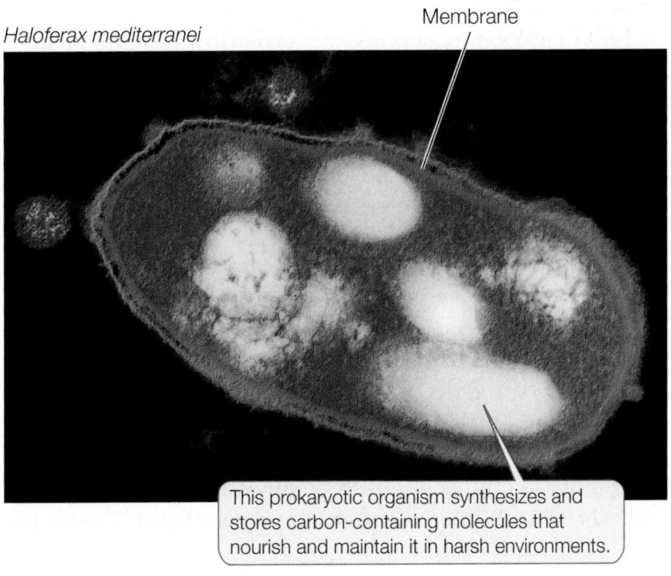

Haloferax mediterranei

Membrane

This prokaryotic organism synthesizes and stores carbon-containing molecules that nourish and maintain it in harsh environments.

FIGURE 1.2 The Basic Unit of Life Is the Cell The concentration of reactions within the enclosing membrane of a cell allowed the evolution of integrated organisms. Today all organisms, even the largest and most complex, are made up of cells. Unicellular organisms such as this one, however, remain the most abundant living organisms (in absolute numbers) on Earth.

(A)

(B)

FIGURE 1.3 Photosynthetic Organisms Changed Earth's Atmosphere Cyanobacteria were the first photosynthetic organisms on Earth. (A) Colonies of cyanobacteria called stromatolites are known from the ancient fossil record. (B) Living stromatolites are still found in appropriate environments on Earth today.

Eukaryotic cells evolved from prokaryotes

Another important step in the history of life was the evolution of cells with membrane-enclosed compartments called **organelles**, within which specialized cellular functions could be performed away from the rest of the cell. The first organelles probably appeared about 2.5 billion years after the appearance of life on Earth (about day 20 on Figure 1.1). One of these organelles, the *nucleus*, came to contain the cell's genetic information. The nucleus (Latin *nux*, "nut" or "core") gives these cells their name: **eukaryotes** (Greek *eu*, "true"; *karyon*, "kernel" or "core"). The eukaryotic cell is completely distinct from the cells of prokaryotes (*pro*, "before"), which lack nuclei and other internal compartments.

Some organelles are hypothesized to have originated by **endosymbiosis** ("living inside another") when larger cells ingested smaller ones. The *mitochondria* that generate a cell's energy probably evolved from engulfed prokaryotic organisms. And *chloroplasts*—the organelles specialized to conduct

photosynthesis—could have originated when photosynthetic prokaryotes were ingested by larger eukaryotes. If the larger cell failed to break down this intended food object, a partnership could have evolved in which the ingested prokaryote provided the products of photosynthesis and the host cell provided a good environment for its smaller partner.

Multicellularity allowed specialization of tissues and functions

For the first few billion years of life, all the organisms that existed—whether prokaryotic or eukaryotic—were unicellular. At some point, the cells of some eukaryotes failed to separate after cell division, remaining attached to each other. Such permanent colonial aggregations of cells made it possible for some of the associated cells to specialize in certain functions, such as reproduction, while other cells specialized in other functions, such as absorbing nutrients. **Cellular specialization** enabled multicellular eukaryotes to increase in size and become more efficient at gathering resources and adapting to specific environments.

Biologists can trace the evolutionary tree of life

If all the organisms on Earth today are the descendants of a single kind of unicellular organism that lived almost 4 billion years ago, how have they become so different? Organisms reproduce by replicating their genomes, as we will discuss shortly. This replication process is not perfect, however, and changes, called **mutations**, are introduced almost every time a genome is replicated. Some mutations give rise to structural and functional changes in organisms. As individuals mate with one another, these changes can spread within a population, but the population will remain one species. However, if something happens to isolate some members of a population from the others, structural and functional differences between the two groups will accumulate over time. The two groups may diverge to the point where their members can no longer reproduce with each other; thus the two populations become distinct species.

Tens of millions of species exist on Earth today. Many times that number lived in the past but are now extinct. Biologists give each of these species a distinctive scientific name formed from two Latinized words—a **binomial**. The first name identifies the species' *genus*—a group of species that share a recent common ancestor. The second is the name of the species. For example, the scientific name for the human species is *Homo sapiens*: *Homo* is our genus and *sapiens* our species. *Homo* is Latin for "man"; *sapiens* is from the Latin word for "wise" or "rational." Our closest relatives in the genus *Homo* (the Neanderthals) are now extinct and are known only from fossil remains.

Much of biology is based on comparisons among species, and these comparisons are useful precisely because we can place species in an evolutionary context relative to one another. Our ability to do this has been greatly enhanced in recent decades by our ability to sequence and compare the genomes

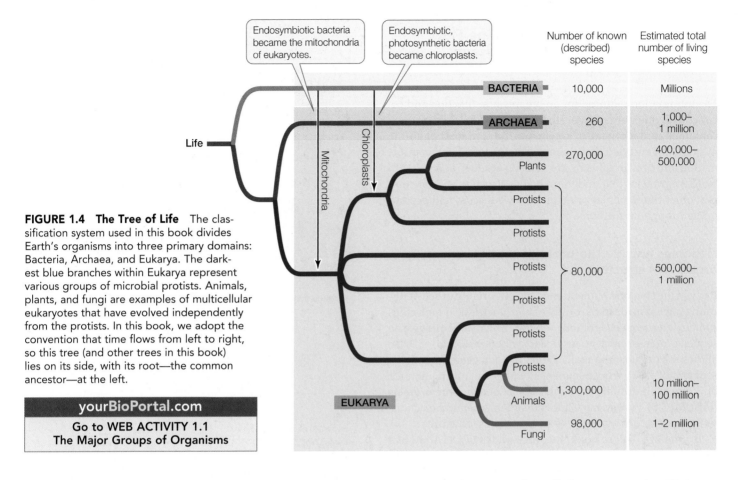

FIGURE 1.4 The Tree of Life The classification system used in this book divides Earth's organisms into three primary domains: Bacteria, Archaea, and Eukarya. The darkest blue branches within Eukarya represent various groups of microbial protists. Animals, plants, and fungi are examples of multicellular eukaryotes that have evolved independently from the protists. In this book, we adopt the convention that time flows from left to right, so this tree (and other trees in this book) lies on its side, with its root—the common ancestor—at the left.

yourBioPortal.com

**Go to WEB ACTIVITY 1.1
The Major Groups of Organisms**

of different species. Genome sequencing and other molecular techniques have allowed biologists to augment evolutionary knowledge based on the fossil record with a vast array of molecular evidence. The result is the ongoing compilation of *phylogenetic trees* that document and diagram evolutionary relationships as part of an overarching tree of life, the broadest categories of which are shown in **FIGURE 1.4**. (The tree is expanded in Appendix A; you can also explore the tree interactively at http://tolweb.org/tree.)

Although many details remain to be clarified, the broad outlines of the tree of life have been determined. Its branching patterns are based on a rich array of evidence from fossils, structures, metabolic processes, behavior, and molecular analyses of genomes. Molecular data in particular have been used to separate the tree into three major **domains**: Archaea, Bacteria, and Eukarya. The organisms of each domain have been evolving separately from those in the other domains for more than a billion years.

Organisms in the domains **Archaea** and **Bacteria** are single-celled prokaryotes. However, members of these two groups differ so fundamentally in their metabolic processes that they are believed to have separated into distinct evolutionary lineages very early. Species belonging to the third domain—**Eukarya**—have eukaryotic cells whose mitochondria and chloroplasts originated from endosymbioses of bacteria.

Plants, fungi, and animals are examples of familiar multicellular eukaryotes that evolved from different groups of unicellular eukaryotes, informally known as *protists*. We know that these three groups (as well as others) had independent origins of multicellularity because they are each most closely related to different groups of unicellular protists, as can be seen from the branching pattern of Figure 1.4.

Discoveries in biology can be generalized

Because all life is related by descent from a common ancestor, shares a genetic code, and consists of similar molecular building blocks, knowledge gained from investigations of one type of organism can, with care, be generalized to other organisms. Biologists use **model systems** for research, knowing they can extend their findings to other organisms, including humans. Our basic understanding of the chemical reactions in cells came from research on bacteria but is applicable to all cells, including those of humans. Similarly, the biochemistry of photosynthesis—the process by which plants use sunlight to produce sugars—was largely worked out from experiments on *Chlorella*, a unicellular green alga. Much of what we know about the genes that control plant development is the result of work on *Arabidopsis thaliana*, a relative of the mustard plant. Knowledge about how animals develop has come from work on sea urchins, frogs, chickens, roundworms, and fruit flies. And recently, the discovery of a major gene controlling human skin color came from work on zebrafish. Being able to generalize from model systems is a powerful tool in biology.

concept
1.2 Genetic Systems Control the Flow, Exchange, Storage, and Use of Information

The information required for an organism to function and interact with other organisms—the "blueprint" for existence—is contained in the organism's **genome**, the sum total of all the information encoded by its genes. The study of genetic information and how organisms are able to "decode" and use it to build the proteins that underlie a body's structure and function is another fundamental principle that we discuss and expand upon throughout the book.

Genomes encode the proteins that govern an organism's structure

Early in the chapter we mentioned the importance of self-replicating nucleic acids in the origin of life. Nucleic acid molecules contain long sequences of four subunits called **nucleotides**. The sequence of these nucleotides in **deoxyribonucleic acid**, or **DNA**, allows the organism to make proteins. Each gene is a specific segment of DNA whose sequence carries the information for building or controlling the expression of one or more **proteins** (**FIGURE 1.5**). Protein molecules govern the chemical reactions within cells and form much of an organism's structure.

By analogy with a book, the nucleotides of DNA are like the letters of an alphabet. The sentences of the book are genes that describe proteins, or provide instructions for making the proteins at a particular time or place. If you were to write out your own genome using four letters to represent the four DNA nucleotides, you would write more than 3 billion letters. Using the size type you are reading now, your genome would fill more than a thousand books the size of this one.

All the cells of a given multicellular organism contain the same genome, yet the different cells have different functions and form different proteins—hemoglobin forms in red blood cells, gut cells produce digestive proteins, and so on. Therefore, different types of cells in an organism must express different parts of the genome. How any given cell controls which genes it expresses (and which genes it suppresses) is a major focus of current biological research.

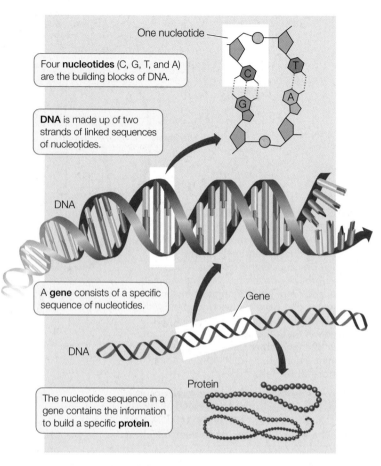

FIGURE 1.5 DNA Is Life's Blueprint The instructions for life are contained in the sequences of nucleotides in DNA molecules. Specific DNA nucleotide sequences comprise genes. The average length of a single human gene is 27,000 nucleotides. The information in each gene provides the cell with the information it needs to manufacture molecules of a specific protein.

The genome of an organism contains thousands of genes. If mutations alter the nucleotide sequence of a gene, the protein that the gene encodes is often altered as well. Mutations may occur spontaneously, as happens when mistakes occur during replication of DNA. Mutations can also be caused by certain chemicals (such as those in cigarette smoke) and radiation

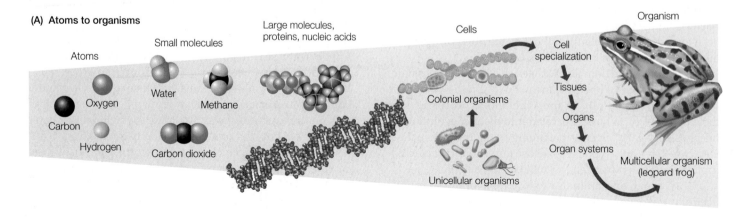

(A) Atoms to organisms

(including UV radiation from the sun). Most mutations are either harmful or have no effect, but occasionally a mutation improves the functioning of the organism under the environmental conditions the individual encounters. Mutations are the raw material of evolution.

Genomes provide insights into all aspects of an organism's biology

The first complete DNA sequence of an organism's genome was determined in 1976. This first sequence was that of a virus, and viral genomes are very small compared with those of most cellular organisms. It was another two decades before the first bacterial genome was sequenced, in 1995. The first animal genome (a relatively small one, that of a roundworm) was determined in late 1998. A massive effort to sequence the complete human genome began in 1990 and culminated 13 years later. Since then, the methods developed in these pioneering projects (as well as new DNA sequencing technologies that appear each year) have resulted in the sequencing of genomes from hundreds of species. As methods have improved, the cost and time for sequencing a complete genome have dropped dramatically. Many biologists expect the routine sequencing of genomes from individual organisms to be commonplace in biological applications of the near future.

What are we learning from genome sequencing? One surprise came when some genomes turned out to contain many fewer genes than expected—for example, there are only about 24,000 different genes that encode proteins in a human genome, whereas most biologists had expected many times more. Gene sequence information is a boon for many areas of biology, making it possible to study the genetic basis of everything from physical structure to the basis of inherited diseases. Biologists can also compare genomes from many species to learn how and why one species differs from another. Such studies allow biologists to trace the evolution of genes through time and to document how particular changes in gene sequence result in changes in structure and function.

The vast amount of information being collected from genome studies has led to rapid development of the field of *bioinformatics*, or the study of biological information. In this emerging field, biologists and computer scientists work in close association to develop new computational tools to organize, process, and study comparative genomic databases.

---

concept 1.3   Organisms Interact with and Affect Their Environments

Another pervasive theme of biology relates to the concepts of *hierarchy* and *integration* of biological systems. Biological systems are organized in a hierarchy from basic building blocks to complete functioning ecosystems of living and nonliving components (**FIGURE 1.6**). Traditionally, each biologist concentrated on understanding a particular level of this hierarchy. Today, however, much of biology involves integrating investigations across many of the hierarchical levels.

> **yourBioPortal.com**
> Go to WEB ACTIVITY 1.2
> The Hierarchy of Life

Organisms use nutrients to supply energy and to build new structures

Living organisms acquire *nutrients* from their environments. Life depends on thousands of biochemical reactions that occur inside cells, and nutrients supply the organism with the energy and

FIGURE 1.6   Biology Is Studied at Many Levels of Organization   (A) Life's properties emerge when DNA and other molecules are organized in cells, which form building blocks for organisms. (B) Organisms exist in populations and interact with other populations to form communities, which interact with the physical environment to make up ecosystems.

(B) Organisms to ecosystems

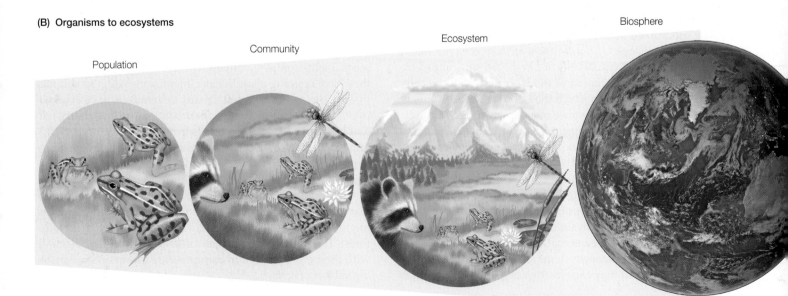

Population   Community   Ecosystem   Biosphere

raw materials to carry out these chemical transformations. Some of the reactions break down nutrient molecules into smaller chemical units, and in the process some of the energy contained in the chemical bonds of the nutrients is captured by molecules that can be used to do different kinds of cellular work.

The most basic cellular work is the building, or *synthesis*, of new complex molecules and structures from smaller chemical units. For example, we are all familiar with the fact that carbohydrates eaten today may be deposited in the body as fat tomorrow. Another kind of work that cells do is mechanical—moving molecules from one cellular location to another, moving whole cells or tissues, or even moving the organism itself, as the proteins in muscle cells do. Still another kind of work is the electrical work that is the essence of information processing in nervous systems, such as vision (recall that you are using captured solar energy to read this book).

The sum total of all the chemical transformations and other work done in all the cells of an organism is called **metabolism**. The many biochemical reactions constantly taking place in cells are integrally linked in that the products of one are the raw materials of the next. Let's consider how these networks of reactions are integrated and controlled.

Organisms regulate their internal environment

The cells of multicellular organisms are specialized, or *differentiated*, to contribute in some way to maintaining the internal environment. With the evolution of specialization, differentiated cells lost many of the functions carried out by single-celled organisms. To accomplish their specialized tasks, assemblages of differentiated cells are organized into *tissues*. For example, a single muscle cell cannot generate much force, but when many cells combine to form the tissue of a working muscle, considerable force and movement can be generated. Different tissue types are organized to form *organs* that accomplish specific functions. The heart, brain, and stomach are each constructed of several types of tissues, as are the roots, stems, and leaves of plants. Organs whose functions are interrelated can be grouped into *organ systems*; the stomach, intestine, and esophagus are parts of the digestive system. The functions of cells, tissues, organs, and organ systems are all integral to the multicellular organism.

The specialized organ systems of multicellular organisms exist in an *internal environment* that is *acellular* (i.e., not made up of cells). The individual cells of a body are surrounded by an extracellular environment of fluids, from which the cells receive nutrients and into which they excrete waste products of metabolism. The maintenance of a narrow range of conditions in this internal environment is known as **homeostasis**. A relatively stable internal (but extracellular) environment means that cells can function efficiently even when conditions outside the organism's body become unfavorable for cellular processes. The organism's *regulatory systems* obtain information from *sensors*, process and integrate this information, and issue instructions to components of physiological systems that produce changes in the organism's internal environment.

Physiological regulatory systems are especially well developed in animals, but they exist in other organisms as well. For example, when conditions are hot and dry, plants can close the small pores (called *stomata*) on the surfaces of their leaves, thereby reducing moisture loss. When external conditions become favorable again, the plants open their stomata, allowing carbon dioxide—which is necessary for photosynthesis—to enter the leaf. This regulatory system is a simple example of a *feedback loop*: cells in the root of the plant (the sensors) release a chemical when they become dehydrated, and this chemical causes the cells around the stomata to shrink, thereby closing the pores. If external conditions improve and the cells in the root become hydrated again, the sensor cells stop releasing the chemical and the stomata open.

The concept of homeostasis extends beyond the regulation of the internal, acellular environment of multicellular organisms. Individual cells (in both unicellular and multicellular organisms) also regulate their internal environment through the actions of a *plasma membrane*, which forms the outer surface of the cell. Thus self-regulation of a more or less constant internal environment is a general attribute of living organisms.

Organisms interact with one another

Organisms do not live in isolation, and the internal hierarchy of the individual organism is matched by the external hierarchy of the biological world (see Figure 1.6). A group of individuals of the same species that interact with one another is a *population*, and populations of all the species that live and interact in the same area are called a *community*. Communities together with their abiotic (nonliving) environment constitute an *ecosystem*.

Individuals in a population interact in many different ways. Animals eat plants and other animals (usually members of another species) and compete with other species for food and other resources. Some animals will prevent other individuals of their own species from exploiting a resource, be it food, nesting sites, or mates. Animals may also *cooperate* with members of their species, forming social units such as a termite colony or a flock of birds. Such interactions have resulted in the evolution of social behaviors such as communication and courtship displays.

Plants also interact with their external environment, which includes other plants, fungi, animals, and microorganisms. All terrestrial plants depend on partnerships with fungi, bacteria, and animals. Some of these partnerships are necessary to obtain nutrients, some to produce fertile seeds, and still others to disperse seeds. Plants compete with each other for light and water, and they have ongoing evolutionary interactions with the animals that eat them. Through time, many adaptations have evolved in plants that protect them from predation (such as thorns) or that help then attract the animals that assist in their reproduction (such as sweet nectar or colorful flowers). The interactions of populations of plant and animal species in a community are major evolutionary forces that produce specialized adaptations.

Communities interacting over a broad geographic area with distinguishing physical features form ecosystems; examples might include the Arctic tundra, a coral reef, or a tropical rainforest. The ways in which species interact with one another and with their environment in communities and in ecosystems is the subject of *ecology*.

concept **1.4** Evolution Explains Both the Unity and Diversity of Life

Evolution—change in the genetic makeup of biological populations through time—is a major unifying principle of biology. A common set of evolutionary mechanisms applies to populations of all organisms. The constant change that occurs among these populations gives rise to all the diversity we see in life. These two themes—unity and diversity—provide a framework for organizing and thinking about biological systems. The similarities of life allow us to make comparisons and predictions from one species to another, and the differences are what make biology such a rich and exciting field for investigation and discovery.

Natural selection is an important mechanism of evolution

Charles Darwin compiled factual evidence for evolution in his 1859 book *On the Origin of Species*. Since then, biologists have gathered massive amounts of data supporting Darwin's idea that all living organisms are descended from a common ancestor. Darwin also proposed one of the most important processes that produce evolutionary change. He argued that differential survival and reproduction among individuals in a population, which he termed **natural selection**, could account for much of the evolution of life.

Although Darwin proposed that living organisms are descended from common ancestors and are therefore related to one another, he did not have the advantage of understanding the mechanisms of genetic inheritance. But he knew that offspring differed from their parents, even though they showed strong similarities. Any population of a plant or animal species displays variation, and if you select breeding pairs on the basis of some particular trait, that trait is more likely to be present in their offspring than in the general population. Darwin himself bred pigeons, and was well aware of how pigeon fanciers selected breeding pairs to produce offspring with unusual feather patterns, beak shapes, or body sizes. He realized that if humans could select for specific traits, the same process could operate in nature; hence the term *natural selection* as opposed to the artificial (human-imposed) selection that has been practiced on crop plants and domesticated animals since the dawn of human civilization.

How does natural selection function? Darwin postulated that different probabilities of survival and reproductive success could account for evolutionary change. He reasoned that the reproductive capacity of plants and animals, if unchecked, would result in unlimited growth of populations, but we do not observe such growth in nature; in most species, only a small percentage of offspring survive to reproduce. Thus any trait that confers even a small increase in the probability that its possessor will survive and reproduce will spread in the population.

Because organisms with certain traits survive and reproduce best under specific sets of conditions, natural selection leads to **adaptations**: structural, physiological, or behavioral traits that enhance an organism's chances of survival and reproduction in its environment (**FIGURE 1.7**). Consider the feet of the frog shown in the opening photograph of this chapter. The toes of the frog's foot are greatly expanded compared with those of frog species that do not live in trees. Expanded toes increase the ability of tree frogs to climb trees, which allows them to seek insects for food in the forest canopy and to escape terrestrial predators. Thus the expanded toe pads of tree frogs are an adaptation to arboreal life.

(A) *Dyscophus guineti*

(B) *Xenopus laevis*

(C) *Agalychnis callidryas*

(D) *Rhacophorus nigropalmatus*

FIGURE 1.7 Adaptations to the Environment The limbs of frogs show adaptations to the different environments of each species. (A) This terrestrial frog walks across the ground using its short legs and peglike digits (toes). (B) Webbed rear feet are evident in this highly aquatic species of frog. (C) This arboreal species has toe pads, which are adaptations for climbing. (D) A different arboreal species has extended webbing between the toes, which increases surface area and allows the frog to glide from tree to tree.

In addition to natural selection, evolutionary processes such as sexual selection (selection due to mate choice) and genetic drift (the random fluctuation of gene frequencies in a population due to chance events) contribute to the rise of biological diversity. These processes operating over evolutionary history have led to the remarkable array of life on Earth.

Evolution is a fact, as well as the basis for broader theory

The famous biologist Theodosius Dobzhansky once wrote that "Nothing in biology makes sense except in the light of evolution." Dobzhansky was emphasizing the need to integrate an evolutionary perspective and approach into all aspects of biological study. Everything in biology is a product of evolution, and biologists need to incorporate a perspective of change and adaptation to fully understand biological systems.

You may have heard it said that evolution is "just a theory," thereby implying that there is some question about whether or not biological populations evolve. This is a common misunderstanding that originates in part from the different meanings of the word "theory" in everyday language and in science. In everyday speech, some people use the word "theory" to mean "hypothesis" or even—disparagingly—"a guess." In science, however, a **theory** is *a body of scientific work in which rigorously tested and well-established facts and principles are used to make predictions about the natural world.* In short, evolutionary theory is (1) a body of knowledge supported by facts and (2) the resulting understanding of the various mechanisms by which biological populations have changed and diversified over time, and by which Earth's populations continue to evolve.

Evolution can be observed and measured directly, and many biologists conduct experiments on evolving populations. We constantly observe changes in the genetic composition of populations over relatively short-term time frames. In addition, we can directly observe a record of the history of evolution in the fossil record over the almost unimaginably long periods of geological time. Exactly *how* biological populations change through time is something that is subject to testing and experimentation. The fact that biological populations evolve, however, is not disputed among biologists

You will see evolution and the other major principles of life described in this chapter at work in each part of this book. In Part I you will learn about the molecular structure of life. We will discuss the origin of life, the energy inherent in atoms and molecules, and how proteins and nucleic acids became the self-replicating cellular systems of life. Part II will describe how these self-replicating systems work and the genetic principles that explain heredity and mutation, which are the building blocks of evolution. In Part III we will describe the mechanisms of evolution and go into greater detail about how evolution works. Part IV will examine the products of evolution: the vast diversity of life and the many different ways organisms solve some common problems such as reproducing, defending themselves, and obtaining nutrients. Parts V and VI will explore the physiological adaptations that allow plants and animals to survive and function in a wide range of physical environments.

Finally, in Part VII we will discuss these environments and the integration of individual organisms, populations, and communities into the interrelated ecosystems of the biosphere.

You may enjoy returning to this chapter occasionally as the course progresses; the necessarily terse explanations given here should begin to cohere and make more sense as you read about the facts and phenomena that underlie the principles. Our knowledge of the "facts" of biology, however, is not based just on reading, contemplation, or discussion—although all of these activities are useful, even necessary. Scientific knowledge is based on active and always-ongoing research.

concept 1.5 Science Is Based on Quantifiable Observations and Experiments

Regardless of the many different tools and methods used in research, all scientific investigations are based on *observation* and *experimentation*. In both, scientists are guided by established principles of a set of scientific methods that allow us to discover new aspects about the structure, function, and history of the natural world.

Observing and quantifying are important skills

Many biologists are motivated by their observations of the living world. Learning *what to observe* in nature is a skill that develops with experience in biology. An intimate understanding of the **natural history** of a group of organisms—how the organisms get their food, reproduce, behave, regulate their internal environments (their cells, tissues, and organs), and interact with other organisms—facilitates observations and leads biologists to ask questions about those observations. The more a biologist knows about general principles, the more he or she is likely to gain new insights from observing nature.

Biologists have always observed the world around them, but today our ability to observe is greatly enhanced by technologies such as electron microscopes, rapid genome sequencing, magnetic resonance imaging, and global positioning satellites. These technologies allow us to observe everything from the distribution of molecules in the body to the daily movement of animals across continents and oceans.

Observation is a basic tool of biology, but as scientists we must also be able to **quantify** our observations. Whether we are testing a new drug or mapping the migrations of the great whales, mathematical and statistical calculations are essential. For example, biologists once classified organisms based entirely on qualitative descriptions of the physical differences among them. There was no way of objectively determining evolutionary relationships of organisms, and biologists had to depend on the fossil record for insight. Today our ability to quantify the molecular and physical differences among species, combined with explicit mathematical models of the evolutionary process, enables quantitative analyses of evolutionary history. These mathematical calculations, in turn, facilitate comparative investigations of all other aspects of an organism's biology.

Scientific methods combine observation, experimentation, and logic

Often, science textbooks describe "*the* scientific method," as if there is a single, simple flow chart that all scientists follow. This is an oversimplification. Although such flow charts incorporate much of what scientists do, you should not conclude that scientists necessarily progress through the steps of the process in one prescribed, linear order.

Observations lead to questions, and scientists make additional observations and often do experiments to answer those questions. This approach, called the *hypothesis–prediction method*, has five steps: (1) making *observations*; (2) asking *questions*; (3) forming *hypotheses*, or tentative answers to the questions; (4) making *predictions* based on the hypotheses; and (5) *testing* the predictions by making additional observations or conducting experiments. These are the steps seen in traditional flow charts such as the one shown in **FIGURE 1.8**.

After posing a question, a scientist often uses *inductive logic* to propose a tentative answer. Inductive logic involves taking observations or facts and creating a new proposition that is compatible with those observations or facts. Such a tentative proposition is called a **hypothesis**. In formulating a hypothesis, scientists put together the facts they already know to formulate one or more possible answers to the question.

The next step in the scientific method is to apply a different form of logic—*deductive logic*—to make predictions based on the hypothesis. Deductive logic starts with a statement believed to be true and goes on to predict what facts would also have to be true to be compatible with that statement.

Getting from questions to answers

Let's consider an example of how scientists can start with a general question and work to find answers. Amphibians—such as the frog in the opening photograph of this chapter—have been around for a long time. They watched the dinosaurs come and go. But today amphibian populations around the world are in dramatic decline, with more than a third of the world's amphibian species threatened with extinction. Why?

Biologists work to answer general questions like this by making observations and doing experiments. There are probably multiple reasons that amphibian populations are declining, but scientists often break up a large problem into many smaller problems and investigate them one at a time. One hypothesis is that frog populations have been adversely affected by agricultural insecticides and herbicides (weed-killers). Several studies have shown that many of these chemicals tested at realistic concentrations do not kill amphibians. But Tyrone Hayes, a biologist at the University of California at Berkeley, probed deeper.

yourBioPortal.com

Go to ANIMATED TUTORIAL 1.1
Using Scientific Methodology

Hayes focused on atrazine, the most widely used herbicide in the world and a common contaminant in fresh water. More than 70 million pounds of atrazine are applied to farmland in the United States every year, and it is used in at least 20 countries. Atrazine kills several types of weeds that can choke fields of important crops such as corn. The chemical is usually applied before weeds emerge in the spring—at the same time many amphibians are breeding and thousands of tadpoles swim in the ditches, ponds, and streams that receive runoff from farms.

In his laboratory, Hayes and his associates raised frog tadpoles in water containing no atrazine and in water with concentrations ranging from 0.01 parts per billion (ppb) up to 25 ppb. The U.S. Environmental Protection Agency considers environmental levels of atrazine of 10–20 ppb of no concern; it considers 3 ppb a safe level in drinking water. Rainwater in Iowa has been measured to contain 40 ppb. In Switzerland, where the use of atrazine is illegal, the chemical has been measured at approximately 1 ppb in rainwater.

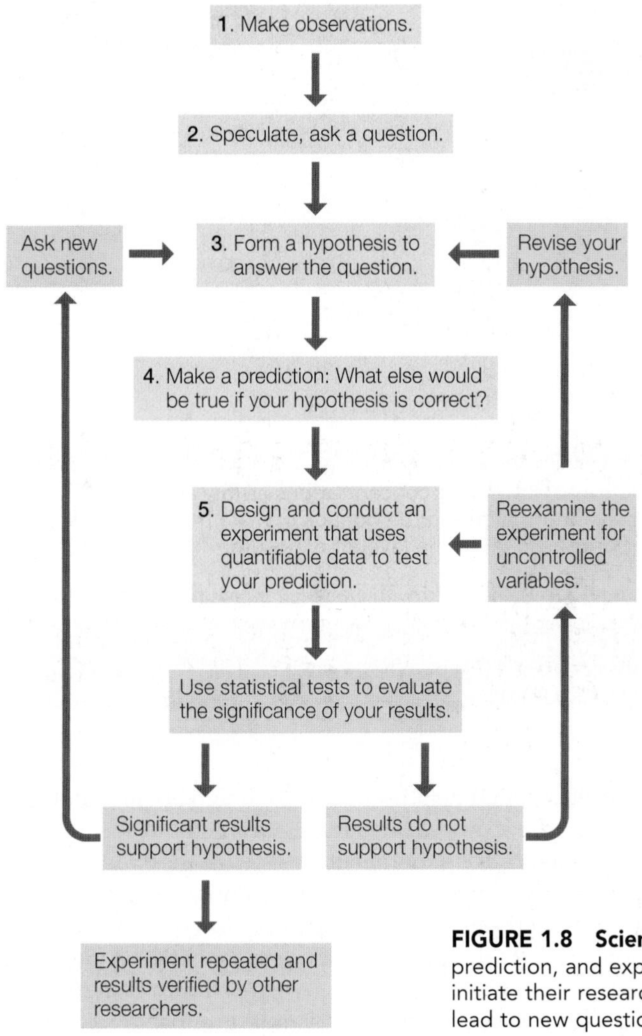

FIGURE 1.8 Scientific Methodology The process of observation, speculation, hypothesis, prediction, and experimentation is a cornerstone of modern science, although scientists may initiate their research at several different points. Answers gleaned through experimentation lead to new questions, more hypotheses, further experiments, and expanding knowledge.

In the Hayes laboratory, an atrazine concentration as low as 0.1 ppb had a dramatic effect on tadpole development: it feminized the males. In some of the adult males that developed from these larvae, the vocal structures used in mating calls were smaller than normal, female sex organs developed, and eggs were found growing in the testes. In other studies, normal adult male frogs exposed to 25 ppb had a tenfold reduction in testosterone levels and did not produce sperm. You can imagine the disastrous effects these developmental and hormonal changes could have on the capacity of frogs to breed and reproduce.

But Hayes's experiments were performed in the laboratory, with a species of frog bred for laboratory use. Would his results be the same in nature? To find out, he and his students traveled across the middle of North America, sampling water and collecting frogs. They analyzed the water for atrazine and examined the frogs. In the only site where atrazine was undetectable in the water, the frogs were normal; in all the other sites, male frogs had abnormalities of the sex organs.

Like other biologists, Hayes made observations. He then made predictions based on those observations, and designed and carried out experiments to test his predictions. Some of the conclusions from his experiments, described below, could have profound implications not only for amphibians but also for other animals, including humans.

Good experiments have the potential to falsify hypotheses

Once predictions are made from a hypothesis, experiments can be designed to test those predictions. The most informative experiments are those that have the ability to show that the prediction is wrong. If the prediction is wrong, the hypothesis must be questioned, modified, or rejected.

There are two general types of experiments, both of which compare data from different groups or samples. A *controlled* experiment manipulates one or more of the factors being tested; *comparative* experiments compare unmanipulated data gathered from different sources.

In a **controlled experiment**, we start with groups or samples that are as similar as possible. We predict on the basis of our hypothesis that some critical factor, or **variable**, has an effect on the phenomenon we are investigating. We devise some method to manipulate *only that variable* in an "experimental" group and compare the resulting data with data from an unmanipulated "control" group. If the predicted difference occurs, we then apply statistical tests to ascertain the probability that the manipulation created the difference (as opposed to the difference being the result of random chance). **FIGURE 1.9** describes one of the many controlled experiments performed by the Hayes laboratory to quantify the effects of atrazine on male frogs.

The basis of controlled experiments is that one variable is manipulated while all others are held constant. The variable that is manipulated is called the *independent variable*, and the response that is measured is the *dependent variable*. A good

---

INVESTIGATION

FIGURE 1.9 Controlled Experiments Manipulate a Variable
The Hayes laboratory created controlled environments that differed only in the concentrations of atrazine in the water. Eggs from leopard frogs (*Rana pipiens*) raised specifically for laboratory use were allowed to hatch and the tadpoles were separated into experimental tanks containing water with different concentrations of atrazine.

HYPOTHESIS
Exposure to atrazine during larval development causes abnormalities in the reproductive tissues of male frogs.

METHOD
1. Establish 9 tanks in which all attributes are held constant except the water's atrazine concentration. Establish 3 atrazine conditions (3 replicate tanks per condition): 0 ppb (control condition), 0.1 ppb, and 25 ppb.
2. Place *Rana pipiens* tadpoles from laboratory-reared eggs in the 9 tanks (30 tadpoles per replicate).
3. When tadpoles have transitioned into adults, sacrifice the animals and evaluate their reproductive tissues.
4. Test for correlation of degree of atrazine exposure with the presence of abnormalities in the gonads (testes) of male frogs.

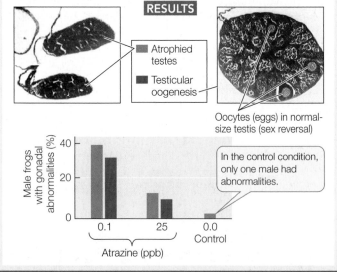

RESULTS
- Atrophied testes
- Testicular oogenesis

Oocytes (eggs) in normal-size testis (sex reversal)

In the control condition, only one male had abnormalities.

Male frogs with gonadal abnormalities (%)

CONCLUSION
Exposure to atrazine at concentrations as low as 0.1 ppb induces abnormalities in the gonads of male frogs. The effect is not proportional to the level of exposure.

For more, go to Working with Data 1.1 at **yourBioPortal.com**.

Go to **yourBioPortal.com** for original citations, discussions, and relevant links for all INVESTIGATION figures.

---

controlled experiment is not easy to design because biological variables are so interrelated that it is difficult to alter just one.

A **comparative experiment** starts with the prediction that there will be a difference between samples or groups based on the hypothesis. In comparative experiments, however, we cannot control the variables; often we cannot even identify all the variables that are present. We are simply gathering and comparing data from different sample groups.

When his controlled experiments indicated that atrazine indeed affects reproductive development in frogs, Hayes and his colleagues performed a comparative experiment. They collected frogs and water samples from eight widely separated sites across the United States and compared the incidence of abnormal frogs from environments with very different levels of atrazine (**FIGURE 1.10**). Of course, the sample sites differed in many ways besides the level of atrazine present.

The results of experiments frequently reveal that the situation is more complex than the hypothesis anticipated, thus raising new questions. There are no "final answers" in science.

Investigations consistently reveal more complexity than we expect. As a result, biologists often develop new questions, hypotheses, and experiments as they collect more data.

Statistical methods are essential scientific tools

Whether we do comparative or controlled experiments, at the end we have to decide whether there is a difference between the samples, individuals, groups, or populations in the study. How do we decide whether a measured difference is enough to support or falsify a hypothesis? In other words, how do we decide in an unbiased, objective way that the measured difference is significant?

Significance can be measured with statistical methods. Scientists use statistics because they recognize that variation is always present in any set of measurements. Statistical tests calculate the probability that the differences observed in an experiment could be due to random variation. The results of statistical tests are therefore probabilities. A statistical test starts with a **null hypothesis**—the premise that any observed differences are simply the result of random differences that arise from drawing two finite samples from the same population. When quantified observations, or **data**, are collected, statistical methods are applied to those data to calculate the likelihood that the null hypothesis is correct.

More specifically, statistical methods tell us the probability of obtaining the same results by chance even if the null hypothesis were true. We need to eliminate, insofar as possible, the chance that any differences showing up in the data are merely the result of random variation in the samples tested. Scientists generally conclude that the differences they measure are significant if statistical tests show that the *probability of error* (that is, the probability that a difference as large as the one observed could be obtained by mere chance) is 5 percent or lower, although more stringent levels of significance may be set for some problems. Appendix B of this book is a short primer on statistical methods that you can refer to as you analyze data that will be presented throughout the text.

INVESTIGATION

FIGURE 1.10 Comparative Experiments Look for Differences among Groups
To see whether the presence of atrazine correlates with testicular abnormalities in male frogs, the Hayes lab collected frogs and water samples from different locations around the U.S. The analysis that followed was "blind," meaning that the frogs and water samples were coded so that experimenters working with each specimen did not know which site the specimen came from.

HYPOTHESIS

Presence of the herbicide atrazine in environmental water correlates with gonadal abnormalities in frog populations.

METHOD

1. Based on commercial sales of atrazine, select 4 sites (sites 1–4) less likely and 4 sites (sites 5–8) more likely to be contaminated with atrazine.
2. Visit all sites in the spring (i.e., when frogs have transitioned from tadpoles into adults); collect frogs and water samples.
3. In the laboratory, sacrifice frogs and examine their reproductive tissues, documenting abnormalities.
4. Analyze the water samples for atrazine concentration (the sample for site 7 was not tested).
5. Quantify and correlate the incidence of reproductive abnormalities with environmental atrazine concentrations.

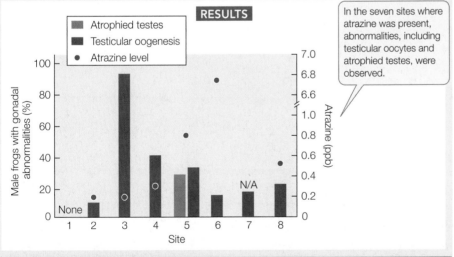

In the seven sites where atrazine was present, abnormalities, including testicular oocytes and atrophied testes, were observed.

CONCLUSION

Reproductive abnormalities exist in frogs from environments in which aqueous atrazine concentration is 0.2 ppb or above. The incidence of abnormalities does not appear to be proportional to atrazine concentration at the time of transition to adulthood.

Go to **yourBioPortal.com** for original citations, discussions, and relevant links for all INVESTIGATION figures.

Not all forms of inquiry into nature are scientific

Science is a unique human endeavor that is bounded by certain standards of practice. Other areas of scholarship share with science the practice of making observations and asking questions, but scientists are distinguished by what they do with their observations and how they answer their questions. Data, subjected to appropriate statistical analysis, are critical in the testing of hypotheses. Science is the most powerful approach humans have devised for learning about the world and how it works.

Scientific explanations for natural processes are objective and reliable because the hypotheses proposed *must be testable* and *must have the potential of being rejected* by direct observations and experiments. Scientists must clearly describe the methods they use to test hypotheses so that other scientists can repeat their results. Not all experiments are repeated, but surprising or controversial results are always subjected to independent verification. Scientists worldwide share this process of testing and rejecting hypotheses, contributing to a common body of scientific knowledge.

If you understand the methods of science, you can distinguish science from non-science. Art, music, and literature all contribute to the quality of human life, but they are not science. They do not use scientific methods to establish what is fact. Religion is not science, although religions have historically attempted to explain natural events ranging from unusual weather patterns to crop failures to human diseases. Most such phenomena that at one time were mysterious can now be explained in terms of scientific principles. Fundamental tenets of religious faith, such as the existence of a supreme deity or deities, cannot be confirmed or refuted by experimentation and are thus outside the realm of science. The power of science derives from the uncompromising objectivity and absolute dependence on evidence that comes from *reproducible and quantifiable observations*. A religious or spiritual explanation of a natural phenomenon may be coherent and satisfying for the person holding that view, but it is not testable and therefore it is not science. To invoke a supernatural explanation (such as a "creator" or "intelligent designer" with no known bounds) is to depart from the world of science. Science does not say that religious beliefs are necessarily wrong; they are just not part of the world of science, and are untestable using scientific methods. In other words, science and religion are nonoverlapping approaches to inquiry.

Science describes how the world works; it is silent on the question of how the world "ought to be." Many scientific advances that contribute to human welfare also raise major ethical issues. Recent developments in genetics and developmental biology may enable us to select the sex of our children, to use stem cells to repair our bodies, and to modify the human genome. Although scientific knowledge allows us to do these things, science cannot tell us whether or not we should do so, or if we choose to do them, how we should regulate them. Such issues are as crucial to human society as the science itself, and a responsible scientist does not lose sight of these questions or neglect the contributions of the humanities in attempting to come to grips with them.

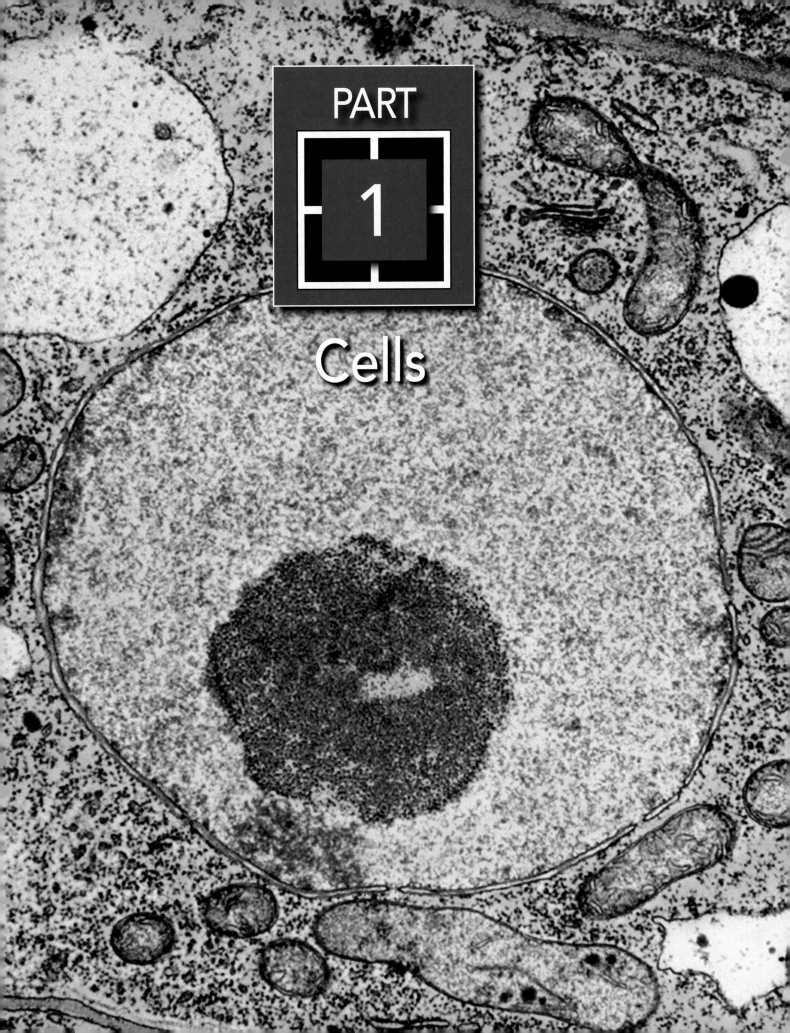

PART

1

Cells

Life Chemistry and Energy

A major discovery of biology was that living things are composed of the same chemical elements as the vast nonliving portion of the universe. This mechanistic view—that life is chemically based and obeys the universal laws of chemistry and physics—is relatively new in human history. Until the nineteenth century, many scientists thought that a "vital force," distinct from the forces governing the inanimate world, was responsible for life. Many people still assume that such a vital force exists, but the mechanistic view of life has led to great advances in biological science, and it underpins many applications of biology to medicine and agriculture. We assume a mechanistic view throughout this book.

Among the most abundant chemical elements in the universe are hydrogen and oxygen, and life as we know it requires the presence of these elements as water (H_2O). Water makes up about 70 percent of the bodies of most organisms, and those that live on land have evolved elaborate ways to retain the water in their bodies. Aquatic organisms do not need these water-retention mechanisms; thus biologists think that life originated in a watery environment.

Life has been found in some surprising places, often in extreme conditions. There are organisms living in hot springs at temperatures above the boiling point of water, beneath the Antarctic ice, 5 kilometers below Earth's surface, at the bottom of the ocean, in extremely acid or salty conditions, and even inside nuclear reactors.

With trillions of galaxies in the universe, each with billions or trillions of stars, there are many planets out there, and if our

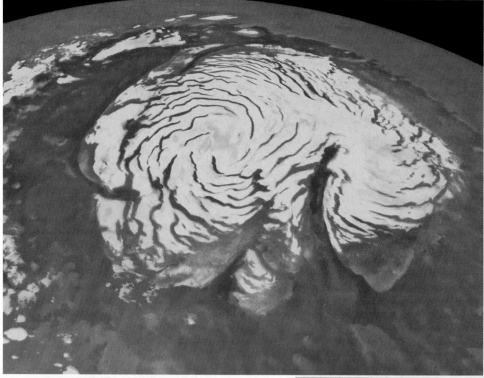

solar system is typical, some of them have the water needed for life. Indeed, space probes have detected water on the moons surrounding Saturn, at the poles and warmer mid-latitudes of Mars, and all over the surface of our own moon. The amount of water on these planetary bodies is not a lot by Earth standards—on the moon, there is about a liter of water per 1,000 kilograms of soil, which is less than the amount in the driest desert on Earth. But given that organisms are found on Earth in extreme environments, the existence of water outside of Earth makes extraterrestrial life seem possible.

Polar ice caps, as shown here, have been observed on Mars for a long time, but recent evidence also shows water at the milder mid-latitudes of Mars.

 QUESTION Why is the search for water important in the search for life?*

*You will find the answer to this question on page 31.

KEY CONCEPTS

2.1 Atomic Structure Is the Basis for Life's Chemistry

2.2 Atoms Interact and Form Molecules

2.3 Carbohydrates Consist of Sugar Molecules

2.4 Lipids Are Hydrophobic Molecules

2.5 Biochemical Changes Involve Energy

concept
2.1 Atomic Structure Is the Basis for Life's Chemistry

Living and nonliving matter is composed of **atoms**. Each atom consists of a dense, positively charged **nucleus**, with one or more negatively charged **electrons** moving around it. The nucleus contains one or more positively charged **protons**, and may contain one or more **neutrons** with no electrical charge:

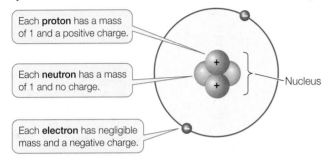

Each **proton** has a mass of 1 and a positive charge.

Each **neutron** has a mass of 1 and no charge.

Each **electron** has negligible mass and a negative charge.

Nucleus

Charges that are different (+/−) attract each other, whereas charges that are alike (+/+, −/−) repel each other. Most atoms are electrically neutral because the number of electrons in an atom equals the number of protons.

The mass of a proton serves as a standard unit of measure called the *dalton* (named after the English chemist John Dalton). A single proton or neutron has a mass of about 1 dalton (Da), which is 1.7×10^{-24} grams, but an electron is even tinier, at 9×10^{-28} g (0.0005 Da). Because the mass of an electron is only about 1/2,000th of the mass of a proton or neutron, the contribution of electrons to the mass of an atom can usually be ignored when chemical measurements and calculations are made.

An element consists of only one kind of atom

An **element** is a pure substance that contains only one kind of atom. The element hydrogen consists only of hydrogen atoms, the element gold only of gold atoms. The atoms of each element have certain characteristics and properties that distinguish them from the atoms of other elements.

There are 94 elements in nature, and at least another 24 have been made in physics laboratories. Almost all of the 94 natural elements have been detected in living organisms, but just a few predominate. About 98 percent of the mass of every living organism (bacterium, turnip, or human) is composed of just six elements:

Carbon (symbol C)   Hydrogen (H)   Nitrogen (N)

Oxygen (O)   Phosphorus (P)   Sulfur (S)

The chemistry of these six elements will be our primary concern in this chapter, but other elements found in living organisms are important as well. Sodium and potassium, for example, are essential for nerve function; calcium can act as a biological signal; iodine is a component of a human hormone; and magnesium is bound to chlorophyll in green plants. The physical and chemical (reactive) properties of atoms depend on the numbers of protons, neutrons, and electrons they contain.

The atoms of an element differ from those of other elements by the number of protons in their nuclei. The number of protons is called the **atomic number**, and it is unique to and characteristic of each element. The atomic number of carbon is 6, and a carbon atom always has six protons; the atomic number of oxygen is always 8. For electrical neutrality, each atom has the same number of electrons as protons, so a carbon atom has six electrons, and an oxygen atom has eight.

Along with a definitive number of protons, every element except hydrogen has one or more neutrons. The *mass number* of an atom is the total number of protons and neutrons in its nucleus. A carbon nucleus contains six protons and six neutrons and has a mass number of 12. Oxygen has eight protons and eight neutrons and has a mass number of 16.

Electrons determine how an atom will react

The **Bohr model** for atomic structure (see diagram at left) provides a concept of an atom that is largely empty space, with a central nucleus surrounded by electrons in orbits, or **electron shells**, at various distances from the nucleus. This model is much like our solar system, with planets orbiting around the sun. Although highly oversimplified (you will learn about the reality of atomic structure in physical chemistry courses), the Bohr model is useful for describing how atoms behave. Specifically, *the behaviors of electrons determine whether a chemical bond will form and what shape the bond will have*. These are two key properties for determining biological changes and structure.

In the Bohr model, each electron shell is a certain distance from the nucleus. Since electrons are negatively charged and protons are positive, an electron needs energy to escape from the attraction of the nucleus. The further away an electron shell is from the nucleus, the more energy the electron must have. We will return to this topic when we discuss biological energetics in Chapter 6. The electron shells, in order of their distance from the nucleus, can be filled with electrons as follows:

- First shell: two electrons
- Second and subsequent shells: eight electrons

FIGURE 2.1 illustrates the electron shell configurations for the six major elements found in living systems.

Atoms with unfilled outer shells (such as oxygen, which has six electrons in its outermost shell) tend to undergo chemical reactions to fill their outer shells. In the case of oxygen, adding two electrons to its outer shell will make a total of eight. These reactive atoms can attain stability either by sharing electrons with other atoms or by losing or gaining one or more electrons. In either case, the atoms involved are *bonded* together into stable associations called **molecules**. The tendency of atoms with at least two electron shells to form stable molecules so they have eight electrons in their outermost shells is known as the *octet rule*. Most atoms in biologically important molecules—for example, carbon (C) and nitrogen (N)—follow this rule.

FIGURE 2.1 Electron Shells Each shell can hold a specific maximum number of electrons and must be filled before electrons can occupy the next shell. The energy level of an electron is higher in a shell farther from the nucleus. An atom with less than the full complement (2 or 8) electrons in its outermost shell can react (bond) with other atoms.

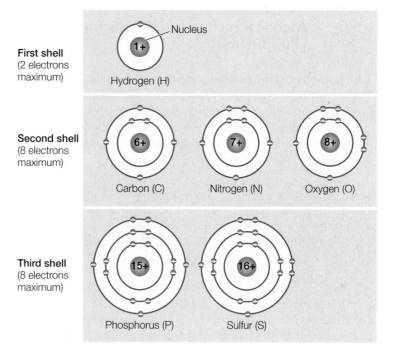

First shell
(2 electrons maximum)
Hydrogen (H)

Second shell
(8 electrons maximum)
Carbon (C) Nitrogen (N) Oxygen (O)

Third shell
(8 electrons maximum)
Phosphorus (P) Sulfur (S)

Do You Understand Concept 2.1?

- What is the arrangement of protons, neutrons, and electrons in an atom?

- Sketch the electron shell configuration of a sodium atom (symbol Na), which has 11 protons. According to the octet rule, what would be the simplest way for a sodium atom to achieve electron stability?

- Many elements have isotopes, which are rare variants of the element with additional neutrons in the nucleus. Deuterium is an isotope of hydrogen that has one neutron (normal hydrogen has no neutrons). Does the neutron change the chemical reactivity of deuterium, compared with normal hydrogen? Explain why or why not.

We have introduced the individual elements that make up all living organisms—the atoms. We have shown how the energy levels of electrons drive an atomic quest for stability. Next we will describe the different types of chemical bonds that can lead to stability, joining atoms together into molecular structures with different properties.

concept 2.2 Atoms Interact and Form Molecules

A **chemical bond** is an attractive force that links two atoms together in a molecule. There are several kinds of chemical bonds (**TABLE 2.1**). In this section we will begin with *ionic bonds*, which form when atoms gain or lose one or more electrons

TABLE 2.1 Chemical Bonds and Interactions

NAME	BASIS OF INTERACTION	STRUCTURE	BOND ENERGY[a]
Ionic attraction	Attraction of opposite charges		3–7
Covalent bond	Sharing of electron pairs		50–110
Hydrogen bond	Sharing of H atom		3–7
Hydrophobic interaction	Interaction of nonpolar substances in the presence of polar substances (especially water)		1–2
van der Waals interaction	Interaction of electrons of nonpolar substances		1

[a]Bond energy is the amount of energy (Kcal/mol) needed to separate two bonded or interacting atoms under physiological conditions.

to achieve stability. Then we will turn to *covalent bonds*—the strong bonds that form when atoms share electrons. We will then consider weaker interactions, including *hydrogen bonds*, which are enormously important to biology. Finally, we will see how atoms are bonded to make *functional groups*—groups of atoms that give important properties to biological molecules.

yourBioPortal.com

Go to ANIMATED TUTORIAL 2.1
Chemical Bond Formation

Ionic bonds form by electrical attraction

In some cases, an atom can transfer or accept a few electrons to complete the octet in its outer shell. Consider sodium (11 protons) and chlorine (17 protons). A sodium atom has only one electron in its outermost shell; this condition is unstable. A chlorine atom has seven electrons in its outermost shell—another unstable condition. The most straightforward way for both atoms to achieve stability is to transfer an electron from sodium's outermost shell to that of chlorine (**FIGURE 2.2**). This reaction makes the two atoms more stable because they both have eight electrons in their outer shells. The result is two *ions*.

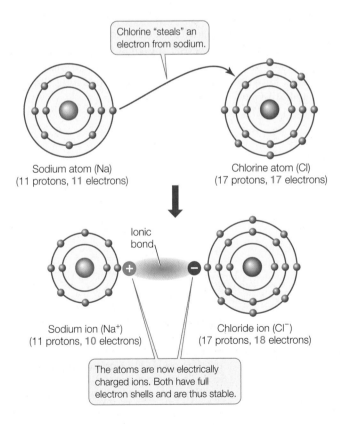

Chlorine "steals" an electron from sodium.

Sodium atom (Na)
(11 protons, **11** electrons)

Chlorine atom (Cl)
(17 protons, **17** electrons)

Ionic bond

Sodium ion (Na⁺)
(11 protons, **10** electrons)

Chloride ion (Cl⁻)
(17 protons, **18** electrons)

The atoms are now electrically charged ions. Both have full electron shells and are thus stable.

FIGURE 2.2   Ionic Bond between Sodium and Chlorine   When a sodium atom reacts with a chlorine atom, the chlorine fills its outermost shell by "stealing" an electron from the sodium. In so doing, the chlorine atom becomes a negatively charged chloride ion (Cl⁻). With one less electron, the sodium atom becomes a positively charged sodium ion (Na⁺).

An **ion** is an electrically charged particle that forms when an atom gains or loses one or more electrons:

- The sodium ion (Na^+) in our example has a charge of +1 because it has one less electron than it has protons. The outermost electron shell of the sodium ion is full, with eight electrons, so the ion is stable. Positively charged ions are called **cations**.

- The chloride ion (Cl^-) has a charge of –1 because it has one more electron than it has protons. This additional electron gives Cl^- a stable outermost shell with eight electrons. Negatively charged ions are called **anions**.

Ionic bonds are formed as a result of the electrical attraction between ions bearing opposite charges. Ionic bonds result in stable molecules that are often referred to as *salts*. An example is sodium chloride (NaCl; table salt), where cations and anions are held together by ionic bonds. While ionic bonds in salts may be stronger, attractions between ions in solution, as occur in living systems, are typically weak (see Table 2.1).

Given that most organisms consist of about 70 percent water, as we described in the opening of this chapter, most of biology (and biochemistry) occurs in the presence of water. Because ionic attractions are weak, salts dissolve in water; the ions separate from one another and become surrounded by water molecules. The water molecules are oriented with their negative poles nearest to the cations and their positive poles nearest to the anions:

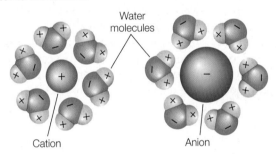

Water molecules

Cation

Anion

Covalent bonds consist of shared pairs of electrons

A **covalent bond** forms when two atoms attain stable electron numbers in their outermost shells by *sharing* one or more pairs of electrons. In this case, each atom contributes one member of each electron pair. Consider two hydrogen atoms coming into close proximity, each with an unpaired electron in its single shell (**FIGURE 2.3**). When the electrons pair up, a stable association is formed, and this links the two hydrogen atoms in a covalent bond, forming the molecule H_2.

Let's see how covalent bonds are formed in the somewhat more complicated methane molecule (CH_4). The carbon atom has six electrons: two electrons fill its inner shell, and four electrons are in its outer shell. Because its outer shell can hold up to eight electrons, carbon can share electrons with up to four other atoms—*it can form four covalent bonds* (**FIGURE 2.4A**). Methane forms when an atom of carbon reacts with four hydrogen atoms. As a result of electron sharing, the outer shell of the carbon atom is now filled with eight electrons—a stable configuration. The outer shell of each hydrogen atom is also

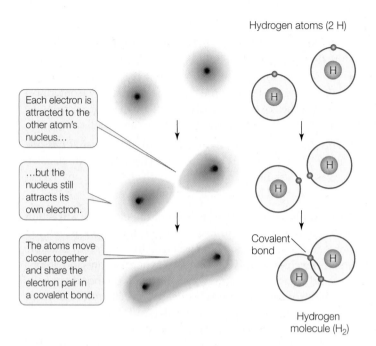

Hydrogen atoms (2 H)

Each electron is attracted to the other atom's nucleus…

…but the nucleus still attracts its own electron.

The atoms move closer together and share the electron pair in a covalent bond.

Covalent bond

Hydrogen molecule (H₂)

FIGURE 2.3 Electrons Are Shared in Covalent Bonds Two hydrogen atoms can combine to form a hydrogen molecule. A covalent bond forms when the electron shells of the two atoms overlap in an energetically stable manner.

ORIENTATION For a given pair of elements, such as carbon bonded to hydrogen, the length of the covalent bond is always the same. And for a given atom within a molecule, the angle of each covalent bond with respect to the others is generally the same. This is true regardless of the type of larger molecule that contains the atom. For example, the four covalent bonds formed by the carbon atom in methane are always distributed in space so that the bonded hydrogens point to the corners of a regular tetrahedron, with the carbon in the center (see Figure 2.4B). Even when carbon is bonded to four atoms other than hydrogen, this three-dimensional orientation is more or less maintained. As you will see, the orientations of covalent bonds in space give molecules their three-dimensional geometry, and the shapes of molecules contribute to their biological functions.

> **FRONTIERS** The activities of biological molecules depend largely on their shapes. As chemists learn more about the geometry of covalent bonds and the forces that affect them, it may be possible to predict the structures of molecules based on their atomic compositions. For a simple molecule like water with only two bonds, this is relatively straightforward. But for complex biological molecules with hundreds or thousands of atoms (like a protein or a new drug), this becomes a subject for modeling by sophisticated computer programs.

filled. Four covalent bonds—four shared electron pairs—hold methane together. **FIGURE 2.4B** shows several different ways to represent the molecular structure of methane.

The properties of molecules are influenced by the characteristics of their covalent bonds. Four important aspects of covalent bonds are their orientation, their strength and stability, multiple covalent bonds, and the degree of sharing of electrons.

STRENGTH AND STABILITY Covalent bonds are very strong (see Table 2.1), meaning it takes a lot of energy to break them. At the temperatures at which life exists, the covalent bonds of biological molecules are quite stable, as are their three-dimensional structures. However, this stability does not preclude change, as we will discover.

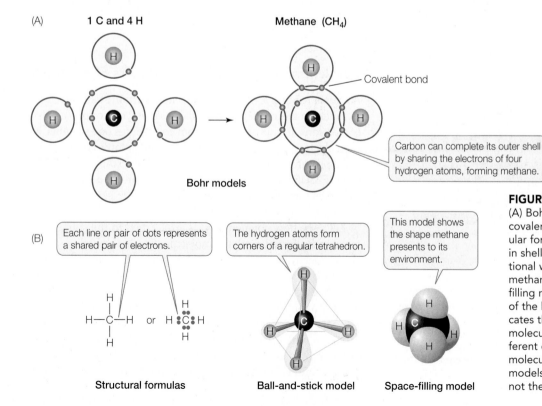

(A)   1 C and 4 H   Methane (CH₄)

H

H   C   H   →   H   C   H

Covalent bond

H

H

Bohr models

Carbon can complete its outer shell by sharing the electrons of four hydrogen atoms, forming methane.

(B) Each line or pair of dots represents a shared pair of electrons.

The hydrogen atoms form corners of a regular tetrahedron.

This model shows the shape methane presents to its environment.

H   H
| ..
H—C—H or H :C: H
| ..
H   H

Structural formulas

H

H   C   H

H

Ball-and-stick model

H

H   C   H

H

Space-filling model

FIGURE 2.4 Covalent Bonding
(A) Bohr models showing the formation of covalent bonds in methane, whose molecular formula is CH₄. Electrons are shown in shells around the nuclei. (B) Three additional ways of representing the structure of methane. The ball-and-stick and the space-filling models show the spatial orientations of the bonds. The space-filling model indicates the overall shape and surface of the molecule. In the chapters that follow, different conventions will be used to depict molecules. Bear in mind that these are models to illustrate certain properties, and not the most accurate portrayal of reality.

MULTIPLE COVALENT BONDS As shown in Figure 2.4B, covalent bonds can be represented by lines between the chemical symbols for the linked atoms:

- A *single bond* involves the sharing of a single pair of electrons (for example, H—H or C—H).
- A *double bond* involves the sharing of four electrons (two pairs; C=C).
- *Triple bonds*—six shared electrons—are rare, but there is one in nitrogen gas (N≡N), which is the major component of the air we breathe.

UNEQUAL SHARING OF ELECTRONS If two atoms of the same element are covalently bonded, there is an equal sharing of the pair(s) of electrons in their outermost shells. However, when the two atoms are different, the sharing is not necessarily equal. One nucleus may exert a greater attractive force on the electron pair than the other nucleus, so that the pair tends to be closer to that atom.

The attractive force that an atomic nucleus exerts on electrons in a covalent bond is called its **electronegativity**. The electronegativity of a nucleus depends on how many positive charges it has (nuclei with more protons are more positive and thus more attractive to electrons) and on the distance between the electrons in the bond and the nucleus (the closer the electrons, the greater the electronegative pull). **TABLE 2.2** shows the electronegativities (which are calculated to produce dimensionless quantities) of some elements important in biological systems.

If two atoms are close to each other in electronegativity, they will share electrons equally in what is called a *nonpolar covalent bond*. Two oxygen atoms, for example, each with an electronegativity of 3.4, will share electrons equally. So will two hydrogen atoms (each with an electronegativity of 2.2). But when hydrogen bonds with oxygen to form water, the electrons involved are *unequally shared*: they tend to be nearer to the oxygen nucleus because it is more electronegative than hydrogen. When electrons are drawn to one nucleus more than to the other, the result is a **polar covalent bond**:

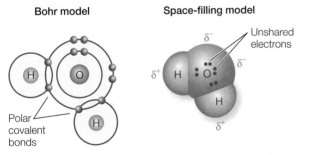

Bohr model   **Space-filling model**

Polar covalent bonds

Unshared electrons

Because of this unequal sharing of electrons, the oxygen end of the bond has a slightly negative charge (symbolized by δ^- and spoken of as "delta negative," meaning a partial unit of charge), and the hydrogen end has a slightly positive charge (δ^+). The bond is *polar* because these opposite charges are separated at the two ends, or poles, of the bond. The partial charges that result from polar covalent bonds produce polar molecules or polar regions of large molecules. Polar bonds within molecules greatly influence the interactions they have with other polar

TABLE 2.2	Some Electronegativities
ELEMENT	**ELECTRONEGATIVITY**
Oxygen (O)	3.4
Chlorine (Cl)	3.2
Nitrogen (N)	3.0
Carbon (C)	2.6
Phosphorus (P)	2.2
Hydrogen (H)	2.2
Sodium (Na)	0.9
Potassium (K)	0.8

molecules. The polarity of the water molecule has significant effects on its physical properties and chemical reactivity, as we will see in later chapters.

Hydrogen bonds may form within or between molecules with polar covalent bonds

In liquid water, the negatively charged oxygen (δ^-) atom of one water molecule is attracted to the positively charged hydrogen (δ^+) atoms of other water molecules (**FIGURE 2.5A**). The bond resulting from this attraction is called a **hydrogen bond**. These bonds are not restricted to water molecules. A hydrogen bond may also form between a strongly electronegative atom and a hydrogen atom that is covalently bonded to another electronegative atom (oxygen or nitrogen), as shown in **FIGURE 2.5B**.

A hydrogen bond is much weaker than a covalent bond (see Table 2.1). Although individual hydrogen bonds are weak, many of them can form within one molecule or between two molecules. In these cases, the hydrogen bonds together have considerable strength and can greatly influence the structure and properties of the substances. Hydrogen bonds play

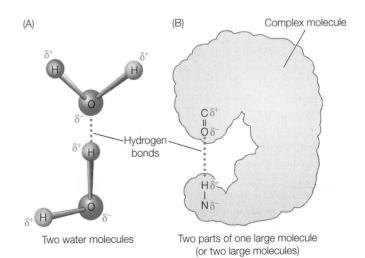

(A) Two water molecules

(B) Two parts of one large molecule (or two large molecules)

Complex molecule

Hydrogen bonds

FIGURE 2.5 Hydrogen Bonds Can Form between or within Molecules (A) A hydrogen bond forms between two molecules because of the attraction between a negatively charged atom on one molecule and a positively charged hydrogen atom on a second molecule. (B) Hydrogen bonds can form between different parts of the same large molecule.

important roles in determining and maintaining the three-dimensional shapes of giant molecules such as DNA and proteins (see Chapter 3). Hydrogen bonding between water molecules also contributes to two properties of water of great significance for living systems: heat capacity and cohesion.

HEAT CAPACITY In liquid water, at any given time, a water molecule forms an average of 3.4 hydrogen bonds (dotted red lines below) with other water molecules:

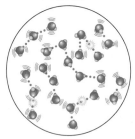

Liquid water

These multiple hydrogen bonds contribute to the high *heat capacity* of water. Raising the temperature of liquid water takes a lot of heat, because much of the heat energy is used to break the hydrogen bonds that hold the liquid together (indicated by the yellow energy bursts above). Think of what happens when you apply heat to a pan of water on the stove: it takes a while for the water to begin boiling. The same happens with an organism—the overwhelming presence of water in living tissues shields them from fluctuations in environmental temperature.

Hydrogen bonding also gives water a high **heat of vaporization**, which means that a lot of heat is required to change water from its liquid to its gaseous state (the process of *evaporation*). Once again, much of the heat energy is used to break the many hydrogen bonds between the water molecules. This heat must be absorbed from the environment in contact with the water. Evaporation thus has a cooling effect on the environment—whether a leaf, a forest, or an entire land mass. This effect explains why sweating cools the human body: as sweat evaporates from the skin, it transforms some of the adjacent body heat.

> **LINK** Evaporation is important in the physiology of both plants and animals; see Concepts 25.3 and 29.4

COHESION The numerous hydrogen bonds that give water a high heat capacity and high heat of vaporization also explain the *cohesive strength* of liquid water. This cohesive strength, or **cohesion**, is defined as the capacity of water molecules to resist coming apart from one another when placed under tension. Water's cohesive strength permits narrow columns of liquid water to move from the roots to the leaves of tall trees. When water evaporates from the leaves, the entire column moves upward in response to the pull of the molecules at the top.

Polar and nonpolar substances: Each interacts best with its own kind

Just as water molecules can interact with one another through hydrogen bonds, any polar molecule can interact with any other

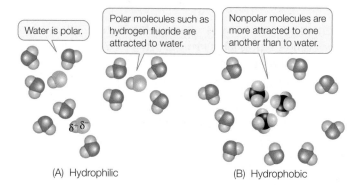

(A) Hydrophilic   (B) Hydrophobic

FIGURE 2.6 Hydrophilic and Hydrophobic (A) Molecules with polar covalent bonds are attracted to polar water (they are hydrophilic). (B) Molecules with nonpolar covalent bonds show greater attraction to one another than to water (they are hydrophobic). The color convention in the models shown here (gray, H; red, O; black, C) is often used.

polar molecule through the weak (δ^+ to δ^-) attractions of hydrogen bonds. Polar molecules interact with water in this way and are called **hydrophilic** ("water-loving"). In aqueous (watery) solutions, these molecules become separated and surrounded by water molecules (**FIGURE 2.6A**).

Nonpolar molecules tend to interact with other nonpolar molecules. For example, molecules containing only hydrogen and carbon atoms—called *hydrocarbon molecules*—are nonpolar. (Compare the electronegativities of hydrogen and carbon in Table 2.2 to see why.) In water these molecules tend to aggregate with one another rather than with the polar water molecules. Therefore, nonpolar molecules are known as **hydrophobic** ("water-hating"), and the interactions between them are called *hydrophobic interactions* (**FIGURE 2.6B**). Hydrophobic substances do not really "hate" water—they can form weak interactions with it, since the electronegativities of carbon and hydrogen are not exactly the same. But these interactions are far weaker than the hydrogen bonds between the water molecules, so the nonpolar substances tend to aggregate.

APPLY THE CONCEPT

Atoms interact and form molecules

The concepts of chemical bonding and electronegativity (see Table 2.2) allow us to predict whether a molecule will be polar or nonpolar, and how it will interact with water. Typically, a difference in electronegativity greater than 0.5 will result in polarity. For each of the bonds below, indicate:

1. Whether the bond is polar or nonpolar
2. If polar, which is the δ^+ end
3. How a molecule with the bond will interact with water (hydrophilic or hydrophobic).

N—H   C—H   C=O   C—N

O—H   C—C   H—H   O—P

Functional groups confer specific properties to biological molecules

Certain small groups of atoms, called **functional groups**, are consistently found together in very different biological molecules. You will encounter several functional groups repeatedly in your study of biology (**FIGURE 2.7**). Each functional group has specific chemical properties, and when attached to a larger molecule, it confers those properties on the larger molecule. One of these properties is polarity. Can you determine which functional groups in Figure 2.7 are the most polar? The consistent chemical behavior of functional groups helps us understand the properties of the molecules that contain them.

Biological molecules often contain many different functional groups. A single large protein may contain hydrophobic, polar, and charged functional groups. Each group gives a different specific property to its local site on the protein, and it may interact with another group on the same protein or with another molecule. Thus, the functional groups determine molecular shape and reactivity.

Large molecules called **macromolecules** are formed by covalent linkages of smaller molecules. Four kinds of macromolecules are characteristic of living things: proteins, carbohydrates, nucleic acids, and lipids.

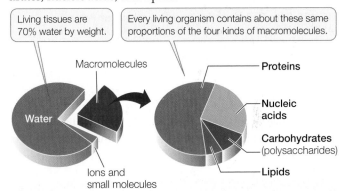

With the exception of lipids, these *biological molecules* are **polymers** (*poly*, "many"; *mer*, "unit") constructed by the covalent bonding of smaller molecules called **monomers**.

- *Proteins* are formed from different combinations of 20 *amino acids*, all of which share chemical similarities.

- *Carbohydrates* can be giant molecules, and are formed by linking together chemically similar sugar monomers (*monosaccharides*) to form *polysaccharides*.

- *Nucleic acids* are formed from four kinds of *nucleotide* monomers linked together in long chains.

- *Lipids* also form large structures from a limited set of smaller molecules, but in this case noncovalent forces maintain the interactions between the lipid monomers.

Polymers are both formed and broken down by a series of reactions involving water (**FIGURE 2.8**):

- In **condensation**, the removal of water links monomers together.

- In **hydrolysis**, the addition of water breaks a polymer into monomers.

Functional group	Class of compounds and an example	Properties
Hydroxyl R—OH	**Alcohols** Ethanol	Polar. Hydrogen bonds with water to help dissolve molecules. Enables linkage to other molecules by condensation.
Aldehyde	**Aldehydes** Acetaldehyde	C=O group is very reactive. Important in building molecules and in energy-releasing reactions.
Keto	**Ketones** Acetone	C=O group is important in carbohydrates and in energy reactions.
Carboxyl	**Carboxylic acids** Acetate	Acidic. Ionizes in living tissues to form —COO⁻ and H⁺. Enters into condensation reactions by giving up —OH. Some carboxylic acids important in energy-releasing reactions.
Amino	**Amines** Methylamine	Basic. Accepts H⁺ in living tissues to form —NH₃⁺. Enters into condensation reactions by giving up H⁺.
Phosphate	**Organic phosphates** 3-Phosphoglycerate	Negatively charged. Enters into condensation reactions by giving up —OH. When bonded to another phosphate, hydrolysis releases much energy.
Sulfhydryl	**Thiols** Mercaptoethanol	By giving up H, two —SH groups can react to form a disulfide bridge (S—S), thus stabilizing protein structure.

FIGURE 2.7 Functional Groups Important to Living Systems Highlighted in yellow are the seven functional groups most commonly found in biological molecules. "R" is a variable chemical grouping.

(A) Condensation

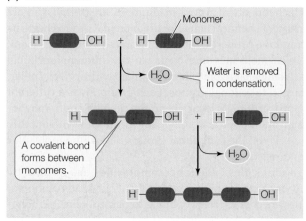

(B) Hydrolysis

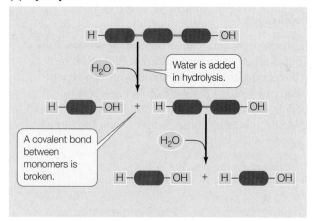

FIGURE 2.8 Condensation and Hydrolysis of Polymers
(A) Condensation reactions link monomers into polymers and produce water. (B) Hydrolysis reactions break polymers into individual monomers and consume water.

How the macromolecules function and interact with other molecules depends on the properties of the functional groups in their monomers.

---

yourBioPortal.com

Go to ANIMATED TUTORIAL 2.2
Macromolecules: Carbohydrates and Lipids

Do You Understand Concept 2.2?

- Compare electron behavior in ionic, covalent, and hydrogen bonds. Which is strongest, and why?
- How do variations in electronegativity result in the unequal sharing of electrons in polar molecules?
- Consider the molecule carbon dioxide (CO_2). Are the bonds between the C and the O atoms ionic or covalent? Is this molecule hydrophobic or hydrophilic? Explain your answers.
- Here is the structure of the molecule glycine:

  a. Is this molecule hydrophilic or hydrophobic? Explain.
  b. Draw two glycine molecules and show how they can be linked by a condensation reaction.

We will begin our discussion of the molecules of life with carbohydrates, as they exemplify many of the chemical principles we have outlined so far.

---

concept
2.3 Carbohydrates Consist of Sugar Molecules

Carbohydrates are a large group of molecules that all have similar atomic compositions but differ greatly in size, chemical properties, and biological functions. Carbohydrates have the general formula $C_n(H_2O)_n$, which makes them appear to be hydrates of carbon—associations between water molecules and carbon; hence their name. However, when their molecular structures are examined, one sees that the carbon atoms are actually bonded with hydrogen atoms (—H) and hydroxyl groups (—OH), rather than with intact water molecules. Carbohydrates have four major biochemical roles:

- They are a source of stored energy that can be released in a form usable by organisms.
- They are used to transport stored energy within complex organisms.
- They function as structural molecules that give many organisms their shapes.
- They serve as recognition or signaling molecules that can trigger specific biological responses.

Some carbohydrates are relatively small, such as the simple sugars (for example, glucose) that are the primary energy source for many organisms. Others are large polymers of simple sugars, such as starch, which is stored in seeds.

Monosaccharides are simple sugars

Monosaccharides (*mono*, "one") are relatively simple molecules with up to seven carbon atoms. They differ in their arrangements of carbon, hydrogen, and oxygen atoms (**FIGURE 2.9**).

Pentoses (*pente*, "five") are five-carbon sugars. Two pentoses are of particular biological importance: the backbones of the nucleic acids RNA and DNA contain ribose and deoxyribose, respectively.

LINK For a description of the nucleic acids RNA and DNA see Concept 3.1

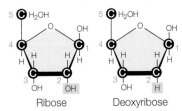

Ribose

Deoxyribose

Ribose and deoxyribose each have five carbons, but very different chemical properties and biological roles.

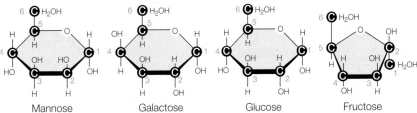

Mannose

Galactose

Glucose

Fructose

These hexoses all have the formula $C_6H_{12}O_6$, but each has distinct biochemical properties.

FIGURE 2.9 Monosaccharides Monosaccharides are made up of varying numbers of carbons. Many have the same kind and number of atoms, but the atoms are arranged differently.

The *hexoses* (*hex*, "six") all have the formula $C_6H_{12}O_6$. They include glucose, fructose (so named because it was first found in fruits), mannose, and galactose.

---

yourBioPortal.com

**Go to WEB ACTIVITY 2.2
Forms of Glucose**

Glycosidic linkages bond monosaccharides

The *disaccharides*, *oligosaccharides*, and *polysaccharides* are all constructed from monosaccharides that are covalently bonded by condensation reactions that form **glycosidic linkages**. A single glycosidic linkage between two monosaccharides forms a **disaccharide**. For example, sucrose—common table sugar—is a major disaccharide formed in plants from a glucose and a fructose:

Glucose   Fructose   Formation of linkage   Glucose   Fructose

Sucrose

Another disaccharide is maltose, formed from two glucose units, which is a product of starch digestion (and an important carbohydrate for making beer).

Oligosaccharides contain several monosaccharides bound together by glycosidic linkages. Many oligosaccharides have additional functional groups, which give them special properties. Oligosaccharides are often covalently bonded to proteins and lipids on the outer surfaces of cells, where they serve as recognition signals. For example, the different human blood groups (the ABO blood types) get their specificity from oligosaccharide chains.

Polysaccharides store energy and provide structural materials

Polysaccharides are large polymers of monosaccharides connected by glycosidic linkages (**FIGURE 2.10**). Polysaccharides are not necessarily linear chains of monomers. Each monomer unit has several sites that are capable of forming glycosidic linkages, and thus branched molecules are possible.

Starches comprise a family of giant molecules that are all polysaccharides of glucose. The different starches can be distin-

guished by the amount of branching in their polymers. Starch is the principal energy storage compound of plants.

Glycogen is a water-insoluble, highly branched polymer of glucose that is the major energy storage molecule in mammals. It is produced in the liver and transported to the muscles. Both glycogen and starch are readily hydrolyzed into glucose monomers, which in turn can be broken down to liberate their stored energy.

If glucose is the major source of fuel, why store it in the form of starch or glycogen? The reason is that 1,000 glucose molecules would exert 1,000 times the *osmotic pressure* of a single glycogen molecule, causing water to enter the cells (see Concept 5.2). If it were not for polysaccharides, many organisms would expend a lot of energy expelling excess water from their cells.

As the predominant component of plant cell walls, *cellulose* is by far the most abundant carbon-containing (*organic*) biological compound on Earth. Like starch and glycogen, cellulose is a polysaccharide of glucose, but its glycosidic linkages are arranged in such a way that it is a much more stable molecule. Whereas starch is easily broken down by chemicals or enzymes to supply glucose for energy-producing reactions, cellulose is an excellent structural material that can withstand harsh environmental conditions without substantial change.

FRONTIERS Cellulose is the most abundant carbon-based material in the living world and is attractive as a source of biofuels, which are plant-derived alternatives to petroleum. It is a significant challenge, however, to find ways to efficiently break down this very stable molecule into simpler fuel molecules.

Do You Understand Concept 2.3?

- Draw the chemical structure of a disaccharide formed by two glucose monosaccharides.
- Examine the glucose molecule shown in Figure 2.9. Identify the functional groups on the molecule.
- Can you see where a large number of hydrogen bonding groups are present in the linear structure of cellulose (see Figure 2.10)? Why is this structure so strong?
- Some sugars have other functional groups in addition to those typically present. Draw the structure of the amino sugar glucosamine, which has an amino group bonded at carbon #2 of glucose. Would this molecule be more or less polar than glucose? Explain why.

(A) Molecular structure

Cellulose

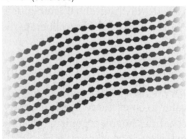

Hydrogen bonding to other cellulose molecules can occur at these points.

Cellulose is an unbranched polymer of glucose with linkages that are chemically very stable.

Starch and glycogen

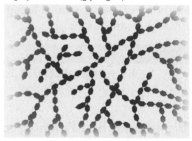

Branching occurs here.

Glycogen and starch are polymers of glucose, with branching at carbon 6 (see Figure 2.9).

(B) Macromolecular structure

Linear (cellulose)

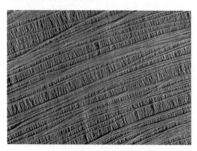

Parallel cellulose molecules form hydrogen bonds, resulting in thin fibrils.

Branched (starch)

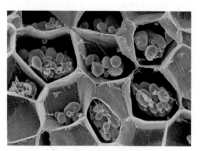

Branching limits the number of hydrogen bonds that can form in starch molecules, making starch less compact than cellulose.

Highly branched (glycogen)

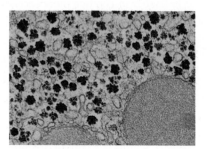

The high amount of branching in glycogen makes its solid deposits more compact than starch.

(C) Polysaccharides in cells

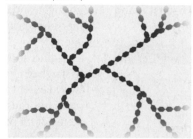

Layers of cellulose fibrils, as seen in this scanning electron micrograph, give plant cell walls great strength.

Within these potato cells, starch deposits (colored purple in this scanning electron micrograph) have a granular shape.

The dark clumps in this electron micrograph are glycogen deposits in a monkey liver cell.

FIGURE 2.10 Polysaccharides Cellulose, starch, and glycogen are all composed of long chains of glucose but with different levels of branching and compaction.

We have seen that carbohydrates are examples of the monomer–polymer theme in biology. Now we will turn to lipids, which are unusual among the four classes of biological macromolecules in that they are not, strictly speaking, polymers.

<table>
<tr><td>concept
2.4</td><td>**Lipids Are Hydrophobic Molecules**</td></tr>
</table>

Lipids—colloquially called *fats*—are hydrocarbons (composed of C and H atoms) that are insoluble in water because of their many nonpolar covalent bonds. As you have seen, nonpolar molecules are hydrophobic and preferentially aggregate together, away from polar water (see Figure 2.6). When nonpolar hydrocarbons are sufficiently close together, weak but additive *van der Waals interactions* (see Table 2.1) hold them together. The huge macromolecular aggregations that can form are not polymers in a strict chemical sense, because the individual lipid molecules are not covalently bonded.

Lipids play several roles in living organisms, including the following:

- They store energy in the C—C and C—H bonds.

- They play important structural roles in cell membranes and on body surfaces, largely because their nonpolar nature makes them essentially insoluble in water.

- Fat in animal bodies serves as thermal insulation.

Fats and oils are triglycerides

The most common units of lipids are **triglycerides**, also known as *simple lipids*. Triglycerides that are solid at room temperature (around 20°C) are called *fats*; those that are liquid at room temperature are called *oils*. A triglyceride contains three *fatty acid* molecules and one *glycerol* molecule. **Glycerol** is a small molecule with three hydroxyl (—OH) groups; thus it is an alcohol. A **fatty acid** consists of a long nonpolar hydrocarbon chain attached to the polar carboxyl (—COOH) group, and it is therefore a carboxylic acid. The long hydrocarbon chain is very hydrophobic because of its abundant C—H and C—C bonds.

Synthesis of a triglyceride involves three condensation reactions (**FIGURE 2.11**). The resulting molecule has very little polarity and is extremely hydrophobic. That is why fats and oils do not mix with water but float on top of it in separate globules or layers. The three fatty acids in a single triglyceride molecule need not all have the same hydrocarbon chain length or structure; some may be saturated fatty acids, while others may be unsaturated:

- In a **saturated fatty acid**, all the bonds between the carbon atoms in the hydrocarbon chain are single; there are no double bonds. That is, all the available bonds are saturated with hydrogen atoms (**FIGURE 2.12A**). These fatty acid molecules are relatively rigid and straight, and they pack together tightly, like pencils in a box.

- In an **unsaturated fatty acid**, the hydrocarbon chain contains one or more double bonds. Linoleic acid is an example of a *polyunsaturated* fatty acid that has two double bonds near the middle of the hydrocarbon chain, causing kinks in the chain (**FIGURE 2.12B**). Such kinks prevent the unsaturated molecules from packing together tightly.

The kinks in fatty acid molecules are important in determining the fluidity and melting point of the lipid. The triglycerides of animal fats tend to have many long-chain saturated fatty acids, which pack tightly together; these fats are usually solid at room temperature and have a high melting point. The triglycerides of plants, such as corn oil, tend to have short or unsaturated fatty acids. Because of their kinks, these fatty acids pack together poorly, have a low melting point, and are usually liquid at room temperature.

FRONTIERS   Plant oils can be artificially hydrogenated to make the fatty acids saturated and the lipids less fluid—desirable qualities for cooking certain foods. However, the process also causes double bonds in the "trans" configuration as a side effect: the resulting trans fats have straight-chain, unsaturated fatty acids that for reasons not fully understood lead to coronary artery blockage and heart attacks. While the food industry is racing to improve the hydrogenation process or change formulations to avoid trans fats, many restaurants and cities have banned food containing them as a public health measure.

Fats and oils are excellent storehouses for chemical energy. As you will see in Chapter 6, when the C—H bond is broken, it releases energy that an organism can use for other purposes, such as movement or to build up complex molecules. On a per weight basis, broken-down lipids yield more than twice as much energy as degraded carbohydrates.

Phospholipids form biological membranes

We have mentioned the hydrophobic nature of the many C—C and C—H bonds in a fatty acid. But what about the carboxyl functional group at the end of the molecule? When it ionizes and forms COO⁻, it is strongly hydrophilic. So a fatty acid is a molecule with a hydrophilic end and a long hydrophobic tail. It has two opposing chemical properties; the technical term for this is **amphipathic**.

In triglycerides, a glycerol molecule is bonded to three fatty acid chains and the resulting molecule is entirely hydrophobic. **Phospholipids** are like triglycerides in that they contain fatty acids bound to glycerol. However, in phospholipids, a phosphate-containing compound replaces one of the fatty acids, giving these molecules amphipathic properties (**FIGURE 2.13A**). The phosphate functional group (there are several different kinds in different phospholipids) has a negative electric charge, so this portion of the molecule is hydrophilic, attracting polar water molecules. But the two fatty acids are

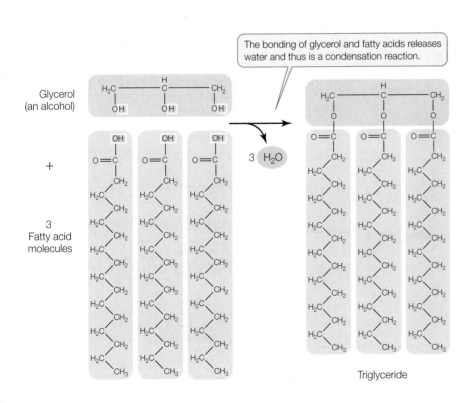

Glycerol (an alcohol)

3 Fatty acid molecules

The bonding of glycerol and fatty acids releases water and thus is a condensation reaction.

3 H₂O

Triglyceride

FIGURE 2.11   Synthesis of a Triglyceride   In living things, the reaction that forms a triglyceride is more complex than the single step shown here.

(A) Palmitic acid

(B) Linoleic acid

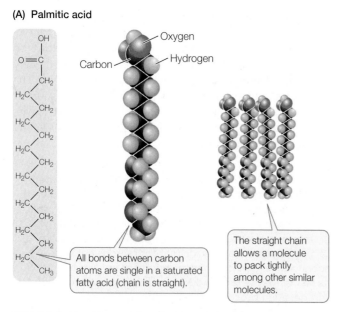

Oxygen

Carbon

Hydrogen

All bonds between carbon atoms are single in a saturated fatty acid (chain is straight).

The straight chain allows a molecule to pack tightly among other similar molecules.

Kinks prevent close packing.

Double bonds between two carbons make an unsaturated fatty acid (carbon chain has kinks).

FIGURE 2.12 Saturated and Unsaturated Fatty Acids (A) The straight hydrocarbon chain of a saturated fatty acid allows the molecule to pack tightly with other, similar molecules. (B) In unsaturated fatty acids, kinks in the chain prevent close packing.

hydrophobic, so they tend to avoid water and aggregate together or with other hydrophobic substances.

In an aqueous environment, phospholipids line up in such a way that the nonpolar, hydrophobic "tails" pack tightly together and the phosphate-containing "heads" face outward, where they interact with water. The phospholipids thus form a **bilayer**: a sheet two molecules thick, with water excluded from the core (**FIGURE 2.13B**). Although no covalent bonds link individual lipids in these large aggregations, such stable aggregations form readily in aqueous conditions. Biological membranes have this kind of **phospholipid bilayer** structure, and we will devote Chapter 5 to their biological functions.

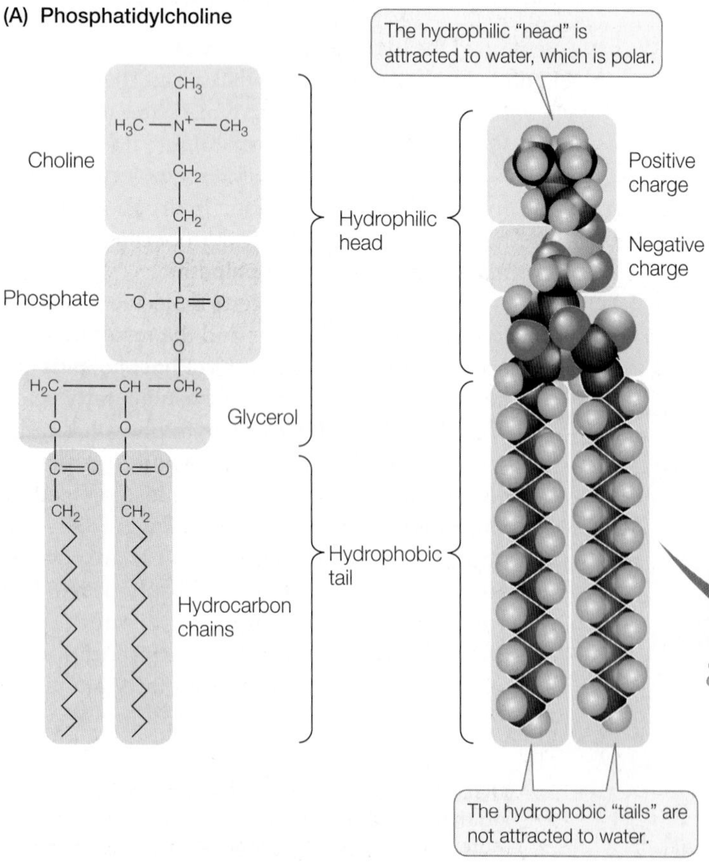

(A) Phosphatidylcholine

Choline

Phosphate

Glycerol

Hydrocarbon chains

The hydrophilic "head" is attracted to water, which is polar.

Hydrophilic head

Positive charge

Negative charge

Hydrophobic tail

The hydrophobic "tails" are not attracted to water.

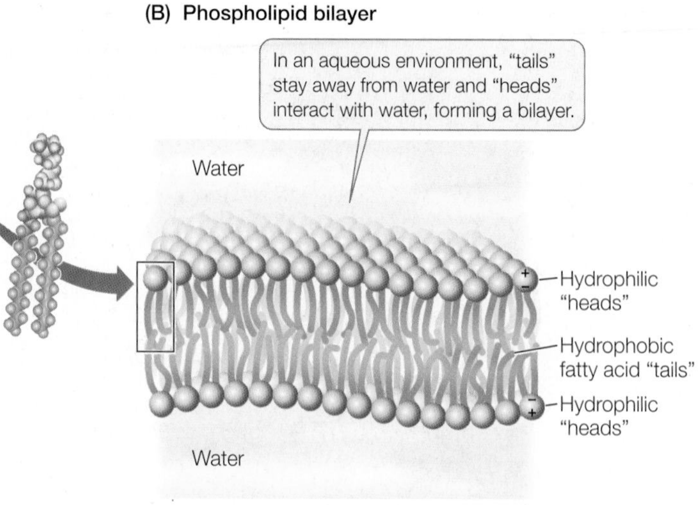

FIGURE 2.13 Phospholipids (A) Phosphatidylcholine (lecithin) is an example of a phospholipid molecule. In other phospholipids, the amino acid serine, the sugar alcohol inositol, or another compound replaces choline. (B) In an aqueous environment, hydrophobic interactions bring the "tails" of phospholipids together in the interior of a bilayer. The hydrophilic "heads" face outward on both sides of the bilayer, where they interact with the surrounding water molecules.

(B) Phospholipid bilayer

In an aqueous environment, "tails" stay away from water and "heads" interact with water, forming a bilayer.

Water

Hydrophilic "heads"

Hydrophobic fatty acid "tails"

Hydrophilic "heads"

Water

Do You Understand Concept 2.4?

- What is the difference between fats and oils?
- Why are phospholipids amphipathic, and how does this result in a lipid bilayer membrane?
- If fatty acids are carefully put onto the surface of water, they form a single molecular layer. If the mixture is then shaken vigorously, the fatty acids will form hollow, round structures called micelles. Explain these observations.

Molecules such as carbohydrates and lipids are not always stable in living systems. Rather, a hallmark of life is its *ability to transform molecules*. This involves making and breaking covalent bonds, as atoms are removed and others are attached. As part of our introduction to biochemical concepts, we will now turn to these processes of chemical change.

concept 2.5 Biochemical Changes Involve Energy

A **chemical reaction** occurs when atoms have sufficient energy to combine, or change their bonding partners. Consider the hydrolysis of the disaccharide sucrose to its component monomers, glucose and fructose (see p. 25 for the chemical structures). We can express this reaction by a chemical equation:

$$\text{sucrose} + \text{H}_2\text{O} \rightarrow \text{glucose} + \text{fructose}$$
$$(\text{C}_{12}\text{H}_{22}\text{O}_{11}) \qquad (\text{C}_6\text{H}_{12}\text{O}_6) \; (\text{C}_6\text{H}_{12}\text{O}_6)$$

In this equation, sucrose and water are the **reactants**, and glucose and fructose are the **products**. Electrons and protons are transferred from the reactants to the products. The products of this reaction have very different properties from the reactants. The sum total of all the chemical reactions occurring in a biological system at a given time is called **metabolism**. Metabolic reactions involve energy changes; for example, the energy contained in the chemical bonds of sucrose (reactants) is greater than the energy in the bonds of the two products, glucose and fructose.

What is energy? Physicists define it as the capacity to do work, which occurs when a force operates on an object over a distance. In biochemistry, it is more useful to consider energy as *the capacity for change*. In biochemical reactions, energy changes are usually associated with changes in the chemical composition and properties of molecules.

There are two basic types of energy

Energy comes in many forms: chemical, electrical, heat, light, and mechanical. But all forms of energy can be considered as one of two basic types:

- *Potential energy* is the energy of state or position—that is, stored energy. It can be stored in many forms: in chemical bonds, as a concentration gradient, or even as an electric charge imbalance.
- *Kinetic energy* is the energy of movement—that is, the type of energy that does work, that makes things change. For example, heat causes molecular motions and can even break chemical bonds.

Potential energy can be converted into kinetic energy and vice versa, and the form that the energy takes can also be converted. Think of reading this book: light energy is converted to chemical energy in your eyes, and then is converted to electrical energy in the nerve cells that carry messages to your brain. When you decide to turn a page, the electrical and chemical energy of nerves and muscles are converted to kinetic energy for movement of your hand and arm.

There are two basic types of metabolism

Energy changes in living systems usually occur as chemical changes, in which energy is stored in or released from chemical bonds.

Anabolic reactions (collectively *anabolism*) link simple molecules to form more complex molecules (for example, the synthesis of sucrose from glucose and fructose). Anabolic reactions require an input of energy—chemists call them *endergonic* or *endothermic* reactions (**FIGURE 2.14A**)—and capture the energy in the chemical bonds that are formed (for example, the glycosidic bond between the two monosaccharides).

Catabolic reactions (collectively *catabolism*) break down complex molecules into simpler ones and release the energy stored in the chemical bonds. Chemists call such reactions *exergonic* or *exothermic* (**FIGURE 2.14B**). For example, when sucrose is hydrolyzed, energy is released.

Catabolic and anabolic reactions are often linked. The energy released in catabolic reactions is often used to drive anabolic reactions—that is, to do biological work. For example, the energy released by the breakdown of glucose (catabolism) is used to drive anabolic reactions such as the synthesis of triglycerides. That is why fat accumulates if you eat food in excess of your energy needs.

Biochemical changes obey physical laws

Recall from the opening of this chapter that we described the mechanistic view of life, whereby living systems obey the same rules that govern the nonliving world. The **laws of thermodynamics** (thermo, "energy"; dynamics, "change") were derived from studies of the fundamental properties of energy, and the ways energy interacts with matter. These laws apply to all matter and all energy transformations in the universe. Their application to living systems helps us understand how organisms and cells harvest and transform energy to sustain life.

The first law of thermodynamics: Energy is neither created nor destroyed. The first law of thermodynamics states that in any conversion, energy is neither created nor destroyed. Another way of stating this is that the total energy before and after an energy conversion is the same (**FIGURE 2.15A**). [Similarly,

(A) Endergonic reaction

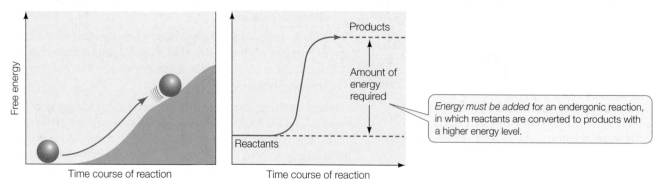

Energy must be added for an endergonic reaction, in which reactants are converted to products with a higher energy level.

(B) Exergonic reaction

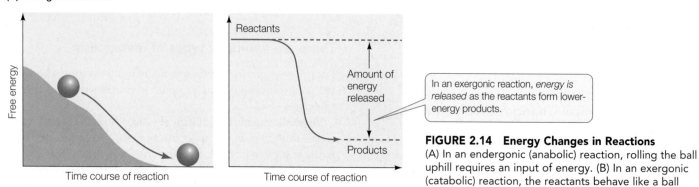

In an exergonic reaction, energy is released as the reactants form lower-energy products.

FIGURE 2.14 Energy Changes in Reactions
(A) In an endergonic (anabolic) reaction, rolling the ball uphill requires an input of energy. (B) In an exergonic (catabolic) reaction, the reactants behave like a ball rolling down a hill, and energy is released.

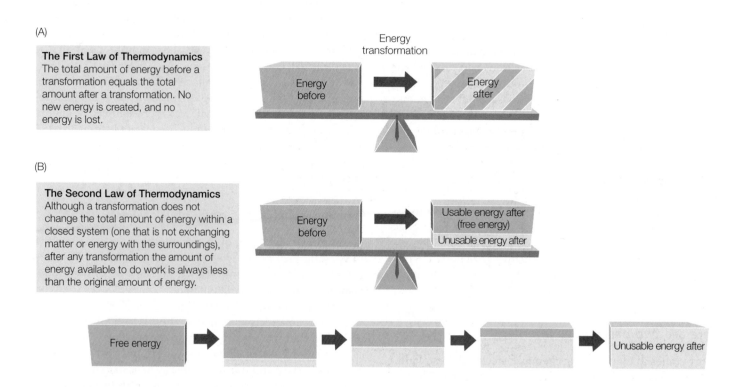

(A)

The First Law of Thermodynamics
The total amount of energy before a transformation equals the total amount after a transformation. No new energy is created, and no energy is lost.

Energy transformation

Energy before

Energy after

(B)

The Second Law of Thermodynamics
Although a transformation does not change the total amount of energy within a closed system (one that is not exchanging matter or energy with the surroundings), after any transformation the amount of energy available to do work is always less than the original amount of energy.

Energy before

Usable energy after (free energy)
Unusable energy after

Free energy

Unusable energy after

Another statement of the second law is that in a closed system, with repeated energy transformations, free energy decreases and unusable energy (disorder) increases—a phenomenon known as the increase in **entropy**.

FIGURE 2.15 The Laws of Thermodynamics (A) The first law states that energy cannot be created or destroyed. (B) The second law states that after energy transformations, some energy becomes unavailable to do work.

APPLY THE CONCEPT

Biochemical changes involve energy

Chemical reactions in living systems involve changes in energy. These can be expressed as changes in available energy, called free energy (designated G, for Gibbs—the scientist who first described this parameter). The overall direction of a spontaneous chemical reaction is from higher to lower free energy. In other words, if the $G_{reactants}$ is greater than the $G_{products}$ (negative ΔG), the reaction will be spontaneous; it will tend to go in the direction from reactants to products, and release free energy in the process. Reactions where the $G_{reactants}$ is less than the $G_{products}$ (positive ΔG) will occur only if additional free energy is supplied.

REACTION	REACTANTS		PRODUCTS	ΔG
Hydrolysis of sucrose:	sucrose + H_2O	→	glucose + fructose	7.0
Triglyceride attachment:	glycerol + fatty acid	→	monoglyceride	3.5
Photosynthesis:	6 CO_2 + 6 H_2O	→	glucose + 6 O_2	686

The table shows some reactions and the absolute values of their associated free energy changes (ΔG).

1. For each reaction, would you expect ΔG to be positive or negative?

2. Which reactions will be spontaneous? Explain your answers.

matter is also conserved: in the hydrolysis of sucrose (see p. 29), there are 12 carbons, 24 hydrogens, and 12 oxygens on both sides of the equation.]

Although the total amount of energy is conserved, chemical reactions involve changes in the amount of (potential) energy stored in chemical bonds. If energy is released during the reaction, it is available to do work—for example, to drive another chemical reaction. In general, reactions that release energy (catabolic, or exergonic reactions) can occur spontaneously.

The second law of thermodynamics: Disorder tends to increase. Although energy cannot be created or destroyed, the second law of thermodynamics implies that when energy is converted from one form to another, some of that energy becomes unavailable for doing work (**FIGURE 2.15B**). In other words, no physical process or chemical reaction is 100 percent efficient; some of the released energy is lost in a form associated with disorder. Think of disorder as a kind of randomness caused by the thermal motion of particles; this energy is so dispersed that it is unusable. Entropy is a measure of the disorder in a system.

If a chemical reaction increases entropy, its products are more disordered or random than its reactants. The disorder in a solution of glucose and fructose is greater than that in a solution of sucrose, where the glycosidic bond between the two monosaccharides prevents free movement. Conversely, if there are fewer products and they are more restrained in their movements than the reactants, the disorder is reduced. But this requires an energy input to achieve.

The second law of thermodynamics predicts that, as a result of energy transformations, disorder tends to increase; some energy is always lost to random thermal motion (entropy). Chemical changes, physical changes, and biological processes all tend to increase entropy (see Figure 2.15B), and this tendency gives direction to these processes. Changes in entropy are mathematically related to changes in free energy, and thus the second law helps to explain why some reactions proceed in one direction rather than another.

How does the second law of thermodynamics apply to organisms? Consider the human body, with its highly organized tissues and organs composed of large, complex molecules. This level of complexity appears to be in conflict with the second law, but for two reasons, it is not. First, the construction of complex molecules also generates disorder. The anabolic reactions needed to construct 1 kg of an animal body require the catabolism of about 10 kg of food. So metabolism creates far more disorder (more energy is lost to entropy) than the amount of order stored in flesh. Second, life requires a constant input of energy to maintain order. Without this energy, the complex structures of living systems would break down. Because energy is used to generate and maintain order, there is no conflict with the second law of thermodynamics.

Do You Understand Concept 2.5?

- Describe the forms of energy and changes involved in reading this book.

- What is the difference between potential energy and kinetic energy? Between anabolism and catabolism? Between endergonic and exergonic reactions?

- Predict whether these situations are endergonic or exergonic and explain your reasoning:
  a. The formation of a lipid bilayer membrane
  b. Turning on a TV set

 QA QUESTION Why is the search for water important in the search for life?

ANSWER You have seen throughout this chapter that water is essential for the chemistry of life. Water is composed of two of the most abundant elements (Concept 2.1), and it is a polar molecule (Concept 2.2). This allows biologically important polar molecules such as monosaccharides (Concept 2.3) to dissolve in water. Because of their hydrophobicity, lipids interact with water to form important biological structures (Concept 2.4). Water molecules participate directly in the formation and breakdown of polymers (Concept 2.2). In short, all of the processes of life as we know it require water.

In the opening essay of this chapter, we described recent evidence for the presence of water on other bodies in our solar system. Could this water harbor life, now or in the past? One way to investigate this possibility is to study how life on Earth may have originated in an aqueous environment. Geological evidence suggests that Earth was formed about 4.5 billion years ago, and that life arose about 3.8 billion years ago. During the time when life originated, there was apparently little oxygen gas (O_2) in the atmosphere. In the 1950s, Stanley Miller and Harold Urey at the University of Chicago set up an experimental "atmosphere" containing various gases thought to be present in Earth's early atmosphere. Among them were ammonia (NH_3), hydrogen (H_2), methane (CH_4), and (importantly) water vapor (H_2O). Miller and Urey passed an electric spark over the mixture to simulate lightning, providing a source of energy for covalent bond formation. Then they cooled the system so the gases would condense and collect in a watery solution, or "ocean" (**FIGURE 2.16**). Note that water was essential for this experiment as a source of oxygen atoms.

After several days of continuous operation, the system contained numerous complex molecules, including amino acids, nucleotides, and sugars—the building blocks of life. In later experiments the researchers added other gases, such as carbon dioxide (CO_2), nitrogen (N_2), and sulfur dioxide (SO_2). This resulted in the formation of functional groups such as carboxylic acids, fatty acids, and pentose sugars. Taken together, these data suggest a plausible mechanism for the formation of life's chemicals in the aqueous environment of early Earth.

yourBioPortal.com

Go to ANIMATED TUTORIAL 2.3 Synthesis of Prebiotic Molecules

INVESTIGATION

FIGURE 2.16 Synthesis of Prebiotic Molecules in an Experimental Atmosphere With an increased understanding of the atmospheric conditions that existed on primitive Earth, the researchers devised an experiment to see if these conditions could lead to the formation of organic molecules.

HYPOTHESIS

Organic chemical compounds can be generated under conditions similar to those that existed in the atmosphere of primitive Earth.

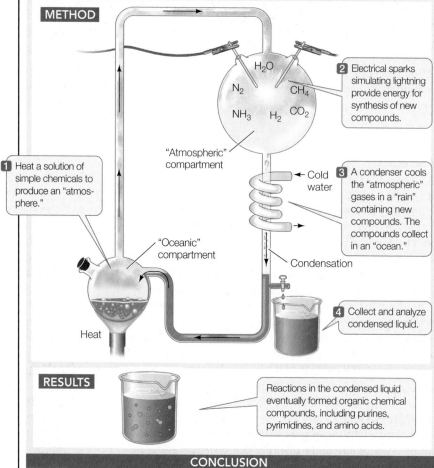

METHOD

H_2O
N_2 CH_4
NH_3 H_2 CO_2

2 Electrical sparks simulating lightning provide energy for synthesis of new compounds.

"Atmospheric" compartment

1 Heat a solution of simple chemicals to produce an "atmosphere."

Cold water

3 A condenser cools the "atmospheric" gases in a "rain" containing new compounds. The compounds collect in an "ocean."

"Oceanic" compartment

Condensation

Heat

4 Collect and analyze condensed liquid.

RESULTS

Reactions in the condensed liquid eventually formed organic chemical compounds, including purines, pyrimidines, and amino acids.

CONCLUSION

The chemical building blocks of life could have been generated in the probable atmosphere of early Earth.

ANALYZE THE DATA

The following data show the amount of energy impinging on Earth in different forms.

Source	Energy (cal cm^{-2} yr^{-1})
Total radiation from sun	260,000
Ultraviolet light	
Wavelength <250 nm	570
Wavelength <200 nm	85
Wavelength <150 nm	3.5
Electric discharges	4
Cosmic rays	0.0015
Radioactivity	0.8
Volcanoes	0.13

A. Only a small fraction of the sun's energy is ultraviolet light (less than 250 nm). What is the rest of the solar energy?

B. The molecules CH_4, H_2O, NH_3, and CO_2 absorb light at wavelengths less than 200 nm. What fraction of total solar radiation is in this range?

C. Instead of electric discharges, what other sources of energy could be used in these experiments?

For more, go to Working with Data 2.1 at **yourBioPortal.com**.

Go to **yourBioPortal.com** for original citations, discussions, and relevant links for all INVESTIGATION figures.

SUMMARY

concept 2.1 Atomic Structure Is the Basis for Life's Chemistry

- Matter is composed of atoms. Each **atom** consists of a positively charged **nucleus** made up of **protons** and **neutrons**, surrounded by **electrons** bearing negative charges.

- The number of protons in the nucleus defines an **element**. There are many elements in the universe, but only a few of them (C, H, O, P, N, and S) make up the bulk of living organisms.

- Electrons are distributed in **electron shells** at varying energy levels away from the nucleus. The first shell has a maximum of two electrons, and subsequent shells have maxima of eight electrons. **Review Figure 2.1**

concept 2.2 Atoms Interact and Form Molecules

- A **chemical bond** is an attractive force that links two atoms together in a molecule. **Review ANIMATED TUTORIAL 2.1**

- **Ions** are electrically charged bodies that form when atoms gain or lose one or more electrons in order to form more stable electron configurations. **Anions** and **cations** are negatively and positively charged ions, respectively. **Ionic bonds** form when ions with opposite charges attract. **Review Figure 2.2**

- A **covalent bond** is a strong bond formed when two atoms share one or more pairs of electrons. **Review Figures 2.3 and 2.4**

- When two atoms of unequal electronegativity bond with each other, a **polar covalent bond** is formed. The two ends, or poles, of the bond have partial charges (δ^+ or δ^-). A **hydrogen bond** is a weak electrical attraction that forms between a δ^+ hydrogen atom in one molecule and a δ^- atom in another molecule (or in another part of a large molecule). Hydrogen bonds are abundant in water. **Review Figure 2.5**

- **Functional groups** are covalently bonded groups of atoms that confer specific properties on biological molecules.

- **Macromolecules** are formed via condensation reactions that link together monomers containing particular functional groups. **Review Figure 2.7, ANIMATED TUTORIAL 2.2, and WEB ACTIVITY 2.1**

concept 2.3 Carbohydrates Consist of Sugar Molecules

- **Carbohydrates** contain carbon bonded to hydrogen and oxygen.

- **Monosaccharides** include pentoses (with five carbons) and hexoses (with six carbons). **Review Figure 2.9 and WEB ACTIVITY 2.2**

- Glycosidic linkages are covalent bonds between saccharides. **Disaccharides** such as sucrose each contain two monosaccharides, whereas **polysaccharides** such as starch and cellulose contain long chains of monomers. **Review Figure 2.10**

concept 2.4 Lipids Are Hydrophobic Molecules

- Fats and oils are **triglycerides**, composed of three **fatty acids** covalently linked to glycerol. **Review Figure 2.11**

- **Saturated fatty acids** have hydrocarbon chains with no double bonds. **Unsaturated fatty acids** contain double bonds in their hydrocarbon chains. **Review Figure 2.12**

- **Phospholipids** contain two fatty acids and a hydrophilic, phosphate-containing polar group attached to glycerol. They are **amphipathic**, with both polar and nonpolar ends. They form into a structural bilayer in water. **Review Figure 2.13**

concept 2.5 Biochemical Changes Involve Energy

- A **chemical reaction** occurs when atoms have sufficient energy to combine or change their bonding partners.

- **Anabolic reactions** are endergonic and require energy. **Catabolic reactions** release energy and are exergonic. **Review Figure 2.14**

- The **laws of thermodynamics** govern biochemical reactions. The first law states that in any transformation, energy is neither created nor destroyed. The second law states that disorder tends to increase. **Review Figure 2.15**

See ANIMATED TUTORIAL 2.3 and WORKING WITH DATA 2.1

Nucleic Acids, Proteins, and Enzymes

Despite suffering from the "ague," the Reverend Edward Stone went walking in the English countryside. Feverish, tired, with aching muscles and joints, he came across a willow tree. Although apparently unaware that many ancient healers used willow bark extracts to reduce fever, the clergyman knew of the tradition of natural remedies for various diseases. The willow reminded him of the bitter extracts from the bark of South American trees then being sold (at high prices) to treat fevers. Removing some willow bark, Stone sucked on it and found it did indeed taste bitter—and that it relieved his symptoms.

Later he gathered a pound of willow bark and ground it into a powder, which he gave to about 50 people who complained of pain; all said they felt better. Stone reported the results of this "clinical test" in a letter to the Royal Society, England's most respected scientific body. Stone had discovered salicylic acid, the basis of the most widely used drug in the world. The date of his letter (which still exists) was April 25, 1763.

The chemical structure of salicylic acid (named for *Salix*, the willow genus) was worked out about 70 years later, and soon chemists could synthesize it in the laboratory. Although the compound alleviated pain, its acidity irritated the digestive system. In the late 1890s, the German chemical company Bayer synthesized a milder yet equally effective form, acetylsalicylic acid, which it marketed as aspirin. The new medicine's success launched Bayer to world prominence as a pharmaceutical company, a position it maintains today.

In the 1960s and 1970s, aspirin use declined when two alternative medications, acetaminophen (Tylenol®) and ibuprofen (Motrin® and Advil®), became widely available. But over this same time, clinical studies revealed a new use for aspirin: it is an effective anticoagulant, shown to prevent heart attacks and strokes caused by blood clots. Today many people take a daily low dose of aspirin as a preventive against clotting disorders.

Fever, joint pain, headache, blood clots. What do these symptoms have in common? They all are mediated by fatty acid products called prostaglandins and molecules derived from them. Salicylic acid blocks the synthesis of the primary prostaglandin. The exact biochemical mechanism by which aspirin works was described in 1971, and as we will see, it requires understanding protein and enzyme function—two subjects of this chapter.

The bark of the willow tree (*Salix alba*) was the original source of salicylic acid, later modified to aspirin.

KEY CONCEPTS

3.1 Nucleic Acids Are Informational Macromolecules

3.2 Proteins Are Polymers with Important Structural and Metabolic Roles

3.3 Some Proteins Act as Enzymes to Speed up Biochemical Reactions

3.4 Regulation of Metabolism Occurs by Regulation of Enzymes

How does an understanding of proteins and enzymes help to explain how aspirin works?

Nucleic Acids Are Informational Macromolecules

Nucleic acids are polymers specialized for the storage, transmission, and use of genetic information. There are two types of nucleic acids: **DNA** (*deoxyribonucleic acid*) and **RNA** (*ribonucleic acid*). DNA encodes hereditary information, and through RNA intermediates, the information encoded in DNA is used to specify the amino acid sequences of proteins. As you will see later in this chapter, proteins are essential in metabolism and structure. So ultimately, *DNA and the proteins encoded by DNA determine metabolic functions.*

Nucleotides are the building blocks of nucleic acids

Nucleic acids are polymers composed of monomers called nucleotides. A **nucleotide** consists of three components: a nitrogen-containing **base**, a pentose sugar, and one to three phosphate groups (**FIGURE 3.1**). Molecules consisting of a pentose sugar and a nitrogenous base—but no phosphate group—are called *nucleosides.* The nucleotides that make up nucleic ac-

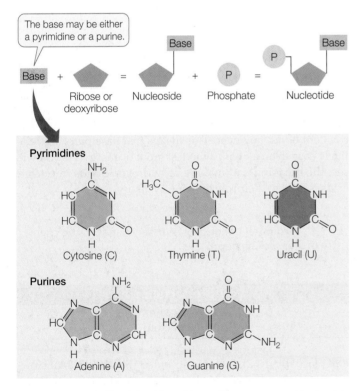

Pyrimidines

Cytosine (C) Thymine (T) Uracil (U)

Purines

Adenine (A) Guanine (G)

FIGURE 3.1 Nucleotides Have Three Components Nucleotide monomers are the building blocks of DNA and RNA polymers. Nucleotides may have one to three phosphate groups; those in DNA and RNA have one.

yourBioPortal.com

Go to WEB ACTIVITY 3.1
Nucleic Acid Building Blocks

TABLE 3.1	Distinguishing RNA from DNA		
NUCLEIC ACID	**SUGAR**	**BASES**	**STRANDS**
RNA	Ribose	Adenine	Single
		Cytosine	
		Guanine	
		Uracil	
DNA	Deoxyribose	Adenine	Double
		Cytosine	
		Guanine	
		Thymine	

ids contain just one phosphate group—they are nucleoside *mono*phosphates.

The bases of the nucleic acids take one of two chemical forms: a six-membered single-ring structure called a **pyrimidine**, or a fused double-ring structure called a **purine** (see Figure 3.1). In DNA, the pentose sugar is **deoxyribose**, which differs from the **ribose** found in RNA by the absence of one oxygen atom (see Figure 2.9).

During the formation of a nucleic acid, new nucleotides are added to an existing chain one at a time. The pentose sugar and phosphate provide the hydroxyl functional groups for the linkage of one nucleotide to the next. This is done through a condensation reaction (see Figure 2.8), and the resulting bond is called a **phosphodiester linkage**. The linkage reaction always occurs between the phosphate on the new nucleotide (which is located at the 5'-carbon atom on the sugar) and the carbon at the 3' position on the last sugar in the existing chain. Thus, *nucleic acids grow in the 5' to 3' direction* (**FIGURE 3.2**).

Nucleic acids can be oligonucleotides, with about 20 nucleotide monomers, or longer polynucleotides:

- *Oligonucleotides* include RNA molecules that function as "primers" to begin the duplication of DNA; RNA molecules that regulate the expression of genes; and synthetic DNA molecules used for amplifying and analyzing other, longer nucleotide sequences.

- *Polynucleotides*, more commonly referred to as nucleic acids, include DNA and most RNA. Polynucleotides can be very long, and indeed are the longest polymers in the living world. Some DNA molecules in humans contain hundreds of millions of nucleotides.

Base pairing occurs in both DNA and RNA

DNA and RNA differ somewhat in their sugar groups, bases, and general structures (**TABLE 3.1**). Four bases are found in DNA: **adenine (A)**, **cytosine (C)**, **guanine (G)**, and **thymine (T)**. RNA is also made up of four different monomers, but its nucleotides have **uracil (U)** instead of thymine. The sugar in DNA is deoxyribose, whereas the sugar in RNA is ribose. The lack of a hydroxyl group at the 2' position in DNA makes its structure less flexible than that of RNA, which, unlike DNA, can form a variety of structures.

FIGURE 3.2 Linking Nucleotides Together
Growth of a nucleic acid (RNA in this figure) from its monomers occurs in the 5′ (phosphate) to 3′ (hydroxyl) direction.

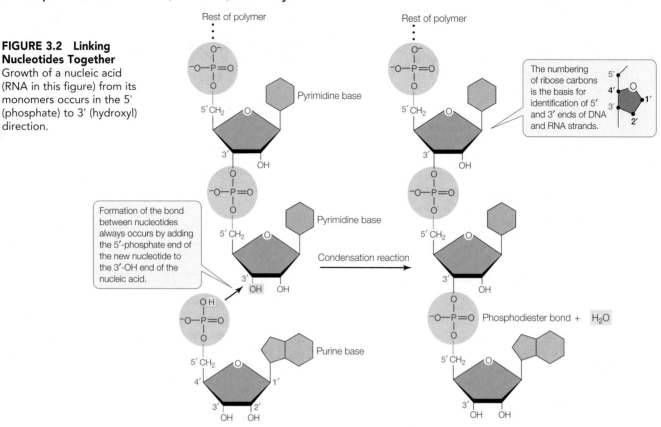

The key to understanding the structure and function of nucleic acids is the principle of **complementary base pairing**. In DNA, adenine and thymine always pair (A-T), and cytosine and guanine always pair (C-G):

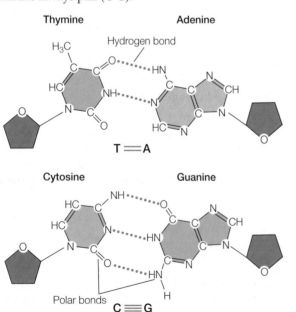

In RNA, the base pairs are A-U and C-G. Base pairs are held together primarily by *hydrogen bonds*. As you can see, there are polar C═O and N—H covalent bonds in the bases; these can form hydrogen bonds between the δ⁻ on an oxygen or nitrogen of one base and the δ⁺ on a hydrogen of another base.

Individual hydrogen bonds are relatively weak, but there are so many of them in a DNA or RNA molecule that collectively they provide a considerable force of attraction, which can bind together two polynucleotide strands, or a single strand that folds back onto itself. This attraction is not as strong as a covalent bond, however. This means that base pairs are relatively easy to break with a modest input of energy. As you will see, the breaking and making of hydrogen bonds in nucleic acids is vital to their role in living systems.

> **LINK** Hydrogen bonds play an essential role in the chemistry of life; see Concept 2.2

RNA Usually, RNA is single-stranded (**FIGURE 3.3A**). However, many single-stranded RNA molecules fold up into three-dimensional structures, because of hydrogen bonding between ribonucleotides in separate portions of the molecules (**FIGURE 3.3B**). This results in a three-dimensional surface for the bonding and recognition of other molecules. It is important to realize that this folding occurs by complementary base pairing, and the structure is thus determined by the particular order of bases in the RNA molecule.

(A)

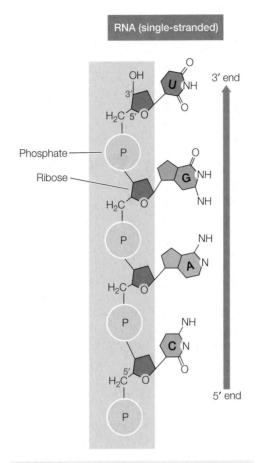

RNA (single-stranded)

3' end

Phosphate

Ribose

5' end

In RNA, the bases are attached to ribose. The bases in RNA are the purines adenine (A) and guanine (G) and the pyrimidines cytosine (C) and uracil (U).

(B)

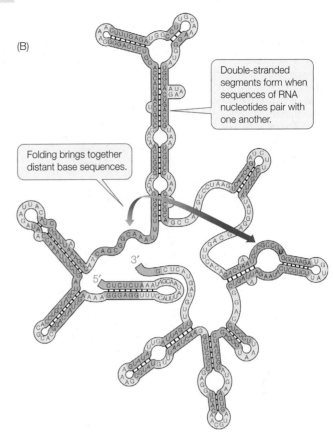

Double-stranded segments form when sequences of RNA nucleotides pair with one another.

Folding brings together distant base sequences.

FIGURE 3.3 RNA (A) RNA is usually a single strand. (B) When a single-stranded RNA folds in on itself, hydrogen bonds between complementary sequences can stabilize it into a three-dimensional shape with complicated surface characteristics.

DNA Usually, DNA is double-stranded; that is, it consists of two separate polynucleotide strands of the same length (**FIGURE 3.4A**). In contrast to RNA's diversity in three-dimensional structure, DNA is remarkably uniform. The A-T and G-C base pairs are about the same size (each is a purine paired with a pyrimidine), and the two polynucleotide strands form a "ladder" that twists into a double helix (**FIGURE 3.4B**). The sugar-phosphate groups form the sides of the ladder, and the bases with their hydrogen bonds form the rungs on the inside. DNA carries genetic information in its sequence of base pairs rather than in its three-dimensional structure. The key differences among DNA molecules are manifest in their different nucleotide base sequences.

DNA carries information and is expressed through RNA

DNA is a purely *informational* molecule. The information is encoded in the sequence of bases carried in its strands. For example, the information encoded in the sequence TCAGCA is different from the information in the sequence CCAGCA. DNA has two functions in terms of information:

- DNA can be reproduced exactly. This is called *DNA replication*. It is done by polymerization using an existing strand as a base-pairing template.

- Some DNA sequences can be copied into RNA, in a process called *transcription*. The nucleotide sequence in the RNA can then be used to specify a sequence of amino acids in a polypeptide chain. This process is called *translation*. The overall process of transcription and translation is called *gene expression*:

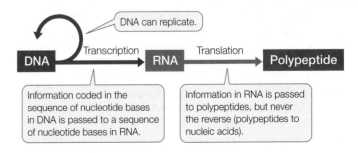

DNA can replicate.

DNA → Transcription → RNA → Translation → Polypeptide

Information coded in the sequence of nucleotide bases in DNA is passed to a sequence of nucleotide bases in RNA.

Information in RNA is passed to polypeptides, but never the reverse (polypeptides to nucleic acids).

FIGURE 3.4 DNA
(A) DNA usually consists of two strands running in opposite directions that are held together by hydrogen bonds between purines on one strand and pyrimidines on the opposing strand. (B) The two strands in a DNA molecule are coiled in a double helix.

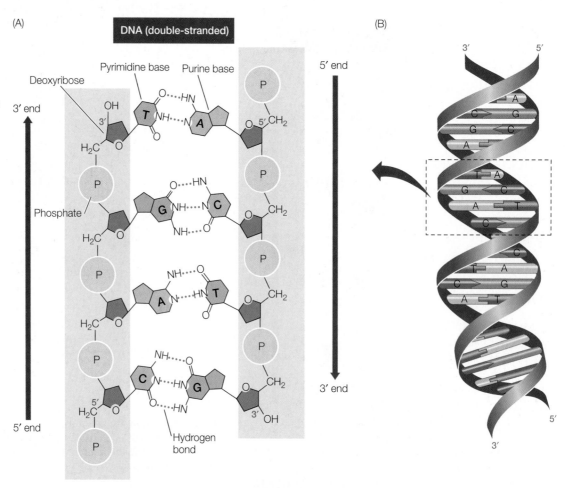

In DNA, the bases are attached to deoxyribose, and the base thymine (T) is found instead of uracil. Hydrogen bonds between purines and pyrimidines hold the two strands of DNA together.

The details of these important processes are described in later chapters, but it is important to realize two things at this point:

1. *DNA replication and transcription depend on the base pairing properties of nucleic acids.* Recall that the hydrogen-bonded base pairs are A-T and G-C in DNA and A-U and G-C in RNA. Consider this double-stranded DNA region:

5'-TCAGCA-3'

3'-AGTCGT-5'

Transcription of the lower strand will result in a single strand of RNA with the sequence 5'-UCAGCA-3'. Can you figure out what the top strand would produce?

2. *DNA replication usually involves the entire DNA molecule.* Since DNA holds essential information, it must be replicated completely so that each new cell or new organism receives a complete set of DNA from its parent (**FIGURE 3.5A**).

The complete set of DNA in a living organism is called its **genome**. However, not all of the information in the genome is needed at all times and in all tissues, and only small sections of the DNA are transcribed into RNA molecules. The sequences of DNA that encode specific proteins and are transcribed into RNA are called **genes** (**FIGURE 3.5B**). In humans, the gene that encodes the major protein in hair (keratin) is expressed only in skin cells. The genetic information in the keratin-encoding gene is transcribed into RNA and then translated into a keratin polypeptide. In other tissues such as the muscles, the keratin gene is not transcribed, but other genes are—for example, the genes that encode proteins present in muscles but not in skin.

The DNA base sequence reveals evolutionary relationships

Because DNA carries hereditary information from one generation to the next, a theoretical series of DNA molecules stretches back through the lineage of every organism to the beginning of biological evolution on Earth, about 3.8 billion years ago. The genomes of organisms gradually accumulate changes in their DNA base sequences over evolutionary time. Therefore, closely related living species should have more similar base sequences than species that are more distantly related.

LINK The use of DNA sequences to reconstruct the evolutionary history of life is described in Concept 16.2

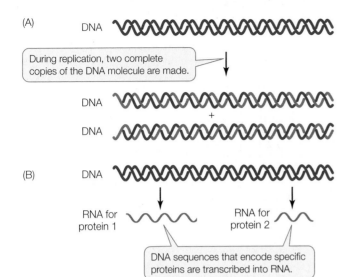

(A)

DNA

During replication, two complete copies of the DNA molecule are made.

DNA

+

DNA

(B)

DNA

RNA for protein 1

RNA for protein 2

DNA sequences that encode specific proteins are transcribed into RNA.

FIGURE 3.5 DNA Replication and Transcription DNA is completely replicated during cell reproduction (A), but it is only partially transcribed (B). In transcription, the DNA code is copied as RNA, which encodes the genes for specific proteins. Transcription of the many different proteins is activated at different times and, in multicellular organisms, in different cells of the body.

Remarkable developments in sequencing and computer technology have enabled scientists to determine the entire DNA base sequences of whole organisms, including the human genome, which contains about 3 billion base pairs. These studies have confirmed many of the evolutionary relationships that were inferred from more traditional comparisons of body structure, biochemistry, and physiology. Traditional comparisons had indicated that the closest living relative of humans (*Homo sapiens*) is the chimpanzee (genus *Pan*). In fact, the chimpanzee genome shares more than 98 percent of its DNA base sequence with the human genome. Increasingly, scientists turn to DNA analyses to elucidate evolutionary relationships when other comparisons are not possible or are not conclusive. For example, DNA studies revealed a close relationship between starlings and mockingbirds that was not expected on the basis of their anatomy or behavior.

Do You Understand Concept 3.1?

- List the key differences between DNA and RNA and between purines and pyrimidines.

- What are the differences between DNA replication and transcription?

- If one strand of a DNA molecule has the sequence 5'-TTCCGGAT-3', what is the sequence of the other strand of DNA? If RNA is transcribed from the 5'-TTCCGGAT-3' strand, what would be its sequence? And if RNA is transcribed from the other DNA strand, what would be its sequence?

- How can DNA molecules be so diverse when they appear to be structurally similar?

Nucleic acids are largely informational molecules that encode proteins. We now turn to a discussion of proteins—the most structurally and functionally diverse class of macromolecules.

concept 3.2 Proteins Are Polymers with Important Structural and Metabolic Roles

Proteins are the fourth and final type of biological macromolecule we will discuss, and in terms of structural diversity and function, they are at the top of the list. Here are some of the major functions of proteins in living organisms:

- *Enzymes* are catalytic proteins that speed up biochemical reactions.

- *Defensive proteins* such as antibodies recognize and respond to substances or particles that invade the organism from the environment.

- *Hormonal and regulatory proteins* such as insulin control physiological processes.

- *Receptor proteins* receive and respond to molecular signals from inside and outside the organism.

- *Storage proteins* store chemical building blocks—amino acids—for later use.

- *Structural proteins* such as collagen provide physical stability and movement.

- *Transport proteins* such as hemoglobin carry substances within the organism.

- *Genetic regulatory proteins* regulate when, how, and to what extent a gene is expressed.

Clearly, the biochemistry of proteins warrants our attention!

Amino acids are the building blocks of proteins

As we noted in Chapter 2, proteins are polymers made up of monomers called **amino acids**. As their name suggests, the amino acids all contain two functional groups: the nitrogen-containing amino group and the carboxylic acid group.

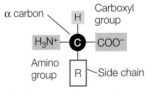

The amino and carboxylic acid groups shown in the diagram are charged. How does this happen? Under the conditions that exist in most living systems, the carboxylic acid group releases a H^+ (a cation), leaving the rest of the group as an anion:

$$-COOH \rightarrow -COO^- + H^+$$

From your studies of chemistry, you may recognize this as an *acid* (hence the name). Conversely, under the same conditions the amino group tends to form a bond with H^+:

$$-NH_2 + H^+ \rightarrow -NH_3^+$$

TABLE 3.2 The Twenty Amino Acids in Proteins

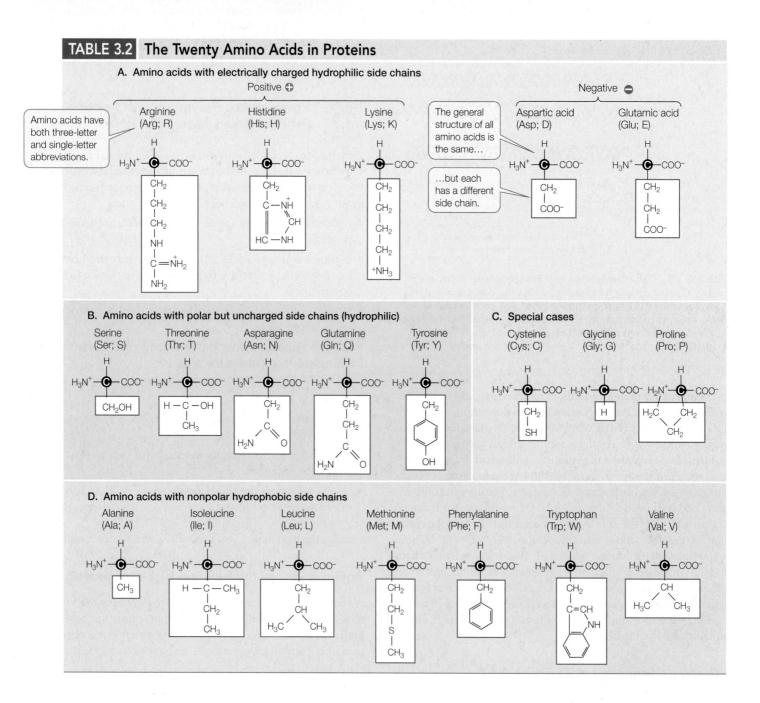

Your chemistry knowledge should tell you that this is a *base*. (We will discuss acids and bases in more detail later in this chapter.)

The central carbon atom of an amino acid—the α carbon—has four available electrons for covalent bonding. In all amino acids, two of the electrons are occupied by the two functional groups noted above, and a third is occupied by a hydrogen atom. The fourth bonding electron is shared with a group that differs in each amino acid. This is often referred to as the **R group**, or *side chain*, and is designated by the letter R. Each amino acid is identified by its R group.

There are hundreds of amino acids known in nature, and many of these occur in plants. But *only 20 amino acids* (listed in **TABLE 3.2**) *occur extensively in the proteins of all organisms.* These 20 amino acids can be grouped according to the properties conferred by their side chains (R groups):

- Five amino acids have electrically charged side chains (+1 or –1), attract water (are hydrophilic), and attract oppositely charged ions of all sorts.

- Five amino acids have polar side chains (δ+, δ−) and tend to form hydrogen bonds with water and other polar or charged substances. These amino acids are also hydrophilic.

- Seven amino acids have side chains that are nonpolar hydrocarbons or very slightly modified hydrocarbons. In the watery environment of the cell, these hydrophobic side chains may cluster together in the interior of the protein. These amino acids are hydrophobic.

Three amino acids—cysteine, glycine, and proline—are special cases, although the side chains of the latter two generally are hydrophobic:

- The *cysteine* side chain, which has a terminal —SH group, can react with another cysteine side chain to form a covalent bond called a **disulfide bridge**, or *disulfide bond* (—S—S—). Disulfide bridges help determine how a polypeptide chain folds.

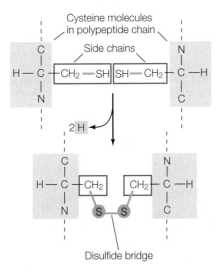

- The *glycine* side chain consists of a single hydrogen atom and is small enough to fit into tight corners in the interior of a protein molecule, where a larger side chain could not fit.

- *Proline* possesses a modified amino group that lacks a hydrogen and instead forms a covalent bond with the hydrocarbon side chain, resulting in a ring structure. This limits both its hydrogen-bonding ability and its ability to rotate. Thus proline often functions to stabilize bends or loops in proteins.

---

yourBioPortal.com

Go to WEB ACTIVITY 3.3
Features of Amino Acids

Amino acids are bonded to one another by peptide linkages

Like nucleotides, amino acids can form short polymers of 20 or fewer amino acids, called *oligopeptides* or simply *peptides*. These include some hormones and other molecules involved in signaling from one part of an organism to another. Even with their relatively short chains of amino acids, oligopeptides have distinctive three-dimensional structures.

More common are the longer polymers called *polypeptides* or *proteins*. Each protein has its own unique proportion and sequence of the 20 amino acids. Proteins range in size from small ones such as insulin, which has a molecular weight of 5,808 daltons and 51 amino acids, to huge molecules such as the muscle protein titin, with a molecular weight of 3,816,188 daltons and 34,350 amino acids. (See p. 17 for a definition of daltons.)

Like nucleic acids, proteins and peptides form via the sequential addition of new amino acids to the ends of existing chains. The amino group of the new amino acid reacts with the carboxyl group of the amino acid at the end of the chain. This condensation reaction forms a **peptide linkage** (also called a *peptide bond*;

FIGURE 3.6). Note that there is directionality here, just as with the nucleic acids. In this case, polymerization takes place in the *amino to carboxyl direction*.

The precise sequence of amino acids in a polypeptide chain constitutes the **primary structure** of a protein. Scientists have determined the primary structures of many proteins. The single-letter abbreviations for amino acids (see Table 3.2) are used to record the amino acid sequences of proteins. Here, for example, are the first 20 amino acids (out of a total of 1,827) in the human protein sucrase:

MARKKFSGLEISLIVLFVIV

The theoretical number of different proteins is enormous. Since there are 20 different amino acids, there could be 20 × 20 = 400 distinct dipeptides (two linked amino acids), and 20 × 20 × 20 = 8,000 different tripeptides (three linked amino acids). So for even a small polypeptide of 100 amino acids there are 20^{100} possible sequences, each with its own distinctive primary structure. How large is the number 20^{100}? Physicists tell us there aren't that many electrons in the entire universe.

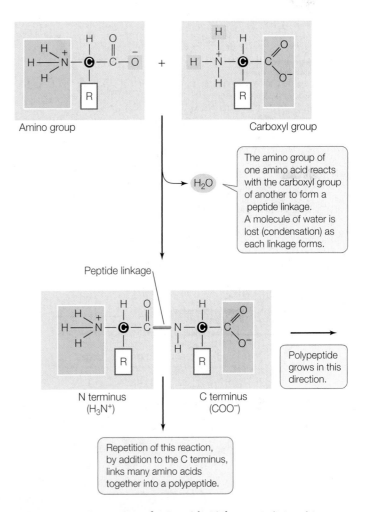

FIGURE 3.6 Formation of a Peptide Linkage In living things, the reaction leading to a peptide linkage (also called a peptide bond) has many intermediate steps, but the reactants and products are the same as those shown in this simplified diagram.

Higher-level protein structure is determined by primary structure

The primary structure of a protein is established by covalent bonds, but higher levels of structure are determined largely by weaker forces, including hydrogen bonds and hydrophobic and hydrophilic interactions. Follow **FIGURE 3.7** as we describe how a protein chain becomes a three-dimensional structure.

SECONDARY STRUCTURE A protein's **secondary structure** consists of regular, repeated spatial patterns in different regions of a polypeptide chain. There are two basic types of secondary structure, both determined by hydrogen bonding between the amino acids that make up the primary structure:

* The **α (alpha) helix** is a right-handed coil that turns in the same direction as a standard wood screw (see Figure 3.7B). The R groups extend outward from the peptide backbone of

FIGURE 3.7 The Four Levels of Protein Structure The primary structure (A) of a protein determines what its secondary (B and C), tertiary (D), and quaternary (E) structures will be.

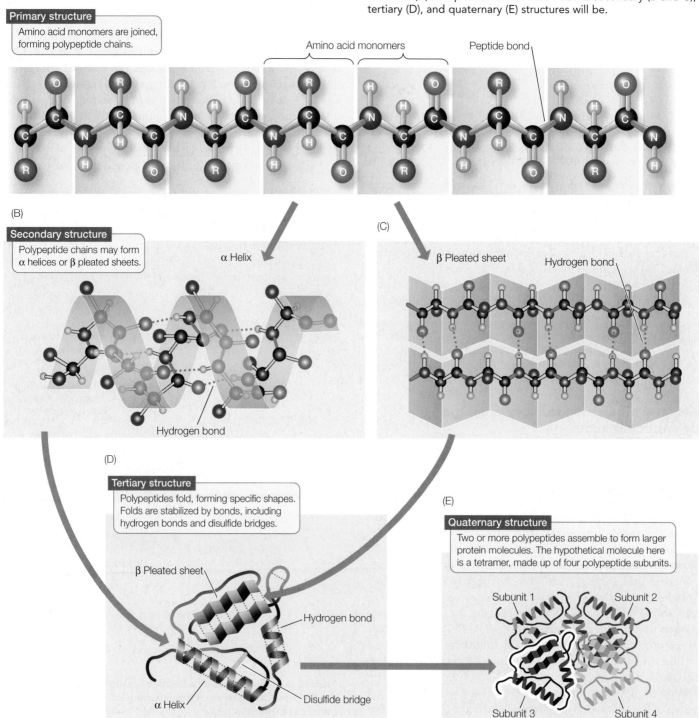

(A)

Primary structure
Amino acid monomers are joined, forming polypeptide chains.

Amino acid monomers

Peptide bond

(B)

Secondary structure
Polypeptide chains may form α helices or β pleated sheets.

α Helix

Hydrogen bond

(C)

β Pleated sheet

Hydrogen bond

(D)

Tertiary structure
Polypeptides fold, forming specific shapes. Folds are stabilized by bonds, including hydrogen bonds and disulfide bridges.

β Pleated sheet

Hydrogen bond

α Helix

Disulfide bridge

(E)

Quaternary structure
Two or more polypeptides assemble to form larger protein molecules. The hypothetical molecule here is a tetramer, made up of four polypeptide subunits.

Subunit 1
Subunit 2
Subunit 3
Subunit 4

the helix. The coiling results from hydrogen bonds that form between the N—H group on one amino acid and the C=O group on another within the same turn of the helix.

- The β (**beta**) **pleated sheet** is formed from two or more polypeptide chains that are extended and aligned. The sheet is stabilized by hydrogen bonds between the N—H groups and the C=O groups on the two chains (see Figure 3.7C). A β pleated sheet may form between separate polypeptide chains or between different regions of a single polypeptide chain that is bent back on itself. Many proteins contain both α helices and β pleated sheets in different regions of the same polypeptide chain.

TERTIARY STRUCTURE In many proteins, the polypeptide chain is bent at specific sites and then folded back and forth, resulting in the **tertiary structure** (see Figure 3.7D). Tertiary structure results in the polypeptide's definitive three-dimensional shape, including a buried interior as well as a surface that is exposed to the environment. The protein's exposed outer surfaces present functional groups capable of interacting with other molecules in the cell. These molecules might be other proteins or smaller chemical reactants (as in enzymes; see below).

Whereas hydrogen bonding between the N—H and C=O groups within and between chains is responsible for a protein's secondary structure, it is the interactions between R groups—the amino acid side chains—that determine tertiary structure (**FIGURE 3.8**):

- Covalent *disulfide bridges* can form between specific cysteine side chains, holding a folded polypeptide together.

- *Hydrogen bonds* between side chains also stabilize folds in proteins.

- *Hydrophobic* side chains can aggregate together in the interior of a protein, away from water, folding the polypeptide in the process.

- *van der Waals interactions* can stabilize close associations between hydrophobic side chains.

- *Ionic interactions* can form between positively and negatively charged side chains, forming *salt bridges* between amino acids. Ionic bonds can also be buried deep within a protein, away from water.

LINK Review the strong and weak interactions that can occur between atoms, described in Concept 2.2

A complete description of a protein's tertiary structure would specify the location of every atom in the molecule in three-dimensional space, relative to all the other atoms. Many such descriptions are available, including one for the human protein sucrase (**FIGURE 3.9**).

Remember that both secondary and tertiary structure derive from primary structure. If a protein is heated slowly, the heat energy will disrupt only the weaker interactions, causing the secondary and tertiary structure to break down. The protein is then said to be **denatured**. But in many cases the protein can return to its normal tertiary structure when it cools, demonstrating that all the information needed to specify its unique shape is contained in its primary structure. This was first shown (using chemicals instead of heat to denature the protein) by biochemist Christian Anfinsen for the protein ribonuclease (**FIGURE 3.10**).

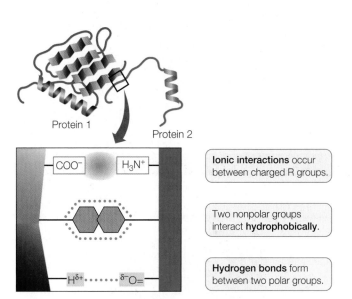

Ionic interactions occur between charged R groups.

Two nonpolar groups interact **hydrophobically**.

Hydrogen bonds form between two polar groups.

FIGURE 3.8 Noncovalent Interactions between Proteins and Other Molecules Noncovalent interactions allow a protein (brown) to bind tightly to another protein (green) with specific properties. Noncovalent interactions also allow regions within the same protein to interact with one another.

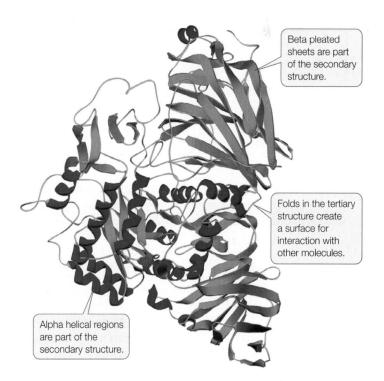

Beta pleated sheets are part of the secondary structure.

Folds in the tertiary structure create a surface for interaction with other molecules.

Alpha helical regions are part of the secondary structure.

FIGURE 3.9 The Structure of a Protein Sucrase has a specific three-dimensional structure, determined by its primary structure. Sucrase plays a role in digestion in humans.

INVESTIGATION

FIGURE 3.10   Primary Structure Specifies Tertiary Structure   Using the protein ribonuclease, Christian Anfinsen showed that proteins spontaneously fold into a functionally correct three-dimensional configuration. As long as the primary structure is not disrupted, the information for correct folding under the right conditions is retained.

HYPOTHESIS

Under controlled conditions that simulate normal cellular environment in the laboratory, the primary structure of a denatured protein can reestablish the protein's three-dimensional structure.

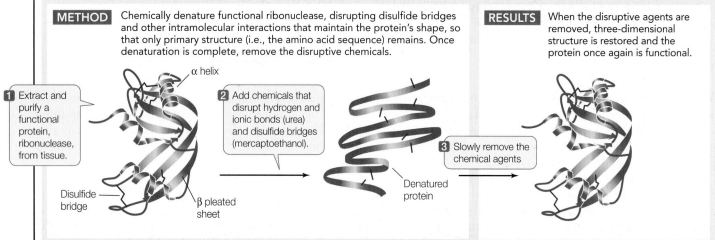

METHOD   Chemically denature functional ribonuclease, disrupting disulfide bridges and other intramolecular interactions that maintain the protein's shape, so that only primary structure (i.e., the amino acid sequence) remains. Once denaturation is complete, remove the disruptive chemicals.

RESULTS   When the disruptive agents are removed, three-dimensional structure is restored and the protein once again is functional.

1 Extract and purify a functional protein, ribonuclease, from tissue.

α helix

Disulfide bridge

β pleated sheet

2 Add chemicals that disrupt hydrogen and ionic bonds (urea) and disulfide bridges (mercaptoethanol).

Denatured protein

3 Slowly remove the chemical agents

CONCLUSION

In normal cellular conditions, the primary structure of a protein specifies how it folds into a functional, three-dimensional structure.

ANALYZE THE DATA

Initially, disulfide bonds (S—S) in RNase A were eliminated because the sulfur atoms in cysteine were reduced (—SH). At time 0, reoxidation began and at various times, the amount of disulfide bond re-formation (blue circles) and the function of ribonuclease (enzyme activity; red circles) were measured by chemical methods. Here are the data:

A. At what time did disulfide bonds begin to form?
B. At what time did enzyme activity begin to appear?
C. Explain the difference between your answers for the times of (A) and (B).

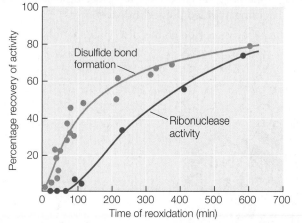

For more, go to Working with Data 3.1 at **yourBioPortal.com**.

Go to **yourBioPortal.com** for original citations, discussions, and relevant links for all INVESTIGATION figures.

QUATERNARY STRUCTURE   Many functional proteins contain two or more polypeptide chains, called subunits, each folded into its own unique tertiary structure. The protein's **quaternary structure** results from the ways in which these subunits bind together and interact (see Figure 3.7E). Hemoglobin (at right) is an example of a protein with multiple subunits.

Hydrophobic interactions, hydrogen bonds, and ionic interactions all help hold the four subunits together to form a hemoglobin macromolecule. The weak nature of these forces permits small changes in the quaternary structure to aid the

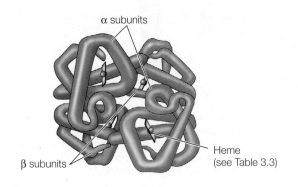

α subunits

β subunits

Heme (see Table 3.3)

protein's function—which is to carry oxygen in red blood cells. As hemoglobin binds one O_2 molecule, the four subunits shift their relative positions slightly, changing the quaternary structure. Ionic interactions are broken, exposing buried side chains that enhance the binding of additional O_2 molecules. The quaternary structure changes again when hemoglobin releases its O_2 molecules to the cells of the body.

FRONTIERS Spiderwebs are composed of a protein that is very strong—possibly the strongest material in the living world—because the web must support the weight of the spider and its prey and stretch without breaking. The protein's strength comes from its multiple interlocking β pleated sheets. It is difficult to collect this protein from spiders, so biologists are using genetic engineering to produce it in more tractable organisms such as goats. The mass-produced spider silk will have uses ranging from surgical threads to bulletproof vests.

Environmental conditions affect protein structure

Because they are held together by weak forces, the three-dimensional structures of proteins are influenced by environmental conditions. Conditions that would not break covalent bonds can disrupt the weaker, noncovalent interactions that determine secondary and tertiary structure. Such alterations may affect a protein's shape and thus its function. Various conditions can alter the weak, noncovalent interactions:

- *Increases in temperature* cause more rapid molecular movements and thus can break hydrogen bonds and hydrophobic interactions.

- *Alterations in the concentration of H+* can change the patterns of ionization of the exposed carboxyl and amino groups, thus disrupting the patterns of ionic attractions and repulsions.

- *High concentrations of polar substances* such as urea can disrupt the hydrogen bonding that is crucial to protein structure.

- *Nonpolar substances* may also denature a protein in cases where hydrophobic groups are essential for maintaining the protein's structure.

Denaturation can be irreversible when amino acids that were buried in the interior of the protein become exposed at the surface, or vice versa, causing a new structure to form, or causing different molecules to bind to the protein. Boiling an egg denatures its proteins and is, as you know, not reversible.

Do You Understand Concept 3.2?

- What attributes of an amino acid's R group would make it hydrophobic? Hydrophilic?

- Sketch the bonding of two amino acids, glycine and leucine, by a peptide linkage. Now add a third amino acid, alanine, in the position it would have if added within a biological system. What is the directionality of this process?

- Examine the structure of sucrase (see Figure 3.9). Where in the protein might you expect to find the following amino acids: valine, proline, glutamic acid, and threonine? Explain your answers.

- Detergents disrupt hydrophobic interactions by coating hydrophobic molecules with a molecule that has a hydrophilic surface. When hemoglobin is treated with a detergent, the four polypeptide chains separate and become random coils. Explain these observations.

We have discussed the remarkable diversity in protein structures. These structures expose atoms that can allow the proteins to interact with other molecules. In the next section we will see how these interactions can result in catalysis, the remarkable speeding up of biochemical reactions.

APPLY THE CONCEPT

Proteins are polymers with important structural and metabolic roles

Biological systems contain "supermolecular complexes" (for example, the ribosome; see Chapter 4), which are composed of individual molecules of RNA and protein that fit together noncovalently. These complexes can be split apart with detergents that disrupt hydrophobic interactions. Based on the concepts discussed in this chapter, fill in the table at right to indicate which of the observations are characteristic of RNA, which are characteristic of protein, and which are characteristic of both. Explain your answers.

OBSERVATION	CHARACTERISTIC OF: PROTEIN	RNA
Has three-dimensional (3-D) structure		
3-D structure destroyed by heat		
Monomers connected by N—C bonds		
Contains sulfur atoms		
Contains phosphorus atoms		

concept
3.3 Some Proteins Act as Enzymes to Speed up Biochemical Reactions

In Chapter 2 we introduced the concepts of biological energetics. We showed that some metabolic reactions are exergonic and some are endergonic, and that biochemistry obeys the laws of thermodynamics (see Figures 2.14 and 2.15). Knowing whether energy is supplied or released in a particular reaction tells us whether the reaction *can* occur in a living system. But it does not tell us *how fast* the reaction will occur.

Living systems depend on reactions that occur spontaneously, but at such slow rates the cells would not survive without ways to speed them up. That is the role of **catalysts**: substances that speed up reactions without themselves being permanently altered. A catalyst does not cause a reaction to occur that would not proceed without it, *but it increases the rate of the reaction*. This is an important point: *No catalyst makes a reaction occur that cannot otherwise occur.*

Most biological catalysts are proteins called *enzymes*. Although we will focus here on proteins, a few important catalysts are RNA molecules called *ribozymes*. A biological catalyst, whether protein or RNA, provides a molecular structure that binds the reactants and can participate in the reaction itself. However, this participation does not permanently change the enzyme. At the end of the reaction, the catalyst is unchanged and available to catalyze additional, similar reactions.

To speed up a reaction, an energy barrier must be overcome

An exergonic reaction may release free energy, but without a catalyst it will take place very slowly. This is because there is an *energy barrier* between reactants and products. Think about the hydrolysis of sucrose, which we described in Concept 2.5.

$$sucrose + H_2O \rightarrow glucose + fructose$$

In humans, this reaction is part of the process of digestion. Even if water is abundant, the sucrose molecule will not bind the H atom and –OH group of water at the appropriate locations to break the covalent bond between glucose and fructose *unless there is an input of energy to initiate the reaction*. Such an input of energy will place the sucrose into a reactive mode called the **transition state**. The energy input required for sucrose to reach this state is called the **activation energy (E_a)**. The following example will help illustrate the ideas of activation energy and transition state:

$$fireworks + O_2 \rightarrow CO_2 + H_2O + energy \text{ (heat and light)}$$

A spark is needed to excite the molecules in the fireworks so they will react with oxygen in the air. Once the transition state is reached, the reaction occurs (**FIGURE 3.11**).

Where does the activation energy come from? In any collection of reactants at room or body temperature, the molecules are moving around. A few are moving fast enough that their kinetic energy can overcome the energy barrier, enter the transition state, and react. So the reaction takes place—but very

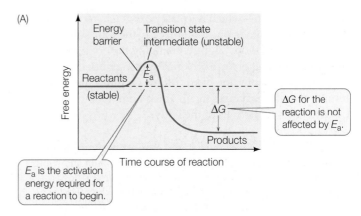

(A)

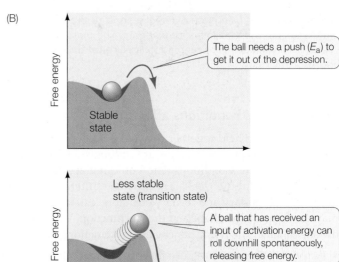

(B)

FIGURE 3.11 Activation Energy Initiates Reactions (A) In any chemical reaction, an initial stable state must become less stable before change is possible. (B) A ball on a hillside provides a physical analogy to the biochemical principle graphed in A. Although these graphs show an exergonic reaction, activation energy is needed for endergonic reactions as well.

slowly. If the system is heated, all the reactant molecules move faster and have more kinetic energy, and the reaction speeds up. You have probably used this technique in the chemistry laboratory.

Adding enough heat to increase the average kinetic energy of the molecules would not work in living systems, however. Such a nonspecific approach would accelerate all reactions, including destructive ones such as the denaturation of proteins.

An enzyme lowers the activation energy for the reaction—it offers the reactants an easier path so they can come together and react more easily (**FIGURE 3.12**). In this way, an enzyme can change the rate of a reaction substantially. For example, if a molecule of sucrose just sits in solution, hydrolysis may occur in about 15 days; with sucrase present, the same reaction occurs in 1 second!

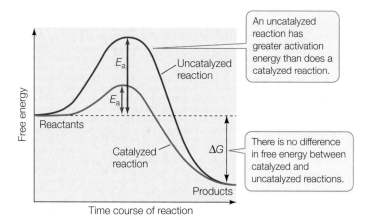

FIGURE 3.12 Enzymes Lower the Energy Barrier Although the activation energy (E_a) is lower in an enzyme-catalyzed reaction than in an uncatalyzed reaction, the energy released is the same with or without catalysis. A lower activation energy means the reaction will take place at a faster rate.

> **yourBioPortal.com**
>
> **Go to WEB ACTIVITY 3.4**
> **Free Energy Changes**

Enzymes bind specific reactants at their active sites

Catalysts increase the rates of chemical reactions. Most nonbiological catalysts are nonspecific. For example, powdered platinum catalyzes virtually any reaction in which molecular hydrogen (H_2) is a reactant. In contrast, most biological catalysts are highly specific. An enzyme usually recognizes and binds to only one or a few closely related reactants, and it catalyzes only a single chemical reaction.

In an enzyme-catalyzed reaction, the reactants are called **substrates**. Substrate molecules bind to a particular site on the enzyme, called the **active site**, where catalysis takes place (**FIGURE 3.13**). The specificity of an enzyme results from the exact three-dimensional shape and chemical properties of its active site. Only a narrow range of substrates, with specific shapes, functional groups, and chemical properties, can fit properly and bind to the active site. The names of enzymes reflect their functions and often end with the suffix "ase." For example, the enzyme sucrase catalyzes the hydrolysis of sucrose, and we write the reaction as follows:

$$\text{sucrose} + H_2O \xrightarrow{\text{Sucrase}} \text{glucose} + \text{fructose}$$

The binding of a substrate to the active site of an enzyme produces an **enzyme–substrate complex (ES)** that is held together by one or more means, such as hydrogen bonding, electrical attraction, or temporary covalent bonding. The enzyme–substrate complex gives rise to product and free enzyme:

$$E + S \rightarrow ES \rightarrow E + P$$

where E is the enzyme, S is the substrate, P is the product, and ES is the enzyme–substrate complex. The free enzyme (E) is

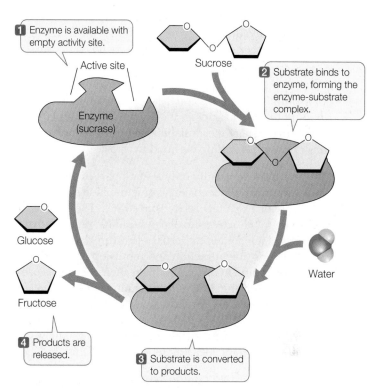

FIGURE 3.13 Enzyme Action Sucrase catalyzes the hydrolysis of sucrose. After the reaction, the enzyme is unchanged and is ready to accept another substrate molecule.

in the same chemical form at the end of the reaction as at the beginning. While bound to the substrate, it may change chemically, but by the end of the reaction it has been restored to its initial form and is ready to bind more substrate (see Figure 3.13).

HOW ENZYMES WORK During and after the formation of the enzyme–substrate complex, chemical interactions occur. These interactions contribute directly to the breaking of old bonds and the formation of new ones. In catalyzing a reaction, an enzyme may use one or more mechanisms:

- *Inducing strain*: Once the substrate has bound to the active site, the enzyme causes bonds in the substrate to stretch, putting it in an unstable transition state:

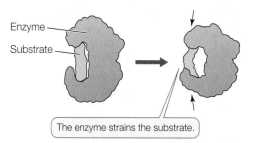

- *Substrate orientation*: When free in solution, substrates are moving from place to place randomly while at the same time vibrating, rotating, and tumbling. They only rarely have the proper orientation to react when they collide. The enzyme lowers the activation energy needed to start the reaction, by bringing together specific atoms so that bonds can form.

- *Adding chemical groups:* The side chains (R groups) of an enzyme's amino acids may be directly involved in the reaction. For example, in *acid–base catalysis*, the acidic or basic side chains of the amino acids in the active site transfer H^+ ions to or from the substrate, destabilizing a covalent bond in the substrate and permitting the bond to break.

The active site is usually only a small part of the enzyme protein. But its three-dimensional structure is so specific that it binds only one or a few related substrates. The binding of the substrate to the active site depends on the same relatively weak forces that maintain the tertiary structure of the enzyme: hydrogen bonds, the attraction and repulsion of electrically charged groups, and hydrophobic interactions. Scientists used to think of substrate binding as being similar to a lock and key fitting together. Actually, for most enzymes and substrates the relationship is more like a baseball and a catcher's mitt: the substrate first binds, and then the active site changes slightly to make the binding tight. **FIGURE 3.14** illustrates this "induced fit" phenomenon.

Induced fit at least partly explains why enzymes are so large. The rest of the macromolecule has at least three roles:

- It provides a framework so the amino acids of the active site are properly positioned in relation to the substrate(s).
- It participates in the changes in protein shape and structure that result in induced fit.
- It provides binding sites for regulatory molecules (see Concept 3.4).

TABLE 3.3	Some Examples of Nonprotein "Partners" of Enzymes	
TYPE OF MOLECULE		**ROLE IN CATALYZED REACTIONS**
Cofactors		
Iron (Fe^{2+} or Fe^{3+})		Oxidation/reduction
Copper (Cu^+ or Cu^{2+})		Oxidation/reduction
Zinc (Zn^{2+})		Helps bind NAD
Coenzymes		
Biotin		Carries —COO^-
Coenzyme A		Carries —CO—CH_3
NAD		Carries electrons
FAD		Carries electrons
ATP		Provides/extracts energy
Prosthetic groups		
Heme		Binds ions, O_2, and electrons; contains iron cofactor
Flavin		Binds electrons
Retinal		Converts light energy

NONPROTEIN PARTNERS FOR ENZYMES Some enzymes require ions or other molecules in order to function (**TABLE 3.3**):

- *Cofactors* are inorganic ions such as copper, zinc, and iron that bind to certain enzymes. For example, the cofactor zinc binds to the enzyme alcohol dehydrogenase.

- A *coenzyme* is a carbon-containing molecule that is required for the action of one or more enzymes. It is usually relatively small compared with the enzyme to which it temporarily binds, and it adds or removes chemical groups from the substrate. A coenzyme is like a substrate in that it does not permanently bind to the enzyme; it binds to the active site, changes chemically during the reaction, and then separates from the enzyme to participate in other reactions. A coenzyme differs from a substrate in that it can participate in many different reactions, with different enzymes.

- *Prosthetic groups* are distinctive, non–amino acid atoms or molecular groupings that are permanently bound to their enzymes. An example is a flavin nucleotide, which binds to succinate dehydrogenase, an important enzyme in energy metabolism.

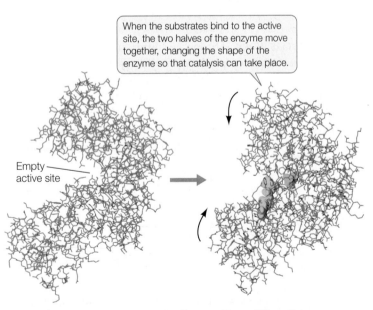

When the substrates bind to the active site, the two halves of the enzyme move together, changing the shape of the enzyme so that catalysis can take place.

Empty active site

FIGURE 3.14 Some Enzymes Change Shape When Substrate Binds to Them Shape changes result in an induced fit between enzyme and substrate, improving the catalytic ability of the enzyme. Induced fit can be observed in the enzyme hexokinase, seen here with and without its substrates, glucose (green) and ATP (yellow).

FRONTIERS Just as proteins can act as catalysts by providing a surface (the active site) that has a particular arrangement of chemical groups, so too can nucleic acids, particularly RNA. Because RNA can also carry genetic information, this has led to the idea that in the evolution of life on Earth, RNA preceded proteins: there was an "RNA world." Recent research supports this theory by showing that both pyrimidines and purines (attached to ribose) can arise from conditions thought to have existed billions of years ago on Earth.

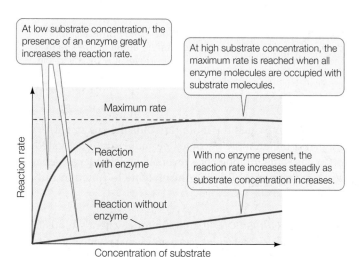

At low substrate concentration, the presence of an enzyme greatly increases the reaction rate.

At high substrate concentration, the maximum rate is reached when all enzyme molecules are occupied with substrate molecules.

Maximum rate

Reaction with enzyme

With no enzyme present, the reaction rate increases steadily as substrate concentration increases.

Reaction without enzyme

Reaction rate

Concentration of substrate

FIGURE 3.15 Catalyzed Reactions Reach a Maximum Rate Because there is usually less enzyme than substrate present, the reaction rate levels off when the enzyme becomes saturated.

RATE OF REACTION The rate of an uncatalyzed reaction is directly proportional to the concentration of the substrate. The higher the concentration, the more reactions per unit of time. As we have seen, the addition of the appropriate enzyme speeds up the reaction, but it also changes the shape of the plot of rate versus substrate concentration (**FIGURE 3.15**). For a given concentration of enzyme, the rate of the enzyme-catalyzed reaction initially increases as the substrate concentration increases from zero, but then it levels off.

Why does this happen? The concentration of an enzyme is usually much lower than that of its substrate and does not change as substrate concentration changes. When all the enzyme molecules are bound to substrate molecules, the enzyme is working at its maximum rate. Under these conditions the active sites are said to be *saturated*.

The maximum rate of a catalyzed reaction can be used to measure how efficient the enzyme is—that is, how many molecules of substrate are converted into product by an individual enzyme molecule per unit of time, when there is an excess of substrate present. This *turnover number* ranges from 1 molecule every second for sucrase to an amazing 40 million molecules per second for the liver enzyme catalase.

Do You Understand Concept 3.3?

- Explain how the structure of an enzyme makes that enzyme specific.

- What is activation energy? How does an enzyme lower the activation energy needed to start a reaction?

- Compare coenzymes with substrates. How do they work together in enzyme catalysis?

- Compare the state of an enzyme active site at a low substrate concentration and at a high substrate concentration. How does this affect the rate of the reaction?

Now that you understand more about how enzymes function, let's see how different enzymes work in the metabolism of living organisms.

concept 3.4 **Regulation of Metabolism Occurs by Regulation of Enzymes**

The enzyme-catalyzed reactions we have been discussing operate within *metabolic pathways* in which the product of one reaction is a substrate for the next. For example, the pathway for the catabolism of sucrose begins with sucrase and ends many reactions later with the production of CO_2 and H_2O. Energy is released along the way. Each step of this catabolic pathway is catalyzed by a specific enzyme:

$$\text{sucrose} + H_2O \xrightarrow{\text{Sucrase}} \text{glucose} + \text{fructose} \longrightarrow$$
$$\xrightarrow{\text{Many enzymes}} \longrightarrow \longrightarrow \longrightarrow \longrightarrow CO_2 + H_2O$$

Other enzymes participate in anabolic pathways, which produce relatively complex molecules from simpler ones. A typical cell contains hundreds of enzymes, which are part of many interconnecting metabolic pathways.

A major characteristic of life is *homeostasis*—the maintenance of stable internal conditions (see Chapter 29). How does a cell maintain a relatively constant internal environment while thousands of chemical reactions are going on?

One way a cell can regulate metabolism is to control the *amount* of an enzyme. For example, the product of a metabolic pathway may be available from the cell's environment in adequate amounts. In this case, it would be energetically wasteful for the cell to continue making large proteins (as most enzymes are) that it doesn't need. For this reason, cells often have the ability to turn off the synthesis of certain enzymes.

LINK The regulation of enzyme synthesis is described in Chapter 11

The consequences of *too little* enzyme can be significant. For example, in humans sucrase is important in digestion. If the enzyme is not present, as in rare cases of infants with congenital sucrase deficiency, the pathway that begins with sucrose is essentially blocked. If such infants ingest fruits or juices containing sucrose, then the sucrose accumulates rather than being catabolized and the infant gets diarrhea and stomach cramps. In some cases, this leads to slower growth. Treatment for sucrase deficiency is to limit sucrose consumption or use a tablet that contains the enzyme at every meal.

Cells can also maintain homeostasis by regulating the *activity* of enzymes. An enzyme protein may be present continuously, but it may be active or inactive depending on the circumstances. Regulation of enzyme activity allows cells to fine-tune metabolism relatively quickly in response to changes in their environment by regulating the functions of particular enzymes. In this section, we will describe how enzyme regulation occurs.

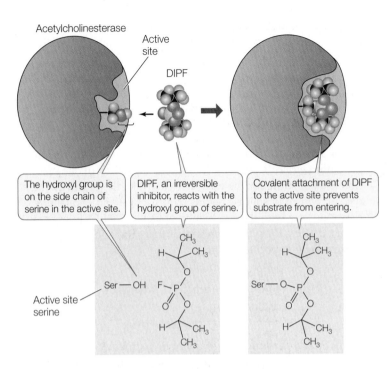

The hydroxyl group is on the side chain of serine in the active site.

DIPF, an irreversible inhibitor, reacts with the hydroxyl group of serine.

Covalent attachment of DIPF to the active site prevents substrate from entering.

FIGURE 3.16 Irreversible Inhibition DIPF forms a stable covalent bond with the amino acid serine at the active site of the enzyme acetylcholinesterase, thus irreversibly disabling the enzyme.

work. In some cases the inhibitor binds the enzyme irreversibly, and the enzyme becomes permanently inactivated. In other cases the inhibitor has reversible effects; it can separate from the enzyme, allowing the enzyme to function fully as before.

IRREVERSIBLE INHIBITION If an inhibitor covalently binds to an amino acid side chain *at the active site* of an enzyme, the enzyme is permanently inactivated because it cannot interact with its substrate. An example of an irreversible inhibitor is DIPF (diisopropyl phosphorofluoridate), which reacts with serine (**FIGURE 3.16**). DIPF is an irreversible inhibitor of acetylcholinesterase, an important enzyme that functions in the nervous system. The widely used insecticide malathion is a derivative of DIPF that inhibits only insect acetylcholinesterase, not the mammalian enzyme. The irreversible inhibition of enzymes is of practical use to humans, but this form of regulation is not common in the cell, because the enzyme is permanently inactivated and cannot be recycled. Instead, cells use reversible inhibition.

Enzymes can be regulated by inhibitors

Various chemical inhibitors can bind to enzymes, slowing down the rates of the reactions they catalyze. Some inhibitors occur naturally in cells; others can be made in laboratories. Naturally occurring inhibitors regulate metabolism; artificial ones can be used to treat disease, kill pests, or study how enzymes

REVERSIBLE INHIBITION In some cases, an inhibitor is similar enough to a particular enzyme's natural substrate that it can bind noncovalently to the active site, yet different enough that no chemical reaction occurs. This is analogous to a key that inserts into a lock but does not turn it. When such a molecule is bound to the enzyme, the natural substrate cannot enter the active site and the enzyme is unable to function. Such a molecule is called a **competitive inhibitor** because it competes with the natural substrate for the active site (**FIGURE 3.17A**). In this case, the inhibition is reversible. When the concentration of the competitive inhibitor is reduced, the active site is less likely to be occupied by the inhibitor, and the enzyme regains activity.

A **noncompetitive inhibitor** binds to an enzyme at a site distinct from the active site. This binding causes a change in the shape of the enzyme that alters its activity (**FIGURE 3.17B**). The active site may no longer bind the substrate, or if it does, the rate of product formation may be reduced. Like competitive inhibitors, noncompetitive inhibitors can become unbound, so their effects are reversible.

(A) Competitive inhibition

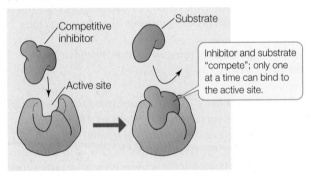

Inhibitor and substrate "compete"; only one at a time can bind to the active site.

(B) Noncompetitive inhibition

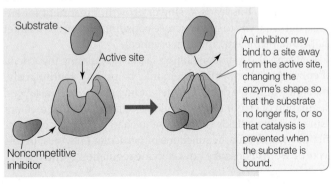

An inhibitor may bind to a site away from the active site, changing the enzyme's shape so that the substrate no longer fits, or so that catalysis is prevented when the substrate is bound.

FIGURE 3.17 Reversible Inhibition (A) A competitive inhibitor binds temporarily to the active site of an enzyme. (B) A noncompetitive inhibitor binds temporarily to the enzyme at a site away from the active site. In both cases, the enzyme's function is disabled for only as long as the inhibitor remains bound.

yourBioPortal.com
Go to ANIMATED TUTORIAL 3.2
Enzyme Catalysis

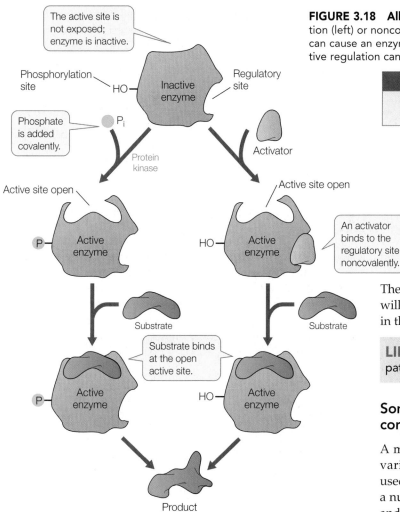

The active site is not exposed; enzyme is inactive.

Phosphorylation site

Regulatory site

Inactive enzyme

HO

Phosphate is added covalently.

P_i

Protein kinase

Activator

Active site open

P

Active enzyme

Active site open

HO

Active enzyme

An activator binds to the regulatory site noncovalently.

Substrate

Substrate

Substrate binds at the open active site.

P

Active enzyme

HO

Active enzyme

Product

FIGURE 3.18 Allosteric Regulation of Enzyme Activity Covalent modification (left) or noncovalent binding of a regulator (in this case an activator; right) can cause an enzyme to change shape and expose an active site. Note that negative regulation can work this way as well, with the active site becoming hidden.

yourBioPortal.com

Go to ANIMATED TUTORIAL 3.3
Allosteric Regulation of Enzymes

An example of allosteric regulation is the activation of *protein kinases*, an important class of enzymes that regulate responses to the environment by organisms. Protein kinases can have profound effects on cell metabolism and are therefore subject to tight allosteric regulation. The active form of a protein kinase in turn regulates the activity of other enzymes, by phosphorylating allosteric or active sites on the other enzymes. There are hundreds of different protein kinases in humans. We will return to the exact functions of protein kinases many times in this book.

LINK The role of protein kinases in intracellular signaling pathways is described in Chapter 5

Some metabolic pathways are usually controlled by feedback inhibition

A metabolic pathway typically involves a starting material, various intermediate products, and an end product that is used for some purpose by the cell. In each pathway there are a number of reactions, each forming an intermediate product and each catalyzed by a different enzyme. In many pathways the first step is the *commitment step*, meaning that once this enzyme-catalyzed reaction occurs, the "ball is rolling," and the other reactions happen in sequence, leading to the end product. But as we pointed out earlier, it is energetically wasteful for the cell to make something it does not need.

One way to regulate a metabolic pathway is by having the final product inhibit the enzyme that catalyzes the commitment step (**FIGURE 3.19**). When the end product is present at a high concentration, some of it binds to a site on the commitment step enzyme, thereby causing it to become inactive. The end product may bind to the active site on the enzyme (as a competitive inhibitor) or an allosteric site (as a noncompetitive inhibitor). This mechanism is known as **feedback inhibition** or *end-product inhibition*. We will describe many other examples of such inhibition in later chapters.

An allosteric enzyme is regulated via changes in its shape

The change in enzyme shape that is due to noncompetitive inhibitor binding is an example of allostery (*allo*, "different"; *stereos*, "shape"). **Allosteric regulation** occurs when a non-substrate molecule binds or modifies *a site other than the active site* of an enzyme (called the *allosteric site*), inducing the enzyme to change its shape. The change in shape alters the chemical attraction (affinity) of the active site for the substrate, and so the rate of the reaction is changed. Allosteric regulation can result in the activation of a formerly inactive enzyme, or the inactivation of an enzyme (as in the case of the noncompetitive inhibitor).

An enzyme can have more than one allosteric site, and these may be modified by either covalent or noncovalent binding (**FIGURE 3.18**):

- *Covalent modification*: For example, an amino acid residue can be covalently modified by the addition of phosphate (in a process called *phosphorylation*). If this occurs in a hydrophobic region of the enzyme, it makes that region hydrophilic, because phosphate carries a negative charge. The protein twists, and this can expose or hide the active site.

- *Noncovalent binding*: A regulatory molecule may bind noncovalently to an allosteric site, causing the enzyme to change shape. This can either activate or inhibit the enzyme's function.

FRONTIERS When a plant is subjected to adverse conditions such as drought, it makes a hormone called abscisic acid, which directs the plant's adaptive responses. For example, it closes tiny holes (stomata) in the plant's leaves to reduce water loss. Key to this drought response is the enzyme that catalyzes the commitment step of the pathway for abscisic acid synthesis. By understanding the regulation of this enzyme, scientists hope to develop crop plants that can tolerate dry conditions.

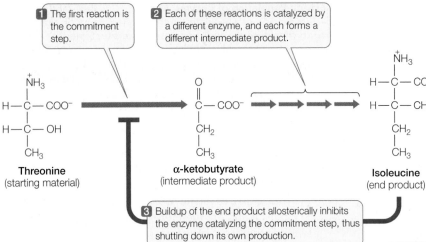

1 The first reaction is the commitment step.

2 Each of these reactions is catalyzed by a different enzyme, and each forms a different intermediate product.

Threonine
(starting material)

α-ketobutyrate
(intermediate product)

Isoleucine
(end product)

3 Buildup of the end product allosterically inhibits the enzyme catalyzing the commitment step, thus shutting down its own production.

FIGURE 3.19 Feedback Inhibition of Metabolic Pathways The first reaction in a metabolic pathway is referred to as the commitment step. Often the end product of the pathway can inhibit the enzyme that catalyzes the commitment step. The specific pathway shown here is the synthesis of isoleucine from threonine in bacteria. It is typical of many enzyme-catalyzed biosynthetic pathways.

Enzymes are affected by their environments

The specificity and activity of an enzyme depend on its three-dimensional structure, and this in turn depends on weak forces such a hydrogen bonds (see Figure 3.7). In living systems, two environmental factors can change protein structure and thereby enzyme activity.

pH AFFECTS ENZYME ACTIVITY We introduced the concept of acids and bases when we discussed amino acids. Some amino acids have side chains that are acidic or basic (see Table 3.2). That is, they either generate H^+ and become anions, or attract H^+ and become cations. These reactions are often reversible. For example:

$$\text{glutamic acid—COOH} \rightleftharpoons \text{glutamic acid—COO}^- + H^+$$

The ionic form of this amino acid (right) is far more hydrophilic than the nonionic form (left).

From your studies of chemistry, you may recall the *law of mass action*. In this case the law implies that the higher the H^+ concentration, the more the reaction will be driven to the left (to the nonionic form of glutamic acid). Therefore, changes in the H^+ concentration can alter the level of hydrophobicity of some regions of a protein and thus affect its shape. To generalize, protein tertiary structure, and therefore enzyme activity, is very sensitive to the concentration of H^+ in the aqueous environment. You may also recall that H^+ concentration is measured by pH (the negative logarithm of the H^+ concentration).

Although the water inside cells is generally at a neutral pH of 7, this can change, and different biological environments have different pH values. Each enzyme has a tertiary structure and amino acid sequence that make it optimally active at a particular pH; its activity decreases as the solution is made more acidic or more basic than this ideal (optimal) pH (**FIGURE 3.20A**). As an example, consider the human digestive system (see Concept 39.3). The pH inside the human stomach is highly acidic, about pH 1.5. Many enzymes that hydrolyze macromolecules in the intestine, such as proteases, have pH optima in the neutral range. So when food enters the small intestine, a buffer (bicarbonate) is secreted into the intestine to raise the pH to 6.5. This allows the hydrolytic enzymes to be active and digest the food.

TEMPERATURE AFFECTS ENZYME ACTIVITY In general, warming increases the rate of a chemical reaction because a greater proportion of the reactant molecules have enough kinetic energy to provide the activation energy for the reaction. Enzyme-catalyzed reactions are no different (**FIGURE 3.20B**). However, temperatures that are too high inactivate enzymes, because at high temperatures the enzyme molecules vibrate and twist so rapidly that some of their noncovalent bonds break. When an enzyme's tertiary structure is changed by heat, the enzyme loses its function. Some enzymes denature at temperatures only slightly above that of the human

(A)

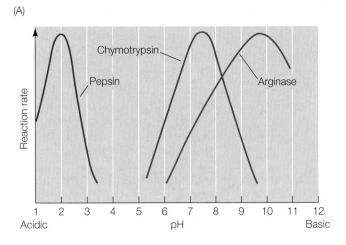

(B)

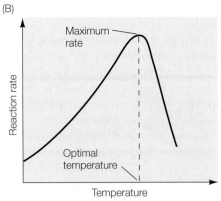

FIGURE 3.20 Enzyme Activity Is Affected by the Environment (A) The activity curve for each enzyme peaks at its optimal pH. For example, pepsin is active in the acidic environment of the stomach, whereas chymotrypsin is active in the small intestine. (B) Similarly, there is an optimal temperature for each enzyme. At higher temperatures the enzyme becomes denatured and inactive; this explains why the activity curve falls off abruptly at temperatures that are above optimal.

APPLY THE CONCEPT

Regulation of metabolism occurs by regulation of enzymes

The concept of enzymes as biological catalysts has many applications. In a pile of clothes in your garage, you notice there are bacteria growing on some socks made of this synthetic polymer:

$$[—CO—(CH_2)_4—CO—NH—(CH_2)_4—NH—]_n$$

You make a protein extract from the bacteria and isolate what you think is an enzyme that can cleave the monomers from the polymer. You also synthesize the dipeptide glycine-glycine (see Table 3.2) to test as a possible inhibitor of the enzyme. The table shows the results from several of your experiments.

EXPERIMENT	CONDITION	RATE OF POLYMER CLEAVAGE
1	No enzyme	0.505
2	Enzyme	825.0
3	Enzyme pre-boiled at 100°C	0.520
4	Enzyme + RNA	799.0
5	Enzyme + dipeptide	0.495

1. Explain the results of each experiment.
2. How do you think the dipeptide works? How would you test your hypothesis?

body, but a few are stable even at the boiling point (or freezing point) of water. All enzymes have an optimal temperature for activity.

Individual organisms adapt to changes in the environment in many ways, one of which is based on groups of enzymes called *isozymes* that catalyze the same reaction but have different chemical compositions and physical properties. Different isozymes within a given group may have different optimal temperatures. The rainbow trout, for example, has several isozymes of the enzyme acetylcholinesterase. If a rainbow trout is transferred from warm water to near-freezing water (2°C), the fish produces a different isozyme of acetylcholinesterase. The new isozyme has a lower optimal temperature, allowing the fish's nervous system to perform normally in the colder water.

In general, enzymes adapted to warm temperatures do not denature at those temperatures because their tertiary structures are held together largely by covalent bonds such as disulfide bridges, instead of the more heat-sensitive weak chemical interactions.

Do You Understand Concept 3.4?

- Explain and give examples of irreversible and reversible enzyme inhibitors.

- The amino acid glutamic acid (see Table 3.2) is at the active site of an enzyme. Normally the enzyme is active at pH 7. At pH 4 (higher concentration of H+), the enzyme is inactive. Explain these observations.

- An enzyme is subject to allosteric regulation. How would you design an inhibitor of the enzyme that was competitive? Noncompetitive? Irreversible?

- Some organisms thrive at a pH of 2; other organisms thrive at a temperature of 65°C. Yet mammals cannot tolerate either environment in their tissues. Explain.

QA **QUESTION** How does an understanding of proteins and enzymes help to explain how aspirin works?

ANSWER The mechanism by which aspirin works exemplifies many of the concepts introduced in this chapter. Robert Vane showed that aspirin binds to a protein with a specific three-dimensional structure (Concept 3.2), and that target is the enzyme cyclooxygenase (Concept 3.3). Aspirin acts as an irreversible inhibitor of the enzyme (Concept 3.4), thus shutting down the metabolic pathway for which cyclooxygenase is a commitment step. Follow the description below carefully, as it illustrates these important concepts.

Cyclooxygenase catalyzes the conversion of a fatty acid with 20 carbon atoms, arachidonic acid, to a structure with a ring (thus the "cyclo" in the name of the enzyme). O_2 is a cofactor (thus the "oxygen"; **FIGURE 3.21**). The product of

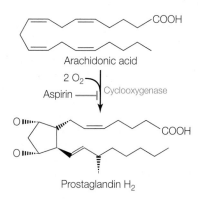

FIGURE 3.21 Aspirin: An Enzyme Inhibitor Aspirin inhibits a key enzyme in the metabolic pathways leading to inflammation and blood clotting.

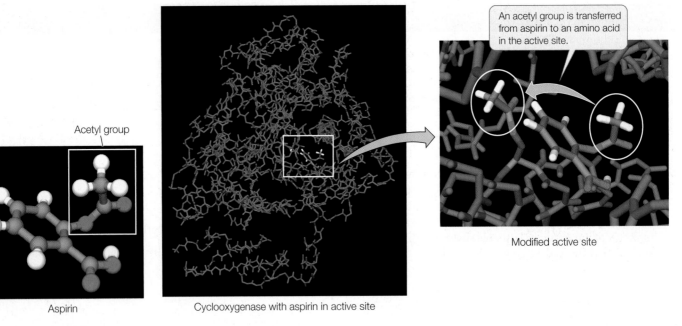

Acetyl group

Aspirin

Cyclooxygenase with aspirin in active site

An acetyl group is transferred from aspirin to an amino acid in the active site.

Modified active site

FIGURE 3.22   Inhibition by Covalent Modification   Aspirin inhibits cyclooxygenase by covalent modification of an amino acid at the active site of the enzyme.

this reaction (prostaglandin H$_2$) is the starting material for the biochemical pathways that produce two types of molecules:

- *Prostaglandins*, which are involved in inflammation and pain, and
- *Thromboxanes*, which stimulate blood clotting and constriction of blood vessels.

Aspirin binds at the active site of cyclooxygenase and transfers an acetyl group to the exposed hydroxyl group of a nearby serine residue (**FIGURE 3.22**):

$$\text{(cyclooxygenase)–serine–OH} \rightarrow$$
$$\text{(cyclooxygenase)–serine–O–CH}_2\text{–CH}_3$$

This covalent modification changes the exposed, polar serine to a less polar molecule, and it becomes slightly more hydrophobic. The shape of the active site changes and becomes inaccessible to the substrate, arachidonic acid. The enzyme is inhibited, and the pathways leading to prostaglandins and thromboxanes are shut down. Less pain, inflammation, and blood clotting are the result. Small wonder that aspirin is taken as a pain reliever and a preventive medicine for heart attacks and strokes. It has come a long way from Edward Stone's walk in the woods.

---

3   SUMMARY

concept 3.1   Nucleic Acids Are Informational Macromolecules

- The nucleic acids—DNA and RNA—are used mainly for information storage and expression.

- Nucleic acids are polymers of nucleotides. A **nucleotide** consists of one to three phosphate groups, a pentose sugar (**ribose** in RNA and **deoxyribose** in DNA), and a nitrogen-containing **base**. Review Figure 3.1 and WEB ACTIVITY 3.1

- In DNA, the nucleotide bases are **adenine** (**A**), **guanine** (**G**), **cytosine** (**C**), and **thymine** (**T**). **Uracil** (**U**) replaces thymine in RNA. The nucleotides are joined by **phosphodiester linkages** between the sugar of one and the phosphate of the next. RNA is usually single-stranded, whereas DNA is double-stranded. Review Figure 3.2

- **Complementary base pairing**, based on hydrogen bonds between A and T, A and U, and G and C, occurs in RNA and DNA. In RNA the hydrogen bonds result in a folded molecule;

in DNA the hydrogen bonds connect two strands into a double helix. **Review Figures 3.3 and 3.4 and WEB ACTIVITY 3.2**

- DNA is expressed as RNA in the process of **transcription**. RNA can then specify the amino acid sequence of a protein in the process of **translation**.
See ANIMATED TUTORIAL 3.1

concept 3.2   Proteins Are Polymers with Important Structural and Metabolic Roles

- The functions of proteins include support, protection, catalysis, transport, defense, regulation, storage, and movement.

- **Amino acids** are the monomers from which polymeric proteins are made by **peptide linkages**. There are 20 different amino acids in proteins, each distinguished by a **side chain (R group)** that confers specific properties. **Review Table 3.2 and WEB ACTIVITY 3.3**

- The **primary structure** of a protein is the sequence of amino acids in the polypeptide chain. This chain is folded into a **secondary structure**, which in different parts of the protein may take the form of an **α helix** or a **β pleated sheet**. Review Figure 3.7

- **Disulfide bridges** and noncovalent interactions between amino acids cause polypeptide chains to fold into three-dimensional **tertiary structures**. Multiple polypeptides can interact to form **quaternary structures**. A protein's unique shape and chemical structure allow it to bind specifically to other molecules.

- Heat and certain chemicals can result in a protein becoming **denatured**, which involves the loss of tertiary or secondary structure. Review Figure 3.10 and WORKING WITH DATA 3.1

concept 3.3 Some Proteins Act as Enzymes to Speed up Biochemical Reactions

- A chemical reaction must overcome an energy barrier to get started. An enzyme is a protein catalyst that affects the rate of a biological reaction by lowering the **activation energy** needed to initiate the reaction. Review Figure 3.12 and WEB ACTIVITY 3.4

- A **substrate** binds to the enzyme's **active site**—the site of catalysis—forming an **enzyme–substrate complex**. Enzymes are highly specific for their substrates.

- At the active site, a substrate enters its **transition state**, and the reaction proceeds.

- Substrate binding causes many enzymes to change shape, exposing their active site(s) and allowing catalysis. Review Figure 3.14

- Some enzymes require nonprotein "partners" to carry out catalysis. Review Table 3.3

- Substrate concentration affects the rate of an enzyme-catalyzed reaction. At the maximum rate, the enzyme is saturated with substrate. Review Figure 3.15

concept 3.4 Regulation of Metabolism Occurs by Regulation of Enzymes

- Metabolism is organized into pathways in which the product of one reaction is a substrate for the next reaction. Each reaction in the pathway is catalyzed by a specific enzyme.

- Enzyme activity is subject to regulation. Some inhibitors bind irreversibly to enzymes. Other inhibitors bind reversibly. Review Figures 3.16 and 3.17 and ANIMATED TUTORIAL 3.2

- In **allosteric regulation**, a molecule binds to a site on the enzyme other than the active site. This changes the overall structure of the enzyme (including that of its active site) and results in either activation or inhibition of the enzyme's catalytic activity. Review Figure 3.18 and ANIMATED TUTORIAL 3.3

- The end product of a metabolic pathway may inhibit an enzyme that catalyzes the "commitment step" of that pathway. This is called **feedback inhibition**. Review Figure 3.19

- Environmental pH and temperature affect enzyme activity. Review Figure 3.20

In 1818, a 19-year-old London writer, Mary Shelley, published a novel that shocked a society in the midst of the Industrial Revolution. In Shelley's story, Dr. Victor Frankenstein discovers how to use electricity to reanimate dead creatures. Collecting body parts from graves and medical labs, the fictional doctor assembles them into a huge 8-foot-tall body and uses his secret method to bring it to life. The results are disastrous, and the novel became a cautionary tale about the limits of science.

Almost 200 years later, in 2010, biologists Craig Venter and Hamilton Smith also gave new life to an "empty shell." In this case, the "shell" was a cell of the tiny bacterium *Mycoplasma discoides*, in which the DNA had been destroyed by hydrolysis. Without DNA to direct the synthesis of its proteins, the cell would die. The scientists used a computer to design an artificial DNA sequence that had all the genes necessary for bacterial life, plus some unique sequences. Then they went into the chemistry lab and made the DNA from individual nucleotides. They inserted this synthetic genome (similar to the genome of the closely related bacterium *Mycoplasma mycoides*) into the bacterium, where it replaced the bacterium's normal DNA, which was hydrolyzed. The new DNA directed the cell to perform all the biochemical characteristics of life, including cell reproduction. Since the new genome had some distinctive DNA sequences devised by the scientists, it was an entirely new organism, called *Mycoplasma mycoides JVCI-syn1.0.*

Why did Venter and Smith need to start with a pre-existing cell? The chemical reactions of metabolism, polymerization, and replication cannot occur in a dilute aqueous environment; it would be too unlikely for reactants and enzymes to collide with one another. Life requires

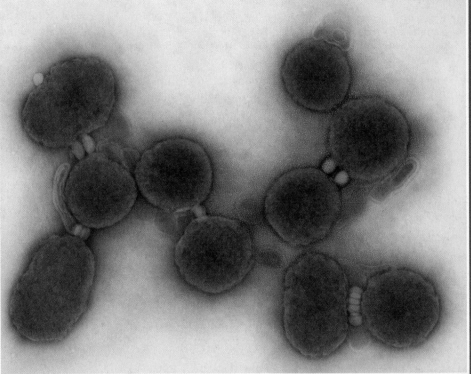

a compartment that brings together and concentrates the molecules involved in these events, which ultimately are directed by the DNA genome.

After several cell divisions (40, by Venter and Smith's estimate), the cells of the new organism no longer had any of the original cell's proteins or small molecules. The cells had used substances in the environment to synthesize their own small and large molecules. They were truly individuals of a new organism, whose "parent" was a computer.

The practical aim of this research is to create cells with new capabilities, such as synthesizing clean-burning fuels. But it also puts cells into broader focus as the basic units of biological structure and function.

Cells of *Mycoplasma mycoides JVCI-syn1.0.* These are the first synthetic cells.

 QUESTION
What do the characteristics of modern cells indicate about how the first cells originated?

KEY CONCEPTS

4.1 Cells Provide Compartments for Biochemical Reactions

4.2 Prokaryotic Cells Do Not Have a Nucleus

4.3 Eukaryotic Cells Have a Nucleus and Other Membrane-Bound Compartments

4.4 The Cytoskeleton Provides Strength and Movement

4.5 Extracellular Structures Allow Cells to Communicate with the External Environment

Cells Provide Compartments for Biochemical Reactions

Cells contain water and other small and large molecules, which we examined in Chapters 2 and 3. Each cell contains at least 10,000 different types of molecules, most of them present in many copies. Cells use these molecules to transform matter and energy, to respond to their environments, and to reproduce. As we mentioned in the essay above, these biological processes would be impossible outside the enclosure of a cell.

The **cell theory**, developed in the nineteenth century, recognizes this basic fact about life. It was the first unifying principle of biology and has three critical components:

• Cells are the fundamental units of life.

• All living organisms are composed of cells.

• All cells come from preexisting cells.

Cell theory has two important conceptual implications:

• *Studying cell biology is in some sense the same as studying life.* The principles that underlie the functions of a single bacterial cell are similar to those governing the approximately 60 trillion cells in an adult human.

• *Life is continuous.* All those human cells came from a single cell, a zygote (or fertilized egg). The zygote was formed when two cells fused: a sperm from the father and an egg from the mother. The cells of the parents' bodies were all derived from their parents, and so on back through the generations—all the way back to the evolution of the first living cells.

Cell size is limited by the surface area-to-volume ratio

Most cells are tiny. Their volumes range from 1 to 1,000 cubic micrometers (**FIGURE 4.1**). There are some exceptions: the eggs of birds are single cells that are, relatively speaking, enormous, and individual cells of several types of algae and bacteria are large enough to be viewed with the unaided eye.

Small cell size is a practical necessity arising from the decrease in the **surface area-to-volume ratio** of any object as it increases in size. As an object increases in volume, its surface area also increases, but not as quickly (**FIGURE 4.2**). This phenomenon has great biological significance for two reasons:

• The *volume* of a cell determines the amount of metabolic activity it carries out per unit of time.

• The *surface area* of a cell determines the amount of substances that can enter it from the outside environment, and the amount of waste products that can exit to the environment.

As a living cell grows larger, its metabolic activity, and thus its need for resources and its rate of waste production, increases faster than its surface area. In addition, substances must move from one location to another within the cell; the smaller the cell, the more easily this is accomplished. The large surface area-to-volume ratio represented by the many small cells of a multicellular organism enables it to carry out the many different functions required for survival.

---

yourBioPortal.com

**Go to WEB ACTIVITY 4.1
The Scale of Life**

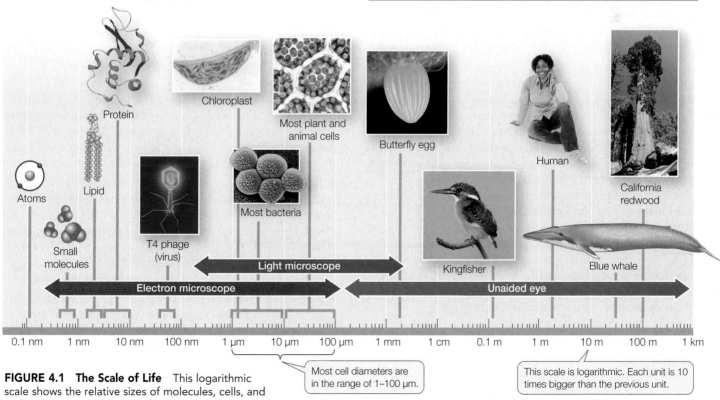

FIGURE 4.1 The Scale of Life This logarithmic scale shows the relative sizes of molecules, cells, and multicellular organisms.

Protein
Chloroplast
Most plant and animal cells
Butterfly egg
Human
California redwood
Atoms
Lipid
Most bacteria
T4 phage (virus)
Kingfisher
Blue whale
Small molecules
Light microscope
Electron microscope
Unaided eye

0.1 nm   1 nm   10 nm   100 nm   1 μm   10 μm   100 μm   1 mm   1 cm   0.1 m   1 m   10 m   100 m   1 km

Most cell diameters are in the range of 1–100 μm.

This scale is logarithmic. Each unit is 10 times bigger than the previous unit.

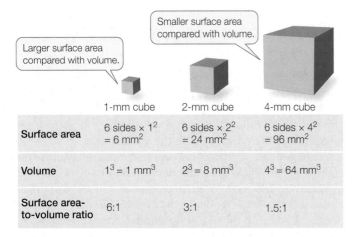

	1-mm cube	2-mm cube	4-mm cube
Surface area	6 sides × 1^2 = 6 mm^2	6 sides × 2^2 = 24 mm^2	6 sides × 4^2 = 96 mm^2
Volume	1^3 = 1 mm^3	2^3 = 8 mm^3	4^3 = 64 mm^3
Surface area-to-volume ratio	6:1	3:1	1.5:1

FIGURE 4.2 Why Cells Are Small As an object grows larger, its volume increases more rapidly than its surface area. Cells must maintain a large surface area-to-volume ratio in order to function. This explains why multicellular organisms must be composed of many small cells rather than a few large ones.

Cells can be studied structurally and chemically

The small sizes of most cells necessitate special instruments to study them and their constituents. For *visualizing cells*, there are two types of microscopes (**FIGURE 4.3**):

- *Light microscopes* use glass lenses and visible light to form images. The smallest detail that can be seen with such a microscope is about 0.2 μm in diameter, which is about 1,000 times smaller than an object the human eye can see. Light microscopes are used to visualize living cells and general cell structure.

- *Electron microscopes* use an electron beam focused by magnets. The size limit is 2 nm, which is 100,000 times smaller than something the human eye can see. Electron microscopes can be used to visualize most structures within preserved cells.

The *chemical analysis* of cells usually begins with breaking them open to make a *cell-free extract*. This can be done physically, using a blender or other homogenizing machinery, or osmotically, by placing the cell in a chemical environment where it swells and bursts (see Figure 5.3). In either case, the resulting extract can be analyzed in terms of its composition and chemical reactions. For example, specific enzyme activities may be measured. If conditions are right in this test tube system, the *properties of the cell-free extract are the same as those inside the cell*. This last statement is of great importance, because it allows biologists to study the chemical processes that occur inside cells in the test tube, so that chemical changes can be easily measured.

A cell's internal structures and even some of its macromolecules can be separated according to their sizes in a centrifuge (**FIGURE 4.4**). Once the subcellular structures are separated from one another, they are much easier to study.

The plasma membrane forms the outer surface of every cell

As we described in the opening of this chapter, a key to the origin of cells was the enclosure of biochemical functions within

FIGURE 4.3 Microscopy Light and electron microscopes are used to examine cell structures.

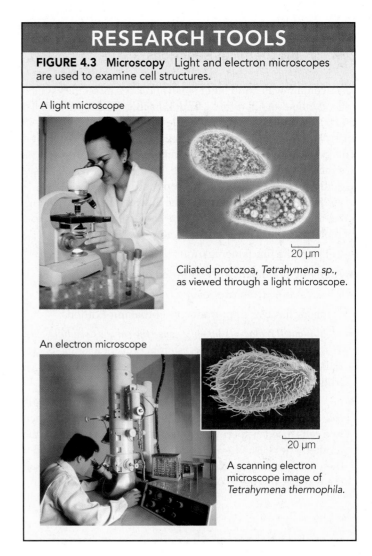

A light microscope

Ciliated protozoa, *Tetrahymena sp.*, as viewed through a light microscope.

20 μm

An electron microscope

A scanning electron microscope image of *Tetrahymena thermophila*.

20 μm

a membrane. We will describe the **plasma membrane** in more detail in Chapter 5, but for now, note that it consists of phospholipid bilayers with proteins, and that there is both compositional and functional diversity within this general framework. Unless stained for light microscopy, the very thin (7 nm) plasma membrane is visible only in electron microscopy. It has several important roles:

- The plasma membrane acts as a *selectively permeable barrier*, preventing some substances from crossing it while permitting other substances to enter and leave the cell. In doing so, it allows the cell to maintain homeostasis, a stable internal environment that is distinct from the surrounding environment. This explains why a red blood cell contains the pigmented molecule hemoglobin but the surrounding blood plasma does not.

- As the cell's boundary with the outside environment, the plasma membrane is important in *communicating* with adjacent cells and receiving signals from the environment.

- The plasma membrane often has proteins protruding from it that are responsible for *binding and adhering* to adjacent cells or to a surface. Thus the plasma membrane plays an important structural role and contributes to cell shape.

RESEARCH TOOLS

FIGURE 4.4 Centrifugation Structures within cells can be separated from one another on the basis of size and density, and the isolated structures can then be analyzed chemically.

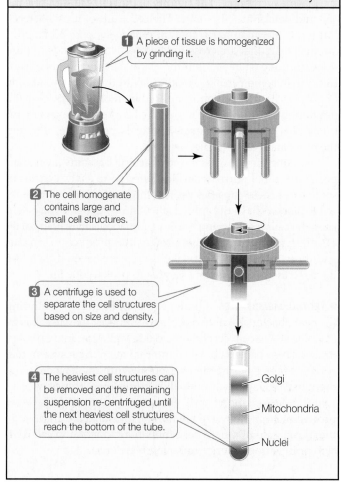

1 A piece of tissue is homogenized by grinding it.

2 The cell homogenate contains large and small cell structures.

3 A centrifuge is used to separate the cell structures based on size and density.

4 The heaviest cell structures can be removed and the remaining suspension re-centrifuged until the next heaviest cell structures reach the bottom of the tube.

Golgi

Mitochondria

Nuclei

Cells are classified as either prokaryotic or eukaryotic

Biologists classify all living things into three domains: Archaea, Bacteria, and Eukarya. The organisms in Archaea and Bacteria are collectively called **prokaryotes** because they have in common a prokaryotic cellular organization. A prokaryotic cell typically does not have membrane-enclosed internal compartments; in particular, it does not have a *nucleus*.

Eukaryotic cell organization is found in members of the domain Eukarya—the **eukaryotes**—which includes the protists (a diverse group of microorganisms), plants, fungi, and animals. In contrast to the prokaryotes, eukaryotes contain membrane-enclosed compartments called **organelles** where specific metabolic functions occur. The most notable of these is the cell **nucleus**, where most of the cell's DNA is located and where gene expression begins:

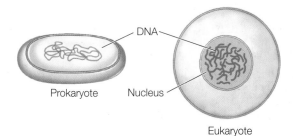

DNA

Prokaryote   Nucleus

Eukaryote

Just as a cell is an enclosed compartment, separating its contents from the surrounding environment, so each organelle provides a compartment that separates certain molecules and biochemical reactions from the rest of the cell. This impressive "division of labor" provides possibilities for regulation and efficiency that were important in the evolution of complex organisms.

LINK Eukaryotes arose from prokaryotes by endosymbiosis; see Concept 20.1

Do You Understand Concept 4.1?

- In considering the origin of cells, why do biologists focus on the origin of the plasma membrane?

- Figure 4.2 shows cube-shaped representations of cells. Many cells, however, are spherical. The surface area of a sphere is $4\pi r^2$ and the volume is $(4/3)(\pi r^3)$. What is the surface area-to-volume ratio of a spherical cell with a diameter of 2 μm and one with a diameter of 3 μm? What are the implications of this difference between the two cells for cell function?

- What evolutionary advantages does a eukaryotic cell have compared with a prokaryotic cell?

As we mentioned in this section, there are two structural themes in cell architecture: prokaryotic and eukaryotic. We'll turn now to the organization of prokaryotic cells.

concept 4.2 Prokaryotic Cells Do Not Have a Nucleus

In terms of sheer numbers and diversity, prokaryotes are the most successful organisms on Earth. As we generalize about the features of these cells, bear in mind that there are vast numbers of prokaryotic species, and that the Bacteria and Archaea can be distinguished from one another in numerous ways. These differences, and the vast diversity of organisms in these two domains, are the subject of Chapter 19.

The volume of a prokaryotic cell is typically about one fiftieth of the volume of a eukaryotic cell. Prokaryotic cells range from about 1 to 10 μm in length, width, and diameter. While some prokaryotes exist as single cells, other types form chains or small clusters of cells, and in some cases certain cells in a group perform specialized functions.

Prokaryotic cells share certain features

All prokaryotes have the same basic structure (**FIGURE 4.5**):

- The plasma membrane encloses the cell, separating its interior from the external environment, and regulates the traffic of materials into and out of the cell.

- The **nucleoid** is a region in the cell where the DNA is located. As we described in Chapter 3, DNA is the hereditary material that controls cell growth, maintenance, and reproduction.

The rest of the material enclosed in the plasma membrane is called the **cytoplasm**. The cytoplasm has two components:

- The **cytosol** consists mostly of water containing dissolved ions, small molecules, and soluble macromolecules such as proteins.
- Within the cytosol are insoluble suspended particles, including ribosomes. **Ribosomes** are complexes of RNA and proteins that are about 25 nm in diameter. They can be visualized only with the electron microscope. They are the sites of protein synthesis, where the information encoded by nucleic acids directs the sequential linking of amino acids to form proteins.

The cytoplasm is not a static region. Rather, the substances in this environment are in *constant motion*. For example, a typical protein moves around the entire cell within a minute, and it collides with many other molecules along the way. This constant motion helps ensure that biochemical reactions proceed at sufficient rates to meet the needs of the cell. Prokaryotes may look simple, but in reality they are functionally complex, carrying out thousands of biochemical reactions.

Specialized features are found in some prokaryotes

As they evolved, some prokaryotes developed specialized structures that gave them a selective advantage in their particular environments. These cells were better able to survive and reproduce than cells lacking the specialized structures.

CELL WALLS   Most prokaryotes have a cell wall located outside the plasma membrane. The rigidity of the cell wall supports the cell and determines its shape. The cell walls of most bacteria, but not those of archaea, contain *peptidoglycan*, a polymer of amino sugars that are cross-linked by covalent bonds to peptides, forming a single giant molecule that surrounds the entire cell. In some bacteria, another layer, the outer membrane (a polysaccharide-rich phospholipid membrane), encloses the peptidoglycan layer (see Figure 4.5). Unlike the plasma membrane, this outer membrane is relatively permeable, allowing the movement of molecules across it.

Enclosing the cell wall of some bacteria is a slimy layer composed mostly of polysaccharides, referred to as the *capsule*. In some cases these capsules protect the bacteria from attack by white blood cells in the animals they infect. Capsules also help keep the cells from drying out, and sometimes they help bacteria attach to other cells. Many prokaryotes produce no capsule, and those that do have capsules can survive even if they lose them, so the capsule is not essential to prokaryotic life.

INTERNAL MEMBRANES   Some groups of bacteria—including the cyanobacteria—carry out photosynthesis: they use energy from the sun to convert carbon dioxide and water into carbohydrates. These bacteria have an internal membrane system that contains molecules needed for photosynthesis. The development of photosynthesis, which requires membranes, was an important event in the early evolution of life on Earth. Other prokaryotes have internal membrane folds that are attached to the plasma membrane. These folds may function in cell division or in various energy-releasing reactions.

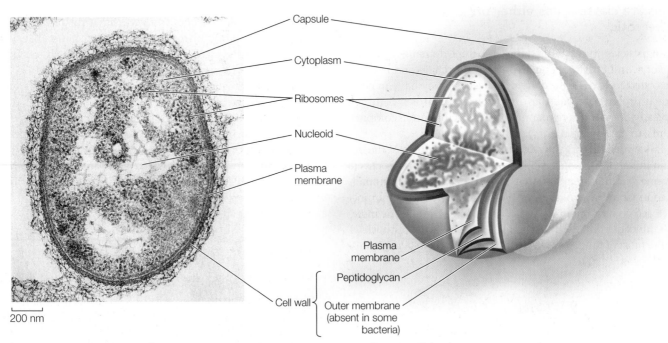

200 nm

FIGURE 4.5   A Prokaryotic Cell   The bacterium *Pseudomonas aeruginosa* illustrates the typical structures shared by all prokaryotic cells. This bacterium also has a protective outer membrane and a capsule, which are not present in all prokaryotes.

(A)

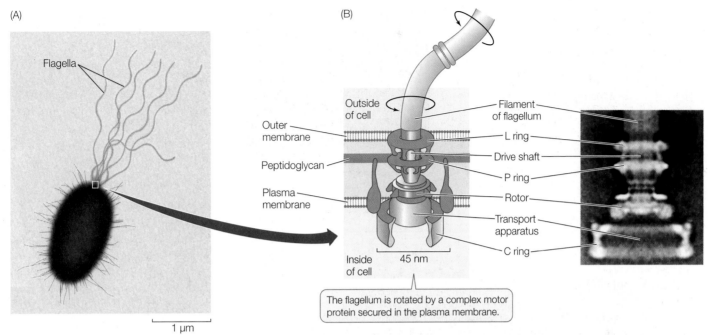

(B)

FIGURE 4.6 Prokaryotic Flagella (A) Flagella contribute to the movement and adhesion of prokaryotic cells. (B) Complex protein ring structures anchored in the plasma membrane form a motor unit that rotates the flagellum and propels the cell.

A bacterium with enclosed compartments would have several evolutionary advantages. Chemicals could be concentrated within particular regions of the cell, allowing chemical reactions to proceed more efficiently. Certain biochemical activities could be segregated within compartments with more favorable conditions for those reactions, such as a different pH from the rest of the cell.

FLAGELLA Some prokaryotes swim by using appendages called **flagella**, which sometimes look like tiny corkscrews (**FIGURE 4.6A**). In bacteria, flagella are made of a protein called *flagellin*. A complex motor protein spins each flagellum on its axis like a propeller, driving the cell along. The motor protein is anchored to the plasma membrane and, in some bacteria, to the outer membrane of the cell wall (**FIGURE 4.6B**). We know that flagella cause the motion of cells because if they are removed, the cells do not move.

CYTOSKELETON Some prokaryotes, especially rod-shaped bacteria, have a helical network of filamentous structures that extend down the length of the cell just inside the plasma membrane. The proteins that make up this structure are similar to actin in eukaryotic cells (which we will discuss next). The helical filaments in these prokaryotes play a role in maintaining their rodlike cell shape.

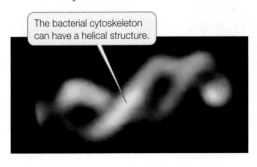

The bacterial cytoskeleton can have a helical structure.

Do You Understand Concept 4.2?

- Describe the structures that are present in all prokaryotic cells.

- Compare the structures and functions of bacterial cell walls with those of bacterial cytoskeletons.

- Flagella are made up of proteins that have other functions, and are present in all prokaryotes. What is the evolutionary advantage of a bacterium that has flagella over bacteria that do not?

As we mentioned earlier, the prokaryotic cell is one of two broad types of cells recognized in cell biology. The other is the eukaryotic cell. Eukaryotic cells, and multicellular eukaryotic organisms, are more structurally and functionally complex than prokaryotic cells.

concept 4.3 Eukaryotic Cells Have a Nucleus and Other Membrane-Bound Compartments

Like prokaryotic cells, eukaryotic cells have a plasma membrane, cytoplasm, and ribosomes. But as you learned earlier in this chapter, eukaryotic cells also have organelles within the cytoplasm whose interiors are separated from the cytosol by membranes (**FIGURE 4.7**).

AN ANIMAL CELL

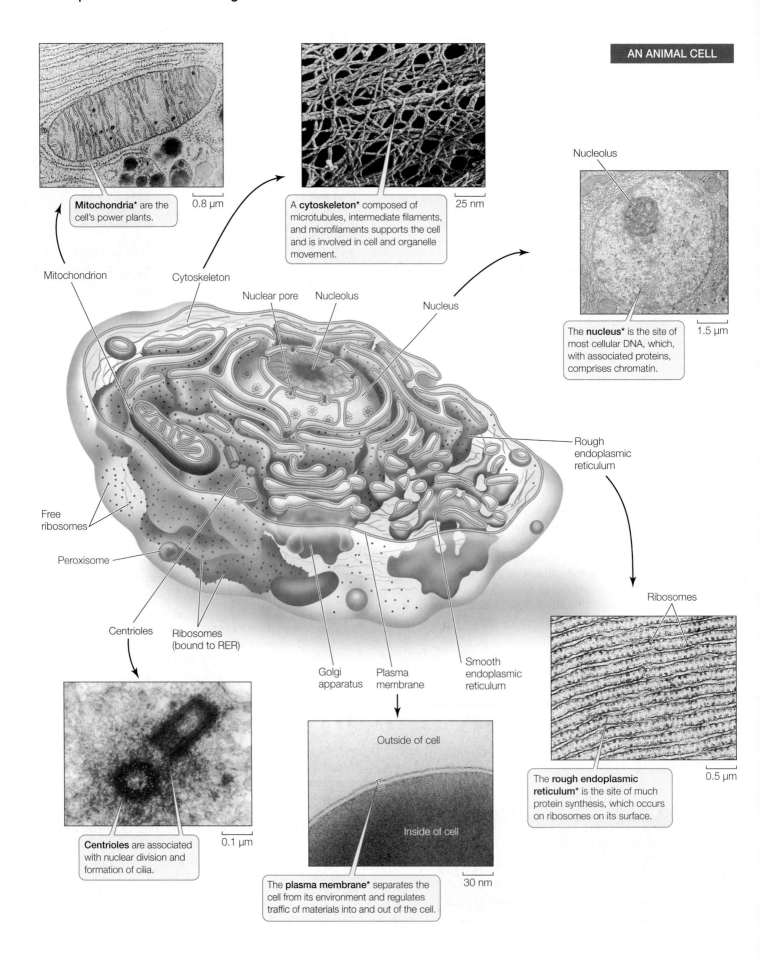

Mitochondria* are the cell's power plants.

0.8 μm

A **cytoskeleton*** composed of microtubules, intermediate filaments, and microfilaments supports the cell and is involved in cell and organelle movement.

25 nm

Nucleolus

1.5 μm

The **nucleus*** is the site of most cellular DNA, which, with associated proteins, comprises chromatin.

Mitochondrion

Cytoskeleton

Nuclear pore   Nucleolus

Nucleus

Rough endoplasmic reticulum

Free ribosomes

Peroxisome

Centrioles

Ribosomes (bound to RER)

Golgi apparatus

Plasma membrane

Smooth endoplasmic reticulum

Ribosomes

Centrioles are associated with nuclear division and formation of cilia.

0.1 μm

Outside of cell

Inside of cell

The **plasma membrane*** separates the cell from its environment and regulates traffic of materials into and out of the cell.

30 nm

The **rough endoplasmic reticulum*** is the site of much protein synthesis, which occurs on ribosomes on its surface.

0.5 μm

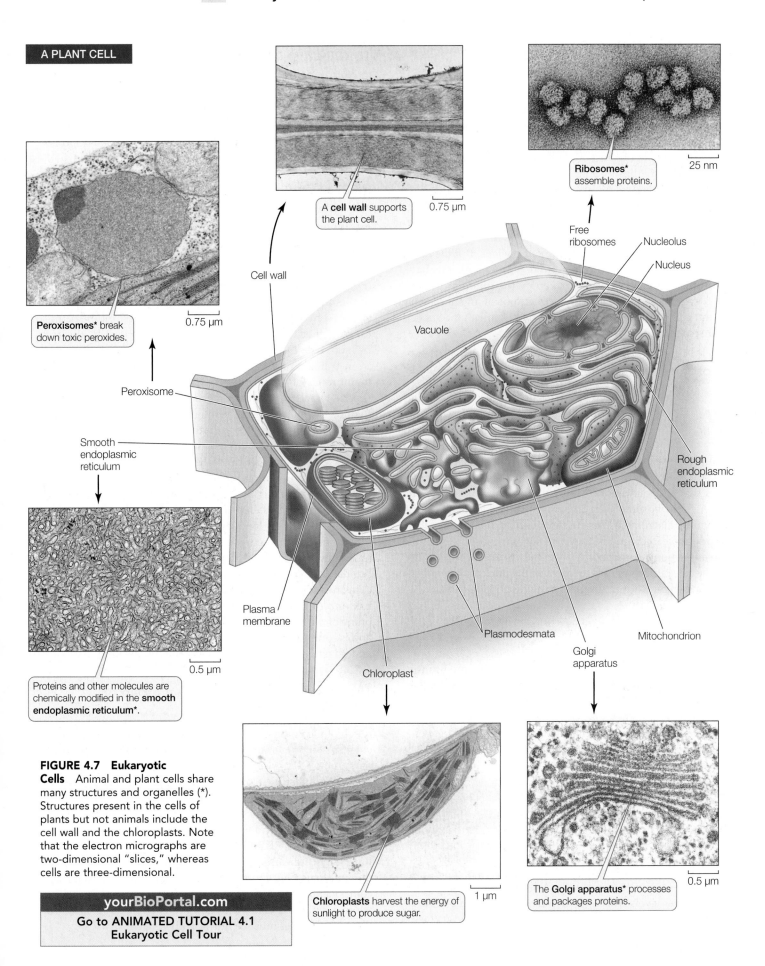

A PLANT CELL

A **cell wall** supports the plant cell.

0.75 μm

Ribosomes* assemble proteins.

25 nm

Peroxisomes* break down toxic peroxides.

0.75 μm

Cell wall

Free ribosomes

Nucleolus

Nucleus

Vacuole

Peroxisome

Smooth endoplasmic reticulum

Rough endoplasmic reticulum

Proteins and other molecules are chemically modified in the **smooth endoplasmic reticulum***.

0.5 μm

Plasma membrane

Plasmodesmata

Mitochondrion

Golgi apparatus

Chloroplast

FIGURE 4.7 Eukaryotic Cells Animal and plant cells share many structures and organelles (*). Structures present in the cells of plants but not animals include the cell wall and the chloroplasts. Note that the electron micrographs are two-dimensional "slices," whereas cells are three-dimensional.

Chloroplasts harvest the energy of sunlight to produce sugar.

1 μm

The **Golgi apparatus*** processes and packages proteins.

0.5 μm

yourBioPortal.com
Go to ANIMATED TUTORIAL 4.1
Eukaryotic Cell Tour

Compartmentalization is the key to eukaryotic cell function

Each type of organelle has a specific role in the cell. Some organelles have been characterized as factories that make specific products. Others are like power plants that take in energy in one form and convert it into a more useful form. In addition to these organelles, eukaryotic cells have some structures that are analogous to those seen in prokaryotes. For example, they have a cytoskeleton composed of protein fibers, and outside the plasma membrane, an extracellular matrix.

When animal and plant cells are examined using electron microscopy, they have many organelles and structures in common—the most obvious is the cell nucleus. But they also have some differences—for example, many plant cells have chloroplasts that are colored green by the pigment used in photosynthesis. Figure 4.7 shows diagrams of an animal and a plant cell, with electron micrographs of some of the subcellular structures.

Ribosomes are factories for protein synthesis

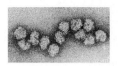

The ribosomes of both prokaryotes and eukaryotes consist of one larger and one smaller subunit; the sizes of the subunits differ between the two cell types. Each subunit consists of one to three large RNA molecules called ribosomal RNA (rRNA) and multiple smaller protein molecules that are bound noncovalently to one another and to the rRNA. The ribosome is an amazingly precise structure. If the individual macromolecules are separated by disruption of their hydrophobic interactions, they will spontaneously reassemble into a functional complex.

The role of ribosomes is to translate the nucleotide sequence of a messenger RNA molecule into a polypeptide chain.

> **LINK**   Protein synthesis is described in more detail in Concept 10.4

Unlike other organelles, ribosomes are not membrane-enclosed compartments. In prokaryotic cells, ribosomes float freely in the cytoplasm. In eukaryotic cells they are found in the cytoplasm, where they may be free or attached to the surface of the endoplasmic reticulum (a membrane-enclosed organelle; see below), and also inside certain organelles—namely the mitochondria and the chloroplasts.

> **FRONTIERS**   Ribosomes are prototypical "macromolecular machines"—large aggregates of macromolecules, bound together noncovalently, that perform complex functions in the cell. Other such aggregates include the chromatin inside the nucleus and the cytoskeleton. Biologists and chemists are currently working to understand how these aggregates fit together, at both the atomic and macromolecular levels. This understanding will lead to greater insights into cell function and into the design of drugs targeted at the aggregates that inhibit protein synthesis.

The nucleus contains most of the DNA

As we noted in Chapter 3, hereditary information is stored in the sequence of nucleotides in DNA molecules. In eukaryotic cells, most of the DNA is in the nucleus. Most cells have a single nucleus, and it is usually the largest organelle; at 5 μm in diameter, the nucleus is substantially larger than most prokaryotic cells. The nucleus has several functions:

- It is the location of the DNA and of DNA replication.
- It is where DNA is transcribed to RNA (see Concept 3.1).
- It contains the **nucleolus**, a region where ribosomes begin to be assembled from RNA and proteins.

If you look carefully at the diagrams and electron micrograph of the nucleus in Figure 4.7, you will see that the nucleus is enclosed by not one but *two membranes*—two lipid bilayers—that together form the *nuclear envelope*. Functionally, this barrier separates DNA transcription (which occurs in the nucleus) from translation (which occurs in the cytoplasm). The two membranes of the nuclear envelope are perforated by thousands of *nuclear pores*, each measuring approximately 9 nm in diameter, which connect the interior of the nucleus to the cytoplasm. The pores regulate the traffic between these two cellular compartments by allowing some molecules to enter or leave the nucleus and by blocking others. This allows the nucleus to regulate its information-processing functions.

Inside the nucleus, DNA is combined with proteins to form a fibrous complex called *chromatin*. Chromatin occurs in the form of exceedingly long, thin threads called *chromosomes*. Different eukaryotic organisms have different numbers of chromosomes (ranging from two in one kind of Australian ant to hundreds in some plants). These DNA–protein complexes become much more compact during cell division, as you will see in Concept 7.2. You can see in Figure 4.7 that the outer membrane of the nuclear envelope folds outward into the cytoplasm and is continuous with the membrane of another organelle, the endoplasmic reticulum, which we will discuss next.

The endomembrane system is a group of interrelated organelles

Much of the volume of many eukaryotic cells is taken up by an extensive **endomembrane system**. This interconnected system of membrane-enclosed compartments includes the nuclear envelope, endoplasmic reticulum, Golgi apparatus, and lysosomes, which are derived from the Golgi. Tiny, membrane-surrounded droplets called **vesicles** shuttle substances between the various components of the endomembrane system, as well as the plasma membrane (**FIGURE 4.8**). In drawings and electron micrographs this system appears static, but in the living cell, membranes and the materials they contain are in constant motion. Membrane components have been observed to shift from one organelle to another within the endomembrane system. This suggests that all these membranes must be functionally related.

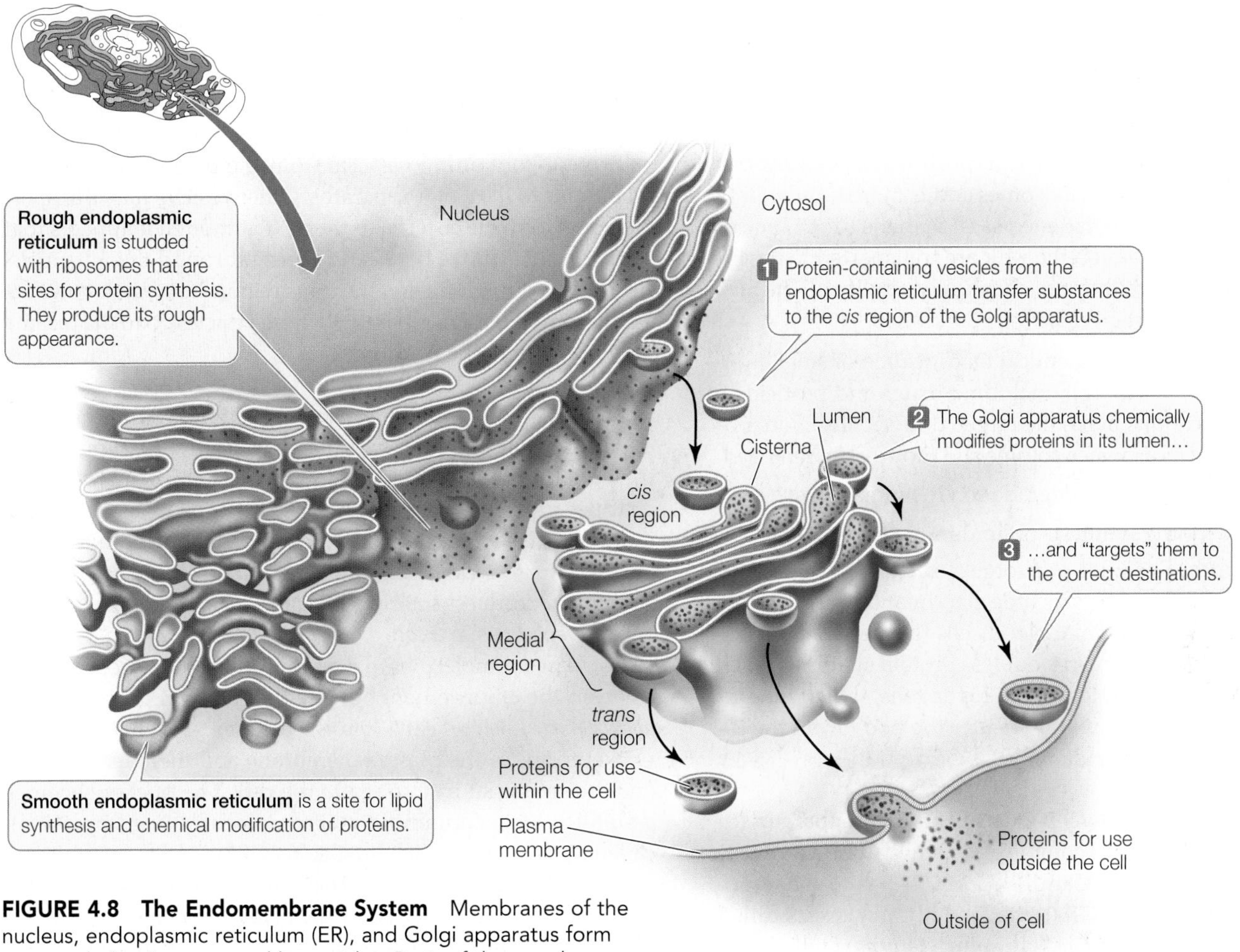

Rough endoplasmic reticulum is studded with ribosomes that are sites for protein synthesis. They produce its rough appearance.

Nucleus

Cytosol

1 Protein-containing vesicles from the endoplasmic reticulum transfer substances to the *cis* region of the Golgi apparatus.

Lumen

Cisterna

cis region

2 The Golgi apparatus chemically modifies proteins in its lumen…

Medial region

3 …and "targets" them to the correct destinations.

trans region

Proteins for use within the cell

Smooth endoplasmic reticulum is a site for lipid synthesis and chemical modification of proteins.

Plasma membrane

Proteins for use outside the cell

Outside of cell

FIGURE 4.8 The Endomembrane System Membranes of the nucleus, endoplasmic reticulum (ER), and Golgi apparatus form a network that is connected by vesicles. Parts of the membrane move between these organelles. Membrane synthesized in the smooth ER becomes sequentially part of the rough ER, then the Golgi, then vesicles formed from the Golgi. These vesicles may eventually fuse with, and become part of, the plasma membrane.

---

yourBioPortal.com

Go to ANIMATED TUTORIAL 4.2
The Golgi Apparatus

ENDOPLASMIC RETICULUM Electron micrographs of eukaryotic cells reveal networks of interconnected membranes branching throughout the cytoplasm, forming tubes and flattened sacs. These membranes are collectively called the **endoplasmic reticulum**, or **ER**. The interior compartment (*lumen*) of the ER is separate and distinct from the surrounding cytoplasm (see Figure 4.8). The ER can enclose up to 10 percent of the interior volume of the cell, and its extensive folding results in a surface area many times greater than that of the plasma membrane. There are two types of ER: rough and smooth.

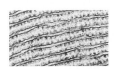 The **rough endoplasmic reticulum (RER)** is called "rough" because of the many ribosomes attached to the outer surface of the membrane, giving it a rough appearance in electron micrographs. These ribosomes are not permanently

attached to the ER but become attached when they begin synthesizing proteins destined for modification within the RER:

- A protein enters the RER only if it contains a specific short sequence of amino acids that signals the ribosome to attach to the RER.
- Once inside the RER, proteins are chemically modified to alter their functions and to chemically "tag" them for delivery to specific cellular destinations.
- The RER participates in transporting these proteins to other locations in the cell. The proteins are transported in vesicles that pinch off from the ER. All secreted proteins pass through the RER.
- Most membrane-bound proteins are made on the RER.

A polypetide that is synthesized on the RER surface is transported across the membrane and into the lumen while it is being translated. Once inside, it undergoes several changes, including the formation of disulfide bridges and folding into its tertiary structure. Many proteins are covalently linked to carbohydrate groups in the RER, thus becoming *glycoproteins*. These carbohydrate groups often have roles in recognition—for example, they identify proteins destined for transfer to specific cellular locations (such as the lysosome; see Figure 4.9). This "addressing" system is very important for ensuring that proteins arrive at their correct destinations. For example, the enzymes within the lysosomes (see below) are highly destructive and could destroy the cell if they were released into the cytosol.

 The **smooth endoplasmic reticulum** (**SER**) is connected to portions of the RER but lacks ribosomes and is more tubular (less like flattened sacs) than the RER. Some of the proteins synthesized on the RER are transported to the lumen of the SER, where they are chemically modified. The SER has three other important roles:

- It is responsible for the chemical modification of small molecules taken in by the cell, including drugs and pesticides. These modifications make the targeted molecules more polar, so they are more water-soluble and more easily removed.
- It is the site for glycogen degradation in animal cells.
- It is the site for the synthesis of lipids and steroids.

Cells that synthesize a lot of protein for export are usually packed with RER. Examples include glandular cells that secrete digestive enzymes and white blood cells that secrete antibodies. In contrast, cells that carry out less protein synthesis (such as storage cells) contain less RER. Liver cells, which modify molecules (including toxins) that enter the body from the digestive system, have abundant SER.

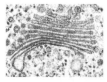 **GOLGI APPARATUS** The **Golgi apparatus** (or Golgi complex) is named after its discoverer, Camillo Golgi. It has two components: flattened membranous sacs called *cisternae* (singular *cisterna*), which are piled up like saucers, and small membrane-enclosed vesicles (see Figure 4.8). The entire apparatus is about 1 μm long.

When protein-containing vesicles from the RER fuse with the Golgi membrane, the proteins are released into the lumen of a Golgi cisterna, where they may be further modified. The Golgi has several roles:

- It concentrates, packages, and sorts proteins before they are sent to their cellular or extracellular destinations.
- It adds some carbohydrates to proteins.
- It is where some polysaccharides for the plant cell wall are synthesized.

FRONTIERS Storage proteins in seeds are a major source of food, but sometimes these proteins are deficient in certain essential amino acids. Crop plants can be genetically engineered to make more nutritious storage proteins; however, the Golgi often packages the altered proteins incorrectly. This makes the proteins unstable, leading to seeds with low protein content. Understanding how proteins are packaged by the Golgi is important for increasing the nutritional value of foods.

The cisternae of the Golgi apparatus have three functionally distinct regions: the *cis* region lies nearest to the nucleus or a patch of RER, the *trans* region lies closest to the plasma membrane, and the *medial* region lies in between (see Figure 4.8). (The terms *cis*, *trans*, and *medial* derive from Latin words meaning "on the same side," "on the opposite side," and "in the middle," respectively.) These three parts of the Golgi apparatus contain different enzymes and perform different functions.

Protein-containing vesicles from the ER fuse with the *cis* membrane of the Golgi apparatus. Other vesicles may transport proteins from one cisterna to the next, although it appears that some proteins move between cisterna through tiny channels. Vesicles budding off from the *trans* region carry their contents away from the Golgi apparatus. These vesicles go to the plasma membrane or to the lysosome.

LYSOSOMES The **primary lysosomes** originate from the Golgi apparatus. They contain hydrolases (digestive enzymes), and they are the sites where macromolecules—proteins, polysaccharides, nucleic acids, and lipids—are hydrolyzed into their monomers (see Chapter 2). A lysosome is about 1 μm in diameter; it is surrounded by a single membrane and has a densely staining, featureless interior (**FIGURE 4.9**). There may be dozens of lysosomes in a cell.

Some macromolecules that are hydrolyzed in lysosomes enter from the environment outside the cell by a process called *phagocytosis* (*phago*, "eat"; *cytosis*, "cellular"). In this process, a pocket forms in the plasma membrane and then deepens and encloses material from outside the cell. The pocket becomes a small vesicle containing macromolecules (e.g., proteins), called a *phagosome*, which breaks free of the plasma membrane to move into the cytoplasm. The phagosome fuses with a primary lysosome to form a **secondary lysosome**, in which hydrolysis occurs. The products of digestion (e.g., amino acids) pass through the membrane of the lysosome, providing monomers for other cellular processes. The "used" secondary lysosome, now containing undigested particles, then moves to the plasma membrane, fuses with it, and releases the undigested contents to the environment.

Phagocytes are specialized cells whose major role is to take in and break down materials; they are found in nearly all animals and many protists. However, lysosomes are active even in cells that do not perform phagocytosis. All cells continually break down some of their components and replace them with new ones. The programmed destruction of cell components is called *autophagy*, and lysosomes are where the cell breaks down its own materials, even entire organelles, hydrolyzing their constituents.

How important is autophagy? An entire class of human diseases called *lysosomal storage diseases* occur when lysosomes fail to digest internal components; these diseases are often very harmful or fatal. An example is Tay-Sachs disease, in which a particular lipid called a ganglioside is not broken down in the lysosomes and instead accumulates in brain cells and damages them. In the most common form of this disease, a baby starts exhibiting neurological symptoms and becomes blind, deaf, and unable to swallow after six months of age. Death occurs before age four.

Plant cells do not appear to contain lysosomes, but the central vacuole of a plant cell (which we will describe below) may

APPLY THE CONCEPT

Eukaryotic cells have a nucleus and other membrane-bound compartments

Proteins can be tagged with radioactivity and specific molecules (antibodies) so they can be detected in particular cell fractions. Liver cells were exposed to radioactive amino acids for 3 minutes, which made all proteins being synthesized during that period radioactive. Then the radioactive amino acids were removed and, at 5-minute intervals, portions of the cells were broken open and fractionated—the organelles were separated from one another as shown in Figure 4.3B. An antibody to the protein lipase was used to distinguish lipase from all the other proteins present. The percentage of radioactively labeled lipase in several cell compartments was determined. The table shows the results.

	PERCENTAGE OF RADIOACTIVE LIPASE			
TIME (MIN)	ER LUMEN	GOLGI	LYSOSOMES	RIBOSOMES
5	5	0	0	95
10	25	10	0	65
15	75	20	5	0
20	25	55	20	0
25	0	65	35	0
30	0	25	75	0
35	0	0	100	0

1. What can you conclude about the pathway of lipase in the cell after it is synthesized?

2. Look up the function of lipase in Concept 39.3. Why is its organelle destination appropriate?

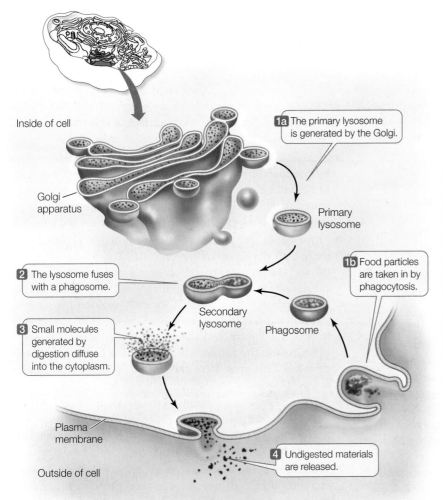

Inside of cell

1a The primary lysosome is generated by the Golgi.

Golgi apparatus

Primary lysosome

2 The lysosome fuses with a phagosome.

1b Food particles are taken in by phagocytosis.

Secondary lysosome

Phagosome

3 Small molecules generated by digestion diffuse into the cytoplasm.

Plasma membrane

4 Undigested materials are released.

Outside of cell

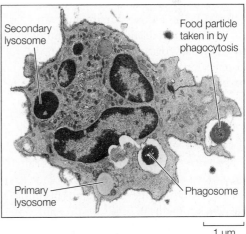

Secondary lysosome

Food particle taken in by phagocytosis

Primary lysosome

Phagosome

1 μm

FIGURE 4.9 Lysosomes Isolate Digestive Enzymes from the Cytoplasm Lysosomes are sites for the hydrolysis of material taken into the cell by phagocytosis.

yourBioPortal.com

Go to WEB ACTIVITY 4.2 Lysosomal Digestion

function in an equivalent capacity because it, like lysosomes, contains many digestive enzymes.

Some organelles transform energy

A cell requires energy to make the molecules it needs for activities such as growth, reproduction, responsiveness, and movement. *Mitochondria* (found in all eukaryotic cells) harvest chemical energy, while *chloroplasts* (found in plants and other photosynthetic cells) harvest energy from sunlight.

 **MITOCHONDRIA** In eukaryotic cells, the breakdown of energy-rich molecules such as the monosaccharide glucose begins in the cytosol. The molecules that result from this partial degradation enter the **mitochondrion** (plural *mitochondria*), whose primary function is to convert the chemical energy of those molecules into a form the cell can use, namely the energy-rich nucleotide ATP (adenosine triphosphate; see Concept 6.1).

A typical mitochondrion is somewhat less than 1.5 μm in diameter and 2–8 μm in length—about the size of many bacteria. It can divide independently of the central nucleus. The number of mitochondria per cell ranges from one gigantic organelle in some unicellular protists to a few hundred thousand in large egg cells. An average human liver cell contains more than 1,000 mitochondria. Cells that are active in movement and growth require the most chemical energy, and these tend to have the most mitochondria per unit of volume.

Mitochondria have two membranes. The outer membrane has large pores, and most substances can pass through it. The inner membrane separates the biochemical processes of the mitochondrion from the surrounding cytosol. The inner membrane is extensively folded into structures called *cristae*, and the fluid-filled region inside the inner membrane is referred to as the mitochondrial *matrix*. The mitochondrion contains many enzymes for energy metabolism, as well as DNA and ribosomes for the synthesis of a small proportion of the mitochondrial proteins.

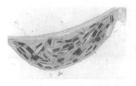 **PLASTIDS** Plastids are present in the cells of plants and algae, and like mitochondria, they can divide autonomously. Plastids can differentiate into a variety of organelles, some of which are used for the storage of pigments (as in flowers), carbohydrates (as in potatoes), lipids, or proteins. An important type of plastid is the **chloroplast**, which contains the green pigment chlorophyll and is the site of photosynthesis (see Concepts 6.5 and 6.6). Photosynthesis is an anabolic process that converts light energy into the chemical energy contained in bonds between the atoms of carbohydrates.

A chloroplast is enclosed within two membranes. In addition, it contains a series of internal membranes that look like stacks of flat, hollow discs, called *thylakoids*. Each stack of thylakoids is called a *granum* (plural *grana*). Light energy is converted to chemical energy on the thylakoid membranes. The aqueous fluid surrounding the thylakoids is called the *stroma*, and it is there that carbohydrates are synthesized. Like the mitochondrial matrix, the chloroplast stroma contains ribosomes and DNA, which are used to synthesize some of the chloroplast proteins.

Several other membrane-enclosed organelles perform specialized functions

There are several other kinds of membrane-bound organelles with specialized functions: peroxisomes, glyoxysomes, and vacuoles, including contractile vacuoles.

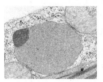 **PEROXISOMES** **Peroxisomes** are small (0.2–1 μm diameter) organelles that accumulate toxic peroxides, such as hydrogen peroxide (H_2O_2), which occur as the by-products of some biochemical reactions in all eukaryotes. These peroxides can be safely broken down inside the peroxisomes without mixing with other components of the cell. A peroxisome has a single membrane and a granular interior containing specialized enzymes.

GLYOXYSOMES **Glyoxysomes** are found only in plants. They are most abundant in young plants and are the locations where stored lipids are converted into carbohydrates for transport to growing cells.

VACUOLES **Vacuoles** occur in many eukaryotic cells, but particularly those of plants and fungi (see Figure 4.6). Plant vacuoles have several functions:

- *Storage*: Like all cells, plant cells produce a variety of toxic by-products and waste products. Plants store many of these in vacuoles. Because they are poisonous or distasteful, these stored materials deter some animals from eating the plants, and may thus contribute to the plants' defenses and survival.

- *Structure*: In many plant cells, enormous vacuoles take up more than 90 percent of the cell volume and grow as the cell grows. The presence of dissolved substances in the vacuole causes water to enter it from the cytoplasm (which in turn takes up water from outside the cell), making the vacuole swell like a water-filled balloon. (This osmotic effect is illustrated in Figure 5.3.) The plant cell wall resists the swelling, causing the cell to stiffen from the increase in water pressure. This pressure is called *turgor pressure*, and it helps support the plant.

- *Reproduction*: Some pigments in the petals and fruits of flowering plants are contained in vacuoles. These pigments—the red, purple, and blue anthocyanins—are visual cues that help attract animals, which assist in pollination and seed dispersal.

- *Catabolism*: In the seeds of some plants, the vacuoles contain enzymes that hydrolyze stored seed proteins into monomers. The developing plant seedling uses these monomers as building blocks and sources of energy.

Many freshwater protists have *contractile vacuoles*. Their function is to get rid of the excess water that rushes into the cell because of the imbalance in solute concentration between the interior of the cell and its freshwater environment. The contractile vacuole enlarges as water enters, and then abruptly contracts, forcing the water out of the cell through a special pore structure.

Do You Understand Concept 4.3?

- Make a table that summarizes eukaryotic cell organelles with regard to size, numbers per cell, and functions.

- What are some functions of the cell nucleus? What are the advantages of confining these functions within the nucleus, separated from the cytoplasm?

- Compare the structural and functional differences between rough and smooth endoplasmic reticulum.

- In I-cell disease, an enzyme in the endomembrane system that normally adds phosphorylated sugar groups to proteins is lacking, and the proteins are not targeted to the lysosomes as they would be in normal cells. The "I" stands for inclusion bodies that appear in the cells. What do you think these inclusions are, and why do they accumulate?

So far, we have discussed numerous membrane-enclosed organelles. Now we'll turn to a group of cytoplasmic structures without membranes.

concept 4.4 The Cytoskeleton Provides Strength and Movement

The interior of the cell has a meshwork of protein filaments. Each type of filament is a polymer, made up of monomers that are proteins (which in turn are polymers of amino acids). This **cytoskeleton** fills several important roles:

- It supports the cell and maintains its shape.
- It controls the positions and movements of organelles within the cell.
- It is involved with bulk movements of the cytoplasm, called *cytoplasmic streaming*.
- It interacts with extracellular structures, helping to anchor the cell in place.

There are three components of the cytoskeleton: microfilaments (smallest diameter), intermediate filaments, and microtubules (largest diameter). These filaments have very different functions.

Microfilaments are made of actin

Microfilaments (FIGURE 4.10A) are usually in bundles. Each filament is about 7 nm in diameter and up to several micrometers long. Microfilaments have two major roles:

- They help the entire cell or parts of the cell to move.
- They determine and stabilize cell shape.

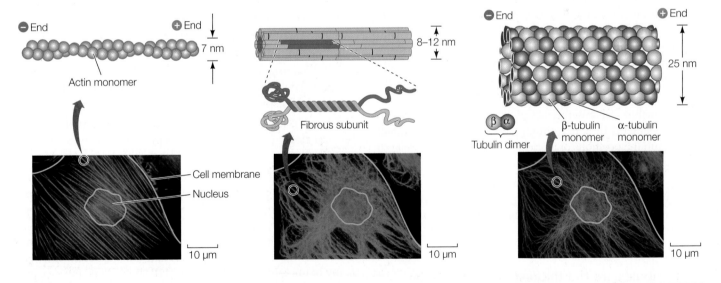

(A) Microfilaments
Made up of strands of the protein actin; often interact with strands of other proteins.

(B) Intermediate filaments
Made up of fibrous proteins organized into tough, ropelike assemblages that stabilize a cell's structure and help maintain its shape.

(C) Microtubules
Long, hollow cylinders made up of many molecules of the protein tubulin. Tubulin consists of two subunits, α-tubulin and β-tubulin.

FIGURE 4.10 The Cytoskeleton Three highly visible and important structural components of the cytoskeleton are shown here in detail. Specific stains were used to visualize them in a single cell. These structures maintain and reinforce cell shape and contribute to cell movement.

Microfilaments are assembled from *actin* monomers that attach to the filament at one end (the "plus end") and detach at the other (the "minus end"). In an intact filament, assembly and detachment are in equilibrium. But sometimes the filaments can shorten (more detachment) or lengthen (more assembly):

actin polymer (filament) ⇌ actin monomers

This property of **dynamic instability** is a hallmark of the cytoskeleton. Portions of it can be made and broken down rather quickly, depending on cell function. Actin-associated proteins work at both ends of the filament to catalyze assembly and disassembly.

In the muscle cells of animals, actin filaments are associated with another protein, the motor protein *myosin*, and the interactions of these two proteins account for the contraction of muscles. In non-muscle cells, actin filaments are associated with localized changes in cell shape. For example, microfilaments are involved in the flowing movement of the cytoplasm called cytoplasmic streaming, in amoeboid movement, and in the "pinching" contractions that divide an animal cell into two daughter cells. Microfilaments are also involved in the formation of cellular extensions called pseudopodia (*pseudo*, "false"; *podia*, "feet") that enable some cells (such as *Amoeba*, see Figure 4.14) to move.

Intermediate filaments are diverse and stable

There are at least 50 different kinds of **intermediate filaments** (**FIGURE 4.10B**), many of them specific to just a few cell types. They generally fall into six molecular classes (based on amino acid sequence) that share the same general structure. One of these classes consists of fibrous proteins of the keratin family, which also includes the proteins in hair and fingernails. The intermediate filaments are tough, ropelike protein assemblages 8–12 nm in diameter. Intermediate filaments are more permanent than the other two types of filaments and do not show dynamic instability. Intermediate filaments have two major structural functions:

- They anchor cell structures in place. In some cells, intermediate filaments radiate from the nuclear envelope and help maintain the positions of the nucleus and other organelles in the cell.

- They resist tension. For example, they maintain rigidity in body surface tissues by extending through the cytoplasm and connecting specialized membrane structures called desmosomes (see Figure 4.18).

Microtubules are the thickest elements of the cytoskeleton

Microtubules are the largest diameter components of the cytoskeletal system (**FIGURE 4.10C**). They are long, hollow, unbranched cylinders about 25 nm in diameter and up to several micrometers long. Microtubules have two roles:

- They form a rigid internal skeleton for some cells or cell regions.

- They act as a framework along which motor proteins can move structures within the cell.

Microtubules are assembled from dimers of the protein *tubulin*. The dimers consist of one molecule each of α-tubulin and β-tubulin. Thirteen chains of tubulin dimers surround the hollow microtubule. Like microfilaments, microtubules show dynamic instability, with (+) and (–) ends and associated proteins.

microtubule ⇌ tubulin monomers

Tubulin polymerization results in a rigid structure, and tubulin depolymerization leads to its collapse. Microtubules often form an interior skeleton for projections that come out of the plasma membrane, such as cilia and flagella (see below).

> **FRONTIERS**   In a search for inhibitors of cell division for treating cancer, the molecule taxol was purified from the Pacific yew, an evergreen tree. Taxol works by blocking the disassembly of microtubules at their minus ends—the exact mechanism remains unclear. Treated microtubules are stable, and this prevents chromosome separation during cell division. Unfortunately, taxol is not very soluble in tissue fluids, and it has adverse side effects. There is a major effort to synthesize and test chemically modified taxol to overcome these problems.

Cilia and flagella provide mobility

Microtubules line movable cell appendages: the **cilia** (singular *cilium*; **FIGURE 4.11**) and the flagella (see Concept 4.2). Many eukaryotic cells have one or both of these appendages, which are projections of the plasma membrane lined with microtubules and their associated proteins:

- Cilia are only 0.25 μm in length. They are present by the hundreds and move stiffly to either propel a cell (for example, in protists) or to move fluid over a stationary cell (as in the human respiratory system).

- Flagella are much longer—100 to 200 μm in length—and occur singly or in pairs. They can push or pull the cell through its aqueous environment.

The microtubules that line cilia and flagella do more than just make them rigid. Microtubules and their associated proteins are responsible for the movement of these organelles by bending.

In cross section, a typical cilium or eukaryotic flagellum is surrounded by the plasma membrane and contains a "9 + 2" array of microtubules. As Figure 4.11B shows, nine fused pairs of microtubules—called doublets—form an outer cylinder, and one pair of unfused microtubules runs up the center. A spoke radiates from one microtubule of each doublet and connects the doublet to the center of the structure. These structures are essential to the bending motions of both cilia and flagella. How does this occur?

The motion of cilia and flagella results from the sliding of the microtubule doublets past each other. This sliding is driven by a motor protein called *dynein*, which can change its three-dimensional shape. (All motor proteins work by undergoing

(A)

The beating of the cilia covering the surface of this unicellular protist propels it through the water of its environment.

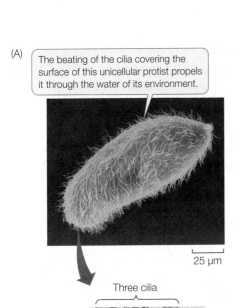

25 µm

Three cilia

250 nm

(B)

Microtubule doublet

Cross section reveals the "9+2" pattern of microtubules, including nine pairs of fused microtubules...

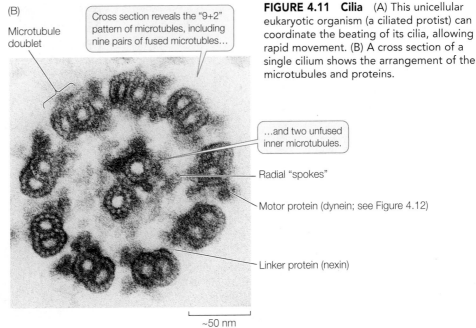

...and two unfused inner microtubules.

Radial "spokes"

Motor protein (dynein; see Figure 4.12)

Linker protein (nexin)

~50 nm

FIGURE 4.11 Cilia (A) This unicellular eukaryotic organism (a ciliated protist) can coordinate the beating of its cilia, allowing rapid movement. (B) A cross section of a single cilium shows the arrangement of the microtubules and proteins.

elles toward the minus end, while kinesin moves them toward the plus end.

Biologists manipulate living systems to establish cause and effect

How do we know that the structural fibers of the cytoskeleton can achieve the dynamic functions described above? We can observe an individual structure under the microscope and a function in a living cell that contains that structure. These simultaneous observations suggest that the structure may carry out that function, but in science, mere *correlation does not establish cause and effect*. For example, light microscopy of living cells reveals the movement of the cytoplasm within the cell. The observed presence of cytoskeletal components *suggests, but does not prove*, their role in this process. Science seeks to show the specific links that relate a structure or molecule ("A") to a

reversible shape changes powered by energy from ATP hydrolysis.) Dynein molecules bind between two neighboring microtubule doublets. As the dynein molecules change shape, they move the doublets past one another (**FIGURE 4.12**). Another protein, *nexin*, can cross-link the doublets and prevent them from sliding past one another; in this case, the cilium bends.

Motor proteins, including *kinesin*, carry protein-laden vesicles from one part of the cell to another (**FIGURE 4.13**). These proteins bind to a vesicle or other organelle, then "walk" it along a microtubule by a repeated series of shape changes. A slightly different form of dynein from the one that moves cilia also performs this function. Recall that microtubules are directional, with a plus end and a minus end. Dynein moves attached organ-

FIGURE 4.12 A Motor Protein Moves Microtubules in Cilia and Flagella The motor protein dynein causes microtubule doublets to slide past one another. If the protein nexin is present to anchor the microtubule doublets together, the flagellum or cilium bends.

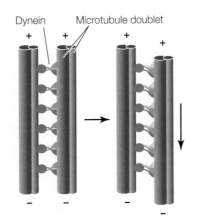

Dynein   Microtubule doublet

In isolated cilia without nexin cross-links, movement of dynein motor proteins causes microtubule doublets to slide past one another.

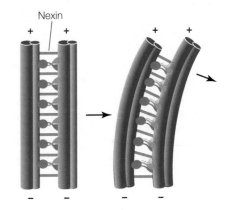

Nexin

When nexin is present to cross-link the doublets, they cannot slide and the force generated by dynein movement causes the cilium to bend.

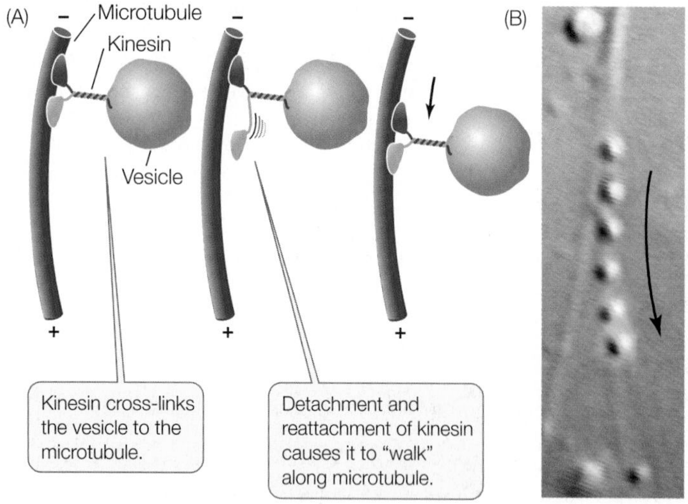

(A)

Microtubule
Kinesin

Vesicle

Kinesin cross-links the vesicle to the microtubule.

Detachment and reattachment of kinesin causes it to "walk" along microtubule.

(B)

FIGURE 4.13 A Motor Protein Drives Vesicles along Microtubules (A) Kinesin delivers vesicles or organelles to various locations in the cell by moving along microtubule "railroad tracks." (B) The process is seen by time-lapse photography at half-second intervals in the protist *Dictyostelium*.

function ("B"). Two manipulative approaches are commonly used in cell biology:

- *Inhibition*: Use a drug that inhibits A, and see if B still occurs. If B does not occur, then A is probably a causative factor for B. **FIGURE 4.14** shows an experiment in which an inhibitor is used to demonstrate cause and effect in the case of the cytoskeleton and cell movement.

- *Mutation*: Examine a cell or organism that lacks the gene (or genes) for A, and see if B still occurs. If it does not, then A is probably a causative factor for B. You will see many examples of this experimental approach later in this book.

Do You Understand Concept 4.4?

- Make a table that compares the three major components of the cytoskeleton with regard to composition, structure, and function.

- The neuron (nerve cell) has a long extension called an axon. Molecules made in the cell's main body must travel a long distance to reach the end of the axon. The axon is lined with microtubules. Explain how motor proteins, vesicles, and microtubules move these molecules along the axon.

- In a dividing cell, the chromosomes become very compact, and then the duplicated sets of chromosomes move along the microtubules to opposite ends of the cell. How would you use an inhibitor to show that microtubules are essential for this chromosomal separation? What control experiments would you suggest?

All cells interact with their environments, and many eukaryotic cells are part of multicellular organisms and must interact with other cells. The plasma membrane plays a crucial role in these interactions, but other structures outside that membrane are

INVESTIGATION

FIGURE 4.14 The Role of Microfilaments in Cell Movement: Showing Cause and Effect in Biology In test tubes, the drug cytochalasin B prevents microfilament formation from monomeric precursors. This led to the question: Will the drug work like this in living cells and inhibit the movement of *Amoeba*?

HYPOTHESIS

Amoeboid cell movements are caused by the cytoskeleton.

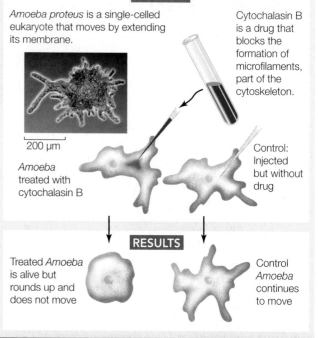

METHOD

Amoeba proteus is a single-celled eukaryote that moves by extending its membrane.

200 μm

Amoeba treated with cytochalasin B

Cytochalasin B is a drug that blocks the formation of microfilaments, part of the cytoskeleton.

Control: Injected but without drug

RESULTS

Treated *Amoeba* is alive but rounds up and does not move

Control *Amoeba* continues to move

CONCLUSION

Microfilaments of the cytoskeleton are essential for amoeboid cell movement

ANALYZE THE DATA

Several important controls were done to validate the conclusions of this experiment. The experiment was repeated in the presence of the following drugs: cycloheximide, which inhibits new protein synthesis; dinitrophenol, which inhibits new ATP formation (energy); and colchicine, which inhibits the polymerization of microtubules. Here are the results:

Condition	Rounded cells (%)
No drug	3
Cytochalasin B	95
Colchicine	4
Cycloheximide	3
Cycloheximide + cytochalasin B	94
Dinitrophenol	5
Dinitrophenol + cytochalasin B	85

Explain each experiment. What can you conclude about *Amoeba* and the cytoskeleton?

Go to **yourBioPortal.com** for original citations, discussions, and relevant links for all INVESTIGATION figures.

involved as well. We now turn to these extracellular structures in animals and plants

Extracellular Structures Allow Cells to Communicate with the External Environment

Although the plasma membrane is the functional barrier between the inside and the outside of a cell, molecules are produced by cells and secreted to the outside of the plasma membrane, where they form structures that play essential roles in protecting, supporting, or attaching cells to each other. Because they are outside the plasma membrane, these structures are said to be *extracellular*. In eukaryotes, these structures are made up of two components:

- A prominent fibrous macromolecule
- A gel-like medium in which the fibers are embedded

The plant cell wall is an extracellular structure

The plant **cell wall** is a semirigid structure outside the plasma membrane (**FIGURE 4.15**). The *fibrous* component is the polysaccharide cellulose (see Figure 2.10), and the *gel-like matrix* contains extensively cross-linked polysaccharides and proteins. The wall has three major roles:

- It provides support for the cell and limits the volume of a mature cell by remaining rigid.
- It acts as a barrier to infection by fungi and other organisms that can cause plant diseases.
- It contributes to plant form by controlling the direction of cell expansion during growth and development.

Because of their thick cell walls, plant cells viewed under a light microscope appear to be entirely isolated from one another. But electron microscopy reveals that this is not the case. The cytoplasms of adjacent plant cells are connected by numerous plasma membrane–lined channels, called **plasmodesmata**, that are about 20–40 nm in diameter and extend through the cell walls (see Figure 4.7). Plasmodesmata allow water, ions, small molecules, hormones, and even some RNA and protein molecules to move between connected cells. In this way, energy-rich molecules such as sugars can be shared among cells, and plant hormones can affect growth at sites far from where they were synthesized. This intercellular communication integrates a plant organ composed of thousands of cells.

> **FRONTIERS** In most plants, cellulose is synthesized by a complex of enzymes at the plasma membrane and then secreted into the cell wall. The orientation of the cellulose fibers determines the final shape of a cell and, in aggregate, the shapes of plant organs such as leaves and roots. Cytoplasmic microtubules appear to be important in orienting the fibers, possibly by interacting with the synthesis complex. However, it is not yet clear how the orientation of the microtubules themselves is determined.

The extracellular matrix supports tissue functions in animals

Animal cells lack the semirigid wall that is characteristic of plant cells, but many animal cells are surrounded by, or in contact with, an **extracellular matrix** (**FIGURE 4.16**). The *fibrous component* of the extracellular matrix is the protein **collagen**, and the gel-like matrix consists of **proteoglycans**, which are glycoproteins with long carbohydrate side chains. A third group of proteins links the collagen and the proteoglycan matrix together.

The extracellular matrices of animal cells have several roles:

- They hold cells together in tissues.
- They contribute to the physical properties of cartilage, skin, and other tissues. For example, the mineral component of bone is laid down on an organized extracellular matrix.
- They help filter materials passing between different tissues. This is especially important in the kidney.
- They help orient cell movements during embryonic development and during tissue repair.

Proteins connect the cell's plasma membrane to the extracellular matrix. These proteins (for example, *integrin*) span the plasma membrane and have two binding sites: one on the interior of the cell, usually to microfilaments in the cytoplasm just

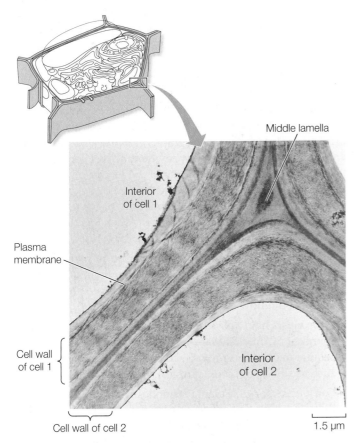

Middle lamella

Interior of cell 1

Plasma membrane

Cell wall of cell 1

Interior of cell 2

Cell wall of cell 2

1.5 µm

FIGURE 4.15 The Plant Cell Wall The semirigid cell wall provides support for plant cells. It is composed of cellulose fibers embedded in a matrix of polysaccharides and proteins.

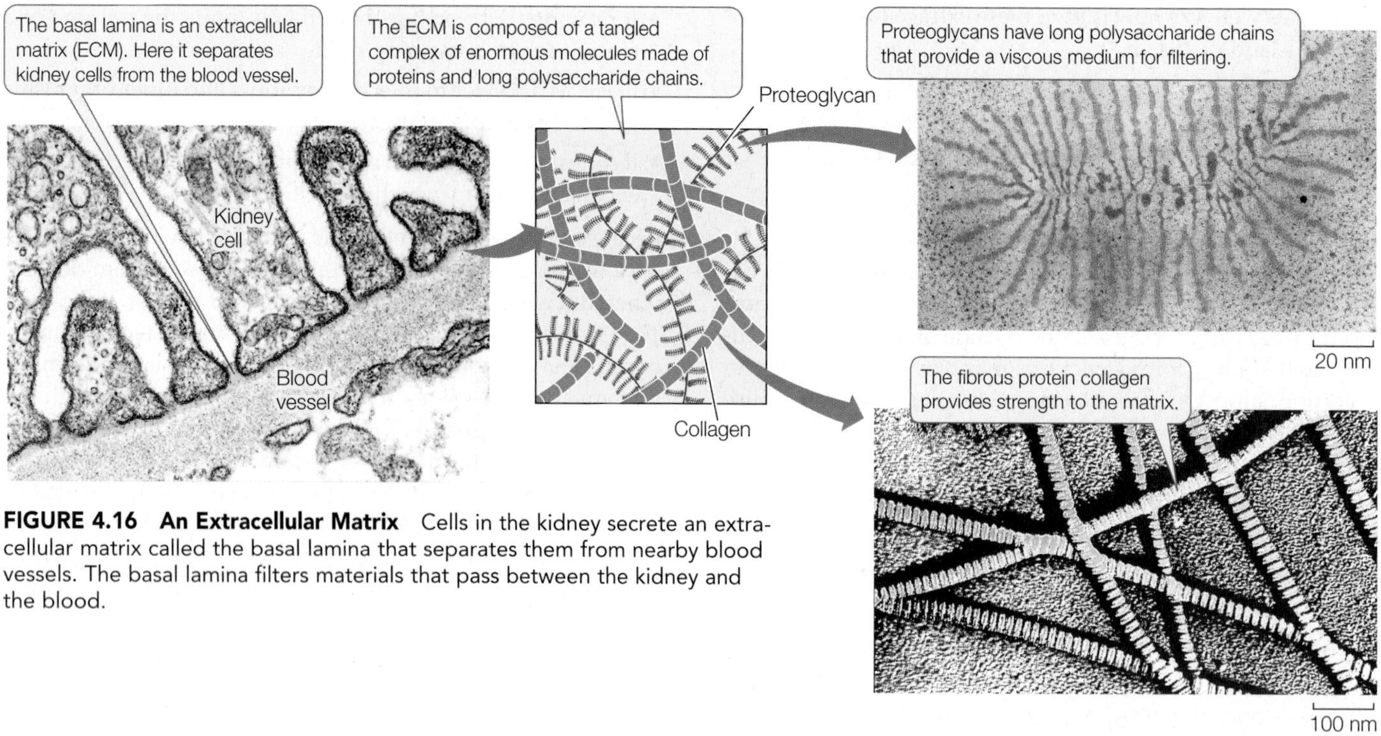

The basal lamina is an extracellular matrix (ECM). Here it separates kidney cells from the blood vessel.

The ECM is composed of a tangled complex of enormous molecules made of proteins and long polysaccharide chains.

Proteoglycans have long polysaccharide chains that provide a viscous medium for filtering.

Proteoglycan

Kidney cell

Blood vessel

Collagen

20 nm

The fibrous protein collagen provides strength to the matrix.

100 nm

FIGURE 4.16 An Extracellular Matrix Cells in the kidney secrete an extracellular matrix called the basal lamina that separates them from nearby blood vessels. The basal lamina filters materials that pass between the kidney and the blood.

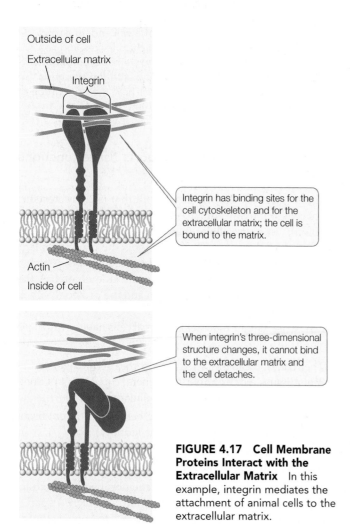

Outside of cell

Extracellular matrix

Integrin

Integrin has binding sites for the cell cytoskeleton and for the extracellular matrix; the cell is bound to the matrix.

Actin

Inside of cell

When integrin's three-dimensional structure changes, it cannot bind to the extracellular matrix and the cell detaches.

FIGURE 4.17 Cell Membrane Proteins Interact with the Extracellular Matrix In this example, integrin mediates the attachment of animal cells to the extracellular matrix.

below the cell surface, and the other to collagen in the extracellular matrix. These binding sites are noncovalent and reversible. When a cell moves its location in an organism, the first step is for integrin to change its three-dimensional structure so that it detaches from the collagen (**FIGURE 4.17**).

Cell junctions connect adjacent cells

In a multicellular animal, specialized structures protrude from adjacent cells to "glue" them together. These **cell junctions** are most evident in electron micrographs of *epithelial tissues*, which are layers of cells that line body cavities or cover body surfaces (examples are skin and the lining of the windpipe leading to the lungs). These surfaces are often exposed to environmental factors that might disrupt the integrity of the tissues, so it is particularly important that their cells stick together tightly. There are three types of junctions (**FIGURE 4.18**):

- *Tight junctions* prevent substances from moving through spaces between cells. For example, the epithelium of the urinary bladder contains tight junctions to prevent urine from leaking out into the body.

- *Desmosomes* hold adjacent cells together with stable protein connections, but materials can still move around in the extracellular matrix. This provides mechanical stability for tissues such as skin that receive physical stress.

- *Gap junctions* are like plant plasmodesmata: they are channels that run between membrane pores in adjacent cells, allowing substances to pass between cells. In the heart, for example, gap junctions allow the rapid spread of electric current mediated by ions so the heart muscle cells can beat in unison.

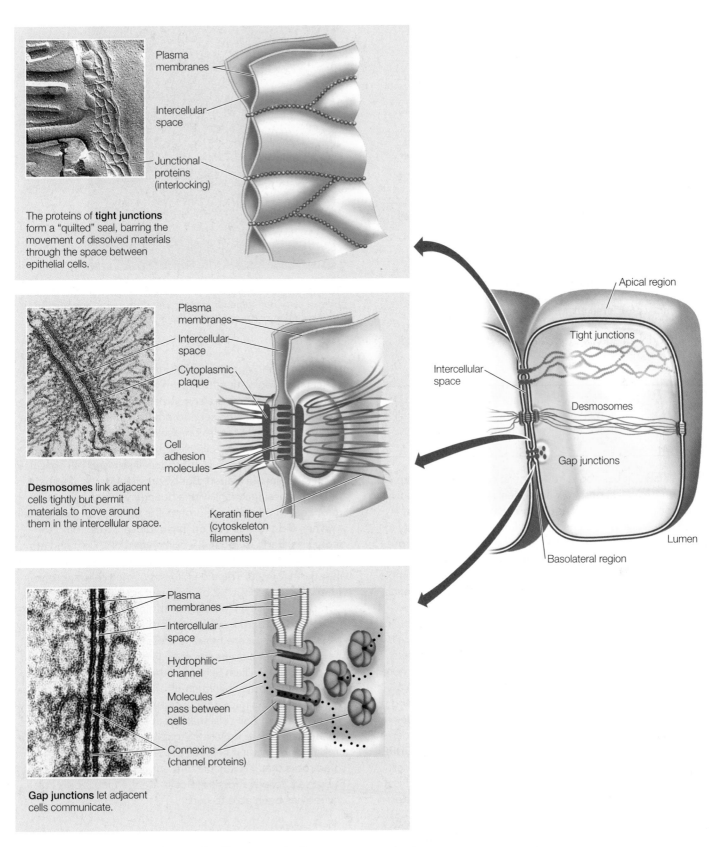

The proteins of **tight junctions** form a "quilted" seal, barring the movement of dissolved materials through the space between epithelial cells.

- Plasma membranes
- Intercellular space
- Junctional proteins (interlocking)

Desmosomes link adjacent cells tightly but permit materials to move around them in the intercellular space.

- Plasma membranes
- Intercellular space
- Cytoplasmic plaque
- Cell adhesion molecules
- Keratin fiber (cytoskeleton filaments)

Gap junctions let adjacent cells communicate.

- Plasma membranes
- Intercellular space
- Hydrophilic channel
- Molecules pass between cells
- Connexins (channel proteins)

- Apical region
- Tight junctions
- Intercellular space
- Desmosomes
- Gap junctions
- Basolateral region
- Lumen

FIGURE 4.18 Junctions Link Animal Cells Although all three types of junctions are shown in the cell at right, they don't necessarily all occur in the same cell.

yourBioPortal.com

Go to WEB ACTIVITY 4.3
Animal Cell Junctions

Do You Understand Concept 4.5?

- Compare the extracellular matrix of plant cells and animal cells.

- What kinds of cell junctions would you expect to find, and why, in the following situations?
  a. In the digestive system, where material must pass through cells and not go through the extracellular material, to get from the intestine to the blood vessels.
  b. In a small animal, where a chemical signal passes rapidly though cells to go from the head to the tail.
  c. In the lining of the intestine, where cells in the lining are constantly jostled by the churning of the underlying muscle

- When cancer spreads from its primary location to other parts of the body (a process called metastasis), tumor cells detach from their original location and then reattach at a different location. How would the integrin-collagen system be involved in this process?

 What do the characteristics of modern cells tell us about how the first cells originated?

ANSWER   Ideas about how the first cells may have formed focus on two questions: how and when. As to how cells could arise from a chemical-rich environment, most biologists assume that a cell membrane formed first and was necessary to provide a compartment for the chemical transformations of life to occur, separated from the environment (Concept 4.1). Biologists also assume that the first cells were relatively simple prokaryotes (Concept 4.2), without the organelles that define eukaryotic cells (Concept 4.3).

Jack Szostak, a Nobel laureate at Harvard University, builds synthetic cell models that give insights into the origin of cells. He and his colleagues make small membrane-lined droplets by putting fatty acids into water and then shaking the mixture. The lipids form water-filled droplets, each surrounded by a lipid bilayer "membrane" (see Figure 2.13). With water (and other molecules of the scientists' choosing) trapped inside, these spheres have many properties characteristic of modern cells—so many that they have been called *protocells* (**FIGURE 4.19**). For example, the membrane barrier determines what goes in and out of a protocell, by excluding macromolecules like RNA but allowing smaller molecules such as nucleotides

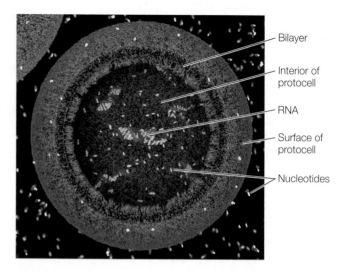

FIGURE 4.19   A Protocell   A protocell can be made in the lab and can carry out some functions of modern cells—in particular, it provides a compartment for biochemical reactions.

to pass through. Moreover, RNA inside the protocell can act as a catalyst, replicating itself from nucleotides that enter the protocell. The spheres are somewhat unstable, and under the microscope they can be seen to grow, elongate, and break, a possible precursor of more precise cell division.

Is this really a cell, possibly like the one where life started? Certainly not: it cannot fully reproduce itself, and its capacities for metabolism are limited. But by providing a compartment for biochemical reactions with a boundary that separates it from the environment, the protocell is a model for the first cell.

When did the first cells on Earth appear? According to geologists, Earth is about 4.5 billion years old. Heat and atmospheric conditions precluded life for at least a half-billion years after Earth formed. The oldest fossils of multicellular organisms date from about 1.2 billion years ago.

In all probability, life began with single-celled organisms resembling modern bacteria. Unfortunately, such cells lack the structures that are typically preserved in fossils, and so they die without a trace. Recently, however, geochemist and paleontologist William Schopf at the University of California, Los Angeles used a new method of microscopy called confocal laser scanning microscopy, combined with chemical analyses, to identify fossil cells that are about 800 million years old. Some of these look like Szostak's protocells. These were probably not the first cells, as there is chemical evidence in some rocks that life was present about 3.8 billion years ago. But so far, Schopf's fossilized cells are the oldest cells that anyone has been able to find.

SUMMARY

See WEB ACTIVITY 4.1

concept 4.1 Cells Provide Compartments for Biochemical Reactions

- Cell theory states that the cell is the fundamental unit of biological structure and function.

- Cells are small because a cell's surface area must be large compared with its volume to accommodate exchanges between the cell and its environment. **Review Figure 4.2**

- All cells are enclosed by a selectively permeable **plasma membrane** that separates their contents from the external environment.

concept 4.2 Prokaryotic Cells Do Not Have a Nucleus

- Prokaryotic cells usually have no internal compartments, but have a **nucleoid** containing DNA, and a **cytoplasm** containing **cytosol**, **ribosomes** (the sites of protein synthesis), proteins, and small molecules. Many have an extracellular **cell wall**. **Review Figure 4.5**

- Some prokaryotes have folded membranes, for example photosynthetic membranes, and some have **flagella** for motility. **Review Figure 4.6**

concept 4.3 Eukaryotic Cells Have a Nucleus and Other Membrane-Bound Compartments

- Eukaryotic cells contain many membrane-enclosed **organelles** that compartmentalize their biochemical functions. **Review Figure 4.7 and ANIMATED TUTORIAL 4.1**

- The **nucleus** contains most of the cell's DNA.

- The **endomembrane system**—consisting of the nuclear envelope, **endoplasmic reticulum**, **Golgi apparatus**, and **lysosomes**—is a series of interrelated compartments enclosed by membranes. It segregates proteins and modifies them. Lysosomes contain many digestive enzymes. **Review Figures 4.8 and 4.9, ANIMATED TUTORIAL 4.2, and WEB ACTIVITY 4.2**

- **Mitochondria** and **chloroplasts** are semiautonomous organelles that process energy.

- A **vacuole** is prominent in many plant cells. It is a membrane-enclosed compartment full of water and dissolved substances.

concept 4.4 The Cytoskeleton Provides Strength and Movement

- The **microfilaments, intermediate filaments**, and **microtubules** of the **cytoskeleton** provide the cell with shape, strength, and movement. **Review Figure 4.10**

- Microfilaments and microtubules have **dynamic instability** and can grow or shrink in length rapidly.

- **Cilia** and **flagella** are microtubule-lined extensions of the plasma membrane that produce movements of cells or their surrounding fluid medium. **Review Figures 4.11 and 4.12**

- Motor proteins move cellular components, such as vesicles, around the cell by "walking" along microtubules. **Review Figure 4.13**

- Biologists establish cause-and-effect relationships by manipulating biological systems. **Review Figure 4.14**

concept 4.5 Extracellular Structures Allow Cells to Communicate with the External Environment

- The plant **cell wall** consists principally of **cellulose**. Cell walls are pierced by **plasmodesmata** that join the cytoplasms of adjacent cells. **Review Figure 4.15**

- In animals, the **extracellular matrix** consists of different kinds of proteins, including collagen and proteoglycans. Integrins connect the cell cytoplasm with the extracellular matrix. **Review Figures 4.16 and 4.17**

- Specialized cell junctions connect cells in animal tissues. These include **tight junctions, desmosomes**, and **gap junctions**. The last are involved in intercellular communication. **Review Figure 4.18 and WEB ACTIVITY 4.3**

If you are like most people, you consume a significant amount of caffeine every day. In fact, more than 90 percent of North Americans and Europeans drink coffee or tea to get their "caffeine fix." Coffee and tea plants contain caffeine as a defense against the insects that eat them. Caffeine acts as an insecticide in plant parts that are particularly vulnerable to insect attacks, such as seeds, young seedlings, and leaves. But it is not toxic to humans.

Legend has it that about 5,000 years ago, a Chinese emperor found out by accident that a pleasant beverage could be made by boiling tea leaves. About 1,000 years ago, monks living in what is now Ethiopia found that roasting coffee seeds ("beans") gave a similarly pleasant effect and that the beverage kept them awake during long periods of prayer. Caffeine is now the most widely consumed psychoactive molecule in the world, but unlike other psychoactive drugs, it is not subject to government regulation.

Most people know from personal experience what caffeine does to the body: because it keeps us awake, it obviously affects the brain. In fact, it is often given to premature babies in the hospital nursery when they stop breathing. But it also affects other parts of the body—for example, it increases urination and speeds up the heart. How does this molecule work?

The key to understanding caffeine's action is to understand how it interacts with the cell membrane. In Chapter 4 we introduce the concept of the membrane as a structural boundary between the inside of a cell and the surrounding environment.

The plasma membrane physically separates the cell cytoplasm from its surroundings and helps maintain chemical differences between these two environments. The same can be said of the membranes that surround cell organelles, separating them from the cytoplasmic environment.

When caffeine arrives at a cell in the body, it first encounters the plasma membrane. The properties of this membrane determine whether and how the cell will react to caffeine. Will it cross the membrane boundary and enter the cell? What determines whether it crosses the membrane? If it does not, how can caffeine's interactions with membrane components lead to changes in cell function?

Q QUESTION What role does the cell membrane play in the body's response to caffeine?

Many people rely on caffeine to wake themselves up and to keep their minds alert.

KEY CONCEPTS

5.1 Biological Membranes Have a Common Structure and Are Fluid

5.2 Some Substances Can Cross the Membrane by Diffusion

5.3 Some Substances Require Energy to Cross the Membrane

5.4 Large Molecules Cross the Membrane via Vesicles

5.5 The Membrane Plays a Key Role in a Cell's Response to Environmental Signals

5.6 Signal Transduction Allows the Cell to Respond to Its Environment

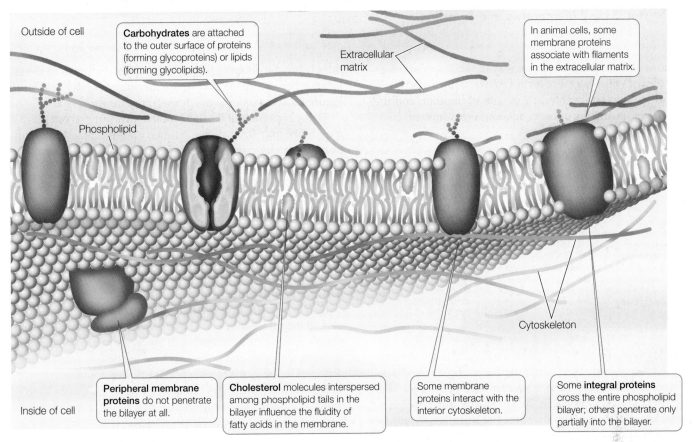

Outside of cell

Carbohydrates are attached to the outer surface of proteins (forming glycoproteins) or lipids (forming glycolipids).

Extracellular matrix

In animal cells, some membrane proteins associate with filaments in the extracellular matrix.

Phospholipid

Cytoskeleton

Inside of cell

Peripheral membrane proteins do not penetrate the bilayer at all.

Cholesterol molecules interspersed among phospholipid tails in the bilayer influence the fluidity of fatty acids in the membrane.

Some membrane proteins interact with the interior cytoskeleton.

Some **integral proteins** cross the entire phospholipid bilayer; others penetrate only partially into the bilayer.

FIGURE 5.1 Membrane Molecular Structure The general molecular structure of biological membranes is a continuous phospholipid bilayer in which proteins are embedded. The phospholipid bilayer separates two aqueous regions, the external environment outside the cell and the cell cytoplasm.

> **yourBioPortal.com**
>
> **Go to WEB ACTIVITY 5.1**
> **Membrane Molecular Structure**

<table>
<tr><td>concept
5.1</td><td>**Biological Membranes Have a Common Structure and Are Fluid**</td></tr>
</table>

The evolution of cellular life required the presence of a boundary, a way to separate the inside of the cell from the surrounding environment. This need was—and still is—fulfilled by biological membranes. Like many of the basic processes of life at the cellular level, the functions of membranes are carried out by a molecular structure shared by most organisms.

A biological membrane's structure and functions are determined by the chemical properties of its constituents: lipids, proteins, and carbohydrates. An important concept that emerges from our consideration of these molecules is polarity and how polarity influences the way a molecule interacts with water. The nonpolar regions of phospholipids and membrane proteins interact to form an insoluble barrier. The phospholipid bilayer serves as a lipid "lake" in which a variety of proteins "float" (**FIGURE 5.1**). This general design is known as the **fluid mosaic model**.

Membranes contain a wide array of proteins, some of which are noncovalently embedded in the phospholipid bilayer. These proteins are held within the membrane by their hydrophobic regions, or *domains*, while the proteins' hydrophilic domains are exposed to the watery conditions on one or both sides of the bilayer. Membrane proteins have several functions: they move materials through the membrane and receive chemical signals from the cell's external environment.

The carbohydrates associated with membranes are attached to either lipids or protein molecules. They are generally located on the outside of the cell, where they interact with substances in the external environment. Like some membrane proteins, carbohydrates are crucial for recognizing specific molecules, such as those on the surfaces of adjacent cells.

Each membrane is made up of constituents that are suitable for the specialized functions of the cell or organelle it surrounds. As you read about the different molecules in membranes, keep in mind that some membranes contain many protein molecules, others are lipid-rich, others have significant amounts of *cholesterol* or other sterols, and still others are rich in carbohydrates.

Lipids form the hydrophobic core of the membrane

The lipids in biological membranes are usually *phospholipids*. Recall from Concept 2.2 that some compounds are hydrophilic ("water-loving") and others are hydrophobic ("water-hating"), and that a phospholipid molecule has regions of both kinds:

- *Hydrophilic regions.* The phosphorus-containing "head" of a phospholipid is electrically charged and therefore associates with polar water molecules.

- *Hydrophobic regions.* The long, nonpolar fatty acid "tails" of a phospholipid associate with other nonpolar materials, but they do not dissolve in water or associate with hydrophilic substances.

Because of these properties, one way in which phospholipids can coexist with water is to form a *bilayer*, with the fatty acid "tails" of the two layers interacting with each other, and the polar "heads" facing the outside, aqueous environment:

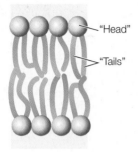

"Head"

"Tails"

LINK The properties of phospholipid bilayers are described in Concept 2.4

The thickness of a biological membrane is about 8 nm (0.008 μm), which is twice the length of a typical phospholipid. This thickness is about 8,000 times thinner than the page you are reading.

As we note in Chapter 4, in the laboratory it is possible to make artificial bilayers with the same organization as natural membranes. Small holes in such bilayers seal themselves spontaneously. The capacity of phospholipids to associate with one another and maintain a bilayer organization helps biological membranes fuse during vesicle formation, phagocytosis, and related processes (see Concept 4.3, especially Figure 4.9).

Although biological membranes all share a similar structure, there are many different kinds of phospholipids, and membranes from different cells or organelles may differ greatly in their *lipid composition*. Not only do membranes contain many different kinds of phospholipids, but also a significant proportion of the lipid content in an animal cell membrane may be cholesterol.

Phospholipids can differ in terms of fatty acid chain length (number of carbon atoms), degree of unsaturation (number of double bonds) in the fatty acids, and the kinds of polar (phosphate-containing) groups present. The most common fatty acids in membranes have chains with 16–18 carbon atoms and 0–2 double bonds. Saturated fatty acid chains (those with no double bonds) allow close packing of phospholipids in the bilayer, whereas the "kinks" in the unsaturated fatty acids (see Figure 2.12) make for a less dense, more fluid packing.

Up to 25 percent of the lipid content of an animal cell plasma membrane may be cholesterol. When present, cholesterol is important for membrane integrity; the cholesterol in your membranes is not hazardous to your health. A molecule of cholesterol is usually situated next to an unsaturated fatty acid.

The fatty acids of the phospholipids make the membrane somewhat fluid—about as fluid as lightweight machine oil. This fluidity permits some molecules to move laterally within the plane of the membrane. A given phospholipid molecule in the plasma membrane can travel from one end of the cell to the other in a little more than one second! However, it is rare for a phospholipid molecule in one half of the bilayer to spontaneously flip over to the other side. For that to happen, the polar part of the molecule would have to move through the hydrophobic interior of the membrane. Since spontaneous flip-flops are rare, the inner and outer halves of the bilayer may be quite different in the kinds of phospholipids they contain.

Membrane fluidity is affected by several factors, two of which are particularly important:

- *Lipid composition*. Cholesterol and long-chain, saturated fatty acids pack tightly together, resulting in less fluid membranes. Unsaturated fatty acids or those with shorter chains tend to increase membrane fluidity. Some anesthetics are nonpolar and act by inserting into the membrane, where they reduce the fluidity of nerve cell membranes, and thereby decrease nerve activity.

- *Temperature*. Membrane fluidity declines under cold conditions because molecules move more slowly at lower temperatures. For example, when your fingers get numb after contact with ice, it is due to a reduction in membrane fluidity in nerve cells. To address this problem, some organisms simply change the lipid composition of their membranes when their environment gets cold, replacing saturated with unsaturated fatty acids and using fatty acids with shorter chains. These changes play a role in the survival of plants, bacteria, and hibernating animals during the winter.

APPLY THE CONCEPT

Biological membranes have a common structure and are fluid

The membrane lipids of a cell can be labeled with a fluorescent tag so the entire surface of the cell will glow evenly under ultraviolet light. If a strong laser light is then shone on a tiny region of the cell, that region gets bleached (the strong light destroys the fluorescent tag) and there is a "hole" in the cell surface fluorescence. After the laser is turned off, the hole gradually fills in with fluorescent lipids that diffuse in from other parts of the membrane. The time it takes for the hole to disappear is a measure of membrane fluidity. The table shows some data for cells with altered membrane compositions. Explain the effect of each alteration.

CONDITION	TIME (SEC) FOR "HOLE" TO BECOME FLUORESCENT
No alteration	65
Decreased length of fatty acid chains	38
Increased cholesterol	88
Increased desaturation of fatty acid chains	42
Increased membrane protein content	90

yourBioPortal.com

Go to INTERACTIVE TUTORIAL 5.1
Lipid Bilayer: Temperature Effects on Composition

Membrane proteins are asymmetrically distributed

All biological membranes contain proteins. Typically, plasma membranes have about 1 protein molecule for every 25 phospholipid molecules. This ratio varies depending on membrane function. In the inner membrane of the mitochondrion, which is specialized for energy processing, there is 1 protein for every 5 lipids. By contrast, myelin—a membrane that encloses portions of some neurons (nerve cells) and acts as an electrical insulator—has only 1 protein for every 70 lipids.

Recall from Table 3.2 that some amino acids contain nonpolar, hydrophobic R groups, while others contain polar (charged), hydrophilic R groups. The arrangement of these amino acids in a membrane protein determines whether the membrane protein will insert into the nonpolar lipid bilayer and how it will be positioned:

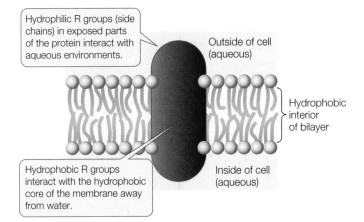

Hydrophilic R groups (side chains) in exposed parts of the protein interact with aqueous environments.

Outside of cell (aqueous)

Hydrophobic interior of bilayer

Hydrophobic R groups interact with the hydrophobic core of the membrane away from water.

Inside of cell (aqueous)

There are two general types of membrane proteins:

- **Peripheral membrane proteins** lack exposed hydrophobic groups and are not embedded in the bilayer. Instead, they have polar or charged regions that interact with exposed parts of integral membrane proteins, or with the polar heads of phospholipid molecules (see Figure 5.1).

- **Integral membrane proteins** are at least partly embedded in the phospholipid bilayer (see Figure 5.1). Like phospholipids, these proteins have both hydrophilic and hydrophobic regions.

Membrane proteins and lipids generally interact only noncovalently. The polar ends of proteins can interact with the polar ends of lipids, and the nonpolar regions of both molecules can interact hydrophobically. However, some membrane proteins have fatty acids or other lipid groups covalently attached to them. Proteins in this subgroup of integral membrane proteins are referred to as *anchored membrane proteins*, because it is their hydrophobic lipid components that anchor them in the phospholipid bilayer.

Proteins are asymmetrically distributed on the inner and outer surfaces of membranes. An integral membrane protein that extends all the way through the phospholipid bilayer and protrudes on both sides is known as a **transmembrane protein**. In addition to one or more *transmembrane domains* that extend through the bilayer, such a protein may have domains with other specific functions on the inner and outer sides of the membrane. Transmembrane proteins are always oriented the same way—domains with specific functions inside or outside the cell are always found on the correct side of the membrane. Peripheral membrane proteins are localized on one side of the membrane or the other. This asymmetrical arrangement gives the two surfaces of the membrane different properties. As we will soon see, these differences have great functional significance.

Like lipids, some membrane proteins move relatively freely within the phospholipid bilayer. Cell fusion experiments illustrate this migration dramatically. When two cells fuse, a single continuous membrane forms around both cells, and some proteins from each cell distribute themselves uniformly around this membrane (**FIGURE 5.2**).

Although some proteins are free to migrate throughout the membrane, others appear to be contained within specific regions. These membrane regions are like a corral of horses on a farm: the horses are free to move around within the fenced area but not outside it. For example, a muscle cell protein that recognizes a chemical signal from a neuron is normally found only within a specific region of the plasma membrane, where the neuron meets the muscle cell.

How does this happen? Proteins inside the cell can restrict the movement of proteins within a membrane. Components of the cytoskeleton may be attached to membrane proteins protruding into the cytoplasm (see Figure 5.1). The stability of the cytoskeleton may thus restrict the movement of attached membrane proteins.

Plasma membrane carbohydrates are recognition sites

In addition to lipids and proteins, the plasma membrane contains carbohydrates. The carbohydrates are located on the outer surface of the plasma membrane and serve as recognition sites for other cells and molecules. Membrane-associated carbohydrates may be covalently bonded to lipids or to proteins:

- A **glycolipid** consists of a carbohydrate covalently bonded to a lipid. Extending outside the cell surface, the carbohydrate may serve as a recognition signal for interactions between cells. For example, the carbohydrates on some glycolipids change when cells become cancerous. This change may allow white blood cells to target cancer cells for destruction.

- A **glycoprotein** consists of a carbohydrate covalently bonded to a protein. The bound carbohydrate is an oligosaccharide of 15 or fewer monosaccharide units (see Concept 2.3). These oligosaccharides often function as signaling sites, as do the carbohydrates attached to glycolipids.

The "alphabet" of monosaccharides on the outer surfaces of membranes can generate a large diversity of messages. There may be linear or branched oligosaccharides with many different three-dimensional shapes. An oligosaccharide of a specific shape on one cell can bind to a complementary shape on an adjacent cell. This binding is the basis of cell–cell adhesion:

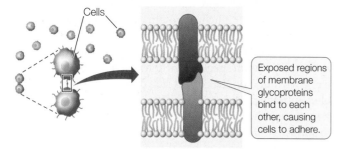

Exposed regions of membrane glycoproteins bind to each other, causing cells to adhere.

FRONTIERS   As an animal embryo develops, cells detach from one region, move, and then reattach at another region. This requires that cell membrane glycoproteins with binding specificities for adjacent cells are replaced by new ones with different specificities, so the cells can bind to their new neighbors. To investigate embryonic development, scientists are making genetically modified organisms whose embryonic cell membranes lack one or another surface binding component.

Membranes are constantly changing

Membranes in eukaryotic cells are constantly forming, transforming from one type to another, fusing with one another, and breaking down. As we discuss in Concept 4.3, fragments of membrane move (in the form of vesicles) from the endoplasmic reticulum (ER) to the Golgi apparatus, and from the Golgi apparatus to the plasma membrane (see Figure 4.7). Secondary lysosomes form when primary lysosomes from the Golgi apparatus fuse with phagosomes from the plasma membrane (see Figure 4.8).

Because all membranes have a similar appearance under the electron microscope, and because they interconvert readily, we might expect all subcellular membranes to be chemically identical. However, that is not the case, for *there are major chemical differences among the membranes of even a single cell.* Membranes are changed chemically when they form parts of certain organelles. In the Golgi apparatus, for example, the membranes of the *cis* face closely resemble those of the endoplasmic reticulum in chemical composition, but those of the *trans* face are more similar to the plasma membrane.

Do You Understand Concept 5.1?

- What are the differences between peripheral and integral membrane proteins?

- Why do phospholipids shaken in a water environment assemble into vesicles surrounded by a lipid bilayer?

- What is the evidence for membrane fluidity?

- If cells that recognize and attach to one another are separated, they reaggregate because of binding between their membrane glycoproteins. What would happen if the same experiment were conducted with cells treated to remove cell surface carbohydrates?

INVESTIGATION

FIGURE 5.2 Rapid Diffusion of Membrane Proteins
A human cell can be fused to a mouse cell in the laboratory, forming a single large cell (heterokaryon). This phenomenon was used to test whether membrane proteins can diffuse independently in the plane of the plasma membrane.

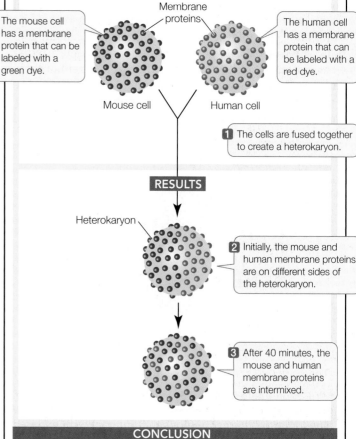

HYPOTHESIS

Proteins embedded in a membrane can diffuse freely within the membrane.

METHOD

The mouse cell has a membrane protein that can be labeled with a green dye.

Membrane proteins

The human cell has a membrane protein that can be labeled with a red dye.

Mouse cell   Human cell

1 The cells are fused together to create a heterokaryon.

RESULTS

Heterokaryon

2 Initially, the mouse and human membrane proteins are on different sides of the heterokaryon.

3 After 40 minutes, the mouse and human membrane proteins are intermixed.

CONCLUSION

Membrane proteins can diffuse rapidly in the plane of the membrane.

ANALYZE THE DATA

The experiment was repeated at various temperatures with the following results:

Temperature (°C)	Cells with mixed proteins (%)
0	0
15	8
20	42
26	77

Plot these data on a graph of Percentage Mixed vs. Temperature. Explain these data, relating the results to the concepts of diffusion and membrane fluidity.

Go to **yourBioPortal.com** for original citations, discussions, and relevant links for all INVESTIGATION figures.

Now that you understand the structure of biological membranes, let's see how their components function. In the sections that follow, we will focus on the plasma membrane (the cell membrane). We'll look at how the plasma membrane regulates the passage of substances that enter or leave a cell. Bear in mind that these principles also apply to the membranes that surround organelles.

concept 5.2 Some Substances Can Cross the Membrane by Diffusion

An important property of all life is the ability to regulate the internal composition of a cell, distinguishing it from the surrounding environment. Biological membranes allow some substances, but not others, to pass through them. This characteristic of membranes is called **selective permeability**.

There are two fundamentally different processes by which substances cross biological membranes:

- The processes of **passive transport** do not require direct input of metabolic energy to drive them.
- The processes of **active transport** require the input of metabolic (chemical) energy from an outside source.

This section focuses on passive transport across membranes. The energy for the passive transport of a substance is found in the difference between its concentration on one side of the membrane and its concentration on the other. Passive transport can occur by either of two types of diffusion: *simple diffusion* through the phospholipid bilayer, or *facilitated diffusion* through channel proteins or by means of carrier proteins.

Diffusion is the process of random movement toward a state of equilibrium

In a solution, there is a tendency for all of the components to be evenly distributed. You can see this when a drop of ink is allowed to fall into a gelatin suspension (a "gel"). Initially the pigment molecules are very concentrated, but they will move about at random, slowly spreading until the intensity of color is exactly the same throughout the gel:

A solution in which the solute molecules are uniformly distributed is said to be at *equilibrium*. This does not mean the molecules have stopped moving; it just means they are moving in such a way that their overall distribution does not change.

Diffusion is the process of random movement toward a state of equilibrium. In effect, it is a net movement from regions of greater concentration to regions of lesser concentration. Diffusion is generally a *very slow process in living tissues*, especially when we consider the gel-like consistency of the cell cytoplasm. For example, it would take about 3 years for a molecule

of oxygen gas (O_2) to diffuse from the human lung to a cell at the fingertip! So it is not surprising that as plants and animals evolved and became larger and multicellular, those with circulatory systems to distribute vital molecules such as O_2 had a distinct advantage over organisms relying on simple diffusion.

How fast a substance diffuses depends on three factors:

- The *diameter* of the molecules or ions: smaller molecules diffuse faster.
- The *temperature* of the solution: higher temperatures lead to faster diffusion because the heat provides more energy for movement.
- The *concentration gradient* in the system—that is, the change in solute concentration with distance in a given direction. The greater the concentration gradient, the more rapidly a substance diffuses.

What does this mean for a cell surrounded by a membrane? The cytoplasm is largely a water-based (aqueous) solution, and so is the surrounding environment. In a complex solution (one with many different solutes), the diffusion of each solute depends only on its own concentration, not the concentrations of other solutes. So one might expect a substance with a higher concentration inside the cell to diffuse out, and one with a higher concentration outside the cell to diffuse in. Indeed, this does occur for some molecules.

Simple diffusion takes place through the phospholipid bilayer

Some small molecules can pass through the phospholipid bilayer of the membrane by **simple diffusion**. A molecule that is hydrophobic and soluble in lipids can enter the membrane readily and pass through it. The more lipid-soluble the molecule is, the more rapidly it diffuses through the lipid bilayer.

In contrast, electrically charged or polar molecules, such as amino acids, sugars, ions, and water, do not pass readily through a membrane, for two reasons. First, these molecules are not very soluble in the hydrophobic interior of the bilayer. Second, charged molecules will form hydrogen bonds with water and ions in the aqueous environment on either side of the membrane. The multiplicity of these hydrogen bonds prevents the substances from moving into the hydrophobic interior of the membrane.

FRONTIERS The effectiveness of many anesthetics in reducing feeling or sensation is directly related to their membrane lipid solubility. But it is not clear exactly what happens after the drug dissolves in the membrane. Scientists are investigating this by measuring the physical properties of membrane components after an anesthetic is added. Understanding how membrane lipids and proteins are affected by anesthetics may help in designing more specific drugs with fewer side effects.

Osmosis is the diffusion of water across membranes

Water molecules pass through specialized channels in membranes (see below) by a diffusion process called **osmosis**. This process depends on the relative concentrations of water

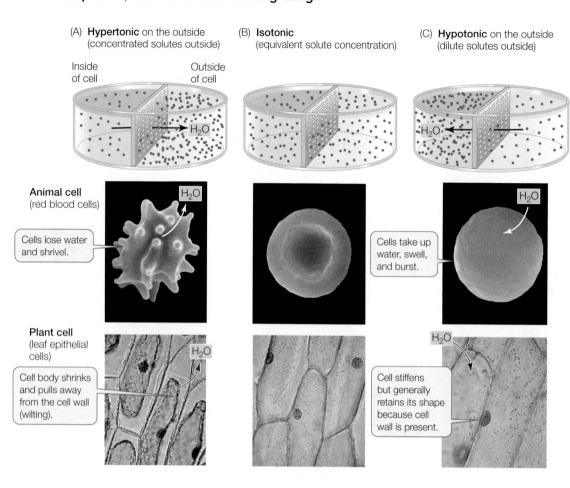

(A) **Hypertonic** on the outside (concentrated solutes outside)

(B) **Isotonic** (equivalent solute concentration)

(C) **Hypotonic** on the outside (dilute solutes outside)

Inside of cell

Outside of cell

H_2O

H_2O

H_2O

Animal cell (red blood cells)

H_2O

Cells lose water and shrivel.

Cells take up water, swell, and burst.

H_2O

Plant cell (leaf epithelial cells)

H_2O

Cell body shrinks and pulls away from the cell wall (wilting).

H_2O

Cell stiffens but generally retains its shape because cell wall is present.

FIGURE 5.3 Osmosis Can Modify the Shapes of Cells
(A) In a solution that is hypertonic to the cytoplasm of a plant or animal cell, water flows out of the cell. (B) In a solution that is isotonic with the cytoplasm, the cell maintains a consistent, characteristic shape because there is no net movement of water into or out of the cell. (C) In a solution that is hypotonic to the cytoplasm, water enters the cell. An animal cell will swell and may burst under these conditions; a plant cell will not swell too much because if its rigid cell wall.

molecules on both sides of the membrane. In a particular solution, the higher the *total* solute concentration, the lower the concentration of water molecules. Consider a situation where a membrane separating two different solutions allows water, *but not solutes*, to pass through. The water molecules will move across the membrane toward the solution with the higher solute concentration and the lower concentration of water molecules.

Here we are referring to the *net* movement of water. Since it is so abundant, water is constantly moving (through protein channels) across the plasma membrane, into and out of cells. But if there is a concentration difference between the two sides of the membrane, the overall movement will be greater in one direction or the other.

Three terms are used to compare the solute concentrations of two solutions separated by a membrane:

- A **hypertonic** solution has a higher solute concentration than the other solution (**FIGURE 5.3A**).

- **Isotonic** solutions have equal solute concentrations (**FIGURE 5.3B**).

- A **hypotonic** solution has a lower solute concentration than the other solution (**FIGURE 5.3C**).

The concentration of solutes in the environment determines the direction of osmosis in all animal cells. A red blood cell takes up water from a solution that is hypotonic to the cell's contents. If this happens, the cell bursts because its plasma membrane cannot withstand the pressure created by the water entry and the resultant swelling (see Figure 5.3C). The integrity of blood

cells is absolutely dependent on the maintenance of a constant solute concentration in the surrounding blood plasma—the plasma must be isotonic to the blood cells. Regulation of the solute concentrations of body fluids is thus an important process for organisms without cell walls.

In contrast to animal cells, the cells of plants, archaea, bacteria, fungi, and some protists have cell walls that limit their volumes and keep them from bursting. Cells with sturdy walls take up a limited amount of water, and in so doing they build up internal pressure against the cell wall, which prevents further water from entering. This pressure within the cell is called **turgor pressure**; it keeps the green parts of plants upright and is the driving force for enlargement of plant cells (see Concept 25.3). It is a normal and essential component of plant growth. If enough water leaves the cells, turgor pressure drops and the plant wilts. Turgor pressure reaches about 100 pounds per square inch (0.7 kg/cm^2), which is many times greater than the pressure in auto tires.

Diffusion may be aided by channel proteins

As we saw earlier, polar or charged substances such as water, amino acids, sugars, and ions do not readily diffuse across membranes. But they can cross the hydrophobic phospholipid bilayer passively (that is, without the input of energy) in one of two ways, depending on the substance:

- **Channel proteins** are integral membrane proteins that form channels across the membrane through which certain substances can pass.

- Some substances can bind to membrane proteins called **carrier proteins** that speed up their diffusion through the phospholipid bilayer.

Both of these processes are forms of **facilitated diffusion**. The substances diffuse according to their concentration gradients, but their diffusion is made easier by channel or carrier proteins. Particular channel or carrier proteins allow diffusion both into and out of a cell or organelle. In other words, they can operate in both directions.

We will focus here on two examples of channel proteins and discuss carrier proteins in the next section.

ION CHANNELS The best-studied channel proteins are the **ion channels**. As you will see in later chapters, the movement of ions across membranes is important in many biological processes, including respiration within the mitochondria, the electrical activity of the nervous system, and the opening of the pores in leaves that allow gas exchange with the environment. Several types of ion channels have been identified, each of them specific for a particular ion. All of them show the same basic structure of a hydrophilic pore that allows a particular ion to move through it.

Just as a fence may have a gate that can be opened or closed, most ion channels are gated: they can be opened or closed to ion passage. A **gated channel** opens when a stimulus causes a change in the three-dimensional shape of the channel. In some cases, this stimulus is the binding of a chemical signal, or **ligand**. Channels controlled in this way are called *ligand-gated channels* (**FIGURE 5.4**). In contrast, a *voltage-gated channel* is stimulated to open or close by a change in the voltage (electrical charge difference) across the membrane.

AQUAPORINS FOR WATER Water crosses membranes at a much faster rate than would be expected if it simply diffused through the phospholipid bilayer. One way it does this is by "hitchhiking" with some ions, such as Na^+, as they pass through ion channels. Up to 12 water molecules may coat an ion as it traverses a channel. But there is an even faster way for water to cross membranes. Plants and some animal cells (such as red blood and kidney cells) have membrane channels called **aquaporins**. These specific channels allow large amounts of water to move along its concentration gradient, as you will see when we discuss water relations in plants (see Chapter 25) and animals (see Chapter 40).

Aquaporins were first identified by Peter Agre at Duke University, who noticed a membrane protein that was present in red blood cells, kidney cells, and plant cells but did not know its function. A colleague suggested that it might be a water channel, because these cell types show rapid diffusion of water across their membranes. Agre tested this idea by creating egg cells (oocytes) with the protein in their membranes. An oocyte membrane does not normally permit much diffusion of water. Agre injected the oocytes with the mRNA for aquaporin; the protein was produced by the cells and inserted into their membranes. Remarkably, the oocytes began swelling immediately after being

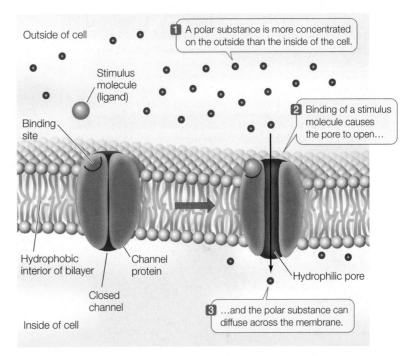

FIGURE 5.4 A Ligand-Gated Channel Protein Opens in Response to a Stimulus The channel protein is anchored in the lipid bilayer by its outer coating of nonpolar (hydrophobic) amino acids. The protein changes its three-dimensional shape when a stimulus molecule (ligand) binds to it, opening a pore lined with polar amino acids. This allows hydrophilic, polar substances to pass through.

transferred to a hypotonic solution, indicating the rapid diffusion of water into the cells (**FIGURE 5.5**).

Carrier proteins aid diffusion by binding substances

Another kind of facilitated diffusion involves the actual binding of the transported substance to a membrane protein called a carrier protein. Carrier proteins transport polar molecules such as sugars and amino acids.

Glucose is the major energy source for most mammalian cells, and they require a great deal of it. Their membranes contain a carrier protein—the glucose transporter—that facilitates glucose uptake into the cell. Binding of glucose to a specific three-dimensional site on one side of the transport protein causes the protein to change its shape and release glucose on the other side of the membrane (**FIGURE 5.6A**). Since glucose is usually broken down as soon as it enters the cell, there is almost always a strong concentration gradient favoring glucose entry (that is, a higher concentration outside the cell than inside). The transporter allows glucose molecules to cross the membrane and enter the cell much faster than they would by simple diffusion through the bilayer. This rapid entry is necessary to ensure that the cell receives enough glucose for its energy needs.

Transport by carrier proteins is different from simple diffusion. In both processes, the rate of movement depends on the concentration gradient across the membrane. However, in carrier-mediated transport, a point is reached at which increases

INVESTIGATION

FIGURE 5.5 Aquaporins Increase Membrane Permeability to Water A protein was isolated from the membranes of cells in which water diffuses rapidly across the membranes. When the protein was inserted into oocytes, which do not normally have it, the water permeability of the oocytes was greatly increased.

HYPOTHESIS

Aquaporin increases membrane permeability to water.

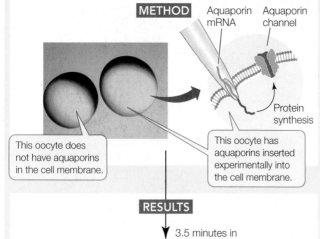

METHOD

Aquaporin mRNA
Aquaporin channel
Protein synthesis

This oocyte does not have aquaporins in the cell membrane.

This oocyte has aquaporins inserted experimentally into the cell membrane.

RESULTS

3.5 minutes in hypotonic solution

Water does not diffuse into the cell, so it does not swell.

Water diffuses into the cell through the aquaporin channels, and it swells.

CONCLUSION

Aquaporin **increases** the rate of water diffusion across the cell membrane.

ANALYZE THE DATA

Oocytes were injected with aquaporin mRNA (red circles) or a solution without mRNA (blue circles). Water permeability was tested by incubating the oocytes in hypotonic solution and measuring cell volume. After time X in the upper curve, intact oocytes were not visible:

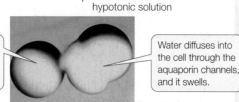

A. Why did the cells with aquaporin mRNA increase in volume?
B. What happened at time X?
C. Calculate the relative rates (volume increase per minute) of swelling in the control and experimental curves. What does this show about the effectiveness of mRNA injection?

For more, go to Working with Data 5.1 at **yourBioPortal.com**.

in the concentration gradient are not accompanied by an increased rate of diffusion. At this point, the facilitated diffusion system is said to be *saturated* (**FIGURE 5.6B**). Because there are only a limited number of carrier protein molecules per unit of membrane area, the rate of diffusion reaches a maximum when all the carrier molecules are fully loaded with solute molecules. This situation is similar to that of enzyme saturation (see Figure 3.15).

Do You Understand Concept 5.2?

- What properties of a substance determine whether, and how fast, it will diffuse across a membrane?

- Compare the process of facilitated diffusion through a channel and by a carrier protein. Which might be faster, and why?

- After celery is stored in an open refrigerator for two days, it is wilted. However, immersing the cut stalk in water for a few hours restores the integrity of the celery. How?

Diffusion tends to equalize the concentrations of substances between the outsides and insides of cells or organelles. However, one hallmark of a living thing is that it can have an internal composition quite different from that of its environment. To achieve this, a cell must sometimes move substances *against their concentration gradients*. This process requires work—the input of energy—and is known as active transport.

concept 5.3 Some Substances Require Energy to Cross the Membrane

In many biological situations, there is a different concentration of a particular ion or small molecule inside compared with outside a cell. In these cases, the concentration imbalance is maintained by a protein in the plasma membrane that moves the substance against its concentration gradient. This is called *active transport*, and because it is acting "against the normal flow," it requires the expenditure of energy. Often the energy source is the nucleotide adenosine triphosphate (ATP). In eukaryotes, ATP is produced in the mitochondria and plastids, and it has chemical energy stored in its terminal phosphate bond. This energy is released when ATP is converted to adenosine diphosphate (ADP) in a hydrolysis reaction that breaks this terminal bond. We will give more details of how ATP provides energy to cells in Concept 6.1.

The differences between diffusion and active transport are summarized in **TABLE 5.1**.

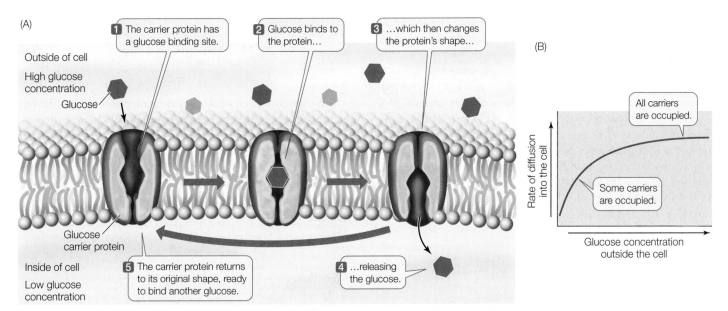

FIGURE 5.6 A Carrier Protein Facilitates Diffusion The glucose transporter is a carrier protein that allows glucose to enter the cell at a faster rate than would be possible by simple diffusion. (A) The transporter binds to glucose, and as it does so, it changes shape, releasing the glucose into the cell cytoplasm. (B) The graph shows the rate of glucose entry via a carrier versus the concentration of glucose outside the cell. As the glucose concentration increases, the rate of diffusion increases until the point at which all the available transporters are being used (the system is saturated).

Active transport is directional

Simple and facilitated diffusion follow concentration gradients and can occur in either direction across a membrane. In contrast, active transport is *directional*, and moves a substance either into or out of a cell or organelle against its concentration gradient. In other words, the substance is transported from a lower to a higher concentration. The direction in which a particular substance is transported usually depends on the cell's needs. As in facilitated diffusion, there is usually a specific carrier protein for each substance that is transported.

Different energy sources distinguish different active transport systems

There are two basic types of active transport:

- **Primary active transport** involves the direct hydrolysis of ATP, which provides the energy required for transport.

- **Secondary active transport** does not use ATP directly. Instead, its energy is supplied by an ion concentration gradient or an electrical gradient, established by primary active transport. This transport system uses the energy of ATP indirectly to set up the gradient.

In primary active transport, energy released by the hydrolysis of ATP drives the movement of specific ions against their concentration gradients. For example, the concentration of potassium ions (K⁺) inside a cell is often much higher than the concentration in the fluid bathing the cell. However, the concentration of sodium ions (Na⁺) is often much higher outside the cell. A protein in the plasma membrane pumps Na⁺ out of the cell and K⁺ into the cell against these concentration gradients, ensuring that the gradients are maintained. This **sodium–potassium (Na⁺–K⁺) pump** is an integral membrane glycoprotein that is found in all animal cells. It breaks down a molecule of ATP to ADP and a free phosphate ion (P_i) and uses the released energy to bring two K⁺ ions into the cell, and export three Na⁺ ions (**FIGURE 5.7**).

FRONTIERS Cancer therapy often fails when the targeted cells become resistant to the drugs used to kill them. The drugs may work initially, only to have the cancer return—this time with a membrane protein that actively transports the drugs out of the cancer cells. Designing ways to block this active transport is a major challenge for effective cancer therapy.

TABLE 5.1	Membrane Transport Mechanisms		
	SIMPLE DIFFUSION	FACILITATED DIFFUSION (CHANNEL OR CARRIER PROTEIN)	ACTIVE TRANSPORT
Cellular energy required?	No	No	Yes
Driving force	Concentration gradient	Concentration gradient	ATP hydrolysis (against concentration gradient)
Membrane protein required?	No	Yes	Yes
Specificity	No	Yes	Yes

In secondary active transport, the movement of a substance against its concentration gradient is accomplished using energy "regained" by letting ions move across the membrane *with* their concentration gradients. For example, once the Na⁺–K⁺ pump establishes a concentration gradient of sodium ions, the passive diffusion of some Na⁺ back into a cell can provide energy for the secondary active transport of glucose into the cell. This occurs when glucose is absorbed into the bloodstream from the digestive tract. Secondary active transport is usually accomplished by a single protein that moves both the ion and the actively transported molecule across the membrane. In some cases, the ion and the transported molecule move in opposite directions, whereas in others they move in the same direction (as for glucose and Na⁺ in the digestive tract). Secondary active transport aids in the uptake of amino acids and sugars, which are essential raw materials for cell maintenance and growth.

Do You Understand Concept 5.3?

- Why is energy required for active transport?

- The drug ouabain inhibits the activity of the Na⁺–K⁺ pump. A nerve cell is incubated in ouabain. Make a table in which you predict what would happen to the concentrations of Na⁺ and K⁺ inside the cell, as a result of the action of ouabain.

- How would you use experiments to distinguish between the following two ways for glucose to enter a cell: facilitated diffusion via a carrier protein and secondary active transport?

yourBioPortal.com
Go to ANIMATED TUTORIAL 5.2
Active Transport

We have examined a number of passive and active ways by which ions and small molecules can enter and leave cells. But what about large molecules such as proteins? Many proteins are so large that they diffuse very slowly, and their bulk makes it difficult for them to pass through the phospholipid bilayer. It takes a completely different mechanism to move intact large molecules across membranes.

concept 5.4   Large Molecules Cross the Membrane via Vesicles

Macromolecules such as proteins, polysaccharides, and nucleic acids are simply too large and too charged or polar to pass through biological membranes. This is a fortunate property—cellular integrity depends on containing these macromolecules in specific locations. However, cells must sometimes take up or *secrete* (release to the external environment) intact large molecules. This is done via vesicles, and the general terms for the mechanisms by which cells take up and secrete large molecules or particles are *endocytosis* and *exocytosis* (**FIGURE 5.8**).

Macromolecules and particles enter the cell by endocytosis

Endocytosis is a general term for a group of processes that bring small molecules, macromolecules, large particles, and

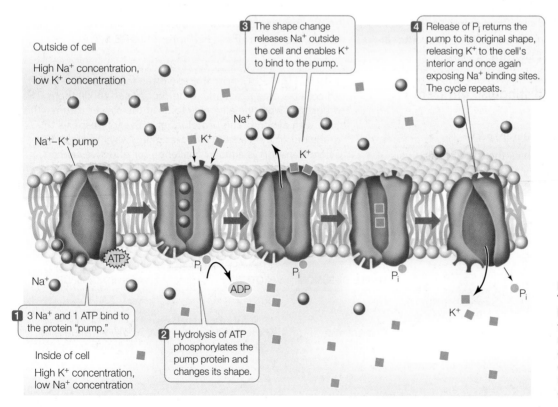

Outside of cell

High Na⁺ concentration, low K⁺ concentration

Na⁺–K⁺ pump

3 The shape change releases Na⁺ outside the cell and enables K⁺ to bind to the pump.

4 Release of P$_i$ returns the pump to its original shape, releasing K⁺ to the cell's interior and once again exposing Na⁺ binding sites. The cycle repeats.

Na⁺

K⁺

K⁺

ATP

P$_i$

P$_i$

P$_i$

P$_i$

ADP

Na⁺

K⁺

1 3 Na⁺ and 1 ATP bind to the protein "pump."

2 Hydrolysis of ATP phosphorylates the pump protein and changes its shape.

Inside of cell

High K⁺ concentration, low Na⁺ concentration

FIGURE 5.7 Primary Active Transport: The Sodium–Potassium Pump In active transport, energy is used to move a solute against its concentration gradient. Here, energy from ATP is used to move Na⁺ and K⁺ against their concentration gradients.

even small cells into the eukaryotic cell (see Figure 5.8A). There are three types of endocytosis: phagocytosis, pinocytosis, and receptor-mediated endocytosis. In all three, the plasma membrane invaginates (folds inward), forming a small pocket around materials from the environment. The pocket deepens, forming a vesicle. This vesicle separates from the plasma membrane and migrates with its contents to the cell's interior.

- In **phagocytosis** ("cellular eating"), part of the plasma membrane engulfs a large particle or even an entire cell. Unicellular protists use phagocytosis for feeding, and some white blood cells use phagocytosis to engulf foreign cells and substances. The food vacuole (phagosome) that forms usually fuses with a lysosome, where its contents are digested.

> **LINK** Review the discussion of phagocytosis in Concept 4.3

- Vesicles also form in **pinocytosis** ("cellular drinking"). However, in this case the vesicles are smaller and they bring fluids and dissolved substances, including proteins, into the cell. Phagocytosis and pinocytosis are relatively nonspecific regarding what they bring into the cell. For example, pinocytosis goes on constantly in the endothelium—the single layer of cells that separates a blood vessel from the surrounding tissue. Pinocytosis allows cells of the endothelium to rapidly acquire fluids and dissolved solutes from the blood.

- In **receptor-mediated endocytosis**, molecules at the cell surface recognize and trigger the uptake of specific materials.

Let's take a closer look at this last process.

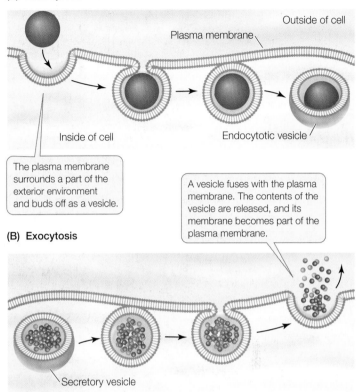

(A) Endocytosis

Outside of cell

Plasma membrane

Inside of cell

Endocytotic vesicle

The plasma membrane surrounds a part of the exterior environment and buds off as a vesicle.

A vesicle fuses with the plasma membrane. The contents of the vesicle are released, and its membrane becomes part of the plasma membrane.

(B) Exocytosis

Secretory vesicle

FIGURE 5.8 Endocytosis and Exocytosis Eukaryotic cells use endocytosis (A) and exocytosis (B) to take up and release large molecules and particles. Even small cells can be engulfed via endocytosis.

Receptor-mediated endocytosis is specific

Receptor-mediated endocytosis is used by animal cells to capture specific macromolecules from the cell's environment. This process depends on **receptors**, which are proteins that bind to specific molecules (their *ligands*) and then set off specific cellular responses. In receptor-mediated endocytosis, the receptors are integral membrane proteins located at particular regions on the extracellular surface of the plasma membrane. These membrane regions are called *coated pits* because they form slight depressions in the plasma membrane, and their cytoplasmic surfaces are coated by another protein (often *clathrin*). The uptake process is similar to that in phagocytosis.

APPLY THE CONCEPT

Some substances require energy to cross the membrane

The liver plays several vital metabolic roles, including protein synthesis, detoxification, and the production of substances necessary for digestion. Liver cells are in contact with the blood and exchange a variety of substances with the blood plasma (the noncellular part of blood). Below and at right is a list of observations about the relative concentrations of various molecules in a liver cell cytoplasm and in the blood plasma. Explain each observation in terms of membrane permeability and transport mechanisms.

1. The concentration of serum albumin, a blood protein, is much higher in the plasma.

2. The concentration of RNA is much higher in the cytoplasm.

3. The concentration of Na^+ is lower in the cytoplasm.

4. The concentration of water is equal in the plasma and the cytoplasm.

5. The concentration of low-density lipoproteins is higher in the cytoplasm.

6. The concentration of glucose is equal in the plasma and the cytoplasm.

7. If K^+ enters the plasma, its concentration rapidly equalizes between the plasma and the cytoplasm.

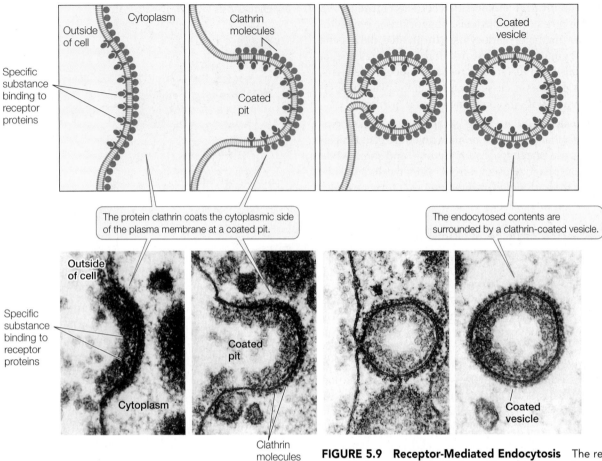

The protein clathrin coats the cytoplasmic side of the plasma membrane at a coated pit.

The endocytosed contents are surrounded by a clathrin-coated vesicle.

FIGURE 5.9 Receptor-Mediated Endocytosis The receptor proteins in a coated pit bind specific macromolecules, which are then carried into the cell by a coated vesicle.

When the receptors in a coated pit bind to their specific ligands (the macromolecules to be taken into the cell), the coated pit invaginates (folds in) and forms a coated vesicle around the bound macromolecules. The clathrin molecules strengthen and stabilize the vesicle, which carries the macromolecules away from the plasma membrane and into the cytoplasm (**FIGURE 5.9**). Once inside, the vesicle loses its clathrin coat and may fuse with a lysosome, where the engulfed material is digested. Because of its specificity for particular macromolecules, receptor-mediated endocytosis is an efficient method for taking up substances that may exist at low concentrations in the cell's environment.

Receptor-mediated endocytosis is the method by which cholesterol is taken up by most mammalian cells. Water-insoluble cholesterol and triglycerides are packaged by liver cells into lipoprotein particles. Most of the cholesterol is packaged into *low-density lipoproteins* (LDLs) and circulated via the bloodstream. When a particular cell requires cholesterol, it produces specific LDL receptors, which are inserted into the plasma membrane in clathrin-coated pits. LDLs bind to the receptors and are taken into the cell via receptor-mediated endocytosis. Within the resulting vesicle, the LDL particles are freed from the receptors. The receptors segregate to a region that buds off and forms a new vesicle, which is recycled to the plasma membrane. The freed LDL particles remain in the original vesicle,

which fuses with a lysosome. There, the LDLs are digested and the cholesterol made available for use by the cell.

In healthy individuals, the liver takes up unused LDLs for recycling. People with the inherited disease *familial hypercholesterolemia* have a deficient LDL receptor in their livers. This prevents receptor-mediated endocytosis of LDLs in the liver, resulting in dangerously high levels of cholesterol in the blood. The cholesterol builds up in the arteries that nourish the heart and causes heart attacks. In extreme cases where only the deficient receptor is present, children and teenagers can have severe cardiovascular disease.

Exocytosis moves materials out of the cell

Exocytosis is the process by which materials packaged in vesicles are secreted from the cell (see Figure 5.8B). When the vesicle membrane fuses with the plasma membrane, an opening is made to the outside of the cell. The contents of the vesicle are released into the environment, and the vesicle membrane is smoothly incorporated into the plasma membrane.

In Chapter 4 we encounter exocytosis as the last step in the processing of material engulfed by phagocytosis—the release of undigested materials back to the extracellular environment (see Figure 4.9). Secreted proteins are transported out of the cell via exocytosis. The proteins are folded and modified in the endoplasmic reticulum, transported in vesicles to the Golgi where they may be further modified, then packaged in new vesicles for secretion (see Figure 4.8).

Exocytosis is important in the secretion of many types of substances, including digestive enzymes from the pancreas, neurotransmitters from neurons, and materials for the construction of the plant cell wall. You will encounter these processes in later chapters.

---

yourBioPortal.com

Go to ANIMATED TUTORIAL 5.3
Endocytosis and Exocytosis

Do You Understand Concept 5.4?

- What is the difference between phagocytosis and pinocytosis?

- Would a small molecule such as an amino acid enter a cell by receptor-mediated endocytosis?

- Exocytosis involves the fusion of the membranes of a vesicle and the plasma membrane. Explain how this might occur.

We have just introduced the concept of a membrane-bound receptor, which is a key factor in a cell's interaction with its environment. Let's look more closely at receptors and how they respond to signals.

concept 5.5 The Membrane Plays a Key Role in a Cell's Response to Environmental Signals

A hallmark of living things is their ability to process information from their environment. This information can be thought of in terms of *signals*. The signal may be a physical stimulus such as light or heat, or a chemical such as a hormone. The mere presence of a signal does not mean a particular cell will respond to it. In order to respond, the cell must have a specific receptor that can detect it. Once the receptor is activated by the signal, it sets off a series of events within the cell. A **signal transduction pathway** is a sequence of molecular events and chemical reactions that lead to a cell's response to a signal. The ability of cells to sense and respond to their environment is key to the maintenance of cellular and organismal *homeostasis*, a theme that recurs throughout our discussion of gene regulation and plant and animal physiology in later chapters of this book.

Cells are exposed to many signals and may have different responses

Inside a large multicellular animal, chemical signals made by the body itself reach a target cell by local diffusion or by circulation within the blood. These signals are usually in tiny concentrations (as low as $10^{-10} M$) and differ in their sources and mode of delivery (**FIGURE 5.10**):

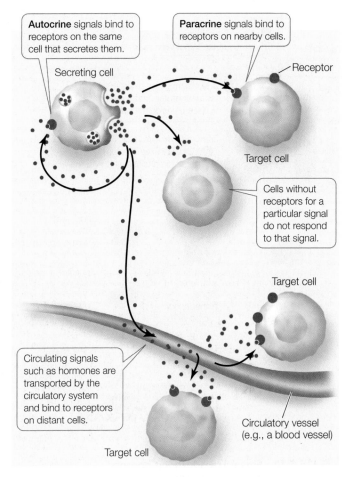

FIGURE 5.10 Chemical Signaling Concepts A signal molecule can act on the cell that produces it, a nearby cell, or be transported by the organism's circulatory system to a distant target cell.

- **Autocrine** signals affect the same cells that release them. For example, many tumor cells reproduce uncontrollably because they self-stimulate cell division by making their own division signals.

- **Paracrine** signals diffuse to and affect nearby cells. An example is a neurotransmitter made by one nerve cell that diffuses to an adjacent cell and stimulates it.

- Signals to distant cells travel through the circulatory system and are called **hormones**.

Chemical signals do not always come from within the multicellular organism—some come from the external environment. For example, specific molecules produced by pathogenic organisms trigger signal transduction pathways in plants, leading to defense responses.

For the information from a signal to be transmitted to a cell, the target cell must be able to *sense* the signal and *respond* to it. In a multicellular animal, all the cells may receive chemical signals that are circulated in the blood, but most body cells are not capable of responding to the signals. *Only the cells with the necessary receptors can respond.*

Typically, a signal transduction pathway involves a signal, a receptor, and a response (**FIGURE 5.11**). These pathways vary

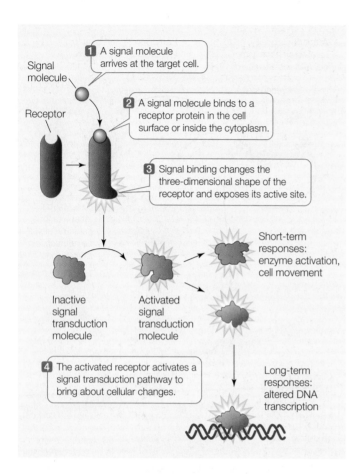

FIGURE 5.11   Signal Transduction Concepts   This general pathway is common to many cells and situations. The ultimate cellular responses are either short-term or long-term.

in their details, but a common mechanism is allosteric regulation. Recall that **allosteric regulation** involves an alteration in the three-dimensional shape of a protein as a result of the binding of another molecule at a site other than the protein's active site (see Figure 3.18). You have already seen an example of allosteric regulation in this chapter when we considered a gated channel, which opens (changes shape) after binding to another molecule (see Figure 5.6).

A signal transduction pathway may end in a response that is short term, such as the activation of an enzyme, or long term, such as an alteration in gene expression.

Membrane proteins act as receptors

A cell must have the appropriate receptor in order to respond to a signal. The signal may be a specific molecule or physical stimulus. In the following discussion, we will focus on chemical signals, such as those described in the previous section.

The signal molecule fits into a three-dimensional site on its corresponding receptor protein (**FIGURE 5.12**). As noted earlier in the chapter, a molecule that binds to a receptor in this way is called a *ligand*. Binding of the ligand causes the receptor to change its three-dimensional shape, and that conformational change initiates a cellular response. In many cases, the receptor has an enzyme function (a catalytic domain), with its active site on the cytoplasmic side of the membrane. When the ligand

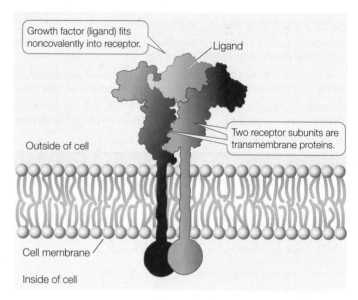

FIGURE 5.12   A Signal Binds to Its Receptor   Human growth factor fits into its membrane-bound receptor (a protein with two subunits) and binds to it noncovalently. There is an equilibrium between bound and unbound receptor.

binds to the receptor (for example, on the outside of the cell), the ligand acts as an allosteric regulator, exposing the active site of the catalytic domain. Generally, the ligand does not contribute further to the cellular response. In fact, the ligand is usually not metabolized (or changed) at all; its role is purely to "knock on the door." (This is in sharp contrast to enzyme–substrate interactions, which we describe in Concept 3.3. The whole purpose of those interactions is to change substrates into useful products.)

Receptors (R) bind to their ligands (L) noncovalently, according to chemistry's *law of mass action*:

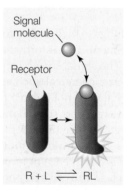

$$R + L \rightleftharpoons RL$$

This means the binding is *reversible*, although for most ligand–receptor complexes, the equilibrium point is far to the right of the above reaction—that is, binding is favored. Reversibility is important, however, because if the ligand were never released, the receptor would be continuously stimulated.

An inhibitor (or *antagonist*) can also bind to a receptor protein, preventing the binding of the normal ligand. This is analogous to the competitive inhibition of enzymes (see Concept 3.4). There are both natural and artificial antagonists of receptor binding. For example, many substances that alter human behavior bind to specific receptors in the brain and prevent the binding of the receptors' specific ligands.

Receptors can be classified by location and function

The chemistry of signal ligands is quite variable, but they can be divided into two groups based on whether or not they can diffuse through membranes. Correspondingly, a receptor can be classified by its location in the cell, which largely depends on the nature of its ligand:

- *Cytoplasmic receptors.* Small or nonpolar ligands can diffuse across the phospholipid bilayer of the plasma membrane and enter the cell. Estrogen, for example, is a lipid-soluble steroid hormone that can easily diffuse across the plasma membrane; it binds to a receptor in the cytoplasm. Many cytoplasmic receptors function as regulators of gene expression; they alter (upregulate or downregulate) the expression of specific genes as a result of ligand binding.

- *Membrane receptors.* Large or polar ligands cannot cross the lipid bilayer. Insulin, for example, is a protein hormone that cannot diffuse through the plasma membrane. Instead, it binds to a transmembrane receptor with an extracellular binding domain.

FRONTIERS Some receptor proteins are concentrated in a particular area of the membrane, trapped in a *lipid raft*—a semisolid region with lipids enriched in cholesterol and long-chain fatty acids. How do these rafts get placed where they are, and how do they trap only certain proteins? Scientists are investigating membrane assembly in the endoplasmic reticulum and Golgi apparatus to find out. This information will be important for understanding how cells optimize functions such as cellular signaling.

In complex eukaryotes such as mammals and higher plants, there are three well-studied categories of plasma membrane receptors, which are grouped according to their activities: ion channels, protein kinase receptors, and G protein–linked receptors. Because you will see these receptors several times later in the text, we describe them in some detail here.

ION CHANNEL RECEPTORS As described in Concept 5.2, the plasma membranes of many cells contain *gated ion channels* for ions such as Na^+, K^+, Ca^{2+}, or Cl^- (see Figure 5.4). The gate-opening mechanism is an alteration in the three-dimensional shape of the channel protein upon ligand binding; thus these proteins function as receptors. An example is the *acetylcholine receptor*, a ligand-gated sodium channel located in the plasma membranes of skeletal muscle cells. The ligand acetylcholine is a neurotransmitter—a chemical signal released from neurons. Opening of the channel allows Na^+, which is more concentrated outside the cell than inside, to diffuse into the cell. This initiates a series of events that result in muscle contraction (see Figure 34.9).

LINK Nerve cells communicate with muscle cells at neuromuscular junctions, which are described in Concept 36.1

PROTEIN KINASE RECEPTORS Like gated channel receptors, protein kinase receptors change shape upon ligand binding. But in

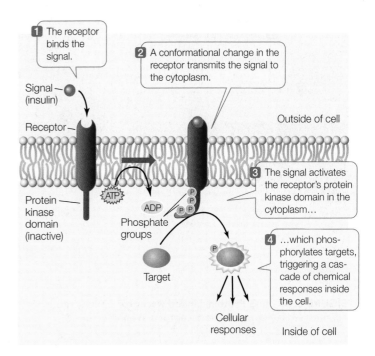

FIGURE 5.13 A Protein Kinase Receptor The mammalian hormone insulin binds to a protein kinase receptor on the outside surface of the cell and initiates a response.

this case, the new conformation exposes or activates a domain on the cytoplasmic side of the transmembrane protein that has catalytic (protein kinase) activity (**FIGURE 5.13**).

In general, **protein kinases** catalyze the following reaction:

$$ATP + protein \rightarrow ADP + phosphorylated\ protein$$

Each protein kinase has specific target(s) in the cell. The reaction results in the covalent modification (phosphorylation) of a target protein, thereby changing its activity. Protein kinases are extraordinarily important in biological signaling: about 1 human gene in 50 is a protein kinase, and there is an even higher proportion of such genes in some plants.

An example of a protein kinase receptor is that for the hormone insulin. The activation of this receptor results in the phosphorylation of target proteins in the cytoplasm. The targeted proteins mediate the cell's response, which includes the insertion of glucose transport proteins into the cell membrane.

It should be noted that not all protein kinases are receptors—many function in later steps of signal transduction pathways. In these cases, the protein kinase is activated by a receptor or other protein, and then phosphorylates the next protein in the pathway (as you will see in Concept 5.6).

G PROTEIN–LINKED RECEPTORS A third category of eukaryotic plasma membrane receptors is the family of **G protein–linked receptors**. In this case, ligand binding on the extracellular domain of the receptor changes the shape of its cytoplasmic region, exposing a site that can bind to a mobile membrane protein called a **G protein**. The G protein is partially inserted in the lipid bilayer and partially exposed on the cytoplasmic surface of the membrane.

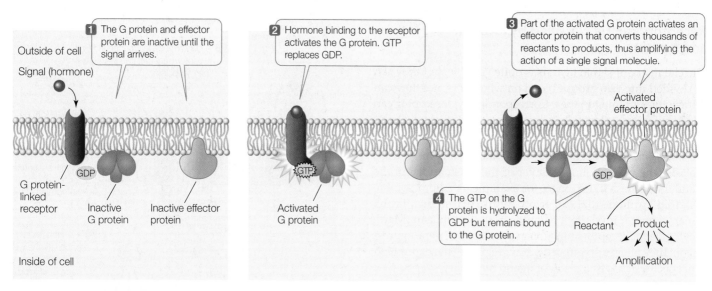

FIGURE 5.14 A G Protein–Linked Receptor The G protein is an intermediary between the receptor and an effector.

Many G proteins have three polypeptide subunits and can bind three different molecules:

- The receptor
- GDP and GTP (guanosine diphosphate and triphosphate, nucleotides used for energy transfer, like ADP and ATP)
- An effector protein (a protein that causes an effect in the cell)

The activated G protein–linked receptor functions as a *guanine nucleotide exchange factor*, which exchanges a GDP nucleotide bound to the G protein for a more energy-rich GTP. The activated G protein in turn activates the effector protein, leading to downstream signal amplification (**FIGURE 5.14**). G protein–linked receptors are especially important in the sensory systems of animals (see Chapter 35).

---

yourBioPortal.com

Go to ANIMATED TUTORIAL 5.4
G Protein–Linked Signal Transduction and Cancer

Do You Understand Concept 5.5?

- What are the three major steps in cell signaling?

- What are the differences and similarities between ion channel receptors and G protein–linked receptors?

- If an intact cell is treated with an enzyme that hydrolyzes proteins, will the cell be able to receive any environmental signals? Explain.

Regardless of the type of receptor, when a signal activates it, a *signal transduction pathway* ensues. This often involves a series of multiple steps, and it leads to a particular cellular response. This could be the expression of a set of genes that were previously silent. The proteins encoded by these genes change the cell in some way so it can respond to the signal.

concept 5.6 Signal Transduction Allows the Cell to Respond to Its Environment

As we mentioned in Concept 5.5, a signal may be a chemical ligand or a physical stimulus such as light or heat. Its effect is to activate a specific receptor—leading to a cellular response, which is mediated by a signal transduction pathway. Typically, signaling at the plasma membrane initiates a *cascade* (series) of events in the cell, in which proteins interact with other proteins until the final responses are achieved. Through such a cascade, an initial signal can be both *amplified* and *distributed* to cause several different responses in the target cell.

Second messengers can stimulate signal transduction

Often there is a small molecule intermediary between the activated receptor and the cascade of events that ensues. In a series of clever experiments, Earl Sutherland and his colleagues at Case Western Reserve University discovered that a small water-soluble chemical messenger can mediate the cytoplasmic events initiated by a plasma membrane receptor. These researchers were investigating the activation of the liver enzyme glycogen phosphorylase by the hormone epinephrine (also called adrenaline)—the "fight-or-flight" hormone. The enzyme is activated when an animal faces life-threatening conditions and needs energy fast for the fight-or-flight response. Glycogen phosphorylase catalyzes the breakdown of glycogen stored in the liver so that the resulting glucose molecules can be released to the blood (see Figure 39.13). The enzyme is present in the liver cell cytoplasm but is inactive in the absence of epinephrine.

The researchers found that epinephrine could activate glycogen phosphorylase in liver cells that had been broken open, but only if the entire cell contents, including plasma membrane fragments, were present. Under these conditions epinephrine was bound to the plasma membrane fragments, but the active phosphorylase was present in the solution. The researchers hypothesized that there must be a second "messenger" that transmits

INVESTIGATION

FIGURE 5.15 The Discovery of a Second Messenger Glycogen phosphorylase is activated in liver cells after epinephrine binds to a membrane receptor. Sutherland and his colleagues observed that this activation could occur in vivo only if fragments of the plasma membrane were present. They designed experiments to show that a second messenger caused the activation of glycogen phosphorylase.

HYPOTHESIS

A second messenger mediates between receptor activation at the plasma membrane and enzyme activation in the cytoplasm.

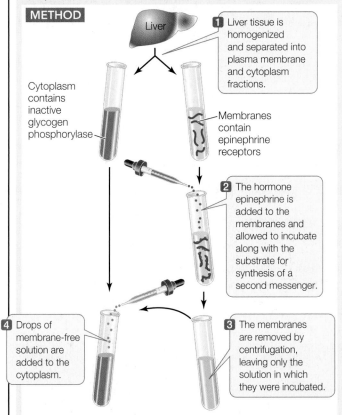

METHOD

Liver

Cytoplasm contains inactive glycogen phosphorylase

1 Liver tissue is homogenized and separated into plasma membrane and cytoplasm fractions.

Membranes contain epinephrine receptors

2 The hormone epinephrine is added to the membranes and allowed to incubate along with the substrate for synthesis of a second messenger.

3 The membranes are removed by centrifugation, leaving only the solution in which they were incubated.

4 Drops of membrane-free solution are added to the cytoplasm.

RESULTS

Active glycogen phosphorylase is present in the cytoplasm.

CONCLUSION

A soluble second messenger, produced by hormone-activated membranes, is present in the solution and activates enzymes in the cytoplasm.

ANALYZE THE DATA

The activity of previously inactive liver glycogen phosphorylase was measured with and without epinephrine incubation, with these results:

Condition	Enzyme activity (units)
Homogenate	0.4
Homogenate + epinephrine	2.5
Cytoplasm fraction	0.2
Cytoplasm + epinephrine	0.4
Membranes + epinephrine	0.4
Cytoplasm + membranes + epinephrine	2.0

A. What do these data show?

B. Propose an experiment to show that the factor that activates the enzyme is stable on heating and give predicted data.

C. Propose an experiment to show that cAMP can replace the membrane fraction and hormone treatment and give predicted data.

For more, go to Working with Data 5.2 at **yourBioPortal.com**.

Go to **yourBioPortal.com** for original citations, discussions, and relevant links for all INVESTIGATION figures.

---

the epinephrine signal (the "first messenger") from the plasma membrane to the phosphorylase in the cytoplasm. They investigated this by separating plasma membrane fragments from the cytoplasmic fractions of broken liver cells and following the sequence of steps described in **FIGURE 5.15**. This experiment confirmed the existence of a second messenger, later identified as **cyclic AMP** (**cAMP**; **FIGURE 5.16**). Second messengers do not have enzymatic activity themselves; rather, they act to regulate target enzymes by binding to them noncovalently.

LINK The different types of enzyme regulation are discussed in Concept 3.4

A second messenger is a small molecule that mediates later steps in a signal transduction pathway. While receptor binding is highly specific, second messengers allow a cell to respond to

FIGURE 5.16 The Formation of Cyclic AMP The formation of cAMP from ATP is catalyzed by adenylyl cyclase, an enzyme that is activated by G proteins.

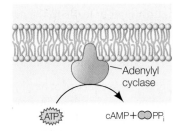

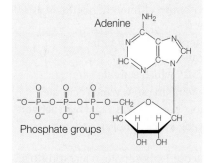

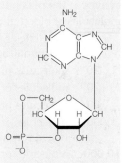

ATP

Cyclic AMP (cAMP)

a single event at the plasma membrane with *many events in-side the cell*—in other words, the second messenger *distributes the initial signal*. Second messengers also serve to *amplify the signal*—for example, the binding of a single epinephrine molecule leads to the production of many molecules of cAMP. In turn, cAMP activates many enzyme targets by binding to them noncovalently. In the case of epinephrine and the liver cell, glycogen phosphorylase is just one of several enzymes that are activated.

A signaling cascade involves enzyme regulation and signal amplification

Signal transduction pathways often involve multiple sequential steps, in which particular enzymes are either activated or inhibited by other enzymes in the pathway. For example, a protein kinase adds a phosphate group to a target protein, and this covalent change alters the protein's conformation and activates or inhibits its function. Cyclic AMP binds noncovalently to a target protein, and this changes the protein's shape, activating or inhibiting its function. In the case of activation, a previously inaccessible active site is exposed, and the target protein goes on to perform a new cellular role.

A good example of a signaling cascade is the G protein–mediated protein kinase pathway stimulated by epinephrine in liver cells (**FIGURE 5.17**). Binding of epinephrine to the membrane receptor results in the activation of a G protein, followed by the production of cAMP, which activates a key signaling molecule, protein kinase A. In turn, protein kinase A phosphorylates two other enzymes, with opposite effects:

- *Inhibition.* Glycogen synthase, which catalyzes the joining of glucose molecules to form the energy-storing molecule glycogen, is inactivated when a phosphate group is added to it by protein kinase A. Thus the epinephrine signal *prevents glucose from being stored* in glycogen (Figure 5.17, step 1).

- *Activation.* Phosphorylase kinase is activated when a phosphate group is added to it. It is part of a cascade of reactions that ultimately leads to the activation of glycogen phosphorylase, another key enzyme in glucose metabolism. This enzyme results in the *liberation of glucose molecules* from glycogen (Figure 5.17, steps 2 and 3).

An important consequence of having multiple steps in a signal transduction cascade is that the signal is *amplified* with each step. The amplification of the signal in the pathway illustrated in Figure 5.17 is impressive. Each molecule of epinephrine that arrives at the plasma membrane ultimately results in 10,000 molecules of blood glucose:

1	molecule of epinephrine bound to the membrane activates
20	molecules of cAMP, which activate
20	molecules of protein kinase A, which activate
100	molecules of phosphorylase kinase, which activate
1,000	molecules of glycogen phosphorylase, which produce
10,000	molecules of glucose 1-phosphate, which produce
10,000	molecules of blood glucose

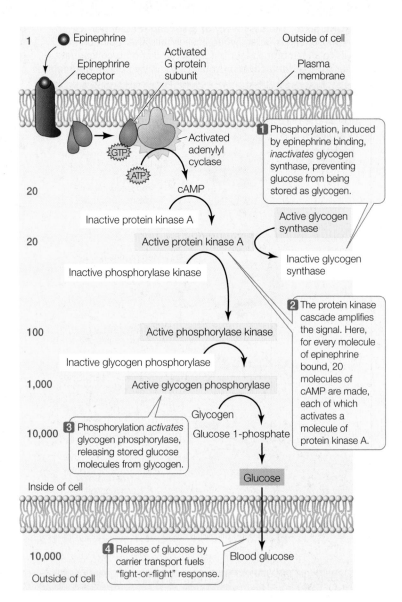

FIGURE 5.17 A Cascade of Reactions Leads to Altered Enzyme Activity Liver cells respond to epinephrine by activating G proteins, which in turn activate the synthesis of the second messenger cAMP. Cyclic AMP initiates a protein kinase cascade, greatly amplifying the epinephrine signal, as indicated by the blue numbers. The cascade both inhibits the conversion of glucose to glycogen and stimulates the release of previously stored glucose.

yourBioPortal.com

Go to **ANIMATED TUTORIAL 5.5**
Signal Transduction Pathway

Signal transduction is highly regulated

Signal transduction is a temporary event in the cell, and gets "turned off" once the cell has responded. To regulate protein kinases, G proteins, and cAMP, there are enzymes that convert each activated transducer back to its inactive precursor (**FIGURE 5.18**). The balance between the activities of these regulating enzymes and the signaling enzymes themselves is what determines the ultimate cellular response to a signal. Cells can alter this balance in several ways:

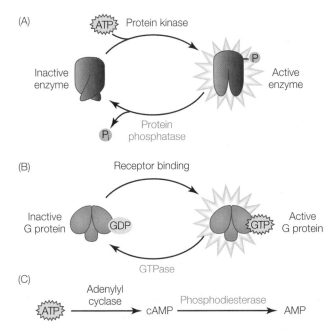

FIGURE 5.18 Signal Transduction Regulatory Mechanisms Some signals lead to the production of active transducers such as (A) protein kinases, (B) G proteins, and (C) cAMP. Other enzymes (red type) inactivate or remove these transducers.

- *Synthesis or breakdown of the enzymes involved.* For example, synthesis of adenylyl cyclase and breakdown of phosphodiesterase (which breaks down cAMP) would tilt the balance in favor of more cAMP in the cell.

- *Activation or inhibition of the enzymes by other molecules.* Examples include the activation of a G protein–linked receptor by ligand binding, and inhibition of phosphodiesterase (which also breaks down cGMP) by sildenafil citrate (Viagra), a drug used to treat erectile dysfunction.

Cell functions change in response to environmental signals

The activation of a receptor by a signal, and the subsequent transduction and amplification of the signal, ultimately leads to changes in cell function. There are many ways in which a cell might respond. Three important types of responses have already been discussed in this chapter:

- *Opening of ion channels* changes the balance of ion concentrations between the outside of the cell and its interior (see Figure 5.4). As you will see in Chapter 34, this results in a change in the electrical potential across the membrane, with important consequences in nerve and muscle cells.

- Many signal transduction pathways lead to *alterations in gene expression.* The expression of some genes may be switched on (upregulated), whereas others are switched off (downregulated). This alters the abundance of the proteins (often enzymes) encoded by the genes, thus changing cell function. You will see many examples that highlight the importance of gene regulation throughout this book.

- A third kind of response involves the *alteration of enzyme activities.* An example is the activation of phosphorylase kinase

and the inhibition of glycogen synthase in liver cells exposed to epinephrine (see Figure 5.17). In this case, the signal transduction pathway does not lead to alterations in the expression of the genes that encode these enzymes. Their *activities* are regulated so that the liver cell can rapidly supply energy-rich glucose for the fight-or-flight response.

Different signal transduction pathways can produce different cellular responses to the same signal-receptor binding event. An example is the responses to epinephrine in heart muscle and in the smooth muscles that line blood vessels. When epinephrine binds to its receptor in heart muscle cells, a G protein is activated that in turn *activates* the signal transduction cascade that results in glucose mobilization for energy and muscle contraction (see Figure 5.17). In smooth muscle cells that line the digestive tract, however, a different G protein gets activated by epinephrine receptor binding, and in this case the G protein *inhibits* a target enzyme, allowing the muscle cells to relax. This increases the diameter of the blood vessels, allowing more nutrients to be carried from the digestive system to the rest of the body. (For more on the flight-or-fight response, see Concept 30.2.) So the same signaling mechanism can lead to different responses in different types of cells.

A great deal has been learned about signal transduction pathways and cellular responses in the past two decades, and there is still much to learn. As biologists tease apart specific pathways, they find that many of them are interconnected: one pathway may be switched on by a particular signal or molecule, and another may be switched off. In this chapter we have concentrated on signaling pathways that occur in animal cells. However, signal transduction pathways are important in the functioning of all living organisms.

Do You Understand Concept 5.6?

- Compare "first messengers" (e.g., hormones) with "second messengers" (e.g., cAMP) with regard to their chemical nature, where and when they are made, and locations of synthesis and activity.

- Outline the steps in the amplification of signaling by epinephrine, resulting in the release of glucose to the bloodstream. At each step, is the amplification due to a covalent or noncovalent interaction?

- What would happen to a liver cell exposed to epinephrine and at the same time to a drug that inhibits protein kinase A? To epinephrine and to a drug that inhibits the hydrolysis of GTP?

- Biochemist Robert Furchgott studied how acetylcholine, released from nerve cells, causes smooth muscles surrounding arteries to relax, dilating the arteries. He found that the smooth muscles in an intact artery section would relax, but if the endothelial cells lining the blood vessel were removed, the addition of acetylcholine would not cause dilation. A second messenger was proposed. How would you investigate this hypothesis? (Hint: see Figure 5.15.)

What role does the cell membrane play in the body's response to caffeine?

ANSWER Caffeine has many effects on the body, but the most noticeable is that it keeps us awake. The caffeine molecule is somewhat large and polar, and it is unlikely to diffuse through the nonpolar lipids of the cell membrane (Concept 5.2). Instead, it binds to receptors on the surfaces of nerve cells in the brain (Concept 5.5).

The nucleoside adenosine (adenine attached to a five-carbon sugar) accumulates in the brain when a person is under stress or has prolonged mental activity. When it binds to a specific receptor in the brain, adenosine sets in motion a signal transduction pathway (Concept 5.6) that results in reduced brain activity, which usually means drowsiness. This membrane-associated signaling by adenosine has evolved as a protective mechanism against the adverse effects of stress.

Caffeine has a three-dimensional structure similar to that of adenosine, and is able to bind to the adenosine receptor (**FIGURE 5.19**). Because its binding does not activate the receptor, caffeine functions as an *antagonist* of adenosine signaling, with the result that the brain stays active and the person remains alert.

When we discussed the interaction between a ligand and its receptor, we noted that this is a reversible, noncovalent interaction. In time, after drinking coffee or tea, the caffeine molecules come off the adenosine receptors in the brain, allowing adenosine to bind once again. Otherwise, coffee drinkers might never get to sleep!

In addition to competing with adenosine for a membrane receptor, caffeine blocks the enzyme cAMP phosphodiesterase. This enzyme acts in signal transduction (Concept 5.6) to break down the second messenger cAMP. Looking at the signal transduction pathway in Figure 5.17, can you explain how caffeine augments the fight-or-flight response, which includes an increase in blood sugar and increased heartbeat?

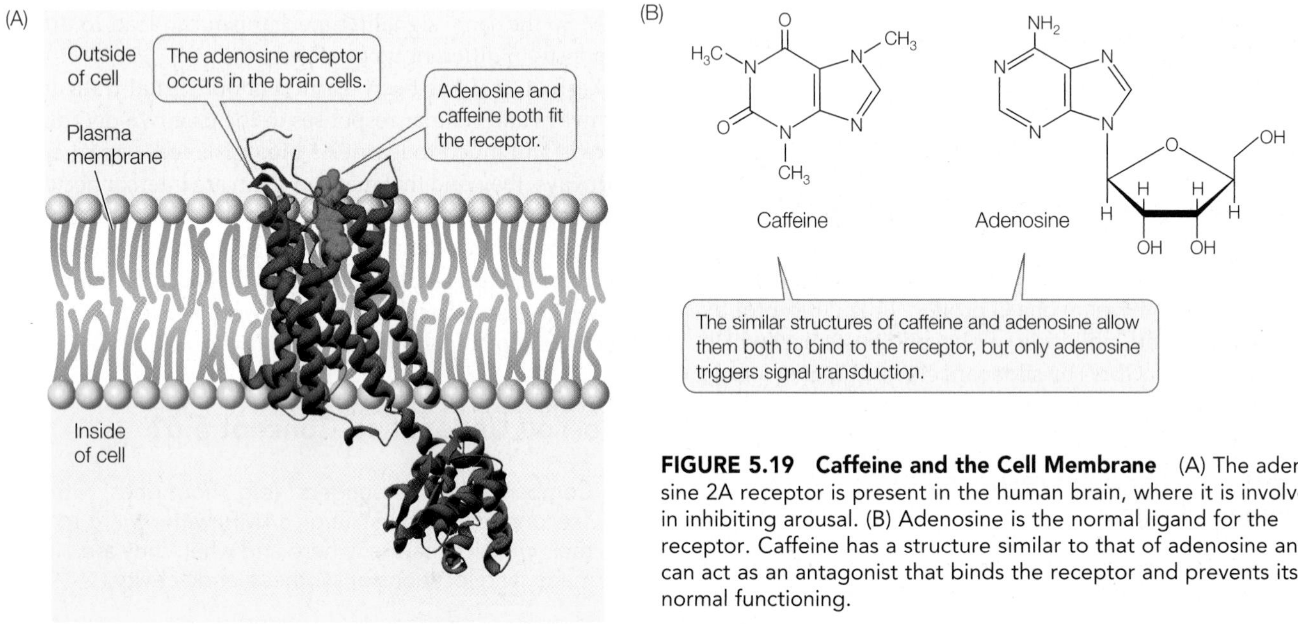

(A)

Outside of cell

The adenosine receptor occurs in the brain cells.

Plasma membrane

Adenosine and caffeine both fit the receptor.

Inside of cell

(B)

Caffeine

Adenosine

The similar structures of caffeine and adenosine allow them both to bind to the receptor, but only adenosine triggers signal transduction.

FIGURE 5.19 Caffeine and the Cell Membrane (A) The adenosine 2A receptor is present in the human brain, where it is involved in inhibiting arousal. (B) Adenosine is the normal ligand for the receptor. Caffeine has a structure similar to that of adenosine and can act as an antagonist that binds the receptor and prevents its normal functioning.

SUMMARY

concept 5.1 Biological Membranes Have a Common Structure and Are Fluid

- Biological membranes consist of lipids, proteins, and carbohydrates. The **fluid mosaic model** of membrane structure describes a phospholipid bilayer in which proteins can move about within the plane of the membrane.

- The two layers of a membrane may have different properties because of their different phospholipid compositions, exposed domains of **integral membrane proteins**, and **peripheral membrane proteins**. **Transmembrane proteins** span the membrane. **Review Figure 5.1, WEB ACTIVITY 5.1, and INTERACTIVE TUTORIAL 5.1**

concept 5.2 Some Substances Can Cross the Membrane by Diffusion

- Membranes exhibit **selective permeability**, regulating which substances pass through them.

- A substance can diffuse passively across a membrane by one of two processes: **simple diffusion** through the phospholipid bilayer or **facilitated diffusion**, either through a channel created by a **channel protein** or by means of a **carrier protein**. In both cases, molecules diffuse down their concentration gradients.

- In **osmosis**, water diffuses from a region of higher water concentration to a region of lower water concentration through membrane channels called aquaporins. Ions diffuse across membranes through ion channels. **Review Figures 5.3 and 5.4, ANIMATED TUTORIAL 5.1, and WORKING WITH DATA 5.1**

- **Carrier proteins** bind to polar molecules such as sugars and amino acids and transport them across the membrane. **Review Figure 5.6**

concept 5.3 Some Substances Require Energy to Cross the Membrane

- **Active transport** requires the use of chemical energy to move substances across membranes against their concentration gradients. The **sodium–potassium (Na^+–K^+) pump** uses energy released from the hydrolysis of ATP to move ions against their concentration gradients. **Review Figure 5.7 and ANIMATED TUTORIAL 5.2**

concept 5.4 Large Molecules Cross the Membrane via Vesicles

- **Endocytosis** is the transport of small molecules, macromolecules, large particles, and small cells into eukaryotic cells via the invagination of the plasma membrane and the formation of vesicles. **Review Figure 5.8A**

- In **receptor-mediated endocytosis**, a specific **receptor** on the plasma membrane binds to a particular macromolecule. **Review Figure 5.9 and ANIMATED TUTORIAL 5.3**

- In **exocytosis**, materials in vesicles are secreted from the cell when the vesicles fuse with the plasma membrane. **Review Figure 5.8B**

concept 5.5 The Membrane Plays a Key Role in a Cell's Response to Environmental Signals

- Cells receive many signals from the physical environment and from other cells. Chemical signals are often at very low concentrations. **Review Figure 5.10**

- A **signal transduction pathway** involves the interaction of a signal (often a chemical **ligand**) with a receptor; the transduction and amplification of the signal via a series of steps within the cell; and a cellular response. The response may be short-term or long-term. **Review Figure 5.11**

- Cells respond to signals only if they have specific receptor proteins that can be activated by those signals. Most receptors are located at the plasma membrane. They include **ion channels**, **protein kinases**, and **G protein–linked receptors**. **Review Figures 5.13 and 5.14 and ANIMATED TUTORIAL 5.4**

concept 5.6 Signal Transduction Allows the Cell to Respond to Its Environment

- A cascade of events, one following another, occurs after a receptor is activated by a signal.

- Often, a soluble second messenger conveys signaling information from the primary messenger (ligand) at the membrane to downstream signaling molecules in the cytoplasm. **Cyclic AMP (cAMP)** is an important second messenger. **Review Figure 5.16 and WORKING WITH DATA 5.2**

- Activated enzymes may in turn activate other enzymes in a signal transduction pathway, leading to impressive amplification of a signal. **Review Figure 5.17 and ANIMATED TUTORIAL 5.5**

- Protein kinases covalently add phosphate groups to target proteins; cAMP binds target proteins noncovalently. Both kinds of binding change the target protein's conformation to expose or hide its active site.

- Signal transduction can be regulated in several ways. The balance between the activation and inactivation of the molecules involved determines the ultimate cellular response to a signal. **Review Figure 5.18**

- The cellular responses to signals may include the opening of ion channels, changes in gene expression, or the alteration of enzyme activities.

See **WEB ACTIVITY 5.2** for a concept review of this chapter.

Pathways that Harvest and Store Chemical Energy

Agriculture was a key invention in the development of human civilizations. The planting and harvesting of seeds began about 10,000 years ago. One of the first plants to be turned into a reliable crop was barley, and one of the first uses of barley was to brew beer. Living in what is now Iraq, ancient Sumerians learned that partly germinated and then mashed-up barley seeds, stored under the right conditions, could produce a potent and pleasant drink. An ancient king, Hammurabi, laid down the oldest known law regarding an alcoholic beverage: the daily beer ration was 2 liters for a normal worker, 3 liters for a civil servant, and 5 liters for a high priest. Alcoholic beverages were not just a diversion to these people, because drinking water from rivers and ponds caused diseases, and whatever caused these diseases was not present in liquids containing alcohol.

Early chemists and biologists were interested in how mashed barley seeds (or grapes, in the case of wine) were transformed into alcoholic beverages. By the nineteenth century there were two theories. Chemists claimed that these transformations were simply chemical reactions, not some special property of the plant material. Biologists, armed with their microscopes and cell theory (see Chapter 4), said that the barley and grape extracts were converted to beer and wine by living cells.

The great French scientist Louis Pasteur tackled the question in the 1860s, responding to a challenge posed by a group of distillers who wanted to use sugar beets to produce alcohol. Pasteur found that (1) nothing happened to beet mash unless microscopic yeast cells were present; (2) in the presence of fresh air, yeast cells grew vigorously on the mash, and bubbles of CO_2 were formed; and (3) without fresh air, the yeast grew slowly, less CO_2 was produced, and alcohol was formed. So the biologists were right: living cells produced alcohol from ground-up, sugary extracts. Later, biochemists broke open yeast cells and unraveled the sequence of chemical transformations from sugar to alcohol. It turned out that the chemists were right too: the production of alcohol was just a series of chemical reactions. These reactions require energy transfers, and the flow of energy in living systems (such as yeast cells) involves the same chemical principles as energy flow in the inanimate world.

These tanks at a winery in California provide conditions without air that are suitable for the production of alcohol by yeast cells.

KEY CONCEPTS

6.1 ATP, Reduced Coenzymes, and Chemiosmosis Play Important Roles in Biological Energy Metabolism

6.2 Carbohydrate Catabolism in the Presence of Oxygen Releases a Large Amount of Energy

6.3 Carbohydrate Catabolism in the Absence of Oxygen Releases a Small Amount of Energy

6.4 Catabolic and Anabolic Pathways Are Integrated

6.5 During Photosynthesis, Light Energy Is Converted to Chemical Energy

6.6 Photosynthetic Organisms Use Chemical Energy to Convert CO_2 to Carbohydrates

Q QUESTION Why does fresh air inhibit the formation of alcohol by yeast cells?

<table>
<tr><td>concept
6.1</td><td>**ATP, Reduced Coenzymes, and Chemiosmosis Play Important Roles in Biological Energy Metabolism**</td></tr>
</table>

In Chapters 2 and 3 we introduce the general concepts of energy, enzymes, and metabolism. Energy is stored in the chemical bonds of molecules, and it can be released and transformed by the metabolic pathways of living cells. There are five *general principles governing metabolic pathways*:

- A complex chemical transformation occurs in a series of separate, intermediate reactions that form a metabolic pathway.
- Each reaction is catalyzed by a specific enzyme.
- Most metabolic pathways are similar in all organisms, from bacteria to plants to humans.
- In eukaryotes, many metabolic pathways are compartmentalized, with certain reactions occurring inside specific organelles.
- Each metabolic pathway is controlled by key enzymes that can be inhibited or activated, thereby determining how fast the reactions will go.

Chemical energy available to do work is termed *free energy* (G). According to the laws of thermodynamics, a biochemical reaction may change the *form* of energy but not the net *amount*. A biochemical reaction is exergonic if it releases energy from the reactants, or endergonic if energy must be added to the reactants.

LINK You can review the principles of energy transformations in Concept 2.5

In the chemistry lab, energy can be released or added in the form of heat. But in cells, energy-transforming reactions are often coupled; that is, an energy-releasing (exergonic) reaction is coupled in time and location to an energy-requiring (endergonic) reaction. Two widely used coupling molecules are the coenzymes ATP and NADH.

ATP hydrolysis releases energy

Cells use adenosine triphosphate (ATP) as a kind of "energy currency." Just as it is more effective, efficient, and convenient for you to trade money for a lunch than to trade your actual labor, it is useful for cells to have a single currency for transferring energy between different reactions and cell processes. Some of the energy that is released in exergonic reactions is captured in chemical bonds when ATP is formed from adenosine diphosphate (ADP) and inorganic phosphate (hydrogen phosphate; commonly abbreviated to P_i). The ATP can then be hydrolyzed at other sites in the cell, releasing free energy to drive endergonic reactions (**FIGURE 6.1**). For example, in Chapter 5 we describe the use of ATP hydrolysis to drive the energy-requiring process of active transport. Later in this chapter you will see how ATP hydrolysis is used in an anabolic (endergonic) pathway—the synthesis of carbohydrates from CO_2 during photosynthesis.

An ATP molecule consists of the nitrogenous base adenine bonded to ribose (a sugar), which is attached to a sequence

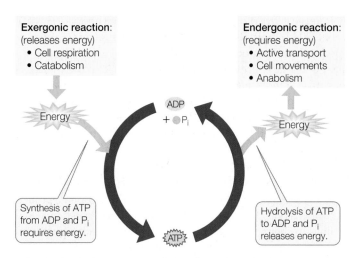

FIGURE 6.1 The Concept of Coupling Reactions Exergonic cellular reactions release the energy needed to make ATP from ADP. The energy released from the conversion of ATP to ADP can be used to drive endergonic reactions.

yourBioPortal.com
Go to WEB ACTIVITY 6.1
ATP and Coupled Reactions

of three phosphate groups (**FIGURE 6.2**). The hydrolysis of a molecule of ATP yields free energy, ADP, and the inorganic phosphate ion (P_i):

$$ATP + H_2O \rightarrow ADP + P_i + \text{free energy}$$

The important property of this reaction is that it is exergonic, releasing free energy. Under standard laboratory conditions,

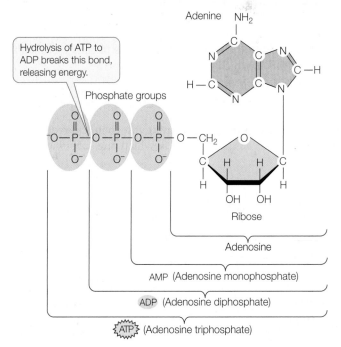

FIGURE 6.2 ATP The nucleotide ATP is built by the addition of terminal phosphate groups onto the nucleoside adenosine.

the *change in free energy* for this reaction (ΔG) is about -7.3 kcal/mol (-30 kJ/mol). Recall that a negative change in free energy means that the product molecules (in this case, ADP and P_i) have less free energy than the reactant (ATP), so the change is negative. A molecule of ATP can also be hydrolyzed to adenosine monophosphate (AMP) and a pyrophosphate ion ($P_2O_7^{4-}$; commonly abbreviated as PP_i). In this case, additional energy may be released by the subsequent conversion of PP_i to two molecules of P_i.

Two characteristics of ATP account for the free energy released by the loss of one or two of its phosphate groups:

- The free energy of the P—O bond between phosphate groups (called a phosphoanhydride bond and often denoted by wavy lines in chemical structures—see below) is much higher than the energy of the O—H bond that forms after hydrolysis. So some usable energy is released by the following hydrolysis:

Adenine—ribose —O—P—O ~ P—O ~ P—O⁻ + H_2O

ATP

Adenine—ribose —O—P—O ~ P—OH + HO—P—O⁻

ADP   **P_i**

- Because phosphate groups are negatively charged and so repel each other, it takes energy to get phosphates near enough to each other to make the covalent bond that links them together in the ATP molecule.

In some reactions, ATP is formed by *substrate-level phosphorylation* because it involves the direct transfer of phosphate to ADP. This is the case for some reactions of glycolysis (see Concept 6.2). But most of the ATP in living cells is formed by *oxidative phosphorylation*, which we will discuss shortly.

Redox reactions transfer electrons and energy

Another way of transferring energy in chemical reactions is to *transfer electrons*. A reaction in which one substance transfers one or more electrons to another substance is called an oxidation–reduction reaction, or **redox** reaction.

- **Reduction** is the gain of one or more electrons by an atom, ion, or molecule.
- **Oxidation** is the loss of one or more electrons.

Oxidation and reduction *always occur together*: as one chemical is oxidized, the electrons it loses are transferred to another chemical, reducing it. Thus, some molecules are called *oxidizing agents* and others are *reducing agents*:

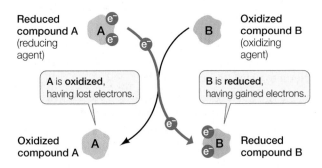

Although oxidation and reduction are defined in terms of traffic in electrons, it is often helpful to think in terms of the *gain or loss of hydrogen atoms*. Transfers of hydrogen atoms involve transfers of electrons ($H = H^+ + e^-$). So *when a molecule loses a hydrogen atom, it becomes oxidized.*

In general, the more reduced a molecule is, the more energy is stored in its covalent bonds (**FIGURE 6.3**). In a redox reaction, some energy is transferred from the reducing agent to the reduced product. Some energy remains in the reducing agent (now oxidized), and some is lost to entropy.

Cells use the coenzyme nicotinamide adenine dinucleotide as an electron carrier in redox reactions (**FIGURE 6.4**). This coenzyme exists in two chemically distinct forms, one oxidized (NAD^+) and the other reduced (NADH). The reduction reaction:

$$NAD^+ + H^+ + 2\,e^- \rightarrow NADH$$

involves the transfer of a proton (the hydrogen ion, H^+) and two electrons, which are released by the accompanying oxidization reaction. This reaction is *highly endergonic*, with a positive ΔG about four times greater than the positive ΔG (free energy change) for ATP formation.

Within the cell, the electrons do not remain with NADH. Oxygen is highly electronegative and readily accepts electrons from the reduced NADH molecule. The oxidation of NADH by O_2 (which occurs in several steps):

$$NADH + H^+ + \tfrac{1}{2}O_2 \rightarrow NAD^+ + H_2O$$

is *highly exergonic*, releasing energy with a ΔG of -52.4 kcal/mol (-219 kJ/mol). Note that the oxidizing agent appears here as "$\tfrac{1}{2}O_2$" instead of "O." This notation emphasizes that it is molecular oxygen, O_2, that acts as the oxidizing agent.

Because the oxidation of NADH releases more energy than the hydrolysis of ATP, NADH can be thought of as a larger package of free energy than ATP (approximately 50 kcal/mol, or 200 kJ/mol). NAD^+ is a common electron carrier in cells, but not

H—C—H   H—C—OH   H—C=O   H—C=O   C=O

Methane   Methanol   Formaldehyde   Formic   Carbon
(CH_4)   (CH_3OH)   (CH_2O)   acid   dioxide
                                      (HCOOH)   (CO_2)

Most reduced state   Most oxidized state
Highest free energy   Lowest free energy

FIGURE 6.3 Oxidation, Reduction, and Energy The more oxidized a carbon atom is, the less its free energy.

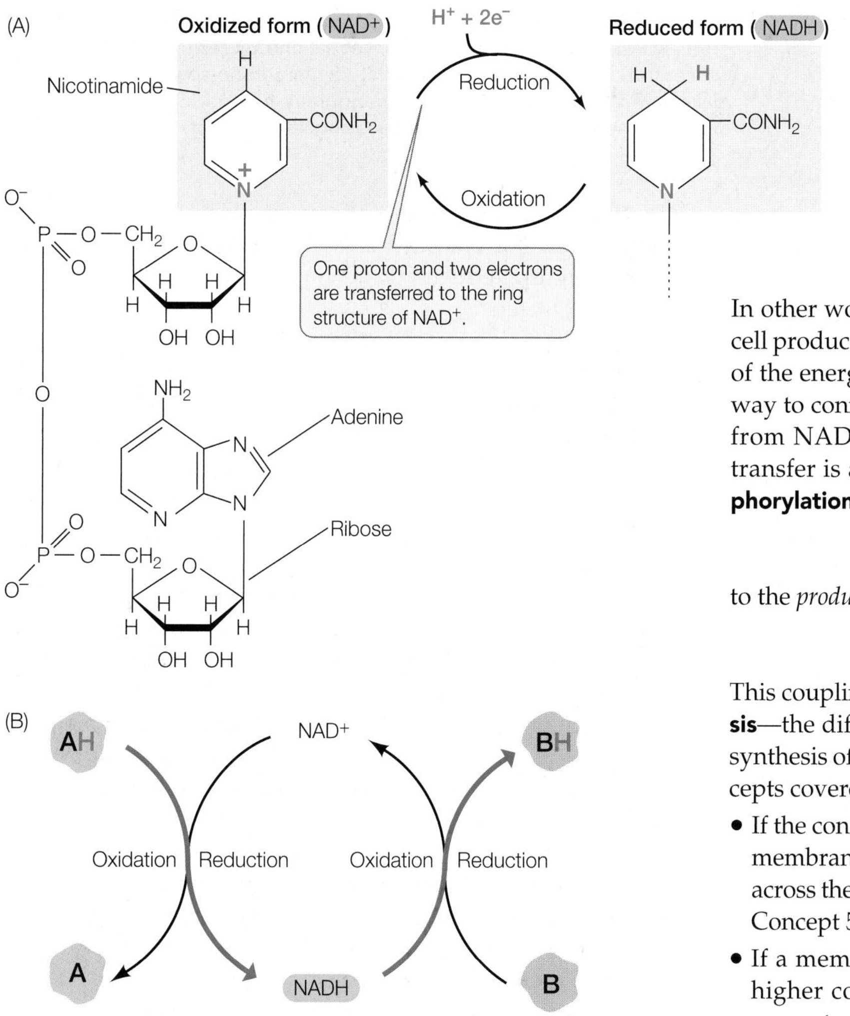

FIGURE 6.4 NAD$^+$/NADH Is an Electron Carrier in Redox Reactions (A) NAD$^+$ is an important electron acceptor in redox reactions, and its reduced form, NADH, is an important energy intermediary in cells. The unshaded portion of the molecule (left) remains unchanged by the redox reaction. (B) Coupling of redox reactions using NAD$^+$/NADH.

the only one. Others include flavin adenine dinucleotide (FAD), which also transfers electrons during glucose metabolism (see Concept 6.2), and nicotinamide adenine dinucleotide phosphate (NADP$^+$), which is used in photosynthesis (see Concept 6.5).

Oxidative phosphorylation couples the oxidation of NADH to the production of ATP

We can summarize the two energy-coupling coenzymes as follows:

- ADP traps chemical energy to make ATP.

- NAD$^+$ traps the energy released in redox reactions to make NADH.

How do these coenzymes participate in the flow of energy within cells? As you will see below, most chemical energy in cells is stored in the C—H bonds of carbohydrates and lipids (see Chapter 2). The release and reuse of this energy can be summarized as follows:

- Energy is released in *catabolism* by oxidation; this energy can be trapped by the reduction of coenzymes such as NADH.

- Energy for many *anabolic* and other energy-requiring processes is supplied by ATP. For example, as we note in Chapter 5, *active transport* requires ATP.

In other words, most of the energy-releasing reactions in the cell produce NADH (or similar reduced coenzymes), but most of the energy-consuming reactions require ATP. Cells need a way to connect the two coenzymes; that is, to transfer energy from NADH to the phosphoanhydride bond of ATP. This transfer is accomplished in a process called **oxidative phosphorylation**—the coupling of the *oxidation of NADH*:

$$NADH \rightarrow NAD^+ + H^+ + 2e^- + energy$$

to the *production of ATP*:

$$energy + ADP + P_i \rightarrow ATP$$

This coupling is achieved via a mechanism called **chemiosmosis**—the diffusion of protons across a membrane, driving the synthesis of ATP (**FIGURE 6.5A**). Chemiosmosis relies on concepts covered in earlier chapters:

- If the concentration of a substance is greater on one side of a membrane than the other, the substance will tend to diffuse across the membrane to its region of lower concentration (see Concept 5.2).

- If a membrane blocks this diffusion, the substance at the higher concentration has potential energy, which can be converted to other forms of energy (see Concept 2.5).

Because the interior of a membrane is nonpolar, protons (H$^+$) cannot readily diffuse across the membrane. Chemiosmosis converts the potential energy of a proton gradient across a membrane into the chemical energy in ATP. This has been called the proton motive force. In prokaryotes, the gradient is set up across the plasma membrane, using energy from various sources. In eukaryotes, chemiosmosis occurs in the mitochondria and chloroplasts. In mitochondria, the H$^+$ gradient is set up across the inner membrane, using energy released by the oxidation of NADH (see Concept 6.2). In chloroplasts, the H$^+$ gradient is set up across the thylakoid membrane using energy from light (see Concept 6.5). Despite these differences in detail, the mechanism of chemiosmosis is the same, and this mechanism is a key to biological energetics in almost all organisms.

A membrane protein called *ATP synthase* uses the potential energy of the H$^+$ gradient to drive ATP synthesis. ATP synthase is a molecular motor composed of two parts: the F$_o$ unit, which is a transmembrane domain that functions as the H$^+$ channel; and the F$_1$ unit, which contains the active sites for ATP synthesis (**FIGURE 6.5B**). The F$_1$ unit consists of six subunits (three each of two polypeptide chains), arranged like the segments of an orange around a central polypeptide. The potential energy set up by the proton gradient drives the passage of protons through the ring of polypeptides that make up

(A)

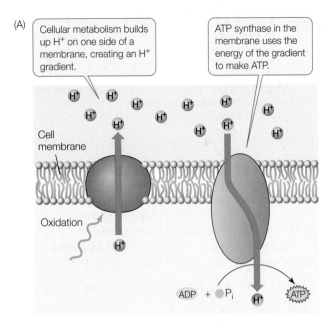

Cellular metabolism builds up H$^+$ on one side of a membrane, creating an H$^+$ gradient.

ATP synthase in the membrane uses the energy of the gradient to make ATP.

Cell membrane

Oxidation

ADP + P$_i$ H$^+$ ATP

(B)

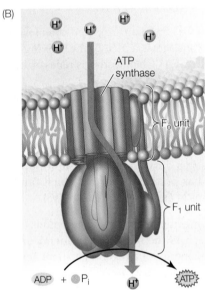

ATP synthase

F$_o$ unit

F$_1$ unit

ADP + P$_i$ H$^+$ ATP

FIGURE 6.5 Chemiosmosis (A) If a cell can generate a proton (H$^+$) gradient across a membrane, the potential energy resulting from the concentration gradient can be used by a membrane-spanning enzyme to make ATP. (B) ATP synthase has a membrane-embedded channel for H$^+$ diffusion and a motor that turns, releasing some energy to produce ATP.

An oxidation reaction is always coupled with a reduction. When NADH is oxidized to NAD$^+$ in the mitochondria, the corresponding reduction reaction (as noted above) is the formation of water:

$$2H^+ + 2e^- + \tfrac{1}{2}O_2 \rightarrow H_2O$$

So the key role of O$_2$ in cells—the reason we breathe and have a blood system to deliver O$_2$ to tissues—is to act as an electron acceptor and become reduced. In choloroplasts, the molecule ultimately reduced is NADP$^+$, a relative of NAD$^+$.

Chemiosmosis can be demonstrated experimentally. If chloroplasts or mitochondria are isolated from cells and put in a test tube, a proton gradient can be introduced artificially. This artificial gradient drives ATP synthesis (**FIGURE 6.6**), but only if ATP synthase, ADP, inorganic phosphate, and the membrane are present.

What happens if the H$^+$ gradient is destroyed by the presence of a membrane channel that is always open to protons? Obviously, ATP cannot be made, but the oxidation of NADH still occurs and O$_2$ is reduced, releasing considerable energy. The released energy forms heat instead to being used to make ATP. In newborn human infants, the membrane protein *thermogenin* disrupts the H$^+$ gradient in fat cell mitochondria, and this results in the release of heat. Because infants lack body hair, this process helps keep them warm.

A popular weight loss drug in the 1930s was the uncoupler molecule dinitrophenol. There were claims of dramatic weight loss when the drug was administered to obese patients. Unfortunately, the heat that was released caused fatally high fevers, and the effective dose and fatal dose were quite close. The use of this drug was discontinued in 1938, but the general strategy of using an uncoupler for weight loss remains a subject of research.

the F$_o$ component. This ring rotates as the protons pass through the membrane, causing the F$_1$ unit to rotate as well. ADP and P$_i$ bind to active sites that become exposed on the F$_1$ unit as it rotates, and ATP is made. The structure and function of ATP synthase are shared by living organisms as diverse as bacteria and humans. These molecular motors make ATP at rates of up to 100 molecules per second.

ATP and reduced coenzymes link catabolism and anabolism: An overview

We have separately described the cycles of biochemical energy storage and release for ATP (see Figure 6.1) and for reduced coenzymes (see Figure 6.4), as well as the role of chemiosmosis in linking the two (see Figure 6.5). Furthermore, we have mentioned how, in living organisms, catabolism releases chemical energy that can be used to fuel anabolism.

LINK See Concept 2.5 to review the principles of catabolism and anabolism

The next two concepts of this chapter deal with a major catabolic pathway called *cellular respiration*, in which a reduced molecule, the carbohydrate glucose, is oxidized, often all the way to CO$_2$:

carbohydrate + 6 O$_2$ → 6 CO$_2$ + 6 H$_2$O + chemical energy

FRONTIERS Scientists have long thought that life originated in a "primordial soup," with the first biochemical reactions driven by energy from space. Recently, however, microscopic geochemical H$^+$ gradients have been discovered in deep-sea vents. These gradients could have provided chemiosmotic energy for the formation of molecules such as carbohydrates, lipids, proteins, and nucleic acids. Scientists are attempting to create such vent systems in the lab to determine if this model is plausible.

INVESTIGATION

FIGURE 6.6 An Experiment Demonstrates the Chemiosmotic Mechanism The chemiosmosis hypothesis was a bold departure from the conventional scientific thinking of the time. It required an intact compartment separated by a membrane. Could a proton gradient drive the synthesis of ATP?

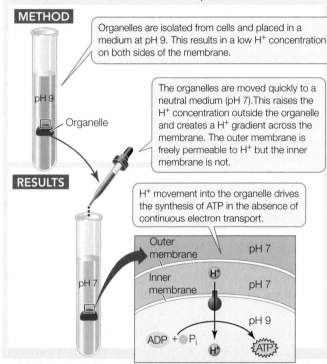

HYPOTHESIS

A H$^+$ gradient can drive ATP synthesis by isolated mitochondria or chloroplasts.

METHOD

Organelles are isolated from cells and placed in a medium at pH 9. This results in a low H$^+$ concentration on both sides of the membrane.

pH 9

Organelle

The organelles are moved quickly to a neutral medium (pH 7).This raises the H$^+$ concentration outside the organelle and creates a H$^+$ gradient across the membrane. The outer membrane is freely permeable to H$^+$ but the inner membrane is not.

RESULTS

H$^+$ movement into the organelle drives the synthesis of ATP in the absence of continuous electron transport.

Outer membrane   pH 7

pH 7

Inner membrane   pH 7

H$^+$

pH 9

ADP + P$_i$ H$^+$ ATP

CONCLUSION

In the absence of electron transport, an artificial H$^+$ gradient is sufficient for ATP synthesis by organelles.

ANALYZE THE DATA

In another experiment, chloroplast thylakoids were added to an acid solution at pH 3.8. After a short time to equilibrate, they were then transferred back to a solution at pH 8 in the presence of ADP, phosphate (P$_i$), and magnesium ions (Mg^{2+}). Some thylakoids were not transferred to pH 3.8, but instead kept at pH 7.0. ATP formation was measured using luciferase, which catalyzes the formation of a luminescent (light-emitting) molecule if ATP is present. Here are the data from the paper:

Reaction mixture	Luciferase activity (light emission)	
	Raw data	Corrected data
Complete, pH 3.8	141	
Complete, pH 7.0	12	
Complete, pH 3.8 – P$_i$	12	
" " – ADP	4	
" " – Mg^{2+}	60	
" " – chloroplasts	7	

A. Which reaction mixture is the control? Use the control data to correct the raw data for the other, experimental reaction mixtures and fill in the table.

B. Why did ATP production go down in the absence of P$_i$?

For more, go to Working with Data 6.1 at **yourBioPortal.com**.

Go to **yourBioPortal.com** for original citations, discussions, and relevant links for all INVESTIGATION figures.

This pathway occurs in some form in most cells. Plant cells and some prokaryotes also carry out *photosynthesis*, a major anabolic pathway. You will see in Concepts 6.5 and 6.6 that photosynthesis converts light energy into chemical energy in the form of carbohydrate:

$$6 \, CO_2 + 6 \, H_2O + \text{light energy} \rightarrow 6 \, O_2 + \text{carbohydrate}$$

How do these pathways fit together? Cellular respiration and photosynthesis are linked not only by their reactants and products (O_2, CO_2, and carbohydrate), but also by the energy "currency" of ATP and reduced coenzymes (**FIGURE 6.7**). The two pathways can occur in the same cell. For example, a cell in the leaf of a green plant carries out both photosynthesis and cellular respiration, often simultaneously. In these cases, the cell must coordinate the activities of the two processes, using the same mechanisms that integrate anabolism and catabolism in all cells (see Concept 6.4).

More often, however, cells do not carry out photosynthesis for themselves. Even in a green plant, a root cell is not photosynthetic and relies on carbohydrates transported from the leaf to carry out cellular respiration. Humans do not carry out photosynthesis anywhere in their bodies; they rely on carbohydrates obtained in their diet (ultimately derived from photosynthesis) to carry out cellular respiration, which provides

yourBioPortal.com

**Go to ANIMATED TUTORIAL 6.1
Two Experiments Demonstrate the Chemiosmotic Mechanism**

the chemical energy for their bodies' activities, such as active transport and anabolism.

Do You Understand Concept 6.1?

- Sketch the coupling of exergonic and endergonic reactions with ATP and NADH as coenzymes.

- What kind of coenzyme would receive energy released from the following reactions? Explain your answer.
  a. Glucose 6-phosphate → glucose
  b. Fatty acid → CO_2 + H_2O

- Hibernating animals have low rates of metabolism. They can keep warm by synthesizing a membrane protein not normally present in their cells. What might that protein do?

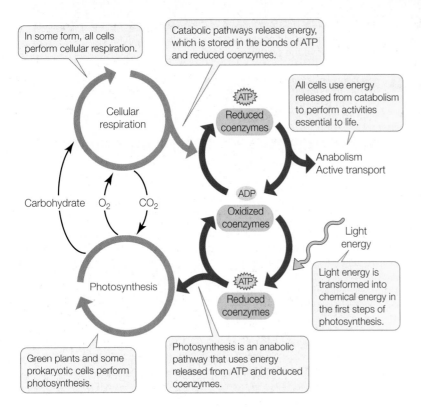

In some form, all cells perform cellular respiration.

Catabolic pathways release energy, which is stored in the bonds of ATP and reduced coenzymes.

All cells use energy released from catabolism to perform activities essential to life.

Light energy is transformed into chemical energy in the first steps of photosynthesis.

Green plants and some prokaryotic cells perform photosynthesis.

Photosynthesis is an anabolic pathway that uses energy released from ATP and reduced coenzymes.

FIGURE 6.7 ATP, Reduced Coenzymes, and Metabolism The major pathways of energy metabolism, cellular respiration and photosynthesis, are related by their use of energy-transferring substrates, ATP and reduced coenzymes. Note that the net result of these pathways is to convert light energy into chemical energy to fuel the processes of life.

with the best motors that humans have devised. The cell can achieve this through the general principles that govern metabolism listed in Concept 6.1. Most notably, the oxidation occurs in a series of small steps (**FIGURE 6.8**).

In the catabolism of glucose under **aerobic** conditions (in the presence of O_2), the small steps can be grouped into three linked biochemical pathways (**FIGURE 6.9**):

- In **glycolysis**, the six-carbon monosaccharide glucose is converted into two three-carbon molecules of pyruvate.
- In **pyruvate oxidation**, two three-carbon molecules of pyruvate are oxidized to two two-carbon molecules of acetyl CoA and two molecules of CO_2.
- In the **citric acid cycle**, two two-carbon molecules of acetyl CoA are oxidized to four molecules of CO_2.

In glycolysis, glucose is partially oxidized and some energy is released

Glycolysis takes place in the cytosol and involves ten enzyme-catalyzed reactions. During glycolysis, some of the covalent bonds between carbon and hydrogen atoms in the glucose molecule are oxidized, releasing some of the

In this concept we saw that in chemiosmosis, a proton gradient across a membrane drives the production of ATP. This oxidative phosphorylation of ADP to produce ATP is linked to NADH oxidation. We will now turn to the biochemical pathway that results in the production of NADH.

---

concept 6.2

Carbohydrate Catabolism in the Presence of Oxygen Releases a Large Amount of Energy

Cellular respiration is the catabolism of organic molecules within cells, and it is one of the key ways in which cells obtain energy. Considerable energy is released when reduced molecules with many C—C and C—H bonds are fully oxidized to CO_2. We will consider in detail only the catabolism of carbohydrates, but bear in mind that cells also obtain energy from the catabolism of other molecules, such as lipids.

The chemical energy released from the complete oxidation of glucose to CO_2 is considerable, and the cell traps the energy by forming ATP:

glucose + 6 O_2 → 6 CO_2 + 6 H_2O + energy

In a chemistry lab this energy is all lost as heat (686 kcal/mol). In the cell, much of the released energy is trapped as ATP (234 kcal/mol). The efficiency of this energy-trapping process is impressive, even when compared

FIGURE 6.8 Energy Metabolism Occurs in Small Steps (A) In living systems, glucose is oxidized via a series of steps, releasing small amounts of energy that can be efficiently trapped by coenzymes. (B) Glucose that is burned releases its energy as heat in one big step.

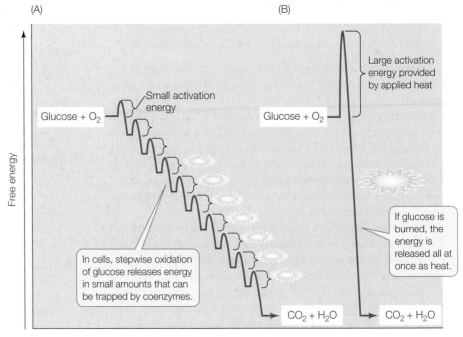

(A)

(B)

Large activation energy provided by applied heat

Small activation energy

Glucose + O_2

Glucose + O_2

In cells, stepwise oxidation of glucose releases energy in small amounts that can be trapped by coenzymes.

If glucose is burned, the energy is released all at once as heat.

Free energy

CO_2 + H_2O

CO_2 + H_2O

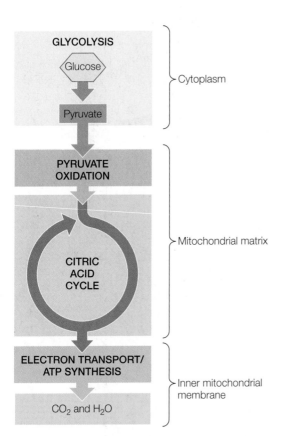

FIGURE 6.9 Energy-Releasing Metabolic Pathways The catabolism of glucose under aerobic conditions occurs in three sequential metabolic pathways: glycolysis, pyruvate oxidation, and the citric acid cycle. The reduced coenzymes are then oxidized by the respiratory chain, and ATP is made.

stored energy. The final products are two molecules of pyruvate (pyruvic acid), two molecules of ATP, and two molecules of NADH. Glycolysis can be divided into two stages: the initial energy-investing reactions that consume chemical energy stored in ATP, and the energy-harvesting reactions that produce ATP and NADH (**FIGURE 6.10**).

To help you understand the process without getting into extensive detail, we will focus on two consecutive reactions in this pathway (steps 6 and 7 in Figure 6.10).

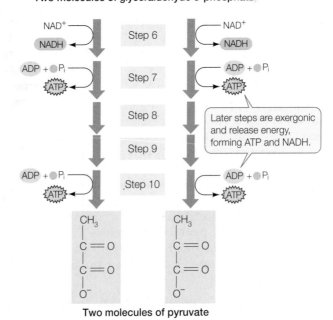

These are examples of two types of reactions that occur repeatedly in glycolysis and in many other metabolic pathways:

1. *Oxidation–reduction*: In this exergonic reaction, more than 50 kcal/mol of energy are released in the oxidation of glyceraldehyde 3-phosphate. (Look at the first carbon atom, where an H is replaced by an O.) The energy is trapped via the reduction of NAD⁺ to NADH.

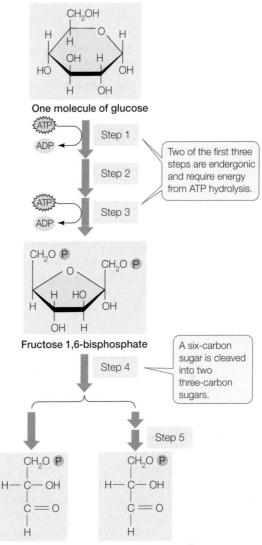

FIGURE 6.10 Glycolysis Converts Glucose into Pyruvate Glucose is converted to pyruvate in ten enzyme-catalyzed steps. Along the way, energy is released to form ATP and NADH.

2. *Substrate-level phosphorylation*: The second reaction in this series is also exergonic, but in this case less energy is released. It is enough to transfer a phosphate from the substrate to ADP, forming ATP.

The end product of glycolysis, pyruvate, is somewhat more oxidized than glucose. In the presence of O_2, further oxidation can occur. In prokaryotes these subsequent reactions take place in the cytosol, but in eukaryotes they take place in the mitochondrial matrix.

Pyruvate oxidation links glycolysis and the citric acid cycle

The next step in the aerobic catabolism of glucose involves the oxidation of pyruvate to a two-carbon acetate molecule and CO_2. The acetate is then bound to **coenzyme A (CoA)**, which is used in various biochemical reactions as a carrier of acyl groups:

This is the link between glycolysis and further oxidative reactions (see Figure 6.9).

The formation of acetyl CoA is a multistep reaction catalyzed by the *pyruvate dehydrogenase complex*, containing 60 individual proteins and 5 different coenzymes. The overall reaction is exergonic, and one molecule of NAD^+ is reduced. But the main role of acetyl CoA is to donate its acetyl group to the four-carbon compound oxaloacetate, forming the six-carbon molecule citrate. This initiates the citric acid cycle, one of life's most important energy-harvesting pathways.

The citric acid cycle completes the oxidation of glucose to CO_2

Acetyl CoA is the starting point for the citric acid cycle. This pathway of eight reactions completely oxidizes the two-carbon acetyl group to two molecules of CO_2. The free energy released from these reactions is captured by ADP and the electron carriers NAD^+ and FAD (**FIGURE 6.11**). This is a cycle because the starting material, oxaloacetate, is regenerated in the last step and is ready to accept another acetate group from acetyl CoA. The citric acid cycle operates twice for each glucose molecule that enters glycolysis (once for each pyruvate that enters the mitochondrion).

Let's focus on the final reaction of the cycle (step 8 in Figure 6.11), as an example of the kind of reaction that occurs:

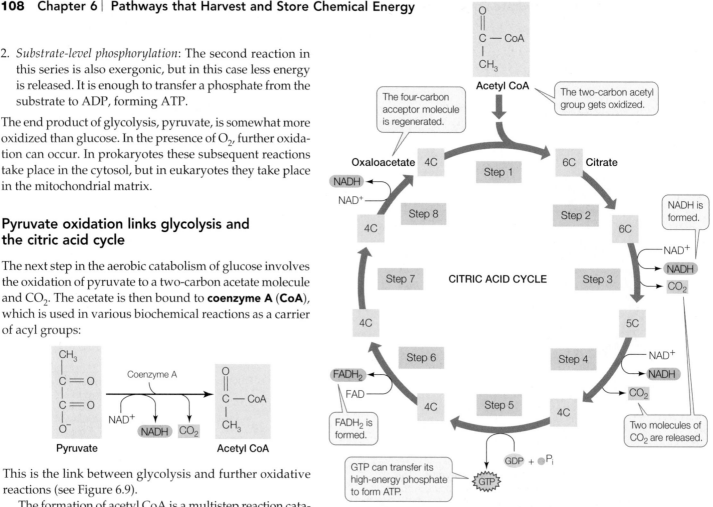

FIGURE 6.11 The Citric Acid Cycle Also called the Krebs cycle for its discoverer, Hans Krebs, the citric acid cycle involves eight steps and fully oxidizes acetyl CoA to CO_2.

This oxidation reaction (see the second carbon atom) is exergonic, and the released energy is trapped by NAD^+, forming NADH. With four such reactions ($FADH_2$ is a reduced coenzyme similar to NADH), the citric acid cycle harvests a great deal of chemical energy from the oxidation of acetyl CoA.

NADH is oxidized by the respiratory chain, and ATP is formed by chemiosmosis

As we discussed in Concept 6.1, the NADH is reoxidized to NAD^+. In the process, O_2 is reduced to H_2O:

$$NADH + H^+ + \tfrac{1}{2} O_2 \rightarrow NAD^+ + H_2O$$

This does not happen in a single step. Rather, there is a series of redox carrier proteins called the *respiratory chain* embedded in the inner membrane of the mitochondrion (**FIGURE 6.12**). The electrons from the oxidation of NADH and $FADH_2$ pass from one carrier to the next in the chain, in a process called **electron transport**. The oxidation reactions are exergonic, and they

Mitochondrion

yourBioPortal.com

Go to ANIMATED TUTORIAL 6.2
Electron Transport and ATP Synthesis
and WEB ACTIVITY 6.3 Respiratory Chain

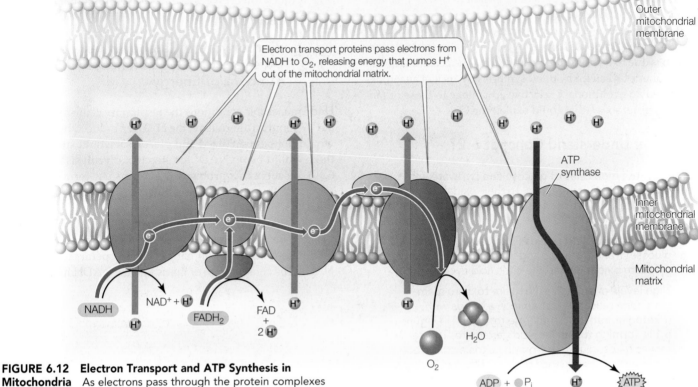

Electron transport proteins pass electrons from NADH to O_2, releasing energy that pumps H^+ out of the mitochondrial matrix.

Cytoplasm

Outer mitochondrial membrane

ATP synthase

Inner mitochondrial membrane

Mitochondrial matrix

NADH

$NAD^+ + H^+$

FADH₂

FAD + 2 H^+

H_2O

O_2

$ADP + P_i$

ATP

FIGURE 6.12 Electron Transport and ATP Synthesis in Mitochondria As electrons pass through the protein complexes of the respiratory chain, protons are pumped from the mitochondrial matrix into the intermembrane space. As the protons return to the matrix through ATP synthase, ATP is formed.

release energy that is used to actively transport H^+ ions out of the mitochondrial matrix. Thus, a proton gradient is set up across the inner membrane. In addition to the electron transport carriers, the inner membrane contains an ATP synthase that uses the H^+ gradient to synthesize ATP by chemiosmosis. The basic principles of chemiosmosis are shown in Figure 6.5. Note that in mitochondria, the inner membrane provides the

compartments needed for separation of H^+ and formation of the H^+ gradient.

Oxidative phosphorylation yields a lot of ATP. For each NADH (or FADH₂) that begins the chain, two to three ATP molecules are formed under the conditions in the cell. Taking 2.5 as an average number, the 4 molecules of reduced coenzyme produced by each turn of the citric acid cycle yield 10 molecules (4 × 2.5) of ATP. Two molecules of acetyl CoA are produced from each glucose, so the total is about 20 ATPs per

APPLY THE CONCEPT

Carbohydrate catabolism in the presence of oxygen releases a large amount of energy

The following reaction occurs in the citric acid cycle:

Succinate → Fumarate

Answer each of the following questions, and explain your answers:

1. Is this reaction an oxidation or reduction?

2. Is the reaction exergonic or endergonic?

3. This reaction requires a coenzyme. What kind of coenzyme?

4. What happens to the fumarate after the reaction is completed?

5. What happens to the coenzyme after the reaction is completed?

molecule of glucose. Add to this the NADH produced by glycolysis and pyruvate oxidation, and the ATP formed by substrate-level phosphorylation during glycolysis and the citric acid cycle, and the total is about *32 molecules of ATP produced per fully oxidized glucose*.

The vital role of O_2 is now clear: most of the ATP produced in cellular respiration is formed by oxidative phosphorylation, which is due to the reoxidation of NADH. The accumulation of atmospheric O_2 caused by photosynthesis by ancient anaerobes provided an evolutionary selective advantage to those organisms—the aerobes—that could exploit the O_2.

Do You Understand Concept 6.2?

- Draw the molecules of glucose and pyruvate and compare them with respect to their oxidation/reduction states. What does this mean in terms of available chemical energy?

- Compare the energy yields (in terms of ATP per glucose) from glycolysis and the combination of pyruvate oxidation and the citric acid cycle.

- High levels of ammonia (NH_3) are toxic to mammals. One reason is that ammonia bonds with α-ketoglutarate (the substrate for step 4 in Figure 6.11), forming the amino acid glutamate. This removes α-ketoglutarate from the citric acid cycle. Explain the consequence of this for the cell.

We have seen that a large amount of energy is released when carbohydrates are catabolized in the presence of O_2. If O_2 is absent—that is, when conditions are **anaerobic**—the yield of ATP is much lower. We will now turn to this situation.

<table>
<tr><td>concept
6.3</td><td>**Carbohydrate Catabolism in the Absence of Oxygen Releases a Small Amount of Energy**</td></tr>
</table>

Under anaerobic conditions the respiratory chain cannot operate. Without an alternative, the NADH produced by glycolysis would not be reoxidized and glycolysis would stop, because there would be no NAD^+ for step 6 of glycolysis (see Figure 6.10). To solve this problem, organisms use **fermentation** to reoxidize the NADH, thus allowing glycolysis to continue (**FIGURE 6.13**).

Like glycolysis, fermentation pathways occur in the cytoplasm. There are many different types of fermentation used by different organisms in nature, but all operate to regenerate NAD^+. The consequence of this is that the NADH made during

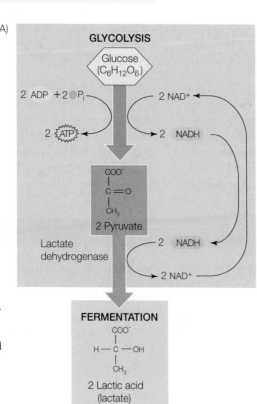

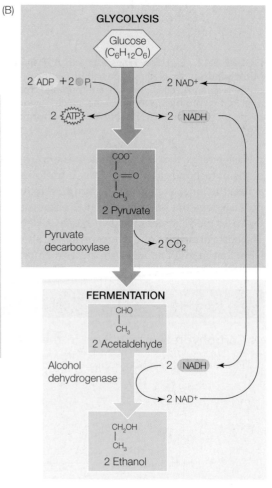

FIGURE 6.13 Fermentation
(A) In lactic acid fermentation, NADH is used to reduce pyruvate to lactic acid, thus regenerating NAD^+ to keep glycolysis operating. (B) In alcoholic fermentation, pyruvate is converted to acetaldehyde, and CO_2 is released. NADH is used to reduce acetaldehyde to ethanol, again regenerating NAD^+ for glycolysis.

yourBioPortal.com

Go to WEB ACTIVITY 6.4
Glycolysis and Fermentation

Summary of reactants and products:
$C_6H_{12}O_6 + 2\ ADP + 2\ P_i \longrightarrow$ 2 lactic acid + 2 ATP

Summary of reactants and products:
$C_6H_{12}O_6 + 2\ ADP + 2\ P_i \longrightarrow$ 2 ethanol + 2 CO_2 + 2 ATP

glycolysis is not available for reoxidation by the respiratory chain to form ATP. Therefore, the overall yield of ATP from fermentation is reduced to only the ATP made in glycolysis (two ATP per glucose).

Two fermentation pathways are found in a wide variety of organisms:

- Lactic acid fermentation, whose end product is lactic acid (lactate)
- Alcoholic fermentation, whose end product is ethyl alcohol (ethanol)

In *lactic acid fermentation*, pyruvate serves as the electron acceptor and lactate is the product (see Figure 6.13A). This process takes place in many microorganisms and complex organisms, including more complex plants and vertebrates. A notable example of lactic acid fermentation occurs in vertebrate muscle tissue. Usually, vertebrates get their energy for muscle contractions aerobically, with the circulatory system supplying O_2 to muscles. This is almost always adequate for small vertebrates, which explains why birds can fly long distances without resting. But in larger vertebrates such as humans, the circulatory system is not up to the task of delivering enough O_2 when the need is great, such as during a long sprint. At this point, the muscle cells break down glycogen (a stored polysaccharide) and undergo lactic acid fermentation. The process is reversible; lactate is converted back to pyruvate once O_2 is available again.

Alcoholic fermentation takes place in certain yeasts (eukaryotic microbes) and some plant cells under anaerobic conditions. In this process, pyruvate is converted to ethanol (see Figure 6.13B). We saw these reactions (as did Pasteur) in the opening story of this chapter. As with lactic acid fermentation, the reactions are essentially reversible.

FRONTIERS Industrial chemicals such as acetone and butyric acid were once made using microbial fermentation. In the twentieth century, fermentation was supplanted by the synthesis of these substances from petroleum. With the higher prices and reduced supply of oil, there is renewed interest in microbial fermentation. The efficiency of the microbes is being improved with biotechnology.

By recycling NAD^+, fermentation allows glycolysis to continue, thus producing small amounts of ATP through substrate-level phosphorylation. The net yield of two ATPs per glucose molecule is much lower than the energy yield from oxidative phosphorylation. For this reason, most organisms existing in anaerobic environments are small microbes that grow relatively slowly.

yourBioPortal.com
Go to WEB ACTIVITY 6.5
Energy Levels

Do You Understand Concept 6.3?

- Why is replenishing NAD^+ crucial to cellular metabolism?
- Compare the sources and total energy yield in terms of ATP per glucose in human cells in the presence versus the absence of O_2.
- Conditions can become anaerobic in a heart muscle cell during a heart attack, because of the inadequate supply of blood. If O_2 is restored, what will happen to the lactate produced by the heart muscle?

You have seen how cells harvest chemical energy in cellular respiration. Now we will see how that energy moves through other metabolic pathways in the cell.

concept 6.4 Catabolic and Anabolic Pathways Are Integrated

Metabolic transformations are a hallmark of life. The pathways we have seen thus far in the chapter, including glycolysis and the citric acid cycle, do not operate in isolation. Rather, there is an interchange of molecules into and out of these pathways, to and from the metabolic pathways for the synthesis and breakdown of amino acids, nucleotides, fatty acids, and other building blocks of life. *Carbon skeletons* (a term describing molecules with covalently linked carbon atoms) can enter catabolic pathways and be oxidized to release their energy, or they can enter anabolic pathways to be used in the formation of the macromolecules that are the major constituents of the cell. These relationships are summarized in **FIGURE 6.14**.

Catabolism and anabolism are linked

A hamburger or veggie burger on a bun contains three major sources of carbon skeletons: carbohydrates, mostly in the form of starch (a polysaccharide); lipids, mostly as triglycerides (three fatty acids attached to glycerol); and proteins (polymers of amino acids). Look at Figure 6.14 to see how each of these three types of macromolecules can be hydrolyzed and used in catabolism or anabolism.

CATABOLIC INTERCONVERSIONS Polysaccharides, lipids, and proteins can all be broken down to provide energy:

- *Polysaccharides* are hydrolyzed to glucose. Glucose then passes through glycolysis, pyruvate oxidation, and the respiratory chain, where its energy is captured in ATP.
- *Lipids* are broken down into their constituents—glycerol and fatty acids. Glycerol is converted into dihydroxyacetone phosphate, an intermediate in glycolysis. Fatty acids are highly reduced molecules that are converted to acetyl CoA in a process called *β-oxidation*. This is carried out by a series of oxidation enzymes inside the mitochondrion. For

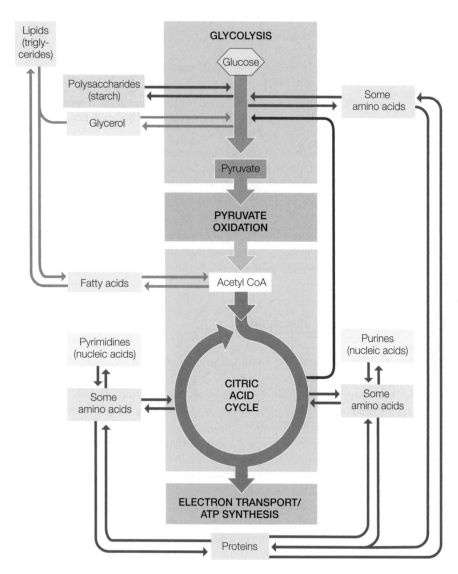

FIGURE 6.14 Relationships among the Major Metabolic Pathways of the Cell Note the central positions of glycolysis and the citric acid cycle in this network of metabolic pathways. Also note that many of the pathways can operate essentially in reverse.

in a process called **gluconeogenesis** (which means "new formation of glucose"). Likewise, acetyl CoA can be used to form fatty acids. The most common fatty acids have even numbers of carbons: 14, 16, or 18. These are formed by the addition of two-carbon acetyl CoA "units" one at a time until the appropriate chain length is reached.

Some intermediates in the citric acid cycle are reactants in pathways that synthesize important components of nucleic acids. For example, α-ketoglutarate and oxaloacetate are starting points for purines and pyrimidines, respectively.

Catabolism and anabolism are integrated

A carbon atom from a protein in your burger can end up in DNA, fat, or CO_2, among other fates. How does the organism "decide" which metabolic pathways to follow, in which cells? With hundreds of enzymes and all the possible interconversions, you might expect that the cellular concentrations of various biochemical molecules would vary widely. Remarkably, the levels of these substances in what is called the *metabolic pool*—the sum total of all the biochemical molecules in a cell—are quite constant. Metabolic changes in the cell are a bit like the changes in traffic patterns in a city: if an accident blocks traffic on a major road, drivers take alternate routes, where the traffic volume consequently changes.

Consider what happens to the starch in your burger bun. In the digestive system, starch is hydrolyzed to glucose, which enters the blood. If it is needed, the glucose is distributed to the rest of the body. But if there is already enough glucose in the blood to supply the body's needs, the excess glucose is converted into glycogen and stored in the liver. If not enough glucose is supplied by food, glycogen is broken down, or other molecules are used to make glucose by gluconeogenesis. The end result is that the level of glucose in the blood is remarkably constant. How does the body accomplish this?

Metabolic enzymes (including those of glycolysis, the citric acid cycle, and the respiratory chain) are subject to regulation, and often the regulatory mechanisms involve allosteric effects. An example is feedback inhibition, illustrated in Figure 3.19. In a metabolic pathway, a high concentration of the final product can inhibit the enzyme that catalyzes the commitment step (see Concept 3.4). By contrast, there can be positive feedback when an excess of the product of one pathway speeds up reactions in another pathway.

example, the β-oxidation of a C_{16} (16-carbon) fatty acid occurs in several steps:

$$C_{16} \text{ fatty acid} + CoA \rightarrow C_{16} \text{ fatty acyl CoA}$$

$$C_{16} \text{ fatty acyl CoA} + CoA \rightarrow C_{14} \text{ fatty acyl CoA} + \text{acetyl CoA}$$

$$\text{repeat 6 times} \rightarrow 8 \text{ acetyl CoA}$$

The acetyl CoA can then enter the citric acid cycle and be catabolized to CO_2.

• *Proteins* are hydrolyzed to their amino acid building blocks. The 20 different amino acids feed into glycolysis or the citric acid cycle at different points. For example, the amino acid glutamate is converted into α-ketoglutarate, an intermediate in the citric acid cycle.

ANABOLIC INTERCONVERSIONS Many catabolic pathways can operate essentially in reverse, with some modifications. Glycolytic and citric acid cycle intermediates, instead of being oxidized to form CO_2, can be reduced and used to form glucose

LINK Review the discussion of enzyme regulation in Concept 3.4

Metabolic enzymes can also be regulated by altering the transcription of the genes that encode them. Excess levels of glucose and other dietary factors can lead to increased transcription of the gene for *fatty acid synthase*, a key enzyme in the synthesis of fatty acids. Excess citrate produced by the citric acid cycle is broken down to acetyl CoA, which in turn is used in fatty acid synthesis. This is one reason why people accumulate fat after eating too much. The fatty acids may be metabolized later to produce more acetyl CoA.

---

yourBioPortal.com

Go to WEB ACTIVITY 6.6
Relationships among Metabolic Pathways

Do You Understand Concept 6.4?

- Give examples of the catabolic conversion of a lipid and the anabolic conversion of a protein.

- Trace the biochemical pathway by which a carbon atom from a starch molecule in rice eaten today can end up in a muscle protein tomorrow.

- Describe what might happen if there were no mechanisms for modulating the level of acetyl CoA.

We have seen how cellular respiration allows organisms to harvest chemical energy from organic molecules. For the rest of the chapter, we'll look at how plants and other photosynthetic organisms produce these organic molecules using energy from light.

concept 6.5 During Photosynthesis, Light Energy Is Converted to Chemical Energy

The energy released by catabolic pathways in all organisms, including animals, plants, and prokaryotes, ultimately comes from the sun. **Photosynthesis** (literally, "synthesis from light") is an anabolic process by which the energy of sunlight is captured and used to convert carbon dioxide (CO_2) and water (H_2O) into carbohydrates (which we represent as a six-carbon sugar, $C_6H_{12}O_6$) and oxygen gas (O_2):

$$6\,CO_2 + 6\,H_2O \rightarrow C_6H_{12}O_6 + 6\,O_2$$

This equation shows a highly endergonic reaction. The net outcome is the reverse of the general equation for glucose catabolism that we discussed in Concept 6.2. Many of the molecular processes of photosynthesis are similar to those for glucose catabolism. For example, both processes involve redox reactions, electron transport, and chemiosmosis. However, the details of photosynthesis are quite different.

Photosynthesis involves two pathways (**FIGURE 6.15**):

- The **light reactions** convert light energy into chemical energy in the form of ATP and the reduced electron carrier NADPH. This molecule is similar to NADH (see Figure 6.4A) but with an additional phosphate group attached to the sugar of its adenosine.

- The **carbon-fixation reactions** do not use light directly, but instead use the ATP and NADPH made by the light reactions, along with CO_2, to produce carbohydrates.

Both the light reactions and the carbon-fixation reactions stop in the dark because ATP synthesis and $NADP^+$ reduction require light. In photosynthetic prokaryotes (e.g., cyanobacteria) the light reactions take place on internal membranes and the carbon-fixation reactions occur in the cytosol. In plants, which will be our focus here, both pathways proceed within the chloroplast, but they occur in different parts of that organelle (see Figure 6.15).

Light energy is absorbed by chlorophyll and other pigments

Light is a form of energy that can be converted to other forms, such as heat or chemical energy. It is helpful here to discuss light in terms of its *photochemistry* and *photobiology*.

FIGURE 6.15 An Overview of Photosynthesis Photosynthesis consists of two pathways: the light reactions and the carbon-fixation reactions. In eukaryotes, these occur in the chloroplast.

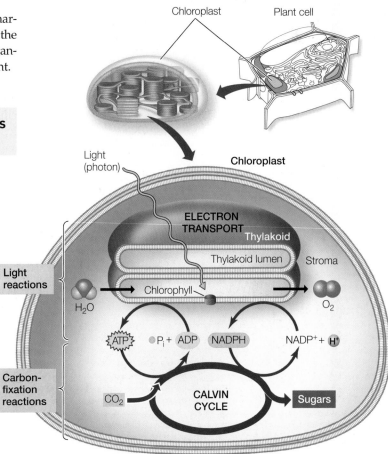

PHOTOCHEMISTRY Light is a form of **electromagnetic radiation**. Electromagnetic radiation is propagated in waves, and the amount of energy in the radiation is inversely proportional to its **wavelength**—the shorter the wavelength, the greater the energy. The visible portion of the electromagnetic spectrum (**FIGURE 6.16**) encompasses a wide range of wavelengths and energy levels. In addition to traveling in waves, light also behaves as particles, called **photons**, which have no mass. In plants and other photosynthetic organisms, receptive molecules absorb photons in order to harvest their energy for biological processes. These receptive molecules absorb only specific wavelengths of light—photons with specific amounts of energy.

When a photon meets a molecule, one of three things can happen:

- The photon may bounce off the molecule—it may be *scattered* or *reflected*.

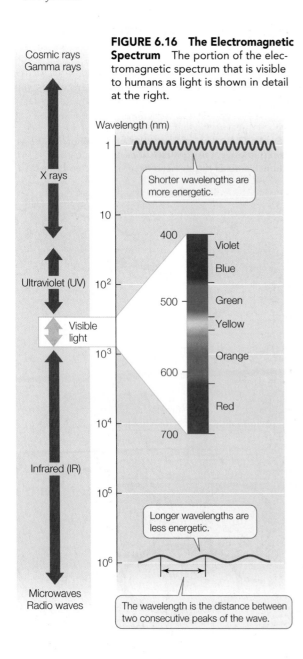

Cosmic rays
Gamma rays

FIGURE 6.16 The Electromagnetic Spectrum The portion of the electromagnetic spectrum that is visible to humans as light is shown in detail at the right.

X rays

Ultraviolet (UV)

Visible light

Infrared (IR)

Microwaves
Radio waves

Wavelength (nm)

Shorter wavelengths are more energetic.

Violet
Blue
Green
Yellow
Orange
Red

Longer wavelengths are less energetic.

The wavelength is the distance between two consecutive peaks of the wave.

- The photon may pass through the molecule—it may be *transmitted*.
- The photon may be *absorbed* by the molecule, adding energy to the molecule.

Neither of the first two outcomes causes any change in the molecule. However, in the case of absorption, the photon disappears and its energy is absorbed by the molecule. The photon's *energy* cannot disappear, because according to the first law of thermodynamics, energy is neither created nor destroyed. When the molecule acquires the energy of the photon it is raised from a ground state (with lower energy) to an excited state (with higher energy):

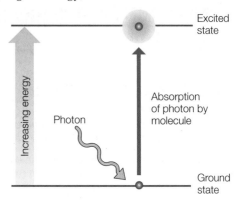

Increasing energy

Photon

Excited state

Absorption of photon by molecule

Ground state

The difference in free energy between the molecule's excited state and its ground state is approximately equal to the free energy of the absorbed photon (a small amount of energy is lost to entropy). The increase in energy boosts one of the electrons within the molecule into a shell farther from its nucleus; this electron is now held less firmly, making the molecule *unstable and more chemically reactive*.

PHOTOBIOLOGY Each type of molecule absorbs light at specific, characteristic wavelengths. Molecules that absorb wavelengths in the visible spectrum are called **pigments**.

When a beam of white light (containing all the wavelengths of visible light) falls on a pigment, certain wavelengths are absorbed. The remaining wavelengths are scattered or transmitted and make the pigment appear to us as colored. For example, the pigment *chlorophyll* absorbs both blue and red light, and we see the remaining light, which is primarily green. If we plot light absorbed by a purified pigment against wavelength, the result is an **absorption spectrum** for that pigment.

In contrast to the absorption spectrum, an **action spectrum** is a plot of the *biological activity* of an organism against the wavelengths of light to which it is exposed. An action spectrum can be determined as follows:

1. Place the organism (for example, a water plant with thin leaves) in a closed container.
2. Expose it to light of a certain wavelength for a period of time.
3. Measure the rate of photosynthesis by the amount of O_2 released.
4. Repeat with light of other wavelengths.

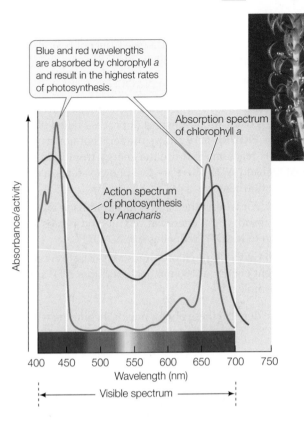

Blue and red wavelengths are absorbed by chlorophyll *a* and result in the highest rates of photosynthesis.

Absorption spectrum of chlorophyll *a*

Action spectrum of photosynthesis by *Anacharis*

Absorbance/activity

Wavelength (nm)

Visible spectrum

Anacharis

FIGURE 6.17 Absorption and Action Spectra The absorption spectrum of the purified pigment chlorophyll *a* from the aquatic plant *Anacharis* is similar to the action spectrum, obtained when different wavelengths of light are shone on the intact plant and the rate of photosynthesis is measured. In the thicker leaves of land plants, the action spectra show less of a dip in the green region (500–650 nm).

are found in red algae and in cyanobacteria, absorb various yellow-green, yellow, and orange wavelengths.

Light absorption results in photochemical change

When chlorophyll absorbs light it enters an excited state. This is an unstable situation, and the chlorophyll rapidly returns to its ground state, releasing most of the absorbed energy. This is an extremely rapid process—measured in picoseconds

FIGURE 6.17 shows the absorption spectrum of the pigment chlorophyll *a*, which was isolated from the leaves of *Anacharis*, a common aquarium plant. Also shown is the action spectrum for photosynthetic activity by the same plant. A comparison of the two spectra shows that the wavelengths at which photosynthesis is highest are the same wavelengths at which chlorophyll *a* absorbs light.

In plants, two chlorophylls absorb light energy to drive the light reactions: **chlorophyll *a*** and **chlorophyll *b***. These two molecules differ only slightly in their molecular structures. Both have a complex ring structure, similar to that of the heme group of hemoglobin, with a magnesium ion at the center (**FIGURE 6.18**). A long hydrocarbon "tail" anchors the chlorophyll molecule to integral proteins in the thylakoid membrane of the chloroplast.

As mentioned above, the chlorophylls absorb blue and red light, which are near the two ends of the visible spectrum (see Figure 6.17). In addition, plants possess accessory pigments that absorb photons intermediate in energy between the red and the blue wavelengths, and then transfer a portion of that energy to the chlorophylls. Among these accessory pigments are *carotenoids* such as β-carotene, which absorb photons in the blue and blue-green wavelengths and appear deep yellow. The *phycobilins*, which

FIGURE 6.18 The Molecular Structure of Chlorophyll Chlorophyll consists of a complex ring structure (green) with a magnesium ion at its center, plus a hydrocarbon "tail." The tail anchors chlorophyll molecules to integral membrane proteins in the thylakoid membrane. Chlorophyll *a* and chlorophyll *b* are identical except for the replacement of a methyl group (—CH_3) with an aldehyde group (—CHO), shown on the upper right side of the ring structure. Most chlorophylls absorb light and pass the energy on to a chlorophyll in the reaction center of a photosystem.

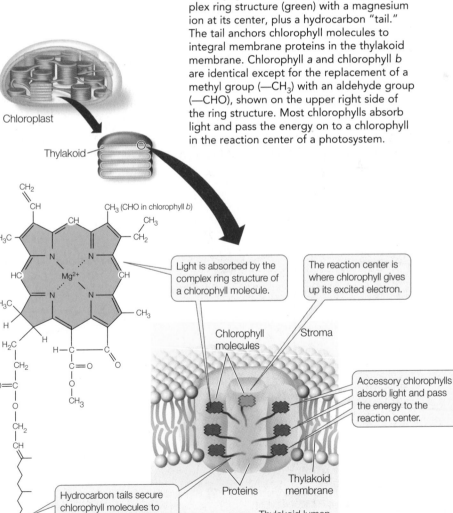

Chloroplast

Thylakoid

CH_2
CH
CH
H_3C
CH_3 (CHO in chlorophyll *b*)
CH_3
CH_2
HC
Mg^{2+}
CH
H_3C
CH_3
H
H
H_2C
H—C—C
CH_2
C=O
O=C
O
O
CH_3
CH_2
CH

Light is absorbed by the complex ring structure of a chlorophyll molecule.

The reaction center is where chlorophyll gives up its excited electron.

Chlorophyll molecules

Stroma

Accessory chlorophylls absorb light and pass the energy to the reaction center.

Hydrocarbon tails secure chlorophyll molecules to hydrophobic proteins inside the thylakoid membrane.

Proteins

Thylakoid membrane

Thylakoid lumen

(trillionths of a second)! For most chlorophyll molecules embedded in the thylakoid membrane, the released energy is absorbed by other, adjacent chlorophyll molecules. The pigments in photosynthetic organisms are arranged into energy-absorbing *antenna systems*, also called **light-harvesting complexes**. These form part of a large multi-protein complex called a **photosystem**. The photosystem spans the thylakoid membrane and consists of multiple antenna systems with their associated pigment molecules, all surrounding a **reaction center** (see Figure 6.18).

A ground-state chlorophyll molecule at the reaction center (symbolized by Chl) absorbs the energy from the adjacent chlorophylls and becomes excited (Chl*), but when this chlorophyll returns to the ground state, something very different occurs. *The reaction center converts the absorbed light energy into chemical energy.* The chlorophyll molecule in the reaction center absorbs sufficient energy that it actually *gives up its excited electron to a chemical acceptor*:

$$\text{Chl*} + \text{acceptor} \rightarrow \text{Chl}^+ + \text{acceptor}^-$$

This, then, is the first consequence of light absorption by chlorophyll: *the reaction center chlorophyll (Chl*) loses its excited electron in a redox reaction and becomes Chl⁺.* As a result of this transfer of an electron, the chlorophyll gets oxidized, while the acceptor molecule is reduced.

Reduction leads to ATP and NADPH formation

The electron acceptor that is reduced by Chl* is the first of a chain of electron carriers in the thylakoid membrane. Electrons are passed from one carrier to another in an energetically "downhill" series of reductions and oxidations. Thus, the thylakoid membrane has an *electron transport* system similar to

the respiratory chain of mitochondria (see Concept 6.2). The final electron acceptor is NADP⁺, which gets reduced:

$$\text{NADP}^+ + \text{H}^+ + 2\,\text{e}^- \rightarrow \text{NADPH}$$

As in mitochondria, ATP is produced chemiosmotically during the process of electron transport (a process called *photophosphorylation*). **FIGURE 6.19** shows the series of noncyclic electron transport reactions that use the energy from light to generate NADPH and ATP. There are two photosystems, each with its own reaction center:

- **Photosystem I** (containing the "P₇₀₀" chlorophylls at its reaction center) absorbs light energy at 700 nm and passes an excited electron to NADP⁺, reducing it to NADPH.

- **Photosystem II** (with "P₆₈₀" chlorophylls at its reaction center) absorbs light energy at 680 nm and produces ATP and oxidizes water molecules.

Let's look in more detail at these photosystems, beginning with photosystem II.

FIGURE 6.19 Noncyclic Electron Transport Uses Two Photosystems As chlorophyll molecules in the reaction centers of photosystems I and II absorb light energy, they pass electrons into a series of redox reactions, ultimately producing NADPH and ATP. The term "Z scheme" describes the path (blue arrows) of electrons as they travel through the two photosystems. In this scheme the vertical positions represent the energy levels of the molecules in the electron transport system.

1. The Chl in the reaction center of photosystem II absorbs light maximally at 680 nm, becoming Chl*. Water gets oxidized.

2. H⁺ from H₂O and electron transport through the electron transport system capture energy for the chemiosmotic synthesis of ATP.

3. The Chl in the reaction center of photosystem I absorbs light maximally at 700 nm, becoming Chl*.

4. Photosystem I reduces an electron carrier, which is used to reduce NADP⁺ to NADPH.

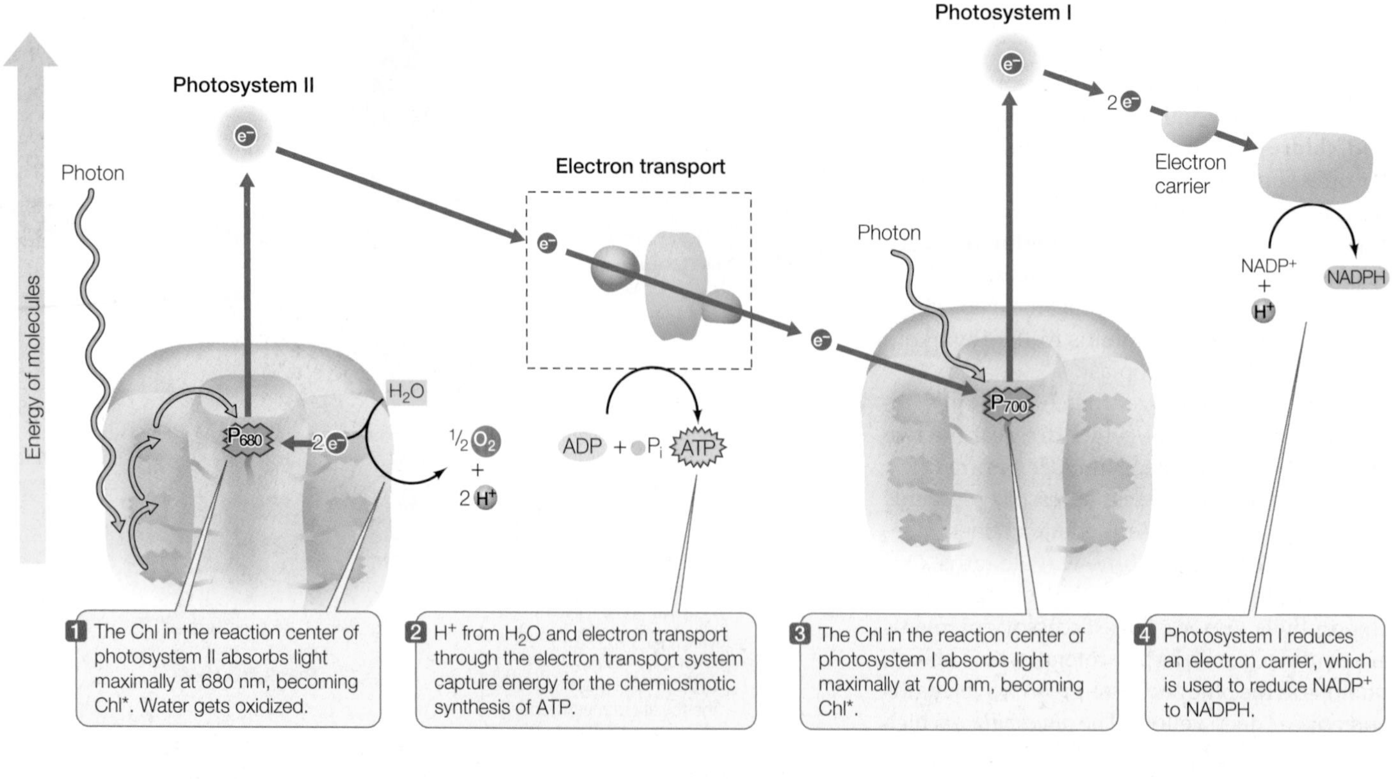

APPLY THE CONCEPT

During photosynthesis, light energy is converted to chemical energy

The key role of water in supplying electrons for reduction of light-activated chlorophyll in the light reactions and in the release of O_2 to the atmosphere in the process of energy conversion has been investigated using isotopes of oxygen. The ^{18}O isotope is heavier than normal oxygen (^{16}O), and a mass spectrometer can be used to detect the difference. Green plant cells were exposed to light, water, and CO_2. (The CO_2 was supplied as the bicarbonate ion HCO_3^-, which forms CO_2 when dissolved in water.) In the first experiment, some of the oxygen atoms in the water molecules were ^{18}O ($H_2^{18}O$), while CO_2 had the normal form of oxygen ($C^{16}O_2$). In the second experiment, the situation was reversed, with $H_2^{16}O$ and $C^{18}O_2$ being supplied to the plants. After 2 hours of photosynthesis, the ratio of ^{18}O to ^{16}O was measured in the O_2 produced by the cells.

EXPERIMENT	ISOTOPE RATIO		
	H_2O	CO_2	O_2
1	0.85	0.31	0.84
2	0.20	0.50	0.20

1. In experiment 1, was the isotopic ratio of O_2 more similar to that of H_2O or CO_2?

2. What about experiment 2? What can you conclude from these data?

PHOTOSYSTEM II After an excited chlorophyll in the reaction center (Chl*) gives up its energetic electron to reduce a chemical acceptor molecule, the chlorophyll lacks an electron and is very unstable. It has a strong tendency to "grab" an electron from another molecule to replace the one it lost—in chemical terms, it is a strong oxidizing agent. The replenishing electrons come from water, splitting the H—O—H bonds:

$$H_2O \rightarrow \tfrac{1}{2} O_2 + 2 H^+ + 2 e^-$$

$$2 e^- + 2 Chl^+ \rightarrow 2 Chl$$

$$\text{Overall:} \quad 2 Chl^* + H_2O \rightarrow 2 Chl + 2 H^+ + \tfrac{1}{2} O_2$$

Notice that *the source of O_2 in photosynthesis is H_2O.*

yourBioPortal.com
Go to ANIMATED TUTORIAL 6.4
The Source of the Oxygen Produced by Photosynthesis

Back to the electron acceptor in the electron transport system: the energetic electrons are passed through a series of membrane-bound carriers to a final acceptor at a lower energy level. As in the mitochondrion, a proton gradient is generated and is used by ATP synthase to store energy in the bonds of ATP.

PHOTOSYSTEM I In photosystem I, an excited electron from the Chl* at the reaction center reduces an acceptor. The oxidized chlorophyll (Chl⁺) now "grabs" an electron, but in this case the electron comes from the last carrier in the electron transport system of photosystem II. This links the two photosystems chemically. They are also linked spatially, with the two photosystems adjacent to one another in the thylakoid membrane. The energetic electrons from photosystem I pass through several molecules and end up reducing NADP⁺ to NADPH.

Next in the process of harvesting light energy to produce carbohydrates is the series of *carbon-fixation* reactions. These reactions require more ATP than NADPH. If the pathway we just described—the *linear* or *noncyclic* pathway—were the only set of light reactions operating, there might not be sufficient ATP for carbon fixation. **Cyclic electron transport** makes up for this imbalance. This pathway uses only photosystem I and produces ATP but not NADPH; it is cyclic because an electron is passed from an excited chlorophyll and recycles back to the same chlorophyll (**FIGURE 6.20**).

FIGURE 6.20 Cyclic Electron Transport Traps Light Energy as ATP Cyclic electron transport produces ATP but no NADPH.

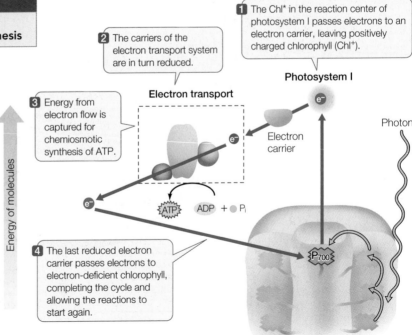

1 The Chl* in the reaction center of photosystem I passes electrons to an electron carrier, leaving positively charged chlorophyll (Chl⁺).

2 The carriers of the electron transport system are in turn reduced.

3 Energy from electron flow is captured for chemiosmotic synthesis of ATP.

4 The last reduced electron carrier passes electrons to electron-deficient chlorophyll, completing the cycle and allowing the reactions to start again.

Electron transport

Photosystem I

Electron carrier

Photon

Energy of molecules

ATP   ADP + Pᵢ

P₇₀₀

Do You Understand Concept 6.5?

- What are the reactants and products of the light reactions of photosynthesis?

- What happens when electromagnetic radiation at the following wavelengths arrives at a leaf: 180 nm; 400 nm; 550 nm; 600 nm; 680 nm; 700 nm; 900 nm?

- Trace the flow of electrons in noncyclic electron transport in the chloroplast and compare it with that of cyclic electron transport.

- Write equations for the production of the following, and indicate whether they are oxidations, reductions, or neither: Chl*; O_2; ATP; NADPH.

We have seen how photosystems I and II absorb light energy, which ultimately ends up as chemical energy in ATP and NADPH. Let's look now at how these two energy-rich mol-ecules are used in the carbon-fixation reactions to reduce CO_2 and thereby form carbohydrates.

concept 6.6 Photosynthetic Organisms Use Chemical Energy to Convert CO_2 to Carbohydrates

The energy in ATP and NADPH is used in the carbon-fixation reactions to "fix" CO_2 into a reduced form and convert it to carbohydrates. Most CO_2 fixation occurs only in the light, when ATP and NADPH are being generated. The metabolic pathway occurs in the stroma, or central region, of the chloroplast (see Figure 6.15) and is called the **Calvin cycle** after one of its discoverers, Melvin Calvin.

Like all biochemical pathways, each reaction in the Calvin cycle is catalyzed by a specific enzyme. The cycle is composed of three distinct processes (**FIGURE 6.21**):

yourBioPortal.com

Go to WEB ACTIVITY 6.7
The Calvin Cycle

FIGURE 6.21 The Calvin Cycle The Calvin cycle uses the ATP and NADPH generated in the light reactions to produce G3P from CO_2. The G3P is used as a starting material for the production of glucose and other carbohydrates. Six turns of the cycle are needed to produce one molecule of the hexose glucose.

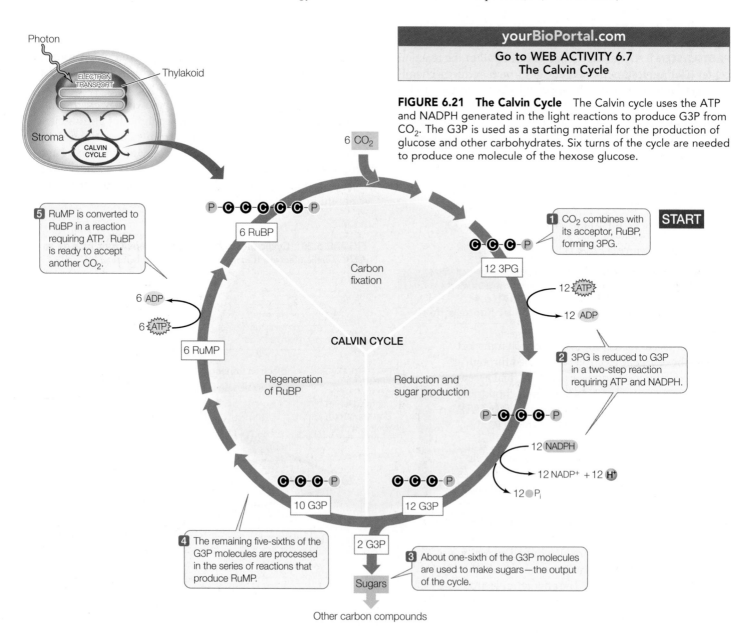

FIGURE 6.22 RuBP Is the Carbon Dioxide Acceptor The enzyme rubisco adds CO_2 to the five-carbon compound RuBP. The resulting six-carbon compound immediately splits into two molecules of 3PG.

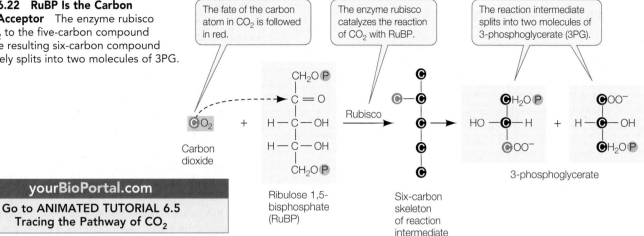

The fate of the carbon atom in CO_2 is followed in red.

The enzyme rubisco catalyzes the reaction of CO_2 with RuBP.

The reaction intermediate splits into two molecules of 3-phosphoglycerate (3PG).

Carbon dioxide

Ribulose 1,5-bisphosphate (RuBP)

Six-carbon skeleton of reaction intermediate

3-phosphoglycerate

yourBioPortal.com

Go to ANIMATED TUTORIAL 6.5
Tracing the Pathway of CO_2

- *Fixation of CO_2.* The initial reaction of the Calvin cycle adds the one-carbon CO_2 to an acceptor molecule, the five-carbon ribulose 1,5-bisphosphate (RuBP). The immediate product is a six-carbon molecule, which quickly breaks down into two three-carbon molecules called 3-phosphoglycerate (3PG; **FIGURE 6.22**). The enzyme that catalyzes this reaction, **ribulose bisphosphate carboxylase/oxygenase** (**rubisco**), is rather sluggish as enzymes go. It typically catalyzes two to three fixation reactions per second. Because of this, plants need a lot of rubisco to perform enough photosynthesis to satisfy the needs of growth and metabolism. Rubisco constitutes about half of all the protein in a leaf, and it is probably the most abundant protein in the world!

- *Reduction of 3PG to form glyceraldehyde 3-phosphate.* This series of reactions involves a phosphorylation (using the high-energy phosphate from an ATP made in the light reactions) and a reduction (using an NADPH made in the light reactions). The product is **glyceraldehyde 3-phosphate** (**G3P**), which is a three-carbon sugar phosphate, also called triose phosphate:

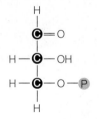

Glyceraldehyde 3-phosphate (G3P)

- *Regeneration of the CO_2 acceptor, RuBP.* Most of the G3P ends up as ribulose monophosphate (RuMP), and ATP is used to convert this compound into RuBP. So for every "turn" of the Calvin cycle, one CO_2 is fixed and the CO_2 acceptor is regenerated.

What happens to the extra G3P made by the Calvin cycle (see Figure 6.21)? It has two fates, depending on the time of day and the needs of different parts of the plant:

- Some of the extra G3P is exported out of the chloroplast to the cytosol, where it is converted to *hexoses* (glucose and fructose). This is the familiar $C_6H_{12}O_6$ from the general equation

for photosynthesis. These molecules may be catabolized for energy; used as carbon skeletons for the synthesis of amino acids and other molecules (see Figure 6.14); or converted to sucrose, which is transported out of the leaf to other organs in the plant.

- Late in the day when glucose has accumulated inside the chloroplast, the glucose units are linked to form the starch polysaccharide. This storage carbohydrate can then be drawn upon during the night so that the photosynthetic tissues can continue to export sucrose to the rest of the plant, even when photosynthesis is not taking place. In addition, starch is abundant in nonphotosynthetic organs such as roots, underground stems, and seeds, where it provides a ready supply of glucose to fuel cellular activities, including plant growth.

FRONTIERS Laboratory and field studies show that increases in CO_2 result in more photosynthesis. With the global increases in atmospheric CO_2, scientists are interested in what the effects will be on plant growth, especially of plants important to humans. By exposing plants in the field to pipes emitting CO_2, biologists are making actual measurements of these possible effects.

The products of the Calvin cycle are of crucial importance to Earth's entire biosphere. The C—H covalent bonds generated by this cycle provide almost all of the energy for life on Earth. Photosynthetic organisms, which are also called **autotrophs** ("self-feeders"), release most of this energy in cellular respiration, and use it to support their own growth, development, and reproduction. But plants are also the source of energy for other organisms. Much plant matter ends up being consumed by **heterotrophs** ("other-feeders"), including animals, which cannot photosynthesize. Heterotrophs depend on autotrophs for chemical energy, which they harvest via cellular respiration. In addition, many heterotrophs rely on plants to produce molecules that the heterotrophs cannot synthesize for themselves.

LINK The roles of autotrophs and heterotrophs in ecosystems are described in Concept 45.3

Do You Understand Concept 6.6?

- What are the three processes of the Calvin cycle?

- If green plant cells are incubated in the presence of CO_2 molecules containing radioactive carbon atoms, the fate of the carbon atoms can be followed. In an experiment, radioactive CO_2 was given for 1 minute to plant cells, and then the cells were examined after 1, 5, 10, 20, and 30 minutes. The following molecules were labeled with radioactive carbon at some point(s): glucose, glyceraldehyde 3-phosphate, glycine (an amino acid), 3-phosphoglycerate, ribulose 1,5-bisphosphate, and sucrose. List these molecules in the order in which they first become labeled.

 QUESTION Why does fresh air inhibit the formation of alcohol by yeast cells?

ANSWER   Armed with our knowledge of metabolism, we can now explain Pasteur's observations of beet sugar and alcohol:

1. *Nothing happened to beet mash unless microscopic yeast cells were present.* Beet sugar is a product of photosynthetic CO_2 fixation (Concept 6.6). Catabolism of carbohydrates is a cellular activity, characteristic of virtually all living things (Concepts 6.2 and 6.3).

2. *In the presence of fresh air, yeast cells grew vigorously on the mash, and bubbles of CO_2 were formed.* Carbohydrate catabolism under aerobic conditions yields a large amount of ATP (Concepts 6.1 and 6.2), which can be used to fuel anabolic pathways for growth (Concept 6.4). Under aerobic conditions, glucose is fully oxidized to CO_2 (Concept 6.2).

3. *Without fresh air, the yeast grew slowly, less CO_2 was produced, and alcohol was formed.* Alcoholic fermentation occurs in anaerobic conditions, and yields far less ATP (for growth) than aerobic metabolism; a small amount of CO_2 is produced (Concept 6.3).

As we note in the opening story, the use by humans of yeasts for fermentation has a long history (**FIGURE 6.23**).

Beer comes from the fermentation of barley seeds. These are soaked to begin the process of germination, which includes the induction of an enzyme that hydrolyzes the starch stored in the seed. The resulting disaccharide, maltose, is what the yeast cells use for energy—they first break the maltose down to its glucose monomers. At a cool temperature and under anaerobic conditions, yeast produces alcohol. An herb called hops is added to impart a distinctive bitter flavor.

Wine grapes are crushed, and the resulting juice contains sugars that are used by yeast for fermentation. There are yeasts growing on the skins of grapes naturally, so the original winemakers did not add yeast. More recently, special yeast strains have been used to control the fermentation process. The longer this process is allowed to proceed, the less sugar remains (resulting in a less sweet taste) and the higher the alcoholic content. After fermentation the wine is stored in wooden casks, typically at cool temperatures. During this time hundreds of molecular transformations occur, giving different wines distinctive properties.

Bread is made from ground-up plant seeds (flour), which contain abundant starch. Moisture activates enzymes inside the seed that catabolize the starch to monosaccharides, which in turn are used by yeast for growth under aerobic conditions. The resulting CO_2 provides bubbles that cause the complex flour mixture to rise to a familiar texture.

FIGURE 6.23   Products of Glucose Metabolism   Beer and wine are made by fermentation, whereas bread rises because of aerobic catabolism of glucose.

SUMMARY

concept 6.1 **ATP, Reduced Coenzymes, and Chemiosmosis Play Important Roles in Biological Energy Metabolism**

- Metabolism is carried out in small steps and involves coenzymes as carriers of chemical energy.

- Adenosine triphosphate (ATP) serves as "energy currency" in the cell. Hydrolysis of ATP releases a large amount of free energy. **Review Figure 6.1 and WEB ACTIVITY 6.1**

- In **oxidation**, a material loses electrons by transfer to another material, which thereby undergoes **reduction**. Such **redox** reactions transfer large amounts of energy.

- The coenzyme nicotinamide adenine dinucleotide is a key electron carrier in biological redox reactions. It exists in two forms, one oxidized (NAD^+) and the other reduced (NADH). **Review Figure 6.4**

- In **oxidative phosphorylation**, ATP is formed with the energy derived from the reoxidation of reduced coenzymes. This depends on the process of **chemiosmosis**, in which a proton gradient across a membrane powers ATP formation. This occurs at the plasma membrane in prokaryotes, and in the mitochondria and chloroplasts in eukaryotes. **Review Figures 6.5 and 6.6, WORKING WITH DATA 6.1, and ANIMATED TUTORIAL 6.1**

concept 6.2 **Carbohydrate Catabolism in the Presence of Oxygen Releases a Large Amount of Energy**

- The sequential pathways of aerobic glucose catabolism are **glycolysis**, **pyruvate oxidation**, and the **citric acid cycle**. **Review Figure 6.9**

- In glycolysis, a series of ten enzyme-catalyzed reactions in the cell cytoplasm converts glucose to two molecules of pyruvate. Some energy is released and captured as ATP and NADH. **Review Figure 6.10**

- The next pathway, pyruvate oxidation, links glycolysis to the citric acid cycle. Pyruvate oxidation converts pyruvate into the two-carbon molecule acetyl CoA.

- In the citric acid cycle, a series of eight enzyme-catalyzed reactions fully oxidizes acetyl CoA to CO_2. Much energy is released, and most is used to form NADH. **Review Figure 6.11 and WEB ACTIVITY 6.2**

- The energy in NADH is used to make ATP via a series of **electron transport** carriers and chemiosmosis. **Review Figure 6.12, ANIMATED TUTORIAL 6.2, and WEB ACTIVITY 6.3**

concept 6.3 **Carbohydrate Catabolism in the Absence of Oxygen Releases a Small Amount of Energy**

- In the absence of O_2, glycolysis is followed by **fermentation**. Together, these pathways partially oxidize pyruvate and generate the end products lactic acid or ethanol. In the process, NAD^+ is regenerated from NADH so that glycolysis can continue, thus generating a small amount of ATP. **Review Figure 6.13 and WEB ACTIVITY 6.4**

- For each molecule of glucose used, fermentation yields 2 molecules of ATP. In contrast, glycolysis, pyruvate oxidation, the citric acid cycle, and oxidative phosphorylation yield up to 32 molecules of ATP per molecule of glucose. **Review WEB ACTIVITY 6.5**

concept 6.4 **Catabolic and Anabolic Pathways Are Integrated**

- The catabolic pathways for the breakdown of carbohydrates, lipids, and proteins feed into the energy-harvesting metabolic pathways. **Review Figure 6.14**

- Anabolic pathways use intermediate components of the energy-harvesting pathways to synthesize fatty acids, amino acids, and other essential building blocks.

- The formation of glucose from intermediates of glycolysis and the citric acid cycle is called **gluconeogenesis**.

- The enzymes of glycolysis and the citric acid cycle are regulated by various mechanisms, including allosteric regulation. Excess acetyl CoA is diverted into fatty acid synthesis. **Review WEB ACTIVITY 6.6**

concept 6.5 **During Photosynthesis, Light Energy Is Converted to Chemical Energy**

- The **light reactions** of photosynthesis convert light energy into chemical energy. They produce ATP and reduce $NADP^+$ to NADPH. **Review Figure 6.15**

- Light is a form of **electromagnetic radiation**. It is emitted in particle-like packets called **photons** but has wavelike properties. Molecules that absorb light in the visible spectrum are called **pigments**. Photosynthetic organisms have several pigments, most notably **chlorophylls**. **Review Figures 6.16 and 6.17**

- The absorption of a photon puts a chlorophyll molecule into an excited state that has more energy than its ground state. This energy can be transferred via other chlorophylls to one in the **reaction center** of a photosystem. **Review Figure 6.18**

- An excited chlorophyll can act as a reducing agent, transferring excited electrons to other molecules. Oxidized chlorophyll regains electrons by the splitting of H_2O.

- In the thylakoid membrane of the chloroplast, photosystems I and II and a noncyclic electron transport system produce ATP via oxidative phosphorylation. NADPH and O_2 are also produced. **Review Figure 6.19 and ANIMATED TUTORIALS 6.3 and 6.4**

- **Cyclic electron transport** uses only photosystem I and produces only ATP. **Review Figure 6.20**

concept 6.6 **Photosynthetic Organisms Use Chemical Energy to Convert CO_2 to Carbohydrates**

- The **Calvin cycle** makes carbohydrates from CO_2. The cycle consists of three processes: fixation of CO_2, reduction and sugar production, and regeneration of RuBP. **Review Figure 6.21 and WEB ACTIVITY 6.7**

- RuBP is the initial CO_2 acceptor, and 3PG is the first stable product of CO_2 fixation. The enzyme **rubisco** catalyzes the reaction of CO_2 and RuBP to form 3PG. **Review Figure 6.22 and ANIMATED TUTORIAL 6.5**

- The ATP and NADPH formed by the light reactions are used to fuel the reduction of 3PG to form **glyceraldehyde 3-phosphate (G3P)**.

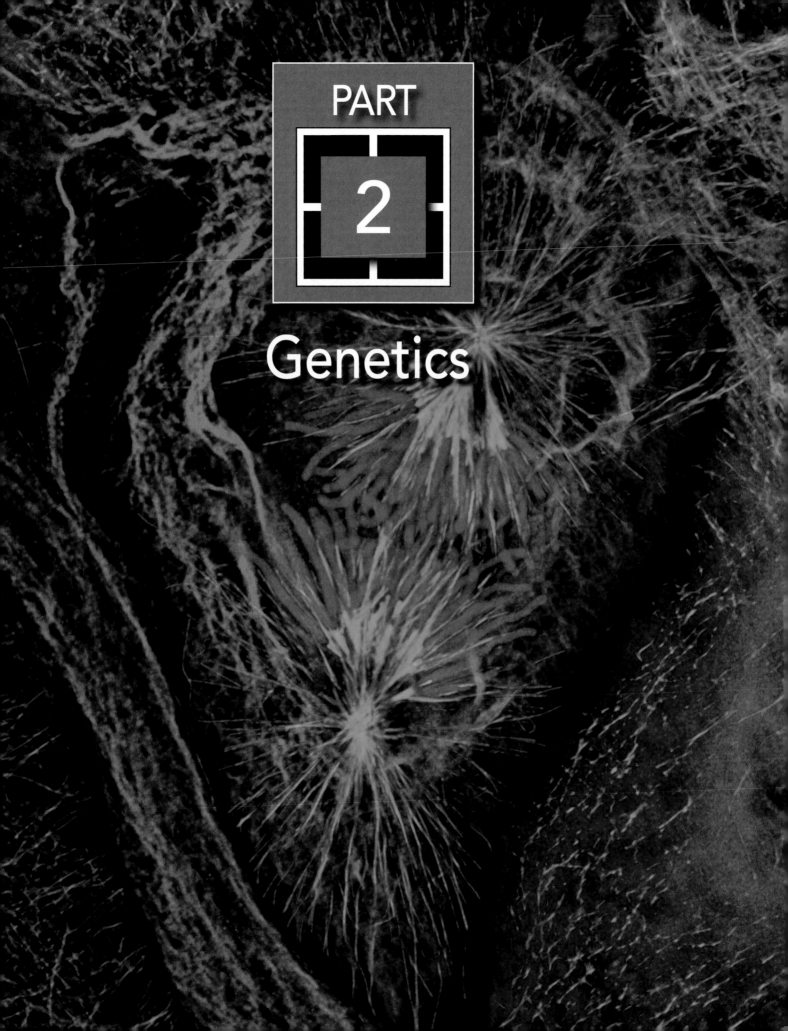

PART

2

Genetics

The Cell Cycle and Cell Division

Ruth felt healthy and was surprised when she was called back to her physician's office a week after her annual checkup. "Your lab report indicates you have early cervical cancer," said the doctor. "I ordered a follow-up test, and it came back positive—at some point, you were infected with HPV."

Ruth felt numb as soon as she heard the word "cancer." Her mother had died of breast cancer in the previous year. The doctor's statement about HPV—human papillomavirus—did not register in her consciousness. Sensing Ruth's discomfort, the doctor quickly reassured her that the cancer was caught at an early stage and that a simple surgical procedure would remove it. Two weeks later, the cancer was removed and Ruth remains cancer-free. She was fortunate that her annual medical exam included a Papanicolau (Pap) test, in which the cells lining the cervix are examined for abnormalities. Since they were begun almost 50 years ago in Europe, Pap tests have resulted in the early detection and removal of millions of early cervical cancers, and the death rate from this potentially lethal disease has plummeted.

Only recently was HPV found to be the cause of most cervical cancers. The German physician Harald Zur-Hausen was awarded the Nobel Prize in 2008 for this discovery, and it has led to a vaccine to prevent future infections. There are many different types of HPV, and many of the ones that infect humans cause warts, which are small, rough growths on the skin. The types of HPV that infect tissues at the cervix get there by sexual transmission, and this is a common infection.

When HPV arrives at the tissues lining the cervix, it has one of two fates. Most of the time it enters the cells and turns them into HPV factories, releasing a lot of

HPV particles into the mucus outside the uterus. These viruses can infect another person during a sexual encounter. But in some cases the virus follows a different, more sinister path. The viral DNA becomes incorporated into the DNA of the cervical cells, and the cells are stimulated to reproduce.

Cell reproduction in healthy humans is tightly controlled by a variety of mechanisms, but the virus-infected cells lose these controls. Understanding how cell division is controlled is clearly an important subject for the development of cancer treatments. But cell division is not just important in medicine. It underlies the growth, development, and reproduction of all organisms.

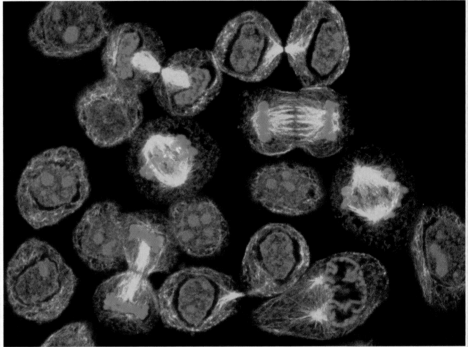

These cervical cancer cells are actively dividing; many of the cells are in various stages of mitosis and cytokinesis.

KEY CONCEPTS

7.1 Different Life Cycles Use Different Modes of Cell Reproduction

7.2 Both Binary Fission and Mitosis Produce Genetically Identical Cells

7.3 Cell Reproduction Is Under Precise Control

7.4 Meiosis Halves the Nuclear Chromosome Content and Generates Diversity

7.5 Programmed Cell Death Is a Necessary Process in Living Organisms

How does infection with HPV result in uncontrolled cell reproduction?*

*You will find the answer to this question on page 141.

Different Life Cycles Use Different Modes of Cell Reproduction

In Chapter 4 we describe cells as the basic compartments of life, where biological processes are separated from the external environment. Cells are also essential for biological reproduction.

The lifespan of an organism, from birth to death, is intimately linked to cell reproduction. (Although you will see that it is somewhat of a misnomer, we will follow convention and *refer to cell reproduction as cell division*.) Cell division plays important roles in the growth and repair of tissues in multicellular organisms, as well as in the reproduction of all organisms (**FIGURE 7.1**). Although the details vary widely, organisms have two basic strategies for reproducing themselves: asexual reproduction and sexual reproduction. These two strategies make use of different types of cell division.

Asexual reproduction by binary fission or mitosis results in genetic constancy

Asexual reproduction is a rapid and effective means of making new individuals, and it is common in nature. The offspring resulting from asexual reproduction are **clones** of the parent organism—they are genetically identical (or virtually identical) to each other and the parent. Any genetic variations among the parent and offspring are due to changes called *mutations*, which are alterations in DNA sequence caused by environmental factors or errors in DNA replication. As you will see, this small amount of variation contrasts with the extensive variation possible in sexually reproducing organisms.

In most cases, single-celled prokaryotes reproduce by *binary fission*, an asexual process that we will discuss in Concept 7.2. A cell of the bacterium *Escherichia coli* is the whole organism, so when it divides to form two new cells, it is reproducing. Similarly, single-celled eukaryotes (such as fission yeast) can reproduce asexually through *mitosis*, a type of cell division that also produces two genetically identical cells.

Many eukaryotes, including fungi and plants, can also reproduce asexually. Perhaps the most dramatic example of this is a forest stand of aspen trees (*Populus tremuloides*) in the Wasatch Mountains of Utah (**FIGURE 7.2**). DNA analyses have shown that these trees are clones—they are virtually identical genetically. Aspen can reproduce sexually, with male and female plants, but in this case all the trees are males. An extensive root system has spread through the soil, and at intervals stems have formed and grown into trees. These trees look like separate organisms, but they are actually clones derived from asexual reproduction.

Sexual reproduction by meiosis results in genetic diversity

Sexual reproduction involves the fusion of two specialized cells called **gametes**, and can result in offspring with considerable

(A) Reproduction

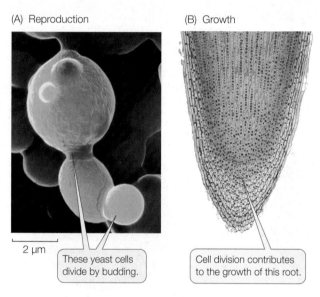

2 μm

These yeast cells divide by budding.

(B) Growth

Cell division contributes to the growth of this root.

(C) Regeneration

Cell division contributes to the regeneration of a lizard's tail.

FIGURE 7.1 The Importance of Cell Division Cell division is the basis for (A) reproduction, (B) growth, and (C) repair and regeneration of tissues.

FIGURE 7.2 Asexual Reproduction on a Large Scale In this forest, the aspen trees arose from asexual reproduction and are virtually identical genetically.

genetic variation. The gametes form by **meiosis**—a process of cell division (described in Concept 7.4) that results in daughter cells with only half the genetic material of the original cell. During meiosis, the genetic material is randomly separated and reorganized so that the daughter cells differ genetically from one another. Because of this genetic variation, some offspring of sexual reproduction may be better adapted than others to survive and reproduce in a particular environment. Meiosis thus generates the genetic diversity that is the raw material for natural selection and evolution.

As we describe in Chapter 4, the DNA in eukaryotic cells is organized into multiple structures called *chromosomes*. Each **chromosome** consists of a single molecule of DNA and associated proteins. In multicellular organisms, the body cells that are *not* specialized for reproduction are called **somatic cells**. Somatic cells each contain *two sets of chromosomes*, and the chromosomes occur in pairs called **homologous pairs**. One chromosome of each pair comes from the organism's female parent, and the other comes from its male parent. For example, in humans with 46 chromosomes, 23 come from the mother and 23 from the father, with, for example, a chromosome 1 from each parent, and so on. The two chromosomes in a homologous pair (called *homologs*) bear corresponding, though not identical, genetic information. For example, a homologous pair of chromosomes in a plant may carry different versions of a gene that controls seed shape. One homolog may carry the version

for wrinkled seeds, while the other may carry the version for smooth seeds.

LINK The inheritance of characteristics such as seed shape is discussed in Chapter 8

Gametes, by contrast, contain only a single set of chromosomes—that is, one homolog from each pair. The number of chromosomes in a gamete is denoted by n, and the cell is said to be **haploid**. During sexual reproduction, two haploid gametes fuse to form a **zygote** in a process called **fertilization**. The zygote thus has two sets of chromosomes, just as the somatic cells do. The chromosome number in the zygote is denoted by $2n$, and the cells are said to be **diploid**.

In many familiar organisms the zygote divides by mitosis, producing a new, mature organism with diploid somatic cells (as we have just described). But some other organisms have haploid stages in their life cycles.

yourBioPortal.com

Go to WEB ACTIVITY 7.1
Sexual Life Cycle

FIGURE 7.3 All Sexual Life Cycles Involve Fertilization and Meiosis In sexual reproduction, haploid (n) cells or organisms alternate with diploid ($2n$) cells or organisms.

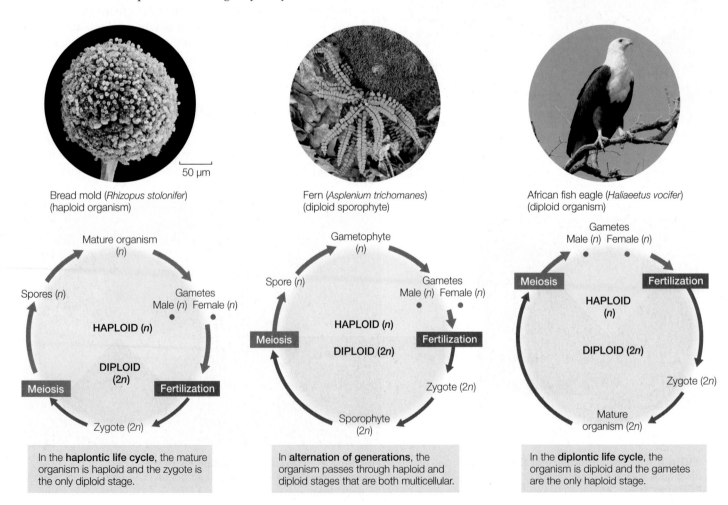

Bread mold (*Rhizopus stolonifer*)
(haploid organism)

Fern (*Asplenium trichomanes*)
(diploid sporophyte)

African fish eagle (*Haliaeetus vocifer*)
(diploid organism)

In the **haplontic life cycle**, the mature organism is haploid and the zygote is the only diploid stage.

In **alternation of generations**, the organism passes through haploid and diploid stages that are both multicellular.

In the **diplontic life cycle**, the organism is diploid and the gametes are the only haploid stage.

All sexual life cycles involve meiosis to produce haploid cells. In some cases, gametes develop immediately after meiosis. In others, each haploid cell divides and develops into a haploid organism—the haploid stage of the life cycle—which eventually produces gametes by mitosis. The fusion of gametes results in a zygote and begins the diploid stage of the life cycle. Since the origin of sexual reproduction, evolution has generated many different versions of the sexual life cycle. **FIGURE 7.3** presents three examples:

- In *haplontic* organisms, including most protists, fungi, and some green algae, the tiny zygote is the only diploid cell in the life cycle. After it is formed it immediately undergoes meiosis to produce more haploid cells. These are usually *spores*, which are the dispersal units for the organism. A spore germinates to form a new haploid organism, which may be single-celled or multicellular. Cells of the mature haploid organism produce gametes that fuse to form the diploid zygote.

- Most plants and some fungi display **alternation of generations**. As in many haplontic organisms, meiosis gives rise to haploid spores, which divide by mitosis to form a haploid life stage called the *gametophyte*. In flowering plants the gametophytes are very small: the male gametophyte is the pollen, and the female gametophyte is the embryo sac—a part of the flower. In mosses and liverworts the gametophytes are larger and multicellular. The gametophyte forms gametes by mitosis, and they fuse to produce the diploid zygote. The zygote divides by mitosis to become the diploid *sporophyte*, which in turn produces haploid spores by meiosis.

- In *diplontic* organisms, which include animals, brown algae, and some fungi, the gametes are the only haploid cells in the life cycle, and the mature organism is diploid.

These life cycles are described in greater detail in Part Four. For now we will focus on the role of sexual reproduction in generating diversity among individual organisms.

The essence of sexual reproduction is the *random selection of half of the diploid chromosome set* to make a haploid gamete, followed by fusion of two haploid gametes from separate parents to produce a diploid cell. As we will see later in this chapter, further diversity is introduced by events that take place during meiosis. All of these steps contribute to a shuffling of genetic information in the population, so that no two individuals have exactly the same genetic constitution. The diversity provided by sexual reproduction opens up enormous opportunities for evolution.

Do You Understand Concept 7.1?

- In terms of the genetic composition of offspring, what is the difference between sexual and asexual reproduction?

- Discuss the advantages and disadvantages of sexual versus asexual reproduction in terms of evolution. Could evolution proceed without sexual reproduction?

We have briefly described the roles that different types of cell division play in the life cycles of organisms. Now let's look in more detail at the processes of cell division, starting with binary fission and mitosis.

concept **7.2** **Both Binary Fission and Mitosis Produce Genetically Identical Cells**

Cell reproduction by either binary fission or mitosis produces two genetically identical cells. For single-celled organisms, this is the basis of asexual reproduction. For multicellular organisms, mitosis is a way to build tissues and organs during development and to repair damaged tissues once development is complete.

In order for any cell to divide, the following events must occur:

- There must be a *reproductive signal*. This signal initiates cell division and may originate from either inside or outside the cell.

- **Replication** of DNA (the genetic material) must occur so that each of the two new cells will have a full complement of genes to complete cell functions.

- The cell must distribute the replicated DNA to each of the two new cells. This process is called **segregation**.

- Enzymes needed for cell division must be synthesized, new organelles must be formed, and new material must be added to the plasma membrane (and the cell wall in organisms that have one). The division of the cytoplasm to form two daughter cells is called **cytokinesis**.

These events proceed somewhat differently in prokaryotes and eukaryotes. We will discuss replication, segregation, and cytokinesis here, and reproductive signals in Concept 7.3.

Prokaryotes divide by binary fission

In prokaryotes, cell division results in the reproduction of the entire single-celled organism. The cell grows in size, replicates its DNA, and then separates the cytoplasm and DNA into two new cells by a process called **binary fission**.

REPLICATION OF DNA In most prokaryotic cells, almost all of the genetic information is carried on one single chromosome. In many cases the ends of the single DNA molecule are covalently joined, making the chromosome circular. Two regions of the prokaryotic chromosome play functional roles in cell reproduction:

- *ori*: the site where replication of the circular chromosome starts (the *ori*gin of replication)

- *ter*: the site where replication ends (the *ter*minus of replication)

Chromosome replication takes place as the DNA is threaded through a "replication complex" of proteins near the center of the cell. Replication begins at the *ori* site and moves toward the *ter* site. When replication is complete, the two daughter DNA molecules separate and segregate from one another at opposite

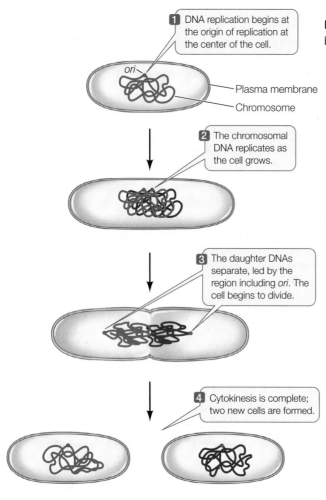

1 DNA replication begins at the origin of replication at the center of the cell.

ori

Plasma membrane

Chromosome

2 The chromosomal DNA replicates as the cell grows.

3 The daughter DNAs separate, led by the region including *ori*. The cell begins to divide.

4 Cytokinesis is complete; two new cells are formed.

FIGURE 7.4 **Prokaryotic Cell Division** The process of cell division in a bacterium involves DNA replication, segregation, and cytokinesis.

ends of the cell. In rapidly dividing prokaryotes, DNA replication occupies the entire time between cell divisions.

SEGREGATION OF DNA Replication begins near the center of the cell, and as it proceeds, the *ori* regions move toward opposite ends of the cell (**FIGURE 7.4**). DNA sequences adjacent to the *ori* region bind proteins that are essential for this segregation. This is an active process, since the binding proteins hydrolyze ATP. Components of the prokaryotic cytoskeleton are involved in the segregation process. In particular, a bacterial protein that is structurally related to actin but functionally related to tubulin provides a filament along which the *ori* regions and their associated proteins move.

LINK Review the description of the cytoskeleton and its components in Concept 4.4

CYTOKINESIS The actual division of a single cell and its contents into two cells begins immediately after chromosome segregation. Initially, there is a pinching in of the plasma membrane caused by the contraction of a ring of fibers on the inside surface of the membrane (similar to a drawstring on shorts being tightened). In this case, the major component of these fibers is structurally similar to eukaryotic tubulin (which makes up microtubules), but its function is analogous to that of actin in the contractile ring of

an animal cell (see below). As the membrane pinches in, new cell wall materials are deposited, which finally separate the two cells.

Eukaryotic cells divide by mitosis followed by cytokinesis

As in prokaryotes, cell reproduction in eukaryotes entails DNA replication, segregation, and cytokinesis. Some of the details, however, are quite different:

- *Replication of DNA.* Unlike prokaryotes, eukaryotes have more than one chromosome. But the replication of each eukaryotic DNA molecule is similar to replication in prokaryotes, in that it is achieved by threading the long strands through replication complexes (see Concept 9.2). DNA replication occurs only during a specific stage of the cell cycle.

- *Segregation of DNA.* When a cell divides, one copy of each chromosome must end up in each of the two new cells—for example, each new somatic cell in a human will have all 46 chromosomes (23 pairs). In eukaryotes, the newly replicated chromosomes are closely associated with each other. They become highly condensed, and a mechanism called **mitosis** segregates them into two new nuclei. The cytoskeleton is involved in this process.

- *Cytokinesis.* Cytokinesis follows mitosis. The process in plant cells (which have cell walls) is different than in animal cells (which do not have cell walls).

These events occur within the context of the **cell cycle**, the period from one division to the next. In eukaryotes, the cell cycle can be divided into mitosis and cytokinesis—referred to as the M phase—and a much longer interphase. During **interphase**, the cell nucleus is visible and typical cell functions occur—including DNA replication in cells that are dividing. This phase of the cell cycle begins when cytokinesis is completed and ends when mitosis begins (**FIGURE 7.5**).

Interphase has three subphases called G1, S, and G2. **G1** (the *G* stands for gap) is quite variable, and a cell may spend a long time in this phase carrying out its specialized functions. The cell's DNA is replicated during **S phase** (the *S* stands for synthesis). During **G2**, the cell makes preparations for mitosis—for example, by synthesizing components of the microtubules that will move the segregating chromosomes to opposite ends of the dividing cell.

yourBioPortal.com

Go to ANIMATED TUTORIAL 7.1
Mitosis

Prophase sets the stage for DNA segregation

In mitosis (the M phase of the cell cycle), *a single nucleus gives rise to two daughter nuclei that each contain the same number of chromosomes as the parent nucleus.* While mitosis is a continuous

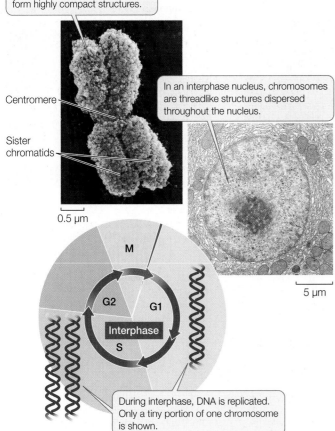

In the M phase cell, the DNA and proteins in each chromosome form highly compact structures.

Centromere

Sister chromatids

In an interphase nucleus, chromosomes are threadlike structures dispersed throughout the nucleus.

0.5 μm

5 μm

M

G2   G1

Interphase

S

During interphase, DNA is replicated. Only a tiny portion of one chromosome is shown.

FIGURE 7.5 The Phases of the Eukaryotic Cell Cycle The eukaryotic cell cycle has several phases. DNA in the interphase nucleus is diffuse and becomes compacted as mitosis begins.

CONDENSED CHROMOSOMES Before the S phase of interphase, each chromosome contains one very long double-stranded DNA molecule. If all of the DNA in a typical human cell were put end to end, it would be nearly 2 meters long. Yet the nucleus is only 5 μm (0.000005 m) in diameter. So even during interphase the eukaryotic DNA is packaged in a highly organized way. The DNA is wound around specific proteins, and other proteins coat the DNA coils (see Concept 11.3). During prophase the chromosomes become much more tightly coiled and condensed.

After DNA replication, each chromosome has *two* DNA molecules, known as **sister chromatids**. The chromatids are held together at a region called the **centromere** until separation during mitosis (see below). During prophase the chromosomes become so compact that they can be seen clearly with a light microscope after staining with special dyes. (The term *chromosomes* means "staining bodies" and was coined in the late nineteenth century before the discovery of DNA and its structure.) Specialized protein structures called **kinetochores** assemble on the centromeres, one on each chromatid. These structures are important for chromosome movement.

For a given organism, the number and sizes of the condensed chromosomes constitute the **karyotype**. Each chromosome has a particular length, and the centromere is located at a particular position along its length. For example, humans have 46 chromosomes that can be distinguished from one another by their sizes and centromere positions. Karyotype analysis used to be a way to identify organisms, and is still used to detect gross chromosomal abnormalities in humans. However, DNA sequence analysis is now used much more commonly to identify and classify organisms.

process in which each event flows smoothly into the next, it is convenient to subdivide it into a series of stages: prophase, prometaphase, metaphase, anaphase, and telophase.

During interphase, only the nuclear envelope and the nucleolus (the region of the nucleus where ribosomes are formed; see Concept 4.3) are visible under the light microscope. The *chromatin* (the DNA with its associated proteins) is not yet condensed, and individual chromosomes cannot be discerned. The appearance of the nucleus changes as the cell enters **prophase**— the beginning of mitosis. Here we describe three structures that appear during prophase and contribute to the orderly segregation of the replicated DNA during mitosis: the condensed chromosomes, the centrosome and the spindle:

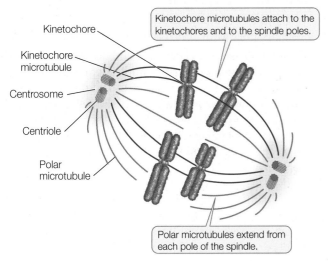

Kinetochore

Kinetochore microtubule

Centrosome

Centriole

Polar microtubule

Kinetochore microtubules attach to the kinetochores and to the spindle poles.

Polar microtubules extend from each pole of the spindle.

CENTROSOME Before the spindle apparatus forms (see below), its orientation is determined. In many cells, this is accomplished by the **centrosome** ("central body"), an organelle in the cytoplasm near the nucleus. The centrosome consists of a pair of **centrioles**, each one a hollow tube formed by nine triplets of microtubules. During S phase the centrosome becomes duplicated, and at the G2-to-M transition, the two centrosomes separate from one another, moving to opposite sides of the nucleus. Eventually these identify "poles" toward which chromosomes move during segregation.

The positions of the centrosomes determine the plane at which the cell divides; therefore they determine the spatial relationship between the two new cells. This relationship may be of little consequence to single free-living cells such as yeasts, but it is important for development in a multicellular organism. For example, during the development of an embryo, the daughter cells from some divisions must be positioned correctly to receive signals to form new tissues. Plant cells lack centrosomes, but distinct microtubule organizing centers at each end of the cell play the same role.

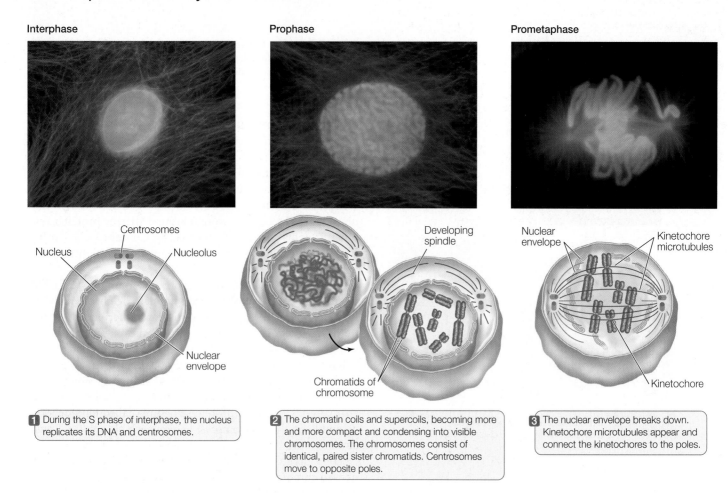

Interphase

Prophase

Prometaphase

1 During the S phase of interphase, the nucleus replicates its DNA and centrosomes.

2 The chromatin coils and supercoils, becoming more and more compact and condensing into visible chromosomes. The chromosomes consist of identical, paired sister chromatids. Centrosomes move to opposite poles.

3 The nuclear envelope breaks down. Kinetochore microtubules appear and connect the kinetochores to the poles.

SPINDLE   Each of the two centrosomes, when positioned on opposite sides of the nucleus, serves as a *pole*, toward which the chromosomes move. Tubulin dimers from around the centrosomes aggregate into microtubules that extend from the poles into the middle region of the cell. Together these microtubules make up a **spindle**. The spindle forms during prophase and prometaphase, when the nuclear envelope breaks down (see below).

There are two types of microtubules in the spindle:

- *Polar microtubules* form the framework of the spindle and run from one pole to the other, keeping the two poles apart.

- *Kinetochore microtubules*, which form later, attach to the kinetochores on the chromosomes. The two sister chromatids in each chromosome become attached to kinetochore microtubules from opposite sides of the cell. This ensures that the two chromatids will move to opposite poles.

The growing end of a developing microtubule is unstable and will fall apart unless it is attached to a kinetochore.

Separation of the chromatids and movement of the **daughter chromosomes** (which the sister chromatids become after separation) is the central feature of mitosis. It accomplishes the segregation that is needed for cell division and completion of the cell cycle.

yourBioPortal.com

**Go to WEB ACTIVITY 7.2
The Mitotic Spindle**

Chromosome separation and movement are highly organized

During the next three phases of mitosis—prometaphase, metaphase, and anaphase—dramatic changes take place in the cell and the chromosomes:

- In **prometaphase** the nuclear envelope breaks down and the compacted chromosomes, each consisting of two chromatids, attach to the kinetochore microtubules.

- In **metaphase** the chromosomes line up at the midline of the cell (the equatorial position).

- In **anaphase** the chromatids separate, and the daughter chromosomes move away from each other toward the poles.

You will find these events depicted and described in **FIGURE 7.6**. The separation of chromatids into daughter chromosomes occurs at the beginning of anaphase. The migration of the daughter chromosomes to the poles of the cell is a highly organized, active process. Two mechanisms operate to move the chromosomes along. First, the kinetochores contain a protein called *cytoplasmic dynein* that acts as a "molecular motor." It hydrolyzes ATP to ADP and phosphate, thus releasing energy that moves the chromosomes along the microtubules toward the poles. This accounts for about 75 percent of the force of motion. Second, the kinetochore microtubules shorten from the poles, drawing the chromosomes toward them, accounting for about 25 percent of the force of motion.

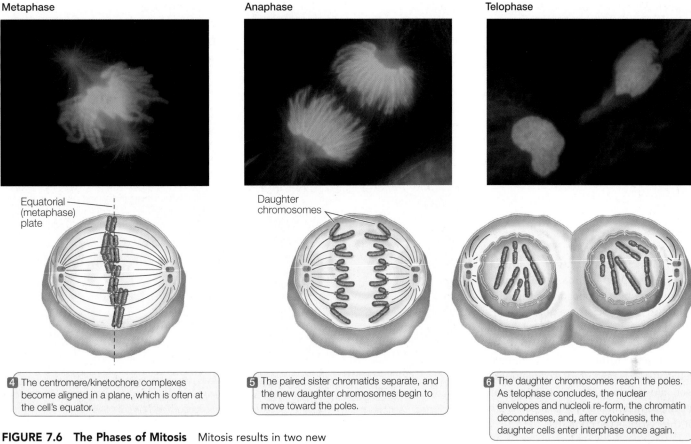

Metaphase   Anaphase   Telophase

Equatorial (metaphase) plate

Daughter chromosomes

4 The centromere/kinetochore complexes become aligned in a plane, which is often at the cell's equator.

5 The paired sister chromatids separate, and the new daughter chromosomes begin to move toward the poles.

6 The daughter chromosomes reach the poles. As telophase concludes, the nuclear envelopes and nucleoli re-form, the chromatin decondenses, and, after cytokinesis, the daughter cells enter interphase once again.

FIGURE 7.6 The Phases of Mitosis Mitosis results in two new nuclei that are genetically identical to each other and to the nucleus from which they were formed. In the micrographs, the green dye stains microtubules (and thus the spindle); the blue dye stains the chromosomes. The chromosomes in the diagrams are stylized to emphasize the fates of the individual chromatids.

yourBioPortal.com

Go to WEB ACTIVITY 7.3
Images of Mitosis

Telophase occurs after the chromosomes have separated and is the last phase of mitosis. During this period, a nuclear envelope forms around each set of new chromosomes, nucleoli appear, and the chromosomes become less compact. The spindle also disappears at this stage. As a result, there are two new nuclei in a single cell.

Cytokinesis is the division of the cytoplasm

Mitosis refers only to the division of the nucleus. Cytokinesis, the division of the cell's cytoplasm, is the final stage of cell reproduction. This process occurs differently in plants and animals.

ANIMAL CELLS Cytokinesis usually begins with a furrowing of the plasma membrane, as if an invisible thread were cinching the cytoplasm between the two nuclei (**FIGURE 7.7A**). This *contractile ring* is composed of microfilaments both actin and myosin, which form a ring on the cytoplasmic surface of the plasma membrane. These two proteins interact to produce a

contraction (just as they do in muscles), pinching the cell in two. The microfilaments assemble rapidly from actin monomers that are present in the interphase cytoskeleton. Their assembly is controlled by calcium ions (commonly used in cellular signaling) that are released from storage sites in the center of the cell.

PLANT CELLS In plant cells, the cytoplasm divides differently because plants have cell walls. As the spindle breaks down after mitosis, membranous vesicles derived from the Golgi apparatus appear along the plane of cell division, roughly midway between the two daughter nuclei. The vesicles are propelled along microtubules by the motor protein *kinesin* and fuse to form a new plasma membrane. At the same time they contribute their contents to a *cell plate*, which is the beginning of a new cell wall that will lie between the two new cells (**FIGURE 7.7B**).

FRONTIERS Orientation of the mitotic spindle and cytokinesis in a dividing cell is usually nonrandom. During embryo development, the plane of cell division and whether the daughter cells get equal amounts of cytoplasm have important consequences for the subsequent fates of the daughter cells (see Concept 14.2). Animal and plant biologists are characterizing the genes and protein products involved in spindle orientation, with the aim of describing how it is involved in the formation of organs.

FIGURE 7.7 Cytokinesis Differs in Animal and Plant Cells (A) A sea urchin zygote (fertilized egg) that has just completed cytokinesis at the end of the first cell division of its development into an embryo. (B) An electron micrograph of a plant cell in late telophase. Plant cells divide differently than animal cells because they have cell walls.

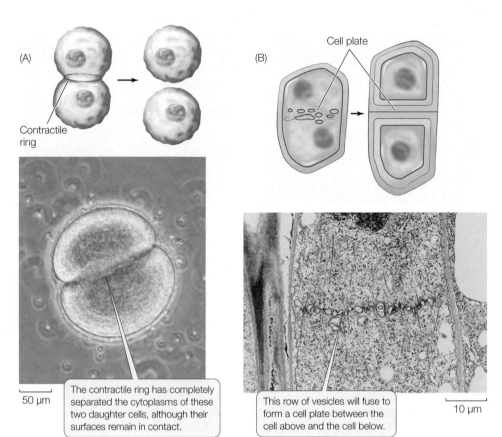

(A)

Contractile ring

50 µm

The contractile ring has completely separated the cytoplasms of these two daughter cells, although their surfaces remain in contact.

(B)

Cell plate

This row of vesicles will fuse to form a cell plate between the cell above and the cell below.

10 µm

Following cytokinesis, *each daughter cell contains all the components of a complete cell*. A precise distribution of chromosomes is ensured by mitosis. In contrast, organelles such as ribosomes, mitochondria, and chloroplasts are not necessarily distributed equally between daughter cells (as long as some of each are present in each cell). Although the *orientation* of cell division is important in development (see above), there does not appear to be a precise mechanism for the distribution of the cytoplasmic contents.

Do You Understand Concept 7.2?

- How does the mitotic spindle ensure that each daughter cell receives a full complement of the genetic material in the cell nucleus?

- Sketch the five stages of mitosis for an organism with four chromosomes (two pairs). Clearly label chromosomes and chromatids and note the number of double-stranded DNA molecules in each structure at each stage.

- The drug cytochalasin B blocks the assembly and function of microfilaments. What would happen if animal cells were treated with this drug after telophase but before cytokinesis?

Light microscopists have studied the dramatic events of mitosis and cell division since the 1880s, and we now have detailed descriptions of these processes. More recently, biologists have focused on the mechanisms controlling cell reproduction, which will be our next topic.

concept 7.3 Cell Reproduction Is Under Precise Control

Cell reproduction cannot go on continuously and indefinitely. If a single-celled species had no control over its reproduction, it would soon overrun its environment and starve to death. In a multicellular organism, cell reproduction must be controlled to maintain the forms and functions of different parts of the body.

The reproductive rates of many prokaryotes respond to general conditions in the environment. For example, the bacterium *Bacillus subtilis* normally divides every 30 minutes. But when nutrients in its environment are low it stops dividing, and then resumes dividing when conditions improve.

Unlike prokaryotes, eukaryotic cells do not constantly divide whenever environmental conditions are adequate. In fact, the specialized cells of a multicellular eukaryotic organism may seldom or never divide. The signals for eukaryotic cell division are related to the needs of the entire organism. Mammals produce a variety of substances called **growth factors** that stimulate cell division and differentiation. For example, if you cut yourself and bleed, a blood clot eventually forms. Cell fragments called platelets in the blood vessels surrounding the clot secrete various growth factors that stimulate nearby skin cells to divide and heal the wound.

The eukaryotic cell division cycle is regulated internally

As we discussed in Concept 7.2, the eukaryotic cell cycle can be divided into four stages: G1, S, G2, and M. Progression through these phases is tightly regulated. For example, at the end of G1, the **G1–S transition** marks a key decision point for the cell: passing this point (called *R*, the *restriction point*) usually means

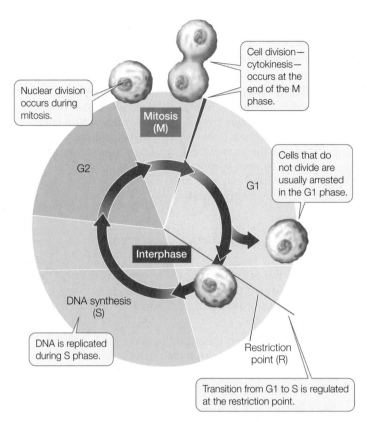

FIGURE 7.8   The Eukaryotic Cell Cycle   The cell cycle consists of a mitotic (M) phase, during which mitosis and cytokinesis take place, and a long period of growth known as interphase. Interphase has three subphases (G1, S, and G2) in cells that divide.

the cell will proceed with the rest of the cell cycle and divide (**FIGURE 7.8**).

What events cause a cell to enter the S or M phases? A first indication that there were substances that control these transitions came from experiments involving *cell fusion*. Polyethylene glycol is a nonpolar solvent that partially dissolves membrane lipids and can be used to make different cells fuse together. Experiments involving the fusion of mammalian cells at different phases of the cell cycle showed that a cell in S phase produces a substance that activates DNA replication (**FIGURE 7.9**). Similar experiments pointed to a molecular activator for entry into M phase.

The cell cycle is controlled by cyclin-dependent kinases

The molecular activators revealed by the cell fusion experiments turned out to be *protein kinases*. These enzymes, common in cell signal transduction (see Concept 5.5), catalyze the phosphorylation of target proteins that regulate the cell cycle:

$$\text{cell cycle regulator} + \text{ATP} \xrightarrow{\text{Protein kinase}} \text{cell cycle regulator-P} + \text{ADP}$$

The class of protein kinases involved in cell cycle regulation is called **cyclin-dependent kinases** (**Cdk's**), and as their name

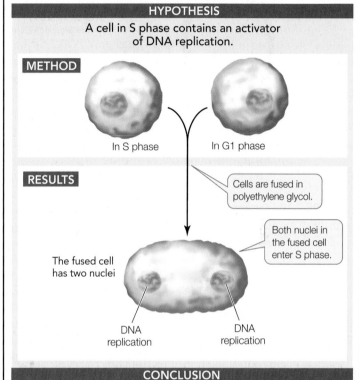

INVESTIGATION

FIGURE 7.9   Regulation of the Cell Cycle   Nuclei in G1 do not undergo DNA replication, but nuclei in S phase do. To determine if there is some signal in the S cells that stimulates G1 cells to replicate their DNA, cells in the G1 and S phases were fused together, creating cells with both G1 and S properties.

HYPOTHESIS

A cell in S phase contains an activator of DNA replication.

METHOD

In S phase   In G1 phase

RESULTS

Cells are fused in polyethylene glycol.

Both nuclei in the fused cell enter S phase.

The fused cell has two nuclei

DNA replication   DNA replication

CONCLUSION

The S phase cell produces a substance that diffuses to the G1 nucleus and activates DNA replication.

ANALYZE THE DATA

The experiment used mammalian cells undergoing the cell cycle synchronously. Radioactive labeling and microscopy were used to determine which nuclei were synthesizing DNA. Here are counts of the cell nuclei that were labeled:

Type of cells	Cells with labeled nuclei/total cells
Unfused G1:	6/300
Unfused S:	435/500
Fused G1 and S cells:	17*/19

*Both nuclei labeled

A. What were the percentages of cells in S phase in each of the three experiments?
B. What does this mean in terms of control of the cell cycle?

Go to **yourBioPortal.com** for original citations, discussions, and relevant links for all INVESTIGATION figures.

implies, Cdk's are activated by binding to the protein **cyclin**. This binding changes the shape of a Cdk such that its active site is exposed, and is an example of *allosteric regulation* (see Concept 3.4).

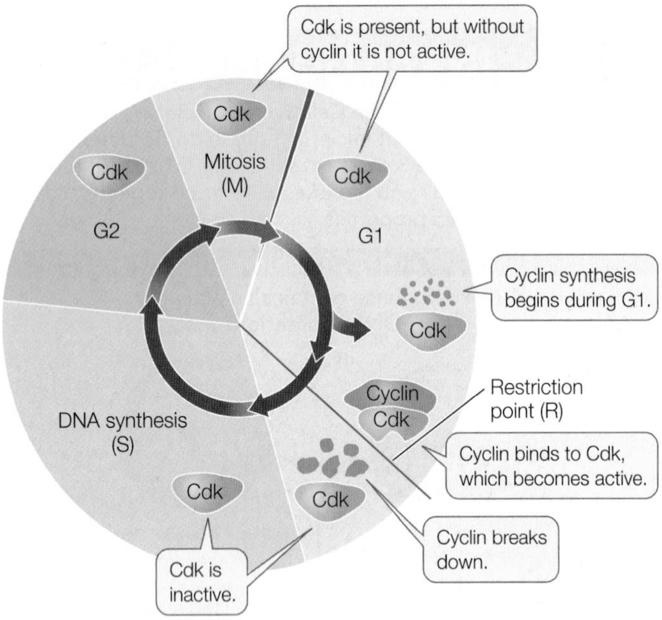

FIGURE 7.10 Cyclins Are Transient in the Cell Cycle Cyclins are made at a particular time and then break down. In this case, the cyclin is present during G1 and activates a Cdk at that time.

There are several Cdk's (which vary in different eukaryotic species) that regulate the cell cycle at specific stages called **cell cycle checkpoints**. Each Cdk has its own cyclin to activate it, and the cyclin is made only at the right time. After the Cdk acts, the cyclin is broken down by a protease (**FIGURE 7.10**). So a key event controlling transition from one cell cycle phase to the next is the synthesis and subsequent breakdown of a cyclin. Cyclins are synthesized in response to various molecular signals, including growth factors. This starts a chain reaction:

$$\text{growth factor} \rightarrow \text{cyclin synthesis} \rightarrow \text{Cdk activation} \rightarrow \text{cell cycle events}$$

To illustrate the concept of cell cycle control by a particular cyclin–Cdk complex, let's take a look at the complex that controls the R point at the G1–S transition (see Figure 7.8).

G1–S cyclin–Cdk catalyzes the phosphorylation of a protein called *retinoblastoma protein* (*RB*). In many cells, RB or a protein like it acts as an *inhibitor of the cell cycle* at the R point. To begin S phase, a cell must overcome the RB block. Here is where G1–S cyclin–Cdk comes in: it catalyzes the addition of a phosphate to RB. This causes a change in the three-dimensional structure of RB, thereby inactivating it. With RB out of the way, the cell cycle can proceed. To summarize:

$$\underset{\text{(active: blocks cell cycle)}}{RB} + ATP \xrightarrow{\text{G1–S cyclin–Cdk}} \underset{\text{(inactive: allows cell cycle)}}{RB\text{-}P} + ADP$$

Now we can be more specific about the chain of events involved in growth factor stimulation of cell division: the specific cyclin whose synthesis is activated is the one that allosterically activates the Cdk that phosphorylates RB, and this allows the cell cycle to exit G1 and begin DNA replication in S phase. This example illustrates how regulation of the cell cycle involves a number of cellular processes that we have examined in this and other chapters: signal transduction (see Chapter 5), gene expression and protein synthesis (see Chapter 3), and cell division.

FRONTIERS Regulation of progress through the cell cycle by Cdk's and cyclins is itself regulated by a host of molecules that "regulate the regulators." These in turn may be regulated by signals from outside the cell. Biologists have found that the constant cell cycling seen in cancer cells can be caused by abnormalities in these regulatory proteins. This new knowledge may lead to cancer therapies that target specific regulatory processes.

Do You Understand Concept 7.3?

- Compare the cell cycles of prokaryotes and eukaryotes with regard to signals for initiation, how many chromosomes are present, and how the replicated DNA segregates.

- Draw a diagram and describe the events that occur in the four stages of the eukaryotic cell cycle.

- Cultures of eukaryotic cells can be synchronized, so they are all at the same phase of the cell cycle at the same time. If you examined a culture at the beginning of G1, would the Cdk that acts at the R point be present? Would it be active? Would its cyclin be present? What would your answers be if the cell was in G2?

Binary fission and mitosis result in daughter cells with the same number of chromosomes as their parent cells. But sexual reproduction requires a process of cell division in which the number of chromosomes in halved. We'll look at this process next.

concept 7.4 Meiosis Halves the Nuclear Chromosome Content and Generates Diversity

In Concept 7.1 we described the role and importance of meiosis in sexual reproduction. Now we will see how meiosis accomplishes the orderly and precise generation of haploid cells.

Meiosis consists of *two* nuclear divisions that reduce the number of chromosomes to the haploid number. Although the *nucleus divides twice* during meiosis, the *DNA is replicated only once*. Unlike the products of mitosis, the haploid cells that are the products of meiosis are genetically different from one another and from the parent cell. **FIGURE 7.11** compares the two processes.

To understand the process of meiosis and its specific details, it is useful to keep in mind the overall functions that meiosis has evolved to serve:

- To reduce the chromosome number from diploid to haploid
- To ensure that each of the haploid products has a complete set of chromosomes
- To generate genetic diversity among the products

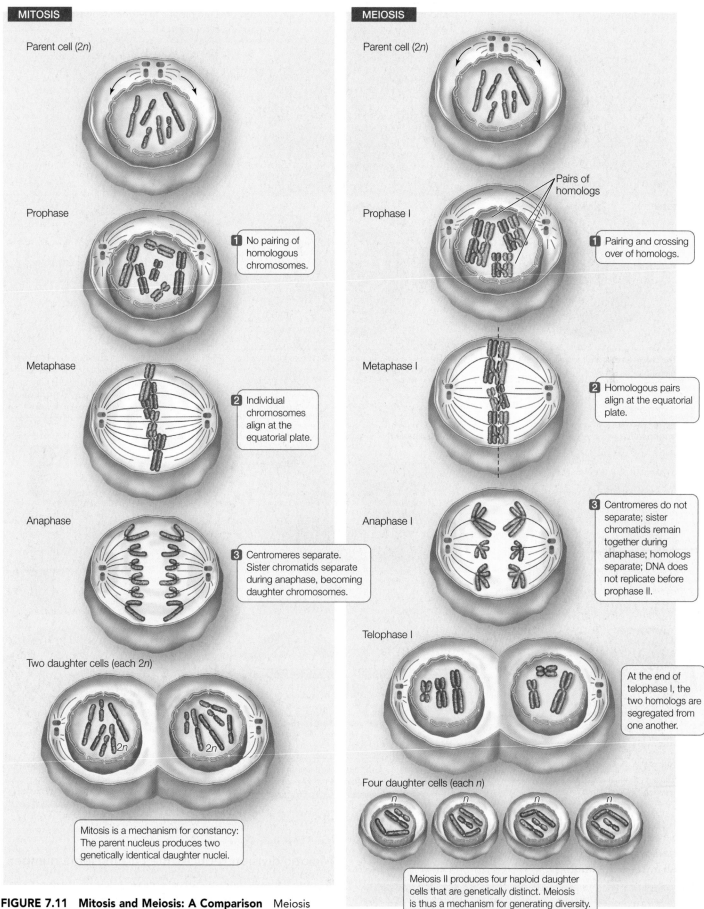

MITOSIS

Parent cell (2n)

Prophase

1 No pairing of homologous chromosomes.

Metaphase

2 Individual chromosomes align at the equatorial plate.

Anaphase

3 Centromeres separate. Sister chromatids separate during anaphase, becoming daughter chromosomes.

Two daughter cells (each 2n)

2n   2n

Mitosis is a mechanism for constancy: The parent nucleus produces two genetically identical daughter nuclei.

MEIOSIS

Parent cell (2n)

Prophase I

Pairs of homologs

1 Pairing and crossing over of homologs.

Metaphase I

2 Homologous pairs align at the equatorial plate.

Anaphase I

3 Centromeres do not separate; sister chromatids remain together during anaphase; homologs separate; DNA does not replicate before prophase II.

Telophase I

At the end of telophase I, the two homologs are segregated from one another.

Four daughter cells (each n)

n   n   n   n

Meiosis II produces four haploid daughter cells that are genetically distinct. Meiosis is thus a mechanism for generating diversity.

FIGURE 7.11 Mitosis and Meiosis: A Comparison Meiosis differs from mitosis chiefly by the pairing of homologs and by the failure of the centromeres to separate at the end of metaphase I.

Early prophase I

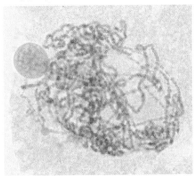

Centrosomes

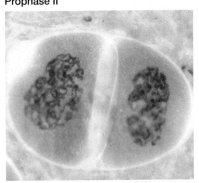

1 The chromatin begins to condense following interphase.

Mid-prophase I

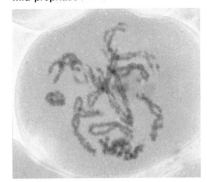

Pairs of homologs

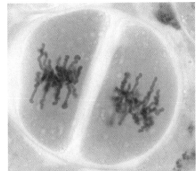

Tetrad

2 Synapsis aligns homologs, and chromosomes condense further.

Late prophase I–Prometaphase

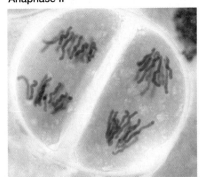

Chiasma

3 The chromosomes continue to coil and shorten. The chiasmata reflect crossing over, the exchange of genetic material between nonsister chromatids in a homologous pair. In prometaphase the nuclear envelope breaks down.

Prophase II

7 The chromosomes condense again, following a brief interphase (interkinesis) in which DNA does not replicate.

Metaphase II

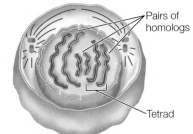

Equatorial plate

8 The centromeres of the paired chromatids line up across the equatorial plates of each cell.

Anaphase II

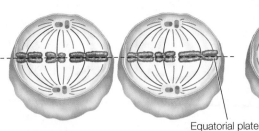

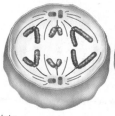
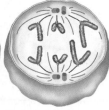

9 The chromatids finally separate, becoming chromosomes in their own right, and are pulled to opposite poles. Because of crossing over and independent assortment, each new cell will have a different genetic makeup.

The events of meiosis are illustrated in **FIGURE 7.12**. In this section, we will discuss some of the key features that distinguish meiosis from mitosis.

yourBioPortal.com

Go to ANIMATED TUTORIAL 7.2
Meiosis

Meiotic division reduces the chromosome number

As noted above, meiosis consists of two nuclear divisions, *meiosis I* and *meiosis II*. Two unique features characterize **meiosis I**:

• *Homologous chromosomes come together and line up* along their entire lengths. No such pairing occurs in mitosis.

Metaphase I

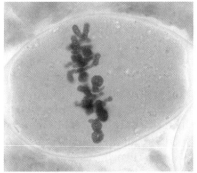

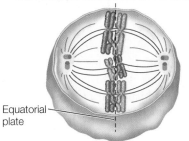

Equatorial
plate

4 The homologous pairs line up on the equatorial (metaphase) plate.

Anaphase I

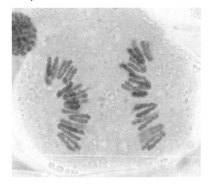

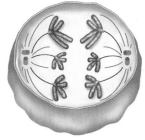

5 The homologous chromosomes (each with two chromatids) move to opposite poles of the cell.

Telophase I

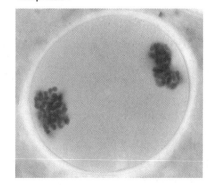

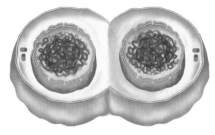

6 The chromosomes gather into nuclei, and the original cell divides.

Telophase II

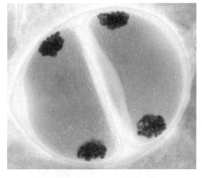

10 The chromosomes gather into nuclei, and the cells divide.

Products

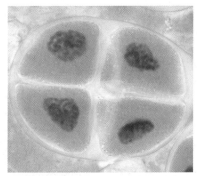

11 Each of the four cells has a nucleus with a haploid number of chromosomes.

FIGURE 7.12 Meiosis: Generating Haploid Cells In meiosis, two sets of chromosomes are divided among four daughter cells, each of which has half as many chromosomes as the original cell. The four haploid cells are the result of two successive nuclear divisions. The micrographs show meiosis in the male reproductive organ of a lily; the diagrams show the corresponding phases in an animal cell. (For instructional purposes, the chromosomes from one parent of the original organism are colored blue and those from the other parent are red.)

yourBioPortal.com

Go to WEB ACTIVITY 7.4 Images of Meiosis

- *The homologous chromosome pairs separate*, but the individual chromosomes, each consisting of two sister chromatids, remain intact. (The chromatids will separate during meiosis II.)

Like mitosis, meiosis I is preceded by an interphase with an S phase, during which each chromosome is replicated. As a result, each chromosome consists of two sister chromatids. At the end of meiosis I two nuclei form, each with half of the original chromosomes (one member of each homologous pair). Since the centromeres did not separate, these chromosomes are still double—composed of two sister chromatids. The sister chromatids are separated during **meiosis II**, which is *not* preceded by DNA replication. As a result, the products of meiosis I and II are four cells, each containing the haploid number of chromosomes. But *these four cells are not genetically identical*.

Crossing over and independent assortment generate diversity

A diploid organism has two sets of chromosomes (*2n*): one set derived from its male parent, the other from its female parent. As the organism grows and develops, its cells undergo mitotic divisions. In mitosis, each chromosome behaves independently of its homolog, and its two chromatids are sent to opposite poles during anaphase. Each daughter nucleus ends up with an identical set of *2n* chromosomes. In meiosis, things are very different (see Figure 7.11).

An important consequence of meiosis is that the four resulting cells differ from one another genetically. The shuffling of genetic material occurs by two processes: crossing over and independent assortment.

CROSSING OVER Meiosis I begins with a long prophase I (the first three panels of Figure 7.12), during which the chromosomes change markedly. The homologous chromosomes pair by adhering along their lengths in a process called *synapsis*. (This does not happen in mitosis.) This pairing process lasts from prophase I to the end of metaphase I. The four chromatids of each pair of homologous chromosomes form a **tetrad**, or *bivalent*. For example, in a human cell at the end of prophase I there are 23 tetrads, each consisting of four chromatids. The four chromatids come from the two partners in each homologous pair of chromosomes.

Throughout prophase I and metaphase I, the chromatin continues to coil and compact and the chromosomes become more condensed. At a certain point, the homologous chromosome pairs appear to repel each other, especially near the centromeres, but they remain attached. The X-shaped attachment points are called *chiasmata* (singular *chiasma*, "cross"):

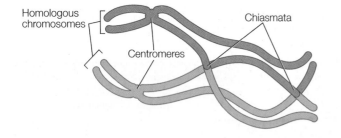

A chiasma is a point where *genetic material is exchanged* between nonsister chromatids on homologous chromosomes—a process called **crossing over** (FIGURE 7.13). Any of the four chromatids in the tedrad can participate in this exchange, and a single chromatid can exchange material at more than one point along its length. Crossing over occurs shortly after synapsis begins, but chiasmata do not become visible until later, when the homologs are repelling each other. Crossing over results in **recombinant** chromatids, and it increases genetic variation among the products of meiosis by reshuffling genetic information between homologous chromosome pairs. In Concept 8.3 we explore further the genetic consequences of crossing over.

Mitosis seldom takes more than an hour or two, but meiosis can take *much* longer. In human males, the cells in the testis that undergo meiosis take about a week for prophase I and about a month for the entire meiotic cycle. In females, prophase I begins long before a woman's birth, during her early fetal development, and ends as much as decades later, during the monthly ovarian cycle.

INDEPENDENT ASSORTMENT In addition to crossing over, meiosis provides a second source of genetic diversity. It is a matter of chance which member of a homologous pair goes to which daughter cell at anaphase I. For example, imagine a diploid organism with two pairs of homologous chromosomes (pairs 1 and 2). One member of each pair came from the male parent of the organism (paternal 1 and 2), and the other member of each pair came from the female parent (maternal 1 and 2). When cells in this organism undergo meiosos, a particular daughter nucleus could receive paternal 1 and maternal 2, paternal 2 and maternal 1, both maternal, or both paternal chromosomes. It all depends on how the homologous pairs line up at metaphase I. This phenomenon is termed **independent assortment**.

Note that of the four possible outcomes just described, only two daughter nuclei receive the full complements of either maternal or paternal chromosomes (apart from the material exchanged by crossing over). *The greater the number of chromosomes, the lower the probability of reestablishing the original parental combinations, and the greater the potential for genetic diversity.* Most species of diploid organisms have more than two pairs of chromosomes. In humans, with 23 chromosome pairs, 2^{23} *(8,388,608) different combinations of maternal and paternal*

APPLY THE CONCEPT

Meiosis halves the nuclear chromosome content and generates diversity

In the anther (male sex organ) of the lily plant, cells undergo mitosis in synchrony. These cells can be removed and studied in the laboratory. An antibody was developed that would bind specifically to a chromosomal protein called B1. A procedure involving antibody binding and staining of the antibody was used to detect the protein in the anther cells. The protein was detected at the centromeres of mitotic chromosomes, and its presence or absence was monitored at different stages of the cell cycle. The results were: G1, absent; early S, absent; late S through metaphase, present; anaphase, absent; telophase, absent.

1. Propose a role for this protein in sister chromatid function.

2. What results would you expect if meiotic cells were examined for this protein? At what stages of meiosis would the protein be present or absent?

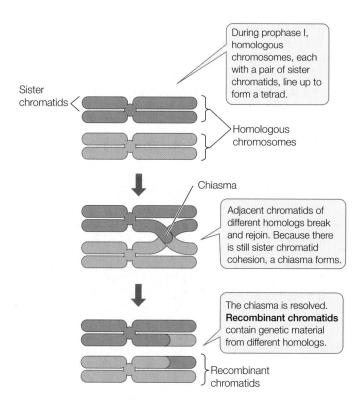

Sister chromatids

During prophase I, homologous chromosomes, each with a pair of sister chromatids, line up to form a tetrad.

Homologous chromosomes

Chiasma

Adjacent chromatids of different homologs break and rejoin. Because there is still sister chromatid cohesion, a chiasma forms.

The chiasma is resolved. **Recombinant chromatids** contain genetic material from different homologs.

Recombinant chromatids

FIGURE 7.13 Crossing Over Forms Genetically Diverse Chromosomes The exchange of genetic material by crossing over results in new combinations of genetic information on the recombinant chromosomes. The two different colors distinguish the chromosomes contributed by the male and female parents of the organism whose cell is undergoing meiosis.

chromosomes can be produced just by the mechanism of independent assortment! Taking the extra genetic shuffling afforded by crossing over into account, the number of possible combinations is virtually infinite. Crossing over and independent assortment, along with the processes that result in mutations, provide the genetic diversity needed for evolution by natural selection.

We have seen how meiosis I is fundamentally different from mitosis. However, meiosis II is similar to mitosis, in that it involves the separation of chromatids into daughter nuclei (see steps 7–11 in Figure 7.12). The final products of meiosis I and meiosis II are four haploid daughter cells, each with one set (*n*) of chromosomes.

Meiotic errors lead to abnormal chromosome structures and numbers

In the complex process of meiosis, things occasionally go wrong. For example, chromosomes may break or homologs may fail to separate at anaphase I or anaphase II. The gametes formed from meiotic errors have abnormal chromosomes, and when they take part in fertilization, the consequences for offspring can be significant.

NONDISJUNCTION Occasionally, a homologous chromosome pair fails to separate (fails to *disjoin*) at anaphase I, or a pair

of chromatids fail to separate at anaphase II. This failure to separate is referred to as **nondisjunction**. If a chromosome pair fails to separate at anaphase I, two of the four daughter nuclei will each end up with both members of that homologous pair, and the other two will have neither member of the pair. If nondisjunction occurs at anaphase II, only two of the four daughter nuclei will be affected: one will have an extra chromosome and the other will have less than the full complement of chromosomes. Using humans as an example, if during anaphase I the two homologs of chromosome 10 fail to separate, half the gametes will have two copies of chromosome 10, with a total of 24 chromosomes instead of 23. If one of these gametes fuses with a normal gamete during fertilization, the zygote will have 47 (23 + 24) chromosomes, with three copies of chromosome 10. The condition of having an extra chromosome (*trisomy*) or missing a chromosome (*monosomy*) is called **aneuploidy**.

For reasons that are unclear, aneuploidy is a common and harmful condition in humans. About 10–30 percent of all conceptions show aneuploidy, but most of the embryos that develop from such zygotes do not survive to birth, and those that do often die before the age of 1 year. At least one-fifth of all recognized human pregnancies are spontaneously terminated (miscarried) during the first 2 months, largely because of trisomies and monosomies. The actual proportion of spontaneously terminated pregnancies is certainly higher, because the earliest ones often go unrecognized. The most common form of aneuploidy in humans is trisomy 16 (three copies of chromosome 16), but almost none of these embryos survive to birth. Among the few aneuploidies that allow survival is Down syndrome—trisomy 21. Such individuals generally have mental retardation, but can lead long and productive lives.

POLYPLOIDY Most organisms are either diploid (for example, most animals) or haploid (for example, most fungi). Under some circumstances, triploid (3*n*), tetraploid (4*n*), or higher-order **polyploid** nuclei may form. This can occur in a variety of ways. For example, there could be an extra round of DNA replication preceding meiosis, or there could be no spindle formed in meiosis II. Polyploidy occurs naturally in some animals and in many plants, and in some cases it has probably led to speciation—the evolution of a new species.

A diploid nucleus can undergo normal meiosis because there are two sets of chromosomes to make up homologous pairs, which separate during anaphase I. Similarly, a tetraploid nucleus has an even number of each kind of chromosome, so each chromosome can pair with its homolog. However, a triploid nucleus cannot undergo normal meiosis because one-third of the chromosomes would lack partners. Polyploidy has implications for agriculture, particularly in the production of hybrid plants. For example, ploidy must be taken into account in wheat breeding because there are diploid, tetraploid, and hexaploid wheat varieties. Polyploidy can be a desirable trait in crops and ornamental plants because it often leads to more robust plants with larger flowers, fruits, and seeds. In addition, triploid fruit varieties are desirable because they are infertile and therefore seedless.

FRONTIERS Polyploid plants, produced either naturally or in the laboratory, often have larger vegetative organs, flowers, or seeds—desirable characteristics for crops and ornamental plants. But for reasons not fully understood, some polyploid plants show no increased vigor. Scientists are analyzing the DNA and proteins of polyploid plants to shed light on the mechanisms of polyploidy vigor.

TRANSLOCATION During crossing over in meiosis I, chromatids from homologous chromosome pairs break and rejoin. Occasionally this can happen between *non-homologous chromosomes*. The result is a **translocation**, and these are quite common, even in mitotic cells. As we point out in our discussion of gene expression and its regulation in Chapters 10 and 11, the location of genes relative to other DNA sequences is important, and translocations can have profound effects on gene expression.

An example of a translocation that is known to occur in humans is the swap of material between chromosomes 9 and 22:

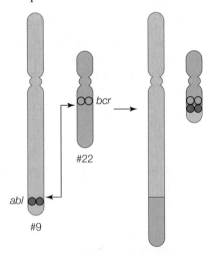

In this case, the DNA sequence *abl* on chromosome 9 comes to lie adjacent to the sequence *bcr* on chromosome 22. If the translocation occurs in a mitotic cell forming white blood cells, the result of this combination is a form of leukemia, a cancer of white blood cells. If a translocation occurs during meiosis, the gametes that result may carry it and pass it on to offspring after fertilization.

Do You Understand Concept 7.4?

- How do crossing over and independent assortment during meiosis result in daughter nuclei that differ genetically?
- What are the differences between meiosis and mitosis?
- An organism has a diploid number of 6. How many chromosomes are present in the following cells: A gamete? A gamete with monosomy of chromosome 2? A liver cell? A sperm cell at meiotic anaphase II?

An essential role of cell division in complex eukaryotes is to replace cells that die. What causes cells to die?

concept 7.5 **Programmed Cell Death Is a Necessary Process in Living Organisms**

Cells die in one of two ways. The first type of cell death, **necrosis**, occurs when cells are damaged by mechanical means or toxins, or are starved of oxygen or nutrients. These cells often swell up and burst, releasing their contents into the extracellular environment. This process often results in inflammation (see Concept 31.1).

APPLY THE CONCEPT

Programmed cell death is a necessary process in living organisms

The DNA content of an individual cell can be measured by applying a DNA-specific dye to the cell and then passing it through an instrument that measures the staining intensity. A new drug was tested on a population of rapidly dividing tumor cells and their DNA contents analyzed and compared with those of untreated cells:

DYE INTENSITY	% OF UNTREATED CELLS	% OF TREATED CELLS
<10	0	20
10	10	5
20	55	60
30	5	5
40	30	10
50	0	0

1. Plot percentage of cells versus DNA content for the untreated and treated cells.
2. Explain the data for the untreated cells. Which cells are in G1? What do the data indicate about how much time cells spend in G1 relative to other phases?
3. Explain the data for treated cells and compare them with untreated cells. At what stage of the cell cycle do you think the new drug acts?

More typically, cell death is due to **apoptosis** (Greek, "falling apart"). Apoptosis is a *genetically programmed series of events that result in cell death*. Why would a cell initiate apoptosis, which is essentially cell suicide? In animals, there are two possible reasons:

- *The cell is no longer needed by the organism.* For example, before birth, a human fetus has weblike hands, with connective tissue between the fingers. As development proceeds, this unneeded tissue disappears as its cells undergo apoptosis in response to specific signals.

- *The longer cells live, the more prone they are to genetic damage that could lead to cancer.* This is especially true of epithelial cells on the surface of an organism, which may be exposed to radiation or toxic substances. Such cells normally die after only days or weeks and are replaced by new cells.

The events of apoptosis are similar in many organisms. The cell becomes detached from its neighbors, hydrolyzes its DNA into small fragments, and forms membranous lobes, or "blebs," that break up into cell fragments (**FIGURE 7.14A**). In a remarkable example of the economy of nature, the surrounding living cells usually ingest the remains of the dead cell by phagocytosis. The remains are digested in the lysosomes, and the digestion products are recycled.

Apoptosis is also used by plant cells in an important defense mechanism called the *hypersensitive response*. Plants can protect themselves from disease by undergoing apoptosis at the site of infection by a fungus or bacterium. With no living tissue to grow in, the invading organism is not able to spread to other parts of the plant. Because of their rigid cell walls, plant cells do not form blebs the way animal cells do. Instead, they digest their own cell contents in the vacuole and then release the digested components into the vascular system.

Despite these differences between plant and animal cells, they share many of the signal transduction pathways that lead to apoptosis. Like the cell division cycle, programmed cell death is controlled by signals, which may come from inside or outside the cell (**FIGURE 7.14B**). Internal signals may be linked to the age of the cell or the recognition of damaged DNA. External signals (or a lack of them) can cause a receptor protein in the plasma membrane to change its shape, and in turn activate a signal transduction pathway. Both internal and external signals can lead to the activation of a class of enzymes called **caspases**. These enzymes are proteases that hydrolyze target molecules in a cascade of events. As a result, the cell dies as the caspases hydrolyze proteins of the nuclear envelope, nucleosomes, and plasma membrane.

Do You Understand Concept 7.5?

- What are some differences between apoptosis and necrosis?

- Give examples of when apoptosis is necessary in animals and in plants.

- In the worm *Caenorhabditis elegans* the fertilized egg divides by mitosis to produce 1,090 somatic cells. But the adult worm has only 959 cells. What happens to the 131 other cells formed during worm embryo development? What might happen if the 131 cells did not undergo this process?

Q&A
QUESTION How does infection with HPV result in uncontrolled cell reproduction?

ANSWER Human papillomavirus (HPV) stimulates the cell cycle when it infects tissues lining the cervix. It does this by "hijacking" the regulatory mechanisms that control the

FIGURE 7.14 Apoptosis: Programmed Cell Death (A) Many cells are programmed to "self-destruct" when they are no longer needed, or when they have lived long enough to accumulate a burden of DNA damage that might harm the organism. (B) Both external and internal signals stimulate caspases, the enzymes that break down specific cell constituents, resulting in apoptosis.

(A)

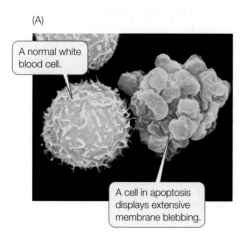

A normal white blood cell.

A cell in apoptosis displays extensive membrane blebbing.

(B)

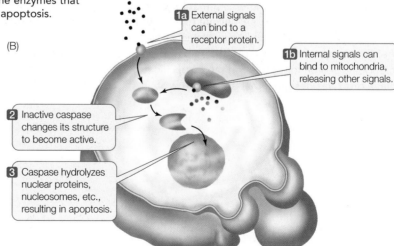

1a External signals can bind to a receptor protein.

1b Internal signals can bind to mitochondria, releasing other signals.

2 Inactive caspase changes its structure to become active.

3 Caspase hydrolyzes nuclear proteins, nucleosomes, etc., resulting in apoptosis.

cell cycle (Concept 7.3). There are two types of proteins that regulate the cell cycle:

- **Oncogene** proteins are positive regulators of the cell cycle in cancer cells. They are derived from normal positive regulators that have become mutated to be overly active, or that are present in excess, and they stimulate the cancer cells to divide more often. An example of an oncogene protein is the growth factor receptor in a breast cancer cell (**FIGURE 7.15A**). Normal breast cells have relatively low numbers of the growth factor receptor *human epidermal growth factor receptor 2* (HER2). So the growth factor does not normally find many breast cell receptors with which to bind and initiate cell division. In about 25 percent of breast cancers, a DNA change results in the increased production of HER2. This results in positive stimulation of the cell cycle, and a rapid proliferation of cells with the altered DNA.

- **Tumor suppressors** are negative regulators of the cell cycle in normal cells, but in cancer cells they are inactive. An example is the retinoblastoma (RB) protein that acts at R (the restriction point) in G1 (see Figure 7.8). When RB is active the cell cycle does not proceed, but it is inactive in cancer cells, allowing the cell cycle to occur (**FIGURE 7.15B**). *This is where HPV hijacks the system.* When it infects cells lining the cervix, HPV causes the synthesis of a protein called E7, which has a three-dimensional shape that just fits into the protein-binding site of RB, thereby inactivating it. With no active RB to prevent it, cell division proceeds. Uncontrolled cell reproduction is a hallmark of cancer—and so cervical cancer begins.

Most tumors are treated by surgery. But when a tumor has spread from its original site (a common occurrence, unfortunately), surgery does not cure it. Instead, drugs—chemotherapy—are used. Generally, these drugs stop cell division by targeting specific cell cycle events (Concepts 7.2 and 7.3). For example, some drugs block DNA replication (e.g., 5-fluorouracil); others damage DNA, stopping the cells at G2 (e.g., etoposide); and still others prevent the normal functioning of the mitotic spindle (e.g., paclitaxel). Many of these drugs do not kill the cell, but they cause the cell cycle to stop, and the damaged cell is stimulated to undergo apoptosis (Concept 7.5).

A major problem with these treatments is that they target normal cells as well as the tumor cells. They are toxic to tissues with large populations of normal dividing cells such as those in the intestine, skin, and bone marrow (producing blood cells). There is an ongoing search for better and more specific drugs. For example, a drug was designed that affects the protein

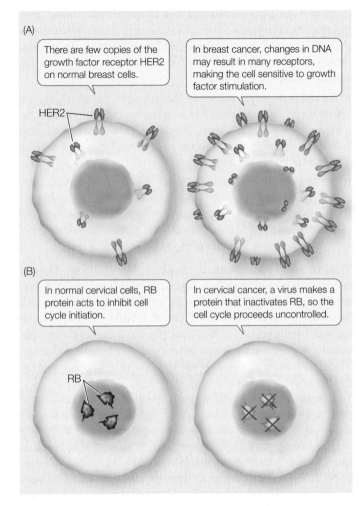

FIGURE 7.15 Molecular Changes Regulate the Cell Cycle in Cancer Cells In cancer cells, oncogene proteins become active (A) and tumor suppressor proteins become inactive (B).

produced from the translocated chromosome (Concept 7.4) that causes leukemia. The drug is rather specific and has been very successful at treating this tumor.

In this chapter we examined the cell cycle and cell division by binary fission and mitosis. We have seen how the normal cell cycle is disrupted in cancer. We also examined meiosis and the production of haploid cells in sexual life cycles. In the coming chapters we examine heredity, genes, and DNA. In Concept 8.1 we discuss Gregor Mendel's studies of heredity, and how the enormous power of his discoveries founded the science of genetics.

7 SUMMARY

concept 7.1 Different Life Cycles Use Different Modes of Cell Reproduction

- Cell division is necessary for the reproduction, growth, and repair of organisms. **Review Figure 7.1**

- **Asexual reproduction** produces clones, new organisms that are virtually identical genetically to the parent. Any genetic variation is the result of mutations.

- In **sexual reproduction**, two **haploid** gametes—one from each parent—unite in **fertilization** to form a genetically unique, **diploid zygote**. There are many different sexual life cycles that can be haplontic, diplontic, or involve **alternation of generations**. **Review Figure 7.3 and WEB ACTIVITY 7.1**

- Diploid cells contain **homologous pairs** of **chromosomes**. In sexually reproducing organisms, certain cells undergo **meiosis**, a process of cell division in which the chromosome number is halved. Each of the haploid daughter cells contains one member of each homologous pair of chromosomes.

concept 7.2 Both Binary Fission and Mitosis Produce Genetically Identical Cells

- Cell division must be initiated by a signal. Before a cell can divide, the genetic material (DNA) must undergo **replication** and **segregation** to separate portions of the cell. **Cytokinesis** then divides the cytoplasm into two cells.

- In prokaryotes, most cellular DNA is a single molecule, usually in the form of a circular chromosome. Prokaryotes reproduce by **binary fission**. **Review Figure 7.4**

- During most of the eukaryotic cell cycle, the cell is in **interphase**, which is divided into three subphases: **S**, **G1**, and **G2**. DNA is replicated during S phase. **Mitosis** (**M** phase) and cytokinesis follow. **Review Figure 7.5 and ANIMATED TUTORIAL 7.1**

- In mitosis, a single nucleus gives rise to two nuclei that are genetically identical to each other and to the parent nucleus.

- At mitosis, the replicated chromosomes, called **sister chromatids**, are held together at the **centromere**. Each chromatid contains one double-stranded DNA molecule. During mitosis, sister chromatids line up at the equatorial plate and attach to the spindle. **Review WEB ACTIVITY 7.2**

- Mitosis can be divided into several phases called **prophase**, **prometaphase**, **metaphase**, **anaphase**, and **telophase**. **Review Figure 7.6 and WEB ACTIVITY 7.3**

- Nuclear division is usually followed by cytokinesis. Animal cell cytoplasms divide via a contractile ring made up of actin microfilaments. In plant cells, cytokinesis is accomplished by vesicles that fuse to form a cell plate. **Review Figure 7.7**

concept 7.3 Cell Reproduction Is Under Precise Control

- Interactions between **cyclins** and **Cdk's** regulate the passage of cells through checkpoints in the cell cycle. External controls such as **growth factors** can stimulate the cell to begin a division cycle. **Review Figure 7.10**

concept 7.4 Meiosis Halves the Nuclear Chromosome Content and Generates Diversity

- Meiosis consists of two nuclear divisions, **meiosis I** and **meiosis II**, which collectively reduce the chromosome number from diploid to haploid. Meiosis results in four genetically diverse haploid cells, often gametes. **Review ANIMATED TUTORIAL 7.2**

- In meiosis I, entire chromosomes, each with two chromatids, migrate to the poles. In meiosis II, the sister chromatids separate. **Review Figures 7.11 and 7.12 and WEB ACTIVITY 7.4**

- During prophase I, homologous chromosomes undergo synapsis to form pairs in a **tetrad**. Chromatids can form junctions called chiasmata, and genetic material may be exchanged between the two homologs by **crossing over**. **Review Figure 7.13**

- Both crossing over during prophase I and **independent assortment** of the homologs as they separate during anaphase I ensure that gametes are genetically diverse.

- Meiotic errors can result in abnormal numbers of chromosomes in the resulting gametes and offspring.

concept 7.5 Programmed Cell Death Is a Necessary Process in Living Organisms

- A cell may die by **necrosis**, or it may self-destruct by **apoptosis**, a genetically programmed series of events that includes the fragmentation of its nuclear DNA.

- Apoptosis is regulated by external and internal signals. These signals result in activation of a class of enzymes called **caspases** that hydrolyze proteins in the cell. **Review Figure 7.14**

Inheritance, Genes, and Chromosomes

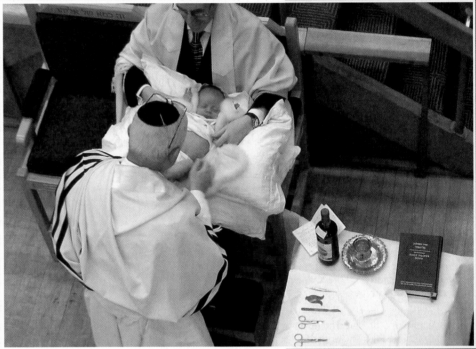

In the Middle Eastern desert 1,800 years ago, a rabbi faced a dilemma. A Jewish woman had given birth to a son. As required by ancient Jewish custom, the mother brought her 8-day-old son to the rabbi for ritual penile circumcision. The rabbi knew that the woman's two previous sons had bled to death when their foreskins were cut. Yet the biblical requirement remained: unless he was circumcised, the boy could not be counted among those with whom their God had made a solemn covenant. After consultation with other rabbis, it was decided to exempt this third son.

Almost 1,000 years later, in the twelfth century, the physician and biblical commentator Moses Maimonides reviewed this and other cases in the rabbinical literature and stated that in such instances the third son should not be circumcised. Furthermore, the exemption should apply whether the mother's son was "from her first husband or from her second husband." The bleeding disorder, he reasoned, was clearly carried by the mother and passed on to her sons. In all cases, the parents did not show any evidence of having the disease.

Without any knowledge of our modern concepts of genes and genetics, the rabbis had linked a human disease with a *pattern of inheritance*. We now have a name for the disease, hemophilia A, which affects about 18,000 people in the United States—mostly males. The bleeding disorder is due to the absence of a specific protein called factor VIII, which is important in the formation of blood clots. When a person without hemophilia gets cut there is usually some bleeding, but then a clot forms to prevent further bleeding. In the case of hemophilia, the bleeding can continue until the person dies. Indeed, well into the twentieth century

the slightest accident could be lethal in such a person. Internal bleeding is also an extremely serious problem for people with this disease, and permanent joint damage due to bleeding in the joints is a common problem for untreated patients.

Treatment of hemophilia A by injection of factor VIII into the bloodstream is now possible, because factor VIII can be isolated from donated blood or made in the laboratory using biotechnology. An issue has been whether people with hemophilia A should be injected with factor VIII all the time as a preventive measure (an expensive proposition) or just take it as needed. Based on reductions in joint damage in children treated by the preventive approach, recent studies have concluded that this approach is best.

A male infant undergoes ritual circumcision in accordance with Jewish laws. Sons of Jewish mothers who carry the gene for hemophilia may be exempted from this ritual.

 QUESTION
How is hemophilia inherited through the mother, and why is it more frequent in males?

KEY CONCEPTS

8.1 Genes Are Particulate and Are Inherited According to Mendel's Laws

8.2 Alleles and Genes Interact to Produce Phenotypes

8.3 Genes Are Carried on Chromosomes

8.4 Prokaryotes Can Exchange Genetic Material

concept 8.1 Genes Are Particulate and Are Inherited According to Mendel's Laws

Genetics, the field of biology concerned with inheritance, has a long history. There is good evidence that people were deliberately breeding animals (horses) and plants (the date palm tree) for desirable characteristics as long as 5,000 years ago. The general idea was to examine the natural variation among the individuals of a species and "breed the best to the best and hope for the best." This was a hit-or-miss method—sometimes the resulting offspring had all the good characteristics of the parents, but often they did not.

By the mid-nineteenth century, two theories had emerged to explain the results of breeding experiments:

- The theory of *blending inheritance* proposed that gametes contained hereditary determinants (what we now call genes) that blended when the gametes fused during fertilization. Like inks of different colors, the two different determinants lost their individuality after blending and could never be separated. For example, if a plant that made smooth, spherical seeds was mated (crossed) with a plant that made wrinkled seeds, the offspring would be intermediate between the two and the determinants for the two parental characteristics would be lost.

- The theory of *particulate inheritance* proposed that each determinant had a physically distinct nature; when gametes fused in fertilization, the determinants remained intact. According to this theory, if a plant that made spherical seeds was crossed with a plant that made wrinkled seeds, the offspring (no matter the shape of their seeds) would still contain the determinants for the two characteristics.

The story of how these competing theories were tested provides a great example of how the *scientific method* can be used to support one theory and reject another. In the following sections we will look in detail at experiments performed in the 1860s by an Austrian monk–scientist, Gregor Mendel, whose work clearly supported the particulate theory.

Mendel used the scientific method to test his hypotheses

After entering the priesthood at a monastery in Brno, in what is now the Czech Republic, Gregor Mendel was sent to the University of Vienna, where he studied biology, physics, and mathematics. He returned to the monastery in 1853 to teach. The abbot in charge had set up a small plot of land to do experiments with plants and encouraged Mendel to continue with them. Over seven years, Mendel made crosses with many thousands of plants. Analysis of his meticulously gathered data suggested to him that inheritance was due to particulate factors.

Mendel presented his theories in a public lecture in 1865 and a detailed written publication in 1866, but his work was ignored by mainstream scientists until 1900. By that time, the discovery of chromosomes had suggested to biologists that genes might be carried on chromosomes. When they read Mendel's work on particulate inheritance, the biologists connected the dots between genes and chromosomes.

Mendel chose to study the common garden pea because of its ease of cultivation and the feasibility of making controlled crosses. Pea flowers have both male and female sex organs: stamens and pistils, which produce gametes contained within the pollen tubes and ovules, respectively:

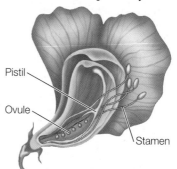

Pea flowers normally self-fertilize. However, the male organs can be removed from a flower so that it can be manually fertilized with pollen from a different flower.

There are many varieties of pea plants with easily recognizable characteristics. A **character** is an observable physical feature, such as seed shape. A **trait** is a particular form of a character, such as spherical or wrinkled seeds. Mendel worked with seven pairs of varieties with contrasting traits for characters such as seed shape, seed color, and flower color. These varieties were *true-breeding*: that is, when he crossed a plant that produced wrinkled seeds with another of the same variety, all of the offspring plants produced wrinkled seeds.

Mendel performed his crosses in the following manner:

- He removed the stamens from (emasculated) flowers of one parental variety so that it couldn't self-fertilize. Then he collected pollen from another parental variety and placed it on the pistils of the emasculated flowers. The plants providing and receiving the pollen were the **parental generation**, designated **P**.

- In due course, seeds formed and were planted. The seeds and the resulting new plants constituted the **first filial generation**, or F_1. (The word "filial" refers to the relationship between offspring and parents, from the Latin *filius*, "son.") Mendel examined each F_1 plant to see which traits it bore and then recorded the number of F_1 plants expressing each trait.

- In some experiments the F_1 plants were allowed to self-pollinate and produce a **second filial generation**, the F_2. Again, each F_2 plant was characterized and counted.

Mendel's first experiments involved monohybrid crosses

The term *hybrid* refers to the offspring of crosses between organisms differing in one or more characters. In Mendel's first experiments, he crossed parental (P) varieties with contrasting traits for a single character, producing monohybrids (from the Greek *monos*, "single") in the F_1 generation. He subsequently planted the F_1 seeds and allowed the resulting plants to self-pollinate to produce the F_2 generation. This technique is referred to as a **monohybrid cross**.

Mendel performed the same experiment for seven pea characters. His method is illustrated in **FIGURE 8.1**, using seed shape as an example. When he crossed a strain that made spherical seeds with one that made wrinkled seeds, the F_1 seeds were spherical—it was as if the wrinkled seed trait had disappeared completely. However, when F_1 plants were allowed to self-pollinate to produce F_2 seeds, about one-fourth of the seeds were wrinkled. These observations were key to distinguishing the two theories noted above:

- The F_1 offspring were not a blend of the two traits of the parents. Only one of the traits was present (in this case, spherical seeds).
- Some F_2 offspring had wrinkled seeds. The trait had not disappeared.

These observations led to a rejection of the blending theory of inheritance and provided support for the particulate theory. We now know that hereditary determinants are not actually "particulate," but they are physically distinct entities: sequences of DNA carried on chromosomes (see Concept 8.3).

All seven crosses between varieties with contrasting traits gave the same kind of data (see Figure 8.1). In the F_1 generation only one of the two traits was seen, but the other one reappeared in about one-fourth of the offspring in the F_2 generation. Mendel called the trait that appeared in the F_1 and was more abundant in the F_2 the **dominant** trait, and the other trait **recessive**.

Mendel went on to expand on the particulate theory. He proposed that hereditary determinants—we will call them *genes* here, though Mendel did not use that term—occur in pairs and segregate (separate) from one another during the formation of gametes. He concluded that each pea plant has two genes for each character (such as seed shape), one inherited from each parent. We now use the term *diploid* to describe the state of having two copies of each gene; *haploids* have just a single copy.

> **LINK** The characteristics of haploid and diploid organisms are discussed in Concept 7.1

Mendel concluded that while each gamete contains one copy of each gene, the resulting zygote contains two copies, because it is produced by the fusion of two gametes. Furthermore, different traits arise from different forms of a gene (now called **alleles**) for a particular character. For example, Mendel studied two alleles for seed shape: one that caused smooth seeds and the other causing wrinkled seeds.

An organism that is **homozygous** for a gene has two alleles that are the same (for example, two copies of the allele for spherical seeds). An organism that is **heterozygous** for a gene has two different alleles (for example, one allele for spherical seeds and one allele for wrinkled seeds). In a heterozygote, one of the two alleles may be dominant (such as spherical, *S*) and the other recessive (wrinkled, *s*). By conven-

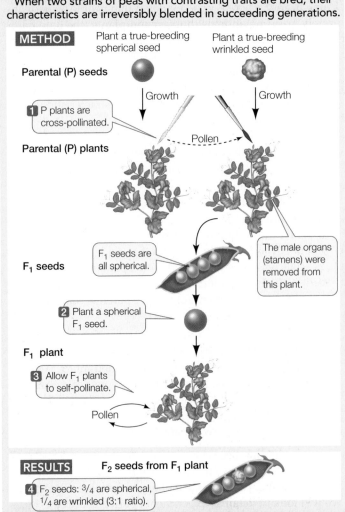

INVESTIGATION

FIGURE 8.1 Mendel's Monohybrid Experiments
Mendel performed crosses with pea plants and carefully analyzed the outcomes to show that genetic determinants are particulate.

HYPOTHESIS

When two strains of peas with contrasting traits are bred, their characteristics are irreversibly blended in succeeding generations.

METHOD
Plant a true-breeding spherical seed
Plant a true-breeding wrinkled seed

Parental (P) seeds

1 P plants are cross-pollinated.

| Growth | | Growth |
Pollen

Parental (P) plants

F_1 seeds

F_1 seeds are all spherical.

The male organs (stamens) were removed from this plant.

2 Plant a spherical F_1 seed.

F_1 plant

3 Allow F_1 plants to self-pollinate.

Pollen

RESULTS F_2 seeds from F_1 plant

4 F_2 seeds: 3/4 are spherical, 1/4 are wrinkled (3:1 ratio).

CONCLUSION

The hypothesis is rejected. No irreversible blending of characteristics, and a recessive trait can reappear in succeeding generations.

ANALYZE THE DATA

Here are Mendel's data (the number of offspring showing each trait) for the F_2 from crosses between plants with contrasting traits:

Characteristic	Dominant	Recessive
Seed shape	5474 spherical	1850 wrinkled
Seed color	6022 yellow	2001 green
Flower color	705 purple	224 white
Pod color	428 green	152 yellow
Stem height	787 tall	277 short

A. Calculate the phenotypic ratio of dominant:recessive in the F_2 offspring.
B. What can you conclude about the behavior of alleles during gamete formation in a plant that is heterozygous for a trait?
C. Perform a chi square analysis to evaluate the statistical significance of these data (refer to Appendix B).

Go to **yourBioPortal.com** for original citations, discussions, and relevant links for all INVESTIGATION figures.

tion, dominant alleles are designated with uppercase letters and recessive alleles with lowercase letters.

The physical appearance of an organism is its **phenotype**. Mendel proposed that the phenotype is the result of the **genotype**, or genetic constitution, of the organism showing the phenotype. Spherical seeds and wrinkled seeds are two phenotypes resulting from three possible genotypes: the wrinkled seed phenotype is produced by the genotype *ss*, whereas the spherical seed phenotype is produced by either of the genotypes *SS* or *Ss* (because the *S* allele is dominant to the *s* allele).

Mendel's first law states that the two copies of a gene segregate

How do Mendel's theories explain the proportions of traits seen in the F_1 and F_2 generations of his monohybrid crosses? Mendel's first law—the **law of segregation**—states that *when any individual produces gametes, the two copies of a gene separate, so that each gamete receives only one copy*. Thus, gametes from a parent with the *SS* genotype will all be *S*; gametes from an *ss* parent will all be *s*; and the progeny derived from a cross between these parents will all be *Ss*, producing seeds with a smooth phenotype (**FIGURE 8.2**).

Now let's consider the composition of the F_2 generation. Because the alleles segregate, half of the gametes produced by the F_1 generation will have the *S* allele and the other half will have the *s* allele. What genotypes are produced when these gametes fuse to form the next (F_2) generation?

The allele combinations that will result from a cross can be predicted using a **Punnett square**, a method devised in 1905 by the British geneticist Reginald Punnett. This device ensures that we consider all possible combinations of gametes when calculating expected genotype frequencies. A Punnett square looks like this:

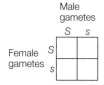

It is a simple grid with all possible male gamete (haploid sperm) genotypes shown along the top and all possible female gamete (haploid egg) genotypes along the left side. The grid is completed by filling in each square with the diploid genotype that can be generated from each combination of gametes. In this example, to fill in the top right square, we put in the *S* from the female gamete (the egg cell) and the *s* from the male gamete (the sperm cell in the pollen tube), yielding *Ss*.

Once the Punnett square is filled in, we can readily see that there are four possible combinations of alleles in the F_2 generation: *SS*, *Ss*, *sS*, and *ss* (see Figure 8.2). Since *S* is dominant, there are three ways to get spherical-seeded plants in the F_2 generation (*SS*, *Ss*, or *sS*), but only one way to get a plant with wrinkled seeds (*ss*). Therefore, we predict a 3:1 ratio of these phenotypes in the F_2 generation, remarkably close to the values Mendel found experimentally for all seven of the traits he compared.

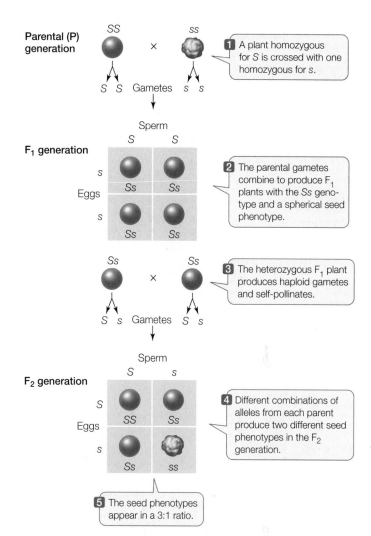

FIGURE 8.2 Mendel's Explanation of Inheritance Mendel concluded that inheritance depends on discrete factors from each parent that do not blend in the offspring.

Mendel did not live to see his theories placed on a sound physical footing with the discoveries of chromosomes and DNA. Genes are now known to be relatively short sequences of DNA (usually one to a few thousand base pairs in length) found on the much longer DNA molecules that make up chromosomes (which are often millions of base pairs long). Today we can picture the different alleles of a gene segregating as chromosomes separate during meiosis I (**FIGURE 8.3**).

LINK The process of meiosis is described in Concept 7.4

Genes determine phenotypes mostly by producing proteins with particular functions, such as enzymes. So in many cases a dominant gene is expressed (transcribed and translated) to produce a functional protein, while a recessive gene is mutated so that it is no longer expressed, or it encodes a mutant protein that is nonfunctional. For example, the molecular nature of the wrinkled pea seed phenotype is the absence of the enzyme SBE1, which is essential for starch synthesis. With less starch, the developing seed has more sucrose and this causes an inflow

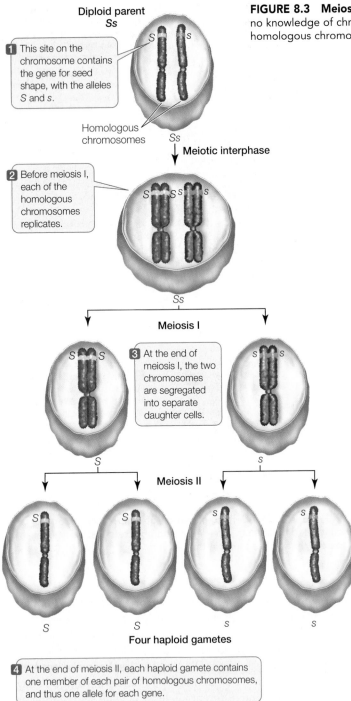

Diploid parent
Ss

1 This site on the chromosome contains the gene for seed shape, with the alleles *S* and *s*.

Homologous chromosomes

Ss

↓ Meiotic interphase

2 Before meiosis I, each of the homologous chromosomes replicates.

Ss

Meiosis I

3 At the end of meiosis I, the two chromosomes are segregated into separate daughter cells.

S | *S*

Meiosis II

S | *S* | *S* | *S*

Four haploid gametes

4 At the end of meiosis II, each haploid gamete contains one member of each pair of homologous chromosomes, and thus one allele for each gene.

FIGURE 8.3 Meiosis Accounts for the Segregation of Alleles Although Mendel had no knowledge of chromosomes or meiosis, we now know that a pair of alleles resides on homologous chromosomes, and that those alleles segregate during meiosis.

the spherical phenotype. Mendel verified this hypothesis by performing test crosses with F_1 seeds derived from a variety of other crosses. A **test cross** is used to determine whether an individual showing a dominant trait is homozygous or heterozygous. The individual in question is crossed with an individual that is homozygous for the recessive trait—an easy individual to identify, because all individuals with the recessive phenotype are homozygous for that trait.

The recessive homozygote for the seed shape gene has wrinkled seeds and the genotype *ss*. The individual being tested may be described initially as *S_* because we do not yet know the identity of the second allele. We can predict two possible results:

- If the individual being tested is homozygous dominant (*SS*), all offspring of the test cross will be *Ss* and show the dominant trait (spherical seeds) (**FIGURE 8.4, LEFT**).
- If the individual being tested is heterozygous (*Ss*), then approximately half of the offspring of the test cross will be heterozygous and show the dominant trait (*Ss*), and the other half will be homozygous for the recessive trait (*ss*) (**FIGURE 8.4, RIGHT**).

Mendel obtained results consistent with both of these predictions; thus his hypothesis accurately predicted the results of his test crosses.

Mendel's second law states that copies of different genes assort independently

Consider an organism that is heterozygous for two genes (*SsYy*). In this example, the dominant *S* and *Y* alleles came from one true-breeding parent, and the recessive *s* and *y* alleles came from the other true-breeding parent. When this organism produces gametes, do the *S* and *Y* alleles always go together in one gamete, and *s* and *y* alleles in another? Or can a single gamete receive one recessive and one dominant allele (*S* and *y* or *s* and *Y*)?

Mendel performed another series of experiments to answer these questions. He began with peas that differed in *two* characters: seed shape and seed color. One parental variety produced only spherical, yellow seeds (*SSYY*), and the other produced only wrinkled, green ones (*ssyy*). A cross between these two varieties produced an F_1 generation in which all the plants were *SsYy*. Because the *S* and *Y* alleles were dominant, the F_1 seeds were all spherical and yellow.

Mendel continued this experiment into the F_2 generation by performing a **dihybrid cross**—a cross between individuals that are identical *double heterozygotes*. In this case, he simply allowed the F_1 plants, which were all double heterozygotes, to self-pollinate (**FIGURE 8.5**). Depending on whether the alleles

of water by osmosis. When the seed matures and dries out, this water is lost, leaving a shrunken seed.

Mendel verified his hypotheses by performing test crosses

As mentioned above, Mendel arrived at his laws of inheritance by developing a series of hypotheses and then designing experiments to test them. One such hypothesis was that there are two possible allele combinations (*SS* or *Ss*) for seeds with

INVESTIGATION

FIGURE 8.4 Homozygous or Heterozygous? An individual with a dominant phenotype may have either a homozygous or a heterozygous genotype. The test cross determines which.

HYPOTHESIS

The progeny of a test cross can reveal whether an organism is homozygous or heterozygous.

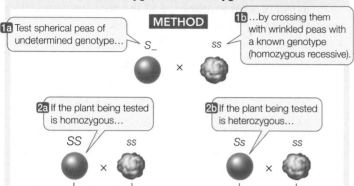

METHOD

1a Test spherical peas of undetermined genotype…

1b …by crossing them with wrinkled peas with a known genotype (homozygous recessive).

$S\_$ × ss

2a If the plant being tested is homozygous…

SS × ss

2b If the plant being tested is heterozygous…

Ss × ss

S S s s Gametes S s s s

RESULTS

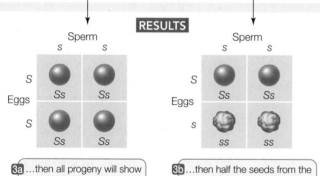

Sperm

	s	s
S	Ss	Ss
S	Ss	Ss

Eggs

Sperm

	s	s
S	Ss	Ss
s	ss	ss

Eggs

3a …then all progeny will show the dominant phenotype (spherical).

3b …then half the seeds from the cross will be wrinkled, and half will be spherical.

CONCLUSION

The plant being tested is homozygous.

CONCLUSION

The plant being tested is heterozygous.

ANALYZE THE DATA

A tall plant was crossed with a short plant to produce three F_1 plants (A, B, and C). Each of the F_1 plants was test crossed with these results:

F_1	Tall progeny	Short progeny
A	34	0
B	18	15
C	21	0

A. For each test cross, what were the genotypes and phenotypes of both parents?
B. Choose and perform a statistical test from Appendix B to evaluate the significance of these data.

Go to **yourBioPortal.com** for original citations, discussions, and relevant links for all INVESTIGATION figures.

yourBioPortal.com

Go to WEB ACTIVITY 8.1
Homozygous or Heterozygous?

---

of the two genes are inherited together or separately, there are two possible outcomes, as Mendel saw:

1. *The alleles could maintain the associations they had in the parental generation—they could be linked.* If this were the case, the F_1 plants would produce two types of gametes (SY and sy). The F_2 progeny resulting from self-pollination of these F_1 plants would consist of *two phenotypes*: spherical yellow and wrinkled green in the ratio of 3:1, just as in the monohybrid cross.

2. *The segregation of S from s could be independent of the segregation of Y from y—the two genes could be unlinked.* In this case, four kinds of gametes would be produced in equal numbers: SY, Sy, sY, and sy. When these gametes combine at random, they should produce an F_2 having nine different genotypes. The nine genotypes would produce *four phenotypes* (spherical yellow, spherical green, wrinkled yellow, wrinkled green). Putting these possibilities into a Punnett square, we can predict that these four phenotypes would occur in a ratio of 9:3:3:1.

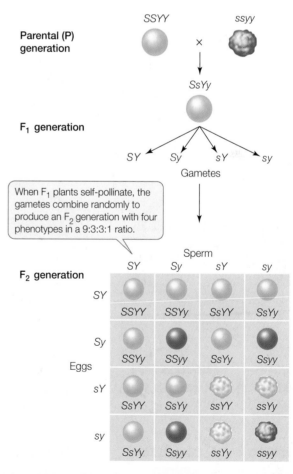

FIGURE 8.5 Independent Assortment The 16 possible combinations of gametes in this dihybrid cross result in nine different genotypes. Because S and Y are dominant over s and y, respectively, the nine genotypes result in four phenotypes in a ratio of 9:3:3:1. These results show that the two genes segregate independently.

Mendel's dihybrid crosses supported the second prediction: four different phenotypes appeared in the F$_2$ generation in a ratio of about 9:3:3:1 (see Figure 8.5). On the basis of such experiments, Mendel proposed his second law—the **law of independent assortment**: *alleles of different genes assort independently of one another during gamete formation*. In the example above, the segregation of the *S* and *s* alleles is independent of the segregation of the *Y* and *y* alleles. As you will see in Concept 8.3, this is not as universal as the law of segregation because it does not apply to genes located near one another on the same chromosome. However, it is correct to say that *chromosomes segregate independently* during the formation of gametes, and so do any two genes located on separate chromosome pairs (**FIGURE 8.6**).

Probability is used to predict inheritance

One of Mendel's major contributions to the science of genetics was his use of a quantitative approach and probability to analyze the large amounts of data he gathered. His mathematical analyses revealed clear patterns that allowed him to formulate his theories. Ever since his work became widely recognized, geneticists have used simple mathematics in the same ways

that Mendel did. Geneticists use probabilities to predict the ratios of genotypes and phenotypes in the progeny of a given cross or mating. They use statistics to determine whether the actual results match the prediction.

You can think of probabilities by considering a coin toss. For any given toss of a fair coin, the probability of heads is independent of what happened in all the previous tosses. The basic conventions of probability are simple:

- If an event is absolutely certain to happen, its probability is 1.
- If it cannot possibly happen, its probability is 0.
- All other events have a probability between 0 and 1.

A coin toss results in heads approximately half the time, so the probability of heads is $^1/_2$—as is the probability of tails.

If two coins (say a penny and a dime) are tossed, each acts independently of the other. What is the probability of both coins coming up heads? In half of the tosses, the penny comes up heads; the dime also comes up heads half of the time. Therefore, the *joint probability* of both coins coming up heads is half of one-half, or $^1/_2 \times ^1/_2 = ^1/_4$. So, to find the joint probability of independent events, we multiply the probabilities of the individual events (**FIGURE 8.7**).

Probability applies to a monohybrid cross (see Figure 8.2) in the same way. After the self-pollination of an *Ss* F$_1$ plant, the probability that an F$_2$ plant will have the genotype *SS* is $^1/_2 \times ^1/_2 = ^1/_4$, because the chance that the sperm will have the genotype *S* is $^1/_2$, and the chance that the egg will have the genotype *S* is also $^1/_2$. Similarly, the probability of *ss* offspring is also $^1/_4$.

You will note in Figures 8.2 and 8.7 that there are *two ways* to get an *Ss* plant or a head and a tail in a coin toss. In the case of the seed shape gene, the *S* allele can come from a sperm and the *s* from an egg (probability $^1/_4$). Or the *S* allele could come from the egg and the *s* from the sperm (probability $^1/_4$). *The probability of an event that can occur in two or more different ways is the sum of the individual probabilities of those ways*. Thus the probability that an F$_2$ plant will be a heterozygote is equal to the sum of the probabilities of the two ways of forming a heterozygote: $^1/_4 + ^1/_4 = ^1/_2$.

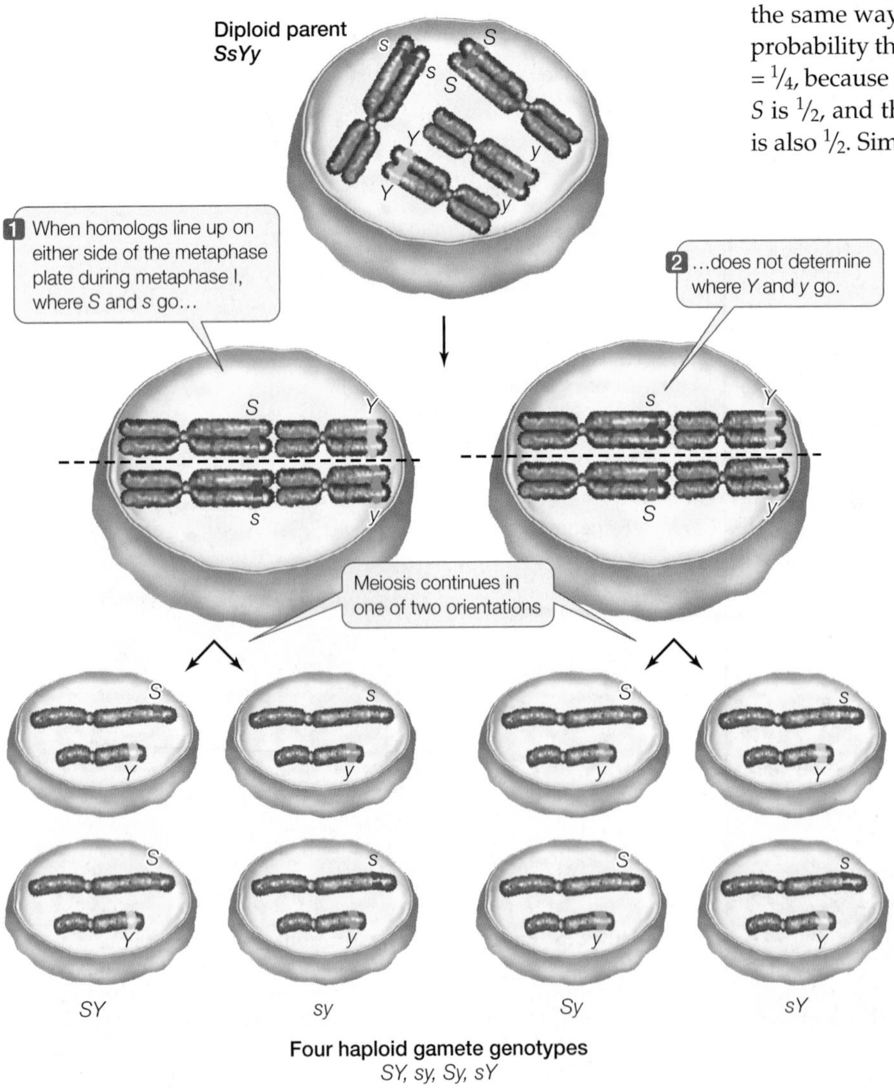

Diploid parent
SsYy

1 When homologs line up on either side of the metaphase plate during metaphase I, where *S* and *s* go...

2 ...does not determine where *Y* and *y* go.

Meiosis continues in one of two orientations

SY   *sy*   *Sy*   *sY*

Four haploid gamete genotypes
SY, sy, Sy, sY

FIGURE 8.6 Meiosis Accounts for Independent Assortment of Alleles We now know that copies of genes on different chromosomes are segregated independently during metaphase I of meiosis. Thus a parent of genotype *SsYy* can form gametes with four different genotypes.

yourBioPortal.com
Go to ANIMATED TUTORIAL 8.1
Independent Assortment of Alleles

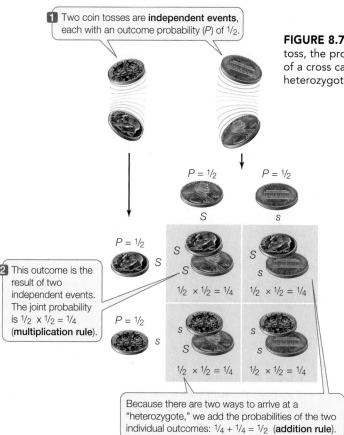

1 Two coin tosses are **independent events**, each with an outcome probability (*P*) of ½.

P = ½   *P* = ½

S   *s*

P = ½

S

2 This outcome is the result of two independent events. The joint probability is ½ × ½ = ¼ (**multiplication rule**).

P = ½

s

$S\,S$ ½ × ½ = ¼ $S\,s$ ½ × ½ = ¼

$S\,s$ ½ × ½ = ¼ $s\,s$ ½ × ½ = ¼

Because there are two ways to arrive at a "heterozygote," we add the probabilities of the two individual outcomes: ¼ + ¼ = ½ (**addition rule**).

FIGURE 8.7 Using Probability Calculations in Genetics Like the results of a coin toss, the probability of any given combination of alleles appearing in the offspring of a cross can be obtained by multiplying the probabilities of each event. Since a heterozygote can be formed in two ways, these two probabilities are added together.

Mendel's laws can be observed in human pedigrees

Mendel developed his theories by performing many planned crosses and counting many offspring. This approach is not possible with humans, so human geneticists rely on **pedigrees**: family trees that show the occurrence of inherited phenotypes in several generations of related individuals (**FIGURE 8.8**).

Because humans have relatively few offspring, human pedigrees do not show the clear proportions of phenotypes that Mendel saw in his pea plants. For example, when a man and a woman who are both heterozygous for a recessive allele (say, *Aa*) have children together, each child has a 25 percent probability of being a recessive homozygote (*aa*). But the offspring of a single couple are likely to be too few to reliably show the one-fourth proportion. In a family with only two children, for example, both could easily be *aa* (or *Aa*, or *AA*).

Despite this limitation, pedigrees do show inheritance patterns that can provide information about the allele(s)

yourBioPortal.com

**Go to INTERACTIVE TUTORIAL 8.1
Pedigree Analysis**

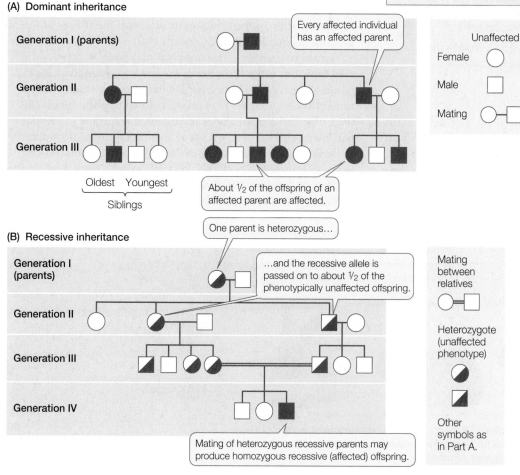

(A) Dominant inheritance

Generation I (parents)

Every affected individual has an affected parent.

Generation II

Generation III

Oldest   Youngest

Siblings

About ½ of the offspring of an affected parent are affected.

(B) Recessive inheritance

One parent is heterozygous...

Generation I (parents)

...and the recessive allele is passed on to about ½ of the phenotypically unaffected offspring.

Generation II

Generation III

Generation IV

Mating of heterozygous recessive parents may produce homozygous recessive (affected) offspring.

Unaffected   Affected

Female

Male

Mating

Mating between relatives

Heterozygote (unaffected phenotype)

Other symbols as in Part A.

FIGURE 8.8 Pedigree Analysis and Inheritance (A) This pedigree represents a family affected by Huntington's disease, which results from a rare dominant allele. Everyone who inherits this allele is potentially affected. (B) The family in this pedigree carries the allele for albinism, a recessive trait. Because the trait is recessive, heterozygotes do not have the albino phenotype, but they can pass the allele on to their offspring. Affected persons must inherit the allele from two heterozygous parents, or (rarely) from one homozygous recessive and one heterozygous parent, or (very rarely) two homozygous recessive parents. In this family, the heterozygous parents in generation III are cousins; however, the same result would occur if the parents were unrelated.

controlling a particular phenotype. For example, it is useful to know whether a particular rare allele that causes an abnormal phenotype is dominant or recessive. Figure 8.8A is a pedigree showing the pattern of inheritance of a rare *dominant* allele. The following are the key features to look for in such a pedigree:

- Every person with the abnormal phenotype (affected) has an affected parent.
- Either all (in the case of a homozygous parent) or about half (in the case of a heterozygous parent) of the offspring in an affected family are affected.

Compare this pattern with the one shown in Figure 8.8B, which is typical for the inheritance of a rare *recessive* allele:

- Affected people often have two parents who are not affected.
- In affected families, about one-fourth of the children of unaffected parents are affected.

Other patterns of inheritance can arise in special cases, such as sex-linked characters (see Concept 8.3).

Do You Understand Concept 8.1?

- What are the differences between genes and alleles? Between homozygous and heterozygous conditions? Between genotype and phenotype? Between Mendel's definition of a gene and the current definition?

- In a monohybrid cross, how do the events of meiosis explain Mendel's first law? In a dihybrid cross, how does meiosis explain Mendel's second law?

- Using the cross shown in Figure 8.5, calculate the probability that an F_2 seed will be spherical and yellow.

The laws of inheritance as articulated by Mendel remain valid today, and his discoveries laid the groundwork for all future studies of genetics. However, as we will see in the next concept, the relationship of one gene to one phenotype turns out to be complicated by interactions between alleles and between genes.

Alleles and Genes Interact to Produce Phenotypes

Phenotypes do not always follow the simple patterns of inheritance shown by the pairs of alleles for seed color and seed shape in peas, described in Concept 8.1. Existing alleles are subject to change by mutation and can give rise to new alleles—in fact, a single gene can have many alleles. In addition, alleles do not always show simple dominant–recessive relationships. A single allele may have multiple phenotypic effects, and a single character may be controlled by multiple genes. The expression of a gene is generally affected by interactions with other genes and with the environment.

New alleles arise by mutation

Genes are subject to **mutations**, which are rare, stable, and inherited changes in the genetic material. In other words, an allele can mutate (change) to become a different allele. For example, we can envision that at one time all pea plants made spherical seeds and had the seed shape allele *S*. At some point a mutation in *S* resulted in a new allele, *s* (wrinkled). If this mutation occurred in a cell that underwent meiosis, some of the resulting gametes would carry the *s* allele, and some offspring of this pea plant would carry the *s* allele in all of their cells. By producing phenotypic variety, mutations are the raw material for evolution. A new allele may become more or less prevalent in a population, depending on its effect on the fitness of the individuals carrying it (see Concept 15.3).

Geneticists usually define one particular allele of a gene as the **wild type**; this is the allele that is present in most individuals in nature ("the wild"). Other alleles of that gene are usually called mutant alleles, and they may produce different phenotypes. The wild-type and mutant alleles are inherited according to Mendelian laws. A gene with a wild-type allele that is present less than 99 percent of the time (the rest of the alleles being mutant) is said to be **polymorphic** (Greek *poly*, "many"; *morph*, "form").

Mendel developed his theories by studying just two alleles of each gene. But often a gene has *multiple alleles*, and the alleles may

Possible genotypes	CC, Cc^{chd}, Cc^h, Cc	$c^{chd}c^{chd}, c^{chd}c$	c^hc^h, c^hc	cc
Phenotype	Dark gray	Chinchilla	Point restricted	Albino

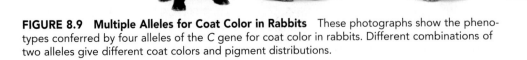

FIGURE 8.9 Multiple Alleles for Coat Color in Rabbits These photographs show the phenotypes conferred by four alleles of the *C* gene for coat color in rabbits. Different combinations of two alleles give different coat colors and pigment distributions.

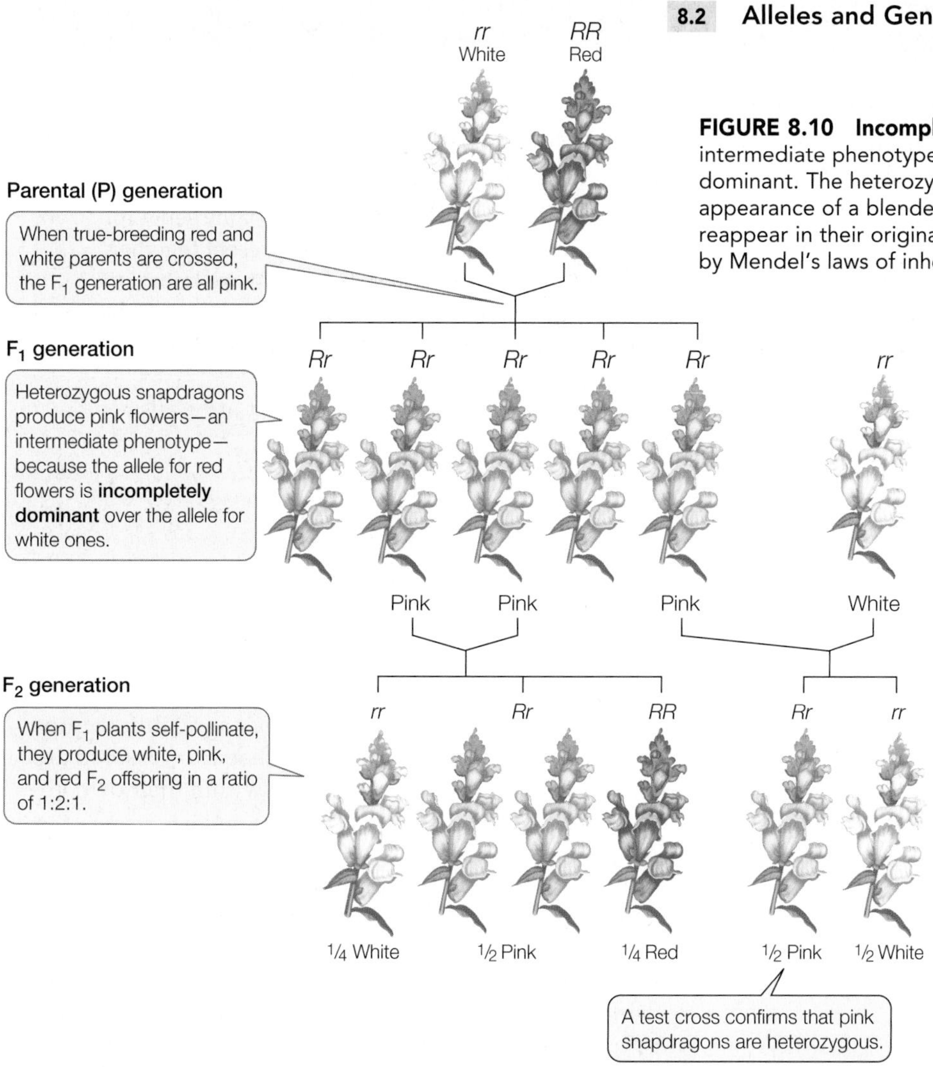

Parental (P) generation

rr White   *RR* Red

When true-breeding red and white parents are crossed, the F$_1$ generation are all pink.

F$_1$ generation

Rr *Rr* *Rr* *Rr* *Rr*   *rr*

Heterozygous snapdragons produce pink flowers—an intermediate phenotype—because the allele for red flowers is **incompletely dominant** over the allele for white ones.

Pink   Pink   Pink   White

F$_2$ generation

rr   *Rr*   *RR*   *Rr*   *rr*

When F$_1$ plants self-pollinate, they produce white, pink, and red F$_2$ offspring in a ratio of 1:2:1.

¼ White   ½ Pink   ¼ Red   ½ Pink   ½ White

A test cross confirms that pink snapdragons are heterozygous.

FIGURE 8.10 Incomplete Dominance Follows Mendel's Laws An intermediate phenotype can occur in heterozygotes when neither allele is dominant. The heterozygous phenotype (here, pink flowers) may give the appearance of a blended trait, but the traits of the parental generation reappear in their original forms in succeeding generations, as predicted by Mendel's laws of inheritance.

with any of the four) is dark gray. The c^{chd} and c^h alleles encode enzymes with specific defects. Both are dominant over the c allele, giving either light gray (c^{chd}) or point restricted (c^h) phenotypes. The homozygous recessive, cc, is albino because there is no tyrosinase for pigment production.

Dominance is not always complete

Many genes have alleles that are neither dominant nor recessive to one another. Instead, the heterozygotes have an *intermediate phenotype* in a situation called **incomplete dominance**. For example, if a true-breeding red snapdragon is crossed with a white one, all the F$_1$ flowers are an intermediate pink. Such cases appear to support the old blending theory of inheritance. However, further crosses indicate that this apparent blending can still be explained in terms of Mendelian genetics (**FIGURE 8.10**). The red and white snapdragon alleles have not disappeared, as those colors reappear in the F$_2$ generation.

Sometimes two alleles of a gene both produce their phenotypes when present in a heterozygote—a phenomenon called **codominance**. For example, in humans the gene I encodes an enzyme involved in attachment of sugars to a glycoprotein on the surface of red blood cells. There are three alleles of the gene: I^A, I^B, and I^O. The I^A and I^B alleles both encode active enzymes, but the enzymes attach different sugars to the glycoprotein; the I^O allele does not encode an active enzyme, so no sugar is attached at that position on the glycoprotein. When two different alleles (e.g., I^A and I^B) are present, both alleles are expressed (both enzymes are made, so both types of glycoproteins are made). The A and B glycoproteins are antigenic: if a red blood cell with the A glycoprotein on its surface gets into the bloodstream of a person who lacks the I^A allele, the recipient mounts an immune response and produces antibodies against the "nonself" cells (**FIGURE 8.11**). While the A and B glycoproteins are antigenic in people who do not have the I^A or I^B alleles, respectively, the O glycoprotein

show a hierarchy of dominance when present in heterozygous individuals. An example is coat color in rabbits (**FIGURE 8.9**):

- *C* determines dark gray
- c^{chd} determines chinchilla, a lighter gray
- c^h determines Himalayan, where pigment is restricted to the extremities (point restricted)
- *c* determines albino, no pigment

The *C* gene encodes the enzyme tyrosinase, which is involved in pigment production. Any rabbit with the *C* allele (paired

Blood type of cells	Genotype	Blood cell types that body rejects	Reaction to added antibodies	
			Anti-A	Anti-B
A	$I^A I^A$ or $I^A I^O$	B		
B	$I^B I^B$ or $I^B I^O$	A		
AB	$I^A I^B$	Neither A nor B		
O	$I^O I^O$	A, B, and AB		

Red blood cells that do not react with antibody remain evenly dispersed.

Red blood cells that react with antibody clump together (speckled appearance).

FIGURE 8.11 ABO Blood Reactions Are Important in Transfusions This table shows the results of mixing red blood cells of types A, B, AB, and O with serum containing anti-A or anti-B antibodies. As you look down the columns, note that each of the types, when mixed separately with anti-A and with anti-B, gives a unique pair of results—this is the basic method by which blood is typed.

Alleles and genes interact to produce phenotypes

1. In the genetic cross $AaBbCcDdEE \times AaBBCcDdEe$ where all the genes are unlinked, what fraction of the offspring will be heterozygous for all of these genes?

2. In a plant species, two alleles control flower color, which can be yellow, blue, or white. Crosses of these plants produce these offspring:

PARENTAL PHENOTYPES	OFFSPRING PHENOTYPES (RATIO)
Yellow × yellow	All yellow
Blue × yellow	Blue or yellow (1:1)
Blue × white	Blue or white (1:1)
White × white	All white

What will be the phenotypes of the offspring and the ratios among them from a cross of blue × blue?

3. In chickens, when the dominant alleles of the genes for rose comb (R) and pea comb (A) are present together, the bird has a walnut comb. Birds that are homozygous recessive for both genes have a single comb. A rose-combed bird mated with a walnut-combed bird, and the offspring were:

3/8 walnut : 3/8 rose : 1/8 pea : 1/8 single

What were the genotypes of the parents?

---

does not provoke an immune response. This makes people who are $I^O I^O$ good blood donors in transfusions. See Chapter 31 for more about the immune system.

Genes interact when they are expressed

Epistasis occurs when the phenotypic expression of one gene is affected by another gene. For example, two genes (B and E) determine coat color in Labrador retrievers:

- Allele B (black pigment) is dominant to b (brown).
- Allele E (pigment deposition in hair) is dominant to e (no deposition, so hair is yellow).

An EE or Ee dog with BB or Bb is black; one with bb is brown. A dog with ee is yellow regardless of the Bb alleles present. Clearly, gene E determines the phenotypic expression of gene B, and is therefore *epistatic* to B (**FIGURE 8.12**).

Perhaps the most dramatic example of interacting genes is **hybrid vigor** (or *heterosis*). In 1876, Charles Darwin reported that when he crossed two different genetic varieties of corn, the offspring were 25 percent taller than either of the parent strains. Darwin's observation was largely ignored for the next 30 years. In 1908, George Shull "rediscovered" this idea, reporting that not just plant height but the weight of the corn grain produced was dramatically higher in the offspring:

Parent: *B73*   Hybrid   Parent: *Mo17*

Agricultural scientists took note, and Shull's paper had a lasting impact on the field of applied genetics. The cultivation of hybrid corn spread rapidly, and the practice of hybridization is now used for many other agricultural crops and animals. For example, beef cattle that are crossbred are larger and live longer than cattle bred within their own genetic strains.

What determines the "vigor" in hybrid vigor? A phenotype such as the amount of grain that a variety of corn produces in a given environment is determined by many genes and their alleles. Presumably, the combinations of different alleles and their products from two different varieties lead to increased grain production. Put more generally, *most complex phenotypes are determined by multiple genes*. Traits conferred by multiple genes are often referred to as **quantitative traits**, because they need to be *measured* rather than assessed qualitatively. For example, grain yield must be measured, whereas a simple trait such as Mendel's pea seed color can be assessed by eye.

> **FRONTIERS** With the sequencing of the human genome and dramatic advances in molecular biology, much has been learned about human genes and what they do. A major new goal is to find out how human gene products interact to form complex phenotypes. A familiar example is height, which is a quantitative trait. A new field of bioinformatics has been developed to analyze the network of interactions between multiple gene products.

The environment affects gene action

The phenotype of an individual does not result from its genotype alone. *Genotype and environment interact to determine the phenotype of an organism.* This is especially important to remember in the era of genome sequencing (see Chapter 12). When the sequence of the human genome was completed in 2003, it was hailed as the "book of life," and public expectations of the benefits gained from this knowledge were (and are) high. But this kind of "genetic determinism"—the idea that an organism's genome sequence determines all of its phenotype—is wrong. Common knowledge tells us that environmental variables such

A dog with alleles *B* and *E* is black.

A dog with alleles *bb* and *E* is brown.

A dog with *ee* is yellow, regardless of its *Bb* alleles.

Black labrador (*B_E_*)

Chocolate labrador (*bbE_*)

Yellow labrador (*_ _ee*)

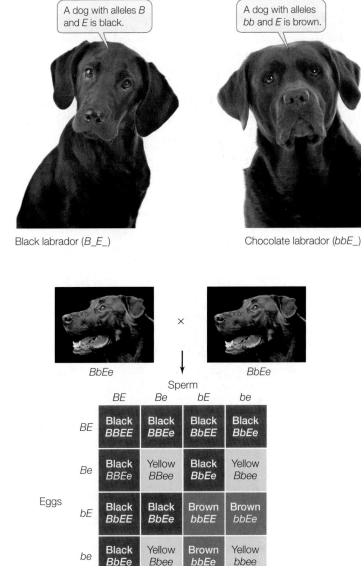

BbEe × *BbEe*

| | Sperm | | | |
	BE	*Be*	*bE*	*be*
Eggs *BE*	Black *BBEE*	Black *BBEe*	Black *BbEE*	Black *BbEe*
Be	Black *BBEe*	Yellow *BBee*	Black *BbEe*	Yellow *Bbee*
bE	Black *BbEE*	Black *BbEe*	Brown *bbEE*	Brown *bbEe*
be	Black *BbEe*	Yellow *Bbee*	Brown *bbEe*	Yellow *bbee*

FIGURE 8.12 Genes Interact Epistatically Epistasis occurs when one gene alters the phenotypic effect of another gene. In Labrador retrievers, the *E* gene determines the expression of the *B* gene.

as light, temperature, and nutrition can affect the phenotypic expression of a genotype. Clearly a person's height is determined not only by multiple genes but also by the quality of nutrition available.

Two parameters describe the effects of genes and environment on phenotype:

- **Penetrance** is the proportion of individuals in a group with a given genotype that actually show the expected phenotype. For example, many people who inherit a mutant allele of the gene *BRCA1* develop breast cancer in their lifetimes. But for reasons that are not yet clear and must involve other genes and/or the environment, some people with the mutation do not develop breast cancer. So the *BRCA1* mutation is said to be incompletely penetrant.

- **Expressivity** is the degree to which a genotype is expressed in an individual. For example, a woman with the *BRCA1* allele may develop both breast and ovarian cancer as part of the phenotype, but another woman with the same mutation may only get breast cancer. So the mutation is said to have variable expressivity.

Do You Understand Concept 8.2?

- What is the difference between incomplete dominance and codominance?

- A point restricted rabbit (see Figure 8.9) was mated with a light gray rabbit. The two offspring were albino and chinchilla. What were the genotypes of the parents?

- If a dominant allele of one gene, *A*, is necessary for hearing in humans, and the dominant allele of another gene, *B*, results in deafness regardless of the presence of other genes, what fraction of offspring in a marriage of *AaBb* and *Aabb* individuals will be deaf?

- Give an example from your own experience of a genotype whose expression is affected by the environment.

So far in this chapter we have considered genes that obey Mendel's law of independent assortment. But many genes are inherited together. This apparent anomaly can be explained by the presence of multiple genes on a single chromosome.

concept **8.3** **Genes Are Carried on Chromosomes**

Genes are parts of chromosomes. More specifically, a gene is a sequence of DNA that resides at a particular site on a chromosome, called a **locus** (plural *loci*). You have seen how the behavior of chromosomes during meiosis can explain Mendel's laws of segregation (see Figure 8.3) and independent assortment (see Figure 8.6). However, the **genetic linkage** of genes on a single chromosome alters their pattern of inheritance.

Genetic linkage was first discovered in the fruit fly *Drosophila melanogaster*. This animal is an attractive experimental subject because it is small, easily bred, and has a short generation time. In fact, the fruit fly has been a *model organism* for experimental genetics for more than a century. In Concept 8.1 we saw how Mendel successfully applied the scientific method to arrive at his laws of inheritance. Now we will examine the work of Thomas Hunt Morgan, who worked at Columbia University early in the twentieth century and used a similar approach to discover genetic linkage.

---

yourBioPortal.com

Go to ANIMATED TUTORIAL 8.2
Alleles That Do Not Assort Independently

Genes on the same chromosome are linked, but can be separated by crossing over in meiosis

Some of the crosses Morgan performed with fruit flies yielded phenotypic ratios that were not in accordance with those predicted by Mendel's law of independent assortment. Morgan did a test cross of *Drosophila* with two known genotypes: *BbVgvg* and *bbvgvg*. The *B* and *Vg* genes control two characters, body color and wing shape:

- *B* (wild-type gray body), is dominant over *b* (black body)
- *Vg* (wild-type wing) is dominant over *vg* (vestigial, or very small, wing)

Morgan expected to see four phenotypes in a ratio of 1:1:1:1, but that is not what he observed. The body color gene and the wing size gene did not assort independently; instead, they were frequently *inherited together*, and most of the progeny showed one or the other of the parental phenotypes (**FIGURE 8.13**).

These results became understandable when Morgan considered the possibility that the two loci were *linked* on the same chromosome. Such genes would not be able to assort independently as predicted by Mendel's second law. In this case, the test cross offspring might be expected to have only the parental phenotypes (gray flies with normal wings or black flies with vestigial wings) in a 1:1 ratio. If linkage were absolute we would *only* see these two types of progeny. However, this did not happen. Why did some of Morgan's flies show phenotypes different from their parents?

Some of Morgan's flies displayed *recombinant phenotypes* because two homologous chromosomes can physically exchange corresponding segments during prophase I of meiosis—that is, by *crossing over* (**FIGURE 8.14**). Each exchange event involves two of the four chromatids in a tetrad—one from each member of the homologous pair—and can occur at any point along the length of the chromosome. The chromosome segments are exchanged reciprocally, so both chromatids become *recombinant*

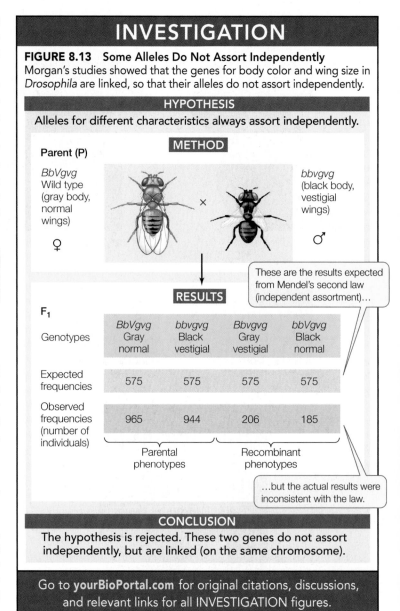

INVESTIGATION

FIGURE 8.13   Some Alleles Do Not Assort Independently
Morgan's studies showed that the genes for body color and wing size in *Drosophila* are linked, so that their alleles do not assort independently.

HYPOTHESIS

Alleles for different characteristics always assort independently.

METHOD

Parent (P)

BbVgvg
Wild type
(gray body,
normal
wings)
♀

bbvgvg
(black body,
vestigial
wings)
♂

RESULTS

These are the results expected from Mendel's second law (independent assortment)…

F₁ Genotypes	*BbVgvg* Gray normal	*bbvgvg* Black vestigial	*Bbvgvg* Gray vestigial	*bbVgvg* Black normal
Expected frequencies	575	575	575	575
Observed frequencies (number of individuals)	965	944	206	185

Parental phenotypes   Recombinant phenotypes

…but the actual results were inconsistent with the law.

CONCLUSION

The hypothesis is rejected. These two genes do not assort independently, but are linked (on the same chromosome).

Go to **yourBioPortal.com** for original citations, discussions, and relevant links for all INVESTIGATION figures.

(that is, each chromatid ends up with genes from both of the organism's parents).

LINK   The process of crossing over is described in Concept 7.4

As a result of crossing over *between* two linked genes, not all the progeny of a cross have the parental phenotypes. Instead, recombinant offspring appear as well, as they did in Morgan's cross. They generally appear in proportions related to the **recombination frequency** between the two genes, which is calculated by dividing the number of recombinant progeny by the total number of progeny (**FIGURE 8.15**). Recombination frequencies are *greater for loci that are farther apart on the chromosome* than for loci that are closer together because crossing over is more likely to occur between genes that are far apart. By

Homologous chromosomes

b b B B
vg vgVg Vg

Meiosis I

Tetrad

Chromatid

Crossover

Genes at different loci on the same chromosome can recombine and separate by crossing over.

b b
vg Vg

B B
vg Vg

Recombinant chromosomes

Meiosis II

b
vg

b
Vg

B
vg

B
Vg

The result is two recombinant gametes from each event of crossing over.

FIGURE 8.14 Crossing Over Results in Genetic Recombination Recombination accounts for why linked alleles are not always inherited together. Alleles at different loci on the same chromosome can be recombined by crossing over and then being separated from one another. Such recombination occurs during prophase I of meiosis.

Recombination between *y* and *v* is more frequent, so they are farther apart. The recombination frequencies are converted to *map units*, which in this case correspond to the distances from *y* ($y = 0$).

The era of gene sequencing has made mapping less important in some areas of genetics research. However, mapping is still one way to verify that a particular DNA sequence corresponds with a particular phenotype. The phenomenon of linkage has allowed biologists to isolate genes and to create genetic markers that are linked to important genes, making it easy to identify individuals carrying particular alleles. This is particularly important in breeding new crops and animals for agriculture, and for identifying humans carrying medically significant mutations.

Linkage is also revealed by studies of the X and Y chromosomes

The fruit fly genome has four pairs of chromosomes: in three pairs, the chromosomes are similar in size to one another and are called *autosomes*. The fourth pair has two chromosomes of different sizes. These determine the sex of the fly and are called the *sex chromosomes*:

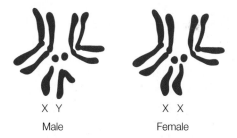

X Y   X X
Male   Female

Note that the female fly has two X chromosomes and that the male has only one, the other being the Y chromosome: females are XX and males are XY. It turns out that in addition to being different sizes, *many genes on the X chromosome are not present on the Y*. The X chromosome was one of the first to have specific genes assigned to it.

Morgan identified a gene that controls eye color in *Drosophila*. The wild-type allele of the gene confers red eyes, whereas a recessive mutant allele confers white eyes. Morgan's experimental crosses with flies carrying the mutant allele demonstrated that this eye color locus is on the X chromosome. If we abbreviate the eye color alleles as *R* (red eyes) and *r* (white eyes), the presence of the alleles on the X chromosome is designated by X^R and X^r.

Morgan crossed a homozygous red-eyed female ($X^R X^R$) with a white-eyed male. The male is designated X^rY because the Y does not carry any allele for this gene. (Any gene that is present as a single copy in a diploid organism is called **hemizygous**.)

calculating recombination frequencies, geneticists can infer the locations of genes along a chromosome and generate a genetic map. Below is a map showing five genes on a fruit fly chromosome. It was constructed using the recombination frequencies generated by test crosses involving various pairs of the genes:

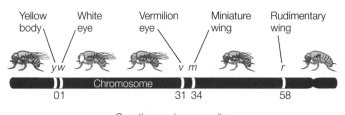

Yellow body | White eye | Vermilion eye | Miniature wing | Rudimentary wing

y w | *v m* | *r*

Chromosome

01   31 34   58

Genetic map in map units

In the chromosome shown above the recombination frequency between *y* and *w* is low, so they are close together on the map.

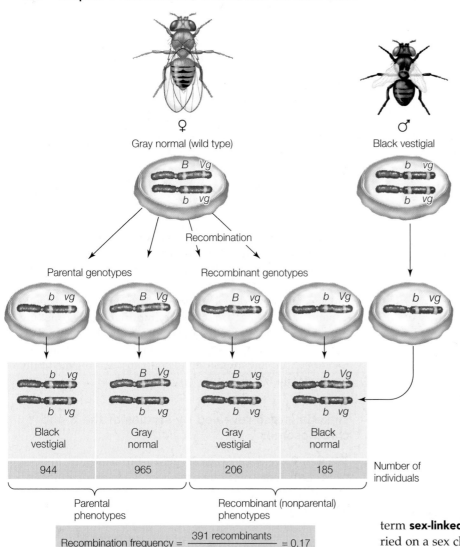

FIGURE 8.15 Recombination Frequencies
The frequency of recombinant offspring (those with a phenotype different from either parent) can be calculated.

♀ Gray normal (wild type)

♂ Black vestigial

Recombination

Parental genotypes   Recombinant genotypes

b *vg*	*B* *Vg*	*B* *vg*	*b* *Vg*
b *vg*	*b* *vg*	*b* *vg*	*b* *vg*
Black vestigial	Gray normal	Gray vestigial	Black normal
944	965	206	185

Number of individuals

Parental phenotypes

Recombinant (nonparental) phenotypes

$$\text{Recombination frequency} = \frac{391 \text{ recombinants}}{2{,}300 \text{ total offspring}} = 0.17$$

All the sons and daughters from this cross had red eyes, because red (*R*) is dominant over white (*r*) and all the progeny had inherited a wild-type X chromosome (X^R) from their mother (**FIGURE 8.16A**). Note that this phenotypic outcome would have occurred even if the *R* gene had been present on

an autosome rather than a sex chromosome. In that case, the male would have been homozygous recessive—*rr*.

When Morgan performed the reciprocal cross, in which a white-eyed female (X^rX^r) was mated with a red-eyed male (X^RY), the results were unexpected: *all the sons were white-eyed and all the daughters were red-eyed* (**FIGURE 8.16B**). The sons from the reciprocal cross inherited their only X chromosome from their white-eyed mother and were therefore hemizygous for the white allele. The daughters, however, got an X chromosome bearing the white allele from their mother and an X chromosome bearing the red allele from their father; therefore they were red-eyed heterozygotes. When these heterozygous females were mated with red-eyed males, half their sons had white eyes but all their daughters had red eyes. Together, these results showed that eye color was carried on the X chromosome and not on the Y.

These and other experiments led to the term **sex-linked inheritance**: inheritance of a gene that is carried on a sex chromosome. (This term is somewhat misleading because "sex-linked" inheritance is not really linked to the sex of an organism—after all, both males and females carry X chromosomes.) In mammals, the X chromosome is larger and carries more genes that the Y. For this reason, most examples of sex-linked inheritance involve genes that are carried on the X chromosome.

Many sexually reproducing species, including humans, have sex chromosomes. As in fruit flies, human males are XY,

APPLY THE CONCEPT

Genes are carried on chromosomes

The pedigree shows the inheritance pattern of a rare mutant phenotype in humans, congenital cataract (filled-in symbols).

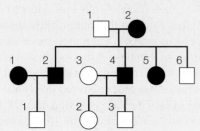

1. Are cataracts inherited as an autosomal dominant? Autosomal recessive? Sex-linked dominant? Sex-linked recessive?

2. Person #5 in the second generation marries a man who does not have cataracts. Two of their four children, a boy and a girl, develop cataracts. What is the probability that their next child will be a girl with cataracts?

(A)

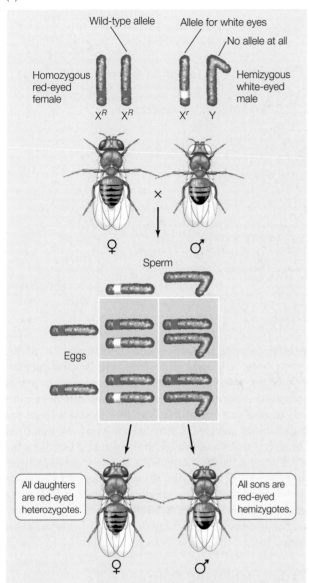

(B)

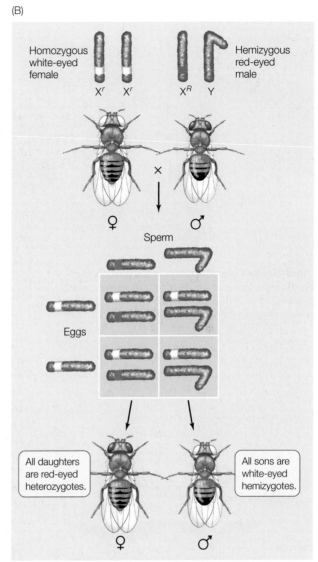

FIGURE 8.16 A Gene for Eye Color Is Carried on the *Drosophila* X Chromosome Morgan demonstrated that a mutant allele that causes white eyes in *Drosophila* is carried on the X chromosome. Note that in this case, the reciprocal crosses do not have the same results.

females are XX, and relatively few of the genes that are present on the X chromosome are present on the Y. Pedigree analyses of X-linked recessive phenotypes like the one in **FIGURE 8.17** reveal the following patterns (compare with the pedigrees of non-X-linked phenotypes in Figure 8.8):

- The phenotype appears much more often in males than in females, because only one copy of the rare allele is needed for its expression in males, whereas two copies must be present in females.

- A male with the mutation can pass it on only to his daughters; all his sons get his Y chromosome.

- Daughters who receive one X-linked mutation are **heterozygous carriers**. They are phenotypically normal, but they can pass the mutant allele to their sons or daughters. On average, half their children will inherit the mutant allele since half of their X chromosomes carry the normal allele.

- The mutant phenotype can skip a generation if the mutation passes from a male to his daughter (who will be phenotypically normal) and then to her son.

Some genes are carried on chromosomes in organelles

The nucleus is not the only organelle in a eukaryotic cell that carries genetic material. Mitochondria and plastids each contain several copies of a small chromosome that carries a small number of genes. For example, in humans there are about 24,000 genes in the nuclear genome and 37 in the mitochondrial genome.

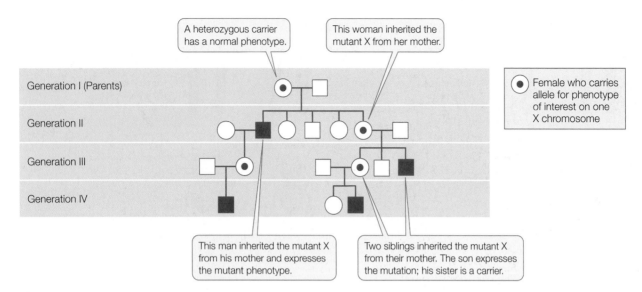

A heterozygous carrier has a normal phenotype.

This woman inherited the mutant X from her mother.

⊙ Female who carries allele for phenotype of interest on one X chromosome

This man inherited the mutant X from his mother and expresses the mutant phenotype.

Two siblings inherited the mutant X from their mother. The son expresses the mutation; his sister is a carrier.

FIGURE 8.17 Red–Green Color Blindness Is Carried on the Human X Chromosome The mutant allele for red–green color blindness is expressed as an X-linked recessive trait, and therefore is always expressed in males when they carry that allele.

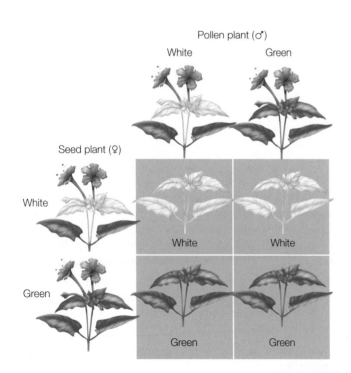

FIGURE 8.18 Cytoplasmic Inheritance In four o'clock plants, leaf color is inherited through the female plant only. In the parent plant with some white leaves, the white leaf color is caused by a chloroplast mutation that occurred during the life of the plant; the leaves that formed before the mutation occurred are green. When this plant is used as a pollen donor in a cross with an all green plant, the offspring are all green. But when the same plant is used as an egg donor, the offspring inherit the mutation cytoplasmically and are entirely white.

The inheritance of organelle genes differs from that of nuclear genes because in most organisms, *mitochondria and plastids are inherited only from the mother*. Most egg cells in plants and animals contain abundant cytoplasm and organelles, but the only part of the sperm that survives to take part in the union of haploid gametes is the nucleus. You have inherited your mother's mitochondria (with their genes) but not your father's. The inheritance of organelles and their genes is therefore non-Mendelian and is described as maternal or *cytoplasmic* inheritance, since the inherited organelles come from the maternal cytoplasm. **FIGURE 8.18** illustrates the cytoplasmic inheritance of a mutant plastid gene that confers white color to leaves.

Do You Understand Concept 8.3?

- Describe the differences in patterns of inheritance between a gene present in the nucleus and a gene present in the mitochondria.

- Explain the concept of linkage. If you performed a test cross with a fruit fly that is heterozygous for two genes, how would you conclude that the two genes are linked?

- Red–green color blindness is inherited as a sex-linked recessive. Two parents with normal color vision have a child who is red–green color-blind. Is it a boy or girl? Draw a pedigree of the family. Draw a Punnett square to show gametes and offspring with regard to the X and Y chromosomes and the normal and color-blind alleles.

Like eukaryotes, prokaryotes contain genes that determine their phenotypes. Sexual reproduction in eukaryotes involves two sets of chromosomes and meiosis, giving rise to haploid gametes. Let's look next at how reproduction and inheritance differ in prokaryotes, which are haploid.

Prokaryotes Can Exchange Genetic Material

As described in Concept 4.2, prokaryotic cells lack nuclei; they contain their genetic material mostly as single chromosomes in central regions of their cells. Prokaryotes reproduce asexually by binary fission, a process that gives rise to progeny that are virtually identical genetically (see Concept 7.2). That is, the offspring of cell reproduction in prokaryotes constitute a clone.

How then do prokaryotes evolve? Mutations occur in prokaryotes just as they do in eukaryotes, and the resulting new alleles increase genetic diversity. You might expect, therefore, that there is no way for individuals of these organisms to exchange genes, as in sexual reproduction. It turns out, however, that prokaryotes do have a process for transferring genes between cells. Along with mutation, this process provides for genetic diversity among prokaryotes. This transfer of genes from one organism to another without sexual reproduction is called horizontal (or lateral) gene transfer, to distinguish it from the vertical gene transfer that occurs from parent to offspring.

LINK The evolutionary consequences of lateral gene transfer are discussed in Concepts 15.6 and 19.1

Bacteria exchange genes by conjugation

To illustrate genetic exchange in bacteria, let's consider two strains of the bacterium *E. coli* with different alleles for each of six genes (each of the genes coding for the synthesis of a certain small molecule). Simply put, the two strains have the following genotypes (remember that bacteria are haploid):

ABCdef and *abcDEF*

where capital letters stand for wild-type alleles and lowercase letters stand for mutant alleles.

When the two strains are grown together in the laboratory, most of the cells produce clones. That is, almost all of the cells that grow have the original genotypes:

ABCdef and *abcDEF*

However, out of millions of bacteria, a few occur that have the genotype: *ABCDEF*

How could these completely wild-type bacteria arise? One possibility is *mutation*: in the *abcDEF* bacteria, the *a* allele could have mutated to *A*, the *b* allele to *B*, and the *c* allele to *C*. The problem with this explanation is that a mutation at any particular point in an organism's DNA sequence is a very rare event (about 1 in a million). The probability of all three events

occurring in the same cell is extremely low—much lower than the actual rate of appearance of cells with the *ABCDEF* genotype. So the mutant cells must have acquired wild-type genes some other way—and this turns out to be the transfer of DNA between cells.

Electron microscopy shows that genetic transfers between bacteria can happen via physical contact between the cells (**FIGURE 8.19A**). Contact is initiated by a thin projection called a **sex pilus** (plural *pili*), which extends from one cell (the donor), attaches to another (the recipient), and draws the two cells together. Genetic material can then pass from the donor cell to the

(A)

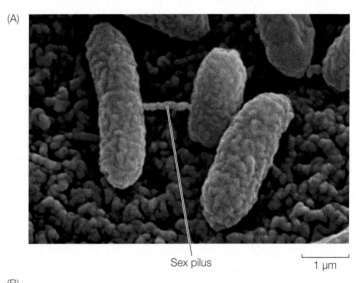

Sex pilus — 1 μm

(B)

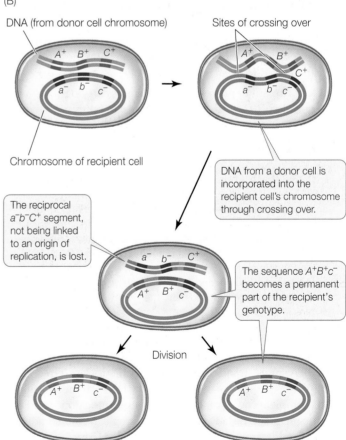

FIGURE 8.19 Bacterial Conjugation and Recombination (A) A sex pilus draws two bacteria into close contact, so that a cytoplasmic bridge (conjugation tube) can form. DNA is transferred from the donor cell to the recipient cell via the conjugation tube. (B) DNA from a donor cell can become incorporated into a recipient cell's chromosome through crossing over.

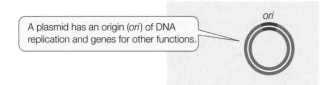

A plasmid has an origin (*ori*) of DNA replication and genes for other functions.

FIGURE 8.20 Gene Transfer by Plasmids When plasmids enter a cell via conjugation, their genes can be expressed in the recipient cell.

recipient through a thin cytoplasmic bridge called a *conjugation tube*. There is no reciprocal transfer of DNA from the recipient to the donor. This process is referred to as **bacterial conjugation**.

Once the donor DNA is inside the recipient cell, it can recombine with the recipient cell's genome. In much the same way that chromosomes pair up, gene for gene, in prophase I of meiosis, the donor DNA can line up beside its homologous genes in the recipient, and crossing over can occur. Gene(s) from the donor can become integrated into the genome of the recipient, thus changing the recipient's genetic constitution (**FIGURE 8.19B**), although only about half the transferred genes become integrated in this way. When the recipient cells proliferate, the integrated donor genes are passed on to all progeny cells, and the other transferred genes are lost.

Plasmids transfer genes between bacteria

In addition to their main chromosome, many bacteria harbor additional smaller, circular DNA molecules called **plasmids**. Plasmids typically contain at most a few dozen genes, which may fall into one of several categories:

- Genes for unusual metabolic capacities, such as the ability to break down hydrocarbons. Bacteria carrying these plasmids can be used to clean up oil spills.

- Genes for antibiotic resistance. Plasmids carrying such genes are called R factors, and since they can be transferred between bacteria via conjugation, they are a major threat to human health.

FRONTIERS Methicillin-resistant *Staphylococcus aureus* (MRSA) is a bacterial strain that is resistant to multiple antibiotics because it contains a chromosome harboring resistance genes. This bacterium causes serious infections in hospital settings. Multidrug resistance is now spreading rapidly through the wider community, because MRSA DNA has been transferred to other bacterial strains. Scientists are trying to fully understand how this DNA transfer occurs and to design new antibiotics to kill the genetically transformed strains.

Plasmids can move between cells during conjugation, thereby transferring new genes to the recipient bacterium (**FIGURE 8.20**). A single strand of the donor plasmid is transferred to the recipient; synthesis of complementary DNA strands results in two complete copies of the plasmid, one in the donor and one in the recipient. Because plasmids can replicate independently of the main chromosome, they do not need to recombine with

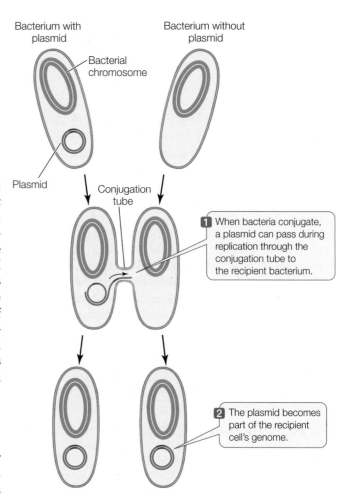

Bacterium with plasmid

Bacterium without plasmid

Bacterial chromosome

Plasmid

Conjugation tube

1 When bacteria conjugate, a plasmid can pass during replication through the conjugation tube to the recipient bacterium.

2 The plasmid becomes part of the recipient cell's genome.

the main chromosome to add their genes to the recipient cell's genome.

Do You Understand Concept 8.4?

- How does recombination occur in prokaryotes?
- What is the evolutionary advantage of recombination in prokaryotes?
- What are the differences between recombination after conjugation in prokaryotes and recombination during meiosis in eukaryotes?

 QUESTION How is hemophilia inherited through the mother, and why is it more frequent in males?

ANSWER The ancient rabbis in the opening story were dealing with babies that had the blood-clotting disease hemophilia A. The mutant allele for factor VIII, the clotting factor missing in these babies, must be recessive since the babies' parents did not suffer from the disease (Concept 8.1). Usually, the disease occurred in boys, and any relatives with the disease were males on the mother's side of the family. This

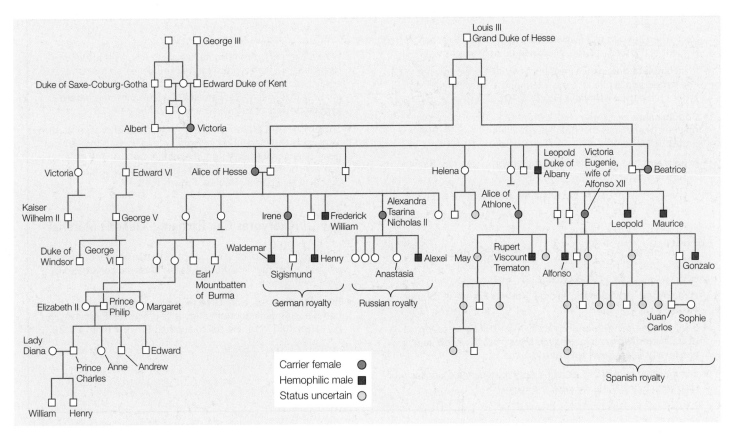

FIGURE 8.21 Sex Linkage in Royal Families of Europe
England's Queen Victoria passed an X chromosome carrying the mutant allele for hemophilia to three of her children.

was because the gene for factor VIII is carried on the human X chromosome and its inheritance is sex-linked, with all boys receiving their single X chromosomes from their mothers (Concept 8.3).

Hemophilia played a role in the history of modern Europe. England's Queen Victoria, who ruled in the nineteenth century, had nine children, and one of them, Leopold, had hemophilia and died after a minor accident at age 31. The queen

did not have the disease, nor was it present in any of her forbears; so it is probable that a mutation arose on Victoria's X chromosome or in that of her father (Concept 8.2). Three of Victoria's grandchildren had hemophilia, indicating that their mothers, two of Victoria's daughters, were carriers of the mutant allele. This spread the mutation to the royal families of Spain, Germany, and Russia in the twentieth century (**FIGURE 8.21**). While there are many descendents of Queen Victoria still living (including Queen Elizabeth II), none have hemophilia; however, it is possible that some of Victoria's female descendents are carriers.

<table>
<tr><td>**8**</td><td>## SUMMARY</td></tr>
</table>

concept 8.1 **Genes Are Particulate and Are Inherited According to Mendel's Laws**

- Mendel's experiments on pea plants supported the particulate theory of inheritance stating that discrete units (now called genes) are responsible for the inheritance of specific traits. **Review Figure 8.1**

- Mendel's first law, the **law of segregation**, states that when any individual produces gametes, the two copies of a gene separate, so that each gamete receives only one member of the pair. **Review Figures 8.2 and 8.3**

- Mendel used a **test cross** to find out whether an individual showing a dominant phenotype was homozygous or heterozygous. **Review Figure 8.4 and WEB ACTIVITY 8.1**

- Mendel's use of **dihybrid crosses** to study the inheritance of two characters led to his second law: the **law of independent assortment**. The independent assortment of genes in meiosis leads to recombinant phenotypes. **Review Figures 8.5 and 8.6 and ANIMATED TUTORIAL 8.1**

- **Pedigree** analysis can determine whether an allele is **dominant** or **recessive**. **Review Figure 8.8 and INTERACTIVE TUTORIAL 8.1**

concept 8.2 **Alleles and Genes Interact to Produce Phenotypes**

- New alleles arise by random mutation. Many genes have multiple alleles. A **wild-type** allele gives rise to the predominant form of a trait. When the wild-type allele is present at a locus less than 99 percent of the time, the locus is said to be polymorphic.

- In **incomplete dominance**, neither of two alleles is dominant. The heterozygous phenotype is intermediate between the homozygous phenotypes. **Review Figure 8.10**

- **Codominance** exists when two alleles at a locus produce two different phenotypes that both appear in heterozygotes. **Review Figure 8.11**

- In **epistasis**, one gene affects the expression of another. **Review Figure 8.12**

- Environmental conditions can affect the expression of a genotype.

concept 8.3 Genes Are Carried on Chromosomes

- Each chromosome carries many genes, and the genes on a single chromosome show **genetic linkage**. **Review Figure 8.13 and ANIMATED TUTORIAL 8.2**

- Genes on the same chromosome can recombine by crossing over. The resulting recombinant chromosomes have new combinations of alleles. **Review Figure 8.14**

- Recombination frequencies can be used to generate a genetic map of a chromosome. **Review Figure 8.15**

- In fruit flies and mammals, the X chromosome carries many genes, but the Y chromosome has only a few. Males have only one allele (are hemizygous) for X-linked genes, so recessive sex-linked mutations are expressed phenotypically more often in males than in females. Females may be unaffected **heterozygous carriers** of such alleles. **Review Figure 8.16**

- Some genes are present on the chromosomes of organelles such as plastids and mitochondria. In many organisms, cytoplasmic genes are inherited only from the mother because the male gamete contributes only its nucleus (i.e., no cytoplasm) to the zygote at fertilization. **Review Figure 8.18**

concept 8.4 Prokaryotes Can Exchange Genetic Material

- Prokaryotes reproduce asexually but can transfer genes from one cell to another in a process called **bacterial conjugation**. **Review Figure 8.19**

- **Plasmids** are small, extra DNA molecules in bacteria that carry genes involved in important metabolic processes. Plasmids can be transmitted from one cell to another. **Review Figure 8.20**

See **WEB ACTIVITIES 8.2 and 8.3** for a concept review of this chapter.

DNA and Its Role in Heredity

Jurassic Park, in both its literary and film incarnations, featured a fictional theme park populated with live dinosaurs. In the story, scientists isolated DNA from dinosaur blood that they found in the digestive tracts of fossil insects. The insects supposedly sucked the reptiles' blood right before being preserved in amber (fossilized tree resin). This DNA, according to the novel, could be manipulated to produce living individuals of long-extinct organisms such as velociraptors and the ever-memorable *Tyrannosaurus rex*.

The late Michael Crichton got the idea for his novel from an actual scientific paper in which the authors cracked open amber that was 40 million years old and extracted DNA from a bee fossil trapped in the amber. Other scientists had reported on ancient DNA from amber-trapped termites and gnats. Then several reports emerged of DNA from 80-million-year-old dinosaur bones. Unfortunately, upon additional study these "preserved" DNAs turned out to be contamination—either from microorganisms living in the surrounding soil or even from the scientists studying the samples. In fact, one of the supposed dinosaur DNAs turned out to be from the human Y chromosome.

It is unlikely that any long DNA polymers would survive over millions of years. The oldest fossilized insects in amber are about 40 million years old, and dinosaurs died out about 65 million years ago. Nevertheless, the huge success of Crichton's book brought ancient DNA to the attention of millions of people, including biologists who study the evolution of life on Earth. DNA samples have been isolated from organisms preserved for many thousands of years in permafrost, and with

improved methods for DNA analysis, large portions of these organisms' genomes are being sequenced.

As methods to replicate tiny amounts of DNA and keep it from contamination have improved, attention has turned to ancient human DNA. DNAs from human remains not overly prone to environmental contamination have been studied. These include people whose bodies have been preserved in ice, such as the "Iceman" who died in the Austrian Alps 5,300 years ago. There is even a Neandertal Genome Project to analyze the DNA from preserved specimens of *Homo neanderthalensis*, a species that lived in Europe at the same time as early humans, between 350,000 and 30,000 years ago.

Michael Crichton's novel *Jurassic Park* was based on the fictional premise that DNA retrieved from fossils could produce living dinosaurs, such as these velociraptors.

Q QUESTION

What can we learn from ancient DNA?

KEY CONCEPTS

9.1 DNA Structure Reflects Its Role as the Genetic Material

9.2 DNA Replicates Semiconservatively

9.3 Mutations Are Heritable Changes in DNA

<table>
<tr><td>concept
9.1</td><td>**DNA Structure Reflects Its Role as the Genetic Material**</td></tr>
</table>

In Concept 8.1 we describe Mendel's experiments in the 1860s demonstrating that genes are physically distinct entities, and Morgan's work in the early twentieth century showing that genes are associated with chromosomes. We now turn to the actual chemical nature of genes, beginning with the evidence that DNA is the carrier of heritable information. Scientists used two types of evidence to show that DNA is the genetic material: circumstantial and experimental. Below we provide examples of both types.

Circumstantial evidence suggested that DNA was the genetic material

Before DNA was accepted as the genetic material, it had to meet certain criteria. Scientists needed to show that eukaryotic DNA:

• was present in the cell nucleus and in condensed chromosomes

• doubled during S phase of the cell cycle

• was twice as abundant in the diploid cells as in the haploid cells of a given organism

• showed the same patterns of transmission as the genetic information it was supposed to carry.

There was circumstantial evidence for all of these criteria well before there was experimental evidence.

DNA IN THE NUCLEUS DNA was first isolated in 1868 by the young Swiss researcher–physician Friedrich Miescher. At that time, dividing cells and chromosomes had been seen under the microscope but genes were not yet described. There is no evidence that Miescher had seen Mendel's paper on inheritance. While still a student, Miescher isolated cell nuclei from white blood cells in pus from the bandages of wounded soldiers. When he treated these nuclei chemically, a fibrous substance came out of solution. He called it "nuclein" and found it contained the elements C, H, O, N, and P. With no evidence except for finding it in the nucleus, Miescher boldly proposed that nuclein was the genetic material. His supervising professor was so astounded by Miescher's work that he repeated it himself in the lab and finally allowed his student to publish it in a scientific journal.

DNA IN THE CHROMOSOMES Dyes were developed that react specifically with DNA, turning color only when they bind to it (**FIGURE 9.1A**). This allowed individual cells to be examined for the location and amount of DNA they contained. When dividing cells were stained with such a dye, only the chromosomes were stained:

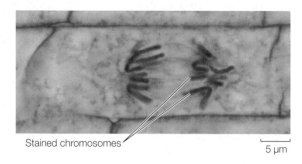

Stained chromosomes

5 μm

DNA IN THE CELLS The amount of dye binding to DNA, and hence the intensity of color observed, was directly related to

(A)

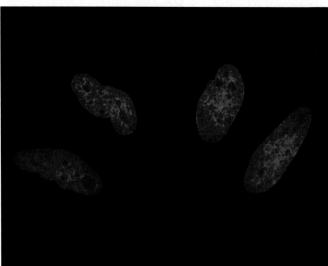

(B)

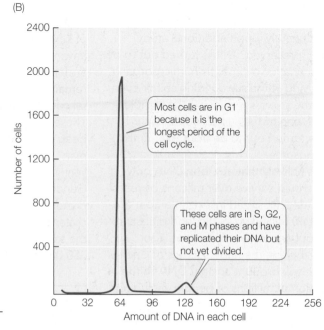

Most cells are in G1 because it is the longest period of the cell cycle.

These cells are in S, G2, and M phases and have replicated their DNA but not yet divided.

FIGURE 9.1 DNA in the Nucleus and in the Cell Cycle (A) A dye—in this case DAPI—can specifically stain DNA. (B) When dividing cells are stained and analyzed by flow cytometry, there are two populations in terms of DNA content.

the amount of DNA present: the greater the intensity, the more DNA. This allowed scientists to analyze DNA amounts in individual cells during the cell cycle (see Concept 7.2).

When a population of actively dividing cells was stained with dye, the amount of DNA in each cell could be quantified by passing the cells one by one through a *flow cytometer*. In general, two populations of cells were seen: most cells were in G1 and contained half the amount of DNA that was in the remaining cells, which were in S, G2, and M (**FIGURE 9.1B**).

Such staining experiments confirmed two other predictions for DNA as the genetic material:

- Virtually all nondividing somatic cells of a particular organism have the same amount of nuclear DNA.

- Similar experiments showed that after meiosis, gametes have half the amount of nuclear DNA as somatic cells.

TRANSMISSION OF DNA Chromosomes in eukaryotic cells contain DNA, but they also contain proteins that are bound to DNA. Therefore, it was difficult for scientists to rule out that genetic information might be carried on proteins. Viruses provided an ideal system to explore this question. Many viruses, including **bacteriophage** (viruses that infect bacteria), are composed of DNA and only one or a few kinds of protein. When a bacteriophage infects a bacterium, it takes about 20 minutes for the virus to hijack the bacterium's metabolic capabilities and turn the bacterium into a virus factory. Minutes later, the bacterium is dead and hundreds of viruses are released.

The transition from bacterium to virus producer is a change in the *genetic program* of the bacterial cell. Careful chemical analyses and observations by electron microscope showed that *only the viral DNA* is injected into the cell during infection (**FIGURE 9.2**). This was further evidence that DNA and not protein was the genetic material.

Experimental evidence confirmed that DNA is the genetic material

Circumstantial evidence can show correlations between two situations. However, *scientists rely on experiments* to provide proof of a cause-and-effect relationship. In order to confirm that DNA was the genetic material, biologists used model organisms such as bacteria in **transformation** experiments. They found that the addition of DNA from one strain of bacterium could genetically transform another strain of bacterium:

bacterium strain A + strain B DNA → bacterium strain B

The transformation of mammalian cells carrying genetic mutations provided another model system for showing that DNA is the genetic material. For example, certain cells were found to lack the gene for thymidine kinase, an enzyme that catalyzes the first step in a pathway that converts thymidine into deoxythymidine triphosphate (dTTP, a nucleotide used in DNA

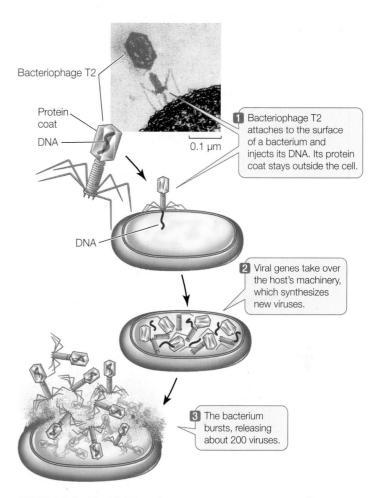

FIGURE 9.2 Viral DNA and Not Protein Enters Host Cells Bacteriophage T2 infects *E. coli* and depends on the bacterium to produce new viruses. The bacteriophage consists entirely of DNA contained within a protein coat. When the virus infects an *E. coli* cell, its DNA, but not its protein coat, is injected into the host bacterium.

synthesis). Such cells cannot grow in a medium that contains thymidine as the only source for dTTP synthesis. However, when the cells were incubated with DNA containing the gene for thymidine kinase (in a solution containing $CaCl_2$; see below), some of the cells became transformed with the TK gene and were able to grow on the thymidine-containing medium (**FIGURE 9.3**).

For successful genetic transformation, DNA must pass through the cell membrane into the cytoplasm and (in most cases) get incorporated into a host cell chromosome. In early transformation experiments, a major stumbling block was the first step, because DNA is negatively charged (because of its phosphate groups) and so are the surfaces of cell membranes (because of their phospholipids). Since like charges repel, DNA does not tend to bind to cell membranes. Biologists found they could circumvent this by incubating the DNA and cells in a calcium salt such as $CaCl_2$. The positively charged Ca^{2+} neutralized the charges, allowing DNA to bind and even pass through the cell membranes. This simple procedure had enormous consequences because it allowed the artificial transformation of bacterial and mammalian cells.

INVESTIGATION

FIGURE 9.3 Transformation of Eukaryotic Cells The use of a marker gene shows that mammalian cells can be genetically transformed by DNA. Usually, the marker gene is carried on a larger DNA molecule.

HYPOTHESIS

DNA can transform eukaryotic cells.

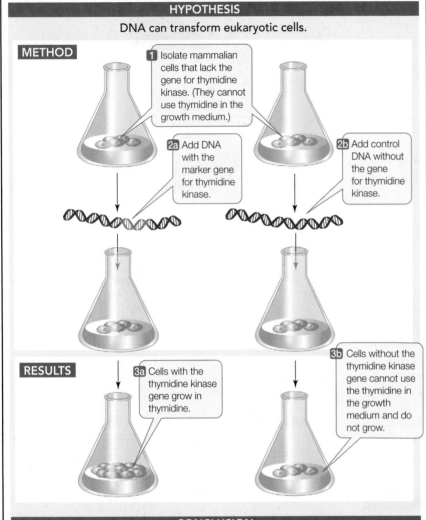

METHOD

1 Isolate mammalian cells that lack the gene for thymidine kinase. (They cannot use thymidine in the growth medium.)

2a Add DNA with the marker gene for thymidine kinase.

2b Add control DNA without the gene for thymidine kinase.

RESULTS

3a Cells with the thymidine kinase gene grow in thymidine.

3b Cells without the thymidine kinase gene cannot use the thymidine in the growth medium and do not grow.

CONCLUSION

The cells were transformed by DNA.

ANALYZE THE DATA

Transformation was achieved by adding the DNA in a solution of calcium chloride ($CaCl_2$) at pH 7. In other experiments, the type or amount of DNA, pH, or $CaCl_2$ concentration was varied. The transformation efficiency was calculated as the percentage of cells that produced colonies on a medium containing thymidine. Explain these data.

Transformation conditions	Efficiency (%)
Mammalian DNA with TK gene	
10 µg	15
20 µg	55
30 µg	10
40 µg	5
20 µg, no $CaCl_2$	10
20 µg, pH 6.5	0
20 µg, pH 7.5	0
Bacterial virus DNA with TK gene	
20 µg	0

Go to **yourBioPortal.com** for original citations, discussions, and relevant links for all INVESTIGATION figures.

Many kinds of cells can be transformed in this way—even egg cells. In this case, a whole new genetically transformed organism can result; such an organism is referred to as **transgenic**. These methods form the basis of much applied research, including biotechnology and genetic engineering. The transformation of multicellular eukaryotes provided powerful experimental evidence for DNA as the genetic material.

The discovery of the three-dimensional structure of DNA was a milestone in biology

Mendel showed that genes are physically distinct entities, and further research identified DNA as the genetic material. The history of how the actual structure of DNA was deciphered is worth considering, as it represents not only brilliant scientists working together, but also a landmark in our understanding of nature.

By the mid-twentieth century, the chemical makeup of DNA, as a polymer made up of nucleotide monomers, had been known for several decades. In determining the structure of DNA, scientists hoped to answer two questions:

1. How is DNA replicated between cell divisions?

2. How does it direct the synthesis of specific proteins?

They were eventually able to answer both questions.

The structure of DNA was deciphered only after many types of experimental evidence were considered together in a theoretical framework. The most crucial evidence was obtained using X-ray crystallography. Some chemical substances, when they are isolated and purified, can be made to form crystals. The positions of atoms in a crystallized substance can be inferred from the diffraction pattern of X rays passing through the substance (**FIGURE 9.4A**). The structure of DNA would not have been characterized without the crystallographs prepared in the early 1950s by the English chemist Rosalind Franklin (**FIGURE 9.4B**). Franklin's work, in turn, depended on the success of the English biophysicist Maurice Wilkins, who prepared samples containing very uniformly oriented DNA fibers. These fibers and the crystallographs Franklin prepared from them suggested a spiral or helical molecule.

(A)

(B)

DNA sample

These spots
are caused
by diffracted
X rays.

Beam of X rays

X ray source   Lead screen

Photographic
plate

**FIGURE 9.4   X-Ray Crystallography Helped Reveal the
Structure of DNA**   (A) The positions of atoms in a crystallized
chemical substance can be inferred by the pattern of diffraction of
X rays passed through it. The pattern of DNA is both highly regular and repetitive. (B) Rosalind Franklin's crystallographs helped
scientists visualize the helical structure of the DNA molecule.

The nucleotide composition of DNA was known

The chemical composition of DNA also provided important
clues to its structure. Biochemists knew that DNA is a polymer
of nucleotides. Each of these nucleotides consists of a molecule
of the sugar deoxyribose, a phosphate group, and a nitrogen-
containing base (see Figure 3.1). The only differences among
the four nucleotides of DNA are their nitrogenous bases: the
purines *adenine* (A) and *guanine* (G), and the pyrimidines *cytosine* (C) and *thymine* (T).

In 1950, biochemist Erwin Chargaff at Columbia University
reported an important observation. He and his colleagues had
found that DNA samples from many different species—and
from different sources within a single organism—exhibited certain regularities. The following rule held for each sample: the
amount of adenine equaled the amount of thymine (A = T), and
the amount of guanine equaled the amount of cytosine (G = C).
As a result, the total abundance of purines (A + G) equaled the
total abundance of pyrimidines (T + C):

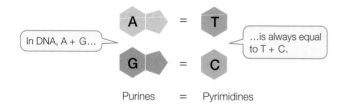

In DNA, A + G…

A   =   T

…is always equal
to T + C.

G   =   C

Purines   =   Pyrimidines

The structure of DNA could not have been worked out without
this observation, now known as *Chargaff's rule*.

Watson and Crick described the double helix

Chemical model building is the assembly of three-dimensional
structures using known relative molecular dimensions and

known bond angles. The English physicist Francis Crick and
the American geneticist James D. Watson (**FIGURE 9.5A**), both
then at the Cavendish Laboratory of Cambridge University,
used model building to solve the structure of DNA.

Watson and Crick attempted to combine all that had been
learned so far about DNA structure into a single coherent
model. Rosalind Franklin's crystallography results convinced
them that the DNA molecule must be **helical**—it must have a
spiral shape like a spring. Density measurements and previous
model building results suggested that there are two polynucleotide chains in the molecule. Modeling studies also showed
that the strands run in opposite directions, that is, they are **antiparallel**. The two strands would not fit together in the model if
they were parallel.

How are nucleotides oriented in DNA chains? Watson and
Crick suggested that:

- the nucleotide bases are on the interior of the two strands,
  with a sugar–phosphate backbone on the outside.
- to satisfy Chargaff's rule (purines = pyrimidines), a purine
  on one strand is always paired with a pyrimidine on the opposite strand. These **base pairs** (A-T and G-C) have the same
  width down the double helix, a uniformity shown by X-ray
  diffraction.

In late February of 1953, Crick and Watson built a model out of
tin that established the general structure of DNA. This structure explained all the known chemical properties of DNA, and
it opened the door to understanding its biological functions.
There have been minor amendments to that first published
structure, but its principal features remain unchanged.

Four key features define DNA structure

Four features summarize the molecular architecture of the
DNA molecule (**FIGURE 9.5B**; also review Figure 3.4):

- It is a *double-stranded helix* of uniform diameter.

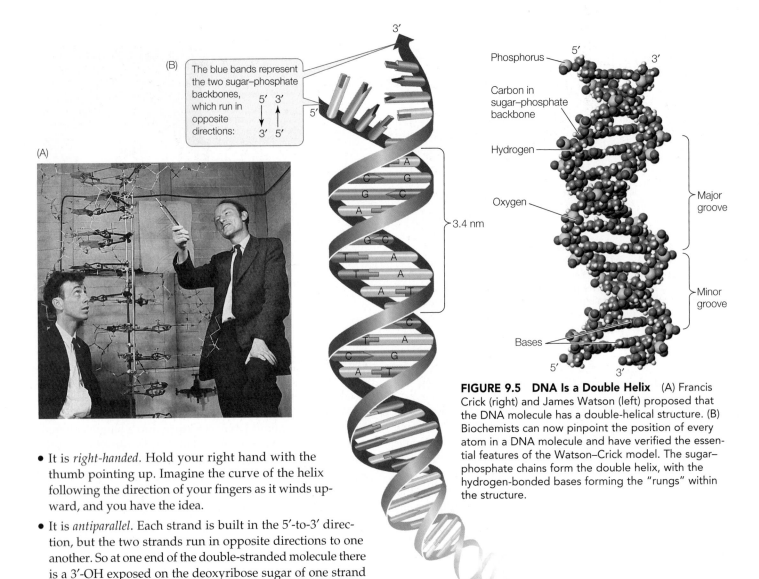

(B) The blue bands represent the two sugar–phosphate backbones, which run in opposite directions:

5′ → 3′
3′ ← 5′

FIGURE 9.5 **DNA Is a Double Helix**   (A) Francis Crick (right) and James Watson (left) proposed that the DNA molecule has a double-helical structure. (B) Biochemists can now pinpoint the position of every atom in a DNA molecule and have verified the essential features of the Watson–Crick model. The sugar–phosphate chains form the double helix, with the hydrogen-bonded bases forming the "rungs" within the structure.

- It is *right-handed*. Hold your right hand with the thumb pointing up. Imagine the curve of the helix following the direction of your fingers as it winds upward, and you have the idea.

- It is *antiparallel*. Each strand is built in the 5′-to-3′ direction, but the two strands run in opposite directions to one another. So at one end of the double-stranded molecule there is a 3′-OH exposed on the deoxyribose sugar of one strand and a 5′-phosphate on the other.

- The outer edges of the nitrogenous bases are exposed in the *major and minor grooves*. These grooves exist because the helices formed by the backbones of the two DNA strands are not evenly spaced relative to one another (see Figure 9.5B). **FIGURE 9.6** shows the four possible configurations of the flat, hydrogen-bonded base pairs within the major and minor grooves. The exposed outer edges of the base pairs are accessible for additional hydrogen bonding. Notice that the arrangements of unpaired atoms and groups differ in the A-T base pairs compared with the G-C base pairs. Thus the *surfaces of the A-T and G-C base pairs are chemically distinct*, allowing other molecules, such as proteins, to recognize specific base-pair sequences and bind to them. The atoms and groups in the major groove are more accessible, and tend to bind other molecules more frequently, than those in the minor groove. *This binding of proteins to specific base-pair sequences is the key to protein–DNA interactions*, which are necessary for the replication and expression of the genetic information in DNA.

While most DNA is a right-handed helix, there are some regions that are left-handed. So-called Z-DNA does not have major and minor grooves and is a tighter helix than normal DNA. Z-DNA appears to form in regions of DNA that are being actively transcribed, and it may play a role in stabilizing the DNA during transcription.

The double-helical structure of DNA is essential to its function

The genetic material performs four important functions, and the DNA structure proposed by Watson and Crick was elegantly suited to three of them.

- *Storage of genetic information*. With its millions of nucleotides, the base sequence of a DNA molecule can encode and store an enormous amount of information. Variations in DNA sequences can account for differences among species and individuals.

- *Precise replication during the cell division cycle*. Replication could be accomplished by complementary base pairing, A with T and G with C. In the original publication of their findings in 1953, Watson and Crick coyly pointed out, "It has not

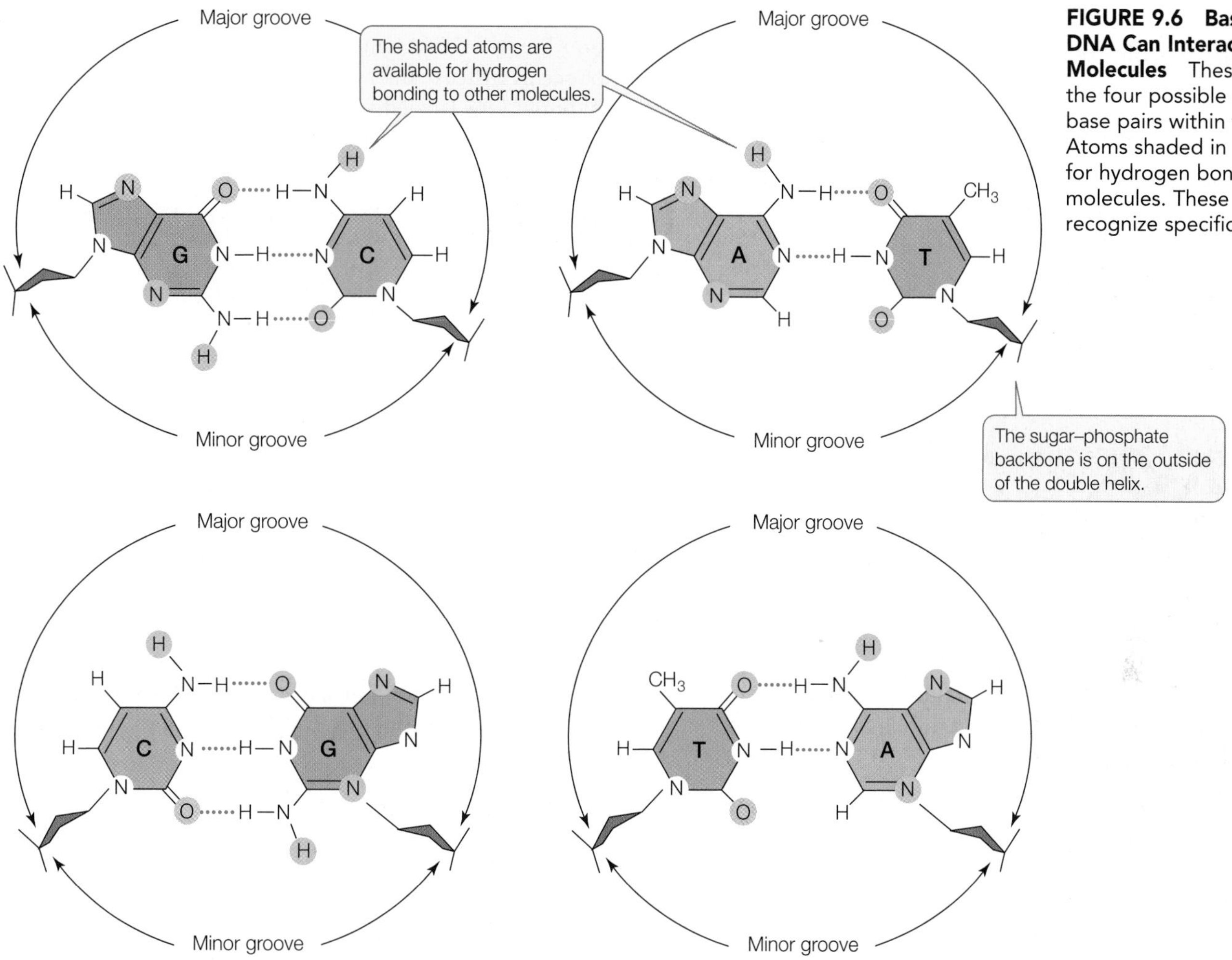

FIGURE 9.6 Base Pairs in DNA Can Interact with Other Molecules These diagrams show the four possible configurations of base pairs within the double helix. Atoms shaded in purple are available for hydrogen bonding with other molecules. These allow proteins to recognize specific DNA sequences.

The shaded atoms are available for hydrogen bonding to other molecules.

The sugar–phosphate backbone is on the outside of the double helix.

escaped our notice that the specific pairing we have postulated immediately suggests a possible copying mechanism for the genetic material." We will turn to DNA replication in Concept 9.2.

- *Susceptibility to mutations.* The structure of DNA suggested an obvious mechanism for mutations: they might be simple changes in the linear sequence of base pairs. We will discuss mutations in Concept 9.3.

- *Expression of the coded information as phenotypes.* The way this function is accomplished is not obvious in the structure of DNA. However, as we describe briefly in Chapter 3 (see Figure 3.5), the nucleotide sequence of DNA is copied (transcribed) into RNA. The linear sequence of nucleotides in RNA is translated into a linear sequence of amino acids—a protein. The folded forms of proteins determine many of the phenotypes of an organism.

LINK Transcription and translation are described in more detail in Concepts 10.2 and 10.4

Do You Understand Concept 9.1?

- Compare circumstantial evidence with experimental evidence in science, using DNA as an example.

- If a DNA molecule is 20 percent G, what are the percentages of the other three bases?

- Describe the evidence that Watson and Crick used to come up with the double-helix model for DNA.

- What would be the consequences if protein and not DNA were the genetic material? What circumstantial and experimental evidence would be acceptable for this conclusion? How would proteins differ from nucleic acids as informational molecules?

We have seen that a DNA molecule consists of long polymers of nucleotides. An individual DNA strand contains thousands or millions (up to about 1 billion) nucleotides in a precise sequence. How is this huge amount of genetic information replicated?

DNA Replicates Semiconservatively

An important requirement for the genetic material is that it replicates *both completely and accurately* during the cell division cycle. The double-helix model of DNA suggested to Watson and Crick how this might be accomplished. **Semiconservative replication** means that each strand of the parental DNA acts as a **template** for a new strand, which is added by base pairing:

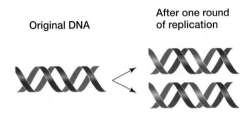

Original DNA   **After one round of replication**

There is abundant evidence supporting this mechanism. In a typical experiment, the parental DNA (represented by the blue strands above) is labeled in some fashion (for example, with a radioactive isotope) and then allowed to replicate in cells for a generation. As the new DNA strands (red in the diagram) are made, they are unlabeled. A *conservative* mode of replication

would show the parental DNA intact with both strands labeled, and the new DNA with both strands unlabeled. This does not occur. Instead, the resulting DNA molecules are always "hybrids" (one labeled strand and one unlabeled strand), providing experimental evidence to support the semiconservative model of replication.

> **yourBioPortal.com**
>
> **Go to ANIMATED TUTORIAL 9.2 Experimental Evidence for Semiconservative DNA Replication**

DNA replication involves a number of different enzymes and other proteins. It takes place in two general steps:

1. The DNA double helix is unwound to separate the two template strands and make them available for new base pairing.

2. As new nucleotides form complementary base pairs with template DNA, they are covalently linked together by phosphodiester bonds, forming a polymer whose base sequence is complementary to the bases in the template strand. The template DNA is read in the 3'-to-5' direction.

During DNA synthesis, *nucleotides are added to the 3' end of the growing new strand*—the end at which the DNA strand has a free hydroxyl (—OH) group on the 3' carbon of its terminal

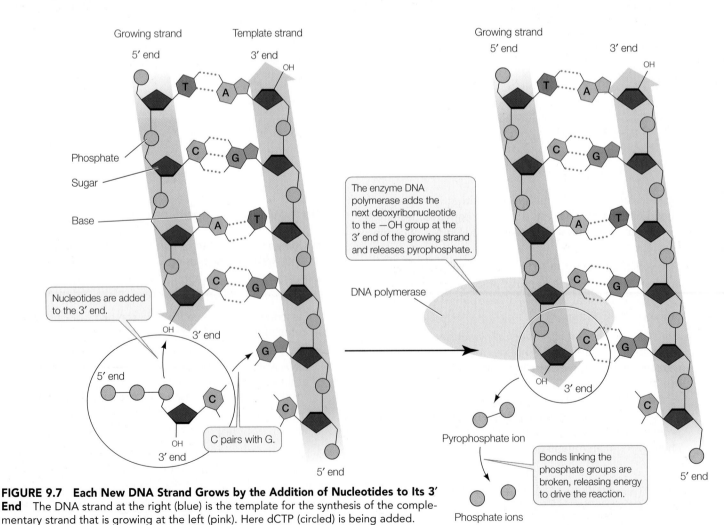

FIGURE 9.7 Each New DNA Strand Grows by the Addition of Nucleotides to Its 3' End The DNA strand at the right (blue) is the template for the synthesis of the complementary strand that is growing at the left (pink). Here dCTP (circled) is being added.

deoxyribose (**FIGURE 9.7**; see also Figure 3.2). As we noted in Concept 3.1, a free nucleotide can have one to three phosphate groups attached to its pentose sugar. The raw materials for DNA synthesis are the four nucleotides deoxyadenosine triphosphate (dATP), deoxythymidine triphosphate (dTTP), deoxycytidine triphosphate (dCTP), and deoxyguanosine triphosphate (dGTP)—collectively referred to as **deoxyribonucleoside triphosphates** (**dNTPs**) or *deoxyribonucleotides*. As their names imply, these nucleotides each carry three phosphate groups. During DNA synthesis, the two outer phosphate groups are released in an exothermic reaction. This provides energy for the formation of a phosphodiester bond between the third phosphate group of the incoming nucleotide and the 3′ position of the sugar at the end of the DNA chain.

> **yourBioPortal.com**
>
> Go to ANIMATED TUTORIAL 9.3 DNA Replication, Part 1: Replication of a Chromosome and DNA Polymerization

DNA polymerases add nucleotides to the growing chain

DNA replication begins with the binding of a large protein complex (the *pre-replication complex*) to a specific site on the DNA molecule. This complex contains several different proteins, among them the enzyme **DNA polymerase**, which catalyzes the addition of nucleotides as the new DNA chain grows. All chromosomes have at least one region called the **origin of replication** (*ori*), to which the pre-replication complex binds.

Binding occurs when proteins in the complex recognize specific DNA sequences within the *ori*.

ORIGINS OF REPLICATION The single circular chromosome of the bacterium *Escherichia coli* has 4×10^6 base pairs (bp) of DNA. The 245 bp *ori* sequence is at a fixed location on the chromosome. Once the pre-replication complex binds to it, the DNA unwinds and replication proceeds in both directions around the circle, forming two **replication forks** (**FIGURE 9.8A**). The replication rate in *E. coli* is approximately 1,000 bp per second, so it takes about 40 minutes to fully replicate the chromosome (with two replication forks). Rapidly dividing *E. coli* cells divide every 20 minutes. In these cells, new rounds of replication begin at the *ori* of each new chromosome before the first chromosome has fully replicated. In this way the cells can divide in less time than the time needed to finish replicating the original chromosome.

Eukaryotic chromosomes are much longer than those of prokaryotes—up to a billion bp—and are linear, not circular. If replication occurred from a single *ori*, it would take weeks to fully replicate a chromosome. So eukaryotic chromosomes have *multiple origins of replication*, scattered at intervals of 10,000–40,000 bp (**FIGURE 9.8B**).

> **FRONTIERS** The organization of multiple DNA growing points along a huge eukaryotic chromosome is not clear. Are the *ori* sequences spaced evenly along the chromosome? How are the many replication forks prevented from interfering with one another? By tagging newly replicating DNA with a visible label, biologists can begin to answer these questions.

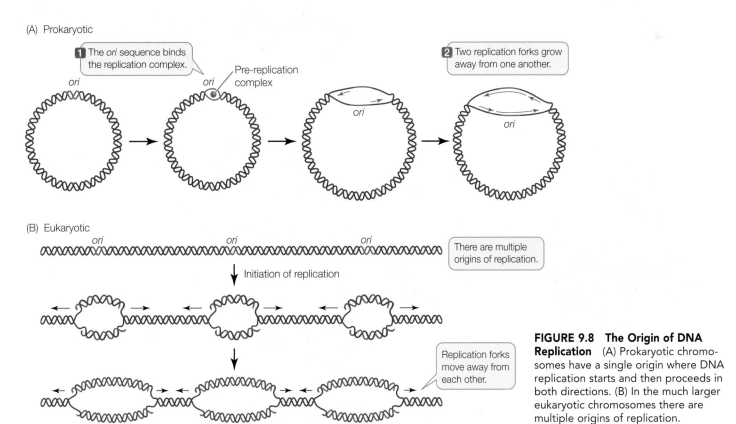

(A) Prokaryotic

1 The *ori* sequence binds the replication complex.

Pre-replication complex

2 Two replication forks grow away from one another.

(B) Eukaryotic

There are multiple origins of replication.

Initiation of replication

Replication forks move away from each other.

FIGURE 9.8 The Origin of DNA Replication (A) Prokaryotic chromosomes have a single origin where DNA replication starts and then proceeds in both directions. (B) In the much larger eukaryotic chromosomes there are multiple origins of replication.

DNA REPLICATION BEGINS WITH A PRIMER DNA polymerase elongates a polynucleotide strand by covalently linking new nucleotides to a preexisting strand. However, it cannot begin this process without a short "starter" strand, called a **primer**. In most organisms this primer is a short single strand of RNA (**FIGURE 9.9**), but in some organisms it is DNA. The primer is complementary to the DNA template and is synthesized one nucleotide at a time by an enzyme called a **primase**. The DNA polymerase then adds nucleotides to the 3′ end of the primer and continues until the replication of that section of DNA has been completed. Then the RNA primer is degraded, DNA is added in its place, and the resulting DNA fragments are connected by the action of other enzymes. When DNA replication is complete, each new strand consists only of DNA.

DNA POLYMERASES ARE LARGE DNA polymerases are much larger than their substrates, the dNTPs, and the template DNA, which is very thin (**FIGURE 9.10A**). Molecular models of the enzyme–substrate–template complex from bacteria show that the enzyme is shaped like an open right hand with a palm, a thumb, and fingers (**FIGURE 9.10B**). Within the "palm" is the active site of the enzyme, which brings together each dNTP substrate and the template. The "finger" regions have precise shapes that can recognize the different shapes of the four nucleotide bases. They bind to the bases by hydrogen bonding and rotate inward.

Most cells contain more than one kind of DNA polymerase, but only one of them is responsible for chromosomal DNA replication. The others are involved in primer removal and DNA repair. Fifteen DNA polymerases have been identified in humans, whereas the bacterium *E. coli* has five DNA polymerases.

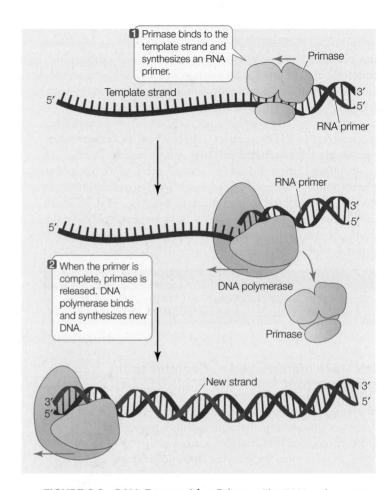

1 Primase binds to the template strand and synthesizes an RNA primer.

2 When the primer is complete, primase is released. DNA polymerase binds and synthesizes new DNA.

FIGURE 9.9 DNA Forms with a Primer The DNA polymerase requires a primer: a "starter" strand of RNA or DNA to which new nucleotides can be added.

yourBioPortal.com

Go to WEB ACTIVITY 9.1
DNA Polymerase

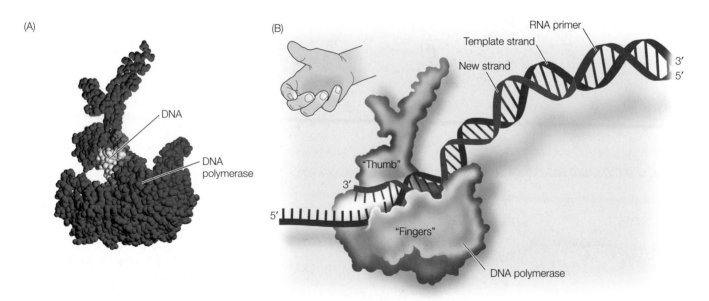

FIGURE 9.10 DNA Polymerase Binds to the Template Strand (A) The DNA polymerase enzyme (blue) is much larger than the DNA molecule (red and white). (B) DNA polymerase is shaped like a hand, and in this side-on view, its "fingers" can be seen curling around the DNA. These fingers can recognize the distinctive shapes of the four bases.

APPLY THE CONCEPT

DNA replicates semiconservatively

Each eukaryotic chromosome is composed of a double-stranded DNA molecule. It is possible to use a microscope to distinguish between a chromosome that has been labeled with radioactivity and one that is unlabeled. Plant cells were grown in the lab in the presence of radioactive thymidine for many generations. The cells were then grown for three cell doublings in nonradioactive thymidine so that any new DNA strands would be unlabeled. These cultured cells had synchronized cell cycles. They were examined when they were in anaphase of mitosis.

DOUBLINGS	LABELED CHROMOSOMES (%)
0	100
1	100
2	50
3	25

1. Use diagrams of double-stranded DNA to explain these data.

2. What would the data and diagrams look like if DNA replicated conservatively?

The two DNA strands grow differently at the replication fork

A single replication fork opens up in one direction. Study **FIGURE 9.11** and try to imagine what is happening over a short period of time. For the purpose of understanding the process, imagine that the DNA opens from one end like a zipper (as we have seen, it actually opens up from within the molecule and replication extends in both directions). Remember two things:

- The two DNA strands are antiparallel—that is, the 3′ end of one strand is paired with the 5′ end of the other.
- DNA replicates in a 5′-to-3′ direction.

One newly synthesized strand—the **leading strand**—is oriented so that it can grow continuously at its 3′ end as the fork opens up. The other new strand—the **lagging strand**—must be synthesized differently because it grows in the direction away from the replication fork.

Synthesis of the lagging strand requires the synthesis of relatively short, *discontinuous stretches* of sequence (100–200 nucleotides in eukaryotes; 1,000–2,000 nucleotides in prokaryotes). These discontinuous stretches are synthesized just as the leading strand is, by the addition of new nucleotides one at a time to the 3′ end, but the new strand grows away from the replication fork. These stretches of new DNA are called **Okazaki fragments** after their discoverer, the Japanese biochemist Reiji Okazaki. To summarize, while the leading strand grows continuously "forward," the lagging strand grows in shorter, "backward" stretches with gaps between them.

yourBioPortal.com

**Go to ANIMATED TUTORIAL 9.4
DNA Replication, Part 2: Coordination of
Leading and Lagging Strand Synthesis**

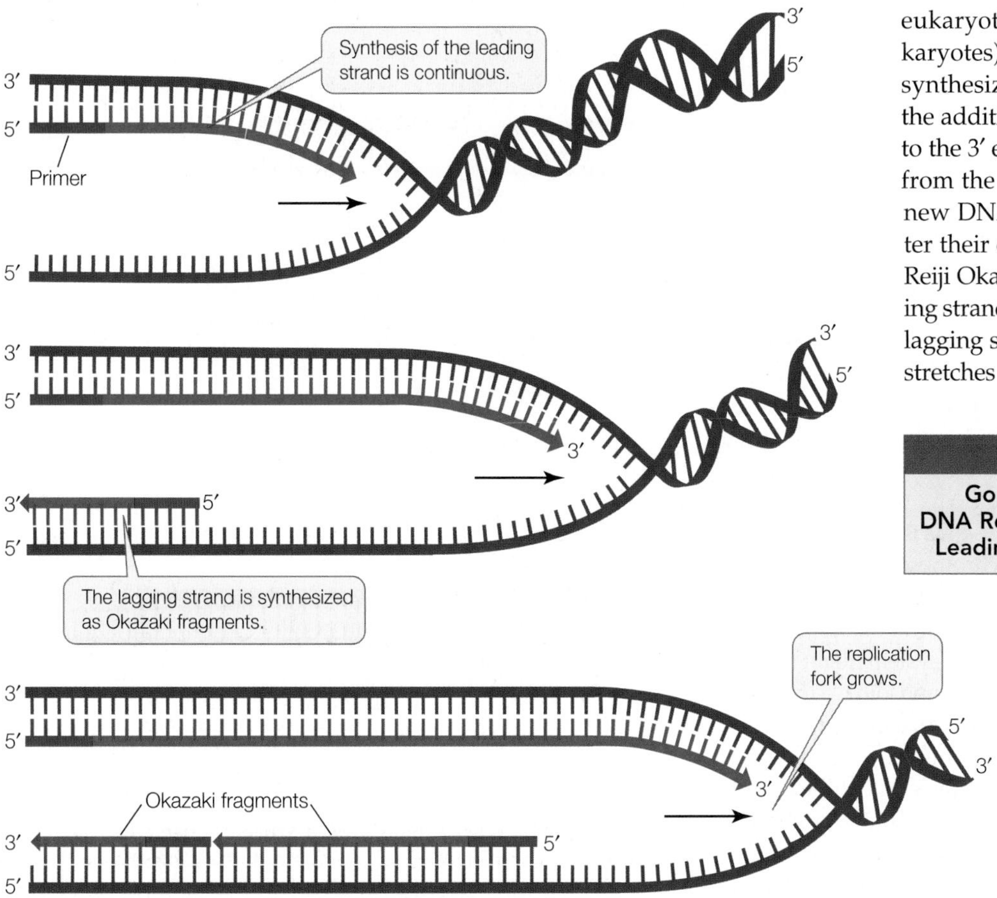

FIGURE 9.11 The Two New Strands Form in Different Ways As the parent DNA unwinds, both new strands are synthesized in the 5′-to-3′ direction, although their template strands are antiparallel. The leading strand grows continuously forward, but the lagging strand grows in short, discontinuous stretches called Okazaki fragments. Okazaki fragments in eukaryotes are 100 to 200 nucleotides long.

A single primer is needed to initiate synthesis of the leading strand, but each Okazaki fragment requires its own primer to be synthesized by the primase. DNA polymerase then synthesizes an Okazaki fragment by adding nucleotides to one primer until it reaches the primer of the previous fragment. At this point, a different DNA polymerase removes the old primer and replaces it with DNA. Left behind is a tiny nick—the final phosphodiester linkage between the adjacent Okazaki fragments is missing. The enzyme **DNA ligase** catalyzes the formation of that bond, linking the fragments and making the lagging strand whole (**FIGURE 9.12**).

DNA replication may appear complex (we have simplified it considerably), but it occurs with astonishing speed and accuracy. As we mentioned earlier, the rate of replication in *E. coli* is about 1,000 base pairs per second, yet the polymerase commits very few errors—less than 1 base in a million. How do DNA polymerases work so fast? We saw in Concept 3.3 that an enzyme catalyzes a chemical reaction through a series of events:

substrate binds to enzyme → one product is formed →
enzyme is released → cycle repeats

DNA replication would not proceed as rapidly as it does if DNA polymerase went through such a cycle for each nucleotide. Instead, DNA polymerase is **processive**—that is, it *catalyzes many sequential polymerization reactions each time it binds to a DNA molecule*:

substrates bind to enzyme → many products are formed →
enzyme is released → cycle repeats

Typically, a DNA polymerase can add thousands of nucleotides before it detaches from DNA.

Telomeres are not fully replicated in most eukaryotic cells

As we have just seen, replication of the lagging strand occurs by the addition of Okazaki fragments to RNA primers. When the terminal RNA primer is removed from the replicating end of a linear eukaryotic chromosome, no DNA can be synthesized to replace it because there is no 3′ end to extend (**FIGURE 9.13A**). So the new chromosome has a bit of single-stranded DNA at each end. This situation activates a mechanism for cutting off the single-stranded region, along with some of the intact double-stranded DNA. Thus the chromosome becomes slightly shorter with each cell division.

Another problem with chromosome ends is that they must be protected from being joined to other chromosomes by the DNA repair system. When DNA is damaged by external or internal agents (e.g., radiation), it is repaired by a combination of DNA polymerase and DNA ligase activities. This system might mistakenly recognize chromosome ends as breaks and join two chromosomes together. This would create havoc with genomic integrity.

To prevent chromosomes from joining, many eukaryotes have strings of repetitive sequences at the ends of their chromosomes called **telomeres**. In humans and other vertebrates, the repeated sequence is TTAGGG, and in humans it is repeated about 2,500 times. These repeats bind special proteins

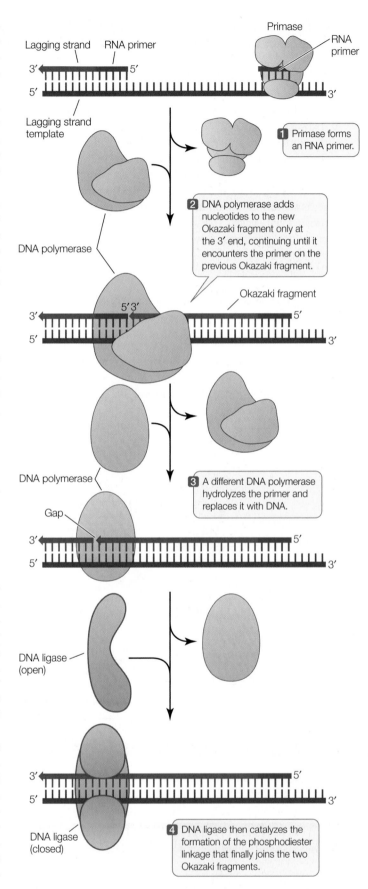

1 Primase forms an RNA primer.

2 DNA polymerase adds nucleotides to the new Okazaki fragment only at the 3′ end, continuing until it encounters the primer on the previous Okazaki fragment.

3 A different DNA polymerase hydrolyzes the primer and replaces it with DNA.

4 DNA ligase then catalyzes the formation of the phosphodiester linkage that finally joins the two Okazaki fragments.

FIGURE 9.12 The Lagging Strand Story In bacteria, DNA polymerases and DNA ligase cooperate to complete the complex task of synthesizing the lagging strand.

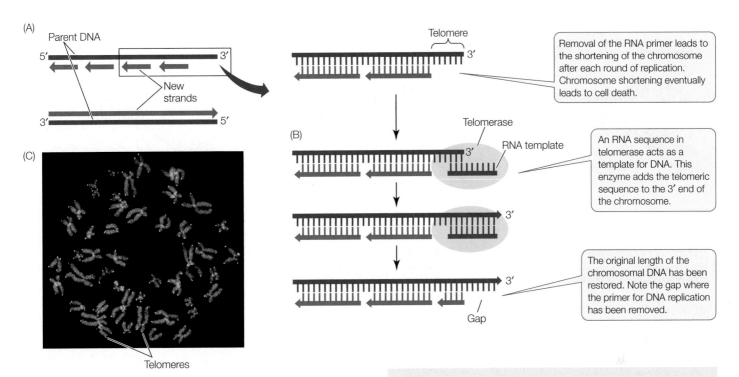

FIGURE 9.13 Telomeres and Telomerase (A) The chromosome ends contain regions of repetitive DNA called telomeres. Removal of the RNA primer at the 3′ end of the template for the lagging strand leaves a short sequence that is unreplicated. (B) In continuously dividing cells, the enzyme telomerase binds to the 3′ end and extends the lagging strand of DNA, so the chromosome does not get shorter. (Although shown here as a single short sequence, actual telomeres are much longer, with many repeats of a short sequence.) (C) Bright fluorescent staining marks the telomeric regions on these blue-stained human chromosomes.

that protect the ends from being joined together by the DNA repair system. In addition, the repeats form loops that have a similar protective role. So telomeres act like the plastic tips of shoelaces to prevent fraying.

Each human chromosome can lose 50–200 bp of telomeric DNA after each round of DNA replication and cell division. After 20–30 cell divisions, the chromosome ends become short enough to lose their protective role, and the chromosomes lose their integrity. Apoptosis (programmed cell death) ensues, and the cell dies. This phenomenon explains, in part, why many cell lineages do not last the entire lifetime of the organism: their telomeres are lost. Yet continuously dividing cells, such as bone marrow stem cells and gamete-producing cells, maintain their telomeric DNA. An enzyme, appropriately called **telomerase**, catalyzes the addition of any lost telomeric sequences in these cells (**FIGURE 9.13B**). Telomerase contains an RNA sequence that acts as a template for the telomeric DNA repeat sequence.

There is a relationship between telomere length and aging: the average telomere length is shorter in older individuals. Furthermore, when a gene expressing high levels of telomerase is added to human cells in culture, their telomeres do not shorten. Instead of living 20–30 cell generations and then dying, the cells become immortal. It remains to be seen how this finding relates to the aging of a whole organism.

FRONTIERS Telomerase is expressed in more than 90 percent of human cancers and may be an important factor in the ability of cancer cells to divide continuously. Since most normal cells do not have this ability, telomerase is an attractive target for drugs designed to attack tumors specifically.

Errors in DNA replication can be repaired

We have stressed that DNA must be accurately replicated; the accurate transmission of genetic information is essential for the proper functioning and even the life of a single cell or multicellular organism. Yet the replication of DNA is not perfectly accurate. DNA polymerases sometimes insert a base that is not complementary to the template (for example, putting an A in the new DNA strand opposite a C in the template strand). In eukaryotes, the error rate is about 1 incorrect base in 100,000 (a 10^{-5} error rate). With a genome size of 4×10^9 bp, this would produce 40,000 errors after every cell division. This is an intolerable mutation rate for survival in the long term. However, if eukaryotic DNA sequences are studied before and after a cell cycle, the actual frequency of DNA errors is 10^{-10} per cell cycle. This means that most of the errors are repaired. There are two major repair mechanisms:

- *Proofreading* occurs right after DNA polymerase inserts a nucleotide (**FIGURE 9.14A**). When a DNA polymerase recognizes a mispairing of bases, it removes the improperly introduced nucleotide and tries again.

- *Mismatch repair* occurs after DNA has been replicated (**FIGURE 9.14B**). A second set of proteins surveys the newly replicated molecule and looks for mismatched base pairs that were missed in proofreading. A portion of the DNA including the incorrect nucleotide is removed and replaced by DNA polymerase.

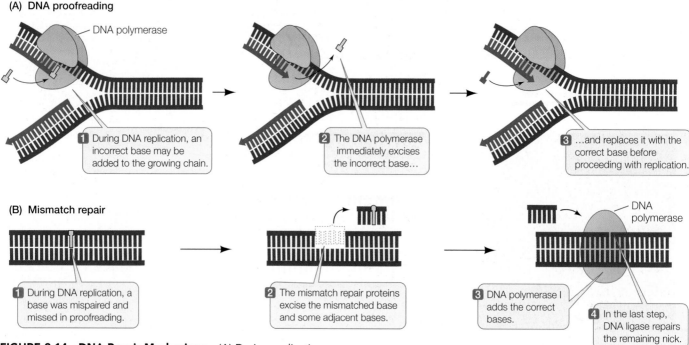

(A) DNA proofreading

DNA polymerase

1 During DNA replication, an incorrect base may be added to the growing chain.

2 The DNA polymerase immediately excises the incorrect base...

3 ...and replaces it with the correct base before proceeding with replication.

(B) Mismatch repair

1 During DNA replication, a base was mispaired and missed in proofreading.

2 The mismatch repair proteins excise the mismatched base and some adjacent bases.

3 DNA polymerase I adds the correct bases.

DNA polymerase

4 In the last step, DNA ligase repairs the remaining nick.

FIGURE 9.14 DNA Repair Mechanisms (A) During replication, the DNA polymerase checks for incorrect bases in the new DNA strand and immediately replaces them with correct ones. This process is called proofreading. (B) After replication, mismatch repair proteins search for incorrect bases that were missed by DNA polymerase and replace them.

The basic mechanisms of DNA replication can be used to amplify DNA in a test tube

The principles underlying DNA replication in cells have been used to develop a laboratory technique that has been vital in analyzing DNA, genes, and genomes. The **polymerase chain reaction** (**PCR**) allows researchers to make multiple copies of short DNA sequences in a test tube—a process referred to as *DNA amplification*. The PCR technique uses:

• a sample of double-stranded DNA, often from a biological sample, to act as the template

• two short, artificially synthesized primers that are complementary to the ends of the sequence to be amplified

• the four dNTPs (dATP, dTTP, dCTP, and dGTP)

• a DNA polymerase that can tolerate high temperatures without becoming denatured

• salts and a buffer to maintain a near-neutral pH.

PCR is a cyclic process in which a sequence of steps is repeated over and over again (**FIGURE 9.15**). Since DNA replication is fast even in a test tube, it takes only a short time to go from 1 to 2 to 4 to million short segments of DNA. The PCR technique requires that the base sequences at each end of the amplified fragment be known ahead of time, so that complementary primers, usually 15–30 bases long, can be made in the laboratory. Because of the uniqueness of DNA sequences, a pair of primers this length will usually bind to only a single region of DNA in an organism's genome. This specificity is a key to the power of PCR to amplify just a small part of a larger DNA molecule. Some of the most striking applications of PCR will be described in Chapters 11 through 13. These applications range from the identification of individuals by their DNA to the detection of diseases.

Do You Understand Concept 9.2?

• What is semiconservative DNA replication?

• Why does the leading strand in DNA replicate continuously and the lagging strand discontinuously?

• Cells from older people have shorter telomeres than cells from younger people. How might this relate to aging?

• If you have a small amount of a large chromosome of 20 million bp and want to amplify a short sequence of 1,000 bp, how would you do it? Explain the role of primers in this process.

We have described how the precise replication of DNA satisfies one of the requirements for proving that DNA is the genetic material. A less obvious requirement is its ability to mutate. Mutation is needed to create variability in DNA, which is the raw material for evolution.

RESEARCH TOOLS

FIGURE 9.15 **The Polymerase Chain Reaction** The steps in this cyclic process are repeated many times to produce millions of identical copies of a DNA fragment. This makes enough DNA for chemical analysis and genetic manipulations.

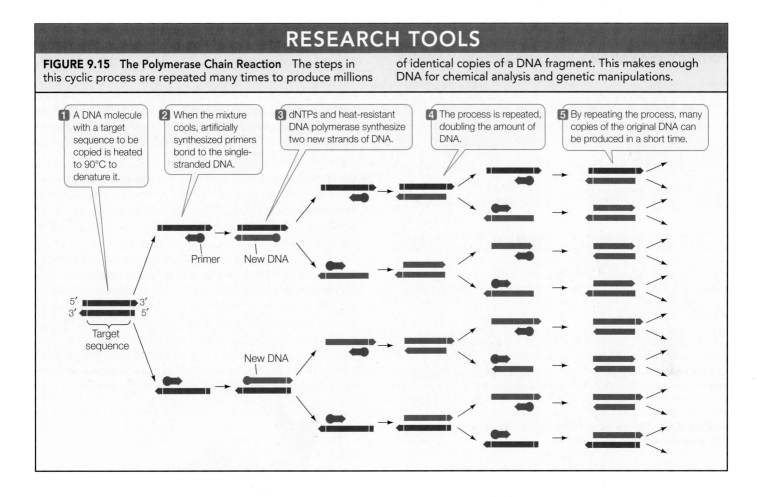

1 A DNA molecule with a target sequence to be copied is heated to 90°C to denature it.

2 When the mixture cools, artificially synthesized primers bond to the single-stranded DNA.

3 dNTPs and heat-resistant DNA polymerase synthesize two new strands of DNA.

4 The process is repeated, doubling the amount of DNA.

5 By repeating the process, many copies of the original DNA can be produced in a short time.

Primer   New DNA

5′ 3′
3′ 5′

Target sequence

New DNA

Mutations Are Heritable Changes in DNA

In Chapter 8 we describe mutations as inherited changes in genes. A mutation may result in a new allele of a gene, and different alleles may produce different phenotypes (for example, pea plants with wrinkled seeds versus spherical seeds). With reference to the chemical nature of genes, we can state that *mutations are changes in the nucleotide sequence of DNA that are passed on from one cell, or organism, to another.*

Mutations occur by a variety of processes. For example, in Concept 9.2 we described how DNA polymerases make errors. Repair systems such as proofreading are in place to correct them, but some errors escape being corrected and are passed on to the daughter cells.

Mutations in multicellular organisms can be divided into two types:

- **Somatic mutations** occur in somatic (body) cells. They are passed on to the daughter cells during mitosis, and in turn to the offspring of those cells. For example, a mutation in a single skin cell could result in a patch of skin cells that all have the same mutation. However, somatic mutations are not passed on to sexually produced offspring. (Exceptions occur in plants, where germline cells can arise from somatic cells and thus pass on somatic mutations.)

- **Germline mutations** occur in the cells of the *germ line*—the specialized cells that give rise to gametes. A gamete with the mutation passes it on to a new organism at fertilization.

In either case, the mutations may or may not have phenotypic effects.

Mutations can have various phenotypic effects

An organism's genome is the total DNA sequence present in all of its chromosomes (or in its single chromosome, in the case of prokaryotes). Depending on the organism, it can consist of millions or billions of base pairs of DNA. As we discuss further in Chapter 12, most genomes include both genes and regions of DNA that are not expressed:

- Genes are transcribed into RNAs. In most cases these RNAs are in turn translated into the amino acid sequences of proteins. Some other genes are transcribed to produce RNA molecules with catalytic functions, such as ribosomal RNAs. All genes consist of the transcribed region and regions at the 5′ and 3′ ends, which are not transcribed but have sequences that bind proteins involved in transcription. A protein-coding gene has sequences within the transcribed region that are translated—these are called the *coding regions*.

- Most genomes also contain extensive regions of DNA that are not expressed (see Chapter 12 for more information on those regions).

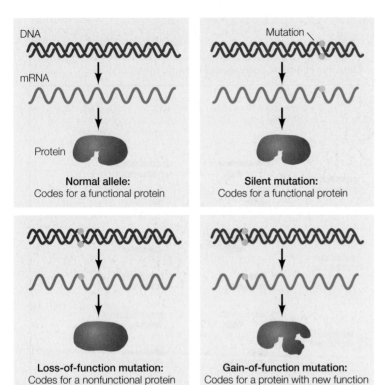

FIGURE 9.16 **Mutation and Phenotype** Mutations may or may not affect the function of a protein.

Mutations are often discussed in terms of their effects on protein-coding genes and their functions (**FIGURE 9.16**):

- **Silent mutations** do not affect gene function. They can be mutations in DNA that is not expressed, or mutations within an expressed region that do not have any effect on the encoded protein. *Most mutations in large genomes are silent.*

> **LINK** Silent mutations are a source of neutral alleles in evolution; see Concept 15.2

- **Loss-of-function mutations** can result in either the loss of expression of a gene or in the production of a nonfunctional protein or RNA. Some loss-of-function mutations prevent a gene from being transcribed or cause transcription to terminate too soon. In other cases the gene is transcribed and translated, but the resulting protein no longer works as a structural protein or enzyme (as illustrated in Figure 9.16). Loss-of-function mutations almost always show recessive inheritance in a diploid organism, because the presence of one wild-type allele usually results in sufficient functional protein for the cell. For example, the wrinkled seed phenotype studied by Mendel (see Figure 8.1) is due to a recessive, loss-of-function mutation in the gene for starch *branching* enzyme 1 (*SBE1*). Even in plants with only one copy of the wild-type allele, there is enough SBE1 enzyme to produce the wild-type, spherical phenotype.

- **Gain-of-function mutations** lead to a protein with an altered function. This kind of mutation usually shows dominant

inheritance, because the presence of the wild-type allele does not prevent the mutant allele from functioning. This type of mutation is common in cancer. For example, a receptor for a growth factor normally requires binding of the growth factor (the ligand) to activate the cell division cycle. Some cancers are caused by mutations in genes that encode these receptors so that the receptors are "always on," even in the absence of their ligands. This leads to the unrestrained cell proliferation that is characteristic of cancer cells.

- **Conditional mutations** cause their phenotypes only under certain *restrictive* conditions. The wild-type phenotype is expressed under other, *permissive* conditions. Many conditional mutants are temperature-sensitive; that is, they show the altered phenotype only at a certain temperature. For example, in cats and rabbits the *C* gene encodes the enzyme tyrosinase, which is involved in pigment production (see Figure 8.9). Some breeds carry the conditional mutant allele c^H, which encodes a tyrosinase protein that denatures easily at temperatures above 35°C. In animals with no dominant wild-type (*C*) allele, the body fur is pale because there is no functional pigment-producing enzyme at body temperature (37°C). However, the extremities—feet, ears, nose, and tail—are cooler, typically less than 30°C, so they are dark:

All mutations are alterations in the nucleotide sequence of DNA. At the molecular level, we can divide mutations into two categories:

- A **point mutation** results from the gain, loss, or substitution of a single nucleotide. After DNA replication, the altered nucleotide becomes a mutant base pair. If a point mutation occurs within a gene, it results in a new allele of that gene. The new allele may or may not result in a new phenotype.

- **Chromosomal mutations** are extensive changes in the chromosomal structure. They may change the position or orientation of a DNA segment without actually removing any genetic information, or they may cause a segment of DNA to be duplicated or irretrievably lost.

Point mutations change single nucleotides

Point mutations result from the addition or subtraction of a nucleotide base, or the substitution of one base for another. Point mutations can arise because of errors in DNA replication that are not corrected during proofreading, or they may be caused by environmental **mutagens**: substances that cause mutations, such as radiation or certain chemicals.

Some point mutations that occur within genes are loss-of-function mutations because they prevent the gene from being properly transcribed. In other cases the gene is transcribed normally. A point mutation in the coding region of a gene may result in changes in the RNA, but changes in the RNA may or may not result in a change in the amino acid sequence of the protein. If the protein is not changed, the mutation is silent.

> **LINK** The genetic code explains why some point mutations in coding regions of DNA are silent; see Figure 10.11

Other mutations result in altered amino acid sequences, and in some cases these changes can have drastic phenotypic effects. An example of a point mutation with a significant effect on phenotype is the one that causes *sickle-cell disease*, a heritable blood disorder. The disease occurs in people who carry two copies of the sickle allele of the gene for human β-globin (a subunit of hemoglobin, the protein in human blood that carries oxygen—see p. 44). The sickle allele differs from the normal allele by one base pair, resulting in a polypeptide that differs by one amino acid from the normal protein. Individuals who are homozygous for this recessive allele have defective, sickle-shaped red blood cells:

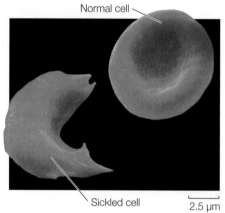

The deformed cells tend to block narrow capillaries, which results in tissue damage. Not all changes in the amino acid sequence of a protein affect its function. For example, a hydrophilic amino acid may be substituted for another hydrophilic amino acid, so that the shape of the protein is unchanged. Or a mutation might result in a protein that has reduced efficiency but is not completely inactivated. Individuals homozygous for a point mutation of this type may survive if enough of the protein's function is retained.

In some cases, gain-of-function point mutations occur. An example is a class of mutations in the human gene *TP53*, which encodes the tumor suppressor protein p53. The p53 protein normally functions to inhibit the cell cycle, but certain mutations cause the protein to promote the cell cycle and prevent programmed cell death. So a p53 protein mutated in this way has a gain of oncogenic (cancer-causing) function.

Chromosomal mutations are extensive changes in the genetic material

The most dramatic changes that can occur in the genetic material are chromosomal mutations. Whole chromosomes can break and rejoin, grossly disrupting the sequences of genes. There are four types of chromosomal mutations: *deletions, duplications, inversions,* and *translocations*. This kind of severe damage to chromosomes can result from mutagens or from drastic errors in chromosome replication.

- **Deletions** result in the removal of part of the genetic material (**FIGURE 9.17A**). Their consequences can be severe or even fatal. It is easy to imagine one mechanism that could produce a deletion: a DNA molecule might break at two points and the two end pieces might rejoin, leaving out the DNA between the breaks.

- **Duplications** can be produced at the same time as deletions (**FIGURE 9.17B**). A duplication would arise if homologous chromosomes broke at different positions and then reconnected to the wrong partners. One of the two chromosomes produced by this mechanism would lack a segment of DNA (it would have a deletion), and the other would have two copies (a duplication) of the segment that was deleted from the first chromosome.

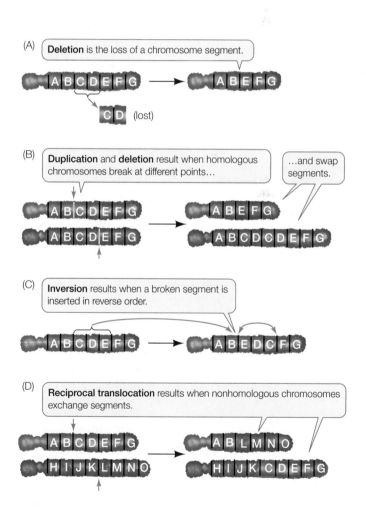

FIGURE 9.17 Chromosomal Mutations Chromosomes may break during replication, and parts of chromosomes may then rejoin incorrectly. The letters on these illustrations represent large chromosomal segments containing anywhere from zero to hundreds or thousands of genes.

- **Inversions** can also result from the breaking and rejoining of chromosomes. A segment of DNA may be removed and reinserted into the same location in the chromosome, but "flipped" end over end so that it runs in the opposite direction (**FIGURE 9.17C**). If either break site occurs within a gene, it is likely to cause a loss-of-function mutation in that gene.

- **Translocations** result when segments of chromosomes break off and become joined to different chromosomes. Translocations may involve reciprocal exchanges of chromosome segments, as in **FIGURE 9.17D**. Translocations often lead to duplications and deletions and may result in sterility if normal chromosome pairing cannot occur during meiosis.

> **LINK** Another way mutations can occur is through mobile DNA elements called transposons; see Figure 12.6

Mutations can be spontaneous or induced

When thinking about the causes of mutations, it is useful to distinguish between mutations that are *spontaneous* and those that are *induced*.

Spontaneous mutations are permanent changes in the genetic material that occur without any outside influence. In other words, they occur simply because cellular processes are imperfect. Spontaneous mutations may occur by several mechanisms:

- *DNA polymerase can make errors in replication.* Most of these errors are repaired by the proofreading function of the replication complex, but some errors escape detection and become permanent.

- *The four nucleotide bases of DNA have alternate structures that affect base pairing.* Each nucleotide can exist in two different forms (called *tautomers*), one of which is common and one rare. When a base temporarily forms its rare tautomer, it can pair with the wrong base (**FIGURE 9.18A, C**).

- *Bases in DNA may change because of spontaneous chemical reactions.* One such reaction is the *deamination* (conversion of an amino group to a keto group) in cytosine to form the base uracil, which pairs with A rather than G. Usually these errors are repaired, but since the repair mechanism is not perfect, the altered nucleotide will sometimes remain and cause a permanent base change after replication.

- *Meiosis is not perfect.* Sometimes errors occur during the complex process of meiosis. This can result in nondisjunction and

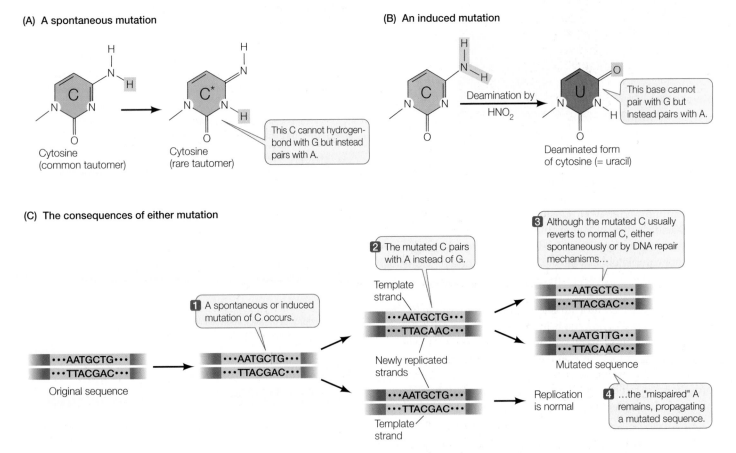

(A) A spontaneous mutation

Cytosine (common tautomer) → Cytosine (rare tautomer)

This C cannot hydrogen-bond with G but instead pairs with A.

(B) An induced mutation

C — Deamination by HNO₂ → U — Deaminated form of cytosine (= uracil)

This base cannot pair with G but instead pairs with A.

(C) The consequences of either mutation

Original sequence
···AATGCTG···
···TTACGAC···

1 A spontaneous or induced mutation of C occurs.
···AATGCTG···
···TTACGAC···

2 The mutated C pairs with A instead of G.

Template strand
···AATGCTG···
···TTACAAC···
Newly replicated strands

3 Although the mutated C usually reverts to normal C, either spontaneously or by DNA repair mechanisms…
···AATGCTG···
···TTACGAC···

···AATGTTG···
···TTACAAC···
Mutated sequence

Template strand
···AATGCTG···
···TTACGAC···

Replication is normal

4 …the "mispaired" A remains, propagating a mutated sequence.

FIGURE 9.18 Spontaneous and Induced Mutations (A) Each nitrogenous base exists in both a common (prevalent) form and a rare form. When a base spontaneously switches to its rare tautomer, it can pair with a different base. (B) Mutagens such as nitrous acid can induce changes in the bases. (C) The results of both spontaneous and induced mutations are permanent changes in the DNA sequence following replication.

aneuploidy (see Concept 7.4) or chromosomal breakage and rejoining (discussed above).

- *Gene sequences can be disrupted.* Random chromosome breakage and rejoining can produce deletions, duplications, inversions, or translocations.

Induced mutations occur when some agent from outside the cell—a mutagen—causes a permanent change in the DNA sequence:

- *Some chemicals alter the nucleotide bases.* For example, nitrous acid (HNO_2) reacts with cytosine and converts it to uracil by deamination (**FIGURE 9.18B**). This alteration has the same result as spontaneous deamination: instead of a G, DNA polymerase inserts an A (see Figure 9.18C).

- *Some chemicals add groups to the bases.* An example is benzopyrene, a component of cigarette smoke that adds a large chemical group to guanine, making it unavailable for base pairing. When DNA polymerase reaches such a modified guanine, it inserts any one of the four bases, resulting in a high frequency of mutations.

- *Radiation damages the genetic material.* Radiation can damage DNA in three ways. First, ionizing radiation (including X rays, gamma rays, and particles emitted by unstable isotopes) can detach electrons from atoms or molecules and produce highly reactive chemicals called *free radicals*. Free radicals can change bases in DNA to forms that are not recognized by DNA polymerase. Second, ionizing radiation can also break the sugar–phosphate backbone of DNA, causing chromosomal abnormalities. And third, ultraviolet radiation (from the sun or a tanning lamp) can cause thymine bases to form covalent bonds with adjacent thymines. This, too, plays havoc with DNA replication by distorting the double helix, and can result in a mutation.

Some base pairs are more vulnerable than others to mutation

DNA sequencing has revealed that mutations occur most often at certain base pairs. These "hotspots" are often located where cytosine has been methylated to 5-methylcytosine. (Methylation of DNA is a normal process in the regulation of chromatin structure and gene expression; see Concept 11.3.)

As we discussed above, unmethylated cytosine can lose its amino group to form uracil (see Figure 9.18B). This error is usually repaired because uracil (which normally occurs in RNA) is recognized by the repair mechanism as an inappropriate component of DNA. When 5-methylcytosine loses its amino group, however, the product is thymine, a natural base in DNA. The DNA repair mechanism ignores the thymine, and in this case, half of the new DNA molecules contain A-T base pairs instead of the original G-C pair (**FIGURE 9.19**). Thus the frequency of mutation is much greater in regions containing 5-methylcytosine.

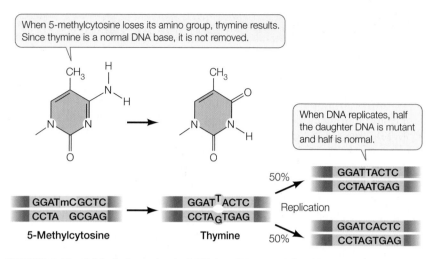

FIGURE 9.19 5-Methylcytosine in DNA Is a "Hotspot" for Mutations If cytosine has been methylated to 5-methylcytosine and then becomes deaminated, the mutation is usually not repaired, and a C-G base pair is replaced with a T-A base pair.

Mutagens can be natural or artificial

Many people associate mutagens with materials made by humans, but there are also many naturally occurring mutagens. Plants (and to a lesser extent animals) make thousands of small molecules that serve a range of purposes, including defense against pathogens (see Concept 28.1). Some of these are mutagenic and potentially carcinogenic. An example of a naturally occurring mutagen is aflatoxin, which is made by the mold *Aspergillus*. When mammals ingest the mold, the aflatoxin is converted into a product that binds to guanine and causes mutations, just as benzopyrene from cigarette smoke does.

Radiation can be human-made or natural. Some of the isotopes made in nuclear reactors and nuclear bomb explosions are certainly harmful—as was shown by the increased mutation rates in survivors of the atom bombs dropped on Japan in 1945. A certain amount of radiation comes from space in the form of cosmic radiation, and as mentioned above, natural ultraviolet radiation in sunlight also causes mutations.

Biochemists have estimated how much DNA damage occurs in the human genome under normal circumstances: among the haploid genome's 3 billion base pairs, there are about 16,000 DNA-damaging events per cell per day, of which 80 percent are repaired.

Mutations have both benefits and costs

What is the overall effect of mutation? For a species as a whole, the evolutionary benefits are clear. But there are costs as well as benefits for individual organisms.

BENEFITS OF MUTATIONS Mutations are the raw material of evolution. As you will see in Part 3 of this book, mutation alone does not drive evolution, but it provides the genetic diversity that makes natural selection possible. A mutation in a germline cell may have no immediate selective advantage to the organism, but it may cause a phenotypic change in the offspring. If

APPLY THE CONCEPT

Mutations are changes in DNA

Nitrosamines ($R-N-NO_2$) are potent mutagens. They can be formed from the reaction of nitrites ($R-NO_2$) with amino groups in proteins. So there are concerns about using nitrites to preserve meats, which contain amino groups. An experiment was performed to test the effect of vitamin C (ascorbate) on mutagenesis by meats cured with nitrites. Bacterial cells were incubated with cured meat extracts in the presence or absence of ascorbate. The rate of mutation (number of mutant bacteria per total bacteria) was determined:

1. What did the experiment with no extract and no ascorbate show?

2. What did the experiments with increasing amounts of extract and no ascorbate show?

3. What was the effect of ascorbate on the mutation rate?

4. In the bacterium tested, the wild-type DNA had the sequence 5′-ACTTAT-3′, and the mutated strain had the sequence 5′-ATTTAT-3′. What does this tell you about the nature of the mutation involved? Clearly outline the steps in mutagenesis, noting DNA replication(s).

MEAT EXTRACT (μg/mL)	AMOUNT OF ASCORBATE (μg/mL)	RATE OF MUTATION ($\times 10^{-5}$)
0	0	2
0	50	2
10	0	5
10	50	2
20	0	14
20	50	2
30	0	35
30	50	2

the environment changes in a later generation, that mutation may be advantageous, enabling the species as a whole to adapt to changing conditions. A mutation in a somatic cell can sometimes benefit the individual organism, particularly if it occurs in a stem cell that produces a large number of offspring cells.

COSTS OF MUTATIONS Mutations can be harmful if they result in the loss of function of genes (and their protein products) or other DNA sequences that are needed for survival. A harmful mutation in a germline cell may be inherited in heterozygous form by the organism's descendants. If two individuals carrying the mutation mate, some of the offspring may be homozygous for the mutation. In their extreme form, such mutations produce phenotypes that are lethal. Lethal mutations can kill an organism during early development, or the organism may die before it matures and reproduces.

In Chapter 7 we described how mutations in somatic cells can lead to cancer. Typically these are mutations in *oncogenes* that result in the stimulation of cell division, or mutations in *tumor suppressor genes* that result in a lack of inhibition of cell division. These mutations can occur by either spontaneous or induced mutagenesis.

MINIMIZING EXPOSURE TO MUTAGENS While spontaneous mutagenesis is not in our control, we can certainly try to avoid mutagenic substances and radiation. Not surprisingly, many things that cause cancer (carcinogens) are also mutagens. A good example is benzopyrene (discussed above), which is found in coal tar, car exhaust fumes, and charbroiled foods, as well as in cigarette smoke.

A major public policy goal is to minimize the effects of both human-made and natural mutagens on human health. For example, the Montreal Protocol is the only international environmental agreement signed and adhered to by all nations. It bans chlorofluorocarbons and other substances that cause depletion of the ozone layer in the upper atmosphere of Earth. The ozone layer is important for screening out ultraviolet radiation from the sun. Ultraviolet radiation can cause somatic mutations that lead to skin cancer. Similarly, bans on cigarette smoking have rapidly spread throughout the world. Cigarette smoking causes cancer because of the increased exposure of lung and other cells to benzopyrene and other carcinogens.

Do You Understand Concept 9.3?

- What are the differences between a germline mutation and a somatic mutation?

- Describe how the same base, cytosine, can be mutated spontaneously or by a mutagen. In which case would the mutation be efficiently repaired?

- The mutation *Bar eyes* in fruit flies was found in genetic crosses to be inherited as an autosomal (non–X chromosome) dominant. Would you expect the mutant allele to be a loss-of-function or a gain-of-function mutation? Explain your answer.

 What can we learn from ancient DNA?

ANSWER As we described in the opening story of this chapter, studying the DNA of fossils is a challenge. DNA is a large polymer that is rather stable (Concept 9.1). However, during the formation of fossils, most of the soft tissues and cells are broken down and their contents consumed by organisms in the environment. So it is not surprising that research on ancient DNA has been focused on places where DNA is likely to be preserved intact, such as frozen specimens and the interior of bones.

The polymerase chain reaction (PCR; Concept 9.2) has been invaluable in amplifying tiny amounts of ancient DNA to be sequenced. However, even a single molecule of contaminating DNA can ruin the experiment, because it too will be amplified. This has been a major challenge in studies of ancient DNA related to humans.

The Neandertal Genome Project involves an international team of scientists who are extracting DNA from the bones of skeletons of Neanderthals who lived in Europe over 50,000 years ago, amplifying it by PCR and then examining the genes. The entire DNA sequence has been completed. It is over 99 percent identical to our human DNA, justifying the classification of Neanderthals as part of the same genus, *Homo*.

Comparisons of humans and Neanderthals with regard to specific genes and mutations (Concept 9.3) is ongoing and has already shown several interesting facts:

- The gene *MC1R* is involved in skin and hair pigmentation. A point mutation found in Neanderthals but not humans caused lower activity of the MC1R protein when it was induced in cell cultures. Such lower activity of MC1R is known to result in fair skin and red hair in humans. So it appears that at least some Neanderthals may have had pale skin and red hair (**FIGURE 9.20**).

- The gene *FOXP2* is involved in vocalization in many organisms, including birds and mammals. Mutations in this gene result in severe speech impairment in humans. The Neanderthal *FOXP2* gene is identical to that of humans, whereas that of chimpanzees is slightly different. This has led to speculation that Neanderthals may have been capable of speech.

- While the two genome sequences are very similar, there are differences in many point mutations and larger

FIGURE 9.20 A Neanderthal Child This reconstruction of a Neanderthal child who lived about 60,000 years ago was made using bones recovered at Gibraltar, as well as phenotypic projections made from DNA analyses.

chromosomal arrangements (Concept 9.3). There are distinctive "human" DNA sequences and also distinctive "Neanderthal" sequences. There is some mixture of the two, suggesting that humans and Neanderthals may have interbred.

These fascinating reconstructions of ancient DNAs and their phenotypic expression are being repeated for many other species. They serve to underline the universality of DNA as the genetic material.

 ## SUMMARY

^{concept} **9.1** **DNA Structure Reflects Its Role as the Genetic Material**

- Circumstantial evidence for DNA as the genetic material includes its presence in the nucleus, its doubling during S phase of the mitotic cell cycle, and its injection into host cells by viruses. **Review Figures 9.1 and 9.2 and ANIMATED TUTORIAL 9.1**

- Experimental evidence for DNA as the genetic material is provided by the **transformation** of one genotype into another by adding DNA. **Review Figure 9.3**

- In DNA, the amount of A equals the amount of T, and the amount of G equals the amount of C. This observation, along with X-ray crystallography data, helped Watson and Crick unravel the **helical** structure of DNA. **Review Figures 9.4 and 9.5**

^{concept} **9.2** **DNA Replicates Semiconservatively**

- DNA exhibits **semiconservative replication**. Each parent strand acts as a **template** for the synthesis of a new strand; thus the two replicated DNA molecules each contain one parent strand and one newly synthesized strand. **Review ANIMATED TUTORIAL 9.2**

- In DNA replication, the enzyme **DNA polymerase** catalyzes the addition of nucleotides to the 3′ end of each new strand. **Review Figure 9.7 and ANIMATED TUTORIAL 9.3**

- Replication proceeds in both directions from the **origin of replication**. The parent DNA molecule unwinds to form a **replication fork**. **Review Figure 9.8**

- **Primase** catalyzes the synthesis of a short RNA **primer** to which nucleotides are added by DNA polymerase. **Review Figure 9.9 and WEB ACTIVITY 9.1**

- The **leading strand** is synthesized continuously. The **lagging strand** is synthesized in pieces called **Okazaki fragments**. The fragments are joined together by **DNA ligase**. **Review Figures 9.11 and 9.12 and ANIMATED TUTORIAL 9.4**

- Eukaryotic chromosomes have repetitive sequences at each end called **telomeres**. DNA replication leaves a short, unreplicated sequence at the 5′ end each new DNA strand. Unless the enzyme **telomerase** is present, the sequence is removed. After multiple cell cycles the telomeres shorten, leading to chromosome instability and cell death. **Review Figure 9.13**

- DNA polymerases make errors, which can be repaired by proofreading and mismatch repair. **Review Figure 9.14**

- The **polymerase chain reaction (PCR)** technique uses DNA polymerase to make multiple copies of DNA in the laboratory. **Review Figure 9.15 and INTERACTIVE TUTORIAL 9.1**

concept 9.3 Mutations Are Changes in DNA

- Mutations are heritable changes in DNA. **Somatic mutations** are passed on to daughter cells; only **germline mutations** are passed on to sexually produced offspring.

- **Point mutations** are alterations in single base pairs of DNA. **Silent mutations** can occur in genes or nontranscribed regions and do not affect the amino acid sequences of proteins. A mutation in a protein-coding region can lead to an alteration in the amino acid sequence of the protein. **Review Figure 9.16**

- Chromosomal mutations (**deletions**, **duplications**, **inversions**, and **translocations**) involve large regions of chromosomes. **Review Figure 9.17**

- **Spontaneous mutations** occur because of instabilities in DNA or chromosomes. **Induced mutations** occur when a mutagen damages DNA. **Review Figure 9.18**

From DNA to Protein: Gene Expression

Humans have more prokaryotic cells on and in their bodies than they have eukaryotic cells of their own. Among the billions of bacteria that inhabit the skin and noses of many people is *Staphylococcus aureus*. Healthy people can carry this bacterium without symptoms, but sometimes, especially when the immune system has been weakened by age or disease, *S. aureus* can cause major skin infections and may even enter the body through the nose or a wound site. In these cases, much more serious infections of organs such as the heart and lungs can occur, and may even result in death.

Until recently, most *S. aureus* infections were successfully treated with penicillin and related drugs, including methicillin. These antibiotics block several related enzymes known as penicillin binding proteins, which are involved in the assembly of bacterial cell walls. Bacteria treated with these antibiotics have defective cell walls, and because of this, new bacterial cells cannot survive after cell division. Unfortunately, some *S. aureus* strains have acquired mutant versions of a penicillin binding protein that can assemble cell walls in the presence of the antibiotics, thus conferring antibiotic resistance to these strains. The mutant penicillin binding protein has an altered shape that doesn't bind the antibiotics. This protein is encoded by the *mecA* gene, which can be passed from one bacterium to another by bacterial conjugation (see Concept 8.4). At a more general level, the mutant phenotype demonstrates that *a gene is expressed as a protein.*

By the late 1990s these bacterial strains were being called "superbugs," with the formal name "methicillin resistant *S. aureus*," or MRSA. The first decade of

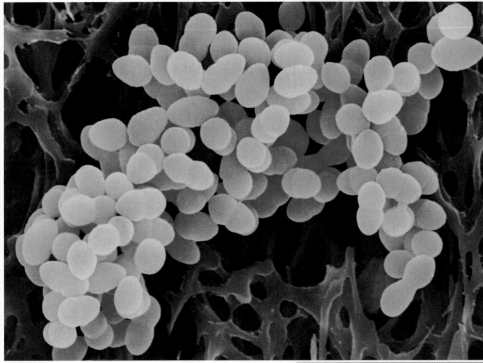

the new millennium saw a dramatic rise in MRSA infections. At first, most cases occurred in hospitals and nursing homes, but more recently MRSA has become rampant in communities as well. Resistant strains have a selective advantage because of the extensive use of antibiotics in health care. With close to 100,000 serious MRSA infections and 20,000 deaths in the United States each year, more people are dying from this infection than from AIDS.

MRSA can be treated if detected early. Antibiotics such as tetracycline, which targets bacterial protein synthesis, are effective in some strains. But there is reasonable concern that MRSA may become resistant to these antibiotics as well.

Q QUESTION How do antibiotics such as tetracycline target bacterial protein synthesis?

Although these *Staphylococcus aureus* cells look like normal bacteria, they have genes for resistance to multiple antibiotics and are difficult to eradicate. This is an increasing challenge to public health.

KEY CONCEPTS

10.1 Genetics Shows That Genes Code for Proteins

10.2 DNA Expression Begins with Its Transcription to RNA

10.3 The Genetic Code in RNA Is Translated into the Amino Acid Sequences of Proteins

10.4 Translation of the Genetic Code Is Mediated by tRNA and Ribosomes

10.5 Proteins Are Modified after Translation

concept 10.1 Genetics Shows That Genes Code for Proteins

Following Mendel's definition of the gene as a physically distinct entity (see Concept 8.1), biologists identified the genetic material as DNA (see Concept 9.1). In this chapter we will show that in most cases, genes code for proteins, and it is proteins that determine phenotypes. The connection between protein and phenotype was made before it was known that DNA is the genetic material.

Observations in humans led to the proposal that genes determine enzymes

The identification of a gene product as a protein began with a mutation. In the early twentieth century, the English physician Archibald Garrod saw several children with a rare disease. One symptom was that the urine turned dark brown or black in air, and for this reason the disease was named *alkaptonuria* ("black urine").

Garrod noticed that the disease was most common in children whose parents were first cousins. Mendelian genetics had just been "rediscovered," and Garrod realized that because first cousins share some alleles from each of their grandparents, the children of first cousins are more likely than other children to inherit rare mutant alleles in the homozygous condition. He proposed that alkaptonuria was a phenotype caused by a recessive, mutant allele.

Garrod took the analysis a step further by identifying the biochemical abnormality in the affected children. He isolated from them an unusual substance, homogentisic acid, which accumulated in the blood, joints (where it crystallized and caused severe pain), and urine (where it turned black when exposed to air).

Enzymes as biological catalysts had just been discovered, and Garrod proposed that in healthy individuals, homogentisic acid might be broken down to a harmless product by an enzyme. He speculated that the synthesis of this enzyme is determined by the wild-type allele of the gene that was mutated in alkaptonuria patients. These and other studies led Garrod to correlate *one gene to one enzyme*, and to coin the term "inborn error of metabolism" to describe this kind of genetically determined biochemical disease.

But Garrod's hypothesis needed direct confirmation by the identification of the specific enzyme and the specific gene mutation involved. In 1958 the enzyme was identified as homogentisic acid oxidase, which breaks down homogentisic acid to a harmless product, just as Garrod predicted (**FIGURE 10.1**). The specific DNA mutation was described in 1996.

Homogentisic acid is part of a biochemical pathway that catabolizes proteins, with the amino acids phenylalanine and tyrosine as intermediate products. Another genetic disease involving this pathway was discovered several decades after Garrod did his work. In *phenylketonuria*, the enzyme that converts phenylalanine to tyrosine is nonfunctional (see Figure 10.1). Untreated, this disease leads to significant mental

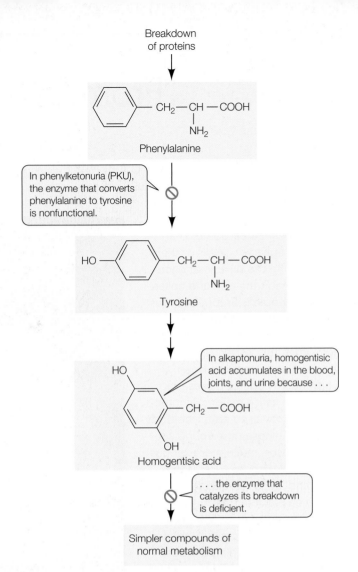

FIGURE 10.1 Metabolic Diseases and Enzymes Both phenylketonuria and alkaptonuria are caused by abnormalities in specific enzymes in a pathway that breaks down proteins.

retardation. Fortunately, the accumulation of phenylalanine can be easily detected in the blood of a newborn infant, and if the child consumes a diet low in proteins containing phenylalanine, mental retardation is avoided.

FRONTIERS Inborn errors of metabolism cause multiple symptoms, which usually vary in severity. While the specific gene mutations and resulting enzyme defects are known for many of these genetic diseases, we don't know why many of them cause mental retardation in infants if untreated, or why some infants are relatively less affected than others. Animal models for these diseases will be useful in answering these questions.

The concept of the gene has changed over time

The phenotypic expression of mutations underlying alkaptonuria and phenylketonuria led to the *one gene–one protein* hypothesis. Once it was known that proteins are polymers of amino acids, and that the sequence of amino acids determines

APPLY THE CONCEPT

Genetics shows that genes code for proteins

Wild-type bacteria can synthesize the amino acid tryptophan (T) using a biochemical pathway that begins with chorismate (C). But strains with mutant alleles for enzymes involved in this pathway cannot synthesize tryptophan, and it must be supplied as a nutrient in the growth medium. The pathway involves four intermediates that we will call D, E, F, and G. In the table are the phenotypes of various mutant strains, each of which has a mutation in a gene for a different enzyme. "+" means the strain grew with the indicated compound added to the medium, and "0" means it did not grow.

Based on these data, order the compounds (C, D, E, F, and T) and enzymes (1, 2, 3, 4, and 5) in a biochemical pathway. Hint: Mutant strain 5 will not grow if any compound

MUTANT STRAIN	ADDITION TO THE MEDIUM					
	C	D	E	F	G	T
1	0	0	0	0	+	+
2	0	+	+	0	+	+
3	0	+	0	0	+	+
4	0	+	+	+	+	+
5	0	0	0	0	0	+

except T is supplied, so it must carry a loss-of-function mutation in the enzyme that transforms another molecule (C, D, E, F, or G) into T. Thus, enzyme 5 is the final enzyme in the pathway.

---

protein function, it became clear that *a mutant phenotype arises from a change in the protein's amino acid sequence*. However, scientists soon realized that the one gene–one protein hypothesis was an oversimplification. Once again, studies of human mutations were a key to this realization.

In humans, the oxygen-carrying protein hemoglobin has a quaternary structure: it is made up of four polypeptide chains—two α-chains and two β-chains (see the illustration in Concept 3.2, p. 44). In Concept 9.3 we introduce sickle-cell disease, which is caused by a point mutation in the gene for β-globin and is inherited as an autosomal recessive (carried on an autosome rather than a sex chromosome). One of the 146 amino acids in the β-globin chain is abnormal: at position 6, the normal glutamic acid has been replaced by valine. This replacement changes the charge of the protein (glutamic acid is negatively charged and valine is neutral), causing it to form long, needlelike aggregates in the red blood cells. The phenotypic result is anemia, an impaired ability of the blood to carry oxygen.

Because hemoglobin is easy to isolate and study, its variations in the human population have been extensively documented (**FIGURE 10.2**). Hundreds of single amino acid alterations in β-globin have been reported. For example, at the same position that is mutated in sickle-cell disease (resulting in hemoglobin S), the normal glutamic acid may be replaced by lysine, causing hemoglobin C disease. In this case, the resulting anemia is usually not severe. Many alterations of hemoglobin do not affect the protein's function. That is fortunate, because about 5 percent of all humans are carriers for one of these variants.

Studies of proteins like hemoglobin, which are made up of multiple polypeptides, resulted in a modification of the one gene–one protein hypothesis. Scientists began to think of the relationship as **one gene–one polypeptide**. This remains a powerful and useful concept today. However, as you will see later in this chapter and in Chapter 11, we are learning that this, too, is an oversimplification.

The mutations we have been discussing result in alterations in amino acid sequences. But not all genes code for polypeptides. As we will see below and in Chapter 11, there are many DNA sequences that are transcribed into RNA molecules that are *not* translated into polypeptides, but instead have other functions. Like any DNA sequences, these RNA genes are also subject to mutations that can affect the functions of the RNAs they produce.

Genes are expressed via transcription and translation

Molecular biology is the study of nucleic acids and proteins, and it often focuses on gene expression. As we describe briefly in

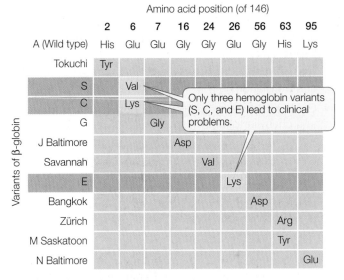

FIGURE 10.2 Gene Mutations and Amino Acid Changes
Each of these mutant alleles (e.g., hemoglobin Tokuchi) codes for a β-globin polypeptide with an alteration in one of its 146 amino acids (e.g., a change from His to Tyr at position 2 of the polypeptide).

Chapter 3, genes are expressed as RNAs, some of which are translated into proteins. This process involves two steps:

- During **transcription**, the information in a DNA sequence (a gene) is copied into a complementary RNA sequence.
- During **translation**, this RNA sequence is used to create the amino acid sequence of a polypeptide.

Here we will consider three types of RNA with regard to their roles in protein synthesis:

- *Messenger RNA and transcription:* When a particular gene is expressed, one of the two DNA strands in the gene is transcribed (copied) to produce a complementary RNA strand, which is then modified to produce a **messenger RNA (mRNA)**. In eukaryotic cells, the mRNA travels from the nucleus to the cytoplasm, where it is translated into a polypeptide (**FIGURE 10.3**). The nucleotide sequence of the mRNA determines the ordered sequence of amino acids in the polypeptide chain, which is built by a ribosome.

- *Ribosomal RNA and translation:* The **ribosome** is essentially a protein synthesis factory with multiple proteins and several **ribosomal RNAs (rRNAs)**. One of the rRNAs catalyzes peptide bond formation between amino acids to form a polypeptide.

- *Transfer RNA mediates between mRNA and protein:* A third RNA called **transfer RNA (tRNA)** can both bind a specific amino acid and recognize a specific sequence of nucleotides in mRNA, by complementary base pairing (A with U, and G with C). It is the tRNA that recognizes which amino acid should be added next to a growing polypeptide chain (see Figure 10.3).

In Chapter 11 we will consider other RNAs that play roles in the regulation of gene expression.

Do You Understand Concept 10.1?

- What is the difference between the "one gene–one enzyme" and "one gene–one polypeptide" hypotheses?
- What is the difference between gene transcription and translation?
- Could a person inherit alleles for both alkaptonuria and phenylketonuria? If so, what would the symptoms be?
- Defining phenotype as the presence of a polypeptide chain of a particular amino acid sequence, would you expect the Zürich variant of β-globin (see Figure 10.2) to be inherited as a dominant, recessive, or codominant? Explain your answer.

In this section we have shown how the connection between genes and phenotypes can be understood in terms of DNA and proteins. We will now turn to some details of the process of gene expression, which is at the heart of what genes do.

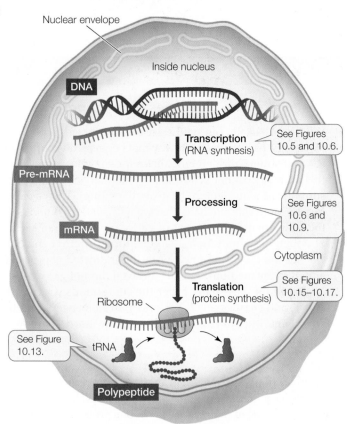

FIGURE 10.3 From Gene to Protein This diagram summarizes the processes of gene expression in eukaryotes.

yourBioPortal.com

Go to WEB ACTIVITY 10.1
Eukaryotic Gene Expression

concept
10.2
DNA Expression Begins with Its Transcription to RNA

Transcription—the formation of a specific RNA sequence from a specific DNA sequence—requires several components:

- A *DNA template* for complementary base pairing
- The appropriate *nucleoside triphosphates* (ATP, GTP, CTP, and UTP) to act as substrates
- An *RNA polymerase* enzyme

In addition to mRNA, other types of RNA are produced by transcription. The same process is responsible for the synthesis of tRNA and rRNA, whose important roles in protein synthesis will be described in Concepts 10.3 and 10.4. Like mRNA, these RNAs are encoded by specific genes. There are in fact many kinds of RNA with other functions in the cell. For example, *small nuclear RNAs* are components of snRNPs (see below); and *microRNAs* play roles in stimulating or inhibiting gene expression (see Concept 11.4).

RNA polymerases share common features

RNA polymerases from both prokaryotes and eukaryotes catalyze the synthesis of RNA from the DNA template. There is only one kind of RNA polymerase in bacteria and archaea,

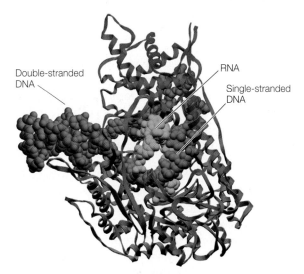

Double-stranded DNA

RNA

Single-stranded DNA

FIGURE 10.4   RNA Polymerase   This enzyme from bacteriophage T7 is smaller than most other RNA polymerases, but has an active site similar to the bacterial and eukaryotic enzymes.

Transcription occurs in three steps

Transcription can be divided into three distinct processes: initiation, elongation, and termination. You can follow these processes in **FIGURE 10.5**.

INITIATION   Transcription begins with initiation, which requires a **promoter**, a special DNA sequence to which the RNA polymerase binds very tightly (see Figure 10.5A). Promoters are important control sequences that "tell" the RNA polymerase two things:

● Where to start transcription

● Which of the two DNA strands to transcribe

The promoter has a nucleotide sequence that can be "read" in a particular direction and orients the RNA polymerase, thus "aiming" it at the correct strand to use as a template. A promoter

whereas there are several kinds in eukaryotes. However, they all share a common structure (**FIGURE 10.4**). Like DNA polymerases, RNA polymerases are *processive*; that is, a single enzyme–template binding event results in the polymerization of hundreds of RNA nucleotides. But unlike DNA polymerases, RNA polymerases do not require a primer.

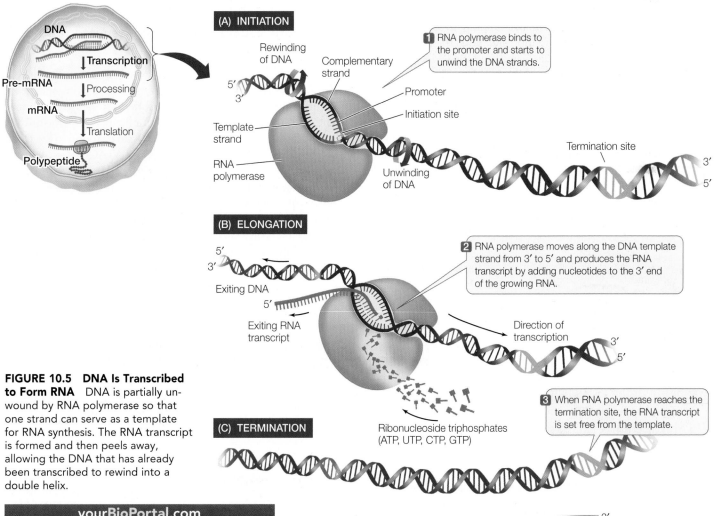

(A) INITIATION

Rewinding of DNA

Complementary strand

5′
3′

Template strand

RNA polymerase

Promoter

Initiation site

Unwinding of DNA

1 RNA polymerase binds to the promoter and starts to unwind the DNA strands.

Termination site

3′

5′

(B) ELONGATION

5′
3′

Exiting DNA

5′

Exiting RNA transcript

2 RNA polymerase moves along the DNA template strand from 3′ to 5′ and produces the RNA transcript by adding nucleotides to the 3′ end of the growing RNA.

Direction of transcription

3′

5′

FIGURE 10.5   DNA Is Transcribed to Form RNA   DNA is partially unwound by RNA polymerase so that one strand can serve as a template for RNA synthesis. The RNA transcript is formed and then peels away, allowing the DNA that has already been transcribed to rewind into a double helix.

Ribonucleoside triphosphates (ATP, UTP, CTP, GTP)

(C) TERMINATION

3 When RNA polymerase reaches the termination site, the RNA transcript is set free from the template.

5′

3′

RNA

DNA

Transcription

Pre-mRNA

Processing

mRNA

Translation

Polypeptide

yourBioPortal.com

Go to ANIMATED TUTORIAL 10.1
Transcription

functions somewhat like the capital letter at the beginning of a sentence, indicating how the sequence of words should be read. Part of each promoter is the **transcription initiation site**, where transcription begins. Groups of nucleotides lying "upstream" from the initiation site (5' on the non-template strand and 3' on the template strand) are bound by other proteins, which help the RNA polymerase bind.

Although every gene has a promoter, not all promoters are identical. Some promoters are more effective at transcription initiation than others. Furthermore, bacteria, archaea, and eukaryotes differ in the details of transcription initiation. But despite these variations, the basic mechanisms of initiation are the same throughout the living world.

> **LINK** The mechanisms of transcription and translation are discussed in an evolutionary context in Chapter 19

ELONGATION Once RNA polymerase has bound to the promoter, it begins the process of **elongation** (see Figure 10.5B). RNA polymerase unwinds the DNA about 13 base pairs at a time and reads the template strand in the 3'-to-5' direction. Like DNA polymerase, RNA polymerase adds new nucleotides to the 3'end of the growing strand, beginning with the first nucleotide at the transcription initiation site. Thus the first nucleotide in the new RNA forms its 5' end, and the RNA transcript is antiparallel to the DNA template strand.

Remember that the nucleotides of RNA contain ribose rather than deoxyribose but are otherwise structurally similar to those of DNA (see Figure 2.9 and Concept 3.1). The RNA polymerase adds new nucleotides to the RNA molecule by complementary base pairing with nucleotides in the template strand of the DNA. This is similar to DNA replication except that in RNA, the base uracil (rather than thymine) in the RNA molecule is paired with adenine in the DNA molecule. In a mechanism

very similar to that used by DNA polymerase, the RNA polymerase uses the ribonucleoside triphosphates ATP, UTP, GTP, and CTP as substrates and catalyzes the formation of phosphodiester bonds between them, releasing pyrophosphate in the process (see Figure 9.7). As transcription progresses, the two DNA strands rewind and the RNA grows as a single-stranded molecule (see Figure 10.5B).

Like DNA polymerases, RNA polymerases and associated proteins have mechanisms for *proofreading* during transcription, but these mechanisms are not as efficient as those for DNA. Transcriptional errors occur at rates of about 1 for every 10^4 to 10^5 bases. Because many copies of RNA are made, however, and because they often have relatively short life spans, these errors are not as potentially harmful as mutations in DNA.

TERMINATION Just as initiation sites in the DNA template strand specify the starting point for transcription, particular base sequences specify its **termination** (see Figure 10.5C). The mechanisms of termination are complex and of more than one kind. In some genes, the newly formed transcript falls away from the DNA template and the RNA polymerase. In others, a helper protein pulls the transcript away.

Eukaryotic coding regions are often interrupted by introns

Coding regions are sequences within a DNA molecule that are expressed as proteins. In prokaryotes, most of the genomic DNA is made up of coding regions, and the mRNAs are collinear with the DNA sequences that code for them: that is, the mRNA sequence (e.g., UAUAUAUCCCC….) can be found in the DNA sequence as its complement (e.g., ATATATAGGGG….). In eukaryotes the situation is different (**TABLE 10.1**).

A diagram of the structure and transcription of a typical eukaryotic gene is shown in **FIGURE 10.6**. In prokaryotes and viruses several adjacent genes sometimes share one promoter, but in eukaryotes each gene has its own promoter. And while the coding region of a prokaryotic gene is usually continuous

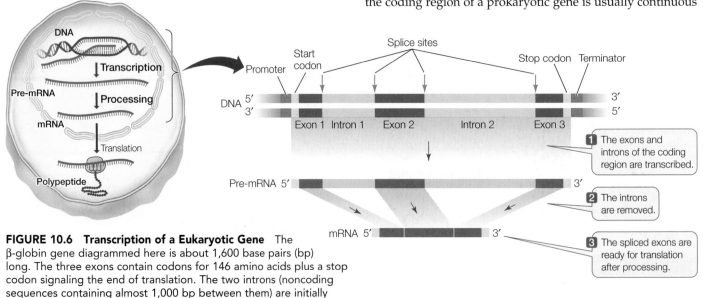

FIGURE 10.6 Transcription of a Eukaryotic Gene The β-globin gene diagrammed here is about 1,600 base pairs (bp) long. The three exons contain codons for 146 amino acids plus a stop codon signaling the end of translation. The two introns (noncoding sequences containing almost 1,000 bp between them) are initially transcribed but then are spliced out of the pre-mRNA transcript.

TABLE 10.1	Differences between Prokaryotic and Eukaryotic Gene Expression	
CHARACTERISTIC	PROKARYOTES	EUKARYOTES
Transcription and translation occurrence	At the same time in the cytoplasm	Transcription in the nucleus, then translation in the cytoplasm
Gene structure	Transcribed regions not interrupted by introns	Transcribed regions often interrupted by non-coding introns
Modification of mRNA after initial transcription but before translation	None	Introns spliced out; 5′ cap and 3′ poly A added

(with no interruptions), a eukaryotic gene may contain noncoding sequences called **introns** (*int*ervening *regions*) that interrupt the coding region. The transcribed regions that are interspersed with the introns are called **exons** (*ex*pressed *regions*). Both introns and exons appear in the primary mRNA transcript, called **pre-mRNA**, but the introns are removed by the time the mature mRNA leaves the nucleus. Pre-mRNA processing involves cutting introns out of the pre-mRNA transcript and splicing together the exon transcripts (see Figure 10.6). If this seems surprising, you are in good company. For scientists who were familiar with prokaryotic genes and gene expression, the discovery of introns in eukaryotic genes was entirely unexpected.

How can we locate introns within a eukaryotic gene? One way is by **nucleic acid hybridization**, the method that originally revealed the existence of introns. This method, outlined in **FIGURE 10.7**, has been crucial for studying the relationship between eukaryotic genes and their transcripts and is widely used in many applications. It involves two steps:

- The DNA to be analyzed is denatured by heat to break the hydrogen bonds between the base pairs and separate the two strands.
- A single-stranded nucleic acid from another source (called a **probe**) is incubated with the denatured DNA. If the probe has a base sequence complementary to the target DNA, a probe–target double helix forms by hydrogen bonding between the bases. Because the two strands are from different sources, the resulting double-stranded molecule is called a hybrid.

Biologists used this technique to examine the β-globin gene (**FIGURE 10.8**). The researchers first denatured DNA containing the gene by heating it slowly, then used previously isolated β-globin mRNA as a probe. They were able to view the hybridized molecules using electron microscopy. As expected, the mRNA bound to the DNA by complementary base pairing. The researchers expected to obtain a linear (1:1) matchup of the mRNA to the coding DNA. That expectation was only partially met: there were indeed stretches of RNA–DNA hybrid, but some *unexpected looped structures* were also visible. These loops turned out to be introns, stretches of DNA that did not have complementary base sequences on the mature mRNA.

When pre-mRNA was used instead of mature mRNA to hybridize to the DNA, there was complete hybridization with *no loops*, revealing that the introns were part of the pre-mRNA transcript. Somewhere on the path from primary transcript (pre-mRNA) to mature mRNA, the introns had been removed, and the exons had been spliced together. We will examine this splicing process in the next section.

Introns *interrupt, but do not scramble*, the DNA sequence of a gene. The base sequences of the exons in the template strand, if joined and taken in order, form a continuous sequence that is complementary to that of the mature mRNA. Most (but not all) eukaryotic genes contain introns, and in rare cases, introns are also found in prokaryotes. The largest human gene encodes a muscle protein called titin; it has 363 exons, which together code for 38,138 amino acids.

RESEARCH TOOLS

FIGURE 10.7   Nucleic Acid Hybridization   Base pairing permits the detection of a sequence that is complementary to the probe.

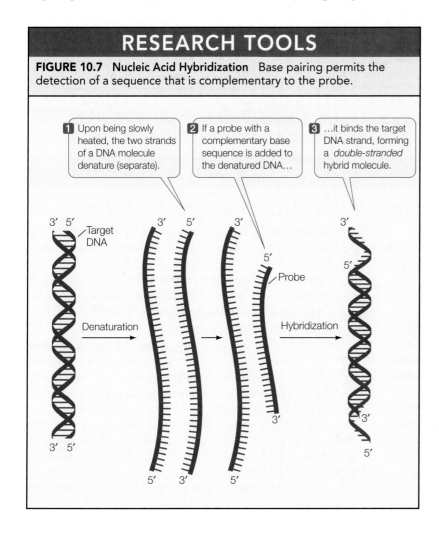

1 Upon being slowly heated, the two strands of a DNA molecule denature (separate).

2 If a probe with a complementary base sequence is added to the denatured DNA…

3 …it binds the target DNA strand, forming a *double-stranded* hybrid molecule.

Target DNA

Denaturation

Probe

Hybridization

INVESTIGATION

FIGURE 10.8 Demonstrating the Existence of Introns When an mRNA transcript of the β-globin gene was hybridized with the double-stranded DNA of that gene, the introns in the DNA "looped out." This demonstrated that the coding region of a eukaryotic gene can contain noncoding DNA that is not present in the mature mRNA transcript.

HYPOTHESIS

All regions within the coding sequence of a gene end up in its mRNA.

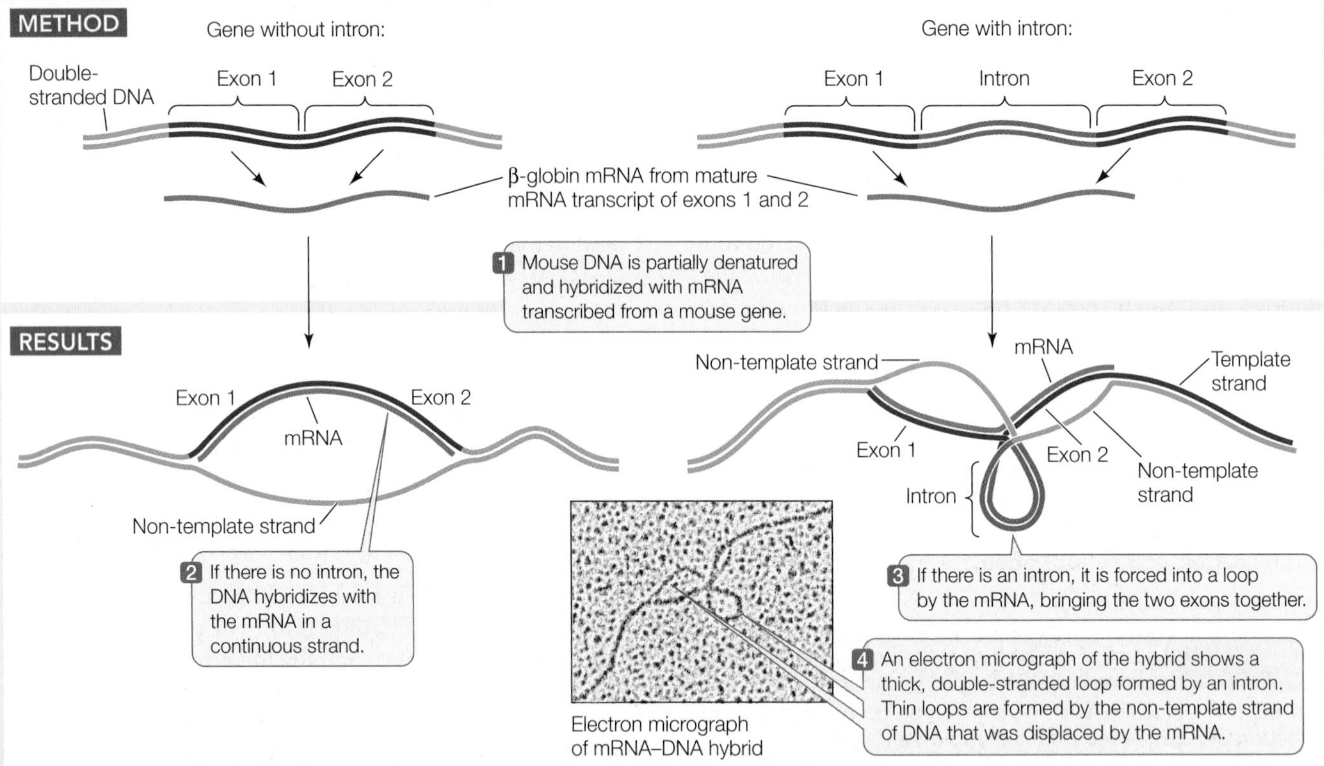

METHOD

Gene without intron:

Gene with intron:

Double-stranded DNA — Exon 1 — Exon 2 — Exon 1 — Intron — Exon 2

β-globin mRNA from mature mRNA transcript of exons 1 and 2

1 Mouse DNA is partially denatured and hybridized with mRNA transcribed from a mouse gene.

RESULTS

Exon 1 — mRNA — Exon 2

Non-template strand

2 If there is no intron, the DNA hybridizes with the mRNA in a continuous strand.

Non-template strand — mRNA — Template strand

Exon 1 — Exon 2 — Non-template strand

Intron

3 If there is an intron, it is forced into a loop by the mRNA, bringing the two exons together.

Electron micrograph of mRNA–DNA hybrid

4 An electron micrograph of the hybrid shows a thick, double-stranded loop formed by an intron. Thin loops are formed by the non-template strand of DNA that was displaced by the mRNA.

CONCLUSION

The DNA contains noncoding regions within the genes that are not present in the mature mRNA.

Go to **yourBioPortal.com** for original citations, discussions, and relevant links for all INVESTIGATION figures.

Eukaryotic gene transcripts are processed before translation

The primary transcript of a eukaryotic gene is modified in several ways before it leaves the nucleus: introns are removed, and both ends of the pre-mRNA are chemically modified.

SPLICING TO REMOVE INTRONS After the pre-mRNA is made, its introns must be removed. If this did not happen, the extra nucleotides in the mRNA would be translated at the ribosome and a nonfunctional protein would result. A process called **RNA splicing** removes the introns and splices the exons together.

At the boundaries between introns and exons are **consensus sequences**—short stretches of DNA that appear with little variation ("consensus") in many different genes. As soon as the pre-mRNA is transcribed, the consensus sequences are bound by several **small nuclear ribonucleoprotein particles (snRNPs)**. The RNA in one of the snRNPs has a stretch of bases complementary to the consensus sequence at the 5′ exon–intron boundary,

and it binds to the pre-mRNA by complementary base pairing. Another snRNP binds to the pre-mRNA near the 3′ intron–exon boundary, and then other proteins accumulate to form a large RNA–protein complex called a **spliceosome**. This complex cuts the pre-mRNA, releases the introns, and joins the ends of the exons together to produce mature mRNA (**FIGURE 10.9**).

Molecular studies of human genetic diseases have provided insights into intron consensus sequences and the splicing machinery. For example, people with the genetic disease β-thalassemia have a defect in the production of one of the hemoglobin subunits. These people suffer from severe anemia because they have an inadequate supply of red blood cells. In some cases, the genetic mutation that causes the disease occurs at an intron consensus sequence in the β-globin gene. Consequently, the β-globin pre-mRNA cannot be spliced correctly, and a defective β-globin mRNA is made. This finding offers another example of how biologists can *use mutations to elucidate cause-and-effect relationships*.

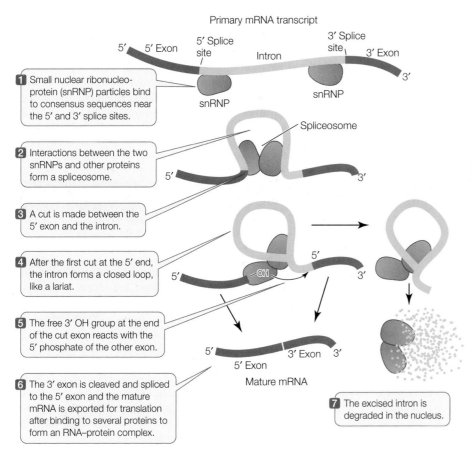

Primary mRNA transcript

1 Small nuclear ribonucleo-protein (snRNP) particles bind to consensus sequences near the 5′ and 3′ splice sites.

2 Interactions between the two snRNPs and other proteins form a spliceosome.

3 A cut is made between the 5′ exon and the intron.

4 After the first cut at the 5′ end, the intron forms a closed loop, like a lariat.

5 The free 3′ OH group at the end of the cut exon reacts with the 5′ phosphate of the other exon.

6 The 3′ exon is cleaved and spliced to the 5′ exon and the mature mRNA is exported for translation after binding to several proteins to form an RNA–protein complex.

7 The excised intron is degraded in the nucleus.

FIGURE 10.9 The Spliceosome: An RNA Splicing Machine Small nuclear ribonucleoproteins (snRNPs) bind to consensus sequences bordering the introns on pre-mRNA transcripts. Other proteins then bind, forming a large complex called a spliceosome. This structure determines the exact position of each cut in the pre-mRNA with great precision.

yourBioPortal.com

Go to ANIMATED TUTORIAL 10.2 RNA Splicing

In Concept 10.1 we mentioned that the one gene–one polypeptide hypothesis is an oversimplification. It has recently become clear that in many cases *alternative splicing* results in *different mRNAs and different polypeptides from a single gene*. This occurs because not all exons may be included in every mRNA: some are spliced out, along with the introns.

LINK See Concept 11.4, especially Figure 11.16, for a discussion of alternative splicing

MODIFICATION AT BOTH ENDS While the pre-mRNA is still in the nucleus it undergoes two processing steps, one at each end of the molecule:

Mature mRNA

- A **5′ cap** (or *G cap*) is added to the 5′ end of the pre-mRNA as it is transcribed. The 5′ cap is a chemically modified molecule of guanosine triphosphate (GTP). It facilitates the binding of mRNA to the ribosome for translation, and it protects the mRNA from being digested by ribonucleases—enzymes that break down RNAs.

- A **poly A tail** is added to the 3′ end of the pre-mRNA at the end of transcription. This sequence of 100 to 300 adenine nucleotides assists in the export of the mRNA from the nucleus and is also important for mRNA stability.

Do You Understand Concept 10.2?

- A DNA template strand has the sequence 5′-ATG-GTGTACG-3′. What will be the sequence of the RNA transcribed from this DNA? (Be careful to specify the 5′ and 3′ ends.)

- What would be the consequences of the following?
  a. A mutation of a promoter such that it is deleted
  b. A mutation of the gene coding for RNA polymerase
  c. Deletion of intron consensus sequences from a gene

- Refer to the experiment shown in Figure 10.8. What would the result have been if there were five exons and four introns? Sketch what this would look like in an electron micrograph.

The transcription of a gene to produce mRNA is only the first step in gene expression. The next step in the pathway from DNA to RNA to protein is translation, the subject of Concepts 10.3 and 10.4. First we will discuss the genetic code, which enables the base sequence in an mRNA to be translated into a specific amino acid sequence in the resulting polypeptide. Then we will look in more detail at the process of translation.

concept 10.3 The Genetic Code in RNA Is Translated into the Amino Acid Sequences of Proteins

The translation of the nucleotide sequence of an mRNA into the amino acid sequence of a polypeptide occurs at the ribosome. In prokaryotes, transcription and translation are coupled: there is no nucleus, and ribosomes often bind to an mRNA as it is being transcribed in the cytoplasm. In eukaryotes, the nuclear envelope separates the locations of mRNA production and translation, the latter occurring at ribosomes in the cytoplasm. In both cases, the key event is the decoding of one chemical "language" (the nucleotide sequence) into another (the amino acid sequence).

The information for protein synthesis lies in the genetic code

The genetic information in an mRNA molecule is a series of sequential, nonoverlapping three-letter "words" called **codons**. Each sequence of three nucleotide bases along the mRNA poly-nucleotide chain is a codon that specifies a particular amino acid. Each codon in the mRNA is complementary to the corresponding triplet of bases in the DNA molecule from which it was transcribed. The genetic code relates codons to their specific amino acids.

CHARACTERISTICS OF THE GENETIC CODE Molecular biologists "broke" the genetic code in the early 1960s. The problem they addressed was perplexing: how could 20 different amino acids be encoded using only four nucleotide bases (A, U, G, and C)? A triplet code with three-letter codons was considered likely because it was the shortest sequence with enough possible variations to encode all the 20 amino acids. With four available bases, a triplet codon could have $4 \times 4 \times 4 = 64$ variations.

Marshall W. Nirenberg and J. H. Matthaei, at the U.S. National Institutes of Health, made the first decoding breakthrough in 1961 when they realized they could use a simple artificial poly-nucleotide instead of a complex natural mRNA as a template for polypeptide synthesis in a test tube. They could then identify the polypeptide that the artificial messenger encoded. This led to the identification of the first three codons (**FIGURE 10.10**).

Other scientists later found that simple artificial mRNAs only three nucleotides long—each amounting to one codon—could bind to a ribosome, and that the resulting complex could then bind to the corresponding tRNA with its specific amino acid. Thus, for example, a simple UUU mRNA caused the tRNA carrying phenylalanine to bind to the ribosome. After this discovery, the complete deciphering of the genetic code was relatively simple.

INVESTIGATION

FIGURE 10.10 Deciphering the Genetic Code Nirenberg and Matthaei used a test-tube protein synthesis system to determine the amino acids specified by particular synthetic mRNAs.

HYPOTHESIS
A triplet codon based on three-base codons specifies amino acids.

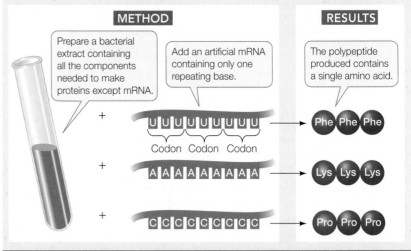

METHOD

Prepare a bacterial extract containing all the components needed to make proteins except mRNA.

Add an artificial mRNA containing only one repeating base.

RESULTS

The polypeptide produced contains a single amino acid.

U U U U U U U U U
Codon Codon Codon → Phe Phe Phe

A A A A A A A A A → Lys Lys Lys

C C C C C C C C C → Pro Pro Pro

CONCLUSION
UUU is an mRNA codon for phenylalanine.
AAA is an mRNA codon for lysine.
CCC is an mRNA codon for proline.

ANALYZE THE DATA
Poly U, an artificial mRNA, was added to a test tube with all other components for protein synthesis ("Complete system"). Other test tubes differed from the complete system as indicated in the table. Samples were tested for radioactive phenylalanine incorporation with these results:

Condition	Radioactivity in protein
Complete system	29,500
Minus poly U mRNA	70
Minus ribosomes	52
Minus ATP	83
Plus RNase (hydrolyzes RNA)	120
Plus DNase	27,600
Radioactive glycine instead of phenylalanine	33
Mixture of 19 radioactive amino acids minus phenylalanine	276

Explain the results for each of the conditions.

For more, go to Working with Data 10.1 at **yourBioPortal.com**.

yourBioPortal.com

Go to ANIMATED TUTORIAL 10.3
Deciphering the Genetic Code

Go to **yourBioPortal.com** for original citations, discussions, and relevant links for all INVESTIGATION figures.

First letter	Second letter: U	Second letter: C	Second letter: A	Second letter: G	Third letter
U	UUU Phenylalanine	UCU Serine	UAU Tyrosine	UGU Cysteine	U
U	UUC Phenylalanine	UCC Serine	UAC Tyrosine	UGC Cysteine	C
U	UUA Leucine	UCA Serine	UAA Stop codon	UGA Stop codon	A
U	UUG Leucine	UCG Serine	UAG Stop codon	UGG Tryptophan	G
C	CUU Leucine	CCU Proline	CAU Histidine	CGU Arginine	U
C	CUC Leucine	CCC Proline	CAC Histidine	CGC Arginine	C
C	CUA Leucine	CCA Proline	CAA Glutamine	CGA Arginine	A
C	CUG Leucine	CCG Proline	CAG Glutamine	CGG Arginine	G
A	AUU Isoleucine	ACU Threonine	AAU Asparagine	AGU Serine	U
A	AUC Isoleucine	ACC Threonine	AAC Asparagine	AGC Serine	C
A	AUA Isoleucine	ACA Threonine	AAA Lysine	AGA Arginine	A
A	AUG Methionine; start codon	ACG Threonine	AAG Lysine	AGG Arginine	G
G	GUU Valine	GCU Alanine	GAU Aspartic acid	GGU Glycine	U
G	GUC Valine	GCC Alanine	GAC Aspartic acid	GGC Glycine	C
G	GUA Valine	GCA Alanine	GAA Glutamic acid	GGA Glycine	A
G	GUG Valine	GCG Alanine	GAG Glutamic acid	GGG Glycine	G

FIGURE 10.11 The Genetic Code Genetic information is encoded in three-letter units—codons—that are read in the 5′-to-3′ direction on the mRNA. To decode a codon, find its first letter in the left column, then read across the top to its second letter, then read down the right column to its third letter. The amino acid the codon specifies is given in the corresponding row. For example, AUG codes for methionine, and GUA codes for valine.

yourBioPortal.com
Go to WEB ACTIVITY 10.2 The Genetic Code

The complete genetic code is shown in **FIGURE 10.11**. Notice that there are many more codons than there are different amino acids in proteins. All possible combinations of the four available bases give 64 (4^3) different three-letter codons, yet these codons determine only 20 amino acids. AUG, which codes for methionine, is also the **start codon**, the initiation signal for translation. Three of the codons (UAA, UAG, and UGA) are **stop codons**, or termination signals for translation. When the translation machinery reaches one of these codons, translation stops and the polypeptide is released from the translation complex.

THE GENETIC CODE IS REDUNDANT BUT NOT AMBIGUOUS The 60 codons that are not start or stop codons are far more than enough to code for the other 19 amino acids—and indeed, there is more than one codon for almost all the amino acids. Thus we say that the genetic code contains redundancies. For example, leucine is represented by six different codons (see Figure 10.11). Only methionine and tryptophan are represented by just one codon each.

A *redundant* code should not be confused with an *ambiguous* code. If the code were ambiguous, a single codon could specify two (or more) different amino acids, and there would be doubt about which amino acid should be incorporated into a growing polypeptide chain. The genetic code is not ambiguous: a given amino acid may be encoded by more than one codon, but each codon encodes only one amino acid.

THE GENETIC CODE IS (NEARLY) UNIVERSAL The same genetic code is used by all the species on our planet. Thus the code must be an ancient one that has been maintained intact throughout the evolution of living organisms. Exceptions are known: within mitochondria and chloroplasts, the code differs slightly from that in prokaryotes and in the nuclei of eukaryotic cells; and in one group of protists, the codons UAA and UAG encode glutamine rather than functioning as stop codons. The significance of these differences is not yet clear. What is clear is that the exceptions are few.

The common genetic code unifies life, and indicates that all life came ultimately from a common ancestor. The genetic code probably originated early in the evolution of life. The common code also has profound implications for genetic engineering, as we can see in Chapter 13, since it means that the code for a human gene is the same as that for a bacterial gene. It is therefore impressive, but not surprising, that a human gene can be expressed in *Escherichia coli* via laboratory manipulations, since these cells speak the same "molecular language."

The codons in Figure 10.11 are mRNA codons. The base sequence of the DNA strand that is transcribed to produce the mRNA is complementary and antiparallel to these codons. Thus, for example:

- 3′-AAA-5′ in the template DNA strand corresponds to phenylalanine (which is encoded by the mRNA codon 5′-UUU-3′).
- 3′-ACC-5′ in the template DNA corresponds to tryptophan (which is encoded by the mRNA codon 5′-UGG-3′).

Point mutations confirm the genetic code

Strong support for the assignment of codons in the genetic code comes from point mutations (changes in single nucleotides within a sequence). These have been studied in a wide variety of organisms. When point mutations within the coding region of a gene are compared with the amino acid sequences in the encoded polypeptide, they are consistent with the genetic code.

In Concept 9.3 we discuss mutations in terms of their effects on phenotypes. We can now define types of mutations in terms of their effects on polypeptide sequences (**FIGURE 10.12**):

- *Silent mutations* can occur because of the redundancy of the genetic code. For example, the codons CCG and CCU are

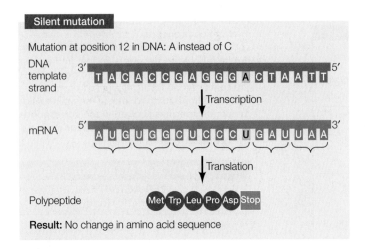

Silent mutation

Mutation at position 12 in DNA: A instead of C

Result: No change in amino acid sequence

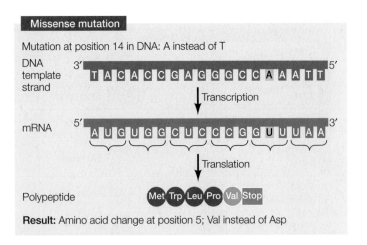

Missense mutation

Mutation at position 14 in DNA: A instead of T

Result: Amino acid change at position 5; Val instead of Asp

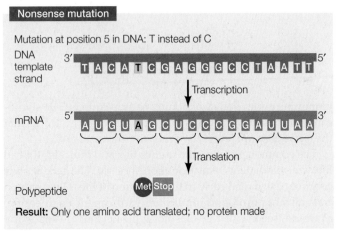

Nonsense mutation

Mutation at position 5 in DNA: T instead of C

Result: Only one amino acid translated; no protein made

FIGURE 10.12 Mutations Changes in a coding region of DNA can have different effects on the protein the DNA encodes.

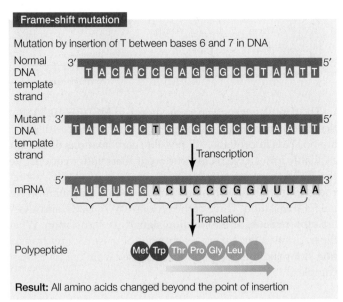

Frame-shift mutation

Mutation by insertion of T between bases 6 and 7 in DNA

Result: All amino acids changed beyond the point of insertion

both translated from mRNA as proline (Pro). So a change in the template strand of the DNA from 3′-GGC to 3′-GGA (a mutation from C to A) will not cause any change in amino acid sequence.

- *Missense mutations* result in a change in the amino acid sequence. For example, GAU in mRNA is translated as aspar-

tic acid (Asp), whereas a mutation that results in GUU is translated as valine (Val).

- *Nonsense mutations* result in a premature stop codon. For example, the codon UGG is translated as the amino acid tryptophan (Trp). A DNA point mutation could convert this to the stop codon UAG, which acts as a translation termination

---

APPLY THE CONCEPT

The genetic code in RNA is translated into the amino acid sequences of proteins

The double-stranded DNA sequence for the coding region of a short peptide is:

```
5′-A T G T T T T C G A C G T G C G A T T G A-3′
3′-T A C A A A A G C T G C A C G C T A A C T-5′
    1       5        10       15       20
```

1. Which strand of DNA (top or bottom) is transcribed into mRNA? Explain.

2. What is the amino acid sequence of the peptide coded for by the DNA?

3. A mutant strain has a C-G base pair at position #5 instead of the T-A pair shown above. What is the amino acid sequence of the peptide? Explain.

4. A mutation at base pair #15 results in a peptide that is not full length. What point mutation would cause this? Explain.

signal. If this occurred, the polypeptide chain would end at the amino acid translated just before the stop codon.

- *Frame-shift mutations* result from the insertion or deletion of one or more base pairs within the coding sequence. Since the genetic code is read as sequential, nonoverlapping triplets, this can cause new triplets to be read, and an altered sequence of amino acids in the resulting polypeptide.

---

yourBioPortal.com

Go to INTERACTIVE TUTORIAL 10.1
Genetic Mutations

Do You Understand Concept 10.3?

- What are the characteristics of the genetic code?

- If the artificial mRNA UAUAUAUAUA... is used in a test tube protein synthesis system, what would be the amino acid sequence of the resulting polypeptide chain? Note that in this system translation can begin anywhere on the mRNA.

- A deletion of two consecutive base pairs in the coding region of DNA causes a frame-shift mutation. But a deletion of three consecutive base pairs causes the deletion of only one amino acid, with the rest of the polypeptide chain intact. Explain.

The mRNA with its coding information is translated into an amino acid sequence at the ribosome. We will now consider this process.

---

concept
10.4 **Translation of the Genetic Code Is Mediated by tRNA and Ribosomes**

The translation of mRNA into proteins requires a molecule that links the information contained in each mRNA codon with a specific amino acid. That function is performed by a set of transfer RNAs (tRNAs). Two key events must take place to ensure that the protein made is the one specified by the mRNA:

- A tRNA must chemically read each mRNA codon correctly.

- The tRNA must deliver the amino acid that corresponds to the mRNA codon.

Once the tRNAs "decode" the mRNA and deliver the appropriate amino acids, components of the ribosome catalyze the formation of peptide bonds between the amino acids.

---

yourBioPortal.com

Go to ANIMATED TUTORIAL 10.4
Protein Synthesis

Transfer RNAs carry specific amino acids and bind to specific codons

There is at least one specific tRNA molecule for each of the 20 amino acids. Each tRNA has three functions that are fulfilled by its structure and base sequence (**FIGURE 10.13**):

- *tRNAs bind to particular amino acids.* Each tRNA binds to a specific enzyme that attaches it to only 1 of the 20 amino acids. This covalent attachment is at the 3′ end of the tRNA. We describe the details of this vital process in the next section.

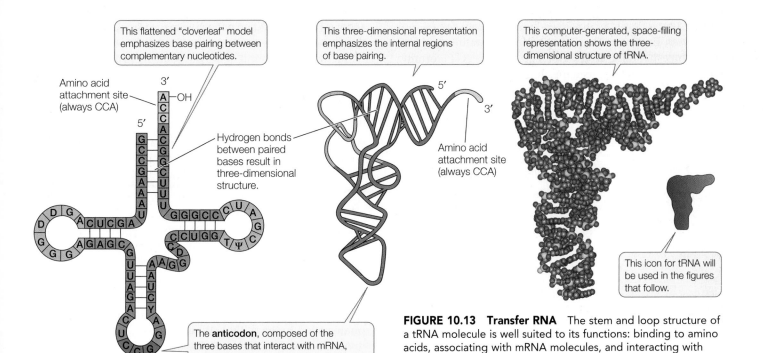

FIGURE 10.13 Transfer RNA The stem and loop structure of a tRNA molecule is well suited to its functions: binding to amino acids, associating with mRNA molecules, and interacting with ribosomes.

When it is carrying an amino acid, the tRNA is said to be "charged."

- *tRNAs bind to mRNA.* At about the midpoint on the tRNA polynucleotide chain there is a triplet of bases called the **anticodon**, which is complementary to the mRNA codon for the particular amino acid that the tRNA carries. Like the two strands of DNA, the codon and anticodon bind together via noncovalent hydrogen bonds. For example, the mRNA codon for arginine is 5′-CGG-3′, and the tRNA anticodon is 3′-GCC-5′.

- *tRNAs interact with ribosomes.* The ribosome has several sites on its surface that just fit the three-dimensional structure of a tRNA molecule. Interaction between the ribosome and the tRNA is noncovalent.

Recall that 61 different codons encode the 20 amino acids in proteins (see Figure 10.11). Does this mean that the cell must produce 61 different tRNA species, each with a different anticodon? No. The cell gets by with about two-thirds of that number of tRNA species because the specificity for the base at the 3′ end of the codon (and the 5′end of the anticodon) is not always strictly observed. This phenomenon is called *wobble*, and it is possible because in some cases unusual or modified nucleotide bases occur in the 5′ position of the anticodon. One such unusual base is inosine (I), which can pair with A, C, and U. For example, its presence allows three of the alanine codons GCA, GCC, and GCU to be recognized by the same tRNA (with the anticodon 3′-CGI-5′). Wobble occurs in some matches but not in others; of most importance, it does not allow the genetic code to be ambiguous. That is, *each mRNA codon binds to just one tRNA species, carrying a specific amino acid.*

Each tRNA is specifically attached to an amino acid

The charging of each tRNA with its correct amino acid is achieved by a family of enzymes known as *aminoacyl-tRNA synthetases*. Each enzyme is specific for one amino acid and for its corresponding tRNA. The reaction uses ATP, forming a high-energy bond between the amino acid and the tRNA:

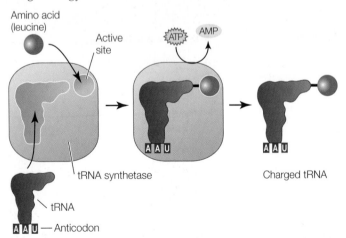

Amino acid (leucine)

Active site

ATP

AMP

tRNA synthetase

Charged tRNA

tRNA

AAU — Anticodon

The energy in this bond is later used in the formation of peptide bonds between amino acids in a growing polypeptide chain. Clearly, the specificity between the tRNA and its corresponding amino acid is extremely important. These reactions, for example, are highly specific:

$$\text{cysteine} + \text{tRNA}_{cys}\ (\text{anticodon ACA}) \xrightarrow{\text{Cys tRNA synthetase}} \text{cys-tRNA}_{cys}$$

$$\text{alanine} + \text{tRNA}_{ala}\ (\text{anticodon CGA}) \xrightarrow{\text{Ala tRNA synthetase}} \text{ala-tRNA}_{ala}$$

A clever experiment by Seymour Benzer and his colleagues at Purdue University demonstrated the importance of this specificity. They took the cys-tRNA molecule (see above) and chemically modified the cysteine, converting it into alanine. Which component—the amino acid or the tRNA—would be recognized when this hybrid charged tRNA was put into a protein-synthesizing system? The answer was the tRNA. Everywhere in the synthesized protein where cysteine was supposed to be, alanine appeared instead. The cysteine-specific tRNA had delivered its cargo (alanine) to every mRNA codon for cysteine. This experiment showed that the protein synthesis machinery recognizes the anticodon of the charged tRNA, not the amino acid attached to it.

Translation occurs at the ribosome

The ribosome is the molecular workbench where the translation of mRNA by tRNA is accomplished. All prokaryotic and eukaryotic ribosomes consist of two subunits (**FIGURE 10.14**). In eukaryotes, the large subunit consists of 3 different ribosomal RNA (rRNA) molecules and about 49 protein molecules arranged in a precise pattern. The small subunit consists of 1 rRNA molecule and about 33 proteins. These two subunits and several dozen other molecules interact noncovalently, fitting together like a jigsaw puzzle. If the hydrophobic interactions between the proteins and RNAs are disrupted, the ribosome falls apart, but it will reassemble perfectly when the disrupting agent is removed. When not active in the translation of mRNA, the ribosome exists as two separate subunits.

On the large subunit of the ribosome there are three sites to which a tRNA can bind, designated the A, P, and E sites (see Figure 10.14). The mRNA and ribosome move in relation to one another, and as they do so, a charged tRNA traverses these three sites in order:

- The *A (amino acid) site* is where the charged tRNA anticodon binds to the mRNA codon, thus lining up the correct amino acid to be added to the growing polypeptide chain.

- The *P (polypeptide) site* is where the tRNA adds its amino acid to the polypeptide chain.

- The *E (exit) site* is where the tRNA, having given up its amino acid, resides before being released from the ribosome and going back to the cytosol to pick up another amino acid and begin the process again.

The ribosome has a *fidelity function*, which ensures that a charged tRNA with the correct anticodon binds to the appropriate codon in the mRNA. When proper binding occurs, hydrogen bonds form between the three base pairs. The rRNA of the small ribosomal subunit plays a role in validating the

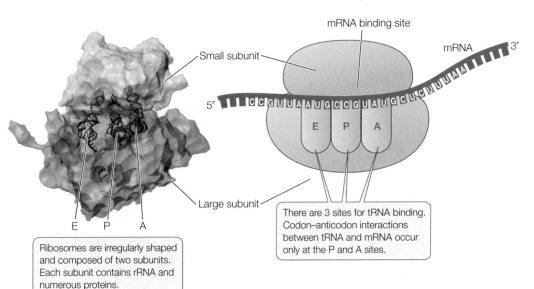

mRNA binding site

Small subunit

mRNA 3′

5′ C C G U U A A U G C C G U A U G C U C U U U A A

E P A

Large subunit

E P A

Ribosomes are irregularly shaped and composed of two subunits. Each subunit contains rRNA and numerous proteins.

There are 3 sites for tRNA binding. Codon–anticodon interactions between tRNA and mRNA occur only at the P and A sites.

FIGURE 10.14 Ribosome Structure Each ribosome consists of a large and a small subunit. The subunits remain separate when they are not in use for protein synthesis.

three-base-pair match. Any tRNA that does *not* form hydrogen bonds with all three bases of the codon is ejected from the ribosome.

Translation takes place in three steps

Like transcription, translation occurs in three steps: initiation, elongation, and termination.

INITIATION The **initiation complex** consists of a charged tRNA and a small ribosomal subunit, both bound to the mRNA (**FIGURE 10.15**). After binding, the small subunit moves along the mRNA until it reaches the start codon. Recall that the mRNA start codon in the genetic code is AUG (see Figure 10.11). The anticodon of a methionine-charged tRNA binds to this start codon by complementary base pairing to complete the initiation complex. Thus the first amino acid in a new polypeptide chain is always methionine. (In bacteria, but not archaea, the first amino acid is a slightly modified form of methionine called formylmethionine.) However, not all mature proteins have methionine as their first amino acid. In many cases, the initiator methionine is removed by an enzyme after translation.

After the methionine-charged tRNA has bound to the mRNA, the large subunit of the ribosome joins the complex. The methionine-charged tRNA now lies in the P site of the ribosome, and the A site is aligned with the second mRNA codon. These ingredients—mRNA, two

FIGURE 10.15 The Initiation of Translation Translation begins with the formation of an initiation complex.

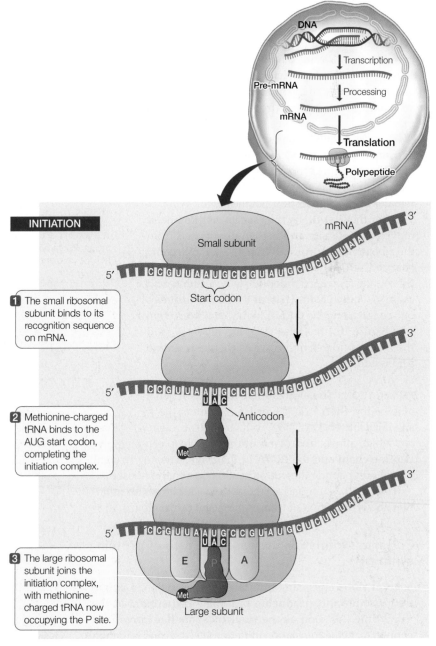

DNA

Transcription

Pre-mRNA

Processing

mRNA

Translation

Polypeptide

INITIATION

Small subunit mRNA 3′

5′ C C G U U A A U G C C G U A U G C U C U U U A A

Start codon

1 The small ribosomal subunit binds to its recognition sequence on mRNA.

3′

5′ C C G U U A A U G C C G U A U G C U C U U U A A
          U A C

2 Methionine-charged tRNA binds to the AUG start codon, completing the initiation complex.

Anticodon

Met

3′

5′ C C G U U A A U G C C G U A U G C U C U U U A A
          U A C

E P A

Met

Large subunit

3 The large ribosomal subunit joins the initiation complex, with methionine-charged tRNA now occupying the P site.

ribosomal subunits, and methionine-charged tRNA—are put together properly by a group of proteins called *initiation factors*.

ELONGATION A charged tRNA whose anticodon is complementary to the second codon of the mRNA now enters the open A site of the large ribosomal subunit (**FIGURE 10.16**). The large subunit then catalyzes two reactions:

- It breaks the bond between the methionine and its tRNA in the P site.
- It catalyzes the formation of a peptide bond between the methionine and the amino acid attached to the tRNA in the A site.

> **LINK** You may review the structure of peptide linkages in Concept 3.2, especially Figure 3.6

Because the large ribosomal subunit performs these two actions, it is said to have **peptidyl transferase** activity. The component with this activity is actually one of the rRNAs in the ribosome, so the catalyst is an example of a *ribozyme* (from *ribonucleic acid* and *enzyme*).

Methionine thus becomes the amino (N) terminus of the new protein (recall that polypeptides grow in the amino to carboxyl direction; see Concept 3.2). The second amino acid is now bound to methionine but remains attached to its tRNA at the A site.

After the first tRNA releases its methionine, it moves to the E site and is then dissociated from the ribosome, returning to the cytosol to become charged with another methionine. The second tRNA, now bearing a dipeptide (a two-amino-acid chain), is shifted to the P site as the ribosome moves one codon along the mRNA in the 5′-to-3′ direction (see Figure 10.16). These steps are repeated, and the polypeptide chain grows as each new amino acid is added.

TERMINATION The elongation cycle ends when a stop codon—UAA, UAG, or UGA—enters the A site (**FIGURE 10.17**). These codons bind a protein *release factor*, which allows hydrolysis of the bond between the polypeptide chain and the tRNA in the P site. The newly completed polypeptide thereupon separates from the ribosome.

TABLE 10.2 summarizes the nucleic acid signals for initiation and termination of transcription and translation.

Polysome formation increases the rate of protein synthesis

Several ribosomes can simultaneously translate a single mRNA molecule, producing multiple polypeptides at the same time. As soon as the first ribosome has moved far enough from the translation initiation site, a second initiation

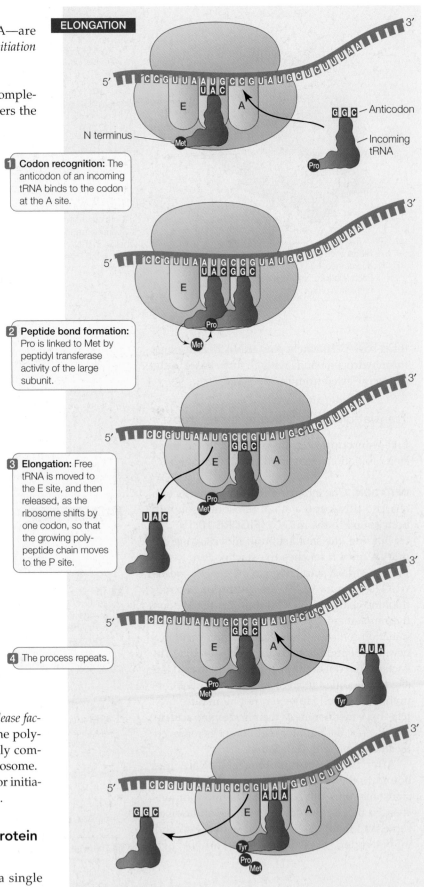

ELONGATION

1 **Codon recognition:** The anticodon of an incoming tRNA binds to the codon at the A site.

2 **Peptide bond formation:** Pro is linked to Met by peptidyl transferase activity of the large subunit.

3 **Elongation:** Free tRNA is moved to the E site, and then released, as the ribosome shifts by one codon, so that the growing polypeptide chain moves to the P site.

4 The process repeats.

FIGURE 10.16 The Elongation of Translation The polypeptide chain elongates as the mRNA is translated.

FIGURE 10.17 The Termination of Translation Translation terminates when the A site of the ribosome encounters a stop codon on the mRNA.

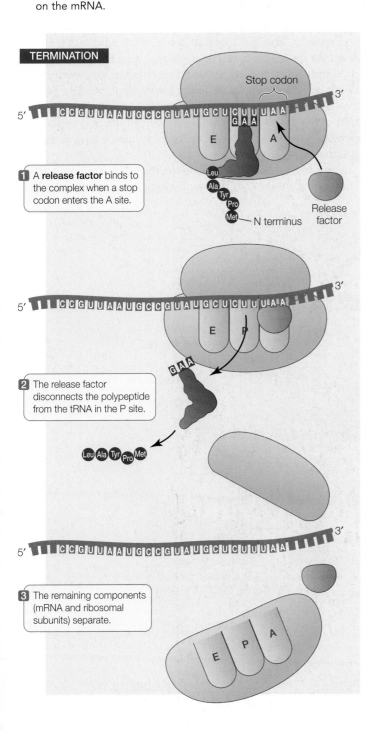

TERMINATION

1 A **release factor** binds to the complex when a stop codon enters the A site.

2 The release factor disconnects the polypeptide from the tRNA in the P site.

3 The remaining components (mRNA and ribosomal subunits) separate.

**FIGURE 10.18
A Polysome**
(A) A polysome consists of multiple ribosomes and their growing polypeptide chains moving along an mRNA molecule.
(B) An electron micrograph of a polysome.

(B)

Ribosome

mRNA

Growing polypeptides

(A)

INITIATION

Small subunit

Large subunit

ELONGATION

Ribosome

mRNA

Direction of translation

Polypeptide chain

TERMINATION

Polypeptides grow longer as each ribosome moves toward the 3′ end of mRNA.

complex can form, then a third, and so on. An assemblage consisting of a strand of mRNA with its beadlike ribosomes and their growing polypeptide chains is called a **polyribosome**, or **polysome** (**FIGURE 10.18**). Cells that are actively synthesizing proteins contain large numbers of polysomes and few free ribosomes or ribosomal subunits.

Do You Understand Concept 10.4?

- Describe the sequence of events in translation that involve tRNA, from charging with an amino acid to initiation, elongation, and termination.

- Imagine a polypeptide whose second amino acid is tryptophan. Sketch a ribosome with the mRNA and the first two tRNAs for this polypeptide, noting their positions in the A, P, and E sites.

- What would happen if a valine tRNA synthetase lost its specificity and attached any of the 20 amino acids to the 3′ end of the valine tRNA?

Usually, the process of protein synthesis does not end with translation. Proteins can undergo covalent modifications both during and after translation—with chemical groups being

TABLE 10.2	Signals that Start and Stop Transcription and Translation

	TRANSCRIPTION	TRANSLATION
Initiation	Promoter DNA	AUG start codon in the mRNA
Termination	Terminator DNA	UAA, UAG, or UGA in the mRNA

added or parts of the polypeptide chains removed. We will now turn to these modifications.

concept 10.5   Proteins Are Modified after Translation

The site of a polypeptide's function in the cell may be far away from its point of synthesis at the ribosome. This is especially true for eukaryotes, where a polypeptide may be moved into an organelle. Furthermore, polypeptides are often modified by the addition of new chemical groups that contribute to the function of the mature protein. In this section we examine these *posttranslational* aspects of protein synthesis.

Signal sequences in proteins direct them to their cellular destinations

Protein synthesis always begins on free ribosomes floating in the cytoplasm, and the "default" location for a protein is the cytosol. As the polypeptide chain emerges from the ribosome it may simply fold into its three-dimensional shape and perform its cellular role. However, a newly formed polypeptide may contain a **signal sequence** (or *signal peptide*)—a short stretch of amino acids that indicates where in the cell the polypeptide belongs. Proteins destined for different locations have different signals.

In the absence of a signal sequence, the protein will remain in the same cellular compartment where it was synthesized.

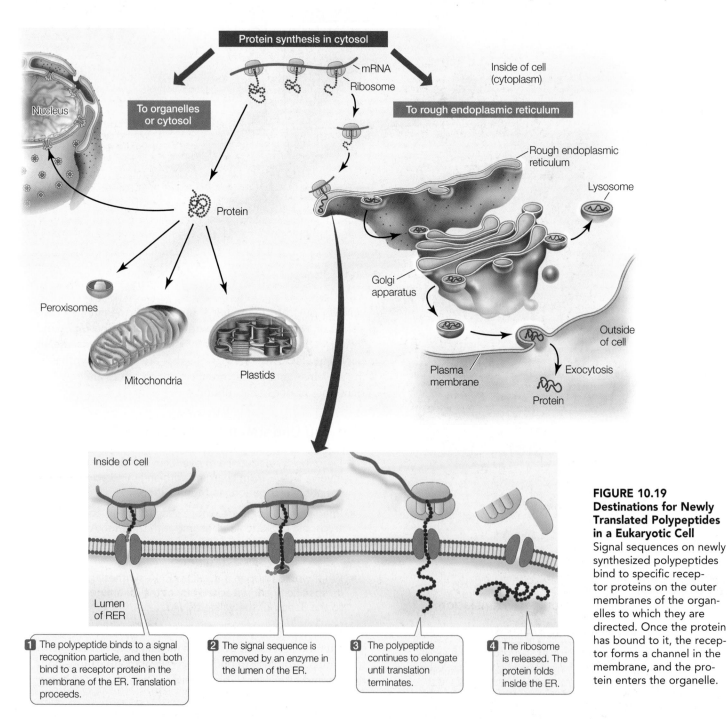

FIGURE 10.19 Destinations for Newly Translated Polypeptides in a Eukaryotic Cell Signal sequences on newly synthesized polypeptides bind to specific receptor proteins on the outer membranes of the organelles to which they are directed. Once the protein has bound to it, the receptor forms a channel in the membrane, and the protein enters the organelle.

1 The polypeptide binds to a signal recognition particle, and then both bind to a receptor protein in the membrane of the ER. Translation proceeds.

2 The signal sequence is removed by an enzyme in the lumen of the ER.

3 The polypeptide continues to elongate until translation terminates.

4 The ribosome is released. The protein folds inside the ER.

Some proteins, however, contain signal sequences that "target" them to the nucleus, mitochondria, plastids, or peroxisomes (**FIGURE 10.19, TOP**). A signal sequence binds to a specific receptor protein at the surface of the organelle. Once it has bound, a channel forms in the organelle membrane, allowing the targeted protein to move into the organelle. For example, here is a nuclear localization signal:

-Pro-Pro-Lys-Lys-Lys-Arg-Lys-Val-

The function of this peptide was established using experiments like the one illustrated in **FIGURE 10.20**. Proteins were made in the laboratory with or without the peptide, and then tested by injecting them into cells. Only proteins with the nuclear localization signal were found in the nucleus.

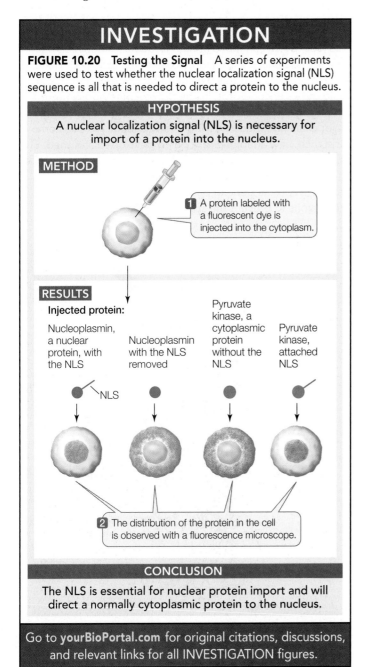

INVESTIGATION

FIGURE 10.20 Testing the Signal A series of experiments were used to test whether the nuclear localization signal (NLS) sequence is all that is needed to direct a protein to the nucleus.

HYPOTHESIS
A nuclear localization signal (NLS) is necessary for import of a protein into the nucleus.

METHOD

1 A protein labeled with a fluorescent dye is injected into the cytoplasm.

RESULTS

Injected protein:

Nucleoplasmin, a nuclear protein, with the NLS

Nucleoplasmin with the NLS removed

Pyruvate kinase, a cytoplasmic protein without the NLS

Pyruvate kinase, attached NLS

NLS

2 The distribution of the protein in the cell is observed with a fluorescence microscope.

CONCLUSION
The NLS is essential for nuclear protein import and will direct a normally cytoplasmic protein to the nucleus.

Go to **yourBioPortal.com** for original citations, discussions, and relevant links for all INVESTIGATION figures.

LINK For a description of how the proteins for this experiment were made, see Concept 13.3

If a polypeptide carries a particular signal sequence of 5–10 hydrophobic amino acids at it its N terminus, it will be directed to the rough endoplasmic reticulum (RER) for further processing (**FIGURE 10.19, BOTTOM**). Translation will pause, and the ribosome will bind to a receptor at the RER membrane. Once the polypeptide–ribosome complex is bound, translation will resume, and as elongation continues, the protein will traverse the RER membrane. Such proteins may be retained in the lumen (the inside) or membrane of the RER, or they may move elsewhere within the endomembrane system (Golgi apparatus, lysosomes, and plasma membrane). If the proteins lack specific signals for destinations within the endomembrane system, they are usually secreted from the cell via vesicles that fuse with the plasma membrane.

Many proteins are modified after translation

Most mature proteins are not identical to the polypeptide chains that are translated from mRNA on the ribosomes. Instead, most polypeptides are modified in any of a number of ways after translation (**FIGURE 10.21**). These modifications are essential to the final functioning of the protein.

- **Proteolysis** is the cutting of a polypeptide chain. For example, the ER signal sequence is cut off from the growing polypeptide chain as it enters the ER. Some mature proteins are actually made from *polyproteins*—long polypeptides containing the primary sequences of multiple distinct proteins—that are cut into final products by enzymes called *proteases*. Proteases are essential to some viruses, including human immunodeficiency virus (HIV), because the large viral polyprotein cannot fold properly unless it is cut. Certain drugs used to treat acquired immune deficiency syndrome (AIDS) work by inhibiting the HIV protease, thereby preventing the formation of proteins needed for viral reproduction.

- **Glycosylation** is the addition of carbohydrates to proteins to form glycoproteins. In both the ER and the Golgi apparatus, resident enzymes catalyze the addition of various oligosaccharides (short chains of monosaccharides; see Concept 2.3) to certain amino acid R groups on proteins. One such type of "sugar coating" is essential for directing proteins to lysosomes. Other types are important for protein conformation and for recognition functions at the cell surface. In many cases the attached oligosaccharides help stabilize proteins, such as those in the extracellular matrix, and those in the storage vacuoles of plants.

- **Phosphorylation** is the addition of phosphate groups to proteins and is catalyzed by *protein kinases*. The charged phosphate groups change the conformation of the protein, often exposing the active site of an enzyme or the binding site for another protein. Phosphorylation is especially important in cell signaling (see Concepts 5.5 and 5.6).

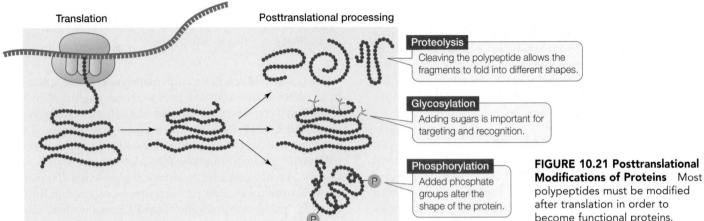

Translation   Posttranslational processing

Proteolysis
Cleaving the polypeptide allows the fragments to fold into different shapes.

Glycosylation
Adding sugars is important for targeting and recognition.

Phosphorylation
Added phosphate groups alter the shape of the protein.

FIGURE 10.21 Posttranslational Modifications of Proteins Most polypeptides must be modified after translation in order to become functional proteins.

FRONTIERS In some proteins, proline is converted to hydroxyproline after translation. This occurs in collagen, a protein that is secreted into the extracellular matrices of animal cells, and extensin, a protein secreted into the plant cell wall. Why is this modification a hallmark of extracellular matrix proteins? Examination of the role of hydroxyproline in the structures of these proteins may provide an answer.

Do You Understand Concept 10.5?

- Describe how signal sequences determine where a protein will go after it is made.

- What are some ways in which posttranslational modifications alter protein structure and function?

- Describe an experiment you would perform to test a proposed chloroplast-targeting signal sequence. Be specific about the type of cell and the proteins you would use.

How do antibiotics such as tetracycline target bacterial protein synthesis?

ANSWER Tetracyclines are widely used antibiotics that are effective against some strains of MRSA and many other bacterial infections. They derive their name from the four hydrocarbon rings that are common to this family of molecules. Tetracyclines kill bacteria by interrupting translation (Concept 10.1). They do this by binding noncovalently to the small subunit of bacterial ribosomes (Concept 10.4), where binding changes ribosome structure such that charged tRNAs can no longer bind to the A site on the ribosome (**FIGURE 10.22**). The specificity of antibiotics for bacterial ribosomes comes from the fact that bacterial and eukaryotic ribosomes have different proteins and RNAs.

Strains of MRSA with resistance to tetracyclines are emerging. The genes that confer resistance to this group of antibiotics are carried on mobile genetic elements such as plasmids, which can move at high frequencies between bacteria, by bacterial conjugation. Some of the resistance genes encode proteins that transfer the tetracyclines out of the cell, while others encode proteins that protect the ribosomes from binding by the antibiotics. These resistance genes present a major challenge, because MRSA can be lethal. To overcome resistance, new antibiotics are being developed and tried. The evolutionary race between genetically caused drug resistance and new therapies continues. In the meantime, health-care providers and the general public are being advised to take precautions to prevent the spread of MRSA.

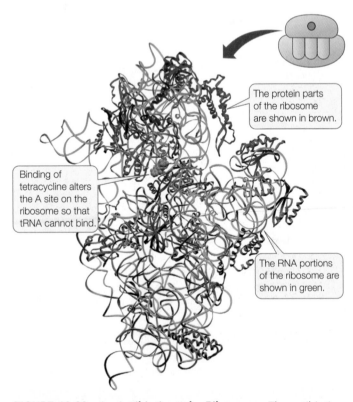

The protein parts of the ribosome are shown in brown.

Binding of tetracycline alters the A site on the ribosome so that tRNA cannot bind.

The RNA portions of the ribosome are shown in green.

FIGURE 10.22 An Antibiotic at the Ribosome The antibiotic tetracycline binds to the small ribosomal subunit of bacteria. This causes a change in the structure of the A site, preventing tRNAs from binding, and protein synthesis stops.

SUMMARY

concept **10.1** ## Genetics Shows That Genes Code for Proteins

- Studies of human genetic diseases such as alkaptonuria linked genes to proteins. **Review Figure 10.1**

- Hemoglobin abnormalities show that mutations can alter the sequence of amino acids in proteins. **Review Figure 10.2**

- Genes are expressed via transcription and translation. During **transcription**, the information in a gene is copied into a complementary RNA sequence. During **translation**, this RNA sequence is used to create the amino acid sequence of a polypeptide. **Review Figure 10.3 and WEB ACTIVITY 10.1**

- The product of transcription is **messenger RNA (mRNA)**. **Transfer RNA (tRNA)** molecules are adapters that translate the genetic information in the mRNA into a corresponding sequence of amino acids.

- **Ribosomal RNA (rRNA)** helps provide structure to the **ribosome** and acts as a ribozyme that catalyzes peptide bond formation between amino acids during protein synthesis.

concept **10.2** ## DNA Expression Begins with Its Transcription to RNA

- In a given gene, only one of the two strands of DNA (the template strand) acts as a template for transcription. **RNA polymerase** is the catalyst for transcription.

- RNA transcription from DNA proceeds in three steps: initiation, **elongation**, and **termination**. Initiation requires a **promoter** to which RNA polymerase binds. Elongation of the RNA molecule proceeds by the addition of nucleotides to the 3′ end of the molecule. **Review Figure 10.5 and ANIMATED TUTORIAL 10.1**

- After transcription, eukaryotic pre-mRNA is spliced to remove **introns**. **Review Figures 10.6 and 10.9 and ANIMATED TUTORIAL 10.2**

- Eukaryotic mRNA is also modified by the addition of a **5′ cap** and a **poly A tail**.

concept **10.3** ## The Genetic Code in RNA Is Translated into the Amino Acid Sequences of Proteins

- Experiments involving synthetic mRNAs and protein synthesis in the test tube established the genetic code. **Review Figure 10.10 and ANIMATED TUTORIAL 10.3, and WORKING WITH DATA 10.1**

- The genetic code consists of triplets of mRNA nucleotide bases (**codons**) that correspond to 20 specific amino acids. There are **start codons** and **stop codons** as well.

- The genetic code is redundant (an amino acid may be represented by more than one codon) but not ambiguous (no single codon represents more than one amino acid). **Review Figure 10.11 and WEB ACTIVITY 10.2**

- Mutations in the coding regions of genes can be silent, missense, nonsense, or frame-shift mutations. **Review Figure 10.12 and INTERACTIVE TUTORIAL 10.1**

concept **10.4** ## Translation of the Genetic Code Is Mediated by tRNA and Ribosomes

Review ANIMATED TUTORIAL 10.4

- Transfer RNA (tRNA) mediates between mRNA and amino acids during translation at the ribosome.

- Each tRNA species has an amino acid attachment site and an **anticodon** that is complementary to a specific mRNA codon. **Review Figure 10.13**

- A specific synthetase enzyme charges each tRNA with its specific amino acid.

- Three sites on the large subunit of the ribosome interact with tRNA anticodons. The A site is where the charged tRNA anticodon binds to the mRNA codon. The P site is where the tRNA adds its amino acid to the growing polypeptide chain. The E site is where the tRNA is released. **Review Figure 10.14**

- Translation occurs in three steps: initiation, **elongation**, and **termination**. **Review Figures 10.15–10.17**

- In a **polyribosome**, or **polysome**, more than one ribosome moves along a strand of mRNA at one time. **Review Figure 10.18**

concept **10.5** ## Proteins Are Modified after Translation

- **Signal sequences** are short sequences of amino acids that direct polypeptides to their cellular destinations.

- These destinations include the nucleus and other organelles, which proteins enter after being recognized and bound by surface receptors.

- If a ribosome begins translating a polypeptide with an N-terminal RER signal sequence, it pauses and then resumes translation after attachment to a receptor in the RER membrane. **Review Figure 10.19**

- Posttranslational modifications of polypeptides include **proteolysis**, in which a polypeptide is cut into smaller fragments; **glycosylation**, in which sugars are added; and **phosphorylation**, in which phosphate groups are added. **Review Figure 10.21**

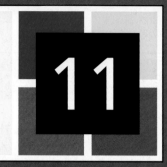
Many people drink alcoholic beverages, but relatively few of them become addicted (alcoholic). Alcoholism is characterized by a compulsion to consume alcohol, tolerance (increasing doses are needed for the same effect), and dependence (abrupt cessation of consumption leads to severe withdrawal symptoms). In most alcoholics, alcohol provides pleasant sensations (positive reinforcement) and alleviates unpleasant ones such as anxiety (negative reinforcement).

Alcoholism is a complex disease. Psychologists sometimes speak of "addictive personalities," and genetic studies indicate there may be inherited factors that predispose people to the disease. One approach to describing the genes involved is to study animal models of alcoholism at the molecular level. James Murphy at Indiana University has bred a genetic strain of alcoholic rats, called P rats, that prefer alcohol when given the choice of alcohol-containing or alcohol-free water. These rats show many of the symptoms of addiction, including compulsive drinking, tolerance, and withdrawal. P rats also appear more anxious than wild-type rats, spending more time in a closed rather than an open environment. Drinking alcohol alters this behavior and seems to relieve their anxiety.

Scientists have found a link between a protein and the genetic mutation(s) leading to alcoholism in rats. CREB (cyclic AMP response element binding protein) is abundant in the brain and regulates the expression of hundreds of genes that are important in metabolism. CREB becomes activated when it is phosphorylated by the enzyme protein kinase A, which in turn is activated by the second messenger cyclic AMP. In an effort to understand the molecular basis of alcoholism and anxiety, neuroscientist Subhash Pandey and his colleagues at the University of Illinois compared CREB levels in the brains of P

rats and wild-type rats. They found that P rats have inherently lower levels of CREB in certain parts of the brain. When these rats consumed alcohol, the total levels of CREB did not increase, but the levels of phosphorylated CREB did. It is the phosphorylated version of CREB that regulates gene transcription.

The prospect that CREB, a molecule that regulates gene expression, is a key element in the genetic propensity for alcoholism is important because it begins to explain the molecular nature of a complex behavioral disease. Such understanding may permit more effective treatment of alcohol abuse, or even its prevention. And with reference to our purpose here, it underscores the importance of the regulation of gene expression in biological processes.

A rat drinks an alcoholic beverage. Some rats are genetically programmed to prefer alcohol over plain water.

How does CREB regulate the expression of many genes?

QUESTION

KEY CONCEPTS

11.1 Several Strategies Are Used to Regulate Gene Expression

11.2 Many Prokaryotic Genes Are Regulated in Operons

11.3 Eukaryotic Genes Are Regulated by Transcription Factors and DNA Changes

11.4 Eukaryotic Gene Expression Can Be Regulated after Transcription

concept
11.1 **Several Strategies Are Used to Regulate Gene Expression**

In Chapter 10 we introduce the concepts of gene expression. DNA is initially expressed as RNA. In many cases the RNA is then translated into protein at the ribosome. Throughout this book we describe instances where gene expression is altered so that the level of protein produced from a particular gene varies. Such variations are influenced by environmental conditions and the developmental stage of the cell or organism. Here are a few examples:

- In Chapter 5: When an extracellular signal binds to its receptor on a eukaryotic cell, it sets in motion a signal transduction pathway that may end with some genes being activated (their expression switched on) or others being repressed (their expression switched off).

- In Chapter 7: During the cell cycle, cyclins are synthesized only at specific points. The genes for cyclins are inactive at other points in the cycle.

- In Chapter 9: When a virus infects a host cell, it can "hijack" the host gene expression machinery and divert it to viral gene expression.

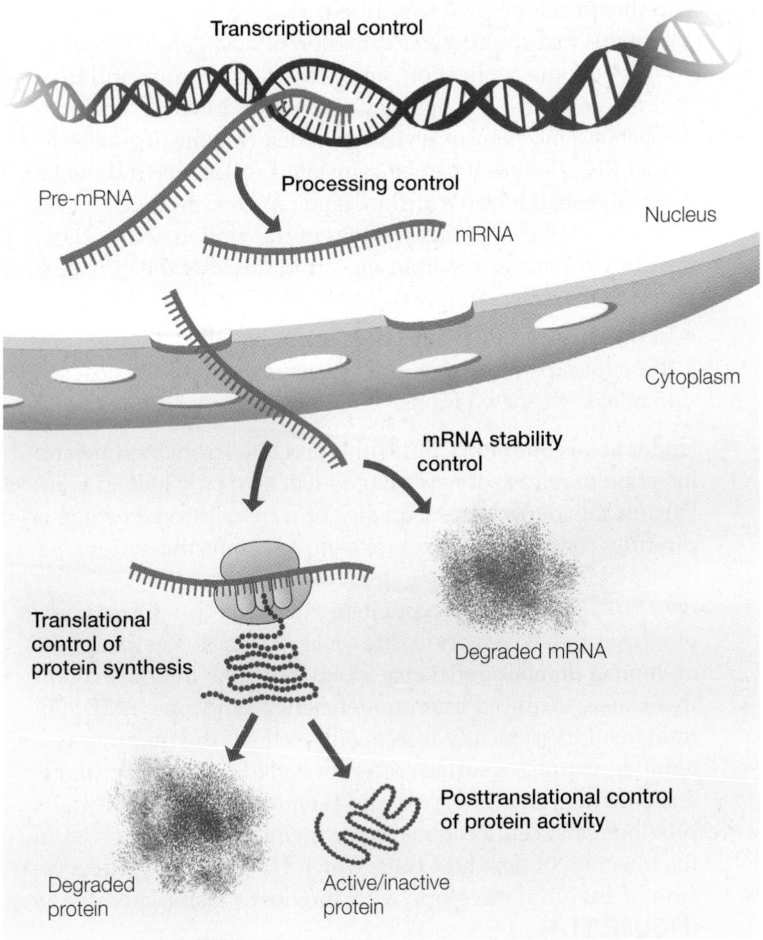

These and other examples indicate that *gene expression is precisely regulated*. In some cases, gene expression is modified to counteract changes in the cell's environment, so that stable conditions are maintained within the cell. In other cases, gene expression changes so that the cell can perform specific functions. For example, all of our cells carry the genes encoding keratin (the protein in our hair and nails) and hemoglobin. Yet keratin is made only by epithelial cells such as skin cells, and hemoglobin is made only by developing red blood cells. In contrast, all human cells express the genes that encode enzymes needed for basic metabolic activities (such as glycolysis), and all cells must synthesize certain structural proteins such as actin (a component of the cytoskeleton). To generalize:

- **Constitutive genes** are actively expressed all the time.

- **Inducible genes** are expressed only when their proteins are needed by the cell.

Our discussion of the regulation of gene expression will focus on inducible genes.

Genes are subject to positive and negative regulation

At every step of the way from DNA to protein that we described in Chapter 10, gene expression can be regulated (**FIGURE 11.1**). As we proceed through this chapter, you will see examples of gene regulation at the transcriptional, posttranscriptional, translational, and posttranslational levels. An important form of gene regulation is at the level of *transcription*.

---

yourBioPortal.com
Go to WEB ACTIVITY 11.1
Eukaryotic Gene Expression Control Points

LINK You may wish to review the processes of transcription described in Concept 10.2

Gene expression begins at the *promoter*, where RNA polymerase binds to initiate transcription. As we mentioned above, not all genes are active (being transcribed) at a given time—there is *selective gene transcription*. Two types of regulatory proteins—also called **transcription factors**—control whether or not a gene is active: repressors and activators. These proteins bind to specific DNA sequences at or near the promoter (**FIGURE 11.2**):

- In negative regulation, a **repressor** binds near the promoter to prevent transcription.

- In positive regulation, the binding of an **activator** stimulates transcription.

FIGURE 11.1 Potential Points for the Regulation of Gene Expression Gene expression can be regulated before transcription, during transcription, after transcription but before translation, at translation, or after translation.

(A) Negative regulation

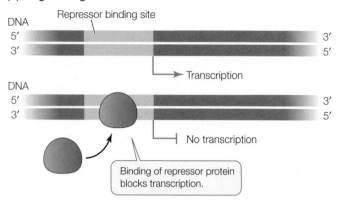

(B) Positive regulation

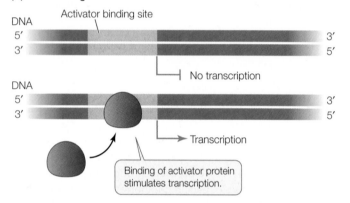

FIGURE 11.2 Positive and Negative Regulation Transcription factors regulate gene expression by binding to DNA and (A) repressing or (B) activating transcription by RNA polymerase.

You will see these mechanisms, or combinations of them, as we examine gene regulation in viruses, bacteria, and eukaryotes.

Viruses use gene regulation strategies to subvert host cells

The immunologist Sir Peter Medawar once described a virus as "a piece of bad news wrapped in protein." As we describe in Concept 9.1, a **virus** injects its genetic material into a host cell, and in many cases it turns that cell into a virus factory (see Figure 9.2). This involves a radical change in gene expression for the host cell, and results in the death of the cell when new viral particles are released. Viral life cycles are very efficient—for example, the poliovirus completes its life cycle (from infection to release of new particles) in 4–6 hours, and each dying host cell can produce up to 10,000 new particles!

Unlike cellular organisms, viruses are *acellular*. They are not cells and do not carry out many of the processes characteristic of life, and they are dependent on living cells to reproduce. Not all viruses use double-stranded DNA as the genetic material that is contained within the viral particle and transmitted from one generation to the next. The viral genome may consist of double-stranded DNA, single-stranded DNA, or double- or single-stranded RNA. But whether the genetic material is DNA

or RNA, the viral genome takes over the host's protein synthetic machinery within minutes of entering the cell.

Typically, the host cell immediately begins to produce new viral particles (*virions*), which are released as the cell breaks open, or *lyses*. This type of viral life cycle is called **lytic**. Some viral life cycles also include a **lysogenic** or dormant phase. In this case the viral genome becomes incorporated into the host cell genome and is replicated along with the host genome. The virus may survive in this way for many host cell generations. Sooner or later, an environmental signal can cause the host cell to begin producing virions—at which point the viral reproductive cycle enters the lytic phase.

By studying the relatively simple reproductive cycles of viruses, biologists have discovered principles of gene regulation that apply to much more complex cellular systems. We will discuss two examples of viruses here: one prokaryotic (a bacteriophage) and the other eukaryotic (the human immunodeficiency virus).

BACTERIOPHAGE Like other viruses, a *bacteriophage* (phage, or bacterial virus) may have a DNA or RNA genome, and its life cycle may or may not include a lysogenic phase. **FIGURE 11.3** illustrates the lytic life cycle of T4, a typical double-stranded DNA phage. At the molecular level, the lytic cycle has two stages, early and late:

- The viral genome contains a promoter that binds host RNA polymerase. In the *early stage*, viral genes that lie adjacent to this promoter are transcribed. These early genes encode proteins that shut down expression of host genes, stimulate viral genome replication, and activate the transcription of viral late genes. The host genes are shut down by a *posttranscriptional* mechanism: a virus-encoded enzyme degrades the host RNA before it can be translated. Another viral nuclease digests the host's chromosome, providing nucleotides for the synthesis of many copies of the viral genome. These processes can occur within a few minutes after the virus first infects the cell.

- In the *late stage*, viral late genes are transcribed; they encode the viral capsid proteins and enzymes that lyse the host cell to release the new virions.

Under ideal conditions, this entire process—from binding and infection to release of new phage—can take only half an hour. During this period, the sequence of transcriptional events is carefully controlled to produce complete, infective virions.

HIV Eukaryotes are susceptible to infections by various kinds of viruses that have various life cycle strategies. We focus here on **human immunodeficiency virus** (**HIV**), the infective agent that causes acquired immunodeficiency syndrome (AIDS) in humans. HIV typically infects only cells of the immune system that express a surface receptor called CD4. The virion is enclosed within a phospholipid membrane derived from its previous host cell. Proteins in the membrane are involved in the infection of new host cells, which HIV enters by direct fusion of the viral envelope with the host plasma membrane (**FIGURE 11.4**).

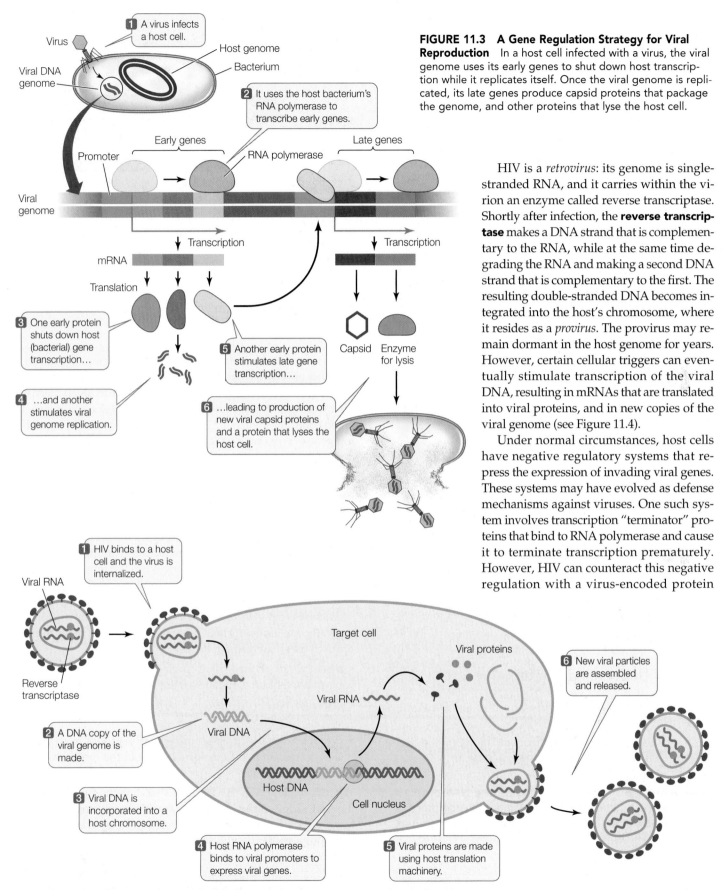

FIGURE 11.3 A Gene Regulation Strategy for Viral Reproduction In a host cell infected with a virus, the viral genome uses its early genes to shut down host transcription while it replicates itself. Once the viral genome is replicated, its late genes produce capsid proteins that package the genome, and other proteins that lyse the host cell.

HIV is a *retrovirus*: its genome is single-stranded RNA, and it carries within the virion an enzyme called reverse transcriptase. Shortly after infection, the **reverse transcriptase** makes a DNA strand that is complementary to the RNA, while at the same time degrading the RNA and making a second DNA strand that is complementary to the first. The resulting double-stranded DNA becomes integrated into the host's chromosome, where it resides as a *provirus*. The provirus may remain dormant in the host genome for years. However, certain cellular triggers can eventually stimulate transcription of the viral DNA, resulting in mRNAs that are translated into viral proteins, and in new copies of the viral genome (see Figure 11.4).

Under normal circumstances, host cells have negative regulatory systems that repress the expression of invading viral genes. These systems may have evolved as defense mechanisms against viruses. One such system involves transcription "terminator" proteins that bind to RNA polymerase and cause it to terminate transcription prematurely. However, HIV can counteract this negative regulation with a virus-encoded protein

FIGURE 11.4 The Reproductive Cycle of HIV This retrovirus enters a host cell via fusion of its envelope with the host's plasma membrane. Reverse transcription of retroviral RNA then produces a complementary DNA that becomes inserted into the host's genome. The inserted viral DNA directs the synthesis of new virus particles.

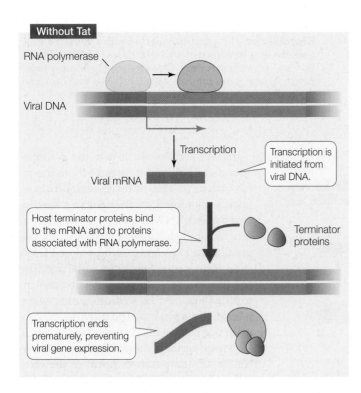

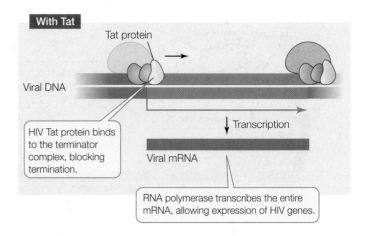

FIGURE 11.5 Regulation of Transcription by HIV The Tat protein acts as an antiterminator, allowing transcription of the HIV genome.

called Tat (*Trans*activator of *t*ranscription), which binds to the viral mRNA along with associated proteins that allow RNA polymerase to transcribe the viral genome (**FIGURE 11.5**).

FRONTIERS Because AIDS is a major challenge worldwide, biologists probably know more about HIV than any other virus. Major efforts are underway to develop drugs targeted at virtually every step in the virus's life cycle. Some of these stop viral transmission and replication without harming human cells.

Do You Understand Concept 11.1?

- What is the difference between positive and negative regulation of gene expression?
- Describe positive and negative regulation of gene expression in bacteriophage and HIV life cycles.
- What would be the effects of the following?
  a. A mutation in the gene that encodes RNA polymerase so that it does not bind to the promoter for late genes in a bacteriophage.
  b. The inhibition of reverse transcriptase in an HIV-infected cell.

We have seen how viruses co-opt the regulatory mechanisms of their host cells in order to express their own genes and reproduce. Now let's turn to a closer examination of gene regulation in prokaryotes.

Many Prokaryotic Genes Are Regulated in Operons

Prokaryotes conserve energy and resources by making certain proteins only when they are needed. Because their environments can change abruptly, prokaryotes have evolved mechanisms to rapidly alter the expression levels of certain genes when conditions warrant. The most efficient means of regulating gene expression is at the level of transcription.

Regulating gene transcription conserves energy

As a normal inhabitant of the human intestine, *Escherichia coli* must be able to adjust to sudden changes in its chemical environment as the foods consumed by its host change (for example, from glucose at one time to lactose at another). In many cases, *E. coli* responds to such changes by changing the expression of its genes. To illustrate this, we will look at the regulation of the pathway for lactose catabolism in *E. coli*.

Lactose is a β-galactoside—a disaccharide containing galactose linked to glucose. Three proteins are involved in the initial uptake and metabolism of lactose by *E. coli*:

- *β-galactoside permease* is a carrier protein in the bacterial plasma membrane that moves the sugar into the cell.
- *β-galactosidase* is an enzyme that hydrolyzes lactose to glucose and galactose.
- *β-galactoside transacetylase* transfers acetyl groups from acetyl CoA to certain β-galactosides. Its role in the metabolism of lactose is not clear.

When *E. coli* is grown on a medium that contains glucose but no β-galactosides, the levels of these three proteins are extremely low—only a few molecules per cell. But if the cells are transferred to a medium with lactose as the predominant sugar, they promptly begin making all three enzymes, and within 10 minutes there are about 3,000 of each of these proteins per cell. Clearly, these are proteins encoded by inducible genes, and their expression is switched on by an **inducer**. In this case the inducer is allolactose, an isomer of lactose.

FIGURE 11.6 Two Ways to Regulate a Metabolic Pathway Feedback from the end product of a metabolic pathway can block enzyme activity (allosteric regulation), or it can stop the transcription of genes that code for the enzymes in the pathway (transcriptional regulation).

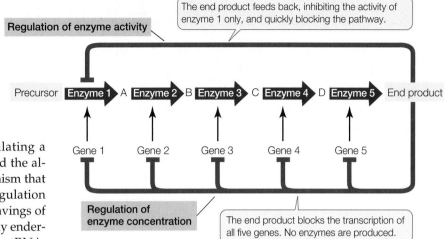

We have now seen two basic ways of regulating a metabolic pathway. In Concept 3.4 we described the allosteric regulation of enzyme activity—a mechanism that allows rapid fine-tuning of metabolism. The regulation of transcription is slower but results in greater savings of energy and resources. Protein synthesis is a highly endergonic process, since assembling mRNA, charging tRNA, and moving the ribosomes along mRNA all require large amounts of energy. **FIGURE 11.6** compares these two modes of regulation.

Operons are units of transcriptional regulation in prokaryotes

The genes that encode the three enzymes for processing lactose in *E. coli* are **structural genes**; they each specify the primary structure (the amino acid sequence) of a protein molecule that is not involved in regulation. The three genes lie adjacent to one another on the *E. coli* chromosome. This arrangement is no coincidence: the genes share a single promoter, and their DNA is transcribed into a single, continuous molecule of mRNA. Because this particular mRNA governs the synthesis of all three lactose-metabolizing enzymes, either all or none of these enzymes are made at any particular time.

A cluster of genes with a single promoter is called an **operon**, and the operon that encodes the three lactose-metabolizing enzymes in *E. coli* is called the *lac* operon. The *lac* operon promoter can be very efficient (the maximum rate of mRNA synthesis can be high), but mRNA synthesis can be shut down when the enzymes are not needed. This example of negative regulation was elegantly worked out by Nobel Prize winners François Jacob and Jacques Monod.

The *lac* operon has another DNA sequence called an **operator**, which is near the promoter and controls transcription of the structural genes (**FIGURE 11.7**). Operators can bind very tightly with repressor proteins, which play different roles in different operons:

- An *inducible operon* is turned *off* unless needed.
- A *repressible operon* is turned *on* unless *not* needed.

Operator–repressor interactions regulate transcription in the *lac* and *trp* operons

In the case of the inducible *lac* operon, a repressor protein prevents transcription until the *lac*-encoded proteins are needed. In contrast, the *trp* operon (described below) is a repressible operon that is turned off by a repressor only under particular circumstances.

***lac* OPERON** As we described above, the *lac* operon is not transcribed unless a β-galactoside (such as lactose) is the predominant sugar available in the cell's environment. A repressor protein is normally bound to the operator, preventing transcription. When lactose is present, the repressor detaches from the operator sequence, allowing RNA polymerase to bind to the promoter and start transcribing the structural genes (**FIGURE 11.8**).

The key to this regulatory system is the repressor protein. Expressed from a constitutive promoter (one that is always active), the repressor is always present in the cell in adequate amounts to occupy the operator and keep the operon turned off. The repressor has a recognition site for the DNA sequence in the operator, and it binds very tightly. However, it also has an allosteric binding site for the inducer. When the inducer (allolactose, an alternate form of lactose) binds to the repressor, the repressor changes shape so that it can no longer bind DNA.

FIGURE 11.7 The *lac* Operon of *E. coli* The *lac* operon of *E. coli* is a segment of DNA that includes a promoter, an operator, and the three structural genes that code for lactose-metabolizing enzymes. In reality, the structural genes are much longer than the short, regulatory sequences.

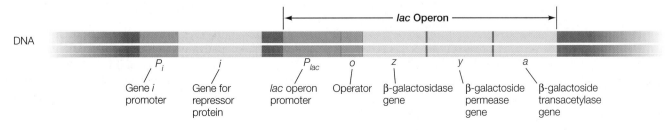

APPLY THE CONCEPT

Many prokaryotic genes are regulated in operons, which include regulatory DNA sequences

Genetic mutations are useful in analyzing the control of gene expression. In the *lac* operon of *E. coli* (see Figure 11.7), gene *i* codes for the repressor protein, P_{lac} is the promoter, *o* is the operator, and *z* is the first structural gene. (+) means wild type; (−) means mutant. Fill in the table, describing the level of transcription in different genetic and environmental conditions.

GENOTYPE	Z TRANSCRIPTION LEVEL	
	LACTOSE PRESENT	LACTOSE ABSENT
$i^- \ P_{lac}^+ \ o^+ \ z^+$		
$i^+ \ P_{lac}^+ \ o^+ \ z^-$		
$i^+ \ P_{lac}^- \ o^+ \ z^+$		
$i^+ \ P_{lac}^+ \ o^- \ z^+$		

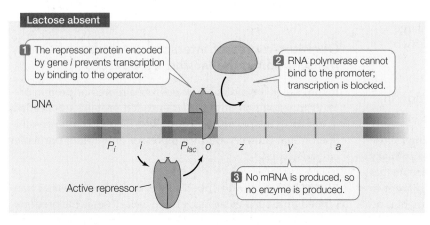

Lactose absent

1 The repressor protein encoded by gene *i* prevents transcription by binding to the operator.

2 RNA polymerase cannot bind to the promoter; transcription is blocked.

DNA

P_i *i* P_{lac} *o*   *z*   *y*   *a*

Active repressor

3 No mRNA is produced, so no enzyme is produced.

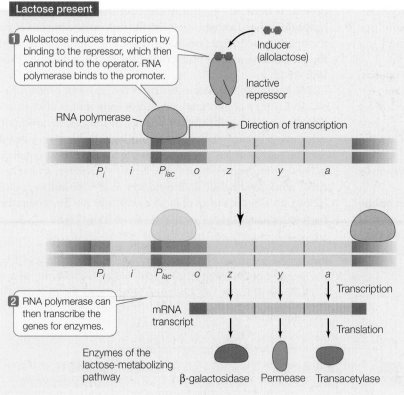

Lactose present

1 Allolactose induces transcription by binding to the repressor, which then cannot bind to the operator. RNA polymerase binds to the promoter.

Inducer (allolactose)

Inactive repressor

RNA polymerase

Direction of transcription

P_i *i* P_{lac} *o*   *z*   *y*   *a*

P_i *i* P_{lac} *o*   *z*   *y*   *a*

2 RNA polymerase can then transcribe the genes for enzymes.

Transcription

mRNA transcript

Translation

Enzymes of the lactose-metabolizing pathway

β-galactosidase   Permease   Transacetylase

FIGURE 11.8 The *lac* Operon: An Inducible System Allolactose (the inducer) leads to synthesis of the enzymes in the lactose-metabolizing pathway by binding to the repressor protein and preventing its binding to the operator.

yourBioPortal.com

Go to ANIMATED TUTORIAL 11.1
The *lac* Operon

***trp* OPERON** Like an inducible operon, a repressible operon is switched off when its repressor is bound to its operator. However in this case, the repressor binds to the DNA only in the presence of a **co-repressor**. The co-repressor is a molecule that binds to the repressor, causing it to change shape and bind to the operator, thereby inhibiting transcription. An example is the operon whose structural genes catalyze the synthesis of the amino acid tryptophan (**FIGURE 11.9**). When tryptophan is present in the cell in adequate concentrations, it is energy efficient to stop making the enzymes for tryptophan synthesis. Therefore, tryptophan itself functions as a co-repressor that binds to the repressor of the *trp* operon, causing the repressor to bind to the *trp* operator to prevent transcription.

To summarize the differences between these two types of operons:

- In *inducible* systems, the substrate of a metabolic pathway (the inducer) interacts with a transcription factor (the repressor), rendering the repressor incapable of binding to the operator and thus allowing transcription.

- In *repressible* systems, the product of a metabolic pathway (the co-repressor) binds to the repressor protein, which is then able to bind to the operator and block transcription.

In general, inducible systems control catabolic pathways (which are turned on only when the substrate is available), whereas repressible systems control anabolic pathways (which are turned on until the concentration of the product becomes excessive).

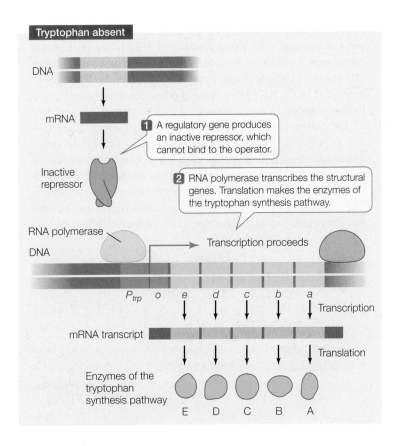

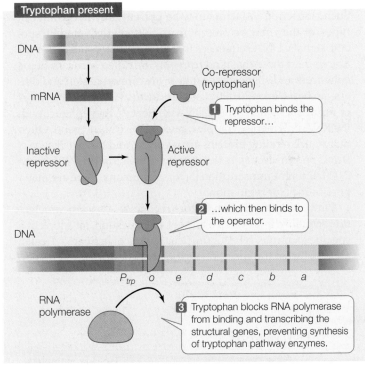

FIGURE 11.9 The *trp* Operon: A Repressible System Because tryptophan activates an otherwise inactive repressor, it is called a co-repressor.

yourBioPortal.com

Go to ANIMATED TUTORIAL 11.2
The *trp* Operon

LINK Review the descriptions of catabolic and anabolic reactions in Concept 2.5

In both of the systems described above, the regulatory protein is a repressor that functions by binding to the operator. Other operons are regulated by activator proteins that bind to DNA elements near the promoter and promote transcription. Like repressors, activators can regulate both inducible and repressible systems. We will discuss transcription factors in more detail in Concept 11.3.

RNA polymerase can be directed to a class of promoters

As noted above and in Chapter 10, RNA polymerase binds to specific DNA sequences at the promoter to initiate transcription. We have just described how repressor proteins can physically block RNA polymerase binding. However, there are other proteins in prokaryotes called **sigma factors** that can bind to RNA polymerase and direct the polymerase to specific promoters.

Genes that encode proteins with related functions may be at different locations in the genome but have the same promoter sequence. This allows them to be expressed at the same time and under the same physiological conditions. For example, some bacteria stop growing when nutrients in their environment are depleted. When this happens, they adopt an alternative lifestyle called *sporulation*—they reduce metabolism and form a tough spore coat. This process involves the sequential expression of specific classes of genes in a manner reminiscent of the early and late genes of bacteriophage infection (see Figure 11.3). Each member of a gene class has a common promoter sequence, and RNA polymerase is directed to the promoter in each case by a specific sigma factor. As we will see in Concept 11.3, this *global gene regulation* by proteins binding to RNA polymerase is also common in eukaryotes.

LINK For more on sporulation as a survival strategy, see Concept 19.2

Do You Understand Concept 11.2?

- Describe the molecular conditions at the *lac* operon promoter in the presence and absence of lactose.

- Describe the molecular events at the *trp* operon promoter in the presence and absence of tryptophan.

- If the *lac* repressor gene is mutated so that the allosteric site on the protein no longer binds allolactose, what would be the effect on transcription of the *lac* operon? What about a similar mutation in the *trp* repressor gene?

Studies of viruses and bacteria provide a basic understanding of the mechanisms that regulate gene expression and of the roles of regulatory proteins in both positive and negative regulation. We will now turn to the control of gene expression in eukaryotes. You will see both negative and positive control of transcription, as well as posttranscriptional mechanisms of regulation.

concept 11.3 Eukaryotic Genes Are Regulated by Transcription Factors and DNA Changes

As we mentioned in Concept 11.1, gene expression can be regulated at a number of different points in the process of transcribing a gene and translating the mRNA into a protein (see Figure 11.1). In this concept we will describe the mechanisms that result in the selective transcription of specific eukaryotic genes. The mechanisms for regulating transcription in eukaryotes have similar themes to those of prokaryotes. Both types of cells use DNA–protein interactions to mediate negative and positive control of gene expression. However, there are significant differences, which generally reflect the greater complexity of eukaryotic organisms (**TABLE 11.1**).

Transcription factors act at eukaryotic promoters

As in bacteria, a eukaryotic promoter is a region of DNA near the 5'-end of a gene where RNA polymerase binds and initiates transcription. Eukaryotic promoters are extremely diverse and difficult to characterize, but they each contain a core promoter sequence to which the RNA polymerase binds. The most common of these is the **TATA box**—so called because it is rich in A-T base pairs.

RNA polymerase II is the polymerase that transcribes the protein-coding genes in eukaryotes. It cannot bind to the promoter and initiate transcription by itself. Rather, it does so only after various **general transcription factors** have bound to the core promoter. General transcription factors bind to most promoters and are distinct from transcription factors that have specific regulatory effects only at certain promoters or classes of promoters. **FIGURE 11.10** illustrates the assembly of the resulting transcription complex at a promoter containing a TATA

box. First, the protein TFIID ("TF" stands for transcription factor) binds to the TATA box. Binding of TFIID changes both its own shape and that of the DNA, presenting a new surface that attracts the binding of other transcription factors. RNA polymerase II binds only after several other proteins have bound to the complex.

The core promoter sequence is bound by general transcription factors that are needed for the expression of all RNA polymerase II–transcribed genes. Other sequences that are (usually) found in or near promoter regions are specific to only a few genes and are recognized by specific transcription factors. These transcription factors may be positive regulators (*activators*) or negative regulators (*repressors*) of transcription:

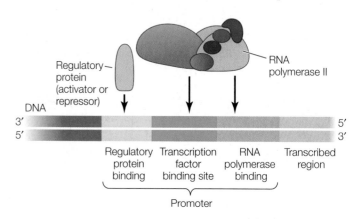

Such transcription factors may be present only in certain cell types, or they may be present in all cells but activated by specific signals. DNA sequences that bind activators are called *enhancers*, and those that bind *repressors* are called silencers. Some enhancers and silencers occur near the core promoter, and others can be as far as 20,000 base pairs away. When the activators or repressors bind to these DNA sequences, they interact with the RNA polymerase complex, causing the DNA to bend. Often many such binding proteins are involved, and *the combination of factors present determines the initiation of transcription*. With about 2,000 different transcription factors in humans, there are many possibilities for regulation.

How do transcription factors recognize a specific nucleotide sequence in DNA? To answer this question, let's look at a specific example. NFATs (*nuclear factors of activated T cells*) are a group of transcription factors that control the expression

TABLE 11.1	Transcription in Bacteria and Eukaryotes	
CHARACTERISTIC	BACTERIA	EUKARYOTES
Locations of functionally related genes	Often clustered in operons	Often distant from one another with separate promoters
RNA polymerases	One	Three: I: transcribes rRNA II: transcribes mRNA III: transcribes tRNA and small RNAs
Promoters and other regulatory sequences	Few	Many
Initiation of transcription	Binding of RNA polymerase	Binding of many proteins, including RNA polymerase, to promoter

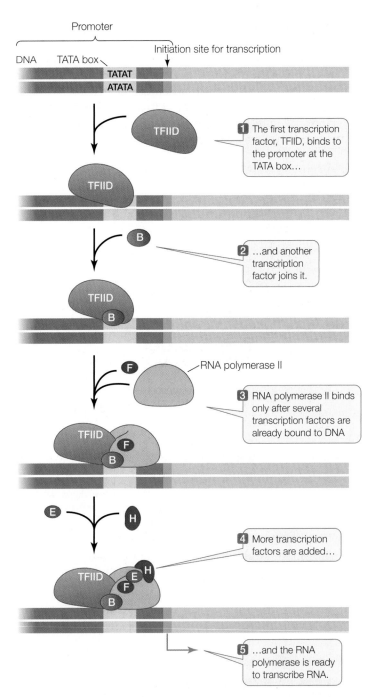

FIGURE 11.10 The Initiation of Transcription in Eukaryotes
Apart from TFIID, which binds to the TATA box, each transcription factor in this transcription complex has binding sites only for the other proteins in the complex, and does not bind directly to DNA. B, E, F, and H are general transcription factors.

> **yourBioPortal.com**
>
> **Go to ANIMATED TUTORIAL 11.3**
> **Initiation of Transcription**

of genes essential for the immune response (see Chapter 31). NFAT proteins bind to a 12-bp recognition sequence near the promoters of these genes, with the sequence CGAG-GAAAATTG (**FIGURE 11.11**). Recall that there are atoms in

the bases of DNA that are available for hydrogen bonding but are not involved in base pairing (see Figure 9.6). These atoms are important in the interactions between an NFAT and the DNA. In addition, there are hydrophobic interactions between the rings in the DNA bases and some amino acid R groups in the protein. As for an enzyme and its substrate (see Concept 3.3), there is an *induced fit* between the NFAT and the DNA, such that the protein undergoes a conformational change after binding begins.

> **FRONTIERS** An important aspect of gene regulation is the specific binding of transcription factors to DNA. Major efforts are underway to understand this binding at the atomic level. The atoms of bases that are exposed within the major or minor grooves of DNA can interact by hydrogen or ionic bonding with the DNA binding domains of transcription factors. Biophysicists are determining the three-dimensional structures of transcription factors so that they can create computer models for how the proteins might interact with DNA.

The expression of sets of genes can be coordinately regulated by transcription factors

We have seen that prokaryotes can coordinate the regulation of several genes by arranging them in an operon. In addition, bacteria can coordinate the expression of groups of genes using sigma factors, which guide RNA polymerase to particular classes of promoters. This latter mechanism is also used in eukaryotes to coordinately regulate genes that may be far apart, even on different chromosomes. The expression of genes can be coordinated if they share regulatory sequences that *bind the same transcription factors.*

This type of coordination is used by organisms to respond to stress—for example, by plants in response to drought.

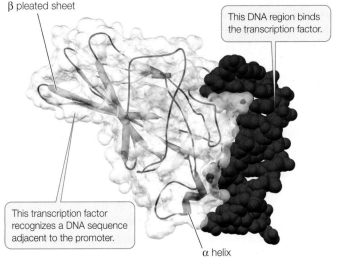

FIGURE 11.11 A Transcription Factor Protein Binds to DNA
The transcription factor NFAT activates genes for the immune response by binding to a specific DNA sequence near the promoters of those genes.

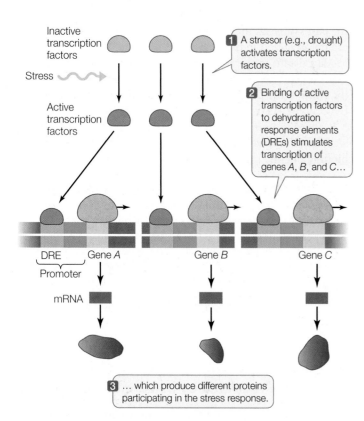

FIGURE 11.12 Coordinating Gene Expression A single environmental signal, such as drought stress, activates a transcription factor that acts on many genes.

Under conditions of drought stress, a plant must simultaneously synthesize a number of proteins whose genes are scattered throughout the genome. The synthesis of these proteins comprises the stress response. To coordinate expression, each of these genes has a specific regulatory sequence near its promoter called the *dehydration response element* (DRE). In response to drought, a transcription factor changes so that it binds to this element and stimulates mRNA synthesis (**FIGURE 11.12**). The dehydration response proteins not only help the plant conserve water, but also protect the plant against freezing or excess salt in the soil. This finding has considerable importance for agriculture because crops are often grown under less than optimal conditions.

Epigenetic changes to DNA and chromatin can regulate transcription

So far we have focused on regulatory events that involve specific DNA sequences at or near a gene's promoter. Eukaryotic cells can also regulate the transcription of large stretches of DNA (containing many genes) by reversible, non-sequence-specific alterations to either the DNA or the chromosomal proteins that package the DNA in the nucleus. These alterations can be passed on to daughter cells after mitosis or meiosis. They are called **epigenetic** changes to distinguish them from mutations, which involve irreversible changes to the DNA's base sequence (see Concept 9.3).

DNA METHYLATION Depending on the organism, from 1 to 5 percent of cytosine residues in the DNA are chemically modified by the addition of a methyl group (—CH₃), to form 5-methylcytosine (**FIGURE 11.13**). This covalent addition is catalyzed by the enzyme **DNA methyltransferase** and, in mammals, usually occurs in C residues that are adjacent to G residues. DNA regions rich in these doublets are called **CpG islands**, and they are especially abundant in promoters.

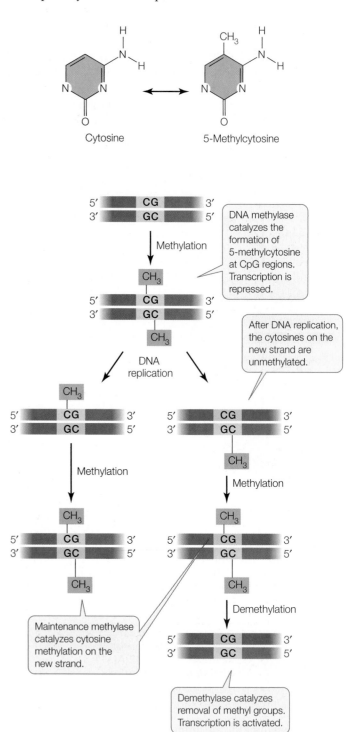

FIGURE 11.13 DNA Methylation: An Epigenetic Change The reversible formation of 5-methylcytosine in DNA can alter the rate of transcription.

This covalent change in DNA is *heritable*: when DNA is replicated, a **maintenance methylase** catalyzes the formation of 5-methylcytosine in the new DNA strand. However, the pattern of cytosine methylation can also be altered, because methylation is reversible: a third enzyme, appropriately called **demethylase**, catalyzes the removal of the methyl group from cytosine (see Figure 11.13).

Methylated DNA binds specific proteins that are involved in the repression of transcription; thus heavily methylated genes tend to be inactive (*silenced*). Sometimes, large stretches of DNA or almost whole chromosomes are methylated. Under a microscope, two kinds of chromatin can be distinguished in the stained interphase nucleus: *euchromatin* and *heterochromatin*. The euchromatin appears diffuse and stains lightly; it contains the DNA that is transcribed into mRNA. Heterochromatin is condensed and stains darkly; any genes it contains are generally not transcribed.

A dramatic example of heterochromatin is the X chromosome in female mammals. A normal female mammal has two X chromosomes, whereas a normal male has an X and a Y (see Concept 8.3). The Y chromosome is smaller and lacks most of the genes present on the X. As a result, females and males differ greatly in the "dosage" of X-linked genes. Because each female cell has two copies of each X chromosome gene, the female should have the potential to produce twice as much of each protein product as the male. Nevertheless, for 75 percent of the genes on the X chromosome, the total amount of mRNA produced is generally the same in males and in females. How does this happen?

In the early female embryo, one copy of X becomes heterochromatic and transcriptionally inactive in each cell, and the same X remains inactive in all of that cell's descendants. In a given female embryo cell, the "choice" of which X to inactivate is random. Recall that one X in a female comes from her father and one from her mother. Thus, in one embryonic cell

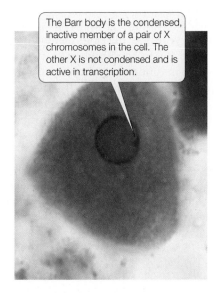

The Barr body is the condensed, inactive member of a pair of X chromosomes in the cell. The other X is not condensed and is active in transcription.

FIGURE 11.14 X Chromosome Inactivation A Barr body in the nucleus of a human female cell is the transcriptionally inactive X chromosome.

the paternal X might be inactivated, but in a neighboring cell the maternal X might be inactive.

The inactive X is identifiable within the nucleus as a heterochromatic *Barr body* (named for its discoverer, Murray Barr) (**FIGURE 11.14**). This clump of heterochromatin consists of heavily methylated DNA. A female with the normal two X chromosomes will have one Barr body, whereas a rare female with three Xs will have two, and an XXXX female will have three. Males that are XXY will have one. These observations suggest that the interphase cells of each person, male or female, have a single active X chromosome, and thus a constant dosage of expressed X chromosome genes.

HISTONE PROTEIN MODIFICATION Another mechanism for epigenetic gene regulation is the alteration of chromatin structure, or **chromatin remodeling**. Large amounts of DNA (nearly 2 meters in humans!) is packed within the nucleus (a 5-μm-diameter organelle). The basic unit of DNA packaging in eukaryotes is the nucleosome, a core of positively charged **histone** proteins around which DNA is wound:

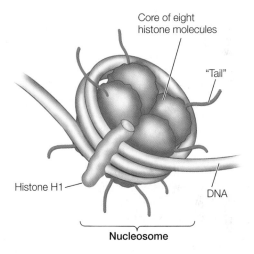

Core of eight histone molecules

"Tail"

Histone H1

DNA

Nucleosome

Nucleosomes can make DNA physically inaccessible to RNA polymerase and the rest of the transcription apparatus. Each histone protein has a "tail" of approximately 20 amino acids at its N terminus that sticks out of the compact structure and contains certain positively charged amino acids (notably lysine). Enzymes called **histone acetyltransferases** can add acetyl groups to these positively charged amino acids, thus neutralizing their charges:

Lysine in histone + Acetyl CoA → Acetyl-lysine + CoA—SH

Ordinarily, there is strong electrostatic attraction between the positively charged histone proteins and DNA, which is negatively charged because of its phosphate groups. Reducing the positive charges of the histone tails reduces the affinity

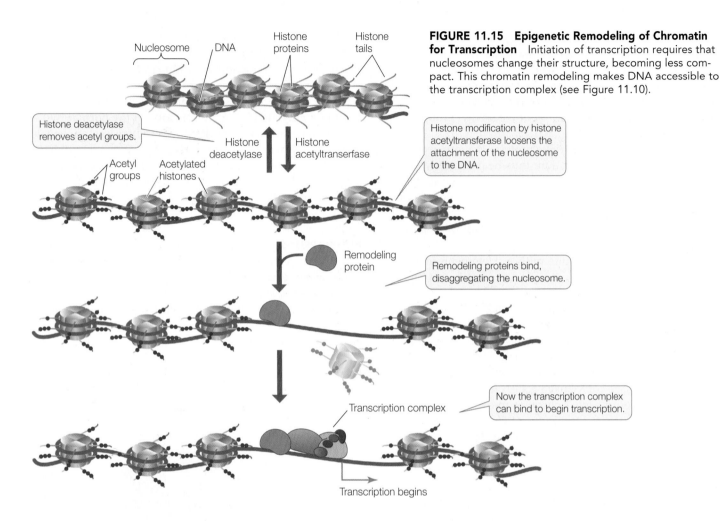

FIGURE 11.15 Epigenetic Remodeling of Chromatin for Transcription Initiation of transcription requires that nucleosomes change their structure, becoming less compact. This chromatin remodeling makes DNA accessible to the transcription complex (see Figure 11.10).

of the histones for DNA, loosening the compact nucleosome. Additional chromatin remodeling proteins can then bind to the nucleosome–DNA complex and open up the DNA for gene expression (**FIGURE 11.15**). Thus, histone acetyltransferases can *activate* transcription. Another kind of chromatin remodeling protein, **histone deacetylase**, can remove the acetyl groups from histones and thereby *repress* transcription.

Other types of histone modification can affect gene activation and repression. For example, histone methylation (not to be confused with the *cytosine* methylation we discussed above) is associated with gene inactivation. Histone phosphorylation also affects gene expression, the specific effect depending on which amino acid of the histone is modified. All of these effects are reversible, and so the transcriptional activity of a eukaryotic gene may be determined by varying patterns of histone modification.

Epigenetic changes can be induced by the environment

Despite the fact that they are reversible, many epigenetic changes such as DNA methylation and histone modification can permanently alter gene expression patterns in a cell. If the cell is a germline cell that forms gametes, the epigenetic changes can be passed on to the next generation. But what de-

termines these epigenetic changes? A clue comes from a recent study of human monozygotic (identical) twins.

Monozygotic twins come from a single fertilized egg that divides to produce two separate cells; each of these develops into a separate individual. Twin brothers or sisters thus have identical genomes. But are they identical in their *epigenomes*? A comparison of DNA in hundreds of such twin pairs shows that in tissues of three-year-olds, the DNA methylation patterns are virtually the same. But by age 50, when the twins have usually been living apart in different environments for decades, the patterns are quite different. This indicates that the *environment plays an important role in epigenetic modifications* and, therefore, in the regulation of genes that these modifications affect.

FRONTIERS Biologists are investigating the inheritance of epigenetic changes by characterizing the epigenetic tags on genes during embryonic development. Early in development, epigenetic tags are removed from almost all genes, so the "epigenome" begins with a largely blank slate. However, some genes escape this process and maintain the epigenetic changes they accumulated while they were in the parents. This is inheritance of an acquired characteristic, and its discovery was a major surprise to biologists—especially geneticists.

What factors in the environment lead to epigenetic changes? One might be stress: when mice are put in a stressful situation, genes that are involved in important brain pathways become heavily methylated (and transcriptionally inactive). Treatment of the stressed mice with an antidepressant drug reverses these changes. Transcription factors such as CREB that mediate addiction (see the opening story of this chapter) are involved in histone acetylation, which leads to subsequent gene activation.

Do You Understand Concept 11.3?

- How do transcription factors regulate gene expression?
- What is the difference between epigenetic regulation and gene regulation by transcription factors?
- How can a pattern of DNA methylation be inherited?
- In colorectal cancer, some tumor suppressor genes are inactive. This is an important factor resulting in uncontrolled cell division. Two of the possible explanations for the inactive genes are: (1) a mutation in the coding region, resulting in an inactive protein, and (2) epigenetic silencing at the promoter of the gene, resulting in reduced transcription. How would you investigate these two possibilities?

Thus far we have examined transcriptional gene regulation in viruses, prokaryotes, and eukaryotes. In the final concept we will focus on the posttranscriptional mechanisms for regulating gene expression in eukaryotes.

Eukaryotic Gene Expression Can Be Regulated after Transcription

Gene expression involves transcription and then translation. So far we have described how eukaryotic gene expression is regulated at the transcriptional level. But as Figure 11.1 shows, there are many points at which regulation can occur after the initial gene transcript is made.

Different mRNAs can be made from the same gene by alternative splicing

Most primary mRNA transcripts in eukaryotes contain several introns (see Figure 10.6). We have seen how the splicing mechanism recognizes the boundaries between exons and introns. What would happen if the β-globin pre-mRNA, which has two introns, were spliced from the start of the first intron to the end of the second? The middle exon would be spliced out along with the two introns. An entirely new protein (certainly not a β-globin) would be made, and the functions of normal β-globin would be lost. Such **alternative splicing** can be a deliberate mechanism for generating a family of different proteins with different activities and functions from a single gene (**FIGURE 11.16**).

Two examples of this mechanism are found in HIV and in the fruit fly (*Drosophila*):

- The HIV genome (see Figure 11.4) encodes nine proteins but is transcribed as a single pre-mRNA. Most of the nine proteins are then generated by alternative splicing of this pre-mRNA.

- In *Drosophila*, sex is determined by the *Sxl* gene. This gene has four exons, which we will designate 1, 2, 3, and 4. In the female embryo, splicing generates two active forms of the Sxl

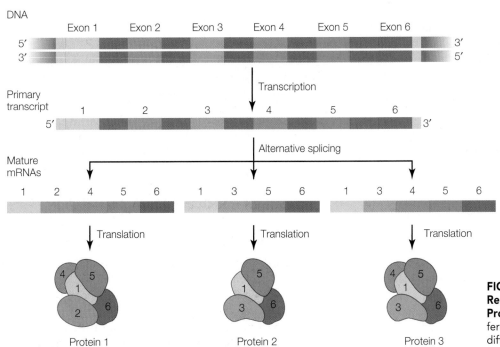

FIGURE 11.16 Alternative Splicing Results in Different Mature mRNAs and Proteins Pre-mRNA can be spliced differently in different tissues, resulting in different proteins.

protein, containing exons 1 and 2, and 1, 2, and 4. However, in the male embryo, the protein contains all four exons (1, 2, 3, and 4) and is inactive.

Before the human genome was sequenced, most scientists estimated that they would find between 80,000 and 150,000 protein-coding genes. You can imagine their surprise when the actual sequence revealed only about 24,000 genes! In fact, there are many more human mRNAs than there are human genes, and most of this variation comes from alternative splicing. Indeed, recent surveys show that more than 80 percent of all human genes are alternatively spliced.

Alternative splicing may be a key to the differences in levels of complexity among organisms. For example, although humans and chimpanzees have similar-sized genomes, there is more alternative splicing in the human brain than in the brain of a chimpanzee.

MicroRNAs are important regulators of gene expression

As we discuss in Concept 12.3, only a fraction of the genome in most plants and animals codes for proteins. Some of the genome encodes ribosomal RNA and transfer RNAs, but until recently biologists thought that the rest of the genome was not transcribed; some even called it "junk." Recent investigations, however, have shown that some of these noncoding regions *are* transcribed. The noncoding RNAs are often very small and therefore difficult to detect. These tiny RNA molecules are called **microRNA (miRNA)**.

The first miRNA sequences were found in the worm *Caenorhabditis elegans*. This model organism, which has been studied extensively by developmental biologists, goes through several larval stages. Victor Ambros at the University of Massachusetts found mutations in two genes that had different effects on progress through these stages:

- *lin-14* mutations (named for abnormal cell *lin*eage) cause the larvae to skip the first stage and go straight to the second stage. Thus the gene's normal role is to facilitate events of the first larval stage.

- *lin-4* mutations cause certain cells in later larval stages to repeat a pattern of development normally observed in the first larval stage. It is as if the cells were stuck in that stage. So the normal role of this gene is to *negatively regulate lin-14*, turning off its expression so the cells can progress to the next stage.

Not surprisingly, further investigation showed that *lin-14* encodes a transcription factor that affects the transcription of genes involved in larval cell progression. It was originally expected that *lin-4*, the negative regulator, would encode a protein that downregulates genes activated by the lin-14 protein. But this turned out to be incorrect. Instead, *lin-4* encodes a 22-base miRNA that inhibits *lin-14* expression *posttranscriptionally by binding to its mRNA*.

Hundreds of miRNAs, in a variety of eukaryotes, have now been described. Each one is about 22 nucletides long and usually has dozens of mRNA targets. Each miRNA is transcribed

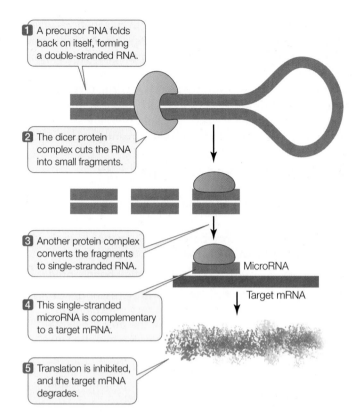

1 A precursor RNA folds back on itself, forming a double-stranded RNA.

2 The dicer protein complex cuts the RNA into small fragments.

3 Another protein complex converts the fragments to single-stranded RNA.

MicroRNA

Target mRNA

4 This single-stranded microRNA is complementary to a target mRNA.

5 Translation is inhibited, and the target mRNA degrades.

FIGURE 11.17 mRNA Degradation Caused by MicroRNAs MicroRNAs inhibit the translation of specific mRNAs by causing their premature degradation.

as a longer precursor that is cleaved through a series of steps to double-stranded miRNAs. A protein complex guides the miRNA to its target mRNA, where translation is inhibited and the mRNA is degraded (**FIGURE 11.17**). The remarkable conservation of this gene-silencing mechanism in eukaryotes indicates that it is evolutionarily ancient and biologically important.

FRONTIERS The patterns of miRNA expression vary in different tissues and at different times. At an early stage of breast cancer, the cancer cells cause a distinctive pattern of miRNAs to appear in blood serum, and this is being investigated as a marker for cancer that might otherwise be undetectable. This may allow earlier detection of breast cancer, which would improve treatment outcomes.

Translation of mRNA can be regulated

The amount of a protein in a cell is not determined simply by the amount of its mRNA. For example, in yeast cells only about a third of the genes show clear correlations in the amounts of mRNA and protein; in these cases, more mRNA leads to more protein. For two-thirds of the genes there is no apparent relationship between the two—there may be lots of mRNA and little or no protein, or lots of protein and little mRNA. The concentrations of these proteins must therefore be determined

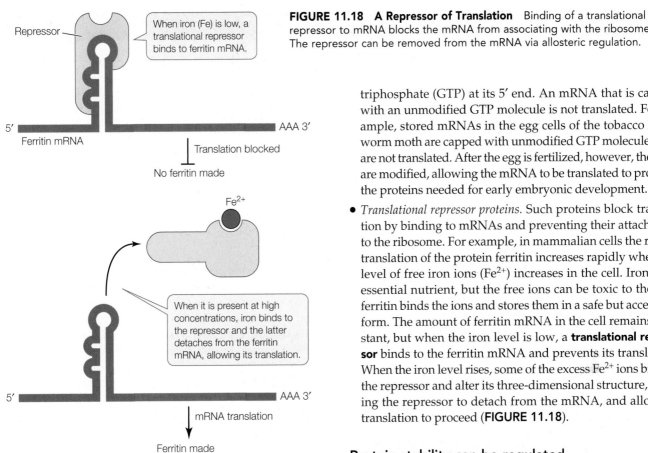

Repressor

When iron (Fe) is low, a translational repressor binds to ferritin mRNA.

5′

Ferritin mRNA

AAA 3′

⊥ Translation blocked

No ferritin made

Fe²⁺

When it is present at high concentrations, iron binds to the repressor and the latter detaches from the ferritin mRNA, allowing its translation.

5′

AAA 3′

↓ mRNA translation

Ferritin made

FIGURE 11.18 A Repressor of Translation Binding of a translational repressor to mRNA blocks the mRNA from associating with the ribosome. The repressor can be removed from the mRNA via allosteric regulation.

triphosphate (GTP) at its 5′ end. An mRNA that is capped with an unmodified GTP molecule is not translated. For example, stored mRNAs in the egg cells of the tobacco hornworm moth are capped with unmodified GTP molecules and are not translated. After the egg is fertilized, however, the caps are modified, allowing the mRNA to be translated to produce the proteins needed for early embryonic development.

• *Translational repressor proteins.* Such proteins block translation by binding to mRNAs and preventing their attachment to the ribosome. For example, in mammalian cells the rate of translation of the protein ferritin increases rapidly when the level of free iron ions (Fe^{2+}) increases in the cell. Iron is an essential nutrient, but the free ions can be toxic to the cell; ferritin binds the ions and stores them in a safe but accessible form. The amount of ferritin mRNA in the cell remains constant, but when the iron level is low, a **translational repressor** binds to the ferritin mRNA and prevents its translation. When the iron level rises, some of the excess Fe^{2+} ions bind to the repressor and alter its three-dimensional structure, causing the repressor to detach from the mRNA, and allowing translation to proceed (**FIGURE 11.18**).

Protein stability can be regulated

The protein content of any cell at a given time is a function of both protein synthesis and protein degradation. Certain proteins can be targeted for destruction in a chain of events that begins when an enzyme attaches a 76–amino acid protein called **ubiquitin** (so named because it is ubiquitous, or widespread) to a lysine residue of the protein to be destroyed. Other ubiquitins then attach to the primary one, forming a polyubiquitin chain. The protein–polyubiquitin complex then binds to a huge protein complex called a **proteasome** (from *protease* and *soma*, "body"; **FIGURE 11.19**). Upon entering the proteasome, the polyubiquitin is removed and ATP energy is used to unfold the target protein. Three different proteases then digest the protein into small peptides and amino acids. You may recall from Chapter 7 that cyclins are proteins that regulate the activities of key enzymes at specific points in the cell cycle. Cyclins must be broken down at just the right time, and this is done by proteasomes.

by factors acting after the mRNA is made. Cells do this in two major ways: by regulating the translation of mRNA or by altering how long proteins persist in the cell.

There are three known ways in which the translation of mRNA can be regulated:

• *Inhibition of translation with miRNAs.* This was discussed in the last section (see above).

• *Modification of the 5′ cap.* As noted in Concept 10.2, an mRNA usually has a chemically modified molecule of guanosine

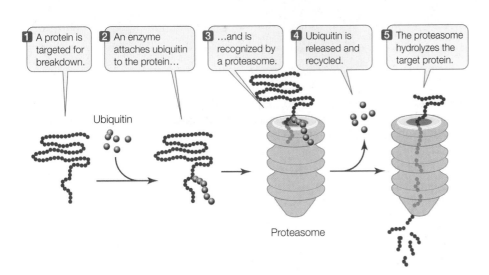

1 A protein is targeted for breakdown.

2 An enzyme attaches ubiquitin to the protein…

3 …and is recognized by a proteasome.

4 Ubiquitin is released and recycled.

5 The proteasome hydrolyzes the target protein.

Ubiquitin

Proteasome

FIGURE 11.19 A Proteasome Breaks Down Proteins Proteins targeted for degradation are bound by ubiquitin, which then directs the targeted protein to a proteasome. The proteasome is a complex structure where proteins are digested by several powerful proteases.

APPLY THE CONCEPT

Eukaryotic gene expression can be regulated transcriptionally and posttranscriptionally

The enzyme HMG CoA reductase (HR) catalyzes an initial step in the synthesis of cholesterol. The table shows the HR levels in liver cells following various treatments.

Explain the results of each treatment.

TREATMENT	AMOUNT OF HR PROTEIN
1. Actinomycin D, a drug that inhibits RNA polymerase II	Reduced
2. Suberoylanilide hydroxamic acid, a histone deacetylase inhibitor	Increased
3. Bortezomib, a proteasome inhibitor	Increased
4. High level of cholesterol	Reduced
5. Azacytidine, inhibitor of DNA methylation	Increased

Do You Understand Concept 11.4?

- How can a single gene transcribed into a single pre-mRNA code for several different proteins?

- Compare inhibition of translation by miRNA with inhibition by a repressor.

- You are studying the enzyme protease in germinating seeds. You find that protease activity increases tenfold after treatment of the seeds with a hormone, gibberellic acid. How would you show that this increase is due to:
  a. release of a translational repressor by the hormone?
  b. allosteric inhibition of a transcriptional repressor by the hormone?

(described in the opening story), CREB has been strongly implicated in long-term memory. Animals with mutations that result in a lack of active CREB can learn a maze test, but they don't remember it later. Similar effects are seen if CREB activity is blocked right after a task is learned. When animals learn a task, imaging studies reveal *CRE*-containing genes becoming active in the hippocampus, a region of the brain that is involved in long-term memory (see Concept 34.5). Thus, CREB provides an insight into the molecular biology of memory, linking learning to the regulation of gene expression.

How does CREB regulate the expression of many genes?

ANSWER   CREB is a family of several closely related transcription factors that can either activate or repress gene expression (Concept 11.1). They bind to the *cAMP response element* (*CRE*), a short DNA sequence of GACGTCA that is found in the promoter regions of many genes (Concept 11.3). CREB proteins have a "leucine zipper" structure that consists of two parallel α helices rich in the amino acid leucine (**FIGURE 11.20**), and fingerlike extensions that fit into the major groove of the DNA double helix.

CREB binding and regulation of gene expression is essential for a number of processes in several organs, including the brain. In addition to its role in drug and alcohol addiction

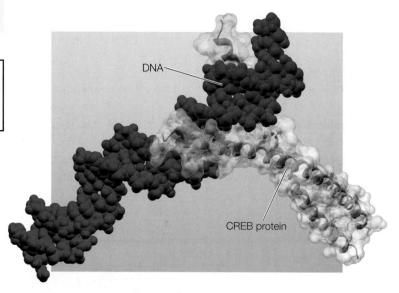

FIGURE 11.20 An Explanation for Alcoholism? The transcription factor CREB binds to DNA and activates the promoters of genes involved in addictive behaviors.

11 SUMMARY

concept 11.1 Several Strategies Are Used to Regulate Gene Expression

- Gene expression can be regulated at the levels of transcription, RNA processing, translation, or posttranslation. **Review Figure 11.1 and WEB ACTIVITY 11.1**

- Some genes are always expressed (**constitutive genes**), whereas others are expressed only at certain times and in certain cells (**inducible genes**).

- **Transcription factors** are regulatory proteins that bind DNA and regulate gene expression. **Activators** positively regulate gene expression. **Repressors** negatively regulate gene expression. **Review Figure 11.2**

- **Viruses** provide examples of gene regulation as they convert the host cell into a virus factory. **Review Figures 11.3 and 11.4**

concept 11.2 Many Prokaryotic Genes Are Regulated in Operons

- A metabolic pathway can be regulated either by allosteric regulation of an enzyme or by regulation of enzyme synthesis. **Review Figure 11.6**

- In prokaryotes, several genes can be part of a single transcriptional unit called an **operon**, which consists of a promoter, an **operator**, and two or more **structural genes**. **Review Figure 11.7**

- An inducible operon is turned off unless its expression is needed, whereas a repressible operon is turned on unless its expression is not needed. When an operon is turned off, it has a repressor protein bound to its operator, preventing transcription.

- The *lac* operon is an example of an inducible system, whereas the *trp* operon is an example of a repressible system. **Review Figures 11.8 and 11.9 and ANIMATED TUTORIALS 11.1 and 11.2**

- **Sigma factors** direct RNA polymerase to specific promoters in prokaryotes.

concept 11.3 Eukaryotic Genes Are Regulated by Transcription Factors and DNA Changes

- Eukaryotic gene expression is regulated both during and after transcription.

- **General transcription factors** bind to the core promoter sequences of protein-coding genes and direct RNA polymerase II to the promoter. **Review Figure 11.10 and ANIMATED TUTORIAL 11.3**

- Specific transcription factors (activators and repressors) bind to specific DNA elements near the promoter and affect the rate of transcription initiation. **Review Figures 11.11 and 11.12**

- The term **epigenetic** refers to changes in gene expression that do not involve changes in DNA sequences.

- Methylation of cytosine residues generally inhibits transcription. **Review Figure 11.13**

- **Chromatin remodeling** via the modification of **histone** proteins in nucleosomes also affects transcription. **Review Figure 11.15**

- Epigenetic changes can be induced by the environment and can be inherited.

concept 11.4 Eukaryotic Gene Expression Can Be Regulated after Transcription

- **Alternative splicing** of pre-mRNA can produce different proteins. **Review Figure 11.16**

- A **microRNA (miRNA)** is a small noncoding RNA that inhibits the translation of specific mRNAs by causing their premature degradation. **Review Figure 11.17**

- The translation of mRNA to proteins can be regulated by **translational repressors**. **Review Figure 11.18**

- A **proteasome** can break down proteins, thus affecting protein longevity. **Review Figure 11.19**

See **WEB ACTIVITY 11.2** for a concept review of this chapter.

Genomes 12

Canis lupus familiaris, the dog, was domesticated by humans from the gray wolf thousands of years ago. There are several kinds of wolves, and they all look more or less the same. Not so with "man's best friend." The American Kennel Club recognizes about 155 different breeds. Dog breeds vary greatly in size, shape, coat color, hair length, and even behavior. For example, an adult Chihuahua weighs just 1.5 kg, whereas a Scottish deerhound weighs 70 kg. No other mammalian species shows such large phenotypic variation. Furthermore, there are hundreds of genetic diseases in dogs, and many of these diseases have counterparts in humans. Biologists are curious about the molecular basis of the phenotypic variation, and they view dogs as models for studying genetic diseases. For these reasons, the Dog Genome Project began in the late 1990s. Since then the sequences of several dog genomes have been published.

Two dogs—a boxer and a poodle—were the first of their species to have their entire genomes sequenced. The dog genome contains 2.8 billion base pairs of DNA in 39 pairs of chromosomes. There are 19,000 protein-coding genes, most of them with close counterparts in other mammals, including humans. The whole genome sequence made it easy to create a map of genetic markers—specific nucleotides or short sequences of DNA at particular locations on the genome that differ between individual dogs or breeds.

Genetic markers are used to map the locations of genes that control particular traits. To do this, scientists must extract

DNA from many individual dogs that vary in just one or a few characters. Taking samples of cells for DNA isolation is relatively easy: a cotton swab is swept over the inside of the dog's cheek. As one scientist conducting genomic analyses of dogs said, the dogs "didn't care, especially if they were going to get a treat or if there was a tennis ball in our other hand."

The molecular methods used to analyze dogs have been applied to many other animals and plants of economic and social importance to humans. And of course, the human genome itself has been sequenced and is being studied intensively.

QUESTION What does genome sequencing reveal about dogs and other animals?

The papillion (left) and the Great Dane (right) are the same species, *Canis lupus familiaris*, and yet they show great variation in size. Genome sequencing has provided insights into how size is controlled by genes.

KEY CONCEPTS

12.1 There Are Powerful Methods for Sequencing Genomes and Analyzing Gene Products

12.2 Prokaryotic Genomes Are Relatively Small and Compact

12.3 Eukaryotic Genomes Are Large and Complex

12.4 The Human Genome Sequence Has Many Applications

There Are Powerful Methods for Sequencing Genomes and Analyzing Gene Products

A major goal in sequencing genomes is to understand genetics—that is, to identify mutations in DNA and relate them to phenotypes. While this had been done for individual genes, the notion of sequencing the entire genome of a complex organism was not contemplated until 1986. The Nobel laureate Renato Dulbecco and others proposed at that time that the world scientific community be mobilized to undertake the sequencing of the entire human genome. One motive was to detect DNA damage in people who had survived the atomic bomb attacks and been exposed to radiation in Japan during World War II. But in order to detect changes in the human genome, scientists first needed to know its normal sequence.

The result was the publicly funded **Human Genome Project**, an enormous undertaking that was successfully completed in 2003. This effort was aided and complemented by privately funded groups. The project benefited from the development of many new methods that were first used in the sequencing of smaller genomes—those of prokaryotes and simple eukaryotes, the model organisms you are familiar with from studies in genetics and cell biology. Many of these methods are still applied widely, and powerful new methods for sequencing genomes have emerged. These are complemented by new ways to examine phenotypic diversity in a cell's proteins and in the metabolic products of the cell's enzymes.

New methods have been developed to rapidly sequence DNA

Many prokaryotes have a single chromosome, whereas eukaryotes have many. Because of their differing sizes, chromosomes can be separated from one another, identified, and experimentally manipulated. It might seem that the most straightforward way to sequence a chromosome would be to start at one end and simply sequence the DNA molecule one nucleotide at a time. The task is somewhat simplified because only one of the two strands needs to be sequenced, the other being complementary. However, this large-polymer approach is not practical, since at most only several hundred base pairs (bp) can be sequenced at a time using current methods.

As you will see, a key to interpreting DNA sequences is to do several separate experiments simultaneously on a given chromosome, first breaking the DNA into *overlapping fragments*.

In the 1970s, the late Frederick Sanger invented a way to sequence DNA by using chemically modified nucleotides that were originally developed to stop cell division in cancer. This method, or a variation of it, was used to obtain the first human genome sequence as well as those of model organisms. However, it was relatively slow, expensive, and labor-intensive. The first decade of the new millennium saw the development of a faster and less-expensive method: **next-generation DNA sequencing** uses miniaturization techniques first developed for the electronics industry, as well as the principles of DNA replication and the polymerase chain reaction (PCR).

LINK DNA replication and PCR can be reviewed in Concept 9.2

There are several variations in the next-generation sequencing techniques. One approach is outlined here and illustrated in **FIGURE 12.1**. First the DNA is prepared for sequencing:

1. A large molecule of DNA is cut into small fragments of about 100 bp each. This can be done by physically breaking up the DNA or by using enzymes that hydrolyze the phosphodiester bonds between nucleotides at intervals in the DNA backbone.

2. The DNA is denatured by heat, breaking hydrogen bonds holding the two strands together. Each single strand acts as a template for the synthesis of new, complementary DNA.

3. Each fragment is attached at each end to short adapter sequences, which are in turn attached to a solid support. The support can be a microbead or a flat surface.

4. Each DNA fragment is amplified by PCR to provide many (approximately 1,000) copies. The multiple copies at a single location allow for easy detection of added nucleotides during the sequencing steps.

Once the DNA has been attached to a solid substrate and amplified, it is ready for sequencing (see Figure 12.1B):

1. At the beginning of each sequencing cycle, the fragments are heated to denature them. A primer, DNA polymerase, and the four nucleotides (each tagged with a different fluorescent dye) are added. A universal primer that is complementary to one of the adaptor sequences is used in the sequencing reactions.

2. The replication process is set up so that DNA (complementary to the template strand) is added one nucleotide at a time. After each addition, the unincorporated nucleotides are removed.

3. The fluorescence of the new nucleotide at each location is detected with a camera. The color of the fluorescence indicates which of the four nucleotides was added.

4. The fluorescent tag is removed from the nucleotide that is already attached, and then the synthesis cycle is repeated. Images are captured after each nucleotide is added. The series of colors at each location indicate the sequence of nucleotides in the growing DNA strand at that location.

The power of this method derives from the fact that:

- It is fully automated and miniaturized.

- Millions of different fragments are sequenced at the same time. This is called *massively parallel sequencing*.

- It is an inexpensive way to sequence large genomes. For example, at the time of this writing, a complete human genome could be sequenced in a few days for several thousand dollars. In contrast, the Human Genome Project took 13 years and $2.7 billion to sequence one genome!

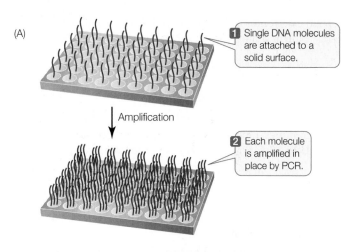

(A)

1 Single DNA molecules are attached to a solid surface.

Amplification

2 Each molecule is amplified in place by PCR.

FIGURE 12.1 DNA Sequencing Next-generation sequencing is faster and cheaper than traditional methods. It involves (A) the chemical amplification of DNA fragments and (B) the synthesis of complementary strands using color-labeled nucleotides.

yourBioPortal.com

Go to ANIMATED TUTORIAL 12.1
Next-Generation Sequencing

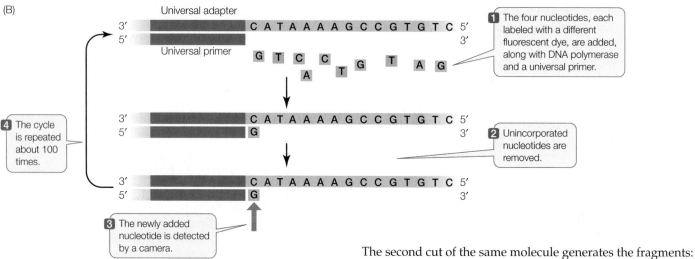

(B)

Universal adapter

3′ CATAAAAGCCGTGTC 5′
5′
Universal primer

GTCCTGTAG
ATG

1 The four nucleotides, each labeled with a different fluorescent dye, are added, along with DNA polymerase and a universal primer.

4 The cycle is repeated about 100 times.

3′ CATAAAAGCCGTGTC 5′
5′ G   3′

2 Unincorporated nucleotides are removed.

3′ CATAAAAGCCGTGTC 5′
5′ G   3′

3 The newly added nucleotide is detected by a camera.

Once the sequences of millions of short fragments have been determined, the problem becomes how to put them together. In other words, how are they arranged in the chromosomes from which they came? Imagine if you cut out every word in this book (there are about 500,000 of them), put them on a table, and tried to arrange them in their original order! The enormous task of determining DNA sequences is possible because the original DNA fragments are *overlapping*.

Let's illustrate the process using a single 10-bp DNA molecule. (This is a double-stranded molecule, but for convenience we show only the sequence of the noncoding strand.) The molecule is cut three ways. The first cut generates the fragments:

TG, ATG, and CCTAC

The second cut of the same molecule generates the fragments:

AT, GCC, and TACTG

The third cut results in:

CTG, CTA, and ATGC

Can you put the fragments in the correct order? (The answer is ATGCCTACTG.) For genome sequencing, the fragments are called "reads" (**FIGURE 12.2**). Of course, the problem of ordering 2.5 million fragments of 100 bp from human chromosome 1 (246 million bp) is more challenging! The field of **bioinformatics** was developed to analyze DNA sequences using complex mathematics and computer programs.

Genome sequences yield several kinds of information

New genome sequences are being published at an accelerating pace, creating a torrent of biological information (**FIGURE 12.3**). This information is used in two related fields of research, both focused on studying genomes. In **functional genomics**, biologists use sequence information to identify the functions of various parts of genomes. These parts include:

• *Open reading frames*, which are the coding regions of genes. For protein-coding genes, these regions can be recognized by the start and stop codons for translation, and by intron

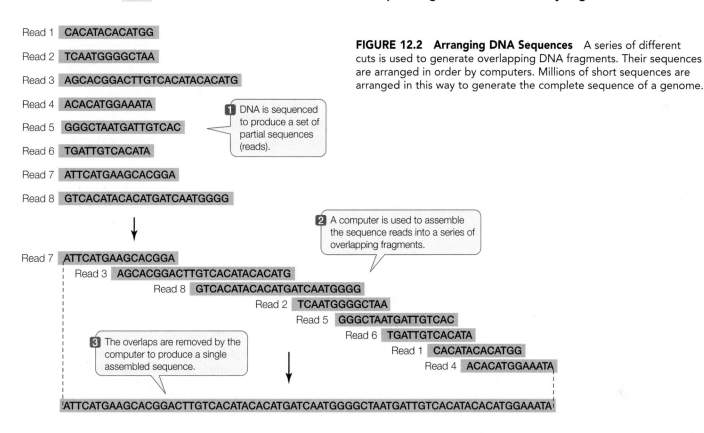

Read 1 `CACATACACATGG`

Read 2 `TCAATGGGGCTAA`

Read 3 `AGCACGGACTTGTCACATACACATG`

Read 4 `ACACATGGAAATA`

Read 5 `GGGCTAATGATTGTCAC`

Read 6 `TGATTGTCACATA`

Read 7 `ATTCATGAAGCACGGA`

Read 8 `GTCACATACACATGATCAATGGGG`

1 DNA is sequenced to produce a set of partial sequences (reads).

2 A computer is used to assemble the sequence reads into a series of overlapping fragments.

Read 7 `ATTCATGAAGCACGGA`
Read 3 `AGCACGGACTTGTCACATACACATG`
Read 8 `GTCACATACACATGATCAATGGGG`
Read 2 `TCAATGGGGCTAA`
Read 5 `GGGCTAATGATTGTCAC`
Read 6 `TGATTGTCACATA`
Read 1 `CACATACACATGG`
Read 4 `ACACATGGAAATA`

3 The overlaps are removed by the computer to produce a single assembled sequence.

`ATTCATGAAGCACGGACTTGTCACATACACATGATCAATGGGGCTAATGATTGTCACATACACATGGAAATA`

FIGURE 12.2 Arranging DNA Sequences A series of different cuts is used to generate overlapping DNA fragments. Their sequences are arranged in order by computers. Millions of short sequences are arranged in this way to generate the complete sequence of a genome.

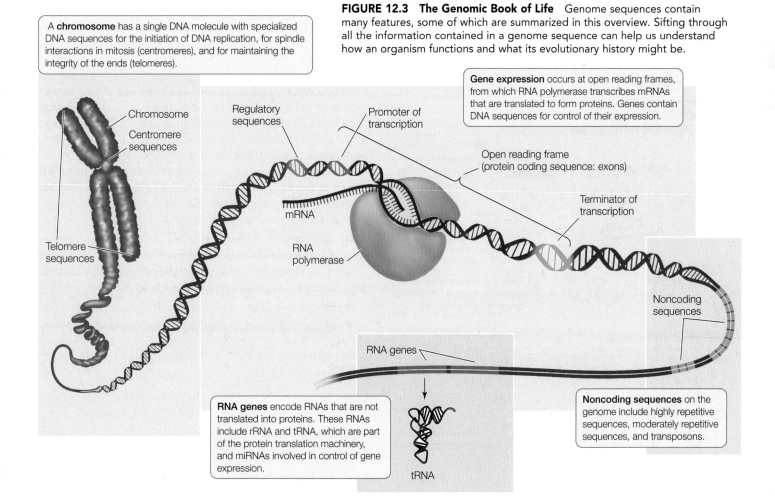

FIGURE 12.3 The Genomic Book of Life Genome sequences contain many features, some of which are summarized in this overview. Sifting through all the information contained in a genome sequence can help us understand how an organism functions and what its evolutionary history might be.

A **chromosome** has a single DNA molecule with specialized DNA sequences for the initiation of DNA replication, for spindle interactions in mitosis (centromeres), and for maintaining the integrity of the ends (telomeres).

Gene expression occurs at open reading frames, from which RNA polymerase transcribes mRNAs that are translated to form proteins. Genes contain DNA sequences for control of their expression.

Chromosome

Centromere sequences

Telomere sequences

Regulatory sequences

Promoter of transcription

Open reading frame (protein coding sequence: exons)

Terminator of transcription

mRNA

RNA polymerase

Noncoding sequences

RNA genes

tRNA

RNA genes encode RNAs that are not translated into proteins. These RNAs include rRNA and tRNA, which are part of the protein translation machinery, and miRNAs involved in control of gene expression.

Noncoding sequences on the genome include highly repetitive sequences, moderately repetitive sequences, and transposons.

consensus sequences that indicate the locations of introns. A major goal of functional genomics is to understand the function of every open reading frame in each genome.

- *Amino acid sequences* of proteins, which can be deduced from the DNA sequences of open reading frames by applying the genetic code.

- *Regulatory sequences*, such as promoters and terminators for transcription.

- *RNA genes*, including rRNA, tRNA, small nuclear RNA, and microRNA genes.

- *Other noncoding sequences* that can be classified into various categories, including centromeric and telomeric regions, transposons, and repetitive sequences.

Sequence information is also used in **comparative genomics**: the comparison of a newly sequenced genome (or parts thereof) with sequences from other organisms. This can provide further information about the functions of sequences and can be used to trace evolutionary relationships among different organisms.

Phenotypes can be analyzed using proteomics and metabolomics

"The human genome is the book of life." Statements like this were common when the human genome sequence was first revealed. They reflect the concept of *genetic determinism*—the idea that a person's phenotype is determined solely by his or her genotype. But is an organism just a product of gene expression? We know that it is not. The proteins and small molecules present in any cell at a given point in time reflect not just gene expression but changes in the proteins by the intracellular and extracellular environments. To take more complete snapshots of cells and organisms, two new fields have emerged to complement genomics: proteomics and metabolomics.

PROTEOMICS Many genes encode more than a single protein (**FIGURE 12.4A**). As we described in Concept 11.4, alternative splicing leads to different combinations of exons in the mature mRNAs transcribed from a single gene. Posttranslational modifications also increase the number and the structural and functional diversity of proteins derived from one gene (see Figures 12.4A and 10.21). The **proteome** is the sum total of the proteins produced by an organism, and it is more complex than its genome.

Two methods are commonly used to analyze proteins and the proteome:

- Because of their unique amino acid compositions (primary structures), most proteins have unique combinations of electric charge and size. On the basis of these two properties, they can be separated by two-dimensional gel electrophoresis. Thus isolated, individual proteins can be analyzed, sequenced, and studied (**FIGURE 12.4B**).

- Mass spectrometry uses electromagnets to identify molecules by the masses of their atoms, and it can also be used to determine the structures of molecules. The results are displayed as peaks on a graph.

While genomics seeks to describe the genome and its expression, **proteomics** seeks to identify and characterize all of the expressed proteins. Its ultimate aim is just as ambitious as that of genomics.

METABOLOMICS Studying genes and proteins gives a limited picture of what is going on in a cell. But as we have seen, both gene function and protein function are affected by a cell's internal and external environments. Many proteins are enzymes, and their activities affect the concentrations of their substrates and products. So as the proteome changes, so do the abundances of these (often small) molecules, called metabolites. The **metabolome** is the quantitative description of all of the metabolites in a cell or organism. These include:

- *Primary metabolites* that are involved in normal processes, such as intermediates in pathways like glycolysis. This category also includes hormones and other signaling molecules.

- *Secondary metabolites*, which are often unique to particular organisms or groups of organisms. They are often involved in special responses to the environment. Examples are antibiotics made by microbes, and the many chemicals made by plants that are used in defense against pathogens and herbivores.

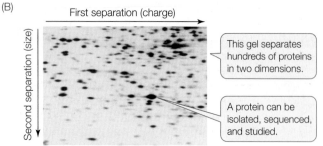

FIGURE 12.4 Proteomics (A) A single gene can code for multiple proteins. (B) A cell's proteins can be separated on the basis of charge and size by two-dimensional gel electrophoresis. The two separations can distinguish most proteins from one another.

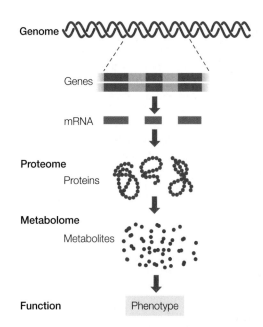

FIGURE 12.5 Genomics, Proteomics, and Metabolomics A combination of these approaches can give more comprehensive information about genotypes and phenotypes.

Metabolomics aims to describe the metabolome of a tissue or organism under particular environmental conditions. Measuring metabolites involves sophisticated analytical instruments. If you have studied organic or analytical chemistry, you may be familiar with gas chromatography and high-performance liquid chromatography which are used to separate molecules with different chemical properties. Mass spectrometry and nuclear magnetic resonance spectroscopy are used to identify molecules. These measurements result in "chemical snapshots" of cells or organisms, which can be related to physiological states.

Taken together, the genome, proteome, and metabolome can move biologists toward a more comprehensive picture of an organism's genotype and phenotype (**FIGURE 12.5**).

Do You Understand Concept 12.1?

- Using a table, compare genomics, proteomics, and metabolomics with regard to the methods used and results obtained.

- A DNA molecule is cut into the following fragments that are sequenced: AGTTT, TAGG, CGAT, and CCT. The same molecule is cut in a different way to produce TTCGA, TCCT, AGT, GG, and TA. A third cut produces TTCGAT, CCTT, AGG, and AGT. What is the sequence of this DNA?

- If you were designing a computer program to recognize important sequences in DNA, what types of sequences would you include? What would you learn from finding these sequences in the DNA? Be as specific as you can.

The first cellular genomes to be fully sequenced were those of prokaryotes. In the next concept we will discuss these relatively small, compact genomes.

concept **12.2**

Prokaryotic Genomes Are Relatively Small and Compact

When DNA sequencing became possible in the late 1970s, the first life forms to be sequenced were the simplest viruses. The sequences quickly provided new information on how viruses infect their hosts and reproduce. The next genomes to be fully sequenced were those of prokaryotes. We now have genome sequences for many microorganisms, to the great benefit of microbiology and medicine.

Prokaryotic genomes are compact

In 1995 a team led by Craig Venter and Hamilton Smith published the first complete genomic sequence of a free-living cellular organism, the bacterium *Haemophilus influenzae*. Many more prokaryotic sequences have followed, revealing not only how prokaryotic genes are organized to perform different cellular functions, but also how certain specialized functions of particular organisms are carried out.

There are several notable features of bacterial and archaeal genomes:

- They are relatively small. Prokaryotic genomes range in size from about 160,000 to 12 million bp and are usually organized into a single circular chromosome.

- They are compact. Typically, over 85 percent of the DNA consists of protein-coding regions or RNA genes, with only short sequences between genes.

- Their genes usually do not contain introns. An exception is the rRNA and tRNA genes of archaea, which are frequently interrupted by introns.

- In addition to the main chromosome, they often carry smaller, circular DNA molecules called plasmids, which may be transferred between cells (see Concept 8.4).

Beyond these broad similarities, there is great diversity among these single-celled organisms, reflecting their huge variety in the environments where they are found.

Let's look in more detail at a few prokaryotic genomes in terms of functional and comparative genomics.

FUNCTIONAL GENOMICS As mentioned above, functional genomics is a biological discipline that assigns functions to the products of genes. This field is less than 20 years old but is now a major occupation of biologists. You can see the various functions encoded by the genomes of three prokaryotes (in this case, all bacteria) in **TABLE 12.1**.

H. influenzae lives in the upper respiratory tracts of humans and can cause ear infections and (more seriously) meningitis. Its single circular chromosome has 1,830,138 bp. In addition

TABLE 12.1 Gene Functions in Three Bacteria

CATEGORY	NUMBER OF GENES IN:		
	E. COLI	*H. INFLUENZAE*	*M. GENITALIUM*
Total protein-coding genes	4,288	1,727	482
Biosynthesis of amino acids	131	68	1
Biosynthesis of cofactors	103	54	5
Biosynthesis of nucleotides	58	53	19
Cell envelope proteins	237	84	17
Energy metabolism	243	112	31
Intermediary metabolism	188	30	6
Lipid metabolism	48	25	6
DNA replication, recombination, and repair	115	87	32
Protein folding	9	6	7
Regulatory proteins	178	64	7
Transcription	55	27	12
Translation	182	141	101
Uptake of molecules from the environment	427	123	34

FRONTIERS Some bacteria, such as *Methanococcus*, produce methane in the stomachs of cattle. Others, such as *Methylococcus*, remove methane from the air and use it as an energy source. The genomes of both of these bacteria have been sequenced. Researchers are analyzing the genes involved in methane production and oxidation to find ways to control atmospheric methane (a powerful greenhouse gas) and thereby slow the progress of global warming.

Some sequences of DNA can move about the genome

Genome sequencing allowed scientists to study more broadly a class of DNA sequences that had been discovered by geneticists decades earlier. Segments of DNA called **transposons** (or *transposable elements*) can move from place to place in the genome and can even move from one piece of DNA (such as a chromosome) to another (such as a plasmid) in the same cell. A transposon might be at one location in the genome of one *E. coli* strain, and at a different location in another strain. The insertion of this movable DNA sequence from elsewhere in the genome into the middle of a protein-coding gene disrupts that gene (**FIGURE 12.6A**). Any mRNA expressed from the disrupted gene will have the extra sequence, and the protein will be abnormal. So transposons can produce significant phenotypic effects by inactivating genes.

Transposons are often short sequences of 1,000–2,000 bp and are found at many sites in prokaryotic genomes. The mechanisms that allow them to move vary. For example, the transposon may be replicated, and then the copy inserted into another site in the genome. Or it might splice out of one location and move to another location.

If a transposon becomes duplicated with two copies separated by one or a few genes, the result may be a single larger transposon (up to about 5,000 bp). In this case, the additional genes can be carried to different locations in the genome (**FIGURE 12.6B**). Some of these transposons carry genes for antibiotic resistance. We will discuss transposons again in Concept 12.3.

Metagenomics allows us to describe new organisms and ecosystems

If you take a microbiology laboratory course, you will learn how to identify various prokaryotes on the basis of their growth in lab cultures. Microorganisms can be identified by their nutritional requirements or the conditions under which they will grow (e.g., aerobic versus anaerobic). For example, staphylococci are a group of bacteria that inhabit skin and nasal passages. Unlike many bacteria, staphylococci can use the sugar alcohol mannitol as an energy source and thus can

to its origin of replication and the RNA genes, this bacterial chromosome has 1,727 open reading frames.

When this sequence was first announced, only 1,007 (58 percent) of the open reading frames encoded proteins with known functions. Since then scientists have identified the role of almost every protein encoded by the *H. influenzae* genome. All of the major biochemical pathways and molecular functions are represented. When the larger *Escherichia coli* genome (4.6 million bp) was determined, it was found that these functions are coded for by more genes (see Table 12.1), indicating that the pathways are more complex in *E. coli*.

COMPARATIVE GENOMICS Soon after the sequence of *E. coli* was announced, the genome of a smaller prokaryote, *Mycoplasma genitalium* (580,073 bp), was completed. Thus began the new era in biology of comparative genomics. Scientists can identify genes that are present in one bacterium and missing in another, allowing them to relate these genes to bacterial function.

For example, *M. genitalium* lacks most of the enzymes needed to synthesize amino acids, which *E. coli* and *H. influenzae* both possess (see Table 12.1). This finding reveals that *M. genitalium* must obtain its amino acids from its environment (usually the human urogenital tract). Furthermore, *E. coli* has dozens of genes for regulatory proteins that encode transcriptional activators or repressors; *M. genitalium* only has 7 such genes. This suggests that the biochemical flexibility of *M. genitalium* is limited by its relative lack of control over gene expression.

LINK For more on the role of activators and repressors in controlling gene expression in bacteria, see Concepts 11.1 and 11.2

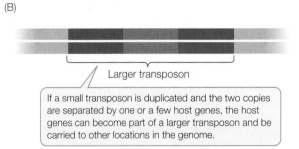

(A)

(B)

Larger transposon

If a small transposon is duplicated and the two copies are separated by one or a few host genes, the host genes can become part of a larger transposon and be carried to other locations in the genome.

FIGURE 12.6 DNA Sequences That Move Transposons (or transposable elements) are DNA sequences that move from one location to another. (A) In one method of transposition, the DNA sequence is replicated and the copy inserts elsewhere in the genome. (B) Transposons can evolve to carry additional genomic sequences.

grow on a special medium containing mannitol. Often a dye is included in the medium, which changes color if the bacteria are pathogenic (disease-causing). Such culture methods have been the mainstay of microbial identification for over a century and are still useful and important. However, scientists can now use PCR and modern DNA analysis techniques to analyze microbes *without* culturing them in the laboratory.

In 1985, Norman Pace, then at Indiana University, came up with the idea of isolating DNA directly from environmental samples. He used PCR to amplify specific sequences from the samples to determine whether particular microbes were present. The PCR products were sequenced to explore their diversity. The term **metagenomics** was coined to describe this approach of analyzing genes without isolating the intact organism. It is now possible to do DNA sequencing with samples from almost any environment. The DNA can be cloned to make "libraries" of sequences (**FIGURE 12.7**), or it can be amplified and sequenced directly, using next-generation sequencing methods. The sequences can be used to detect the presence of previously unidentified organisms as well as known microbes. For example:

- Sequencing of DNA from 200 liters of seawater indicated that it contained 5,000 different viruses and 2,000 different bacteria, many of which had not been described previously.
- One kilogram of marine sediment contained a million different viruses, most of them new.

- Water runoff from a mine contaminated with toxic chemicals contained many new species of prokaryotes thriving in this apparently inhospitable environment. Some of these organisms exhibited metabolic pathways that were previously unknown to biologists. These organisms and their capabilities may be useful in cleaning up pollutants from the water.

- Gut samples from 124 Europeans revealed that each person harbored at least 160 species of bacteria (constituting their "gut microflora"). Many of these species were found in all of the individuals, but the presence of other bacteria varied from person to person. Such variations in gut microflora may be associated with obesity or bowel diseases.

LINK For more on the complex ecosystem inside the human gut, see Figure 42.2

These and other discoveries are truly extraordinary and potentially very important. It is estimated that 90 percent of the microbial world has been invisible to biologists and is only now being revealed by metagenomics. Entirely new ecosystems of bacteria and viruses are being discovered in which, for example, one species produces a molecule that another metabolizes. It is hard to overemphasize the importance of such an increase in our knowledge of the hidden world of microbes. This new knowledge underscores the remarkable diversity among prokaryotic organisms, and will further our understanding of natural ecological processes. Furthermore, it has the potential to help us find better ways to manage environmental catastro-

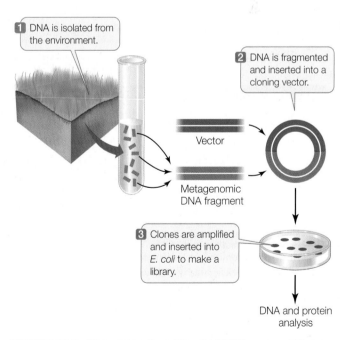

FIGURE 12.7 Metagenomics Microbial DNA extracted from the environment can be amplified and analyzed. This has led to the description of many new genes and species.

INVESTIGATION

FIGURE 12.8   Using Transposon Mutagenesis to Determine the Minimal Genome   *Mycoplasma genitalium* has one of the smallest genomes of any prokaryote. But are all of its genes essential to life? By inactivating the genes one by one, scientists determined which of them are essential for the cell's survival. This research may lead to the construction of artificial cells with customized genomes, designed to perform functions such as degrading oil and making plastics.

HYPOTHESIS

Only some of the genes in a bacterial genome are essential for cell survival.

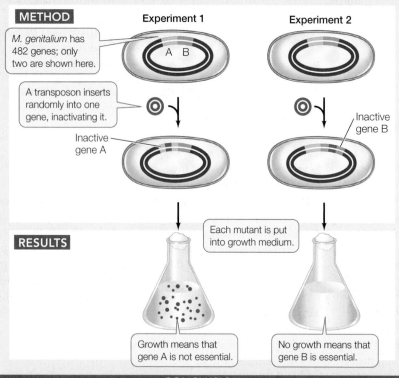

METHOD

M. genitalium has 482 genes; only two are shown here.

A transposon inserts randomly into one gene, inactivating it.

Inactive gene A

Inactive gene B

Each mutant is put into growth medium.

RESULTS

Growth means that gene A is not essential.

No growth means that gene B is essential.

CONCLUSION

If each gene is inactivated in turn, a "minimal essential genome" can be determined.

ANALYZE THE DATA

The growth of *M. genitalium* strains with insertions in genes (intragenic regions) was compared with the growth of strains with insertions in noncoding (intergenic) regions of the genome:

Type of insertion	Number of different genes/regions	Number that grew
Intragenic	482	100
Intergenic	199	184

A. Explain these data in terms of genes essential for growth and survival. Are all of the genes in *M. genitalium* essential for growth? If not, how many are essential? Why did some of the insertions in intergenic regions prevent growth?

B. If a transposon inserts into the following regions of genes, there might be no effect on phenotoype. Explain in each case:
(i) near the 3' end of the coding region
(ii) within a gene coding for rRNA

How does this affect your answer to the first question?

Go to **yourBioPortal.com** for original citations, discussions, and relevant links for all INVESTIGATION figures.

phes such as oil spills, or to remove toxic heavy metals from soil and water.

Will defining the genes required for cellular life lead to artificial life?

When the genomes of prokaryotes and eukaryotes are compared, a striking conclusion arises: certain genes are present in all organisms (universal genes). There are also some (nearly) universal gene segments that are present in many genes in many organisms. One example is a sequence encoding an ATP binding site, which is a domain found in many proteins. These findings suggest that there is some ancient, minimal set of DNA sequences common to all cells. One way to identify these sequences is to look for them in computer analyses of sequenced genomes.

Another way to define the minimal genome is to take an organism with a simple genome and deliberately mutate one gene at a time to see what happens. *M. genitalium* has one of the smallest known genomes—only 482 genes. Even so, some of its genes are dispensable under some circumstances. For example, it has genes for metabolizing both glucose and fructose, but it can survive in the laboratory on a medium containing only one of these sugars.

What about other genes? Researchers addressed this question using transposons as mutagens. When transposons in the bacterium were activated, they inserted themselves into genes at random, mutating and inactivating them (**FIGURE 12.8**). The mutated bacteria were tested for growth and survival, and DNA from interesting mutants was sequenced to find out which genes were mutated. The astonishing result of these studies was that *M. genitalium* could survive in the laboratory with a minimal genome of only 382 functional genes!

Is this really all it takes to make a viable organism? As described in the opening story of Chapter 4, scientists made a synthetic genome based on that of the bacterium *Mycoplasma discoides* and inserted it into an empty cell, creating the first organism with a synthetic genome. This research has promise for making organisms with novel functions, such as the synthesis of plastics polymers or the ability to break down environmental pollutants.

Do You Understand Concept 12.2?

- What are the characteristics of most prokaryotic genomes?

- Examine Table 12.1 and Figure 12.8. What gene functions would you predict are nonessential for *M. genitalium* as determined by transposon-mediated inactivation?

- You want to isolate a prokaryote that can live on discarded Styrofoam cups. Such an organism might live in a landfill where ground-up cups are discarded. How would you use metagenomics to identify such a bacterium? How would you show that its ability to live on Styrofoam is essential, and that it cannot live in another environment?

The methods used to sequence and analyze prokaryotic genomes have also been applied to eukaryotic genomes, which we will examine next.

---

concept 12.3 Eukaryotic Genomes Are Large and Complex

As genomes have been sequenced and described, a number of major differences have emerged between eukaryotic and prokaryotic genomes:

- *Eukaryotic genomes are larger than those of prokaryotes*, and they have more protein-coding genes. This difference is not surprising given that multicellular organisms have many cell types with specific functions that require specialized proteins. A typical virus contains enough DNA (about 10,000 bp) to encode only a few proteins. As we saw above, one of the simplest prokaryotes, *Mycoplasma*, has several hundred protein-coding genes in a genome of about 0.5 million bp. A rice plant, in contrast, has 37,544 genes.

- *Eukaryotic genomes have more regulatory sequences*—and many more regulatory proteins—than prokaryotic genomes. The greater complexity of eukaryotes requires much more regulation, which is evident in the many points of control associated with the expression of eukaryotic genes (see Concepts 11.3 and 11.4).

- *Much of eukaryotic DNA is noncoding*. Distributed throughout many eukaryotic genomes are various kinds of DNA sequences that are not transcribed into functional RNAs (such

as rRNA and tRNA) or mRNA. Some of these sequences are introns or regulatory sequences. In addition, eukaryotic genomes contain various kinds of repeated sequences.

Model organisms reveal many characteristics of eukaryotic genomes

Most of the lessons learned from eukaryotic genomes have come from several model organisms that have been studied extensively: the yeast *Saccharomyces cerevisiae*, the nematode (roundworm) *Caenorhabditis elegans*, the fruit fly *Drosophila melanogaster*, and the plant *Arabidopsis thaliana* (thale cress). Model organisms have been chosen because they are relatively easy to grow and study in a laboratory, their genetics are well studied, and they exhibit characteristics that represent a larger group of organisms. **TABLE 12.2** shows some characteristics of the genomes of these organisms.

YEAST: THE BASIC EUKARYOTIC MODEL Yeasts are single-celled eukaryotes. Like other eukaryotes, they have membrane-enclosed organelles. They can live as either haploid or diploid organisms, and this is usually determined by environmental conditions: under adverse conditions the diploid cells will undergo meiosis and sporulation. While the prokaryote *E. coli* has a single circular chromosome and 4,290 genes, *Saccharomyces cerevisiae* has 16 linear chromosomes and 5,770 genes. The most striking difference between the yeast genome and that of *E. coli* is in the number of genes for targeting proteins to organelles. Both of these single-celled organisms appear to use about the same number of genes to perform the basic functions of cell survival. It is the compartmentalization of the eukaryotic yeast cell into organelles that requires it to have many more genes.

TABLE 12.2	Representative Sequenced Genomes			
ORGANISM	HAPLOID GENOME SIZE (Mb)	NUMBER OF GENES	PROTEIN-CODING SEQUENCE	NOTABLE GENES
Bacteria				
M. genitalium	0.58	482	88%	Minimal genome
H. influenzae	1.8	1,738	89%	
E. coli	4.6	4,377	88%	
Yeasts				Targeting; cell organelles
S. cerevisiae	12.1	5,770	70%	
S. pombe	12.1	4,929	60%	
Plants				Photosynthesis; cell walls
A. thaliana	157	28,000	25%	
Rice	394	37,544	12%	Water tolerance for roots
Soybean	1,115	46,000	7%	Lipid synthesis, storage
Animals				
C. elegans	100	19,427	25%	Tissue formation
D. melanogaster	130	13,379	13%	Embryonic development
Human	3,200	24,000	1.2%	

Mb = millions of base pairs

This finding is direct, quantitative confirmation of something we have known for a century: the eukaryotic cell is structurally and functionally more complex than the prokaryotic cell.

THE NEMATODE: UNDERSTANDING CELL DIFFERENTIATION The 1-mm-long nematode *Caenorhabditis elegans* normally lives in the soil. It can also live in the laboratory, where it has become a favorite model organism of developmental biologists (see Chapter 14). The nematode has a transparent body that develops over 3 days from a fertilized egg to an adult worm that has a nervous system, digests food, and reproduces sexually. Its genome is 8 times larger than that of yeast and has 3.5 times as many protein-coding genes. Gene inactivation studies show that the worm can survive in laboratory cultures with only 10 percent of these genes. But this "minimum genome" is still five times the size of the minimum genome for *Mycoplasma*. The extra genes in the nematode minimal genome encode proteins needed for cell differentiation, for intercellular communication, and for holding cells together to form tissues.

***DROSOPHILA MELANOGASTER*: UNDERSTANDING GENETICS AND DEVELOPMENT** The fruit fly *Drosophila melanogaster* is a famous model organism. Studies of fruit fly genetics resulted in the formulation of many basic principles of genetics (see Concept 8.3). The fruit fly is a much larger organism than *C. elegans* (it has ten times as many cells), and it is much more complex: it undergoes complicated developmental transformations from egg to larva to pupa to adult. These differences are reflected in the fruit fly genome, which has many genes encoding transcription factors needed for complex embryonic development (you will study some of these in Chapter 14). In general, the fruit fly genome has a distribution of coding sequence functions quite similar to those of many other complex eukaryotes (**FIGURE 12.9**).

***ARABIDOPSIS*: STUDYING THE GENOMES OF PLANTS** About 250,000 species of flowering plants dominate the land and fresh water.

Although there is generally more interest in the plants we use for food and fiber, scientists first sequenced the genome of a simpler flowering plant with a relatively small genome. *Arabidopsis thaliana*, thale cress, is a member of the mustard family and has long been a favorite model organism of plant biologists. It is small (hundreds could grow and reproduce in the space occupied by this page) and easy to manipulate. Its genome has about 28,000 protein-coding genes, but many of these are duplicates and probably originated by chromosomal rearrangements. When these duplicate genes are subtracted from the total, about 15,000 unique genes are left—similar to the gene numbers found in fruit flies and nematodes. Indeed, many of the genes found in these animals have **orthologs**—genes with very similar sequences—in *Arabidopsis* and other plants, suggesting that plants and animals have a common ancestor.

FRONTIERS The genome sequences of corn and soybean reveal that many protein-coding genes in these important crop plants are duplicated as a result of chromosome rearrangements. This makes "gene counts" somewhat misleading, since many genes are redundant. More importantly, this redundancy presents a challenge to plant breeders trying to improve the characteristics of these plants: instead of one simple cross to transfer a single allele into a new strain, the breeders may have to do multiple crosses to transfer multiple copies of that allele.

Arabidopsis has some genes, however, that are unique to plants. These include genes involved in photosynthesis, in the transport of water throughout the plant, in the assembly of the cell wall, in the uptake and metabolism of inorganic substances from the environment, and in the synthesis of specific molecules used for defense against microbes and herbivores. These plant defense molecules may be a major reason why the numbers of protein-coding genes in some plants are higher than in many animals. Plants cannot escape their enemies or other adverse conditions as animals can, and they must cope with situations where they are. So they make tens of thousands of molecules to help them fight their enemies and adapt to their changing environments (see Chapter 28).

These plant-specific genes are also found in the genomes of other plants, including rice (*Oryza sativa*), which was the first major crop plant to be fully sequenced. Rice is the world's most important crop—it is a staple in the diets of 3 billion people. The larger genome of rice has a set of genes remarkably similar to that of *Arabidopsis*. The genome of the poplar tree (*Populus trichocarpa*) was also sequenced, to gain insight into the potential for this rapidly growing tree to be used as a source of fuel. A comparison of the three genomes shows many genes in common, which may comprise the *basic plant genome* (**FIGURE 12.10**).

Gene families exist within individual eukaryotic organisms

About half of all eukaryotic protein-coding genes exist as only one copy in the haploid genome (with two alleles in somatic

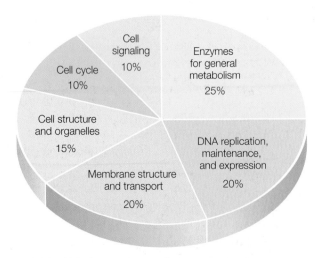

FIGURE 12.9 Functions of the Eukaryotic Genome The distribution of gene functions in *Drosophila melanogaster* shows a pattern that is typical of many complex organisms.

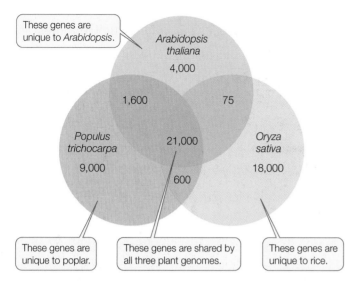

FIGURE 12.10 **Plant Genomes** Three plant genomes share a common set of approximately 21,000 genes that appear to comprise the "minimal" plant genome.

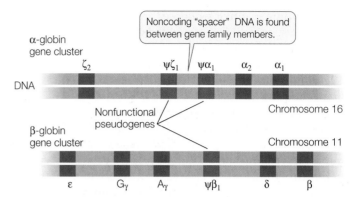

FIGURE 12.11 **The Globin Gene Family** The α-globin and β-globin clusters of the human globin gene family are located on different chromosomes. The genes of each cluster are separated by noncoding "spacer" DNA. The nonfunctional pseudogenes are indicated by the Greek letter psi (ψ). The γ gene has two variants, A$_γ$ and G$_γ$.

cells). The rest are present in multiple copies that arose from gene duplications. Over evolutionary time, different copies of the genes have undergone separate mutations, giving rise to groups of closely related genes called **gene families**. Some gene families, such as those encoding the globin proteins that make up hemoglobin, contain only a few members in a single organism. Other families, such as the genes encoding the immunoglobulins that make up antibodies, have hundreds of members.

> **LINK** Gene duplication is a key mechanism for the evolution of new functions; see Concept 15.6

Within a single organism, the genes in a family are usually slightly different from one another. As long as at least one member encodes a functional protein, the other members may mutate in ways that change the functions of the proteins they encode. For evolution, the availability of multiple copies of a gene allows for selection of mutations that provide advantages under certain circumstances. If a mutated gene is useful, it may be selected for in succeeding generations. If the mutated gene is a total loss, the functional copy is still there to carry out its role.

As an example, let's look at the gene family encoding the globins in vertebrates. These proteins are found in hemoglobin and myoglobin (an oxygen-binding protein present in muscle). The globin genes all arose long ago from a single common ancestral gene (see Figure 15.22). In humans there are three functional members of the α-globin cluster and five in the β-globin cluster (**FIGURE 12.11**). Each hemoglobin molecule in an adult human is a tetramer containing two identical α-globin subunits, two identical β-globin subunits, and four heme pigments (see Chapter 3, p. 44).

During human development, different members of the globin gene cluster are expressed at different times and in different tissues. This differential gene expression has great physiological significance. For example, the hemoglobin in the human

fetus contains γ-globin, which binds O_2 more tightly than adult hemoglobin. This specialized form of hemoglobin ensures that in the placenta, O_2 will be transferred from the mother's blood to the developing fetus's blood. Just before birth the liver stops synthesizing fetal hemoglobin and the bone marrow cells take over, making the adult forms (two α and two β). Thus hemoglobins with different binding affinities for O_2 are provided at different stages of human development.

In addition to genes that encode functional proteins, many gene families include nonfunctional **pseudogenes**, which are designated with the Greek letter psi (ψ; see Figure 12.11). These pseudogenes result from mutations that cause a loss of function rather than an enhanced or new function. The DNA sequence of a pseudogene may not differ greatly from that of other family members. It may simply lack a promoter, for example, and thus fail to be transcribed. Or it may lack a recognition site needed for the removal of an intron, so that the transcript it makes is not correctly processed into a useful mature mRNA. In some gene families pseudogenes outnumber functional genes. Because some members of the family are functional, there appears to be little selection pressure to preserve the functions of redundant genes.

Eukaryotic genomes contain many repetitive sequences

Eukaryotic genomes contain numerous repetitive DNA sequences that do not code for polypeptides. There are highly repetitive sequences and moderately repetitive sequences, which include rRNA genes, tRNA genes, and transposons (**TABLE 12.3**).

Highly repetitive sequences are short (less than 100 bp) sequences that are repeated thousands of times in tandem (side-by-side) arrangements in the genome. They are not transcribed. Their proportion in eukaryotic genomes varies, from 10 percent in humans to about half the genome in some species of fruit flies. Often they are associated with heterochromatin, the densely packed, transcriptionally inactive part of the genome.

Other highly repetitive sequences are scattered around the genome. For example, **short tandem repeats** (**STRs**) of 1–5 bp can be repeated up to 100 times at a particular chromosomal location. The copy number of an STR at a particular location varies between individuals and is inherited.

Moderately repetitive sequences are repeated 10–1,000 times in the eukaryotic genome. These sequences include the genes that are transcribed to produce tRNAs and rRNAs, which are used in protein synthesis. The cell makes tRNAs and rRNAs constantly, but even at the maximum rate of transcription, single copies of the tRNA and rRNA genes would be inadequate to supply the large amounts of these molecules needed by most cells. Thus the genome has multiple copies of these genes, in clusters containing transcribed regions (with introns) and nontranscribed "spacers" between the genes.

TABLE 12.3	Types of Sequences in Eukaryotic Genomes		
CATEGORY		TRANSCRIBED	TRANSLATED
Single-copy genes			
Promoters and expression control sequences		No	No
Introns		Yes	No
Exons		Yes	Yes
Moderately repetitive sequences			
rRNA and tRNA genes		Yes	No
Transposons			
I. Retrotransposons			
LTR retrotransposons		Yes	No
SINEs		Yes	No
LINEs		Yes	Yes
II. DNA transposons		Yes	Yes
Highly repetitive short sequences		No	No

DNA

13,000 bp
Transcribed region

30,000 bp
Nontranscribed spacer region

Most moderately repetitive sequences are not stably integrated into the genome but instead are transposons (see Figure 12.6). Transposons make up over 40 percent of the human genome. There are two main types of transposons in eukaryotes: retrotransposons (Class I transposons) and DNA transposons (Class II; see Table 12.3). **Retrotransposons** make RNA copies of themselves, which are then copied back into DNA before insertion at new locations in the genome. They are divided into two categories:

1. *LTR retrotransposons* have long terminal repeats (LTRs) of DNA sequence (100–5,000 bp) at each end. LTR retrotransposons constitute about 8 percent of the human genome.

2. *Non-LTR retrotransposons* do not have LTR sequences at their ends. They are further divided into two subcategories: SINEs and LINEs. *SINEs* (*s*hort *in*terspersed *e*lements) are up to 500 bp long and are transcribed but not translated. There are about 1.5 million of them scattered over the human genome, making up about 15 percent of the total DNA content. A single type, the 300-bp *Alu* element, accounts for 11 percent of the human genome; it is present in a million copies. *LINEs* (*l*ong *in*terspersed *e*lements) are up to 7,000 bp long, and some are transcribed and translated into proteins. They constitute about 17 percent of the human genome.

DNA transposons do not use RNA intermediates. Like some prokaryotic transposons, they are excised from the original location and become inserted at a new location without being replicated.

APPLY THE CONCEPT

Eukaryotic genomes are large and complex

Repetitive DNA sequences can be classified by nucleic acid hybridization (see Figure 10.7). A genome is initially cut into 300-bp fragments, and these are heated to denature the DNA. If the solution is cooled, the DNA strands will form hydrogen bonds and reassociate into double-stranded structures. If there are many copies of a DNA sequence in the solution (repetitive DNA), it will find its complementary sequence and reassociate faster than if there are only a few copies. The table shows results from equal amounts of DNA from three species.

1. Why do yeast and mouse DNAs reassociate faster than *E. coli* DNA?

REASSOCIATION TIME (MIN)	PERCENTAGE OF DNA REASSOCIATED		
	E. COLI	YEAST	MOUSE
1	0	3	10
10	0	17	35
100	100	100	100

2. Would you expect human DNA to reassociate faster or slower than yeast DNA?

Do You Understand Concept 12.3?

- Compare the general properties of the genomes of prokaryotes and eukaryotes.

- Does the size of a genome determine how much information it contains? Explain in terms of repetitive sequences and protein-coding genes.

- What is the evolutionary role of eukaryotic gene families?

- During transposition, an adjacent gene is sometimes replicated along with a retrotransposon. What would be the consequence of making a new copy of this gene at a new location in the genome?

The analysis of eukaryotic genomes has resulted in an enormous amount of useful information, as we have seen. In the next concept we look more closely at the human genome.

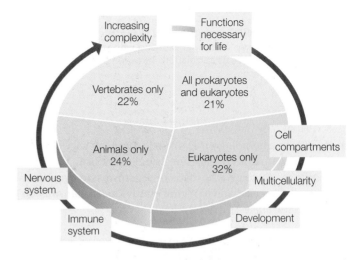

FIGURE 12.12 Evolution of the Genome A comparison of the human and other genomes has revealed how genes with new functions have been added over the course of evolution. Each percentage number refers to genes in the human genome. Thus, 21 percent of human genes have homologs in prokaryotes and other eukaryotes, 32 percent of human genes occur only in other eukaryotes, and so on.

concept 12.4 The Human Genome Sequence Has Many Applications

During the first decade of this millennium the haploid genomes of more than ten individuals were sequenced and published. With the rapid development of new sequencing technologies, the time is approaching when a human genome can be sequenced for less than $1,000.

The human genome sequence held some surprises

The following are just some of the interesting facts we have learned about the human genome:

- Among the 3.2 billion bp in the haploid human genome, there are about 24,000 protein-coding genes. This was a surprise. Before sequencing began, the diversity of human proteins suggested there would be 80,000–150,000 genes. The actual number—not many more than in a fruit fly—means that posttranscriptional mechanisms (such as alternative splicing; see Concept 11.4) must account for the observed number of proteins in humans. It turns out that most human genes encode multiple proteins.

- The average gene spans 27,000 bp. Gene sizes vary greatly, from about 1,000 to 2.4 million bp. Variation in gene size is to be expected given that human proteins vary in size, from about 100 to 5,000 amino acids per polypeptide chain.

- Virtually all human genes have many introns.

- At least 3.5 percent of the genome is functional but noncoding. These sequences have roles in gene regulation (e.g., are transcribed to microRNAs) or in chromosome structure (e.g., telomere DNA).

- Over 50 percent of the genome is made up of transposons and other repetitive sequences.

- Most of the genome (at least 97 percent) is the same in all people. Despite this apparent homogeneity, there are, of course, many individual differences. Scientists have mapped over 7 million single nucleotide polymorphisms (SNPs; see below) in humans, and more will be found as more human genomes are sequenced.

Comparisons between sequenced genomes from prokaryotes and eukaryotes have revealed some of the evolutionary relationships between genes. Some genes are present in both prokaryotes and eukaryotes; others are only in eukaryotes; still others are only in animals, or only in vertebrates (**FIGURE 12.12**).

More comparative genomics is possible now that the genomes of two other primates, the chimpanzee and the rhesus macaque, have been sequenced. The chimpanzee is evolutionarily close to humans and shares 95 percent of the human genome sequence. The more distantly related rhesus macaque shares 91 percent of the human sequence. The search is on for a set of human genes that differ from the other primates and "make humans human."

Human genomics has potential benefits in medicine

Complex phenotypes are determined not by single genes but by multiple genes interacting with the environment. The single-allele models of phenylketonuria and sickle-cell anemia (see Concept 9.3) do not apply to such common disorders as diabetes, heart disease, and Alzheimer's disease. To understand the genetic bases of these diseases, biologists are now using rapid genotyping technologies to create "haplotype maps," which are used to identify genes involved in disease.

Haplotype maps are based on **single nucleotide polymorphisms (SNPs)**—DNA sequence variations that involve single nucleotides. SNPs (pronounced "snips") arise as point mutations (see Concept 9.3). Because of these mutations, a single

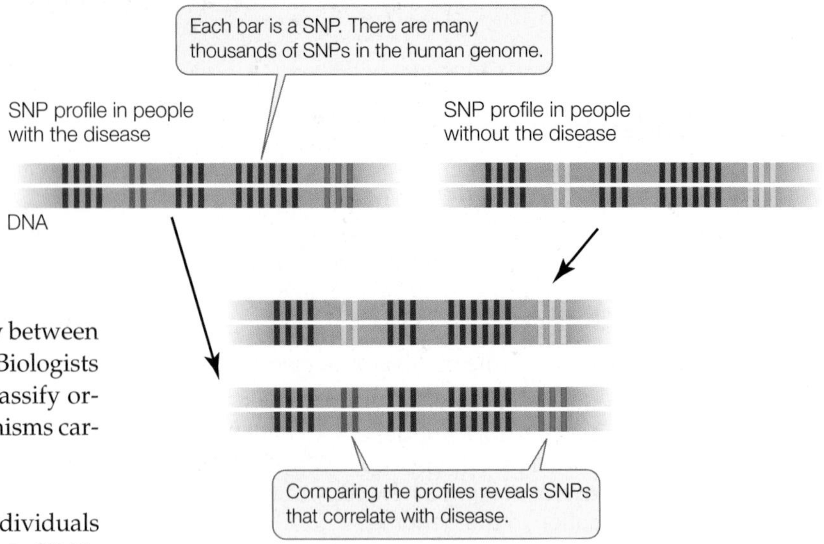

FIGURE 12.13 **SNP Genotyping and Disease**
Scanning the genomes of people with and without particular diseases reveals correlations between SNPs and complex diseases.

Each bar is a SNP. There are many thousands of SNPs in the human genome.

SNP profile in people with the disease

SNP profile in people without the disease

DNA

Comparing the profiles reveals SNPs that correlate with disease.

nucleotide in a homologous DNA sequence may vary between individuals or between alleles in a single organism. Biologists use SNPs to create genetic maps of organisms, to classify organisms and species, and to identify individual organisms carrying specific alleles.

HAPLOTYPE MAPPING The SNPs that differ between individuals are not all inherited as independent alleles. Rather, *a set of SNPs that are close together on a chromosome are inherited as a unit* (they are tightly linked). A piece of chromosome with a set of linked SNPs is called a **haplotype**. You can think of the haplotype as a sentence and the SNP as a word in the sentence. Analyses of haplotypes in humans from all over the world have thus far identified 500,000 common variations.

GENOTYPING TECHNOLOGY AND PERSONAL GENOMICS New technologies are continually being developed to analyze thousands or millions of SNPs in the genomes of individuals. Such technologies include next-generation sequencing methods and DNA microarrays, which depend on hybridization to identify specific SNPs.

A **DNA microarray** is a grid of microscopic spots of oligonucleotides arrayed on a solid surface. It can be "probed" with a complex mixture of DNA or RNA; if the mixture contains a sequence that is complementary to one of the oligonucleotides, the sequence will hybridize to that spot. Colored fluorescent dyes are used to detect hybridizing spots. For example, a microarray of 500,000 SNP-containing oligonucleotides has been used to analyze DNA from thousands of people to find out which SNPs are associated with specific diseases. The aim is to identify particular alleles that contribute (along with particular alleles of other genes) to each complex disease (**FIGURE 12.13**). The amount of data from 500,000 SNPs and thousands of people with thousands of medical records is prodigious. With so much natural variation, statistical measures of association between a haplotype and a disease need to be very rigorous.

These association tests have revealed haplotypes or alleles that are associated with modestly increased risks for such complex diseases as breast cancer, diabetes, arthritis, obesity, and coronary heart disease. For example, 12 sequence variants are associated with increased incidence of heart attacks, and if considered together, the variants can be used to identify individuals who are at increased risk. Indeed, the predictive value of such a genetic test is greater than the widely used test for elevated blood cholesterol level. Private companies will now scan a human genome for SNP alleles, and the price for this service keeps getting lower. However, at this point it is unclear what a person without symptoms should do with the information, since multiple genes, environmental influences, and epigenetic effects all contribute to the development of these diseases.

FRONTIERS Perhaps the most remarkable human genome sequenced thus far is that of an ancient human who lived 4,000 years ago in Greenland. Using a hair sample unearthed from permafrost, scientists extracted the entire genome. Surprisingly, initial SNP analyses show that "Inuk" had blood type A, brown eyes, and dark skin—characteristics shared with modern natives of Siberia. Additional human genome sequences, especially of Arctic peoples, will help researchers reconstruct the human migration that brought Inuk to Greenland.

PHARMACOGENOMICS Genetic variation can affect how an individual responds to a particular drug. For example, consider an enzyme in the liver that catalyzes the following reaction:

active drug → less active drug

A mutation in the gene that encodes this enzyme may make the enzyme less active and reduce the rate at which the active drug is modified to a less active form. For a given dose of the drug, a person with the mutation would have more active drug in his or her bloodstream than a person without the mutation. So the effective dose of the drug would be lower for this person.

Now consider a different case, in which a liver enzyme is needed to make the drug active:

inactive drug → active drug

A person carrying a mutation in the gene encoding this enzyme would not be affected by the drug, since the activating enzyme is not present.

The study of how an individual's genome affects his or her response to drugs or other agents is called **pharmacogenomics**. This type of analysis makes it possible to predict whether or not a drug will be effective. The objective is to *personalize drug treatment* so that a physician can know in advance whether an individual will benefit from a particular drug (**FIGURE 12.14**). This approach might also be used to reduce the incidence of adverse drug reactions in individuals who metabolize particular drugs slowly.

PROTEOMICS Comparisons of the proteomes of humans and other eukaryotic organisms have revealed a common set of

All patients with the same diagnosis

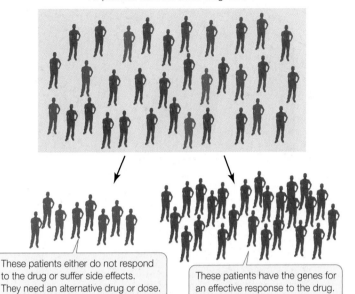

These patients either do not respond to the drug or suffer side effects. They need an alternative drug or dose.

These patients have the genes for an effective response to the drug.

FIGURE 12.14 Pharmacogenomics Correlations between genotypes and responses to drugs will help physicians develop more personalized medical care.

proteins that can be categorized into groups (families) with similar amino acid sequences and similar functions. Forty-six percent of the yeast proteome, 43 percent of the worm proteome, and 61 percent of the fruit fly proteome are shared by the human proteome. Functional analyses indicate that this set of 1,300 proteins provides the basic metabolic functions of a eukaryotic cell—including glycolysis, the citric acid cycle, membrane transport, protein synthesis and targeting, and DNA replication.

Of course, these are not the only human proteins. There are many more, which presumably distinguish us as *human* eukaryotic organisms. As we describe elsewhere, proteins have different functional regions called domains—for example, a protein may have a ligand-binding domain and a membrane-spanning domain. While a particular organism may have many unique proteins, those proteins are often just unique combinations of domains that exist in other organisms. For example, a ligand-gated channel might have its ligand-binding domain

swapped for another with a different specificity, thereby changing the signal that opens the channel. *This reshuffling of the genetic deck is a key to evolution.*

There is considerable interest in using proteomics in the diagnosis of diseases. For diseases caused by a single gene, examining the single protein involved is possible (e.g., the hemoglobins; see Figure 10.2). But for more complex diseases such as diabetes or cancer, many genes and proteins may be involved, and the *pattern* of proteins made in a particular tissue at a certain time might indicate the presence or likelihood of the disease. Proteomic analyses of tissues (including blood) are being developed to provide early warnings of the presence of particular diseases before symptoms occur.

METABOLOMICS There has been some progress in defining the human metabolome. A database created by David Wishart and colleagues at the University of Alberta contains over 6,500 metabolite entries. The challenge now is to relate the levels of these substances to physiology. For example, high levels of glucose in the blood are associated with diabetes, and there may be patterns of metabolites that are diagnostic of other diseases. This could aid in early diagnosis and treatment.

DNA fingerprinting uses short tandem repeats

As noted in Concept 12.3, short tandem repeats (STRs) are blocks of 1–5 bp that can be repeated up to 100 times at particular locations on chromosomes. Since the number of repeats can vary widely, there are usually numerous alleles for a particular STR. For example, at a particular location on human chromosome 15 there might be an STR of "AGG." An individual might inherit an allele with six copies of the repeat (AGGAGGAGGAGGAGGAGG) from her mother and an allele with two copies (AGGAGG) from her father (**FIGURE 12.15A**). PCR can be used to amplify DNA fragments containing these repeat sequences, and the number of copies of the repeat can be determined by sizing the DNA fragments.

DNA fingerprinting refers to a group of techniques used to identify particular individuals by their DNA; the most common of these techniques involves STR analysis. When several different STR loci are analyzed, an individual's unique pattern becomes apparent. The U.S. Federal Bureau of Investigation

APPLY THE CONCEPT

The human genome sequence has many applications

It is the year 2025. You are taking care of a patient who is worried that he may have an early stage of kidney cancer. His mother died from this disease.

1. Assume that the SNPs linked to genes involved in the development of this type of cancer have been identified. How would you determine if this man has a genetic predisposition for developing kidney cancer? Explain how you would do the analysis.

2. How might you develop a metabolomic profile for kidney cancer and then use it to determine whether your patient has kidney cancer?

3. If the patient is diagnosed with the cancer, how would you use pharmacogenomics to choose the right medications to treat his tumor?

(A)

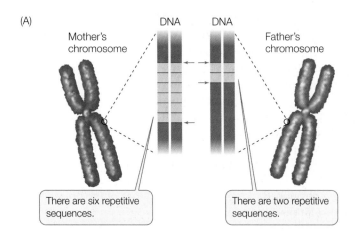

There are six repetitive sequences.

There are two repetitive sequences.

uses 13 STR loci in its Combined DNA Index System (CODIS) database.

DNA fingerprinting is used to resolve questions of paternity, and in forensics (crime investigations) to identify criminals. It also has other uses—for example, it can help in analyses of historical events. In 1918 the Russian tsar Nicholas II and his wife and five children were killed during the Communist revolution. A report that the bodies had been burned was not questioned until 1991, when a shallow grave with several skeletons was discovered a few miles from the presumed execution site. The remains were from a man, a woman, and three female children, and STR analysis indicated that they were all related to one another (**FIGURE 12.15B**). These STR patterns were also related to living descendents of the tsar. The accuracy and specificity of this method gave historical and cultural closure to a major event of the twentieth century.

Do You Understand Concept 12.4?

- The average human gene spans 27,000 bp. The average human polypeptide has 300 amino acids. Explain.

- What is a haplotype with regard to an STR? How can haplotypes be used to relate DNA to a phenotype?

- A person has a rare allele for an STR (STR-1) that has a frequency in the population of 1 percent (0.01). The same person has allele frequencies for other STRs as follows: STR-2, 0.005; STR-3, 0.01; STR-4, 0.05; STR-5, 0.01. What is the probability that an individual will have all of these alleles? What does this mean in terms of identifying an individual with this genotype?

 QA **QUESTION** What does genome sequencing reveal about dogs and other animals?

ANSWER In the opening story of this chapter we described how genome sequencing is being applied to breeds of dogs. For example, next-generation sequencing methods (Concept 12.1) have allowed biologists to collect data on genes that control body size. This has led to the identification

(B)

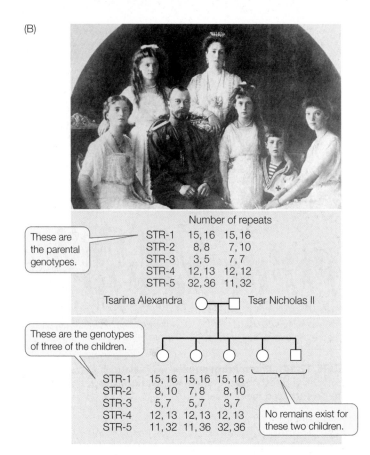

These are the parental genotypes.

	Number of repeats	
STR-1	15, 16	15, 16
STR-2	8, 8	7, 10
STR-3	3, 5	7, 7
STR-4	12, 13	12, 12
STR-5	32, 36	11, 32

Tsarina Alexandra   Tsar Nicholas II

These are the genotypes of three of the children.

STR-1	15, 16	15,16	15, 16
STR-2	8, 10	7, 8	8, 10
STR-3	5, 7	5, 7	3, 7
STR-4	12, 13	12, 13	12, 13
STR-5	11, 32	11, 36	32, 36

No remains exist for these two children.

FIGURE 12.15 DNA Fingerprinting (A) A short tandem repeat (STR) can occur in a specific, inherited pattern. (B) STR analyses were used to determine that bony remains were from one family, and other evidence pointed to the Russian tsar and his family.

of a SNP (Concept 12.4) in the gene for *insulin-like growth factor 1* (IGF-1) that is important in determining size. Large breeds have an allele that encodes an active IGF-1, and small breeds have a different allele that encodes a less active version of the protein. In humans, IGF-1 mediates the overall effects of growth hormone, and people with a mutation in the IGF-1 gene have short stature.

Another gene important to phenotypic variation is found in whippets, sleek dogs that run fast and are often raced. A mutation in the gene for myostatin, a protein that inhibits the overdevelopment of muscles, results in a whippet that is more muscular and runs faster (**FIGURE 12.16A**). Comparative genomics shows that this gene is important in other animals as well. In Belgian Blue cattle, individuals homozygous for a particular myostatin SNP have huge muscles (**FIGURE 12.16B**). There is interest in applying this knowledge to humans. For example, in muscular dystrophy the skeletal muscles waste away, and blocking myostatin could be useful in keeping muscles robust. Athletes anxious to have bulkier muscles have also been focusing on this gene and its protein product.

Inevitably, some scientists have set up companies to test dogs for genetic variations using DNA supplied by anxious owners and breeders. The black dog rescued from the pound that looks like a Labrador retriever may turn out to be a German pointer. Some traditional breeders frown on this practice, but others say it will bring more joy (and prestige) to owners.

(A)

(B)

FIGURE 12.16 Muscular Gene (A) These dogs are both whippets, but the muscle-bound dog (left) has a mutation in a gene that normally limits muscle buildup. (B) A similar mutation in Belgian Blue cattle also leads to overgrown muscles.

12 SUMMARY

concept 12.1 There Are Powerful Methods for Sequencing Genomes and Analyzing Gene Products

- To sequence a genome, the chromosomes are cut into overlapping fragments, which are sequenced. Then the fragment sequences are lined up to assemble the DNA sequence of the chromosome.

- **Next-generation sequencing** involves attaching short, single-stranded DNA fragments to a solid surface. A primer and DNA polymerase are added, and tagged nucleotides are detected by a camera as they are added to the complementary DNA strand. Many sequences can be done in parallel. **Review Figure 12.1 and ANIMATED TUTORIAL 12.1**

- The analysis of DNA sequences is done by computer. Genomic sequences include protein-coding genes, RNA genes, transcription control elements, and repeated sequences. **Review Figure 12.3**

- The **proteome** is the total protein content of an organism. It can be analyzed using chemical methods that separate and identify proteins. These include two-dimensional electrophoresis and mass spectrometry. **Review Figure 12.4**

- The **metabolome** is the total content of small molecules in a tissue under particular conditions. These molecules include intermediates in metabolism, hormones and other signaling molecules, and secondary metabolites. **Review Figure 12.5**

concept 12.2 Prokaryotic Genomes Are Relatively Small and Compact

- Prokaryotic genomes have been studied using **functional genomics** to determine the roles of various parts of the genome, including the protein-coding genes. **Comparative genomics** is used to compare sequences among organisms. **Review Table 12.1**

- **Transposons** are sequences of DNA that can move about the genome. **Review Figure 12.6**

- **Metagenomics** is the identification of DNA sequences in environmental samples without first isolating, growing, and identifying the organisms. **Review Figure 12.7**

- Transposon mutagenesis can be used to inactivate genes one by one. Then the organism can be tested for survival. In this way, a minimal genome can be identified. **Review Figure 12.8**

concept 12.3 Eukaryotic Genomes Are Large and Complex

- Sequences from model organisms have highlighted some common features of eukaryotic genomes. In addition, there are specialized genes such as those for cellular compartmentalization, development, and features unique to plants. **Review Figures 12.9 and 12.10 and Table 12.2**

- Some genes exist as members of **gene families**. Proteins may be made from these closely related genes at different times and in different tissues. **Review Figure 12.11**

- Eukaryotic genomes contain various kinds of repeated sequences. **Review Table 12.3**

concept 12.4 The Human Genome Sequence Has Many Applications

- The haploid human genome has 3.2 billion bp.

- Only 1.5 percent of the genome codes for proteins; much of the rest consists of repeated sequences.

- Virtually all human genes have introns, and alternative splicing leads to the production of more than one protein per gene.

- Genotyping using **single nucleotide polymorphisms** (SNPs) can be used to correlate variations in the genome with diseases or drug sensitivity. It may lead to personalized medicine. **Review Figure 12.13**

- **Pharmacogenomics** is the analysis of how a person's genetic make-up affects his or her drug metabolism. **Review Figure 12.14**

- **Short tandem repeats** (STRs) are DNA sequences that are variable in length. They can be used to identify individuals. **Review Figure 12.15**

See WEB ACTIVITY 12.1 for a concept review of this chapter.

Biotechnology — 13

The United Nations defines **biotechnology** as "any technological application that uses biological systems, living organisms, or derivatives thereof to make or modify products or processes." This broad definition includes major human activities such as brewing beer (see Chapter 6) and the domestication of animals and plants (see Chapter 12). More recently, biotechnology has become associated with the genetic modification of microorganisms for the production of particular substances, and of a variety of plants and animals used in agriculture.

Industrial biotechnology began in a lab in England in 1917 during World War I. The production of cordite, an explosive used to propel a bullet or shell to its target, required the solvent acetone, $(CH_3)_2CO$. But acetone was mostly manufactured by England's enemy, Germany. A microbiologist of the University of Manchester, Chaim Weizmann, found that if the bacterium *Clostridium acetylbutylicum* was grown using starch as an energy source, it produced abundant quantities of acetone. The British government set up a factory to grow large vats of these bacteria, and the cordite shortage was solved.

The contemporary era of biotechnology as a major industry dates from June 16, 1980, on the steps of the U.S. Supreme Court. In this case, scientists were studying bacteria not for their ability to make something, but to break it down. Many bacteria have genes that code for unusual enzymes and biochemical pathways, and they can use all sorts of substances as nutrients, including pollutants. Scientists have discovered these organisms simply by mixing polluted soil with water and seeing what grows. In 1971, Ananda Chakrabarty at the General Electric Research Center in New York used genetic crosses to develop a single strain of the bacterium *Pseudomonas* that carried genes for the

breakdown of various hydrocarbons in oil. He and his company applied for a patent to legally protect their discovery and profit from it. In a landmark case, the U.S. Supreme Court ruled in 1980 that "a live, human-made microorganism is patentable" under the U.S. Constitution.

The Supreme Court ruling came at a time when new laboratory methods were being developed to insert specific DNA sequences into organisms by recombinant DNA technology. Since then, an entirely new biotechnology industry has sprung up. The resulting flood of patents for DNA sequences and genetically modified organisms, some of them developed to improve the environment, continues to this day.

Ananda Chakrabarty received the first patent for a genetically modified organism, a bacterium that breaks down crude oil.

How is biotechnology used to alleviate environmental problems?

KEY CONCEPTS

13.1 Recombinant DNA Can Be Made in the Laboratory

13.2 DNA Can Genetically Transform Cells and Organisms

13.3 Genes and Gene Expression Can Be Manipulated

13.4 Biotechnology Has Wide Applications

concept
13.1 Recombinant DNA Can Be Made in the Laboratory

Biotechnology began with the use of organisms with genetic capabilities that occur in nature. For example, existing biochemical pathways in yeast are used to make alcohol, and as noted in the opening story, bacteria can be grown in large quantities to make acetone and other industrial chemicals. More recently, it has become possible to genetically modify organisms with genes from other, distantly related organisms, to create new combinations of genes that would not otherwise occur in the same cell. This technology involves the use of **recombinant DNA**: single molecules containing DNA sequences from two or more organisms. Biologists have relied on manipulating natural molecules and processes to develop the techniques for making recombinant DNA. Three key tools are widely used:

- Restriction enzymes for cutting DNA into pieces (fragments) that can be manipulated
- Gel electrophoresis for the analysis and purification of DNA fragments
- DNA ligase for joining DNA fragments together in novel combinations

Restriction enzymes cleave DNA at specific sequences

All organisms, including bacteria, must have ways of dealing with their enemies. As we described in Concept 9.1, bacteria are attacked by viruses called bacteriophage (or phage, for short). These viruses inject their genetic material into the host cells and turn them into virus-producing factories, eventually killing the cells. Some bacteria defend themselves against such invasions by producing **restriction enzymes** (also known as restriction endonucleases), which cut double-stranded DNA molecules—such as those injected by bacteriophage—into smaller, noninfectious fragments (**FIGURE 13.1**). These enzymes break the bonds of the DNA backbone between the 3′ hydroxyl group of one nucleotide and the 5′ phosphate group of the next nucleotide.

There are many different restriction enzymes, each of which cleaves DNA at a specific sequence of bases called a **recognition sequence** or a **restriction site**. Most recognition sequences are 4–6 base pairs (bp) long. As examples, here are three enzymes that each recognize a different 6-bp sequence and cleave the DNA as indicated by the red arrows:

$$
\begin{array}{lll}
BamHI & 5'\cdots \text{G\;G\;A\;T\;C\;C} \cdots 3' \\
& 3'\cdots \text{C\;C\;T\;A\;G\;G} \cdots 5'
\end{array}
$$

$$
\begin{array}{lll}
HindIII & 5'\cdots \text{A\;A\;G\;C\;T\;T} \cdots 3' \\
& 3'\cdots \text{T\;T\;C\;G\;A\;A} \cdots 5'
\end{array}
$$

$$
\begin{array}{lll}
EcoRI & 5'\cdots \text{G\;A\;A\;T\;T\;C} \cdots 3' \\
& 3'\cdots \text{C\;T\;T\;A\;A\;G} \cdots 5'
\end{array}
$$

Note that each of these sequences forms a *palindrome*: the opposite strands have the same sequences when they are read

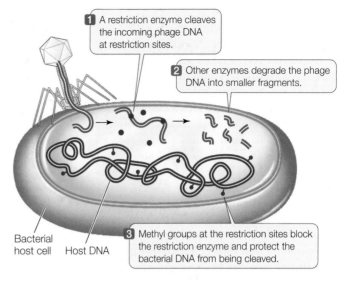

FIGURE 13.1 Bacteria Fight Invading Viruses by Making Restriction Enzymes

1 A restriction enzyme cleaves the incoming phage DNA at restriction sites.

2 Other enzymes degrade the phage DNA into smaller fragments.

3 Methyl groups at the restriction sites block the restriction enzyme and protect the bacterial DNA from being cleaved.

Bacterial host cell   Host DNA

from their 5′ ends. This is similar to a palindromic word such as "racecar," which is the same when read in either direction. Restriction enzymes have two identical active sites on two subunits, which cleave the two strands simultaneously. The names of these enzymes reflect their origins; for example *EcoRI* was found in *E. coli*, and *BamHI* in the bacterium *Bacillus amyloliquifaciens*.

Note also that these enzymes cut the two DNA strands in such a way that there will be a short sequence of single-stranded DNA at each cut end. Many restriction enzymes create these single-stranded overhangs, which are referred to as "sticky ends" because they are able to form hydrogen bonds with complementary sequences on other DNA molecules. Other restriction enzymes make cuts directly opposite one another on the two DNA strands, creating "blunt ends."

The bacterial cell protects itself from its own restriction enzymes by modifying the restriction sites in its own DNA. Specific modifying enzymes called *methylases* add methyl ($-CH_3$) groups to certain bases at the restriction sites after the chromosome has been replicated (see Figure 11.13). The restriction enzymes do not recognize or cut the methylated restriction sites in the host's DNA. But unmethylated phage DNA is efficiently recognized and cleaved.

The *EcoRI* recognition sequence occurs, on average, about once in every 4,000 bp in a typical prokaryotic genome, or about once per four prokaryotic genes. So *EcoRI* can chop a large piece of DNA into smaller pieces containing, on average, just a few genes. When *EcoRI* is used in the laboratory to cut a small genome such as that of a virus with, say, 50,000 bp, only a few fragments are obtained. For a huge eukaryotic chromosome with tens of millions of bp, a very large number of fragments is created.

Of course, "on average" does not mean that the enzyme cuts all stretches of DNA at regular intervals. For example, the *EcoRI* recognition sequence does not occur even once in the 40,000 bp of the T7 phage genome—a fact that is crucial to the survival of this virus, since its host is *E. coli*. Fortunately for *E.*

coli, the *Eco*RI recognition sequence does appear in the DNA of other bacteriophage.

Hundreds of restriction enzymes (all with unique recognition sequences) have been purified from various microorganisms. In the laboratory, different restriction enzymes can be used to cut samples of DNA from the same source. The procedure of using restriction enzymes in the lab is called setting up a *restriction digest*. Thus restriction enzymes can be used to cut a sample of DNA in many different, specific places. When analyzed by gel electrophoresis (see below), the fragments formed can be used to create a physical map of the intact DNA molecule. Before DNA sequencing technology became automated and widely available, this was a way to physically map base sequences on DNA.

Gel electrophoresis separates DNA fragments

After a sample of DNA has been cut with a restriction enzyme, the fragments must be separated from each other to determine the number of fragments and the size (in bp) of each fragment. In this way an individual fragment can be identified, and it can then be purified for further analysis or for use in an experiment.

A convenient way to separate or purify DNA fragments is by **gel electrophoresis**. Samples containing the fragments are placed in wells at one end of a semisolid gel (usually made of agarose or polyacrylamide polymers), and an electric field is applied to the gel (**FIGURE 13.2**). Because of its phosphate groups, DNA is negatively charged at neutral pH; therefore, because opposite charges attract, the DNA fragments move through the gel toward the positive end of the field. Because the spaces between the polymers of the gel are small, small DNA molecules can move through the gel faster than larger ones. Thus, DNA fragments of different sizes separate from one another and can be detected with a dye. This gives us three types of information about a DNA sample:

- *The number of fragments.* The number of fragments produced by digestion of a DNA sample with a given restriction enzyme depends on how many times that enzyme's recognition sequence occurs in the sample. Thus gel electrophoresis can provide some information about the presence of specific DNA sequences in the DNA sample.

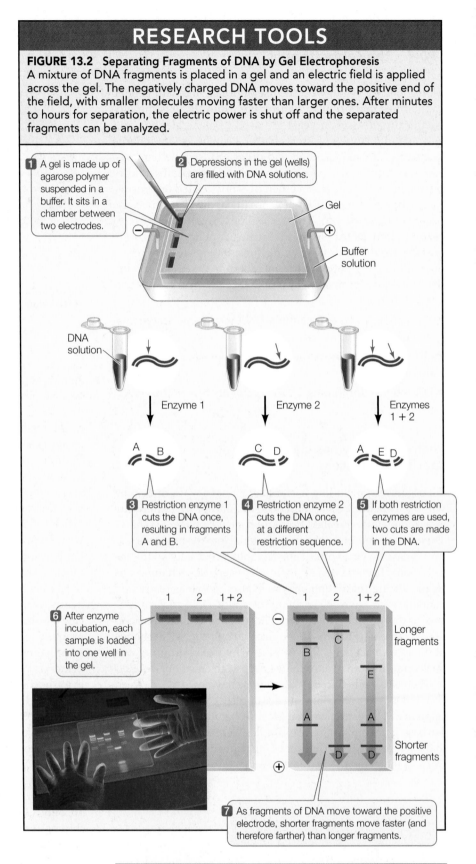

RESEARCH TOOLS

FIGURE 13.2 Separating Fragments of DNA by Gel Electrophoresis
A mixture of DNA fragments is placed in a gel and an electric field is applied across the gel. The negatively charged DNA moves toward the positive end of the field, with smaller molecules moving faster than larger ones. After minutes to hours for separation, the electric power is shut off and the separated fragments can be analyzed.

1 A gel is made up of agarose polymer suspended in a buffer. It sits in a chamber between two electrodes.

2 Depressions in the gel (wells) are filled with DNA solutions.

Gel

Buffer solution

DNA solution

Enzyme 1   Enzyme 2   Enzymes 1 + 2

A B   C D   A E D

3 Restriction enzyme 1 cuts the DNA once, resulting in fragments A and B.

4 Restriction enzyme 2 cuts the DNA once, at a different restriction sequence.

5 If both restriction enzymes are used, two cuts are made in the DNA.

6 After enzyme incubation, each sample is loaded into one well in the gel.

1   2   1+2   1   2   1+2

Longer fragments

Shorter fragments

7 As fragments of DNA move toward the positive electrode, shorter fragments move faster (and therefore farther) than longer fragments.

yourBioPortal.com
Go to ANIMATED TUTORIAL 13.1
Separating Fragments of DNA by Gel Electrophoresis

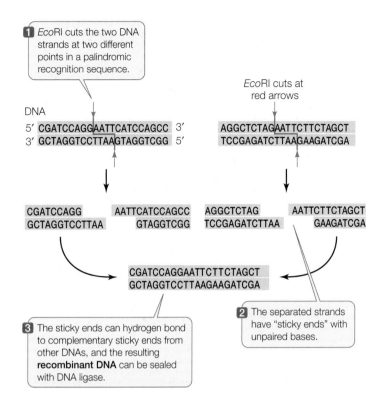

1 EcoRI cuts the two DNA strands at two different points in a palindromic recognition sequence.

DNA

5′ CGATCCAGGAATTCATCCAGCC 3′
3′ GCTAGGTCCTTAAGTAGGTCGG 5′

EcoRI cuts at red arrows

AGGCTCTAGAATTCTTCTAGCT
TCCGAGATCTTAAGAAGATCGA

CGATCCAGG   AATTCATCCAGCC
GCTAGGTCCTTAA   GTAGGTCGG

AGGCTCTAG   AATTCTTCTAGCT
TCCGAGATCTTAA   GAAGATCGA

CGATCCAGGAATTCTTCTAGCT
GCTAGGTCCTTAAGAAGATCGA

2 The separated strands have "sticky ends" with unpaired bases.

3 The sticky ends can hydrogen bond to complementary sticky ends from other DNAs, and the resulting **recombinant DNA** can be sealed with DNA ligase.

FIGURE 13.3 Cutting, Splicing, and Joining DNA Many restriction enzymes (EcoRI is shown here) make staggered cuts in DNA. EcoRI can be used to cut two different DNA molecules (blue and orange). The exposed bases can hydrogen bond with complementary exposed bases on other DNA fragments, forming recombinant DNA. DNA ligase stabilizes the recombinant molecule by forming covalent phosphodiester bonds in the DNA backbone.

After separation on the gel, a slice of gel containing the desired DNA fragment (identified by its size) can be cut out and then be purified by one of a variety of methods. This fragment can then be analyzed to determine its sequence or used to make recombinant DNA.

Recombinant DNA can be made from DNA fragments

Another enzyme that is involved in DNA metabolism in cells is **DNA ligase**, which catalyzes the joining of DNA fragments by making phosphodiester bonds between them. This is the enzyme that joins Okazaki fragments during DNA replication (see Concept 9.2). With restriction enzymes (which *break* bonds) and DNA ligase (which *makes* bonds), scientists can cut DNA into fragments and then splice them together in new combinations as recombinant DNA. As shown in **FIGURE 13.3**, two fragments with complementary sticky ends first join by hydrogen bonding, and then the DNA ligase forms a phosphodiester bond in each strand, making a single intact DNA molecule.

In the early 1970s, Stanley Cohen and Herbert Boyer wondered whether recombinant DNA could be a functional carrier of genetic information. They used restriction enzymes to cut sequences from two *E. coli* plasmids (small circular DNAs; see Concept 8.4) containing different antibiotic resistance genes. Then they used DNA ligase to join the fragments together. The resulting plasmid, when inserted into new *E. coli* cells, gave those cells resistance to both antibiotics (**FIGURE 13.4**). A new era of biotechnology was born.

- *The sizes of the fragments.* DNA fragments of known size are often placed in one well of the gel to provide a standard for comparison. This tells us how large the DNA fragments in the other wells are. By comparing the fragment sizes obtained with two or more restriction enzymes, the locations of their recognition sites relative to one another can be worked out (mapped).

- *The relative abundance of a fragment.* In many experiments, the investigator is interested in how much DNA is present. The relative *intensity* of a band produced by a specific fragment can indicate the amount of that fragment.

APPLY THE CONCEPT

Recombinant DNA can be made in the laboratory

The specificity of restriction enzyme recognition can be used to detect mutations. For example, the enzyme *Mst*II cuts DNA at CCTNAGG, where N is any base. Around the sixth codon in the β-globin gene is the sequence CCTGAGGAG. In addition to this *Mst*II site, there are additional sites on each side of the sequence, such that when *Mst*II is used to cut human DNA in this region, two fragments of 1.15 and 0.20 kilobase (one kilobase = 1,000 bp) are obtained. The sickle allele of the β-globin gene causes sickle-cell anemia when it occurs in the homozygous state. In this allele, the sequence around the sixth codon is mutated to CCTGTGGAG.

1. Identify the point mutation that led to the sickle allele.

2. What fragment(s) would result when *Mst*II is used to cut DNA with the sickle allele?

3. Sketch the patterns of cuts with normal and sickle alleles on a gel.

4. Could an individual have both patterns? Explain.

5. How can this information be used to make a DNA test for these alleles?

INVESTIGATION

FIGURE 13.4 Recombinant DNA With the discovery of restriction enzymes and DNA ligase, it became possible to combine DNA fragments from different sources in the laboratory. But would such "recombinant DNA" be functional when inserted into a living cell? The results of this experiment completely changed the scope of genetic research, increasing our knowledge of gene structure and function, and ushering in the new field of biotechnology.

HYPOTHESIS

Biologically functional recombinant chromosomes can be made in the laboratory.

METHOD *E. coli* plasmids carrying a gene for resistance to either the antibiotic kanamycin (*kan^r*) or tetracycline (*tet^r*) are cut with a restriction enzyme.

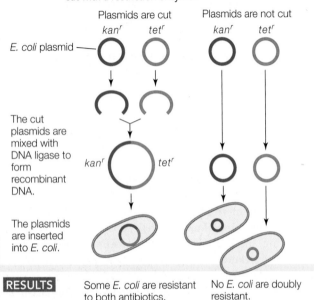

The cut plasmids are mixed with DNA ligase to form recombinant DNA.

The plasmids are inserted into *E. coli*.

RESULTS Some *E. coli* are resistant to both antibiotics. No *E. coli* are doubly resistant.

CONCLUSION

Two DNA fragments with different genes can be joined to make a recombinant DNA molecule, and the resulting DNA is functional.

ANALYZE THE DATA

Two plasmids were used in this study: pSC101 had a gene for resistance to tetracycline and pSC102 had a gene for resistance to kanamycin. Equal quantities of the plasmids—either intact, cut with *Eco*RI, or cut with *Eco*RI and then sealed with DNA ligase—were mixed and incubated with antibiotic-sensitive *E. coli*. The *E. coli* were then grown on various combinations of the antibiotics. Here are the results:

DNA treatment	Number of resistant colonies		
	Tetracycline only	Kanamycin only	Both antibiotics
None	200,000	100,000	200
*Eco*RI cut	10,000	1,100	70
*Eco*RI, then ligase	12,000	1,300	570

A. Did treatment with *Eco*RI affect the transformation efficiency? Explain.
B. Did treatment with DNA ligase affect the transformation efficiency of each cut plasmid? Which quantitative data support your answer?
C. How did doubly antibiotic-resistant bacteria arise in the "none" treatment? (Hint: see Concept 9.3.)
D. Did the *Eco*RI followed by ligase treatment increase the appearance of doubly antibiotic-resistant bacteria? What data support your answer?

For more, go to Working with Data 13.1 at **yourBioPortal.com**.

Go to **yourBioPortal.com** for original citations, discussions, and relevant links for all INVESTIGATION figures.

Do You Understand Concept 13.1?

- Using diagrams of the chemical structure of DNA (see Concepts 3.1 and 9.2), compare the actions of a restriction enzyme and DNA ligase.
- How is recombinant DNA technology different from genetic recombination that occurs in meiosis?
- A DNA molecule of 12,000 bp (12 kilobase, or kb) is cut by restriction enzymes as follows:

CONDITION	SIZES OF FRAGMENTS (KB)
Enzyme A	2, 10
Enzyme B	2, 10
Enzymes A + B	2, 8

  a. Sketch the results of the three cuts on an electrophoresis gel.
  b. Indicate on a linear map where each enzyme cuts the DNA.

With the tools described in this concept—restriction enzymes, gel electrophoresis, and DNA ligase—scientists can cut and rejoin different DNA molecules from any and all sources, including artificially synthesized DNA sequences. In the next concept we will examine some of the ways these recombinant DNA molecules are used.

concept 13.2 DNA Can Genetically Transform Cells and Organisms

One goal of recombinant DNA technology is to **clone**—produce many identical copies of—a particular DNA sequence. Cloning might be done for sequence analysis, to produce a protein product in quantity, or as a step toward creating an organism with a new phenotype. Recombinant DNA can be cloned by inserting it into host cells in a process known as **transformation** (or **transfection** if the host cells are derived from an animal). A host cell or organism that contains recombinant DNA is referred to as a **transgenic** cell or organism. Later in this chapter we will encounter many examples of transgenic cells and organisms, including yeast, mice, rice plants, and even cattle.

Various methods are used to create transgenic cells. Generally, *only a few* of the cells that are exposed to the recombinant DNA actually become transformed with it. In order to grow only the transgenic cells, **selectable marker** genes, such as genes that confer resistance to antibiotics, are often included as part of the recombinant DNA molecule. Antibiotic resistance genes were the markers used in Cohen and Boyer's experiment (see Figure 13.4).

Genes can be inserted into prokaryotic or eukaryotic cells

In theory, any cell or organism can act as a host for the introduction of recombinant DNA. Most research has been done using model organisms:

- *Bacteria* are easily grown and manipulated in the laboratory. Much of their molecular biology is known, especially for well-studied bacteria such as *E. coli*. Furthermore, bacteria contain plasmids, which are easily manipulated to carry recombinant DNA into the cell. Because the processes of transcription and translation proceed differently in prokaryotes than they do in eukaryotes, however, bacteria might not be suitable as hosts to express eukaryotic genes.

- *Yeasts* such as *Saccharomyces* are commonly used as eukaryotic hosts for recombinant DNA studies. The advantages of using yeasts include rapid cell division (a life cycle completed in 2–8 hours), ease of growth in the laboratory, and a relatively small genome size. In addition, yeasts have most of the characteristics of other eukaryotes, except for those characteristics involved in multicellularity.

- *Plant cells* are good hosts because of their ability to make stem cells (unspecialized, totipotent cells) from mature plant tissues. The unspecialized cells can be transformed with recombinant DNA and then studied in culture, or grown into new plants. There are also methods for making whole transgenic plants without going through the cell culture step. These methods result in plants that carry the recombinant DNA in all their cells, including the germline cells.

- *Cultured animal cells* can be used to study expression of human or animal genes, for example for medical purposes. Whole transgenic animals can also be created.

Recombinant DNA enters host cells in a variety of ways

Methods for inserting DNA into host cells vary. The cells may be chemically treated to make their outer membranes more permeable, and then mixed with the DNA so it can diffuse into the cells. Another approach is called *electroporation*: a short electric shock is used to create temporary pores in the membranes through which the DNA can enter. Viruses and bacteria can be altered so that they carry or insert recombinant DNA into cells. Transgenic animals can be produced by injecting recombinant DNA into the nuclei of fertilized eggs. There are even "gene guns," which "shoot" the host cells with tiny particles carrying the DNA.

The challenge of inserting new DNA into a cell lies not just in getting it into the host cell, but in getting it to replicate as the host cell divides. DNA polymerase does not bind to just any sequence. If the new DNA is to be replicated, it must become part of a segment of DNA that contains an origin of replication (see Concept 9.2). Such a DNA molecule is called a **replicon**, or replication unit.

There are two general ways in which the newly introduced DNA can become part of a replicon within the host cell:

- It may be inserted into a host chromosome. Although the site of insertion is usually random, this is nevertheless a common method of integrating new genes into host cells.

- It can enter the host cell as part of a carrier DNA sequence, called a **vector**, and can either integrate into the host chromosome or have its own origin of DNA replication.

Several types of vectors are used to get DNA into cells.

PLASMIDS AS VECTORS As we described in Concept 8.4, plasmids are small, circular DNA molecules that replicate autonomously in many prokaryotic cells. A number of characteristics make plasmids useful as transformation vectors:

- They are relatively small (an *E. coli* plasmid usually has 2,000–6,000 bp) and therefore easy to manipulate in the laboratory.

- A plasmid will usually have one or more restriction enzyme recognition sequences that each occur only once in the plasmid sequence. These sites make it easy to insert additional DNA into the plasmid before it is used to transform host cells.

- Many plasmids contain genes that confer resistance to antibiotics, and thus can serve as selectable markers.

- Plasmids have a bacterial origin of replication (*ori*) and can replicate independently of the host chromosome. It is not uncommon for a bacterial cell to contain hundreds of copies of a recombinant plasmid. For this reason, the power of bacterial transformation to amplify a gene is extraordinary. A 1-liter culture of bacteria harboring the human β-globin gene in a typical plasmid has as many copies of that gene as there are cells in a typical adult human (10^{14}).

The plasmids used as vectors in the laboratory have been extensively altered to include convenient features: multiple cloning sites with 20 or more unique restriction enzyme sites for cloning purposes; origins of replication for a variety of host cells; and various kinds of reporter genes (see below) and selectable marker genes:

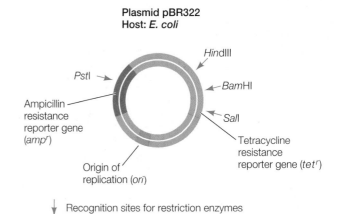

Plasmid pBR322
Host: *E. coli*

*Hind*III
*Pst*I
*Bam*HI
Ampicillin resistance reporter gene (*amp^r*)
*Sal*I
Tetracycline resistance reporter gene (*tet^r*)
Origin of replication (*ori*)

↓ Recognition sites for restriction enzymes

PLASMID VECTORS FOR PLANTS An important vector for carrying new DNA into many types of plants is a plasmid found in the bacterium *Agrobacterium tumefaciens*. This bacterium lives in the soil, infects plants, and causes a disease called crown gall, which is characterized by the presence of growths (or tumors) on the plant. *A. tumefaciens* contains a plasmid called Ti (for *tumor-inducing*):

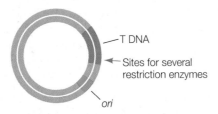

Ti plasmid
Hosts: *Agrobacterium tumefaciens* (plasmid)
and infected plants (T DNA)

The Ti plasmid carries genes that allow the bacterium to infect plant cells and then insert a region of its DNA called the T DNA into the chromosomes of infected cells. The T DNA contains genes that cause the growth of tumors and the production of specific sugars that the bacterium uses as sources of energy. Scientists have exploited this remarkable natural "genetic engineer" to insert foreign DNA into the genomes of plants.

When used as a vector for plant transformation, the tumor-inducing and sugar-producing genes on the T DNA are removed and replaced with foreign DNA. The altered Ti plasmids are first used to transform *Agrobacterium* cells from which the original Ti plasmids have been removed. Then the *Agrobacterium* cells are used to infect plant cells.

VIRUSES AS VECTORS Constraints on plasmid replication limit the size of the new DNA that can be inserted into a plasmid to about 10,000 bp. Although many prokaryotic genes may be smaller than this, most eukaryotic genes—with their introns and extensive flanking sequences—are bigger. A vector that accommodates larger DNA inserts is needed for these genes.

Both prokaryotic and eukaryotic viruses are often used as vectors for eukaryotic DNA. Bacteriophage λ, which infects *E. coli*, has a DNA genome of about 45,000 bp. If the genes that cause the host cell to die and lyse—about 20,000 bp—are eliminated, the virus can still attach to a host cell and inject its DNA. The deleted 20,000 bp can be replaced with DNA from another organism. Because viruses infect cells naturally, they offer a great advantage over plasmids, which often require artificial means to coax them to enter host cells.

FRONTIERS Some human diseases, for example certain cancers and genetic diseases, are caused by deficiencies in the expression of a specific gene. Such diseases could theoretically be treated by gene therapy—the insertion of functional copies of the affected gene into the patient's cells. Viral vectors that infect human cells are being developed for gene therapy. Success has been limited so far, but the research continues.

Reporter genes are used to identify host cells containing recombinant DNA

Even when a population of host cells interacts with an appropriate vector, only a small proportion of the cells actually take up the vector. Furthermore, the process of making recombinant DNA is far from perfect. After a ligation reaction, not all the

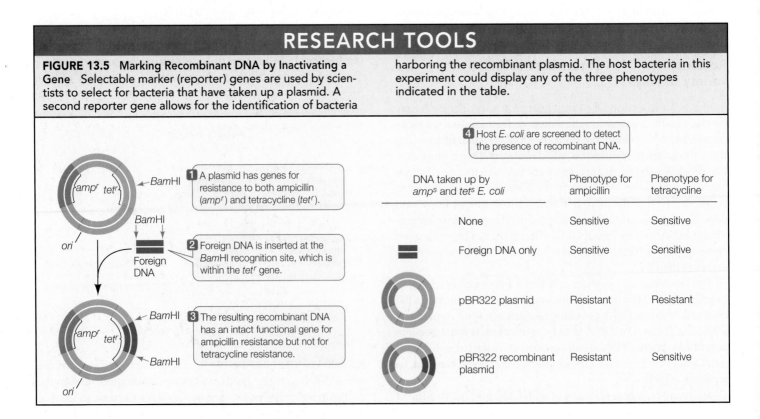

RESEARCH TOOLS

FIGURE 13.5 Marking Recombinant DNA by Inactivating a Gene Selectable marker (reporter) genes are used by scientists to select for bacteria that have taken up a plasmid. A second reporter gene allows for the identification of bacteria harboring the recombinant plasmid. The host bacteria in this experiment could display any of the three phenotypes indicated in the table.

1 A plasmid has genes for resistance to both ampicillin (*amp^r*) and tetracycline (*tet^r*).

2 Foreign DNA is inserted at the *Bam*HI recognition site, which is within the *tet^r* gene.

3 The resulting recombinant DNA has an intact functional gene for ampicillin resistance but not for tetracycline resistance.

4 Host *E. coli* are screened to detect the presence of recombinant DNA.

DNA taken up by *amp^s* and *tet^s* E. coli	Phenotype for ampicillin	Phenotype for tetracycline
None	Sensitive	Sensitive
Foreign DNA only	Sensitive	Sensitive
pBR322 plasmid	Resistant	Resistant
pBR322 recombinant plasmid	Resistant	Sensitive

FIGURE 13.6 Green Fluorescent Protein as a Reporter The presence of a plasmid with the gene for green fluorescent protein is readily apparent in transgenic cells because they glow under ultraviolet light.

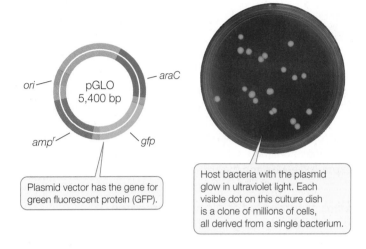

Plasmid vector has the gene for green fluorescent protein (GFP).

Host bacteria with the plasmid glow in ultraviolet light. Each visible dot on this culture dish is a clone of millions of cells, all derived from a single bacterium.

vector copies contain the foreign DNA. How can we identify or select the host cells that contain the vector with foreign DNA?

As we described above, selectable markers such as antibiotic resistance genes can be used to select cells containing those genes. Only cells carrying the antibiotic resistance gene can grow in the presence of that antibiotic. During the process of making recombinant DNA, a second antibiotic resistance gene can be used to identify cells that carry the vector *with the desired insert*. If the second antibiotic resistance gene is inactivated by the insertion of additional DNA, then cells carrying copies of the vector with the inserted DNA can be identified by their sensitivity to that antibiotic (**FIGURE 13.5**). Since the uptake of recombinant DNA is a rare event (only about 1 cell in 10,000 takes up a plasmid in such experiments), it is vital to be able to select the small number of cells harboring the recombinant DNA.

Selectable markers are one type of **reporter gene**, which is any gene whose expression is easily assayed. Other reporter genes code for proteins that can be detected visually. For example, green fluorescent protein, which normally occurs in the jellyfish *Aequorea victoria*, emits green light when exposed to ultraviolet light. The gene for this protein has been isolated and incorporated into vectors and is now widely used as a reporter gene (**FIGURE 13.6**).

Such reporters are not just used to select and identify cells carrying recombinant DNA. They can be attached to promoters in order to study how the promoters function under different conditions or in different tissues of a transgenic multicellular organism. They can also be attached to other proteins, to study how and where those proteins become localized within eukaryotic cells.

Do You Understand Concept 13.2?

- Outline the steps used to create recombinant DNA, transform host cells, and detect cells carrying the recombinant DNA.

- *Shuttle vectors* have the ability to transform both prokaryotic and eukaryotic cells. What sequences would you expect these vectors to have?

- What are the advantages of green fluorescent protein over antibiotic resistance as a marker on a plasmid for genetic transformation?

We have described how DNA can be cut, inserted into a vector, and introduced into host cells. We have also seen how host cells carrying recombinant DNA can be identified. Now let's consider the sources of DNA used for cloning, as well as some molecular methods for manipulating gene expression.

APPLY THE CONCEPT

DNA can genetically transform cells and organisms

The β-galactosidase (*lacZ*) gene in the *E. coli lac* operon (see Concept 11.2) encodes an enzyme that can convert the colorless substrate X-gal into a bright blue product. This product is clearly visible in a colony of bacteria carrying the *lacZ* gene. Colonies that do not have a functional *lacZ* gene are white when exposed to X-gal. A plasmid vector contains the *lacZ* gene with a restriction site for *Eco*RI within its coding sequence, and a gene for resistance (*r*) to the antibiotic ampicillin. In cloning experiments involving this vector, a strain of *E. coli* is used that carries no other functional *lacZ* gene and that is sensitive (*s*) to ampicillin. A biologist performs three transformations in a procedure designed to

clone a gene from a wheat plant into this vector. The *Eco*RI site in the *lacZ* gene is used for inserting the wheat DNA. Fill in the table with the results you would expect from each transformation.

DNA TAKEN UP BY *E. COLI*	PHENOTYPE FOR AMPICILLIN	PHENOTYPE FOR X-GAL (BLUE OR WHITE)
None		
Plasmid only		
Recombinant plasmid		

concept **13.3** ## Genes and Gene Expression Can Be Manipulated

Cloning and the formation of recombinant DNA are applications of the concepts of DNA as a double-stranded, base-paired molecule, and as the genetic material in model organisms such as bacteria and viruses. In this section, you can see these same concepts applied to other aspects of DNA manipulation, such as the selection or creation of DNAs for amplification, the detection of expressed genes, and the artificial regulation of gene expression.

DNA fragments for cloning can come from several sources

The goal of cloning by recombinant DNA techniques is often to elucidate the functions of specific DNA sequences, including regulatory and coding regions, and the proteins they encode. The DNA fragments used in cloning procedures are obtained from a number of sources. In many cases the first step is to create a *library* of DNA fragments: a collection of clones that can be searched for the gene or genes of interest, or analyzed in other ways to learn more about the original source of the DNA fragments.

GENOMIC LIBRARIES A **genomic library** is a collection of DNA fragments that together comprise the genome of an organism. This is the starting point for some methods of genome sequencing. Restriction enzymes or other means, such as mechanical shearing, can be used to break chromosomes into smaller pieces (**FIGURE 13.7A**). Each fragment is inserted into a vector, which is then taken up by a host cell. Proliferation of a single transformed cell in a selective medium (such as for antibiotic resistance) produces a colony of recombinant cells, each of which harbors many copies of the same fragment of DNA. The colonies are grown by spreading the transformed cells over a solid culture medium in petri dishes (small circular plates), which are incubated at a suitable temperature for the host cells to grow.

A single petri dish can hold thousands of bacterial colonies and is easily screened for the presence of a particular DNA sequence. Colonies containing that sequence are identified by DNA hybridization using a probe labeled with complementary fluorescent or radioactive nucleotides. To do this, the petri dish with its bacterial colonies is duplicated, and then the bacteria on one of the plates are treated to expose the DNA for hybridization (see Figure 10.7).

cDNA LIBRARIES A much smaller DNA library—one that includes only the genes transcribed in a particular tissue—can be made from **complementary DNA**, or **cDNA** (**FIGURE 13.7B**). This involves isolating mRNA from cells and making cDNA copies of that mRNA by complementary base pairing. The enzyme *reverse transcriptase* catalyzes this reaction. This collection of cDNAs from a particular tissue at a particular time in the life cycle of an organism is called a **cDNA library**, which is a "snapshot" of the transcription pattern of the cell. cDNA libraries have been invaluable for comparing gene expression in different tissues at different stages of development. For example, if cDNAs derived from developing red blood cells are examined, the globin sequences (encoding the subunits of hemoglobin) are prominent. But a cDNA library derived from hair follicles does not contain those sequences.

SYNTHETIC DNA In Concept 9.2 (see Figure 9.15) we described the polymerase chain reaction (PCR), a method of amplifying DNA in a test tube. PCR can begin with as little as 10^{-12} g (a picogram) of DNA. Any fragment of DNA can be amplified as long as appropriate primers are available. This amplified DNA can then be inserted into a plasmid to create recombinant DNA, and cloned in host cells. The *artificial synthesis* of DNA by organic chemistry

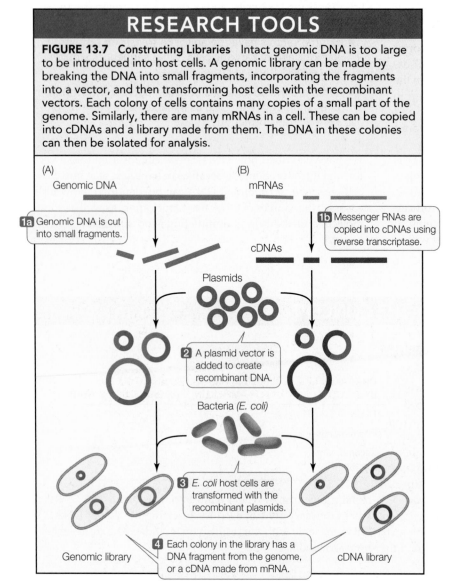

RESEARCH TOOLS

FIGURE 13.7 Constructing Libraries Intact genomic DNA is too large to be introduced into host cells. A genomic library can be made by breaking the DNA into small fragments, incorporating the fragments into a vector, and then transforming host cells with the recombinant vectors. Each colony of cells contains many copies of a small part of the genome. Similarly, there are many mRNAs in a cell. These can be copied into cDNAs and a library made from them. The DNA in these colonies can then be isolated for analysis.

(A) Genomic DNA

(B) mRNAs

1a Genomic DNA is cut into small fragments.

1b Messenger RNAs are copied into cDNAs using reverse transcriptase.

cDNAs

Plasmids

2 A plasmid vector is added to create recombinant DNA.

Bacteria (*E. coli*)

3 *E. coli* host cells are transformed with the recombinant plasmids.

Genomic library

4 Each colony in the library has a DNA fragment from the genome, or a cDNA made from mRNA.

cDNA library

is now fully automated. Synthetic oligonucleotides (single-stranded DNA fragments of 20 to 40 bp) are used as primers in PCR reactions. These primers can be designed to create short new sequences at the ends of the PCR products. This might be done to create a mutation in a recombinant gene, or to add restriction enzyme sites at the ends of the PCR product to aid in ligation reactions. Longer synthetic sequences can be pieced together to construct an *artificial gene* (or even an entire artificial genome; see the opening story of Chapter 4). If the amino acid sequence of a desired protein product is known, the genetic code can be used to determine the corresponding DNA sequence, which can be synthesized in the laboratory.

RESEARCH TOOLS

FIGURE 13.8 Making a Knockout Mouse Animals carrying mutations are rare. Homologous recombination is used to replace a normal mouse gene with an inactivated copy of that gene, thus "knocking out" the gene. Discovering what happens to a mouse with an inactive gene tells us much about the normal role of that gene.

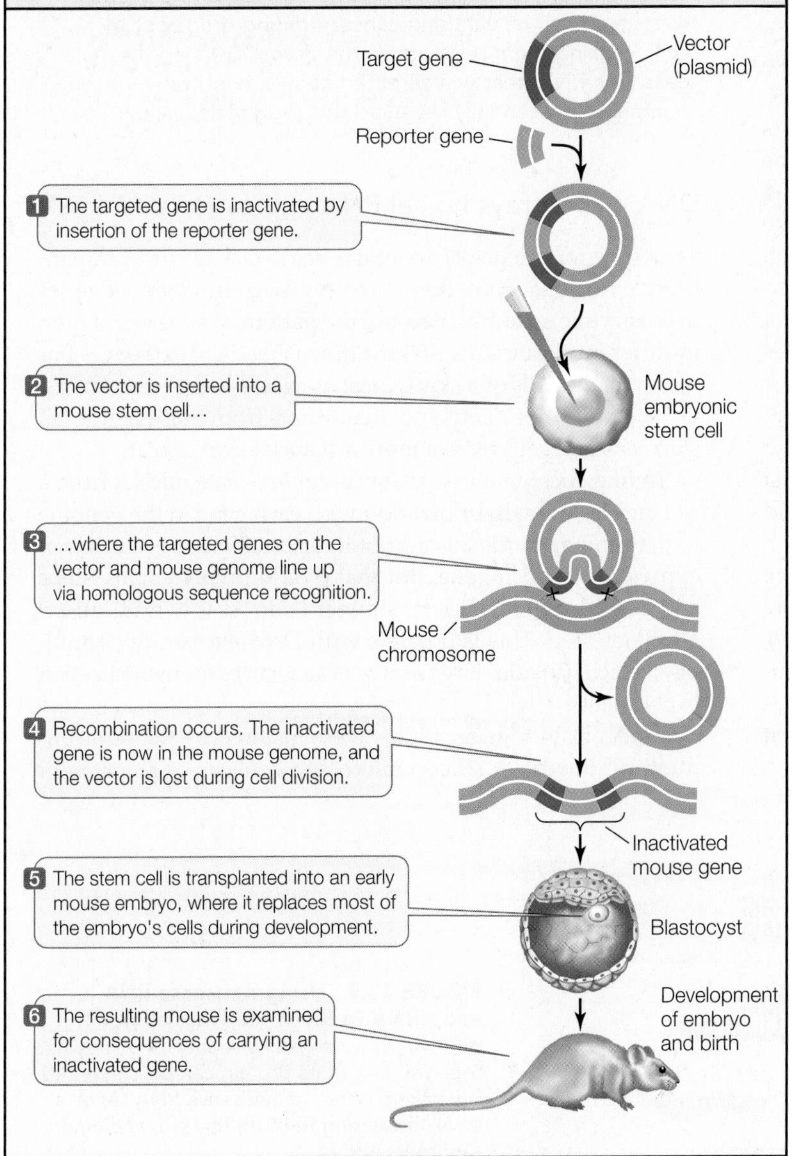

1 The targeted gene is inactivated by insertion of the reporter gene.

2 The vector is inserted into a mouse stem cell...

3 ...where the targeted genes on the vector and mouse genome line up via homologous sequence recognition.

4 Recombination occurs. The inactivated gene is now in the mouse genome, and the vector is lost during cell division.

5 The stem cell is transplanted into an early mouse embryo, where it replaces most of the embryo's cells during development.

6 The resulting mouse is examined for consequences of carrying an inactivated gene.

Target gene — Vector (plasmid)

Reporter gene

Mouse embryonic stem cell

Mouse chromosome

Inactivated mouse gene

Blastocyst

Development of embryo and birth

DNA mutations can be made in the laboratory

Mutations that occur in nature have been important in demonstrating cause-and-effect relationships in biology. However, mutations in nature are rare events. Recombinant DNA technology allows us to ask "what if" questions by creating mutations artificially. Because synthetic DNA can be made with any desired sequence, it can be manipulated to create specific mutations, the consequences of which can be observed when the mutant DNA is expressed in host cells. These mutagenesis techniques have revealed many cause-and-effect relationships.

For example, consider the experiment illustrated in Figure 10.20. Researchers hypothesized that a nuclear localization signal (NLS)—a short sequence of amino acids—is necessary for targeting a protein to the nucleus after it is made at the ribosome. The researchers used recombinant DNA technology to synthesize genes encoding proteins with and without the sequence, which were then used to transform cells. Without the NLS, newly synthesized proteins did not enter the nucleus. Knowing this, the researchers then asked, "Are certain amino acids more functionally important to the NLS than others?" In follow-up experiments, they made a series of mutated genes to test whether certain amino acids were needed at certain locations in the NLS. They found that changing the amino acids at the very beginning or very end of the NLS, but not the middle, abolished its function. This led to a fuller description of the binding of the NLS to its nuclear receptor. Without the ability to generate specific mutations, these experiments would not have been possible.

Genes can be inactivated by homologous recombination

Another way to understand a gene's function is to inactivate it so it is not transcribed and translated into a protein. An example of this approach is the use of transposon mutagenesis in experiments designed to describe the minimal genome (see Figure 12.8). In animals, such manipulation is called a **knockout** experiment. Techniques have been developed for knocking out genes in mice, fruit flies, and the nematode *Caenorhabditis elegans*, all of which are model organisms. We will focus here on the technique used for mice.

In mice, a process called **homologous recombination** is used to target a specific gene (**FIGURE 13.8**). The normal allele of the mouse gene to be tested is inserted into a plasmid. Restriction enzymes are then used to insert a fragment containing a reporter gene or selectable marker into the middle of the normal gene. This addition of extra DNA disrupts the gene's coding region so that it no longer encodes a functional protein product.

Once the recombinant plasmid has been made, it is used to transfect mouse embryonic stem cells. A **stem cell** is an unspecialized cell that divides and differentiates into specialized cells. The sequences of the targeted gene in the plasmid tend to line up with their homologous sequences in the mouse chromosome. If recombination occurs, the disrupted, inactive allele is "swapped" with the functional allele in the host cell.

The knockout technique has been important in assessing the roles of many genes, and is especially valuable in studying human genetic diseases. Many such diseases (including phenylketonuria; see Concept 10.1) have *knockout mouse models*: mouse strains with similar diseases that were produced by homologous recombination. These models can be used to study the diseases and to test potential treatments. Mario Capecchi, Martin Evans, and Oliver Smithies shared a Nobel Prize for developing the knockout mouse technique.

Complementary RNA can prevent the expression of specific genes

Another way to study the expression of a specific gene is to block the translation of its mRNA. This is yet another example of scientists imitating nature. As described in Concept 11.4, gene expression is sometimes controlled in nature by the production of short, single-stranded RNA molecules (*microRNAs*) that are complementary to specific mRNA sequences. Such a complementary molecule is called **antisense RNA** because it binds by base pairing to the "sense" bases on the mRNA. The resulting partially double-stranded RNA hybrid inhibits translation of the mRNA. Although the gene continues to be transcribed, translation does not take place. After determining the sequence of a gene in the laboratory, scientists can make a specific, single-stranded antisense RNA and add it to a cell to prevent translation of that gene's mRNA (**FIGURE 13.9A**).

Several drugs use antisense RNA to reduce the expression of genes involved with cancer. For example, the gene *BCL2* codes for a protein that blocks apoptosis, and in some forms of cancer *BCL2* expression is activated inappropriately through mutation. These cells fail to undergo apoptosis, continue to divide, and form tumors. The drug oblimersen is an antisense RNA that binds to *BCL2* mRNA. It prevents production of the protein and leads to apoptosis of the tumor cells and shrinkage of the tumor.

A related technique takes advantage of **RNA interference** (**RNAi**), a natural mechanism for inhibiting mRNA translation. RNAi occurs via the action of *small interfering RNAs* (siRNAs), which are formed in a process similar to the processing of microRNAs. An siRNA is a short (21–25 bp) double-stranded RNA which is unwound to single strands by a protein complex that guides one strand to a complementary region on an mRNA molecule. The protein complex then catalyzes the breakdown of the targeted mRNA. RNAi was not discovered until the late 1990s, but since then scientists have synthesized double-stranded siRNAs to inhibit the expression of known genes (**FIGURE 13.9B**). This technique has been used extensively to block expression of specific genes in the model animal *C. elegans*.

Because these double-stranded siRNA molecules are more stable than antisense RNA, the use of siRNA is now the preferred approach for blocking translation in both research and medicine. Macular degeneration is an eye disease that results in near blindness when blood vessels proliferate in the eye. The signaling molecule that stimulates vessel proliferation is a growth factor. An RNAi-based therapy is being developed to target this growth factor's mRNA, and the therapy shows promise for stopping and even reversing the progress of the disease.

> **FRONTIERS** RNAi is now the major tool used by biologists to experimentally reduce gene expression. The technique is called "gene knockdown" (as opposed to "knockout"), and it is important in establishing cause-and-effect relationships. For example, a particular membrane protein was suspected of pumping cancer-fighting drugs out of cells, making them resistant to the drugs. Treating these cells with RNAi that was directed against expression of the pump gene essentially reversed this drug resistance.

DNA microarrays reveal RNA expression patterns

The emerging science of genomics has to face two major quantitative realities. First, there are very large numbers of genes in eukaryotic genomes. Second, the pattern of gene expression in different tissues at different times is quite distinctive. For example, the cells of a skin cancer at its early stage may have a unique mRNA "fingerprint" that differs from those of normal skin cells and cells from a more advanced skin cancer.

To find such patterns, scientists could isolate mRNA from a cell and test it by hybridization with each gene in the genome to determine, by the amount of hybridization, the amount of expression of each gene. But that would involve many steps and take a long time. It is far simpler to do these hybridizations all in one step. This is possible with *DNA microarray* technology, which provides large arrays of sequences for hybridization experiments.

DNA arrays ("gene chips") were inspired by the semiconductor industry. A silicon microchip consists of an array of

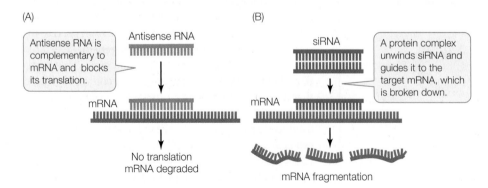

(A)

Antisense RNA is complementary to mRNA and blocks its translation.

Antisense RNA

mRNA

No translation
mRNA degraded

(B)

siRNA

A protein complex unwinds siRNA and guides it to the target mRNA, which is broken down.

mRNA

mRNA fragmentation

FIGURE 13.9 Using Antisense RNA and siRNA to Block the Translation of mRNA Once a gene's sequence is known, the synthesis of its protein can be prevented by making either an antisense RNA (A) or a small interfering RNA (B) that is complementary to its mRNA.

microscopic electric circuits etched onto a tiny silicon base, called a chip. In the same way, a series of DNA sequences can be attached to a glass slide in a precise order. The slide is divided into a grid of microscopic spots, or "wells." Each spot contains thousands of copies of a particular oligonucleotide of 20 or more bases that corresponds to a unique sequence from the genome of an organism. A computer controls the addition of these oligonucleotide sequences in a predetermined pattern. Each oligonucleotide can hybridize with only one DNA or RNA sequence, and thus is a unique identifier of a gene. Many thousands of different oligonucleotides can be placed on a single microarray.

Microarrays can be used to examine patterns of gene expression in different tissues and under different conditions, and they can be used to identify individual organisms with particular mutations (see Concept 12.4). You can see the concept of microarray analysis by following the example illustrated in **FIGURE 13.10**. Most women with breast cancer are treated with surgery to remove the tumor, and then treated with radiation soon afterward to kill cancer cells that the surgery may have missed. But a few cancer cells may still survive in some patients, and these eventually form tumors in the breast or elsewhere in the body. The challenge for physicians is to develop

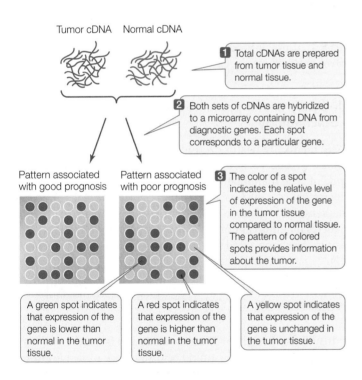

FIGURE 13.10 Using DNA Microarrays for Clinical Decision-Making The pattern of expression of 70 genes in tumor tissues (the pattern of colored spots) indicates whether breast cancer is likely to recur. Actual arrays have more dots than shown here.

---

yourBioPortal.com

Go to ANIMATED TUTORIAL 13.2
DNA Chip Technology

---

criteria to identify patients with surviving cancer cells so they can be treated aggressively with tumor-killing chemotherapy. Scientists at the Netherlands Cancer Institute used medical records to identify patients whose cancer recurred or did not recur. Then they extracted mRNA from the patients' tumors, which had been stored after surgery, and made cDNA from the samples. The cDNAs were hybridized to microarrays containing sequences derived from 1,000 human genes. The scientists found 70 genes whose expression differed dramatically between tumors from patients whose cancers recurred and tumors from patients whose cancers did not recur. From this information the Dutch group identified *gene expression signatures* that are useful in clinical decision-making: patients with a good prognosis can avoid unnecessary chemotherapy, while those with a poor prognosis can receive more aggressive treatment.

Do You Understand Concept 13.3?

- Outline the steps involved in "knocking out" a gene in a bacterium and in an animal. What are the uses for these methods?

- What are the differences between a genomic library and a cDNA library? Between antisense RNA and siRNA?

- You hypothesize that when a corn plant is infected with a fungus, a set of genes is turned on that fights the infection. How would you investigate this using microarray technology?

We have now seen how recombinant DNA is made, how cells and organisms are transformed, and how gene expression can be manipulated. In the final concept we will look at some of the many applications of biotechnology.

---

concept
13.4 **Biotechnology Has Wide Applications**

In the opening story of this chapter we defined biotechnology as the use of cells or whole living organisms to make or modify materials or processes that are useful to people, such as foods, medicines, and chemicals. Bacteria and yeast cells can be transformed with almost any gene, and they can be induced to express that gene at high levels and to export the protein product out of their cells. This technology has turned these microbes into versatile factories for many important products. Today there is interest in producing nutritional supplements and pharmaceuticals in whole transgenic animals and plants and harvesting the products in large quantities—for example, from cow's milk or rice grains. Another goal is to produce animals and plants with improved characteristics, such as increased nutritional value or better tolerance of harsh environments. Key to this boom in biotechnology has been the development of specialized vectors that not only carry genes into animal and plant cells, but also make those cells express the genes at high levels.

Expression vectors can turn cells into protein factories

Many potentially useful proteins come from eukaryotes. But if a eukaryotic gene is inserted into a typical plasmid and used to transform *E. coli*, none of the gene product will be made. Other key prokaryotic DNA sequences must be included with the gene. A bacterial promoter, a signal for transcription termination, and a special sequence that is necessary for ribosome binding on the mRNA must all be included in the transformation vector if the gene is to be expressed in the bacterial cell.

To solve this kind of problem, scientists make **expression vectors** that have all the characteristics of typical vectors, as well as the extra sequences needed for the foreign gene (also called a *transgene*) to be expressed in the host cell. For bacterial hosts, these additional sequences include the elements named above (**FIGURE 13.11**); for eukaryotes, they include the poly A–addition sequence, transcription factor binding sites, and enhancers. An expression vector can include various types of promoters and other features:

- An *inducible promoter*, which responds to a specific signal, can be included. For example, a promoter that responds to hormonal stimulation can be used so that the transgene will be expressed at high levels when the hormone is added.

- A *tissue-specific promoter*, which is expressed only in a certain tissue at a certain time, can be used if localized expression is desired. For example, many seed proteins are expressed only in the plant embryo. Coupling a transgene to a seed-specific promoter will allow it to be expressed only in seeds and not, for example, in leaves.

- *Signal sequences* can be added so that the gene product is directed to an appropriate destination. For example, when a protein is made by yeast or bacterial cells in a liquid medium, it is economical to include a signal directing the protein to be secreted into the extracellular medium for easier recovery.

> **LINK** You may wish to review the mechanisms of transcription, described in Concept 10.2

> **FRONTIERS** Tobacco mosaic virus infects but does not kill tobacco plants. Using the virus as vector, tobacco leaves can become factories for the synthesis of valuable proteins. For example, the gene for an anticancer vaccine has been inserted into the virus, and infected tobacco plants make a great deal of the vaccine protein, which is being tested on patients. An important challenge with this technique is purification of the desired protein, separating it completely from plant proteins.

Medically useful proteins can be made by biotechnology

Some medically useful products are being made by biotechnology (**TABLE 13.1**), and more are in various stages of development. The manufacture of human insulin provides a good illustration of a medical application of biotechnology. Insulin is essential for glucose uptake into cells. People with type I diabetes mellitus cannot make this hormone and must receive insulin injections. Until recently, the insulin used for this purpose came from cattle and pigs.

> **LINK** The role of insulin in regulating glucose metabolism is detailed in Concept 39.4

Insulin is a protein of 51 amino acids and is made up of two polypeptide chains. The amino acid sequence of insulin from cattle differs from the human sequence by three amino acids, and the pig sequence differs from that of humans by just one amino acid. These differences are enough to cause an immune reaction in some patients who are diabetic, and these patients need to be treated with the human protein. Since the hormone is made in tiny amounts, harvesting enough from deceased persons is not practical. Recombinant DNA biotechnology has solved this problem.

FIGURE 13.12 illustrates the strategy used to make human insulin. The two insulin polypeptides are synthesized

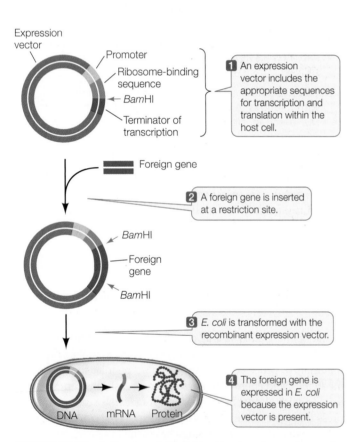

FIGURE 13.11 A Transgenic Cell Can Produce Large Amounts of the Transgene's Protein Product To be expressed in *E. coli*, a gene derived from a eukaryote requires bacterial sequences for transcription initiation (promoter), transcription termination, and ribosome binding. Expression vectors contain these additional sequences, enabling the eukaryotic protein to be synthesized in the prokaryotic cell.

Expression vector
- Promoter
- Ribosome-binding sequence
- *Bam*HI
- Terminator of transcription

1 An expression vector includes the appropriate sequences for transcription and translation within the host cell.

- Foreign gene

2 A foreign gene is inserted at a restriction site.

- *Bam*HI
- Foreign gene
- *Bam*HI

3 *E. coli* is transformed with the recombinant expression vector.

DNA   mRNA   Protein

4 The foreign gene is expressed in *E. coli* because the expression vector is present.

yourBioPortal.com
Go to WEB ACTIVITY 13.1
Expression Vectors

TABLE 13.1 Some Medically Useful Products of Biotechnology

PRODUCT	USE
Erythropoietin	Prevents anemia in patients undergoing kidney dialysis and cancer therapy
Colony-stimulating factor	Stimulates production of white blood cells in patients with cancer and AIDS
Bovine/porcine somatotropin	Stimulates growth and milk production in animals
Tissue plasminogen activator	Dissolves blood clots after heart attacks and strokes
Human growth hormone	Replaces missing hormone in people of short stature
Human insulin	Stimulates glucose uptake from blood in patients with type I diabetes mellitus
Factor VIII	Replaces clotting factor missing in patients with hemophilia A
Platelet-derived growth factor	Stimulates wound healing

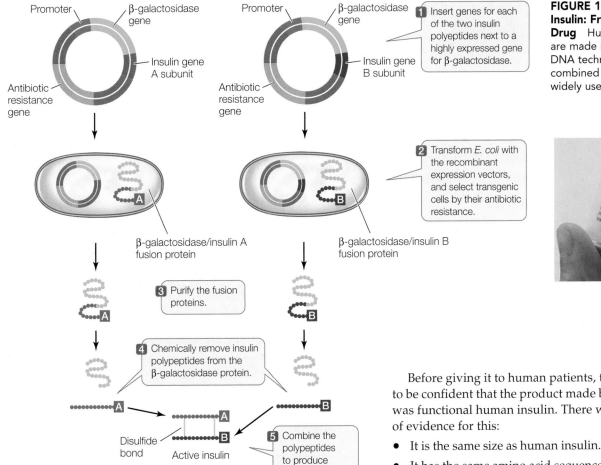

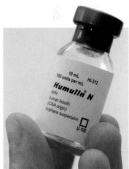

FIGURE 13.12 Human Insulin: From Gene to Drug Human insulin chains are made by recombinant DNA technology and then combined to produce the widely used drug.

Before giving it to human patients, the scientists had to be confident that the product made by biotechnology was functional human insulin. There were several lines of evidence for this:

- It is the same size as human insulin.
- It has the same amino acid sequence.
- It has the same shape, as measured by physical techniques.
- It binds to the insulin receptor on cells and stimulates glucose uptake.

Another way of making medically useful products in large amounts is **pharming**: the production of pharmaceuticals in farm animals or plants. For example, a gene encoding a useful protein might be placed next to the promoter of the gene that encodes lactoglobulin, an abundant milk protein. Transgenic animals carrying this recombinant DNA will secrete large amounts of the foreign protein into their milk. These natural "bioreactors" can produce abundant supplies of the protein,

separately using an expression vector containing the gene for β-galactosidase, which is part of the inducible *lac* operon (see Concept 11.2). Each insulin gene is inserted into the vector in such a way that it is induced, transcribed, and translated along with the β-galactosidase gene. After extraction and purification of the β-galactosidase-insulin fusion proteins, the insulin polypeptides are cleaved off by chemical treatment. The two insulin peptides are then combined to make a complete, functional human insulin molecule.

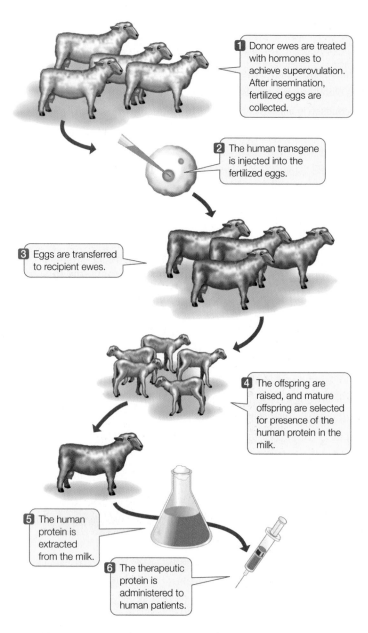

1 Donor ewes are treated with hormones to achieve superovulation. After insemination, fertilized eggs are collected.

2 The human transgene is injected into the fertilized eggs.

3 Eggs are transferred to recipient ewes.

4 The offspring are raised, and mature offspring are selected for presence of the human protein in the milk.

5 The human protein is extracted from the milk.

6 The therapeutic protein is administered to human patients.

FIGURE 13.13   Pharming   An expression vector carrying a desired gene can be put into an animal egg, which is implanted into a surrogate mother. The transgenic offspring produce the new protein in their milk. The milk is easily harvested and the protein isolated, purified, and made clinically available for patients.

which can be separated easily from the other components of the milk (**FIGURE 13.13**).

Human growth hormone, a protein made in the pituitary gland, has many effects, especially in growing children (see Concept 30.3). People with growth hormone deficiencies have short stature as well as other abnormalities. In the past they were treated with protein isolated from the pituitary glands of dead people, but the supply was too limited to meet demand. Recombinant DNA technology was used to coax bacteria into making the protein, but the cost of treatment was high ($30,000 a year). In 2004, a team led by Daniel Salamone at the University of Buenos Aires made a transgenic cow that secretes

human growth hormone in her milk. The yield is prodigious: only 15 such cows are needed to meet the worldwide demand of children suffering from this type of dwarfism.

DNA manipulation is changing agriculture

The cultivation of plants and the husbanding of animals provide the world's oldest examples of biotechnology, dating back more than 10,000 years. Over the centuries, people have adapted crops and farm animals to their needs. Through selective breeding of these organisms, desirable characteristics such as large seeds, high fat content in milk, or resistance to disease have been selected for and improved.

The traditional way to improve crop plants and farm animals was to identify individuals with desirable phenotypes that existed as a result of natural variation. Through deliberate crosses, the genes responsible for the desirable traits could be introduced into widely used varieties or breeds. Despite some spectacular successes, such as the breeding of high-yielding varieties of wheat, rice, and hybrid corn, such deliberate crossing has been a hit-or-miss affair. Many desirable traits are controlled by multiple genes, and it is hard to predict the results of a cross or to maintain a prized combination in a true-breeding variety. In sexual reproduction, combinations of desirable genes are quickly separated by meiosis. Furthermore, traditional breeding takes a long time: many plants and animals take years to reach maturity and then can reproduce only once or twice a year.

Recombinant DNA technology has several advantages over traditional methods of breeding (**FIGURE 13.14**):

- *The ability to identify specific genes.* The development of genetic markers allows breeders to select for specific desirable genes, making the breeding process more precise and rapid.

- *The ability to introduce any gene from any organism into a plant or animal species.* This ability, combined with mutagenesis techniques, vastly expands the range of possible new traits.

- *The ability to generate new organisms quickly.* Manipulating cells in the laboratory and regenerating a whole plant by cloning is much faster than traditional breeding.

Consequently, recombinant DNA technology has found many applications in agriculture (**TABLE 13.2**). We will describe a

TABLE 13.2	Potential Agricultural Applications of Biotechnology	
PROBLEM	**TECHNOLOGY/GENES**
Improving the environmental adaptations of plants	Genes for drought tolerance, salt tolerance
Improving nutritional traits	High-lysine seeds; β-carotene in rice
Improving crops after harvest	Delay of fruit ripening; sweeter vegetables
Using plants as bioreactors	Plastics, oils, and drugs produced in plants

Conventional breeding
Many generations; only gene(s) from same species can be used

Biotechnology
One generation; gene(s) from any organism can be used

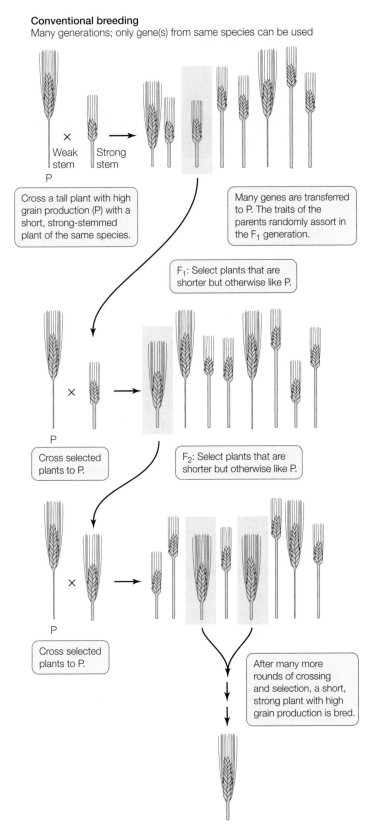

Cross a tall plant with high grain production (P) with a short, strong-stemmed plant of the same species.

Many genes are transferred to P. The traits of the parents randomly assort in the F$_1$ generation.

F$_1$: Select plants that are shorter but otherwise like P.

Cross selected plants to P.

F$_2$: Select plants that are shorter but otherwise like P.

Cross selected plants to P.

After many more rounds of crossing and selection, a short, strong plant with high grain production is bred.

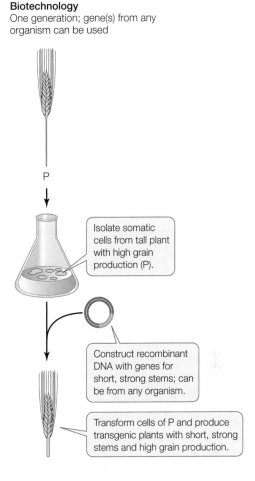

Isolate somatic cells from tall plant with high grain production (P).

Construct recombinant DNA with genes for short, strong stems; can be from any organism.

Transform cells of P and produce transgenic plants with short, strong stems and high grain production.

FIGURE 13.14 Genetic Modification of Plants versus Conventional Plant Breeding Plant biotechnology offers many potential advantages over conventional breeding. In the hypothetical example here, the objective is to transfer gene(s) for short, strong stems into a wheat plant that has high grain production but a tall, weak stem.

few examples to demonstrate the approaches that plant scientists have used to improve crop plants.

PLANTS THAT MAKE THEIR OWN INSECTICIDES From the locusts of biblical (and modern) times to the cotton boll weevil, insects have continually eaten the crops people grow. The development of insecticides has improved the situation somewhat, but insecticides have their own problems. Many are relatively nonspecific and kill beneficial insects in the broader ecosystem as well as crop pests. Some have toxic effects on other groups of organisms, including people, and persist in the environment for a long time.

Some bacteria protect themselves by producing proteins that can kill insects. For example, the bacterium *Bacillus thuringiensis* (Bt) produces a protein that is toxic to the insect larvae that prey on it:

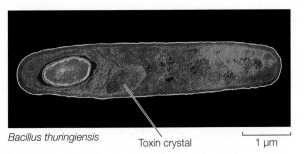

Bacillus thuringiensis   Toxin crystal   1 μm

The toxicity of this protein is 80,000 times greater than that of a typical commercial insecticide. When a hapless larva eats the bacteria, the toxin becomes activated and binds specifically to the insect's gut, producing pores and killing the insect. Dried preparations of Bt have been sold for decades as safe insecticides that break down rapidly in the environment. But the biodegradation of these preparations is their limitation, because it means the dried bacteria must be applied repeatedly during the growing season.

A more permanent approach is to have the crop plants themselves make the toxin. The toxin gene from Bt has been isolated, cloned, and extensively modified by the addition of a plant promoter and other regulatory sequences. Transgenic (also called genetically modified or GM) corn, cotton, soybeans, tomatoes, and other crops are now being grown successfully with this added gene. Pesticide usage by farmers growing these transgenic crops is greatly reduced.

GRAINS WITH IMPROVED NUTRITIONAL CHARACTERISTICS To remain healthy, humans must consume adequate amounts of β-carotene, which the body converts into vitamin A. About 400 million people worldwide suffer from vitamin A deficiency, which makes them susceptible to infections and blindness. One reason is that rice grains, which do not contain β-carotene, make up a large part of their diets. Other parts of the rice plant—and indeed many plants and other organisms—contain enzymes for the biochemical pathway that leads to β-carotene production.

Plant biologists Ingo Potrykus and Peter Beyer isolated one of the genes for the β-carotene pathway from the bacterium *Erwinia uredovora* and another from daffodil plants. They added a promoter and other signals for expression in the developing rice grain, and then transformed rice plants with the two genes. The resulting rice plants produce grains that look yellow because of their high β-carotene content. A newer variety with a corn gene replacing the one from daffodils makes even more β-carotene and is golden in color (**FIGURE 13.15**). A daily intake of about 150 grams of this cooked rice can supply all the β-carotene a person needs. This new transgenic strain has been crossed with strains adapted for various local environments, in the hope of improving the diets of millions of people.

CROPS THAT ADAPT TO THE ENVIRONMENT Agriculture depends on ecological management—tailoring the environment to the needs of crop plants and animals. A farm field is an unnatural, human-designed system that must be carefully managed to maintain optimal conditions for crop growth. For example, repeated irrigation results in increases in soil salinity. The Fertile Crescent, the region between the Tigris and Euphrates rivers in the Middle East where agriculture was practiced 10,000 years ago, is no longer fertile. It is now a desert, largely because the soil has a high salt concentration. Few plants can grow on salty soils, partly because of osmotic effects that result in wilting, and partly because excess salt ions are toxic to plant cells.

Some plants can tolerate salty soils because they have a protein that transports Na$^+$ ions out of the cytoplasm and into the vacuole, where the ions can accumulate without harming plant

Wild type   Golden rice 1   Golden rice 2

FIGURE 13.15 Transgenic Rice Rich in β-Carotene Right and middle: The grains from these transgenic rice strains are colored because they make the pigment β-carotene, which is converted to vitamin A in the human body. Left: Normal (wild-type) rice grains do not contain β-carotene.

growth (see Concept 4.3 for a description of the plant vacuole). In many salt-intolerant plants, including *Arabidopsis thaliana*, the gene for this protein exists but is inactive. Recombinant DNA technology has allowed scientists to create active versions of this gene and to use it to transform crop plants such as rapeseed, wheat, and tomatoes. When this gene was added to tomato plants, they grew in water that was four times as salty as the typical lethal level (**FIGURE 13.16**). This finding raises the prospect of growing useful crops on what were previously unproductive soils.

The example described here illustrates what could become a fundamental shift in the relationship between crop plants and the environment. *Instead of manipulating the environment to suit the plant, biotechnology may allow us to adapt the plant to the environment.* As a result, some of the negative effects of agriculture, such as water pollution, could be lessened.

There is public concern about biotechnology

Concerns have been raised about the safety and wisdom of genetically modifying crops and other organisms. These concerns are centered on three claims:

- Genetic manipulation is an unnatural interference with nature.
- Genetically altered foods are unsafe to eat.
- Genetically altered crop plants are dangerous to the environment.

Advocates of biotechnology tend to agree with the first claim. However, they point out that all crops are unnatural in the sense that they come from artificially bred plants growing in a manipulated environment (a farmer's field). Recombinant DNA technology just adds another level of sophistication to these technologies.

To counter the concern about whether genetically engineered crops are safe for human consumption, biotechnology advocates point out that only single genes are added, and that these genes have specific, known functions. For example, the Bt toxin produced by transgenic plants has no effect on people. However, as plant biotechnology moves from adding genes that improve plant growth to adding genes that affect human nutrition, such concerns will become more pressing.

Various negative environmental impacts have been envisaged. There is concern about the possible "escape" of transgenes from crops to other species. For example, a drought-tolerant crop plant might spread into, and upset the ecology of, a desert. Or beneficial insects could eat plant materials containing Bt toxin and die. Transgenic organisms undergo extensive field-testing before they are approved for use, but the complexity of the biological world makes it impossible to predict all the potential environmental effects of releasing these organisms. In fact, some spreading of transgenes has been detected. Because of the potential benefits of agricultural biotechnology (see Table 13.2), most scientists believe that it is wise to proceed, but with abundant caution

Do You Understand Concept 13.4?

- In addition to the coding sequence for a gene of interest, what other DNA sequences are required for the gene to be expressed in a different host?

- What is pharming, how is it done, and what are its advantages over more conventional biotechnology approaches?

- What are the advantages of using biotechnology for plant breeding compared with traditional methods?

- What are some of the concerns that people might have about agricultural biotechnology?

How is biotechnology used to alleviate environmental problems?

ANSWER Among the thousands of species of bacteria, there are many unique enzymes and biochemical pathways. New pathways are continually being discovered as new bacterial species are found (for example, by metagenomics; see Figure 12.7). Bacteria are nature's recyclers, thriving on many types of nutrients—including what humans refer to as wastes.

Bioremediation is the use by humans of other organisms to remove contaminants from the environment. Two well-known examples of bioremediation are composting and wastewater treatment. *Composting* involves using bacteria and other microbes to break down the large molecules, including carbon-rich polymers and proteins, in waste products such as wood chips, paper, straw, and kitchen scraps. For example, some species of bacteria make cellulase, an enzyme that hydrolyzes cellulose. Bacteria are used in *wastewater treatment* to break down human wastes, paper products, and household chemicals.

In the opening story we described how conventional genetic crosses were used to produce (and patent) bacteria that have the capacity to break down oil. Since that time, scientists have improved on nature. The oil-degrading genes from such bacteria have been isolated and spliced into vectors to make recombinant DNA (Concept 13.1). This DNA has been used to transform several species of bacteria that live in the soils where oil spills have occurred (Concept 13.2), and these transgenic bacteria are used to help clean up oil spills.

In the summer of 1990, soldiers from Iraq invaded neighboring Kuwait. The reason was oil: the Iraqis were angry because Kuwait was pumping too much of it, keeping prices low. Six months later, a United Nations–sponsored coalition army from more than 30 countries drove the Iraqis out of Kuwait and back to their homeland. For Kuwait, the Gulf War was a success, but it left an environmental disaster. As they fled, the Iraqi soldiers set fire to more than 700 oil wells. It took over

FIGURE 13.16 Salt-tolerant Tomato Plants Transgenic plants containing a gene for salt tolerance thrive in salty water (A), whereas plants without the transgene die (B). This technology may allow crops to be grown on salty soils.

(A)

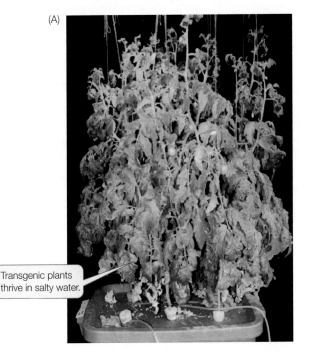

Transgenic plants thrive in salty water.

(B)

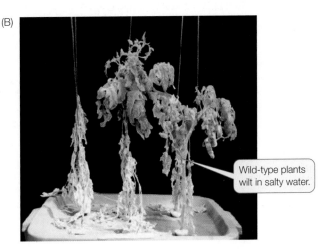

Wild-type plants wilt in salty water.

six months to put out the fires, and in the meantime an astounding 250 million gallons of crude oil were spilled into the desert. Twenty years later, much of the oil remains as a gooey coating, severely affecting the organisms that live there (**FIGURE 13.17**). The government of Kuwait is using a variety of processes to get rid of the contaminating oil. Among them is bioremediation with bacteria that can break down oil and use the hydrocarbons as energy sources for growth.

The Kuwait episode is not the first major use of bacteria for bioremediation. In 2010, about 200 million gallons of oil escaped from an offshore well into the Gulf of Mexico. Although there was limited use of genetically modified microbes near the shore, naturally occurring microbes in the open ocean have been very important in breaking down the oil and reducing its adverse impacts on the environment.

FIGURE 13.17 The Spoils of War Massive oil spills occurred in Kuwait during the 1990–1991 Gulf War.

13 SUMMARY

concept 13.1 Recombinant DNA Can Be Made in the Laboratory

- **Biotechnology** is the use of living cells to make or modify materials and processes useful to people.

- **Restriction enzymes** make cuts in double-stranded DNA, creating fragments of various lengths.

- DNA fragments can be separated by size using gel electrophoresis. **Review Figure 13.2 and ANIMATED TUTORIAL 13.1**

- DNA fragments from different sources can be used to create recombinant DNA by splicing them together using **DNA ligase**. **Review Figures 13.3 and 13.4 and WORKING WITH DATA 13.1**

concept 13.2 DNA Can Genetically Transform Cells and Organisms

- One goal of recombinant DNA technology is to **clone** a particular gene, either for analysis or to produce its protein product in quantity.

- Bacteria, yeasts, and cultured plant and animal cells are commonly used as hosts for recombinant DNA. The insertion of foreign DNA into host cells is called **transformation**, or if the host cells are derived from an animal, **transfection**.

- Various methods are used to get recombinant DNA into cells. These include chemical and physical treatments for plasmids and the use of viral vectors.

- **Selectable markers** such as genes for antibiotic resistance are used to select for host cells that have taken up a foreign gene. **Review Figure 13.5**

- **Reporter genes** (of which selectable markers are one type) are genetic markers with easily identifiable phenotypes. **Review Figure 13.6**

concept 13.3 Genes and Gene Expression Can Be Manipulated

- DNA fragments from a genome can be inserted into host cells to create a **genomic library**. A **cDNA library** is made by reverse transcribing mRNA to make cDNA. **Review Figure 13.7**

- Synthetic DNA containing any desired sequence can be made in the laboratory.

- Manipulating gene expression is one way to study the functions of particular genes.

- Homologous recombination is used to knock out a gene in a living organism. **Review Figure 13.8**

- Gene silencing techniques using **antisense RNA** or siRNA are used to prevent the translation of genes. **Review Figure 13.9**

- DNA microarray technology permits the screening of thousands of cDNA sequences at the same time. **Review Figure 13.10 and ANIMATED TUTORIAL 13.2**

concept 13.4 Biotechnology Has Wide Applications

- **Expression vectors** allow transgenes to be expressed in host cells. **Review Figure 13.11 and WEB ACTIVITY 13.1**

- Recombinant DNA techniques have been used to make medically useful proteins. **Review Figure 13.12**

- **Pharming** is the use of transgenic plants or animals to produce pharmaceuticals. **Review Figure 13.13**

- Transgenic crop plants can be adapted to their environments, rather than vice versa. **Review Figure 13.16**

- There is public concern about the application of recombinant DNA technology to food production.

Genes, Development, and Evolution

After winning a race in 2005, the thoroughbred Greg's Gold was limping because of a shredded tendon in his right front leg. A tendon is like a rubber band connecting muscles and bones, and tendons in the legs store energy when an animal runs. Typically, a damaged tendon is allowed to heal naturally, but scar tissue makes it less flexible, and a horse cannot run as fast as it did before injuring a tendon. It looked as if Greg's Gold might have to retire from racing.

Greg's Gold's trainer, David Hofmans, decided to try a new therapy. A veterinarian removed a small amount of adipose (fatty) tissue from the horse's hindquarters and sent it to a cell biology laboratory, where biologists isolated mesenchymal stem cells. Stem cells are actively dividing, unspecialized cells that have the potential to produce different cell types depending on the signals they receive from the body. Mesenchymal stem cells are able to differentiate into various kinds of connective tissue, including bone, cartilage, blood vessels, tendons, and muscle.

Two days after the tissue was taken, Greg's Gold's veterinarian received the stem cells back from the lab and injected them into the site of the damaged tendon. After several months, the tendon healed with little scar tissue, and Greg's Gold's trainer returned him to the racetrack. Greg's Gold raced for almost two more years, winning over $1 million in purse money before being retired. The mesenchymal stem cell treatment has been used successfully on several thousand horses, and on dogs with arthritis.

The processes by which an unspecialized stem cell proliferates and forms

specialized cells and tissues with distinctive appearances and functions are similar to the developmental processes that occur in the embryo. Much of our knowledge of developmental biology has come from studies on model organisms such as the fruit fly *Drosophila melanogaster*, the nematode worm *Caenorhabditis elegans*, zebrafish, the mouse, and the small flowering plant *Arabidopsis thaliana*. Eukaryotes share many similar genes, and the cellular and molecular principles underlying their development also turn out to be similar. Thus discoveries from one organism can aid us in understanding other organisms, including ourselves.

Fat stem cells helped repair damage to Greg's Gold's tendons, and he was able to race—and win—again.

KEY CONCEPTS

14.1 Development Involves Distinct but Overlapping Processes

14.2 Changes in Gene Expression Underlie Cell Differentiation in Development

14.3 Spatial Differences in Gene Expression Lead to Morphogenesis

14.4 Gene Expression Pathways Underlie the Evolution of Development

14.5 Developmental Genes Contribute to Species Evolution but Also Pose Constraints

QUESTION Why are stem cells so useful?

Development Involves Distinct but Overlapping Processes

Development is the process by which a multicellular organism, beginning with a single cell, goes through a series of changes, taking on the successive forms that characterize its life cycle (**FIGURE 14.1**). After the egg is fertilized, it is called a *zygote*, and in the earliest stages of development a plant or animal is called an **embryo**. A series of embryonic stages precedes emergence of the new, independent organism. Many organisms continue to develop throughout their life cycle, with development ceasing only with death.

Four key processes underlie development

The developmental changes an organism undergoes as it progresses from an embryo to mature adulthood involve four processes:

- **Determination** sets the developmental *fate* of a cell—what type of cell it will become—even before any characteristics of that cell type are observable. For example, the mesenchymal stem cells described in the opening story look unspecialized, but their fate to become connective tissue cells has already been determined.

- **Differentiation** is the process by which different types of cells arise from less specialized cells, leading to cells with specific structures and functions. For example, mesenchymal stem cells differentiate to become muscle, fat, tendon, or other connective tissue cells.

- **Morphogenesis** (Greek for "origin of form") is the organization and spatial distribution of differentiated cells into the multicellular body and its organs. For example, the stem cells used to heal Greg's Gold became part of a tendon with a distinct structure. Morphogenesis can occur by cell division, cell expansion (especially in plants), cell movements, and apoptosis (programmed cell death).

- **Growth** is the increase in size of the body and its organs by cell division and cell expansion. Growth can occur by an increase in the number of cells or by the enlargement of existing cells. Growth continues throughout the individual's life

---

yourBioPortal.com

Go to **WEB ACTIVITY 14.1**
Stages of Development

FIGURE 14.1 Development The stages and processes of development from zygote to maturity are shown for an animal and for a plant. The blastula is a hollow sphere of cells; the gastrula has three cell layers.

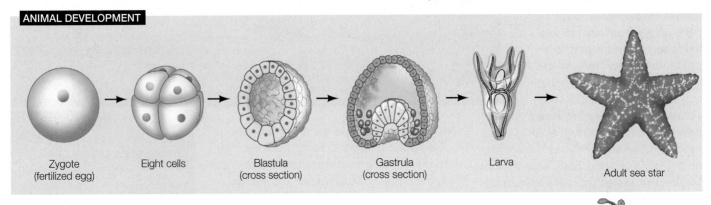

ANIMAL DEVELOPMENT

Zygote (fertilized egg)　　Eight cells　　Blastula (cross section)　　Gastrula (cross section)　　Larva　　Adult sea star

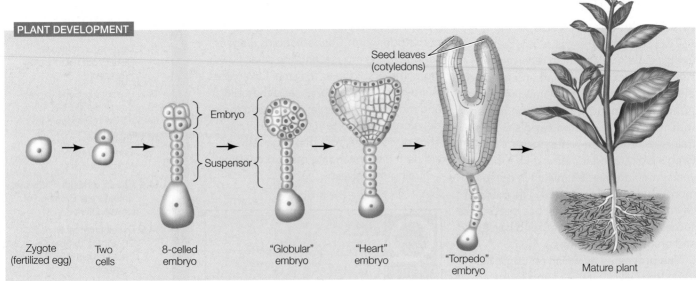

PLANT DEVELOPMENT

Seed leaves (cotyledons)

Embryo

Suspensor

Zygote (fertilized egg)　　Two cells　　8-celled embryo　　"Globular" embryo　　"Heart" embryo　　"Torpedo" embryo　　Mature plant

in some organisms, but reaches a more or less stable end point in others.

Cell fates become progressively more restricted during development

A zygote is a single cell that gives rise to all the cells in the individual organism that will develop from it. As the zygote divides to form a multicellular embryo, each of the embryo's undifferentiated cells are destined to become part of a particular type of tissue—this is referred to as the **cell fate** of that undifferentiated cell.

When is the fate determined? At what point does a cell become committed to a particular fate and no other? One way to find out is to transplant cells from one embryo to a different region of a recipient embryo (**FIGURE 14.2**). A dye is used to mark the transplanted cells so that their subsequent development can be followed. The question is, will the transplanted cells adopt the differentiation pattern of their new surroundings, or will they continue on their own path, with their fate already sealed?

Experiments with frog embryos give an answer for that organism: If the donor tissue is from an early-stage embryo (blastula), it adopts the fate of its new surroundings; its fate *has not* been sealed. But if the donor tissue is from an older embryo (gastrula), it continues on its own path; its fate *has* been sealed. Determination is influenced by changes in gene expression as well as the extracellular environment and is not something that is visible under the microscope—cells do not change their appearance when they become determined. Determination is followed by differentiation—the actual changes in biochemistry, structure, and function that result in cells of different types. *Determination is a commitment; the final realization of that commitment is differentiation.*

> **yourBioPortal.com**
>
> Go to INTERACTIVE TUTORIAL 14.1
> **Cell Fates: Genetic vs. Environmental Influences**

Cell differentiation is not irreversible

Once a cell's fate is determined, the cell differentiates. However, under the right experimental conditions, a determined or differentiated cell can become undetermined again. In some cases, the cell can even become **totipotent**, meaning it is able to form the entire organism, with all of its differentiated cells. Normally, this is a property of only the zygote.

PLANT CELL TOTIPOTENCY A carrot root cell normally faces a dark future. It cannot photosynthesize and generally does not give rise to new carrot plants. However, in 1958 Frederick Steward at Cornell University showed that if he isolated cells from a carrot root and maintained them in a suitable nutrient medium, he could induce them to *dedifferentiate*—to lose their differentiated characteristics. The cells could divide and give rise to masses of undifferentiated cells called *calli* (singular *callus*), which could be maintained in culture indefinitely. If they were provided with the right chemical cues, the cells could develop into embryos and eventually into complete new plants (**FIGURE 14.3**).

Since the new plants in Steward's experiments were genetically identical to the cells from which they came, they were *clones* of the original carrot plant. The ability to produce clones is evidence for the **genomic equivalence** of somatic (body) cells; that is, all cells in a plant have a complete genome and thus have all the genetic information needed to become any cell in the plant.

Many types of cells from other plant species show similar behavior in the laboratory. This ability to generate a whole plant from a single cell has been invaluable in agriculture and forestry. For example, trees from planted forests are used in making paper, lumber, and other products. To replace the trees reliably, forestry companies regenerate new trees from the leaves of selected trees with desirable traits. The characteristics of these clones are more uniform and predictable than those of trees grown from seeds.

NUCLEAR TOTIPOTENCY IN ANIMALS Animal somatic cells cannot be manipulated as easily as plant cells can. Until recently, it was not possible to induce a cell from a fully developed animal to dedifferentiate and then redifferentiate into another cell type. However, nuclear transfer experiments have shown that the genetic information from an animal cell can be used to create cloned animals. The nucleus from an unfertilized egg is removed, forming an *enucleated* egg. A donor nucleus from a differentiated cell

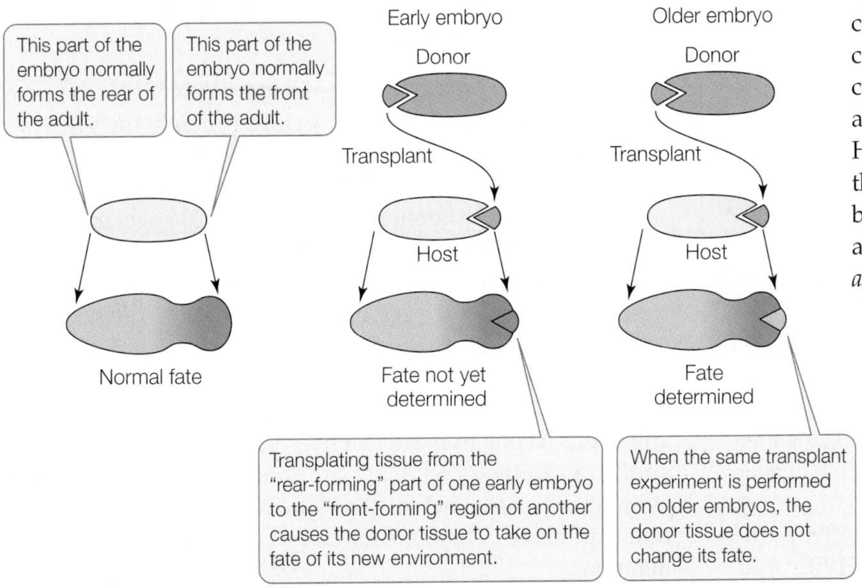

This part of the embryo normally forms the rear of the adult.

This part of the embryo normally forms the front of the adult.

Early embryo

Donor

Transplant

Host

Normal fate

Fate not yet determined

Older embryo

Donor

Transplant

Host

Fate determined

Transplanting tissue from the "rear-forming" part of one early embryo to the "front-forming" region of another causes the donor tissue to take on the fate of its new environment.

When the same transplant experiment is performed on older embryos, the donor tissue does not change its fate.

FIGURE 14.2 A Cell's Fate Is Determined in the Embryo Transplantation experiments using frog embryos show that the fate of cells is determined as the embryo develops.

INVESTIGATION

FIGURE 14.3 Cloning a Plant When cells were removed from a plant and put into a medium with nutrients and hormones, they lost many of their specialized features—in other words, they dedifferentiated and became totipotent.

HYPOTHESIS

Differentiated plant cells can be totipotent and can be induced to generate all types of the plant's cells.

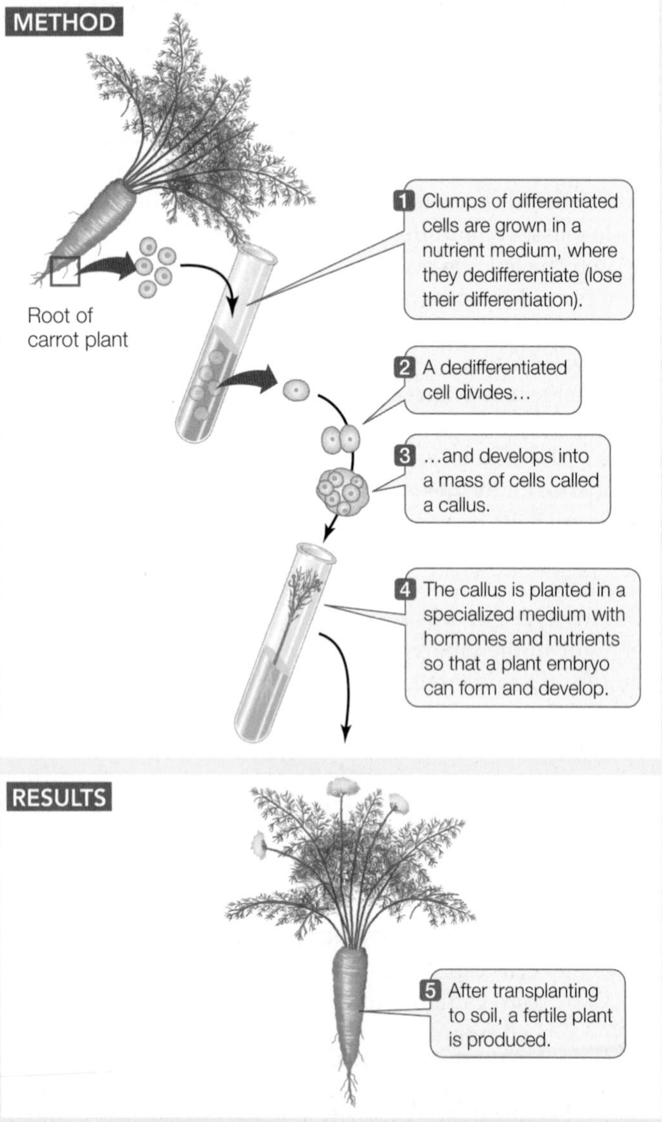

METHOD

Root of carrot plant

1. Clumps of differentiated cells are grown in a nutrient medium, where they dedifferentiate (lose their differentiation).

2. A dedifferentiated cell divides...

3. ...and develops into a mass of cells called a callus.

4. The callus is planted in a specialized medium with hormones and nutrients so that a plant embryo can form and develop.

RESULTS

5. After transplanting to soil, a fertile plant is produced.

CONCLUSION

Differentiated plant cells can be totipotent.

Go to **yourBioPortal.com** for original citations, discussions, and relevant links for all INVESTIGATION figures.

is then introduced into the "empty" egg. If it is then stimulated to divide, the egg forms an embryo and adult with the genetic composition of its nuclear donor. This is the basis of cloning animals such as Dolly, the landmark sheep born in 1996 as the first experimentally produced mammalian clone (**FIGURE 14.4**).

Many other animal species, including cats, dogs, horses, pigs, rabbits, and mice, have since been cloned by nuclear transfer. As in plants, the cloning of animals has shown that their differentiated cells have genomic equivalence. Cloning of animals has practical uses as well:

- *Expansion of the numbers of valuable animals*: One goal of the experiments that resulted in Dolly the sheep was to develop a method of cloning transgenic animals carrying genes with therapeutic properties (see Concept 13.4). For example, a cow that was genetically engineered to make human growth hormone in milk has been cloned to produce two more cows that do the same thing. Only 15 such cows would supply the world's need for this medication, which is used to treat short stature due to growth hormone deficiency.

- *Preservation of endangered species*: The banteng, a relative of the cow, was the first endangered animal to be cloned, using a cow enucleated egg, the nucleus from a banteng cell, and a cow surrogate mother. Cloning may be the only way to save endangered species with low rates of natural reproduction, such as the Sumatran tiger.

- *Preservation of pets*: Many people get great personal benefit from pets, and the death of a pet can be devastating. Companies have been set up to clone cats and dogs from cells provided by their owners. Of course, the behavioral characteristics of the beloved pet, which were certainly determined in part from the environment, may not be the same in the cloned pet as in its genetic parent.

FRONTIERS Most male cattle are castrated as calves (they are referred to as steers) to improve the color and texture of their meat. Some steers may grow up to have superior phenotypes, and a cattle breeder may want to use such a steer for breeding purposes. Unfortunately, the steer is sterile. Cloning can produce a fertile animal, thereby "reversing" castration.

Stem cells differentiate in response to environmental signals

The embryo is not the only time and place where the processes of development occur. In adult plants, the growing regions at the tips of roots and stems contain *meristems*, which are clusters of undifferentiated, rapidly dividing *stem cells*. These cells can differentiate into the 15–20 specialized cell types that make up roots, stems, leaves, and flowers. As you will see in Chapter 26, the plant body undergoes constant growth and renewal, with new organs forming often. (Think of leaves in the spring and flowers.)

In adult mammals, stem cells occur in virtually all tissues, where they are used as a pool of cells that can differentiate and replace cells that are lost by "wear and tear" (necrosis) and programmed cell death (apoptosis). This is especially evident in tissues such as the skin, inner lining of the intestine, and blood. There are about 300 different cell types in a mammal.

INVESTIGATION

FIGURE 14.4 Cloning a Mammal The experimental procedure described here produced the first cloned mammal, a Dorset sheep named Dolly (shown on the left in the photo). As an adult, Dolly mated and subsequently gave birth to a normal offspring (the lamb on the right), thus proving the genetic viability of cloned mammals.

HYPOTHESIS

Differentiated animal cells are totipotent.

METHOD

1. Cells are removed from the udder of a Dorset ewe.

2. An egg is removed from a Scottish blackface ewe.

Dorset sheep (#1)

Scottish blackface sheep (#2)

Nucleus

Micropipette

3. Udder cells are deprived of nutrients in culture to halt the cell cycle prior to DNA replication.

4. The nucleus is removed from the egg.

Donor nucleus (from sheep #1)

5. The udder cell (donor) and enucleated egg are fused.

Enucleated egg (from sheep #2)

6. Mitosis-stimulating inducers cause the cell to divide.

7. An early embryo develops and is transplanted into a receptive ewe.

Scottish blackface sheep (#3)

RESULTS

8. The embryo develops and a Dorset sheep, genetically identical to #1, is born.

CONCLUSION

Differentiated animal cells are totipotent in nuclear transplant experiments.

ANALYZE THE DATA

The team that cloned Dolly the sheep used a nucleus from a mammary epithelium (ME) cell. They also tried cloning by transplanting nuclei from fetal fibroblasts (FB) and embryos (EC), with these results:

Stage	Number of attempts that progressed to each stage		
	ME	FB	EC
Egg fusions	277	172	385
Embryos transferred to recipients	29	34	72
Pregnancies	1	4	14
Live lambs	1	2	4

A. Calculate the percentage survival of eggs from fusion to birth. What can you conclude about the efficiency of cloning?

B. Compare the efficiencies of cloning using different nuclear donors. What can you conclude about the ability of nuclei at different stages to be totipotent?

C. What statistical test would you use to show whether the differences in A and B were significant (see Appendix B)?

For more, go to Working with Data 14.1 at **yourBioPortal.com**.

Go to **yourBioPortal.com** for original citations, discussions, and relevant links for all INVESTIGATION figures.

MULTIPOTENT STEM CELLS Stem cells in particular mammalian tissues are **multipotent**, meaning they can form a limited repertoire of differentiated cells. For example, there are two types of multipotent stem cells in bone marrow. One type (called *hematopoietic stem cells*) produces the various kinds of red and white blood cells, whereas the other type (*mesenchymal stem cells*) produces the cells that make bone and surrounding tissues, such as muscle.

The differentiation of multipotent stem cells is "on demand." The blood cells that differentiate in the bone marrow do so in

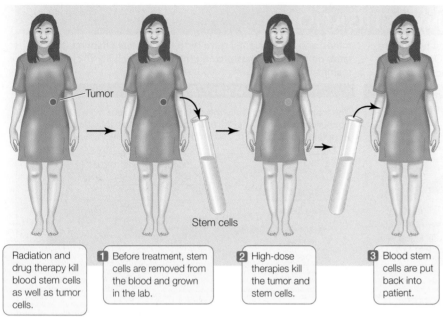

| Radiation and drug therapy kill blood stem cells as well as tumor cells. | **1** Before treatment, stem cells are removed from the blood and grown in the lab. | **2** High-dose therapies kill the tumor and stem cells. | **3** Blood stem cells are put back into patient. |

FIGURE 14.5 Multipotent Stem Cells In hematopoietic stem cell transplantation, blood stem cells are used to replace stem cells destroyed by cancer therapy.

PLURIPOTENT STEM CELLS In mammals, toti-potent stem cells that can individually give rise to an organism are found only in very early embryos. In both mice and humans, the earliest embryonic stage before differentiation occurs is called a *blastocyst* (the term for a mammalian blastula; see Figures 14.1 and 33.4). Although they cannot form an entire embryo, a group of cells in the blastocyst still retains the ability to form all of the cells in the body; these cells are **pluripotent**. These **embryonic stem cells (ESCs)** can be removed from the blastocyst and grown in laboratory culture almost indefinitely if provided with the right conditions. They can also be induced to express appropriate genes and differentiate in a particular way if the right signal is provided (**FIGURE 14.6A**). For example, treatment of mouse ESCs with a derivative of vitamin A causes them to form neurons (nerve cells), while other growth factors induce them to form blood cells. Such experiments demonstrate both the cells'

response to specific signals. These can be from adjacent cells (as in the repairing tendon of Greg's Gold) or from the circulation (as in the differentiation of bone marrow stem cells to form blood cells). This is the basis of an important cancer therapy called *hematopoietic stem cell transplantation* (HSCT; **FIGURE 14.5**). Because some treatments that kill cancer cells also kill other dividing cells, bone marrow stem cells in patients will die if exposed to these treatments. To circumvent this problem, stem cells are removed from the patient's blood before treatments begin, and are given signals to increase their numbers in the laboratory. The cells are stored during treatment, and then added back to populate the depleted bone marrow when treatment is over. The stored stem cells retain their ability to differentiate in the bone marrow environment. By allowing the use of high doses of treatment to kill tumors, bone marrow transplantation saves thousands of lives each year.

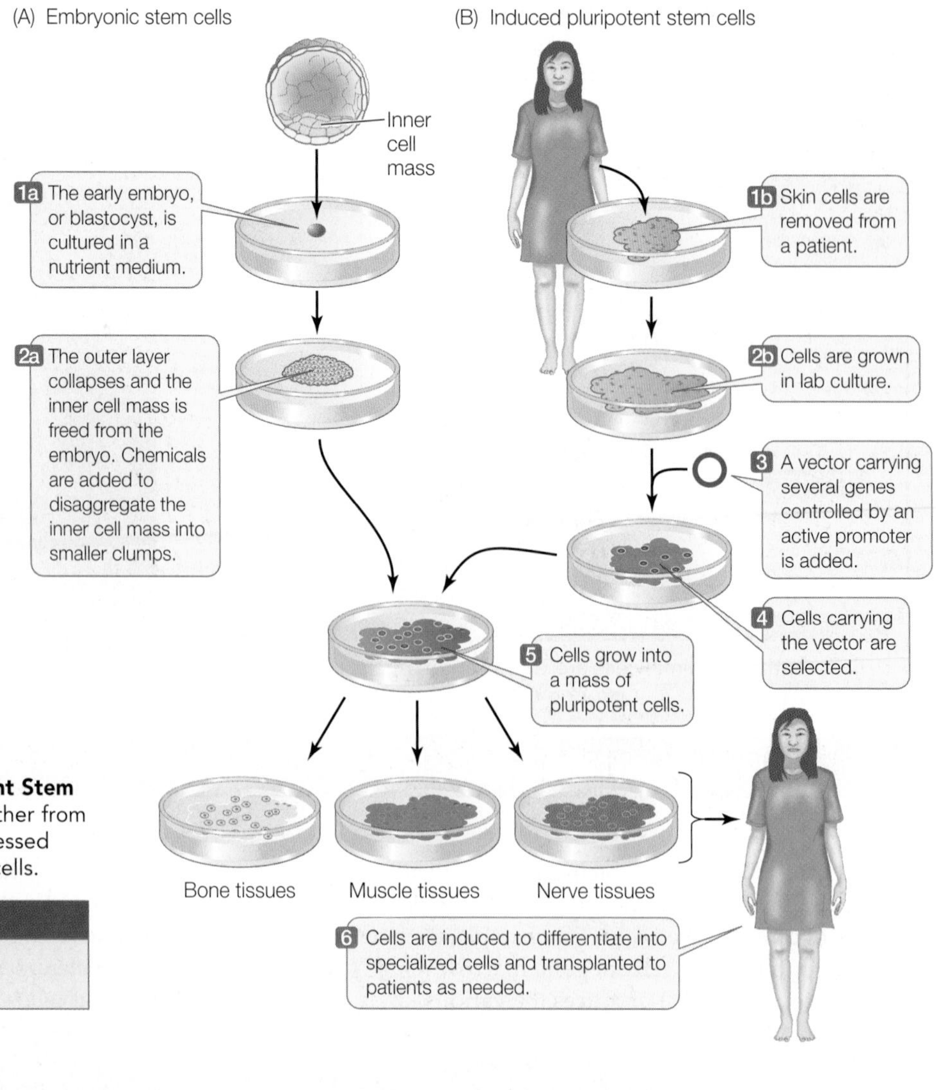

FIGURE 14.6 Two Ways to Obtain Pluripotent Stem Cells Pluripotent stem cells can be obtained either from (A) human embryos or (B) by adding highly expressed genes to skin cells to transform them into stem cells.

(A) Embryonic stem cells

Inner cell mass

1a The early embryo, or blastocyst, is cultured in a nutrient medium.

2a The outer layer collapses and the inner cell mass is freed from the embryo. Chemicals are added to disaggregate the inner cell mass into smaller clumps.

5 Cells grow into a mass of pluripotent cells.

Bone tissues   Muscle tissues   Nerve tissues

(B) Induced pluripotent stem cells

1b Skin cells are removed from a patient.

2b Cells are grown in lab culture.

3 A vector carrying several genes controlled by an active promoter is added.

4 Cells carrying the vector are selected.

6 Cells are induced to differentiate into specialized cells and transplanted to patients as needed.

yourBioPortal.com

Go to ANIMATED TUTORIAL 14.1
Embryonic Stem Cells

developmental potential and the roles of environmental signals. This finding raises the possibility of using ESC cultures as sources of differentiated cells to repair specific tissues, such as a damaged pancreas in diabetes, or a brain that malfunctions in Parkinson's disease.

ESCs can be harvested from human embryos conceived by in vitro ("under glass"—in the laboratory) fertilization, with the consent of the donors. Since more than one embryo is usually conceived in this procedure, embryos not used for reproduction might be available for embryonic stem cell isolation. These cells could then be grown in the laboratory and used as sources of tissues for transplantation into patients with tissue damage. There are two problems with this approach:

- Some people object to the destruction of human embryos for this purpose.
- The stem cells, and tissues derived from them, would provoke an immune response in a recipient (see Chapter 31).

Shinya Yamanaka and coworkers at Kyoto University in Japan and the Gladstone Institute, San Francisco, have developed another way to produce pluripotent stem cells that uses the concepts of gene expression and development (**FIGURE 14.6B**). Instead of extracting ESCs from blastocysts, they make **induced pluripotent stem cells** (**iPS cells**) from skin cells. They developed this method systematically (see Chapter 13 for more information on the techniques discussed here):

1. First, they used microarrays to compare the genes expressed in ESCs with nonstem cells. They found several genes that were uniquely expressed at high levels in ESCs. These genes were believed to be essential to the undifferentiated state and function of stem cells.

2. Next, they isolated the genes and inserted them into a vector for genetic transformation of skin cells. They found that the skin cells now expressed the newly added genes at high levels.

3. Finally, they showed that the transformed cells were pluripotent and could be induced to differentiate into many tissues—they had become iPS cells.

Because iPS cells can be made from skin cells of the individual who is to be treated, an immune response may be avoided. Such cells have already been used for cell therapy in animals for diseases similar to human Parkinson's disease (a brain disorder), diabetes, and sickle-cell anemia. Human uses are sure to follow.

Do You Understand Concept 14.1?

- Describe the four major processes of development and, using Figure 14.1, describe when these processes occur in plant and animal development.

- Not all the DNA in a cell is in the nucleus. What are the genetic differences between cloning in carrot plants and cloning in sheep? How would you show this?

- Identical twins are formed when a zygote divides once by mitosis and then each mitotic product forms an embryo. Are identical twins clones? Explain your answer.

Having considered the general principles of development, we now turn to the mechanisms that govern developmental events. Not surprisingly, these mechanisms have been studied at the molecular level and involve changes in gene expression and the activities of specific proteins.

concept **14.2** **Changes in Gene Expression Underlie Cell Differentiation in Development**

Every cell of an individual organism contains all the genes needed to produce every protein encoded by the organism's genome. Each cell, however, expresses only a subset of these genes; the identity of these genes and the processes by which they are activated form the basis of cell determination and differentiation. For example, certain cells in hair follicles produce keratin, the protein that makes up hair, while other cell types in the body do not. What determines whether a cell will produce keratin? Chapter 11 describes a number of ways in which cells regulate gene expression and protein production—by controlling transcription, translation, and posttranslational protein modifications. The mechanisms that control gene expression resulting in cell differentiation generally work at the level of transcription. So in the case of keratin, skin cells produce keratin mRNA, but other cells, such as liver cells, do not.

Differential gene transcription is a hallmark of cell differentiation

What leads to this differential gene expression? One well-studied example of cell differentiation is the conversion of undifferentiated muscle precursor cells into cells that are destined to form muscle (**FIGURE 14.7**). In the vertebrate embryo, muscle precursor cells come from a tissue layer called the *mesoderm* (see Concept 33.3), and they are multipotent. A key event in the commitment of these cells to become muscle is that they stop dividing. Indeed, in many parts of the embryo, *cell division and cell differentiation are mutually exclusive*. Cell signaling activates the gene for a transcription factor called **MyoD** (*myoblast-determining gene*). Recall that transcription factors are DNA binding proteins that regulate the expression of specific genes. In this case, MyoD activates the gene for p21, an inhibitor of cyclin-dependent kinases (Cdk's) that normally stimulate the cell cycle at G1 (see Figure 7.10). Expression of the *p21* gene causes the cell cycle to stop, and other transcription factors then enter the picture so that differentiation into myoblasts (muscle precursor cells) can begin.

Genes such as *myoD* that direct the most fundamental decisions in development (often by regulating the expression of other genes on other chromosomes) usually encode transcription factors. In some cases, a single transcription factor can cause a cell to differentiate in a certain way. In other cases, complex interactions between genes and proteins determine a sequence of transcriptional events that leads to differential

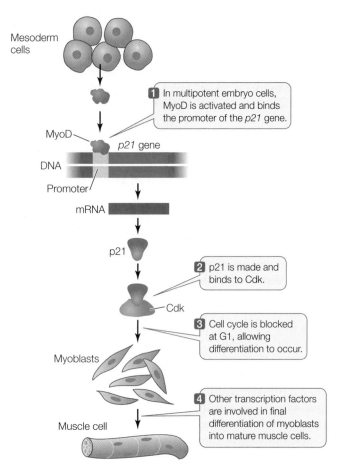

FIGURE 14.7 Transcription and Differentiation in the Formation of Muscle Cells Activation of a transcription factor, MyoD, is important in muscle cell differentiation.

gene expression. There are two ways a cell can be made to transcribe a different set of genes than another cell:

- by asymmetrical distribution of a cytoplasmic factor inside the cell so that its two progeny cells receive unequal amounts of the factor
- by differential exposure of the two cells to an external inducer

Gene expression can be regulated by cytoplasmic polarity

An early event in development is often the establishment of axes that relate to the body plan of the organism. For example, an embryo may develop a distinct "top" and "bottom" corresponding to what will become opposite ends of the mature organism; such a difference is called **polarity**. Many examples of polarity are observed as development proceeds. Our heads are distinct from our rear ends, and the distal (far from the center) ends of our arms and legs (wrists, ankles, fingers, toes) differ from the proximal (near) ends (shoulders and hips).

Polarity may develop early; even within the egg, the yolk and other factors are often distributed asymmetrically. During early development in animals, polarity is specified by an *animal pole* at the top of the zygote and a *vegetal pole* at the bottom. This

polarity can lead to determination of cell fates at a very early stage of development. For example, sea urchin embryos can be bisected at the eight-cell stage in two different ways:

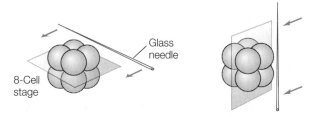

If the two halves (with four cells each) of these embryos are allowed to develop, the results are dramatically different for the two different cuts:

- For an embryo cut into a top half and a bottom half (left, above), the bottom half develops into a small sea urchin and the top half does not develop at all.
- For an embryo cut into two side halves (right, above), both halves develop normally, though smaller.

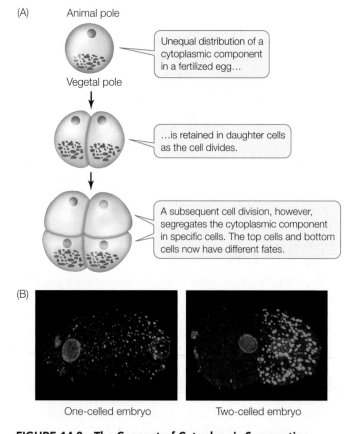

FIGURE 14.8 The Concept of Cytoplasmic Segregation
(A) The unequal distribution of some component in the cytoplasm of a cell may determine the fates of its descendants. (B) The zygote of the nematode worm (left) shows an asymmetrical distribution of cytoplasmic particles (stained green). The progeny of the first cell division (right) receive unequal amounts of the particles.

In sum, the top and bottom halves of an eight-cell sea urchin embryo have already developed distinct fates. These observations led to the model of **cytoplasmic segregation** shown in **FIGURE 14.8A**. This model states that certain materials, called **cytoplasmic determinants**, are distributed unequally in the egg cytoplasm. During the early cell divisions of the embryo's development, the progeny cells receive unequal amounts of these determinants. The amount of determinant received determines each cell's fate and its pattern of gene expression.

Cytoplasmic determinants play roles in directing the embryonic development of many organisms. What is the chemical nature of these determinants, and how is their asymmetrical distribution established? It turns out that the cytoskeleton contributes to the asymmetrical distribution of these determinants in the egg. Recall from Concept 4.4 that an important function of the microtubules and microfilaments in the cytoskeleton is to help move materials in the cell. Two properties allow these structures to accomplish this:

- Microtubules and microfilaments have polarity—they grow by adding subunits to the plus end.
- Cytoskeletal elements can bind specific proteins, which can be used in the transport of mRNA.

For example, in the sea urchin egg, a protein binds to both the growing (+) end of a microfilament and to an mRNA encoding a cytoplasmic determinant. As the microfilament grows toward one end of the cell, it carries the mRNA along with it. The asymmetrical distribution of the mRNA leads to asymmetrical distribution of the protein it encodes, a regulator of gene transcription. A similar process is also seen in other organisms (**FIGURE 14.8B**). So what biologists once called unspecified "cytoplasmic determinants" can now be defined in terms of cellular structures, mRNAs, and proteins.

Inducers passing from one cell to another can determine cell fates

The second way that embryos stimulate differential gene transcription is by **induction**, exemplified here by vulval development in the model organism *Caenorhabditis elegans* (**FIGURE 14.9**). The term "induction" has different meanings in different contexts. In biology it can be used broadly to refer to the initiation of, or cause of, a change or process.

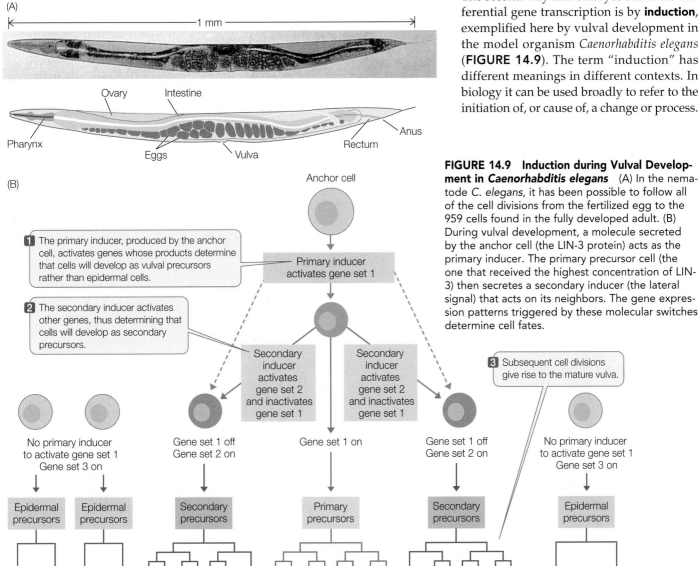

FIGURE 14.9 Induction during Vulval Development in *Caenorhabditis elegans* (A) In the nematode *C. elegans*, it has been possible to follow all of the cell divisions from the fertilized egg to the 959 cells found in the fully developed adult. (B) During vulval development, a molecule secreted by the anchor cell (the LIN-3 protein) acts as the primary inducer. The primary precursor cell (the one that received the highest concentration of LIN-3) then secretes a secondary inducer (the lateral signal) that acts on its neighbors. The gene expression patterns triggered by these molecular switches determine cell fates.

(A)

|← 1 mm →|

Ovary Intestine

Pharynx

Eggs Vulva Rectum Anus

(B)

Anchor cell

1 The primary inducer, produced by the anchor cell, activates genes whose products determine that cells will develop as vulval precursors rather than epidermal cells.

Primary inducer activates gene set 1

2 The secondary inducer activates other genes, thus determining that cells will develop as secondary precursors.

Secondary inducer activates gene set 2 and inactivates gene set 1

Secondary inducer activates gene set 2 and inactivates gene set 1

3 Subsequent cell divisions give rise to the mature vulva.

No primary inducer to activate gene set 1
Gene set 3 on

Gene set 1 off
Gene set 2 on

Gene set 1 on

Gene set 1 off
Gene set 2 on

No primary inducer to activate gene set 1
Gene set 3 on

Epidermal precursors

Epidermal precursors

Secondary precursors

Primary precursors

Secondary precursors

Epidermal precursors

Epidermis

Vulva

Epidermis

But in the context of cellular differentiation, it refers to the signaling events by which cells in a developing embryo communicate and influence one another's developmental fate. Induction involves chemical signals and signal transduction mechanisms (see Concepts 5.5 and 5.6). Exposure to different amounts of such inductive signals can lead to differences in gene expression among cells in a developing organism.

The nematode worm *Caenorhabditis elegans*, whose genome was one of the first eukaryotic genomes to be sequenced (see Chapter 12), develops from fertilized egg to larva in only about 8 hours and reaches the adult stage in just 3.5 days. The process is easily observed using a low-magnification dissecting microscope because the body covering is transparent (see Figure 14.9A). To illustrate the principles of induction, we focus here on the development of one part of the *C. elegans* body: the vulva.

The adult nematode is *hermaphroditic*, containing both male and female reproductive organs. It lays eggs through a pore called the *vulva* on the ventral (lower) surface of the worm. During development, a single cell, called the *anchor cell*, induces the vulva to form from six cells on the worm's ventral surface. In this case, there are two molecular signals, the *primary inducer* and the *secondary inducer*. Each of the six ventral cells has three

possible fates: it may become a primary vulval precursor cell, a secondary vulval precursor cell, or simply become part of the worm's skin—an epidermal cell. You can follow the sequence of events in Figure 14.9B. The concentration gradient of the primary inducer, LIN-3, is key: the anchor cell produces LIN-3, which diffuses out of the cell and forms a concentration gradient with respect to adjacent cells. The three cells that are closest to the anchor cell receive the most LIN-3 and become vulval precursor cells; cells slightly farther from the anchor cell receive less LIN-3 and become epidermal cells. A second induction event results in the two classes of vulval precursor cells: primary and secondary. Induction involves the activation or inactivation of specific sets of genes through signal transduction cascades in the responding cells (**FIGURE 14.10**).

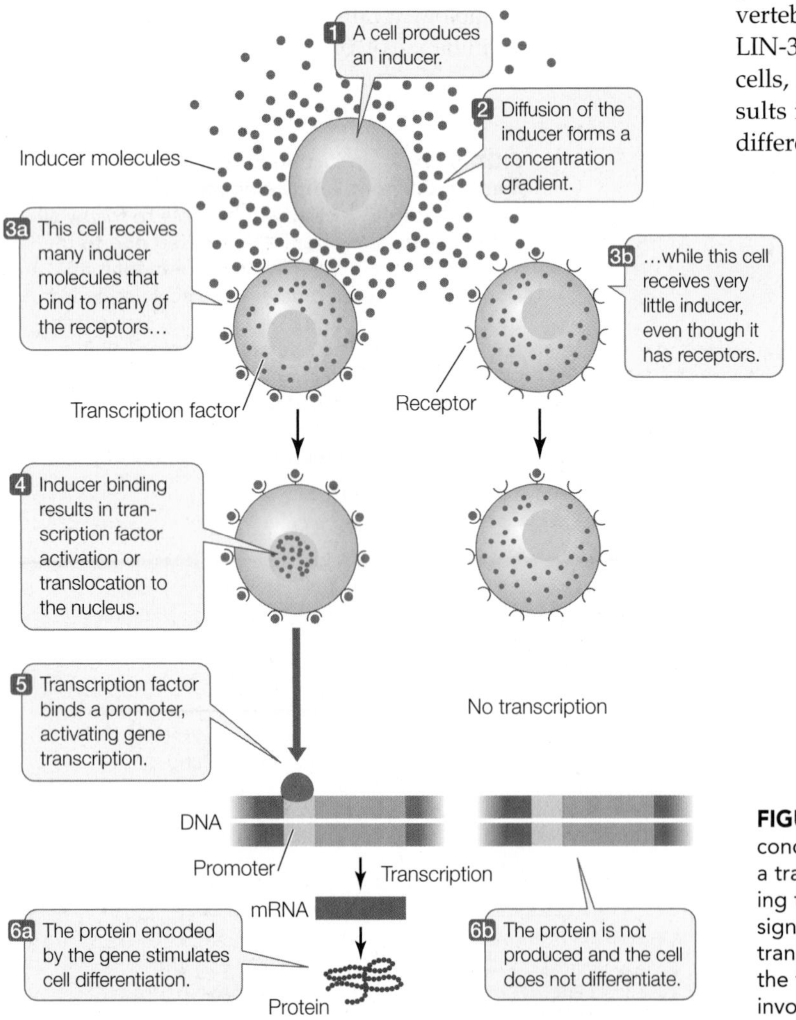

> **LINK** For more on signal transduction cascades, see Concepts 5.5 and 5.6

This example from nematode development illustrates an important observation: *much of development is controlled by molecular switches that allow a cell to proceed down one of two alternative tracks.* One challenge for developmental biologists is to find these switches and determine how they work. The primary inducer, LIN-3, released by the *C. elegans* anchor cell is a growth factor that is orthologous (evolutionarily related) to a vertebrate growth factor called EGF (*epidermal growth factor*). LIN-3 binds to a receptor on the surfaces of vulval precursor cells, setting in motion a signal transduction cascade that results in increased transcription of the genes involved in the differentiation of vulval cells.

1 A cell produces an inducer.

2 Diffusion of the inducer forms a concentration gradient.

3a This cell receives many inducer molecules that bind to many of the receptors...

3b ...while this cell receives very little inducer, even though it has receptors.

Inducer molecules

Transcription factor

Receptor

4 Inducer binding results in transcription factor activation or translocation to the nucleus.

5 Transcription factor binds a promoter, activating gene transcription.

No transcription

DNA

Promoter

Transcription

mRNA

6a The protein encoded by the gene stimulates cell differentiation.

6b The protein is not produced and the cell does not differentiate.

Protein

Do You Understand Concept 14.2?

- What would be the effect of injecting an inhibitor of microtubule polymerization into the fertilized eggs of a sea urchin embryo?

- Compare the internal and external stimuli that lead to differential gene expression in embryonic cells.

- What would be the consequences of a homozygous deletion mutation for *lin-3*?

Cytoplasmic polarity and inducers alter the expression of genes that determine cell fate. We will now see that the expression of genes also results in the next phase of development, the formation of tissues and organs in morphogenesis.

FIGURE 14.10 The Concept of Embryonic Induction The concentration of an inducer directly affects the degree to which a transcription factor is activated. The inducer acts by binding to a receptor on the target cell. This binding is followed by signal transduction involving transcription factor activation or translocation from the cytoplasm to the nucleus. In the nucleus, the transcription factor acts to stimulate the expression of genes involved in cell differentiation.

Spatial Differences in Gene Expression Lead to Morphogenesis

Pattern formation is the developmental process that results in the spatial organization of a tissue or organism. It is inextricably linked to morphogenesis, the creation of body form. Underlying both of these processes are spatial differences in gene expression, which determine whether, for example, a particular piece of tissue will become a leg or a wing or a flower petal. These spatial differences in gene expression, in turn, depend on two cellular processes:

- The cells in the tissue must "know" where they are in relation to rest of the body.
- The cells must activate the pattern of gene expression that is appropriate for their location.

In the sections that follow, we will explore the mechanisms used by various organisms to direct pattern formation and morphogenesis.

Multiple genes interact to determine developmental programmed cell death

You might expect morphogenesis to involve a lot of cell division, followed by differentiation—and it does. But what you might not expect is the amount of programmed cell death—apoptosis—that occurs during morphogenesis. For example, in an early human embryo, the hands and feet look like tiny paddles: the tissues that will become fingers and toes are linked by connective tissue. Between days 41 and 56 of development, the cells between the digits die, freeing the individual fingers and toes:

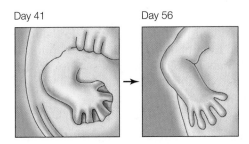

Day 41   Day 56

Many cells and structures form and then disappear during development, in processes involving apoptosis.

LINK Concept 7.5 describes some of the cellular events of apoptosis

Model organisms have been very useful in studying the genes involved in apoptosis. For example, the nematode worm *C. elegans* produces precisely 1,090 somatic cells as it develops from a fertilized egg into an adult, but 131 of those cells die (leaving 959 cells in the adult worm). The sequential expression of two genes called *ced-4* and *ced-3* (for *c*ell *d*eath) is essential to this programmed cell death. The CED-3 protein is a caspase (a protease involved in apoptosis) that turns out to be similar to

the caspase protein encoded by an orthologous gene in humans. Several other genes involved in the nematode apoptosis pathway (including *ced-4*) also have relatives in humans. So humans and nematodes, two species separated by more than 600 million years of evolutionary history, have similar genes controlling programmed cell death. The commonality of this pathway indicates its importance: mutations are harmful and evolution selects against them. We will return to other examples of links between evolution and development in Concept 14.4.

Our example of apoptosis in the development of fingers and toes shows one of the many ways that the behavior of cells can give rise to body form during development. It also illustrates the two cellular processes underlying pattern formation: only cells in a particular place (between the digits) activate a specific pattern of gene expression (to trigger apoptosis).

FRONTIERS The genes that control apoptosis in the worm have orthologs in humans. One of them, *BCL2*, prevents apoptosis. In a type of human cancer called follicular lymphoma, a chromosome translocation occurs in white blood cells such that the *BCL2* gene is next to a highly active promoter. The resulting high levels of BCL2 protein prevent the apoptosis that normally occurs in the white blood cells, and cancer results. Drugs that prevent expression of *BCL2* result in apoptosis of the tumor cells and shrinkage of the tumor. These drugs are being tested for other cancers as well.

Expression of transcription factor genes determines organ placement in plants

Like animals, plants have organs—for example, leaves and roots. Many plants form flowers, and many flowers are composed of four types of organs: sepals, petals, stamens (male reproductive organs), and carpels (female reproductive organs). These floral organs occur in concentric *whorls* (rings), with groups of each organ type encircling a central axis. The sepals are on the outside and the carpels are on the inside (**FIGURE 14.11A**).

In the model plant *Arabidopsis thaliana* (thale cress), the whorls develop from a meristem of about 700 undifferentiated cells arranged in a dome, which is at the growing point on the stem. A group of genes called **organ identity genes** encode proteins that act in combination to produce specific whorl features (**FIGURE 14.11B,C**):

- Genes in class A are expressed in whorls 1 and 2 (which form sepals and petals, respectively).
- Genes in class B are expressed in whorls 2 and 3 (which form petals and stamens).
- Genes in class C are expressed in whorls 3 and 4 (which form stamens and carpels).

Genes in classes A, B, and C code for transcription factors (see Concept 11.3) that are active as dimers, that is, proteins with two polypeptide subunits (see Figure 14.11B). Gene regulation in these cases is *combinatorial*—that is, the composition of the dimer depends on the location of the cell and determines which

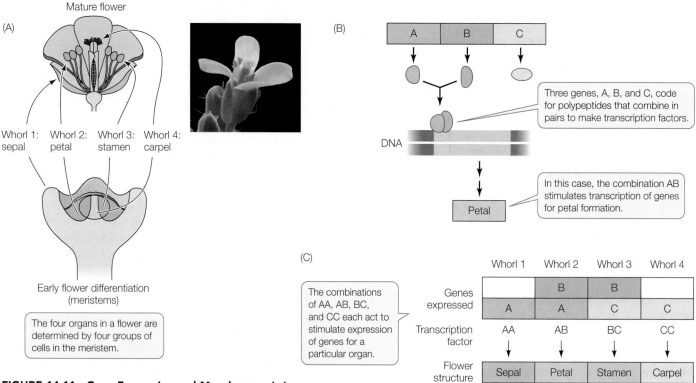

FIGURE 14.11 Gene Expression and Morphogenesis in *Arabidopsis* Flowers (A) The four organs of a flower—carpels (yellow), stamens (green), petals (purple), and sepals (pink)—grow in whorls that develop from the floral meristem. (B) Floral organs are determined by three genes whose polypeptide products combine in pairs to form transcription factors. (C) Combinations of polypeptide subunits in transcription factors activate gene expression for specific organs.

genes will be activated. For example, a dimer made up of two class A monomers, which is found only in cells in whorl 1, activates transcription of the genes that make sepals; a dimer made up of A and B monomers results in petals, and so forth.

Two lines of experimental evidence support this model of organ identity gene function:

- *Loss-of-function mutations*: for example, a mutation in a class A gene results in no sepals or petals. Instead, carpels and stamens form in their place. In any organism, the replacement of one organ for another is called *homeosis*, and this type of mutation is a **homeotic mutation**.

- *Gain-of-function mutations*: for example, a promoter for a class C gene can be artificially coupled to a class A gene that is then placed in a transgenic plant. In this case, class A genes are expressed in all four whorls (in the first two whorls from their natural promoters, and in the second two whorls from the class C gene promoter), resulting in only sepals and petals.

Transcription of the floral organ identity genes is controlled by other gene products, including the LEAFY protein, which is a transcription factor that stimulates the expression of the class A, B, and C genes. This finding has practical applications. It usually takes 6–20 years for a citrus tree to produce flowers and fruits. Scientists have made transgenic orange trees expressing the *LEAFY* gene coupled to a strongly expressed promoter. These trees flower and fruit years earlier than normal trees.

Morphogen gradients provide positional information during development

During development, the key cellular question "What am I (or what will I be)?" is often answered in part by "Where am I?" Think of the cells in the developing nematode, which develop into different parts of the vulva depending on their positions relative to the anchor cell. The same is true for the cells between the digits of a developing hand and in different whorls of a developing flower. This spatial "sense" is called **positional information**.

Positional information often comes in the form of an inducer called a **morphogen**, which diffuses from one group of cells to surrounding cells, setting up a concentration gradient. There are two requirements for a signal to be considered a morphogen:

- It must specifically affect target cells.

- Different concentrations of the signal must cause different effects.

Developmental biologist Lewis Wolpert uses the "French flag model" to explain the action of morphogens (**FIGURE 14.12A**). This model can be applied to the differentiation of the vulva in *C. elegans* (see Figure 14.9), which relies on a gradient of LIN-3. Another example can be seen in the development of vertebrate limbs.

As we described earlier for humans, the vertebrate limb develops from a paddle-shaped *limb bud* (**FIGURE 14.12B**). The cells that develop into different digits must receive positional information; if they do not, the limb will not be organized properly (imagine a hand with only thumbs or only little fingers). How do the cells know where they are? A group of cells at the posterior base of the limb bud, just where it joins the

(A)

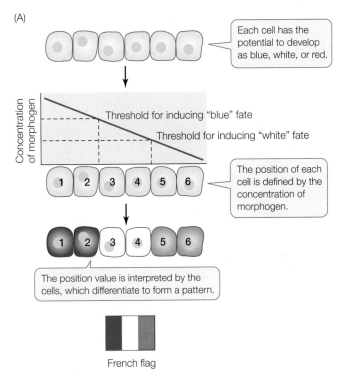

(B)

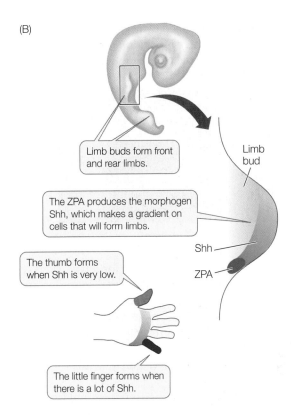

FIGURE 14.12 The French Flag Model (A) In the "French flag" model, a concentration gradient of a diffusible morphogen signals each cell to specify its position. (B) The zone of polarizing activity (ZPA) in the limb bud of the embryo secretes the morphogen Sonic hedgehog (Shh). Cells in the bud form different digits depending on the concentration of Shh.

body wall, is called the *zone of polarizing activity* (ZPA). The cells of the ZPA secrete a morphogen called *Sonic hedgehog* (Shh), which forms a gradient that determines the posterior–anterior (little finger to thumb) axis of the developing limb. The cells getting the highest dose of Shh form the little finger; those getting the lowest dose develop into the thumb. Recall the French flag model when considering the gradient of Shh.

A cascade of transcription factors establishes body segmentation in the fruit fly

A major achievement in studies of developmental biology has been the ever-advancing description of how morphogens act in another model organism, the fruit fly *Drosophila melanogaster*.

As you will see in Concept 14.4, the molecular events that underlie fruit fly development turn out to be similar to events that occur in many other organisms, including ourselves. So they merit examination in some detail.

The insect body is made up of *segments* that differ from one another. The adult fly has an anterior *head* (composed of several fused segments), three different *thoracic* segments, and eight *abdominal* segments at the posterior end. Each segment gives rise to different body parts: for example, antennae and eyes develop from head segments, wings from the thorax, and so on.

The life cycle of *Drosophila* from fertilized egg to adult takes about 2 weeks. The egg hatches into a larva, which then forms a pupa, which finally is transformed into the adult fly. By the time a larva appears—about 24 hours after fertilization—there are recognizable segments. The thoracic and abdominal segments all look similar, but *the fates of the cells to become different adult segments are already determined*.

As with other organisms, fertilization in *Drosophila* leads to a rapid series of mitoses. However, until the 13th division cycle,

APPLY THE CONCEPT

Spatial differences in gene expression lead to morphogenesis

Molecular biologists can attach genes to active promoters and insert them into cells. This results in higher than normal expression (overexpression) of the gene. What would happen if the genes listed at right were inserted and overexpressed? Explain your answers.

1. *ced-3* in embryonic neuron precursors of *C. elegans*
2. *myoD* in undifferentiated myoblasts
3. *Sonic hedgehog* in a chick limb bud
4. *LEAFY* in a leaf bud meristem of *Arabidopsis*

these mitoses are not accompanied by cytokinesis. So instead of a multicellular embryo, there is a *multinucleate embryo* (the nuclei are brightly stained in the micrographs below):

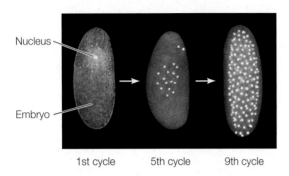

This is an important property, as it allows for relatively easy diffusion of morphogens from one part of the embryo to another (because there are no cell membranes to cross). As you will see, many of these morphogens end up affecting transcription in the nuclei. The determination events in the first 24 hours will be our focus here.

Experimental genetics was used to elucidate these events in *Drosophila*:

- First, developmental mutations were identified. For example, a mutant strain might produce larvae with two heads or no segments.

- Second, each mutant was compared with wild-type flies, and the gene responsible for the developmental mistake, and its protein product (if appropriate), was isolated.

- Finally, experiments with the gene (making transgenic flies) and protein (injecting the protein into an egg or embryo) were done to confirm the proposed developmental pathway.

Together, these approaches revealed a sequential pattern (cascade) of gene expression that results in the determination of each segment within 24 hours after fertilization. Several classes of genes are involved:

- *Maternal effect genes*, which set up the major axes (anterior–posterior and dorsal–ventral) of the egg

- *Segmentation genes*, which determine the boundaries and polarity of each of the segments

- *Hox genes*, which determine what organ will be made at a given location

MATERNAL EFFECT GENES Like the eggs and early embryos of many other organisms (see Figure 14.8), *Drosophila* eggs and larvae are characterized by unevenly distributed cytoplasmic determinants. These molecular determinants, which include both mRNAs and proteins, are the products of specific **maternal effect genes**. These genes are transcribed in the cells of the mother's ovary that surround what will be the anterior portion of the egg. The transcription products are passed to the egg by cytoplasmic bridges. Two maternal effect genes, called *bicoid* and *nanos*, help determine the anterior–posterior axis of the egg. (The dorsal–ventral, or back–belly, axis is determined by other maternal effect genes that will not be described here.)

The mRNAs for *bicoid* and *nanos* diffuse from the mother's cells into what will be the anterior end of the egg. The *bicoid* mRNA is translated to produce Bicoid protein, which diffuses away from the anterior end, establishing a concentration gradient in the egg cytoplasm:

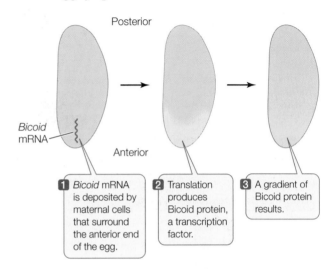

At this point, the egg is in its multinucleate stage.

Where it is present in sufficient concentration, Bicoid acts as a transcription factor to stimulate the transcription of the *hunchback* gene in the early embryo. Consequently, the nuclei nearest the anterior end are most active in the transcription of the *hunchback* gene, and the resulting gradient of Hunchback protein (itself a transcription factor) establishes the head, or anterior, region.

Meanwhile, the egg's cytoskeleton transports the *nanos* mRNA from the anterior end of the egg, where it was deposited by the maternal cells, to the posterior end, where it is translated. This results in a gradient of the Nanos protein, with the highest concentration at the posterior end. At that end, the Nanos protein inhibits the translation of *hunchback* mRNA, preventing accumulation of Hunchback protein. Thus, the actions of both Bicoid and Nanos establish a Hunchback protein gradient, which determines the anterior and posterior ends of the embryo by influencing the gene expression patterns of the nuclei along the gradient.

The events involving *bicoid*, *nanos*, and *hunchback* begin before fertilization and continue after it, during the multinucleate stage, which lasts a few hours. At this stage, the embryo looks like a bunch of indistinguishable nuclei under the light microscope. But the fates of the individual nuclei and the cells they will occupy have already begun to be determined. After the anterior and posterior ends have been established, the next step in pattern formation in fruit flies is the determination of segment number and locations.

SEGMENTATION GENES The number, boundaries, and polarity of the *Drosophila* larval segments are determined by proteins encoded by the **segmentation genes**. These genes are expressed when there are about 6,000 nuclei in the embryo (about 3 hours after fertilization). Three classes of segmentation genes act one after the other to regulate finer and finer

FIGURE 14.13 A Gene Cascade Controls Pattern Formation in the *Drosophila* Embryo
Maternal effect genes induce gap, pair rule, and segment polarity genes—collectively referred to as segmentation genes. By the end of this cascade, a group of nuclei at the anterior of the embryo, for example, is determined to become the first head segment in the adult fly. In the micrographs at left various staining methods have been used to highlight the different gene products.

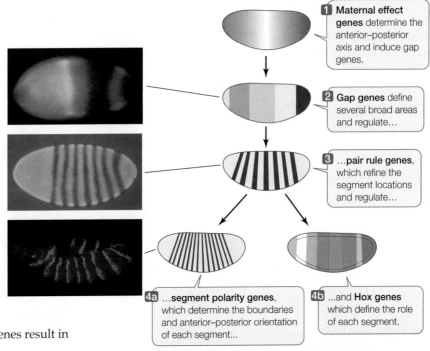

1 **Maternal effect genes** determine the anterior–posterior axis and induce gap genes.

2 **Gap genes** define several broad areas and regulate…

3 …**pair rule genes**, which refine the segment locations and regulate…

4a …**segment polarity genes**, which determine the boundaries and anterior–posterior orientation of each segment…

4b …and **Hox genes** which define the role of each segment.

details of the segmentation pattern (**FIGURE 14.13**):

- **Gap genes** organize broad areas along the anterior–posterior axis. Mutations in gap genes result in gaps in the body plan—the omission of several consecutive larval segments.

- **Pair rule genes** divide the embryo into units of two segments each. Mutations in pair rule genes result in embryos missing every other segment.

- **Segment polarity genes** determine the boundaries and anterior–posterior organization of the individual segments. Mutations in segment polarity genes can result in segments in which posterior structures are replaced by reversed (mirror-image) anterior structures.

By the end of this cascade, nuclei throughout the embryo "know" which segment they will be part of in the adult fly. The next set of genes in the cascade determines the form and function of each segment.

HOX GENES Hox genes encode a family of transcription factors that are expressed in different combinations along the length of the embryo, and help determine cell fates within each segment. Hox gene expression tells certain cells in a head segment to make eyes, other cells in a thorax segment to make wings, and so on. Hox genes are homeotic genes that are shared by all animals, and they are functionally similar to the organ identity genes of plants (see Figure 14.11).

How do we know that the Hox genes determine segment identity? A clue comes from homeotic mutations in *Drosophila*. A mutation in the Hox gene *Antennapedia* causes legs to grow on the head in place of antennae (**FIGURE 14.14**). When another Hox gene, *bithorax*, is mutated, an extra pair of wings grows in a thoracic segment where wings do not normally occur. So the normal (wild-type) functions of the Hox genes must be to "tell" a segment what organ to form. The *Antennapedia* and *bithorax* genes both encode transcription factors and have a common 180-bp sequence called the **homeobox** (from which Hox genes get their name). The homeobox encodes a 60-amino acid sequence called the *homeodomain*. The homeodomain

(A)

Antenna

(B)

Leg where antenna should be

FIGURE 14.14 A Homeotic Mutation in *Drosophila* Mutations of the Hox genes cause body parts to form on inappropriate segments. (A) A wild-type fruit fly. (B) An *Antennapedia* mutant fruit fly. Mutations such as this reveal the normal role of the *Antennapedia* gene in determining segment function.

recognizes and binds to a specific DNA sequence in the promoters of its target genes. As you will see in Concept 14.4, this domain is found in transcription factors that regulate development in many other animals with an anterior–posterior axis.

LINK To review mechanisms of transcriptional regulation, see Concept 11.3

Do You Understand Concept 14.3?

- Outline the steps that determine that a nucleus and cell in the developing *Drosophila* embryo will be part of an antenna.
- Compare the determination of organ identity in *Arabidopsis* and *Drosophila*.
- How does the "French flag" model apply to development of *Drosophila*?
- In the nematode nervous system, 302 neurons come from 405 precursors. How would you investigate the fate of the 103 "missing" cells? What gene(s) might be involved?

We have seen how positional information leads to changes in the expression of key developmental genes, which in turn control morphogenesis. It turns out that there are remarkable similarities in the genes used to guide development in diverse organisms, and this has led to a new way to look at the evolution of development.

concept 14.4 Gene Expression Pathways Underlie the Evolution of Development

The discovery of the genes that control the development of *Drosophila* provided biologists with tools to investigate the development of other organisms. For example, when scientists used homeobox DNA as a hybridization probe (see Figure 10.7) to search for similar genes elsewhere, they found that the homeobox is present in many genes in many other organisms. This, and other astounding discoveries that followed, showed a similarity in the molecular events underlying morphogenesis in organisms ranging from flies to fish to mammals. These results suggested that just as the forms of organisms evolved through descent with modification from a common ancestor, so did the molecular mechanisms that produce those forms. Biologists started to ask new questions about the interplay between evolutionary and developmental processes—a field of study called **evolutionary developmental biology (evo-devo)**. The major ideas of evo-devo are:

- Many groups of animals and plants, even distantly related ones, share similar molecular mechanisms for morphogenesis and pattern formation. These mechanisms can be thought of as "toolkits," in the same sense that a few tools in a carpenter's toolkit can be used to build many different structures.

- The molecular pathways that determine different developmental processes, such as anterior–posterior polarity and organ formation in animals, operate independently from one another. This is called **modularity**.

- Changes in the location and timing of expression of key genes are important in the evolution of new body forms and structures.

- Development produces morphology, and much of morphological evolution occurs by modifications of existing development genes and pathways, rather than the introduction of radically new developmental mechanisms.

yourBioPortal.com

**Go to WEB ACTIVITY 14.2
Plant and Animal Development**

Developmental genes in distantly related organisms are similar

Initially through hybridization with a homeobox probe, and then by genome sequencing and comparative genomics (see Concept 12.3), biologists have found that diverse animals share numerous molecular pathways that govern gene expression during development. For example, fruit fly homeotic genes such as *Antennapedia* and *bithorax* have mouse (and human) orthologs that are also developmental genes. Remarkably, these orthologous genes are arranged along a chromosome in both the fruit fly and mouse in the same order as they are expressed along the anterior–posterior axis of their embryos (**FIGURE 14.15**). This means that the positional information controlled by these genes has been conserved, even as the structures formed at each position have changed. Over the millions of years that have elapsed since these animals diverged from a common ancestor, the genes in question have mostly been maintained, suggesting that their functions were favored under many different conditions.

These and other examples have led biologists to the idea that certain developmental mechanisms, controlled by specific DNA sequences, have been conserved over long periods during the evolution of multicellular organisms. These sequences comprise a **genetic toolkit**, the contents of which have been modified and reshuffled over the course of evolution to produce the remarkable diversity of plants, animals, and other organisms in the world today.

Genetic switches govern how the genetic toolkit is used

The genetic toolkit is also used to generate diverse structures within a single organism. Different structures can evolve within a single organism using a common set of genetic instructions because there are mechanisms called **genetic switches** (also called molecular switches) that control how the genetic toolkit is used. These mechanisms involve promoters and the transcription factors that bind them, as well as enhancers and their associated proteins. The signal cascades that converge on and operate these switches determine when and where genes will

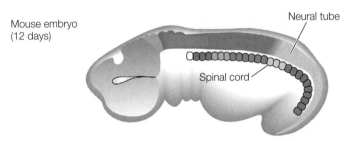

FIGURE 14.15 Regulatory Genes Show Similar Expression Patterns Orthologous genes encoding similar transcription factors are expressed in similar patterns along the anterior–posterior axes of both insects and vertebrates. Orthologous genes and the locations of their expression are indicated by shared colors. The mouse (and human) Hox genes are actually present in multiple copies; this prevents a single mutation from having drastic effects.

Genetic switches translate positional information in the embryo into specific changes in gene expression. In this way, they play key roles in determining the developmental pathways of different modules. For example, each Hox gene codes for a transcription factor that is expressed in a particular segment or appendage of the developing fruit fly. The pattern formation and functioning of each segment depend on the unique Hox gene or combination of Hox genes that is expressed in the segment. Genetic switches determine when and where each Hox gene (each one a tool from the genetic toolkit) is expressed.

Consider the formation of fruit fly wings. *Drosophila* has three thoracic segments, the first of which bears no wings. The second segment bears the large forewings, and the third segment bears small hindwings, called *halteres*, that function as balancing organs. The *Hox* gene *Ultrabithorax* (*Ubx*) is not expressed in forewing cells, but hindwing cells express the *Ubx* gene because a set of genetic switches activates it in the third thoracic segment. The Ubx transcription factor turns off genes that promote the formation of the veins and other structures of the forewing, and it turns on genes that promote the formation of hindwing features (**FIGURE 14.16**).

This same genetic switch turns on different genes in butterflies, where instead of turning *off* wing-forming genes in the third thoracic segment as in fruit flies, Ubx turns them *on*, so that full hindwings develop. Therefore, a simple genetic change in the effect of Ubx on genes that promote wing development results in a major morphological difference in the wings of flies and butterflies. This phenomenon—the same switch having different effects on target genes in different species—is important in evolution.

be turned on and off. Multiple switches control each gene by influencing its expression at different times and in different places. In this way, elements of the genetic toolkit can be involved in multiple developmental processes while still allowing individual modules to develop independently. You have seen several examples of this phenomenon in Concept 14.3. For example, the morphogenesis of flower organs is determined by combinations of transcription factors (the ABC model; see Figure 14.11) acting at specific times and locations.

FRONTIERS Signaling pathways involving transcription factors often regulate numerous pathways in an organism. *Sonic hedgehog* (*Shh*) is a gene in humans that is orthologous to the developmentally important *hedgehog* gene in fruit flies. In both organisms, the product is a secreted protein that activates a signaling pathway involved in several aspects of development. In humans, over-production of Shh occurs in basal cell carcinoma, a tumor that is a common form of skin cancer. Understanding the pathway activated by Shh may lead to better treatment of this tumor.

APPLY THE CONCEPT

Gene expression pathways underlie the evolution of development

Control of eye formation during development of many animals is under the control of a genetic switch involving a transcription factor. Here are partial DNA sequences for the control gene from two organisms:

Mouse
Pax6 gene: 5′-GTATCCAACGGTTGTGTGAGTAAAATT-3′

Fruit fly
eyeless gene: 5′-GTATCAAATGGATGTGTGAGCAAAATT-3′

1. Calculate the percentage of identity between the two DNA sequences.

2. Use the genetic code (see Figure 10.11) to determine the amino acid sequences encoded by the two regions, and calculate their percentage of similarity.

3. The fruit fly and mouse evolved from a common ancestor about 500 million years ago. Comment on your answers to 1 and 2 in terms of the evolution of developmental pathways.

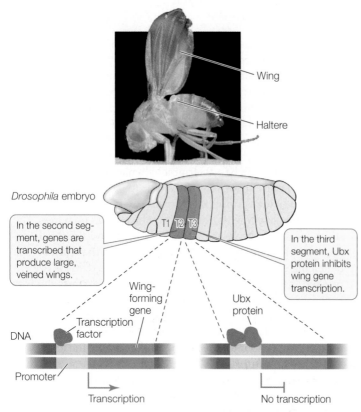

Drosophila embryo

In the second segment, genes are transcribed that produce large, veined wings.

In the third segment, Ubx protein inhibits wing gene transcription.

Wing-forming gene

Ubx protein

Transcription factor

DNA

Promoter

Transcription

No transcription

FIGURE 14.16 Segments Differentiate under Control of Genetic Switches   The binding of a single protein, Ultrabithorax (Ubx), determines whether a thoracic segment in *Drosophila* produces full wings or halteres (balancers).

Modularity allows for differences in the pattern of gene expression among organisms

The modularity of development means that the molecular pathways for developmental processes such as organ formation operate independently from one another. For example, an *Antennapedia* mutant grows a leg where an antenna should be, but all of the mutant's other organs develop normally and in their proper places. On an evolutionary time scale, modularity means that the timing and position of a particular developmental process can change without disrupting the whole organism.

yourBioPortal.com

Go to ANIMATED TUTORIAL 14.4
Modularity

TIMING DIFFERENCES   The genes regulating the development of a module may be expressed at different developmental stages or for different durations in different species, a phenomenon called **heterochrony**. An example is the evolution of the giraffe's neck. As in virtually all mammals (with the exception of manatees and sloths), there are seven vertebrae in the neck of the giraffe. So the giraffe did not get a longer neck than other mammals by adding vertebrae. Instead, each of the cervical (neck) vertebrae of the giraffe is much longer than those of other mammals (**FIGURE 14.17**). How does this happen?

Bones grow because of the proliferation of cartilage-producing cells called *chondrocytes*. Bone growth is stopped by a signal that results in death of the chondrocytes and calcification of the bone matrix. In giraffes this signaling process is delayed in the cervical vertebrae, with the result that these vertebrae grow longer. Thus, the evolution of longer necks occurred through *changes in the timing of expression* of the genes that control bone formation.

SPATIAL DIFFERENCES   Changes in the *spatial expression pattern* of a developmental gene can also result in evolutionary change. For example, the difference in foot webbing in ducks versus chickens is determined by an alteration in the spatial expression of a single gene. The feet of all bird embryos have webs of skin that connect their toes. This webbing is retained in adult ducks (and other aquatic birds) but not in adult chickens (and other non-aquatic birds). The loss of webbing is caused by a

(A) Giraffe

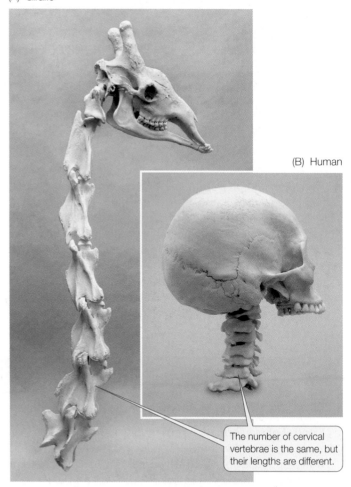

(B) Human

The number of cervical vertebrae is the same, but their lengths are different.

FIGURE 14.17 Heterochrony in the Development of a Longer Neck   There are seven vertebrae in the neck of the giraffe (left) and human (right; not to scale). But the vertebrae of the giraffe are much longer (25 cm compared with 1.5 cm) because during development, growth continues for a longer period of time. This timing difference is called heterochrony.

signaling protein called bone morphogenetic protein 4 (BMP4) that instructs the cells in the webbing to undergo apoptosis. The death of these cells destroys the webbing between the toes.

Embryonic duck and chicken hindlimbs both express the *BMP4* gene in the webbing between the toes, but they differ in expression of a gene called *Gremlin*, which encodes a BMP *inhibitor* protein (**FIGURE 14.18**). In ducks, but not chickens, the *Gremlin* gene is expressed in the webbing cells. The Gremlin protein inhibits the BMP4 protein from signaling for apoptosis, and the result is a webbed foot.

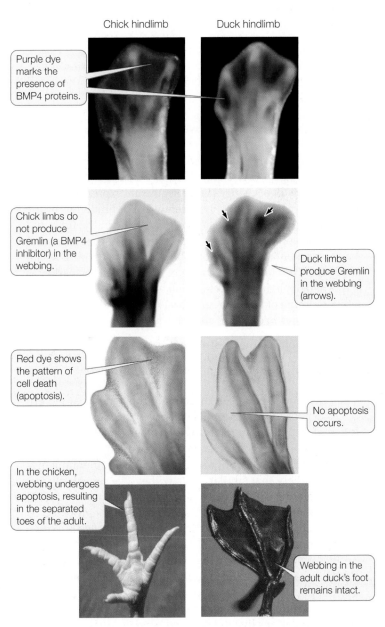

Chick hindlimb   Duck hindlimb

Purple dye marks the presence of BMP4 proteins.

Chick limbs do not produce Gremlin (a BMP4 inhibitor) in the webbing.

Duck limbs produce Gremlin in the webbing (arrows).

Red dye shows the pattern of cell death (apoptosis).

No apoptosis occurs.

In the chicken, webbing undergoes apoptosis, resulting in the separated toes of the adult.

Webbing in the adult duck's foot remains intact.

FIGURE 14.18 Changes in Gremlin Expression Correlate with Changes in Hindlimb Structure The left column of photos shows the development of a chicken's foot; the right column shows foot development in a duck. Gremlin protein in the webbing of the duck foot inhibits BMP4 signaling, thus preventing the embryonic webbing from undergoing apoptosis.

Do You Understand Concept 14.4?

- Describe the major ideas of evolutionary developmental biology.

- What is the evidence that there was a common ancestor for the developmental pathways leading to segment identity in insects and the organization of the spinal cord in mice?

- What is the evidence that changes in the transcription of a single gene can lead to differences in morphogenesis between different regions of an embryo?

- Examine Figure 14.18 and the related text. If Gremlin protein were added to the webbed region between the developing toes of a chicken, what would be the result?

The tools in the genetic toolkit guide morphogenesis of an individual organism in accordance with the individual's species. In the next concept, we turn to the roles some of these same tools play in the evolution of new forms and new species.

concept 14.5 Developmental Genes Contribute to Species Evolution but Also Pose Constraints

The genetic switches that allow different structures to develop in different regions of an embryo can also give rise to major morphological differences among species. We have already seen examples of this in the development of cervical vertebrae in giraffes versus other mammals, and in the differences in Gremlin expression that determine whether a bird's foot will be webbed or not. Thus changes in the timing and position of genetic switch activity can generate morphological variation, which then can be acted on by natural selection.

At the same time, the reliance of development on a genetic toolkit with a limited set of tools places constraints on how radically organisms can differ from one another. Four decades ago, the French geneticist François Jacob made the analogy that evolution works like a tinker, assembling new structures by *combining and modifying the available materials*, and not like an engineer, who is free to develop dramatically different designs (say, a jet engine to replace a propeller-driven engine). The evolution of morphology has not been governed by the appearance of radically new genes, but by modifications of existing genes and their regulatory pathways. Thus, developmental genes and their expression constrain evolution in two major ways:

- Nearly all evolutionary innovations are modifications of previously existing structures.

- The genes that control development are highly conserved; that is, the regulatory genes themselves change slowly over the course of evolution.

FIGURE 14.19   A Mutation in a Hox Gene Changed the Number of Legs in Insects   In the insect lineage (blue box) of the arthropods, a change to the *Ubx* gene resulted in a protein that inhibits the *Dll* gene, which is required for legs to form. Because insects express this modified *Ubx* gene in their abdominal segments, no legs grow from these segments. Other arthropods, such as centipedes, do grow legs from their abdominal segments.

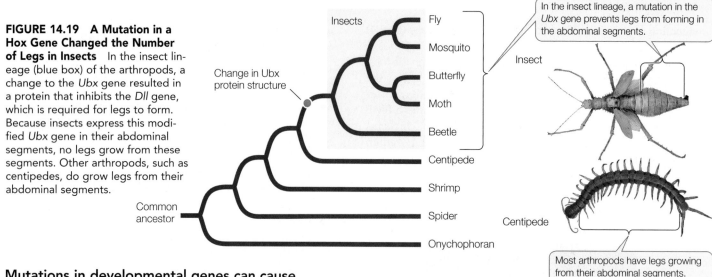

In the insect lineage, a mutation in the *Ubx* gene prevents legs from forming in the abdominal segments.

Most arthropods have legs growing from their abdominal segments.

Mutations in developmental genes can cause major evolutionary changes

The arthropods (which include crustaceans, centipedes, spiders, and insects) are segmented, with head, thoracic, and abdominal segments. In centipedes, both thoracic and abdominal segments form legs; but in insects, only the thoracic segments do. Arthropods express a gene called *Distal-less* (*Dll*) that causes legs to form from segments. What shuts down *Dll* expression in insect abdominal segments? The product of the Hox gene *Ubx* (which by now should be a familiar tool from the genetic toolkit) is produced in arthropod abdominal segments. But it has very different effects in different organisms. In centipedes, the Ubx protein apparently activates expression of the *Dll* gene to promote the formation of legs. But during the evolution of insects, a change in the *Ubx* gene sequence resulted in a modified Ubx protein that *represses Dll* expression in abdominal segments, so that leg formation is inhibited. A phylogenetic tree of arthropods shows that this change in *Ubx* occurred in the ancestor of insects at the same time that abdominal legs were lost (**FIGURE 14.19**).

LINK   Arthropod evolution and diversity are discussed in Concept 23.4

Evolution proceeds by changing what's already there

The features of organisms almost always evolve from preexisting features in their ancestors. New "wing genes" did not suddenly appear in birds and bats; instead, wings arose as modifications of existing structures (**FIGURE 14.20**). *In vertebrates, the wings are modified limbs*.

Although the wings of birds and bats look different, they are made from the same basic parts. Like limbs, wings have a common structure: a humerus that connects to the body; two longer bones, the radius and ulna, that project away from the humerus; and then metacarpals and phalanges (digits). During development these bones take on different lengths and weights in different organisms. For example, the phalanges are relatively short in birds and relatively long in bats. These differences arise from changes in the molecular mechanisms

that control development, as we saw for cervical vertebrae in giraffes.

Developmental controls also influence how organisms *lose structures*. The ancestors of present-day snakes lost their forelimbs as a result of changes in the segmental expression of Hox genes. The snake lineage subsequently lost its hindlimbs by the loss of expression of the *Sonic hedgehog* gene in the limb bud tissue. But some snake species such as boas and pythons still have rudimentary pelvic bones and upper leg bones. Recall that we encountered an ortholog of Sonic hedgehog earlier as a morphogen in hand development (see p. 273). This is yet another example of how the same basic genetic tools are used in different ways in different species.

Conserved developmental genes can lead to parallel evolution

The nucleotide sequences of many of the genes that govern development have been highly conserved throughout the evolution of multicellular organisms—in other words, these genes exist in similar form across a broad spectrum of species.

The existence of highly conserved developmental genes makes it likely that similar traits will evolve repeatedly, especially among closely related species—a phenomenon called **parallel phenotypic evolution**. A good example is provided by a small fish, the three-spined stickleback (*Gasterosteus aculeatus*: "bony stomach with spines").

Sticklebacks are widely distributed across the Atlantic and Pacific oceans and are also found in many freshwater lakes. Marine populations of this species spend most of their lives at sea but return to fresh water to breed. Members of freshwater populations live in lakes and never journey to salt water. Genetic evidence shows that freshwater populations have arisen independently from marine populations many times, most recently at the end of the last ice age. Marine sticklebacks have several structures that protect them from predators: well-developed pelvic bones with pelvic spines, and bony plates. In the freshwater populations descended from them, this body

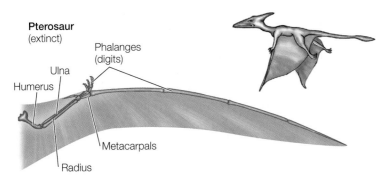

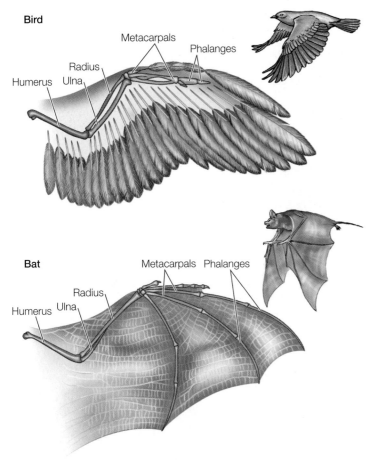

FIGURE 14.20 Wings Evolved Three Times in Vertebrates
The wings of pterosaurs (the earliest flying vertebrates, which lived from 265 to 220 million years ago), birds, and bats are all modified forelimbs constructed from the same skeletal components. However, the components have different forms in the different groups of vertebrates.

armor is greatly reduced, and dorsal and pelvic spines are much shorter or even lacking (**FIGURE 14.21**).

The difference between marine and freshwater sticklebacks is not induced by environmental conditions. Marine species that are reared in fresh water still grow spines. Not surprisingly, the difference is due to a gene that affects development. The *Pituitary homoeobox transcription factor 1* (*Pitx1*) gene codes for a transcription factor that is normally expressed in regions of the developing embryo that form the head, trunk, tail, and pelvis of the marine stickleback. However, David Kingsley and his colleagues

at Stanford University have found that in independent populations from Japan, British Columbia, California, and Iceland, an enhancer sequence that drives the expression of *Pitx1* has been deleted, so that the gene has evolved so that it is no longer expressed in the pelvis, and the spines do not develop. *This same gene sequence has evolved to produce similar phenotypic changes in several independent populations*, and is thus a good example of parallel evolution. What could be the common selective mechanism in these cases? Possibly, the decreased predation pressure in the freshwater environment allows for increased reproductive success in animals that invest less energy in the development of unnecessary protective structures.

Do You Understand Concept 14.5?

- How have diverse body forms such as wings evolved by means of modifications in the functioning of existing genes?

- What would happen at the molecular and phenotypic levels if a *Ubx* gene from a spider replaced the *Ubx* gene in a fertilized insect egg?

- When several freshwater populations of stickleback fish were compared, the *Pitx1* gene coding region was identical to that of marine populations. But in every case, the freshwater fish had mutations in noncoding regions of *Pitx1* that led to reduced expression. What might these noncoding region mutations be?

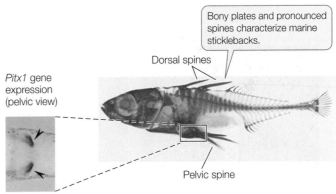

Bony plates and pronounced spines characterize marine sticklebacks.

Pitx1 gene expression (pelvic view)

Dorsal spines

Pelvic spine

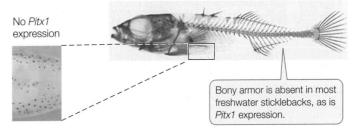

No *Pitx1* expression

Bony armor is absent in most freshwater sticklebacks, as is *Pitx1* expression.

FIGURE 14.21 Parallel Phenotypic Evolution in Sticklebacks
A developmental gene, *Pitx1*, encodes a transcription factor that stimulates the production of plates and spines. This gene is active in marine sticklebacks but is mutated and inactive in various freshwater populations of the fish. The fact that this mutation is found in geographically distant and isolated freshwater populations is evidence for parallel evolution.

Q&A QUESTION — Why are stem cells so useful?

ANSWER The use of stem cells to help heal injuries in horses applies many of the concepts of developmental biology. The zygote and early embryonic cells are totipotent and can give rise to an entire new organism. But as development proceeds, cells become restricted in their fate (Concept 14.1). In adult mammals, however, not all the cells are fully differentiated. Some are rapidly dividing, undifferentiated stem cells, as occur in fat tissue.

Gene expression underlies the differentiation of stem cells (Concept 14.2). In the case of fat stem cells, specific proteins are formed as they differentiate into bone, cartilage, muscle, fat, organ, blood vessel, and nerve cells (**FIGURE 14.22**). In the clinical setting where fat stem cells are injected into a damaged tissue, as in the case of Greg's Gold's damaged tendon, the environment around the stem cells largely determines the products of cell differentiation, probably by inducers (Concept 14.3).

Fat stem cells are being used therapeutically in humans. As with horses, human fat stem cells are easily isolated, and indeed a procedure has been developed to isolate them in large quantities in the operating room, where they can then be used to repair tissues, for example, after surgery for breast cancer.

FIGURE 14.22 Differentiation Potential of Stem Cells from Fat Stem cells isolated from adipose tissue can differentiate into a variety of cell types.

Adipose stem cells

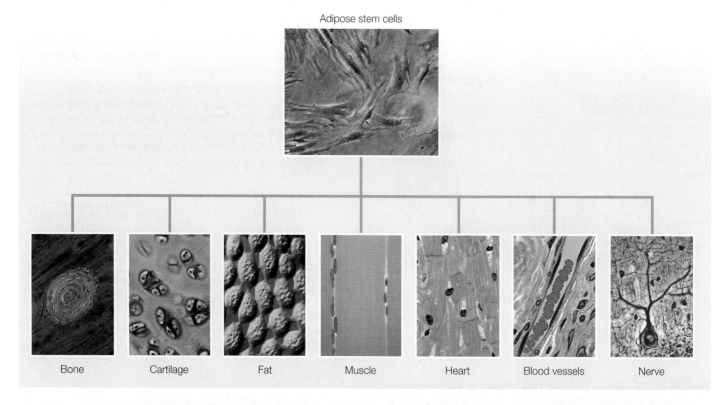

Bone   Cartilage   Fat   Muscle   Heart   Blood vessels   Nerve

14 SUMMARY

concept 14.1 Development Involves Distinct but Overlapping Processes

- A multicellular organism begins its development as an embryo, and several embryonic stages precede the birth of an independent organism. **Review Figure 14.1 and WEB ACTIVITY 14.1**

- The processes of development are **determination**, **differentiation**, **morphogenesis**, and **growth**.

- The zygote is **totipotent**; it is capable of producing an entire new organism, with every type of cell in the adult body. **Review INTERACTIVE TUTORIAL 14.1**

- The ability to create clones from differentiated cells demonstrates the principle of **genomic equivalence**. **Review Figures 14.3 and 14.4 and WORKING WITH DATA 14.1**

- **Multipotent** stem cells occur in the growing regions of plants and in most tissues of mammals. They constantly divide and form a pool of cells that can be used for differentiation to specialized cells. **Review Figure 14.5**

- **Pluripotent** stem cells can form every cell type of a mammal, but not an entire organism. They occur in the embryo and can be induced to form in the laboratory. They may have medical uses. **Review Figure 14.6 and ANIMATED TUTORIAL 14.1**

concept 14.2 Changes in Gene Expression Underlie Cell Differentiation in Development

- Differential gene expression results in cell differentiation. Transcription factors are especially important in regulating gene expression during differentiation.

- Cytoplasmic segregation—the unequal distribution of **cytoplasmic determinants** in the egg, zygote, or early embryo—can establish **polarity** and lead to cell fate determination. **Review Figure 14.8 and ANIMATED TUTORIAL 14.2**

- **Induction** is a process by which embryonic animal tissues direct the development of neighboring cells and tissues by secreting chemical signals called inducers. **Review Figure 14.10**

concept 14.3 Spatial Differences in Gene Expression Lead to Morphogenesis

- During development, selective elimination of cells by apoptosis results from the expression of specific genes.

- In plants, **organ identity genes** encode polypeptides that associate to form transcription factors. These proteins determine the formation of flower organs. **Review Figure 14.11**

- Both plants and animals use **positional information** in the form of a signal called a **morphogen** to stimulate cell determination. **Review Figure 14.12**

- In the fruit fly *Drosophila melanogaster*, a cascade of transcriptional activation sets up the axes of the embryo, the development of the segments, and the determination of cell fate in each segment. **Review Figure 14.13 and ANIMATED TUTORIAL 14.3**

- **Hox genes** determine cell fate in the embryos of many animals. The **homeobox** is a DNA sequence found in Hox genes and other genes that code for transcription factors. The sequence of amino acids encoded by the homeobox is called the homeodomain.

concept 14.4 Gene Expression Pathways Underlie the Evolution of Development

- **Evolutionary developmental biology (evo-devo)** is the modern study of the evolutionary aspects of development, and it focuses on molecular mechanisms. **Review WEB ACTIVITY 14.2**

- Hox genes have evolved from a common ancestor. **Review Figure 14.15**

- Genes such as Hox genes underlie evolutionary changes in morphology that produce major differences in body forms.

- Evolutionary diversity is produced using a modest number of regulatory genes. **Review Figure 14.16**

- The transcription factors and chemical signals that govern pattern formation in the bodies of multicellular organisms, and the genes that encode them, can be thought of as a **genetic toolkit**.

- The bodies of developing and mature organisms are organized into self-contained units that can be modified independently in space and time. **Review ANIMATED TUTORIAL 14.4**

- Changes in **genetic switches** that determine where and when a set of genes will be expressed underlie the transformation of an individual from egg to adult.

concept 14.5 Developmental Genes Contribute to Species Evolution but Also Pose Constraints

- Evolutionary innovations are modifications of preexisting structures. **Review Figure 14.20**

- Because many genes that govern development have been highly conserved, similar traits are likely to evolve repeatedly, especially among closely related species. This process is called **parallel phenotypic evolution**. **Review Figure 14.21**

PART

3

Evolution

Mechanisms of Evolution

15

On November 11, 1918, an armistice agreement signed in France signaled the end of World War I. But the death toll from four years of war was soon surpassed by the casualties of a massive influenza epidemic that began in the spring of 1918 among soldiers in a U.S. Army barracks. Over the next 18 months, this particular strain of flu virus spread across the globe, killing more than 50 million people worldwide—more than twice the number of World War I-related combat deaths.

The 1918–1919 pandemic was noteworthy because the death rate among young adults—who are usually less likely to die from influenza than are the elderly or the very young—was 20 times higher than in flu epidemics before or since. Why was that particular virus so deadly, especially to typically hardy individuals? The 1918 flu strain triggered an especially intense reaction in the human immune system. This overreaction meant that people with strong immune systems were likely to be more severely affected.

In most cases, however, our immune system helps us fight viruses; this response is the basis of *vaccination*. Since 1945, programs to administer flu vaccines have helped keep the number and severity of influenza outbreaks in check. Last year's vaccine, however, will probably not be effective against this year's virus. New strains of flu virus are evolving continuously, ensuring genetic variation in the population. If these viruses did not evolve, we would become resistant to them and annual vaccination would become unnecessary. But because they do evolve, biologists must develop a new and different flu vaccine each year.

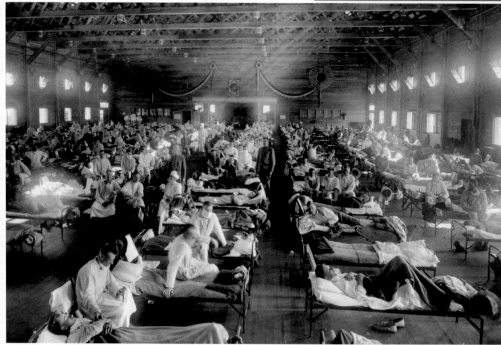

Vertebrate immune systems recognize proteins on the viral surface, and changes in these proteins mean that the virus can escape immune detection. Virus strains with the greatest number of changes to their surface proteins are most likely to avoid detection and infect their hosts, and thus have an advantage over other strains. Biologists can observe evolution in action by following changes in influenza virus proteins from year to year.

We learn a great deal about the processes of evolution by examining rapidly evolving organisms such as viruses, and these studies contribute to the development of evolutionary theory. Evolutionary theory, in turn, is put to practical uses, such as the development of better strategies for combating deadly diseases.

How do biologists use evolutionary theory to develop better flu vaccines?

You will find the answer to this question on page 312.

Flu victims are treated at a U.S. Army hospital in 1918.

KEY CONCEPTS

15.1 Evolution Is Both Factual and the Basis of Broader Theory

15.2 Mutation, Selection, Gene Flow, Genetic Drift, and Nonrandom Mating Result in Evolution

15.3 Evolution Can Be Measured by Changes in Allele Frequencies

15.4 Selection Can Be Stabilizing, Directional, or Disruptive

15.5 Genomes Reveal Both Neutral and Selective Processes of Evolution

15.6 Recombination, Lateral Gene Transfer, and Gene Duplication Can Result in New Features

15.7 Evolutionary Theory Has Practical Applications

concept 15.1 Evolution Is Both Factual and the Basis of Broader Theory

That biological populations change over time, or **evolve**, is a fact that is not disputed by scientists. We can, and do, observe evolutionary change on a regular basis, both in laboratory experiments and in natural populations. We measure the rate at which new mutations arise, observe the spread of new genetic variants through a population, and see the effects of genetic change on the form and function of organisms. In the fossil record, we observe the long-term morphological changes (which are the result of underlying genetic changes) that have occurred among living organisms. These underlying changes in the genetic makeup of populations (sometimes referred to as *microevolution*) drive the origin and extinction of species and fuel the diversification of life (*macroevolution*).

In addition to observing and recording physical changes over evolutionary time, biologists have accumulated a large body of evidence about *how* these changes occur, and about *what* evolutionary changes have occurred in the past. The resulting understanding and application of the mechanisms of evolutionary change to biological problems is known as **evolutionary theory**.

Evolutionary theory has many useful applications. We constantly apply it to the study and treatment of diseases; to the development of better agricultural crops and practices; and to the development of industrial processes that produce new molecules with useful properties. At a more basic level, knowledge of evolutionary theory allows biologists to understand how life diversified and has provided insight into how species interact. It also helps us to make predictions about the biological world.

In everyday speech, people tend to use the word "theory" to mean an untested hypothesis, or even a guess. But *evolutionary theory* does not refer to any single hypothesis, and it certainly is not guesswork. The concept of evolutionary change among living organisms was present among a few scientists even before Charles Darwin so clearly described his observations, presented his conclusions, and articulated the premise of natural selection in *The Origin of Species*. The rediscovery of Mendel's experiments and the subsequent establishment of the principles of genetic inheritance early in the 1900s set the stage for vast amounts of research. By the end of the twentieth century, findings from many fields of biology firmly upheld Darwin's basic premises about the common ancestry of life and the role of natural selection as an important mechanism of evolution. Today a vast and rich array of geological, morphological, and molecular data all support and expand the factual basis of evolution.

When we refer to evolutionary theory, we are referring to our understanding of the mechanisms that result in genetic changes in populations over time and to our use of that understanding to interpret changes in and interactions among living organisms. We can directly observe the evolution of influenza viruses, but it is evolutionary theory that allows us to apply our observations to the task of developing more effective vaccines. Several mechanisms of evolutionary change are recognized, and the scientific community is continually using evolutionary theory to expand its understanding of how and when these mechanisms apply to particular biological problems.

Darwin and Wallace introduced the idea of evolution by natural selection

In the early 1800s, it was not yet evident to many people that life evolves. But several biologists had suggested that the species living on Earth had changed over time—that is, that evolution had taken place. Jean-Baptiste Lamarck, for one, presented strong evidence for the fact of evolution in 1809, but his ideas about *how* it occurred were not convincing. At that time, no one had yet envisioned a viable mechanism for evolution.

Charles Robert Darwin

In the 1820s, a young Charles Darwin became passionately interested in the subjects of geology (with its new sense of Earth's great age) and *natural history* (the scientific study of how different organisms function and carry out their lives in nature). Despite these interests, he planned, at his father's behest, to become a doctor. But surgery conducted without anesthesia nauseated Darwin, and he gave up medicine to study at Cambridge University for a career as a clergyman in the Church of England. Always more interested in science than in theology, he gravitated toward scientists on the faculty, especially the botanist John Henslow. In 1831, Henslow recommended Darwin for a position on HMS *Beagle*, a Royal Navy vessel that was preparing for a survey voyage around the world.

HMS *Beagle*

Whenever possible during the 5-year voyage (**FIGURE 15.1**), Darwin went ashore to study rocks and to observe and collect plants and animals. He noticed striking differences between the species he saw in South America and those of Europe. He

FIGURE 15.1 The Voyage of the *Beagle* The mission of HMS *Beagle* was to chart the oceans and collect oceanographic and biological information from around the world. The world map indicates the ship's path; the inset map shows the Galápagos Islands, whose organisms were an important source of Darwin's ideas on natural selection.

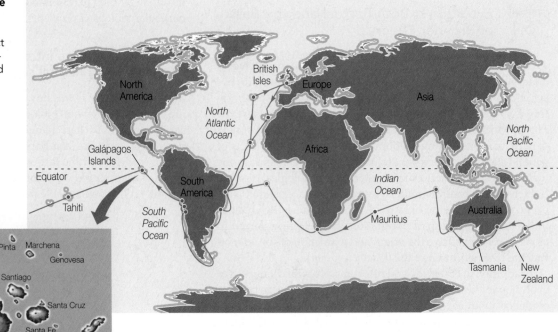

observed that the species of the temperate regions of South America (Argentina and Chile) were more similar to those of tropical South America (Brazil) than they were to temperate European species. When he explored the islands of the Galápagos archipelago west of Ecuador, he noted that most of the animals were *endemic* to the islands (that is, unique and found nowhere else), although they were similar to animals found on the mainland of South America. Darwin also observed that the fauna of the Galápagos differed from island to island. He postulated that some animals had come to the archipelago from mainland South America and had subsequently undergone different changes on each of the islands. He wondered what might account for these changes.

When he returned to England in 1836, Darwin continued to ponder his observations. His ruminations were strongly influenced by the geologist Charles Lyell, who had recently popularized the idea that Earth had been shaped by slow-acting forces that are still at work today. Darwin reasoned that similar thinking could be applied to the living world. Within a decade, he had developed the framework of an explanatory theory for evolutionary change based on three major propositions:

• Species are not immutable; they change over time.

• Divergent species share a common ancestor.

• The mechanism that produces changes in species is **natural selection**: the differential survival and reproduction of individuals in a population based on variation in their traits.

The first of these propositions was not unique to Darwin; several earlier authors had argued for the fact of evolution. A more revolutionary idea was his second proposition, that *divergent species are related to one another through common descent*. In 1844, Darwin wrote a long essay on his third proposition, describing natural selection as the mechanism of evolution, but he was reluctant to publish it, preferring to assemble more evidence first.

Darwin's hand was forced in 1858, when he received a letter and manuscript from another traveling English naturalist, Alfred Russel Wallace, who was studying the biota of the Malay Archipelago. Wallace asked Darwin to evaluate his manuscript, which included an explanation of natural selection almost identical to Darwin's. Darwin was at first dismayed, believing Wallace to have preempted his idea. Parts of Darwin's 1844 essay, together with Wallace's manuscript, were presented to the Linnaean Society of London on July 1, 1858, thereby crediting both men for the idea of natural selection. Darwin then worked quickly to finish his own book, *The Origin of Species*, which was published the following year.

yourBioPortal.com

Go to ANIMATED TUTORIAL 15.1
Natural Selection

Although Darwin and Wallace independently articulated the concept of natural selection, Darwin developed his ideas first. Furthermore, *The Origin of Species* proved to be a stunning work of scholarship that provided exhaustive evidence from many fields supporting both the premise of evolution itself and the notion of natural selection as a mechanism of evolution. Thus both concepts are more closely associated with Darwin than with Wallace.

The publication of *The Origin of Species* in 1859 stirred considerable interest (and controversy) among scientists and the public alike. Scientists spent much of the rest of the nineteenth century amassing biological and paleontological data to test evolutionary ideas and document the history of life on Earth. By 1900, the fact of biological evolution (by then defined as change in the physical characteristics of populations over time) was established beyond any reasonable doubt. But the *genetic* basis of evolutionary change was not yet understood.

Evolutionary theory has continued to develop over the past century

In 1900, several individuals rediscovered the work of Gregor Mendel (which had been published in 1866 but rarely read or cited), and the basic mechanisms of genetic inheritance began to be unraveled. In the first decades of the twentieth century, Thomas Hunt Morgan's studies on fruit flies led to his discovery of the role of chromosomes in inheritance. In the 1920s and early 1930s, the major principles of population genetics were established, the genetic basis of new variation (i.e., mutations) began to be understood, and mechanisms of evolution such as genetic drift were described (see Concept 15.2). This work set the stage for a "modern synthesis" of genetics and evolution that took place over the period 1936–1947. Some of the major contributors to this synthesis and a few of their books are listed in **FIGURE 15.2**.

Although chromosomes were now understood to be the basis of genetic transmission in eukaryotes, their molecular structure remained a mystery until soon after the modern synthesis. Then, in 1953, Watson and Crick published their paper on the

structure of DNA, opening the door to our current detailed understanding of molecular evolutionary mechanisms. By the 1960s, biologists could study and document changes in allele frequencies in populations over time (see Concept 15.3). Most of this early work necessarily focused on variants of proteins that differed within and between populations and species; even

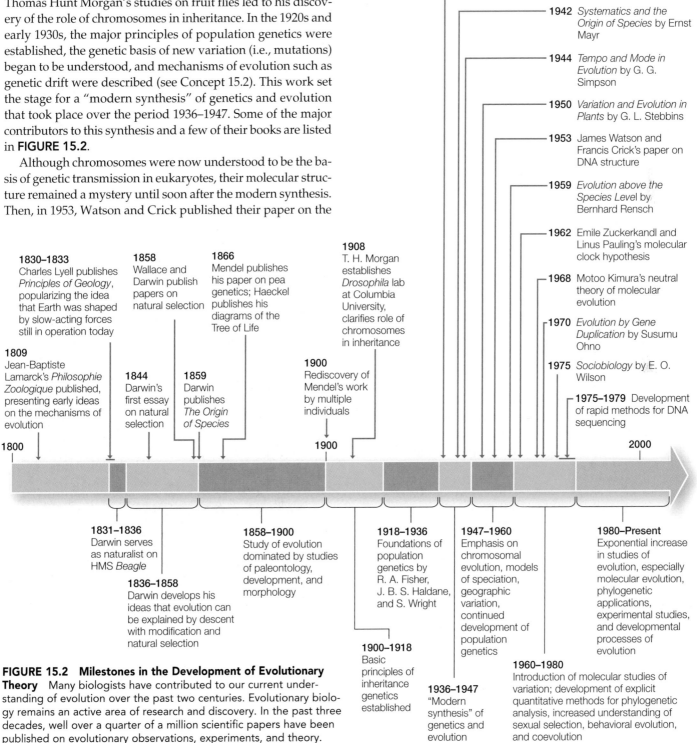

FIGURE 15.2 Milestones in the Development of Evolutionary Theory Many biologists have contributed to our current understanding of evolution over the past two centuries. Evolutionary biology remains an active area of research and discovery. In the past three decades, well over a quarter of a million scientific papers have been published on evolutionary observations, experiments, and theory.

though the molecular structure of DNA was known, it was not yet practical to sequence long stretches of DNA. Nonetheless, many important advances occurred in evolutionary theory during this time (see Figure 15.2), and these advances were not focused solely on a genetic understanding of evolution. E. O. Wilson's 1975 book *Sociobiology*, for example, invigorated studies of the evolution of behavior (a subject that had fascinated Darwin).

In the late 1970s, several techniques were developed that allowed the rapid sequencing of long stretches of DNA, which in turn allowed researchers to ascertain the amino acid sequences of proteins. This ability opened a new door for evolutionary biologists, who can now explore the structure of genes and proteins and document evolutionary changes within and between species in ways never before possible.

Do You Understand Concept 15.1?

- Why do biologists speak of "evolutionary theory" if the facts of evolution are not in doubt?

- Why do you think Darwin and Wallace formulated their ideas on natural selection at about the same time?

- Discuss the significance of each of the following scientific advances for evolutionary theory:
  a. Elucidation of the principles of chromosomal inheritance
  b. The discovery of DNA, its structure, and the universal genetic code
  c. Technology that allows us to sequence long segments of DNA

  Keep your discussion in mind as you continue reading this chapter.

Darwin had important insights into the mechanisms of evolution, even though he had a poor understanding of genetic transmission. Next we'll consider the primary mechanisms of evolution in light of our current understanding of genetics.

concept 15.2 Mutation, Selection, Gene Flow, Genetic Drift, and Nonrandom Mating Result in Evolution

Although the word "evolution" is often used in a general sense to mean simply "change," in a biological context **evolution** refers specifically to changes in the genetic makeup of populations over time. Developmental changes that occur in a single individual over the course of the life cycle are not the result of evolutionary change. Evolution is genetic change occurring in a **population**—a group of individuals of a single species that live and interbreed in a particular geographic area at the same time. It is important to remember that *individuals do not evolve; populations do.*

The premise of natural selection was one of Darwin's principal insights and has been demonstrated to be an important

mechanism of evolution, but natural selection does not act alone. Three additional processes—gene flow, genetic drift, and nonrandom mating—affect the genetic makeup of populations over time. Before we consider how these processes change the frequencies of gene variants in a population, however, we need to understand how mutation brings such variants into existence.

Mutation generates genetic variation

The origin of genetic variation is mutation. As described in Concept 9.3, a *mutation* is any change in the nucleotide sequences of an organism's DNA. The process of DNA replication is not perfect, and some changes appear almost every time a genome is replicated. Mutations occur randomly with respect to an organism's needs; it is natural selection acting on this random variation that results in adaptation. Most mutations are either harmful to their bearers (*deleterious mutations*) or have no effect (*neutral mutations*). But a few mutations are *beneficial*, and even previously deleterious or neutral alleles may become advantageous if environmental conditions change. In addition, mutation can restore genetic variation that other evolutionary processes have removed. Thus mutation both creates and helps maintain genetic variation in populations.

Mutation rates can be high, as we saw in the case of the influenza viruses described at the opening of this chapter, but in many organisms the mutation rate is very low (on the order of 10^{-8} to 10^{-9} changes per base pair of DNA per generation). Even low overall mutation rates, however, create considerable genetic variation, because each of a large number of genes may change, and populations often contain

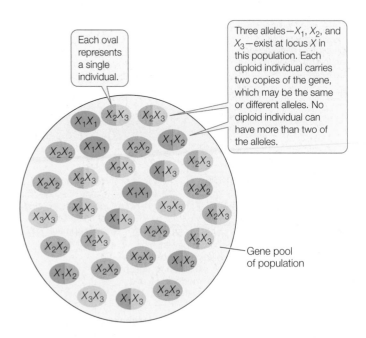

FIGURE 15.3 A Gene Pool A gene pool is the sum of all the alleles found in a population or at a particular locus. This figure shows the gene pool for one locus, X. The allele frequencies in this case are 0.20 for X_1, 0.50 for X_2, and 0.30 for X_3 (see Figure 15.10).

large numbers of individuals. For example, if the probability of a point mutation (an addition, deletion, or substitution of a single base) were 10^{-9} per base pair per generation, then each human gamete—the DNA of which contains 3×10^9 base pairs—would average three new point mutations ($3 \times 10^9 \times 10^{-9} = 3$), and each zygote would carry an average of six new mutations. The current human population of about 7 billion people would thus be expected to carry about 42 billion new mutations (i.e., changes in the nucleotide sequences of their DNA that were not present one generation earlier). So even though the mutation rate in humans is low, human populations still contain enormous genetic variation on which other evolutionary mechanisms can act.

As a result of mutation, different forms of a gene, known as *alleles*, may exist at a particular chromosomal locus. At any particular locus, a single diploid individual has no more than two of the alleles found in the population to which it belongs. The sum of all copies of all alleles at all loci found in a population constitutes its **gene pool (FIGURE 15.3)**. (We can also refer to the gene pool for a particular chromosomal locus or loci.) The gene pool is the sum of the genetic variation in the population. The proportion of each allele in the gene pool is the **allele frequency.** Likewise, the proportion of each genotype among individuals in the population is the **genotype frequency**.

> **LINK** Review the nature of alleles and genetic inheritance in Concepts 8.1 and 8.2

Selection on genetic variation leads to new phenotypes

As a result of mutation, the gene pools of nearly all populations contain variation for many characters. Selection on different characters in a single European species of wild mustard produced many important crop plants (**FIGURE 15.4**). Agriculturalists were able to achieve these results because the original mustard population had genetic variation for the characters of interest (such as stem thickness or number of leaves).

Darwin compared this **artificial selection** by animal and plant breeders with natural selection. Many of Darwin's observations on the nature of variation and selection came from domesticated plants and animals. Darwin bred pigeons and thus knew firsthand the astonishing diversity in color, size, form, and behavior that breeders could achieve (**FIGURE 15.5**). He recognized close parallels between selection by breeders and selection in nature. Natural selection resulted in traits that helped organisms survive and reproduce more effectively; artificial selection resulted in traits that were preferred by the human breeders, for whatever reason.

Laboratory experiments also demonstrate the existence of considerable genetic variation in populations. In one such experiment, investigators bred populations of the fruit fly *Drosophila melanogaster* with high or low numbers of bristles on their abdomens from an initial population with intermediate numbers of bristles. After 35 generations, all flies in both the high- and low-bristle lineages had bristle numbers that fell well outside the range found in the original population (**FIGURE 15.6**). Selection for high and low bristle numbers resulted in new combinations of the many different genes that were present in the original population, so that the phenotypic variation seen in subsequent generations fell outside the phenotypic variation seen in the original population.

Natural selection increases the frequency of beneficial mutations in populations

Darwin knew that far more individuals of most species are born than survive to reproduce. He also knew that, although offspring tend to resemble their parents, the offspring of most organisms are not identical either to their parents or to one another. He suggested that slight differences among individuals affect the chance that a given individual will survive and

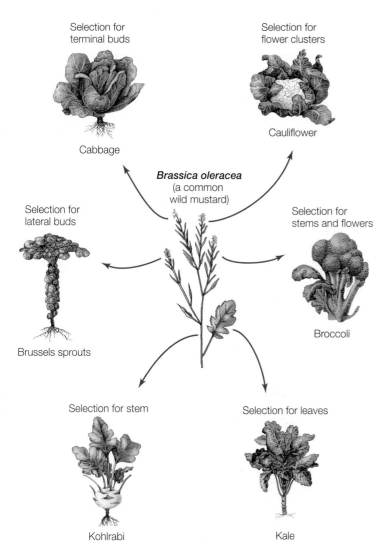

Selection for terminal buds
Cabbage

Selection for flower clusters
Cauliflower

Selection for lateral buds
Brussels sprouts

Brassica oleracea
(a common wild mustard)

Selection for stems and flowers
Broccoli

Selection for stem
Kohlrabi

Selection for leaves
Kale

FIGURE 15.4 Many Vegetables from One Species All of the crop plants shown here derive from a single wild mustard species. European agriculturalists produced these crop species by selecting and breeding plants with unusually large buds, stems, leaves, or flowers. The results substantiate the vast amount of variation present in a gene pool.

FIGURE 15.5 Artificial Selection Charles Darwin raised pigeons as a hobby and noted similar forces at work in artificial and natural selection. The "fancy" pigeons shown here represent three of the more than 300 varieties derived from the wild rock pigeon (*Columba livia*; left) by artificial selection for character traits such as color and feather distribution.

reproduce, which increases the frequency of the favored trait in the next generation. A favored trait that evolves through natural selection is known as an **adaptation**; this word is used to describe both the trait itself and the process that produces the trait.

Biologists regard an organism as being *adapted* to a particular environment when they can demonstrate that a slightly different organism reproduces and survives less well in that environment. To understand adaptation, biologists compare the performances of individuals that differ in their traits.

Natural selection also acts to remove deleterious mutations from populations. Individuals with deleterious mutations are less likely to survive and reproduce, so they are less likely to pass their alleles on to the next generation.

Gene flow may change allele frequencies

Few populations are completely isolated from other populations of the same species. Migration of individuals and movements of gametes between populations—a phenomenon called **gene flow**—can change allele frequencies in a population. If the arriving individuals survive and reproduce in their new location, they may add new alleles to the population's gene pool, or they may change the frequencies of alleles present in the original population.

LINK If gene flow between two populations stops, those populations may diverge and become different species; see Concept 17.2

Genetic drift may cause large changes in small populations

In small populations, **genetic drift**—random changes in allele frequencies from one generation to the next—may produce large changes in allele frequencies over time. Harmful alleles may increase in frequency, and rare advantageous alleles may be lost. Even in large populations, genetic drift can influence the frequencies of neutral alleles (which do not affect the survival and reproductive rates of their bearers).

As an example, suppose there are only two females in a small population of mice, and one of these females carries a newly arisen dominant allele that produces black fur. Even in the absence of any selection, it is unlikely that the two females will produce exactly the same number of offspring. Even if they do produce identical litter sizes and identical numbers of litters, chance events that have nothing to do with genetic

FIGURE 15.6 Artificial Selection Reveals Genetic Variation When investigators subjected *Drosophila melanogaster* to artificial selection for abdominal bristle number, that trait evolved rapidly. The graph shows the number of flies with different numbers of bristles in the original population and after 35 generations of artificial selection. The bristle numbers of the selected lineages clearly diverged from those of the original population.

characteristics are likely to result in differential mortality among their offspring. If each female produces one litter, but a flood envelops the black female's nest and kills all of her offspring, the novel allele could be lost from the population in just one generation. In contrast, if the wild-type female's litter is lost, then the frequency of the newly arisen allele (and phenotype) for dark fur will rise dramatically in just one generation.

Genetic drift also operates when a population is reduced dramatically in size. Even populations that are normally large may occasionally pass through environmental events that only a small number of individuals survive, a situation known as a **population bottleneck**. The effect of genetic drift in such a situation is illustrated in **FIGURE 15.7**, in which red and yellow beans represent two alleles of a gene. Most of the beans in the small sample of the "population" that "survives" the bottleneck event are, just by chance, red, so the new population has a much higher frequency of red beans than the previous generation had. In a real population, the red and yellow allele frequencies would be described as having "drifted."

A population forced through a bottleneck is likely to lose much of its genetic variation. For example, when Europeans first arrived in North America, millions of greater prairie-chickens (*Tympanuchus cupido*) inhabited the midwestern prairies. As a result of hunting and habitat destruction by the new settlers, the Illinois population of this species plummeted from about 100 million birds in 1900 to fewer than 50 individuals in the 1990s. A comparison of DNA from birds collected in Illinois during the middle of the twentieth century with DNA from the surviving population in the 1990s showed that Illinois prairie-chickens have lost most of their genetic diversity. Loss of genetic variation in small populations is one of the problems facing biologists who attempt to protect endangered species.

Genetic drift can have similar effects when a few pioneering individuals colonize a new region. Because of its small size, the colonizing population is unlikely to possess all of the alleles found in the gene pool of its source population. The resulting change in genetic variation, called a **founder effect**, is equivalent to that in a large population reduced by a bottleneck.

Nonrandom mating can change genotype or allele frequencies

Mating patterns often alter genotype frequencies because the individuals in a population do not choose mates at random. For example, self-fertilization (*selfing*) is common in many groups of organisms, especially plants. Any time individuals mate preferentially with other individuals of the same genotype (including themselves), homozygous genotypes will increase in frequency and heterozygous genotypes will decrease in frequency over time. The opposite effect (more heterozygotes, fewer homozygotes) is expected when individuals mate primarily or exclusively with individuals of different genotypes.

Sexual selection results from a specific type of nonrandom mating in which an organism's phenotype influences its ability to attract mates. For example, female peacocks may choose their male mates on the basis of his bright tail feathers and associated mating display. Males with brighter feathers are more likely to attract females. The higher reproductive success of colorful males results in an increase in the frequency of the alleles associated with colorful tail feathers in the next generation.

In *The Origin of Species*, Darwin devoted a few pages to sexual selection, but in 1871 he wrote an entire book about it: *The Descent of Man, and Selection in Relation to Sex*. Sexual selection was Darwin's explanation for the evolution of conspicuous characters that would appear to inhibit survival, such as bright colors, long tails, and elaborate courtship displays in males of many species. He hypothesized that these features either improved the ability of their bearers to compete for access to mates (*intrasexual selection*) or made their bearers more attractive to members of the opposite sex (*intersexual selection*). The concept of sexual selection was either ignored or questioned for many decades, but recent investigations have demonstrated its importance.

Whereas Darwin associated natural selection with traits that enhance the survival of their bearers or their bearers' descendants, sexual selection is primarily about successful reproduction. Of course, an animal must survive long enough to reproduce, but if it survives and fails to reproduce, it makes no contribution to the next generation. Thus sexual selection may favor traits that enhance an individual's chances of reproduction even when these traits reduce its chances of survival. For example, females may be more likely to see or hear males with a given trait (and thus be more likely to mate with those males), even though the favored trait also increases the chances that the male will be seen or heard by a predator.

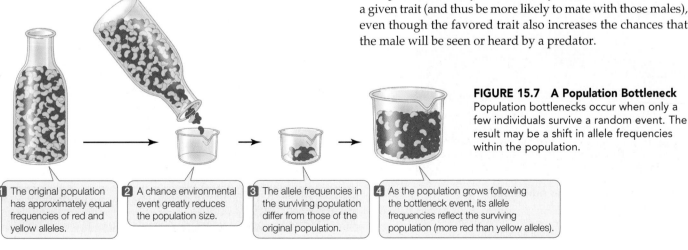

FIGURE 15.7 A Population Bottleneck Population bottlenecks occur when only a few individuals survive a random event. The result may be a shift in allele frequencies within the population.

1 The original population has approximately equal frequencies of red and yellow alleles.

2 A chance environmental event greatly reduces the population size.

3 The allele frequencies in the surviving population differ from those of the original population.

4 As the population grows following the bottleneck event, its allele frequencies reflect the surviving population (more red than yellow alleles).

In other cases, a male's sexual signal directly indicates a successful genotype. In many species of frogs, for example, females prefer males with low-frequency calls. Males' calls vary with body size, and a low-frequency call is indicative of a large-bodied frog. Frogs exhibit *indeterminate growth*—that is, they continue to grow throughout their lives—so a large frog is a long-lived frog, and size is an indication of survivorship. In this case, the sexual signal represents what is known as an *honest signal* of the male's ability to survive in the local environment.

> **LINK** Some of the animal behaviors that have evolved in response to sexual selection are described in Concepts 41.5 and 41.6

One example of a trait that Darwin attributed to sexual selection is the remarkable tail of the male African long-tailed widowbird (*Euplectes progne*), which is longer than the bird's head and body combined (**FIGURE 15.8**). Male widowbirds normally select, and defend from other males, a territory where they perform courtship displays to attract females. To investigate whether sexual selection drove the evolution of widowbird tails, Malte Andersson, a behavioral ecologist at Gothenburg University in Sweden, clipped the tails of some captured male widowbirds and lengthened the tails of others by gluing on additional feathers. He then cut and reglued the tail feathers of still other males, which served as controls. Both short- and long-tailed males successfully defended their display territories, indicating that

Euplectes progne

FIGURE 15.8 What Is the Advantage? The extensive tail of the male African long-tailed widowbird actually inhibits its ability to fly. Darwin attributed the evolution of this seemingly nonadaptive trait to sexual selection.

INVESTIGATION

FIGURE 15.9 Sexual Selection in Action Behavioral ecologist Malte Andersson tested Darwin's hypothesis that excessively long tails evolved in male widowbirds because female preference for longer-tailed males increased their mating and reproductive success.

HYPOTHESIS

Female widowbirds prefer to mate with the male that displays the longest tail; longer-tailed males thus are favored by sexual selection because they will father more offspring.

METHOD

1. Capture males and artificially lengthen or shorten tails by cutting or gluing on feathers. In a control group, cut and replace tails to their normal length (to control for the effects of tail-cutting).
2. Release the males to establish their territories and mate.
3. Count the nests with eggs or young on each male's territory.

RESULTS

Male widowbirds with artificially shortened tails established and defended display sites sucessfully but fathered fewer offspring than did control or unmanipulated males. Males with artificially lengthened tales fathered the most offspring.

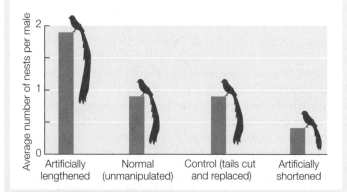

CONCLUSION

Sexual selection in *Euplectes progne* has favored the evolution of long tails in the male.

ANALYZE THE DATA

Are the differences plotted above significantly different? See Working with Data 15.1 at **yourBioPortal.com** for a simple method to test the statistical significance of the differences using the following data.

	Number of nests per male		
Group	Shortened tail	Control	Elongated tail
1	0	0	2
2	0	0	2
3	2	3	5
4	1	2	4
5	0	1	2
6	0	1	2
7	0	1	0
8	0	0	0
9	1	0	0

Go to **yourBioPortal.com** for original citations, discussions, and relevant links for all INVESTIGATION figures.

a long tail does not confer an advantage in male–male competition. However, males with artificially elongated tails attracted about four times more females than did males with shortened tails (**FIGURE 15.9**).

Why do female widowbirds prefer males with long tails? One possibility is that the ability to grow and maintain a costly feature such as a long tail may indicate that the male bearing it is vigorous and healthy, even though the tail impairs his ability to fly. If so, then females that are attracted to long tails are indirectly attracted to vigorous, healthy males, which are likely to carry beneficial genes that will lead to higher survivorship of their offspring.

Do You Understand Concept 15.2?

- How do deleterious, neutral, and beneficial mutations differ?
- Can you explain how natural selection results in an increase in the frequency of beneficial alleles in a population over time, and a decrease in the frequency of deleterious alleles?
- How can genetic drift cause large changes in small populations?
- How do selfing and sexual selection differ in their expected effects on genotype and allele frequencies over time?

The mechanisms of mutation, selection, gene flow, genetic drift, and nonrandom mating can all result in evolutionary change. We will consider next how evolutionary change that results from these mechanisms is measured.

concept 15.3   Evolution Can Be Measured by Changes in Allele Frequencies

Much of evolution occurs through gradual changes in the relative frequencies of different alleles in a population from one generation to the next. Major genetic changes can also be sudden, as happens when two formerly separated populations merge and hybridize, or when genes within a population are duplicated within the genome (see Concept 15.6). We can measure evolution by looking at changes in allele frequencies in populations.

Allele frequencies are usually estimated in locally interbreeding populations. To measure allele frequencies in a population precisely, we would need to count every allele at every locus in every individual in the population. Fortunately, we do not need to make such complete measurements because we can reliably estimate allele frequencies for a given locus by counting alleles in a sample of individuals from the population. The sum of all allele frequencies at a locus is equal to 1, so measures of allele frequency range from 0 to 1.

An allele's frequency is calculated using the following formula:

$$\frac{\text{number of copies of the allele in the population}}{\text{total number of copies of all alleles in the population}}$$

If only two alleles (we'll call them A and a) for a given locus are found among the members of a diploid population, those alleles can combine to form three different genotypes: AA, Aa, and aa (see Figure 15.3). A population with more than one allele at a locus is said to be *polymorphic* ("many forms") at that locus. Applying the formula above as shown in **FIGURE 15.10**, we can calculate the relative frequencies of alleles A and a in a population of N individuals as follows:

- Let N_{AA} be the number of individuals that are homozygous for the A allele (AA).
- Let N_{Aa} be the number that are heterozygous (Aa).
- Let N_{aa} be the number that are homozygous for the a allele (aa).

Note that $N_{AA} + N_{Aa} + N_{aa} = N$, the total number of individuals in the population, and that the total number of copies of both

RESEARCH TOOLS

FIGURE 15.10   Calculating Allele and Genotype Frequencies Allele and genotype frequencies for a gene locus with two alleles in the population can be calculated using the equations in panel 1. When the equations are applied to two populations (panel 2), we find that the *frequencies of alleles A and a in the two populations are the same*, but the alleles are distributed differently between heterozygous and homozygous genotypes.

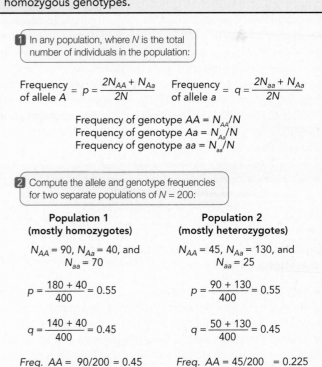

1 In any population, where N is the total number of individuals in the population:

$$\text{Frequency of allele } A = p = \frac{2N_{AA} + N_{Aa}}{2N} \qquad \text{Frequency of allele } a = q = \frac{2N_{aa} + N_{Aa}}{2N}$$

Frequency of genotype $AA = N_{AA}/N$
Frequency of genotype $Aa = N_{Aa}/N$
Frequency of genotype $aa = N_{aa}/N$

2 Compute the allele and genotype frequencies for two separate populations of $N = 200$:

Population 1 (mostly homozygotes)	Population 2 (mostly heterozygotes)
$N_{AA} = 90$, $N_{Aa} = 40$, and $N_{aa} = 70$	$N_{AA} = 45$, $N_{Aa} = 130$, and $N_{aa} = 25$
$p = \dfrac{180 + 40}{400} = 0.55$	$p = \dfrac{90 + 130}{400} = 0.55$
$q = \dfrac{140 + 40}{400} = 0.45$	$q = \dfrac{50 + 130}{400} = 0.45$
Freq. $AA = 90/200 = 0.45$	Freq. $AA = 45/200 = 0.225$
Freq. $Aa = 40/200 = 0.20$	Freq. $Aa = 130/200 = 0.65$
Freq. $aa = 70/200 = 0.35$	Freq. $aa = 25/200 = 0.125$

alleles present in the population is 2*N*, because each individual is diploid. Each *AA* individual has two copies of the *A* allele, and each *Aa* individual has one copy of the *A* allele. Therefore, the total number of *A* alleles in the population is $2N_{AA} + N_{Aa}$. Similarly, the total number of *a* alleles in the population is $2N_{aa} + N_{Aa}$. If *p* represents the frequency of *A*, and *q* represents the frequency of *a*, then

$$p = \frac{2N_{AA} + N_{Aa}}{2N}$$

and

$$q = \frac{2N_{aa} + N_{Aa}}{2N}$$

Figure 15.10 applies these formulas to calculate the allele and genotype frequencies in two hypothetical populations, each containing 200 diploid individuals. The calculations in Figure 15.10 demonstrate two important points. First, notice that for each population, *p* + *q* = 1, which means that *q* = 1 − *p*. So when there are only two alleles at a given locus in a population, we can calculate the frequency of one allele and obtain the second allele's frequency by subtraction. If there is only one allele at a given locus in a population, its frequency is 1: the population is then *monomorphic* at that locus, and the allele is said to be *fixed*.

The second thing to notice is that population 1 (consisting mostly of homozygotes) and population 2 (consisting mostly of heterozygotes) have the same allele frequencies for *A* and *a*. Thus they have the same gene pool for this locus. Because the alleles in the gene pool are distributed differently among individuals, however, the *genotype frequencies* of the two populations differ.

The frequencies of the different alleles at each locus and the frequencies of the different genotypes in a population describe that population's **genetic structure**. Allele frequencies measure the amount of genetic variation in a population; genotype frequencies show how a population's genetic variation is distributed among its members. Other measures, such as the proportion of loci that are polymorphic, are also used to measure variation in populations. With these measurements, it becomes possible to consider how the genetic structure of a population changes or remains the same over generations—that is, to measure evolutionary change.

---

yourBioPortal.com

Go to ANIMATED TUTORIAL 15.2
Hardy–Weinberg Equilibrium

Evolution will occur unless certain restrictive conditions exist

In 1908, the British mathematician Godfrey Hardy and the German physician Wilhelm Weinberg independently deduced the conditions that must prevail if the genetic structure of a population is to remain the same over time. If the conditions they identified do not exist, then evolution will occur. The resulting principle is known as **Hardy–Weinberg equilibrium**. Hardy–Weinberg equilibrium describes a model in which allele frequencies do not change across generations and

genotype frequencies can be predicted from allele frequencies (**FIGURE 15.11**). The principles of Hardy–Weinberg equilibrium apply only to sexually reproducing organisms. Several conditions must be met for a population to be at Hardy–Weinberg equilibrium (which, you should notice, correspond precisely to the five principal mechanisms of evolution discussed in Concept 15.2):

- *There is no mutation.* The alleles present in the population do not change, and no new alleles are added to the gene pool.

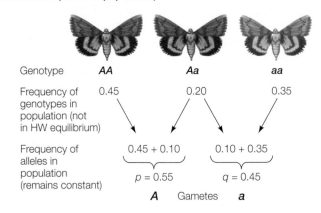

Generation I (Founder population)

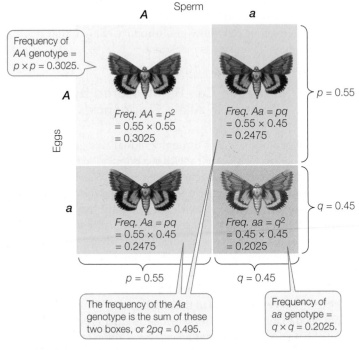

FIGURE 15.11 One Generation of Random Mating Restores Hardy–Weinberg Equilibrium Generation I of this population is made up of migrants from several source populations, and so is not initially in Hardy–Weinberg equilibrium. After one generation of random mating, the allele frequencies are unchanged, and the genotype frequencies return to Hardy–Weinberg expectations. The lengths of the sides of each rectangle are proportional to the allele frequencies in the population; the areas of the rectangles are proportional to the genotype frequencies.

- *There is no differential selection among genotypes.* Individuals with different genotypes have equal probabilities of survival and equal rates of reproduction.

- *There is no gene flow.* There is no movement of individuals into or out of the population or reproductive contact with other populations.

- *Population size is infinite.* The larger a population, the smaller will be the effect of genetic drift.

- *Mating is random.* Individuals do not preferentially choose mates with certain genotypes.

If these "ideal" conditions hold, two major consequences follow. First, the frequencies of alleles at a locus remain constant from generation to generation. Second, following one generation of random mating, the genotype frequencies occur in the following proportions:

Genotype	AA	Aa	aa
Frequency	p^2	$2pq$	q^2

To understand why these consequences are important, start by considering a population that is *not* in Hardy–Weinberg equilibrium, such as generation I in Figure 15.11. This could occur, for example, if the initial population is founded by migrants from several other populations, thus violating the Hardy–Weinberg assumption of no gene flow. In this example, generation I has more homozygous individuals and fewer heterozygous individuals than would be expected under Hardy–Weinberg equilibrium (a condition known as *heterozygote deficiency*).

Even with a starting population that is not in Hardy–Weinberg equilibrium, we can predict that after a single generation of random mating, and if the other Hardy–Weinberg assumptions are not violated, the *allele frequencies* will remain unchanged, but the *genotype frequencies* will return to Hardy–Weinberg expectations. Let's explore why this is true.

In generation I of Figure 15.11, the frequency of the A allele (p) is 0.55. Because we assume that individuals select mates at random, without regard to their genotype, gametes carrying A or a combine at random—that is, as predicted by the allele frequencies p and q. Thus, in this example, the probability that a particular sperm or egg will bear an A allele is 0.55. In other words, 55 out of 100 randomly sampled sperm or eggs will bear an A allele. Because $q = 1 - p$, the probability that a sperm or egg will bear an a allele is $1 - 0.55 = 0.45$.

LINK You may wish to review the discussion of probability and inheritance in Concept 8.1

To obtain the probability of two A-bearing gametes coming together at fertilization, we multiply the two independent probabilities of their occurrence:

$$p \times p = p^2 = (0.55)^2 = 0.3025$$

Therefore, 0.3025, or 30.25 percent, of the offspring in generation II will have homozygous genotype AA. Similarly, the probability of two a-bearing gametes coming together is

$$q \times q = q^2 = (0.45)^2 = 0.2025$$

Thus 20.25 percent of generation II will have the aa genotype.

There are two ways of producing a heterozygote: an A sperm may combine with an a egg, the probability of which is $p \times q$; or an a sperm may combine with an A egg, the probability of which is $q \times p$. Consequently, the overall probability

---

APPLY THE CONCEPT

Evolution can be measured by changes in allele frequencies

Imagine you have discovered a new population of curly-tailed lizards established on an island after immigrants have arrived from several different source populations during a hurricane. You collect and tabulate genotype data (right) for the lactate dehydrogenase gene (*Ldh*) for each of the individual lizards. Use the table to answer the following questions.

1. Calculate the allele and genotype frequencies of *Ldh* in this newly founded population.

2. Is the population in Hardy–Weinberg equilibrium? If not, which genotypes are over- or underrepresented? Given the population's history, what is a likely explanation of your answer?

3. Under Hardy–Weinberg assumptions, what allele and genotype frequencies do you predict for the next generation?

4. Imagine that you are able to continue studying this population and determine the next generation's actual allele and genotype frequencies. What are some of the

INDIVIDUAL NUMBER	SEX	INDIVIDUAL GENOTYPE FOR *Ldh*
1	Male	*Aa*
2	Male	*AA*
3	Female	*AA*
4	Male	*aa*
5	Female	*aa*
6	Female	*AA*
7	Male	*aa*
8	Male	*aa*
9	Female	*Aa*
10	Male	*AA*

principal reasons you might expect the observed allele and genotype frequencies to differ from the Hardy–Weinberg expectations you calculated in question 3?

of obtaining a heterozygote is $2pq$, or 0.495. The frequencies of the AA, Aa, and aa genotypes in generation II of Figure 15.11 now meet Hardy–Weinberg expectations, and the frequencies of the two alleles (p and q) have not changed from generation I.

Under the assumptions of Hardy–Weinberg equilibrium, allele frequencies p and q remain constant from generation to generation. If Hardy–Weinberg assumptions are violated and the genotype frequencies in the parental generation are altered (say, by the loss of a large number of AA individuals from the population), then the allele frequencies in the next generation will be altered. However, based on the new allele frequencies, another generation of random mating will be sufficient to restore the genotype frequencies to Hardy–Weinberg equilibrium.

Deviations from Hardy–Weinberg equilibrium show that evolution is occurring

You probably have realized that populations in nature never meet the stringent conditions necessary to be at Hardy–Weinberg equilibrium—which explains why all biological populations evolve. Why, then, is this model considered so important for the study of evolution? There are two reasons. First, the equation is useful for predicting the approximate genotype frequencies of a population from its allele frequencies. Second—and crucially—the model allows biologists to evaluate which mechanisms are acting on the evolution of a particular population. The specific patterns of deviation from Hardy–Weinberg equilibrium can help us identify the various mechanisms of evolutionary change.

Do You Understand Concept 15.3?

- Why is the concept of Hardy–Weinberg equilibrium important even though the assumptions on which it is based are never completely met in nature?

- Although the stringent assumptions of Hardy–Weinberg equilibrium are never met completely in real populations, the genotype frequencies of many populations do not deviate significantly from Hardy–Weinberg expectations. Can you explain why?

- Suppose you examine a population of toads breeding in a single pond and find that heterozygous genotypes at several different loci are present at significantly lower frequencies than predicted by Hardy–Weinberg equilibrium. What are some possible explanations?

Our discussion so far has focused on changes in allele frequencies at a single gene locus. Genes do not exist in isolation, however, but interact with one another (and with the environment) to produce an organism's phenotype. What effects can these interactions have on selection?

Selection Can Be Stabilizing, Directional, or Disruptive

Until now, we have only discussed traits influenced by alleles at a single locus. Such traits are often distinguished by discrete qualities (black versus white, or smooth versus wrinkled), and so are called *qualitative traits*. Many traits, however, are influenced by alleles at more than one locus. Such traits are likely to show continuous quantitative variation rather than discrete qualitative variation, and so are known as *quantitative traits*. For example, the distribution of body sizes of individuals in a population, a trait that is influenced by genes at many loci as well as by the environment, is likely to resemble a continuous bell-shaped curve.

Natural selection can act on characters with quantitative variation in any one of several different ways, producing quite different results (**FIGURE 15.12**):

(A) Stabilizing selection

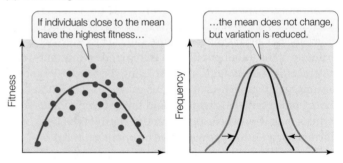

(B) Directional selection

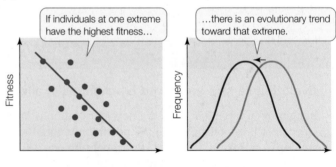

(C) Disruptive selection

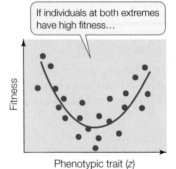

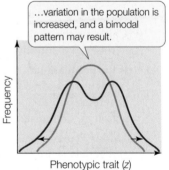

FIGURE 15.12 Natural Selection Can Operate in Several Ways The graphs in the left-hand column show the fitness of individuals with different phenotypes of the same trait. The graphs on the right show the distribution of the phenotypes in the population before (light green) and after (dark green) the influence of selection.

- **Stabilizing selection** preserves the average characteristics of a population by favoring average individuals.

- **Directional selection** changes the characteristics of a population by favoring individuals that vary in one direction from the mean of the population.

- **Disruptive selection** changes the characteristics of a population by favoring individuals that vary in both directions from the mean of the population.

Stabilizing selection reduces variation in populations

If the smallest and largest individuals in a population contribute fewer offspring to the next generation than do individuals closer to the average size, then stabilizing selection is operating on size (see Figure 15.12A). Stabilizing selection reduces variation in populations, but it does not change the mean. Natural selection frequently acts in this way, countering increases in variation brought about by sexual recombination, mutation, or gene flow. Rates of phenotypic change in many species are slow because natural selection is often stabilizing. Stabilizing selection operates, for example, on human birth weight. Babies who are lighter or heavier at birth than the population mean die at higher rates than babies whose weights are close to the mean (**FIGURE 15.13**). In discussions of specific genes, stabilizing selection is often called *purifying selection* because there is selection against any deleterious mutations to the usual gene sequence.

Directional selection favors one extreme

Directional selection is operating when individuals at one extreme of a character distribution contribute more offspring to the next generation than other individuals do, shifting the

FIGURE 15.14 Long Horns Are the Result of Directional Selection Long horns were advantageous for defending young calves from attacks by predators, so horn length increased in feral herds of Spanish cattle in the American Southwest between the early 1500s and the 1860s. The result was the familiar Texas Longhorn breed. This evolutionary trend has been maintained in modern times by ranchers practicing artificial selection.

average value of that character in the population toward that extreme. In the case of a single gene locus, directional selection may result in favoring a particular genetic variant—referred to as *positive selection* for that variant. By favoring one phenotype over another, directional selection results in an increase of the frequencies of alleles that produce the favored phenotype (as with the surface proteins of influenza discussed in the opening of this chapter).

If directional selection operates over many generations, an *evolutionary trend* is seen in the population (see Figure 15.12B). Evolutionary trends often continue for many generations, but they can be reversed if the environment changes and different phenotypes are favored, or halted when an optimal phenotype is reached or trade-offs between different adaptational advantages oppose further change. The character then undergoes stabilizing selection.

Many cases of directional selection have been observed directly, and long-term examples abound in the fossil record. The long horns of Texas Longhorn cattle (**FIGURE 15.14**) are an example of a trait that has evolved through directional selection. Texas Longhorns are descendants of cattle brought to the New World by Christopher Columbus, who picked up a few cattle in the Canary Islands and brought them to the island of Hispaniola in 1493. The cattle multiplied, and their descendants were taken to the mainland of Mexico. Spaniards exploring what would become Texas and the southwestern United States brought these cattle with them, some of which escaped and formed feral herds. Populations of feral cattle increased greatly over the next few hundred years, but there was heavy predation from bears, mountain lions, and wolves, especially on the young calves. Cows with longer horns were more successful in protecting their calves against attacks, and over a few hundred years the average horn length in the feral herds increased considerably. In addition, the cattle evolved resistance to endemic diseases of the Southwest, as well as higher fecundity and longevity. Texas Longhorns often live and produce calves

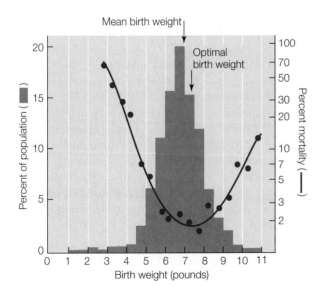

FIGURE 15.13 Human Birth Weight Is Influenced by Stabilizing Selection Babies that weigh more or less than average are more likely to die soon after birth than babies with weights close to the population mean.

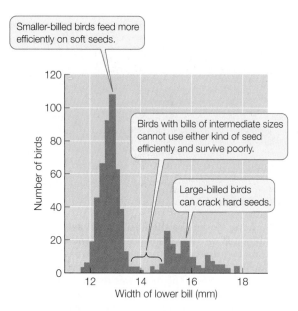

FIGURE 15.15 Disruptive Selection Results in a Bimodal Character Distribution The bimodal distribution of bill sizes in the black-bellied seedcracker of West Africa is a result of disruptive selection, which favors individuals with larger and smaller bill sizes over individuals with intermediate-sized bills.

well into their twenties—about twice as long as many breeds of cattle that have been artificially selected by humans for traits such as high fat content or high milk production (which are examples of artificial directional selection).

Disruptive selection favors extremes over the mean

When disruptive selection operates, individuals at opposite extremes of a character distribution contribute more offspring to the next generation than do individuals close to the mean, which increases variation in the population (see Figure 15.12C).

The strikingly *bimodal* (two-peaked) distribution of bill sizes in the black-bellied seedcracker (*Pyrenestes ostrinus*), a West African finch (**FIGURE 15.15**), illustrates how disruptive selection can influence populations in nature. The seeds of two types of sedges (marsh plants) are the most abundant food source for these finches during part of the year. Birds with large bills can readily crack the hard seeds of the sedge *Scleria verrucosa*. Birds with small bills can crack *S. verrucosa* seeds only with difficulty; however, they feed more efficiently on the soft seeds of *S. goossensii* than do birds with larger bills. Young finches whose bills deviate markedly from the two predominant bill sizes do not survive as well as finches whose bills are close to one of the two sizes represented by the distribution peaks. Because there are few abundant food sources in the finches' environment, and because the seeds of the two sedges do not overlap in hardness, birds with intermediate-sized bills are less efficient in using either one of the species' principal food sources. Disruptive selection therefore maintains a bimodal bill size distribution.

Do You Understand Concept 15.4?

- What are the different expected outcomes of stabilizing, directional, and disruptive selection?
- Why would you expect selection on human birth weight to be stabilizing rather than directional?
- Can you think of examples of extreme phenotypes in animal or plant populations that could be explained by directional selection?

Our discussion so far has largely focused on the evolution of phenotypes (what organisms look like and how they behave). We will now consider the specific mechanistic processes that operate at the level of genes and genomes.

concept 15.5 Genomes Reveal Both Neutral and Selective Processes of Evolution

Most natural populations harbor far more genetic variation than we would expect to find if genetic variation were influenced by natural selection alone. This discovery, combined with the knowledge that many mutations do not change molecular function, provided a major stimulus to the development of the field of *molecular evolution*.

To discuss the evolution of genes, we need to consider the specific types of mutations that are possible. A *nucleotide substitution* is a change in a single nucleotide in a DNA sequence (a type of point mutation). Many nucleotide substitutions have no effect on phenotype, even if the change occurs in a gene that encodes a protein, because most amino acids are specified by more than one codon. A substitution that does not change the encoded amino acid is known as a *silent substitution* or **synonymous substitution** (**FIGURE 15.16A**). Synonymous substitutions do not affect the functioning of a protein (although they may have other effects, such as changes in mRNA stability or translation rates) and are therefore less likely to be influenced by natural selection.

A nucleotide substitution that *does* change the amino acid sequence encoded by a gene is known as a *missense substitution* or **nonsynonymous substitution** (**FIGURE 15.16B**). In general, nonsynonymous substitutions are likely to be deleterious to the organism. But not every amino acid replacement alters a protein's shape and charge (and hence its functional properties). Therefore, some nonsynonymous substitutions are selectively neutral, or nearly so. A third possibility is that a nonsynonymous substitution alters a protein in a way that confers an advantage to the organism, and is therefore favored by natural selection.

LINK The genetic code determines the amino acid that is encoded by each codon; see Figure 10.11

The rate of synonymous substitutions in most protein-coding genes is much higher than the rate of nonsynonymous substitutions. In other words, *substitution rates are highest at nucleotide*

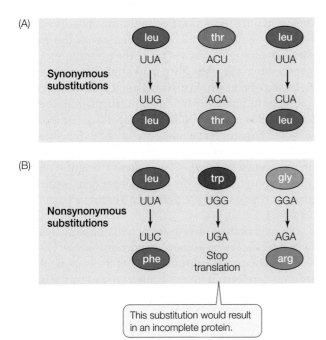

FIGURE 15.16 When One Nucleotide Changes (A) Synonymous substitutions do not change the amino acid specified and do not affect protein function. Such substitutions are less likely to be subject to natural selection, although they contribute greatly to the buildup of neutral genetic variation in a population. (B) Nonsynonymous substitutions do change the amino acid sequence and are likely to have an effect (often deleterious, but sometimes beneficial) on protein function. Such nucleotide substitutions are targets for natural selection.

positions that do not change the amino acid being expressed (**FIGURE 15.17**). The rate of substitution is even higher in **pseudogenes**, which are copies of genes that are no longer functional.

Insertions, deletions, and rearrangements of DNA sequences are all mutations that may affect a larger portion of the gene or genome than do point mutations (see Concept

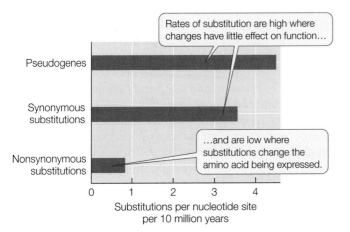

FIGURE 15.17 Rates of Substitution Differ Rates of nonsynonymous substitution are typically much lower than rates of synonymous substitution, and much lower than substitution rates in pseudogenes. This pattern reflects stronger stabilizing selection in functional genes than in pseudogenes.

9.3). Insertions and deletions of nucleotides in a protein-coding sequence interrupt its reading frame, unless they occur in multiples of three nucleotides (the length of one codon). Rearrangements may merely change the order of whole genes along chromosomes, or they may rearrange functional domains among individual genes.

When biologists began to examine the details of genetic variation of populations, they soon discovered many gene variants that had little or no effect on function. This gave rise to new ideas about how these *neutral variants* arise and spread in populations.

Much of molecular evolution is neutral

Motoo Kimura proposed the **neutral theory** in 1968. He suggested that, at the molecular level, the majority of variants found in most populations are selectively neutral. That is, most gene variants confer neither an advantage nor a disadvantage on their bearers. Therefore, these neutral variants must accumulate through genetic drift rather than through positive selection.

We saw in Concept 15.2 that genetic drift of existing gene variants tends to be greatest in small populations. However, the *rate of fixation* of neutral mutations by genetic drift is independent of population size. To see why this is so, consider a population of size N and a neutral mutation rate of μ (mu) per gamete per generation at a particular locus. The number of new mutations would be, on average, $\mu \times 2N$, because $2N$ gene copies are available to mutate in a population of diploid organisms. The probability that a given mutation will be fixed by drift alone is its frequency, which equals $1/(2N)$ for a newly arisen mutation. We can multiply these two terms to get the rate of fixation of neutral mutations in a given population of N individuals:

$$2N\mu \frac{1}{2N} = \mu$$

Therefore, the rate of fixation of neutral mutations depends only on the neutral mutation rate μ and is independent of population size. Any given mutation is more likely to appear in a large population than in a small one, but any mutation that does appear is more likely to become fixed in a small population. These two influences of population size cancel each other out. Therefore, the rate of fixation of neutral mutations is equal to the mutation rate.

As long as the underlying mutation rate is constant, macromolecules evolving in different populations should diverge from one another in neutral changes at a constant rate. The rate of evolution of particular genes and proteins is indeed often relatively constant over time, and therefore can be used as a "molecular clock" to calculate evolutionary divergence times between species (see Concept 16.3).

Although much of the genetic variation present in a population is the result of neutral evolution, the neutral theory does not imply that most mutations have no effect on the individual organism. Many mutations are never observed in populations because they are lethal or strongly detrimental, and the individuals that carry them are quickly removed from the population through natural selection. Similarly, because mutations that confer a selective advantage tend to be quickly fixed in populations, they also do not result in significant variation at the population level. Nonetheless, if we compare homologous proteins from different populations or species, some amino acid positions will remain constant under purifying selection, others will vary through neutral genetic drift, and still others will differ between species as a result of positive selection for change. How can these evolutionary processes be distinguished?

Positive and purifying selection can be detected in the genome

Positive and purifying selection are defined with respect to the **fitness** of the genotype, or change in the relative frequency of the genotype in the population from one generation to the next. Genotypes of higher fitness increase in frequency over time; those of lower fitness decrease over time.

Relative rates of synonymous and nonsynonymous substitution differ among codons of protein-coding genes as a function of selection. Each codon specifies an amino acid residue in the encoded protein. Changes in some amino acid residues have a large effect on protein function, whereas other changes have little or no effect on function. The nature and rates of substitutions in the corresponding genes can identify codons and genes that are evolving under neutral or selective processes:

- If the rates of synonymous and nonsynonymous substitution at a codon position are very similar (that is, the ratio of the two rates is close to 1), then the corresponding amino acid residue is likely drifting neutrally among states.

- If the rate of nonsynonymous substitution exceeds the rate of synonymous substitution at a codon position, then positive selection likely accounts for change in the corresponding amino acid residue.

- If the rate of synonymous substitution exceeds the rate of nonsynonymous substitution at a codon position, then purifying selection is likely resisting change in the corresponding amino acid residue.

FRONTIERS Biologists have compared the complete genomes of humans and our closest living relatives, chimpanzees. Analysis of ratios of rates of nonsynonymous to synonymous substitution reveals hundreds of genes that are evolving under positive selection in one or both lineages. Further analysis of these genes is expected to provide insights into the major selective changes that have occurred in humans and chimpanzees since our most recent common ancestor.

The evolution of lysozyme illustrates how and why particular codons in a gene sequence might be under different modes of selection. Lysozyme is an enzyme that is found in almost all animals; it is produced in the tears, saliva, and milk of mammals and in the albumen (whites) of bird eggs. Lysozyme digests the cell walls of bacteria, rupturing and killing them. Most animals defend themselves against bacteria by digesting them, which is probably why most animals have lysozyme. Some animals, however, also use lysozyme to digest their food.

Among mammals, a mode of digestion called *foregut fermentation* has evolved twice. In mammals with this mode of digestion, the foregut—consisting of the posterior esophagus and/or the stomach—has been converted into a chamber in which bacteria break down ingested plant matter by fermentation. Foregut fermenters can obtain nutrients from the otherwise indigestible cellulose that makes up a large proportion of plant tissue. Foregut fermentation evolved independently in *ruminants* (a group of hoofed mammals that includes cattle) and in certain leaf-eating monkeys, such as langurs. We know that these evolutionary events were independent because both langurs and ruminants have close relatives that are not foregut fermenters.

In both mammalian foregut-fermenting lineages, lysozyme has been modified to play a new, nondefensive role. The modified lysozyme enzyme ruptures some of the bacteria that live in the foregut, releasing nutrients metabolized by the bacteria, which the mammal then absorbs. How many changes in the lysozyme molecule were needed to allow it to perform this function amid the digestive enzymes and acidic conditions of the mammalian foregut? To answer this question, biologists compared the lysozyme-coding sequences in foregut fermenters with those in several of their nonfermenting relatives. They determined which amino acids differed and which were shared among the species (**FIGURE 15.18A**), as well as the rates of synonymous and nonsynonymous substitution in lysozyme genes across the evolutionary history of the sampled species.

The researchers found that the rate of synonymous substitution within the gene that codes for lysozyme was much higher than the rate of nonsynonymous substitution. This observation indicates that many of the amino acids that make up lysozyme are evolving under purifying selection. In other words, there is selection against change in the lysozyme protein at these positions, and the encoded amino acids must therefore be critical for lysozyme function. At other positions, several different amino acids function equally well, and the corresponding codons have similar rates of synonymous and nonsynonymous substitution.

The most striking finding was that amino acid replacements in lysozyme happened at a much higher rate in the lineage leading to langurs than in any other primates. The high rate of nonsynonymous substitution in the langur lysozyme gene shows that lysozyme went through a period of rapid change in adapting to the stomachs of langurs. Moreover, the lysozymes of langurs and cattle share five convergent amino acid replacements, all of which lie on the surface of the lysozyme molecule, well away from the enzyme's active site. Several of these shared replacements are changes from arginine to lysine, which make the protein more resistant to degradation by the stomach enzyme pepsin. By understanding the functional significance of amino acid replacements, biologists can explain the observed changes in amino acid sequences in terms of changes in the functioning of the protein.

A large body of fossil, morphological, and molecular evidence shows that langurs and cattle do not share a recent common ancestor. However, langur and ruminant *lysozymes* share several amino acids that neither mammal shares with the

(A) *Semnopithecus* sp.   *Bos taurus*

(B) *Opisthocomus hoazin*

The lysozymes of langurs and cattle are convergent for 5 amino acid residues, indicative of the independent evolution of foregut fermentation in these two species.

	Langur	Baboon	Human	Rat	Cattle	Horse
Langur		14	18	38	32	65
Baboon	0		14	33	39	65
Human	0	1		37	41	64
Rat	0	0	0		55	64
Cattle	5	0	0	0		71
Horse	0	0	0	0	1	

FIGURE 15.18 Convergent Molecular Evolution of Lysozyme
(A) The numbers of amino acid differences in the lysozymes of several pairs of mammals are shown above the diagonal line; the numbers of similarities that arose from convergence between species are shown below the diagonal. The two foregut-fermenting species (cattle and langur) share five convergent amino acid replacements related to this digestive adaptation. (B) The hoatzin—the only known foregut-fermenting bird species—has been evolving independently from mammals for hundreds of millions of years but has independently evolved modifications to lysozyme similar to those found in cattle and langurs.

lysozymes of its own closer relatives. The lysozymes of these two mammals have undergone *convergent evolution* of some amino acid residues despite their very different ancestry. The amino acids they share give these lysozymes the ability to lyse the bacteria that ferment plant material in the foregut.

The hoatzin, an unusual leaf-eating South American bird (**FIGURE 15.18B**) and the only known avian foregut fermenter, offers another remarkable example of the convergent evolution of lysozyme. Many birds have an enlarged esophageal chamber called a *crop*. The hoatzin crop contains lysozyme and bacteria and acts as a fermentation chamber. Many of the amino acid replacements that occurred in the adaptation of hoatzin lysozyme are identical to those that evolved in ruminants and langurs. Thus, even though the hoatzin and foregut-fermenting mammals have not shared a common ancestor in hundreds of millions of years, similar adaptations have evolved in their lysozyme enzymes, enabling both groups to recover nutrients from fermenting bacteria.

APPLY THE CONCEPT

Genomes reveal both neutral and selective processes of evolution

Analysis of synonymous and nonsynonymous substitutions in protein-coding genes can be used to detect neutral evolution, positive selection, and purifying selection. An investigator compared many gene sequences that encode the protein hemagglutinin (a surface protein of influenza virus) sampled over time, and collected the data at right. Use the table to answer the following questions.

1. Which codon positions encode amino acids that have probably changed as a result of positive selection? Why?

2. Which codon position is most likely to encode an amino acid that drifts neutrally among states? Why?

3. Which codon positions encode amino acids that have probably changed as a result of purifying selection? Why?

CODON POSITION	NUMBER OF SYNONYMOUS SUBSTITUTIONS IN CODON	NUMBER OF NONSYNONYMOUS SUBSTITUTIONS IN CODON
12	0	7
15	1	9
61	0	12
80	7	0
137	12	1
156	24	2
165	3	4
226	38	3

Heterozygote advantage maintains polymorphic loci

In many cases, different alleles of a particular gene are advantageous under different environmental conditions. Most organisms, however, experience a wide diversity of environments. A night is dramatically different from the preceding day. A cold, cloudy day differs from a clear, hot one. Day length and temperature change seasonally. For many genes, a single allele is unlikely to perform well under all these conditions. In such situations, a heterozygous individual (with two different alleles) is likely to outperform individuals that are homozygous for either one of the alleles.

Colias butterflies of the Rocky Mountains live in environments where dawn temperatures often are too cold, and afternoon temperatures too hot, for the butterflies to fly. Populations of these butterflies are polymorphic for the gene that encodes phosphoglucose isomerase (PGI), an enzyme that influences how well an individual flies at different temperatures. Butterflies with certain PGI genotypes can fly better during the cold hours of early morning; those with other genotypes perform better during midday heat. The optimal body temperature for flight is 35°C–39°C, but some butterflies can fly with body temperatures as low as 29°C or as high as 40°C. Heat-tolerant genotypes are favored during spells of unusually hot weather; during spells of unusually cool weather, cold-tolerant genotypes are favored.

Heterozygous *Colias* butterflies can fly over a greater temperature range than homozygous individuals because they produce two different forms of PGI. This greater range of activity should give them an advantage in foraging and finding mates. A test of this prediction did find a mating advantage in heterozygous males, and further found that this mating advantage maintains the polymorphism in the population (**FIGURE 15.19**). The heterozygous condition can never become fixed in the population, however, because the offspring of two heterozygotes will always include both classes of homozygotes in addition to heterozygotes.

Genome size and organization also evolve

We know that genome size varies tremendously among organisms. Across broad taxonomic categories, there is some correlation between genome size and organismal complexity. The genome of the tiny bacterium *Mycoplasma genitalium* has only 470 genes. *Rickettsia prowazekii*, the bacterium that causes typhus, has 634 genes. *Homo sapiens*, by

---

INVESTIGATION

FIGURE 15.19   A Heterozygote Mating Advantage   Among butterflies of the genus *Colias*, males that are heterozygous for two alleles of the PGI enzyme can fly farther under a broader range of temperatures than males that are homozygous for either allele. Does this ability give heterozygous males a mating advantage?

HYPOTHESIS

Heterozygous male *Colias* will have proportionally greater mating success than homozygous males.

METHOD

1. For each of two *Colias* species, capture butterflies in the field. In the laboratory, determine their genotypes and allow them to mate.
2. Determine the genotypes of the offspring, thus revealing paternity and mating success of the males.

RESULTS

For both species, the proportion of heterozygous males that mated successfully was higher than the proportion of all males seeking females ("flying").

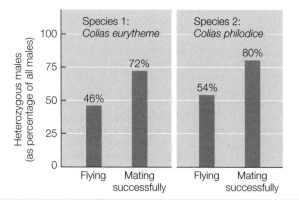

CONCLUSION

Heterozygous *Colias* males have a mating advantage over homozygous males.

ANALYZE THE DATA

Analyze these sampling data collected during the experiment (only one of several samples is shown for each species).

Species	All viable males* Heterozygous/total	All viable males* % heterozygous	Mating males Heterozygous/total	Mating males % heterozygous
C. philodice	32/74	43.2	31/50	62.0
C. eurytheme	44/92	47.8	45/59	76.3

*"Viable males" are all males captured flying with females (hence with the potential to mate)

A. Under the assumption that the proportions of each genotype (heterozygotes and homozygotes) of mating males are the same as the proportions seen among all viable males, calculate the number of *mating males* expected to be heterozygous and the number expected to be homozygous.

B. Use a chi-square test (see Appendix B) to evaluate the significance of the difference in your expected numbers in (A) and the observed percentages of heterozygous mating males. The critical value ($P = 0.05$) of the chi-square distribution with one degree of freedom is 3.841. Are the observed and expected numbers of heterozygotes and homozygotes among mating males significantly different in these samples?

For more, go to Working with Data 15.2 at **yourBioPortal.com**.

Go to **yourBioPortal.com** for original citations, discussions, and relevant links for all INVESTIGATION figures.

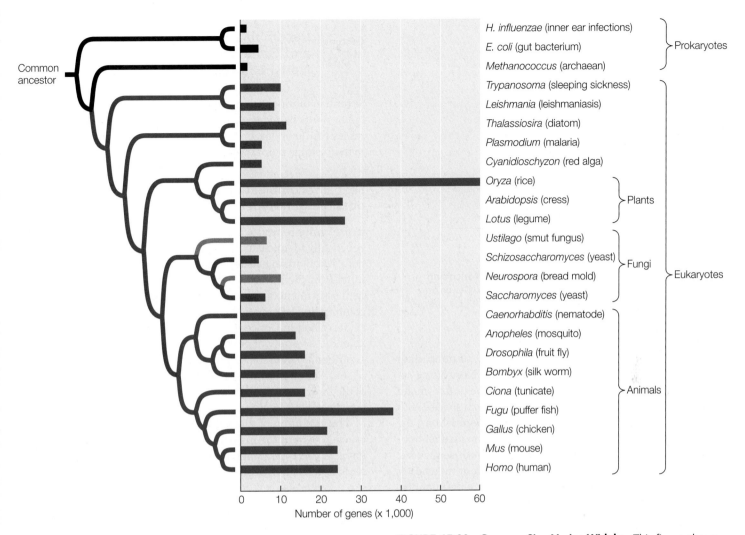

Number of genes (x 1,000)

FIGURE 15.20   Genome Size Varies Widely   This figure shows the number of genes from a sample of organisms whose genomes have been fully sequenced, arranged by their evolutionary relationships. Bacteria and archaea (black branches) typically have fewer genes than most eukaryotes. Among eukaryotes, multicellular organisms with tissue organization (plants and animals; blue branches) have more genes than single-celled organisms (red branches) or multicellular organisms that lack pronounced tissue organization (green branches).

contrast, has about 23,000 protein-coding genes. **FIGURE 15.20** shows the number of genes from a sample of organisms whose genomes have been fully sequenced, arranged by their evolutionary relationships. As this figure reveals, however, a larger genome does not always indicate greater complexity (compare rice with the other plants, for example). It is not surprising that more complex genetic instructions are needed for building and maintaining a large, multicellular organism than a small, single-celled bacterium. What is surprising is that some organisms, such as lungfishes, some salamanders, and lilies, have about 40 times as much DNA as humans do (**FIGURE 15.21**). Structurally, a lungfish or a lily is not 40 times more complex than a human. So why does genome size vary so much?

Differences in genome size are not so great if we take into account only the portion of DNA that actually encodes RNAs or proteins. The organisms with the largest total amounts of nuclear DNA (some ferns and flowering plants) have 80,000 times as much DNA as do the bacteria with the smallest genomes, but no species has more than about 100 times as many protein-coding genes as a bacterium. Therefore, much of the variation in genome size lies not in the number of functional genes, but in the amount of noncoding DNA (see Figure 15.21).

Why do the cells of most eukaryotic organisms have so much noncoding DNA? Does this noncoding DNA have a function? Although some of this DNA does not appear to have a direct function, it can alter the expression of the genes surrounding it. The degree or timing of gene expression can vary dramatically depending on the gene's position relative to noncoding sequences. Other regions of noncoding DNA consist of pseudogenes that are simply carried in the genome because the cost of doing so is very small. These pseudogenes may become the raw material for the evolution of new genes with novel functions. Some noncoding sequences function in maintaining chromosomal structure. Still others consist of parasitic transposable elements that spread through populations because they reproduce faster than the host genome.

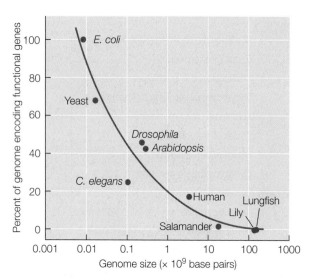

FIGURE 15.21 A Large Proportion of DNA Is Noncoding
Most of the DNA of bacteria and yeasts encodes RNAs or proteins, but a large percentage of the DNA of multicellular species is noncoding.

Another hypothesis is that the proportion of noncoding DNA is related primarily to population size. Noncoding sequences that are only slightly deleterious to the organism are likely to be purged by selection most efficiently in species with large population sizes. In species with small populations, the effects of genetic drift can overwhelm selection against noncoding sequences that have small deleterious consequences. Therefore, selection against the accumulation of noncoding sequences is most effective in species with large populations, so such species (such as bacteria or yeasts) have relatively little noncoding DNA compared with species with small populations (see Figure 15.21).

Do You Understand Concept 15.5?

- How can the ratio of synonymous to nonsynonymous substitutions be used to determine whether a particular gene is evolving neutrally, under positive selection, or under stabilizing selection?

- Why is the rate of fixation of neutral mutations independent of population size?

- Why do heterozygous individuals sometimes have an advantage over homozygous individuals?

- Why can a mutation that results in the replacement of one amino acid by another be a neutral event in some cases and in other cases be detrimental or beneficial? (Hint: Review the information about amino acids in Table 3.2 and the details of protein structure in Concept 3.2.)

- Postulate and contrast two hypotheses for the wide diversity of genome sizes among different organisms.

Most of our discussion so far has centered on changes in existing genes and phenotypes. Next we consider how new genes with novel functions arise in populations in the first place.

Recombination, Lateral Gene Transfer, and Gene Duplication Can Result in New Features

Several evolutionary processes can result in the acquisition of major new characteristics in populations. Each of these processes results in larger and more rapid evolutionary changes than do single point mutations.

Sexual recombination amplifies the number of possible genotypes

In asexually reproducing organisms, each new individual is genetically identical to its parent unless there has been a mutation. When organisms reproduce sexually, however, offspring differ from their parents because of crossing over and independent assortment of chromosomes during meiosis, as well as the combination of genetic material from two different gametes, as described in Concept 7.4. Sexual recombination generates an endless variety of genotype combinations that increase the evolutionary potential of populations—a long-term advantage of sex. Although some species may reproduce asexually most of the time, most asexual species have some means of achieving genetic recombination.

The evolution of meiosis and sexual recombination was a crucial event in the history of life. Exactly how these mechanisms arose is puzzling, however, because in the short term, sex has at least three striking disadvantages:

- Recombination breaks up adaptive combinations of genes.
- Sex reduces the rate at which females pass genes on to their offspring.
- Dividing offspring into separate genders greatly reduces the overall reproductive rate.

To see why this last disadvantage exists, consider an asexual female that produces the same number of offspring as a sexual female. Assume that both females produce two offspring, but that half of the sexual female's offspring are males. In the next (F_1) generation, then, each of the two asexual F_1 females will produce two more offspring—but there is only one sexual F_1 female to produce offspring. Thus, the effective reproductive rate of the asexual lineage is twice that of the sexual lineage. The evolutionary problem is to identify the advantages of sex that can overcome such short-term disadvantages.

A number of hypotheses have been proposed to explain the existence of sex, none of which are mutually exclusive. One is that sexual recombination facilitates repair of damaged DNA, because breaks and other errors in DNA on one chromosome can be repaired by copying the intact sequence from the homologous chromosome.

Another advantage of sexual reproduction is that it permits the elimination of deleterious mutations through recombination followed by selection. As Concept 9.2 described, DNA replication is not perfect, and many replication errors result in lower fitness. Meiotic recombination distributes these deleterious mutations unequally among gametes. Sexual reproduction then produces some individuals with more deleterious mutations and some with fewer. The individuals with fewer deleterious mutations are more likely to survive. Therefore, sexual reproduction allows natural selection to eliminate particular deleterious mutations from the population over time.

In asexual reproduction, deleterious mutations can be eliminated only by the death of the lineage or by a rare back mutation (that is, when a subsequent mutation returns a mutated sequence to its original DNA sequence). Hermann J. Muller noted that deleterious mutations in a non-recombining genome accumulate— "ratchet up" —at each replication. Mutations occur and are passed on each time a genome replicates, and these mutations accumulate with each subsequent generation. This accumulation of deleterious mutations in lineages that lack genetic recombination is known as *Muller's ratchet*.

Another explanation for the existence of sex is that the great variety of genetic combinations created in each generation can itself be advantageous. For example, genetic variation can be a defense against pathogens and parasites. Most pathogens and parasites have much shorter life cycles than their hosts and can rapidly evolve counteradaptations to host defenses. Sexual recombination might give the host's defenses a chance to keep up.

Sexual recombination does not directly influence the frequencies of alleles. Rather, *it generates new combinations of alleles on which natural selection can act*. It expands variation in quantitative characters by creating new genotypes. That is why artificial selection for bristle number in *Drosophila* (see Figure 15.6) resulted in flies that had either more or fewer bristles than the flies in the initial population had.

Lateral gene transfer can result in the gain of new functions

The tree of life is usually visualized as a branching diagram, with each lineage diverging into two (or more) lineages over time, from one common ancestor to the millions of species that are alive today. Ancestral lineages divide into descendant lineages, and it is those speciation events that the tree of life captures. However, there are also processes of **lateral gene transfer**, which allow individual genes, organelles, or fragments of genomes to move horizontally from one lineage to another. Some species may pick up fragments of DNA directly from the environment. A virus may pick up some genes from one host and transfer them to a new host when the virus becomes integrated into the new host's genome. Hybridization between species also results in the lateral transfer of large numbers of genes.

Lateral gene transfer can be highly advantageous to the species that incorporates novel genes from a distant relative. Genes that confer antibiotic resistance, for example, are commonly transferred among different species of bacteria. Lateral gene transfer is another way, in addition to mutation and recombination, that species can increase their genetic variation.

The degree to which lateral gene transfer events occur in various parts of the tree of life is a matter of considerable current investigation and debate. Lateral gene transfer appears to be relatively uncommon among most eukaryote lineages,

although the two major endosymbioses that gave rise to mitochondria and chloroplasts involved lateral transfers of entire bacterial genomes to the eukaryote lineage. Some groups of eukaryotes, most notably some plants, are subject to relatively high levels of hybridization among closely related species. Hybridization leads to the exchange of many genes among recently separated lineages of plants. The greatest degree of lateral transfer, however, appears to occur among bacteria. Many genes have been transferred repeatedly among bacteria, to the point that relationships and boundaries among species of bacteria are sometimes hard to decipher.

Many new functions arise following gene duplication

Gene duplication is yet another way in which genomes can acquire new functions. When a gene is duplicated, one copy of that gene is potentially freed from having to perform its original function. The identical copies of a duplicated gene can have any one of four different fates:

- Both copies of the gene may retain their original function (which can result in a change in the amount of gene product that is produced by the organism).

- Both copies of the gene may retain the ability to produce the original gene product, but the expression of the genes may diverge in different tissues or at different times in development.

- One copy of the gene may be incapacitated by the accumulation of deleterious mutations and become a functionless pseudogene.

- One copy of the gene may retain its original function while the second copy changes and evolves a new function.

How often do gene duplications arise, and which of these four outcomes is most likely? Investigators have found that rates of gene duplication are fast enough for a yeast or *Drosophila* population to acquire several hundred duplicate genes over the course of a million years. They have also found that most of the duplicated genes that are still present in these organisms are very young. Many duplicated genes are lost from a genome within 10 million years—an eyeblink on an evolutionary time scale.

Many gene duplications affect only one or a few genes at a time, but in some cases entire genomes may be duplicated. When all the genes are duplicated, there are massive opportunities for new functions to evolve. That is exactly what seems to have happened during the course of vertebrate evolution. The genomes of the jawed vertebrates have four diploid sets of many major genes, which leads biologists to conclude that two genome-wide duplication events occurred in the ancestor of these species. These duplications allowed considerable specialization of individual vertebrate genes, many of which are now highly tissue-specific in their expression.

LINK   See Concept 14.4 for a discussion of the role of duplicated Hox genes in vertebrate evolution

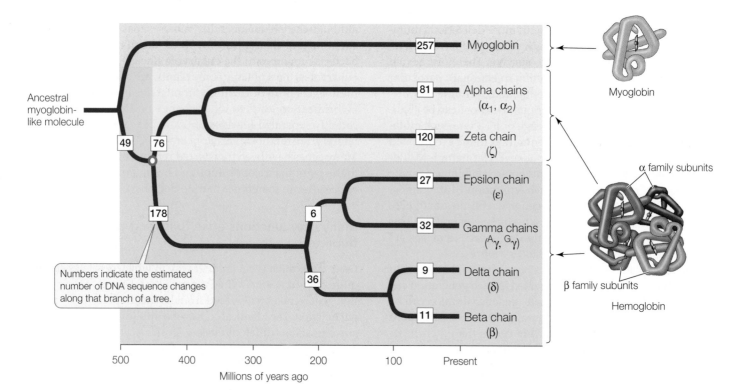

FIGURE 15.22 A Globin Family Gene Tree This gene tree suggests that the α-globin and β-globin gene clusters diverged about 450 million years ago (open circle), soon after the origin of the vertebrates.

Several successive rounds of duplication and sequence evolution may result in a *gene family*, a group of homologous genes with related functions, often arrayed in tandem along a chromosome. An example of this process is provided by the globin gene family (**FIGURE 15.22**). Comparisons of the amino acid sequences among globins strongly suggest that this family of proteins arose via gene duplications.

Hemoglobin is a tetramer (four-subunit molecule) consisting of two α-globin and two β-globin polypeptide chains. It carries oxygen in the blood. Myoglobin, a monomer, is the primary O_2 storage protein in muscle. Myoglobin's affinity for O_2 is much higher than that of hemoglobin, but hemoglobin has evolved to be more diversified in its role. Hemoglobin binds O_2 in the lungs or gills, where the O_2 concentration is relatively high, transports it to deep body tissues, where the O_2 concentration is low, and releases it in those tissues. With its more complex tetrameric structure, hemoglobin is able to carry four molecules of O_2, as well as hydrogen ions and carbon dioxide, in the blood. Hemoglobin and myoglobin are estimated to have arisen through gene duplication about 500 million years ago.

Do You Understand Concept 15.6?

- What are some of the potential advantages of lateral gene transfer to the organisms that gain new genes by this mechanism?
- Why is gene duplication considered important for long-term evolutionary change?
- Why is sexual reproduction so prevalent in nature, despite its having at least three short-term evolutionary disadvantages?

The development of evolutionary theory has helped reveal how biological molecules function, how genetic diversity is created and maintained, and how organisms develop new features. Next we will see how biologists put this theory into practice.

concept 15.7 Evolutionary Theory Has Practical Applications

Evolutionary theory has many practical applications across biology, and new ones are being developed every day. Here we'll discuss a few of these applications to fields such as agriculture, industry, and medicine.

Knowledge of gene evolution is used to study protein function

Earlier in this chapter we discussed the ways in which biologists can detect codons or genes that are under positive selection for change. These methods have greatly increased our understanding of the functions of many genes. Consider, for example, the gated sodium channel genes. Sodium channels have many functions, including the control of nerve impulses in the nervous system (see Concept 34.2). Sodium channels can become blocked when they bind certain toxins, one of which is the tetrodotoxin (TTX) present in puffer fishes and many other animals. A human who eats puffer fish tissues that contain TTX can become paralyzed and die because the toxin-blocked sodium channels prevent nerves and muscles from functioning properly.

But puffer fish themselves have sodium channels, so why doesn't the TTX in their system paralyze them? Nucleotide substitutions in the puffer fish genome have resulted in structural changes in the proteins that form the sodium channels, and those changes prevent TTX from binding to the channel pore. Several different substitutions that result in such resistance have evolved in the various duplicated sodium channel genes of the many species of puffer fish. Many other changes that have nothing to do with the evolution of tetrodotoxin resistance have occurred in these genes as well.

So how does what we have learned about the evolution of TTX-resistant sodium channels affect our lives? Mutations in human sodium channel genes are responsible for a number of neurological pathologies. By studying the function of sodium channels and understanding which changes have produced tetrodotoxin resistance, we are learning a great deal about how these crucial channels work and how various mutations affect them. Biologists do this by comparing rates of synonymous and nonsynonymous substitutions across sodium channel genes in various animals that have evolved TTX resistance. In a similar manner, molecular evolutionary principles are used to understand function and diversification of function in many other proteins.

In vitro evolution produces new molecules

Living organisms produce thousands of compounds that humans have found useful. The search for naturally occurring compounds that can be used for pharmaceutical, agricultural, or industrial purposes has been termed *bioprospecting*. These compounds are the result of millions of years of molecular evolution across millions of species of living organisms. Yet biologists can imagine molecules that could have evolved but have not, in the absence of the right combination of selection pressures and opportunities.

For instance, we might want to find a molecule that binds a particular environmental contaminant so that the contaminant can be isolated and extracted from the environment. But if the contaminant is synthetic (not produced naturally), then it is unlikely that any living organism would have evolved a molecule with the function we desire. This problem was the inspiration for the field of **in vitro evolution**, in which new molecules are produced in the laboratory to perform novel and useful functions.

The principles of in vitro evolution are based on principles of molecular evolution that we have learned from the natural world. Consider a new RNA molecule that was produced in

the laboratory using the principles of mutation and selection. The new molecule's intended function was to join two other RNA molecules (acting as a ribozyme with a function similar to that of the naturally occurring DNA ligase described in Concept 9.2, but for RNA molecules). The process started with a large pool of random RNA sequences (10^{15} different sequences, each about 300 nucleotides long), which were then selected for displaying any ligase activity (**FIGURE 15.23**). None were very effective ligases, but some were slightly better than others. The most functional of the ribozymes were selected and reverse-transcribed into cDNA (using the enzyme reverse transcriptase). The cDNA molecules were then amplified using the polymerase chain reaction (PCR; see Figure 9.15). PCR amplification is not perfect, and it introduced many new mutations

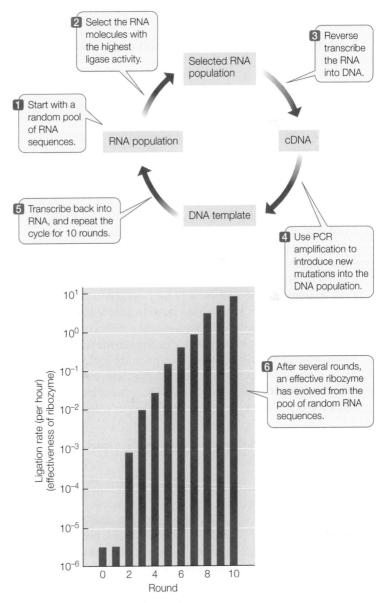

FIGURE 15.23 In Vitro Evolution Starting with a large pool of random RNA sequences, David Bartel and Jack Szostak of Massachusetts General Hospital produced a new ribozyme through rounds of mutation and selection for the ability to ligate RNA sequences.

into the pool of RNA sequences. These sequences were then transcribed back into RNA molecules using RNA polymerase, and the process was repeated.

The ligase activity of the RNAs evolved quickly; after 10 rounds of in vitro evolution, it had increased by about 7 million times. Similar techniques have been used to create a wide variety of molecules with novel enzymatic and binding functions.

FRONTIERS In vitro evolution of a bacterial enzyme has identified molecular changes that give rise to antibiotic resistance in bacteria. This research may lead to the design of new antibiotics that are harder for bacteria to evolve resistance against. Developing new, effective antibiotics is critical to human health, since many of the existing antibiotic drugs are rapidly losing their effectiveness against bacterial pathogens.

Evolutionary theory provides multiple benefits to agriculture

Well before humans had a clear understanding of evolution, they were selecting beneficial traits in the plants and animals they used for food. Modern agricultural practices have benefitted from a clearer understanding of evolutionary principles. Agriculturists have also used knowledge of evolutionary relationships and principles to incorporate beneficial genes into our food crops from many wild species.

Evolutionary theory has also proved important for understanding how to reduce the threats of pesticide and herbicide resistance. When farmers use the same pesticide over many seasons, the pests they are trying to kill gradually evolve resistance to the pesticide. Each year, a few pest individuals are slightly better at surviving in the presence of the pesticide, and those individuals produce most of the next generation of crop pests. Because their genes allow them to survive at a higher rate, and because they pass these resistant genes on to their offspring, pesticide resistance quickly evolves in the entire population. To combat this problem, evolutionary biologists have devised pesticide application and rotation schemes to reduce the rate of evolution of pesticide resistance, thus allowing farmers to use pesticides more effectively for longer periods of time.

Knowledge of molecular evolution is used to combat diseases

Many of the most problematic human diseases are caused by living, evolving organisms that present a moving target for modern medicine, as we described for influenza at the start of this chapter. The control of these and many other human diseases depends on techniques that can track the evolution of pathogenic organisms over time.

During the past century, transportation advances have allowed humans to move around the world with unprecedented speed and increasing frequency. Unfortunately, this mobility has increased the rate at which pathogens are transmitted among human populations, leading to the global emergence of many "new" diseases. Most of these emerging diseases are

caused by viruses, and virtually all new viral diseases have been identified by evolutionary comparison of their genomes with those of known viruses. In recent years, rodent-borne hantaviruses have been identified as the source of widespread respiratory illnesses, and the virus that causes sudden acute respiratory syndrome (SARS) has been identified, as has its host, using evolutionary comparisons of genes. Studies of the origins, timing of emergence, and global diversity of many human pathogens (including HIV, the human immunodeficiency virus) depend on evolutionary principles and methods, as do efforts to develop effective vaccines against these pathogens.

At present, it is difficult to identify many common infections (the viral strains that cause "colds," for instance). As genomic databases increase, however, automated methods of sequencing and making evolutionary comparisons of sequences will allow us to identify and treat a much wider array of human (and other) diseases. Once biologists have collected genome data for enough infectious organisms, it will be possible to identify an infection by sequencing a portion of the pathogen's genome and comparing this sequence with other sequences on an evolutionary tree.

Do You Understand Concept 15.7?

- How can gene evolution be used to study protein function?
- How are principles of evolutionary biology used to identify emerging diseases?
- What are the key elements of in vitro evolution, and how do these elements correspond to natural evolutionary processes?

The mechanisms of evolution have produced a remarkable variety of organisms, some of which are adapted to most environments on Earth. In the next chapter, we will describe how biologists study the evolutionary relationships across the great diversity of life.

 QA **QUESTION** How do biologists use evolutionary theory to develop better flu vaccines?

ANSWER Many different strains of influenza virus circulate among human populations and other vertebrate hosts each year, but only a few of those strains survive to leave descendants. Selection among these circulating influenza strains results in rapid evolution of the viral genome. One of the ways that influenza strains differ is in the configuration of proteins on their surface. These surface proteins are the targets of recognition by the host immune system (**FIGURE 15.24**).

When changes occur in the surface proteins of an influenza virus, the host immune system may no longer detect the invading virus, so the virus is more likely to replicate

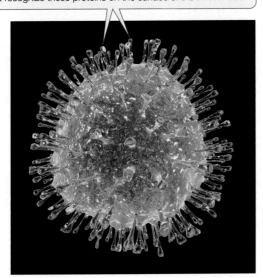

A vaccine stimulates our immune system to produce antibodies that recognize these proteins on the surface of the H1N1 virus.

FIGURE 15.24 Evolutionary Analysis of Surface Proteins Leads to Improved Flu Vaccines This computer-generated image is of the H1N1 virus that was the target of a 2009–2010 flu vaccine. Rapidly evolving surface proteins ("spikes" in this illustration) allow flu viruses to escape detection by the host's immune system. Analyzing the surface proteins among current strains of the virus can help biologists anticipate which strains are most likely to be the cause of future epidemics.

successfully. The viral strains with the greatest number of changes to their surface proteins are most likely to escape detection by the host immune system, and are therefore most likely to spread among the host population and result in future flu epidemics. In other words, there is positive selection for change in the surface proteins of influenza.

By comparing the survival and proliferation rates of virus strains that have different gene sequences coding for their surface proteins, biologists can study adaptation of the viruses over time (Concept 15.2). If biologists can predict which of the currently circulating flu virus strains are most likely to escape host immune detection, then they can identify the strains of that are most likely to be involved in upcoming influenza epidemics and can target those strains for vaccine production.

How can biologists make such predictions? By examining the ratio of synonymous to nonsynonymous substitutions in genes that encode viral surface proteins, biologists can detect which codon changes (i.e., mutations) are under positive selection (Concept 15.5). They can then assess which of the currently circulating flu strains show the greatest number of changes in these positively selected codons. It is these flu strains that are most likely to survive and lead to the flu epidemics of the future, so they are the best targets for new vaccines. This practical application of evolutionary theory leads to more effective flu vaccines—and thus fewer illnesses and influenza-related deaths each year.

---

15 SUMMARY

concept 15.1 Evolution Is Both Factual and the Basis of Broader Theory

- **Evolution** is genetic change in populations over time. Evolution can be observed directly in living populations as well as in the fossil record of life.

- **Evolutionary theory** refers to our understanding and application of the mechanisms of evolutionary change.

- Charles Darwin in best known for his ideas on the common ancestry of divergent species and on **natural selection** as a mechanism of evolution. **See ANIMATED TUTORIAL 15.1**

- Since Darwin's time, many biologists have contributed to the development of evolutionary theory, and rapid progress in our understanding continues today. **Review Figure 15.2**

concept 15.2 Mutation, Selection, Gene Flow, Genetic Drift, and Nonrandom Mating Result in Evolution

- Mutation produces new genetic variants (**alleles**).

- Within populations, natural selection acts to increase the frequency of beneficial alleles and decrease the frequency of deleterious alleles.

- **Adaptation** refers both to a trait that evolves through natural selection and to the process that produces such traits.

- Migration or mating of individuals between populations results in **gene flow**.

- In small populations, **genetic drift**—the random loss of individuals and the alleles they possess—may produce large changes in allele frequencies from one generation to the next and greatly reduce genetic variation.

- **Population bottlenecks** occur when only a few individuals survive a random event, resulting in a drastic shift in allele frequencies within the population and the loss of variation. Similarly, a population established by a small number of individuals colonizing a new region may lose variation via a **founder effect**. **Review Figure 15.7**

- **Nonrandom mating** may result in changes in genotype frequencies in a population.

- **Sexual selection** results from differential mating success of individuals based on their phenotype. **Review Figure 15.9 and WORKING WITH DATA 15.1**

concept 15.3 Evolution Can Be Measured by Changes in Allele Frequencies

- Allele frequencies measure the amount of genetic variation in a population. Genotype frequencies show how a population's genetic variation is distributed among its members. Together, allele and genotype frequencies describe a population's **genetic structure**. Review Figure 15.10 and INTERACTIVE TUTORIAL 15.1

- **Hardy–Weinberg equilibrium** predicts genotype frequencies from allele frequencies in the absence of evolution. Deviation from these frequencies indicates that evolutionary mechanisms are at work. **Review Figure 15.11 and ANIMATED TUTORIAL 15.2**

concept 15.4 Selection Can Be Stabilizing, Directional, or Disruptive

- Natural selection can act on characters with quantitative variation in three different ways. **Review Figure 15.12**

- **Stabilizing selection** acts to reduce variation without changing the mean value of a trait.

- **Directional selection** acts to shift the mean value of a trait toward one extreme.

- **Disruptive selection** favors both extremes of trait values, resulting in a bimodal character distribution.

concept 15.5 Genomes Reveal Both Neutral and Selective Processes of Evolution

- **Nonsynonymous substitutions** of nucleotides result in amino acid replacements in proteins, but **synonymous substitutions** do not. **Review Figure 15.16**

- Rates of synonymous substitution are typically higher than rates of nonsynonymous substitution in protein-coding genes (a result of stabilizing selection). **Review Figure 15.17**

- Much of the change in nucleotide sequences over time is a result of neutral evolution. The rate of fixation of neutral mutations is independent of population size and is equal to the mutation rate.

- Positive selection for change in a protein-coding gene may be detected by a higher rate of nonsynonymous than synonymous substitution.

- Specific codons within a given gene sequence can be under different modes of selection. **Review Figure 15.19 and WORKING WITH DATA 15.2**

- The total size of genomes varies much more widely across multicellular organisms than does the number of functional genes. **Review Figures 15.20 and 15.21**

- Even though many noncoding regions of the genome may not have direct functions, these regions can affect the phenotype of an organism by influencing gene expression.

- Functionless **pseudogenes** can serve as the raw material for the evolution of new genes.

concept 15.6 Recombination, Lateral Gene Transfer, and Gene Duplication Can Result in New Features

- Despite its short-term disadvantages, sexual reproduction generates countless genotype combinations that increase genetic variation in populations.

- **Lateral gene transfer** can result in the rapid acquisition of new functions from distantly related species.

- **Gene duplications** can result in increased production of the gene's product, in divergence of the duplicated genes' expression, in pseudogenes, or in new gene functions. Several rounds of gene duplication can give rise to multiple genes with related functions, known as a gene family. **Review Figure 15.22 and WEB ACTIVITY 15.1**

concept 15.7 Evolutionary Theory Has Practical Applications

- Protein function can be studied by examining gene evolution. Detection of positive selection can be used to identify molecular changes that have resulted in functional changes.

- Agricultural applications of evolution include the development of new crop plants and domesticated animals, as well as a reduction in the rate of evolution of pesticide resistance.

- **In vitro evolution** is used to produce synthetic molecules with particular desired functions. **Review Figure 15.23**

- Many diseases are identified, studied, and combated through molecular evolutionary investigations.

Reconstructing and Using Phylogenies

Green fluorescent protein (GFP) was discovered in 1962 when Osamu Shimomura, an organic chemist and marine biologist, led a team that was able to purify the protein from the tissues of the bioluminescent jellyfish *Aequorea victoria*. Some 30 years after its initial discovery, Martin Chalfie had the idea (and the technology) to link the gene for GFP to other protein-coding genes, so that the expression of specific genes of interest could be visualized in glowing green within cells and tissues of living organisms (see Figure 13.6). This work was extended by Roger Tsien, who changed some of the amino acids within GFP to create proteins of several distinct colors. Different colored proteins meant that the expression of a number of different proteins could be visualized and studied in the same organism at the same time. These three scientists were awarded the 2008 Nobel Prize in Chemistry for the isolation and development of GFP for visualizing gene expression.

Although Tsien was able to produce different colored proteins, he could not produce a *red* protein. This was frustrating; a red fluorescent protein would be particularly useful to biologists because red light penetrates tissues more easily than do other colors. Tsien's work stimulated Mikhail Matz to look for new fluorescent proteins in corals (which are relatives of the jellyfishes). Among the different species he studied, Matz found coral proteins that fluoresced in various shades of green, cyan (blue-green)—and red.

How had fluorescent red pigments evolved among the corals, given that the necessary molecular changes had eluded Tsien? To answer this question, Matz sequenced the genes of the fluorescent proteins and used these sequences to reconstruct the evolutionary history of the amino acid changes that produced different colors in different species of corals.

Matz's work showed that the ancestral fluorescent protein in corals was green, and that red fluorescent proteins evolved in a series of gradual steps. His analysis of evolutionary relationships allowed him to retrace these steps. Such an evolutionary history, as depicted in a tree of relationships among lineages, is called a phylogeny.

The evolution of many aspects of an organism's biology can be studied using phylogenetic methods. This information is used in all fields of biology to understand the structure, function, and behavior of organisms.

The reef-building coral *Acropora millepora* shows cyan and red fluorescence. This photograph was taken under a fluorescent microscope that affects the colors we see; the colors are perceived differently by marine animals in their natural environment.

KEY CONCEPTS

16.1 All of Life Is Connected through Its Evolutionary History

16.2 Phylogeny Can Be Reconstructed from Traits of Organisms

16.3 Phylogeny Makes Biology Comparative and Predictive

16.4 Phylogeny Is the Basis of Biological Classification

Q QUESTION
How are phylogenetic methods used to resurrect protein sequences from extinct organisms?

All of Life Is Connected through Its Evolutionary History

The sequencing of complete genomes from many diverse species has confirmed what biologists have long suspected: all of life is related through a common ancestor. The common ancestry of life explains why the general principles of biology apply to all organisms. Thus we can learn much about how the human genome works by studying the biology of model organisms because we share a common evolutionary history with those organisms. The evolutionary history of these relationships is known as **phylogeny**, and a **phylogenetic tree** is a diagrammatic reconstruction of that history.

Phylogenetic trees are commonly used to depict the evolutionary history of species, populations, and genes. For many years such trees have been constructed based on physical structures, behaviors, and biochemical attributes. Now, as genomes are sequenced for more and more organisms, biologists are able to reconstruct the history of life in ever greater detail.

In Chapter 15, we discussed why we expect populations of organisms to evolve over time. Such a series of ancestor and descendant populations forms a **lineage**, which we can depict as a line drawn on a time axis:

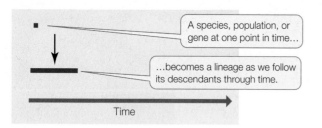

What happens when a single lineage divides into two? For example, a geographic barrier may divide an ancestral population into two descendant populations that no longer interact with one another. We depict such an event as a split, or *node*, in a phylogenetic tree. Each of the descendant populations give rise to a new lineage, and as these independent lineages evolve, new traits arise in each:

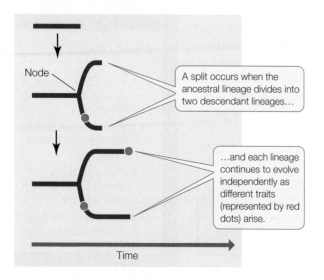

As the lineages continue to split over time, this history can be represented in the form of a branching tree that can be used to trace the evolutionary relationships from the ancient common ancestor of a group of species, through the various lineage splits, up to the present populations of the organisms:

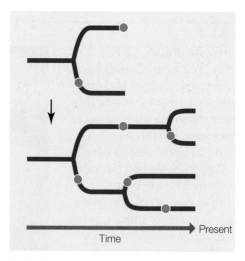

A phylogenetic tree may portray the evolutionary history of all life forms; of a major evolutionary group (such as the insects); of a small group of closely related species; or in some cases, even the history of individuals, populations, or genes within a species. The common ancestor of all the organisms in the tree forms the *root* of the tree. The depictions of phylogenetic trees in this book are rooted at the left, with time flowing from left (earliest) to right (most recent):

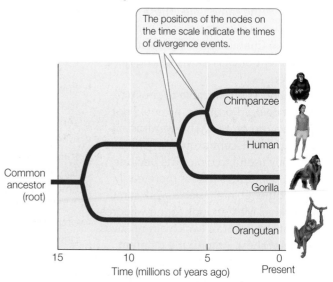

The timing of splitting events in lineages is shown by the position of nodes on a time axis. These splits represent events where one lineage diverged into two, such as a *speciation event* (for a tree of species), a *gene duplication event* (for a tree of genes), or a *transmission event* (for a tree of viral lineages transmitted through a host population). The time axis may have an explicit scale, or it may simply show the relative timing of divergence events.

In this book's illustrations, the order in which nodes are placed along the horizontal (time) axis has meaning, but the vertical distance between the branches does not. Vertical distances have been adjusted for legibility and clarity of presentation; they do not correlate with the degree of similarity or difference between groups. Note too that lineages can be rotated around nodes in the tree, so the vertical order of lineages is also largely arbitrary:

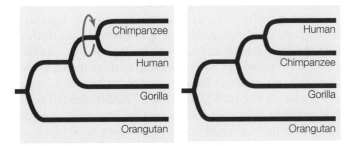

Any group of species that we designate with a name is a **taxon** (plural *taxa*). Examples of familiar taxa include humans, primates, mammals, and vertebrates; in this series, each taxon is also a member of the next, more inclusive taxon. Any taxon that consists of all the evolutionary descendants of a common ancestor is called a **clade**. Clades can be identified by picking any point on a phylogenetic tree and from that point tracing all the descendant lineages to the tips of the terminal branches (**FIGURE 16.1**). Two species that are each other's closest relatives are called **sister species**; similarly, any two clades that are each other's closest relatives are **sister clades**.

Before the 1980s, phylogenetic trees tended to be seen only in the literature on evolutionary biology, especially in the area of **systematics**—the study and classification of biodiversity. But almost every journal in the life sciences published during the last few years contains phylogenetic trees. Trees are widely used in molecular biology, biomedicine, physiology, behavior, ecology, and virtually all other fields of biology. Why have phylogenetic studies become so widespread?

Phylogenetic trees are the basis of comparative biology

In biology, we study life at all levels of organization—from genes, cells, organisms, populations, and species to the major divisions of life. In most cases, however, no individual gene or organism (or other unit of study) is exactly like any other gene or organism that we investigate.

Consider the individuals in your biology class. We recognize each person as an individual human, but we know that no two are exactly alike. If we knew everyone's family tree in detail, the genetic similarity of any pair of students would be more predictable. We would find that more closely related students have many more traits in common (from the color of their hair to their susceptibility or resistance to diseases). Likewise, biologists use phylogenies to make comparisons and predictions about shared traits across genes, populations, and species.

The evolutionary relationships among species, as represented in the tree of life, form the basis for biological classification. Biologists estimate that there are tens of millions of species on Earth. So far, however, only about 1.8 million species have been *classified*—that is, formally described and named. New species are being

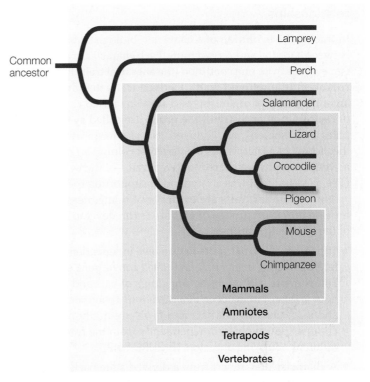

FIGURE 16.1 Clades Represent All the Descendants of a Common Ancestor All clades are subsets of larger clades, with all of life as the most inclusive taxon. In this example, the groups called mammals, amniotes, tetrapods, and vertebrates represent successively larger clades. Only a few species within each clade are represented on this tree.

discovered all the time and phylogenetic analyses are constantly reviewed and revised, so our knowledge of the tree of life is far from complete. Yet knowledge of evolutionary relationships is essential for making comparisons in biology, so biologists build phylogenies for groups of interest as the need arises. The tree of life's evolutionary framework allows us to make many predictions about the behavior, ecology, physiology, genetics, and morphology of species that have not yet been studied in detail.

When biologists compare species, they observe traits that differ within the group of interest and try to ascertain when these traits evolved. In many cases, investigators are interested in how the evolution of a trait relates to environmental conditions or selective pressures. For instance, scientists have used phylogenetic analyses to discover changes in the genome of human immunodeficiency virus that confer resistance to particular drug treatments. The association of a particular genetic change in HIV with a particular treatment provides a hypothesis about the evolution of resistance that can be tested experimentally.

Any features shared by two or more species that have been inherited from a common ancestor are said to be **homologous**. Homologous features may be any heritable traits, including DNA sequences, protein structures, anatomical structures, and even some behavior patterns. For example, all living vertebrates have a vertebral column, as did the ancestral vertebrate. Therefore, the vertebral column is judged to be homologous in all vertebrates.

Derived traits provide evidence of evolutionary relationships

In tracing the evolution of a character, biologists distinguish between *ancestral* and *derived* traits. Each character of an organism evolves from one condition (the **ancestral trait**) to another condition (the **derived trait**). Derived traits that are shared among a group of organisms and are also viewed as evidence of the common ancestry of the group are called **synapomorphies** (*syn*, "shared"; *apo*, "derived"; *morph*, "form," referring to the "form" of a trait). Thus the vertebral column is considered a synapomorphy—a shared, derived trait—of the vertebrates. (The ancestral trait was an undivided supporting rod.)

Not all similar traits are evidence of relatedness. Similar traits in unrelated groups of organisms can develop for either of the following reasons:

- Superficially similar traits may evolve independently in different lineages, a phenomenon called **convergent evolution**. For example, although the *wing bones* of bats and birds are homologous, having been inherited from a common tetrapod ancestor, the *wings* of bats and birds are not homologous because they evolved independently from the forelimbs of different nonflying ancestors (**FIGURE 16.2**).

- A character may revert from a derived state back to an ancestral state in an event called an **evolutionary reversal**. For example, the derived limbs of terrestrial tetrapods evolved from the ancestral fins of their aquatic ancestors. Then, within the mammals, the ancestors of modern cetaceans (whales and dolphins) returned to the ocean, and cetacean limbs evolved to once again resemble their ancestral state—fins. The superficial similarity of cetacean and fish fins does not suggest a close relationship between these groups; the similarity arises from evolutionary reversal.

Similar traits generated by convergent evolution and evolutionary reversals are called *homoplastic traits* or **homoplasies**.

A particular trait may be ancestral or derived, depending on our point of reference. For example, all birds have feathers. We infer from this that feathers (which are highly modified scales) were present in the common ancestor of modern birds. Therefore, we consider the presence of feathers to be an *ancestral* trait for any particular group of modern birds, such as the songbirds. However, feathers are not present in any other living animals. In reconstructing a phylogeny of all living vertebrates, the presence of feathers is a *derived* trait found only among birds, and thus is a synapomorphy of the birds.

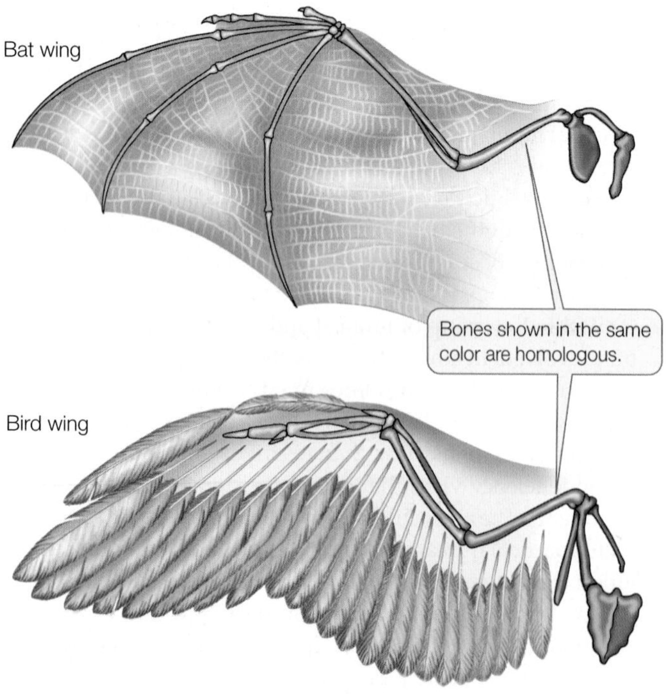

FIGURE 16.2 The Bones Are Homologous, the Wings Are Not The supporting bone structures of both bat wings and bird wings are derived from a common tetrapod (four-limbed) ancestor and are thus homologous. However, the wings themselves—an adaptation for flight—evolved independently in the two groups.

> Bat wing
>
> Bird wing
>
> Bones shown in the same color are homologous.

Do You Understand Concept 16.1?

- What biological processes are represented in a phylogenetic tree?
- Why is it important to consider only homologous characters in reconstructing phylogenetic trees?
- What are some reasons that similar traits might arise independently in species that are only distantly related? Can you think of examples among familiar organisms?

Phylogenetic analyses of evolutionary history have become increasingly important to many types of biological research in recent years, and they are the basis for the comparative nature of biology. For the most part, however, evolutionary history cannot be observed directly. How, then, do biologists reconstruct the past?

concept 16.2 Phylogeny Can Be Reconstructed from Traits of Organisms

Consider the eight vertebrate animals listed in **TABLE 16.1**: lamprey, perch, salamander, lizard, crocodile, pigeon, mouse, and chimpanzee. To illustrate how the phylogenetic tree in **FIGURE 16.3** is constructed, we assume initially that a given derived trait arose only once during the evolution of these animals (there has been no convergent evolution), and that no derived traits were lost from any of the descendant groups (there has been no evolutionary reversal). For simplicity, we have selected traits that are either present (+) or absent (−).

In a phylogenetic study, the group of organisms of primary interest is called the **ingroup**. As a point of reference, an ingroup is compared with an **outgroup**: a species or group known to be closely related to but phylogenetically outside the group of interest. If the outgroup is known to have diverged before the ingroup, the outgroup can be used to determine which traits of the ingroup are derived (i.e., evolved within the

TABLE 16.1 Eight Vertebrates and the Presence or Absence of Some Shared Derived Traits

| | | | | DERIVED TRAIT | | | | |
TAXON	JAWS	LUNGS	CLAWS OR NAILS	GIZZARD	FEATHERS	FUR	MAMMARY GLANDS	KERATINOUS SCALES
Lamprey (outgroup)	–	–	–	–	–	–	–	–
Perch	+	–	–	–	–	–	–	–
Salamander	+	+	–	–	–	–	–	–
Lizard	+	+	+	–	–	–	–	+
Crocodile	+	+	+	+	–	–	–	+
Pigeon	+	+	+	+	+	–	–	+
Mouse	+	+	+	–	–	+	+	–
Chimpanzee	+	+	+	–	–	+	+	–

ingroup) and which are ancestral (i.e., evolved before the origin of the ingroup). The lamprey belongs to a group of jawless fishes thought to have separated from the lineage leading to the other vertebrates before the jaw arose. Therefore, we have specified the lamprey as the outgroup for our analysis. Because derived traits were acquired by other members of the vertebrate lineage *after* they diverged from the outgroup, any trait that is present in both the lamprey and the other vertebrates is judged to be ancestral.

We begin by noting that the chimpanzee and mouse share two traits—mammary glands and fur—that are absent in both the outgroup and in the other species of the ingroup. Therefore, we infer that mammary glands and fur are derived traits that evolved in a common ancestor of chimpanzees and mice after that lineage separated from the lineages leading to the other vertebrates. These characters are synapomorphies that unite chimpanzees and mice (as well as all other mammals, although we have not included other mammalian species in

yourBioPortal.com

Go to WEB ACTIVITY 16.1
Constructing a Phylogenetic Tree

FIGURE 16.3 Inferring a Phylogenetic Tree This phylogenetic tree was constructed from the information given in Table 16.1 using the parsimony principle. Each clade in the tree is supported by at least one shared derived trait, or synapomorphy.

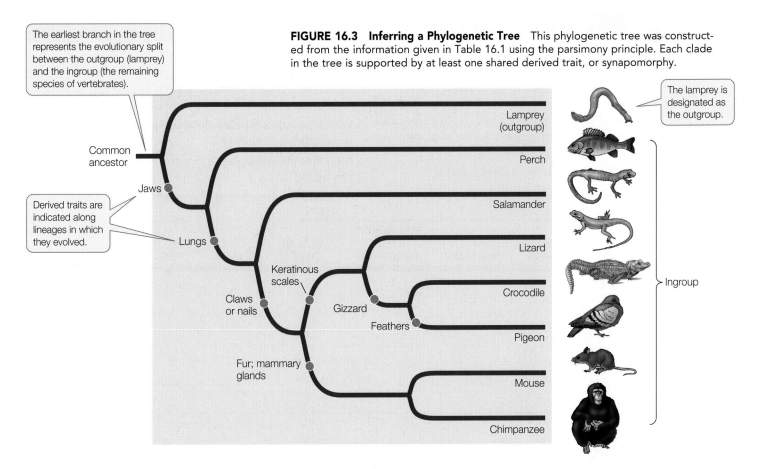

The earliest branch in the tree represents the evolutionary split between the outgroup (lamprey) and the ingroup (the remaining species of vertebrates).

The lamprey is designated as the outgroup.

Common ancestor

Derived traits are indicated along lineages in which they evolved.

Jaws

Lungs

Keratinous scales

Claws or nails

Gizzard

Feathers

Fur; mammary glands

Lamprey (outgroup)

Perch

Salamander

Lizard

Crocodile

Pigeon

Mouse

Chimpanzee

Ingroup

this example). By the same reasoning, we can infer that the other shared derived traits are synapomorphies for the various groups in which they are expressed. For instance, keratinous scales are a synapomorphy of the lizard, crocodile, and pigeon.

Table 16.1 also tells us that, among the animals in our ingroup, the pigeon has a unique trait: the presence of feathers. Feathers are a synapomorphy of birds and their extinct relatives. However, because we only have one bird in this example, the presence of feathers provides no clues concerning relationships among these eight species of vertebrates. However, gizzards are found in both birds and crocodiles, so this trait is evidence of a close relationship between birds and crocodilians.

By combining information about the various synapomorphies, we can construct the phylogenetic tree in Figure 16.3. We infer from our information that mice and chimpanzees—the only two animals that share fur and mammary glands—share a more recent common ancestor with each other than they do with pigeons and crocodiles. Otherwise, we would need to assume that the ancestors of pigeons and crocodiles also had fur and mammary glands but subsequently lost them; these additional assumptions are unnecessary in this case.

This particular tree was easy to construct because it is based on a very small sample of traits, and the derived traits we examined evolved only once and were never lost after they appeared. Had we included a snake in the group, our analysis would not have been as straightforward. We would need to examine additional characters to determine that snakes evolved from a group of lizards that had limbs. In an evolutionary

reversal, limbs were lost in the ancestors of snakes as an adaptation to a subterranean existence.

Typically, biologists construct phylogenetic trees using hundreds or thousands of traits. With larger data sets, we would expect to observe traits that have changed more than once, and thus would expect to see convergence and evolutionary reversal. How do we determine which traits are synapomorphies and which are homoplasies? One way is to invoke the principle of *parsimony*.

Parsimony provides the simplest explanation for phylogenetic data

In its most general form, the **parsimony principle** states that the preferred explanation of observed data is the simplest explanation. Applying the principle of parsimony to the reconstruction of phylogenies entails minimizing the number of evolutionary changes that need to be assumed over all characters in all groups in the tree. In other words, the best hypothesis under the parsimony principle is one that requires the fewest homoplasies. This application of parsimony is a specific case of a general principle of reasoning called *Occam's razor*: The best explanation is the one that best fits the data while making the fewest assumptions. More complicated explanations are accepted only when the evidence requires them. Phylogenetic trees represent our best estimates about evolutionary relationships, given the evidence available. As with all analyses in science, phylogenetic trees are continually modified as additional information is obtained.

APPLY THE CONCEPT

Phylogeny can be reconstructed from traits of organisms

The matrix below supplies data for seven land plants and an outgroup (an aquatic plant known as a stonewort). Each trait is scored as either present (+) or absent (–) in each of the plants. Use this data matrix to reconstruct the phylogeny of land plants and answer the questions.

1. Which two of these taxa are most closely related?

2. Plants that produce seeds are known as seed plants. What is the sister group to the seed plants among these taxa?

3. Which two traits evolved along the same branch of your reconstructed phylogeny?

4. Are there any homoplasies in your reconstructed phylogeny?

	TRAIT						
TAXON	PROTECTED EMBRYOS	TRUE ROOTS	PERSISTENTLY GREEN SPOROPHYTE	VASCULAR CELLS	STOMATA	MEGAPHYLLS (TRUE LEAVES)	SEEDS
Stonewort (outgroup)	–	–	–	–	–	–	–
Liverwort	+	–	–	–	–	–	–
Pine tree	+	+	+	+	+	+	+
Bracken fern	+	+	+	+	+	+	–
Club moss	+	+	+	+	+	–	–
Sphagnum moss	+	–	–	–	+	–	–
Hornwort	+	–	+	–	+	–	–
Sunflower	+	+	+	+	+	+	+

Phylogenies are reconstructed from many sources of data

Naturalists have constructed various forms of phylogenetic trees for more than 150 years. In fact, the only figure in the first edition of Darwin's *Origin of Species* was a conceptual diagram of a phylogenetic tree. Phylogenetic tree construction has been revolutionized by the advent of computer software that allows us to consider far more data and analyze far more traits than could ever before be processed. Combining these advances in methodology with the massive comparative data sets being generated through genome sequencing and other molecular studies, biologists are learning details about the tree of life at a remarkable pace.

Any trait that is genetically determined—and therefore heritable—can be used in a phylogenetic analysis. Evolutionary relationships can be revealed through studies of morphology, development, the fossil record, behavioral traits, and molecular traits such as DNA and protein sequences.

---

yourBioPortal.com

Go to INTERACTIVE TUTORIAL 16.1
Phylogeny and Molecular Evolution

MORPHOLOGY An important source of phylogenetic information is *morphology*: the presence, size, shape, and other attributes of body parts. Since living organisms have been observed, depicted, and studied for millenia, we have a wealth of recorded morphological data as well as extensive museum and herbarium collections of organisms whose traits can be measured. New technological tools, such as the electron microscope and computed tomography (CT) scans, enable systematists to examine and analyze the structures of organisms at much finer scales than was formerly possible.

Most species are described and known primarily by their morphology, and morphology still provides the most comprehensive data set available for many taxa. The morphological features that are important for phylogenetic analysis are often specific to a particular group. For example, the presence, development, shape, and size of various features of the skeletal system are important in vertebrate phylogeny, whereas the structures of the floral organs (petals, carpels, stamens and sepals; see p. 557) are important for studying the relationships among flowering plants.

Reconstructing the evolutionary relationships of most extinct species depends almost exclusively on morphological comparisons. Fossils show us where and when organisms lived in the past and give us an idea of what they looked like. Fossils provide important evidence that helps distinguish ancestral from derived traits. The fossil record can also reveal when lineages diverged and began their independent evolutionary histories. Furthermore, in groups with few species that have survived to the present, information on extinct species is often critical to an understanding of the large divergences among the surviving species. The fossil record is limited, however; for some groups, few or no fossils have been found, and for others the fossil record is fragmentary.

Although useful, morphological approaches to phylogenetic analysis do have limitations. Some taxa exhibit little morphological diversity, despite great species diversity. For example, the phylogeny of the leopard frogs of North and Central America would be difficult to infer from morphological differences alone, because the many species look very similar, despite important differences in their behavior and physiology. At the other extreme, few morphological traits can be compared across distantly related species (an earthworm and a mammal, for instance). Furthermore, some morphological variation has an environmental rather than a genetic basis and so must be excluded from phylogenetic analyses. For these reasons, an accurate phylogenetic analysis often requires information beyond that supplied by morphology.

DEVELOPMENT Observations of similarities in developmental patterns may reveal evolutionary relationships. Some organisms exhibit similarities in early developmental stages only. The larvae of marine creatures called sea squirts, for example, have a flexible gelatinous rod in the back—the *notochord*—that disappears as the larvae develop into adults. All vertebrate animals also have a notochord at some time during their development (**FIGURE 16.4**). This shared structure is one of the reasons for inferring that sea squirts are more closely related to vertebrates than would be suspected if only adult sea squirts were examined.

LINK For more on the role of developmental processes in evolution, see Concepts 14.4 and 14.5

BEHAVIOR Some behavioral traits are culturally transmitted (learned from other individuals); others have a genetic basis (see Chapter 41). If a particular behavior is culturally transmitted, it may not accurately reflect evolutionary relationships (but may nonetheless reflect cultural connections). Bird songs, for instance, are often learned and may be inappropriate traits for phylogenetic analysis. Frog calls, however, are genetically determined and appear to be acceptable sources of information for reconstructing phylogenies.

MOLECULAR DATA All heritable variation is encoded in DNA, and so the complete genome of an organism contains an enormous set of traits (the individual nucleotide bases of DNA) that can be used in phylogenetic analyses. In recent years, DNA sequences have become among the most widely used sources of data for constructing phylogenetic trees. Comparisons of nucleotide sequences are not limited to the DNA in the cell nucleus. Eukaryotes have genes in their mitochondria as well as in their nuclei; plant cells have genes in their chloroplasts as well.

The chloroplast genome (cpDNA) is used extensively in phylogenetic studies of plants because it has changed slowly over evolutionary time and can thus be used to study relatively ancient phylogenetic relationships. Most animal mitochondrial DNA (mtDNA) has changed more rapidly, so mitochondrial genes are used to study evolutionary relationships among closely related animal species (the mitochondrial genes of plants evolve more slowly). Many nuclear gene sequences are also commonly analyzed, and now that entire genomes have been sequenced from many species, they too are used to construct phylogenetic trees. Information on gene products (such as the amino acid sequences of proteins) is also widely used for phylogenetic analyses.

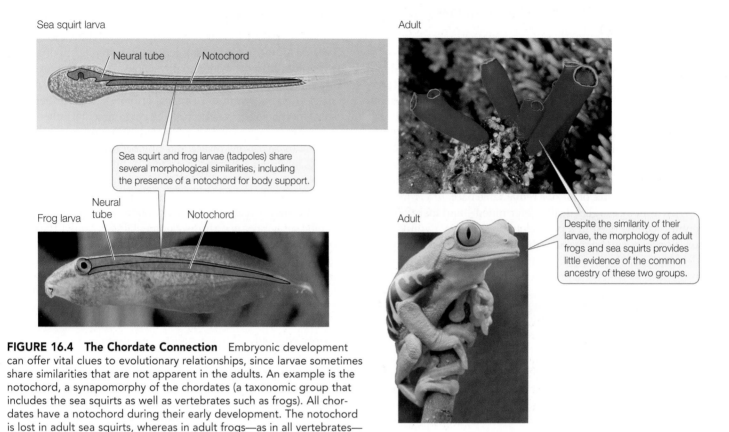

Sea squirt larva

Neural tube   Notochord

Sea squirt and frog larvae (tadpoles) share several morphological similarities, including the presence of a notochord for body support.

Neural tube

Frog larva   Notochord

Adult

Adult

Despite the similarity of their larvae, the morphology of adult frogs and sea squirts provides little evidence of the common ancestry of these two groups.

FIGURE 16.4 The Chordate Connection Embryonic development can offer vital clues to evolutionary relationships, since larvae sometimes share similarities that are not apparent in the adults. An example is the notochord, a synapomorphy of the chordates (a taxonomic group that includes the sea squirts as well as vertebrates such as frogs). All chordates have a notochord during their early development. The notochord is lost in adult sea squirts, whereas in adult frogs—as in all vertebrates—the vertebral column replaces the notochord as the body's support structure.

Mathematical models expand the power of phylogenetic reconstruction

As biologists began to use DNA sequences to infer phylogenies, they developed explicit mathematical models describing how DNA sequences change over time. These models account for multiple changes at a given position in a DNA sequence. They also take into account different rates of change at different positions in a gene, at different positions in a codon, and among different nucleotides. For example, *transitions* (changes between two purines or between two pyrimidines) are usually more likely than are *transversions* (changes between a purine and pyrimidine).

Mathematical models can be used to compute how a tree might evolve given the observed data. A **maximum likelihood** method will identify the tree that most likely produced the observed data, given the assumed model of evolutionary change. Maximum likelihood methods can be used for any kind of characters, but they are most often used with molecular data, for which explicit mathematical models of evolutionary change are easier to develop. The principal advantages to maximum likelihood analyses are that they incorporate more information about evolutionary change than do parsimony methods, and they are easier to treat in a statistical framework. The principal disadvantages are that they are computationally intensive and require explicit models of evolutionary change (which may not be available for some kinds of character change).

The accuracy of phylogenetic methods can be tested

If phylogenetic trees represent reconstructions of past events, and if many of these events occurred before any humans were around to witness them, how can we test the accuracy of phylogenetic methods? Biologists have conducted experiments both in living organisms and with computer simulations that have demonstrated the effectiveness and accuracy of phylogenetic methods.

In one experiment designed to test the accuracy of phylogenetic analysis, a single viral culture of bacteriophage T7 was used as a starting point, and lineages were allowed to evolve from this ancestral virus in the laboratory (**FIGURE 16.5**). The initial culture was split into two separate lineages, one of which became the ingroup for analysis and the other of which became the outgroup for rooting the tree. The lineages in the ingroup were split in two after every 400 generations, and samples of the virus were saved for analysis at each branching point. The lineages were allowed to evolve until there were eight lineages in the ingroup. Mutagens were added to the viral cultures to increase the mutation rate so that the amount of change and the degree of homoplasy would be typical of the organisms analyzed in average phylogenetic analyses. The investigators then sequenced samples from the end points of the eight lineages, as well as from the ancestors at the branching points. They then gave the sequences from the end points of the lineages to other

INVESTIGATION

FIGURE 16.5 The Accuracy of Phylogenetic Analysis
To test whether analysis of gene sequences can accurately reconstruct evolutionary phylogeny, we must have an unambigu- ously known phylogeny to compare against the reconstruction. Will the observed phylogeny match the reconstruction?

HYPOTHESIS

A phylogeny reconstructed from analysis of the DNA sequences of living organisms can accurately match the known evolutionary history of the organisms.

METHOD

In the laboratory, one group of investigators produced an unambiguous phylogeny of 9 viral lineages, enhancing the mutation rate to increase variation among the lineages.

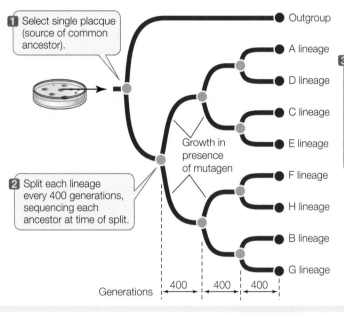

1 Select single placque (source of common ancestor).

2 Split each lineage every 400 generations, sequencing each ancestor at time of split.

Growth in presence of mutagen

3 Present final genes (blue dots) to a second group of investigators who are unaware of the history of the lineages or the gene sequences of the ancestral viruses. These "blind" investigators then determine the sequences of the descendant genes and use these sequences to reconstruct the evolution of these lineages in the form of a phylogenetic tree.

Outgroup
A lineage
D lineage
C lineage
E lineage
F lineage
H lineage
B lineage
G lineage

Generations ← 400 → ← 400 → ← 400 →

RESULTS

The true phylogeny and ancestral DNA sequences were accurately reconstructed solely from the DNA sequences of the viruses at the tips of the tree.

CONCLUSION

Phylogenetic analysis of DNA sequences can accurately reconstruct evolutionary history.

ANALYZE THE DATA

The full DNA sequences for the T7 strains in this experiment are thousands of nucleotides long. The nucleotides ("characters") at 23 DNA positions are given in the table.

	Character at position																						
	1	2	3	4	5	6	7	8	9	10	11	12	13	14	15	16	17	18	19	20	21	22	23
Outgroup	C	C	G	G	G	C	C	T	C	C	T	C	G	A	C	C	G	G	C	A	C	G	G
A	T	C	G	G	G	C	C	C	C	C	C	C	A	A	C	C	G	A	T	A	C	A	A
B	C	C	G	G	G	T	C	C	C	T	C	C	G	A	T	T	A	G	C	G	T	G	G
C	C	C	G	G	G	C	C	T	C	C	T	A	A	C	C	G	G	T	A	C	A	A
D	T	C	A	G	G	C	C	C	C	C	C	A	A	C	C	G	A	T	A	C	A	A
E	C	T	G	G	G	C	C	C	C	C	T	A	A	C	C	G	G	T	A	C	A	A
F	C	T	G	A	A	C	C	C	C	C	C	G	A	C	T	G	G	C	G	C	G	G
G	C	C	G	G	G	T	T	C	C	T	C	C	G	A	T	T	A	G	C	G	C	G	G
H	C	C	G	G	A	C	C	C	C	C	C	C	G	C	C	T	G	G	C	G	C	G	G

A. Construct a phylogenetic tree from these DNA positions using the parsimony method. Use the outgroup to root your tree. Assume that all changes among nucleotides are equally likely.

B. Using your tree, reconstruct the DNA sequences of the ancestral lineages.

For more, go to Working with Data 16.1 at **yourBioPortal.com**.

Go to **yourBioPortal.com** for original citations, discussions, and relevant links for all INVESTIGATION figures.

investigators to analyze, without revealing the known history of the lineages or the sequences of the ancestral viruses.

After the phylogenetic analysis was completed, the investigators asked two questions. Did phylogenetic methods reconstruct the known history correctly? And, were the sequences of the ancestral viruses reconstructed accurately? The answer in both cases was "yes." The branching order of the lineages was reconstructed exactly as it had occurred; more than 98% of the nucleotide positions of the ancestral viruses were reconstructed correctly; and 100% of the amino acid changes in the viral proteins were reconstructed correctly.

---

yourBioPortal.com

Go to ANIMATED TUTORIAL 16.1
Using Phylogenetic Analysis to Reconstruct Evolutionary History

---

The experiment shown in Figure 16.5 demonstrated that phylogenetic analysis was accurate under the conditions tested, but it did not examine all possible conditions. Other experimental studies have taken other factors into account, such as the sensitivity of phylogenetic analysis to convergent environments and highly variable rates of evolutionary change. In addition, computer simulations based on evolutionary models have been used extensively to study the effectiveness of phylogenetic analysis. These studies have also confirmed the accuracy of phylogenetic methods and have been used to refine those methods and extend them to new applications.

Do You Understand Concept 16.2?

- How is the parsimony principle used in reconstructing evolutionary history?

- Why is it useful to consider the entire life cycle when reconstructing an organism's evolutionary history?

- What are some comparative advantages and disadvantages of morphological and molecular approaches for reconstructing phylogenetic trees?

- Contrast experimental and simulation approaches for testing the accuracy of phylogenetic reconstructions of evolutionary history. Can you think of some aspects of phylogenetic accuracy that might be more practical to test using computer simulation than with experimental studies of viruses?

Why do biologists expend the time and effort necessary to reconstruct phylogenies? In fact, information about the evolutionary relationships among organisms is a useful source of data for scientists investigating a wide variety of biological questions. Next we will describe how phylogenetic trees are used to answer questions about the past, and to predict and compare traits of organisms in the present.

concept 16.3 Phylogeny Makes Biology Comparative and Predictive

Once a phylogeny is reconstructed, what do we do with it? What beyond an understanding of evolutionary history does phylogeny offer us?

Reconstructing the past is important for understanding many biological processes

Phylogeny often clarifies the origin and evolution of traits that are of great interest in understanding fundamental biological processes. This information is then widely applied in diverse fields in the life sciences, including agriculture and medicine.

SELF-COMPATIBILITY Like most animals, flowering plants (angiosperms) often reproduce by mating with another individual of the same species. But in many angiosperm species, the same individual produces both male and female gametes (contained within pollen and ovules, respectively). *Self-incompatible* species have mechanisms to prevent fertilization of the ovule by the individual's own pollen, and so must reproduce by *outcrossing* with another individual. Individuals of some species, however, regularly fertilize their ovules using their own pollen; they are self-fertilizing or *selfing* species, and their gametes are *self-compatible*.

The evolution of angiosperm fertilization mechanisms was examined in *Leptosiphon*, a genus in the phlox family that exhibits a diversity of mating systems and pollination mechanisms. The self-incompatible (outcrossing) species of *Leptosiphon* have long petals and are pollinated by long-tongued flies. In contrast, self-pollinating species have short petals and do not require insect pollinators to reproduce successfully. Using nuclear ribosomal DNA sequences, investigators reconstructed a phylogeny this genus (**FIGURE 16.6**). They then determined whether each species was self-compatible by artificially pollinating flowers with the plant's own pollen or with pollen from other individuals and observing whether viable seeds formed.

The reconstructed phylogeny suggests that self-incompatibility is the ancestral state and that self-compatibility evolved three times within this group of *Leptosiphon*. The change to self-compatibility eliminated the plants' dependence on an outside pollinator and has been accompanied by the evolution of reduced petal size. Indeed, the striking morphological similarity of the flowers in the self-compatible groups once led to their being classified as members of a single species. Phylogenetic analysis, however, shows them to be members of three distinct lineages.

LINK Some mechanisms of self-incompatibility are discussed in Concept 27.1

ZOONOTIC DISEASES Many infectious pathogens, particularly viruses, affect individuals of only one species. However, *zoonotic diseases*, or *zoonoses*, are caused by infectious organisms transmitted from an infected animal of a different species. Approximately 150 zoonotic diseases are known to affect humans;

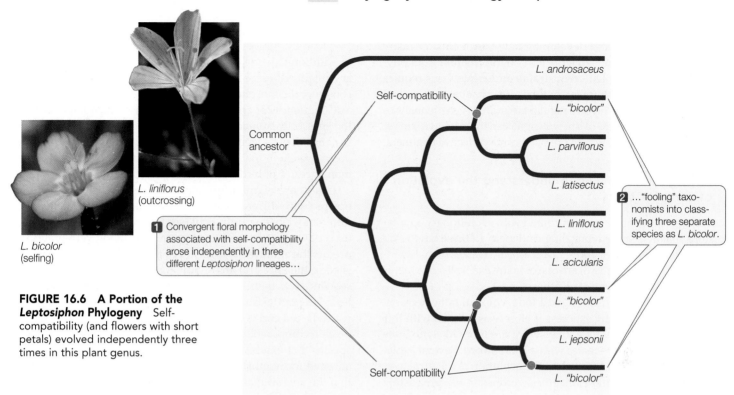

FIGURE 16.6 A Portion of the Leptosiphon Phylogeny Self-compatibility (and flowers with short petals) evolved independently three times in this plant genus.

L. liniflorus (outcrossing)

L. bicolor (selfing)

1 Convergent floral morphology associated with self-compatibility arose independently in three different *Leptosiphon* lineages...

2 ..."fooling" taxonomists into classifying three separate species as *L. bicolor*.

rabies and bubonic plague are historically well-known examples. Lyme disease and acquired immunodeficiency syndrome (AIDS) are more recent examples of zoonotic transfers.

In dealing with zoonotic diseases, it is important to understand when, where, and how the disease first entered a human population. Phylogenetic analyses have become important for studying the transmission of zoonotic pathogens, including the human immunodeficiency virus (HIV) that causes AIDS. Phylogenies are also important for understanding the present global diversity of HIV and for determining the virus's origins in human populations. A broader phylogenetic analysis of immunodeficiency viruses shows that humans acquired these viruses from two different hosts: HIV-1 from chimpanzees, and HIV-2 from sooty mangabeys (**FIGURE 16.7**).

HIV-1 is the common form of the virus in human populations in central Africa, where chimpanzees are hunted for food. HIV-2 is the common form in human populations in western Africa, where sooty mangabeys are hunted for food. These viruses apparently entered human populations through hunters who cut themselves while skinning their primate prey. The

relatively recent global pandemic of AIDS occurred when these infections in local African populations spread through human populations around the world. By understanding how such viruses first entered human populations, health care workers can know what steps need to be taken to prevent other zoonotic diseases from spreading into human populations.

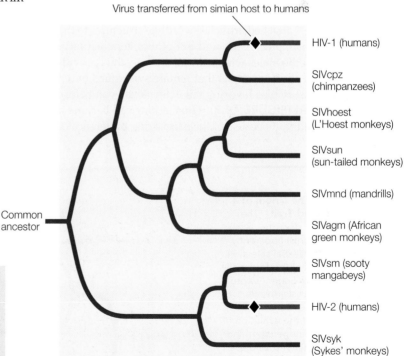

Virus transferred from simian host to humans

FIGURE 16.7 Phylogenetic Tree of Immunodeficiency Viruses
Immunodeficiency viruses have been transmitted to humans from two different simian hosts: HIV-1 from chimpanzees and HIV-2 from sooty mangabeys. SIV stands for "simian immunodeficiency virus."

FRONTIERS Using phylogenetics, it is often possible to trace rare viral infections in humans from specific sources, as in cases of forensic investigations of rape, aggravated assault, and even attempted murder. Biologists are now considering the possibility of tracing the transmission paths of common viral diseases among individuals. This information would allow epidemiologists to identify how these diseases spread, which could lead to prevention of future epidemics.

Understanding and developing new treatments for many viral diseases depends in large part on information from phylogenetic analyses. Different strains of dengue virus require different forms of treatment, but the virus changes quickly and new variants are found on a regular basis. Phylogenetic analyses show the relationships of a new dengue variant to known strains, which is an effective way to predict the most effective treatment.

Phylogenies allow us to understand the evolution of complex traits

Biologists are constantly confronted with evolutionary adaptations that at first seem puzzling, and many of these have to do with mating behavior and *sexual selection*. The tails of male widowbirds (see Figure 15.8) are one of many examples; another is found among the swordtail fishes.

Male swordtail fishes have a long, colorful tail extension, and their reproductive success is closely associated with this appendage. Males with a long sword are more likely to mate successfully than are males with a short sword. Several explanations have been advanced for the evolution of this structure, including the hypothesis that the sword exploits a preexisting bias in the sensory system of the female fish. This *sensory exploitation hypothesis* suggests that female swordtails had a bias to prefer males with long tails even before this trait evolved (perhaps because females assess the size of males by their total body length—including the tail—and prefer larger males).

To test the sensory exploitation hypothesis, phylogenetic analysis identified the swordtail relatives that had split most recently from their lineage before the evolution of sword extensions. These closest relatives turned out to be the platyfishes. Even though male platyfishes do not normally have swords, when researchers attached artificial swordlike structures to the tails of some male platyfishes and not others, female platyfishes preferred the males with artificial swords, thus providing support for the hypothesis that female swordtails had a preexisting sensory bias favoring tail extensions even before the trait evolved (**FIGURE 16.8**). Thus, a long tail became a sexually selected trait because of the preexisting preference of the females.

Ancestral states can be reconstructed

In addition to using phylogenetic methods to infer evolutionary relationships, biologists can use these techniques to reconstruct the morphology, behavior, or nucleotide and amino acid sequences of ancestral species (as was demonstrated for the ancestral sequence of bacteriophage T7 in Figure 16.5). For instance, a phylogenetic analysis was used to reconstruct an opsin protein in the ancestral archosaur (the most recent common ancestor of birds, dinosaurs, and crocodiles).

Opsins are pigment proteins involved in vision; different opsins (with different amino acid sequences) are excited by different wavelengths of light (see Figure 35.17). Investigators used phylogenetic analysis of opsin from living vertebrates to predict the amino acid sequence of the visual pigment that existed in the ancestral archosaur. A protein with the predicted sequence was then constructed in the laboratory. Investigators tested the reconstructed opsin and found a significant shift toward the red end of the spectrum in the light sensitivity of this protein compared with that of most modern opsins. Modern species that exhibit similar red (and infrared) sensitivity are adapted for nocturnal vision; thus the investigators inferred that the ancestral archosaur was likely to have been active at night. These findings may remind you of the movie *Jurassic Park*, although here the extinct species are being "brought back to life" one protein at a time.

Molecular clocks help date evolutionary events

For many applications, biologists want to know not only the order in which evolutionary lineages split but also the timing of those splits. In 1965, Emile Zuckerkandl and Linus Pauling hypothesized that rates of molecular change were constant enough that they could be used to predict evolutionary divergence times—an idea that has become known as the *molecular clock hypothesis*.

Different genes evolve at different rates. In addition, there are differences in evolutionary rates among species related to differing generation times, environments, efficiencies of DNA repair systems, and other biological factors. Nonetheless, among closely related species, a given gene usually evolves at

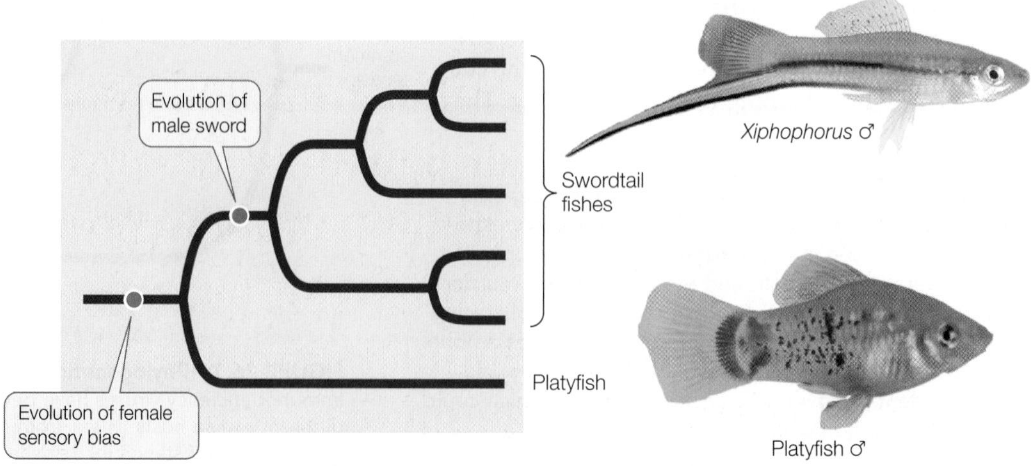

FIGURE 16.8 The Origin of a Sexually Selected Trait The large tail of male swordtail fishes (genus *Xiphophorus*) apparently evolved through sexual selection, with females mating preferentially with males with a longer "sword." Phylogenetic analysis reveals that the platyfishes split from the swordtails before the evolution of the sword. The independent finding that female platyfishes prefer males with an artificial sword further supports the idea that this appendage evolved as a result of a preexisting preference in the females.

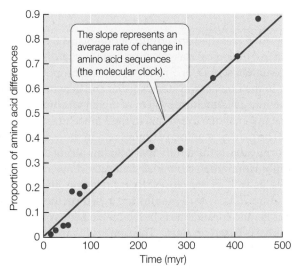

FIGURE 16.9 A Molecular Clock of the Protein Hemoglobin Amino acid replacements in hemoglobin have occurred at a relatively constant rate over nearly 500 million years of evolution. The graph shows the relationship between time of divergence and proportion of amino acid change for 13 pairs of vertebrate hemoglobin proteins. The average rate of change represents the molecular clock for hemoglobin in vertebrates.

a reasonably constant rate. Therefore, the protein encoded by the gene accumulates amino acid replacements at a relatively constant rate (**FIGURE 16.9**). A **molecular clock** uses the average rate at which a given gene or protein accumulates changes to gauge the time of divergence for a particular split in the phylogeny. Molecular clocks must be calibrated using independent data such as the fossil record (see Concept 18.1), known times of divergence, or biogeographic dates (e.g., the time of separations of continents; see Concept 18.3). Using such calibrations, times of divergence have been estimated for many groups of species that have diverged over millions of years.

Besides being useful for dating ancient events, molecular clocks can also be used to study the timing of comparatively recent events. Although HIV-1 samples have generally been collected from humans only since the early 1980s, a few isolates from medical biopsies are available from as early as the 1950s. But biologists can use the observed changes in HIV-1 over the past three decades to project back to the common ancestor of all HIV-1 isolates, and thus can estimate when HIV-1 first entered human populations from chimpanzees (**FIGURE 16.10**). The clock can be calibrated using the samples from the 1980s and 1990s, then tested using the samples from the 1950s. As shown in Figure 16.10C, a sample from a 1959 biopsy is dated by molecular clock analysis at 1957 ± 10 years. The molecular clock was also used to project back to the common ancestor of this group of HIV-1 samples. Extrapolation suggests a date of origin for this group of viruses of about 1930. Although AIDS was unknown to Western medicine until the 1980s, this analysis shows that HIV-1 was present (probably at very low frequency) in human populations in Africa for at least 50 years before its emergence as a global pandemic. Biologists have used similar analyses to conclude that immunodeficiency viruses have been transmitted repeatedly into human populations from multiple primates for more than a century.

FIGURE 16.10 Dating the Origin of HIV-1 in Human Populations (A) A phylogenetic analysis of the main group of HIV-1 viruses. The dates indicate the years in which samples were taken. (For clarity, only a small fraction of the samples that were examined in the original study are shown.) (B) A plot of year of isolation versus genetic divergence from the common ancestor provides an average rate of divergence, or a molecular clock. (C) The molecular clock is used to date a sample taken in 1959 (as a test of the clock) and the unknown date of origin of the HIV-1 main group (about 1930).

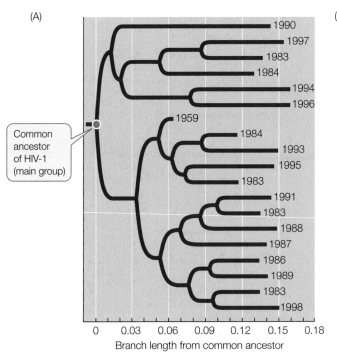

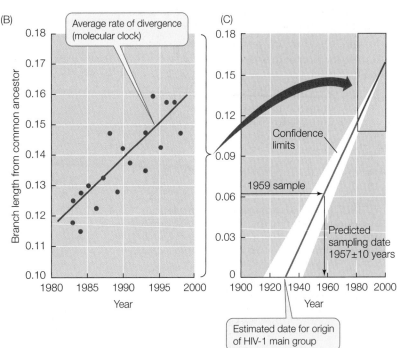

Do You Understand Concept 16.3?

- How can phylogenetic trees help determine the number of times a particular trait evolved?
- How does the reconstruction of ancestral traits help biologists explain the biology of extinct species?
- What is the importance of adding a time dimension to phylogenetic trees, and how do biologists accomplish this?

All of life is connected through evolutionary history, and the relationships among organisms provide a natural basis for making biological comparisons. For these reasons, biologists use phylogenetic relationships as the basis for organizing life into a coherent classification system.

concept 16.4 Phylogeny Is the Basis of Biological Classification

The biological classification system in widespread use today is derived from that developed by the Swedish biologist Carolus Linnaeus in the mid-1700s. Linnaeus developed a system of **binomial nomenclature**. Linnaeus gave each species two names, one identifying the species itself and the other the group of closely related species, or **genus** (plural, *genera*) to which it belongs. Optionally, the name of the taxonomist who first proposed the species name may be added at the end. Thus *Homo sapiens* Linnaeus is the name of the modern human species. *Homo* is the genus, *sapiens* identifies the particular species in the genus *Homo*, and Linnaeus is the person who proposed the name *Homo sapiens*.

You can think of *Homo* as equivalent to your surname and *sapiens* as equivalent to your first name. The first letter of the genus name is capitalized, and the specific name is lowercase. Both of these formal designations are italicized. Rather than repeating the name of a genus when it is used several times in the same discussion, biologists often spell it out only once and abbreviate it to the initial letter thereafter (e.g., *D. melanogaster* rather than *Drosophila melanogaster*).

As we noted earlier, any group of organisms that is treated as a unit in a biological classification system, such as all species in the genus *Drosophila*, or all insects, or all arthropods, is called a *taxon*. In the Linnaean system, species and genera are further grouped into a hierarchical system of higher taxonomic categories. The taxon above the genus in the Linnaean system is the **family**. The names of animal families end in the suffix "-idae." Thus Formicidae is the family that contains all ant species, and the family Hominidae contains humans and our recent fossil relatives, as well as our closest living relatives, the chimpanzees and gorillas. Family names are based on the name of a member genus; Formicidae is based on the genus *Formica*, and Hominidae is based on *Homo*. The same rules are used in classifying plants, except that the suffix "-aceae" is used for plant family names instead of "-idae." Thus Rosaceae is the family that includes the genus *Rosa* (roses) and its relatives.

In the Linnaean system, families are grouped into **orders**, orders into **classes**, and classes into **phyla** (singular *phylum*), and phyla into **kingdoms**. However, the ranking of taxa within Linnaean classification is subjective. Whether a particular taxon is considered, say, an order or a class is informative only with respect to the *relative* ranking of other related taxa. Although families are always grouped within orders, orders within classes, and so forth, there is nothing that makes a "family" in one group equivalent (in number of genera or in evolutionary age, for instance) to a "family" in another group.

Today, the Linnaean terms above the genus level are used largely for convenience. Linnaeus recognized the overarching hierarchy of life, but he developed his system before evolutionary thought had become widespread. Biologists today recognize the tree of life as the basis for biological classification and often name taxa without placing them into the various Linnaean ranks. But, regardless of whether they rank organisms into Linnaean categories or use unranked taxon names,

FIGURE 16.11 Monophyletic, Polyphyletic, and Paraphyletic Groups Monophyletic groups are the basis of biological taxa in modern classifications. Polyphyletic and paraphyletic groups do not accurately reflect evolutionary history.

yourBioPortal.com

**Go to WEB ACTIVITY 16.2
Types of Taxa**

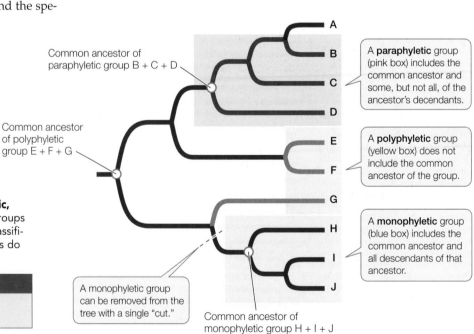

Common ancestor of paraphyletic group B + C + D

Common ancestor of polyphyletic group E + F + G

A **paraphyletic** group (pink box) includes the common ancestor and some, but not all, of the ancestor's decendants.

A **polyphyletic** group (yellow box) does not include the common ancestor of the group.

A **monophyletic** group (blue box) includes the common ancestor and all descendants of that ancestor.

A monophyletic group can be removed from the tree with a single "cut."

Common ancestor of monophyletic group H + I + J

modern biologists use evolutionary relationships as the basis for distinguishing, naming, and classifying biological groups.

Evolutionary history is the basis for modern biological classification

Today's biological classifications express the evolutionary relationships of organisms. Taxa are expected to be **monophyletic**, meaning that the taxon contains an ancestor and all descendants of that ancestor, and no other organisms. In other words, a *monophyletic taxon* is a historical group of related species, or a complete branch on the tree of life. As noted earlier, this is also the definition of a *clade*. A true monophyletic group can be removed from a phylogenetic tree by a single "cut" in the tree, as shown in **FIGURE 16.11**.

Note that there are many monophyletic groups on any phylogenetic tree, and that these groups are successively smaller subsets of larger monophyletic groups. This hierarchy of biological taxa, with all of life as the most inclusive taxon and many smaller taxa within larger taxa, down to the individual species, is the modern basis for biological classification.

Although biologists seek to describe and name only monophyletic taxa, the detailed phylogenetic information needed to do so is not always available. A group that does not include its common ancestor is **polyphyletic**; a group that does not include all the descendants of a common ancestor is referred to as **paraphyletic** (see Figure 16.11). Virtually all taxonomists now agree that polyphyletic and paraphyletic groups are inappropriate as taxonomic units because they do not correctly reflect evolutionary history. Some classifications still contain such groups because some organisms have not been evaluated phylogenetically. As mistakes in prior classifications are detected, taxonomic names are revised and polyphyletic and paraphyletic groups are eliminated from the classifications.

Several codes of biological nomenclature govern the use of scientific names

Several sets of explicit rules govern the use of scientific names. Biologists around the world follow these rules voluntarily to facilitate communication and dialogue. There may be dozens of common names for an organism in many different languages, and the same common name may refer to more than one species (**FIGURE 16.12**). The rules of biological nomenclature are designed so that there is only one correct scientific name for any single recognized taxon and (ideally) a given scientific name applies only to a single taxon (that is, each scientific name is unique). Sometimes the same species is named more than once (when more than one taxonomist has taken up the task); the rules specify that the valid name is the first name that was proposed. If the same name is inadvertently given to two different species, then a replacement name must be given to the species that was named second.

Because of the historical separation of the fields of zoology, botany (which originally included mycology, the study of

(A) *Asclepias tuberosa*

(B) *Castilleja coccinea*

(C) *Hieracium aurantiacum*

FIGURE 16.12 Same Common Name, Not the Same Species All three of these distinct plant species are commonly called "Indian paintbrush" in North America. Unique scientific binomials allow biologists to communicate clearly about each species. (A) *Asclepias tuberosa* is a perennial milkweed native to eastern North America. (B) *Castilleja coccinea* is also native to eastern North America, but is a member of a very different group of plants called scrophs. (C) *Hieracium aurantiacum* is a European species of aster that has been widely introduced into North America.

fungi), and microbiology, different sets of taxonomic rules were developed for each of these groups. Yet another set of rules emerged later for classifying viruses. This separation of fields resulted in duplicated taxon names in groups governed by the different sets of rules; *Drosophila*, for instance, is both a genus of fruit flies and a genus of fungi, and some species in both groups have identical names. Until recently these duplicated

FRONTIERS Biologists are working on a universal code of nomenclature that can be applied to all organisms, so that every species will have a unique identifying name or registration number. This will assist efforts to build an online *Encyclopedia of Life* that links all the information for all the world's species.

APPLY THE CONCEPT

Phylogeny is the basis of biological classification

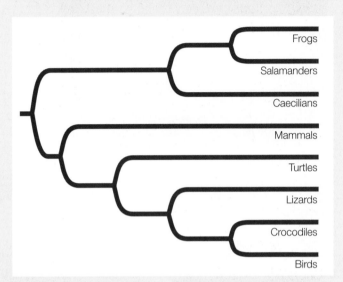

Frogs
Salamanders
Caecilians
Mammals
Turtles
Lizards
Crocodiles
Birds

Classification One:

Named group	Included taxa
Amphibia	Frogs, salamanders, and caecilians
Mammalia	Mammals
Reptilia	Turtles, lizards, and crocodiles
Aves	Birds

Classification Two:

Named group	Included taxa
Amphibia	Frogs, salamanders, and caecilians
Mammalia	Mammals
Reptilia	Turtles, lizards, crocodiles, and birds

Classification Three:

Named group	Included taxa
Amphibia	Frogs, salamanders, and caecilians
Homothermia	Mammals and birds
Reptilia	Turtles, lizards, and crocodiles

Consider the above phylogeny and three possible classifications of the living taxa.

1. Which of these classifications contains a paraphyletic group?

2. Which of these classifications contains a polyphyletic group?

3. Which of these classifications is consistent with the goal of including only monophyletic groups in a biological classification?

4. Starting with the classification you named in Question 3, how many additional group names would you need to include all the clades shown in this phylogenetic tree?

names caused little confusion, since traditionally biologists who studied fruit flies were unlikely to read the literature on fungi (and vice versa). Today, given the prevalence of large, universal biological databases (such as GenBank, which includes DNA sequences from across all life), it is increasingly important that each taxon have a unique and unambiguous name.

Do You Understand Concept 16.4?

- What is the difference between monophyletic, paraphyletic, and polyphyletic groups?

- Why do biologists prefer monophyletic groups in formal classifications?

- What advantages or disadvantages do you see to having separate sets of taxonomic rules for animals, plants, bacteria, and viruses?

Having described some of the mechanisms by which evolution occurs and how phylogenies can be used to study evolutionary relationships, we are now ready to consider the subject of speciation. Speciation is the process that leads to the splitting events (nodes) on the tree of life and eventually results in the millions of distinctive species that constitute Earth's biodiversity.

QA QUESTION How are phylogenetic methods used to resurrect protein sequences from extinct organisms in the laboratory?

ANSWER Most genes and proteins of organisms that lived millions of years ago have decomposed in the fossil remains of these species. Nonetheless, sequences of many ancient genes and proteins can be reconstructed. As we discussed in Concept 16.3, just as we can reconstruct the morphological features of a clade's ancestors, we can also reconstruct their DNA and protein sequences—if we have enough information about the genomes of their descendants.

Biologists have now reconstructed gene sequences from many species that have been extinct for millions of years. Using this information, a laboratory can reconstruct real proteins that correspond to the proteins that were present in long-extinct species. This is how Mikhail Matz and his colleagues were able to resurrect fluorescent proteins from the extinct ancestors of modern corals and then visualize the colors produced by these proteins in the laboratory and recreate the probable evolutionary path of the different pigments (**FIGURE 16.13**).

Biologists have even used phylogenetic analysis to reconstruct some protein sequences that were present in the

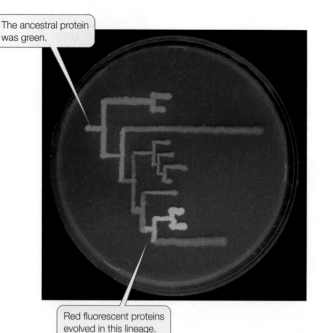

The ancestral protein was green.

Red fluorescent proteins evolved in this lineage.

FIGURE 16.13 Evolution of Fluorescent Proteins of Corals Mikhail Matz and his colleagues used phylogenetic analysis to reconstruct the sequences of extinct fluorescent proteins that were present in the ancestors of modern corals. They then expressed these proteins in bacteria and plated the bacteria in the form of a phylogenetic tree to show how the colors evolved over time.

common ancestor of life described in Concept 16.1. These hypothetical protein sequences can then be resurrected into actual proteins in the laboratory. When biologists measured the temperature optima for these resurrected proteins, they found the proteins functioned best in the range of 55°C–65°C. This analysis is consistent with hypotheses that life evolved in a high-temperature environment.

To reconstruct protein sequences from species that have been extinct for millions or even billions of years, biologists use detailed mathematical models that take into account much of what we have learned of molecular evolution, as described in Concept 16.2. These models incorporate information on rates of replacement among different amino acid residues in proteins, information on different substitution rates among nucleotides, and changes in the rate molecular evolution among the major lineages of life.

16 SUMMARY

concept 16.1 All of Life Is Connected through Its Evolutionary History

- **Phylogeny** is the history of descent of organisms from their common ancestor. Groups of evolutionarily related species are represented as related branches in a **phylogenetic tree**.

- A group of species that consists of a common ancestor and all its evolutionary descendants is called a **clade**. Named clades and species are called **taxa**. Review Figure 16.1

- **Homologies** are similar traits that have been inherited from a common ancestor.

- A trait that is shared by two or more taxa and is derived through evolution from a common ancestral form is called a **synapomorphy**.

- Similar traits may occur among species that do not result from common ancestry. **Convergent evolution** and **evolutionary reversals** can give rise to such traits, which are called **homoplasies**. Review Figure 16.2

concept 16.2 Phylogeny Can Be Reconstructed from Traits of Organisms

- Phylogenetic trees can be inferred from synapomorphies and using the **parsimony principle**. Review Figure 16.3 and WEB ACTIVITY 16.1

- Sources of phylogenetic information include morphology, patterns of development, the fossil record, behavioral traits, and molecular traits such as DNA and protein sequences. Review INTERACTIVE TUTORIAL 16.1

- Phylogenetic trees can be inferred with **maximum likelihood** methods, which calculate the probability that a particular tree

will have generated the observed data. **See WORKING WITH DATA 16.1 and ANIMATED TUTORIAL 16.1**

concept 16.3 Phylogeny Makes Biology Comparative and Predictive

- Phylogenetic trees are used to reconstruct the past and understand the origin of traits. **Review Figure 16.6**

- Phylogenetic trees are used to make appropriate evolutionary comparisons among living organisms.

- Biologists can use phylogenetic trees to reconstruct ancestral states. **Review Figure 16.7**

- Phylogenetic trees may include estimates of times of divergence of lineages determined by **molecular clock** analysis. **Review Figure 16.10**

concept 16.4 Phylogeny Is the Basis of Biological Classification

- Taxonomists organize biological diversity on the basis of evolutionary history.

- Taxa in modern classifications are expected to be clades, or **monophyletic** groups. **Paraphyletic** and **polyphyletic** groups are not considered appropriate taxonomic units. **Review Figure 16.11 and WEB ACTIVITY 16.2**

- Several sets of rules govern the use of scientific names, with the goal of providing unique and universal names for biological taxa.

Speciation 17

Not quite 2 million years ago, a tectonic split in the Great Rift Valley of East Africa led to the formation of Lake Malawi, which lies between the modern countries of Malawi, Tanzania, and Mozambique. A few fish species entered the new lake, including a type known as a haplochromine cichlid. Today the descendants of this early invader include nearly a thousand species of haplochromine cichlids. All of them are *endemic* to Lake Malawi—they are found nowhere else. This vast array of cichlid species makes this the most diverse lake in the world in terms of its fish fauna. How did so many different species arise from a single ancestral species in less than 2 million years?

As we noted in Chapter 16, *speciation* is the process that produces the splits among lineages in the tree of life. Biologists have studied the history and timing of speciation events in Lake Malawi and have pieced together some of the processes that led to so many cichlid species. The earliest haplochromine cichlids to enter the new lake encountered diverse habitats in Lake Malawi, as some shores were rocky and others were sandy. Cichlid populations quickly adapted to these distinct habitat types. Fish in rocky habitats adapted to breeding and living in rocky conditions, and those in sandy habitats evolved specializations for life over sand. These changes resulted in an early speciation event.

Within each of these major habitat types, there were numerous opportunities for diet specialization. Various populations of cichlids became rock scrapers, bottom feeders, fish predators, scale biters, pelagic zooplankton eaters, and plant specialists. Each of these feeding specializations requires different mouth morphology. The offspring of fish that bred with fish of similar morphology were more

likely to survive than were fish with two very different parents. These differences in fitness led to the formation of many more new species, each adapted to a different feeding mode.

But still the Lake Malawi cichlids continued to diverge and form new species. Male cichlids compete for the attention of females through their bright body colors. Diversification of the body colors of males, and of the preferences of females for different body colors, led to many more new species of cichlids, each isolated from the other by their sexual preferences. Now biologists are studying the genomes of these Lake Malawi cichlids to understand the details of the genetic changes that have given rise to so many species over so little time.

This composite photograph shows several of the hundreds of species of diverse haplochromine cichlids that are endemic to Lake Malawi.

KEY CONCEPTS

17.1 Species Are Reproductively Isolated Lineages on the Tree of Life

17.2 Speciation Is a Natural Consequence of Population Subdivision

17.3 Speciation May Occur through Geographic Isolation or in Sympatry

17.4 Reproductive Isolation Is Reinforced When Diverging Species Come into Contact

 QUESTION Can biologists study the process of speciation in the laboratory?

17.1 Species Are Reproductively Isolated Lineages on the Tree of Life

Although "species" is a useful and common term in biology, its usage varies among biologists who are interested in different aspects of **speciation**—the divergence of biological lineages and the emergence of reproductive isolation between lineages. Different biologists think about species differently because they ask different questions: How can we recognize and identify species? How do new species arise? How do different species remain separate? Why do rates of speciation differ among groups of organisms? In answering these questions, biologists focus on different attributes of species, leading to several different ways of thinking about what species are and how they form. Most of the various *species concepts* proposed by biologists are simply different ways of approaching the question "What are species?"

We can recognize many species by their appearance

Biological diversity does not vary in a smooth, incremental way. People have long recognized groups of similar organisms that mate with one another, and they have noticed that there are usually distinct morphological breaks between these groups. Groups of organisms that mate with one another are commonly called *species* (note that this is both the plural and singular form of the word). Someone who is knowledgeable about a group of organisms, such as birds or flowering plants, can usually distinguish the different species found in a particular area simply by looking at them. Standard field guides to birds, mammals, insects, and wildflowers are possible only because many species change little in appearance over large geographic distances (**FIGURE 17.1A**).

More than 250 years ago, Carolus Linnaeus developed the system of binomial nomenclature by which species are named today

(see Concept 16.4). Linnaeus described and named thousands of species, but because he knew nothing about genetics or the mating behavior of the organisms he was naming, he classified them on the basis of their appearance alone. In other words, Linnaeus used a **morphological species concept**, a construct that assumes a species comprises individuals that "look alike," and that individuals that do not look alike belong to different species. Although Linnaeus did not know it, the members of most of the groups he classified as species look alike because they share many alleles of genes that code for morphological features.

Using morphology to define species has limitations. Members of the same species do not always look alike. For example, males, females, and young individuals do not always resemble one another closely (**FIGURE 17.1B**). Furthermore, morphology is of little use in the case of *cryptic species*—instances in which two or more species are morphologically indistinguishable but do not interbreed (**FIGURE 17.2**). Biologists therefore cannot rely on appearance alone in determining whether individual organisms are members of the same or different species. Today, biologists use several additional types of information—especially behavioral and genetic data—to differentiate species.

Reproductive isolation is key

The most important factor in the long-term isolation of sexually reproducing lineages from one another is the evolution of **reproductive isolation**, a state in which two groups of organisms can no longer exchange genes. If individuals of group "A" mate and reproduce only with one another, group "A" constitutes a distinct species within which genes recombine. In other words, group "A" is an independent evolutionary lineage—a separate branch on the tree of life.

It was his recognition of the importance of reproductive isolation that brought Ernst Mayr to propose the **biological species concept**: "*Species are groups of actually or potentially interbreeding natural populations which are reproductively isolated from other such groups.*" The phrase "actually or potentially" is an important

(A)

(B)

Aix sponsa
Male, Florida

Aix sponsa
Male, California

Aix sponsa
Female

FIGURE 17.1 Members of the Same Species Look Alike—or Not (A) It is easy to identify these two male wood ducks as members of the same species, even though they are found on opposite coasts 2,000 miles apart. Despite their geographic separation, the two individuals are morphologically very similar. (B) Wood ducks are sexually dimorphic, which means the female's appearance is quite different from that of the male.

(A)

(B)

Hyla versicolor

Hyla chrysoscelis

FIGURE 17.2 Cryptic Species Look Alike but Do Not Interbreed These two species of gray treefrogs (*Hyla versicolor* and *H. chrysoscelis*) cannot be distinguished by their external morphology, but they do not interbreed even when they occupy the same geographic range. *Hyla versicolor* is a tetraploid species, whereas *H. chrysoscelis* is diploid. Although they look alike, the males have distinctive mating calls, and based on these calls, the females recognize and mate with males of their own species.

element of this definition. "Actually" says that the individuals live in the same area and interbreed with one another. "Potentially" says that even though the individuals do not live in the same area, and therefore do not interbreed, other information suggests that they *would* do so if they were able to get together. This widely used species concept does not apply to organisms that reproduce asexually, and it is limited to a single point in evolutionary time.

The lineage approach takes a long-term view

Evolutionary biologists often think of species as branches on the tree of life. This idea can be termed a **lineage species concept**. In this framework for thinking about species, one species splits into two or more daughter species, which thereafter evolve as distinct lineages. A lineage concept allows biologists to consider species over evolutionary time.

Recall from Chapter 16 that a *lineage* is an ancestor–descendant series of populations followed over time. Each species has a history that starts with a speciation event by which one lineage on the tree is split into two, and ends either at extinction or at another speciation event, at which time the species produces two daughter species. The process of lineage splitting may be gradual, taking thousands of generations to complete. At the other extreme, an ancestral lineage may be split in two within a few generations (as happens with polyploidy, which we'll discuss in Concept 17.3). The gradual nature of some splitting events means that at a single point in time, the final outcome of the process may not be clear. In these cases, it may be difficult to predict whether the incipient species will continue to diverge and become fully isolated from its sibling species, or if they will merge again in the future.

The different species concepts are not mutually exclusive

Many named variants of these three major classes of species concepts exist. The various concepts are not entirely

incompatible; they simply emphasize different aspects of species or speciation. Morphological species concepts emphasize the practical aspects of recognizing species, although sometimes underestimating or overestimating the actual number of species. Mayr's biological species concept emphasizes that reproductive isolation is what allows sexual species to evolve independently of one another. Lineage species concepts embrace the idea that sexual species are maintained by reproductive isolation, but extend the concept of a species as a lineage over evolutionary time. The species-as-lineage concept is also able to accommodate species that reproduce asexually.

Virtually all species exhibit some degree of genetic recombination among individuals, even if recombination events are relatively rare. Significant reproductive isolation between species is therefore necessary for lineages to remain distinct over evolutionary time. Furthermore, reproductive isolation is responsible for the morphological distinctiveness of most species, because mutations that result in morphological changes cannot spread between reproductively isolated species. Therefore, no matter which species concept we emphasize, the evolution of reproductive isolation is important for understanding the origin of species.

Do You Understand Concept 17.1?

- Why do different biologists emphasize different attributes of species in formulating species concepts?
- What makes reproductive isolation such an important component of each of the species concepts discussed here?
- Why is the biological species concept not applicable to asexually reproducing organisms? Do you think this limits its applicability?

Although Charles Darwin titled his groundbreaking book *The Origin of Species*, in fact it included very little about speciation as we understand it today. Darwin devoted most of his attention to demonstrating that individual species are altered over time by natural selection. The remaining sections discuss the many aspects of speciation that biologists have learned about since Darwin's time.

concept
17.2
Speciation Is a Natural Consequence of Population Subdivision

Not all evolutionary changes result in new species. A single lineage may change over time without giving rise to a new species. Speciation requires the interruption of gene flow within a species whose members formerly exchanged genes. But if a genetic change prevents reproduction between individuals of a species, how can such a change spread through a species in the first place?

Incompatibilities between genes can produce reproductive isolation

If a new allele that causes reproductive incompatibility arises in a population, it cannot spread through the population because no other individuals will be reproductively compatible with the individual that carries the new allele. So how can one reproductively cohesive lineage ever split into two reproductively isolated species? Several early geneticists, including Theodosius Dobzhansky and Hermann Joseph Muller, developed a genetic model to explain this apparent conundrum (**FIGURE 17.3**).

The Dobzhansky–Muller model is quite simple. First, assume that a single ancestral population is subdivided into two daughter populations (by the formation of a new mountain range, for instance), which then evolve as independent lineages. In one of the descendant lineages, a new allele (*A*) arises and becomes fixed (see Figure 17.3). In the other population, another new allele (*B*) becomes fixed *at a different gene locus*. Neither new allele at either locus results in any loss of reproductive compatibility. However, the two new forms of these two different genes have never occurred together in the same individual or population. Recall that the products of many genes must work together in an organism. It is possible that the new protein forms encoded by the two new alleles will not be compatible with each other. If individuals from the two lineages come back together after these genetic changes, they may still be able to interbreed, or *hybridize*. However, the hybrid offspring may have a new combination of genes that is functionally inferior, or even lethal. This will not happen with all new combinations of genes, but over time, isolated lineages will accumulate many allele differences at many gene loci. Some combinations of these differentiated genes will not function well together in hybrids. Thus genetic incompatibility between the two isolated populations will develop over time.

> **FRONTIERS**   Although the Dobzhansky–Muller model is based on a minimum of two gene loci, reproductive incompatibility between pairs of closely related species may involve many more genes. The number of genes involved in known cases of reproductive incompatibility has been estimated for more than 20 such species pairs and ranges from 2 to nearly 200 genes.

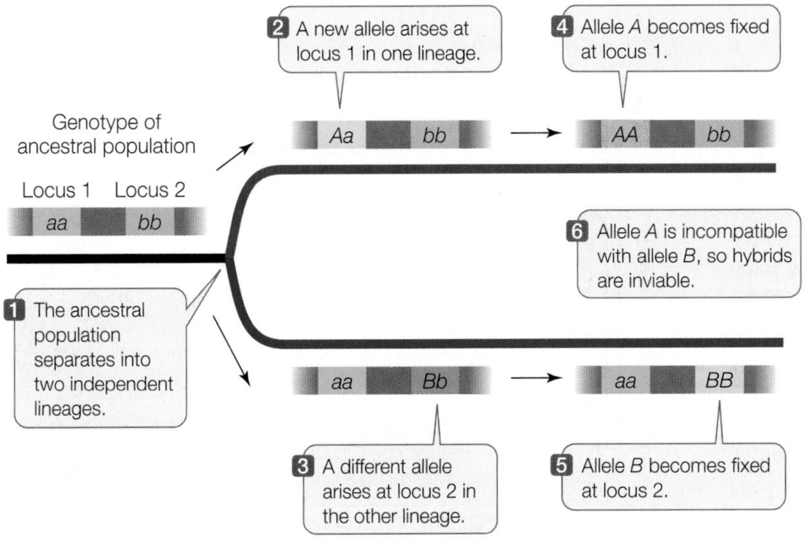

Many empirical examples support the Dobzhansky–Muller model. This model works not only for pairs of individual genes but also for some kinds of chromosomal rearrangements. Bats of the genus *Rhogeessa*, for example, exhibit considerable variation in *centric fusions* of their chromosomes. The chromosomes of the various species contain the same basic chromosomal arms, but in some species two *acrocentric* (one-armed) chromosomes have fused at the centromere to form larger, *metacentric* (two-armed) chromosomes. A polymorphism in centric fusion causes few, if any, problems in meiosis because the respective chromosomes can still align and assort normally. Therefore, a given centric fusion can become fixed in a lineage. However, if a *different* centric fusion becomes fixed in a second lineage, then hybrids between individuals of each lineage will not be able to produce normal gametes in meiosis (**FIGURE 17.4**). Most of the closely related species of *Rhogeessa* display different combinations of these centric fusions and are thereby reproductively isolated from one another.

FIGURE 17.3   The Dobzhansky–Muller Model   In this simple two-locus version of the model, two lineages from the same ancestral population become separated from each other and evolve independently. A new allele becomes fixed in each descendant lineage, but at two different genes. Neither of the new alleles is incompatible with the ancestral alleles, but the two new alleles in the two different genes are incompatible with each other. Thus the two descendant lineages are reproductively incompatible.

yourBioPortal.com
Go to INTERACTIVE TUTORIAL 17.1
Speciation: Trends

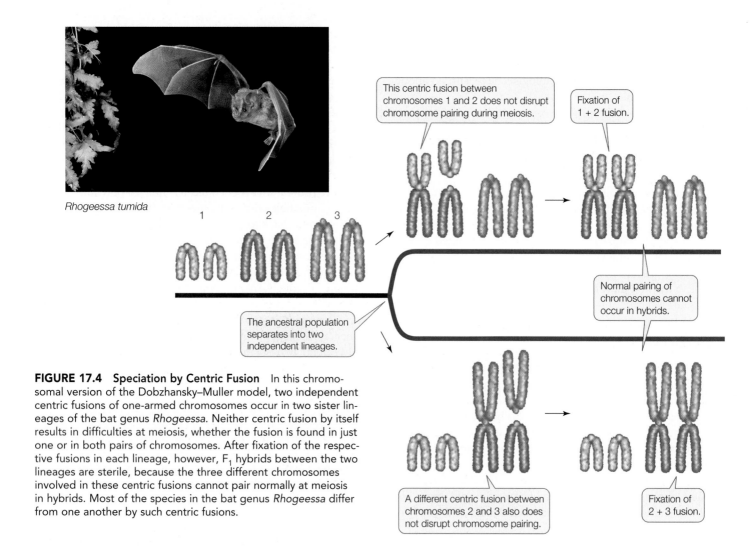

Rhogeessa tumida

FIGURE 17.4 Speciation by Centric Fusion In this chromosomal version of the Dobzhansky–Muller model, two independent centric fusions of one-armed chromosomes occur in two sister lineages of the bat genus *Rhogeessa*. Neither centric fusion by itself results in difficulties at meiosis, whether the fusion is found in just one or in both pairs of chromosomes. After fixation of the respective fusions in each lineage, however, F_1 hybrids between the two lineages are sterile, because the three different chromosomes involved in these centric fusions cannot pair normally at meiosis in hybrids. Most of the species in the bat genus *Rhogeessa* differ from one another by such centric fusions.

Reproductive isolation develops with increasing genetic divergence

As pairs of species diverge genetically, they become increasingly reproductively isolated (**FIGURE 17.5**). Both the rate at which reproductive isolation develops and the mechanisms that produce it vary from group to group. Reproductive incompatibility has been shown to develop gradually in many groups of plants, animals, and fungi, reflecting the slow pace at which incompatible genes accumulate in each lineage. In some cases, complete reproductive isolation may take millions of years. In other cases (as with the chromosomal fusions of *Rhogeessa* described above), reproductive isolation can develop over just a few generations.

Partial reproductive isolation has evolved in strains of *Phlox drummondii* artificially isolated by humans. In 1835, Thomas Drummond, after whom this species of garden plant is named, collected seeds in Texas and distributed them to nurseries in Europe. The European nurseries established more than 200 true-breeding strains of *P. drummondii* that differed in flower size, flower color, and plant growth form. The breeders did not select directly for reproductive incompatibility between strains, but in subsequent experiments in which strains were crossed and seed production was measured and compared,

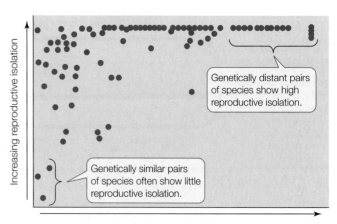

Increasing genetic divergence

FIGURE 17.5 Reproductive Isolation Increases with Genetic Divergence Among pairs of *Drosophila* species, the more the species differ genetically, the greater their reproductive isolation from each other. Each dot represents a comparison of one species pair. Such positive relationships between genetic distance and reproductive isolation have been observed in many groups of plants, animals, and fungi.

biologists found that reproductive compatibility between strains had been reduced by 14 to 50 percent, depending on the cross—even though the strains had been isolated from one another for less than two centuries.

Do You Understand Concept 17.2?

- The Dobzhansky–Muller model suggests that divergence among alleles at *different* gene loci leads to genetic incompatibility between species. Why is genetic incompatibility between two alleles at the *same* locus considered less likely?

- Why do some combinations of chromosomal centric fusions cause problems in meiosis? Can you diagram what would happen at meiosis in a hybrid of the divergent lineages shown in Figure 17.4?

- Assume that the reproductive isolation seen in *Phlox* strains results from lethal combinations of incompatible alleles at several loci among the various strains. Given this assumption, why might the reproductive isolation seen among these strains be partial rather than complete?

We have now seen how splitting an ancestral population leads to genetic divergence in the two descendant lineages. Next we will consider ways in which the descendant lineages could have become separated in the first place.

concept 17.3 Speciation May Occur through Geographic Isolation or in Sympatry

Many scientists who study speciation have concentrated on geographic processes that result in the splitting of an ancestral species. Splitting the range of a species is one obvious way of achieving such a division, but it is not the only way.

Physical barriers give rise to allopatric speciation

Speciation that results when a population is divided by a physical barrier is known as **allopatric speciation** (Greek *allos*, "other"; *patria*, "homeland") or *geographic speciation* (**FIGURE 17.6**). Allopatric speciation is thought to be the dominant mode of speciation in most groups of organisms. The physical barrier that divides the range of a species may be a body of water, a mountain range, or other inhospitable habitat for terrestrial organisms, or dry land for aquatic organisms. Such barriers can form when continents drift, sea levels rise and fall, glaciers advance and retreat, or climates change. These processes continue to generate physical barriers today. The populations separated by such barriers are often, but not always, initially large. The lineages that descend from these founding populations evolve differences for a variety of reasons, including genetic drift and adaptation to the different environments in the two areas.

Allopatric speciation also may result when some members of a population cross an existing barrier and establish a new,

(A) Pliocene

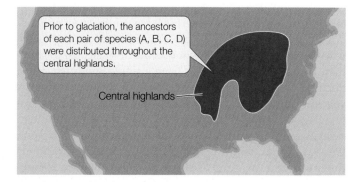

Prior to glaciation, the ancestors of each pair of species (A, B, C, D) were distributed throughout the central highlands.

Central highlands

(B) Pleistocene

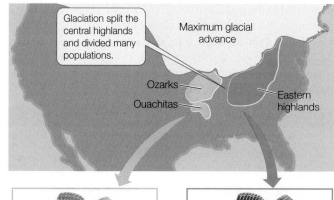

Glaciation split the central highlands and divided many populations.

Maximum glacial advance

Ozarks

Ouachitas

Eastern highlands

A₁ Missouri saddled darter
Etheostoma tetrazonum

A₂ Variegated darter
E. variatum

B₁ Bleeding shiner
Luxilus zonatus

B₂ Warpaint shiner
L. coccogenis

C₁ Ozark minnow
Notropis nubilus

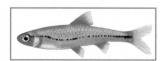

C₂ Tennessee shiner
N. leuciodus

D₁ Ozark madtom
Noturus albater

D₂ Elegant madtom
N. elegans

FIGURE 17.6 Allopatric Speciation Allopatric speciation may result when an ancestral population is divided into two separate populations by a physical barrier, and the lineages that descend from these populations then diverge. (A) Many species of freshwater stream fishes were distributed throughout the central highlands of North America in the Pliocene epoch (about 3–5 million years ago). (B) During the Pleistocene, glaciers advanced and isolated fish populations in the Ozark and Ouachita mountains to the west from fish populations in the highlands to the east. Numerous species diverged as a result of this separation, including the ancestors of the four species pairs shown here.

isolated population. The 14 species of finches found in the Galápagos, an archipelago some 1,000 km off the coast of Ecuador, are the result of this process. Darwin's finches (as they are usually called, because Darwin was the first scientist to study them) arose in the Galápagos from a single South American finch species that colonized the islands. Today the Galápagos species differ strikingly from their closest mainland relative and from one another (**FIGURE 17.7**). The islands of the archipelago are sufficiently far apart that the finches move among them only infrequently. In addition, environmental conditions differ widely from island to island. Some islands are relatively flat and arid; others have forested mountain slopes. Over millions of years, finch lineages on the different islands have differentiated to the point that when occasional immigrants arrive from other islands, they either do not breed with the residents or, if they do, the resulting offspring do not survive as well as the offspring of established residents. The genetic distinctness of each finch species from the others and the genetic cohesiveness of the individual species are thus maintained.

---

yourBioPortal.com

Go to ANIMATED TUTORIAL 17.1
Founder Events and Allopatric Speciation

Sympatric speciation occurs without physical barriers

Although geographic isolation is usually required for speciation, under some circumstances speciation can occur in the absence of a physical barrier. Speciation without physical isolation is called **sympatric speciation** (Greek *sym*, "together with"). But how can such speciation happen? Given that speciation is usually a gradual process, how can reproductive isolation develop when individuals have frequent opportunities to mate with one another?

Sympatric speciation may occur with some forms of disruptive selection (see Concept 15.4) in which individuals with certain genotypes have a preference for distinct microhabitats where mating takes place. For example, sympatric speciation via disruptive selection appears to be taking place in the apple maggot fly (*Rhagoletis pomonella*) of eastern North America. Until the mid-1800s, *Rhagoletis* flies courted, mated, and deposited their eggs only on hawthorn fruits. About 150 years ago, some flies began to lay their eggs on apples, which European immigrants had introduced into eastern North America. Apple trees are closely related to hawthorns, but the smell of the fruits differs, and the apple fruits appear earlier than those of hawthorns. Some early-emerging female *Rhagoletis* laid their eggs on apples, and over time, a genetic preference for the smell of apples evolved among early-emerging insects. When the offspring of these flies sought out apple trees for mating and egg deposition, they mated with other flies reared on apples, which shared the same preferences.

Today the two groups of *Rhagoletis pomonella* in the eastern U.S. appear to be on the way to becoming distinct species. One group mates and lays eggs primarily on hawthorn fruits, the other on apples. The incipient species are partially reproductively isolated because they mate primarily with individuals raised on the same fruit and because they emerge from their pupae at different times of the year. In addition, the apple-feeding flies now grow more rapidly on apples than they originally did. Sympatric speciation that arises from such host-plant specificity may be widespread among insects, many of which feed only on a single plant species.

APPLY THE CONCEPT

Speciation may occur through geographic isolation

The different species of Darwin's finches shown in the phylogeny in Figure 17.7 have all evolved on islands of the Galápagos archipelago within the past 3 million years. Molecular clock analysis (see Concept 15.5) has been used to determine the dates of the various speciation events in that phylogeny. Geological techniques for dating rock samples (see Concept 18.1) have been used to determine the ages of the various Galápagos islands. The table shows the number of species of Darwin's finches and the number of islands that have existed in the archipelago at several times during the past 4 million years.

TIME (MYA)	NUMBER OF ISLANDS	NUMBER OF FINCH SPECIES
0.25	18	14
0.50	18	9
0.75	9	7
1.00	6	5
2.00	4	3
3.00	4	1
4.00	3	0

1. Plot the number of species of Darwin's finches and the number of islands in the Galápagos archipelago (dependent variables) against time (independent variable).

2. Are the data consistent with the hypothesis that isolation of populations on newly formed islands is related to speciation in this group of birds? Why or why not?

3. If no more islands form in the Galápagos archipelago, do you think that speciation by geographic isolation will continue to occur among Darwin's finches? Why or why not? What additional data could you collect to test your hypothesis (without waiting to see if speciation occurs)?

FIGURE 17.7 Allopatric Speciation among Darwin's Finches The descendants of the ancestral finch that colonized the Galápagos archipelago several million years ago evolved into at least 14 different species whose members are variously adapted to feed on seeds, buds, and insects.

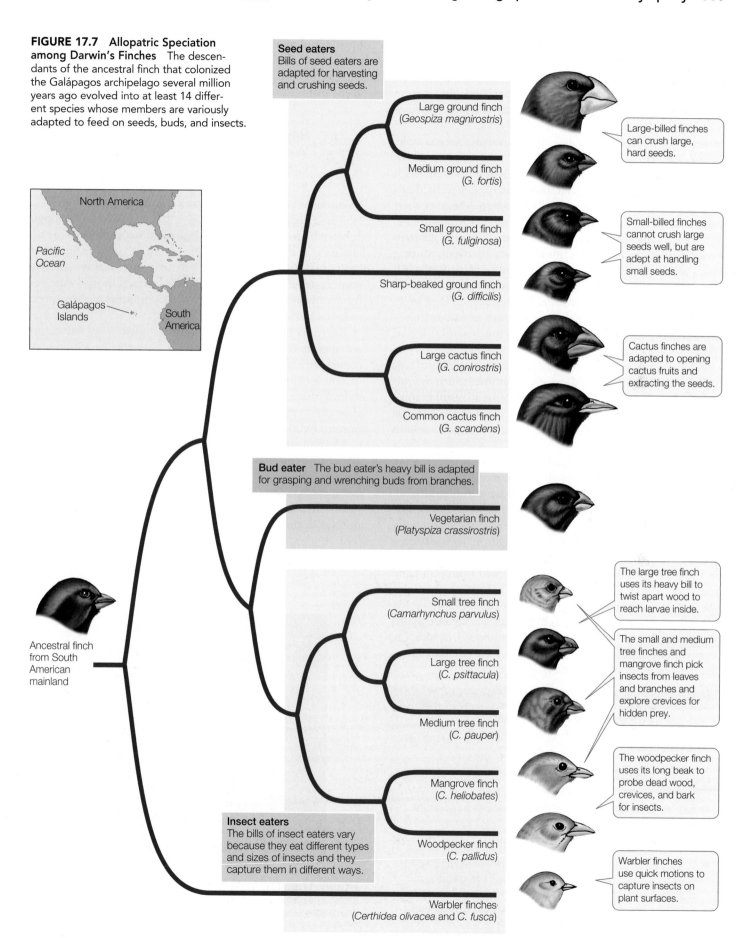

Seed eaters Bills of seed eaters are adapted for harvesting and crushing seeds.

Large ground finch
(*Geospiza magnirostris*)

Medium ground finch
(*G. fortis*)

Small ground finch
(*G. fuliginosa*)

Sharp-beaked ground finch
(*G. difficilis*)

Large cactus finch
(*G. conirostris*)

Common cactus finch
(*G. scandens*)

Large-billed finches can crush large, hard seeds.

Small-billed finches cannot crush large seeds well, but are adept at handling small seeds.

Cactus finches are adapted to opening cactus fruits and extracting the seeds.

Bud eater The bud eater's heavy bill is adapted for grasping and wrenching buds from branches.

Vegetarian finch
(*Platyspiza crassirostris*)

Small tree finch
(*Camarhynchus parvulus*)

Large tree finch
(*C. psittacula*)

Medium tree finch
(*C. pauper*)

Mangrove finch
(*C. heliobates*)

Woodpecker finch
(*C. pallidus*)

The large tree finch uses its heavy bill to twist apart wood to reach larvae inside.

The small and medium tree finches and mangrove finch pick insects from leaves and branches and explore crevices for hidden prey.

The woodpecker finch uses its long beak to probe dead wood, crevices, and bark for insects.

Insect eaters The bills of insect eaters vary because they eat different types and sizes of insects and they capture them in different ways.

Warbler finches
(*Certhidea olivacea* and *C. fusca*)

Warbler finches use quick motions to capture insects on plant surfaces.

North America

Pacific Ocean

Galápagos Islands

South America

Ancestral finch from South American mainland

The most common means of sympatric speciation, however, is **polyploidy**, or the duplication of sets of chromosomes within individuals. Polyploidy can arise either from chromosome duplication in a single species (**autopolyploidy**) or from the combining of the chromosomes of two different species (**allopolyploidy**).

An autopolyploid individual originates when, for example, two accidentally unreduced diploid gametes (with two sets of chromosomes) combine to form a tetraploid individual (with four sets of chromosomes). Tetraploid and diploid individuals of the same species are reproductively isolated because their hybrid offspring are triploid. Even if these offspring survive, they are usually sterile; they cannot produce normal gametes because their chromosomes do not segregate evenly during meiosis. So a tetraploid individual cannot produce fertile offspring by mating with a diploid individual—but it *can* do so if it self-fertilizes or mates with another tetraploid. Thus polyploidy can result in complete reproductive isolation in two generations—an important exception to the general rule that speciation is a gradual process.

Allopolyploids may be produced when individuals of two different (but closely related) species interbreed. Such hybridization often disrupts normal meiosis, which can result in chromosomal doubling. Allopolyploids are often fertile because each of the chromosomes has a nearly identical partner with which to pair during meiosis.

Speciation by polyploidy has been particularly important in the evolution of plants, although it has contributed to speciation in animals as well (such as the example in Figure 17.2). Botanists estimate that about 70 percent of flowering plant species and 95 percent of fern species are the result of recent polyploidization. Some of these species arose from hybridization between two species followed by chromosomal duplication and self-fertilization. Other species diverged from polyploid ancestors, so that the new species shared their ancestors' duplicated sets of chromosomes. New species arise by polyploidy more easily among plants than among animals because plants of many species can reproduce by self-fertilization. In addition, if polyploidy arises in several offspring of a single parent, the siblings can fertilize one another.

---

yourBioPortal.com

Go to ANIMATED TUTORIAL 17.2
Speciation Mechanisms

Do You Understand Concept 17.3?

- Explain how speciation via polyploidy can happen in only two generations.

- What are some obstacles to sympatric speciation?

- If allopatric speciation is the most prevalent mode of speciation, what do you predict about the geographic distributions of many closely related species? Does your answer differ for species that are sedentary versus species that are highly mobile?

Most populations separated by a physical barrier become reproductively isolated only slowly and gradually. If two incipient species once again come into contact with each other, what keeps them from merging back into a single species?

concept 17.4 Reproductive Isolation Is Reinforced When Diverging Species Come into Contact

As discussed in Concept 17.2, once a barrier to gene flow is established, reproductive isolation will begin to develop through genetic divergence. Over many generations, differences accumulate in the isolated lineages, reducing the probability that individuals from each lineage will mate successfully with one another when they come back into contact. In this way, reproductive isolation can evolve as a by-product of the genetic changes in the two diverging lineages.

Reproductive isolation may be incomplete when the incipient species come back into contact, however, in which case some hybridization will occur. If hybrid individuals are less fit than non-hybrids, selection favors parents that do not produce hybrid offspring. Under these conditions, selection results in strengthening, or **reinforcement**, of isolating mechanisms that prevent hybridization.

Mechanisms that prevent hybridization from occurring are called **prezygotic isolating mechanisms**. Mechanisms that reduce the fitness of hybrid offspring are called **postzygotic isolating mechanisms**. Postzygotic mechanisms result in selection against hybridization, which leads to the reinforcement of the prezygotic mechanisms.

FIGURE 17.8 Mechanical Isolation through Mimicry Many orchid species maintain reproductive isolation by means of flowers that look and smell like females of a specific bee or wasp species. Shown here are an Australian orchid (*Cryptostylis* sp.) and its pollinator, a male wasp of the genus *Lissopimpla*.

(A) Allopatric populations

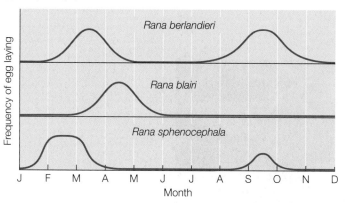

(B) Sympatric populations

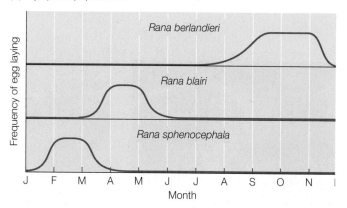

FIGURE 17.9 Temporal Isolation of Breeding Seasons (A) The peak breeding seasons of three species of *Rana* overlap when the species are physically separated (allopatry). (B) Where two or more species of *Rana* live together (sympatry), overlap between their peak breeding seasons is greatly reduced or eliminated. Selection against hybridization in areas of sympatry helps reinforce this prezygotic isolating mechanism.

Prezygotic isolating mechanisms prevent hybridization between species

Prezygotic isolating mechanisms, which come into play before fertilization, can prevent hybridization in several ways.

MECHANICAL ISOLATION Differences in the sizes and shapes of reproductive organs may prevent the union of gametes from different species. With animals, there may be a match between the shapes of the reproductive organs of males and females of the same species, so that reproduction between species with mismatched structures is not physically possible. In plants, mechanical isolation may involve a pollinator. For example, orchids of the genus *Cryptostylis* produce flowers that look and smell like the females of particular species of wasps (**FIGURE 17.8**). When a male wasp visits and attempts to mate with the flower (thinking it is a female wasp of his species), his mating behavior results in the transfer of pollen to and from his body by appropriately configured anthers and stigmas on the flower. Insects that visit the flower but do not attempt to mate with it do not trigger the transfer of pollen between the insect and the flower.

TEMPORAL ISOLATION Many organisms have distinct mating seasons. If two closely related species breed at different times of the year (or different times of day), they may never have an opportunity to hybridize. For example, in sympatric populations of three closely related leopard frog species, each species breeds at a different time of year (**FIGURE 17.9**). Although there is some overlap in the breeding seasons, the opportunities for hybridization are minimized.

BEHAVIORAL ISOLATION Individuals may reject, or fail to recognize, individuals of other species as potential mating partners. For example, the mating calls of male frogs of related species diverge quickly (**FIGURE 17.10**). Female frogs respond to mating calls from males of their own species but ignore the calls of other species, even closely related ones. The evolution of female preferences for certain male coloration patterns among the cichlids of Lake Malawi, described at the opening of this chapter, is another example of behavioral isolation.

Sometimes the mate choice of one species is mediated by the behavior of individuals of other species. For example, whether

Gastrophryne olivacea ■

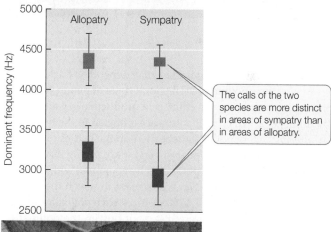

The calls of the two species are more distinct in areas of sympatry than in areas of allopatry.

Gastrophryne carolinensis ■

FIGURE 17.10 Behavioral Isolation in Mating Calls The males of most frog species produce species-specific calls. The calls of the two closely related frog species shown here differ in their dominant frequency (a high-frequency sound wave results in a high-pitched sound; a low frequency results in a low-pitched sound). Female frogs are attracted to the calls of males of their own species.

(A)

(B)

(C) *Aquilegia formosa*

(D) *A. pubescens*

FIGURE 17.11 Floral Morphology Is Associated with Pollinator Morphology (A) This hummingbird's morphology and behavior are adapted for feeding on nectar from flowers that are pendant. (B) The nectar-extracting proboscis of this hawkmoth is adapted to flowers that grow upright. (C) *Aquilegia formosa* flowers are normally pendant and are pollinated by hummingbirds. (D) Flowers of *A. pubescens* are normally upright and are pollinated by hawkmoths. In addition, their long floral spurs appear to restrict access by some other potential pollinators.

or not two plant species hybridize may depend on the food preferences of their pollinators. The floral traits of plants, including their color and shape, can enhance reproductive isolation either by influencing which pollinators are attracted to the flowers or by altering where pollen is deposited on the bodies of pollinators. A plant whose flowers are pendant (hanging downward; **FIGURE 17.11A**) will be pollinated by an animal with different physical characteristics than will a plant in which the flowers grow upright (**FIGURE 17.11B**). Because each pollinator prefers (and is adapted to) a different type of flower, the pollinators rarely transfer pollen from one plant species to the other.

Such isolation by pollinator behavior is seen in the mountains of California in two sympatric species of columbines (*Aquilegia*) that have diverged in flower color, structure, and orientation. *Aquilegia formosa* (**FIGURE 17.11C**) has pendant flowers with short spurs (spikelike, nectar-containing structures) and is pollinated by hummingbirds. *A. pubescens* (**FIGURE 17.11D**) has upright, lighter-colored flowers with long spurs and is pollinated by hawkmoths. The difference in pollinators means that these two species are effectively reproductively isolated even though they populate the same geographic range.

HABITAT ISOLATION When two closely related species evolve preferences for living or mating in different habitats, they may never come into contact during their respective mating periods.

The *Rhagoletis* flies discussed in Concept 17.3 experienced such habitat isolation, as did the cichlid fishes that first adapted to rocky and sandy habitats upon entering Lake Malawi, as described at the opening of this chapter.

LINK Some plants and their pollinators become so tightly adapted to each other that they develop mutually dependent relationships, as described in Concept 21.5

GAMETIC ISOLATION The sperm of one species may not attach to the eggs of another species because the eggs do not release the appropriate attractive chemicals, or the sperm may be unable to penetrate the egg because the two gametes are chemically incompatible. Thus, even though the gametes of two species may come into contact, the gametes never fuse into a zygote.

Gametic isolation is extremely important for many aquatic species that *spawn* (release their gametes directly into the environment); it has been extensively studied in sea urchins. A protein known as *bindin* is found in sea urchin sperm and functions in attaching ("binding") the sperm to eggs. All sea urchin species studied produce this egg-recognition protein, but the bindin gene sequence diverges so rapidly that it becomes species-specific. Since sperm can only attach to eggs of the same species, no interspecific hybridization occurs.

Postzygotic isolating mechanisms result in selection against hybridization

Genetic differences that accumulate between two diverging lineages may reduce the survival and reproductive rates of hybrid offspring in any of several ways:

- *Low hybrid zygote viability.* Hybrid zygotes may fail to mature normally, either dying during development or developing phenotypic abnormalities that prevent them from becoming reproductively capable adults.
- *Low hybrid adult viability.* Hybrid offspring may have lower survivorship than non-hybrid offspring.
- *Hybrid infertility.* Hybrids may mature into infertile adults. For example, the offspring of matings between horses and donkeys—mules—are sterile; although otherwise healthy, they produce no descendants.

Natural selection does not directly favor the evolution of postzygotic isolating mechanisms. But if hybrids are less fit, individuals that breed only within their own species will leave more surviving offspring than will individuals that interbreed with another species. Therefore, individuals that can avoid interbreeding with members of other species will have a selective advantage, and any trait that contributes to such avoidance will be favored.

Donald Levin of the University of Texas has studied reinforcement of prezygotic isolating mechanisms in flowers of the genus *Phlox.* Levin noticed that individuals of *Phlox drummondii* in most of the range of the species in Texas usually have pink flowers. However, where *P. drummondii* is sympatric with its close relative the pink-flowered *P. cuspidata,* *P. drummondii* usually has red flowers. No other *Phlox* species has red flowers. Levin performed an experiment whose results showed that reinforcement may explain why red flowers are favored where the two species are sympatric (**FIGURE 17.12**).

Likely cases of reinforcement are often detected by comparing sympatric and allopatric populations of potentially hybridizing species, as in the case of *Phlox.* If reinforcement is occurring, then

INVESTIGATION

FIGURE 17.12 Flower Color and Reproductive Isolation Most *Phlox drummondii* individuals have pink flowers, but in regions where the species is sympatric with *P. cuspidata*—which is always pink—most *P. drummondii* individuals have red flowers. Most pollinators preferentially visit flowers of one color or the other. In this experiment, Donald Levin explored whether flower color acts as a prezygotic isolating mechanism that lessens the chances of hybridization between the two species.

HYPOTHESIS

Red-flowered *P. drummondii* are less likely to hybridize with *P. cuspidata* than are pink-flowered *P. drummondii.*

METHOD

1. Introduce equal numbers of red-flowered and pink-flowered *P. drummondii* individuals into an area where many pink-flowered *P. cuspidata* are growing.

P. cuspidata

P. drummondii

2. After the flowering season ends, measure hybridization by assessing the genetic composition of the seeds produced by *P. drummondii* plants of both colors.

RESULTS

Of the seeds produced by pink-flowered *P. drummondii*, 38% were hybrids with *P. cuspidata.* Only 13% of the seeds produced by red-flowered *P. drummondii* were hybrids with *P. cuspidata*

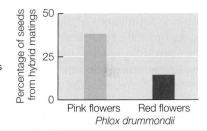

CONCLUSION

P. drummondii and *P. cuspidata* are less likely to hybridize if the flowers of the two species differ in color.

ANALYZE THE DATA

Data from Levin's experiment show that the frequency of hybridization between *Phlox drummondii* and *P. cuspidata* was strongly dependent on the former's flower color.

P. drummondii flower color	Number of progeny (seeds)		
	P. drummondii	Hybrid	Total
Red	181 (87%)	27 (13%)	208
Pink	86 (62%)	53 (38%)	139

A. The data reveal that red-flowered *P. drummondii* produced more seeds (208) than pink-flowered plants did (139), even though equal numbers of individuals of the two flower types were used. Does this difference influence your interpretation of Levin's results? Why or why not?

B. How would you improve the experimental design of this study? Should replicate or control sites be added? What kinds of additional test sites and conditions would you add?

For more, go to Working with Data 17.1 at **yourBioPortal.com.**

Go to **yourBioPortal.com** for original citations, discussions, and relevant links for all INVESTIGATION figures.

APPLY THE CONCEPT

Reproductive isolation is reinforced when diverging species come into contact

As shown in Figure 17.9, the leopard frogs *Rana berlandieri* and *R. sphenocephala* usually have non-overlapping breeding seasons in areas of sympatry, but where the species are allopatric, both species breed in both spring and fall. But when new ponds are created where the ranges of the two species come close together, frogs from previously allopatric populations may colonize the new ponds and hybridize during their overlapping breeding seasons.

Imagine you have collected and tabulated data on hybridization between these two frog species. You have sampled various life stages of frogs and their tadpoles for two years after an initial spring breeding season at a newly established pond. Use the data (above right) to answer the following questions.

1. Create four pie charts (one for each life stage) showing the percentage of each species and the percentage of hybrids at each stage.

LIFE STAGE	R. BERLANDIERI	R. SPHENOCEPHALA	F₁ HYBRIDS
Recently hatched tadpoles (spring, year 1)	155	125	238
Late-stage tadpoles (summer, year 1)	45	55	64
Newly metamorphosed froglets (fall, year 1)	32	42	15
Adult frogs (year 2)	10	15	1

2. What are some possible reasons for the differences in the percentages of hybrids found at each life stage? Suggest some postzygotic isolating mechanisms that are consistent with your data.

3. Over time, what changes might you expect in the breeding seasons of the two species at this particular pond, and why? How would future pie charts change if your predictions about breeding seasons are correct?

sympatric populations of closely related species are expected to evolve more effective prezygotic reproductive barriers than do allopatric populations of the same species. As Figure 17.8 shows, the breeding seasons of sympatric populations of different leopard frog species overlap much less than do those of the corresponding allopatric populations. Similarly, the frequencies of the frog mating calls illustrated in Figure 17.10 are more divergent in sympatric populations than in allopatric populations. In both cases, there appears to have been natural selection against hybridization in areas of sympatry.

Hybrid zones may form if reproductive isolation is incomplete

Unless reproductive isolation is complete, closely related species may hybridize in areas where their ranges overlap, resulting in the formation of a **hybrid zone**. When a hybrid zone first forms, most hybrids are offspring of crosses between purebred individuals of the two species. However, subsequent generations will include a variety of individuals with varying proportions of their genes derived from the original two species. Thus hybrid zones often contain recombinant individuals resulting from many generations of hybridization.

Detailed genetic studies can tell us much about why narrow hybrid zones may persist for long periods between the ranges of two species. In Europe, the hybrid zone between two toad species of the genus *Bombina* has been studied intensively. The fire-bellied toad (*B. bombina*) lives in eastern Europe; the closely related yellow-bellied toad (*B. variegata*) lives in western and southern Europe. The ranges of the two

FRONTIERS *Homo neanderthalensis* coexisted with modern *Homo sapiens* for tens of thousands of years before Neanderthals became extinct. Anthropologists have speculated as to whether the two groups, apparently genetically and morphologically quite distinct, were capable of interbreeding. Improvements in genome sequencing have now permitted biologists to sequence much of the Neanderthal genome from fossil remains, revealing that 1–4 percent of the genes present in modern European and Asian (but not African) human populations appears to have come from Neanderthals. This evidence suggests that *Homo sapiens* interbred with Neanderthals about 50,000 years ago in the Middle East—after humans left Africa but before they spread through the rest of the world.

species overlap in a long but very narrow zone stretching 4,800 kilometers from eastern Germany to the Black Sea (**FIGURE 17.13**). Hybrids between the two species suffer from a range of defects, many of which are lethal. Those hybrids that survive often have skeletal abnormalities, such as misshapen mouths, ribs that are fused to vertebrae, and a reduced number of vertebrae. By following the fates of thousands of toads from the hybrid zone, investigators found that a hybrid toad, on average, is only half as fit as a purebred individual of either species. The hybrid zone remains narrow because there is strong selection against hybrids and because adult toads do not move over long distances. The zone has persisted for hundreds of years, however, because individuals of both species continue to move short distances into it, continually replenishing the hybrid population.

B. bombina
(fire-bellied toad)

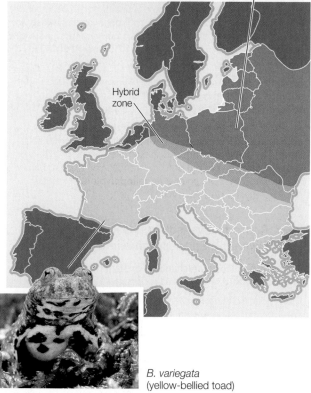

Hybrid
zone

B. variegata
(yellow-bellied toad)

FIGURE 17.13 **A Hybrid Zone** The long, narrow zone where fire-bellied toads meet and hybridize with yellow-bellied toads has been stable for hundreds of years.

Do You Understand Concept 17.4?

- Why would reinforcement of prezygotic isolating mechanisms be expected even if postzygotic isolating mechanisms already exist?

- Why would you expect to find different types of prezygotic isolating mechanisms among different groups of organisms? Describe some specific examples.

- Why don't most narrow hybrid zones, such as the one between *Bombina bombina* and *Bombina variegata*, get wider over time?

The result of 4 billion of years of evolution has been many millions of species, each adapted to live in a particular environment and to use environmental resources in a particular way. In the next chapter we will describe how geological and biological processes interacted over the course of Earth's history to produce this vast biodiversity.

Q&A
QUESTION
Can biologists study the process of speciation in the laboratory?

ANSWER Although speciation usually takes thousands or millions of years, and although it is typically studied in natural settings such as Lake Malawi, some aspects of speciation can be studied and observed in controlled laboratory experiments. Most such experiments use organisms with short generation times, in which evolution is expected to be relatively rapid.

William Rice and George Salt conducted an experiment in which fruit flies were allowed to choose food sources in different habitats, where mating also took place. The habitats were vials in different parts of an experimental cage (**FIGURE 17.14**). The vials differed in three environmental factors: (1) light; (2) the direction (up or down) in which the fruit flies had to move to reach food; and (3) the concentrations of two aromatic chemicals, ethanol and acetaldehyde. In just 35 generations, the two groups of flies that chose the most divergent habitats had become reproductively isolated from each other, having evolved distinct preferences for the different habitats.

The experiment by Rice and Salt demonstrated an example of habitat isolation evolving to function as a prezygotic isolating mechanism. Even though the different habitats were in the same cage, and individual fruit flies were capable of flying from one habitat to the other, habitat preferences were inherited by offspring from their parents, and populations from the two divergent habitats did not interbreed. Similar habitat selection is thought to have resulted in the early split in cichlids that preferred the rocky versus the sandy shores of Lake Malawi.

In controlled experiments like this one, biologists can observe many aspects of the process of speciation directly.

FIGURE 17.14 **Evolution in the Laboratory** For their experiments on the evolution of prezygotic isolating mechanisms in *Drosophila melanogaster*, Rice and Salt built an elaborate system of varying habitats contained within vials inside a large fly enclosure. Some groups of flies developed preferences for widely divergent habitats and became reproductively isolated within 35 generations.

SUMMARY

17

concept
17.1 **Species Are Reproductively Isolated Lineages on the Tree of Life**

- **Speciation** is the process by which one species splits into two or more daughter species, which thereafter evolve as distinct lineages.

- The **morphological species concept** distinguishes species on the basis of physical similarities; it often underestimates or overestimates the actual number of reproductively isolated species.

- The **biological species concept** distinguishes species on the basis of **reproductive isolation**.

- **Lineage species concepts**, which recognize independent evolutionary lineages as species, allow biologists to consider species over evolutionary time.

concept
17.2 **Speciation Is a Natural Consequence of Population Subdivision**

- Genetic divergence results from the interruption of gene flow within a population.

- The Dobzhansky–Muller model describes how reproductive isolation between two descendant lineages can develop through the accumulation of incompatible genes or chromosomal arrangements. **Review Figure 17.3 and INTERACTIVE TUTORIAL 17.1**

- Reproductive isolation increases with increasing genetic divergence between populations.

concept
17.3 **Speciation May Occur through Geographic Isolation or in Sympatry**

- **Allopatric speciation**, which results when populations are separated by a physical barrier, is the dominant mode of speciation. This type of speciation may follow founder events, in which some members of a population cross a barrier and found a new, isolated population. **Review Figure 17.6 and ANIMATED TUTORIAL 17.1**

- **Sympatric speciation** results when two species diverge in the absence of geographic isolation. It can result from disruptive selection in two or more distinct microhabitats.

- Sympatric speciation can occur within two generations via **polyploidy**, an increase in the number of chromosomes sets. Polyploidy may arise from chromosome duplications within a species (**autopolyploidy**) or from hybridization that combines the chromosomes of two species (**allopolyploidy**). **Review ANIMATED TUTORIAL 17.2**

concept
17.4 **Reproductive Isolation Is Reinforced When Diverging Species Come into Contact**

- **Prezygotic isolating mechanisms** prevent hybridization; **postzygotic isolating mechanisms** reduce the fitness of hybrids.

- Postzygotic isolating mechanisms lead to **reinforcement** of prezygotic isolating mechanisms by natural selection. **Review Figures 17.8 and 17.10**

- **Hybrid zones** may persist between species with incomplete reproductive isolation. **Review Figure 17.12 and WORKING WITH DATA 17.1**

See **WEB ACTIVITY 17.1** for a concept review of this chapter

The History of Life on Earth

Almost anyone who has spent time around freshwater ponds is familiar with dragonflies. Their bright colors and transparent wings stimulate our visual senses on bright summer afternoons as they fly about their business of devouring mosquitoes, mating, and laying their eggs. The largest dragonflies alive today have wingspans that can be covered by a human hand. Three hundred million years ago, however, dragonflies such as *Meganeuropsis permiana* had wingspans of more than 70 centimeters—well over 2 feet, matching or exceeding the wingspans of many modern birds of prey. These dragonflies were the largest flying predators of their time.

No flying insects alive today are anywhere near this size. But during the Carboniferous and Permian geological periods, 350–250 million years ago, many groups of flying insects contained gigantic members. *Meganeuropsis* probably ate huge mayflies and other giant flying insects that shared their home in the Permian swamps. These enormous insects were themselves eaten by giant amphibians.

None of these insects or amphibians would be able to survive on Earth today. The oxygen concentrations in Earth's atmosphere were about 50 percent higher then than they are now, and those high oxygen concentrations are thought to have been necessary to support giant insects and their huge amphibian predators.

Paleontologists have uncovered fossils of *Meganeuropsis permiana* in the rocks of Kansas. How do we know the age of these fossils, and how can we know how much oxygen that long-vanished atmosphere contained? The layering of the rocks allows us to tell their ages relative to one another, but it does not by itself indicate a given layer's absolute age.

One of the remarkable achievements of twentieth-century scientists was the development of sophisticated techniques that use the decay rates of various radioisotopes, changes in Earth's magnetic field, and the ratios of certain molecules to infer conditions and events in the remote past and to date them accurately. It is those methods that allow us to age the fossils of *Meganeuropsis* and to calculate the concentration of oxygen in Earth's atmosphere at the time.

Earth is about 4.5 billion years old, and life has existed on it for about 3.8 billion of those years. That means human civilizations have occupied Earth for less than 0.0003 percent of the history of life. Discovering what happened before humans were around is an ongoing and exciting area of science.

Meganeuropsis permiana, shown in a reconstruction from fossils. Except for its size, this giant from the Permian period was similar to modern dragonflies (shown in the inset at the same scale).

Q QUESTION
Can modern experiments test hypotheses about the evolutionary impact of ancient environmental changes?

KEY CONCEPTS

18.1 Events in Earth's History Can Be Dated

18.2 Changes in Earth's Physical Environment Have Affected the Evolution of Life

18.3 Major Events in the Evolution of Life Can Be Read in the Fossil Record

concept 18.1 **Events in Earth's History Can Be Dated**

Many evolutionary changes happen rapidly enough to be studied directly and manipulated experimentally. Plant and animal breeding by agriculturalists and evolution of resistance to pesticides are examples of rapid, short-term evolution that we saw in Chapter 15. Other evolutionary changes, such as the appearance of new species and evolutionary lineages, usually take place over much longer time scales.

To understand long-term patterns of evolutionary change, we must not only think in time scales spanning many millions of years, but also consider events and conditions very different from those we observe today. Earth of the distant past was so unlike our present Earth that it would seem like a foreign planet inhabited by strange organisms. The continents were not where they are now, and climates were sometimes dramatically different from those of today. We know this because much of Earth's history is recorded in its rocks.

We cannot tell the ages of rocks just by looking at them, but we can visually determine the ages of rocks *relative to one another*.

The first person to formally recognize this fact was the seventeenth-century Danish physician Nicolaus Steno. Steno realized that in undisturbed **sedimentary rocks** (rocks formed by the accumulation of sediments), the oldest layers of rock, or **strata** (singular *stratum*), lie at the bottom, and successively higher strata are progressively younger.

Geologists subsequently combined Steno's insight with their observations of fossils contained in sedimentary rocks. They developed the following principles of *stratigraphy*:

- Fossils of similar organisms are found in widely separated places on Earth.
- Certain fossils are always found in younger strata, and certain other fossils are always found in older strata.
- Organisms found in younger strata are more similar to modern organisms than are those found in older strata.

These patterns revealed much about the relative ages of sedimentary rocks and the fossils they contain, as well as patterns in the evolution of life. But the geologists still could not tell how old particular rocks were. A method of dating rocks did not become available until after radioactivity was discovered at the beginning of the twentieth century.

TABLE 18.1 Earth's Geological History

RELATIVE TIME SPAN	ERA	PERIOD	ONSET	MAJOR PHYSICAL CHANGES ON EARTH
	Cenozoic	Quaternary	2.6 mya	Cold/dry climate; repeated glaciations
		Tertiary	65 mya	Continents near current positions; climate cools
	Mesozoic	Cretaceous	145 mya	Northern continents attached; Gondwana begins to drift apart; meteorite strikes near present Yucatán Peninsula
		Jurassic	200 mya	Two large continents form: Laurasia (north) and Gondwana (south); climate warm
		Triassic	251 mya	Pangaea begins to slowly drift apart; hot/humid climate
	Paleozoic	Permian	297 mya	Extensive lowland swamps; O_2 levels 50% higher than present; by end of period continents aggregate to form Pangaea, and O_2 levels begin to drop rapidly
		Carboniferous	359 mya	Climate cools; marked latitudinal climate gradients
		Devonian	416 mya	Continents collide at end of period; meteorite probably strikes Earth
		Silurian	444 mya	Sea levels rise; two large land masses emerge; hot/humid climate
		Ordovician	488 mya	Massive glaciation, sea level drops 50 meters
		Cambrian	542 mya	O_2 levels approach current levels
Precambrian	Precambrian		900 mya	O_2 level at ≈5% of current level
			1.5 bya	O_2 level at ≈1% of current level
			3.8 bya	O_2 first appears in atmosphere
			4.5 bya	

Note: mya, million years ago; bya, billion years ago.

Radioisotopes provide a way to date rocks

Radioactive isotopes of atoms—*radioisotopes*—decay in a predictable pattern over long periods. Over a specific time interval, known as a **half-life**, half of the atoms in a radioisotope decay to become a different, stable (nonradioactive) isotope (**FIGURE 18.1A**). The use of this knowledge to date fossils and rocks is known as **radiometric dating**.

To use a radioisotope to date a past event, we must know or estimate the concentration of that isotope at the time of that event, and we must know the radioisotope's half-life. In the case of carbon-14, a radioisotope of carbon, the production of new carbon-14 (^{14}C) in the upper atmosphere—by the reaction of neutrons with nitrogen-14—just balances the natural radioactive decay of ^{14}C into ^{14}N. Therefore, the ratio of ^{14}C to the more common stable isotope of carbon, carbon-12 (^{12}C), is relatively constant in living organisms and in their environment. As soon as an organism dies, however, it ceases to exchange carbon compounds with its environment. Its decaying ^{14}C is no longer replenished, and the ratio of ^{14}C to ^{12}C in its remains decreases over time. Paleontologists can use the ratio of ^{14}C to ^{12}C in fossil material to date fossils that are less than 60,000 years

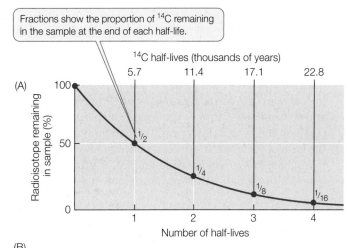

FIGURE 18.1   Radioactive Isotopes Allow Us to Date Ancient Rocks   The decay of radioactive isotopes into stable isotopes happens at a steady rate. A half-life is the time it takes for half of the remaining atoms to decay in this way. (A) The graph demonstrates the principle of half-life using carbon-14 (^{14}C) as an example. The half-life of ^{14}C is 5,700 years. (B) Different radioisotopes have different characteristic half-lives that allow us to measure how much time has elapsed since the rocks containing them were laid down.

Radioisotope	Half-life (years)	Decay product	Useful dating range (years)
Carbon-14 (^{14}C)	5,700	Nitrogen-14 (^{14}N)	100 – 60,000
Potassium-40 (^{40}K)	1.3 billion	Argon-40 (^{40}Ar)	10 million – 4.5 billion
Uranium-238 (^{238}U)	4.5 billion	Lead-206 (^{206}Pb)	10 million – 4.5 billion

MAJOR EVENTS IN THE HISTORY OF LIFE

Humans evolve; many large mammals become extinct

Diversification of birds, mammals, flowering plants, and insects

Dinosaurs continue to diversify; mass extinction at end of period (≈76% of species disappear)

Diverse dinosaurs; radiation of ray-finned fishes; first fossils of flowering plants

Early dinosaurs; first mammals; marine invertebrates diversify; mass extinction at end of period (≈65% of species disappear)

Reptiles diversify; giant amphibians and flying insects present; mass extinction at end of period (≈96% of species disappear)

Extensive "fern" forests; first reptiles; insects diversify

Fishes diversify; first insects and amphibians; mass extinction at end of period (≈75% of marine species disappear)

Jawless fishes diversify; first ray-finned fishes; plants and animals colonize land

Mass extinction at end of period (≈75% of species disappear)

Rapid diversification of multicellular animals; diverse photosynthetic protists

Earliest fossils of multicellular animals

Eukaryotes evolve

Origin of life; prokaryotes flourish

old (and thus the sedimentary rocks that contain those fossils). If fossils are older than that, so little ^{14}C remains that the limits of detection using this particular isotope are reached.

Radiometric dating methods have been expanded and refined

Sedimentary rocks are formed from materials that existed for varying lengths of time before being weathered, fragmented, and transported, sometimes over long distances, to the site of their deposition. Therefore, the radioisotopes in sedimentary rock do not contain reliable information about the date of its formation. Radiometric dating of rocks older than 60,000 years requires estimating radioisotope concentrations in *igneous* rocks, which are formed when molten material cools. To date sedimentary strata, geologists search for places where volcanic ash or lava flows have intruded into the sedimentary rock.

A preliminary estimate of the age of an igneous rock determines which radioisotopes can be used to date it (**FIGURE 18.1B**). The decay of potassium-40 (which has a half-life of 1.3 billion years) to argon-40, for example, has been used to date many of the ancient events in the evolution of life. Fossils in the adjacent sedimentary rock that are similar to those in other rocks of known ages provide additional clues to the rock's age.

APPLY THE CONCEPT

Events in Earth's history can be dated

Imagine you have been assigned the job of producing a geological map of volcanic rocks that were formed between 400 and 600 million years ago. You collect samples from ten sites (1–10 on the map at lower right). Then you can determine the ratio of ^{206}Pb to ^{238}U for each sample and use these ratios to estimate the ages of the rock samples, as given in the table below. Use the data to answer the following questions.

SITE	^{206}Pb/^{238}U RATIO	ESTIMATED AGE (MYA)
1	0.076	474
2	0.077	479
3	0.069	431
4	0.081	505
5	0.076	474
6	0.070	435
7	0.089	550
8	0.080	500
9	0.079	495
10	0.077	479

1. Use the table shown here and Table 18.1 to assign each sample to a geological period.
2. Use these estimated ages and geological periods of the samples to mark rough boundaries between the geological periods among the sample locations on the map below.
3. If you wanted to refine the boundary between the Ordovician and Silurian on your map, which of three new sampling sites—*x*, *y*, or *z*—would you add to your analysis next?

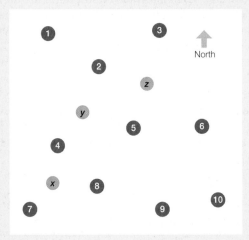

Scientists have used several methods to construct a geological time scale

Radiometric dating of rocks, combined with fossil analysis, is the most powerful method of determining geological age. But in places where sedimentary rocks do not contain suitable igneous intrusions and few fossils are present, paleontologists turn to other dating methods.

One method, known as **paleomagnetic dating**, relates the ages of rocks to patterns in Earth's magnetism, which change over time. Earth's magnetic poles move and occasionally reverse themselves. Both sedimentary and igneous rocks preserve a record of Earth's magnetic field at the time they were formed, and that paleomagnetism can be used to determine the ages of those rocks. Other dating methods use information about continental drift, information about sea level changes, and molecular clocks (the last of which was described in Concept 16.3).

Using all of these methods, geologists developed a geological time scale. They divided the history of life into *eras*, which in turn are subdivided into *periods* (**TABLE 18.1**). They based the boundaries between these divisions of time on the striking differences they observed in the assemblages of fossil organisms contained in successive strata. Geologists defined and named these divisions before they were able to establish the ages of rocks or fossils, and refined the geological time scale as new methods for geological dating were developed.

Do You Understand Concept 18.1?

- What observations about fossils suggested to geologists that they could be used to determine the relative ages of rocks?
- Why are radioisotopes not measured directly from sedimentary rocks to determine their ages?
- Given the problems of dating sedimentary rocks using radioisotopes, what other methods can geologists use to date sedimentary rocks?

The relative time spans at the left of Table 18.1 give us a sense of the scope of geological time—especially the vast expanse of Precambrian time during which, as we describe next, early life evolved amid stupendous physical changes in Earth and its atmosphere.

concept 18.2 Changes in Earth's Physical Environment Have Affected the Evolution of Life

Physical changes to Earth did not end in the Precambrian. Earth has continued to undergo massive physical changes throughout its history that have influenced the evolution of life. And, in its turn, life has influenced Earth's physical environment.

The continents have not always been where they are today

The globes and maps that adorn our walls, shelves, and books give an impression of a static Earth. It would be easy for us to assume that the continents have always been where they are. But we would be wrong. The idea that Earth's land masses have changed position over the millennia, and that they continue to do so, was first put forth in 1912 by the German meteorologist and geophysicist Alfred Wegener. His idea was initially met with skepticism and resistance. By the 1960s, however, physical evidence and increased understanding of **plate tectonics**—the geophysics of the movement of major land masses—had convinced virtually all geologists of the reality of Wegener's vision.

Earth's crust consists of several solid plates approximately 40 kilometers thick, which collectively make up the *lithosphere.* The lithospheric plates float on a fluid layer of molten rock, or *magma* (**FIGURE 18.2**). Heat produced by radioactive decay deep in Earth's core sets up convection currents in the fluid magma, which rises and exerts tremendous pressure on the solid plates. Where the pressure of the rising magma pushes plates apart, ocean basins may form. Where it pushes plates together, the plates may move sideways past each other. Alternatively, one plate may slide under the other, which can lead to the formation of mountain ranges and volcanoes. The movement of the lithospheric plates and the continents they carry is known as **continental drift**.

Many physical conditions on Earth have oscillated in response to plate tectonic processes such as continental drift and volcanic activity. We now know that the movement of the plates has brought continents together at times and at other times has pushed them apart (see the maps across the top of

Figure 18.12). The positions and sizes of the continents influence oceanic circulation patterns, global climates, and sea levels. Sea level is influenced directly by plate tectonic processes (which can influence the depth of ocean basins) and indirectly by oceanic circulation patterns, which affect patterns of glaciation. As climates cool, glaciers tie up water over land masses; as climates warm, they melt and release water.

Some of these dramatic changes in Earth's physical parameters resulted in **mass extinctions**, during which a large proportion of the species living at the time disappeared. These mass extinctions are the cause of the striking differences in fossil assemblages that geologists used to divide the units of the geological time scale. After each mass extinction, the diversity of life rebounded, but recovery took millions of years.

Earth's climate has shifted between hot and cold conditions

Through much of its history, Earth's climate was considerably warmer than it is today, and temperatures decreased more gradually toward the poles. At other times, Earth was colder than it is today. Rapid drops in sea level near the ends of the Ordovician, Devonian, Permian, Triassic, and Cretaceous periods, and most recently in the Quaternary period, resulted mainly from increased global glaciation (**FIGURE 18.3**). Many of these drops in sea level were accompanied by mass extinctions—particularly of marine organisms, which could not survive the disappearance of the shallow seas that covered vast areas of the continental shelves.

Earth's cold periods were separated by long periods of milder climates. Because we are living in one of the colder periods, it is difficult for us to imagine the mild climates that were found at high latitudes during much of the history of life. The Quaternary period has been marked by a series of glacial advances, interspersed with warmer *interglacial* intervals during which the glaciers retreated.

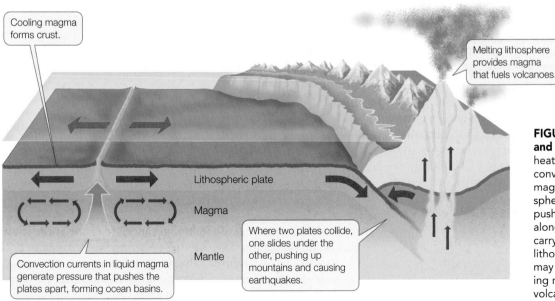

Cooling magma forms crust.

Melting lithosphere provides magma that fuels volcanoes.

Lithospheric plate

Magma

Mantle

Convection currents in liquid magma generate pressure that pushes the plates apart, forming ocean basins.

Where two plates collide, one slides under the other, pushing up mountains and causing earthquakes.

FIGURE 18.2 Plate Tectonics and Continental Drift The heat of Earth's core generates convection currents in the fluid magma underlying the lithospheric plates. Those currents push the lithospheric plates, along with the land masses they carry, together or apart. Where lithospheric plates collide, one may slide under the other, creating mountain ranges and often volcanoes.

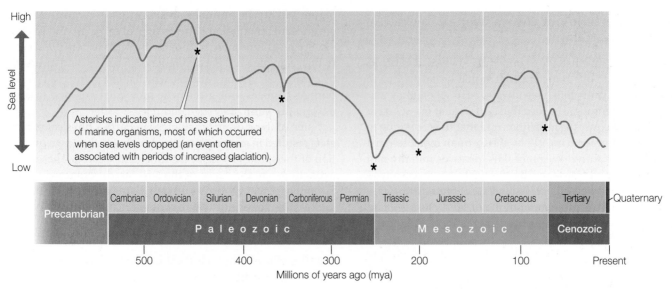

FIGURE 18.3 Sea Levels Have Changed Repeatedly Most mass extinctions of marine organisms (indicated by asterisks) have coincided with low sea levels. Rapid drops in sea level are associated with periods of increased glaciation.

Weather refers to the daily events at a given location, such as individual storms and the high and low temperatures on a given day. *Climate* refers to long-term average expectations over the various seasons at a given location. Weather often changes rapidly; climates typically change slowly. However, major climatic shifts have taken place over periods as short as 5,000 to 10,000 years, primarily as a result of changes in Earth's orbit around the sun. A few climatic shifts have been even more rapid; during one Quaternary interglacial period, the ice-locked Antarctic Ocean became nearly ice-free in less than 100 years. Some climate changes have been so rapid that the extinctions caused by them appear to be nearly instantaneous in the fossil record. Such rapid changes are usually caused by sudden shifts in ocean currents.

We are currently living in a time of rapid climate change thought to be caused by a buildup of atmospheric CO_2, primarily from the burning of fossil fuels by human populations. We are reversing the energy transformations accrued in the burial and decomposition of organic material that occurred (especially) in the Carboniferous and the Permian, but we are doing so over a few hundred years, rather than the many millions of years over which those deposits accumulated. The current rate of increase of atmospheric CO_2 is unprecedented in Earth's history. A doubling of the atmospheric CO_2 concentration—which may happen during the current century—is expected to increase the average temperature of Earth, change rainfall patterns, melt glaciers and ice caps, and raise sea level.

LINK The consequences of today's rapid climate changes will be discussed in Concept 46.5

Volcanoes have occasionally changed the history of life

Most volcanic eruptions produce only local or short-lived effects, but a few large volcanic eruptions have had major consequences for life. When Krakatau (a volcanic island in the Sunda Strait off Indonesia) erupted in 1883, it ejected more than 25 cubic kilometers of ash and rock, as well as large quantities of sulfur dioxide gas (SO_2). The SO_2 was ejected into the stratosphere and carried by high-altitude winds around the planet. Its presence led to high concentrations of sulfurous acid (H_2SO_3) in high-altitude clouds, creating a "parasol effect" so that less sunlight reached Earth's surface. Global temperatures dropped by 1.2°C in the year following the eruption, and global weather patterns showed strong effects for another 5 years. More recently, the eruption of Mount Pinatubo in the Philippines in 1991 (**FIGURE 18.4**) temporarily reduced global temperatures by about 0.5°C. Although these individual volcanoes had only relatively short-term effects on global temperatures, they suggest that the simultaneous eruption of many volcanoes could have a much stronger effect on Earth's climate.

FRONTIERS The temporary cooling of our planet that has taken place after volcanic eruptions led some scientists to propose a controversial plan for slowing or even reversing the recent pattern of global warming. Clouds of small reflective particles injected high into the stratosphere (using high-altitude planes, or ground-to-sky pipelines supported by balloons) could reflect small amounts of sunlight and thereby keep Earth from warming too quickly. Although such a scheme might have unintended consequences, nations may begin to consider such radical solutions to counter the deadly consequences of global warming.

What would cause many volcanoes to erupt at the same time? The collision of continents during the Permian period (about 275 mya) formed a single, gigantic land mass and caused a multitude of massive volcanic eruptions as the continental

FIGURE 18.4 Volcanic Eruptions Can Cool Global Temperatures When Mount Pinatubo erupted in 1991, it increased the concentrations of sulfurous acid in high-altitude clouds, which temporarily lowered global temperatures by about 0.5°C.

plates overrode one another (see Figure 18.2). Emissions from these eruptions blocked considerable sunlight, contributing to the advance of glaciers and a consequent drop in sea level (see Figure 18.3). Thus volcanoes were probably responsible, at least in part, for the greatest mass extinction in Earth's history.

Extraterrestrial events have triggered changes on Earth

At least 30 meteorites of sizes between tennis and soccer balls strike Earth each year. Collisions with larger meteorites or comets are rare, but such collisions have probably been responsible for several mass extinctions. Several types of evidence tell us about these collisions. Their craters, and the dramatically disfigured rocks that result from their impact, are found in many places. Geologists have discovered compounds in these rocks that contain helium and argon with isotope ratios characteristic of meteorites, which are very different from the ratios found elsewhere on Earth.

A meteorite caused or contributed to a mass extinction at the end of the Cretaceous period (about 65 mya). The first clue that a meteorite was responsible came from the abnormally high concentrations of the element iridium found in a thin layer separating rocks deposited during the Cretaceous from rocks deposited during the Tertiary (**FIGURE 18.5**). Iridium is abundant in some meteorites, but it is exceedingly rare on Earth's surface. When scientists then discovered a circular crater 180 km in diameter buried beneath the northern coast of the Yucatán Peninsula of Mexico, they constructed the following scenario. When it collided with Earth, the meteorite released energy equivalent to that of 100 million megatons of high explosives, creating great tsunamis. A massive plume of debris rose into the atmosphere, spread around Earth, and descended. The descending debris heated the atmosphere to several hundred degrees and ignited massive fires. It also blocked the sun, preventing plants from photosynthesizing. The settling debris formed the iridium-rich layer. About a billion tons of soot with a composition matchng that of smoke from forest fires was also deposited. These events had devastating effects on biodiversity. Many fossil species (including non-avian dinosaurs) that are found in Cretaceous rocks are not found in the Tertiary rocks of the next stratum.

Oxygen concentrations in Earth's atmosphere have changed over time

As the continents have moved over Earth's surface, the world has experienced other physical changes, including large increases and decreases in atmospheric oxygen concentrations. The atmosphere of early Earth probably contained little or no free oxygen gas (O_2). The increase in atmospheric O_2 came in two big steps more than a billion years apart.

Iridium-rich layer at the Cretaceous-Tertiary (K/T) boundary

FIGURE 18.5 Evidence of a Meteorite Impact The white layers of rock are Cretaceous in age; the layers at the upper left were deposited in the Tertiary. Between the two is a thin, dark layer of clay that contains large amounts of iridium, a metal common in some meteorites but rare on Earth. Its high concentration in these sediments, deposited about 65 million years ago, suggests the impact of a large meteorite at that time.

(A)

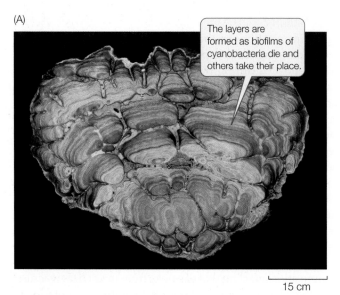

The layers are formed as biofilms of cyanobacteria die and others take their place.

15 cm

(B)

Living cyanobacteria are found in the upper parts of these stromatolites.

30 cm

FIGURE 18.6 Stromatolites (A) A vertical section through a fossil stromatolite. (B) These rocklike structures are living stromatolites that thrive in the very salty waters of Shark Bay in Western Australia.

The first step occurred at least 2.4 billion years ago (bya), when certain bacteria evolved the ability to use water as the source of hydrogen ions for photosynthesis. By chemically splitting H_2O, these bacteria generated O_2 as a waste product. They also made electrons available for reducing CO_2 to form the carbohydrate end-products of photosynthesis (see Concepts 6.5 and 6.6). The second step occurred about a billion years later, when some of these bacteria became symbiotic within eukaryote cells, leading to the evolution of chloroplasts in photosynthetic plants and other eukaryotes.

One group of O_2-generating bacteria, the *cyanobacteria*, formed rocklike structures called *stromatolites*, which are abundantly preserved in the fossil record. To this day, cyanobacteria still form stromatolites in a few very salty places (**FIGURE 18.6**). Cyanobacteria liberated enough O_2 to open the way for the evolution of oxidation reactions as the energy source for the synthesis of ATP.

The evolution of life irrevocably changed the physical nature of Earth. Those physical changes, in turn, influenced the evolution of life. When it first appeared in the atmosphere, O_2 was poisonous to most of the anaerobic prokaryotes that inhabited Earth at the time. Over millennia, however, prokaryotes that evolved the ability to tolerate and use O_2 not only survived but gained the advantage. Aerobic metabolism proceeds more rapidly, and harvests energy more efficiently, than anaerobic metabolism. Organisms with aerobic metabolism replaced anaerobes in most of Earth's environments.

An atmosphere rich in O_2 also made possible larger and more complex organisms. Small single-celled aquatic organisms can obtain enough oxygen by simple diffusion even when dissolved oxygen concentrations in the water are very low. Larger single-celled organisms, however, have lower surface area-to-volume ratios; to obtain enough oxygen by simple diffusion, they must live in an environment with a relatively high oxygen concentration. Bacteria can thrive at 1 percent of the current oxygen concentration; eukaryotic cells require levels that are at least 2–3 percent of the current concentration. (For concentrations of dissolved oxygen in the oceans to have reached these levels, much higher atmospheric concentrations were needed.)

Probably because it took many millions of years for Earth to develop an oxygenated atmosphere, only single-celled prokaryotes lived on Earth for more than 2 billion years. About 1.5 bya, atmospheric O_2 concentrations became high enough for larger eukaryotic cells to flourish (**FIGURE 18.7**). Further increases in atmospheric O_2 concentrations in the late Precambrian enabled several groups of multicellular organisms to evolve.

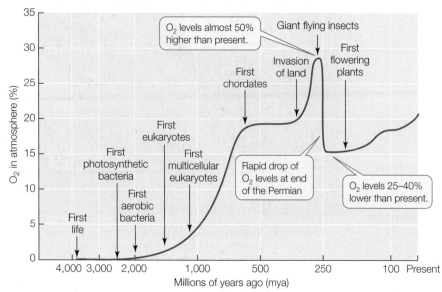

FIGURE 18.7 Larger Cells and Organisms Need More Oxygen Changes in atmospheric oxygen concentrations have strongly influenced, and been influenced by, the evolution of life. (Note that the horizontal axis of the graph is on a logarithmic scale.)

INVESTIGATION

FIGURE 18.8 Atmospheric Oxygen Concentrations and Body Size in Insects C. Jaco Klok and his colleagues asked whether insects raised in hyperoxic conditions would evolve to be larger than their counterparts raised under today's atmospheric conditions. They raised strains of fruit flies (*Drosophila melanogaster*) under both conditions to test the effects of increased O_2 concentrations on the evolution of body size.

HYPOTHESIS

In hyperoxic conditions, increased partial pressure of oxygen results in evolution of increased body size in flying insects.

METHOD

1. Separate a population of fruit flies into multiple lines.
2. Raise half the lines in current atmospheric (control) conditions; raise the other lines in hyperoxic (experimental) conditions. Continue all lines for seven generations.
3. Raise the F_8 individuals of all lines under identical (current) atmospheric conditions.
4. Weigh 50 flies from each of the replicate lines and test for statistical differences in body weight.

RESULTS

The average body mass of F_8 individuals of both sexes raised under hyperoxic conditions was significantly (p < 0.001) greater than that of individuals in the control lines:

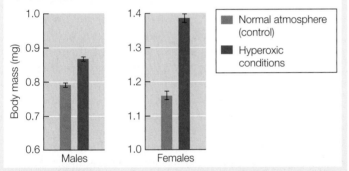

Legend:
- Normal atmosphere (control)
- Hyperoxic conditions

CONCLUSION

Increased O_2 concentrations led to evolution of larger body size in fruit flies, consistent with the trends seen among other flying insects in the fossil record.

ANALYZE THE DATA

The table shows the average body masses of the flies raised in hyperoxic conditions in the F_0 (i.e, before the first generation in hyperoxia), F_1, F_2, and F_8 generations.

Generation	Average body mass (mg)	
	Males	Females
F_0	0.732	1.179
F_1	0.847	1.189
F_2	0.848	1.254
F_8	0.878	1.392

A. Graph body mass versus generation for males and females.
B. Do the rates of evolution of larger body size appear to be constant throughout the experiment?
C. If you doubled the number of generations in the experiment, would you expect the increase in body mass seen under the hyperoxic conditions to double? Why or why not?

Go to **yourBioPortal.com** for original citations, discussions, and relevant links for all INVESTIGATION figures.

O_2 concentrations increased again during the Carboniferous and Permian periods because of the evolution of large vascular plants. These plants lived in the expansive lowland swamps that existed at the time (see Table 18.1). Massive amounts of organic material were buried in these swamps as the plants died, leading to the formation of Earth's vast coal deposits. Because the buried organic material was not subject to oxidation as it decomposed, and because the living plants were producing large quantities of O_2, atmospheric O_2 increased to concentrations that have not been reached again in Earth's history (see Figure 18.7). As mentioned at the opening of this chapter, these high concentrations of atmospheric O_2 allowed the evolution of giant flying insects and amphibians that could not survive in today's atmosphere.

The drying of the lowland swamps at the end of the Permian reduced burial of organic matter as well as the production of O_2, so atmospheric O_2 concentrations dropped rapidly. Over the past 200 million years, with the diversification of flowering plants, O_2 concentrations have again increased, but not to the levels that characterized the Carboniferous and Permian periods.

Biologists have conducted experiments that demonstrate the changing selection pressures that can accompany changes in atmospheric O_2 concentrations. When fruit flies (*Drosophila*) are raised in *hyperoxic* conditions (i.e., with artificially increased atmospheric concentrations of O_2), they evolve larger body sizes in just a few generations (**FIGURE 18.8**). The present atmospheric O_2 concentrations appear to constrain body size in these flying insects; increases in O_2 appear to relax those constraints. This experiment demonstrates that the *stabilizing selection* on body size at present O_2 concentrations can quickly switch to *directional selection* for a change in body size in response to a change in O_2 concentrations (see Concept 15.4).

Do You Understand Concept 18.2?

- How have volcanic eruptions and meteorite strikes influenced the course of life's evolution?
- Explain why an occasional major winter blizzard is irrelevant to discussions of global climatic warming.
- Research and explain the different climatic responses to the "parasol effect" of major volcanic eruptions and the "greenhouse effect" of long-term CO_2 emissions. What is the fundamental difference in the two effects?
- How have increases in atmospheric concentrations of O_2 affected the evolution of multicellular organisms?

The many dramatic physical events of Earth's history have influenced the nature and timing of evolutionary changes among Earth's living organisms. We will now look more closely at some of the major events that characterize the history of life on Earth.

APPLY THE CONCEPT

Changes in Earth's atmosphere have affected the evolution of life

In the experiment shown in Figure 18.8, body mass of individuals in the experimental population of *Drosophila* increased (on average) about 2% per generation in the high-oxygen environment (although the rate of increase was not constant over the experiment). What rate of increase in body mass per generation would be sufficient to account for the giant dragonflies of the Permian?

1. Assume that the average rate of increase in dragonfly size during the Permian was much slower than the rate observed in the experiment in Figure 18.8. We'll assume that the actual rate of increase for dragonflies was just 0.01% per generation, rather than the 2% observed over a few generations for *Drosophila*. We'll assume further that dragonflies complete just one generation per year (as opposed to 40 or more generations for *Drosophila*). Starting with an average body mass of 1 gram, calculate the projected increase in body mass over 50,000 years.*

2. What percentage of the Permian period does 50,000 years represent? Use Table 18.1 for your calculation.

3. Given your calculations, do you think that increased oxygen concentrations during the Permian were sufficient to account for the evolution of giant dragonflies? Why or why not?

*This calculation is similar to computing compound interest for a savings account. Use the formula $W = S(1 + R)^N$, where W = the final mass, S = the starting mass (1 gram), R = the rate of increase per generation (0.0001 in this case), and N = the number of generations.

<table>
<tr><td>concept
18.3</td><td>**Major Events in the Evolution of Life Can Be Read in the Fossil Record**</td></tr>
</table>

How do we know about the physical changes in Earth's environment and their effects on the evolution of life? To reconstruct life's history, scientists rely heavily on the fossil record. Geologists divided Earth's history into eras and periods based on their distinct fossil assemblages (see Table 18.1). Biologists refer to the assemblage of all organisms of all kinds living at a particular time or place as a **biota**. All of the plants living at a particular time or place are its **flora**; all of the animals are its **fauna**.

About 300,000 species of fossil organisms have been described, and the number steadily grows. The number of named species, however, is only a tiny fraction of the species that have ever lived. We do not know how many species lived in the past, but we have ways of making reasonable estimates. Of the present-day biota, nearly 1.8 million species have been named. The actual number of living species is probably well over 10 million, and possibly much higher, because many species have not yet been discovered and described by biologists. So the number of described fossil species is only about 3 percent of the estimated minimum number of living species. Life has existed on Earth for about 3.8 billion years. Many species last only a few million years before undergoing speciation or going extinct; therefore, Earth's biota must have turned over many times during geological history. So the total number of species that have lived over evolutionary time must vastly exceed the number living today. Why have only about 300,000 of these tens of millions of species been described from fossils to date?

Most organisms live and die in oxygen-rich environments in which they quickly decompose. They are not likely to become fossils unless they are transported by wind or water to sites that lack oxygen, where decomposition proceeds slowly or not at all. Furthermore, geological processes transform many rocks, destroying the fossils they contain, and many fossil-bearing rocks are deeply buried and inaccessible. Paleontologists have studied only a tiny fraction of the sites that contain fossils, although they find and describe many new ones every year.

The fossil record is most complete for marine animals that had hard skeletons (which resist decomposition). Among the nine major animal groups with hard-shelled members, approximately

Solenopsis sp.

FIGURE 18.9 Insect Fossils Chunks of amber—fossilized tree resin—often contain insects that were preserved when they were trapped in the sticky resin. This fire ant fossil is some 30 million years old.

Several processes contribute to the paucity of fossils

Only a tiny fraction of organisms ever become fossils, and only a tiny fraction of fossils are ever discovered by paleontologists.

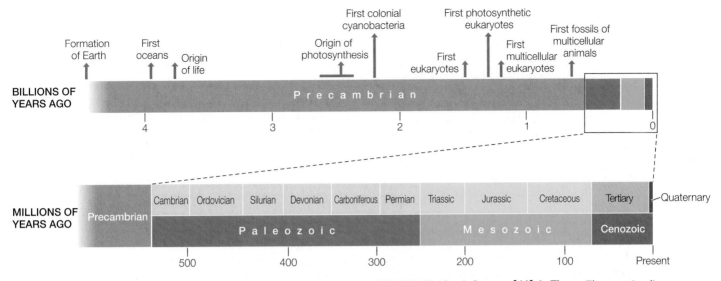

First colonial
cyanobacteria

First photosynthetic
eukaryotes

First fossils of
multicellular
animals

Formation
of Earth

First
oceans

Origin
of life

Origin of
photosynthesis

First
eukaryotes

First
multicellular
eukaryotes

BILLIONS OF
YEARS AGO

P r e c a m b r i a n

4   3   2   1   0

| Cambrian | Ordovician | Silurian | Devonian | Carboniferous | Permian | Triassic | Jurassic | Cretaceous | Tertiary | Quaternary |

MILLIONS OF
YEARS AGO   Precambrian

P a l e o z o i c   M e s o z o i c   Cenozoic

500   400   300   200   100   Present

FIGURE 18.10 A Sense of Life's Time The top timeline shows the 4.5-billion-year history of Earth. Most of this history is accounted for by the Precambrian, a 3.4-billion-year time span that saw the origin of life and the evolution of cells, photosynthesis, and multicellularity. The final 600 million years are expanded in the bottom timeline and detailed in Figure 18.12.

200,000 species have been described from fossils—roughly twice the number of living marine species in these same groups. Paleontologists lean heavily on these groups in their interpretations of the evolution of life. Insects and spiders are also relatively well represented in the fossil record because they are numerically abundant and have hard exoskeletons (**FIGURE 18.9**). The fossil record, though incomplete, is good enough to document clearly the factual history of the evolution of life.

By combining information about physical changes during Earth's history with evidence from the fossil record, scientists have composed portraits of what Earth and its inhabitants may have looked like at different times. We know in general where the continents were and how life changed over time, but many of the details are poorly known, especially for events in the more remote past.

Precambrian life was small and aquatic

Life first appeared on Earth about 3.8 bya. By about 1.5 bya, eukaryotic organisms had evolved. The fossil record of organisms that lived prior to 550 mya is fragmentary, but it is good enough to establish that the total number of species and individuals increased dramatically in the late Precambrian.

For most of its history, life was confined to the oceans, and all organisms were small. Over the long Precambrian—more than

3 billion years—the shallow seas slowly began to teem with life. For most of the Precambrian, life consisted of microscopic prokaryotes; eukaryotes evolved about three-quarters of the way through the era (**FIGURE 18.10**). Unicellular eukaryotes and small multicellular animals fed on floating photosynthetic microorganisms. Small floating organisms, known collectively as *plankton*, were strained from the water and eaten by slightly larger *filter-feeding* animals. Other animals ingested sediments on the seafloor and digested the remains of organisms within them. By the late Precambrian (630–542 mya), many kinds of multicellular soft-bodied animals had evolved. Some of them were very different from any animals living today and may be members of groups that have no living descendants (**FIGURE 18.11**).

FIGURE 18.11 Precambrian Life These fossils of soft-bodied marine invertebrates, excavated at Ediacara in southern Australia, were formed about 600 million years ago. These organisms, which are very different from later life forms, illustrate the diversity of life at the end of the Precambrian.

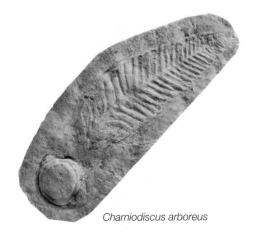

Charniodiscus arboreus

Tribrachidium heraldicum

Dickinsonia costata

FRONTIERS Some Precambrian-age tracks preserved in ancient seabeds were once attributed to large multicellular animals that left no direct fossil remains. But recent deep-sea studies of the relatively large (but single-celled) spherical protist *Gromia sphaerica* have shown that these organisms can pull themselves along the seafloor using pseudopods, thereby creating tracks similar to those preserved in the Precambrian rocks. The grape-sized *Gromia* "roll" along the sea bottom, even pulling themselves up moderate inclines. This finding has stimulated paleontologists to reconsider the source of some Precambrian tracks.

Life expanded rapidly during the Cambrian period

The Cambrian period (542–488 mya) marks the beginning of the Paleozoic era. The O_2 concentration in the Cambrian atmosphere was approaching its current level, and the land masses had come together to form several large continents. A geologically rapid diversification of life took place that is sometimes referred to as the **Cambrian explosion** (although in fact it began before the Cambrian, and the "explosion" took millions of years). Many of the major animal groups represented by species alive today first appeared during the Cambrian. Such periods of rapid diversification are known as **evolutionary radiations**. FIGURE 18.12 provides an overview of the continental and biotic shifts that characterized the Cambrian and subsequent periods.

For the most part, fossils tell us only about the hard parts of organisms, but in three known Cambrian fossil beds—the Burgess Shale in British Columbia, Sirius Passet in northern Greenland, and the Chengjiang site in southern China—the soft parts of many animals were preserved. Crustacean arthropods (crabs, shrimps, and their relatives) are the most diverse group in the Chinese fauna; some of them were large carnivores. Multicellular life was largely or completely aquatic during the Cambrian. If there was life on land at this time, it was probably restricted to microorganisms.

Many groups of organisms that arose during the Cambrian later diversified

Geologists divide the remainder of the Paleozoic era into the Ordovician, Silurian, Devonian, Carboniferous, and Permian periods. Each period is characterized by the diversification of specific groups of organisms. Mass extinctions marked the ends of the Ordovician, Devonian, and Permian.

THE ORDOVICIAN (488–444 MYA) During the Ordovician period, the continents, which were located primarily in the Southern Hemisphere, still lacked multicellular plants. Evolutionary radiation of marine organisms was spectacular during the early Ordovician, especially among animals, such as brachiopods and mollusks, that lived on the seafloor and filtered small prey from the water. At the end of the Ordovician, as massive glaciers formed over the southern continents, sea levels dropped about 50 meters, and ocean temperatures dropped. About 75 percent of all animal species became extinct, probably because of these major environmental changes.

THE SILURIAN (444–416 MYA) During the Silurian period, the continents began to merge together. Marine life rebounded from the mass extinction at the end of the Ordovician. Animals able to swim in open water and feed above the ocean floor appeared for the first time. Jawless fishes diversified, and the first fishes with supporting rays in their fins evolved (see Figure 23.38). The tropical sea was uninterrupted by land barriers, and most marine organisms were widely distributed. On land, the first vascular plants evolved late in the Silurian (about 420 mya). The first terrestrial arthropods—scorpions and millipedes—evolved at about the same time.

THE DEVONIAN (416–359 MYA) Rates of evolutionary change accelerated in many groups of organisms during the Devonian period. The major land masses continued to move slowly toward each other. In the oceans there were great evolutionary radiations of corals and of shelled, squidlike cephalopod mollusks. Fishes diversified as jawed forms replaced jawless ones (see Figure 23.39) and as bony armor gave way to the less rigid scales of modern fishes.

Terrestrial communities changed dramatically during the Devonian. Club mosses, horsetails, and tree ferns became common; some attained the size of large trees. Their roots accelerated the weathering of rocks, resulting in the development of the first forest soils. The first plants to produce seeds appeared in the Devonian. The earliest fossil centipedes, spiders, mites, and insects date to this period, as do the earliest fossil terrestrial amphibians.

A massive extinction of about 75 percent of all marine species marked the end of the Devonian. Paleontologists are uncertain about its cause, but two large meteorites collided with Earth at about that time (one in present-day Nevada, the other in western Australia) and may have been responsible, or at least a contributing factor. The continued coalescence of the continents, with the corresponding reduction in the area of continental shelves, may have also contributed to this mass extinction.

THE CARBONIFEROUS (359–297 MYA) Large glaciers formed over high-latitude portions of the southern land masses during the Carboniferous period, but extensive swamp forests grew on the tropical continents. These forests were dominated by giant tree ferns and horsetails with small leaves. Their fossilized remains formed the coal we now mine for energy. In the seas, crinoids (sea lilies and feather stars) reached their greatest diversity, forming "meadows" on the seafloor.

FIGURE 18.12 A Brief History of Multicellular Life on Earth ▶ The geologically rapid "explosion" of life during the Cambrian saw the rise of several animal groups that have representatives surviving today. The following three pages depict life's history from the Cambrian forward. The movements of the major continents during the past half-billion years are shown in the maps of Earth, and associated biotas for each time period are depicted. The artists' reconstructions are based on fossils such as those shown in the photographs.

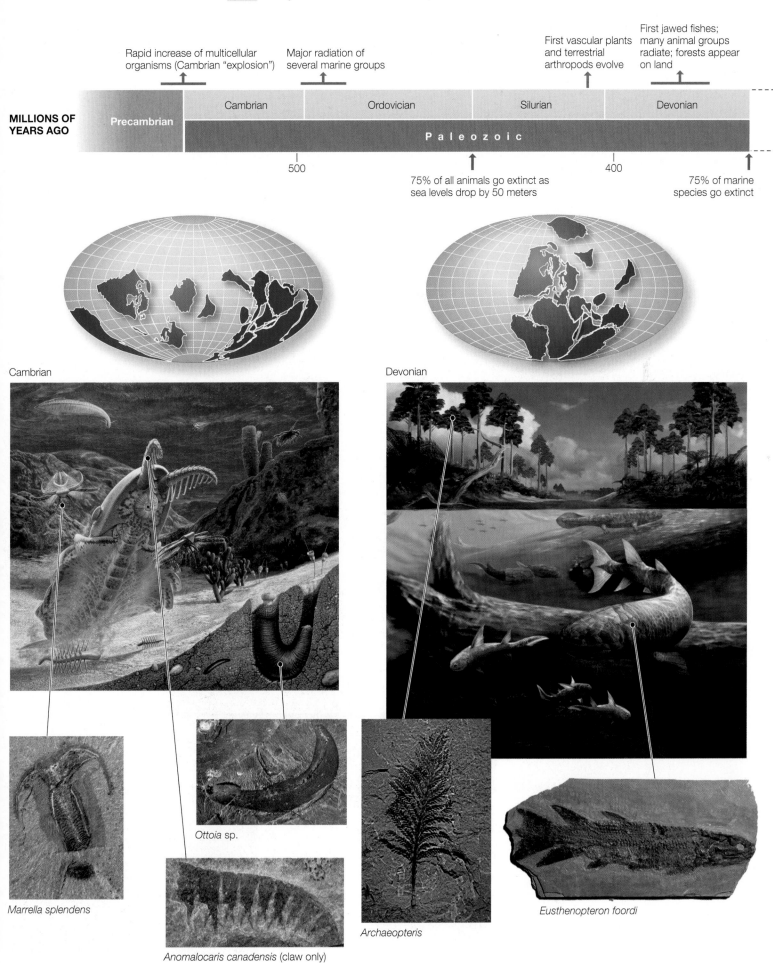

MILLIONS OF YEARS AGO

Rapid increase of multicellular organisms (Cambrian "explosion")

Major radiation of several marine groups

First vascular plants and terrestrial arthropods evolve

First jawed fishes; many animal groups radiate; forests appear on land

Precambrian | Cambrian | Ordovician | Silurian | Devonian

Paleozoic

500

400

75% of all animals go extinct as sea levels drop by 50 meters

75% of marine species go extinct

Cambrian

Devonian

Ottoia sp.

Marrella splendens

Anomalocaris canadensis (claw only)

Archaeopteris

Eusthenopteron foordi

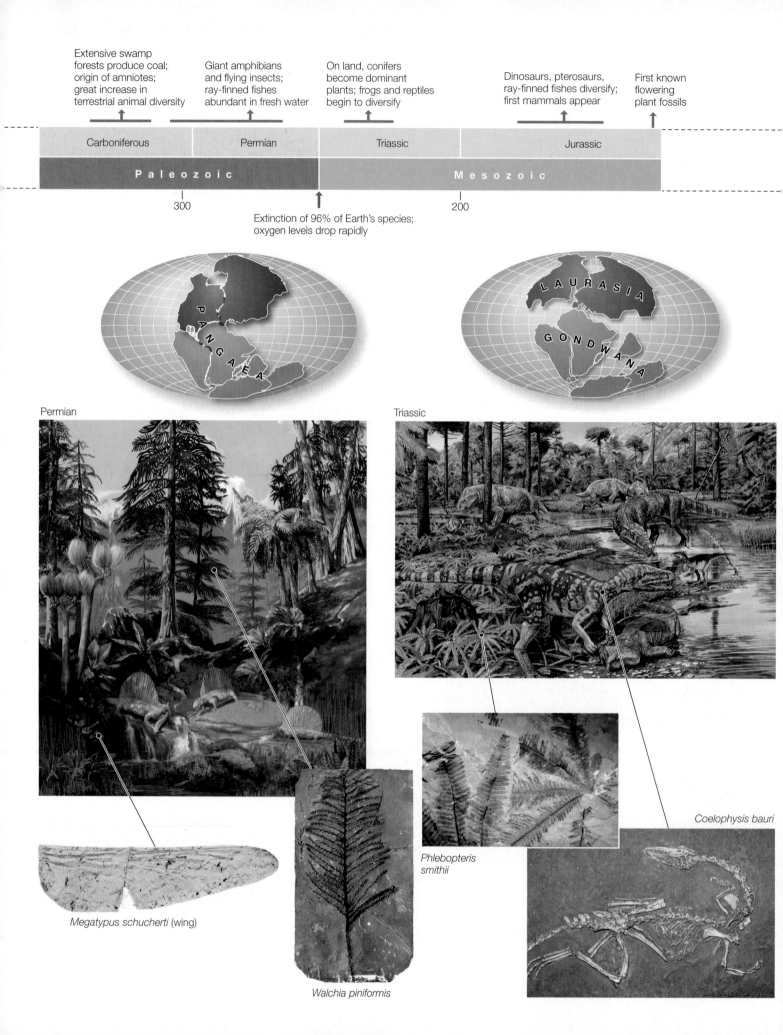

Extensive swamp forests produce coal; origin of amniotes; great increase in terrestrial animal diversity

Giant amphibians and flying insects; ray-finned fishes abundant in fresh water

On land, conifers become dominant plants; frogs and reptiles begin to diversify

Dinosaurs, pterosaurs, ray-finned fishes diversify; first mammals appear

First known flowering plant fossils

| Carboniferous | Permian | Triassic | Jurassic |

Paleozoic — Mesozoic

300

200

Extinction of 96% of Earth's species; oxygen levels drop rapidly

PANGAEA

LAURASIA

GONDWANA

Permian

Triassic

Megatypus schucherti (wing)

Walchia piniformis

Phlebopteris smithii

Coelophysis bauri

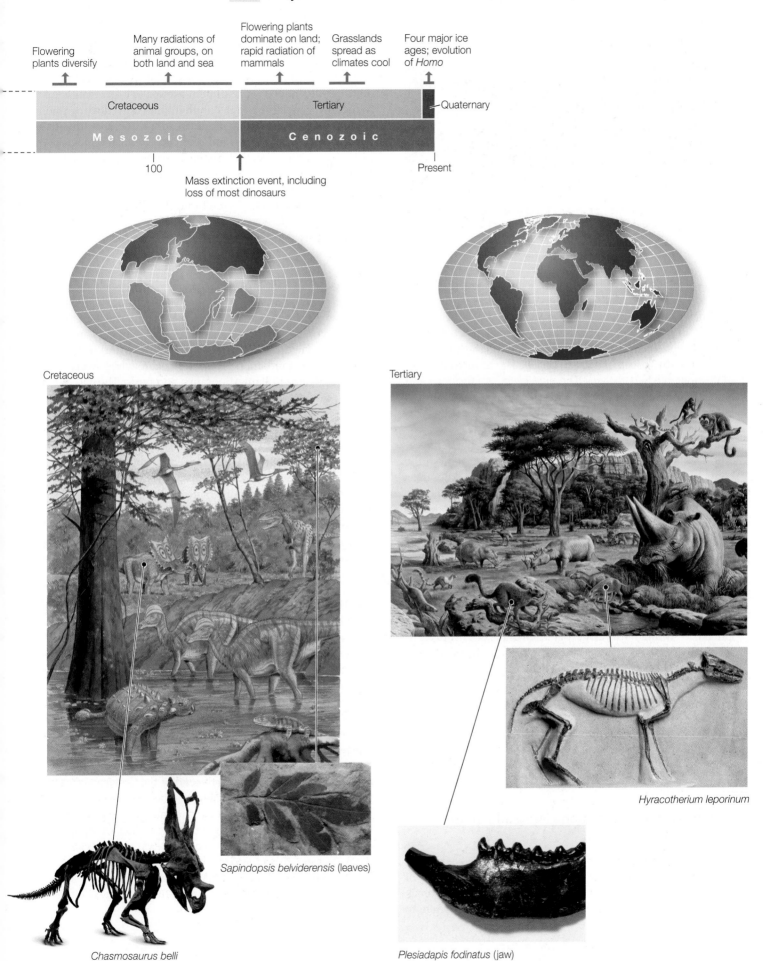

Flowering plants diversify

Many radiations of animal groups, on both land and sea

Flowering plants dominate on land; rapid radiation of mammals

Grasslands spread as climates cool

Four major ice ages; evolution of *Homo*

Cretaceous

Tertiary

Quaternary

Mesozoic

Cenozoic

100

Present

Mass extinction event, including loss of most dinosaurs

Cretaceous

Tertiary

Chasmosaurus belli

Sapindopsis belviderensis (leaves)

Plesiadapis fodinatus (jaw)

Hyracotherium leporinum

FIGURE 18.13 Evidence of Insect Diversification The margins of this fossil fern leaf from the Carboniferous have been chewed by insects.

The diversity of terrestrial animals increased greatly during the Carboniferous. Snails, scorpions, centipedes, and insects were abundant and diverse. Insects evolved wings, becoming the first animals to fly. Flight gave herbivorous insects easy access to tall plants; plant fossils from this period show evidence of chewing by insects (**FIGURE 18.13**). The terrestrial vertebrates split into two lineages. The *amphibians* became larger and better adapted to terrestrial existence, while the sister lineage led to the *amniotes*—vertebrates with well-protected eggs that can be laid in dry places (see Concept 23.6).

THE PERMIAN (297–251 MYA) During the Permian period, the continents coalesced into a single supercontinent, known as **Pangaea**. Permian rocks contain representatives of many of the major groups of insects we know today. By the end of the period the amniotes had split into two lineages: the reptiles, and a second lineage that would lead to the mammals. Ray-finned fishes became common in the fresh waters of Pangaea.

Toward the end of the Permian, conditions for life deteriorated. Massive volcanic eruptions resulted in outpourings of lava that covered large areas of Earth. The ash and gases produced by the volcanoes blocked sunlight and cooled the climate, resulting in the largest glaciers in Earth's history. Atmospheric oxygen concentrations gradually dropped from about 30 percent to 15 percent. At such low concentrations, most animals would have been unable to survive at elevations above 500 meters; thus about half of the land area would have been uninhabitable at the end of the Permian. The combination of these changes resulted in the most drastic mass extinction in Earth's history. Scientists estimate that about 96 percent of all multicellular species became extinct at the end of the Permian.

Geographic differentiation increased during the Mesozoic era

The few organisms that survived the Permian mass extinction found themselves in a relatively empty world at the start of the Mesozoic era (251 mya). As Pangaea slowly began to break apart, the biotas of the newly separated continents began to diverge. The oceans rose and once again flooded the continental shelves, forming huge, shallow inland seas. Atmospheric oxygen concentrations gradually rose. Life once again proliferated and diversified, but different groups of organisms came to the fore. The three groups of phytoplankton (floating photosynthetic organisms) that dominate today's oceans—dinoflagellates, coccolithophores, and diatoms—became ecologically important at this time; their remains are the primary origin of the world's oil deposits. Seed-bearing plants replaced the trees that had ruled the Permian forests.

The Mesozoic era is divided into three periods: the Triassic, Jurassic, and Cretaceous. The Triassic and Cretaceous were terminated by mass extinctions, probably caused by meteorite impacts.

THE TRIASSIC (251–200 MYA) Pangaea began to break apart during the Triassic period. Many invertebrate groups diversified, and many burrowing animals evolved from groups living on the surfaces of seafloor sediments. On land, conifers and seed ferns were the dominant trees. The first frogs and turtles appeared. A great radiation of reptiles began, which eventually gave rise to crocodilians, dinosaurs, and birds. The end of the Triassic was marked by a mass extinction that eliminated about 65 percent of the species on Earth.

THE JURASSIC (200–145 MYA) During the Jurassic period, Pangaea became fully divided into two large continents: **Laurasia**, which drifted northward, and **Gondwana**, which drifted southward. Ray-finned fishes rapidly diversified in the oceans. The first lizards appeared, and flying reptiles (pterosaurs) evolved. Most of the large terrestrial predators and herbivores of the period were dinosaurs. Several groups of mammals made their first appearance, and the earliest known fossils of flowering plants are from late in this period.

THE CRETACEOUS (145–65 MYA) By the early Cretaceous period, Laurasia and Gondwana had begun to break apart into the continents we know today. A continuous sea encircled the tropics. Sea levels were high, and Earth was warm and humid. Life proliferated both on land and in the oceans. Marine invertebrates increased in diversity. On land, the reptile radiation continued as dinosaurs diversified further and the first snakes appeared. Early in the Cretaceous, flowering plants began the radiation that led to their current dominance of the land. By the end of the period, many groups of mammals had appeared.

As described in Concept 18.2, another meteorite-caused mass extinction took place at the end of the Cretaceous. In the seas, many planktonic organisms and bottom-dwelling invertebrates became extinct. On land, almost all animals larger than about 25 kg in body weight became extinct. Many species of insects died out, perhaps because the growth of their food plants was greatly reduced following the impact. Some species in northern North America and Eurasia survived in areas that were not subjected to the devastating fires that engulfed most low-latitude regions.

Modern biotas evolved during the Cenozoic era

By the early Cenozoic era (65 mya), the positions of the continents resembled those of today, but Australia was still attached to Antarctica, and the Atlantic Ocean was much narrower. The Cenozoic was characterized by an extensive radiation of mammals, but other groups were also undergoing important changes.

Flowering plants diversified extensively and came to dominate world forests, except in the coolest regions, where the forests were composed primarily of gymnosperms. Mutations of two genes in one group of plants (the legumes) allowed them to use atmospheric nitrogen directly by forming symbioses with a few species of nitrogen-fixing bacteria. The evolution of this symbiosis was the first "green revolution" and dramatically increased the amount of nitrogen available for terrestrial plant growth; this symbiosis remains fundamental to the ecological base of life as we know it today (see Chapter 25).

The Cenozoic era is divided into the Tertiary and the Quaternary periods. Because both the fossil record and our subsequent knowledge of evolutionary history become more extensive as we approach our own time, paleontologists have subdivided these two periods into *epochs* (**TABLE 18.2**).

THE TERTIARY (65–2.6 MYA) During the Tertiary period, Australia began its northward drift. By 20 mya it had nearly reached its current position. The early Tertiary was a hot and humid time, and the ranges of many plants shifted latitudinally. The tropics were probably too hot to support rainforest vegetation and instead were clothed in low-lying vegetation. In the middle of the Tertiary, however, Earth's climate became considerably cooler and drier. Many lineages of flowering plants evolved herbaceous (nonwoody) forms, and grasslands spread over much of Earth.

By the start of the Cenozoic era, invertebrate faunas had already come to resemble those of today. It is among the terrestrial vertebrates that evolutionary changes during the Tertiary were most rapid. Frogs, snakes, lizards, birds, and mammals all underwent extensive radiations during this period. Three waves of mammals dispersed from Asia to North America across one of the several land bridges that have intermittently connected the two continents during the past 55 million years. Rodents, marsupials, primates, and hoofed mammals appeared in North America for the first time.

THE QUATERNARY (2.6 MYA TO PRESENT) We are living in the Quaternary period. It is subdivided into two epochs, the Pleistocene and the Holocene (the Holocene is also known as the Recent).

The Pleistocene was a time of drastic cooling and climate fluctuations. During 4 major and about 20 minor "ice ages," massive glaciers spread across the continents, and the ranges of animal and plant populations shifted toward the equator. The last of these glaciers retreated from temperate latitudes less than 15,000 years ago. Organisms are still adjusting to this change. Many high-latitude ecological communities have occupied their current locations for no more than a few thousand years.

It was during the Pleistocene that divergence within one group of mammals, the primates, resulted in the evolution of the hominoid lineage. Subsequent hominoid radiation eventually led to the species *Homo sapiens*—modern humans. Many large bird and mammal species became extinct in Australia and in the Americas when *H. sapiens* arrived on those continents about 45,000 and 15,000 years ago, respectively. Many paleontologists believe these extinctions were the result of hunting and other influences of *Homo sapiens*.

LINK The evolution of modern humans and their close relatives during the Pleistocene is discussed in Concept 23.7

The tree of life is used to reconstruct evolutionary events

The fossil record reveals broad patterns in life's evolution. To reconstruct major events in the history of life, biologists also rely on the phylogenetic information in the tree of life. We can use phylogeny, in combination with the fossil record, to reconstruct the timing of such major events as the acquisition of mitochondria in the ancestral eukaryotic cell, the several independent origins of multicellularity, and the movement of life onto dry land. We can also follow major changes in the genomes of organisms, and we can even reconstruct many gene sequences of species that are long extinct.

Changes in Earth's physical environment have clearly influenced the diversity of organisms we see on the planet today. To study the evolution of that diversity, biologists examine the evolutionary relationships among species. Deciphering phylogenetic relationships is an important step in understanding how life has diversified on Earth. The next part of this book explores the major groups of life and the different solutions these groups have evolved to major challenges such as reproduction, energy acquisition, dispersal, and escape from predation.

PERIOD	EPOCH	ONSET (MYA)
Quaternary	Holocene (Recent)	0.01 (~10,000 years ago)
	Pleistocene	2.6
Tertiary	Pliocene	5.3
	Miocene	23
	Oligocene	34
	Eocene	55.8
	Paleocene	65

TABLE 18.2 Subdivisions of the Cenozoic Era

Do You Understand Concept 18.3?

- Why have so few of the organisms that have existed over Earth's history become fossilized?
- What do we mean by the "Cambrian explosion"? How long did it last, and in what sense was it an "explosion"?
- What are some of the ways in which continental drift has affected the evolution of life?

QUESTION

Can modern experiments test hypotheses about the evolutionary impact of ancient environmental changes?

ANSWER Several experiments have been conducted to test the link between O_2 concentrations and evolution of body size in flying insects. One of these is discussed in Figure 18.8. The results of these experiments are consistent with the evolution of larger body size in flying insects in hyperoxic (high-oxygen) environments.

Experiments have also been conducted under *hypoxic* (low-oxygen) conditions, as existed at the end of the Permian. These experiments suggest that the evolution of body size is constrained under hypoxic conditions, even under strong artificial selection for larger body size. These latter results are consistent with the extinction of many large flying insects at the end of the Permian as a result of rapidly decreasing O_2 concentrations. Giant flying insects simply could not have survived the lower O_2 concentrations that existed at that time. The mass extinction at the end of the Permian is the only known mass extinction that involved considerable loss of insect diversity.

---

18 SUMMARY

concept 18.1 Events in Earth's History Can Be Dated

- The relative ages of organisms can be determined by the dating of fossils and the **strata** of **sedimentary rocks** in which they are found.

- **Radiometric dating** techniques use a variety of radioisotopes with different **half-lives** to date events in the remote past. **Review Figure 18.1**

- Geologists divide the history of life into eras and periods, based on major differences in the fossil assemblages found in successive strata. **Review Table 18.1**

concept 18.2 Changes in Earth's Physical Environment Have Affected the Evolution of Life

- Earth's crust consists of solid lithospheric plates that float on fluid magma. **Continental drift** is caused by convection currents in the magma, which move the plates and the continents that lie on top of them. **Review Figure 18.2 and ANIMATED TUTORIAL 18.1**

- Major physical events on Earth, such as continental collisions and volcanic eruptions, have affected Earth's climate, atmosphere, and sea levels. In addition, extraterrestrial events such as meteorite strikes have created sudden and dramatic environmental shifts. All of these changes affected the history of life. **Review Figure 18.3 and Table 18.1**

- Oxygen-generating cyanobacteria liberated enough O_2 to open the door to oxidation reactions in metabolic pathways. Aerobic prokaryotes were able to harvest more energy than anaerobic organisms and began to proliferate. Increases in atmospheric O_2 concentrations supported the evolution of large eukaryotic cells and, eventually, multicellular organisms. **Review Figures 18.7 and 18.8**

concept 18.3 Major Events in the Evolution of Life Can Be Read in the Fossil Record

- Paleontologists use fossils and evidence of geological changes to determine what Earth and its biota may have looked like at different times. **Review Figure 18.12**

- During the Precambrian, which accounts for most of life's history, life was confined to the oceans. Multicellular life diversified extensively during the **Cambrian explosion**, a prime example of an **evolutionary radiation**.

- The periods of the Paleozoic era were each characterized by the diversification of specific groups of organisms.

- During the Mesozoic era, distinct terrestrial **biotas** evolved on each continent.

- Five episodes of **mass extinction** punctuated the history of life in the Paleozoic and Mesozoic eras.

- Earth's **flora** has been dominated by flowering plants since the Cenozoic era began.

- The tree of life can be used to reconstruct the timing of evolutionary events.

See WEB ACTIVITY 18.1 for a concept review of this chapter.

PART

4

Diversity

Bacteria, Archaea, and Viruses   19

On the night of January 25, 1995, the British merchant vessel *Lima* was off the coast of Somalia, near the Horn of Africa. This area is infamous for bands of pirates, so the crew was keeping a watchful eye on the seas. On the horizon, they spotted an eerie whitish glow. It was directly in their path, and there was no way to avoid it. Was the glow the result of some strange trick of piracy?

Within 15 minutes of first spotting the glow, the *Lima* was surrounded by glowing waters for as far as her crew could see. As the ship's log recorded, "it appeared as though the ship was sailing over a field of snow or gliding over the clouds." Fortunately for the crew, the glow had nothing to do with pirates.

For centuries, mariners in this part of the world had reported occasional "milky seas" in which the sea surface produced a strange glow at night, extending from horizon to horizon. Scientists to that point had never been able to confirm the reality or the cause of such phenomena. It was well established, however, that many organisms can emit light by *bioluminescence*—a complex, enzyme-catalyzed biochemical reaction that results in the emission of light but not heat.

But what kind of organisms could cause the vast expanse of bioluminescence observed by the *Lima*? Some protists are known to emit flashes of light when disturbed, but they could not produce the sustained and uniform glow seen in milky seas. The only organisms known to produce the quality of bioluminescence consistent with milky seas are prokaryotes, such as bacteria of the genus *Vibrio*.

Using information supplied by the *Lima*, biologists scanned satellite images of the Indian Ocean for the specific light wavelengths emitted by *Vibrio*. The satellite images clearly showed thousands of square kilometers of *Vibrio*-produced milky seas.

Why does *Vibrio* produce bioluminescence? These bacteria can live freely, but they thrive inside the guts of fish. Inside a fish, they may attach themselves to food particles, including phytoplankton, and may then be expelled as waste. At low densities, *Vibrio* populations do not glow. But as the bacteria reproduce on the phytoplankton, their population increases. When it reaches a certain density, the bacteria begin to glow. The glow attracts other fish, which ingest the bacteria along with the phytoplankton—giving the bacteria a new home and food source for a while.

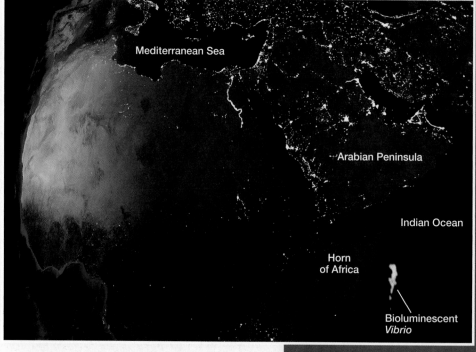

Mediterranean Sea

Arabian Peninsula

Indian Ocean

Horn of Africa

Bioluminescent *Vibrio*

A satellite image reveals thousands of square kilometers of "milky seas" in the Indian Ocean. This expanse of bioluminescence is produced by *Vibrio* bacteria.

 QUESTION How do *Vibrio* populations detect when they are dense enough to produce bioluminescence?*

*You will find the answer to this question on page 386.

KEY CONCEPTS

19.1 Life Consists of Three Domains That Share a Common Ancestor

19.2 Prokaryote Diversity Reflects the Ancient Origins of Life

19.3 Ecological Communities Depend on Prokaryotes

19.4 Viruses Have Evolved Many Times

Life Consists of Three Domains That Share a Common Ancestor

You may think that you have little in common with a bacterium. But all multicellular eukaryotes—including you—share many attributes with bacteria and archaea, together called *prokaryotes*. For example, all organisms, whether eukaryotes or prokaryotes,

- have plasma membranes and ribosomes (see Chapter 4).

- have a common set of metabolic pathways, such as glycolysis (see Chapter 6).

- replicate DNA semiconservatively (see Chapter 9).

- use DNA as the genetic material to encode proteins, and use a similar genetic code to produce those proteins by transcription and translation (see Chapter 10).

These shared features support the conclusion that all living organisms are related. If life had multiple origins, there would be little reason to expect all organisms to use overwhelmingly similar genetic codes or to share structures as unique as ribosomes. Furthermore, similarities in the DNA sequences of universal genes (such as those that encode the structural components of ribosomes) confirm the monophyly of life.

Despite these commonalities, major differences have also evolved across the diversity of life. Based on the significant distinctions in cell structure and biochemical functioning, many biologists now recognize three *domains* of life, two prokaryotic and one eukaryotic (**FIGURE 19.1**). Note that "domain" is a subjective term used for the largest divisions of life.

All prokaryotic organisms are unicellular, although they may form large coordinated colonies or biofilms consisting of many individuals. The domain Eukarya, by contrast, encompasses both unicellular and multicellular life forms. As was described in Chapter 4, prokaryotic cells differ from eukaryotic cells in some important ways:

- *Prokaryotic cells do not divide by mitosis.* Instead, after replicating their DNA, prokaryotic cells divide by their own method, *binary fission*.

- *The organization of the genetic material differs.* The DNA of the prokaryotic cell is not organized within a membrane-enclosed nucleus. DNA molecules in prokaryotes are often circular. Many (but not all) prokaryotes have only one main chromosome and are effectively haploid, although many have additional smaller DNA molecules, called *plasmids*.

- *Prokaryotes have none of the membrane-enclosed cytoplasmic organelles—mitochondria, Golgi apparatus, and others—that are found in most eukaryotes.* However, the cytoplasm of a prokaryotic cell may contain a variety of infoldings of the plasma membrane and photosynthetic membrane systems not found in eukaryotes.

Although the study and classification of eukaryotic organisms goes back centuries, much our knowledge of the evolutionarily ancient prokaryotic domains is extremely recent. Not until the final quarter of the twentieth century did advances in molecular genetics and biochemistry reach a point that enabled research that revealed deep-seated distinctions between the domains Bacteria and Archaea.

The two prokaryotic domains differ in significant ways

A glance at **TABLE 19.1** will show you that there are major differences (most of which cannot be seen even under an electron microscope) between the two prokaryotic domains. In some

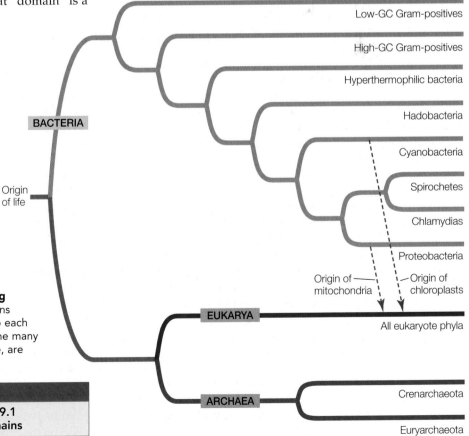

FIGURE 19.1   The Three Domains of the Living World   This summary classification of the domains Bacteria and Archaea shows their relationships to each other and to Eukarya. The relationships among the many clades of bacteria, not all of which are listed here, are incompletely resolved at this time.

Low-GC Gram-positives
High-GC Gram-positives
Hyperthermophilic bacteria
Hadobacteria
Cyanobacteria
Spirochetes
Chlamydias
Proteobacteria
Origin of mitochondria
Origin of chloroplasts
All eukaryote phyla
BACTERIA
Origin of life
EUKARYA
ARCHAEA
Crenarchaeota
Euryarchaeota

TABLE 19.1	The Three Domains of Life on Earth		
		DOMAIN	
CHARACTERISTIC	BACTERIA	ARCHAEA	EUKARYA
Membrane-enclosed nucleus	Absent	Absent	Present
Membrane-enclosed organelles	Few	Absent	Many
Peptidoglycan in cell wall	Present	Absent	Absent
Membrane lipids	Ester-linked	Ether-linked	Ester-linked
	Unbranched	Branched	Unbranched
Ribosomes[a]	70S	70S	80S
Initiator tRNA	Formylmethionine	Methionine	Methionine
Operons	Yes	Yes	Rare
Plasmids	Yes	Yes	Rare
RNA polymerases	One	One[b]	Three
Ribosomes sensitive to chloramphenicol and streptomycin	Yes	No	No
Ribosomes sensitive to diphtheria toxin	No	Yes	Yes

[a] 70S ribosomes are smaller than 80S ribosomes.
[b] Archaeal RNA polymerase is similar to eukaryotic polymerases.

ways archaea are more like eukaryotes; in other ways they are more like bacteria. (Note that we use lowercase when referring to the members of these domains and uppercase when referring to the domains themselves.) The basic unit of an *archaeon* (the term for a single archaeal organism) or *bacterium* (a single bacterial organism) is the prokaryotic cell. Each single-celled organism contains a full complement of genetic and protein-synthesizing systems, including DNA, RNA, and all the enzymes needed to transcribe and translate the genetic information into proteins. The prokaryotic cell also contains at least one system for generating the ATP it needs.

Genetic studies clearly indicate that all three domains had a single common ancestor. Across a major portion of their genome, eukaryotes share a more recent common ancestor with Archaea than they do with Bacteria (see Figure 19.1). However, the mitochondria of eukaryotes (as well as the chloroplasts of photosynthetic eukaryotes, such as plants) originated through endosymbiosis with a bacterium. Some biologists prefer to view the origin of eukaryotes as a fusion of two equal partners (one ancestor that was related to modern archaea, and another that was more closely related to modern bacteria). Others view the divergence of the early eukaryotes from the archaea as a separate and earlier event than the later endosymbioses. In either case, some eukaryote genes are most closely related to those of archaea, while others are most closely related to those of bacteria. The tree of life therefore contains some merging of lineages as well as the predominant diverging of lineages.

LINK   The origin of mitochondria and chloroplasts by endosymbiosis is described in Concept 20.1

The last common ancestor of the three domains probably lived about 3 billion years ago. It probably had DNA as its genetic material, and its machinery for transcription and translation probably produced RNAs and proteins, respectively. This ancestor probably had a circular chromosome. Archaea, Bacteria, and Eukarya are all the products of billions of years of mutation, natural selection, and genetic drift, and they are all well adapted to present-day environments. The earliest prokaryote fossils, which date back at least 3.5 billion years, indicate that there was considerable diversity among the prokaryotes even during those earliest days of life.

The small size of prokaryotes has hindered our study of their evolutionary relationships

Until about 300 years ago, nobody had even *seen* an individual prokaryote; these organisms remained invisible to humans until the invention of the first simple microscope. Prokaryotes are so small, however, that even the best light microscopes don't reveal much about them. It took advanced microscopic equipment and modern molecular techniques to open up the microbial world. (Microscopic organisms—both prokaryotes and eukaryotes—are often collectively referred to as *microbes*.)

Before DNA sequencing became practical, taxonomists based prokaryote classification on observable phenotypic characters such as shape, color, motility, nutritional requirements, and sensitivity to antibiotics. One of the characters most widely used to classify prokaryotes is the structure of their cell walls.

The cell walls of almost all bacteria contain **peptidoglycan**, a cross-linked polymer of amino sugars that produces a meshlike structure around the cell. Peptidoglycan is a substance unique to bacteria; its absence from the cell walls of archaea is a key difference between the two prokaryotic domains. Peptidoglycan is also an excellent target for combating pathogenic (disease-causing) bacteria because it has no counterpart in eukaryotic cells. Antibiotics such as penicillin and ampicillin, as well as other agents that specifically interfere with the synthesis of peptidoglycan-containing cell walls, tend to have little, if any, effect on the cells of humans and other eukaryotes.

(A)

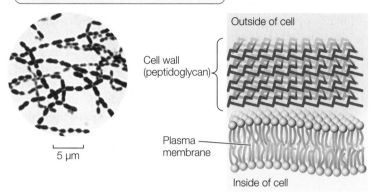

Gram-positive bacteria have a uniformly dense cell wall consisting primarily of peptidoglycan.

Outside of cell

Cell wall (peptidoglycan)

Plasma membrane

Inside of cell

5 μm

(B)

Gram-negative bacteria have a very thin peptidoglycan layer and an outer membrane.

Outside of cell

Outer membrane of cell envelope

Periplasmic space

Peptidoglycan layer

Periplasmic space

Plasma membrane

Inside of cell

5 μm

FIGURE 19.2 The Gram Stain and the Bacterial Cell Wall When treated with Gram stain, the cell walls of bacteria react in one of two ways. (A) Gram-positive bacteria have a thick peptidoglycan cell wall that retains the violet dye and appears deep blue or purple. (B) Gram-negative bacteria have a thin peptidoglycan layer that does not retain the violet dye, but picks up the counterstain and appears pink to red.

The **Gram stain** is a technique that separates most types of bacteria into two distinct groups, Gram-positive and Gram-negative. A smear of bacterial cells on a microscope slide is soaked in a violet dye and treated with iodine; it is then washed with alcohol and counterstained with a red dye (safranine). **Gram-positive bacteria** retain the violet dye and appear blue to purple (**FIGURE 19.2A**). The alcohol washes the violet stain out of **Gram-negative bacteria**, which then pick up the safranine counterstain and appear pink to red (**FIGURE 19.2B**). For most bacteria, the effect of the Gram stain is determined by the chemical structure of the cell wall.

- *A Gram-negative cell wall* usually has a thin peptidoglycan layer, which is surrounded by a second, outer membrane quite distinct in chemical makeup from the

plasma membrane (see Figure 19.2B). Between the plasma membrane and the outer membrane is a *periplasmic space*. This space contains proteins that are important in digesting some materials, transporting others, and detecting chemical gradients in the environment.

- *A Gram-positive cell wall* usually has about five times as much peptidoglycan as a Gram-negative cell wall. Its thick peptidoglycan layer is a meshwork that may serve some of the same purposes as the periplasmic space of the Gram-negative cell envelope (cell wall plus outer membrane).

Shape is another phenotypic characteristic that is useful for the basic identification of bacteria. Three major shapes are common—spheres, rods, and spiral forms (**FIGURE 19.3**). Many bacterial names are based on these shapes. A spherical bacterium is called a **coccus** (plural *cocci*). Cocci may live singly or may associate in two- or three-dimensional arrays as chains, plates, blocks, or clusters of cells. A rod-shaped bacterium is called a **bacillus** (plural *bacilli*). The spiral form (like a corkscrew), or **helix** (plural *helices*), is the third main bacterial shape. Bacilli and helices may be single, form chains, or gather in regular clusters. Among the other bacterial shapes are long filaments and branched filaments.

Less is known about the shapes of archaea because many of these organisms have never been seen. Many archaea are known only from samples of DNA from the environment. However, the species whose morphologies are known include cocci, bacilli, and even triangular and square-shaped species; the last grow on surfaces, arranged like sheets of postage stamps.

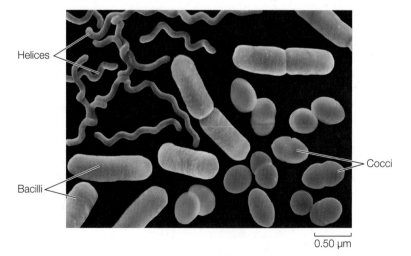

Helices

Bacilli

Cocci

0.50 μm

FIGURE 19.3 Bacterial Cell Shapes This composite, colorized micrograph shows three common bacterial shapes. Spherical cells are called cocci; these are a species of *Enterococcus* from the mammalian gut. Rod-shaped cells are called bacilli; *Escherichia coli* also reside in the gut. The spiral-shaped helices are *Leptospira interrogans*, a human pathogen.

The nucleotide sequences of prokaryotes reveal their evolutionary relationships

Analyses of the nucleotide sequences of ribosomal RNA (rRNA) genes provided the first comprehensive evidence of evolutionary relationships among prokaryotes. For several reasons, rRNA is particularly useful for phylogenetic studies of living organisms:

- rRNA was present in the common ancestor of all life and is therefore evolutionarily ancient.
- No free-living organism lacks rRNA, so rRNA genes can be compared throughout the tree of life.
- rRNA plays a critical role in translation in all organisms, so lateral transfer of rRNA genes among distantly related species is unlikely.
- rRNA has evolved slowly enough that gene sequences from even distantly related species can be aligned and analyzed.

Comparisons of rRNA genes from a great many organisms have revealed the probable phylogenetic relationships throughout the tree of life. Databases such as GenBank contain rRNA gene sequences from hundreds of thousands of species—more than any other type of gene sequence.

Although studies of rRNA genes reveal much about the evolutionary relationships of prokaryotes, they don't always reveal the entire evolutionary history of these organisms. In some groups of prokaryotes, analyses of multiple gene sequences have suggested several different phylogenetic patterns. How could such differences among different gene sequences arise? Studies of whole prokaryotic genomes have revealed that even distantly related prokaryotes sometimes exchange genetic material.

Lateral gene transfer can lead to discordant gene trees

As noted earlier, prokaryotes reproduce by binary fission. If we could follow these divisions back through evolutionary time, we would be tracing the path of the complete tree of life for bacteria and archaea. This underlying tree of relationships (represented in highly abbreviated form in Appendix A) is called the *organismal* (or *species*) *tree*. Because whole genomes are replicated during binary fission, we would expect phylogenetic trees constructed from most gene sequences (see Chapter 16) to reflect these same relationships.

Even though binary fission is an asexual process, there are other processes—including transformation, conjugation, and transduction—that allow the exchange of genetic information between some prokaryotes without reproduction. Thus prokaryotes can exchange and recombine their DNA with that of other individuals (this is sex in the genetic sense of the word), but this genetic exchange is not directly linked to reproduction, as it is in most eukaryotes.

> **LINK** Prokaryote exchange of genetic material by transformation, conjugation, and transduction is described in Concept 8.4

From early in evolution to the present day, some genes have been moving "sideways" from one prokaryote species to another, a phenomenon known as **lateral gene transfer**. Lateral gene transfers are well documented, especially among closely related species; some have been documented even across the domains of life.

Consider, for example, the genome of *Thermotoga maritima*, a bacterium that can survive extremely high temperatures. In comparing the 1,869 gene sequences of *T. maritima* against sequences encoding the same proteins in other species, investigators found that some of this bacterium's genes have their closest relationships not with the genes of other bacterial

FIGURE 19.4 Lateral Gene Transfer Complicates Phylogenetic Relationships (A) The phylogeny of four hypothetical prokaryote species, two of which have been involved in a lateral transfer of gene *x*. (B) A tree based only on gene *x* shows the phylogeny of the laterally transferred gene, rather than the organismal phylogeny. (C) In many cases, a consensus tree based on a "stable core" of genes accurately reflects the organismal phylogeny.

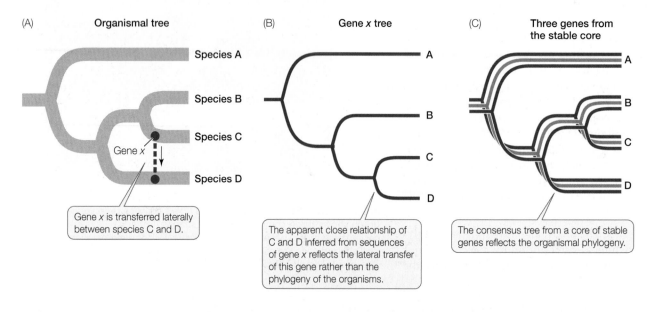

(A) Organismal tree

Species A
Species B
Species C
Species D

Gene *x*

Gene *x* is transferred laterally between species C and D.

(B) Gene *x* tree

A
B
C
D

The apparent close relationship of C and D inferred from sequences of gene *x* reflects the lateral transfer of this gene rather than the phylogeny of the organisms.

(C) Three genes from the stable core

A
B
C
D

The consensus tree from a core of stable genes reflects the organismal phylogeny.

species, but with the genes of archaea that live in similar extreme environments.

When genes involved in lateral transfer events are sequenced and analyzed, the resulting individual *gene trees* will not match the organismal tree in every respect (**FIGURE 19.4**). The individual gene trees will vary because the history of lateral transfer events is different for each gene. Biologists can reconstruct the underlying organismal phylogeny by comparing multiple genes (to produce a *consensus tree*) or by concentrating on genes that are unlikely to be involved in lateral gene transfer events. For example, genes that are involved in fundamental cellular processes (such as the rRNA genes discussed above) are unlikely to be replaced by the same genes from other species because functional, locally adapted copies of these genes are already present.

What kinds of genes are most likely to be involved in lateral gene transfer? Genes that result in a new adaptation that confers higher fitness on a recipient species are most likely to be transferred repeatedly among species. For example, genes that produce antibiotic resistance are often transferred among bacterial species on plasmids, especially under the strong selection pressure such as that imposed by modern antibiotic medications. Improper or overly frequent use of antibiotics can select for resistant strains of bacteria that are much harder to treat. This selection for antibiotic resistance explains why informed physicians have become more careful in prescribing antibiotics.

It is debatable whether lateral gene transfer has seriously complicated our attempts to resolve the tree of prokaryotic life. Recent work suggests that it has not; although it complicates studies in some individual species, it need not present problems at higher taxonomic levels. It is now possible to make nucleotide sequence comparisons involving entire genomes, and these studies are revealing a *stable core* of crucial genes that are uncomplicated by lateral gene transfer. Gene trees based on this stable core more accurately reveal the organismal phylogeny (see Figure 19.4). The problem remains, however, that only a very small proportion of the prokaryotic world has been described and studied.

The great majority of prokaryote species have never been studied

Most prokaryotes have defied all attempts to grow them in pure culture, causing biologists to wonder how many species, and possibly even clades, we might be missing. A window onto this problem was opened with the introduction of a new way of examining nucleic acid sequences. When biologists are unable to work with the whole genome of a single prokaryote species, they can instead examine individual genes collected from a random sample of the environment (see Figure 12.7).

Norman Pace of the University of Colorado isolated individual rRNA gene sequences from extracts of environmental samples such as soil and seawater. Comparing such sequences with previously known ones revealed that an extraordinary number of the sequences were new, implying that they came from previously unrecognized species. Biologists have described

only about 10,000 species of bacteria and only a few hundred species of archaea (see Figure 1.4). The results of Pace's and similar studies suggest that there may be millions—perhaps hundreds of millions—of prokaryote species on Earth. Other biologists put the estimate much lower, arguing that the high dispersal ability of many bacterial species greatly reduces local endemism (i.e., the number of species restricted to a small geographic area). Only the magnitude of these estimates differs, however; all sides agree that we have just begun to uncover Earth's bacterial and archaeal diversity.

Do You Understand Concept 19.1?

- Why were all prokaryotes once considered "equal," and what findings led to the establishment of Bacteria and Archaea as separate domains?
- How did biologists classify bacteria before it became possible to determine nucleotide sequences?
- Why are nucleotide sequences of rRNA genes particularly useful for evolutionary studies?
- How does lateral gene transfer complicate evolutionary studies?

Despite the challenges of reconstructing the phylogeny of prokaryotes, taxonomists are beginning to establish evolutionary classification systems for these organisms. With a full understanding that new information requires periodic revisions in these classifications, we next apply a current system of classification to organize our survey of prokaryote diversity.

concept 19.2 Prokaryote Diversity Reflects the Ancient Origins of Life

The prokaryotes were alone on Earth for a very long time, adapting to new environments and to changes in existing environments. They have survived to this day, in massive numbers and incredible diversity, and they are found everywhere. If success is measured by numbers of individuals, the prokaryotes are the most successful organisms on Earth. Individual bacteria and archaea in the oceans number more than 3×10^{28}—perhaps 100 million times greater than the number of stars in the visible universe. Closer to home, the individual bacteria living in your intestinal tract outnumber all the humans who have ever lived.

Given our still fragmentary knowledge of prokaryote diversity, it is not surprising that there are several different hypotheses about the relationships of the major groups of prokaryotes. In this book we use a widely accepted classification system that has considerable support from nucleotide sequence data. We will discuss the eight bacterial groups that have the broadest phylogenetic support and have received the most study: the low-GC Gram-positives, high-GC Gram-positives, hyperthermophilic bacteria, hadobacteria, cyanobacteria, spirochetes, chlamydias, and proteobacteria (see Figure 19.1). We will then describe the archaea, whose diversity is even less well studied than that of the bacteria.

The low-GC Gram-positives include the smallest cellular organisms

The **low-GC Gram-positives**, also known as *Firmicutes*, derive the first part of their name from the relatively low ratio of G-C to A-T nucleotide base pairs in their DNA. The second part of their name is less accurate: some of the low-GC Gram-positives are in fact Gram-negative, and some have no cell wall at all. Despite these differences, phylogenetic analyses of DNA sequences support the monophyly of this bacterial clade.

Some low-GC Gram-positives can produce heat-resistant resting structures called **endospores** (**FIGURE 19.5**). When a key nutrient such as nitrogen or carbon becomes scarce, the bacterium replicates its DNA and encapsulates one copy, along with some of its cytoplasm, in a tough cell wall heavily thickened with peptidoglycan and surrounded by a spore coat. The parent cell then breaks down, releasing the endospore. Endospore production is not a reproductive process; the endospore merely replaces the parent cell. The endospore, however, can survive harsh environmental conditions that would kill the parent cell, such as high or low temperatures or drought, because it is *dormant*—its normal metabolic activity is suspended. Later, if it encounters favorable conditions, the endospore becomes metabolically active and divides, forming new cells that are like the parent cells. Members of this endospore-forming group of low-GC Gram-positives include the many species of *Clostridium* and

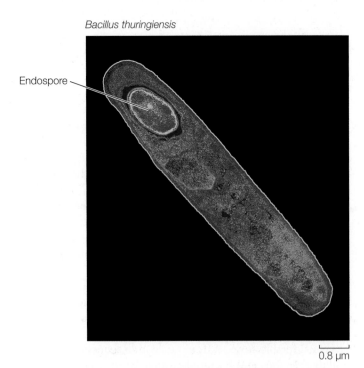

Bacillus thuringiensis

Endospore

0.8 μm

FIGURE 19.5 A Structure for Waiting Out Bad Times Some low-GC Gram-positive bacteria can replicate their DNA and encase it in a resistant endospore. In harsh conditions, the parent cell breaks down and the endospore survives in a dormant state until conditions improve.

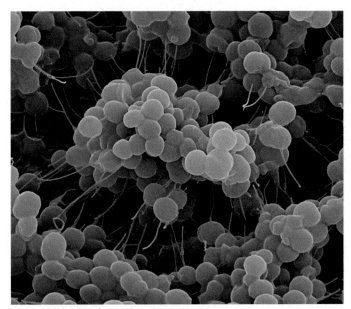

Staphylococcus epidermis

1.2 μm

FIGURE 19.6 Staphylococci "Grape clusters" are the usual arrangement of these low-GC Gram-positive coccal bacteria, often the cause of skin or wound infections.

Bacillus. Some of their endospores can be reactivated after more than 1,000 years of dormancy. There are even credible claims of reactivation of *Bacillus* endospores after millions of years.

Dormant endospores of *Bacillus anthracis* are the source of an *exotoxin* (a toxin produced by the bacteria's metabolism and released into the environment) that causes anthrax. Anthrax is primarily a disease of cattle and sheep, but it can also be fatal in humans. The endospores reactivate when they sense macrophages in the cytoplasm of mammalian blood cells. *Bacillus anthracis* has been used as a bioterrorism agent because large quantities of its endospores are relatively easy to transport and spread among human populations, where they may be inhaled or ingested.

Low-GC Gram-positives of the genus *Staphylococcus*—the **staphylococci** (**FIGURE 19.6**)—are abundant on the human body surface; they are responsible for boils and many other skin problems. *Staphylococcus aureus* is the best-known human pathogen in this genus; it is present in 20 to 40 percent of normal adults (and in 50 to 70 percent of hospitalized adults). In addition to skin diseases, *S. aureus* can cause respiratory, intestinal, and wound infections.

Another interesting group of low-GC Gram-positives, the **mycoplasmas**, lack cell walls, although some have a stiffening material outside the plasma membrane. The mycoplasmas are among the smallest cellular organisms known (**FIGURE 19.7**). The smallest mycoplasmas capable of multiplying by fission have a diameter of about 0.2 μm. They are small in another crucial sense as well: they have less than half as much DNA as most other prokaryotes. It has been speculated that the DNA in a mycoplasma, which codes for fewer than 500 proteins, may be close to the minimum amount required to encode the essential properties of a living cell (see Figure 12.8).

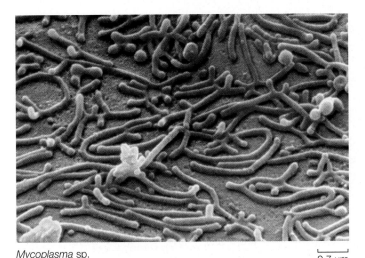

Mycoplasma sp.

0.7 μm

FIGURE 19.7 Tiny Cells With about one-fifth as much DNA as *E. coli*, mycoplasmas are among the smallest known bacteria.

Some high-GC Gram-positives are valuable sources of antibiotics

High-GC Gram-positives, also known as *Actinobacteria*, have a higher ratio of G-C to A-T nucleotide base pairs than do the low-GC Gram-positives. These bacteria develop an elaborately branched system of filaments (**FIGURE 19.8**) that resembles the filamentous growth habit of fungi, albeit at a reduced scale. Some high-GC Gram-positives reproduce by forming chains of spores at the tips of the filaments. In species that do not form spores, the branched, filamentous growth ceases and the structure breaks up into typical cocci or bacilli, which then reproduce by binary fission.

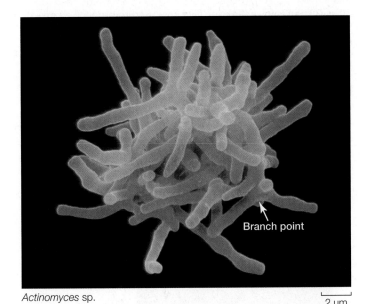

Actinomyces sp.

Branch point

2 μm

FIGURE 19.8 Actinomycetes Are High-GC Gram-Positives The tangled, branching filaments seen in this scanning electron micrograph are typical of this medically important bacterial group.

The high-GC Gram-positives include several medically important bacteria. *Mycobacterium tuberculosis* causes tuberculosis, which kills 3 million people each year. Genetic data suggest that this bacterium arose 3 million years ago in East Africa, making it the oldest known human bacterial pathogen. The genus *Streptomyces* produces streptomycin as well as hundreds of other antibiotics. We derive most of our antibiotics from members of the high-GC Gram-positives.

Hyperthermophilic bacteria live at very high temperatures

Several lineages of bacteria and archaea are **extremophiles**: they thrive under extreme conditions that would kill most other organisms. The **hyperthermophilic bacteria**, for example, are *thermophiles* (Greek, "heat-lovers"). Genera such as *Aquifex* live near volcanic vents and in hot springs, sometimes at temperatures near the boiling point of water. Some species of *Aquifex* need only hydrogen, oxygen, carbon dioxide, and mineral salts to live and grow. Species of the genus *Thermotoga* live deep underground in oil reservoirs, as well as in other high-temperature environments.

Biologists have hypothesized that high temperatures characterized the ancestral conditions for life on Earth, given that most environments on early Earth were much hotter than those of today. Reconstructions of ancestral bacterial genes have supported this hypothesis by showing that the ancestral sequences functioned best at elevated temperatures. The monophyly of the hyperthermophilic bacteria, however, is not well established.

Hadobacteria live in extreme environments

The **hadobacteria**, including such genera as *Deinococcus* and *Thermus*, are another group of thermophilic extremophiles. The group's name is derived from Hades, the ancient Greek name for the underworld. *Deinococcus* are resistant to radiation and can consume nuclear waste and other toxic materials. They can also survive extremes of cold as well as hot temperatures. Another member of this group, *Thermus aquaticus*, was the source of the thermally stable DNA polymerase that was critical for the development of the polymerase chain reaction. *Thermus aquaticus* was originally isolated from a hot spring, but it can be found wherever hot water occurs (including many residential hot water heaters).

Cyanobacteria were the first photosynthesizers

Cyanobacteria, sometimes called *blue-green bacteria* because of their pigmentation, are photosynthetic bacteria that require only water, nitrogen gas, oxygen, a few mineral elements, light, and carbon dioxide to survive. They use chlorophyll *a* for photosynthesis and release oxygen gas (O_2); many species also fix nitrogen. Photosynthesis by these bacteria was the basis of the "green revolution" that transformed Earth's atmosphere (see Concept 18.2).

Cyanobacteria carry out the same type of photosynthesis that is characteristic of eukaryotic photosynthesizers. They

(A) *Anabaena* sp.

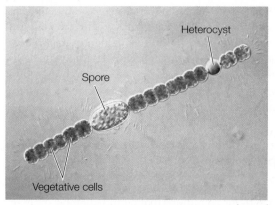

Heterocyst

Spore

Vegetative cells

4 μm

(B) *Nostoc punctiforme*

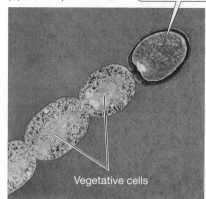

A thick wall separates the cytoplasm of the nitrogen-fixing heterocyst from the surrounding environment.

Vegetative cells

0.4 μm

FIGURE 19.9 Cyanobacteria
(A) Some cyanobacteria form filamentous colonies containing three cell types, including reproductive spores and photosynthesizing vegetative cells. (B) Heterocysts are specialized for nitrogen fixation and may serve as a breaking point when filaments reproduce. (C) This pond in Germany has experienced eutrophication: phosphorus and other nutrients generated by human activity have accumulated, feeding an immense green mat (commonly referred to as "pond scum") that is made up of several species of free-living cyanobacteria.

(C)

colonies of cyanobacteria differentiate into three specialized cell types: vegetative cells, spores, and heterocysts (**FIGURE 19.9**). **Vegetative cells** photosynthesize, **spores** are resting stages that can survive harsh environmental conditions and eventually develop into new filaments, and **heterocysts** are cells specialized for nitrogen fixation. All of the known cyanobacteria with heterocysts fix nitrogen. Heterocysts also have a role in reproduction: when filaments break apart to reproduce, the heterocyst may serve as a breaking point.

Spirochetes move by means of axial filaments

Spirochetes are Gram-negative, motile bacteria characterized by unique structures called *axial filaments* (**FIGURE 19.10A**),

contain elaborate and highly organized internal membrane systems called *photosynthetic lamellae*. As mentioned in Concept 19.1, the chloroplasts of photosynthetic eukaryotes are derived from an endosymbiotic cyanobacterium.

Cyanobacteria may live free as single cells or associate in multicellular colonies. Depending on the species and on growth conditions, these colonies may range from flat sheets one cell thick to filaments to spherical balls of cells. Some filamentous

FIGURE 19.10 Spirochetes Get Their Shape from Axial Filaments (A) A spirochete from the gut of a termite, seen in cross section, shows the axial filaments used to produce a corkscrew-like movement of these helical prokaryotes. (B) This spirochete species causes syphilis in humans.

(B)

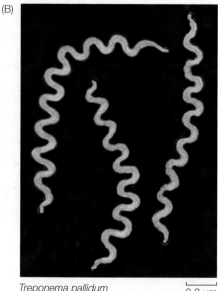

(A)

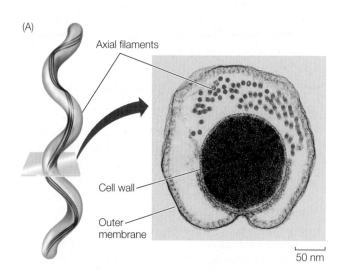

Axial filaments

Cell wall

Outer membrane

50 nm

Treponema pallidum

0.8 μm

which are modified flagella running through the periplasmic space. The cell body is a long cylinder coiled into a helix (**FIGURE 19.10B**). The axial filaments begin at either end of the cell and overlap in the middle. Motor proteins connect the axial filaments to the cell wall, enabling the corkscrew-like movement of the bacterium. Many spirochetes are parasites of humans; a few are pathogens, including those that cause syphilis and Lyme disease. Others live free in mud or water.

Chlamydias are extremely small parasites

Chlamydias are among the smallest bacteria (0.2–1.5 μm in diameter). They can live only as parasites in the cells of other organisms. It was once believed that their obligate parasitism resulted from an inability to produce ATP—that chlamydias were "energy parasites." However, genome sequencing indicates that chlamydias have the genetic capacity to produce at least some ATP. They can augment this capacity by using an enzyme called a translocase, which allows them to take up ATP from the cytoplasm of their host in exchange for ADP from their own cells.

These tiny, Gram-negative cocci are unique among prokaryotes because of a complex life cycle that involves two different forms of cells, *elementary bodies* and *reticulate bodies* (**FIGURE 19.11**). Various strains of chlamydias cause eye infections (especially trachoma), sexually transmitted diseases, and some forms of pneumonia in humans.

The proteobacteria are a large and diverse group

By far the largest group of bacteria, in terms of number of described species, is the **proteobacteria**. The proteobacteria include

Salmonella typhimurium

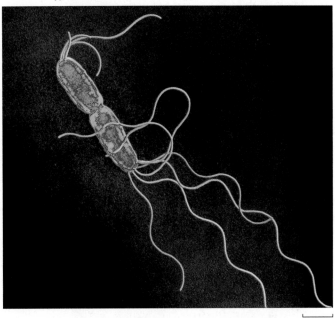

1.5 μm

FIGURE 19.12 Proteobacteria Include Human Pathogens These conjugating cells of *Salmonella typhimurium* are exchanging genetic material (see Concept 8.4). Representing only one group of the highly diverse proteobacteria, this pathogen causes a wide range of gastrointestinal illnesses in humans.

many species of Gram-negative photoautotrophs that use light-driven reactions to metabolize sulfur, as well as dramatically diverse bacteria that bear no phenotypic resemblance to the photoautotrophic species. Genetic and morphological evidence indicates that the mitochondria of eukaryotes were derived from a proteobacterium by endosymbiosis.

Among the proteobacteria are some nitrogen-fixing genera, such as *Rhizobium*, and other bacteria that contribute to the global nitrogen and sulfur cycles. *Escherichia coli*, one of the most studied organisms on Earth, is a proteobacterium. So, too, are many of the most famous human pathogens, such as *Yersinia pestis* (which causes bubonic plague), *Vibrio cholerae* (cholera), and *Salmonella typhimurium* (gastrointestinal disease; **FIGURE 19.12**). The bioluminescent *Vibrio* we discussed at the start of the chapter are also members of this group.

> **FRONTIERS** There are many potential applications of the genes that encode bioluminescent proteins in bacteria. Already, these genes are inserted into the genomes of other species where the resulting bioluminescence is used as a marker of gene expression. Proposals for making use of bioluminescence in bioengineered organisms include crop plants that glow when they become water-stressed and need to be irrigated, and glowing trees that could light highways at night in place of electric lights.

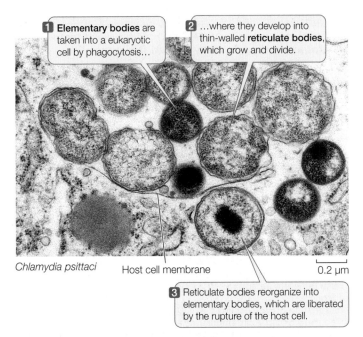

1 **Elementary bodies** are taken into a eukaryotic cell by phagocytosis...

2 ...where they develop into thin-walled **reticulate bodies**, which grow and divide.

Chlamydia psittaci   Host cell membrane   0.2 μm

3 Reticulate bodies reorganize into elementary bodies, which are liberated by the rupture of the host cell.

FIGURE 19.11 Chlamydias Change Form during Their Life Cycle Elementary bodies and reticulate bodies are the two major phases of the chlamydia life cycle.

Although fungi cause most plant diseases, and viruses cause others, about 200 known plant diseases are of bacterial origin. *Crown gall*, with its characteristic tumors (**FIGURE 19.13**), is one of the most striking. The causal agent of crown gall is

FIGURE 19.13 Crown Gall Crown gall, a type of tumor shown here growing on the trunk of a white oak, is caused by the proteobacterium *Agrobacterium tumefaciens*.

Agrobacterium tumefaciens, a proteobacterium that harbors a plasmid used in recombinant DNA studies as a vehicle for inserting genes into new plant hosts.

Gene sequencing enabled biologists to differentiate the domain Archaea

The separation of Archaea from Bacteria and Eukarya was originally based on phylogenetic relationships determined from sequences of rRNA genes. It was supported when biologists sequenced the first archaeal genome. That genome consisted of 1,738 genes, more than half of which were unlike any genes ever found in the other two domains.

Archaea are well known for living in extreme habitats such as those with high salinity (salt content), low oxygen concentrations, high temperatures, or high or low pH (**FIGURE 19.14**). Many archaea are not extremophiles, however, but live in moderate habitats—they are common in soil, for example. Perhaps the largest numbers of archaea live in the ocean depths.

One current classification scheme divides Archaea into two principal groups, **Crenarchaeota** and **Euryarchaeota**. Less is known about two more recently discovered groups, **Korarchaeota** and **Nanoarchaeota**. In fact, we know relatively little about the phylogeny of archaea, in part because the study of these prokaryotes is still in its early stages.

Two characteristics shared by all archaea are the absence of peptidoglycan in their cell walls and the presence of lipids of distinctive composition in their cell membranes (see Table 19.1). The unusual lipids in the membranes of

INVESTIGATION

FIGURE 19.14 What Is the Highest Temperature Compatible with Life? Can any organism thrive at temperatures above 120°C? This is the temperature used for sterilization, known to destroy all previously described organisms. Kazem Kashefi and Derek Lovley isolated an unidentified prokaryote from water samples taken near a hydrothermal vent and found it survived and even multiplied at 121°C. The organism was dubbed "Strain 121," and its gene sequencing results indicate that it is an archaeal species.

HYPOTHESIS

Some prokaryotes can survive at temperatures above the 120°C threshold of sterilization.

METHOD

1. Seal samples of unidentified, iron-reducing, thermal vent prokaryotes in tubes with a medium containing Fe^{3+} as an electron acceptor. Control tubes contain Fe^{3+} but no organisms.

2. Hold both tubes in a sterilizer at 121°C for 10 hours. If the iron-reducing organisms are metabolically active, they will reduce the Fe^{3+} to Fe^{2+} (as magnetite, which can be detected with a magnet).

RESULTS

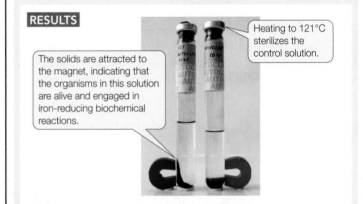

The solids are attracted to the magnet, indicating that the organisms in this solution are alive and engaged in iron-reducing biochemical reactions.

Heating to 121°C sterilizes the control solution.

CONCLUSION

This thermal vent organism (Strain 121) can survive at temperatures above the previously defined sterilization limit.

ANALYZE THE DATA

After Strain 121 was isolated, its growth was examined at various temperatures. The table shows generation time (the time between cell divisions) at nine temperatures.

Temperature (°C)	Generation time (hr)
85	10
90	4
95	3
100	2.5
105	2
110	4
115	6
120	20
130	No growth, but cells not killed

A. Make a graph showing generation time as a function of temperature.
B. Which temperature appears to be closest to the optimum for growth of Strain 121?
C. Note that no growth occurred at 130°C, but that the cells were not killed. How would you demonstrate that these cells were still alive?

Go to **yourBioPortal.com** for original citations, discussions, and relevant links for all INVESTIGATION figures.

archaea are found in all archaea and in no bacteria or eukaryotes. Most lipids in bacterial and eukaryotic membranes contain unbranched long-chain fatty acids connected to glycerol molecules by *ester linkages*:

$$
\begin{matrix}
& O & & H \\
& \| & & | \\
-C & -O & -C- \\
& & & | \\
& & & H
\end{matrix}
$$

In contrast, some lipids in archaeal membranes contain long-chain hydrocarbons connected to glycerol molecules by *ether linkages*:

$$
\begin{matrix}
H & & H \\
| & & | \\
-C & -O & -C- \\
| & & | \\
H & & H
\end{matrix}
$$

These ether linkages are a synapomorphy of archaea.

In addition, the long-chain hydrocarbons of archaea are branched. One class of archaeal lipids, with hydrocarbon chains 40 carbon atoms in length, contains glycerol at *both* ends of the hydrocarbons (**FIGURE 19.15**). These lipids form a *lipid monolayer* structure that is unique to archaea. They still fit into a biological membrane because they are twice as long as the typical lipids in the bilayers of other membranes. Lipid monolayers and bilayers are both found among the archaea. The effects, if any, of these structural features on membrane performance are unknown. In spite of this striking difference in their lipids, the membranes of all three domains have similar overall structures, dimensions, and functions.

Most crenarchaeotes live in hot or acidic places

Most known crenarchaeotes are either thermophilic, acidophilic (acid loving), or both. Members of the genus *Sulfolobus* live in hot sulfur springs at temperatures of 70°C to 75°C. They become metabolically inactive at 55°C (131°F). Hot sulfur springs are also extremely acidic. *Sulfolobus* grows best in the

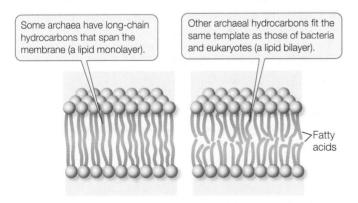

FIGURE 19.15 Membrane Architecture in Archaea The long-chain hydrocarbons of many archaeal lipids have glycerol molecules at both ends, so that the membranes they form consist of a lipid monolayer. In contrast, the membranes of other archaea, bacteria, and eukaryotes consist of a lipid bilayer.

Some archaea have long-chain hydrocarbons that span the membrane (a lipid monolayer).

Other archaeal hydrocarbons fit the same template as those of bacteria and eukaryotes (a lipid bilayer).

Fatty acids

FIGURE 19.16 Crenarchaeotes Like It Hot Thermophilic archaea can thrive in the intense heat of volcanic hot sulfur springs such as those in Wyoming's Yellowstone National Park.

range from pH 2 to pH 3, but some members of this genus readily tolerate pH values as low as 0.9. Most acidophilic thermophiles maintain an internal pH of 5.5 to 7 (close to neutral) in spite of their acidic environment. These and other thermophiles thrive where very few other organisms can even survive (**FIGURE 19.16**).

Euryarchaeotes are found in surprising places

Some species of Euryarchaeota are **methanogens**: they produce methane (CH_4) by reducing carbon dioxide, and this is the key step in their energy metabolism. All of the methanogens are *obligate anaerobes* (see Concept 19.3). Comparison of their rRNA gene sequences has revealed a close evolutionary relationship among these methanogenic species, which were previously assigned to several different groups of bacteria.

Methanogenic euryarchaeotes release approximately 2 billion tons of methane gas into Earth's atmosphere each year, accounting for 80 to 90 percent of the methane that enters the atmosphere, including that produced in some mammalian digestive systems. Approximately a third of this methane comes from methanogens living in the guts of ruminants such as cattle, sheep, and deer, and another large fraction comes from methanogens living in the guts of termites and cockroaches. Methane is increasing in Earth's atmosphere by about 1 percent per year and contributes to the greenhouse effect (see Concept 46.4). Part of that increase is due to increases in cattle and rice farming and the methanogens associated with both.

Another group of euryarchaeotes, the **extreme halophiles** (salt lovers), lives exclusively in very salty environments. Because they contain pink carotenoid pigments, these archaea are sometimes easy to see (**FIGURE 19.17**). Extreme halophiles grow in

FIGURE 19.17 Extreme Halophiles Highly saline environments such as these commercial seawater evaporating ponds in San Francisco Bay are home to extreme halophiles. The archaea are easily visible here because of the rich red coloration from their carotenoid pigments.

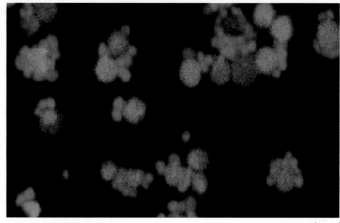

1 μm

FIGURE 19.18 A Nanoarchaeote Growing in Mixed Culture with a Crenarchaeote *Nanoarchaeum equitans* (red), discovered living near deep-sea hydrothermal vents, is the only representative of the nanoarchaeote group so far identified. This tiny organism lives attached to cells of the crenarchaeote *Ignicoccus* (green). For this confocal laser micrograph, the two species were visually differentiated by fluorescent dye "tags" that are specific to their distinct gene sequences.

the Dead Sea and in brines of all types; the reddish pink spots that can occur on pickled fish are colonies of halophilic archaea. Few other organisms can live in the saltiest of the homes that the extreme halophiles occupy; most would "dry" to death, losing too much water to the hypertonic environment. Extreme halophiles have been found in lakes with pH values as high as 11.5—the most alkaline environment inhabited by living organisms, and almost as alkaline as household ammonia.

Some of the extreme halophiles have a unique system for trapping light energy and using it to form ATP—without using any form of chlorophyll—when oxygen is in short supply. They use the pigment *retinal* (also found in the vertebrate eye) combined with a protein to form a light-absorbing molecule called *microbial rhodopsin*.

Another member of the Euryarchaeota, *Thermoplasma*, has no cell wall. It is thermophilic and acidophilic, its metabolism is aerobic, and it lives in coal deposits. Its genome of 1,100,000 base pairs is among the smallest (along with the mycoplasmas) of any free-living organism, although some parasitic organisms have even smaller genomes.

FRONTIERS The halophilic euryarchaeote *Haloferax mediterranei* (see Figure 1.2) accumulates large quantities of a white polyester for carbon and energy storage. The structure and synthesis of this molecule is being studied for possible use of the polyester as a substitute for petroleum-based products.

Korarchaeotes and nanoarchaeotes are less well known

The korarchaeotes are known only from DNA isolated directly from hot springs. No korarchaeote has been successfully grown in pure culture.

Another distinctive archaeal lineage has been discovered at a deep-sea hydrothermal vent off the coast of Iceland. It is the first representative of the group christened Nanoarchaeota because of their minute size (Greek *nanos*, "dwarf"). This organism is a parasite that lives on cells of *Ignicoccus*, a crenarchaeote. Because of their association, the two species can be grown together in culture (**FIGURE 19.18**).

Do You Understand Concept 19.2?

- How does the diversity of environments occupied by prokaryotes compare with the diversity of environments occupied by multicellular organisms you may be familiar with?
- Explain why Gram staining is of limited use in understanding the evolutionary relationships of bacteria.
- What makes the membranes of archaea unique?
- Given that all species of life have evolved for the same amount of time since their common origin, how would you respond to someone who characterizes prokaryotes as "primitive"?

Prokaryotes are found almost everywhere on Earth and live in a wide variety of ecosystems. Next we examine the contributions of prokaryotes to the functioning of those ecosystems.

concept 19.3 Ecological Communities Depend on Prokaryotes

Prokaryotic cells and their associations do not usually live in isolation. Rather, they live in communities of many different species, often including microscopic eukaryotes. While some

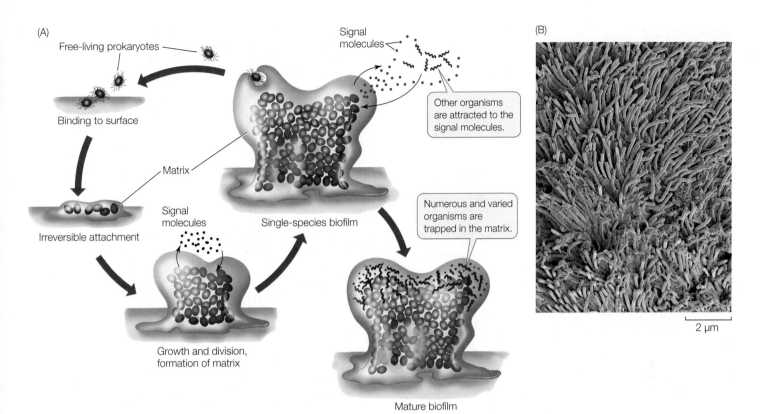

FIGURE 19.19 Forming a Biofilm (A) Free-living prokaryotes readily attach themselves to surfaces and form films that are stabilized and protected by a surrounding matrix. Once the population is large enough, the developing biofilm can send chemical signals that attract other microorganisms. (B) Scanning electron micrography reveals a biofilm of dental plaque. The bacteria (red) are embedded in a matrix consisting of proteins from both bacterial secretions and saliva.

microbial communities are harmful to humans, others provide important services. They help us digest our food, break down municipal waste, and recycle organic matter and chemical elements in the environment.

Many prokaryotes form complex communities

Some microbial communities form layers in sediments; others form clumps a meter or more in diameter. Many microbial communities tend to form dense **biofilms**. Upon contacting a solid surface, the cells bind to that surface and secrete a sticky, gel-like polysaccharide matrix that traps other cells (**FIGURE 19.19**). Once this biofilm forms, it is difficult to kill the cells.

Pathogenic bacteria are difficult for the immune system—and modern medicine—to combat once they form a biofilm. For example, the film may be impermeable to antibiotics. Worse, some drugs stimulate the bacteria in a biofilm to lay down more matrix, making the film even more impermeable. Biofilms may form on just about any available surface, including contact lenses and artificial joint replacements. They foul metal pipes and cause corrosion, a major problem in steam-driven electricity generation plants. The material on our teeth that we call dental plaque is a biofilm. Fossil stromatolites—large, rocky structures made up of alternating layers of fossilized microbial biofilm and calcium carbonate—are among the oldest remnants of life on Earth (see Figure 18.6A).

Biofilms are the subject of much current research. For example, some biologists are studying the chemical signals that prokaryotes in biofilms use to communicate with one another. By blocking the signals that lead to the production of the matrix polysaccharides, researchers may be able to prevent biofilms from forming.

Prokaryotes have amazingly diverse metabolic pathways

Bacteria and archaea outdo the eukaryotes in terms of metabolic diversity. Although they are much more diverse in size and shape, eukaryotes draw on fewer metabolic mechanisms for their energy needs. In fact, much of the eukaryotes' energy metabolism is carried out in organelles—mitochondria and chloroplasts—that are endosymbiotic descendants of bacteria. The long evolutionary history of bacteria and archaea, during which they have had time to explore a wide variety of habitats, has led to the extraordinary diversity of their metabolic "lifestyles"—their use or nonuse of oxygen, their energy sources, their sources of carbon atoms, and the materials they release as waste products.

ANAEROBIC VERSUS AEROBIC METABOLISM Some prokaryotes can live only by anaerobic metabolism because oxygen is poisonous to them. These oxygen-sensitive organisms are called **obligate anaerobes**. Other prokaryotes can shift their metabolism between anaerobic and aerobic modes and are thus called **facultative anaerobes**. Many facultative anaerobes alternate between anaerobic metabolism (such as fermentation) and cellular

TABLE 19.2	How Organisms Obtain Their Energy and Carbon	
NUTRITIONAL CATEGORY	**ENERGY SOURCE**	**CARBON SOURCE**
Photoautotrophs (found in all three domains)	Light	Carbon dioxide
Photoheterotrophs (some bacteria)	Light	Organic compounds
Chemolithotrophs (some bacteria, many archaea)	Inorganic substances	Carbon dioxide
Chemoheterotrophs (found in all three domains)	Organic compounds	Organic compounds

respiration as conditions dictate. **Aerotolerant anaerobes** cannot conduct cellular respiration, but they are not damaged by oxygen when it is present. By definition, an *anaerobe* does not use oxygen as an electron acceptor for its respiration.

At the other extreme from the obligate anaerobes, some prokaryotes are **obligate aerobes**, unable to survive for extended periods in the *absence* of oxygen. They require oxygen for cellular respiration.

NUTRITIONAL CATEGORIES All living organisms face the same nutritional challenges: they must synthesize energy-rich compounds such as ATP to power their life-sustaining metabolic reactions, and they must obtain carbon atoms to build their own organic molecules. Biologists recognize four broad nutritional categories of organisms: photoautotrophs, photoheterotrophs, chemolithotrophs, and chemoheterotrophs. Prokaryotes are represented in all four groups (**TABLE 19.2**).

Photoautotrophs perform photosynthesis. They use light as their energy source and carbon dioxide (CO_2) as their carbon source. The cyanobacteria, like green plants and other

photosynthetic eukaryotes, use chlorophyll *a* as their key photosynthetic pigment and produce oxygen gas (O_2) as a by-product of noncyclic electron transport.

There are other photoautotrophs among the bacteria, but these organisms use *bacteriochlorophyll* as their key photosynthetic pigment, and they do not produce O_2. Instead, some of these photosynthesizers produce particles of pure sulfur, because hydrogen sulfide (H_2S), rather than H_2O, is their electron donor for photophosphorylation. Many proteobacteria fit into this category. Bacteriochlorophyll molecules absorb light of longer wavelengths than the chlorophyll molecules used by other photosynthesizing organisms. As a result, bacteria using this pigment can grow in water under fairly dense layers of algae, using light of wavelengths that are not absorbed by the algae (**FIGURE 19.20**).

Photoheterotrophs use light as their energy source but must obtain their carbon atoms from organic compounds made by other organisms. Their "food" consists of organic compounds such as carbohydrates, fatty acids, and alcohols. For example, compounds released from plant roots (as in rice paddies) or from decomposing photosynthetic bacteria in hot springs are taken up by photoheterotrophs and metabolized to form building blocks for other compounds. Sunlight provides the ATP necessary for metabolism through photophosphorylation.

Chemolithotrophs (also called *chemoautotrophs*) obtain their energy by oxidizing inorganic substances, and they use some of that energy to fix carbon. Some chemolithotrophs use reactions identical to those of the typical photosynthetic cycle, but others use alternative pathways for carbon fixation. Some bacteria oxidize ammonia or nitrite ions to form nitrate ions. Others oxidize hydrogen gas, hydrogen sulfide, sulfur, and other materials. Many archaea are chemolithotrophs.

Finally, **chemoheterotrophs** obtain both energy and carbon atoms from one or more complex organic compounds that have been synthesized by other organisms. Most known bacteria and archaea are chemoheterotrophs—as are all animals and fungi and many protists.

Prokaryotes play important roles in element cycling

The metabolic diversity of the prokaryotes makes them key players in the cycles that keep elements moving through ecosystems. Many prokaryotes are *decomposers*: organisms that metabolize organic compounds in dead organic material and return the

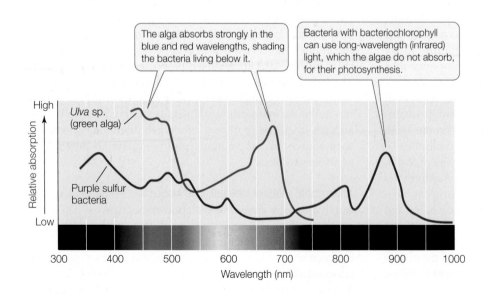

The alga absorbs strongly in the blue and red wavelengths, shading the bacteria living below it.

Bacteria with bacteriochlorophyll can use long-wavelength (infrared) light, which the algae do not absorb, for their photosynthesis.

Ulva sp. (green alga)

Purple sulfur bacteria

Relative absorption — High / Low

Wavelength (nm) — 300 400 500 600 700 800 900 1000

FIGURE 19.20 Bacteriochlorophyll Absorbs Long-Wavelength Light The green alga *Ulva* contains chlorophyll, which absorbs no light of wavelengths longer than 750 nm. Purple sulfur bacteria, which contain bacteriochlorophyll, can conduct photosynthesis using longer wavelengths. As a result, these bacteria can grow under layers of algae.

products to the environment as inorganic substances. Prokaryotes, along with fungi, return tremendous quantities of carbon to the atmosphere as carbon dioxide, thus carrying out a key step in the carbon cycle.

The key metabolic reactions of many prokaryotes involve nitrogen or sulfur. For example, some bacteria carry out respiratory electron transport without using oxygen as an electron acceptor. These organisms use oxidized inorganic ions such as nitrate, nitrite, or sulfate as electron acceptors. Examples include the **denitrifiers**, which release nitrogen to the atmosphere as nitrogen gas (N_2). These normally aerobic bacteria, mostly species of the genera *Bacillus* and *Pseudomonas*, use nitrate (NO_3^-) as an electron acceptor in place of oxygen if they are kept under anaerobic conditions:

$$2\ NO_3^- + 10\ e^- + 12\ H^+ \rightarrow N_2 + 6\ H_2O$$

Denitrifiers play a key role in the cycling of nitrogen through ecosystems. Without denitrifiers, which convert nitrate ions back into nitrogen gas, all forms of nitrogen would leach from the soil and end up in lakes and oceans, making life on land much more difficult.

Nitrogen fixers convert atmospheric nitrogen gas into a chemical form (ammonia) that is usable by the nitrogen fixers themselves as well as by other organisms:

$$N_2 + 6\ H \rightarrow 2\ NH_3$$

All organisms require nitrogen in order to build proteins, nucleic acids, and other important compounds. Nitrogen fixation is thus vital to life as we know it. This all-important biochemical process is carried out by a wide variety of archaea and bacteria (including cyanobacteria) but by no other organisms, so we depend on these prokaryotes for our very existence.

LINK For descriptions of the role of nitrogen in plant nutrition and of the global nitrogen cycle, see Concepts 25.2 and 46.3

Ammonia is oxidized to nitrate in soil and in seawater by chemolithotrophic bacteria called **nitrifiers**. Bacteria of two genera, *Nitrosomonas* and *Nitrosococcus*, convert ammonia (NH_3) to nitrite ions (NO_2^-), and *Nitrobacter* oxidize nitrite to nitrate (NO_3^-), the form of nitrogen most easily used by many plants. What do the nitrifiers get out of these reactions? Their metabolism is powered by the energy released by the oxidation of ammonia or nitrite. For example, by passing the electrons from nitrite through an electron transport system, *Nitrobacter* can make ATP, and using some of this ATP, can also make NADH. With this ATP and NADH, the bacterium can convert CO_2 and H_2O into glucose.

We have already seen the importance of the cyanobacteria in the cycling of oxygen: in ancient times, the oxygen generated by their photosynthesis converted Earth's atmosphere from an anaerobic to an aerobic environment. Other prokaryotes—both bacteria and archaea—contribute to the cycling of sulfur. Deepsea hydrothermal vent ecosystems depend on chemolithotrophic prokaryotes that are incorporated into large communities of crabs, mollusks, and giant worms, all living at a depth of 2,500 meters—below any hint of sunlight. These bacteria obtain energy by oxidizing hydrogen sulfide and other substances released in the near-boiling water flowing from volcanic vents in the ocean floor.

Prokaryotes live on and in other organisms

Prokaryotes work together with eukaryotes in many ways. As we have seen, the mitochondria and chloroplasts of eukaryotes are descended from what were once free-living bacteria. Much later in evolutionary history, some plants became associated with bacteria to form cooperative nitrogen-fixing nodules on their roots.

Many animals harbor a variety of bacteria and archaea in their digestive tracts. Cattle depend on prokaryotes to perform important steps in digestion. Like most animals, cattle cannot produce cellulase, the enzyme needed to start the digestion of the cellulose that makes up the bulk of their plant food. However, bacteria living in a special section of the gut, called the rumen, produce enough cellulase to process the daily diet for the cattle.

FRONTIERS Humans use some of the metabolic products—especially vitamins B_{12} and K—produced by bacteria living in our large intestine. These prokaryotes line our intestines with a dense biofilm that is in intimate contact with the mucosal lining of the gut. This biofilm facilitates nutrient transfer from the intestine into the body, functioning like a specialized "tissue" that is essential to our health. This biofilm has a complex ecology that scientists have just begun to explore in detail—including the possibility that the species composition of an individual's prokaryote gut fauna may contribute to obesity (or the resistance to it).

Humans are heavily populated both inside and out by bacteria. A 2009 study identified more than 1,000 species of bacteria that live on human skin. Many of these bacteria are thought to be critical to maintaining the skin's health.

Of the tiny percentage of all prokaryotes that are pathogens, all those that are known are in the domain Bacteria. Although only a few bacterial species are pathogens, popular notions of bacteria as "germs" and fear of the consequences of infection arouse our curiosity about those few.

A small minority of bacteria are pathogens

The late nineteenth century was a productive era in the history of medicine—a time when bacteriologists, chemists, and physicians proved that many diseases are caused by microbial agents. During this time, the German physician Robert Koch laid down a set of four rules for establishing that a particular microorganism causes a particular disease:

1. The microorganism is always found in individuals with the disease.

2. The microorganism can be taken from the host and grown in pure culture.

Marshall and Warren set out to satisfy Koch's postulates:

Test 1

The microorganism must be present in every case of the disease.

Results: Biopsies from the stomachs of many patients revealed that the bacterium was always present if the stomach was inflamed or ulcerated.

Test 2

The microorganism must be cultured from a sick host.

Results: The bacterium was isolated from biopsy material and eventually grown in culture media in the laboratory.

Test 3

The isolated and cultured bacteria must be able to induce the disease.

Results: Marshall was examined and found to be free of bacteria and inflammation in his stomach. After drinking a pure culture of the bacterium, he developed stomach inflammation (gastritis).

Test 4

The bacteria must be recoverable from newly infected individuals.

Results: Biopsy of Marshall's stomach 2 weeks after he ingested the bacteria revealed the presence of the bacterium, now christened *Helicobacter pylori*, in the inflamed tissue.

Conclusion

Antibiotic treatment eliminated the bacteria and the inflammation in Marshall. The experiment was repeated on healthy volunteers, and many patients with gastric ulcers were cured with antibiotics. Thus Marshall and Warren demonstrated that the stomach inflammation leading to ulcers is caused by *H. pylori* infections in the stomach.

3. A sample of the culture produces the same disease when injected into a new, healthy host.

4. The newly infected host yields a new, pure culture of microorganisms identical to those obtained in the second step.

These rules, called **Koch's postulates**, were important tools in a time when it was not widely understood that microorganisms cause disease. Although modern medical science has more powerful diagnostic tools, Koch's postulates remain useful. For

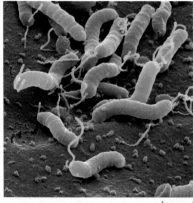

FIGURE 19.21 Satisfying Koch's Postulates Robin Warren and Barry Marshall of the University of Western Australia won the 2005 Nobel Prize in Medicine for showing that ulcers are caused not by the action of stomach acid but by infection with the bacterium *Helicobacter pylori*.

Helicobacter pylori   1.5 μm

example, physicians were taken aback in the 1990s when stomach ulcers—long accepted and treated as the result of excess stomach acid—were shown by Koch's postulates to be caused by the bacterium *Helicobacter pylori* (**FIGURE 19.21**).

For an organism to be a successful pathogen, it must:

- arrive at the body surface of a potential host;
- enter the host's body;
- evade the host's defenses;
- reproduce inside the host; and
- infect a new host.

Failure to complete any of these steps ends the reproductive career of a pathogenic organism. Yet in spite of the many defenses available to potential hosts (see Chapter 31), some bacteria are very successful pathogens. Pathogenic bacteria are often surprisingly difficult to combat, even with today's arsenal of antibiotics. One source of this difficulty is their ability to form biofilms.

For the host, the consequences of a bacterial infection depend on several factors. One is the **invasiveness** of the pathogen: its ability to multiply in the host's body. Another is its **toxigenicity**: its ability to produce *toxins*, chemical substances that are harmful to the host's tissues. *Corynebacterium diphtheriae*, the agent that causes diphtheria, has low invasiveness and multiplies only in the throat, but its toxigenicity is so great that the entire body is affected. In contrast, *Bacillus anthracis*, which

APPLY THE CONCEPT

A small minority of bacteria are pathogens

Imagine you are in charge of maintaining a trout hatchery. Some trout are exhibiting loss of tissue at the tips of their fins, and you suspect a bacterial infection. You isolate and culture two species of bacteria from a trout with affected fins.

1. How would you satisfy the first of Koch's postulates?

2. Imagine that bacterium 1 satisfies the first of Koch's postulates, but bacterium 2 does not. What presence or

absence of data from the cultures from many samples of infected fish would lead you to this conclusion?

3. You already know that bacterium 1 satisfies the second of Koch's postulates. You decide to conduct a test of the third postulate. To your surprise, the test animals are all healthy and show no sign of disease. What are some possible explanations? How would you test your hypotheses?

causes anthrax, has low toxigenicity but is so invasive that the entire bloodstream ultimately teems with the bacteria.

There are two general types of bacterial toxins: exotoxins and endotoxins. **Endotoxins** are released when certain Gram-negative bacteria grow or lyse (burst). Endotoxins are lipopoly-saccharides (complexes consisting of a polysaccharide and a lipid component) that form part of the outer bacterial membrane. Endotoxins are rarely fatal to the host; they normally cause fever, vomiting, and diarrhea. Among the endotoxin producers are some strains of the proteobacteria *Salmonella* and *Escherichia*.

Exotoxins are soluble proteins released by living, multiplying bacteria. They are highly toxic—often fatal—to the host. Human diseases induced by bacterial exotoxins include tetanus (*Clostridium tetani*), cholera (*Vibrio cholerae*), and bubonic plague (*Yersinia pestis*). Anthrax is caused by three exotoxins produced by *Bacillus anthracis*. Botulism is caused by exotoxins produced by *Clostridium botulinum*, which are among the most poisonous ever discovered. The lethal dose for humans of one exotoxin of *C. botulinum* is about one-millionth of a gram. Nonetheless, much smaller doses of this exotoxin, marketed under various trade names (e.g., Botox), are used to treat muscle spasms and for cosmetic purposes (temporary wrinkle reduction in the skin).

Do You Understand Concept 19.3?

- How do biofilms form, and why are they of special interest to researchers?
- How are the four nutritional categories of prokaryotes distinguished?
- How is nitrogen metabolism in the prokaryotes vital to other organisms? Given the roles of bacteria in the nitrogen cycle, how would you answer people who consider all bacteria to be "germs" and dangerous?

Before moving on to discuss the diversity of eukaryotic life, it is appropriate to consider another category of life that includes some pathogens: the viruses. Although they are not cellular, viruses are numerically among the most abundant forms of life on Earth. Their effects on other organisms are enormous. Where did viruses come from, and how do they fit into the tree of life? Biologists are still working to answer these questions.

concept 19.4   Viruses Have Evolved Many Times

Some biologists do not think of viruses as living organisms, primarily because they are not cellular and must depend on cellular organisms for basic life functions such as replication and metabolism. But viruses are derived from the cells of living organisms. They use the same essential forms of genetic information storage and transmission as do cellular organisms. Viruses infect all cellular forms of life—bacteria, archaea, and eukaryotes. They replicate, mutate, evolve, and interact with other organisms, often causing serious diseases in their hosts. Finally, viruses clearly evolve independently of other organisms, so it is almost impossible not to treat them as a part of life.

Several factors make virus phylogeny difficult to resolve. The tiny size of many virus genomes restricts the phylogenetic analyses that can be conducted to relate viruses to cellular organisms. Their rapid mutation rate, which results in rapid evolution of virus genomes, tends to cloud evolutionary relationships over long periods. There are no known fossil viruses (viruses are too small and delicate to fossilize), so the paleontological record offers no clues to virus origins. Finally, viruses are highly diverse (**FIGURE 19.22**). Several lines of evidence support the hypothesis that viruses have evolved repeatedly within each of the major groups of life. The difficulty in resolving deep evolutionary relationships of viruses makes a phylogeny-based classification difficult. Instead, viruses are placed in one of several functionally similar groups on the basis of the structure of their genomes (for example, whether the genomes are composed of RNA or DNA, and are double- or single-stranded). Most of these defined groups are not thought to represent monophyletic taxa, however.

Many RNA viruses probably represent escaped genomic components

Although viruses are now obligate parasites of cellular species, many viruses may once have been cellular components involved in basic cellular functions—that is, they may be "escaped" components of cellular life that now evolve independently of their hosts.

NEGATIVE-SENSE SINGLE-STRANDED RNA VIRUSES   A case in point is a class of viruses whose genome is composed of single-stranded *negative-sense* RNA: RNA that is the complement of the mRNA needed for protein translation. Many of these *negative-sense single-stranded RNA viruses* have only a few genes, including one for an RNA-dependent RNA polymerase that allows them to make mRNA from their negative-sense RNA genome. Modern cellular organisms cannot generate mRNA in this manner (at least in the absence of viral infections), but scientists speculate that single-stranded RNA genomes may have been common in the distant past, before DNA became the primary molecule for genetic information storage.

A self-replicating RNA polymerase gene that began to replicate independently of a cellular genome could conceivably acquire a few additional protein-coding genes through recombination with its host's DNA. If one or more of these genes were to foster the development of a protein coat, the virus might then survive outside the host and infect new hosts. It is believed that this scenario has been repeated many times independently across the tree of life, given that many of the negative-sense single-stranded RNA viruses that infect organisms from bacteria to humans are not closely related to one another. In other words, negative-sense single-stranded RNA viruses do not represent a distinct taxonomic group, but rather exemplify a particular process of cellular escape that probably happened many different times.

(A)

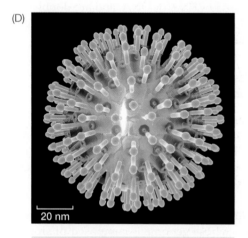

A negative-sense single-stranded RNA virus: The H1N1 influenza A virus prevalent in 2009–2010. Surface view.

(B)

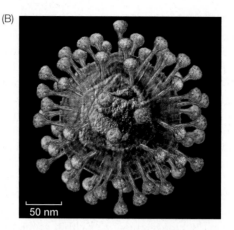

A positive-sense single-stranded RNA virus: Coronavirus of a type thought to be responsible for severe acute respiratory syndrome (SARS). Surface view.

(C)

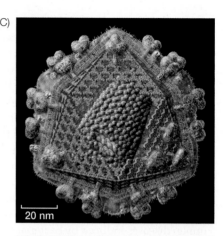

An RNA retrovirus: One of the human immunodeficiency viruses (HIV) that causes AIDS. Cutaway view.

(D)

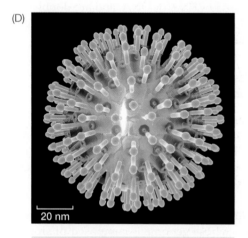

A double-stranded DNA virus: One of the many herpes viruses (Herpesviridae). Different herpes viruses are responsible for many human infections, including chicken pox, shingles, cold sores, and genital herpes (HSV1/2). Surface view.

(E)

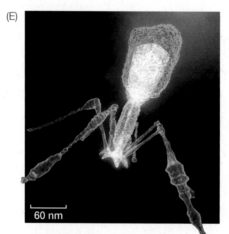

A double-stranded DNA virus: Bacteriophage T4. Viruses that infect bacteria are referred to as bacteriophage (or simply phage). T4 attaches leglike fibers to the outside of its host cell and injects its DNA into the cytoplasm through its "tail" (pink structure in this rendition).

(F)

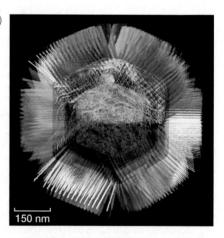

A double-stranded DNA mimivirus: This *Acanthamoeba polyphaga* mimivirus (APMV) has the largest diameter of all known viruses and a genome larger than some prokaryote genomes. Cutaway view.

FIGURE 19.22 Viruses Are Diverse Relatively small genomes and rapid evolutionary rates make it difficult to reconstruct phylogenetic relationships among some classes of viruses. Instead, viruses are classified largely by general characteristics of their genomes. The images here are computer artists' reconstructions based on cryoelectron micrographs.

Familiar examples of negative-sense single-stranded RNA viruses include the viruses that cause measles, mumps, rabies, and influenza (see Figure 19.22A).

POSITIVE-SENSE SINGLE-STRANDED RNA VIRUSES The genome of another type of single-stranded RNA virus is composed of positive-sense RNA. Positive-sense genomes are already set for translation; no replication of the genome to form a complement strand is needed before protein translation can take place.

Positive-sense single-stranded RNA viruses (see Figure 19.22B) are the most abundant and diverse class of viruses. Most of the viruses that cause diseases in crop plants are members of this group. These viruses kill patches of cells in the leaves or stems of plants, leaving live cells amid a patchwork of discolored dead tissue (giving them the name of *mosaic* or *mottle viruses*; **FIGURE 19.23**). Other viruses in this group infect bacteria, fungi, and animals. Human diseases caused by positive-sense single-stranded RNA viruses include polio, hepatitis C, and the common cold. As is true of the other functionally defined groups of viruses, these viruses appear to have evolved multiple times across the tree of life from different groups of cellular ancestors.

APPLY THE CONCEPT

Viruses have evolved many times

When Gram-staining revealed the first mimivirus (discovered living inside an amoeba), it was mistakenly identified as a parasitic Gram-positive bacterium. It was soon discovered that this species was not cellular, however, and it was reclassified as a virus. The graph below shows the genome size and number of protein-coding genes for various species of bacteria. Use the graph to answer the following questions.

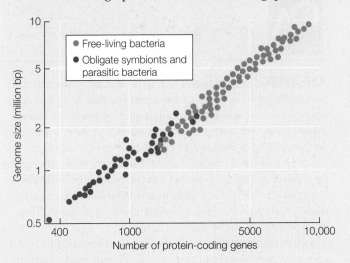

1. What is a likely explanation for the generally smaller genome size and smaller number of protein-coding genes in parasitic bacteria than in most free-living species? Can you form a hypothesis about which species of parasitic bacteria are likely to have existed for the longest time as parasites?

2. The genome of *Acanthamoeba polyphaga mimivirus* (the first discovered mimivirus) is 1,181,404 base pairs, encompassing 911 protein-coding genes. Plot the relevant data for this mimivirus on the graph. How does your result fit with the hypothesis that the mimivirus evolved from a parasitic bacterium?

3. Earlier in this chapter, we described a minute, parasitic group of archaea, the Nanoarchaeota (see Figure 19.18). The genome of one species, *Nanoarchaeum equitans*, is 490,885 base pairs and contains 536 protein-coding genes. Plot the relevant data for this species on the graph. How does the genome of this parasitic archaea species differ from the genomes of the parasitic bacteria? From the genome of the mimivirus?

Yellow areas are dead leaf cells, killed by the mosaic virus.

FIGURE 19.23 Mosaic Viruses Are a Problem for Agriculture Mosaic, or "mottle," viruses are the most diverse class of viruses. This leaf is from an apple tree infected with mosaic virus.

RNA RETROVIRUSES The *RNA retroviruses* are best known as the group that includes the human immunodeficiency viruses (HIV; see Figure 19.22C). Like the previous two categories of viruses, RNA retroviruses have genomes composed of single-stranded RNA and probably evolved as escaped cellular components.

Retroviruses are so named because they regenerate themselves by reverse transcription. When the retrovirus enters the nucleus of its vertebrate host, viral reverse transcriptase produces complementary DNA (cDNA) from the viral RNA genome and then replicates the single-stranded cDNA to produce double-stranded DNA. Another virally encoded enzyme called integrase catalyzes the integration of the new piece of double-stranded viral DNA into the host's genome. The viral genome is then replicated along with the host cell's DNA; the integrated retroviral DNA is known as a *provirus*. Many components of the genomes of cellular species (such as retrotransposons; see Concept 12.3) resemble components of retroviruses.

Retroviruses are only known to infect vertebrates, although genomic elements that resemble portions of these viruses are a component of the genomes of a wide variety of organisms, including bacteria, plants, and many animals. Several retroviruses are associated with the development of various forms of cancer, as cells infected with these viruses are likely to undergo uncontrolled replication.

DOUBLE-STRANDED RNA VIRUSES   *Double-stranded RNA viruses* may have evolved repeatedly from single-stranded RNA ancestors—or perhaps vice versa. These viruses, which are not closely related to one another, infect organisms from throughout the tree of life. Many plant diseases are caused by double-stranded RNA viruses. Other viruses of this type cause many cases of infant diarrhea in humans.

Some DNA viruses may have evolved from reduced cellular organisms

Another class of viruses is composed of viruses that have a double-stranded DNA genome (see Figure 19.22D–F). This group is also almost certainly polyphyletic (with many independent origins). Many of the common phage that infect bacteria (*bacteriophage*) are double-stranded DNA viruses, as are the viruses that cause smallpox and herpes in humans.

Some biologists think that at least some of the DNA viruses may represent highly reduced parasitic organisms that have lost their cellular structure as well as their ability to survive as free-living species. For example, the *mimiviruses*, which are some of the largest DNA viruses (see Figure 19.22F), have a genome in excess of a million base pairs of DNA that encode over 900 proteins. This genome is similar in size to the genomes of many parasitic bacteria and about twice as large as the genome of the smallest bacteria. Phylogenetic analyses of these DNA viruses suggest that they have evolved repeatedly from cellular organisms. Furthermore, recombination among different viruses may have allowed the exchange of various genetic modules, further complicating the history and origins of these viruses.

Do You Understand Concept 19.4?

- Why is it difficult to place viruses precisely within the tree of life?
- What are the two main hypotheses of virus origins?

It appears that the enormous diversity of viruses is, at least in part, a result of their multiple origins from many different cellular organisms. It may be best to view viruses as spinoffs from the various branches on the tree of life—sometimes evolving independently of cellular genomes, sometimes recombining with them. One way to think of viruses is as the "bark" on the tree of life: certainly an important component all across the tree, but not quite like the main branches.

Q&A QUESTION   How do *Vibrio* populations detect when they are dense enough to produce bioluminescence?

ANSWER   Bacteria release chemical substances that are sensed by other bacteria of the same species. They can announce their availability for conjugation by means of such signals. They can also monitor the density of their population. As the density of bacteria in a particular region increases, the concentration of a chemical signal builds up. When the bacteria sense that their population has become sufficiently dense, they can commence activities that smaller densities could not manage, such as forming a biofilm (see Figure 19.19). This phenomenon is called *quorum sensing*.

In the case of *Vibrio*, bioluminescence requires a critical concentration of a chemical signal produced by the bacteria. At low densities, there is not enough of this substance to produce bioluminescence, so free-living bacteria do not glow. When a colony of bacteria is growing on phytoplankton, however, concentrations of this substance are high enough to produce light at a rate of about 10^3 photons per second per cell. This light does more than amaze human observers: by attracting fish that consume the phytoplankton on which the bacteria are growing, it gets the bacteria into a new host.

---

19   SUMMARY

concept 19.1   Life Consists of Three Domains That Share a Common Ancestor

- Two of life's three domains, Bacteria and Archaea, are prokaryotic. They are distinguished from Eukarya in several ways, including their lack of a nucleus and of membrane-enclosed organelles. **Review Table 19.1**

- Eukaryotes are related to both Archaea and Bacteria and appear to have formed through endosymbiosis between members of these two lineages. The last common ancestor of all three domains probably lived about 3 billion years ago. **Review Figure 19.1 and ANIMATED TUTORIAL 19.1**

- Early attempts to classify prokaryotes were hampered by the organisms' small size and the difficulties of growing them in pure culture.

- The cell walls of almost all bacteria contain **peptidoglycan**, whereas the cell walls of archaea and eukaryotes lack peptidoglycan.

- Bacteria can be differentiated into two groups by the **Gram stain**. **Gram-negative bacteria** have a periplasmic space between the plasma membrane and a distinct outer membrane. **Gram-positive bacteria** have a thick cell wall containing about five times as much peptidoglycan as a Gram-negative wall. **Review Figure 19.2 and WEB ACTIVITY 19.1**

- The three most common bacterial morphologies are **cocci** (spheres), **bacilli** (rods), and **helices** (spirals). **Review Figure 19.3**

- The cells of some bacteria aggregate, forming multicellular colonies.

- Phylogenetic classification of prokaryotes is now based principally on the nucleotide sequences of rRNA and other genes involved in fundamental cellular processes.

- Prokaryotes reproduce asexually by binary fission, but may exchange genetic material. Reproduction and genetic exchange are not directly linked in prokaryotes.

- Although lateral gene transfer has occurred throughout prokaryotic evolutionary history, elucidation of prokaryote phylogeny is still possible. **Review Figure 19.4**

concept 19.2 Prokaryote Diversity Reflects the Ancient Origins of Life

- Prokaryotes are the most numerous organisms on Earth, only a small fraction of prokaryote diversity has been characterized to date.

- Some **high-GC Gram-positives** produce important antibiotics.

- The **low-GC Gram-positives** include the **mycoplasmas**, which are among the smallest cellular organisms ever discovered.

- The photosynthetic **cyanobacteria** release oxygen into the atmosphere. Cyanobacteria may live free as single cells or associate in multicellular colonies. **Review Figure 19.9**

- **Spirochetes** have unique structures called axial filaments that allow them to move in a corkscrew-like manner. **Review Figure 19.10**

- The **proteobacteria** embrace the largest number of known species of bacteria. Smaller groups include the **hyperthermophilic bacteria**, **hadobacteria**, and **chlamydias**.

- The best-studied groups of Archaea are **Crenarchaeota** and **Euryarchaeota**.

- Many archaea are **extremophiles**. Most known crenarchaeotes are thermophilic, acidophilic, or both. Some euryarchaeotes are **methanogens**; **extreme halophiles** are also found among the euryarchaeotes. **Review Figure 19.14**

- Ether linkages in the branched long-chain hydrocarbons of the lipids that make up their cell membranes are a synapomorphy of Archaea. **Review Figure 19.15**

concept 19.3 Ecological Communities Depend on Prokaryotes

- Prokaryotes form complex communities, of which **biofilms** are one example. **Review Figure 19.19**

- Prokaryote metabolism is very diverse. Some prokaryotes are anaerobic, others are aerobic, and still others can shift between these modes.

- Prokaryotes fall into four broad nutritional categories: **photoautotrophs**, **photoheterotrophs**, **chemolithotrophs**, and **chemoheterotrophs**. **Review Table 19.2**

- Prokaryotes play key roles in the cycling of elements such as nitrogen, oxygen, sulfur, and carbon. One such role is as decomposers of dead organisms.

- Some prokaryotes metabolize sulfur or nitrogen. **Nitrogen fixers** convert nitrogen gas into a form that organisms can metabolize. **Nitrifiers** convert that nitrogen into forms that can be used by plants, and **denitrifiers** return nitrogen to the atmosphere.

- Prokaryotes inhabit the guts of many animals (including humans) and help them digest food.

- **Koch's postulates** establish the criteria by which an organism may be classified as a pathogen. Relatively few bacteria—and no archaea—are known to be pathogens. **Review Figure 19.21**

concept 19.4 Viruses Have Evolved Many Times

- Viruses have evolved many times from many different groups of cellular organisms. They do not represent a single taxonomic group.

- Some viruses are probably derived from escaped genetic elements of cellular species; others are thought to have evolved as highly reduced parasites.

- Viruses are categorized by the nature of their genomes.

The Origin and Diversification of Eukaryotes

In summer 2005, a devastating red tide crippled the shellfish industry along the Atlantic coast of North America from Canada to Massachusetts. This red tide was produced by a bloom of dino-flagellates of the genus *Alexandrium*. These protists produce a powerful toxin that accumulates in clams, mussels, and oysters. A person who eats a mollusk contaminated with the toxin can experience a syndrome known as paralytic shellfish poisoning. The losses to the shellfish industry in 2005 were estimated at $50 million, and many people were sickened by eating mollusks that were harvested before the problem was diagnosed.

Several species of dinoflagellates produce toxic red tides in many parts of the world. Along the Gulf of Mexico, red tides caused by dinoflagellates of the genus *Karenia* produce a neurotoxin that affects the central nervous systems of fish, which become paralyzed and cannot respire effectively. Huge numbers of dead fish wash up on Gulf Coast beaches during a *Karenia* red tide. In addition, wave action can produce aerosols of the *Karenia* toxin, and these aerosols often cause asthma-like symptoms in humans on shore.

After the losses that resulted from the 2005 red tide, biologists at the Woods Hole Oceanographic Institution (WHOI) on Cape Cod began to monitor and model dinoflagellate populations off the New England coast. If biologists could accurately forecast future blooms, people in the area could be made aware of the problem in advance and adjust the shellfish harvest (and their eating habits) accordingly.

Biologists from WHOI monitored counts of dinoflagellates in the water and in seafloor sediments. They also monitored river runoff, water currents, water temperature and salinity, winds, and tides. An additional environmental factor was the "nor'easter" storms common along the New England coast. By correlating their measurements of these environmental factors with dinoflagellate counts, biologists produced a model that predicted growth of dinoflagellate populations.

In spring 2008, the WHOI team determined that all the factors were in place to produce another red tide like the one of 2005—if a nor'easter occurred to blow the dinoflagellates toward the coast. A nor'easter did occur at just the wrong time, and another red tide materialized in summer 2008, just as predicted. But this time, people were warned. Shellfish harvesters adjusted their harvest, and many fewer people were harmed by eating toxic mollusks.

Blooms of dinoflagellates can cause toxic red tides, such as this one in the Bountiful Islands off the coast of Queensland, Australia.

KEY CONCEPTS

20.1 Eukaryotes Acquired Features from Both Archaea and Bacteria

20.2 Major Lineages of Eukaryotes Diversified in the Precambrian

20.3 Protists Reproduce Sexually and Asexually

20.4 Protists Are Critical Components of Many Ecosystems

QUESTION

Red tides are harmful, but can dinoflagellates also be beneficial to marine ecosystems?

concept 20.1 Eukaryotes Acquired Features from Both Archaea and Bacteria

We easily recognize trees, mushrooms, and insects as plants, fungi, and animals, respectively. But there is a dazzling assortment of other eukaryotic organisms—mostly microscopic—that do not fit into these three groups. Eukaryotes that are not plants, animals, or fungi have traditionally been called **protists**. But phylogenetic analyses reveal that many of the groups we commonly refer to as protists are not, in fact, closely related. Thus the term "protist" does not describe a formal taxonomic group, but is a convenience term for "all the eukaryotes that are not plants, animals, or fungi."

The unique characteristics of the eukaryotic cell lead scientists to conclude that the eukaryotes are monophyletic, and that a single eukaryotic ancestor diversified into the many different protist lineages as well as giving rise to the plants, fungi, and animals. As we saw in Concept 19.1, eukaryotes are generally thought to be more closely related to Archaea than to Bacteria. The mitochondria and chloroplasts of eukaryotes, however, are clearly derived from bacterial lineages (see Figure 19.1).

Traditionally, biologists have hypothesized that the *split* of Eukarya from Archaea was followed by the endosymbioses with bacterial lineages that led to the origin of mitochondria and chloroplasts. Some biologists prefer to view the origin of eukaryotes as the *fusion* of lineages from the two prokaryote groups. This difference is largely a semantic one that hinges on the subjective point at which we deem the eukaryote lineage to have become definitively "eukaryotic." In either case, we can make some reasonable inferences about the events that led to the evolution of a new cell type, bearing in mind that the environment underwent an enormous change—from low to high availability of free atmospheric oxygen—during the course of these events.

The modern eukaryotic cell arose in several steps

Several events were important in the origin of the modern eukaryotic cell (**FIGURE 20.1**):

- The origin of a flexible cell surface
- The origin of a cytoskeleton

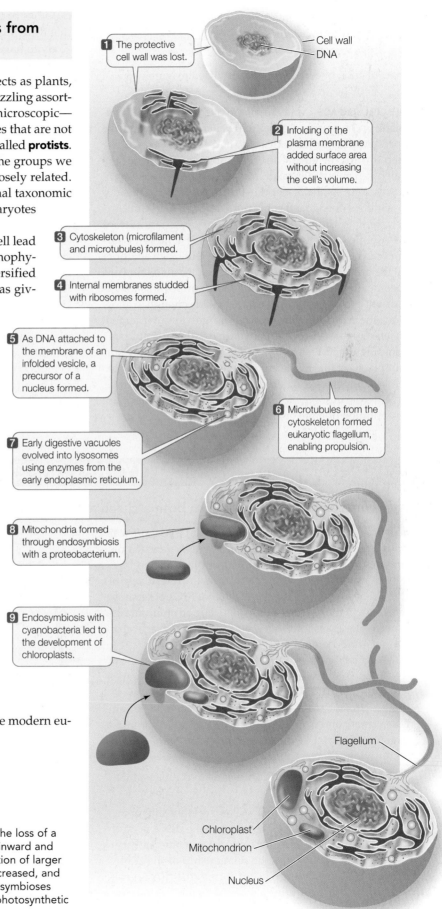

1 The protective cell wall was lost.

Cell wall
DNA

2 Infolding of the plasma membrane added surface area without increasing the cell's volume.

3 Cytoskeleton (microfilament and microtubules) formed.

4 Internal membranes studded with ribosomes formed.

5 As DNA attached to the membrane of an infolded vesicle, a precursor of a nucleus formed.

6 Microtubules from the cytoskeleton formed eukaryotic flagellum, enabling propulsion.

7 Early digestive vacuoles evolved into lysosomes using enzymes from the early endoplasmic reticulum.

8 Mitochondria formed through endosymbiosis with a proteobacterium.

9 Endosymbiosis with cyanobacteria led to the development of chloroplasts.

Flagellum

Chloroplast
Mitochondrion

Nucleus

FIGURE 20.1 Evolution of the Eukaryotic Cell The loss of a rigid cell wall allowed the plasma membrane to fold inward and create more surface area, which facilitated the evolution of larger cells. As cells grew larger, cytoskeletal complexity increased, and the cell became increasing compartmentalized. Endosymbioses involving bacteria gave rise to mitochondria and (in photosynthetic eukaryotes) to chloroplasts.

- The origin of a nuclear envelope, which enclosed a genome organized into chromosomes
- The appearance of *digestive vacuoles*
- The acquisition of certain organelles via endosymbiosis

FLEXIBLE CELL SURFACE Many fossil prokaryotes look like modern bacilli, and we presume that these ancient organisms, like most present-day prokaryotic cells, had firm cell walls. The first step toward the eukaryotic condition was the loss of the cell wall by an ancestral prokaryotic cell. This wall-less condition is present in some present-day prokaryotes. Let's consider the possibilities open to a flexible cell without a firm wall.

First, think of cell size. As a cell grows larger, its surface area-to-volume ratio decreases (see Figure 4.2). Unless the surface area can be increased, the cell volume will reach an upper limit. If the cell's surface is flexible, however, it can fold inward and become more elaborate, creating more surface area for gas and nutrient exchange. With a surface flexible enough to allow infolding, the cell can exchange materials with its environment rapidly enough to sustain a larger volume and more rapid metabolism (Figure 20.1, steps 1–2). Furthermore, a flexible surface can pinch off bits of the environment, bringing them into the cell by endocytosis. These infoldings of the cell surface, which also exist in some modern prokaryotes, were important for the evolution of large eukaryotic cells.

CHANGES IN CELL STRUCTURE AND FUNCTION Other early steps that were important for the evolution of the eukaryotic cell are likely to have included three advances: the formation of ribosome-studded internal membranes, some of which surrounded the DNA; development of a more complex cytoskeleton; and the evolution of digestive vacuoles (Figure 20.1, steps 3–7).

Until a few years ago, biologists thought that cytoskeletons were restricted to eukaryotes. Improved imaging technology and molecular analyses have now revealed homologs of many cytoskeletal proteins in prokaryotes, so simple cytoskeletons evolved before the origin of eukaryotes. The cytoskeleton of a eukaryote, however, is much more developed and complex than that of a prokaryote. This greater development of microfilaments and microtubules supports the eukaryotic cell and allows it to manage changes in shape, to distribute daughter chromosomes, and to move materials from one part of its larger cell to other parts. In addition, the presence of microtubules in the cytoskeleton could have given rise in some cells to the characteristic eukaryotic flagellum.

The DNA of a prokaryotic cell is attached to a site on its plasma membrane. If that region of the plasma membrane were to fold into the cell, the first step would be taken toward the evolution of a *nucleus*, a primary feature of the eukaryotic cell.

The nuclear envelope appeared early in the eukaryote lineage. The next step was probably *phagocytosis*—the ability to engulf and digest other cells. Early eukaryotes may also have

APPLY THE CONCEPT

Eukaryotes acquired features from both archaea and bacteria

Ribosomal RNA (rRNA) genes are present in the nuclear genome of eukaryotes. There are also rRNA genes in the genomes of mitochondria and chloroplasts. Therefore, photosynthetic eukaryotes have three different sets of rRNA genes, which encode the structural RNA of separate ribosomes in the nucleus, mitochondria, and chloroplasts, respectively. Translation of each genome takes place on its own set of ribosomes.

The gene tree shows the evolutionary relationships of rRNA gene sequences isolated from the nuclear genomes of humans, yeast, and corn; from an archaeon (*Halobacterium*), a proteobacterium (*E. coli*), and a cyanobacterium (*Chlorobium*); and from the mitochondrial and chloroplast genomes of corn. Use the gene tree to answer the following questions.

1. Why aren't the three rRNA genes of corn one another's closest relatives?

2. How would you explain the closer relationship of the mitochondrial rRNA gene of corn to the rRNA gene of *E. coli* than to the nuclear rRNA genes of other eukaryotes? Can you explain the relationship of the rRNA gene from the chloroplast of corn to the rRNA gene of the cyanobacterium?

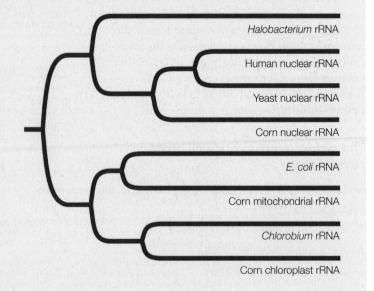

3. If you were to sequence the rRNA genes from human and yeast mitochondrial genomes, where would you expect these two sequences to fit on the gene tree?

had an associated endoplasmic reticulum and Golgi apparatus, and perhaps one or more flagella of the eukaryotic type.

ENDOSYMBIOSIS At the same time the processes outlined above were taking place, cyanobacteria were generating O_2 as a product of photosynthesis. The increasing concentrations of O_2 in the atmosphere had disastrous consequences for most organisms of the time, which were unable to tolerate the newly oxidizing environment. But some prokaryotes managed to cope with these changes, and—fortunately for us—so did some of the new phagocytic eukaryotes.

At about this time, endosymbioses began to play a role in eukaryote evolution (Figure 20.1, steps 8–9). The theory of endosymbiosis proposes that certain organelles are the descendants of prokaryotes engulfed, but not digested, by ancient eukaryotic cells. One crucial event in the history of eukaryotes was the incorporation of a proteobacterium that evolved into the mitochondrion. Initially, the new organelle's primary function was probably to detoxify O_2 by reducing it to water. Later, this reduction became coupled with the formation of ATP in cellular respiration. Upon completion of this step, the essential modern eukaryotic cell was complete.

> **LINK** You may wish to review the reactions of cellular respiration in Concept 6.2

Photosynthetic eukaryotes are the result of yet another endosymbiotic step: the incorporation of a prokaryote related to today's cyanobacteria, which became the chloroplast.

Chloroplasts have been transferred among eukaryotes several times

Eukaryotes in several different groups possess chloroplasts, and groups with chloroplasts appear in several distantly related eukaryote clades. Some of these groups differ in the photosynthetic pigments their chloroplasts contain. And not all chloroplasts are limited to a pair of surrounding membranes—in some microbial eukaryotes, they are surrounded by *three or more* membranes. We now view these observations as evidence of a remarkable series of endosymbioses. This conclusion is supported by extensive evidence from electron microscopy and nucleic acid sequence comparisons.

All chloroplasts trace their ancestry back to the engulfment of one cyanobacterium by a larger eukaryotic cell. This event, the step that first gave rise to the photosynthetic eukaryotes, is known as **primary endosymbiosis** (**FIGURE 20.2A**). The cyanobacterium, a Gram-negative bacterium, had both an inner and an outer membrane (see Figure 19.2B). Thus the original chloroplasts had two surrounding membranes: the inner and outer membranes of the cyanobacterium. Remnants of the peptidoglycan-containing cell wall of the bacterium are present

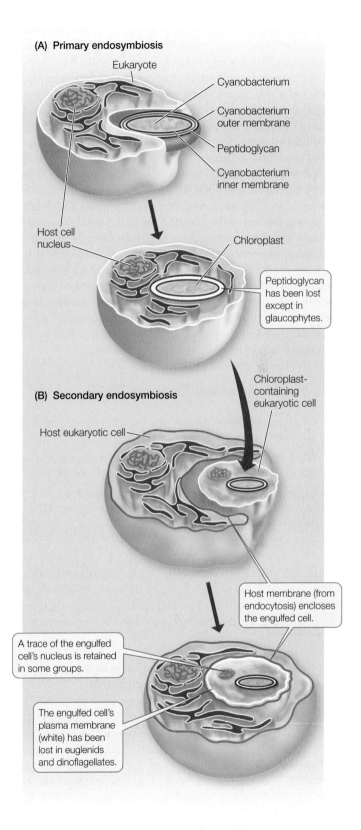

(A) Primary endosymbiosis

- Eukaryote
- Cyanobacterium
- Cyanobacterium outer membrane
- Peptidoglycan
- Cyanobacterium inner membrane
- Host cell nucleus
- Chloroplast

Peptidoglycan has been lost except in glaucophytes.

(B) Secondary endosymbiosis

Chloroplast-containing eukaryotic cell

Host eukaryotic cell

Host membrane (from endocytosis) encloses the engulfed cell.

A trace of the engulfed cell's nucleus is retained in some groups.

The engulfed cell's plasma membrane (white) has been lost in euglenids and dinoflagellates.

FIGURE 20.2 Endosymbiotic Events in the Evolution of Chloroplasts (A) A single instance of primary endosymbiosis ultimately gave rise to all of today's chloroplasts. (B) Secondary endosymbiosis—the uptake and retention of a chloroplast-containing cell by another eukaryotic cell—took place several times, independently.

in the form of a bit of peptidoglycan between the chloroplast membranes of *glaucophytes*, the first eukaryote group to branch off following primary endosymbiosis (as we will see in Chapter 21). Primary endosymbiosis also gave rise to the chloroplasts of the red algae, green algae, and land plants. The red algal chloroplast retains certain pigments of the original cyanobacterial endosymbiont that are absent in green algal chloroplasts.

Almost all remaining photosynthetic eukaryotes are the result of additional rounds of endosymbiosis. For example, the photosynthetic *euglenids* derived their chloroplasts from **secondary endosymbiosis (FIGURE 20.2B)**. Their ancestor took up a unicellular green alga, retaining its chloroplast and eventually losing the rest of the constituents of the alga. This history explains why the photosynthetic euglenids have the same photosynthetic pigments as the green algae and land plants. It also accounts for the third membrane of the euglenid chloroplast, which is derived from the euglenid's plasma membrane (as a result of endocytosis). An additional round—**tertiary endosymbiosis**—occurred when a dinoflagellate apparently lost its chloroplast and took up another protist that had acquired its chloroplast through secondary endosymbiosis.

Do You Understand Concept 20.1?

- Why was the development of a flexible cell surface a key event for eukaryote evolution?
- What do you consider the most critical events that led to the evolution of the eukaryotic cell? Why?
- Explain how increased availability of atmospheric oxygen (O_2) could have impacted the evolution of the eukaryotic cell.

The features that eukaryotes gained from archaea and bacteria have allowed them to exploit many different environments. This led to the evolution of great diversity among eukaryotes, beginning with a radiation that started in the Precambrian.

concept 20.2 | Major Lineages of Eukaryotes Diversified in the Precambrian

Most eukaryotes can be classified in one of eight major clades that began to diversify about 1.5 billion years ago: alveolates, excavates, stramenopiles, plants, rhizaria, amoebozoans, fungi, and animals (**FIGURE 20.3**). Plants, fungi, and animals each have close protist relatives (such as the choanoflagellate relatives of animals), which we will discuss along with those major multicellular eukaryote groups in Chapters 21–23.

Each of the five major groups of protistan eukaryotes covered in this chapter consists of organisms with enormously diverse body forms and nutritional lifestyles. Some protists are motile, whereas others do not move; some are photosynthetic, others heterotrophic; most are unicellular, but some are multicellular. Most are microscopic, but a few are huge (giant kelps, for example, can grow to half the length of a football field). We refer to the unicellular species of protists as *microbial eukaryotes*, but keep in mind that there are large, multicellular protists as well.

Biologists used to classify protists largely on the basis of their life histories and reproductive features (see Concept 20.3). In recent years, however, electron microscopy and gene sequencing have revealed many new patterns of evolutionary relatedness among these groups. Analyses of slowly evolving gene sequences are making it possible to explore evolutionary relationships among eukaryotes in ever greater detail and with greater confidence. Nonetheless, some substantial areas of uncertainty remain, and lateral gene transfer may complicate efforts to reconstruct the evolutionary history of protists (as was also true for prokaryotes; see Concept 19.1). Today we recognize great diversity among the many distantly related protist clades.

Alveolates have sacs under their plasma membranes

Alveolates are so named because they possess sacs called *alveoli* just beneath their plasma membranes, which may play a role in supporting the cell surface. All alveolates are unicellular, and most are photosynthetic, but they are diverse in body form. The groups considered in detail here are the *dinoflagellates*, *apicomplexans*, and *ciliates*.

DINOFLAGELLATES Most **dinoflagellates** are marine and photosynthetic; they are important primary producers of organic matter in the oceans. The dinoflagellates are of great ecological, evolutionary, and morphological interest. A distinctive mixture of photosynthetic and accessory pigments gives their chloroplasts a golden brown color. Some dinoflagellate species cause red tides, as discussed at the start of this chapter. Other species are photosynthetic endosymbionts that live within the cells of other organisms, including invertebrate animals (such as corals) and other marine protists (see Figures 20.13A and 20.19). Some are nonphotosynthetic and live as parasites within other marine organisms.

Dinoflagellates have a distinctive appearance. They generally have two flagella, one in an equatorial groove around the cell, the other starting near the same point as the first and passing down a longitudinal groove before extending into the surrounding medium (**FIGURE 20.4**). Some dinoflagellates can take on different forms, including amoeboid ones, depending on environmental conditions. It has been claimed that the dinoflagellate *Pfiesteria piscicida* can occur in at least two dozen distinct forms, although this claim is highly controversial. In any case, this remarkable dinoflagellate is harmful to fish and can, when present in great numbers, both stun and feed on them.

APICOMPLEXANS The exclusively parasitic **apicomplexans** derive their name from the *apical complex*, a mass of organelles contained in the *apical* end (the tip) of the cell. These organelles help the apicomplexan invade its host's tissues. For example, the apical complex enables *Plasmodium*, the causative agent of malaria, to enter its target cells in the human body after transmission by a mosquito (see p. 403).

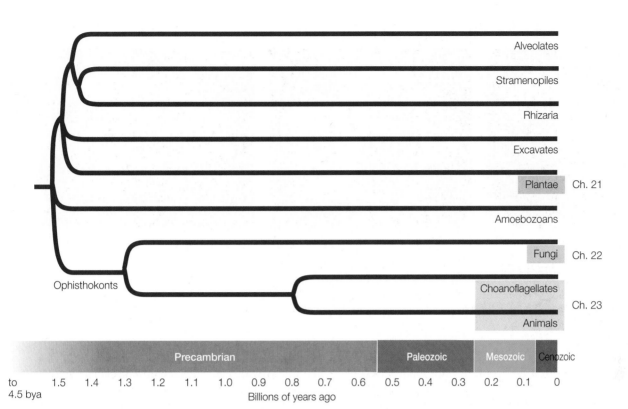

FIGURE 20.3 Precambrian Divergence of Major Eukaryote Groups A phylogenetic tree shows one current hypothesis and estimated time line for the origin of the major groups of eukaryotes. The rapid divergence of major lineages between 1.5 and 1.4 bya makes reconstruction of their precise relationships difficult. The major multicellular groups (tinted boxes) will be covered in subsequent chapters.

Like many obligate parasites, apicomplexans have elaborate life cycles featuring asexual and sexual reproduction by a series of very dissimilar life stages. In many species, these life stages are associated with two different types of host organisms, as is the case with *Plasmodium*. Another apicomplexan, *Toxoplasma*, alternates between cats and rats to complete its life cycle. A rat infected with *Toxoplasma* loses its fear of cats, which makes it more likely to be eaten by, and thus transfer the parasite to, a cat.

FRONTIERS Apicomplexans contain a much-reduced chloroplast, derived from secondary endosymbiosis of a red alga, that no longer has a photosynthetic function. Researchers are targeting this chloroplast in *Plasmodium* as the site of attack for a future antimalarial drug. Its discovery is viewed as an important advance in the effort to treat malaria victims, as existing medications only reduce the chance of initial infection and do not cure infected individuals.

CILIATES The **ciliates** are named for their numerous hairlike cilia, which are shorter than, but otherwise identical to, eukaryotic flagella. The ciliates are much more complex in body form than are most other unicellular eukaryotes (**FIGURE 20.5**). Their definitive characteristic is the possession of two types of nuclei (whose roles we will describe in Concept 20.3 when we discuss protist reproduction). Almost all ciliates

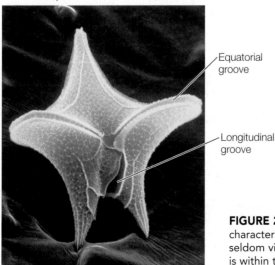

Peridinium sp.

Equatorial groove

Longitudinal groove

20 μm

FIGURE 20.4 A Dinoflagellate The presence of two flagella is characteristic of many dinoflagellates, although these appendages are seldom visible, being contained within deep grooves. One flagellum is within the equatorial groove and provides forward thrust and spin to the organism. The second flagellum originates in the longitudinal groove and acts like the rudder of a boat.

(A) *Paramecium* sp.

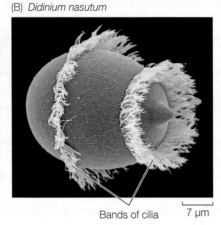

Cilia   10 μm

(B) *Didinium nasutum*

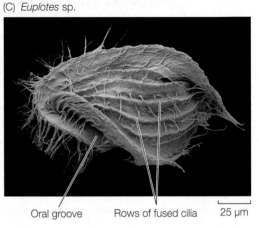

Bands of cilia   7 μm

(C) *Euplotes* sp.

Oral groove   Rows of fused cilia   25 μm

FIGURE 20.5 Diversity among the Ciliates   (A) A free-swimming organism, this *Para-mecium* belongs to a ciliate group whose members have many cilia of uniform length. (B) The barrel-shaped *Didinium nasutum* feeds on other ciliates, including *Paramecium*. Their cilia occur in two separate bands. (C) Some of the cilia in *Euplotes* fuse into flat sheets that direct food particles into an oral groove.

are heterotrophic, although a few contain photosynthetic endosymbionts.

Paramecium, a frequently studied ciliate genus, exemplifies the complex structure and behavior of ciliates (**FIGURE 20.6**). The slipper-shaped cell is covered by an elaborate *pellicle*, a structure composed principally of an outer membrane and an inner layer of closely packed, membrane-enclosed sacs (the alveoli) that surround the bases of the cilia. Defensive organelles called *trichocysts* are also present in the pellicle. In response to a threat, a microscopic explosion expels the trichocysts in a few milliseconds, and they emerge as sharp darts, driven forward at the tip of a long, expanding filament.

The cilia provide *Paramecium* with a form of locomotion that is generally more precise than locomotion by flagella or pseudopods. A *Paramecium* can coordinate the beating of its cilia to propel itself either forward or backward in a spiraling manner. It can also back off swiftly when it encounters a barrier or a negative stimulus. The coordination of ciliary beating is probably the result of a differential distribution of ion channels in the plasma membrane near the two ends of the cell.

Organisms living in fresh water are hypertonic to their environment. Many freshwater protists, including *Paramecium*, address this problem by means of specialized **contractile vacuoles** that excrete the excess water the organisms constantly take in by osmosis. The excess water collects in the contractile vacuoles, which then contract and expel the water from the cell.

Paramecium and many other protists engulf solid food by endocytosis, forming a **digestive vacuole** within which the food is digested. Smaller vesicles containing digested food pinch away from the digestive vacuole and enter the cytoplasm. These tiny vesicles provide a large surface area across which the products of digestion can be absorbed by the rest of the cell.

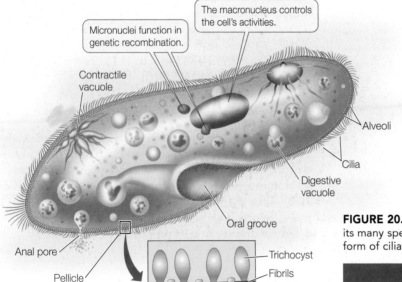

Micronuclei function in genetic recombination.

The macronucleus controls the cell's activities.

Contractile vacuole

Alveoli

Cilia

Digestive vacuole

Oral groove

Anal pore

Pellicle

Trichocyst

Fibrils

Alveolus

Cilium

FIGURE 20.6 Anatomy of a *Paramecium*   A *Paramecium*, with its many specialized organelles, exemplifies the complex body form of ciliates.

yourBioPortal.com

Go to WEB ACTIVITY 20.1
Anatomy of a *Paramecium*

and ANIMATED TUTORIAL 20.2
Digestive Vacuoles

(A) *Giardia muris*

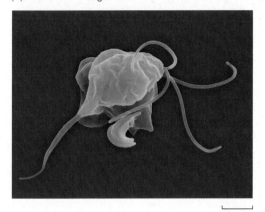

2 μm

(B) *Trichomonas vaginalis*

2 μm

FIGURE 20.7 Some Excavate Groups Lack Mitochondria (A) *Giardia*, a diplomonad, has flagella and two nuclei. (B) *Trichomonas*, a parabasalid, has flagella and undulating membranes. Neither of these organisms possesses mitochondria.

> **FRONTIERS** *Giardia* infections usually result from drinking contaminated water. In the United States, these infections are most common in hikers and campers using spring or stream water in recreational areas. Recent research shows that *Giardia* can switch surface proteins to avoid detection by the host immune system, which explains why the human body has a difficult time clearing *Giardia* infections without drug treatment.

Excavates began to diversify about 1.5 billion years ago

The **excavates** include a number of diverse groups that began to split from one another soon after the origin of eukaryotes. Several groups of excavates lack mitochondria, an absence that once led to the view that these groups might represent early-diverging eukaryotes that diversified before the evolution of mitochondria. However, the discovery of nuclear genes normally associated with mitochondria in these organisms suggests that the absence of mitochondria is a derived condition. In other words, ancestors of these excavate groups probably possessed mitochondria that were lost or reduced over the course of evolution. The existence of these organisms today shows that eukaryotic life is possible without mitochondria.

DIPLOMONADS AND PARABASALIDS The **diplomonads** and **parabasalids** are unicellular and lack mitochondria. *Giardia lamblia*, a diplomonad, is a familiar parasite that contaminates water supplies and causes the intestinal disease giardiasis. This tiny organism contains two nuclei bounded by nuclear envelopes, and it has a cytoskeleton and multiple flagella (**FIGURE 20.7A**).

In addition to flagella and a cytoskeleton, the parabasalids have undulating membranes that also contribute to the cell's locomotion. *Trichomonas vaginalis* (**FIGURE 20.7B**) is a parabasalid responsible for a sexually transmitted disease in humans. Infection of the male urethra, where it may occur without symptoms, is less common than infection of the vagina.

HETEROLOBOSEANS The amoeboid body form appears in several protist groups that are only distantly related to one another. The body forms of **heteroloboseans**, for example, resemble those of loboseans, an amoebozoan group that is not at all closely related to heteroloboseans (see p. 399). Amoebas of the free-living heterolobosean genus *Naegleria*, some of which can enter the human body and cause a fatal disease of the nervous system, usually have a two-stage life cycle, in which one stage has amoeboid cells and the other flagellated cells.

EUGLENIDS AND KINETOPLASTIDS The **euglenids** and **kinetoplastids** together constitute a clade of unicellular excavates with flagella. Their mitochondria contain distinctive, disc-shaped cristae, and their flagella contain a crystalline rod not found in other organisms. They reproduce primarily asexually by binary fission.

The flagella of euglenids arise from a pocket at the anterior end of the cell. Spiraling strips of proteins under the plasma membrane control the cell's shape. Some euglenids are photosynthetic.

FIGURE 20.8 depicts a cell of the genus *Euglena*. Like most other euglenids, this common freshwater organism has a

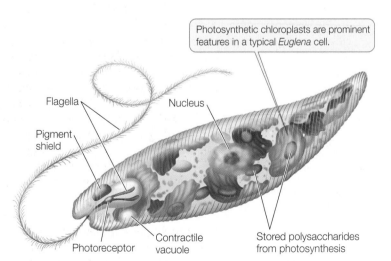

Photosynthetic chloroplasts are prominent features in a typical *Euglena* cell.

Flagella

Nucleus

Pigment shield

Photoreceptor

Contractile vacuole

Stored polysaccharides from photosynthesis

FIGURE 20.8 A Photosynthetic Euglenid In the *Euglena* species illustrated in this drawing, the second flagellum is rudimentary. Note that the primary flagellum originates at the anterior of the organism and trails toward its posterior.

TABLE 20.1	A Comparison of Three Kinetoplastid Trypanosomes		
	TRYPANOSOMA BRUCEI	*TRYPANOSOMA CRUZI*	*LEISHMANIA MAJOR*
Human disease	Sleeping sickness	Chagas' disease	Leishmaniasis
Insect vector	Tsetse fly	Assassin bug	Sand fly
Vaccine or effective cure	None	None	None
Strategy for survival	Changes surface recognition molecules frequently	Causes changes in surface recognition molecules on host cell	Reduces effectiveness of macrophage hosts
Site in human body	Bloodstream; attacks nerve tissue in final stages	Enters cells, especially muscle cells	Enters cells, primarily macrophages
Approximate number of deaths per year	50,000	43,000	60,000

complex cell structure. It propels itself through the water with the longer of its two flagella, which may also serve as an anchor to hold the organism in place. The second flagellum is often rudimentary.

The euglenids have diverse nutritional requirements. Many species are always heterotrophic. Other species are fully autotrophic in sunlight, using chloroplasts to synthesize organic compounds through photosynthesis. When kept in the dark, these euglenids lose their photosynthetic pigment and begin to feed exclusively on dissolved organic material in the water around them. Such a "bleached" *Euglena* resynthesizes its photosynthetic pigment when it is returned to the light and becomes autotrophic again. But *Euglena* cells treated with certain antibiotics or mutagens lose their photosynthetic pigment completely; neither they nor their descendants are ever autotrophs again. However, those descendants function well as heterotrophs.

The kinetoplastids are unicellular parasites with two flagella and a single, large mitochondrion. That mitochondrion contains a *kinetoplast*, a unique structure housing multiple circular DNA molecules and associated proteins. Some of these DNA molecules encode "guide proteins" that edit mRNA within the mitochondrion.

The kinetoplastids include several medically important species of pathogenic *trypanosomes* (**TABLE 20.1**). Some of these organisms are able to change their cell surface recognition molecules frequently, allowing them to evade our best attempts to kill them and thus eradicate the diseases they cause.

> **LINK** The role of a pathogen's cell surface recognition molecules in the mammalian immune response is covered in Chapter 31

Stramenopiles typically have two unequal flagella, one with hairs

A morphological synapomorphy of most **stramenopiles** is the possession of rows of tubular hairs on the longer of their two flagella. Some stramenopiles lack flagella, but they are descended from ancestors that possessed flagella. The stramenopiles include the diatoms and the brown algae, which are photosynthetic, and the oomycetes, which are not.

DIATOMS All of the **diatoms** are unicellular, although some species associate in filaments. Many have sufficient carotenoids in their chloroplasts to give them a yellow or brownish color. All of them synthesize carbohydrates and oils as photosynthetic storage products. Diatoms lack flagella except in male gametes.

Architectural magnificence on a microscopic scale is the hallmark of the diatoms. Almost all diatoms deposit silica (hydrated silicon dioxide) in their cell walls. The cell wall of a diatom is constructed in two pieces, with the top overlapping the bottom like the top of a petri dish. The silica-impregnated walls have intricate patterns unique to each species (**FIGURE 20.9**). Despite their remarkable morphological diversity, all diatoms are symmetrical—either bilaterally (with "right" and

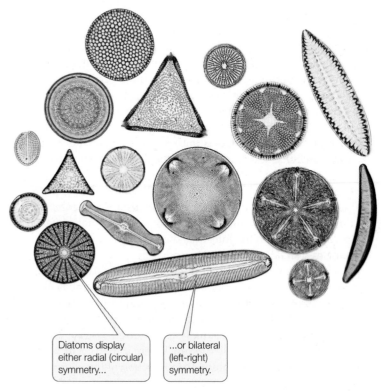

Diatoms display either radial (circular) symmetry...

...or bilateral (left-right) symmetry.

FIGURE 20.9 Diatom Diversity This bright-field micrograph illustrates the variety of species-specific forms found among the diatoms.

(A) *Himanthalia elongata*

(B) *Postelsia palmaeformis*

Holdfasts

FIGURE 20.10 Brown Algae (A) This seaweed illustrates the filamentous growth form of the brown algae. (B) Sea palms exemplify the leaflike growth form of brown algae. Sea palms and many other brown algal species are "glued" to the rocks by tough, branched structures called holdfasts that can withstand the pounding of the surf.

"left" halves) or radially (with the type of symmetry possessed by a circle).

Diatoms reproduce both sexually and asexually. Asexual reproduction by binary fission is somewhat constrained by the stiff cell wall. Both the top and bottom of the "petri dish" become tops of new "dishes" without changing appreciably in size; as a result, the new cell made from the former bottom is smaller than the parent cell. If this process continued indefinitely, one cell line would simply vanish, but sexual reproduction largely solves this potential problem. Gametes are formed, shed their cell walls, and fuse. The resulting zygote then grows substantially in size before a new cell wall is laid down.

Diatoms are found in all the oceans and are frequently present in great numbers. They are major photosynthetic producers in coastal waters and are among the dominant organisms in the dense "blooms" of phytoplankton that occasionally appear in the open ocean (see Concept 20.4). Diatoms are also common in fresh water and even occur on the wet surfaces of terrestrial mosses.

BROWN ALGAE The **brown algae** obtain their namesake color from the carotenoid *fucoxanthin*, which is abundant in their chloroplasts. The combination of this yellow-orange pigment with the green of chlorophylls *a* and *c* yields a brownish tinge. All brown algae are multicellular, and some are extremely large. Giant kelps, such as those of the genus *Macrocystis*, may be up to 60 meters long.

The brown algae are almost exclusively marine. They are composed either of branched filaments (**FIGURE 20.10A**) or of leaflike growths (**FIGURE 20.10B**). Some float in the open ocean; the most famous example is the genus *Sargassum*, which

forms dense mats in the Sargasso Sea in the mid-Atlantic. Most brown algae, however, attach themselves to rocks near the shore. A few thrive only where they are regularly exposed to heavy surf. All of the attached forms develop a specialized structure, called a *holdfast*, that literally glues them to the rocks. The "glue" of the holdfast is *alginic acid*, a gummy polymer of sugar acids found in the walls of many brown algal cells. In addition to its function in holdfasts, it cements algal cells and filaments together. It is harvested and used by humans as an emulsifier in ice cream, cosmetics, and other products.

OOMYCETES The **oomycetes** consist in large part of the water molds and their terrestrial relatives, such as the downy mildews. Water molds are filamentous and stationary. They are *absorptive heterotrophs*—that is, they secrete enzymes that digest large food molecules into smaller molecules that they can absorb. If you have seen a whitish, cottony mold growing on dead fish or dead insects in water, it was probably a water mold of the common genus *Saprolegnia* (**FIGURE 20.11**).

Oomycetes were once classified as fungi. However, we now know that their similarity to fungi is only superficial, and that the oomycetes are more distantly related to the fungi than are many other eukaryote groups, including humans (see Figure 20.3). For example, the cell walls of oomycetes are typically made of cellulose, whereas those of fungi are made of chitin.

The water molds, such as *Saprolegnia*, are all aquatic and **saprobic**—meaning they feed on dead organic matter. Some other oomycetes are terrestrial. Although most of the terrestrial oomycetes are harmless or helpful decomposers of dead

Saprolegnia sp.

2 mm

FIGURE 20.11 An Oomycete The filaments of a water mold radiate from the carcass of an insect.

Elphidium crispum

200 μm

FIGURE 20.12 Building Blocks of Limestone Some foraminiferans secrete calcium carbonate to form shells. The shells of different species have distinctive shapes. Over millions of years, the shells of foraminiferans have accumulated to form limestone deposits.

matter, a few are plant parasites that attack crops such as avocados, grapes, and potatoes.

Rhizaria typically have long, thin pseudopods

The three primary groups of **Rhizaria**—cercozoans, foraminiferans, and radiolarians—are unicellular and mostly aquatic. These organisms typically have long, thin pseudopods that contrast with the broader, lobelike pseudopods of the more familiar amoebozoans. The rhizaria have contributed to ocean sediments, some of which have become terrestrial features in the course of geological history.

CERCOZOANS The **cercozoans** are a diverse group with many forms and habitats. Some are aquatic; others live in soil. One group of cercozoans possesses chloroplasts derived from a green alga by secondary endosymbiosis—and that chloroplast contains a trace of the alga's nucleus.

FORAMINIFERANS Some **foraminiferans** secrete external shells of calcium carbonate (**FIGURE 20.12**). These shells have accumulated over time to produce much of the world's limestone. Some foraminiferans live as plankton; others live on the seafloor. Living foraminiferans have been found at the deepest point in the world's oceans—10,896 meters down in the Challenger Deep in the western Pacific. At that depth, however, they cannot secrete normal shells because the surrounding water is too poor in calcium carbonate.

In living planktonic foraminiferans, long, threadlike, branched pseudopods extend through numerous microscopic apertures in the shell and interconnect to create a sticky, reticulated net, which the foraminiferans use to catch smaller plankton. In some foraminiferan species, the pseudopods provide locomotion.

RADIOLARIANS The **radiolarians** are recognizable by their thin, stiff pseudopods, which are reinforced by microtubules (**FIGURE 20.13A**). These pseudopods greatly increase the surface area of the cell, and they help the cell stay afloat in its marine environment.

Radiolarians are immediately recognizable by their distinctive radial symmetry. Almost all radiolarian species secrete glassy *endoskeletons* (internal skeletons). The skeletons of the different species are as varied as snowflakes, and many have elaborate geometric designs (**FIGURE 20.13B**). A few radiolarians are among the largest of the unicellular eukaryotes, measuring several millimeters across.

Amoebozoans use lobe-shaped pseudopods for locomotion

Amoebozoans appear to have diverged from other eukaryotes about 1.5 billion years ago (see Figure 20.3). It is not yet clear if they are more closely related to opisthokonts (including fungi and animals) or to other major groups of eukaryotes.

The lobe-shaped pseudopods of amoebozoans (**FIGURE 20.14**) are a hallmark of the amoeboid body form. Amoebozoan pseudopods differ in form and function from the slender

(A)

350 μm

(B) *Hexacontium* sp.

50 μm

FIGURE 20.13 Radiolarians Exhibit Distinctive Pseudopods and Radial Symmetry (A) The radiolarians are distinguished by their thin, stiff pseudopods and by their radial symmetry. The green pigmentation seen at the center of this radiolarian's glassy endoskeleton is imparted by endosymbiotic dinoflagellates. (B) The endoskeleton secreted by a radiolarian.

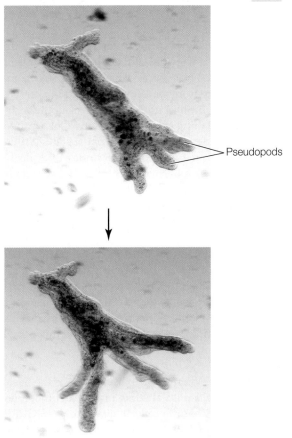

Pseudopods

Nebela collaris

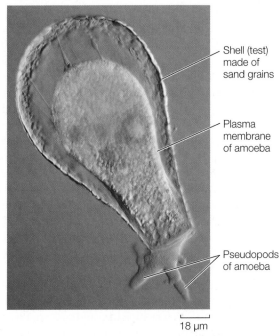

Shell (test) made of sand grains

Plasma membrane of amoeba

Pseudopods of amoeba

18 μm

FIGURE 20.15 Life in a Glass House This testate amoeba has built a lightbulb-shaped shell, or *test*, by gluing sand grains together. Its pseudopods extend through the single aperture in the test.

120 μm

FIGURE 20.14 An Amoeba in Motion The flowing pseudopods of this "chaos amoeba" are constantly changing shape as it moves and feeds.

relatively rich supply of sedentary organisms or organic particles. Most loboseans exist as predators, parasites, or scavengers. Members of one group of loboseans, the testate amoebas, live inside shells. Some of these amoebas produce casings by gluing sand grains together (**FIGURE 20.15**); other testate amoebas have shells secreted by the organism itself.

pseudopods of rhizaria. We consider three amoebozoan groups here: the loboseans and two groups known as slime molds.

LOBOSEANS Loboseans are small amoebozoans that feed on other small organisms and particles of organic matter by phagocytosis, engulfing them with pseudopods. Many loboseans are adapted for life on the bottoms of lakes, ponds, and other bodies of water. Their creeping locomotion and their manner of engulfing food particles fit them for life close to a

PLASMODIAL SLIME MOLDS If the nucleus of an amoeba began rapid mitotic division, accompanied by a tremendous increase in cytoplasm and organelles but no cytokinesis, the resulting organism would resemble the multinucleate mass of a **plasmodial slime mold**. During its vegetative (feeding, nonreproductive) stage, a plasmodial slime mold is a wall-less mass of cytoplasm with numerous diploid nuclei. This mass streams very slowly over its substrate in a remarkable network of strands called a *plasmodium* (**FIGURE 20.16A**). The plasmodium of such a slime mold is an example of a **coenocyte**: many nuclei enclosed in a

(A)

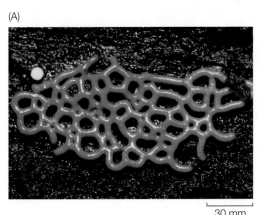

30 mm

(B)

1.5 mm

FIGURE 20.16 Plasmodial Slime Molds (A) The plasmodial form of the slime mold *Hemitrichia serpula* covers rocks, decaying logs, and other objects as it engulfs bacteria and other food items; it is also responsible for its common name of "pretzel mold." (B) Fruiting structures of *Hemitrichia*.

single plasma membrane. The outer cytoplasm of the plasmodium (closest to the environment) is normally less fluid than the interior cytoplasm and thus provides some structural rigidity.

Plasmodial slime molds provide a dramatic example of movement by cytoplasmic streaming. The outer cytoplasmic region of the plasmodium becomes more fluid in places, and cytoplasm rushes into those areas, stretching the plasmodium. This streaming somehow reverses its direction every few minutes as cytoplasm rushes into a new area and drains away from an older one, moving the plasmodium over its substrate. Sometimes an entire wave of plasmodium moves across a surface, leaving strands behind. Microfilaments and a contractile protein called *myxomyosin* interact to produce the streaming movement. As it moves, the plasmodium engulfs food particles by endocytosis—predominantly bacteria, yeasts, spores of fungi, and other small organisms as well as decaying animal and plant remains.

A plasmodial slime mold can grow almost indefinitely in its plasmodial stage as long as the food supply is adequate and other conditions, such as moisture and pH, are favorable. If conditions become unfavorable, however, one of two things can happen. First, the plasmodium can form an irregular mass of hardened cell-like components called a *sclerotium*. This resting structure rapidly becomes a plasmodium again when favorable conditions are restored.

Alternatively, the plasmodium can transform itself into spore-bearing *fruiting structures* (**FIGURE 20.16B**). These stalked or branched structures rise from heaped masses of plasmodium. They derive their rigidity from walls that form and thicken between their nuclei. The diploid nuclei of the plasmodium divide by meiosis as the fruiting structure develops. One or more knobs, called *sporangia*, develop on the end of the stalk. Within a sporangium, haploid nuclei become surrounded by walls to form *spores*. Eventually, as the fruiting structure dries, it sheds its spores.

The spores germinate into wall-less, haploid cells called *swarm cells*, which can either divide mitotically to produce more haploid swarm cells or function as gametes. Swarm cells can live as separate individual cells that move by means of flagella or pseudopods, or they can become walled and resistant *resting cysts* when conditions are unfavorable; when conditions improve again, the cysts release swarm cells. Two swarm cells can also fuse to form a diploid zygote, which divides by mitosis (but without a wall forming between the nuclei) and thus forms a new, coenocytic plasmodium.

CELLULAR SLIME MOLDS Whereas the plasmodium is the basic vegetative unit of the plasmodial slime molds, an amoeboid cell is the vegetative unit of the **cellular slime molds** (**FIGURE 20.17**). Large numbers of cells called *myxamoebas*, which have

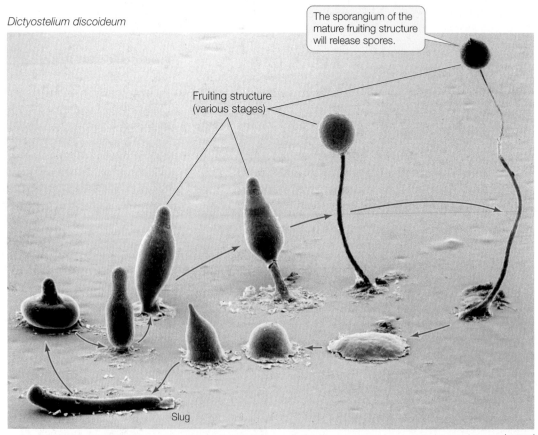

Dictyostelium discoideum

The sporangium of the mature fruiting structure will release spores.

Fruiting structure (various stages)

Slug

FIGURE 20.17 A Cellular Slime Mold This composite micrograph shows the life cycle of the slime mold *Dictyostelium*.

0.25 mm

single haploid nuclei, engulf bacteria and other food particles by endocytosis and reproduce by mitosis and fission. This simple life cycle stage, consisting of swarms of independent, isolated cells, can persist indefinitely as long as food and moisture are available.

When conditions become unfavorable, the cellular slime molds aggregate and form fruiting structures, as do their plasmodial counterparts. The individual myxamoebas aggregate into a mass called a *slug* or *pseudoplasmodium*. Unlike the true plasmodium of the plasmodial slime molds, however, this structure is not simply a giant sheet of cytoplasm with many nuclei; the individual myxamoebas retain their plasma membranes and, therefore, their identity.

A slug may migrate over a substrate for several hours before becoming motionless and reorganizing to construct a delicate, stalked fruiting structure. Cells at the top of the fruiting structure develop into thick-walled spores, which are eventually released. Later, under favorable conditions, the spores germinate, releasing myxamoebas.

The cycle from myxamoebas through slug and spores to new myxamoebas is asexual. Cellular slime molds also have a sexual cycle, in which two myxamoebas fuse. The product of this fusion develops into a spherical structure that ultimately germinates, releasing new haploid myxamoebas.

Do You Understand Concept 20.2?

- Explain why the term "protists" does not refer to a formal taxonomic group.
- Contrast the major distinctive features of alveolates, excavates, stramenopiles, rhizaria, and amoebozoans. Which of these groups is most closely related to fungi and animals? What morphological evidence supports this relationship?
- The fossil record of eukaryotes in the Precambrian is poor compared with those of the Cambrian and later geological periods, even though eukaryotes were diversifying for the last billion years of the Precambrian. Can you think of some possible reasons for the better fossil record beginning in the Cambrian?

The ancient origins of major eukaryote lineages, and the adaptation of these lineages to a wide variety of lifestyles and environments, resulted in enormous protist diversity. It is not surprising, then, that reproductive modes among protists are also highly diverse.

concept 20.3 Protists Reproduce Sexually and Asexually

Although most protists engage in both asexual and sexual reproduction, sexual reproduction has yet to be observed in some groups. In some protists, as in all prokaryotes, the acts of sex and reproduction are not directly linked.

Several asexual reproductive processes have been observed among the protists:

- *Binary fission* is the equal splitting of one cell into two by mitosis followed by cytokinesis.
- *Multiple fission* is the splitting of one cell into multiple (i.e., more than two) cells.
- *Budding* refers to the outgrowth of a new cell from the surface of an old one.
- *Sporulation* is the formation of specialized cells (spores) that are capable of developing into new individuals.

Asexual reproduction results in offspring that are genetically nearly identical to their parents (they only differ by new mutations that may arise during DNA replication). Such asexually reproduced groups of nearly identical organisms are known as **clonal lineages**.

Sexual reproduction among the protists takes various forms. In some protists, as in animals, the gametes are the only haploid cells. In others, the zygote is the only diploid cell. In still others, both diploid and haploid cells undergo mitosis, giving rise to alternating multicellular diploid and haploid life stages.

Some protists have reproduction without sex and sex without reproduction

Members of the genus *Paramecium* are ciliates, which as we noted in Concept 20.2 are characterized by the possession of two types of nuclei in a single cell: commonly one *macronucleus* and from one to several *micronuclei*. The micronuclei, which are typical eukaryotic nuclei, are essential for genetic recombination. Each macronucleus contains many copies of the genetic information, packaged in units containing very few genes each. The macronuclear DNA is transcribed and translated to regulate the life of the cell.

When paramecia reproduce asexually, all of the nuclei are copied before the cell divides. Paramecia also have an elaborate sexual behavior called **conjugation**, in which two individuals line up tightly against each other and fuse in the oral groove region of the body. Nuclear material is extensively reorganized and exchanged over the next several hours (**FIGURE 20.18**). Each cell ends up with two haploid micronuclei, one of its own and one from the other cell, which fuse to form a new diploid micronucleus. A new macronucleus develops from the micronucleus through a series of dramatic chromosomal rearrangements. The exchange of nuclei is fully reciprocal: each of the two paramecia gives and receives an equal amount of DNA. The two organisms then separate and go their own ways, each equipped with new combinations of alleles.

Conjugation in *Paramecium* is a *sexual* process of genetic recombination, but it is not a *reproductive* process. Two cells begin the process, and two cells are there at the end, so no new cells are created. As a rule, each asexual clone of paramecia must periodically conjugate. Experiments have shown that if some species are not permitted to conjugate, the clones can live through no more than approximately 350 cell divisions before they die out.

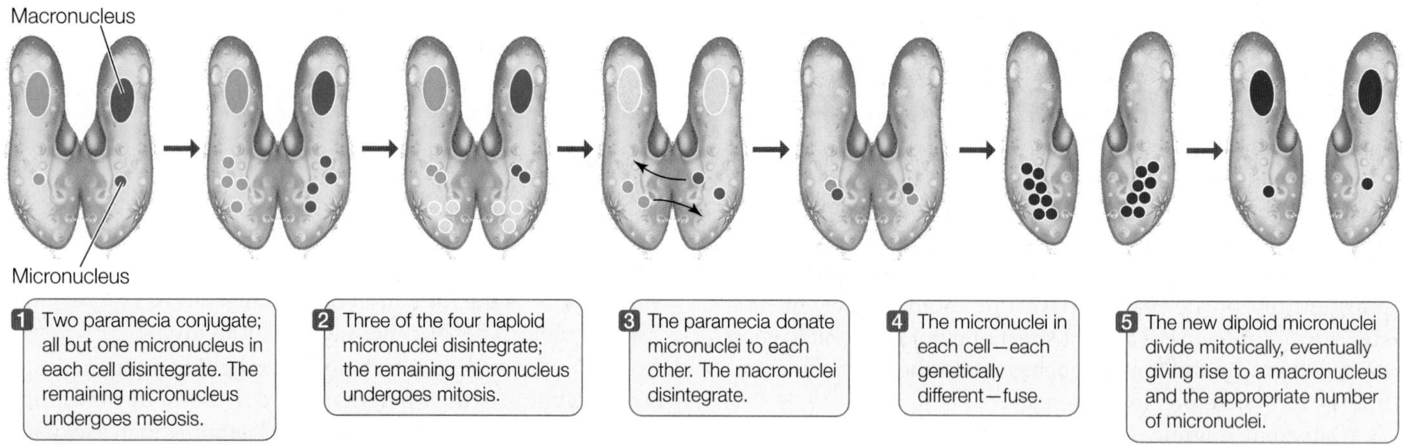

Macronucleus

Micronucleus

1 Two paramecia conjugate; all but one micronucleus in each cell disintegrate. The remaining micronucleus undergoes meiosis.

2 Three of the four haploid micronuclei disintegrate; the remaining micronucleus undergoes mitosis.

3 The paramecia donate micronuclei to each other. The macronuclei disintegrate.

4 The micronuclei in each cell—each genetically different—fuse.

5 The new diploid micronuclei divide mitotically, eventually giving rise to a macronucleus and the appropriate number of micronuclei.

FIGURE 20.18 Paramecia Achieve Genetic Recombination by Conjugating The exchange of micronuclei by two conjugating *Paramecium* individuals results in genetic recombination. After conjugation, the cells separate and continue their lives as two individuals.

Some protist life cycles feature alternation of generations

Alternation of generations is a type of life cycle found in many multicellular protists, all land plants, and some fungi. A multicellular, diploid, spore-producing organism gives rise to a multicellular, haploid, gamete-producing organism. When two haploid gametes fuse, a diploid organism is produced. The haploid organism, the diploid organism, or both may also reproduce asexually.

The two alternating (spore-producing and gamete-producing) generations differ genetically (one has diploid cells, the other haploid cells), but they may or may not differ morphologically. In **heteromorphic** alternation of generations, the two generations differ morphologically; in **isomorphic** alternation of generations, they do not. Examples of both heteromorphic and isomorphic alternation of generations are found among the brown algae.

The gamete-producing generation does not produce gametes by meiosis because the gamete-producing organism is already haploid. Instead, specialized cells of the diploid spore-producing organism, called **sporocytes**, divide meiotically to produce four haploid *spores*. The spores may eventually germinate and divide mitotically to produce the multicellular haploid generation, which then produces gametes by mitosis and cytokinesis.

Gametes, unlike spores, can produce new organisms only by fusing with other gametes. The fusion of two gametes produces a diploid zygote, which then undergoes mitotic divisions to produce a diploid organism. The diploid organism's sporocytes then undergo meiosis and produce haploid spores, starting the cycle anew.

Do You Understand Concept 20.3?

- Why is conjugation between paramecia considered a sexual process but not a reproductive process?

- Why do you think paramecia that are not allowed to conjugate begin to die out after about 350 rounds of asexual reproduction?

- Although most diploid animals have haploid stages (for example, eggs and sperm), their life cycles are not considered alternation of generations. Why not?

concept 20.4 Protists Are Critical Components of Many Ecosystems

Some protists are food for marine animals, while others poison those animals. Some are packaged as nutritional supplements for humans, and some are human pathogens. The remains of some form the sands of many modern beaches, and others are a major source of the oil that sometimes fouls those beaches.

Phytoplankton are primary producers

A single protist clade, the diatoms, performs about one-fifth of all photosynthetic carbon fixation on Earth—about the same amount as all of Earth's rainforests. These spectacular unicellular organisms (see Figure 20.9) are the predominant component of the phytoplankton, but the phytoplankton includes many other protists that contribute heavily to global photosynthesis. Like green plants on land, these "floating photosynthesizers" are the gateway for energy from the sun into the rest of the living world; in other words, they are *primary producers* that are eaten by heterotrophs, including animals and many other protists. Those consumers are, in turn, eaten by other consumers. Most aquatic heterotrophs (with the exception of some species in the deep sea) depend on photosynthesis performed by phytoplankton.

Some microbial eukaryotes are deadly

Some microbial eukaryotes are pathogens that cause serious diseases in humans and other vertebrates. The best-known pathogenic protists are members of the genus *Plasmodium*, a highly specialized group of apicomplexans that spend part of their life cycle as parasites in human red blood cells, where they are the cause of malaria. In terms of the number of people affected, malaria is one of the world's three most serious infectious diseases: it infects over 350 million people, and kills over 1 million people, each year. On average, about two people die from malaria every minute of every day—most of them in sub-Saharan Africa, although malaria occurs in more than 100 countries.

Mosquitoes of the genus *Anopheles* transmit *Plasmodium* to humans. The parasites enter the human circulatory system when an infected female *Anopheles* mosquito penetrates the skin in search of blood. The parasites find their way to cells in the liver and the lymphatic system, change their form, multiply, and re-enter the bloodstream, where they invade red blood cells.

The parasites multiply inside the red blood cells, which then burst, releasing new swarms of parasites. These episodes of bursting red blood cells coincide with the primary symptoms of malaria, which include fever, shivering, vomiting, joint pains, and convulsions. If another *Anopheles* bites the victim, the mosquito takes in *Plasmodium* cells along with blood. Some of the ingested cells develop into gametes that unite in the mosquito, forming zygotes. The zygotes lodge in the mosquito's gut, divide several times, and move into its salivary glands, from which they can be passed on to another human host. Thus

Plasmodium is an extracellular parasite in the mosquito vector and an intracellular parasite in the human host.

---

yourBioPortal.com

Go to ANIMATED TUTORIAL 20.3
Life Cycle of the Malarial Parasite

Plasmodium has proved to be a singularly difficult pathogen to attack. The complex *Plasmodium* life cycle is best broken by the removal of stagnant water, in which mosquitoes breed. Using insecticides to reduce the *Anopheles* population can also be effective, but the benefits must be weighed against the ecological, economic, and health risks posed by the insecticides themselves.

Even some of the phytoplankton that are such important primary producers can be deadly, as described in this chapter's opening story. Some diatoms and dinoflagellates reproduce in enormous numbers when environmental conditions are favorable for their growth. In the resulting "red tides," the concentration of dinoflagellates may reach 60 million per liter of ocean water and produce potent nerve toxins that harm or kill many vertebrates, especially fish.

Some microbial eukaryotes are endosymbionts

Endosymbiosis is common among the microbial eukaryotes, many of which live within the cells of animals. Many radiolarians harbor photosynthetic endosymbionts (see Figure 20.13A). As a result, these radiolarians, which are not photosynthetic themselves, appear greenish or golden, depending on the type of endosymbiont they contain. This arrangement is often

---

APPLY THE CONCEPT

Protists are critical components of many ecosystems

In most temperate regions of the oceans, there is a spring bloom of phytoplankton. Although the red tide blooms described in this chapter's opening story are harmful, phytoplankton blooms can also be beneficial for marine communities. In fact, many species of marine life depend on these blooms for their survival.

The dates of spring phytoplankton blooms near the coast of Nova Scotia, Canada, were determined by examining remote satellite images. The table at right presents these dates as deviations from the mean date of the spring bloom in this region. The table also gives the survival index for larval haddock (an important commercial fish) for the year after each bloom. The survival index is the ratio of the mass of juvenile fish to the mass of mature fish; higher values indicate better survival of larval fish.

1. Plot the survival index of larval haddock against the deviation in the date of the spring phytoplankton bloom. Calculate a correlation coefficient for their relationship (see Appendix B).

YEAR	DEVIATION IN BLOOM DATE* (DAYS)	SURVIVAL INDEX
1	+5	1.9
2	+11	2.2
3	−15	6.8
4	+5	1.9
5	−4	4.9
6	−20	10.3
7	+6	2.1
8	+14	1.9

*Negative values indicate blooms occurring earlier than the mean date; positive values indicate later blooms.

2. Formulate one or more hypotheses to explain your results. Keep in mind that larval haddock include phytoplankton in their diet, and that phytoplankton blooms also provide some cover in which larval fish can hide from potential predators.

mutually beneficial: the radiolarian can make use of the carbon compounds produced by its photosynthetic guest, and the guest may in turn make use of metabolites made by the host or receive physical protection. In some cases, the guest is exploited for its photosynthetic products while receiving little or no benefit itself.

Dinoflagellates are also common endosymbionts and can be found in both animals and other protists. Most, but not all, dinoflagellate endosymbionts are photosynthetic. Some dinoflagellates live endosymbiotically in the cells of corals, contributing the products of their photosynthesis to the partnership. Their importance to the corals is demonstrated when the dinoflagellates die or are expelled by the corals as a result of changing environmental conditions such as rising water temperatures or increased water turbidity. This phenomenon is known as *coral bleaching*. Unless the corals can acquire new endosymbionts, they are ultimately damaged or destroyed as a result of their reduced food supply (**FIGURE 20.19**).

INVESTIGATION

FIGURE 20.19 Can Corals Reacquire Dinoflagellate Endosymbionts Lost to Bleaching? Some corals lose their chief nutritional source when their photosynthetic endosymbionts die, often as a result of changing environmental conditions. This experiment by Cynthia Lewis and Mary Alice Coffroth investigated the ability of corals to acquire new endosymbionts after bleaching.

HYPOTHESIS

Bleached corals can acquire new photosynthetic endosymbionts from their environment.

METHOD

1. Count numbers of *Symbiodinium*, a photosynthetic dinoflagellate, living symbiotically in samples of a coral (*Briareum* sp.).
2. Stimulate bleaching by maintaining all *Briareum* colonies in darkness for 12 weeks.
3. After 12 weeks of darkness, count numbers of *Symbiodinium* in the coral samples; then return all colonies to light.
4. In some of the bleached colonies (the experimental group), introduce *Symbiodinium* strain B211—dinoflagellates that contain a unique molecular marker. A control group of bleached colonies is not exposed to strain B211. Maintain both groups in the light for 6 weeks.

RESULTS

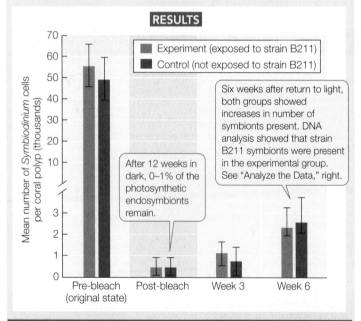

■ Experiment (exposed to strain B211)
■ Control (not exposed to strain B211)

After 12 weeks in dark, 0–1% of the photosynthetic endosymbionts remain.

Six weeks after return to light, both groups showed increases in number of symbionts present. DNA analysis showed that strain B211 symbionts were present in the experimental group. See "Analyze the Data," right.

CONCLUSION

Corals can acquire new strains of endosymbionts from their environment following bleaching.

ANALYZE THE DATA

These data—the results of DNA analysis of the *Symbiodinium* endosymbionts—reveal that many of the experimental colonies took up strain B211 from their environment. The control colonies recovered their native *Symbiodinium*, except in colonies in which endosymbionts were completely lost. Use these data to answer the questions below.

	Symbiodinium strain present (% of colonies)			
	Non-B211	B211	None*	Colony died
Experimental colonies (strain B211 added)				
Pre-bleach	100	0	0	0
Post-bleach	58	0	42	0
Week 3	0	92	0	8
Week 6	8	58	8	25
Control colonies (no strain B211)				
Pre-bleach	100	0	0	0
Post-bleach	67	0	33	0
Week 3	67	0	33	0
Week 6	67	0	17	17

*Colonies remained alive but no *Symbiodinium* were detected.

A. Are new strains of *Symbiodinium* taken up only by coral colonies that have lost all their original endosymbionts?

B. Does the acquisition of a new *Symbiodinium* strain always result in survival of a recovering *Briareum* colony?

C. In week 3, only strain B211 was detected in the experimental colonies, but in week 6, non-B211 *Symbiodinium* were detected in 8% of the experimental colonies. Can you suggest an explanation for this observation?

Pre-bleach

Post-bleach

Go to **yourBioPortal.com** for original citations, discussions, and relevant links for all INVESTIGATION figures.

We rely on the remains of ancient marine protists

Diatoms are lovely to look at, but their importance to us goes far beyond aesthetics, and even beyond their role as primary producers. Diatoms store oil as an energy reserve and to keep themselves afloat at the correct depth in the ocean. Over millions of years, diatoms have died and sunk to the ocean floor, where they have undergone chemical changes. In this way, they have become a major source of petroleum and natural gas, two of our most important energy supplies and political concerns.

Because the silica-containing cell walls of dead diatoms resist decomposition, some sedimentary rocks are composed almost entirely of diatom skeletons that sank to the seafloor over time. Diatomaceous earth, which is obtained from such rocks, has many industrial uses, such as insulation, filtration, and metal polishing. It has also been used as an "Earth-friendly" insecticide that clogs the tracheae (breathing structures) of insects.

Other ancient marine protists have also contributed to today's world. Some foraminiferans, as we have seen, secrete shells of calcium carbonate. After they reproduce (by mitosis and cytokinesis), the daughter cells abandon the parent shell and make new shells of their own. The discarded shells of ancient foraminiferans make up extensive limestone deposits in various parts of the world, forming a layer hundreds to thousands of meters deep over millions of square kilometers of ocean bottom. Foraminiferan shells also make up much of the sand of some beaches. A single gram of such sand may contain as many as 50,000 foraminiferan shells and shell fragments.

The shells of individual foraminiferans are easily preserved as fossils in marine sediments. Each geological period has a distinctive assemblage of foraminiferan species. Because the shells of foraminiferan species have distinctive shapes (see Figure 20.12), and because they are so abundant, the remains of foraminiferans are especially valuable in classifying and dating sedimentary rocks. In addition, analyses of the chemical makeup of foraminiferan shells can be used to estimate the global temperatures prevalent at the time when the shells were formed.

Do You Understand Concept 20.4?

- What is the role of female *Anopheles* mosquitoes in the transmission of malaria?
- Explain the role of dinoflagellates in the two very different phenomena of coral bleaching and red tides.
- What are some of the ways in which diatoms are important to human society?

The next three chapters will explore the three major groups of multicellular eukaryotes, along with the protist ancestors from which they arose. Chapter 21 will describe the origin and diversification of the plants, Chapter 22 will present the fungi, and Chapter 23 will describe the animals.

QUESTION Red tides are harmful, but can dinoflagellates also be beneficial to marine ecosystems?

ANSWER Not all dinoflagellate blooms produce problems for other species. Dinoflagellates are important components of many ecosystems, as we have seen throughout this chapter. Corals and many other species depend on symbiotic dinoflagellates for photosynthesis (see Figure 20.19). In addition, as photosynthetic organisms, free-living planktonic dinoflagellates are among the most important primary producers in aquatic food webs. They are a major component of the phytoplankton and provide an important food source for many species (see Concept 20.4). Photosynthetic dinoflagellates also produce much of the atmospheric oxygen that most animals need to survive.

Some dinoflagellates produce a beautiful bioluminescence (**FIGURE 20.20**). Unlike the bioluminescent bacteria discussed at the start of Chapter 19, however, dinoflagellate bioluminescence is not steady. These protists produce flashes of light when disturbed, as people who swim in the ocean at night in certain regions often observe.

FIGURE 20.20 Light Up the Sea
Bioluminescent dinoflagellates flash as an outrigger disturbs the ocean surface off the island of Bali.

SUMMARY

Eukaryotes Acquired Features from Both Archaea and Bacteria

- Early events in the evolution of the eukaryotic cell probably included the loss of the firm cell wall and infolding of the plasma membrane. Such infolding probably led to segregation of the genetic material in a membrane-enclosed nucleus. **Review Figure 20.1**

- Some organelles were acquired by endosymbiosis. Mitochondria evolved by endosymbiosis with a proteobacterium.

- **Primary endosymbiosis** of a eukaryote and a cyanobacterium gave rise to the first chloroplasts. **Secondary endosymbiosis** and **tertiary endosymbiosis** between chloroplast-containing eukaryotes and other eukaryotes gave rise to the distinctive chloroplasts of euglenids, dinoflagellates, and other groups. **Review Figure 20.2 and ANIMATED TUTORIAL 20.1**

concept
20.2 **Major Lineages of Eukaryotes Diversified in the Precambrian**

- Most eukaryotes can be placed in one of eight major clades that originated in the Precambrian: alveolates, excavates, stramenopiles, rhizaria, amoebozoans, Plantae, fungi, and animals. The first five of these clades are collectively referred to as **protists**. **Review Figure 20.3**

- The term "protist" does not describe a formal taxonomic group, but is shorthand for "all eukaryotes that are not plants, animals, or fungi." Most, but not all, protists are unicellular.

- **Alveolates** are unicellular organisms with sacs (alveoli) beneath their plasma membranes. Alveolate clades include the marine **dinoflagellates**, the parasitic **apicomplexans**, and the diverse, highly motile **ciliates**. See **WEB ACTIVITY 20.1** and **ANIMATED TUTORIAL 20.2**

- The **excavates** include a wide variety of symbiotic as well as free-living species. The **diplomonads** and **parabasalids** lack mitochondria, having apparently lost them during the course of their evolution. **Heteroloboseans** are amoebas with a two-stage life cycle. **Euglenids** are often photosynthetic and have anterior flagella and spiraling strips of protein that support their cell surface. The **kinetoplastids**, which include several human pathogens, have a single, large mitochondrion.

- **Stramenopiles** typically have two flagella of unequal length, the longer one bearing rows of tubular hairs. Among the stramenopiles are the unicellular **diatoms**, the multicellular **brown algae**, and the nonphotosynthetic **oomycetes**, many of which are **saprobic**.

- **Rhizaria** are unicellular and aquatic. They include the **foraminiferans**, whose shells have contributed to great limestone deposits; the **radiolarians**, which have thin, stiff pseudopods and glassy endoskeletons; and the **cercozoans**, which take many forms and live in diverse habitats.

- The **amoebozoans** move by means of lobe-shaped pseudopods. A **lobosean** consists of a single amoeboid cell. **Plasmodial slime molds** are amoebozoans whose vegetative stage is a **coenocyte** that moves by cytoplasmic streaming. In **cellular slime molds**, the individual cells maintain their identity at all times but aggregate to form fruiting structures.

concept
20.3 **Protists Reproduce Sexually and Asexually**

- Asexual reproduction gives rise to **clonal lineages** of organisms.

- **Conjugation** in *Paramecium* is a sexual process but not a reproductive one. **Review Figure 20.18**

- **Alternation of generations**, which includes a multicellular diploid phase and a multicellular haploid phase, is a feature of many multicellular protist life cycles (as well as of some fungi and all land plants). The alternating generations may be **heteromorphic** or **isomorphic**.

- In alternation of generations, specialized cells of the diploid organism, called **sporocytes**, divide meiotically to produce haploid spores. Spores give rise to the multicellular, haploid, gamete-producing generation through mitosis. Gametes fuse and give rise to the diploid organism.

concept
20.4 **Protists Are Critical Components of Many Ecosystems**

- The diatoms are responsible for about one-fifth of the photosynthetic carbon fixation on Earth. They and other members of the phytoplankton are the primary producers in the marine environment.

- Endosymbiotic relationships are common among microbial protists and typically benefit both the endosymbionts and their protist or animal partners. **Review Figure 20.19**

- Some protists are pathogens of humans and other vertebrates. See **ANIMATED TUTORIAL 20.3**

- Ancient diatoms are the major source of today's petroleum and natural gas deposits.

The Evolution of Plants

In the early 1860s, while the United States was entangled in its tragic civil war, much of middle- and upper-class England was caught up in an orchid frenzy. Amateur plant breeders and professional botanists alike were enchanted with raising the beautiful flowers. After *The Origin of Species* appeared in 1859, Charles Darwin wrote his next book on this group of plants, publishing *Fertilisation of Orchids* in 1862.

There are more than 25,000 species of orchids, which makes them one of the most diverse plant groups. Darwin wanted to know why orchids had experienced such rapid diversification and was particularly impressed with the role that insect pollinators might have played in this process. He wanted examples to demonstrate the power of natural selection; he found such examples in abundance among the orchids.

Orchids show an impressive variety of specialized pollination mechanisms, many of which demonstrate that they have coevolved with their pollinators. For example, Darwin observed a South American orchid of the genus *Catasetum* shooting a packet of pollen at an insect that landed on its flower. When he was shown *Angraecum sesquipedale*, an orchid from Madagascar with a nectar tube over a foot long, Darwin hypothesized that there must be a moth with a proboscis of unprecedented length that fed from and pollinated that flower. Many people scoffed at his vision, but the moth he described was eventually discovered—21 years after his death.

In 1836 the explorer Robert Schomburgk shook the botanical world with a report that he had seen flowers described as belonging to three different genera of orchids—*Catasetum*, *Monachanthus*, and *Myanthus*—growing together on a single plant. The English botanist John Lindley remarked that this observation would "shake to the foundation all our ideas of the stability of genera and species." Orchid enthusiasts were befuddled by their efforts to grow specimens of *Myanthus*, only to have them flower with the more common blooms of *Catasetum*. Darwin knew that he needed to find the explanation for these odd observations, for otherwise he would have to conclude that individual plants were able to change their specific identity, something that did not fit with his explanations of the evolution of diversity.

 QUESTION What was Darwin's explanation for the three distinct flowers growing on a single orchid plant?

On examining a specimen of the orchid *Angraecum sesquipedale*, Charles Darwin predicted the existence of a pollinator with an exceptionally long proboscis. This pollinator, the sphinx moth *Xanthopan morgani*, was not discovered until after Darwin's death.

KEY CONCEPTS

21.1 Primary Endosymbiosis Produced the First Photosynthetic Eukaryotes

21.2 Key Adaptations Permitted Plants to Colonize Land

21.3 Vascular Tissues Led to Rapid Diversification of Land Plants

21.4 Seeds Protect Plant Embryos

21.5 Flowers and Fruits Increase the Reproductive Success of Angiosperms

(A)

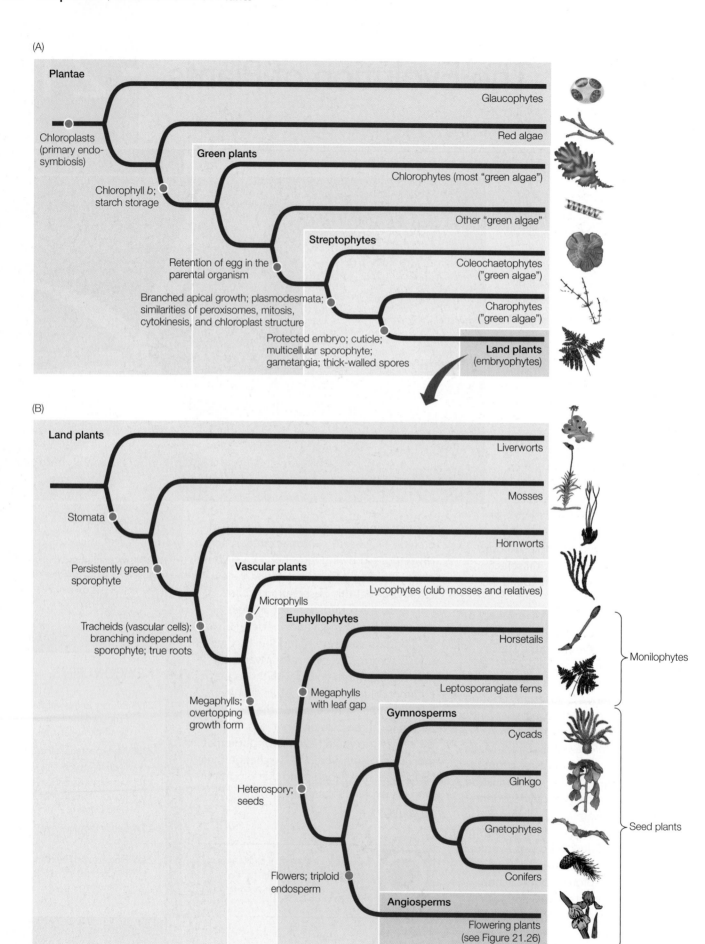

Plantae

Chloroplasts (primary endo-symbiosis)

Glaucophytes

Red algae

Chlorophyll *b*; starch storage

Green plants

Chlorophytes (most "green algae")

Other "green algae"

Streptophytes

Retention of egg in the parental organism

Coleochaetophytes ("green algae")

Branched apical growth; plasmodesmata; similarities of peroxisomes, mitosis, cytokinesis, and chloroplast structure

Charophytes ("green algae")

Protected embryo; cuticle; multicellular sporophyte; gametangia; thick-walled spores

Land plants (embryophytes)

(B)

Land plants

Liverworts

Stomata

Mosses

Persistently green sporophyte

Hornworts

Vascular plants

Lycophytes (club mosses and relatives)

Microphylls

Tracheids (vascular cells); branching independent sporophyte; true roots

Euphyllophytes

Horsetails

Megaphylls; overtopping growth form

Megaphylls with leaf gap

Leptosporangiate ferns

Monilophytes

Gymnosperms

Cycads

Heterospory; seeds

Ginkgo

Gnetophytes

Conifers

Flowers; triploid endosperm

Angiosperms

Flowering plants (see Figure 21.26)

Seed plants

Primary Endosymbiosis Produced the First Photosynthetic Eukaryotes

Over a billion years ago, when a cyanobacterium was first engulfed by an early eukaryote, the history of life was altered radically. The chloroplasts that resulted from primary endosymbiosis of this cyanobacterium (see Figure 20.2) were obviously important for the evolution of plants and other photosynthetic eukaryotes, but they were also critical to the evolution of all life on land. Until photosynthetic plants were able to move onto land, there was very little on land to support multicellular animals or fungi, and almost all life was restricted to the oceans and fresh waters.

Primary endosymbiosis is a shared derived trait, or *synapomorphy*, of the group known as **Plantae** (**FIGURE 21.1**). Although *Plantae* is Latin for "plants," throughout this book and in everyday language the unmodified common name "plants" is usually used to refer only to the land plants. The first several clades to branch off the tree of life after primary endosymbiosis, however, are all aquatic. Aquatic photosynthetic eukaryotes (other than those secondarily derived from land plants) are known by the common name **algae**. This name, however, is just a convenient way to refer to all such groups, which are not all closely related; in other words, the algae are polyphyletic.

Several distinct clades of algae were among the first photosynthetic eukaryotes

The ancestor of Plantae was unicellular and may have been similar in general form to the modern **glaucophytes**. These microscopic, freshwater algae are thought to be the sister group to the rest of Plantae (see Figure 21.1A). The chloroplast of glaucophytes is unique in containing a small amount of peptidoglycan between its inner and outer membranes—the same arrangement found in cyanobacteria. Peptidoglycan has been lost from the remaining photosynthetic eukaryotes.

In contrast to the glaucophytes, almost all **red algae** are multicellular (**FIGURE 21.2**). Their characteristic color is a result of the accessory photosynthetic pigment *phycoerythrin*, which is found in relatively large amounts in the chloroplasts of many red algae. In addition to phycoerythrin, red algal chloroplasts contain the accessory pigments phycocyanin, carotenoids, and chlorophyll *a*.

The red algae include species that grow in the shallowest tide pools as well as the photosynthesizers found deepest in the ocean (as deep as 260 meters if nutrient conditions are right

◀ **FIGURE 21.1   The Evolution of Plants**   In its broadest definition, the term "plant" includes the green plants, red algae, and glaucophytes—all the groups descended from a common ancestor with chloroplasts. Some biologists restrict the term "plant" to the green plants (those with chlorophyll *b*) or, even more narrowly, to the land plants. Three key characteristics that emerged during the evolution of land plants—protected embryos, vascular tissues, and seeds—led to their success in the terrestrial environment. (See Working with Data 21.1 at **yourBioPortal.com**.)

Nemalion helminthoides

FIGURE 21.2   A Red Alga   The photosynthetic pigment phycoerythrin gives this red alga its rich hue, showing up vividly against a background of green algae.

and the water is clear enough to permit light to penetrate). A few red algae inhabit fresh water. Most grow attached to a substrate by a holdfast.

Despite their name, red algae don't always appear red in color. The ratio of two pigments—phycoerythrin (red) and chlorophyll *a* (green)—depends largely on the intensity of light that reaches the alga. In deep water, where light is dim, algae accumulate large amounts of phycoerythrin and appear red. But many species growing near the surface contain a higher concentration of chlorophyll *a* and thus appear bright green.

The remaining algal groups in Plantae are the various "green algae." Like land plants, the green algae contain both chlorophylls *a* and *b* and store their reserve of photosynthetic products as starch in chloroplasts. All the groups that share these features are commonly called **green plants** because of the strong green color of chlorophyll *b*.

Three important clades of green algae are the **chlorophytes**, **coleochaetophytes**, and **charophytes** (**FIGURE 21.3**). The latter two groups are most closely related to the land plants, and together with the land plants form a group known as **streptophytes**.

FRONTIERS   Some human pathologies, such as polycystic kidney disease, are caused by defects in cilia (hence their name, *ciliopathies*). Recent genome analyses have revealed remarkable similarities between the human and green algal proteins involved in cilium assembly and function. The discovery of how little these proteins have changed throughout evolution suggested the use of green algae as experimental organisms for research on diseases that affect human cilia. The single-celled chlorophyte *Chlamydomonas* is now being used for that purpose.

One of the key synapomorphies of the **land plants** is development from an embryo that is protected by tissues of the parent plant. For this reason, land plants are sometimes called

(A) *Ulva rigida*

(B) *Coleochaete* sp.

(C) *Chara* sp.

FIGURE 21.3 "Green Algae" Consist of Several Distantly Related Groups (A) Sea lettuce, a chlorophyte, grows in ocean tidewaters. (B) The coleochaetophytes are thought to be the sister group of charophytes plus land plants. (C) The land plants probably evolved from a common ancestor shared with charophytes such as this one, which displays the branching pattern we associate with land plants.

embryophytes (*phyton*, "plant"). The green plants, the streptophytes, and the land plants each have been called "the plant kingdom" by different authorities; others take an even broader view and include red algae and glaucophytes as "plants." To avoid confusion in this chapter, we will use modifying terms ("land plants" or "green plants," for example) to refer to the various clades of Plantae shown in Figure 21.1.

There are ten major groups of land plants

The land plants that exist today fall naturally into ten major clades (**TABLE 21.1**). Members of seven of those clades possess well-developed vascular systems that transport materials throughout the plant body. We call these seven groups, collectively, the **vascular plants**, or **tracheophytes**, because they all possess fluid-conducting cells called *tracheids*.

The remaining three clades (liverworts, hornworts, and mosses) lack tracheids. We refer to these three clades collectively as **nonvascular land plants**. Note, however, that *these three groups do not form a clade*. Some nonvascular land plants have conducting cells, but none have tracheids.

Do You Understand Concept 21.1?

- Explain the different possible uses of the term "plant."
- What are some of the key differences between glaucophytes, red algae, and the various clades of green algae?
- What evidence supports the phylogenetic relationship between land plants and green algae? Why doesn't the name "algae" designate a formal taxonomic group?

TABLE 21.1   Classification of Land Plants

GROUP	COMMON NAME	CHARACTERISTICS
Nonvascular Land plants		
Hepatophyta	Liverworts	No filamentous stage; gametophyte flat
Anthocerophyta	Hornworts	Embedded archegonia; sporophyte grows basally (from the ground)
Bryophyta	Mosses	Filamentous stage; sporophyte grows apically (from the tip)
Vascular plants		
Lycopodiophyta	Lycophytes: Club mosses and allies	Microphylls in spirals; sporangia in leaf axils
Monilophyta	Horsetails, ferns	Megaphylls with a leaf gap (a space in the stem from which the leaf emerges)
SEED PLANTS		
Gymnosperms		
Cycadophyta	Cycads	Compound leaves; swimming sperm; seeds on modified leaves
Ginkgophyta	Ginkgo	Deciduous; fan-shaped leaves; swimming sperm
Gnetophyta	Gnetophytes	Vessels in vascular tissue; opposite, simple leaves
Coniferophyta	Conifers	Seeds in cones; needlelike or scalelike leaves
Angiosperms	Flowering plants	Endosperm; carpels; gametophytes much reduced; seeds within fruit

The green algal ancestors of the land plants lived at the margins of ponds or marshes, ringing them with a mat of dense green. It was from such a marginal habitat, which was sometimes wet and sometimes dry, that early plants made the transition onto land.

concept 21.2 Key Adaptations Permitted Plants to Colonize Land

How did the land plants arise? To address this question, we can compare land plants with their closest relatives among the green algae. The features that differ between the two groups include the adaptations that allowed the first land plants to survive in the terrestrial environment.

Two groups of green algae share many features with land plants

Several microscopic structural features, backed by clear-cut evidence from molecular studies, indicate that the coleochaetophytes and the charophytes are the closest relatives of the land plants. Both of these green algal groups retain their female gametes (eggs) within the parental organism, as do land plants. Of these two candidates, charophytes are thought to be the sister group of land plants. That conclusion is based on the following synapomorphies (as well as studies of DNA sequences):

- Plasmodesmata that join the cytoplasm of adjacent cells
- Growth that is branching and apical (from the tip)
- Similar peroxisome contents, mechanics of mitosis and cytokinesis, and chloroplast structure

Both coleochaetophytes and charophytes have some general features that are similar to those of some groups of land plants. The flattened growth form of coleochaetophytes, represented by the genus *Coleochaete* (see Figure 21.3B), is like the growth form of early land plants such as liverworts. But the charophytes, as represented by stoneworts of the genus *Chara* (see Figure 21.3C), exhibit the branching growth form found among most land plants.

Adaptations to life on land distinguish land plants from green algae

Land plants, or their immediate ancestors in those ancient green mats, first appeared in the terrestrial environment between 400 and 500 million years ago. How did they survive in an environment that differed so dramatically from the aquatic environment of their ancestors? While the water essential for life is everywhere in the aquatic environment, water is difficult to obtain and retain in the terrestrial environment.

No longer bathed in fluid, organisms on land faced potentially lethal desiccation (drying). Large terrestrial organisms had to develop ways to transport water to body parts distant from the source of the water. And whereas water provides aquatic organisms with support against gravity, a plant living on land must either have some other support system or sprawl unsupported on the ground. A land plant must also use different mechanisms for dispersing its gametes and progeny than its aquatic relatives, which can simply release them into the water.

Survival on land was facilitated by the evolution among plants of numerous adaptations:

- The *cuticle*, a coating of waxy lipids that retards water loss
- *Stomata*, small closable openings in leaves and stems that are used to regulate gas exchange
- *Gametangia*, multicellular organs that enclose plant gametes and prevent them from drying out
- *Embryos*, young plants contained within a protective structure
- Certain *pigments* that afford protection against the mutagenic ultraviolet radiation that bathes the terrestrial environment
- Thick *spore walls* containing *sporopollenin*, a polymer that protects the spores from desiccation and resists decay
- A *mutually beneficial association with fungi* that promotes nutrient uptake from the soil

The **cuticle** may be the most important—and earliest—of these features. Composed of several unique waxy lipids that coat the leaves and stems of land plants, the cuticle has several functions, the most obvious and important of which is to keep water from evaporating from the plant body.

As ancient plants colonized the land, they not only adapted to the terrestrial environment—they modified it by contributing to the formation of soil. Acids secreted by plants help break down rock, and the organic compounds produced by the breakdown of dead plants contribute nutrients to the soil. Such effects are repeated today as plants grow in new areas.

Life cycles of land plants feature alternation of generations

A universal feature of the life cycles of land plants is alternation of generations. Recall from Concept 20.3 the two hallmarks of alternation of generations:

- The life cycle includes both a multicellular diploid stage and a multicellular haploid stage.
- Gametes are produced by mitosis, not by meiosis. Meiosis produces *spores* that develop into multicellular haploid organisms.

If we begin looking at the land plant life cycle at the single-cell stage—the diploid zygote—then the first phase of the cycle is the formation, by mitosis and cytokinesis, of a multicellular embryo, which eventually grows into a mature diploid plant. This multicellular diploid plant is called the **sporophyte** ("spore plant").

Cells contained within specialized reproductive organs of the sporophyte, called **sporangia** (singular *sporangium*), undergo meiosis to produce haploid, unicellular spores. By

FIGURE 21.4 Alternation of Generations in Land Plants A multicellular diploid sporophyte generation that produces spores by meiosis alternates with a multicellular haploid gametophyte generation that produces gametes by mitosis.

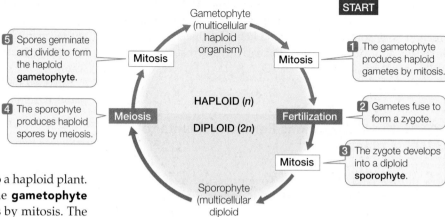

mitosis and cytokinesis, a spore develops into a haploid plant. This multicellular haploid plant, called the **gametophyte** ("gamete plant"), produces haploid gametes by mitosis. The fusion of two gametes (*syngamy*, or *fertilization*) forms a single diploid cell—the zygote—and the cycle is repeated (**FIGURE 21.4**).

The *sporophyte generation* extends from the zygote through the adult multicellular diploid plant and sporangium formation; the *gametophyte generation* extends from the spore through the adult multicellular haploid plant to the gametes. The transitions between the generations are accomplished by fertilization and by meiosis. In all land plants, the sporophyte and the gametophyte differ genetically: the sporophyte has diploid cells, and the gametophyte has haploid cells.

There is a trend toward reduction of the gametophyte generation in plant evolution. In the nonvascular land plants, the gametophyte is larger, longer-lived, and more self-sufficient than the sporophyte. In those groups that appeared later in plant evolution, however, the sporophyte generation is the larger, more conspicuous, longer-lived, and more self-sufficient one.

Nonvascular land plants live where water is readily available

The living species of nonvascular land plants are the liverworts, mosses, and hornworts. These three groups are thought to be similar in many ways to the earliest land plants. Most of these plants grow in dense mats, usually in moist habitats. Even the largest of these species are only about half a meter tall, and most are only a few centimeters tall or long. Why have they not evolved to be taller? The probable answer is that they lack an efficient vascular system for transporting water and minerals from the soil to distant parts of the plant body.

The nonvascular land plants lack the true leaves, stems, and roots that characterize the vascular plants, although they have structures analogous to each. Their growth form allows water to move through the mats of plants by capillary action. They have leaflike structures that readily catch and hold any water that splashes onto them. They are small enough that minerals can be distributed throughout their bodies by diffusion. As in all land plants, layers of maternal tissue protect their embryos from desiccation. Nonvascular land plants also have a cuticle, although it is often very thin (or even absent in some species) and thus is not highly effective in retarding water loss.

Most nonvascular land plants live on the soil or on vascular plants, but some grow on bare rock, on dead and fallen tree

trunks, and even on buildings. Their ability to grow on such marginal surfaces results from a mutualistic association with fungi. The earliest association of land plants with fungi dates back at least 460 million years. This mutualism probably facilitated the absorption of water and minerals, especially phosphorus, from the first soils.

LINK Land plants of many groups have mutualistic associations with fungi, as described in Concept 22.2

Nonvascular land plants are widely distributed over six continents and even exist (albeit very locally) on the coast of the seventh, Antarctica. Most are terrestrial. Although a few species live in fresh water, these aquatic species are descended from terrestrial ones. None live in the oceans.

LIVERWORTS There are about 9,000 species of **liverworts**. Most liverworts have green, leaflike gametophytes that lie close to or flat on the ground (**FIGURE 21.5A**). The simplest liverworts are flat plates of cells a centimeter or so long with structures that produce sperm or eggs on their upper surfaces and rootlike filaments on their lower surfaces. The sporophyte remains attached to the larger gametophyte and rarely exceeds a few millimeters in length. Most liverworts can reproduce asexually (through simple division of the gametophyte) as well as sexually.

MOSSES The most familiar of the nonvascular land plants are the **mosses** (**FIGURE 21.5B**). These hardy little plants, of which there are about 15,000 species, are found in almost every terrestrial environment. They are often found on damp, cool ground, where they form thick mats.

The mosses are the sister lineage to the vascular plants plus the hornworts (see Figure 21.1). They share with those lineages an advance over the liverworts in their adaptation to life on land: they have stomata, which are important for both water retention and gas exchange.

Some moss gametophytes are so large that they cannot transport enough water throughout their bodies solely by diffusion. Gametophytes and sporophytes of many mosses contain a type of cell called a *hydroid*, which dies and leaves a

(A) *Marchantia* sp.

Spore case

Sporophytes

Gametophytes

(B) *Polytrichum* sp.

FIGURE 21.5 Diversity among Nonvascular Land Plants (A) Gametophytes of a liverwort. (B) The sporophytes and gametophytes are easily distinguished in this moss. (C) The sporophytes of many hornworts resemble little horns.

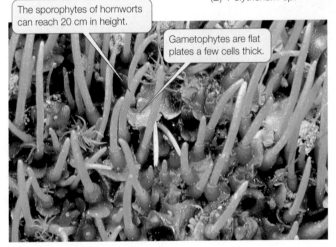

The sporophytes of hornworts can reach 20 cm in height.

Gametophytes are flat plates a few cells thick.

(C) *Anthoceros* sp.

all three groups, those of the hornworts come closest to being capable of growth without a set limit. Liverwort and moss sporophytes have a stalk that stops growing as the spore-producing structure matures, so elongation of the sporophyte is strictly limited. The hornwort sporophyte, however, has no stalk. Instead, a basal region remains capable of indefinite cell division, continuously producing new spore-bearing tissue above. The sporophytes of some hornworts growing in mild and continuously moist conditions can grow as tall as 20 centimeters. Eventually, however, the sporophyte's growth is limited by the lack of a transport system.

Hornworts have evolved a symbiotic relationship that promotes their growth by providing them with greater access to nitrogen, which is often a limiting resource for plants. These plants have internal cavities filled with mucilage, and the cavities are often populated by cyanobacteria that convert atmospheric nitrogen gas into a form of nitrogen usable by their host plant.

The sporophytes of nonvascular land plants are dependent on the gametophytes

In the nonvascular land plants, the conspicuous green structure visible to the naked eye is the gametophyte (see Figure 21.6). The gametophyte is photosynthetic and is therefore nutritionally independent; the sporophyte may or may not be photosynthetic, but it is always nutritionally dependent on the gametophyte and remains permanently attached to it.

FIGURE 21.6 illustrates a moss life cycle that is typical of the life cycles of nonvascular land plants. A sporophyte produces unicellular haploid spores as products of meiosis within a sporangium. When a spore germinates, it gives rise to a multicellular haploid gametophyte whose cells contain chloroplasts and are thus photosynthetic. Eventually gametes form within specialized sex organs, called the **gametangia**. The **archegonium** is a multicellular, flask-shaped female sex organ with a long neck and a swollen base; it produces a single egg. The **antheridium** is a male sex organ in which sperm, each bearing two flagella, are produced in large numbers. Both archegonia and antheridia are produced on the same individual, so each individual has both male and female reproductive structures. Adjacent individuals often fertilize one another's gametes, however, which helps maintain genetic diversity in the population.

tiny channel through which water can travel. The hydroid is functionally similar to the tracheid, the characteristic water-conducting cell of vascular plants, but it lacks the lignin and the cell-wall structure that characterize tracheids. The possession of hydroids shows that the term "nonvascular land plant" is somewhat misleading when applied to mosses. Despite their simple systems of internal transport, however, the mosses are not considered vascular plants because of their lack of a well-developed vascular system.

HORNWORTS The **hornworts** are so named because their sporophytes often look like little horns (**FIGURE 21.5C**). The approximately 100 species of hornworts have two characteristics that distinguish them from liverworts and mosses. First, the cells of hornworts each contain a single large, platelike chloroplast, whereas the cells of the other two groups contain numerous small, lens-shaped chloroplasts. Second, of the sporophytes in

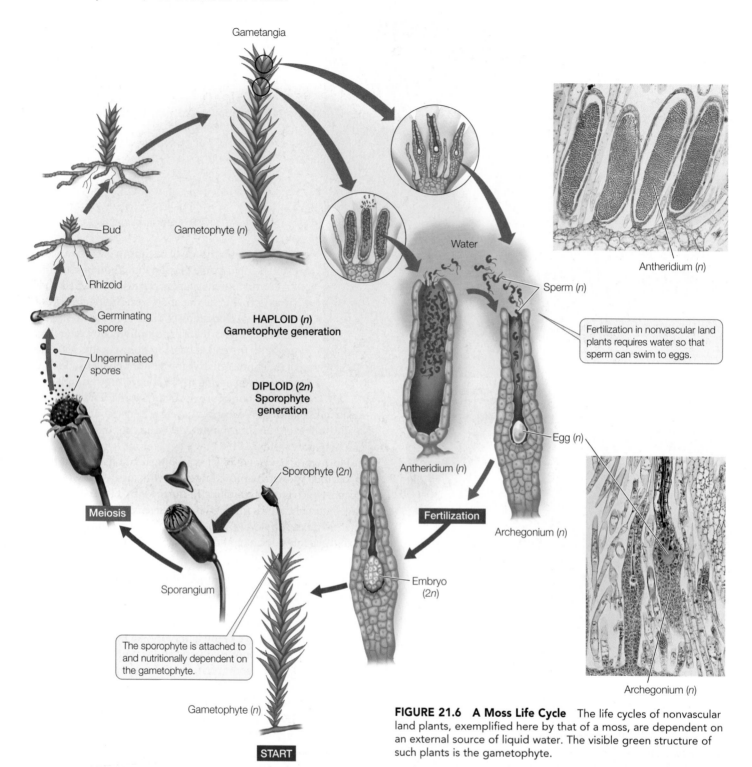

HAPLOID (*n*)
Gametophyte generation

DIPLOID (2*n*)
Sporophyte generation

Gametangia

Gametophyte (*n*)

Bud

Rhizoid

Germinating spore

Ungerminated spores

Meiosis

Sporangium

Sporophyte (2*n*)

Antheridium (*n*)

Archegonium (*n*)

Water

Sperm (*n*)

Fertilization in nonvascular land plants requires water so that sperm can swim to eggs.

Antheridium (*n*)

Egg (*n*)

Fertilization

Embryo (2*n*)

The sporophyte is attached to and nutritionally dependent on the gametophyte.

Gametophyte (*n*)

START

Archegonium (*n*)

FIGURE 21.6 A Moss Life Cycle The life cycles of nonvascular land plants, exemplified here by that of a moss, are dependent on an external source of liquid water. The visible green structure of such plants is the gametophyte.

Once released from the antheridium, the sperm must swim or be splashed by raindrops to a nearby archegonium on the same or a neighboring plant—a constraint that reflects the aquatic origins of the nonvascular land plants' ancestors. The sperm are aided on their journey by chemical attractants released by the egg or the archegonium. Before sperm can enter the archegonium, however, certain cells in the neck of the archegonium must break down, leaving a water-filled canal through which the sperm can swim to complete their journey. Notice that *all of these events require liquid water.*

yourBioPortal.com

Go to ANIMATED TUTORIAL 21.1
Life Cycle of a Moss

Once sperm arrive at an egg, the nucleus of a sperm fuses with the egg nucleus to form a diploid zygote. Mitotic divisions of the zygote produce a multicellular, diploid sporophyte embryo. The sporophyte matures and produces a sporangium, within which meiotic divisions produce spores and thus the next gametophyte generation.

Do You Understand Concept 21.2?

- Describe several adaptations of plants to the terrestrial environment, and describe the distribution of those adaptations in liverworts, mosses, and hornworts.
- Explain what is meant by alternation of generations.
- What aspects of the reproductive cycle of nonvascular land plants appear to have been retained from their aquatic ancestors?

Further adaptations to the terrestrial environment appeared as plants continued to evolve. One of the most important of these later adaptations was the appearance of vascular tissues—the characteristic that defines the vascular plants.

concept 21.3 Vascular Tissues Led to Rapid Diversification of Land Plants

The first plants possessing vascular tissues did not arise until tens of millions of years after the earliest plants had colonized the land. But once they arose, their ability to transport water and food throughout their bodies allowed them to spread to new environments and diversify rapidly.

Vascular tissues transport water and dissolved materials

Vascular plants differ from the other land plants in crucial ways, one of which is the possession of a well-developed **vascular system** consisting of tissues specialized for the transport of materials from one part of the plant to another. One type of vascular tissue, the *xylem*, conducts water and minerals from the soil to aerial parts of the plant. Because some of its cell walls contain a stiffening substance called lignin, xylem also provides support against gravity in the terrestrial environment. The other type of vascular tissue, the *phloem*, conducts the products of photosynthesis from sites where they are produced or released to sites where they are used or stored.

LINK The vascular tissues of plants are described in detail in Chapter 24

Although the vascular plants are an extraordinarily large and diverse group, a particular event was critical to their evolution. Sometime during the Paleozoic era, probably in the mid-Silurian (430 mya), a new cell type called a **tracheid** evolved in sporophytes of the earliest vascular plants. The tracheid is the principal water-conducting element of the xylem in all vascular plants except the *angiosperms* (flowering plants), and even in the angiosperms, tracheids persist along with a more specialized and efficient system derived from them.

The evolution of tracheids set the stage for the complete and permanent invasion of land by plants. First, these cells provided a pathway for the transport of water and mineral nutrients from a source of supply to regions of need in the plant body. And second, the cell walls of tracheids, stiffened by lignin, provided rigid structural support. This support is a crucial factor in a terrestrial environment because it allows plants to grow upward and thus compete for sunlight. A taller plant can intercept more direct sunlight (and thus conduct photosynthesis more readily) than a shorter plant, which may be shaded by the taller one. Increased height also improves the dispersal of spores.

The vascular plants featured another evolutionary novelty: a branching, independent sporophyte. A branching sporophyte body can produce more spores than an unbranched body, and it can develop in complex ways. The sporophyte of a vascular plant is nutritionally independent of the gametophyte at maturity. Among the vascular plants, the sporophyte is the large and obvious plant that one normally pays attention to in nature, in contrast to the relatively small, dependent sporophytes typical of most nonvascular land plants.

Vascular plants have been evolving for almost half a billion years

The evolution of an effective cuticle and protective layers for the gametangia helped make the first vascular plants successful, as did the initial absence of herbivores (plant-eating animals) on land. By the late Silurian period (about 425 mya), vascular plants were being preserved as fossils that we can study today. Their proliferation made the terrestrial environment more hospitable to animals. Arthropods, vertebrates, and other animals moved onto land only after vascular plants became established there.

Trees of various kinds appeared in the Devonian period and dominated the landscape of the Carboniferous period (359–297 mya). Forests of lycophytes (club mosses) up to 40 meters tall, along with horsetails and tree ferns, flourished in the tropical swamps of what would become North America and Europe (**FIGURE 21.7**). Plant parts from those forests sank into the swamps and were gradually covered by layers of sediment. Over millions of years, as the buried plant material was subjected to intense pressure and elevated temperatures, it was transformed into coal. Today that coal provides over half of our electricity. The world's coal deposits, although huge, are not infinite, and they cannot be renewed because the conditions that created coal no longer exist.

In the subsequent Permian period, when the continents came together to form the single gigantic land mass of Pangaea, the continental interior became warmer and drier, but late in the period glaciation was extensive. The 200-million-year reign of the lycophyte–fern forests came to an end as they were replaced by forests of early *gymnosperms*, which were prevalent until angiosperms became the dominant vegetation over much of the landscape about 65 million years ago.

The earliest vascular plants lacked roots

The earliest known vascular plants belonged to a now-extinct group called the **rhyniophytes**. The rhyniophytes were one of

FIGURE 21.7 Reconstruction of an Ancient Forest Forests of the Carboniferous period were characterized by abundant vascular plants such as club mosses, ferns, and horsetails, some of which reached heights of 40 meters. Huge flying insects (see p. 347) thrived in these forests, which are the source of modern coal deposits.

a very few types of vascular plants in the Silurian period. The landscape at that time probably consisted mostly of bare ground, with stands of rhyniophytes in low-lying moist areas. Early versions of the structural features of all the other vascular plant groups appeared in the rhyniophytes of that time. These shared features strengthen the case for the origin of all vascular plants from a common nonvascular land plant ancestor.

Rhyniophytes did not have roots. Like most modern ferns and lycophytes, they were apparently anchored in the soil by horizontal portions of stem, called **rhizomes**, which bore water-absorbing unicellular filaments called **rhizoids**. These rhizomes also bore aerial branches, and sporangia—homologous to the sporangia of mosses—were found at the tips of those branches. Their branching pattern was *dichotomous*; that is, the apex (tip) of the shoot divided to produce two equivalent new branches, each pair diverging at approximately the same angle from the original stem.

The lycophytes are sister to the other vascular plants

The club mosses and their relatives, the spike mosses and quillworts, are collectively called **lycophytes**. The lycophytes are the sister group to the remaining vascular plants (see Figure 21.1B). There are relatively few surviving species of lycophytes—just over 1,200.

The lycophytes have true roots that branch dichotomously. The arrangement of vascular tissue in their stems is simpler than that in other vascular plants. They bear simple leaflike structures called **microphylls**, which are arranged spirally on the stem. Growth in lycophytes comes entirely from apical cell division. Branching in the stems, which is also dichotomous, occurs by division of an apical cluster of dividing cells.

The sporangia of many club mosses are aggregated in conelike structures called *strobili* (singular *strobilus*; **FIGURE 21.8A**). The strobilus of a club moss is a cluster of spore-bearing leaves inserted on an axis (a linear supporting structure). Other club mosses lack strobili and bear their sporangia on (or adjacent to) the upper surfaces of specialized leaves.

Horsetails and ferns constitute a clade

The horsetails and ferns were once thought to be only distantly related. As a result of genomic analyses, we now know that they form a clade: the **monilophytes**, or "ferns and fern allies." Within the monilophytes, the horsetails are monophyletic, but

the ferns are not. However, most ferns do belong to a single clade, the *leptosporangiate ferns*. In the monilophytes—as in the *seed plants*, to which they are the sister group (see Table 21.1)—there is differentiation between a main stem and side branches. This pattern contrasts with the dichotomous branching characteristic of the lycophytes and rhyniophytes, in which each split gives rise to two branches of similar size. A synapomorphy of the monilophytes is the presence of openings in the stem, called *leaf gaps*, from which the leaves emerge.

Today there are only about 15 species of **horsetails**, all in the genus *Equisetum*. The horsetails have reduced true leaves that form in distinct whorls (circles) around the stem (**FIGURE 21.8B**). Horsetails are sometimes called "scouring rushes" because rough silica deposits found in their cell walls once made them useful for cleaning. They have true roots that branch irregularly. Horsetails have a large sporophyte and a small gametophyte, both independent.

The first **leptosporangiate ferns** appeared during the Devonian period; today this group comprises more than 12,000 species. The sporangia of leptosporangiate ferns have walls only one cell thick. The name "leptosporangiate" refers to these thin-walled sporangia (*lepton*, "thin"). Three other small groups that are also called "ferns" evolved independently, even though some are superficially similar to the leptosporangiate ferns.

Although most ferns are terrestrial, a few species live in shallow fresh water (**FIGURE 21.8C**). Terrestrial ferns are

(A) *Lycopodium annotinum*

(B) *Equisetum pratense*

(D) *Dicksonia antarctica*

(C) *Marsilea* sp.   *Salvinia* sp.

FIGURE 21.8 Lycophytes and Monilophytes (A) A strobilus is visible at the tip of this club moss. Club mosses have microphylls arranged spirally on their stems. (B) Horsetails have a distinctive growth pattern in which the stem grows in segments above each whorl of leaves. These are fertile shoots with sporangia-bearing structures at the apex. (C) The leaves of two species of water ferns. (D) Tree ferns dominate this forest on the island of Tasmania, Australia.

characterized by large leaves with branching vascular strands (**FIGURE 21.8D**). Some fern leaves become climbing organs and may grow to be as long as 30 meters.

In the alternating generations of a fern, the gametophyte is small, delicate, and short-lived, but the sporophyte can be very large and can sometimes survive for hundreds of years (**FIGURE 21.9**). Ferns require liquid water for the transport of the male gametes to the female gametes, so most ferns inhabit shaded, moist woodlands and swamps. The sporangia of ferns typically are borne on a stalk in clusters called *sori*. The sori are found on the undersurfaces of the leaves, sometimes covering the entire undersurface and sometimes located at the edges.

The vascular plants branched out

Several features that were new to the vascular plants evolved in lycophytes and monilophytes. Roots probably had their evolutionary origins as a branch, either of a rhizome or of the aboveground portion of a stem. That branch presumably penetrated the soil and branched further. The underground portion could anchor the plant firmly, and even in this primitive condition, it could absorb water and minerals.

The microphylls of lycophytes were probably the first leaflike structures to evolve among the vascular plants.

Microphylls are usually small and only rarely have more than a single vascular strand, at least in existing species. Some biologists believe that microphylls had their evolutionary origins as sterile sporangia (**FIGURE 21.10A**). The principal characteristic of a microphyll is a vascular strand that departs from the vascular system of the stem in such a way that the structure of the stem's vascular system is scarcely disturbed. This pattern was evident even in the lycophyte trees of the Carboniferous period, many of which had microphylls many centimeters long.

The monilophytes and seed plants constitute a clade called the **euphyllophytes** (*eu*, "true"; *phyllos*, "leaf"). An important synapomorphy of the euphyllophytes is **overtopping**, a growth pattern in which one branch differentiates from and grows beyond the others (**FIGURE 21.10B**). Overtopping would have given these plants an advantage in the competition for light, enabling them to shade their dichotomously branching competitors. The overtopping growth of the euphyllophytes also allowed a new type of leaflike structure to evolve. This larger, more complex leaf is called a **megaphyll**. The megaphyll is thought to have arisen from the flattening of a portion of a branching stem system that exhibited overtopping growth. This change was followed by the development of photosynthetic tissue between the members of overtopped groups of

FIGURE 21.9 **Life Cycle of a Fern** The most conspicuous stage in the fern life cycle is the mature diploid sporophyte, shown at the bottom of this diagram. The inset shows sori on the underside of a fern leaf. Each sorus contains many spore-producing sporangia.

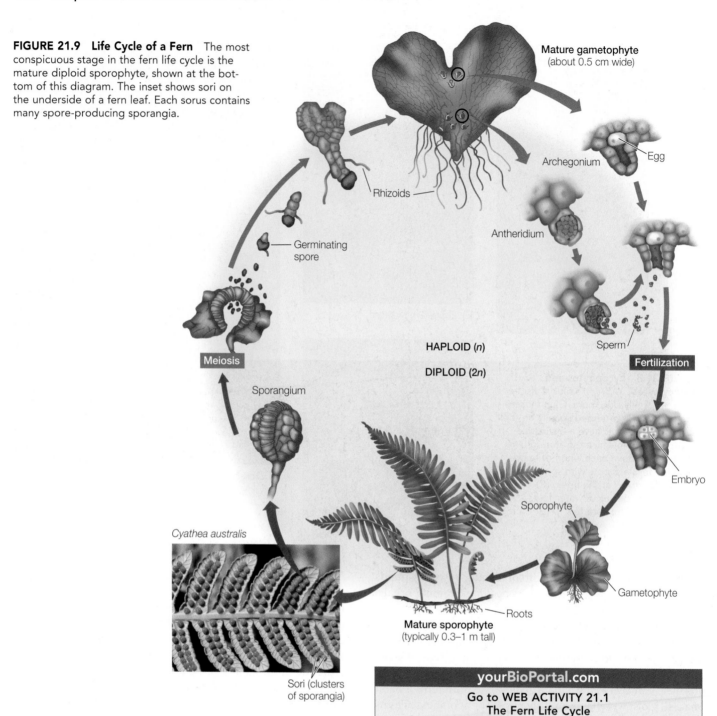

Mature gametophyte
(about 0.5 cm wide)

Archegonium

Egg

Antheridium

Rhizoids

Germinating spore

Sperm

HAPLOID (*n*)

DIPLOID (2*n*)

Meiosis

Fertilization

Sporangium

Embryo

Cyathea australis

Sporophyte

Gametophyte

Roots

Mature sporophyte
(typically 0.3–1 m tall)

Sori (clusters of sporangia)

yourBioPortal.com

Go to WEB ACTIVITY 21.1
The Fern Life Cycle

branches, which had the advantage of increasing the photosynthetic surface area of those branches.

Heterospory appeared among the vascular plants

In the lineages of present-day vascular plants that are most similar to their ancestors, the gametophyte and the sporophyte are independent, and both are usually photosynthetic. The spores produced by the sporophyte are of a single type and develop into a single type of gametophyte that bears both female and male reproductive organs (see Figure 21.9). Such plants, which bear a single type of spore, are said to be **homosporous** (FIGURE 21.11A).

A system with two distinct types of spores evolved somewhat later. Plants of this type are said to be **heterosporous** (FIGURE 21.11B). In heterospory, one type of spore—the **megaspore**—develops into a specifically female gametophyte (a **megagametophyte**) that produces only eggs. The other type, the **microspore**, is smaller and develops into a male gametophyte (a **microgametophyte**) that produces only sperm. The sporophyte produces megaspores in small numbers in *megasporangia* and microspores in large numbers in *microsporangia*. Heterospory affects not only the spores and the gametophytes, but also the sporophyte plant itself, which must develop two types of sporangia.

(A)

Vascular tissue

Sporangia

Sporangium

Microphyll

A sporangium evolved into a simple leaflike structure.

Time

Lycopodium (club moss)

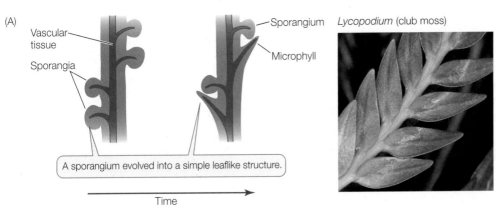

FIGURE 21.10 Evolution of Leaves (A) Microphylls are thought to have evolved from sterile sporangia. (B) The megaphylls of monilophytes and seed plants may have arisen as photosynthetic tissue developed between branch pairs that were "left behind" as dominant branches overtopped them.

(B)

A branching stem system became progressively reduced and flattened.

Overtopping

Flat plates of photosynthetic tissue developed between branches.

Megaphyll

The end branches evolved into the veins of leaves.

Time

Adiantum (fern)

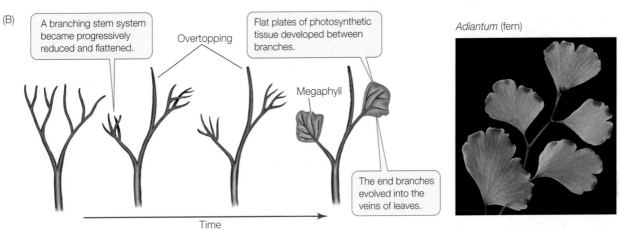

(A) **Homospory**

Homosporous plants produce a single type of spore.

The spores of homosporous plants produce a single type of gametophyte with both male and female reproductive organs.

Gametophyte (n)

Spore (n)

Antheridium (♂) (n)

Archegonium (♀) (n)

Sperm (n)

Egg (n)

Meiosis

HAPLOID (n)

DIPLOID (2n)

Fertilization

Spore mother cell (2n)

Zygote (2n)

Sporangium (2n)

Sporophyte (2n)

Embryo (2n)

(B) **Heterospory**

Heterosporous plants produce two types of spores: a larger megaspore and a smaller microspore.

The spores of heterosporous plants produce separate male and female gametophytes.

Megagametophyte (♀) (n)

Microgametophyte (♂) (n)

Megaspore (n)

Microspore (n)

Egg (n)

Sperm (n)

Meiosis

HAPLOID (n)

DIPLOID (2n)

Fertilization

Spore mother cell (2n)

Spore mother cell (2n)

Zygote (2n)

Megasporangium (2n)

Microsporangium (2n)

Sporophyte (2n)

Embryo (2n)

FIGURE 21.11 Homospory and Heterospory (A) Homosporous plants bear a single type of spore. Each gametophyte has two types of sex organs, antheridia (male) and archegonia (female). (B) Heterosporous plants bear two types of spores that develop into distinctly male and female gametophytes.

The earliest vascular plants were all homosporous, but heterospory evidently evolved several times independently among later groups of vascular plants. The fact that heterospory evolved repeatedly suggests that it affords selective advantages. Subsequent evolution in the land plants featured ever greater specialization of the heterosporous condition.

Do You Understand Concept 21.3?

- How do the vascular tissues xylem and phloem serve the vascular plants?

- Describe the evolution and distribution of different kinds of leaves and roots among the vascular plants.

- Explain the concept of heterospory. Why is heterospory thought to provide selective advantages over homospory?

All of the vascular plant groups we have discussed thus far disperse by means of spores. The embryos of these seedless vascular plants develop directly into sporophytes, which either survive or die, depending on environmental conditions. The spores of some seedless plants may remain dormant and viable for long periods, but the embryos of seedless plants are relatively unprotected (see Figure 21.9). Greater protection of the embryo evolved in the seed plants.

concept 21.4 Seeds Protect Plant Embryos

The **seed plants**—the gymnosperms and the angiosperms—provide a secure and lasting dormant stage for the embryo, known as a seed. In seeds, embryos may safely lie dormant, in some cases for many years (even centuries), until conditions are right for germination. How is this protection provided?

Features of the seed plant life cycle protect gametes and embryos

In Concept 21.2 we described a trend in plant evolution: the sporophyte became less dependent on the gametophyte, which became smaller in relation to the sporophyte. This trend continued with the seed plants, whose gametophyte generation is reduced even further than it is in the ferns (**FIGURE 21.12**). The haploid gametophyte develops partly or entirely while attached to and nutritionally dependent on the diploid sporophyte.

Among the seed plants, only the earliest diverging groups of gymnosperms (including modern cycads and ginkgos) have swimming sperm. Other groups of gymnosperms and the angiosperms evolved other means of bringing eggs and sperm together. The culmination of this striking evolutionary trend in seed plants was independence from the liquid water that earlier plants needed to assist sperm in reaching the egg. This adaptation, along with the advent of the seed, gave seed plants the opportunity to colonize drier areas and spread over the terrestrial environment.

Seed plants are heterosporous (see Figure 21.11B); that is, they produce two types of spores, one that becomes a microgametophyte and one that becomes a megagametophyte. They form separate microsporangia and megasporangia on structures that are grouped on short axes, such as the stamens and pistils of an angiosperm flower.

FIGURE 21.12 The Relationship between Sporophyte and Gametophyte In the course of plant evolution, the gametophyte (brown) has been reduced and the sporophyte (blue) has become more prominent.

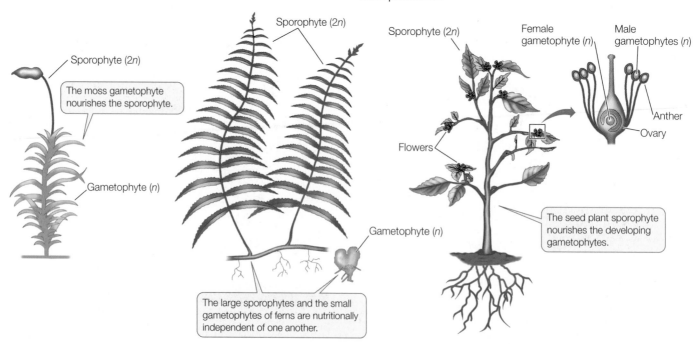

Sporophyte (2n)

Sporophyte (2n)

The moss gametophyte nourishes the sporophyte.

Gametophyte (n)

The large sporophytes and the small gametophytes of ferns are nutritionally independent of one another.

Sporophyte (2n)

Female gametophyte (n)

Male gametophytes (n)

Anther

Ovary

Flowers

Gametophyte (n)

The seed plant sporophyte nourishes the developing gametophytes.

Within the microsporangium, the meiotic products are microspores, which divide mitotically within the spore wall one or a few times to form a multicellular male gametophyte called a **pollen grain**. Pollen grains are released from the microsporangium to be distributed by wind or by an animal pollinator (**FIGURE 21.13**). The spore wall that surrounds the pollen grain contains *sporopollenin*, the most chemically resistant biological compound known, which protects the pollen grain against dehydration and chemical damage—another advantage in terms of survival in the terrestrial environment.

In contrast to the pollen grains produced by the microsporangia, the megaspores of seed plants are not shed. Instead, they develop into female gametophytes within the megasporangia. These megagametophytes are dependent on the sporophyte for food and water.

In most seed plant species, only one of the meiotic products in a megasporangium survives. The surviving haploid nucleus divides mitotically, and the resulting cells divide again to produce a multicellular female gametophyte. The megasporangium is surrounded by sterile sporophytic structures, which form an **integument** that protects the megasporangium and its contents. Together, the megasporangium and integu-

Salix cinerea

FIGURE 21.13 Pollen Grains Pollen grains are the male gametophytes of seed plants. The pollen of pussy willow is dispersed by the wind. The grains may land near female gametophytes of the same or other willow plants.

ment constitute the **ovule**, which will develop into a seed after fertilization (**FIGURE 21.14**).

The arrival of a pollen grain at an appropriate landing point, close to a female gametophyte on a sporophyte of the same species, is called **pollination**. A pollen grain that reaches this point develops further. It produces a slender **pollen tube** that

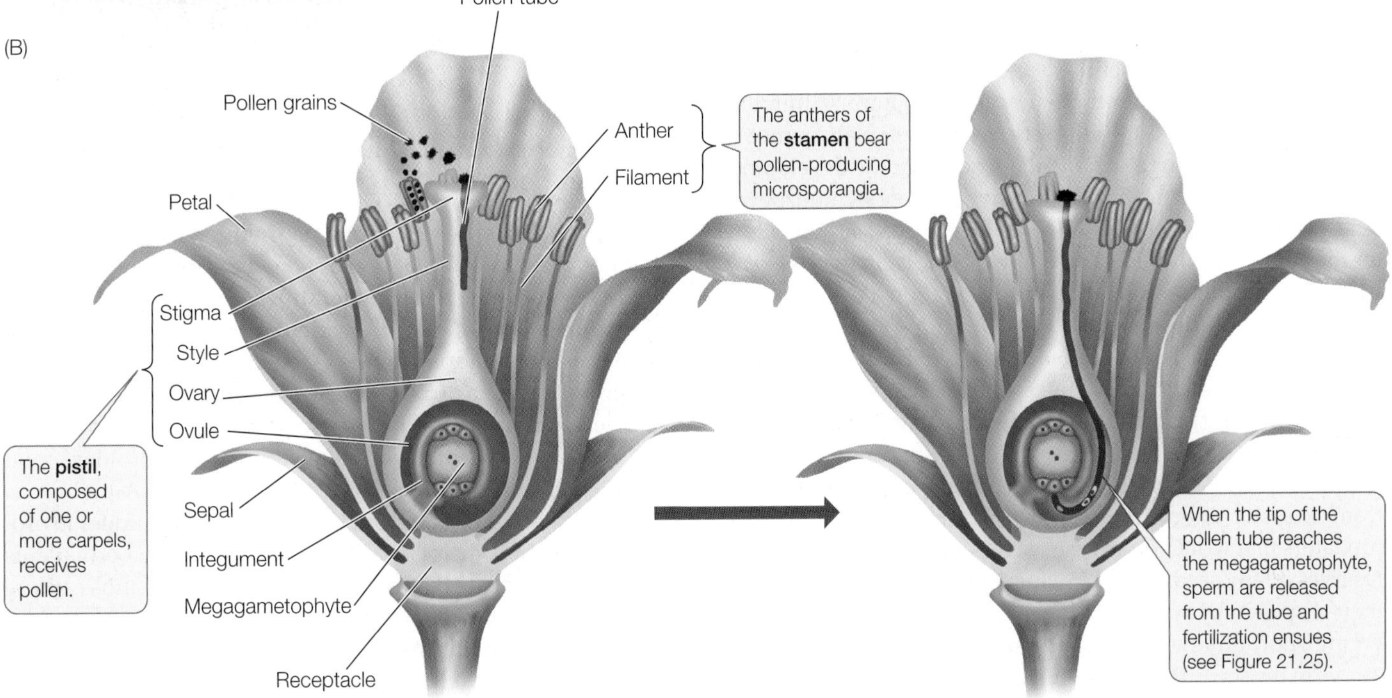

FIGURE 21.14 Pollination in Seed Plants In all seed plants, a pollen tube grows from the pollen grain to the megagametophyte, where sperm are released. (A) Scanning electron micrograph of a pollen tube growing in a prairie gentian flower. (B) The process of pollination is diagrammed for a generalized angiosperm flower.

(A)

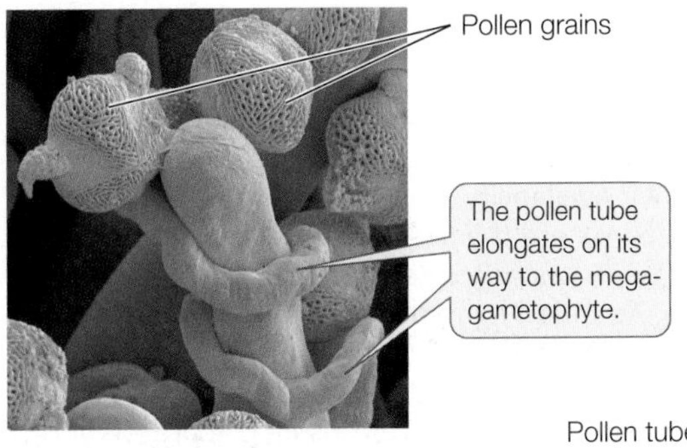

Pollen grains

The pollen tube elongates on its way to the megagametophyte.

(B)

Pollen tube

Pollen grains

Anther

Filament

The anthers of the **stamen** bear pollen-producing microsporangia.

Petal

Stigma

Style

Ovary

Ovule

The **pistil**, composed of one or more carpels, receives pollen.

Sepal

Integument

Megagametophyte

Receptacle

When the tip of the pollen tube reaches the megagametophyte, sperm are released from the tube and fertilization ensues (see Figure 21.25).

APPLY THE CONCEPT

Seeds protect plant embryos

In 1879, W. J. Beal began an experiment that he could not hope to finish in his lifetime. He prepared 20 lots of seeds for long-term storage. Each lot consisted of 50 seeds from each of 23 species. He mixed each lot of seeds with sand and placed the mixture in an uncapped bottle, then buried all the bottles on a sandy knoll. At regular intervals over the next century, several different biologists have excavated a bottle and checked the viability of its contents. The table shows the number of germinating seeds (of the original 50) from three species in years 50–100 of this ongoing experiment.

SPECIES	YEARS AFTER BURIAL					
	50	60	70	80	90	100
Oenothera biennis (Evening primrose)	19	12	7	5	0	0
Rumex crispus (Curly dock)	26	2	7	1	0	0
Verbascum blattaria (Moth mullein)	31	34	37	35	10	21

1. Calculate the percentage of surviving seeds for these three species in years 50–100 and graph seed survivorship as a function of time buried.

2. No seeds of the first two species were viable after 90 years of the experiment. Assume 100% seed viability at the start of the experiment (year 0), and predict from your graph the approximate year when you think the last of the *Verbascum blattaria* seeds will germinate.

3. What factors do you think might influence the differences in long-term seed viability among the species?

---

elongates and digests its way toward the megagametophyte (see Figure 21.14). When the tip of the pollen tube reaches the megagametophyte, sperm are released from the tube and fertilization occurs.

The resulting diploid zygote divides repeatedly, forming an embryonic sporophyte. After a period of embryonic development, growth is temporarily suspended (the embryo enters a *dormant* stage). The end product at this stage is the multicellular seed.

The seed is a complex, well-protected package

A **seed** contains tissues from three generations. A *seed coat* develops from the integument—the tissues of the diploid sporophyte parent that surround the megasporangium. Within the megasporangium is haploid tissue from the female gametophyte, which contains a supply of nutrients for the developing embryo. (This tissue is fairly extensive in most gymnosperm seeds. In angiosperm seeds it is greatly reduced, and nutrition for the embryo is supplied instead by a tissue called *endosperm*.) In the center of the seed is the third generation, the embryo of the new diploid sporophyte.

The seed is a well-protected resting stage. The seeds of some species may remain dormant but stay *viable* (capable of growth and development) for many years, germinating only when conditions are favorable for the growth of the sporophyte.

During the dormant stage, the seed coat protects the embryo from excessive drying and may also protect it against potential predators that would otherwise eat the embryo and its nutrient reserves. Many seeds have structural adaptations that promote their dispersal by wind or, more often, by animals. When the young sporophyte resumes growth, it draws on the food reserves in the seed. The possession of seeds is a major reason for the enormous evolutionary success of the seed plants, which are the dominant life forms of most modern terrestrial floras.

FRONTIERS Seeds of many plant species important to humans are stored at low temperatures in seed banks to ensure their availability for future human populations. These seed banks are themselves subject to destruction by humans, however, as happened in Iraq and Afghanistan in recent wars. To guard against permanent losses of plant species needed for human survival, a multinational global seed vault has been constructed in Norway, deep inside a sandstone mountain. Its features ensure the preservation of the seeds inside in all but the most extreme of possible future scenarios.

A change in anatomy enabled seed plants to grow to great heights

The earliest fossil seed plants were the seed ferns found in late Devonian rocks. These plants had extensively thickened woody stems, which developed through the proliferation of xylem. This type of growth, which increases the diameter of stems and roots in some modern seed plants, is called **secondary growth**, and its product is called secondary xylem, or *wood*.

The younger portion of the wood produced by secondary growth is well adapted for water transport, but older wood becomes clogged with resins or other materials. Although no longer functional in transport, the older wood continues to provide support for the plant. This support allows woody plants to grow taller than other plants around them and thus capture more light for photosynthesis.

Not all seed plants are woody. In the course of seed plant evolution, many groups lost the woody growth habit; however, other advantageous attributes helped them become established in an astonishing variety of places.

Gymnosperms have naked seeds

The two major groups of living seed plants are the **gymnosperms** (such as pines and cycads) and the **angiosperms** (flowering plants; see Figure 21.1B). We'll discuss the flowering plants in Concept 21.5 and examine the gymnosperms here.

The gymnosperms are seed plants that do not form flowers or fruits. Gymnosperms (which means "naked-seeded") are so named because their ovules and seeds, unlike those of angiosperms, are not protected by ovary or fruit tissue. Although there are probably fewer than 1,200 living species of gymnosperms, these plants are second only to the angiosperms in their dominance of the terrestrial environment. The gymnosperms can be divided into four major groups:

- **Cycads** are palmlike plants of the tropics and subtropics (**FIGURE 21.15A**). Of the present-day gymnosperms, the cycads are probably the earliest-diverging clade. There are about 300 species, some of which grow as tall as 20 meters. The tissues of many species are highly toxic to humans if ingested.

- **Ginkgos**, common during the Mesozoic era, are represented today by a single genus and species: *Ginkgo biloba*, the maidenhair tree (**FIGURE 21.15B**). There are both male (microsporangiate) and female (megasporangiate) maidenhair trees. The difference is determined by X and Y sex chromosomes, as in humans; few other plants have distinct sex chromosomes.

- **Gnetophytes** number about 90 species in three very different genera, which share certain characteristics analogous to ones found in the angiosperms. One of the gnetophytes is *Welwitschia* (**FIGURE 21.15C**), a long-lived desert plant with just two straplike leaves that sprawl on the sand and can grow as long as 3 meters.

- **Conifers** are by far the most abundant of the gymnosperms. There are about 700 species of these cone-bearing plants, including the pines and redwoods (**FIGURE 21.15D**).

With the exception of the gnetophytes, the living gymnosperm groups have only tracheids as water-conducting and support cells within the xylem; they lack the cells called *vessel*

(A) *Encephalartos transvenosus*

(B) *Ginkgo biloba*

(D) *Pinus longaeva*

(C) *Welwitschia mirabilis*

FIGURE 21.15 Diversity among the Gymnosperms (A) Many cycads have growth forms that resemble both ferns and palms, although cycads are not closely related to either group. (B) The characteristic broad leaves of the maidenhair tree. (C) The straplike leaves of *Welwitschia*, a gnetophyte, grow throughout the life of the plant, breaking and splitting as they grow. (D) Conifers dominate many types of landscapes in the Northern Hemisphere. Bristlecone pines such as these are the longest-lived individual trees known.

elements and *fibers* (specialized for water conduction and support, respectively) that are found in angiosperms. While the gymnosperm water-transport and support system may thus seem somewhat less efficient than that of the angiosperms, it serves some of the largest trees known. The coastal redwoods of California are the tallest gymnosperms; the largest are well over 100 meters tall.

During the Permian, as environments became warmer and drier, the conifers and cycads flourished. Gymnosperm forests

(A) *Pinus contorta*

Woody scales are modifications of branches

Seed

Central axis

Cross section of a megastrobilus

Female (seed-bearing) cones, or megastrobili

(B) *Pinus contorta*

Herbaceous scales are modifications of leaves

Microsporangia bearing pollen

Central axis

Cross section of a microstrobilus

Male (pollen-bearing) cones, or microstrobili

FIGURE 21.16 Female and Male Cones   (A) The scales of female cones (megastrobili) are modified branches. (B) The scales of male cones (microstrobili) are modified leaves.

everywhere, rank among the great forests of the world. All these trees belong to one group of gymnosperms: the conifers, or cone-bearers.

Male and female **cones** contain the reproductive structures of conifers. The female (seed-bearing) cone is known as a **megastrobilus** (plural *megastrobili*); an example is the familiar woody cone of pine trees. The seeds in a megastrobilus are protected by a tight cluster of woody scales, which are modifications of branches extending from a central axis (**FIGURE 21.16A**). The typically much smaller male (pollen-bearing) cone is known as a **microstrobilus**. The microstrobilus is typically herbaceous rather than woody, as its scales are composed of modified leaves, beneath which are the pollen-bearing microsporangia (**FIGURE 21.16B**).

The life cycle of a pine illustrates reproduction in gymnosperms (**FIGURE 21.17**). The production of male gametophytes in the form of pollen grains frees the plant completely from its dependence on liquid water for fertilization. Wind, rather than water, assists conifer pollen grains in their first stage of travel from the microstrobilus to the female gametophyte inside a cone. A pollen tube provides the sperm with the means for the last stage of travel by elongating through maternal sporophytic tissue. When the pollen tube reaches the female gametophyte, it releases two sperm, one of which degenerates after the other unites with an egg. Union of sperm and egg results in a zygote; mitotic divisions and further development of the zygote result in an embryo.

The megasporangium, in which the female gametophyte will form, is enclosed in a layer of sporophytic tissue—the integument—that will eventually develop into the seed coat that protects the embryo. The integument, the megasporangium inside it, and the tissue attaching it to the maternal sporophyte constitute the ovule. The pollen grain enters through a small opening in the integument at the tip of the ovule, the **micropyle**.

Most conifer ovules (which will develop into seeds after fertilization) are borne exposed on the upper surfaces of the scales of the cone (megastrobilus). The only protection of the ovules comes from the scales, which are tightly pressed against one another within the cone. Some pines, such as the lodgepole pine, have such tightly closed cones that only fire suffices to split them open and release the seeds. These species are said to be *fire-adapted*, and fire is essential to their reproduction.

About half of all conifer species have soft, fleshy modifications of cones that envelop their seeds; examples are the fruitlike cones ("berries") of juniper and yew. Animals may eat these tissues and disperse the seeds in their feces, often carrying them considerable distances from the parent plant.

changed over time as the gymnosperm groups evolved. Gymnosperms dominated the Mesozoic era, during which the continents drifted apart and large dinosaurs lived. Gymnosperms were the principal trees in all forests until about 65 million years ago, and even today conifers are the dominant trees in many forests, especially at high latitudes and altitudes. The oldest living single organism on Earth today is a gymnosperm in California—a bristlecone pine that germinated about 4,800 years ago, at about the time the ancient Egyptians were starting to develop writing (see Figure 21.15D).

Conifers have cones but no motile gametes

The great Douglas fir and cedar forests found in the northwestern United States and the massive boreal forests of pine, fir, and spruce of the northern regions of Eurasia and North America, as well as on the upper slopes of mountain ranges

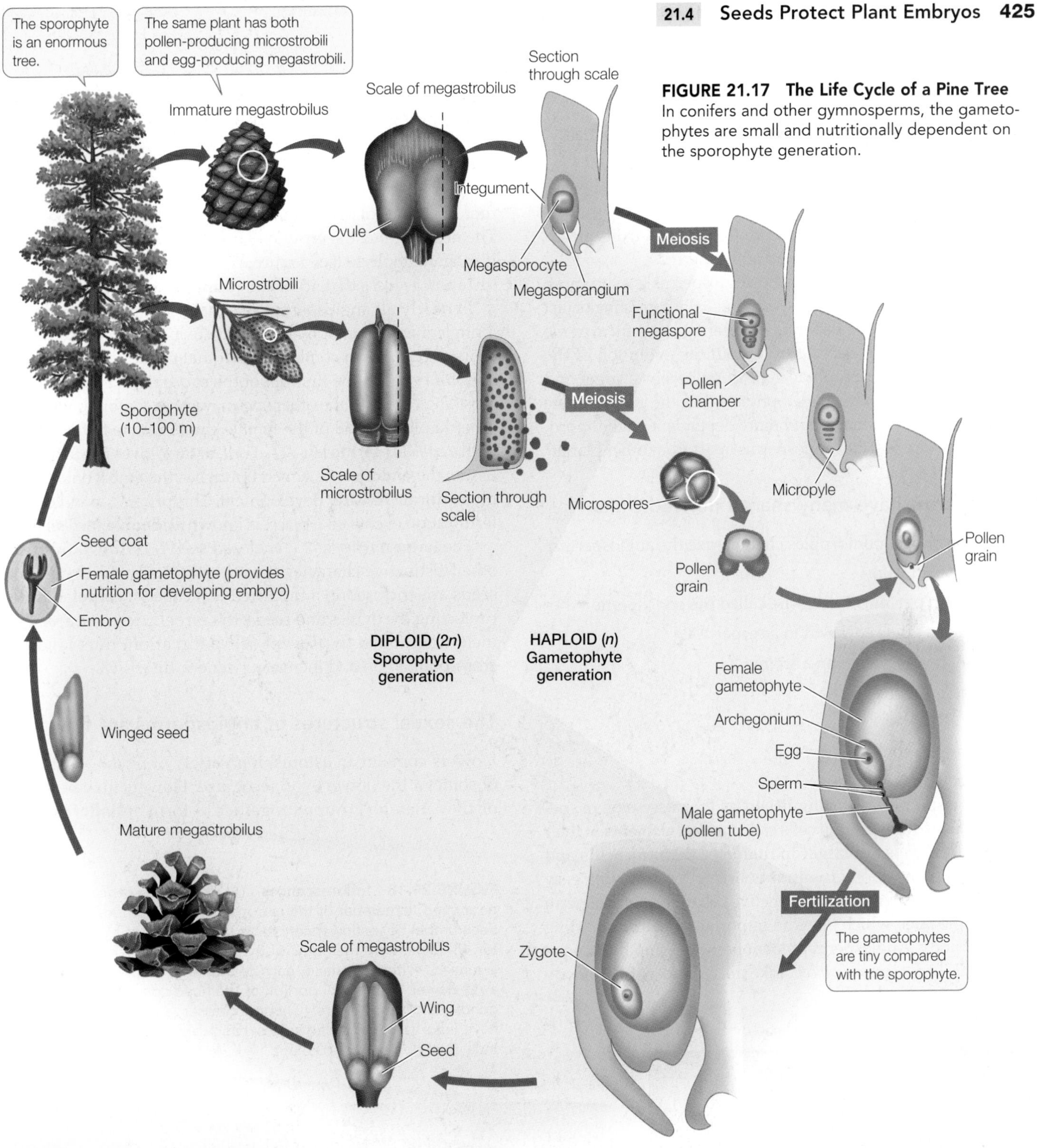

The sporophyte is an enormous tree.

The same plant has both pollen-producing microstrobili and egg-producing megastrobili.

Immature megastrobilus

Scale of megastrobilus

Section through scale

FIGURE 21.17 The Life Cycle of a Pine Tree
In conifers and other gymnosperms, the gametophytes are small and nutritionally dependent on the sporophyte generation.

Integument

Ovule

Megasporocyte

Megasporangium

Meiosis

Functional megaspore

Pollen chamber

Microstrobili

Sporophyte (10–100 m)

Scale of microstrobilus

Section through scale

Meiosis

Microspores

Micropyle

Pollen grain

Pollen grain

Seed coat

Female gametophyte (provides nutrition for developing embryo)

Embryo

DIPLOID (2n) Sporophyte generation

HAPLOID (n) Gametophyte generation

Female gametophyte

Archegonium

Egg

Sperm

Male gametophyte (pollen tube)

Pollen grain

Winged seed

Mature megastrobilus

Scale of megastrobilus

Zygote

Fertilization

The gametophytes are tiny compared with the sporophyte.

Wing

Seed

yourBioPortal.com
Go to WEB ACTIVITY 21.4 and ANIMATED TUTORIAL 21.2
Life Cycle of a Conifer

Do You Understand Concept 21.4?

- Distinguish between the roles of the megagametophyte and the pollen grain.
- Explain the importance of pollen in freeing seed plants from dependence on liquid water.
- What are some of the advantages afforded by seeds? By wood?
- Do you understand how fire can be necessary for the survival of some plant species?

The "berries" on some gymnosperms such as juniper and yew are not true fruits but rather are fleshy cones. As we will see next, true fruits are the ripened ovaries of plants. Ovaries are absent in gymnosperms but are characteristic of the plant group that is dominant today: the angiosperms, or flowering plants, which include more than 250,000 species.

Flowers and Fruits Increase the Reproductive Success of Angiosperms

Production of fruits is a unique characteristic of angiosperms. Their most obvious feature, however, is of course their sexual organs: **flowers**. As we will see, both flowers and fruits afford major reproductive advantages to angiosperms.

The female gametophyte of the angiosperms is even more reduced than that of the gymnosperms, usually consisting of just seven cells. Thus the angiosperms represent the current extreme of the trend we have traced throughout the evolution of the vascular plants: the sporophyte generation becomes larger and more independent of the gametophyte, while the gametophyte generation becomes smaller and more dependent on the sporophyte. What else sets the angiosperms apart from other plants?

Angiosperms have many shared derived traits

Several major synapomorphies characterize the angiosperms:

- Double fertilization
- Production of a nutritive tissue called the endosperm
- Ovules and seeds enclosed in a carpel
- Germination of pollen on a stigma
- Flowers
- Fruits
- Phloem with companion cells
- Reduced gametophytes

Most angiosperms are distinguished by the possession of specialized water-transporting cells called **vessel elements** in their xylem. These cells are larger in diameter than tracheids and connect with one another without obstruction, allowing easy water movement. A second distinctive cell type in angiosperm xylem is the **fiber**, which plays an important role in supporting the plant body. Angiosperm phloem possesses another unique cell type, called a *companion cell*. Like the gymnosperms, woody angiosperms show secondary growth, increasing in diameter by producing secondary xylem and secondary phloem.

Pollination in the angiosperms consists of the arrival of a microgametophyte—a pollen grain—on a receptive surface in a flower (the *stigma*). As in the gymnosperms, pollination is the first in a series of events that result in the formation of a seed. The next event is the growth of a pollen tube extending to the megagametophyte (see Figure 21.14). The third event is a fertilization process that, in detail, is unique to the angiosperms.

In nearly all angiosperms, *two* male gametes, contained in a single microgametophyte, participate in fertilization. The nucleus of one sperm combines with that of the egg to produce a diploid zygote, the first cell of the sporophyte generation. In most angiosperms, the other sperm nucleus combines with two other haploid nuclei of the female gametophyte to form a cell with a *triploid* (3n) nucleus. That cell, in turn, gives rise to triploid tissue, the **endosperm**, which nourishes the embryonic sporophyte during its early development. This process, in which two fertilization events take place, is known as **double fertilization**.

The name *angiosperm* ("enclosed seed") is drawn from another distinctive characteristic of these plants: the ovules and seeds are enclosed in a modified leaf called a **carpel**. Besides protecting the ovules and seeds, the carpel often interacts with incoming pollen to prevent self-pollination, thus favoring cross-pollination and increasing genetic diversity.

The sexual structures of angiosperms are flowers

Flowers come in an astonishing variety of forms—just think of some of the flowers you recognize. Flowers may be single, or they may be grouped together to form an **inflorescence**.

FIGURE 21.18 Inflorescences (A) The inflorescence of bishop's goutweed, a member of the carrot family, is a compound umbel. Each umbel bears flowers on stalks that arise from a common center. (B) Zinnias are members of the aster family; their inflorescence is a head. Within the head, each of the long, petal-like structures is a ray flower; the central portion of the head consists of dozens to hundreds of disc flowers. (C) Some grasses, such as quack grass, have inflorescences called spikes, which are composed of many individual flowers, or spikelets.

(A) *Aegopodium podagraria*   Umbel

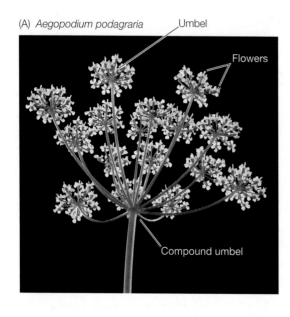

Flowers

Compound umbel

(B) *Zinnia elegans*

Ray flowers   Disc flowers

(C) *Agropyron repens*

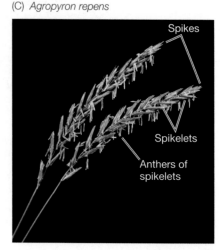

Spikes

Spikelets

Anthers of spikelets

Different families of flowering plants have characteristic types of inflorescences, such as the compound umbels of the carrot family (**FIGURE 21.18A**), the heads of the aster family (**FIGURE 21.18B**), and the spikes of many grasses (**FIGURE 21.18C**).

If you examine any familiar flower, you will notice that the outer parts look somewhat like leaves. In fact, all the parts of a flower *are* modified leaves.

The diagram in Figure 21.14B represents a generalized flower (for which there is no exact counterpart in nature). The structures bearing microsporangia are called **stamens**. Each stamen is composed of a **filament** bearing an **anther** that contains the pollen-producing microsporangia. The structures bearing megasporangia are the carpels. A structure composed of one carpel or two or more fused carpels is called a **pistil**. The swollen base of the pistil, containing one or more ovules (each containing a megasporangium surrounded by its protective integument), is called the **ovary**. The apical stalk of the pistil is the **style**, and the terminal surface that receives pollen grains is the **stigma**.

In addition, many flowers have specialized sterile (non-spore-bearing) leaves. The inner ones are called **petals** (collectively, the **corolla**) and the outer ones **sepals** (collectively, the **calyx**). The corolla and calyx (collectively, the *perianth*) can be quite showy and often play roles in attracting animal pollinators to the flower. The calyx more commonly protects the immature flower in bud. From base to apex, these floral organs—sepals, petals, stamens, and carpels—are usually positioned in circular arrangements or whorls and attached to a central stalk (called the *receptacle*).

The generalized flower in Figure 21.14B has functional megasporangia and microsporangia; such flowers are referred to as **perfect** (or hermaphroditic). Many angiosperms produce two types of flowers, one with only megasporangia and the other with only microsporangia. Consequently, either the stamens or the carpels are nonfunctional or absent in a given flower, and the flower is referred to as **imperfect**.

Species such as corn or birch, in which both megasporangiate (female) and microsporangiate (male) flowers occur on the same plant, are said to be **monoecious** ("one-housed"—but, it must be added, one house with separate rooms). Complete separation of imperfect flowers occurs in some other angiosperm species, such as willows and date palms; in these species, an individual plant produces either flowers with stamens or flowers with carpels, but never both. Such species are said to be **dioecious** ("two-housed").

Flower structure has evolved over time

The flowers of the earliest-diverging clades of angiosperms have a large and variable number of *tepals* (undifferentiated sepals and petals), carpels, and stamens (**FIGURE 21.19A**). Evolutionary change within the angiosperms has included some striking modifications of this early condition: reductions in the number of each type of floral organ to a fixed number, differentiation of petals from sepals, and changes in symmetry from radial (as in a lily or magnolia) to bilateral (as in a sweet pea or orchid), often accompanied by an extensive fusion of parts (**FIGURE 21.19B**).

According to one hypothesis, the first carpels to evolve were leaves with marginal sporangia, folded but incompletely closed. Early in angiosperm evolution, the carpels fused and became progressively more buried in receptacle tissue, forming the ovary (**FIGURE 21.20A**). In some flowers, the other floral organs are attached at the top of the ovary, rather than at the bottom as in Figure 21.14B. The stamens of the most ancient flowers may have appeared leaflike (**FIGURE 21.20B**), little resembling those of the generalized flower in Figure 21.14B.

Why do so many flowers have pistils with long styles and anthers with long filaments? Natural selection has favored length in both of these structures, probably because length increases the likelihood of successful pollination. Long filaments may bring the anthers into contact with insect bodies, or they may place the anthers in a better position to catch the wind. Similar arguments apply to long styles.

A perfect flower represents a compromise of sorts. On the one hand, by attracting a pollinating bird or insect, the plant is attending to both its female and male functions with a single flower type, whereas plants with imperfect flowers must create that attraction twice—once for each type of flower. On the other hand, the perfect flower can favor self-pollination, which is usually disadvantageous. Another potential problem is that the female and male functions might interfere with each other—for example, the stigma might be so placed as to make it

(A) *Nymphaea* sp.

(B) *Viola tricolor*

FIGURE 21.19 Flower Form and Evolution
(A) A water lily shows the major features of early flowers: it is radially symmetrical, and the individual tepals, stamens, and carpels are separate, numerous, and attached at their bases. (B) Violets such as this "Johnny jump-up" have a bilaterally symmetrical structure that evolved much later than radial flower symmetry.

(A) Carpel evolution

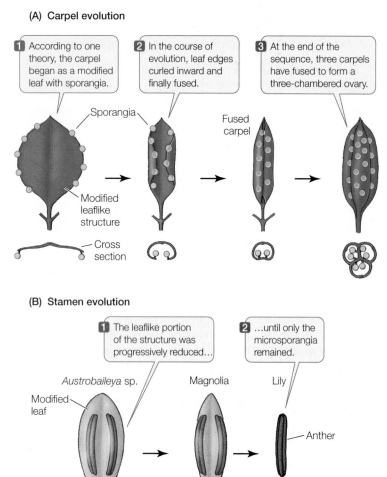

1 According to one theory, the carpel began as a modified leaf with sporangia.

2 In the course of evolution, leaf edges curled inward and finally fused.

3 At the end of the sequence, three carpels have fused to form a three-chambered ovary.

Sporangia

Fused carpel

Modified leaflike structure

Cross section

(B) Stamen evolution

1 The leaflike portion of the structure was progressively reduced...

2 ...until only the microsporangia remained.

Austrobaileya sp.

Magnolia

Lily

Modified leaf

Anther

Sporangia

Cross section

Filament

FIGURE 21.20 Carpels and Stamens Evolved from Leaflike Structures (A) Possible stages in the evolution of a carpel from a more leaflike structure. (B) The stamens of three modern plants show three possible stages in the evolution of that organ. (It is *not* implied that these species evolved from one another; their structures simply illustrate the possible stages.)

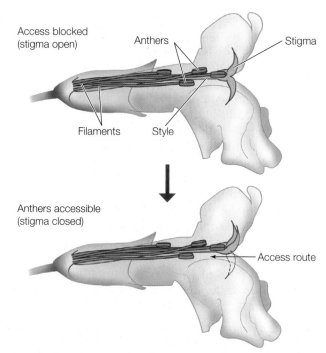

Access blocked (stigma open)

Anthers

Stigma

Filaments

Style

Anthers accessible (stigma closed)

Access route

FIGURE 21.21 An Unusual Way to Prevent Selfing Both long stamens and long styles facilitate cross-pollination, but if these male and female structures are too close to each other, the likelihood of (disadvantageous) self-pollination increases. In *Mimulus aurantiacus*, the stigma is initially open, blocking access to the anthers. A hummingbird's touch as it deposits pollen on the stigma causes one lobe of the stigma to retract, creating a path to the anthers and allowing pollen dispersal.

difficult for pollinators to reach the anthers, thus reducing the export of pollen to other flowers.

Might there be a way around these problems? One solution is seen in the bush monkeyflower (*Mimulus aurantiacus*), which is pollinated by hummingbirds. Its flower has a stigma that initially serves as a screen, hiding the anthers (**FIGURE 21.21**). Once a hummingbird touches the stigma, one of the stigma's two lobes is retracted, so that subsequent hummingbird visitors pick up pollen from the previously screened anthers. Thus the first bird to visit the flower transfers pollen from another plant to the stigma, eventually leading to fertilization. Later visitors pick up pollen from the now-accessible anthers, fulfilling the flower's male function. **FIGURE 21.22** describes the experiment that revealed the function of this mechanism.

Angiosperms have coevolved with animals

Whereas many gymnosperms are pollinated by wind, most angiosperms are pollinated by animals. The many different pollination mutualisms between plants and animals are vital to both parties. We discussed coevolution of insects and orchids briefly in the opening of this chapter, but we'll consider a few additional aspects of plant–pollinator coevolution here.

Many flowers entice animals to visit them by providing food rewards. Some flowers produce a sugary fluid called nectar, and some of these flowers have specialized structures to store and distribute it, as we saw at the opening of this chapter. Pollen grains themselves sometimes serve as food for animals. In the process of visiting flowers to obtain nectar or pollen, animals often carry pollen from one flower to another or from one plant to another. Thus, in their quest for food, the animals contribute to the genetic diversity of the plant population. Insects, especially bees, are among the most important pollinators; birds and some species of bats are also major pollinators.

For more than 150 million years, angiosperms and their animal pollinators have coevolved in the terrestrial environment. The animals have affected the evolution of the plants, and the plants have affected the evolution of the animals. Flower structure has become incredibly diverse under these selection pressures. Some of the products of coevolution are highly specific; for example, some yucca species are pollinated by only one

INVESTIGATION

FIGURE 21.22 The Effect of Stigma Retraction in Monkeyflowers Elizabeth Fetscher's experiments showed that the unusual stigma retraction response to pollination in monkeyflowers (illustrated in Figure 21.21) enhances the dispersal of pollen to other flowers.

HYPOTHESIS

The stigma-retraction response in *M. aurantiacus* increases the likelihood than an individual flower's pollen will be exported to another flower once pollen from another flower has been deposited on its stigma.

METHOD

1. Set up three groups of monkeyflower arrays. Each array consists of one pollen-donor flower and multiple pollen-recipient flowers (with the anthers removed to prevent pollen donation).
2. In control arrays, the stigma of the pollen donor is allowed to function normally.
3. In one set of experimental arrays, the stigma of the pollen donor is permanently propped open.
4. In a second set of experimental arrays, the stigma of the pollen donor is artificially sealed closed.
5. Allow hummingbirds to visit the arrays, then count the pollen grains transferred from each donor flower to the recipient flowers in the same array.

RESULTS

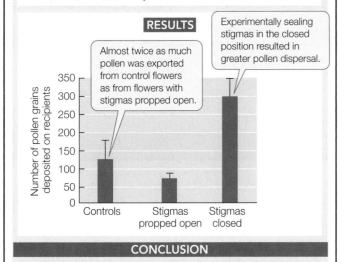

Almost twice as much pollen was exported from control flowers as from flowers with stigmas propped open.

Experimentally sealing stigmas in the closed position resulted in greater pollen dispersal.

y-axis: Number of pollen grains deposited on recipients — 0, 50, 100, 150, 200, 250, 300, 350

x-axis: Controls, Stigmas propped open, Stigmas closed

CONCLUSION

The stigma-retraction response enhances the male function of the flower (dispersal of pollen) once the female function (receipt of pollen) has been performed.

Go to **yourBioPortal.com** for original citations, discussions, and relevant links for all INVESTIGATION figures.

species of yucca moth, and that moth may exclusively pollinate just one species of yucca. Such specific relationships provide plants with a reliable mechanism for transferring pollen only to members of their own species.

Most plant–pollinator interactions are much less specific; that is, many different animal species pollinate the same plant species, and the same animal species pollinates many different plant species. However, even these less specific interactions have developed some specialization. Bird-pollinated flowers

Rubeckia fulgida

FIGURE 21.23 See Like a Bee To normal human vision (above), the petals of a black-eyed Susan appear solid yellow. Ultraviolet photography reveals patterns that attract bees to the central region, where pollen and nectar are located.

are often red and odorless. Many insect-pollinated flowers have characteristic odors, and bee-pollinated flowers may have conspicuous markings, or *nectar guides*, that may be visible only in the ultraviolet region of the spectrum, which is visible to bees (**FIGURE 21.23**).

Fruits aid angiosperm seed dispersal

The ovary of a flower (together with the seeds it contains) develops into a **fruit** after fertilization. The fruit protects the seeds and can also promote seed dispersal by becoming attached to or being eaten by an animal. Fruits are not necessarily fleshy; they can be hard and woody, or small and have modified structures that allow the seeds to be dispersed by wind or water (**FIGURE 21.24**).

A fruit may consist of only the mature ovary and its seeds, or it may include other parts of the flower or structures associated with it. A *simple fruit* is one that develops from a single carpel or several fused carpels, such as a plum or peach. A raspberry is an example of an *aggregate fruit*—one that develops from several separate carpels of a single flower. Pineapples and figs are examples of *multiple fruits*, formed from a cluster of flowers (an inflorescence). Fruits derived from parts in addition to the carpel and seeds are called *accessory fruits*; examples are apples, pears, and strawberries.

FIGURE 21.24 Fruits Come in Many Forms (A) The single seeds inside the simple fruits of peaches are dispersed by animals. (B) Each macadamia seed is covered by a hard, woody fruit that allows it to survive drought. (C) The highly reduced simple fruits of dandelions are dispersed by wind. (D) A multiple fruit, the jackfruit (*Artocarpus heterophyllus*) of tropical Asia, is the largest tree-borne fruit in the world. (E) An aggregate fruit (blackberry). (F) An accessory fruit (pear).

APPLY THE CONCEPT

Fruits increase the reproductive success of angiosperms

Many fleshy fruits attract animals, which eat the fruit and then disperse the seeds in their feces. If all other factors are equal, large seeds have a better chance of producing a successful seedling than small seeds. So why isn't there selection for larger seeds in the fruits of all plants?

In one study in Peru, the feces of the spider monkey *Ateles paniscus* were found to contain seeds from 71 species of plants. After eating fruit, the monkeys usually travel some distance before defecating, thus dispersing any undigested seeds.

If monkey feces are left undisturbed on the forest floor, rodents eat and destroy the vast majority of the seeds in the feces. To germinate successfully, the seeds in spider monkey feces need to be buried by dung beetles, which makes the discovery and destruction of seeds by rodents much less likely.

Ellen Andresen hypothesized that dung beetles were more likely to remove larger than smaller seeds from spider monkey dung before burying the dung. She added plastic beads of various diameters to spider monkey dung (to simulate seeds) and measured the percentage of beads buried with the dung by the beetles. Use her data to answer the questions at right.

Bead diameter (mm)	2	4	6	8	10	12
Percentage buried	100	76	52	39	20	4

1. Plot bead size (the independent variable) versus percentage of beads buried by dung beetles (the dependent variable).

2. Calculate a regression line for the relationship shown in your graph (see Appendix B). Approximately what percentage of beads with a diameter of 5 mm would you predict would be buried by the beetles? What about beads 14 mm in diameter?

3. What other factors besides size might influence the probability of seed burial by dung beetles? Can you design an experiment to test your hypotheses?

4. Describe how changes in the population sizes of spider monkeys, rodents, and dung beetles would affect the reproductive success of various plant species.

FIGURE 21.25   The Life Cycle of an Angiosperm
Triploid endosperm is produced among many species of angiosperms. One sperm nucleus fertilizes the egg to form the zygote, while the other combines with the two polar nuclei to form the endosperm.

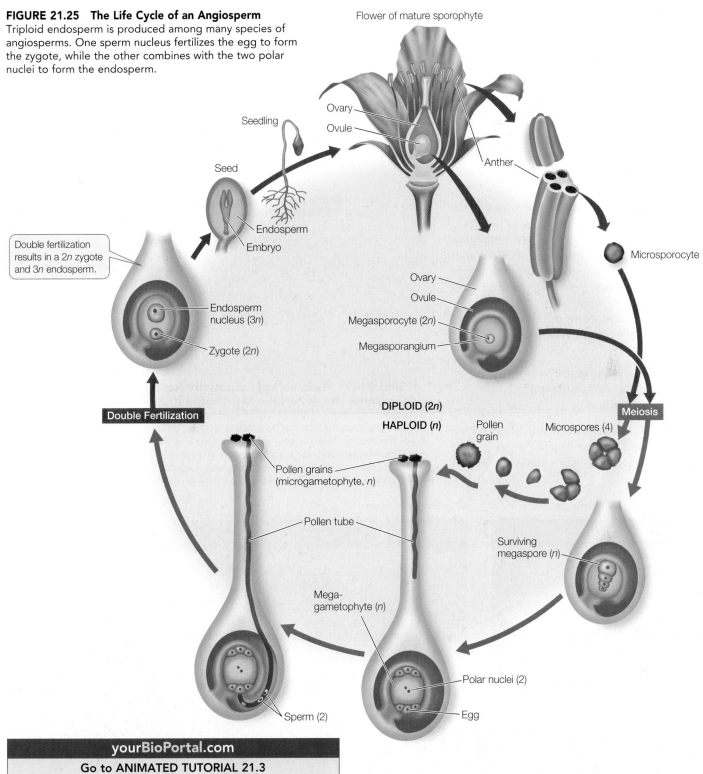

yourBioPortal.com
Go to ANIMATED TUTORIAL 21.3
Life Cycle of an Angiosperm

The angiosperm life cycle produces diploid zygotes nourished by triploid endosperms

Like all seed plants, angiosperms are heterosporous. As we have seen, their ovules are contained within carpels rather than being exposed on the surfaces of scales, as in most gymnosperms. The male gametophytes, as in the gymnosperms, are pollen grains. (Compare the conifer life cycle in Figure 21.17 with Figure 21.25.)

As **FIGURE 21.25** shows, the zygote develops into an embryo, which consists of an embryonic axis (the "backbone" that will become a stem and a root) and one or two **cotyledons**, or "seed leaves." The cotyledons have different fates in different plants. In many, they serve as absorptive organs that take up and digest the endosperm. In others, they enlarge and become

FIGURE 21.26 Evolutionary Relationships among the Angiosperms Recent analyses of many angiosperm genes have clarified the relationships among the major groups.

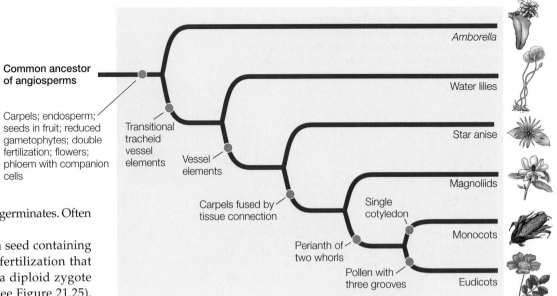

photosynthetic when the seed germinates. Often they play both roles.

The ovule develops into a seed containing the products of the double fertilization that characterizes angiosperms: a diploid zygote and a triploid endosperm (see Figure 21.25). The endosperm serves as storage tissue for starch or lipids, proteins, and other substances that will be needed by the developing embryo.

Recent analyses have revealed the phylogenetic relationships of angiosperms

FIGURE 21.26 shows the relationships among the major angiosperm clades. The two largest clades—the **monocots** and the **eudicots**—include

FIGURE 21.27 Monocots and Eudicots Are Not the Only Surviving Angiosperms (A) *Amborella*, a shrub, is sister to the remaining extant angiosperms. Of interest in this photo of a female flower is the pair of false anthers, possibly serving to lure insects that are searching for pollen. (B) The water lily clade was the next to diverge after *Amborella* (see also Figure 21.19A). (C) Star anise and its relatives belong to another early-diverging angiosperm clade. (D,E) The largest clade other than the monocots and eudicots is the magnoliid complex, which includes magnolias and the group known as "Dutchman's pipe."

(A)

(B) *Saccharum* sp.

(C) *Posidonia oceanica*

(D) *Phoenix dactylifera*

FIGURE 21.28 Monocots   (A) Monocots include many popular garden flowers such as these hyacinths (*Muscari armeniacum*; blue), tulips (*Tulipa* sp.; red) and daffodils (*Narcissus* sp.; yellow). (B) Monocot grasses such as sugarcane feed the world; wheat, rice, and maize (corn) are also grasses. (C) Seagrasses such as this Neptune's grass form "meadows" in the shallow, sunlit waters of the world's oceans. (D) Palms are among the few monocot trees. Date palms like these are a major food source in some areas of the world.

the great majority of angiosperm species. The monocots are so called because they have a single embryonic cotyledon; the eudicots have two. Some familiar angiosperms belong to other clades, among them the water lilies, star anise and its relatives, and the magnoliids (**FIGURE 21.27**). The **magnoliids** are the sister group to the monocots and eudicots. Although less numerous than the latter two clades, the magnoliids include many familiar and useful plants, such as avocados, cinnamon, black pepper, and magnolias.

The root of the evolutionary tree of flowering plants was once a matter of great controversy. A fundamental challenge was identifying the group that is sister to the remaining angiosperms, and the magnoliid clade was a leading candidate for the position. At the close of the twentieth century, however, an impressive convergence of molecular and morphological evidence led to the conclusion that the sister group to the remaining flowering plants is a clade that today consists of a single species of the genus *Amborella* (see Figure 21.27A). This woody shrub, with cream-colored flowers, lives only on New

Caledonia, an island in the South Pacific. Its 5 to 8 carpels are in a spiral arrangement, and it has 30 to 100 stamens. The xylem of *Amborella* lacks vessel elements, which evolved after this deepest split in the angiosperm evolutionary tree.

Representatives of the two largest angiosperm clades are everywhere. The monocots (**FIGURE 21.28**) include grasses, cattails, lilies, orchids, and palms. The eudicots (**FIGURE 21.29**) include the vast majority of familiar seed plants, including most herbs (i.e., nonwoody plants), vines, trees, and shrubs. Among the eudicots are such diverse plants as oaks, willows, beans, snapdragons, roses, and sunflowers.

(A) *Laburnum* sp.

(B) *Passiflora caerulea*

Filaments of corona

(C) *Rafflesia arnoldii*

(D) *Echinocereus reichenbachii*

FIGURE 21.29 Eudicots (A) This golden chain tree is a member of the legume clade, an important plant group with a large number of economically important species. (B) Passionflower vines are found throughout the tropics and subtropics. Their flowers are distinguished by elaborate coronas (purple filaments on this flower). (C) The largest flower in the world is *Rafflesia arnoldii*, found in the rainforests of Indonesia. The flower lives as a parasite on tropical vines and has lost its leaf, stem, and even root structures. (D) Cacti comprise a large group of eudicots, with about 1,500 species in the Americas. Many, such as this black lace cactus, bear large flowers for a brief period of each year.

Do You Understand Concept 21.5?

- Explain the difference between pollination and fertilization.
- Give some examples of how animals have affected the evolution of the angiosperms.
- What are the respective roles of the two sperm in double fertilization?
- What are the different functions of flowers, fruits, and seeds?

Once life moved onto land, it was plants that shaped the terrestrial environment. Terrestrial ecosystems could not function without the foods and habitats provided by plants. Plants produce oxygen and remove carbon dioxide from the atmosphere. They play important roles in forming soils and renewing their fertility. Plant roots help hold soil in place, providing protection against erosion by wind and water. Plants also moderate local climates in various ways, such as by increasing humidity, providing shade, and blocking wind. All of these ecosystem services permit a great diversity of fungi and animals to exist on land.

What was Darwin's explanation for the three distinct flowers growing on a single orchid plant?
QUESTION

ANSWER Darwin obtained specimens of the plants in question and dissected the flowers. He was able to demonstrate that the orchid was a single species (*Catasetum macrocarpum*) that bore three distinct types of flowers: megasporangiate (female), microsporangiate (male), and perfect (hermaphroditic). The three types of flowers were remarkable in their morphological differences, which had misled botanists into describing the different flower types as species in different genera. Most plants were either male (specimens identified as *Catasetum*) or female (specimens identified as *Monachanthus*), but some individuals that bore predominately male or female flowers also produced perfect flowers (specimens identified as *Myanthus*).

The case of *C. macrocarpum* demonstrates that some plants blur the lines between strict dioecy (male and female flowers in separate individuals), monoecy (male and female flowers in the same individual), and perfect flowers (flowers with both male and female parts). The flowers on a *C. macrocarpum* are either male or female at any one time, except that plants can bear some perfect flowers as well.

Why do the male and female flowers of *C. macrocarpum* look so different? Part of the explanation is their different roles in pollination. Recall Darwin's observation of a *Catasetum*

flower shooting a packet of pollen at an insect that landed on its flower. The pollinia (pollen packets) and associated structures in male flowers of *Catasetum* are coiled like springs and are released suddenly when disturbed by an insect. This release forcefully propels the pollinia precisely into position on the back of the insect. The insect pollinator of *C. macrocarpum* is a specific bee species, the males of which are attracted to the odor of the flowers. The flowers produce no nectar reward, but the male bee does gather the chemical that produces the scent. The bee then moves on to another flower. When the bee visits a female flower on a different *C. macrocarpum* individual (again attracted by the same scent), no such "loaded spring" awaits. Instead, the morphology of the female flower enhances the removal of the pollinia from the insect's body. In this way, floral morphology makes cross-fertilization more likely and reduces the chances of self-pollination.

Orchids were important in forming Darwin's ideas about the mechanisms of evolution, for they showed that even aspects of the coloration and form of flowers evolved in response to natural selection. This conclusion ran counter to the thinking of the day. For example, Thomas Huxley, one of Darwin's earliest and strongest supporters, doubted that the beauty of color in plants and animals could be explained on the basis of their importance to function. Darwin showed that the beauty of flowers is indeed connected to their reproductive success and is a key element in explaining the great diversity of plants.

21 SUMMARY

concept 21.1 Primary Endosymbiosis Produced the First Photosynthetic Eukaryotes

- Primary endosymbiosis gave rise to chloroplasts and the subsequent diversification of the **Plantae**. The descendants of the first photosynthetic eukaryote include **glaucophytes**, **red algae**, several groups of green algae, and **land plants**, all of which contain chlorophyll *a*. Review Figure 21.1 and WORKING WITH DATA 21.1

- **Streptophytes** include the land plants and two groups of green algae. **Green plants**, which include the streptophytes and the remaining green algae, are characterized by the presence of chlorophyll *b* (in addition to chlorophyll *a*). Review Figure 21.1

- Land plants, also known as **embryophytes**, arose from an aquatic green algal ancestor related to today's **charophytes**. Land plants develop from embryos that are protected by parental tissue. Review Figure 21.1

concept 21.2 Key Adaptations Permitted Plants to Colonize Land

- The acquisition of a **cuticle**, stomata, **gametangia**, a protected embryo, protective pigments, thick spore walls with a protective polymer, and a mutualistic association with a fungus were all adaptations of land plants to terrestrial life.

- All land plant life cycles feature alternation of generations, in which a multicellular diploid **sporophyte** alternates with a multicellular haploid **gametophyte**. Review Figure 21.4

- The **nonvascular land plants** comprise the **liverworts**, **hornworts**, and **mosses**. These groups lack specialized vascular tissues for the conduction of water or nutrients through the plant body.

- The life cycles of nonvascular land plants depend on liquid water. The sporophyte is usually smaller than the gametophyte and depends on it for water and nutrition.

- In many land plants, spores form in structures called **sporangia** and gametes form in structures called **gametangia**. Female and male gametangia are, respectively, an **archegonium** and an **antheridium**. Review Figure 21.6 and ANIMATED TUTORIAL 21.1

concept 21.3 Vascular Tissues Led to Rapid Diversification of Land Plants

- The **vascular plants** have a **vascular system** consisting of xylem and phloem that conducts water, minerals, and products of photosynthesis through the plant body. The vascular system includes cells called **tracheids**.

- The **rhyniophytes**, the earliest known vascular plants, are known to us only in fossil form. They lacked true roots and leaves but apparently possessed **rhizomes** and **rhizoids**.

- Among living vascular plant groups, the **lycophytes** (club mosses and relatives) have only small, simple leaflike structures (**microphylls**). True leaves (**megaphylls**) are found in **monilophytes** (which include **horsetails** and **leptosporangiate ferns**). The monilophytes and the seed plants are collectively called **euphyllophytes**. Review WEB ACTIVITY 21.1

- Roots may have evolved either from rhizomes or from stems. Microphylls probably evolved from sterile sporangia, and megaphylls may have resulted from the flattening and reduction of a portion of a stem system with **overtopping** growth. **Review Figure 21.10**

- The earliest-diverging groups of vascular plants are **homosporous**, but **heterospory**—the production of distinct **megaspores** and **microspores**—has evolved several times. Megaspores develop into female **megagametophytes**; microspores develop into male **microgametophytes**. **Review Figure 21.11 and WEB ACTIVITIES 21.2 and 21.3**

concept 21.4 Seeds Protect Plant Embryos

- All **seed plants** are heterosporous, and their gametophytes are much smaller than (and dependent on) their sporophytes. **Review Figure 21.12**

- Seed plants do not require liquid water for fertilization. **Pollen grains**, the microgametophytes of seed plants, are carried to a megagametophyte by wind or by animals. Following **pollination**, a **pollen tube** emerges from the pollen grain and elongates to deliver gametes to the megagametophyte. **Review Figure 21.14**

- An **ovule** consists of the seed plant megagametophyte and the **integument** of sporophytic tissue that protects it. The ovule develops into a **seed**. **Review Figure 21.14B**

- Seeds are well protected, and they are often capable of long periods of dormancy, germinating when conditions are favorable.

- Fossils of woody seed ferns are the earliest evidence of seed plants. The surviving groups of seed plants are the **gymnosperms** and **angiosperms**. **Review Figure 21.1**

- The gymnosperms produce ovules and seeds that are not protected by ovary or fruit tissues. The major gymnosperm groups are the **cycads**, **ginkgos**, **gnetophytes**, and **conifers**. Review Figure 21.15

- The megaspores of conifers are produced in woody **cones** called **megastrobili**; the microspores are produced in herbaceous cones called **microstrobili**. Pollen reaches the megagametophyte by way of the **micropyle**, an opening in the integument of the ovule. Review Figures 21.16 and 21.17, WEB ACTIVITY 21.4, and ANIMATED TUTORIAL 21.2

concept 21.5 Flowers and Fruits Increase the Reproductive Success of Angiosperms

- **Flowers** and **fruits** are unique to the angiosperms, distinguishing them from the gymnosperms.

- The xylem of angiosperms is more complex than that of the gymnosperms. It contains two specialized cell types: **vessel elements**, which function in water transport, and **fibers**, which play an important role in structural support.

- The ovules and seeds of angiosperms are enclosed in and protected by **carpels**.

- The floral organs, from the base to the apex of the flower, are the **sepals**, **petals**, **stamens**, and **pistil**. Stamens bear microsporangia in **anthers**. The pistil (consisting of one or more carpels) includes an **ovary** containing ovules. The **stigma** is the receptive surface of the pistil. **Review Figure 21.14B and WEB ACTIVITY 21.5**

- The structure of flowers has evolved over time. A flower with both megasporangia and microsporangia is referred to as **perfect**; a flower with only one or the other is **imperfect**. Some plants with perfect flowers have adaptations to prevent self-fertilization. **Review Figures 21.21 and 21.22**

- A **monoecious** species has megasporangiate and microsporangiate flowers on the same plant. A **dioecious** species is one in which megasporangiate and microsporangiate flowers occur on different plants.

- Flowers may be pollinated by wind or by animals. Many angiosperms have coevolved with their animal pollinators.

- Nearly all angiosperms exhibit **double fertilization**, resulting in the production of a diploid zygote and an **endosperm** (which is triploid in most species). **Review Figure 21.25 and ANIMATED TUTORIAL 21.3**

- The oldest evolutionary split among the angiosperms is between the clade represented by the single species in the genus *Amborella* and all the remaining flowering plants. **Review Figure 21.26**

- The most species-rich angiosperm clades are the **monocots** and the **eudicots**. The **magnoliids** are the sister group to the monocots and eudicots.

The Evolution and Diversity of Fungi

Alexander Fleming was already a famous scientist in 1928, but his laboratory was often a mess. That year he was studying the properties of *Staphylococcus* bacteria, the agents of dangerous staph infections. In August, he took a long vacation with his family. When he returned in early September, he found that some of his petri dishes of *Staphylococcus* had become infested with a fungus that killed many of the bacteria.

Many scientists would have sighed at the loss, thrown out the petri dishes, and started new cultures of bacteria. But when Fleming looked at the dishes, he saw something exciting. Around each colony of fungi was a ring within which all the bacteria were dead.

Fleming hypothesized that the bacteria-free rings around the fungal colonies were produced by a substance excreted from the fungi, which he initially called "mould juice." He identified the fungi as members of the genus *Penicillium* and eventually named the antibacterial substance produced by these fungi penicillin. Fleming published his discovery in 1929, but initially, the finding received very little attention.

Over the next decade, Fleming produced small quantities of penicillin for testing as an antibacterial agent. Some of the tests showed promise, but many were inconclusive, and eventually Fleming gave up on the research. But his tests had shown enough promise to attract the attention of several chemists, who worked out the practical problems of producing a stable form of the substance. Clinical trials of this stable form of penicillin were extremely successful, and by 1945 it was being produced and distributed as an antibiotic on a large scale. That same year, Fleming and two of the chemists, Howard Florey and Ernst Chain, won the Nobel Prize in Medicine for their work on penicillin.

The development of penicillin was one of the most important achievements in modern medicine. Until the introduction of modern antibiotics, the most widespread agents of human death included bacterial infections such as gangrene, tuberculosis, and syphilis. Penicillin proved to be highly effective in curing such infections, and its success led to the creation of the modern pharmaceutical industry. Soon many additional antibiotic compounds were isolated from other fungi or synthesized in the laboratory, leading to a "golden age" of human health.

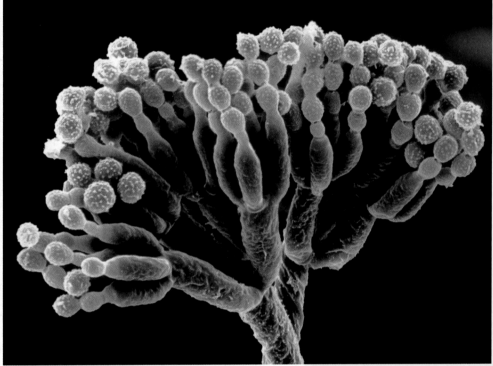

All species of the fungus *Penicillium* are recognizable by their dense spore-bearing structures (see Figure 22.17).

Q **QUESTION** Have antibiotics derived from fungi eliminated the danger of bacterial diseases in human populations?

KEY CONCEPTS

22.1 Fungi Live by Absorptive Heterotrophy

22.2 Fungi Can Be Saprobic, Parasitic, Predatory, or Mutualistic

22.3 Major Groups of Fungi Differ in Their Life Cycles

22.4 Fungi Can Be Sensitive Indicators of Environmental Change

concept 22.1   Fungi Live by Absorptive Heterotrophy

Fungi are organisms that digest their food outside their bodies. They secrete digestive enzymes to break down large food molecules in the environment, then absorb the breakdown products through the plasma membranes of their cells in a process known as **absorptive heterotrophy**. This mode of nutrition is successful in a wide variety of environments. Many fungi are *saprobes*, which absorb nutrients from dead organic matter. Others are *parasites*, which absorb nutrients from living hosts. Still others are *mutualists* living in intimate associations with other organisms that benefit both partners.

Modern fungi are believed to have evolved from a unicellular protist ancestor that had a flagellum. The probable common ancestor of the animals was also a flagellated protist much like the living choanoflagellates (see Figure 23.2). Current evidence, including the sequences of many genes, suggests that the fungi, choanoflagellates, and animals share a common ancestor not shared by other eukaryotes. These three lineages are often grouped together as the *opisthokonts* (**FIGURE 22.1**). A synapomorphy of the opisthokonts is a flagellum that, if present, is posterior, as in animal sperm. The flagella of all other eukaryotes are anterior.

Synapomorphies that distinguish the fungi as a group among the opisthokonts include absorptive heterotrophy and the presence of chitin in their cell walls. The fungi represent one of four large, independent evolutionary origins of multicellular organisms (plants, brown algae, and animals are the other three).

Unicellular yeasts absorb nutrients directly

Most fungi are multicellular, but single-celled species are found in most fungal groups. Unicellular, free-living fungi are referred to as **yeasts** (**FIGURE 22.2**). Some fungi that have yeast life stages also have multicellular life stages. Thus the term "yeast" does not refer to a single taxonomic group, but rather to a lifestyle that has evolved multiple times. Yeasts live in liquid or moist environments and absorb nutrients directly across their cell surfaces.

The ease with which many yeasts can be cultured, combined with their rapid growth rates, has made them ideal model

FIGURE 22.2 Yeasts Unicellular, free-living fungi are known as yeasts. Many yeasts reproduce by budding—mitosis followed by asymmetrical cell division—as those shown here are doing.

organisms for study in the laboratory. They present many of the same advantages to laboratory investigators as do many bacteria, but because they are eukaryotes, their genome structures and cells are much more like those of humans and other eukaryotes than are those of bacteria.

Multicellular fungi use hyphae to absorb nutrients

The body of a multicellular fungus is called a **mycelium** (plural *mycelia*). A mycelium is composed of a mass of individual tubular filaments called **hyphae** (singular *hypha*; **FIGURE 22.3A,B**), in which absorption of nutrients takes place. The cell walls of the hyphae are greatly strengthened by microscopic fibrils of *chitin*, a nitrogen-containing structural polysaccharide. In some species of fungi, the hyphae are subdivided into cell-like compartments by *incomplete* cross-walls called **septa** (singular *septum*); these hyphae are referred to as **septate**. Septa do not completely close off compartments in the hyphae. Gaps at the centers of the septa known as *pores* allow organelles—sometimes even nuclei—to move in a controlled way between compartments (**FIGURE 22.3C**). In other species of fungi, the hyphae lack septa, but may contain hundreds of nuclei; these hyphae are referred to as **coenocytic**. The coenocytic condition results from repeated nuclear divisions without cytokinesis.

The total hyphal growth of a fungal mycelium (not the growth of an individual hypha) may exceed 1 kilometer a day! The hyphae may be widely dispersed to forage for nutrients over a large area, or they may clump together in a cottony mass to exploit a rich nutrient source. Some species of fungi produce sexual spores, in which case portions of the mycelium become reorganized into a reproductive *fruiting structure*, such as a mushroom. The mycelial mass is often far larger than the mushroom alone. The mycelium of one individual fungus in Michigan covers 15 hectares underground and weighs more than a blue whale (the largest animal). Aboveground, this individual is evident only as isolated clumps of mushrooms.

Certain modified hyphae, called *rhizoids*, anchor some fungi to their substrate (i.e., the dead organism or other matter on which they feed). These rhizoids are not homologous to the

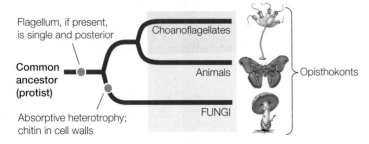

FIGURE 22.1 Fungi in Evolutionary Context Absorptive heterotrophy and the presence of chitin in their cell walls distinguish the fungi from other opisthokonts.

(A)

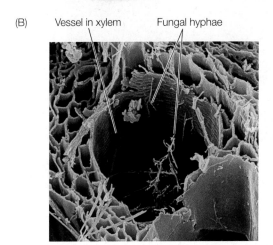

Fruiting structure (mushroom)

The main (vegetative) portion of the mycelium is typically much more extensive than the fruiting structure (only a small portion is shown in this figure).

(B)   Vessel in xylem   Fungal hyphae

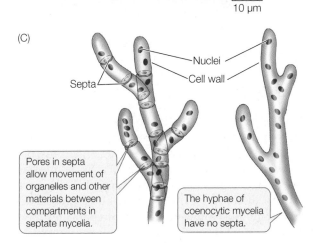

10 μm

(C)

Septa

Nuclei
Cell wall

Pores in septa allow movement of organelles and other materials between compartments in septate mycelia.

The hyphae of coenocytic mycelia have no septa.

FIGURE 22.3 Mycelia Are Made Up of Hyphae (A) The fruiting structure of a club fungus is short-lived, but the filamentous, nutrient-absorbing mycelium can be long-lived and cover large areas. (B) The minute individual hyphae of fungal mycelia can penetrate small spaces. In this artificially colored micrograph, hyphae (yellow structures) of a dry-rot fungus are penetrating the xylem tissues of a log. (C) The hyphae of septate fungal species are divided into organelle-containing compartments by porous septa, while coenocytic hyphae have no septa.

The downside of the large surface area-to-volume ratio of the mycelium is its tendency to lose water rapidly in a dry environment. Thus fungi are most common in moist environments. You have probably observed the tendency of molds, toadstools, and other fungi to appear in damp places.

Another characteristic of some fungi is a tolerance for highly hypertonic environments (those with a solute concentration higher than their own; see Concept 5.2). Many fungi are more resilient than bacteria in hypertonic surroundings. Jelly in the refrigerator, for example, will not become a growth medium for bacteria because it is too hypertonic to those organisms, but it may eventually harbor mold colonies. Mold in the refrigerator illustrates yet another trait of many fungi: tolerance of temperature extremes. Many fungi grow in temperatures as low as –6°C, and some tolerate temperatures above 50°C.

Do You Understand Concept 22.1?

- Describe the relationship between fungal structure and absorptive heterotrophy.
- What are the advantages and disadvantages to multicellular fungi of the large surface-to-volume ratio of the mycelium?

Fungi are important components of healthy ecosystems. They interact with other organisms in many ways, some of which are harmful and some beneficial to those other organisms.

concept 22.2 Fungi Can Be Saprobic, Parasitic, Predatory, or Mutualistic

Without the fungi, our planet would be very different. Picture Earth with only a few stunted plants and watery environments choked with the remains of dead organisms. Fungi do much of Earth's garbage disposal. Fungi not only help clean up the landscape and form soil, but also play a key role in recycling mineral nutrients. Furthermore, the colonization of the terrestrial environment was made possible in large part by associations fungi formed with land plants and other organisms.

Saprobic fungi are critical to the planetary carbon cycle

Saprobic fungi, along with bacteria, are the major decomposers on Earth, contributing to the decay of nonliving organic

rhizoids of plants, and they are not specialized to absorb nutrients and water.

Fungi are in intimate contact with their environment

The filamentous hyphae of a fungus give it a unique relationship with its physical environment. The fungal mycelium has an enormous surface area-to-volume ratio compared with that of most large multicellular organisms. This large ratio is a marvelous adaptation for absorptive heterotrophy. Throughout the mycelium (except in fruiting structures), all of the hyphae are very close to their food source.

matter and thus to recycling of the elements used by living things. In forests, for example, the mycelia of fungi absorb nutrients from fallen trees, thus decomposing their wood. Fungi are the principal decomposers of cellulose and lignin, the main components of plant cell walls (most bacteria cannot break down these materials). Other fungi produce enzymes that decompose keratin and thus break down animal structures such as hair and nails.

Were it not for the fungal decomposers, Earth's carbon cycle would fail: great quantities of carbon atoms would remain trapped forever on forest floors and elsewhere. Instead, those carbon atoms are returned to the atmosphere in the form of CO_2 by fungal respiration, where they are again available for photosynthesis by plants.

> **LINK** Earth's carbon cycle is described in Concept 46.3

In fact, there was a time in Earth's history when populations of saprobic fungi declined dramatically. Vast tropical swamps existed during the Carboniferous period, as we saw in Chapter 21. When plants in these swamps died, they began to form peat. Peat formation led to acidification of the swamps; that acidity, in turn, drastically reduced the fungal population. The result? With the decomposers largely absent, large quantities of peat remained on the swamp floor and over time were converted into coal.

In contrast to their decline during the Carboniferous, fungi did very well at the end of the Permian, a quarter of a billion years ago, when the aggregation of continents produced volcanic eruptions that triggered a planetwide mass extinction of many other organisms (see Chapter 18). The fossil record shows that even as 96 percent of all multicellular species became extinct, fungi flourished—demonstrating both their hardiness and their role in recycling the elements in dead plants and animals.

Simple sugars and the breakdown products of complex polysaccharides are the favored source of carbon for saprobic fungi. Most fungi obtain nitrogen from proteins or the products of protein breakdown. Many fungi can use nitrate (NO_3^-) or ammonium (NH_4^+) ions as their sole source of nitrogen. No known fungus can get its nitrogen directly from inorganic nitrogen gas, however, as can some bacteria and plant–bacteria associations (that is, fungi cannot fix nitrogen; see Concept 19.3).

What happens when a fungus faces a dwindling food supply? A common strategy is to reproduce rapidly and abundantly. When conditions are good, fungi produce great quantities of spores, but the rate of spore production is commonly even higher when nutrient supplies go down. The spores may then remain dormant until conditions improve, or they may be dispersed to areas where nutrient supplies are higher.

Not only are fungal spores abundant in number, but they are extremely tiny and easily spread by wind or water (**FIGURE 22.4**). These attributes virtually ensure that they will be scattered over great distances, and that at least some of them will find conditions suitable for growth. No wonder we find fungi just about everywhere.

Lycoperdon perlatum

FIGURE 22.4 Spores Galore Puffballs (a type of club fungus) disperse trillions of spores in great bursts. Few of the spores travel very far, however; some 99 percent of them fall within 100 meters of the parent puffball.

Some fungi engage in parasitic or predatory interactions

Whereas saprobic fungi obtain their energy, carbon, and nitrogen directly from dead organic matter, other species of fungi obtain their nutrition from parasitic—and even predatory—interactions.

PARASITIC FUNGI Mycologists distinguish between two classes of parasitic fungi based on their degree of dependence on their host. *Facultative* parasites can grow on living organisms, but can also grow by themselves (including on artificial media). *Obligate* parasites can grow only on their specific living host, often a plant or insect species. The fact that their growth depends on a living host shows that obligate parasites have specialized nutritional requirements.

The filamentous structure of fungal hyphae is especially well suited to a life of absorbing nutrients from living plants. The slender hyphae of a parasitic fungus can invade a plant through stomata, through wounds, or in some cases, by direct penetration of epidermal cell walls (**FIGURE 22.5A**). Once inside the plant, the hyphae branch out to expand the mycelium. Some hyphae produce **haustoria**, branching projections that push through cell walls into living plant cells, absorbing the nutrients within those cells. The haustoria do not break through the plasma membranes inside the cell walls; they simply invaginate into the membranes, so that the plasma membrane fits them like a glove (**FIGURE 22.5B**). Fruiting structures may form, either within the plant body or on its surface. Some parasitic fungi live in a close physical (*symbiotic*) relationship with their host that is usually not lethal to the plant. Others, however, are *pathogenic*, sickening or even killing the host from which they derive nutrition.

PATHOGENIC FUNGI Although most human diseases are caused by bacteria or viruses, fungal pathogens are a major cause of death among people with compromised immune systems. Most people with AIDS die of fungal diseases, such as the pneumonia caused by *Pneumocystis jirovecii* or incurable diarrhea caused by other fungi. *Candida albicans* and certain other yeasts also cause severe diseases, such as esophagitis (which impairs swallowing), in individuals with AIDS and in individuals taking immunosuppressive drugs. Fungal diseases are a growing international health problem, requiring vigorous research. Our limited understanding of the basic biology of these fungi still hampers our ability to treat the diseases they cause. Various fungi cause other, less threatening human diseases, such as ringworm and athlete's foot.

The worldwide decline of amphibian species has been linked to the spread of a chytrid fungus, *Batrachochytrium dendrobatidis*. Genetic analyses indicate that the fungus populations attacking amphibian populations around the world are genetically almost identical, which suggests a recent introduction of the fungus across the globe. This chytrid appears to be endemic to southern Africa, and its spread around the world may have been initiated in the 1930s with exports of the African clawed frog (*Xenopus laevis*), which was once widely used in human pregnancy tests.

Fungi are by far the most important plant pathogens, causing crop losses amounting to billions of dollars. Bacteria and viruses are less important than fungi as plant pathogens. Major fungal diseases of crop plants include black stem rust of wheat and other diseases of wheat, corn, and oats. The agent of black stem rust is *Puccinia graminis*, which has a complicated life cycle that involves two plant hosts (wheat and barberry). In an epidemic in 1935, *P. graminis* was responsible for the loss of about one-fourth of the wheat crop in Canada and the United States.

FRONTIERS Although many pathogenic fungi cause problems when they attack agricultural crops, pathogenic fungi of the genus *Fusarium* can benefit agriculture by killing certain weed species, such as witchweed, a serious pest of cereal crops. Another strain of *Fusarium* has been proposed as a tool in the war against cocaine. The fungus could be applied to kill coca plants, the source of the drug. The use of *Fusarium* to kill coca plants is highly controversial, however, as the fungus may not be specific to this target species.

PREDATORY FUNGI Some fungi have adaptations that enable them to function as active predators, trapping nearby microscopic protists or animals. The most common predatory strategy seen in fungi is to secrete sticky substances from the hyphae so that passing organisms stick to them. The hyphae then quickly invade the trapped prey, growing and branching within it, spreading through its body, absorbing nutrients, and eventually killing it.

A more dramatic adaptation for predation is the constricting ring formed by some species of soil fungi (**FIGURE 22.6**). When nematodes (tiny roundworms) are present in the soil, these fungi form three-celled rings with a diameter that just fits a nematode. A nematode crawling through one of these rings stimulates the fungus, causing the cells of the ring to swell and trap the worm. Fungal hyphae quickly invade and digest the unlucky victim.

Mutualistic fungi engage in relationships beneficial to both partners

Certain kinds of relationships between fungi and other organisms have nutritional consequences for both partners. Two relationships of this type are highly specific and are **symbiotic** (the

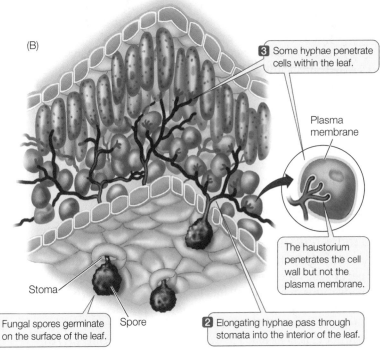

(A) Hyphae of fungal mycelium — Leaf cells — Stoma of leaf

This hypha is penetrating the leaf's interior through a stoma.

2 μm

(B) 3 Some hyphae penetrate cells within the leaf.

Plasma membrane

The haustorium penetrates the cell wall but not the plasma membrane.

Stoma — Spore

1 Fungal spores germinate on the surface of the leaf.

2 Elongating hyphae pass through stomata into the interior of the leaf.

FIGURE 22.5 Invading a Leaf (A) Hyphae of the mildew *Phyllactinia guttata* growing on the surface of a hazel leaf. (B) Haustoria are fungal hyphae that push into the living cells of plants, from which they absorb nutrients.

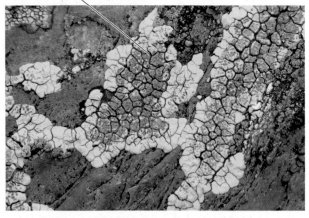

Nematode   Fungal hyphae

20 µm

FIGURE 22.6 Fungus as Predator A nematode is trapped by hyphal rings of the soil-dwelling fungus *Arthrobotrys dactyloides*.

(B) Fruticose (*Cladonia* sp.)

Foliose (*Parmotrema* sp.)

FIGURE 22.7 Lichen Body Forms Lichens fall into three principal categories—crustose, fruticose, or foliose—based on their body form. (A) A crustose lichen growing on the surface of an exposed rock. (B) Fruticose lichens (above) appear "shrubby," whereas foliose lichens (below) look "leafy."

partners live in close, permanent contact with each other) as well as **mutualistic** (the relationship benefits both partners; see Chapter 44).

Lichens are associations of a fungus with a cyanobacterium, a unicellular photosynthetic alga, or both. **Mycorrhizae** (singular *mycorrhiza*) are associations between fungi and the roots of plants. In these associations, the fungus obtains organic compounds from its photosynthetic partner and provides it with minerals and water in return, so that the partner's nutrition is also promoted.

LICHENS A lichen is not a single organism, but rather a meshwork of two radically different species: a fungus and a photosynthetic microorganism. Together the organisms that constitute a lichen can survive some of the harshest environments on Earth (although they are sensitive to poor air quality; see Concept 22.4). The biota of Antarctica, for example, features more than a hundred times as many species of lichens as of plants. Relatively little experimental work has focused on lichens, perhaps because they grow so slowly—typically less than 1 centimeter in a year.

There are nearly 30,000 described "species" of lichens, each of which is assigned the name of its fungal component. These fungal components may constitute as many as 20 percent of all fungal species. Most of them are sac fungi (Ascomycota). Some of them are able to grow independently without a photosynthetic partner, but most have never been observed in nature other than in a lichen association. The photosynthetic component of a lichen is most often a unicellular green alga, but it can be a cyanobacterium, or may even include both.

Lichens are found in all sorts of exposed habitats: on tree bark, on open soil, and on bare rock. Reindeer moss (not a moss at all, but the lichen *Cladonia subtenuis*) covers vast areas in Arctic, sub-Arctic, and boreal regions, where it is an important part of the diets of reindeer and other large mammals. The body forms of lichens fall into three principal categories: *Crustose* (crustlike) lichens look like colored powder dusted over their substrate (**FIGURE 22.7A**). *Foliose* (leafy) and *fruticose* (shrubby) lichens may have complex forms (**FIGURE 22.7B**).

Visible in a cross section of a typical foliose lichen are a tight upper region of fungal hyphae, a layer of photosynthetic cyanobacteria or algae, a looser hyphal layer, and finally hyphal rhizoids that attach the entire structure to its substrate (**FIGURE 22.8**). The meshwork of fungal hyphae takes up some mineral nutrients needed by the photosynthetic cells and provides a suitably moist environment for them by holding water tenaciously. The fungi obtain fixed carbon from the photosynthetic products of the algal or cyanobacterial cells.

Within the lichen, the fungal hyphae are tightly pressed against the algal or cyanobacterial cells and sometimes even invade them without breaching the plasma membrane (as we described earlier for haustoria in parasitic fungi; see Figure 22.5). The bacterial or algal cells not only survive these indignities, but continue their growth and photosynthesis. In fact, the algal cells in a lichen "leak" photosynthetic products at a greater rate than do similar cells growing on their own, and photosynthetic cells taken from lichens grow more rapidly on their own than when associated with a fungus. On this basis, we could consider lichen fungi to be parasitic on their photosynthetic

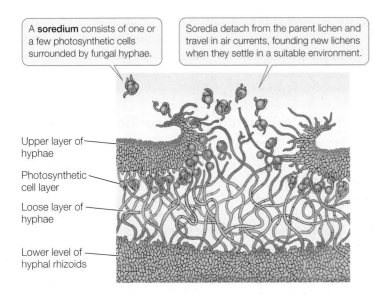

A **soredium** consists of one or a few photosynthetic cells surrounded by fungal hyphae.

Soredia detach from the parent lichen and travel in air currents, founding new lichens when they settle in a suitable environment.

Upper layer of hyphae

Photosynthetic cell layer

Loose layer of hyphae

Lower level of hyphal rhizoids

FIGURE 22.8 Lichen Anatomy Cross section showing the layers of a foliose lichen and the release of soredia.

partners. In many places where lichens grow, however, the photosynthetic cells could not grow at all on their own.

Lichens can reproduce simply by fragmentation of the vegetative body (the *thallus*), or by means of specialized structures called *soredia* (singular *soredium*). Soredia consist of one or a few photosynthetic cells bound by fungal hyphae. The soredia become detached from the lichen, are dispersed by air currents, and upon arriving at a favorable location, develop into a new lichen thallus. Alternatively, the fungal partner may go through its sexual cycle, producing haploid spores. When these spores are discharged, however, they disperse alone, unaccompanied by the photosynthetic partner.

Lichens are often the first colonists on new areas of bare rock. They get most of the mineral nutrients they need from the air and rainwater, augmented by minerals absorbed from dust. A lichen begins to grow shortly after a rain, as it begins to dry. As it grows, the lichen acidifies its environment slightly, and this acidity contributes to the slow breakdown of rocks, an early step in soil formation. With further drying, the lichen's photosynthesis ceases. The water content of the lichen may drop to less than 10 percent of its dry weight, at which point it becomes highly insensitive to extremes of temperature.

MYCORRHIZAE Many vascular plants depend on a symbiotic association with fungi. Unassisted, the root hairs of such plants often do not take up enough water or minerals to sustain growth. However, their roots usually do become infected with fungi, forming an association called a mycorrhiza. Mycorrhizae are of two types, distinguished by whether or not the fungal hyphae penetrate the plant cell walls.

In *ectomycorrhizae*, the fungus wraps around the root, and its mass is often as great as that of the root itself

(**FIGURE 22.9A**). The fungal hyphae wrap around individual cells in the root, but do not penetrate the cell walls. An extensive web of hyphae penetrates the soil in the area around the root, so that up to 25 percent of the soil volume near the root may be fungal hyphae. The hyphae attached to the root increase the surface area for the absorption of water and minerals, and the mass of hyphae in the soil acts like a sponge to hold water in the neighborhood of the root. Infected roots are short, swollen, and club-shaped, and they lack root hairs.

The fungal hyphae of *arbuscular mycorrhizae* enter the root and penetrate the cell walls of the root cells, forming arbuscular (treelike) structures inside the cell wall but outside the plasma membrane. These structures, like the haustoria of parasitic fungi and the contact regions of fungal hyphae and photosynthetic cells in lichens, become the primary site of exchange between plant and fungus (**FIGURE 22.9B**). As in the ectomycorrhizae, the fungus forms a vast web of hyphae leading from the root surface into the surrounding soil.

LINK Concept 25.2 contains a detailed description of the role of arbuscular mycorrhizae in plant nutrition

The mycorrhizal association is important to both partners. The fungus obtains needed organic compounds, such as sugars and amino acids, from the plant. In return, the fungus, because of its very high surface area-to-volume ratio and its ability to penetrate the fine structure of the soil, greatly increases the plant's ability to absorb water and minerals (especially phosphorus). The fungus may also provide the plant with certain growth hormones and may protect it against attack

(A)

(B) Arbuscule of *Glomus mosseae*

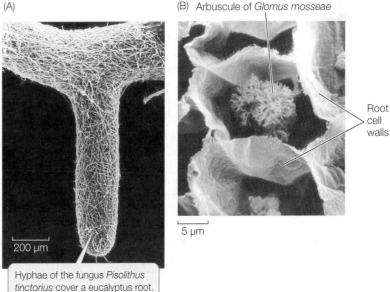

Root cell walls

200 μm

Hyphae of the fungus *Pisolithus tinctorius* cover a eucalyptus root.

5 μm

FIGURE 22.9 Mycorrhizal Associations (A) Ectomycorrhizal fungi wrap themselves around a plant root, increasing the area available for absorption of water and minerals. (B) Hyphae of arbuscular mycorrhizal fungi infect the root internally and penetrate the root cell walls, branching within the cells and forming a treelike structure, the arbuscule. (For purposes of this scanning electron micrograph, the cell cytoplasm was removed to better visualize the arbuscule.)

by disease-causing microorganisms. Plants that have active arbuscular mycorrhizae typically are a deeper green and may resist drought and temperature extremes better than plants of the same species that have little mycorrhizal development.

Attempts to introduce some plant species to new areas have failed until a bit of soil from the native area (presumably containing the fungus necessary to establish mycorrhizae) was provided. Trees without ectomycorrhizae do not grow well in the absence of abundant nutrients and water, so the health of our forests depends on the presence of ectomycorrhizal fungi. Many agricultural crops require inoculation of seeds with appropriate mycorrhizal fungi prior to planting. Without these fungi, the plants are unlikely to grow well, or in some cases at all. Certain plants that live in nitrogen-poor habitats, such as cranberry bushes and orchids, invariably have mycorrhizae. Orchid seeds will not germinate in nature unless they are already infected by the fungus that will form their mycorrhizae. Plants that lack chlorophyll always have mycorrhizae, which they often share with the roots of green, photosynthetic plants. In effect, these plants without chlorophyll are feeding on nearby green plants, using the fungus as a bridge.

Endophytic fungi protect some plants from pathogens, herbivores, and stress

In a tropical rainforest, 10,000 or more fungal spores may land on a single leaf each day. Some are plant pathogens, some do not affect the plant at all, and some invade the plant in a beneficial way. Fungi that live within aboveground parts of plants without causing obvious deleterious symptoms are called **endophytic fungi**. Recent research has shown that endophytic fungi are abundant in plants in all terrestrial environments.

Among the grasses, individual plants with endophytic fungi are more resistant to pathogens and to insect and mammalian herbivores than are plants lacking endophytes. The fungi produce alkaloids (nitrogen-containing compounds) that are toxic to animals. The alkaloids do not harm the host plant; in fact, some plants produce alkaloids (such as nicotine) themselves. The fungal alkaloids also increase the ability of grasses to resist stress of various types, including drought (water shortage) and salty soils. Such resistance is useful in agriculture.

The role, if any, of endophytic fungi in most broad-leaved plants is unclear. They may convey protection against pathogens, or they may simply occupy space within leaves without conferring any benefit, but also without doing harm. The benefit, in fact, might be all for the fungus.

Do You Understand Concept 22.2?

- What is the role of fungi in Earth's carbon cycle?
- Describe the nature and benefits of the lichen association.
- Why do plants grow better when infected with mycorrhizal fungi?

Before molecular techniques clarified the phylogenetic relationships among fungi, one criterion used for assigning fungi to taxonomic groups was the nature of their life cycles. The next section takes a closer look at life cycles in the six major groups of fungi.

concept 22.3 Major Groups of Fungi Differ in Their Life Cycles

Major fungal groups were originally defined by their structures and processes for sexual reproduction and also, to a lesser extent, by other morphological differences. Although fungal life cycles are even more diverse than was once realized, specific types of life cycles generally distinguish the six major groups of fungi—microsporidia, chytrids, zygospore fungi (Zygomycota), arbuscular mycorrhizal fungi (Glomeromycota), sac fungi (Ascomycota), and club fungi (Basidiomycota). **FIGURE 22.10** diagrams the evolutionary relationships of these groups as they are understood today.

The chytrids and the zygospore fungi may not represent monophyletic groups, as they each consist of several distantly

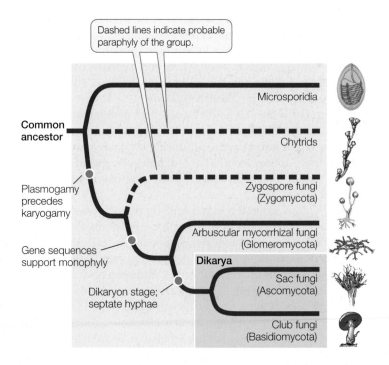

FIGURE 22.10 A Phylogeny of the Fungi Microsporidia are reduced, parasitic fungi whose relationships among the fungi are uncertain. They may be the sister group of most other fungi or more closely related to particular groups of chytrids or zygospore fungi. The dashed lines indicate that chytrids and zygospore fungi are thought to be paraphyletic; the relationships of the lineages within these two informal groups (see Table 22.1) are not yet well resolved. The sac fungi and club fungi together form the clade Dikarya.

---

yourBioPortal.com

Go to WEB ACTIVITY 22.1
Fungal Phylogeny

related lineages that retain some ancestral features. The clades that are thought to be monophyletic within these two informal groupings are listed in **TABLE 22.1**. Recent evidence from DNA analyses has established the placement of the microsporidia among the fungi, the likely paraphyly of the chytrids and the zygospore fungi, the independence of arbuscular mycorrhizal fungi from the other fungal groups, and the monophyly of sac fungi and club fungi.

Fungi reproduce both sexually and asexually

Both asexual and sexual reproduction occur among the fungi (**FIGURE 22.11**). Asexual reproduction takes several forms:

- The production of (usually) haploid spores within structures called *sporangia*
- The production of haploid spores (not enclosed in sporangia) at the tips of hyphae; such spores are called *conidia* (Greek *konis*, "dust")
- Cell division by unicellular fungi—either a relatively equal division of one cell into two (*fission*) or an asymmetrical division in which a smaller daughter cell is produced (*budding*)
- Simple breakage of the mycelium

Asexual reproduction in fungi can be spectacular in terms of spore quantity. A 2.5-centimeter colony of *Penicillium*, the mold that produces the antibiotic penicillin, can produce as many as 400 million conidia. The air we breathe contains as many as 10,000 fungal spores per cubic meter.

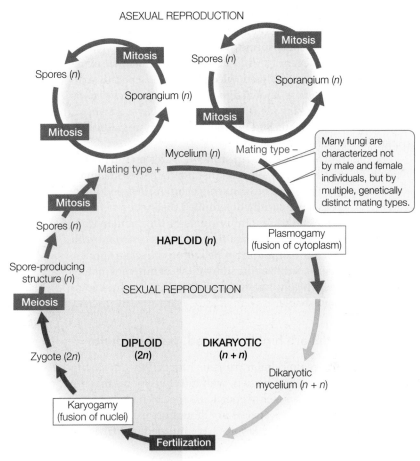

FIGURE 22.11 A Fungal Life Cycle Environmental conditions may determine which mode of reproduction—sexual or asexual—takes place at a given time.

TABLE 22.1	Classification of the Fungi	
GROUP	**COMMON NAME**	**FEATURES**
Microsporidia	Microsporidia	Intracellular parasites of animals; greatly reduced, among smallest eukayotes known; polar tube used to infect hosts
Chytrids (paraphyletic)[a] Chytridiomycota Neocallimastigomycota Blastocladiomycota	Chytrids	Mostly aquatic and microscopic; zoospores have flagella
Zygomycota (paraphyletic)[a] Entomophthoromycotina Kickxellomycotina Mucoromycotina Zoopagomycotina	Zygospore fungi	Reproductive structure is a unicellular zygospore with many diploid nuclei in a zygosporangium; hyphae coenocytic; usually no fleshy fruiting body
Glomeromycota	Arbuscular mycorrhizal fungi	Form arbuscular mycorrhizae on plant roots; only asexual reproduction is known
Ascomycota	Sac fungi	Sexual reproductive saclike structure known as an ascus, which contains haploid ascospores; hyphae septate; dikaryon
Basidiomycota	Club fungi	Sexual reproductive structure is a basidium, a swollen cell at the tip of a specialized hypha that supports haploid basidiospores; hyphae septate; dikaryon

[a]The formally named groups within the chytrids and Zygomycota are each thought to be monophyletic, but their relationships to one another (and to Microsporidia) are not yet well resolved.

Sexual reproduction is rare (or even unknown) in some groups of fungi, but common in others. Sexual reproduction may not occur in some species, or it may occur so rarely that biologists have never observed it. Species in which no sexual stage has been observed were once placed in a separate taxonomic group because knowledge of the sexual life cycle was considered necessary for classifying fungi. Now, however, these species can be related to other species of fungi through analysis of their DNA sequences.

Sexual reproduction in some fungi features an interesting twist: There is no morphological distinction between female and male structures, or between female and male individuals. Rather, there is a genetically determined distinction between two *or more* **mating types**. Individuals of the same mating type cannot mate with each another, but they can mate with individuals of another mating type within the same species, thus avoiding self-fertilization. Individuals of different mating types differ genetically, but are often visually and physiologically indistinguishable.

Microsporidia are highly reduced, parasitic fungi

Microsporidia are unicellular parasitic fungi. They are among the smallest eukaryotes known, with infective spores that are only 1–40 μm in diameter. About 1,500 species have been described, but many more are thought to exist. Their relationships among the eukaryotes have puzzled biologists for many decades.

Microsporidia lack true mitochondria, although they have reduced structures known as *mitosomes* that are derived from mitochondria. Unlike mitochondria, however, mitosomes contain no DNA; the mitochondrial genome has been completely transferred to the nucleus. Because microsporidia lack mitochondria, biologists initially suspected that they represented an early lineage of eukaryotes that diverged before the endosymbiotic event from which mitochondria evolved. The presence of mitosomes, however, indicates that this hypothesis is incorrect. DNA sequence analysis, along with the fact that their cell walls contain chitin, has confirmed that the microsporidia are in fact highly reduced, parasitic fungi, although their exact placement among the fungal lineages is still being investigated.

Microsporidia are obligate intracellular parasites of animals, especially of insects, crustaceans, and fishes. Some species are known to infect mammals, including humans. Most infections by microsporidia cause chronic diseases in the host, with effects that include weight loss, reduced fertility, and shortened life span. The host cell is penetrated by a *polar tube* that grows from the microsporidian spore, and the contents of the spore, called the *sporoplasm*, are injected into the host (**FIGURE 22.12**). The sporoplasm then replicates within the host cell and produces new infective spores. The life cycle of some species is complex and involves multiple hosts, whereas other species infect a single host. In some insects, parasitic microsporidia are transmitted vertically (i.e., from parent to offspring). Reproduction is thought to be strictly asexual in some microsporidians, but includes poorly understood asexual and sexual cycles in other species.

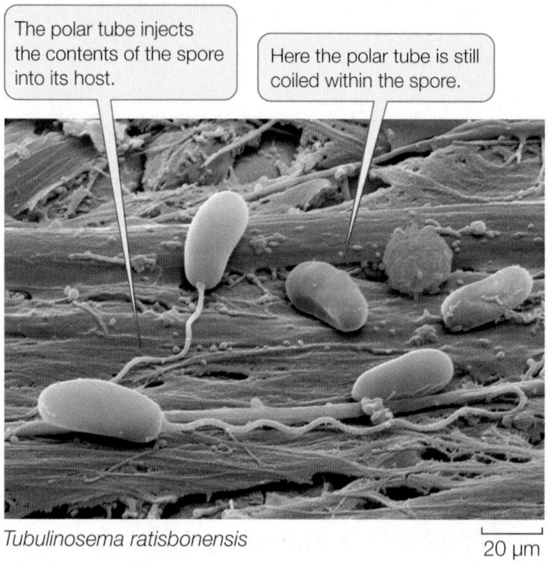

The polar tube injects the contents of the spore into its host.

Here the polar tube is still coiled within the spore.

Tubulinosema ratisbonensis

20 μm

FIGURE 22.12 Invasion of the Microsporidia Spores The polar tubes of microsporidian spores transfer the contents of the spores into the host's cells. The species shown here infects many animals, including humans.

Most chytrids have an aquatic life cycle

The **chytrids** include several distinct lineages of aquatic microorganisms once classified with the protists. However, morphological evidence (cell walls that consist primarily of chitin) and molecular evidence support their classification as early-diverging fungi. In this book we use the term "chytrid" to refer to all three of the formally named clades listed as chytrids in Table 22.1, but some mycologists use this term to refer to only one of those clades, the Chytridiomycota. There are fewer than one thousand described species among the three groups of chytrids.

Like the animals (and many other eukaryotes), most chytrids possess flagellated gametes. The retention of this trait reflects the aquatic environment in which fungi first evolved. Chytrids are the only fungi that have flagella at any life cycle stage.

Chytrids reproduce both sexually and asexually. The alternation between multicellular haploid (n) and multicellular diploid ($2n$) generations that evolved in plants and certain protist groups is seen in some chytrids as well (**FIGURE 22.13A**). Alternation of generations is not usual in the life cycles of other fungal groups. These chytrids have flagellated spores as well as flagellated gametes.

Chytrids may be parasitic (on organisms such as algae, mosquito larvae, nematodes, and amphibians) or saprobic. Some have complex mutualistic relationships with foregut-fermenting animals such as cattle and deer. Many chytrids live in freshwater habitats or in moist soil, but some are marine. Some chytrids are unicellular, others have rhizoids, and still others have coenocytic hyphae.

Some fungal life cycles feature separate fusion of cytoplasms and nuclei

Most members of the remaining four groups of fungi are terrestrial. Although the terrestrial fungi grow in moist places, they do not have motile gametes, and they do not release gametes

(A) Chytrids

The life cycle of some aquatic chytrids features alternation of generations.

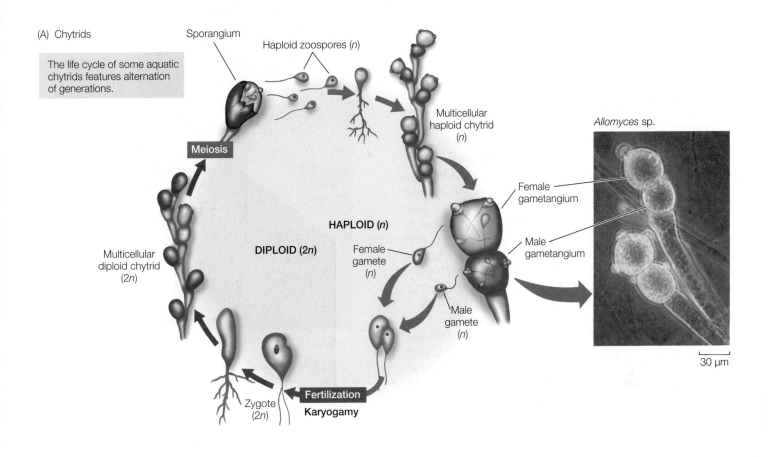

Sporangium

Haploid zoospores (n)

Meiosis

Multicellular haploid chytrid (n)

Multicellular diploid chytrid (2n)

HAPLOID (n)

DIPLOID (2n)

Female gamete (n)

Male gamete (n)

Female gametangium

Male gametangium

Allomyces sp.

30 μm

Zygote (2n)

Fertilization

Karyogamy

(B) Zygospore fungi (Zygomycota)

The sporangium of zygospore fungi contains haploid nuclei that are incorporated into spores.

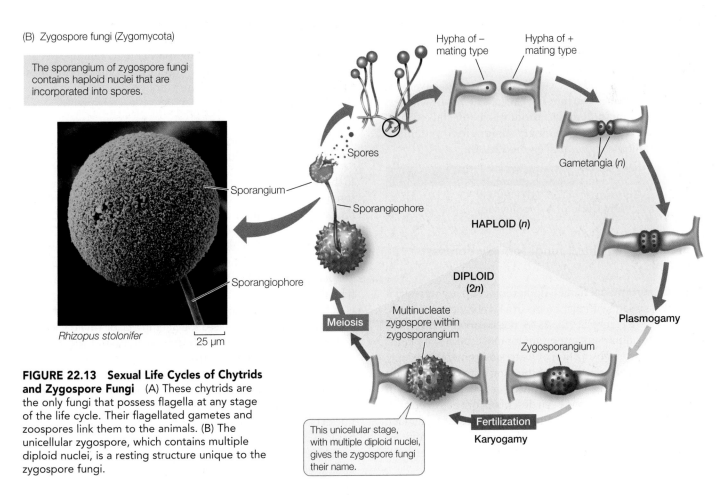

Hypha of – mating type

Hypha of + mating type

Spores

Sporangium

Gametangia (n)

Sporangiophore

HAPLOID (n)

DIPLOID (2n)

Plasmogamy

Multinucleate zygospore within zygosporangium

Zygosporangium

Meiosis

Fertilization

Karyogamy

This unicellular stage, with multiple diploid nuclei, gives the zygospore fungi their name.

Rhizopus stolonifer

Sporangium

Sporangiophore

25 μm

FIGURE 22.13 Sexual Life Cycles of Chytrids and Zygospore Fungi (A) These chytrids are the only fungi that possess flagella at any stage of the life cycle. Their flagellated gametes and zoospores link them to the animals. (B) The unicellular zygospore, which contains multiple diploid nuclei, is a resting structure unique to the zygospore fungi.

into the environment. Instead, the cytoplasms of two individuals of different mating types fuse (a process called **plasmogamy**) before their nuclei fuse (a process called **karyogamy**). Therefore, liquid water is not required for fertilization. Sexual species of terrestrial fungi include some zygospore fungi, sac fungi, and club fungi.

Zygospore fungi reproduce sexually when adjacent hyphae of two different mating types release chemical signals called *pheromones*, which cause them to grow toward each other. These hyphae produce gametangia, in which the nuclei replicate without cell division, resulting in multiple haploid nuclei in both gametangia. The two gametangia then fuse to form a *zygosporangium* with many haploid nuclei from each mating type (**FIGURE 22.13B**). Pairs of haploid nuclei then form multiple diploid nuclei within a unicellular **zygospore**, which is the basis of the name of the zygospore fungi. A thick, multilayered cell wall develops to protect the zygospore. The zygospore is a resting stage that may remain dormant for months before its nuclei undergo meiosis and a stalked **sporangiophore** sprouts, bearing one or many *sporangia*. Each sporangium contains the products of meiosis: haploid nuclei that are incorporated into spores. These spores disperse and germinate to form a new generation of haploid hyphae.

The zygospore fungi include four major lineages of terrestrial fungi that live on soil as saprobes, as parasites of insects and spiders, or as mutualists of other fungi and invertebrate animals. They produce no cells with flagella, and only one diploid cell—the zygospore—appears in the entire life cycle. Their hyphae are coenocytic. Most species do not form a fleshy fruiting structure; rather, the hyphae spread in a radial pattern from the spore, with occasional stalked sporangiophores reaching up into the air (**FIGURE 22.14**).

More than a thousand species of zygospore fungi have been described. One species you may have seen is *Rhizopus stolonifer*, the black bread mold. *Rhizopus* produces many stalked sporangiophores, each bearing a single sporangium containing hundreds of minute spores (see Figure 22.13B).

---

yourBioPortal.com

Go to ANIMATED TUTORIAL 22.1
Life Cycle of a Zygomycete

Arbuscular mycorrhizal fungi form symbioses with plants

Arbuscular mycorrhizal fungi (Glomeromycota) are terrestrial fungi that associate with plant roots in a symbiotic, mutualistic relationship (see Figure 22.9B). As we noted earlier in this chapter, these associations are important for most species of plants, which benefit from absorption of water and mineral nutrients through the large surface area of the fungal mycelium. Many of the fungi found in soils are arbuscular mycorrhizal fungi. Fewer than 200 species have been described, but 80–90 percent of all plants have associations with them. Molecular systematic studies have suggested that arbuscular mycorrhizal fungi are the sister group to the Dikarya (sac fungi and club fungi).

Pilobolus sp.   150 μm

FIGURE 22.14 Zygospore Fungi Produce Sporangiophores The transparent structures are sporangiophores growing on decomposing animal dung. The sporangiophores grow toward the light and end in tiny sporangia, which the stalked sporangiophores can eject as far as 2 meters. Animals ingest sporangia that land on grass and then disseminate the spores in their feces.

The hyphae of arbuscular mycorrhizal fungi are coenocytic. These fungi use glucose from their plant partners as their primary energy source, converting it into other, fungus-specific sugars that cannot return to the plant. Arbuscular mycorrhizal fungi reproduce asexually; there is not yet any direct evidence that they reproduce sexually.

The dikaryotic condition is a synapomorphy of sac fungi and club fungi

In the two remaining groups of fungi—the sac fungi and the club fungi—certain hyphae have a nuclear configuration other than the familiar haploid or diploid states (**FIGURES 22.15A, B**). In these fungi, sexual reproduction begins in two distinct steps: karyogamy (fusion of nuclei) occurs long after plasmogamy (fusion of cytoplasm), so that *two genetically different haploid nuclei co-exist and divide within the each cell of the mycelium.* This stage of the fungal life cycle is called a **dikaryon** ("two nuclei") and its ploidy is indicated as *n* + *n*. The dikaryon is a synapomorphy of these two groups, which are placed together in a clade called Dikarya.

Eventually, specialized fruiting structures form, within which pairs of genetically dissimilar nuclei—one from each parent—fuse, giving rise to zygotes long after the original "mating." The diploid zygote nucleus undergoes meiosis, producing four haploid nuclei. The mitotic descendants of those nuclei become spores, which germinate to give rise to the next haploid generation.

A life cycle with a dikaryon stage has several unusual features. First, there are no gamete *cells*, only gamete *nuclei*. Second, the

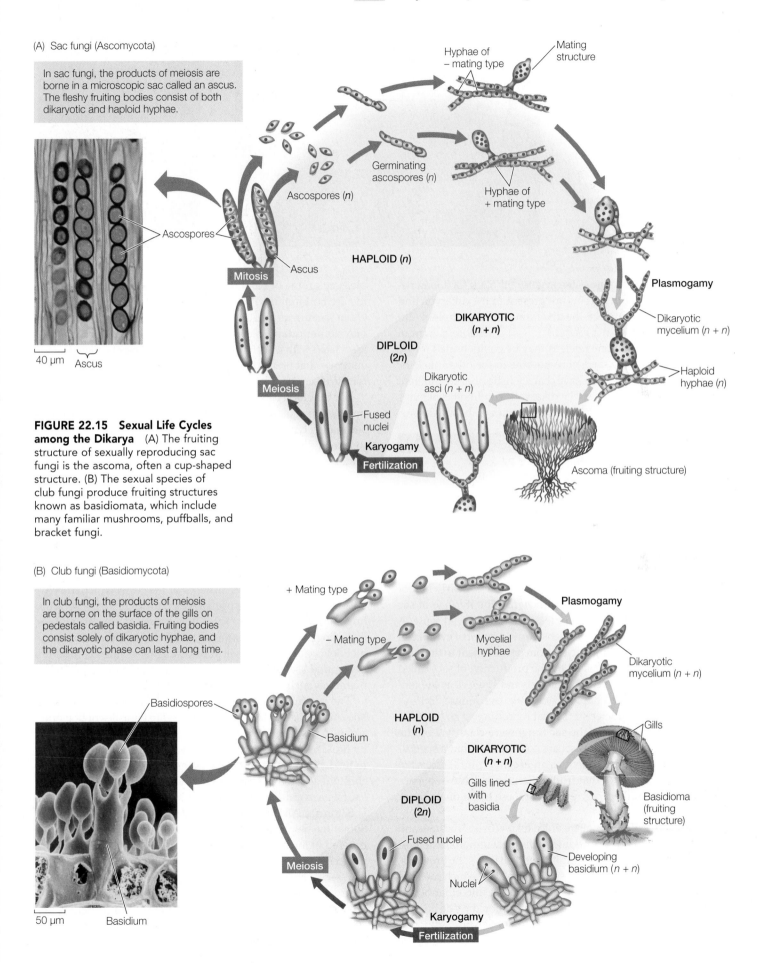

(A) Sac fungi (Ascomycota)

In sac fungi, the products of meiosis are borne in a microscopic sac called an ascus. The fleshy fruiting bodies consist of both dikaryotic and haploid hyphae.

Hyphae of – mating type

Mating structure

Germinating ascospores (n)

Ascospores (n)

Hyphae of + mating type

Ascospores

Ascus

HAPLOID (n)

Mitosis

Plasmogamy

Dikaryotic mycelium (n + n)

40 µm   Ascus

Meiosis

DIKARYOTIC (n + n)

DIPLOID (2n)

Dikaryotic asci (n + n)

Haploid hyphae (n)

Fused nuclei

Karyogamy

Fertilization

Ascoma (fruiting structure)

FIGURE 22.15 Sexual Life Cycles among the Dikarya (A) The fruiting structure of sexually reproducing sac fungi is the ascoma, often a cup-shaped structure. (B) The sexual species of club fungi produce fruiting structures known as basidiomata, which include many familiar mushrooms, puffballs, and bracket fungi.

(B) Club fungi (Basidiomycota)

In club fungi, the products of meiosis are borne on the surface of the gills on pedestals called basidia. Fruiting bodies consist solely of dikaryotic hyphae, and the dikaryotic phase can last a long time.

+ Mating type

Plasmogamy

– Mating type

Mycelial hyphae

Dikaryotic mycelium (n + n)

Basidiospores

Basidium

HAPLOID (n)

Gills

DIKARYOTIC (n + n)

Gills lined with basidia

DIPLOID (2n)

Basidioma (fruiting structure)

Fused nuclei

Developing basidium (n + n)

Meiosis

Nuclei

50 µm   Basidium

Karyogamy

Fertilization

(A) *Aleuria aurantia*

(B) *Morchella esculenta*

FIGURE 22.16 Sac Fungi
(A) These brilliant red cups are the ascomata of a cup fungus. (B) Morels, which have a spongelike ascoma and a subtle flavor, are considered a culinary delicacy by humans.

only true diploid structure is the zygote, although for a long period the genes of both parents are present in the dikaryon and can be expressed. In effect, the dikaryon is neither diploid (2*n*) nor haploid (*n*); rather, it is *dikaryotic* (*n* + *n*). Therefore, a harmful recessive mutation in one nucleus may be compensated for by a normal allele on the same chromosome in the other nucleus, and dikaryotic hyphae often have characteristics that are different from their *n* or 2*n* products. The dikaryotic condition is perhaps the most distinctive of the genetic peculiarities of the fungi.

> **yourBioPortal.com**
>
> Go to **WEB ACTIVITY 22.2**
> **Life Cycle of a Dikaryotic Fungus**

The sexual reproductive structure of sac fungi is the ascus

The **sac fungi** (Ascomycota) are a large and diverse group of fungi found in marine, freshwater, and terrestrial habitats. There are approximately 64,000 known species, nearly half of which are the fungal partners in lichens. The hyphae of sac fungi are segmented by more or less regularly spaced septa. A pore in each septum permits extensive movement of cytoplasm and organelles (including nuclei) from one segment to the next.

Sac fungi are distinguished by the production of sacs called **asci** (singular *ascus*), which after meiosis and spore cleavage contain sexually produced haploid *ascospores* (see Figure 22.15A). The ascus is the characteristic sexual reproductive structure of the sac fungi. In the past, the sac fungi were classified on the basis of whether or not the asci are contained within a specialized fruiting structure known as an **ascoma** (plural *ascomata*) and the morphology of that fruiting structure. DNA sequence analyses have resulted in a revision of these traditional groupings, however.

Some species of sac fungi are unicellular yeasts. The thousand or so species in this group are among the most important domesticated fungi. Perhaps the best known is baker's, or brewer's, yeast (*Saccharomyces cerevisiae*; see Figure 22.2), which metabolizes glucose obtained from its environment into ethanol and carbon dioxide by fermentation. It forms carbon dioxide bubbles in bread dough and gives baked bread its light texture. Although they are baked away in bread making (which produces the pleasant aroma of baking bread), the ethanol and carbon dioxide are both retained when the yeast ferments grain into beer. Other sac fungus yeasts live on fruits such as figs and grapes and play an important role in the making of wine. Many others are associated with insects; in the guts of some insects, they provide enzymes that break down materials that are otherwise difficult for the insects to digest, especially cellulose.

Sac fungus yeasts reproduce asexually by budding. Sexual reproduction takes place when two adjacent haploid cells of opposite mating types fuse. In some species, the resulting zygote buds to form a diploid cell population. In others, the zygote nucleus undergoes meiosis immediately; when this happens, the entire cell becomes an ascus. Depending on whether the products of meiosis then undergo mitosis, a yeast ascus usually contains either eight or four ascospores. Ascospores germinate to become haploid cells. The sac fungus yeasts have lost the dikaryon stage.

Most sac fungi are filamentous species, such as the cup fungi (**FIGURE 22.16**), in which the ascomata are cup-shaped and can be as large as several centimeters across (although most are much smaller). The inner surfaces of the cups, which are covered with a mixture of specialized hyphae and asci, produce huge numbers of spores. The edible ascomata of some species, including morels and truffles, are regarded by humans as gourmet delicacies (and can sell at prices higher than gold). The underground ascomata of truffles have a strong odor that attracts mammals such as pigs, which then eat and disperse the fungus.

The sac fungi also include many of the filamentous fungi known as *molds*. Many of these species are parasites of flowering plants. Chestnut blight and Dutch elm disease are both caused by molds. The chestnut blight fungus, which was introduced to the United States in the 1890s, had destroyed the American chestnut as a commercial species by 1940. Before the blight, this species accounted for more than half the trees in the eastern U.S. forests. Another familiar story is that of the American elm. Sometime before 1930, the Dutch elm disease fungus (first discovered in the Netherlands, but native to Asia) was introduced into the U.S. on infected elm logs from Europe. Spreading rapidly—sometimes by way of connected root systems—the fungus destroyed great numbers of American elm trees.

Other plant pathogens among the sac fungi include the powdery mildews that infect cereal crops, lilacs, and roses, among many other plants. Mildews can be a serious problem

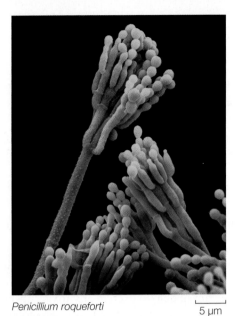

Penicillium roqueforti

5 μm

FIGURE 22.17 Conidia Chains of conidia (yellow) are developing at the tips of specialized hyphae arising from a *Penicillium* mold; compare the conidia of this species (which is used to produce "bleu" cheese) with the species shown on p. 437.

to farmers and gardeners, and a great deal of research has focused on ways to control these agricultural pests.

Brown molds of the genus *Aspergillus* are important in some human diets. *A. tamarii* acts on soybeans in the production of soy sauce, and *A. oryzae* is used in brewing the Japanese alcoholic beverage sake. Some species of *Aspergillus* that grow on grains and on nuts such as peanuts and pecans produce extremely carcinogenic (cancer-inducing) compounds called *aflatoxins*. In the United States and most other industrialized countries, moldy grain infected with *Aspergillus* is thrown out. In Africa, where food is scarcer, the grain gets eaten, moldy or not, and causes severe health problems, including high levels of certain cancers.

Penicillium is a genus of green molds, of which some species produce the antibiotic penicillin, presumably for defense against competing bacteria. Two species, *P. camembertii* and *P. roquefortii*, are the organisms responsible for the characteristic strong flavors of Camembert and Roquefort cheeses, respectively.

The filamentous sac fungi reproduce asexually by means of conidia that form at the tips of specialized hyphae (**FIGURE 22.17**). Small chains of conidia are produced by the millions and can survive for weeks in nature. The conidia are what give molds their characteristic colors.

The sexual reproductive cycle of filamentous sac fungi includes the formation of a dikaryon, although this stage is relatively brief compared with that in club fungi. Many filamentous sac fungi form multinucleate mating structures (see Figure 22.15A). Mating structures of two different mating types fuse and produce a dikaryotic mycelium, with nuclei from both mating types. The dikaryotic mycelium typically forms a cup-shaped ascoma, which bears the asci. Only with the formation of asci do the nuclei from the two mating types finally fuse. Both nuclear fusion and the subsequent meiosis that produces haploid ascospores take place within individual asci. The ascospores are ultimately released (sometimes shot off forcefully) by the ascus to begin the new haploid generation.

The sexual reproductive structure of club fungi is the basidium

Club fungi (Basidiomycota) produce some of the most spectacular fruiting structures found among the fungi. These fruiting structures, called **basidiomata** (singular *basidioma*), include puffballs (see Figure 22.4), some of which may be more than half a meter in diameter, mushrooms of all kinds, and the bracket fungi often encountered on trees and fallen logs in a damp forest. About 30,000 species of club fungi have been described. They include about 4,000 species of mushrooms, including both poisonous and edible species (**FIGURE 22.18A**). Bracket fungi (**FIGURE 22.18B**) do great damage to both cut

(A) *Armillaria* sp.

(B) *Coriolus versicolor*

FIGURE 22.18 Club Fungus Basidiomata (A) These edible mushrooms are the fruiting structures of a honey fungus; mycelia of this genus form connected underground masses that cover many hectares and, if considered as one organism, are among the largest and longest-lived organisms in the world. (B) Saprophytic bracket fungi are agents of decay on dead wood.

lumber and timber stands; by breaking down wood, they also play an important role in the carbon cycle. Some of the most damaging plant pathogens are club fungi, including the rust fungi and smut fungi that parasitize cereal grains. In contrast, other club fungi contribute to the survival of plants as fungal partners in ectomycorrhizae.

The hyphae of club fungi characteristically have septa with small, distinctive pores. The **basidium** (plural *basidia*), a swollen cell at the tip of a specialized hypha, is the characteristic sexual reproductive structure of the club fungi. In mushroom-forming club fungi, the basidia typically form on specialized structures known as *gills*. The basidium is the site of nuclear fusion and meiosis and thus plays the same role in the club fungi as the ascus does in the sac fungi and the zygosporangium does in the zygospore fungi.

After nuclei fuse in the basidium, the resulting diploid nucleus undergoes meiosis, and the four resulting haploid nuclei are incorporated into haploid *basidiospores*, which form on tiny stalks on the outside of the basidium (see Figure 22.15B). A single basidioma of the common bracket fungus *Ganoderma applanatum* can produce as many as 4.5 *trillion* basidiospores in one growing season. Basidiospores typically are forcibly discharged from their basidia and then germinate, giving rise to hyphae with haploid nuclei. As these hyphae grow, haploid hyphae of different mating types meet and fuse, forming dikaryotic hyphae, each cell of which contains two nuclei, one from each parent hypha. The dikaryotic mycelium grows and eventually, when triggered by rain or another environmental cue, produces a basidioma. The dikaryon stage may persist for years—some club fungi live for decades or even centuries. This pattern contrasts with the life cycle of the sac fungi, in which the dikaryon is found only in the stages leading up to formation of the asci.

Do You Understand Concept 22.3?

- Explain the concept of mating types. How are they different from male and female sexes?
- Explain how the microsporidia infect the cells of their animal hosts.
- What feature of the chytrids suggests that the fungi had an aquatic ancestor?
- What is the role of the zygospore in the life cycle of zygospore fungi?
- What distinguishes the fruiting structures of sac fungi from those of club fungi?

Fungi are of interest to biologists because the roles they play in interactions with other organisms. But they can also be useful as tools for studying environmental problems and for finding solutions to those problems.

concept 22.4 Fungi Can Be Sensitive Indicators of Environmental Change

We've already noted the important roles that fungi play in ecosystems, from decomposers to pathogens to plant mutualists. These diverse ecological roles have led to their use in studies of environmental change and in environmental remediation.

Lichen diversity and abundance indicate air quality

Lichens can live in many harsh environments where few other species can survive, as we saw in Concept 22.2. In spite of their hardiness, however, lichens are highly sensitive to air pollution because they are unable to excrete any toxic substances they absorb. This sensitivity means that lichens are good biological indicators of air pollution levels. It also explains why they are not commonly found in heavily industrialized regions or in large cities.

Monitoring the diversity and abundance of lichens growing on trees is a practical and inexpensive system for gauging air quality around cities (**FIGURE 22.19**). Maps of lichen diversity provide environmental biologists with a tool for tracking the distribution of air pollutants and their effects. Sensitive biological indicators of pollution, such as lichen growth, allow biologists to monitor air quality without the use of specialized equipment. Lichens are naturally distributed across the environment, and they can also provide a long-term measure of the effects of air pollution across many seasons and years.

Fungi contain historical records of pollutants

Biologists deposit samples of many groups of organisms in the collections of natural history museums. These museum collections serve many purposes. Biologists borrow specimens from these museums to study many aspects of evolution and ecology, and the collections document changes in the biota of our planet over time.

Collections of fungi made over many decades or centuries provide a record of the environmental pollutants that were present when the fungi were growing. Biologists can analyze these historical samples to see how different sources of pollutants were affecting our environment before anyone thought to take direct measures. These long-term records are also useful for analyzing the effectiveness of cleanup efforts and regulatory programs for controlling environmental pollutants.

Reforestation may depend on mycorrhizal fungi

When a forest is cut down, it is not just the trees that are lost. A forest is an ecosystem that depends on the interaction of many species. As we have discussed, many plants depend on close relationships with mycorrhizal fungal partners. When trees are removed from a site, the populations of mycorrhizal fungi

(A)

(B)

FIGURE 22.19 More Lichens, Better Air Lichen abundance and diversity are excellent indicators of air quality. (A) Many lichen species show luxuriant growth on trees in suitable environments with few pollutants in the air. (B) As air quality declines, so do the number and diversity of lichens.

APPLY THE CONCEPT

Fungi can be sensitive indicators of environmental change

Biologists analyzed museum samples of lace lichens (*Ramalina menziesii*) collected near San Francisco, California from 1892 to 2006 for evidence of lead contamination. They measured concentrations of lead (Pb) as well as the ratios of its isotopes ^{206}Pb and ^{207}Pb. The latter measurement was used to determine the source of lead contamination. Possible sources included a lead smelter that operated in the area from 1885 to 1971 (which produced emissions with a ^{206}Pb/^{207}Pb ratio of about 1.15–1.17); leaded gasoline in use from the 1930s to the early 1980s, peaking in 1970 (with a ^{206}Pb/^{207}Pb ratio of 1.18–1.23); and resuspension of historic lead contamination as atmospheric aerosols in recent decades (with an intermediate ^{206}Pb/^{207}Pb ratio of about 1.16–1.19).

Before analyzing the data, use the information provided above to formulate hypotheses about these questions: What trends in atmospheric lead concentrations would you expect to see? What ^{206}Pb/^{207}Pb ratios would you expect to find at different times from the late 1800s to the early 2000s?

1. Plot lead concentration in the lichen samples against year of sample collection. Make a second plot of ^{206}Pb/^{207}Pb ratio against year of sample collection.

2. Do your analyses support the hypotheses you formulated? Are your hypotheses consistent with your analyses of both lead concentrations and ^{206}Pb/^{207}Pb ratios through time? If not, how would you modify your hypotheses, and what additional tests can you design to test your ideas?

SAMPLE	YEAR COLLECTED	LEAD CONCENTRATION (µg OF Pb/g OF LICHEN)	^{206}Pb/^{207}Pb RATIO
1	1892	11.9	1.165
2	1894	4.0	1.155
3	1906	13.7	1.154
4	1907	22.9	1.157
5	1945	49.9	1.187
6	1957	34.2	1.185
7	1978	50.9	1.221
8	1982	10.0	1.215
9	1983	4.6	1.224
10	1987	1.0	1.198
11	1988	1.3	1.199
12	1995	1.9	1.202
13	2000	0.4	1.184
14	2006	1.8	1.184

there decline rapidly. If we wish to restore the forest on the site, we cannot simply replant it with trees and other plants and expect them to survive. The mycorrhizal fungal community must be reestablished as well. For large forest restoration projects, a planned succession of plant growth and soil improvement is often necessary before forest trees can be replanted. As the community of soil fungi gradually recovers, trees that have been inoculated with appropriate mycorrhizal fungi in tree nurseries can be planted to reintroduce greater diversity to the soil fungal community.

Do You Understand Concept 22.4?

- What are some advantages of using surveys of lichen diversity and museum collections of lichens to measure long-term changes in air quality, compared with direct measurements of atmospheric pollutants?

- Can you develop a strategy for tree harvest that would ease the difficulty and expense of reforestation projects by retaining viable communities of mycorrhizal fungi?

Whether living on their own or in symbiotic associations, fungi have spread successfully over much of Earth since their origin from a protist ancestor. An earlier ancestor of fungi also gave rise to the choanoflagellates and the animals, as we will describe in Chapter 23.

| | Have antibiotics derived from fungi eliminated the danger of bacterial diseases in human populations? |

ANSWER Beginning in the 1940s, antibiotics derived from fungi ushered in a "golden age" of freedom from bacterial infections. Today, however, that golden age may be coming to an end. Many antibiotics are losing their effectiveness as pathogenic bacteria evolve resistance to these drugs (**FIGURE 22.20**). Some bacterial diseases, such as tuberculosis, are increasingly serious health problems because of the evolution of new strains that are resistant to most classes of antibiotics.

Why do bacteria evolve resistance to antibiotics? Mutations that allow bacteria to survive in the presence of an antibiotic are favored by selection in a bacterial population whenever an antibiotic is used. Such mutations often carry a cost to the bacteria, so they may be selected against in the absence of regular antibiotic use. To reduce the rate of evolution of antibiotic resistance, antibiotics should be used only for the

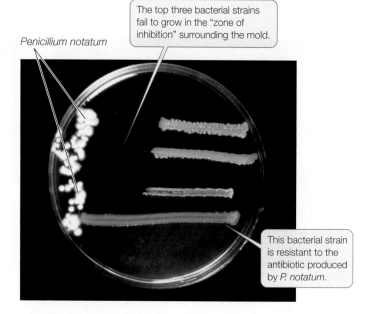

FIGURE 22.20 **Penicillin Resistance** In a petri dish similar to those in Alexander Fleming's lab, four strains of a pathogenic bacterium have been cultured along with *Penicillium* mold. One strain is resistant to the mold's antibiotic substance, as is evidenced by its growth up to the mold.

treatment of appropriate bacterial diseases, and then used to completely clear the bacterial infection.

Most medical antibiotics are semisynthetic, meaning that they are chemically modified forms of the substances that are found naturally in fungi and other organisms. Fungi naturally produce antibiotic compounds to defend themselves against bacterial growth and to reduce competition from bacteria for nutritional resources. These naturally occurring compounds are usually chemically modified to increase their stability, to improve their effectiveness, and to facilitate synthetic production.

From the late 1950s to the late 1990s, no new major classes of antibiotics were discovered. In recent years, however, three new classes of antibiotics have been synthesized based on the information learned from naturally occurring antibiotics, leading to improved treatment of some formerly resistant strains of bacteria.

Fungi have also been used to combat non-bacterial diseases. One of the more unusual applications of fungi is in the war against malaria, one of the biggest killers of humans in sub-Saharan Africa. Biologists have discovered that two species of fungi, *Beauveria bassiana* and *Metarhizium anisopliae*, can kill malaria-causing mosquitoes when applied to mosquito netting. Mosquitoes have not yet shown evidence of developing resistance to these biological pathogens, unlike most chemical pesticides.

SUMMARY

concept 22.1 Fungi Live by Absorptive Heterotrophy

- Fungi are opisthokonts with **absorptive heterotrophy** and with chitin in their cell walls. Fungi have various nutritional modes: some are saprobes, others are parasites, and some are mutualists. **Yeasts** are unicellular, free-living fungi. **Review Figure 22.1**

- The body of a multicellular fungus is a **mycelium**—a meshwork of filaments called **hyphae**. Hyphae may be **septate** (having **septa**) or **coenocytic** (multinucleate). **Review Figure 22.3**

- Fungi are often tolerant of hypertonic environments, and many are tolerant of extreme temperatures.

concept 22.2 Fungi Can Be Saprobic, Parasitic, Predatory, or Mutualistic

- Saprobic fungi, as decomposers, make crucial contributions to the recycling of elements, especially carbon.

- Many fungi are parasites of plants, harvesting nutrients from plant cells by means of **haustoria**. **Review Figure 22.5**

- Certain fungi have relationships with other organisms that are both **symbiotic** and **mutualistic**.

- Some fungi associate with unicellular green algae, cyanobacteria, or both to form **lichens**, which live on exposed surfaces of rocks, trees, and soil. **Review Figure 22.8**

- **Mycorrhizae** are mutualistic associations of fungi with plant roots. They improve a plant's ability to take up nutrients and water.

- **Endophytic fungi** live within plants and may provide protection to their hosts from herbivores and pathogens.

concept 22.3 Major Groups of Fungi Differ in Their Life Cycles

- The microsporidia, chytrids, and zygospore fungi diversified early in fungal evolution. The mycorrhizal fungi, sac fungi, and club fungi form a monophyletic group, and the latter two groups form the clade Dikarya. **Review Figure 22.10 and Table 22.1; see WEB ACTIVITY 22.1**

- Many species of fungi reproduce both sexually and asexually. In many fungi, sexual reproduction occurs between individuals of different **mating types**. **Review Figure 22.11**

- The **microsporidia** are highly reduced unicellular fungi. They are obligate intracellular parasites that infect several animal groups.

- The three distinct lineages of **chytrids** all have flagellated gametes. **Review Figure 22.13A**

- In the sexual reproduction of terrestrial fungi, hyphae fuse, allowing "gamete" nuclei to be transferred. **Plasmogamy** (fusion of cytoplasm) precedes **karyogamy** (fusion of nuclei).

- **Zygospore fungi** have a resting stage with many diploid nuclei, known as a **zygospore**. Their fruiting structures are simple stalked **sporangiophores**. **Review Figure 22.13B and ANIMATED TUTORIAL 22.1**

- **Arbuscular mycorrhizal fungi** form symbiotic associations with plant roots. They are only known to reproduce asexually. Their hyphae are coenocytic.

- In sac fungi and club fungi, a mycelium containing two genetically different haploid nuclei, called a **dikaryon**, is formed. The dikaryotic ($n + n$) condition is unique to the fungi. **Review Figures 22.15 and WEB ACTIVITY 22.2**

- **Sac fungi** have septate hyphae; their sexual reproductive structures are **asci**. Many sac fungi are partners in lichens. Filamentous sac fungi produce fleshy fruiting structures called **ascomata**. The dikaryon stage in the sac fungus life cycle is relatively brief. **Review Figure 22.15A**

- **Club fungi** have septate hyphae. Many club fungi are plant pathogens, although mushroom-forming species are more familiar to most people. Their fruiting structures are called **basidiomata**, and their sexual reproductive structures are **basidia**. The dikaryon stage may last for years. **Review Figure 22.15B**

concept 22.4 Fungi Can Be Sensitive Indicators of Environmental Change

- The diversity and abundance of lichen growth on trees provide sensitive indications of air quality. **Review Figure 22.19**

- Museum collections of fungi provide a historical record of atmospheric pollutants that were present when the fungi were growing.

- Reforestation projects require restoration of the mycorrhizal fungal community.

Animal Origins and Diversity

Almost 1.5 million species of animals have been discovered and named by biologists. One group of animals, the insects, accounts for about 1 million of these species, or more than half of all known species of living organisms. Although these numbers may seem incredibly large, they represent a relatively small fraction of the total animal diversity that is thought to exist on Earth.

As recently as the 1980s, many biologists thought that about half of existing insect species had been described, but today they think that the number of described insect species may be a much smaller fraction of the total number of living species. Why did they change their minds?

A simple but important field study published in 1988 suggested that the number of existing insect species had been significantly underestimated. Knowing that the insects of tropical forests—the most species-rich habitat on Earth—were poorly known, entomologist Terry Erwin made a comprehensive sample of one group of insects, the beetles, in the canopies of a single species of tropical forest tree, *Luehea seemannii*, in Panama. Erwin fogged the canopies of 19 large *L. seemannii* trees with a pesticide and collected the insects that fell from the trees in collection nets. His sample contained about 1,200 species of beetles—many of them undescribed—from this one species of tree.

Erwin then used a set of assumptions to estimate the total number of insect species in tropical evergreen forests. His assumptions included estimates of the number of species of host trees in these forests; the proportion of beetles that specialized on a specific species of host tree; the relative proportion of beetles to other

insect groups; and the proportion of beetles that live in trees versus leaf litter. From this and similar studies, Erwin estimated that there may be 30 million or more species of insects on Earth. Although recent tests of Erwin's assumptions suggest 30 million was an overestimate, it is clear that the vast majority of insect species remain to be discovered by biologists.

Erwin's pioneering study highlighted the fact that we live on a poorly known planet, most of whose species have yet to be named and described. Much of the undiscovered diversity occurs among several groups of physically small yet biologically diverse animal groups, although even many larger species of animals continue to be discovered.

Entomologists collecting insects from a tree canopy in the Yungas rainforest of Argentina. Such research illuminates the vast number of insect species that remain undescribed by science.

KEY CONCEPTS

23.1 Distinct Body Plans Evolved among the Animals

23.2 Some Animal Groups Fall Outside the Bilateria

23.3 There Are Two Major Groups of Protostomes

23.4 Arthropods Are Diverse and Abundant Animals

23.5 Deuterostomes Include Echinoderms, Hemichordates, and Chordates

23.6 Life on Land Contributed to Vertebrate Diversification

23.7 Humans Evolved among the Primates

 QUESTION Besides the insects, which other groups of animals are thought to contain many more species than are known at present?

Distinct Body Plans Evolved among the Animals

How do we recognize an animal? The answer may seem obvious for familiar animals, but consider the sponges, which were once thought to be plants, or the rotifers, which are smaller than some protists.

The general characteristics often used to recognize animals include multicellularity, heterotrophy (using nutrients produced by other organisms), internal digestion, and motility (independent movement). Although these features are useful, none is diagnostic for all animals. Some animals do not move independently (at least during certain life stages), and some plants and fungi do have limited movement. Many parasitic animals lack a gut for internal digestion, and many other organisms (notably the fungi) are multicellular and heterotrophic. So what evidence leads us to group the animals in a monophyletic clade?

Animal monophyly is supported by gene sequences and cellular morphology

The most convincing evidence that all the organisms considered to be animals share a common ancestor comes from phylogenetic analyses of their gene sequences. Relatively few complete animal genomes are available, but more are being sequenced each year. Analyses of these genomes and of many individual gene sequences have shown that all animals are indeed monophyletic. A currently well-supported phylogenetic tree of the animals is shown in **FIGURE 23.1**.

Surprisingly few morphological features are shared across all animal species. The morphological *synapomorphies* that are present are evident primarily at the cellular level. They include

FIGURE 23.1 Animal Phylogeny This tree presents the best-supported current hypotheses of the evolutionary relationships among major groups of animals. The traits highlighted by red circles will be explained as you read this chapter. (See Working with Data 23.1 at **yourBioPortal.com**.)

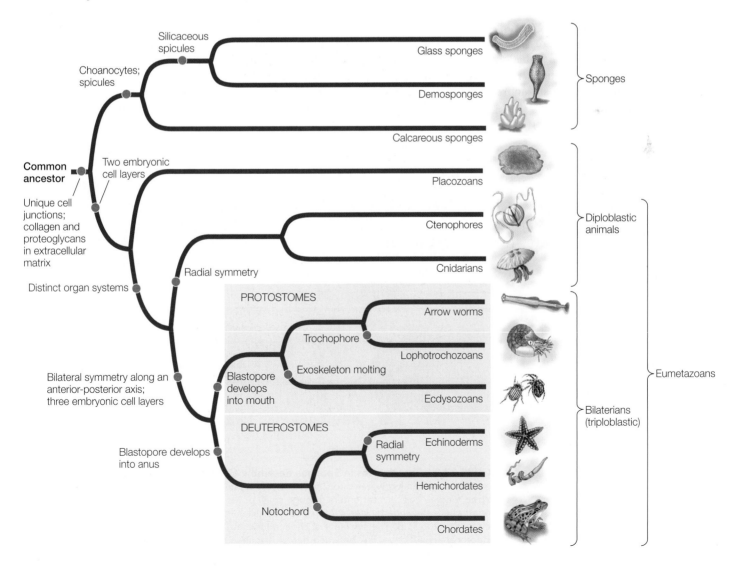

several unique features of the junctions between animal cells and a common set of extracellular matrix molecules, including collagen and proteoglycans. Although some animals in a few groups lack one or another of these cellular characteristics, it is believed that these traits were possessed by the ancestor of all animals and were subsequently lost in those groups.

> **LINK** Animal cell junctions and extracellular matrix molecules are described in Concept 4.5

Similarities in the organization and function of Hox and other developmental genes provide additional evidence of developmental mechanisms shared by a common animal ancestor. The Hox genes specify body pattern and axis formation, leading to developmental similarities across animals (see Chapter 14).

The common ancestor of animals was probably a colonial flagellated protist similar to existing colonial choanoflagellates, which have similarities to the multicellular sponges (**FIGURE 23.2**). The best-supported hypothesis postulates a choanoflagellate lineage in which certain cells in the colony began to be specialized—some for movement, others for nutrition, others for reproduction, and so on. Once this *functional specialization* had begun, cells could have continued to differentiate. Coordination among groups of cells could have improved by means of specific regulatory and signaling molecules that guided differentiation and migration of cells in developing embryos. Such coordinated groups of cells eventually evolved into the larger and more complex organisms that we call animals.

Nearly 80 percent of the 1.8 million named species of living organisms are animals, and millions of additional animal species await discovery. Evidence about the evolutionary relationships among animal groups can be found in fossils, in patterns of embryonic development, in the morphology and physiology of living animals, in the structure of animal proteins, and in gene sequences. Increasingly, studies of higher-level relationships have come to depend on genomic sequence comparisons, as genomes are ultimately the source of all inherited trait information.

Basic developmental patterns and body plans differentiate major animal groups

Distinct layers of cells form during the early development of most animals. These cell layers differentiate into specific organs and organ systems as development continues. The embryos of **diploblastic** animals have two cell layers: an outer *ectoderm* and an inner *endoderm*. Embryos of **triploblastic** animals have, in addition to ectoderm and endoderm, a third distinct cell layer, *mesoderm*, between the ectoderm and the endoderm. The existence of three cell layers in embryos is a synapomorphy of triploblastic animals, whereas the diploblastic animals (placozoans, ctenophores, and cnidarians) exhibit the ancestral condition. Some biologists consider sponges to be diploblastic, but since they do not have clearly differentiated tissue types or embryonic cell layers, the term is not usually applied to them.

During early development in many animals, in a process known as *gastrulation*, a hollow ball one cell thick indents to

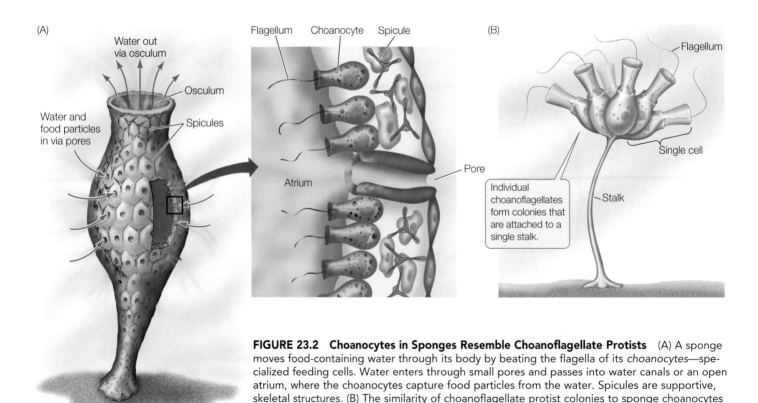

FIGURE 23.2 Choanocytes in Sponges Resemble Choanoflagellate Protists (A) A sponge moves food-containing water through its body by beating the flagella of its *choanocytes*—specialized feeding cells. Water enters through small pores and passes into water canals or an open atrium, where the choanocytes capture food particles from the water. Spicules are supportive, skeletal structures. (B) The similarity of choanoflagellate protist colonies to sponge choanocytes supports an evolutionary link between this protist lineage and the animals.

form a cup-shaped structure. The opening of the cavity formed by this indentation is called the *blastopore*:

Blastopore

Gastrulation is covered in detail in Chapter 33; the point to remember here is that the *overall pattern* of gastrulation immediately after the blastopore forms divides the triploblastic animals into two major groups:

- In the **protostomes** (Greek, "mouth first"), the mouth arises from the blastopore, and the anus forms later.
- In the **deuterostomes** ("mouth second"), the blastopore becomes the anus, and the mouth forms later.

Although the developmental patterns of animals are more varied than suggested by this simple dichotomy, phylogenies based on DNA sequences indicate that the protostomes and deuterostomes are distinct clades. These groups are known as the *bilaterians* (named for their usual bilateral symmetry), and they account for the vast majority of animal species.

The general structure of an animal, the arrangement of its organ systems, and the integrated functioning of its parts are referred to as its **body plan**. As Chapter 14 describes, the regulatory and signaling genes that govern the development of body symmetry, body cavities, segmentation, and appendages are widely shared among the different animal groups. Although the myriad animal body plans cover a wide range of morphologies, they can be seen as variations on four key features:

- The *symmetry* of the body
- The structure of the *body cavity*
- The *segmentation* of the body
- The existence and location of *external appendages* that are used for sensing, chewing, locomotion, mating, and other functions

Each of these features affects how an animal interacts with its environment.

Most animals are symmetrical

The overall shape of an animal can be described by its **symmetry**. An animal is said to be *symmetrical* if it can be divided along at least one plane into similar halves. Animals that have no plane of symmetry are *asymmetrical*. Placozoans and many sponges are asymmetrical, but most other animals have some kind of symmetry, which is governed by the expression of regulatory genes during development.

In organisms with **radial symmetry**, body parts are arranged around a single axis at the body's center:

Radial symmetry

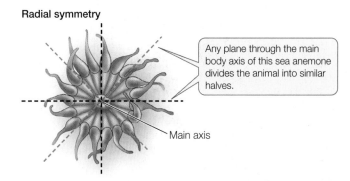

Any plane through the main body axis of this sea anemone divides the animal into similar halves.

Main axis

A perfectly radially symmetrical animal can be divided into similar halves by any plane that contains the central axis. However, most radially symmetrical animals are slightly modified, so fewer planes can divide them into identical halves. Some radially symmetrical animals are sessile (sedentary) or drift with water currents. Others move slowly but can move equally well in any direction.

Bilateral symmetry is characteristic of animals that have a distinct front end, which typically precedes the rest of the body as the animal moves. A bilaterally symmetrical animal can be divided into mirror-image (left and right) halves by a single plane that passes through the midline of its body. This plane runs from the front, or *anterior*, end of the body, to the rear, or *posterior*, end:

Bilateral symmetry

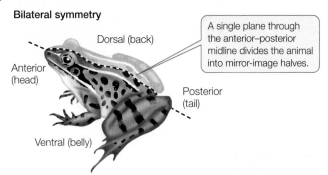

Dorsal (back)

Anterior (head)

A single plane through the anterior–posterior midline divides the animal into mirror-image halves.

Posterior (tail)

Ventral (belly)

A plane at right angles to the midline divides the body into two dissimilar sides, its *dorsal* and *ventral* surfaces.

Bilateral symmetry is strongly correlated with **cephalization**, which is the concentration of sensory organs and nervous tissues in an anterior head. Cephalization has been evolutionarily favored because the anterior end of a bilaterally symmetrical animal typically encounters new environments first.

The structure of the body cavity influences movement

Animals with three embryonic cell layers (see Figure 23.1) can be divided into three types—*acoelomate, pseudocoelomate,* and *coelomate*—based on the presence and structure of an internal, fluid-filled body cavity.

- **Acoelomate** animals such as flatworms lack an enclosed, fluid-filled body cavity. Instead, the space between the gut (derived from endoderm) and the muscular body wall (derived from mesoderm) is filled with masses of cells called *mesenchyme* (**FIGURE 23.3A**). These animals typically move by beating cilia.

- **Pseudocoelomate** animals have a body cavity called a *pseudocoel*, a fluid-filled space in which many of the internal organs are suspended. A pseudocoel is enclosed by muscles

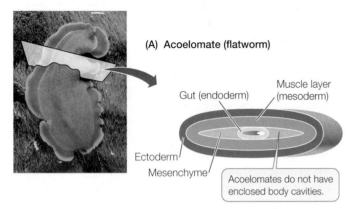

(A) Acoelomate (flatworm)

Gut (endoderm)
Muscle layer (mesoderm)
Ectoderm
Mesenchyme

Acoelomates do not have enclosed body cavities.

(B) Pseudocoelomate (roundworm)

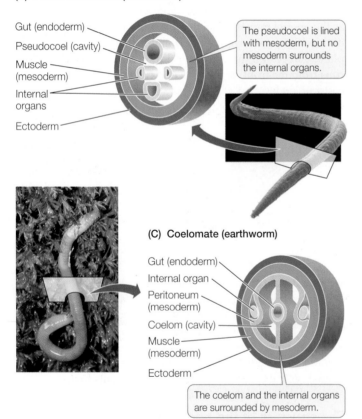

Gut (endoderm)
Pseudocoel (cavity)
Muscle (mesoderm)
Internal organs
Ectoderm

The pseudocoel is lined with mesoderm, but no mesoderm surrounds the internal organs.

(C) Coelomate (earthworm)

Gut (endoderm)
Internal organ
Peritoneum (mesoderm)
Coelom (cavity)
Muscle (mesoderm)
Ectoderm

The coelom and the internal organs are surrounded by mesoderm.

FIGURE 23.3 Animal Body Cavities (A) Acoelomates do not have enclosed body cavities. (B) Pseudocoelomates have a body cavity bounded by endoderm and mesoderm. (C) Coelomates have a peritoneum surrounding the internal organs in a region bounded by mesoderm.

yourBioPortal.com
Go to WEB ACTIVITY 23.1
Animal Body Cavities

(mesoderm) only on its outside; there is no inner layer of mesoderm surrounding the internal organs (**FIGURE 23.3B**).

- **Coelomate** animals have a *coelom*, a body cavity that develops within the mesoderm. It is lined with a layer of muscular tissue called the *peritoneum*, which also surrounds the internal organs. The coelom is thus enclosed on both the inside and the outside by mesoderm (**FIGURE 23.3C**). A coelomate animal has more refined control over the movement of the fluids in its body cavity than a pseudocoelomate animal does.

The structure of an animal's body cavity strongly influences the ways in which it can move, and the body cavities of many animals function as *hydrostatic skeletons* (see Figure 36.12). Fluids are relatively incompressible, so when the muscles surrounding them contract, fluids shift to another part of the cavity. If the body tissues around the cavity are flexible, fluids squeezed out of one region can cause another region to expand. The moving fluids can thus move specific body parts. (You can see how a hydrostatic skeleton works by watching a snail emerge from its shell.) An animal with both *circular muscles* (encircling the body cavity) and *longitudinal muscles* (running along the length of the body) has even greater control over its movement.

In terrestrial environments, the hydrostatic function of fluid-filled body cavities applies mostly to relatively small, soft-bodied organisms. Most larger animals (as well as many smaller ones) have hard skeletons that provide protection and facilitate movement. Muscles are attached to those firm structures, which may be inside the animal or on its outer surface (in the form of a shell or cuticle).

Segmentation improves control of movement

Many animal bodies are divided into segments. **Segmentation** facilitates specialization of different body regions. It also allows an animal to alter the shape of its body in complex ways and to control its movements precisely. If an animal's body is segmented, muscles in each individual segment can change the shape of that segment independently of the others. In only a few segmented animals is the body cavity separated into discrete compartments, but even partly separated compartments allow better control of movement. Segmentation occurs in several groups of protostomes and deuterostomes.

In some animals, segments are not apparent externally (e.g., the segmented vertebrae of vertebrates). In other animals, similar body segments are repeated many times (**FIGURE 23.4A**). And in yet other animals, including most arthropods, the visible segments differ strikingly (**FIGURE 23.4B**). The dramatic evolutionary radiation of the arthropods (including the insects, spiders, centipedes, and crustaceans) was based on changes in a segmented body plan that features muscles attached to the inner surface of an external skeleton, including a variety of external *appendages* that move these animals.

Appendages have many uses

Getting around under their own power allows animals to obtain food, to avoid predators, and to find mates. Even some

(A) *Hermodice carunculata*

The fireworm is a marine worm that displays an evenly segmented body plan.

Protective bristles

Tail segments are modified for hunting and defense.

Anterior segments have fused and bear appendages for locomotion and feeding.

Abdominal segments are modified for digestion and reproduction.

(B) *Ophistothalmus walberghi*

FIGURE 23.4 Body Segmentation (A) All of the segments of this marine fireworm, an annelid, are similar. Its appendages are tipped with bristles that are used for locomotion and (in this species) for protection, since they contain a noxious toxin. (B) Segmentation allows the evolution of differentiation among the segments. The segments of this scorpion, an arthropod, differ in their form, function, and the appendages they bear.

sedentary species, such as sea anemones, have larval stages that use cilia to swim, thus increasing the animal's chances of finding a suitable habitat. Appendages that project externally from the body greatly enhance an animal's ability to move around.

Highly controlled, rapid movement is greatly enhanced in animals whose appendages have become modified into specialized *limbs*. In the arthropods and vertebrates, the presence of *jointed limbs* has been a prominent factor in their evolutionary success. In four independent instances—among the arthropod insects and among the vertebrate pterosaurs, birds, and bats—body plans emerged in which limbs were modified into wings, allowing these animals to take to the air.

Appendages also include many structures that are not used for locomotion. Many animals have antennae, which are specialized appendages used for sensing the environment. Other appendages (such as claws and mouthparts of many arthropods) are adaptations for capturing prey or chewing food. In some species, appendages are used for reproductive purposes, such as sperm transfer or egg incubation.

Do You Understand Concept 23.1?

- Describe the difference between diploblastic and triploblastic embryos, and between protostomes and deuterostomes.
- Describe the main types of symmetry found in animals. How can an animal's symmetry influence the way it moves?
- Explain several ways in which body cavities and segmentation improve control over movement.

Variations in body symmetry, body cavity structure, life cycles, patterns of development, and survival strategies differentiate millions of animal species. **TABLE 23.1** summarizes the living members of the major animal groups we will describe in the rest of this chapter.

concept 23.2 Some Animal Groups Fall Outside the Bilateria

Looking at Figure 23.1, you can see that the protostome and deuterostome animals together comprise a monophyletic group known as the **Bilateria**. Some major traits that support the monophyly of bilaterians (in addition to genomic analyses) are strong bilateral symmetry, the presence of three distinct cell layers in embryos (*triploblasty*), and the presence of at least seven Hox genes (see Chapter 14). There are, however, four animal groups—the sponges, placozoans, ctenophores, and cnidarians—that are not bilaterians.

The simplest animals, the **sponges**, have no distinct tissue types. The **placozoans** have four cell types and weakly differentiated tissue layers. All other animals groups, including the bilaterians, are known as **eumetazoans**; eumetazoans have obvious body symmetry, a gut, a nervous system, and tissues organized into distinct organs (although there have been secondary losses of some of these structures in some eumetazoans). Sponges and placozoans lack all of these features.

yourBioPortal.com

Go to WEB ACTIVITY 23.2
Sponge and Diploblast Classification

Sponges and placozoans are weakly organized animals

Sponges have hard skeletal elements called **spicules**, which may be small and simple or large and complex. The three major groups of sponges separated soon after the split between sponges and the rest of the animals. Members of two groups (*glass sponges* and *demosponges*) have skeletons composed of spicules made of hydrated silicon dioxide (**FIGURE 23.5A,B**). These siliceous spicules have greater flexibility and toughness than synthetic glass rods of similar length. Members of the third group, the *calcareous sponges*, take their name from their

FIGURE 23.5 Sponge Diversity (A) The great majority of sponge species are demosponges, such as these Pacific barrel sponges. (B) Like the demosponges, the supporting structures of glass sponges are siliceous spicules. The pores and water canals "typical" of the sponge body plan are apparent in both (A) and (B). (C) The spicules of calcareous sponges, prominent in this photo, are made of calcium carbonate.

(A) *Xestospongia testudinaria*

(B) *Staurocalyptus* sp.

(C) *Sycon* sp.

TABLE 23.1 Summary of Living Members of the Major Animal Groups

	APPROXIMATE NUMBER OF LIVING SPECIES DESCRIBED	MAJOR GROUPS		APPROXIMATE NUMBER OF LIVING SPECIES DESCRIBED	MAJOR GROUPS
Sponges	9,000	Demosponges, glass sponges, calcareous sponges	**Ecdysozoans**		
			Kinorhynchs	150	
Placozoans	2		Loriciferans	100	
Ctenophores	150		Priapulids	16	
			Horsehair worms	320	
Cnidarians	11,000	Anthozoans: Corals, sea anemones	Nematodes	25,000	
			Onychophorans	150	
		Hydrozoans: Hydras and hydroids	Tardigrades	800	
		Scyphozoans: Jellyfishes	Arthropods:		
			Crustaceans	52,000	Crabs, shrimps, lobsters, barnacles, copepods
PROTOSTOMES					
Arrow worms	100		Hexapods	1,000,000	Insects and relatives
Lophotrochozoans			Myriapods	14,000	Millipedes, centipedes
Bryozoans	4,500		Chelicerates	98,000	Horseshoe crabs, arachnids (scorpions, harvestmen, spiders, mites, ticks)
Flatworms	25,000	Free-living flatworms; flukes and tapeworms (all parasitic); monogeneans (ectoparasites of fishes)			
			DEUTEROSTOMES		
Rotifers	1,800		Echinoderms	7,000	Crinoids (sea lilies and feather stars); brittle stars; sea stars; sea urchins; sea cucumbers
Ribbon worms	1,000				
Phoronids	20				
Brachiopods	335		Hemichordates	100	Acorn worms and pterobranchs
Annelids	16,500	Polychaetes (all marine)	Urochordates	3,000	Ascidians (sea squirts)
		Clitellates: Earthworms, freshwater worms, leeches	Cephalochordates	30	Lancelets
Mollusks	100,000	Monoplacophorans	Vertebrates	62,000	Hagfish; lampreys
		Chitons			Cartilaginous fishes
		Bivalves: Clams, oysters, mussels			Ray-finned fishes
		Gastropods: Snails, slugs, limpets			Coelacanths; lungfishes
		Cephalopods: Squids, octopuses, nautiloids			Amphibians
					Reptiles (including birds)
					Mammals

(A)

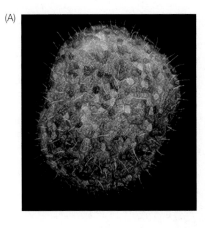

(B)

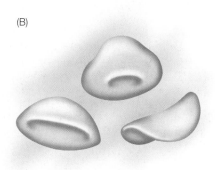

FIGURE 23.6 Placozoan Simplicity (A) As seen in this artist's rendition, adult placozoans are tiny, flattened, asymmetrical animals with only four cell types. (B) Recent studies have found a weakly swimming pelagic stage of placozoan to be abundant in many warm tropical and subtropical seas.

skeletons composed of calcium carbonate spicules (**FIGURE 23.5C**).

The body plan of all sponges—even large ones, which may reach a meter or more in length—is an aggregation of cells built around a water canal system. Water, along with any food particles it contains, enters the sponge by way of small pores and passes into the water canals or a central atrium, where choanocytes capture food particles (see Figure 23.2).

A skeleton of simple or branching spicules, and often a complex network of elastic fibers, supports the body of most sponges. Sponges also have an extracellular matrix, composed of collagen, adhesive glycoproteins, and other molecules, that holds the cells together. Most species are filter feeders; a few species are carnivores that trap prey on hook-shaped spicules that protrude from the body surface.

Most of the 9,000 species of sponges are marine; only about 50 species live in fresh water. The wide varieties of sponge sizes and shapes are adapted to different movement patterns of water. Sponges living in intertidal or shallow subtidal environments are subject to strong wave action and attach firmly to the substrate. Most sponges that live in slowly flowing water are flattened and are oriented at right angles to the direction of current flow, allowing them to intercept water and the food it contains as it flows past them.

Sponges reproduce both sexually and asexually. In most species, a single individual produces both eggs and sperm, but individuals do not self-fertilize. Water currents carry sperm from one individual to another. Asexual reproduction is by budding and regeneration (see Concept 32.1).

Placozoans are structurally very simple animals with only four distinct cell types (**FIGURE 23.6A**). Individuals in the mature life stage are usually observed adhering to surfaces (such as the glass of aquariums, where they were first discovered, or to rocks and other hard substrata in nature). Their structural simplicity—they have no mouth, gut, or nervous system—initially led biologists to suspect they might be the sister group of all other animals. Most phylogenetic analyses have not supported this hypothesis, however, and some aspects of the placozoans' structural simplicity may be secondarily derived. They are generally considered to have a diploblastic body plan, with upper and lower epithelial (surface) layers that sandwich a layer of contractile fiber cells.

Recent studies have found that placozoans have a pelagic (open-ocean) stage that is capable of swimming (**FIGURE**

23.6B), but the life history of placozoans is incompletely known. It is known that placozoans can reproduce both asexually as well as sexually, although we know almost none of the details of their sexual reproduction. Most studies have focused on the larger adherent stages that can be found in aquariums, where they appear after being inadvertently collected with other marine organisms. The transparent nature and small size of placozoans make them very difficult to observe in nature.

Ctenophores are radially symmetrical and diploblastic

Ctenophores, also known as *comb jellies*, lack most of the Hox genes found in all other eumetazoans. Ctenophores have a radially symmetrical, diploblastic body plan. The two cell layers are separated by an inert, gelatinous extracellular matrix called **mesoglea**. Ctenophores have a *complete gut*: food enters through a mouth, and wastes are eliminated through two anal pores.

Ctenophores move by beating cilia rather than by muscular contractions. Most of the 150 known species have eight comb-like rows of cilia-bearing plates, called **ctenes** (**FIGURE 23.7**). The feeding tentacles of ctenophores are covered with cells that discharge adhesive material when they contact prey. After capturing its prey, a ctenophore retracts its tentacles to bring the food to its mouth. In some species, the entire surface of the body is coated with sticky mucus that captures prey. Most ctenophores eat small planktonic organisms, although some eat other ctenophores. Ctenophores are common in open seas and can become abundant in protected bodies of water, where large populations can damage local ecosystems.

Ctenophore life cycles are uncomplicated. Gametes are released into the body cavity and then discharged through the mouth or the anal pores. Fertilization takes place in open seawater. In nearly all species, the fertilized egg develops directly into a miniature ctenophore that gradually grows into an adult.

Cnidarians are specialized carnivores

The **cnidarians** (jellyfishes, sea anemones, corals, and hydrozoans) may be the sister group of the ctenophores, although some biologists think they are more closely related to the bilaterians. The mouth of a cnidarian is connected to a blind sac called the **gastrovascular cavity** (a cnidarian thus does not have a complete gut). The gastrovascular cavity functions in digestion, circulation, and gas exchange, and it also acts as a hydrostatic skeleton. The single opening serves as both mouth and anus.

(A)

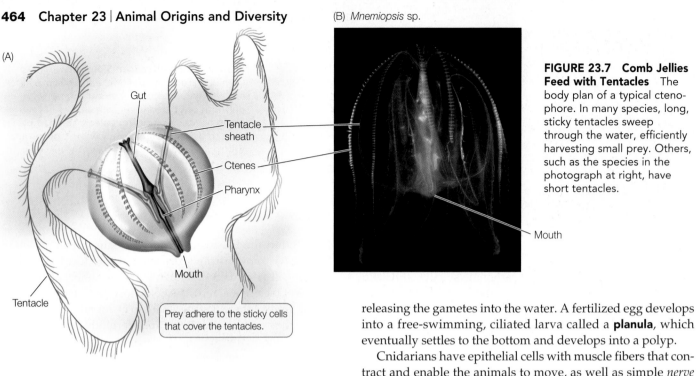

(B) *Mnemiopsis* sp.

Gut

Tentacle
sheath

Ctenes

Pharynx

Mouth

Tentacle

Prey adhere to the sticky cells
that cover the tentacles.

**FIGURE 23.7 Comb Jellies
Feed with Tentacles** The
body plan of a typical ctenophore. In many species, long,
sticky tentacles sweep
through the water, efficiently
harvesting small prey. Others,
such as the species in the
photograph at right, have
short tentacles.

Mouth

The life cycle of many cnidarians has two distinct stages,
one sessile and the other motile (**FIGURE 23.8**), although one
or the other of these stages is absent in some groups. In the
sessile **polyp** stage, a cylindrical stalk is attached to the substrate. The motile **medusa** (plural *medusae*) is a free-swimming
stage shaped like a bell or an umbrella. It typically floats with
its mouth and feeding tentacles facing downward. Mature
polyps produce medusae by asexual budding. Medusae then
reproduce sexually, producing eggs or sperm by meiosis and

releasing the gametes into the water. A fertilized egg develops
into a free-swimming, ciliated larva called a **planula**, which
eventually settles to the bottom and develops into a polyp.

Cnidarians have epithelial cells with muscle fibers that contract and enable the animals to move, as well as simple *nerve
nets* that integrate their body activities. They are specialized
carnivores, using the toxin in harpoonlike structures called *nematocysts* to capture relatively large and complex prey. Some
cnidarians, including many corals and anemones, gain additional nutrition from photosynthetic endosymbionts that live
in their tissues.

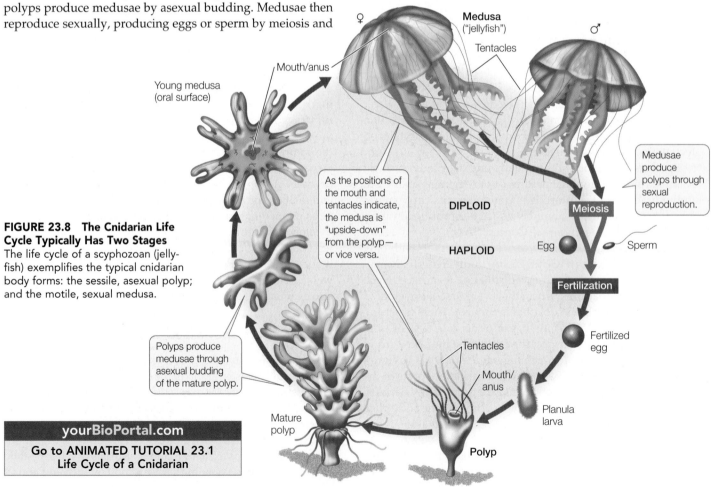

**FIGURE 23.8 The Cnidarian Life
Cycle Typically Has Two Stages**
The life cycle of a scyphozoan (jellyfish) exemplifies the typical cnidarian
body forms: the sessile, asexual polyp;
and the motile, sexual medusa.

Young medusa
(oral surface)

Mouth/anus

Medusa
("jellyfish")

Tentacles

Medusae
produce
polyps through
sexual
reproduction.

As the positions of
the mouth and
tentacles indicate,
the medusa is
"upside-down"
from the polyp—
or vice versa.

DIPLOID

HAPLOID

Meiosis

Egg

Sperm

Fertilization

Polyps produce
medusae through
asexual budding
of the mature polyp.

Mature
polyp

Tentacles

Mouth/
anus

Fertilized
egg

Planula
larva

Polyp

yourBioPortal.com

**Go to ANIMATED TUTORIAL 23.1
Life Cycle of a Cnidarian**

(A) *Sagartia modesta*

(B) *Gonionemus vertens*

FIGURE 23.9 Diversity among Cnidarians (A) Sea anemones are sessile, living attached to a marine substrate. Water currents carry prey into the netmatocyst-studded tentacles. (B) A jellyfish illustrates the complexity of a scyphozoan medusa. (C) Numerous corals species on a reef in Fiji.

(C)

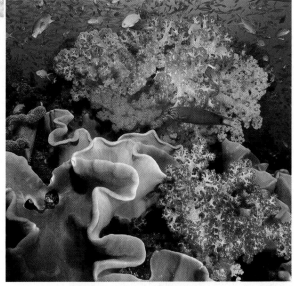

> **LINK** The importance of protist endosymbionts in coral colonies is discussed in Concept 20.4

Of the roughly 11,000 living cnidarian species, all but a few live in the oceans. The smallest cnidarians can hardly be seen without a microscope. The largest known jellyfish is 2.5 meters in diameter, and some colonial species can reach lengths in excess of 30 meters. Three clades of cnidarians have many species, some of which are shown in **FIGURE 23.9**. The *anthozoans* include sea anemones, sea pens, and corals; the *scyphozoans* are commonly known as jellyfishes or sea jellies. The less familiar *hydrozoans* include both freshwater and marine species.

Do You Understand Concept 23.2?

- Why are sponges and placozoans considered to be animals, even though they lack the complex body structures found among most other animal groups?
- How does feeding differ between ctenophores and cnidarians?
- Can you think of any advantages to the biphasic life cycle (sessile polyp and motile medusa) of many cnidarians?

In Concept 23.1 we noted that the name protostome means "mouth first," a developmental synapomorphy that links all these animals. The protostomes can be divided into two major clades—the lophotrochozoans and the ecdysozoans—largely on the basis of DNA sequence analysis.

concept 23.3 There Are Two Major Groups of Protostomes

The protostomes are a highly diverse group of animals. Nearly all protostomes can be readily classified as either *lophotrochozoans* or *ecdysozoans*. Although their body plans are extremely varied, they are all bilaterally symmetrical animals whose bodies exhibit two major derived traits:

- An anterior *brain* that surrounds the entrance to the digestive tract
- A ventral *nervous system* consisting of paired or fused longitudinal nerve cords

Other aspects of protostome body organization differ widely from group to group. Although the common ancestor of the protostomes had a coelom, subsequent modifications of the coelom distinguish many protostome lineages. In at least one protostome lineage (the flatworms), the coelom has been lost (that is, the flatworms reverted to an acoelomate state). Some lineages are characterized by a pseudocoel (see Figure 23.3B). In two of the most prominent protostome groups, the coelom has been highly modified.

- The *arthropods* lost the ancestral condition of the coelom over the course of evolution. Their internal body cavity has become a *hemocoel*, or "blood chamber," in which fluid from an open circulatory system bathes the internal organs before returning to blood vessels.
- Most *mollusks* have an open circulatory system with some of the attributes of the hemocoel, but they retain vestiges of an enclosed coelom around their major organs.

The evolutionary relationships of one small group of protostomes, the **arrow worms** (see Figure 23.1), have been debated for many years. Although recent gene sequence studies clearly identify arrow worms as protostomes, there remains some question as to whether they are the closest relatives of the lophotrochozoans, or possibly the sister group of all other protostomes.

The 100 or so living species of arrow worms are small (3 mm–12 cm) marine predators of planktonic protists and small fish.

Cilia-bearing lophophores and trochophore larvae evolved among the lophotrochozoans

Lophotrochozoans derive their name from two different features that involve cilia: a feeding structure known as a *lophophore* and a free-living larva known as a *trochophore*. Neither feature is universal for all lophotrochozoans, however.

Several distantly related groups of lophotrochozoans (including bryozoans, brachiopods, and phoronids) have a **lophophore**, a circular or U-shaped ring of ciliated, hollow tentacles around the mouth (**FIGURE 23.10A**). This complex structure is an organ for both food collection and gas exchange. The lophophore appears to have evolved independently several times, although it may be an ancestral feature that has been lost in many groups. Nearly all animals with a lophophore are sessile as adults, using the lophophore's tentacles and cilia to capture small floating organisms from the water.

Some lophotrochozoans, especially in their larval form, use cilia for locomotion. The larval form known as a **trochophore** moves by beating a band of cilia:

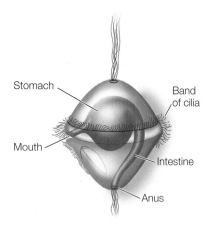

This movement of cilia also brings plankton closer to the larva, where the plankton can be captured and ingested (similar in function to the cilia of the lophophore). Trochophore larvae are found among many of the major groups of lophotrochozoans, including the mollusks, annelids, ribbon worms, and bryozoans. This larval form was probably present in the common ancestor of lophotrochozoans, although it has been subsequently lost in several lineages.

Lophotrochozoans range from relatively simple animals with a blind gut (that is, a gut with a single opening that both takes in food and expels wastes) and no internal transport system to animals with a complete gut (having separate entrance and exit openings) and a complex internal transport system. A number of these groups exhibit wormlike bodies, but the lophotrochozoans encompass a wide diversity of morphologies, including a few groups with external shells. Included among the lophotrochozoans are species-rich groups such as flatworms, annelids, and mollusks, along with many less well known groups, some of which have only recently been discovered.

BRYOZOANS The 4,500 species of **bryozoans** ("moss animals") are colonial animals that live in a "house" made of material secreted by the external body wall. Almost all bryozoans are marine, although a few species occur in fresh or brackish water. A bryozoan colony consists of many small (1–2 mm) individuals connected by strands of tissue along which nutrients can be moved (**FIGURE 23.10B**). The colony is created by the asexual reproduction of its founding member, and a single colony may contain as many as 2 million individuals. Rocks in coastal regions in many parts of the world are covered with luxuriant growths of bryozoans. Some bryozoans create miniature reefs in shallow waters. In some species, the individual colony members are differentially specialized for feeding, reproduction, defense, or support.

(A) *Cristatella mucedo*

Mouth

Lophophore

Tentacles of lophophore

(B) *Cabera boryi*

FIGURE 23.10 Bryozoans Form Colonies (A) The extended lophophores of its individual members dominate the anatomy of this colonial bryozoan. (B) The rigid orange tissue of this marine bryozoan colony connects and supplies nutrients to thousands of individual animals.

FIGURE 23.11 **Flatworms** (A) The fluke diagrammed here is representative of many parasitic flatworms. Absorbing nutrition from the host animal's gut, these internal parasites do not require elaborate feeding or digestive organs and can devote most of their bodies to reproduction. (B) Some flatworms species, such as this Hawaiian marine species, are free-living; this one was photographed on the surface of a sea star.

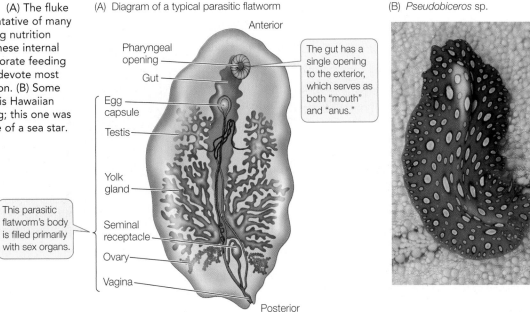

(A) Diagram of a typical parasitic flatworm

Anterior

Pharyngeal opening

Gut

The gut has a single opening to the exterior, which serves as both "mouth" and "anus."

Egg capsule

Testis

Yolk gland

This parasitic flatworm's body is filled primarily with sex organs.

Seminal receptacle

Ovary

Vagina

Posterior

(B) *Pseudobiceros* sp.

Bryozoans can also reproduce sexually by releasing sperm into the water, which carries the sperm to other individuals. Eggs are fertilized internally; developing embryos are brooded before they exit as larvae to seek suitable sites for attachment to the substrate.

FLATWORMS Most of the 25,000 species of **flatworms** are *tapeworms* and *flukes*; members of these two groups are internal parasites, particularly of vertebrates (**FIGURE 23.11A**). Because they absorb digested food from the guts of their hosts, many parasitic flatworms lack digestive tracts of their own. Some cause serious human diseases, such as schistosomiasis, which is common in parts of Asia, Africa, and South America. The species that causes this devastating disease has a complex life cycle involving both freshwater snails and mammals as hosts. Other flatworms are external parasites of fishes and other aquatic vertebrates. The *turbellarians* include most of the free-living species (**FIGURE 23.11B**).

Flatworms lack specialized organs for transporting oxygen to their internal tissues. Lacking a gas transport system, each cell must be near a body surface, a requirement met by the dorsoventrally flattened body form. In flatworms that have a digestive tract, this consists of a mouth opening into a blind sac. The sac is often highly branched, forming intricate patterns that increase the surface area available for the absorption of nutrients. Some small free-living flatworms are cephalized, with a head bearing chemoreceptor organs, two simple eyes, and a tiny brain composed of anterior thickenings of the longitudinal nerve cords. Free-living flatworms glide over surfaces, powered by broad bands of cilia.

ROTIFERS Most **rotifers** are tiny—50–500 μm long, smaller than some ciliate protists—but they have

specialized internal organs (**FIGURE 23.12**). A complete gut passes from an anterior mouth to a posterior anus; the body cavity is a pseudocoel that functions as a hydrostatic skeleton. Rotifers typically propel themselves through the water by means of rapidly beating cilia rather than by muscular contraction.

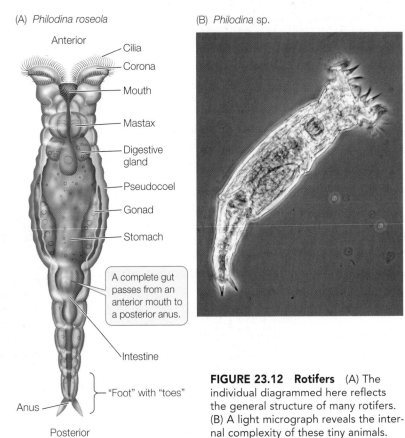

(A) *Philodina roseola*

Anterior

Cilia

Corona

Mouth

Mastax

Digestive gland

Pseudocoel

Gonad

Stomach

A complete gut passes from an anterior mouth to a posterior anus.

Intestine

"Foot" with "toes"

Anus

Posterior

(B) *Philodina* sp.

FIGURE 23.12 **Rotifers** (A) The individual diagrammed here reflects the general structure of many rotifers. (B) A light micrograph reveals the internal complexity of these tiny animals.

(A)

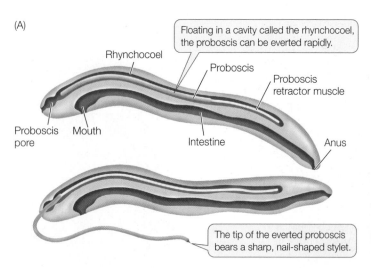

Rhynchocoel

Floating in a cavity called the rhynchocoel, the proboscis can be everted rapidly.

Proboscis

Proboscis retractor muscle

Proboscis pore   Mouth

Intestine

Anus

The tip of the everted proboscis bears a sharp, nail-shaped stylet.

(B) *Tubulanus annulatus*

FIGURE 23.13 **Ribbon Worms** (A) The proboscis is the ribbon worm's feeding organ. (B) Although most nemerteans are small, some marine species such as this one can reach several meters in length.

The most distinctive organ of rotifers is a conspicuous ciliated organ called the *corona*, which surmounts the head of many species. Coordinated beating of the cilia sweeps particles of organic matter from the water into the animal's mouth and down to a complicated structure called the *mastax*, in which food is ground into small pieces. By contracting muscles around the pseudocoel, a few rotifer species that prey on protists and small animals can protrude the mastax through their mouth and seize small objects with it.

Most of the 1,800 known species of rotifers live in fresh water. Some species rest on the surfaces of mosses or lichens in a desiccated, inactive state until it rains. When rain falls, they absorb water and become mobile, feeding in the films of water that temporarily cover the plants. Most rotifers live no longer than a few weeks.

Both males and females are found in some species of rotifers, but only females are known among the *bdelloids* (the *b* is silent). Biologists have concluded that the bdelloid rotifers may have existed for tens of millions of years without regular sexual reproduction. In general, lack of genetic recombination leads to the buildup of deleterious mutations, so long-term asexual reproduction typically leads to extinction (see Concept 15.6). However, recent studies indicate that bdelloid rotifers may avoid this problem by taking up fragments of genes directly from the environment during the desiccation–rehydration cycle; such a mechanism would allow genetic recombination among individuals in the absence of direct sexual exchange.

RIBBON WORMS The *nemerteans*, or **ribbon worms**, have simple nervous and excretory systems similar to those of flatworms. Unlike flatworms, however, they have a complete digestive tract with a mouth at one end and an anus at the other. Small ribbon worms move slowly by beating their cilia. Larger ones employ waves of muscle contraction to move over the surface of sediments or to burrow into them.

Within the body of nearly all of the 1,000 species of ribbon worms is a fluid-filled cavity called the *rhynchocoel*, within which lies a hollow, muscular *proboscis*. The proboscis, which is the worm's feeding organ, may extend much of the length of the body. Contraction of the muscles surrounding the rhynchocoel causes the proboscis to evert explosively through an anterior pore (**FIGURE 23.13A**). The proboscis may be armed with sharp stylets that pierce prey and discharge paralytic toxins into the wound.

Ribbon worms are largely marine, although there are some freshwater and terrestrial species. Most ribbon worms are less than 20 centimeters long, but individuals of some species

(B)   Anterior

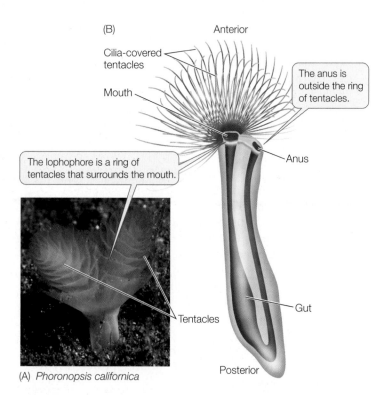

Cilia-covered tentacles

Mouth

The anus is outside the ring of tentacles.

Anus

The lophophore is a ring of tentacles that surrounds the mouth.

Tentacles

Gut

Posterior

(A) *Phoronopsis californica*

FIGURE 23.14 **Phoronids** (A) The tentacles of this species' lophophores form two spirals. (B) The phoronid gut is U-shaped, as seen in this generalized diagram.

reach 20 *meters* or more. Some genera feature species that are conspicuous and brightly colored (**FIGURE 23.13B**).

PHORONIDS AND BRACHIOPODS Like bryozoans (see Figure 23.10A), phoronids and brachiopods also feed using a lophophore, but this structure may have evolved more than once among these groups. Although neither the phoronids nor the brachiopods are represented by many living species, the brachiopods—which have shells and thus leave an excellent fossil record—are known to have been much more abundant during the Paleozoic and Mesozoic eras.

The 20 known species of **phoronids** are small (5–25 cm long), sessile worms that live in muddy or sandy sediments or attached to rocky substrata. Phoronids are found in marine waters, from the intertidal zone to about 400 meters deep. They secrete tubes made of chitin, within which they live (**FIGURE 23.14**). Their cilia drive water into the top of the lophophore, and the water exits through the narrow spaces between the tentacles. Suspended food particles are caught and transported to the mouth by ciliary action. Eggs are fertilized internally, and the embryos are either released into the water or, in species with large embryos, retained in the parent's body, where they are brooded until they hatch.

Brachiopods are solitary marine animals with a rigid shell that is divided into two parts connected by a ligament (**FIGURE**

Terebraulina septentrionalis — Lophophore rings — Tentacles

FIGURE 23.15 Brachiopods The lophophores of these North Atlantic brachiopods can be seen between their shells; although their shells resemble those of the bivalve mollusks (see Figure 23.18D), they evolved independently.

23.15). Although brachiopods superficially resemble bivalve mollusks, shells evolved independently in the two groups. The two halves of the brachiopod shell are dorsal and ventral, rather than lateral as in bivalves. The lophophore is located within the shell. Most brachiopods are 4–6 centimeters long. More than 26,000 fossil brachiopod species have been described, but only about 335 species survive.

ANNELIDS As discussed in Concept 23.1, segmentation allows an animal to move different parts of its body independently of one another, giving it much better control of its movement. A clear and obvious example of segmentation is seen in the body plan of the **annelids** (**FIGURE 23.16**; see also Figure 23.4A).

In most large annelids, the coelom in each segment is isolated from those in other segments. A separate nerve center called a *ganglion* controls each segment; nerve cords that connect the ganglia coordinate their functioning. Most annelids lack a rigid external protective covering; instead, they have a thin, permeable body wall that serves as a general surface for gas

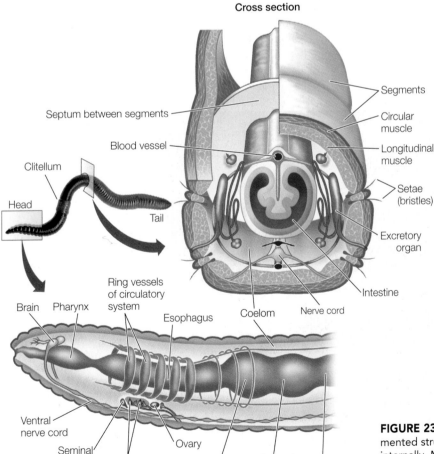

Cross section

Septum between segments — Blood vessel — Clitellum — Head — Tail — Segments — Circular muscle — Longitudinal muscle — Setae (bristles) — Excretory organ — Intestine — Brain — Pharynx — Ring vessels of circulatory system — Esophagus — Coelom — Nerve cord — Ventral nerve cord — Seminal receptacle — Testes and sperm sacs — Ovary — Crop — Gizzard — Intestine

FIGURE 23.16 Annelids Have Many Body Segments The segmented structure of the annelids is apparent both externally and internally. Many organs of this earthworm, a common annelid, are repeated serially.

exchange. These animals are thus restricted to moist environments because they lose body water rapidly in dry air. The approximately 16,500 described annelid species live in marine, freshwater, and moist terrestrial environments.

Many annelids have one or more pairs of eyes and one or more pairs of tentacles (with which they capture prey or filter food from the surrounding water) at the anterior end of the body (**FIGURE 23.17A**). In some species, the body wall of most segments extends laterally as a series of thin outgrowths called *parapodia*. The parapodia function in gas exchange, and some species use them to move. Stiff bristles called *setae* protrude from each parapodium, forming temporary contact with the substrate and preventing the animal from slipping backward when its muscles contract.

Members of one annelid group, the *pogonophorans*, secrete tubes made of chitin and other substances, in which they live (**FIGURE 23.17B**). Pogonophorans have lost their digestive tract (they have no mouth or gut). So how do they obtain nutrition? Part of the answer is that pogonophorans can take up dissolved organic matter directly from the sediments in which they live or from the surrounding water. Much of their nutrition, however, is provided by endosymbiotic bacteria that live in a specialized organ known as the *trophosome*. These bacteria oxidize hydrogen sulfide and other sulfur-containing compounds, fixing carbon from methane in the process. The uptake of the hydrogen sulfide, methane, and oxygen used by the bacteria is facilitated by hemoglobin in the pogonophorans' tentacles. It is this hemoglobin that gives the tentacles their red color (see Figure 23.17B).

Oligochaetes have no parapodia, eyes, or anterior tentacles, and they have only four pairs of setae bundles per segment. Earthworms—the most familiar oligochaetes—burrow in and ingest soil, from which they extract food particles. All oligochaetes are *hermaphroditic*; that is, each individual is both male and female. Sperm are exchanged simultaneously between two copulating individuals. Eggs and sperm are deposited outside the adult's body, in a cocoon secreted by the *clitellum* (see Figure 23.16). Fertilization occurs within the cocoon after it is shed, and when development is complete, miniature worms emerge and immediately begin independent life.

Leeches, like oligochaetes, lack parapodia and tentacles. They live in freshwater or terrestrial habitats. The leech coelom is not divided into compartments, and the coelomic space is largely filled with undifferentiated tissue. Groups of segments at each end of the body are modified to form suckers, which serve as temporary anchors that help the leech move. With its posterior sucker attached to a substrate, the leech extends its body by contracting its circular muscles. The anterior sucker is then attached, the posterior one detached, and the leech shortens itself by contracting its longitudinal muscles.

A leech feeds by making an incision in its vertebrate host, from which blood flows; it then secretes an anticoagulant into the wound to keep the blood flowing. For centuries, medical practitioners used leeches to treat diseases they believed were caused by an excess of blood or by "bad blood." Although most leeching practices (such as inserting a leech in a person's throat to alleviate swollen tonsils) have been abandoned, *Hirudo medicinalis* (the medicinal leech; **FIGURE 23.17C**) is used medically even today to reduce fluid pressure and prevent blood clotting in damaged tissues, to eliminate pools of coagulated blood, and to prevent scarring. The anticoagulants of certain other leech species contain anesthetics and blood vessel dilators and are being studied for possible medical uses.

(A) *Spirographis spallanzanii*

(B) *Riftia* sp.

(C) *Hirudo medicinalis*

FIGURE 23.17 Diversity among the Annelids (A) "Fan worms," or "feather duster worms," are sessile marine annelids that grow in chitinous tubes, from which their tentacles extend and filter food from the water. (B) Pogonophorans live around hydrothermal vents deep in the ocean. Like fan worms, their tentacles protrude from chitinous tubes. They do not possess a digestive tract, however, and obtain most of their nutrition from endosymbiotic bacteria. (C) The medicinal leech has been a tool of physicians and healers for centuries.

MOLLUSKS The most diverse group of lophotrochozoans are the **mollusks**, with about 100,000 species that inhabit a wide array of aquatic and terrestrial environments. There are four major clades of mollusks: chitons, gastropods, bivalves, and cephalopods. Although these groups differ dramatically in morphology, they all share the same three major body components: a foot, a visceral mass, and a mantle (**FIGURE 23.18A**).

- The molluscan *foot* is a muscular structure that originally was both an organ of locomotion and a support for internal organs. In cephalopods such as squids and octopuses, the foot has been modified to form arms and tentacles borne on a head with complex sensory organs. In other groups, such as clams, the foot is a burrowing organ. In some groups the foot is greatly reduced.

- The heart and the digestive, excretory, and reproductive organs are concentrated in a centralized, internal *visceral mass*.

- The *mantle* is a fold of tissue that covers the organs of the visceral mass. The mantle secretes the hard, calcareous shell that is typical of many mollusks.

In most mollusks, the mantle extends beyond the visceral mass to form a *mantle cavity*. Within this cavity lie *gills* that are used for gas exchange. When cilia on the gills beat, they create a current of water. The gill tissue, which is highly vascularized (contains many blood vessels), takes up oxygen from the water and releases carbon dioxide. Many mollusks use their gills as filter-feeding devices; others feed using a rasping structure known as the *radula* to scrape algae from rocks. In some mollusks, such as the marine cone snails, the radula has been modified into a drill or poison dart.

Molluscan blood vessels do not form a closed circulatory system. Blood and other fluids empty into a large, fluid-filled hemocoel, through which fluids move around the animal and deliver oxygen to the internal organs. Eventually, the fluids reenter the blood vessels and are moved by a heart.

The approximately 1,000 living species of **chitons** (**FIGURE 23.18B**) are characterized by eight overlapping calcareous plates, surrounded by a structure known as the girdle. These plates and girdle protect the chiton's internal organs and muscular foot. The chiton body is bilaterally symmetrical, and the internal organs, particularly the digestive and nervous systems, are relatively simple. Most chitons are marine omnivores that scrape algae, bryozoans, and other organisms from rocks with their sharp radula. An adult chiton spends most of its life clinging to rock surfaces with its large, muscular, mucus-covered foot. It moves slowly by means of rippling waves of muscular contraction in the foot. Fertilization in most chitons takes place in the water, but in a few species fertilization is internal and embryos are brooded within the body.

Gastropods are the most species-rich and widely distributed mollusks, with nearly 70,000 living species. Snails, whelks, limpets, slugs, nudibranchs (sea slugs), and abalones are all gastropods. Most species move by gliding on their muscular foot, but in a few species—the sea butterflies and heteropods—the foot is a swimming organ with which the animal moves through open ocean waters. The only mollusks that live in terrestrial environments—land snails and slugs—are gastropods

(**FIGURE 23.18C**). In these terrestrial species, the mantle tissue is modified into a highly vascularized lung.

Clams, oysters, scallops, and mussels are all familiar **bivalves**. The 30,000 living species are found in both marine and freshwater environments. Bivalves have a very small head and a hinged, two-part shell that extends over the sides of the body as well as the top (**FIGURE 23.18D**). Many clams use their foot to burrow into mud and sand. Bivalves feed by taking in water through an opening called an *incurrent siphon* and filtering food from the water with their large gills, which are also the main sites of gas exchange. Water and gametes exit through the *excurrent siphon*. Fertilization takes place in open water in most species.

There are about 800 living species of **cephalopods**—squids, octopuses, and nautiluses. Their excurrent siphon is modified to allow the animal to control the water content of the mantle cavity. The modification of the mantle into a device for forcibly ejecting water from the cavity through the siphon enables these animals to move rapidly by "jet propulsion" through the water. With their greatly enhanced mobility, cephalopods (which first appeared early in the Cambrian) became the major predators in the open waters of the Devonian oceans. They remain important marine predators today.

Cephalopods capture and subdue prey with their tentacles (see Figure 23.19B). As is typical of active, rapidly moving predators, cephalopods have a head with complex sensory organs, most notably eyes that are comparable to those of vertebrates in their ability to resolve images. The head is closely associated with a large, branched foot that bears the tentacles and a siphon (**FIGURE 23.18E**). The large, muscular mantle provides an external supporting structure. The gills hang in the mantle cavity. Many cephalopods have elaborate courtship behavior, which can involve striking color changes.

Many early cephalopods had a chambered external shell divided by partitions penetrated by tubes through which gases and liquids could be moved to control the animal's buoyancy. Nautiluses are the only surviving cephalopods that have such external chambered shells, although squids and cuttlefish retain internal shells.

Shells have been lost several times among the mollusks, as in several groups of gastropods, including the slugs and nudibranchs (**FIGURE 23.19A**). These shell-less gastropods gain some protection from predation by being distasteful or toxic to many species. The often brilliant coloration of nudibranchs is *aposematic*, meaning it serves to warn potential predators of toxicity. Among the cephalopods, the octopuses have lost both external and internal shells (**FIGURE 12.19B**). Their lack of shells allows octopuses to escape predators by squeezing into small crevices.

Ecdysozoans must shed their cuticles

The distinguishing characteristic of **ecdysozoans** is their stiff external covering, or **cuticle**, which is secreted by the underlying *epidermis* (the outermost cell layer). The cuticle provides these animals with both protection and support. Once formed, however, the cuticle cannot grow. How, then, can ecdysozoans increase in size? They do so by shedding, or **molting**, the cuticle

(A) Generalized molluscan body plan

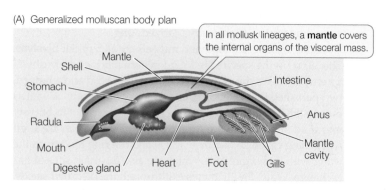

In all mollusk lineages, a **mantle** covers the internal organs of the visceral mass.

Mantle
Shell
Stomach
Intestine
Radula
Anus
Mouth
Mantle cavity
Digestive gland
Heart
Foot
Gills

FIGURE 23.18 Organization and Diversity of Molluscan Bodies (A) The major molluscan groups display different variations on a general body plan that includes three major components: a foot, a visceral mass of internal organs, and a mantle. In many species, the mantle secretes a calcareous shell. (B) Chitons have eight overlapping calcareous plates surrounded by a girdle. (C) Most gastropods have a single dorsal shell, into which they can retreat for protection. (D) Bivalves get their name from their two hinged shells, which can be tightly closed. (E) Cephalopods are active predators; they use their arms and tentacles to capture prey. This cuttlefish has an internal shell but no external shell.

(B) Chitons

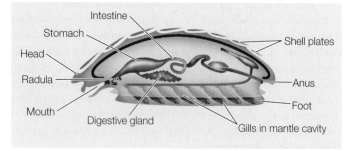

Intestine
Stomach
Shell plates
Head
Radula
Anus
Foot
Mouth
Digestive gland
Gills in mantle cavity

Chaetopleura angulata

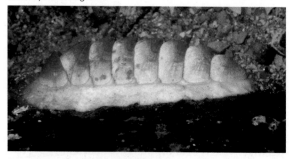

(C) Gastropods

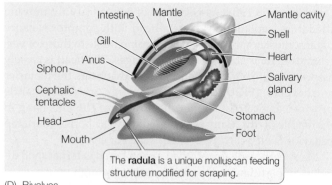

Intestine
Mantle
Mantle cavity
Gill
Shell
Anus
Heart
Siphon
Salivary gland
Cephalic tentacles
Head
Stomach
Mouth
Foot

The **radula** is a unique molluscan feeding structure modified for scraping.

Helix pomatia

(D) Bivalves

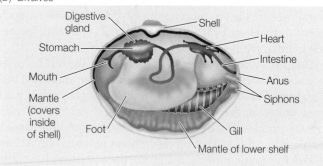

Digestive gland
Shell
Stomach
Heart
Mouth
Intestine
Mantle (covers inside of shell)
Anus
Siphons
Foot
Gill
Mantle of lower shelf

Chlamys varia

(E) Cephalopods

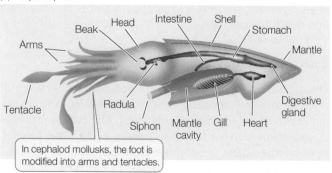

Head
Beak
Intestine
Shell
Stomach
Arms
Mantle
Tentacle
Radula
Digestive gland
Siphon
Mantle cavity
Gill
Heart

In cephalod mollusks, the foot is modified into arms and tentacles.

Sepia sp.

(A) *Hermissenda crassicornis*

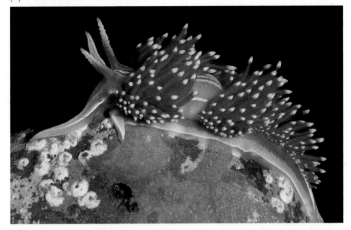

(B) *Octopus macropus*

FIGURE 23.19 Mollusks in Some Groups Have Lost Their Shells (A) Nudibranchs ("naked gills), also called sea slugs, are shell-less gastropods. This species is brightly colored, alerting potential predators of its toxicity. (B) Octopuses have lost both the external and internal shell, which allows these cephalopods to squeeze through tight spaces.

and replacing it with a new, larger one. This molting process gives the clade its name (Greek *ecdysis*, "to get out of").

A soft-bodied arthropod from the Cambrian, fossilized in the process of molting, shows that molting evolved more than 500 million years ago (**FIGURE 23.20A**). An increasingly rich array of molecular and genetic evidence, including a set of Hox genes shared by all ecdysozoans, suggests they have a single common ancestor. Thus molting of a cuticle is a trait that may have evolved only once during animal evolution.

Before an ecdysozoan molts, a new cuticle is already forming underneath the old one. Once the old cuticle is shed, the new one expands and hardens. Until it has hardened, though, the animal is vulnerable to its enemies, both because its outer surface is easy to penetrate and because an animal with a soft cuticle moves slowly or not at all (**FIGURE 23.20B**).

In many ecdysozoans that have wormlike bodies, the cuticle is relatively thin and flexible; it offers the animal some protection but provides only modest body support. A thin cuticle

allows the exchange of gases, minerals, and water across the body surface, but it restricts the animal to moist habitats. Many species of ecdysozoans with thin cuticles live in marine sediments from which they obtain prey. Some freshwater species absorb nutrients directly through their thin cuticles, as do parasitic species that live within their hosts.

The cuticles of other ecdysozoans, notably the arthropods, function as external skeletons, or **exoskeletons**. These exoskeletons are thickened by layers of protein and a strong,

(A)

Molted exoskeleton

Emerging animal

(B) *Heterophrynus batesi*   Molted exoskeleton

The newly emerged whip scorpion's body is still soft and vulnerable.

FIGURE 23.20 Molting, Past and Present (A) A 500-million-year-old fossil from the Cambrian captured an individual of a long-extinct arthropod species in the process of molting, demonstrating that molting is an evolutionarily ancient trait. (B) This tailless whip scorpion has just emerged from its discarded exoskeleton and will be highly vulnerable until its new cuticle has hardened.

waterproof polysaccharide called **chitin**. An animal with a rigid, chitin-reinforced exoskeleton can neither move in a wormlike manner nor use cilia for locomotion. A hard exoskeleton also impedes the passage of oxygen and nutrients into the animal, presenting new challenges in other areas besides growth. New mechanisms of locomotion and gas exchange evolved in those ecdysozoans with hard exoskeletons.

To move rapidly, an animal with a rigid exoskeleton must have body extensions that can be manipulated by muscles. Such *appendages* evolved in the late Precambrian and led to the **arthropod** (Greek, "jointed foot") clade. Arthropod appendages exist in an amazing variety of forms. They serve many functions, including walking and swimming, gas exchange, food capture and manipulation, copulation, and sensory perception. Arthropods grasp food with their mouths and associated appendages and digest it internally. Their muscles are attached to the inside of the exoskeleton.

The arthropod exoskeleton has had a profound influence on the evolution of these animals. Encasement within a rigid body covering provides support for walking on dry land, and the waterproofing provided by chitin keeps the animal from dehydrating in dry air. These features have allowed arthropods to invade terrestrial environments several times.

The evolution of the diverse arthropods and their relatives will be described in greater detail in Concept 23.4. Here we describe the other major groups of ecdysozoans.

NEMATODES **Nematodes**, or *roundworms*, have a thick, multilayered cuticle that gives their unsegmented body its shape. As a nematode grows, it sheds its cuticle four times. Nematodes exchange oxygen and nutrients with their environment through both the cuticle and the gut, which is only one cell

layer thick. Materials are moved through the gut by rhythmic contraction of a highly muscular organ, the *pharynx*, at the worm's anterior end. Nematodes move by contracting their longitudinal muscles.

Nematodes are probably the most abundant and universally distributed of all major animal groups. About 25,000 species have been described, but the actual number of living species may be in the millions. Many are microscopic; the largest known nematode, which reaches a length of 9 meters, is a parasite in the placentas of sperm whales. Countless nematodes live as scavengers in the upper layers of the soil, on the bottoms of lakes and streams, and in marine sediments. The topsoil of rich farmland may contain 3–9 billion nematodes per acre. A single rotting apple may contain as many as 90,000 individuals.

One soil-inhabiting nematode, *Caenorhabditis elegans*, serves as a model organism in the laboratories of geneticists and developmental biologists. It is ideal for such research because it is easy to cultivate, matures in 3 days, and has a fixed number of body cells. Its genome has been completely sequenced.

Many nematodes are predators, feeding on protists and small animals (including other roundworms). Most significant to humans, however, are the many species that parasitize plants and animals (**FIGURE 23.21A**). The nematodes that parasitize humans (causing serious diseases such as trichinosis, filariasis, and elephantiasis), domesticated animals, and economically important plants have been studied intensively in an effort to find ways of controlling them.

The structure of parasitic nematodes is similar to that of free-living species, but the life cycles of many parasitic species have special stages that facilitate the transfer of individuals among hosts. *Trichinella spiralis*, the species that causes the human disease trichinosis, has a relatively simple life cycle. A person may become infected by eating the flesh of an animal (usually a pig) that has *Trichinella* larvae encysted in its muscles (**FIGURE 23.21B**). The larvae are activated in the person's digestive tract,

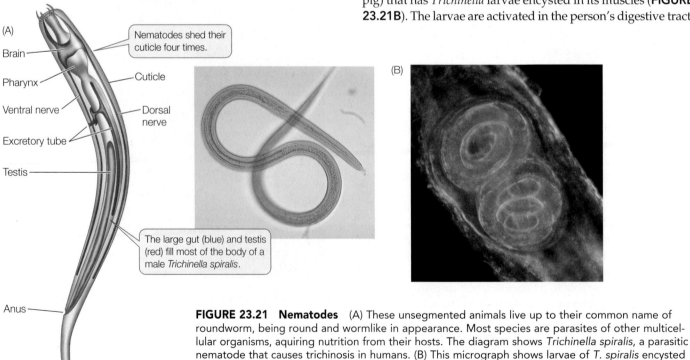

(A)

Brain

Pharynx

Ventral nerve

Excretory tube

Testis

Anus

Nematodes shed their cuticle four times.

Cuticle

Dorsal nerve

The large gut (blue) and testis (red) fill most of the body of a male *Trichinella spiralis*.

(B)

FIGURE 23.21 Nematodes (A) These unsegmented animals live up to their common name of roundworm, being round and wormlike in appearance. Most species are parasites of other multicellular organisms, aquiring nutrition from their hosts. The diagram shows *Trichinella spiralis*, a parasitic nematode that causes trichinosis in humans. (B) This micrograph shows larvae of *T. spiralis* encysted in the muscle tissue of a host swine.

emerge from their cysts, and attach to the intestinal wall, where they feed. Later, they bore through the intestinal wall and are carried in the bloodstream to muscles, where they form new cysts. If present in great numbers, these cysts can cause severe pain or death.

FRONTIERS Although nematodes that infect plants are pests of many crop species, parasitic and predatory nematodes of insects are now being selected for use in controlling crop pests. The larva of the beetle *Diabrotica virgifera*, known as western corn rootworm, devastates untreated corn fields in large areas of North America and Europe, causing over $1 billion a year in losses in the United States alone. Pesticides that kill the beetle larvae also harm many beneficial insects. New strains of a nematode that kills the corn rootworm have been selected by researchers for attraction to the "alarm molecules" produced by infected corn plants. These modified strains of nematodes show great promise for effectively controlling corn rootworms without using nonspecific pesticides.

HORSEHAIR WORMS About 320 species of the unsegmented **horsehair worms** have been described. As their name implies, these animals are extremely thin in diameter; they range from a few millimeters up to a meter in length. Most adult horsehair worms live in fresh water among the leaf litter and algal mats that accumulate near the shores of streams and ponds. A few species live in damp soil.

Horsehair larvae are internal parasites of freshwater crayfish and of terrestrial and aquatic insects (**FIGURE 23.22**). An adult horsehair worm has no mouth, and its gut is greatly reduced and probably nonfunctional. Some species feed only as larvae, absorbing nutrients from their hosts across the body wall. But other species continue to grow and shed their curticles even after they have left their hosts, suggesting that some adult worms may be able to absorb nutrients from their environment.

An adult horsehair worm exits the wood cricket it parasitized during its larval development.

FIGURE 23.22 Horsehair Worm Larvae Are Parasitic The larvae of this horsehair worm (*Paragordius tricuspidatus*) can manipulate its host's behavior. The hatching worm causes the cricket to jump into water, where the worm will continue its life cycle as a free-living adult. The insect, having delivered its parasitic burden, drowns.

SMALL MARINE CLADES There are several small, relatively poorly known groups of benthic marine ecdysozoans. The 16 species of **priapulids** are cylindrical, unsegmented, wormlike animals with a three-part body plan consisting of a proboscis, trunk, and caudal appendage ("tail"). It should be clear from their appearance why they were named after the Greek fertility god Priapus (**FIGURE 23.23A**). The 150 species of **kinorhynchs** live in marine sands and muds and are virtually microscopic; no kinorhynchs are longer than 1 millimeter. Their bodies are divided into 13 segments, each with a separate cuticular plate (**FIGURE 23.23B**). The minute (less than 1 mm long) **loriciferans** were not discovered until 1983. About 100 living species are known to exist, although many of these are still being

---

APPLY THE CONCEPT

There are two major groups of protostomes

The table shows a small sample of amino acid residues that have been used to reconstruct the relationships of some major animal groups. (The full dataset includes data on 11,234 amino acid positions across 77 species of animals.)

1. Construct a phylogenetic tree of these seven animals using the parsimony method (see Chapter 16). Use the sponge as your outgroup to root the tree. Assume that all changes among amino acids are equally likely.

2. How many changes (from one amino acid residue to another) occur along each branch on your tree?

3. Which branch on your tree represents the eumetazoans? The protostomes? The lophotrochozoans? The ecdysozoans? The deuterostomes?

ANIMAL	AMINO ACID AT POSITION											
	1	2	3	4	5	6	7	8	9	10	11	12
Sponge	P	P	M	S	V	R	Q	N	L	V	I	L
Clam	E	A	M	R	I	K	L	S	I	V	I	L
Earthworm	E	P	M	R	I	K	L	S	I	V	M	L
Tardigrade	E	K	M	R	I	R	L	N	L	V	L	L
Fruit fly	E	K	M	R	I	S	L	D	L	V	L	L
Sea urchin	E	P	M	R	V	R	Q	N	L	T	V	K
Human	E	P	I	R	V	R	Q	N	L	T	V	K

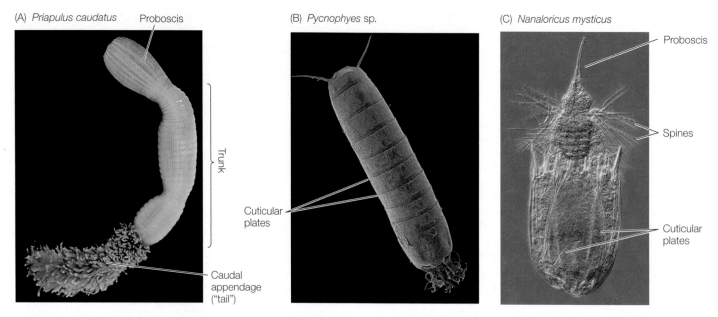

(A) *Priapulus caudatus* — Proboscis, Trunk, Caudal appendage ("tail")

(B) *Pycnophyes* sp. — Cuticular plates

(C) *Nanaloricus mysticus* — Proboscis, Spines, Cuticular plates

FIGURE 23.23 Benthic Marine Ecdysozoans Members of these groups are marine bottom-dwellers. (A) Most priapulid species live in burrows on the ocean floor, extending the proboscis for feeding. (B) Kinorhynchs are virtually microscopic. The cuticular plates that cover their bodies are molted periodically. (C) Six cuticular plates form a "corset" around the minute loriciferan body.

described. The body is divided into a head, neck, thorax, and abdomen and is covered by six plates, from which the loriciferans get their name (Latin *lorica*, "corset"; **FIGURE 23.23C**).

Do You Understand Concept 23.3?

- How does an animal's body covering influence the way it breathes, feeds, and moves?
- Why are most annelids restricted to moist environments?
- Briefly describe how the basic body organization of mollusks has been modified to yield a wide diversity of animals.
- What are three ways in which nematodes have a significant impact on humans?

We turn now to the arthropods. Members of the four arthropod subgroups not only dominate the ecdysozoan clade but are also among the most diverse animals on Earth.

concept
23.4 ## Arthropods Are Diverse and Abundant Animals

Arthropods and their relatives are ecdysozoans with paired appendages. Arthropods are an extremely diverse group of animals in numbers of species. Furthermore, the number of individual arthropods alive at any one time is estimated to be about 10^{18}, or a billion billion. Among the animals, only the nematodes are thought to exist in greater numbers.

Several key features have contributed to the success of the arthropods. Their bodies are segmented, and their muscles are attached to the inside of their rigid exoskeletons. Each segment has muscles that operate that segment and the jointed appendages attached to it. Jointed appendages permit complex movements, and different appendages are specialized for different functions. Encasement of the body within a rigid exoskeleton provides the animal with support for walking in the water or on dry land and provides some protection against predators. The waterproofing provided by chitin keeps the animal from dehydrating in dry air.

Representatives of the four major arthropod groups living today are all species-rich: the *chelicerates* (including the arachnids—spiders, scorpions, mites, and their relatives), *myriapods* (millipedes and centipedes), *crustaceans* (including shrimps, crabs, and barnacles), and *hexapods* (insects and their relatives). Phylogenetic relationships among arthropod groups are currently being reexamined in light of a wealth of new information, much of it based on gene sequences. These studies suggest that the chelicerates are the sister group to the remaining arthropods, and that the crustaceans may be paraphyletic with respect to the hexapods. There is strong support for the monophyly of arthropods as a whole.

The jointed appendages of arthropods gave the clade its name, from the Greek words *arthron*, "joint," and *podos*, "foot" or "limb." Arthropods evolved from ancestors with simple, unjointed appendages. The exact forms of those ancestors are unknown, but some arthropod relatives with segmented bodies and unjointed appendages survive today. Before we describe the modern arthropods, we will discuss those arthropod relatives.

Arthropod relatives have fleshy, unjointed appendages

Until fairly recently, biologists debated whether the **onychophorans** (velvet worms) were more closely related to annelids or arthropods, but molecular evidence clearly links them to the

(A) *Peripatus* sp.

(B) *Echiniscus* sp.

FIGURE 23.24 Arthropod Relatives with Unjointed Appendages (A) Onychophorans have unjointed legs and use the body cavity as a hydrostatic skeleton. (B) Tardigrades can be abundant on the wet surfaces of mosses and plants and in temporary pools of water.

lack both a circulatory system and gas-exchange organs. The 800 known species live in marine sands and on temporary water films on plants. When these films dry out, the animals also lose water and shrink to small, barrel-shaped objects that can survive for at least a decade in a dormant state. Tardigardes have been found in densities as high as 2 million per square meter of moss.

Chelicerates are characterized by pointed, nonchewing mouthparts

In the **chelicerates**, the head bears two pairs of pointed appendages modified to form mouthparts, called *chelicerae*, which are used to grasp (rather than chew) prey. In addition, many chelicerates have four pairs of walking legs. The 98,000 described species are placed in three major clades: pycnogonids, horseshoe crabs, and arachnids.

The *pycnogonids*, or sea spiders, are a poorly known group of about 1,000 marine species (**FIGURE 23.25A**). Most are small, with leg spans less than 1 cm, but some deep-sea species have leg spans up to 60 cm. A few pycnogonids eat algae, but most are carnivorous, eating a variety of small invertebrates.

There are four living species of *horseshoe crabs*, but many close relatives are known from fossils. Horseshoe crabs, which have changed very little morphologically during their long fossil history, have a large horseshoe-shaped covering over most of the body. They are common in shallow waters along the eastern coast of North America and the southern and eastern coasts of Asia, where they scavenge and prey on bottom-dwelling animals. Periodically they crawl into the intertidal zone in large numbers to mate and lay eggs (**FIGURE 23.25B**).

Arachnids are abundant in terrestrial environments. Most arachnids have a simple life cycle in which miniature adults hatch from internally fertilized eggs and begin independent lives almost immediately. Some arachnids retain their eggs during development and give birth to live young.

arthropods. Indeed, with their soft, fleshy, unjointed, claw-bearing legs, onychophorans may be similar in appearance to the arthropod ancestor (**FIGURE 23.24A**). The 150 species of onychophorans live in leaf litter in humid tropical environments. Their soft, segmented bodies are covered by a thin, flexible cuticle that contains chitin. They use their fluid-filled body cavities as hydrostatic skeletons. Fertilization is internal, and large, yolky eggs are brooded within the female's body.

Tardigrades (water bears) also have fleshy, unjointed legs and use their fluid-filled body cavities as hydrostatic skeletons (**FIGURE 23.24B**). Tardigrades are tiny (0.5–1.5 mm long) and

(A) *Anoplodactylus* sp.

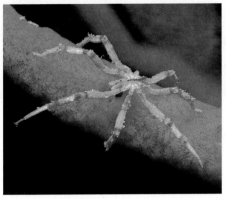

(B) *Limulus polyphemus*

FIGURE 23.25 Two Small Chelicerate Groups (A) Although they are not spiders, it is easy to see why sea spiders were given their common name. (B) Horseshoe crabs are an ancient group that has changed very little in morphology over time; such species are sometimes referred to as "living fossils."

FIGURE 23.26 Arachnid Diversity
(A) Known for their web-spinning abilities, spiders are expert predators. This green lynx spider has captured a fly. (B) A desert scorpion assumes the animal's aggressive attack position. These arachnids are fearsome predators, and their sting can be dangerous to humans. (C) Harvestmen, also called daddy longlegs, are scavengers. (D) Mites include many free-living species as well as blood-sucking, external parasites.

(A) *Peucetia* sp.

(B) *Opistophthalmus carinatus*

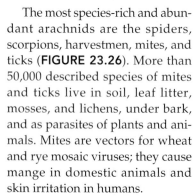

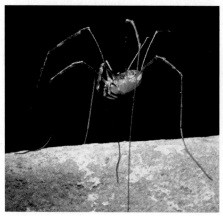

(D) *Lorryia formosa*

(C) *Leiobunum rotundum*

The most species-rich and abundant arachnids are the spiders, scorpions, harvestmen, mites, and ticks (**FIGURE 23.26**). More than 50,000 described species of mites and ticks live in soil, leaf litter, mosses, and lichens, under bark, and as parasites of plants and animals. Mites are vectors for wheat and rye mosaic viruses; they cause mange in domestic animals and skin irritation in humans.

Spiders, of which 38,000 species have been described, are important terrestrial predators with hollow chelicerae, which they use to inject venom into their prey. Some have excellent vision that enables them to chase and seize their prey. Others spin elaborate webs made of protein threads in which they snare prey. The threads are produced by modified abdominal appendages connected to internal glands that secrete the proteins, which solidify on contact with air. The webs of different groups of spiders are strikingly varied, and this variation enables the spiders to position their snares in many different environments for many different types of prey.

Mandibles and antennae characterize the remaining arthropod groups

The remaining arthropods have *mandibles*, rather than chelicerae, as mouthparts, so they are together called **mandibulates**. Mandibles are often used for chewing as well as for biting and holding food. Another distinctive characteristic of the mandibulates is the presence of sensory antennae on the head.

The **myriapods** comprise the centipedes, millipedes, and their close relatives. Centipedes and millipedes have a well-formed head with the mandibles and antennae characteristic of mandibulates. Their distinguishing feature is a long, flexible, segmented trunk that bears many pairs of legs (**FIGURE 23.27**). Centipedes, which have one pair of legs per segment, prey on insects and other small animals. In millipedes, two adjacent segments are fused so that each fused segment has two pairs of legs. Millipedes scavenge and eat plants. More than 3,000 species of centipedes and 11,000 species of millipedes have been

(A) *Scolopendra angulata*

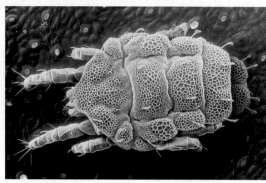

(B) *Sigmoria trimaculata*

FIGURE 23.27 Myriapods (A) Centipedes have modified appendages that function as poisonous fangs for capturing active prey. They have one pair of legs per segment. (B) Millipedes, which are scavengers and plant eaters, have smaller jaws and legs than centipedes do. They have two pairs of legs per segment.

(A) *Grapsus grapsus*

(B) *Armadillium vulgare*

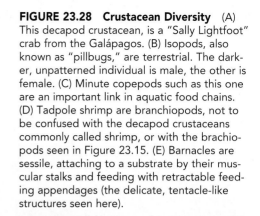

FIGURE 23.28 Crustacean Diversity (A) This decapod crustacean, is a "Sally Lightfoot" crab from the Galápagos. (B) Isopods, also known as "pillbugs," are terrestrial. The darker, unpatterned individual is male, the other is female. (C) Minute copepods such as this one are an important link in aquatic food chains. (D) Tadpole shrimp are branchiopods, not to be confused with the decapod crustaceans commonly called shrimp, or with the brachiopods seen in Figure 23.15. (E) Barnacles are sessile, attaching to a substrate by their muscular stalks and feeding with retractable feeding appendages (the delicate, tentacle-like structures seen here).

(C) *Eudiaptomus gracilis*

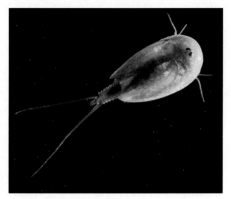

(D) *Lepidurus* sp.

(E) *Lepas anatifera*

described; many more species probably remain unknown. Although most myriapods are less than a few centimeters long, some tropical species are ten times that size.

Crustaceans are the dominant marine arthropods today, and they are also common in freshwater and some terrestrial environments. The most familiar crustaceans are the shrimps, lobsters, crayfishes, and crabs (all *decapods*; **FIGURE 23.28A**) and the sow bugs (*isopods*; **FIGURE 23.28B**). Additional species-rich groups include the *amphipods, ostracods, copepods* (**FIGURE 23.28C**), and *branchiopods* (**FIGURE 23.28D**), all of which are found in freshwater and marine environments.

Barnacles are unusual crustaceans that are sessile as adults (**FIGURE 23.28E**). Adult barnacles look more like mollusks than like other crustaceans, but as the zoologist Louis Agassiz remarked more than a century ago, a barnacle is "nothing more than a little shrimp-like animal, standing on its head in a limestone house and kicking food into its mouth."

Most of the 52,000 described species of crustaceans have a body that is divided into three regions: head, thorax, and abdomen (**FIGURE 23.29A**). The segments of the head are fused together, and the head bears five pairs of appendages. Each of the multiple thoracic and abdominal segments usually bears one pair of appendages. The appendages on different parts of the body are specialized for different functions, such as gas exchange, chewing, capturing food, sensing, walking, and swimming. In many species, a fold of the exoskeleton, the *carapace*, extends dorsally and laterally back from the head to cover and protect some of the other segments.

The fertilized eggs of most crustacean species are attached to the outside of the female's body, where they remain during their early development. At hatching, the young of some species are released as larvae; those of other species are released as juveniles that are similar in form to the adults. Still other species release eggs into the water or attach them to an object in the environment.

Over half of all described species are insects

During the Devonian period, more than 400 million years ago, some mandibulates colonized terrestrial environments. Of the several groups (including some crustacean isopods and decapods) that successfully colonized the land, none is more prominent today than the six-legged **hexapods**: the insects and their relatives. Insects are abundant and diverse in terrestrial and freshwater environments; only a few live in salt water.

The wingless relatives of the insects—the springtails, two-pronged bristletails, and proturans—are probably the most similar of living forms to insect ancestors. These insect relatives have a simple life cycle; they hatch from eggs as miniature adults. They differ from insects in having internal mouthparts. Springtails can be extremely abundant (up to 200,000 per square meter) in soil, leaf litter, and on vegetation and are the most abundant hexapods in the world in terms of number of individuals (rather than number of species).

About 1 million of the 1.8 million described living species of life are insects. As we discussed at the beginning of this chapter, biologists have estimated that many millions of insect species remain to be discovered. Like crustaceans, insects have a body with three regions—head, thorax, and abdomen. They have a single pair of antennae on the head and three pairs of legs attached to the thorax. In most groups of insects, the thorax also bears two pairs of wings. Unlike other arthropods, insects have no appendages growing from their abdominal segments (**FIGURE 23.29B**).

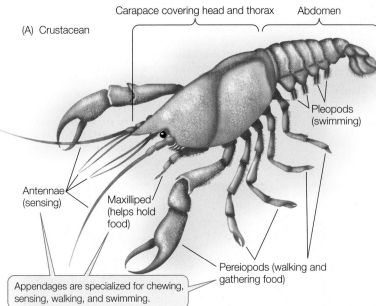

(A) Crustacean

Carapace covering head and thorax   Abdomen

Pleopods (swimming)

Antennae (sensing)

Maxilliped (helps hold food)

Pereiopods (walking and gathering food)

Appendages are specialized for chewing, sensing, walking, and swimming.

FIGURE 23.29 Two Segmented Body Plans The bodies of crustaceans and insects are divided into three regions, the head, thorax, and abdomen. (A) In crustaceans, each body region bears specialized appendages, and a shell-like carapace covers the head and thorax. (B) The thorax of insects bears three pairs of legs, and in most groups, two pairs of wings. Unlike other arthropods, insects have no appendages growing from their abdominal segments (see Figure 14.19).

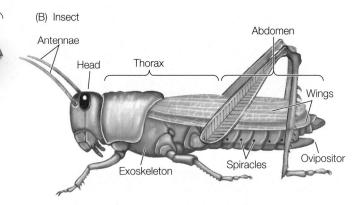

(B) Insect

Antennae

Head

Thorax

Abdomen

Wings

Ovipositor

Spiracles

Exoskeleton

Insects can be distinguished from springtails and other hexapods by their external mouthparts and by antennae that contain a motion-sensitive receptor called Johnston's organ. In addition, insects have a derived mechanism for gas exchange in air: a system of air sacs and tubular channels called *tracheae* (singular *trachea*) that extend from external openings called *spiracles* inward to tissues throughout the body (see Figure 37.2). Insects use nearly all species of plants and many species of animals as food.

TABLE 23.2 lists the major insect groups. Two groups—the jumping bristletails and silverfish—are wingless and have simple life cycles, like the springtails and other close insect

APPLY THE CONCEPT

Over half of all described species are insects

The following data were used by entomologist Terry Erwin to estimate the undescribed diversity of insects. Review the design of Erwin's experiment in the opening story of this chapter. Then use Erwin's data to answer the questions and consider how changes in the assumptions may affect the estimates.

Approximate number of beetle species collected from *Luehea seemannii* trees:	1,200
Estimated number of host-specific beetles in this sample:	163
Number of species of canopy trees per hectare of forest:	70
Percentage of beetle species living in canopy (as opposed to ground-dwelling species):	75%
Percentage of beetles among all insect species:	40%

1. From the data above, estimate the number of insect species in an average hectare of Panamanian forest. Assume that the data for beetles on *L. seemanii* are representative of the other tree species, and that all the species of beetles that are *not* host-specific were collected in the original sample. Remember to sum your estimates of the number of (a) host-specific

beetle species in the forest canopy; (b) non-host-specific beetle species in the forest canopy; (c) beetle species on the forest floor; and (d) species of all insects other than beetles.

2. Use the following information to estimate the number of insect species on Earth. There are about 50,000 species of tropical forest trees. Assume that the data for beetles on *L. seemanii* are representative for other species of tropical trees, and calculate the number of host-specific beetles found on these trees. Add an estimated 1 million species of beetles that are expected across different species of trees (including temperate regions). Estimate the number of ground-dwelling beetle species based on the percentage used in question 1. Now estimate the number of species of insects found worldwide, based on the percentage of beetles among all insect species.

3. The estimates in questions 1 and 2 are based on many assumptions. Do you think these assumptions are reasonable? Why or why not? Would you argue for a different set of assumptions? How do you think these changes in assumptions would affect your calculations? Can you think of ways to test your assumptions?

TABLE 23.2 The Major Insect Groups[a]

GROUP	APPROXIMATE NUMBER OF DESCRIBED LIVING SPECIES
Jumping bristletails (Archaeognatha)	300
Silverfish (Thysanura)	370
PTERYGOTE (WINGED) INSECTS (PTERYGOTA)	
Mayflies (Ephemeroptera)	2,000
Dragonflies and damselflies (Odonata)	5,000
Neopterans (Neoptera)[b]	
Ice-crawlers (Grylloblattodea)	25
Gladiators (Mantophasmatodea)	15
Stoneflies (Plecoptera)	1,700
Webspinners (Embioptera)	300
Angel insects (Zoraptera)	30
Earwigs (Dermaptera)	1,800
Grasshoppers and crickets (Orthoptera)	20,000
Stick insects (Phasmida)	3,000
Cockroaches (Blattodea)	3,500
Termites (Isoptera)	2,750
Mantids (Mantodea)	2,300
Booklice and barklice (Psocoptera)	3,000
Thrips (Thysanoptera)	5,000
Lice (Phthiraptera)	3,100
True bugs, cicadas, aphids, leafhoppers (Hemiptera)	80,000
Holometabolous neopterans (Holometabola)[c]	
Ants, bees, wasps (Hymenoptera)	125,000
Beetles (Coleoptera)	375,000
Twisted-wing parasites (Strepsiptera)	600
Lacewings, ant lions, dobsonflies (Neuropterida)	4,700
Scorpionflies (Mecoptera)	600
Fleas (Siphonaptera)	2,400
True flies (Diptera)	120,000
Caddisflies (Trichoptera)	5,000
Butterflies and moths (Lepidoptera)	250,000

[a] The hexapod relatives of insects include the springtails (Collembola; 3,000 spp.), two-pronged bristletails (Diplura; 600 spp.), and proturans (Protura; 10 spp.). All are wingless and have internal mouthparts.

[b] Neopteran insects can tuck their wings close to their bodies.

[c] Holometabolous insects are neopterans that undergo complete metamorphosis.

insects between molts are called **instars**. A substantial change that occurs between one developmental stage and another is called **metamorphosis**. If the changes between its instars are gradual, an insect is said to have **incomplete metamorphosis**. If the change between at least some instars is dramatic, an insect is said to have **complete metamorphosis** (see Figure 30.3). In many insects with complete metamorphosis, the different life stages are specialized for different environments and use different food sources. In many species, the larvae are adapted for feeding and growing, whereas the adults are specialized for reproduction and dispersal.

> **LINK** Insect metamorphosis is under the control of hormones, as described in Concept 30.1

Pterygote insects were the first animals in evolutionary history to achieve the ability to fly. Flight opened up many new lifestyles and feeding opportunities that only the insects could exploit, and it is almost certainly one of the reasons for the remarkable numbers of insect species and individuals, and for their unparalleled evolutionary success.

Molecular data suggest that insects began to diversify about 450 million years ago, about the time of the appearance of the first land plants. These early hexapods evolved in a terrestrial environment that lacked any similar organisms, which in part accounts for their remarkable success. But the success of the insects is also due to their wings. Homologous genes control the development of insect wings and crustacean appendages, suggesting that the insect wing evolved from a dorsal branch of a crustacean-like limb:

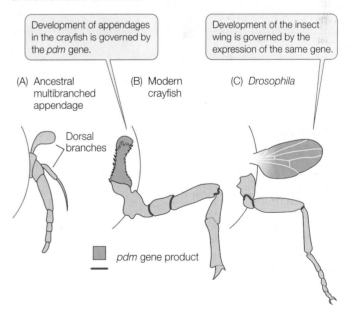

Development of appendages in the crayfish is governed by the *pdm* gene.

Development of the insect wing is governed by the expression of the same gene.

(A) Ancestral multibranched appendage

(B) Modern crayfish

(C) *Drosophila*

Dorsal branches

■ *pdm* gene product

The dorsal limb branch of crustaceans is used for gas exchange. Thus the insect wing probably evolved from a gill-like structure that had a gas exchange function.

The adults of most flying insects have two pairs of stiff, membranous wings attached to the thorax. True flies, however, have one pair of wings and a pair of stabilizers called *haltares*. In winged beetles, one pair of wings—the forewings—forms

relatives. The remaining groups are all *pterygote insects*. Pterygotes have two pairs of wings, except in some groups where one or both pairs of wings have been secondarily lost. Secondarily wingless groups include the parasitic lice and fleas, some beetles, and the worker individuals in many ants.

Hatchling pterygotes do not look like adults, and they undergo substantial changes at each molt. The immature stages of

(A) *Libellula saturata*, Odonata

(B) *Melanoplus differentialis*, Orthoptera

(C) *Anisocelis flavolineata*, Hemiptera

(D) *Limnephilus* (larva), Trichoptera

(E) *Trachelophorus giraffa*, Coleoptera

(F) *Junonia coenia*, Lepidoptera

(G) *Scathophaga stercoraria*, Diptera

(H) *Polistes gallicus*, Hymenoptera

FIGURE 23.30 Diverse Winged Insects (A) Unlike most flying insects, a dragonfly cannot fold its wings over its back. (B) Orthopteran insects such as grasshoppers undergo several larval molts (instars), but the juveniles resemble small adults (incomplete metamorphosis). (C) Hemipterans are "true" bugs. The leaflike extensions on this Costa Rican bug are cryptic (camouflaging) in the leaf litter the bug normally inhabits. (D–H) Holometabolous insects undergo complete metamorphosis. (D) A larval caddisfly (right) emerges from its dark pupal case. (E) Coleoptera is the largest insect group; beetles such as this Madagascan giraffe weevil account for more than half of all holometabolous species. (F) The buckeye butterfly is found in most parts of North America. (G) Many fly species feed on carrion and excrement, as does this golden dung fly. (H) This European paper wasp is a hymenopteran, a group in which most members display social behaviors.

This is the ancestral condition for pterygote insects, and the mayflies and dragonflies are not closely related to one another. Members of these groups have predatory or herbivorous aquatic larvae that transform into flying adults after they crawl out of the water. Dragonflies (and their relatives the damselflies) are active predators as adults. In contrast, adult mayflies lack functional digestive tracts. Mayflies live only about a day, just long enough to mate and lay eggs.

All other pterygote insects—the *neopterans*—can tuck their wings out of the way upon landing and crawl into crevices and other tight places. Some neopteran groups undergo incomplete metamorphosis, so hatchlings of these insects are sufficiently similar in form to adults to be recognizable. Examples include the grasshoppers (**FIGURE 23.30B**), roaches, mantids, stick insects, termites, stoneflies, earwigs, thrips, true bugs (**FIGURE 23.30C**), aphids, cicadas, and leafhoppers. They acquire adult organ systems, such as wings and compound eyes, gradually through several juvenile instars.

More than 80 percent of all insects belong to a subgroup of the neopterans called the *holometabolous* insects (see Table 23.2), which undergo complete metamorphosis (**FIGURE 23.30D**). The many species of beetles account for almost half of this group (**FIGURE 23.30E**). Also included are lacewings and their relatives; caddisflies; butterflies and moths (**FIGURE 23.30F**); sawflies; true flies (**FIGURE 23.30G**); and bees, wasps, and ants, some species of which display unique and highly specialized social behaviors (**FIGURE 23.30H**).

heavy, hardened wing covers. Flying insects are important pollinators of flowering plants.

Two groups of pterygotes, the mayflies and dragonflies (**FIGURE 23.30A**), cannot fold their wings against their bodies.

Do You Understand Concept 23.4?

- What features have contributed to making arthropods among the most abundant animals on Earth, both in number of species and number of individuals?
- Describe the difference between incomplete and complete metamorphosis.
- Can you think of some possible reasons why there are so many species of insects?

The majority of Earth's animal species are protostomes, so it is not surprising that the protostome groups display a huge variety of different body plans and other characteristics. We will now consider the diversity of the deuterostomes, a group that contains far fewer species than the protostomes but which is even more intensively studied and, not incidentally, is the group to which humans belong.

concept 23.5 Deuterostomes Include Echinoderms, Hemichordates, and Chordates

It may surprise you to learn that you and a sea urchin are both deuterostomes. Adult sea stars, sea urchins, and sea cucumbers—the most familiar echinoderms—look so different from adult vertebrates (fishes, frogs, lizards, birds, and mammals) that it is difficult to believe all these animals are closely related. Two major pieces of evidence indicate that the deuterostomes share a common ancestor not shared with the protostomes:

- Deuterostomes share a pattern of early development in which the mouth forms at the opposite end of the embryo from the blastopore, and the blastopore develops into the anus (this is opposed to the protostomes, in which the blastopore becomes the mouth).
- Recent phylogenetic analyses of the DNA sequences of many different genes offer strong support for the shared evolutionary relationships of deuterostomes.

Note that neither of the above factors is apparent in the morphology of the adult animals.

Although there are far fewer species of deuterostomes than of protostomes (see Table 23.1), the deuterostomes are of special interest because they include many large animals—including humans—that strongly influence the characteristics of ecosystems. Many deuterostome species have been intensively studied in all fields of biology. Complex behaviors are especially well developed among some deuterostomes and are a vast and fascinating field of study in themselves (see Chapter 41).

There are three major clades of living deuterostomes:

- **Echinoderms**: sea stars, sea urchins, and their relatives
- **Hemichordates**: acorn worms and pterobranchs
- **Chordates**: sea squirts, lancelets, and vertebrates

All deuterostomes are triploblastic and coelomate (see Figure 23.3C). Skeletal support features, where present, are internal rather than external. Some species have segmented bodies, but the segments are less obvious than those of annelids and arthropods.

The earliest deuterostomes were bilaterally symmetrical, segmented animals with a pharynx that had slits through which water flowed. Echinoderms evolved their adult forms with unique symmetry (in which the body parts are arranged along five radial axes) much later, whereas other deuterostomes retained the ancestral bilateral symmetry.

The echinoderms and hemichordates (together known as *ambulacrarians*) have a bilaterally symmetrical, ciliated larva (**FIGURE 23.31A**). Adult hemichordates also are bilaterally symmetrical. Echinoderms, however, undergo a radical change in form as they develop into adults (**FIGURE 23.31B**), changing from a bilaterally symmetrical larva to an adult with **pentaradial symmetry** (symmetry in five or multiples of five). As is typical of animals with radial symmetry, echinoderms have no head, and they move equally well (but usually slowly) in many directions. Rather than having an anterior–posterior (head–tail) and dorsal–ventral (back–belly) body organization, echinoderms have an *oral* side (containing the mouth) and an opposite *aboral* side (containing the anus).

Echinoderms have unique structural features

About 13,000 species of echinoderms in 23 major groups have been described from fossil remains. They are probably only a small fraction of those that actually lived. Only 6 of the 23 major groups known from fossils are represented by species that survive today; many clades became extinct during the periodic mass extinctions that have occurred throughout Earth's history (see Table 18.1). Nearly all of the 7,000 extant species of echinoderms live only in marine environments.

In addition to having pentaradial symmetry, adult echinoderms have two unique structural features. One is a system of calcified internal plates covered by thin layers of skin and some muscles. The calcified plates of most echinoderms are thick, and they fuse inside the entire body, forming an *internal skeleton*. The other unique feature is a **water vascular system**, a network of water-filled canals leading to extensions called **tube feet**. This system functions in gas exchange, locomotion, and feeding (see Figure 23.31B).

Members of one major extant clade, the *crinoids* (sea lilies and feather stars), were more abundant and species-rich 300 to 500 million years ago than they are today. There are some 80 described living sea lily species, most of which are sessile organisms attached to a substrate by a stalk. Feather stars (**FIGURE 23.32A**) grasp the substrate with flexible appendages that allow for limited movement. About 600 living species of feather stars have been described.

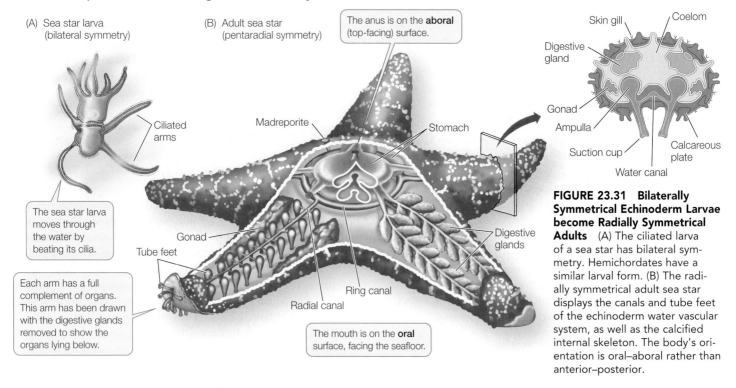

(A) Sea star larva (bilateral symmetry)

The sea star larva moves through the water by beating its cilia.

Ciliated arms

(B) Adult sea star (pentaradial symmetry)

The anus is on the **aboral** (top-facing) surface.

Madreporite

Stomach

Gonad

Tube feet

Each arm has a full complement of organs. This arm has been drawn with the digestive glands removed to show the organs lying below.

Radial canal

Ring canal

Digestive glands

The mouth is on the **oral** surface, facing the seafloor.

Skin gill

Coelom

Digestive gland

Gonad

Ampulla

Suction cup

Water canal

Calcareous plate

FIGURE 23.31 Bilaterally Symmetrical Echinoderm Larvae become Radially Symmetrical Adults (A) The ciliated larva of a sea star has bilateral symmetry. Hemichordates have a similar larval form. (B) The radially symmetrical adult sea star displays the canals and tube feet of the echinoderm water vascular system, as well as the calcified internal skeleton. The body's orientation is oral–aboral rather than anterior–posterior.

Unlike the crinoids, most of the other surviving echinoderms are motile. The two main groups of motile echinoderms are the *echinozoans* (sea urchins and sea cucumbers; **FIGURES 23.32B,C**) and *asterozoans* (sea stars and brittle stars; **FIGURE 23.32D,E**).

The tube feet of the different echinoderm groups have been modified in a great variety of ways to capture prey. Sea lilies, for example, feed by orienting

FIGURE 23.32 Echinoderm Diversity (A) The flexible arms of this feather star are clearly visible. (B) Sea urchins are important grazers on algae in the intertidal zones of the world's oceans. (C) Sea cucumbers are unique among echinderms in having an anterior–posterior rather than an oral–aboral orientation of the mouth and anus. This individual is in the spawning position; normally it would travel across the seafloor on its tube feet. (D) Sea stars are important predators on bivalve mollusks such as mussels and clams. Suction tips on its tube feet allow a sea star to grasp both shells of the bivalve and pull them open. (E) The arms of the brittle star are composed of hard but jointed plates.

(A) *Comanthina schlegeli*

(B) *Sphaerechinus granularis*

(C) *Thelenota* sp.

(D) *Marthasterias glacialis*

(E) *Ophiopholis aculeata*

their arms in passing water currents. Food particles strike and stick to their tube feet, which are covered with mucus-secreting glands. The tube feet transfer food particles to grooves in the arms, where ciliary action carries the particles to the mouth. Sea cucumbers capture food with their anterior tube feet, which are modified into large, feathery, sticky tentacles that can be protruded from the mouth. Periodically, a sea cucumber withdraws the tentacles, wipes off the material that has adhered to them, and digests it.

Many sea stars use their tube feet to capture large prey such as annelids, gastropod and bivalve mollusks, small crustaceans such as crabs, and fishes. With hundreds of tube feet acting simultaneously, a sea star can grasp a bivalve in its arms, anchor the arms with its tube feet, and by steady contraction of the muscles in its arms, gradually exhaust the muscles the bivalve uses to keep its shell closed (see Figure 23.32D). To feed on a bivalve, a sea star can push its stomach out through its mouth and then through the narrow space between the two halves of the bivalve's shell. The sea star's stomach then secretes enzymes that digest the prey.

Most sea urchins eat algae, which they catch with their tube feet from the plankton or scrape from rocks with a complex rasping structure. Most of the 2,000 species of brittle stars ingest particles from the upper layers of sediments and assimilate the organic material from them, although some species filter suspended food particles from the water, and others capture small animals.

Hemichordates are wormlike marine deuterostomes

The roughly 100 species of hemichordates—acorn worms and pterobranchs—have a body organized in three major parts, consisting of a *proboscis*, a *collar* (which bears the mouth), and a *trunk* (which contains the other body parts). The 70 known species of acorn worms range up to 2 meters in length (**FIGURE 23.33A**). They live in burrows in muddy and sandy marine sediments. The digestive tract of an acorn worm consists of a mouth behind which are a muscular *pharynx* and an *intestine*. The pharynx opens to the outside through a number of *pharyngeal slits* through which water can exit. Highly vascularized tissue surrounding the pharyngeal slits serves as a gas-exchange apparatus. Acorn worms breathe by pumping water into the mouth and out through the pharyngeal slits. They capture prey with the large proboscis, which is coated with sticky mucus to which small organisms in the sediment stick. The mucus and its attached prey are conveyed by cilia to the mouth. In the esophagus, the food-laden mucus is compacted into a ropelike mass that is moved through the digestive tract by ciliary action.

The 30 living species of pterobranchs are sedentary marine animals up to 12 millimeters long that live in a tube secreted by the proboscis. Some species are solitary; others form colonies of individuals joined together (**FIGURE 23.33B**). Behind the proboscis is a collar with anywhere from one to nine pairs of arms. The arms bear long tentacles that capture prey and function in gas exchange.

Chordate characteristics are most evident in larvae

As mentioned earlier, it is not obvious from examining the morphology of adult animals that echinoderms and chordates share a common ancestor. The evolutionary relationships among some chordate groups are not immediately apparent either. The features that reveal these evolutionary relationships are seen primarily in the larvae—in other words, it is during the early developmental stages that their evolutionary relationships are evident.

There are three principal chordate clades: the **cephalochordates**, **urochordates**, and **vertebrates**. There are about 3,000 living species of urochordates and 62,000 living species of vertebrates, but only about 30 living species of cephalochordates.

(A)

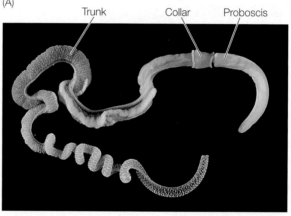

Trunk   Collar   Proboscis

Saccoglossus kowalevskii

(B)

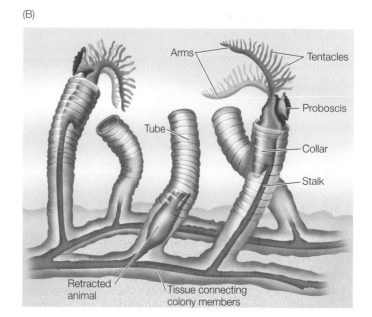

Arms   Tentacles   Tube   Proboscis   Collar   Stalk   Retracted animal   Tissue connecting colony members

FIGURE 23.33 Hemichordates (A) The proboscis of an acorn worm is modified for burrowing. (B) The structure of a colonial pterobranch.

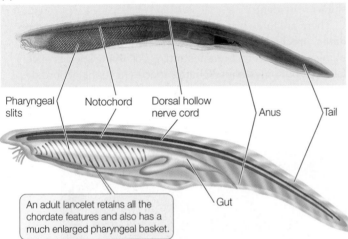

FIGURE 23.34 **The Key Features of Chordates Are Most Apparent in Early Developmental Stages** The pharyngeal slits of both the urochordate ascidian and the cephalochordate lancelet develop into a pharyngeal basket. (A) The ascidian larva (but not the adult) has all three chordate features: a dorsal hollow nerve cord, postanal tail, and notochord. (B) All three chordate synapomorphies are retained in the adult lancelet.

Adult chordates vary greatly in form, but all chordates display the following derived structures at some stage in their development (**FIGURE 23.34**):

- A dorsal hollow nerve cord
- A tail that extends beyond the anus
- A dorsal supporting rod called the notochord

The **notochord** is the most distinctive derived chordate trait. It is composed of a core of large cells with turgid fluid-filled vacuoles, which make it rigid but flexible. In most urochordates the notochord is lost during metamorphosis to the adult stage. In most vertebrate species, it is replaced during development by vertebrae that provide support for the body.

The ancestral pharyngeal slits (not a derived feature of this group) are present at some developmental stage of chordates but are often lost in adults. The pharynx, which develops around the pharyngeal slits, functioned in chordate ancestors as the site for oxygen uptake and the elimination of carbon

dioxide and water (as in acorn worms). The pharynx is much enlarged in some chordate species (as in the *pharyngeal basket* of the lancelet in Figure 23.34B).

Adults of most cephalochordates and urochordates are sessile

The 30 species of cephalochordates, or *lancelets*, are small animals that rarely exceed 5 centimeters in length. The notochord, which provides body support, extends the entire length of the body throughout their lives (see Figure 23.34B). Lancelets are found in shallow marine and brackish waters worldwide. Most of the time they lie covered in sand with their head protruding above the sediment, but they can swim. They filter prey from the water with their pharyngeal basket.

All urochordates live in marine environments. More than 90 percent of the known species are *ascidians* (sea squirts). Individual ascidians range in length from less than 1 millimeter to 60 centimeters. Some ascidians form colonies by asexual budding from a single founder. Colonies may measure several meters across. The baglike body of an adult ascidian is enclosed in a tough tunic, leading to its alternate name of "tunicate" (**FIGURE 23.35**). The tunic is composed of proteins and a complex polysaccharide secreted by epidermal cells. The ascidian pharynx is enlarged into a pharyngeal basket that filters prey from the water passing through it.

In addition to its pharyngeal slits, an ascidian larva has a dorsal hollow nerve cord and a notochord that is restricted mostly to the tail region (see Figure 23.34A). Bands of muscle that surround the notochord provide support for the body. After a short time swimming in the plankton, the larvae of most species settle on the seafloor and transform into sessile adults. The swimming,

Polycarpa aurata

FIGURE 23.35 **An Adult Urochordate** The colorful tunic of this adult ascidian (also known as a tunicate) has led to the species' two common names, "golden sea squirt" and "ink spot sea squirt."

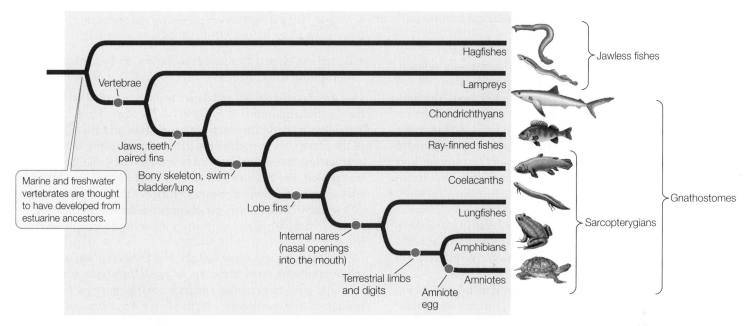

FIGURE 23.36 **Phylogeny of the Living Vertebrates** This phylogenetic tree shows the evolution of some of the key innovations among the major groups of vertebrates.

tadpolelike larvae suggest a close evolutionary relationship between ascidians and vertebrates (see Figure 16.4).

A dorsal supporting structure replaces the notochord in vertebrates

The vertebrates take their name from the unique, jointed, dorsal **vertebral column** that replaces the notochord during early development as their primary supporting structure (**FIGURE 23.36**).

As we have discussed, the hemichordates, echinoderms, cephalochordates, and urochordates are marine animals. The lineage that led to the vertebrates is also thought to have evolved in the oceans, although probably in an estuarine environment (where fresh water meets salt water). Vertebrates have since radiated into marine, freshwater, terrestrial, and aerial environments worldwide.

The *hagfishes* are thought by many biologists to be the sister group to the remaining vertebrates (as shown in Figure 23.36). Hagfishes (**FIGURE 23.37A**) have a weak circulatory system, with three small accessory hearts (rather than a single, large heart), a partial *cranium* or skull (containing no *cerebrum* or *cerebellum*, two main regions of the brain of other vertebrates), and no jaws or stomach. They also lack separate, jointed vertebrae and have a skeleton composed of cartilage. Thus, some biologists do not consider hagfishes vertebrates, and use instead the term *craniates* to refer collectively to the hagfishes and the vertebrates. Some analyses of gene sequences suggest, however, that hagfishes may be more closely related to the vertebrate lampreys (**FIGURE 23.37B**); in this phylogenetic arrangement, the two groups are collectively called the *cyclostomes* ("circle mouths"). If in fact the hagfishes and lampreys do form a monophyletic group, then hagfishes must have secondarily

Eptatretus stoutii

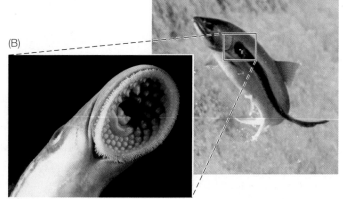
Petromyzon marinus

FIGURE 23.37 **Modern Jawless Fishes** (A) Hagfishes burrow in the ocean mud, from which they extract small prey. They also scavenge on dead or dying fish. Their degenerate eyes led to their being miscalled "blind eels." (B) Sea lampreys are ectoparasites that attach to the bodies of living fish and use their large, jawless mouths to suck blood and flesh. They can survive in both fresh and salt water, as this individual attached to a salmon returning to its spawning ground will do.

lost many of the major vertebrate morphological features during their evolution.

The 58 known species of hagfishes are unusual marine animals that produce copious quantities of slime as a defense. They are virtually blind and rely largely on the four pairs of sensory tentacles around their mouth to detect food. Although they have no jaws, hagfishes have a tonguelike structure equipped with toothlike rasps that they can use to tear apart dead organisms and to capture their principal prey, annelid worms. Hagfishes have direct development (no larvae), and individuals may actually change sex from year to year (from male to female and vice versa).

The nearly 50 species of lampreys either live in fresh water, or they live in coastal salt water and move into fresh water to breed. Although the lampreys and hagfishes may look superficially similar (with elongate eel-like bodies and no paired fins), they differ greatly in their biology.

Lampreys have a complete braincase and distinct and separate (although rudimentary) vertebrae, all cartilaginous rather than bony. Lampreys undergo a complete metamorphosis from filter-feeding larvae known as *ammocoetes*, which are morphologically quite similar in general structure to adult lancelets. The adults of many species of lampreys are parasitic, although several lineages of lampreys evolved to become nonfeeding as adults. These nonfeeding adult lampreys survive only a few weeks after metamorphosis—just long enough to breed. In the species that are parasitic as adults, the round mouth is a rasping and sucking organ that is used to attach to their prey and rasp at the flesh (see Figure 23.37B). Some species of lampreys are critically endangered because of recent habitat changes and losses.

The vertebrate body plan can support large, active animals

Four key features characterize the vertebrates:

- An anterior *skull* with a large brain
- A rigid internal *skeleton* supported by the vertebral column
- Internal organs *suspended in a coelom*
- A well-developed *circulatory system*, driven by contractions of a ventral *heart*

This organization of the vertebrate body is exemplified by the bony fish diagrammed in **FIGURE 23.38**. Many kinds of jawless fishes were found in the seas, estuaries, and fresh waters of the Devonian period, but hagfishes and lampreys are the only jawless fishes that survived beyond the Devonian. During that period, the *gnathostomes* (Greek, "jaw mouths") evolved jaws via modifications of the skeletal arches that supported the gills (**FIGURE 23.39**). Jaws greatly improved feeding efficiency, as an animal with jaws can grasp, subdue, and swallow large prey.

The earliest jaws were simple, but the evolution of *teeth* made predators more effective. In predators, teeth function crucially both in grasping and in breaking up prey. In both predators and herbivores, teeth enable an animal to chew both soft and hard body parts of their food. Chewing also aids chemical digestion and improves an animal's ability to extract nutrients from its food.

Fins and swim bladders improved stability and control over locomotion

Paired fins stabilize the position of jawed fishes in water (and in some cases, help propel them). Most aquatic gnathostomes have a pair of pectoral fins just behind the gill slits, and a pair of pelvic fins anterior to the anal region (see Figures 23.38 and 23.40). Median dorsal and anal fins stabilize the fish, or may be used for propulsion in some species. In many fishes, the caudal (tail) fin helps propel the animal and enables it to turn rapidly.

Several groups of gnathostomes became abundant during the Devonian. Among them were the **chondrichthyans**—sharks, skates, and rays (940 living species) and chimaeras (40 living

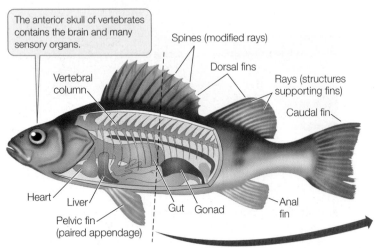

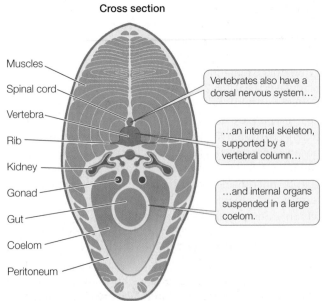

FIGURE 23.38 The Vertebrate Body Plan A ray-finned fish illustrates the structural elements common to all vertebrates. In addition to the paired pelvic fins, these fishes have paired pectoral fins on the sides of their bodies (not seen in this cutaway view).

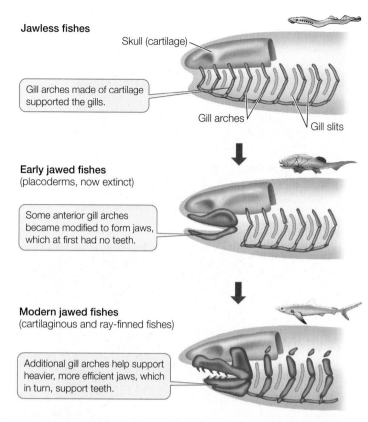

Jawless fishes

Skull (cartilage)

Gill arches made of cartilage supported the gills.

Gill arches

Gill slits

Early jawed fishes
(placoderms, now extinct)

Some anterior gill arches became modified to form jaws, which at first had no teeth.

Modern jawed fishes
(cartilaginous and ray-finned fishes)

Additional gill arches help support heavier, more efficient jaws, which in turn, support teeth.

FIGURE 23.39 Verterbrate Jaw Evolution Jaws of vertebrates are derived from modifications of the anterior gill arches of jawless fishes.

species). Like hagfishes and lampreys, these fishes have a skeleton composed entirely of firm but pliable cartilage. Their skin is flexible and leathery, sometimes bearing scales that give it the consistency of sandpaper. Sharks move forward by means of lateral undulations of their body and caudal fin (**FIGURE 23.40A**). Skates and rays propel themselves by means of vertical undulating movements of their greatly enlarged pectoral fins (**FIGURE 23.40B**).

Most sharks are predators, but some feed by straining plankton from the water. Most skates and rays live on the ocean floor, where they feed on mollusks and other animals buried in the sediments. Nearly all cartilaginous fishes live in the oceans, but a few are estuarine or migrate into lakes and rivers. One group of stingrays is found in river systems of South America. The less familiar chimaeras (**FIGURE 23.40C**) live in deep-sea or cold waters.

In the ancestor of the bony vertebrates, gas-filled sacs that extended from the digestive tract supplemented the gas-exchange function of the gills by giving the animals access to atmospheric oxygen. These features enabled those fishes to live where oxygen was periodically in short supply, as it often is in freshwater environments. These lunglike sacs evolved into *swim bladders* (organs of buoyancy), as well as into the lungs of tetrapods. By adjusting the amount of gas in its swim bladder, a fish can control the depth at which it remains suspended in the water while expending very little energy to maintain its position.

Ray-finned fishes, and most remaining groups of vertebrates, have internal skeletons of calcified, rigid *bone* rather than flexible cartilage. The outer body surface of most species of ray-finned fishes is covered with flat, thin, lightweight scales that provide some protection or enhance movement through the water. The gills of ray-finned fishes open into a single chamber covered by a hard flap, called an *operculum*. Movement of the operculum improves the flow of water over the gills, where gas exchange takes place.

(A) *Carcharhinus galapagensis*

Dorsal fin

Caudal fin   Pelvic fin   Pectoral fins

FIGURE 23.40 Chondrichthyans The fins of chondrichthyans lack supportive rays. (A) Most sharks, such as this sandbar shark, are active marine predators. (B) The fins of skates and rays, represented here by a ribbontail stingray, are modified for feeding on the ocean bottom. (C) A chimaera, or ratfish.

(B) *Taeniura lymma*

(C) *Hydrolagus colliei*

The pectoral fins of stingrays are modified for propulsion; their other fins are greatly reduced.

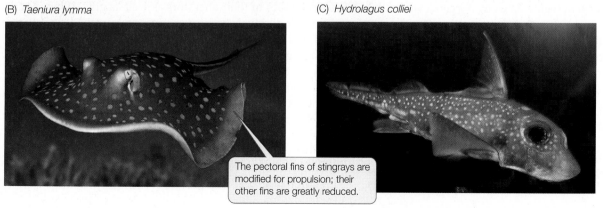

FIGURE 23.41 Ray-Finned Fishes (A) The two ray-finned species shown here are mutualists in their coral reef ecosystem. The larger sweetlips illustrates the body plan most commonly associated with this highly diverse clade. (B) Eels such as this moray have the large teeth and powerful jaws typical of predatory fishes. Their elongated, virtually finless bodies are adapted to hunting and hiding in narrow crevices.

Ray-finned fishes radiated extensively in the Tertiary. Today there are about 30,000 known living species, encompassing a remarkable variety of sizes, shapes, and lifestyles (**FIGURE 23.41**). The smallest are less than 1 centimeter long as adults; the largest weigh as much as 900 kilograms. Ray-finned fishes exploit nearly all types of aquatic food sources. In the oceans they filter plankton from the water, rasp algae from rocks, eat corals and other soft-bodied colonial animals, dig animals from soft sediments, and prey on virtually all kinds of other fishes. In fresh water they eat plankton, devour insects, eat fruits that fall into the water in flooded forests, and prey on other aquatic vertebrates and, occasionally, terrestrial vertebrates.

Do You Understand Concept 23.5?

- What are three developmental patterns the earliest deuterostomes had in common?
- What are some of the ways that echinoderms use their tube feet to obtain food?
- What synapomorphies respectively characterize the chordates and the vertebrates?
- How do the hagfishes differ from the lampreys in morphology and life history? Why do some biologists not consider the hagfishes to be vertebrates?

In some fishes, the lunglike sacs that gave rise to swim bladders became specialized for another purpose: breathing air. That adaptation set the stage for the vertebrates to move onto the land.

concept **23.6** **Life on Land Contributed to Vertebrate Diversification**

The evolution of lunglike sacs in fishes set the stage for the invasion of the land. Some early ray-finned fishes probably used those sacs to supplement their gills when oxygen levels in the water were low, as lungfishes and many groups of ray-finned fishes do today. But with their unjointed fins, those fishes could only flop around on land. Changes in the structure of the fins first allowed some fishes to support themselves better in shallow water and, later, to move better on land.

Jointed fins enhanced support for fishes

Two pairs of muscular, jointed fins evolved in the ancestor of the **sarcopterygians**, which include coelacanths, lungfishes, and tetrapods. Each of the jointed appendages of sarcopterygians is

(A)

> Cleaner wrasse (*Labroides dimidiatus*) feed on parasites off the body of a much larger ribbon sweetlips (*Plectorhincus polytaenia*).

(B) *Gymnothorax meleagris*

joined to the body by a single enlarged bone. The coelacanths flourished from the Devonian until about 65 million years ago, when they were thought to have become extinct. However, in 1938 a commercial fisherman caught a living coelacanth off South Africa. Since that time, hundreds of individuals of this extraordinary fish, *Latimeria chalumnae* (**FIGURE 23.42A**), have been collected. A second species, *L. menadoensis*, was discovered in 1998 off the Indonesian island of Sulawesi.

Lungfishes, which also have jointed fins that are connected to the body by a single enlarged bone, were important predators in shallow-water habitats in the Devonian, but most lineages died out. The six surviving species live in stagnant swamps and muddy waters in South America, Africa, and Australia (**FIGURE 23.42B**). Lungfishes have lungs derived from the lunglike sacs of their ancestors as well as gills. When ponds dry up, individuals of most species can burrow deep into the mud and survive for many months in an inactive state while breathing air.

Some early aquatic sarcopterygians began to use terrestrial food sources, became more fully adapted to life on land, and eventually evolved to become ancestral **tetrapods** (four-legged vertebrates). The earliest tetrapod limbs appear to have functioned in holding the animals upright in shallow water, allowing them to hold their head above the water's surface. These same structures were then co-opted for movement on land, at first probably for foraging on brief trips out of water.

(A) *Latimeria chalumnae*

(B) *Protopterus annectens*

FIGURE 23.42 The Closest Relatives of Tetrapods (A) The African coelacanth, discovered in deep waters of the Indian Ocean, represents one of two surviving species of a group that was once thought to be extinct. (B) All surviving lungfish species, such as this African lungfish, live in the Southern Hemisphere.

Amphibians adapted to life on land

Most modern **amphibians** are confined to moist environments because they lose water rapidly through the skin when exposed to dry air. In addition, their eggs are enclosed within delicate membranous envelopes that cannot prevent water loss in dry conditions. In some amphibian species, adults live mostly on land but return to fresh water to lay their eggs, which are usually fertilized outside the body (**FIGURE 23.43**). The fertilized eggs give rise to larvae that live in water until they undergo metamorphosis to become terrestrial adults. However, many amphibians (especially those in tropical and subtropical areas) have evolved a wide diversity of additional reproductive modes and types of parental care. Internal fertilization evolved many times among the amphibians. Many species develop directly into adultlike forms from fertilized eggs laid on land or carried by the parents. Other species of amphibians are entirely aquatic, never leaving the water at any stage of their lives, and many of these species retain a larval-like morphology.

The more than 6,500 known species of amphibians living on Earth today belong to three major groups: the wormlike, limbless, tropical, burrowing or aquatic *caecilians* (**FIGURE 23.44A**), the tail-less frogs and toads (collectively called *anurans*; **FIGURE 23.44B**), and the tailed *salamanders* (**FIGURE 23.44C,D**).

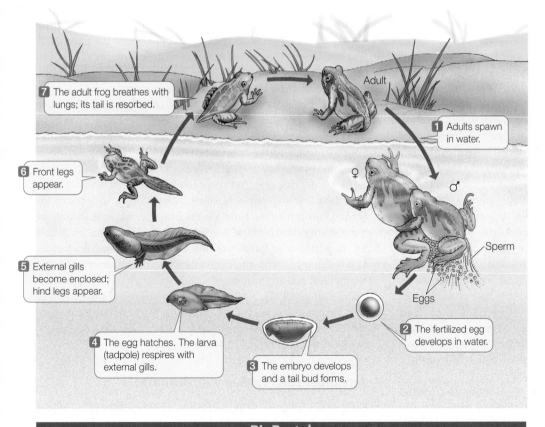

7 The adult frog breathes with lungs; its tail is resorbed.

Adult

1 Adults spawn in water.

6 Front legs appear.

♀

♂

Sperm

5 External gills become enclosed; hind legs appear.

Eggs

2 The fertilized egg develops in water.

4 The egg hatches. The larva (tadpole) respires with external gills.

3 The embryo develops and a tail bud forms.

yourBioPortal.com

Go to ANIMATED TUTORIAL 23.2
Life Cycle of a Frog

FIGURE 23.43 In and Out of the Water Most early stages in the life cycle of many amphibians take place in water. The aquatic tadpole transforms into a terrestrial adult through metamorphosis. Some species of amphibians, however, have direct development (with no aquatic larval stage), and others are aquatic throughout life.

FIGURE 23.44 Diversity among the Amphibians
(A) Burrowing caecilians superficially look more like worms than amphibians. (B) Male golden toads in the cloud forest of Monteverde, Costa Rica. This species has recently become extinct, one of many amphibian species to do so in the past few decades. (C) An adult Chinese salamander. (D) This Austin blind salamander's life cycle remains aquatic; it has no adult terrestrial stage. The eyes of this cave dweller have become greatly reduced.

(A) *Gymnophis multiplicata*

(B) *Bufo periglenes*

(C) *Tylototriton verrucosus*

(D) *Eurycea waterlooensis*

Anurans are most diverse in wet tropical and warm temperate regions, although a few are found at very high latitudes. There are far more anurans—about 5,800 described species, with more being discovered every year—than any other amphibians. Some anurans have tough skins and other adaptations that enable them to live for long periods in very dry deserts, whereas others live in moist terrestrial and arboreal environments. Some species are completely aquatic as adults. All anurans have a very short vertebral column, with a strongly modified pelvic region that is adapted for leaping, hopping, or propelling the body through water by kicking the hind legs.

The approximately 600 described species of salamanders are most diverse in temperate regions of the Northern Hemisphere, but many species are also found in cool, moist environments in the mountains of Central America, and a few species penetrate into the South American tropics. Many salamanders live in rotting logs or moist soil. One major group has lost lungs, and these species exchange gases entirely through the skin and mouth lining—body parts that all amphibians use in addition to their lungs. Through *paedomorphosis* (retention of the juvenile state), a completely aquatic lifestyle has evolved several times among the salamanders (see Figure 23.44D). Most species of salamanders have internal fertilization, which is usually achieved through the transfer of a small jellylike, sperm-embedded capsule called a *spermatophore*.

Many amphibians have complex social behaviors. Most male anurans utter loud, species-specific calls to attract females of their own species (and sometimes to defend breeding territories), and they compete for access to females that arrive at the breeding sites. Many amphibians lay large numbers of eggs, which they abandon once they are deposited and fertilized. Some amphibians lay only a few eggs, which are fertilized and then guarded in a nest or carried on the backs, in the vocal pouches, or even in the stomachs of one of the parents. A few species of frogs, salamanders, and caecilians are *viviparous*, meaning they give birth to well-developed young that have received nutrition from the female during gestation.

Amphibians are the focus of much attention today because populations of many species are declining rapidly, especially in mountainous regions of western North America, Central and South America, and northeastern Australia. Worldwide, about one-third of amphibian species are now threatened with extinction or have disappeared completely in the last few decades. Scientists are investigating several hypotheses to account for these population declines, including the adverse effects of habitat alteration by humans, increased solar radiation caused by destruction of Earth's ozone layer, pollution from urban and industrial areas and airborne agricultural pesticides and herbicides, and the spread of a pathogenic chytrid fungus that attacks amphibians. Scientists have documented the spread of the chytrid fungus through Central America, where many species of amphibians have become extinct (including Costa Rica's golden toad; see Figure 23.44B).

Amniotes colonized dry environments

Several key innovations contributed to the ability of members of one clade of tetrapods to exploit a wide range of terrestrial habitats. The animals that evolved these water-conserving traits are called **amniotes**.

The **amniote egg** (which gives the group its name) is relatively impermeable to water and allows the embryo to develop in a contained aqueous environment:

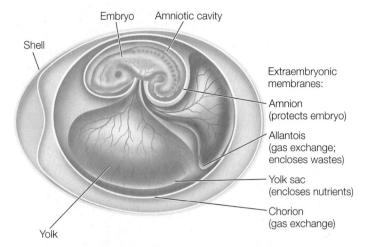

The leathery or brittle, calcium-impregnated shell of the amniote egg retards evaporation of the fluids inside but permits passage of oxygen and carbon dioxide. The egg also stores large quantities of food in the form of *yolk*, allowing the embryo to attain a relatively advanced state of development before it hatches. Within the shell are *extraembryonic membranes* that protect the embryo from desiccation and assist its gas exchange and excretion of waste nitrogen.

---

yourBioPortal.com

Go to WEB ACTIVITY 23.6
The Amniote Egg

In several different groups of amniotes, the amniote egg became modified, allowing the embryo to grow inside (and receive nutrition from) the mother. For instance, the mammalian egg lost its shell while the functions of the extraembryonic membranes were retained and expanded.

> **LINK** The roles of the extraembryonic membranes of the amniote egg are discussed in Concept 33.5

Other innovations evolved in the organs of terrestrial adults. A tough, impermeable skin, covered with scales or modifications of scales such as hair and feathers, greatly reduced water loss. Adaptations of the vertebrate excretory organs, the kidneys, allowed amniotes to excrete concentrated urine, ridding the body of waste nitrogen without losing a large amount of water in the process.

During the Carboniferous, amniotes split into two major groups, the mammals and reptiles (**FIGURE 23.45**). More than 18,700 species of **reptiles** exist today, over half of which are *birds*. Birds are the only living members of the otherwise extinct *dinosaurs*, the dominant terrestrial predators of the Mesozoic.

Reptiles adapted to life in many habitats

The lineage leading to modern reptiles began to diverge from other amniotes about 250 million years ago. One reptilian group that has changed very little over the intervening millennia is the *turtles*. The dorsal and ventral bony plates of turtles form a shell into which the head and limbs can be withdrawn in many species (**FIGURE 23.46A**). The dorsal shell is an expansion of

FIGURE 23.45 Phylogeny of Amniotes This tree of amniote relationships shows the primary split between mammals and reptiles. The reptile portion of the tree shows a lineage leading to the turtles, another to the lepidosaurs (snakes, lizards, and tuataras), and a third branch that includes all the archosaurs (crocodiles, several extinct groups, and the birds). There is some uncertainty in the placement of the turtle lineage; some data support a relationship between turtles and archosaurs.

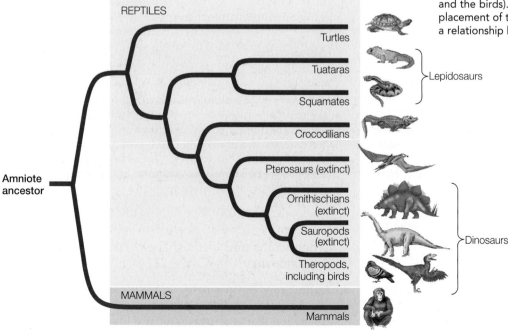

FIGURE 23.46 Reptilian Diversity (A) Turtles are unique shelled vertebrates. The dorsal shell, or carapace, is an adaptation of the rib bones. (B) The tuatara represents one of only two surviving species in a lineage that diverged long ago. (C,D) Lizards and snakes are the major squamate groups. (C) The leopard gecko is a desert lizard native to Afghanistan, Pakistan, and northwestern India. (D) The nonvenomous Mexican milk snake has a color pattern similar to that of the deadly coral snake. Such evolutionary adaptation, in which the harmless species gains protection from its resemblance to a deadly animal, is known as Batesian mimicry.

(A) *Emydoidea blandingii*

(B) *Sphenodon punctatus*

(C) *Eublepharis macularius*

(D) *Lampropeltis triangulum*

the ribs, and it is a mystery how the pectoral girdles evolved to be inside the ribs of turtles, making them unlike any other vertebrates. Most turtles live in aquatic environments, but several groups, such as tortoises and box turtles, are terrestrial. Sea turtles spend their entire lives at sea except when they come ashore to lay eggs. Human exploitation of sea turtles and their eggs has resulted in worldwide declines of these species, all of which are now endangered. A few species of turtles are strict herbivores or carnivores, but most species are omnivores that eat a variety of aquatic and terrestrial plants and animals.

The *lepidosaurs* constitute the second-most species-rich clade of living reptiles. This group is composed of the *squamates* (lizards, snakes, and amphisbaenians—the last a group of mostly legless, wormlike, burrowing reptiles with greatly reduced eyes) and the *tuataras*, which superficially resemble lizards but differ from them in tooth attachment and several internal anatomical features. Many species related to the tuataras lived during the Mesozoic era, but today only two species, restricted to a few islands off New Zealand, survive (**FIGURE 23.46B**).

The skin of a lepidosaur is covered with horny scales that greatly reduce loss of water from the body surface. These scales, however, make the skin unavailable as an organ of gas exchange. Gases are exchanged almost entirely via the lungs, which are proportionally much larger in surface area than those of amphibians. A lepidosaur forces air into and out of its lungs by bellowslike movements of its ribs. The three-chambered lepidosaur heart partially separates oxygenated blood from the lungs from deoxygenated blood returning from the body. With this type of heart, lepidosaurs can generate high blood pressure and can sustain a relatively high metabolism.

Most lizards are insectivores, but some are herbivores; a few prey on other vertebrates. The largest lizard, which grows as long as 3 meters and can weigh more than 150 kilograms, is the predaceous Komodo dragon of the East Indies. Most lizards walk on four limbs (**FIGURE 23.46C**), although limblessness has evolved repeatedly in the group, especially in burrowing and grassland species. One major group of limbless squamates is the snakes (**FIGURE 23.46D**). All snakes are carnivores, and many can swallow objects much larger than themselves. Several snake groups evolved venom glands and the ability to inject venom rapidly into their prey.

FRONTIERS One group of lizards, the geckoes, are well known for their ability to climb smooth vertical surfaces. Biologists recently discovered that they accomplish this task using microscopically branched elastic fibers on their toe pads, which grip surfaces using atomic-scale attractive forces. These microscopic structures of gecko feet are now the basis for the development of new dry adhesives based on similar carbon "nanotubes." These new adhesives are expected to have many applications, from connecting materials in the vacuum of space to the assembly of electronic microdevices.

Crocodilians and birds share their ancestry with the dinosaurs

Another reptilian group, the *archosaurs*, includes the crocodilians, dinosaurs, and birds. *Dinosaurs* rose to prominence about 215 million years ago and dominated terrestrial environments

(A) *Crocodylus niloticus*

(B) *Rhea americana*

FIGURE 23.47 Archosaurs (A) Crocodiles, alligators, and their relatives live in tropical and warm temperate climates. The Nile crocodile is commonly found over much of the African continent. (B) Birds are the other living archosaur group. This South American rhea is a palaeognath, a group that includes several flightless and weakly flying birds of the Southern Hemisphere. Flightless members of this early-diverging group lack the sternum keel structure that anchors the flight muscles of other bird groups.

for about 150 million years; only one group of dinosaurs, the *birds*, survived the mass extinction event at the Cretaceous–Tertiary boundary. During the Mesozoic, most terrestrial animals more than a meter long were dinosaurs. Many were agile and could run rapidly; they had special muscles that enabled the lungs to be filled and emptied while the limbs moved. We can infer the existence of such muscles in extinct dinosaurs from the structure of the vertebral column in fossils. Some of the largest dinosaurs weighed as much as 70,000 kilograms.

Modern *crocodilians*—crocodiles, caimans, gharials, and alligators—are confined to tropical and warm temperate environments (**FIGURE 23.47A**). Crocodilians spend much of their time in water, but they build nests on land or on floating piles of vegetation. The eggs are warmed by heat generated by decaying organic matter that the female places in the nest. Typically, the female guards the eggs until hatching, and she often facilitates hatching. In some species, the female continues to guard and communicate with her offspring after they hatch. All crocodilians are carnivorous. They eat vertebrates of all kinds, including large mammals.

Biologists have long accepted the phylogenetic position of birds among the reptiles, although birds clearly have many unique, derived morphological features. In addition to the strong morphological evidence for the placement of birds among the reptiles, fossil and molecular data emerging over the last few decades have provided definitive supporting evidence. Birds are thought to have emerged among the *theropods*, a group of predatory dinosaurs that share such traits as bipedal stance, hollow bones, a *furcula* ("wishbone"), elongated

metatarsals with three-fingered feet, elongated forelimbs with three-fingered hands, and a pelvis that points backward.

The living bird species fall into two major groups that diverged during the late Cretaceous, about 80 to 90 million years ago, from a flying ancestor. The few modern descendants of one lineage include a group of secondarily flightless and weakly flying birds, some of which are very large. This group, called the *palaeognaths*, includes the South and Central American tinamous and several large, flightless birds of the southern continents; these include the rheas (**FIGURE 23.47B**), emu, kiwis, cassowaries, and the world's largest bird, the ostrich. The second lineage, the *neognaths*, has left a much larger number of descendants, most of which have retained the ability to fly.

The evolution of feathers allowed birds to fly

During the Mesozoic era, about 175 million years ago, a lineage of theropods gave rise to the birds. Recent fossil discoveries show that the scales of some theropod dinosaurs were modified to form *feathers*. The feathers of many of these dinosaurs were structurally similar to those of modern birds (**FIGURE 23.48**).

The oldest known avian fossil, *Archaeopteryx*, which lived about 150 million years ago, had teeth, but it was covered with feathers that are virtually identical to those of modern birds (see Figure 23.48B). It also had well-developed wings, a long tail, and a furcula to which some of the flight muscles were probably attached. *Archaeopteryx* had clawed fingers on its forelimbs, but it also had typical perching bird claws on its hindlimbs. It probably lived in trees and shrubs and used the fingers to assist it in clambering over branches.

The evolution of feathers was a major force for diversification. Feathers are lightweight but are strong and structurally complex. The large quills of the flight feathers on the wings arise from the skin of the forelimbs to create the flying surfaces. Other strong feathers sprout like a fan from the shortened tail and serve as stabilizers during flight. The feathers that cover the body, along with an underlying layer of down feathers, provide birds with insulation that helps them survive in virtually all of Earth's climates.

The bones of theropod dinosaurs, including birds, are hollow with internal struts that increase their strength. Hollow bones would have made early theropods lighter and more mobile; later they facilitated the evolution of flight. The sternum (breastbone) of flying birds forms a large, vertical keel to which the flight muscles are attached.

Flight is metabolically expensive. A flying bird consumes energy at a rate 15 to 20 times faster than a running lizard of

(A)

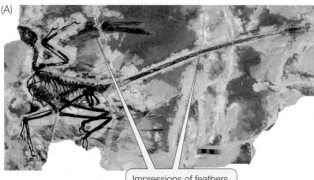

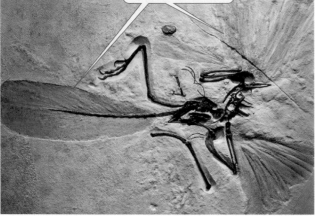

Impressions of feathers can be seen around the fossilized skeletons.

(B)

FIGURE 23.48 Mesozoic Bird Relatives The fossil record supports the evolution of birds from other dinosaurs. (A) *Microraptor gui* was a feathered dinosaur from the early Cretaceous (about 140 mya). (B) Dating from roughly the same time frame, *Archaeopteryx* is the oldest known birdlike fossil.

and other vertebrates. By eating the fruits and seeds of plants, birds serve as major agents of seed dispersal. **FIGURE 23.49** shows representatives of a few of the major groups of birds.

Mammals radiated as nonavian dinosaurs declined in diversity

Small and medium-sized **mammals** coexisted with the dinosaurs for millions of years. After the nonavian dinosaurs disappeared during the mass extinction at the close of the Mesozoic era, mammals increased dramatically in numbers, diversity, and size. Today mammals range in size from tiny shrews and bats weighing only about 2 grams to the blue whale, the largest animal on Earth, which measures up to 33 meters long and can weigh as much as 160,000 kilograms. Mammals have far fewer, but more highly differentiated, teeth than do fishes, amphibians, or reptiles. Differences among mammals in the number, type, and arrangement of teeth reflect their varied diets.

the same weight. Because birds have such high metabolic rates, they generate large amounts of heat. They control the rate of heat loss using their feathers, which may be held close to the body or elevated to alter the amount of insulation they provide (see Concept 29.4). The lungs of birds allow air to flow through unidirectionally rather than by pumping air in and out (see Figure 37.5). This flow-through structure of the lungs increases the efficiency of gas exchange and thereby supports an increased metabolic rate.

There are about 10,000 species of birds alive today. They range in size from the 150-kilogram ostrich to a hummingbird weighing only 2 grams. The teeth—so prominent among other dinosaurs—were secondarily lost in the ancestral birds, but birds nonetheless eat almost all types of animal and plant material. Insects and fruits are the most important dietary items for terrestrial species. Birds also eat seeds, nectar and pollen, leaves and buds, carrion, and fish

(B) *Megascops asio*

(A) *Balearica regulorum gibbericeps*

FIGURE 23.49 Some Diverse Birds (A) The East African crowned crane feeds on insects, small reptiles, and even small mammals. Like other crane species, its mating and courtship behavior includes an elaborate dance display. (B) Owls such as this screech owl are nighttime predators that can locate prey using their sensitive auditory systems. (C) Perching, or passeriform, birds such as this European goldfinch comprise the most species-rich of all bird groups.

(C) *Carduelis carduelis*

(A) *Ornithorhynchus anatinus*

(B) *Didelphis virginiana*

(C) *Spermophilus mexicanus*

(E)

(F) *Megatera novaeangliae*

(D) *Macroderma gigas*

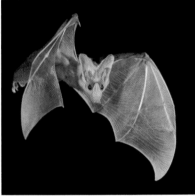

Four key features distinguish the mammals:

- *Sweat glands,* which secrete sweat that evaporates and thereby cools an animal (see Concept 29.4).

- *Mammary glands*, which in females secrete a nutritive fluid (milk) on which newborn individuals feed.

- *Hair*, which provides a protective and insulating covering.

- A *four-chambered heart* that completely separates the oxygenated blood coming from the lungs from the deoxygenated blood returning from the body (this last characteristic is convergent with the archosaurs, including modern birds and crocodiles; see Concept 38.2).

Mammalian eggs are fertilized within the female's body, and the embryos undergo a period of development in the female's body in an organ called the *uterus* prior to being born. Most mammals have a covering of hair (fur), which is luxuriant in some species but has been greatly reduced in others, including the cetaceans (whales and dolphins) and humans. Thick layers of insulating fat (blubber) replace hair as a heat-retention mechanism in the cetaceans; humans learned to use clothing for this purpose when they dispersed from warm tropical areas.

FIGURE 23.50 Mammalian Diversity (A) The duck-billed platypus is one of the five surviving species of prototherians—mammals that lay eggs. (B) Female marsupial mammals have a ventral pouch in which they nurture their offspring, which are extremely small at birth. The young of this opossum have grown large enough to leave her pouch. (C–F) Eutherian mammals. (C) Almost half of all eutherians are rodents, such as this Mexican ground squirrel. (D) Flight evolved in the common ancestor of bats. Virtually all bat species are nocturnal. (E) Many large mammals are important herbivores in terrestrial environments. Nowhere are these assemblages more spectacular than on the grass and brushlands of the African continent. (F) Humpback whales are cetaceans, a cetartiodactyl group that returned to the marine environment. A mother and calf are seen here.

The approximately 5,000 species of living mammals are divided into two primary groups: the *prototherians* and the *therians*. Only five species of prototherians are known, and they are found only in Australia and New Guinea. These mammals, the duck-billed platypus and four species of echidnas, differ from other mammals in lacking a placenta, laying eggs, and having sprawling legs (**FIGURE 23.50A**). Prototherians supply milk for their young, but they have no nipples on their mammary glands; the milk simply oozes out and is lapped off the fur by the offspring.

TABLE 23.3	Major Groups of Living Eutherian Mammals	
GROUP	**APPROXIMATE NUMBER OF LIVING SPECIES**	**EXAMPLES**
Gnawing mammals (Rodentia)	2,300	Rats, mice, squirrels, woodchucks, ground squirrels, beaver, capybara
Flying mammals (Chiroptera)	1,100	Bats
Soricomorph insectivores (Soricomorpha)	430	Shrews, moles
Even-toed hoofed mammals and cetaceans (Cetartiodactyla)	320	Deer, sheep, goats, cattle, antelopes, giraffes, camels, swine, hippopotamus, whales, dolphins
Carnivores (Carnivora)	290	Wolves, dogs, bears, cats, weasels, pinnipeds (seals, sea lions, walruses)
Primates (Primates)	235	Lemurs, monkeys, apes, humans
Lagomorphs (Lagomorpha)	80	Rabbits, hares, pikas
African insectivores (Afrosoricida)	50	Tenrecs, golden moles
Spiny insectivores (Erinaceomorpha)	24	Hedgehogs
Armored mammals (Cingulata)	21	Armadillos
Tree shrews (Scandentia)	20	Tree shrews
Odd-toed hoofed mammals (Perissodactyla)	20	Horses, zebras, tapirs, rhinoceroses
Long-nosed insectivores (Macroscelidea)	16	Elephant shrews
Pilosans (Pilosa)	10	Anteaters, sloths
Pholidotans (Pholidota)	8	Pangolins
Sirenians (Sirenia)	5	Manatees, dugongs
Hyracoids (Hyracoidea)	4	Hyraxes, dassies
Elephants (Proboscidea)	3	African and Indian elephants
Dermopterans (Dermoptera)	2	Flying lemurs
Aardvark (Tubulidentata)	1	Aardvark

Most mammals are viviparous

Members of the viviparous *therian* clade are further divided into the *marsupials* and the *eutherians*. Females of most marsupial species have a ventral pouch in which they carry and feed their offspring (**FIGURE 23.50B**). Gestation (pregnancy) in marsupials is brief; the young are born tiny but with well-developed forelimbs, with which they climb to the pouch. They attach to a nipple but cannot suck. The mother ejects milk into the tiny offspring until they grow large enough to suckle. Once her offspring have left the uterus, a female marsupial may become sexually receptive again. She can then carry fertilized eggs that are capable of initiating development and can replace the offspring in her pouch should something happen to them.

Eutherians include the majority of mammals. Eutherians are sometimes called *placental mammals,* but this name is inappropriate because some marsupials also have placentas. Eutherians are more developed at birth than are marsupials; no external pouch houses them after they are born.

The approximately 5,000 living species of morphologically diverse eutherians are divided into 20 major groups (**TABLE 23.3**). The largest group is the rodents (**FIGURE 23.50C**), with about 2,300 species. Rodents are traditionally defined by the unique morphology of their teeth, which are adapted for gnawing through substances such as wood. The next largest group comprises the approximately 1,100 bat species (**FIGURE 23.50D**)—the flying mammals. The bats are followed by the

moles and shrews, with about 430 species. The relationships of the major groups of eutherians to one another have been difficult to determine because most of the major groups diverged in a short period of time during an explosive adaptive radiation.

Grazing and *browsing* by members of several eutherian groups helped transform the terrestrial landscape. Herds of grazing herbivores feed on open grasslands, whereas browsers feed on shrubs and trees. The effects of herbivores on plant life favored the evolution of the spines, tough leaves, and difficult-to-eat growth forms found in many plants. In turn, adaptations to the teeth and digestive systems of many herbivore lineages allowed these species to consume many plants despite such defenses—a striking example of coevolution. A large animal can survive on food of lower quality than a small animal can, and large size evolved in several groups of grazing and browsing mammals (**FIGURE 23.50E**). The evolution of large herbivores, in turn, favored the evolution of large carnivores able to attack and overpower them.

Several lineages of terrestrial eutherians subsequently returned to the aquatic environments their ancestors had left behind (**FIGURE 23.50F**). The completely aquatic cetaceans—whales and dolphins—evolved from artiodactyl ancestors (whales are closely related to the hippopotamuses). The seals, sea lions, and walruses also returned to the marine environment, and their limbs became modified into flippers. Weasel-like otters retain their limbs but have also returned to aquatic environments, colonizing both fresh and salt water. The manatees and dugongs colonized estuaries and shallow seas.

Do You Understand Concept 23.6?

- Many amphibians have a biphasic life cycle that includes an aquatic larva and a terrestrial adult. What are some exceptions to this common pattern?

- In the not-too-distant past, the idea that birds were reptiles met with skepticism. Explain how fossils, morphology, and molecular evidence now support the position of birds among the reptiles.

- Contrast the reproductive modes of prototherians, marsupials, and eutherians.

The evolutionary history of one eutherian group—the primates—is of special interest to us because it includes the human lineage. The primates have been the subject of extensive research in most aspects of their biology, including behavior, ecology, physiology, and molecular biology.

concept 23.7 Humans Evolved among the Primates

The **primates** underwent extensive evolutionary radiation from an ancestral small, arboreal, insectivorous mammal. Grasping limbs with opposable digits are one of the major adaptations to arboreal life that distinguish primates from other mammals.

Early in their evolutionary history, the primates split into two main clades, the prosimians and the anthropoids (**FIGURE 23.51**). *Prosimians*—lemurs, lorises, and their close relatives—once lived on all continents, but today they are restricted to Africa, Madagascar, and tropical Asia. A second primate lineage, the *anthropoids*—tarsiers, New World monkeys, Old World monkeys, and apes—evolved about 65 million years ago in Africa or Asia. New World monkeys diverged from Old World monkeys and apes at a slightly later date, but at an early enough date that they might have reached South America from Africa when those two continents were still close to each other. All New World monkeys are arboreal, and many have long, prehensile tails with which they can grasp branches. Many Old World monkeys are arboreal as well, but a number of species are terrestrial. No Old World primate has a prehensile (grasping) tail.

About 35 million years ago, a lineage that led to the apes separated from the Old World monkeys. Between 22 and 5.5 million years ago, dozens of species of apes lived in Europe, Asia, and Africa. The Asian apes (gibbons and orangutans), African apes (gorillas and chimpanzees), and humans are their modern descendants.

Human ancestors evolved bipedal locomotion

About 6 million years ago in Africa, a lineage split occurred that would lead to the chimpanzees on the one hand and to the *hominid* clade on the other. It is the latter that includes modern humans and their extinct close relatives.

The earliest known protohominids, the *ardipithecines*, had distinct morphological adaptations for *bipedal locomotion* (walking on two legs). Bipedal locomotion frees the forelimbs to manipulate objects and to carry them while walking. It also elevates the eyes, enabling the animal to see over tall vegetation to spot predators and prey. Bipedal locomotion is also energetically more economical than quadrupedal locomotion. All three advantages were probably important for the ardipithecines and their descendants, the australopithecines.

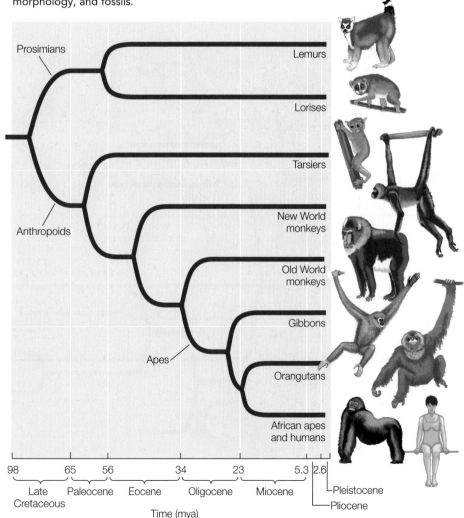

FIGURE 23.51 Phylogeny of the Primates The phylogeny of primates is among the best studied of any major group of mammals. This tree is based on evidence from many genes, morphology, and fossils.

The first *australopithecine* skull was found in South Africa in 1924. Since then australopithecine fossils have been found at many African sites. The most complete fossil skeleton of an australopithecine yet found was discovered in Ethiopia in 1974. The skeleton was approximately 3.5 million years old and was that of a young female who has since become known to the world as "Lucy." Lucy was assigned to the species *Australopithecus afarensis*, and her discovery captured worldwide interest. Fossil remains of more than a hundred *A. afarensis* individuals have since been discovered, and recent discoveries have unearthed fossils of other australopithecines that lived in Africa 4–5 million years ago.

Experts disagree over how many species are represented by australopithecine fossils, but it is clear that multiple species of hominids lived together over much of eastern Africa several million years ago (**FIGURE 23.52**). A lineage of larger species (weighing about 40 kilograms) is represented by *Paranthropus robustus* and *P. boisei*, both of which died out between 1 and 1.5 million years ago. Members of a smaller lineage of australopithecines gave rise to the genus *Homo*.

Early members of the genus *Homo* lived contemporaneously with *Paranthropus* in Africa for about a million years. Some 2-million-year-old fossils of an extinct species called *H. habilis* were discovered in the Olduvai Gorge, Tanzania. Other fossils of *H. habilis* have been found in Kenya and Ethiopia.

Associated with the fossils are tools that these early hominids used to obtain food.

Another extinct hominid species, *Homo erectus*, evolved in Africa about 1.6 million years ago. Soon thereafter it had spread as far as eastern Asia, becoming the first hominid to leave Africa. Members of *H. erectus* were nearly as large as modern people, but their brains were smaller and they had comparatively thick skulls. *Homo erectus* used fire for cooking and for hunting large animals, and made characteristic stone tools that have been found in many parts of the Old World. Populations of *H. erectus* survived until at least 250,000 years ago, although more recent fossils may also be attributable to this species. In 2004 some 18,000-year-old fossil remains of a small *Homo* were found on the island of Flores in Indonesia. Since then, numerous additional fossils of this diminutive hominid have been found on Flores, dating from 95,000 to 17,000 years ago. Many anthropologists think that this small species, named *H. floresiensis*, was most closely related to *H. erectus*.

FIGURE 23.52 A Phylogenetic Tree of Hominids At times in the past, more than one species of hominid lived on Earth at the same time. Originating in Africa, hominids spread to Europe and Asia multiple times. All but one of these closely related species are now extinct; modern *Homo sapiens* have colonized nearly every corner of the planet.

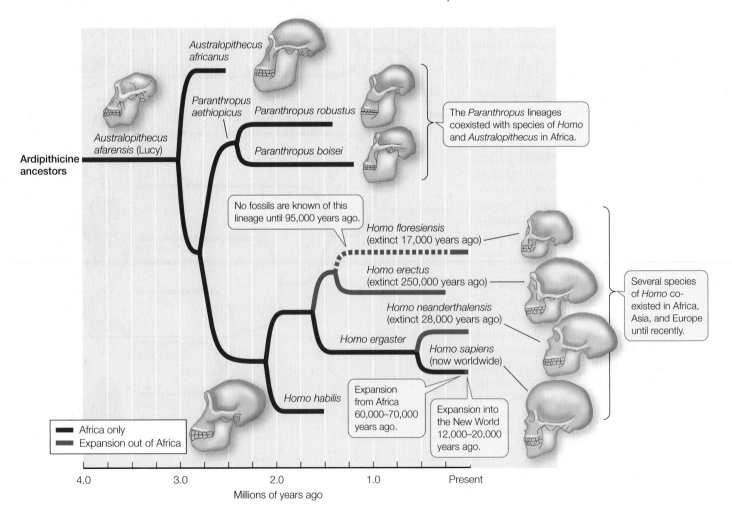

Africa only
Expansion out of Africa

4.0   3.0   2.0   1.0   Present
Millions of years ago

Human brains became larger as jaws became smaller

In the hominid lineage leading to *Homo sapiens* and *H. neanderthalensis*, the brain increased rapidly in size. At the same time, the powerful jaw muscles of our ancestors dramatically decreased in size. These two changes were simultaneous, suggesting they might have been functionally correlated. A mutation in a myosin gene that is expressed in jaw muscles may have removed a barrier that had previously prevented this remodeling of the human cranium.

The striking enlargement of the brain relative to body size in the hominid lineage was probably favored by an increasingly complex social life. Any features that allowed group members to communicate more effectively with one another would have been valuable in cooperative hunting and gathering and for improving one's status in the complex social interactions that must have characterized early human societies, just as they do in ours today.

Several *Homo* species coexisted during the mid-Pleistocene epoch, from about 1.5 million to about 250,000 years ago. All were skilled hunters of large mammals, but plants were important components of their diets. During this period another distinctly human trait emerged: rituals and a concept of life after death. Deceased individuals were buried with tools and clothing, supplies for their presumed existence in the next world.

One species, *Homo neanderthalensis*, was widespread in Europe and Asia between about 500,000 and 28,000 years ago. Neanderthals were short, stocky, and powerfully built. Their massive skull housed a brain somewhat larger than our own. They manufactured a variety of tools and hunted large mammals, which they probably ambushed and subdued in close combat. Early modern humans (*H. sapiens*) expanded out of Africa between 70,000 and 60,000 years ago. Then about 35,000 years ago, *H. sapiens* moved into the range of *H. neanderthalensis* in Europe and western Asia, so the two species must have interacted with one another. Neanderthals abruptly disappeared about 28,000 years ago. Many anthropologists believe it is likely that the Neanderthals were exterminated by these early modern humans. Scientists have been able to isolate large parts of the genome of *H. neanderthalensis* from recent fossils and to compare it with our own. These studies suggest that a small amount of interbreeding between the two species occurred shortly after *H. sapiens* expanded out of Africa.

Early modern humans made and used a variety of sophisticated tools. They created the remarkable paintings of large mammals, many of them showing scenes of hunting, found in European caves. The animals depicted were characteristic of the cold steppes and grasslands that occupied much of Europe during periods of glacial expansion. Early modern humans also spread across Asia, reaching North America perhaps as early as 20,000 years ago, although the date of their arrival in the New World is still uncertain. Within a few thousand years, they had spread southward through North America to the southern tip of South America.

Do You Understand Concept 23.7?

- Describe the differences between Old World and New World monkeys.
- What leads biologists to conclude that the evolution of brain size and jaw size may be functionally linked?
- If interbreeding between *Homo sapiens* and *H. neanderthalensis* took place after *H. sapiens* expanded out of Africa, what would you predict about the distribution of Neanderthal genes in modern human populations?

 QUESTION Besides the insects, which other groups of animals are thought to contain many more species than are known at present?

ANSWER It is perhaps easier to list the groups of animals for which a nearly complete inventory of living species has been completed than to list all the groups for which many new species remain to be described. Among the insects, the best-studied group in terms of species is the butterflies, which are widely collected and studied. There are still many species of other lepidopterans (such as moths), however, remaining to be discovered. Most other major insect groups contain many undescribed species. Among the vertebrates, most species of birds and mammals have been described, although several new species of mammals are still discovered and named each year. Species discovery and description remains high for almost all other major groups of animals.

After insects, and perhaps even rivaling the insects in undiscovered diversity, are the nematodes. Although known nematode diversity is only about 1/40th of the known insect diversity (in terms of number of described species), the taxonomy of nematodes is much more poorly studied than that of insects. Some biologists think there are likely to be species-specific parasitic nematodes of most other species of multicellular organisms. If so, then there may be as many species of nematodes as there are of plants, fungi, and other animals combined.

Most of the other diverse groups of animals contain many as yet undetected species, judging from the rate of new species descriptions. In particular, flatworms (especially the parasitic flukes and tapeworms), marine annelids, mollusks, crustaceans, myriapods, and chelicerates all contain large numbers of undescribed species. Among the vertebrates, the rate of discovery of new species remains particularly high for the fishes and amphibians.

SUMMARY

concept 23.1 Distinct Body Plans Evolved among the Animals

- Animals share a set of derived traits not found in other groups of organisms, including similarities in the sequences of many of their genes, the structure of their cell junctions, and the components of their extracellular matrix. **Review Figure 23.1 and WORKING WITH DATA 23.1**

- Patterns of embryonic development provide clues to the evolutionary relationships among animals. **Diploblastic** animals develop two embryonic cell layers; **triploblastic** animals develop three cell layers.

- Differences in their patterns of early development characterize two major clades of triploblastic animals, the **protostomes** and the **deuterostomes**.

- Animal **body plans** can be described in terms of **symmetry**, **body cavity** structure, **segmentation**, and type of appendages.

- Most animals have either **radial symmetry** or **bilateral symmetry**. Many bilaterally symmetrical animals exhibit **cephalization**, with sensory and nervous tissues in an anterior head.

- On the basis of their body cavity structure, animals can be described as **acoelomates**, **pseudocoelomates**, or **coelomates**. **Review Figure 23.3 and WEB ACTIVITY 23.1**

- Segmentation takes many forms and improves control of movement, especially if the animal also has appendages.

concept 23.2 Some Animal Groups Fall Outside the Bilateria

- All animals other than **sponges**, **placozoans**, **ctenophores**, and **cnidarians** belong to a large, monophyletic group called the **Bilateria**. **Eumetazoans**, which have tissues organized into distinct organs, include all animals other than sponges and placozoans. **See WEB ACTIVITY 23.2**

- Sponges and placozoans are simple, asymmetrical animals that lack differentiated cell layers and true organs. Sponges have skeletons made up of silicaceous or calcareous **spicules**. They create water currents and capture food with flagellated feeding cells called choanocytes. Choanocytes are an evolutionary link between the animals and the choanoflagellate protists.

- The two cell layers of the radially symmetrical ctenophores are separated by an inert extracellular matrix called **mesoglea**. Ctenophores move by beating fused plates of cilia called **ctenes**. **Review Figure 23.7**

- The life cycle of most cnidarians has two distinct stages: a sessile **polyp** stage and a motile **medusa**. A fertilized egg develops into a free-swimming larval **planula**, which settles to the bottom and develops into a polyp. **Review Figure 23.8 and ANIMATED TUTORIAL 23.1**

concept 23.3 There Are Two Major Groups of Protostomes

- Protostomes ("mouth first") are bilaterally symmetrical animals that have an anterior brain surrounding the entrance to the digestive tract and a ventral nervous system. Protostomes comprise two major clades, the lophotrochozoans and the ecdysozoans. **See WEB ACTIVITIES 23.3 and 23.4**

- **Lophotrochozoans** include a wide diversity of animals. Within this group evolved **lophophores** (a complex organ for food collection and gas exchange) and free-living **trochophore** larvae. These features were subsequently lost in some lineages. **Arrow worms** may be related to lophotrochozoans, or they may be the sister group to all other protostomes.

- Lophophores, wormlike body forms, and external shells are each found in many distantly related groups of lophotrochozoans. The most species-rich groups of lophotrochozoans are the **flatworms**, **annelids**, and **mollusks**.

- Annelids are a diverse group of segmented worms that live in moist terrestrial and aquatic environments. **Review Figure 23.16**

- Mollusks underwent a dramatic evolutionary radiation based on a body plan consisting of three major components: a foot, a mantle, and a visceral mass. The four major living molluscan clades—**chitons**, **bivalves**, **gastropods**, and **cephalopods**—demonstrate the diversity that evolved from this three-part body plan. **Review Figure 23.18**

- **Ecdysozoans** have a **cuticle** covering their body, which they must **molt** in order to grow. Some ecdysozoans, notably the **arthropods**, have a rigid cuticle reinforced with **chitin** that functions as an **exoskeleton**. New mechanisms of locomotion and gas exchange evolved among the arthropods.

- **Nematodes**, or roundworms, have a thick, multilayered cuticle. Nematodes are among the most abundant and universally distributed of all animal groups. **Review Figure 23.21**

- **Horsehair worms** are extremely thin; their larvae are internal parasites.

- Many ecdysozoan groups are wormlike in form. Members of several species-poor groups of wormlike marine ecdysozoans—**priapulids**, **kinorhynchs**, and **loriciferans**—have thin cuticles.

- One major ecdysozoan clade, the arthropods, has evolved jointed, paired appendages that have a wide diversity of functions.

concept 23.4 Arthropods Are Diverse and Abundant Animals

- Arthropods are the dominant animals on Earth in number of described species, and among the most abundant in number of individuals.

- Encasement within a rigid exoskeleton provides arthropods with support for walking as well as some protection from predators. The waterproofing provided by chitin keeps arthropods from dehydrating in dry air.

- Jointed appendages permit complex movements. Each arthropod segment has muscles attached to the inside of the exoskeleton that operate that segment and the appendages attached to it.

- Two groups of arthropod relatives, the **onychophorans** and the **tardigrades**, have simple, unjointed appendages.

- **Chelicerates** have a two-part body and pointed mouthparts than grasp prey; most chelicerates have four pairs of walking legs.

- Mandibles and antennae are synapomorphies of the **mandibulates**, which include **myriapods**, **crustaceans**, and **hexapods**.

- **Crustaceans** are the dominant marine arthropods and are also found in many freshwater and some terrestrial environments. Their segmented bodies are divided into three regions (head, thorax, and abdomen) with different, specialized appendages in each region.

- Hexapods—insects and their relatives—are the dominant terrestrial arthropods. They have the same three body regions as crustaceans, but no appendages form in their abdominal segments. Wings and the ability to fly first evolved among the insects, allowing them to exploit new lifestyles.

concept 23.5 Deuterostomes Include Echinoderms, Hemichordates, and Chordates

- Deuterostomes vary greatly in adult form, but based on the distinctive patterns of early development they share and on phylogenetic analyses of gene sequences, they represent a monophyletic group. There are far fewer species of deuterostomes than of protostomes, but many deuterostomes are large and ecologically important. **See WEB ACTIVITY 23.5**

- Echinoderms and hemichordates both have bilaterally symmetrical, ciliated larvae.

- Most adult echinoderms have **pentaradial symmetry**. Echinoderms have an internal skeleton of calcified plates and a unique **water vascular system** connected to extensions called **tube feet**. Review Figure 23.31

- Hemichordate adults are bilaterally symmetrical and have a three-part body that is divided into a proboscis, collar, and trunk. They include the acorn worms and the pterobranchs. Review Figure 23.33

- Chordates fall into three principal subgroups: **cephalochordates**, **urochordates**, and **vertebrates**.

- At some stage in their development, all chordates have a dorsal hollow nerve cord, a postanal tail, and a **notochord**. Review Figure 23.34

- Urochordates include the ascidians (sea squirts), which are sessile filter feeders as adults. Cephalochordates are the lancelets, which live buried in the sand of shallow marine and brackish waters.

- The vertebrate body is characterized by a rigid internal skeleton, which is supported by a **vertebral column** that replaces the notochord, internal organs suspended in a coelom, a ventral heart, and an anterior skull with a large brain. Review Figure 23.38

- The evolution of jaws from gill arches enabled individuals to grasp large prey and, together with teeth, cut them into small pieces.

- **Chondrichthyans** have skeletons of cartilage; almost all species are marine. The skeletons of **ray-finned fishes** are made of bone; these fishes have colonized most aquatic environments.

concept 23.6 Life on Land Contributed to Vertebrate Diversification

- Lungs and jointed appendages enabled vertebrates to colonize the land. The earliest split in the **tetrapod** tree is between the **amphibians** and the **amniotes** (reptiles and mammals).

- Most modern amphibians are confined to moist environments because they and their eggs lose water rapidly. **See ANIMATED TUTORIAL 23.2**

- An impermeable skin, efficient kidneys, and an egg that could resist desiccation evolved in the amniotes. **See WEB ACTIVITY 23.6**

- The major living **reptile** groups are the turtles, the lepidosaurs (tuataras, lizards, snakes, and amphisbaenas), and the archosaurs (crocodilians and birds). Review Figure 23.45

- **Mammals** are unique among animals in supplying their young with a nutritive fluid (milk) secreted by mammary glands. There are two primary mammalian clades: the prototherians (of which there are only five species) and the species-rich therians. The therian clade is subdivided into the marsupials and the eutherians. Review Table 23.3

concept 23.7 Humans Evolved among the Primates

- Grasping limbs with opposable digits distinguish **primates** from other mammals. The prosimian clade includes the lemurs and lorises; the anthropoid clade includes monkeys, apes, and humans. Review Figure 23.51

- Hominid ancestors developed efficient bipedal locomotion. In the lineage leading to *Homo*, brains became larger as jaws became smaller; the two events may have been developmentally linked. Several species of *Homo* coexisted in parts of the world until recently. Review Figure 23.52

Plant Form and Function

The Plant Body

People use plants and substances made from them not only for food, but also for fiber, fuel, shelter, and medicines. The clothes you are wearing probably contain cotton, and the paper this book is printed on comes from wood pulp. About 5,000 years ago, Egyptians found that when stems of the papyrus plant (*Cyperus papyrus*) were peeled and dried, they could be formed into sheets that were easier to write on and store than parchment made from animal skins. About 3,000 years later, the Chinese invented a new method of papermaking that persists today: they mixed fibrous plant materials with water, mashed the mixture into a pulp, and pressed it into sheets, which they then dried.

With the advent of computers, economists predicted a lower demand for paper and therefore for the trees that provide the wood pulp used in papermaking. But this has not happened. Every year, U.S. residents use about 500 kilograms of paper per person, and paper constitutes about one-third of all solid waste. In both Europe and the United States, 90 percent of the forest cover that existed when papermaking was invented has disappeared because of the demand for fiber for paper as well as wood for energy production, heating, and construction. To regenerate these forests and to provide a sustainable resource for papermaking, there is considerable interest in growing fibrous crops that can substitute for trees.

Kenaf (*Hibiscus cannabinus*) is an angiosperm that grows readily and rapidly in Asia and parts of Africa. Although it does not produce wood, kenaf has stems rich in fiber. It has been cultivated for

three millennia in Asia, where its fiber is used to make bags, rope, dry bedding for animals, and even sails. There is now interest in using kenaf fiber for papermaking. Compared with southern pine trees, the current main source of pulp for paper, kenaf grows more rapidly (it matures in months rather than years), produces more fiber per hectare of land, and requires less water and fewer chemicals when used in paper production. Newspapers and greeting cards made from kenaf paper are stronger, yet thinner, than products made from typical wood pulp. The future of this plant as a resource for papermaking looks bright.

Kenaf is being grown as an alternative source of pulp for paper production

 QUESTION

What are the properties of the kenaf plant that make it suitable for papermaking?*

*You will find the answer to this question on page 519.

KEY CONCEPTS

24.1 The Plant Body Is Organized and Constructed in a Distinctive Way

24.2 Meristems Build Roots, Stems, and Leaves

24.3 Domestication Has Altered Plant Form

The Plant Body Is Organized and Constructed in a Distinctive Way

Plants must harvest energy from sunlight and mineral nutrients from the soil. Because they are stationary (*sessile*), they cannot move to find a more favorable environment or to avoid predators. Their body plan and physiology allow plants to respond to these challenges:

• Stems, leaves, and roots have structural adaptations that enable plants anchored to one spot to capture scarce resources effectively, both above and below the ground.

• Their ability to grow throughout their lifetimes enables plants to respond to environmental cues by redirecting their growth to exploit environmental opportunities, as when roots grow toward a water supply.

In Chapter 21 we see how modern plants arose from aquatic ancestors, giving rise to simple land plants and then vascular plants. Despite their obvious differences in size and form, all vascular plants have essentially the same simple structural organization. This chapter describes the basic architecture of the largest group of vascular plants, the angiosperms (flowering plants), and shows how so much diversity (there are more than 250,000 species of angiosperms) can literally grow out of such a simple basic form. In this chapter we focus on the angiosperms' three kinds of *vegetative* (nonsexual) organs: roots, stems, and leaves.

Plant organs are organized into two systems (**FIGURE 24.1**):

• The **root system** anchors the plant in place. **Roots** absorb water and dissolved minerals and store the products of photosynthesis. The extreme branching of roots and their high surface area-to-volume ratios allow them to absorb water and mineral nutrients from the soil efficiently.

• The **shoot system** consists of the stems, leaves, and flowers. The **leaves** are the chief organs of photosynthesis. The **stems** hold and display the leaves to the sun and provide connections for the transport of materials between roots and leaves. The flowers are reproductive organs.

The leaves arise from *nodes* on the stem, and the repeating unit between the nodes is called a *phytomer*.

As noted in Chapter 21, most angiosperms belong to one of two major clades. *Monocots* are generally narrow-leaved plants such as grasses, lilies, orchids, and palms. *Eudicots* are generally broad-leaved plants such as soybeans, roses, sunflowers, and maples. These two clades, which account for 97 percent of angiosperm species, differ in several important structural characteristics, such as the organization of the root system and the structure of the leaves (see Figure 24.1).

Plants develop differently than animals

The four processes that govern the development of all organisms, whether plant or animal, are *determination* (the commitment of an embryonic cell to its ultimate fate in the organism); *differentiation* (the specialization of a cell); *morphogenesis* (the organization and spatial distribution of cells into tissues and organs); and *growth* (increase in body size). In plants, these processes are influenced by three unique properties: apical meristems, cell walls, and the totipotency of most cells.

LINK The processes of development are described in Concept 14.1

APICAL MERISTEMS Plants grow throughout their lifetimes. Whereas adult humans, for example, have stem cells to replace tissues lost through damage or apoptosis, plants have *apical meristems* that are "always embryonic," producing new roots, stems, leaves, and flowers throughout the plant's life. We will discuss apical meristems in more detail below and in Concept 24.2.

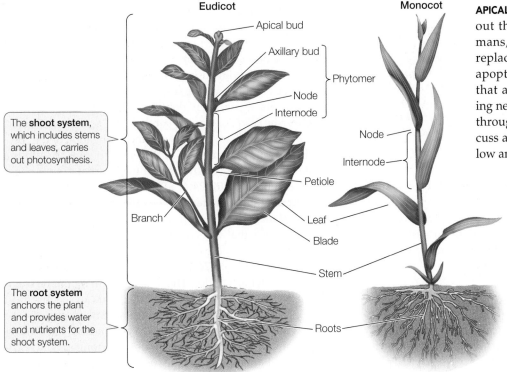

Eudicot

Apical bud

Axillary bud

Phytomer

Node

Internode

The **shoot system**, which includes stems and leaves, carries out photosynthesis.

Branch

Petiole

Leaf

Blade

Stem

The **root system** anchors the plant and provides water and nutrients for the shoot system.

Monocot

Node

Internode

Roots

FIGURE 24.1 Vegetative Plant Organs and Systems The basic plant body plan, with root and shoot systems, and the principal vegetative organs are similar in eudicots and monocots, although there are also some structural differences between the two clades.

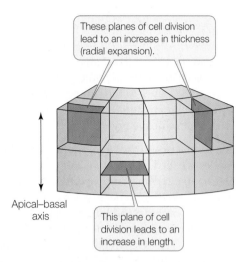

FIGURE 24.2 **Cytokinesis and Morphogenesis** The plane of cell division can determine the growth pattern of a plant organ, as in this section of a shoot.

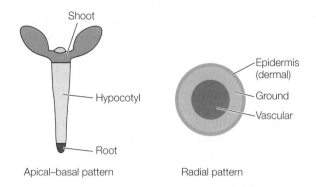

FIGURE 24.3 **Two Patterns for Plant Morphogenesis** (A) The apical–basal pattern is the arrangement of cells and tissues along the main axis from root to shoot. (B) The radial pattern determines the concentric arrangement of tissues when organs grow in thickness.

CELL WALLS Plant cells are surrounded by a cell wall. This rigid extracellular matrix makes it impossible for cells to move from place to place as they do in animal development. Instead, plant morphogenesis occurs through alterations in the plane of cell division at cytokinesis, which in turn change the direction in which a piece of tissue grows (**FIGURE 24.2**). In addition, unequal cell division can occur when the cytoplasm contains differentiation signals (cytoplasmic determinants) that are localized in one part of a cell (see Concept 14.2). Plant cytokinesis occurs along a *cell plate* laid down by Golgi vesicles (see Figure 7.7B). Unlike animal cells, in which the location of cytokinesis depends on the location of the middle of the mitotic spindle, the location of the plant cell plate is determined earlier—as early as mitotic prophase.

> **FRONTIERS** Because the plane of cell division is so important in plant morphogenesis, the mechanism by which a cell determines where the cell plate will form is under investigation. By screening for mutations leading to abnormal cell plate orientation in the model organism *Arabidopsis thaliana*, biologists have isolated several genes that are involved in determining the plane of cell division. How these genes work remains to be discovered.

TOTIPOTENCY In animals, only early embryonic cells are totipotent, although some stem cell populations, such as those in bone marrow, are multipotent. In contrast, in plants even differentiated cells are pluripotent, and most are totipotent (see Figure 14.3). This means a plant can readily repair damage wrought by the environment or herbivores.

Apical–basal polarity and radial symmetry are characteristic of the plant body

Two basic patterns are established early in plant embryogenesis (embryo formation) (**FIGURE 24.3**):

- The *apical–basal* axis: the arrangement of cells and tissues along the main axis from root to shoot
- The *radial* axis: the concentric arrangement of the *tissue systems* (which we will discuss in the next section)

Both axes are best understood in developmental terms. We focus here on embryogenesis in the model eudicot *Arabidopsis thaliana*, in which the process has been most intensively studied.

The first step in the formation of a plant embryo is a mitotic division of the zygote that gives rise to two daughter cells (**FIGURE 24.4, STEP 1**). An asymmetrical plane of cell division results in an uneven distribution of cytoplasm between these two cells, which face different fates. Signals in the smaller, apical (upper) daughter cell induce it to produce the embryo proper, while the other, larger daughter cell produces a supporting structure, the **suspensor** (**FIGURE 24.4, STEP 2**). This division not only establishes the apical–basal axis of the new plant but also determines its polarity (which end is the tip, or apex, and which is the base). A long, thin suspensor and a more spherical or globular embryo are distinguishable after just four mitotic divisions (see Figure 14.1). The suspensor soon ceases to elongate.

In eudicots, the initially globular embryo develops into the characteristic *heart stage* as the **cotyledons** ("seed leaves") start to grow (**FIGURE 24.4, STEP 3**). Further elongation of the cotyledons and of the main axis of the embryo gives rise to the *torpedo stage*, during which some of the internal tissues begin to differentiate (**FIGURE 24.4, STEP 4**). Between the cotyledons is the **shoot apical meristem**; at the other end of the axis is the **root apical meristem**. Each of these meristems contains undifferentiated cells that will continue to divide to give rise to the organs that will develop over the life of the plant.

As shown in Figure 24.4, step 2, the plant embryo is first a sphere and later a cylinder. The root and stem retain the cylindrical shape throughout the plant's life. You can see this most easily in the trunk of a tree. By the end of embryogenesis, the radial symmetry of the plant has been established. The embryonic plant contains three tissue systems, arranged concentrically, that will give rise to the tissues of the adult plant body.

FIGURE 24.4 Plant Embryogenesis The basic body plan of the model eudicot *Arabidopsis thaliana* is established in several steps. By the heart stage, the three tissue systems are established: the dermal (gold), ground (light green), and vascular (blue) tissue systems.

The plant body is constructed from three tissue systems

A *tissue* is an organized group of cells that have features in common and that work together as a structural and functional unit. In plants, tissues, in turn, are grouped into **tissue systems**. Despite their structural diversity, all vascular plants are constructed from three tissue systems: *dermal*, *vascular*, and *ground*. These three tissue systems are established during embryogenesis and ultimately extend throughout the plant body in a concentric arrangement (**FIGURE 24.5**). Each tissue system has distinct functions and is composed of different mixtures of cell types.

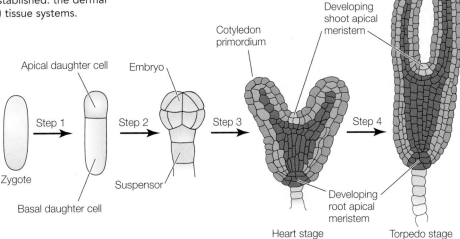

DERMAL TISSUE SYSTEM The **dermal tissue system** forms the *epidermis*, or outer covering, of a plant, which usually consists of a single cell layer. Epidermal cells are typically small. Some of them differentiate to form one of three specialized structures:

- *Stomata*, which are pores for gas exchange in leaves
- *Trichomes*, or leaf hairs, which provide protection against insects and damaging solar radiation

- *Root hairs*, which greatly increase root surface area, thus providing more surface for the uptake of water and mineral nutrients

Aboveground epidermal cells secrete a protective extracellular **cuticle** made of *cutin* (a polymer composed of long chains of fatty acids), a complex mixture of waxes, and cell wall polysaccharides. The cuticle limits water loss, reflects potentially damaging solar radiation, and serves as a barrier against pathogens.

The stems and roots of woody plants develop a dermal tissue called *periderm*. As we will see in Concept 24.2, this tissue ultimately forms *cork*, an external layer that is impermeable to water and gases.

GROUND TISSUE SYSTEM Virtually all the tissue lying between dermal tissue and vascular tissue in both shoots and roots is part of the **ground tissue system**, which therefore makes up most of the plant body. Ground tissue contains three cell types.

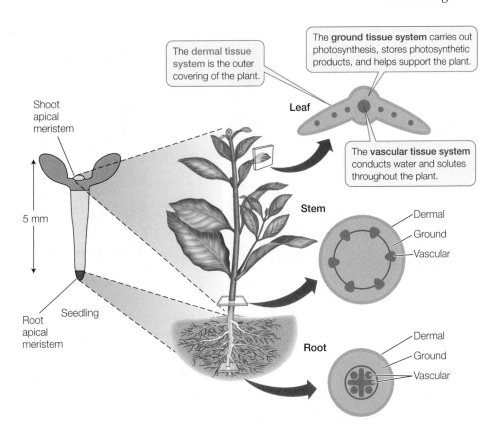

FIGURE 24.5 Three Tissue Systems Extend throughout the Plant Body The arrangement shown here is typical of eudicots, but the three tissue systems are continuous in the bodies of all vascular plants.

Parenchyma cells are the most abundant ground tissue cells. They have large vacuoles and relatively thin cell walls. They perform photosynthesis (in the shoot) and store protein (in fruits) and starch (in roots).

Parenchyma cells

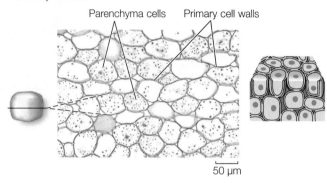

Parenchyma cells   Primary cell walls

50 μm

Collenchyma cells are elongated and have thick cell walls. They provide support for growing tissues such as stems. The familiar "strings" in celery consist primarily of collenchyma cells.

Collenchyma cells

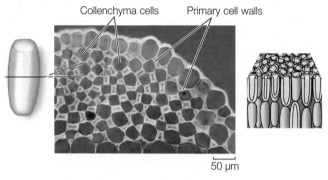

Collenchyma cells   Primary cell walls

50 μm

Sclerenchyma cells have very thick walls reinforced with the polyphenol polymer *lignin*. Most sclerenchyma cells undergo programmed cell death (apoptosis), but their strong cell walls remain to provide support for the plant. There are two types of sclerenchyma cells: fibers and sclereids. Elongated **fibers** provide relatively rigid support to wood and other parts of the plant, within which they are often organized into bundles.

Fibers

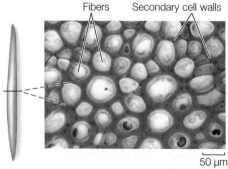

Fibers   Secondary cell walls

50 μm

The bark of trees owes much of its mechanical strength to long fibers. **Sclereids** occur in various shapes and may pack together densely, as in a nut's shell or in some seed coats.

Sclereids

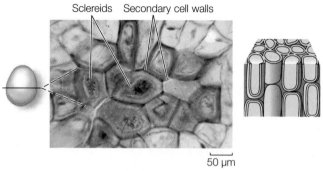

Sclereids   Secondary cell walls

50 μm

Isolated clumps of sclereids, called *stone cells*, in pears and some other fruits give them their characteristic gritty texture.

VASCULAR TISSUE SYSTEM   The **vascular tissue system** is the plant's plumbing or transport system—the distinguishing feature of vascular plants. Its two constituent tissues, the *xylem* and *phloem*, distribute materials throughout the plant.

Xylem distributes water and mineral ions taken up by the roots to all the cells of the roots, stems, and leaves. Like most sclerenchyma, xylem is made up of dead cells. It contains two types of conducting cells. **Tracheids** are spindle-shaped cells, with thinner regions in the cell wall called *pits* through which water can move with little resistance from one tracheid to its neighbors. **Vessel elements** are larger in diameter than tracheids. They meet end-to-end and partially break down their end walls, forming an open pipeline for water conduction.

Tracheids   Vessel elements

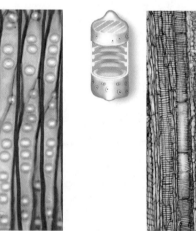

50 μm   100 μm

Xylem

Phloem generally consists of living cells. It transports carbohydrates (primarily sugars) from sites where they are produced by photosynthesis (called *sources*; primarily leaves) to sites where they are used or stored (called *sinks*, such as growing tissues, roots, and developing flowers). The characteristic cells of the phloem are **sieve tube elements**, which, like vessel elements, meet end-to-end, forming *sieve tubes*. Instead of the ends of the cells breaking down, however, the sieve tube elements are connected by plasmodesmata (see Figure 4.7), which form a set of pores called a *sieve plate*. Adjacent to the sieve tubes are metabolically more active *companion cells*.

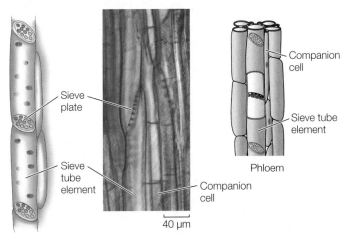

Sieve tube elements

Sieve plate

Sieve tube element

Sieve tube element

Companion cell

Companion cell

Sieve tube element

Phloem

40 μm

Xylem and phloem tissues may also contain fibers, which provide structural support.

Do You Understand Concept 24.1?

- Make a table listing the three tissue systems of plants and their functions.

- What are the similarities and differences between plant and animal development?

- What are the similarities and differences between a root and a shoot?

- A mutant strain of *Arabidopsis thaliana* lacks apical–basal polarity in the embryo. What would you expect the phenotype of the plant to be?

By the end of embryogenesis, the plant embryo is encased in a seed and is ready to germinate. We will return to seeds and their germination in subsequent chapters. For now, let's see how the cells and tissues we have just described allow the embryo to build an adult plant body.

concept 24.2 Meristems Build Roots, Stems, and Leaves

All plants experience **primary growth**, which is characterized by the lengthening of roots and shoots and by the proliferation of new roots and shoots through branching. Primary growth occurs throughout the life of the plant. In addition, many gymnosperms (e.g., pine trees) and eudicots, especially trees, experience **secondary growth**, by which they increase in thickness. Tissues produced by primary growth are referred to as *primary tissues* and comprise what is called the primary plant body. Tissues produced by secondary growth are referred to as *secondary tissues* and comprise the secondary plant body. A plant that

results entirely from primary growth consists of nonwoody tissues and is referred to as *herbaceous*. *Woody* plants, such as trees and shrubs, have, in addition to the primary plant body, a secondary plant body consisting of wood and bark.

As we discuss how vegetative plant organs develop and grow, you will find examples of organs whose functions deviate from the norm, such as roots and stems that are used to store water and nutrients, and leaves that are used for protection or support. The evolution of these novel roles for established organs is an example of natural selection working with what is already present, as well as an example of the interaction between evolution and development—the topic of evolutionary developmental biology (evo-devo).

LINK See Concepts 14.4 and 14.5 for a discussion of evolutionary developmental biology

A hierarchy of meristems generates the plant body

Even before seed germination, the plant embryo has two meristems: a shoot apical meristem at the end of the embryonic shoot and a root apical meristem near the end of the embryonic root (see Figure 24.4). The dividing cells of the meristems are comparable to animal stem cells: after their division, one daughter cell is capable of differentiating, and the other retains its undifferentiated phenotype.

In the adult plant, two types of meristems contribute to growth and development (**FIGURE 24.6**):

- **Apical meristems** orchestrate primary growth, giving rise to the primary plant body. This growth is characterized by cell division followed by cell expansion (elongation).

- **Lateral meristems** orchestrate secondary growth. Two lateral meristems, *vascular cambium* and *cork cambium*, contribute to the secondary plant body.

Because apical meristems can perpetuate themselves indefinitely, a shoot or root can continue to lengthen and grow indefinitely; in other words, the growth of a shoot or root is **indeterminate**. Primary growth leads to the elongation of shoots and roots and the formation of organs. All plant organs arise ultimately from cell divisions in apical meristems, followed by cell expansion and cell differentiation. You will see this pattern of

$$\text{division} \rightarrow \text{expansion} \rightarrow \text{differentiation}$$

in several types of meristems.

As they do in the embryo, apical meristems in the adult plant form root and shoot organs. They do so by producing three **primary meristems**. From the outside to the inside of the root or shoot, which are both cylindrical, the primary meristems are the **protoderm**, the **ground meristem**, and the **procambium**. These meristems, in turn, give rise to the three tissue systems:

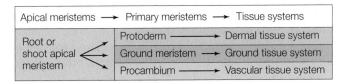

Apical meristems →	Primary meristems →	Tissue systems
Root or shoot apical meristem	Protoderm →	Dermal tissue system
	Ground meristem →	Ground tissue system
	Procambium →	Vascular tissue system

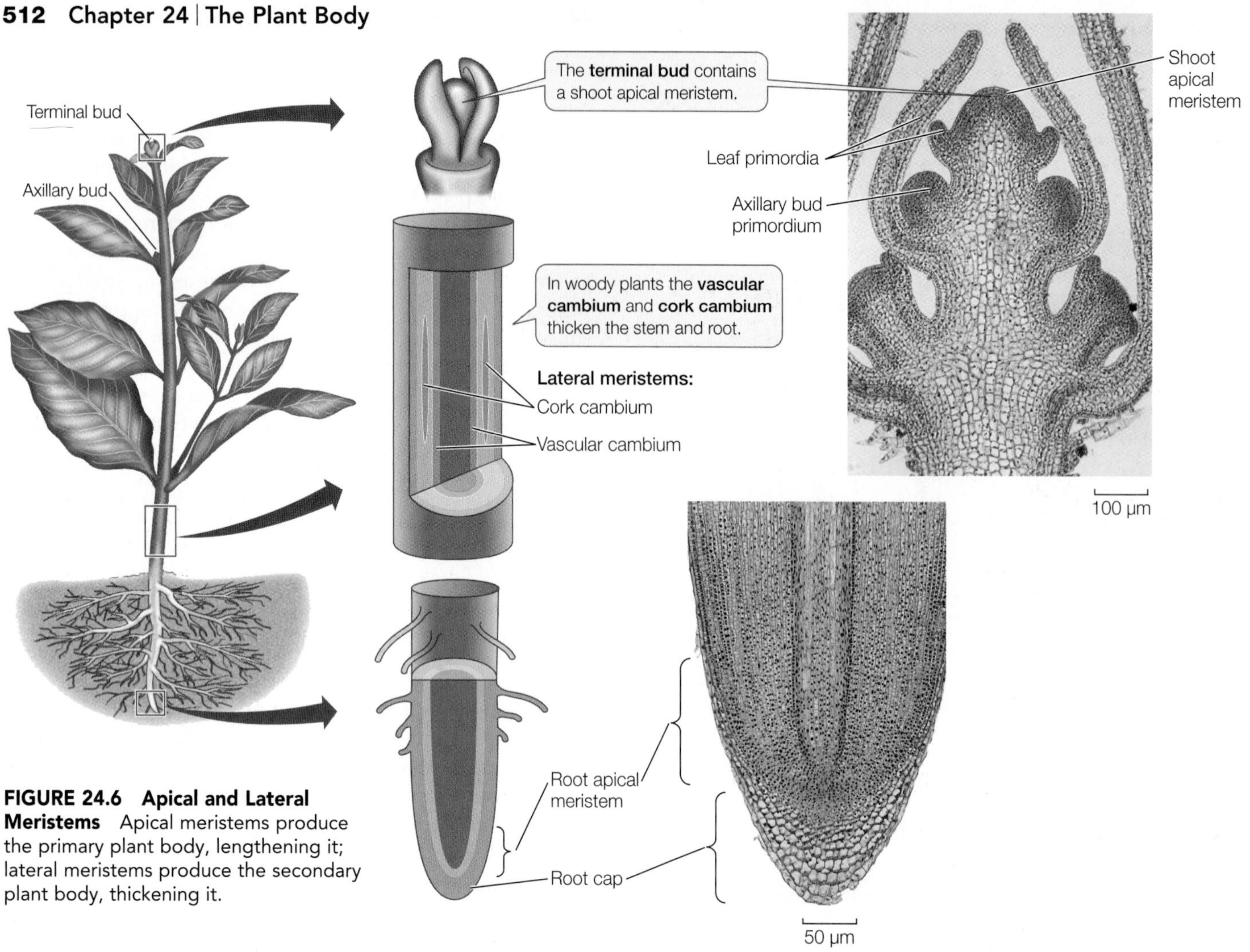

Terminal bud

Axillary bud

The **terminal bud** contains a shoot apical meristem.

In woody plants the **vascular cambium** and **cork cambium** thicken the stem and root.

Lateral meristems:
Cork cambium
Vascular cambium

Shoot apical meristem

Leaf primordia

Axillary bud primordium

100 µm

Root apical meristem

Root cap

50 µm

FIGURE 24.6 Apical and Lateral Meristems Apical meristems produce the primary plant body, lengthening it; lateral meristems produce the secondary plant body, thickening it.

Let's look now at how these meristems function in the root and shoot.

FRONTIERS The relative activities of meristems that produce vegetative growth and those that produce reproductive growth are important measures for plant breeders. Plant breeders are interested in knowing what genes regulate the ratio between these two types of growth. For example, such knowledge might allow them to increase the amount of seeds that a crop plant makes by increasing (within limits) the ratio of reproductive to vegetative growth. Analyses of mutants and of the genomes of crops and model organisms may reveal which genes control meristem activity.

The root apical meristem gives rise to the root cap and the root primary meristems

The root apical meristem produces all the cells that contribute to growth in the length of a root (**FIGURE 24.7A**). Some of the daughter cells from the apical (tip) end of the root apical meristem contribute to a **root cap**, which protects the delicate growing region of the root as it pushes through the soil. The root cap secretes a mucopolysaccharide (slime) that acts as a lubricant. Even so, the cells of the root cap are often damaged or scraped away and must therefore be replaced constantly. The root cap is also the structure that detects the pull of gravity and thus controls the downward growth of roots.

Above the root cap, the cells in the root tip are arranged in several zones. The apical and primary meristems constitute the **zone of cell division**, the source of all the cells of the root's primary tissues. The daughter cells of the divisions that occur in this zone become the three cylindrical primary meristems: the protoderm, the ground meristem, and the procambium. Just above this zone is the **zone of cell elongation**, where the newly formed cells are elongating and thus pushing the root farther into the soil. Above that zone is the **zone of cell maturation**, where the cells are differentiating. These three zones grade imperceptibly into one another; there is no abrupt line of demarcation.

The products of the root's primary meristems become root tissues

The products of the three primary meristems (the protoderm, ground meristem, and procambium) are the tissue systems of the mature root: the *epidermis* (dermal tissue), *cortex* (ground tissue), and *stele* (vascular tissue).

The protoderm gives rise to the **epidermis**, an outer layer of cells that protect the root and absorb mineral ions and water. Many of the epidermal cells produce long, delicate root hairs (**FIGURE 24.7B**).

Internal to the protoderm, the ground meristem gives rise to a region of ground tissue that is many cells thick, called the

APPLY THE CONCEPT

Meristems build roots

A research team investigated how long root cells continued to divide after being produced by the root apical meristem. They incubated germinating bean seeds for 2 hours with their roots suspended in water containing radioactive thymidine, which is taken up by dividing cells and incorporated into their DNA as it replicates (making the DNA radioactive).

After this incubation period, they removed some of the seeds (time 0 in the table) and took longitudinal slices from the roots (see Figure 24.7A). They transferred the rest of the seeds to water containing nonradioactive thymidine and left them there for various times (12, 36, and 72 hours) before taking root slices. They tallied the number of radioactive cells found in each of the four regions of each root and compared those numbers with the total number of cells in each region. The table shows the results.

TIME AFTER TRANSFER (HR)	RADIOACTIVE CELLS/TOTAL CELLS			
	ROOT CAP	ROOT APICAL MERISTEM	ZONE OF CELL ELONGATION	ZONE OF CELL MATURATION
0	0/63	90/115	0/32	0/22
12	2/58	95/110	16/29	0/27
36	36/60	45/119	14/30	12/31
72	12/24	0/116	2/31	30/32

1. At 0 hours, why did only root apical meristem cells contain radioactive thymidine?

2. For each region, calculate the percentages of radioactive cells and plot those percentages over time.

3. Explain the rise and fall in the percentage of radioactive cells in each region.

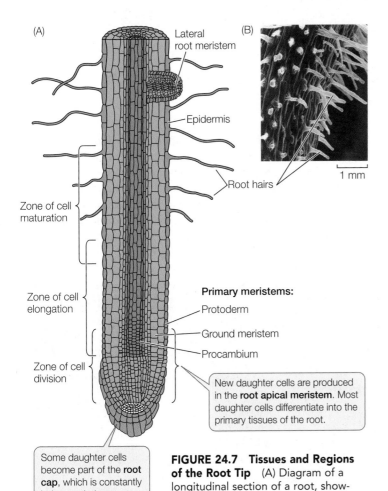

(A)

Lateral root meristem

Epidermis

Root hairs

(B)

1 mm

Zone of cell maturation

Zone of cell elongation

Zone of cell division

Primary meristems:

Protoderm

Ground meristem

Procambium

New daughter cells are produced in the **root apical meristem**. Most daughter cells differentiate into the primary tissues of the root.

Some daughter cells become part of the **root cap**, which is constantly being eroded away.

FIGURE 24.7 Tissues and Regions of the Root Tip (A) Diagram of a longitudinal section of a root, showing how cell division creates the complex structure of the root. (B) Root hairs, seen with a scanning electron microscope.

cortex. The parenchyma cells of the cortex are relatively unspecialized and often serve as storage depots. The innermost layer of the cortex is the **endodermis**. Unlike those of other cortical cells, the cell walls of the endodermal cells contain *suberin*, a waterproof substance. Strategic placement of suberin in only certain parts of the cell wall enables the cylindrical ring of endodermal cells to control the movement of water and dissolved mineral ions into the vascular tissue system.

LINK The role of the endodermis is described in more detail in Concept 25.3

Interior to the endodermis is the *vascular cylinder*, or **stele**, produced by the procambium. The stele consists of three types of cells: pericycle, xylem, and phloem. The **pericycle** consists of one or more layers of parenchyma cells. It has three important functions:

- It is the tissue within which *lateral roots* arise.

- It can contribute to secondary growth by giving rise to lateral meristems that thicken the root.

- The plasma membranes of its cells contain membrane transport proteins that export nutrient ions into the cells of the xylem.

At the very center of the root of a eudicot lies the xylem. Seen in cross section, it typically has the shape of a star with a variable number of points (**FIGURE 24.8A**). Between the points are bundles of phloem. In monocots, a region of parenchyma cells, called the **pith**, typically lies in the center of the root, surrounded by xylem and phloem (**FIGURE 24.8B**). Pith, which often stores carbohydrate reserves, is also found in the stems of eudicots.

FIGURE 24.8 Products of the Root's Primary Meristems The protoderm gives rise to the epidermis, the outermost layer of cells. The ground meristem produces the cortex, the innermost layer of which is the endodermis. The vascular tissues of the root are found in the stele, which is the product of the procambium. The arrangement of tissues in the stele differs in the roots of (A) eudicots and (B) monocots.

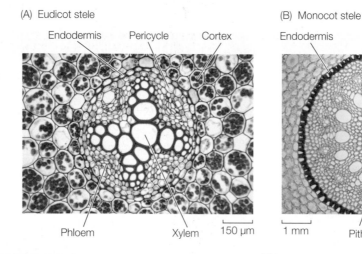

(A) Eudicot stele

Endodermis Pericycle Cortex

Phloem Xylem 150 µm

(B) Monocot stele

Endodermis Phloem Xylem

1 mm Pith Cortex

yourBioPortal.com

**Go to WEB ACTIVITIES 24.1 and 24.2
Eudicot Root • Monocot Root**

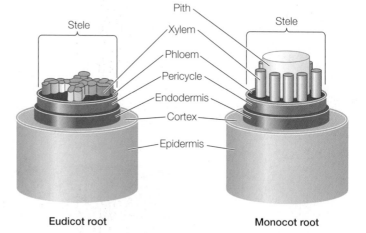

Stele Pith Stele
Xylem
Phloem
Pericycle
Endodermis
Cortex
Epidermis

Eudicot root Monocot root

The root system anchors the plant and takes up water and dissolved minerals

Water and mineral nutrients enter most plants through the root system, which is located in the soil. Although hidden from view, the root system is often larger than the visible shoot system. For example, the root system of a 4-month-old winter rye plant (*Secale cereale*) was found to be 130 times larger than the shoot system, with almost 13 million branches that had a cumulative length of over 500 kilometers!

The root system of angiosperms originates in an embryonic root called the *radicle*. In most eudicots the radicle develops as a primary root (called the **taproot**) that extends downward by tip growth and outward by initiating **lateral roots** (see Figure 24.7A). The taproot itself often functions as a nutrient storage organ, as in carrots, sugar beets, and sweet potato (**FIGURE 24.9A**).

In monocots, the radicle-derived initial root is short-lived. Monocots form a **fibrous root system** composed of numerous thin roots that are all roughly equal in diameter (**FIGURE 24.9B**). Because they originate from the stem at ground level or below, the roots of a typical monocot are called **adventitious** ("arriving from outside") **roots**. Many fibrous root systems have a large surface area for the absorption of water and minerals, and they cling to soil very well. The fibrous root systems of

FIGURE 24.9 Root Systems of Eudicots and Monocots (A) The taproot systems of some eudicots, such as carrots (*Daucus carota*) and sugar beets (*Beta vulgaris*), are nutrient storage organs. (B) These leeks (*Allium ampeloprasum*), which are monocots, have fibrous root systems. (C) The stems of this hala tree (*Pandanus tectorius*), another monocot, are supported by prop roots.

(A) Taproots

(B) Fibrous root system

(C) Prop roots

grasses, for example, may protect steep hillsides where runoff from rain would otherwise cause erosion.

In some monocots—corn, banyan trees, and some palms, for example—adventitious roots function as props to help support the shoot (**FIGURE 24.9C**). **Prop roots** are critical to these plants, which, unlike most eudicot tree species, are unable to support aboveground growth through the thickening of their stems.

The products of the stem's primary meristems become stem tissues

The shoot is composed of repeating modules called *phytomers*. Each phytomer consists of a *node* with its attached leaf or leaves; an *internode*, or section of stem between nodes; and one or more *axillary buds* in the angle between each leaf and the stem (see Figure 24.1). The shoot grows by adding new phytomers. Those new phytomers originate from shoot apical meristems, which are formed at the tips of stems and in axillary buds.

The shoot apical meristem, like the root apical meristem, forms three primary meristems: protoderm, ground meristem, and procambium. These primary meristems, in turn, give rise to the three shoot tissue systems. The shoot apical meristem repetitively lays down the beginnings of leaves and axillary buds. Leaves arise from bulges called *leaf primordia*, which form

as cells divide on the sides of the shoot apical meristem (see Figure 24.6). *Bud primordia* form above the bases of the leaf primordia, where they may become new apical meristems and initiate new shoots. The growing stem has no protective structure analogous to the root cap, but the leaf primordia can act as a protective covering for the shoot apical meristem.

The vascular tissue of young stems is divided into discrete **vascular bundles** (**FIGURE 24.10**). Each vascular bundle contains both xylem and phloem. In eudicots, the vascular bundles are arranged along the periphery and appear in a circle in cross section, but in monocots they are seemingly scattered throughout the stem.

The stem supports leaves and flowers

The central function of stems is to elevate and support the photosynthetic organs (leaves) as well as the reproductive organs (flowers).

Various modifications of stems are seen in nature. The *tuber* of a potato, for example—the part of the plant eaten by

yourBioPortal.com

**Go to WEB ACTIVITIES 24.3 and 24.4
Eudicot Stem • Monocot Stem**

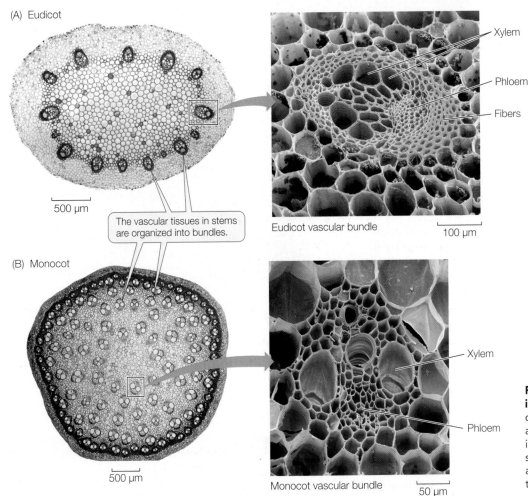

(A) Eudicot

500 μm

The vascular tissues in stems are organized into bundles.

Eudicot vascular bundle

Xylem

Phloem

Fibers

100 μm

(B) Monocot

500 μm

Xylem

Phloem

Monocot vascular bundle

50 μm

FIGURE 24.10 Vascular Bundles in Stems (A) In herbaceous eudicot stems, the vascular bundles are arranged in a cylinder, with pith in the center and the cortex outside the cylinder. (B) A scattered arrangement of vascular bundles is typical of monocot stems.

FIGURE 24.11 Modified Stems
(A) A potato (*Solanum tuberosum*) is a modified stem called a tuber; the sprouts that grow from its eyes are shoots, not roots. (B) The stem of this barrel cactus (*Echinocactus grusoni*) is enlarged to store water. Its highly modified leaves serve as thorny spines. Most of this plant's photosynthesis occurs in the stem. (C) The runners of beach strawberry (*Fragaria chiloensis*) are horizontal stems that produce roots and shoots at intervals. Rooted portions of the plant can live independently if the runner is cut.

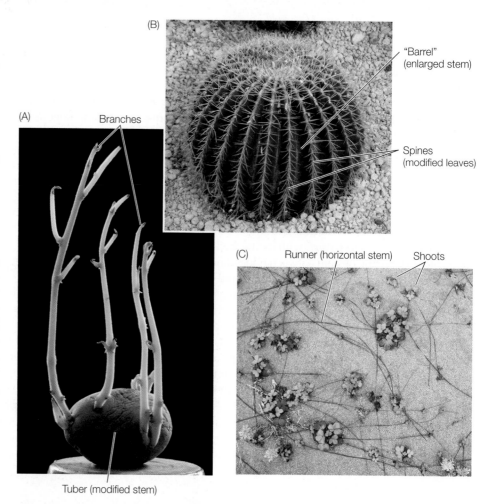

humans—is not a root but rather an underground stem. The "eyes" of a potato are depressions containing axillary buds—in other words, a sprouting potato is just a branching stem (**FIGURE 24.11A**). Many desert plants have enlarged, water-retaining stems (**FIGURE 24.11B**). The *runners* of strawberry plants are horizontal stems from which roots grow at frequent intervals (**FIGURE 24.11C**). If the links between the rooted portions are broken, independent plants can develop on each side of the break—a form of vegetative (asexual) reproduction (see Concept 7.1).

Leaves are the primary site of photosynthesis

For most of its life, a plant produces leaves from shoot apical meristems. As shown in Figure 24.6, leaves originate from leaf primordia at the sides of the shoot apical meristem. Unlike the growth of roots and stems, the growth of leaves is finite, or **determinate**: leaves generally stop growing once they reach a predetermined mature size.

Typically, the **blade** of a leaf is a thin, flat structure attached to the stem by a stalk called a **petiole** (see Figure 24.1). In many plants, the leaf blade is held by its petiole at an angle almost perpendicular to the rays of the sun. This orientation, with the leaf surface facing the sun, maximizes the amount of light available for photosynthesis. Some leaves track the sun over the course of the day, moving so that they constantly face it.

Leaf anatomy is beautifully adapted to carry out photosynthesis and to support that process by exchanging the gases O_2 and CO_2 with the environment, limiting evaporative water loss, and exporting the products of photosynthesis to the rest of the plant. **FIGURE 24.12** shows a section of a typical eudicot leaf in three dimensions.

Most eudicot leaves have two zones of photosynthetic parenchyma tissue called **mesophyll** (which means "middle of the leaf"). The upper layer or layers of mesophyll, which consist of elongated cells, constitute a zone called *palisade mesophyll*. The lower layer or layers, which consist of irregularly shaped cells, constitute a zone called *spongy mesophyll*. Within the mesophyll is a network of air spaces through which CO_2 can diffuse to

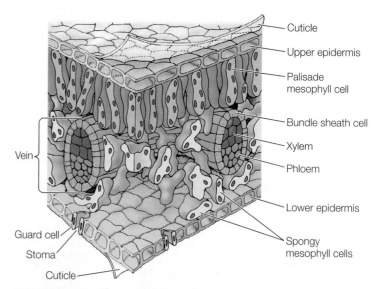

FIGURE 24.12 Eudicot Leaf Anatomy This three-dimensional diagram shows a section of a eudicot leaf.

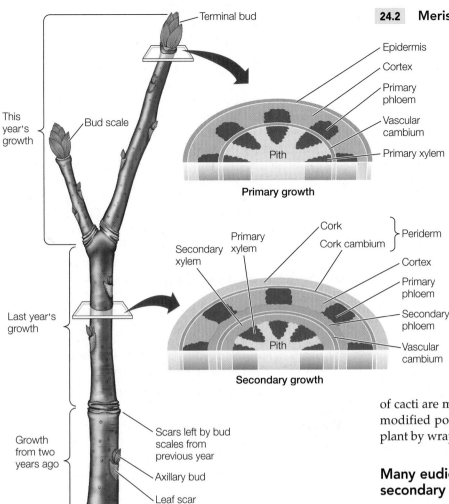

Primary growth

- Terminal bud
- Bud scale
- Epidermis
- Cortex
- Primary phloem
- Vascular cambium
- Primary xylem
- Pith

This year's growth

Last year's growth

Secondary growth

- Secondary xylem
- Primary xylem
- Cork
- Cork cambium
- Periderm
- Cortex
- Primary phloem
- Secondary phloem
- Vascular cambium
- Pith

Growth from two years ago

- Scars left by bud scales from previous year
- Axillary bud
- Leaf scar

FIGURE 24.13 A Woody Twig Has Both Primary and Secondary Tissues The shoot apical meristems inside the buds of this dormant twig will produce primary growth in spring. Lateral meristems in the wood will produce secondary growth.

---

yourBioPortal.com

Go to ANIMATED TUTORIAL 24.1
Secondary Growth: The Vascular Cambium

CO_2—the other raw material of photosynthesis. Therefore, water and gases are exchanged through pores in the leaf called *stomata* (see Concept 25.3).

In some plant species, leaves are highly modified for special functions. For example, some modified leaves serve as storage depots for energy-rich molecules, as in the bulbs of onions. In other species, such as succulents, the leaves store water. The protective spines of cacti are modified leaves. Other plants, such as peas, have modified portions of leaves called *tendrils* that support the plant by wrapping around other structures or plants.

Many eudicot stems and roots undergo secondary growth

As we have seen, the roots and stems of some eudicots develop a secondary plant body, the tissues of which we commonly refer to as *wood* and *bark*. These tissues are derived by secondary growth from the two lateral meristems:

- The **vascular cambium** produces cells that make up *secondary xylem* (wood) and *secondary phloem* (inner bark).

- The **cork cambium** produces mainly waxy-walled protective cells. It supplies some of the cells that become outer bark.

Each year, deciduous trees lose their leaves, leaving bare branches and twigs in winter. These twigs illustrate both primary and secondary growth (**FIGURE 24.13**). The apical meristems of the twigs are enclosed in buds protected by *bud scales*. When the buds begin to grow in spring, the scales fall away, leaving scars, which show us where the bud was and allow us to identify each year's growth. The woody portions of the twig are the product of primary and secondary growth. Only the buds and the newest growth from the apical meristem consist entirely of primary tissues.

In the stem, the vascular cambium is initially a single layer of cells lying between the primary xylem and the primary phloem within the vascular bundles. A stem or root increases in diameter when the cells of the vascular cambium divide, producing secondary xylem cells toward the inside of the stem or root and secondary phloem cells toward the outside (see Figure 24.13).

Secondary xylem conducts water and minerals and continues to do so for many years. Secondary phloem, however, is active in transport of sugars for only a year. As secondary growth of a stem or root continues, the expanding vascular

---

photosynthesizing cells. Vascular tissue branches extensively throughout the leaf, forming a network of *veins* that extend to within a few cell diameters of all its cells, ensuring that the mesophyll cells are well supplied with water and mineral nutrients. The products of photosynthesis are loaded into the veins for export to the rest of the plant.

LINK Photosynthesis is described in Concepts 6.5 and 6.6

FRONTIERS What directs a cell in a leaf primordium to differentiate appropriately for its eventual position on the upper or lower side of the mature leaf, so that the cell becomes, for example, a palisade mesophyll cell or a spongy mesophyll cell? Some sort of positional information is presumably involved. The nature of this information is being investigated in *Arabidopsis thaliana*. Mutational analyses indicate that both transcription factors and microRNAs are involved in the differentiation process. Further studies may reveal exactly how positional information is transmitted during leaf development.

Covering virtually the entire leaf on both its upper and lower surfaces is a layer of nonphotosynthetic epidermal cells. These cells secrete a cuticle that is impermeable to water. Although the epidermis keeps water in the leaf, it also keeps out

tissue stretches and breaks the epidermis and the outer layers of the cortex, which ultimately flake away. Tissue derived from the secondary phloem then becomes the outermost part of the stem. Before the dermal tissues are broken away, however, cells lying near the surface of the secondary phloem begin to divide, forming a cork cambium. This meristematic tissue produces layers of **cork**, a protective tissue composed of cells with thick walls waterproofed with suberin. The cork soon becomes the outermost tissue of the stem or root. Without the activity of the cork cambium, the sloughing off of the outer primary tissues would expose the plant to potential damage, such as excessive water loss or invasion by microorganisms. Sometimes the cork cambium produces cells to the inside as well as to the outside; these inside cells constitute a tissue known as the *phelloderm*.

The cork cambium, cork, and phelloderm constitute a secondary dermal tissue called **periderm**. As the vascular cambium continues to produce secondary vascular tissue, these corky layers are lost, but the continuous formation of new cork cambia in the underlying secondary phloem gives rise to new corky layers. The periderm and the secondary phloem—that is, all the tissues external to the vascular cambium—constitute the **bark**.

Cross sections of most tree trunks (which are mature stems) in temperate-zone forests reveal annual rings in the wood (**FIGURE 24.14**), which result from seasonal changes in environmental conditions. In spring, when water is relatively plentiful, the tracheids or vessel elements produced by the vascular cambium tend to be large in diameter and thin-walled. Such wood is well adapted for transporting large quantities of water and mineral nutrients. As water becomes less available in summer, narrower cells with thicker walls are produced, which make this summer wood darker and sometimes more dense than the wood formed in spring. Thus each growing season is recorded by a clearly visible annual ring. Trees in the moist tropics do not experience these seasonal changes, so they do not lay down such obvious regular rings. In temperate-zone trees, variations in temperature or water supply can lead to the formation of more than one "annual" ring in a single year, but most years bring a new annual ring and a new batch of leaves.

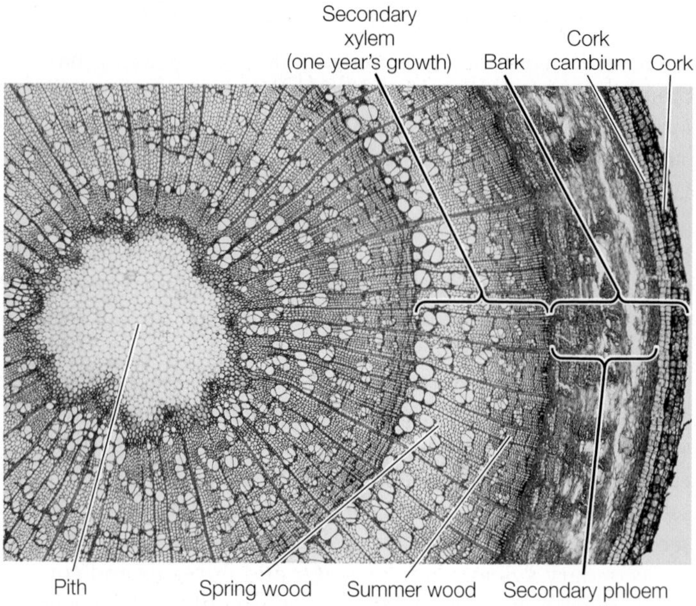

Only eudicots and other non-monocot angiosperms, along with many gymnosperms, have a vascular cambium and a cork cambium and thus undergo secondary growth. The few monocots that form thickened stems—palms, for example—do so without secondary growth. Palms have a very wide apical meristem that produces a wide stem, and dead leaf bases add to the diameter of the stem. All monocots grow in essentially this way, as do other angiosperms that lack secondary growth.

Do You Understand Concept 24.2?

- Compare primary and secondary growth, creating a table with the following headings: the plant groups that undergo them; the organs that are involved during the plant life cycle; the meristems that are involved; and the products of the meristems.

- Why are annual rings used to determine the age of a tree trunk?

- What is the difference between determinate and indeterminate growth? What plant organs undergo each type of growth?

- What are the selective advantages of indeterminate growth?

Natural selection has acted on the plant body plan to produce a remarkable diversity of plant forms. Artificial selection by humans on the plants that provide us with food and fiber has greatly altered plant form as well.

concept 24.3 Domestication Has Altered Plant Form

We have seen in this chapter that a very simple plant body plan—with roots, stems, leaves, meristems, and relatively few tissue and cell types—underlies the diversity of the flowering plants that cover our planet. Although differences in form among species are expected, members of the same plant species can be remarkably diverse in form as well. From a genetic perspective, this observation suggests that minor differences in genomes or gene regulation can underlie dramatic differences in plant form. (Nevertheless, some plant species do differ greatly in genome size, content, and organization.)

The form of the plant body is subject to natural selection. Phenotypes that can outcompete their peers for limiting resources have the highest fitness. Some plants have developed a viny, climbing phenotype that gives them access to light in crowded conditions. Others have a branched phenotype that puts neighboring plants in the shade. Still others have a deep and extensive root system that gives them access to water. Clearly, the vegetative body plan is as important in terms of

FIGURE 24.14 Annual Rings Rings of secondary xylem are a noticeable feature of this cross section from a tree trunk.

Teosinte   Corn

FIGURE 24.15 Corn Was Domesticated from the Wild Grass Teosinte Beginning more than 8,000 years ago in Mexico, farmers favored plants with minimal branching. Reducing the number of branches results in fewer ears per plant but allows each ear to grow larger and produce more seeds.

evolution as are phenotypes relating directly to reproduction, such as flower morphology.

When humans domesticated crops and chose from nature those phenotypes most suitable for cultivation, they set in motion a long experiment in artificial selection of the plant body. For example, in the case of corn (maize), farmers selected plants with tall, compact growth and few branches, minimizing competition among individual plants. Corn was domesticated from the wild grass teosinte, which still grows in the hills of Mexico (**FIGURE 24.15**). One of the most conspicuous differences between teosinte and domesticated corn is that teosinte, like other wild grasses, is highly branched, with many shoots, while domesticated corn has a single shoot. This morphological difference is due in large part to the activity of a single gene called *teosinte branched 1* (*tb1*). The protein product of *tb1* regulates the growth of axillary buds (see Figure 24.1). The allele of *tb1* in domesticated corn represses branching, while the allele in teosinte permits branching. Similarly, in the domestication of rice, farmers selected for a vertical orientation of the upper leaves so that sunlight could reach all the leaves and photosynthesis was maximized.

It may be hard to believe that a single species, *Brassica oleracea* (wild mustard), is the ancestor of so many familiar and morphologically diverse crops, including kale, broccoli, brussels sprouts, and cabbage (see Figure 15.4). An understanding of how the basic body plan of plants arises

makes it possible to appreciate how each of these crops was domesticated. Starting with morphologically diverse populations of wild mustard, humans selected and planted the seeds from variants with the trait they found desirable. Many generations of such artificial selection produced the crops that fill the produce section of the supermarket and the stands of the farmers' market.

Q&A QUESTION What are the properties of the kenaf plant that make it suitable for papermaking?

ANSWER Kenaf is a biennial, herbaceous plant that grows rapidly from very active shoot apical meristems (Concept 24.1), reaching a height of 4–5 meters in 5 months. There are over 500 known genetic strains of kenaf. Over the centuries since its domestication, kenaf has been selected to grow taller and to branch less (Concept 24.3). In addition, the plant's adventitious roots have become longer, at about 2 meters (Concept 24.2), and more numerous, promoting the spread of the plant into dense stands in the field.

Kenaf stems are harvested for fiber (**FIGURE 24.16**). The phloem fibers are 3 millimeters long, somewhat longer than the fibers in wood; longer fibers are desirable for stronger products. In addition, the cell walls in kenaf stems are rich in cellulose but relatively poor in lignin. This makes kenaf easier to pulp than wood, which requires strong chemical treatments to hydrolyze the lignin in its cell walls. Furthermore, 1 hectare of kenaf produces 11–17 tons of fiber suitable for pulp and paper production; this is at least three times more than a hectare of southern pine trees produces. And bear in mind that it takes 5 months to grow the kenaf and 20 years to grow the pines!

The main uses of kenaf today are for the manufacture of rope, fiber-based bags, particleboard, and most recently, paper. The last use looks increasingly viable economically.

FIGURE 24.16 Kenaf Stems The stem of kenaf plants is rich in long phloem fibers. Their low lignin content makes the fibers attractive for making paper as well as other products. Paper derived from kenaf fibers has been used for printing newspapers.

SUMMARY

concept 24.1 The Plant Body Is Organized and Constructed in a Distinctive Way

- The vegetative organs of flowering plants are **roots**, which form a **root system**, and **stems** and **leaves**, which (together with flowers, which are sexual organs) form a **shoot system**. Review Figure 24.1

- Plant development is influenced by three unique properties of plants (compared to animals): apical meristems, the presence of cell walls, and the totipotency of most plant cells. **Review Figure 24.2**

- During embryogenesis, the apical–basal axis and the radial axis of the plant body are established, as are the **shoot apical meristem** and the **root apical meristem**. Review Figures 24.3 and 24.4

- Three **tissue systems**, arranged concentrically, extend throughout the plant body: the **dermal tissue system**, **ground tissue system**, and **vascular tissue system**. Review Figure 24.5

- The vascular tissue system includes **xylem**, which conducts water and mineral ions absorbed by the roots to the shoot, and **phloem**, which conducts the products of photosynthesis throughout the plant body.

concept 24.2 Meristems Build Roots, Stems, and Leaves

- **Primary growth** is characterized by the lengthening of roots and shoots and by the proliferation of new roots and shoots through branching. Some plants also experience **secondary growth**, by which they increase in thickness.

- **Apical meristems** generate primary growth, and **lateral meristems** generate secondary growth. Review Figure 24.6

- Apical meristems at the tips of shoots and roots give rise to three **primary meristems (protoderm, ground meristem,** and **procam-**

bium), which in turn produce the three tissue systems of the plant body.

- The root apical meristem gives rise to the **root cap** and the three primary meristems. The cells in the root tip are arranged in three zones that grade into one another: the **zone of cell division**, **zone of cell elongation**, and **zone of cell maturation**. Review Figure 24.7

- The vascular tissue of roots is contained within the **stele**. It is arranged differently in eudicot and monocot roots. Review Figure 24.8 and WEB ACTIVITIES 24.1 and 24.2

- In stems, the vascular tissue is divided into **vascular bundles**, which containing both xylem and phloem. Review Figure 24.10 and WEB ACTIVITIES 24.3 and 24.4

- Eudicot leaves have two zones of photosynthetic **mesophyll** cells that are supplied by veins with water and minerals. Review Figure 24.12 and WEB ACTIVITY 24.5

- Two lateral meristems, the **vascular cambium** and **cork cambium**, are responsible for secondary growth. The vascular cambium produces secondary xylem (wood) and secondary phloem (inner bark). The cork cambium produces a protective tissue called **cork**. Review Figures 24.13 and 24.14 and ANIMATED TUTORIAL 24.1

concept 24.3 Domestication Has Altered Plant Form

- Although the plant body plan is simple, it can be changed dramatically by minor genetic differences, as evidenced by the natural diversity of wild plants.

- Crop domestication involves artificial selection of certain desirable traits found in wild populations. As a result of artificial selection over many generations, the body forms of crop plants are very different from those of their wild relatives. Review Figure 24.15

Plant Nutrition and Transport

The 2010 earthquake was a catastrophe for Haiti, the poorest country in the Western Hemisphere. Over 200,000 people died, and over a million people were left homeless. The urban infrastructure in the capital city was destroyed. Amid the outpouring of help from the rest of the world came a realization that another disaster has been occurring in Haiti for decades: the destruction of its forests and farms. Plants need fertile soil to grow, and in Haiti that has been in increasingly short supply.

Haiti shares the island of Hispaniola with the Dominican Republic. The two countries have had dramatically different policies with regard to land conservation, with equally dramatic results. While the Dominicans have carefully managed their land, the story of Haiti is one of overexploitation. First, Spanish colonists planted sugarcane, a crop that depleted the soil. Then the French colonists that followed them cut down forests to plant coffee and tobacco for export to Europe. After the former slaves revolted and expelled the Europeans in the early 1800s, land speculators chased out indigenous farmers from fertile valleys and planted export crops such as corn and beans. Today only 4 percent of the forests that existed when the Europeans arrived in 1492 remain. As Haitians have said, "Te a fatige (The earth is tired)." Because there are few plant roots to stabilize the soil, especially in Haiti's many mountainous regions, rain washes the soil into the sea. From an airplane window it is easy to see the 120-mile-long border between the two countries. The Dominican side is

verdant and forested, while the Haitian side is devoid of plant life and has lost most of its soil.

As human land use intensifies, ecological disasters like Haiti's are becoming all too common. Today, parts of sub-Saharan Africa's farmland are losing their topsoil as a result of poor land management, swelling populations, and a challenging climate. Crop failures, starvation, and large-scale human displacements are inevitable consequences. The need for a better understanding of how plants interact with the soil is a pressing human concern.

An aerial view of the border between Haiti (left) and the Dominican Republic (right) provides a dramatic illustration of the extent of soil degradation in Haiti.

KEY CONCEPTS

25.1 Plants Acquire Mineral Nutrients from the Soil

25.2 Soil Organisms Contribute to Plant Nutrition

25.3 Water and Solutes Are Transported in the Xylem by Transpiration–Cohesion–Tension

25.4 Solutes Are Transported in the Phloem by Pressure Flow

How can soil be managed for optimal plant growth?

<table>
<tr><td>concept
25.1</td><td>**Plants Acquire Mineral Nutrients from the Soil**</td></tr>
</table>

Like all organisms, plants are made up primarily of water and molecules that contain the major elements of life. Plants are *autotrophs* that obtain carbon from atmospheric carbon dioxide through the carbon-fixation reactions of photosynthesis. They obtain hydrogen and oxygen mainly from water, so these elements are plentiful as long as plants have an adequate water supply.

> **LINK**   The major elements of life are described in Chapter 2, and the chemical reactions of photosynthesis in Concepts 6.5 and 6.6

The plant body contains other elements as well. Nitrogen, for example, is essential for the synthesis of proteins and nucleic acids. Plants obtain nitrogen, directly or indirectly, through the activities of bacteria, as we will see later in this chapter. In addition, proteins contain sulfur (S), nucleic acids contain phosphorus (P), and chlorophyll contains magnesium (Mg). Plants obtain these and other mineral nutrients mainly from the soil. Mineral nutrients dissolve in water in the soil as ions, forming a solution—called the **soil solution**—that surrounds the roots of plants.

Nutrients can be defined by their deficiency

A plant nutrient is called an **essential element** if its absence causes severe disruption of normal plant growth and reproduction. An essential element cannot be replaced by any other element. There are two categories of essential elements:

- **Macronutrients** have concentrations of at least 1 gram per kilogram of the plant's dry matter. There are six macronutrients:

  Nitrogen (taken up as NO_3^-)
  Phosphorus (taken up as PO_4^{3-})
  Potassium (taken up as K^+)
  Sulfur (taken up as SO_4^{2-})
  Calcium (taken up as Ca^{2+})
  Magnesium (taken up as Mg^{2+})

- **Micronutrients** have concentrations of less than $0.1\,g/kg$ of the plant's dry matter. They include iron (Fe^{2+}), chlorine (Cl^-), manganese

(Mn^{2+}), zinc (Zn^{2+}), copper (Cu^{2+}), nickel (Ni^{2+}), and molybdenum ($Mo(O_4)^{2-}$).

Plants show characteristic deficiency symptoms, most noticeable in leaves, that can be used to diagnose deficiencies of essential elements (**FIGURE 25.1**). With proper diagnosis, the missing nutrient(s) can be provided in the form of a supplement, or **fertilizer**.

Experiments using hydroponics have identified essential elements

The essential elements for plants were identified by growing plants **hydroponically**—that is, with their roots suspended in nutrient solutions instead of soil. Growing plants in this manner allows for greater control of nutrient availability than is possible in a complex medium like soil.

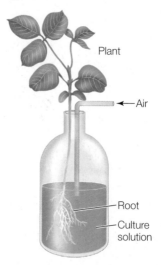

In the first successful experiments of this type, performed a century and a half ago, plants grew seemingly normally in solutions containing only calcium nitrate ($Ca(NO_3)_2$), magnesium sulfate ($MgSO_4$), and potassium phosphate (KH_2PO_4). A solution missing any of these compounds could not support normal plant growth. Tests with other compounds that included

FIGURE 25.1   Mineral Nutrient Deficiency Symptoms   The plants on the left were grown with a full complement of essential nutrients. The center plants were deprived of iron, while the plants on the right were deprived of nitrogen.

yourBioPortal.com

Go to ANIMATED TUTORIAL 25.1
Nitrogen and Iron Deficiencies

various combinations of these same elements soon established the six macronutrients listed above.

Identifying micronutrients by this experimental approach proved to be more difficult because of the small amounts involved. Sufficient amounts of a micronutrient can be present in the environment used to grow plants. Therefore, nutrition experiments must be performed in tightly controlled laboratories with special air filters that exclude microscopic mineral particles, and they must use only the purest available chemicals. Furthermore, a seed may contain enough of a micronutrient to supply the embryo and the resultant second-generation plant throughout its lifetime. There might even be enough left over to pass on to the next generation of plants. Thus, even if a plant is completely deprived of a micronutrient, it may take a long time for deficiency symptoms to appear. The most recent micronutrient to be listed as essential was nickel, in 1983 (**FIGURE 25.2**).

Soil provides nutrients for plants

Most terrestrial plants grow in soil. Plants derive a number of benefits from soil:

- Anchorage for mechanical support of the shoot
- Mineral nutrients and water from the soil solution
- O_2 for root respiration from air spaces between soil particles
- The services of other soil organisms, including bacteria, fungi, protists, and animals such as earthworms and arthropods

Although soils vary greatly, almost all of them have a *soil profile* consisting of several recognizable horizontal layers, called **horizons**, lying on top of one another. Soil scientists recognize three major horizons—termed A, B, and C—in a typical soil (**FIGURE 25.3**):

- The **A horizon** is the *topsoil* that supports the plant's mineral nutrient needs. It contains most of the soil's living and dead organic matter.

INVESTIGATION

FIGURE 25.2 Nickel Is an Essential Element for Plants Using highly purified minerals in a hydroponic solution, Patrick Brown and his colleagues tested whether barley can complete its life cycle in the absence of nickel. Other investigators showed that no other element could substitute for nickel.

HYPOTHESIS
Nickel is an essential element for plants.

METHOD
1. Grow barley plants for three generations in hydroponic solutions containing 0, 0.6, and 1.0 µM nickel sulfate ($NiSO_4$).
2. Harvest seeds from 5–6 third-generation plants in each of the groups.
3. Determine the nickel concentration in seeds from each plant.
4. Germinate other seeds from the same plants in a nickel-free hydroponic solution and plot the success of germination against nickel concentration.

RESULTS
There was a positive correlation between seed germination and nickel concentration. There was significantly less germination at the lowest nickel concentrations.

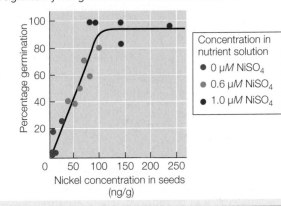

Concentration in nutrient solution
- 0 µM $NiSO_4$
- 0.6 µM $NiSO_4$
- 1.0 µM $NiSO_4$

CONCLUSION
Barley seeds require nickel in order to germinate and complete their life cycle.

ANALYZE THE DATA
A. What do the blue points have in common? Why are there five blue points instead of just one?
B. What is the biological meaning of the zero/zero point?
C. Why didn't the investigators examine only the plants grown without nickel in their hydroponic solution?
D. Plant biologists define the *critical value* for a mineral nutrient as that concentration in plant tissue that results in a 15 percent reduction in the optimal yield of the plant. Using maximum germination percentage in this experiment as the optimal yield, calculate the critical value for nickel.

For more, go to Working with Data 25.1 at **yourBioPortal.com**.

Go to **yourBioPortal.com** for original citations, discussions, and relevant links for all INVESTIGATION figures.

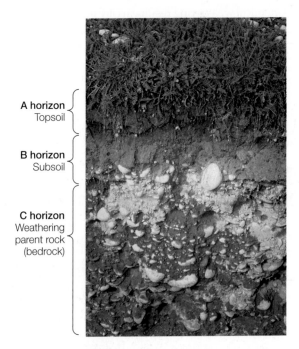

A horizon
Topsoil

B horizon
Subsoil

C horizon
Weathering
parent rock
(bedrock)

FIGURE 25.3 A Soil Profile The A, B, and C horizons of a soil can sometimes be seen in road cuts such as this one in Australia. The dark upper layer (the A horizon) is home to most of the living organisms in the soil.

- The **B horizon** is the *subsoil*, which accumulates materials from the topsoil above it and the parent rock below.

- The **C horizon** is the *parent rock* from which the soil arises.

Rocks are broken down into soil particles (*weathered*) in two ways. *Mechanical weathering* is the physical breakdown of materials by wetting, drying, and freezing. *Chemical weathering* is the alteration of the chemistry of the materials in the rocks by oxidation, water, and acids.

The soil's *fertility* (its ability to support plant growth) is determined by several factors. The size of the soil particles in the A horizon, for example, is important in determining nutrient retention and other soil characteristics. Mineral nutrients in the soil solution can be **leached** (washed away) from the A horizon downward by rain (or over-irrigation) and thereby become unavailable to plants. Tiny *clay* particles bind a lot of water, and mineral nutrients are not readily leached from them, but clay particles pack tightly and leave little room for air. Large *sand* particles present the opposite problems. A **loam** is a soil that is an optimal mixture of sand, silt, and clay, and thus has sufficient supplies of air, water, and nutrients for plants.

In addition to mineral particles, soils contain dead organic matter, largely from plants. Soil organisms break down dead leaves and other organic materials that fall to the ground into a substance called **humus**. This material is used as a food source by microbes that break down complex organic molecules and release simpler molecules into the soil solution. Humus also improves the texture of soil and provides air spaces that increase oxygen availability to plant roots.

Ion exchange makes nutrients available to plants

Chemical weathering results in clay particles covered with negatively charged chemical groups (**FIGURE 25.4**). These negatively charged clay particles bind the positively charged ions (cations) of many minerals that are important for plant nutrition, such as potassium (K^+), magnesium (Mg^{2+}), and calcium (Ca^{2+}). This binding prevents the cations from being leached out of the soil, but to become available to plants, they must be detached from the clay particles.

Recall that the root surface is covered with epidermal cells bearing root hairs (see Figure 24.7B). Proteins in the plasma membrane of these cells actively pump protons (H^+) out of the cells, as we'll describe in Concept 25.3. In addition, cellular respiration in the roots releases CO_2, which dissolves in the soil water and reacts with it to form carbonic acid. This acid ionizes to form bicarbonate and free protons:

$$CO_2 + H_2O \rightleftharpoons H_2CO_3 \rightleftharpoons H^+ + HCO_3^-$$

Proton pumping by the root cells and ionization of carbonic acid both act to increase the proton concentration in the soil surrounding the root. The protons bind more strongly to clay particles than do mineral cations; in essence, the protons trade places with the cations in a process called **ion exchange** (see Figure 25.4). Ion exchange releases mineral nutrient cations into the soil solution, where they are available to be taken up by roots. Soil fertility is determined in part by the soil's ability to provide nutrients in this manner.

> **FRONTIERS** Some mineral ions are toxic to plants. At low soil pH, acids cause clay particles to release aluminum oxide (Al_2O_3), which is poisonous to the roots of many species. Its presence can severely reduce crop growth in areas with naturally acidic soils. In Colombia, plant breeders have produced several strains of crop plants that grow well in acidic soils. These plants are now being used in breeding programs to introduce their characteristics into commercial varieties of the same crops.

There is no comparable mechanism for binding and releasing negatively charged ions. As a result, important anions such as nitrate (NO_3^-) and sulfate (SO_4^{2-}) may be leached from the A horizon.

Fertilizers can be used to add nutrients to soil

Leaching and the harvesting of crops may deplete a soil of its nutrients, in which case the growth of subsequent crops on that soil will be reduced. There are three ways to restore the nutrient content of such a soil: shifting agriculture, organic fertilizers, and chemical fertilizers.

SHIFTING AGRICULTURE In the past, when the soil could no longer support a level of plant growth sufficient for agricultural purposes, people simply moved to another location. This *nomadic* existence allowed the soil to replenish its nutrients naturally, by weathering of the parent rock and gradual accumulation and breakdown of organic matter, building back up the soil's humus. Weathering is

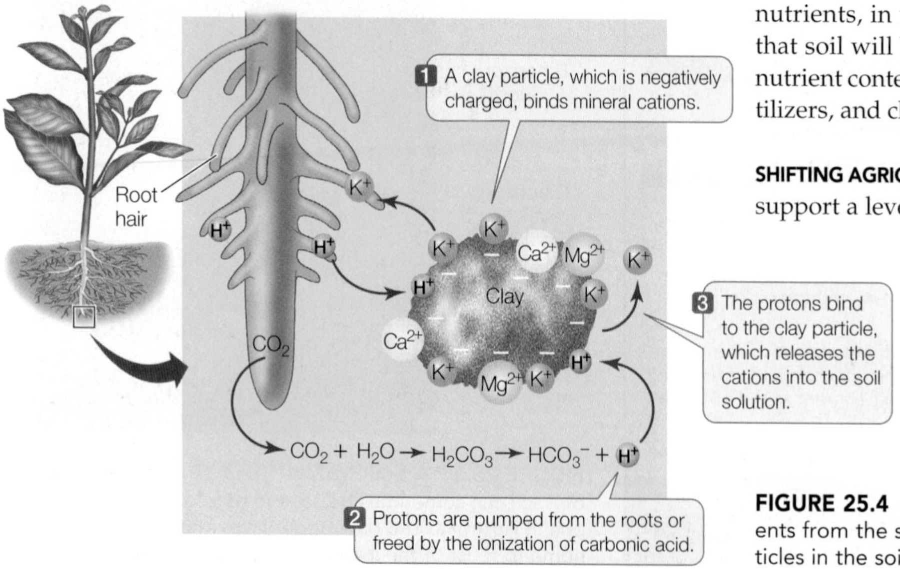

1 A clay particle, which is negatively charged, binds mineral cations.

2 Protons are pumped from the roots or freed by the ionization of carbonic acid.

3 The protons bind to the clay particle, which releases the cations into the soil solution.

FIGURE 25.4 Ion Exchange Plants obtain some mineral nutrients from the soil in the form of positive ions bound to clay particles in the soil.

a geological process that takes a long time, but that was not a problem as long as a lot of land was available. After many years in other locations, the people could return to their original land, which was again able to support farming. Today, the food needs of a large human population are too great to allow land to be left vacant for a long time; in addition, most people live more settled lives now and do not want to move.

ORGANIC FERTILIZERS Humus is used as a food source by soil organisms, which in turn release simpler molecules to the soil solution. For example:

$$\text{leaf proteins} \xrightarrow{\text{Bacteria}} NH_4^+ \xrightarrow{\text{Bacteria}} NO_3^-$$

These simpler molecules can dissolve in soil water and enter plant roots.

Farmers can increase the nutrient content of soil by adding organic materials such as *compost* (partially decomposed plant material). Another way to increase the fertility of soil is to add the waste from farm animals (*manure*), which is a particularly good source of nitrogen. In either case, these **organic fertilizers** add nutrients to the soil much more rapidly than natural weathering, but still allow for a slow release of ions, with little leaching, as the materials decompose.

INORGANIC FERTILIZERS Because organic fertilizers may still act too slowly to restore fertility if a soil is to be used every year, **inorganic fertilizers** are used to supply mineral nutrients directly in forms that can be immediately taken up by plants. Inorganic fertilizers are characterized by their "N-P-K" percentages. A 5-10-10 fertilizer, for example, contains 5 percent nitrogen, 10 percent phosphate (P_2O_5), and 10 percent potash (K_2O) by weight (of the nutrient-containing compound, not as weights of the elements N, P, and K). Sulfur, in the form of ammonium sulfate (($NH_4)_2SO_4$), is also occasionally added to soils.

To return to our example of nitrogen, inorganic fertilizers reach plants by this process:

$$\text{chemical manufacture} \to NH_4^+ \xrightarrow{\text{Bacteria}} NO_3^-$$

The manufacture of nitrogen fertilizer begins with an energy-intensive industrial process in which N_2 is combined with H_2 at high temperatures and pressures in the presence of a catalyst to form ammonia (NH_3). While organic and inorganic fertilizers have the same effects on crop nutrition, their costs are very different: animal manure and plant waste are free, whereas NH_4^+ made in a factory is very expensive because the manufacturing process requires large amounts of energy.

FRONTIERS About 100 million metric tons of ammonia are made industrially for use as fertilizer every year. The process of manufacturing ammonia, invented in 1909, combines N_2 and H_2 gases and is highly energy dependent. Ammonia production accounts for about 2 percent of the world's energy use, mostly in the form of natural gas. As fossil fuel prices rise, so does the cost of inorganic fertilizer. Chemists are exploring ways to use renewable energy sources to lower the cost.

Do You Understand Concept 25.1?

- What are the differences between:
  a. macronutrients and micronutrients?
  b. organic and inorganic fertilizers?
  c. the availability to plant roots of a cation and an anion?
- Outline an experimental method for determining whether an element is essential to a plant.
- A label on a bin in a market states "Organic apples, grown without chemical fertilizers." Are these apples different in terms of plant molecules than they would be if they were grown with chemical fertilizers?

We have just described some of the physical and chemical aspects of soils that are vital to plant nutrition. But many other organisms besides plants inhabit soils. We will turn now to the effects of those organisms on plants.

concept 25.2 Soil Organisms Contribute to Plant Nutrition

One gram of soil contains from 6,000 to 50,000 bacterial species and up to 200 meters of fungal hyphae, although both are largely invisible to the naked eye. Many of these soil organisms consume living and dead plant materials. Plants have an array of defenses they can use against potentially harmful soil organisms (which we describe in Chapter 28). In a few cases, however, plants *actively encourage* fungi and bacteria to infect their roots. Here we describe the resulting "intercellular trading posts," where products are exchanged to the mutual benefit of plants and a few very special soil microbes. In these interactions, the plants usually "trade" the products of photosynthesis, which the microbes use as an energy source, for other essential elements, such as phosphorus and nitrogen, which the microbes can supply. We'll begin by describing how these associations form; then we'll see how they benefit the plants and their microbial partners.

Plants send signals for colonization

In Concept 22.2 we described the association of fungi with plant roots, an interaction that occurs in over 90 percent of terrestrial plants. Our focus here is on arbuscular mycorrhizae, in which fungal hyphae penetrate roots (see Figure 22.9B). In Chapter 19 we introduced the nitrogen-fixing bacteria that convert atmospheric N_2 into ammonia. Here we also describe the close association between certain nitrogen-fixing bacteria called rhizobia and the roots of some plants. In the case of both arbuscular mycorrhizae and rhizobia, the plant roots send signals that attract the soil organisms, and the development of the association involves similar genes and cellular pathways.

FORMATION OF MYCORRHIZAE The events in the formation of arbuscular mycorrhizae are shown in **FIGURE 25.5A**. Plant roots produce compounds called **strigolactones** that stimulate rapid growth of fungal hyphae toward the root. In response, the fungi produce signals that stimulate expression of plant symbiosis-related genes. The products of some of these genes give rise to the *prepenetration apparatus* (PPA), which guides the

growth of the fungal hyphae into the root cortex. The sites of nutrient exchange between fungus and plant are the *arbuscules*, which form at root cortical cells. Despite the intimacy of this association, the plant and fungal cytoplasms never mix—they are separated by two membranes, the fungal plasma membrane and the *periarbuscular membrane* (PAM), which is continuous with the plant plasma membrane.

FORMATION OF NITROGEN-FIXING NODULES A group of plants called *legumes* can form symbioses with several species of soil bacteria collectively known as *rhizobia*, which live in nodules that form on the legumes' roots. The roots of these plants release flavonoids and other chemical signals that attract the rhizobia to the vicinity of the roots (**FIGURE 25.5B**). The flavonoids also trigger the transcription of bacterial *nod* genes, the products of which synthesize Nod (nodulation) factors. These factors, when secreted by the bacteria, cause cells in the root

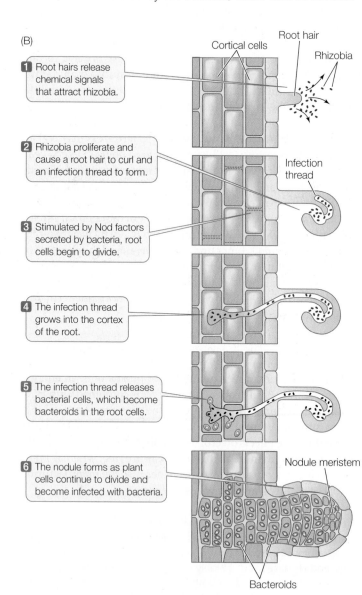

(A)

Cortical cells
Epidermis
Strigolactones
Spore
Root tip

1 Plant roots produce strigolactones that stimulate rapid growth of fungal hyphae toward the root.

Hypha

Fungal signal

PPA

2 Fungal signal stimulates plant to produce a pre-penetration apparatus (PPA).

3 Fungal hypha enters the PPA and is guided to the root cortex through the apoplast.

4 Fungus grows along the root length.

PAM

5 Hyphae induce formation of new PPA structures inside cortical cells.

6 Hyphae enter PPAs and branch to form arbuscules, where nutrients are exchanged.

(B)

Cortical cells
Root hair
Rhizobia

1 Root hairs release chemical signals that attract rhizobia.

2 Rhizobia proliferate and cause a root hair to curl and an infection thread to form.

Infection thread

3 Stimulated by Nod factors secreted by bacteria, root cells begin to divide.

4 The infection thread grows into the cortex of the root.

5 The infection thread releases bacterial cells, which become bacteroids in the root cells.

6 The nodule forms as plant cells continue to divide and become infected with bacteria.

Nodule meristem

Bacteroids

FIGURE 25.5 Roots Send Signals for Colonization Plant roots send chemical signals to arbuscular mycorrhizal fungi (A) and nitrogen-fixing bacteria (B) to stimulate colonization.

APPLY THE CONCEPT

Soil organisms contribute to plant nutrition

Researchers grew lemon seedlings in soils that contained phosphate fertilizer, arbuscular mycorrhizal fungi, or both. The table shows the mean dry weight of the seedlings after 6 months.

Use the data in the table to answer the following questions:

1. What do these results tell us about the role of phosphate fertilizer in the growth of lemon plants?

2. Why didn't the mean weight of the seedlings increase when the amount of phosphate fertilizer, in the absence of mycorrhizae, was doubled?

PHOSPHATE FERTILIZER ADDED (g)	MYCORRHIZAE PRESENT	SEEDLING DRY WEIGHT (g)
0	No	1
0	Yes	10
12	No	28
12	Yes	166
24	No	20
24	Yes	210

cortex to divide, leading to the formation of a *primary nodule meristem*. This meristem gives rise to the plant tissue that constitutes the root nodule. Bacteria enter the root via an infection thread, analogous to the PPA in mycorrhizal associations, and eventually reach cells inside the root nodule. There, the bacteria are released into the cytoplasm of the nodule cells and are enclosed in membrane vesicles similar to the PAM. Inside the vesicles, the bacteria differentiate into **bacteroids**—the form of the bacteria that can fix nitrogen.

Mycorrhizae expand the root system

In most cases, the roots of vascular plants cannot optimally support plant growth alone—they simply cannot reach all the nutrients available in the soil. Mycorrhizae expand the root surface area 10-fold to 1,000-fold, increasing the amount of soil the plant can explore for nutrients. In addition, because fungal hyphae are much finer than root hairs, they can get into pores in the soil that are inaccessible to roots. In this way, mycorrhizae probe a vast expanse of soil for nutrients and deliver them into root cortical cells.

The primary nutrient that the plant obtains from a mycorrhizal interaction is phosphorus. In exchange, the fungus obtains an energy source: the products of photosynthesis. In fact, up to 20 percent of the photosynthate of terrestrial plants is directed to and consumed by arbuscular mycorrhizal fungi. Such

associations are excellent examples of *mutualism*, an interaction between two species in which both species benefit (which is further discussed in Chapter 44). They are also examples of *symbiosis*, in which two different species live in close contact for a significant portion of their life cycles.

Rhizobia capture nitrogen from the air and make it available to plant cells

Although nitrogen is essential for their survival, plants cannot use atmospheric N_2 because the triple bond that links the two N atoms in N_2 requires a large amount of energy to break. Some bacteria, however, have an enzyme called **nitrogenase** that enables them to convert N_2 into a more reactive and biologically useful form, ammonia (NH_3), by a process called **nitrogen fixation**. This process is by far the most important route by which nitrogen enters biological systems. Biological nitrogen fixation is the *reduction* of nitrogen gas, and it requires a great deal of energy. It proceeds by the stepwise addition of three pairs of hydrogen atoms to N_2 (**FIGURE 25.6**).

FIGURE 25.6 Nitrogenase Fixes Nitrogen Throughout the chemical reactions of nitrogen fixation, the reactants are bound to the enzyme nitrogenase. A reducing agent transfers hydrogen atoms to a molecule of nitrogen gas (N_2), and eventually the final product—ammonia (NH_3)—is released. This reaction requires a large input of energy: about 16 ATPs.

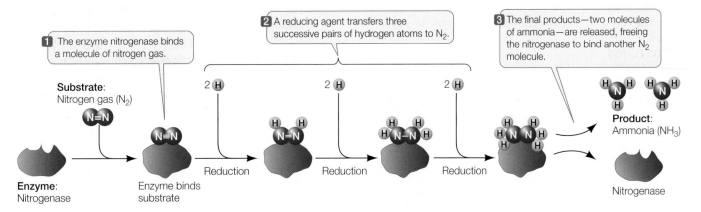

Nitrogenase is strongly inhibited by oxygen. Many free-living (nonsymbiotic) nitrogen-fixing bacteria are anaerobes that live in environments that naturally contain little or no O_2. In the aerobic environment of plant roots, the nodules provide the rhizobia with the conditions necessary for nitrogen fixation (see Figure 25.5B). Within a nodule, O_2 is maintained at a low level that is sufficient to support aerobic respiration by the bacteria (which is necessary to supply energy for the fixation reaction) but not so high as to inactivate nitrogenase. The O_2 level is regulated by a plant-produced protein called **leghemo-globin**, which is an O_2 carrier. Leghemoglobin is a close relative of hemoglobin, the red, oxygen-carrying pigment of animals, and is thus an evolutionarily ancient molecule.

> **LINK** The cycling of nitrogen through biological systems is described in Chapter 46, and some of the prokaryotes that participate in that cycling are described in Concept 19.3

As with mycorrhizae, attracting and maintaining nitrogen-fixing bacteria is costly to the plant, consuming as much as 20 percent of all of the energy stored in photosynthate. Relatively few plant species are legumes, but those species are important to humans because they can be grown in nitrogen-poor soil. They include peas, soybeans, clover, alfalfa, and some tropical shrubs and trees. Most other plants depend on nitrogen fixed by free-living bacteria in the soil or released by the breakdown of proteins present in dead organic matter.

Some plants obtain nutrients directly from other organisms

Although the majority of plants obtain most of their mineral nutrients from the soil solution (with the help of fungi), some, such as carnivorous and parasitic plants, use other sources.

CARNIVOROUS PLANTS There are about 500 carnivorous plant species that obtain some of their nutrients by digesting arthropods. These plants typically grow in boggy soils where little nitrogen or phosphorus is available. Digestion (hydrolysis) of arthropod prey helps provide those missing nutrients. A well-known example of a carnivorous plant is the Venus flytrap (genus *Dionaea*; **FIGURE 25.7A**), which has a modified leaf with two halves that fold together. When an insect touches trigger hairs on the leaf, its two halves quickly come together, and their spiny margins interlock and trap the insect before it can escape. The leaf then secretes enzymes that digest the prey.

PARASITIC PLANTS Approximately 1 percent of flowering plant species derive some or all of their water, nutrients, and even photosynthate from other plants. **Hemiparasites** can photosynthesize but derive water and mineral nutrients from the living bodies of other plants. Mistletoes, for example, contain chlorophyll and carry out some photosynthesis, but they parasitize other plants for water and mineral nutrients and may derive some photosynthetic products from them as well. Dwarf mistletoe (*Arceuthobium americanum*) is a serious parasite in forests

FIGURE 25.7 Nutrients from Other Organisms (A) The Venus flytrap (*Dionaea muscipula*) obtains nitrogen from the bodies of arthropods trapped inside the plant when its specialized leaves snap shut. (B) Purple witchweed (*Striga hermonthica*) parasitizes sorghum plants.

of the western United States, destroying more than 3 billion board feet of lumber per year.

Holoparasites are completely parasitic and do not perform photosynthesis. Witchweed (*Striga*) is a serious parasite of crops, causing over $1 billion a year in corn losses in Africa (**FIGURE 25.7B**). Interestingly, strigolactone, the molecule made by plant roots that attracts arbuscular mycorrhizal fungi, is closely related to the molecule that *Striga* responds to as well. Scientists hypothesize that a mechanism evolved in the ancestors of modern *Striga* to take advantage of a compound that was already produced by plants to attract soil microbes.

> ## Do You Understand Concept 25.2?
>
> - Compare the events that occur when arbuscular mycorrhizal fungi and rhizobia infect a plant root. What is the nutritional advantage to the plant in each case?
> - How might you turn wheat, a plant that does not have symbiotic relationships with nitrogen-fixing bacteria, into a nitrogen-fixing plant? What biochemical challenges would you face?
> - How do the nutritional needs of holoparasitic plants differ from those of carnivorous plants?
> - What characteristics are shared among nonparasitic plant–parasitic plant, plant–mycorrhizal fungus, and plant–rhizobia associations?

Now that we have learned about plant nutrients in the soil, let's consider how soil water, with its dissolved nutrients, enters the plant root, and how it moves from the root to the rest of the plant body.

concept 25.3 Water and Solutes Are Transported in the Xylem by Transpiration–Cohesion–Tension

Nutrients, as we have seen, enter plants as ions dissolved in soil water. But plants need that water as more than a solvent for nutrients. Plants require water to carry out photosynthesis, to transport solutes between plant organs, to cool their bodies by evaporation, and to maintain the internal pressure that supports their bodies.

Differences in water potential govern the direction of water movement

To enter a root cell, a solution must pass through the cell's plasma membrane. In Concept 5.2 we described *osmosis*, the movement of water through a selectively permeable membrane toward a region of higher solute concentration (that is, lower water concentration). Plant biologists define **water potential (psi, Ψ)** as the tendency of a solution (water plus solutes) to take up water from pure water across a membrane. By definition, the water potential of pure water is zero. Any solution with a water potential of less than zero has a tendency to take up water from pure water, and the lower (the more negative) the water potential, the greater the driving force for water movement across the membrane.

Water potential has two components:

• **Solute potential (Ψ_s):** Solutes affect the osmotic behavior of a solution. As solutes are added, the concentration of free water is reduced; the more solutes, the lower the water potential. So solute potential is usually negative.

• **Pressure potential (Ψ_p):** As plant cells take up water, they tend to swell. The presence of the cell wall, however, provides resistance to that swelling (see Figure 5.3C). The result is an increase in internal pressure in the cell (*turgor pressure*), which decreases the tendency of the cell to take up more water.

The water potential is the sum of the solute potential and the pressure potential:

$$\Psi = \Psi_s + \Psi_p$$

We can measure all three potentials in *megapascals* (MPa), a unit of pressure. Atmospheric pressure—"one atmosphere"—is about 0.1 MPa, or 14.7 pounds per square inch (a typical pressure in an automobile tire is about 0.2 MPa).

Whenever water moves across a selectively permeable membrane by osmosis, it moves toward the region of lower (more negative) water potential (**FIGURE 25.8A**).

In a plant cell (**FIGURE 25.8B**), turgor pressure is equivalent to the pressure potential (Ψ_p) exerted by the piston in Figure 25.8A. Water will enter plant cells by osmosis until the pressure

FIGURE 25.8 Water Potential, Solute Potential, and Pressure Potential (A) A theoretical illustration of water potential. (B) The effect of differences in water potential on a plant cell.

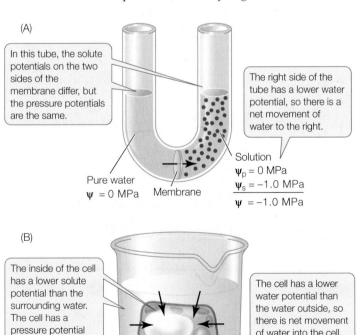

(A)

In this tube, the solute potentials on the two sides of the membrane differ, but the pressure potentials are the same.

The right side of the tube has a lower water potential, so there is a net movement of water to the right.

Pure water
$\Psi = 0$ MPa

Membrane

Solution
$\Psi_p = 0$ MPa
$\Psi_s = -1.0$ MPa
$\Psi = -1.0$ MPa

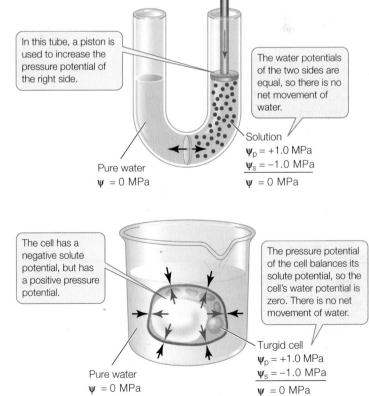

In this tube, a piston is used to increase the pressure potential of the right side.

The water potentials of the two sides are equal, so there is no net movement of water.

Pure water
$\Psi = 0$ MPa

Solution
$\Psi_p = +1.0$ MPa
$\Psi_s = -1.0$ MPa
$\Psi = 0$ MPa

(B)

The inside of the cell has a lower solute potential than the surrounding water. The cell has a pressure potential of zero.

The cell has a lower water potential than the water outside, so there is net movement of water into the cell.

Pure water
$\Psi = 0$ MPa

Flaccid cell
$\Psi_p = 0$ MPa
$\Psi_s = -1.0$ MPa
$\Psi = -1.0$ MPa

The cell has a negative solute potential, but has a positive pressure potential.

The pressure potential of the cell balances its solute potential, so the cell's water potential is zero. There is no net movement of water.

Pure water
$\Psi = 0$ MPa

Turgid cell
$\Psi_p = +1.0$ MPa
$\Psi_s = -1.0$ MPa
$\Psi = 0$ MPa

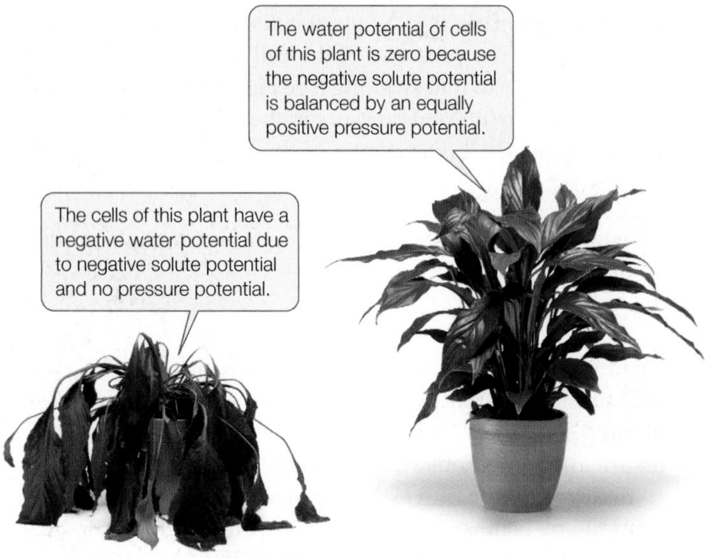

> The water potential of cells of this plant is zero because the negative solute potential is balanced by an equally positive pressure potential.

> The cells of this plant have a negative water potential due to negative solute potential and no pressure potential.

FIGURE 25.9 A Wilted Plant A plant wilts when the pressure potential of its cells drops.

potential exactly balances the solute potential. At this point, the cell is *turgid*; that is, it has a significant positive pressure potential. The physical structure of many plants is maintained by the (positive) pressure potential of their cells; if the pressure potential drops (for example, if the plant does not have enough water), the plant *wilts* (**FIGURE 25.9**).

Water and ions move across the root cell plasma membrane

The movement of a soil solution containing mineral ions across a root cell plasma membrane faces two major challenges:

- The membrane is hydrophobic, whereas water and mineral ions are polar.
- Some mineral ions must be moved against their concentration gradient.

These two challenges are met by membrane proteins.

> **LINK** Review the structure and functions of biological membranes in Chapter 5

AQUAPORINS *Aquaporins* are membrane channels through which water can diffuse (see Figure 5.5). Although aquaporins were first discovered in animal cells, they are also abundant in plant plasma membranes. Their concentration in the plasma membrane and their function are both regulated so that the rate of osmosis can be changed. But the direction of osmosis always remains the same: water moves to the region of more negative water potential.

ION CHANNELS AND PROTON PUMPS When the concentration of an ion in the soil solution is greater than that in the root, transport proteins can move the ion into the plant by facilitated diffusion. The concentrations of many mineral ions in the soil solution, however, are lower than those required inside the plant. In addition, electric charge differences play a role in the uptake of mineral ions. For example, a negatively charged ion that moves into a negatively charged cellular compartment is moving against an *electrical gradient*, and this movement requires energy. Concentration and electrical gradients combine to form an *electrochemical gradient*. Uptake against an electrochemical gradient requires *active transport*.

> **LINK** Electrochemical gradients are discussed in more detail in Concept 34.2

Animals use a $Na^+–K^+$ pump to generate a Na^+ gradient that can be used to drive the uptake of other solutes (see Figure 5.7). Instead of a $Na^+–K^+$ pump, plants have a **proton pump**, which uses energy from ATP to move protons out of the cell against a *proton concentration gradient* (**FIGURE 25.10, STEP 1**). Because protons (H^+) are positively charged, their accumulation outside the cell has two results:

- An electrical gradient is created such that the region just outside the cell becomes more positively charged than the inside of the cell.
- A proton concentration gradient develops, with more protons just outside the cell than inside the cell.

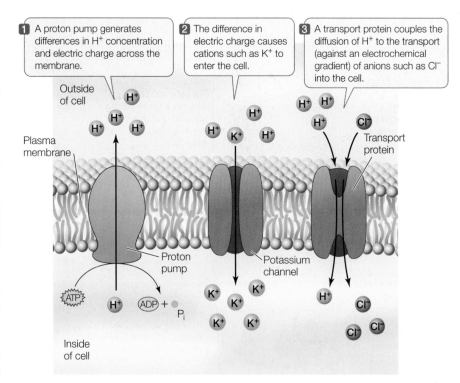

1 A proton pump generates differences in H^+ concentration and electric charge across the membrane.

2 The difference in electric charge causes cations such as K^+ to enter the cell.

3 A transport protein couples the diffusion of H^+ to the transport (against an electrochemical gradient) of anions such as Cl^- into the cell.

FIGURE 25.10 Ion Transport into Plant Cells The active transport of protons (H^+) out of the cell by a proton pump (step 1) drives the movement of both cations (step 2) and anions (step 3) into the cell.

Each of these results has consequences for the movement of other ions. Because the inside of the cell is now more negative than the outside, cations such as potassium (K^+) move into the cell by facilitated diffusion through their specific membrane channels (**FIGURE 25.10, STEP 2**). In addition, the proton concentration gradient can be harnessed to drive active transport. Anions such as chloride (Cl^-) are moved into the cell against an electrochemical gradient by a membrane transport protein that couples their movement with that of H^+ (**FIGURE 25.10, STEP 3**). This process is an example of secondary active transport.

Water and ions pass to the xylem by way of the apoplast and symplast

Water and ions from the soil solution move through the roots to the xylem by one of two pathways: the *apoplast* or the *symplast* (**FIGURE 25.11**).

- The **apoplast** (Greek *apo*, "away from"; *plast*, "living material") consists of the cell walls, which lie outside the plasma

membranes, and intercellular spaces (spaces between cells), which are common in many plant tissues. The apoplast is a continuous meshwork through which water and solutes can flow without ever having to cross a membrane.

- The **symplast** (Greek *sym*, "together with") passes through the continuous cytoplasm of the living cells connected by plasmodesmata. The selectively permeable plasma membranes of the root cells control access to the symplast, so movement of water and solutes into the symplast is tightly regulated.

Water and minerals that pass from the soil solution through the apoplast can travel freely as far as the endodermis, the

FIGURE 25.11 Apoplast and Symplast Plant cell walls and intercellular spaces constitute the apoplast. The symplast consists of the cytoplasm of living cells, which are connected by plasmodesmata. To enter the symplast, water and solutes must pass through a plasma membrane. No such selective barrier limits movement through the apoplast.

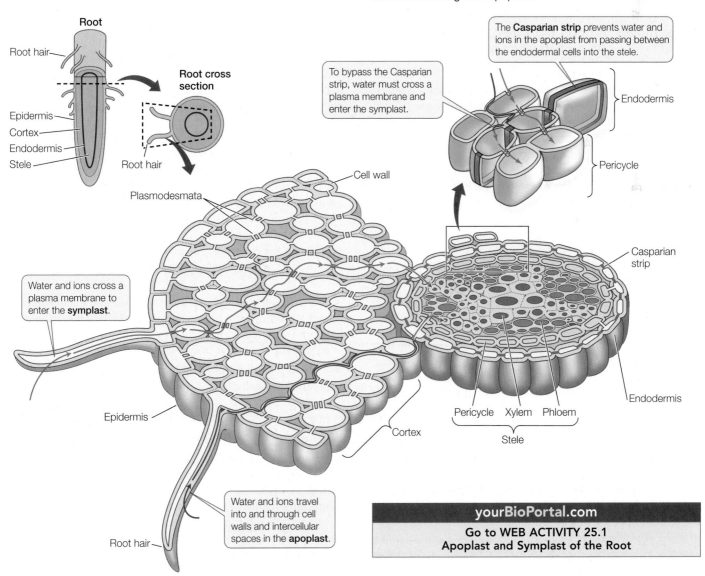

Root

Root hair

Root cross section

Epidermis
Cortex
Endodermis
Stele

Root hair

The **Casparian strip** prevents water and ions in the apoplast from passing between the endodermal cells into the stele.

To bypass the Casparian strip, water must cross a plasma membrane and enter the symplast.

Cell wall

Plasmodesmata

Endodermis

Pericycle

Casparian strip

Water and ions cross a plasma membrane to enter the **symplast**.

Epidermis

Cortex

Pericycle Xylem Phloem

Stele

Endodermis

Water and ions travel into and through cell walls and intercellular spaces in the **apoplast**.

Root hair

innermost layer of the root cortex. The endodermis is distinguished from the rest of the ground tissue by the presence of the **Casparian strip**. This waxy, suberin-impregnated region of the endodermal cell wall forms a hydrophobic belt around each endodermal cell where it is in contact with other endodermal cells. The Casparian strip acts as a seal that prevents water and ions from moving through spaces between the endodermal cells; instead, water and ions enter the cytoplasm of the endodermal cells. (The Casparian strip is functionally equivalent to the tight junctions found in some animal tissues; see Figure 4.18.) Water and ions already in the symplast can enter the endodermal cells through plasmodesmata.

Once they have passed the endodermal barrier, water and minerals remain in the symplast until they reach parenchyma cells in the pericycle or xylem. These cells then actively transport mineral ions into the apoplast of the stele. As ions are transported into the solution in the cell walls of the stele, water potential in the apoplast becomes more negative; consequently, water moves out of the cells and into the apoplast by osmosis. In other words, ions are transported actively, and water follows passively. The end result is that water and minerals end up in the xylem, where they constitute the *xylem sap*.

Water moves through the xylem by the transpiration–cohesion–tension mechanism

Once water has arrived in the xylem, it is all uphill from there! Consider the magnitude of what xylem accomplishes. A single maple tree 15 m tall has been estimated to have some 177,000 leaves, with a total leaf surface area of 675 m^2—1.5 times the area of a basketball court. During a summer day, that tree loses 220 liters of water *per hour* to the atmosphere by evaporation from its leaves. So, to prevent wilting, the xylem must transport 220 liters of water 15 m from the roots up to the leaves every hour.

How is the xylem able to move so much water to such great heights? Part of the answer lies in the structure of the xylem, which we described in Chapter 24. Recall that xylem vessels consist of a long "straw" of cell walls, which provide both structural support and the rigidity needed to maintain a pressure gradient within the xylem sap. Early experiments ruled out two possible ways that had been hypothesized for moving water up through these vessels from roots to leaves: upward pressure and capillary action.

Upward pressure would move water by "pushing the bottom." A simple experiment in 1893 ruled out this mechanism: A tree was cut at its base, and the sawed-off part was placed in a vat of poison that killed living cells. (Recall that mature xylem is made up of dead cells.) The poison continued to rise through the trunk, even though cells were killed along the way. This observation showed that a living "pump" is not required to push the xylem sap up a tree. Furthermore, the roots were absent, yet upward flow in the trunk continued, demonstrating that the roots do not provide upward pressure. When the poison reached the leaves and killed them, however, all further movement up the trunk stopped, demonstrating that leaves must be alive for water to move in the xylem. *Capillary action* was also ruled out as the mechanism that makes water rise in

the xylem. Water molecules have strong cohesion, so a column of water can indeed rise in a thin tube (e.g., a straw or a xylem vessel) by capillary action, but calculations have shown that xylem vessels, at 100 μm in diameter, are simply too wide to get water to the top of a 15-m tree in this fashion. In fact, the maximum height for a water column raised by capillary action alone in a 100-μm-diameter vessel is only 0.15 m.

LINK Review the properties of water in Chapter 2

Almost a century ago, the **transpiration–cohesion–tension** theory was proposed to explain water transport in the xylem (**FIGURE 25.12**). According to this theory, the key elements of xylem transport are:

- *Transpiration:* evaporation of water from cells within the leaves
- *Cohesion* of water molecules in the xylem sap due to hydrogen bonding
- *Tension* on the xylem sap resulting from transpiration

The concentration of water vapor in the atmosphere is lower than that in the apoplast of a leaf. Because of this difference, water vapor diffuses from the intercellular spaces of the leaf and through the *stomata* to the outside air in a process called **transpiration**. Within the leaf, water evaporates from the moist walls of the mesophyll cells and enters the intercellular spaces. As water evaporates from the aqueous film coating each cell, the film shrinks back into tiny spaces in the cell walls, increasing the curvature of the water surface and thus increasing its surface tension. Because of hydrogen bonding, water molecules have cohesion, and therefore the increased *tension* (negative pressure potential) in the surface film draws more water into the cell walls, replacing that which was lost by evaporation. The resulting tension in the mesophyll draws water from the xylem of the nearest vein into the apoplast surrounding the mesophyll cells. The removal of water from the veins, in turn, establishes tension on the entire column of water contained in the xylem, so that the column is drawn upward all the way from the roots.

Each part of this theory is supported by evidence:

- The difference in water potential between the soil solution and air is huge, on the order of –100 MPa. This difference should generate more than enough tension to pull a water column up the tallest tree.
- There is a continuous column of water in the xylem, which is caused by cohesion.
- Measurements of xylem pressures in cut stems show a negative pressure potential, which indicates considerable tension.

In addition to its role in the transpiration–cohesion–tension mechanism, transpiration has the added benefit of cooling a plant's leaves (much as humans sweat to cool off). The evaporation of water from mesophyll cells consumes heat, thereby decreasing the leaf temperature. A farmer can hold a leaf between thumb and forefinger to estimate its temperature; if the leaf doesn't feel cool, that means transpiration is not occurring and it must be time to water.

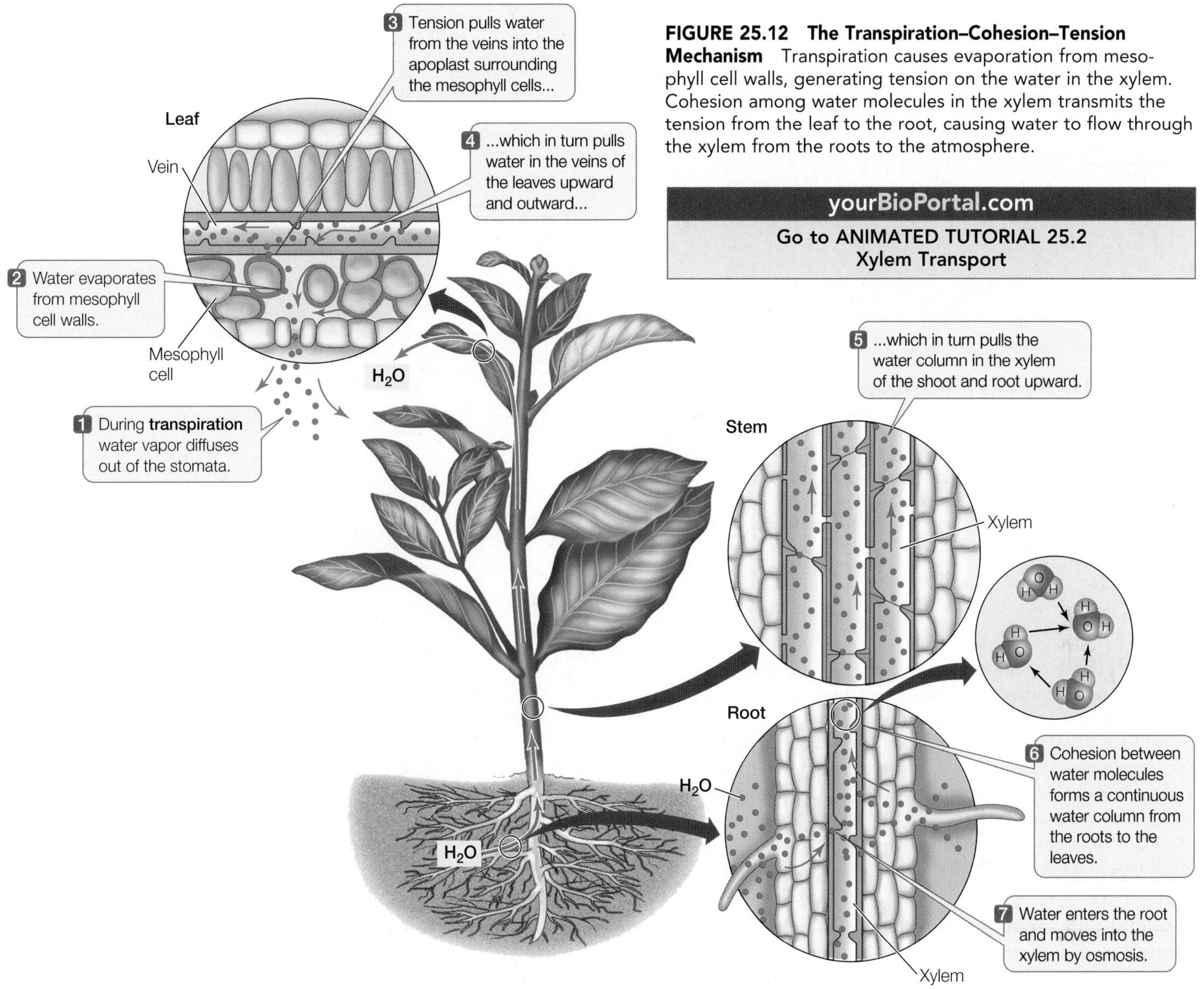

3 Tension pulls water from the veins into the apoplast surrounding the mesophyll cells...

4 ...which in turn pulls water in the veins of the leaves upward and outward...

Leaf

Vein

2 Water evaporates from mesophyll cell walls.

Mesophyll cell

H_2O

1 During **transpiration** water vapor diffuses out of the stomata.

FIGURE 25.12 The Transpiration–Cohesion–Tension Mechanism Transpiration causes evaporation from mesophyll cell walls, generating tension on the water in the xylem. Cohesion among water molecules in the xylem transmits the tension from the leaf to the root, causing water to flow through the xylem from the roots to the atmosphere.

yourBioPortal.com
Go to ANIMATED TUTORIAL 25.2
Xylem Transport

5 ...which in turn pulls the water column in the xylem of the shoot and root upward.

Stem

Xylem

Root

H_2O

6 Cohesion between water molecules forms a continuous water column from the roots to the leaves.

7 Water enters the root and moves into the xylem by osmosis.

H_2O

Xylem

Although transpiration provides the driving force for the transport of water and minerals in the xylem, it also results in the loss of tremendous quantities of water from the plant. How do plants control this loss?

Stomata control water loss and gas exchange

In Chapter 6 we describe photosynthesis, which has the general equation

$$CO_2 + H_2O \rightarrow carbohydrate + O_2$$

Leaves are the chief organs of photosynthesis, so they must have a large surface area across which to exchange the gases O_2 and CO_2. This surface area is provided by the mesophyll cells being surrounded by abundant air spaces inside the leaf (see Figure 25.12), which are also the location of the evaporation that drives water movement in the xylem. The epidermis of leaves and stems minimizes transpirational water loss by secreting a waxy cuticle, which is impermeable to water. However, the cuticle is also impermeable to CO_2 and O_2. How can

the plant balance its need to retain water with its need to obtain CO_2 for photosynthesis?

Plants have met this challenge with the evolution of **stomata** (singular *stoma*): pores in the leaf epidermis, typically on the underside of the leaf (**FIGURE 25.13A**). There are up to 1,000 stomata per square millimeter in some leaves, constituting up to 3 percent of the total surface area of the shoot. Such an abundance of stomata could lead to excessive water loss. But stomata are not static structures. A pair of specialized epidermal cells, called **guard cells**, controls the opening and closing of each stoma.

When the stomata are open, CO_2 can enter the leaf by diffusion, but water vapor diffuses out of the leaf at the same time. Closed stomata prevent water loss but also exclude CO_2 from the leaf. Most plants open their stomata when the light intensity is sufficient to maintain a moderate rate of photosynthesis. At night, when darkness precludes photosynthesis, their stomata are closed; no CO_2 is needed at this time, and water is conserved. Even during the day, the stomata close if water is being lost at too great a rate.

(A)

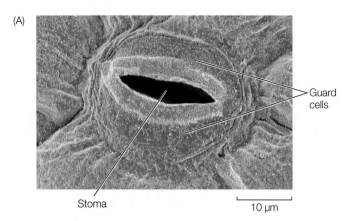

Guard cells

Stoma

10 μm

FIGURE 25.13 Stomata Regulate Gas Exchange and Transpiration (A) Scanning electron micrograph of an open stoma formed by two sausage-shaped guard cells. (B) Potassium ion concentrations affect the water potential of the guard cells, controlling the opening and closing of stomata. Negatively charged ions (e.g., Cl^-) that accompany K^+ maintain electrical balance and contribute to the changes in water potential that open and close the stomata.

(B)

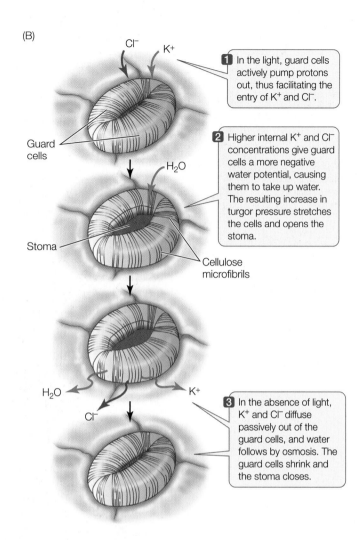

Guard cells

Stoma

Cl^- K^+

H_2O

Cellulose microfibrils

H_2O K^+ Cl^-

1 In the light, guard cells actively pump protons out, thus facilitating the entry of K^+ and Cl^-.

2 Higher internal K^+ and Cl^- concentrations give guard cells a more negative water potential, causing them to take up water. The resulting increase in turgor pressure stretches the cells and opens the stoma.

3 In the absence of light, K^+ and Cl^- diffuse passively out of the guard cells, and water follows by osmosis. The guard cells shrink and the stoma closes.

FRONTIERS Water-use efficiency is a measure of how much water is consumed (in kilograms) per kilogram of plant material produced. This measure varies among crop species, from water-hungry soybeans at 650 kg/kg to drought-adapted sorghum at 300 kg/kg. A single transcription factor, first found in *Arabidopsis*, improves water-use efficiency dramatically. Scientists are now trying to find out if genes encoding this transcription factor can be used to produce crop plants that need less water.

Guard cells can respond to changes in light and CO_2 concentration in a matter of minutes by changing their solute potential. These changes are driven by changes in the K^+ concentration in the guard cells. The absorption of light by a pigment in the guard cell plasma membrane activates a proton pump (see Figure 25.10), which actively transports H^+ out

of the guard cells and into the apoplast of the surrounding epidermis. The resulting electrochemical gradient drives K^+ into the guard cells, where it accumulates (**FIGURE 25.13B**). The increased internal concentration of K^+ makes the solute potential of the guard cells more negative. Negatively charged chloride ions and organic ions also move into and out of the

APPLY THE CONCEPT

Water and solutes are transported in the xylem by transpiration–cohesion–tension

Here are measurements of water potential (Ψ) in a 100-m-tall tree and its surroundings:

REGION	Ψ (MPa)
Soil water	−0.3
Xylem of root	−0.6
Xylem of trunk	−1.2
Inside of leaf	−2.0
Outside air	−58.5

Gravity exerts a force of −0.01 MPa per meter of height above ground.

1. Is the water potential in the leaf sufficiently low to draw water to the top of the tree?

2. Would transpiration continue if soil water potential decreased to −1.0?

3. What would you expect to happen to the xylem water potential if all of the stomata closed?

guard cells along with the K^+ ions, maintaining electrical balance and contributing to the change in the solute potential of the guard cells. Water then enters by osmosis (guard cell membranes are particularly rich in aquaporins), increasing the turgor pressure of the guard cells. The guard cells change their shape in response to the increase in pressure potential, so that a space—the stoma—appears between them. The stoma closes in the absence of light: the proton pump becomes less active, K^+ ions diffuse passively out of the guard cells, water follows by osmosis, the pressure potential decreases, and the guard cells sag together and seal off the stoma.

Do You Understand Concept 25.3?

- What distinguishes water potential, solute potential, and pressure potential?

- What are the roles of transpiration, cohesion, and tension in xylem transport?

- If *Arabidopsis* is exposed to high CO_2 concentrations, the new leaves that form on the plant have fewer stomata than they would have had under normal conditions. Why do you think this might be advantageous?

- When water is in short supply, some plants shed their leaves. Explain why this occurs.

Now that we understand how leaves obtain the water and CO_2 required for photosynthesis, let's examine how the products of photosynthesis are transported to other parts of the plant where they are needed.

concept 25.4 Solutes Are Transported in the Phloem by Pressure Flow

As photosynthesis occurs in the leaf, the carbohydrate products of photosynthesis (mainly sucrose) diffuse to the nearest small vein, where they are actively transported into sieve tube elements. The movement of carbohydrates and other solutes through the plant in the phloem is called **translocation**. These solutes are translocated from *sources* to *sinks*:

- A **source** is an organ (such as a mature leaf or a storage root) that *produces* (by photosynthesis or by digestion of stored reserves) more photosynthate than it requires.

- A **sink** is an organ (such as a root, flower, developing fruit or tuber, or immature leaf) that *consumes* photosynthate for its own growth and storage needs.

Sources and sinks can change roles. For example, storage roots (such as sweet potatoes) are sinks when they accumulate photosynthate, but they are sources when their stored reserves are needed to nourish other organs in the plant.

LINK The anatomy of sieve tubes is described in Concept 24.1

Sucrose and other solutes are carried in the phloem

Evidence that the phloem carries sucrose and other solutes initially came in the 1600s, when Marcello Malpighi removed a ring of bark (containing the phloem) from the trunk of a tree, while leaving the xylem intact—that is, he *girdled* the tree. Over time, the bark in the region above the girdle swelled:

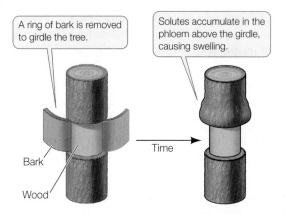

A ring of bark is removed to girdle the tree.

Solutes accumulate in the phloem above the girdle, causing swelling.

Bark

Wood

Time

Malpighi correctly concluded that the solution coming from the leaves above the girdle was trapped in the bark. Later, the bark below the girdle died because it was no longer receiving sugars from the leaves. Eventually the roots, and then the entire tree, died.

In the twentieth century, plant biologists analyzed the contents of phloem with the help of aphids. These insects feed on plants by drilling into sieve tube elements with a specialized organ, the *stylet*. The pressure potential in the sieve tube is higher than that outside the plant, so the nutritious phloem sap is forced through the stylet and into the aphid's digestive tract. So great is the pressure that sugary liquid is forced through the insect's body and out its anus:

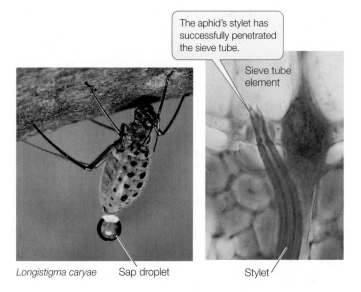

The aphid's stylet has successfully penetrated the sieve tube.

Sieve tube element

Longistigma caryae   Sap droplet   Stylet

If an aphid is frozen in the act of feeding, its body can be chopped off the plant stem, leaving the stylet intact. Phloem sap continues to flow out of the stylet for hours.

Several observations came from analyzing phloem sap in this way, as well as from other experiments:

- Ninety percent of the phloem solutes consist of sucrose, with the rest hormones, other small molecules such as mineral nutrients and amino acids, and viruses.

- The flow rate can be very high, as much as 100 centimeters per hour.

- Different sieve tube elements conduct their contents in different directions (up or down the stem, for example). Movement in the phloem as a whole is bidirectional, but it is unidirectional within an individual sieve tube.

- Movement of sap in the phloem requires living cells, in contrast to movement in the xylem.

The pressure flow model describes the movement of fluid in the phloem

The observations described above led to the **pressure flow model** as an explanation for translocation in the phloem (**FIGURE 25.14**). According to this model, sucrose is actively transported at a source into companion cells, from which it flows through plasmodesmata into the sieve tube elements. This gives those cells a higher sucrose concentration than the surrounding cells (a more negative solute potential), and water therefore enters the sieve tube elements from the xylem by osmosis. The entry of this water causes turgor pressure (a more positive pressure potential) at the source end of the sieve tube, so that the entire fluid content of the sieve tube is pushed toward the sink end of the tube—in other words, the sap moves in response to a pressure gradient. In the sink, the sucrose is unloaded both passively and by active transport, and water moves back into the xylem. In this way, the gradient of solute potential and pressure potential needed for the movement of phloem sap (translocation) is maintained.

Two steps in phloem translocation require metabolic energy:

- Transport of sucrose and other solutes from sources into companion cells and then into the sieve tubes; called *loading*

- Transport of solutes from the sieve tubes into sinks; called *unloading*

The need for metabolic energy is the reason why phloem transport, unlike xylem transport, requires living cells.

Two general routes can be taken by sucrose and other solutes as they move from the mesophyll cells

into the phloem. In many plants, sucrose and other solutes follow an *apoplastic pathway*: they leave the mesophyll cells and enter the apoplast before they reach the sieve tube elements. Specific sugars and amino acids are then actively transported into cells of the phloem. Because the solutes cross at least one selectively permeable membrane in the apoplastic pathway, selective transport can be used to regulate which specific substances enter the phloem. In other plants, solutes follow a *symplastic pathway*: the solutes remain within the symplast all the way from the mesophyll cells to the sieve tube elements. Because no membranes are crossed in the symplastic pathway, a mechanism that does not involve membrane transport is used to load sucrose into the phloem.

In sink regions, the solutes are actively transported *out* of the sieve tube elements and into the surrounding tissues. This unloading serves two purposes: it helps maintain the gradient of solute potential, and hence of pressure potential, in the sieve tubes; and it helps build up high concentrations of carbohydrates in storage organs, such as developing fruits and seeds.

Do You Understand Concept 25.4?

- Deer sometimes graze on trees by eating bark during the winter when other food is scarce. By spring, those trees may be dead. How does this happen?

- What would happen if the ends of sieve tubes were blocked and did not allow phloem sap to flow through them?

- Make a table that compares the flow of fluids in xylem and in phloem with respect to the driving force for movement; where flow occurs and whether the tubes are alive; and whether the driving pressure is positive or negative.

FIGURE 25.14 The Pressure Flow Model Differences in water potential produce a pressure gradient that moves phloem sap from sources to sinks.

yourBioPortal.com

Go to ANIMATED TUTORIAL 25.3
The Pressure Flow Model

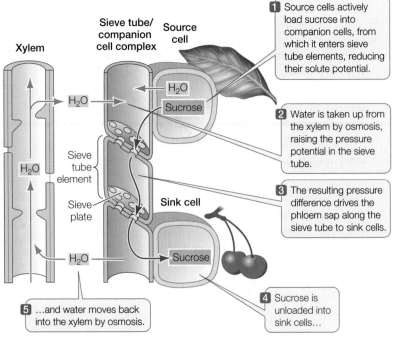

1 Source cells actively load sucrose into companion cells, from which it enters sieve tube elements, reducing their solute potential.

2 Water is taken up from the xylem by osmosis, raising the pressure potential in the sieve tube.

3 The resulting pressure difference drives the phloem sap along the sieve tube to sink cells.

4 Sucrose is unloaded into sink cells…

5 …and water moves back into the xylem by osmosis.

QA
QUESTION

How can soil be managed for optimal plant growth?

ANSWER   Every year, 25 billion tons of topsoil (A horizon) are lost worldwide from potentially arable land. This loss reduces the nutrient content (Concept 25.1) and water-holding capacity (Concept 25.3) of the remaining soil. It therefore creates added demand for fertilizers and irrigation to restore the soils, but these inputs are too expensive in most areas of the world, and they can have adverse effects on the rest of the environment. For example, adding fertilizer to soils can pollute nearby waters, and huge irrigation projects destroy natural habitats and can lead to flooding.

Faced with these challenges, scientists and farmers (often the same people!) have invented methods of *sustainable farming*. In the United States, for example, a branch of the federal government called the Natural Resource Conservation Service (formerly the Soil Conservation Service) was established in response to the country's experiences with soil degradation during the Dust Bowl of the 1930s.

There are several strategies to sustain the structure, nutrient content, and water-holding capacity of soil. One of these strategies is *crop rotation*. Different plants have different nutritional requirements and so remove different nutrients from the soil as they grow and are harvested. For example, corn requires large amounts of nitrogen, whereas soybeans harbor nitrogen-fixing bacteroids in their roots (Concept 25.2). In the midwestern United States, corn and soybeans are planted in alternate growing seasons.

Another strategy is *conservation tillage*. This refers to methods such as plowing that are used to prepare soil for planting. Conservation tillage can help maintain soil and reduce soil losses. If you fly over any heavily farmed area of the

Seedlings of current crop   Residue from previous crop

FIGURE 25.15 Conservation Tillage   These rye plants are growing amidst the residue of corn plants that were harvested for their grain. The crop residues help conserve the soil.

United States, look out the window and you will see contour plowing patterns; these patterns are used to control runoff of the topsoil during heavy rains. More recently, it has become increasingly common for farmers to leave behind on the field up to half of the stems and leaves of plants that are harvested for grain (**FIGURE 25.15**). The farmers plow these crop residues back into the soil, where they help maintain soil structure, bind water, and provide nutrients because they are organic matter that builds up humus. Over one-third of the farmland in the United States is now being farmed in this way.

Our increasing understanding of plants and their relationship to soils will doubtless lead to more sustainable practices in the future.

---

25 SUMMARY

concept 25.1 Plants Acquire Mineral Nutrients from the Soil

- Plants are photosynthetic autotrophs that require water and certain mineral nutrients to survive. They obtain most of these mineral nutrients as ions from the **soil solution**.

- The **essential elements** for plants include six **macronutrients** and several **micronutrients**. Plants that lack a particular nutrient show characteristic deficiency symptoms. **Review Figure 25.1 and ANIMATED TUTORIAL 25.1**

- The essential elements were discovered by growing plants **hydroponically** in solutions that lacked individual elements. **Review Figure 25.2 and WORKING WITH DATA 25.1**

- Soils supply plants with mechanical support, water and dissolved ions, air, and the services of other organisms. **Review Figure 25.3**

- Protons take the place of mineral nutrient cations bound to clay particles in soil in a process called **ion exchange**. **Review Figure 25.4**

- Farmers may use shifting agriculture or **fertilizer** to make up for nutrient deficiencies in soil.

concept 25.2 Soil Organisms Contribute to Plant Nutrition

- Signaling molecules called **strigolactones** induce the hyphae of arbuscular mycorrhizal fungi to invade root cortical cells and form arbuscules, which serve as sites of nutrient exchange between fungus and plant. **Review Figure 25.5A**

- Legumes signal nitrogen-fixing bacteria (rhizobia) to form **bacteroids** within nodules that form on their roots. **Review Figure 25.5B**

- In nitrogen fixation, nitrogen gas (N_2) is reduced to ammonia in a reaction catalyzed by **nitrogenase**. **Review Figure 25.6**

- Carnivorous plants supplement their nutrient supplies by trapping and digesting arthropods. Parasitic plants obtain minerals, water, or products of photosynthesis from other plants.

concept 25.3 Water and Solutes Are Transported in the Xylem by Transpiration–Cohesion–Tension

- Water moves through biological membranes by osmosis, always moving toward regions with a more negative water potential. The **water potential** (Ψ) of a cell or solution is the sum of its **solute potential** (Ψ_s) and its **pressure potential** (Ψ_p). **Review Figure 25.8 and INTERACTIVE TUTORIAL 25.1**

- The physical structure of many plants is maintained by the positive pressure potential of their cells (turgor pressure); if the pressure potential drops, the plant wilts.

- Water moves into root cells by osmosis through aquaporins. Mineral ions move into root cells through ion channels, by facilitated diffusion, and by secondary active transport. **Review Figure 25.10**

- Water and ions may pass from the soil into the root by way of the **apoplast** or the **symplast**, but they must pass through the symplast to cross the endodermis and enter the xylem. The **Casparian strip** in the endodermis blocks the movement of water and ions through the apoplast. **Review Figure 25.11 and WEB ACTIVITY 25.1**

- Water is transported in the xylem by the **transpiration–cohesion–tension** mechanism. Evaporation from the leaf produces tension in the mesophyll, which pulls a column of water—held together by cohesion—up through the xylem from the root. **Review Figure 25.12 and ANIMATED TUTORIAL 25.2**

- **Stomata** allow a balance between water retention and CO_2 uptake. Their opening and closing is regulated by **guard cells**. **Review Figure 25.13**

concept 25.4 Solutes Are Transported in the Phloem by Pressure Flow

- **Translocation** is the movement of the products of photosynthesis, as well as some other small molecules, through sieve tubes in the phloem. The solutes move from **sources** to **sinks**.

- Translocation is explained by the **pressure flow model**: the difference in solute potential between sources and sinks creates a difference in pressure potential that pushes phloem sap along the sieve tubes. **Review Figure 25.14 and ANIMATED TUTORIAL 25.3**

Plant Growth and Development

In their constant search for ways to help farmers produce more food for a growing population, biologists have developed cereal crops whose physiology allows them to produce more grain per plant (a higher yield). When a plant produces a lot of seeds, however, the sheer weight of the seeds may cause the stem to bend over or even break. This makes harvesting the seeds impossible: think of how hard it would be to get enough food for your family if you had to pick up seeds on the ground, some of which had already sprouted.

In 1945, the U.S. Army temporarily occupied Japan, which it had defeated in World War II. During the war, Japan, an island nation with a limited amount of land suitable for farming, was blockaded and could not import food. How had Japan been able to grow enough grain to feed its people? The answer lay in the fields: the Japanese had bred genetic strains of rice and wheat with short, strong stems that could bear a high yield of grain without bending or breaking. This innovation made an impression on an agricultural advisor who happened to be among the first wave of U.S. occupiers, and seeds of the Japanese strains were sent back to the United States.

A decade later, Norman Borlaug, a plant geneticist who was working in Mexico at the time, began genetic crosses of what were known as semi-dwarf wheat plants from Japan with wheat varieties that had genes conferring rapid growth, adaptability to varying climates, and resistance to fungal diseases. The results were new genetic strains of wheat that produced record yields, first in Mexico and then in India and Pakistan in the 1960s. At about the same time, and using a similar

strategy, scientists in the Philippines developed new semi-dwarf strains of rice with equally spectacular results. People who had lived on the edge of starvation now produced enough food. Countries that had been relying on food aid from other countries were now growing so much grain that they could export the surplus. The development of these new semi-dwarf strains began what was called the "Green Revolution." Borlaug was awarded the Nobel Peace Prize for his research on wheat, which is estimated to have saved a billion lives.

QUESTION What changes in their growth patterns made the new strains of cereal crops produced by the Green Revolution so successful?

Plant geneticist Norman Borlaug, seen here in a field of semi-dwarf wheat, carried out a program of genetic crosses that led to high-yielding varieties of wheat and saved millions of people from starvation.

KEY CONCEPTS

26.1 Plants Develop in Response to the Environment

26.2 Gibberellins and Auxin Have Diverse Effects but a Similar Mechanism of Action

26.3 Other Plant Hormones Have Diverse Effects on Plant Development

26.4 Photoreceptors Initiate Developmental Responses to Light

Plants Develop in Response to the Environment

Plants are sessile organisms that must seek out resources above and below the ground. To maximize their access to the resources they need to grow and reproduce, plants have constantly dividing meristems that form new organs throughout their lives. Thus they can respond to an ever-changing environment by selective growth of organs (see Chapter 24).

The *development* of a plant—the series of progressive changes that take place throughout its life—is regulated in many ways. Several key factors are involved in regulating plant growth and development:

- *Environmental cues*, such as day length
- *Receptors* that allow a plant to sense environmental cues, such as photoreceptors that absorb light
- *Hormones*—chemical signals that mediate the effects of environmental cues, including those sensed by receptors
- The plant's *genome*, which encodes regulatory proteins and enzymes that catalyze the biochemical reactions of development

We will explore these factors in more detail later in this chapter. But first let's look at the initial steps of plant development—from seed to seedling—and the environmental cues and internal responses that guide them.

The seed germinates and forms a growing seedling

Concepts 24.1 and 27.1 describe the events of plant reproduction and development that lead to the formation of seeds. Here we begin with the seed, the structure that contains the early embryo. Unlike most animal embryos, plant seeds may be held in "suspended animation," with development of the embryo halted, for long periods. If development stops, even when external conditions (such as water supply) are adequate for growth, the seed is said to be **dormant**.

DORMANCY Seed dormancy may last for weeks, months, years, or even centuries: in 2005, a botanist was able to germinate a date palm seed recovered from a 2,000-year-old storage bin at Masada in Israel. Plants use several mechanisms to maintain dormancy:

- *Exclusion of water or oxygen* from the embryo by an impermeable seed coat
- *Mechanical restraint* of the embryo by a tough seed coat
- *Chemical inhibition* of germination

Dormancy can be broken by factors that overcome these mechanisms. For example, the seed coat may be damaged by passage through an animal's digestive system, or heavy rains may wash away chemical inhibitors.

Seed dormancy is a common phenomenon, so it must have selective advantages for plants. Dormancy ensures survival during unfavorable conditions and results in germination when conditions are most favorable for growth. For example, to avoid germination in the dry days of late summer, some seeds require exposure to a long cold period (winter) before they will germinate. Dormancy also helps seeds survive long-distance dispersal, allowing plants to colonize new territory.

GERMINATION Seeds begin to **germinate**, or sprout, when dormancy is broken and environmental conditions are satisfactory. For example, a seed may have a germination inhibitor that gets washed away in heavy rain, which means the soils will have plenty of water and germination will be possible. The first step in germination is the uptake of water, called **imbibition** (from *imbibe*, "to drink in"). A dormant seed contains very little water: only 5–15 percent of its weight is water, compared with 80–95 percent for most other plant parts. Seeds also contain polar macromolecules, such as cellulose and starch, that attract and bind polar water molecules. Consequently, a seed has a very negative water potential (see Concept 25.3) and will take up water if the seed coat is permeable. The force exerted by imbibing seeds, which expand severalfold in volume, demonstrates the magnitude of their water potential. Imbibing cocklebur seeds can exert a pressure of up to 1,000 atmospheres (approximately 100 kilopascals, or 15,000 pounds per square inch).

As a seed takes up water, it undergoes metabolic changes: enzymes are activated upon hydration, RNA and then proteins are synthesized, the rate of cellular respiration increases, and other metabolic pathways are activated. In many seeds, cell division is not initiated during the early stages of germination. Instead, growth results solely from the expansion of small preformed cells.

As the seed begins to germinate, the growing embryo obtains chemical building blocks for its development—carbohydrates, amino acids, and lipid monomers—by hydrolyzing starch, proteins, and lipids stored in the seed. These reserves are stored in the *cotyledons* or in the *endosperm*. Germination is completed when the **radicle** (embryonic root) emerges from the seed coat. The plant is then called a **seedling**.

If the seed germinates underground, the new seedling must elongate rapidly (in the right direction!) and cope with a period of life in darkness or dim light. A series of photoreceptors direct this stage of development and prepare the seedling for growth in the light.

The pattern of early shoot development varies among the flowering plants. **FIGURE 26.1** shows the shoot development patterns of monocots and eudicots. In monocots, the growing shoot is protected by a cylindrical sheath of cells called the **coleoptile** as it pushes its way through the soil. In eudicots, the shoot is protected by the cotyledons.

Several hormones and photoreceptors help regulate plant growth

This survey of the early stages of plant development illustrates the many environmental cues that influence plant growth. A plant's responses to these cues are initiated and maintained by two types of regulators: *hormones* and *photoreceptors*. Both types of regulators act through signal transduction pathways

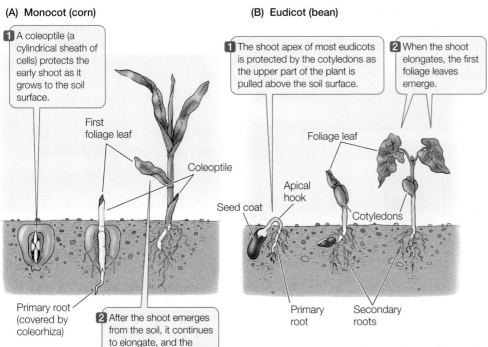

(A) Monocot (corn)

1 A coleoptile (a cylindrical sheath of cells) protects the early shoot as it grows to the soil surface.

First foliage leaf

Coleoptile

Seed coat

Primary root (covered by coleorhiza)

2 After the shoot emerges from the soil, it continues to elongate, and the leaves emerge.

(B) Eudicot (bean)

1 The shoot apex of most eudicots is protected by the cotyledons as the upper part of the plant is pulled above the soil surface.

2 When the shoot elongates, the first foliage leaves emerge.

Foliage leaf

Apical hook

Cotyledons

Primary root

Secondary roots

FIGURE 26.1 Patterns of Early Shoot Development (A) In grasses and some other monocots, growing shoots are protected by a coleoptile until they reach the soil surface. (B) In most eudicots, the growing point of the shoot is protected within the cotyledons.

yourBioPortal.com

Go to
**WEB ACTIVITIES 26.1 and 26.2
Monocot Shoot Development
Eudicot Shoot Development**

(series of biochemical steps within a cell that occur between a stimulus and a response).

LINK The general characteristics of signal transduction pathways are discussed in Concepts 5.5 and 5.6

Hormones are chemical signals that act at very low concentrations at sites often distant from where they are produced. Plant hormones are very different from animal hormones (**TABLE 26.1**; see also Chapter 30). Each plant hormone plays multiple regulatory roles, and interactions among them can be complex. Several hormones regulate the growth and development of plants from seedling to adult (**TABLE 26.2**). Other hormones are involved in the plant's defenses against herbivores and microorganisms (which we discuss in Chapter 28).

Photoreceptors are pigments associated with proteins. Light acts directly on photoreceptors, which in turn regulate developmental processes that need to be responsive to light, such as the many changes that occur as a seed germinates and a seedling emerges from the soil.

Genetic screens have increased our understanding of plant signal transduction

In Chapter 14 we describe how genetic studies can be used to identify the steps along a developmental pathway. You will recall the reasoning behind these experiments: if a mutation of a gene involved in a certain biochemical process disrupts a developmental event, then that biochemical process must be essential for that developmental event. Similarly, genetic studies can be used to analyze pathways of receptor activation and signal transduction in plants: if proper signaling does not occur in a mutant strain, then the mutant gene must be involved in the signal transduction process. Mapping the mutant gene and identifying its function are starting points for understanding the signaling pathway. *Arabidopsis thaliana*, a member of the mustard family, has been a major model organism for plant biologists investigating signal transduction.

One technique for identifying the genes involved in a plant signal transduction pathway is illustrated in **FIGURE 26.2**. This technique, called a **genetic screen**, involves creating a large collection of randomly mutated plants and identifying those individuals that are likely to have a defect in the pathway of interest. Plant genes can be randomly mutated in two ways: by insertion of transposons (see Concept 12.3) and by treatment of seeds with chemical mutagens, most commonly ethyl methane sulfonate. In both cases, the treated plants are grown and then examined for a specific phenotype, usually a characteristic that is known to be influenced by the pathway of interest. Once mutant plants have been selected, their genotypes are compared with those of wild-type plants. *Arabidopsis* mutants with altered developmental patterns have provided a wealth of new information about the hormones present in plants and the mechanisms of hormone and photoreceptor action.

TABLE 26.1	Comparison of Plant and Animal Hormones	
CHARACTERISTIC	**PLANT HORMONES**	**ANIMAL HORMONES**
Size, chemistry	Small organic molecules	Peptides, proteins, small molecules
Site of synthesis	Throughout the plant	Specialized glands or cells
Site of action	Local or distant	Distant, transported
Effects	Diverse	Often specific
Regulation	Decentralized	By central nervous system

RESEARCH TOOLS

FIGURE 26.2 A Genetic Screen Genetics of the model plant *Arabidopsis thaliana* can be used to identify the steps of a signal transduction pathway. If a mutant strain does not respond to a hormone (in this case, ethylene), the corresponding wild-type gene must be essential for the pathway (in this case, ethylene response). This method has been instrumental to scientists in understanding plant growth regulation.

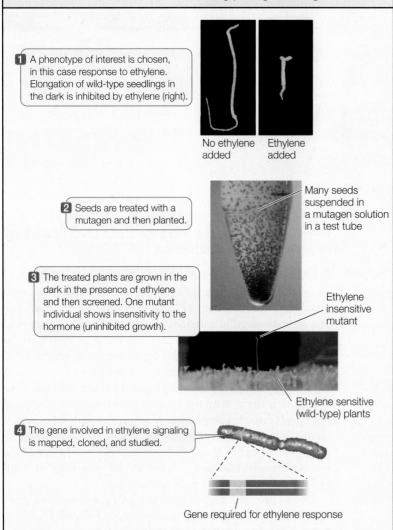

1 A phenotype of interest is chosen, in this case response to ethylene. Elongation of wild-type seedlings in the dark is inhibited by ethylene (right).

No ethylene added   Ethylene added

2 Seeds are treated with a mutagen and then planted.

Many seeds suspended in a mutagen solution in a test tube

3 The treated plants are grown in the dark in the presence of ethylene and then screened. One mutant individual shows insensitivity to the hormone (uninhibited growth).

Ethylene insensitive mutant

Ethylene sensitive (wild-type) plants

4 The gene involved in ethylene signaling is mapped, cloned, and studied.

Gene required for ethylene response

Do You Understand Concept 26.1?

- Describe how monocots and eudicots differ in their early development.

- Seeds of the Australian shrub *Acacia myrtifolia* require extreme heat (fire) to germinate. How do you think fire might work as an environmental cue to break dormancy? What would be the selective advantage of this mechanism for breaking dormancy?

- Seedlings grown in the dark grow thin and tall, a phenomenon called *etiolation*. How would you set up a genetic screen using *Arabidopsis* to investigate the signaling pathway that controls etiolation?

With this overview of plant development, we will begin our examination of the specific mechanisms that underlie these events. Two hormones—gibberellins and auxin—have prominent roles in plant development.

concept 26.2 **Gibberellins and Auxin Have Diverse Effects but a Similar Mechanism of Action**

The discovery of two key plant hormones exemplifies the experimental approaches that plant biologists have used to investigate the mechanisms of plant development. **Gibberellins** (of which there are several active forms) and **auxin** were the first plant hormones to be identified, early in the twentieth century. In both cases, the discoveries came from observations of natural phenomena:

- *Gibberellins*: In rice plants, a disease caused by the fungus *Gibberella fujikuroi* resulted in plants that grew overly tall and spindly.

- *Auxin*: Charles Darwin and his son Francis noted that canary grass seedlings would bend toward the light when placed near a light source.

In both cases, a chemical substance was isolated that could cause the phenomenon:

- Gibberellic acid (see Table 26.2) made by the fungus caused rice plants to overgrow. Later, it was found that plants also make gibberellic acid, and that applying it to plants causes growth.

- Auxin made by the growing tip of the canary grass coleoptile diffused asymmetrically, causing cell elongation on the side away from the light, which resulted in the coleoptile bending toward the light.

In each case, mutant plants that do not make the hormone exhibit a phenotype expected in the absence of the hormone, and adding the hormone reverses that phenotype (**FIGURE 26.3**):

- Tomato plants that do not make gibberellic acid are very short; supplying them with the hormone results in normal growth.

- *Arabidopsis thaliana* individuals that do not make auxin are also short; supplying them with that hormone reverses that phenotype.

Note that the phenotype involved—short stature, or *dwarfism*—is similar in both cases. This observation exemplifies a concept that is important to keep in mind when studying plant hormones: their actions are not unique and specific, as is the case with animal hormones (see Table 26.1).

The three-part approach outlined above for gibberellins and auxin—observation, hormone isolation, and analysis of mutants—has been used to identify other plant hormones.

TABLE 26.2 Plant Growth Hormones

HORMONE	STRUCTURE	TYPICAL ACTIVITIES
Abscisic acid		Maintains seed dormancy; closes stomata
Auxin (indole-3-acetic acid)		Promotes stem elongation, lateral root initiation, and fruit development; inhibits axillary bud outgrowth, leaf abscission, and root elongation
Brassinosteroids		Promote stem and pollen tube elongation; promote vascular tissue differentiation
Cytokinins		Inhibit leaf senescence; promote cell division and axillary bud outgrowth; affect root growth
Ethylene		Promotes fruit ripening and leaf abscission; inhibits stem elongation
Gibberellins		Promote seed germination, stem growth, and fruit development; break winter dormancy; mobilize nutrient reserves in grass seeds

Gibberellins have many effects on plant growth and development

The functions of gibberellins can be inferred from the effects of experimentally decreasing concentrations of gibberellins or blocking their action at various times in plant development. Such experiments reveal that gibberellins have multiple roles in regulating plant growth.

STEM ELONGATION The effects of gibberellins on wild-type plants are not as dramatic as their effects on dwarf plants. We know, however, that gibberellins are indeed active in wild-type plants because inhibitors of gibberellin synthesis cause a reduction in stem elongation. Such inhibitors can be put to practical uses. For example, plants such as chrysanthemums that are grown in greenhouses tend to get tall and spindly, but such plants do not appeal to consumers. Flower growers spray such plants with gibberellin synthesis inhibitors to control their height.

FRUIT GROWTH Gibberellins can regulate the growth of fruit. Grapevines that produce seedless grapes develop smaller fruit than varieties that produce seed-bearing grapes. Biologists wanting to explain this

(A)

Twenty-two days after being sprayed with a dilute gibberellin solution, this plant reached the size of a nondwarf plant.

This untreated mutant plant remained a dwarf.

(B)

Supplying auxin to this mutant plant made it grow.

This untreated mutant plant does not make auxin.

FIGURE 26.3 Hormones Reverse a Mutant Phenotype (A) The two mutant dwarf tomato plants in this photograph were the same size when the one on the right was treated with gibberellins. (B) The short phenotype of this *Arabidopsis* mutant was reversed in the plant on the right by supplying auxin.

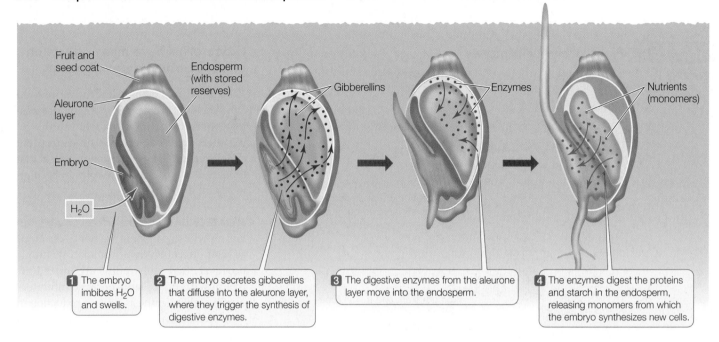

1. The embryo imbibes H_2O and swells.

2. The embryo secretes gibberellins that diffuse into the aleurone layer, where they trigger the synthesis of digestive enzymes.

3. The digestive enzymes from the aleurone layer move into the endosperm.

4. The enzymes digest the proteins and starch in the endosperm, releasing monomers from which the embryo synthesizes new cells.

phenomenon removed seeds from immature seeded grapes and found that this prevented normal fruit growth. This suggested that the seeds are sources of a growth regulator. Biochemical studies showed that developing seeds produce gibberellins, which diffuse out into the immature fruit tissue. Spraying young seedless grapes with a gibberellin solution, which causes them to grow as large as seeded ones, is now a standard commercial practice.

MOBILIZATION OF SEED RESERVES As noted in Concept 26.1, an important event early in seed germination is the hydrolysis of stored starch, proteins, and lipids. In germinating seeds of barley and other cereal crops, the embryo secretes gibberellins just after imbibition. The hormones diffuse through the endosperm to a surrounding tissue called the **aleurone layer**, which lies underneath the seed coat. The gibberellins trigger a cascade of events in the aleurone layer, causing it to synthesize and secrete enzymes that digest proteins and starch stored in the endosperm (**FIGURE 26.4**). These observations have practical importance: in the beer-brewing industry, gibberellins are used to enhance the "malting" (germination) of barley and the

FIGURE 26.4 Gibberellins and Seed Germination During seed germination in cereal crops, gibberellins trigger a cascade of events that result in the conversion of starch and protein stores in the endosperm into monomers that can be used by the developing embryo.

---

yourBioPortal.com
Go to WEB ACTIVITY 26.3
Events of Seed Germination

---

breakdown of its endosperm, producing glucose that is fermented into alcohol.

The transport of auxin mediates some of its effects

As we noted above, auxin was discovered in the context of **phototropism**: a response to light in which plant stems bend toward a light source. Auxin is made in the shoot apex and diffuses down the shoot in a *polar* (unidirectional) fashion, stimulating cell elongation. Unidirectional longitudinal movement of auxin occurs in other plant organs as well. For example, in a leaf petiole, which connects the leaf blade to the stem, auxin moves from the leaf blade end toward the stem. In roots, auxin moves unidirectionally toward the root tip.

Polar transport of auxin depends on four biochemical processes that may be familiar from earlier chapters (**FIGURE 26.5**):

- *Diffusion across a plasma membrane.* Polar (hydrophilic) molecules diffuse across plasma membranes less readily than nonpolar (hydrophobic) molecules (see Concept 5.2).

- *Membrane protein asymmetry.* Active transport, or efflux, carriers (see Concept 5.3) for auxin are located only in the portion of the plasma membrane at the basal (bottom) end of the cell.

- *Proton pumping/chemiosmosis.* Proton pumps (see Concept 25.3) remove H^+ from the cell, thereby increasing the intracellular pH and decreasing the pH in the cell wall. Proton

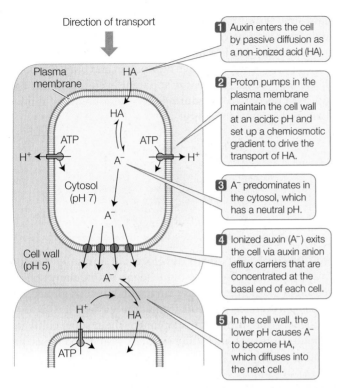

Direction of transport

1 Auxin enters the cell by passive diffusion as a non-ionized acid (HA).

2 Proton pumps in the plasma membrane maintain the cell wall at an acidic pH and set up a chemiosmotic gradient to drive the transport of HA.

3 A⁻ predominates in the cytosol, which has a neutral pH.

4 Ionized auxin (A⁻) exits the cell via auxin anion efflux carriers that are concentrated at the basal end of each cell.

5 In the cell wall, the lower pH causes A⁻ to become HA, which diffuses into the next cell.

FIGURE 26.5 Polar Transport of Auxin Proton pumps set up a chemiosmotic gradient that drives the uptake of non-ionized auxin (HA) and the efflux of ionized auxin (A⁻) through the basally placed auxin anion efflux carriers, leading to a net movement of auxin in a basal direction.

pumping also sets up an electrochemical gradient, which provides potential energy to drive the transport of auxin by the carriers mentioned above.

- *Ionization of a weak acid.* Indole-3-acetic acid (the chemical name for auxin; see Table 26.2) is a weak acid; it forms ions (H⁺ and A⁻, where A⁻ stands for indole acetate) in solution that also tend to recombine to form the acid (HA):

$$A^- + H^+ \rightleftharpoons HA$$

When the pH is low, the increased H⁺ concentration drives this reaction to the right, and HA (non-ionized auxin) is the predominant form. When the pH is higher, there is more A⁻ (ionized auxin).

Whereas polar auxin transport distributes the hormone along the longitudinal axis of the plant, *lateral* (side-to-side) redistribution of auxin is responsible for directional plant growth. This redistribution is carried out by auxin efflux carriers that move from the base of the cell to one side; because of this, auxin exits the cell only on that side of the cell, rather than at the base, and moves sideways within the tissue.

This lateral movement of auxin explains the bending of canary grass seedlings toward light. When light strikes a canary grass coleoptile on one side, auxin at the tip moves laterally toward the shaded side. The asymmetry thus established is maintained as polar transport moves auxin down the coleoptile, so that in the growing region below, the auxin concentration is

(A) Phototropism

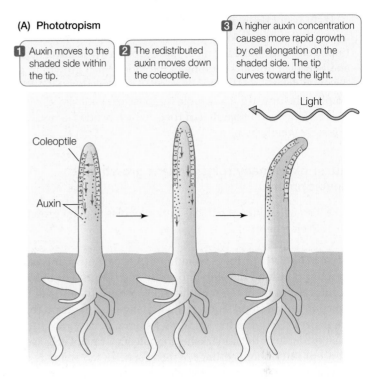

1 Auxin moves to the shaded side within the tip.

2 The redistributed auxin moves down the coleoptile.

3 A higher auxin concentration causes more rapid growth by cell elongation on the shaded side. The tip curves toward the light.

(B) Negative gravitropism of shoot

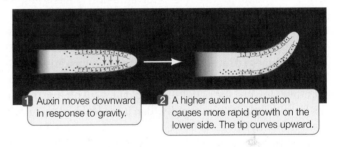

1 Auxin moves downward in response to gravity.

2 A higher auxin concentration causes more rapid growth on the lower side. The tip curves upward.

FIGURE 26.6 Plants Respond to Light and Gravity Phototropism (A) and gravitropism (B) occur in shoot apices in response to a redistribution of auxin.

yourBioPortal.com
Go to ANIMATED TUTORIAL 26.1
Tropisms

highest on the shaded side. Cell elongation is thus speeded up on that side, causing the coleoptile to bend toward the light (**FIGURE 26.6A**). The same type of bending also occurs in eudicots:

Light

Light is not the only signal that can cause redistribution of auxin. Auxin moves to the lower side of a shoot that has been tipped sideways, causing more rapid growth in the lower side and hence an upward bending of the shoot. Such growth in a direction determined by gravity is called **gravitropism (FIGURE 26.6B)**. The upward gravitropic response of shoots is defined as *negative gravitropism*; that of roots, which bend downward, is *positive gravitropism*.

Auxin plays many roles in plant growth and development

Like the gibberellins, auxin has multiple roles in plant growth and development.

ROOT INITIATION Cuttings from the shoots of some plants can produce roots and develop into entire new plants. For this to occur, cells in the interior of the shoot, originally destined to function only in food storage, must change their fate and become organized into the apical meristem of a new root (this is an example of the remarkable totipotency of plant cells; see Concept 14.1). Shoot cuttings of many species can be made to develop roots by dipping the cut surfaces into an auxin solution. These observations suggest that in an intact plant, the plant's own auxin plays a role in the initiation of lateral roots.

LEAF ABSCISSION In contrast to its stimulatory effect on root initiation, auxin inhibits the detachment of old leaves from stems. This detachment process, called **abscission**, is the cause of autumn leaf fall. Recall that the blade of a leaf produces auxin that moves toward the petiole. If the blade is cut off, the petiole falls from the plant more rapidly than if the leaf had remained intact. If the cut surface is treated with an auxin solution, however, the petiole remains attached to the plant, often longer than an intact leaf would have. The timing of leaf abscission in nature appears to be determined in part by a decrease in the movement of auxin through the petiole.

APICAL DOMINANCE Auxin helps maintain **apical dominance**, a phenomenon in which apical buds inhibit the growth of axillary buds (see Figure 24.1), resulting in the growth of a single main stem with minimal branching. A diffusion gradient of auxin from the apex of the shoot down the stem results in lower branches receiving less auxin and therefore branching more, while higher branches receive more auxin and branch less. The effect of the auxin gradient is apparent in conifers: next time you see a decorated tree during the winter holidays, think of auxin and apical dominance.

FRUIT DEVELOPMENT Fruit development normally depends on prior fertilization of the ovule (egg), but in many species, treatment of an unfertilized ovary with auxin or gibberellins causes **parthenocarpy**: fruit formation without fertilization. Parthenocarpic fruits form spontaneously in some cultivated varieties of plants, including seedless grapes, bananas, and some cucumbers.

CELL ELONGATION The elongation of plant cells is a key process in plant growth. Because the plant cell wall normally prevents expansion of the cell, it plays a key role in controlling the rate and direction of plant cell growth. Auxin acts on the cell wall to regulate this process.

The expansion of a plant cell is driven primarily by the uptake of water, which enters the cell by osmosis. As it expands, the cell contents press against the cell wall, which resists this force (producing turgor pressure). The cell wall is an extensively cross-linked network of polysaccharides and proteins, dominated by cellulose fibrils. If the cell is to expand, some adjustments must be made in the wall structure to allow the wall to "give" under turgor pressure. Think of a balloon (the cell surrounded by a membrane) inside a box (the cell wall).

LINK Water movement and turgor pressure in plant cells are discussed in Concept 25.3

APPLY THE CONCEPT

Gibberellins and auxin have diverse effects but a similar mechanism of action

Oat seedlings with coleoptiles 20 mm long were exposed to light from one side and checked for bending after 6 hours. Various regions of the seedlings were covered with foil to block the light. The results varied depending on which region of the coleoptile was covered (see table).

1. Which part of the coleoptile senses the light?
2. For each treatment, what would happen if supplementary auxin were applied to the coleoptiles on the same side as the light? On the side away from the light?

REGION COVERED	BENDING
None	Yes
All	No
Apical 5 mm	No
Apical 10 mm	No
Apical 5–10 mm	Yes
Apical 10–15 mm	Yes

The **acid growth hypothesis** explains auxin-induced cell expansion (**FIGURE 26.7**). This hypothesis proposes that protons (H^+) are pumped from the cytoplasm into the cell wall, lowering the pH of the wall and activating enzymes called *expansins*, which catalyze changes in the cell wall structure such that the polysaccharides adhere to each other less strongly. This change loosens the cell wall, allowing it to stretch as the cell expands. Auxin is believed to have two roles in this process: to increase the synthesis of proton pumps, and to guide their insertion into the plasma membrane. Several lines of evidence support the acid growth hypothesis. For example, adding acid to the cell wall to lower the pH stimulates cell expansion even in the absence of auxin. Conversely, when a buffer is used to prevent the wall from becoming more acidic, auxin-induced cell expansion is blocked.

At the molecular level, auxin and gibberellins act similarly

The molecular mechanisms underlying both auxin and gibberellin action have been worked out with the help of genetic screens (see Figure 26.2). Biologists started by identifying mutant plants whose growth and development are *insensitive* to the hormones; that is, plants that are *not* affected by added hormone. Such mutants fall into two general categories:

- *Excessively tall plants.* These plants resemble wild-type plants given an excess of hormone, and they grow no taller when given extra hormone. They grow tall even when treated with inhibitors of hormone synthesis. Their hormone response is always "on," even in the absence of the hormone. It is presumed that the normal allele for the mutant gene codes for an *inhibitor* of the hormone signal transduction pathway. In wild-type plants, that pathway is "off," but in the mutant plants the pathway is "on," and the plant grows tall.

- *Dwarf plants.* These plants resemble dwarf plants that are deficient in hormone synthesis (see Figure 26.3), but they *do not respond* to added hormone. In these mutants the hormone response is always "off," regardless of the presence of the hormone.

Remarkably, some mutations of both types *affect the same protein*, which turns out to be a *repressor* of a transcription factor that stimulates the expression of growth-promoting genes. The repressor protein has two important domains, which explains how mutations in the same protein can have seemingly opposite effects:

- *One region of the repressor protein binds to the transcription complex to inhibit transcription.* This is the mutant region in the excessively tall plants: the growth-promoting genes are always "on" because the repressor does not bind to the transcription complex.

  - *Another region of the repressor protein causes it to be removed from the transcription complex.* This is the mutant region in the dwarf plants: the growth-promoting genes are always "off" because the repressor is always bound to the complex.

> **LINK**   Review the process of transcription initiation in Concept 11.3

These observations allowed biologists to figure out how auxin and gibberellins work in wild-type plants. Of course, the actual proteins involved with the two hormones are different, but the actions of both

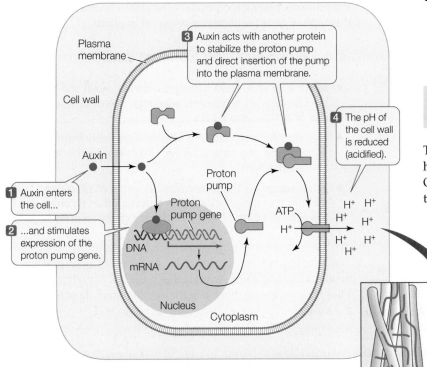

1 Auxin enters the cell...

2 ...and stimulates expression of the proton pump gene.

3 Auxin acts with another protein to stabilize the proton pump and direct insertion of the pump into the plasma membrane.

4 The pH of the cell wall is reduced (acidified).

Plasma membrane

Cell wall

Auxin

Proton pump

Proton pump gene

DNA

mRNA

ATP

Nucleus

Cytoplasm

FIGURE 26.7   Auxin and Cell Expansion   The plant cell wall is an extensive network of cross-linked polymers. Auxin induces loosening of the cell wall by activating proton pumps that reduce pH in the cell wall.

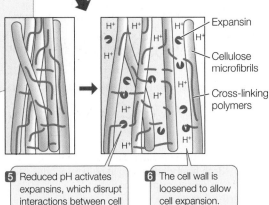

Expansin

Cellulose microfibrils

Cross-linking polymers

5 Reduced pH activates expansins, which disrupt interactions between cell wall polymers.

6 The cell wall is loosened to allow cell expansion.

yourBioPortal.com
Go to ANIMATED TUTORIAL 26.2
Auxin Affects Cell Walls

APPLY THE CONCEPT

Gibberellins and auxin have diverse effects but a similar mechanism of action

The aleurone layer of germinating barley can be isolated and studied for the induction of α-amylase, the enzyme that catalyzes starch hydrolysis. Predict the amount of α-amylase activity in aleurone layers subjected to the following treatments:

1. Incubation with gibberellic acid

2. Incubation with auxin

3. Incubation with gibberellic acid in the presence of an inhibitor of DNA transcription to mRNA

4. Incubation without gibberellic acid

5. Incubation without gibberellic acid in the presence of an inhibitor of transcription

6. Incubation with gibberellic acid in the presence of an inhibitor of the proteasome

7. Incubation without gibberellic acid in the presence of an inhibitor of the proteasome

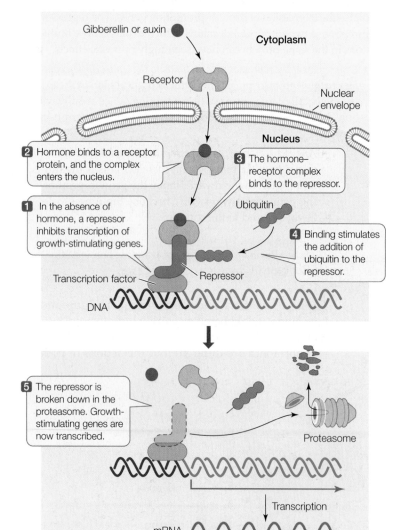

FIGURE 26.8 Gibberellins and Auxin Have Similar Signal Transduction Pathways Both hormones act to stimulate gene transcription by inactivating a repressor protein.

hormones are similar: they *act by removing the repressor from the transcription complex* (**FIGURE 26.8**). The hormones do this by binding to a receptor protein, which in turn binds to the repressor. Binding of the hormone–receptor complex stimulates polyubiquitination of the repressor, targeting it for breakdown in the proteasome (see Figure 11.19). The receptors for both auxin and gibberellins contain a region called an **F-box** that facilitates protein–protein interactions necessary for protein breakdown. While animal genomes have few F-box–containing proteins, plant genomes have hundreds, an indication that this type of gene regulation is common in plants.

Do You Understand Concept 26.2?

- Make a table for two activities each for auxin and gibberellin with three column headings: site of synthesis, site of action, and effect.

- Explain why, even though auxin moves *away* from the lighted side of a coleoptile tip, the coleoptile bends *toward* the light.

- What would be the phenotype, in terms of seed germination, of a plant that has a mutation that disables the F-box region of the receptor for gibberellin?

The same approaches that led to the discovery of auxin and gibberellins have revealed several other classes of plant hormones. These hormones are diverse in their actions and affect many of the same developmental processes as auxin and gibberellins.

concept 26.3 Other Plant Hormones Have Diverse Effects on Plant Development

The discoveries of gibberellins and auxin have led to the isolation of additional classes of plant hormones (see Table 26.2). These hormones are diverse in their actions, and they interact with one another in many cases.

Ethylene is a gaseous hormone that promotes senescence

Whereas the cytokinins delay senescence, another plant hormone promotes it: the gas **ethylene** (see Table 26.2), which is sometimes called the senescence hormone. When streets were lit by gas rather than by electricity, leaves on trees near street lamps dropped earlier in the fall than those on trees farther from the lamps. We now know why: ethylene, a combustion product of the illuminating gas, caused the early abscission. While auxin inhibits leaf abscission, ethylene strongly promotes it; thus the balance of auxin and ethylene controls abscission.

FRUIT RIPENING By promoting senescence, ethylene speeds the ripening of fruit. As a fruit ripens, it loses chlorophyll and its cell walls break down, making the fruit softer; ethylene promotes both of these processes. Ethylene also causes an increase in its own production. Thus once ripening begins, more and more ethylene forms, and because it is a gas, it diffuses readily throughout the fruit and even to neighboring fruits on the same or other plants. The old saying that "one rotten apple spoils the barrel" is true. That rotten apple is a rich source of ethylene, which speeds the ripening and subsequent rotting of the other fruit in a barrel or other confined space.

LINK The action of ethylene in fruit ripening is an example of positive feedback; see Concept 29.2

Farmers in ancient times poked holes in developing figs to make them ripen faster. Wounding causes an increase in ethylene production by the fruit, and that increase promotes ripening. Today commercial shippers and storers of fruit hasten ripening by adding ethylene to storage chambers. This use of ethylene is the single most important use of a natural plant hormone in agriculture and commerce. Conversely, ripening can be delayed by the use of "scrubbers" and adsorbents that remove ethylene from the atmosphere in fruit storage chambers.

These stored tomatoes were treated with ethylene to promote ripening.

As flowers senesce, their petals may abscise, which decreases their value in the cut-flower industry. Growers and florists often immerse the cut stems of ethylene-sensitive flowers in dilute solutions of silver thiosulfate before selling them. Silver salts inhibit ethylene action by interacting directly with the ethylene receptor—thus they delay senescence, keeping flowers "fresh" for a longer time.

STEM GROWTH The stems of many eudicot seedlings (which lack protective coleoptiles) form an **apical hook** that protects the delicate shoot apex while the stem grows through the soil (see Figure 26.1B). The apical hook is maintained through an asymmetrical production of ethylene, which inhibits the elongation of cells on the inner surface of the hook. Once the seedling breaks through the soil surface and is exposed to light, ethylene synthesis stops, and the cells of the inner surface are no longer inhibited. Those cells now elongate, and the hook unfolds, raising the shoot apex and the expanding leaves into the sun. Ethylene also inhibits stem elongation in general, promotes lateral swelling of stems (as do the cytokinins), and decreases the sensitivity of stems to gravitropic stimulation. Together, these three phenomena constitute the *triple response*, a well-characterized stunted growth habit observed when plants are treated with ethylene.

FRONTIERS Ethylene is a gas, and its ready diffusion in plant tissues aids the hormone's signal transduction. The receptor for ethylene is a protein on the endoplasmic reticulum. Because ethylene is hydrophobic, binding to the receptor occurs at the interior of the protein. How does a dissolved gas that diffuses rapidly (an unusual type of receptor ligand) bind to a receptor that initiates further events in the cell? Mutants for the ethylene signal transduction pathway may be useful in answering this question.

Cytokinins are active from seed to senescence

Like bacteria and yeasts, some plant cells, such as parenchyma cells, can be grown in a liquid or solidified medium containing sugars and salts. For continuous cell division, cultured plant cells also require a type of hormone called a **cytokinin**, first isolated from corn. Over 150 different cytokinins have been isolated, and most are derivatives of adenine.

Cytokinins have a number of different effects (see Table 26.2), in many cases by interacting with auxin:

- Adding an appropriate combination of auxin and cytokinins to a growth medium induces rapid proliferation of cultured plant cells.

- In cell cultures, a high cytokinin-to-auxin ratio promotes the formation of shoots; a low ratio promotes the formation of roots.

- Cytokinins can cause certain light-requiring seeds to germinate even when they are kept in constant darkness.

- Cytokinins usually inhibit the elongation of stems, but they cause lateral swelling of stems and roots (the fleshy roots of radishes are an extreme example).

- Cytokinins stimulate axillary buds to grow into branches; the auxin-to-cytokinin ratio controls the extent of branching (bushiness) of a plant.

- Cytokinins delay the senescence of leaves. If leaf blades are detached from a plant and floated on water or a nutrient solution, they gradually turn yellow and show other signs of senescence. If instead they are floated on a solution containing a cytokinin, they remain green and senesce much more

Outside of cell

Cytokinin

1 As long as cytokinin is not bound to its receptor (AHK)...

Receptor (AHK)

Cytoplasm

Nuclear envelope

Nucleus

Genes that respond to cytokinin

DNA

Promoter

2 ...cytokinin response genes are not expressed.

Outside of cell

3 Binding of cytokinin to its receptor causes a conformational change that exposes the active site for the protein kinase domain of the receptor.

Cytokinin

Receptor

Cytoplasm

AHP

4 An intermediate protein (AHP) is phosphorylated...

Nuclear envelope

Nucleus

Transcription factor (ARR)

5 ...enters the nucleus, and phosphorylates a transcription factor (ARR).

DNA

Promoter

Transcription

mRNA

6 The phosphorylated transcription factor binds to promoters of over 20 cytokinin response genes, initiating their transcription.

FIGURE 26.9 The Cytokinin Signal Transduction Pathway
Plant cells respond to cytokinins through a two-component signal transduction pathway involving a receptor and a target protein.

slowly. Roots contain abundant cytokinins, and cytokinin transport to the leaves delays senescence.

This leaf was treated with cytokinin, which delays senescence.

This detached leaf was not treated.

Cytokinin appears to act through a pathway that includes proteins with amino acid sequences similar to those of proteins in *two-component systems*, a type of signal transduction pathway commonly found in bacteria. Indeed, this pathway was one of the first of its kind discovered in eukaryotes. The two components in such a system are:

- a *receptor* that can act as a protein kinase, phosphorylating itself as well as a target protein

- a *target protein*, generally a transcription factor, that regulates the response

Genetic screens in *Arabidopsis* for abnormalities in the response to cytokinin have identified the receptor (AHK; *Arabidopsis h*istidine *k*inase) and the target protein (ARR; *Arabidopsis r*esponse *r*egulator) involved. The target protein acts as a transcription factor when phosphorylated. The cytokinin signal transduction pathway also includes a third protein (AHP; *Arabidopsis h*istidine *p*hosphotransfer protein), which transfers phosphates from the receptor to the target protein (**FIGURE 26.9**). The plant genome has over 20 genes that are expressed in response to this signaling pathway.

Brassinosteroids are plant steroid hormones

In animals, steroid hormones such as cortisol and estrogen, which are formed from cholesterol, have been well studied for decades (see Chapter 30). In the 1970s, biologists isolated a steroid from the pollen of rape (*Brassica napus*), a member of the mustard family. When applied to various plant tissues, this hormone stimulated cell elongation, pollen tube elongation, and vascular tissue differentiation, but it inhibited root elongation. Since then, dozens of chemically related, growth-affecting **brassinosteroids** have been found in plants. Like other plant hormones, brassinosteroids have diverse effects (see Table 26.2):

- They enhance cell elongation and cell division in shoots.
- They promote xylem differentiation.
- They promote growth of pollen tubes during reproduction.
- They promote seed germination.
- They promote apical dominance and leaf senescence.

The signaling pathway for these plant steroids differs sharply from those for steroid hormones in animals. In animals, steroids diffuse through the plasma membrane and bind to receptors in the cytoplasm. In contrast, the receptor for brassinosteroids is an integral protein in the plasma membrane.

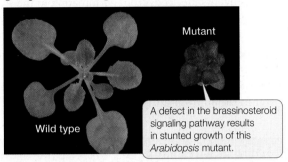

Mutant

A defect in the brassinosteroid signaling pathway results in stunted growth of this *Arabidopsis* mutant.

Wild type

Abscisic acid acts by inhibiting development

The hormone **abscisic acid** is found in most plant tissues. Many of its actions involve inhibition of other hormones' actions in the plant (see Table 26.2):

- *Prevention of seed germination.* Abscisic acid prevents seeds from germinating on the parent plant before the seeds are dry. Premature germination, termed *vivipary*, is undesirable in cereal crops because the grain is ruined if it has started to sprout. Viviparous seedlings are unlikely to survive if they remain attached to the parent plant and are unable to establish themselves in the soil.

- *Promotion of seed dormancy.* Abscisic acid promotes seed dormancy and inhibits the initiation of germination events, such as the hydrolysis of stored macromolecules, by gibberellins.

- *Reaction to stress.* Abscisic acid is a "stress hormone" that mediates a number of plant responses to pathogens and environmental stresses. We describe some of these effects in Chapter 28. For now, consider that when the air is dry, abscisic acid stimulates the closure of stomata to prevent water loss by transpiration (see Concept 25.3).

FRONTIERS Abscisic acid provides an example of the complex interactions between plant hormones. Whereas abscisic acid promotes seed dormancy, gibberellic acid inhibits it; whereas abscisic acid inhibits seed germination, both gibberellic acid and ethylene promote it. The exact mechanism of this hormonal cross-talk is not clear. Molecular studies of the signal transduction pathways for these hormones may reveal common intermediates where the pathways intersect and influence one another.

Do You Understand Concept 26.3?

- Which plant hormones affect cell elongation, and what is the effect in each case?

- At the grocery store, you can buy plastic bags that keep fruit fresh for a much longer period than other containers. How do you think they work?

- A mutant strain of corn shows vivipary. Why is this a disadvantage for the plant and the crop, and what gene might be involved?

As we have seen, plant development is strongly influenced by the environment. For example, in Concept 26.1 we alluded to the role of photoreceptors whose responses to light guide the early growth of seedlings. We will now turn to the molecular nature of these and other plant photoreceptors.

concept
26.4

Photoreceptors Initiate Developmental Responses to Light

Plants are completely dependent on sunlight for energy, so it is not surprising that they have multiple ways to sense light and respond appropriately. A number of physiological and developmental events in plants are controlled by light, a process called **photomorphogenesis**. Plants respond to two aspects of light: its *quality*—that is, the wavelengths of light to which the plant is exposed; and its *quantity*—that is, the intensity and duration of light exposure.

Chapter 6 describes how chlorophyll and other photosynthetic pigments absorb light of certain wavelengths (see Figure 6.17). Here we consider how light affects plant development. Light influences seed germination, phototropism, shoot elongation, initiation of flowering, and many other important aspects of plant development. Several photoreceptors take part in these processes.

Phototropin, cryptochromes, and zeaxanthin are blue-light receptors

An action spectrum for the phototropic response of a coleoptile (**FIGURE 26.10**) shows that blue light at 436 nm is the most

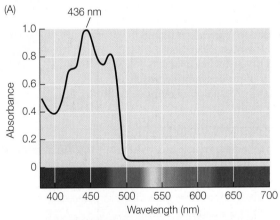

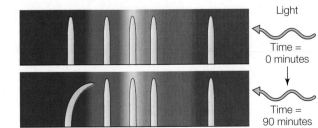

FIGURE 26.10 Action Spectrum for Phototropism The absorption spectrum for phototropin (A) is similar to the action spectrum for the bending of a coleoptile toward light (B). After 90 minutes, only the coleoptiles exposed to blue light bend.

effective in stimulating that response. To identify the blue-light-absorbing molecule involved, biologists have taken a genetic approach using the model plant *Arabidopsis*.

> **LINK** Action spectra are discussed in Concept 6.5; see Figure 6.17

By recovering blue-light-insensitive *Arabidopsis* mutants from a genetic screen, researchers were able to identify the gene for a blue-light receptor protein located in the plasma membrane, called **phototropin**. This protein has a flavin mononucleotide associated with it that absorbs blue light at 436 nm. Light absorption leads to a change in the shape of the protein that exposes an active site for a protein kinase, which in turn initiates a signal transduction cascade that ultimately results in stimulation of cell elongation by auxin. Phototropin also participates with another type of blue-light receptor, the plastid pigment **zeaxanthin**, in the light-induced opening of stomata.

> **FRONTIERS** In bright sunlight, chloroplasts in leaves move to the edges of cells, making the leaves look pale green. When the weather is cloudy, chloroplasts move toward the center of cells to maximize light exposure. Genetic screens of *Arabidopsis* show that these responses are mediated by phototropin. Studying this photoreceptor's signal transduction pathway may allow scientists to manipulate pigment-mediated chloroplast movements. This ability could be useful in increasing the efficiency of photosynthesis, which often limits the production of crops.

A third class of blue-light receptors is the **cryptochromes**, which are yellow pigments that absorb blue and ultraviolet light. These pigments are located primarily in the plant cell nucleus and affect seedling development and flowering. The exact mechanism of cryptochrome action is not yet known. Strong blue light inhibits cell elongation through the action of cryptochromes, although the most rapid responses are mediated by phototropin.

Phytochrome mediates the effects of red and far-red light

Plants also have responses that are specific to light in the red region of the spectrum. Here are two examples:

- Lettuce seeds spread on soil will germinate only in response to light. Even a brief flash of dim light will suffice. This response prevents the seeds from germinating in the shade of other plants or so deep in the soil that the seedlings will not be able to reach the light before their seed reserves run out.

- Adult cocklebur plants flower when they are exposed to long nights. If there is a brief flash of light in the middle of the night, they do not flower. This response enables plants to flower in the appropriate season.

INVESTIGATION

FIGURE 26.11 Photomorphogenesis and Red Light Lettuce seeds will germinate if exposed to a brief period of light. The action spectrum for germination indicated that red light was most effective in promoting it, but far-red light would reverse the effect if presented right after a red-light flash. Harry Borthwick and his colleagues asked what the effect of repeated alternating flashes of red and far-red light would be. In each case, the final exposure determined the germination response. This observation led to the conclusion that a single, photo-reversible molecule was involved. That molecule turned out to be phytochrome.

HYPOTHESIS
The effects of red and far-red light on lettuce seed germination are mutually reversible.

METHOD Expose lettuce seeds to alternating periods of red light R for 1 minute and far-red light FR for 4 minutes.

RESULTS

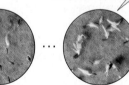
Seeds germinate if the final exposure is to red R …

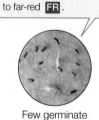
…and remain dormant if the final exposure is to far-red FR.

Most germinate   Few germinate   …   Most germinate   Few germinate

CONCLUSION
Red light and far-red light reverse each other's effects.

ANALYZE THE DATA
Seven groups of 200 lettuce seeds each were incubated in water for 16 hours in the dark. One group was then exposed to white light for 1 min. A second group (controls) remained in the dark. Five other groups were exposed to red (R) and/or far-red (FR) light. All the seeds were then returned to darkness for 2 more days. Germination was then observed.

Condition	Seeds germinated
1. White light	199
2. Dark	17
3. R	196
4. R then FR	108
5. R then FR then R	200
6. R then FR then R then FR	86
7. R then FR then R then FR then R	198

A. Calculate the percentage of seeds that germinated in each case.

B. What can you conclude about the photoreceptors involved?

Go to **yourBioPortal.com** for original citations, discussions, and relevant links for all INVESTIGATION figures.

The action spectra of these processes show that they are induced by red light (650–680 nm). This indicates that plants must have a photoreceptor that absorbs red light.

What is especially remarkable about these red light responses is that *they are reversible by far-red light* (710–740 nm). For example, if lettuce seeds are exposed to brief, alternating periods of red and far-red light in close succession, they respond only to the final exposure. If it is red, they germinate; if it is far-red, they remain dormant (**FIGURE 26.11**). This reversibility of the effects of red and far-red light regulates many other aspects of plant development, including flowering and shoot development.

The basis for the effects of red and far-red light is a bluish photoreceptor in the cytosol of plants called **phytochrome**. Phytochrome exists in two interconvertible *isoforms*, or states. The molecule undergoes a conformational change when it absorbs light at particular wavelengths. The default or "ground" state, which absorbs principally red light, is called P_r. When P_r absorbs a photon of red light, it is converted into the other isoform, P_{fr}. The P_{fr} isoform preferentially absorbs far-red light; when it does so, it is converted back into P_r:

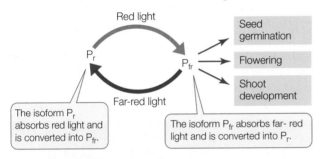

The isoform P_r absorbs red light and is converted into P_{fr}.

The isoform P_{fr} absorbs far-red light and is converted into P_r.

Red light

Far-red light

Seed germination

Flowering

Shoot development

P_{fr}, not P_r, is the active isoform of phytochrome—the form that triggers important biological processes in various plants. These processes include seed germination, shoot development after etiolation (growth in the dark), and flowering.

The ratio of red to far-red light determines whether a phytochrome-mediated response will occur. For example, during daylight, the ratio is about 1.2:1; because there is more red than far-red light, the P_{fr} isoform predominates. But for a plant growing in the shade of other plants, the ratio may be as low as 0.13:1, and most phytochrome is in the P_r isoform. The low ratio of red to far-red light in the shade results from absorption of red light by chlorophyll in the leaves overhead, so less of the red light gets through to the plants below. Shade-intolerant species respond by stimulating cell elongation in the stem and thus growing taller to escape the shade (similar to the behavior of etiolated seedlings grown in the dark). Similarly, shade cast by other plants prevents germination of seeds that require red light to germinate (see Figure 26.11).

Phytochrome stimulates gene transcription

How does phytochrome—or more specifically, P_{fr}—work? Phytochrome is a cytoplasmic protein composed of two subunits (**FIGURE 26.12**). Each subunit contains a protein chain and a nonprotein pigment called a *chromophore*. In *Arabidopsis*, there is a gene family that encodes five slightly different phytochromes, each functioning in different photomorphogenic responses.

The transcription of genes involved in phytochrome responses is changed when P_r is converted into the P_{fr} isoform. When P_r absorbs red light, the chromophore changes shape, which leads to a change in the conformation of the phytochrome protein from the P_r isoform to the P_{fr} isoform. Conversion to the P_{fr} isoform exposes two important regions of the protein (see Figure 26.12), both of which affect transcriptional activity:

- Exposure of a *nuclear localization signal sequence* (see Figure 10.20) results in movement of P_{fr} from the cytosol to the nucleus. Once in the nucleus, P_{fr} binds to transcription factors and thereby stimulates expression of genes involved in photomorphogenesis.

- Exposure of a *protein kinase* domain causes the P_{fr} protein to phosphorylate itself and other proteins involved in red-light signal transduction. Those proteins change the activity of other transcription factors.

FIGURE 26.12 Phytochrome Stimulates Gene Transcription Phytochrome is composed of two subunits, each containing a protein chain and a chromophore. When the chromophore absorbs red light, phytochrome is converted into the P_{fr} isoform, which activates transcription of phytochrome-responsive genes.

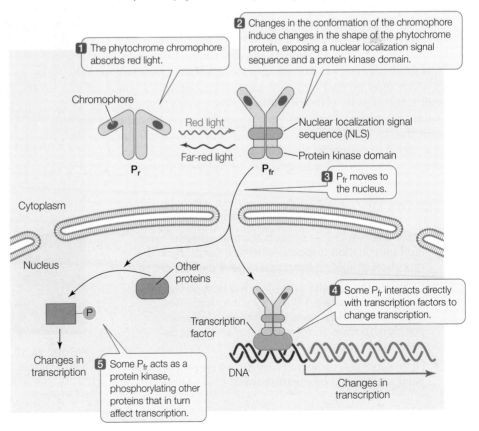

1 The phytochrome chromophore absorbs red light.

2 Changes in the conformation of the chromophore induce changes in the shape of the phytochrome protein, exposing a nuclear localization signal sequence and a protein kinase domain.

Chromophore

Red light

Far-red light

P_r

P_{fr}

Nuclear localization signal sequence (NLS)

Protein kinase domain

3 P_{fr} moves to the nucleus.

Cytoplasm

Nucleus

Other proteins

Transcription factor

4 Some P_{fr} interacts directly with transcription factors to change transcription.

P

Changes in transcription

5 Some P_{fr} acts as a protein kinase, phosphorylating other proteins that in turn affect transcription.

DNA

Changes in transcription

The effect of this activation of transcription factors is quite large: in *Arabidopsis*, phytochrome affects an amazing *2,500 genes* (10 percent of the entire genome) by either increasing or decreasing their expression. Some of these genes are related to other hormones. For example, when P_{fr} is formed at the time of seed germination, genes for gibberellin synthesis are activated and genes for gibberellin breakdown are repressed. As a result, gibberellins accumulate and seed reserves are mobilized (see Figure 26.4).

Circadian rhythms are entrained by photoreceptors

The timing and duration of biological activities in living organisms are governed in all eukaryotes and some prokaryotes by what is commonly called a "biological clock": an oscillator within cells that alternates back and forth between two states at roughly 12-hour intervals. The major outward manifestations of this clock are known as **circadian rhythms** (Latin *circa*, "about," and *dies*, "day"). In plants, circadian rhythms influence, for example, the opening (during the day) and closing (at night) of stomata in *Arabidopsis* and the elevation toward the sun (during the day) and lowering (at night) of leaves in bean plants.

LINK For a description of the role of circadian rhythms in animal behavior, see Concept 41.4

The circadian rhythms of plants can be reset, or *entrained*, within limits, by light–dark cycles that change. Consider what happens when plants adapt to day length as the seasons change during the year. The action spectrum for plant entrainment indicates that phytochrome (and to a lesser extent, blue-light receptors) is very likely involved. At sundown, phytochrome is mostly in the active P_{fr} isoform. But as the night progresses, P_{fr} is gradually converted back into the inactive P_r isoform. By dawn, most phytochrome is in the P_r state, but as daylight begins, it is rapidly converted into P_{fr}. The switch to the P_{fr} isoform resets the plant's biological clock. However long the night, the clock is still reset at dawn every day. Thus while the total period measured by the clock does not change, the clock adjusts to changes in day length over the course of the year.

Do You Understand Concept 26.4?

- Trace the events that occur in the phytochrome-mediated germination response of lettuce seeds, from the reception of light to alteration in gene expression.

- Compare the wavelengths of light absorbed by chlorophyll (see Figure 6.17) and by phototropin (see Figure 26.10). How does the difference influence a plant's responses to light when a plant grows beneath a forest canopy?

- How would you show that a physiological process in a plant is mediated by phytochrome?

What changes in their growth patterns made the new strains of cereal crops produced by the Green Revolution so successful?

ANSWER   It is only recently that plant biologists have discovered why semi-dwarf wheat and rice have short stems (**FIGURE 26.13**). This knowledge has come from an understanding of the effects of various hormones on plant growth and development (Concept 26.1), the signal transduction pathways for response to these hormones (Concepts 26.2 and 26.3), and the biochemical pathways for the synthesis of these hormones. It turns out that the genes involved in wheat and in rice are different.

In semi-dwarf wheat, the mutant allele that produces short plants is involved in signal transduction in response to gibberellin. The gene involved is called *Rht* ("reduced height"). The mutant allele was first seen in Japanese semi-dwarf varieties. It is dominant, so its phenotypes are readily seen. Norman Borlaug used conventional plant breeding to transfer the *Rht* mutant allele into wheat strains in Mexico. In addition to having short stature and not falling over when bearing a high grain yield, *Rht* mutant plants put a greater proportion of their photosynthate into making seeds than wild-type plants do. This higher *harvest index* is also a major reason why most wheat varieties today carry an *Rht* mutation.

FIGURE 26.13 Semi-Dwarf Rice   This short variety of rice (left) produces higher yields of grain than its taller counterpart (right). The semi-dwarf rice produces less than normal amounts of gibberellins.

A hint that *Rht* might be involved in hormone regulation came when semi-dwarf plants harboring *Rht* mutations were found to be insensitive to added gibberellin and did not grow to normal height when the hormone was applied. *Rht* encodes the transcriptional repressor in the gibberellin signal transduction pathway (see Figure 26.8). Mutations in semi-dwarf plants are in the region that binds the gibberellin receptor and that signals polyubiquitination and degradation in the proteasome. Thus in the mutant plants, the repressor is always bound to the transcription factor, inhibiting its activity and keeping the gibberellin response "off." This mutation therefore results in shorter stature.

In semi-dwarf rice, the mutant allele that produces short plants is involved with gibberellin *synthesis*. The gene involved is called *sd-1* ("semi-dwarf"). In contrast to semi-dwarf wheat, rice plants with mutations in *sd-1* are sensitive to gibberellin—that is, the semi-dwarf phenotype can be corrected by supplying gibberellin. This observation indicates that the mutant plants have the signal transduction machinery intact but lack the hormone. Indeed, *sd-1* encodes an enzyme that catalyzes one of the last steps in gibberellin synthesis.

SUMMARY

concept 26.1 — Plants Develop in Response to the Environment

- Plant development is regulated by environmental cues, receptors, hormones, and the plant's genome.

- Seed **dormancy**, which has adaptive advantages, is maintained by a variety of mechanisms. When dormancy ends, the seed **imbibes** water, **germinates**, and develops into a **seedling**. **Review Figure 26.1 and WEB ACTIVITIES 26.1 and 26.2**

- **Hormones** and **photoreceptors** act through signal transduction pathways to regulate plant growth and development.

- **Genetic screens** using the model organism *Arabidopsis thaliana* have contributed greatly to our understanding of signal transduction pathways in plants. **Review Figure 26.2**

concept 26.2 — Gibberellins and Auxin Have Diverse Effects but a Similar Mechanism of Action

- **Gibberellins** stimulate growth of stems and fruits as well as mobilization of seed reserves in cereal crops. **Review Figure 26.4 and WEB ACTIVITY 26.3**

- **Auxin** is made in cells at the shoot apex and moves down to the growing region in a polar manner. **Review Figure 26.5**

- Lateral movement of auxin, mediated by auxin efflux carriers, is responsible for **phototropism** and **gravitropism**. **Review Figure 26.6 and ANIMATED TUTORIAL 26.1**

- Auxin plays roles in lateral root formation, leaf **abscission**, and **apical dominance**.

- The **acid growth hypothesis** explains how auxin promotes cell expansion by increasing proton pumps in the plasma membrane, which loosens the cell wall. **Review Figure 26.7 and ANIMATED TUTORIAL 26.2**

- Both auxin and gibberellins act by binding to their respective receptors, which then bind to a transcriptional repressor, leading to the repressor's breakdown in the proteasome. **Review Figure 26.8**

concept 26.3 — Other Plant Hormones Have Diverse Effects on Plant Development

- **Cytokinins** are adenine derivatives that often interact with auxin. They promote plant cell division, promote seed germination in some species, and inhibit stem elongation, among other activities.

- Cytokinins act on plant cells through a two-component signal transduction pathway. **Review Figure 26.9**

- A balance between auxin and **ethylene** controls leaf abscission. Ethylene promotes senescence and fruit ripening. It causes the stems of eudicot seedlings to form a protective **apical hook**. In stems, it inhibits elongation, promotes lateral swelling, and decreases sensitivity to gravitropic stimulation.

- **Brassinosteroids** promote cell elongation, pollen tube elongation, and vascular tissue differentiation but inhibit root elongation. Unlike animal steroids, these hormones act at a plasma membrane receptor.

- **Abscisic acid** inhibits seed germination, promotes dormancy, and stimulates stomatal closing in response to dry conditions in the environment.

concept 26.4 — Photoreceptors Initiate Developmental Responses to Light

- **Phototropin** is a blue-light receptor protein involved in phototropism. **Zeaxanthin** acts in conjunction with phototropin to mediate the light-induced opening of stomata. **Cryptochromes** are blue-light receptors that affect seedling development and flowering and inhibit cell elongation. **Review Figure 26.10**

- **Phytochrome** is a photoreceptor that exists in the cytosol in two interconvertible isoforms, P_r and P_{fr}. The relative amounts of these two isoforms are a function of the ratio of red to far-red light. Phytochrome plays a number of roles in **photomorphogenesis**. **Review Figure 26.11**

- The phytochrome signal transduction pathway affects transcription in two ways: the P_{fr} isoform interacts directly with some transcription factors and influences transcription indirectly by phosphorylating other proteins. **Review Figure 26.12**

- **Circadian rhythms** are changes that occur on a daily cycle. Light can entrain circadian rhythms through photoreceptors such as phytochrome.

Dairy farmer Albert Ecke was fascinated by the red and green shrubs that grew all over southern California and were used by Mexican-Americans for red dye. The shrub, *Euphorbia pulcherrima*, got the name "poinsettia" from the man who first brought it to the United States: the first U.S. ambassador to Mexico, Joel Roberts Poinsett. In the early 1900s, Ecke started selling the plants at his farm in Hollywood. But two challenges stood in the way of his making this plant a commercial success.

First, although poinsettias generally bloomed in the fall and early winter in the mild climate of southern California, when Ecke grew them in fields in Hollywood, the formation of flowers was unreliable. Biologists later found that the flowering of poinsettias required at least a 14-hour night for several weeks. Any interruption in the long night—by passing cars or street lamps, for example—inhibited flowering. So Ecke's son Paul moved the growing operation south, to isolated fields far from Los Angeles. A second challenge was that the plants were tall and gangly. Although pretty, they were hard to transport and not attractive indoors. Paul Ecke found a variety of poinsettia that was much more compact. He propagated this variety and eventually developed it into the short potted plant that is so popular today. The Eckes decided that the time of flowering was just right for making the poinsettia a "Christmas flower," so they promoted this now portable plant as a holiday decoration, blanketing live television shows with free plants between Thanksgiving and Christmas. The campaign was successful: over 100 million poinsettias are

now sold in the United States during the winter holidays every year, making it the nation's best-selling potted plant.

Breeding more attractive flowering plants that are easier to grow is an ongoing part of *floriculture*, the industry involved with the production of floral crops. You may be surprised to learn that the brightly colored "flowers" of poinsettias are not flowers at all. The red parts of the plant that we most notice and appreciate are actually specialized leaves called *bracts*. The poinsettia has a single tiny yellow female flower, without petals, surrounded by male flowers.

The poinsettia is a popular Christmas flower. But the red "flowers" are not flowers at all; instead, they are leaves (bracts). The flower is the yellow structure in the center of a group of red leaves.

Q **QUESTION** How did an understanding of angiosperm reproduction allow floriculturists to develop a commercially successful poinsettia?

KEY CONCEPTS

27.1 Most Angiosperms Reproduce Sexually

27.2 Hormones and Signaling Determine the Transition from the Vegetative to the Reproductive State

27.3 Angiosperms Can Reproduce Asexually

concept 27.1 Most Angiosperms Reproduce Sexually

Most angiosperms (flowering plants) have evolved to reproduce sexually because this strategy results in the genetic diversity that is the raw material for evolution. Sexual reproduction in angiosperms involves mitosis, meiosis, and the alternation of haploid and diploid generations (see Figure 7.3). There are several important differences between sexual reproduction in angiosperms and in vertebrate animals (see Chapter 32):

- Meiosis in plants produces spores, after which mitosis produces gametes; in animals, meiosis usually produces gametes directly.

- In most plants, there are multicellular diploid (*sporophyte*) and haploid (*gametophyte*) life stages (alternation of generations); in animals, there is no multicellular haploid stage.

- In plants, the cells that will form gametes are determined in the adult organism, usually in response to environmental conditions; in animals, the germline cells are determined before birth.

LINK The life cycles of plants have many unique characteristics, which are detailed in Chapter 21

The flower is the reproductive organ of angiosperms

In angiosperms, male or female gametophytes (and sometimes both) are contained in flowers. As we saw in Concept 21.5, a complete flower consists of four concentric groups of organs arising from modified leaves: the *carpels*, *stamens*, *petals*, and *sepals*.

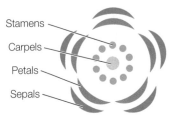

The parts of the flower are all derived from a modified shoot apical meristem.

- The carpels are the female sex organs that contain the developing female gametophytes.

- The stamens are the male sex organs that contain the developing male gametophytes.

The differentiation of these organs is controlled by specific transcription factors (see Figure 14.11).

As you know, flowers come in many shapes, sizes, and forms. Some have both stamens (male sex organs) and carpels (female sex organs); such flowers are termed *perfect* (**FIGURE 27.1A**). *Imperfect* flowers, by contrast, are those with only male or only female sex organs. Male flowers have stamens but not carpels, and female flowers have carpels but not stamens. Some plants, such as corn, bear both male and female flowers on

(A) Perfect: lily (*Lilium* sp.)

Carpel

Stamens

FIGURE 27.1 Perfect and Imperfect Flowers (A) A lily is an example of a perfect flower, meaning one that has both male and female sex organs. (B) Imperfect flowers are either male or female. Corn is a monoecious species: both types of imperfect flowers are borne on the same plant. (C) American holly is a dioecious species; some American holly plants bear male imperfect flowers, and others bear female imperfect flowers.

(B) Imperfect monoecious: corn (*Zea mays*)

Male flower with stamens

Female flower with carpels

(C) Imperfect dioecious: American holly (*Ilex opaca*)

Male flower with stamens

Female flower with carpels

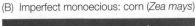

the same individual plant; such species are called **monoecious** ("one housed"; **FIGURE 27.1B**). In **dioecious** ("two-housed") species, individual plants bear either male-only or female-only flowers; an example is American holly (**FIGURE 27.1C**).

LINK Review the diverse structures and functions of flowers in Concept 21.5

Angiosperms have microscopic gametophytes

FIGURE 27.2 offers a detailed look at angiosperm gametophytes. These haploid, microscopic gamete-producing structures develop from haploid spores in the flower:

- Female gametophytes (*megagametophytes*), which are also called **embryo sacs**, develop in megasporangia. The meiotic products that give rise to megagametophytes are called *megaspores*.

- Male gametophytes (*microgametophytes*), which are also called **pollen grains**, develop in microsporangia. The meiotic products that give rise to microgametophytes are called *microspores*.

FEMALE GAMETOPHYTE Of the four haploid megaspores resulting from meiosis, three undergo apoptosis (programmed cell death). Typically, the remaining megaspore undergoes three mitotic divisions without cytokinesis, producing eight haploid nuclei, all initially contained within a single cell—three nuclei at one end, three at the other, and two in the middle. Subse-

quent cell wall formation leads to an elliptical, seven-celled megagametophyte with a total of eight nuclei:

- At one end of the megagametophyte are three small cells: the *egg* cell and two cells called *synergids*. The egg cell is the female gamete, and the synergids participate in fertilization by attracting the *pollen tube* and receiving the sperm nuclei prior to their movement to the egg cell and the central cell.

- At the opposite end of the megagametophyte are three *antipodal* cells, which eventually degenerate.

- In the large central cell are two **polar nuclei**.

The megagametophyte, or embryo sac, is the entire seven-cell, eight-nucleus structure.

MALE GAMETOPHYTE The four haploid products of meiosis in the microsporangia each develop a cell wall and undergo a single mitotic division, producing four two-celled pollen grains that are released into the environment. The two cells in a pollen grain have different roles:

- The *generative cell* divides by mitosis to form two *sperm* cells that participate in fertilization.

- The *tube cell* forms the elongating pollen tube that delivers the sperm to the embryo sac.

FIGURE 27.2 Sexual Reproduction in Angiosperms The embryo sac is the female gametophyte; the pollen grain is the male gametophyte. The male and female nuclei meet and fuse in the embryo sac. Angiosperms have double fertilization, in which a zygote and an endosperm nucleus form from separate fusion events—the zygote from one sperm and the egg, and the endosperm from the other sperm and the two polar nuclei.

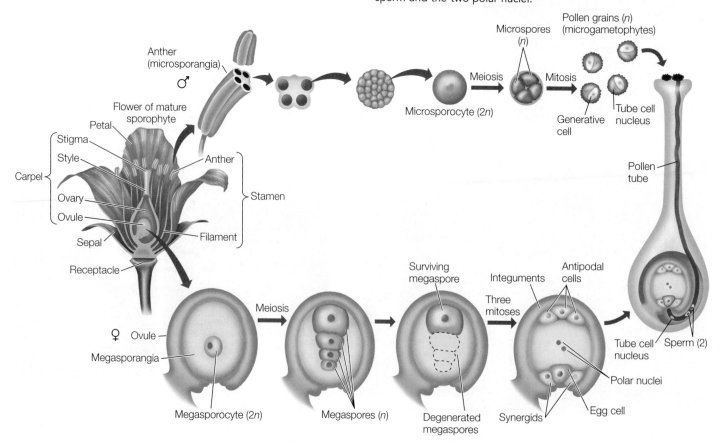

These events occur after the pollen grain is transferred to a stigma (part of the female reproductive organ)—a process called *pollination*.

Angiosperms have mechanisms to prevent inbreeding

As we discuss in Concept 21.4, the evolution of pollen in the seed plants made it possible for male gametes to reach the female gametophyte in the absence of liquid water. The transfer of pollen from plant to plant presents its own challenges, however. In the first seed plants to evolve, tiny pollen grains were carried to stigmas by wind. Wind-pollinated flowers have sticky or featherlike stigmas, and they produce a great number of pollen grains because of the low probability of an individual grain encountering a suitable stigma. More recently in evolutionary time, most angiosperms have come to rely on animals for pollination. Pollen transport by animals greatly increases the probability that pollen will get to a female gametophyte of the same species.

> **LINK** The evolution of plant–pollinator interactions and their effects on plant speciation are described in Concepts 17.4 and 21.5

Some plants *self-pollinate*: that is, they are fertilized by pollen that lands on the stigma of a flower on the same plant, or even the same flower. Mendel's garden peas, for example, reproduce in this way, so he had to remove stamens to control the parentage of his genetic crosses (see Figure 8.1). Self-pollination mitigates the difficulties of wind or animal pollination of a distant flower. However, "selfing" leads to homozygosity, which can reduce the reproductive fitness of offspring (known as inbreeding depression). Most plants have evolved mechanisms that prevent self-pollination and self-fertilization.

SEPARATION OF MALE AND FEMALE GAMETOPHYTES Self-fertilization is not possible in dioecious species, which bear only male or female flowers on an individual plant. Pollination in dioecious species is accomplished only when one plant pollinates another. In monoecious plants, which bear both male and female flowers on the same plant, the physical separation of male and female flowers is often sufficient to prevent self-fertilization. Some monoecious species prevent self-fertilization by staggering the development of male and female flowers so they do not bloom at the same time, making these species functionally dioecious.

GENETIC SELF-INCOMPATIBILITY A pollen grain that lands on a stigma of the same plant will fertilize the female gamete only if the plant is *self-compatible* (capable of self-fertilization). To prevent self-fertilization, many plants are **self-incompatible**. This mechanism depends on the ability of a plant to determine whether pollen is genetically similar or genetically different from "self." Rejection of "same-as-self" pollen prevents self-fertilization.

Self-incompatibility in plants is controlled by a cluster of tightly linked genes called the *S* locus (for self-incompatibility). The *S* locus encodes proteins in the pollen and style that interact during the recognition process. A self-incompatible species typically has many alleles of the *S* locus, and when the pollen carries an allele that matches one of the alleles of the recipient pistil, the pollen is rejected. Depending on the type of self-incompatibility system, the rejected pollen either fails to germinate or the pollen tube is prevented from growing through the style (**FIGURE 27.3**); either way, self-fertilization is prevented.

A pollen tube delivers sperm cells to the embryo sac

When a genetically appropriate pollen grain lands on the stigma of a compatible pistil, it germinates. As with seed germination, a key event is water uptake by pollen, this time from the stigma. Germination involves the development of a pollen tube. The pollen tube either traverses the spongy tissue of the style or, if the style is hollow, grows on the inner surface of the style until it reaches an ovule. The pollen tube typically grows at the rate of 1.5–3 millimeters per hour, taking just an hour or two to reach its destination, the female gametophyte.

The growth of the pollen tube may be guided in part by a chemical signal produced by the synergids in the ovule. If one synergid is destroyed, the ovule still attracts pollen tubes, but destruction of both synergids renders the ovule unable to attract pollen tubes, and fertilization does not occur. The attractant appears to be species-specific: in some cases, isolated

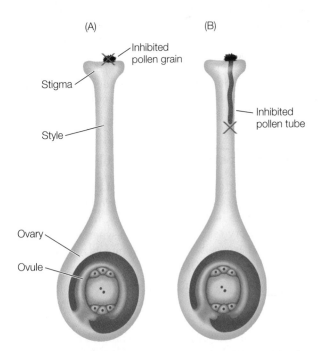

FIGURE 27.3 Self-incompatibility In a self-incompatible plant, pollen is rejected if it expresses an *S* allele that matches one of the *S* alleles of the stigma and style. There are two mechanisms of rejection: (A) the pollen may fail to germinate, or (B) the pollen tube may die before reaching an ovule. Both mechanisms prevent the egg from being fertilized by a sperm from the same plant.

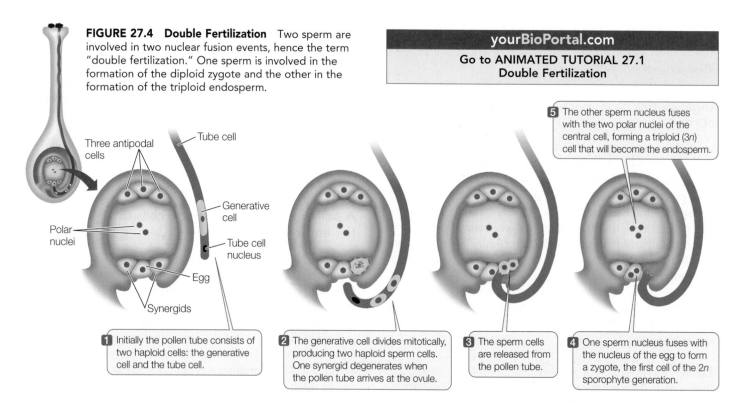

FIGURE 27.4 Double Fertilization Two sperm are involved in two nuclear fusion events, hence the term "double fertilization." One sperm is involved in the formation of the diploid zygote and the other in the formation of the triploid endosperm.

Three antipodal cells

Tube cell

Generative cell

Tube cell nucleus

Polar nuclei

Egg

Synergids

1 Initially the pollen tube consists of two haploid cells: the generative cell and the tube cell.

2 The generative cell divides mitotically, producing two haploid sperm cells. One synergid degenerates when the pollen tube arrives at the ovule.

3 The sperm cells are released from the pollen tube.

4 One sperm nucleus fuses with the nucleus of the egg to form a zygote, the first cell of the 2*n* sporophyte generation.

5 The other sperm nucleus fuses with the two polar nuclei of the central cell, forming a triploid (3*n*) cell that will become the endosperm.

yourBioPortal.com

Go to ANIMATED TUTORIAL 27.1
Double Fertilization

female gametophytes attract only pollen tubes of the same species.

Angiosperms perform double fertilization

The pollen tube grows down the style, under the direction of the haploid tube cell. Meanwhile, the generative cell divides once to form two haploid sperm cells. Once the pollen tube reaches the embryo sac, one of the two synergids degenerates, and the two sperm cells are released into its remains. *Two fertilization events occur* (**FIGURE 27.4**):

- One sperm cell fuses with the egg cell, producing the diploid zygote, which forms the new sporophyte embryo.

- The other sperm cell fuses with the two polar nuclei in the central cell of the embryo sac. This fusion forms a triploid nucleus that undergoes rapid mitosis to form a specialized nutritive tissue, the *endosperm*. The endosperm contains the reserves that will provide chemical building blocks for the developing embryo while it is underground and cannot perform photosynthesis.

This process of *double fertilization* is a characteristic feature of angiosperm reproduction. The remaining cells of the female gametophyte—the antipodal cells and the remaining synergid—eventually degenerate, as does the pollen tube nucleus.

Embryos develop within seeds contained in fruits

Fertilization initiates the highly coordinated growth and development of the embryo, endosperm, integuments, and carpel. The *integuments*—tissue layers immediately surrounding the megasporangium—develop into the seed coat, and the carpel ultimately becomes the wall of the fruit that encloses the seed (see Figure 27.2).

In Chapter 26 we described the events in plant embryonic development and its hormonal control. As seeds develop, they prepare for dispersal and dormancy by losing up to 95 percent of their water content. You can see this desiccation by comparing corn grains (e.g., popcorn) with ripe corn from the cob or a can. A dry seed is still alive; it has protective proteins that keep its cells in a viscous state.

In angiosperms, the ovary—together with the seeds it contains—develops into a fruit after fertilization has occurred. Fruits have two main functions:

- They protect the seed from damage by animals and infection by microbial pathogens.

- They aid in seed dispersal.

A **fruit** may consist of only the mature ovary and seeds, or it may include other parts of the flower. Some species produce fleshy, edible fruits, such as peaches and tomatoes, whereas the fruits of other species are dry or inedible (**FIGURE 27.5**; see also Figure 21.24).

(A)   Seed with embryo   Fruit (ovary wall)

(B)

(C)

Seed with embryo

Seed with embryo   Fruit (ovary wall)   Fruit (ovary wall)

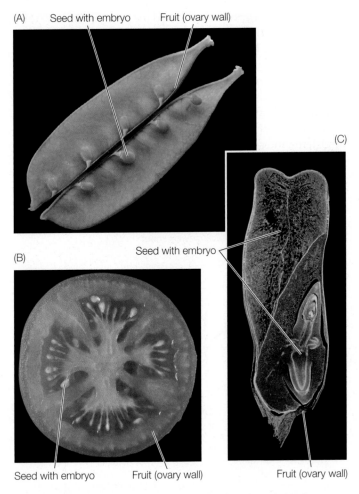

FIGURE 27.5 Angiosperm Fruits There are a variety of fruits, but all have seed containing the embryo, surrounded by a fruit that derives from the wall of the ovary. (A) Garden pea. (B) Tomato. (C) Corn.

germination and growth to sexual maturity. Some, like those of the thistle, are carried to new locations by wind:

Others attach themselves to animals (or to your clothes and shoes):

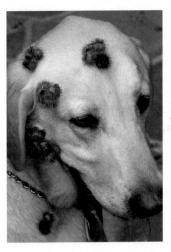

Water disperses some fruits; coconuts have been known to float thousands of miles between islands. Seeds swallowed whole by an animal along with fruits such as berries travel through the animal's digestive tract and are deposited some distance from the parent plant (see Apply the Concept in Chapter 21, p.430).

The diverse forms of fruits reflect the varied strategies plants use to disperse their progeny. Because plants cannot move, their progeny need some mechanism for separating themselves from their parent. Fruits are clearly an important way to move seeds away from the parent plant. Wide dispersal of progeny may not always be advantageous, however. If a plant has successfully grown and reproduced, its location is likely to be favorable for the next generation too. Some offspring do indeed stay near the parent, as is the case in many tree species, whose seeds simply fall to the ground. This strategy, however, has several potential disadvantages. If the species is a *perennial*, offspring that germinate near their parent will be competing with their parent for resources, which may be too limited to support a dense population. Furthermore, even though local conditions were good enough for the parent to produce at least some seeds, there is no guarantee that conditions will still be good the next year, or that they won't be better elsewhere. Thus in many cases, seed dispersal is vital to a species' survival.

Some fruits help disperse seeds over substantial distances, increasing the probability that at least a few of the many seeds produced by a plant will find suitable conditions for

Do You Understand Concept 27.1?

- Make a table that compares the haploid and diploid life stages of a human and a corn plant with regard to (1) the relative sizes of the haploid and diploid stages, (2) how many cells are in the haploid stage, (3) the names of the haploid cells, (4) the time in the life cycle that the haploid stage occurs, (5) the location of the haploid cells, and (6) how the haploid stages develop in males and females.

- Describe the processes that angiosperms use to ensure a genetically diverse population.

- The two sperm that participate in plant double fertilization are genetically identical. Are the two products of fertilization genetically identical?

Now that we have described the cellular aspects of angiosperm reproduction, let's turn to the events that signal a transition from vegetative growth of the shoot to the formation of flowers.

concept
27.2 **Hormones and Signaling Determine the Transition from the Vegetative to the Reproductive State**

Flowering represents a reallocation of energy and materials away from making more roots, stems, and leaves (*vegetative growth*) to making flowers and gametes (*reproductive growth*). Flowering can happen at maturity as part of a predetermined developmental program (as in a dandelion plant in summer) or in response to environmental cues such as light or temperature (as with the poinsettias we described at the opening of this chapter). In either case, flowering is initiated through a cascade of gene expression.

Plants fall into three categories based on when they mature and initiate flowering and what happens after they flower:

- **Annuals** complete their lives within a year. Annuals include many crops important to humans, such as corn, wheat, rice, and soybeans. When the environment is suitable, annuals grow rapidly. After flowering, they channel most of their energy into the development of seeds and fruits. Then the rest of the plant withers and dies.

- **Biennials** take two years to complete their lives. They are much less common than annuals and include carrots, cabbage, and onions. Typically, biennials produce only vegetative growth during their first year and store carbohydrates in underground roots (carrots) or stems (onions). In their second year, they use most of the stored carbohydrates to produce flowers and seeds rather than vegetative growth, and the plant dies after seeds form.

- **Perennials** live three or more—sometimes many more—years. Maple trees, for example, can live up to 400 years. Perennials include many wildflowers as well as trees and shrubs. Typically these plants flower each year but stay alive and keep growing the next seasons.

Shoot apical meristems can become inflorescence meristems

The transition from purely vegetative growth to the flowering state requires a change in one or more apical meristems in the shoot system. As we described in Chapter 24, shoot apical meristems continually produce leaves, axillary buds, and stem tissues (**FIGURE 27.6A**) in a kind of unrestricted growth called *indeterminate growth*.

A shoot apical meristem becomes an **inflorescence meristem** (recall that an *inflorescence* is a cluster of flowers) when it ceases production of leaves and stems and produces other structures: smaller leafy structures called *bracts* (such as the red leaves in poinsettias), as well as new meristems

in the angles between the bracts and the stem (**FIGURE 27.6B**). These new meristems may also be inflorescence meristems, or they may be **floral meristems**, each of which gives rise to a single flower.

Each floral meristem typically produces four consecutive *whorls*, or spirals, of organs—the sepals, petals, stamens, and carpels—separated by very short internodes, which keep the flower compact (**FIGURE 27.6C**). In contrast to shoot apical meristems and some inflorescence meristems, floral meristems are responsible for *determinate growth*—growth of limited extent, like that of leaves.

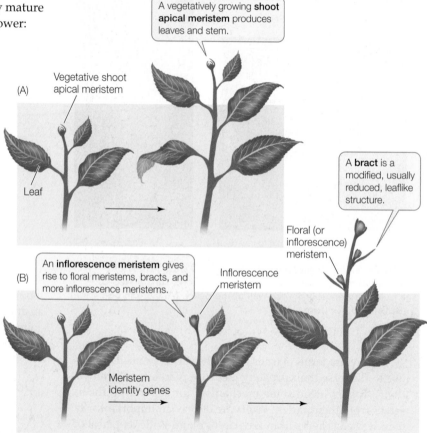

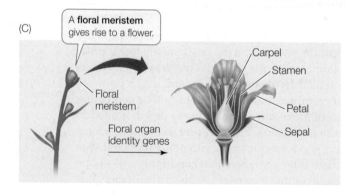

FIGURE 27.6 The Transition to Flowering (A) A shoot apical meristem produces vegetative organs without producing flowers. (B) A shoot apical meristem becomes an inflorescence meristem when it ceases production of leaves and stems and produces other structures. (C) A floral meristem becomes a single flower.

A cascade of gene expression leads to flowering

The genes that determine the transition from shoot apical meristems to inflorescence meristems and from inflorescence meristems to floral meristems have been studied in model organisms such as *Arabidopsis*.

MERISTEM IDENTITY GENES Expression of two **meristem identity genes** initiates a cascade of further gene expression that leads to flower formation. Expression of the genes *LEAFY* and *APETALA1* is both necessary and sufficient for flowering. Evidence for the role of these genes is both genetic and molecular. For example, a mutant allele of *APETALA1* leads to continued vegetative growth, even if conditions are suitable for flowering. However, if the wild-type *APETALA1* gene is coupled to an active promoter and introduced into a shoot apical meristem, the plant will flower regardless of environmental conditions. This is powerful evidence that *APETALA1* plays a role in switching shoot apical meristem cells from a vegetative to a reproductive fate (see Figure 27.6B).

FLORAL ORGAN IDENTITY GENES The products of the meristem identity genes trigger the expression of **floral organ identity genes**, which work in concert to specify the successive whorls of the flower (see Figure 27.6C). Floral organ identity genes are homeotic genes whose products are transcription factors that determine whether cells in the floral meristem will be sepals, petals, stamens, or carpels. An example is the gene *AGAMOUS*, a class C gene that causes florally determined cells to form stamens and carpels in the "ABC" system described in Chapter 14 (see Figure 14.11).

Depending on the species, plants initiate these gene expression changes, and the events that follow, in response to either internal or external cues. Among external clues, the best studied are *photoperiod* (day length) and temperature. We will begin with photoperiod.

Photoperiodic cues can initiate flowering

The study of how light affects the transition to flowering began with two observations in the early twentieth century:

- A mutant tobacco strain appropriately called Maryland Mammoth grew 5 m tall (normal tobacco is about 1.5 m tall). Instead of flowering in summer, it continued to grow until the late fall frost killed it. Farmers in Virginia were frustrated because they could not get seed of this luxuriant plant for the next year's crop.

- Because of improvements in agricultural techniques, soybean yields became so great that it was hard for farmers to harvest all the plants at once. So they tried planting the seeds in groups, several weeks apart. Unfortunately, the resulting plants all formed flowers and seeds at the same time.

The explanation for both of these observations was the same: the signal that set the plants' shoot apical meristems on the path to flowering was the length of daylight, or **photoperiod**. When soybeans experienced days of a certain length, they flowered,

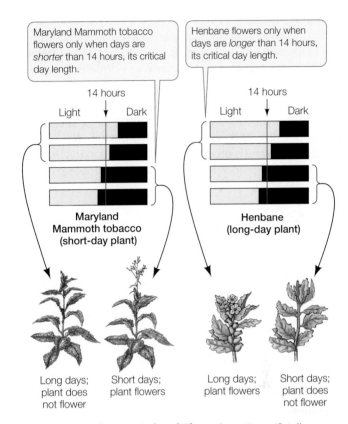

FIGURE 27.7 Photoperiod and Flowering By artificially varying the length of daylight in a 24-hour period, biologists showed that the flowering of Maryland Mammoth tobacco is initiated when the days become shorter than a critical length. Maryland Mammoth tobacco is thus called a short-day plant. Henbane, a long-day plant, shows an inverse pattern of flowering.

regardless of how "old" they were. Maryland Mammoth tobacco *could* flower, but it did not do so in Virginia because it died when the weather got cold. When the plant was grown in a greenhouse to prevent freezing, however, it flowered in December, when the days were short. Maryland Mammouth is now grown commercially in Florida.

Greenhouse experiments measured the day length required for different plant species to flower. Maryland Mammoth tobacco did not flower if exposed to more than 14 hours of light per day; flowering was only initiated once day length became shorter than 14 hours, as it does in December. Other plants (such as soybeans and henbane) flowered only when the days were long (**FIGURE 27.7**).

Plants vary in their responses to photoperiodic cues

Plants that flower in response to photoperiodic stimuli fall into two main classes, although there are variations on these patterns:

- **Short-day plants** (SDPs) flower only when the day is *shorter* than a critical maximum. They include poinsettias and chrysanthemums as well as Maryland Mammoth tobacco. Thus, for example, we see chrysanthemums in nurseries in fall and poinsettias in winter (as noted at the opening of this chapter).

- **Long-day plants (LDPs)** flower only when the day is *longer* than a critical minimum. Spinach and clover are examples of LDPs. Spinach, for example, tends to flower and become bitter in the summer and is therefore normally planted in early spring.

Photoperiodic control of flowering serves an important role: it synchronizes the flowering of plants of the same species in a local population. This synchronization promotes cross-pollination and successful reproduction. It also means that floriculturists can vary light exposures in greenhouses to produce flowers at any time of year.

Night length is the key photoperiodic cue that determines flowering

The terms "short-day plant" and "long-day plant" assume that the key events that induce flowering occur in the daytime. What about the night? After all, short-day plants could just as easily be called "long-night" plants, and long-day plants could be called "short-night plants." As it turns out, *night length is indeed the critical factor that induces flowering*, as a series of greenhouse experiments confirmed (**FIGURE 27.8**). In a greenhouse, the overall length of a day or night can be varied irrespective of the 24-hour natural cycle. For example, if cocklebur, a SDP, is exposed to several long periods of light (16 hours each), it will still flower as long as the dark period between them is 9 hours or longer. This 9-hour inductive dark period also induces flowering even if the light period varies from 8 hours to 12 hours.

Biologists noticed that when the inductive dark period was interrupted by a brief period of light, the flowering signal generated by the long night disappeared. It took several days of long nights for the plant to recover and initiate flowering. Interrupting the day with a dark period had no effect on flowering. A clue as to what occurred in the plant when the flash of light was given came when biologists determined the action spectrum for the wavelengths of light that were effective. As with lettuce seed germination (see Figure 26.11), red light was most effective at breaking the "night" stimulus, and its effect was

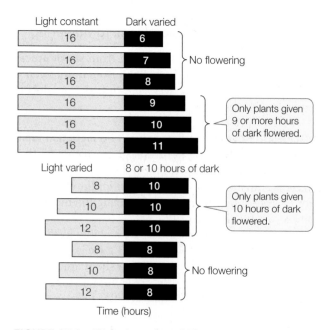

FIGURE 27.8 Night Length and Flowering Greenhouse experiments using cocklebur, a SDP, demonstrated that night length, not day length, is the environmental cue that initiates flowering.

reversible by far-red light. This indicated that the photoreceptor involved in flowering is phytochrome.

> **LINK** The properties of phytochrome as a photoreversible photoreceptor are described in Concept 26.4

---

yourBioPortal.com

Go to ANIMATED TUTORIAL 27.2
The Effect of Interrupted Days and Nights

Plants sense night length by measuring the ratio of the P_{fr} and P_r isoforms of phytochrome. During the day, there is more light in the red than in the far-red range. By the end of the day,

---

APPLY THE CONCEPT

Hormones and signaling determine the transition from the vegetative to the flowering state

The experiments reported on in Figure 27.8 were repeated using a long-day plant (LDP) that normally flowers with a 16-hour photoperiod. The plant was subjected to the light regimes listed in the table.

1. List the light regimes that should have resulted in flowering.

2. Predict the effect of adding a brief period of far-red light in the middle of the dark period of each light regime.

3. Predict the effect of adding a brief period of red light in the middle of the dark period of each light regime.

A. 14 hours light + 6 hours dark	G. 8 hours light + 8 hours dark
B. 14 hours light + 7 hours dark	H. 10 hours light + 8 hours dark
C. 14 hours light + 8 hours dark	I. 12 hours light + 8 hours dark
D. 14 hours light + 10 hours dark	J. 8 hours light + 10 hours dark
E. 14 hours light + 12 hours dark	K. 10 hours light + 10 hours dark
F. 14 hours light + 14 hours dark	L. 12 hours light + 10 hours dark

much of the P_r has absorbed red light and been converted to P_{fr}. At night, there is a gradual, spontaneous conversion of this P_{fr} back to P_r. The longer the night, the more P_r there is at the beginning of the next day. A SDP flowers when the ratio of P_{fr} to P_r is low at the end of the night, whereas a LDP flowers when this ratio is high. You can compare this mechanism to an hourglass egg timer. The sand in the top chamber is analogous to P_{fr} and gradually runs down (is converted to P_r) over the course of the night; the shorter the night, the more sand remains at the top at dawn. To "reset" an hourglass timer, you simply turn it over; this would be analogous to dawn of the next day, when P_r begins being converted back to P_{fr}.

The flowering stimulus originates in the leaf

The photoperiodic receptor, phytochrome, is located in the leaf, not in the shoot apical meristem. "Masking" experiments can show this: if a single leaf of a SDP growing under short-night conditions is covered to simulate a long-night exposure, the shoot apical meristems will transition to flowering as if the entire plant were exposed to long nights (**FIGURE 27.9**). Reversing the masking (leaves exposed, buds covered) does not induce flowering.

Because the receptor of the photoperiodic stimulus (phytochrome in the leaf) is physically separated from the tissue on which the stimulus acts (the shoot apical meristem), the inference can be drawn that a signal travels from the leaf through the plant's tissues to the shoot apical meristem. Unlike animals, plants do not have a nervous system, so the signal must be a diffusible chemical. Biologists have found additional evidence that a diffusible signal travels from the leaf to the shoot apical meristem:

- If a photoperiodically induced leaf is immediately removed from a plant after the inductive dark period, the plant does not flower. If the induced leaf remains attached to the plant for several hours, however, the plant flowers. This result suggests that something is synthesized in the leaf in response to the inductive dark period, and then moves out of the leaf to induce flowering.

- If two or more cocklebur plants are grafted together, and if one plant is exposed to inductive long nights and its graft partners are exposed to noninductive short nights, all of the plants flower.

- In several species, if an induced leaf from one species is grafted onto another, noninduced plant of a different species, the recipient plant flowers. This indicates that the same diffusible chemical signal is used by both species.

Although the diffusible signal was given a name, **florigen** ("flower inducing"), decades ago, the nature of the signal has only recently been explained.

Florigen is a small protein

The characterization of the protein florigen (FT) was made possible by genetic and molecular studies of the model

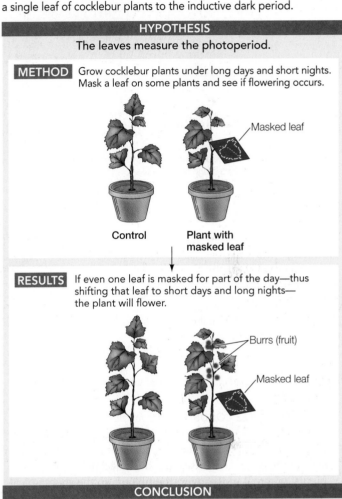

INVESTIGATION

FIGURE 27.9 The Flowering Signal Moves from Leaf to Bud
Phytochrome, the receptor for photoperiod, is in the leaf, but flowering occurs in the shoot apical meristem. To investigate whether there is a diffusible substance that travels from leaf to bud, James Knott exposed a single leaf of cocklebur plants to the inductive dark period.

HYPOTHESIS
The leaves measure the photoperiod.

METHOD Grow cocklebur plants under long days and short nights. Mask a leaf on some plants and see if flowering occurs.

Masked leaf

Control   Plant with masked leaf

RESULTS If even one leaf is masked for part of the day—thus shifting that leaf to short days and long nights— the plant will flower.

Burrs (fruit)

Masked leaf

CONCLUSION
The leaves measure the photoperiod. Therefore, some signal must move from the induced leaf to the flowering parts of the plant.

ANALYZE THE DATA
In related experiments, leaves were removed from plants before the plants were exposed to the inductive dark period. There were six plants in each condition.

Condition	Number of plants that flowered
No inductive dark period, intact plant	0
Inductive dark period, intact plant	6
Inductive dark period, all leaves removed	0
Inductive dark period, all but one leaf removed	6

A. Based on these data, which part of the plant senses photoperiod?

B. What do these data tell you about the signal that is generated by the plant in response to the photoperiod and that induces the shoot apical meristem to produce flowers?

Go to **yourBioPortal.com** for original citations, discussions, and relevant links for all INVESTIGATION figures.

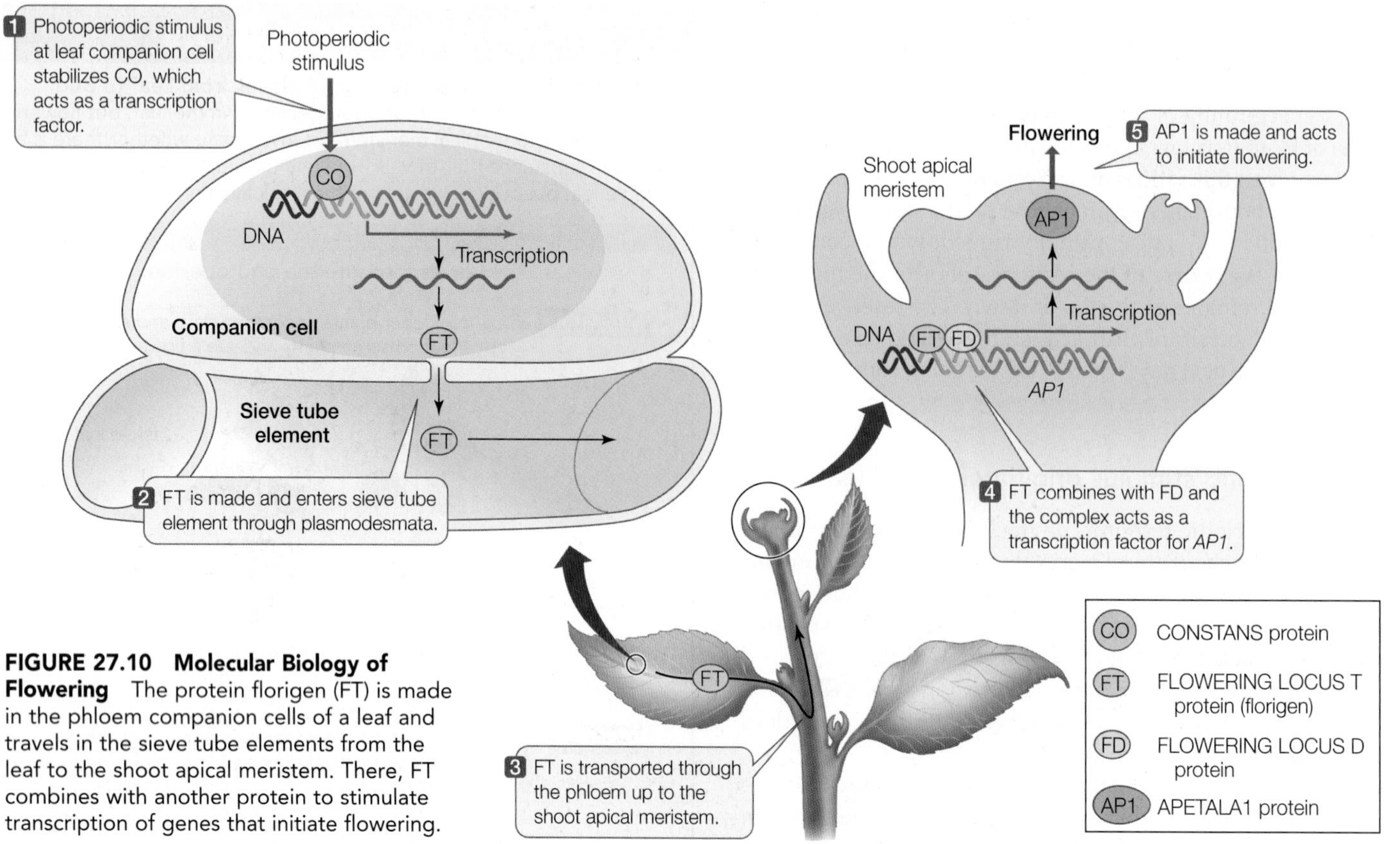

FIGURE 27.10 Molecular Biology of Flowering The protein florigen (FT) is made in the phloem companion cells of a leaf and travels in the sieve tube elements from the leaf to the shoot apical meristem. There, FT combines with another protein to stimulate transcription of genes that initiate flowering.

organism *Arabidopsis*, a LDP. Three genes are involved (**FIGURE 27.10**):

- *FT (FLOWERING LOCUS T) codes for florigen.* FT is a small protein (20 kDa molecular weight) that can travel through plasmodesmata. It is synthesized in phloem companion cells in the leaf and diffuses into the adjacent sieve tube elements, where it moves with the phloem to the shoot apical meristem.

- *CO (CONSTANS) codes for a transcription factor that activates the synthesis of FT.* Like *FT*, *CO* is expressed in phloem companion cells in the leaf. CO protein is expressed all the time but is unstable; an appropriate photoperiodic stimulus stabilizes CO so that there is enough to turn on FT synthesis.

- *FD (FLOWERING LOCUS D) codes for a protein that binds to FT protein when it arrives in the shoot apical meristem.* The FD protein is a transcription factor that forms a complex with FT protein; the complex activates promoters for meristem identity genes, such as *APETALA1*.

Genetic studies have shown that the *FT* gene is involved in photoperiod signaling in many species:

- Transgenic plants (e.g., tobacco and tomato) that express the *Arabidopsis FT* gene at high levels flower regardless of day length.

- Transgenic *Arabidopsis* plants that express high levels of *FT* orthologs from other plants (e.g., rice and tomato) flower regardless of day length.

FRONTIERS The events that occur between signal reception by phytochrome and the molecular events of flowering are not clear. Researchers are using molecular and genetic approaches to determine whether altered gene transcription caused by P_r or P_{fr} directly or indirectly affects expression of the *FT* gene. The identification of all the photoperiodic signal transduction steps will advance our understanding of the mechanisms that control flowering.

Whereas photoperiod induces flowering via the FT–FD pathway in some plants, in many other species flowering is induced by other stimuli through other pathways.

Flowering can be induced by temperature or gibberellins

TEMPERATURE In some plant species, notably certain cereal crops, the environmental signal for flowering is exposure to cold temperatures, or **vernalization** (Latin *vernus*, "spring"). In both wheat and rye, we can distinguish two categories of flowering behavior. Spring wheat, for example, is a typical annual plant: it is sown in the spring and flowers in the same year. Winter wheat is sown in the fall, grows into a seedling, overwinters (often covered by snow), and flowers the following summer. If winter wheat is not exposed to cold in its first year, it will not flower normally the next year.

Some strains of *Arabidopsis* require vernalization, and those strains have been useful in elucidating the molecular pathway

APPLY THE CONCEPT

Hormones and signaling determine the transition from the vegetative to the flowering state

Describe the genes and mutations that could be involved in each of the following observations:

1. A mutant plant flowers without its normal inductive dark period. When a leaf from the mutant plant is grafted onto an unexposed wild-type plant, the recipient plant flowers.

2. A mutant plant does not flower when exposed to the normal inductive dark period. When a leaf from a mutant plant that has been exposed to the inductive dark period is grafted onto an unexposed wild-type plant, the recipient plant flowers.

3. A plant flowers only after exposure to cold.

4. If a gene is coupled to an active promoter and expressed at high levels in the shoot apical meristem, flowering is induced even in the absence of an appropriate photoperiodic stimulus.

5. If a gene is experimentally overexpressed in the leaf, flowering is induced. Overexpression of the gene in the shoot apical meristem does not, however, induce flowering.

---

involved (**FIGURE 27.11**). The gene *FLC* (*FLOWERING LOCUS C*) encodes a transcription factor that blocks the FT–FD pathway by inhibiting expression of FT and FD. Cold temperatures inhibit the synthesis of FLC protein, allowing FT and FD proteins to be expressed and flowering to proceed. Similar proteins control some steps in vernalization in cereals.

GIBBERELLINS *Arabidopsis* plants do not flower if they are genetically deficient in gibberellin synthesis or if they are treated with an inhibitor of gibberellin synthesis. These observations implicate gibberellins in flowering. Direct application of gibberellins to buds in *Arabidopsis* results in activation of the meristem identity gene *LEAFY*, which in turn promotes the transition to flowering (see Concept 14.3).

Some plants do not require an environmental cue to flower

Some plants flower in response to cues from an "internal clock." For example, flowering in some strains of tobacco is initiated in the apical bud when the stem has grown four phytomers in length (recall that stems are composed of repeating

units called *phytomers*; see Figure 24.1). If a terminal bud and a single adjacent phytomer are removed from a plant this size and planted, the cutting will flower because the bud has already received the cue for flowering. But the rest of the shoot below the bud that has been removed will not flower because it is only three phytomers long. After it grows an additional phytomer, it will flower. These results suggest that there is something about the *position* of the bud (atop four phytomers of stem) that determines its transition to flowering.

> **LINK** The role of positional information in morphogenesis in both plants and animals is described in Concept 14.3

The bud might "know" its position by the concentration of some substance that forms a positional gradient along the apical–basal axis of the plant. Such a gradient could be formed, for example, if the root makes a diffusible inhibitor of flowering whose concentration diminishes with plant height. When the plant reached a certain height, the concentration of the inhibitor would become sufficiently low at the tip of the shoot to allow flowering. What this inhibitor might be is unclear, but there is evidence that it acts by decreasing the amount of FLC, allowing the FT–FD pathway to proceed (just as cold acts on FLC in vernalization). A positional gradient that acts on FLC would be consistent with other mechanisms affecting

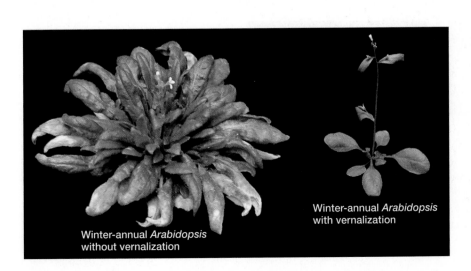

Winter-annual *Arabidopsis* without vernalization

Winter-annual *Arabidopsis* with vernalization

FIGURE 27.11 Vernalization A genetic strain of *Arabidopsis* (winter-annual *Arabidopsis*) requires vernalization for flowering. Without it, the plant is large and vegetative (left), but when exposed to a cold period, it is smaller and flowers (right).

flowering, which all converge on the meristem identity genes, *LEAFY* and *APETALA1*:

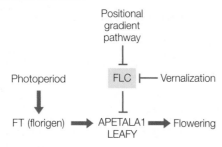

Do You Understand Concept 27.2?

- How do shoot apical meristems, inflorescence meristems, and floral meristems differ, and what are the genes that control the transitions between them?
- Explain why "short-day plant" is a misleading term.
- If a LDP is exposed to its inductive dark period and one of its leaves is removed and grafted onto a SDP that has not been exposed to its inductive dark period, what will happen?
- How would you show that vernalization acts through the *FLC* gene?

Sexual reproduction is not the only way in which angiosperms give rise to more plants. We will now turn to some ways in which plants can reproduce asexually.

---

concept
27.3 Angiosperms Can Reproduce Asexually

A plant that reproduces asexually produces progeny genetically identical to itself (*clones*). This mode of reproduction has the disadvantage of not generating the diversity of offspring that sexual reproduction can produce. But if a plant is well adapted to its environment, the parent plant can pass on to all its progeny a combination of alleles that function well in that environment, and which might otherwise be separated by sexual recombination. Furthermore, the plant avoids the costs of producing flowers and the potentially unreliable processes of cross-pollination and seed germination.

Angiosperms use many forms of asexual reproduction

Asexual reproduction is often accomplished through the modification of vegetative organs such as roots, stems, or leaves, which is why the term **vegetative reproduction** is sometimes used to describe asexual reproduction in plants. Another type of asexual reproduction, **apomixis**, involves flowers but no fertilization.

VEGETATIVE REPRODUCTION The totipotency of plant cells is dramatically shown when a cell from a fully differentiated organ gives rise to an entire new plant. Strawberries, for example, produce horizontal stems, called *stolons* or runners, that grow along the soil surface, form roots at intervals, and establish potentially independent plants (see Figure 24.11C). Another example is bamboo, which has underground stems called *rhizomes* that can form an entire forest (**FIGURE 27.12A**). Fleshy underground stems, such as potato tubers, produce multiple plants from "eyes" (see Figure 24.11A). Garlic can produce multiple plants from bulbs, which are also modified stems (**FIGURE 27.12B**). Even leaves can be the source of new plants, as in the succulent *Kalanchoe* (**FIGURE 27.12C**).

Plants that reproduce vegetatively often grow in physically unstable environments, such as eroding hillsides, or places where conditions for seed germination are unreliable. Plants with stolons or rhizomes, such as beach grasses, rushes, and

FIGURE 27.12 Vegetative Reproduction (A) The rhizomes of bamboo are underground stems that produce plants at intervals. (B) Bulbs are short stems with large leaves that store nutrients and can give rise to new plants. (C) In *Kalanchoe*, new plantlets can form on leaves.

(A)

These bamboo shoots all arise from the same underground stem.

(B)

Each clove of garlic is a bulb that can give rise to a new plant.

(C)

The plantlets forming on the margin of this *Kalanchoe* leaf will fall to the ground and become independent plants.

sand verbena, are common on coastal sand dunes. Rapid vegetative reproduction enables these plants, once introduced, not only to multiply but also to stay ahead of the shifting sand; in addition, the dunes are stabilized by the extensive network of rhizomes or stolons that develops. Vegetative reproduction is also common in some deserts, where for much of the time the environment is unsuitable for seed germination and the establishment of seedlings.

APOMIXIS Some plants produce flowers but use them to reproduce asexually rather than sexually. Dandelions, blackberries, some citrus trees, and some other plants reproduce by the asexual production of seeds, called *apomixis*. In sexual reproduction, seeds form from the union of haploid gametes in the embryo sac. But in apomixis, one of two other things happens:

- The megasporocyte in the ovule that is supposed to undergo meiosis fails to do so, resulting in a diploid egg cell, which then goes on to form an embryo and seed.
- Diploid cells from the integument surrounding the embryo sac form a diploid embryo sac, which goes on to form an embryo and seed.

In both cases, seed and fruit development proceed normally. But the genetic consequences are profound: *apomixis produces clones*.

Some agriculturally important plants are hybrids that are derived from the crossing of two genetically different species. Because these hybrids have two sets of chromosomes that are not homologous to one another, they cannot undergo meiosis and are sterile. This means they cannot form the seeds and

fruits that are desired by people. In citrus, apomixis occurs naturally, which gets around this problem. Kentucky bluegrass, a mainstay of lawns, reproduces in this manner as well.

Many important crops, such as corn, are grown as hybrids because the progeny of a cross between two inbred, homozygous genetic strains are often superior to either of their parents, a phenomenon called *hybrid vigor* (see Concept 8.2). Unfortunately, once farmers have obtained a hybrid with desirable characteristics, they cannot use those plants for further crosses with themselves (selfing) to get more seeds for the next generation. You can imagine the genetic chaos when a hybrid, which is heterozygous at many of its loci (e.g., *AaBbCcDdEe*, etc.), is crossed with itself: there will be many new combinations of alleles (e.g., *AabbCCDdee*, etc.), resulting in highly variable progeny. The only way to reliably reproduce the hybrid is to maintain populations of the original parents to cross again each year. That is exactly what seed companies do. For farmers, it means buying new seeds every year—an expensive proposition.

If a hybrid carried a gene for apomixis, however, it could reproduce asexually, and its offspring would be genetically identical to itself (**FIGURE 27.13**). New hybrids could be developed that would be adapted to specific environments and could be propagated by the farmers on the spot. So an intensive search is on for genes for apomixis that could be introduced into desirable crops and allow them to be propagated indefinitely. Such a gene has been identified in corn, but the yield of the variety that contains it is low.

> **FRONTIERS** In *Arabidopsis* a gene called *dyad* has been identified that causes apomixis when mutated, because the dyad protein is necessary for chromosome segregation in meiosis I. Plant scientists are searching for an analogous gene in crop plants and for ways of transferring the mutant *Arabidopsis dyad* gene to crop plants to produce apomictic hybrids.

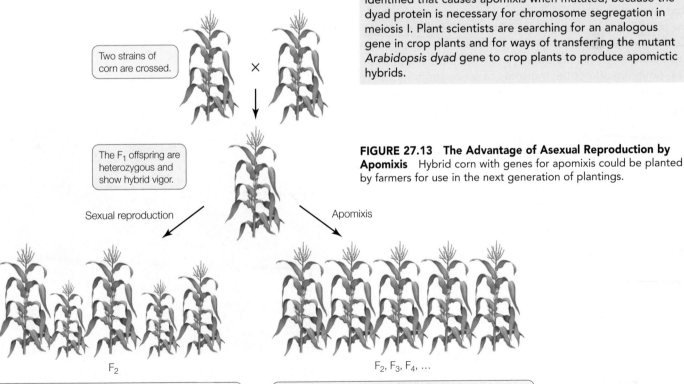

FIGURE 27.13 The Advantage of Asexual Reproduction by Apomixis Hybrid corn with genes for apomixis could be planted by farmers for use in the next generation of plantings.

Two strains of corn are crossed.

×

The F₁ offspring are heterozygous and show hybrid vigor.

Sexual reproduction

Apomixis

F₂

F₂, F₃, F₄, ...

Sexual reproduction by crossing of F₁ plants results in many genotypes and phenotypes, and overall hybrid vigor is lost.

Apomixis of F₁ plants produces offspring genetically identical to the F₁ plants, and hybrid vigor is retained.

In grafting, the scion is aligned so that its vascular cambium is adjacent to the vascular cambium in the stock.

Scion

Stock

FIGURE 27.14 Grafting Grafting—attaching a piece of a plant to the root or root-bearing stem of another plant—is a common horticultural technique. The "host" root or stem is the stock; the upper grafted piece is the scion. In the photo, an avocado scion is being grafted onto a stock of a different variety of avocado.

Vegetative reproduction is important in agriculture

One of the oldest methods of vegetative reproduction used in agriculture consists of simply making cuttings of stems, inserting them in soil, and waiting for them to form roots and become autonomous plants. The cuttings are usually encouraged to root by treatment with auxin, as described in Concept 26.2.

Woody plants can be propagated asexually by **grafting**: attaching a bud or a piece of stem from one plant to a root or root-bearing stem of another plant. The part of the resulting plant that comes from the root-bearing "host" is called the **stock**; the part grafted on is the **scion** (**FIGURE 27.14**). The vascular cambium of the scion associates with that of the stock, forming a continuous cambium that produces xylem and phloem. The cambium allows the transport of water and minerals to the scion and of photosynthate to the stock. Much of the fruit grown for market in the United States is produced on grafted trees, as are wine grapes.

Another method widely used for asexual plant propagation is **meristem culture**, in which pieces of shoot apical meristem are cultured on growth media to generate plantlets, which can then be planted in the field. This strategy is vital when uniformity is desired, as in forestry, or when virus-free plants are the goal, as with strawberries and potatoes.

Do You Understand Concept 27.3?

- What are the advantages and disadvantages of asexual as compared with sexual reproduction in plants?

- In addition to the *dyad* mutation, what other kinds of apomixis genes would you search for?

- Most of the navel orange trees that once grew all over southern California are derived by grafting and apomixis from a single tree that still exists in Riverside, CA. How would you prove this?

QA QUESTION How did an understanding of angiosperm reproduction allow floriculturists to develop a commercially successful poinsettia?

ANSWER The poinsettia illustrates well how the concepts of plant physiology and reproduction are used in floriculture. As Albert Ecke found, poinsettias are short-day plants, using the long nights of winter as a signal for flowering (Concept 27.2). In fact, the poinsettia's inductive dark period of 14 hours was among the first described in a flowering plant. The Eckes' plants did not flower consistently until they moved their growing operations from urban Los Angeles to an isolated rural region with less light pollution. Today the plants are grown in greenhouses, where the photoperiod is carefully regulated.

A second important factor in the poinsettia's commercial success was the plant's growth habit. In Mexico, the wild relatives of cultivated varieties grow up to 3 meters tall, with few branches (**FIGURE 27.15**; compare these plants with the 0.5-meter-tall plants in the photograph at the opening of this chapter). Paul Ecke found a variety that was much shorter and formed an attractive branching plant with axillary shoots. These compact poinsettias were initially propagated asexually by grafting to native plants (Concept 27.3).

The cause of the compact phenotype was assumed to be a dwarfing gene, but several observations cast doubt on this assumption. First, heat treatment of short, branching plants changed their phenotype to the tall, less branched growth habit. Second, heat-treated plants would regain their short, branching growth habit if they were grafted onto untreated short, branching plants. These and other results pointed to a *transmissible infective agent* as the cause of the short, branching phenotype.

FIGURE 27.15 A Wild Relative of Poinsettia Poinsettias growing in the wild in Mexico have an entirely different growth habit from the commercially propagated plants sold for holiday decorations.

Dr. Ing-Ming Lee at the U.S. Department of Agriculture isolated a bacterium called *Phytoplasma* from the short, branching plants and proved by experiments that it was the causative agent of the compact growth habit: when young plants with the tall growth habit were inoculated with the bacteria, they grew up short and branching. It is suspected that the bacteria change cytokinin levels in the plants, since this hormone regulates branching (see Concept 26.3).

Many new varieties of poinsettias have been generated by conventional sexual reproduction (Concept 27.1). You can look for the phenotypes of these varieties next holiday season. They include stiff stems, new colors (such as white), and longer lifetimes for the colorful bracts.

 # SUMMARY

concept 27.1 Most Angiosperms Reproduce Sexually

- Sexual reproduction promotes genetic diversity in a population. The flower is an angiosperm's structure for sexual reproduction.

- Flowering plants have microscopic gametophytes. The megagametophyte is the **embryo sac**, which typically contains eight nuclei in seven cells. The microgametophyte is the two-celled **pollen grain**. **Review Figure 27.2**

- Following pollination, the pollen grain delivers sperm cells to the embryo sac by means of a pollen tube.

- Angiosperms exhibit double fertilization, forming a diploid zygote that becomes the embryo and a triploid endosperm that stores reserves. **Review Figure 27.4 and ANIMATED TUTORIAL 27.1**

concept 27.2 Hormones and Signaling Determine the Transition from the Vegetative to the Reproductive State

- In **annuals** and **biennials**, flowering and seed formation are followed by the death of the rest of the plant. **Perennials** live longer and reproduce repeatedly.

- For a vegetatively growing plant to flower, a shoot apical meristem must become an **inflorescence meristem**, which in turn must give rise to one or more **floral meristems**. These events are determined by specific genes. **Review Figure 27.6**

- Some plants flower in response to **photoperiod. Short-day plants (SDPs)** flower when nights are longer than a critical length specific to each species; **long-day plants (LDPs)** flower when nights are shorter than a critical length. **Review Figures 27.7 and 27.8 and ANIMATED TUTORIAL 27.2**

- The mechanism of photoperiodic control of flowering involves phytochromes and a diffusible protein signal, **florigen (FT)**, which is formed in the leaf and is translocated to the shoot apical meristem. **Review Figures 27.9 and 27.10**

- In some angiosperms, exposure to cold—called **vernalization**—is required for flowering. In others, internal signals (such as gibberellin) induce flowering. All of these stimuli converge on the **meristem identity genes**.

concept 27.3 Angiosperms Can Reproduce Asexually

- Asexual reproduction allows rapid multiplication of organisms that are well suited to their environment.

- **Vegetative reproduction** involves the modification of a vegetative organ for reproduction. **Review Figure 27.12**

- Some plant species produce seeds asexually by **apomixis**. **Review Figure 27.13**

- Woody plants can be propagated asexually by **grafting**.

In late 1998, William Wagoire, a plant geneticist in Uganda, was astounded when he noticed some red blotches on the stems of wheat plants he was breeding. Wheat rust (*Puccinia graminis*), the fungus that causes the blotches, had supposedly been rendered almost extinct 25 years previously by the "Green Revolution," when a gene from rye plants had been crossed into wheat, making it resistant to this fungal disease. This landmark achievement had protected the crop that provides one-third of the human diet from a mold that can devastate it.

Two generations of farmers worldwide had never seen an epidemic of wheat rust, but its presence in history is vivid. Even the ancient Romans feared this disease, worshipping a god, "Robigus," who was thought to help their crop ward off the mold. In the seventeenth century, the early colonists in Massachusetts almost starved because the wheat they planted got infected with wheat rust. In 1917, an epidemic of wheat rust in the United States reduced the crop by one-third, leading to widespread panic.

Wagoire used DNA markers to identify the strain of wheat rust he had found. When he compared it with known strains, which are stored in a few laboratories, he found that his strain was new and unique. Clearly, Ug99 (for Uganda 1999) had evolved a way to get around the resistance genes in modern wheat.

When the blotches of wheat rust on a stem burst, they release thousands of spores, any one of which can be carried by the wind to a susceptible plant. When you consider that 1 hectare of an infected

wheat field can release over 10 billion spores, the possibility of epidemic spread becomes apparent. Ug99 has begun a relentless path of infection, carried by prevailing winds. By 2001 it had wreaked havoc on wheat in Kenya; in 2003 it was in Ethiopia; in 2006 it crossed the Red Sea to Yemen; and in 2009 it was in Iran and Iraq. Biologists now fear that Ug99 could reach North America by the "747 route," accidentally dusting a traveler's clothes somewhere in the now widespread infective region. A Global Rust Initiative has been set up to try to use knowledge of plant and fungal biology to stop the spread of this disease before it is too late.

A field of wheat infected with the wheat rust fungus (right) has a much reduced growth rate and produces much less grain than an uninfected field of wheat that is resistant to the fungus (left).

Q QUESTION

How can knowledge of plant and fungal biology be used to prevent the spread of wheat rust?

KEY CONCEPTS

28.1 Plants Have Constitutive and Induced Responses to Pathogens

28.2 Plants Have Mechanical and Chemical Defenses against Herbivores

28.3 Plants Adapt to Environmental Stresses

concept 28.1 Plants Have Constitutive and Induced Responses to Pathogens

Botanists know of dozens of diseases that can kill a wheat plant, each of them caused by a different pathogen, each of which in turn has many different genetic strains. Plant pathogens—which include fungi, bacteria, protists, and viruses—are part of nature, and for that reason alone they merit our study in biology. Because we humans depend on plants for our food, however, the stakes in our effort to understand plant pathology are especially high. That is why, just as medical schools have departments of pathology, universities in agricultural regions have departments of plant pathology.

Successful infection by a pathogen can have significant effects on a plant, reducing photosynthesis and causing massive cell and tissue death. Plants and pathogens have evolved together in a continuing "arms race": pathogens have evolved mechanisms with which to attack plants, and plants have evolved mechanisms for defending themselves against those attacks. Like the responses of the human immune system, the responses by which plants fight off infection can be either **constitutive**—always present in the plant—or **induced**—produced in reaction to the presence of a pathogen.

LINK The human immune system, with its constitutive and induced defenses, is described in Chapter 31

Physical barriers form constitutive defenses

As with humans and their skin, a plant's first line of defense is its outer surfaces, which can prevent the entry of pathogens. The parts of stems and leaves exposed to the outside environment are covered with cutin, suberin, and waxes. These substances not only prevent water loss by evaporation but can also prevent fungal spores and bacteria from entering the underlying tissues.

In addition to using physical barriers to prevent infection, plants use chemical warfare. Unlike animals, plants cannot flee from animals that eat them, so they make molecules that deter or destroy herbivores, as we will see in Concept 28.2. Some of these constitutively synthesized molecules also inhibit pathogens. For example, in plantains (*Plantago*) and some other plants, *iridoid glycosides*, which are glucose molecules covalently linked to short-chain fatty acids, inhibit the growth of fungal pathogens.

When constitutive defenses fail to deter a pathogen, plants initiate *induced resistance mechanisms*. As we discuss these mechanisms, refer to the overview in **FIGURE 28.1**.

Induced responses to pathogens may be genetically determined

Plant pathogens *induce* the host plant to activate various chemical defense responses. Many distinctive molecules called

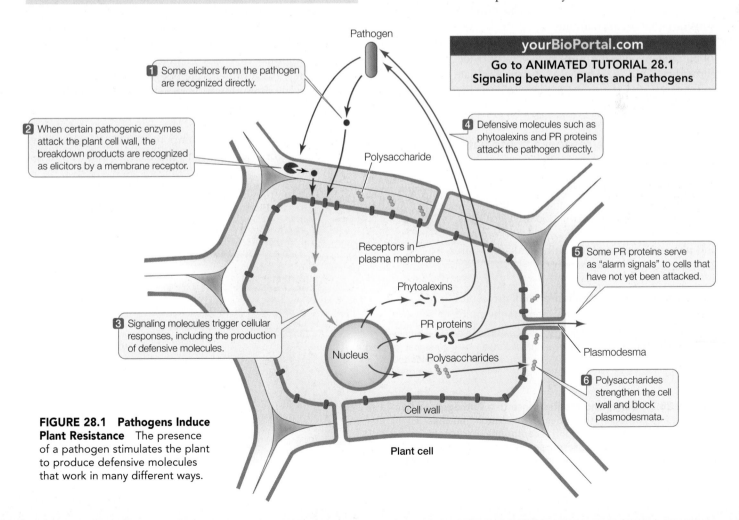

yourBioPortal.com
Go to ANIMATED TUTORIAL 28.1
Signaling between Plants and Pathogens

1 Some elicitors from the pathogen are recognized directly.

2 When certain pathogenic enzymes attack the plant cell wall, the breakdown products are recognized as elicitors by a membrane receptor.

3 Signaling molecules trigger cellular responses, including the production of defensive molecules.

4 Defensive molecules such as phytoalexins and PR proteins attack the pathogen directly.

5 Some PR proteins serve as "alarm signals" to cells that have not yet been attacked.

6 Polysaccharides strengthen the cell wall and block plasmodesmata.

Pathogen
Polysaccharide
Receptors in plasma membrane
Phytoalexins
PR proteins
Polysaccharides
Nucleus
Cell wall
Plasmodesma
Plant cell

FIGURE 28.1 Pathogens Induce Plant Resistance The presence of a pathogen stimulates the plant to produce defensive molecules that work in many different ways.

elicitors have been identified that trigger these plant defenses. These molecules vary in character, from peptides made by bacteria to cell wall fragments from fungi. Elicitors can also be derived from fragments of plant cell wall components broken down by pathogens (see Figure 28.1, steps 1 and 2). Pathogen genes that code for elicitors are called **avirulence (Avr) genes**; there are hundreds of such genes, and they vary among pathogen species and strains.

When an elicitor meets a plant cell, it may encounter a membrane receptor protein encoded by a **resistance (R) gene**. There are hundreds of *R* genes, each encoding a receptor that is specific for one or a few elicitors. If a plant receptor binds to an elicitor, a signal transduction pathway is set in motion that leads to the plant's defensive response. This type of resistance is called **gene-for-gene resistance** because it involves a recognition event between two specific molecules: one determined by an *Avr* gene in the pathogen and the other determined by an *R* gene in the plant (**FIGURE 28.2**).

If the plant has no receptor to bind to the elicitor made by a particular species or strain of pathogen, the plant does not recognize the pathogen and does not turn on its inducible defenses. The lack of a receptor makes the plant especially susceptible to that pathogen. There is an ongoing coevolutionary "arms race" between the ability of pathogens to evade detection by plants and the ability of plants to detect them. A major goal of plant breeders for the past 50 years has been to identify *R* and *Avr* genes and to breed new *R* genes into crops to make them more resistant to pathogens. This effort speeds up the arms race that occurs naturally.

The signal transduction pathway started by receptor–elicitor binding involves the production of nitric oxide and the toxic peroxide H_2O_2. Together, these substances initiate *local defenses* and, later, *systemic defenses* (defenses in parts of the plant distant from the attack site).

The hypersensitive response fights pathogens at the site of infection

The **hypersensitive response** is a local defense induced by pathogen invasion. This response has three components: the production of *phytoalexins*, the synthesis of *pathogenesis-related (PR) proteins*, and *physical isolation* of the pathogen (see Figure 28.1, steps 3 and 4).

PHYTOALEXINS Phytoalexins are small molecules produced within hours of an infection by plant cells near the infection site. They are antibiotics, toxic to pathogens. An example is *camalexin*, a phy-

toalexin made by the model organism *Arabidopsis thaliana* from the amino acid tryptophan:

Tryptophan → Camalexin

> **FRONTIERS** Why phytoalexins such as camalexin are toxic to pathogens is not clear. However, plant biologists are interested in developing phytoalexins as fungicides that could be sprayed on crops to protect them, especially strains that are not resistant to fungal infection by other means such as *R* genes. To make phytoalexin fungicides work, biologists will have to determine how to get plants to absorb the chemicals and how to prevent possible adverse environmental effects.

PATHOGENESIS-RELATED PROTEINS Plants produce several types of **pathogenesis-related**, or **PR**, **proteins**. Some are enzymes that break down the cell walls of pathogens. Chitinase, for example, is a PR protein that breaks down chitin, a distinctive component of fungal cell walls. In some cases, the breakdown products of the pathogen's cell walls serve as elicitors that trigger further defensive responses. Other PR proteins may serve as alarm signals to plant cells that have not yet been attacked, so they can initiate defensive responses of their own (see Figure 28.1, step 5). PR proteins are not rapid-response weapons; rather, they act more slowly, perhaps after other mechanisms have blunted the pathogen's attack.

PHYSICAL ISOLATION A third component of the hypersensitive response seals off the damaged plant tissue and the pathogen

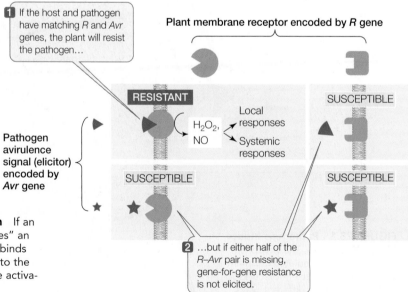

1 If the host and pathogen have matching *R* and *Avr* genes, the plant will resist the pathogen…

Plant membrane receptor encoded by *R* gene

RESISTANT   SUSCEPTIBLE

H_2O_2, NO → Local responses / Systemic responses

Pathogen avirulence signal (elicitor) encoded by *Avr* gene

SUSCEPTIBLE   SUSCEPTIBLE

2 …but if either half of the *R*–*Avr* pair is missing, gene-for-gene resistance is not elicited.

FIGURE 28.2 Genes and the Response to a Pathogen If an *Avr* gene in a pathogen that codes for an elicitor "matches" an *R* gene in a plant that codes for a receptor, the receptor binds the elicitor, and a defensive response results. In addition to the production of H_2O_2 and NO, responses may include gene activation and modifications of signaling proteins.

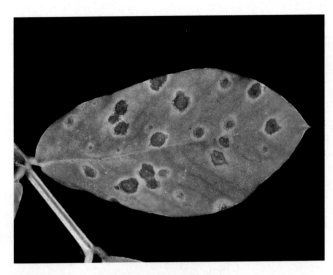

FIGURE 28.3 Sealing Off the Pathogen and the Damage
These necrotic lesions on a peanut leaflet are a response to infection by a strain of early leaf spot fungus (*Cercospora arachidicola*).

from the rest of the plant. Cells around the site of infection undergo apoptosis, preventing the spread of the pathogen by depriving it of nutrients. Some of these cells produce phytoalexins and other toxic chemicals before they die. Others produce cell wall polysaccharides to seal off plasmodesmata and prevent the spread of the pathogen to healthy tissues (see Figure 28.1, step 6). The dead tissue, called a *necrotic lesion*, contains and isolates what is left of the infection (**FIGURE 28.3**). The rest of the plant remains free of the infecting pathogen.

Plants can develop systemic resistance to pathogens

Systemic acquired resistance is a *general increase in the resistance of the entire plant* to a wide range of pathogens. It is not localized and not limited to the pathogen that originally triggered it, and its effect may last as long as an entire growing season.

This defensive response is initiated by salicylic acid, a chemical produced during the local hypersensitive response:

Salicylic acid

Systemic acquired resistance is accompanied by the synthesis of PR proteins. Infection in one part of a plant can lead to the export of salicylic acid to other parts, where it triggers the production of PR proteins ahead of the spread of infection. Infected plant parts also produce the closely related compound methyl salicylate (also known as oil of wintergreen). This volatile substance travels to other plant parts through the air and may trigger the production of PR proteins in neighboring plants that have not yet been infected.

Another type of systemic acquired resistance is a more specific defense against viruses whose genomes are made of RNA. The plant uses its own enzymes to convert some of the single-stranded RNA of the invading virus into double-stranded RNA (dsRNA) and to chop that dsRNA into small pieces called small interfering RNAs (siRNAs). The siRNAs interact with another cellular component to degrade viral mRNAs, blocking viral replication. This phenomenon is an example of *RNA interference* (RNAi) (see Figure 13.9). The siRNAs spread rapidly through the plant by way of plasmodesmata, providing systemic resistance.

Do You Understand Concept 28.1?

- Compare constitutive and induced responses to pathogens.
- Explain how inherited traits are involved in both constitutive and induced responses to pathogens.
- If a group of plants is sprayed with oil of wintergreen, they all show resistance to a variety of pathogens. Explain how this occurs.

APPLY THE CONCEPT

Plants have constitutive and induced responses to pathogens

Because you are a biology student, your nonscientist neighbors are confident that you know all about plants. A neighbor brings you a potted plant he bought 3 weeks ago from a nursery. The leaves have brown spots. He bought another plant of the same species from a different nursery 3 weeks ago, and that plant has no spots.

1. You suspect that the spots are caused by either a disease or a nutritional deficiency. Given a large supply of the plants and soils involved, how would you distinguish between these possibilities?

2. Assuming that the spots are caused by a fungal disease, how would you investigate at the molecular level the mechanism for formation of the brown leaf spots?

3. You go to the nursery where the neighbor bought the infected plant and find several pots of plants with no brown spots on their leaves. What plant characteristics might account for the absence of spots?

Because they are sessile, plants must deal with herbivores, as they do with pathogens, where they are. But that does not mean they are defenseless against their would-be consumers, as we will see next.

concept 28.2 ## Plants Have Mechanical and Chemical Defenses against Herbivores

Herbivores—animals that eat plants—depend on plants for energy and nutrients. Their foraging activities cause physical damage to plants, and they often spread pathogens among plants as well. While the majority of herbivores are insects (**FIGURE 28.4**), most other groups of animals also include some herbivores. Plants have both constitutive and induced mechanisms to protect themselves from herbivory.

Constitutive defenses are physical and chemical

Plants have a number of *constitutive anatomical features*, such as trichomes (leaf hairs), thorns, and spines that may be specialized for defense. In addition, normal plant features, including thick cell walls and tree bark, may deter some herbivores.

Plants revert to "chemical warfare" to repel or inhibit other herbivores, using special chemicals called **secondary metabolites**. These substances are referred to as "secondary" because, unlike proteins and nucleic acids, they are not used for basic cellular processes. The more than 10,000 known secondary metabolites range in molecular mass from about 70 to more than 390,000 daltons, but most are small (**TABLE 28.1**). Some are produced by only a single plant species, whereas others are characteristic of entire genera or even families. The effects of defensive secondary metabolites on animals are diverse. Some act on the nervous systems of herbivorous insects, mollusks, or mammals. Others mimic the natural hormones of insects, causing some larvae to fail to develop into adults. Still others damage the digestive tracts of herbivores, and others are toxic to fungal pathogens.

An example of a secondary metabolite is *canavanine*, an amino acid that is similar to the amino acid arginine:

(A) *Locusta migratoria*

(B) *Manduca sexta*

FIGURE 28.4 Insect Herbivores The great majority of herbivores are insects. (A) Some herbivores, such as this locust, are generalists that will attack nearly any plant. (B) Others are specialists, like this tobacco hornworm (the caterpillar of the Carolina sphinx moth), which feeds only on tobacco plants and a few closely related species.

When an insect larva consumes canavanine-containing plant tissue, the canavanine is incorporated into the insect's proteins in some of the places where the insect's mRNA codes for arginine. The enzyme that charges the tRNA specific for arginine fails to discriminate accurately between the two amino acids. The structure of canavanine, however, is different enough from that of arginine that some of the resulting proteins end up with a modified tertiary structure and hence reduced biological activity. These defects in protein structure and function lead to developmental abnormalities that kill the insect.

LINK The charging of tRNA is described in Concept 10.4

While the mechanisms of action of secondary metabolites and their presence in plant tissues are circumstantial evidence of their role in defense, experiments can be done to provide cause-and-effect evidence. Nicotine, for example, kills insects by acting as an inhibitor of nervous system function, and its function in tobacco is presumed to be insecticidal. Yet commercial varieties of tobacco and related plants that produce nicotine are still attacked, with moderate damage, by pests such as the tobacco hornworm (see Figure 28.4B). Given that observation, does nicotine really deter herbivores? Biologists answered this question conclusively with a study that used tobacco plants in which an enzyme involved in nicotine biosynthesis had been

Arginine

Canavanine

A seemingly slight chemical difference…

…produces an inactive protein.

TABLE 28.1 Secondary Metabolites Used in Plant Defense

CLASS	TYPE	ROLE	EXAMPLE
Nitrogen-containing	Alkaloids	Neurotoxin	Nicotine in tobacco
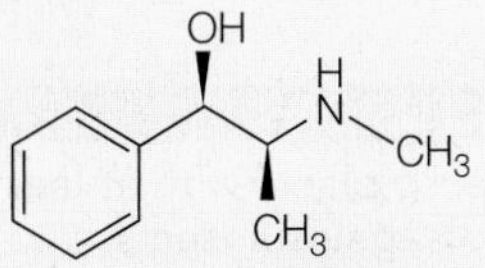	Glycosides	Inhibit electron transport	Dhurrin in sorghum
	Nonprotein amino acids	Disrupt protein structure	Canavanine in jack bean

Ephedrine (an alkaloid)

Nitrogen- and sulfur-containing	Glucosinolates	Inhibit respiration	Methylglucosinolate in cabbage

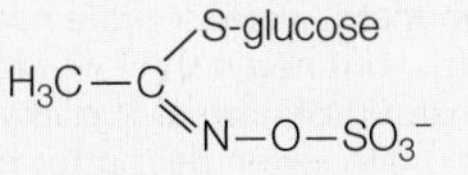

Methylglucosinolate

Phenolics	Coumarins	Block cell division	Umbelliferone in carrots
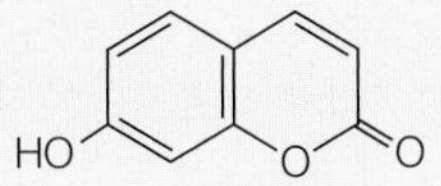	Flavonoids	Phytoalexins	Capsidol in peppers
	Tannins	Inhibit enzymes	Gallotannin in oak trees

Umbelliferone

Terpenes	Monoterpenes	Neurotoxins	Pyrethrin in chrysanthemums
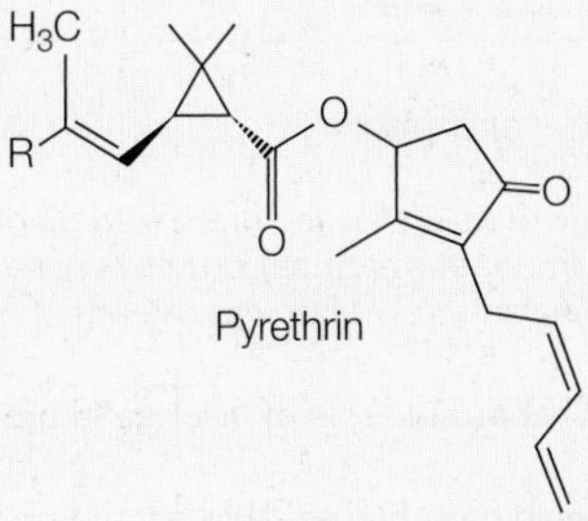	Diterpenes	Disrupt reproduction and muscle function	Gossypol in cotton
	Triterpenes	Inhibit ion transport	Digitalis in foxglove
	Sterols	Block animal hormones	Spinasterol in spinach
	Polyterpenes	Deter feeding	Latex in *Euphorbia*

Pyrethrin

inhibited, lowering the nicotine concentration in the plants by more than 95 percent. These low-nicotine plants suffered much more damage from insect herbivory than normal plants did (**FIGURE 28.5**).

FRONTIERS Secondary metabolites have long been used by people as medicines. Salicylic acid, for example, is the basis of aspirin (see Chapter 3). Screening metabolites for pharmaceutical activity is a major activity in drug development. The foods we eat also contain many secondary metabolites, and scientists are trying to determine their effects on human health. Some cause diseases such as cancer, and others prevent them.

Plants respond to herbivory with induced defenses

The first step in a plant's response to herbivory is to sense the attack, either by means of changes in membrane potential or by means of chemical signals. The detection of an herbivore attack then triggers signal transduction pathways that induce plant defenses.

MEMBRANE SIGNALING The plasma membrane is the part of the plant cell that is in contact with the environment. Within the first minute after an herbivore strikes, changes in the electric potential of the plasma membrane occur in the damaged area. The continuity of the *symplast* (see Concept 25.3) ensures that the signal travels over much of the plant within 10 minutes.

LINK In animals, direct electrical coupling between cells occurs only in specific tissues, such as cardiac muscle; see Concept 36.1

CHEMICAL SIGNALING When some insects (such as caterpillars) chew on a plant, substances in the insect's saliva combine with fatty acids derived from the consumed plant tissue. The resulting compounds act as elicitors to trigger both local and systemic responses to the herbivore. In corn, the elicitor produced by one particular moth larva has been named *volicitin* for its ability to induce production of volatile signals that can travel to other plant parts—and to neighboring corn plants—and simulate their defensive responses.

INVESTIGATION

FIGURE 28.5 Nicotine Is a Defense against Herbivores
The secondary metabolite nicotine, made by tobacco plants, is an insecticide, yet most commercial varieties of tobacco are suscep- tible to insect attack. Ian Baldwin demonstrated that a tobacco strain with a reduced nicotine concentration was much more susceptible to insect damage.

HYPOTHESIS
Nicotine helps protect tobacco plants against insects.

CONCLUSION
Nicotine provides tobacco plants with at least some protection against insects.

METHOD

Create a strain of low-nicotine tobacco plants.

↓

Plant normal (control) and low-nicotine (mutant) plants together in a field where they are accessible to insects.

↓

Assess the extent of leaf damage at 2-day intervals.

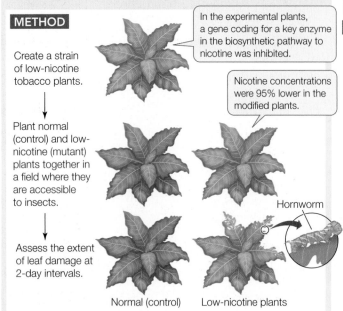

In the experimental plants, a gene coding for a key enzyme in the biosynthetic pathway to nicotine was inhibited.

Nicotine concentrations were 95% lower in the modified plants.

Hornworm

Normal (control)   Low-nicotine plants

RESULTS
The low-nicotine plants suffered more than twice as much leaf damage as did the normal controls.

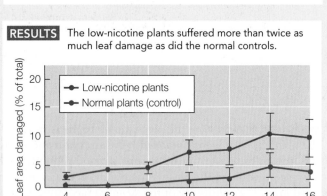

ANALYZE THE DATA

In a separate experiment, Baldwin and his colleagues showed that treatment with jasmonic acid (jasmonate) increased the concentration of nicotine in normal tobacco plants, but not in the low-nicotine plants. The researchers planted a group of normal and low-nicotine plants and treated them with jasmonate seven days after planting. The plants were assessed for herbivore damage every 2 days after being planted. Here are the results:

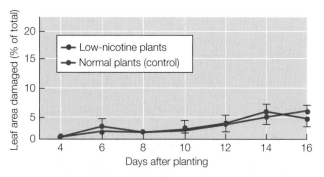

A. Compare these data to those at left for the untreated plants. What was the effect of jasmonate treatment on the resistance of normal plants to herbivore damage? What was the effect on low-nicotine plants?

B. What do these data reveal about the role of nicotine in preventing herbivore damage?

C. Explain how jasmonate could have had the effect it did on the low-nicotine plants, even though their nicotine levels were still low after jasmonate treatment.

D. The error bars at each data point are the standard error of the mean (SEM). What statistical test would you use to determine the possible significance of differences between the jasmonate-treated and untreated plants? At 10 days, the mean damage ± SEMs for untreated low-nicotine plants was 6.0 ± 1.5 percent ($n = 36$), and for treated low-nicotine plants it was 2.2 ± 0.6 percent ($n = 28$). Run a statistical test comparing the two results, calculate the P-value, and comment on significance.

For more, go to Working with Data 28.1 at **yourBioPortal.com**.

Go to **yourBioPortal.com** for original citations, discussions, and relevant links for all INVESTIGATION figures.

SIGNAL TRANSDUCTION PATHWAY The perception of damage from herbivory can initiate a signal transduction pathway in the plant that involves the plant hormone *jasmonic acid* (**jasmonate**) (**FIGURE 28.6**; also see Figure 28.5, Analyze the Data):

Jasmonic acid

When the plant senses an herbivore-produced elicitor, it makes jasmonate, which triggers many plant defenses, including the synthesis of a protease inhibitor. The inhibitor, once in an insect's gut, interferes with the digestion of proteins and thus stunts the insect's growth. Jasmonate also "calls for help" by triggering the formation of volatile compounds that attract insects that prey on the herbivores attacking the plant. Similarly, the volatile compounds induced by volicitin in corn attract wasps that parasitize herbivorous caterpillars, so the "call for help" is a recurrent theme in plant–insect interactions.

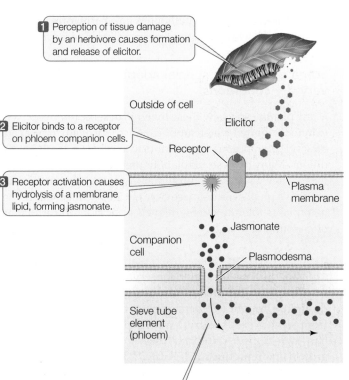
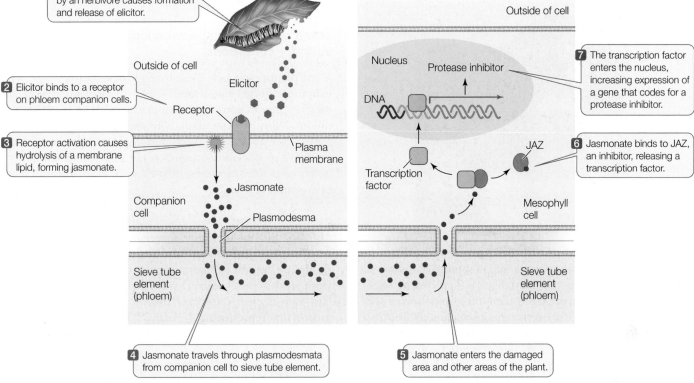

1. Perception of tissue damage by an herbivore causes formation and release of elicitor.

Outside of cell

2. Elicitor binds to a receptor on phloem companion cells.

Elicitor

Receptor

3. Receptor activation causes hydrolysis of a membrane lipid, forming jasmonate.

Plasma membrane

Jasmonate

Companion cell

Plasmodesma

Sieve tube element (phloem)

Outside of cell

Nucleus   Protease inhibitor

DNA

7. The transcription factor enters the nucleus, increasing expression of a gene that codes for a protease inhibitor.

JAZ

6. Jasmonate binds to JAZ, an inhibitor, releasing a transcription factor.

Transcription factor

Mesophyll cell

Sieve tube element (phloem)

4. Jasmonate travels through plasmodesmata from companion cell to sieve tube element.

5. Jasmonate enters the damaged area and other areas of the plant.

FIGURE 28.6 A Signaling Pathway for Induced Defenses against Herbivory   The chain of events initiated by herbivory that leads to the production of a defensive chemical can consist of many steps. These steps may include the synthesis of one or two hormones, binding of receptors, gene activation, and finally, synthesis of the defensive chemical.

Why don't plants poison themselves?

Why don't the constitutive defensive chemicals that are so toxic to herbivores and pathogens kill the plants that produce them? Plants that produce toxic defensive chemicals generally use one of several measures to protect themselves.

COMPARTMENTALIZATION   Isolation of the toxic substance is the most common means of avoiding exposure. Plants store their toxins in vacuoles if the toxins are water-soluble. If they are hydrophobic, the toxins may be dissolved in latex (a complex mixture of dissolved molecules and suspended particles) and stored in a specialized compartment, or they may be dissolved in waxes on the epidermal surface. Such compartmentalized storage keeps the toxins away from the mitochondria, chloroplasts, and other parts of the plant's metabolic machinery.

STORAGE OF PRECURSORS   Some plants store the precursors of toxic substances in one type of tissue, such as the epidermis, and store the enzymes that convert those precursors into the active toxin in another type, such as the mesophyll. When an herbivore chews part of the plant, cells are ruptured, the enzymes come into contact with the precursors, and the toxin is

produced. The only part of the plant that is damaged by the toxin is that which was already damaged by the herbivore. Plants such as sorghum and some legumes, which respond to herbivory by producing cyanide (a potent inhibitor of cellular respiration), are among those that use this type of protective measure.

MODIFIED PROTEINS   The plant has modified proteins that do not react with the toxin. As noted above, canavanine resembles arginine and therefore plays havoc with protein synthesis in insect larvae. In plants that make canavanine, the plant enzyme that charges the arginine tRNA discriminates correctly between arginine and canavanine, so canavanine is not incorporated into the plant's proteins.

Plants don't always win the arms race

Milkweeds such as *Asclepias syriaca* store their defensive chemicals in latex in specialized tubes called **laticifers**, which run alongside the veins in the leaves. When damaged, a milkweed releases copious amounts of toxic latex from its laticifers. Field studies have shown that most insects that feed on neighboring plants of other species do not attack laticiferous plants, but there are exceptions. One population of beetles that feeds on *A. syriaca* exhibits a remarkable prefeeding behavior: these beetles cut a few veins in the leaves before settling down to dine. Cutting the veins causes massive latex leakage from the adjacent laticifers and interrupts the latex supply to a downstream portion of the leaf. The beetles then move to the relatively latex-free portion and eat their fill.

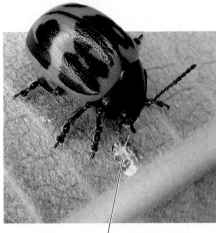

Latex from laticifer

This is just one of many ways that herbivores circumvent plant defenses. A successful plant defense exerts strong selection pressure on herbivores to get around it somehow; a successful herbivore, in turn, exerts strong selection pressure on plants to develop new defensive strategies. One can imagine, for example, that over time, *A. syriaca* might evolve to have thicker walls around the base of its laticifers, such that the beetles can no longer cut them, or to produce a different toxin that does not depend on laticifers.

Do You Understand Concept 28.2?

- Using a plant growing near you as an example, describe anatomical features that help plants avoid herbivores.

- A mutant strain of a plant makes jasmonate constitutively. What would be the phenotype of this plant?

- You have isolated a secondary metabolite from apple tree leaves. What circumstantial and experimental evidence would you use to show that the metabolite prevents the herbivorous larva of the moth *Spilonota ocellana* from damaging apple orchards?

In addition to coping with their biological enemies, plants must survive in their physical environment. Plants have a number of adaptations that allow them to cope with stressful environmental conditions.

concept 28.3 Plants Adapt to Environmental Stresses

In an ever-changing environment, plants face a number of potential stressors: drought, submersion, heat, cold, and high concentrations of salt and heavy metals in the soil. Here we will focus on the overall mechanisms by which plants deal with extremes of climate and soil conditions. You will see that in these cases,

as with defenses against pathogens and herbivores, there are constitutive defenses (adaptations) as well as induced responses.

Some plants have special adaptations to live in very dry conditions

Many plants, especially those living in deserts, must cope with extremely limited water supplies. A variety of anatomical and life-cycle adaptations allow plants to survive under these conditions. Many of these adaptations are ways to avoid or reduce the inevitable water loss through transpiration that occurs during active photosynthesis (see Concept 25.3). Other adaptations help plants tolerate the high levels of light and heat that are often found in deserts.

DROUGHT AVOIDERS Some desert plants have no special structural adaptations for water conservation. Instead, these desert annuals, called *drought avoiders*, simply evade periods of drought. Drought avoiders carry out their entire life cycle—from seed to seed—during a brief period in which rainfall has made the surrounding desert soil sufficiently moist for growth and reproduction (**FIGURE 28.7**). A different drought

FIGURE 28.7 Desert Annuals Avoid Drought The seeds of many desert annuals lie dormant for long periods, awaiting conditions appropriate for germination. When they do receive enough moisture to germinate, they grow and reproduce rapidly before the short wet season ends. During the long dry spells, only dormant seeds remain alive.

avoidance strategy is seen in deciduous perennial plants, particularly in Africa and South America, that shed their leaves in response to drought as a way to conserve water. These plants remain dormant until conditions are again favorable for growth, much as deciduous trees in temperate climates shed their leaves in fall and are dormant until the following spring.

LEAF STRUCTURES Most desert plants are not drought avoiders, but rather grow in their dry environment year-round. Plants adapted to dry environments are called **xerophytes** (Greek *xeros*, "dry"). Three structural adaptations are found in the leaves of many xerophytes:

- Specialized leaf anatomy that reduces water loss
- A thick cuticle and a profusion of trichomes over the leaf epidermis, which retard water loss
- Trichomes that diffract and diffuse sunlight, thereby decreasing the intensity of light impinging on the leaves and the risk of damage to the photosynthetic apparatus by excess light

In some xerophytes, the stomata are strategically located in sunken cavities below the leaf surface (known as *stomatal crypts*), where they are sheltered from the drying effects of air currents (**FIGURE 28.8**). Trichomes surrounding the stomata slow air currents further. Cacti and similar plants have spines rather than typical leaves, and photosynthesis is confined to the fleshy stems. The spines may help the plants cope with desert conditions by reflecting solar radiation or by dissipating heat. The spines also deter herbivores.

WATER-STORING STRUCTURES Succulence—the possession of fleshy, water-storing leaves or stems—is another adaptation to dry environments (**FIGURE 28.9**). This adaptation allows plants to take up large amounts of water when it is available (such as after a brief thunderstorm) and then draw on the stored water during subsequent dry periods. Other

FIGURE 28.9 Succulence *Aloe vera* stores water in its fleshy leaves.

adaptations of succulents include a reduced number of stomata and a variant form of photosynthesis, both of which reduce water loss.

ROOT SYSTEMS THAT MAXIMIZE WATER UPTAKE Roots may also be adapted to dry environments. Cacti have shallow but extensive fibrous root systems that effectively intercept water at the soil surface following even light rains. The tamarugo tree (**FIGURE 28.10**) obtains water through taproots that grow to great depths, reaching water supplies far underground, as well as from condensation on its leaves. The Atacama Desert in northern Chile often goes several years without measurable rainfall, but the landscape there has many surprisingly large shrubs.

SOLUTE ACCUMULATION Xerophytes and other plants that must cope with inadequate water supplies may accumulate high concentrations of the amino acid proline or of secondary metabolites in their vacuoles. This solute accumulation lowers the water potential in the plant's cells below that in the soil, which allows the plant to take up water via osmosis. Plants living in saline environments share this and several other adaptations with xerophytes, as we will see shortly.

LINK Review the principles of water potential and water movement in Concept 25.3

Some plants grow in saturated soils

For some plants, the environmental challenge is the opposite of that faced by xerophytes: too much water. They live in environments so wet that the diffusion of oxygen to their roots is severely limited. These plants have shallow root systems that grow slowly; oxygen levels are likely to be highest near the surface of the soil, and slow growth decreases the roots' need for oxygen.

The root systems of some plants adapted to swampy environments, such as cypresses and some plants that grow in

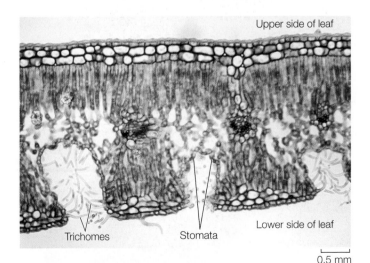

Upper side of leaf

Lower side of leaf

Trichomes   Stomata

0.5 mm

FIGURE 28.8 Stomatal Crypts Stomata in the leaves of some xerophytes, in this case oleander (*Nerium oleander*), are located in sunken cavities called stomatal crypts. The trichomes (hairs) covering these crypts trap moist air.

Plants can respond to drought stress

The adaptations of xerophytes for coping with dry environments are generally constitutive—they are always present—and under normal conditions they prevent the plants from experiencing drought stress. When conditions become so dry that even xerophytes are stressed, however, the plants turn to inducible responses. The same responses are found in many other plants, including those that are not adapted to grow in dry climates.

When the weather is abnormally dry, the water content of the soil is reduced, and less water is available to plants. Water deficits in plant cells have two major biochemical effects: a reduction in membrane integrity as the polar–nonpolar forces that orient the lipid bilayer are reduced, and changes in the three-dimensional structures of proteins. Plant growth is reduced when the structure of plant cells is compromised in these ways. Indeed, inadequate water supply is the single most important factor that limits production of our most important food crops.

coastal mangrove habitats, have **pneumatophores**, which are extensions that grow out of the water and up into the air (**FIGURE 28.11A**). Pneumatophores contain *lenticels* (openings) that allow oxygen to diffuse through them, aerating the submerged parts of the root system.

Many submerged or partly submerged aquatic plants have large air spaces in the leaf and stem parenchyma and in the petioles. Tissue containing such air spaces is called **aerenchyma** (**FIGURE 28.11B**). Aerenchyma stores oxygen produced by photosynthesis and permits its ready diffusion to parts of the plant where it is needed for cellular respiration. Aerenchyma also imparts buoyancy. Furthermore, because aerenchyma contains far fewer cells than most other plant tissues, metabolism in aerenchyma proceeds at a lower rate, so the need for oxygen is much reduced.

FIGURE 28.11　Plant Adaptations to Saturated Habitats (A) The submerged roots of mangrove trees obtain oxygen through pneumatophores. (B) This cross section of a petiole of the yellow water lily shows the air-filled channels of aerenchyma tissue.

(A)

Pneumatophores are root extensions that grow out of the water, under which the rest of the roots are submerged.

(B)

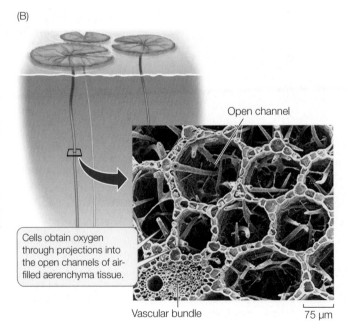

Open channel

Cells obtain oxygen through projections into the open channels of air-filled aerenchyma tissue.

Vascular bundle

75 μm

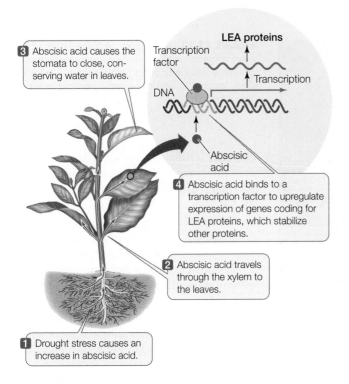

3 Abscisic acid causes the stomata to close, conserving water in leaves.

LEA proteins

Transcription factor

Transcription

DNA

Abscisic acid

4 Abscisic acid binds to a transcription factor to upregulate expression of genes coding for LEA proteins, which stabilize other proteins.

2 Abscisic acid travels through the xylem to the leaves.

1 Drought stress causes an increase in abscisic acid.

FIGURE 28.12 A Signaling Pathway in Response to Drought Stress Acclimation to drought stress begins in the root with the production of the hormone abscisic acid.

These hydrophobic proteins also accumulate in maturing seeds as they dry out (hence their name). The LEA proteins bind to membrane proteins and other cellular proteins to stabilize them, preventing their aggregation during desiccation.

Plants can cope with temperature extremes

Temperatures that are too high or too low can stress plants and even kill them. Plant species differ in their sensitivity to heat and cold, but all plants have their limits. Any temperature extreme can damage cellular membranes.

- High temperatures destabilize membranes and denature many proteins, especially some of the enzymes of photosynthesis.
- Low temperatures cause membranes to lose their fluidity and alter their permeabilities to solutes.
- Freezing temperatures may cause ice crystals to form, damaging membranes.

Plants have both constitutive adaptations and inducible responses for coping with temperature extremes.

ANATOMICAL ADAPTATIONS Many plants living in hot environments have constitutive adaptations similar to those of xerophytes. These adaptations include hairs and spines that dissipate heat and leaf forms that intercept less direct sunlight.

HEAT SHOCK RESPONSE The plant inducible response to heat stress is similar to the response to drought stress in that new proteins are made, often under the direction of an abscisic acid–mediated signaling pathway. Within minutes of experimental

When plants sense a water deficit in their roots, a signaling pathway is set in motion that initiates several measures to conserve water and maintain cellular integrity. This pathway begins with the production of the hormone abscisic acid (see Concept 26.3) in the roots. This hormone travels from the roots to the shoot, where it causes stomatal closure and initiates gene transcription that leads to other physiological events that conserve water and cellular integrity (**FIGURE 28.12**).

Many plant genes whose expression is altered by drought stress have been identified, largely through research using DNA microarrays, proteomics, and other molecular approaches (see Chapters 12 and 13). One group of proteins whose production is upregulated during drought stress is the *late embryogenesis abundant* (LEA; pronounced "lee-yuh") proteins.

APPLY THE CONCEPT

Plants respond to environmental stresses

The *Arabidopsis* gene *cor15a* is under the control of a promoter that responds to environmental conditions. *Arabidopsis* plants were genetically modified with a copy of the *cor15a* promoter fused to a reporter gene that is expressed as an enzyme that is readily detected in leaf extracts. The transgenic plants were transferred to environments kept at two different temperatures (19°C and 2°C), and the enzyme activity in their leaves was measured at different times after the transfer. The results are shown in the table.

1. Plot these data with hours after transfer on the *x* axis and enzyme activity on the *y* axis. What can you conclude about the role of *cor15a*?

2. In a separate experiment, plants were kept at 19°C for 36 hours in the presence of abscisic acid, and the

HOURS AFTER TRANSFER	ENZYME ACTIVITY (UNITS/g LEAF PROTEIN)	
	19°C	2°C
12	0.35	3.21
24	0.51	5.66
36	0.45	9.65
48	0.60	10.10
72	0.49	11.33

enzyme activity was 8.99. What can you conclude from this result?

3. How would you investigate the possible role of *cor15a* in the plant's response to drought? Why would you expect that there might be such a role?

FIGURE 28.13 Salty Soil Accumulation of salt from irrigation water in an area with inadequate drainage has caused this soil in central California (San Joaquin Valley) to become unsuitable for most plant growth.

exposure to raised temperatures (typically a 5°C–10°C increase), plants synthesize several kinds of **heat shock proteins**. Among these proteins are *chaperonins*, which help other proteins maintain their structures and avoid denaturation. Threshold temperatures for the production of heat shock proteins vary, but 39°C is sufficient to induce them in most plants.

COLD-HARDENING Low temperatures above the freezing point can cause *chilling injury* in many plants, including crops such as rice, corn, and cotton as well as tropical plants such as bananas. Many plant species can acclimate to cooler temperatures through a process called **cold-hardening**, which requires repeated exposure to cool temperatures over many days. A key change during the hardening process is an increase in the proportion of unsaturated fatty acids in cell membranes, which allows them to retain their fluidity and function normally at cooler temperatures (see Figure 2.12). Plants have a greater ability to modify the degree of saturation of their membrane lipids than animals do. In addition, low temperatures induce the formation of proteins similar to heat shock proteins, which protect against chilling injury.

If ice crystals form within plant cells, they can kill the cells by puncturing organelles and plasma membranes. Furthermore, the growth of ice crystals outside the cells can draw water from the cells and dehydrate them. Freeze-tolerant plants have a variety of adaptations to cope with these problems, including the production of *antifreeze proteins* that slow the growth of ice crystals.

Some plants can tolerate soils with high salt concentrations

Salty, or *saline*, environments (high in Na^+, K^+, Ca^{2+}, and Cl^-) are found in nature in diverse locales, from hot, dry deserts to moist, cool coastal marshes. Much of the land in Australia is naturally saline. In addition, agricultural land can become increasingly saline as a result of irrigation and the application of

chemical fertilizers. This *salinization*, which can eventually make land unsuitable for farming, is an increasing problem worldwide (**FIGURE 28.13**).

Because of its high salt concentration, a saline environment has a very negative water potential. To obtain water from such an environment, a plant must have an even more negative water potential (see Concept 25.3); otherwise water will diffuse out of its cells, and the plant will wilt and die. Plants in saline environments are also challenged by the potential toxicity of sodium ions, which inhibit enzymes and protein synthesis.

Plants that are adapted for survival in saline soils are called **halophytes**. Most halophytes take up Na^+, and many take up Cl^-, into their roots and transport those ions to their leaves, where they accumulate in the central vacuoles of leaf cells, away from more sensitive parts of the cells. The accumulated salts in the tissues of halophytes make their water potential more negative than the soil solution and allow them to take up water from their saline environment.

> **FRONTIERS** Knowledge about the mechanisms by which plants accumulate salt is being applied to plant breeding. Selection of mutant *Arabidopsis* plants that can tolerate high salt concentrations has led to the isolation of a gene for salt tolerance and the creation of transgenic crop plants that can tolerate saline conditions. Agricultural scientists and farmers hope that these transgenic crops will allow farming in regions where salinization has made agriculture unfeasible.

Some halophytes have **salt glands** in their leaves. These glands excrete salt, which collects on the leaf surface until it is removed by rain or wind (**FIGURE 28.14**). This adaptation, which reduces the danger of poisoning by accumulated salt, is found in some desert plants and in some plants growing in mangrove habitats. The negative water potential in the salt-laden leaves also promotes water flow from the roots up through the xylem by *transpiration–cohesion–tension* (see Concept 25.3).

Some plants can tolerate heavy metals

High concentrations of heavy metal ions are toxic to most plants. Some soils are naturally rich in heavy metals as a result of normal geological processes or acid rain. The mining of metallic ores also leaves localized areas with high concentrations of heavy metals and low concentrations of nutrients.

FIGURE 28.14 Excreting Salt This saltwater mangrove plant has specialized glands that excrete salt, which appears here as crystals on the leaves.

Such sites are hostile to most plants, and seeds falling on them generally do not produce adult plants.

Some plants can survive in these environments, however, by accumulating concentrations of heavy metals that would kill most plants. Over 200 plant species have been identified as **hyperaccumulators** that store large quantities of metals such as arsenic (As), cadmium (Cd), nickel (Ni), aluminum (Al), and zinc (Zn).

Perhaps the best-studied hyperaccumulator is alpine pennycress (*Thlaspi caerulescens*). Before the advent of chemical analysis, miners used the presence of this plant as an indicator of mineral-rich deposits. A *Thlaspi* plant may accumulate as much as 30 g/kg Zn (most plants contain 0.1 g/kg) and 1.5 g/kg Cd (most plants contain 0.001 g/kg). Studies of *Thlaspi* and other hyperaccumulators have revealed the presence of several common adaptations:

- Increased ion transport into the roots
- Increased rates of translocation of ions to the leaves
- Accumulation of ions in vacuoles in the shoot
- Resistance to the ions' toxicity

Knowledge of these hyperaccumulation mechanisms and the genes underlying them has led to the emergence of **phytoremediation**, a form of bioremediation (see Chapter 13) that uses plants to clean up environmental pollution in soils. Some phytoremediation projects use natural hyperaccumulators, while others use genes from hyperaccumulators

FIGURE 28.15 Phytoremediation Plants that accumulate heavy metals can be used to clean up contaminated soils. Here, poplars are being used to remove contaminants from an air base.

to create transgenic plants that grow more rapidly in and are better adapted to a particular polluted environment. In either case, the plants are grown in the contaminated soil, where they act as natural "vacuum cleaners" by taking up the contaminants (**FIGURE 28.15**). The plants are then harvested and disposed of to remove the contaminants. Perhaps the most dramatic use of phytoremediation occurred after an accident at the nuclear power plant at Chernobyl, Ukraine (then part of the Soviet Union), in 1986, when sunflower plants were used to remove uranium from the nearby soil. Phytoremediation is now widely used to clean up land after strip mining.

Do You Understand Concept 28.3?

- Compare the constitutive and induced mechanisms that allow plants to cope with drought conditions.
- Describe a mechanism shared by plants that tolerate high concentrations of salts and those that tolerate heavy metals.
- If an inhibitor of abscisic acid synthesis was sprayed on a plant and the plant was exposed to drought conditions, what would happen? Give details.

QA
QUESTION

How can knowledge of plant and fungal biology be used to prevent the spread of wheat rust?

ANSWER Today, *Puccinia graminis* strain Ug99 continues its spread and threatens wheat crops worldwide. In the Green Revolution of the 1960s to 1980s, wheat plants were bred to have the gene complexes *Sr24* and *Sr31*, *R* genes that conferred resistance to existing strains of wheat rust fungus (Concept 28.1). The first strain of Ug99 found in Uganda had *Avr* genes that form elicitor proteins that do not bind to Sr31 and

therefore overcome part of the plant's resistance. Then, in 2006, a new genetic variant of Ug99 was found in Kenya with additional *Avr* genes that allow the pathogen to overcome the *Sr24* gene resistance as well (**FIGURE 28.16**).

Because 90 percent of the wheat grown in the world has no resistance to this new strain of Ug99, an intensive search is under way for additional wheat resistance genes. In the regions where Ug99 infections are now raging, occasional wheat plants grow that are resistant. Samples of both wheat plants and fungus are sent to a U.S. government laboratory in Minnesota, where, under high security and sterility, they are tested and examined for *Avr* and *R* genes. In addition, seeds from thousands of varieties of wheat, collected from all over the world, are being grown into seedlings and examined for *R* genes.

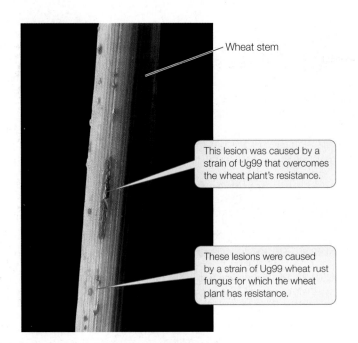

Wheat stem

This lesion was caused by a strain of Ug99 that overcomes the wheat plant's resistance.

These lesions were caused by a strain of Ug99 wheat rust fungus for which the wheat plant has resistance.

FIGURE 28.16 Overcoming Resistance to Pathogens Blotches caused by wheat rust are visible on the stem of a wheat plant. This plant has a gene called *Sr24* that confers resistance to the wheat rust pathogen, but it has been infected by a strain of the pathogen, Ug99, that can overcome that resistance.

28 SUMMARY

concept 28.1 Plants Have Constitutive and Induced Responses to Pathogens

- Plants and pathogens have evolved together in a continuing "arms race": pathogens have evolved mechanisms for attacking plants, and plants have evolved mechanisms for defending themselves against those attacks.

- Some of the responses by which plants fight off pathogens are **constitutive**—always present in the plant—whereas others are **induced**—produced in reaction to the presence of a pathogen. **Review Figure 28.1 and ANIMATED TUTORIAL 28.1**

- Plants use physical barriers to block pathogen entry and seal off infected regions.

- **Gene-for-gene resistance** depends on a match between a plant's **resistance (R) genes** and a pathogen's **avirulence (Avr) genes**. **Review Figure 28.2**

- In the **hypersensitive response** to infection, cells produce **phytoalexins** and **pathogenesis-related (PR) proteins**, and the plant isolates the area of infection by forming necrotic lesions.

- The hypersensitive response may be followed by another defensive reaction, **systemic acquired resistance**, in which salicylic acid activates further synthesis of defensive compounds throughout the plant.

concept 28.2 Plants Have Mechanical and Chemical Defenses against Herbivores

- Physical structures such as spines and thick cell walls deter some herbivores.

- Plants produce **secondary metabolites** as defenses against herbivores. **Review Table 28.1, Figure 28.5, and WORKING WITH DATA 28.1**

- Hormones, including **jasmonate**, participate in signaling pathways leading to the production of defensive compounds. **Review Figure 28.6**

- Plants protect themselves against their own toxic defensive chemicals by compartmentalizing those chemicals, by storing their precursors separately, or through modifications of their own proteins.

concept 28.3 Plants Adapt to Environmental Stresses

- **Xerophytes** are plants adapted to dry environments. Their structural adaptations include thickened cuticles, specialized trichomes, **stomatal crypts**, **succulence**, and long taproots.

- Some plants accumulate solutes in their cells, which lowers their water potential so they can more easily take up water.

- Adaptations to water-saturated habitats include **pneumatophores**, extensions of roots that allow oxygen uptake from the air, and **aerenchyma**, tissue in which oxygen can be stored and can diffuse throughout the plant. **Review Figure 28.11**

- A signaling pathway involving abscisic acid initiates a plant's response to drought stress. **Review Figure 28.12**

- Plants respond to high temperatures by producing **heat shock proteins**. Low temperatures can result in **cold-hardening**.

- Plants that are adapted for survival in saline soils are called **halophytes**. Most halophytes accumulate salt. Some have **salt glands** that excrete salt to the leaf surface.

- Some plants living in soils that are rich in heavy metals are **hyperaccumulators** that take up and store large amounts of those metals into their tissues.

- **Phytoremediation** is the use of hyperaccumulating plants or their genes to clean up environmental pollution in soils.

See WEB ACTIVITY 28.1 for a concept review of this chapter.

PART

6

Animal Form and Function

Physiology, Homeostasis, and Temperature Regulation

Ground squirrels live in high mountains such as California's Sierra Nevada, where the ground is covered by snow for more than 5 months of the year. How do these animals, which derive all of their food from sources on or near the ground, survive the winter? One strategy is to store enough energy to last through the time of scarcity.

Belding's ground squirrel (*Citellus beldingi*) lives in alpine meadows and, in summer, feeds on grass and other green plants. As fall approaches, these animals store energy in the form of body fat, growing to about three times their lean body mass. An individual that weighed 160 g in the spring may put on over 300 g of fat before snow begins to fall and it retires to its burrow system. In contrast, the golden-mantled ground squirrel (*Citellus lateralis*) lives on forested slopes during the summer, eating plants, berries, seeds, and fungi. Its lean body mass is similar to that of Belding's ground squirrel, but it does not put on nearly as much pre-winter fat. Instead, *C. lateralis* stores energy in the form of food in its burrow. The success of each of these strategies depends on the balance between supply and demand. How much energy is available, and how much is needed to get through the winter?

The resting metabolic rate of a 160-g ground squirrel is about 1.2 Kcal/hr. But to maintain the animal's body temperature in a cold burrow under snow, the metabolic rate rises to about 4.0 Kcal/hr. Over 5 months, this amounts to 14,400 Kcal. Three hundred grams of fat yield

A golden-mantled ground squirrel feeds in the summer forest of the Sierra Nevada.

2,700 Kcal—which would mean a serious shortfall for the Belding's ground squirrel.

How about the food hoarder? If we assume that the golden-mantled ground squirrel stores fir and pine seeds with an average energy content of 5 Kcal/g, an individual would have to store about 3 kg of seeds—an enormous amount. (See Figure 21.16, or better yet, try prying apart one scale of a pine cone and removing the seed; it takes a lot of seeds to make 3 kg.)

The supply side of the winter survival equation does not explain how ground squirrels survive the months they cannot be active above ground. What about the demand side? How can the animal reduce its energetic needs?

Q QUESTION

What can ground squirrels do to lower the metabolic demands of surviving through the winter?*

*You will find the answer to this question on page 601.

KEY CONCEPTS

29.1 Multicellular Animals Require a Stable Internal Environment

29.2 Physiological Regulation Achieves Homeostasis of the Internal Environment

29.3 Living Systems Are Temperature-Sensitive

29.4 Animals Control Body Temperature by Altering Rates of Heat Gain and Loss

29.5 A Thermostat in the Brain Regulates Mammalian Body Temperature

concept
29.1 ## Multicellular Animals Require a Stable Internal Environment

Animal cells must take in nutrients and oxygen from the environment and eliminate carbon dioxide and other waste products of metabolism to the environment. The cells of tiny or thin aquatic animals are so close to the surrounding water that they can meet their needs by diffusional exchange. The cells of larger animals are too far from the outside environment to depend on diffusional exchange of nutrients, oxygen, and wastes.

Multicellular animals have an internal environment of extracellular fluid

The cellular needs of multicellular animals are served by a fluid environment that is internal to the animal. That internal fluid environment mediates the exchange of substances between the cells of the body and the external environment. Much of physiology—the study of how organisms function—involves analyzing exchanges of energy and materials that take place between the cells and the internal environment, and between the internal environment and the external environment.

Animals are mostly water (you, for example, are about 60 percent water). That water is distributed in different body compartments. Most of the water in an animal's body is located within its cells; this is *intracellular fluid*. The rest is the *extracellular fluid*. For all animals with blood circulatory systems, a portion of the extracellular water is the liquid portion of the blood, or *plasma*. The majority of extracellular fluid, however, is found between the cells of the body and is called **interstitial fluid** (**FIGURE 29.1**). Most cells exchange materials and nutrients with the interstitial fluid that surrounds them.

Maintainence of the internal environment protects the cells of the body from external changes and harsh conditions, making it possible for an animal to occupy habitats that would kill its cells if they were exposed to the environment directly. The constancy of the internal environment is critical, and any change in its composition can threaten the health of cells. The activities of the cells can deplete the nutrients present in the

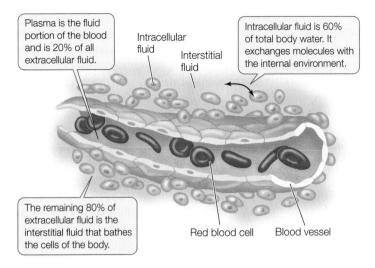

Plasma is the fluid portion of the blood and is 20% of all extracellular fluid.

Intracellular fluid

Interstitial fluid

Intracellular fluid is 60% of total body water. It exchanges molecules with the internal environment.

The remaining 80% of extracellular fluid is the interstitial fluid that bathes the cells of the body.

Red blood cell   Blood vessel

FIGURE 29.1 The Internal Environment Extracellular fluid is the body's internal environment. Its components include the liquid portion of the blood (plasma) and the interstitial fluid that bathes all the cells of the body. By exchanging molecules with the intracellular fluid, the internal environment serves the needs of all the body's cells.

interstitial fluid; furthermore, cells are constantly dumping their waste products into the interstitial fluid and can poison this critical environment.

The physical and chemical nature of an animal's habitat, along with the metabolic activities of its cells, constantly threaten to change the composition of its internal environment. How is the internal environment kept constant?

Homeostasis is the process of maintaining stable conditions in the internal environment

As multicellular organisms evolved, cells became specialized for maintaining specific aspects of the internal environment—among them temperature, pH, ion concentrations, O_2 and CO_2 levels, and levels of fuel molecules such as glucose—within specific limits. The maintenance of such stability is known as **homeostasis**.

APPLY THE CONCEPT

Multicellular animals have an internal environment of extracellular fluid

Different substances injected into the bloodstream will distribute into the body's intracellular and extracellular fluid compartments. Their distribution pattern depends on the chemical properties of the injected substances. In three experiments, 10 ml of 10% solutions of each of three compounds (^2H_2O, or heavy water; inulin, a polysaccharide; and Evan's Blue, a dye) are injected into the bloodstream of a 70-kg man. After 1 hour, a blood sample is taken from the subject and the concentrations of the substances are measured. From the results shown and the information in Figure 29.1, what would

SUBSTANCE	CONCENTRATION INJECTED	FINAL CONCENTRATION IN PLASMA
Heavy water	10%	0.0024%
Inulin	10%	0.006%
Evan's Blue	10%	0.03%

you conclude about how these three substances are distributed among the fluid compartments of the body?

The presence of an internal environment facilitated the evolution of specializations of cells, tissues, and organs. When each cell no longer had to carry out all exchanges with the external environment for itself, cells could adapt to contribute a variety of specialized functions to the maintenance of the internal environment, all the while benefitting from that internal environment. Some cells became the boundary between the internal and the external environments, transporting nutrients in, moving wastes out, and maintaining appropriate ion concentrations. Others provided specialized internal functions such as storing energy, processing information, and moving the body. As complexity of animals evolved, these specialized cells evolved into tissues, organs, and ultimately organ systems. These systems function to maintain homeostasis of the internal environment. Examples are the digestive, respiratory, and circulatory systems. Adaptations of these systems made it possible for animals to maintain homeostasis and survive in many different habitats.

Cells, tissues, and organs serve homeostatic needs

Each physiological system is composed of discrete **organs**, such as the liver, heart, lungs, and kidneys, that serve specific functions in the body. These organs are made up of tissues. A **tissue** is an assemblage of cells; however, although there are a multitude of different cell types, there are only four kinds of tissue: *epithelial*, *connective*, *nervous*, and *muscle*.

- **Epithelial tissues** are sheets of densely packed, interconnected *epithelial cells* that cover inner and outer body surfaces. Some epithelial cells secrete substances such as hormones, mucus, digestive enzymes, or sweat. Some, such as the lining of blood vessels, serve as selective barriers. Some serve transport functions, such as absorption of nutrients from the gut. Some, like the lining of the respiratory passages, have cilia that move mucus and detritus over their surfaces. Some epithelial tissues serve sensory functions, including smell, taste, and touch.

- **Connective tissues** are dispersed populations of cells embedded in an *extracellular matrix* that the cells themselves secrete. The composition and properties of the matrix differentiate the types of connective tissues. A matrix of protein fibers (collagen and elastin) provides the structural strength and elasticity that characterizes cartilage (the substance that lines the joints between bones and gives structure to the nose and the external ears). The extracellular matrix of bone is mineralized to give it rigid structural strength. In contrast, the extracellular matrix of blood cells—the plasma—is liquid so blood can flow through vessels. Adipose tissue is a connective tissue of energy-storing fat cells with little extracellular matrix.

- **Nervous tissues** consist of neurons and glial cells. *Neurons* are individual cells that generate and conduct electrical signals, or *nerve impulses*, around the body. They are the foundational units of the *central* and *peripheral nervous systems* that transmit information from both the internal and external environments to the other cells of the body, as we will detail in Chapters 34 and 35. Neurons communicate with each other using chemical signals called *neurotransmitters*. Glial cells do not fire nerve impulses, but they do release chemical signals, and they support the information-processing functions of neurons in various ways.

- **Muscle tissues** account for most of the body mass of vertebrates. They consist of elongated cells that contract to generate forces that result in movement, as we will describe in Chapter 36. *Skeletal muscle* (so named because it is usually attached to bones) is responsible for locomotion and other body movements. The heart is made up of *cardiac muscle* that generates the heartbeat and pumps blood. *Smooth muscle* generates forces in many hollow internal organs, including the gut, bladder, and blood vessels.

Organs include more than one kind of tissue, and most organs include all four (**FIGURE 29.2**). The wall of the stomach or the small intestine is a good example. Its inner surface is lined with a sheet of columnar epithelial cells. Different epithelial cells secrete mucus, enzymes, or stomach acid. Beneath the epithelial lining is connective tissue. Within this connective tissue are blood vessels, neurons, and glands (clusters of secretory epithelial cells). Concentric layers of smooth muscle tissue enable the stomach to contract to mix food with digestive juices. A network of neurons between the muscle layers controls these movements. Surrounding the stomach is a sheath of connective tissue and an epithelial lining.

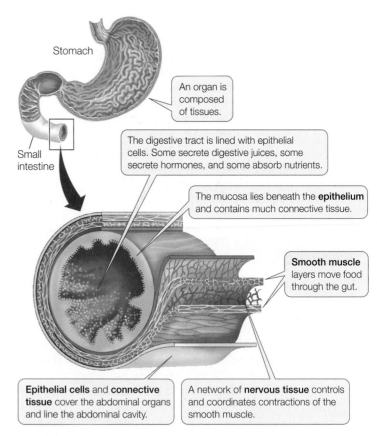

Stomach

Small intestine

An organ is composed of tissues.

The digestive tract is lined with epithelial cells. Some secrete digestive juices, some secrete hormones, and some absorb nutrients.

The mucosa lies beneath the **epithelium** and contains much connective tissue.

Smooth muscle layers move food through the gut.

Epithelial cells and **connective tissue** cover the abdominal organs and line the abdominal cavity.

A network of **nervous tissue** controls and coordinates contractions of the smooth muscle.

FIGURE 29.2 Tissues Form Organs Most organs contain more than one of the four types of tissue; many organs are made up of all four types.

An individual organ is usually part of an **organ system**—a group of organs that work together to carry out certain functions. The stomach is part of the digestive system. If an organ fails to function properly, homeostasis is compromised and cells may be damaged and die.

---

yourBioPortal.com

Go to WEB ACTIVITY 29.1
Tissues and Cell Types

Do You Understand Concept 29.1?

- What are the roles of the blood plasma and the interstitial fluid in serving the needs of the body, and how do these two extracellular fluids interrelate?

- What are the properties and functions of each of the four tissue types?

- How do organ systems contribute to the maintenance of the internal environment? Describe an example.

To maintain homeostasis, the activities of each organ and organ system must be matched to the demands of the body's cells. This matching requires that physiological systems be controlled and regulated in response to changes in both the external and internal environments.

FIGURE 29.3   Control, Regulation, and Feedback   The animal body uses information and control mechanisms to maintain homeostasis, just as a driver uses them to regulate the speed of a car.

---

concept
29.2

Physiological Regulation Achieves Homeostasis of the Internal Environment

As we will detail in subsequent chapters, the activities of an animal's physiological systems are controlled—speeded up or slowed down—by actions of the nervous and endocrine systems. But to *regulate* the internal environment, information is required.

Regulating physiological systems requires feedback information

Think of control and regulation in the context of driving a car. You *control* the speed of your car with the accelerator and the brakes; but to use the accelerator and the brakes to *regulate* the speed of your car, you have to know both how fast you are going and how fast you want to go. The desired speed is a **set point**; the reading on your speedometer is **feedback information**. When the set point and feedback information are compared, any difference between them is an **error signal**. Error signals suggest corrective actions, which you make by using the accelerator or brake (**FIGURE 29.3**).

Regulatory systems obtain, process, and integrate feedback information and then issue commands. An important component of any regulatory system is a *sensor*, which provides the feedback information that is compared with the set point. (Sensors will be described in Chapter 35.) Once sensory information is processed, the regulatory system issues commands to *effectors*, so named

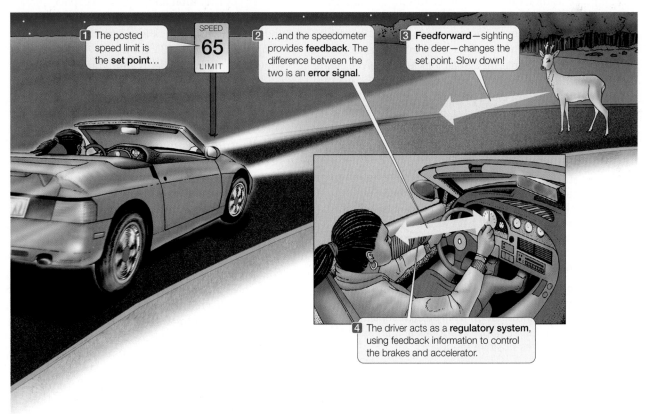

1 The posted speed limit is the **set point**...

SPEED **65** LIMIT

2 ...and the speedometer provides **feedback**. The difference between the two is an **error signal**.

3 **Feedforward**—sighting the deer—changes the set point. Slow down!

4 The driver acts as a **regulatory system**, using feedback information to control the brakes and accelerator.

because they effect changes in the internal—or external—environment. The accelerator and brake act as effectors in your car; muscles and glands are important effectors in the animal body (see Chapter 36). Effectors are **controlled systems** because their activities are controlled by commands from regulatory systems.

Feedback information can be negative or positive

A fundamental way to study a physiological regulatory system is to identify the information to which it responds. What detects the information? How is the information from the sensors used? **Negative feedback** is the most common use of sensory information in regulatory systems. The word "negative" indicates that this feedback information is used to counteract the influence that created an error signal. Whatever force is pushing the system away from its set point is "negated" by negative feedback, resulting in a movement back to the set point. In our car analogy, the recognition that you are going too fast is negative feedback that causes you to slow down. Negative feedback is a stabilizing influence in physiological systems.

Rather than returning a system to a set point, **positive feedback** amplifies a response (i.e., it *increases* the deviation from the set point). Although it is not as common as negative feedback, examples of positive feedback are seen in regulation of the responses that empty body cavities, such as urination and defecation. Another example is sexual behavior, in which a little stimulation increases the behavioral response, which increases stimulation, and so on. Positive feedback responses tend to reach a limit and then terminate rapidly.

The function of **feedforward information** is to change the set point. Seeing a deer ahead on the road when you are driving is an example of feedforward information; this information takes precedence over the posted speed limit, and you change your set point to a slower speed. In physiology, feedforward information is predictive of a change in the internal environment before that change occurs. For example, the climatological signs of oncoming winter (shorter days, dying vegetation, cooler temperatures) are feedforward information that stimulates physiological changes in the ground squirrels mentioned at the start of this chapter.

Do You Understand Concept 29.2?

- A regulatory system you encounter often is the thermostatically controlled heat and air conditioning system that keeps a building's temperature constant in changing external conditions. Describe such a system in terms of the regulating system, the controlled system, the set points, feedback information, and effector commands.

- A common type of sensor is a nerve cell that increases its activity when it is stretched. Describe a situation where information from stretch-sensitive neurons is used as negative feedback and an example where it is used as positive feedback.

The principles of control and regulation help organize our thinking about physiological systems. Once we understand

how a system works, we can then ask how it is regulated. For each of the various animal organ systems described in the subsequent chapters of this unit, we will consider the challenges imposed by the environment, some of the physiological adaptations that have evolved to meet those challenges, and how those physiological mechanisms are regulated. The remainder of this chapter applies that approach to the regulation of body temperature.

---

concept 29.3	Living Systems Are Temperature-Sensitive

Temperatures over the face of Earth vary enormously, from the boiling hot springs of Yellowstone National Park to the interior of Antarctica, where temperatures can fall below –80°C. But most eukaryotic cells can function only over a relatively narrow range of temperatures. If cells cool below 0°C, ice crystals can form and damage their structures. Some animals have adaptations (such as "antifreeze molecules" in their blood) that help them resist freezing; others are able to survive freezing. Generally, however, cells must remain above 0°C to stay alive.

The upper temperature limit for survival of most cells is about 45°C, and proteins begin to denature and lose their function as soon as temperatures rise above 40°C. Therefore, normal cellular functions are limited to the range between 0°C and 40°C, which approximates the thermal limits for life of cells. A particular species, however, usually has much narrower limits. Humans face heat exhaustion and other serious heat-induced illness as body temperatures rise above 40°C, and we begin to lose coordination and cognitive capacity when body temperature falls below 35°C. Thus the homeostatic mechanisms of the human thermoregulatory system must maintain the body's temperature between 35°C and 40°C—no matter what the *ambient* (external environment) temperature may be.

Q_{10} is a measure of temperature sensitivity

Most biochemical reactions and physiological processes are temperature-sensitive, and they usually proceed faster at higher temperatures (see Figure 3.20). The temperature sensitivity of a reaction or process can be described in terms of **Q_{10}**, a factor calculated by dividing the rate of a process or reaction at a certain temperature, R_T, by the rate of that process or reaction at a temperature 10°C lower, R_{T-10}:

$$Q_{10} = R_T / R_{T-10}$$

Q_{10} can be measured for a simple enzymatic reaction or for a complex physiological process, such as rate of oxygen consumption. If a reaction or process is not temperature-sensitive, it has a Q_{10} of 1. Most biological Q_{10} values are between 2 and 3. A Q_{10} of 2 means that the reaction rate doubles as temperature increases by 10°C, and a Q_{10} of 3 indicates a tripling of the rate (**FIGURE 29.4**).

Changes in body temperature can disrupt an animal's physiology because not all of the biochemical reactions that comprise

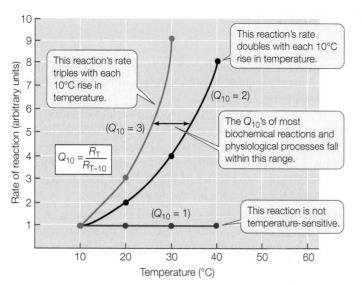

FIGURE 29.4 Q_{10} **and Reaction Rate** The larger the Q_{10} of a reaction or process, the faster its rate rises in response to an increase in temperature.

the metabolism of an animal have the same Q_{10}. These biochemical reactions are linked together in complex networks in which the products of one reaction are the reactants for other reactions. Because different reactions have different Q_{10} values, changes in tissue temperature will shift the rates of some reactions more than others, disrupting the overall network. To maintain homeostasis, organisms must be able to compensate for or prevent changes in body temperature.

Animals can acclimatize to seasonal temperature changes

The body temperatures of some animals (especially aquatic animals) are tightly coupled to the environmental temperature. The body temperature of a fish in a pond, for example, will always be the same as the water temperature, which might range more than 20°C seasonally. Nevertheless, the fish can maintain similar levels of activity in summer and winter because, over time, it can acclimatize. Important means of acclimatization are to produce more or fewer enzymes or enzymes with different temperature optima. The ability to acclimatize means that metabolic functions are less sensitive to long-term changes in temperature than they are to short-term changes.

Animals can regulate body temperature

The body temperature of a fish, amphibian or reptile depends on the temperature of its environment, so they are called **ectotherms** ("heat from the outside"). In contrast, birds and mammals are referred to as **endotherms** ("heat from the inside") because they maintain high and rather constant body temperatures over a wide range of ambient temperatures, largely because of their ability to generate internal heat metabolically.

FIGURE 29.5 shows the differences in the effects of ambient temperature on the metabolism of endotherms and ectotherms. In the experiments summarized here, an endotherm (mouse) and an ectotherm (lizard) of the same body size are put in closed chambers. Air is circulated through the chambers,

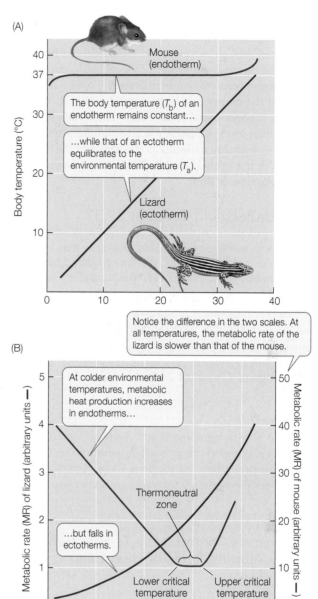

FIGURE 29.5 Ectotherms and Endotherms React Differently to Environmental Temperatures (A) At the same ambient temperature, an ectotherm and an endotherm of approximately the same body size (here, a lizard and a mouse) have different body temperatures. (B) The metabolic rate of the lizard drops as the environmental temperature falls, but below a lower critical temperature the mouse curve rises as ambient temperature falls. The mouse curve turns up again at an upper critical temperature because it takes metabolic energy to actively dissipate heat.

yourBioPortal.com

Go to WEB ACTIVITY 29.2
Thermoregulation in an Endotherm

and the animals' rates of oxygen consumption (a measure of their metabolic rate, MR) are measured at different ambient temperatures (T_a). The first thing we notice is that the body temperature (T_b) of the mouse remains constant at 37°C over a broad range of T_a, whereas the T_b of the lizard equilibrates with T_a (see Figure 29.5A).

In the lizard, MR shows a typical biological temperature sensitivity (see Figure 29.5B) while T_b follows a Q_{10} curve. The MR of the mouse responds quite differently; over a range of T_a, the MR of the mouse remains low and constant. We call this range the **thermoneutral zone** (**TNZ**). Within the TNZ, MR is at the minimal level compatible with all of the physiological functions essential to maintain homeostasis of the internal environment in the resting animal; this level is known as the **basal metabolic rate**, or **BMR**.

Do You Understand Concept 29.3?

- What are some cellular processes or events that would lead to the survival of cells being limited to an approximately 40°C range of temperatures?
- Explain how a physiological process can be disrupted by changes in its temperature when its component biochemical reactions have different Q_{10}'s.
- Why does MR of a mammal increases as T_a decreases, whereas MR of a reptile decreases as T_a decreases, in the range of T_a's below 25 C in the example in the figure?"

We have seen how living systems are affected by temperature and how they respond to changes in temperature. Next we examine how the properties and adaptations of animals interact with the thermal environment to determine body temperature.

concept 29.4 Animals Control Body Temperature by Altering Rates of Heat Gain and Loss

An animal's body temperature is the result of the balance between thermal energy flowing into the body from the environment and from metabolism (heat$_{in}$), and thermal energy leaving the animal (heat$_{out}$). The two sides of this **heat budget equation** must be equal if the body temperature of the animal is to remain constant. In other words, unless heat$_{in}$ = heat$_{out}$, body temperature changes.

If we expand both sides of the heat budget equation to describe all of the pathways of thermal energy exchange between the animal and the environment, we can see all of the possible ways an animal can control its body temperature (**FIGURE 29.6**). All gains and losses of thermal energy that an animal experiences occur by the following mechanisms:

- **Metabolism**: All energy conversions are inefficient and produce heat as a by-product. Thus, the conversion of chemical bond energy in nutrients to the chemical bond energy in ATP, and the use of ATP to do work result in the production of heat in the animal.
- **Radiation**: Heat transfers from warmer objects to cooler ones via the exchange of infrared radiation (e.g., what you feel when you stand in front of a fire). Incoming radiation is that which is absorbed, R_{abs}, and outgoing radiation is that which is emitted, R_{out}.
- **Convection**: Heat transfers to a surrounding medium such as air or water as that medium flows over a surface (the "wind chill" factor).
- **Conduction**: Heat transfers directly when objects of two different temperatures come into contact (think of putting an icepack on a sprained ankle).
- **Evaporation**: Heat transfers away from a surface when water evaporates on that surface (the effect of sweating).

Under most circumstances, heat$_{in}$ for an animal is generated by metabolism and thermal radiation (MR and R_{abs}). The other terms in the above list are mostly avenues of heat loss and are on the right side of the heat budget equation (heat$_{out}$). "Under most circumstances" is important, because the direction of heat flow through radiation, conduction, and convection depends on the difference between the temperature of the environment and the temperature of the animal's surface. Thus,

$$\underbrace{\text{heat}_{in}}_{\text{metabolism} + R_{abs}} = \underbrace{\text{heat}_{out}}_{R_{out} + \text{convection} + \text{conduction} + \text{evaporation}}$$

APPLY THE CONCEPT

Living systems are temperature-sensitive

Neurons convey information from one part of the body to another as streams of electrical signals called nerve impulses. Information is coded as the frequency of nerve impulses (impulses/sec). Thus, changes in the temperature of neurons can alter the information they are transmitting. A neurobiologist recorded the impulse frequencies of three different types of neurons in the skin of an animal at different skin temperatures. Use the data in the table to answer the questions at right.

FIRING RATE (IMPULSES/SEC)	SKIN TEMPERATURE (°C)			
	25	30	35	40
Neuron A	1	2	6	18
Neuron B	12	6	3	1
Neuron C	8	7	9	8

1. Calculate the Q_{10} for each of the three neuron types.
2. On the basis of your calculations, what type of information do you think each different neuron might be conveying from the skin to the brain?

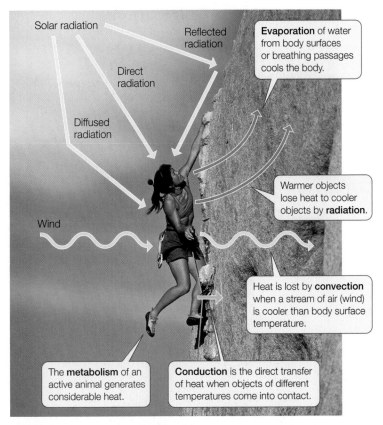

FIGURE 29.6 Animals Exchange Heat with the Environment
An animal's body temperature is determined by the balance between internal heat production—metabolism—and four avenues of heat exchange with the environment.

Any adaptation that serves to regulate an animal's body temperature must influence one or more terms of this equation. Moreover, if we expanded each term on the right side of the equation and looked at them as a physicist would, we would see that recurrent factors are surface temperature and surface area. Thus, an important insight we gain from the heat budget equation is the central importance of *surface temperature* and *surface area* for heat exchange. Those two terms are key determinants of radiative, conductive, and convective heat loss.

Mammals and birds have high rates of metabolic heat production

A close look at Figure 29.5B reveals that even the highest MR of the resting lizard is lower than the BMR of a mouse. Most of the energy a resting endotherm expends goes into pumping ions across membranes. To the extent that cell membranes are permeable to ions, the ions diffuse down their concentration gradients (from high concentration to low). To maintain proper concentrations inside and outside cells, the ions must be transported back "uphill," which requires an expenditure of energy. While this is true for both ectotherms and endotherms, the cells of endotherms are leakier to ions than those of ectotherms. Thus endotherms must expend more energy (and thus release more heat) to maintain ion concentration gradients. This is akin to running

on a treadmill: the faster the treadmill goes (analogous to leaking ions), the faster you have to run (analogous to pumping ions) to remain in the same position.

LINK Review the mechanisms of ion transport described in Concepts 5.2 and 5.3

When T_a falls below an endotherm's thermoneutral zone—the lower end of which is called the *lower critical temperature*—the animal must increase its metabolic heat production or its T_b will fall. One adaptation that produces metabolic heat in all endotherms is shivering. Skeletal muscles are organized in antagonistic pairs, one of which will flex a joint and the other of which will extend it. If both muscles contract at the same time, they consume energy and produce heat, but they will not produce major movements other than a tremor that is experienced as shivering. Such *shivering thermogenesis* allows a mammal or a bird to increase its MR by about four times above its BMR (see Figure 29.5B).

Some mammals also are capable of *non-shivering thermogenesis*. Tissues other than skeletal muscle can produce metabolic heat by uncoupling oxidative phosphorylation (see Chapter 6, p. 104). These tissues burn fuel without producing ATP, and therefore the animal's metabolic rate is not limited by the buildup of ATP. The major evolutionary adaptation for non-shivering thermogenesis is **brown fat** (**FIGURE 29.7**). This adipose tissue looks brown because it has a high density of mitochondria and blood vessels. It is found in the newborns of many species of mammals, including humans. For mammals that hibernate, it is an important source of heat as they cycle out of hibernation.

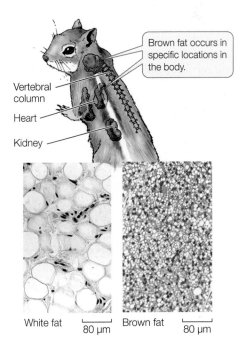

FIGURE 29.7 Brown Fat In many mammals, specialized brown fat tissue produces heat. White fat cells (left) contain large droplets of lipid but have few organelles and limited blood supply. Brown fat cells (right) are packed with mitochondria and richly supplied with blood.

Basal metabolic rates are correlated with body size

As you would expect, the total metabolic expenditure of a resting elephant (its basal metabolic rate) is greater than that of a mouse. After all, the elephant is more than 100,000 times more massive than the mouse. The elephant's BMR, however, is only about 7,000 times greater than that of the mouse. That means that 1 g of mouse tissue uses energy at a rate about 15 times greater than does 1 g of elephant tissue (**FIGURE 29.8**). Across all endotherms, BMR per gram of tissue increases as animals get smaller.

Why should this disproportionate difference exist? Several reasons have been suggested, but none of these is sufficient by itself. As animals get bigger, they have a smaller ratio of surface area to volume (see Figure 4.2). Because heat production is a function of an animal's volume, or mass, but its capacity to dissipate heat is a function of its surface area, it has been reasoned that larger animals evolved lower metabolic rates to avoid overheating. This explanation is insufficient because the relationship between body mass and metabolic rate also holds for small organisms and for ectotherms, in which overheating is not a problem. Another hypothesis is that larger animals have a greater proportion of support tissues (notably skin and bone) that are not very active metabolically. The real answer is probably a mixture of different factors.

Insulation is the major adaptation of endotherms to cold climates

Endotherms from all climates, tropical to arctic, fall on the same line when it comes to relating their BMRs to their body sizes; thus we can conclude that BMR is not an adaptation to different climates. When we compare the MR versus T_a curves for different endotherms, a decreased response to declines in environmental temperature indicates that an ability to conserve heat is an important adaptation for cold climate endotherms. How do you conserve body heat? You put on more or heavier

FIGURE 29.9 Anatomical Adaptations to Climate (A) The desert fox has short fur and limited body insulation. Its large ears and long limbs allow heat to radiate from these extended body surfaces. (B) The Arctic fox has a thick coat of insulating fur. Its small ears, short limbs, and rounded body shape give it a smaller surface-to-volume ratio than its desert-dwelling relative—in other words, there is less surface area over which heat can be lost to the environment.

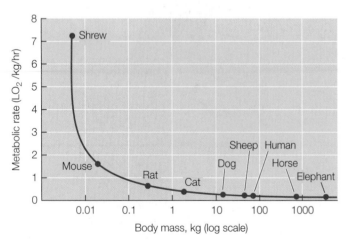

FIGURE 29.8 The Mouse-to-Elephant Curve On a weight-specific basis, the basal metabolic rates of small endotherms are much greater those that pf of larger endotherms. This graph plots O_2 consumption per kg of body weight (a measure of metabolic rate) against a logarithmic plot of body mass.

clothing—that is, you increase your thermal insulation. What you do behaviorally, evolution has provided anatomically for many species (**FIGURE 29.9**).

How can we measure and compare the thermal insulation of animals? Referring back to Figure 29.5B, the rising portion of the graph that occurs in the mouse when T_a falls below the lower critical temperature is described by the equation

$$MR = K(T_b - T_a)$$

where K is the slope of the curve, or *thermal conductance*—a measure of how readily the animal loses heat. The inverse of thermal conductance is *insulation*, which is a measure of how effectively the animal conserves heat. Note that this equation predicts that when $T_a = T_b$, MR will be 0. But, of course, MR is never 0 because there is always a BMR necessary for maintaining the homeostatic functions of life. Nevertheless, the equation tells us that the curve describing the effect of low T_a on heat production (i.e., on MR) will intersect with the x axis at T_a equal to T_b.

Above lower critical temperatures, the MR versus T_a curves for endotherms show a region of thermoneutral zones over which MR remains at basal levels. How can the MR be at the BMR level over such a wide range of ambient temperatures? Surely, heat loss continues to be a function of the difference between T_b and T_a within the thermoneutral zone. The answer is that, above the lower critical temperature, the animal can change the value of K. It can smooth or fluff its fur or feathers. It can send more or less blood to skin surfaces that dissipate heat to the environment. And it can change its posture and positioning to expose parts of its body that are not as well insulated with fur or feathers.

Some fish can conserve metabolic heat

Active fish produce substantial amounts of metabolic heat, but they have difficulty retaining any of that heat. Blood pumped from the fish heart goes directly to the gills for respiratory gas exchange. The gills are thin and have a huge surface area, so the blood comes into thermal equilibrium with the surrounding water. Thus any heat the blood picks up from the metabolically active muscles is lost to the cold water flowing over the gills. But some large, rapidly swimming fishes, such as bluefin tuna and great white sharks, can maintain body temperatures 10°C–15°C higher than the temperature of the surrounding water. The heat comes from their powerful swimming muscles, and the ability

FRONTIERS   Covered with insulation, how can a heavily furred or feathered endotherm dissipate the heat generated by high levels of activity? Such animals have heat-loss portals—body areas that are not furred or feathered, and which are underlain by special blood vessels that bring large volumes of blood close to the skin surface (thereby acting as radiators). These body areas include the palms of the hands, soles of the feet, and parts of the face. Humans inherited these special blood vessels in non-hairy skin from our primate ancestors. Scientists exploited this adaptation with a new technology: a portable device that uses a partial vacuum to increase blood flow through the heat-dissipating vessels and a heat sink to efficiently transfer the heat from the skin surface. Using this technology, scientists found that muscle fatigue is a function of rising muscle temperature, and that extracting that heat could increase an individual's ability to sustain high levels of physical effort, even in hot environments.

of these "hot" fish to conserve that heat is based on the arrangement of their blood vessels.

In the typical ("cold") fish circulatory system, cold oxygenated blood from the gills travels through the center of the fish in a large artery that distributes blood to all regions of the body (**FIGURE 29.10A**). "Hot" fish have a smaller central artery, and most of their cold oxygenated blood is transported in large vessels just under the skin (**FIGURE 29.10B**). Thus, the cold blood from the gills is kept close to the surface of the body. Smaller vessels transporting this cold blood into the muscle mass run

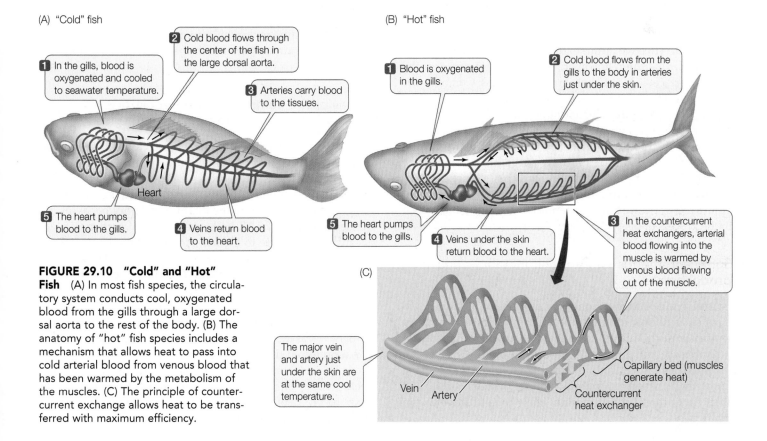

(A) "Cold" fish

1 In the gills, blood is oxygenated and cooled to seawater temperature.

2 Cold blood flows through the center of the fish in the large dorsal aorta.

3 Arteries carry blood to the tissues.

Heart

5 The heart pumps blood to the gills.

4 Veins return blood to the heart.

(B) "Hot" fish

1 Blood is oxygenated in the gills.

2 Cold blood flows from the gills to the body in arteries just under the skin.

5 The heart pumps blood to the gills.

4 Veins under the skin return blood to the heart.

3 In the countercurrent heat exchangers, arterial blood flowing into the muscle is warmed by venous blood flowing out of the muscle.

(C)

The major vein and artery just under the skin are at the same cool temperature.

Vein   Artery

Capillary bed (muscles generate heat)

Countercurrent heat exchanger

FIGURE 29.10   "Cold" and "Hot" Fish   (A) In most fish species, the circulatory system conducts cool, oxygenated blood from the gills through a large dorsal aorta to the rest of the body. (B) The anatomy of "hot" fish species includes a mechanism that allows heat to pass into cold arterial blood from venous blood that has been warmed by the metabolism of the muscles. (C) The principle of countercurrent exchange allows heat to be transferred with maximum efficiency.

parallel to vessels transporting warm blood from the muscle mass back toward the heart. Heat flows from the warm to the cold blood and is therefore retained in the muscle mass.

Because heat is exchanged between blood vessels carrying blood in opposite directions, this adaptation is called a **countercurrent heat exchanger** (**FIGURE 29.10C**). Countercurrent heat exchange enables these fish to have an internal body temperature considerably above the water temperature. Why is this advantageous for the fish? Each 10°C rise in muscle temperature increases the fish's sustainable power output almost threefold, giving it a faster swimming speed that is useful in capturing prey.

> **LINK** The principle of countercurrent exchange is seen in other physiological systems, including oxygen transport (see Concept 37.2) and the ability of the kidney to concentrate urine (see Concept 40.4)

Evaporation is an effective but expensive avenue of heat loss

Blood flow to the skin maximizes passive heat loss only when there is a substantial difference between body temperature and ambient temperature. Therefore, high ambient temperatures as well as high metabolic rates threaten to raise body temperature. Endotherms live close to their upper lethal temperatures, so even small increases in body temperature can be dangerous. Many endotherms counter this increase with evaporative cooling by sweating or panting. You probably have "worked up a sweat" in a hot environment or when exercising.

When a gram of water evaporates, it absorbs about 580 calories of heat. If that evaporation takes place on the skin, much of that 580 calories is removed from the body. Evaporation is therefore an effective means of dissipating heat. During heavy sweating or panting, however, any water that drips off the body's surface does not contribute to heat loss. Combine this with the fact that animals carry little excess water and can die rapidly from dehydration, and it becomes clear that evaporative heat loss is a physiologically expensive and potentially dangerous thermoregulatory adaptation. In addition, sweating is an *active* process that requires the expenditure of metabolic energy. That is why the metabolic rate increases above the upper critical temperature (the rising curve on the right side of Figure 29.5B). A sweating or panting animal is generating heat in the process of dissipating heat.

Some ectotherms elevate their metabolic heat production

Although elevated metabolic heat production is characteristic of endotherms, some ectotherms can also raise their body temperatures in this way. For example, the flight muscles of many insects must reach 35°C–40°C before the insects can fly, and they must maintain these high temperatures during flight. Such insects produce the required heat by contracting their flight muscles in a manner analogous to shivering in mammals.

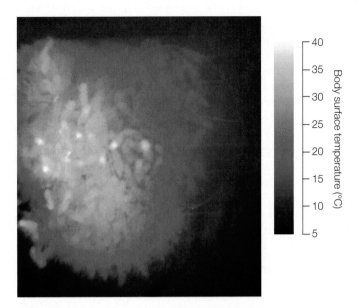

FIGURE 29.11 Bees Keep Warm in Winter Honey bee colonies survive winter cold because workers generate metabolic heat. In this infrared photograph of the center of an overwintering hive, individual bees are discernible by the heat their bodies produce as they cluster around their queen as she lays eggs.

Honey bees regulate temperature as a group. They live in large colonies consisting mostly of female worker bees that maintain the hive and rear the larval offspring of the queen bee. During winter, worker bees cluster around the brood (eggs and larvae). They adjust their individual metabolic heat production and density of clustering so that the brood temperature remains remarkably constant, at about 34°C, even as the outside air temperature drops below freezing (**FIGURE 29.11**).

Behavior is a common thermoregulatory adaptation in ectotherms and endotherms

The lizard in a metabolism chamber (see Figure 29.5) can't do much to regulate its body temperature, yet in its natural desert environment it does remarkably well. When it is active during the day, it maintains its body temperature in the 35°C–40°C range. It achieves this by the thermoregulatory behavior described in **FIGURE 29.12**. In the morning, the lizard maximizes its surface exposure to the sun and absorbs solar radiation. Through the middle of the day, it maintains its body temperature by moving between sun and shade and by climbing rocks and brush to be in convective air streams. Only when the lizard retires to its underground nightly refuge does its body temperature equilibrate with the ambient.

Behavioral thermoregulation is not limited to ectotherms; it is a preferred thermoregulatory strategy of endotherms as well. The elephant sprays itself with water or dust, the lion seeks midday shade, and you would rather put on a sweater than shiver. All of these behavioral strategies alter components of the heat budget equation.

Do You Understand Concept 29.4?

- Compare the two foxes in Figure 29.9 and explain how the differences you observe influence each element of the heat balance equation for these two species.

- Many newborn mammals have brown fat even if brown fat is not seen in adults of that species. Why is brown fat adaptive for newborns?

- For both hot and cold stress, explain how behavioral adaptations can alter heat exchange through all of the components of the heat budget equation.

- If sweating and panting are such costly avenues of cooling (due to water loss and metabolic energy expenditure), why haven't evolutionary mechanisms eliminated them as regulatory mechanisms?

There are many adaptations for altering heat exchange between animals and their environments. Regulation of body temperature, however, requires that these various adaptations be controlled in a way that results in the ability of the animal to regulate its body temperature. We next explore how mammals achieve this feat of homeostatic regulation.

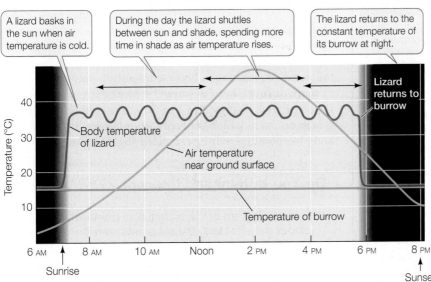

| A lizard basks in the sun when air temperature is cold. | During the day the lizard shuttles between sun and shade, spending more time in shade as air temperature rises. | The lizard returns to the constant temperature of its burrow at night. |

FIGURE 29.12 Ectotherms Can Use Behavior to Regulate Body Temperature
The body temperature of a lizard depends on the heat of its environment, but the lizard can regulate its temperature by moving from place to place within that environment. The behavior of a desert species, the spiny lizard *Sceloporus serrifer*, is illustrated here.

concept 29.5 A Thermostat in the Brain Regulates Mammalian Body Temperature

Hormonal and neural mechanisms are responsible for controlling thermoregulatory adaptations, whether these changes involve the involuntary constriction or dilation of blood vessels, changes in shivering or non-shivering thermogenesis, sweating or panting, or changes in behavior. What sets the level of the neural and hormonal regulatory commands to the effector mechanisms? To achieve a constant, optimal body temperature, a regulatory system must be able to receive information about actual body temperature, compare it to a set point, and issue appropriate commands to effector mechanisms. In the case of thermoregulation, we can call such a regulatory system a *thermostat*. We will focus here on the thermostat of mammals.

The mammalian thermostat uses feedback information

In vertebrates, the major integrative center of the thermostat is at the bottom of the brain, in a structure called the **hypothalamus** (see Figure 30.7A). The hypothalamus is a key part of many homeostatic regulatory systems we will discuss in the following chapters.

In many species, the temperature of the hypothalamus itself is the major source of feedback information to the thermostat. Cooling the hypothalamus causes fish and reptiles to seek a warmer environment, and warming the hypothalamus causes them to seek a cooler environment. In mammals, cooling the hypothalamus stimulates constriction of the blood vessels supplying the skin and increases metabolic heat production. Because it activates these thermoregulatory responses, cooling the hypothalamus causes the body temperature to rise. Conversely, mild warming of the hypothalamus stimulates dilation of the blood vessels supplying the skin, while stronger hypothalamic heating stimulates sweating or panting. Consequently, heating the hypothalamus can actually cause the overall body temperature to fall (**FIGURE 29.13**).

The hypothalamus generates a set point like a setting on a home thermostat. When the temperature of the hypothalamus exceeds or drops below that set point, thermoregulatory responses are activated to reverse the direction of temperature change. Hence, hypothalamic temperature is a negative feedback signal.

The mammalian thermoregulatory system has adjustable set points and integrates sources of information in addition to hypothalamic temperature. For example, temperature sensors in the skin register environmental temperature; change in skin temperature is feedforward information that shifts the hypothalamic set point for thermoregulatory responses. The set point for metabolic heat production is higher when skin is cold and lower when skin is warm.

INVESTIGATION

FIGURE 29.13 The Hypothalamus Regulates Body Temperature
A mammal's hypothalamus was subjected directly to temperature manipulation. The body's responses to the manipulations were as expected if the hypothalamus is the mammalian "thermostat."

HYPOTHESIS

Heating or cooling the mammalian hypothalamus results in corresponding and predictable changes in body temperature.

METHOD

1. Implant a probe into the hypothalamus of a living ground squirrel's brain. Use the probe to heat or cool the hypothalamus directly (i.e., without affecting the ambient temperature).
2. Manipulate the hypothalamic temperature T_H.

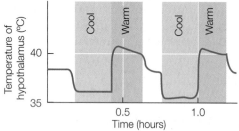

3. Measure the animal's metabolic rate and body temperature throughout the period of hypothalamic manipulation.

1 When the hypothalamus was cooled, metabolic heat production increased and the animal's body temperature rose.

RESULTS

2 When the hypothalamus was heated, the squirrel's metabolic rate and body temperature fell.

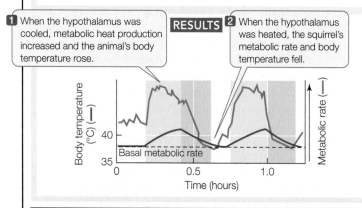

CONCLUSION

The ground squirrel's hypothalamus acts as a thermostat. When cooled it activates metabolic heat production; when warmed, it suppresses metabolic heat production and favors heat loss.

ANALYZE THE DATA

The data below were recorded from an experiment in which the hypothalamus of a ground squirrel was randomly cooled (T_H) while the animal's metabolic rate (MR) was measured. T_H is given in °C, MR is given in calories per gram of body mass per minute.

T_H	MR	T_H	MR	T_H	MR
39.5	0.040	36.5	0.038	37.5	0.041
39.0	0.041	36.0	0.040	37.0	0.039
38.5	0.040	35.5	0.060	34.5	0.110
38.0	0.038	35.0	0.080	34.0	0.140

A. Plot the data and describe what it tells you about the properties of the hypothalamic thermostat.

B. What is the threshold temperature for the metabolic heat production response?

For more, go to Working with Data 29.1 at **yourBioPortal.com**.

Go to **yourBioPortal.com** for original citations, discussions, and relevant links for all INVESTIGATION figures.

Other factors can shift hypothalamic set points for thermoregulatory responses. Set points are higher during wakefulness than during sleep, and they are higher during the active part of the daily cycle than during the inactive part, even if the animal is awake at both times.

yourBioPortal.com
Go to ANIMATED TUTORIAL 29.1
The Hypothalamus

The thermostat can be adjusted up and down

Fever is an adaptive response that helps the body fight pathogens. A *fever* is a regulated rise in body temperature caused by a rise in the hypothalamic set point for the metabolic heat production response. As a result, you shiver and crawl under a blanket, and your body temperature rises until it matches a new and higher set point. At the higher body temperature you no longer feel cold, and you may not feel warm, but someone touching your forehead will say that you are "burning up." Taking aspirin lowers your set point to normal.

FRONTIERS Although a fever that raises body temperature above 42°C can cause brain damage and even death, medical evidence also points to the possibility that fever helps fight against bacterial and viral infections. Many pathogenic microbes grow best at normal body temperature, and their growth is inhibited at higher temperatures. In addition, the production of certain white blood cells that fight microbes goes up with a rise in body temperature. Fever also decreases the levels of iron and zinc in the blood, both of which are necessary for the proliferation of the infecting organisms.

The regulated body temperature can also be lowered. This happens every day as the body temperature cycles from a high during the active phase of the daily rhythm to a low during the rest phase of the daily cycle. There is both a daily rhythm of the set point for body temperature and a further lowering of regulated body temperature during sleep. In some temperate-climate animals, lowering body temperature during a portion of the day has evolved into an important energy saving adaptation called *daily torpor*. Body temperature may decline 15°C or more during a bout of torpor. A more extreme adaptation for conserving energy and surviving the cold winter season is *hibernation*, in which bouts of torpor last multiple days and body temperature may be regulated slightly above freezing.

Do You Understand Concept 29.5?

- Draw a diagram of the mammalian thermostat that includes the effectors, the set points, and the negative feedback.

- Explain at least four sources of feedforward information for the mammalian thermostat that can either keep body temperature constant in spite of changes in the environment, or can cause body temperature to change.

What can ground squirrels do to lower the metabolic demands of surviving through the winter?

ANSWER   Being mammals, ground squirrels have high basal metabolic rates, and in a cold winter environment (below their lower critical temperatures), the metabolic cost of maintaining a normal mammalian body temperature is much higher. Neither the enormous fat reserves of Belding's ground squirrel nor the food stores of the golden-mantled ground squirrel can fuel this metabolic demand for an average 5 months of underground, snow-shrouded existence.

One possible physical-chemical basis for an adaptation was discussed in Concept 29.3. Lowering the temperature of biological systems lowers their metabolic demand, with a Q_{10} of about 2.5. Because there is a large range of temperatures between normal mammalian body temperature and the lower limit for cell survival, the range for lowering the temperature of mammalian metabolism extends over more than 30°C. Lowering body temperature to the lowest possible level would result in a greater than 16-fold decrease in metabolic demand. Ground squirrels take advantage of these metabolic savings to stretch their fat or food reserves; they achieve this by hibernating.

Hibernation is a regulated lowering of body temperature over long periods when food is not available (**FIGURE 29.14A**). Typically a ground squirrel will retire to its burrow when snow falls, and for the rest of the winter it will undergo repeated bouts of hibernation during which its body temperature will drop to a level close to that of the environment. A bout of hibernation will last about a week, and then the squirrel will return to a normal mammalian body temperature for a day or less before entering the next bout (**FIGURE 29.14B**). The periodic arousals are the major energy expenditure of these animals over the winter, yet we still do not know their function.

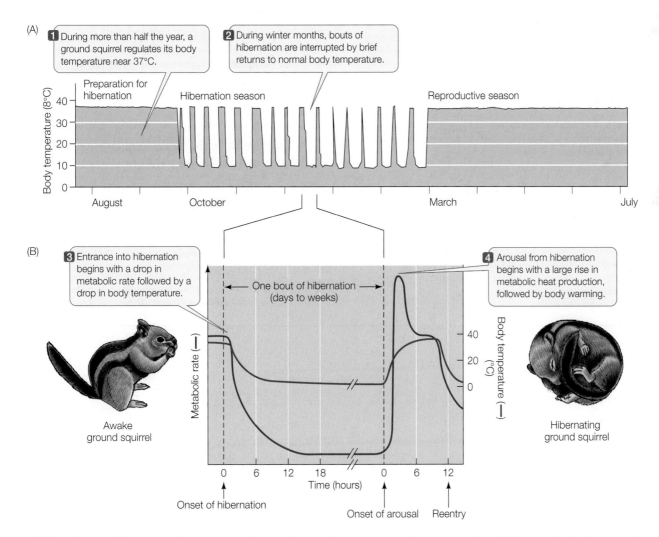

FIGURE 29.14   Hibernation Patterns in a Ground Squirrel
(A) During most of the year, the ground squirrel regulates its body temperature around 37°C, as other mammals do. But when snow falls and temperatures plunge, these animals retire to underground burrows and spend the winter living off stored fuel, in the form of either food or fat. (B) The metabolic demands for these stored fuels are decreased by repeated bouts of torpor, during which the regulated body temperature drops close to that of the environment for long periods of time.

SUMMARY

concept 29.1 Multicellular Animals Require a Stable Internal Environment

- Multicellular animals provide for the needs of all of their cells by maintaining a stable internal environment, which consists of the extracellular fluid. **Review Figure 29.1**

- **Tissues** are assemblages of cells. Although there are many cell types, there are only four types of tissues: epithelial, muscle, connective, and nervous. **Review WEB ACTIVITY 29.1**

- **Organs** are made up of tissues; most organs contain all four tissue types. Organs are grouped into **organ systems**. **Review Figure 29.2**

concept 29.2 Physiological Regulation Achieves Homeostasis of the Internal Environment

- Physiological systems are regulated primarily through **negative feedback**, which keeps the system stable by returning it toward a **set point**.

- **Positive feedback** increases the deviation from a set point. **Feedforward information** changes a system's set point. **Review Figure 29.3**

concept 29.3 Living Systems Are Temperature-Sensitive

- Life is sustained within a narrow range of ambient (environmental) temperatures. Q_{10} is a measure of the sensitivity of a life process to temperature. **Review Figure 29.4**

- An animal's **metabolic rate (MR)** is a measure of the energy turnover of its cells. The MR is often measured in terms of the rate of O_2 consumption.

- The body temperature of an ectotherm depends on heat from the environment.

- **Endotherms** regulate body temperature independent of the environment by metabolically producing heat internally.

- The minimal metabolic rate at which an endotherm can maintain homeostasis of its internal environment in a resting state is its **basal metabolic rate (BMR)**. **Review Figure 29.5 and WEB ACTIVITY 29.2**

concept 29.4 Animals Control Body Temperature by Altering Rates of Heat Gain and Loss

- A **heat budget** describes all pathways for heat exchange between an organism and its environment. These avenues of heat exchange are **metabolism, radiation, convection, conduction**, and **evaporation**. **Review Figure 29.6**

- The BMRs of endothermic animals are higher than the resting metabolic rates of ectotherms when at the same body temperature.

- The basal metabolic rate is a function of body size. **Review Figure 29.8**

- Surface insulation decreases heat loss.

- Adaptations such as **countercurrent heat exchange** can conserve metabolic heat. **Review Figure 29.10**

- Behavioral thermoregulation is the first line of defense against changes in the thermal environment. **Review Figure 29.11**

- Evaporative water loss is an effective but expensive means of heat loss.

concept 29.5 A Thermostat in the Brain Regulates Mammalian Body Temperature

- In mammals, the control of thermoregulatory effectors relies on commands from a regulatory center in the **hypothalamus** of the brain.

- The hypothalamus uses its own temperature as a negative feedback signal and external (skin) temperature as a feedforward signal. **Review Figure 29.13, WORKING WITH DATA 29.1, and ANIMATED TUTORIAL 29.1**

- Fever is a regulated increase in body temperature. Hibernation is a regulated decrease. **Review Figure 29.14**

Animal Hormones

In humans, genes on the Y chromosome are essential for development of the male body plan, including the formation of the testes—the primary male sex organs. Early in a male's development, the testes start to produce the *steroid hormone* testosterone. Testosterone and other masculinizing steroid hormones are known as *androgens* (Greek, "male-makers").

Many characteristics of the male phenotype are the direct result of testosterone and other androgens. The changes in these characteristics at puberty are due to a large increase in testosterone production. The hormone enters cells, binds to receptors, and alters gene expression throughout the body, promoting the mature male phenotype: deep voice, facial and body hair, and increased muscle and bone mass.

The bodies of both sexes respond to testosterone, but women normally have lower concentrations of this hormone than men do. The extreme transformative powers of testosterone can be seen in female bodybuilders who take high doses of synthetic androgens, a class of drugs known as anabolic steroids. These women develop male muscle patterns, a deep voice, and body and facial hair. Because male sex steroids inhibit the hormones controlling female reproductive physiology, the women's breast tissue diminishes, menstruation stops, and they become infertile. Analogous negative feedback in males who take steroids also causes infertility. Other dangerous side effects for both men and women are increased risks of cancer and of heart, liver, and kidney diseases.

The transformative power of testosterone is also seen in certain *intersex* conditions in which individuals' physical phenotypes are discordant with their chromosomal sex. In androgen insensitivity

syndrome (AIS), for example, a person's cells do not respond to testosterone. An individual with AIS who is a genetic male (XY) will have internal testes that produce testosterone, but that person develops the external sex organs and secondary sexual characteristics of a female: breasts, widening of hips, and subcutaneous fat deposition.

A rare, inherited intersex condition is "guevedoces," or "testes-at-twelve," first studied in a small community in the Dominican Republic. Children with this particular genetic mutation all appear to be female until puberty. But at puberty, those with the XY genotype suddenly develop the external sex organs of males, and they function as males. Both of these intersex conditions offer insights into the mechanisms of action of *hormones*.

QUESTION How do the transformative effects of testosterone exemplify the way many hormones work?

Synthetic androgens known as anabolic steroids mimic the male hormone testostorone. This action can result in enhanced muscle development in both males and females.

KEY CONCEPTS

30.1 Hormones Are Chemical Messengers

30.2 Hormones Act by Binding to Receptors

30.3 The Pituitary Gland Links the Nervous and Endocrine Systems

30.4 Hormones Regulate Mammalian Physiological Systems

concept
30.1 Hormones Are Chemical Messengers

In any multicellular body, the individual cells must be able to communicate with one another. Most such intercellular communication is by means of chemical signals that are released from one cell and travel to another (the *target cell*). The signal molecules bind to receptors on the target cell, triggering a response in it; these responses can result in profound effects on the organism's physiology, anatomy, and behavior. Although details vary, this general pattern of chemical signaling, which we described in Chapter 5 of this book, is seen in physiological systems across all multicellular phyla. In this chapter we will focus on the *endocrine system*, which consists of those cells that produce and release the chemical signals we call hormones.

Endocrine signals can act locally or at a distance

Cells can secrete substances either into the extracellular fluid (*endocrine secretion*) or into a duct or an internal body cavity that communicates to the external world (*exocrine secretion*). For example, endocrine cells in the digestive tract secrete signaling molecules that diffuse into the blood and are transported to other organs of the digestive system to control their activities. Exocrine cells in the digestive tract secrete enzymes into the digestive tract that break down the complex molecules of food. Other exocrine secretions include saliva, sweat, and pheromones (chemical signals that an animal releases into the environment and that communicate information to other individuals of the same species). You can think of "endocrine" as referring to "into the body" and "exocrine" as referring to "out of the body" or at least into a body cavity lined by epithelial cells.

Cells that secrete endocrine signals are called **endocrine cells**. Some endocrine cells exist as single cells scattered within a larger organ, as is the case with the digestive tract endocrine cells mentioned above. In other cases, endocrine cells are aggregated into secretory organs called **endocrine glands**. The testes and thyroid are examples of endocrine glands.

As noted in Chapter 5 (see Figure 5.10), endocrine signaling molecules have traditionally been classified as paracrine signals, autocrine signals, or hormones, according to the distance over which they operate. While paracrine signals act on nearby cells and autocrine signals stimulate the same cell that produces them, **hormones** are "long-distance" endocrine signals that are released into the bloodstream and circulate through the entire body.

Whether the endocrine signal is a hormone, paracrine, or autocrine, the cell that responds to it is the target cell. **Target cells** have receptor proteins that recognize and bind to a signal (even when that signal is present only at extremely low concentrations, as is usually the case with hormones) and can trigger appropriate responses. The same hormone can influence the activities of a variety of different target cells, all of which are distant from the cells that originally released it.

The dividing lines between hormones, paracrine signals, and autocrine signals have blurred somewhat as researchers have discovered endocrine signals that operate over a variety of distances. In this chapter we will focus on the "classic" hormones that are released into the bloodstream and circulate widely.

Hormones can be divided into three chemical groups

There is enormous diversity in the chemical structure of hormones, but most hormones can be classified into three groups:

- **Peptide and protein hormones** are relatively large molecules (**FIGURE 30.1A**) that are water-soluble and therefore easily transported in the blood. They are packaged in vesicles within the cells that synthesize them and are released by exocytosis. Since water-soluble molecules cannot cross cell membranes easily, the receptors for peptide hormones must be on the exterior of the target cell. Most hormones are peptides: there are dozens, perhaps hundreds, of such hormones, with diverse structures and effects.

- **Steroid hormones** are synthesized from cholesterol molecules and share a similar structure of four interlinked rings (**FIGURE 30.1B**). Steroids are lipid-soluble and readily pass through the membranes of the cells that synthesize them, but they do not dissolve well in blood plasma and thus are usually bound to carrier proteins for transport in the blood. Once they reach a target cell they easily pass through its membrane; thus most steroid receptors are found inside target cells.

- **Amine hormones** are small molecules synthesized from single amino acids. For example, *thyroxine* and *epinephrine* are made from tyrosine (**FIGURE 30.1C**), and melatonin is made from tryptophan. Depending on whether the modified amino acid is polar or nonpolar, an amine hormone may be water-soluble (can't cross membranes) or lipid-soluble (can cross membranes).

Hormonal communication has a long evolutionary history

Intercellular chemical signaling was critical for the evolution of multicellularity. The most primitive of the multicellular animals—the sponges—do not have nervous systems, but they do have intercellular chemical communication. Plants and fungi also use a wide variety of hormones.

> **LINK** The study of plant growth and development has contributed greatly to our knowledge of hormones and how they work; see Chapter 26

Studying the evolution of hormonal signaling reveals an interesting generalization: the signal molecules themselves are highly conserved over evolutionary time, although their functions may change. For example, the hormone prolactin was first shown to be a signal stimulating milk production in the mammary glands of female mammals. Prolactin (but not milk

(A) Protein hormones

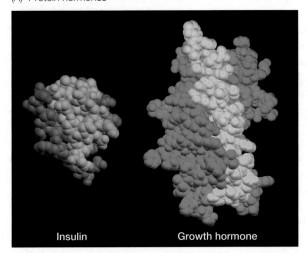

Insulin   Growth hormone

FIGURE 30.1 Three Classes of Hormones (A) The largest hormone molecules are peptides or proteins. This class includes insulin and growth hormone. (B) Steroid hormones are modified from cholesterol molecules. They include the corticosteroids produced by the adrenal gland and the sex steroids produced primarily by the gonads. (C) Amine hormones are small molecules synthesized from a single amino acid. Thyroxine and epinephrine are both made from tyrosine, but thyroxine is lipid-soluble and epinephrine is water-soluble; their modes of release, dynamics of transport, and locations of receptor binding differ accordingly.

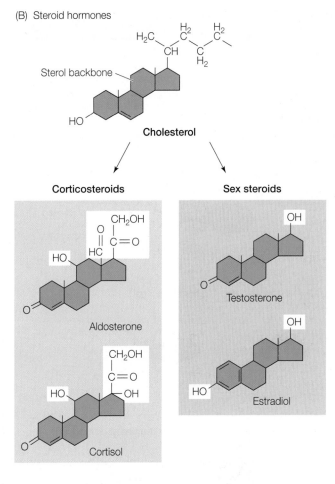

or mammary glands) has since been found in all the vertebrate phyla. In salmon, prolactin is involved in triggering the animals' migrations from salt water back to their freshwater spawning grounds. In birds, prolactin can stimulate nest building and parental care. Many of prolactin's functions relate to reproduction, but significant differences in the specifics of this hormone's action have evolved in the different vertebrate groups.

The ubiquity and shared evolutionary history of the hormones means we can study their actions in organisms from across the tree of life. An example is the well-documented hormonal control of molting and metamorphosis in arthropods.

Insect studies reveal hormonal control of development

The many arthropod classes, including the insects, are all characterized by having a rigid exoskeleton. Their growth is episodic, punctuated with *molts* (shedding of the exoskeleton). The hormonal mechanisms that control insect molting and metamorphosis are a longstanding area of research.

CONTROL OF MOLTING A classic series of experiments on the bloodsucking bug *Rhodnius prolixus* demonstrated that molting in this insect is controlled by hormones. A juvenile *Rhodnius* bug molts five times before developing into a mature adult; the growth stages between the molts are called *instars*. Each episode of molting is triggered by a blood meal.

Rhodnius is an amazingly hardy experimental animal—it can survive for more than a week after researchers cut off its head. A bug decapitated within an hour after a blood meal never

molts; however, if it is decapitated a week after its blood meal, it *does* molt. The British physiologist Sir Vincent Wigglesworth hypothesized that this time lag meant that some substance that triggers molting diffuses slowly from the head. He tested his hypothesis with the experiment described in **FIGURE 30.2**.

We now know that two diffusible hormones working in sequence regulate molting in arthropods: *prothoracicotropic hormone* (PTTH, a peptide) and *ecdysone* (a steroid related to the vertebrate hormone testosterone). Cells in the brain produce PTTH, which is why it is sometimes called "brain hormone." PTTH is transported to and stored in paired structures—the *corpora cardiaca*—attached to the brain. After appropriate

INVESTIGATION

FIGURE 30.2 A Diffusible Substance Triggers Molting
The bloodsucking bug *Rhodnius prolixus* develops from hatchling to adult in a series of five molts (instars) that are triggered by ingesting blood. Sir Vincent Wigglesworth's experiments demonstrated that a blood meal stimulates production of some molt-inducing substance in the insect's head.

HYPOTHESIS

The substance that controls molting in *R. prolixus* is produced in the head segment and diffuses slowly through the body.

OBSERVATION

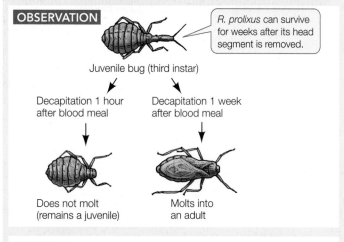

R. prolixus can survive for weeks after its head segment is removed.

Juvenile bug (third instar)

Decapitation 1 hour after blood meal

Decapitation 1 week after blood meal

Does not molt (remains a juvenile)

Molts into an adult

METHOD

1. Decapitate third-instar juveniles at different times after blood meal.

1 hour after blood meal   1 week after blood meal

2. Join bugs with glass tube

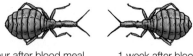

Tubing allows body fluids to pass from one bug to another

RESULTS

Both bugs molt into adults

CONCLUSION

A blood meal stimulates production of some substance within the insect's head that then diffuses slowly through the body, triggering a molt.

Go to **yourBioPortal.com** for original citations, discussions, and relevant links for all INVESTIGATION figures.

stimulation (which for *Rhodnius* bugs is a blood meal), PTTH is released and diffuses through the extracellular fluid to an endocrine gland, the prothoracic gland. PTTH stimulates the prothoracic gland to release ecdysone, which in turn diffuses to target tissues and stimulates molting.

The control of molting by PTTH and ecdysone exemplifies not only Wigglesworth's "diffusible substance," but also how the endocrine system works with the nervous system to integrate diverse information and induce long-term effects. The nervous system of an arthropod receives many types of information about the animal's environment (such as day length, temperature, social cues, and nutrition) that help determine the optimal timing for the stages of growth and development. When conditions are right, the brain (nervous system) signals the prothoracic gland (endocrine system), which produces ecdysone, which in turn orchestrates the processes of development and molting. Later in this chapter we will see similar links between the nervous system and endocrine glands in vertebrates.

JUVENILE HORMONE AND METAMORPHOSIS PTTH and ecdysone determine whether an insect will molt. But what determines whether the molting insect becomes just a slightly larger juvenile or a dramatically different adult? A third hormone, **juvenile hormone (JH)**, determines the final outcome of the molt. In immature insects, JH is produced continuously from the *corpora allata* (structures attached to the corpora cardiaca). Further experiments with *Rhodnius* demonstrated that if JH is present in high concentration during a molt, the molting bug becomes a slightly larger juvenile; if JH levels are low, the bug becomes an adult.

Some insects, such as butterflies, undergo a particularly dramatic developmental change called *complete metamorphosis*. A fertilized egg hatches into a *larva*, which feeds and molts several times, becoming bigger each time. After a fixed number of molts, it enters an inactive stage called *pupation*. The pupa undergoes major body reorganization and finally emerges as an adult. Studies in silkworm moths have shown that larvae will molt into larger larvae as long as juvenile hormone is present in high concentrations. It is only when the level of JH falls that the larvae spin cocoons, pupate, and metamorphose into adults (**FIGURE 30.3**).

yourBioPortal.com
Go to ANIMATED TUTORIAL 30.1
Complete Metamorphosis

Do You Understand Concept 30.1?

- How do hormone and paracrine signals differ?

- Compare the different methods by which peptide hormones and steroid hormones are released, travel through the blood, and reach their receptors. What is the underlying chemical reason for these differences?

- Suppose a newly hatched butterfly larva (i.e., a first-instar caterpillar) has a mutation such that its corpora allata cannot produce any hormones. Will this caterpillar ever molt? If it does, what will happen?

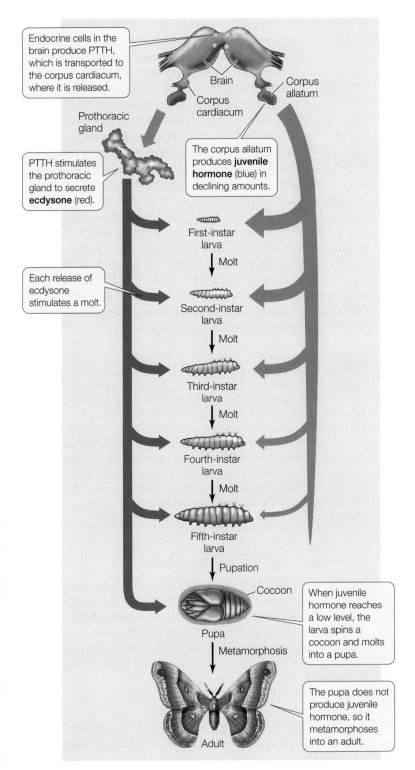

Endocrine cells in the brain produce PTTH, which is transported to the corpus cardiacum, where it is released.

Brain

Corpus cardiacum

Corpus allatum

Prothoracic gland

PTTH stimulates the prothoracic gland to secrete **ecdysone** (red).

The corpus allatum produces **juvenile hormone** (blue) in declining amounts.

First-instar larva

Molt

Each release of ecdysone stimulates a molt.

Second-instar larva

Molt

Third-instar larva

Molt

Fourth-instar larva

Molt

Fifth-instar larva

Pupation

Cocoon

When juvenile hormone reaches a low level, the larva spins a cocoon and molts into a pupa.

Pupa

Metamorphosis

The pupa does not produce juvenile hormone, so it metamorphoses into an adult.

Adult

FIGURE 30.3 Hormonal Control of Metamorphosis Three hormones control molting and metamorphosis in the silkworm moth *Hyalophora cecropia*.

Dozens of hormone molecules are circulating in your blood right now, conveying messages to virtually every organ in your body. But each hormone molecule is only a message. To have a physiological effect, the messages must be "read" and acted upon, which involves receptor molecules.

Hormones Act by Binding to Receptors

A hormone can affect only those cells with the appropriate receptor proteins that bind to the specific hormone and trigger specific responses in the target cell. A hormone released into the blood circulates throughout the body. If and when it contacts a cell that has the right receptor, the hormone binds (usually temporarily) to that receptor; the receptor then does the rest. If the cell does not have a receptor protein for a given hormone, that hormone will have no effect on that cell.

LINK Review the mechanisms of receptor protein actions in Concepts 5.5 and 5.6

Hormone receptors can be membrane-bound or intracellular

Peptide and some amine hormones, being water-soluble, cannot pass readily through plasma membranes, so they must bind to receptors on the surfaces of target cells. Peptide receptors are large transmembrane complexes with three domains: a *binding domain* that projects outside the plasma membrane; a *transmembrane domain* that anchors the receptor in the membrane; and a *cytoplasmic domain* that extends into the cytoplasm of the cell. When a hormone binds, it changes the conformation of the receptor, and the cytoplasmic domain initiates the target cell's response, often through a second messenger cascade that eventually activates protein kinases or protein phosphatases (see Figure 5.17). In most cases, these kinases and phosphatases activate or inactivate enzymes in the cytoplasm, which leads to the cell's response.

Steroid and some amine hormones are lipid-soluble and thus can diffuse through plasma membranes; their receptors are usually in the target cell's cytoplasm. When the steroid binds to its receptor in the cytoplasm, the hormone–receptor complex usually moves to the target nucleus, where it alters gene expression. For example, when testosterone binds to androgen receptors in the cytoplasm of skeletal muscle cells, testosterone–receptor complexes move into the nuclei and activate several genes, including genes for the proteins that form contractile filaments (see Concept 36.1). Thus, one effect of testosterone is an increase in muscle mass.

Hormone action depends on the nature of the target cell and its receptors

The same hormone can cause very different responses in different types of cells, depending on which receptors are present in those cells and which cellular responses are triggered by the receptors. For example, the amine hormone epinephrine is involved in the rapid physiological response to sudden danger or highly stressful situations known as the **fight-or-flight response** (FIGURE 30.4). Suppose you are hiking and almost

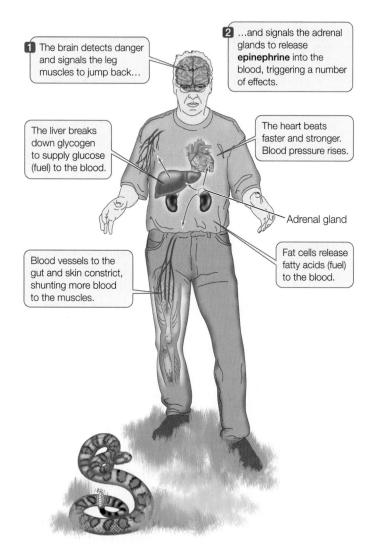

1 The brain detects danger and signals the leg muscles to jump back...

2 ...and signals the adrenal glands to release **epinephrine** into the blood, triggering a number of effects.

The liver breaks down glycogen to supply glucose (fuel) to the blood.

The heart beats faster and stronger. Blood pressure rises.

Adrenal gland

Blood vessels to the gut and skin constrict, shunting more blood to the muscles.

Fat cells release fatty acids (fuel) to the blood.

FIGURE 30.4 The Fight-or-Flight Response When a person is suddenly faced with a threatening situation, the brain sends a signal to the adrenal glands, which almost instantaneously release epinephrine. Epinephrine circulating through the body induces the different components of the fight-or-flight response.

step on a rattlesnake. You jump back, and your heart starts to thump. Both these actions are driven by your nervous system, which reacts very quickly. Simultaneously, your nervous system stimulates endocrine cells in the *adrenal glands* located just above your kidneys (see Figures 30.6 and 30.12) to secrete **epinephrine** and a related hormone, **norepinephrine**. Within seconds, these hormones are circulating through your body to activate the many components of the fight-or-flight response.

Epinephrine and norepinephrine (sometimes known as *adrenaline* and *noradrenaline*) bind to a set of **adrenergic receptors**. We know of at least five types of these receptors, all of which are G protein-linked receptors (see Concept 5.5). They are grouped into two basic categories—α-adrenergic and β-adrenergic—with different affinities for epinephrine and norepinephrine (**FIGURE 30.5**). Stimulation of one type of β-adrenergic receptor in heart cells causes a faster and stronger heartbeat (thus pumping more blood). Stimulation

of α-adrenergic receptors causes blood vessels in the skin to constrict, causing pallor and cold hands and feet. These same receptors in the digestive system shut off the secretion of digestive enzymes and decrease blood flow through your gut (digesting lunch can wait). Meanwhile, β receptors cause arterioles in skeletal muscle to dilate (enlarge), so increased blood flow is diverted from your gut to your skeletal muscles (which you need to run). In the liver β receptors cause the breakdown of glycogen into glucose for a quick energy supply (see Chapter 39), while in fatty tissue β receptors stimulate the breakdown of fats to yield fatty acids—another source of energy.

These diverse actions are triggered by the same two hormones. The variation in cellular responses arises because, first, cells of different tissues can have different receptors; and second, even when different types of cells have the same receptor, they may respond very differently to activation of that receptor. Pharmacologists have exploited this diversity in the adrenergic system by developing drugs called "beta blockers" that selectively inhibit the activation of β-adrenergic receptors, thereby taking the edge off anxiety reactions.

FRONTIERS The genes for hormone receptors frequently have similar sequences and can be grouped into gene families. Armed with this information, investigators have scanned the human genome sequence looking for homologies to known receptor gene sequences. This method has resulted in the discovery of many "orphan" hormone receptors—receptors that do not appear to bind any known hormone. The existence of orphan receptors raises the possibility that there may be many as-yet undiscovered human hormones, and that determining the effects of these hormones could have profound implications for health.

Hormone receptors are regulated

Like the hormones themselves, hormone receptors are subject to feedback control. In some cases, continuous high concentrations of a hormone can cause a decrease in the number of its receptors in the target cell's cytoplasm or nucleus, a process known as *downregulation*. Downregulation makes the cells less sensitive to the hormone, in essence causing the cells to ignore the excess hormone. An example of downregulation occurs in type II diabetes mellitus, in which chronically high levels of the hormone insulin (usually caused by excessive carbohydrate intake) downregulate production of insulin receptors throughout the body. The result is that cells become less sensitive to insulin.

LINK The crucial role of insulin in regulating glucose metabolism is detailed in Concept 39.4

Conversely, *upregulation* of receptors can occur if hormone secretion is chronically low. Cells often respond to low signal levels by making more receptors for that hormone. An example of upregulation is seen in people who have been on a regular dose of the beta blocker medications mentioned in the previous section. If the activity of β-adrenergic receptors is blocked

β-Adrenergic receptors

α-Adrenergic receptors

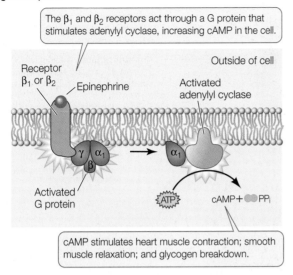

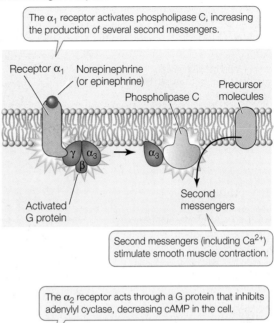

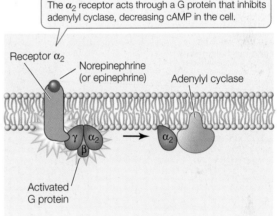

FIGURE 30.5 Adrenergic Receptors Receptors that respond to epinephrine and norepinephrine are G protein-linked (see Figure 5.14) and fall into two broad categories. Epinephrine acts on both α- and β-adrenergic receptors, whereas norepinephrine acts mostly on α-adrenergic receptors. The effect produced when the hormone binds depends on the type of cell.

for some length of time, more receptors are produced. If the person then goes off the medication suddenly, the effects of the receptors are amplified, resulting in heightened anxiety and blood pressure. This explains why changes in dosage for medications that cause up- or downregulation of receptors are usually gradual and carefully supervised.

Do You Understand Concept 30.2?

- Growth hormone is a large, water-soluble peptide. Given these characteristics, how would you expect this hormone to act on its target cells?

- What do you think might be the cause of androgen insensitivity syndrome, one of the intersex conditions described on page 603?

- When you are suddenly alarmed, your adrenal glands release epinephrine, norepinephrine, and the steroid hormone cortisol. The effects of epinephrine and norepinephrine are seen within seconds and decrease within minutes; cortisol's effects are not seen for many minutes and last for hours. Explain some reasons for these differences in time course of action.

Hundreds of animal and plant hormones are known, and more are being discovered each year. The rest of this chapter will focus primarily on the human endocrine system (**FIGURE 30.6**), but the same hormonal systems are acting in other species. We look first at the hormones involved in coordinating the endocrine system with the nervous system.

concept
30.3

The Pituitary Gland Links the Nervous and Endocrine Systems

The nervous and endocrine systems represent two means of communication using molecular signals. Whereas the nervous system communicates via molecules called *neurotransmitters* that are released at the junction between nerve cells (see Concept 34.3), the endocrine system communicates by releasing molecular signals into the blood, which carries them to target cells throughout the body. These two methods of communication complement each other, with nerve cells often mediating rapid, highly specific responses and hormones mediating longer-term, more broadly distributed responses.

In addition to complementing each other, the nervous and endocrine systems also interact. The nervous system directly or indirectly controls the activity of many endocrine glands—for example, the production of sex steroids by the gonads. Some neurons even secrete hormones directly; hormones secreted by neurons are known as *neurohormones*. Conversely, hormones

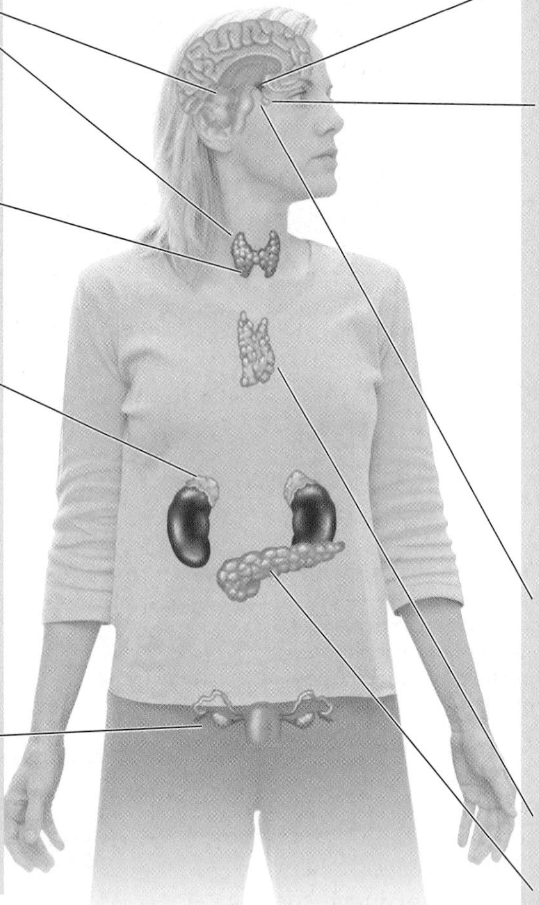

Pineal gland
Melatonin: regulates daily rhythms

Thyroid gland (see Figure 30.11)
Thyroxine (T_3 and T_4): increases cell metabolism; essential for growth and neural development
Calcitonin: lowers blood calcium levels, stimulates incorporation of calcium into bone

Parathyroid glands (on posterior surface of thyroid; see Figure 30.11)
Parathyroid hormone (PTH): stimulates release of calcium from bone and absorption of calcium by gut and kidney

Adrenal gland (see Figure 30.12)
Cortex
Cortisol: mediates long-term metabolic responses to stress
Aldosterone: involved in salt and water balance
Sex steroids

Medulla
Epinephrine (adrenaline) and *norepinephrine* (noradrenaline): stimulate immediate fight-or-flight reactions

Gonads (see Chapter 32)
Testes (male)
Androgens (esp. testosterone): development and maintenance of male sexual characteristics
Ovaries (female)
Estrogens: development and maintenance of female sexual characteristics
Progesterone: supports pregnancy

Hypothalamus (see Figure 30.7)
Releasing and release-inhibiting neuro-hormones control the anterior pituitary; *ADH* and *oxytocin* are transported to and released from the posterior pituitary

Anterior pituitary (see Figure 30.8)
Thyroid-stimulating hormone (TSH): activates the thyroid gland; also called *thyrotropin*
Follicle-stimulating hormone (FSH): in females, stimulates maturation of ovarian follicles; in males, stimulates spermatogenesis
Luteinizing hormone (LH): in females, triggers ovulation and ovarian production of estrogens and progesterone; in males, stimulates production of testosterone
Adrenocorticotropic hormone (ACTH): stimulates growth and secretory activity of the adrenal cortex; also called corticotropin
Growth hormone (GH): stimulates protein synthesis and growth
Prolactin: stimulates milk production
Melanocyte-stimulating hormone (MSH): controls skin pigmentation
Endorphins and *enkephalins*: pain control

Posterior pituitary (see Figure 30.7)
Receives and releases two hypothalamic neurohormones:
Oxytocin: stimulates contraction of uterus, flow of milk, interindividual bonding
Antidiuretic hormone (ADH; also known as vasopressin): promotes water conservation by kidneys

Thymus (diminishes in adults)
Thymosin: activates immune system T cells

Pancreas (islets of Langerhans)
Insulin: stimulates cells to take up and use glucose
Glucagon: stimulates liver to release glucose
Somatostatin: slows release of insulin and glucagon and digestive tract functions

Other organs include cells that produce and secrete hormones:

Organ	Hormone
Adipose tissue	Leptin
Heart	Atrial natriuretic peptide
Kidney	Erythropoietin
Stomach	Gastrin
Intestine	Secretin, cholecystokinin
Skin	Vitamin D (calciferol)
Liver	Somatomedins, insulin-like growth factors

FIGURE 30.6 The Human Endocrine System Cells that produce and secrete hormones may be organized into discrete endocrine glands, or they may be embedded in the tissues of other organs, such as the digestive tract or kidneys. Although the hypothalamus is part of the brain, it includes cells that secrete neurohormones into the extracellular fluid.

yourBioPortal.com
Go to WEB ACTIVITY 30.1
The Human Endocrine Glands

can influence the activity of the nervous system, as when exposure of the brain to sex steroid hormones stimulates the development and expression of sexual behavior.

The key meeting point for these two communication networks is the **pituitary**, an endocrine gland that serves as an interface between the nervous and endocrine systems and is involved in the hormonal control of many physiological processes.

yourBioPortal.com
Go to ANIMATED TUTORIAL 30.2
The Hypothalamus–Pituitary–Endocrine Axis

The pituitary gland has two parts

The pituitary sits in a depression at the bottom of the skull, just over the back of the roof of the mouth (**FIGURE 30.7A**). It is attached by a stalk to a region of the brain called the **hypothalamus**, which is itself involved in many physiological regulatory systems. The pituitary has two parts with different developmental origins: the **anterior pituitary** (derived from gut epithelium) and the **posterior pituitary** (derived from neural tissue).

THE POSTERIOR PITUITARY Long processes of hypothalamic neurons extend into the posterior pituitary. The ends (terminals)

(A)

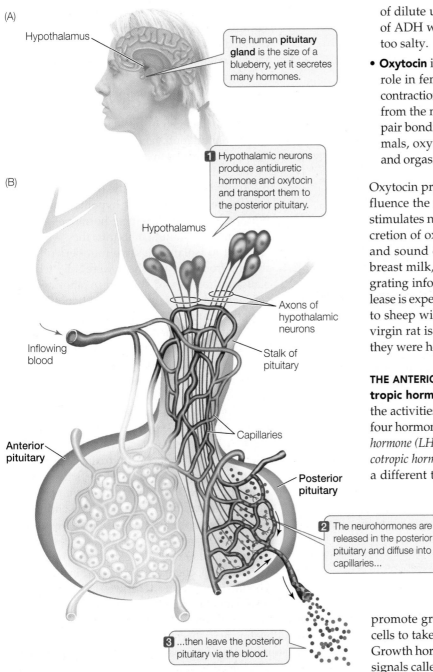

Hypothalamus

The human **pituitary gland** is the size of a blueberry, yet it secretes many hormones.

(B)

1 Hypothalamic neurons produce antidiuretic hormone and oxytocin and transport them to the posterior pituitary.

Hypothalamus

Axons of hypothalamic neurons

Inflowing blood

Stalk of pituitary

Anterior pituitary

Capillaries

Posterior pituitary

2 The neurohormones are released in the posterior pituitary and diffuse into capillaries...

3 ...then leave the posterior pituitary via the blood.

FIGURE 30.7 The Posterior Pituitary Neurons in the hypothalamus produce two peptide neurohormones, which are stored and released by the posterior pituitary.

of those axons release two neurohormones produced by the hypothalamic neurons, antidiuretic hormone (also called *vasopressin*) and oxytocin (**FIGURE 30.7B**).

- The main action of **antidiuretic hormone (ADH)** in mammals and birds is to increase the amount of water conserved by the kidneys. When ADH secretion is high, the kidneys produce only a small volume of highly concentrated urine. When ADH secretion is low, the kidneys produce a large volume

of dilute urine. The posterior pituitary increases its release of ADH when blood pressure falls or if the blood becomes too salty.

- **Oxytocin** is produced in both sexes but is best known for its role in female reproduction, in which it stimulates uterine contractions during birth and also stimulates the flow of milk from the mother's breasts. In both sexes, oxytocin promotes pair bonding and trust. In humans and in some other mammals, oxytocin secretion rises with intimate sexual contact and orgasm and has thus been called the "cuddle hormone."

Oxytocin provides an excellent example of how hormones influence the nervous system. After birth, the baby's suckling stimulates neurons in the mother's brain that cause further secretion of oxytocin and release of breast milk. Even the sight and sound of a baby can cause a nursing mother to release breast milk, an example of the sensory nervous system integrating information to affect hormone release. If oxytocin release is experimentally blocked, mammalian mothers from rats to sheep will reject their newborn offspring; conversely, if a virgin rat is given oxytocin, she will adopt strange pups as if they were her own.

THE ANTERIOR PITUITARY The anterior pituitary releases four **tropic hormones**, which are hormones that direct and control the activities of other endocrine glands (**FIGURE 30.8**). These four hormones are *thyroid-stimulating hormone (TSH)*, *luteinizing hormone (LH)*, *follicle-stimulating hormone (FSH)*, and *adrenocorticotropic hormone (ACTH)*. Each tropic hormone is produced by a different type of pituitary cell. We will say more about the tropic hormones later in this chapter when we describe their target glands (thyroid, testes, ovaries, and adrenal cortex, respectively).

The anterior pituitary also produces several other peptide hormones, including prolactin (see pp. 604–605) and growth hormone. **Growth hormone (GH;** see Figure 30.1A) acts on a wide variety of tissues to promote growth. One of its important effects is to stimulate cells to take up the amino acids needed for protein synthesis. Growth hormone also stimulates the liver to produce chemical signals called *somatomedins*, or *insulin-like growth factors (IGFs)*, which stimulate the growth of bone and cartilage. Overproduction of growth hormone causes *gigantism*, in which affected individuals may grow to nearly 8 feet tall. Underproduction causes *pituitary dwarfism*, in which individuals fail to reach normal adult height.

The neurohormones of the anterior pituitary are produced in minute amounts

Whereas the posterior pituitary secretes neurohormones produced by the hypothalamus, the anterior pituitary makes and secretes its own hormones—but its production of hormones is under the *control* of hypothalamic neurohormones. The hypothalamus receives information about conditions in the body

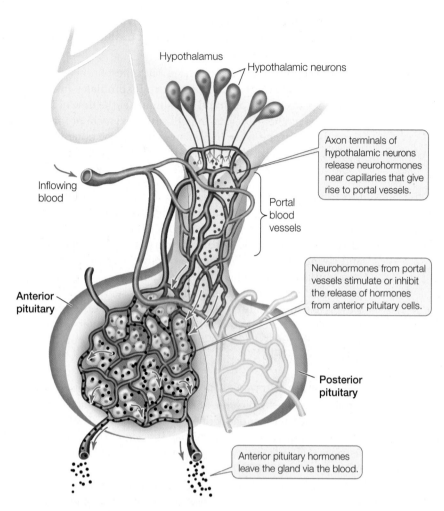

Hypothalamus

Hypothalamic neurons

Inflowing blood

Axon terminals of hypothalamic neurons release neurohormones near capillaries that give rise to portal vessels.

Portal blood vessels

Neurohormones from portal vessels stimulate or inhibit the release of hormones from anterior pituitary cells.

Anterior pituitary

Posterior pituitary

Anterior pituitary hormones leave the gland via the blood.

FIGURE 30.8 The Anterior Pituitary Cells of the anterior pituitary produce four tropic hormones that control other endocrine glands, as well as several other peptide hormones. These cells are controlled by neurohormones produced in the hypothalamus and delivered through portal blood vessels that run between the hypothalamus and the anterior pituitary through the pituitary stalk.

and the external environment, and it communicates that information to the anterior pituitary by secreting neurohormones.

FRONTIERS In the 1960s, some children with severe growth hormone deficiencies were treated with GH extracted from the pituitaries of cadavers. The treatments were successful in stimulating growth, but the availability of human pituitaries was a limiting factor. In 1981, the biotechnology pioneer Genentech used recombinant DNA technology to produce bacteria that synthesized human growth hormone. Today GH treatment for pituitary dwarfism is common, but its wider availability has engendered controversy. Is it valid, for example, to administer the hormone to children merely because they are shorter than average? The use of GH as a performance-enhancing drug is also under scrutiny.

The hypothalamic neurohormones that control the anterior pituitary are carrried in a special type of blood vessel, the *portal blood vessels*. Portal blood vessels begin in capillaries in the hypothalamus and then run down the pituitary stalk, ending in capillaries within the anterior pituitary (see Figure 30.8). Because the portal vessels deliver molecules directly to the anterior pituitary and not into the general circulation, neurohormones can be produced and secreted in extremely small quantities.

In the 1960s, two teams of scientists, one led by Roger Guillemin and the other by Andrew Schally, began to search for hypothalamic neurohormones, but because these hormones are produced in such minute amounts, they were very difficult to isolate. Massive numbers of hypothalami from pigs and sheep were collected from slaughterhouses and shipped to laboratories in refrigerated trucks. In one extraction effort, the hypothalami from 270,000 sheep yielded only 1 milligram of purified **thyrotropin-releasing hormone** (**TRH**), the first *releasing hormone* (that is, release-stimulating hormone) to be isolated and characterized. TRH turned out to be a small peptide of only three amino acids that, when secreted by the hypothalamus, causes certain anterior pituitary cells to release **thyroid-stimulating hormone** (**TSH**). TSH then stimulates the thyroid gland to releases a third hormone, thyroxine (see p. 614).

Frequently in science, giant steps forward are made possible by the invention of a new technique. Research on the hypothalamic hormones was largely dependent on Rosalyn Yalow's invention of the *immunoassay*, a technique that allowed accurate measurement of extremely low concentrations of biochemical substances, including antibodies (see Chapter 31) as well as hormones and other signaling molecules. In 1977, Yalow shared the Nobel Prize in Medicine with Guillemin and Schally. Many other hypothalamic neurohormones, including both releasing and release-inhibiting hormones, are now known. In most cases they are involved in stimulating or inhibiting signaling cascades like that for TRH → TSH → thyroxine.

Negative feedback loops also regulate the pituitary hormones

As well as being controlled by hypothalamic hormones, the endocrine cells of the anterior pituitary are under negative feedback control by the hormones of the target glands they stimulate (**FIGURE 30.9**). As an example, consider the cortisol cascade, in which corticotropin-releasing hormone (CRH) from the hypothalamus stimulates release of adrenocorticotropic hormone (ACTH) from the anterior pituitary, which in turn stimulates release of the hormone cortisol from the adrenal glands. Once cortisol is released, it circulates throughout the body—including back to the hypothalamus and pituitary

APPLY THE CONCEPT

The pituitary gland links the nervous and endocrine systems

Once released from the adrenal gland into the bloodstream, cortisol diffuses from blood plasma into saliva, enabling the collection of measureable amounts of hormone from saliva samples. The table gives cortisol concentrations in saliva samples taken from two female Asian elephants over a 6-day period during which the animals were introduced to each other at a Berlin zoo. Given that (a) baseline salivary cortisol in elephants is 4–5 nanomoles per liter and (b) elephant 1 received an immunization shot on day 1, graph the data in the table and answer the following questions.

1. Do the data indicate that the elephants experienced an adrenal stress response on meeting each other?

2. What physiological effects would elevated cortisol levels cause in the elephants?

3. Do these data indicate communication between the nervous and endocrine systems? Explain your answer.

	CORTISOL CONCENTRATION (NMOL/L)	
DAY	ELEPHANT 1	ELEPHANT 2
1	10	4
2	6	5
3*	21	4
4	33	22
5	5	4
6	4	5

*Elephants were introduced to each other on day 3. The cortisol sample from elephant 1 was taken after the introduction; that from elephant 2 was taken before the introduction.

gland, where it directly inhibits further release of both CRH and ACTH, thereby helping to maintain a constant level of cortisol secretion.

Do You Understand Concept 30.3?

- What are the developmental, anatomical, and functional relationships between the brain and the pituitary gland?

- Why are so many anterior pituitary hormones called tropic hormones?

- The hypothalamus and the pituitary gland both have cortisol receptors. If they lose their cortisol receptors, these glands will immediately begin secreting unusually large amounts of CRH and ACTH. Why?

The hypothalamus and pituitary control the secretion of hormones from several other endocrine glands. What hormones do these other glands release, and what are their functions?

concept 30.4 Hormones Regulate Mammalian Physiological Systems

Hormones are involved in regulating most mammalian physiological systems. Some regulatory systems are controlled by pituitary hormones, and all are controlled by feedback information from the physiological functions they regulate. We will examine a few examples of important mammalian hormones, but you should be aware that subsequent chapters will describe many additional examples of regulatory hormones.

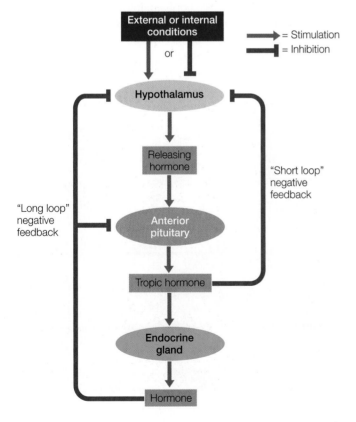

FIGURE 30.9 Multiple Feedback Loops Control Hormone Secretion Multiple negative feedback loops regulate the chain of command from hypothalamus to anterior pituitary to endocrine glands.

Thyroxine helps regulate metabolic rate and body temperature

The **thyroid gland** wraps around the front of the windpipe (*trachea*) and expands into a lobe on either side (see Figure 30.6). Two different cell types in the thyroid gland produce two different hormones, thyroxine and *calcitonin*. In addition, embedded in or near the thyroid gland (depending on the species) are the **parathyroid glands**, which produce *parathyroid hormone*. Calcitonin and parathyroid hormone are involved in calcium regulation and will be discussed in the next section. Here we focus on thyroxine.

Thyroxine is an amine hormone that is synthesized from the amino acid tyrosine. A single thyroxine molecule consists of two tyrosine residues with four iodine atoms covalently attached (two to each tyrosine residue). Thyroxine is also known as T_4, the subscript signifying that it carries the full complement of four iodine atoms:

The thyroid also releases a nearly identical hormone, T_3, that has only three iodine atoms:

The thyroid usually releases about four times as much T_4 as T_3; however, T_3 is more active in the cells of the body. Circulating T_4 can be converted to T_3 by an enzyme within target cells. Each target cell can set its own sensitivity to thyroid hormones by controlling the conversion of T_4 to the more active T_3. Although we use the general term thyroxine in this discussion, keep in mind that most of the hormonal activity is due to T_3.

ACTION AND CONTROL OF THYROXINE In birds and mammals, thyroxine raises metabolic rate. Thyroxine is lipid-soluble and binds to an intracellular receptor that stimulates transcription of numerous genes. These different genes encode proteins involved in energy pathways as well as transport and structural proteins—all of which leads to a general increase in cell metabolism. Thyroxine is especially crucial during development and growth, as it promotes amino acid uptake and protein synthesis. Insufficient thyroxine in a human fetus or growing child greatly retards physical and mental development, resulting in a condition known as cretinism.

As described in Concept 30.3, thyroxine levels are controlled by the hypothalamus and the anterior pituitary. The hypothalamus releases thyrotropin-releasing hormone (TRH), which stimulates the anterior pituitary to release thyroid-stimulating hormone (TSH). TSH stimulates the thyroid gland to produce and release thyroxine. This sequence of steps is regulated by a negative feedback loop like the one described earlier for cortisol (see Figure 30.9).

GOITERS A *goiter* is an enlarged thyroid gland (**FIGURE 30.10**) that can be associated with either *hyperthyroidism* (excess production of thyroxine) or *hypothyroidism* (thyroxine deficiency). The negative feedback loop whereby thyroxine controls TSH release explains how two very different conditions can result in the same symptom.

The most common cause of hyperthyroid goiter is an autoimmune condition involving an antibody that binds to and activates TSH receptors, causing uncontrolled production and release of thyroxine. Blood levels of TSH are low because of negative feedback from high levels of thyroxine, but the thyroid gland itself remains maximally stimulated and grows bigger. People with hyperthyroidism have a high metabolic rate, tend to be nervous and "jumpy," usually feel hot, and may develop a buildup of fat behind the eyeballs that causes the eyes to bulge.

Hypothyroid goiter results when there is not enough circulating thyroxine to turn off TSH production. The most common cause of this condition is too little iodine in the diet; without dietary iodine, the thyroid gland cannot make functional thyroxine. Without sufficient thyroxine, TSH levels remain high, stimulating the thyroid to become bigger. The symptoms of hypothyroidism are low metabolism, intolerance of cold, and general physical and mental sluggishness. The occurrence of hypothyroid goiter has been greatly reduced by the widespread use of iodized table salt.

FIGURE 30.10 Goiter Iodine deficiency can result in hypothyroid goiter. In this condition, a lack of functional thyroxine results in enlargement of the thyroid gland.

Hormonal regulation of calcium concentration is vital

Calcium ions (Ca^{2+}) must be maintained at a precise concentration in the blood plasma because Ca^{2+} flow across cell membranes is the trigger for several critical physiological processes. Neurons use calcium to trigger neurotransmitter release, and muscles use calcium to trigger muscle contraction. Shifts in blood calcium concentration above or below a certain narrow range can cause serious problems. Concentrations that are too low can cause muscle spasms and even seizures; excessively high levels of Ca^{2+} can cause the entire nervous system to become depressed, and muscles—including the heart—to weaken or even to stop completely.

Almost 99 percent of the calcium in the human body is located in our bones. About 1 percent is found in cells, and only about 0.1 percent Ca^{2+} is in the extracellular fluid. It is essential that this tiny pool of circulating Ca^{2+} be maintained at precise concentrations. Because the circulating Ca^{2+} pool can be affected greatly by relatively small shifts in the much larger pools in the cells and bones, and because maintaining blood calcium concentration within narrow limits is vital, the body has multiple mechanisms for adjusting and regulating Ca^{2+} levels. These mechanism include:

- Deposition or resorption of bone
- Excretion or retention of calcium by the kidneys
- Absorption of calcium from the digestive tract

These mechanisms are controlled by the hormones **calcitonin, calcitriol** (a hormone synthesized from vitamin D), and **parathyroid hormone** (**PTH**).

CALCITONIN   Calcitonin is released by the thyroid gland and lowers the concentration of calcium in the blood, mainly by regulating bone turnover (**FIGURE 30.11**). Bone is continuously remodeled through a dynamic process that involves both resorption of old bone and synthesis of new bone (see Concept 36.3). Cells called *osteoclasts* break down bone and release calcium into the blood, whereas cells called *osteoblasts* take up calcium from the blood and deposit it in new bone. Calcitonin decreases the activity of osteoclasts and thereby favors removal of calcium from the blood and its deposition in bone by osteoblasts. The turnover of bone in adult humans is not high, so calcitonin probably does not play a major role in calcium homeostasis in adults. It is likely to be considerably more important in young individuals whose bones are actively growing.

VITAMIN D AND CALCITRIOL   As we will discuss in Chapter 39, a *vitamin* is a substance that the body requires in small quantities but cannot synthesize and must therefore obtain from the diet. By this definition, vitamin D (also known as *calciferol*) is not a vitamin, because the body can and does synthesize it—assuming the skin gets adequate exposure to sunlight. The ultraviolet wavelengths in sunlight convert cholesterol present in the skin cells into vitamin D. If sun exposure is inadequate (as in high-latitude regions during winter months), vitamin D must be obtained from the diet or from supplements. Once present in the body, vitamin D is converted by the liver and kidneys into calcitriol, a hormone whose actions include stimulating the cells of the digestive tract to absorb calcium from ingested food.

---

yourBioPortal.com

Go to ANIMATED TUTORIAL 30.3
Hormonal Regulation of Calcium

PARATHYROID HORMONE   The single most important hormone in the regulation of blood calcium levels is parathyroid hormone (PTH), whose synthesis is triggered when blood calcium levels fall below a set point. PTH increases the concentration of calcium in the blood by three different mechanisms:

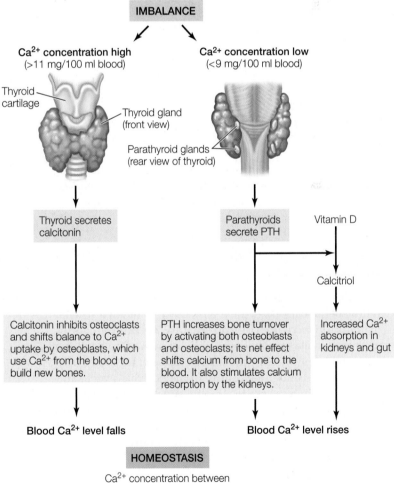

FIGURE 30.11 Hormonal Regulation of Calcium   Calcitonin, parathyroid hormone (PTH), and vitamin D regulate calcium levels in the blood.

APPLY THE CONCEPT

Hormones regulate mammalian physiological systems

The time courses of action for different hormones vary widely. Some hormones are released rapidly, establish their effects almost immediately, and then are cleared from the bloodstream within minutes. Others are released slowly and remain in the blood for many hours or even days.

One way of characterizing the time course of a hormone is to measure its *half-life* in the blood: the length of time it takes for the blood level of a given hormone to fall from its maximum level following its release (or injection) halfway back to baseline values. The table gives blood concentrations of thyroxine (T_4) following a 600-µg injection. Plot these data.

1. Before assessing your graph, consider what you know about the functioning of T_4. Would you expect this hormone to have a short or a long half-life? Why?

2. Use the data in the table to estimate the actual half-life of 600 µg of T_4 in the bloodstream. (Time 0 is the baseline.)

TIME (HRS)	CONCENTRATION (µg/DECILITER)
0	7.5
6	13.7
12	12.3
24	11.1
36	10.7
48	10.3
60	9.9
72	9.5
84	9.3
96	9.1

3. If you were trying to correct a hormonal deficit by administering hormone therapy and needed to maintain a relatively constant hormone level in the blood (i.e., minimizing peaks and troughs), how would your dosing differ for hormones with different half-lives?

- It stimulates the bone-turnover activities of osteoclasts and osteoblasts.

- It stimulates the kidneys to reabsorb calcium.

- It increases the body's absorption of calcium from food by activating the synthesis of calcitriol from vitamin D.

Circulating calcium activates receptors in the plasma membrane of the parathyroid cells, inhibiting the synthesis and release of PTH. A fall in blood calcium removes this inhibition so that PTH can be synthesized and released.

The two segments of the adrenal gland coordinate the stress response

The **adrenal glands** sit above the kidneys, just below the middle of the back (see Figure 30.6). Each of the two adrenal glands is a gland within a gland (**FIGURE 30.12**). The core, or **adrenal medulla**, produces the neurohormones epinephrine and norepinephrine. The medulla's release of these two neurohormones is under the control of the nervous system and is very rapid in response to stressful situations, arousing the body to action as detailed in Concept 30.2. Epinephrine and (to a lesser degree) norepinephrine increase heart rate and blood pressure and divert blood flow to active muscles and away from the gut and skin. They also cause fat and liver cells to release metabolic fuels (see Figure 30.4).

Surrounding the adrenal medulla is the **adrenal cortex**. Under the control of the hypothalamus and anterior pituitary, the adrenal cortex produces a few sex steroids and, importantly, two classes of *corticosteroid hormones*:

- The **mineralocorticoids** influence salt and water balance of the extracellular fluid. The mineralocorticoid *aldosterone*

stimulates the kidneys to conserve sodium and excrete potassium (see Concept 40.5).

- The **glucocorticoids** increase blood glucose concentrations and affect many other aspects of fat, protein, and carbohydrate metabolism. The primary glucocorticoid in most vertebrates is *corticosterone*; in humans and a few other mammals it is **cortisol**.

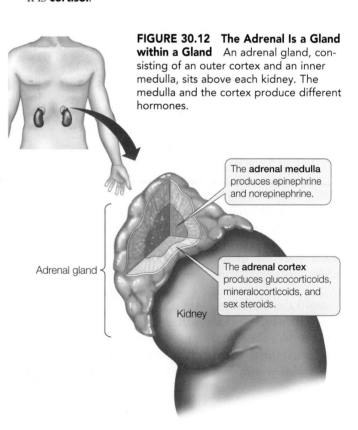

FIGURE 30.12 The Adrenal Is a Gland within a Gland An adrenal gland, consisting of an outer cortex and an inner medulla, sits above each kidney. The medulla and the cortex produce different hormones.

The **adrenal medulla** produces epinephrine and norepinephrine.

The **adrenal cortex** produces glucocorticoids, mineralocorticoids, and sex steroids.

Adrenal gland

Kidney

Corticosterone and cortisol are critical for mediating long-term (hours to days) metabolic responses to stress. Within minutes of a stressful stimulus, blood cortisol levels begin to rise. Cells not critical for fight-or-flight responses are stimulated by cortisol to decrease their use of blood glucose and to shift instead to using fats and proteins for energy. This is no time to feel sick, have allergic reactions, or heal wounds, so cortisol also blocks immune system reactions—which is why cortisol and drugs that mimic its action are useful for reducing inflammation and allergic responses. If glucocorticoids remain elevated for too long, however, they increase the risk of several stress-related conditions, including cardiovascular disorders, gastrointestinal problems, infertility, and lowered immune responses.

Sex steroids from the gonads control reproduction

The gonads—the testes of the male and the ovaries of the female—produce hormones as well as the gametes (sperm and ova). The male steroids are collectively called **androgens**, and the dominant one in humans is *testosterone*. The female sex steroids are **estrogens** and **progesterone**. It is important to note that both sexes make and use androgens and estrogens; however, the relative levels vary in the two sexes.

In embryonic life, the sex steroids determine whether a human embryo develops into a phenotypic female or male. At puberty, they trigger maturation of the reproductive organs and the development of *secondary sexual characteristics*, such as breasts and facial hair. They continue to play essential roles in reproductive cycles and sexual functioning throughout adulthood, as described in Chapter 32.

SEXUAL DIFFERENTIATION IN EMBRYOS If a Y chromosome is present, the undifferentiated embryonic gonads begin producing testosterone in the seventh week of development, along with a peptide hormone called *MIS (Müllerian-inhibiting substance)*. Testosterone causes testes and the male reproductive ducts to develop, and MIS induces apoptosis of the cells making up the female reproductive (Müllerian) ducts, so they disappear. If no androgens or MIS are produced, female reproductive structures develop (**FIGURE 30.13**). In other words, androgens are required to trigger male development in humans, and in their absence the fetus develops as a female.

PUBERTY Sex steroids are produced at extremely low levels by the juvenile gonads, but their production increases rapidly at **puberty**—a time of sexual maturation and dramatic physical transformation that occurs around the age of 12 or 13. What triggers this sudden increase?

The production of sex steroids by the gonads is controlled by the anterior pituitary hormones **luteinizing hormone** (**LH**) and **follicle-stimulating hormone** (**FSH**), which together are called the **gonadotropins**. Gonadotropin production is in turn under the control of the hypothalamic neurohormone **gonadotropin-releasing hormone** (**GnRH**). Before puberty, the hypothalamus produces only very low levels of GnRH. Puberty is initiated by a reduction in the sensitivity of hypothalamic

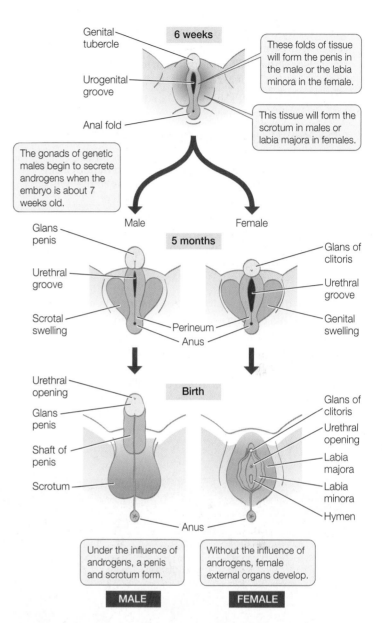

FIGURE 30.13 Sex Steroids Direct the Development of Human Sex Organs The sex organs of early human embryos are undifferentiated. Androgens promote the development of male sex organs. In the absence of androgens, female sex organs form.

GnRH-producing cells to negative feedback from the sex steroids and gonadotropins. This reduced sensitivity overcomes the negative feedback, so GnRH release increases, stimulating increased production of gonadotropins and hence increased production of sex steroids.

In females, increased levels of LH stimulate the ovaries to increase estrogen production. Increased circulating levels of estrogen initiate development of the traits of a sexually mature woman: enlarged breasts, vagina, and uterus; broadened hips; increased subcutaneous fat; pubic hair; and the menstrual cycle. In males, increasing LH stimulates cells in the testes to produce testosterone, which initiates the physiological, anatomical, and psychological changes associated with male adolescence. The voice deepens, hair begins to grow on the face and body, muscle mass increases, and the testes and penis grow larger. At this time the anterior pituitary also increases FSH production; in females this stimulates the maturation of ovarian follicles, and in males it stimulates the production of sperm.

Do You Understand Concept 30.4?

- Explain how both hyperthyroidism and hypothyroidism can cause goiter. Refer to the roles of the hypothalamus and the pituitary in your answer.

- How does the hypothalamus affect the endocrine activity of the gonads and the adrenal cortex? What is similar about the way the gonads and the adrenal cortex are regulated?

We have discussed just a few of the glands and their hormones in this chapter—enough to give an overview of how hormones act and the types of physiological functions they regulate. Many more hormones exist, and new hormones are discovered almost every year. As we discuss the organ systems of the body in the chapters that follow, we will see many other examples of hormones regulating other physiological processes.

QUESTION How do the transformative effects of testosterone exemplify the way many hormones work?

ANSWER The effects of testosterone start when this lipid-soluble hormone binds to its receptor in the target cell cytoplasm (Concept 30.1). The hormone–receptor complex then moves into the target cell nucleus, where it stimulates the expression of different genes in different cells; its actions differ depending on the type of target cell it binds to (Concept 30.2). In muscle cells, for example, testosterone induces expression of the genes for the contractile proteins, so muscle mass increases. In hair cells, the genes for the protein keratin are activated.

Androgen insensitivity syndrome (AIS) demonstrates the essential role of hormone receptors, since its cause is not a lack of testosterone but a lack of functional testosterone receptors. Without these receptors, testosterone cannot stimulate development of the male reproductive ducts, male genitalia, or male secondary sexual characteristics. However, the peptide hormone MIS is produced normally and inhibits the development of internal female reproductive ducts. The result is a genetically XY individual whose body appears female on the outside but lacks internal female sex organs (**FIGURE 30.14**).

The rare condition "testes-at-twelve" exemplifies another aspect of hormonal control. It turns out that, like the T_3 and T_4 forms of thyroid hormone (Concept 30.4), testosterone can be modified into an even more potent form known as dihydrotestosterone, or DHT. Guevedoces lack the gene for the enzyme that converts testosterone to DHT, and DHT is required to masculinize the external genitalia during early development. If DHT is not present, the penis and scrotum do not form in the infant, and the young child is perceived to be female. When much higher levels of testosterone are triggered by puberty, however, testosterone stimulates formation of the penis and scrotum, and thus at puberty XY individuals are revealed to be male.

FIGURE 30.14 Real People, Real Lives This photo shows a group of women with AIS and similar intersex conditions at a support group meeting.

30 SUMMARY

concept 30.1 Hormones Are Chemical Messengers

- **Endocrine cells** secrete chemical signals that induce responses in other cells—called **target cells**—that have receptors for those molecules. In some cases endocrine cells are aggregated into **endocrine glands**. See WEB ACTIVITY 30.1

- **Hormones** are endocrine signals that are secreted from a cell, circulate in the blood, and bind to target cells distant from the secreting cell.

- Most hormones are either **peptide hormones**, **steroid hormones**, or **amine hormones**. Peptides and some amines are water-soluble; steroids and some amines are lipid-soluble. **Review Figure 30.1**

- Two hormones, PTTH and ecdysone, control molting in arthropods. A third hormone, **juvenile hormone (JH)**, prevents maturation. When an insect stops producing juvenile hormone, it molts into an adult. **Review Figure 30.3 and ANIMATED TUTORIAL 30.1**

concept 30.2 Hormones Act by Binding to Receptors

- Receptors for water-soluble hormones are located on the cell surface. Receptors for most lipid-soluble hormones are inside the cell.

- Hormones cause different responses in different target cells. **Review Figure 30.5**

- The sensitivity of a cell to hormones can be altered by down-regulation or upregulation of the receptors in that cell.

concept 30.3 The Pituitary Gland Links the Nervous and Endocrine Systems

- The **pituitary gland** is the interface between the nervous and endocrine systems.

- The posterior pituitary secretes two peptide neurohormones: **antidiuretic hormone (ADH)** and **oxytocin**. Review Figure 30.7

- The anterior pituitary is controlled by neurohormones produced by cells in the **hypothalamus**. The anterior pituitary secretes four **tropic hormones** as well as **growth hormone**, prolactin, and a few other hormones. **Review Figure 30.8 and ANIMATED TUTORIAL 30.2**

- Hormone release is controlled by negative feedback loops. **Review Figure 30.9**

concept 30.4 Hormones Regulate Mammalian Physiological Systems

- The **thyroid gland** is controlled by **thyroid-stimulating hormone (TSH)** and secretes **thyroxine**, which controls cell metabolism.

- **Calcitonin** from the thyroid lowers blood calcium by promoting bone deposition. **Parathyroid hormone (PTH)** raises blood calcium by promoting bone turnover and decreasing calcium excretion. **Calcitriol**, synthesized from vitamin D, promotes calcium absorption from the digestive tract. **Review Figure 30.11 and ANIMATED TUTORIAL 30.3**

- The **adrenal gland** has two parts. The inner **adrenal medulla** releases **epinephrine** and **norepinephrine** in a rapid response to stress. The **adrenal cortex** produces steroids, including sex steroids, **glucocorticoids**, and **mineralocorticoids**. **Review Figure 30.12**

- Sex steroids (**androgens** in males, **estrogens** and **progesterone** in females) are produced by the gonads in response to tropic hormones. Sex steroids control sexual development, secondary sexual characteristics, and reproductive functions. **Review Figure 30.13**

See WEB ACTIVITY 30.2 for a concept review of this chapter

Immunology: Animal Defense Systems

As a rule, however, there was no ostensible cause; but people in good health were all of a sudden attacked by violent heats in the head, and redness and inflammation in the eyes, the inward parts, such as the throat or tongue, becoming bloody and emitting an unnatural and fetid breath.

The Athenian historian Thucydides wrote these words in 430 B.C.E. He was describing a rapidly spreading infectious disease, or plague, that was sweeping the city of Athens in ancient Greece, and would ultimately kill about one-third of its inhabitants.

One year earlier, in 431 B.C.E., the Spartans had surrounded Athens. Sparta was a rival state with a highly disciplined army, while Athens was near a port and had a powerful navy. The Athenian leader Pericles decided to take advantage of his naval supremacy. He instructed his soldiers to abandon the countryside to the Spartans, relying on the navy to keep the enemies out of the city and to keep the Athenians supplied with food. People living in the countryside poured into Athens.

As Athens became more crowded, deteriorating sanitation and close living conditions resulted in the city becoming an incubator for infectious diseases. In his famous *History of the Peloponnesian War*, Thucydides speculated that this disease came to Athens from Africa. His precise clinical description has provoked great debate among medical historians regarding its nature. In 1994, a burial ground was found near Athens that apparently contained the remains of several dozen people who died of the disease. Samples of dental pulp from the skeletons were examined using the polymerase chain reaction (PCR), and the results revealed the

presence of DNA sequences from the bacterium that causes typhoid fever. Based on these results, it was proposed that the disease that killed so many Athenians was typhoid fever. Others disagree, using arguments based on a clinical description of typhoid fever today.

Whatever the cause, another passage from Thucydides (who was one of the few lucky survivors of the disease) is an interesting observation about immunity—the subject of this chapter:

Yet it was with those who had recovered from the disease that the sick and the dying found most compassion. These knew what it was from experience and now had no fear for themselves; for the same man was never attacked twice, never at least fatally.

The siege of Athens by Sparta and its Peloponnesian allies made the city crowded and vulnerable to the spread of an infectious disease.

QUESTION How can a person survive an infection and be resistant to further infection?

KEY CONCEPTS

31.1 Animals Use Innate and Adaptive Mechanisms to Defend Themselves against Pathogens

31.2 Innate Defenses Are Nonspecific

31.3 The Adaptive Immune Response Is Specific

31.4 The Adaptive Humoral Immune Response Involves Specific Antibodies

31.5 The Adaptive Cellular Immune Response Involves T Cells and Their Receptors

concept 31.1 Animals Use Innate and Adaptive Mechanisms to Defend Themselves against Pathogens

Animals have a number of ways of defending themselves against **pathogens**—harmful organisms and viruses that can cause disease. There are two general types of defense mechanisms that can provide **immunity**—the ability to avoid disease when invaded by a pathogen:

- **Innate immunity** is nonspecific in that it is deployed against a wide variety of invasive organisms. Innate immunity systems include barriers such as the skin and molecules that are toxic to or destroy invaders; these provide the first line of defense for the body. A second line of innate defense includes phagocytic cells (*phagocytes*), which ingest foreign cells and other particles. These defenses may be present all the time, or they may be activated in response to an injury or invasion by a pathogen. Such inducible innate defenses are typically activated very rapidly. Most animals have innate immunity.

LINK You may review the description of phagocytosis in Concept 4.3 (see Figure 4.9)

- **Adaptive immunity** is specific. It distinguishes between substances that are made by the organism (*self*) and substances that are not part of the organism (*nonself*). Adaptive immunity involves antibody proteins and other proteins that recognize, bind to, and aid in the destruction of specific viruses and bacteria. Adaptive immunity is typically slow to develop and long lasting, and it is found only in vertebrate animals.

Mammals have both innate and adaptive immunity and are the focus of this chapter. In mammals and other vertebrates, these nonspecific and specific mechanisms operate together as a *coordinated* defense system, usually in sequence. **TABLE 31.1** gives an overview of these processes during the course of an infection. Innate immunity is the body's first line of defense; adaptive immunity often requires days or even weeks to become effective.

The major players in immunity are specific cells and proteins. These are produced in the blood and lymphoid tissues (see Concept 38.4) and are circulated throughout the body where they interact with almost all of its other tissues and organs. We will describe these cells and proteins briefly here and then in more detail later in the chapter.

White blood cells play many roles in immunity

One milliliter of human blood typically contains about *5 billion* red blood cells and *7 million* white blood cells. While the main function of red blood cells is to carry oxygen throughout the body, **white blood cells** are specialized for various functions in the immune system. Examine **FIGURE 31.1** and you will see that there are two major kinds of white blood cells (also called *leukocytes*): phagocytes and lymphocytes. **Phagocytes** (such as *macrophages*) are large cells that engulf pathogens and other substances by phagocytosis. They are involved in both innate and adaptive immunity. **Lymphocytes**, which include *B cells* and *T cells*, are involved in adaptive immunity.

> **yourBioPortal.com**
> Go to ANIMATED TUTORIAL 31.1
> Cells of the Immune System

Immune system proteins bind pathogens or signal other cells

The cells of the adaptive immune system work together, interacting with one another and with the cells of invading pathogens. These cell–cell interactions depend on a variety of key proteins, including receptors, other cell surface proteins, and signaling molecules. Four of the major players are listed here and will be discussed in more detail later:

- **Antibodies** are proteins that bind specifically to substances identified by the immune system as nonself. This binding can inactivate and destroy microorganisms and toxins, and it can act as a tag on nonself cells, making them easier for the immune system cells to attack. Antibodies are produced by B cells.

- **Major histocompatibility complex** (**MHC**) proteins are found in two classes (I and II). MHC I proteins are found on the surfaces of most cells in the mammalian body. MHC II proteins are found on most immune system cells. MHC proteins are important self-identifying labels and play a major role in coordinating interactions between lymphocytes and macrophages.

- **T cell receptors** are integral membrane proteins on the surfaces of T cells. They recognize and bind to nonself substances presented by MHC proteins on the surfaces of other cells.

TABLE 31.1	Innate and Adaptive Immune Responses to an Infection	
RESPONSE (TIME AFTER INFECTION BY A PATHOGEN)	**SYSTEM**	**MECHANISMS**
Early (0–4 hr)	Innate, nonspecific (first line)	Barrier (skin and lining of organs)
		Dryness, low pH
		Mucus
		Lysozyme, defensins
Middle (4–96 hr)	Innate, nonspecific (second line)	Inflammation
		Phagocytosis
		Natural killer cells
		Complement system
		Interferons
Late (>96 hr)	Adaptive, specific	Humoral immunity (B cells, antibodies)
		Cellular immunity (T cells)

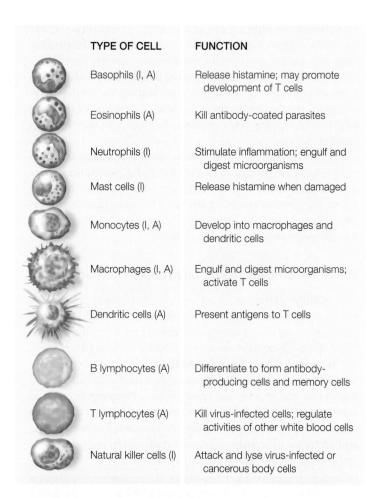

TYPE OF CELL	FUNCTION
Basophils (I, A)	Release histamine; may promote development of T cells
Eosinophils (A)	Kill antibody-coated parasites
Neutrophils (I)	Stimulate inflammation; engulf and digest microorganisms
Mast cells (I)	Release histamine when damaged
Monocytes (I, A)	Develop into macrophages and dendritic cells
Macrophages (I, A)	Engulf and digest microorganisms; activate T cells
Dendritic cells (A)	Present antigens to T cells
B lymphocytes (A)	Differentiate to form antibody-producing cells and memory cells
T lymphocytes (A)	Kill virus-infected cells; regulate activities of other white blood cells
Natural killer cells (I)	Attack and lyse virus-infected or cancerous body cells

FIGURE 31.1 White Blood Cells White blood cells have key roles in both innate (I) and adaptive (A) immunity. The lymphocytes are the B cells and T cells; the other cell types are phagocytes.

- **Cytokines** are soluble signaling proteins released by many cell types. They bind to cell surface receptors and alter the behavior of their target cells. Various cytokines activate or inactivate B cells, macrophages, and T cells.

Do You Understand Concept 31.1?

- List the differences between innate and adaptive immunity.
- Rarely, a person is born with a genetic disease and has no lymphocytes. What would be the consequences of this situation for the immune system? Explain why such a person would have to be very careful about going outdoors.

Now that we've seen a brief overview of innate and adaptive immunity, let's look at some of the innate defenses that animals have against invading organisms.

concept 31.2 Innate Defenses Are Nonspecific

Innate, nonspecific defenses are general protection mechanisms that attempt to either stop pathogens from invading the body or quickly eliminate pathogens that do manage to invade. They are genetically programmed (innate) and "ready to go," in contrast to adaptive, specific immunity, which takes time to develop after a pathogen or toxin has been recognized as nonself. In mammals, innate defenses include physical barriers as well as cellular and chemical defenses (**FIGURE 31.2**).

Barriers and local agents defend the body against invaders

The first line of innate defense is encountered by a potential pathogen as soon as it lands on the surface of the animal. Consider a pathogenic bacterium that lands on human skin. The challenges faced by the bacterium just to reach its target are formidable:

- *The physical barrier of the skin*: Bacteria rarely penetrate intact skin (which explains why broken skin increases the risk of infection).
- *The saltiness of the skin*: This condition is usually not hospitable to the growth of the bacterium.
- *The presence of normal flora*: Bacteria and fungi that normally live in great numbers on body surfaces without causing disease will compete with potential pathogens for space and nutrients.

If a pathogen lands inside the nose or another internal organ, it faces other innate defenses:

- **Mucus** is a slippery secretion produced by *mucous membranes*, which line various body cavities that are exposed to the

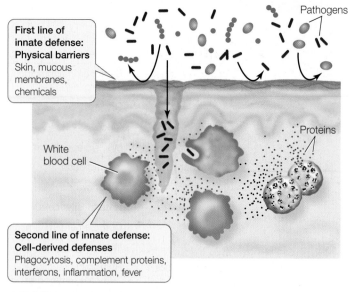

First line of innate defense: Physical barriers
Skin, mucous membranes, chemicals

Pathogens

Proteins

White blood cell

Second line of innate defense: Cell-derived defenses
Phagocytosis, complement proteins, interferons, inflammation, fever

FIGURE 31.2 Innate Immunity Physical barriers, cells, and proteins (complement and interferons) provide nonspecific defenses against invading pathogens.

external environment. Mucus traps microorganisms so they can be removed by the beating of *cilia* (see Concept 4.4, p. 70), which continuously move the mucus and its trapped debris away.

- **Lysozyme**, an enzyme made by mucous membranes, cleaves bonds in the cell walls of many bacteria, causing them to *lyse* (burst open).

- **Defensins**, also made by mucous membranes, are peptides of 18–45 amino acids. They contain hydrophobic domains and are toxic to a wide range of pathogens, including bacteria, microbial eukaryotes, and enveloped (membrane-enclosed) viruses. Defensins insert themselves into the plasma membranes of these organisms and make the membranes freely permeable to water and all solutes, thus killing the invaders:

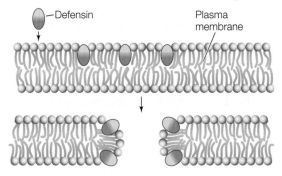

Defensins are also produced inside phagocytes, where they kill pathogens trapped by phagocytosis. Plants also produce defensins in response to pathogens (see Concept 28.1).

- Harsh conditions in the internal environment can kill pathogens. For example, the gastric juice in the stomach is a deadly environment for many bacteria because of the hydrochloric acid and proteases that are secreted into it.

Other innate defenses include specialized proteins and cellular processes

Pathogens that penetrate the body's outer and inner surfaces encounter a second line of innate defenses. These include the activation of defensive cells such as phagocytes and natural killer cells, and the secretion of various defensive proteins such as *complement* and *interferon* proteins.

PHAGOCYTES Pathogenic cells, viruses, or fragments of these invaders can be recognized by phagocytes, which then ingest them by phagocytosis:

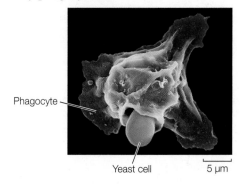

Phagocyte

Yeast cell   5 μm

Once inside the phagocyte, the invader is killed either by hydrolysis within the lysosomes (see Figure 4.9) or by defensins.

NATURAL KILLER CELLS One class of lymphocytes, known as **natural killer cells**, can distinguish between healthy body cells and those that are infected by viruses or have become cancerous. In the latter cases they initiate apoptosis (programmed cell death; see Concept 7.5) in the target cells. In addition to this nonspecific action, natural killer cells interact with the adaptive defense mechanisms by lysing antibody-labeled target cells.

COMPLEMENT PROTEINS Vertebrate blood contains more than 20 different proteins that make up the antimicrobial **complement system**. Once this system has been activated, the proteins function in a characteristic sequence, or cascade, with each protein activating the next:

1. One complement protein binds to components on the surface of an invading cell. This binding helps phagocytes recognize and destroy the invading cell.

2. Another protein activates the inflammation response (see below) and attracts phagocytes to the site of infection.

3. Finally, other proteins lyse the invading cell.

The complement system can be activated by various mechanisms, including innate (nonspecific) and adaptive (specific) immune responses. In the latter case, the complement cascade is initiated by the binding of a complement protein to an antibody bound to the surface of the invading cell.

INTERFERONS When a cell is infected by a viral pathogen, the cell produces small amounts of signaling proteins called **interferons** that help increase the resistance of neighboring cells to infection. Interferons are a class of cytokines and are found in many vertebrates. Various molecules induce the production of interferons. One such inducer is double-stranded viral RNA; thus interferons are particularly important in defense against viruses. Interferons bind to receptors on the plasma membranes of uninfected cells, stimulating a signaling pathway that inhibits viral reproduction if the cells are subsequently infected. In addition, interferons stimulate the cells to hydrolyze bacterial or viral proteins to peptides, an initial step in adaptive immunity.

Inflammation is a coordinated response to infection or injury

When a tissue is damaged because of infection or injury, the body responds with **inflammation**: redness, swelling, and heat near the damaged site, which can become painful. This response can happen almost anywhere in the body, internally as well as on the outer surface. Inflammation is an important phenomenon: it isolates the area to stop the spread of the damage; it recruits cells and molecules to the damaged location to kill any pathogens that might be present; and it promotes healing.

Among the first responders to tissue damage are **mast cells**, which adhere to the skin and the linings of organs and release numerous chemical signals, including:

- **tumor necrosis factor**, a cytokine protein that kills target cells and activates immune cells.

- **prostaglandins**, fatty acid derivatives that play roles in various responses, including the initiation of inflammation in nearby tissues.

- **histamine**, an amino acid derivative that increases the permeability of blood vessels to white blood cells and molecules so they can act in nearby tissues.

The redness and heat of inflammation result from the dilation of blood vessels in the infected or injured area (**FIGURE 31.3**). Phagocytes enter the inflamed area, where they engulf pathogens and dead tissue cells. Phagocytes and neutrophils are responsible for most of the healing associated with inflammation. They produce several cytokines, which (among other functions) can signal the brain to produce a fever. This rise in body temperature accelerates lymphocyte production and phagocytosis, thereby speeding the immune response. In some cases pathogens are temperature-sensitive, and their growth is inhibited by the fever. The pain of inflammation results from increased pressure due to swelling,

the action of leaked enzymes on nerve endings, and the action of prostaglandins, which increase the sensitivity of the nerve endings to pain. (Recall that in Chapter 3, we describe aspirin and how it alleviates pain by blocking the synthesis of prostaglandins.)

Following inflammation, pus may accumulate. *Pus* is a mixture of leaked fluid and dead cells: bacteria, white blood cells, and damaged body cells. Pus is a normal result of the body's fighting a disease and/or inflammation and is gradually removed by macrophages. If there is a wound, *platelets*—small, irregularly shaped cell fragments that are present in the blood—aggregate at the wound site. The platelets produce growth factors that stimulate nearby skin cells to divide and heal the wound.

Inflammation can cause medical problems

While inflammation is generally a good thing, sometimes the inflammatory response is inappropriately strong:

- In an **allergic** reaction, a nonself molecule that is normally harmless binds to mast cells, causing the release of histamine and subsequent inflammation (along with itchy, watery eyes, and rashes in some cases). The nonself molecule may come from food or from the environment—for example, on the

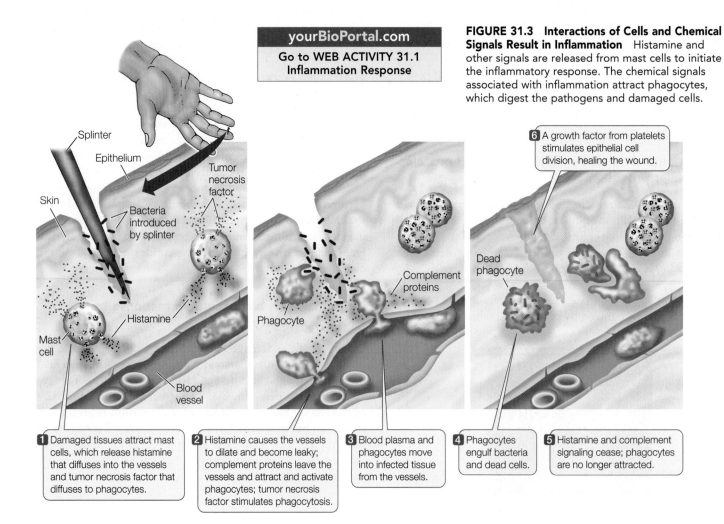

yourBioPortal.com

Go to WEB ACTIVITY 31.1
Inflammation Response

FIGURE 31.3 Interactions of Cells and Chemical Signals Result in Inflammation Histamine and other signals are released from mast cells to initiate the inflammatory response. The chemical signals associated with inflammation attract phagocytes, which digest the pathogens and damaged cells.

1 Damaged tissues attract mast cells, which release histamine that diffuses into the vessels and tumor necrosis factor that diffuses to phagocytes.

2 Histamine causes the vessels to dilate and become leaky; complement proteins leave the vessels and attract and activate phagocytes; tumor necrosis factor stimulates phagocytosis.

3 Blood plasma and phagocytes move into infected tissue from the vessels.

4 Phagocytes engulf bacteria and dead cells.

5 Histamine and complement signaling cease; phagocytes are no longer attracted.

6 A growth factor from platelets stimulates epithelial cell division, healing the wound.

surface of a plant pollen grain (as in hay fever) or a particle in dust.

- In **autoimmune diseases** such as rheumatoid arthritis, the immune system fails to distinguish between self and non-self, and attacks tissues in the organism's own body. In these cases the inflammation is somewhat more general, affecting tissues throughout the body.

- In **sepsis**, the inflammation due to a bacterial infection does not remain local. Instead it extends throughout the body, with the dilation of blood vessels throughout the body. The resulting drop in blood pressure is a medical emergency and can be lethal.

FRONTIERS Anti-inflammatory drugs are among the most popular medications. Many of the commonly used ones block prostaglandins and have side effects because of their inhibition of prostaglandin-related processes elsewhere in the body. Knowledge of the role of tumor necrosis factor is leading to the development of more specific agents to block inflammation in rheumatoid arthritis.

Do You Understand Concept 31.2?

- Outline the sequence of innate defenses encountered by a pathogenic bacterial cell if it is ingested in food. Similarly, outline the defenses encountered by a bacterial cell that lands in the nose.

- Antihistamines are used to treat the symptom of sneezing due to inflammation caused by irritants in the airways. How do you think antihistamines might work?

- A massive inflammation due to a food allergy can be treated with an injection of epinephrine. Refer to the description of epinephrine's effects in Concept 30.2. How do you think this hormone relieves the inflammation symptoms?

In most instances innate immunity is sufficient to block a pathogen from affecting the body. But many pathogens are present in huge numbers (think of a viral infection), and some may escape the innate responses and begin to proliferate in the body. In these cases, adaptive immunity takes over.

concept 31.3 The Adaptive Immune Response Is Specific

Long before the twentieth century, scientists had suspected that blood was somehow involved in immunity against pathogens, but they did not have definitive experimental evidence to support their suspicions. Over a century ago, Emil von Behring and Shibasaburo Kitasato at the University of Marburg in Germany performed a key experiment that pointed to blood as an important factor in adaptive immunity (**FIGURE 31.4**).

They showed that guinea pigs injected with a sublethal dose of diphtheria toxin developed in their blood serum (the non-cellular fluid that remains after blood is clotted) a factor that when injected into other guinea pigs was able to protect them from a lethal dose of the same toxin. In other words, the serum recipients had somehow acquired immunity. The response of the donor guinea pigs to the diphtheria toxin is an example of *adaptive immunity*: after exposure to the toxin, they made a protective factor that was not present before their exposure. Moreover, the immunity was specific: the factor made by the first guinea pig protected the others only against the specific toxin produced by the strain of bacteria with which the first guinea pigs had been injected.

Based on the animal model, Behring realized that serum protection might work for human diseases as well. It did, and he won the Nobel Prize for his efforts in protecting children against diphtheria. Later, the agent of this immunity was identified as an antibody protein, and the process of acquiring immunity from antibodies received from another individual was called *passive immunity*.

LINK Many bacteria produce toxins of various types; see Concept 19.3

Adaptive immunity has four key features

As we explore adaptive immunity in more detail in this chapter, it is important to keep in mind the key features of the adaptive immune system that enable it to protect the body:

- It is *specific*, which allows it to focus its responses on pathogens that are actually present.

- It is *diverse*, which enables it to respond to novel pathogens.

- It *distinguishes self from nonself*, which prevents it from destroying self cells.

- It has *immunological memory*, which allows it to respond more effectively in later exposures to the same pathogen.

SPECIFICITY B and T lymphocytes are crucial for the specificity of adaptive immune responses. T cell receptors and the antibodies produced by B cells recognize and bind to specific nonself substances called **antigens**, and this interaction initiates a specific immune response. Each T cell and each antibody-producing B cell is specific for a single antigen.

The sites on antigens that the immune system recognizes are called **antigenic determinants** or *epitopes*:

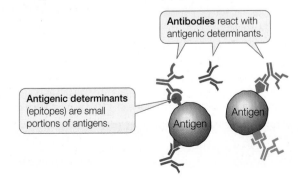

Antibodies react with antigenic determinants.

Antigenic determinants (epitopes) are small portions of antigens.

Antigen

INVESTIGATION

FIGURE 31.4 The Discovery of Specific Immunity Until the twentieth century, most people did not survive an attack of diphtheria, but a few did. Emil von Behring and Shibasaburo Kitasato performed a key experiment using an animal model. They demonstrated that the factor(s) responsible for immunity against diphtheria was in the blood serum.

HYPOTHESIS

Serum from guinea pigs injected with a sublethal dose of diphtheria toxin protects other guinea pigs that are exposed to a lethal dose of the same toxin.

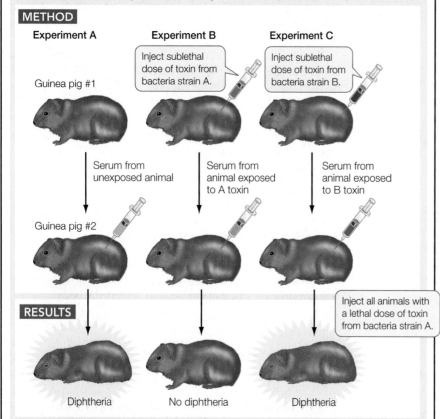

METHOD

Experiment A **Experiment B** **Experiment C**

Inject sublethal dose of toxin from bacteria strain A.

Inject sublethal dose of toxin from bacteria strain B.

Guinea pig #1

Serum from unexposed animal

Serum from animal exposed to A toxin

Serum from animal exposed to B toxin

Guinea pig #2

Inject all animals with a lethal dose of toxin from bacteria strain A.

RESULTS

Diphtheria No diphtheria Diphtheria

CONCLUSION

Serum of toxin-exposed guinea pigs is protective against later exposure to a lethal dose of toxin from the same genetic strain of bacteria, but not a different strain.

ANALYZE THE DATA

In their experiments, Behring and Kitasato used different doses of toxin to test the immunity of guinea pig #2 in each experiment. These are the results:

Experiment	Symptoms	
	0.5 ml dose	10 ml dose
A	Diphtheria	Diphtheria
B	No diphtheria	No diphtheria
C	Diphtheria	Diphtheria

A. Explain these data in terms of the level of protection afforded by the serum.

B. These experiments could be performed with either intact bacteria that cause diphtheria, or a bacteria-free filtrate of a 10-day-old culture of the bacteria. Explain.

Go to **yourBioPortal.com** for original citations, discussions, and relevant links for all INVESTIGATION figures.

Antigens are usually proteins or polysaccharides, and there can be multiple antigens on a single invading bacterium. An antigenic determinant is a specific portion of an antigen, such as a certain sequence of amino acids that may be present in a protein. A single antigenic molecule can have multiple different antigenic determinants. For the remainder of the chapter, we will refer to antigenic determinants simply as "antigens."

DIVERSITY Pathogens take many forms: viruses, bacteria, protists, fungi, and multicellular parasites, and toxins made by these organisms. Furthermore, each pathogenic species usually exists as many subtly different genetic strains, and each strain possesses multiple surface features. Estimates vary, but a reasonable guess is that humans can respond specifically to 10 million different antigens. Upon recognizing an antigen, the immune system responds by activating lymphocytes of the appropriate specificity.

To have the ability to respond to the large number of potential pathogens, the body needs to generate a vast diversity of lymphocytes that are specific for different antigens. These lymphocytes represent a pool from which specific cells are selected when needed.

- *Diversity is generated primarily by DNA changes*—chromosomal rearrangements and other mutations—that occur just after the B and T cells are formed in the bone marrow. Each B cell is able to produce only one kind of antibody; thus there are millions of different B cells. Similarly, there are millions of different T cells with specific T cell receptors. The adaptive immune system is "predeveloped"—*all of the machinery available to respond to an immense diversity of antigens is already there, even before the antigens are encountered.*

- *Antigen binding "selects" a particular B or T cell for proliferation.* For example, when an antigen fits the surface receptor on a B cell and binds to it, that B cell is *activated*. It divides to form a clone of cells (a genetically identical group derived from a single cell), all of which produce and/or secrete antibodies with the same specificity as the receptor (**FIGURE 31.5**). Binding, activation, and proliferation also apply to T lymphocytes. A particular lymphocyte is selected via binding and activation, and then it proliferates to generate a clone—hence the name **clonal selection** for this mechanism of producing an immune response.

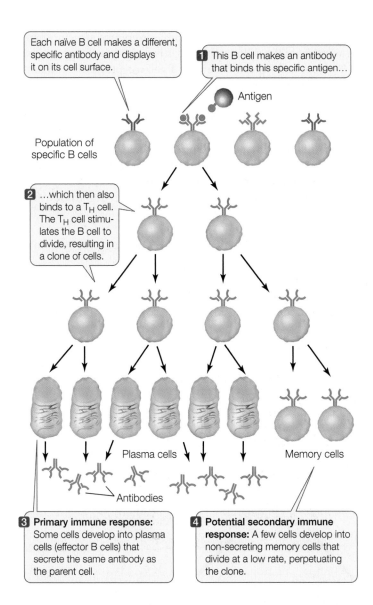

Each naïve B cell makes a different, specific antibody and displays it on its cell surface.

1 This B cell makes an antibody that binds this specific antigen…

Antigen

Population of specific B cells

2 …which then also binds to a T_H cell. The T_H cell stimulates the B cell to divide, resulting in a clone of cells.

Plasma cells

Memory cells

Antibodies

3 Primary immune response: Some cells develop into plasma cells (effector B cells) that secrete the same antibody as the parent cell.

4 Potential secondary immune response: A few cells develop into non-secreting memory cells that divide at a low rate, perpetuating the clone.

FIGURE 31.5 Clonal Selection in B Cells The binding of an antigen to a specific receptor on the surface of a B cell stimulates that cell to divide, producing a clone of genetically identical cells to fight that invader.

DISTINGUISHING SELF FROM NONSELF The human body contains tens of thousands of different molecules, each with a specific three-dimensional structure capable of generating immune responses. Thus every cell in the body bears a tremendous number of antigens. A crucial requirement of an individual's immune system is that it recognize the body's own antigens and not attack them.

Normally, the body is *tolerant* of its own molecules—the same molecules that would generate an immune response in another individual. This occurs primarily during the early differentiation of T and B cells, when they encounter self antigens. Any immature B or T cell that shows the potential to mount a strong immune response against self-antigens undergoes apoptosis within a short time. This process is called **clonal deletion**.

A failure of clonal deletion leads to an immune response within an individual to self-antigens, or **autoimmunity**. Autoimmunity does not always result in disease, but a number of autoimmune diseases are common:

- People with *systemic lupus erythematosis* (SLE) have antibodies to many cellular components. These antibodies can cause serious damage when they bind to normal tissue antigens and form large circulating antigen–antibody complexes. These can become stuck in tissues and provoke inflammation.

- *Hashimoto's thyroiditis* is the most common autoimmune disease in women over age 50. Immune cells attack thyroid tissue, resulting in fatigue, depression, weight gain, and other symptoms associated with impaired thyroid function.

FRONTIERS One possible mechanism for inducing auto-immunity is *molecular mimicry*, in which exposure to an antigen similar to a self antigen provokes an immune response that cross-reacts with the self antigen. There is a search underway for antigens that cause this response, which can lead to autoimmune diseases. The antigens may be in the environment, foods, or microbes.

IMMUNOLOGICAL MEMORY After responding to a particular type of pathogen once, the immune system "remembers" that pathogen and can usually respond more rapidly and powerfully to the same threat in the future. This **immunological memory** usually saves us from repeats of childhood infectious diseases.

The first time a vertebrate animal is exposed to a particular antigen, it takes several days before the adaptive immune system produces antibodies and T cells specific for the antigen. This is the **primary immune response**. Immunological memory arises from the fact that activated lymphocytes divide and differentiate to produce *two types* of daughter cells: effector cells and memory cells (see Figure 31.5).

- **Effector cells** carry out the attack on the antigen. Effector B cells, called **plasma cells**, secrete antibodies. Effector T cells release cytokines and other molecules that initiate reactions that destroy nonself or altered cells. Effector cells live for only a few days.

- **Memory cells** are long-lived cells that retain the ability to start dividing on short notice to produce more effector and more memory cells. Memory B and T cells may survive in the body for decades, rarely dividing.

After a primary immune response to a particular antigen, subsequent encounters with the same antigen will trigger a much more rapid and powerful **secondary immune response**. The memory cells that bind with that antigen proliferate, launching a huge army of plasma cells and effector T cells. The principle behind vaccination is to trigger a primary immune response that prepares the body to mount a stronger, quicker secondary response if it encounters the actual pathogen again.

Two types of adaptive immune responses interact

The adaptive immune response involves three phases (**FIGURE 31.6**):

- *Recognition phase.* The organism discriminates between self and nonself to detect a pathogen.

- *Activation phase.* The recognition event leads to a mobilization of cells and molecules to fight the invader.

- *Effector phase.* The mobilized cells and molecules destroy the invader.

Depending on the tissue and antigen, these three phases can occur in either of two types of responses: the *humoral immune response* and the *cellular immune response.*

B cells that make antibodies are the workhorses of the **humoral immune response**, and **cytotoxic T (T_C) cells** are the workhorses of the **cellular immune response**. A key event early in these two processes is the exposure or *presentation* of the antigen to the immune system, where recognition can occur. In humoral immunity, this primarily occurs when an antigen binds to a B cell that has on its surface an antibody that is specific for that antigen. In cellular immunity, an antigen is inserted into the plasma membrane of an antigen-presenting cell, with the unique epitope structure protruding from the cell membrane. This antigen is recognized by a **T-helper cell (T_H cell)** bearing a T cell receptor protein that is specific for that antigen. In both cases, binding is highly specific and initiates the activation phase.

Antigen binding readies a B cell for division. Some antigen is internalized by the B cell after binding and digested into fragments. When an antigenic fragment binds to the B cell MHC protein and is presented on the B cell surface, a T_H cell whose receptor is specific for that antigen binds to the B cell. The binding of the T cell stimulates the B cell to divide and

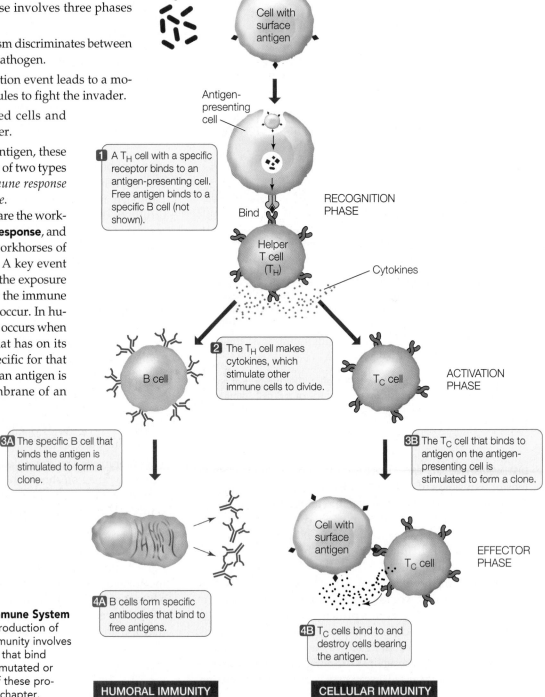

FIGURE 31.6 The Adaptive Immune System Humoral immunity involves the production of antibodies by B cells. Cellular immunity involves the activation of cytotoxic T cells that bind to and destroy self cells that are mutated or infected by pathogens. Details of these processes are described later in this chapter.

APPLY THE CONCEPT

The adaptive immune response is specific

In humans, the dominant gene *D* codes for the RhD (Rhesus) protein on the surfaces of red blood cells. The recessive allele *d* does not encode a cell surface protein. If cells from an individual who is *DD* or *Dd* (Rh⁺) enter the bloodstream of an individual who is *dd* (Rh⁻), the Rh⁻ individual makes antibodies that react with the Rh⁺ cells. During development, the blood vessels of a mother and her fetus are near to each other but do not mix. However, antibodies can pass through capillary walls between the two blood supplies.

MOTHER		FETUS		
GENOTYPE	Rh PHENOTYPE	GENOTYPE	Rh PHENOTYPE	RESULT
Dd		*Dd*		
dd		*Dd*		

1. Hemolytic disease of the newborn (sometimes resulting in stillbirth) can occur when antibodies from the mother pass into the fetal circulatory system. Fill in the table, which predicts RhD incompatibility.

2. During birth there is mixing of the blood supplies of mother and fetus. In a situation of RhD incompatibility, the first birth does not result in a clinical problem. But a second RhD incompatible birth with this mother results in a hemolytic disease. Explain how this can happen (hint: immunological memory).

3. Today, RhD incompatibility does not result in stillbirths. The mother is treated with serum containing anti-RhD antibodies late in the first and subsequent pregnancies. How can this prevent the immunological reactions?

---

form a clone. In the cellular immune response, binding of the T_H cell to the antigen-presenting cell causes the T_H cell to release cytokines that stimulate T_C cells bearing the same T cell receptor to divide. So the result of activation is the formation of two clones of cells:

• A clone of B cells that can produce antibodies specific for the antigen

• A clone of T_C cells that express a T cell receptor that can bind to any cell expressing the antigen on its surface

Once formed, the two clones perform their roles in the effector phase. Cells of the B clone produce antibodies that bind to free antigen in the bloodstream. Such binding results in inactivation and destruction of the antigen. Cells of the T_C clone bind to cells bearing the antigen and destroy them.

Do You Understand Concept 31.3?

• Make a table describing the four key features of the adaptive immune system.

• In 2009, the H1N1 strain of influenza was a worldwide epidemic. Notably, very old people, who had been alive during the 1918 flu outbreak, had low rates of H1N1 infection. Explain this in terms of immunological memory.

• Insulin-dependent diabetes (Type 1) results from a destruction of the cells in the pancreas that make the hormone insulin (see Concept 39.4). One hypothesis for this disease is that it is caused by an autoimmune reaction. Explain how this might happen and how you would investigate your hypothesis.

With this overview of the basic characteristics of adaptive immunity and the events of the immune response, we will now turn to some of the cellular and molecular details of how these events occur.

concept 31.4 The Adaptive Humoral Immune Response Involves Specific Antibodies

Every day in the human body, billions of B cells survive the test of clonal deletion and are released from the bone marrow into the circulatory system. B cells are the basis for the humoral immune response.

yourBioPortal.com
Go to ANIMATED TUTORIAL 31.2
Humoral Immune Response

Plasma cells produce antibodies

A B cell begins life as a "naïve" B cell with a receptor protein on its cell surface that is specific for a particular antigen. The cell is activated by antigen binding to this receptor, and, after stimulation by a T_H cell, it produces a clone of plasma cells that make antibodies as well as memory cells (see Figure 31.5). In addition to this stimulation by antigen binding, a B cell can be stimulated to divide by a T_H cell binding to the exposed antigen on the B cell surface. The specific T_H cell may come from a clone that was activated by the cellular immune response (see Concept 31.5). This type of interaction between B cells and T_H cells provides a connection between the cellular and humoral systems. The T_H cell bound to the B cell secretes cytokines that stimulate the B cell to divide.

All the plasma cells arising from a given B cell produce antibodies that are specific for the antigen that originally bound to the parent B cell. Thus antibody specificity is maintained as B cells proliferate.

Antibodies share a common overall structure

There are several classes of antibodies (also called **immunoglobulins**), but all contain a tetramer with four polypeptide chains (**FIGURE 31.7**). In each immunoglobulin molecule, two of the polypeptides are identical *light chains*, and two are identical

heavy chains. Disulfide bonds hold the chains together. Each polypeptide chain has a constant region and a variable region:

- The amino acid sequence of the **constant region** determines the general structure and function (the *class*) of an immunoglobulin. All immunoglobulins in a particular class have a similar constant region. When an antibody acts as a B cell receptor, the constant region inserts into the plasma membrane.

- The amino acid sequence of the **variable region** is different for each specific immunoglobulin. The variable region's three-dimensional antigen-binding site is determined by its secondary structure, and is responsible for antibody specificity.

The two antigen-binding sites on each immunoglobulin molecule are identical, making the antibody *bivalent* (*bi*, "two"; *valent*, "binding"). This ability to bind two antigen molecules at once, along with the existence of multiple epitopes on each antigen, permits antibodies to form large complexes with the antigens. For example, one antibody might bind two molecules of an antigen. Another antibody might bind the same antigen at a different epitope. It may bind one of the antigen molecules that is already bound to the first antibody, along with a third antigen molecule:

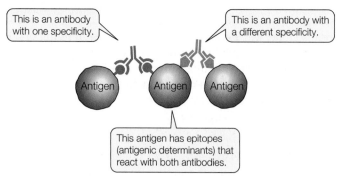

This binding of multiple antigens and multiple antibodies can result in large complexes that are easy targets for ingestion and breakdown by phagocytes.

There are five classes of immunoglobulins (Ig), and they differ in function and in the type of heavy chain:

- IgG is secreted by B cells and constitutes about 80 percent of circulating antibodies.

- IgD is the cell surface receptor on a B cell.

- IgM is the initial surface and circulating antibody released by a B cell.

- IgA protects mucosa on epithelia exposed to the environment.

- IgE binds to mast cells and is involved with inflammation.

We will focus on IgG antibodies in this chapter.

Antibody diversity results from DNA rearrangements and other mutations

Each mature B cell makes one—and only one—specific antibody targeted to a single antigen. And there are millions of possible antigens to which a human can be exposed. How can

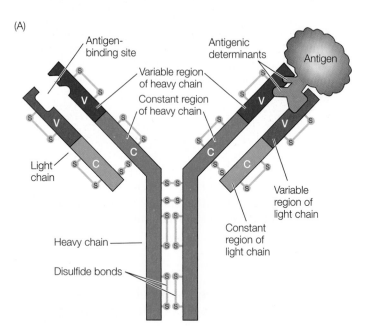

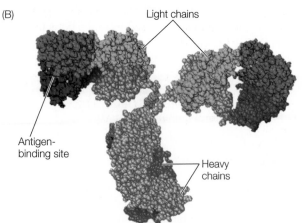

FIGURE 31.7 The Structure of an Immunoglobulin Four polypeptide chains (two light, two heavy) make up an immunoglobulin molecule. Here we show both diagrammatic (A) and space-filling (B) representations of an immunoglobulin.

yourBioPortal.com

Go to WEB ACTIVITY 31.2
Immunoglobulin Structure

the genome encode enough different antibodies to protect the body against all the possible pathogens? With millions of possible amino acid sequences in immunoglobulins, one explanation might be that there are millions of genes, each one encoding one antibody molecule. A simple calculation using approximate numbers shows that this is impossible:

one IgG heavy chain = 500 amino acids coded for by 1,500 bp of DNA

one IgG light chain = 200 amino acids coded for by 600 bp of DNA

Therefore,

one antibody (two identical heavy and two identical light chains) is encoded by 2,100 bp of DNA

10 million different antibodies = 21 billion bp of DNA

This is 6.5 times the size of the entire human genome! There must be another way to generate antibody diversity.

It turns out that instead of a single gene encoding each complete immunoglobulin, the genome of the differentiating B cell *has a number of different coding regions for each domain* of the protein, and that *diversity is generated by putting together different combinations of these regions.* Shuffling of this genetic deck generates the enormous immunological diversity that characterizes each individual mammal.

Each gene encoding an immunoglobulin chain is in reality a "supergene" assembled by means of genetic recombination from several clusters of smaller genes scattered along part of a chromosome (**FIGURE 31.8**). Every cell in the body has

hundreds of immunoglobulin genes located in separate clusters that are potentially capable of participating in the synthesis of both the variable and constant regions of immunoglobulin chains. In most body cells and tissues, these genes remain intact and separated from one another. But during B cell development, these genes are cut out, rearranged, and joined together in DNA recombination events. One gene from each cluster is chosen randomly for joining, and the others are deleted. In the case of one multigene set, the J genes, the extra sequences are removed in RNA splicing (**FIGURE 31.9**).

In this manner, a unique immunoglobulin supergene is assembled from randomly selected "parts." Each B cell precursor assembles two supergenes, one for a specific heavy chain and the other, assembled independently, for a specific light chain. This remarkable example of irreversible cell differentiation generates an enormous diversity of immunoglobulins from the same genome. It is a major exception to the generalization that all somatic cells derived from the fertilized egg have identical DNA.

In both humans and mice, the gene clusters encoding immunoglobulin heavy chains are on one pair of chromosomes and those for the light chains are on two other pairs. Three families of genes (V, D, and J) encode the variable region of the heavy chain. There are two classes of light chain, and the variable regions in both classes are made up of two domains, encoded by two gene families (V and J).

Figure 31.8 illustrates the gene families that encode the constant and variable regions of the heavy chain in mice. There are multiple genes that encode each of the three parts of the variable region: 100 V, 30 D, and 6 J genes. Each B cell randomly selects one gene from each of these clusters to make the final

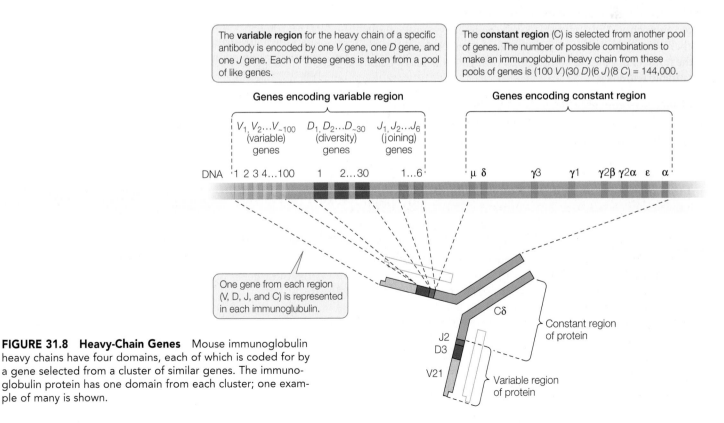

The **variable region** for the heavy chain of a specific antibody is encoded by one V gene, one D gene, and one J gene. Each of these genes is taken from a pool of like genes.

The **constant region** (C) is selected from another pool of genes. The number of possible combinations to make an immunoglobulin heavy chain from these pools of genes is (100 V)(30 D)(6 J)(8 C) = 144,000.

Genes encoding variable region

Genes encoding constant region

$V_1, V_2...V_{\sim100}$ (variable) genes $D_1, D_2...D_{\sim30}$ (diversity) genes $J_1, J_2...J_6$ (joining) genes

DNA 1 2 3 4...100   1   2...30   1...6   μ δ   γ3   γ1   γ2β γ2α ε   α

One gene from each region (V, D, J, and C) is represented in each immunoglubulin.

Cδ

J2
D3

Constant region of protein

V21

Variable region of protein

FIGURE 31.8 Heavy-Chain Genes Mouse immunoglobulin heavy chains have four domains, each of which is coded for by a gene selected from a cluster of similar genes. The immunoglobulin protein has one domain from each cluster; one example of many is shown.

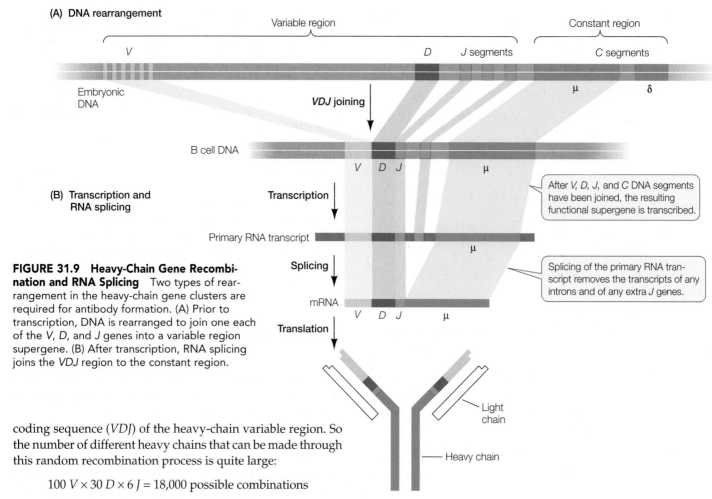

(A) DNA rearrangement

(B) Transcription and RNA splicing

After *V, D, J*, and *C* DNA segments have been joined, the resulting functional supergene is transcribed.

Splicing of the primary RNA transcript removes the transcripts of any introns and of any extra *J* genes.

FIGURE 31.9 Heavy-Chain Gene Recombination and RNA Splicing Two types of rearrangement in the heavy-chain gene clusters are required for antibody formation. (A) Prior to transcription, DNA is rearranged to join one each of the *V, D*, and *J* genes into a variable region supergene. (B) After transcription, RNA splicing joins the *VDJ* region to the constant region.

coding sequence (*VDJ*) of the heavy-chain variable region. So the number of different heavy chains that can be made through this random recombination process is quite large:

$$100 \; V \times 30 \; D \times 6 \; J = 18{,}000 \text{ possible combinations}$$

Now consider that the light chains are similarly constructed, with a similar amount of diversity made possible by random recombination. If we assume that the degree of potential light-chain diversity is the same as that for heavy-chain diversity, the number of possible combinations of light- and heavy-chain variable regions is:

18,000 different light chains × 18,000 different heavy chains = 324 million possibilities!

Other mechanisms generate even more diversity:

- When the DNA sequences that encode the *V, D*, and *J* regions are rearranged so that they are next to one another, the recombination event is not precise, and errors occur at the junctions. This *imprecise recombination* can create frameshift mutations, generating new codons at the junctions, with resulting amino acid changes.
- After the DNA sequences are cut and before they are rejoined, the enzyme *terminal transferase* often adds some

APPLY THE CONCEPT

The adaptive humoral immune response involves specific antibodies

Mammals have two kinds of light chains: the kappa (κ) and lambda (λ) chains, and each variable region consists of a V domain and a J domain. Imagine a strain of mouse that has a family of 350 genes encoding the κ V domain and 10 encoding the κ J domain, while the λ variable regions are encoded by 300 V and 5 J genes. The IgG heavy chain is encoded by 350 V, 8 J, and 5 D genes.

1. How many different types of antibodies can be made from these genes by recombination?
2. What are two other ways to generate antibody diversity from these genes?

nucleotides to the free ends of the DNA pieces. These additional bases create insertion mutations.

- There is a relatively high *spontaneous mutation rate* in immunoglobulin genes. Once again, this process creates many new alleles and adds to antibody diversity.

When we include these possibilities with the millions of combinations that can be made by random DNA rearrangements, it is not surprising that the immune system can mount a response to almost any natural or artificial substance.

Once the DNA rearrangements are completed, each supergene is transcribed and then translated to produce an immunoglobulin light chain or heavy chain. These chains combine to form an active immunoglobulin protein.

---

yourBioPortal.com

Go to ANIMATED TUTORIAL 31.3
A B Cell Builds an Antibody

Antibodies bind to pathogens on cells or in the bloodstream

Recall that antibodies have two roles in B cells after they undergo DNA rearrangements and RNA splicing. First, by being expressed on the cell surface, a unique antibody can act as a receptor for an antigen in the recognition phase of the humoral response. Second, in the effector phase of the humoral response, specific antibodies are produced in large amounts by a clone of B cells. These antibodies are secreted from the B cells and enter the bloodstream where they act in either of two ways:

- Some antibodies bind to the antigen that is expressed on the surface of a pathogen. This can stimulate macrophages to ingest, or natural killer cells to destroy, the pathogen.

- If the antigen is free in the bloodstream, antibodies may bind to it using their cross-linking function (see p. 630) to form large, insoluble antibody–antigen complexes. These are ingested and destroyed by phagocytic cells.

Do You Understand Concept 31.4?

- Sketch an IgG antibody, identifying the variable and constant regions, light and heavy chains, and antigen-binding sites.

- The bacterium that causes diphtheria (see Figure 31.4) synthesizes a toxic protein. You have probably not been exposed to this bacterium or its toxin. At the present time, are you making B cells and antibodies that bind specifically to diphtheria toxin? Explain your answer.

- When an antigenic protein such as diphtheria toxin (see above) enters the bloodstream, numerous clones of B cells are activated. Explain.

The humoral immune response works in concert with the cellular immune response in adaptive immunity. Now let's turn to a closer examination of the cellular response.

---

concept 31.5 The Adaptive Cellular Immune Response Involves T Cells and Their Receptors

Two types of effector T cells (T-helper cells and cytotoxic T cells) are involved in the cellular immune response, along with proteins of the major histocompatibility complex (MHC proteins), which underlie the immune system's tolerance for the body's own cells.

---

yourBioPortal.com

Go to ANIMATED TUTORIAL 31.4
Cellular Immune Response

T cell receptors specifically bind to antigens on cell surfaces

Like B cells, T cells possess specific membrane receptors. The T cell receptor is not an immunoglobulin, however, but a glycoprotein with a molecular weight of about half that of an IgG. It is made up of two polypeptide chains, each encoded by a separate gene (**FIGURE 31.10**). The two chains have distinct regions with constant and variable amino acid sequences. As in the immunoglobulins, the variable regions provide the site for specific binding to antigens. T cell receptors typically bind to a piece of an antigen, such as a peptide from a protein, displayed

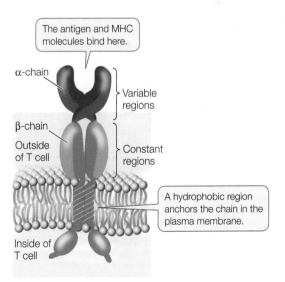

FIGURE 31.10 A T Cell Receptor The receptors on T lymphocytes are smaller than those on B lymphocytes, but their two polypeptides contain both variable and constant regions. As with the B cell receptors, the constant regions fix the receptor in the plasma membrane, while the variable regions establish the specificity for binding to antigen.

Receptor   Antibody

1 Binding of antibody to a receptor activates phagocytosis.

Antigen

Macrophage

Antigen

Class II MHC protein

2 The macrophage takes up the antigen by phagocytosis.

3 The macrophage breaks down the antigen into fragments in the lysosome.

Lysosome

Macrophage

4 A class II MHC protein binds an antigen fragment.

5 The MHC presents the antigen to a T_H cell.

T cell receptor   **T_H cell**

FIGURE 31.11 Macrophages Are Antigen-Presenting Cells
A fragment of an antigen is displayed by MHC II on the surface of a macrophage. T cell receptors on a specific T-helper cell can then bind to and interact further with the antigen–MHC II complex.

on the surface of an antigen-presenting cell in the presence of an MHC protein.

MHC proteins present antigen to T cells and result in recognition

Recall that so far we have described two types of T cells: T_H and T_C cells. Both cell types express T cell receptors than bind to antigen on the cell surface. But the response of each cell type to binding is quite different. T_H binding results in activation of the adaptive immune response, whereas T_C binding results in death of the cell carrying the antigen in the adaptic response. The MHC proteins form complexes with antigens on cell surfaces and assist with recognition by the T cells, so that the appropriate type of T cell binds.

The MHC proteins are plasma membrane glycoproteins. Two types of MHC proteins function to present antigens to the two different types of T lymphocytes:

- **Class I MHC** proteins are present on the surface of every nucleated cell in the mammalian body. They present antigens to T_C cells. These antigens can be fragments of virus proteins in virus-infected cells or abnormal proteins made by cancer cells as a result of somatic mutations.

- **Class II MHC** proteins are on the surfaces of macrophages, B cells, and dendritic cells. They present antigens to T_H cells. The three cell types ingest antigens and break them down; one of the fragments then binds to MHC II for presentation (**FIGURE 31.11**).

In humans, there are three genetic loci for class I MHC proteins and three for class II MHC proteins. Each of these six loci has as many as 100 different alleles. With so many possible allele combinations, it is not surprising that different people are very likely to have different MHC genotypes. MHC proteins are "self" markers. To accomplish its role in antigen presentation, an MHC protein has an antigen-binding site that can hold a peptide of about 10–20 amino acids. The T cell receptor recognizes not just the antigenic fragment but also the class I or II MHC molecule to which the fragment is bound.

Information on MHC proteins, the cellular origins of antigens, and T lymphocytes is summarized in **TABLE 31.2**.

Activation of the cellular response results in death of the targeted cell

Activation of a T_H cell results in the stimulation of B cells to propagate and produce antibodies against that antigen. In the case of binding of a T_C cell, activation results in the production of a clone of T_C cells with the specific T cell receptor. These T_C cells bind to cells carrying the antigen–MHC I protein complex (e.g., virus-infected cells). When bound, the T_C cells do two things to eliminate the antigen-carrying cell:

- They produce *perforin*, which lyses the bound target cell.
- They stimulate apoptosis in the target cell.

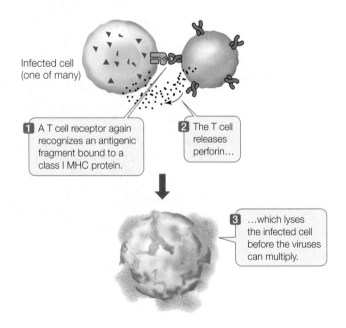

Infected cell (one of many)

1 A T cell receptor again recognizes an antigenic fragment bound to a class I MHC protein.

2 The T cell releases perforin...

3 ...which lyses the infected cell before the viruses can multiply.

TABLE 31.2	The Interaction between T Cells and Antigen-Presenting Cells		
PRESENTING CELL TYPE	**ANTIGEN PRESENTED**	**MHC CLASS**	**T CELL TYPE**
Any cell	Intracellular protein fragment	Class I	Cytotoxic T cell (T$_C$)
Macrophages and B cells	Fragments from extracellular proteins	Class II	Helper T cell (T$_H$)

FRONTIERS Cancer cells express specific molecules on their surfaces. When an antibody binds to such a molecule, the targeted cell dies by a mechanism that is not well understood. Nevertheless, several antibodies to cancer cell membrane proteins are in widespread clinical use for breast and colon cancer.

Regulatory T cells suppress the humoral and cellular immune responses

A third class of T cells called **regulatory T cells** (**Tregs**) ensures that the immune response does not spiral out of control. Like T$_H$ and T$_C$ cells, Tregs are made in the thymus gland, express the T cell receptor, and become activated if they bind to antigen–MHC complexes. But Tregs are different in one important way: the antigens that Tregs recognize are *self antigens*. The activation of Tregs causes them to secrete the cytokine *interleukin-10*, which blocks T cell activation and leads to apoptosis of the T$_C$ and T$_H$ cells that are bound to the same antigen-presenting cell (**FIGURE 31.12**).

The important role of Tregs is to mediate tolerance to self antigens. Thus they constitute one of the mechanisms for distinguishing self from nonself. There are two lines of experimental evidence for the role of Tregs:

• If Tregs are experimentally destroyed during development in the thymus of a mouse, the mouse grows up with an out-of-control immune system, mounting strong immune responses to self antigens—autoimmunity.

• In humans, a rare X-linked inherited disease occurs when a gene critical to Treg function is mutated. An infant with this disease, called IPEX (*i*mmune dysregulation, *p*olyendocrinopathy and *e*nteropathy, *X*-linked), mounts an immune response that attacks the pancreas, thyroid, and intestine. Most affected individuals die within the first few years of life.

AIDS is an immune deficiency disorder

There are a number of inherited and acquired *immune deficiency disorders*. In some individuals, T or B cells never form; in others, B cells lose the ability to give rise to plasma cells. In either case, the affected individual is unable to mount an immune response and thus lacks a major line of defense against pathogens. A

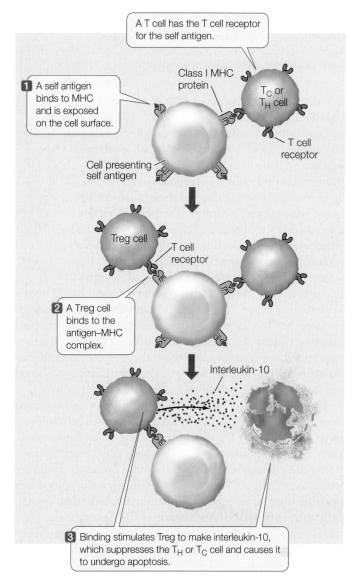

FIGURE 31.12 Tregs and Tolerance A special class of T cells called regulatory T cells (Tregs) inhibits the activation of the immune system in response to self antigens.

more common disease that was first detected in the early 1980s is **acquired immune deficiency syndrome** (**AIDS**), which results from infection by **human immunodeficiency virus** (**HIV**).

The course of HIV infection provides a good review of adaptive immunity (see Figure 31.6). HIV is transmitted from person to person in blood, semen, vaginal fluid, or breast milk; the recipient tissue is either blood (by transfusion or injection) or a mucous membrane lining an organ. HIV initially infects macrophages, T$_H$ cells, and antigen-presenting dendritic cells (another type of white blood cell) in the blood and tissues. At first there is an immune response to the viral infection, and some T$_H$ cells are activated. But because HIV infects the T$_H$ cells, they are killed both by HIV itself and by T$_C$ cells that lyse infected T$_H$ cells. Consequently, T$_H$ cell numbers decline after the first month or so of infection. Meanwhile, the extensive production of HIV by infected cells activates the humoral immune system. Antibodies bind to HIV and the complexes are removed by phagocytes. The HIV level in blood goes down. There is still a low level of

FIGURE 31.13 The Course of an HIV Infection
An HIV infection may be carried, unsuspected, for many years before the onset of symptoms.

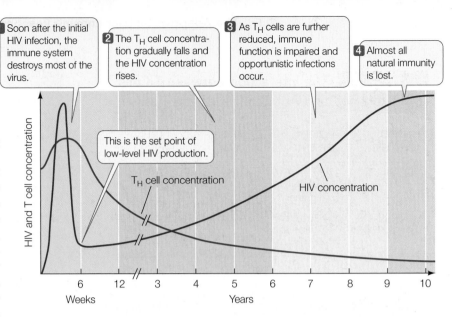

1 Soon after the initial HIV infection, the immune system destroys most of the virus.

2 The T$_H$ cell concentration gradually falls and the HIV concentration rises.

3 As T$_H$ cells are further reduced, immune function is impaired and opportunistic infections occur.

4 Almost all natural immunity is lost.

This is the set point of low-level HIV production.

T$_H$ cell concentration

HIV concentration

infection, however, because of the depletion of T$_H$ cells (**FIGURE 31.13**).

During this dormant period, people carrying HIV generally feel fine, and their T$_H$ cell levels are adequate for them to mount immune responses against other infections. Eventually, however, the virus destroys the T$_H$ cells, and their numbers fall to the point where the infected person is susceptible to infections that the T$_H$ cells would normally eliminate. These infections result in conditions such as Kaposi's sarcoma, a skin tumor caused by a herpesvirus; pneumonia caused by the fungus *Pneumocystis jirovecii*; and lymphoma tumors caused by the Epstein–Barr virus. These conditions result from *opportunistic infections* because the pathogens take advantage of the crippled immune system of the host. In addition, HIV infection somehow stimulates Treg cells, and these also downregulate the immune response. The combination of infections and a weakened immune response can lead to death within a year or two.

The molecular biology of HIV and its life cycle have been intensively studied. Drug treatments are focused on inhibiting processes necessary for viral entry, assembly, and replication, such as *reverse transcriptase* that makes cDNA from the viral RNA, and *viral protease* that cuts the large precursor viral protein into its final active proteins. Combinations of such drugs result in long-term survival. Unfortunately, like many medical treatments, HIV drugs are not available to all who need them—particularly in poor regions of the world where AIDS is prevalent. As a result, there are about 2 million deaths per year worldwide from AIDS.

LINK For more on the life cycle of HIV, see Concept 11.1

Do You Understand Concept 31.5?

- Compare the T cell receptor and B cell receptor in terms of structure, diversity, and function.

- What are the similarities and differences in function between class I and II MHC proteins?

- What are the roles of T$_H$ cells in cellular and humoral immunity?

- Since MHC proteins are highly variable and almost always differ between unrelated people, an organ transplant between such people will generally provoke a cellular immune response, and the organ will be rejected. Patients receiving organ transplants are treated with cyclosporin, a drug that inhibits T cell development. How do you think cyclosporin prevents rejection? What side effects might you expect in treated people?

Q&A QUESTION How can a person survive an infection and be resistant to further infection?

ANSWER Almost 2,500 years after Thucydides's observations, we have some answers to this question. During the Athenian plague, innate immunity (Concept 31.2) kept most of the disease agent out of the body. Any agent that penetrated the innate defenses was attacked by the adaptive immune system (Concepts 31.3 and 31.4). The agent was engulfed by macrophages, which broke it up and presented fragments on their cell surfaces, in a complex with class II MHC proteins. These complexes bound to T$_H$ cells with the appropriate T cell receptors. A clone of these T cells ensued, which bound to B cells expressing the appropriate receptor. The B cells then formed clones and made antibodies that bound to the agent; the antibody-bound agent was then destroyed by phagocytes. Meanwhile, if cells were infected by the agent, a parallel series of events occurred in the cellular immune response, resulting in a clone of T$_C$ cells that killed the infected cells (Concept 31.5). These two systems allowed Thucydides to survive the infection that killed so many of his fellow Athenians.

The T and B cells activated by the two adaptive immune responses also formed smaller clones of memory cells. When Thucydides took care of others in Athens who had the disease, the re-entry of the pathogen into his body provoked a rapid, massive T and B cell response. This response prevented the pathogen from proliferating and making him sick again.

An important application of immunological memory is the use of **vaccines**. Exposure to an antigen in a form that does not cause disease can still initiate a primary immune response, generating memory cells without making the person ill. Later, if a pathogen carrying the same antigen attacks, specific memory cells already exist. They recognize the antigen and quickly overwhelm the invaders with a massive production of lymphocytes and antibodies (**FIGURE 31.14**).

Because the antigens used for immunization or vaccination are produced by pathogenic organisms, they must be altered so they cannot cause disease. This is achieved in a variety of

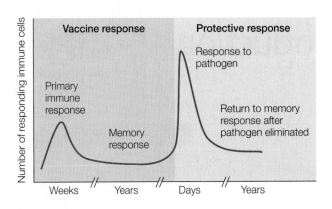

FIGURE 31.14 Vaccination Immunological memory from exposure to an antigen that does not cause disease can result in a massive response to the disease agent when it appears later.

ways, including inactivation of the pathogen by chemicals or heat, mutations that render the pathogen inactive, or recombinant DNA technology to make nonharmful peptides derived from the pathogen. The use of vaccines, especially in children, has been a great success in public health. In industrialized countries, widespread vaccination has completely or almost completely wiped out some deadly diseases, including smallpox, diphtheria, and polio.

31 SUMMARY

concept 31.1 Animals Use Innate and Adaptive Mechanisms to Defend Themselves against Pathogens

- Animal defenses against **pathogens** are based on the body's ability to distinguish between self and nonself.

- **Innate immunity** is a set of nonspecific, inherited mechanisms that protect the body from many kinds of pathogens. These defenses typically act rapidly.

- **Adaptive immunity** is a set of specific mechanisms that respond to specific pathogens. These defenses develop more slowly than nonspecific defenses but are long lasting.

- Many of these defenses are implemented by cells and proteins carried in the circulatory and lymphatic systems. **Review Figure 31.1 and ANIMATED TUTORIAL 31.1**

concept 31.2 Innate Defenses Are Nonspecific

- Innate defenses include physical barriers such as the skin, mucous membranes, and competing resident microorganisms. **Review Figure 31.2**

- Circulating defensive cells, such as **phagocytes** and **natural killer cells**, work to eliminate invaders.

- The **complement system** consists of more than 20 different antimicrobial proteins that act to alter membrane permeability and kill targeted cells.

- The **inflammation** response activates several types of cells and proteins that act against invading pathogens. **Mast cells** release **histamine**, which increases the permeability of blood vessels to aid in inflammation. **Review Figure 31.3 and WEB ACTIVITY 31.1**

concept 31.3 The Adaptive Immune Response Is Specific

- The adaptive immune system recognizes specific **antigens**, responds to an enormous diversity of **antigenic determinants** (also called epitopes), distinguishes self from nonself, and remembers the antigens it has encountered.

- **Clonal selection** accounts for the specificity and diversity of the immune response and for **immunological memory**. **Review Figure 31.5**

- Adaptive immunity includes the **humoral immune response**, which involves antibody production by B cells, and the **cellular immune response**, mediated by T cells. Both require specific receptors to bind to each antigen. **Review Figure 31.6**

concept 31.4 The Adaptive Humoral Immune Response Involves Specific Antibodies

Go to ANIMATED TUTORIAL 31.2

- Naïve B cells are activated by binding of the antigen and by stimulation from T_H cells with the same specificity, and then form **plasma cells**. These cells synthesize and secrete specific antibodies.

- The basic unit of an **immunoglobulin** is a tetramer of four polypeptides: two identical light chains and two identical heavy chains, each consisting of a **constant region** and a **variable region**. The variable regions determine the specificity of an immunoglobulin, and the constant regions of the heavy chain determine its class. **Review Figure 31.7 and WEB ACTIVITY 31.2**

- There are five classes of immunoglobulins, differing in function and in the type of heavy chain.

- B cell genomes undergo recombination events in which the genes that encode specific domains of the immunoglobulin variable regions are randomly selected from large clusters of genes. This DNA rearrangement yields millions of different immunoglobulin proteins. **Review Figures 31.8 and 31.9 and ANIMATED TUTORIAL 31.3**

concept 31.5 The Adaptive Cellular Immune Response Involves T Cells and Their Receptors

Go to ANIMATED TUTORIAL 31.4

- T cell receptors are somewhat similar in structure to the immunoglobulins, having variable and constant regions. **Review Figure 31.10**

- There are several types of T cells. **Cytotoxic T cells (T_C)** recognize and kill virus-infected cells or mutated cells. **T-helper cells (T_H)** direct both the cellular and humoral immune responses.

- The genes of the major histocompatibility complex (MHC) encode membrane proteins that bind antigenic fragments and present them to T cells. **Review Figure 31.11**

- **Regulatory T cells (Tregs)** inhibit the other T cells from mounting an immune response to self antigens. **Review Figure 31.12**

- **Acquired immune deficiency syndrome (AIDS)** arises from depletion of the T_H cells as a result of infection with **human immunodeficiency virus (HIV)**. **Review Figure 31.13**

Animal Reproduction

In 1839 Charles Goodyear discovered a method of curing rubber that made it stronger and more elastic. Soon thereafter, this improved material was used to make condoms. The spread of these contraceptive devices led to fears about licentiousness in American society, and in 1873 Congress passed the Comstock Law outlawing contraceptive devices and the dissemination of information for preventing pregnancy as obscene and immoral. Ignorance of reproductive biology and the lack of birth control technologies at the time were at the root of widespread venereal disease and high rates of infant and maternal mortality.

Margaret Higgins Sanger was born 1879, the middle of her mother's 11 children from 18 pregnancies. Margaret cared for her mother as she was dying from tuberculosis, cervical cancer, and physical exhaustion. Later, as a nurse, Margaret worked among the poor in New York City. Witnessing the misery of so many women and children and the deaths from self-induced and illegal abortions led her to write a newspaper column called "What Every Girl Should Know." She also coined the term "birth control" in her monthly newsletter, "The Woman Rebel." For her efforts, she was indicted under the Comstock Law. Undeterred, in 1916 Margaret opened the first U.S. family planning and birth control clinic, located in Brooklyn; within days she was jailed. An appeal of her sentence, however, led to a landmark ruling that allowed physicians to prescribe contraceptive devices. Margaret's movement for reproductive education and rights grew steadily (and she was arrested seven more times).

Among Margaret's friends was Katherine McCormick, heiress to a considerable fortune. With Margaret's zeal and Katherine's money, they convinced reproductive biologist Gregory Pincus to undertake development of a birth control pill. Pincus showed that high doses of the hormone progesterone prevented ovulation in rabbits, but the hormone had to be isolated from animals and was costly. The creation of synthetic progestins by chemists dramatically lowered the costs, and in 1957 "the pill" was released to treat "gynecological disorders." In 1960 it gained FDA approval for use as an oral contraceptive. However, the Comstock Law was not overturned until 1965, when the U.S. Supreme Court ruled that it violated the individual's right to privacy.

Condoms and "the pill" are widely used means of preventing unwanted pregnancies. See Table 32.1.

QUESTION How does "the pill" prevent conception?

KEY CONCEPTS

32.1 Reproduction Can Be Sexual or Asexual

32.2 Gametogenesis Produces Haploid Gametes

32.3 Fertilization Is the Union of Sperm and Ovum

32.4 Human Reproduction Is Hormonally Controlled

32.5 Humans Use a Variety of Methods to Control Fertility

concept 32.1 Reproduction Can Be Sexual or Asexual

Sexual reproduction is a nearly universal trait in animals. However, a variety of animals, mostly invertebrates, also reproduce asexually, and some species only reproduce asexually. Asexual reproduction is efficient because it does not require a mate, and because all members of a population can reproduce (not just the females). Because offspring produced asexually are genetically identical to the parent that produced them, asexual reproduction is likely to be observed in relatively constant environments where genetic diversity is not critical for species success. In fact, asexual reproduction is a good way to preserve a genotype that is successful in a particular environment—a good strategy as long as that environment does not change.

Three common modes of asexual reproduction are *budding*, *regeneration*, and *parthenogenesis*.

Budding and regeneration produce new individuals by mitosis

Some simple animals produce offspring by **budding**, where new individuals simply form as outgrowths, or buds, from the bodies of mature animals. A bud grows by mitotic cell division and eventually breaks away from the parent (**FIGURE 32.1A**). The bud may grow as large as the parent before it becomes independent.

Some species can reproduce by **regeneration**, the development of a complete individual from a piece of an organism. Echinoderms (sea stars), for example, have remarkable abilities to regenerate. If sea stars are cut into pieces, each piece that includes an arm and a portion of the central disc can grow into a new animal (**FIGURE 32.1B**). In the early 1900s, oyster fishermen in Rhode Island's Narragansett Bay tried to eliminate the sea stars that were preying on their oysters. Whenever they encountered

a sea star, they chopped it up and threw it back into the water. As a result, the sea star population increased explosively.

Parthenogenesis is the development of unfertilized eggs

Offspring can sometimes develop from unfertilized eggs, a mode of asexual reproduction known as **parthenogenesis**. Parthenogenesis is common in arthropods and also occurs in some species of fish, amphibians, and reptiles.

Parthenogenesis in some species requires sexual behavior even though no fertilization occurs. In a parthenogenetic species of whiptail lizard, stimulation from sexual activity triggers the release of eggs from the ovaries. This species has no males, but some of the females act as males in mating behavior, though they do not produce or transfer sperm (**FIGURE 32.1C**). Whether a specific female acts as a female or male depends on cyclical hormonal states. When her estrogen levels are high, she acts as a female. When her progesterone level peaks, she acts as a male.

In some species, parthenogenesis may also be part of the mechanism that determines sex. In honey bees (and in most ants and wasps), males develop from unfertilized eggs and are haploid, whereas females develop from fertilized eggs and are diploid.

Most animals reproduce sexually

Most animals reproduce by sexual reproduction—the joining of two haploid sex cells, called *gametes*, to form a diploid

FIGURE 32.1 Three Forms of Asexual Reproduction (A) Budding. A new individual forms as an outgrowth from an adult hydra. (B) Regeneration. A single severed arm and a piece of the central disc of a mature sea star can regenerate into an entire animal. (C) Parthenogenesis. Whiptail lizards are all females. Depending on their hormonal status, they act as males or females in reproductive behavior. That behavior stimulates ovulation in the lizard behaving as the female, but her ova develop without being fertilized.

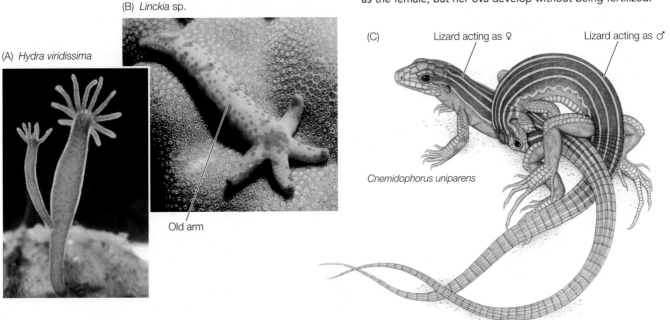

(A) *Hydra viridissima*

(B) *Linckia* sp.

Old arm

(C) Lizard acting as ♀   Lizard acting as ♂

Cnemidophorus uniparens

offspring. Sexual reproduction in animals consists of three fundamental steps:

- *Gametogenesis*: production of haploid gametes via meiotic cell divisions
- *Spawning* or *mating*: behavior that brings gametes together
- *Fertilization*: fusion of gametes to form a diploid zygote

Sexual reproduction involves costs that asexual reproduction does not. Half of the individuals in a sexually reproducing population—the males—do not bear young even though they might help rear them. Potential mates have to find each other, and they must frequently engage in complex reproductive behaviors that have metabolic and opportunity costs and that expose them to predation. Nevertheless, sexual reproduction has one big advantage: the generation of genetic diversity. Two events in meiosis contribute to diversity: *crossing over* between homologous chromosomes and the *independent assortment* of chromosomes. The resulting genetic variation among the gametes of an individual parent and between any two parents produces an enormous potential for genetic variation in the offspring, an evolutionary advantage that can exceed the cost of sex.

Do You Understand Concept 32.1?

- What are the adaptive advantages and disadvantages of asexual reproduction?
- In spring, newly formed rosebuds sometimes become covered with aphids within days. All of these aphids are females and carry preformed female offspring within their bodies. In fall, however, the aphid population includes males, and the females produce eggs rather than live offspring. The eggs survive the winter and hatch into females the following spring. What forms of reproduction are the aphids using? Propose an evolutionary explanation for this strategy.

The next two concepts will examine the fundamental steps of sexual reproduction in more detail. How exactly are gametes produced, and how does fertilization occur?

concept 32.2 Gametogenesis Produces Haploid Gametes

Gametogenesis occurs in the gonads—the *testes* in males and *ovaries* in females. The tiny male gametes, the **sperm**, move by beating their flagella. The larger gametes of females, called *eggs* or **ova** (singular *ovum*), are nonmotile.

Gametes are produced from **germ cells**, which have their origin in the earliest cell divisions of the embryo and remain distinct from all other (*somatic*) cells of the body. The embryonic germ cells move into the gonads, where they proliferate by mitosis, producing diploid **spermatogonia** (singular *spermatogonium*) in males and **oogonia** (singular *oogonium*) in females. The diploid spermatogonia and oogonia then multiply

by mitosis, producing **primary spermatocytes** and **primary oocytes** that enter meiosis (see Figure 7.11) and produce the haploid gametes (sperm and ova). The production of sperm is called **spermatogenesis**, and the production of ova is called **oogenesis**.

Spermatogenesis produces four sperm from one parent cell

As illustrated in **FIGURE 32.2A**, a primary spermatocyte undergoes the first meiotic division to form two secondary spermatocytes. The second meiotic division produces four haploid **spermatids** (two from each secondary spermatocyte) which bear little resemblance to mature sperm. Through further differentiation, each spermatid becomes compact, streamlined, and grows a flagellum to become a motile sperm. We will look at the production of human sperm in detail in Concept 32.4.

Oogenesis produces one large ovum from one parent cell

Primary oocytes immediately enter prophase of the first meiotic division. In many species, including humans, the oocyte then undergoes developmental arrest and may remain in this stage for days, months, or years (**FIGURE 32.2B**). In the human female, this period of arrest may be at least 10 years (i.e., until puberty), and some primary oocytes may remain in prophase I for up to 50 years (i.e., until menopause). In contrast, spermatogenesis continues, uninterrupted, to completion once the primary spermatocyte has differentiated.

Near the end of the oocyte's prolonged prophase I, it grows larger and acquires all the nutrients, raw materials, and RNA that the ovum will need to survive its first few cell divisions after fertilization. The ovum must contain sufficient nutrients to maintain the embryo until it is either nourished by the maternal circulatory system or can feed on its own.

When a primary oocyte resumes meiosis, its nucleus completes the first meiotic division near the surface of the cell. The daughter cells of this division receive grossly unequal shares of cytoplasm. This asymmetry represents another major difference from spermatogenesis, in which cytoplasm is apportioned equally. The daughter cell that receives almost all the cytoplasm becomes the **secondary oocyte**, and the one that receives almost none forms the **first polar body**.

A second period of arrested development occurs after formation of the secondary oocyte. The egg may be expelled from the ovary in this condition. In many species, including humans, the second meiotic division is not completed until the egg is fertilized by a sperm.

The second meiotic division of the secondary oocyte is again characterized by an asymmetrical division of the cytoplasm. One daughter cell is a large **ootid**, which immediately differentiates to become a mature ovum. The other daughter cell forms the **second polar body**. Polar bodies degenerate, so the end result of oogenesis is only one mature ovum from each primary oocyte that enters meiosis. However, that ovum is a large, well-provisioned cell.

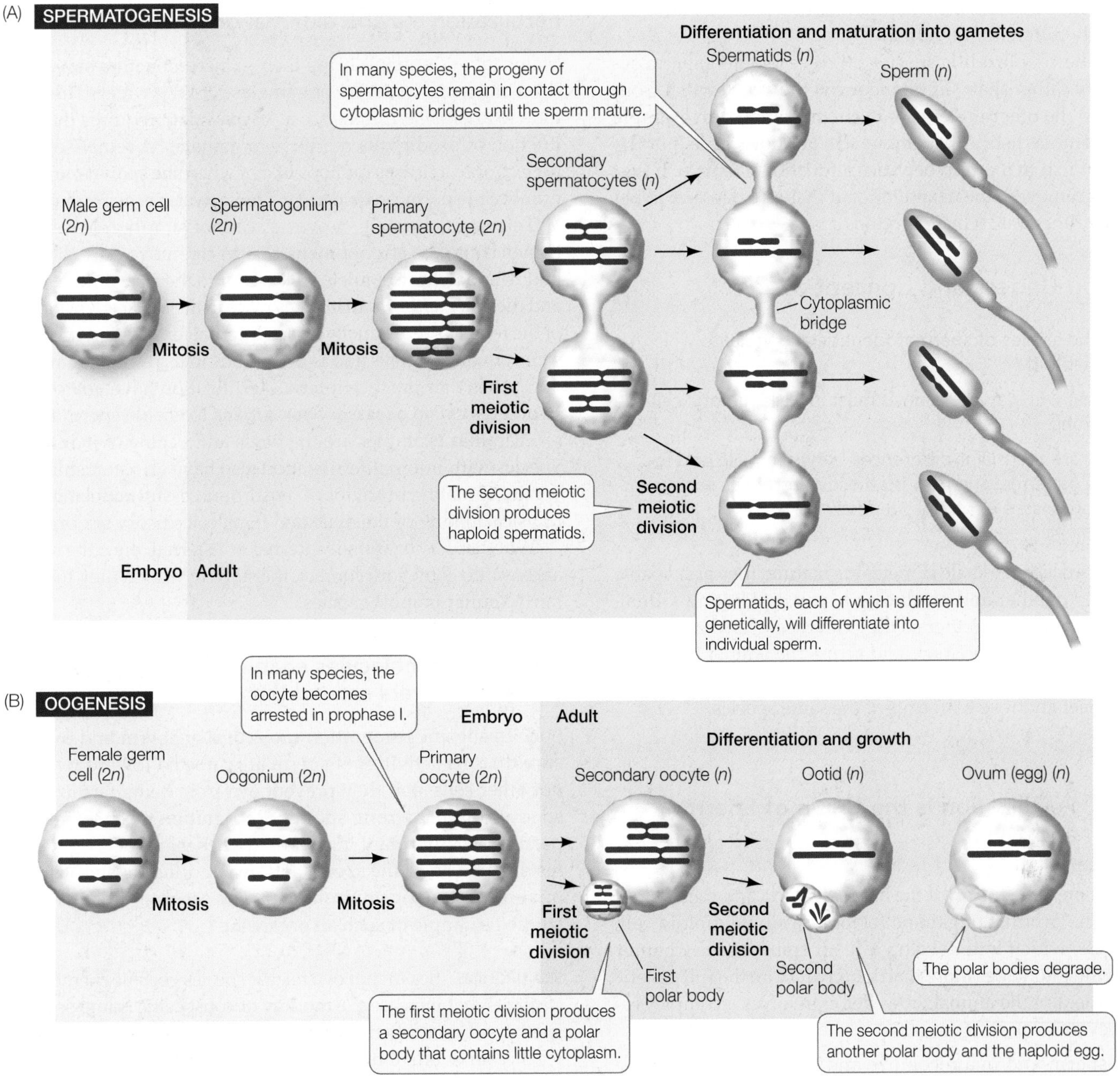

(A) **SPERMATOGENESIS**

Differentiation and maturation into gametes

Spermatids (*n*)

Sperm (*n*)

In many species, the progeny of spermatocytes remain in contact through cytoplasmic bridges until the sperm mature.

Secondary spermatocytes (*n*)

Male germ cell (2*n*)

Spermatogonium (2*n*)

Primary spermatocyte (2*n*)

Mitosis

Mitosis

First meiotic division

Cytoplasmic bridge

The second meiotic division produces haploid spermatids.

Second meiotic division

Embryo   Adult

Spermatids, each of which is different genetically, will differentiate into individual sperm.

(B) **OOGENESIS**

In many species, the oocyte becomes arrested in prophase I.

Embryo   Adult

Differentiation and growth

Female germ cell (2*n*)

Oogonium (2*n*)

Primary oocyte (2*n*)

Secondary oocyte (*n*)

Ootid (*n*)

Ovum (egg) (*n*)

Mitosis

Mitosis

First meiotic division

Second meiotic division

First polar body

Second polar body

The polar bodies degrade.

The first meiotic division produces a secondary oocyte and a polar body that contains little cytoplasm.

The second meiotic division produces another polar body and the haploid egg.

FIGURE 32.2 Gametogenesis Male and female germ cells proliferate by mitosis to produce diploid spermatogonia and oogonia, which mature into primary spermatocytes and oocytes before entering meiosis. (A) Spermatogonia continue to divide by mitosis in adult males, producing a steady supply of primary spermatocytes. The primary spermatocytes divide meiotically to produce haploid spermatids, which differentiate into sperm. (B) In many species, primary oocytes remain arrested in prophase I of meiosis for a long period. Each primary oocyte produces one haploid ovum.

Very few primary oocytes complete all the meiotic stages shown in Figure 32.2B, and as a result, females produce far fewer gametes than do males. For example, a newborn girl has about a million primary oocytes in each ovary. By the time she reaches puberty, she has about 200,000; the rest have degenerated. The average woman goes through about 450 menstrual cycles during her fertile years, releasing one ovum each cycle. Around the age of 50, a woman reaches **menopause**—the end of fertility—at which time she may have few, if any, oocytes left. Thus a woman produces a total of only a few hundred ova in a lifetime, far below a man's production of over 100 million sperm per *day*.

Hermaphrodites can produce both sperm and ova

In most species, gametes are produced by individuals that are either male or female. In **hermaphroditic** species, however, a single individual may produce both sperm and ova; such species are found among both invertebrates and vertebrates. A mature earthworm, for example, produces both sperm and ova simultaneously. When two earthworms mate, they exchange sperm, and as a result, the ova of each are fertilized by another, genetically different individual.

Some hermaphroditic species produce ova and sperm sequentially rather than simultaneously; they can switch from one sex to the other under certain circumstances. An example is the anemone fish. All anemone fish are born male, but the dominant fish in a group becomes a functional female. If that female is removed, the next dominant male changes sex to become the functional female producing ova.

Do You Understand Concept 32.2?

- List the stages of spermatogenesis and indicate which are haploid.
- Explain why gametogenesis must include meiotic cell divisions.
- What are the major differences between male and female gametes, and how do these differences arise in the processes of gametogenesis?

Once the male and female gametes are mature, they are capable of joining together and producing a new diploid individual. Bringing gametes together so they can fuse—fertilization—is crucial to life's continuation, and across the animal kingdom, surprisingly similar mechanisms have evolved to ensure that sperm meet and fuse with eggs of the same species.

concept 32.3 Fertilization Is the Union of Sperm and Ovum

Fertilization is the fusion of the haploid sperm and the haploid ovum. The product of gamete fusion is a single diploid cell, the *zygote*, which will develop into an embryo (see Chapter 33). Fertilization does more than simply restore the full genetic complement of the animal, however. It involves a complex series of events:

- The sperm and ova recognize and bind to each other by means of species-specific molecules.
- The sperm is *activated*, enabling it to gain access to the plasma membrane of a single ovum.
- The plasma membrane of the ovum fuses with the plasma membrane of a single sperm; the ovum then blocks entry of any additional sperm.
- The ovum is metabolically activated and stimulated to start development.
- The ovum and sperm nuclei fuse to create the diploid nucleus of the zygote.

Before the events of fertilization can begin, however, the male and female gametes must meet.

Fertilization may be external or internal

Many aquatic animals bring their gametes together simply by releasing large numbers of gametes into the water. This prelude to *external fertilization*, called **spawning**, requires the production of prodigious numbers of gametes. A female oyster, for example, releases millions of ova when she spawns, and the number of sperm produced by a male oyster is astronomical.

Terrestrial animals, however, cannot simply release their gametes into the environment. Sperm can move only through liquid, and delicate gametes released into the air would dry out and die. *Internal fertilization* takes place in the moist environment of the female's reproductive tract. Internal fertilization occurs in some aquatic animals and is ubiquitous in terrestrial animals.

Species that practice internal fertilization have anatomical structures called **accessory sex organs** to enable sperm transfer. Familiar examples are the penis and vagina of mammals. Species with internal fertilization also have an astonishing diversity of mating behaviors to ensure successful **copulation**: the physical joining of the male and female accessory sex organs.

Whether fertilization is internal or external, once the sperm and ova do come into contact, the steps by which they fuse are fairly similar in most species.

Recognition molecules enable sperm to penetrate protective layers around the ovum

Species-specific recognition molecules of sperm and ova ensure that the activities of sperm are directed toward ova and not other cells, and they prevent ova from being fertilized by sperm from the wrong species. Recognition mechanisms are particularly important in aquatic species that practice external fertilization since the sperm and ova of different species may mix in the aquatic environment. The sea urchin is the best-studied example of such an organism.

SEA URCHINS Sea urchin ova release species-specific chemical attractants that increase the motility of conspecific (same-species) sperm and cause them to swim toward the ova. When a sperm reaches an ovum, it must get through two protective layers—a **jelly coat** and a proteinaceous **vitelline envelope**—before it comes into contact with the ovum's plasma membrane (**FIGURE 32.3A**). The success of a sperm's assault on the ovum's protective layers depends on the **acrosome**, a structure at the tip of the sperm head that contains enzymes and other proteins enclosed by a membrane. When the sperm makes contact with the protective layers surrounding the egg, substances in those layers trigger an *acrosomal reaction* in the sperm that begins with the breakdown of the plasma membrane covering the sperm head (**FIGURE 32.3B**). The acrosomal enzymes are then released and digest a hole through the jelly coat.

A structure called the *acrosomal process*, produced by polymerization of the protein actin, then extends out of the head of the sperm. The sea urchin acrosomal process is coated with species-specific recognition molecules known as *bindins*. Bindins recognize and bind to receptors on the vitelline envelope and on the egg membrane. This final recognition process brings about

(A)

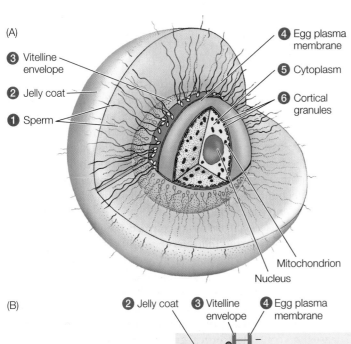

❸ Vitelline envelope
❷ Jelly coat
❶ Sperm
❹ Egg plasma membrane
❺ Cytoplasm
❻ Cortical granules
Mitochondrion
Nucleus

the slow block to polyspermy and egg activation. Plasma membranes of sperm and ovum fuse, forming a *fertilization cone* that engulfs the sperm head and brings it into the egg cytoplasm.

MAMMALS In animals in which fertilization is internal (i.e., occurs within the female's body), mating behaviors can guarantee species specificity, but ova–sperm recognition mechanisms still exist. The mammalian ovum is protected by a thick gelatinous layer called the **cumulus** surrounding a glycoprotein envelope called the **zona pellucida** (plural *zonae pellucidae*), or simply the *zona* (**FIGURE 32.4**). When sperm make contact with the zona, species-specific glycoproteins bind to recognition molecules on the sperm. This binding triggers the acrosomal reaction, releasing acrosomal enzymes that digest a path through the zona. When the sperm head reaches the ovum plasma membrane, other proteins facilitate the fusion.

(B)

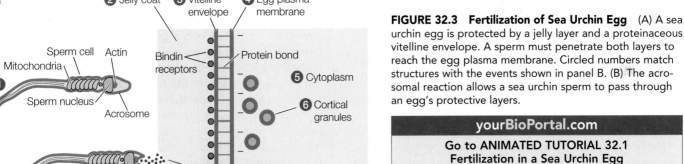

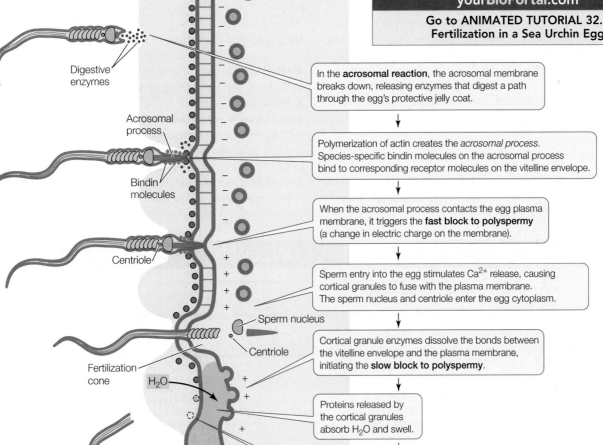

❷ Jelly coat ❸ Vitelline envelope ❹ Egg plasma membrane

Sperm cell Actin
Mitochondria
Sperm nucleus
Acrosome
❶

Bindin receptors
Protein bond
❺ Cytoplasm
❻ Cortical granules

Digestive enzymes

Acrosomal process

Bindin molecules

Centriole

Sperm nucleus
Centriole

Fertilization cone
H₂O

Fertilization envelope

FIGURE 32.3 Fertilization of Sea Urchin Egg (A) A sea urchin egg is protected by a jelly layer and a proteinaceous vitelline envelope. A sperm must penetrate both layers to reach the egg plasma membrane. Circled numbers match structures with the events shown in panel B. (B) The acrosomal reaction allows a sea urchin sperm to pass through an egg's protective layers.

yourBioPortal.com

**Go to ANIMATED TUTORIAL 32.1
Fertilization in a Sea Urchin Egg**

In the **acrosomal reaction**, the acrosomal membrane breaks down, releasing enzymes that digest a path through the egg's protective jelly coat.

Polymerization of actin creates the *acrosomal process.* Species-specific bindin molecules on the acrosomal process bind to corresponding receptor molecules on the vitelline envelope.

When the acrosomal process contacts the egg plasma membrane, it triggers the **fast block to polyspermy** (a change in electric charge on the membrane).

Sperm entry into the egg stimulates Ca²⁺ release, causing cortical granules to fuse with the plasma membrane. The sperm nucleus and centriole enter the egg cytoplasm.

Cortical granule enzymes dissolve the bonds between the vitelline envelope and the plasma membrane, initiating the **slow block to polyspermy**.

Proteins released by the cortical granules absorb H₂O and swell.

Enzymes remove sperm-binding receptors. The vitelline envelope hardens, forming a fertilization envelope.

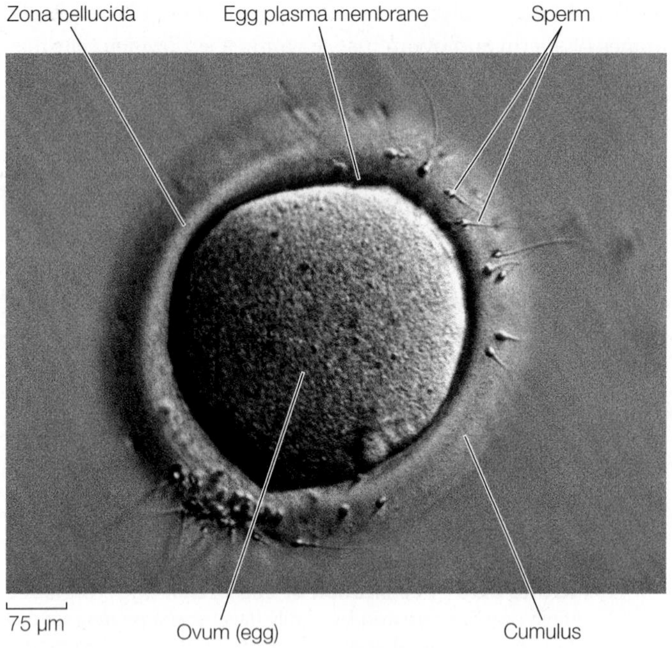

FIGURE 32.4 **Barriers to Mammalian Sperm** This human ovum, like other mammalian ova, is surrounded by a cumulus and a zona pellucida. A single sperm must penetrate both layers to fertilize the ovum.

The importance of these species-specific recognition mechanisms was revealed in experiments in which human ova were stripped of their zonae pellucidae and exposed to hamster sperm. Fertilization took place, resulting in a hamster–human hybrid zygote. The zygote did not survive its first cell division, but hybrid fertilization would never have occurred had the zona pellucida been intact.

Only one sperm can fertilize an ovum

The entry of a sperm into the ovum initiates a programmed sequence of events, the first of which are **blocks to polyspermy**— that is, mechanisms that prevent additional sperm from entering the ovum.

SEA URCHINS Blocks to polyspermy have been studied extensively in sea urchin ova. Within seconds after the sperm and ovum plasma membranes make contact, an influx of sodium ions (Na^+) changes the electric charge difference across the ovum's plasma membrane. This *fast block to polyspermy* prevents the fusion of any other sperm with the ovum's plasma membrane. The change in membrane charge lasts only about a minute, but that is enough time to allow an additional block to sperm entry to develop.

The *slow block to polyspermy* converts the vitelline envelope into a physical barrier that sperm cannot penetrate. Just under the ovum's plasma membrane are vesicles called *cortical granules* (see Figure 32.3) that contain enzymes and other proteins. Sperm entry stimulates the release of calcium ions (Ca^{2+}) from the ovum's endoplasmic reticulum; this wave of Ca^{2+} causes the cortical granules to fuse with the plasma membrane and release their contents. Cortical granule enzymes break the bonds between the vitelline envelope and plasma membrane, and cortical granule proteins attract water into the space between the two layers. As a result, the vitelline envelope swells and rises from the surface of the ovum to form a *fertilization envelope*. Cortical granule enzymes also degrade the sperm-binding molecules on the surface of the fertilization envelope and cause it to harden. All of these mechanisms prevent additional sperm from reaching the surface of the ovum.

MAMMALS The fast block to polyspermy does not occur in mammals, but something similar to the slow block does. As in sea urchins, a release of Ca^{2+} results in the fusion of cortical granules with the ovum plasma membrane. Although no fertilization envelope forms, the cortical granule enzymes destroy the sperm-binding glycoproteins in the zona pellucida. The rise in Ca^{2+} in its cytosol also signals the ovum to complete meiosis. The stage is thus set for the fusion of the gametic nuclei of the sperm and ovum, followed by the first cell division.

APPLY THE CONCEPT

Fertilization is the union of sperm and ovum

The zona pellucida of mammalian ova is a glycoprotein matrix secreted by the developing oocyte. Several glycoproteins have been isolated from mouse zonae pelluicidae and named ZP1, ZP2, and ZP3. To identify which, if any, of these glycoproteins bind to and activate sperm, investigators collected mouse sperm and incubated them in media with different glycoprotein concentrations (here listed as 1, 3, and 5, in terms of their relative glycoprotein levels). Sperm were then exposed to mouse zonae to assess their binding capacity.

1. Plot these data using percentage of sperm binding as your y axis and glycoprotein concentration as your x axis.

2. Which glycoprotein stimulates the acrosomal reaction?

3. Design an additional experiment to test your conclusion.

	CONTROL (NO GLYCOPROTEIN)	ZP1			ZP2			ZP3		
		1	3	5	1	3	5	1	3	5
Sperm binding (%)	100	95	98	92	88	94	93	54	42	30

Fertilized ova may be released into the environment or retained in the mother's body

The development of the embryo beyond the zygote stage is the subject of Chapter 33. However, we note here that among the animals, two patterns of care and nurture of the embryo have evolved: oviparity (egg laying) and viviparity (live bearing).

Oviparous animals deposit their fertilized ova in the environment, and their embryos develop outside the mother's body. Oviparous terrestrial animals such as insects, reptiles, and birds protect their ova with membranes or shells that deter some potential predators and decrease water loss.

Viviparous animals retain the embryo within the mother's body during its early developmental stages. Examples of viviparity exist in all vertebrate groups except the crocodiles, turtles, and birds. Viviparity is especially well developed in mammals. Almost all mammals have a specialized portion of the female reproductive tract, the **uterus** or *womb*, that holds the embryo and interacts with it to produce a **placenta**. The placenta enables the exchange of nutrients and wastes between the blood of the mother and that of the embryo.

Do You Understand Concept 32.3?

- What two layers surround a mammalian ovum? Where in these layers, and how, is the acrosomal reaction initiated?

- Describe the slow and fast blocks to polyspermy in sea urchin ova and explain how they prevent polyspermy.

- Suppose an ovum has a metabolic defect such that it contains very few calcium ions. How might this affect that ovum's capability to undergo fertilization?

Having covered the general aspects of animal gametogenesis and fertilization, we will next consider the male and female reproductive systems in mammals, using our own species as the primary example.

concept 32.4   Human Reproduction Is Hormonally Controlled

We learned in Chapter 30 that the gonads produce sex steroids as well as gametes, and we saw that these hormones play key roles in the development of the reproductive system. We will now see how these hormones regulate every aspect of mammalian sexual reproduction.

Male sex organs produce and deliver semen

The male reproductive organs are diagrammed in **FIGURE 32.5**. Sperm are produced in the testes (the paired male gonads). The testes of most

FIGURE 32.5 Reproductive Organs of the Human Male The organs of the male reproductive tract are shown (A) from the rear and (B) from the side.

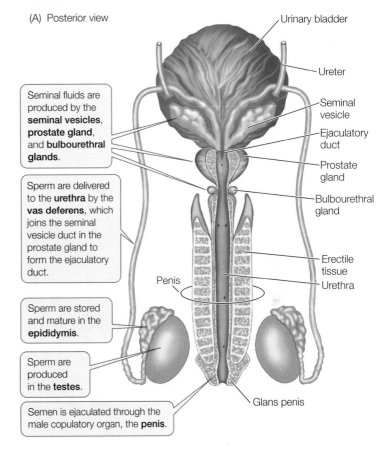

(A) Posterior view

Urinary bladder
Ureter
Seminal vesicle
Ejaculatory duct
Prostate gland
Bulbourethral gland
Erectile tissue
Urethra
Penis
Glans penis

Seminal fluids are produced by the **seminal vesicles**, **prostate gland**, and **bulbourethral glands**.

Sperm are delivered to the **urethra** by the **vas deferens**, which joins the seminal vesicle duct in the prostate gland to form the ejaculatory duct.

Sperm are stored and mature in the **epididymis**.

Sperm are produced in the **testes**.

Semen is ejaculated through the male copulatory organ, the **penis**.

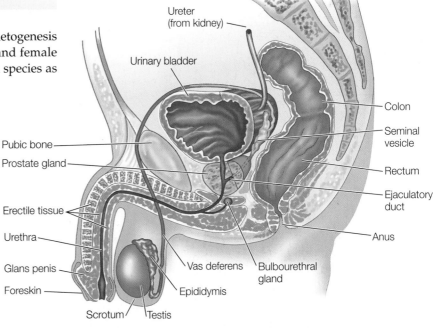

(B) Side view

Ureter (from kidney)
Urinary bladder
Colon
Seminal vesicle
Rectum
Ejaculatory duct
Anus
Pubic bone
Prostate gland
Erectile tissue
Urethra
Glans penis
Foreskin
Scrotum
Testis
Vas deferens
Bulbourethral gland
Epididymis

yourBioPortal.com

Go to WEB ACTIVITY 32.1
The Human Male Reproductive Tract

mammals are located outside the body cavity in a pouch of skin called the **scrotum**. Why should the testes be located outside the body cavity? In most mammals, the optimal temperature for the health of the testes and for spermatogenesis is slightly lower than the normal body temperature. The scrotum keeps the testes at this optimum by expanding or contracting, thus placing the testes closer to or farther from the heat of the body.

Sperm are the product of the male reproductive system. In addition to sperm, **semen** contains a complex mixture of substances that support the sperm and facilitate fertilization. Sperm make up less than 5 percent of the volume of semen.

SPERMATOGENESIS Spermatogenesis begins at puberty and continues throughout a man's life. It takes place within the **seminiferous tubules**, great lengths of which are tightly coiled in each testis (**FIGURE 32.6A**). Between the seminiferous tubules are clusters of *Leydig cells*, which produce testosterone (**FIGURE 32.6B**). Spermatogonia reside in the outermost regions of the tubules (**FIGURE 32.6C**) and are intimately associated with *Sertoli cells*, which provide nutrients for the developing sperm.

Each primary spermatocyte gives rise to four spermatids (see Figure 32.2A) that develop into sperm as they migrate toward the lumen of the seminiferous tubule. The nucleus

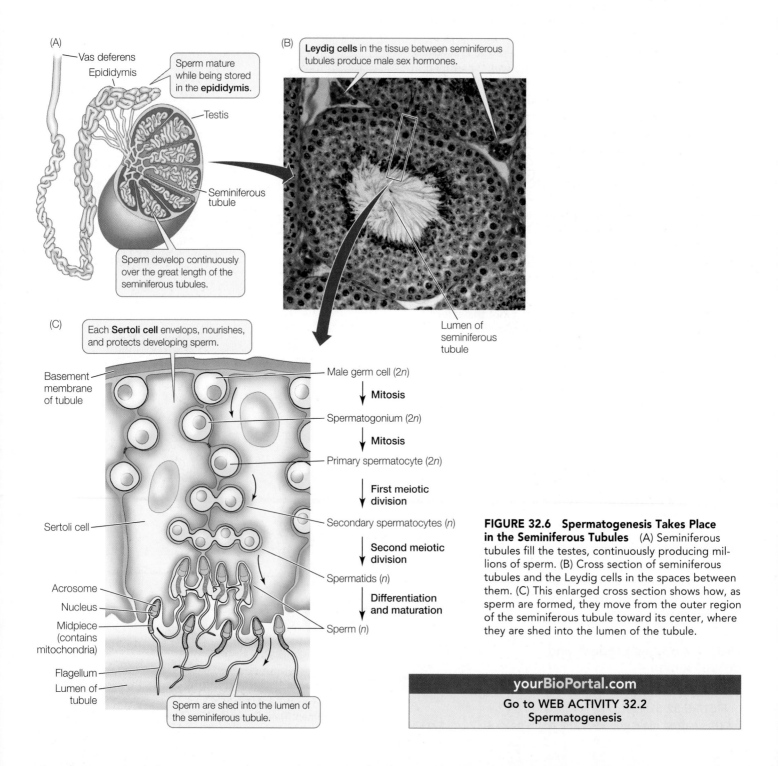

FIGURE 32.6 Spermatogenesis Takes Place in the Seminiferous Tubules (A) Seminiferous tubules fill the testes, continuously producing millions of sperm. (B) Cross section of seminiferous tubules and the Leydig cells in the spaces between them. (C) This enlarged cross section shows how, as sperm are formed, they move from the outer region of the seminiferous tubule toward its center, where they are shed into the lumen of the tubule.

yourBioPortal.com

Go to WEB ACTIVITY 32.2
Spermatogenesis

becomes compact, and the surrounding cytoplasm is lost. A flagellum develops, and mitochondria that will provide energy for its motility are packed into a midpiece between the head and the flagellum. An acrosome forms over the nucleus in the head of the sperm.

Immature sperm are shed into the lumen of the seminiferous tubules. From there, they move into the *epididymis*, where they mature and become motile. From the epididymis, the sperm travel in the *vas deferens* (plural *vasa deferentia*) to a point where the vas deferens joins the duct from the seminal vesicles and becomes the semen-carrying *ejaculatory duct*, which joins the *urethra* at the base of the penis. The urethra—the common final duct for both the urinary and reproductive systems—originates in the bladder, runs through the penis, and opens to the outside of the body at the tip of the penis (see Figure 32.5).

SEMINAL FLUID The components of the semen other than sperm are referred to as *seminal fluid* and are the products of several accessory glands: the paired *seminal vesicles*, the *prostate gland*, and the *bulbourethral glands* (see Figure 32.5). Seminal fluid is thick because it contains mucus and fibrinogen. It also contains fructose, an energy source for the sperm.

The *prostate gland* produces an alkaline fluid that neutralizes the acidity in the male and female reproductive tracts and makes those environments more hospitable to sperm. Prostate fluid also contains a clotting enzyme that causes fibrinogen to convert the semen into a gelatinous mass, facilitating its ejaculation into and retention in the upper regions of the female reproductive tract. Another enzyme in the prostate fluid, fibrinolysin, is activated shortly after it enters the female reproductive tract, where it dissolves clotted semen and liberates the sperm.

The *bulbourethral glands* produce a small volume of an alkaline secretion that helps neutralize acidity in the urethra and lubricates it to facilitate ejaculation. Secretions of the bulbourethral glands precede ejaculation and can carry with them residual sperm from prior sexual activity. Thus it is possible for pregnancy to occur even if the penis is withdrawn from the female just before ejaculation (a rather ineffective birth control practice known as *coitus interruptus*).

MALE SEXUAL FUNCTION The **penis** is the male copulatory organ. The shaft of the penis is covered with normal skin, but the highly sensitive tip, the *glans penis*, is covered with thinner, more sensitive skin. A fold of skin called the *foreskin* (or prepuce) covers the glans penis. The procedure known as *circumcision* removes a portion of the foreskin.

Sexual stimulation triggers responses in the nervous system that result in penile **erection**. Nerve endings release a gaseous neurotransmitter, nitric oxide (NO), onto blood vessels leading into the penis. NO stimulates production of the second messenger cGMP, which causes the vessels to dilate, filling shafts of spongy *erectile tissue* along the length of the penis with blood and compressing the vessels that normally carry blood out of the penis. As a result, the penis becomes stiff and engorged with blood.

At the climax of copulation, 2 to 6 milliliters of semen are *ejaculated*—propelled via smooth muscle contractions through the vasa deferentia and the urethra. The muscle contractions are accompanied by feelings of intense pleasure known as *orgasm*. After ejaculation, NO decreases and enzymes break down cGMP, causing the blood vessels flowing into the penis to constrict. The blood pressure in the erectile tissue decreases, relieving the compression of the blood vessels leaving the penis, and the erection declines. Following an ejaculation, there is a *refractory period*, during which sexual stimulation cannot generate an erection.

Erectile dysfunction (ED), or *impotence*, is the inability to achieve or sustain an erection. Drugs used to treat ED act by inhibiting the breakdown of cGMP, thus enhancing the effect of NO.

Male sexual function is controlled by hormones

Like the maintenance of the male secondary sexual characteristics described in Chapter 30 (body form, facial hair, and a deep voice), spermatogenesis depends on the hormone testosterone, which is produced by the Leydig cells of the testes. Regular pulses of gonadotropin-releasing hormone (GnRH) secreted by the hypothalamus stimulate anterior pituitary cells to increase their secretion of luteinizing hormone (LH) and follicle-stimulating hormone (FSH) (**FIGURE 32.7**). LH then stimulates the Leydig cells to secrete testosterone. FSH and testosterone together stimulate the Sertoli cells to support spermatogenesis. The Sertoli cells also produce a hormone called *inhibin*, which exerts negative feedback on the anterior pituitary cells that produce and secrete FSH.

LINK For more information on the gonadal hormones, or sex steroids, see Concept 30.4

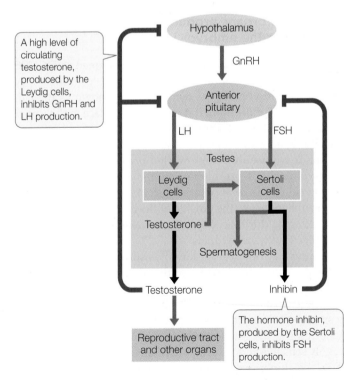

A high level of circulating testosterone, produced by the Leydig cells, inhibits GnRH and LH production.

The hormone inhibin, produced by the Sertoli cells, inhibits FSH production.

FIGURE 32.7 Male Reproductive Hormones The male reproductive system is under hormonal control by the hypothalamus and the anterior pituitary.

APPLY THE CONCEPT

Human reproduction is hormonally controlled

In human males the testes develop in the abdomen over the course of gestation and normally descend into the scrotum shortly before birth. If they do not descend—a condition known as cryptorchidism—the functional cells of the testes undergo changes that can result in reduced fertility. In most cases, the testes descend on their own during an infant's first year, but if they do not, surgical intervention is common. A medical question is whether it is advantageous for the surgery to take place at a very early age.

ADULT FUNCTIONAL MEASURE[a]	AGE AT SURGERY				
	0–2 YRS	2–5 YRS	5–8 YRS	8–11 YRS	P-VALUE[b]
Sperm count (x 10⁶/mL)	60 ± 42	48 ± 47	46 ± 34	53 ± 52	<0.85
Testosterone (ng/dL)	653 ± 118	656 ± 171	570 ± 143	557 ± 193	<0.03
Inhibin (pg/mL)	159 ± 60	106 ± 54	121 ± 59	103 ± 36	<0.03
FSH (units/L)	4.4 ± 3.2	6.1 ± 3.3	6.5 ± 6.4	7.4 ± 4.3	<0.09
LH (units/L)	4.8 ± 3.0	4.0 ± 1.4	4.9 ± 5.5	4.2 ± 1.8	<0.81

[a] Measures are given as the mean value ± 1 standard deviation (see Appendix B).
[b] P-values indicate the level of significance of the correlation between age at surgery and the adult testicular function measured.

The data in the table represent measures of testicular function (sperm count and blood levels of four reproductive hormones) for 85 young adult men who underwent surgery at different ages.

1. Explain why we can claim significant correlations with inhibin and testosterone levels, but only a trend for FSH levels.

2. Referring to Figure 32.7, explain the possible relationship between the inhibin and FSH results.

3. Given the observed differences in inhibin, testosterone, and FSH, suggest why there is no significant difference in sperm counts.

4. Based on the results shown here, would you advise the parents of an infant with cryptorchidism to have the procedure done as soon as possible or to postpone surgery until middle childhood?

Female sex organs produce ova, receive sperm, and nurture the embryo

The female reproductive organs are diagrammed in **FIGURE 32.8**. When an ovum matures, it is released from the ovary and moves into one of the **oviducts** (also known as the *Fallopian tubes*). Fertilization, if it occurs, takes place in the upper region of the oviduct. Whether or not the ovum is fertilized, cilia lining the oviduct propel it slowly toward the uterus. The uterus is a muscular, thick-walled organ within which the embryo develops. At the bottom, the uterus narrows into a region called the **cervix** (Latin, "neck"), which leads into the **vagina** that opens to the outside of the body. Sperm deposited in the vagina during copulation move up through the cervix and uterus into the oviducts.

In humans, two sets of skin folds surround the opening of the vagina and the opening of the urethra. The inner, more delicate folds are the *labia minora*; the outer, thicker folds are the *labia majora*. At the anterior tip of the labia minora is the *clitoris*, a small bulb of erectile tissue that has the same developmental origins as the glans penis (see Figure 30.13). The clitoris is highly sensitive and plays an important role in sexual response. The labia minora and the clitoris become engorged with blood in response to sexual stimulation. Unlike men, women do not have a well-defined refractory period, and some women can experience several orgasms in rapid succession.

The female reproductive cycle is controlled by hormones

Women undergo a regularly repeating reproductive cycle that is about 28 days long. In the first half of the cycle, an ovum matures and is released from the ovary. In the second half of the cycle, the ovary ceases growth and maturation of ova, and the uterus prepares for the possible arrival of an embryo. If no embryo appears, the uterus sheds its lining, and the cycle begins again. Thus the female reproductive cycle consists of two simultaneous cycles: an *ovarian cycle* that produces mature ova and hormones; and a *uterine*, or *menstrual*, *cycle* that prepares the uterus for the arrival of an embryo.

In the ovary, each primary oocyte is surrounded by a layer of ovarian cells. An oocyte and its surrounding cells constitute a **follicle** (**FIGURE 32.9**). At the beginning of a cycle, the anterior pituitary increases its secretion of FSH and LH. In response, some 6 to 12 follicles begin to grow in the ovary. Thus the first 2 weeks of the ovarian cycle are called the *follicular phase*.

As the follicles grow, they synthesize and secrete increasing amounts of estrogen. After about a week, one follicle is larger than the rest; it continues to grow while the others cease to develop and die (they undergo *atresia*). At first, estrogen exerts negative feedback control on gonadotropin release by the anterior pituitary. Then, on about day 12, estrogen exerts positive feedback control on the pituitary, resulting in a surge

(A) Front view

Eggs mature in and are released by the **ovaries**.

Oviduct (Fallopian tube)

Fimbria

Ovary

Ligament

Endometrium (lines uterus)

Cervix

Sperm are deposited in the **vagina** during copulation. The vagina is also the birth canal.

Eggs are taken into the **oviducts**, where they travel to the **uterus**. Fertilization occurs in the upper regions of the oviduct, where development begins.

The blastocyst implants in the **endometrium** of the uterus, where embryonic development continues.

The neck of the uterus is the **cervix**, which remains closed during pregnancy and dilates to allow childbirth.

FIGURE 32.8 Reproductive Organs of the Human Female The female reproductive organs are shown (A) from the front and (B) from the side.

yourBioPortal.com

Go to WEB ACTIVITY 32.3 The Human Female Reproductive Tract

(B) Side view

Ovary

Oviduct

Urinary bladder

Pubic bone

Urethra

Clitoris

Labia majora

Labia minora

Vagina

Colon

Uterus

Cervix

Rectum

Anus

FIGURE 32.9 The Ovarian Cycle The ovarian cycle progresses from the development of a follicle to ovulation and finally to growth and degeneration of the corpus luteum. The micrograph at right shows a mature mammalian follicle; the oocyte is in the center.

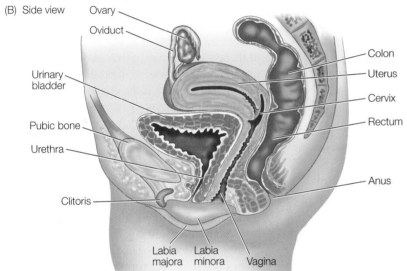

(A)

5 The remaining follicle cells form the **corpus luteum**, which produces progesterone and estrogen.

6 If pregnancy does not occur, the corpus luteum degenerates.

Primary oocytes

Ligament (holds ovary in place in the abdomen)

Ruptured follicle

START

1 About once a month between puberty and menopause, 6–12 primary oocytes begin to mature. A primary oocyte and its surrounding cells constitute a **follicle**.

Follicle

(B)

4 At ovulation, the follicle ruptures, releasing an oocyte.

Ovary

3 After 1 week, usually only one primary oocyte continues to develop. A meiotic division just before ovulation creates the secondary oocyte (n).

2 The surrounding follicular cells nourish the developing oocyte and release estrogen.

Primary oocyte

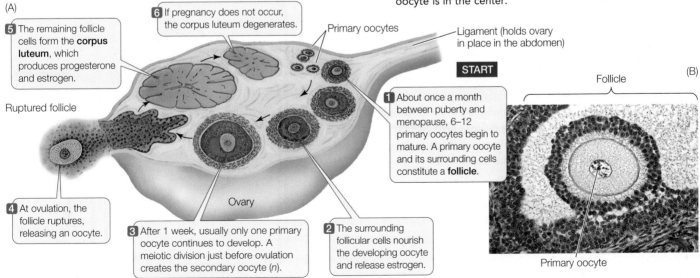

(A) Gonadotropins (from anterior pituitary)

FSH and LH secretion are under control of GnRH from the hypothalamus and the ovarian hormones estrogen and progesterone (part C).

Estrogen inhibits LH and FSH release | Estrogen stimulates LH and FSH release | Estrogen inhibits LH and FSH release

LH surge triggers ovulation.

Luteinizing hormone (LH)

Follicle-stimulating hormone (FSH)

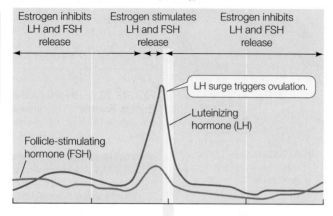

(B) Events in ovary (ovarian cycle)

FSH stimulates the development of follicles; the LH surge causes ovulation and then the development of the corpus luteum.

Oocyte maturation | Developing follicle | Ovulation (day 14) | Developing oocyte | Corpus luteum

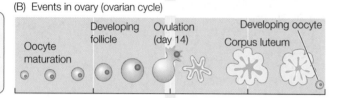

(C) Ovarian hormones and the uterine cycle

Estrogen and progesterone stimulate the development of the endometrium in preparation for pregnancy.

Estrogen

Progesterone

(D) Endometrium of uterus

Highly proliferated and vascularized endometrium

Bleeding and sloughing (menstruation)

Thickness of endometrium

0   7   14   21   28
Day of uterine cycle

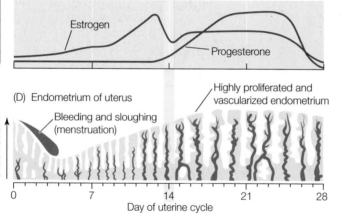

FIGURE 32.10   The Ovarian and Uterine Cycles   During a woman's ovarian and uterine cycles, coordinated changes occur in (A) gonadotropin release by the anterior pituitary, (B) the ovary, (C) the release of female sex steroids, and (D) the uterus. The cycles begin with the onset of menstruation; ovulation is at midcycle (yellow bar).

yourBioPortal.com

**Go to ANIMATED TUTORIAL 32.2
The Ovarian and Uterine Cycles**

the granulosa cells to produce more FSH receptors (positive feedback). The increasing levels of estrogen in the circulation have a negative feedback effect on the hypothalamus and the anterior pituitary (**FIGURE 32.11**). As a result, FSH and LH levels fall. Negative feedback on FSH production also derives from *inhibin* released from granulosa cells (similar to the release of inhibin by Sertoli cells in the testes). As a result of these sources of negative feedback, gonadotropin levels decline, so the follicle that has the most FSH receptors is the one that survives.

At the time of ovulation, most follicle cells remain behind in the ovary, where they proliferate to form a mass of yellow tissue called the *corpus luteum* (Latin, "yellow body"). The corpus luteum secretes high levels of progesterone, along with some estrogen, for about 2 weeks (the *luteal phase* of the ovarian cycle). Progesterone causes the epithelial lining of the uterus, called the **endometrium**, to thicken in preparation for a possible pregnancy. Progesterone and estrogen, along with inhibin, exert negative feedback control on the pituitary, inhibiting gonadotropin release and thus preventing new follicles from beginning to mature.

of LH and a lesser surge of FSH. The LH surge triggers *ovulation*—rupture of the follicle and release of the oocyte from the ovary (**FIGURE 32.10**).

Why is it that usually only one follicle reaches the stage of ovulation? The follicle consists of two types of cells. Those immediately surrounding the oocyte are *granulosa cells*, and those enclosing the whole follicle are *thecal cells*. Thecal cells (like Leydig cells in the testes) produce androgens when stimulated by LH. Granulosa cells (like Sertoli cells in the testes) support the developing oocyte and are stimulated by FSH. Androgens diffuse from the thecal cells to the granulosa cells, which produce the enzyme *aromatase* that converts androgens to estrogen. Estrogen released from the granulosa cells also stimulates

FIGURE 32.11   Hormones Control the Female Reproductive Cycles   The ovarian and uterine cycles are under a complex series of positive and negative feedback controls involving several hormones, as described in the text. Estrogen has a positive feedback effect on the granulosa cells by stimulating the expression of FSH receptors. The follicle with the most FSH receptors survives declining FSH production and goes on to mature and release its ovum.

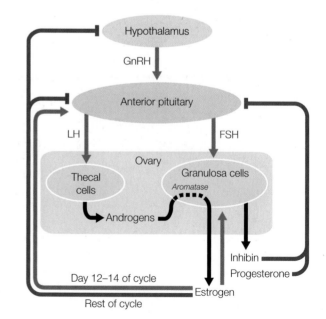

Hypothalamus

GnRH

Anterior pituitary

LH | FSH

Ovary

Thecal cells | Granulosa cells *Aromatase*

Androgens

Inhibin

Progesterone

Day 12–14 of cycle

Estrogen

Rest of cycle

If an embryo does not arrive in the uterus within 2 weeks after ovulation, the corpus luteum degenerates and stops secreting estrogen, progesterone, and inhibin. The falling progesterone causes the endometrium to break down, and the sloughed-off tissue, including blood, flows from the body through the vagina—the process of **menstruation** (Latin *menses*, "month").

The decrease in estrogen, progesterone, and inhibin also releases the hypothalamus and pituitary from negative feedback control, so GnRH, FSH, and LH all begin to increase. The increase in these hormones induces the next round of follicle development, and the ovarian cycle begins again.

The uterine cycles of most mammals other than humans do not include menstruation; instead, the uterine lining typically is reabsorbed. In these species, the most obvious correlate of the ovarian cycle is a state of sexual receptivity called *estrus* at about the time of ovulation. When the female mammal comes into estrus, or "heat," she actively solicits male attention and may be aggressive to other females. Humans are unusual among mammals in that females are potentially sexually receptive throughout their ovarian cycles and at all seasons of the year.

In pregnancy, tissues derived from the embryo produce hormones

If fertilization occurs, the zygote undergoes its first few cell divisions as it moves down the oviduct to the uterus, where it attaches itself to the endometrium as a blastocyst. It then burrows into the endometrium—a process called *implantation*—and interacts with it to form the placenta (see Figure 33.4). Once implantation has occurred, a layer of cells covering the embryo begins to secrete a new hormone, **human chorionic gonadotropin (hCG)**. This hormone stimulates the corpus luteum to continue to produce progesterone, along with some estrogen, to support the endometrium and thereby prevent menstruation. Pregnancy tests use an antibody to detect hCG in urine.

The placenta forms from a combination of embryonic and maternal tissues (see Figure 33.16). The placenta produces estrogen and progesterone, eventually replacing the corpus luteum as the most important source of these hormones. Continued high levels of progesterone maintain the endometrial lining, and both hormones prevent the pituitary from secreting gonadotropins; thus the ovarian cycle ceases for the duration of pregnancy.

Hormonal and mechanical stimuli trigger childbirth

A normal human pregnancy lasts about 38 weeks. Both hormonal and mechanical stimuli contribute to the onset of labor (the uterine muscle contractions that push the child through the birth canal). Progesterone inhibits and estrogen stimulates contractions of uterine muscle. Toward the last month of pregnancy, the estrogen–progesterone ratio shifts in favor of estrogen. The onset of labor is marked by increased secretion of the hormone oxytocin—a powerful stimulant of uterine muscle contraction—by the posterior pituitaries of both mother and fetus.

Mechanical stimuli, from the stretching of the uterus and pressure on the cervix by the growing fetus, increase the release of oxytocin by the mother's posterior pituitary, which in turn increases the activity of uterine muscle, which causes even more pressure on the cervix. This positive feedback loop results in stronger contractions. The hormonal changes and the pressure created by the contractions cause the cervix to dilate (expand) until it is large enough to allow the baby to pass through. When the cervix is fully dilated (to a diameter of about 10 centimeters), the baby's head can move into the vagina; its passage through the vagina is assisted by the mother's bearing down ("pushing") with her abdominal and other muscles. Once the baby's head and shoulders clear the cervix, the rest of its body eases out rapidly.

If the baby suckles at the breast immediately following birth, its suckling stimulates additional secretion of oxytocin, which augments uterine contractions that reduce the size of the uterus and help stop bleeding. Oxytocin also promotes bonding between the mother and infant.

Do You Understand Concept 32.4?

- Diagram the hormonal controls of male and female reproduction. What roles do the same hormones play in each sex? Given that the female system cycles and the male system does not, what phase of the female cycle is most similar to the state of the male system?
- Suppose a mutation prevented the production of normal hCG. What problems would this create for a pregnancy?
- Describe two examples of positive feedback control in the human reproductive system.

Our growing understanding of the physiology of human reproduction has led to numerous methods and technologies for controlling it, either to prevent unwanted pregnancies or to overcome infertility.

concept 32.5 Humans Use a Variety of Methods to Control Fertility

According to a recent study, almost half of the more than 6 million pregnancies that occur in the United States each year are unintended. For women of college age, a single act of unprotected sexual intercourse in the two days prior to ovulation carries a chance of conception as high as 50 percent.

Many contraceptive methods are available

The only failure-proof methods of preventing pregnancy are complete abstinence from sexual activity or the surgical removal of the gonads. There are, however, other, less drastic methods, many of which prevent fertilization or implantation (*conception*) and are therefore referred to as **contraception**. **TABLE 32.1** lists the most commonly used contraceptive methods and their relative failure rates; note that the different methods vary enormously in their effectiveness.

TABLE 32.1　Methods of Contraception

METHOD	MODE OF ACTION	FAILURE RATE[a]	COMMENTS
Unprotected	No form of birth control	85	High risk of pregnancy, especially for women 15–30.
Nontechnological methods			
Rhythm method	The couple abstains from intercourse between days 10 and 20 of the ovarian cycle (peak fertility).	15–35	High failure rate due to miscalculation and/or variation of individual cycles.
Coitus interruptus	The man withdraws his penis prior to ejaculation with the intention of not depositing sperm into the vagina.	20–40	Requires self-control, especially by the man. Very high failure rate.
Barrier methods[b]			
Condom	A sheath of impermeable material (often latex) is fitted over the erect penis. Semen is trapped in the condom, so no sperm are deposited in the vagina.	15	If fitted correctly, an intact condom can prevent pregnancy and provide protection against sexually transmitted diseases (STDs), including HIV (AIDS).
Spermicidal jellies	Applied inside the vagina, these chemical compounds kill or immobilize sperm.	25	Used alone, spermicidal compounds have a fairly high failure rate.
Diaphragms, cervical caps	Inserted by the woman prior to intercourse, these devices work by blocking the cervix so that sperm cannot pass into the uterus.	10–15	Approximately the same failure rate as condom use by men, but do not protect against STDs. Can be used in conjunction with spermicidal jelly for extra protection.
Hormone-based contraceptives			
Oral hormones ("the pill")	A daily pill for women containing a combination of synthetic estrogens and progesterone (progestin). These hormones mimic pregnancy to the extent that the ovarian cycle and ovulation are suspended. The uterine cycle is allowed to continue by including a week of non-hormone administration every 21–28 days.	0–3	Requires medical consultation and prescription. Taken correctly, oral contraceptives are extremely effective. In the U.S., more than 12 million women use them each year; they are sometimes prescribed to treat menstrual disorders.
Non-orally administered hormones	Making use of same hormonal actions as the pill, these methods include long-acting injections, patches that release hormones transdermally (through the skin), and a hormone-containing vaginal ring.	<1	Same as oral hormones. A slightly lower failure rate because the woman does not have to remember to take a daily pill.
Progestin-only pill	An oral contraceptive meant to be taken within 72 hours after unprotected sex. A high dose of progestin in two pills prevents ovulation in the same manner birth control pills do.	5–40[c]	Not an "abortion pill," this drug will not terminate an existing pregnancy. Currently available to women over 17 without a prescription.
Implantation blockers			
Intrauterine device (IUD)	A medical professional inserts a small plastic or metal device into the uterus. The resulting inflammation reaction (see Chapter 31) releases prostaglandins, which prevent implantation of the fertilized egg.	0.5–5	A highly effective contraceptive, it is the most widely used birth control device in China (and hence the world). With medical monitoring, can remain in place for several years.
Mifepristone (RU-486)	This drug blocks progesterone receptors necessary to maintain the endometrium during implantation and pregnancy.	0.5–6	Prevents implantation when taken up to several days after unprotected intercourse. Can terminate a pregnancy up to the time of the first missed menstrual period. In the U.S., available from specialized providers.
Sterilization			
Vasectomy	The vasa deferentia (see Figure 32.5A) are cut and tied off so that sperm can no longer pass into the urethra. Sperm continue to be produced but are reabsorbed by the man's body. Male hormone levels and sexual responses are not affected.	0–0.15	A simple surgical procedure performed under local anesthetic in a doctor's office. Although it can theoretically be reversed, vasectomy should be considered permanent.
Tubal ligation	The oviducts (see Figure 32.8A) are tied off so that eggs cannot reach the uterus and sperm cannot reach the egg. As with vasectomy, hormone levels and sexual responses are not affected.	0–0.05	This surgical procedure is somewhat more complex than vasectomy. It is often performed in conjunction with childbirth when a woman has decided that her family is complete.

[a] "Failure rate" refers to the number of pregnancies per 100 women per year.

[b] All of these barrier methods are routinely available without medical prescription.

[c] Failure rate varies widely depending on when taken.

Once a fertilized egg is successfully implanted in the uterus, any termination of the pregnancy is called an **abortion**. A *spontaneous abortion* is the medical term for what is commonly called a miscarriage. Spontaneous abortions are quite frequent early in pregnancy and are usually the result of either a chromosomal abnormality in the embryo or a breakdown in the process of implantation. Many spontaneous abortions occur before a woman even realizes she is pregnant.

Abortions that result from medical intervention may be performed either for therapeutic purposes or for fertility control. A therapeutic abortion may be necessary to protect the health of the mother, or it may be performed because prenatal testing reveals that the fetus has a severe defect. In a medical abortion, the cervix is dilated and some of the endometrium, along with the implanted embryo, is removed from the uterus. When performed in the first trimester (the first 12 weeks) of a pregnancy, a medical abortion carries less risk of death to the mother than a full-term pregnancy. The risk rises after the first trimester but still remains less than that of a full-term pregnancy through the second trimester.

Reproductive technologies help solve problems of infertility

There are many reasons for infertility—usually defined as the persistent inability of a couple to conceive a child—and they are equally distributed between men and women. A number of technologies have been developed to overcome barriers to both conceiving and bearing a child.

The simplest treatment available is **artificial insemination**, in which the physician positions sperm within the woman's reproductive tract. This technique is useful if the male partner's sperm count is low or his sperm lack motility, or if certain conditions in the woman's reproductive tract prevent the normal progress of sperm to the ovum. It is also useful if a woman does not have a male partner.

FRONTIERS Most domestic cattle are produced through artificial insemination. Dairy farming is far more efficient and profitable when cows produce female rather than male offspring, and dairy farmers preferentially produce female offspring based on the DNA content of the sperm they use to inseminate their cows. Sperm sorting uses an instrument called a flow cytometer to separate sperm based on the density of a fluorescent dye marker. X-bearing (female-producing) sperm have more DNA than Y-bearing sperm and thus can be distinguished by the greater amount of dye they absorb. Human sperm can be sorted in a similar way, raising controversy over the desirability of allowing parents to choose the sex of their offspring.

For situations in which artificial insemination does not work, there are more complex procedures called **assisted reproductive technologies**, or **ARTs**. These procedures involve removing unfertilized eggs from the ovary, combining them with sperm outside the body, and then placing fertilized eggs or egg–sperm mixtures in the appropriate location in the woman's reproductive tract for implantation. The first successful ART was *in vitro fertilization* (IVF). In IVF, the woman is treated with hormones that stimulate many follicles in her ovaries to mature. Eggs are collected from these follicles and combined with sperm collected from the man. The resulting embryos can be injected into the mother's uterus or kept frozen for implantation later. The first "test-tube baby" resulting from IVF was Louise Brown, born in England in 1978. Since then, more than 3 million babies have been produced using this ART. For creating this technology, Robert Edwards received the 2010 Nobel Prize in Physiology and Medicine.

Do You Understand Concept 32.5?

- According to the data in Table 32.1, what are the two least effective methods of contraception? Which is the only method that protects against sexually transmitted diseases?
- What are some possible reasons that a sperm might be unable to fertilize an ovum?
- Explain how reproductive technologies such as artificial insemination and IVF make it possible for parents to choose the sex of their offspring. What are some of the arguments in favor of and against doing so?

The fertilized ovum, or zygote, of a sexually reproducing organism is a single cell containing all the genetic information needed to create a new organism. The emergence of the new adult organism from a zygote is the process of *development*, the subject of Chapter 33.

QA **QUESTION** How does "the pill" prevent conception?

ANSWER The basic mechanism of the birth control pill is to create in the female a hormonal condition similar to that of pregnancy. Most birth control pills are the so-called combined pill that contains synthetic hormones resembling both estrogen and progesterone. Some pills, sometimes called "minipills," contain only synthetic progesterone. Daily doses of either pill type mimic the secretion of the hormones of pregnancy by the corpus luteum and the placenta, as described in Concept 32.4. Heightened levels of these hormones in the circulation exert negative feedback on the hypothalamus and the anterior pituitary so that release of GnRH by the hypothalamus and of FSH and LH by the anterior pituitary remains low. As a result, follicles do not mature and ovulation does not occur (review Figures 32.9 and 32.10).

Minipills are taken every day. Combined pills are taken once a day for 21 days, followed by 7 days with either no pills or dummy pills. The drop in hormone levels during that 7 days allows menstruation to occur.

32 SUMMARY

concept 32.1 Reproduction Can Be Sexual or Asexual

- Asexual reproduction produces offspring that are genetically identical to their parent and to one another; it produces no genetic diversity.

- Modes of asexual reproduction include **budding**, **regeneration**, and **parthenogenesis**. Review Figure 32.1

- Sexual reproduction consists of three basic steps: **gametogenesis**, spawning or mating, and **fertilization**.

concept 32.2 Gametogenesis Produces Haploid Gametes

- Gametogenesis occurs in **testes** and **ovaries**. In **spermatogenesis** (the production of **sperm**) and **oogenesis** (the production of **ova**), the **germ cells** proliferate mitotically, undergo meiosis, and mature into gametes. Review Figure 32.2

- Each **primary spermatocyte** produces four haploid sperm through the two cell divisions of meiosis.

- **Primary oocytes** immediately enter prophase of the first meiotic division, and in many species, including humans, their development is arrested at this point. Each primary oocyte produces only one ovum.

- In **hermaphroditic** species, the same individual can produce both sperm and ova, either simultaneously or sequentially.

concept 32.3 Fertilization Is the Union of Sperm and Ovum

- External fertilization is common in aquatic species. **Spawning** is the release of large numbers of gametes into the water.

- Internal fertilization is necessary in terrestrial species. It usually requires **copulation**, the physical joining of the male and female **accessory sex organs**.

- Fertilization involves species-specific binding of sperm to ovum, the acrosomal reaction, the sperm's passage through the protective layers covering the ovum, and fusion of sperm and ovum plasma membranes. Review Figure 32.3 and ANIMATED TUTORIAL 32.1

- The entry of the sperm into the ovum triggers **blocks to polyspermy**, which prevent additional sperm from entering the ovum, and in mammals, signal the ovum to complete meiosis.

concept 32.4 Human Reproduction Is Hormonally Controlled

- The male sex organs produce and deliver **semen**. Semen consists of sperm suspended in seminal fluid, which nourishes the sperm and facilitates fertilization. Review Figure 32.5 and WEB ACTIVITY 32.1

- Sperm are produced in the **seminiferous tubules** of the testes, mature in the epididymis, and are delivered to the urethra through the vasa deferentia. Review Figure 32.6 and WEB ACTIVITY 32.2

- Spermatogenesis depends on testosterone secreted by the Leydig cells of the testes, which are under the control of hormones produced in the anterior pituitary and hypothalamus. Review Figure 32.7

- Ova mature in the female's ovaries and are released into the **oviducts**. Sperm deposited in the **vagina** during copulation move up through the **cervix** and **uterus** into the oviducts, where fertilization occurs. Review Figure 32.8 and WEB ACTIVITY 32.3

- The maturation and release of ova constitute an ovarian cycle. This cycle takes about 28 days. The uterine cycle prepares the **endometrium** of the uterus for receipt of an embryo. If no embryo arrives, the endometrium sloughs off in the process of **menstruation**. Review Figures 32.9 and 32.10 and ANIMATED TUTORIAL 32.2

- Both the ovarian and the uterine cycles are under the control of hypothalamic and pituitary hormones, which in turn are under the feedback control of estrogen and other hormones. Review Figure 32.11

concept 32.5 Humans Use a Variety of Methods to Control Fertility

- Methods of **contraception** include abstention from intercourse and the use of technologies that decrease the probability of fertilization and implantation. Review Table 32.1

- A number of **assisted reproductive technologies** (ARTs) have been developed to treat infertility.

Animal Development

In Homer's *Odyssey*, the mythic hero Odysseus visits the island of the Cyclopes, a race of one-eyed giants. He and his men come across an enormous cave inhabited by one of these beings. After feasting on the cyclops's stored food, they hide as the monster returns: "The cyclops … blotted out the light in the doorway. He was as tall and rugged as an alp. One huge eye glared out of the center of his forehead."

The one-eyed monster of Greek mythology has a basis in reality. A common birth defect in humans is a condition known as holoprosencephaly (Greek, "one forebrain"). In its extreme form, as the name implies, the forebrain does not develop as two hemispheres but as one, and the face can develop with a single, centrally located eye. These extreme cases of holoprosencephaly invariably lead to death of the fetus and a miscarriage. Less extreme cases, however, result in infants born with cleft lips and cleft palates.

A number of factors have been implicated in holoprosencephaly, most of them genetic. Mutations in several genes that are part of the hedgehog signaling factor family can cause this condition. The first *hedgehog* gene was discovered in fruit flies, where it plays a role in determining the differences in the body segments of the fly. The name arises from the fact that flies with a defect in this gene are hunched up and have a continuous covering of tiny projections, so they resemble a hedgehog. The homologous vertebrate gene, *Sonic hedgehog* (named for the video game character), is critical in central nervous system development.

Environmental factors can induce holoprosencephaly. A particularly powerful mutagen is a molecule found in the corn lily, *Veratrum californicum*. Cows and ewes grazing in fields where this plant grows have high incidences of stillbirths with extreme forms of holoprosencephaly; in fact, when the molecule was extracted and purified from corn lily, it was named cyclopamine. In humans, the ingestion early in pregnancy of a class of drugs called statins has been associated with these birth defects. Learning how these environmental factors cause holoprosencephlic development helps explain why Sonic hedgehog is critical in limb and nervous system development.

QUESTION How does the Sonic hedgehog pathway control development of the vertebrate brain and eyes?

In mammals, normal *Sonic hedgehog* gene expression over the course of development leads to the formation of two separate eye fields.

KEY CONCEPTS

33.1 Fertilization Activates Development

33.2 Cleavage Repackages the Cytoplasm of the Zygote

33.3 Gastrulation Creates Three Tissue Layers

33.4 Neurulation Creates the Nervous System

33.5 Extraembryonic Membranes Nourish the Growing Embryo

Fertilization Activates Development

In sexually reproducing animals, fertilization is the joining of sperm and egg to form a diploid zygote, as described in Concept 32.3. You might think of fertilization as the event that begins development. However, it is preceded by critical events in the maturing egg that influence subsequent development. Here we take a closer look at the cellular and molecular interactions of sperm and egg that activate the continuing processes of development.

The sperm and ovum make different contributions to the zygote

In most species, eggs are much larger than sperm. The sperm contains almost no cytoplasm and is little more than a DNA delivery vehicle. In addition to providing its haploid nucleus, however, the sperm makes an additional, and vital, contribution to the zygote in most species: a *centriole*. The centrosome of the egg degrades during oogenesis, so the sperm centriole becomes the centrosome of the zygote and plays essential roles in microtubule organization and formation of the primary cilium (see Concept 7.2).

In contrast to sperm, the egg has copious cytoplasm stocked with organelles, nutrients, and a variety of molecules, including transcription factors and mRNAs. Nearly everything the embryo needs during its first stages of development comes from the mother. Cytoplasmic factors in the egg play important roles in setting up the signaling cascades that orchestrate the major processes of development described in Chapter 14: *determination, differentiation, morphogenesis,* and *growth* (see p. 264). Throughout this chapter, we will see how these processes underlie the development of a multicellular individual from a single cell.

> **LINK** Concepts 14.1 through 14.3 explain how differential gene expression acts during early development

Rearrangements of egg cytoplasm set the stage for determination

Certain molecules in the egg's cytoplasm are not homogeneously distributed even before fertilization, and the entry of the sperm into the egg stimulates further rearrangements. These rearrangements establish the *polarity* of the zygote, and when cell divisions begin, the informational molecules that will guide development are not divided equally among daughter cells.

Cytoplasmic rearrangement following fertilization is easily observed in some frog species because their eggs have pigments in the cytoplasm. The nutrients in an unfertilized frog egg are dense yolk granules that are concentrated by gravity in the lower half of the egg, or **vegetal hemisphere**. The haploid nucleus of the egg is located at the opposite end, in the **animal hemisphere**. The outermost (*cortical*) cytoplasm of the animal hemisphere is heavily pigmented, and the underlying

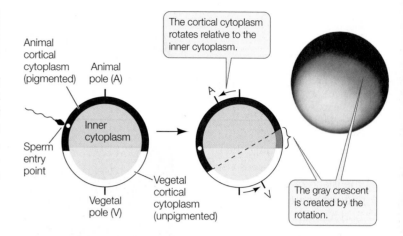

FIGURE 33.1 The Gray Crescent Rearrangement of the cytoplasm of frog eggs after fertilization creates the gray crescent and imposes an anterior–posterior axis on the egg.

cytoplasm has more diffuse pigmentation. The vegetal hemisphere is not pigmented.

Sperm-binding sites are localized on the surface of the animal hemisphere, so that is where the sperm enters the egg. When a sperm enters the egg, cortical cytoplasm rotates toward the site of sperm entry. This rotation brings different regions of cytoplasm into contact with each other on opposite sides of the egg, producing a band of diffusely pigmented cytoplasm on the side opposite the site of sperm entry. This band, called the **gray crescent**, marks the location of important developmental events in some amphibians (**FIGURE 33.1**). These rearrangements also impose an anterior–posterior axis on the egg, so that the zygote now has bilateral symmetry as well as a prospective head and tail.

Cytoplasmic reorganization is initiated by the one non-nuclear organelle that the sperm contributes to the egg: the centriole. The centriole organizes the microtubules in a parallel array extending through the vegetal hemisphere cytoplasm. This array guides the movement of the cortical cytoplasm. These microtubules also appear to be directly responsible for movement of specific organelles and proteins that travel from the vegetal hemisphere to the gray crescent region faster than the cortical cytoplasm rotates.

The movement of cytoplasm, proteins, and organelles changes the distribution of critical developmental signals. For example, the transcription factor β-catenin, necessary for specifying and positioning the cells of the three embryonic *germ layers* (see Concept 33.3), is produced from maternal mRNA (i.e., mRNA produced and stored in the egg while it was maturing in the ovary). Both β-catenin mRNA and a protein kinase called glycogen synthase kinase-3 (GSK-3) are found throughout the cytoplasm; GSK-3 targets β-catenin for degradation. However, an *inhibitor* of GSK-3 is segregated in the egg's vegetal cortex. After sperm entry, this inhibitor migrates along microtubules to the gray crescent, where it prevents the degradation of β-catenin. As a result, the concentration of β-catenin is higher on the dorsal than on the ventral side of the developing embryo (**FIGURE 33.2**).

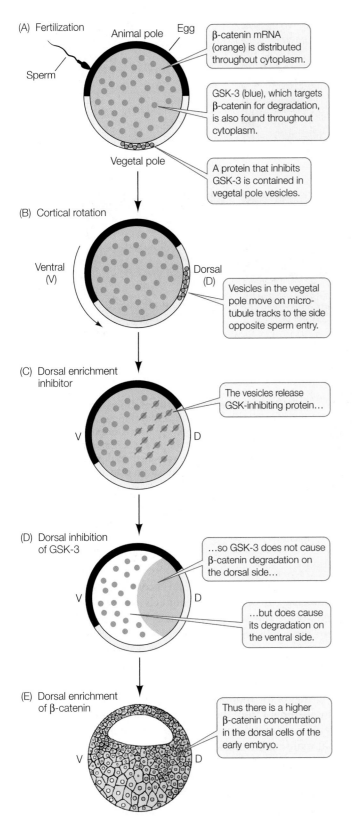

FIGURE 33.2 Cytoplasmic Factors Set Up Signaling Cascades Cytoplasmic movement changes the distributions of critical developmental signals. In the frog zygote, the interaction of the protein kinase GSK-3, its inhibitor, and the protein β-catenin are crucial in specifying the dorsal–ventral axis of the embryo.

Do You Understand Concept 33.1?

- What organelles does the sperm contribute to the zygote?
- Describe how cytoplasmic rearrangements in the frog egg set up the anterior–posterior and dorsal–ventral body axes.
- In most species, no genes are transcribed immediately following fertilization and during the early cell divisions. What sources of information determine the spatial organization of an early-stage embryo?

β-Catenin plays a major role in the signaling cascade that begins the process of cell specification and determination. But before cell–cell signaling can occur, there must be multiple cells. We turn next to the early series of cell divisions that produces a multicellular embryo.

concept 33.2 Cleavage Repackages the Cytoplasm of the Zygote

The diploid zygote becomes a mass of cells through a rapid series of cell divisions called **cleavage**. Because the cytoplasm of the zygote is not homogeneous (see Concept 33.1), these first cell divisions result in the differential distribution of nutrients and cytoplasmic determinants (mRNA and proteins).

Cleavage increases cell number without cell growth

In most animals, cleavage proceeds with rapid DNA replication and mitosis but with no cell growth and little gene expression. The embryo becomes a ball of ever-smaller cells. Eventually, this ball forms a central fluid-filled cavity called a **blastocoel**, at which point the embryo is called a **blastula**. Its individual cells are called **blastomeres**. The pattern of cleavage in different species influences the form of their blastulas.

- **Complete cleavage** occurs in most eggs that have little *yolk* (stored nutrients). In this pattern, early cleavage furrows divide the egg completely and all the blastomeres are of similar size. The frog egg undergoes complete cleavage, but because its vegetal pole contains more yolk, the division of the cytoplasm is unequal and the blastomeres in the animal hemisphere are smaller than those in the vegetal hemisphere (**FIGURE 33.3A**).

- **Incomplete cleavage** occurs in many species in which the egg contains a lot of yolk and the cleavage furrows do not penetrate it all. **Discoidal cleavage** is a type of incomplete cleavage that is common in fishes, reptiles, and birds, the eggs of which contain a dense yolk mass. The embryo forms as a disc of cells, called a **blastodisc**, that sits on top of the yolk mass (**FIGURE 33.3B**).

- Cleavage in mammals combines characteristics of both of the above types. Mammalian eggs contain little yolk and

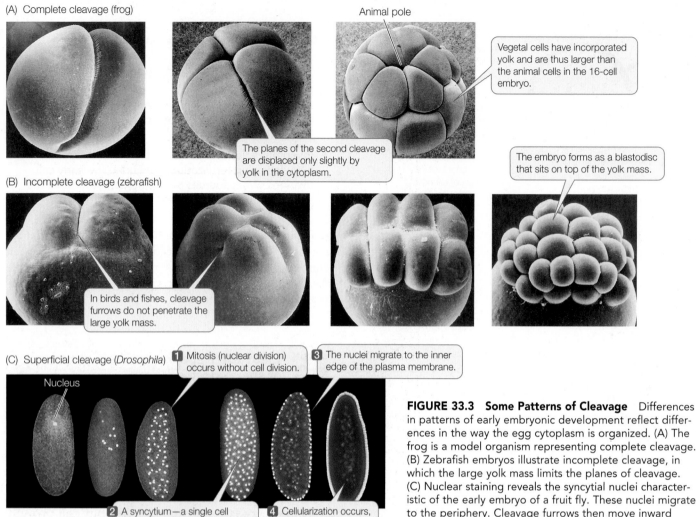

(A) Complete cleavage (frog)

Animal pole

Vegetal cells have incorporated yolk and are thus larger than the animal cells in the 16-cell embryo.

The planes of the second cleavage are displaced only slightly by yolk in the cytoplasm.

The embryo forms as a blastodisc that sits on top of the yolk mass.

(B) Incomplete cleavage (zebrafish)

In birds and fishes, cleavage furrows do not penetrate the large yolk mass.

(C) Superficial cleavage (*Drosophila*)

Nucleus

1 Mitosis (nuclear division) occurs without cell division.

3 The nuclei migrate to the inner edge of the plasma membrane.

2 A syncytium—a single cell with many nuclei—is produced.

4 Cellularization occurs, creating a blastoderm.

FIGURE 33.3 Some Patterns of Cleavage Differences in patterns of early embryonic development reflect differences in the way the egg cytoplasm is organized. (A) The frog is a model organism representing complete cleavage. (B) Zebrafish embryos illustrate incomplete cleavage, in which the large yolk mass limits the planes of cleavage. (C) Nuclear staining reveals the syncytial nuclei characteristic of the early embryo of a fruit fly. These nuclei migrate to the periphery. Cleavage furrows then move inward to separate the nuclei into individual cells, forming the blastoderm.

undergo complete cleavage, yet the blastula develops as a blastodisc—most likely a carryover from shared ancestry with reptiles and birds.

- **Superficial cleavage** is a variation of incomplete cleavage that occurs in insects such as the fruit fly (*Drosophila*). Early in development, cycles of mitosis occur without cell division, producing a *syncytium*—a single cell with many nuclei (**FIGURE 33.3C**). The nuclei eventually migrate to the periphery of the egg, after which the plasma membrane of the egg grows inward, partitioning the nuclei into individual cells surrounding a core of yolk.

Cleavage in mammals is unique

Several features of early cell divisions in placental mammals are different from those seen in other animals. Early cell divisions in mammals are very slow—some 12 to 24 hours apart, compared with tens of minutes to a few hours in non-mammalian species. In mammals, genes are transcribed earlier in cleavage than in most other species.

Another unique feature of mammalian cleavage occurs during the fourth division, when the cells separate into two groups. The **inner cell mass** will become the embryo, while the surrounding outer cells become an encompassing sac called the **trophoblast**. The cells of the inner cell mass are not yet determined but are *pluripotent* ("capable of much"). When grown in culture these cells are the *embryonic stem cells* (ESCs) that are the subject of intense research on their therapeutic potential (see Concept 14.1). The trophoblast cells secrete fluid, creating a blastocoel with the inner cell mass at one end (**FIGURE 33.4A**). At this stage, the mammalian embryo is called a **blastocyst**, distinguishing it from the *blastulas* of other animal groups.

Why is mammalian cleavage so distinctive? A key factor is that mammalian eggs contain almost no yolk; the embryo develops within the mother's uterus and derives its nutrients from the mother's bloodstream. Thus, the mammalian blastocyst produces both an embryo (from the inner cell mass) and the support structures necessary to attach the embryo to the mother's uterine wall (from the trophoblast).

Fertilization in mammals occurs in the upper reaches of the mother's oviduct, and cleavage occurs as the zygote travels down the oviduct to the uterus. When the blastocyst arrives in the uterus, the zygote hatches out of the zona pellucida (see Concept 32.3). The trophoblast then adheres to the lining of the uterus (the *endometrium*), beginning the process

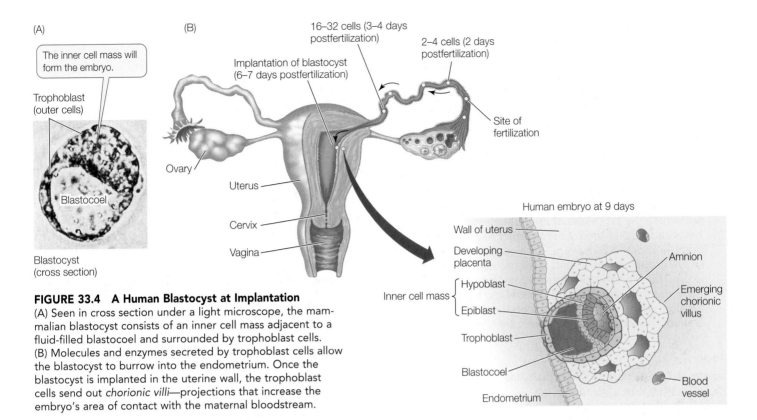

FIGURE 33.4 A Human Blastocyst at Implantation
(A) Seen in cross section under a light microscope, the mammalian blastocyst consists of an inner cell mass adjacent to a fluid-filled blastocoel and surrounded by trophoblast cells. (B) Molecules and enzymes secreted by trophoblast cells allow the blastocyst to burrow into the endometrium. Once the blastocyst is implanted in the uterine wall, the trophoblast cells send out *chorionic villi*—projections that increase the embryo's area of contact with the maternal bloodstream.

of **implantation**—burrowing into the uterine wall. In humans, implantation begins about 6 days after fertilization (**FIGURE 33.4B**).

Specific blastomeres generate specific tissues and organs

During the next stage of development, the cells of the blastula will move around and come into new associations with one another, communicate instructions to one another, and begin to differentiate. In many animals, these blastomere movements are so regular and well orchestrated that it is possible to label a specific blastomere (e.g., with a dye) and identify the tissues and organs that form from its progeny. Such labeling experiments produce **fate maps** of the blastula (**FIGURE 33.5**).

Blastomeres become *determined*—irreversibly committed to specific fates—at different times in different species. In some species, such as roundworms, blastomere fates are restricted as early as the two-cell stage. If one of these blastomeres is experimentally removed, a particular portion of the embryo will not form. This type of development has been called **mosaic development** because each blastomere appears to contribute a specific set of "tiles" to the final "mosaic" that is the adult animal.

In contrast to mosaic development, the loss of some cells during cleavage in **regulative development** does not affect the

developing embryo, because the remaining cells compensate for the loss. Regulative development is typical of many vertebrate species, including humans.

If some blastomeres can change their fate to compensate for the loss of other cells during cleavage and blastula formation, can those cells form an entire embryo? To a certain extent, yes. During cleavage or early blastocyst formation in mammals, for example, if the blastomeres are physically separated into two groups, both groups can produce complete embryos. Since the two embryos come from the same zygote, they will be genetically identical—*monozygotic twins*. If the inner cell mass splits only partially, the result may be *conjoined twins*—monozygotic twins who are physically attached and usually share some organs and/or limbs. *Dizygotic twins* occur when two separate eggs are fertilized by two separate sperm. Sometimes referred

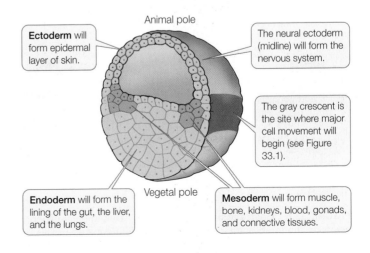

FIGURE 33.5 Fate Map of a Frog Blastula Colors indicate the portions of the blastula that will form the three germ layers and subsequently the frog's tissues and organs.

to as "fraternal twins," dizygotic twins are genetically nonidentical full siblings.

Do You Understand Concept 33.2?

- How does early cell division differ in a mammal and a frog?
- Characterize complete, incomplete, and discoidal cleavage and explain where mammals fit into this scheme.
- An important function of early cell divisions is to differentially distribute cytoplasmic determinants that will initiate the processes of differentiation of cell types and localize them to the correct region of the body. In superficial cleavage, such early partitioning of cytoplasmic determinants does not occur. How do you think positional information is organized in early insect development? (Consider the fact that the anterior–posterior and dorsal–ventral axes reflect the position that the egg had in the ovary.)

Of the next stage of development—gastrulation—the developmental biologist Louis Wolpert once said, "It is not birth, marriage, or death, but gastrulation which is the most important time in your life." During gastrulation, cell movements create the three-layered body plan and set the stage for development of the first organs.

concept 33.3   Gastrulation Creates Three Tissue Layers

Gastrulation is the process whereby the blastula is transformed by massive movements of cells into an embryo with multiple tissue layers and distinct body axes. The resulting spatial relationships between tissues make possible the interactions between cells that trigger differentiation and organ formation.

During gastrulation, three **germ layers** (also called *cell layers* or *tissue layers*) form (see Figure 33.5):

- The innermost germ layer, the **endoderm**, forms from blastomeres that migrate to the inside of the embryo. The endoderm produces the linings of internal spaces such as those of the digestive and respiratory tracts and the urinary bladder. It also contributes to the structure of some internal organs, including the endocrine glands, pancreas, and liver.
- The **ectoderm** is the outer germ layer, formed from those cells remaining on the outside of the embryo. The ectoderm gives rise to the nervous system, including the eyes and ears, and produces the epidermal layer of the skin (and structures derived from skin, such as hair, feathers, and teeth).
- The middle layer, the **mesoderm**, is made up of cells that migrate between the endoderm and the ectoderm. The mesoderm gives rise to the heart, blood vessels, muscles, bones, and several other organs.

What directs the cell movements of gastrulation, and what is responsible for the resulting patterns of cell differentiation and organ formation? We will begin with sea urchin gastrulation because it is the simplest to conceptualize in spatial terms. We will then describe the more complex pattern of gastrulation in frogs, which in turn will help elucidate the still more complex patterns in reptiles, birds, and mammals.

> **LINK**   Early development and the germ layers reflect phylogenetic relationships among the animals; see Concept 23.1

Invagination of the vegetal pole characterizes gastrulation in the sea urchin

The sea urchin blastula is a hollow ball of cells only one cell thick. The beginning of gastrulation is marked by a flattening of the vegetal hemisphere as the individual blastomeres change shape (**FIGURE 33.6**). These cells, which are originally rather cuboidal, become wedge-shaped, with smaller outer edges and larger inner edges. As a result, the vegetal pole bulges inward, or

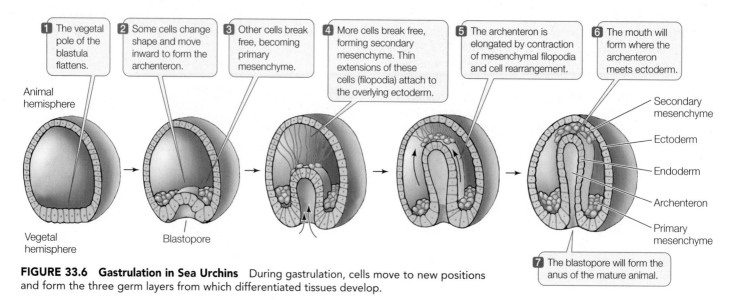

FIGURE 33.6 Gastrulation in Sea Urchins During gastrulation, cells move to new positions and form the three germ layers from which differentiated tissues develop.

invaginates, as if someone were poking a finger into a hollow ball (see Figure 33.6). The invaginating cells become endoderm and form a primitive gut called the **archenteron**. Meanwhile, some cells of the vegetal pole break away from neighboring cells and migrate into the central cavity. These cells become **mesenchyme** —cells of the middle germ layer. Mesenchymal cells act as independent units, migrating into and among the other tissue layers.

Changes in cell shapes cause the initial invagination of the archenteron, but eventually it is pulled by additional mesenchyme cells that form at the tip of the archenteron and send out extensions called *filopodia* that adhere to the overlying ectoderm. When the filopodia contract, they pull the archenteron toward the ectoderm at the opposite end of the embryo from where the invagination began. The mouth of the animal forms where the archenteron makes contact with this overlying ectoderm. The opening created by the invagination of the vegetal pole is called the **blastopore**; it will become the anus of this deuterostome animal (see p. 459).

Only cells from the vegetal pole are capable of bulging inward to initiate gastrulation, probably because of uneven distribution of regulatory proteins in the egg cytoplasm. As cleavage progresses, the regulatory proteins are localized in different groups of cells. Thereafter, specific sets of genes are activated in different cells, determining their different developmental capacities.

We turn now to gastrulation in the frog, in which several key signaling molecules have been identified.

Gastrulation in the frog begins at the gray crescent

Amphibian blastulas have considerable yolk and, unlike sea urchin blastulas, are more than one cell thick. Although the

yourBioPortal.com

Go to ANIMATED TUTORIAL 33.1
Gastrulation

details vary among species, in general amphibian gastrulation begins when certain cells in the gray crescent region change their shapes and cell adhesion properties. These cells bulge inward toward the blastocoel while remaining attached to the outer surface of the blastula by slender necks. Because of their shape, these are called *bottle cells*.

Bottle cells mark the spot where the **dorsal lip** of the blastopore will form (**FIGURE 33.7**). As the bottle cells move inward, the dorsal lip is created, and a sheet of cells moves over it into the blastocoel. This process is called **involution**. One group of involuting cells is the prospective endoderm; these cells form the primitive gut, or archenteron. Another group will move between the endoderm and the outermost cells to form the mesoderm.

As gastrulation proceeds, cells from the animal hemisphere flatten and move toward the site of involution, a process known as **epiboly**. The blastopore lip widens and eventually forms a complete circle surrounding a "plug" of yolk-rich cells. As cells continue to move inward through the blastopore, the archenteron grows, gradually displacing the blastocoel.

As gastrulation comes to an end, the amphibian embryo consists of three germ layers: ectoderm on the outside, endoderm on the inside, and mesoderm in between. The embryo also has a dorsal–ventral and anterior–posterior organization. Most importantly, the fates of specific regions of the endoderm, mesoderm, and ectoderm have been determined. The experiments that revealed how determination takes place in the amphibian embryo are an old but exciting story.

The dorsal lip of the blastopore organizes amphibian embryo formation

In the early 1900s, the German biologist Hans Spemann was studying the development of salamander eggs. With great patience and dexterity, he formed loops from single hairs taken

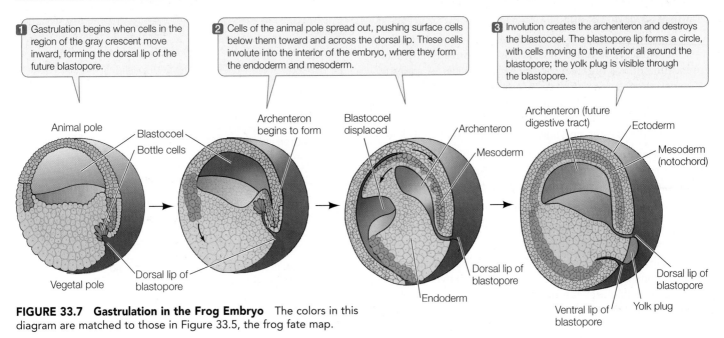

1 Gastrulation begins when cells in the region of the gray crescent move inward, forming the dorsal lip of the future blastopore.

2 Cells of the animal pole spread out, pushing surface cells below them toward and across the dorsal lip. These cells involute into the interior of the embryo, where they form the endoderm and mesoderm.

3 Involution creates the archenteron and destroys the blastocoel. The blastopore lip forms a circle, with cells moving to the interior all around the blastopore; the yolk plug is visible through the blastopore.

Animal pole
Blastocoel
Bottle cells
Vegetal pole
Dorsal lip of blastopore

Archenteron begins to form
Blastocoel displaced

Archenteron
Mesoderm
Dorsal lip of blastopore
Endoderm

Archenteron (future digestive tract)
Ectoderm
Mesoderm (notochord)
Dorsal lip of blastopore
Ventral lip of blastopore
Yolk plug

FIGURE 33.7 Gastrulation in the Frog Embryo The colors in this diagram are matched to those in Figure 33.5, the frog fate map.

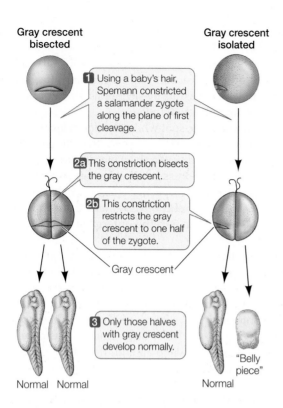

Gray crescent bisected

Gray crescent isolated

1 Using a baby's hair, Spemann constricted a salamander zygote along the plane of first cleavage.

2a This constriction bisects the gray crescent.

2b This constriction restricts the gray crescent to one half of the zygote.

Gray crescent

3 Only those halves with gray crescent develop normally.

Normal Normal

"Belly piece"

Normal

FIGURE 33.8 Gastrulation and the Gray Crescent Spemann's research revealed that gastrulation and subsequent normal development in salamanders depend on cytoplasmic determinants localized in the gray crescent.

(**FIGURE 33.8**). Spemann hypothesized that cytoplasmic factors unequally distributed in the fertilized egg were necessary for gastrulation and the development of a normal salamander.

To test the hypothesis that cells receiving different complements of cytoplasmic factors had different developmental fates, Spemann transplanted pieces of gastrulas to various locations on other gastrulas. When he performed these experiments in early gastrulas, the transplanted pieces always developed into tissues that were appropriate for their new locations. In late gastrulas, however, the transplanted cells became tissues appropriate to their original location. For example, if late-gastrula cells were transplanted from a region that would become skin to a region that would become nervous system, the transplanted cells produced patches of skin cells within the nervous system. At some point during gastrulation, the fates of the embryonic cells had become determined.

Spemann's next experiment, done with his student Hilde Mangold, was to transplant the dorsal blastopore lip. The results were momentous. When this small piece of tissue was transplanted into the presumptive belly area of another gastrula, it stimulated a second site of gastrulation, and a second complete embryo formed belly-to-belly with the original embryo (**FIGURE 33.9**). Because the dorsal blastopore lip of amphibians was apparently capable of inducing host tissue to form an entire embryo, Spemann and Mangold dubbed this tissue the *primary embryonic organizer* or, simply, the **organizer**.

from his baby daughter and tied them around fertilized eggs, effectively dividing the eggs in half. When his loops bisected the gray crescent, both halves of the zygote developed into a complete embryo. When he tied the loops so the gray crescent was on only one side of the constriction, the half lacking gray crescent material underwent cell division but became a clump of undifferentiated cells that Spemann called a "belly piece"

APPLY THE CONCEPT

Gastrulation creates three tissue layers

What about the amphibian dorsal blastopore lip makes it so crucial to "organizing" the three germ layers? The distribution of β-catenin in the late blastula corresponds to the location of this organizer in the early gastrula. A series of experiments endeavored to ascertain whether β-catenin is indeed "necessary and sufficient" for subsequent functions of the organizer. Review the results of these experiments in the table at right, and answer the questions that follow.

1. Which experiment shows that β-catenin is *necessary* for induction of the organizer? Explain.

2. Which experiment shows that β-catenin is *sufficient* for induction of the organizer? Explain.

3. What was the purpose of experiment B? Of experiment D?

4. The series of experiments above demonstrates what is sometimes called the "find it, lose it, move it" approach to understanding a gene's function. Explain why this catchy phrase is appropriate.

EXPERIMENT	RESULT
A. "Knock out" β-catenin by injecting β-catenin antisense mRNA into frog egg (see Figure 13.9).	Dorsal lip of blastopore does not form.
B. Repeat experiment A, but follow injection of antisense mRNA with an injection of exogenous (i.e., from an external source) β-catenin protein into the egg.	Gastrulation is normal.
C. Inject exogenous β-catenin into ventral region of an early gastrula (where β-catenin does not normally occur).	A second site of gastrulation is stimulated, producing results similar to those of Spemann and Mangold (see Figure 33.9).
D. Repeat experiment C, but follow exogenous β-catenin with β-catenin antisense mRNA.	Gastrulation is normal.

For more than 80 years, the organizer has been an active area of research. Molecular biologists are making great strides at elucidating the molecular mechanisms of the organizer, and the transcription factor β-catenin (see Concept 33.1) has been suspected of being a key inductive signal.

The distribution of β-catenin in the late blastula (see Figure 33.2) corresponds to the location of the organizer in the early gastrula, but this observation is only a correlation. To prove that a protein is an inductive signal, it has to be shown that its presence is both *necessary* and *sufficient* for the proposed inductive effect. In other words, the induction should not occur if the candidate protein is not present (necessity), and the candidate protein should be capable of inducing the induction where it would otherwise not occur (sufficiency). Thanks to a series of experiments in the 1980s and 1990s (see Apply the Concept, previous page), we now know that β-catenin triggers a complex series of interactions between various transcription factors and growth factors that control gene expression and determine which cells become the primary organizer. Several of these transcription factors have been identified.

Transcription factors underlie the organizer's actions

The first organizer cells to enter the embryo migrate anteriorly and *induce* formation of the structures of the head—that is, they cause neighboring cells to form head structures. Organizer cells that migrate into the embryo later will induce structures of the trunk, and the last of the organizer cells to migrate inward will induce structures of the tail. How do these organizer cells induce these different structures?

As organizer cells move forward in the blastocoel, they come into contact with new populations of cells that produce a number of different growth factors. For head structures to form, certain of these growth factors have to be suppressed. The anteriormost organizer cells release *antagonists* to those growth factors, suppressing them and allowing head structures to form (**FIGURE 33.10**).

The induction of trunk structures requires suppression of a different set of growth factors. In organizer cells that involute later than the head organizers, different growth factor antagonists are expressed. The induction of tail structures requires still different activities of the organizer cells that involute last. Thus, the organizer cells express appropriate sets of growth factor antagonists at the right times to achieve different patterns of differentiation on the anterior–posterior axis.

Reptilian and avian gastrulation is an adaptation to yolky eggs

The eggs of reptiles and birds contain a mass of yolk, and the blastulas of these groups develop as a disc of cells on top of the yolk (see Figure 33.3B).

INVESTIGATION

FIGURE 33.9 The Dorsal Lip Induces Embryonic Organization In a classic experiment, Hans Spemann and Hilde Mangold transplanted the dorsal blastopore lip mesoderm of an early gastrula stage salamander embryo. The results showed that the cells of this embryonic region, which they dubbed "the organizer," could direct the formation of an entire embryo.

HYPOTHESIS

Cytoplasmic factors in the early dorsal blastopore lip organize cell differentiation in amphibian embryos.

METHOD

1. Excise a patch of mesoderm tissue from above the dorsal blastopore lip of an early gastrula stage salamander embryo (the donor).
2. Transplant the donor tissue onto a recipient embryo at the same stage. The donor tissue is transplanted onto a region of ectoderm that should become epidermis (skin).

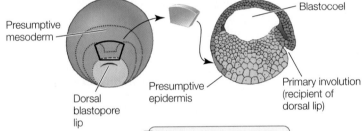

RESULTS

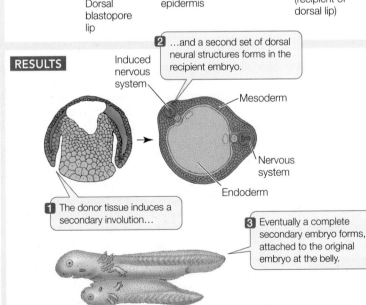

2 ...and a second set of dorsal neural structures forms in the recipient embryo.

1 The donor tissue induces a secondary involution...

3 Eventually a complete secondary embryo forms, attached to the original embryo at the belly.

CONCLUSION

The cells of the dorsal blastopore lip can induce other cells to change their developmental fates.

INVESTIGATION

FIGURE 33.10 Differentiation Can Be Due to Inhibition of Transcription Factors When organizer cells involute to underlie dorsal ectoderm along the embryo midline, that overlying ectoderm becomes neural tissue rather than skin (epidermis). But do the organizer cells *cause* dorsal ectoderm to become neural tissue, or do they *prevent* this ectoderm from becoming skin?

HYPOTHESIS

The default state of amphibian dorsal ectoderm is neural; it is induced by underlying mesoderm to become epidermis.

METHOD

1. Excise animal caps (presumptive ectoderm) from amphibian blastulas. Culture presumptive mesodermal cells from early gastrulas and extract BMP4. BMP4 is a growth factor released from the notochord that induces overlying ectoderm to become skin.

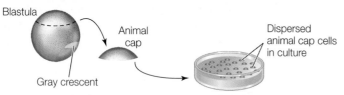

2. Prepare four separate cultures of embryonic ectodermal cells. Incubate with no additions (control); with BMP4 from step 1; with a BMP4 inhibitor; and with both molecules.

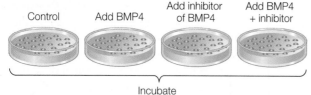

Incubate

3. After incubation, extract mRNAs from the cultured ectodermal cells and run on gels to reveal expression of NCAM (a neural tissue marker) and keratin (an epidermal tissue marker).

RESULTS

Gels reveal that different treatments with BMP4 and its inhibitor alter patterns of gene expression in cultured ectodermal cells.

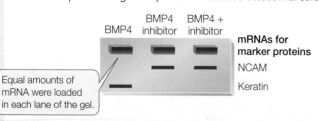

CONCLUSION

The organizer cells secrete an inhibitor of BMP4.

ANALYZE THE DATA

Use the gel results shown above to answer the questions.

A. Does BMP4 induce expression of neural specific or epidermal specific genes?

B. Does BMP4 block any gene expression in the ectodermal cells?

C. What is the evidence that BMP4 has an inductive and/or an inhibitory effect on gene expression in ectodermal cells?

D. Do these results support the hypothesis?

Go to **yourBioPortal.com** for original citations, discussions, and relevant links for all INVESTIGATION figures.

We will use the chicken egg to show how gastrulation proceeds in such eggs.

Cleavage in the chick results in a flat, circular layer of cells called a blastodisc (**FIGURE 33.11**). Between the blastodisc and the yolk mass there is a fluid-filled space. Some cells from the blastodisc break free and move into this space. These cells come together to form a continuous layer called the **hypoblast**, which will later contribute to *extraembryonic membranes* that will support and nourish the developing embryo. The overlying cells make up the **epiblast**, from which the embryo proper will form. Thus, the avian blastula is a flattened structure consisting of an upper epiblast and a lower hypoblast that are joined at the margins of the blastodisc.

Gastrulation begins with a thickening in the posterior region of the epiblast, caused by the movement of cells toward the midline and then forward along the midline (see Figure 33.11). The result is a midline ridge called the *primitive streak*. A depression called the *primitive groove* forms along the length of the primitive streak. The primitive groove functions as the blastopore, and cells migrate through it into the blastocoel to become endoderm and mesoderm.

In the chick embryo, no archenteron forms, but the endoderm and mesoderm migrate forward to form the gut and other structures. At the anterior end of the primitive groove is a thickening called **Hensen's node**, which in birds, reptiles, and mammals is the equivalent of the dorsal lip of the amphibian blastopore. Many signaling molecules that have been identified in the frog organizer are also expressed in Hensen's node.

FRONTIERS Early development explains why we are bilaterally symmetrical—two ears, arms, lungs, and so on. Despite this symmetry, the human heart and stomach are usually positioned toward the left side of the body, the liver and appendix to the right. Where does left–right positional information come from? At the anterior end of the primitive groove, cells with motile cilia move extracellular fluid through the groove. Other cells have nonmotile stereocilia. Movement of fluid exposes the nonmotile stereocilia to signaling molecules and bending forces that induce signaling cascades that affect organ development. Because the flow of extracellular fluid has a left–right directionality, the cascade introduces a left–right pattern into the process of organogenesis. There are individuals whose organ symmetry is reversed, and they may be unaware of it unless the reversal is revealed by imaging or surgery.

Placental mammals retain the avian/reptilian gastrulation pattern but lack yolk

Mammals, reptiles, and birds are amniotes—they produce eggs that develop extraembryonic membranes—so it is not surprising that they share certain patterns

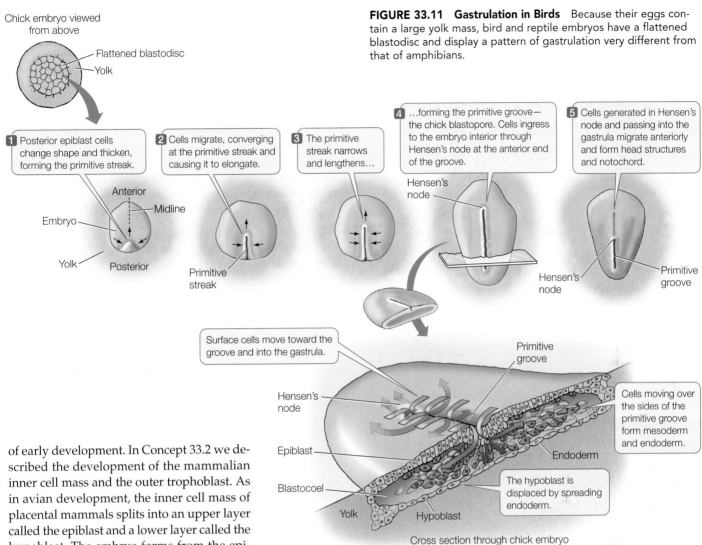

Chick embryo viewed from above

Flattened blastodisc
Yolk

FIGURE 33.11 Gastrulation in Birds Because their eggs contain a large yolk mass, bird and reptile embryos have a flattened blastodisc and display a pattern of gastrulation very different from that of amphibians.

1 Posterior epiblast cells change shape and thicken, forming the primitive streak.

2 Cells migrate, converging at the primitive streak and causing it to elongate.

3 The primitive streak narrows and lengthens…

4 …forming the primitive groove—the chick blastopore. Cells ingress to the embryo interior through Hensen's node at the anterior end of the groove.

5 Cells generated in Hensen's node and passing into the gastrula migrate anteriorly and form head structures and notochord.

Anterior
Midline
Embryo
Yolk
Posterior
Primitive streak

Hensen's node

Hensen's node

Hensen's node
Primitive groove

Surface cells move toward the groove and into the gastrula.

Primitive groove

Hensen's node

Cells moving over the sides of the primitive groove form mesoderm and endoderm.

Epiblast
Endoderm

Blastocoel

The hypoblast is displaced by spreading endoderm.

Yolk
Hypoblast

Cross section through chick embryo

of early development. In Concept 33.2 we described the development of the mammalian inner cell mass and the outer trophoblast. As in avian development, the inner cell mass of placental mammals splits into an upper layer called the epiblast and a lower layer called the hypoblast. The embryo forms from the epiblast, while the hypoblast contributes to the extraembryonic membranes that will encase the developing embryo and help form the *placenta* (see Figure 33.4). Gastrulation occurs in the mammalian epiblast just as it does in the avian epiblast. A primitive groove forms, and epiblast cells migrate through the groove to become layers of endoderm and mesoderm.

Do You Understand Concept 33.3?

- What are the major similarities and differences between sea urchin gastrulation and frog gastrulation?
- Explain how Spemann's tissue transplant experiments on gastrulas disproved his initial hypothesis and generated a new hypothesis.
- Explain why Hensen's node in the chick is the equivalent of the dorsal lip of the amphibian blastopore.

Gastrulation produces an embryo with three germ layers that are positioned to influence one another through inductive tissue interactions. During the next phase of development, called **organogenesis**, the major organs and organ systems develop.

concept 33.4 Neurulation Creates the Nervous System

The first major event of vertebrate organogenesis, and the only one we will describe in detail here, is **neurulation**—the formation of the nervous system. We will detail neurulation in the amphibian embryo, but it occurs in a similar fashion in reptiles, birds, and mammals.

The notochord induces formation of the neural tube

As we learned in Concept 33.3, one group of cells that passes over the dorsal lip of the blastopore moves anteriorly and becomes the endodermal lining of the digestive tract. The other group of cells that involutes over the dorsal lip becomes *chordamesoderm*, so named because it forms a rod of mesoderm—the **notochord**—that extends down the center of the embryo. The notochord gives structural support to the developing embryo and also plays a critical role in inducing neurulation.

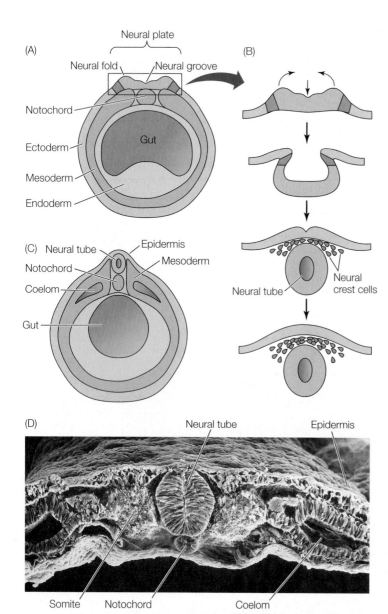

(A) Neural plate

Neural fold | Neural groove

Notochord

Gut

Ectoderm

Mesoderm

Endoderm

(B)

(C) Neural tube | Epidermis

Notochord | Mesoderm

Coelom

Gut

Neural tube | Neural crest cells

(D) Neural tube | Epidermis

Somite | Notochord | Coelom

FIGURE 33.12 Neurulation in a Vertebrate (A) At the start of neurulation, the ectoderm of the neural plate (green) is flat. (B) The neural plate invaginates and folds, forming a tube. (C,D) The completely formed neural tube seen in (C) diagrammatic form and (D) in a scanning electron micrograph of a chick embryo.

Neurulation involves the formation of an internal neural tube from an external sheet of cells. The first signs of neurulation are flattening and thickening of the ectoderm overlying the notochord; this thickened area forms the *neural plate* (**FIGURE 33.12A**). The edges of the neural plate continue to thicken to form folds. Between these neural folds, a groove forms and deepens. The folds move toward each other to fuse in the midline, forming a hollow cylinder, the **neural tube**, and a continuous overlying layer of epidermal ectoderm (**FIGURE 33.12B–D**). Signaling molecules from the notochord are involved in the differentiation of the neural tube into the various structures of the nervous system. Two of these signaling molecules are appropriately named Noggin and Chordin. Another important

factor controlling differentiation of the neural tube is Sonic hedgehog (Shh), the transcription factor mentioned at the start of this chapter. Shh is released by the notochord and diffuses into the ventral region of the neural tube, where is directs the development of this region into the ventral structures and circuits of the spinal cord. These spinal cord circuits will carry the major output from the motor pathways of the nervous system.

Cells from the most lateral portions of the neural plate do not become part of the neural tube, but disassociate from it and come to lie between the neural tube and the overlying epidermis. These **neural crest cells** migrate outward to lead the development of the connections between the central nervous system (brain and spinal cord) and the rest of the body.

The central nervous system develops from the embryonic neural tube

At its anterior end, the vertebrate neural tube forms three swellings that become the major divisions of the adult brain: the **hindbrain**, **midbrain**, and **forebrain**. The rest of the neural tube becomes the spinal cord. This general pattern holds for all vertebrates, although the complexity of certain structures, notably the cerebral hemispheres, is greatest among the mammals. Here we detail the development of the human brain.

Although the developing brain folds and becomes an astonishingly complex structure, information flow in the mature nervous system (which will be the subject of Chapters 34 and 35) follows paths that can be traced back to the simple, linear neural tube of the early embryo (**FIGURE 33.13A,B**). The hindbrain and midbrain give rise to structures that collectively are called the **brain stem** and that govern critical physiological functions such as heartbeat and breathing; the hindbrain also produces the **cerebellum**, a structure that governs motor control and some cognitive functions. The forebrain develops into the **cerebral hemispheres**—the major information-processing areas of the brain—plus the underlying *thalamus, hypothalamus,* and *pituitary* (**FIGURE 33.13C**) The thalamus is the major relay station for sensory information coming into the cerebral hemispheres, and the hypothalamus receives information relevant to the regulation of the internal environment such as temperature and hormone levels, as described in Chapters 29 and 30.

Information traveling between the forebrain and other regions of the body passes through the brain stem. Structures in the brain stem make use of this information to regulate many basic functions such as breathing, circulation, and motor coordination. We will discuss the organization of the nervous system in more detail in Chapter 34, and the functions of specific regions in controlling body functions in the chapters devoted to those physiological systems.

Body segmentation develops during neurulation

The vertebrate body plan, like that of arthropods, consists of repeating segments that are modified during development. These segments are most evident as the repeating patterns of vertebrae, ribs, nerves, and muscles along the anterior–posterior axis.

(A)

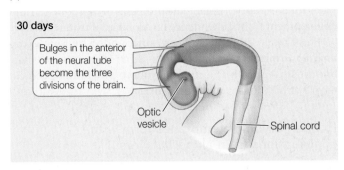

30 days

Bulges in the anterior of the neural tube become the three divisions of the brain.

Optic vesicle

Spinal cord

(B)

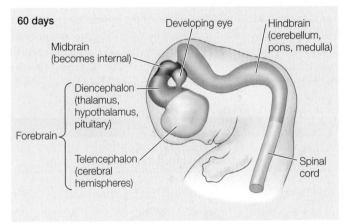

60 days

Developing eye

Hindbrain (cerebellum, pons, medulla)

Midbrain (becomes internal)

Diencephalon (thalamus, hypothalamus, pituitary)

Forebrain

Telencephalon (cerebral hemispheres)

Spinal cord

(C)

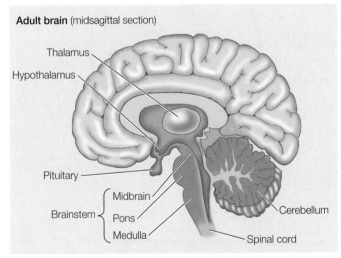

Adult brain (midsagittal section)

Thalamus

Hypothalamus

Pituitary

Brainstem
Midbrain
Pons
Medulla

Cerebellum

Spinal cord

FIGURE 33.13 Development of the Central Nervous System In vertebrate embryos, the anterior end of the hollow neural tube differentiates into forebrain, midbrain, and hindbrain. Each of these regions develops into several structures in the adult brain. The remainder of the neural tube becomes the spinal cord.

(A)

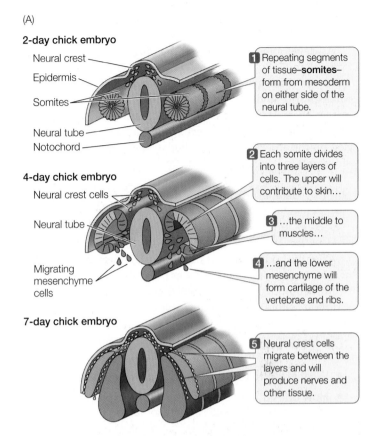

2-day chick embryo

Neural crest

Epidermis

Somites

Neural tube

Notochord

1 Repeating segments of tissue–**somites**–form from mesoderm on either side of the neural tube.

4-day chick embryo

Neural crest cells

Neural tube

Migrating mesenchyme cells

2 Each somite divides into three layers of cells. The upper will contribute to skin…

3 …the middle to muscles…

4 …and the lower mesenchyme will form cartilage of the vertebrae and ribs.

7-day chick embryo

5 Neural crest cells migrate between the layers and will produce nerves and other tissue.

(B)

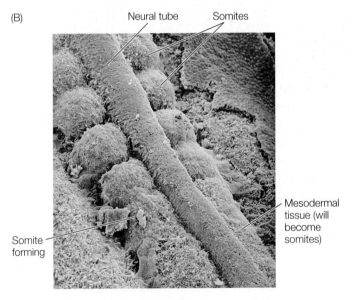

Neural tube   Somites

Somite forming

Mesodermal tissue (will become somites)

FIGURE 33.14 Body Segmentation (A) Repeating blocks of tissue called somites form on either side of the neural tube. Muscle, cartilage, bone, and the lower layer of the skin form from the somites. (B) In this scanning electron micrograph of somite formation in a chick embryo, the overlying ectoderm has been removed and the neural tube and somites are seen from above.

As the neural tube forms, mesodermal tissues gather along the sides of the notochord to form separate, segmented blocks of cells called **somites** (**FIGURE 33.14**). The somites produce cells that will become the vertebrae, ribs, muscles of the trunk and limbs, and lower layer of the skin. Again, tissue interactions and signaling molecules direct these various processes

of differentiation. For example, the transcription factor Sonic hedgehog, mentioned above as being released from the notochord and diffusing to the neural tube, also diffuses laterally to reach the ventral region of the somites. Shh directs cells in that region into the developmental pathways that form skeletal structures.

As development progresses, the different segments of the body change. Regions of the vertebral column differ in that some vertebrae grow ribs of various sizes and others do not, forelegs arise in the anterior part of the embryo, and hind legs arise in the posterior region. Gene expression patterns (including those of *Hox genes*) give cells information about their position on the anterior–posterior body axis, causing appropriate structures to develop.

> **LINK** Gene expression patterns and morphogenic gradients that lead to segmentation during development are described in Concept 14.3

After the body segments have been specified, the formation of other organs and organ systems progresses rapidly, directed by further inductive tissue interactions involving a host of transcription factors and other molecules.

Do You Understand Concept 33.4?

- What is the role of the notochord in neurulation?
- How does the process of neurulation explain why the spinal cord has a central cavity, or lumen?
- The epidermis of the skin forms from ectoderm. Under the epidermis are muscle tissues derived from mesoderm. Within and under the muscle tissues are nerves derived from ectoderm. Explain how ectodermally derived tissues come to lie under mesodermally derived tissues.

In mammals, the circulatory systems of the fetus and mother are separate, yet the fetus derives nutrition and metabolic support from the maternal circulation. The next concept examines the placenta—the structure that makes these exchanges between the maternal and fetal blood possible.

concept 33.5 Extraembryonic Membranes Nourish the Growing Embryo

Reptiles, birds, and mammals are all *amniotes*, meaning they produce amniote eggs—or in the case of placental mammals, structures that are homologous with those found in the amniote egg. We saw in Concept 23.6 the evolutionary significance

FIGURE 33.15 The Extraembryonic Membranes of Amniotes In birds, reptiles, and mammals, the embryo constructs four extraembryonic membranes. The yolk sac encloses the yolk, and the amnion and chorion enclose the embryo. Fluids secreted by the amnion fill the amniotic cavity, providing an aqueous environment for the embryo. The chorion, along with the allantoic membrane, mediates gas exchange between the embryo and its environment. The allantois stores the embryo's waste products.

of the amniote egg: the contained aqueous environment it provides for the embryo, freeing the processes of reproduction and development from dependence on an external water supply.

Embryos in amniote eggs are surrounded by four **extraembryonic membranes** that function in nutrition, gas exchange, and waste removal. The chicken embryo provides a good example of how these membranes form.

Extraembryonic membranes form with contributions from all germ layers

In the chick, four membranes form—the *yolk sac*, the *allantoic membrane*, the *amnion*, and the *chorion* (**FIGURE 33.15**). The **yolk sac** is the first to form, by extension of the hypoblast layer along with some adjacent mesoderm. The yolk sac grows to enclose the yolk in the egg and provides nutrients to the embryo. It constricts at the top to create a tube that is continuous with the gut of the embryo. The **allantoic membrane** is also an outgrowth

> **yourBioPortal.com**
>
> Go to WEB ACTIVITY 33.1
> **Extraembryonic Membranes**

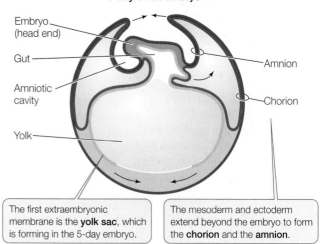

5-day chick embryo

Embryo (head end)
Gut
Amniotic cavity
Yolk
Amnion
Chorion

The first extraembryonic membrane is the **yolk sac**, which is forming in the 5-day embryo.

The mesoderm and ectoderm extend beyond the embryo to form the **chorion** and the **amnion**.

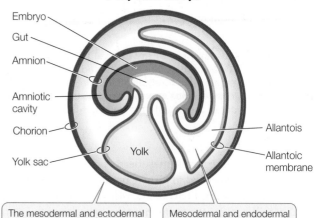

9-day chick embryo

Embryo
Gut
Amnion
Amniotic cavity
Chorion
Yolk sac
Yolk
Allantois
Allantoic membrane

The mesodermal and ectodermal layers fuse below the yolk so that the chorion lines the shell.

Mesodermal and endodermal tissues form the **allantois**, a sac for metabolic wastes.

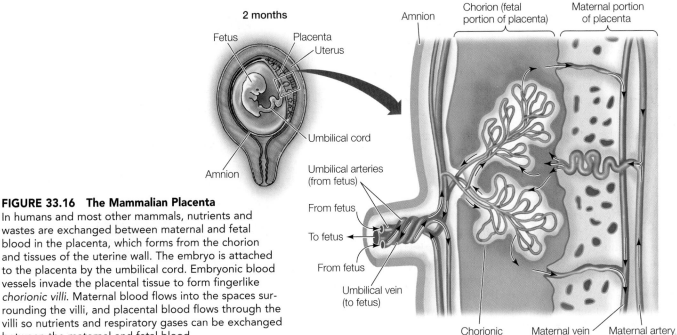

FIGURE 33.16 The Mammalian Placenta
In humans and most other mammals, nutrients and wastes are exchanged between maternal and fetal blood in the placenta, which forms from the chorion and tissues of the uterine wall. The embryo is attached to the placenta by the umbilical cord. Embryonic blood vessels invade the placental tissue to form fingerlike *chorionic villi*. Maternal blood flows into the spaces surrounding the villi, and placental blood flows through the villi so nutrients and respiratory gases can be exchanged between the maternal and fetal blood.

of the extraembryonic endoderm plus adjacent mesoderm. It forms the allantois, a sac for storage of metabolic wastes.

The other two membranes are formed by ectoderm and mesoderm that extend all along the inside of the eggshell, both over the embryo and below the yolk sac. Where they meet, they fuse, forming two membranes, the inner amnion and the outer chorion. The **amnion** surrounds the embryo, forming the amniotic cavity. The amnion secretes fluid into the cavity, providing a protective environment for the embryo. The **chorion**, located just beneath the egg's shell, forms a continuous membrane that limits water loss from the egg and works with the allantoic membrane to exchange respiratory gases between the embryo and the outside world.

Extraembryonic membranes in mammals form the placenta

In placental mammals, the trophoblast adheres to the uterine wall, secretes proteolytic enzymes, and burrows into the endometrium, beginning the process of implantation (see Figure 33.4). Eventually, the entire trophoblast (and the enclosed embryo) is embedded within the wall of the uterus. Meanwhile, the hypoblast cells proliferate to form what in the bird would be the yolk sac. But there is no yolk in eggs of placental mammals, so the yolk sac contributes mesodermal tissues that interact with trophoblast tissues to form the chorion. The chorion, along with maternal tissues of the uterine wall, produces the placenta (**FIGURE 33.16**). Meanwhile, the epiblast produces the amnion, which grows to enclose the entire embryo in a fluid-filled amniotic cavity. (A pregnant woman's "water breaks" when the amnion bursts at the onset of or during labor and releases the amniotic fluid.)

An allantois also develops in mammals, but its importance depends on how well nitrogenous wastes can be transferred across the placenta. In humans the allantois is minor; in pigs it is important. In humans and other placental mammals, allantoic tissues contribute to the formation of the umbilical cord, by which the embryo is attached to the placenta.

FRONTIERS Oxygen and nutrients are not the only things a fetus obtains from its mother. As we will see in Chapter 39, the human gut is richly populated with bacteria that are necessary for normal digestion. Where do these bacteria come from? A recent study shows that the gut flora of children resemble the species assemblage found in their mothers. In adults, the bacterial assemblages in siblings are more similar to each other than to that of their marriage partner. Thus, early transmission of bacteria from mother to infant is indicated. Further studies show that children who were vaginally delivered have gut bacteria more similar to their mothers' than do children born by Caesarean section, so it appears that one route of transmission is via the vaginal canal during birth.

Human gestation is traditionally divided into three periods of roughly 12 weeks each, called *trimesters*. In the first trimester, gastrulation occurs, tissues differentiate, and the placenta forms. This is a time of rapid cell division and tissue differentiation, and the embryo is very sensitive to damage from radiation, drugs, chemicals, and pathogens that can cause birth defects. By the end of the first trimester, most organs have started to form, and the embryo is about 8 centimeters long. At about this time, the human embryo is medically and legally referred to as a **fetus**. During the second and third trimesters, the fetus grows rapidly. Toward the end of the third trimester the internal organs mature. A human infant is born as soon as the last of its critical organs matures—the lungs.

Do You Understand Concept 33.5?

- Name the four extraembryonic membranes of a chick embryo and describe their functions.

- Would you expect a frog embryo to have the same extraembryonic membranes as a lizard? Why or why not?

- Compare the mammalian placenta with a chicken egg. What are the similarities? What do you think is the evolutionary reason for the similarities?

Development does not end with birth. Growth continues until adult size is reached, and even when growth stops, organs of the body continue to repair and renew themselves. In humans, the brain in particular continues to develop and mature long after birth. The mechanisms of the vertebrate (and particularly the human) nervous system, the end results of which are remarkable levels of sensation and perception as well as extensive motor control, will be detailed in the next three chapters.

FIGURE 33.17 Environmentally Induced Holoprosencephaly A stillborn lamb born of a ewe that grazed on corn lily leaves. Alkaloids in the plant interfered with the Sonic hedgehog signaling pathway in the embryo, resulting in fusion of the cerebral hemispheres and the formation of a single eye.

 QUESTION How does the Sonic hedgehog pathway control development of the vetebrate brain and eyes?

ANSWER As we learned in Concept 33.4, Sonic hedgehog (Shh) is a transcription factor secreted by the notochord that diffuses to the overlying neural tube, where it induces differentiation of the ventral neural tube into motor neurons and other features of the ventral regions of the spinal cord. Shh also plays different inductive roles at other locations. At the anterior end of the neural tube where the forebrain will develop, the closed neural tube undergoes reciprocal inductive tissue interactions with the overlying ectoderm. Signaling molecules from the neural tissue induce the ectoderm to form the lens of the eye, and the developing lens produces signaling molecules that induce the underlying neural tissue to form an optic cup that will differentiate into the retina. But why do these events happen bilaterally?

Mesoderm that is anterior to the notochord produces and releases Shh, as does the notochord. Shh induces the ventral region of the anterior neural tube to differentiate into the structures of the forebrain. One of the actions of Shh is to inhibit a transcription factor called Pax6 that is expressed by the neural tissue. Pax6 is essential for determination of the eye field (i.e., tissues that are competent to undergo the inductive interactions with the neural tissue to form eyes). The midline expression of Shh splits the eye field into two and also enables the forebrain to develop into left and right hemispheres.

When environmental factors affect the action of Shh, holoprosencephaly can result. When pregnant cows or ewes graze on *Veratrum californicum* at certain critical points in gestation, the embryo is exposed to cyclopamine and the Shh signaling pathway is blocked. Depending on when the exposure occurs, bilateral development of the forebrain may not take place, resulting in holoprosencephaly and cyclopia (**FIGURE 33.17**). Another environmental factor of great interest for humans is the fact that Shh requires cholesterol to be an effective signal, and abnormally low levels of cholesterol during pregnancy (such as might occur with the use of certain statin drugs) could cause some of the aspects of holoprosenecephaly.

33 | SUMMARY

concept 33.1 Fertilization Activates Development

- The sperm and egg make different contributions to the zygote. The egg contributes a haploid nucleus, nutrients, ribosomes, mitochondria, mRNAs, and proteins. The sperm contributes a haploid nucleus and, in most species, a centriole.

- In amphibians, the cytoplasmic contents of the egg are not distributed homogeneously, and they are rearranged after fertilization to set up the major axes of the future embryo. The nutrient molecules are generally found in the **vegetal hemisphere**, whereas the nucleus is found in the **animal hemisphere**. Review Figures 33.1 and 33.2

concept 33.2 Cleavage Repackages the Cytoplasm of the Zygote

- **Cleavage** is a period of rapid cell division without cell growth. Except in mammals, little if any gene expression occurs during cleavage. Cleavage can be complete or incomplete. The result of cleavage is a ball or mass of cells called a **blastula**. Review Figure 33.3

- Early cell divisions in mammals are unique in being slow and allowing for gene expression. These cell divisions produce a **blastocyst** composed of an **inner cell mass** that becomes the embryo and an outer layer of cells that becomes the **trophoblast**. The trophoblast helps the blastocyst implant in the uterine wall. Review Figures 33.4 and 33.5

- Some species undergo **mosaic development**, in which the fate of each cell is determined during early divisions. Other species, including vertebrates, undergo **regulative development**, in which remaining cells can compensate for cells lost in early cleavages.

concept 33.3 Gastrulation Creates Three Tissue Layers

- **Gastrulation** involves massive cell movements that produce three **germ layers** and place cells from various regions of the blastula into new associations with one another. Review Figure 33.6

- The initial step of sea urchin and amphibian gastrulation is inward movement of certain blastomeres. The site of inward movement becomes the **blastopore**. Cells that move into the blastula become the **endoderm** and **mesoderm**; cells remaining on the outside become the **ectoderm**. Review Figures 33.6 and 33.7 and ANIMATED TUTORIAL 33.1

- The **dorsal lip** of the amphibian blastopore is a critical site for cell determination. It has been called the **organizer** because it induces determination in cells that pass over it during gastrulation. **Review Figures 33.7, 33.8, 33.9 and ANIMATED TUTORIAL 33.2**

- The protein β-catenin activates a signaling cascade that induces the primary embryonic organizer and sets up the anterior–posterior body axis.

- Gastrulation in reptiles and birds differs from that in sea urchins and frogs because the large amount of yolk in reptile and bird eggs causes the blastula to form a flattened disc of cells. **Review Figure 33.11**

- Although their eggs have no yolk, placental mammals have a pattern of gastrulation similar to that of reptiles and birds.

- Gastrulation is followed by **organogenesis**, the process whereby tissues interact to form organs and organ systems.

concept 33.4 Neurulation Creates the Nervous System

- In the formation of the vertebrate nervous system, one group of cells that migrates over the blastopore lip becomes the **notochord**. The notochord organizes the overlying ectoderm to form a **neural tube**. Review Figure 33.12

- The anterior region of the neural tube becomes the brain. Different regions differentiate into the **forebrain**, **midbrain**, and **hindbrain**. The posterior region of the neural tube becomes the spinal cord. Review Figure 33.13

- The notochord and **neural crest cells** participate in the segmental organization of mesoderm into structures called **somites** along the body axis. Rudimentary organs and organ systems form during these stages. Review Figure 33.14

concept 33.5 Extraembryonic Membranes Nourish the Growing Embryo

- Amniote eggs contain four **extraembryonic membranes**. In a bird or reptile, the **yolk sac** surrounds the yolk and provides nutrients to the embryo, the **chorion** lines the eggshell and participates in gas exchange, the **amnion** surrounds the embryo and encloses it in an aqueous environment, and the **allantoic membrane** forms the allantois, a storage sac for metabolic wastes. **Review Figure 33.15 and WEB ACTIVITY 33.1**

- In placental mammals, the chorion interacts with maternal uterine tissues to form a placenta, which provides the embryo with nutrients and gas exchange. The amnion encloses the embryo in an aqueous environment. **Review Figure 33.16**

- Development continues throughout life.

Neurons and Nervous Systems

Fear is a strong emotion that you probably would not immediately think of as pleasant. Yet many of us seek opportunities to experience fear. Riding roller coasters, sky diving, driving fast, surfing, skiing, even watching horror movies—all of these activities can produce a thrill of fear that many people find enjoyable.

Physiologically, fear is a protective emotion that comes in two forms, reactive and proactive. A sudden and unexpected threat generates reactive fear that is associated with the fight-or-flight response—its function is to get you out of danger quickly. Proactive fear gets you ready for an anticipated threat by sharpening your senses, putting you on alert, and preparing you to take action. In short, fear stimulates the nervous system, and that can be an experience that is sought after, as well as being something that can save your life.

Fear can also be pathological. Extreme, irrational fears are described as *phobias*; claustrophobia (fear of enclosed spaces), agoraphobia (fear of public or unfamiliar places), and xenophobia (fear of strangers) are a few common examples. Such extreme fears can be incapacitating and can drive people to irrational behaviors.

Both adaptive fear and pathological fear are functions of the nervous system. How can we understand the neural basis for this emotional reaction and develop effective therapies when it is abnormal, such as in posttraumatic stress disorder? Sometimes case histories are instructive.

Charles Whitman was a normal and responsible child. He became one of the youngest Eagle Scouts in the country. He was a fine son and husband and received commendations as a U.S. Marine. While in the service, however, he began having unexplained fits of anger and other personality disorders. He was discharged from the Marines and entered the University of Texas. Several times he visited campus doctors and complained about having violent and irrational thoughts. Then, in the early morning hours of August 1, 1966, he killed his mother and his wife, leaving notes of love and regret. He then packed weapons and supplies and went to the campus where he barricaded himself inside the top floor of the clock tower. Using high-powered rifles, he killed 14 people and wounded 38 others before being shot and killed by Austin police.

Charles Whitman left a suicide note requesting that proceeds from his insurance be donated to a mental health foundation. He also requested an autopsy, and that autopsy revealed a small brain tumor.

Some humans enjoy stimulating their fear response through risk-taking behavior.

How can a small brain tumor so dramatically affect personality and behavior?

KEY CONCEPTS

34.1 Nervous Systems Consist of Neurons and Glia

34.2 Neurons Generate and Transmit Electrical Signals

34.3 Neurons Communicate with Other Cells at Synapses

34.4 The Vertebrate Nervous System Has Many Interacting Components

34.5 Specific Brain Areas Underlie the Complex Abilities of Humans

concept 34.1 Nervous Systems Consist of Neurons and Glia

Nervous systems are composed of two types of cells: *nerve cells*, or **neurons**, and *glial cells*, or **glia**. We will describe each of these cell types separately before considering how they are linked together to form nervous systems.

Neurons transmit electrical and chemical signals

Most neurons have four regions—a *cell body*, *dendrites*, one or more *axons*, and *axon terminals* (**FIGURE 34.1**). The *cell body* contains the nucleus and most of the cell's organelles. Extending from the cell body are many shrublike **dendrites** (from the Greek *dendron*, "tree"), which bring information from other neurons or sensory cells to the cell body. The degree of branching of the dendrites differs among different types of neurons. Neurons are *excitable*, meaning they generate and transmit electrical signals. These signals are called *nerve impulses*, or *action potentials*, and we will describe them in detail in Concept 34.2.

In most neurons, one projection—the **axon**—is much longer than the others, and may extend long distances such as from your spinal cord to your toes. Axons are the "telephone lines" of the nervous system. Acting on information received by dendrites, an axon generates action potentials that are conducted down the axon toward a target cell. A **nerve** (as distinct from a neuron) is a bundle of axons that come from many different neurons.

At the target cell, the axon divides into a spray of fine nerve endings, each with a swelling called an **axon terminal** that comes very close to the membrane of the target cell to form a synapse. A **synapse** is a tiny gap across which two neurons communicate, either with electrical signals or with chemical signaling molecules called *neurotransmitters*. The neuron sending the information is the **presynaptic neuron**, and the neuron receiving the information is the **postsynaptic neuron** (see Concept 34.3).

FRONTIERS Most neurons do not undergo mitosis and the cycle of cell death and replacement typical of other cells. Until recently, it was believed that an animal is born with all the neurons it will ever have, and that neurons that die are not replaced. That was until scientists discovered that songbirds produce a complement of new neurons each year when the birds come into reproductive condition and begin to sing. Subsequent research on mammals has shown that, in at least two regions of the brain, new neurons are produced regularly. This generation of neurons is believed to be stimulated by new experiences and by exercise. The study of mammalian neurogenesis is an exciting and expanding area of research.

Glia support, nourish, and insulate neurons

There are many more glia than neurons in the human brain. Glia do not generate or conduct action potentials, but they can release neurotransmitter molecules. During embryonic development, some glia physically support developing neurons and orient

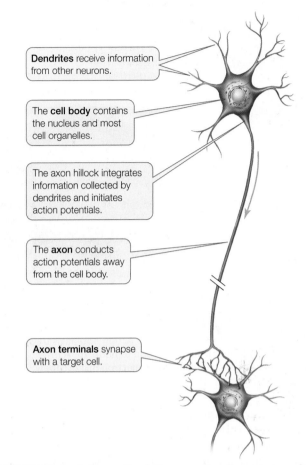

FIGURE 34.1 A Generalized Neuron The diagram shows the features typical of most neurons. The form of these features, including axon length and the density and branching patterns of dendrites, can vary greatly among the many different types of neurons.

Labels in figure:
- **Dendrites** receive information from other neurons.
- The **cell body** contains the nucleus and most cell organelles.
- The axon hillock integrates information collected by dendrites and initiates action potentials.
- The **axon** conducts action potentials away from the cell body.
- **Axon terminals** synapse with a target cell.

them toward their targets. In the postembryonic brain, glia provide homeostatic support for neurons by maintaining the local extracellular environment and provide energy substrates when needed. They help clear the synapse of neurotransmitters and modulate neurotransmission by releasing their own chemicals. Glia play important roles in synapse modification, a phenomenon known as *neuroplasticity*. They also assist in neuronal repair and remove debris such as dead neurons.

Glia called **astrocytes** surround the smallest, most permeable blood vessels in the brain, contributing to a **blood–brain barrier** that prevents toxic chemicals (and most water-soluble or large molecules) from reaching the brain. The barrier is not perfect, however. The blood–brain barrier consists of plasma membranes and is permeable to fat-soluble substances such as anesthetics and alcohol; thus, these substances have rapid and marked effects on the nervous system. The blood–brain barrier usually prevents antibodies from entering the brain. To provide the brain with immune defenses, glia called **microglia** act as macrophages and mediators of inflammatory responses.

One crucial function of glia is to electrically insulate axons by wrapping around them, covering them with concentric layers of insulating plasma membrane. In the brain and spinal cord, the membranes of glia called **oligodendrocytes** wrap around axons; in nerves outside the brain and spinal cord, a

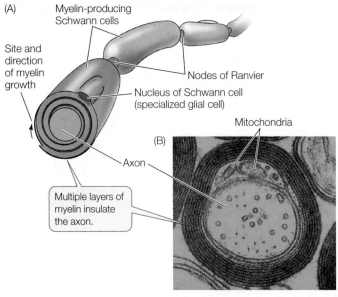

FIGURE 34.2 Wrapping Up an Axon (A) Schwann cells produce layers of myelin, a type of plasma membrane that provides electrical insulation to the axon. At the intervals between Schwann cells—the nodes of Ranvier—the axon is exposed. Action potentials travel along the axon by "jumping" from node to node. (B) A myelinated axon, seen in cross section through an electron microscope.

different type of glia called **Schwann cells** perform this function (**FIGURE 34.2**). This multilayered wrap of glial membranes on axons forms a lipid-rich, nonconductive sheath called **myelin**. Parts of the nervous system consisting mostly of myelinated axons have a glistening white appearance, giving rise to the term *white matter*. Areas of the nervous system that are rich in cell bodies do not appear white and are called *gray matter*. Not all axons are myelinated, but those that are can conduct action potentials more rapidly than unmyelinated axons.

Diseases that affect myelin impair conduction of action potentials. The most common demyelinating disease is *multiple sclerosis*, an autoimmune disease in which antibodies attack proteins in the myelin. Symptoms can vary dramatically, depending on which part of the nervous system is affected. Unfortunately, there are no known cures for demyelinating diseases.

Neurons are linked into information-processing networks

Neurons are organized into information-processing **neural networks**. These networks include three functional categories of neurons, which can be thought of as being involved with input, output, and integration, respectively:

- **Afferent neurons** carry sensory information into the nervous system. That information comes from specialized **sensory cells** that transduce (convert) sensory stimuli (e.g., light, pressure) into action potentials.

- **Efferent neurons** carry commands to physiological and behavioral *effectors* such as muscles and glands. **Motor neurons** are a type of effector neuron that carry commands to muscle cells.

- **Interneurons** integrate and store information and communicate between afferent and efferent neurons. Most neurons in the human brain are interneurons.

Neural networks can be simple, like the reflex that causes your leg to kick when a physician taps your knee (see Figure 34.14); or exceedingly complex, like the network that enables you to read, comprehend, and remember the words and ideas in this chapter. Stationary, simple animals such as cnidarians (e.g., sea anemones) generally process information with *nerve nets*—neural networks that do little more than provide direct lines of communication from sensory cells to effectors (**FIGURE 34.3A**). There is little or no integration or processing of signals.

Animals that actively move about need to process and integrate larger amounts of information. The neurons of earthworms, for example, are organized into clusters called **ganglia** (singular *ganglion*; **FIGURE 34.3B**). In bilaterally symmetrical animals, ganglia frequently come in pairs, with one on each side of the body. In animals with complex behavior, some ganglia

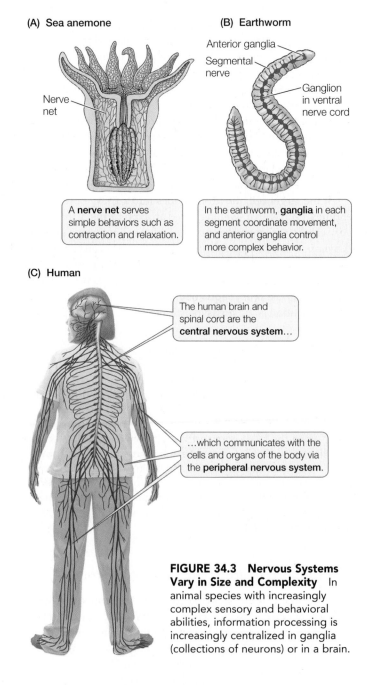

FIGURE 34.3 Nervous Systems Vary in Size and Complexity In animal species with increasingly complex sensory and behavioral abilities, information processing is increasingly centralized in ganglia (collections of neurons) or in a brain.

are enlarged or fused together at the anterior end, forming a larger, centralized integrative center—in other words, a **brain** (**FIGURE 34.3C**).

Do You Understand Concept 34.1?

- Describe how the different parts of a neuron function in the flow of information through the neuron.
- What are the roles of the different types of glia?
- Propose, in terms of simple neural networks, how input coming from different sources can be integrated to produce a single output in an efferent neuron.

The one feature common to all nervous systems is that they encode and transmit information in the form of action potentials. In the next concept we will focus on how action potentials are generated and transmitted by neurons.

concept 34.2 Neurons Generate and Transmit Electrical Signals

Neurons generate changes in *membrane potential*—the difference in electrical charge across their plasma membranes. These electrical changes generate the nerve impulses, or *action potentials*, that carry information along neurons. **Action potentials** are sudden, large, transient changes in membrane potential that travel along axons and cause the release of chemical signals at the axon terminals. Here we will cover the generation of action potentials in detail, starting with a review of some basic electrical principles.

Simple electrical concepts underlie neural function

Voltage is a measure of the *difference in electrical charge between two points*. Voltage represents potential energy (see Concept 2.5) because opposite charges will move together if given a chance. For example, when the negative and positive poles of a battery are connected by a wire, an electric current flows through the wire because there is a voltage difference between the two poles. This flow of electric current can be used to do work, just as the force of a water current can be used to do work.

In wires, electric current is carried by electrons; in solutions and across cell membranes, electric current is carried by ions. The major ions that carry electric charges across the membranes of neurons are sodium (Na^+), potassium (K^+), calcium (Ca^{2+}), and chloride (Cl^-). In cells, these ions are kept at different concentrations inside and outside the cell. The result of these differing concentrations is a voltage across the cell membrane, known as a **membrane potential**.

Membrane potentials exist and can be measured in all cells (**FIGURE 34.4**). In an inactive neuron, this membrane potential is referred to as the **resting potential**, which is typically between –60 and –70 millivolts (mV). The minus sign indicates that the inside of the cell is electrically *negative* compared with the outside. Action potentials are generated when the membrane

RESEARCH TOOLS

FIGURE 34.4 Measuring the Membrane Potential
An electrode can be made from a glass pipette with a very sharp tip filled with a solution that conducts electric charges. If one electrode is placed inside the plasma membrane of an axon and another is placed just outside the axon, the difference in voltage can be measured.

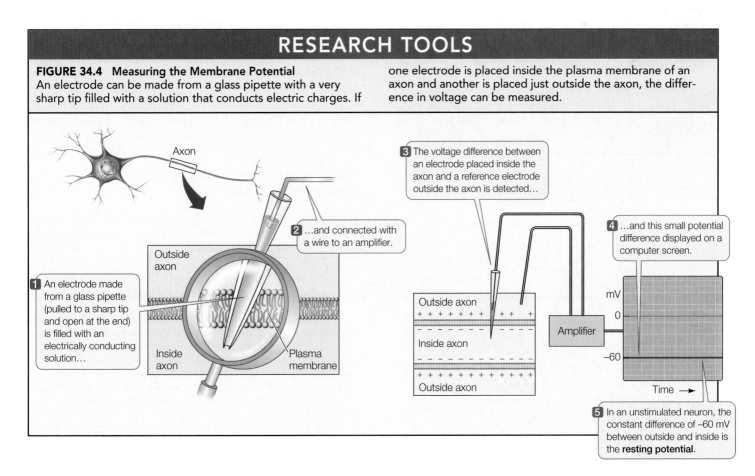

1 An electrode made from a glass pipette (pulled to a sharp tip and open at the end) is filled with an electrically conducting solution...

2 ...and connected with a wire to an amplifier.

Axon

Outside axon

Inside axon

Plasma membrane

3 The voltage difference between an electrode placed inside the axon and a reference electrode outside the axon is detected...

Outside axon
+ + + + + + + + + +
– – – – – – – – – –

Inside axon
– – – – – – – – – –
+ + + + + + + + + +

Outside axon

Amplifier

4 ...and this small potential difference displayed on a computer screen.

mV
0
–60
Time →

5 In an unstimulated neuron, the constant difference of –60 mV between outside and inside is the **resting potential**.

potential changes suddenly to become more *positive* inside than outside. How do the movements of ions establish the resting potential and generate action potentials?

The sodium–potassium pump sets up concentration gradients of Na⁺ and K⁺

The plasma membranes of neurons, like those of all other cells, are lipid bilayers that are impermeable to ions but contain ion transporters and channels. A major ion transporter in the plasma membranes of neurons (and all other cells) is the **sodium–potassium pump** (also known as *sodium–potassium ATPase*), which uses the energy of ATP to actively expel Na⁺ ions from inside the cell, exchanging them for K⁺ ions from the outside (**FIGURE 34.5A**).

> **LINK** The energetics of the sodium–potassium pump are described in Concept 5.3

Because Na⁺ and Cl⁻ are the predominant ions in the extracellular fluid, and the sodium–potassium pump is constantly pumping Na⁺ out and K⁺ in, the concentration of Na⁺ is higher outside the cell than inside, and the K⁺ concentration is higher

inside than outside. In other words, the pump creates *concentration gradients* for both ions. If the ions could suddenly move freely down their concentration gradients, K⁺ would diffuse out of the cell and Na⁺ would diffuse in. As we will see, these concentration gradients can be used to generate the resting potential. Importantly, they can also generate *changes* in the resting potential.

The resting potential is mainly caused by K⁺ leak channels

If you placed two electrodes at different locations in a resting neuron, you would not record a voltage difference between the electrodes. Similarly, if you placed two electrodes outside a resting neuron, you would not record a voltage difference between them. However, if you placed *one electrode outside and one inside the cell*, you would note a voltage difference, with the inside being negative to the outside (see Figure 34.4). Why is this so?

The voltage difference is the result of *leak currents*. Leak currents occur because there are open channels that allow only certain ions to "leak" passively across the cell membrane (**FIGURE 34.5B**). Potassium channels are the most common open, or *leak*, channels in the membranes of resting neurons. Because there are more potassium ions inside than outside, K⁺ diffuses out of the cell through the open channels, down its concentration gradient. But this leakage of K⁺ leaves behind an unbalanced negative charge that tends to pull K⁺ back into the cell. These two motive forces—concentration gradient and overall electrical gradient—constitute an ion's **electrochemical gradient**.

Potassium ions diffuse through K⁺ channels until the concentration gradient pushing K⁺ out exactly equals the electrical charge pulling K⁺ back in. At this equilibrium point the electrochemical gradient is zero, and there is no net movement of K⁺. The membrane potential at this point is the **equilibrium potential** of K⁺. Simply put, if potassium is the only ion that is free to move, it will drive the cell's membrane potential toward the potassium equilibrium potential; thus K⁺ leak channels are largely responsible for the membrane's resting potential.

The Nernst equation can predict a neuron's membrane potential

The equilibrium potential for potassium, E_K, can be calculated from the concentrations of K⁺ on the two sides of the membrane using the **Nernst equation**:

$$E_{ion} = 2.3 \frac{RT}{zF} \log \frac{[ion]_o}{[ion]_i}$$

(A) Na⁺–K⁺ pump (ATPase)

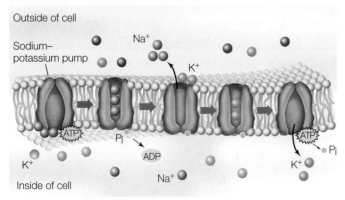

(B) Na⁺–K⁺ channels

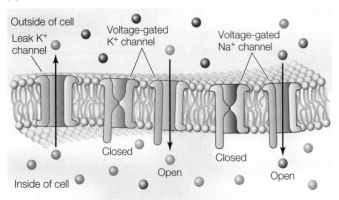

FIGURE 34.5 Ion Transporters and Channels (A) The sodium–potassium pump actively moves K⁺ to the inside of a neuron and Na⁺ to the outside. (B) Ion channels allow specific ions to diffuse down their concentration gradients; K⁺ tends to leave neurons when potassium channels are open, and Na⁺ tends to enter neurons when sodium channels are open. Leak channels like the K⁺ channel shown are always open and create the resting membrane potential. Gated channels like the Na⁺ channels shown are opened by chemical or electrical stimulation.

In this equation, R is the universal gas constant, T is the absolute temperature, F is the Faraday constant, and z is the charge of the ion. Plugging in the constant values, and assuming that reactions proceed at "room temperature" (20°C), $2.3\, RT/F = 58$. Thus the equation simplifies as follows:

$$E_{ion} = \frac{58\ \text{mV}}{z} \log \frac{[\text{ion}]_o}{[\text{ion}]_i}$$

The practical utility of the Nernst equation is that it can predict a neuron's membrane potential if one ion is significantly freer to move than other ions. In resting neurons, K^+ is freer to move than other ions, so the actual resting potential is quite close to the potassium equilibrium potential predicted by the Nernst equation. That the resting potential is in fact slightly less negative than E_K is because resting neurons are slightly permeable to other ions besides K^+ (such as Na^+ and Cl^-; see Apply the Concept below).

---

yourBioPortal.com

Go to WEB ACTIVITY 34.1
The Nernst Equation

Gated ion channels can alter membrane potential

Potassium leak channels are always open, but other ion channels behave as if they contain "gates": they are open under some conditions and closed under other conditions. *Gating provides a means for neurons to change their membrane potentials in response to a stimulus.* **Voltage-gated channels** open or close in response to local changes in voltage across the membrane. **Chemically gated channels** open or close in response to binding a specific molecule. **Mechanically gated channels** open or close in response to mechanical force applied to the cell membrane.

Openings and closings of gated channels can alter the membrane potential. For example, imagine what happens if sodium channels in the plasma membrane open, given that Na^+ ions are more abundant outside the cell than inside, and that the inside of the resting cell is negatively charged. If Na^+ channels open, some Na^+ ions enter and the inside of the cell becomes less negative. When the inside of a neuron becomes less negative (or, alternatively, more positive) in comparison to its resting condition, its plasma membrane is **depolarized** (**FIGURE 34.6**). An opposite change in polarity occurs if gated K^+ channels open. When K^+ leaving the neuron surpasses the normal leak current, the membrane potential becomes even more negative than its resting state, and the membrane is **hyperpolarized**.

The openings and closings of ion channels that result in changes in the voltage across the plasma membrane are the basic mechanisms by which neurons respond to stimuli, be they electrical, chemical, or mechanical.

Graded changes in membrane potential spread to nearby parts of the neuron

If a gated ion channel is stimulated by the appropriate stimulus, it can open or close to cause a local change in membrane potential. These small, local changes in membrane potential are called **graded membrane potentials**, so named because they can vary in magnitude. For example, if a single chemically gated Na^+ channel on a dendrite is stimulated by a neurotransmitter molecule and opens, a small amount of Na^+ will enter the cell, causing a local graded potential in that dendrite (in this case, depolarizing the cell slightly). If more neurotransmitter molecules open more gated Na^+ channels, more Na^+ enters and the graded potential becomes larger, depolarizing the cell even further. Graded potentials are a means of integrating inputs to a cell: the degree of membrane depolarization or hyperpolarization is determined by summing all of the inputs.

Graded potentials spread quickly to nearby areas of the neuron. However, this local flow of electric current decays as it spreads, like water flowing through a leaky hose. Graded potentials do not spread far, and they die out entirely before they get very far along the axon. To carry a signal to the end of an axon, the cell must generate an action potential.

APPLY THE CONCEPT

Neurons generate and transmit electrical signals

The data below were recorded from the large axon of a squid (see pp. 679–680). They show the concentrations of four ions both inside the axon's cytoplasm and outside the cell, in a seawater bath.

ION	ION CONCENTRATION (m*M*)	
	IN SQUID CYTOPLASM	IN SEAWATER
K^+	400	20
Na^+	50	460
Ca^{2+}	0.5	10
Cl^-	50	560

1. Use the Nernst equation to predict the equilibrium potential for each of the four ions.

2. The measured resting potential of this axon is –66 mV. How can you explain that resting potential on the basis of the equilibrium potentials you calculated?

3. Another equation—the Goldman–Hodgkin–Katz equation—includes a relative permeability of the membrane for each ion. Why is this necessary for accurately predicting the membrane potential?

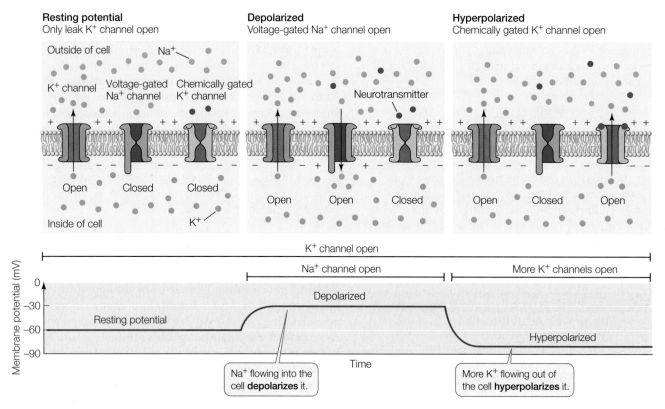

FIGURE 34.6 Membranes Can Be Depolarized or Hyperpolarized The resting potential is produced by leak K⁺ channels. A shift from the resting potential to a less negative membrane potential, as occurs when Na⁺ enters the cell through a gated sodium channel, is called depolarization. Hyperpolarization occurs when the membrane potential becomes more negative, as when additional K⁺ leaves the cell through gated K⁺ channels.

Sudden changes in Na⁺ and K⁺ channels generate action potentials

If we place a pair of electrodes on either side of the plasma membrane of a resting axon, the voltage reading might be about –60 mV. If these electrodes are in place when an action potential travels down the axon, they register a rapid depolarization of the membrane potential, from –60 mV to about +50 mV. The membrane potential returns to its resting level of –60 mV soon after the action potential passes (**FIGURE 34.7**). Unlike graded potentials, action potentials always produce the same amount of depolarization, and they can travel a long distance.

The action potential is generated by voltage-gated Na⁺ and K⁺ channels in the plasma membrane of the axon. At the resting potential, most of these channels are closed (step 1 in Figure 34.7). Local depolarization caused by gated channels in the dendrites produces a graded potential that spreads to the **axon hillock**, the region of the cell body at the base of the axon (see Figure 34.1). Voltage-gated Na⁺ channels are concentrated in the axon hillock. If the plasma membrane in this area depolarizes to 5–15 mV above the resting potential, a **threshold** is reached at which many voltage-gated Na⁺ channels open rapidly (step 3 in Figure 34.7). Enough Na⁺ enters that the

membrane potential becomes positive. Within 1–2 milliseconds, the membrane potential once again becomes negative.

Why does the axon return to resting potential? There are two contributing factors: first, voltage-gated K⁺ channels open, and then with a slight delay the inactivation gates of the Na⁺ channels close (step 4 in Figure 34.7). The voltage-gated K⁺ channels open more slowly than the Na⁺ channels and stay open longer, allowing K⁺ to carry excess positive charges out of the axon. As a result, the membrane potential returns to a negative value (and in fact usually becomes even more negative than the resting potential until the voltage-gated K⁺ channels close; step 5 in Figure 34.7).

Another feature of the voltage-gated Na⁺ channels is that once they have opened and closed, they have a *refractory period* during which they cannot open again. Thus once an action potential has occurred at a particular region of membrane, that region cannot generate another action potential for a few milliseconds.

Action potentials are conducted along axons without loss of signal

Action potentials can travel over long distances with no loss of signal strength. If we place pairs of electrodes at two different locations along an axon, we can record the action potential at those two locations as it travels along the axon. The magnitude of the action potential does not change between the two recording sites. This constancy is possible because an action potential is an all-or-nothing, self-regenerating event.

- An action potential is *all-or-nothing* because of the interaction between the voltage-gated Na⁺ channels and the membrane

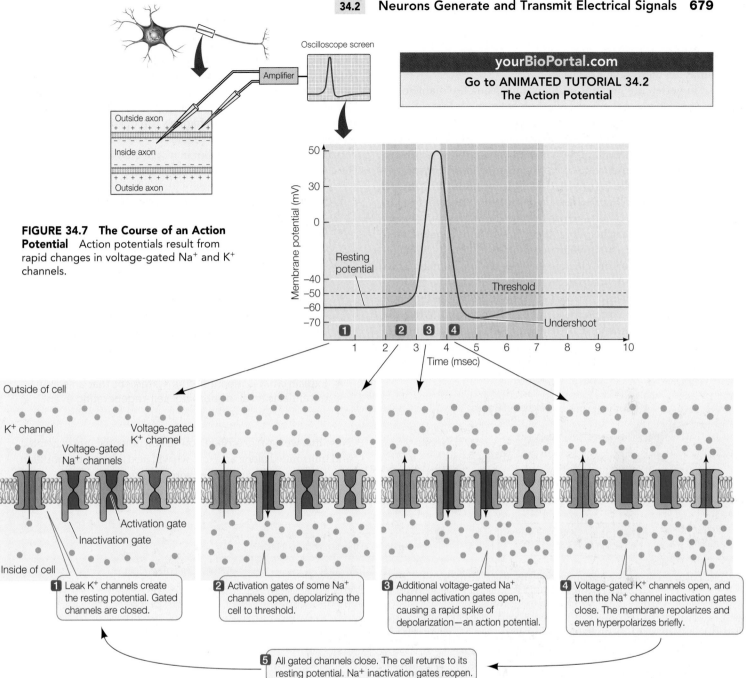

FIGURE 34.7 The Course of an Action Potential Action potentials result from rapid changes in voltage-gated Na⁺ and K⁺ channels.

1 Leak K⁺ channels create the resting potential. Gated channels are closed.

2 Activation gates of some Na⁺ channels open, depolarizing the cell to threshold.

3 Additional voltage-gated Na⁺ channel activation gates open, causing a rapid spike of depolarization—an action potential.

4 Voltage-gated K⁺ channels open, and then the Na⁺ channel inactivation gates close. The membrane repolarizes and even hyperpolarizes briefly.

5 All gated channels close. The cell returns to its resting potential. Na⁺ inactivation gates reopen.

potential. If the membrane is depolarized slightly, some voltage-gated Na⁺ channels open. Some sodium ions cross the plasma membrane and depolarize it even more, opening more voltage-gated Na⁺ channels, and so on, until the membrane reaches threshold and generates an action potential. This positive feedback mechanism ensures that action potentials always rise to their maximum value.

- An action potential is *self-regenerating* because it spreads by local current flow to adjacent regions of the plasma membrane. The resulting depolarization brings those neighboring areas of membrane to threshold. So when an action potential occurs at one location on an axon, it stimulates the adjacent region of axon to generate an action potential, and so on down the length of the axon.

Normally, an action potential is propagated in only one direction—away from the cell body. It cannot reverse itself because

the voltage-gated Na⁺ channels in the region of the membrane it came from are in their refractory period.

Action potentials travel faster in large axons and in myelinated axons

Action potentials do not travel along all axons at the same speed. They travel faster in large-diameter than in small-diameter axons because the resistance to current flow decreases as an axon's diameter gets bigger. Many invertebrate animals rely on increased axon diameter to increase the speed of action potential conduction. In squid, neurons involved in escape responses have exceptionally large-diameter axons, allowing the mollusk to respond rapidly to threats. These squid axons are so large—up to 1 millimeter in diameter—that they are relatively easy to manipulate and study. Many of the basic principles of neuronal and action potential function were discovered by studying these "giant axons."

In vertebrate nervous systems, increasing the diameter of axons to increase the speed of action potentials is less common because of the huge number of axons involved. Vertebrates have evolved a different way of increasing conduction velocity of axons, and that adaptation is *myelination*. When glia wrap themselves around axons, covering them with concentric layers of myelin (see Figure 34.2), they leave **nodes of Ranvier**: regularly spaced gaps where the axon is not covered. Myelin reduces leakage of ions across the membrane, so electric current can spread farther along the inside of a myelinated axon than it can along a nonmyelinated axon. Additionally, voltage-gated ion channels are clustered at the nodes of Ranvier. Thus an axon can fire action potentials only at nodes. The positive charges that flow into the axon at the node move rapidly down the inside of the axon in the form of electric current. When the current reaches the next node, the plasma membrane at that node is depolarized to threshold and fires another action potential. Action potentials therefore appear to "jump" from node

to node (**FIGURE 34.8**). This form of rapid impulse propagation is called **saltatory conduction** (Latin *saltare*, "to jump").

Saltatory conduction by myelinated axons is extremely fast because electric current flows much faster through the cytoplasm than membrane ion channels can open and close. In unmyelinated axons, action potentials can be conducted at speeds up to 2 meters per second, but in myelinated axons the conduction velocity can be 100 meters per second.

Do You Understand Concept 34.2?

- Give a specific example of each of the following types of ion transporters: a leak channel, an ion pump, a voltage-gated channel, and a chemically gated channel.
- Describe two different evolutionary adaptations to speed the conduction of action potentials.
- Why do neurons generate action potentials at all? That is, why don't they just use graded potentials to carry information to the end of the axon?
- Explain what is meant by the statement that an action potential is an all-or-nothing, self-regenerating event.

Now that we have described how action potentials are generated and transmitted along an axon, we will next address the question of what happens when the action potential arrives at the axon terminal. How is the information communicated to the next cell?

Nodes of Ranvier

Myelin-encased Schwann cells

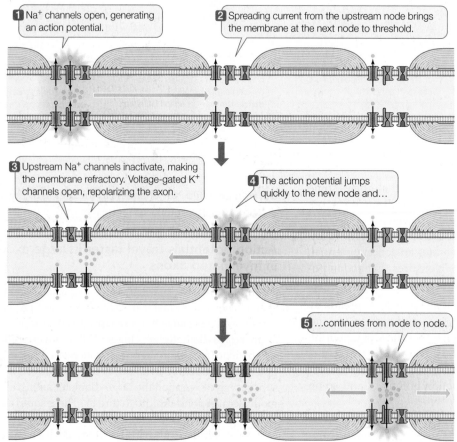

1 Na⁺ channels open, generating an action potential.

2 Spreading current from the upstream node brings the membrane at the next node to threshold.

3 Upstream Na⁺ channels inactivate, making the membrane refractory. Voltage-gated K⁺ channels open, repolarizing the axon.

4 The action potential jumps quickly to the new node and…

5 …continues from node to node.

FIGURE 34.8 Saltatory Action Potentials Action potentials appear to jump from node to node in myelinated axons.

concept 34.3 Neurons Communicate with Other Cells at Synapses

Neurons communicate with each other and with other cells at synapses. The number of synapses can make a nervous system extraordinarily complex. A neuron in a brain may have 1,000 or more synapses. The human brain contains roughly 10^{14} synapses (10^{11} neurons, $\times 10^3$ synapses per neuron).

The most common type of synapse is the **chemical synapse**, in which neurotransmitters released from a *presynaptic cell* bind to receptors in the membrane of a *postsynaptic cell*, inducing changes in the postsynaptic cell. At chemical synapses, a space about 25 nanometers wide—called the **synaptic cleft**—separates the presynaptic and postsynaptic cells.

Electrical synapses physically join the cytoplasms of pre- and postsynaptic cells through *gap junctions*. As we saw in Figure 4.18, a gap junction is made up of proteins (*connexins*) that create channels between cells. Ions can flow through these channels, allowing the action potential to spread passively from the presynaptic to the postsynaptic cell. The advantage of electrical synapses is that they enable fast transmission of information; their disadvantage is that they do not allow complex integration of inputs from multiple sources—which, as we will see, are important properties of chemical synapses. We will focus here on chemical synapses that are characteristic of most parts of the vertebrate nervous system.

The neuromuscular junction is a model chemical synapse

Neuromuscular junctions are synapses between motor neurons and skeletal muscle cells. They are excellent examples of how chemical synaptic transmission works (**FIGURE 34.9**). The axon of a motor neuron branches into numerous axon terminals that may synapse with many muscle cells. Each axon terminal ends in a buttonlike structure (*bouton*) containing vesicles filled with neurotransmitter molecules. The neurotransmitter used by all vertebrate neuromuscular synapses is *acetylcholine* (ACh).

FIGURE 34.9 Chemical Synaptic Transmission The neuromuscular junction depicted here is a typical chemical synapse. The events shown for acetylcholine release are similar for other neurotransmitters, and are precipitated by the arrival of an action potential.

yourBioPortal.com
Go to **ANIMATED TUTORIAL 34.3**
Synaptic Transmission and
INTERACTIVE TUTORIAL 34.1
Neurons: Electrical and Chemical Conduction

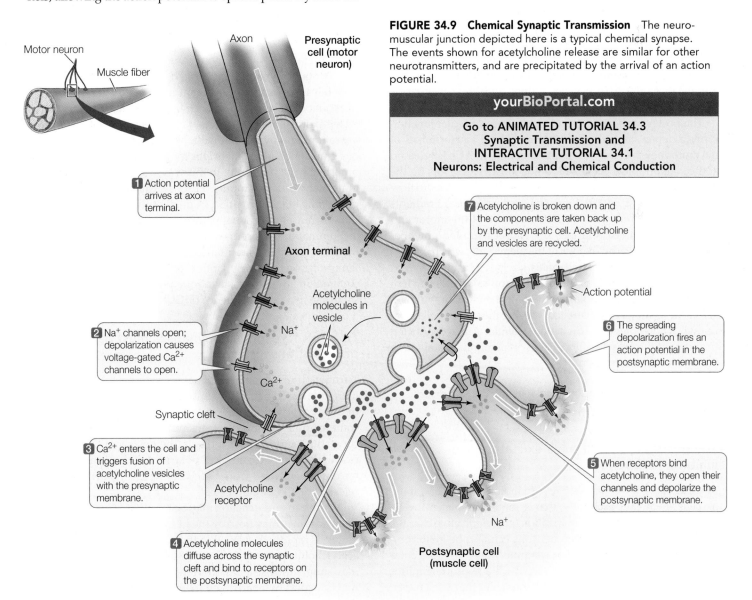

Motor neuron
Muscle fiber

Axon
Presynaptic cell (motor neuron)

1 Action potential arrives at axon terminal.

Axon terminal

Acetylcholine molecules in vesicle

2 Na$^+$ channels open; depolarization causes voltage-gated Ca^{2+} channels to open.

Na$^+$

Ca^{2+}

Synaptic cleft

3 Ca^{2+} enters the cell and triggers fusion of acetylcholine vesicles with the presynaptic membrane.

Acetylcholine receptor

4 Acetylcholine molecules diffuse across the synaptic cleft and bind to receptors on the postsynaptic membrane.

7 Acetylcholine is broken down and the components are taken back up by the presynaptic cell. Acetylcholine and vesicles are recycled.

Action potential

6 The spreading depolarization fires an action potential in the postsynaptic membrane.

5 When receptors bind acetylcholine, they open their channels and depolarize the postsynaptic membrane.

Na$^+$

Postsynaptic cell (muscle cell)

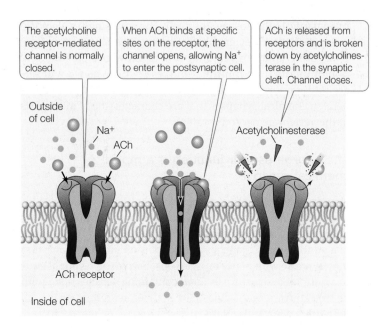

The acetylcholine receptor-mediated channel is normally closed.

When ACh binds at specific sites on the receptor, the channel opens, allowing Na⁺ to enter the postsynaptic cell.

ACh is released from receptors and is broken down by acetylcholinesterase in the synaptic cleft. Channel closes.

Outside of cell

Na^+
ACh

Acetylcholinesterase

ACh receptor

Inside of cell

FIGURE 34.10 Chemically Gated Channels The motor end plate contains acetylcholine (ACh) receptors, which are chemically gated ion channels. When one of these receptors binds ACh, its channel pore opens and Na^+ ions move into the postsynaptic cell, depolarizing its plasma membrane. The enzyme acetylcholinesterase breaks down ACh in the synapse, closing the channel; the breakdown products (acetate and choline) are then taken up by the presynaptic membrane and resynthesized into more ACh.

ACh is released when an action potential arrives at the axon terminal. The action potential causes the opening of voltage-gated Ca^{2+} channels in the presynaptic membrane. Because Ca^{2+} concentration is greater outside the cell than inside, Ca^{2+} enters the axon terminal. Increased Ca^{2+} inside the axon terminal causes the ACh vesicles to fuse with the presynaptic membrane, releasing ACh via exocytosis into the synaptic cleft.

The postsynaptic membrane of the neuromuscular junction forms a depression called a **motor end plate**. The terminals of the motor neuron sit in this depression. When ACh is released at the synaptic cleft, some of it binds to ACh receptors on the motor end plate. These receptors allow both Na^+ and K^+ to flow through (**FIGURE 34.10**). Since the electrochemical gradients favor a greater influx of Na^+, the response of the motor end plate to ACh is to depolarize. This causes a graded potential that spreads to nearby regions of the membrane, which contain voltage-gated Na^+ and K^+ channels.

If the axon terminal of a motor neuron releases sufficient ACh to adequately depolarize the motor end plate, voltage-gated Na^+ channels are activated, causing the muscle cell to fire its own action potential. This action potential is then conducted throughout the muscle cell's system of membranes, activating the contractile mechanism.

LINK The physical and molecular interactions of muscle contraction are described in Concept 36.1

The postsynaptic cell sums excitatory and inhibitory input

In vertebrates, the synapses between motor neurons and muscle cells are always **excitatory**; that is, motor end plates always respond to ACh with a depolarization (greater positive charge). However, synapses between neurons can also be **inhibitory**, causing hyperpolarization. For example, there are more chloride ions (Cl^-) outside cells than inside. If a neurotransmitter causes opening of a Cl^- channel, Cl^- ions enter the postsynaptic

cell and hyperpolarize it (make it more negative). Hyperpolarization takes the postsynaptic cell farther from threshold, making it less likely that the cell will fire action potentials.

FRONTIERS Brain function depends on a balance between excitation and inhibition. Too much excitation can result in psychotic behavior and even seizures. Too much inhibition results in mental impairment and even coma. The major inhibitory neurotransmitter in the human brain is γ-aminobutyric acid (GABA), which controls Cl^- channels. Recent research has shown that low doses of drugs that block GABA activity can greatly reduce the learning disabilities associated with Down syndrome.

A single postsynaptic neuron may receive over 1,000 synapses from presynaptic neurons. The different presynaptic neurons release a variety of different neurotransmitters that may cause depolarization or hyperpolarization at a given synapse. At the axon hillock of the postsynaptic neuron, all of these depolarizations and hyperpolarizations are summed up in the form of a graded potential that brings the axon hillock either closer to or farther from threshold. In this way, a single neuron can integrate many different inputs, a process called *summation*. At any given time, the information from all of the inputs is translated into the rate at which that neuron generates action potentials.

Summation can occur over space and over time. **Spatial summation** adds up the simultaneous influences of synapses at different sites on the postsynaptic cell (**FIGURE 34.11A**). **Temporal summation** adds up postsynaptic potentials generated at the same site in a rapid sequence (**FIGURE 34.11B**).

To turn off responses, synapses must be cleared of neurotransmitter

Turning off neurotransmitter action is as important as turning it on. If released neurotransmitter molecules simply remained in the synaptic cleft, the postsynaptic membrane would become saturated with neurotransmitter, and receptors would be constantly activated. As a result, the postsynaptic cell would remain hyperpolarized or depolarized. Thus neurotransmitter must be cleared from the synaptic cleft shortly after it is released by the axon terminal.

Neurotransmitter action may be terminated in several ways. First, enzymes may destroy the neurotransmitter. ACh, for example, is rapidly destroyed in the synaptic cleft by the enzyme acetylcholinesterase (see Figure 34.10). Neurotransmitter also may simply diffuse away from the cleft, or be taken up via active transport by nearby cell membranes of neurons and glia.

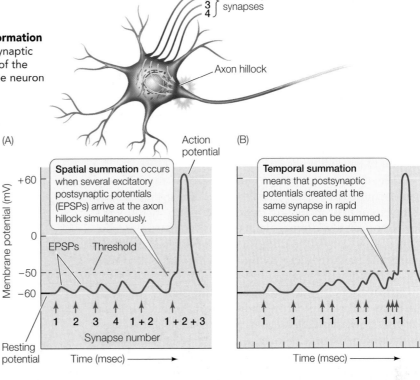

FIGURE 34.11 The Postsynaptic Neuron Sums Information

Individual neurons sum excitatory and inhibitory postsynaptic potentials over (A) space and (B) time. When the sum of the potentials depolarizes the axon hillock to threshold, the neuron generates an action potential.

Some antidepressant drugs (e.g., Prozac) slow reuptake of the neurotransmitter serotonin, thus enhancing its activity at the synapse.

There are many types of neurotransmitters

More than 50 neurotransmitters are now recognized, and more will surely be discovered. In addition to being used by vertebrate motor neurons, acetylcholine is also used in the brain, but the brain also has many other neurotransmitters, including simple amino acids such as glutamate and γ-aminobutyric acid (GABA); modified amino acid derivatives such as

(A) Spatial summation occurs when several excitatory postsynaptic potentials (EPSPs) arrive at the axon hillock simultaneously.

(B) Temporal summation means that postsynaptic potentials created at the same synapse in rapid succession can be summed.

APPLY THE CONCEPT

Neurons communicate with other cells at synapses

How do we know that Ca^{2+} influx into the presynaptic nerve ending causes the release of neurotransmitter? Because the squid giant axon and its nerve endings are so large, they are a convenient system for experiments. It is possible to inject substances into both the presynaptic and postsynaptic cells near the synapse. Some of the substances that can be injected are Ca^{2+} ions and BAPTA, a substance that binds Ca^{2+} ions. Also, channel blockers can be added to the culture medium. For example, cadmium blocks Ca^{2+} channels. Shown below are the results of a series of experiments using these substances.

1. What is happening during the delay between the pre- and postsynaptic membrane events in the control condition?

2. Explain the postsynaptic response in the absence of a presynaptic response in experiment 1.

3. Explain why there is a presynaptic but no postsynaptic response in experiment 2.

4. Why are there no pre- or postsynaptic responses in experiment 3?

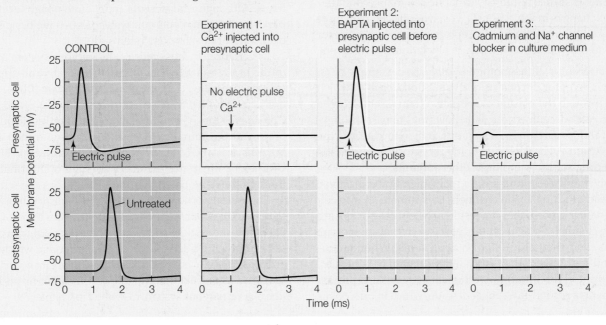

dopamine, norepinephrine, and serotonin; and peptides such as endorphins.

In addition to the multitude of neurotransmitters, each neurotransmitter may have multiple receptor types, which may have different effects on postsynaptic cells. For example, ACh has two receptor types: *nicotinic receptors*, which tend to be excitatory, and *muscarinic receptors*, which tend to be inhibitory. The action of a neurotransmitter ultimately depends on the receptor to which it binds.

Where do neurotransmitter molecules come from? The enzymes required for ACh synthesis are produced in the cell body of the motor neuron and are transported along microtubules down the axon to the terminals. In other types of neurons, neurotransmitters may be produced in the cell body, packaged into membrane-bound vesicles by the Golgi apparatus, and then transported down the axon.

Certain drugs can interfere with neurotransmitter release, as can many naturally occurring toxins. Toxins from *Clostridium* bacteria, for example, destroy several of the proteins necessary for the binding of vesicles to the presynaptic membrane. These toxins cause botulism and tetanus, frequently fatal diseases that involve muscle impairment due of loss of neurotransmitter release.

Synapses can be fast or slow depending on the nature of receptors

Most neurotransmitter receptors induce changes in postsynaptic cells by opening or closing ion channels. How they do so is the basis for grouping receptors into two general categories:

* **Ionotropic receptors** are ion channels themselves. Neurotransmitter binding to an ionotropic receptor causes a direct change in ion movement across the plasma membrane of the postsynaptic cell. These proteins enable fast, short-lived responses.

* **Metabotropic receptors** are not ion channels, but they induce signaling cascades in the postsynaptic cell that secondarily lead to changes in ion channels. Postsynaptic cell responses mediated by metabotropic receptors are generally slower and longer-lived than those induced by ionotropic receptors.

The ACh receptor of the motor end plate is an example of an ionotropic receptor. It consists of five subunits, each of which extends through the plasma membrane. When assembled, the subunits create a central pore that allows ions to pass through. Of several different kinds of subunits, only one kind has the ability to bind ACh. Each functional receptor has two of the ACh-binding subunits and three other subunits.

Metabotropic receptors are also transmembrane proteins, but instead of acting as ion channels, they initiate an intracellular signaling process that can result in the opening or closing of an ion channel. These receptors have seven transmembrane domains, and they are linked to G proteins. When a neurotransmitter binds to the extracellular domain of a metabotropic receptor, the intracellular domain activates a G protein that initiates a second-messenger cascade, which in turn opens

an ion channel. Because metabotropic receptors act through a sequence of cellular reactions, their responses are slower, but they can last longer.

Do You Understand Concept 34.3?

* List the sequence of events at a neuromuscular junction, starting with the firing of an action potential of the presynaptic neuron and ending with contraction of the muscle cell.

* Compare chemical and electrical synapses. Why do you think electrical synapses are more predominant in invertebrates than in vertebrates?

* Dendrites can extend quite a distance from the cell body. Synapses can form at all locations on the dendrites and on the cell body itself. Which synapses do you think will have a stronger effect on the firing rate of the cell? Why?

* Some forms of learning involve physical and chemical changes in the postsynaptic cells. Do you think such synapses are more likely to be ionotropic or metabotropic, and why?

We have seen how neurons use ion pumps and ion channels to generate action potentials and how action potentials trigger neurotransmitter release, allowing neurons to communicate with each other and with other cells. Next we'll look at how neurons form a complex nervous system.

concept 34.4 The Vertebrate Nervous System Has Many Interacting Components

As we saw in Concept 34.1, nervous systems range from simple to highly complex. Here we focus on the complex system of humans. Although it represents an extremely high level of complexity, the human nervous system shares the basic structure of all vertebrate nervous systems.

In vertebrates, most cells of the nervous system are found in the brain and the **spinal cord**. The brain and spinal cord together are called the **central nervous system** (**CNS**). Information is transmitted from sensory cells to the CNS, and from the CNS to effectors, via neurons that extend or reside outside the brain and the spinal cord; these neurons and their supporting cells are called the **peripheral nervous system** (**PNS**). FIGURE 34.12 illustrates the major avenues of information flow through the nervous system.

The *afferent* portion of the PNS carries sensory information to the CNS. The *efferent* portion of the PNS carries information from the CNS to the muscles and glands of the body. Efferent pathways can be further divided into a voluntary division, which executes our conscious movements, and an involuntary, or *autonomic*, division. An additional division of the nervous system exists in the gut (the enteric nervous system, discussed in Chapter 39).

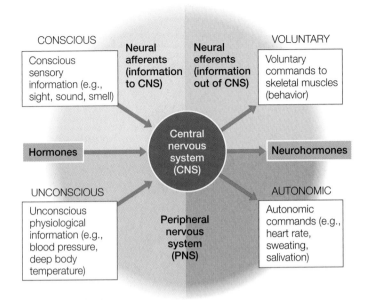

FIGURE 34.12 Organization of the Nervous System The peripheral nervous system (beige and green) carries information both to (afferent) and from (efferent) the central nervous system (purple). The CNS also receives hormonal inputs and produces hormonal outputs (lavender).

The autonomic nervous system controls involuntary physiological functions

The **autonomic nervous system**, or **ANS**, comprises the output pathways of the CNS that control involuntary functions that are crucial to homeostasis, such as heart rate, sweating, and some functions of the gut. The ANS has two divisions, **sympathetic** and **parasympathetic**, that frequently work in opposition to each other (**FIGURE 34.13**). Generally, the sympathetic division produces the fight-or-flight response: increased heart rate, blood pressure, and cardiac output (see Concept 30.2). The sympathetic division prepares the body for emergencies and reduces less urgent activities such as digestion. In contrast, the parasympathetic division slows the heart, lowers blood pressure, and increases digestion; its actions have been characterized as "rest and digest."

Whether sympathetic or parasympathetic, every autonomic efferent pathway begins with a *preganglionic neuron* that has its cell body in the CNS (see Figure 34.13). Axons of preganglionic neurons lead to a ganglion (a collection of neurons) outside the CNS, where they synapse with *postganglionic neurons*. Most of the ganglia of the sympathetic division are lined up in two chains, one on either side of the spinal cord. In contrast, the parasympathetic ganglia are close to the target organs. In both divisions, axons of the postganglionic neurons lead to the target organs, where they synapse with target cells.

The postganglionic neurons of the sympathetic division are *noradrenergic*, meaning they release norepinephrine as their neurotransmitter. In contrast, the postganglionic neurons of the parasympathetic division are *cholinergic* neurons, meaning they release acetylcholine. In organs that receive both sympathetic and parasympathetic input, the target cells respond in

opposite ways to norepinephrine and to ACh. This happens, for example, in a region of the heart called the *pacemaker*, which generates the heartbeat. Stimulating the sympathetic nerve to the heart or dripping norepinephrine onto the pacemaker cells increases their firing rate and causes the heart to beat faster. In contrast, stimulating the parasympathetic nerve to the heart or dripping acetylcholine onto the pacemaker cells decreases their firing rate and causes the heart to beat more slowly.

The sympathetic and parasympathetic divisions of the ANS can also be distinguished by anatomy. The preganglionic neurons of the parasympathetic division come from the brainstem and the *sacral* region of the spinal cord; those of the sympathetic division come from the *thoracic* and *lumbar* regions of the spinal cord (see Figure 34.13).

The spinal cord transmits and processes information

The spinal cord has afferent and efferent tracts of axons communicating between the brain and the body's organs. It also integrates much of the information coming from the PNS and responds to that information by issuing motor commands.

A cross section of the spinal cord reveals a central area of gray matter in the shape of a butterfly, surrounded by an area of white matter (see Figure 34.14). Remember that **gray matter** is rich in neural cell bodies, whereas **white matter** is rich in myelinated axons. The gray matter of the spinal cord contains the cell bodies of the spinal neurons; the white matter contains the axons that conduct information up and down the spinal cord.

Spinal nerves extend from the spinal cord at regular intervals on each side. Each spinal nerve has two roots, one connecting with the *dorsal horn* of the gray matter, the other with the *ventral horn*. The afferent (sensory) axons in a spinal nerve enter the spinal cord through the *dorsal root*, and the efferent (motor) axons leave through the *ventral root*.

The spinal cord is capable of some simple reactions to certain stimuli without involving the brain. These reactions are called **spinal reflexes**. The simplest type of spinal reflex circuit has only two neurons with one synapse between them, and is therefore called a **monosynaptic reflex**. An example is the knee-jerk reflex (**FIGURE 34.14**). When a physician taps your knee with a little hammer, the tap stretches the patellar tendon connecting the extensor muscle (quadriceps) in your upper leg to the femur in your lower leg. Modified muscle fibers in the quadriceps called "muscle spindles" form stretch receptors that are innervated by sensory neurons. Stretching the muscle spindle generates action potentials in that sensory neuron. The axon of that sensory neuron travels to the spinal cord, enters the dorsal horn, and continues to the ventral horn where it synapses onto a motor neuron. This synapse is excitatory and causes the motor neuron to fire action potentials that travel in the axon of that motor neuron back out to the quadriceps, causing it to contract. The result is that your lower leg kicks forward.

Most spinal reflexes are more complex than this monosynaptic reflex, and may involve additional muscles and more neurons. But the basic principle is the same: information enters

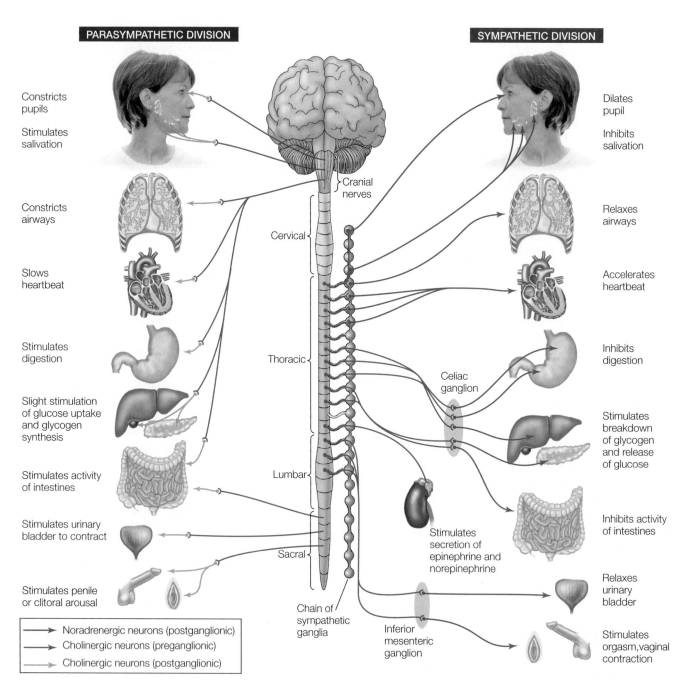

PARASYMPATHETIC DIVISION

Constricts pupils

Stimulates salivation

Constricts airways

Slows heartbeat

Stimulates digestion

Slight stimulation of glucose uptake and glycogen synthesis

Stimulates activity of intestines

Stimulates urinary bladder to contract

Stimulates penile or clitoral arousal

→ Noradrenergic neurons (postganglionic)
→ Cholinergic neurons (preganglionic)
→ Cholinergic neurons (postganglionic)

SYMPATHETIC DIVISION

Dilates pupil

Inhibits salivation

Relaxes airways

Accelerates heartbeat

Inhibits digestion

Stimulates breakdown of glycogen and release of glucose

Inhibits activity of intestines

Relaxes urinary bladder

Stimulates orgasm, vaginal contraction

Cranial nerves

Cervical

Thoracic

Lumbar

Sacral

Chain of sympathetic ganglia

Celiac ganglion

Stimulates secretion of epinephrine and norepinephrine

Inferior mesenteric ganglion

FIGURE 34.13 The Autonomic Nervous System The autonomic nervous system is divided into the sympathetic and parasympathetic divisions. The two divisions work in opposition to each other in their effects on most organs; one results in an increase and the other a decrease in activity.

the spinal cord from sensory neurons, and commands leave the spinal cord through motor neurons — all without any involvement of the brain. Spinal reflexes allow rapid responses to certain simple stimuli and are important in maintaining posture and balance.

Interneurons coordinate polysynaptic reflexes

Even a monosynaptic reflex involves additional neurons. Muscles of the limbs are organized in antagonistic pairs of *flexors* and *extensors* (see Concept 36.3). Thus if a muscle such as the quadriceps is going to contract, its antagonistic muscle has to relax. The sensory neurons coming from the quadriceps stretch

receptors also synapse onto *interneurons* in the spinal cord (see Figure 34.14). Those interneurons make *inhibitory* synapses onto the motor neurons of the flexor of the lower leg, allowing it to move forward when the quadriceps contracts.

Now think about your reflex movements when you step on something sharp. You immediately withdraw the foot generating the painful sensation, and at the same time shift your weight to your other leg, making many small adjustments in

FIGURE 34.14 The Spinal Cord Coordinates the Knee-jerk Reflex Sensory (afferent) information enters the spinal cord through the dorsal horns (red pathway), and motor (efferent) output leaves it via the ventral horns (blue pathways). Information travels to the brain in white matter tracts. Interneurons make connections in the spinal cord that result in a complex, coordinated behavior pattern.

yourBioPortal.com
**Go to ANIMATED TUTORIAL 34.4
Information Processing in the Spinal Cord**

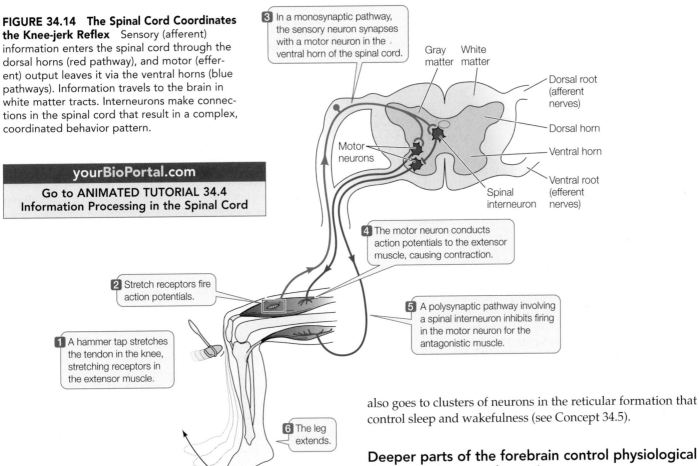

3 In a monosynaptic pathway, the sensory neuron synapses with a motor neuron in the ventral horn of the spinal cord.

Gray matter White matter

Dorsal root (afferent nerves)

Dorsal horn

Ventral horn

Ventral root (efferent nerves)

Motor neurons

Spinal interneuron

4 The motor neuron conducts action potentials to the extensor muscle, causing contraction.

2 Stretch receptors fire action potentials.

5 A polysynaptic pathway involving a spinal interneuron inhibits firing in the motor neuron for the antagonistic muscle.

1 A hammer tap stretches the tendon in the knee, stretching receptors in the extensor muscle.

6 The leg extends.

muscles to maintain your balance. What you experience is a *polysynaptic reflex* mediated by interneurons in the spinal cord.

The brainstem transfers information between the brain and spinal cord

The embryonic neural tube in vertebrates forms three anterior swellings that become the *hindbrain, midbrain,* and *forebrain* (see Figure 33.13). From the hindbrain come the *medulla, pons,* and *cerebellum.* The medulla and pons contain distinct groups of neurons involved in controlling physiological functions such as breathing and circulation, while the cerebellum is involved in coordinating muscle activity and maintaining balance. The midbrain gives rise to certain structures that process aspects of visual and auditory information.

All information traveling between the spinal cord and higher brain areas must pass through the pons, the medulla, and the midbrain, which are collectively known as the **brainstem**. Many sensory axons give off branches in the brainstem that form synapses with a network of brainstem neurons called the **reticular system**. Information from joints and muscles, for example, is directed to areas of the reticular system that are involved in balance and coordination. Sensory information

also goes to clusters of neurons in the reticular formation that control sleep and wakefulness (see Concept 34.5).

Deeper parts of the forebrain control physiological drives, instincts, and emotions

The forebrain includes a central region called the **diencephalon** and a surrounding structure called the **telencephalon**. The diencephalon, in turn, consists of an upper structure, the *thalamus,* and a lower structure, the *hypothalamus.* The thalamus is the final relay station for sensory information going to the telencephalon. The hypothalamus regulates many physiological functions and biological drives such as hunger and thirst; it receives a lot of physiological information of which we are not conscious.

LINK The roles of the hypothalamus in homeostatic regulation and sensory integration are detailed in Concepts 29.5 and 30.3

Surrounding the diencephalon of all vertebrates are phylogenetically older structures of the telencephalon called the **limbic system** (**FIGURE 34.15**). The limbic system is responsible for many basic functions such as instinctive reactions, fear, pleasure, pain, rage, and memory formation. If a rat is given the opportunity to stimulate the pleasure centers of the limbic system by pressing a switch, it will ignore food, water, and even sex, pushing the switch until it is exhausted. An oval-shaped structure called the **amygdala** (Latin, "almond"), part of the limbic system, is the brain's center for the fear reactions and fear memories. Another limbic structure, the **hippocampus**, is necessary for the transfer of short-term memory to long-term memory, as we will see in Concept 34.5.

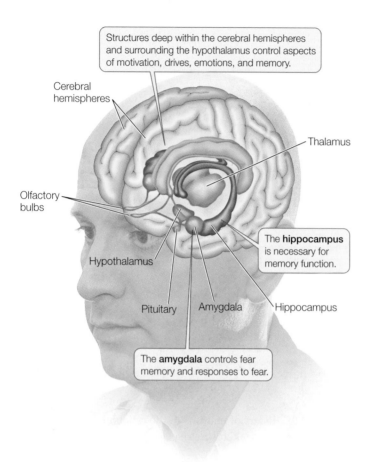

Structures deep within the cerebral hemispheres and surrounding the hypothalamus control aspects of motivation, drives, emotions, and memory.

Cerebral hemispheres

Thalamus

Olfactory bulbs

The **hippocampus** is necessary for memory function.

Hypothalamus

Pituitary   Amygdala   Hippocampus

The **amygdala** controls fear memory and responses to fear.

FIGURE 34.15   The Limbic System   The evolutionarily primitive parts of the telencephalon are referred to as the limbic system. The hippocampus is involved in forming long-term memory. The amygdala triggers fear emotions and fear memories (see p. 693).

the **cerebral cortex**, a thin layer rich in cell bodies. The cerebral cortex is only about 4 millimeters thick, but it is folded into ridges, or *convolutions*, that increase its surface area.

The cerebral cortex plays a major role in sensory perception, learning, memory, and conscious behavior. Different regions of the cerebral cortex have specific functions (**FIGURE 34.16B**). Some of those functions are easily defined, such as receiving and processing sensory information or generating motor commands, but most of the cortex is involved in higher-order information processing that is less easy to define. These latter areas are given the general name of **association cortex**, so named because they integrate, or *associate*, information from different sensory modalities and from memory.

Humans (and, interestingly, dolphins) have much bigger brains than would be predicted from their body size, primarily because of an unusually large telencephalon. The telencephalon is by far the largest part of the human brain. Humans also have the greatest degree of convolution of the cerebral cortex. The percentage of human cerebral cortex that is association cortex (devoted to the integration of information) is greater than in other species. It is these evolutionary changes, primarily in the cortex, that provide the resources for the great intellectual capacity of humans—a topic to which we will return at the end of the chapter.

Regions of the telencephalon interact to produce consciousness and control behavior

The outer part of the telencephalon—and the dominant structure of the mammalian brain—is the **cerebrum**, with its left and right **cerebral hemispheres** (**FIGURE 34.16A**). In humans, the cerebral hemispheres cover all other parts of the brain except the cerebellum. The outermost layer of the cerebrum is

FIGURE 34.16   The Human Cerebrum   (A) Each cerebral hemisphere is divided into four lobes. (B) Different functions are localized in particular areas of the cerebral lobes.

yourBioPortal.com

Go to WEB ACTIVITY 34.2
The Human Cerebrum

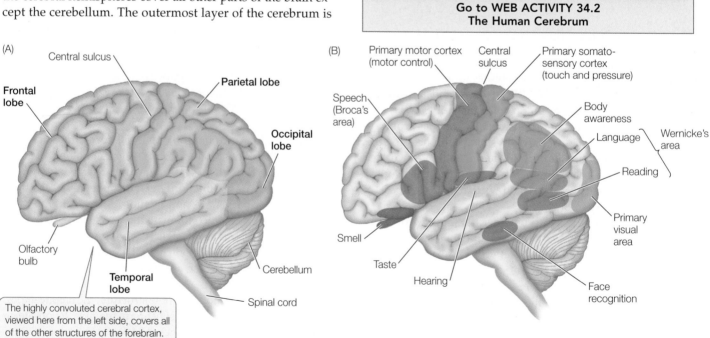

(A)

Central sulcus

Parietal lobe

Frontal lobe

Occipital lobe

Olfactory bulb

Temporal lobe

Cerebellum

Spinal cord

The highly convoluted cerebral cortex, viewed here from the left side, covers all of the other structures of the forebrain.

(B)

Primary motor cortex (motor control)   Central sulcus   Primary somatosensory cortex (touch and pressure)

Speech (Broca's area)

Body awareness

Language   Wernicke's area

Reading

Smell

Primary visual area

Taste

Hearing

Face recognition

A curious feature of the human nervous system is that the left side of the body is served (in both sensory and motor aspects) mostly by the right side of the brain, and vice versa. Thus sensory input from the right hand goes to the left cerebral hemisphere, and sensory input from the left hand goes to the right cerebral hemisphere. The exception is the head, where the left side is controlled by the left cerebral hemisphere and the right side by the right cerebral hemisphere.

Each cerebral hemisphere has four lobes

Each cerebral hemisphere consists of four lobes: the **temporal lobe**, **frontal lobe**, **parietal lobe**, and **occipital lobe** (see Figure 34.16A).

THE TEMPORAL LOBE The upper region of the temporal lobe receives and processes auditory information. The lower regions are involved in visual processing. The association areas of this lobe are involved in recognizing, identifying, and naming objects. Damage to the temporal lobe results in disorders called *agnosias*, in which the individual is aware of an object but cannot identify it. For example, damage to a certain area of the temporal lobe results in the inability to recognize faces. The temporal lobe also contains areas involved in language (see Concept 34.5).

THE FRONTAL LOBE This is the largest of the brain lobes in humans. Its association areas are involved with feeling and planning, and contribute significantly to what we consider our "personality." People with frontal lobe damage have drastic alterations of personality and difficulty planning future events.

The strip of the frontal lobe cortex that is just anterior to the parietal lobe is called the **primary motor cortex** (**FIGURE 34.17A**). The neurons in this region control muscles in specific parts of the body. Regions of the body with fine motor control, such as the face and hands, have disproportionate representation in the primary motor cortex.

THE PARIETAL LOBE A major association function of the parietal lobe is attending to complex stimuli. The parietal lobe also helps translate visual information into a perception of objects being located in a three-dimensional space surrounding the body. Just behind the primary motor cortex of the frontal lobe is a similar strip of parietal lobe called the **primary somatosensory cortex** (**FIGURE 34.17B**). This area receives touch and pressure information relayed from the body through the thalamus. Just as with the primary motor cortex, the entire body surface can be mapped onto the primary somatosensory cortex. Areas of the body that are capable of making fine discriminations in touch (such as the lips and fingers) have disproportionately large representation.

THE OCCIPITAL LOBE The occipital lobe receives and processes visual information. The association areas of the occipital cortex are essential for making sense of the visual world and translating visual experience into language. For example, in one case a woman with limited damage to the occipital lobe was unable to see motion. Her vision was intact, but she could see a waterfall only as a still image, and an approaching car only as a series of stationary objects at different distances.

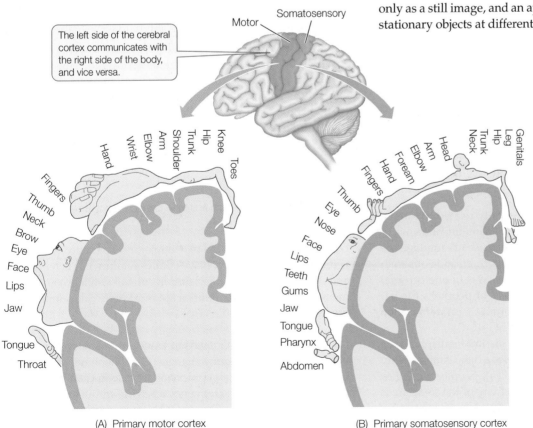

(A) Primary motor cortex   (B) Primary somatosensory cortex

FIGURE 34.17 The Body Is Represented in Primary Motor and Primary Somatosensory Cortexes Neurons in the primary motor cortex (A) control muscles in specific parts of the body, while neurons in the primary somatosensory cortex (B) receive information from specific parts of the body. The locations of these neurons in the cortex correspond to "maps" on which regions of the body are represented in proportion to the amount of primary cortical area devoted to them.

Do You Understand Concept 34.4?

- Describe the sympathetic, parasympathetic, central, autonomic, and peripheral nervous systems and explain how they relate to one another.

- What senses do you think would be most affected if a person's occipital and temporal lobes were damaged? Where would you expect to find the major cortical area involved in reading?

- The activity of different brain regions can be imaged as a person is shown different pictures. When a picture of a familiar, pleasant place is viewed, the activity of the hippocampus increases. When a picture of a place associated with a traumatic event is shown, the activities of the hippocampus and the amygdala increase. How would you interpret these findings?

So far this chapter has described the cellular components of the nervous system and how these cells are organized into networks that send, receive, and process information. We have described regions of the human brain that epitomize the information-processing ability of the vertebrate nervous system. Next let's see how these regions are responsible for the most complex of abilities and behaviors.

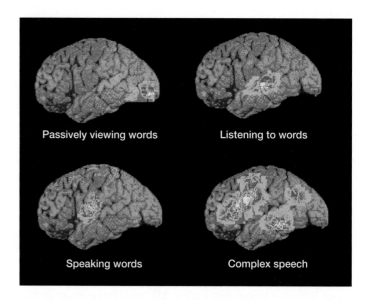

FIGURE 34.18 Imaging Techniques Reveal Active Parts of the Brain Positron emission tomography (PET) scanning reveals the brain regions activated by different aspects of language use. Radioactively labeled glucose is given to the subject. Brain areas take up radioactivity in proportion to their metabolic use of glucose. The PET scan visualizes levels of radioactivity in specific brain regions when a particular activity is performed. The red and white areas are the most active; green and blue the least active.

concept 34.5 Specific Brain Areas Underlie the Complex Abilities of Humans

The following discussion addresses several of the most complex functions of the human brain, including our ability to extensively use and manipulate language; our ability to learn and remember, both of which significantly surpass similar abilities in other primates; and finally, research into the functions of sleep and dreaming, as well as the nature of consciousness. These areas are still not completely understood, but neuroscientists are making fascinating progress.

Language abilities are localized in the left cerebral hemisphere

No aspect of brain function is as integrally related to human consciousness and intellect as language. Therefore, brain mechanisms that underlie the acquisition and use of language are extremely interesting to neuroscientists. A curious observation about language ability is that it resides in one cerebral hemisphere—which in about 85 percent of people is the left hemisphere. This phenomenon is referred to as **lateralization** of language functions.

Individuals who suffer damage to the left hemisphere are frequently left with some form of **aphasia**, a deficit in the ability to use or understand words. PET scans of people reading or speaking have revealed heightened activity of the same areas in the left hemisphere (**FIGURE 34.18**), thus identifying several *language areas*:

- **Broca's area**, located in the frontal lobe (see Figure 34.16B), is essential for production of language. Damage to Broca's area results in halting, poorly articulated speech or complete loss of speech, as well as loss of the ability to write. Patients can still read and can understand speech.

- **Wernicke's area**, located in the temporal lobe (see Figure 34.16B), is essential for understanding language. Damage to Wernicke's area results in loss of the ability both to read and to understand speech. Patients may still produce fluent-sounding, but nonsensical, speechlike sounds.

Normal language ability depends on the flow of information among these and several other areas of the left cerebral cortex.

Some learning and memory can be localized to specific brain areas

Learning is the modification of behavior by experience. *Memory* is the ability of the nervous system to retain what is learned and experienced. Even very simple animals can learn and remember, but these two abilities are most highly developed in humans.

LEARNING Learning that leads to long-term memory must involve long-lasting synaptic changes. One phenomenon that is involved in long-term synaptic changes is **long-term potentiation (LTP)**. In LTP, high-frequency electrical stimulation of certain neuronal circuits makes these circuits more sensitive to subsequent stimulation.

Several kinds of learning exist. A form that is widespread among animal species is **associative learning**, in which two unrelated stimuli become linked to the same response. The simplest example of associative learning is the **conditioned reflex**, described by the Russian physiologist Ivan Pavlov. Pavlov observed that a dog salivates at the sight or smell of food—a simple autonomic reflex. He discovered that if he rang a bell just before food was presented to the dog, after a few trials the dog would salivate at the sound of the bell, even if no food followed. The salivation reflex was "conditioned" to be associated with the sound of a bell, a stimulus that normally is unrelated to feeding and digestion.

More complex forms of learning, often referred to as *observational learning*, are the foundation of human intelligence. The general pattern of successful observational learning has three elements:

- We pay attention to another person's behavior.
- We retain a memory of what we have observed.
- We try to copy or use that information.

A key to this scheme of learning is the way in which we create and recall memories.

MEMORY You experience several forms of memory everyday. Memory of facts—people, places, and events that you can consciously recall and describe—is **declarative memory**. **Procedural memory**, in contrast, is the memory of how to perform a motor task. When you learn to ride a bicycle or use a computer keyboard, you form procedural memories.

Declarative memory lasts varying lengths of time. You have *immediate memory* for events that are happening right now. Immediate memory is virtually photographic, but it is retained for only seconds. *Short-term memory* contains less information but lasts longer—10 to 15 minutes. If you are introduced to a group of new people, you may remember most of their names for a few minutes, but you will have forgotten them in an hour or so if you have not written them down or used them in a conversation. Repetition facilitates the transfer of short-term memory to *long-term memory*, which can last for days, months, or a lifetime.

Where in the brain does memory reside? Electrical stimulation of small areas of association cortex (those areas devoted to the integration of sensory information) elicits recall of vivid memories. Destroying a small area does not completely erase a memory, however, so memory may be distributed over many brain regions. Additional clues come from observations of persons who have lost parts of the limbic system, notably the hippocampus. A famous case is that of the man identified as H.M., whose hippocampus on both sides of the brain was removed in 1953 in an effort to control severe epilepsy. After the surgery, H.M. was unable to transfer information to long-term memory. If someone was introduced to him and then left the room for several minutes, when that person returned, H.M did not recognize him. Up until his death 55 years later, H.M. remembered events that happened before his surgery but could not remember postsurgery events for more than a few minutes.

Although H.M. was incapable of forming declarative memories, he could form procedural memories. When taught a motor task day after day, he could not consciously recall the lessons of the previous day, yet his performance steadily improved. Thus procedural memory must involve mechanisms different from those used in declarative memory.

There are two different states of sleep

All vertebrates (and many invertebrates) sleep, even those that must take considerable risks to do so, such as prey animals and marine mammals (which risk drowning when they sleep, yet sleep anyway). We humans spend one-third of our lives sleeping. Lack of sleep results in a progressive decline in alertness and function, impaired learning, immune system suppression, hormonal changes, hypothermia, and many other pathological changes.

How we fall asleep and what happens in our brains while we sleep have been the subject of a great deal of research in recent years. A primary tool of sleep researchers is the *electroencephalogram*, or EEG, where electrodes are placed at different locations on the subject's scalp. Rather than recording the activity of single neurons, the EEG characterizes the electrical activity of entire brain regions, primarily those of the cerebral cortex. EEG studies have revealed that, in birds and mammals, there are two very different sleep states: **rapid eye movement (REM) sleep** and **non-REM sleep (FIGURE 34.19A)**.

When we are awake, clusters of neurons with specific neurochemical characteristics (called *nuclei*, but unrelated to the nuclei of cells) in different regions of the brainstem are continuously active. Their axons project broadly to the forebrain, where they release depolarizing neurotransmitters. These neurotransmitters keep the resting potentials of the neurons of the thalamus and cortex close to threshold and sensitive to synaptic inputs, thereby maintaining the responsiveness of the brain that characterizes being awake.

With the onset of sleep, these brainstem neurons release less neurotransmitter. As a result, the resting potentials of the cells of the thalamus and cortex become more negative (hyperpolarized) and are less sensitive to excitatory synaptic input. Their processing of information is inhibited, and consciousness is lost. This state is non-REM sleep. As non-REM sleep deepens, neurons over broad areas of the cerebral cortex begin to fire action potentials in slow, synchronized bursts (see Figure 34.19A, Stages 3 and 4). Because an EEG records these bursts as large deflections with a slow wavelength, this stage of non-REM sleep is called *slow-wave sleep*.

After some time in non-REM sleep, a dramatic change occurs. Some of the brainstem neurons that were inactive suddenly become active again, causing a general depolarization of cortical neurons. The synchronized bursts of firing cease, and the EEG resembles that of the awake brain (see Figure 34.19A, REM). Because the resting potentials of the neurons return to near-threshold levels, the cortex can again process information. This is the state of REM sleep, so named because the eyes—even behind closed eyelids—move rapidly during this stage. Vivid dreams occur during REM sleep.

During REM sleep, the brain inhibits both afferent (sensory) and efferent (motor) pathways. In other words, during REM

(A)

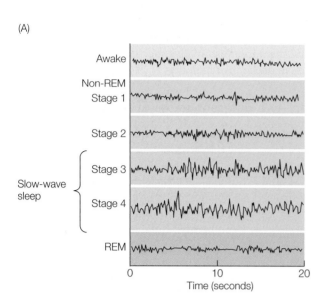

Slow-wave sleep

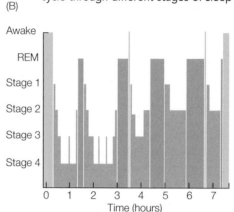

(B)

FIGURE 34.19 Stages of Sleep (A) Electrical activity in the cerebral cortex is detected by electrodes that are placed on the scalp and record changes in voltage between the electrodes through time. The resulting record is an electroencephalogram, or EEG. (B) Humans cycle through different stages of sleep throughout the night.

sleep our active brains are isolated from the external sensory world, and we are paralyzed (which is why we may have trouble awakening ourselves from a nightmare even when we realize we are dreaming). The bizarre nature of dreams may be due to the lack of sensory information—the cortex is functioning but is out of touch with reality. The muscle paralysis prevents the acting out of dreams. (Sleepwalking occurs during non-REM sleep.)

After a period of REM sleep, we return to non-REM sleep. We cycle repeatedly between non-REM and REM sleep, and similar sleep cycles occur in other mammals (**FIGURE 34.19B**).

Why do we have two sleep states with very different neurophysiological characteristics? Why does non-REM sleep always occur first? Why do the two states cycle? Understanding the cellular mechanisms of sleep has not yet answered these questions. We know sleep is essential for life, but we don't know why. Various hypotheses hold that sleep may be necessary for proper immune function, for the maintenance and repair of neural connections, and for the neural changes involved in learning and memory. The last hypothesis is supported by experiments showing that performance of a learned task is impaired if sleep is prevented, and is enhanced following a period of sound sleep.

We still cannot answer the question "What is consciousness?"

Even with all of our current knowledge of the human brain, we still cannot answer the question "What is consciousness?" Consciousness is the sense of being mentally aware of yourself, your environment, events going on around you, and your past, in such a way that you can plan for future events and make decisions.

The central requirement for consciousness is a perception of self that can be integrated with information from the environment and from past experience. A perception of self could, in theory, be derived from the afferent (sensory) information that comes from all parts of the body. Generally, this afferent

information goes to appropriate areas in the brainstem and forebrain. Visual information goes to the visual cortex, somatosensory information goes to the somatosensory cortex, and so on. However, all sensory information also goes to an area deep within the forebrain called the *insular cortex*, or *insula*. The insula appears to integrate information from all over the body to create a sensation of how the body "feels."

In humans, great apes, and a few other mammals (e.g., dolphins and elephants), the insula is greatly expanded and has types of neurons not seen in other animals. The circuitry of the insula has also expanded to communicate with parts of the brain that are involved in planning and decision making. In imaging studies, the insula is seen to be active in a great diversity of situations that involve strong feelings, be they pleasure, disgust, humor, pain, lust, guilt, or empathy. Damage to the insula results in apathy, loss of ability to enjoy music or good food, and loss of sexual response. Fascinatingly, the few species that have expanded insulas and the unusual neurons are also the only species that are able to recognize themselves in a mirror. Could it be that this tiny part of our brains contains the neurobiological bases for self-awareness and conscious experience?

Do You Understand Concept 34.5

- Read this sentence out loud. Then name the brain areas you used to perform the feat of seeing a printed sentence, understanding it, saying it out loud, and finally hearing your own voice.

- Compare and contrast slow-wave sleep and REM sleep. Which one is more similar to the "awake" state?

- Birds that store seeds for the winter in hidden locations often have a very well developed hippocampus. Why?

The next two chapters will encompass two major aspects of the nervous system: the sensory neurons and connections that receive and process information from the environment; and the musculoskeletal system that makes behavior possible.

QUESTION

How can a small brain tumor so dramatically affect personality and behavior?

ANSWER Charles Whitman's brain tumor was just below his thalamus and was about the size of a walnut. It was pressing on his hypothalamus and parts of his limbic system, notably the *amygdala* (**FIGURE 34.20**). The amygdala is involved with intense emotions such as fear. When the neurons of the amygdala are activated and fire action potentials, you experience fear. If you are faced with a threatening situation, your amygdala is activated.

People with a damaged amygdala frequently have trouble engaging in normal social relationships. They cannot "read" the nature, mood, or intentions of other people by looking at their faces. This loss of the ability to assess the possible negative consequences of one's actions is effectively the loss of the ability to perceive fear. Without an amygdala, you would never be frightened—and *not* being frightened could be hazardous to your health. Conversely, studies in which the amygdalas of animals and humans have been stimulated electrically produce intense expressions of fear and rage.

FIGURE 34.20 Source of the Fear Response Frightening situations—or even memories of such a situation—activate a brain region called the amygdala, as shown in this functional magnetic resonance image (fMRI) of the brain of a person experiencing fear.

The pressure on Charles Whitman's amygdala might have been a factor in the emotions that drove him to mass murder. This sad case is a vivid example of how the actions of neurons in certain parts of the brain underlie the core of who we are—our individual behaviors and personalities. Understanding these brain systems and how they function in health and disease is critical to the development of improved therapies for psychiatric illnesses.

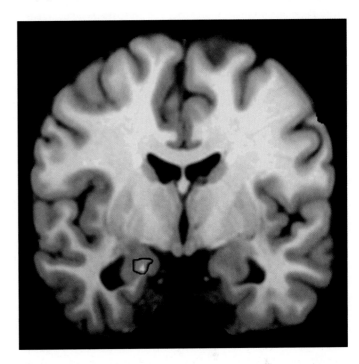

34 SUMMARY

- The cells of the nervous system are either **neurons** or **glia**. Neurons generally receive information via their **dendrites** and transmit information via their **axon**. Review Figure 34.1

- Glia include **Schwann cells** and **oligodendrocytes**, both of which generate **myelin** sheets on axons. Glia also include **astrocytes**, which contribute to the **blood–brain barrier**. Review Figure 34.2

- Neurons are organized in networks with sensory inputs (**afferent neurons**), outputs (**efferent neurons**), and integration (**interneurons**). These networks may be simple or complex. Review Figure 34.3

concept **34.2** Neurons Generate and Transmit Electrical Signals

- Neurons have an electric charge difference across their plasma membranes, called the **membrane potential**. The membrane potential is created by ion transporters and channels. In inactive neurons, the membrane potential is called the **resting potential**.

Review Figures 34.4 and 34.5 and ANIMATED TUTORIAL 34.1

- The **sodium–potassium pump** concentrates K$^+$ on the inside of a neuron and Na$^+$ on the outside. Potassium channels allow K$^+$ to diffuse out, causing the resting potential to be negative. Review Figure 34.5

- The **Nernst equation** can be used to calculate the equilibrium potential of a single ion. Review WEB ACTIVITY 34.1

- When ion channels open or close, the plasma membrane can become **depolarized** or **hyperpolarized**. This causes a **graded membrane potential**. Review Figure 34.6

- An **action potential** is a rapid reversal in charge across a portion of the plasma membrane resulting from the opening and closing of **voltage-gated channels** of Na$^+$ and K$^+$. These voltage-gated channels open when the plasma membrane depolarizes to a **threshold** level. Review Figure 34.7 and ANIMATED TUTORIAL 34.2

- Action potentials are all-or-none, self-regenerating events. They are conducted down axons because local current flow depolarizes adjacent regions of membrane and brings them to threshold.

- In myelinated axons, action potentials jump between **nodes of Ranvier**, patches of membrane that are not covered by myelin. **Review Figure 34.8**

Neurons Communicate with Other Cells at Synapses

- Neurons communicate with each other and with other cells by transmitting information over **chemical synapses** (with neurotransmitters) or **electrical synapses**.

- The **neuromuscular junction** is a well-studied chemical synapse between a motor neuron and a skeletal muscle cell. Its neurotransmitter is acetylcholine (ACh). **Review Figure 34.9**

- When an action potential reaches an axon terminal, it causes the release of neurotransmitters, which diffuse across the **synaptic cleft** and bind to receptors on the postsynaptic membrane. **Review Figures 34.9 and 34.10, ANIMATED TUTORIAL 34.3, and INTERACTIVE TUTORIAL 34.1**

- Synapses between neurons can be either **excitatory** or **inhibitory**. A postsynaptic neuron integrates information by summation of excitatory and inhibitory postsynaptic potentials in both space (**spatial summation**) and time (**temporal summation**). **Review Figure 34.11**

- There are many different neurotransmitters and types of receptors. The action of a neurotransmitter depends on the receptor to which it binds.

- Synapses can be fast or slow, depending on the nature of their receptors. **Ionotropic receptors** are ion channels and generate fast, short-lived responses. **Metabotropic receptors** initiate second-messenger cascades that lead to slower, more sustained responses.

The Vertebrate Nervous System Has Many Interacting Components

- The brain and **spinal cord** make up the **central nervous system** (**CNS**); neurons that extend or reside outside the brain and the spinal cord, together with their supporting cells, make up the **peripheral nervous system** (**PNS**). **Review Figure 34.12**

- The **autonomic nervous system** (**ANS**) is the part of the PNS that controls involuntary physiological functions. Its **sympathetic** and **parasympathetic** divisions differ in anatomy, neurotransmitters, and effects on target tissues. **Review Figure 34.13**

- The spinal cord communicates information between the brain and the rest of the body. It can issue some commands to the body

without input from the brain (reflexes). **Review Figure 34.14 and ANIMATED TUTORIAL 34.4**

- The embryonic brain consists of a hindbrain, midbrain, and forebrain. The forebrain develops into the **cerebral hemispheres** (the **telencephalon**, or **cerebrum**) and the underlying thalamus and hypothalamus (which together compose the **diencephalon**). The midbrain and hindbrain develop into the **brainstem** and the **cerebellum**.

- The **reticular system** is a complex network in the brainstem that controls various autonomic functions and transmits sensory information to the forebrain.

- The **limbic system** is an evolutionarily primitive part of the telencephalon that is involved in emotions, physiological drives, instincts, and memory. **Review Figure 34.15**

- The cerebral hemispheres are the dominant structures of the human brain. Their surfaces are layers of neurons called the **cerebral cortex**. **Review Figure 34.16**

- Each cerebral hemisphere can be divided into temporal, frontal, parietal, and occipital lobes. Many motor functions are localized in the **frontal lobe**. Information from many sensory receptors projects to the **parietal lobe**. Visual information goes to the **occipital lobe**, and auditory and visual information goes to the **temporal lobe**. **Review Figure 34.16 and WEB ACTIVITY 34.2**

Specific Brain Areas Underlie the Complex Abilities of Humans

- Language abilities are localized mostly in the left cerebral hemisphere, a phenomenon known as **lateralization**. Different areas of the left hemisphere are responsible for the production and understanding of language. **Review Figure 34.18**

- Complex memories can be elicited by stimulating small regions of association cortex. Damage to the hippocampus can destroy the ability to form long-term **declarative memory** but not **procedural memory**.

- Most animals, including humans, have a daily cycle of sleep and waking. Sleep can be divided into **rapid-eye-movement** (**REM**) **sleep** and **non-REM sleep**. **Review Figure 34.19**

- A sense of the physiological state of the body may be created in the insular cortex. Evolution of this integrative function could be the basis for conscious experience.

See **WEB ACTIVITY 34.3** for a concept review of this chapter.

Sensors  35

On moonless, pitch-black nights in the desert, kangaroo rats move silently about to forage and to seek mates. When two kangaroo rats meet, they immediately identify each other's gender and behave accordingly—males fight with other males and try to court females. We may not be able to observe their behavior in total darkness, but they can easily identify each other, interact, and also gather food. However, other denizens of the desert can observe the kangaroo rats and attempt to capture them for food.

If a kangaroo rat chances to hop near a coiled rattlesnake, the snake can detect the rat, accurately fix its location, and with a lightning fast strike attempt to nail the rat. But even as the strike of the rattlesnake is in progress, the kangaroo rat senses it and can make a giant leap that usually saves it from the deadly fangs. Another predator of kangaroo rats comes silently from the air. Also in total darkness, an owl can swoop down, flare its wings and extend its talons just above the ground, and accurately seize an unlucky kangaroo rat.

Similar dramas play out overhead. Moths fly about seeking mates as well as night-blooming, nectar-producing plants. Bats pursue the moths and are amazingly accurate in detecting and capturing them in flight. Frequently, however, the moths use a very specific behavior to escape. As a bat is closing in, the moth suddenly dives straight down, and the bat misses its prey.

These feats of detection and identification of individuals of the same or different species in total darkness, and these feats of prey capture and predator avoidance, are remarkable. The kangaroo rat distinguishes when another kangaroo rat is a male or a female. The rattlesnake and the

owl detect potential prey at a distance and accurately attack. The kangaroo rat senses the lightning fast strike of the snake and avoids it. The bats accurately home in on fluttering moths, and the moths sense the incoming bat and take evasive action.

A human being would not even be able to observe any of these nighttime dramas without the benefit of special technologies. We may have impressive sensory capabilities of our own, such as excellent color vision during the daytime, but in a nighttime desert, we would be humbled by the remarkable sensory capabilities of the animals that live there.

The acute senses of this barn owl (*Tyto alba*) have enabled it to detect and capture its kangaroo rat prey.

 QUESTION How do kangaroo rats, rattlesnakes, owls, bats, and moths "see in the dark"?

KEY CONCEPTS

35.1 Sensory Systems Convert Stimuli into Action Potentials

35.2 Chemoreceptors Detect Specific Molecules or Ions

35.3 Mechanoreceptors Detect Physical Forces

35.4 Photoreceptors Detect Light

Sensory Systems Convert Stimuli into Action Potentials

Sensory receptor cells, sometimes simply called *sensors* or *receptors*, transduce (convert) sensory stimuli such as light, sound, and touch into changes in membrane potential that cause in these cells or in neighboring cells the generation of action potentials (see Concept 34.2). These action potentials convey the sensory information to the central nervous system for processing and interpretation. In the first step in this process, a sensory cell must somehow respond to a certain type of stimulus with a change in membrane potential.

Sensory transduction involves changes in membrane potentials

Sensory transduction typically begins with a **receptor protein** that opens or closes ion channels in response to a specific stimulus, such as heat or light. The resulting change in ion flow alters the receptor cell's membrane potential. A change in the membrane potential of a receptor cell in response to a stimulus is called a **receptor potential**. Receptor potentials are *graded membrane potentials* that spread over only short distances. To travel long distances in the nervous system, receptor potentials must generate action potentials, which they can do in two ways:

- The receptor potential may trigger action potentials in the receptor cell itself.

- The receptor potential may cause the receptor cell to release neurotransmitters that can induce a postsynaptic neuron to generate action potentials.

A good model of how a receptor cell generates action potentials can be seen in the *stretch receptor* of a crayfish (**FIGURE 35.1**). Stretching of the muscle to which the stretch receptor is attached causes receptor potentials. These receptor potentials spread to the base of the cell's axon, where they generate action potentials that travel down the axon to the CNS. The rate at which action potentials are fired depends on the magnitude of the receptor potential; that magnitude, in turn, depends on how much the muscle is stretched.

In receptor cells that do not fire action potentials (such as the photoreceptors in the retina), the receptor potential induces the release of neurotransmitter. The intensity of the stimulus influences how much neurotransmitter is released.

LINK Review the mechanics of graded membrane potentials and action potentials in Concept 34.2 and of synaptic transmission in Concept 34.3

Different sensory receptors detect different types of stimuli

Different types of sensory receptor proteins respond to particular types of environmental energy. *Mechanoreceptors* detect physical forces, such as pressure (touch) and variations

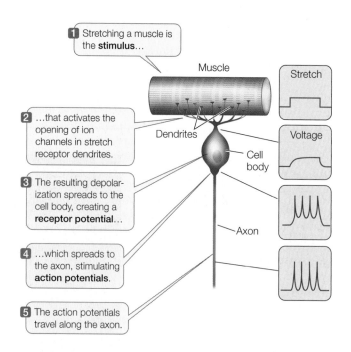

1 Stretching a muscle is the **stimulus**...

2 ...that activates the opening of ion channels in stretch receptor dendrites.

3 The resulting depolarization spreads to the cell body, creating a **receptor potential**...

4 ...which spreads to the axon, stimulating **action potentials**.

5 The action potentials travel along the axon.

Muscle · Stretch · Dendrites · Voltage · Cell body · Axon

FIGURE 35.1 Stimulating a Sensory Cell Produces a Receptor Potential Signal transduction in the stretch receptor of a crayfish can be investigated by measuring the membrane potential at different places on the stretch receptor neuron while stretching the muscle innervated by that sensory neuron.

in pressure (sound waves); the crayfish cell described above contains mechanoreceptors in its dendrites. *Thermoreceptors* respond to temperature. *Electrosensors* are directly sensitive to changes in membrane potential. *Chemoreceptors* respond to the presence or absence of certain specific chemicals, and *photoreceptors* detect light. All of these receptor proteins respond to the appropriate stimulus by causing changes in ion flow (i.e., by directly or indirectly causing ion channels to open or close; **FIGURE 35.2**). These changes in ion flow result in a receptor potential.

Some sensory receptor cells are assembled with other types of cells into *sensory organs*, such as eyes, ears, and noses, that enhance the ability of the sensory cells to collect, filter, and amplify stimuli. A **sensory system** consists of sensory cells, the associated structures, and the neural networks that process the information.

Sensation depends on which neurons receive action potentials from sensory cells

All sensory systems, no matter what type of stimulus they detect, convey information in the form of action potentials. But the sensations we perceive—such as heat, pressure, light, smell, and sound—differ because the messages from different kinds of sensory cells arrive at different places in the CNS. Action potentials arriving in the visual cortex of the brain are interpreted as light, in the auditory cortex as sound, in the olfactory bulb as odors, and so forth. Action potentials arriving from the various

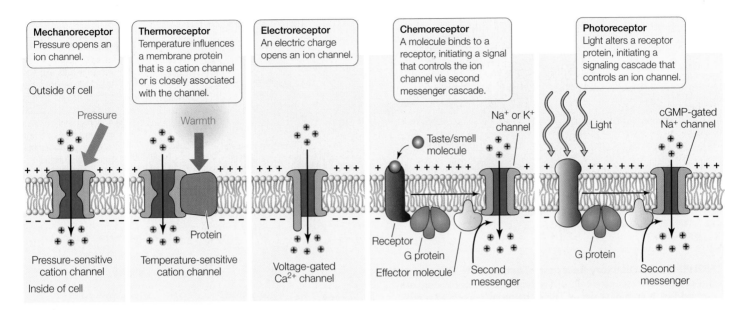

FIGURE 35.2 Sensory Receptor Proteins Respond to Stimuli by Opening or Closing Ion Channels Some receptor proteins are themselves ion channels. Others initiate signal transduction cascades that eventually open or close ion channels.

sensory neurons in a patch of skin may be perceived as heat, cold, itch, pressure, or pain, depending on which neuron is involved and which pathways of the CNS are stimulated. The intensity of the sensation, meanwhile, is coded as the frequency of action potentials.

Some sensory cells transmit information about internal conditions in the body, but we may not be consciously aware of that information. The brain continuously receives information about body temperature, blood carbon dioxide and oxygen concentrations, arterial pressure, muscle tension, and the positions of the limbs—all of which are important for homeostasis. All of this sensory information goes to the CNS, but it does not always result in conscious sensation.

Many receptors adapt to repeated stimulation

Some sensory cells give gradually diminishing responses to maintained or repeated stimulation. This phenomenon is known as **adaptation**, and it enables an animal to ignore background or unchanging conditions while remaining sensitive to changes or to new information. (Note that this use of the term "adaptation" is different from its application in an evolutionary context.) When you dress, you feel each item of clothing touch your skin, but the sensation of clothes touching your skin soon fades from awareness. You are immediately aware, however, of new sensations, such as someone touching your back.

Some sensory cells adapt very little or very slowly, such as pain receptors and the mechanoreceptors for balance. You shouldn't ignore pain, because it is signaling that something is wrong in your body, and to maintain equilibrium you must continuously know (albeit unconciously) the tensions and forces on all of your joints and muscles.

Do You Understand Concept 35.1?

- An ice cube held in your fingertips causes the sensation of cold, whereas a pinprick to the same finger causes the quite different sensation of pain. How can action potentials from the same patch of skin result in such different sensations?

- We have stretch receptors in our skeletal muscles (see Figure 34.14). Would you expect them to adapt to a prolonged stretch? Why or why not?

- A cell on a mosquito's foot responds to heat, letting the mosquito know if it has landed on a warm body. Another cell on its antennae detects the CO_2 exhaled by other animals. What types of sensory receptor proteins do these cells probably contain?

The remainder of this chapter will discuss specific types of sensory receptors and their associated sensory systems. We will begin with chemosensation, the basis of smell and taste.

---

concept 35.2 Chemoreceptors Detect Specific Molecules or Ions

A colony of corals responds to a small amount of meat extract in seawater by extending bodies and tentacles and searching for food; a solution of a single amino acid can stimulate this response. Conversely, a small amount of seawater in which corals were crushed will stimulate a defensive retraction of the coral polyps. Humans react strongly to certain chemical stimuli. When we smell freshly baked bread, we salivate and feel hungry; when we smell rotting meat, we feel nauseated.

All animals receive information about chemical stimuli through **chemoreceptors**, which are receptor proteins that bind to certain molecules. Chemoreceptors are also responsible for

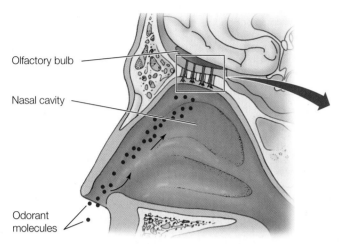

FIGURE 35.3 Olfactory Receptors Communicate Directly with the Brain The receptor cells of the human olfactory system are embedded in epithelial tissues lining the nasal cavity and send their axons to the olfactory bulb of the brain.

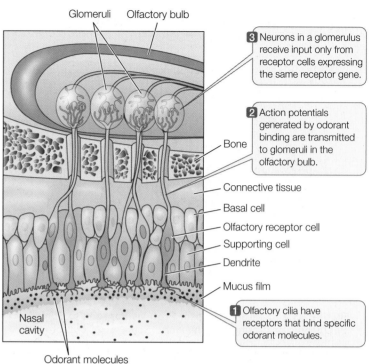

monitoring some aspects of the internal environment, such as the level of carbon dioxide in the blood.

Olfaction is the sense of smell

In vertebrates, the sense of smell, **olfaction**, depends on olfactory sensors—chemoreceptive neurons—embedded in a layer of epithelial tissue at the top of the nasal cavity (**FIGURE 35.3**). The axons from these neurons extend to the olfactory integration area of the brain (the *olfactory bulb*), whereas their dendrites end in olfactory cilia on the surface of the nasal epithelium. A protective layer of mucus covers the epithelium. Molecules from the environment must diffuse through the mucus to reach the receptor proteins on the olfactory cilia. When you have a cold, the amount of mucus in your nose increases and the epithelial tissue swells; this is why you can temporarily lose your sense of smell.

The olfactory world has an enormous number of odors and a correspondingly large number of olfactory receptor proteins. In the 1990s, Linda Buck and Richard Axel discovered a family of about 1,000 genes in mice (about 3 percent of the mouse genome) that code for olfactory receptor proteins (a discovery that won them a Nobel Prize). Each receptor protein is found in a limited number of receptor cells in the olfactory epithelium, and each cell expresses just one type of receptor protein. Each receptor can bind to a set of similar molecules called **odorants**.

When an odorant binds to a receptor, it activates a G protein. The G protein in turn activates an enzyme that causes an increase of a second messenger (cAMP in vertebrates) in the cytoplasm (see the chemoreceptor panel in Figure 35.2). The second messenger causes an influx of Na^+, depolarizing the olfactory neuron.

LINK G protein–linked receptors are involved in many sensory mechanisms and are described in Concept 5.5; see especially Figure 5.14

The next stage of processing olfactory information is in the olfactory bulb, where axons from neurons expressing the same receptor protein cluster together on the dendrites of olfactory bulb neurons, forming structures called *glomeruli* (singular *glomerulus*; see Figure 35.3). A complex odorant molecule may bind to several different types of receptor proteins, and thus can activate a unique combination of glomeruli in the olfactory bulb. Therefore an olfactory system with hundreds of different receptor proteins can discriminate an astronomically large number of smells. The more odorant molecules that bind to receptors, the greater the frequency of action potentials and the greater the intensity of the perceived smell.

Humans are unusual among mammals in that we depend more on vision than on olfaction. About two-thirds of the mammalian genes for olfactory receptor proteins are thought to be nonfunctional in humans, and our nasal anatomy reflects the reduced importance of olfaction. A typical dog's nasal epithelium is 15–20 times larger than a human's and has about 1 *billion* olfactory receptors, compared with about 20 million in humans. Dogs can detect some scents at concentrations 100 million times lower than humans can. The olfactory worlds of dogs, and of most mammals, are much richer than ours.

Some chemoreceptors detect pheromones

A specialized type of chemical signal used for communication among members of the same species is a **pheromone**. Many animals use pheromones to attract mates, and often these are detected by specialized chemoreceptors that are separate from the primary olfactory system.

Male moths have chemoreceptors in their antennae, which are used to detect the presence of female moths (**FIGURE 35.4**).

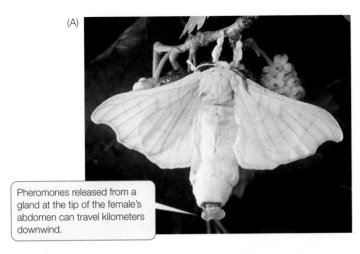

Pheromones released from a gland at the tip of the female's abdomen can travel kilometers downwind.

A male moth detects this pheromone in the air passing over his antennae, which are covered with chemo-sensitive hairs.

FIGURE 35.4 Some Scents Travel Great Distances Mating in silkworm moths of the genus *Bombyx* is coordinated by pheromones released from the female.

In the silkworm moth *Bombyx mori*, for example, a female releases a blend of pheromone molecules from a gland at the tip of her abdomen; the male silkworm moth has receptors for these molecules on his antennae. Each feathery antenna carries about 10,000 pheromone-sensitive hairs. A single molecule of pheromone may be sufficient to generate action potentials in the antennal nerve. When approximately 200 hairs per second are activated, the male orients upwind in search of the female. Because the rate of firing in the male's sensory nerves is proportional to the pheromone concentration in the air, he can follow the airborne concentration gradient and home in on the signaling female.

Many vertebrates use pheromones, often detecting them with a specialized olfactory organ called the **vomeronasal organ**, or **VNO**. The VNO is a small, paired tubular structure embedded deep within the nasal epithelium. When the animal sniffs, the VNO draws a sample of nasal fluid over the chemoreceptors embedded in its walls. The information from these chemoreceptors goes to an accessory olfactory bulb in the brain, and from there to brain regions involved in sexual and other instinctive behaviors. Some humans have a very small, vestigial VNO that is probably nonfunctional.

Gustation is the sense of taste

In terrestrial vertebrates, the sense of taste, or **gustation**, depends on clusters of chemoreceptors in **taste buds** in the mouth. Some fishes have taste buds in their skin that allow them to find food without the use of vision. The duck-billed platypus, a prototherian mammal (see Figure 23.50A), has similar abilities as a result of taste buds on the sensitive skin of its bill.

APPLY THE CONCEPT

Chemoreceptors detect specific molecules or ions

If the vomeronasal organ of a male rodent is surgically removed, both his mating behavior toward females and aggressive behavior toward other males are impaired. One hypothesis to explain this finding is that the VNO is essential for detecting pheromones given off by sexually receptive females, and that these pheromones stimulate male mating behavior and aggression. Catherine Dulac and her students genetically engineered male mice so they lacked one of the VNO receptor ion channels, called TRPC2. Normal (TRPC2$^{+/+}$) and receptor-deficient (TRPC2$^{-/-}$) male mice were tested by placing castrated male mice or female mice into their cages and observing the behavior of the normal (control) and receptor-deficient (experimental) male mice. The experiments were repeated with the castrated male mice being swabbed with urine from intact males. Use the results shown in the table to answer the questions.

RESIDENT MOUSE GENOTYPE	INTRUDER MOUSE[a]			
	CASTRATED ♂	FEMALE	CASTRATED ♂, URINE-SWABBED	FEMALE, URINE-SWABBED
TRPC2$^{+/+}$	N	M	A	A
TRPC2$^{-/-}$	M	M	M	M

[a]The resident's behavior when confronted with different intruders was scored as neutral (N), aggressive (A), or mating attempted (M).

1. Do the results of these experiments support the hypothesis that TRPC2 receptors are essential for males to respond to female pheromones? Why or why not?

2. What function(s) can you hypothesize for the TRPC2 receptor?

The human tongue has approximately 10,000 taste buds, mostly embedded in the epithelium on the sides of papillae (**FIGURE 35.5**). (Look at your tongue in a mirror—the papillae make it look fuzzy.) Each taste bud has a pore that exposes the tips of sensory receptor cells. These sensory cells generate action potentials when they are exposed to certain chemicals. They synapse at their bases with sensory neurons that convey the information to the CNS.

Humans perceive at least five tastes—sweet, salty, sour, bitter, and *umami* (a savory, meaty taste). Much of our experience of taste is actually due to our olfactory receptors, which is why you lose much of your sense of taste when you have a cold.

Research has identified the mechanism of action of all five taste perceptions. "Salty" receptors allow sodium to diffuse through open Na^+ channels, directly depolarizing the sensory cell. "Sour" receptors are in fact detecting acidity—that is, the concentration of H^+. "Sweet" receptors bind to a variety of different sugars, and the umami receptor detects the presence of amino acids, such as the glutamate in monosodium glutamate (MSG).

Bitterness is the most complicated taste, involving at least 30 different taste receptors. Each of these can detect a different molecule at very low concentration. Many different bitter receptors may be present on the same cell, so that all of these different molecules evoke the same taste sensation. These sensitive "bitter" taste receptors probably evolved in response to toxic plant compounds such as quinine, caffeine, and nicotine. Plants produce many such toxic molecules because they repel herbivorous predators; the ability to detect these potentially dangerous molecules in the plants we consume is thus adaptive.

Do You Understand Concept 35.2

- Which sense, olfaction or gustation, has more types of chemoreceptor proteins?

- Explain why we have evolved dozens of different bitterness taste receptors but only one type of saltiness receptor.

- Most dogs can distinguish millions of different odors. Explain how this is possible when a dog's nose contains only about 1,000 different types of olfactory receptors.

We've seen how chemoreceptors can give rise to the senses of smell and taste. We will turn next to mechanoreceptors, the sensory cells that respond to mechanical forces.

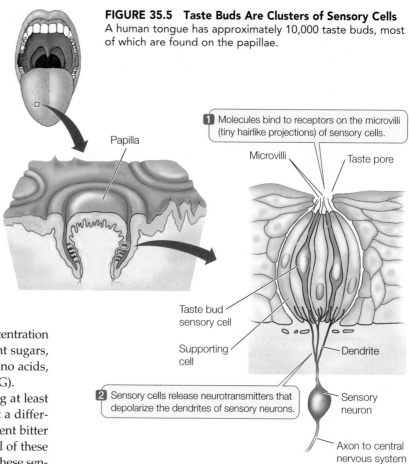

FIGURE 35.5 Taste Buds Are Clusters of Sensory Cells
A human tongue has approximately 10,000 taste buds, most of which are found on the papillae.

Papilla

1 Molecules bind to receptors on the microvilli (tiny hairlike projections) of sensory cells.

Microvilli

Taste pore

Taste bud sensory cell

Supporting cell

Dendrite

2 Sensory cells release neurotransmitters that depolarize the dendrites of sensory neurons.

Sensory neuron

Axon to central nervous system

concept 35.3 Mechanoreceptors Detect Physical Forces

Mechanoreceptors are cells that detect a variety of physical forces. Physical distortion of a mechanoreceptor's plasma membrane causes ion channels to open, creating a receptor potential that in turn can lead to the release of neurotransmitter or the generation of action potentials. A considerable diversity of mechanosensory cells and mechanisms has evolved. Involved in many sensory systems, their functions range from interpreting skin sensations to sensing blood pressure to hearing and maintaining balance.

Many different cells respond to touch and pressure

The most obvious sensory receptors in your skin are the many *free nerve endings* that detect heat, cold, and pain. Mammalian skin is also packed with diverse mechanoreceptors that generate varied sensations (**FIGURE 35.6**). The most important tactile receptors are *Merkel's discs*, which adapt rather slowly and provide continuous information about things touching the skin. *Meissner's corpuscles* are very sensitive but adapt rapidly, so they provide information about *changes* in objects contacting the skin. The rapid adaptation of Meissner's corpuscles is why you roll a small object between your fingers (rather than

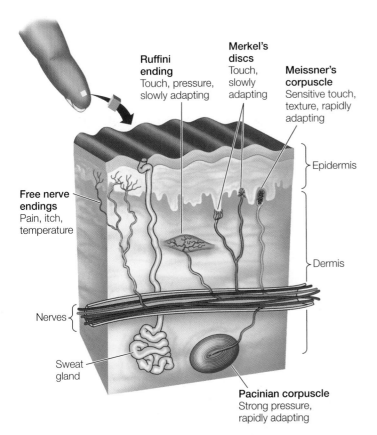

Ruffini ending
Touch, pressure, slowly adapting

Merkel's discs
Touch, slowly adapting

Meissner's corpuscle
Sensitive touch, texture, rapidly adapting

Epidermis

Dermis

Free nerve endings
Pain, itch, temperature

Nerves

Sweat gland

Pacinian corpuscle
Strong pressure, rapidly adapting

FIGURE 35.6 The Skin Feels Many Sensations Even a very small patch of skin contains a variety of sensory cells, making the skin a multimodal receptor that can sense temperature, pressure, texture, pain, touch, and itch.

yourBioPortal.com
Go to INTERACTIVE TUTORIAL 35.1
Sensory Receptors

holding it still) to discern its shape and texture: as you roll it, the object continues to stimulate Meissner's corpuscles.

Two other kinds of mechanoreceptors are found deeper in the skin. *Ruffini endings* adapt slowly and are good at providing information about vibrating stimuli of low frequencies. *Pacinian corpuscles*, which adapt rapidly, provide information about high-frequency alternating stimuli, such as when you

run your fingers over a rough surface. Even deeper in the skin, dendrites of sensory neurons wrap around hair follicles. When the surface hairs are displaced, as when a mosquito lands on your skin, those neurons are stimulated.

Mechanoreceptors are found in muscles, tendons, and ligaments

Mechanoreceptors also provide information about the position of the limbs and the stresses on muscles and joints. Though we are not consciously aware of this information, it is essential for postural control and the coordination of movements.

For example, skeletal muscle contains mechanoreceptor organs called *muscle spindles*. These contain **stretch receptors**, modified muscle cells that are innervated by sensory neurons. Whenever the muscle is stretched, the muscle spindles are also stretched, and the neurons transmit action potentials to the CNS. The CNS uses this information to adjust the strength of the muscle contraction to match the load put on the muscle (**FIGURE 35.7**). This enables you to hold a glass in a steady position as someone else pours liquid into it.

Another type of mechanoreceptor, the *Golgi tendon organ*, provides information about the force being applied to tendons and ligaments. If a tendon or ligament is loaded to an unusual degree, action potentials from the Golgi tendon organ inhibit the spinal cord motor neurons innervating that muscle, causing it to relax and protecting it from tearing.

Hair cells are mechanoreceptors of the auditory and vestibular systems

The mechanoreceptors of the auditory system are the highly specialized **hair cells**. **Stereocilia**—fingerlike extensions of the cell membrane stiffened by cross-linked actin filaments—project from the surface of each hair cell like a set of organ pipes (**FIGURE 35.8A**). Stereocilia bend in response to waves of pressure; bending of the stereocilia in one direction depolarizes the hair cell, and bending in the other direction hyperpolarizes it (**FIGURE 35.8B**).

FIGURE 35.7 Stretch Receptors Stretch receptors provide information about the stresses on muscles and joints in an animal's limbs. In vertebrates, sensory organs called muscle spindles send information to the CNS about how much the muscle is being stretched.

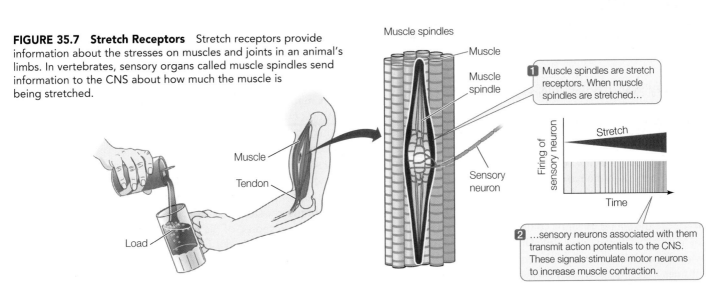

Muscle spindles

Muscle

Muscle spindle

Muscle

Tendon

Load

Sensory neuron

1 Muscle spindles are stretch receptors. When muscle spindles are stretched…

Firing of sensory neuron

Stretch

Time

2 …sensory neurons associated with them transmit action potentials to the CNS. These signals stimulate motor neurons to increase muscle contraction.

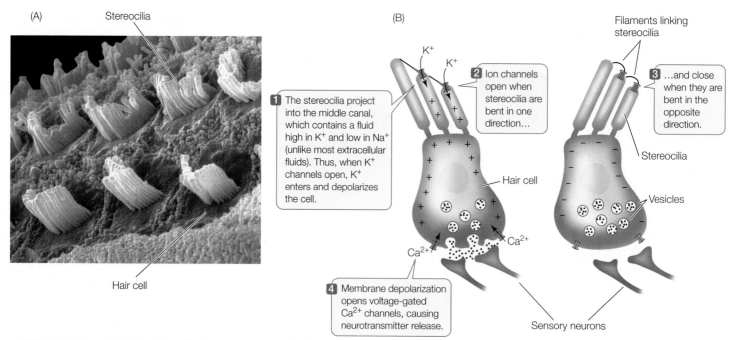

FIGURE 35.8 Hair Cells Have Mechanosensors on Their Stereocilia (A) Hair cells have stereocilia that are connected with each other by small filaments. (B) When a cilium is bent in one direction, the filament opens a mechanosensory ion channel at the tip of the neighboring cilium. The depolarized hair cell releases neurotransmitter onto a sensory neuron.

Measurements with microelectrodes have shown that bending of stereocilia causes local electric currents near their tip, indicating that ion channels near the tips must be opening or closing. Electron microscope images reveal minute filaments that connect the tip of each stereocilium to its taller neighbor. It is hypothesized that these filaments are fine molecular attachments to the ion channels, and that they act like springs that open the channels. If the taller neighboring stereocilium is bent away, the spring tightens and the ion channel is opened. If the taller neighbor bends toward its shorter neighbor, the spring is relaxed and the channel closes (see Figure 35.8B).

Hair cells are the mechanoreceptors for the vertebrate auditory (sound-perceiving) and vestibular (equilibrium-maintaining) systems. Both of these systems are housed in the complex structures of the vertebrate ear.

Auditory systems use hair cells to sense sound waves

Auditory systems use hair cells to convert pressure waves in air into receptor potentials that initiate the perception of sound. Special structures in the ear gather and amplify sound waves and direct them to the hair cells.

OUTER EAR The two prominent structures on the sides of our heads are the *pinnae*. A pinna collects sound waves and directs them into the *auditory canal* (**FIGURE 35.9A**). The eardrum, or **tympanic membrane**, covers the end of the auditory canal and vibrates in response to sound waves.

MIDDLE EAR The middle ear is an air-filled cavity containing three delicate bones called **ossicles**, individually named the *malleus* (Latin, "hammer"), *incus* ("anvil"), and *stapes* ("stirrup") (**FIGURE 35.9B**). Together, the ossicles act as a lever that amplifies vibrations of the tympanic membrane into much more forceful vibrations of another flexible membrane called the **oval window**. Behind the oval window lies the fluid-filled inner ear.

The middle ear is connected to the throat via the **eustachian tube**, allowing air pressure to equilibrate between the middle ear and the outside world. If a eustachian tube becomes blocked because of a respiratory infection, changes in external air pressure can cause the tympanic membrane to stretch or even rupture. If you have ever taken an airplane flight when you had a bad cold, you might have experienced this painful phenomenon.

INNER EAR The inner ear consists of two sets of fluid-filled canals. One is the organ of balance, the **vestibular system**, and the other is the organ of hearing, the **cochlea**. The cochlea (Latin and Greek, "snail" or "spiral shell") is a long, tapered, coiled canal. A cross section of the cochlea reveals that it is composed of three parallel canals separated by two membranes: **Reissner's membrane** and the **basilar membrane** (**FIGURE 35.9C**). Sitting on the basilar membrane is the **organ of Corti**, which transduces pressure waves into action potentials. The organ of Corti contains hair cells with stereocilia. The tips of the longest stereocilia are embedded in a gelatinous overhanging shelf called the *tectorial membrane* (**FIGURE 35.9D**).

Because their tips are attached to the more rigid tectorial membrane, the stereocilia are bent when the basilar membrane flexes. The response of the hair cell is a graded membrane potential. Hair cells do not fire action potentials themselves, but they release neurotransmitter onto sensory neurons, which transmit action potentials to the brain.

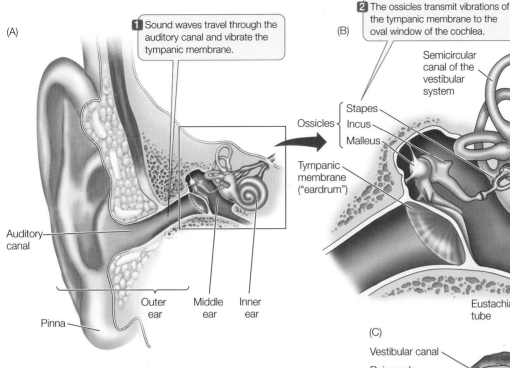

(A)

1 Sound waves travel through the auditory canal and vibrate the tympanic membrane.

Auditory canal

Pinna

Outer ear | Middle ear | Inner ear

FIGURE 35.9 Structures of the Human Ear (A) The pinnae direct sound waves down the auditory canal to impinge on the tympanic membrane. The tympanic membrane mechanically transmits these pressure waves into movements of the ossicles in the middle ear. (B) The ossicles transmit their movement into pressure waves in the fluid of the cochlea at the oval window. (C) The cochlea is divided into fluid-filled chambers; pressure waves from the ossicles cause the membranes between the chambers to flex. (D) Flexing of the basilar membrane bends stereocilia on hair cells in the organ of Corti.

yourBioPortal.com
Go to WEB ACTIVITY 35.1
Structures of the Human Ear

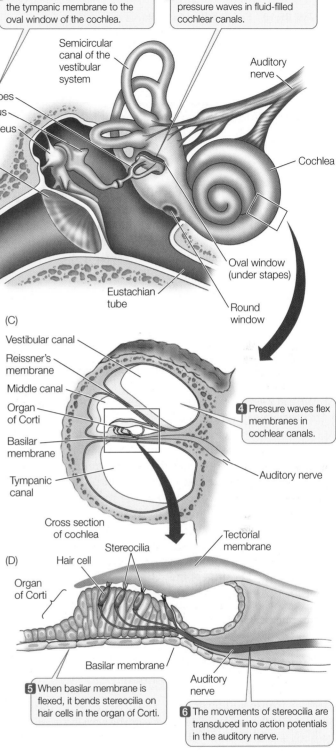

(B)

2 The ossicles transmit vibrations of the tympanic membrane to the oval window of the cochlea.

3 Vibrations at oval window create pressure waves in fluid-filled cochlear canals.

Semicircular canal of the vestibular system

Ossicles { Stapes, Incus, Malleus }

Auditory nerve

Tympanic membrane ("eardrum")

Cochlea

Eustachian tube

Oval window (under stapes)

Round window

(C)

Vestibular canal
Reissner's membrane
Middle canal
Organ of Corti
Basilar membrane
Tympanic canal

4 Pressure waves flex membranes in cochlear canals.

Auditory nerve

Cross section of cochlea

(D)

Organ of Corti

Stereocilia
Hair cell
Tectorial membrane

Basilar membrane
Auditory nerve

5 When basilar membrane is flexed, it bends stereocilia on hair cells in the organ of Corti.

6 The movements of stereocilia are transduced into action potentials in the auditory nerve.

Flexion of the basilar membrane is perceived as pitch

What causes the basilar membrane to flex, and how does this mechanism distinguish sounds of different pitches (frequencies)? **FIGURE 35.10** shows the cochlea uncoiled, to make it easier to understand its structure and function. The upper and lower canals separated by the basilar membrane are joined at the distal end of the cochlea (the end farthest from the oval window), making one continuous canal that turns back on itself. Just as the oval window is a flexible membrane at the beginning of the upper canal of the cochlea, the **round window** is a flexible membrane at the end of the lower canal.

Air is highly compressible but liquids are not. Therefore, a pressure wave can travel through air without much displacement of the air, whereas a pressure wave in an aqueous fluid displaces that fluid. When the stapes pushes on the oval window, the fluid in the upper canal of the cochlea is displaced. If this movement of the oval window occurs slowly, the cochlear fluid pressure wave travels down the upper canal, around the bend, and back through the lower canal. At the end of the

lower canal, the displacement pressure is dissipated by the outward bulging of the round window.

If the oval window vibrates in and out rapidly, the waves of fluid pressure create traveling waves in the basilar membrane. The basilar membrane is not uniform; it is thicker and stiffer at its base and wider and thinner at its apical end. Pressure waves in the cochlear fluid have different frequencies and set up different patterns of traveling waves. High-frequency waves cause maximal flexion at the basal end of the basilar membrane, whereas low-frequency pressure waves result in maximal flexion at its

FIGURE 35.10 Sensing Pressure Waves in the Inner Ear
Pressure waves of different frequencies flex the basilar membrane at different locations. Information about sound frequency is specified by which hair cells are activated. For simplicity, this representation illustrates the cochlea as uncoiled, and leaves out the middle ear.

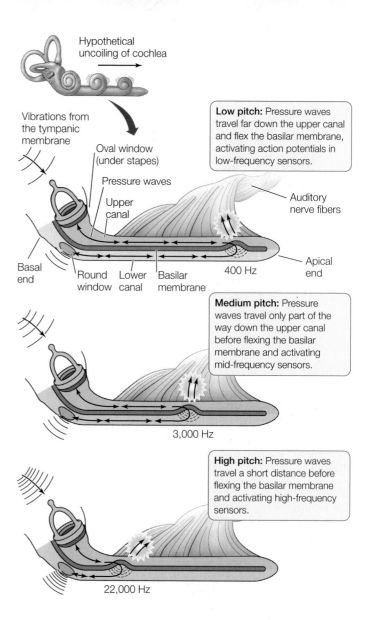

apical end. Thus different pitches of sound flex the basilar membrane at different locations and activate different sets of hair cells. Action potentials stimulated by the mechanoreceptors at different positions along the organ of Corti travel in the auditory nerve to different regions of the *auditory cortex*.

Various types of damage can result in hearing loss

There are two general types of hearing loss, or deafness. *Conduction deafness* is caused by the loss of function of the tympanic membrane or the ossicles of the middle ear. For example, the ossicles often stiffen with increasing age, resulting in a gradual loss of the ability to hear high-frequency sounds.

Nerve deafness is caused by damage to the inner ear or the auditory pathways. A common cause of nerve deafness is damage to the hair cells of the delicate organ of Corti by exposure to loud sounds. Consistent exposure to sounds above 85 decibels can damage hearing; this damage is cumulative and irreversible. For example, personal stereo earphones can reach 120 decibels, and people commonly use them at 100 decibels (equivalent to being at a rock concert).

FRONTIERS A recent study at Brigham and Woman's Hospital in Boston examined hearing in 5,000 adolescents ages 12 to 19. One fifth of the subjects had significant hearing loss. This represents a 30 percent increase in the incidence of hearing loss in this age group compared with an identical study done about 15 years earlier.

The vestibular system uses hair cells to detect forces of gravity and momentum

The mammalian vestibular system is a function of several bony structures in the cochlea of the inner ear. The working units of the vestibule are three **semicircular canals** and two chambers, the *saccule* and the *utricle*. Hair cells in the vestibular system detect the position and movement of the head—information that is essential for maintaining balance (equilibrium) and for certain eye reflexes.

The canals and chambers of the vestibular system are filled with a fluid called endolymph. In the semicircular canals, the endolymph shifts when the head turns. Projecting into the base of each canal is a gelatinous swelling called a *cupula* (plural *cupulae*) that encloses a cluster of hair cell stereocilia. When the shifting endolymph pushes on the cupulae, it bends the stereocilia (**FIGURE 35.11A**). Since the three semicircular canals have different orientations, together they detect rotational movement in any plane.

The stereocilia in the saccule and utricle are bent in a different way. These stereocilia are embedded in otolithic membranes, gelatinous structures covered with a layer of calcium carbonate crystals called *otoliths* (Latin, "ear stones"). When the head changes position or when it accelerates or decelerates, gravitational forces are exerted on the otoliths and bend the stereocilia (**FIGURE 35.11B**).

Do You Understand Concept 35.3?

- Explain how very tiny vibratory movements of the tympanum are translated into forces that bend the basilar membrane at different locations.

- Fish have lateral lines that are fluid-filled channels running down the sides of their bodies. In the channels are hair cells. What do you think the functions of these lateral-line hair cells are?

- Hair cell membrane potentials can respond to stimuli much faster than chemosensors can. What are the cellular mechanisms responsible for this difference?

- How is your auditory system able to distinguish a high-pitched sound from a low-pitched one?

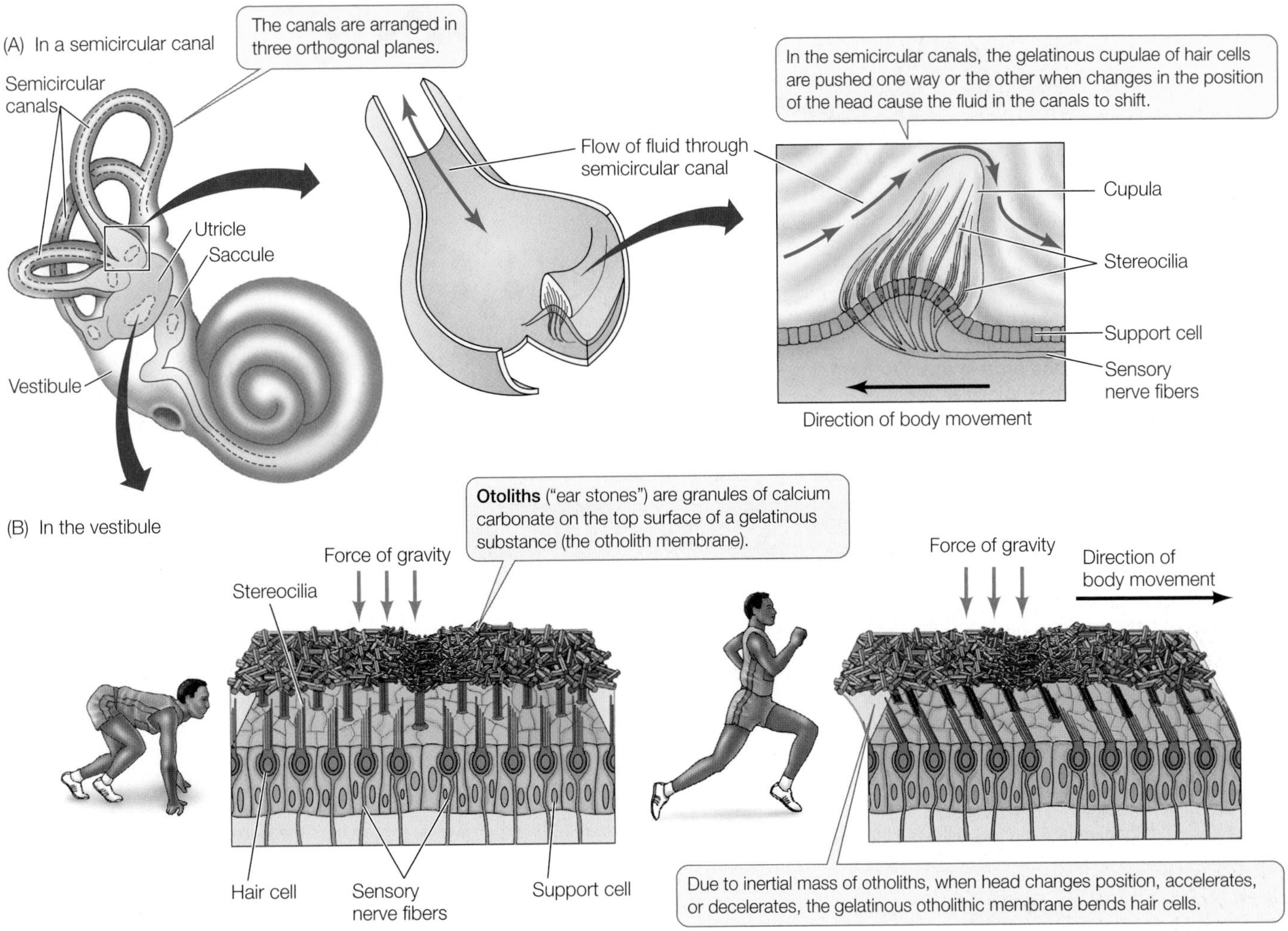

(A) In a semicircular canal

Semicircular canals

The canals are arranged in three orthogonal planes.

Utricle

Saccule

Vestibule

Flow of fluid through semicircular canal

In the semicircular canals, the gelatinous cupulae of hair cells are pushed one way or the other when changes in the position of the head cause the fluid in the canals to shift.

Cupula

Stereocilia

Support cell

Sensory nerve fibers

Direction of body movement

(B) In the vestibule

Otoliths ("ear stones") are granules of calcium carbonate on the top surface of a gelatinous substance (the otolith membrane).

Force of gravity

Stereocilia

Force of gravity

Direction of body movement

Hair cell

Sensory nerve fibers

Support cell

Due to inertial mass of otoliths, when head changes position, accelerates, or decelerates, the gelatinous otolithic membrane bends hair cells.

FIGURE 35.11 Organs of Equilibrium The vestibular system consists of bony chambers and fluid-filled canals. (A) Each semicircular canal has a cupula containing stereocilia. When fluid moves against the cupula, the stereocilia bend. (B) In the saccule and utricle, stereocilia are bent by gravitational forces on the otoliths.

yourBioPortal.com

Go to ANIMATED TUTORIAL 35.2
Photosensitivity

We have seen how pressure and chemicals can be detected by sensory cells that give us the senses of smell, taste, touch, hearing, and balance. Now let's turn our attention to vision, which in humans is perhaps the most elaborate of the senses.

concept
35.4 Photoreceptors Detect Light

Sensitivity to light—**photosensitivity**—confers on the simplest animals the ability to orient to the sun and sky, and gives more complex animals rapid and extremely detailed information about objects in their environment. It is not surprising that both simple and complex animals can sense and respond to light. What is remarkable is that across the entire range of animal species, evolution has conserved the same basis for photosensitivity: a family of pigments called **rhodopsins**.

Rhodopsins are responsible for photosensitivity

Photosensitivity depends on the ability of rhodopsins to absorb photons of light and to undergo a change in conformation. A rhodopsin molecule consists of a protein, **opsin** (which alone is not photosensitive), and an associated nonprotein light-absorbing group, **11-cis-retinal**, cradled in the center of the opsin and bound covalently to it. The entire rhodopsin molecule sits within the plasma membrane of a photoreceptor cell.

When 11-cis-retinal absorbs a photon of light energy, it changes into a different isomer of retinal, called all-trans-retinal. This change puts a strain on the bonds between retinal and opsin, changing the conformation of opsin. This change signals the detection of light. In vertebrate eyes, the retinal and the opsin eventually separate from each other—a process called bleaching, which causes the molecule to lose its photosensitivity. A series of enzymatic reactions is then required to return the all-trans-retinal to the 11-cis isomer, which then recombines with opsin so that it once again becomes the photosensitive pigment rhodopsin (**FIGURE 35.12**).

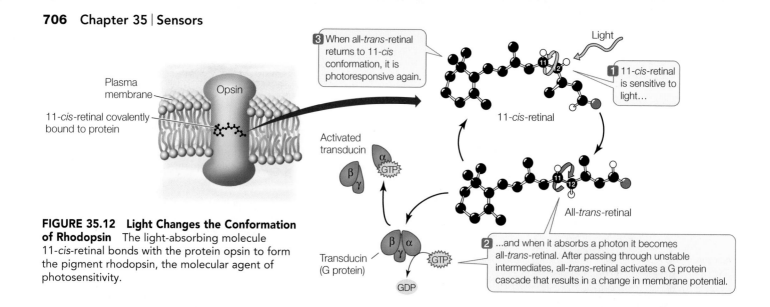

FIGURE 35.12 Light Changes the Conformation of Rhodopsin The light-absorbing molecule 11-*cis*-retinal bonds with the protein opsin to form the pigment rhodopsin, the molecular agent of photosensitivity.

Plasma membrane

11-*cis*-retinal covalently bound to protein

Opsin

3 When all-*trans*-retinal returns to 11-*cis* conformation, it is photoresponsive again.

Light

1 11-*cis*-retinal is sensitive to light…

11-*cis*-retinal

Activated transducin

All-*trans*-retinal

2 …and when it absorbs a photon it becomes all-*trans*-retinal. After passing through unstable intermediates, all-*trans*-retinal activates a G protein cascade that results in a change in membrane potential.

Transducin (G protein)

GDP

INVESTIGATION

FIGURE 35.13 A Rod Cell Responds to Light The plasma membrane of a rod cell hyperpolarizes—becomes more negative—in response to a flash of light. Rod cells do not fire action potentials, but in response to the absorption of light energy, the neuron experiences a change in membrane potential.

HYPOTHESIS

When a rod cell absorbs photons (light energy), its membrane potential changes in proportion to the strength of the light stimulus.

CONCLUSION

The membrane potential of rod cells is depolarized in the dark and hyperpolarizes (becomes more negative) in response to light.

METHOD
1. Record membrane potentials from the inner segment of a rod cell.
2. Stimulate the rod cells with light flashes of varying intensity and record the results.

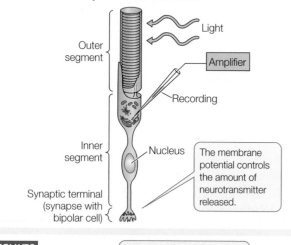

Light

Outer segment

Amplifier

Recording

Inner segment

Nucleus

The membrane potential controls the amount of neurotransmitter released.

Synaptic terminal (synapse with bipolar cell)

ANALYZE THE DATA

In related experiments, researchers measured the effect of light on the current across the rod cell membrane. The figure shows a series of recordings of membrane currents (inward currents of positive ions) in rod cells when they are illuminated with lights of varying intensities. The initial values on the graph represent the condition of the cell when it is in total darkness. The light flash is given at time 0, and the intensity of the flashes is indicated on the right side of the response curves.

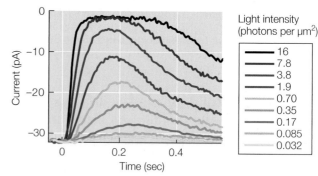

Light intensity (photons per μm^2)

| 16 |
| 7.8 |
| 3.8 |
| 1.9 |
| 0.70 |
| 0.35 |
| 0.17 |
| 0.085 |
| 0.032 |

A. If instead of an inward current, you measured membrane potential, how would the resulting recordings differ from this one?

B. Why is there no difference between the currents induced by flashes of light at 7.8 and 16 photons per square micrometer?

C. Why does a rod maintain its minimum current for longer in response to a flash of light at 16 photons per square micrometer than it does to one at 7.8 photons per square micrometer?

RESULTS

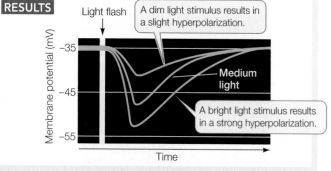

Light flash

A dim light stimulus results in a slight hyperpolarization.

Medium light

A bright light stimulus results in a strong hyperpolarization.

For more, go to Working with Data 35.1 at **yourBioPortal.com**.

Go to **yourBioPortal.com** for original citations, discussions, and relevant links for all INVESTIGATION figures.

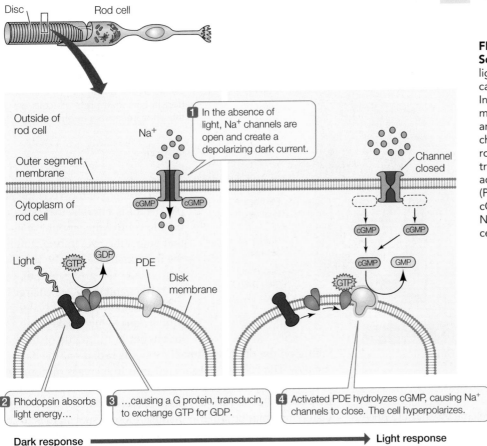

FIGURE 35.14 Light Absorption Closes Sodium Channels The absorption of light by rhodopsin initiates a signaling cascade that hyperpolarizes the rod cell. In the dark, Na⁺ channels in the plasma membrane of the rod cell's outer segment are held open by cGMP, allowing positive charges to enter the cell (step 1). When the rod cell is stimulated by light, it activates transducin (steps 2 and 3). Transducin activates a molecule of phosphodiesterase (PDE), which catalyzes the breakdown of cGMP to GMP and results in closure of the Na⁺ channels and hyperpolarization of the cell (step 4).

Figure labels:
- Disc / Rod cell
- Outside of rod cell
- Na⁺
- Outer segment membrane
- Cytoplasm of rod cell
- Channel closed
- cGMP
- cGMP / GMP
- Light / GTP / GDP / PDE
- Disk membrane
- GTP
- **1** In the absence of light, Na⁺ channels are open and create a depolarizing dark current.
- **2** Rhodopsin absorbs light energy...
- **3** ...causing a G protein, transducin, to exchange GTP for GDP.
- **4** Activated PDE hydrolyzes cGMP, causing Na⁺ channels to close. The cell hyperpolarizes.
- Dark response → Light response

How does the conformational change of rhodopsin transduce light into a cellular response? After retinal is converted from the 11-*cis* to the all-*trans* form, it triggers a cascade of reactions involving a G protein signaling mechanism that results in the alteration of membrane potential. As an example, let's look at one type of vertebrate photoreceptor cell, the rod cell.

Rod cells respond to light

The **rod cell**, named for its shape, is a modified neuron that has an outer segment, an inner segment, and a synaptic terminal (**FIGURE 35.13**). The outer segment contains a stack of discs of plasma membrane densely packed with rhodopsin. The function of the discs is to capture photons of light passing through the rod cell. The inner segment contains the cell nucleus, mitochondria, and other organelles. Rod cells do not generate their own action potentials, but they can alter how much neurotransmitter they release at the synaptic terminal, where they synapse with other neurons.

From what we have learned about other types of sensory receptors, we might expect that stimulation of the rod cell by light would make its membrane potential *less* negative (i.e., depolarized). But the opposite is true—it becomes *more* negative (hyperpolarized).

When a rod cell is kept in the dark, Na⁺ continually enters the outer segment of the cell—a flow of Na⁺ ions called the *dark current* (**FIGURE 35.14, STEP 1**). This keeps the rod cell depolarized, and it continually releases large amounts of neurotransmitter. When light is flashed on the dark-adapted rod cell, its membrane potential becomes more negative—that is, it

hyperpolarizes—and its release of neurotransmitter decreases.

How does light cause hyperpolarization? Light photoexcites rhodopsin, which activates a G protein called *transducin* (**FIGURE 35.14, STEPS 2 AND 3**; see Figure 5.14). Activated transducin in turn activates a phosphodiesterase (PDE; see Figure 5.18). Activated PDE converts cyclic GMP (cGMP) to GMP, which causes the Na⁺ channels to close, and the cell hyperpolarizes (**FIGURE 35.14, STEP 4**).

This mechanism may seem like a roundabout way of doing business, but its advantage is its enormous amplification ability. Each molecule of photoexcited rhodopsin can activate several hundred transducin molecules, thus activating a large number of PDE molecules, each of which can hydrolyze several hundred molecules of cGMP per second. The bottom line is that a single photon of light can cause a huge number of Na⁺ channels to close.

FRONTIERS Algae have a form of rhodopsin (called channel rhodopsin) that opens a cation channel when it is stimulated by blue light. Neurobiologists use channel rhodopsin in a new technique called optogenetics. A channel rhodopsin gene is combined with the promotor for a gene—such as a gene encoding a neurotransmitter—that is uniquely expressed in a particular class of neurons. This genetic construct is then transferred into a target tissue, such as a region of the brain. The channel rhodopsin is expressed only in the cells that normally activate the promotor used in the construct. Those cells can then be excited by blue light, giving the scientists the ability to turn specific types of neurons on and off at will.

Animals have a variety of visual systems

Photoreceptors using rhodopsin are incorporated into a variety of visual systems, from simple to complex. One of the simplest systems is the paired bilateral **eye cups** of flatworms. Eye cups cannot produce detailed images, but the arrangement of

(A)

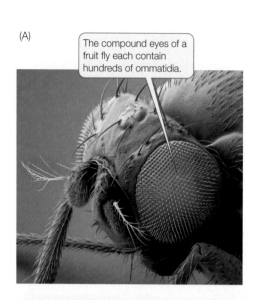

The compound eyes of a fruit fly each contain hundreds of ommatidia.

(B)

Light

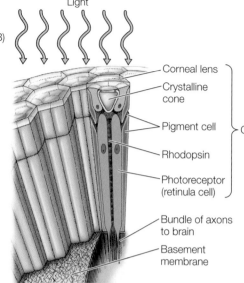

Corneal lens

Crystalline cone

Pigment cell

Rhodopsin

Photoreceptor (retinula cell)

Ommatidium

Bundle of axons to brain

Basement membrane

FIGURE 35.15 Ommatidia: The Functional Units of Insect Eyes (A) The micrograph shows the compound eye of a fruit fly. (B) The rhodopsin-containing retinula cells are the photoreceptors in ommatidia.

yourBioPortal.com

Go to INTERACTIVE TUTORIAL 35.2
Visual Receptive Fields

pigments and photoreceptors in the eyes means that the photoreceptors on the two sides of the animal are unequally stimulated unless the animal is facing directly toward or away from a light source. The resulting directional information enables the flatworm to move away from light.

Arthropods have elaborate **compound eyes**, so called because each eye consists of many optical units called **ommatidia** (singular *ommatidium*), each with its own narrow-angle lens that directs light onto photoreceptor (retinula) cells (**FIGURE 35.15**). The inner borders of the photoreceptors are covered with microvilli (tiny hairlike projections) that contain rhodopsin and trap light. Axons from the photoreceptors send the light information to the nervous system.

Each ommatidium of a compound eye is directed at a slightly different part of the visual world. The more ommatidia in the eye, the higher the resolution of the image compiled by the CNS, rather like the pixel density in a computer monitor. The number of ommatidia in a compound eye varies from only a few in some ants to 800 in fruit flies and to 30,000 in some dragonflies.

Visual information is processed by the retina and the brain

Vertebrates have *image-forming eyes*. Like a camera, an image-forming eye has a lens that focuses images on an internal surface that is sensitive to light. (Cephalopod mollusks, including squid and octopus, also have image-forming eyes, which evolved independently from the vertebrate eye.)

The vertebrate eye—exemplified by the human eye illustrated in **FIGURE 35.16A**—is a fluid-filled structure bounded by a tough connective tissue layer called the *sclera*. At the front of the eye, the sclera forms the transparent **cornea**, through which light enters the eye. Just inside the cornea is the colored **iris**. The

central opening of the iris is the **pupil**. The iris is a muscle that can contract to control the amount of light that enters the eye. In bright light, the iris constricts and the pupil is very small. As light levels fall, the iris relaxes and the pupil enlarges.

Behind the iris is the crystalline protein **lens**, which makes fine adjustments in the focus of images falling on the photosensitive layer—the **retina**—at the back of the eye. The retina is organized into five layers of neurons, including the photoreceptors (rods and cones) at the back (**FIGURE 35.16B**). The various cells of the retina receive and process visual information, then transduce these visual signals to the brain. The layer of photoreceptor cells sends information to a layer of *bipolar cells*, which sends information to a layer of *ganglion cells*. The axons of ganglion cells conduct information to the brain.

The other two cell layers, the *horizontal cells* and *amacrine cells*, communicate laterally across the retina, so that photoreceptors and bipolar cells can communicate with their neighbors. Horizontal cells help sharpen the perception of contrast between light and dark. Amacrine cells have several functions, including detecting motion and adjusting visual sensitivity to the overall level of light. Ultimately, all of this information converges onto the ganglion cells.

The function of these retinal cross-connections was explored in a series of classic experiments by Stephen Kuffler in 1953. Kuffler used electrodes to record the activity in the axons of single ganglion cells of cat eyes. His experiments revealed that each ganglion cell has a well-defined circular **receptive field** composed of a group of photoreceptors. Individual photoreceptors contribute to the receptive fields of multiple ganglion cells, so receptive fields overlap. Some ganglion cells are excited by light falling on the center of the receptive field and are inhibited if light falls in the periphery (surround) of the receptive field. Other ganglion cells are the reverse; they are inhibited by light falling in the center and are stimulated by light falling on the surround. The result of this complex circuitry is that ganglion cells can communicate information to the brain about simple visual patterns such as spots, edges and areas of contrast.

What further processing happens in the brain itself? In the 1960s, another series of classic experiments by David Hubel and Torsten Wiesel demonstrated that neurons in the visual cortex, like retinal ganglion cells, have receptive fields. These neurons each receive input from multiple ganglion cells, in

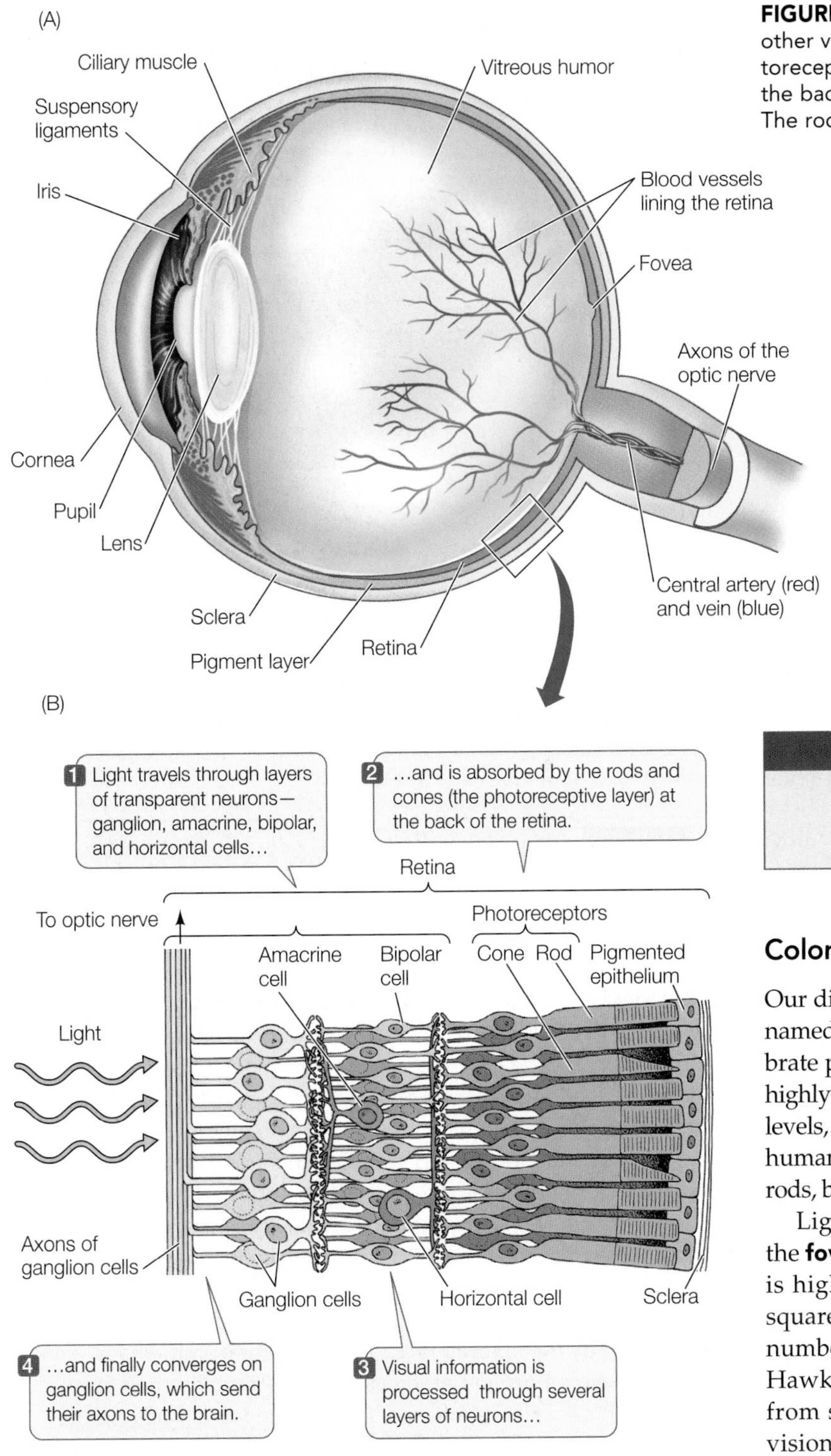

(A)

FIGURE 35.16 The Human Eye (A) The human eye, like that of other vertebrates, has a lens that focuses images on layers of photoreceptor cells. (B) The retina consists of five layers of neurons at the back of the eye that receive and process visual information. The rods and cones are photoreceptors.

(B)

analyzing edges in patterns of light and dark. Each retina sends a million axons to the brain, but there are *hundreds of millions* of neurons in the visual cortex. The action potentials from one retinal ganglion cell are received by hundreds of cortical neurons, each responsive to a different combination of orientation, position, color, and movement of contrasting lines in the patterns of light and dark falling on the retina. The end result is that before visual information is sent on to different parts of the brain (for example, the parts that can recognize faces, written words, or the shapes of objects), the information has already been highly analyzed by both the retina and the visual cortex to highlight patterns such as spots, lines, and certain shapes and motions.

---

yourBioPortal.com

Go to WEB ACTIVITIES 35.2 and 35.3
Structure of the Human Eye
Structure of the Human Retina

Color vision is due to cone cells

Our discussion so far has focused on rod cells. **Cone cells**, also named for their shape, constitute the other major class of vertebrate photoreceptors (**FIGURE 35.17A**). Whereas rod cells are highly sensitive to light and are responsible for vision at low light levels, cone cells are responsible for high-acuity color vision. The human retina has about 5 million cones and about 100 million rods, but their densities are not even across the entire retina.

Light coming from the center of the visual field falls on the **fovea** (see Figure 35.16A), where the density of cone cells is highest. The human fovea has about 160,000 cones per square millimeter. The fovea of a hawk has almost twice that number, making the hawk's vision much sharper than ours. Hawks also have *two* foveae in each eye. One receives light from straight ahead, the other from a more lateral field of vision. The hawk's forward-looking foveae make binocular vision possible, while the lateral-looking foveae provide high-acuity vision that is directed toward the sides.

Cones have low sensitivity to light and contribute little to night vision. Night vision depends mostly on rod cells; vision in dim light is mostly in shades of gray, and acuity is low. You may have trouble seeing a small object such as a keyhole at night when you are looking straight at it—that is, when its image is falling on your fovea. If you look a little to the side, so that the image falls on a rod-rich area of the retina, you see the object better (astronomers looking for faint objects in the night sky learned this trick a long time ago).

such a way that they are maximally stimulated by bars of light with a particular orientation. Cortical receptive fields from the center of the retina are small and contribute to high-acuity vision, while receptive fields from the periphery of the retina are large and provide low-acuity vision. Also, most cells in the primary visual cortex respond well to motion, and many are directionally selective. Hubel and Wiesel received a Nobel Prize for their work in 1981.

The concept that emerges from these experiments is that the brain assembles a mental image of the visual world by

(A)

(B)

FIGURE 35.17 Rods and Cones (A) This scanning electron micrograph of photoreceptors in the retina of a mud puppy (an amphibian) shows cylindrical rods and tapered cones. This micrograph has been artificially colored and does not show the actual color of the rod and cone cells. (B) The three types of cone cells in the human retina can detect different wavelengths of light. The dashed line shows the spectral sensitivity of rod cells.

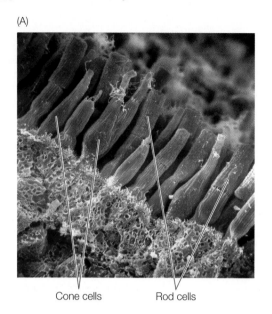

Cone cells   Rod cells

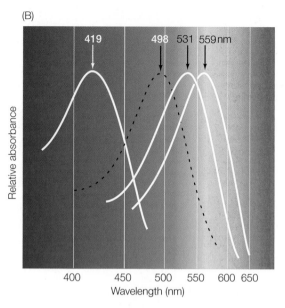

The human retina has three kinds of cone cells, each containing slightly different opsin molecules. Although all opsins contain the same 11-*cis*-retinal group, the three cone opsins differ in the wavelengths of light they absorb best (**FIGURE 35.17B**), because of molecular interactions between the retinal and the opsin. Therefore the brain can interpret the relative inputs from the three classes of cones to determine what wavelength—that is, what color—is being viewed.

Most mammals have two classes of cones, whereas most other vertebrates have at least four. Mammals tend to be nocturnal, and it appears that during their long evolutionary history of nocturnal living, they have lost much of the color vision capabilities found in other vertebrates.

In humans and a few other primates, a third opsin has re-evolved, by means of gene duplication of one of the opsin genes. This duplicated gene is nonfunctional in a substantial number of human males (about 10 percent of men of European descent), resulting in an inability to distinguish red from green. The gene in question is located on the X chromosome, and the condition, known as red–green color blindness, is inherited as a sex-linked trait.

Do You Understand Concept 35.4?

- Would you expect a nocturnal animal such as a flying squirrel to have a high or low density of cones in its retinas? How about a diurnal animal such as a gray squirrel? Explain.

- How do opsins produce light-sensitive responses when opsin molecules themselves are not affected directly by light?

- If you record action potentials from single retinal ganglion cells, they only respond optimally to circular spots of light. However, if you record from neurons in the visual cortex, they can respond optimally to bars of light in specific orientations. Explain how information coming to the cortex from the retinal ganglion cells could result in a linear receptive field.

We have seen how sensory receptors detect a wide variety of sensory information and transmit it to the central nervous system. In response, the CNS analyzes the sensory information and sends out an appropriate command to a muscle, gland, or other *effector cell*. In the next chapter, we will discuss how these effectors work.

> **QA QUESTION** How do kangaroo rats, rattlesnakes, owls, bats, and moths "see in the dark"?

ANSWER Kangaroo rats have large eyes that give them maximal visual sensitivity even in very dim light, but that is not the only sense that enables them to function on moonless nights. Their long whiskers enable them to use tactile information to explore their environments. Their olfactory sensitivity enables them to locate and identify other kangaroo rats, both as to the identity of familiar individuals and the gender of unfamiliar ones. The acute auditory sensitivity of the kangaroo rat enables it to hear the air disturbance caused by the strike of the rattlesnake and to jump reflexively.

The rattlesnake can "see" the kangaroo rat and other small mammals by the infrared radiation (heat) they emit. These snakes have pit organs lateral to their nostrils, and their eyes have high densities of infrared-sensitive neurons. Because the pits have overlapping fields of "view," they give the snake a three-dimensional perspective. Information from the pit organs goes to the same region of the brain as information from the eyes, so rattlesnakes actually do see the world in infrared. Their prey literally glows in the dark.

Bats detect their flying prey with their voices and ears. Bats emit remarkably loud, brief sound pulses at higher sound frequencies than we can hear. Between the pulses, the bat listens for echoes of these sounds bouncing off objects in the environment. The bat's brain analyzes the delays of the echoes with such accuracy that it gives the bat information about the distance to the moth, the moth's trajectory, and even the moth's size, shape, and wingbeat frequency. The moth, of course, can also hear the bat's sound pulses and can respond with a reflexive avoidance behavior that sometimes allows it to escape.

Our senses are our windows on the world, but we sense only a limited range of the information available. Animals with different ranges of sensitivity perceive the world quite differently.

35 SUMMARY

concept 35.1 Sensory Systems Convert Stimuli into Action Potentials

- Sensory receptor cells, also known as sensors or receptors, transduce information about an animal's external and internal environment into action potentials.

- Receptor potentials can spread to regions of the cell's plasma membrane that generate action potentials. Some sensors do not fire action potentials but release neurotransmitter onto sensory neurons that do fire action potentials. **Review Figure 35.1**

- Sensory receptor cells have **receptor proteins** that cause ion channels to open or close, affecting the cells' membrane potential. This causes a graded membrane potential called a receptor potential. **Review Figure 35.2**

- The interpretation of action potentials as particular sensations depends on which neurons in the CNS receive them.

- **Adaptation** enables the nervous system to ignore irrelevant or continuous stimuli while remaining responsive to relevant or new stimuli.

concept 35.2 Chemoreceptors Detect Specific Molecules or Ions

- **Chemoreceptors** are responsible for **olfaction**, **gustation**, and the sensing of **pheromones**.

- Mammalian olfactory sensors project directly to the olfactory bulb of the brain. Sensors for the same **odorant** project to the same area of the olfactory bulb.

- Each olfactory receptor cell expresses one receptor protein that can bind to a set of similar odorant molecules. **Review Figure 35.3**

- Pheromones are detected with chemoreceptors that are usually separate from the main olfactory system. In vertebrates, pheromones are detected with the **vomeronasal organ**. **Review Figure 35.4**

- In vertebrates, **taste buds** in the mouth cavity are responsible for gustation. The five basic tastes are sweet, salty, sour, bitter, and umami. **Review Figure 35.5**

concept 35.3 Mechanoreceptors Detect Physical Forces

- The skin contains a variety of **mechanoreceptors** that respond to touch and pressure. **Review Figure 35.6**

- **Stretch receptors** in muscle spindles and Golgi tendon organs inform the CNS of the loads on muscles and tendons. **Review Figure 35.7 and INTERACTIVE TUTORIAL 35.1**

- **Hair cells** are mechanoreceptors in the mammalian inner ear. Physical bending of their **stereocilia** alters their membrane potentials. Hair cells are found in the **cochlea** and **semicircular canals** and function in the auditory and vestibular systems. **Review Figures 35.8 and 35.9**

- In mammalian auditory systems, ear pinnae collect and direct sound waves into the auditory canal, where the **tympanic membrane** vibrates in response to sound waves. The movements of the tympanic membrane are amplified through a chain of **ossicles** that conduct the vibrations to the **oval window**. Movements of the oval window create pressure waves in the fluid-filled **cochlea**. **Review Figure 35.9 and WEB ACTIVITY 35.1**

- The **basilar membrane** running down the center of the cochlea is distorted by pressure waves at specific locations that depend on the frequency of the waves, generating information we interpret as the pitch of sound. These distortions cause the bending of hair cells in the **organ of Corti**. **Review Figures 35.9 and 35.10 and ANIMATED TUTORIAL 35.1**

concept 35.4 Photoreceptors Detect Light

- **Photosensitivity** depends on the absorption of photons of light by **rhodopsin**, which consists of an **opsin** protein and a nonprotein light-absorbing group called **11-cis-retinal**. Absorption of light by 11-cis-retinal leads to a change in the membrane potential of the photoreceptor cell. **Review Figure 35.12 and ANIMATED TUTORIAL 35.2**

- When excited by light, vertebrate photoreceptor cells hyperpolarize and release less neurotransmitter. They do not fire action potentials. **Review Figures 35.13 and 35.14 and WORKING WITH DATA 35.1**

- Visual systems range from the simple **eye cups** of flatworms, which sense the direction of a light source, to the **compound eyes** of arthropods, which detect shapes and patterns, to the image-forming eyes of vertebrates and cephalopods, which focus detailed images onto dense arrays of photoreceptors. **Review Figure 35.15**

- The vertebrate **retina** consists of five layers of neurons lining the back of the eye, including the light-absorbing photoreceptor cells at the back. Extensive processing of visual information occurs in the retina and also in the brain. **Review Figure 35.16, WEB ACTIVITIES 35.2 and 35.3, and INTERACTIVE TUTORIAL 35.2**

- Vertebrates have two types of photoreceptors, **rod cells** and **cone cells**. In humans, the **fovea** contains almost exclusively cone cells, which are responsible for color vision but are not very sensitive in dim light. Color vision in humans arises from three types of cone cells with different spectral absorption properties. **Review Figure 35.17**

Musculoskeletal Systems

The Olympic record for the women's long jump is 7.4 meters, set in 1988 by Jackie Joyner-Kersee. Another world-record long jump that still stands was set two years earlier by Rosie the Ribeter, who jumped 6.5 meters. Rosie was a frog competing in the Calaveras County Jumping Frog Contest. In some ways, Rosie's jump is more impressive: while Jackie's jump was about 5 times her body length (i.e., her height), Rosie's was about *20 times* her body length.

Both jumps were powered by skeletal muscle. Muscles respond to commands from the nervous system. The cellular mechanisms of muscle contraction are essentially the same in the frog and the human, and in both the muscles accomplish their tasks by pulling on bones, so why is the frog's jump so much more impressive?

Another measure of performance is endurance. It would be hard to match the endurance capacity of a camel—the ship of the desert. Carrying a load that can be almost as heavy as its body mass, the camel can go for days and cover hundreds of kilometers of desert without food or water. The British Camel Corps recommends 200 kilograms as a load for a camel, but much greater loads are routinely imposed. In one study comparing the metabolic rates of walking camels with and without loads that amounted to a third of their body weights, there was only a slight but not significant difference in the cost of locomotion. Part of the camel's endurance is due to its adaptations to conserve water, but its highly efficient mode of locomotion is also important. What adaptations make the camel such an efficient long-distance traveler?

Still another measure of performance is efficiency. The camel's remarkable endurance is partly due to its efficiency of locomotion, but the kangaroo also puts in an impressive performance. Your experience

Jackie Joyner-Kersee set an Olympic record for the women's long jump at the Seoul Olympics in 1988.

will tell you that as you run faster, the number of strides you take per minute, and therefore the energy you expend, increases. Neither is true for the kangaroo. When moving at speeds from about 5 to 25 kilometers per hour, the kangaroo takes the same number of strides per minute, and its metabolic rate does not increase. How does the kangaroo achieve such remarkable speeds at such low metabolic cost?

Adaptations of musculoskeletal systems, the subject of this chapter, underlie the exceptional but different performance capabilities of humans, frogs, camels, and kangaroos.

QUESTION How have musculoskeletal systems evolved to maximize force generation and do so at minimal metabolic cost?

KEY CONCEPTS

36.1 Cycles of Protein–Protein Interactions Cause Muscles to Contract

36.2 The Characteristics of Muscle Cells Determine Muscle Performance

36.3 Muscles Pull on Skeletal Elements to Generate Force and Cause Movement

concept 36.1 Cycles of Protein–Protein Interactions Cause Muscles to Contract

Most behavior and many physiological actions, such as beating of the heart and moving of food through the digestive tract, depend on muscle contraction. There are three types of vertebrate muscle:

- **Skeletal muscle** is responsible for all voluntary movements, such as running or playing a piano. It is also involved in some involuntary actions, such as breathing, shivering, and maintaining posture (**FIGURE 36.1**).

- **Cardiac muscle** is responsible for the beating of the heart.

- **Smooth muscle** is responsible for the movement in many hollow internal organs, such as the gut, bladder, and blood vessels, and is under the control of the autonomic (involuntary) nervous system.

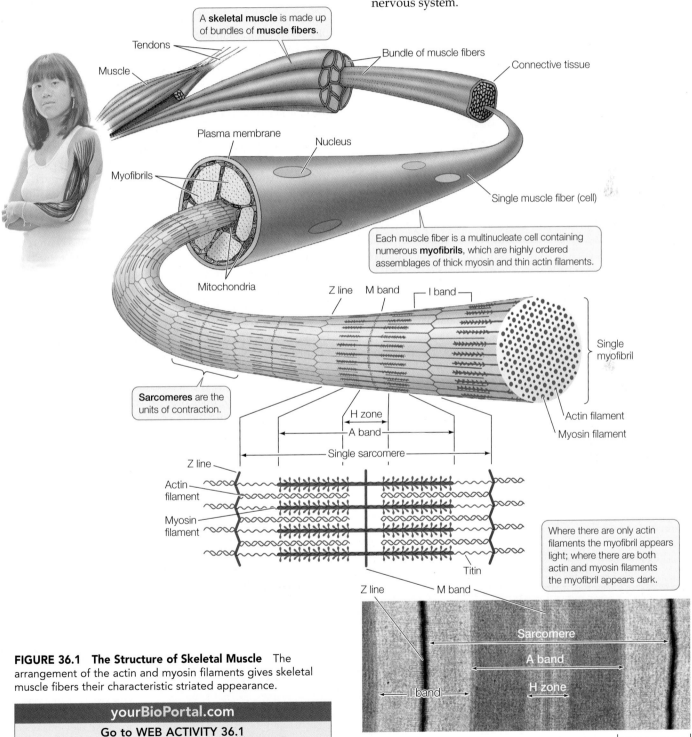

FIGURE 36.1 The Structure of Skeletal Muscle The arrangement of the actin and myosin filaments gives skeletal muscle fibers their characteristic striated appearance.

yourBioPortal.com

Go to WEB ACTIVITY 36.1
The Structure of a Sarcomere

All three muscle types use the same contractile mechanism, and we begin our study of musculoskeletal movement by describing its underlying molecular mechanisms. We will use vertebrate skeletal muscle as our primary example. Later we will discuss the differences in cardiac and smooth muscle that adapt them to their particular functions.

Sliding filaments cause skeletal muscle to contract

Skeletal muscle is also called *striated muscle* because of its striped appearance:

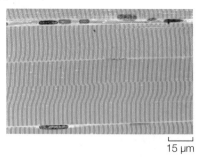

15 µm

Skeletal muscle cells, called **muscle fibers**, are large and have many nuclei. These multinucleate cells form during development through the fusion of many individual embryonic muscle cells called *myoblasts*. A muscle such as your biceps (which bends your arm) is composed of hundreds or thousands of muscle fibers bundled together by connective tissue (see Figure 36.1).

Muscle contraction is due to the interaction between the contractile proteins **actin** and **myosin**. Within muscle cells, actin and myosin molecules are organized into filaments consisting of many molecules. Actin filaments are also called *thin filaments*, and myosin filaments are *thick filaments*. The two kinds of filaments lie parallel to each other. When muscle contraction is triggered, the actin and myosin filaments slide past each other in a telescoping fashion.

> **LINK** Actin and myosin are part of the cytoskeleton, which is responsible for other types of movement in cells; see Concept 4.4

What is the relationship between a skeletal muscle fiber (cell) and the actin and myosin filaments responsible for its contraction? Each muscle fiber is packed with **myofibrils**—bundles of thin actin and thick myosin filaments arranged in an orderly fashion. In most regions of the myofibril, each thick myosin filament is surrounded by six thin actin filaments, and each thin actin filament sits within a triangle of three thick myosin filaments (see coss section of myofibril in Figure 36.1).

A longitudinal view of a myofibril shows that it consists of repeating units called **sarcomeres**. Each sarcomere is made of overlapping filaments of actin and myosin, creating a distinct banding pattern (see Figure 36.1). Before the molecular nature of the muscle banding pattern was known, the bands were given names that are still used today. Each sarcomere is bounded by *Z lines* that anchor the thin actin filaments (Z is for German *zwishen*, "between"). Centered in the sarcomere is the *A band*, which contains all the myosin filaments. The *H zone* (H is for German *hell*, "bright") and the *I band* appear light because the actin and myosin filaments do not overlap in these regions when the muscle is relaxed. The dark stripe in the H zone is called the *M band* (for middle); it contains proteins that help hold the myosin filaments in their regular arrangement.

The bundles of myosin filaments are held in a centered position within the sarcomere by a protein called **titin**. Titin is the largest protein in the body; it runs the full length of the sarcomere from Z line to Z line. Each titin molecule runs right through a myosin bundle. Between the ends of the myosin bundles and the Z lines, titin molecules are very stretchable, like bungee cords. In a relaxed skeletal muscle, resistance to stretch is mostly due to the elasticity of the titin molecules.

As the muscle contracts, the sarcomeres shorten and the band pattern changes. The H zone and the I band become much narrower, and the Z lines move toward the A band as the actin filaments slide into the H zone, the region occupied by the myosin filaments (**FIGURE 36.2**). This shortening of the sarcomeres is the basis of the *sliding filament contractile mechanism* of muscle contraction.

Muscle relaxed

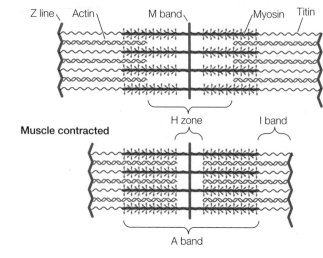

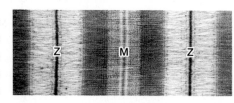

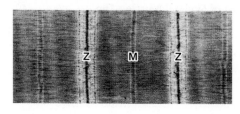

FIGURE 36.2 Sliding Filaments The banding pattern of the sarcomere changes as it shortens. This shortening results from the sliding filament contractile mechanism of muscle contraction.

FIGURE 36.3 Actin and Myosin Filaments Overlap in Myofibrils Myosin filaments are bundles of molecules with globular heads and polypeptide tails; the protein titin holds these filaments centered within the sarcomeres. Actin filaments consist of two chains of actin monomers twisted together, along with two other proteins: tropomyosin and troponin.

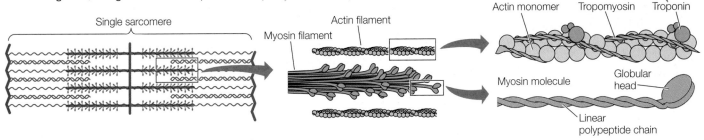

yourBioPortal.com

Go to ANIMATED TUTORIAL 36.1
Molecular Mechanisms of Muscle Contraction

Actin–myosin interactions cause filaments to slide

To understand how the sarcomere contracts, we must examine the structures of actin and myosin (**FIGURE 36.3**). A myosin molecule consists of two long polypeptide chains coiled together, each ending in a large globular head. A myosin filament is made up of many myosin molecules arranged in parallel, with their heads projecting sideways at each end of the filament.

An actin filament consists of actin monomers polymerized into long chains that look like two strands of pearls twisted together. Twisting around the actin chains is another protein, *tropomyosin*, and attached to tropomyosin at intervals are molecules of *troponin*. These molecules control the cycles of contraction and relaxation, as we will see in the next section.

The myosin heads can bind specific sites on actin, forming cross-bridges between the myosin and the actin filaments. Moreover, when a myosin head binds to an actin filament, the head's conformation changes, and it bends and exerts a tiny force that causes the actin filament to move 5–10 nanometers relative to the myosin filament (see Figure 36.6). The myosin heads also have ATPase activity; when they are bound to actin, they can bind and hydrolyze ATP. The energy released when this happens changes the conformation of the myosin head, causing it to release the actin and return to its extended position, from which it can bind to actin again.

Together, these details explain the cycle of events that cause the actin and myosin filaments to slide past each other and shorten the sarcomere. They also explain *rigor mortis*—the stiffening of muscles soon after death. ATP is needed to break the actin–myosin bonds, so when ATP production ceases with death, the actin–myosin bonds cannot be broken and the muscles stiffen. Eventually, however, the proteins begin to lose their integrity, and the muscles soften. The timing of these events helps a medical examiner estimate the time of death.

We have been discussing the cycle of contraction in terms of a single myosin head. Remember that each myosin filament has many myosin heads at both ends and is surrounded by six actin filaments; thus the contraction of the sarcomere involves a great many cycles of interaction between actin and myosin

molecules. These cycles of binding and release are not synchronized, so when a single myosin head breaks its contact with actin, other bridges are still connected and flexing, and the actin filaments do not slip backward.

Actin–myosin interactions are controlled by calcium ions

Like neurons, muscle cells are *excitable*: their plasma membranes can generate and conduct action potentials. In skeletal muscle fibers, action potentials are initiated by acetylcholine (ACh) released by motor neuron terminals at the *neuromuscular junction*. The axon terminals of motor neurons are generally highly branched and form synapses with hundreds of muscle fibers (**FIGURE 36.4**). A motor neuron and all of the fibers with which it forms synapses constitute a **motor unit**. The fibers of a motor unit contract simultaneously when its motor neuron fires. A muscle can consist of many motor units. Thus there are two ways to increase a muscle's strength of contraction: increase the firing rate of an individual motor neuron, or activate more motor neurons.

As described in Concept 34.3, when an action potential arrives at a neuromuscular junction, the neurotransmitter ACh is released from the motor neuron terminals, diffuses across the

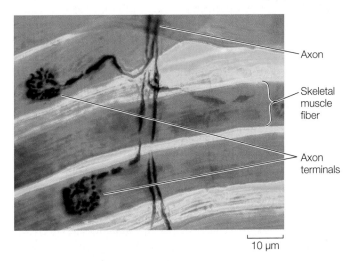

FIGURE 36.4 The Neuromuscular Junction Axons branching from a single motor neuron end in terminals that innervate multiple skeletal muscle fibers.

APPLY THE CONCEPT

Cycles of protein–protein interactions cause muscles to contract

The ability of a muscle fiber to generate force depends on how many actin–myosin cross-bridges can form. This in turn depends on the degree of overlap between the actin and myosin filaments in the sarcomeres, which can vary considerably, as shown in Figure 36.2. When a muscle is stretched from its resting length, there is less overlap between the filaments; as the muscle contracts, there is more overlap between the filaments. We can hypothesize that the maximum force a muscle can produce during a given contraction will be a function of its length at the beginning of the contraction relative to its resting length. In an experiment, a bundle of muscle fibers can be attached at either end to an apparatus that can stretch the fibers and measure the force they generate when the motor neuron to those muscle fibers is stimulated. The results are shown in the table.

Plot these results with force generated on the y axis, and length of fibers at beginning of contraction on the x axis.

1. At what fiber length is a muscle's force generating capacity greatest?

LENGTH OF FIBERS AT BEGINNING OF CONTRACTION AS PERCENTAGE OF RESTING LENGTH	FORCE GENERATED AS PERCENTAGE OF MAXIMUM FORCE
35%	10%
75%	50%
100%	100%
115%	100%
130%	80%
150%	50%
175%	10%

2. At what percentages of the resting length do you predict that the muscle will not be able to generate any force?

3. Why does the force generating capacity of the muscle decrease at shorter lengths?

4. When you do a pull-up, what stages of the pull-up are most difficult?

synaptic cleft, binds to receptors in the postsynaptic membrane, and causes ion channels in the motor end plate to open (see Figures 34.9 and 34.10). Most of the ions that flow through these channels are Na^+, and therefore the motor end plate is depolarized. The depolarization spreads to the surrounding plasma membrane of the muscle fiber, which contains voltage-gated sodium channels. When threshold is reached, the plasma membrane fires an action potential that is conducted rapidly to all points on the surface of the muscle fiber.

An action potential in a muscle fiber also travels deep within the cell because the plasma membrane is continuous with a distribution system of tubules that descend into the muscle fiber cytoplasm (also called the **sarcoplasm**). The action potential that spreads over the plasma membrane also spreads through this system of transverse tubules, or **T tubules** (**FIGURE 36.5**).

The T tubules are very close to the endoplasmic reticulum (ER) of the muscle cells. In muscle cells, the ER is called the **sarcoplasmic reticulum**, and it is a closed compartment surrounding every myofibril. Calcium pumps in the sarcoplasmic reticulum take up Ca^{2+} ions from the sarcoplasm. Therefore, when the muscle fiber is at rest, there is a higher concentration of Ca^{2+} in the sarcoplasmic reticulum and a lower concentration in the sarcoplasm.

Spanning the space between the membranes of the T tubules and the membranes of the sarcoplasmic reticulum are two proteins. One protein, the *dihydropyridine (DHP) receptor*, is located in the T tubule membrane; it is voltage-sensitive and changes its conformation when an action potential reaches it. The other protein, the *ryanodine receptor*, is located in the sarcoplasmic

reticulum membrane; it is a Ca^{2+} channel. These two proteins are physically connected. When the DHP receptor is activated by an action potential, it changes conformation, causing the ryanodine receptor to allow Ca^{2+} to leave the sarcoplasmic reticulum. Ca^{2+} ions diffuse into the sarcoplasm surrounding the actin and myosin filaments and trigger the interaction of actin and myosin and the sliding of the filaments. How do the Ca^{2+} ions do this?

An actin filament, as we have seen, is a helical arrangement of actin monomers. Twisted around the actin filament are two strands of the protein **tropomyosin** (**FIGURE 36.6**; see also Figure 36.3). At regular intervals, the filament also includes a globular protein, **troponin**. The troponin molecule has three subunits: one binds actin, one binds tropomyosin, and one binds Ca^{2+}.

When the muscle is at rest, the tropomyosin strands are positioned so that they block the sites on the actin filament where myosin heads can bind. When Ca^{2+} is released into the sarcoplasm, it binds to troponin, changing its conformation. Because the troponin is bound to the tropomyosin, this conformational change twists the tropomyosin enough to expose the actin–myosin binding sites. Thus the cycle of making and breaking actin–myosin bonds is initiated, the filaments are pulled past each other, and the muscle fiber contracts. When the calcium pumps remove the Ca^{2+} ions from the sarcoplasm, the conformation of the tropomyosin returns to the state in which it blocks the binding of myosin heads to actin, and the muscle fiber returns to its resting condition. Figure 36.6 summarizes this cycle.

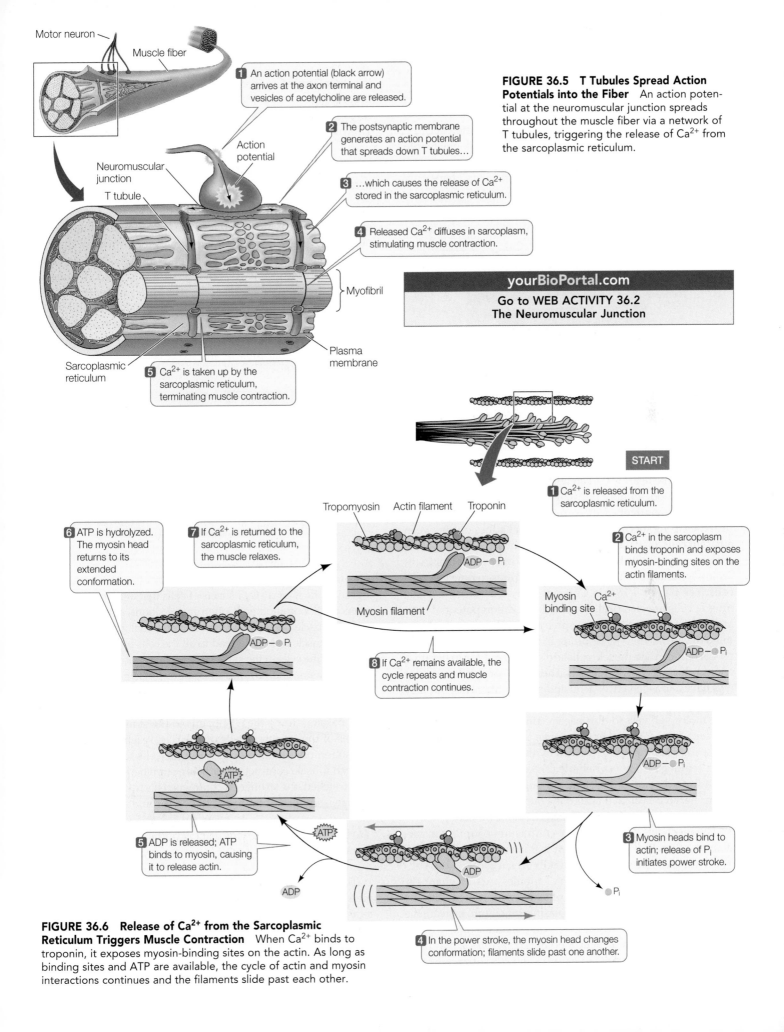

Motor neuron

Muscle fiber

1 An action potential (black arrow) arrives at the axon terminal and vesicles of acetylcholine are released.

FIGURE 36.5 T Tubules Spread Action Potentials into the Fiber An action potential at the neuromuscular junction spreads throughout the muscle fiber via a network of T tubules, triggering the release of Ca^{2+} from the sarcoplasmic reticulum.

Action potential

Neuromuscular junction

T tubule

2 The postsynaptic membrane generates an action potential that spreads down T tubules…

3 …which causes the release of Ca^{2+} stored in the sarcoplasmic reticulum.

4 Released Ca^{2+} diffuses in sarcoplasm, stimulating muscle contraction.

Myofibril

yourBioPortal.com

Go to WEB ACTIVITY 36.2 The Neuromuscular Junction

Plasma membrane

Sarcoplasmic reticulum

5 Ca^{2+} is taken up by the sarcoplasmic reticulum, terminating muscle contraction.

START

1 Ca^{2+} is released from the sarcoplasmic reticulum.

6 ATP is hydrolyzed. The myosin head returns to its extended conformation.

7 If Ca^{2+} is returned to the sarcoplasmic reticulum, the muscle relaxes.

Tropomyosin   Actin filament   Troponin

ADP—P_i

Myosin filament

2 Ca^{2+} in the sarcoplasm binds troponin and exposes myosin-binding sites on the actin filaments.

ADP—P_i

Myosin binding site Ca^{2+}

ADP—P_i

8 If Ca^{2+} remains available, the cycle repeats and muscle contraction continues.

ADP—P_i

5 ADP is released; ATP binds to myosin, causing it to release actin.

ATP

ATP

ADP

ADP

3 Myosin heads bind to actin; release of P_i initiates power stroke.

P_i

4 In the power stroke, the myosin head changes conformation; filaments slide past one another.

FIGURE 36.6   Release of Ca^{2+} from the Sarcoplasmic Reticulum Triggers Muscle Contraction When Ca^{2+} binds to troponin, it exposes myosin-binding sites on the actin. As long as binding sites and ATP are available, the cycle of actin and myosin interactions continues and the filaments slide past each other.

Cardiac muscle is similar to and different from skeletal muscle

Like skeletal muscle, cardiac muscle appears striated because of the regular arrangement of actin and myosin filaments into sarcomeres. The difference between cardiac and skeletal muscle is that cardiac muscle cells are much smaller and have only one nucleus each (*uninucleate*). Cardiac muscle cells branch, and the branches of adjoining cells interdigitate into a meshwork that is resistant to tearing:

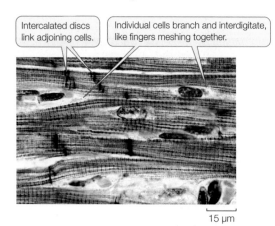

Intercalated discs link adjoining cells.

Individual cells branch and interdigitate, like fingers meshing together.

15 μm

As a result, the heart walls can withstand high pressures while pumping blood, without the danger of developing leaks. Adding to the strength of cardiac muscle are *intercalated discs* that provide strong mechanical and functional adhesions between adjacent cells. Gap junctions—protein structures that allow cytoplasmic continuity between cells—in the intercalated discs offer low-resistance pathways for ionic currents to flow between cells (see Figure 4.18 and Concept 34.3). Therefore, cardiac muscle cells are electrically coupled. An action potential initiated at one point in the heart spreads rapidly through a large mass of cardiac muscle, resulting in an integrated contraction.

Certain cardiac muscle cells are specialized for generating and conducting electrical signals. These *pacemaker* and *conducting cells* have a low density of actin and myosin filaments, but they initiate and coordinate the rhythmic contractions of the heart. (The molecular basis for this pacemaking function is covered in Concept 38.3.) Pacemaker cells make the vertebrate heartbeat *myogenic*, meaning it is generated by the heart muscle itself. A heart removed from a vertebrate can continue to beat with no input from the nervous system; although input from the autonomic nervous system modifies the *rate* of the pacemaker cells, it is not essential for their continued rhythmic function.

The mechanism of excitation–contraction coupling in cardiac muscle cells is different from that in skeletal muscle cells. The T tubules in cardiac muscle cells are larger, and the voltage-sensitive DHP proteins in the T tubules are Ca^{2+} channels. These T tubule proteins are not physically connected with the ryanodine receptors in the sarcoplasmic reticulum. Instead, the ryanodine receptors are ion-gated Ca^{2+} channels that are sensitive to Ca^{2+}. When an action potential spreads down the T tubules, it causes the voltage-gated channels to open, allowing extracellular Ca^{2+} to flow into the sarcoplasm. This slight rise in sarcoplasmic Ca^{2+} concentration opens the Ca^{2+} channels in the sarcoplasmic reticulum, which in turn causes a huge rise in sarcoplasmic Ca^{2+} concentration, resulting in fiber contraction. This mechanism is called *Ca^{2+}-induced Ca^{2+} release*.

Smooth muscle causes slow contractions of many internal organs

Smooth muscle provides the contractile force for most of our internal organs, which are under the control of the autonomic nervous system. Smooth muscle moves food through the digestive tract, controls the flow of blood through blood vessels, and empties the urinary bladder. Structurally, smooth muscle cells are the simplest muscle cells. They are smaller than skeletal muscle cells, usually long and spindle-shaped, and each has a single nucleus. They are "smooth" because the actin and myosin filaments are not as regularly arranged as they are in skeletal and cardiac muscle, and so the cells do not have a striated appearance:

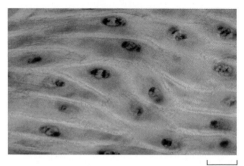

30 μm

Some smooth muscle tissue, such as that from the wall of the digestive tract, consists of sheets of cells that are all in electrical contact with one another through gap junctions, as the cells are in cardiac muscle. As a result, an action potential generated in one smooth muscle cell can spread to all the cells in the sheet of tissue. Thus the cells in the sheet contract in a coordinated fashion.

The plasma membranes of smooth muscle cells are sensitive to stretch, with important consequences. If the wall of the digestive tract is stretched in one location (as by a mouthful of food passing down the esophagus to the stomach), the membranes of the stretched cells depolarize, reach threshold, and fire action potentials, which cause the cells to contract. Thus smooth muscle contracts after being stretched, and the harder it is stretched, the stronger it contracts. This behavior of smooth muscle is important for moving food through the gut.

LINK The structure and function of the gut are detailed in Concept 39.3

The neural influences on smooth muscle come from the two divisions of the autonomic nervous system. The neurotransmitters of the sympathetic and parasympathetic postganglionic cells alter the membrane potential of smooth muscle cells. For example, in the digestive tract, acetylcholine causes smooth muscle cells to depolarize, making them more likely to fire action potentials and contract. Antagonistically, norepinephrine causes

INVESTIGATION

FIGURE 36.7 Neurotransmitters and Stretch Alter the Membrane Potential of Smooth Muscle Cells   Several factors influence the motility of the gut. Motility increases after a meal when the ingested food stretches the walls of the gut. Activity of the autonomic nervous system also controls gut motility; the sympathetic nervous system inhibits gut motility, and the parasympathetic nervous system stimulates it. The experimental setup below was used to study how stretching and autonomic neurotransmitters influence gut smooth muscle activity.

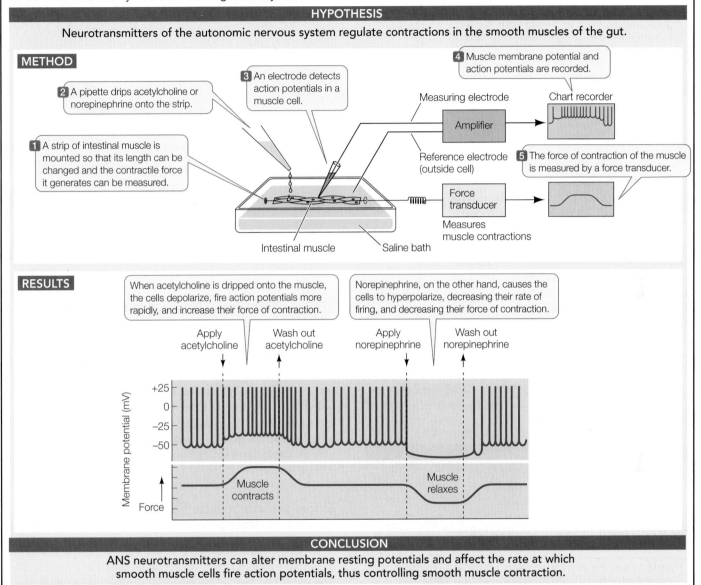

HYPOTHESIS

Neurotransmitters of the autonomic nervous system regulate contractions in the smooth muscles of the gut.

METHOD

2 A pipette drips acetylcholine or norepinephrine onto the strip.

3 An electrode detects action potentials in a muscle cell.

4 Muscle membrane potential and action potentials are recorded.

Measuring electrode

Chart recorder

Amplifier

1 A strip of intestinal muscle is mounted so that its length can be changed and the contractile force it generates can be measured.

Reference electrode (outside cell)

5 The force of contraction of the muscle is measured by a force transducer.

Force transducer

Measures muscle contractions

Intestinal muscle   Saline bath

RESULTS

When acetylcholine is dripped onto the muscle, the cells depolarize, fire action potentials more rapidly, and increase their force of contraction.

Norepinephrine, on the other hand, causes the cells to hyperpolarize, decreasing their rate of firing, and decreasing their force of contraction.

Apply acetylcholine   Wash out acetylcholine   Apply norepinephrine   Wash out norepinephrine

Membrane potential (mV): +25, 0, −25, −50

Muscle contracts   Muscle relaxes

Force

CONCLUSION

ANS neurotransmitters can alter membrane resting potentials and affect the rate at which smooth muscle cells fire action potentials, thus controlling smooth muscle contraction.

ANALYZE THE DATA

In another experiment, the muscle strip was pulled to lengthen it, and the following data were collected:

Length of strip	Membrane potential	Firing rate	Contractile force
10 mm	−50 mv	0.8 hz	5 g
20 mm	−40 mv	1.2 hz	10 g
30 mm	−35 mv	1.6 hz	15 g

A. What is the relationship between the amount the muscle was stretched and its membrane potential?
B. What is the relationship between membrane potential and firing rate?
C. How does firing rate influence the force of contraction?
D. What would happen to your stomach motility if following a large meal you were confronted by a robber?

Go to **yourBioPortal.com** for original citations, discussions, and relevant links for all INVESTIGATION figures.

these muscle cells to hyperpolarize and thus be less likely to fire action potentials and contract (**FIGURE 36.7**). In contrast, norepinephrine causes the smooth muscle in arteries serving the gut to contract. Remember that the action of the neurotransmitters depends on the receptors and response mechanisms in the target tissues.

FIGURE 36.8 The Role of Ca²⁺ in Smooth Muscle Contraction When a smooth muscle cell is stimulated by neurotransmitter, Ca²⁺ enters the sarcoplasm and binds to calmodulin, which in turn activates an enzyme that phosphorylates the myosin heads, causing them to bind to actin. As long as the myosin remains phosphorylated, actin and myosin go through cycles of binding and release. Thus in smooth muscle, the Ca²⁺-mediated change is on myosin, whereas in skeletal and cardiac muscle it is on the actin-tropomyosin filament.

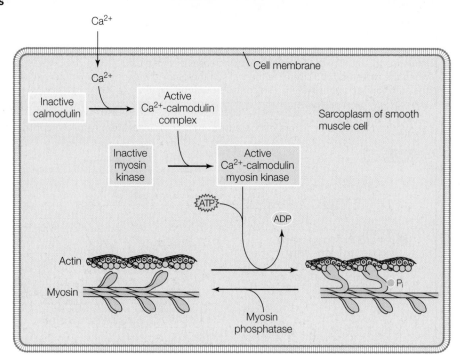

Smooth muscle cell contraction is not controlled by the troponin–tropomyosin mechanism, but calcium still plays a critical role (**FIGURE 36.8**). A Ca²⁺ influx into the sarcoplasm of a smooth muscle cell can be stimulated by neurotransmitters, action potentials spreading from neighboring cells, hormones, or stretching. The Ca²⁺ that enters the sarcoplasm combines with a protein called *calmodulin*. The Ca²⁺–calmodulin complex activates an enzyme called myosin kinase that phosphorylates myosin heads. When the myosin heads in smooth muscle are phosphorylated, they undergo cycles of binding and releasing actin, causing muscle contraction. As Ca²⁺ is removed from the sarcoplasm, it dissociates from calmodulin, and the activity of myosin kinase falls. An additional enzyme, myosin phosphatase, dephosphorylates the myosin to help reduce actin–myosin interactions.

---

yourBioPortal.com

**Go to ANIMATED TUTORIAL 36.2
Smooth Muscle Action**

Do You Understand Concept 36.1?

- Explain how the cellular and subcellular structures of skeletal muscle cause muscle contraction. What role does ATP play?

- Explain the role of Ca²⁺ in the contractile mechanism of skeletal, cardiac, and smooth muscle.

- How does an action potential arriving at the neuromuscular junction cause the muscle to contract?

- A calcium ionophore is a chemical that allows calcium ions to cross membranes freely, without having to pass through a membrane channel. What do you think would happen in the experiment shown in Figure 36.7 if a calcium ionophore were added to the saline bath?

Now that we understand how the components of muscles generate force, let's look at what determines the characteristics of a muscle as a whole, and how individual muscles can change their characteristics with regular use and conditioning.

**concept
36.2** **The Characteristics of Muscle Cells Determine Muscle Performance**

The functions that different muscles perform place different demands on them. Some muscles, such as postural muscles, must sustain a load continuously over long periods of time. Other muscles, such as those that control your fingers, generally do not have to sustain long contractions, but they must be able to contract quickly. And, of course, muscles have to be able to contract in a graded manner to adjust to different loads placed on them.

Single skeletal muscle fibers can generate graded contractions

In skeletal muscle, the arrival of an action potential at a neuromuscular junction causes an action potential in a muscle fiber. The spread of that action potential through the muscle fiber's T tubule system causes a minimum unit of contraction, called a **twitch**. A twitch can be measured in terms of the *tension*, or force, it generates (**FIGURE 36.9A**). A single action potential stimulates a single twitch, but the ultimate force generated by an action potential can vary enormously depending on how many muscle fibers are in the motor unit it innervates. The level of tension an entire muscle generates depends on two factors:

- the number of motor units activated, and

- the frequency at which the motor units fire.

In muscles responsible for fine movements, such as those of the fingers, a motor neuron may innervate only one or a few muscle fibers, but in a muscle that produces large forces, such as the biceps, a motor neuron innervates a large number of muscle fibers.

At the level of a muscle fiber, a single action potential stimulates a single twitch. If action potentials reaching the muscle

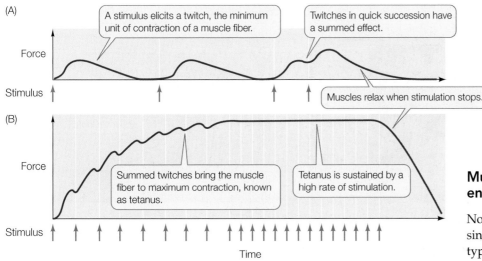

FIGURE 36.9 Twitches and Tetanus (A) Action potentials from a motor neuron cause a muscle fiber to twitch. Twitches in quick succession can be summed. (B) Summation of many twitches can bring the muscle fiber to tetanus.

fiber are adequately separated in time, each twitch is a discrete, all-or-none phenomenon. If action potentials are fired more rapidly, however, new twitches are triggered before the myofibrils have a chance to return to their resting condition. As a result, the twitches sum, and the tension generated by the fiber increases and becomes more sustained. Thus an individual muscle fiber can show a graded response to increased levels of stimulation by its motor neuron.

Twitches sum at high levels of stimulation because the calcium pumps in the sarcoplasmic reticulum are not able to clear the Ca^{2+} ions from the sarcoplasm between action potentials. Eventually a stimulation frequency can be reached that results in continuous presence of Ca^{2+} in the sarcoplasm at high enough levels to cause continuous activation of the contractile machinery—a condition known as **tetanus** (**FIGURE 36.9B**). (Do not confuse this condition with the disease called tetanus, which is caused by a bacterial toxin and is characterized by spastic contractions of skeletal muscles.)

How long a muscle fiber can maintain a tetanic contraction depends on its supply of ATP. Eventually the fiber will become fatigued. It may seem paradoxical that the *lack* of ATP causes fatigue, since the action of ATP is to break actin–myosin bonds. But remember that the energy released from the hydrolysis of ATP "re-cocks" the myosin heads, allowing them to cycle through another power stroke. When a muscle is contracting against a load, the cycle of making and breaking actin–myosin bonds must continue, to prevent the load from stretching the muscle. The situation is like rowing a boat upstream: you cannot maintain your position relative to the stream bank by just holding the oars out against the current; you have to keep rowing. Likewise, actin–myosin bonds have to keep cycling to maintain tension in the muscle.

Many muscles of the body maintain a low level of tension even when the body is at rest. For example, the muscles of the neck, trunk, and limbs that maintain our posture against the pull of gravity are always working, even when we are standing or sitting still. *Muscle tone* comes from the activity of a small but changing number of motor units in a muscle; at any one time, some of the muscle's fibers are contracting and others are relaxed. The nervous system is constantly readjusting muscle tone.

Muscle fiber types determine endurance and strength

Not all skeletal muscle fibers are alike, and a single muscle often contains more than one type of fiber. The two major types of skeletal muscle fibers express different genes for different myosin molecules, and these myosin variants have different rates of ATPase activity. Those with high ATPase activity can recycle their actin–myosin cross-bridges rapidly and are therefore called fast-twitch fibers. Slow-twitch fibers have lower ATPase activity; they develop tension more slowly but can maintain it longer.

Slow-twitch fibers are also called *oxidative* or *red muscle* because they contain the oxygen-binding protein *myoglobin*, have many mitochondria, and are well supplied with blood vessels. These characteristics both increase the fibers' capacity for oxidative metabolism and result in their red appearance. The maximum tension a slow-twitch fiber produces is low and develops slowly but is highly resistant to fatigue. Slow-twitch fibers have substantial reserves of fuel (glycogen and fat), so they can maintain steady, prolonged production of ATP as long as oxygen is available. Muscles with high proportions of slow-twitch fibers are good for long-term *aerobic* work (that is, work that requires oxygen). Long-distance runners, swimmers, cyclists, and other athletes whose activities require endurance have leg and arm muscles consisting mostly of slow-twitch fibers (**FIGURE 36.10**).

Some **fast-twitch fibers** are also called *glycolytic* or *white muscle* because, compared with slow-twitch fibers, they have few mitochondria, little or no myoglobin, and fewer blood vessels; thus they look pale. Fast-twitch glycolytic fibers can develop maximum tension more rapidly than slow-twitch fibers can, and that maximum tension is greater. However, fast-twitch fibers fatigue rapidly. The myosin of these fibers puts the energy of ATP to work very rapidly, but the fibers cannot replenish ATP quickly enough to sustain contraction for a long time. Fast-twitch fibers are especially good for short-term work that requires maximum strength. Weight lifters and sprinters have leg and arm muscles with high proportions of fast-twitch fibers. Rosie the Ribbeter in the opener to this chapter was using fast-twitch fibers in her jump. In fish, dark trunk muscle (slow-twitch) is used for steady, continuous swimming, whereas fast-twitch muscle is used for rapid spurts of activity such as escape or prey capture. When rapid contractions are necessary, as in hummingbird flight muscles or in sound generation by insects, fast-twitch muscles are involved.

(A) Cross sections of leg muscles

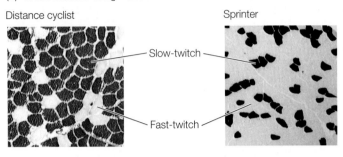

(B)

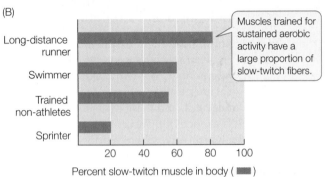

Muscles trained for sustained aerobic activity have a large proportion of slow-twitch fibers.

Percent slow-twitch muscle in body (■)

FIGURE 36.10 Slow- and Fast-twitch Muscle Fibers (A) The skeletal muscles in the micrographs were stained with a reagent that shows slow-twitch fibers as dark and fast-twitch fibers as light. (B) Athletes in different sports show different percentages of slow-twitch muscles.

Muscle ATP supply limits performance

Muscles have three systems for supplying the ATP they need for contraction:

- The *immediate system* uses preformed ATP and a storage compound called *creatine phosphate* (*CP*).
- The *glycolytic system* metabolizes carbohydrates to lactate and pyruvate.
- The *oxidative system* metabolizes carbohydrates or fats all the way to H_2O and CO_2.

The capacity of these three systems and the rates at which they can produce ATP determine both work capacity and endurance (**FIGURE 36.11**).

ATP is present in muscles in very small amounts. However, muscle fibers also contain CP. This molecule stores energy in a phosphate bond, which it can transfer to ADP:

$$CP + ADP \rightarrow creatine + ATP$$

The total energy available in all the muscles of your body in the form of ATP and CP—the immediate energy system—is only about 10 kilocalories. When at rest, you metabolize 1 kilocalorie of energy in less than a minute. Even though the energy available from ATP and CP is limited, it is available immediately, and it enables fast-twitch fibers to generate a lot of force quickly. During burst activity, the immediate system is exhausted in seconds, but that is enough for a 100-meter sprint.

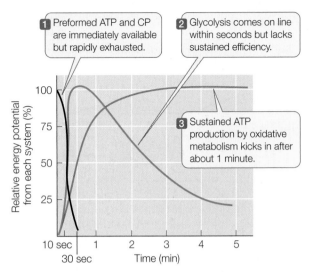

FIGURE 36.11 Supplying Fuel for High Performance Muscles have three systems for obtaining the ATP they need for contraction during exertion such as running.

The glycolytic system activates within a few seconds to replace the ATP depleted at the onset of muscle activity. The glycolytic enzymes are located in the cytoplasm of the muscle fiber, and therefore the ATP they generate is rapidly available to the myosin filaments. However, as noted in Chapter 6, glycolysis alone is an inefficient way to produce ATP, and it leads to the accumulation of lactic acid, which slows the process. Thus, the glycolytic system and the immediate system together can provide most of the energy for active muscles for less than a minute (see Figure 36.11).

Oxidative metabolism (which involves *oxidative phosphorylation*) becomes fully active in about a minute, producing relatively huge amounts of ATP because it can completely metabolize carbohydrates and fats. However, it requires many reactions (see Chapter 6), and it takes place in the mitochondria, so O_2 and substrate must diffuse into the mitochondria, and the formed ATP must diffuse from the mitochondria to the myosin filaments in the muscle. These processes are not instantaneous, so the rate at which oxidative metabolism can make ATP available to do work is slower than the rate at which the other two systems can supply ATP.

Do You Understand Concept 36.2?

- How do muscle fibers generate graded forces?
- Describe the differences between slow-twitch and fast-twitch fibers.
- It has recently been reported that pyruvate kinase, the enzyme that catalyzes the final step of glycolysis, is temperature sensitive. It is inactivated at high temperatures that can occur in muscles during heavy exercise. What would be the effect of inactivation of pyruvate kinase on muscle performance?

Regardless of how much force a muscle can generate, how long it can sustain a work load, or how fast it can contract and relax, a muscle needs something to pull on; otherwise it would just be a lump of pulsating or quivering tissue. Let's look now at how skeletal systems help generate movement.

concept 36.3 Muscles Pull on Skeletal Elements to Generate Force and Cause Movement

Muscles can contract and exert force, or they can relax. To create significant movement, they must have something to pull on and something that stretches the muscle back to a longer position. In some cases, muscles pull on each other, as in the trunk of an elephant or the arms of an octopus. In most cases, however, **skeletal systems** are the rigid supports against which muscles pull to create directed movement. In this section we examine the three types of skeletal systems: hydrostatic skeletons, exoskeletons, and endoskeletons.

A hydrostatic skeleton consists of fluid in a muscular cavity

Cnidarians, annelids, and other soft-bodied invertebrates have **hydrostatic skeletons** consisting of a volume of fluid enclosed in a body cavity surrounded by muscle (see Concept 23.1). When muscles oriented in one direction contract, the fluid-filled body cavity bulges out in the opposite direction.

An earthworm uses its hydrostatic skeleton to crawl. The earthworm's body cavity is divided into many separate segments, each of which contains a compartment filled with extracellular fluid. The body wall surrounding each segment has two muscle layers: a circular layer and a longitudinal layer. Alternating contractions of the earthworm's circular and longitudinal muscles create waves of narrowing and widening, lengthening and shortening, that travel down the body (**FIGURE 36.12**). Bulging, shortened segments serve as anchors as

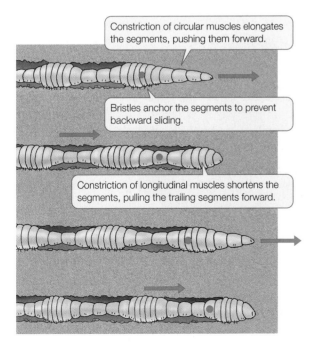

Constriction of circular muscles elongates the segments, pushing them forward.

Bristles anchor the segments to prevent backward sliding.

Constriction of longitudinal muscles shortens the segments, pulling the trailing segments forward.

long, narrow segments project forward and longitudinal contractions pull other segments forward. Bristles help the widest parts of the body hold firm against the substrate.

Exoskeletons are rigid outer structures

An *exoskeleton* is a hardened, rigid outer surface to which muscles can be attached. Contractions of the muscles cause jointed segments of the exoskeleton to move relative to each other. Exoskeletons range from the relatively simple shells of mollusks (for example, clams and snails) to the complex, jointed *cuticles* of arthropods (for example, ants, spiders, and lobsters). The arthropod cuticle contains stiffening materials everywhere except at the joints, where flexibility must be retained. Muscles attached to the inner surfaces of the arthropod exoskeleton move its parts around the joints.

LINK The diverse structures and compositions of exoskeletons are detailed in Concept 23.3

A rigid exoskeleton can provide significant protection for the soft tissues of the body. However, a drawback of the arthropod exoskeleton is that it cannot expand. Therefore, if the animal is to become larger, it must *molt*, shedding its exoskeleton and forming a new, larger one.

Vertebrate endoskeletons consist of cartilage and bone

The **endoskeleton** of vertebrates is an internal scaffolding. Muscles are attached to it and pull against it. Endoskeletons are composed of rodlike, platelike, and tubelike bones connected to one another at a variety of joints that allow a wide range of movements.

The human skeleton consists of 206 bones, some of which are shown in **FIGURE 36.13**. It can be divided into an *axial skeleton*, which includes the skull, vertebral column, sternum, and ribs; and an *appendicular skeleton*, which includes the pectoral girdle, pelvic girdle, and bones of the arms, legs, hands, and feet.

The vertebrate endoskeleton consists of two kinds of connective tissue, *cartilage* and *bone*, which are produced by two kinds of connective tissue cells. *Cartilage cells* produce an extracellular matrix that is a tough, rubbery mixture of polysaccharides and proteins—mainly fibrous collagen. Collagen fibers run in all directions like reinforcing cords through the gel-like matrix and give it the well-known strength and resiliency of "gristle." This matrix, called **cartilage**, is found in parts of the endoskeleton where both stiffness and resiliency are required, such as on the surfaces of joints where bones move against one another. Cartilage is also the supportive tissue in stiff but flexible structures such as the larynx (voice box), nose, and ear

FIGURE 36.12 A Hydrostatic Skeleton Alternating waves of muscle contraction move the earthworm through the soil. The red dot enables you to follow the changes in one segment as the worm moves forward.

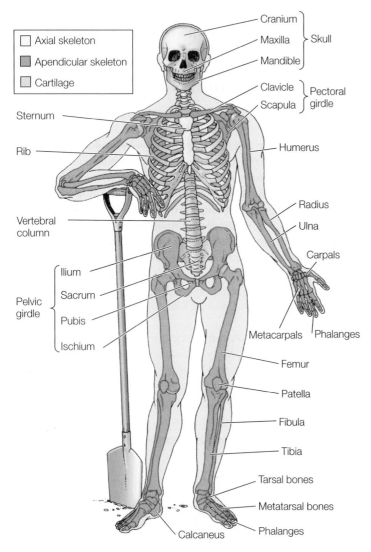

FIGURE 36.13 The Human Endoskeleton Cartilage and bone make up the internal skeleton of a human being.

pinnae. Sharks and rays are called *cartilaginous fishes* because their skeletons are composed entirely of cartilage. In most other vertebrates, cartilage is the principal component of the embryonic skeleton, but during development most of it is gradually replaced by bone.

Bone also contains collagen fibers, but it gets its rigidity and hardness from an extracellular matrix of insoluble calcium phosphate crystals. Bone serves as a reservoir of calcium for the rest of the body and is in dynamic equilibrium with soluble calcium in the extracellular fluids of the body. This equilibrium is under the control of calcitonin, vitamin D, and parathyroid hormone (see Figure 30.11). If too much calcium is taken from the skeleton, the bones are seriously weakened.

The living cells of bone—*osteoblasts, osteocytes,* and *osteoclasts*—are responsible for the constant dynamic remodeling of bone (**FIGURE 36.14**). **Osteoblasts** lay down new matrix material on bone surfaces. These cells gradually become surrounded by matrix and eventually become enclosed within the bone, at which point they cease laying down matrix but

continue to exist within small lacunae (cavities) in the bone. In this state they are called **osteocytes**. Despite the vast amounts of matrix between them, osteocytes remain in contact with one another through long cellular extensions that run through tiny channels in the bone. Communication between osteocytes is important in controlling the activities of the cells that are laying down or removing bone.

The cells that resorb bone are the **osteoclasts**. They are derived from the same cell lineage that produces white blood cells. Osteoclasts erode bone, forming cavities and tunnels. Osteoblasts follow osteoclasts, depositing new bone. Thus the interplay of osteoblasts and osteoclasts constantly replaces and remodels the bones, allowing a bone to recover from damage and adjust to the forces placed on it.

FRONTIERS When astronauts spend long periods in a zero gravity environment, their bones decalcify and weaken. Conversely, in athletes, certain bones thicken during training, but if a bone is broken and immobilized in a cast, it thins. In some way that is not understood, placing stress on bones is necessary to keep them healthy. To learn more, animal physiologists are studying black bears; their bones do not thin as much as expected during their long hibernations, and they rebuild rapidly in the spring. An understanding of the mechanisms underlying the black bear's ability to maintain bone mass could help in the development of treatments for bone thinning caused by aging or inactivity.

Bones develop from connective tissues

Bones are divided into two types on the basis of how they develop. **Membranous bone** forms on a scaffold of connective tissue membrane. **Cartilage bone** forms first as a cartilaginous structure resembling the future mature bone, then gradually hardens, or

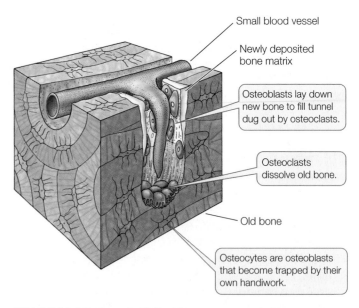

FIGURE 36.14 Bone Is Living Tissue Bones are constantly being remodeled by osteoblasts, which lay down bone, and osteoclasts, which resorb bone.

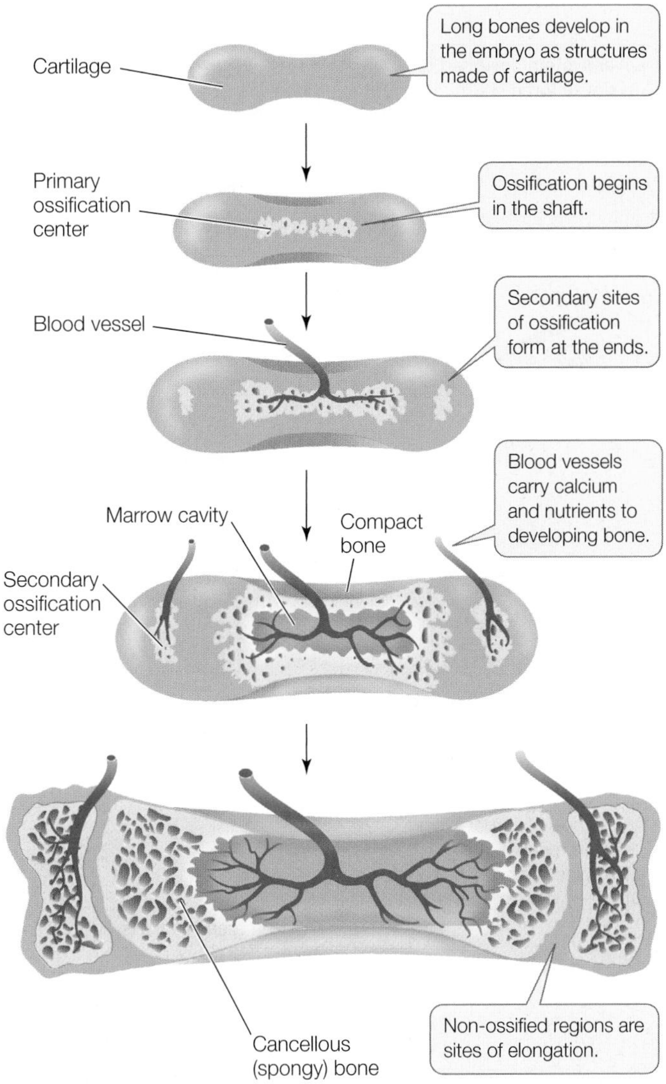

Cartilage

Long bones develop in the embryo as structures made of cartilage.

Primary ossification center

Ossification begins in the shaft.

Blood vessel

Secondary sites of ossification form at the ends.

Blood vessels carry calcium and nutrients to developing bone.

Marrow cavity

Compact bone

Secondary ossification center

Cancellous (spongy) bone

Non-ossified regions are sites of elongation.

FIGURE 36.15 The Growth of Long Bones In the long bones of human limbs, ossification occurs first at the centers and later at each end.

ossifies, to become bone. The outer bones of the skull are membranous bones; the bones of the limbs are cartilage bones.

Cartilage bones can grow throughout the ossification process. The long bones of the legs and arms, for example, ossify first at the centers and later at each end (**FIGURE 36.15**). Growth can continue until these areas of ossification join. The membranous bones forming the skull cap grow until their edges meet. The soft spot on the top of a baby's head (the fontanelle) is the point at which the skull bones have not yet joined.

The structure of bone may be **compact** (solid and hard) or **cancellous** (having numerous internal cavities that make it appear spongy, although it is rigid). Most bones have both compact and cancellous regions. The shafts of the long bones of the limbs, for example, are cylinders of compact bone surrounding central cavities that contain the bone marrow, where the cellular elements of the blood are made. The ends of the long bones are cancellous (see Figure 36.15). Cancellous bone is lightweight because of its numerous cavities, but it is also strong because its internal meshwork constitutes a support system. It

can withstand considerable forces of compression. The rigid, tubelike shaft of compact bone can withstand compression and bending forces. Architects and nature alike use hollow tubes as lightweight structural elements.

Bones that have a common joint can work as a lever

Muscles and bones work together around **joints**, where two or more bones come together. Different kinds of joints allow motion in different directions (**FIGURE 36.16**), but muscles can exert force in only one direction. Therefore, muscles create movement around joints by working in antagonistic pairs: when one muscle contracts, the other relaxes. When both contract, the joint becomes rigid (which is important for maintaining posture, for example).

With respect to a particular joint, such as the knee, we refer to the muscle that bends, or flexes, the joint as the **flexor**, and the muscle that straightens, or extends, the joint as the **extensor**. The bones that meet at the joint are held together by **ligaments**, which are flexible bands of connective tissue. Other

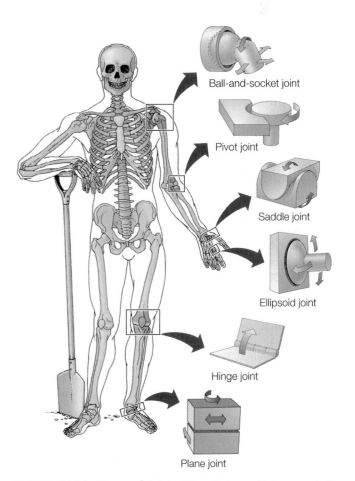

Ball-and-socket joint

Pivot joint

Saddle joint

Ellipsoid joint

Hinge joint

Plane joint

FIGURE 36.16 Types of Joints The designs of joints are similar to mechanical counterparts and enable a variety of movements.

yourBioPortal.com
Go to WEB ACTIVITY 36.3
Joints

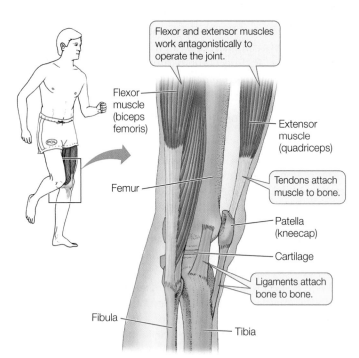

FIGURE 36.17 Joints, Ligaments, and Tendons A side view of the knee shows the interactions of muscle, bone, cartilage, ligaments, and tendons at this crucial and vulnerable human joint.

straps of connective tissue, called **tendons**, attach the muscles to the bones (**FIGURE 36.17**). In many kinds of joints, only the tendon spans the joint, sometimes moving over the surfaces of the bones like a rope over a pulley. The tendon of the quadriceps muscle traveling over the knee joint is what is tapped to elicit the knee-jerk reflex (see Figure 34.14).

Bones constitute a system of levers that are moved around joints by the muscles. A lever has an *effort arm* and a *load arm* that work around a *fulcrum* (pivot). The length ratio of the two arms determines whether a particular lever can exert a lot of force over a short distance or is better at translating force into large or fast movements. Compare the jaw and knee joints, for example (**FIGURE 36.18**). The effort arm of the jaw is long relative to the load arm, allowing the jaw to apply great force over a small distance. Think of the powerful jaws of carnivores that can easily crack bones. The effort arm of the lower leg, by contrast, is short relative to the load arm, so you can run fast, jump high, and deliver swift kicks.

Do You Understand Concept 36.3?

- How do the muscles and fluid-filled body cavity of an earthworm interact to enable the animal to crawl?

- Describe the differences between membranous and cartilaginous bone and between compact and cancellous bone.

- Wombats (groundhog-sized marsupials) are powerful digging animals, and kangaroos are powerful jumping animals. How do you think their leg structures compare in terms of their designs as lever systems?

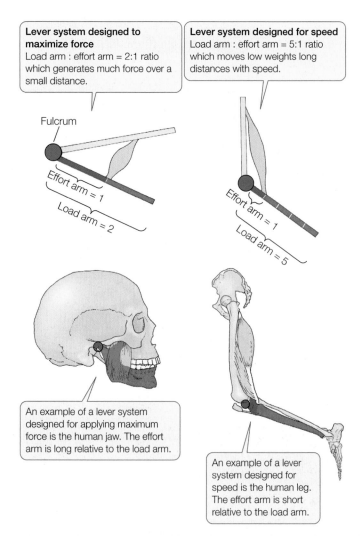

FIGURE 36.18 Bones and Joints Work Like Systems of Levers A lever system can be designed for maximizing either force or speed.

  How have musculoskeletal systems evolved to maximize force generation and do so at minimal metabolic cost?

ANSWER Presumably both Jackie and Rosie the Ribeter have leg muscles that are rich in fast-twitch fibers, so that cannot explain the difference in their jumping abilities. The answer is to be found in the last topic covered in this chapter—leverage. The ratio of leg length to body mass is much greater in the frog than in the human (**FIGURE 36.19**). Thus frog legs are better at propelling a small mass a long distance, and human legs are better at propelling a large mass a short distance.

The camel has an amazing set of adaptations for efficient locomotion. The muscles that move its long legs are actually very small, and they reside mostly within or very close to the main mass of the body. Long tendons travel down the legs and insert on the bones of the legs and feet. The leverage that results is such that very small contractions of

FIGURE 36.19 Champion Jumpers Relative to their size, many animals have more impressive jumping skills than humans. This leopard frog (*Rana pipiens*) can leap distances up to 20 times its body length.

the muscles result in large-dimension movements of the legs. You might ask what about strength to carry such heavy loads? The camel does not need a lot of strength to move its legs because they have very little mass. And the gait of the camel is unusual—it paces. That means the two legs on one side of the body move forward at the same time while the weight of the camel is supported by the legs on the opposite side of the body. Thus the camel rocks from side to side as it walks, rather than moving the mass of its body up and down as most other mammals do.

The kangaroo takes advantage of elastic recoil of its tendons. When it hits the ground with its strong hindlimbs, the tendons in those limbs stretch and store part of the kinetic energy of the jump. The elasticity of those tendons then returns their stored energy to the next jump. As the kangaroo jumps faster, it takes the same number of jumps per minute, but they are longer. The longer jumps stretch the tendons more, creating more elastic energy that can help propel the next jump.

36 SUMMARY

concept 36.1 Cycles of Protein–Protein Interactions Cause Muscles to Contract

- **Skeletal muscle** consists of bundles of **muscle fibers**. Each skeletal muscle fiber is a large cell containing multiple nuclei.

- Skeletal muscles contain numerous **myofibrils**, which are bundles of **actin** and **myosin** filaments. The regular, overlapping arrangement of the actin and myosin filaments into **sarcomeres** gives skeletal muscle its striated appearance. **Review Figures 36.1 and 36.2 and WEB ACTIVITY 36.1**

- The molecular mechanism of muscle contraction involves the binding of the globular heads of myosin molecules to actin. Upon binding, the myosin head changes its conformation, causing the two filaments to move past each other (the sliding filament contractile mechanism). Release of the myosin heads from actin and their return to their original conformation requires ATP. **Review ANIMATED TUTORIAL 36.1**

- All the fibers activated by a single motor neuron constitute a **motor unit**. Each nerve ending of the motor neuron forms a synapse with the muscle cell membrane. When neurotransmitter is released at the synapse, the muscle cell membrane is depolarized and action potentials are generated. Action potentials spread across the plasma membrane and through the **T tubules**, causing Ca^{2+} to be released from the **sarcoplasmic reticulum**. **Review Figure 36.5 and WEB ACTIVITY 36.2**

- Ca^{2+} binds to **troponin** and changes its conformation, pulling the **tropomyosin** strands away from the myosin-binding sites on the actin filament. The muscle fiber continues to contract until the Ca^{2+} is returned to the sarcoplasmic reticulum. **Review Figure 36.6**

- **Cardiac muscle** cells are striated, uninucleate, branching, and electrically connected by gap junctions, so that action potentials spread rapidly throughout sheets of cardiac muscle and cause coordinated contractions. Some cardiac muscle cells are pacemaker cells that generate and conduct electrical signals.

- **Smooth muscle** provides contractile force for internal organs. Smooth muscle cells respond to stretch, action potentials spreading from neighboring cells, hormones, or neurotransmitters from the autonomic nervous system. **Review Figure 36.7 and ANIMATED TUTORIAL 36.2**

concept 36.2 The Characteristics of Muscle Cells Determine Muscle Performance

- In skeletal muscle, a single action potential causes a minimum unit of contraction called a **twitch**. Twitches occurring in rapid succession can be summed, thus increasing the strength of contraction. Maximum sustained tension is called **tetanus**. **Review Figure 36.9**

- **Slow-twitch fibers** facilitate extended, aerobic work; **fast-twitch fibers** generate maximum forces for short periods of time. **Review Figure 36.10**

- Muscle performance depends on a supply of ATP. Available ATP and creatine phosphate (CP) can fuel maximum tension instantaneously but are exhausted within seconds. Glycolysis can regenerate ATP rapidly but is limited by accumulation of lactic acid. Oxidative metabolism delivers ATP more slowly but can continue to do so for a long time. **Review Figure 36.11**

---

concept
36.3
Muscles Pull on Skeletal Elements to Generate Force and Cause Movement

- **Skeletal systems** provide supports against which muscles can pull.

- **Hydrostatic skeletons** are fluid-filled body cavities that can be squeezed by muscles. **Review Figure 36.12**

- Exoskeletons are hardened outer surfaces to which internal muscles are attached.

- **Endoskeletons** are internal systems of rigid rodlike, platelike, and tubelike supports, consisting of **bone** and **cartilage** to which muscles are attached. **Review Figure 36.13**

- Bone is continually remodeled by **osteoblasts**, which lay down new bone, and **osteoclasts**, which erode bone. **Review Figure 36.14**

- Bones develop from connective tissue membranes (**membranous bone**) or from cartilage (**cartilage bone**) through ossification. Cartilage bone can grow until the centers of ossification meet. **Review Figure 36.15**

- Bone can be **compact** (solid and hard) or **cancellous** (containing numerous internal spaces).

- **Joints** enable muscles to power movements in different directions. Muscles and bones work together around joints as systems of levers. **Review Figures 36.16 and 36.18 and WEB ACTIVITY 36.3**

- **Tendons** connect muscle to bone; **ligaments** connect bones to each other. **Review Figure 36.17**

Gas Exchange in Animals

And so back up the gradual slopes, the wind behind me. A much greater effort this, stopping every few yards with a slight anxiety lest I should not make the distance. As I approached the tents, I was astonished to see a bird, a chough, strutting about on the stones near me.... During this day, too, Charles Evans saw what must have been a migration of small grey birds. Neither of us had thought to find any signs of life as high as this.

Sir John Hunt,
Ascent of Everest, 1953

The passage above comes from a book about the first successful climbing of Mount Everest, 8,850 meters high, and it describes an observation Hunt made at the last camp before the summit attempt. At that altitude, experienced and acclimated climbers are practically incapacitated if they do not breathe supplemental oxygen from pressurized bottles. Prior to the moment described in the passage, Hunt had gone a short distance downhill from his tent without supplemental oxygen.

Why do people have so much difficulty at high altitude? The atmosphere has the same percentage of oxygen at sea level and on top of Mount Everest—but there is less atmosphere the higher you go. A cubic meter of air atop Mount Everest has about one-third as much oxygen as an equivalent volume of air at sea level. Birds, however, are able to function normally and even fly under those conditions.

The champion for high-altitude flight is the bar-headed goose, which migrates over the Himalayan mountains. Migrating groups of bar-headed geese have been observed at 10,175 meters. There is so little oxygen at that altitude that kerosene will not burn. Yet many species migrate at these high altitudes, as evidenced by radar signals, plane sightings (and collisions), and eyewitness accounts. These feats are all the more remarkable because the energetic costs of flapping flight are high—about the equivalent metabolic output of a well-trained human athlete engaged in a high-intensity activity that can be sustained for only minutes—yet the migrating birds can sustain it for hours and even days. Furthermore, the lungs of a bird are smaller than the lungs of a similar-sized mammal. Bird lungs are also less compliant—they do not inflate as much. Long-distance bird flight at high altitudes is the ultimate athletic performance.

Bar-headed geese have adaptations that enable them to fly over high mountains.

QUESTION

How are bar-headed geese able to sustain the high metabolic cost of flight at altitudes higher than Mount Everest?

KEY CONCEPTS

37.1 Fick's Law of Diffusion Governs Respiratory Gas Exchange

37.2 Respiratory Systems Have Evolved to Maximize Partial Pressure Gradients

37.3 The Mammalian Lung Is Ventilated by Pressure Changes

37.4 Respiration Is under Negative Feedback Control by the Nervous System

37.5 Respiratory Gases Are Transported in the Blood

Fick's Law of Diffusion Governs Respiratory Gas Exchange

The **respiratory gases** that animals must exchange are oxygen (O_2) and carbon dioxide (CO_2). Cells need O_2 from the environment to produce ATP by cellular respiration (see Chapter 6), and they must get rid of CO_2, the end product of cellular respiration. Diffusion is the only way respiratory gases can be exchanged between an animal and the environment (air or water); there are no active transport mechanisms to move respiratory gases across biological membranes. Diffusion is a physical process, and knowing the physical factors that influence rates of diffusion helps us understand the diverse adaptations of gas exchange systems. In this concept we will discuss environmental factors that influence diffusion rates; we will then consider adaptations that facilitate the exchange of respiratory gases.

LINK Concept 5.2 discusses the diffusion of molecules in solution, and Concept 6.2 describes aerobic cellular respiration and the production of ATP

Diffusion is driven by concentration differences

Because diffusion results from the random motion of molecules, the net movement of molecules is always from an area of higher concentration to an area of lower concentration. Biologists express the concentrations of different gases in a mixture as the *partial pressures* of those gases. For air-breathing animals, the total pressure of the gas mixture they breathe is atmospheric pressure, or **barometric pressure**, and it is measured with a *barometer*. A classic barometer is a glass tube closed at one end, filled with mercury, and inverted over a pool of mercury with the open end under the surface of the mercury. In a total vacuum such as outer space, the mercury would just flow out of the tube, but on Earth it does not because of the pressure the atmosphere exerts on the pool of mercury. At sea level, the pressure exerted by the atmosphere supports, and is therefore equal to, a column of mercury about 760 mm high (depending on the weather):

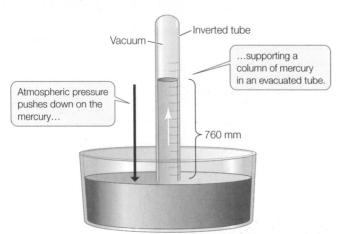

Vacuum — Inverted tube

...supporting a column of mercury in an evacuated tube.

Atmospheric pressure pushes down on the mercury...

760 mm

Therefore barometric pressure (atmospheric pressure) at sea level is 760 mm of mercury (mm Hg). Because dry air is 20.9

percent O_2, the **partial pressure of oxygen** (P_{O_2})—that is, the contribution of O_2 to the total air pressure—at sea level is 20.9 percent of 760 mm Hg, or about 159 mm Hg.

Describing the concentration of respiratory gases in a liquid such as water is a little more complicated because another factor is involved—the solubility of the gas in the liquid. The actual amount of a gas in a liquid depends on the partial pressure of that gas in the gas phase in contact with the liquid, as well as on the solubility of that gas in that liquid. So for fresh water close to 0°C in equilibrium with the atmosphere at 760 mm Hg, the partial pressure of O_2 in both air and water is 159 mm Hg, but whereas the volume of O_2 in the air is about 200 ml/L, the volume of O_2 dissolved in the water is only 10 ml/L. The diffusion of the gas between the gaseous phase and the liquid depends on the partial pressures of the gas in the two phases, but the volume of gas that enters the liquid depends on its solubility.

Fick's law applies to all systems of gas exchange

Diffusion is a physical phenomenon that can be described with a simple equation called **Fick's law of diffusion**. All environmental variables that limit respiratory gas exchange and all adaptations that maximize respiratory gas exchange are reflected in one or more components of this equation. Fick's law is written as

$$Q = DA\frac{P_1 - P_2}{L}$$

where

- Q is the rate at which a gas such as O_2 diffuses between two locations.
- D is the *diffusion coefficient*, which is a characteristic of the diffusing substance, the medium, and the temperature. (For example, perfume has a higher D than motor oil vapor, and all substances diffuse faster at higher temperatures, as well as diffusing faster in air than in water.)
- A is the cross-sectional area across which the gas is diffusing.
- P_1 and P_2 are the partial pressures of the gas at the two locations.
- L is the path length, or distance, between the two locations.

Therefore $(P_1 - P_2)/L$ is a *partial pressure gradient*.

Animals can maximize D for respiratory gases by using air rather than water as their gas exchange medium whenever possible. All other adaptations for maximizing respiratory gas exchange must influence the surface area (A) for gas exchange or the partial pressure gradient $[(P_1 - P_2)/L]$ across that surface area.

Air is a better respiratory medium than water

Oxygen can be obtained more easily from air than from water for several reasons:

- The O_2 content of air is much higher than the O_2 content of an equal volume of water. A cascading freshwater stream has an O_2 content of less than 10 ml of O_2 per liter of water, whereas the air over the stream has an O_2 content of about 200 ml of O_2 per liter of air.

- O_2 diffuses about 8,000 times more rapidly in air than in water.

- When an animal breathes, it does work to move water or air over its specialized gas exchange surfaces. More energy is required to move water than to move air because water is much more dense and viscous than air.

Even in air-breathing animals, the slow rate of O_2 diffusion in water limits the efficiency of O_2 distribution from gas exchange surfaces to the sites of cellular respiration. Eukaryotic cells carry out cellular respiration in their mitochondria, which are surrounded by aqueous cytoplasm. The cells themselves are bathed in extracellular fluid—also an aqueous medium. In addition, all respiratory surfaces must be protected from desiccation by a thin film of fluid. O_2 breathed in must diffuse through all of these aqueous compartments to get to the mitochondria.

Diffusion of O_2 in water is so slow that even animal cells with low rates of metabolism can be no more than a few millimeters away from a good source of environmental O_2. Therefore most invertebrates that lack internal systems for distributing O_2 are very small or, like flat worms, have evolved a flat, thin body that maximizes surface area:

Pseudoceros ferrugineus

Others, such as sponges and cnidarians, have a thin body structure with a central cavity through which water circulates:

Callyspongia plicifera

A critical factor enabling larger, more complex animal bodies has been the evolution of specialized respiratory systems with large surface areas for enhancing respiratory gas exchange.

O_2 availability is limited in many environments

Environmental temperature can compromise the O_2 available for water-breathers. Warm water holds less O_2 per liter than does cold water. Furthermore, most animals that live in water are ectothermic, meaning their body temperature is the same as the temperature of the surrounding environment. This means that as water warms up, a fish warms up too, and that means its metabolic rate increases. As metabolic rate increases, the fish needs more and more O_2—just when there is less and less O_2 available in the water. If the surrounding water becomes too warm, a fish will soon reach a point when it simply cannot obtain enough O_2 from the water to survive.

As discussed in the opening section of this chapter, O_2 availability for air-breathers decreases as altitude increases. Since O_2's rate of diffusion depends on the P_{O_2} difference between the air and the body fluids, the drastically reduced P_{O_2} in the air at

---

APPLY THE CONCEPT

Fick's law of diffusion governs respiratory gas exchange

Water contains much less O_2 than an equal volume of air, and it requires more effort to move it over respiratory exchange surfaces. Other factors that affect respiration in water-breathing animals are changes in solubility of O_2 in water and the effect of temperature on their O_2 needs. Answer each of the following questions by constructing a graph from the data in the table. Use water temperature as your x axis.

1. How does water temperature influence the O_2 available to a water-breathing animal?

2. How does water temperature affect the O_2 demand of a water-breathing animal?

3. From the data, what is the evidence that O_2 availability is a limiting factor for metabolism in warm water?

WATER TEMPERATURE (°C)	O_2 CONTENT OF WATER (ML/L)	METABOLIC RATE OF INACTIVE FISH (ML O_2/KG/HR)	METABOLIC RATE OF ACTIVE FISH (ML O_2/KG/HR)
5	9.0	8	30
15	7.0	50	110
25	5.8	140	255
35	5.0	225	285

4. From this study, what general argument would you make about the effect of environmental warming on the viability of aquatic species?

high altitudes sharply reduces the rate of diffusion of O_2 from air to blood.

CO_2 is easily lost by diffusion

CO_2 diffuses out of the body as O_2 diffuses in. Since the amount of CO_2 in the atmosphere is extremely low (0.03%), air-breathing animals always have a large concentration gradient for diffusion of CO_2 from their bodies.

Getting rid of CO_2 is also not a problem for water-breathing animals. CO_2 is much more soluble in water than is O_2, so even in stagnant water, where the partial pressure of carbon dioxide (P_{CO_2}) is higher than in moving water, the lack of O_2 becomes a problem for an animal before CO_2 exchange difficulties arise.

Do You Understand Concept 37.1?

- Explain the phenomenon of diffusion and how it is influenced by each component of Fick's law.

- In many fishes, lungs or lunglike structures have evolved that allow the fish to gulp mouthfuls of air from the surface to supplement their O_2 intake. Predict whether this air-gulping ability has evolved more often in warm-water fishes or in cold-water fishes, and explain your reasoning.

- For each component of Fick's law, propose an adaptation that would improve respiratory gas exchange for an air-breathing and a water-breathing animal.

Now that we have an understanding of the physical factors that influence diffusion rates of respiratory gases between animals and their environments, let's look at some of the corresponding adaptations that have evolved for maximizing respiratory gas exchange.

---

concept **37.2** **Respiratory Systems Have Evolved to Maximize Partial Pressure Gradients**

Gas exchange systems are made up of gas exchange surfaces and the mechanisms that *ventilate* those surfaces with air or water, and *perfuse* those surfaces with blood or other body fluids. As you might expect from the components of Fick's law of diffusion, adaptations to maximize respiratory gas exchange can be categorized as those that:

- increase the surface area for gas exchange
- maximize the partial pressure difference driving diffusion
- minimize the diffusion path length
- minimize the diffusion processes that take place in an aqueous medium

Respiratory organs have large surface areas

A variety of anatomical adaptations maximize the specialized body surface area (A) over which respiratory gases can diffuse. Water-breathing animals generally have *gills*, and air-breathing animals have *tracheae* or *lungs*. **External gills** are highly branched extensions of the body surface that provide a large surface area for gas exchange with water (**FIGURE 37.1A**). External gills are found in larval amphibians and in many insect larvae. Because they consist of thin, delicate tissues, external gills minimize the path length (L) traversed by diffusing molecules of O_2 and CO_2. In many animals, protective body cavities for gills have evolved. Such **internal gills** are found in shelled mollusks such as clams, aquatic arthropods, and all fishes (**FIGURE 37.1B**).

Lungs are internal cavities for respiratory gas exchange with air (**FIGURE 37.1C**). Lungs have a large surface area because they are highly divided; and because they are elastic, they can be inflated with air and deflated.

Instead of lungs, insects have a network of air-filled tubes called **tracheae** that branch through all tissues of the insect's body (**FIGURE 37.1D**).

Partial pressure gradients can be optimized in several ways

Diffusion is driven by partial pressure gradient [($P_1 - P_2$)/L], so the larger the gradient, the greater the rate of gas exchange. Partial pressure gradients can be maximized in several ways:

- *Minimization of path length*: Very thin tissues in gills and lungs reduce the diffusion path length (L).

- *Ventilation*: Actively moving the external medium over the gas exchange surfaces (i.e., breathing) regularly exposes those surfaces to fresh respiratory medium containing maximum O_2 and minimum CO_2 concentrations. This maximizes $P_1 - P_2$.

- *Perfusion*: Actively moving the internal medium (e.g., blood) over the internal side of the exchange surfaces transports

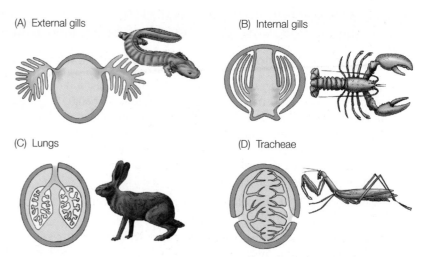

(A) External gills   (B) Internal gills

(C) Lungs   (D) Tracheae

FIGURE 37.1 Gas Exchange Systems Large surface areas (blue in these diagrams) for the diffusion of respiratory gases are common features of animals. External (A) and internal (B) gills are adaptations for gas exchange with water. Lungs (C) and tracheae (D) are organs for gas exchange with air.

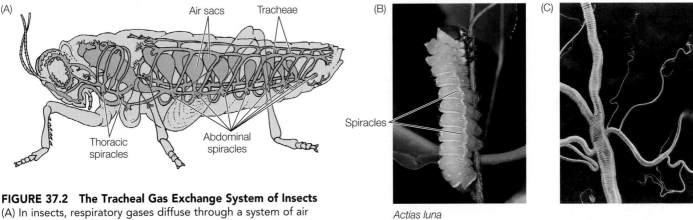

FIGURE 37.2 The Tracheal Gas Exchange System of Insects (A) In insects, respiratory gases diffuse through a system of air tubes (tracheae) that open to the external environment through holes called spiracles. (B) The spiracles of a luna moth larva run down its sides. (C) A scanning electron micrograph shows an insect trachea dividing into smaller tracheoles and still finer air capillaries.

CO_2 to those surfaces and O_2 away from them. This also maximizes $P_1 - P_2$.

All gas exchange organs in animals rely on these basic principles, but the details vary in different animal groups. Below we will look at three examples: insect tracheae, fish gills, and lungs—in particular, avian lungs, which differ from those of other terrestrial vertebrates.

Insects have airways throughout their bodies

Rather than having a central respiratory organ like a lung or gill, insects have an extensive networking system of air tubes that bring air close to every cell. Thus respiratory gases diffuse through air most of the way to and from each cell. Air enters through openings called *spiracles* in the sides of the abdomen and thorax (**FIGURE 37.2A AND B**). The spiracles can close to decrease water loss. They connect to tubes called *tracheae* that branch into even finer tubes, or *tracheoles*, that end in tiny *air capillaries*, which are the gas exchange surfaces (**FIGURE 37.2C**). The terminal branches of these tubes are so numerous that they have an enormous surface area. In the insect's flight muscles and other highly active tissues, every mitochondrion is close to an air capillary. Rhythmic body contractions that promote ventilation throughout this network can be observed in some insects.

Fish gills use countercurrent flow to maximize gas exchange

The most efficient way for one solution (e.g., blood) to take up O_2 from another solution (e.g., an aqueous environment) is by **countercurrent exchange** (**FIGURE 37.3**). The structure and function of fish gills exemplify this principle. The internal gills of fishes are supported by *gill arches* that lie between the mouth cavity and the protective *opercular flaps* on the sides of the fish

just behind the eyes (**FIGURE 37.4A**). Water flows *unidirectionally* into the fish's mouth, over the gills, and out from under the opercular flaps. Thus the gills are continuously bathed with fresh water. This constant, one-way flow of water over the gills maximizes the P_{O_2} on the external gill surfaces. On the internal side of the gill membranes, the circulation of blood minimizes the P_{O_2} by sweeping O_2 away as rapidly as it diffuses across. Both of these features help maximize the partial pressure gradient for O_2 uptake.

Gills have an enormous surface area for gas exchange because they are so highly divided. Each gill consists of hundreds of ribbonlike *gill filaments* (**FIGURE 37.4B**). The upper and lower flat surfaces of each gill filament are covered with rows of evenly spaced folds, or *lamellae*. The lamellae are the actual gas exchange surfaces. Because the lamellae are exceedingly

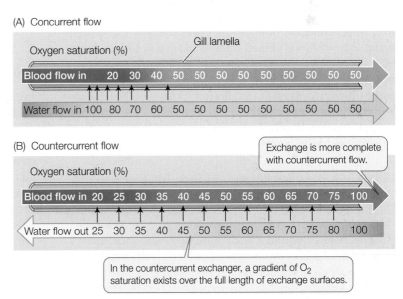

FIGURE 37.3 Countercurrent Exchange Is More Efficient In these models of concurrent and countercurrent gas exchange, the numbers represent the O_2 saturation percentages of blood and water. The thin black arrows represent the diffusion of O_2 molecules. (A) In a concurrent exchanger, the saturation percentages of blood and water reach equilibrium halfway across the exchange surface. (B) A countercurrent exchanger allows more complete gas exchange because the water is always more O_2-saturated than the blood; thus a gradient of O_2 saturation is maintained.

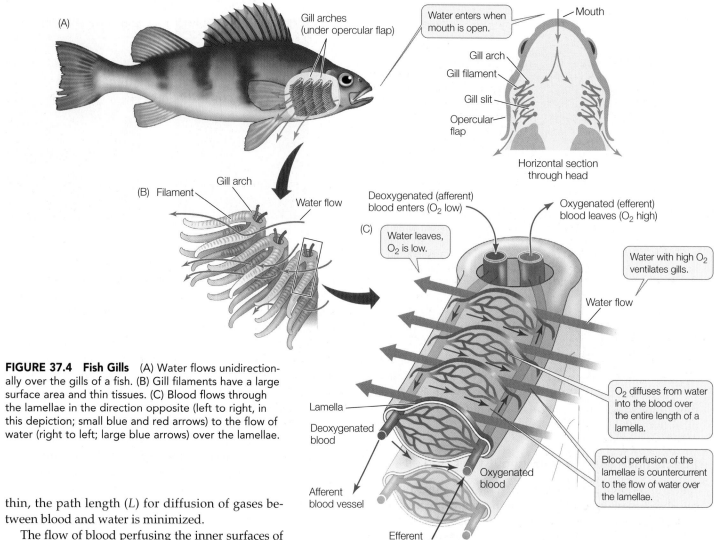

FIGURE 37.4 Fish Gills (A) Water flows unidirectionally over the gills of a fish. (B) Gill filaments have a large surface area and thin tissues. (C) Blood flows through the lamellae in the direction opposite (left to right, in this depiction; small blue and red arrows) to the flow of water (right to left; large blue arrows) over the lamellae.

thin, the path length (L) for diffusion of gases between blood and water is minimized.

The flow of blood perfusing the inner surfaces of the lamellae, like the flow of water over the gills, is unidirectional. *Afferent* blood vessels bring deoxygenated blood to the gills, while *efferent* blood vessels take oxygenated blood away from the gills (**FIGURE 37.4C**). Blood flows through the lamellae in the direction opposite to the flow of water over the lamellae. This *countercurrent flow* optimizes the P_{O_2} gradient between water and blood, making gas exchange more efficient than it would be in a system using concurrent (parallel) flow (see Figure 37.3).

> **LINK** Countercurrent flow is also important for thermoregulation in some animals and in mammalian kidney function; see Figures 29.10 and 40.8

Some fishes, including anchovies, tuna, and some sharks, ventilate their gills by swimming almost constantly, thus forcing water into their open mouths and over their gills. Most fishes, however, ventilate their gills by means of a two-pump mechanism. Closing and contracting the mouth cavity pushes water over the gills, and the expansion of the opercular cavity prior to opening of the opercular flaps pulls water over the gills.

These adaptations for maximizing the surface area (A) for diffusion, minimizing the path length (L) for diffusion, and maximizing the P_{O_2} gradient allow fishes to extract an adequate supply of O_2 from meager environmental sources.

Most terrestrial vertebrates use tidal ventilation

Lungs evolved in the first "air-gulping" vertebrates as outpocketings of the digestive tract. Although their structure has evolved considerably, lungs remain dead-end sacs in all air-breathing vertebrates except birds. Because of this, ventilation cannot be constant and unidirectional but must be **tidal**: fresh air flows in and exhaled gases flow out by the same route. Because of this two-way tidal flow, lungs and airways are never completely emptied of air. At the end of an exhalation they always contain some *dead space*, or **residual volume (RV)**, containing "stale" (low-O_2) air. On the next inhalation, fresh air entering the lungs mixes with the stale air that was already in the lungs. Therefore, the P_{O_2} in the mixed air that reaches the gas exchange surfaces is less than the P_{O_2} of the outside air. Tidal ventilation also precludes countercurrent exchange as a means of maximizing the partial pressure gradients. Amphibians, reptiles, and mammals all use tidal ventilation.

Mammals have high metabolic rates, and mammalian lungs have some evolved features that maximize the rate of gas exchange: an enormous surface area, a very short path length for diffusion, and a high ventilatory capacity (an ability to move a

(A) Avian air sacs and lungs

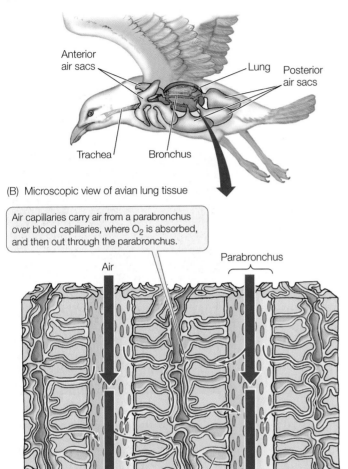

(B) Microscopic view of avian lung tissue

Air capillaries carry air from a parabronchus over blood capillaries, where O_2 is absorbed, and then out through the parabronchus.

Air

Parabronchus

Blood capillary   Air capillaries

FIGURE 37.5 The Respiratory System of a Bird (A) Air sacs and air spaces in the bones are unique to birds. (B) Air flows through bird lungs unidirectionally in parabronchi. Air capillaries, the site of gas exchange, branch off the parabronchi.

yourBioPortal.com

Go to ANIMATED TUTORIAL 37.1
Airflow in Birds and
WORKING WITH DATA 37.1

lung in a series of tubes called *parabronchi*. Branching off the parabronchi are tiny *air capillaries*, which are the gas exchange surfaces. The air capillaries are so numerous that they provide an enormous surface area for gas exchange. The parabronchi coalesce into larger bronchi that take the air out of the anterior end of the lung. Since air leaves the lung by a different path than it entered, bird lungs have very little dead space, and fresh incoming air is not mixed with stale air. In this way, a high P_{O_2} gradient is maintained between the incoming air and the blood.

Bird lungs also receive fresh air continuously, even during exhalation. This is made possible by multiple **air sacs**, thin-walled air storage spaces that are not gas exchange organs. The air sacs are interconnected with each other, with the lungs, and with air spaces in some of the bones (see Figure 37.5A), and they can be divided into two groups: anterior and posterior. During inhalation the posterior air sacs receive the incoming air, and the anterior air sacs receive the air that was in the lungs. During exhalation the air in the posterior air sacs flows into the lungs, and the air in the anterior air sacs leaves the bird. Thus the air sacs work like bellows to create a continuous flow of fresh air through the lungs.

Do You Understand Concept 37.2?

- As we will discuss in the Chapter 38, insects do not have closed blood circulatory systems that could transport respiratory gases around their bodies, yet some insects are able to sustain high metabolic rates. Explain how this is possible.

- Compare the adaptations of fishes and birds that optimize the terms of Fick's law of diffusion. In what ways are they similar and in what ways different?

- Present-day reptiles are good at burst activity but not at long, sustained, high-metabolic efforts. Birds and other dinosaurs share a common reptilian ancestor (see Chapter 23). What feature would you look for in dinosaur fossils to support the hypothesis that dinosaurs were capable of sustained, high levels of metabolic activity like present-day birds?

large volume of air in and out of the lungs). These adaptations help mammals obtain enough O_2 to support their high metabolic rates. But what about the group of vertebrates that have the highest metabolic rates of all—the birds?

Birds have air sacs that supply a continuous unidirectional flow of fresh air

As discussed in the chapter opener, the ability of birds to sustain high metabolic rates at high altitudes is remarkable. Yet the lungs of birds are smaller than the lungs of similar-sized mammals, and during the breathing cycle, bird lungs do not change in volume as much as mammalian lungs do. And the most unusual feature of bird lungs is that they decrease in volume during inhalation and increase in volume during exhalation. How are birds able to fly in environments where the P_{O_2} is so low that a comparably sized mammal would pass out?

A lung structure evolved in birds that allows air to flow unidirectionally through the lungs (**FIGURE 37.5**). Air enters a bird lung from the posterior end and flows through the

Let's take a closer look now at the anatomy and function of mammalian lungs, focusing on the human respiratory system as an example.

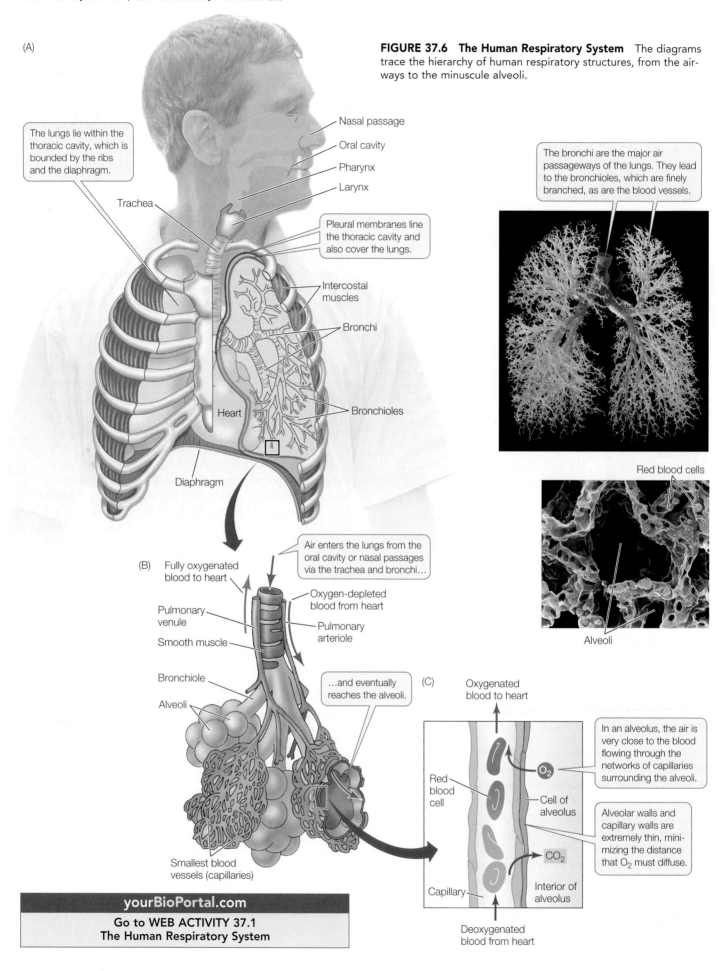

(A)

The lungs lie within the thoracic cavity, which is bounded by the ribs and the diaphragm.

Nasal passage
Oral cavity
Pharynx
Larynx

Trachea

Pleural membranes line the thoracic cavity and also cover the lungs.

Intercostal muscles

Bronchi

Heart

Bronchioles

Diaphragm

FIGURE 37.6 The Human Respiratory System The diagrams trace the hierarchy of human respiratory structures, from the airways to the minuscule alveoli.

The bronchi are the major air passageways of the lungs. They lead to the bronchioles, which are finely branched, as are the blood vessels.

Red blood cells

Alveoli

(B)

Fully oxygenated blood to heart

Air enters the lungs from the oral cavity or nasal passages via the trachea and bronchi…

Oxygen-depleted blood from heart

Pulmonary venule

Smooth muscle

Pulmonary arteriole

Bronchiole

Alveoli

…and eventually reaches the alveoli.

Smallest blood vessels (capillaries)

(C)

Oxygenated blood to heart

Red blood cell

Cell of alveolus

In an alveolus, the air is very close to the blood flowing through the networks of capillaries surrounding the alveoli.

O_2

CO_2

Capillary

Interior of alveolus

Alveolar walls and capillary walls are extremely thin, minimizing the distance that O_2 must diffuse.

Deoxygenated blood from heart

yourBioPortal.com

Go to WEB ACTIVITY 37.1
The Human Respiratory System

The Mammalian Lung Is Ventilated by Pressure Changes

In mammals, air enters the lungs through the oral cavity and the nasal passage, which join together in the *pharynx* (**FIGURE 37.6A**). Below the pharynx, the *esophagus* directs food to the stomach, and the *trachea* leads to the lungs. At the beginning of the trachea is the *larynx*, or voice box, which houses the vocal cords. The larynx is the "Adam's apple" that you can see or feel on the front of your neck. The trachea's thin walls are prevented from collapsing by bands of cartilage. If you run your fingers down the front of your neck just below your larynx, you can feel a few of these bands.

The trachea branches into two *bronchi*, one leading to each lung. Each bronchus branches repeatedly to generate a treelike structure of progressively smaller airways extending to all regions of the lungs. The smallest airways have no cartilage supports and are called **bronchioles**. The bronchioles branch a few more times, eventually leading to tiny, thin-walled air sacs, the **alveoli**. Alveoli are the sites of gas exchange (**FIGURE 37.6B**).

Human lungs have about 300 million alveoli. Although each alveolus is very small, their combined surface area for diffusion of respiratory gases is about 70 square meters—about one-fourth the size of a basketball court. Each alveolus is made of very thin cells. Between and surrounding the alveoli are networks of capillaries whose walls are also made up of exceedingly thin cells. Where capillary meets alveolus, very little tissue separates them (**FIGURE 37.6C**), so the length of the diffusion path between air and blood is less than 2 micrometers.

Secretions from lung tissues are not directly involved in gas exchange but are important for lung function. *Mucus* secreted in the airways helps capture inhaled particles and pathogens. Cilia in the bronchioles and bronchi continuously sweep the mucus and its trapped debris up to the pharynx where it can be swallowed. Nicotine in tobacco paralyzes these cilia, making it difficult to clear the airways and resulting in smoker's cough.

Specialized cells in the alveoli secrete a fatty substance that serves as a *surfactant* to reduce the surface tension of the thin film of fluid that lines the inner surfaces of the alveoli. The work that must be done to inflate the lungs includes overcoming both the elasticity of the alveolar tissues and the surface tension. By reducing that surface tension, the surfactant decreases the effort required for inhalation.

> **FRONTIERS** A baby that is born extremely prematurely—at 30 weeks of gestation or sooner—has lungs that are not fully developed and may not be producing lung surfactant. This makes it very difficult for the baby to inflate its lungs, resulting in Respiratory Distress Syndrome. If untreated, the baby gradually becomes exhausted from the breathing effort and can die in 2 to 7 days. A very successful treatment available today is application of artificial lung surfactant as an aerosol on the first day after birth. This treatment has reduced mortality from about 100 percent to 10 percent of cases. Among those 10 percent, however, are infants who are not extremely premature but are, for unknown reasons, resistant to treatment. Additional causes of failure of the lung surfactant system are being sought in the genes involved in surfactant synthesis and secretion.

At rest, only a small portion of the lung's volume is exchanged

The amount of air that moves in and out per breath when a person is at rest is called the **tidal volume** (**TV**). When we breathe in as much as possible, the additional volume is the **inspiratory reserve volume** (**IRV**). Conversely, if we forcefully exhale as much air as possible, the additional amount of air expelled is the **expiratory reserve volume** (**ERV**). The maximum capacity for air exchange in one breath, or the **vital capacity** (**VC**), is the sum of TV + IRV + ERV (**FIGURE 37.7**). The vital capacity of an athlete is generally greater than that of a nonathlete, and vital capacity decreases with age because of stiffening of the lung tissues.

As noted in Concept 37.2, tidal ventilation involves a residual volume (RV) because the lungs and airways can never be completely emptied of air. The O_2-poor RV dilutes the O_2 in the inhaled air. In most adult humans, TV averages about 500 ml, ERV is about 1,000 ml, and RV is 1,000 ml (see Figure 37.7). Thus the

---

RESEARCH TOOLS

FIGURE 37.7 Measuring Lung Ventilation A spirometer is a device that measures the volume of air a person breathes through a mouthpiece. The combined tidal volume, inspiratory reserve volume, and expiratory reserve volume are the lungs' vital capacity.

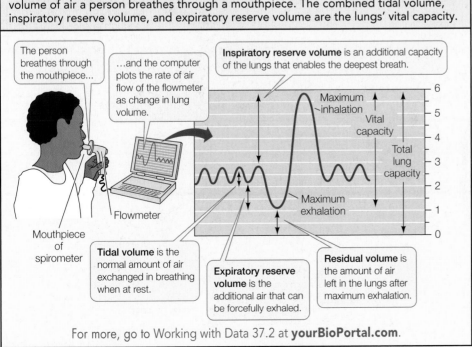

The person breathes through the mouthpiece...

...and the computer plots the rate of air flow of the flowmeter as change in lung volume.

Inspiratory reserve volume is an additional capacity of the lungs that enables the deepest breath.

Maximum inhalation

Vital capacity

Total lung capacity

Maximum exhalation

Flowmeter

Mouthpiece of spirometer

Tidal volume is the normal amount of air exchanged in breathing when at rest.

Expiratory reserve volume is the additional air that can be forcefully exhaled.

Residual volume is the amount of air left in the lungs after maximum exhalation.

For more, go to Working with Data 37.2 at **yourBioPortal.com**.

air that reaches the alveoli with each breath consists of only 500 ml of fresh air diluted by 2,000 ml of stale air. The maximum P_{O_2} in this mixed air is thus much below the P_{O_2} of the outside air.

Any disease or condition that increases the RV compromises a person's respiratory ability. For example, *emphysema* is a condition in which inflammation damages and eventually destroys the walls of the alveoli, resulting in fewer but larger alveoli. The result is that RV increases. This diminishes the partial pressure gradient of O_2, resulting in less O_2 diffusing to the blood. Emphysema is the fourth leading cause of death in the United States; the principal cause is smoking.

Lungs are ventilated by pressure changes in the thoracic cavity

Mammalian lungs are suspended in the **thoracic cavity**, which is bounded on the top by the shoulder girdle, on the sides by the rib cage, and on the bottom by a sheet of muscle called the **diaphragm**. Each lung is covered by a sheet of tissue called the **pleural membrane**, and the thoracic cavity is also lined by pleural membrane (see Figure 37.6A). There is a thin film of fluid in the pleural cavity between the pleural membranes of the lung and the thoracic cavity. This fluid lubricates the inner surfaces of the pleural membranes so they can slip and slide against each other during breathing movements. However, because of the cohesive forces (hydrogen bonding) between water molecules, it is difficult to pull the pleural membranes apart. Think of two wet panes of glass or two wet microscope slides; you can slide them past each other, but it is difficult to separate them. Because the pleural membranes covering the cavity wall and the lung surface are stuck to each other by cohesive forces, any increase in volume of the thoracic cavity increases the tension between the pleural membranes, causing inflation of the lungs.

At rest, inhalation is initiated by contraction of the muscular diaphragm (**FIGURE 37.8A**). As the domed diaphragm contracts, it pulls down, expanding the thoracic cavity and pulling on the pleural membranes. The pleural membranes pull on the lungs, and the elastic lungs expand, drawing air in from outside. Exhalation, in contrast, is usually a passive

FIGURE 37.8 Into the Lungs and Out Again (A) Inhalation is an active process driven by contraction of the diaphragm. (B) Exhalation generally is a passive process as the diaphragm relaxes.

(A)

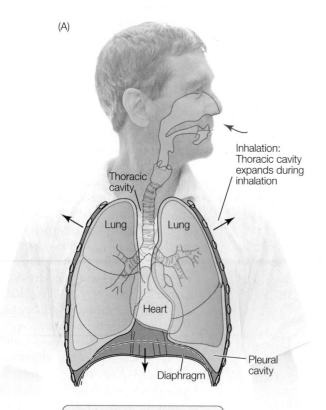

During **inhalation**:
• Diaphragm contracts
• Thoracic cavity expands
• Pleural membranes pull on lungs
• Lungs expand
• Air rushes in

(B)

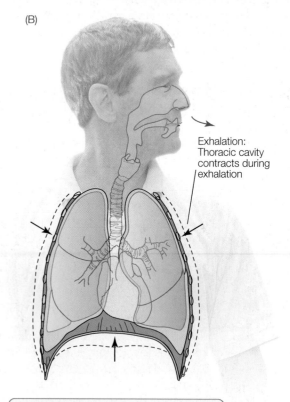

During **exhalation**:
• Diaphragm relaxes
• Thoracic cavity contracts
• Pleural membranes pull less strongly on lungs
• Lungs contract
• Gases in lungs are expelled

process requiring no muscular activity. The diaphragm simply relaxes, and the elastic recoil of the lung tissues pulls the diaphragm up and pushes air out through the airways (**FIGURE 37.8B**).

During strenuous exercise, additional muscles come into play. *External intercostal muscles* between the ribs expand the thoracic cavity by lifting the ribs up and outward, expanding the lungs more than usual and thus making use of the inspiratory reserve volume. Exhalation becomes an active process, with *internal intercostal muscles* pulling the ribs down and inward, and contraction of abdominal muscles pushing the abdominal contents up on the diaphragm. Together, the internal intercostal muscles and the abdominal muscles empty the lungs more than usual, making use of the expiratory reserve volume.

Do You Understand Concept 37.3?

- Name, in order, the anatomical structures that a molecule of O_2 passes through as it travels from the outside air to the site of gas exchange in the human lung. How many membranes must the O_2 molecule diffuse through in going from the alveolus to the inside of the red blood cell?

- Recall that Fick's law of diffusion describes all the variables that can affect the rate of diffusion of O_2 from air to blood. What aspects of the structure and anatomy of the human lung relate to (1) maximizing A, (2) maximizing $P_1 - P_2$, and (3) minimizing L?

- A puncture wound to the chest wall can cause a collapsed lung. In terms of pressures involved in the breathing cycle, explain how this condition occurs, and once a lung collapses, why is it difficult to reinflate it.

We have seen how the respiratory muscles cause pressure changes in the thoracic cavity that cause the mammalian lung to inflate and deflate. Let's now see how the respiratory muscles are controlled.

---

concept **37.4** **Respiration Is under Negative Feedback Control by the Nervous System**

Breathing is an involuntary function of the central nervous system. The medulla oblongata in the brainstem controls the respiratory muscles and sets the respiratory rate. Groups of respiratory motor neurons in the medulla increase their firing rates just before an inhalation begins. As more and more of these neurons fire—and fire faster and faster—the diaphragm contracts. All of a sudden, the neurons stop firing, the diaphragm relaxes, and exhalation begins. Brain areas above the medulla can modify the breathing pattern to allow for eating, drinking, and vocalizing.

The depth and frequency of breathing are modulated to meet the body's demands for O_2 supply and CO_2 elimination.

What determines the respiratory rate? What cues can cause the medulla to increase or decrease the rate of breathing?

Respiratory rate is primarily regulated by CO_2

One might expect that respiratory rate is regulated by the O_2 content of the blood. In fishes, blood O_2 is indeed the primary feedback stimulus for gill ventilation. However, humans (and other mammals) turn out to be remarkably insensitive to falling levels of O_2 in arterial blood, and are instead extremely sensitive to increases in CO_2. That is, arterial P_{O_2} can deviate considerably from normal without causing much of an increase in ventilation rate, but even a small rise in arterial P_{CO_2} causes a large increase in ventilation.

Is P_{CO_2}, then, the primary determinant of the increase in respiratory rate when we exercise? As the experiment in **FIGURE 37.9** demonstrates, when dogs run on treadmills at a steady speed but with the slope gradually increased or decreased to change the workload, their respiratory rate does indeed track the P_{CO_2} of their blood quite closely. When an animal simply increases its speed, however, its respiratory rate increases immediately before any change in P_{CO_2} occurs (see Analyze the Data in Figure 37.9). You may have noticed the same phenomenon in yourself; when you start running, your respiratory rate jumps immediately. During exercise, information from receptors in muscles and joints changes the respiratory center's sensitivity to CO_2, in anticipation of the rise in P_{CO_2} that will follow. This is an example of using *feedforward* information to maintain homeostasis.

LINK Review the discussion of feedforward information in Concept 29.2

CO_2 affects the medulla indirectly via pH changes

The major site of P_{CO_2} sensitivity is an area on the ventral surface of the medulla, not far from the groups of neurons that generate the breathing rhythm (**FIGURE 37.10**). These cells are not directly sensitive to CO_2, however. Rather, they are stimulated by H^+ ions. The pH in the environment of these cells is a direct reflection of the P_{CO_2} of the blood. When the P_{CO_2} of the blood is higher than that of the extracellular fluid, CO_2 diffuses out of the blood. That CO_2 interacts with H_2O to form carbonic acid (H_2CO_3), which dissociates into H^+ ions and bicarbonate ions (HCO_3^-).

$$CO_2 + H_2O \rightleftharpoons H_2CO_3 \rightleftharpoons H^+ + HCO_3^-$$

The H^+ ions stimulate the chemosensitive cells that increase respiratory gas exchange. Thus, blood P_{CO_2} indirectly controls breathing, via changes in pH.

O_2 is also monitored

Though respiratory rate is controlled primarily by blood P_{CO_2} (via pH), blood P_{O_2} is monitored as well. Minor fluctuations in P_{O_2} do not affect respiratory rate very much, but a large drop in P_{O_2} (such as occurs at high altitude) will stimulate an

INVESTIGATION

FIGURE 37.9 Sensitivity of the Respiratory Control System Changes with Exercise What is the metabolic feedback signal that controls ventilation rate during exercise? In experiments with dogs running on treadmills with different slopes, the ventilation rate increases as the P_{CO_2} in the arterial blood increases. When the dogs run at different speeds instead, their ventilation rates are different but their arterial P_{CO_2}'s are the same. How can P_{CO_2} be the metabolic stimulus for breathing?

HYPOTHESIS

P_{CO_2} is the feedback stimulus controlling ventilation.

METHOD

1. Dogs are trained to run on a treadmill.
2. The dogs are equipped with instruments that measure respiratory rate and with arterial catheters that enable sampling of blood.
3. As a dog runs, either the slope of the treadmill or its speed is changed to increase the metabolic workload.
4. Ventilation rate (V; L/min) is plotted as a function of arterial P_{CO_2} (mm Hg).

Catheter for taking blood samples

To flowmeter and respiratory analyzer

RESULTS

When the workload is altered by slowly changing the slope of the treadmill (no change in speed), the ventilation rate is a function of P_{CO_2}.

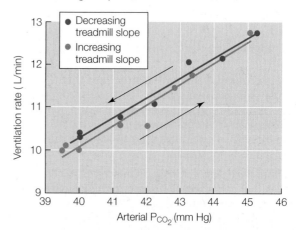

CONCLUSION

Arterial P_{CO_2} can be the metabolic feedback signal controlling ventilation in response to changes in workload.

ANALYZE THE DATA

Additional experiments were done in which the ventilation rate and arterial P_{CO_2} were measured when the treadmill speed was changed. The average values when the dog was running at 3 mph were P_{CO_2} = 39.0 and V = 9. The data for the first seven breaths after the treadmill speed was increased to 6 mph were:

Breath	P_{CO_2}	V		Breath	P_{CO_2}	V
1	38.0	13.0		5	37.2	12.0
2	37.0	14.0		6	36.8	13.0
3	36.5	11.2		7	37.2	13.0
4	37.2	11.2				

After the dog had run at 6 mph for several minutes, the values averaged P_{CO_2} = 41.0 and V = 15.0. Plot these data and use the information to answer the following questions:

A. Do these data support the hypothesis that P_{CO_2} is the feedback signal controlling ventilation rate? Why or why not?
B. How do you explain the average values after the dog had been running at the higher speed for a few minutes?
C. Relate these results to those obtained in the experiment in which the slope of the treadmill was gradually raised and lowered. Explain the differences between the results in terms of the P_{CO_2} sensing mechanism.

Go to **yourBioPortal.com** for original citations, discussions, and relevant links for all INVESTIGATION figures.

increase in respiratory rate. The P_{O_2} detectors reside in nodes of neural tissue on the large blood vessels leaving the heart, the aorta and the carotid arteries (see Figure 37.10). The nodes, called **carotid** and **aortic bodies**, are chemosensors sensitive to O_2. If their blood supply decreases, or if the blood P_{O_2} falls dramatically, the chemosensors are activated and send nerve impulses to the breathing control center.

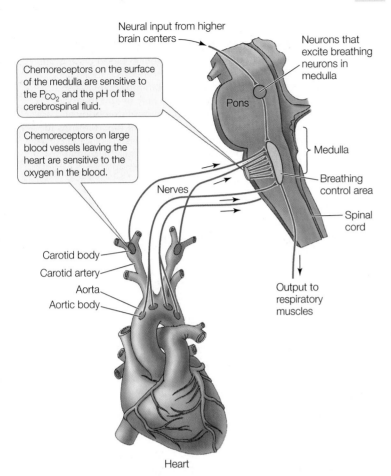

Neural input from higher brain centers

Neurons that excite breathing neurons in medulla

Chemoreceptors on the surface of the medulla are sensitive to the P_{CO_2} and the pH of the cerebrospinal fluid.

Pons

Chemoreceptors on large blood vessels leaving the heart are sensitive to the oxygen in the blood.

Medulla

Nerves

Breathing control area

Spinal cord

Carotid body

Carotid artery

Aorta

Aortic body

Output to respiratory muscles

Heart

FIGURE 37.10 Feedback Information Controls Breathing The body uses feedback information from chemosensors in the heart and brain to match breathing rate to metabolic demand.

Do You Understand Concept 37.4?

- In 1875 three French physiologists went up in a hot air balloon to see what would happen to their respiration at high altitude. Without registering any unusual physiology or sense of danger in their notebooks, they all went unconscious and two died. Explain this event from what you know about regulation of respiration.

- An untreated diabetic who lacks insulin will metabolize fats for energy and produce an excess of acidic end products, lowering the blood pH. How would you expect this condition to affect the individual's breathing?

- People who have recently climbed to high altitude often show dramatic oscillations in respiratory rate. When they rest or sleep, their respiratory rates alternate between periods of rapid ventilation and periods when they stop breathing. Propose an explanation for this phenomenon, which is called Cheyne–Stokes respiration.

Now that we have discussed how respiratory gases move from the environment into the bloodstream, we can look at how these gases are transported to the rest of the body.

concept 37.5 Respiratory Gases Are Transported in the Blood

One of the main functions of the circulatory system is to transport respiratory gases to and from the lungs. O_2, however, is a nonpolar molecule that does not dissolve well in aqueous solutions, such as the *blood plasma*, the liquid part of the blood. Human blood contains only about 0.3 ml of dissolved O_2 per 100 ml of plasma; this is inadequate to support even basal metabolism. Therefore most animals, vertebrate and invertebrate, have transport molecules in their blood that reversibly bind O_2 and thus augment the blood's transport capacity. These molecules pick up O_2 where P_{O_2} is high and release it where P_{O_2} is lower. In vertebrates this role is played by **hemoglobin**, a protein contained in red blood cells. Hemoglobin increases the capacity of blood to transport O_2 by about 60-fold, making high metabolic rates possible.

> **LINK** The systems that circulate blood throughout the bodies of animals are described in Concepts 38.1 and 38.2

Hemoglobin combines reversibly with O_2

Hemoglobin consists of four polypeptide subunits called *globins* (see Concept 3.2, p. 44), each of which surrounds a *heme group*—an iron-containing ring structure that can reversibly bind a molecule of O_2. Thus each hemoglobin molecule can bind up to four O_2 molecules.

Hemoglobin's ability to pick up or release O_2 depends on the P_{O_2} in its environment. When the P_{O_2} of the blood plasma is high, as it usually is in the lung capillaries, each hemoglobin molecule can carry its maximum load of four O_2 molecules. As the blood circulates through the rest of the body, it releases some of the O_2 it is carrying when it encounters lower P_{O_2} values (**FIGURE 37.11**).

The relation between P_{O_2} and the amount of O_2 bound to hemoglobin is not linear but S-shaped (sigmoidal). The O_2 binding/dissociation properties of hemoglobin that are reflected in this sigmoidal curve help get O_2 to the tissues that need it most. In the lungs, where the P_{O_2} is about 100 mm Hg, hemoglobin is 100 percent saturated with O_2. The P_{O_2} in blood returning to the heart from the body is usually about 40 mm Hg. You can see in Figure 37.11 that at this P_{O_2}, the hemoglobin is still about 75 percent saturated. This means that as the blood circulates around the body, it releases only about one in four of the O_2 molecules it carries. This system seems inefficient, but it is really quite adaptive, because the hemoglobin keeps 75 percent of its O_2 in reserve to meet peak demands of highly active tissues.

If a tissue becomes starved of O_2 and its local P_{O_2} falls below 40 mm Hg, the hemoglobin flowing through that tissue is on the steep portion of the sigmoidal binding/dissociation curve. That means relatively small decreases in P_{O_2} below 40 mm Hg will result in the release of lots of O_2 to the tissue. Thus hemoglobin is very effective in making O_2 available to tissues precisely when and where it is needed most.

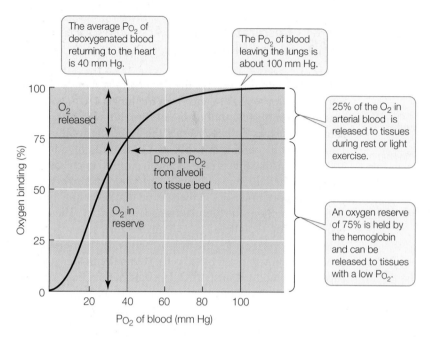

The average P_{O_2} of deoxygenated blood returning to the heart is 40 mm Hg.

The P_{O_2} of blood leaving the lungs is about 100 mm Hg.

O_2 released

25% of the O_2 in arterial blood is released to tissues during rest or light exercise.

Drop in P_{O_2} from alveoli to tissue bed

O_2 in reserve

An oxygen reserve of 75% is held by the hemoglobin and can be released to tissues with a low P_{O_2}.

FIGURE 37.11 Binding of O_2 to Hemoglobin Depends on P_{O_2} Hemoglobin in blood leaving the lungs is 100 percent saturated (four O_2 molecules are bound to each hemoglobin molecule). Most hemoglobin molecules drop only one of their four O_2 molecules as they circulate through the body and are still 75 percent saturated when the blood returns to the lungs. The steep portion of this O_2-binding curve comes into play when tissue P_{O_2} falls below the normal 40 mm Hg, at which point hemoglobin "unloads" its O_2 reserves.

The O_2 transport function of hemoglobin can rapidly and tragically be disrupted by carbon monoxide (CO). CO binds to hemoglobin with a 240-fold higher affinity than O_2 and therefore prevents hemoglobin from binding and transporting O_2. If CO accumulates in a closed space—often from a faulty heating system or engine exhaust—the results can be deadly. In the United States, more than 5,000 people die each year from CO poisoning.

Myoglobin holds an O_2 reserve

Muscle cells have their own O_2-binding molecule, *myoglobin*. Myoglobin consists of just one polypeptide chain associated with an iron-containing ring structure that can bind one O_2 molecule (see Figure 15.22). Myoglobin has a higher affinity for O_2 than hemoglobin does, so it picks up and holds O_2 at P_{O_2}

values at which hemoglobin is releasing its bound O_2 (**FIGURE 37.12**).

Myoglobin facilitates the diffusion of O_2 in muscle cells and provides an O_2 reserve for times when metabolic demands are high. When tissue P_{O_2} values are low and hemoglobin can no longer supply more O_2, myoglobin releases its bound O_2. Diving mammals such as seals have high concentrations of myoglobin in their muscles, which is one reason they can stay underwater for so long. Even in nondiving animals, muscles called on for extended periods of work frequently have more myoglobin than muscles that are used for short, intermittent periods.

Various factors influence hemoglobin's affinity for O_2

Many factors can influence hemoglobin's affinity for O_2, thereby influencing O_2 delivery to tissues. Here we examine three of those factors: the particular globin chains present in the hemoglobin, the blood pH, and the amount of 2,3-bisphosphoglyceric acid (BPG) in red blood cells.

HEMOGLOBIN COMPOSITION Recall that hemoglobin is composed of four globin chains. Mammals have several globin genes that code for different types of globin, which can be assembled in various combinations to form different versions of hemoglobin (see Figure 15.22). For example, the hemoglobin of adult humans has two α-globin chains and two β-globin chains. Before birth, the human fetus has a different form of hemoglobin, consisting of two α-globin and two γ-globin

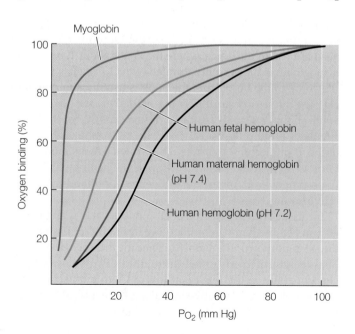

Myoglobin

Human fetal hemoglobin

Human maternal hemoglobin (pH 7.4)

Human hemoglobin (pH 7.2)

FIGURE 37.12 Oxygen-binding Adaptations Myoglobin and the different hemoglobins have different O_2-binding properties adapted to different circumstances. Human fetal hemoglobin has a higher affinity for O_2 than does maternal hemoglobin, facilitating O_2 transfer in the placenta. When high metabolism lowers the pH of the blood, hemoglobin releases more of its O_2.

yourBioPortal.com

Go to WEB ACTIVITY 37.2
Oxygen-Binding Curves and
INTERACTIVE TUTORIAL 37.1
Hemoglobin: Loading and Unloading

chains. The functional difference between fetal and adult hemoglobin is that fetal hemoglobin has a higher affinity for O_2. Therefore, the fetal hemoglobin–oxygen binding/dissociation curve is shifted to the left compared with that for adults (see Figure 37.12). You can see from these curves that if both types of hemoglobin are at the same P_{O_2} (as they are in the placenta), fetal hemoglobin has a greater affinity for O_2 than the maternal hemoglobin and will therefore pick up O_2 released from the maternal hemoglobin. The difference in O_2 affinities of fetal and maternal hemoglobin enables the efficient transfer of O_2 from the mother's blood to the fetus's blood.

HEMOGLOBIN AND pH As blood passes through metabolically active tissue such as exercising muscle, it picks up acidic metabolites such as lactic acid, fatty acids, and CO_2. As a result, blood pH falls. The excess H^+ ions bind to hemoglobin and change its affinity for O_2, shifting the binding/dissociation curve to the right. (This is an example of *allosteric regulation*, which is described in the context of enzyme regulation in Concept 3.4.) This shift, known as the **Bohr effect**, means the hemoglobin will release more O_2 in tissues where pH is low—another way that O_2 is supplied where and when it is most needed.

2,3-BISPHOSPHOGLYCERIC ACID BPG is a metabolite of glycolysis (see Concept 6.2). Mammalian red blood cells respond to low P_{O_2} by increasing their rate of glycolysis and producing more BPG. BPG, like excess H^+, reversibly combines with deoxygenated hemoglobin and lowers its affinity for O_2. The result is that at any P_{O_2}, hemoglobin releases more of its bound O_2 than it

otherwise would. In other words, BPG shifts the O_2 binding/ dissociation curve to the right.

Going to high altitude or beginning to exercise after a sedentary lifestyle increases the production of BPG in the red blood cells, making it possible for hemoglobin to deliver more O_2 to the tissues. The reason fetal hemoglobin has a left-shifted O_2 binding/dissociation curve (a relatively high affinity for O_2) is that its γ-globin chains have a lower affinity for BPG than do the β-globin chains of adult hemoglobin.

CO_2 is transported primarily as bicarbonate ions in the blood

Delivering O_2 to tissues is only half the respiratory function of blood. Blood also takes CO_2, a metabolic waste product, away from tissues (**FIGURE 37.13**). CO_2 is highly soluble and readily diffuses through cell membranes, moving from its site of production in the tissues into the blood. A small amount of CO_2 is then transported in the blood dissolved in the blood plasma, and a slightly greater amount is transported bound to hemoglobin. But most CO_2 produced by the tissues is transported to the lungs in the form of **bicarbonate ions**, HCO_3^-. As mentioned in Concept 37.4, when CO_2 dissolves in water, some of it reacts with the water to form carbonic acid (H_2CO_3), which dissociates into H^+ and HCO_3^-. This is a reversible reaction.

In the endothelial cells of the capillaries and in the red blood cells, the enzyme *carbonic anhydrase* speeds up the first step of this process, the conversion of CO_2 to H_2CO_3. The newly formed H_2CO_3 dissociates, and the resulting bicarbonate ions enter the plasma in exchange for chloride ions (Cl^-; see Figure 37.13).

In the lungs, these reactions are reversed. Remember that an enzyme such as carbonic anhydrase only speeds up a reversible reaction; it does not determine its direction. The direction is determined by concentrations of reactants and products.

FIGURE 37.13 Carbon Dioxide Is Transported as Bicarbonate Ions Carbonic anhydrase in capillary endothelial cells and in red blood cells facilitates conversion of CO_2 produced by tissues into bicarbonate ions carried by the plasma. In the lungs, the process is reversed as CO_2 is exhaled.

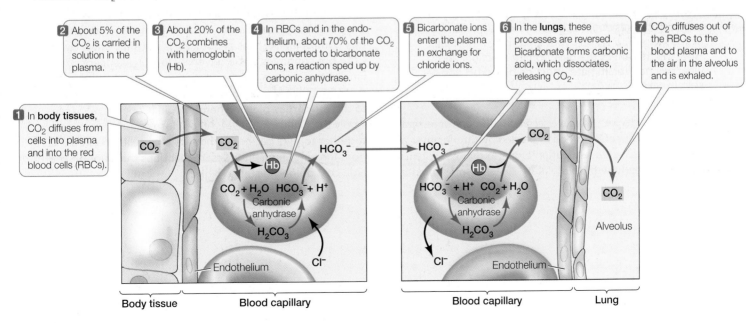

APPLY THE CONCEPT

Respiratory gases are transported in the blood

The transfer of O_2 from maternal to fetal blood is facilitated by the higher affinity of fetal hemoglobin for O_2 compared with that of maternal hemoglobin. South American llamas live at high altitude, and their hemoglobin has a higher affinity for O_2 than does the hemoglobin of other mammals that live at low altitude (e.g., humans). The O_2 affinities of the hemoglobins of fetal llamas and fetal humans are about the same. If we compare the affinities of the various hemoglobins for BPG, we find the following relationships: the affinity of human adult hemoglobin for BPG is about six times greater than that of human fetal hemoglobin, and the affinity of llama

adult hemoglobin for BPG is about three times greater than that of llama fetal hemoglobin.

1. Why is it significant that human fetal hemoglobin and llama adult hemoglobin have higher affinities for O_2 than human adult hemoglobin does?

2. The difference between fetal and adult hemoglobins' affinities for BPG is greater in humans than in llamas. Explain why.

3. Explain the differences in the way llamas and humans adapt to high altitude.

Ventilation keeps the CO_2 concentration in the alveoli low, so CO_2 diffuses from the blood plasma into the alveoli, lowering the CO_2 concentration in the blood, which favors the conversion of HCO_3^- into CO_2.

> **LINK** To review the properties of enzymes and the rules that determine the direction of chemical reactions, see Concepts 3.3 and 3.4

Do You Understand Concept 37.5?

- Why do most animals have a transport protein in their blood that can bind to O_2?

- Draw the O_2 dissociation curve of fetal hemoglobin compared with that of adult hemoglobin. Explain why they differ chemically and the functional consequences of the difference.

- If you hyperventilate (pant deeply and rapidly), you can flush the residual volume in your lungs with fresh air, reducing the P_{CO_2} in your lungs below normal. How will this affect the bicarbonate reaction, and what will be the ultimate effect on blood pH and respiratory rate?

Throughout this chapter, we have seen that diffusion of respiratory gases in and out of the blood is affected by simple physical laws that have driven the evolution of gas exchange systems. We have also seen how the circulatory system works hand in

hand with the gas exchange system, perfusing the gas exchange organs with blood and then transporting the respiratory gases to and from the rest of the body. In the next chapter we will turn to the many other functions of the circulatory system.

 QA QUESTION How are bar-headed geese able to sustain the high metabolic cost of flight at altitudes higher than Mount Everest?

ANSWER Bar-headed geese are exquisitely adapted to function at high altitude by virtue of adaptations in every aspect of respiratory gas exchange we have discussed in this chapter. Most importantly, because they are birds they have a respiratory system that maintains a continuous and unidirectional flow of air across the respiratory exchange surfaces (Concept 37.2). In addition, bar-headed geese have greater ventilatory capacity (L/min) than do geese that live at low altitudes. When the geese are exposed to low O_2, this ventilatory capacity increases even more. Their respiratory control (unlike that of humans) is not shut down by low CO_2, but continues to be driven by low O_2 (Concept 37.4). Finally, they have a point mutation in their hemoglobin gene that gives their hemoglobin a higher affinity for O_2—similar to the properties of llama hemoglobin (Concept 37.5; see Apply the Concept above). All of these adaptations work together to give bar-headed geese the ability to extract O_2 from the air at high altitudes in quantities that support the high cost of long-distance flight.

SUMMARY

concept 37.1 Fick's Law of Diffusion Governs Respiratory Gas Exchange

- Most cells require a constant supply of O_2 and continuous removal of CO_2. These **respiratory gases** are exchanged between an animal and its environment by diffusion.

- **Fick's law of diffusion** shows how various physical factors influence the diffusion rate of gases. Adaptations to maximize respiratory gas exchange influence one or more variables of Fick's law.

- In water, gas exchange is limited by the low diffusion rate and low solubility of O_2 in water. O_2 becomes less soluble in water as water temperature increases.

- In air, the **partial pressure of oxygen** (P_{O_2}) decreases as altitude increases.

concept 37.2 Respiratory Systems Have Evolved to Maximize Partial Pressure Gradients

- Adaptations to maximize gas exchange include increasing the surface area for gas exchange and maximizing partial pressure gradients by ventilating the outer surface with the respiratory medium and perfusing the inner surface with blood.

- Insects distribute air throughout their bodies in a system of **tracheae**, tracheoles, and air capillaries. **Review Figure 37.2**

- The **internal gills** of fishes have large gas exchange surface areas that are ventilated continuously and unidirectionally with water. The **countercurrent flow** of blood helps increase the efficiency of gas exchange. **Review Figures 37.3 and 37.4**

- In all air-breathing vertebrates except birds, breathing is **tidal**, in which inhaled air is always mixed with some stale air. This is a less efficient form of gas exchange than that of fishes and birds.

- The gas exchange system of birds includes **air sacs** that supply fresh air to the lungs even during exhalation. Air flows unidirectionally and continuously through bird lungs. **Review Figure 37.5, ANIMATED TUTORIAL 37.1, and WORKING WITH DATA 37.1**

concept 37.3 The Mammalian Lung Is Ventilated by Pressure Changes

- In mammalian lungs, the gas exchange surface area provided by the millions of **alveoli** is enormous, and the diffusion path length is short. **Review Figure 37.6 and WEB ACTIVITY 37.1**

- Inhalation occurs when contractions of the **diaphragm** pull on the **pleural membranes** and reduce the pressure in the **thoracic cavity**. Relaxation of the diaphragm increases pressure in the thoracic cavity and results in exhalation. **Review Figure 37.7, ANIMATED TUTORIAL 37.2, and WORKING WITH DATA 37.2**

- During strenuous exercise, the intercostal muscles and abdominal muscles increase the volume of air inhaled and exhaled.

concept 37.4 Respiration Is under Negative Feedback Control by the Nervous System

- The breathing rhythm is generated by neurons in the medulla and modulated by higher brain centers.

- The most important feedback stimulus for breathing is the level of CO_2 in the blood, which is detected by the medulla indirectly via changes in the blood pH. **Review Figure 37.9**

- A large change in blood O_2 level can also affect respiration rate. O_2 is detected by chemosensors in the **carotid** and **aortic bodies** on the large vessels leaving the heart. **Review Figure 37.10**

concept 37.5 Respiratory Gases Are Transported in the Blood

- O_2 is reversibly bound to **hemoglobin** in red blood cells. Each hemoglobin molecule can carry a maximum of four O_2 molecules.

- Hemoglobin's affinity for O_2 depends on the P_{O_2} to which the hemoglobin is exposed. Therefore, hemoglobin picks up O_2 as it flows through respiratory exchange structures and gives up O_2 in metabolically active tissues. The affinity of hemoglobin for O_2 is altered by the structure of the hemoglobin molecules, by the presence of H^+ ions, and by the concentration of BPG in the red blood cells. **Review Figure 37.11 and INTERACTIVE TUTORIAL 37.1**

- Myoglobin serves as an O_2 reserve in muscle.

- Fetal hemoglobin has a higher affinity for O_2 than does maternal hemoglobin, allowing fetal blood to pick up O_2 from the maternal blood in the placenta. **Review Figure 37.12 and WEB ACTIVITY 37.2**

- CO_2 is transported in the blood principally as **bicarbonate ions** (HCO_3^-). **Review Figure 37.13**

See **WEB ACTIVITY 37.3** for a concept review of this chapter.

Circulatory Systems

38

Cardiovascular disease is responsible for about one-third of all deaths each year in the United States and Europe. The immediate cause of most of these deaths is heart attack or stroke, but those events are frequently the end result of arterial disease that begins years before symptoms are detected—hence the designation of cardiovascular disease as a "silent killer."

Healthy arteries have a smooth internal lining of endothelial cells that can be damaged by chronic high blood pressure, smoking, a high-fat diet, or infection. A process of plaque formation occurs at these sites of damage. The damaged endothelial cells attract certain white blood cells to the site. These cells are joined by smooth muscle cells migrating from the deeper layers of the arterial wall. Lipids, especially cholesterol, are deposited in these cells, so that the developing plaque becomes fatty. Fibrous connective tissue made by the invading smooth muscle cells and deposits of calcium make the artery wall stiffer—hence "hardening of the arteries." The growing plaque deposit narrows the artery and causes turbulence in the blood flow. Blood platelets stick to the plaque and initiate the formation of a blood clot (thrombus) that further occludes the artery or breaks loose and blocks a smaller downstream artery.

Coronary arteries supply blood to the heart muscle. These vessels are particularly susceptible to plaque formation, and as they narrow, blood flow to the heart muscle decreases, causing chest pain and shortness of breath during mild exertion. If a thrombus forms or lodges in one of these arteries, blood flow to a portion of the heart muscle is cut off, and the result is a heart attack. The weakened heart cannot pump enough blood to meet the demands of the body, and the victim

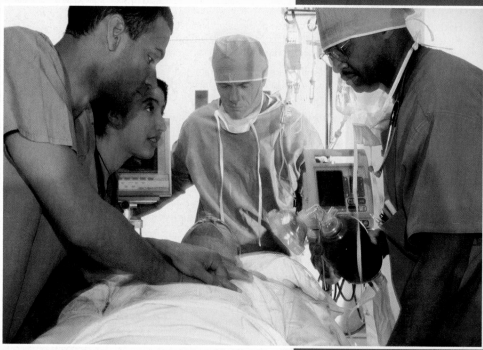

typically collapses. The events that follow are hardly "silent": 911, sirens, flashing lights, paramedics, the emergency room, and the frantic treatments commonly seen on TV medical dramas.

Quick treatment is critical if the patient is to survive the initial consequences of a heart attack. The brain cannot be without blood flow for more than a few minutes, and the functions of other organs are compromised and the heart muscle itself will die if blood flow is not restored. First priorities are to get blood flowing by means of CPR, get the heart beating if it has stopped, and clear the blocked artery. These actions may help the patient survive, but they may not be enough to ensure recovery; for this, the responses of the patient over the next few days are critical.

Q **QUESTION**
What are the critical factors that determine whether a person recovers from a heart attack?

Emergency medical workers need to work quickly in cases of heart attack to ensure the patient's survival and recovery.

KEY CONCEPTS

38.1 Circulatory Systems Can Be Open or Closed

38.2 Circulatory Systems May Have Separate Pulmonary and Systemic Circuits

38.3 A Beating Heart Propels the Blood

38.4 Blood Consists of Cells Suspended in Plasma

38.5 Blood Circulates through Arteries, Capillaries, and Veins

38.6 Circulation Is Regulated by Autoregulation, Nerves, and Hormones

concept 38.1 Circulatory Systems Can Be Open or Closed

The function of a **circulatory system** is to transport substances around the body—hormones, respiratory gases, blood cells, nutrients, waste products, and more. A circulatory system consists of a muscular pump (the heart), a fluid (blood), and a series of conduits (blood vessels) through which the fluid can be circulated.

Single-celled organisms and some very small multicellular organisms do not need a circulatory system because nutrients, respiratory gases, and wastes diffuse directly between the cells of their bodies and the environment. Some larger aquatic multicellular animals have highly branched central cavities called *gastrovascular systems* that essentially bring the external environment into close contact with every cell. All the cells of a sponge, for example, are in contact with, or very close to, the surrounding seawater. However, this system only works because sponges are not very active. Large, active animals need a true circulatory system.

Open circulatory systems move extracellular fluid

In **open circulatory systems**, the circulatory fluid (called *hemolymph*) leaves the vessels of the circulatory system, flows slowly between cells and through tissues, and then flows back into the heart or vessels of the circulatory system. Open circulatory systems are found in arthropods, mollusks (**FIGURE 38.1A**), and several other invertebrate groups.

Closed circulatory systems circulate blood through a system of blood vessels

In **closed circulatory systems**, blood vessels keep the circulating fluid (*blood*) separate from the fluid around the cells (*interstitial fluid*). The blood consists of a liquid *blood plasma* (containing dissolved solutes) and *blood cells*. Blood cells and large molecules stay within the circulatory system, but water and low-molecular-weight solutes leak out of the smallest vessels, the *capillaries*, which are highly permeable. Together, the blood plasma and the interstitial fluid make up the *extracellular fluid*.

Closed circulatory systems are found in annelids, vertebrates, and a few other groups. The earthworm (an annelid) provides a good example of a simple closed circulatory system. This animal has a large ventral blood vessel that carries blood from the worm's anterior end to its posterior end (**FIGURE 38.1B**). Small vessels branch off the ventral blood vessel and transport the blood to tiny capillaries, where respiratory gases, nutrients, and wastes diffuse between the blood and interstitial fluid. The blood then flows into larger vessels that lead into one large dorsal vessel, which carries the blood back to the anterior end of the body. Five pairs of muscular vessels connect the dorsal and ventral vessels. The dorsal vessel and the five connecting vessels also serve as simple hearts, contracting to keep the blood circulating. One-way valves keep the blood flowing in the right direction.

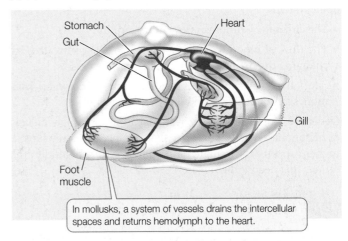

(A) Open circulatory system: mollusk

Stomach
Gut
Heart
Gill
Foot muscle

In mollusks, a system of vessels drains the intercellular spaces and returns hemolymph to the heart.

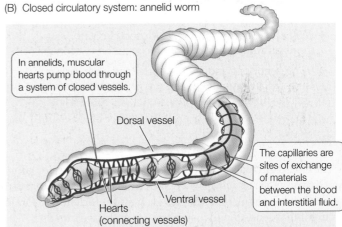

(B) Closed circulatory system: annelid worm

In annelids, muscular hearts pump blood through a system of closed vessels.

Dorsal vessel

The capillaries are sites of exchange of materials between the blood and interstitial fluid.

Ventral vessel

Hearts (connecting vessels)

FIGURE 38.1 Circulatory Systems (A) Open circulatory systems are found in mollusks (illustrated here by a clam) and arthropods. Hemolymph is pumped by a tubular heart to different regions of the body, where it leaves the circulatory vessels and flows through intercellular spaces. (B) Annelids (illustrated by an earthworm) and vertebrates have a closed circulatory system, in which the blood is confined within a continous system of vessels. The blood is pumped through those vessels by one or more muscular hearts.

Vertebrates have a closed circulatory system with a similar basic arrangement but with a more elaborate heart that has two or more chambers. In vertebrates, a multichambered **heart** pumps blood into large vessels called **arteries** that carry blood away from the heart. Arteries branch into smaller vessels called **arterioles**, which feed blood into **capillary beds**, networks of minute, thin-walled **capillaries** where materials are exchanged between the blood and the interstitial fluid. Small vessels called **venules** drain capillary beds. The venules join to form large **veins** that deliver blood back to the heart. Valves in the heart and in some of the veins prevent the backflow of blood.

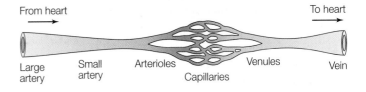

From heart

To heart

Large artery Small artery Arterioles Capillaries Venules Vein

Closed circulatory systems have several advantages over open systems:

- The circulatory fluid can flow more rapidly.
- The flow of blood to specific tissues can be controlled, by changing the diameter (hence resistance) of vessels.
- Specialized cells and molecules that transport oxygen, hormones, and nutrients can be kept in the vessels.

Generally, closed circulatory systems can support higher levels of metabolic activity than open systems because they have a higher capacity to circulate blood and can direct blood to the tissues that need it the most.

Do You Understand Concept 38.1?

- What characteristics of a sponge enable it to function without a circulatory system?
- What are the advantages of closed circulatory systems as compared with open circulatory systems?
- A grasshopper and a clam both have open circulatory systems. What enables the grasshopper to be more active and sustain a higher metabolic rate? (Hint: review the respiratory systems of insects in Concept 37.2, especially Figure 37.2.)

The closed circulatory system clearly has advantages for large, active animals. However, there is a further refinement that can make closed circulatory systems even more efficient: a separate circuit that sends blood just to the gas exchange organs.

concept 38.2 Circulatory Systems May Have Separate Pulmonary and Systemic Circuits

Closed circulatory systems in vertebrates move blood through the respiratory gas exchange organs and through the rest of the body. In fish there is a single circuit in which blood goes from the heart to the respiratory organ (gills) and then directly to the rest of the body and back to the heart. In birds, crocodilians, and mammals, there are separate circuits: the **pulmonary circuit** takes blood from the heart to the gas exchange organs (lungs) and back to the heart, and the **systemic circuit** takes blood from the heart to the rest of the body and back to the heart. In amphibians and all reptiles except birds and crocodilians, these two circuits are not completely separate. Here we will examine the anatomy and evolution of these different vertebrate circulatory systems.

yourBioPortal.com
Go to WEB ACTIVITY 38.1
Vertebrate Circulatory Systems

Most fishes have a two-chambered heart and a single circuit

Most fishes have a simple two-chambered heart. The first chamber of the heart is a thin-walled **atrium** that receives blood from the body and pumps it into a more muscular chamber, the **ventricle**. The ventricle pumps blood to the gills, which are arranged on supportive gill arches (see Figure 37.4). Blood flows into each gill arch in an *afferent arteriole* and leaves the gill in an *efferent arteriole*. The efferent arterioles join together into a single dorsal aorta that carries blood to capillaries in the rest of the body tissues.

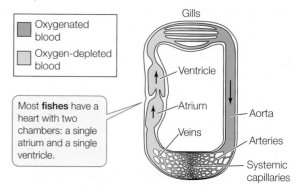

This simple design places a limit on an important feature of the circulatory system—the pressure that propels blood through the non-respiratory tissues and organs. Most of the pressure imparted to the blood by the contraction of the ventricle is lost as the blood travels through the narrow gill lamellae. Therefore, blood entering the aorta is under low pressure and moves relatively slowly as it travels to the rest of the body. This limits the maximum capacity of the fish circulatory system to supply tissues with oxygen and nutrients.

Lungfishes evolved a partially separate circuit that serves the lung

The evolutionary transition from breathing water to breathing air had important consequences for the vertebrate circulatory system. The African lungfishes (several species in the genus *Protopterus*) show how the circulatory system changed to serve a primitive lung.

African lungfish have gills and also a simple lung that forms as an outpocketing of the gut, near the gills. The blood vessels in the posterior pair of gill arteries have been modified to be a low-resistance path for blood to the lung. A new vessel carries oxygenated blood from the lung back to the heart. In addition, two anterior gill arches have lost their gills, and their blood vessels deliver blood from the heart directly to the dorsal aorta. The lungfish heart has a partially divided atrium, with the left side receiving oxygenated blood from the lungs, and the right side receiving oxygen-depleted blood from the other tissues. These two bloodstreams stay mostly separate as they flow through the ventricle and the large vessel leading to the gill arches. As a result, oxygenated blood goes mostly to the anterior gill arteries leading to the dorsal aorta, and oxygen-depleted blood goes mostly to those arteries that lead to the gills and to the lung.

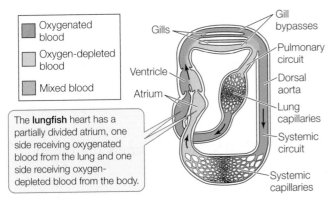

The **lungfish** heart has a partially divided atrium, one side receiving oxygenated blood from the lung and one side receiving oxygen-depleted blood from the body.

The lung probably evolved as a means of supplementing the oxygen uptake from the gills. Lungfishes are periodically exposed to water with low oxygen content and to situations in which their aquatic environment dries up. When the water is fully oxygenated, lungfishes can depend on their gills; but in oxygen-depleted water, they can augment their oxygen intake by gulping air. The associated modifications of the lungfish vascular system set the stage for the evolution of separate pulmonary and systemic circulations in terrestrial vertebrates.

Amphibians and most reptiles have partially separated circuits

In adult amphibians and most reptiles, the pulmonary and systemic circuits are even more fully separated—each circuit has its own set of arteries, capillaries, and veins. The heart has two separate atria and thus is three-chambered. One atrium receives oxygenated blood from the lungs, and the other receives oxygen-depleted blood from the rest of the body. However, the oxygen-depleted blood and oxygenated blood mix to some degree as they pass into a single or only partially divided ventricle.

In adult amphibians, the ventricle has anatomical features that keep the oxygenated and oxygen-depleted blood somewhat separate. Compared with the circulatory system of fishes, this system can deliver blood at higher pressure into the systemic circuits, but the blood may not be totally oxygenated.

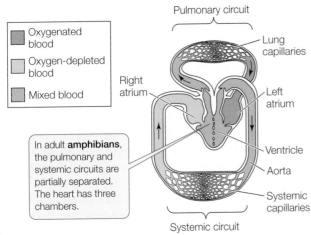

In adult **amphibians**, the pulmonary and systemic circuits are partially separated. The heart has three chambers.

Turtles, lizards, and snakes have taken the amphibian design one step further by evolving a partially divided ventricle, so that oxygen-depleted and oxygenated blood are kept almost entirely separate. The partial connection between the two sides of the ventricle is highly adaptive for the lifestyles of these animals. Some reptiles have burst activity but low resting metabolic rates, whereas others, such as sea turtles, do not breathe for long periods when they are underwater. If an animal doesn't need more oxygen (e.g., when it is resting) or isn't breathing, it is a waste of energy to pump blood through the lungs. These reptiles increase the efficiency of their circulatory system by being able to shunt blood from the pulmonary to the systemic circuit, and that ability depends on the anatomy of their heart and the arteries leaving it.

During development, all vertebrates have two systemic arteries leaving their heart—a left and a right aorta. Birds lose the left aorta during fetal life, while mammals lose the right. In turtles, snakes, and lizards, both aortas remain, with the right aorta connected to the heart close to the location where the two sides of the ventricle are in communication. When the animal is breathing, the resistance in the pulmonary circuit is lower than the resistance in the systemic circuit, so blood from the right side of the ventricle mainly flows into the pulmonary artery that leaves the right side of the heart; blood from the left side of the ventricle mainly flows into the systemic circuit through the left and the right aortas. When the animal is not breathing, the resistance in the pulmonary circuit goes up. This higher resistance limits blood flow in the pulmonary circuit, and therefore blood from the right side of the ventricle flows primarily into the right aorta and into the systemic circuit.

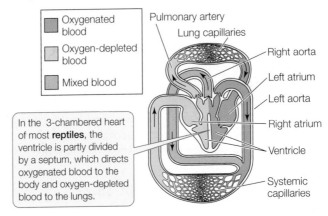

In the 3-chambered heart of most **reptiles**, the ventricle is partly divided by a septum, which directs oxygenated blood to the body and oxygen-depleted blood to the lungs.

Crocodilians, birds, and mammals have fully separated pulmonary and systemic circuits

Crocodilians, birds, and mammals have a four-chambered heart and completely separate pulmonary and systemic circuits. Oxygen-depleted blood from the body arrives in the right atrium of the heart and flows from there into the right ventricle. The right ventricle pumps the blood through **pulmonary arteries** to the lungs, where the blood picks up oxygen and sheds carbon dioxide. The oxygenated blood then returns to the heart in **pulmonary veins**, arriving in the left atrium of the heart from where it flows into the powerful left ventricle. The left ventricle pumps the oxygenated blood into a massive artery called the **aorta** that sends blood at high pressure to capillary beds in the rest of the body, where the blood delivers oxygen and picks up carbon dioxide. Oxygen-depleted blood then returns in the **vena cavae**, large veins that empty into the right atrium of the heart.

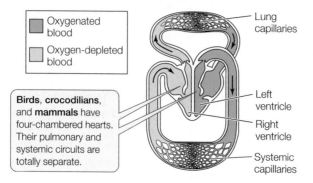

The pulmonary arteries, lung capillaries, and pulmonary veins make up the pulmonary circuit. The aorta, the capillary beds in the rest of the body, and the associated veins make up the systemic circuit. The right side of the heart sends blood to the pulmonary circuit; the left side sends blood to the systemic circuit. But, since these two circuits are connected in series, each side of the heart is pumping the same volume of blood.

Fully separate circuits have several advantages:

- Oxygenated blood can be distributed to the tissues of the body at higher pressure and flow than is possible in fishes.
- Oxygenated and oxygen-depleted blood cannot mix. The systemic circuit always receives blood with the highest oxygen content, and the pulmonary circuit always receives blood with the lowest oxygen content, maximizing respiratory gas exchange.
- The pressures in the systemic and pulmonary circuits are different; low pressure in the pulmonary circuit protects the delicate lung tissues, and high pressure in the systemic circuit effectively delivers blood throughout the body.

Fully separate pulmonary and systemic circuits enable crocodilians, birds, and mammals to maintain a continuously high metabolic rate and to direct blood flow to where it is most needed.

Do You Understand Concept 38.2?

- Draw sketches of circulatory systems showing a single and a double circuit design. Label all of the parts.
- Why can a mammalian circulatory system have different pressures in the pulmonary and the systemic circuits, and why do you think this is important?
- Crocodiles have a fully separated four-chambered heart, but like reptiles they have two aortas and a low resting metabolic rate, and they spend time underwater. Thus they do not breathe continuously. Their right aorta leaves their right ventricle, but just outside the heart there is a connection between the left and the right aortas. Knowing that pressure is higher in the systemic circuit than in the pulmonary circuit and that the resistance in the pulmonary circuit goes up when an animal is not breathing, how do you think the crocodilian system functions to decrease blood flow to the lungs when these animals are not breathing?

All of these circulatory systems, whether single circuit or double circuit, require a steadily beating muscular pump—a heart—to propel the blood. In the next concept we will study the structure and function of the heart of a mammal—the human.

---

concept
38.3   **A Beating Heart Propels the Blood**

Your heart is a muscular pump that, at rest, beats an average of 60–70 times per minute. With each beat, it circulates about 70 milliliters of blood throughout the body. Without taking work or exercise into account, that is 300 liters per hour, 7,200 liters per day, 2.6 million liters per year—without a break. How does the heart accomplish this never-ending task?

Blood flows from right heart to lungs to left heart to body

The structure of the four-chambered human heart is shown in **FIGURE 38.2**. The right atrium receives oxygen-depleted blood from the venae cavae, and also from veins coming from the heart muscle itself. From the right atrium, the blood flows through an *atrioventicular (AV) valve* into the right ventricle. Most of the filling of the ventricle results from passive flow while the heart is relaxed between beats. Just at the end of this interbeat interval, the atrium contracts and adds a little more blood to the ventricle. The right ventricle then contracts, causing the one-way flaps of the AV valve to close (preventing backflow into the atrium) and pumping blood into the pulmonary artery leading to the lungs. Once the ventricle relaxes, the *pulmonary valve* at the base of the pulmonary artery closes to prevent backflow into the heart.

Pulmonary veins return oxygenated blood from the lungs to the left atrium. From here blood flows into the left ventricle through another AV valve. As on the right side of the heart, most left ventricular filling occurs while the ventricle is relaxing, but at the end of this interbeat interval, atrial contraction adds a little more blood to the ventricle.

When the left ventricle contracts, it pumps the blood into the aorta to begin its circulation through the body. Since the systemic circuit has many more kilometers of blood vessels than does the pulmonary circuit, it presents a higher resistance to flow. To meet this challenge, the left ventricle has thicker walls than the right ventricle, contracts with more force, and generates higher pressure. When the left ventricle relaxes, the *aortic valve* at the base of the aorta closes to prevent backflow.

During exercise, a larger volume of blood returns to the heart between beats, and the ventricles must contract much more forcefully to move this increased volume. This adjustment in force of contraction occurs automatically because of a property of cardiac muscle cells known as the **Frank–Starling law**: if the cardiac muscle cells are stretched, as they are when the volume of returning blood increases, they contract more forcefully.

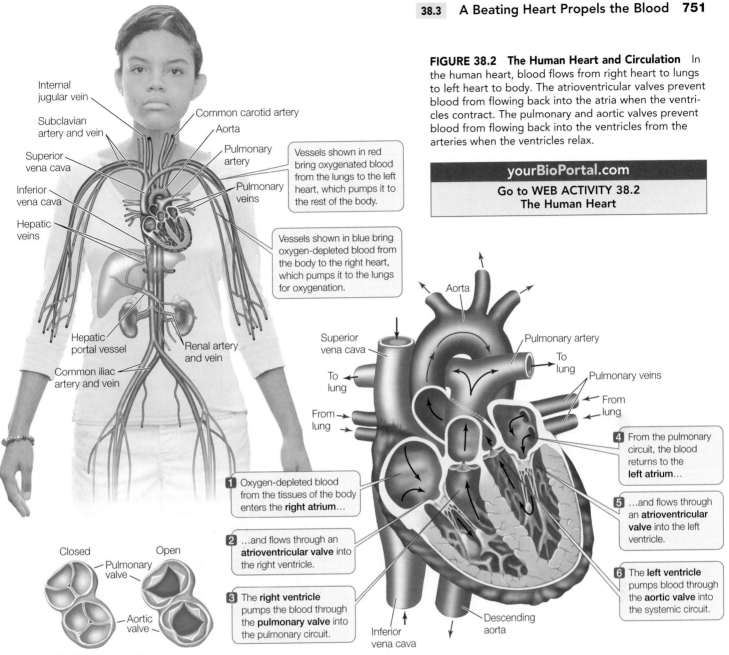

FIGURE 38.2 The Human Heart and Circulation In the human heart, blood flows from right heart to lungs to left heart to body. The atrioventricular valves prevent blood from flowing back into the atria when the ventricles contract. The pulmonary and aortic valves prevent blood from flowing back into the ventricles from the arteries when the ventricles relax.

yourBioPortal.com

Go to WEB ACTIVITY 38.2
The Human Heart

Vessels shown in red bring oxygenated blood from the lungs to the left heart, which pumps it to the rest of the body.

Vessels shown in blue bring oxygen-depleted blood from the body to the right heart, which pumps it to the lungs for oxygenation.

Internal jugular vein
Subclavian artery and vein
Superior vena cava
Inferior vena cava
Hepatic veins
Hepatic portal vessel
Common iliac artery and vein
Common carotid artery
Aorta
Pulmonary artery
Pulmonary veins
Renal artery and vein

Superior vena cava
To lung
From lung
Inferior vena cava
Aorta
Pulmonary artery
To lung
Pulmonary veins
From lung
Descending aorta

1 Oxygen-depleted blood from the tissues of the body enters the **right atrium**…

2 …and flows through an **atrioventricular valve** into the right ventricle.

3 The **right ventricle** pumps the blood through the **pulmonary valve** into the pulmonary circuit.

4 From the pulmonary circuit, the blood returns to the **left atrium**…

5 …and flows through an **atrioventricular valve** into the left ventricle.

6 The **left ventricle** pumps blood through the **aortic valve** into the systemic circuit.

Closed
Open
Pulmonary valve
Aortic valve

The cycle of contraction and relaxation of the heart is called the **cardiac cycle**. The cardiac cycle is divided into two phases: **systole** (pronounced sís-toll-ee), when the ventricles contract, and **diastole** (die-ás-toll-ee), when the ventricles relax (**FIGURE 38.3**). The atria contract a moment before the ventricles do, just at the end of diastole, to top off the volume of blood in the ventricles.

The sounds of the cardiac cycle, the "lub-dup" heard through a stethoscope, are created by the heart valves slamming shut. These valves open and close passively because of pressure differences. As the ventricles begin to contract, the pressure in them rises above the pressure in the atria, so the AV valves snap closed ("lub"). When the ventricles begin to relax, the high pressure in the aorta and pulmonary artery closes the aortic and pulmonary valves ("dup").

Defective valves that do not close completely produce turbulent blood flow and the sounds known as *heart murmurs*. For

example, if an AV valve does not close completely, blood will flow back into the atrium with a "whoosh" following the "lub."

The heartbeat originates in the cardiac muscle

Cardiac muscle has unique adaptations that enable it to function as a pump. First, cardiac muscle cells are in electrical contact with one another through gap junctions that enable *action potentials* to spread rapidly from cell to cell. Because a spreading action potential stimulates contraction, large groups of cardiac muscle cells contract in unison. This coordinated contraction is essential for pumping blood effectively.

LINK Action potentials are described in Concept 34.2; Concept 36.1 provides more details about the special characteristics of cardiac muscle

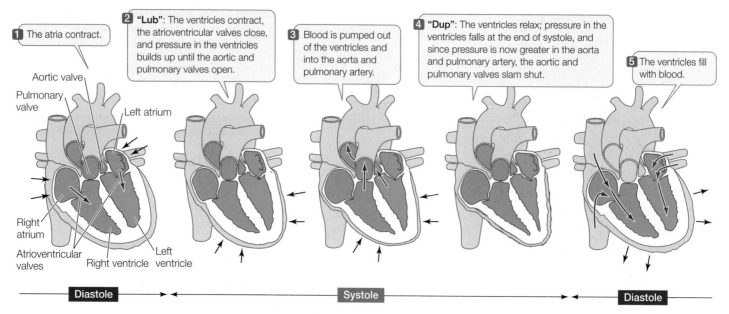

FIGURE 38.3 The Cardiac Cycle The rhythmic contraction (systole) and relaxation (diastole) of the ventricles is called the cardiac cycle.

---

yourBioPortal.com

Go to ANIMATED TUTORIAL 38.1
The Cardiac Cycle

Second, some cardiac muscle cells are **pacemaker cells** that can initiate action potentials without stimulation from the nervous system. When they fire action potentials, they stimulate neighboring cells to contract. The primary pacemaker of the heart is a group of modified cardiac muscle cells, the **sinoatrial node**, located at the junction of the superior vena cava and right atrium (see Figure 38.5). The electrical properties of pacemaker

cells are different from those of neurons and other muscle cells because of the types of ion channels in their membranes.

The action potential of the pacemaker cells is generated by voltage-gated Ca^{2+} channels. This is unlike the voltage-gated Na^+ channel mechanism in neurons and other muscle cells, and therefore the shape of the action potential of pacemaker cells is different (compare Figure 38.4 and Figure 34.7). Most important, however, is that the resting potential of these pacemaker cells is not stable, but slowly drifts upward. There are two reasons for this unstable resting potential. First, these cells are more permeable to Na^+ than are other cardiac muscle cells, as indicated by their higher resting potential; and second, their K^+ channels that open to repolarize the membrane following an action potential gradually close. As these channels close, the balance between Na^+ influx and K^+ efflux changes, and the membrane potential

---

APPLY THE CONCEPT

A beating heart propels the blood

During the cardiac cycle, each chamber of the heart goes through characteristic changes in pressure and volume. The graph shows measurements of the volume of the left ventricle and the pressure in the left ventricle during a single cardiac cycle. The sequence of these measurements goes from A to B to C to D to A.

1. Systole is between which two points?

2. Diastole is between which two points?

3. How much blood does the heart pump with each cycle?

4. Where on the curve would the first ("lub") and second ("dup") heart sounds occur?

5. Where on the curve would the aortic valve open?

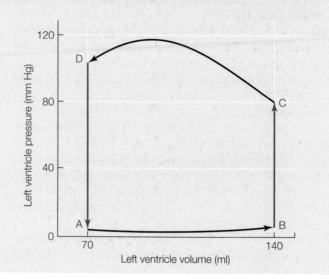

INVESTIGATION

FIGURE 38.4   The Autonomic Nervous System Controls Heart Rate   The membrane potentials of pacemaker cells spontaneously depolarize until action potential threshold is reached. Neurotransmitter signals from the two divisions of the autonomic nervous system speed up and slow down the rate at which the pacemaker membrane potential drifts upward, thereby controlling the rate at which pacemaker cells fire action potentials.

HYPOTHESIS

The ANS neurotransmitters norepinephrine (NE) and acetylcholine (ACh) influence the membrane potentials of pacemaker cells by altering the properties of the ion channels that determine membrane potential.

METHOD
1. Culture living sinoatrial node tissue in a dish. Insert an intracellular recording electrode into pacemaker cells.
2. Measure the membrane potential of pacemaker cells during a resting heartbeat (the control) and after applications of the ANS neurotransmitters NE (sympathetic) and ACh (parasympathetic).

RESULTS

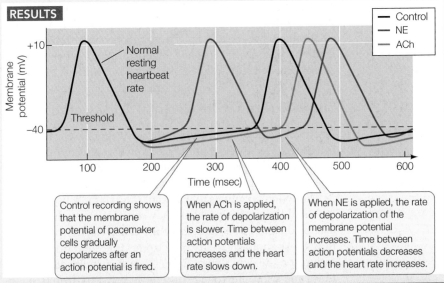

Control recording shows that the membrane potential of pacemaker cells gradually depolarizes after an action potential is fired.

When ACh is applied, the rate of depolarization is slower. Time between action potentials increases and the heart rate slows down.

When NE is applied, the rate of depolarization of the membrane potential increases. Time between action potentials decreases and the heart rate increases.

CONCLUSION

The ANS neurotransmitters NE and ACh influence heart rate by altering the membrane resting potentials of pacemaker cells.

ANALYZE THE DATA

Additional experiments were performed using specific Ca^{2+}, Na^+, and K^+ channel blockers to investigate the roles of these ion channels in generating action potentials (spikes) in pacemaker cells. The table shows the results.

Treatment	Spike height	Spike width	Interspike intervals
Control	+10 mv	100 msec	300 msec
Add Ca^{2+} channel blocker		Action potentials cease	
Add NE	+10 mv	100 msec	200 msec
Add ACh	+10 mv	100 msec	350 msec
Add Na^+ channel blocker and NE	+10 mv	100 msec	250 msec
Add K^+ channel blocker	+10 mv	100 msec	250 msec

A. What ion channel is responsible for the spikes?

B. How do the ANS neurotransmitters alter the characteristics of the pacemaker membrane potential?

C. What ion channels are likely involved in the actions of the ANS neurotransmitters?

Go to **yourBioPortal.com** for original citations, discussions, and relevant links for all INVESTIGATION figures.

---

drifts up until the threshold that opens the voltage-gated Ca^{2+} channels is reached.

Because of the pacemaker cells, the heart initiates its own contractions; that is, it does not need repetitive signals from nerves to beat. Instead, the pacemaker cells automatically depolarize, eventually reaching a threshold that triggers an action potential, repolarize, and repeat the entire process continuously. The pacemaker action potential rapidly spreads to other cardiac cells, triggering systole.

The autonomic nervous system does control the *rate* of the heartbeat, however, by speeding up or slowing down the pacemaker cells' rate of depolarization (**FIGURE 38.4**). *Norepinephrine* released onto pacemaker cells by sympathetic nerves increases the permeability of the Na^+ channels and the Ca^{2+} channels, causing the pacemaker cells to depolarize more rapidly. As a result, the heart beats faster. Conversely, the parasympathetic neurotransmitter acetylcholine has the opposite effect, slowing the rate of depolarization and slowing down the heartbeat.

A conduction system coordinates the contraction of heart muscle

A normal heartbeat begins with an action potential in the sinoatrial node (**FIGURE 38.5**). This action potential spreads rapidly throughout the electrically coupled cells of the atria, causing them to contract. However, there are no gap junctions between the cells of the atria and those of the ventricles, so the action potential does not spread directly to the ventricles. Therefore, the ventricles do not contract in unison with the atria.

How does the action potential move from the atria to the ventricles? Situated at the junction of the atria and the ventricles is a nodule of modified cardiac muscle cells—the **atrioventricular node**—which is stimulated by the depolarization of the atria. With a slight delay, it generates action potentials that are conducted to the ventricles via the **bundle of His**, which consists of modified cardiac muscle fibers that do not contract. These fibers divide into right and left *bundle branches* that run to the very bottom of the ventricles (the apex of the heart) and then spread throughout the ventricular

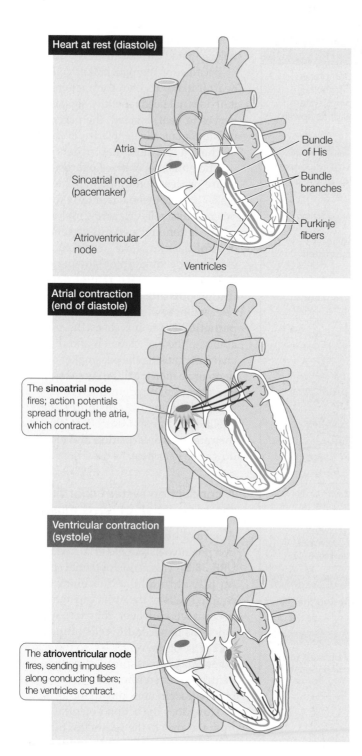

Heart at rest (diastole)

Atria

Sinoatrial node
(pacemaker)

Atrioventricular
node

Ventricles

Bundle
of His

Bundle
branches

Purkinje
fibers

**Atrial contraction
(end of diastole)**

The **sinoatrial node**
fires; action potentials
spread through the atria,
which contract.

**Ventricular contraction
(systole)**

The **atrioventricular node**
fires, sending impulses
along conducting fibers;
the ventricles contract.

FIGURE 38.5 The Heartbeat Pacemaker cells in the sinoatrial node initiate the heartbeat by firing action potentials that spread through the electrically coupled atrial muscle. The atrial action potential eventually spreads to the atrioventricular node, which, with a delay, conducts it through the bundle of His and Purkinje fibers to the cells of the ventricles.

imposed by the atrioventricular node ensures that the atria contract before the ventricles do.

Electrical properties of ventricular muscles sustain heart contraction

The muscle fibers of the ventricles contract for about 300 milliseconds—much longer than the contractions of skeletal muscle fibers. As in neuronal and skeletal muscle fibers, the rising phase of the ventricular muscle cell action potential is due to the opening of voltage-gated Na^+ channels. Unlike neurons and skeletal muscle fibers, however, ventricular muscle cells remain depolarized for a long time. This extended plateau of the action potential is due to sustained opening of voltage-gated Ca^{2+} channels. As in other muscle, cardiac muscle contraction is stimulated by the release of Ca^{2+} from the sarcoplasmic reticulum and the binding of that Ca^{2+} to troponin (see Figure 36.6). As long as the Ca^{2+} channels remain open and Ca^{2+} remains in the sarcoplasm, the ventricular muscle cells continue to contract. The sustained contraction of ventricular muscle fibers facilitates the emptying of ventricles.

The ECG records the electrical activity of the heart

Electrical events in the cardiac muscle during the cardiac cycle can be recorded by electrodes placed on the surface of the body. Such a recording is an **electrocardiogram** (**ECG**), an important tool for diagnosing heart problems (**FIGURE 38.6**).

The action potentials that sweep through the muscles of the atria and the ventricles when they contract are such massive, localized electrical events that they cause electric currents to flow throughout the body. Electrodes placed at different locations on the skin detect those electric currents at different times and register a voltage difference between them (see Figure 38.6B). The wave patterns of the ECG are designated P, Q, R, S, and T, each letter representing a particular event in the cardiac muscle, as shown in the figure.

Do You Understand Concept 38.3?

- A red blood cell is flowing through the superior vena cava. List all of the heart chambers, valves, vessels, and organs it will pass through before it leaves the heart in the aorta.

- How do pacemaker cells generate the cardiac cycle?

- Why do the atria contract before the ventricles? What would happen if they contracted at the same time?

- The drug digitalis impedes the Ca^{2+} transporter that moves Ca^{2+} into the sarcoplasmic reticulum. Explain why digitalis is effective in strengthening weakened hearts.

muscle mass as **Purkinje fibers**. These conducting fibers ensure that the cardiac action potential spreads rapidly throughout the ventricles. The short delay in the spread of the action potential

Let's turn now to the fluid that the heart is pumping: the blood. What are the major components of blood, and what are their major functions?

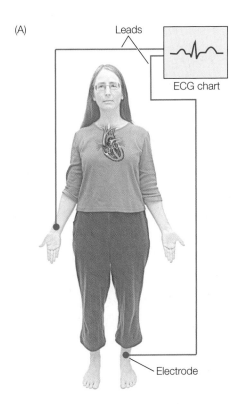

(A) Leads

ECG chart

Electrode

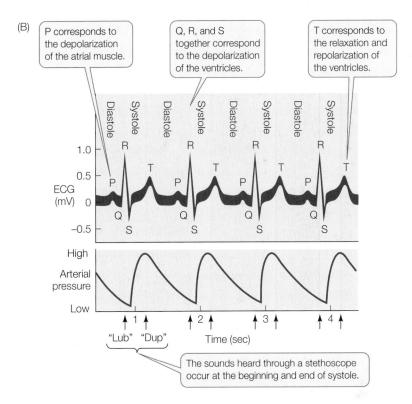

(B)

P corresponds to the depolarization of the atrial muscle.

Q, R, and S together correspond to the depolarization of the ventricles.

T corresponds to the relaxation and repolarization of the ventricles.

ECG (mV)

Diastole | Systole | Diastole | Systole | Diastole | Systole | Diastole | Systole

High
Arterial pressure
Low

"Lub" "Dup"   Time (sec)

The sounds heard through a stethoscope occur at the beginning and end of systole.

FIGURE 38.6 The Electrocardiogram An ECG records heart function. Electrodes attached to various parts of the body record electrical events that correspond to the sequence of contraction and relaxation of the atria and the ventricles.

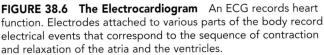

concept 38.4 Blood Consists of Cells Suspended in Plasma

Blood consists of cells suspended in a liquid extracellular matrix called **plasma**. The plasma contains many solutes, such as glucose and other nutrients, ions, waste products, hormones, respiratory gases, clotting proteins, and many other components (**FIGURE 38.7**). Most of the cells suspended in the plasma are **erythrocytes** (red blood cells, or RBCs), which carry respiratory gases. Blood also contains a smaller number of **leukocytes** (white blood cells)—cells of the immune system—as well as **platelets** (pinched-off fragments of cells), which are involved in blood clotting. Here we discuss the red blood cells and platelets.

LINK The functions of leukocytes in the immune system are discussed in Chapter 31

Red blood cells transport respiratory gases

Mature RBCs are biconcave, flexible discs packed with hemoglobin. Their function is to transport respiratory gases. Their shape gives them a large surface area for gas exchange, and their flexibility enables them to squeeze through narrow capillaries (see Figure 38.10).

RBCs and all the other cellular components of blood are generated by stem cells in the bone marrow, particularly in the ribs, breastbone, pelvis, vertebrae, and the long bones of the limbs. RBC production is controlled by a hormone, **erythropoietin**, which is released by cells in the kidneys in response to insufficient oxygen—**hypoxia**. Increased circulating erythropoietin extends the lives of mature RBCs and stimulates production of new RBCs in the bone marrow, thereby increasing the blood's capacity for carrying oxygen.

FRONTIERS Hypoxia inducible factor (HIF) is a transcription factor that was first discovered as part of the mechanism whereby low oxygen levels stimulate kidney cells to produce erythropoietin. This mechanism is now known to be highly conserved. Transcription factors related to HIF are found in animals ranging from worms to humans and are involved in many developmental and physiological processes. HIF is constantly being produced and broken down, and the effect of hypoxia is to stabilize HIF, thereby increasing its activity. The many roles of HIF in health and disease are just beginning to be discovered.

Under normal conditions, your bone marrow produces about 2 million RBCs every second. Immature RBCs in the bone marrow produce a large amount of hemoglobin, storing it for later use. When an immature RBC nears maturity, the nucleus, endoplasmic reticulum, Golgi apparatus, and mitochondria of the cell begin to break down. In most mammals, the nucleus breaks down too, though in all other vertebrates the nucleus is retained. The new RBC—containing hemoglobin, a few enzymes, and not much else—then enters the circulation.

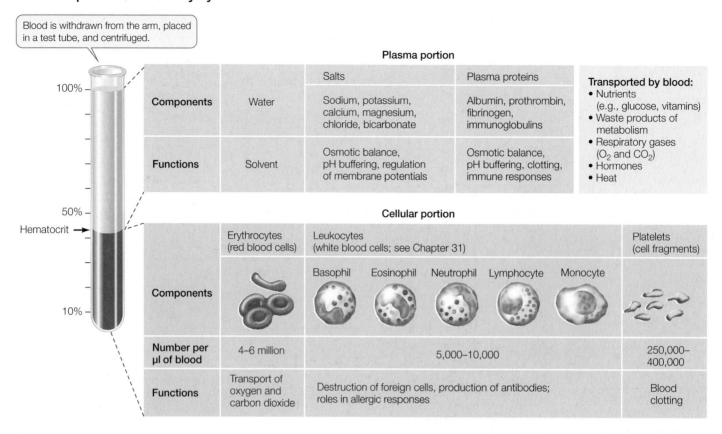

FIGURE 38.7 The Composition of Blood   Blood consists of a complex aqueous solution (the plasma) and numerous cell types and cell fragments. The hematocrit (arrow) is a measure of the cellular portion as a percentage of total blood volume.

In humans, each RBC circulates for about 120 days. As it gets older, its membrane becomes less flexible, and eventually an old red blood cell will rupture as it bends to fit through narrow capillaries. The major site of destruction of old RBCs is the **spleen**, an organ just below the stomach. In the spleen, the blood passes through narrow spaces that cause the less flexible RBCs to rupture. The remnants are taken up and degraded by macrophages (leukocytes that ingest debris and foreign materials). The macrophages recycle the iron from the hemoglobin, sending the iron back into circulation so that new RBCs in the bone marrow can take up the iron to make new hemoglobin.

The RBC content of blood can be quickly assessed by spinning a blood sample in a centrifuge (see Figure 4.4). The cells are spun to the bottom of the tube, leaving the lighter, straw-colored plasma on top (see Figure 38.7). The *hematocrit* is the percentage of blood volume made up by cells, which are primarily RBCs. Normal hematocrit is about 42 percent for women and 46 percent for men, with some variation. A substantially lower number may indicate *anemia*—an insufficient number of RBCs. Anemia can cause body-wide oxygen deprivation, with symptoms such as fatigue, breathlessness, and fainting.

Platelets are essential for blood clotting

Besides producing erythrocytes and leukocytes, the bone marrow stem cells also produce cells called *megakaryocytes.*

Megakaryocytes are large cells that remain in the bone marrow and release platelets into the circulation. A platelet is just a tiny fragment of a cell; it has no organelles but is packed with enzymes and chemicals necessary for its function: sealing leaks in blood vessels and initiating **blood clotting** (**FIGURE 38.8**).

Clotting is triggered when damage to a blood vessel exposes collagen fibers. A platelet that encounters collagen becomes *activated*—it swells, becomes irregularly shaped and sticky, and releases chemicals that activate other platelets and initiate the clotting of blood. The sticky platelets also form a plug at the damaged site.

Blood clotting requires many *clotting factors*, most of which are proteins that circulate in the blood in an inactive form. Activated platelets trigger a complex cascade of biochemical reactions in which many of these clotting proteins are activated in succession. The end result of this cascade is to convert an inactive circulating enzyme, **prothrombin**, to its active form, **thrombin**. Thrombin cleaves molecules of **fibrinogen**, a plasma protein, forming persistant threads of **fibrin**. The fibrin quickly forms a meshwork—a clot—that binds platelets, seals the vessel, and provides a scaffold for the formation of scar tissue.

The absence of any one of the clotting factors can impair clotting, causing excessive bleeding. Because the liver produces most of the clotting factors, liver diseases such as *hepatitis* and *cirrhosis* can result in excessive bleeding. People with *hemophilia* have a genetic inability to produce one of the clotting factors, and if not treated with replacement clotting factor, they can experience uncontrolled bleeding from very minor injuries. A hemophiliac can bleed for weeks from a minor paper cut.

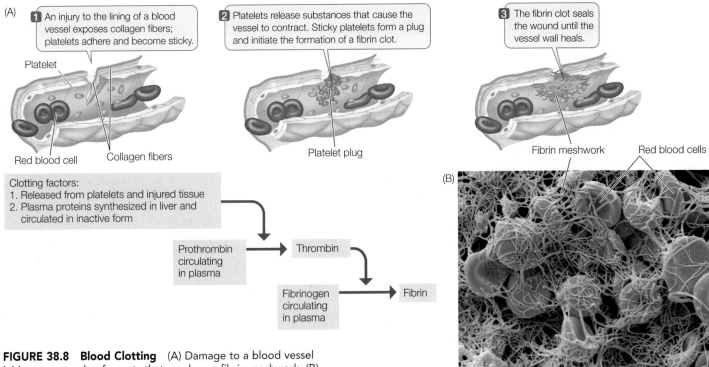

(A)

1 An injury to the lining of a blood vessel exposes collagen fibers; platelets adhere and become sticky.

2 Platelets release substances that cause the vessel to contract. Sticky platelets form a plug and initiate the formation of a fibrin clot.

3 The fibrin clot seals the wound until the vessel wall heals.

Platelet

Red blood cell

Collagen fibers

Platelet plug

Fibrin meshwork

Red blood cells

(B)

Clotting factors:
1. Released from platelets and injured tissue
2. Plasma proteins synthesized in liver and circulated in inactive form

Prothrombin circulating in plasma → Thrombin

Fibrinogen circulating in plasma → Fibrin

FIGURE 38.8 Blood Clotting (A) Damage to a blood vessel initiates a cascade of events that produce a fibrin meshwork. (B) As the meshwork forms, red blood cells are enmeshed in the fibrin threads, forming a clot, as shown in this color-enhanced electron micrograph.

Do You Understand Concept 38.4?

- List the major components of blood (liquid and cellular) and state their functions.

- An excess of clotting components can be as dangerous as a deficiency. How can excessive clotting be dangerous?

- Some competitive cyclists have admitted using erythropoietin as an illegal performance-enhancing drug. How can erythropoietin enhance performance?

The RBCs, white blood cells, and platelets, along with the many dissolved substances in the plasma, circulate continuously around the vertebrate body through a closed network of vessels. Let's turn our attention now to the vessels that carry the blood.

---

concept
38.5 Blood Circulates through Arteries, Capillaries, and Veins

Recall that blood leaves the heart in *arteries* and moves to *arterioles* that feed *capillary beds*, where nutrients and gases are exchanged. Blood leaving capillary beds collects in *venules* that empty into *veins* that conduct the blood back to the heart. These different types of blood vessels all have distinct characteristics that help control the distribution and flow of blood.

Arteries have elastic, muscular walls that help propel and direct the blood

Arteries must withstand much higher pressure than veins. Pressure in the arteries reaches a peak during systole when the ventricles contract, then drops during diastole when the ventricles relax. The walls of the large arteries have thick layers of extracellular collagen and elastin fibers, which enable them to withstand the high blood pressure of systole (**FIGURE 38.9A**). During systole, the pressure wave created by the contraction of the left ventricle surges through the arteries and can be felt as the pulsing of the arteries in the wrists, neck, and elsewhere.

The elastic tissues in artery walls also have another important function: because they are stretched during systole, they store some of the energy imparted to the blood by the heart. Elastic recoil during diastole returns this energy to the blood by squeezing it and pushing it forward. As a result, even though pressure in the arteries drops somewhat during diastole, it never drops all the way to zero. This means that blood continues moving forward, even during diastole when the heart is not contracting.

Arteries and arterioles also contain a layer of smooth muscle that can contract or relax, thereby constricting or dilating the vessel. This change in an artery's diameter in turn changes its resistance to blood flow, and that changes the volume of blood flowing through the artery. As we will see in the next concept, neural and hormonal mechanisms can control the flow of blood to certain tissues by acting on these smooth muscle cells.

APPLY THE CONCEPT

Blood circulates in arteries, capillaries, and veins

A physician measures your blood pressure with an inflatable pressure cuff and a pressure gauge, together called a *sphygmomanometer*, and a stethoscope. Typically the pressure cuff is placed around your upper arm, and the stethoscope is used to listen to the sounds of blood flowing through the artery at the crook of your elbow. When the cuff is inflated to a high enough pressure, only a background rumble is heard in the stethoscope. Now the physician slowly lowers the pressure in the cuff and makes the observations shown at the right.

1. Why is there no sound at the high cuff pressure, and why does the tapping start as the pressure is lowered?

2. Why does the tapping sound disappear as the cuff pressure is further lowered?

CUFF PRESSURE (mm Hg)	SOUND
160	None
140	None
130	None
120	Tapping
100	Tapping
80	Tapping
70	Tapping
60	None
50	None

3. What pressure reading is closest to the systolic pressure? To the diastolic pressure?

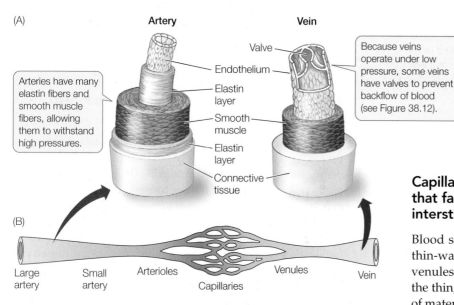

(A)

Artery

Vein

Arteries have many elastin fibers and smooth muscle fibers, allowing them to withstand high pressures.

Valve
Endothelium
Elastin layer
Smooth muscle
Elastin layer
Connective tissue

Because veins operate under low pressure, some veins have valves to prevent backflow of blood (see Figure 38.12).

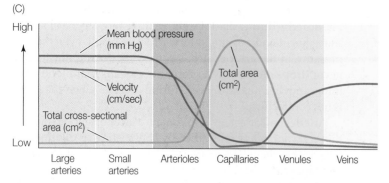

(B)

Large artery · Small artery · Arterioles · Capillaries · Venules · Vein

(C)

High

Mean blood pressure (mm Hg)

Velocity (cm/sec)

Total area (cm²)

Total cross-sectional area (cm²)

Low

Large arteries · Small arteries · Arterioles · Capillaries · Venules · Veins

FIGURE 38.9 Anatomy of Blood Vessels (A) The different anatomical characteristics of arteries and veins match their functions. (B) Blood from the arterial system feeds into capillary beds, where exchanges with the interstitial fluid occur. The venous system returns blood to the heart. (C) The area encompassed by each vessel type is graphed along with the pressure and velocity of the blood within it.

Capillaries have thin, permeable walls that facilitate exchange of materials with interstitial fluid

Blood slows down dramatically when it reaches the thin-walled capillaries that lie between arterioles and venules (**FIGURE 38.9B**). The slow flow of blood and the thin, permeable capillary walls facilitate exchanges of materials between the blood and the interstitial fluid surrounding nearby cells. Most cells lie within a few cell diameters of a capillary.

Why does blood slow down so dramatically in capillaries? Imagine a single garden hose spraying water out at high velocity. If we connected this hose to a large number of lawn sprinklers, the pressure and velocity in each sprinkler would be low. Similarly, a small number of arteries supplies a huge number of capillaries. Even though each capillary has a very small diameter, there are so many capillaries that their total cross-sectional area is much greater than that of the arteries, so the blood pressure and rate of flow decrease greatly when blood reaches the capillaries (**FIGURE 38.9C**).

Capillaries are extremely small vessels that require the RBCs to flow through in single file (**FIGURE 38.10**). Their walls consist of a single layer of thin endothelial cells that are permeable to oxygen, carbon dioxide,

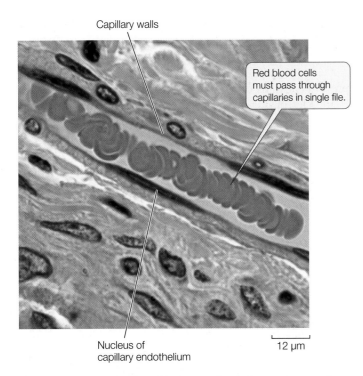

Capillary walls

Red blood cells must pass through capillaries in single file.

Nucleus of capillary endothelium

12 μm

FIGURE 38.10 A Narrow Lane Capillaries have a very small diameter, and blood flows through them slowly.

glucose, lactate, and small ions such as Na^+ and Cl^-. However, capillaries vary in their permeability to larger molecules. In most tissues other than the brain, capillaries have tiny holes, or fenestrations (singular fenestra; "windows"). This allows some larger molecules to pass through, though the largest proteins are still retained in the capillaries. Capillaries are especially permeable in the digestive tract, where nutrients are absorbed, and in the kidneys, where wastes are filtered. In contrast, the capillaries of the brain do not have fenestrations, and therefore not much can pass through them other than lipid-soluble substances, such as alcohol and anesthetics. This low permeability of brain capillaries is known as the blood–brain barrier, and it helps protect the brain from toxins that are not lipid soluble (see Concept 34.1).

Most capillaries are so permeable that at their arterial (high-pressure) end, blood pressure squeezes water and some small solutes out of the capillaries and into the surrounding intercellular spaces. Why don't water and small-molecular-weight solutes collect in the intercellular spaces? How is the blood volume maintained if fluid is continuously leaking out of the capillaries?

More than 100 years ago, the physiologist E. H. Starling suggested that water movement across capillary walls is a result of two opposing forces, which are now known as **Starling's forces**:

- Blood pressure forces water and small solutes out of the capillaries.
- Osmotic pressure pulls water back into the capillaries.

Blood pressure is high at the arterial end of a capillary bed and steadily drops as blood moves toward the venous end (**FIGURE 38.11**). The osmotic pressure is due to the large protein molecules that cannot leave the capillaries, and is assumed to be relatively constant along a capillary. As long as the blood pressure is above the osmotic pressure, fluid leaves the capillary. But at the venule end, blood pressure falls below the osmotic pressure, so fluid returns to the capillary.

Overall, there is still a *slight* net loss of fluid to the intercellular spaces. This loss, about 4 liters per day, percolates between cells and then enters lymphatic vessels, which we will discuss shortly.

Several observations support Starling's model. In people with severe liver disease or protein starvation, a fall in blood protein concentration leads to an accumulation of fluid in the extracellular spaces, which results in tissue swelling, or *edema*. Edema is also characteristic of the inflammation response accompanying tissue damage or allergic responses (see Figure 31.3). *Histamine*, a mediator of inflammation, increases capillary permeability and relaxes the smooth muscles of the arterioles, raising blood pressure in the capillaries and leading to fluid leakage into tissues. The result is the typical swelling associated with inflammation.

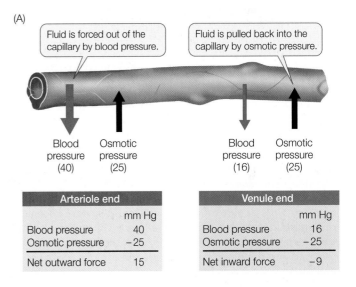

(A)

Fluid is forced out of the capillary by blood pressure.

Fluid is pulled back into the capillary by osmotic pressure.

Blood pressure (40) Osmotic pressure (25) Blood pressure (16) Osmotic pressure (25)

Arteriole end	
	mm Hg
Blood pressure	40
Osmotic pressure	−25
Net outward force	15

Venule end	
	mm Hg
Blood pressure	16
Osmotic pressure	−25
Net inward force	−9

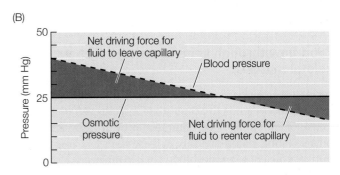

(B)

Net driving force for fluid to leave capillary

Blood pressure

Osmotic pressure

Net driving force for fluid to reenter capillary

Pressure (mm Hg)

FIGURE 38.11 Starling's Forces Starling's model explains how blood volume is maintained in the capillary beds. (A) When blood pressure is greater than osmotic pressure, fluid leaves the capillaries; when blood pressure falls below osmotic pressure, fluid returns to the capillaries. (B) The balance of these two forces changes over the capillary bed as blood pressure falls.

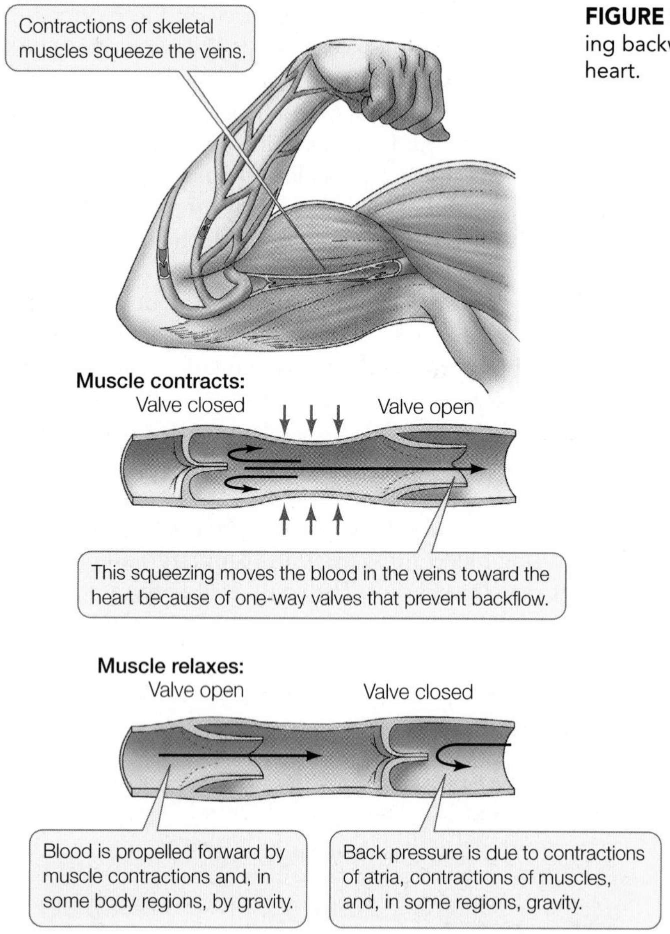

Contractions of skeletal muscles squeeze the veins.

FIGURE 38.12 One-Way Flow Veins have valves that prevent blood from flowing backward, and contractions of skeletal muscle help move blood toward the heart.

Muscle contracts:
Valve closed · Valve open

This squeezing moves the blood in the veins toward the heart because of one-way valves that prevent backflow.

Muscle relaxes:
Valve open · Valve closed

Blood is propelled forward by muscle contractions and, in some body regions, by gravity.

Back pressure is due to contractions of atria, contractions of muscles, and, in some regions, gravity.

FRONTIERS Starling's model does not explain certain puzzling phenomena. During strenuous exercise in humans, blood pressure can be very high, but edema does not occur. Birds have high blood pressure and low osmotic pressure, yet birds do not suffer from edema. Why not? Recent research suggests that bicarbonate ions (HCO_3^-), produced from carbon dioxide in the capillary endothelial cells and released into the blood, can increase the osmotic pressure and thereby serve as a force that pulls water back into the capillary. This hypothesis still requires further research.

Blood flow through veins is assisted by skeletal muscles

The walls of veins are more expandable than the walls of arteries, so blood tends to accumulate in veins. As much as 60 percent of your total blood volume may be in your veins when you are resting. Blood pressure in these veins is so low that it is often insufficient to propel blood back to the heart. How, then, does the blood get back to the heart?

Blood flow through veins that are above the heart is assisted by gravity. Below the level of the heart, however, venous return is against gravity. The most important force propelling blood from these regions is the squeezing of the veins by the contractions of surrounding skeletal muscles. One-way valves in the veins

prevent backflow of blood. Thus whenever a vein is squeezed by muscles, blood is propelled toward the heart (**FIGURE 38.12**).

If the muscles are inactive for a long period of time, gravity causes backpressure in capillary beds of the lower body. This shifts the balance between blood pressure and osmotic pressure, causing increased loss of fluid to the intercellular spaces. That is why your feet swell during a long airline flight.

Two other forces also assist with returning blood to the heart. The muscles involved in inhalation create negative pressure that pulls air into the lungs (see Figure 37.8), and this negative pressure also pulls blood toward the chest. Some of the largest veins contain smooth muscle that contracts at the onset of exercise, thereby increasing venous return to the heart.

Lymphatic vessels return interstitial fluid to the blood

Capillary beds experience a slight but steady loss of fluid to the surrounding interstitial fluid. A separate system of vessels—the **lymphatic system**—collects this excess fluid and returns it to the blood.

Each capillary bed contains at least one blind-ended *lymph capillary* that continuously takes up the excess fluid. Once in a lymphatic vessel, the interstitial fluid is called **lymph**. Fine lymphatic capillaries merge into progressively larger vessels and ultimately into two lymphatic vessels—the **thoracic ducts**—that empty into large veins at the base of the neck (**FIGURE 38.13**). Lymph, like blood, is propelled toward the heart by skeletal muscle contractions, breathing movements, and one-way valves.

Mammals and birds have **lymph nodes** along the major lymphatic vessels. Lymph nodes produce and house lymphocytes that continuously screen the lymph fluid for pathogens, removing microorganisms and other foreign material by phagocytosis.

Do You Understand Concept 38.5?

- In which vessels of the circulatory system does the blood move most rapidly? In which does it move most slowly? Why?

- Compare and contrast the structure of the walls of arteries, capillaries, and veins. How does the structure of the wall relate to each vessel's function?

- Certain parasitic worms can block lymph vessels. Suppose the lymph vessels draining a leg were completely blocked by a parasite infestation. What do you think would happen to the leg? Explain your answer.

- Occasionally soldiers or guards standing at attention and not moving will faint. Why do you think that occurs? How could they avoid that embarrassing situation?

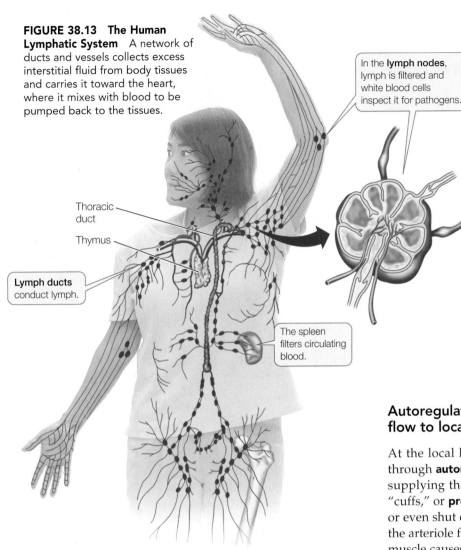

FIGURE 38.13 The Human Lymphatic System A network of ducts and vessels collects excess interstitial fluid from body tissues and carries it toward the heart, where it mixes with blood to be pumped back to the tissues.

In the **lymph nodes**, lymph is filtered and white blood cells inspect it for pathogens.

Thoracic duct

Thymus

Lymph ducts conduct lymph.

The spleen filters circulating blood.

Every tissue in the body requires an adequate flow of blood. Blood flow depends on the maintenance of an appropriate blood pressure, which thus is carefully regulated, as we will see next.

---

concept 38.6 Circulation Is Regulated by Autoregulation, Nerves, and Hormones

Blood flow through a vessel or a system of vessels depends on the pressure difference and the resistance. The relationship is analogous to Ohm's law for electrical current in a circuit: current equals the voltage difference divided by the resistance. In the circulatory system, the heart creates the driving pressure and the characteristics and actions of the vessels determine the resistance. If the pressure imparted to the blood by the heart is too high, vessels may rupture. If the pressure is too low, tissues will not be adequately served. Therefore, arterial blood pressure is closely monitored and regulated.

Blood pressure is determined by heart rate, stroke volume, and peripheral resistance

If blood pressure deviates from normal, what can be done to correct the situation? The most important determinant of blood flow in the entire circulatory system is the *mean arterial pressure* (MAP)

in the aorta. MAP is largely determined by two factors: the volume of blood that the heart pumps per minute, or *cardiac output* (CO), and the resistance to flow in the "downstream" blood vessels, or *total peripheral resistance* (TPR). Thus:

$$MAP = CO \times TPR$$

Cardiac output is the product of beats per minute, or *heart rate* (HR), and the volume of blood per beat, or *stroke volume* (SV). So:

$$MAP = HR \times SV \times TPR$$

Thus a deviation in blood pressure can be corrected by adjusting the heart rate, stroke volume, or total peripheral resistance. All three of these factors are controlled by multiple neural and hormonal mechanisms, at both the local and systemic levels.

Autoregulation matches local blood pressure and flow to local need

At the local level, each tissue controls its own blood flow through **autoregulatory mechanisms** that cause the arterioles supplying that tissue to constrict or dilate. Smooth muscle "cuffs," or **precapillary sphincters**, on an arteriole can reduce or even shut off the supply of blood to the capillary bed that the arteriole feeds (**FIGURE 38.14**). Relaxation of this smooth muscle causes the arteriole to dilate, increasing blood flow to the capillary bed.

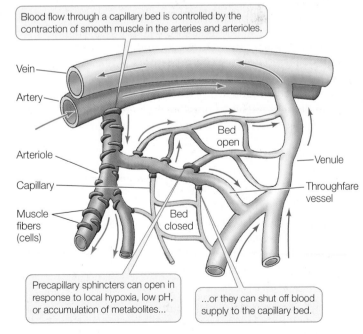

Blood flow through a capillary bed is controlled by the contraction of smooth muscle in the arteries and arterioles.

Vein

Artery

Arteriole

Capillary

Muscle fibers (cells)

Bed open

Bed closed

Venule

Throughfare vessel

Precapillary sphincters can open in response to local hypoxia, low pH, or accumulation of metabolites...

...or they can shut off blood supply to the capillary bed.

FIGURE 38.14 Local Control of Blood Flow Low O_2 concentrations or high levels of metabolic by-products cause the smooth muscle of the arteries and arterioles to relax, thus increasing the supply of blood to the capillary bed.

Autoregulation depends on the sensitivity of the smooth muscle to its local chemical environment. Low O_2 concentrations and high CO_2 concentrations cause the smooth muscle to relax, thus increasing the supply of blood, which brings in more O_2 and carries away CO_2. Other by-products of metabolism, such as lactate, hydrogen ions, potassium, and adenosine (all of which increase in exercising muscle), also cause arterioles to dilate. Hence activities that increase the metabolic rate of a tissue also increase blood flow to that tissue.

The collective autoregulatory actions in the capillary beds in all tissues of the body determine TPR and therefore affect MAP. If many arterioles suddenly dilate, TPR goes down and MAP falls. If many arterioles constrict, TPR goes up and MAP goes up.

Sympathetic and parasympathetic nerves affect blood pressure

The diameter of arteries and arterioles is affected by the autonomic nervous system and by hormones. Arterioles especially are innervated by the sympathetic division of the autonomic nervous system. Sympathetic nerves release norepinephrine onto smooth muscle cells of arterioles, causing constriction of the arterioles. The result is reduction of blood flow through these arterioles and an elevation in MAP. The overall result is a diversion of blood flow from "nonessential" tissues (such as the gut) to skeletal muscle.

As we discussed earlier in this chapter, increased sympathetic activity also increases heart rate, and by increasing the strength of the cardiac muscle contraction, it also increases stroke volume. Thus sympathetic activation sharply raises blood pressure by all three mechanisms (HR, SV, and TPR), enabling rapid delivery of oxygen and fuel to those tissues that need it, principally the skeletal muscles.

Activation of the sympathetic nerves is coordinated by the medulla of the brainstem. The medulla receives information on blood pressure from **baroreceptors** (stretch receptors) in the major arteries and then issues appropriate commands via parasympathetic and sympathetic nerves (**FIGURE 38.15**). For example, suppose blood pressure becomes too high. Increased activity in baroreceptors of the large arteries signals this change to the medulla. This triggers two sets of commands from the medulla: (1) sympathetic nerves are inhibited, causing arterioles in most tissues to dilate; and (2) parasympathetic nerves slow the heart's pacemaker (see Figure 38.5). Both of these responses help reduce blood pressure.

The medulla also monitors blood concentrations of certain key chemicals (such as oxygen and H^+) via *chemoreceptors* in the walls of the large arteries leading to the brain, and integrates this information with other inputs to adjust blood pressure.

LINK For more information about how the medulla monitors oxygen and pH, which also influence respiratory rate, see Concept 37.4

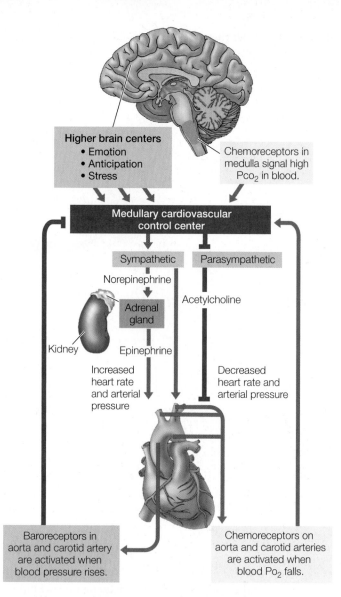

FIGURE 38.15 Regulating Cardiac Output The autonomic nervous system controls heart rate in response to information about blood pressure and blood composition originating in baroreceptors and chemosensors, as shown at the bottom of the figure. Information from these sensors goes to the cardiovascular control center in the medulla, where it is integrated with other information. The medullary center generates responses in the sympathetic and parasympathetic nervous systems that control cardiac output. Green lines indicate excitatory connections and red lines indicate inhibitory connections.

Many hormones affect blood pressure

By affecting the diameter of arteries and arterioles, hormones also play roles in regulating arterial pressure. Epinephrine, like norepinephrine, causes constriction of arterioles in some tissues, such as those of the digestive tract, and as a result diverts blood to other tissues, such as brain and skeletal muscle. Epinephrine is released from the adrenal gland in response to a fall in arterial pressure or by activation of the fight-or-flight response.

Another hormone, *angiotensin*, is produced when blood pressure to the kidneys falls (**FIGURE 38.16**). Angiotensin

FIGURE 38.16   Influences of Local and Systemic Mechanisms on Blood Pressure   A drop in arterial pressure reduces blood flow to tissues, resulting in local accumulation of metabolic wastes. This change in the extracellular environment stimulates autoregulatory dilation of arterioles that causes a further drop in arterial pressure (inner circle). Several systemic responses to a fall in arterial pressure serve to raise and maintain arterial pressure (outer circle).

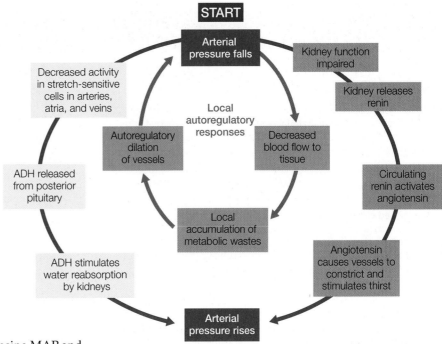

constricts arterioles throughout the body, increasing MAP and blood flow to essential organs such as the heart, brain, and kidneys.

A third hormone, antidiuretic hormone (ADH, also called vasopressin), is secreted by the posterior pituitary when blood pressure drops. ADH causes the kidneys to reabsorb more water, thereby maintaining blood volume, which helps prevent blood pressure from dropping further (see Concept 40.5).

Overall, all of these hormonal and neural signals interact to maintain blood pressure near the ideal level necessary to maintain the flow of oxygen and nutrients to those tissues that need it.

Do You Understand Concept 38.6?

- What three factors determine the mean arterial pressure?
- If blood pressure gets too high, will angiotensin levels go up or down? What about ADH levels? Explain your answer.
- Suppose you suffer a bad cut and lose a lot of blood. The pressure in your aorta starts to drop precipitously. How will your body detect this change, and how it will it attempt to restore homeostasis?

In this chapter we've learned how the circulatory system continuously circulates blood to all the cells of the body. The blood circulates so that it can transport substances such as respiratory gases (discussed in Chapter 37), nutrients, and waste products. In the next chapters we will turn to the nutrients and the waste products, and to the two physiological systems concerned with them: the digestive system (Chapter 39) and the excretory system (Chapter 40).

 QUESTION   What are the critical factors that determine whether a person recovers from a heart attack?

ANSWER   The critical factor for *surviving* a heart attack is rapid restoration of an adequate supply of blood to the brain, but the critical factor for *recovering* from a heart attack is sustaining a high enough cardiac output to all of the organs of the body. The immediate (within seconds) reaction of the body to a heart attack is massive sympathetic activity that increases cardiac output by increasing the rate and strength of contraction of the damaged heart (Concept 38.6). If cardiac output does not reach a critical level, which is about 5 liters of blood per minute for an adult human, many dangerous consequences begin to occur. Because the heart cannot pump the blood returning to it, blood slowly accumulates in the veins, causing a back pressure that increases leakage of fluids from the capillaries into the extracellular space (Concept 38.5). In the case of the lungs, this can compromise gas exchange, and therefore the blood being pumped by the heart may contain less oxygen, making the body even more hypoxic. Low cardiac output also has a negative effect on the kidneys, which as we will see in Chapter 40 depend on blood pressure for their function of filtering blood and eliminating fluid and wastes. The resulting fluid accumulation in the body puts an even greater load on the weakened heart. To recover from a heart attack, it is essential for the kidneys to resume functioning, and that requires the cardiac output to reach the life-sustaining level of about 5 liters per minute.

38 SUMMARY

concept 38.1 Circulatory Systems Can Be Open or Closed

- The metabolic needs of the cells of many small animals are met by direct exchange of materials with the external medium. Larger, active animals need a **circulatory system** that transports nutrients, respiratory gases, and metabolic wastes throughout the body.

- In **open circulatory systems**, found in mollusks and arthropods, extracellular fluid leaves vessels and percolates through tissues. **Review Figure 38.1A**

- In **closed circulatory systems**, found in annelids and vertebrates, the blood is contained in a system of vessels. Closed circulatory systems have several advantages, including faster transport and the ability to selectively direct blood to specific tissues, and the compartmentalization of functional elements such as red blood cells. **Review Figure 38.1B**

- The circulatory system of vertebrates consists of a **heart**, **arteries**, and **arterioles** that carry blood from the heart; **capillaries**, where materials are exchanged between blood and interstitial fluid; and **venules** and **veins** that carry blood back to the heart.

concept 38.2 Circulatory Systems May Have Separate Pulmonary and Systemic Circuits

- The vertebrate heart evolved from two chambers in fishes; to three in lungfishes, amphibians, and most reptiles; to four in crocodilians, birds, and mammals. This evolutionary progression has led to an increasing separation of blood that flows to the gas exchange organs and blood that flows to the rest of the body. **Review WEB ACTIVITY 38.1**

- The simplest (two-chambered heart) has an **atrium** that receives blood from the body and a **ventricle** that pumps blood out of the heart. An aorta distributes blood to arteries.

- In crocodilians, birds, and mammals, blood circulates through two completely separate circuits: the **pulmonary circuit**, which transports oxygen-depleted blood to the lungs, and the **systemic circuit**, which transports oxygenated blood to the rest of the body.

concept 38.3 A Beating Heart Propels the Blood

- The human heart has four chambers, with valves to prevent backflow. Blood flows from the right atrium and ventricle to the lungs, then to the left atrium and ventricle, and then to the rest of the body. **Review Figure 38.2 and WEB ACTIVITY 38.2**

- The **cardiac cycle** has two phases: **systole**, when the ventricles contract, and **diastole**, when the ventricles relax. The heart sounds ("lub-dup") are made by the closing of the heart valves. **Review Figure 38.3 and ANIMATED TUTORIAL 38.1**

- The heartbeat is generated by **pacemaker cells** in the **sinoatrial node** that spontaneously depolarize, triggering an action potential that spreads to the rest of the heart through the **atrioventricular node**, **bundle of His**, and **Purkinje fibers**. **Review Figure 38.5**

- The autonomic nervous system controls heart rate by changing the rate of depolarization of the pacemaker cells. Norepinephrine from sympathetic nerves increases heart rate, and acetylcholine from parasympathetic nerves decreases it. **Review Figure 38.4**

concept 38.4 Blood Consists of Cells Suspended in Plasma

- Blood consists of a liquid **plasma** and cellular components (**erythrocytes**, or red blood cells, **platelets**, and **leukocytes**, or white blood cells). All of the cellular components are produced in the bone marrow. **Review Figure 38.7**

- Erythrocytes transport oxygen. Their production in the bone marrow is stimulated by **erythropoietin**, which is produced in response to **hypoxia** (low oxygen levels) in the tissues.

- Platelets, along with circulating proteins, are involved in **blood clotting**, which results in a meshwork of **fibrin** threads that help seal vessels. **Review Figure 38.8**

concept 38.5 Blood Circulates through Arteries, Capillaries, and Veins

- Arteries and arterioles contain elastic fibers that enable them to withstand high pressure, and smooth muscle that can change their diameters. **Review Figure 38.9 and WEB ACTIVITY 38.3**

- Capillaries have thin, permeable walls that allow exchange of materials with the interstitial fluid. Capillaries in most tissues other than the brain have pores that allow many large molecules to pass through. **Review Figures 38.9 and 38.10**

- **Starling's forces** suggest that blood volume is maintained in the capillary beds by an exchange of fluids driven by both blood pressure and osmotic pressure. **Review Figure 38.11**

- Veins have a high capacity for storing blood. Return of blood to the heart is aided by gravity, skeletal muscles, and breathing. **Review Figure 38.12**

- The **lymphatic system** returns interstitial fluid to the blood. **Review Figure 38.13**

concept 38.6 Circulation Is Regulated by Autoregulation, Nerves, and Hormones

- Deviations in blood pressure can be corrected by altering total peripheral resistance (TPR), heart rate (HR), or stroke volume (SV). **Review INTERACTIVE TUTORIAL 38.1**

- Autoregulation causes arterioles to dilate in response to low oxygen, high carbon dioxide, and certain metabolic waste products. This increases blood flow to the area. **Review Figure 38.14**

- Blood pressure and heart rate are also regulated by the autonomic nervous system, which responds to information about both blood pressure and blood composition that is integrated by regulatory centers in the medulla. **Review Figure 38.15**

- Many hormones affect blood pressure, including epinephrine, angiotensin, and ADH. **Review Figure 38.16**

Nutrition, Digestion, and Absorption

There is an obesity epidemic in the United States. About 75 percent of the adult U.S. population is overweight, and more than one-third of those individuals are obese. This problem is relatively recent in origin; the obesity percentage increased from 19 percent to 27 percent in only 10 years.

How do we define obesity, what causes it, and what are the consequences? An objective measure of body composition is the Body Mass Index. Your BMI is your mass in kilograms divided by your height in meters squared. A BMI of 18.5 to 24.9 is considered normal, 25 to 29.9 overweight, and 30 or above, obese. A simple statement of what causes obesity is the consumption of more calories than the body burns. But the problem is much more complicated. Multiple factors determine how many calories the body burns; the caloric density of foods varies enormously; and appetite is not a simple calorie meter. To find answers to questions about the causes and consequences of obesity, it is valuable to have homogeneous populations in which the problems can be studied.

For about 2,000 years the Pima people were hunters and gatherers who supplemented their diet with subsistence agriculture. Their home is the arid environment that is now the southwestern United States and northern Mexico. Archeological studies have revealed vast, sophisticated networks of irrigation canals that enabled the Pima to raise plants in that arid environment. These peoples frequently encountered drought and subsequent starvation.

Today most individuals of the ethnic Pima population in the southwestern U.S. are clinically obese. As a population, they are one of the heaviest in the world. With obesity come health problems such as adult onset diabetes, and obesity is associated with high blood pressure and heart disease. In the 1960s, epidemiologists first noticed a disturbing incidence of type II diabetes mellitus among the Pima in Arizona. Recently, the incidence of diabetes among adult Pima was claimed to be 50 percent, and 95 percent of those individuals were obese. Furthermore, diabetes is occurring in younger Pima than ever before.

Another population of Pima lives in northern Mexico. Genetically, the Pima in Mexico and Arizona are nearly identical. Yet somehow the Mexican Pima have avoided the diabetes that plagues their northern kin; they have a diabetes incidence of only 6 percent. Even this, however, is higher than that of non-Pima Mexicans, who almost never have diabetes. The two populations of Pima people have provided researchers with an opportunity to investigate the causes and consequences of obesity.

 Why are some ethnic groups of humans much more prone to obesity and diabetes than other groups?

The American Pima are an example of a human population that has a high incidence of obesity.

KEY CONCEPTS

39.1 Food Provides Energy and Nutrients

39.2 Digestive Systems Break Down Macromolecules

39.3 The Vertebrate Digestive System Is a Tubular Gut with Accessory Glands

39.4 Food Intake and Metabolism Are Regulated

concept
39.1 **Food Provides Energy and Nutrients**

Animals are **heterotrophs**—they derive their nutrition from eating other organisms. In contrast, **autotrophs** (most plants, some bacteria, some archaea, and some protists) can use solar energy or inorganic chemical energy to serve all of their energy needs. Directly and indirectly, heterotrophs take advantage of—indeed, depend on—the organic synthesis carried out by autotrophs (**FIGURE 39.1**). But what, exactly, do heterotrophs need from their food? First, heterotrophs need energy. Second, they also need certain *nutrients* from their food—specific molecules and ions required for optimal health.

Food provides energy

One of the primary reasons that heterotrophs must eat is to obtain energy. *Energy is the capacity to do work*, and it comes in different forms: electrical energy, chemical energy, and thermal energy (see Concept 2.5). The energy in food is in the form of chemical energy, stored in chemical bonds of carbon compounds.

The energy content of any food item can be measured by the amount of heat energy the food produces if it is completely burned. Specifically, a *calorie* (note the small *c*) is the amount of heat necessary to raise the temperature of 1 gram of water 1°C. Since this is a tiny amount of energy, physiologists commonly use the **kilocalorie (kcal)** as a unit of measure (1 kcal = 1,000 calories). Nutritionists refer to the kilocalorie as the **Calorie (Cal)**, always capitalized to distinguish it from the single calorie. For example, a pat of butter has about 35,000 calories =

35 kilocalories = 35 Calories. Physiologists also use a unit called the *joule*, which is the unit of energy measurement in the metric system. One joule equals 0.239 calories.

How many calories, or joules, an animal needs per day depends on its *metabolic rate*, a measure of the overall rate of chemical reactions throughout its entire body. *Basal metabolic rate* (BMR) is the minimum energy necessary to sustain life when an animal is resting quietly and is not expending any extra energy on growth, thermoregulation, or digestion. The basal metabolic rate of a human is about 1,400 Cal/day for an adult female and 1,700 Cal/day for an adult male. Any physical activity adds to this basal energy requirement. For example, a woman doing a day of hard physical labor may need 3,000 Cal/day. Professional athletes may need even more; cyclists in the Tour de France can eat an astonishing 9,000 Cal/day during the toughest part of the race.

Generally, fats yield 9.5 Cal/gram, carbohydrates 4.2 Cal/gram, and proteins about 4.1 Cal/gram. Thus fats store far more energy per gram than do carbohydrates or proteins.

Excess energy is stored as glycogen and fat

Catabolism of food molecules results in energy transfers needed for life. The cells of the body use energy continuously, but most animals do not eat continuously. Therefore, animals must store fuel molecules that can be released as needed between meals. A small portion of fuel, about a day's worth of energy requirements, is stored in the form of the carbohydrate *glycogen* in liver and muscle cells. But animals store the majority of their fuel—usually several weeks' worth—in the form of fat. (Proteins are not usually used for energy storage, although body protein can be broken down for energy as a last resort.)

(A)

(B)

FIGURE 39.1 Heterotrophs Get Energy from Autotrophs (A) Some heterotrophs—herbivores—get their energy directly from autotrophs. (B) Other heterotrophs—carnivores—obtain energy indirectly from autotrophs, since the energy stored in their prey was originally obtained from autotrophs.

APPLY THE CONCEPT

Food provides energy and nutrients

MET, the *metabolic equivalent of task*, is an index of the intensity of physical activity. Exercise physiologists define the resting metabolic rate as 1 MET and increasingly energetic activities as multiples of 1 MET. Thus walking has a value of about 5 METs (i.e., it uses five times as much energy as resting quietly), jogging about 8 METs, and cross-country skiing about 10 METs. Assuming that your basal metabolic rate is 1,600 Cal/day, fill in the table with the time it would take you to burn off the calories from the listed foods while resting, jogging, or cross-country skiing.

		TIME REQUIRED TO BURN OFF CALORIES		
FOOD	CALORIES	RESTING	JOGGING	CROSS-COUNTRY SKIING
Fruit yogurt (6 oz)	130			
Turkey sandwich on whole wheat bread	215			
Sports energy bar	250			
Small cheeseburger	530			
Chicago-style cheese pizza	1,300			

Not only does fat have more energy per gram than glycogen, but it can be stored with little associated water, making it more compact. If migrating birds had to store all their energy as glycogen instead of fat, they would be too heavy to fly.

Food provides essential molecular building blocks

In addition to obtaining energy from their food, animals also need to ingest specific nutrients for the biosynthesis and function of essential molecular structures. If a nutrient cannot be synthesized in the body but is absolutely required, it is an *essential nutrient*. Nutrients required in large amounts are called **macronutrients**; those required in only tiny amounts (generally less than 100 mg/day) are called **micronutrients**.

Macronutrients generally are the building blocks of molecules found in large amounts throughout the body, such as proteins, nucleic acids, sugars, fats, and heme groups. Animals typically cannot make all parts of many of these molecules from scratch, so their food must contain specific compounds whose carbon skeletons are used as building blocks.

Amino acids, for example, are the building blocks of proteins. Most animals can synthesize some of the amino acids they need, but not all of them. Thus each species must eat certain **essential amino acids**. Essential amino acids vary by species. Adult humans need eight essential amino acids in their diets: isoleucine, leucine, lysine, methionine, phenylalanine, threonine, tryptophan, and valine. (Babies require four additional amino acids: histidine, tyrosine, cysteine, and arginine.)

Milk, eggs, meat, and soybeans contain all eight essential amino acids, but most plants do not. A *complementary* diet of plant foods, however, supplies all eight essential amino acids (**FIGURE 39.2**). In general, grains (such as rice, wheat, and corn) complement legumes (such as beans and peas). Long before the chemical basis for this complementarity was understood, societies with little access to meat developed complementary diets. Many Central and South American peoples traditionally eat beans with corn, and the native peoples of North America complemented their beans with squash.

We also need certain **essential fatty acids**. Humans can synthesize most lipids from other compounds, but we must have a dietary source of certain fatty acids, especially linoleic acid. Linoleic acid is an unsaturated fatty acid needed to synthesize membrane phospholipids, signaling molecules such as prostaglandins, and some other molecules. A deficiency of linoleic acid causes symptoms that include dermatitis, infertility, and impaired lactation. Linoleic acid is common in vegetable oils, so deficiencies are rare.

Food provides essential minerals

A chemical element required in the diet is known as an **essential mineral**. Familiar examples are calcium, iron, and potassium (**TABLE 39.1**). Most minerals are needed only in minute quantities and hence are considered micronutrients, but a few qualify as macronutrients.

Calcium, for example, is the fifth most abundant element in the body; a 70-kg person contains about 1.2 kg of calcium.

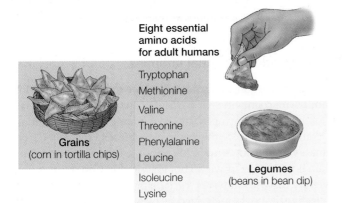

FIGURE 39.2 A Strategy for Vegetarians By combining cereal grains with legumes, an adult vegetarian can obtain all eight essential amino acids.

TABLE 39.1 Mineral Elements Required by Animals

ELEMENT	SOURCE IN HUMAN DIET	MAJOR FUNCTIONS
MACRONUTRIENTS		
Calcium (Ca)	Dairy foods, eggs, green leafy vegetables, whole grains, legumes, nuts, meat	Found in bones and teeth; blood clotting; nerve and muscle action; enzyme activation
Chlorine (Cl)	Table salt (NaCl), meat, eggs, vegetables, dairy foods	Water balance; digestion (as HCl); principal negative ion in extracellular fluid
Magnesium (Mg)	Green vegetables, meat, whole grains, nuts, milk	Required by many enzymes; found in bones and teeth
Phosphorus (P)	Dairy, eggs, meat, whole grains, legumes, nuts	Found in nucleic acids, ATP, and phospholipids; bone formation; buffers; metabolism of sugars
Potassium (K)	Meat, whole grains, fruits, vegetables	Nerve and muscle action; protein synthesis; principal positive ion in cells
Sodium (Na)	Table salt, dairy foods, meat, eggs	Nerve and muscle action; water balance; principal positive ion in extracellular fluid
Sulfur (S)	Meat, eggs, dairy foods, nuts, legumes	Found in proteins and coenzymes; detoxification of harmful substances
MICRONUTRIENTS		
Chromium (Cr)	Meat, dairy, whole grains, legumes, yeast	Glucose metabolism
Cobalt (Co)	Meat, tap water	Found in vitamin B_{12}; formation of red blood cells
Copper (Cu)	Liver, meat, fish, shellfish, legumes, whole grains, nuts	Found in active site of many redox enzymes and electron carriers; production of hemoglobin; bone formation
Fluorine (F)	Most water supplies	Found in teeth; helps prevent decay
Iodine (I)	Fish, shellfish, iodized salt	Found in thyroid hormones
Iron (Fe)	Liver, meat, green vegetables, eggs, whole grains, legumes, nuts	Found in active sites of many redox enzymes and electron carriers, hemoglobin, and myoglobin
Manganese (Mn)	Organ meats, whole grains, legumes, nuts, tea, coffee	Activates many enzymes
Molybdenum (Mo)	Organ meats, dairy, whole grains, green vegetables, legumes	Found in some enzymes
Selenium (Se)	Meat, seafood, whole grains, eggs, milk, garlic	Fat metabolism
Zinc (Zn)	Liver, fish, shellfish, and many other foods	Found in some enzymes and some transcription factors; insulin physiology

Calcium is a component of calcium phosphate (the principal structural material in bones and teeth) and is also necessary for nerve function and muscular contraction (see Concepts 34.2 and 36.1). Calcium is lost from the body in urine, sweat, and feces, so it must be replaced regularly. Humans require 800–1,000 mg of calcium per day in their diet, and hence calcium is a macronutrient. Insufficient calcium can cause *osteoporosis*, a progressive thinning and weakening of the bones.

Iron is an example of a micronutrient. It is the oxygen-binding atom in hemoglobin and myoglobin and is also a component of enzymes in the respiratory chain. Nevertheless, the total amount of iron in a 70-kg person is only about 4 g. Men require only about 8 mg per day of iron, since iron is recycled efficiently in the body and is not lost in the urine. However, women lose some iron every month during menstruation, so adult women require about 18 mg per day. This is still a tiny amount, yet many adult women do not get sufficient iron in their diets. Insufficient dietary iron causes *iron-deficiency anemia*, too few red blood cells in the blood, with symptoms including fatigue, lightheadedness, and fainting. Iron-deficiency anemia is the most common mineral nutrient deficiency in the world.

yourBioPortal.com

Go to WEB ACTIVITY 39.1
Mineral Elements Required by Animals

Food provides essential vitamins

Vitamins are carbon compounds that are micronutrients. Most vitamins are coenzymes or parts of coenzymes (see Table 3.3). Different species may need different vitamins. Most mammals, for example, can make their own ascorbic acid (*vitamin C*), but primates (including humans) cannot synthesize it and must obtain it in the diet. If we do not get vitamin C in our food, we develop a disease known as *scurvy*, a progressive breakdown of connective tissue characterized by bleeding gums, loss of teeth, extensive bruising, and slow wound healing. The disease was a frequently fatal problem for sailors on long voyages until a Scottish physician, James Lind, discovered that scurvy could be prevented if the sailors ate citrus fruit, a good source of vitamin C. The British Admiralty made limes standard provisions for its ships (and British sailors have been called "limeys" ever since). When the active ingredient in limes was isolated, it was named *ascorbic* ("without scurvy") *acid*.

Humans require 13 vitamins (**TABLE 39.2**). They are divided into two groups: water-soluble and fat-soluble. Excess water-soluble vitamins are simply eliminated in the urine. (This is the fate of much of the excess vitamin C that people take.) Fat-soluble vitamins, however, can accumulate in body fat and may build up to toxic levels in the liver if taken in excess.

The fat-soluble vitamin D, which is actually a hormone that stimulates calcium absorption (see Figure 30.11), is a special case because the body can synthesize it if the skin is exposed to sunlight. Thus vitamin D must be obtained in the diet only if there is inadequate sun exposure. The need for vitamin D may be the reason that human races differ in skin color. Races that live in low latitudes are exposed frequently to strong UV light and have no trouble manufacturing plenty of vitamin D, but they run the risk of UV damage. The dark skin common in these latitudes is protection against the damage of UV radiation. Races from higher latitudes have less risk of UV damage but a much greater risk of vitamin D insufficiency, and this is likely the reason they evolved pale skin. The Inuit of the Arctic are an exception, but these dark-skinned people obtain ample vitamin D from the abundant whale blubber and fish oils in their diet.

yourBioPortal.com

Go to WEB ACTIVITY 39.2
Vitamins in the Human Diet

Nutrient deficiencies result in diseases

The lack of any essential nutrient in the diet produces a state of deficiency called **malnutrition**, and chronic malnutrition leads to a characteristic **deficiency disease** (see Table 39.2). We have already mentioned scurvy, caused by vitamin C deficiency. Another deficiency disease, *beriberi*, was directly involved in the discovery of vitamins.

Beriberi is a nervous system disorder characterized by extreme fatigue. It became prevalent in Asia in the nineteenth century when people began to eat white rice (discarding the brown hulls). Interestingly, chickens and pigeons developed beriberi-like symptoms when fed only white rice. In 1912, Casimir Funk cured pigeons of beriberi by feeding them rice hulls. Funk proposed that beriberi and some other diseases are caused by dietary deficiencies of what he called "vitamines" (or "vital amines"—he mistakenly thought all vitamins had amino groups). In 1926, thiamin (vitamin B_1)—the substance present in rice hulls—was the first vitamin to be isolated.

Lack of minerals can also cause deficiency diseases. Insufficient iron causes anemia; insufficient iodine causes goiter and hypothyroidism (see Concept 30.4). Goiter is rare today because of the addition of iodide to salt. However, it is still a

TABLE 39.2 Vitamins in the Human Diet

VITAMIN	SOURCE	FUNCTION	DEFICIENCY SYMPTOMS
WATER-SOLUBLE			
B_1 (thiamin)	Liver, legumes, whole grains	Coenzyme in cellular respiration	Beriberi, loss of appetite, fatigue
B_2 (riboflavin)	Dairy, meat, eggs, green leafy vegetables	Coenzyme in FAD	Lesions in corners of mouth, eye irritation, skin disorders
Niacin	Meat, fowl, liver, yeast	Coenzyme in NAD and NADP	Pellagra, skin disorders, diarrhea, mental disorders
B_6 (pyridoxine)	Liver, whole grains, dairy foods	Coenzyme in amino acid metabolism	Anemia, slow growth, skin problems, convulsions
Pantothenic acid	Liver, eggs, yeast	Found in acetyl CoA	Adrenal problems, reproductive problems
Biotin	Liver, yeast, bacteria in gut	Found in coenzymes	Skin problems, loss of hair
B_{12} (cobalamin)	Liver, meat, dairy foods, eggs	Formation of nucleic acids, proteins, and red blood cells	Pernicious anemia
Folic acid	Vegetables, eggs, liver, whole grains	Coenzyme in formation of heme and nucleotides	Anemia
C (ascorbic acid)	Citrus fruits, tomatoes, potatoes	Formation of connective tissues; antioxidant	Scurvy, slow healing, poor bone growth
FAT-SOLUBLE			
A (retinol)	Fruits, vegetables, liver, dairy	Found in visual pigments	Night blindness
D (cholecalciferol)	Fortified milk, fish oils, sunshine	Absorption of calcium and phosphate	Rickets
E (tocopherol)	Meat, dairy foods, whole grains	Muscle maintenance, antioxidant	Anemia
K (menadione)	Intestinal bacteria, liver	Blood clotting	Blood-clotting problems

major health problem in some areas of the world. Probably the single least expensive action to improve global health would be to provide iodized salt for everyone.

Do You Understand Concept 39.1?

- Describe three deficiency diseases and the nutrients that can cure them.
- Suppose a friend of yours is so concerned about obesity in children that she puts her baby on an entirely fat-free diet. Is this wise?
- As you have probably noticed, two people of identical weight can spend identical amounts of time per day exercising, and yet often one person can eat more food than the other without gaining weight. Propose at least two explanations for this phenomenon.

Heterotrophs need food that contains sufficient energy and certain essential nutrients. However, food usually consists of complex macromolecules that cannot be used as is; the food must be broken down into smaller components for absorption.

concept 39.2 Digestive Systems Break Down Macromolecules

The function of the digestive system, or the *gut*, is to break down large pieces of food into small molecules that can be absorbed—providing the animal with the energy and the essential nutrients it needs. Proteins, carbohydrates, and fats are all broken down into their constituent monomers by hydrolytic enzymes produced by the digestive system. All of these enzymes cleave the chemical bonds of macromolecules through *hydrolysis*, a reaction that splits a bond by adding a water molecule, leaving one product with an additional H^+ and the other with an additional OH^- (see Figure 2.8). Digestive enzymes are classified according to the substances they hydrolyze: *proteases* hydrolyze the bonds between adjacent amino acids in proteins; *carbohydrases* hydrolyze the bonds between adjacent sugar units of carbohydrates; *peptidases* hydrolyze peptides; *lipases*, fats; and *nucleases*, nucleic acids.

> **LINK** For more on the reactions that interconvert macromolecules and their constituent monomers, see Concepts 2.2–2.4, 3.1, and 3.2

Some heterotrophs such as sea stars and spiders secrete hydrolytic enzymes onto or into food sources in the external environment and then take in the resulting monomers. However, most animals *ingest* (take in) the food into a digestive tract, consisting of a body cavity that is continuous with the outside environment.

Simple digestive systems are cavities with one opening

The simplest digestive systems are **gastrovascular cavities**, which connect to the outside world through a single opening. A jellyfish, for example, captures its prey with stinging organs called *nematocysts* and uses its tentacles to pull the prey into a gastrovascular cavity. Enzymes in the gastrovascular cavity digest the prey, and cells lining the cavity take in small food particles by endocytosis. Since there is no *anus*, any undigested pieces are released through the same opening through which the prey entered. Most cnidarians and flatworms have a gastrovascular cavity.

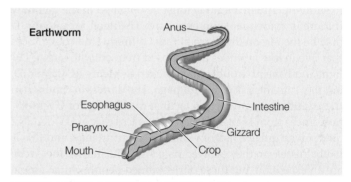

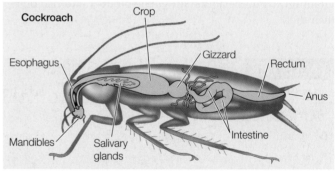

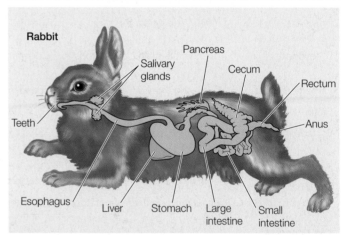

FIGURE 39.3 Compartments for Digestion and Absorption Most invertebrates and all vertebrates have a tubular gut that begins with a mouth, which takes in food, and ends in an anus, which eliminates wastes. Between these two structures are specialized regions for digestion and nutrient absorption; the structures in these regions are adapted to different diets and vary from species to species.

Tubular guts have an opening at each end

Other animals have a tubular gut with two ends. A **mouth** takes in food, molecules are digested (broken down into monomers) and then absorbed throughout the length of the gut, and undigested wastes are eliminated through an anus. Different regions in the gut are specialized for particular functions (**FIGURE 39.3**). In most animals the gut can be roughly divided into three sections: foregut, midgut, and hindgut.

FOREGUT The foregut is where food is physically broken up into small pieces, preparing it for enzymatic digestion. The foregut begins with a mouth cavity. In many animals, sharks for instance, the mouth is simply where food is grasped and swallowed, with little processing. In others, the mouth contains tools for fragmenting or liquefying the food, such as the *radula* in snails, *mandibles* in many arthropods, and specialized *teeth* in vertebrates.

From the mouth, food moves through a tube, the **esophagus**, to a storage sac, usually called a **stomach**. The stomach physically (and sometimes enzymatically) breaks down the food, producing a slurry of small particles that is then delivered to the midgut, ready for digestion. Many animals, among them reptiles (including birds), earthworms, and various insects, have two stomachlike organs in a row: the first is a simple storage sac called a **crop** that just holds food, and the second is a muscular organ called a **gizzard** that grinds up the food (see Figure 39.3).

MIDGUT After leaving the stomach (or gizzard), food travels through the midgut. A long, thin midgut is often called a **small intestine** ("small" referring to its diameter, not its length). The midgut is the primary site of digestion and absorption. Here hydrolytic enzymes are secreted by intestinal cells and associated glands, and macromolecules are broken down into monomers that are absorbed into the blood. In many vertebrates the midgut is highly convoluted or has fingerlike projections called **villi** (singular *villus*) to increase the surface area of contact between the gut wall and the food (**FIGURE 39.4**). The villi in turn have microscopic projections called **microvilli**, which provide even more surface area for absorbing nutrients.

HINDGUT The final segment of the gut, called the hindgut or **large intestine**, recovers water and ions and stores undigested wastes, or **feces**, until they can be expelled. The end of the

FIGURE 39.4 Intestinal Surface Area and Nutrient Absorption Maximizing the surface area of the gut increases an animal's ability to absorb nutrients.

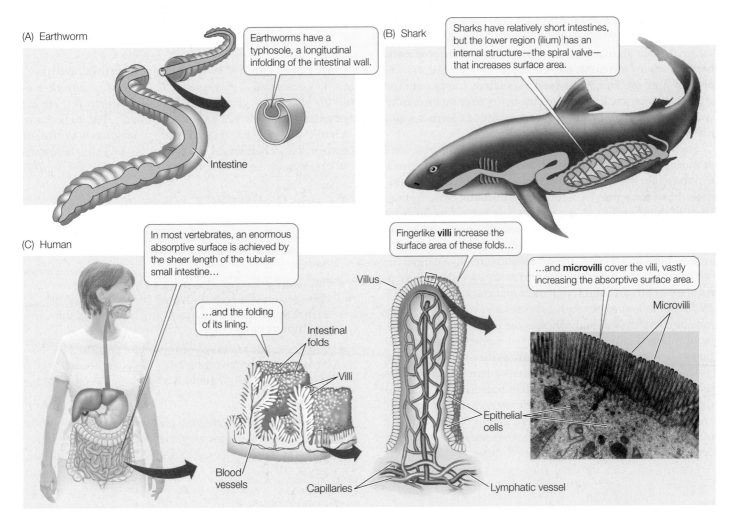

(A) Earthworm

Earthworms have a typhosole, a longitudinal infolding of the intestinal wall.

Intestine

(B) Shark

Sharks have relatively short intestines, but the lower region (ilium) has an internal structure—the spiral valve—that increases surface area.

(C) Human

In most vertebrates, an enormous absorptive surface is achieved by the sheer length of the tubular small intestine…

…and the folding of its lining.

Intestinal folds

Villi

Blood vessels

Fingerlike **villi** increase the surface area of these folds…

Villus

Capillaries

Epithelial cells

Lymphatic vessel

…and **microvilli** cover the villi, vastly increasing the absorptive surface area.

Microvilli

digestive tract is usually called the **anus**. If urinary wastes and digestive wastes are expelled through the same opening, as is the case in most amphibians and in birds and other reptiles, the opening is called a **cloaca** (Latin, "sewer").

SYMBIOTIC BACTERIA The digestive tracts of most animals contain symbiotic bacteria. The human gut contains roughly 10^{14} bacteria—10 times more than the number of human cells in the body, and representing at least 500 different species. In many species, gut bacteria assist in the digestive process. For example, leeches of the genus *Hirudo* feed on vertebrate blood, and they depend on their gut bacteria to produce the enzymes needed to digest the blood proteins. The resulting amino acids are subsequently used by both the leech and the bacteria, so the arrangement is mutually beneficial. Similarly, many plant-eating animals, from cows to termites, house large colonies of symbiotic gut bacteria that can break down the plant polysaccharide *cellulose* in their food, producing sugars that are used both by the bacteria and by the host. We will see some examples of this in Concept 39.3.

Heterotrophs may specialize in different types of food

Heterotrophs can be classified by how they acquire their food. **Saprobes** (or *decomposers*) are organisms—mostly protists and fungi—that absorb nutrients from dead organic matter. **Detritivores**, such as earthworms and crabs, feed on dead organic material. Animals that feed on living organisms are **predators**: **herbivores** prey on plants, **carnivores** prey on animals, and **omnivores** prey on both. **Filter feeders**, such as clams and blue whales, prey on small aquatic organisms by filtering them from water. **Fluid feeders** include mosquitoes, aphids, leeches, vampire bats, and hummingbirds.

Each of these dietary specializations has its distinctive suite of morphological, physiological, and behavioral adaptations. Compare, for example, carnivorous and herbivorous mammals. Most mammals have four types of **teeth**: *incisors* at the front of the jaws for cutting; *canines* just behind the incisors for stabbing; and *premolars* and *molars* (distinguished by the number of roots) in the back for grinding. In many carnivores such as lions and wolves, the canines are greatly enlarged for seizing prey, and the premolars and molars are short, powerful shears that can cut through meat and crack bone (**FIGURE 39.5**). But in herbivores such as horses and cows, the canines may be completely absent, whereas the premolars and molars have enlarged into a long array of flat-topped grinding teeth that grow continuously out of the jaw as they are worn down.

Diet affects the rest of the digestive system too. Carnivores tend to have very short digestive tracts, because meat is easy to digest. Plant material is harder to digest and also has a lower energy content, so herbivores have longer digestive tracts (relative to body size), often with large compartments for the cellulose-digesting bacteria mentioned above.

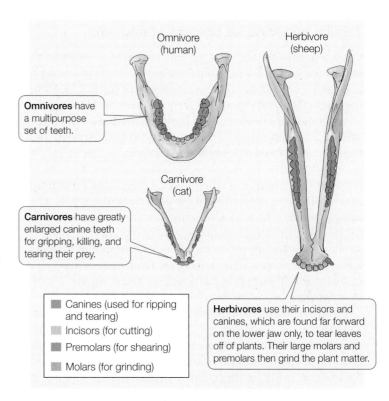

FIGURE 39.5 Mammalian Teeth The teeth of different mammalian species are specialized for different diets. This illustration depicts the teeth of the lower jaw, viewed from above.

What about humans? We have canine teeth (like carnivores), but they are small (see Figure 39.5). Our molars and premolars are shaped for grinding (like those of herbivores) but are not enlarged and do not grow continuously. We do not have cellulose-digesting bacteria, and our gut length is intermediate between that of herbivores and carnivores. Overall, these traits indicate that we are omnivores.

Do You Understand Concept 39.2?

- What are the respective terms for the enzymes that break down proteins, fats, carbohydrates, and nucleic acids?

- How would the digestive system of a fluid feeder differ from that of a detritivore? For instance, which animal would have the longer gut? How would the mouthparts differ?

- Very few birds are herbivores. Why? (Hint: The few birds that are herbivorous, such as geese, are so heavy that they sometimes have trouble taking flight.)

We have seen that the structural and functional details of the digestive system vary from species to species. Let's take a closer look at the processes of digestion in the vertebrate gut.

concept
39.3

The Vertebrate Digestive System Is a Tubular Gut with Accessory Glands

The vertebrate digestive system is a tubular gut running from mouth to anus, with several accessory glands, including the liver and pancreas (**FIGURE 39.6**). The digestive system functions like an assembly line, or rather a *dis*-assembly line that takes food apart. Different processes occur sequentially in different sections. We will first consider the basic structure common to all regions of the tubular gut, and then we will examine the functions of each region. Our focus will be the human system.

The vertebrate gut consists of concentric tissue layers

The tissues of the vertebrate gut are arranged in concentric layers (**FIGURE 39.7**). Surrounding the gut cavity, or **lumen**, is the innermost layer of the gut, the **mucosa**. The mucosa consists

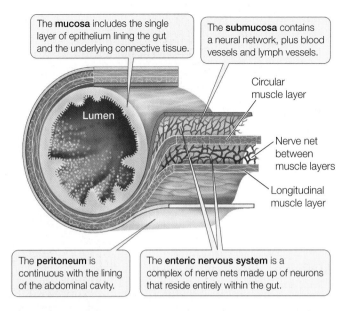

The **mucosa** includes the single layer of epithelium lining the gut and the underlying connective tissue.

The **submucosa** contains a neural network, plus blood vessels and lymph vessels.

Circular muscle layer

Nerve net between muscle layers

Longitudinal muscle layer

The **peritoneum** is continuous with the lining of the abdominal cavity.

The **enteric nervous system** is a complex of nerve nets made up of neurons that reside entirely within the gut.

FIGURE 39.7 Tissue Layers of the Vertebrate Gut The organization of tissue layers is the same in all compartments of the gut, but specialized adaptations of specific tissues characterize different regions.

of delicate epithelial cells that secrete mucus, enzymes, and hormones. Outside the mucosa is the **submucosa** tissue layer, containing blood and lymph vessels that carry absorbed nutrients to the rest of the body. The submucosa also contains a network of sensory and motor nerves that mostly control the secretory functions of the gut.

Surrounding the submucosa is a *circular muscle layer* that constricts the gut, then another nerve net, and over that a *longitudinal muscle layer* that shortens the gut. These smooth muscle layers contract automatically in response to being stretched, such as when food passes through the lumen. The nerve net coordinates these contractions so that the contents of the gut are pushed along by a rippling muscular action called **peristalsis**. The passage of the contents through the gut is further controlled by sphincter muscles at the entrance to the stomach, the small intestine, the large intestine, and at the anus.

The nerve nets in the submucosa and between the smooth muscle layers are together called the **enteric nervous system**. They are unusual because most of their nerves do not synapse with the central nervous system (CNS). Instead they synapse with other neurons in their network. Although the CNS can influence the activity of the enteric nervous system and can receive information from it, the gut truly has a "mind" of its own.

Finally, the outermost layer of the gut is a membrane called the visceral **peritoneum** that surrounds the gut and other abdominal organs. The parietal peritoneum lines the wall of the abdominal cavity and is continuous with the visceral peritoneum. The peritoneum secretes a fluid that lubricates the organs so they can easily slide against each other.

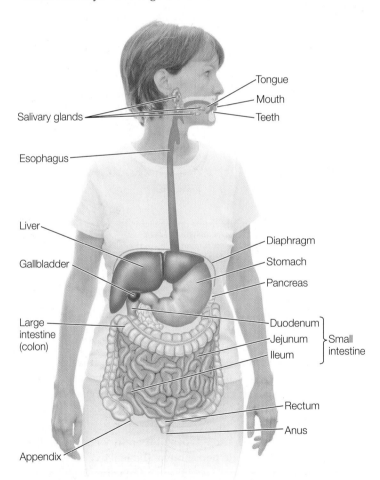

Tongue
Mouth
Teeth
Salivary glands
Esophagus
Liver
Gallbladder
Diaphragm
Stomach
Pancreas
Large intestine (colon)
Duodenum
Jejunum
Ileum
Small intestine
Rectum
Anus
Appendix

FIGURE 39.6 The Vertebrate Digestive System Different compartments in the long tubular gut specialize in digesting food, absorbing nutrients, and storing and expelling wastes. Accessory organs provide secretions containing enzymes and other molecules.

Digestion begins in the mouth

The digestive tract begins with the mouth. Most vertebrates, such as fishes, amphibians, and birds and other reptiles, simply

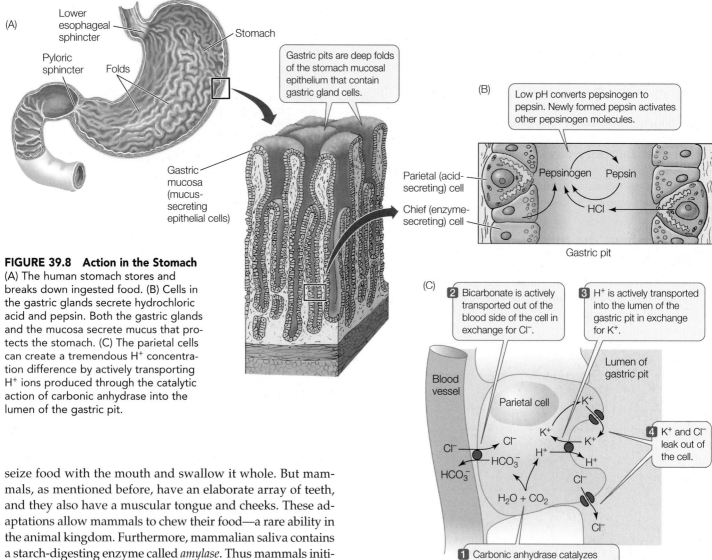

FIGURE 39.8 Action in the Stomach
(A) The human stomach stores and breaks down ingested food. (B) Cells in the gastric glands secrete hydrochloric acid and pepsin. Both the gastric glands and the mucosa secrete mucus that protects the stomach. (C) The parietal cells can create a tremendous H^+ concentration difference by actively transporting H^+ ions produced through the catalytic action of carbonic anhydrase into the lumen of the gastric pit.

seize food with the mouth and swallow it whole. But mammals, as mentioned before, have an elaborate array of teeth, and they also have a muscular tongue and cheeks. These adaptations allow mammals to chew their food—a rare ability in the animal kingdom. Furthermore, mammalian saliva contains a starch-digesting enzyme called *amylase*. Thus mammals initiate both physical and chemical digestion while food is still in the mouth.

A mouthful of food is called a *bolus*, and when that bolus is *swallowed*, it is propelled through the pharynx (where the mouth and nasal passages join), over the opening of the trachea (windpipe), and into the esophagus. To prevent food from entering the trachea, a flap of tissue called the *epiglottis* moves down to cover the entrance to the larynx (voice box). Swallowing involves many muscles contracting in a precise sequence. That sequence is an autonomic reflex that is stimulated by contact of the bolus with the tissue at the back of the mouth. If you take a cotton swab and carefully touch the back of your mouth, you will see that you cannot do it without triggering a swallowing reflex.

The bolus moves toward the stomach via peristalsis, aided by gravity. The enteric nervous system helps coordinate peristalsis by creating an *anticipatory wave of relaxation*. As one region of the gut contracts, the smooth muscles of the region just beyond that point relax to accommodate the bolus. A sphincter at the end of the esophagus is normally closed to prevent the contents of the stomach from going back into the esophagus. The anticipatory wave of relaxation allows the sphincter to open so the bolus can pass into the stomach.

The stomach breaks up food and begins protein digestion

The vertebrate stomach has several functions. It holds large batches of food, physically breaks up the food, and begins protein digestion. It also kills many of the microorganisms that inevitably enter with food.

Deep infoldings in the walls of the stomach called **gastric pits** are lined with three types of secretory cells (**FIGURE 39.8A**). *Chief cells* secrete **pepsinogen**, the inactive form of the protease **pepsin**. Pepsin is secreted in an inactive form so that it will not digest the very cells that secrete it. Another cell type, *parietal cells*, secrete hydrochloric acid (HCl), which kills most ingested microorganisms, helps break down food, and also converts the pepsinogen to active pepsin. Once some pepsin is activated, it activates other pepsin molecules—a process called autocatalysis (**FIGURE 39.8B**). This nasty mix of substances could damage the stomach wall, but a third cell type secretes mucus that provides a protective coating for the stomach.

If the mucus lining of the stomach is damaged, an *ulcer* can develop—a place where the acid and pepsin begin to eat through the stomach wall. Pioneering research by Robin Warren showed that most stomach ulcers are caused by a bacterium, *Helicobacter pylori*, that has adapted to the acidic environment of the stomach. *H. pylori* infection causes inflammation of the stomach lining, which in turn causes thinning of the mucus layer, allowing stomach juices to eat into the stomach lining. Dr. Warren memorably tested his theory by drinking a glass of *H. pylori* giving himself a pre-ulcerous inflammation and then curing himself with antibiotics. In 2005 Dr. Warren and his colleague Barry Marshall received a Nobel Prize for their work on bacterial causes of ulcers, and the practical result is that many ulcers are now treated successfully with antibiotics.

The parietal cells lining the gastric pits produce about 2 liters of a concentrated HCl solution per day, enough to bring the stomach contents below a pH of 1—the same as that of battery acid and 10 times more acidic than lemon juice. This means that parietal cells can create a H^+ concentration difference of 3 million-fold across their plasma membranes. Such a feat of ion transport is not seen anywhere else in the body. How do they do it?

Parietal cells contain the enzyme *carbonic anhydrase*, which catalyzes the formation of H_2CO_3 from CO_2 and water. H_2CO_3 immediately dissociates into H^+ and bicarbonate ion (HCO_3^-). On the blood side of the parietal cells, an *antiporter* protein (a type of secondary active transport protein; see Concept 5.3) exports HCO_3^- into the blood, exchanging it for Cl^-. On the gastric side, a different antiporter pumps H^+ into the stomach cavity, exchanging it for K^+ (**FIGURE 39.8C**). However, this K^+ can leak out again down its concentration gradient. Thus the inward transport of K^+ acts like an endless conveyer belt moving H^+ out of the parietal cells and into the stomach lumen.

LINK Carbonic anhydrase is also important for CO_2 transport in blood; see Figure 37.13

Muscle contractions mix the stomach's contents and push them into the small intestine

The contractions of the smooth muscles in the wall of most of the stomach do not contract and relax in a clear peristaltic fashion as in the esophagus. Instead, they create mixing motions that turn over the contents of the stomach and create a slurry of partly digested food, acid, and enzymes called **chyme**. In the lower regions of the stomach, the peristaltic contractions resume and push the chyme up against the pyloric sphincter leading to the small intestine (see Figure 39.8A). Anticipatory waves of relaxation cause the pyloric sphincter to open and allow small squirts of chyme to enter the small intestine. This measured introduction of food into the small intestine enables it to continue the digestive processes efficiently. If the contents pass through too quickly, digestion and absorption will not be complete; if the contents pass through too slowly, the metabolic needs of the animal will not be met.

Ruminants have a specialized four-chamber stomach

In **ruminants** (cud chewers), the stomach is highly specialized to house cellulose-digesting bacteria (**FIGURE 39.9**). The ruminant stomach has four chambers. The first two, the **rumen** and the **reticulum**, are packed with bacteria. The ruminant periodically regurgitates the contents of the rumen (the *cud*) into the mouth for rechewing. Enormous numbers of microorganisms leave the rumen along with the partially digested food. This mass is concentrated by water absorption in the **omasum** before it enters the true stomach, the **abomasum**, which functions much like our own stomach—secreting acid and enzymes, and killing the bacteria. This system provides a cow not only with glucose from cellulose breakdown, but also with more than 100 grams of protein per day from digestion of its own bacterial culture.

The small intestine continues digestion and does most absorption

In the small intestine, the digestion of carbohydrates and proteins continues, the digestion of fats begins, and nutrients of all types are absorbed. The small intestine is actually a very

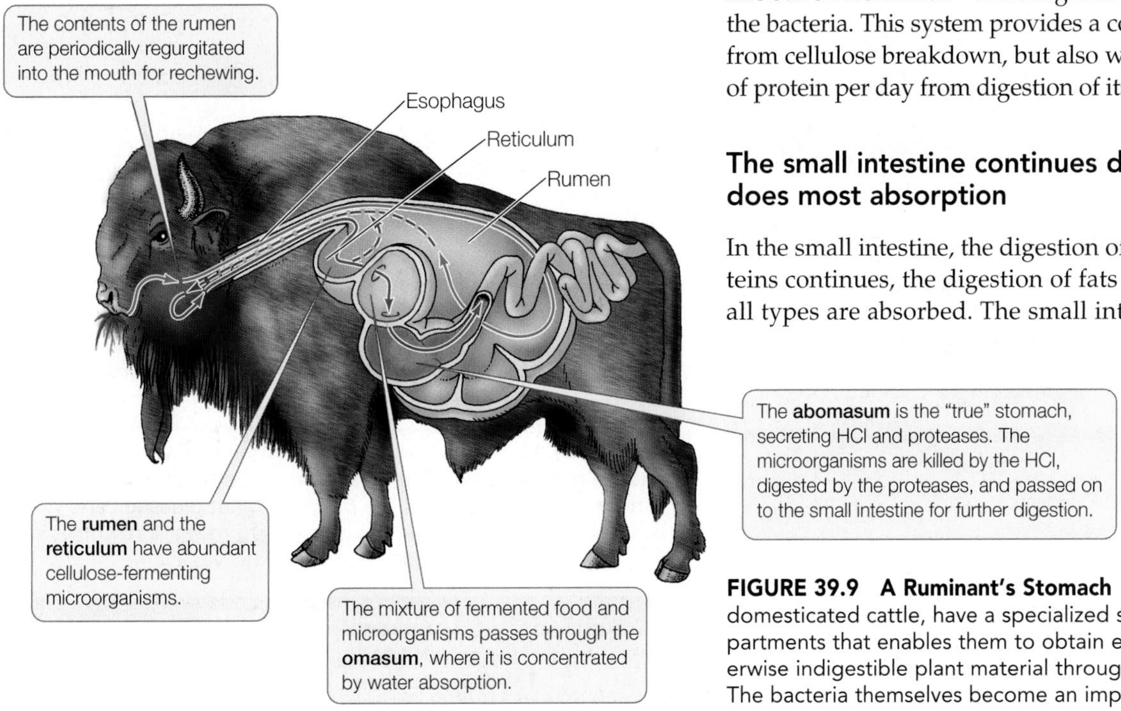

The contents of the rumen are periodically regurgitated into the mouth for rechewing.

Esophagus

Reticulum

Rumen

The **rumen** and the **reticulum** have abundant cellulose-fermenting microorganisms.

The mixture of fermented food and microorganisms passes through the **omasum**, where it is concentrated by water absorption.

The **abomasum** is the "true" stomach, secreting HCl and proteases. The microorganisms are killed by the HCl, digested by the proteases, and passed on to the small intestine for further digestion.

FIGURE 39.9 A Ruminant's Stomach Bison, like their relatives domesticated cattle, have a specialized stomach with four compartments that enables them to obtain energy from coarse, otherwise indigestible plant material through bacterial fermentation. The bacteria themselves become an important source of nutrition.

long organ, about 6 meters in an adult human. The mucosal epithelium in the small intestine has an enormous surface area for absorption because of its many villi and microvilli.

The small intestine of humans has three sections. A short initial section called the **duodenum** carries out most digestion. This requires many specialized enzymes and other secretions, some from the duodenum itself and some from the liver and pancreas. Most of the nutrients are then absorbed in the later sections of the small intestine, the **jejunum** and **ileum** (see Figure 39.6).

DUODENUM In the duodenum, mucosal epithelial cells produce peptidases that cleave small peptides into absorbable amino acids. These epithelial cells also produce the enzymes maltase, lactase, and sucrase that cleave the common disaccharides into absorbable monosaccharides—glucose, galactose, and fructose. The epithelial cells also produce some **lipases**— enzymes that digest fats.

> **FRONTIERS** Most mammals, including most humans, stop producing the enzyme *lactase* after weaning. Without lactase, adult mammals have difficulty digesting the disaccharide *lactose*, which is abundant in milk. Undigested lactose is metabolized by bacteria in the large intestine, causing gas, diarrhea, and abdominal cramps. Why, then, can some people drink milk as adults? Recent research indicates that about 10,000 years ago, several human ethnic groups in northern Europe and Africa gained the ability to produce lactase throughout adulthood. This evolutionary event coincided with the domestication of cattle in each of those ethnic groups, a striking example of culture driving evolution.

LIVER The liver produces and secretes a green liquid known as **bile** that is stored temporarily in the **gallbladder** before being sent to the duodenum through the *common bile duct* (**FIGURE 39.10**). Bile's main function is to *emulsify* fats—break up large fat droplets so they will mix more readily in water. Bile includes molecules called **bile salts** that have one lipophilic end (soluble in fat) and one hydrophilic end (soluble in water). The lipophilic ends merge with the fat droplets, leaving their hydrophilic ends sticking out. As a result, bile salts prevent the fat droplets (micelles) from sticking together (**FIGURE 39.11A**). This greatly increases the surface area of the fats that can be exposed to lipases. Soon, the fats are digested into their component fatty acids, after which they move into the plasma membranes of the intestinal cells.

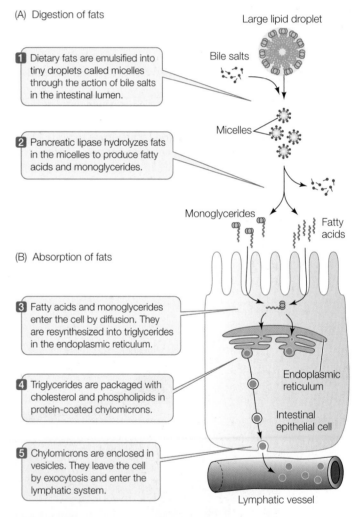

(A) Digestion of fats

1 Dietary fats are emulsified into tiny droplets called micelles through the action of bile salts in the intestinal lumen.

Large lipid droplet
Bile salts
Micelles

2 Pancreatic lipase hydrolyzes fats in the micelles to produce fatty acids and monoglycerides.

Monoglycerides
Fatty acids

(B) Absorption of fats

3 Fatty acids and monoglycerides enter the cell by diffusion. They are resynthesized into triglycerides in the endoplasmic reticulum.

4 Triglycerides are packaged with cholesterol and phospholipids in protein-coated chylomicrons.

5 Chylomicrons are enclosed in vesicles. They leave the cell by exocytosis and enter the lymphatic system.

Endoplasmic reticulum
Intestinal epithelial cell
Lymphatic vessel

FIGURE 39.11 Digestion and Absorption of Fats (A) Dietary fats are broken up by bile into small micelles that present a large surface area to lipases. (B) The products of fat digestion are absorbed by intestinal mucosal cells, where they are resynthesized into triglycerides and exported to lymphatic vessels.

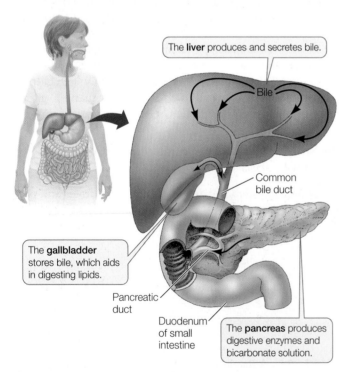

The **liver** produces and secretes bile.

Bile

Common bile duct

The **gallbladder** stores bile, which aids in digesting lipids.

Pancreatic duct

Duodenum of small intestine

The **pancreas** produces digestive enzymes and bicarbonate solution.

FIGURE 39.10 The Liver, Gallbladder, and Pancreas The liver and pancreas produce enzymes and other secretions that aid digestion and are released into the duodenum through the common bile duct.

> **yourBioPortal.com**
>
> Go to ANIMATED TUTORIAL 39.1
> The Digestion and Absorption of Fats

PANCREAS The **pancreas** is a large gland that lies just behind and below the stomach (see Figures 39.6 and 39.10). It is both an endocrine gland (secreting hormones into the blood) and an exocrine gland (secreting digestive juices to the gut lumen). The endocrine cells of the pancreas produce the glucose-regulating hormones *insulin* and *glucagon*, which we will discuss in the next concept. The exocrine cells of the pancreas produce a host of digestive enzymes, including lipases, amylases, proteases, and nucleases (**TABLE 39.3**). They also secrete bicarbonate ions (HCO_3^-) that neutralize the highly acidic pH of the chyme that enters the duodenum from the stomach. The pancreatic enzymes cannot function in an acidic solution.

Pancreatic enzymes are produced and secreted into the pancreatic duct in inactive forms called zymogens. An enzyme called *enterokinase*, secreted by the duodenal epithelium, initiates the activation of these zymogens once they leave the common bile duct that conducts the secretions of the gallbladder and the pancreas into the duodenum. One of the zymogens, trypsinogen, is converted to its active form *trypsin* by enterokinase. Trypsin can then activate other trypsinogen molecules—autocatalysis, another example of positive feedback—as well as the other zymogens.

JEJUNUM AND ILEUM The breakdown of complex food molecules into their constituent units by pancreatic enzymes

TABLE 39.3	Major Digestive Enzymes of Humans
SOURCE/ENZYME	ACTION
SALIVARY GLANDS	
Salivary amylase	Starch → Maltose
STOMACH	
Pepsin	Proteins → Peptides; autocatalysis
PANCREAS	
Pancreatic amylase	Starch → Maltose
Lipase	Fats → Fatty acids and glycerol
Nuclease	Nucleic acids → Nucleotides
Trypsin	Proteins → Peptides; zymogen activation
Chymotrypsin	Proteins → Peptides
Carboxypeptidase	Peptides → Shorter peptide and amino acids
SMALL INTESTINE	
Aminopeptidase	Peptides → Shorter peptides and amino acids
Dipeptidase	Dipeptides → Amino acids
Enterokinase	Trypsinogen → Trypsin
Nuclease	Nucleic acids → Nucleotides
Maltase	Maltose → Glucose
Lactase	Lactose → Galactose and glucose
Sucrase	Sucrose → Fructose and glucose

continues in the jejunum. In addition, dipeptidases and disaccharidases produced by the cells lining the jejunum complete the breakdown of proteins and carbohydrates into the basic monomeric units of amino acids and six-carbon sugars. This final step is accomplished on the surfaces of the cells, and therefore these final breakdown products can be absorbed immediately and not contribute to a drastic rise in the osmotic potential of the gut contents. Throughout the length of the jejunum and ileum, the small molecules that are the final products of digestion—amino acids, sugars, nucleic acids, ions, and so forth—are absorbed into the bloodstream across the microvilli-covered surfaces of the intestinal epithelial cells.

Nutrients and ions are absorbed by diverse methods, including diffusion, facilitated diffusion, osmosis, active transport, and secondary active transport. Fructose, for example, is transported by facilitated diffusion, moving down its concentration gradient from the gut lumen into the cell. This mechanism works because once fructose enters an intestinal cell, it is converted to glucose. Thus the concentration of fructose in the cell is always low and the concentration gradient is maintained.

In contrast, glucose and many amino acids are transported against a concentration gradient by secondary active transporters that exploit the concentration gradient of Na^+ between the outside and the inside of the epithelial cells that is maintained by the Na^+–K^+ pump common to all animal cells (see Figure 5.7). As Na^+ is pulled down its concentration gradient into the cell, the "hitchhiking" molecules (e.g., glucose) are carried along with it.

LINK Active transport mechanisms and the Na^+–K^+ pump are described in Concept 5.3

Many inorganic ions such as sodium, calcium, and iron are also absorbed by active transport. For example, Na^+ pumps move Na^+ from the interior of the epithelial cells to the blood. This maintains a low concentration of Na^+ in the epithelial cells. About 30 grams of Na^+ are transported this way every day, and Cl^- follows. The transport of Na^+ and other ions is also important for water absorption because it creates an osmotic concentration gradient across the epithelial lining of the gut. At least 7 to 8 liters of water per day are absorbed because of this osmotic gradient, and it is mostly through the tiny spaces between the individual epithelial cells.

FRONTIERS Because water moves out of the gut through *spaces* between the epithelial cells and not through the cells themselves, it can carry with it nutrients that are in solution. Since calculations of the maximum capacity of the transport mechanisms of the epithelial cells fall short of the actual amounts of nutrients absorbed, it has been proposed that absorption of dissolved nutrients by hydraulic flow (solvent drag) through the intercellular spaces is a major mechanism of absorption. More research is needed to test this hypothesis.

Fats, as we have seen, pose a problem for the digestive process because of their poor solubility in water. The same problem applies to their transport in the blood stream. The intestinal epithelial cells solve this problem by packaging fats into water-soluble particles called **chylomicrons**. The products of fat digestion (fatty acids and monoglycerides) move readily through the plasma membranes of the intestinal epithelium. These cells then resynthesize triglycerides from these fatty acids and monoglycerides, package them with cholesterol and phospholipids, and coat them with protein to form the water-soluble chylomicrons (**FIGURE 39.11B**). Rather than enter the blood directly, chylomicrons pass into blind-ended lymph vessels that are in each villus. They then flow through the lymphatic system, finally entering the bloodstream through the thoracic ducts at the base of the neck (see Figure 38.13).

Absorbed nutrients go to the liver

In most organs, blood leaving a capillary bed flows directly back to the heart. The digestive system is an exception. Blood leaving the capillary beds of the digestive tract flows in the *hepatic portal vein* to the liver. There it circulates through small spaces between groups of liver cells. Thus the liver is able to process, store, and screen the newly absorbed materials before they are sent on to the rest of the body. The liver cells replenish their store of glycogen from glucose, sucrose, and fructose; build proteins from amino acids; and break down various toxins and drugs that are ingested.

The large intestine absorbs water and ions

The small intestine gradually empties into the *large intestine*, or **colon**. Most of the available nutrients have been absorbed by now, leaving a lot of water, inorganic ions, and any indigestible components. The colon absorbs the water and ions and packs the remaining material into semisolid *feces*. Feces are stored in the final portion of the colon, the **rectum**, until elimination.

The colon's reabsorption of water is important. In humans, the small intestine adds about 8 liters of water per day into the gut lumen during the digestive process. If this water is not reabsorbed, the resulting *diarrhea* will rapidly cause serious dehydration. Cholera toxin, for example, stimulates excessive ion and fluid secretion from the small intestine and can cause such extreme diarrhea that death may occur in hours.

The colon often houses symbiotic bacteria. In humans, colon bacteria produce most of our vitamin K, essential for blood clotting. In most vertebrates, there is a dead-end sac at the junction of the small and large intestines, called the *cecum* (plural *ceca*), that houses cellulose-digesting bacteria in many herbivores. Horses, rabbits, geese, and iguanas are among those herbivores that have enormous ceca packed with bacteria. Some of these animals can absorb in the colon carbohydrates and other nutrients produced by bacterial action, and others re-ingest feces that contain these nutrients (coprophagy). In humans, however, the cecum is a relatively short sac that ends in a small vestigial stub called the *appendix*.

LINK For more on the diverse bacteria in the human gut and their possible effects on human health, see Concept 42.1

Do You Understand Concept 39.3?

- Where in the digestive tract does digestion of proteins, carbohydrates, and fats begin, and how are those processes initiated?
- If a person's stomach is surgically removed, what will be the consequences?
- Sometimes a gallbladder has to be removed. How does this affect digestion?
- A blow to the abdomen can damage the common bile duct and block the flow of pancreatic zymogens. If in this condition, a molecule of trypsinogen is spontaneously converted to trypsin, what could be the consequences?

We have seen how the vertebrate digestive system digests a single meal. But what happens in between meals, and what triggers the drive to begin eating the next meal? Food intake and nutrient flow are both carefully regulated.

concept 39.4 Food Intake and Metabolism Are Regulated

Most animals do not eat continuously. Instead, they cycle between an **absorptive state** (food in the gut) and a **postabsorptive state** (no food in the gut). Nutrients must be stored while they are abundant (during the absorptive state) and then released gradually from those stores in between meals (during the postabsorptive state). Furthermore, hunger must be regulated, to ensure that animals search for food when they need to eat and stop eating when the gut is full. These processes are critical for homeostasis and are regulated by multiple neuronal and hormonal systems.

Neuronal reflexes control many digestive functions

As already mentioned, the digestive tract contains its own *enteric nervous system*. The enteric nervous system communicates some information to and from the CNS, but its most important role is to coordinate the digestive tract. In particular, the enteric nervous system is able to coordinate, largely on its own, the movement of food through the gut.

A few digestive reflexes do involve the CNS. Swallowing is one example; salivation at the sight or smell of food is another, and defecation is a third.

Hormones regulate many digestive functions

Over two dozen hormones regulate the activities of the digestive tract and its accessory organs (**FIGURE 39.12**). Several of

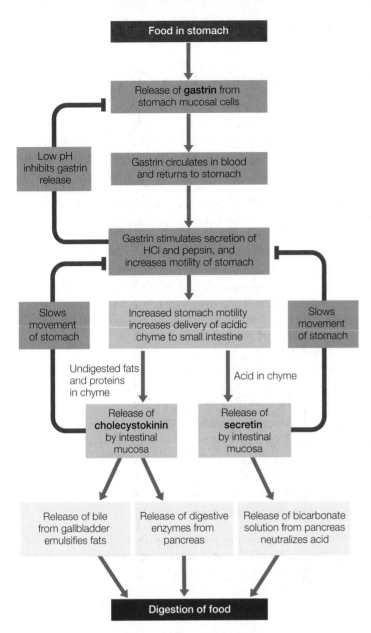

FIGURE 39.12 Hormones Control Digestion The hormones gastrin, cholecystokinin, and secretin are involved in feedback loops that control the sequential processing of food in the digestive tract. Red lines indicate inhibitory actions; green lines indicate stimulatory actions.

these hormones are released when food arrives in the stomach and duodenum. The presence of food in the stomach causes cells in the wall of the stomach to secrete the hormone **gastrin**. Gastrin circulates in the blood and returns to the stomach, causing an increase in motility and the secretion of digestive juices. Gastrin release begins to be inhibited when the pH of the stomach contents falls below 3—an example of negative feedback (see Concept 29.2).

Acidic chyme arriving in the duodenum stimulates cells in the duodenal epithelium to release **secretin** (the first hormone ever discovered, hence its name). Secretin stimulates the pancreas to produce and secrete bicarbonate ions that neutralize the acidic chyme. Similarly, the presence of fats and proteins in the chyme stimulates duodenal cells to release **cholecystokinin (CCK)**. CCK then stimulates the delivery of bile into the

common bile duct by increasing peristalsis of the gallbladder and the duct. CCK also stimulates the pancreas to release digestive enzymes. Both secretin and CCK are negative feedback signals to the stomach to slow the delivery of chyme into the duodenum (see Figure 39.12).

Insulin and glucagon regulate blood glucose

Two of the most important hormones from the digestive system, *insulin* and *glucagon*, regulate glucose metabolism. Glucose is a fuel that can be used by all cells of the body, and it gets into cells by diffusion. Thus it is necessary to have an adequate blood glucose concentration to provide a concentration gradient across cell membranes. When blood glucose levels fall, most cells can switch to alternative fuels, mostly fats, but the nervous system tissues cannot—they require glucose. **Insulin** is the major hormone that maintains blood glucose levels.

Insulin regulates fuel metabolism by controlling the presence of the transporters in the cell membrane that facilitate the diffusion of glucose into cells. When an animal is in the absorptive state (food in gut), the rise in blood glucose as carbohydrates are digested stimulates the β cells of the *islets of Langerhans* in the pancreas to secrete insulin (**FIGURE 39.13**).

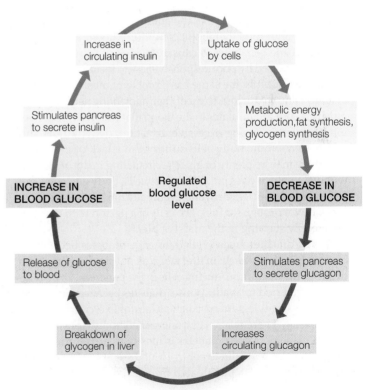

FIGURE 39.13 Regulating Glucose Levels in the Blood Insulin (blue) and glucagon (brown) interactions maintain the homeostasis of circulating glucose. It is important for blood glucose to remain stable because it is the essential source fuel for the nervous system.

yourBioPortal.com

Go to ANIMATED TUTORIAL 39.2
Insulin and Glucose Regulation

Circulating insulin causes the insertion of glucose transporters into cell membranes so that more glucose can diffuse into the cells, and the cells use the glucose for energy and for the synthesis of energy storage molecules—glycogen and fat.

In the postabsorptive state, blood glucose levels decline and insulin secretion stops. The lack of insulin decreases the uptake of glucose into cells, forcing them to use alternative energy supplies. The lack of insulin also changes enzymatic activity in key tissues, enabling the body to maintain blood glucose levels and to shift fuel usage to fats. In the liver, the enzymes of glycogen synthesis become less active, and the enzymes that break down glycogen and provide glucose to the blood are activated. In fat cells, lipases are activated. The cells of the nervous system do not require insulin to take up glucose, so they continue to take up glucose during the postabsorptive phase. Thus the lack of insulin protects glucose supplies for the nervous system.

The metabolism of fuel molecules during the postabsorptive state is mostly controlled by the lack of insulin, but if blood glucose falls too low, another pancreatic hormone, **glucagon**, is released. Glucagon's effect is opposite that of insulin: it stimulates liver cells to break down glycogen and to carry out gluconeogenesis—the synthesis of glucose from some amino acids and from lactate. Thus, under the influence of glucagon, the liver produces glucose and releases it into the blood.

Lack of insulin, or lack of the ability to respond to insulin, causes a dangerous metabolic condition known as **diabetes mellitus**. Type I diabetes is caused by lack of insulin; type II diabetes is caused by poor responsiveness to insulin. Both types of diabetes result in the same basic problem: most cells cannot take up enough glucose to meet their metabolic needs, and so much glucose accumulates in the blood that it starts to spill over into the urine, causing excessive urination, dehydration, and thirst. Meanwhile, body cells suffer from a lack of metabolic fuel; there may be plenty of glucose circulating in the blood, but without insulin it does not diffuse into cells. Body glycogen and fat stores become depleted, and eventually muscle cells begin to break down protein for fuel. The cells of a person with diabetes are literally starving in the midst of plenty.

Type I diabetes is also called juvenile diabetes because its onset is frequently early in life, when an autoimmune reaction destroys the insulin-producing cells in the pancreas. Type II diabetes is referred to as adult-onset diabetes because it is mostly caused by lifestyle patterns such as lack of exercise, smoking, and obesity. There is also a strong genetic predisposition for type II diabetes. Type I diabetes is mostly treated through use of supplemental insulin to control blood sugar levels, whereas type II diabetes is mostly treated through changes in lifestyle.

The liver directs the traffic of the molecules that fuel metabolism

The liver coordinates many of the changes necessary to switch between absorptive and postabsorptive states. During the absorptive state the liver (under the influence of insulin) stores fuel in the form of glycogen and fats. The liver also synthesizes blood plasma proteins from circulating amino acids. Later, when fuel molecule levels in the blood decline, the liver taps its reserves and delivers nutrients into the blood.

The liver is also a major controller of fat metabolism through its production of **lipoproteins**. Lipoproteins are particles of fat and cholesterol with a covering of protein that allow them to be suspended in water. The chylomicrons made by the intestine are a type of lipoprotein. As chylomicrons travel through the bloodstream, their triglyceride and cholesterol cargo is delivered to the liver or to fat cells. The liver repackages the dietary lipids into various other types of lipoproteins that shuttle the lipids to and from the tissues of the body. The "good cholesterol" and "bad cholesterol" that you have probably heard of are actually types of lipoproteins, categorized by density:

- **High-density lipoproteins** (**HDLs**) remove cholesterol from tissues and carry it to the liver, where it is used to synthesize bile. These are the "good" lipoproteins, and their levels are higher in people who exercise.

- **Low-density lipoproteins** (**LDLs**) transport cholesterol around the body for use in biosynthesis and for storage. These are the "bad" lipoproteins associated with a high risk for cardiovascular disease.

- **Very low-density lipoproteins** (**VLDLs**) contain mostly triglyceride fats, which they transport to fat cells. These are especially "bad"; they are associated with excessive fat deposition as well as a high risk for cardiovascular disease.

Many hormones affect food intake

Obesity is a major health issue in the United States. A simple rule—take in fewer calories than your body burns while eating a balanced diet—should solve the problem, but it doesn't. Why not? Lifestyle plays a major role in obesity, but genetic and regulatory factors "weigh in" as well. Ultimately, the amount of food an animal eats is governed by its sensations of hunger and satiety. These sensations are influenced by the hypothalamus (see Concept 30.3). But how does the hypothalamus know whether the body has adequate (or even excessive) stores of fuel? It appears that the hunger and satiety centers in the hypothalamus are powerfully affected by several digestive hormones, among them insulin, CCK, and **leptin**.

Leptin was discovered in a strain of mice that tend to eat enormous amounts of food (**FIGURE 39.14**). Given free access to food, these mice always become obese. The mice have a mutation in the gene for leptin, now known to be a peptide hormone produced by adipose cells. The more fat an adipose cell contains, the more leptin it produces. Thus the overall level of leptin in the bloodstream is the body's "fat report"—the higher the leptin, the more fat the body has. The hypothalamic centers that control hunger and satiety contain leptin receptors. Thus leptin is probably providing feedback information to the hypothalamus about the status of the body's fat reserves.

yourBioPortal.com

Go to INTERACTIVE TUTORIAL 39.1
Parabiotic Mice: Regulation of Food Intake

INVESTIGATION

FIGURE 39.14 A Single-Gene Mutation Leads to Obesity in Mice In mice the *Ob* gene codes for the protein leptin, a satiety factor that signals the brain when enough food has been consumed. The recessive *ob* allele is a loss-of-function allele, so *ob/ob* mice do not produce leptin; they do not experience satiety and become obese. The *Db* gene encodes the leptin receptor, so mice homozygous for the recessive loss-of-function allele *db*, even if they produce leptin, cannot use it and so become obese.

HYPOTHESIS

Mice who cannot produce the satiety signal protein leptin will not become obese if they are able to obtain leptin from an outside source.

METHOD
1. Create two strains of genetically obese laboratory mice, one of which lacks functional leptin (genotype *ob/ob*) and one which lacks the receptor for leptin (genotype *db/db*).
2. Create parabiotic pairs by surgically joining the circulatory systems of a non-obese (wild-type) mouse with a partner from one of the obese strains.
3. Allow mice to feed at will.

Parabiotic pair

Wild-type mouse   Genetically obese mouse
(*Ob/–* and *Db/–*)   (either *ob/ob* or *db/db*)

RESULTS Parabiotic *ob/ob* mice obtain leptin from the wild-type partner and lose fat. Parabiotic *db/db* mice remain obese because they lack the leptin receptor and thus the leptin they obtain from their partner has no effect.

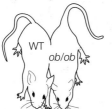 

CONCLUSION

The protein leptin is a satiety signal that acts to prevent overeating and resultant obesity.

ANALYZE THE DATA

In a further experiment, leptin was purified and injected into *ob/ob* and *Ob/Ob* (wild-type) mice daily. The data in the table were collected before the injections began (baseline) and 10 days later.

Parameter	Baseline		Day 10	
	ob/ob	*Ob/Ob*	*ob/ob*	*Ob/Ob*
Food intake (g/day)	12.0	5.5	5.0	6.0
Body mass (g)	64	35	50	38
Metabolic rate (ml O_2/kg/hr)	900	1150	1100	1150
Body temperature (°C)	34.8	37.0	37.0	37.0

A. Do the data support the hypothesis that leptin is a satiety signal? Why or why not?
B. Discuss what factors might explain the loss of body mass in the *ob/ob* mice.

Go to **yourBioPortal.com** for original citations, discussions, and relevant links for all INVESTIGATION figures.

When leptin was injected into the obese mice, they ate much less and lost body fat. This led to great excitement that leptin injections might be able to "cure" obesity in humans. In a few obese people, leptin injections did cause dramatic weight loss, but unfortunately, leptin injections had no effect on most obese people. Most obese people actually turn out to have *higher* levels of leptin than normal; it appears they are unable to respond normally to leptin (this is analogous to type II diabetes; the hormone is being produced, but the body cannot respond).

Many other hormones and feedback signals are now known to involved in regulating food intake in mammals. A hormone called *peptide YY* from the gut reduces appetite. The hormone *ghrelin*, produced by the stomach when it is empty, strongly stimulates appetite. Over a dozen other hormones and neurotransmitters are also known that affect the hunger and satiety centers of the hypothalamus. The search for an "obesity cure" continues. If one thing is clear, it is that the control of food intake involves more than just willpower; powerful physiological feedback systems are also at work.

FRONTIERS Recent research has identified the enzyme AMP-activated protein kinase (AMPK) as a possible signal of nutrient insufficiency that may drive hunger. When most cells are nutrient deprived, they produce AMPK, which stimulates the catabolism of substrates to replenish ATP. In the hypothalamus, increased levels of AMPK stimulate food intake, and lowered levels of AMPK inhibit food intake. In addition, AMPK activity in the hypothalamus is inhibited by leptin and insulin and is stimulated by ghrelin. Thus AMPK may sit at a point where several signaling pathways involved in controlling food intake converge into a single pathway. If true, this would make AMPK an attractive target for therapies to treat obesity and other types of eating disorders.

Do You Understand Concept 39.4?

- Briefly describe the anatomical origins, the factors that stimulate release, and the effects of the hormones gastrin, secretin, cholecystokinin, insulin, and leptin.

- Which condition do you think would more rapidly cause death: complete pancreatic failure or complete liver failure? Why?

- This concept described a strain of obese mice that could not produce leptin. A different strain of obese mice also has a defect in the leptin system, yet these mice did not respond to leptin injections—they continued eating and remained obese. Can you figure out why?

 Why are some ethnic groups of humans much more prone to obesity and diabetes than other groups?

ANSWER   At least two interacting factors are involved in the Pimas' struggle with diabetes and obesity: genetics

and lifestyle. Geneticists hypothesize that recurring episodes of starvation produced strong selective pressure for "thrifty genes"—alleles that result in greater-than-average efficiency in storing fat. Thrifty genes would give individuals a strong selective advantage when food is scarce. The Pima, for example, turn out to have a very low resting metabolic rate (Concept 39.1). They also release a large amount of insulin in response to glucose (recall that insulin promotes the synthesis of fat) (Concept 39.4). A typical Pima responds to a standard amount of glucose with a rise in insulin levels that is three times higher than in Americans of European ancestry.

The other major factor in the Pima obesity epidemic is an abrupt change in their traditional lifestyle. Pima living in the United States today eat a typical "American" diet of high-fat, high-calorie fast foods, and they also engage in less physical activity than their ancestors did. In contrast, the Mexican Pima still live a traditional lifestyle and eat traditional foods. Whereas the Arizona Pima engage in an average of only 2 hours of physical work per week, the Mexican Pima average 23 hours per week.

A high-calorie diet and sedentary lifestyle affect not just the Pima but undoubtedly contribute to the alarming increase in obesity that has occurred throughout the U.S. population in recent decades.

---

39 SUMMARY

concept 39.1 Food Provides Energy And Nutrients

- Animals are **heterotrophs** that derive their energy and molecular building blocks, directly or indirectly, from **autotrophs**. **Review Figure 39.1**

- Carbohydrates, fats, and proteins in food supply animals with metabolic energy. A measure of the energy content of food is the **Calorie (Cal)**. Excess energy is stored as glycogen and fat.

- Food also provides **macronutrients**, which are needed in large quantities, and **micronutrients**, which are needed in small amounts.

- Food provides essential carbon skeletons that animals cannot synthesize themselves.

- Adult humans require eight **essential amino acids** and at least one **essential fatty acid**. **Review Figure 39.2**

- **Essential minerals** are chemical elements that are required in the diet. **Review Table 39.1 and WEB ACTIVITY 39.1**

- **Vitamins** are small organic molecules needed as micronutrients, usually functioning as enzymes or coenzymes. **Review Table 39.2 and WEB ACTIVITY 39.2**

- **Malnutrition** results when any essential nutrient is lacking from the diet. Chronic malnutrition causes **deficiency diseases**.

concept 39.2 Digestive Systems Break Down Macromolecules

- Digestion involves the breakdown of complex food molecules into monomers that can be absorbed and used by cells. Hydrolytic enzymes break down proteins, carbohydrates, fats, and nucleic acids into their monomeric units.

- In most animals, digestion takes place in a tubular gut that has two openings. The tubular gut can be divided into a foregut that physically breaks up food, a midgut that digests and absorbs the food, and a hindgut that packages wastes. **Review Figure 39.3**

- In most vertebrates, absorptive areas of the gut are characterized by a large surface area produced by extensive folding and numerous **villi** and **microvilli**. **Review Figure 39.4**

- Animals can be characterized by how they acquire their food: **saprobes** and **detritivores** eat dead organic matter, **filter feeders** strain the aquatic environment for small food items, **herbivores** eat plants, and **carnivores** eat animals.

concept 39.3 The Vertebrate Digestive System Is a Tubular Gut with Accessory Glands

- The vertebrate gut can be divided into several compartment with different functions. **Review Figure 39.6 and WEB ACTIVITY 39.3**

- The gut consists of an inner **mucosa** that secretes mucus, enzymes, and hormones and absorbs nutrients; a **submucosa** containing blood and lymph vessels; two layers of smooth muscle; and a semi-independent nerve network called the **enteric nervous system**. **Review Figure 39.7**

- **Peristalsis** moves food throughout the length of the gut. Sphincters block the gut at certain locations, but they relax as a wave of peristalsis approaches.

- Digestion begins in the **mouth**. Mammals chew their food, and their saliva contains the starch-digesting enzyme amylase.

- The **stomach** breaks up food, begins the process of protein digestion, controls the flow of chyme to the small intestine, and kills many microorganisms. Stomach cells secrete HCl, **pepsinogen** (the inactive form of **pepsin**), and mucus that protects the stomach wall. **Review Figure 39.8**

- In the **duodenum**, pancreatic enzymes carry out most of the digestion of food. Bile from the liver and **gallbladder** emulsifies fats for digestion. Bicarbonate ions from the **pancreas** neutralize the pH of the **chyme**. **Review Figure 39.10**

- The **jejunum** and **ileum** absorb most of the products of digestion, including amino acids, monosaccharides, and inorganic ions. Specific transporter proteins are sometimes involved.

- Fats broken down by **lipases** are resynthesized into triglycerides within intestinal cells, combined with cholesterol and phospholipids, and coated with protein to form **chylomicrons**, which move into lymph vessels. **Review Figure 39.11 and ANIMATED TUTORIAL 39.1**

- Water and ions are absorbed in the large intestine as waste matter and consolidated into **feces**, which are periodically eliminated.

- Microorganisms in some compartments of the gut digest materials that their host cannot. **Review Figure 39.9**

concept 39.4 Food Intake and Metabolism Are Regulated

- Animals cycle between an **absorptive state** (food in the gut) and a **postabsorptive state** (no food in the gut).

- Neuronal reflexes coordinate much of the activity of the digestive tract.

- The actions of the stomach and small intestine are largely controlled by the hormones **gastrin**, **secretin**, and **cholecystokinin** (**CCK**). **Review Figure 39.12**

- **Insulin** from the pancreas largely controls glucose metabolism. In the postabsorptive state, lack of insulin blocks the uptake and use of glucose by most cells of the body except neurons. If blood glucose levels fall, **glucagon** secretion increases, stimulating the liver to break down glycogen and release glucose to the blood. **Review Figure 39.13 and ANIMATED TUTORIAL 39.2**

- The liver plays a central role in directing the traffic of fuel molecules. In the absorptive state, the liver takes up and stores fats, carbohydrates, and proteins. In the postabsorptive state, the liver can produce glucose, either via breakdown of glycogen or via gluconeogenesis from proteins or other molecules.

- Fat and cholesterol are shipped out of the liver as **low-density lipoproteins** (**LDLs**). High-density lipoproteins (**HDLs**) act as acceptors of cholesterol and are believed to bring fat and cholesterol back to the liver.

- Food intake is governed by sensations of hunger and satiety, which are determined by brain mechanisms responding to feedback signals such as insulin, **leptin**, and ghrelin. **Review INTERACTIVE TUTORIAL 39.1**

Salt and Water Balance and Nitrogen Excretion

Blood, sweat, and tears taste salty; their ionic concentrations are similar to those of the interstitial fluids of the body. The volume and solute composition of the interstitial fluids must remain within certain limits and be kept relatively free of wastes. Maintaining homeostasis of the interstitial fluids is the job of the excretory system, and it can be challenging. The challenges depend on an animal's environment and lifestyle. Some desert animals rarely encounter free water. Like all animals, they derive water from the metabolism of food, but how can they survive with only a meager water budget?

Animals such as fish, invertebrates, and even some terrestrial animals that live in fresh water, such as beavers, have the opposite problem; water continuously enters their bodies by osmosis or with the food they consume, so they must excrete water while conserving important ions that are scarce in their food. Some marine animals such as albatrosses have a still different problem: they take in enormous amounts of ions with their food and have to excrete them without losing much water. Some animals face dramatic changes in their salt-water-waste balance within a single day. Consider vampires—the bats, not the horror-film kind.

Vampire bats are small tropical mammals that feed on the blood of large mammals such as cattle. The bat lands on a sleeping victim and uses its sharp incisor teeth to knick a blood vessel close to the skin surface. The bat laps up the blood that wells out. Blood contains nutritious protein, but it is mostly water.

Feeding may be cut short by the victim waking up, so the bat needs to quickly consume as much as it can. To maximize

its protein intake and keep its weight low enough to fly, it rapidly eliminates the water from its meal. Within minutes of starting to feed, the bat is producing copious dilute urine. The warm trickle down the neck of the victim is not blood!

Between meals, the bat must conserve water. Metabolizing protein produces nitrogenous wastes that have to be excreted even though water may be scarce. Within a short time, the excretory system of the bat switches from producing abundant, dilute urine to producing a tiny amount of concentrated urine. Rapidly, the bat cycles from an excretory physiology typical of a mammal such as a beaver to that of a desert rodent.

The feeding behavior of this Common Vampire Bat (*Desmodus rotundus*) poses extreme challenges for the animal's excretory system.

Q QUESTION How do excretory systems of animals maintain homeostasis of the interstitial fluid in the face of extreme challenges?

KEY CONCEPTS

40.1 Excretory Systems Maintain Homeostasis of the Extracellular Fluid

40.2 Excretory Systems Eliminate Nitrogenous Wastes

40.3 Excretory Systems Produce Urine by Filtration, Reabsorption, and Secretion

40.4 The Mammalian Kidney Produces Concentrated Urine

40.5 The Kidney Is Regulated to Maintain Blood Pressure, Blood Volume, and Blood Composition

The concentration of solutes in extracellular fluid determines the water balance of the cells and thus influences the health and functions of all the organs of the body. In most marine vertebrates and in all freshwater and terrestrial animals, the composition of the extracellular fluid differs considerably from the external environment. These animals must depend on **excretory systems** to maintain the volume, concentration, and composition of their extracellular fluid.

Excretory systems have four basic functions:

- Regulating the volume of fluid in the body
- Regulating the overall solute concentration—the *osmolarity*—of the extracellular fluid within a narrow range
- Maintaining individual solutes such as Ca^{2+}, H^+, Na^+, and glucose at appropriate concentrations
- Eliminating the toxic *nitrogenous wastes* produced by protein and nucleic acid catabolism

The excreted water, solutes, and nitrogenous wastes form the liquid waste product called **urine**.

Excretory systems maintain osmotic equilibrium

The **osmolarity** of a solution is the number of osmoles of solute particles per liter of solvent. For example, seawater has an osmolarity of about 1 mole of solute particles, or 1 *osmole*, per liter of water (1 osm/L). In vertebrates, the extracellular fluid has a much lower osmolarity of about 0.3 moles of solute particles per liter of water, or 300 *milliosmoles* per liter (300 mosm/L).

Why is it so important to regulate the osmolarity of the extracellular fluid? Any difference in osmolarity between the extracellular fluid and the intracellular fluid could cause dangerous changes in cell volume. If the osmolarity of the extracellular fluid is less than that of the cytoplasm, water moves into the cells via *osmosis*, causing them to swell and possibly burst (see Figure 5.3). If the osmolarity of the extracellular fluid is greater than that of the cytoplasm, the cells lose water, shrink, and can ultimately die.

Note that osmolarity can be different from the *molarity* of a solution because some molecules dissociate in solution into more than one particle. Thus a 1 molar solution of glucose is also a 1 osmolar (1 osmole per liter) solution, but a 1 molar solution of sodium chloride (NaCl) is a 2 osmolar solution, because each NaCl molecule dissociates into two ions, Na^+ and Cl^-.

Animals can be osmoconformers or osmoregulators

Animals that live in terrestrial, freshwater, or marine environments face different osmolarity problems. On land, both salts and water are usually scarce and must be conserved, so terrestrial animals are **osmoregulators**; they actively regulate the osmolarity of their extracellular fluid.

In fresh water, water is plentiful but salts are scarce, so freshwater animals have to conserve salts and excrete the water

that continuously invades their bodies through osmosis. So freshwater animals also are osmoregulators.

Seawater, however, has abundant salts. Most marine invertebrates simply allow their extracellular fluid to equilibrate with seawater and are therefore called **osmoconformers**. But most marine vertebrates (except sharks and rays) are osmoregulators that maintain extracellular fluid osmolarities much lower than seawater.

Even animals that can osmoconform over a wide range of osmolarities must osmoregulate in extreme environments. *Artemia* brine shrimp, for example are found in huge numbers in the saltiest environments known, such as Utah's Great Salt Lake (**FIGURE 40.1**). They can survive in fluid with an osmolarity of an astonishing 2,500 mosm/L. In these extremely salty environments, *Artemia* actively transport Na^+ (with Cl^- following) from their extracellular fluid across their gill membranes to the environment. *Artemia* can also survive in dilute seawater (though not in pure fresh water) by reversing the direction of transport of ions across their gill membranes. Because of their excellent control of sodium transport across the gills, *Artemia* can live in environments of almost any osmolarity.

Animals can be ionic conformers or ionic regulators

Even if the extracellular fluid is at a proper osmolarity, the concentrations of certain ions may still need to be adjusted. Most animals are thus *ionic regulators*, regulating the ionic composition of their extracellular fluid. Even many osmoconformers are ionic regulators—although they may not regulate overall

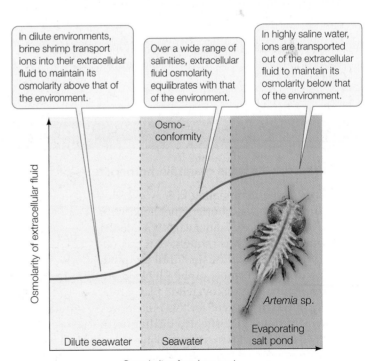

In dilute environments, brine shrimp transport ions into their extracellular fluid to maintain its osmolarity above that of the environment.

Over a wide range of salinities, extracellular fluid osmolarity equilibrates with that of the environment.

In highly saline water, ions are transported out of the extracellular fluid to maintain its osmolarity below that of the environment.

Osmo-conformity

Osmolarity of extracellular fluid

Artemia sp.

Dilute seawater | Seawater | Evaporating salt pond

Osmolarity of environment

FIGURE 40.1 Osmoconformity Has Limits *Artemia* brine shrimp are osmoconformers over a wide range of salinities. In extremely low- or high-salt environments, however, the shrimp osmoregulate by actively transporting ions across their gill membranes—in either direction, depending on the situation.

osmolarity, they usually do regulate the concentrations of certain ions. Some osmoconformers, however, are *ionic conformers*, allowing the ionic composition, as well as the osmolarity, of their extracellular fluid to match that of the environment.

Some commonly regulated ions are Na^+, Cl^-, K^+, Ca^{2+}, H^+, HCO_3^- (bicarbonate), and several others. Many of these ions are critical for cellular transport, cellular signaling, or protein function. For example, H^+ concentration (i.e., pH) is usually closely regulated because pH changes can drastically influence the structure and function of proteins.

LINK   For information on the effects of pH on protein structure, see Concept 3.4

One way to minimize pH changes in a solution is to add a *buffer*—a substance that can either absorb or release hydrogen ions. In mammals, the major buffer in the blood is bicarbonate ions (HCO_3^-; see Figure 37.13). HCO_3^- is indirectly formed from CO_2, as follows:

$$CO_2 + H_2O \rightleftharpoons H_2CO_3 \rightleftharpoons H^+ + HCO_3^-$$

From this equation, you can see that if excess hydrogen ions are added to this reaction mixture, the reaction will move to the left as bicarbonate absorbs the excess H^+ ions. If hydrogen ions are removed from the reaction mixture, however, the reaction will move to the right and supply more H^+. The HCO_3^- buffer system is important for controlling pH, because the reaction can be pushed to the right or the left physiologically: vertebrate lungs can eliminate CO_2 from the blood, and vertebrate kidneys can reabsorb HCO_3^- from the urine and excrete H^+ into the urine.

LINK   Blood pH is an important input in the control of respiratory rate and thus the elimination of CO_2; see Concept 37.4

In mammals, for example, kidney tubule cells reabsorb bicarbonate ions by transporting H^+ into the urine in exchange for Na^+. The H^+ combines with HCO_3^- in the urine, producing H_2CO_3, which disassociates into H_2O and CO_2. The CO_2 diffuses into the tubule cell, where the enzyme carbonic anhydrase catalyzes the combination of CO_2 with water to produce HCO_3^-, which is transported into the interstitial fluid and thence to the blood, and a H^+ ion that is transported into the urine; thus leading to the excretion of acid.

Do You Understand Concept 40.1?

- Would you expect a freshwater snail to be an osmoconformer or an osmoregulator? Explain.

- Salmon hatch in fresh water and migrate to the ocean where they spend several years before returning to the spawning grounds where they hatched. Describe the osmotic challenges salmon have to solve during their lifetimes.

- When a person first climbs to high altitude, he or she will hyperventilate (breathe rapidly) because of the low oxygen content of the air. This causes blood pH to rise, often causing headache and nausea. These symptoms resolve in a week or two, because of an adjustment made by the kidney. First, explain why hyperventilation causes blood pH to rise (you may find it helpful to review Concepts 37.4 and 37.5); then develop a hypothesis about what the kidney is doing to correct the pH problem.

In addition to maintaining salt and water balance and regulating ion concentrations, excretory systems must also eliminate the waste products of metabolism. The major problem is nitrogen.

APPLY THE CONCEPT

Excretory systems maintain homeostasis of the extracellular fluid

One way the mammalian kidney excretes hydrogen ions and buffers the blood with HCO_3^- is to secrete ammonia (NH_3) into the tubular fluid. It was thought that this occurred by simple diffusion until a protein called Rhcg, a member of the Rh blood antigen family, was discovered to play a role. Mice in which the gene for this protein was knocked out had difficulty dealing with NH_4Cl (a mild acid) in their drinking water. NH_4Cl was administered for 6 days to wild-type and knockout mice, and the effects on blood pH, NH_4^+ in the urine, and HCO_3^- in the blood plasma were recorded (see table).

1. What is the effect of ingestion of NH_4Cl on wild-type mice, and how do the knockout mice differ?

DAY	BLOOD pH		NH$_4^+$ IN URINE (μMOL/DAY)		HCO$_3^-$ IN BLOOD PLASMA (mM)	
	WILD TYPE	KNOCKOUT	WILD TYPE	KNOCKOUT	WILD TYPE	KNOCKOUT
0	7.28	7.31	20	15	18.0	17.9
2	7.12	7.14	265	145	12.3	12.1
6	7.25	7.13	225	145	16.5	11.0

2. What is the evidence that the response to ingestion of NH_4Cl is the excretion of NH_3 from the renal tubules? Are the knockout mice impaired in this regard?

3. What is the relationship between the excretion of NH_4^+ and blood buffering mechanisms (see Figure 37.13)?

Excretory Systems Eliminate Nitrogenous Wastes

The end products of the catabolism of carbohydrates and fats are water and carbon dioxide, which are not difficult to eliminate. Proteins and nucleic acids, however, contain nitrogen, so their catabolism produces not only water and carbon dioxide, but also **nitrogenous wastes**, which can be toxic and must be excreted from the body.

Animals excrete nitrogen in a number of forms

The most common nitrogenous waste is **ammonia** (NH_3). Because it is highly toxic, ammonia must be either excreted continuously to prevent its accumulation, or detoxified by conversion into **urea** or **uric acid** (**FIGURE 40.2**).

AMMONIA Ammonia is highly soluble in water and diffuses rapidly, so its excretion is relatively simple for many water-breathing animals; they continuously lose ammonia from their blood to the environment by diffusion across their gill membranes. Animals that excrete ammonia, such as aquatic invertebrates and bony fishes, are **ammonotelic**.

UREA Ureotelic animals, such as mammals, most amphibians, and cartilaginous fishes (sharks and rays), excrete urea as their principal nitrogenous waste product. Urea is quite soluble in water, but its excretion still requires a large loss of water that many animals can ill afford. As we will see later in this chapter, excretory systems that conserve water have evolved in mammals. A different strategy has evolved in sharks and

rays, which can tolerate unusually high concentrations of urea in their body tissues.

URIC ACID Uricotelic animals, which include insects, some amphibians, and reptiles (including birds), excrete uric acid. Uric acid is not very soluble in water, so it forms a colloidal suspension in the urine and is excreted as a semisolid, such as the whitish paste in bird droppings. A uricotelic animal loses very little water in excreting its nitrogenous wastes because precipitated uric acid does not exert an osmotic force in the urine, allowing water to be reabsorbed.

Most species produce more than one nitrogenous waste

Most species produce primarily one nitrogenous waste but also produce small amounts of the others. For example, humans are ureotelic, but we also excrete some uric acid, largely from the metabolism of nucleic acids and caffeine. If uric acid levels in the extracellular fluids rise too high, uric acid crystals can precipitate in joints, causing the age-old malady called gout. As we saw earlier, mammals can also excrete ammonia, which is one of our mechanisms for regulating pH of the extracellular fluid (see Apply the Concept on the opposite page).

Species that live in different habitats at different developmental stages may use more than one mechanism of nitrogen excretion. The tadpoles of frogs and toads, for example, excrete ammonia across their gill membranes, but adult frogs and toads generally excrete urea.

Do You Understand Concept 40.2?

- Name the three major nitrogenous wastes. Which one is the most toxic?
- Some desert-dwelling amphibians excrete uric acid as adults, instead of urea as other adult amphibians do. What selective advantage would this adaptation have had for the early ancestors of these desert species?
- Urea excretion generally rises after a period of heavy physical exercise. Why?

Whichever nitrogenous waste an animal excretes, and whatever osmotic challenge it might be facing in its environment, all excretory systems use the same basic processes to produce urine: filtration, reabsorption, and secretion.

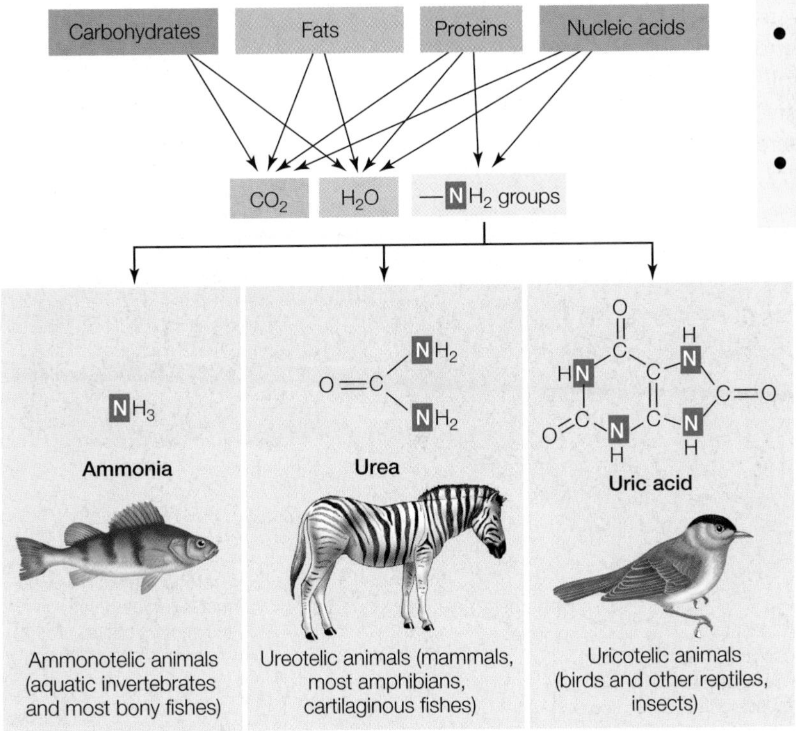

FIGURE 40.2 Waste Products of Metabolism The metabolism of proteins and nucleic acids produces nitrogenous wastes. Many aquatic animals, including most fishes, excrete nitrogenous wastes as ammonia, which is highly diffusible and soluble in an aqueous environment. Most terrestrial animals and some aquatic animals excrete either urea or uric acid.

concept
40.3 **Excretory Systems Produce Urine by Filtration, Reabsorption, and Secretion**

Most animal excretory systems produce urine by first *filtering* the extracellular fluid into some sort of cavity or tubule. In animals with a closed circulatory system, blood pressure drives the filtration of blood through the capillary walls and into the excretory tubule. The resulting fluid, called a *filtrate*, is initially similar in composition to the blood plasma, except that it contains no cells and few large molecules. As the filtrate flows through the tubules, its composition and concentration are modified by *reabsorption* of any valuable solutes that need to be retained (such as glucose) and by *secretion* of certain solutes that the body tries to excrete (such as some drugs, e.g., penicillin). This modified filtrate becomes the urine that leaves the body.

The processing of the filtrate into urine involves movement of water into and out of the tubules. However, recall that there is no mechanism for active transport of water. Instead, water must be moved either by a pressure difference (as in filtration) or by a difference in osmolarity (via osmosis). Water always flows down a pressure gradient or up a solute concentration gradient.

In this concept we will explore how filtration (driven by hydrostatic pressure) and osmosis are used in three different excretory systems: the *metanephridia* of annelids, the *Malpighian tubules* of insects, and the *kidneys* of vertebrates.

The metanephridia of annelids process coelomic fluid

Annelids, such as the earthworm, are segmented worms. In each segment they have a fluid-filled body cavity called a *coelom* (see Figure 23.3). Annelids have a closed circulatory system through which blood is pumped under pressure. The pressure causes the blood to be filtered across the thin, permeable capillary walls into the coelom. Some waste products, such as ammonia, diffuse directly from the tissues into the coelom. Where does this coelomic fluid go?

Each segment of the earthworm contains a pair of **metanephridia** (singular *metanephridium*; Greek *meta*, "akin to," and *nephros*, "kidney").

FIGURE 40.3 Metanephridia in Earthworms
The metanephridia of annelids are arranged segmentally. Coelomic fluid enters the nephrostome and flows through tubules leading to the nephridiopore. A close association of the tubules and blood capillaries facilitates the active exchange of substances between the blood and the tubular fluid.

Each metanephridium begins as a ciliated, funnel-like opening called a *nephrostome* in one segment, continues as a tubule in the next segment, and ends in a pore called a *nephridiopore* that opens to the outside of the animal (**FIGURE 40.3**). Coelomic fluid is swept into the metanephridia through the ciliated nephrostomes. As the fluid passes through the tubules, the tubule cells actively reabsorb certain molecules and actively secrete others. Eventually a dilute urine containing nitrogenous wastes and other solutes leaves the nephridiopores.

The Malpighian tubules of insects depend on active transport

The insect excretory system consists of **Malpighian tubules**. An individual insect has from 2 to more than 100 of these blind-ended tubules that open into the gut between the midgut and hindgut (**FIGURE 40.4**). Insects have an open circulatory system and therefore cannot use a pressure difference to filter extracellular fluids into the Malpighian tubules. Instead, the cells of the tubules actively transport uric acid, potassium ions, and sodium ions from the extracellular fluid into the tubules. The high concentration of solutes in the tubules causes water to follow osmotically, which flushes the tubule contents toward the gut. However, the water must then be recovered. Epithelial cells of the hindgut and rectum actively transport sodium and potassium ions from the gut back into the extracellular fluid, creating an osmotic gradient that pulls water out of the rectal contents. The final waste product is a semisolid mixture of uric acid mixed with other wastes. The Malpighian tubule system is highly effective at excreting nitrogenous wastes and some salts without losing much water. As a result, insects can live in the driest habitats on Earth.

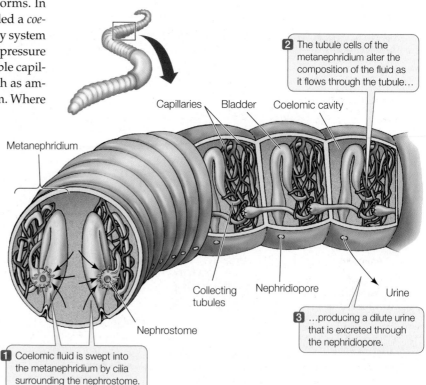

2 The tubule cells of the metanephridium alter the composition of the fluid as it flows through the tubule...

Capillaries   Bladder   Coelomic cavity

Metanephridium

Collecting tubules   Nephridiopore   Urine

3 ...producing a dilute urine that is excreted through the nephridiopore.

Nephrostome

1 Coelomic fluid is swept into the metanephridium by cilia surrounding the nephrostome.

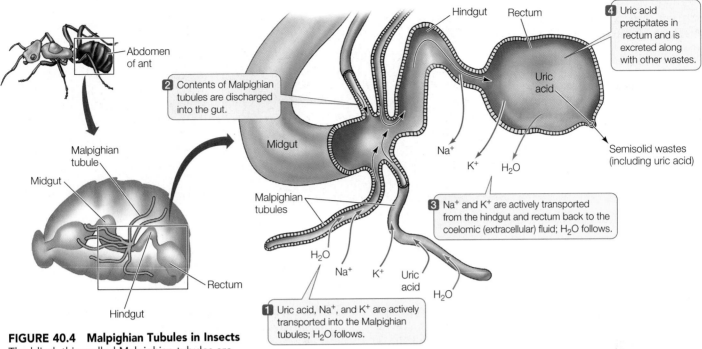

FIGURE 40.4 Malpighian Tubules in Insects
The blind, thin-walled Malpighian tubules are attached to the junction of the insect's midgut and hindgut and project into the spaces containing extracellular fluid. This system makes it possible to excrete wastes with very little loss of water.

The vertebrate kidney is adapted for excretion of excess water

The main excretory organ of vertebrates is the **kidney**, and the functional unit of the kidney is the **nephron**. A nephron consists of a **renal tubule** and its surrounding blood vessels (**FIGURE 40.5**). Nephrons can filter large volumes of blood and can reabsorb large amounts of salts and glucose, making the vertebrate kidney well adapted for the excretion of excess water. Each human kidney has about 1 million nephrons.

A nephron begins with a ball-shaped structure called **Bowman's capsule** that encloses a dense ball of capillaries called a **glomerulus** (plural *glomeruli*; **FIGURE 40.6**). The glomerulus sits within Bowman's capsule much like a fist pushed into an inflated balloon. Blood enters the glomerulus via an *afferent arteriole* and leaves via an *efferent arteriole*. The glomerulus is highly permeable to water, ions, and small molecules but impermeable to cells and large molecules such as proteins.

Filtration takes place as blood pressure drives water and small solutes through *fenestrations* of the glomerular

FIGURE 40.5 The Vertebrate Nephron The vertebrate nephron consists of a renal tubule closely associated with two capillary beds, the glomerulus and the peritubular capillaries.

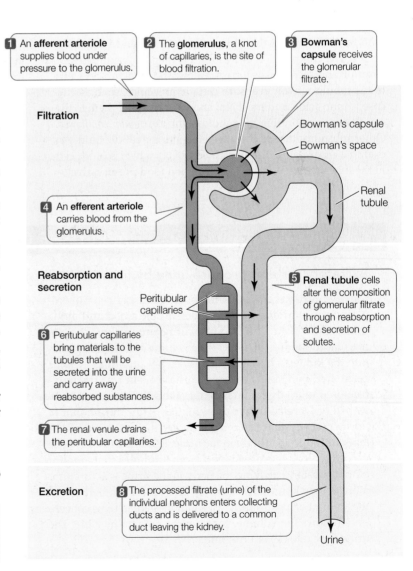

FIGURE 40.6 A Tour of the Nephron Scanning electron micrographs illustrate the anatomical basis for blood filtration by the kidneys. (A) In a preparation showing only the blood vessels (tubular tissue has been digested away), the glomeruli appear as balls of capillaries served by arterioles. (B) This image of an intact glomerulus shows how it fits within Bowman's capsule.

(A) Arterioles

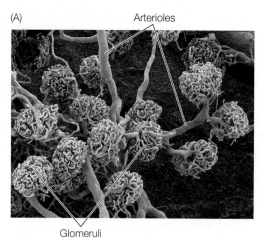

Glomeruli

(B) Capillaries covered by podocytes

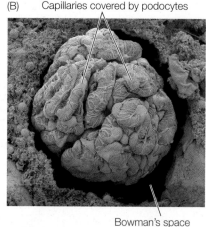

Bowman's space

capillaries (see Concept 38.5), through the basal lamina under the capillary cells, through slits between the cells of Bowman's capsule, and into the capsule's lumen (Bowman's space). *Filtration slits* in Bowman's capsule are formed by *podocytes*, specialized cells with numerous fingerlike projections that wrap around the glomerular capillaries. The rate at which the filtered fluid flows into Bowman's capsule is called the **glomerular filtration rate (GFR)**.

The filtrate that enters Bowman's capsule is similar to the blood plasma except that it lacks blood cells and large molecules. The composition of this fluid is then adjusted by reabsorption and secretion as it passes along the renal tubule. Cells of the tubule actively reabsorb certain molecules, such as glucose, amino acids, and most Na^+ and Cl^-, returning them to the capillaries surrounding the tubules—the *peritubular capillaries*. Other substances are actively secreted into the tubule fluid. An example is para-aminohippuric acid (PAH), which is produced in the liver from benzoic acid, a common food preservative.

Mechanisms to conserve water have evolved in several groups of vertebrates

Vertebrates probably evolved in fresh water, so it makes sense that they have an excretory system that can filter the blood, reabsorb essential molecules, and excrete a dilute urine to get rid of excess water. How, then, have modern vertebrates adapted to environments where water must be conserved and salts excreted? The answer to this question differs among vertebrate groups.

MARINE BONY FISHES Marine bony fishes maintain an osmolarity of one-third to one-half the osmolarity of seawater, which is around 1,000 mosm. Their only source of water is the sea around them, so they must conserve water and excrete excess solutes. Because marine bony fishes cannot produce urine that is more concentrated than their extracellular fluids (unlike mammals; see Concept 40.4), they minimize water loss by producing very little urine. How, then, do they get rid of the excess salts that they ingest with food? First, marine bony fishes do

APPLY THE CONCEPT

Excretory systems produce urine by filtration, reabsorption, and secretion

Kidney function can be quantitatively characterized by comparing concentrations of substances in the blood and in the urine, and by knowing the rate of urine production. The glomerular filtration rate (GFR)—the volume of fluid that passes into the kidney's Bowman's capsules per unit time—can be calculated by using a marker substance (such as the polyfructose inulin) that is filtered out of the blood, but not reabsorbed or secreted by the renal tubules. Thus the amount of the marker substance entering Bowman's capsules equals the amount leaving the body in the urine. If a different substance is secreted or reabsorbed by the renal tubules, the amount of that substance leaving the body in the urine will be either more than or less than the amount filtered. Remembering that the amount of a substance in a solution is equal to its concentration times the volume, consider the data in the table. The urine flow rate (V) is 1 ml/min.

SUBSTANCE	CONCENTRATION IN BLOOD	CONCENTRATION IN URINE
Inulin	1 mg/ml	125 mg/ml
Para-aminohippuric acid (PAH)	0.01 mg/ml	5.85 mg/ml
Na^+	140 mosm/L	70 mosm/L

1. What is the glomerular filtration rate (in ml/min)?
2. Is PAH secreted or reabsorbed, and at what rate (in mg/min)?
3. Is Na^+ secreted or reabsorbed, and at what rate (in mosm/min)?

not absorb from their guts some of the ions they ingest, especially divalent ions such as Mg^{2+} or SO_4^{2-} that are excreted in the feces. Second, bony fishes have an additional excretory organ besides the kidney: the gills. Na^+ is actively excreted across the gill membranes, with Cl^- following. As mentioned earlier, bony fishes can also lose their nitrogenous waste, ammonia, by diffusion across their gill membranes.

CARTILAGINOUS FISHES The cartilaginous fishes (sharks and rays) use a different strategy. They are osmoconformers but not ionic conformers. They convert nitrogenous wastes to urea and trimethylamine oxide, which they retain in large amounts in their body fluids. These species have adapted to a concentration of urea in the body fluids that would be toxic to other vertebrates. As a result, their body fluids have an osmolarity close to that of seawater, so they do not lose body water to the environment by osmosis. Excess NaCl is eliminated by a rectal gland that actively secretes Na^+ (with Cl^- following) into the feces.

AMPHIBIANS Most amphibians live in fresh water or in humid environments. Like freshwater fishes, they produce large amounts of dilute urine and conserve salts. But a few amphibians have adaptations that allow them to live in dry habitats. For example, some of these amphibians have waxy substances in their skin that reduce evaporative water loss. Some frogs burrow into the ground during dry seasons, entering a state of low metabolic activity called **estivation**. These frogs have an enormous urinary bladder that they fill with dilute urine just before digging their burrows. This serves as a reservoir of water that is gradually reabsorbed into the blood during the long period of estivation.

REPTILES Many reptiles, including birds, have successfully adapted to extremely hot, dry habitats. In fact, snakes, lizards, and birds are among the most prominent members of many desert faunas. Three major adaptations have freed reptiles from a close association with water. First, reptiles are *amniotes* and do not need fresh water to reproduce; they use internal fertilization and lay eggs with shells that retard evaporative water loss. Second, they have a dry epidermis (skin) that also minimizes evaporative water loss. Third, they excrete nitrogenous wastes as uric acid semisolids, losing little water in the process.

LINK See Concept 23.6 for more on the adaptations that allowed amniotes to colonize dry environments

MAMMALS Mammals, like reptiles, are amniotes that do not need to deposit their eggs in external bodies of water. They also lose little water across the skin unless they are heat stressed and need to employ evaporative cooling. However, mammals excrete urea instead of uric acid, and since urea must be flushed out of the kidney with water, mammals tend to lose large amounts of water in their urine. As a result, the most challenging environments for mammals are those in which water is severely limited. A kidney has evolved in mammals, however, that enables them to produce urine several times more concentrated than their extracellular fluids.

> ### Do You Understand Concept 40.3?
>
> - Name the type of excretory system found in a fish, an earthworm, and a grasshopper. In each one, what is the site where extracellular fluid enters the excretory system?
> - Comparing the three excretory systems that you named above, which one empties into the gut? Which filters extracellular fluid into the coelom? Which contains nephrons?
> - You compare the kidneys of a marine and a freshwater bony fish species and discover that there is a huge difference in the numbers of nephrons in their kidneys. How would you explain this finding?

A variety of unique adaptations to conserve water have evolved in animals in terrestrial and saltwater habitats. In the next concept we will take a closer look at one of these adaptations: the ability of the mammalian kidney to produce urine that is more concentrated than extracellular fluids.

concept 40.4 The Mammalian Kidney Produces Concentrated Urine

Like other vertebrates, mammals have kidneys consisting of nephrons where the extracellular fluid is filtered into Bowman's capsules, followed by tubular reabsorption and secretion to form urine. However, mammals (and also birds) have an additional refinement: a long loop in the nephron called the **loop of Henle** that dips into the interior region of the kidney. Ions are transported in and out of this loop in such a way that a high-osmolarity region is created at the base of the loop. This high-osmolarity region allows water to be reabsorbed from the urine.

To understand this process, we first need to learn about the anatomy of the mammalian kidney and how the kidney transports ions. We will use the human kidney as an example.

A mammalian kidney has a cortex and a medulla

Humans (and all mammals) have two kidneys behind the abdominal cavity (**FIGURE 40.7A**). Each kidney filters blood, processes the filtrate into urine, and releases that urine into a duct called the **ureter**. The ureter of each kidney leads to the **urinary bladder**, where the urine is stored until it is excreted through the **urethra**, a short tube that opens to the outside of the body.

A kidney is shaped like a kidney bean (**FIGURE 40.7B**), with the ureter, a **renal artery**, and a **renal vein** entering or leaving on its concave side. If you slice a kidney open, you will see that it has two distinctly different regions: a granular outer **cortex** covering an interior **medulla** with a very different appearance. All of the glomeruli and Bowman's capsules are in the cortex. Leaving Bowman's capsule, the renal tubule is called the **proximal convoluted tubule**—"proximal" because it is closest to the

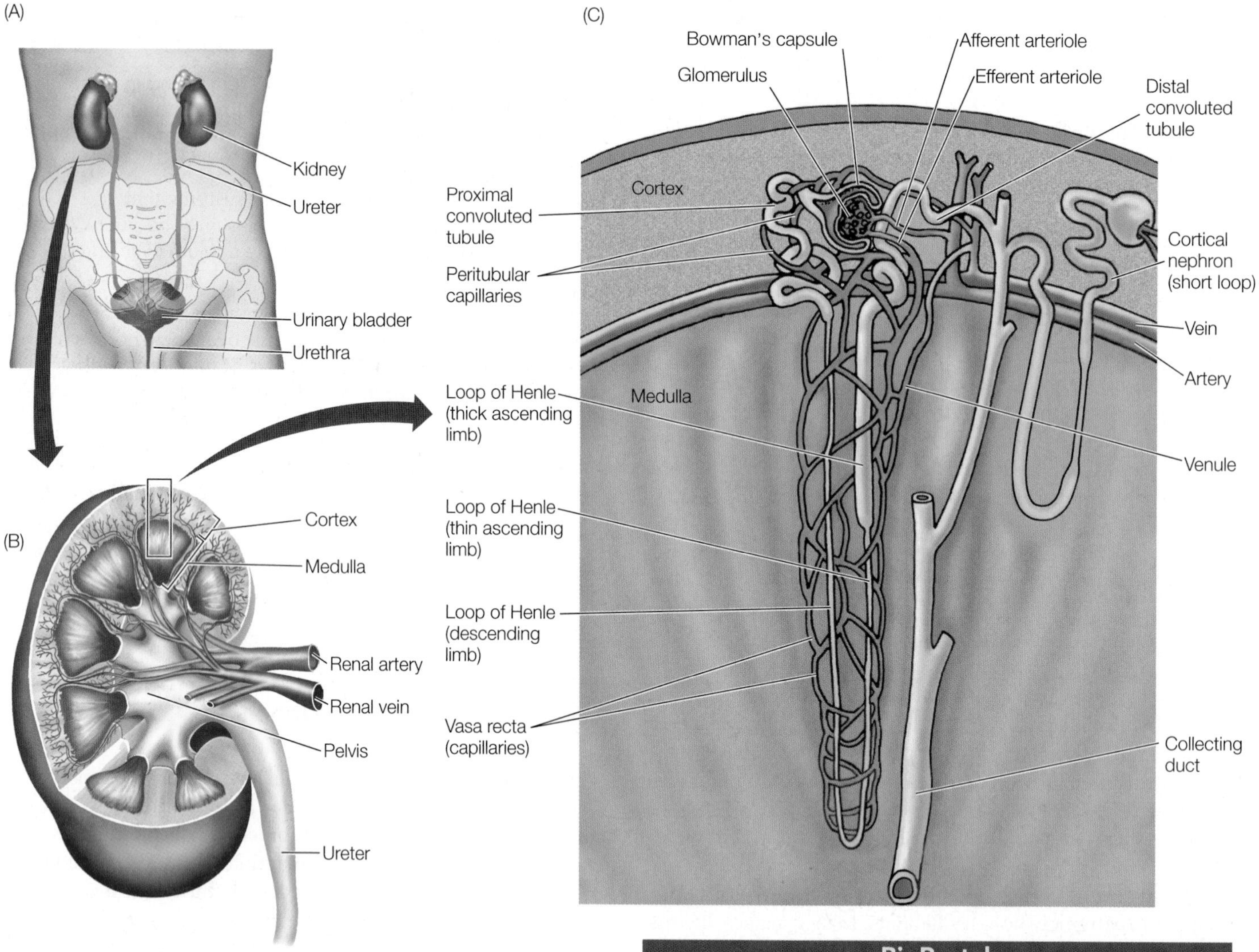

(A)

(B)

(C)

FIGURE 40.7 The Human Excretory System (A) The human kidneys lie behind the abdominal cavity, in the region of the middle back. (B) Certain parts of the nephrons are in the organ's outer region, called the cortex; other parts are in the internal region, called the medulla. (C) The glomeruli and the proximal and distal convoluted tubules are located in the cortex of the kidney. The loops of Henle run in parallel as straight sections down into the renal medulla and back up to the cortex. Collecting ducts run from the cortex to the inner surface of the medulla, where they open into the ureter. The vasa recta are peritubular capillaries that parallel the loops of Henle.

glomerulus, and "convoluted" because it is twisted—is also in the cortex (**FIGURE 40.7C**).

The tubule then straightens, descends directly down into the medulla, makes a hairpin turn, and ascends back to the cortex. This forms the loop of Henle. Some nephrons have loops that descend deep into the medulla; others have shorter loops (see the cortical nephron at the right in Figure 40.7C). The loop of Henle leads to the **distal convoluted tubule** in the cortex ("distal" because it is farther from the glomerulus). The distal convoluted tubules of many nephrons then join together to form a **collecting duct**. There are many collecting ducts, all

yourBioPortal.com

Go to WEB ACTIVITY 40.3
The Human Excretory System

of which descend straight down through the medulla, emptying the urine into a funnel-shaped space called a *pelvis*, which drains into the ureter (see Figure 40.7B).

The blood plasma that does not get filtered into Bowman's capsule (about 80%), along with the red blood cells, leaves the glomerulus in the *efferent arteriole* that feeds the peritubular capillaries surrounding the renal tubules. The peritubular capillaries run into the medulla in parallel with the loops of Henle and the collecting ducts, forming a vascular network called the **vasa recta** (see Figure 40.7C). These blood vessels play an important role in secretion and reabsorption and in maintaining the high-osmolarity region created by the loop of Henle.

yourBioPortal.com

Go to ANIMATED TUTORIAL 40.1
The Mammalian Kidney and
WORKING WITH DATA 40.1

Most of the glomerular filtrate is reabsorbed by the proximal convoluted tubule

The kidneys receive an impressive blood flow of about 1,500 liters of blood per day, more than many larger organs receive. This means that the entire blood volume of the body (about 5 liters in an average man) circulates through the kidneys over 300 times a day. Of this large volume, about 12 percent, or 180 liters of plasma per day, moves into Bowman's capsules. However, we only urinate about 2 to 3 liters per day, so the other 178 or so liters—98 percent of the original filtrate volume—must be returned to the blood somewhere along the length of the nephrons. Where and how is this enormous fluid volume reabsorbed?

The proximal convoluted tubule (PCT) is responsible for about 75 percent of the reabsorption. The cells of the PCT actively transport Na⁺ (with Cl⁻ following), glucose, and amino acids out of the tubule and into the interstitial fluid. Water follows osmotically. The water and solutes are then taken up by the peritubular capillaries.

The PCT changes the solute composition of the tubule fluid and reduces its volume but does not change its osmolarity; the fluid leaving the PCT still has the same osmolarity as the blood plasma. If urine were excreted at this concentration, it would result in a large loss of water, which most terrestrial vertebrates can ill afford. The next steps in urine processing in mammalian kidneys are (1) to reabsorb salts, leaving urea as the major solute in the urine, and (2) to set up the conditions that enable the production of a hypertonic urine.

The loops of Henle create a concentration gradient in the renal medulla

The concentrating ability of the mammalian kidney arises from a **countercurrent multiplier** mechanism made possible by the anatomical arrangement of the loops of Henle (**FIGURE 40.8**).

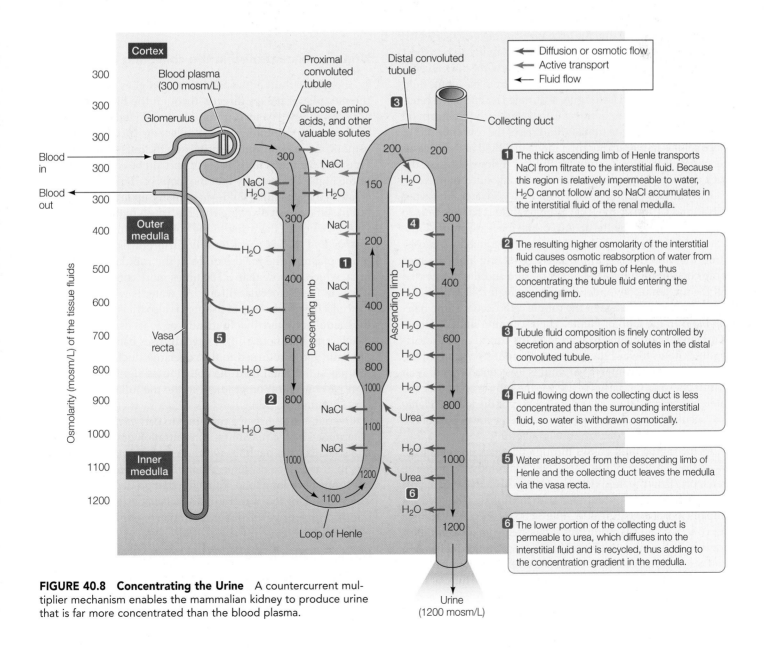

FIGURE 40.8 Concentrating the Urine A countercurrent multiplier mechanism enables the mammalian kidney to produce urine that is far more concentrated than the blood plasma.

The term "countercurrent" refers to the opposing directions of flow of the two limbs of a loop of Henle. The term "multiplier" refers to the ability of this system to create a solute concentration gradient.

> **LINK** Countercurrent mechanisms also contribute to the efficiency of gas exchange in fish gills; see Figure 37.3

The loops of Henle do not concentrate the urine that is passing through them; rather, they increase the osmolarity of the extracellular fluid in the medulla. In humans, for example, near the top of the medulla, next to the cortex, the extracellular fluid is about 300 mosm/L (the same as blood plasma). But deeper in the medulla, the extracellular fluid becomes more and more concentrated. At the hairpin turn of the loop of Henle, the osmolarity of the extracellular fluid is 1,200 mosm/L—*four times* more concentrated than the blood plasma. How do the loops generate this large concentration gradient running from the cortex to the bottom of the medulla?

The loop of Henle has three regions: the *thin descending limb*, the *thin ascending limb*, and the *thick ascending limb*, each of which differs in its permeability to water and its ability to transport NaCl. Most critical are the actions of the cells of the thick ascending limb that actively transport Na$^+$ and Cl$^-$ from the tubule fluid to the interstitial fluid (see Figure 40.8, note 1). The thick ascending limb is relatively impermeable to water, so the reabsorption of Na$^+$ and Cl$^-$ from the tubular fluid raises the concentration of those solutes in the surrounding interstitial fluid and decreases the concentration of the tubular fluid as it flows through the thick ascending limb and into the distal convoluted tubule.

The thin descending limb is permeable to water but less so to Na$^+$ and Cl$^-$. Since the local interstitial fluid has been made more concentrated by the actions of the thick ascending limb, water is drawn out from the thin descending limb by osmosis. Thus, the fluid flowing in the thin descending limb becomes more concentrated than the blood plasma and the filtrate that left the proximal convoluted tubule (see Figure 40.8, note 2).

An important change in the tubule cells occurs at the hairpin bend in the loop of Henle. The ascending limb of Henle is no longer permeable to water. As the fluid in the ascending limb flows toward the cortex, it encounters regions of the medulla where the interstitial fluid is less concentrated. As a result, Na$^+$ and Cl$^-$ passively leave the tubule and enter the interstitial space. Thus, the overall action of the loop of Henle is to accumulate more and more solute in the lower part of the medulla, creating a concentration gradient.

As the tubule fluid passes through the thick portion of the ascending limb of Henle, more NaCl is removed by active transport. Thus the fluid flowing into the distal convoluted tubule is actually less concentrated than the fluid that entered the loop of Henle (see Figure 40.8, note 3). So the loop of Henle has not concentrated the urine, but it has created a concentration gradient in the interstitial fluid of the medulla. And it has had another important effect: because of the removal of much NaCl from the tubular fluid, the dominant solute in the tubular fluid entering the distal convoluted tubule is urea.

The distal convoluted tubule fine-tunes the composition of the urine

Next, fluid moves from the loop of Henle to the distal convoluted tubule. The urine entering the distal tubule is less concentrated than the surrounding interstitial fluid. If water is not reabsorbed as the urine flows through the distal tubule and the collecting duct, a hypotonic urine is created. But, if the distal tubule and collecting duct are permeable to water, water will be reabsorbed into the extracellular fluid, producing a hypertonic urine.

The distal tubule also fine-tunes the ionic composition of the urine (see Figure 40.8, note 3). For example, if a person is potassium depleted, potassium is reabsorbed in the distal convoluted tubule. But if a person has an excess of potassium, potassium is secreted in the distal convoluted tubule. Other ions also are adjusted in this way. The distal convoluted tubule contains dozens of different types of ion transporters. The fluid that leaves the distal convoluted tubule may be hypotonic or isotonic with the plasma, and its major solute is urea.

Urine is concentrated in the collecting duct

As the tubule fluid in the collecting duct goes down through the medulla, it passes directly through the high-osmolarity area created by the surrounding loops of Henle. Depending on the water permeability of the collecting duct, water may be osmotically drawn from the urine into the interstitial fluid (see Figure 40.8, notes 4 and 5). The urine can be concentrated up to the highest osmolarity that the loops of Henle were able to create in the interstitial fluid—in humans, up to 1,200 mosm/L, but in some desert dwelling rodents, much higher.

As we noted, the urine in the collecting duct has a solute *composition* that is considerably different from that of the blood plasma. The major solute in the tubular fluid is now urea, which is more than a 100 times more concentrated than it was in the glomerular filtrate. As water is withdrawn from the collecting duct, some of the urea is also transported out into the medullary interstitial fluid, adding to its osmotic potential (see Figure 40.8, note 6). This urea diffuses back into the loop of Henle and is returned to the collecting duct. The recycling of urea in the renal medulla contributes significantly to the osmotic concentration gradient in the medulla and assists in concentrating the urine.

Overall, the ability of a mammal to concentrate its urine is determined by the maximum concentration gradient it can establish in its renal medulla, which in turn is determined by the length of the loops of Henle. About 20 to 30 percent of human nephrons have long loops of Henle that go deep into the medulla; these long loops are the ones that enable us to concentrate our urine to four times the concentration of our blood plasma.

Kidney failure is treated with dialysis

Loss of kidney function, or *renal failure*, results in the retention of salts and water (hence high blood pressure), retention of urea (uremic poisoning), and a decreasing pH (acidosis). A person who suffers complete renal failure will die within 2 weeks if not treated. The best treatment is kidney transplant, but few kidneys are available. So most patients with renal failure must undergo *dialysis* several times a week—cleansing of the blood with a machine referred to as an artificial kidney but that works completely differently than the kidney.

In a dialysis machine, the patient's blood flows through many small tubes made of semipermeable membranes. A dialysis solution flows on the other side of these membranes, through which small molecules can diffuse. Solutes always diffuse from higher concentration to lower concentration, so the composition of the dialysis fluid is crucial. Solutes that need to be conserved must be at the same concentration in the dialysis fluid as they are in the blood, while solutes that need to be removed from the blood must be near zero in the dialysis fluid. The total osmolarity of the dialysis fluid must equal that of the plasma, so that the patient does not lose or gain water.

FRONTIERS   Researchers at the University of California in San Francisco, working with collaborators at 10 other institutions, recently built the first prototype implantable artificial kidney that would run off the body's own circulatory system and therefore not require electricity or pumps. The artificial kidney uses nanotechnology to filter the blood; the filtrate then flows over cultured kidney cells that can reabsorb glucose and essential salts before allowing the processed filtrate to leave the body. Nanotechnology has made it possible to shrink the functions of a huge dialysis machine into a very small package, but there is still the challenge to bioengineer the prototype for testing in animals and eventual implantation in humans.

Do You Understand Concept 40.4?

- Explain what is meant by a countercurrent multiplier in the kidney and how it enables the production of a hypertonic urine.

- Many diseases, including high blood pressure and heart failure, result in water accumulation in the body. Drugs called diuretics are frequently used to increase the loss of water in these patients. One drug, bumetanide, decreases the activity of the Na^+ transporters in the thick ascending limb of Henle. How would this drug cause increased water loss?

- Explain the difference in how the kidney excretes metabolic wastes and how that function is carried out by a dialysis machine.

The kidney contributes to homeostasis in many critical ways, such as regulating the water content of the body, as well as the solute concentration and composition of the extracellular fluid. As we will see next, the kidneys also play a major role in regulating blood pressure.

concept 40.5   The Kidney Is Regulated to Maintain Blood Pressure, Blood Volume, and Blood Composition

The ability of the kidneys to filter blood depends on a constant glomerular filtration rate (GFR), which in turn depends on an adequate blood supply to the kidneys at an adequate blood pressure. If blood pressure in the glomeruli falls, several *autoregulatory mechanisms* (i.e., initiated by the kidneys themselves) activate to correct the problem. In addition, several hormones released by other organs also help regulate the kidneys to maintain normal blood pressure, blood volume, and blood composition. We will discuss these mechanisms separately, but keep in mind that they are always working together.

The renin-angiotensin-aldosterone system raises blood pressure

If GFR begins to fall, the first autoregulatory response is dilation (expansion) of the afferent renal arterioles, decreasing resistance and increasing local blood pressure in the glomerulus. But if arteriole dilation does not bring the GFR back to normal, the kidneys release an enzyme, **renin**, into the blood. Renin, along with an enzyme in the lungs called ACE (for *a*ngiotensin *c*onverting *e*nzyme), converts an inactive protein in the blood into the hormone **angiotensin** (**FIGURE 40.9**), which increases blood pressure by several mechanisms:

- It constricts the efferent renal arterioles, elevating local blood pressure in the glomerular capillaries.

- It constricts peripheral blood vessels all over the body, elevating blood pressure throughout the body.

- It stimulates thirst. Increased water intake then increases blood volume and increases blood pressure.

- It stimulates the adrenal cortex to release the steroid hormone **aldosterone**. Aldosterone stimulates Na^+ reabsorption by the kidneys. Water follows the Na^+, so more water is reabsorbed, helping maintain blood volume and therefore blood pressure.

Thus the renin-angiotensin-aldosterone system is a powerful regulatory mechanism that raises blood pressure throughout the body, not just in the kidneys (see Figure 38.16).

ADH decreases excretion of water

When serum osmolality rises, hypothalamic osmoreceptors are stimulated to produce, among other responses, thirst. In addition, the release of *antidiuretic hormone* (ADH, also called vasopressin) is triggered by osmoreceptor stimulation. An antidiuretic is any substance that reduces the volume of urine. ADH causes cells of the collecting duct to insert *aquaporins* (water channels; see Concept 5.2, especially Figure 5.5) into their plasma membranes, increasing their permeability to water (**FIGURE 40.10**). Therefore, more water is reabsorbed

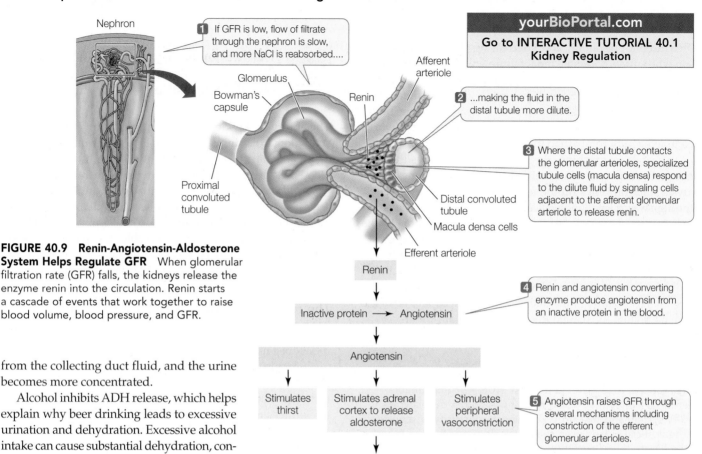

Nephron

1 If GFR is low, flow of filtrate through the nephron is slow, and more NaCl is reabsorbed....

Glomerulus

Bowman's capsule

Renin

Afferent arteriole

yourBioPortal.com

Go to INTERACTIVE TUTORIAL 40.1
Kidney Regulation

2 ...making the fluid in the distal tubule more dilute.

3 Where the distal tubule contacts the glomerular arterioles, specialized tubule cells (macula densa) respond to the dilute fluid by signaling cells adjacent to the afferent glomerular arteriole to release renin.

Proximal convoluted tubule

Distal convoluted tubule

Macula densa cells

Efferent arteriole

FIGURE 40.9 Renin-Angiotensin-Aldosterone System Helps Regulate GFR When glomerular filtration rate (GFR) falls, the kidneys release the enzyme renin into the circulation. Renin starts a cascade of events that work together to raise blood volume, blood pressure, and GFR.

Renin

Inactive protein → Angiotensin

4 Renin and angiotensin converting enzyme produce angiotensin from an inactive protein in the blood.

Angiotensin

Stimulates thirst

Stimulates adrenal cortex to release aldosterone

Stimulates peripheral vasoconstriction

5 Angiotensin raises GFR through several mechanisms including constriction of the efferent glomerular arterioles.

Aldosterone increases Na⁺ reabsorption

from the collecting duct fluid, and the urine becomes more concentrated.

Alcohol inhibits ADH release, which helps explain why beer drinking leads to excessive urination and dehydration. Excessive alcohol intake can cause substantial dehydration, contributing to the symptoms of a hangover.

The heart produces a hormone that helps lower blood pressure

You may not think of the heart as an endocrine organ, but it is. When there is excess body fluid volume, the heart must contract with greater force, and that puts strain on the heart. Under these conditions, the increased venous return stretches the atria of the heart. This causes atrial cells to release a peptide hormone called **atrial natriuretic peptide** (**ANP**). This peptide hormone enters the circulation, and in the kidneys it decreases the reabsorption of Na⁺. If less Na⁺ is reabsorbed, less water is reabsorbed, and a more voluminous and dilute urine is produced. Thus ANP has the effect of lowering blood volume and therefore blood pressure.

Do You Understand Concept 40.5?

- Name four hormones involved in kidney function. Briefly describe the function of each one.

- Angiotensin converting enzyme (ACE) inhibitors are used to treat high blood pressure. By what actions would these drugs have their effect?

- Suppose you sweat a great deal on a hot day, losing both salt and water, so that your blood volume, and hence your blood pressure, decrease sharply without your blood osmolarity changing much. Which hormone levels will change in response to this situation, and why?

How do excretory systems of animals maintain homeostasis of the interstitial fluid in the face of extreme challenges?

ANSWER The active transport of Na⁺ plays many roles in excretory systems. The transport of Na⁺, usually with Cl⁻ following, creates osmotic concentration gradients that move water across the membranes of tubules. In the evolution of the desert rodents, selection exploited the role of NaCl transport in the countercurrent multiplier to the extreme (Concept 40.4). The loops of Henle in these species are so long relative to the size of the kidney that the medulla sticks way out of the kidney and into the ureter and produces a huge concentration gradient. These animals can produce urine with a concentration over 5,000 mosm/L. Crystals appear in the urine as it leaves the body of the animal.

Freshwater animals use the transport ability of their renal tubules to conserve NaCl and to provide secondary active transport to reabsorb other solutes. They have short loops of Henle and produce only dilute urine.

The albatross and many other marine birds that are mostly at sea and take in large amounts of salt with their food have a special salt-secreting organ above the eyes that uses NaCl transporters. These salt glands consist of tubules that take up NaCl from the interstitial fluid (similar to our tear glands). The

INVESTIGATION

FIGURE 40.10 ADH Induces Insertion of Aquaporins into Plasma Membranes Aquaporin proteins make some regions of renal tubules permeable to water. One aquaporin, AQP-2, is responsible for the permeability of the collecting duct cells. How does antidiuretic hormone (ADH) act on these proteins to control the level of permeability in renal cells?

HYPOTHESIS

ADH controls permeability by changing the location of aquaporin proteins.

METHOD

1. Isolate collecting ducts from rat kidney.
2. Use immunochemical staining to localize AQP-2 in collecting duct cells with and without ADH, and after ADH is applied and then washed away.
3. Measure the water permeability of the collecting duct cells under the same three conditions.

RESULTS

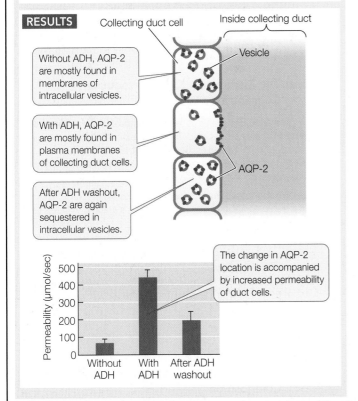

Without ADH, AQP-2 are mostly found in membranes of intracellular vesicles.

With ADH, AQP-2 are mostly found in plasma membranes of collecting duct cells.

After ADH washout, AQP-2 are again sequestered in intracellular vesicles.

Collecting duct cell

Inside collecting duct

Vesicle

AQP-2

The change in AQP-2 location is accompanied by increased permeability of duct cells.

CONCLUSION

In the absence of ADH, AQP-2 is sequestered intracellularly. When ADH is present, these water channels are inserted into the plasma membranes, making the cells more permeable to water.

ANALYZE THE DATA

The results showed that ADH controls the location of AQP-2 in collecting duct cells, but does it also affect the amount of AQP-2? The Brattleboro strain of rats does not produce ADH, and the DI strain of mice has a defect in the signaling pathway of the ADH receptor. The amount of AQP-2 produced in response to ADH and an ADH receptor antagonist was studied in these animals. Values were normalized to baseline amounts in normal animals.

Treatment	Amount of AQP-2			
	Normal rat	Brattleboro rat	Normal mouse	DI mouse
Baseline	1.0	0.5	1.0	0.1
ADH	1.0	1.5	1.0	0.1
ADH receptor antagonist	0.5	0.5	0.5	0.1

A. How does ADH affect the amount of AQP-2 in normal and Brattleboro rats? Explain the difference in the responses of the two rat strains.

B. Why does ADH not have an effect on the amount of AQP-2 in DI mice?

C. Considering these data as a whole, construct a hypothesis to explain why Brattleboro mice can maintain AQP-2 levels at 50% of normal. How would you test your hypothesis?

Go to **yourBioPortal.com** for original citations, discussions, and relevant links for all INVESTIGATION figures.

tubules coalesce into a central canal that drains into the nasal cavity. Drops of highly salty fluid are usually dripping from their bills (**Figure 40.11**).

The vampire bat uses its regulatory mechanisms to turn on or turn off its capacity to concentrate its urine. Its long loops of Henle use NaCl transport to establish a large concentration gradient in its renal medulla, but by changing its levels of ADH release (Concept 40.5), it can turn on or turn off the reabsorption of water from its collecting ducts.

A salty fluid drips from the bird's bill.

FIGURE 40.11 Salt Excretion in a Marine Bird This northern fulmar rids itself of excess NaCl by excreting it in drops of fluid produced by a special organ above its eyes.

40 SUMMARY

concept 40.1 Excretory Systems Maintain Homeostasis of the Extracellular Fluid

- **Excretory systems** maintain the volume, concentration, and composition of the extracellular fluids and eliminate nitrogenous wastes. **Urine** is the output of excretory systems.

- **Osmolarity** is the overall solute concentration of a fluid. Animals must maintain the osmolarity of their extracellular fluids within an acceptable range.

- Animals can be **osmoconformers** or **osmoregulators**. Terrestrial and freshwater animals must be osmoregulators. Most marine invertebrates are osmoconformers, but most marine vertebrates are osmoregulators. **Review Figure 40.1**

- In addition to regulating the overall osmolarity of the extracellular fluid, most animals also regulate the concentrations of specific ions.

- The mammalian kidney reabsorbs bicarbonate ions from the urine and returns them to the blood, where they act as a pH buffer to absorb excess H^+ ions.

concept 40.2 Excretory Systems Eliminate Nitrogenous Wastes

- Metabolism of proteins and nucleic acids produces toxic **nitrogenous wastes**, which must be eliminated from the body. The most common nitrogenous waste is **ammonia**. **Review Figure 40.2**

- **Ammonotelic** animals produce ammonia as their primary nitrogenous waste. They are typically water-breathing aquatic animals that eliminate ammonia by diffusion across their gill membranes.

- **Ureotelic** animals detoxify ammonia by converting it to **urea** before excretion, and then flush the urea out with water. These animals include mammals, most amphibians, and cartilaginous fishes.

- **Uricotelic** animals convert ammonia to **uric acid**. They include insects, some amphibians, and birds and other reptiles. Uric acid can be excreted without much loss of water.

- Most species produce more than one kind of nitrogenous waste.

concept 40.3 Excretory Systems Produce Urine by Filtration, Reabsorption, and Secretion

- Excretory systems produce urine through the processes of filtration, reabsorption, and secretion.

- Water cannot be actively transported against its concentration gradient, so it must be moved across membranes by a difference in either osmolarity or pressure.

- In annelid worms, blood pressure causes filtration of the blood across capillary walls and into the coelom, where it is taken up by **metanephridia**, which adjust the composition of the filtrate by active transport. **Review Figure 40.3 and WEB ACTIVITY 40.1**

- The **Malpighian tubules** of insects receive ions and nitrogenous wastes by active transport across the tubule cells. Water follows by osmosis. Ions and water are reabsorbed from the hindgut and rectum, so the insect excretes semisolid wastes. **Review Figure 40.4**

- The functional unit of the vertebrate **kidney** is the **nephron**. A nephron consists of a ball of capillaries called a **glomerulus**, from which the blood is filtered, a **renal tubule**, which processes the

filtrate into urine to be excreted, and a system of peritubular capillaries, which surround the tubule and assist with secretion and reabsorption. The vertebrate kidney is adapted for ridding the body of excess water. **Review Figures 40.5 and 40.6 and WEB ACTIVITY 40.2**

- The rate at which the filtrate flows into Bowman's capsule is called the **glomerular filtration rate (GFR)**.

- Several groups of vertebrates have evolved different mechanisms for conserving water. Marine bony fishes produce little urine. Cartilaginous fishes retain urea so that the osmolarity of their body fluids remains close to that of seawater. Most amphibians produce large amounts of dilute urine. Reptiles lose little water in urine because they excrete nitrogenous wastes as uric acid. Mammals must lose some water to flush out urea, but they can produce urine more concentrated than their extracellular fluids.

concept 40.4 The Mammalian Kidney Produces Concentrated Urine

- The glomeruli, Bowman's capsules, and **proximal** and **distal convoluted tubules** are located in the **cortex** of the kidney. Straight sections of renal tubules called **loops of Henle** and **collecting ducts** are arranged in parallel in the **medulla** of the kidney. **Review Figure 40.7 and WEB ACTIVITY 40.3**

- The proximal convoluted tubule reabsorbs water, NaCl, and certain solutes, reducing the renal filtrate's volume without appreciably changing its osmolarity.

- The loops of Henle create a concentration gradient in the interstitial fluid of the renal medulla by a **countercurrent multiplier** mechanism. Urine flowing down the collecting ducts is concentrated by the osmotic reabsorption of water caused by the concentration gradient in the surrounding interstitial fluid. **Review Figure 40.8, ANIMATED TUTORIAL 40.1, and WORKING WITH DATA 40.1**

concept 40.5 The Kidney Is Regulated to Maintain Blood Pressure, Blood Volume, and Blood Composition

- Kidney function in mammals is controlled by autoregulatory mechanisms that maintain a constant high GFR even if blood pressure varies.

- The kidneys release **renin** if GFR falls. Renin activates **angiotensin**, which causes the constriction of efferent glomerular arterioles and peripheral blood vessels, causes the release of **aldosterone** (which enhances water reabsorption), and stimulates thirst. **Review Figure 40.9 and INTERACTIVE TUTORIAL 40.1**

- A fall in blood pressure or a rise in blood osmolarity increases the release of antidiuretic hormone (ADH) from the posterior pituitary. ADH increases the permeability of the collecting duct to water and therefore increases the reabsorption of water from the urine.

- If the volume of blood returning to the heart increases and stretches the atrial walls, heart cells release **atrial natriuretic peptide (ANP)**. ANP increases excretion of salt and water, reducing blood volume and blood pressure.

See **WEB ACTIVITY 40.4** for a review of the major human organ systems.

Many behaviors of many species are so stereotypic that they can frequently be used to identify the species. An example is spider webs; different species of spiders spin webs of different designs. Web spinning involves thousands of movements performed in the right sequence, and each time a spider builds a web, the sequence is largely the same, even though it has to accommodate different physical environments. Yet in many species there is no opportunity for a spider to learn how to build a web from another spider because the adults die before their eggs hatch. How do spiders know how to build a complicated and precise web? It has to be programmed into their genome.

Circumstances favoring genetic programming of behavior are easy to understand: (1) there may be no opportunities to learn (no adults to teach offspring), (2) mistakes may be costly, as in predator avoidance (no second chances), and (3) potential mates of the correct species must be identified.

The mating behavior of fruit flies (*Drosophila*) exemplifies all three of these reasons. Flies emerging from pupation in a piece of rotting fruit are ready to mate in about 8 hours. There may be no adult flies around to provide lessons in mating behavior. There may be other species hatching out from that piece of rotting fruit. Making a mistake in selecting a partner can be costly in terms of energy investment (especially for the females) and loss of opportunity (especially for the males, as the availability of virgin females falls rapidly).

Stereotypical, complex courtship rituals guarantee that mating pairs are of the same species. The male fruit fly chases after the female and taps her body with his forelegs. The common female response is to run away, but if she does not, the male will extend one wing and vibrate it to produce a courtship song, and if the female

still does not run away, he will lick her genitals and attempt to copulate.

Mutations of single genes can disrupt the fruit fly courting ritual. The gene *per* (for period) was discovered because it altered the daily rhythms of the flies, but it also altered the frequency of the wing beat and therefore the courtship song. That was enough to cause receptive females to break off courtship. Another gene mutation called *fru* (for fruitless) results in the inability of males to discriminate between males and females. Groups of males with the *fru* mutation form mating circles, with each male trying to mate with the one in front of it. Obviously none has mating success even though their displays are perfect.

How can a single gene be responsible for a complex behavior?
QUESTION

In fruit flies and many other species, mating is the culmination of an elaborate, genetically determined series of behaviors.

KEY CONCEPTS

41.1 Behavior Has Proximate and Ultimate Causes

41.2 Behaviors Can Have Genetic Determinants

41.3 Developmental Processes Shape Behavior

41.4 Physiological Mechanisms Underlie Behavior

41.5 Individual Behavior Is Shaped by Natural Selection

41.6 Social Behavior and Social Systems Are Shaped by Natural Selection

Behavior Has Proximate and Ultimate Causes

The study of behavior involves all biology subject areas covered in this book, from the molecular to the ecological and evolutionary. Molecular, cellular, and physiological mechanisms underlie specific behaviors. Behaviors develop, they have genetic determinants, and they evolve in response to selective pressures imposed by the social and physical environment. Scientists have taken different approaches to studying behavior, depending on their orientations: laboratory versus field, genetic determinism versus learning, physiology versus ecology, and so on. In this chapter we will explore several behaviors from these different perspectives.

Biologists ask four questions about a behavior

Niko Tinbergen, a prominent naturalist, ecologist, and animal behavioral biologist in the mid-1900s, proposed that scientists investigate behavior as four questions:

- *Causation:* What is the immediate stimulus for the behavior?
- *Development:* How does the behavior change with age and with learning, and what experiences are necessary for it to be displayed?
- *Function:* How does the behavior affect the animal's chances for survival and reproduction?
- *Evolution:* How does the behavior compare with similar behaviors in related species, and how might it have evolved?

The first two questions refer to the **proximate causes** of behavior: the immediate genetic, physiological, neurological, and developmental mechanisms that determine how an individual is behaving at time scales ranging from milliseconds to the entire life span. The third and fourth questions refer to the **ultimate causes** of behavior: the evolutionary processes that produced the animal's capacity and tendency to behave in particular ways over many generations of natural selection.

Questions about proximate causes lead to mechanistic approaches

Two classical schools of animal behavior focused on proximate causes. One, **behaviorism**, derived from the famous discoveries of the physiologist Ivan Pavlov, who discovered that neural reflexes controlling digestive processes (including salivation) could be modified by experience to respond to an unnatural stimulus rather than the sight or smell of food. This, of course, was the famous conditioned reflex (**FIGURE 41.1**) that became a paradigm for laboratory studies of learning and memory, mostly in rats.

Another classical school of animal behavior, **ethology**, focused on behavior patterns ethologists called "fixed

action patterns." These are not learned, and in fact are resistant to modification by learning. Fixed action patterns are characteristically triggered by very simple stimuli such as color, smell, or sound. Such triggers are called **releasers**. For example, Tinbergen studied the begging behavior of gull chicks. In some species of gulls, adults have a red dot on their bills. When a parent gull returns to the nest to feed its chicks, the chicks peck on the red dot, which stimulates the parent to regurgitate food (**FIGURE 41.2A**). To determine what stimulated chicks to peck their parents' bills, experimenters showed to newly hatched, naïve gull chicks models of gull heads of different shapes and colors (**FIGURE 41.2B**), as well as models of a beak without a head. The results showed that the red dot was necessary for the release of chick pecking behavior. In fact, a pencil with a

(A) Before conditioning

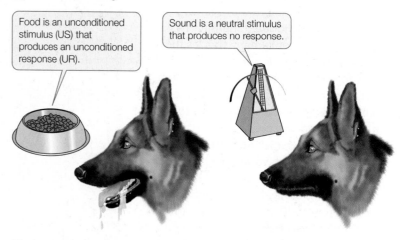

Food is an unconditioned stimulus (US) that produces an unconditioned response (UR).

Sound is a neutral stimulus that produces no response.

(B) Conditioning

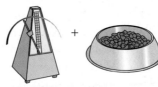

Conditioning repeatedly presents the US immediately following presentation of the neutral stimulus.

(C) After conditioning

The neutral stimulus has become a conditioned stimulus (CS) that by itself produces the conditioned response (CR).

FIGURE 41.1 The Conditioned Reflex Ivan Pavlov discovered that when a normal response is paired with an artificial stimulus, an animal learns to produce the response even when only the artificial stimulus is presented.

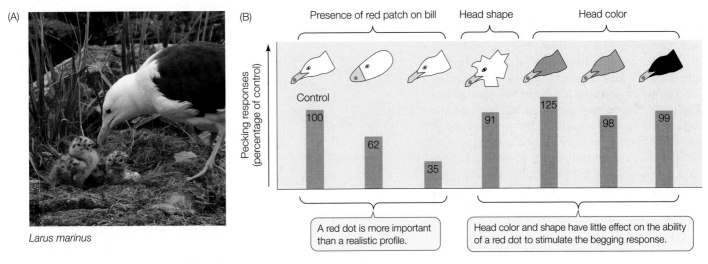

Larus marinus

FIGURE 41.2 Releasing a Fixed Action Pattern
(A) Gull chicks instinctively peck at the red dot on the parent's bill, a behavior that induces the parent to regurgitate food into the chick's mouth. (B) Tinbergen's work showed that the red dot on the parent's bill is the critical component that releases the pecking response.

red eraser elicited a more robust pecking response than did an accurate model of a gull head without a red dot.

Ethologists concluded that fixed action patterns are adaptive in situations where learning is not possible or where mistakes could be costly. But to become genetically programmed, releasers have to be rather simple subsets of information encountered by the animal.

Modern studies of the proximate causes of animal behavior cover a wide range of approaches ranging from molecular genetics to neurophysiology. It is recognized that most behaviors have genetic determinants as well as the capacity to be modified by experience; however, the balance differs for different species and different behaviors.

Questions about ultimate causes lead to ecological/evolutionary approaches

Behavioral ecologists seek answers to why particular behaviors have evolved in a species by trying to understand how those behaviors affect the reproductive fitness of the individuals in the population. To do this, they have to understand the selective pressures placed on the individuals by the environment. Obvious environmental conditions that animals face include availability of food, mates, and good nesting sites, as well as challenges from predators, competitors, and weather. Many behaviors involve making choices: where to build a nest, whom to choose as a mate, when to aggressively defend a territory or a food source, and so on. Experiments frequently involve studying animals under different conditions or manipulating conditions to see how the animals respond.

In the concepts that follow we will see many examples of how the evolution of behaviors is investigated by efforts to understand how the behaviors contribute to the reproductive fitness of the individuals of the population.

Do You Understand Concept 41.1?

- Questions about proximate and ultimate causes of behavior are called how and why questions, respectively. Explain.

- When a male robin is in reproductive condition, it has bright red breast feathers. It aggressively attacks other male robins, but it will also attack a simple tuft of red feathers on a stick. How does this seemingly maladaptive behavior relate to concepts of fixed action patterns and releasers?

- African sunbirds, which are nectar feeders, defend their feeding territories in some habitats but not in others. What selective pressures might have shaped these behavioral choices?

Early work on animal behavior left no doubt that behavior can be genetically determined. But genes code for proteins, and behaviors are highly complex traits involving sensory inputs and intricate patterns of responses. How are genes and behavior connected?

---

concept 41.2 Behaviors Can Have Genetic Determinants

Behaviors are complex traits. Nevertheless, evidence from multiple approaches shows that alterations in single genes can result in discrete behavioral phenotypes subject to natural selection.

Breeding experiments can reveal genetic determinants of a behavior

Breeding experiments can be used to reveal whether or not a specific behavior has a genetic determinant. If a behavior is variably expressed in a population and follows a Mendelian inheritance pattern (see Concept 8.1), then that trait is likely to be controlled by a small number of genes. A classic example is the genetics of hive cleaning by honey bees.

A honey bee colony consists of a reproductive queen and a huge number of nonreproductive workers. The worker bees build the honeycomb, and the queen lays eggs in cells of that honeycomb. The workers feed the larvae that hatch from those eggs. When the larvae pupate, the workers cap and seal their cells. If a pupa dies, workers uncap the cell and drag the carcass outside the hive. This *hygienic* behavior decreases the probability of pathogens spreading to other larvae in the hive. *Nonhygienic* hives do not show this behavior and are more susceptible to the spread of the disease.

When a nonhygienic queen bee was crossed with a hygienic male, the offspring (the F₁ generation) were all nonhygienic, indicating that the genetic determinant of nonhygienic behavior is dominant (**FIGURE 41.3**). (Note that in bees, only females are diploid; the males are haploid and thus have only one copy of each gene instead of two.) When these nonhygienic F₁ females were backcrossed to hygienic males, however, four phenotypes were produced: one hygienic, one nonhygienic, and two intermediates that were very interesting. Workers with one intermediate phenotype opened the cells of dead pupae but did not remove them. Workers with the other intermediate phenotype did not open cells, but if a cell was already open, they would dispose of the dead pupa. Thus it was concluded that the hygienic behavior had at least two components controlled by separate genes. (Compare these experiments with those of Mendel's on the color and shape of pea seeds; see Figure 8.5.) But breeding experiments alone cannot identify which genes influence behavior and how they do so. Molecular genetic approaches are used to answer these questions.

Studies of mutants can reveal the roles of specific genes

Molecular genetic studies of a behavior can begin with naturally occurring mutants or mutants created in the laboratory through exposure to chemicals or radiation. If a behavioral trait is altered in the mutant, then efforts can be made to localize and identify the gene involved.

An example is the *Drosophila* mutant *fruitless* (*fru*), described in the chapter opener. Mutant males do not have the ability to discriminate males from females and are just as likely to court either sex. Cloning and sequencing the *fruitless* gene led to many studies advancing knowledge of its regulation and its roles in male sexual behavior. Targeted alterations in the gene resulted in a variety of effects on the sexual behavior of the male *Drosophila*. Some of these engineered mutants had subtle differences in their courtship displays; others showed no courtship behavior at all. It was found that the *fru* gene is selectively expressed in cells in the fly's nervous system that are involved in sexual differentiation and sexual behavior, and that the gene

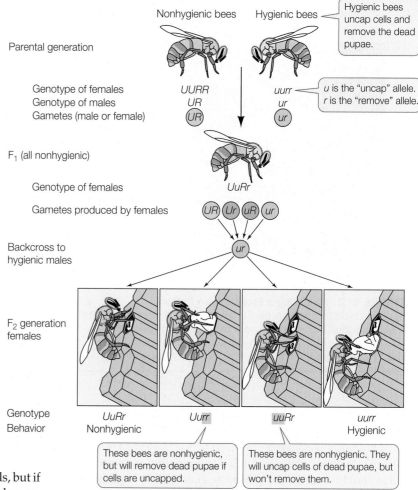

FIGURE 41.3 Genes and Hygienic Behavior Some worker honey bees remove carcasses of dead pupae from their hive. Two components of this behavior—cell uncapping and removal of the carcass—are under the control of separate recessive alleles, designated *u* and *r*.

product is a transcription factor that can control the expression of many other genes. Modifications in any one of those genes or their expression can alter behavior. Thus, even though no behavior is coded for by a single gene, alterations in single genes can influence behavior in ways that affect an animal's fitness. Genes frequently function in gene cascades, which creates many opportunities for a change in a single gene to alter even complex behaviors.

LINK Because of the gene cascades involved in development, single genes can also cause profound changes in morphology; see Concepts 14.4 and 14.5

Gene knockouts can reveal the roles of specific genes

Once a gene has been identified, molecular genetic methods can be used to eliminate it or silence it to see what effects the loss of that gene has on behavior. One method is to genetically engineer an animal that lacks the gene—a gene knockout (see Figure 13.8). Knocking out genes involved in sensory pathways can have pronounced effects on behavior. One example is a

INVESTIGATION

FIGURE 41.4 The Mouse Vomeronasal Organ Identifies Gender The mouse VNO is located adjacent to the nasal passages. It contains pheromone receptors whose input travels to a specific region of the olfactory bulb (the accessory olfactory bulb). Dulac used knockout mice to demonstrate that pheromone receptors in the VNO are involved in stimulating mating behaviors in male mice (pursuit of females and aggression toward other males).

HYPOTHESIS

When the receptors in a male's VNO bind to sex pheromones produced by other mice, the receptors stimulate mating behavior.

METHOD

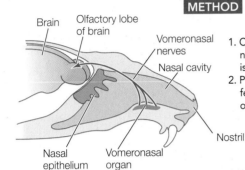

1. Create knockout mice in which a gene necessary for VNO receptor signaling is inactivated.
2. Place knockout male mice with wild-type females and wild-type males and observe their behavior.

RESULTS

Knockout males pursued and mated with wild-type females

Knockout males also pursued and attempted to mate with wild-type males

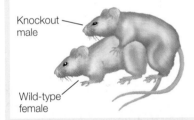

CONCLUSION

Properly functioning VNO receptors appear to be essential not for sexual attraction, but for gender identification.

ANALYZE THE DATA

Mice have two pheromone receptors in the VNO, VN1 and VN2. Knockout males in which either the VN1 or VN2 receptor was inactivated were exposed to either a wild-type male, a castrated male, or a female. The results are shown in the table. M indicates that the test male tried to mate with the introduced individual, O means that he did not react to the individuals, and A indicates that he was aggressive towards that individual.

Introduced individual	Genotype of resident male		
	Wild type	VN1 knockout	VN2 knockout
Male	A	O	M
Castrated male	O	O	M
Female	M	M	M

A. What behavioral trait is dependent on a functional VN1 receptor?

B. What behavioral trait is dependent on a functional VN2 receptor?

C. If a male mouse lacked both a VN1 receptor and a VN2 receptor, how would he react to an introduced wild-type male?

Go to **yourBioPortal.com** for original citations, discussions, and relevant links for all INVESTIGATION figures.

gene for a specific olfactory receptor in mice.

Mice have two olfactory organs: the nasal olfactory epithelium common to all mammals, and a small organ adjacent to the nasal passages, called the *vomeronasal organ*, or VNO. Catherine Dulac at Harvard University discovered that pheromone receptors were expressed in the VNO and used gene knockouts to demonstrate that the receptors are involved in stimulating mating behavior (**FIGURE 41.4**). *Pheromones* are signaling molecules released into the environment; see Concept 35.2.)

Do You Understand Concept 41.2?

- How did breeding experiments on honey bees reveal that two genes play major roles in hygienic behavior? Would you expect the two genes to be linked or to be inherited by independent assortment? Why?

- After the *fru* gene was cloned and sequenced in *Drosophila*, what evidence was obtained to support the conclusion that the product of this gene controlled gene expression cascades specifically involved in sexual behavior?

- How was Catherine Dulac able to conclude that the pheromone receptor she was investigating in mice was specifically responsible for gender identification? What results would have been obtained if her original hypothesis was correct that the receptor controlled mating behavior?

How can the genetic cascades that underlie complex behaviors be programmed to respond selectively to specific sets of stimuli? How can their expression be limited to appropriate times in an animal's life? The answers to these questions can by found by studying how behaviors develop over the life span.

<table>
<tr><td>concept
41.3</td><td>**Developmental Processes Shape
Behavior**</td></tr>
</table>

The emergence of behavior as an animal develops and matures depends on the development of the nervous system as well as on the growth and maturation of other body systems. A bird cannot fly until its wings grow and its muscles and flight feathers mature. But even with anatomical and physiological competence, specific behaviors may not be expressed. Behaviors that are adaptive at one stage in an animal's life may not be adaptive at other stages. Behaviors typical of juvenile animals, such as begging for food, may disappear and new behavior patterns of a mature individual, such as courtship displays, appear.

Hormones can determine behavioral potential and timing

Hormones can determine the development of a behavioral potential at an early age and the expression of that behavior at a later age. An excellent example of this is sexual behavior in rats (**FIGURE 41.5**). Normally, adult male and female rats exhibit different patterns of sexual behavior: females adopt a sexually receptive posture, called *lordosis*, in the presence of males, and males copulate with receptive females. Neither sex, however, expresses these behaviors until the animals have reached adulthood. Experiments in which newborn and adult rats were neutered (to remove the influence of sex hormones naturally produced by their gonads) and artificially treated with hormones led to the following conclusions:

- Development of male sexual behavior requires that the brain of the newborn rat be exposed to testosterone, but development of female sexual behavior does not require exposure to estrogen.

- Testosterone masculinizes the nervous systems of both genetic males and genetic females.

- Exposure to sex steroids in adulthood is necessary for the expression of sexual behavior, but testosterone produces male sexual behavior only in adult rats whose brains were masculinized when they were newborns, and estrogen produces female sexual behavior only in adult rats whose brains were *not* masculinized when they were newborns.

Thus the sex steroid hormones that are present around the time of birth determine which pattern of behavior develops, and the sex steroids that are present in adulthood determine when that pattern is expressed.

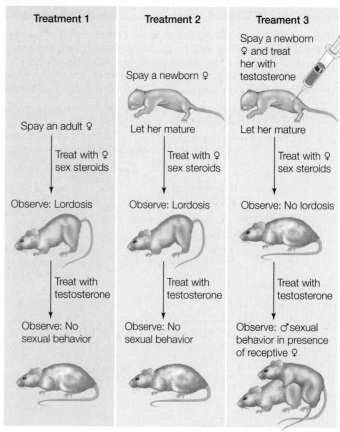

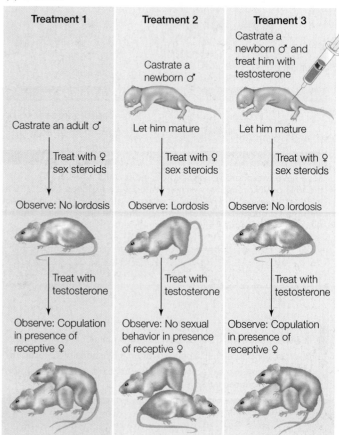

FIGURE 41.5 Hormonal Control of Sexual Behavior
Experimental hormonal treatments of rats demonstrated that the sex steroids present during early development determine what sexual behavior patterns develop, but the sex steroids present in adulthood control the expression of those patterns.

Some behaviors can be acquired only at certain times

Responsiveness to simple releasers is sufficient for certain behaviors such as begging behavior in gull chicks, but more complete information that cannot be genetically programmed is required for other behaviors. An example is parent–offspring recognition. When animals live in close proximity to other individuals, as in a herd or a nesting colony, it is important for a mother and her offspring to learn each other's identity soon after birth so they will be able to find each other in a crowded situation. In many such cases, a parent–offspring bond is formed by **imprinting**. What characterizes imprinting is that an animal learns a specific set of stimuli during a limited time called a **critical period** or **sensitive period**.

Imprinting requires only a brief exposure, but its effects are strong and long-lasting. Emperor penguins reproduce during the coldest, darkest time of year in Antarctica. The parents walk up to 150 kilometers inland to form a dense colony, where the female lays her egg. She then walks back to the ocean to feed while her mate incubates the egg. By the time she returns, the chick has hatched. She then takes over its care and feeding, and the father walks back to the ocean to feed. Generally, he is away so long that the mother must leave to find food as well to avoid starvation. Thus after being away for weeks, the father must find his chick in a crowded, milling colony of chicks, all calling for their parents. Yet he can unerringly locate his own offspring by recognizing its call, which he learned before he left to feed.

The critical, or sensitive, period for imprinting may be determined by a brief developmental or hormonal state. For example, if a mother goat does not nuzzle and lick her newborn within 10 minutes after its birth, she will not recognize it as her own offspring later. For goats, the sensitive period is associated with peaking levels of the hormone oxytocin in the mother's circulatory system at the time she gives birth and is sensing the olfactory cues emanating from her newborn kid. A female goat rendered incapable of smelling before giving birth is unable to differentiate between her own kid and other kids after giving birth.

Bird song learning involves genetics, imprinting, and hormonal timing

Male songbirds use species-specific song to claim and advertise a breeding territory, compete with other males, and declare dominance. They also use song to attract females, which recognize the song of their species even though they do not sing it. For males of many species, learning is an essential step in the acquisition of song. *What* they learn seems to be influenced by genes, and *when* they learn by a limited *developmental time frame* for learning. For example, a male white-crowned sparrow hatchling hears his father and other white-crowned sparrows singing. He also hears the songs of many other bird species. But he does not sing until he approaches sexual maturity almost a year later, and when he does, he sings his father's type of song.

Studies of song learning in white-crowned sparrows were initiated in the 1960s by Peter Marler, who hand-reared hatchlings in the laboratory, where he could expose them to different recorded songs at different times in their development. He discovered that male birds cannot produce their species-specific song as adults unless they hear it as nestlings in the first 2 months of their lives (**FIGURE 41.6**). The hatchlings also have to be able to hear themselves as they approach sexual maturity and begin practicing their singing. If a bird is deafened just before he begins to sing, he will not be able to match his own song with his stored song memory. If he is deafened *after* he sings his correct species-specific song, however, he will continue to sing like a normal bird. We say that at this point the song behavior pattern is *crystalized*. Thus there are two critical periods for song learning: the first in the nestling stage, when a song memory is imprinted; the second as the bird approaches sexual maturity, when he learns to match that song memory.

To learn why birds learn only the song of their own species, Marler played tape recordings of other species' songs to

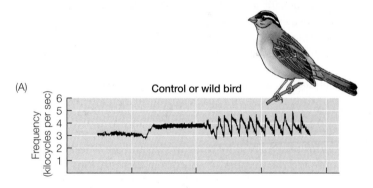

(A)
Control or wild bird

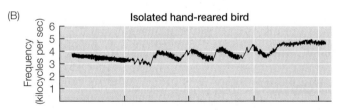

(B)
Isolated hand-reared bird

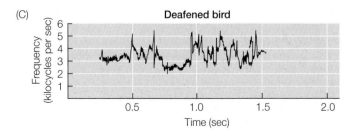

(C)
Deafened bird

FIGURE 41.6 Sensitive Periods for Song Learning (A) Sonogram showing the species-specific song of an adult male white-crowned sparrow (*Zonotrichia leucophrys*). (B) Song of an adult male raised in isolation (never having heard the song as a nestling). (C) Song of an adult male that heard the song as a nestling but was deafened prior to ever singing himself. Marler's experiments showed the bird must first acquire a song memory by hearing the song as a nestling, and must then be able to hear himself as he attempts to match his singing to that song memory.

the white-crowned sparrow hatchlings in his laboratory. The young male sparrows did not learn the songs of other species, even if they heard them many times, but hearing songs of their own species just a few times was sufficient for imprinting. Thus the male sparrows seem to have a genetic predisposition to learn their own song and not the songs of other species. Marler called this phenomenon "an instinct to learn."

Since Marler's pioneering work, more investigations have revealed additional complexity in white-crowned sparrow song learning capacity. In nature, for example, the sparrows are occasionally heard singing the songs of other species. Further laboratory experiments suggested why. Hand-reared nestlings were exposed to the sight and sound of a related species in an adjacent cage while recordings of their own species' song were played in the background (sound only). When these sparrows matured, they sang the song of the other species. Thus social experience (sight plus sound) has a powerful effect on what the young bird can and will learn.

The timing and expression of bird song are under hormonal control

As we have seen, both male and female songbirds hear their species-specific song as nestlings, but only the males of most species sing as adults, and they do so only in spring. Hormones underlie both the difference in song expression between male and female songbirds and the timing of song expression.

When investigators injected adult female songbirds with testosterone in spring, the females sang a rough version of their species-specific song. Apparently females form a memory of their species-specific song and have the physical capacity to sing, but under normal circumstances they lack the hormonal stimulation. This memory, however, may be important in their mate selection.

How does testosterone cause a songbird to sing? A remarkable study revealed that each spring, an increase in circulating testosterone levels causes certain parts of the male's brain necessary for singing to grow larger. Individual neurons in those regions of the brain increase in size and grow longer extensions, and the number of neurons in those regions increases. These results demonstrated that, contrary to the once widely held belief that new neurons are not produced in the brains of adult vertebrates, hormones can control behavior by changing brain structure as well as brain function, both developmentally and in response to environmental cues.

FRONTIERS To determine what triggers the release of testosterone in birds in response to the onset of spring, Takashi Yoshimura conducted a DNA microarray analysis of Japanese quail brains. He examined the responses of 32,000 transcripts to increases in day length. Fourteen hours after dawn on the first day that reached the critical day length for induction of singing, genes in the brain that code for thyroid-stimulating hormone were switched on, initiating a cascade of hormonal controls that stimulate the growth of the testes and the production of testosterone.

Do You Understand Concept 41.3?

- We say that female reproductive behavior is the default developmental pattern in rats. Explain what that means and why it happens.
- What is the adaptive value of imprinting?
- Describe the series of events necessary for a male white-crowned sparrow to sing its species-specific song in spring.

Complex behaviors result from the interactions of genetic, physiological, and environmental factors. Studying these proximate causes of behavior reveals the mechanisms responsible for the behavioral abilities of animals and the differences between species.

concept 41.4 Physiological Mechanisms Underlie Behavior

Control of behavior involves the nervous and endocrine systems. Execution of most behavior involves the musculoskeletal system. All of the physiological systems we have covered in this book have behavioral aspects. Therefore, the field of behavioral physiology is enormous; here we will dig deeper into just three different phenomena studied by behavioral physiologists: the timing of behavior, navigation, and communication.

Biological rhythms coordinate behavior with environmental cycles

Earth turns on its axis once every 24 hours, generating daily cycles of light and dark, temperature, humidity, and tides. In addition, Earth is tilted on its axis, so the light–dark cycle changes as Earth revolves around the sun. These daily and seasonal cycles profoundly influence the physiology and behavior of animals. Animals tend to be either day-active (diurnal) or night-active (nocturnal), and have appropriate sensory capabilities. Therefore it is adaptive to organize behavior on a cycle that corresponds with the environmental cycle of light and dark. Similarly, a behavior that is adaptive at one time of year may not be adaptive at other times. Thus it is important for animals to be able to time their behavior to appropriate times of the day or year and to be able to anticipate those times.

Experimental animals kept in constant darkness and at a constant temperature, with food and water available all the time, still demonstrate daily cycles of activities such as sleeping, eating, drinking, and just about anything else that can be measured. The persistence of these daily cycles in the absence of environmental time cues suggests that animals have an internal clock. Because these daily cycles are not exactly 24 hours long, they are known as **circadian rhythms** (*circa*, "about," and *dies*, "day").

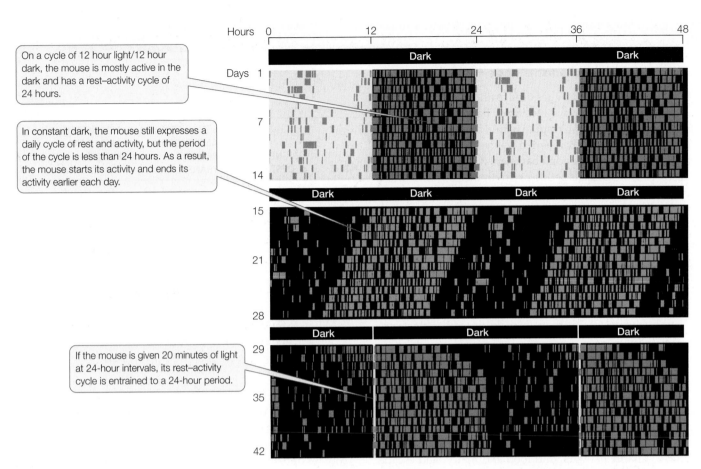

On a cycle of 12 hour light/12 hour dark, the mouse is mostly active in the dark and has a rest–activity cycle of 24 hours.

In constant dark, the mouse still expresses a daily cycle of rest and activity, but the period of the cycle is less than 24 hours. As a result, the mouse starts its activity and ends its activity earlier each day.

If the mouse is given 20 minutes of light at 24-hour intervals, its rest–activity cycle is entrained to a 24-hour period.

FIGURE 41.7 Circadian Rhythms Are Entrained by Environmental Cues The activity–rest cycle of a laboratory mouse (a nocturnal animal) responds to the light–dark cycle under which it is kept. The gray bars indicate times when the mouse is running on an activity wheel. Two days of activity are recorded on each horizontal line; the data for each day are plotted twice—once on the *right* half of each line (hours 24–48) and again on the *left* half of the line below it (hours 0–24). This double plotting is merely to make the pattern easier to see.

yourBioPortal.com

Go to ANIMATED TUTORIAL 41.1
Circadian Rhythms

As described in Concept 26.4, any biological rhythm can be viewed as a series of cycles, and the length of one of those cycles is the *period* of the rhythm. Any point in the cycle is a *phase* of that cycle. Hence, when two rhythms completely match, they are *in phase*, and if a rhythm is shifted (as in the resetting of a clock) it is *phase-advanced* or *phase-delayed*. Because the period of a circadian rhythm is not exactly 24 hours, it must be phase-advanced or phase-delayed each day to remain in phase with the daily cycle of the environment. In other words, the rhythm has to be **entrained** to the cycle of light and dark in the environment.

An animal kept under constant conditions will not be entrained to the light–dark cycle of the environment, and its circadian clock will run according to its natural period—it will be *free-running*. If the period is less than 24 hours, the animal will begin its activity a little earlier each day (**FIGURE 41.7**). The period of the free-running circadian rhythm is under genetic control. Different species may have different average periods, and within a species, mutations can lead to different period lengths.

Under natural conditions, environmental time cues, such as the onset of light or dark, entrain the free-running rhythm to the light–dark cycle of the environment. In the laboratory, it is possible to entrain the circadian rhythms of free-running animals with short pulses of light or dark administered every 24 hours (see the bottom panel of Figure 41.7).

In mammals, the master circadian "clock" consists of two clusters of neurons just above the optic chiasm (the area of the brain where the optic nerves come together and some fibers cross to the opposite side of the brain). These structures are called the **suprachiasmatic nuclei** (**SCN**). If they are destroyed, the animal becomes *arrhythmic*: it is just as likely to eat, drink, sleep, or wake at any time of day.

The molecular mechanism of the circadian clock involves negative feedback loops. When certain *clock genes* are expressed in SCN cells, the mRNA enters the cytoplasm and is translated. The resulting proteins dimerize and reenter the nucleus, where they act as a transcription factor that shuts off the expression of the clock genes. The period of this cycle of gene activation and inactivation is about a day.

APPLY THE CONCEPT

Physiological mechanisms underlie behavior

Martin Ralph and colleagues used artificial selection to produce two strains of hamsters: one with a short circadian period and one with a long circadian period. When the SCNs in some of these adult hamsters were destroyed, the animals became arrhythmic. After several weeks, the scientists transplanted SCN tissue from fetal hamsters into the brains of the adult hamsters whose own SCNs had been destroyed. Long-period adult hamsters received tissue from short-period fetuses, and short-period adults received tissue from long-period fetuses. The effects of these treatments on the hamsters' circadian periods are shown in the table.

1. Why did destroying the SCN make the adult hamsters arrhythmic?

| | TREATMENT | | | |
RECIPIENT	NONE	SCN DESTROYED	SCN DESTROYED AND SHORT-PERIOD TRANSPLANT	SCN DESTROYED AND LONG-PERIOD TRANSPLANT
Short-period adult	Short-period	Arrhythmic	Not done	Long-period
Long-period adult	Long-period	Arrhythmic	Short-period	Not done

2. What result would you expect if SCN tissue from a long-period fetal hamster were transplanted into a long-period adult whose own SCN was destroyed?

3. What do the results of the transplantation experiments tell you about where in the body the circadian rhythm phenotype (long versus short) is expressed?

Animals must find their way around their environment

To locate suitable habitats, find food and mates, and escape from predators and bad weather, an animal needs to be able to find its way around its environment. Within its local habitat, an animal can organize its behavior spatially by orienting to landmarks. But what about destinations farther away?

PILOTING: ORIENTATION BY LANDMARKS Most animals find their way by knowing and remembering the structure of their environment. This form of navigation is called **piloting**. Gray whales, for example, migrate seasonally between the Bering Sea and the coastal lagoons of Mexico (**FIGURE 41.8**). They find their way in part by following the west coast of North America. Coastlines, mountain chains, rivers, water currents, and wind patterns can all serve as piloting cues. But some remarkable cases of long-distance orientation and movement cannot be explained by piloting.

HOMING: RETURN TO A SPECIFIC LOCATION The ability to return to a nest site, burrow, or other specific location is called **homing**. Homing can be accomplished by piloting in a known environment, but some animals that travel long distances through unfamiliar territory perform much more sophisticated homing. The ability of pigeons to return to their home loft even after being transported to remote sites is well known. Pigeons use the sun as a compass (see Figure 41.9), but they can still find their way home when the sun is not visible. Other experiments have shown that pigeons equipped with frosted contact lenses can find their way home, suggesting that visual cues, though sufficient, are not necessary. Most amazing is that pigeons can detect Earth's magnetic field and orient to it much as a human

orients with a compass. Taken together, the studies of homing by pigeons suggest that they can use multiple, redundant sources of directional information and can switch among those sources depending on the circumstances.

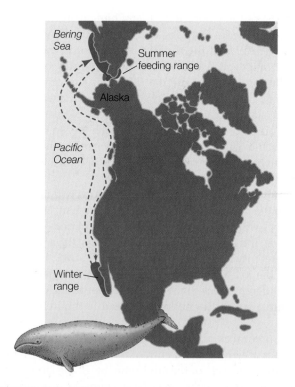

FIGURE 41.8 Piloting Gray whales (*Eschrichtius robustus*) migrate south in winter from the Bering Sea to the coast of Baja California by piloting, in part by following the western coast of North America.

MIGRATION AND NAVIGATION OVER GREAT DISTANCES Many homing and migrating species are able to take direct routes to their destinations through environments they have never experienced because they use mechanisms of navigation other than piloting. Humans use two major forms of navigation:

- *Distance–direction navigation* (dead reckoning) requires knowing in what direction and how far away the destination is. With a compass to determine direction and a means of measuring distance, humans can navigate.

- *Bicoordinate navigation*, also known as *true navigation*, requires knowing the latitude and longitude (the map coordinates) of both the current position and the destination, as well as a compass to determine direction.

Animals, of course, do not carry compasses or GPS receivers, but many of them have a *compass sense* that allows them to use environmental cues to determine direction, and some appear to have a *map sense* that allows them to determine their position.

A compass sense is not difficult to explain. If you know what time it is—remember the circadian clock described above—you can use the position of the sun to indicate compass direction. A *time-compensated solar compass* was demonstrated in pigeons by placing them in an outdoor circular cage that enabled them to see the sun and sky, but no other visual cues (**FIGURE 41.9**). Food bins were arranged around the sides of the cage, and the birds were trained to expect food in the bin at one particular compass point in the cage—south, in this case. The birds were then moved indoors and placed in rooms with controlled photoperiods that were different than in the outside environment. In a week or so, their circadian rhythms entrained to the new photoperiods—just like you recover from jet lag. When the birds were then returned to the outdoor cage, they remembered to seek food at a particular compass point in their cage, but they showed an error in that directional selectivity that matched their shifted circadian rhythms. For example, if the birds were shifted by 6 hours so their internal clocks said noon when it was really 6:00 A.M., when they were returned to the outdoor cage they searched for food in the direction of the sun. But at sunrise that was in the east, not the south. Similar experiments have shown that many species can orient by means of a time-compensated solar compass.

Birds that migrate at night could use a time-compensated star map for directional orientation, but a simpler source of directional information is available in the night time sky. Because of the rotation of Earth, the star pattern rotates around fixed points in the sky that lie over the North or South Poles. Steve Emlen at Cornell University raised birds in a planetarium, rotated the sky around different constellations, and showed that birds from the Northern Hemisphere learn to identify the fixed point in the sky as north. It did not matter which star in which constellation was used as that fixed point.

There is no clear demonstration of true bicoordinate navigational mechanisms in animals, but many behave as if they have such capabilities. For example, gray-headed albatrosses breed on small islands in the Southern Hemisphere. After young birds leave their nests, they wander widely over the southern oceans for 8 or 9 years before reaching reproductive maturity (**FIGURE 41.10**). Then they fly back to the island where they hatched to find a mate and breed. It is difficult to imagine how they could locate such a specific locality in the vast southern oceans after years of wandering unless they were able to know the map coordinates of where they are and where they need to go.

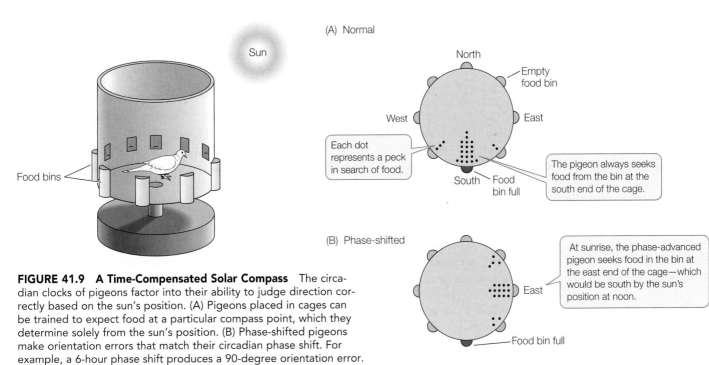

FIGURE 41.9 A Time-Compensated Solar Compass The circadian clocks of pigeons factor into their ability to judge direction correctly based on the sun's position. (A) Pigeons placed in cages can be trained to expect food at a particular compass point, which they determine solely from the sun's position. (B) Phase-shifted pigeons make orientation errors that match their circadian phase shift. For example, a 6-hour phase shift produces a 90-degree orientation error.

(A)

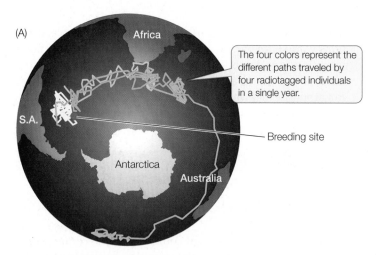

The four colors represent the different paths traveled by four radiotagged individuals in a single year.

Breeding site

(B) *Thalassarche chrysostoma*

FIGURE 41.10 Coming Home (A) Gray-headed albatrosses are born on islands in the subantarctic oceans. Young birds roam widely over the southern oceans for 8 or 9 years. (B) Once they reach maturity, the birds return to the island where they were born to mate and raise their own young. A courting couple is shown here.

Animals use multiple modalities to communicate

As individual animals interact, they exchange information; therefore, animal behaviors can evolve into systems of information exchange, or **communication**. Specific behaviors can evolve into communication systems if they convey information that benefits both the sender and the receiver. For example, consider bird song, which we discussed in Concept 41.3. Natural selection has favored male singing displays because they communicate the male's sex, his reproductive condition, and even his possession of a territory that would provide a good nest site and food supply. Natural selection has favored females' responses to singing displays that convey information about the male's qualities as a potential mate and father. Thus this communication signal contributes to the reproductive success of both the male and the female.

Animals communicate using a variety of sensory modalities that vary in the nature of the signal produced, the specificity of the information conveyed, the speed and persistence of the signal, and its suitability in different environments.

PHEROMONES Odorants used for communication are called *pheromones*. Pheromones vary in their volatility (ease of vaporization) and diffusibility, which in turn determine their range and persistance. Pheromones that act as alarm signals, for example, are highly volatile and diffusible, so their messages spread rapidly but disappear rapidly. Territory-marking and trail-marking pheromones have low volatilities and diffusibilities and stay effective for a long time. Sex pheromones, such as that of the silkworm moths (see Figure 35.4), are intermediate in these properties; they spread a long distance but do not disappear rapidly.

Because of the potential diversity of their molecular structures, pheromones can communicate very specific, information-rich messages. Mammals that mark their territories with pheromones reveal a great deal of information about

themselves: species, individual identity, reproductive status, size (indicated by the height of the marking), and how recently the animal has been in the area (indicated by the strength of the scent). Pheromones are especially effective for exchanging species-specific information because the recipient must have the proper receptor to detect the pheromone; thus predators do not easily intercept the signal.

VISUAL SIGNALS Visual signals offer the advantage of rapid delivery of information over considerable distances (depending on the environment and the visual acuity of the receiver). Visual signals convey without ambiguity the position of the signaler. Visual signals also can provide information about the signaler's sex, species, vigor, health, social status, and reproductive state. Unlike pheromones, effective visual signals require sufficient light, and the receiver must be looking directly at the signaler. Thus visual communication is not particularly useful at night or in dark environments, such as caves and ocean depths. Some species have overcome this constraint with light-emitting mechanisms. Fireflies, for example, use an enzymatic mechanism to create flashes of light. By emitting flashes in species-specific patterns, fireflies advertise for mates at night.

Another drawback of visual signals is that they can be intercepted by other species. There are predatory firefly species, for example, that mimic the flash pattern of females of other species. A male that approaches the mimicking "female" becomes a meal rather than a mate.

ACOUSTIC SIGNALS Sound cannot convey complex information as rapidly as visual signals can. But acoustic signals, unlike visual signals, can be used in dark environments. They can also be transmitted in complex environments, such as forests, that might hinder visual signals. Acoustic signals are often better than visual signals at getting the attention of a receiver because the receiver does not have to be looking at the signaler. Sounds are also useful for communicating over long distances—loud sounds can transmit information over longer distances than visual signals can. The complex songs of humpback whales, for example, can be heard up to hundreds of kilometers away. In this way, humpback whales can locate one another across vast expanses of ocean.

The information content of acoustic signals can be increased by varying their frequency, as in the species-specific songs of

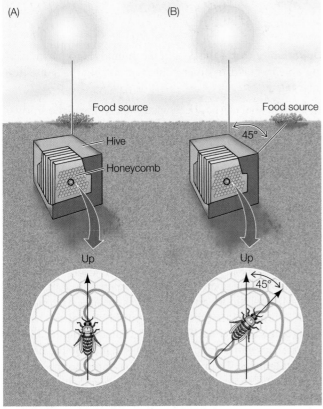

(A)

(B)

Food source

Food source

45°

Hive

Honeycomb

Up

Up

45°

Pattern of waggle dance

Pattern of waggle dance

FIGURE 41.11 The Waggle Dance of the Honey Bee (A) A honey bee runs straight up on the vertical surface of the honeycomb in the dark hive while wagging her abdomen to tell her hive-mates that there is a food source in the direction of the sun. (B) When the waggle runs are at an angle from the vertical, the other bees know that the same angle separates the direction of the food source from the direction of the sun.

yourBioPortal.com

Go to WEB ACTIVITY 41.1
Honey Bee Dance Communication

birds shown in Figure 41.6 (and in our own speech). However, acoustic signals place the signaler at risk for detection by predators, which can be minimized by adjustments of frequency and signal structure that decrease the directional information the receiver can extract from the signal. Alarm calls tend to be pure tones (a single frequency) without much temporal structure, which are difficult to localize. By contrast, territorial calls or mating calls tend to cover a broad frequency range and have temporal structure. These calls are easy to localize. Acoustic signals are also adapted to specific habitats: pure tones at lower frequencies carry better in forests, and more complex calls at higher frequencies carry well in open habitats.

MECHANOSENSORY SIGNALS Animals in close contact with one another communicate by touch, especially under conditions where visual communication is difficult. The best-studied use of mechanosensory communication is the dance of honey bees, first described by Karl von Frisch. Honey bees can reliably communicate the location of food sources as far as 2.5 kilometers away from their hive. When a forager bee finds

food, she returns to the hive and communicates her discovery to her hive-mates by dancing on the vertical surface of the honeycomb.

The returning forager performs a *waggle dance* (**FIGURE 41.11**), a repeating figure-eight pattern on the honeycomb. She alternates half-circles to the left and right with vigorous wagging of her abdomen in a short, straight run between turns. The angle of the straight run indicates the direction of the food source relative to the sun's azimuth (the position of the sun projected down to the horizon). The directional information in the straight runs becomes more accurate as the distance to the food increases. Researchers used to think that the bees had a separate dance to indicate nearby food, but it turned out to be just a larger variation of the waggle dance. Even under partially cloudy conditions that obscure the sun, the forager can provide directions to the food source because bees can see polarized light in a clear patch of sky and infer the position of the sun from the pattern of polarization. The honey bee's internal clock allows the dancer to adjust her dance to take into account the sun's movement during her return flight—another example of a time-compensated solar compass. The dance also indicates the distance to the food source: the farther away it is, the longer the duration of each waggle run. The odor on the dancer's body indicates the flower to be looked for, and she shares some nectar with hive-mates to demonstrate the quality of the food source. When the food source is close, the distance and direction information is less critical because other workers excited by the food smell and the dancer's level of arousal can simply follow her back to the food source.

COMMUNICATION IN MULTIPLE SENSORY MODALITIES Specificity is a high priority in any signaling system and can be enhanced by using multiple sensory modalities. Courtship behavior in fruit flies, as we saw in the opener to this chapter, involves tactile, chemical, visual, and acoustic signals. The male fruit fly identifies the female by her pheromones (chemical), taps her body with his foreleg (touch), and vibrates his wing to produce his "courtship song" (acoustic). The male then extends his mouthparts to taste the female's genitalia (chemical and tactile); if she is receptive, he initiates copulation. If at any point sensory feedback indicates to either the male or the female that their pairing is inappropriate, the courtship abruptly ends.

Do You Understand Concept 41.4?

- What is meant by a "free-running" rhythm? How is it entrained to the 24-hour environmental cycle?

- Experimenters captured young starlings in the Netherlands and transported them to Switzerland. Normally these young starlings would migrate to western France, but instead they flew to Spain. Adult starlings used in the same experiment flew correctly from Switzerland to western France. Explain the different navigational systems that could have been used by juvenile and adult birds to yield these results.

- Draw a table that lists at least one advantage and one disadvantage of visual, acoustic, and chemical communication.

We have just explored three fields of behavior in which researchers primarily focus on physiological mechanisms underlying behaviors and also seek to understand how those mechanisms are adapted to specific environmental variables. Another approach to studying behavior is to investigate how the environment influences the fitness of individuals and thereby shapes the evolution of their behavior.

concept 41.5 Individual Behavior Is Shaped by Natural Selection

Environmental conditions are highly variable over both time and space. Animal behaviors are also variable within and between species. Behavioral ecologists strive to discover the relationships between behavior and environment that lead to the evolution of specific behaviors.

Animals must make choices

Over an animal's lifetime, its behavior is largely a series of choices: where and when to move, where to build a nest, what to eat, when to fight and when to flee, with whom to associate, with whom to mate. Making wrong choices reduces fitness.

The choice of a place to live, for example, is one that has many consequences. In most cases a **habitat**—the environment in which an animal lives—provides not only shelter but also food and access to mates. The environmental cues animals use to make their habitat choices can be quite simple. For example, seabirds select cliffs or offshore rocks for nesting because those sites offer protection from predators. Animals with very specialized food requirements obviously select habitats where those foods are abundant. The general hypothesis that guides behavioral ecologists is that the cues animals use for habitat selection are those that over evolutionary time have been reliable correlates of good fitness outcomes.

Behaviors have costs and benefits

Behavioral ecologists often use a cost–benefit approach to investigate the relationship between behavior, environment, and fitness. A **cost–benefit approach** assumes that an animal has only a limited amount of time and energy, and therefore cannot afford to engage in behaviors that cost more to perform than they bring in benefits.

LINK A cost–benefit approach can also be used to analyze life history trade-offs during evolution; see Concept 43.3

The benefits of a behavior can be measured in terms of the enhancement in fitness an animal accrues by performing the behavior. The cost of a behavior typically has three components:

- **Energetic cost** is the energy the animal expends performing the behavior.

- **Risk cost** is the increased chance of being injured or killed as a result of performing the behavior.

- **Opportunity cost** is the benefit the animal forgoes by not being able to perform other behaviors during the same time interval.

Cost–benefit analysis has been used extensively in the study of **territorial behavior**: aggressive behavior used by an animal to actively deny other animals access to a habitat or resource. Optimal habitats and resources are frequently in short supply, so individuals of the same species must compete for them. Many individual animals—usually males—defend all-purpose territories that provide shelter, food, and access to mates. Through territorial behavior, the males obtain the resources they need for reproductive success, but they also pay a price.

Territorial behavior requires considerable expenditure of energy, makes a male more vulnerable to predation, and detracts from time spent feeding or engaging in parental behavior. Michael Moore and Catherine Marler at Arizona State University investigated the costs incurred by males when defending a territory (**FIGURE 41.12**). Yarrow's spiny lizards (*Sceloporus jarrovii*) defend territories that include the habitats of several females. Their territorial behavior is normally most intense during September and October, when their circulating testosterone levels are high and the females are most receptive to mating. The researchers varied the intensity of the lizards' territorial behavior by implanting testosterone capsules in some males in summer, when they are not normally highly territorial.

Testosterone-treated males spent more time patrolling their territories, performed more displays, and expended about one-third more energy than control males (an energetic cost). As a result, they had less time to feed (an opportunity cost), captured fewer insects, stored less energy, and had a higher death rate (a risk cost). In summer, when females are not normally receptive, these high costs of vigorous territorial defense outweigh the reproductive benefits of territoriality. Thus natural selection has favored seasonal variation in the level of the hormone controlling territorial behavior in this species.

The cost–benefit approach explains the diversity of territorial behaviors seen in different species. Even if a resource is absolutely essential to an animal, if it cannot be defended economically, the animal will not engage in territorial behavior. For example, the open ocean where seabirds feed cannot be defended. But safe nest sites on islands or rocky cliffs are in short supply, and they can be defended. Thus the territories of seabirds may be no larger than the distance the birds can reach while sitting on their nests and are defended only while offspring are present.

In some cases the resource that is defended is the mate itself. Elephant seals spend most of their lives and collect all of their food at sea, but females come to land at traditional beach sites to give birth to their pups. Male elephant seals arrive at these sites ahead of time and stake out territories through vigorous fighting. When the females arrive, they enter the territories of the males. As long as the male territory holder can fend off challengers, he will be able to mate with all the females using his piece of the beach.

INVESTIGATION

FIGURE 41.12 The Costs of Defending a Territory
By using testosterone implants to increase territorial behavior, Moore and Marler measured the costs to male Yarrow's spiny lizards (*Sceloporus jarrovii*) of defending a territory during the summer, when they do not normally do so.

HYPOTHESIS

Yarrow's spiny lizards do not defend a territory during summer because the energetic costs of territorial behavior in that season outweigh the benefits.

METHOD

1. During the summer, when female lizards are not sexually receptive, insert testosterone capsules under the skin of some males; leave other males untreated as controls.
2. Observe the patterns of territorial behavior and the survival rate of the two groups of males.

RESULTS

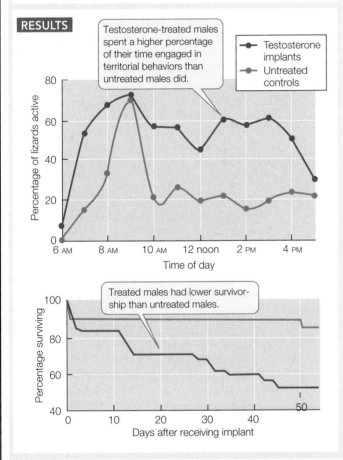

Testosterone-treated males spent a higher percentage of their time engaged in territorial behaviors than untreated males did.

Treated males had lower survivorship than untreated males.

CONCLUSION

For these lizards, the cost of defending territories during summer significantly reduces their survival rate without increasing their reproductive success.

ANALYZE THE DATA

Given the trade-off between time spent in territorial behavior and time available for foraging, researchers hypothesized that the cost of territoriality should be reflected in decreased growth rates. Two measures of body size in a lizard are length (tip of snout to urogenital vent) and weight. Lizards were captured and implanted with testosterone capsules or empty capsules. The individual lizards varied widely in size at the time of capture, but the average size was the same for the two treatment groups. Three months later the lizards were recaptured and their sizes measured (see table).

Controls		Testosterone implants	
Weight (g)	Length (mm)	Weight (g)	Length (mm)
3.9	73	3.1	73
4.9	78	3.2	76
5.1	78	3.2	79
4.3	80	4.2	78
4.8	80	4.1	79
4.6	82	4.8	80
5.8	84	3.5	82
5.2	85	4.5	84
8.2	86	4.2	87
5.5	90	7.5	102
6.9	90		
6.4	91		
8.1	93		
7.0	94		
7.0	96		

A. For each treatment, calculate the mean and standard deviation for the two size measures.

B. Using statistical tests from Appendix B, what can you conclude from these data?

C. Since the lizards varied greatly in size at the beginning of the experiment, how could you use regression analysis to compare the two experimental groups?

Go to **yourBioPortal.com** for original citations, discussions, and relevant links for all INVESTIGATION figures.

The most unusual form of male territorial behavior arises in a situation where neither food, nest sites, nor females are defended. A **lek** is an area where males gather to engage in communal displays of their territorial prowess for the purpose of attracting mates. Even though space is not limited, each male defends a small piece of real estate on which he performs a display. Those pieces of territory closest to the center of the lek are the prime sites, and males compete intensely for those locations. The females stroll into the lek, observe the males, and generally mate with the males holding the prime sites. The benefit of this system to the female is that she is inseminated by a successful competitor, and therefore her offspring will carry the genes that contributed to his success. The costs of lekking to males are high, as they engage in continuous, intense territorial behavior that precludes eating, drinking, and sleeping until they are displaced. The benefit is the chance to maximize their fitness by mating with many females.

Cost–benefit analysis can be applied to many aspects of behavior. For example, foraging behavior can be analyzed in terms of energy and time expended versus energy obtained. Is it more beneficial to exploit a readily abundant food source with low energy content or a more dispersed food source with higher energy content? Is it more beneficial to pursue a high-quality food source for which there is fierce competition or a low-quality food source that is free for the taking?

Of course, there are other considerations than energy in the evolution of feeding behavior. Essential minerals, for example, are in short supply in some animals' diets, and those animals may incur large energetic costs and risks to obtain them. Foods may also be chosen for their medicinal value. Chimpanzees, for example, have been observed eating the pith of a plant called *Vernonia amygdalina*. The pith contains very small quantities of a compound called vernonioside B1, which is toxic to chimps at high concentrations, but at low concentrations it can kill their intestinal parasites. Chimps that consume this pith have fewer parasites. Thus the behavioral ecologist must be cognizant of all aspects of an animal's behavior that could influence its reproductive fitness.

Do You Understand Concept 41.5?

- List at least five types of behavioral choices that an animal must make in its life.

- Some bird species defend all-purpose territories that provide nest sites and all the food needed to raise their young, but other species defend much smaller territories that cannot provide all of the food they will need. What factors could influence the evolution of these different territorial strategies?

- In lek species the male provides no resources or help to the female with which he mates. What are the advantages to the female of selecting a mate that is in the center of the lek?

In this concept we focused on the behavioral choices made by individual animals and how those behaviors can influence the fitness of the individual. However, in nature, species can display complex social behaviors that are not as easy to analyze in terms of the reproductive success of the individual.

The evolution of social behavior became a field of study in 1975, with the publication of E. O. Wilson's landmark book, *Sociobiology*. A minimal social system is the reproductive behavior of a single male and female, but even here there is diversity. Species differ in the number of mates each individual has (ranging from monogamous to promiscuous); the amount of parental care they give their young; and the degree to which the male contributes to raising the young. Beyond these relatively simple mating systems, larger numbers of reproductive individuals may associate in *polygynous* mating systems, in which a male has more than one mate, or *polyandrous* mating systems, in which a female has more than one mate. Even more complex interactions include extended families that participate in raising young, and societies such as honey bee colonies in which large numbers of nonreproductive individuals assist a single reproductive individual. Sociobiology proposes that the evolution of all these variations can be understood by asking how social behavior contributes to the fitness of the *individuals* involved.

Mating systems maximize the fitness of both partners

Voles are small rodents, and there are many species of voles living in very different habitats. Prairie voles (*Microtus ochrogaster*) are *monogamous*; they form strong pair bonds that last for life, and both parents participate in rearing the young. In contrast, montane voles (*M. montanus*) are *promiscuous*: males mate with many females, and the young are raised by the females alone. The proximate mechanisms behind these stark behavioral differences involve differences in the release of oxytocin during mating and the distribution of the oxytocin receptors in the brains of the two species. Because of this role of oxytocin in pair bonding, it has been called the "cuddle hormone." The ultimate question is why two such different mating systems evolved in two species that are so closely related.

To answer this question, we begin with the premise that there is an asymmetry in the contributions of male and female animals to their offspring at the time of fertilization. As we learned in Concept 33.1, females produce a limited number of eggs, and each egg is generously stocked with resources. Males produce an almost infinite number of sperm that contain almost no resources. So the energetic and opportunity costs of reproduction are greater for the female than for the male. In mammals, this asymmetry increases throughout gestation as the female bears most of the costs. Under these circumstances, the main way for the female to maximize her fitness is to make sure her young are healthy and survive to pass on her genes. Females that invest heavily in current offspring may be favored over those that "skimp" and produce weaker offspring, because the weaker young are unlikely to be as successful in their reproductive efforts.

The male has different options for maximizing his fitness. He can simply move on after inseminating the female and seek additional mates as a means of maximizing his reproductive success—as in the case of the montane vole. Or he can stay with the female he inseminated, protect her, and help care for their young—as in the case of the prairie vole. Which strategy maximizes his fitness depends on a number of factors that are influenced by the species' environment, such as the likelihood that a female and her offspring will survive without a male's help, and a male's likelihood of finding another fertile female. Sociobiologists seek to quantify these factors in nature as a means of explaining observed differences in mating systems.

POLYGYNY **Polygynous** mating systems, in which a male has more than one mate, involve a different asymmetry. If a male can sequester a group of females from other males, he can increase his fitness by increasing the number of females in his group. As we mentioned in Concept 41.5, male elephant seals accomplish this by protecting an area of beach where females give birth. Male baboons do so by herding females. Male red-winged blackbirds may acquire more than one mate by defending high-quality nesting territories, building more than one nest, and attracting females to that desirable real estate. Since sex ratios in all these species are close to 50:50, a large differential in male fitness is established: some males have high reproductive success while many males have none. Selection favors males that are successful in competing with other males to obtain and protect access to many females. In general, bigger, stronger males are the winners, and sexual dimorphism in body size evolves. The elephant seal is an extreme example: males may weigh more than three times as much as females. When species with polygynous mating systems are compared, there is a strong correlation between the number of females a male controls and the degree of sexual dimorphism.

Why do females participate in these polygynous mating systems? In some cases they have no choice. A female elephant seal is most likely to successfully raise her pup if she gives birth on a safe beach, and to do so she must enter the territory of a male. A female red-winged blackbird is more likely to have a successful brood if she accepts a mate that defends a nesting site that is safest from predators and provides access to a good food supply. To achieve this end, she may have to share the attentions of the territory's owner with other females. Even if a female has a choice of mates, she can maximize her fitness by mating with a strong dominant male so her sons are also likely to be strong and dominant, thereby maximizing her number of grandchildren. Strong dominant males may be able to protect more than one female.

POLYANDRY **Polyandry**—one female with multiple mates—is a relatively rare mating system seen in some birds and a few mammals in which paternal care for the young can have a large effect on fitness. An example is the golden lion tamarin (*Leontopithecus rosalia*), a primate native to the Brazilian tropical forests (**FIGURE 41.13**). Compared with other primates, adult tamarins are tiny—under 1 kilogram—so they face high predation

Leontopithecus rosalia

FIGURE 41.13 Polyandry in a Small Primate The endangered golden lion tamarins of Brazil are small primates whose unique life history has given rise to polyandry in some groups, with males playing a major role in rearing the young.

pressure. Females usually give birth to twins and thus newborns constitute a higher percentage of maternal weight than in other primates. They also grow faster, so nursing costs are high. For all these reasons, young tamarins cared for by their mother alone are unlikely to survive.

A male tamarin helps guarantee his reproductive success by watching out for predators and gathering food for the female and her young. Tamarin mothers with twins in tow spend 92 percent of their time resting, compared with only 58 percent of the time when they are not carrying young. When a male is present, however, he carries the young about one-third of the time, so the mother has much more time for foraging and feeding.

If one male is helpful in protecting and raising young, then two should be even more helpful. Some females can attract a second mate by being sexually receptive to him. Neither male can be sure that any eventual offspring are his, so it is in the best interest of both to help in their rearing. Of the social groups of golden lion tamarins observed in field studies, only 22 percent had one male and one female, whereas 61 percent had multiple males and one female.

Fitness can be enhanced through the reproductive success of related individuals

As humans, we readily understand the concept of extended family—brothers, sisters, aunts, uncles, nieces, nephews, and so forth. Extended families are a form of social organization in other species as well. Members of these families may cooperate in territory defense, predator avoidance, foraging, and rearing of young. If behavior is favored when it increases the fitness of the individual performing it, how can we explain the evolution

FIGURE 41.14 Helpers at the Nest Young Florida scrub jays (*Aphelocoma coerulescens*) often forego reproduction in their first few years of adulthood to help their parents raise their siblings. These young birds help their parents feed the nestlings, defend the territory, and protect the nest from predators.

of social behaviors that do not lead to the performer having more offspring? How can we explain behaviors that appear to be **altruistic**—benefiting another individual at a cost to the performer?

Having offspring increases an individual's fitness because those offspring carry the parent's genes into the next generation. Fitness gained by producing offspring is referred to as **individual fitness**. However, an individual's genes can be carried into the next generation by other routes. In diploid organisms, two offspring of the same parents share, on average, 50 percent of the same alleles, and an individual is likely to share 25 percent of its alleles with its sibling's offspring. Therefore, by helping parents and other relatives raise their offspring, an individual increases the transmission of those shared alleles to the next generation. **Inclusive fitness** is the fitness derived from an individual's own reproductive success plus that derived through the reproductive success of its relatives.

Maximization of inclusive fitness is the mechanism behind **kin selection**: selection for behaviors that increase the reproductive success of relatives even when they come at a cost to the performer. An example is the phenomenon of *helping at the nest*, which was extensively studied in Florida scrub jays by Glen Woolfenden. Scrub jay pairs mate for life and establish large territories, which they defend aggressively. The mating pair may be assisted in rearing their young by three to five helpers (**FIGURE 41.14**). The helpers guard against predators, feed the young, clean the nest, and fly with fledglings to protect them. Why are these birds helping others rather than rearing their own young? Through a long-term study, Woolfenden was able to establish a number of important facts:

- The helpers are prior offspring of the mating pair and are usually 1 to 3 years old.
- Young birds that attempt to set up their own territories and breed have almost zero reproductive success.
- Mating pairs with helpers have approximately three times the reproductive success of those without helpers.
- When birds that have been helpers eventually establish their own nests, they fledge more offspring than birds that have never been helpers. Thus experience pays off.

These results support the conclusion that helper scrub jays are maximizing their inclusive fitness by helping their parents

raise siblings until the helpers are mature enough to have a reasonable probability of successfully raising their own offspring.

The concept of kin selection was formalized by W. D. Hamilton in what has become known as **Hamilton's rule**. He argued that for an apparent altruistic behavior to be adaptive, the cost to the performer must be less than the benefit to the recipient times the degree of relatedness between the performer and the recipient. This relationship was clearly stated years before by the eminent geneticist J. B. S. Haldane, who said during an argument about altruism that he would not be willing to risk his life to save his brother—but for two brothers or eight cousins, he would consider it.

Eusociality is the extreme result of kin selection

The social groups of **eusocial** species include nonreproductive workers. Prime examples are wasps, bees, and ants—members of the large insect order Hymenoptera. In a honey bee colony, a single reproductive female—the queen—occasionally produces a few male offspring, but most of the thousands of other individuals in the colony— all her offspring—are sterile female workers. One theory for explaining the evolution of eusociality in hymenopterans focuses on their sex determination mechanism, called **haplodiploidy** because diploid individuals are female and haploid individuals are male. The queen carries a lifetime supply of sperm obtained during her single mating flight, and she controls whether her eggs are fertilized or not. An unfertilized egg develops into a male, a fertilized egg develops into a female.

If a queen copulates with only one male, all the sperm she receives are identical because a haploid male has only one set of chromosomes, all of which are transmitted to every sperm cell. Therefore, the queen's daughters share all of their father's genes and, on average, half of their mother's genes. As a result, the workers in the hive—all sisters—share, on average, 75 percent of their alleles. Were they to reproduce, they would share only 50 percent of their alleles with their own female offspring. Thus they can potentially increase their inclusive fitness more per individual raised by caring for their sisters than by producing and caring for their own offspring.

Heterocephalus glaber

Haplodiploidy is not essential for eusociality to evolve. This social system may also arise if it is costly or dangerous to establish new colonies. Nearly all eusocial animals construct elaborate nests or burrow systems within which their offspring are reared. Such a structure represents an enormous investment of resources. Naked mole-rats—the most eusocial mammals—live in elaborate underground tunnel systems that can extend as much as 5 kilometers in cumulative length (**FIGURE 41.15**). A colony includes 70 to 80 individuals but only 1 reproductive female and a few reproductive males. The other colony members—all offspring of the single reproductive female—are sterile workers that dig and maintain the tunnels, guard against intruders, harvest food (tubers), and use their feces to feed

FIGURE 41.15 A Eusocial Mammal Naked mole-rats live in a large colony with only one reproductive female (shown here nursing her young) and a few reproductive males. They live in an elaborate tunnel system excavated by the colony over time.

the queen and her offspring. Individuals attempting to found new colonies have a high risk of failing or being captured by predators. When chances of individual reproductive success are practically zero, an individual can best maximize its inclusive fitness by staying with and helping maintain the colony.

Group living has benefits and costs

The cost–benefit approach of behavioral ecology is relevant to understanding the evolution of social behavior. An obvious benefit of group living is improved foraging efficiency. By hunting in packs, for example, wolves employ cooperative strategies that enable them to bring down larger prey than could a single wolf. Living in a group can also reduce the risk of the members' becoming prey themselves. One investigator found that a trained goshawk's success in capturing a pigeon in a flock decreased as the number of pigeons in the flock increased (**FIGURE 41.16A**). The larger the flock, the sooner some individual in the flock spotted the hawk and flew away. This escape behavior stimulated other individuals in the flock to take flight as well.

Alarm calling is another means of reducing predation risk, but the caller incurs a risk cost by calling attention to itself. Belding's ground squirrels live in large colonies in open meadows. When one squirrel announces the presence of a predator with loud, sharp barks, all the nearby squirrels dive into their burrows (**FIGURE 41.16B**). Paul Sherman and others showed that this altruistic behavior, which doubles the caller's risk of

(A)

Goshawk

Wood pigeon

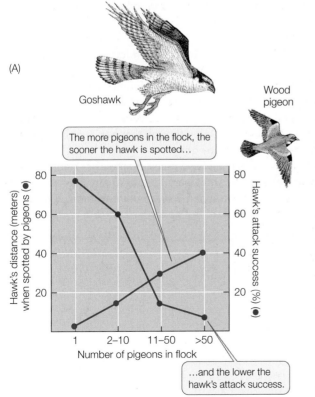

The more pigeons in the flock, the sooner the hawk is spotted...

...and the lower the hawk's attack success.

(B) *Spermophilus beldingi*

FIGURE 41.16 Group Living Provides Protection from Predators Animals that live in groups can spread the cost of looking out for predators. (A) The larger the number of pigeons in a flock, the greater the chances that one of the pigeons will spot a hawk before it attacks, and the lower the chances that the hawk will capture one of the pigeons. (B) A young male Belding's ground squirrel gives an alarm call upon spotting a predator. Although this behavior increases the squirrel's individual risk of becoming prey, it increases the survival chances of many of his close relatives.

being preyed on, is a product of kin selection. In this polygynous species, males establish large territories in the spring that include the territories of several females, which they inseminate. Female offspring settle near their mothers, so neighboring females in a colony tend to be sisters, and they defend each other's young. Sherman showed that males are much less likely to give alarm calls than females, and that females are more likely to give alarm calls when related individuals are nearby.

Social behavior has many costs as well as benefits. Foraging in a group may reduce the amount of food available to each individual, and the foraging individuals may interfere with one another's foraging activities. Individuals living in groups may face more competition for mates, as well as for food, than solitary individuals would. A large group may actually attract the attention of predators. And living at high population densities can increase the risk of disease transmission. The behavioral ecologist employing cost–benefit analysis must take these various factors into consideration.

Do You Understand Concept 41.6?

- Explain how differential parental investment can favor the evolution of either monogamy or promiscuity.
- Explain the evolutionary relationship between sexual dimorphism in body size and degree of polygyny (female to male ratio) in a species.
- Describe the differences in the selective pressures producing eusocial behavior in bees and naked mole-rats.

Knowledge of the behavior of particular species—how they use the environment, how they obtain food, how they organize their activities spatially and temporally—is essential for understanding how species interact in nature. These interactions are one focus of the science of ecology, the subject of Part 7 of this book.

QUESTION How can a single gene be responsible for a complex behavior?

ANSWER A gene cannot code for a behavior. Genes code for proteins, and proteins play many roles—structural, functional, and signaling. By interfering with any of these roles, a gene can have an effect on a behavior (Concept 41.2). We saw a simple example in the gene critical for the function of a pheromone receptor in mice. An alteration in this single gene destroyed the ability of the male mouse to tell a male from a female—an essential component of territorial and mating behaviors. Another example is the prairie (monogamous) and the montane (polygynous) voles (Concept 41.6). Both species of voles release oxytocin when copulating, but the expression pattern of the oxytocin receptor is different in the brains of the two species. Notably, the receptor is heavily expressed in a particular area of the brain (the nucleus accumbens) of the prairie vole, but not in the montane vole. The nucleus accumbens is associated with sensations of pleasure and reward, so we can hypothesize that its stimulation could play a role in forming a strong bond between two young voles mating for the first time. The role of hormones in stimulating neurogenesis in songbirds or in causing the development of neural circuitry responsible for masculinization of the newborn rat brain are other examples where genes can influence developmental and functional processes that underlie behavior (Concept 41.3). A signaling role for a gene in behavior was seen in the fruit fly example in the chapter opener. The development and expression of male and female sexual behavior in fruit flies is under control of a gene cascade that differs in males and females because of different splicing patterns of gene products early in the cascade. As a result, one gene, *fruitless*, produces a slightly different protein in males and females. This protein serves as a transcription factor and controls the expression of downstream genes that make the difference between male and female behavior. It is easy to see how elimination or alteration of this one gene could have large effects on courtship behavior of this species even though that behavior is really under the control of many genes.

It is fair to say that we have many examples of single genes affecting behavior and that we understand why that result occurs. However, we are a long way from understanding how spider webs are coded in the genome.

41 | SUMMARY

concept 41.1 Behavior Has Proximate and Ultimate Causes

- Questions about animal behavior can be grouped into four inter-related categories: causation, development, function, and evolution. The first two refer to **proximal causes** of behavior and the later two to **ultimate causes** of behavior.

- **Behaviorism** originated in the study of conditioned reflexes and focused on this phenomenon as a mechanism of learning. By contrast, **ethology** focused on fixed action patterns and the nature of stimuli, called **releasers**, that trigger those genetically inherited behavior patterns. **Review Figures 41.1 and 41.2**

concept 41.2 Behaviors Can Have Genetic Determinants

- Breeding experiments can reveal whether a behavioral phenotype is inherited. A simple Mendelian ratio of traits in offspring indicates a small number of responsible genes for a behavioral difference. **Review Figure 41.3**

- Induced mutations can reveal genetic determinants of a behavior. However, a mutation may affect a behavior because it is a transcription factor influencing the expression of many genes, or because it is a gene in a cascade of gene expression that underlies a behavior.

- Gene knockout experiments can reveal the roles of specific genes underlying a behavior. An example is the knockout of genes responsible for pheromone detection in mice. **Review Figure 41.4**

concept 41.3 Developmental Processes Shape Behavior

- Hormones can determine the pattern of behavior that develops and the timing of its expression. **Review Figure 41.5**

- **Imprinting** is a process by which an animal learns a specific set of stimuli during a limited **critical**, or **sensitive**, **period**. That critical period may be determined by hormones.

- The development and expression of song in white-crowned sparrows involves a genetic predisposition to learn the species-specific song, a critical period for imprinting of a song memory, and hormonally controlled timing of song expression. Social interactions may also play a role. **Review Figure 41.6**

concept 41.4 Physiological Mechanisms Underlie Behavior

- **Circadian rhythms** control the daily cycle of behavior. Without environmental time cues, circadian rhythms free-run with a period that is genetically programmed. They are normally **entrained** to the light–dark cycle by environmental cues. **Review Figure 41.7 and ANIMATED TUTORIAL 41.1**

- Two forms of navigation used by animals to find their way in the environment are piloting (orienting to landmarks) and distance–direction navigation. Bicoordinate navigation may also be used by some animals. Navigation mechanisms include celestial navigation and a time-compensated solar compass. **Review Figures 41.8–41.10, ANIMATED TUTORIAL 41.2, and INTERACTIVE TUTORIAL 41.1**

- The behaviors of individuals may become communication systems if the transmission of information benefits both the sender and the receiver.

- Chemical communication signals (pheromones) can be highly specific and have different time courses depending on their volatility and diffusibility. Visual signals can convey complex messages rapidly, but the recipient must be looking at the sender. Acoustic signals travel well over distances, do not require a focused recipient, and can be modified to reveal or conceal directional information. Tactile signals are used by animals in close proximity and can convey complex messages. **Review Figure 41.11 and WEB ACTIVITY 41.1**

concept 41.5 Individual Behavior Is Shaped by Natural Selection

- An animal's behavior is a series of choices that influence its fitness. To make these choices, animals use environmental cues that are reliable predictors of the potential effects of their choices on their fitness.

- The **cost–benefit approach** can be used to investigate the fitness value of specific behaviors. The cost of a behavior typically has three components: **energetic cost**, **risk cost**, and **opportunity cost**.

- Different types of territorial behavior can be understood in terms of the costs and benefits of defending resources, among them nesting sites, food, and even mates. **Review Figure 41.12 and ANIMATED TUTORIALS 41.3 and 41.4**

concept 41.6 Social Behavior and Social Systems Are Shaped by Natural Selection

- The evolution of mating systems can be understood by looking at the fitness costs and benefits incurred by each partner in the species' environment. **Polygynous** mating systems, in which one male controls and mates with many females, can result in great variation in male reproductive success. **Polyandry**—a female mating with multiple males—can evolve in circumstances in which a male can make a substantial contribution to the survival of his offspring. **Review Figure 41.13**

- The fitness an individual gains by producing offspring (**individual fitness**) plus the fitness it gains by increasing the reproductive success of relatives with whom it shares alleles is called **inclusive fitness**. **Kin selection** may favor **altruistic** behavior toward relatives, despite its cost to the performer, if it increases the performer's inclusive fitness. **Review Figure 41.14**

- As a result of **haplodiploidy**, the sex determination mechanism of the Hymenoptera, nonreproductive workers share more alleles with one another than females share with their own offspring. Haplodiploidy has probably facilitated the evolution of **eusocial behavior** in this group through kin selection. Eusocial behavior has also risen in diploid species in which chances of individual reproductive success are extremely low. **Review Figure 41.15**

- Group living confers benefits such as greater foraging efficiency and protection from predators, but it also has costs, such as increased competition for food and ease of transmission of diseases. **Review Figure 41.16**

See **WEB ACTIVITY 41.2** for a concept review of this chapter.

PART

7

Ecology

Organisms in Their Environment

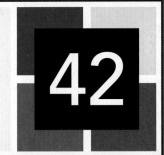

The U.S.–Mexico Borderlands of southern Arizona have a colorful "Wild West" history peopled by Spanish explorers and missionaries, cowboys and Indians, cavalry, outlaws and sheriffs, miners and ranchers, cattle rustlers, and railroad tycoons. Today most of this history is seen only in a few monuments and historic buildings. One period, however, left a lasting legacy on the grasslands of the Borderlands: the cattle boom of the 1880s.

The first herds of cattle appeared in the Borderlands in 1687, when Spanish missions and associated settlements were established. Livestock numbers fluctuated as conflicts between settlers and Indians waxed and waned, but remained below 40,000 head until the 1880s. At that time, several events—the end of the Indian Wars, the completion of transcontinental railroads that opened up huge markets for beef, and an influx of investment capital—combined to bring 1.5 million animals to fatten on the free grass of Arizona's "open range." Within a few years, the cattle grazed to the ground the sea of grass that had once grown "as high as the belly of a horse."

The cattle boom did not last long. A drought in 1892 and 1893 killed 75 percent of the animals, leaving such a density of carcasses that one could walk across the San Simon Valley, it is said, without setting foot on the ground. Prickly, unpalatable shrubs such as mesquite (*Prosopis* spp.) gained a toehold in the denuded landscape when torrential rains returned in the late 1890s and washed away the soil. Rich grassland had been turned into unprofitable shrubland.

Soon thereafter, state and federal governments closed the "open range," substituting a lease system that would allow them to better manage public rangelands. They tried to restore the grass using a method that had worked for thousands of years in the pastures of Europe: adjusting the number of cattle up or down depending on pasture condition. This technique assumes that the primary effect of grazing is to deplete palatable grasses and favor unpalatable ones, and that reducing the number of cattle will allow the palatable species to recover and restore the productive capacity of the pasture. Alas, "resting the range" did not work in the arid Borderlands: shrubs continued to spread, and grasses did not increase measurably.

QUESTION

Why did the rangeland restoration method that worked in Europe fail to work in the Borderlands?

*You will find the answer to this question on page 840.

A cow grazes in the mesquite-dominated "pasture" of southern Arizona. Modern range management techniques have failed to restore this region's grassland ecosystem.

KEY CONCEPTS

42.1 Ecological Systems Vary in Space and over Time

42.2 Climate and Topography Shape Earth's Physical Environments

42.3 Physical Geography Provides the Template for Biogeography

42.4 Geological History Has Shaped the Distributions of Organisms

42.5 Human Activities Affect Ecological Systems on a Global Scale

42.6 Ecological Investigation Depends on Natural History Knowledge and Modeling

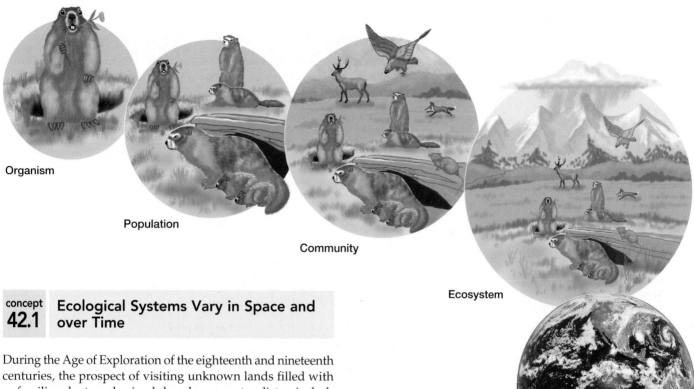

Organism

Population

Community

Ecosystem

Biosphere

concept 42.1 Ecological Systems Vary in Space and over Time

During the Age of Exploration of the eighteenth and nineteenth centuries, the prospect of visiting unknown lands filled with unfamiliar plants and animals lured many naturalists—including Charles Darwin—to travel the world. The environments they saw on their journeys were very different not only from those they were familiar with at home, but from one another. As the early explorer–naturalists observed and catalogued this variation, they soon realized that organisms and their environments are closely linked. Their work laid the foundations for two new fields of study: **physical geography**, the study of the distributions of Earth's climates and surface features; and **biogeography**, the study of the distributions of organisms.

FIGURE 42.1 The Hierarchy of Ecological Systems Ecological systems can include any part of the biological hierarchy from the individual to the biosphere. Each successive level brings in new interacting biotic and abiotic elements and involves progressively larger spatial scales.

Ecological systems comprise organisms plus their external environment

All organisms, humans included, are surrounded by and interact with their external environment. That external environment is made up of nonliving, or **abiotic**, components and living, or **biotic**, components. We can think of one or more organisms plus the external environment with which they interact as an **ecological system**.

The term "ecological" derives from **ecology**, a word that the German biologist Ernst Haeckel constructed in 1866 from the Greek roots *oikos*, "house," and *logia*, "study of." By "house" he meant "the total relations of the animal both to its inorganic [abiotic] and its organic [biotic] environment." Haeckel's new word referred to a very old subject, for even the earliest humans knew a great deal about their food, their enemies, and natural hazards—indeed, the earliest available religious texts and scholarly writings are full of ecological topics. By coining the term "ecology," however, Haeckel helped to establish "all those complex interrelations referred to by Darwin as 'the conditions of the struggle for existence'" as a legitimate subject for scientific study. Furthermore, he emphasized that ecology is relevant to evolution because ecological interactions drive natural selection.

The term "system" derives from the Greek *systema*, meaning "organized whole." Nicolas Léonard Sadi Carnot, a nineteenth-century French physicist who is considered the father of thermodynamics, first introduced the idea of a system as a whole comprising a set of interacting parts in which neither the parts nor the whole can be understood without taking account of the interactions.

Ecological systems can be small or large

A system is defined by the interacting parts it contains. But scientists have a great deal of freedom to choose the set of interacting parts they wish to study (see Figure 1.6). In the case of ecological systems, a boundary can be drawn around any part of the biological hierarchy from the individual on up (**FIGURE 42.1**). Each successive level of the hierarchy brings in new interacting parts at progressively larger spatial scales.

At the smallest scale is the fundamental ecological unit: the individual organism and its immediate environment. Individuals remove materials and energy from the environment, convert those materials and energy into forms that can be used by other organisms, and by their presence and activities, modify the physical environment. In aggregate, these humble exchanges between individuals and their immediate surroundings drive vast fluxes of energy and materials, as we will see in Chapter 46.

The individual organism is part of a **population**: a group of individuals of the same species that live, interact, and interbreed in a particular geographic area at the same time. When individuals reproduce or die, they contribute to changes in the size (number of individuals) of the population of which they are a part. The role a population plays in its ecological system is determined by the average individual's rate of interaction or exchange with its environment, multiplied by the population size.

> **LINK** The population is the unit of evolution; review Concepts 15.2 and 15.3

The assemblage of interacting populations of different species within a particular geographic area forms a biological **community**. As we expand the geographic area we include within the boundary of our ecological system, we include multiple communities in **landscapes** until, at the largest scale, we include all the organisms and environments of the planet—the entire **biosphere**.

In 1935 the British ecologist Sir Arthur Tansley first introduced the idea of an ecological system. He called it an **ecosystem**. Although Tansley's original use of this term referred to all levels of the biological hierarchy from individuals to communities plus their abiotic environment, in modern ecology this term is mostly used to refer to communities plus their physical environment. It was at the community level that the concept was first applied in a concrete way by Raymond Lindeman. In an influential paper published in 1942, Lindeman showed that properties of the community in a freshwater lake resulted from exchanges of matter and energy within and between the biotic and abiotic components of the system. In this chapter we use "ecosystem" more in Tansley's broad sense, as a contraction of "ecological system." In Chapter 46 we consider ecosystems in the modern sense: as communities plus their abiotic environments.

Large ecological systems tend to be more complex than small ones, not only because they contain more interacting parts, but also because those parts interact over a greater range of spatial and temporal scales. The components of the biosphere, for example, are linked not only by the rapid and localized exchanges between individual organisms and their surroundings, but also by slower and more extensive migrations of organisms, great cycles of air and water movement, and the geological churning of Earth's crust.

INVESTIGATION

FIGURE 42.2 The Microbial Community of the Human Gut Depends on the Host's Diet To treat certain digestive disorders, doctors try to increase the relative abundance of beneficial *Bifidobacterium* bacteria in the human gut. This manipulation of the nutritional environment as a way of adjusting the species composition of the gut community is known as *prebiotic therapy*.

HYPOTHESIS

The abundance of beneficial *Bifidobacterium* bacteria in the human gut can be increased relative to pathogenic bacteria by including oligofructose in the diet.

METHOD

1. Feed 8 healthy human subjects a standardized experimental diet for 45 days. On days 1–15 supplement the diet with 15 g sucrose/day ("Sucrose 1"); on days 16–30, substitute 15 g oligofructose for the sucrose; on days 31–45 return to 15 g sucrose ("Sucrose 2").
2. Collect stool samples during each 15-day period and assay for bacterial composition and energy content.

RESULTS

The number of bifidobacteria was highest on the oligofructose diet.

The numbers of these potentially pathogenic bacteria were lowest on the oligofructose diet.

Type of bacteria	Number of bacteria ($\log_{10}$/g stool; mean ± SD)		
	Sucrose 1	Oligofructose	Sucrose 2
Bifidobacteria	**8.8 ± 0.5**	**9.5 ± 0.7**[a]	**8.9 ± 0.9**[b]
Lactobacilli	6.8 ± 1.2	7.0 ± 1.4	7.1 ± 1.0
Coliforms	6.0 ± 1.2	5.9 ± 0.7	5.8 ± 1.0
Gram-positive cocci	5.8 ± 1.0	5.8 ± 0.9	5.5 ± 0.8
Bacteroides	9.4 ± 0.8	8.8 ± 1.1[a]	8.9 ± 0.9[c]
Fusobacteria	8.5 ± 0.7	7.7 ± 0.9[a]	8.1 ± 0.8[c]
Clostridia	8.0 ± 1.2	7.5 ± 0.9[c]	7.7 ± 0.7

[a]Significantly different from sucrose 1 ($P < 0.01$).
[b]Significantly different from oligofructose ($P < 0.01$).
[c]Significantly different from sucrose 1 ($P < 0.05$).

CONCLUSION

Oligofructose stimulated the growth of bifidobacteria at the expense of potentially pathogenic bacteria.

ANALYZE THE DATA

Researchers measured the energy content of the subjects' stools and compared it with the energy content of the ingested oligofructose. The stools contained no oligofructose because it had been fermented—converted into bacterial biomass and waste heat.

	Energy (kJ/day)
Ingested as oligofructose	240
Excreted in stools	77

A. Was all of the oligofructose energy excreted?
B. What happened to the energy that wasn't excreted?
C. What do these observations suggest about the role of gut bacteria in host nutrition?
D. Why did the researchers follow the oligofructose diet with "Sucrose 2"? (Hint: If the microbial community in Sucrose 2 had not moved back toward its composition in Sucrose 1, would the Conclusion be different?)

Go to **yourBioPortal.com** for original citations, discussions, and relevant links for all INVESTIGATION figures.

> **LINK** Scientists can analyze complex microbial ecosystems by sequencing DNA present in environmental samples; see Figure 12.7

Even very small systems, however, can be complex. Consider the ecosystem within each of us. The human large intestine is one of the most densely populated ecosystems on Earth. It contains hundreds of microbial species, and on the order of 10^{12} individual microbes per gram of fecal material. The bacterial cells in our bodies vastly outnumber our own trillion or so human cells, and their combined metabolism rivals that of an organ such as the liver.

It's no surprise that bacteria thrive in the mammalian gut. That environment is regulated within narrow physiological limits, and there is a steady input of nutrients, in the form of food ingested by the host. The gut bacteria interact with their environment as any other organism does. They metabolize food materials that the host cannot digest, and they excrete waste products, some of which are used by other bacterial species or the host. The gut bacteria modify the water and acid content of their environment, provide their host with products such as B vitamins, suppress populations of other, potentially pathogenic bacteria, and interact with epithelial cells to "educate" the immune system.

Each ecological system at each time is potentially unique

The properties of ecological systems depend on their interacting components—biotic and abiotic—and the patterns, strengths, and rates of the interactions among them. Because these components and interactions are distributed unevenly in space and change over time, each system, at any given time, is potentially unique.

The ecosystem of the human gut exemplifies this uniqueness. Even though our knowledge of this particular system is in its infancy, what is emerging from recent studies is that the species of bacteria in the gut are highly diverse (hundreds of species) and vary from person to person and with diet. These observations suggest that both the host's genotype and inputs from the host's diet affect the gut environment from the bacterial point of view. The gut bacteria, in turn, influence their environment, which includes the host. The composition of the gut bacterial community is associated with the presence or absence of various health problems in the host, such as gastric ulcers, irritable bowel syndrome, and obesity. These interactions between the bacterial community and the host affect the properties of the entire system. For example, bacteria in the genus *Bifidobacterium* are human gut microbes that are particularly adept at fermenting oligofructose—a carbohydrate that humans can't digest. Bifidobacteria are also beneficial because they suppress pathogenic bacteria and supply nutrients.

Researchers are actively investigating whether certain health disorders can be treated by manipulating the "ecosystem within." Administering live bifidobacteria by mouth—a strategy called *probiotic therapy*—met with limited success because few bacteria survived the low pH of the stomach to reach the gut. Scientists then tested an alternative strategy known as *prebiotic therapy*: stimulating growth of an already-resident *Bifidobacterium* population by including oligofructose in the patient's diet (**FIGURE 42.2**).

Do You Understand Concept 42.1?

- What two components make up the external environment of organisms?
- Explain why ecological systems at large scales tend to be more complex than smaller-scale systems.
- How might advances in medicine arise from considering the human gut as an ecosystem?

Variation over space and time can be seen in ecosystems at all scales, from the human gut to the entire biosphere. Next we turn to some of the physical factors that underlie this heterogeneity at the biosphere scale: Earth's climates and topography.

concept 42.2 Climate and Topography Shape Earth's Physical Environments

Variation in Earth's physical environments results from patterns of circulation in the atmosphere and the oceans and from the geological processes that sculpt the surface of the planet. It is these processes that determine whether a particular place is hot or cold, moist or dry, aquatic or terrestrial.

Latitudinal gradients in solar energy input drive climate patterns

Atmospheric conditions—temperature, humidity, precipitation (rainfall or snowfall), and wind direction and speed—are important features of terrestrial environments. We refer to the state of atmospheric conditions in a particular place at a particular time as **weather**, and to their average state and pattern of variation over longer periods as **climate**. In other words, climate is what you expect; weather is what you get. The responses of organisms to weather are usually short-term; for example, animals may seek shelter from a sudden rainstorm, and plants may close their stomata on a hot summer day. Responses to climate, on the other hand, tend to involve adaptations that prepare organisms for expected weather patterns. Organisms in hot climates have high thermal tolerances; those that inhabit seasonal climates use cues such as day length to prepare for an impending cold or dry season.

Globally, air temperatures decrease from low latitudes (near the equator) to high latitudes (near the poles). Averaged over the course of the year, air temperature decreases by about 0.8°C with every degree of latitude (about 110 km) north or south of the equator. High latitudes also experience more pronounced **seasonality**—that is, greater fluctuations in temperature over the course of a year—than low latitudes. These global climatic patterns are the result of uneven input of solar radiation. That uneven input is caused by the shape of Earth and by the orientation of the axis around which it spins relative to Earth's orbit around the sun.

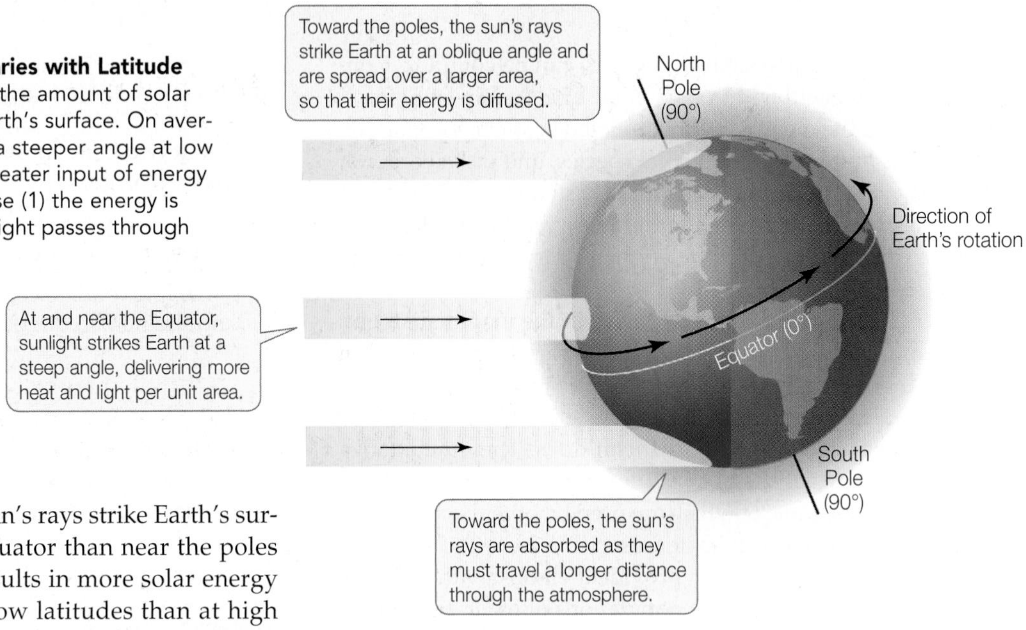

FIGURE 42.3 Solar Energy Input Varies with Latitude
The angle of incoming sunlight affects the amount of solar energy that reaches a given area of Earth's surface. On average, sunlight strikes Earth's surface at a steeper angle at low than at high latitudes. This results in greater input of energy at low (i.e., equatorial) latitudes because (1) the energy is spread over a smaller area and (2) sunlight passes through less atmosphere.

> Toward the poles, the sun's rays strike Earth at an oblique angle and are spread over a larger area, so that their energy is diffused.

North Pole (90°)

Direction of Earth's rotation

Equator (0°)

South Pole (90°)

> At and near the Equator, sunlight strikes Earth at a steep angle, delivering more heat and light per unit area.

> Toward the poles, the sun's rays are absorbed as they must travel a longer distance through the atmosphere.

Because Earth is spherical, the sun's rays strike Earth's surface at a steeper angle near the equator than near the poles (**FIGURE 42.3**). This difference results in more solar energy input per unit of surface area at low latitudes than at high latitudes, for two reasons. First, the energy of perpendicular rays of sunlight is spread over a smaller surface area. Second, perpendicular rays of sunlight pass through less of Earth's atmosphere, so less of their energy is absorbed and reflected by the atmosphere before it reaches the ground.

Seasonality occurs because Earth's axis of rotation is tilted at an angle of approximately 23.5 degrees relative to its orbit around the sun. This tilt causes different latitudes to receive their greatest solar energy input at different times during the

yearly orbit, in a yearly cycle of solstice and equinox (**FIGURE 42.4**). In late June of every year, the North Pole tilts toward the sun, and the sun is directly over the Tropic of Cancer (latitude 23.5°N). At this time the Northern Hemisphere is basking in the long, warm days of summer, and the Southern Hemisphere is shivering in the long, cold nights of winter. In late December of every year, the North Pole points away from the sun, and the Tropic of Capricorn (latitude 23.5°S) receives the greatest solar input; then it is winter in the Northern Hemisphere and summer in the Southern Hemisphere. In late March and late September, the sun is directly above the equator, and both hemispheres experience equal lengths of day and night.

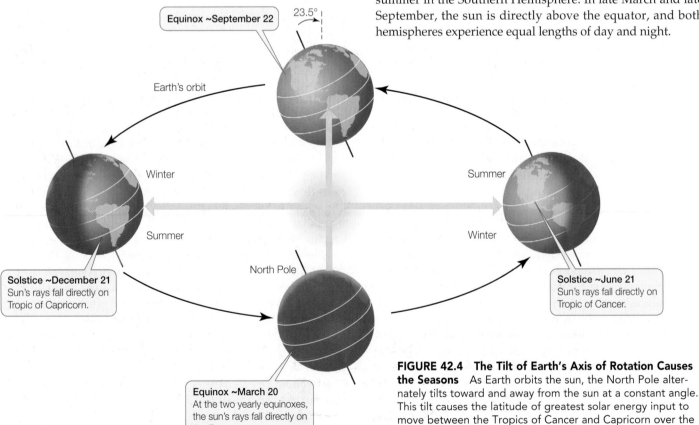

Equinox ~September 22

23.5°

Earth's orbit

Winter

Summer

Summer

Winter

North Pole

Solstice ~December 21
Sun's rays fall directly on Tropic of Capricorn.

Solstice ~June 21
Sun's rays fall directly on Tropic of Cancer.

Equinox ~March 20
At the two yearly equinoxes, the sun's rays fall directly on the Equator.

FIGURE 42.4 The Tilt of Earth's Axis of Rotation Causes the Seasons As Earth orbits the sun, the North Pole alternately tilts toward and away from the sun at a constant angle. This tilt causes the latitude of greatest solar energy input to move between the Tropics of Cancer and Capricorn over the course of a year.

Solar energy drives global air circulation patterns

Regardless of seasonal latitudinal shifts in solar input, the total input of solar energy is always greatest between the two Tropics (see Figure 42.3). This solar energy drives a vast global pattern of atmospheric circulation that defines broad latitudinal patterns of surface winds and precipitation.

As the sun beats down on the tropics, the air there absorbs radiant energy and heats up—that is, its molecules move faster (recall that temperature is a measure of disordered molecular motion), colliding and pushing one another farther apart. The heated air expands, becoming less dense and more buoyant, and rises. As it rises, it continues to expand and cool **adiabatically** (this is the principle on which an air conditioner works: when compressed gas is allowed to expand, it cools), until it reaches the same low temperature and density as the surrounding air, high in the atmosphere. Because tropical air is the warmest, it boils up to greater altitudes before it stops rising than does high-latitude air.

This rising tropical air is replaced by cooler, denser surface air that flows in from the north and south. As that air heats, expands, and rises, it pushes the cool air high above the tropics away from the equator. No longer buoyed by rising tropical air, the air aloft eventually sinks back toward Earth. As it descends, it is compressed by overlying atmosphere and warms adiabatically. It reaches Earth's surface at latitudes of roughly 30°N and 30°S, at which point some of it flows back toward the equator to replace air that is rising in the tropics. Thus it completes two patterns of vertical atmospheric circulation—one north and one south of the equator—that we call **Hadley cells**. The Hadley cells give rise to similar air circulation patterns at higher latitudes (**FIGURE 42.5**). Some of the air that descends at 30°N and 30°S flows poleward. At about 60°N and 60°S, this poleward-flowing surface air rises again and flows aloft either toward the equator or poleward. At the poles, where there is little solar energy input, cold air descends and flows toward

the equator. Because of these three latitudinal circulation cells, surface air flows toward the equator between 30°N and 30°S latitudes, toward the poles between 30° and 60°N and S, and again toward the equator above 60°N and 60°S latitude.

These vertical atmospheric circulation cells influence latitudinal patterns of precipitation as well as surface winds. As water molecules absorb the intense tropical sunlight, they evaporate and the tropical air becomes rich in the resulting water vapor. As this warm, moist tropical air expands, rises, and cools, fewer water molecules bounce apart after they collide. Instead they form droplets of liquid water (or ice crystals), which grow and—when they become massive enough—fall to the ground as precipitation. Thus the warm, rising air of the Hadley cells drops most of its moisture on the tropics as rain. The high-altitude air that eventually descends at about 30°N and 30°S is largely depleted of water vapor; moreover, because it warms as it descends and is compressed, the little water that remains stays in a gaseous state. Earth's great deserts are located at these dry subtropical latitudes.

The prevailing wind patterns at Earth's surface reflect Earth's counterclockwise rotation as seen from the North Pole (see Figure 42.3), as well as the north–south atmospheric circulation (see Figure 42.5). Because Earth is a sphere, the rotation of its surface is fastest at the equator, where its circumference is greatest, and slowest at the poles. Air that is not moving north or south has the same speed of eastward rotation as the surface beneath it. But as an air mass moves toward the equator, it lags behind the faster-moving surface beneath it and is therefore deflected to the west. Conversely, the rotational movement of an air mass moving poleward is faster than that of Earth beneath it, and it is deflected to the east. This *Coriolis effect* causes prevailing surface winds to blow from east to west in the tropics

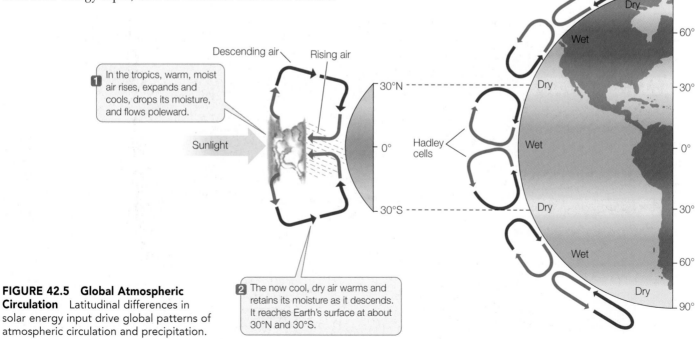

FIGURE 42.5 Global Atmospheric Circulation Latitudinal differences in solar energy input drive global patterns of atmospheric circulation and precipitation.

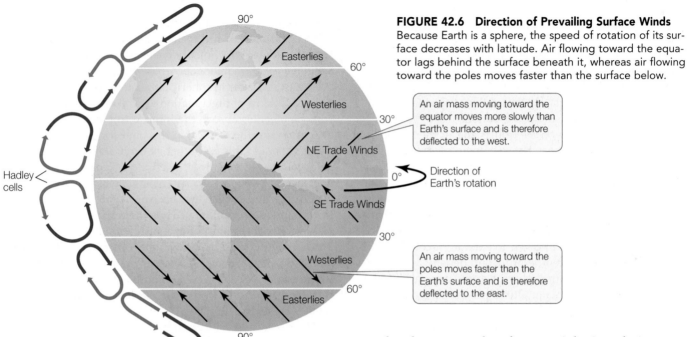

FIGURE 42.6 Direction of Prevailing Surface Winds
Because Earth is a sphere, the speed of rotation of its surface decreases with latitude. Air flowing toward the equator lags behind the surface beneath it, whereas air flowing toward the poles moves faster than the surface below.

An air mass moving toward the equator moves more slowly than Earth's surface and is therefore deflected to the west.

Direction of Earth's rotation

An air mass moving toward the poles moves faster than the Earth's surface and is therefore deflected to the east.

(the northeast or southeast *trade winds*), from west to east in the mid-latitudes (the *westerlies*), and from east to west again above 60°N or 60°S latitude (the *easterlies*; **FIGURE 42.6**).

The spatial arrangement of continents and oceans influences climate

The prevailing winds drive massive circulation patterns in the surface waters of the oceans, known as **currents** (**FIGURE 42.7**). Winds move the water over which they blow by frictional drag. In the tropics, for example, the northeast trade winds drag surface waters to the west. When this westward-moving water

reaches the western edge of an oceanic basin and encounters a continent, most of it flows northward until the westerlies of the northern temperate zone drag it eastward again, completing a circuit. The water that doesn't flow north sloshes back to the east near the equator because the trade winds have pushed it to a higher sea level, and it flows downhill. The pattern in the Southern Hemisphere is essentially a mirror image of that just described.

Surface currents are not the whole story. Gradients in water density caused by latitudinal variation in temperature and salinity cause surface waters to sink in certain regions, forming deep currents. These deep currents regain the surface in areas

FIGURE 42.7 Ocean Currents Ocean surface currents are driven by prevailing winds (see Figure 42.6) and are deflected by continents.

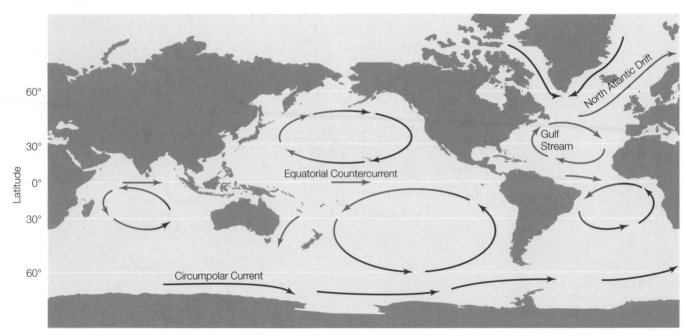

of upwelling, completing a vertical ocean circulation. Thus oceanic circulation is complex and three-dimensional.

Oceans and large lakes moderate Earth's terrestrial climates because water has a high heat capacity, as we saw in Chapter 2. This means that the temperature of water changes relatively slowly as it exchanges heat with the air it contacts and as it absorbs or emits radiant energy. As a result, water temperatures fluctuate less with the seasons and with the day–night cycle than do land temperatures, and the air over land close to oceans or lakes also shows less daily and seasonal temperature fluctuation.

Furthermore, poleward-flowing ocean currents carry heat from the tropics toward the poles, moderating the climates of higher latitudes. The Gulf Stream and North Atlantic Drift, for example, bring warm water from the tropical Atlantic Ocean and the Gulf of Mexico north and east across the Atlantic, warming the air above the North Atlantic (see Figure 42.7). Prevailing westerlies then carry this warmed air across northern Europe, moderating temperatures there.

Walter climate diagrams summarize climate in an ecologically relevant way

The climate at any location can be summarized in a **climate diagram**. Modern climate diagrams were devised by the German biogeographer Heinrich Walter as a way of graphically summarizing the climate in a given location. Recognizing that ecological processes are jointly influenced by temperature and moisture, Walter superimposed graphs of average monthly temperature and precipitation through the year. He scaled the axes of the two graphs to incorporate the rule of thumb that plant growth requires at least 20 mm of precipitation per 10°C of temperature above 0°C. This scaling makes it easy to see when conditions favor terrestrial plant growth—when temperature is greater than 0°C and the precipitation line is above the temperature line—and when they do not (**FIGURE 42.8**).

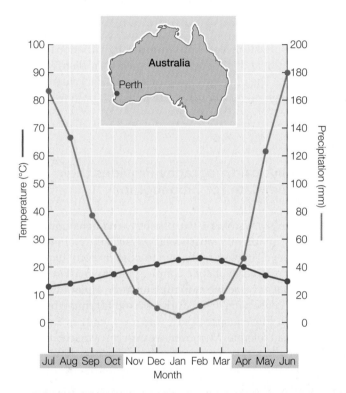

FIGURE 42.8 Walter Climate Diagrams Summarize Climate in an Ecologically Relevant Way Walter diagrams traditionally combine average temperature (left-hand axis) and precipitation (right-hand axis) throughout the year, as shown here for Perth, Western Australia. The summer solstice (December in Perth) is placed at the center of the diagram. The axes are scaled so that precipitation favors plant growth when the precipitation line is above the temperature line. Months highlighted in green typically are favorable for plant growth— temperatures are above-freezing and precipitation is adequate.

APPLY THE CONCEPT

Climate shapes Earth's physical environments

Shown below are the Walter climate diagrams for three different locations. Using the information in these diagrams and what you've learned about the various factors that determine a location's climate, answer the following questions and explain the logic that led to your answer.

1. Which location is near the equator?

2. Which location is close to 30°N?

3. Which location has the greatest seasonality?

4. Which location has the longest period favorable for plant growth?

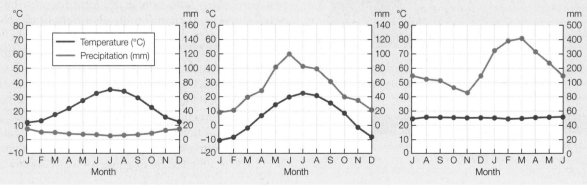

Topography produces additional environmental heterogeneity

Earth's surface is not flat; it is rumpled and crumpled into mountains, valleys, and ocean basins. This variation in the elevation of Earth's surface, known as **topography**, affects the physical environment.

Mountains, for example, are "weather formers." Because rising air cools adiabatically, temperatures become cooler as elevation (distance above sea level) increases. If you climb a mountain, you will find that the temperature drops by about 1°C for each 220 m of elevation gained. Where prevailing winds blow over a mountain range, the mountains force the air to rise, cool, and drop its moisture as precipitation. Abundant precipitation on the windward side (the side facing into the wind) of a mountain range may lead to the growth of lush vegetation. On the leeward side (the side away from the wind), air descends and warms, and little precipitation falls (**FIGURE 42.9**). The resulting dry area on the leeward side of a mountain range is known as a "rain shadow."

Topography also influences physical conditions in aquatic environments. The velocity of water flow, for example, is dictated by topography: water flows rapidly down steep slopes, more slowly down gradual slopes, and pools as lakes, ponds, or oceans in depressions. The depth of a water-filled depression determines gradients of many abiotic factors, including temperature, pressure, light penetration, and water movement (see Figure 42.13).

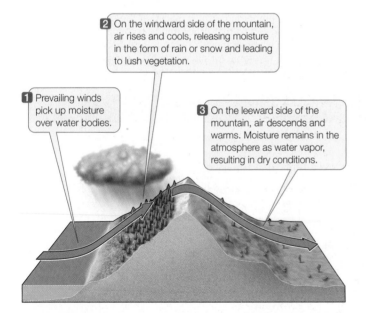

2 On the windward side of the mountain, air rises and cools, releasing moisture in the form of rain or snow and leading to lush vegetation.

1 Prevailing winds pick up moisture over water bodies.

3 On the leeward side of the mountain, air descends and warms. Moisture remains in the atmosphere as water vapor, resulting in dry conditions.

FIGURE 42.9 A Rain Shadow Precipitation tends to be lower on the leeward side of a mountain range than on the windward side, particularly when oceans lie upwind of mountain ranges.

yourBioPortal.com

Go to ANIMATED TUTORIAL 42.1
Rain Shadow

How do the physical conditions imposed by variations in solar energy input and in topography influence the distributions of organisms? The physical conditions in any particular place, and their range of variation, determine which organisms can live there.

concept 42.3 Physical Geography Provides the Template for Biogeography

An organism's physiology, morphology, and behavior affect how well it can tolerate a particular physical environment. A lizard that needs to bask in the sun to raise its body temperature to an operative level, for example, cannot persist in shady environments or be active at night. Similarly, a plant that has no means of conserving water during droughts cannot thrive in a desert. As a consequence, species are found only in environments they can tolerate. Moreover, we expect species that occur in the same physical environment to have evolved means of coping with the challenges that environment presents.

LINK Some adaptations of plants to challenging climatic conditions are described in Concept 28.3

Similarities in terrestrial vegetation led to the biome concept

The naturalist–explorers of the eighteenth and nineteenth centuries laid the foundations for understanding how the distribution of Earth's physical environments shapes the distribution of organisms. The German explorer Alexander von Humboldt was particularly influential in this regard because he explicitly related his observations on the distributions of plants to his painstaking measurements of physical conditions, including latitudinal and elevational gradients in temperature. Humboldt's observations, and those of other naturalists, eventually revealed a striking convergence in the characteristics of the vegetation found in similar climates around the world.

This convergence led to the concept of a **biome**: a distinct physical environment that is inhabited by ecologically similar organisms with similar adaptations. Often the species that occupy the same biome in geographically separate regions

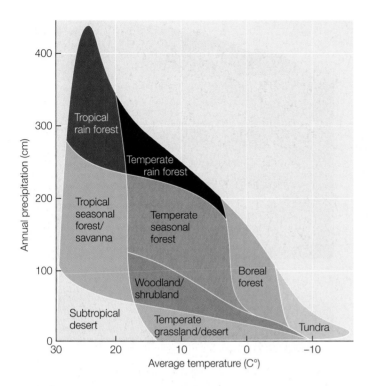

FIGURE 42.10 Temperature and Precipitation Gradients Determine Terrestrial Biomes Average annual temperature and precipitation determine terrestrial vegetation structure of a given locality.

The distribution of terrestrial biomes is broadly determined by annual patterns of temperature and precipitation (**FIGURE 42.10**). Latitudinal gradients in average temperature and its seasonal fluctuation separate tropical, subtropical, temperate, boreal, and polar biomes. Within those broad temperature zones, average annual precipitation and its seasonal fluctuation define biome types. In tropical biomes, for example, the length of the dry season determines whether tropical rainforest, tropical seasonal forest, or savanna develops. Elevational gradients in temperature also determine the distribution of biomes; for example, tundra can be found on mountaintops even in the tropics (**FIGURE 42.11**).

Since biomes were first recognized, ecologists have compared the attributes of organisms within and among biomes to understand how the physical environment influences their adaptations. This approach has led to the discovery that features of the physical environment other than climate—particularly soil characteristics—interact with climate to influence the character of vegetation. Southwestern Australia, for example, is one of a number of places around the world with a Mediterranean

are not closely related phylogenetically. Their morphological, physiological, or behavioral similarities therefore reflect *convergent evolution* in response to similar natural selection (see Concept 16.1). Terrestrial biomes are generally distinguished by the characteristics of their vegetation—whether the dominant plants are woody trees or shrubs, grasses, or broad-leaved herbaceous plants, and whether their leaves are big or small, deciduous (dropped seasonally) or evergreen. The attributes of other organisms vary among biomes as well, however, since other organisms are directly affected by climate, as well as manifesting indirect effects in response to the climatic adaptations of vegetation.

FIGURE 42.11 Global Terrestrial Biomes The global distribution of terrestrial biomes reflects latitudinal and elevational gradients in temperature and precipitation. The same biomes can be found in widely separated locations.

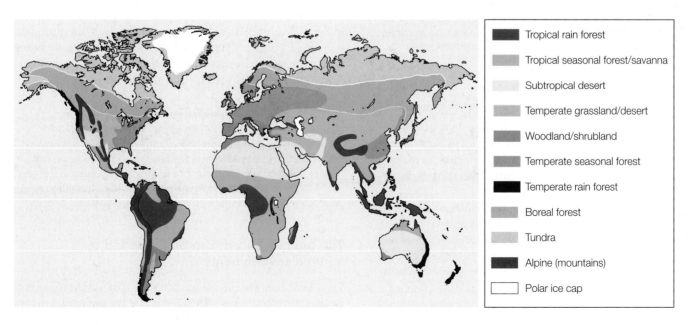

(A)

(B)

(C)

(D)

FIGURE 42.12 Same Biome, Different Continents The Mediterranean climate (described by the climate diagram in Figure 42.8) produces a convergent woodland/shrubland vegetation that is found throughout the world. (A) Southwestern Australia's Fitzgerald River National Park. (B) "Maritime" shrubland on the windward side of a California mountain, North America. (C) The region bordering the Mediterranean Sea gave this climate type its name; this "bloom of broom" is in Croatia. (D) A unique flora known as *fynbos* is found in the Cape region of South Africa.

climate in which summers are hot and dry and winters are cool and moist (see Figure 42.8). This climate supports woodland/shrubland vegetation (**FIGURE 42.12**). Succulent plants, which are well adapted to summer drought, are common in many of these places but not in southwestern Australia. What is responsible for this difference?

Southwestern Australia has extremely nutrient-poor soils. The lack of nitrogen in these soils makes it expensive for plants to construct the nitrogen-rich photosynthetic machinery of new leaves. As a result, the leaves of many plants in the region are long-lived, and they are defended against herbivores by chemical compounds that do not contain nitrogen. These defensive compounds include tough, indigestible components of plant cell walls such as lignin or silica, compounds such as tannins that reduce the nutritional quality of leaves, and toxic oils and

resins (see Concept 28.2). These defensive compounds make plant material not only very slow to decompose, so that it accumulates in the ecosystem, but also highly flammable. Plant material produced during the winter rains dries out during the summer drought and fuels intense fires that periodically sweep across the landscape. As a result, succulent plants, which are easily killed by fires, are rare in southwestern Australia.

FRONTIERS Plant scientists have formed a global consortium called Glopnet to gather and make digitally available new information about patterns in the chemical, structural, and physiological traits of leaves. The hope is that understanding how climate interacts with the adaptive characteristics of leaves will allow them to predict how biome boundaries will change with changing land use and climate.

The biome concept can be extended to aquatic environments

The biome concept can be applied to aquatic environments as well as terrestrial ones (**TABLE 42.1**). Like terrestrial organisms, aquatic organisms show convergent adaptations to similar physical environments. In aquatic environments, however,

TABLE 42.1 Major Aquatic Biomes

BIOME[a]	DESCRIPTION
FRESHWATER	
Rivers and streams	Flowing water. Many fast-flowing, small *source streams* form on high ground, feeding into networks of ever larger, slower-flowing streams and rivers. Biota adapted to constantly moving water.
Wetlands	Glades, swamps, and marshes. Rich biota adapted to water-saturated soil and/or standing fresh water.
Ponds and lakes	Significant bodies of standing fresh water. Ponds are smaller and shallower, subject to drying. Biotic zones determined by distance from shore and light penetration (see Figure 42.13A).
ESTUARIES[b]	
Salt marshes	Cool-temperate stands of salt-tolerant grasses, herbaceous plants, and low-growing shrubs. Crucial to nutrient cycling and coastal protection; rich habitat supporting diverse aquatic and terrestrial life.
Mangrove forests	Tropical and warm subtropical coasts and river deltas. Dominated by mangrove trees with aerial roots (see Figure 28.11A). Rich in animal life; protect against coastal erosion.
MARINE	
Intertidal	Sandy or rocky coastlines subject to rising and falling tides; organisms adapted to withstand both submerged and dry conditions, as well as the force of waves and moving water.
Kelp forests	Found in shallow coastal waters of temperate and cold regions. Dominated by large, leaflike brown algae (kelp) that support a wide variety of marine life.
Seagrass beds	"Meadows" of monocot grasses (see Figure 21.29C) found in shallow, light-filled temperate and tropical waters.
Coral reefs	Rich, highly endangered ecosystems of shallow tropical waters. Dependent on cnidarian corals (see Figure 23.9C) and their photosynthetic endosymbionts (see Concept 20.4).
Open ocean	The *pelagic zone* (see Figure 42.13B) is rich in photosynthetic planktonic organisms that support a host of marine animals. Below the level of light penetration, the *abyssal zone* supports a fauna largely dependent on detritus that sinks down from pelagic regions.
Hydrothermal vents	Abyssal ecosystems warmed by volcanic emissions. Chemolithotrophic prokaryotes (see Concept 19.3) nourish large annelid worms (see Figure 23.17B) and other invertebrates.

[a] A *benthic* region—silt, sand, or other substrate and the organisms encompassed there—occurs in all three biome types.

[b] *Estuaries* are coastal biomes where the water is brackish (i.e., fresh and salt water mix).

there is no structurally dominant group of organisms like terrestrial plants that can be used to distinguish biomes, and climate is less of a distinguishing physical factor than are other abiotic factors such as water depth and movement, temperature, pressure, salinity, and characteristics of the substrate.

The primary distinction among aquatic biomes is salinity, which distinguishes freshwater (streams, ponds, and lakes), saltwater (salt lakes and oceans), and estuarine (at river mouths where fresh and salt water mix) biomes. Salinity determines what organisms are found in a biome because the salt concentration in an aquatic organism's external environment strongly affects its ability to osmoregulate (see Concept 40.1).

Freshwater biomes can be categorized by water movement. Streams form wherever precipitation exceeds evaporation, as water that does not penetrate the soil flows downhill. These streams may be small initially, but they grow as they join other streams, forming rivers that eventually reach a lake or ocean. Within streams, the velocity of the water's flow determines the strength of the current against which fish must swim and the amount of force that can dislodge bottom-dwelling organisms from the streambed. It also affects the characteristics of the stream bottom. Rapid water flow scours away sediment, exposing rocky surfaces to which organisms cling. As water slows, it deposits sediment, forming a soft bottom in which organisms can burrow.

Still-water biomes such as lakes and oceans can be divided into water-depth zones (**FIGURE 42.13**). The nearshore regions of lakes (**littoral zone**) and oceans (littoral or **intertidal zone**) are shallow, affected by wave action, and periodically exposed to air by fluctuations in water level. There is often conspicuous vertical zonation of species here according to their tolerance of heat, desiccation, and wave energy.

Because only the surface water is in contact with air, dissolved oxygen concentrations are highest in surface waters. Light, too, penetrates only a short distance into water. Photosynthetic organisms are confined to this zone of light penetration, called the **photic zone**. Aquatic vascular plants and their multicellular algal equivalents (as well as corals with their photosynthetic algal symbionts) grow where the photic zone extends to the bottom, sometimes forming communities with a complex vertical structure much like that of terrestrial plant communities. In the open-water **limnetic zone** of lakes and the **pelagic zone** of oceans beyond the continental shelf, the prominent photosynthesizers are phytoplankton.

No photosynthetic organisms occur in the **aphotic zone**, below the reach of light, which as a consequence is sparsely populated. The lake bottom or ocean floor is called the **benthic zone**. Water is heavy, and pressure increases with water depth. Temperature generally decreases with depth, because cold water is denser than warm water and sinks. Organisms that dwell

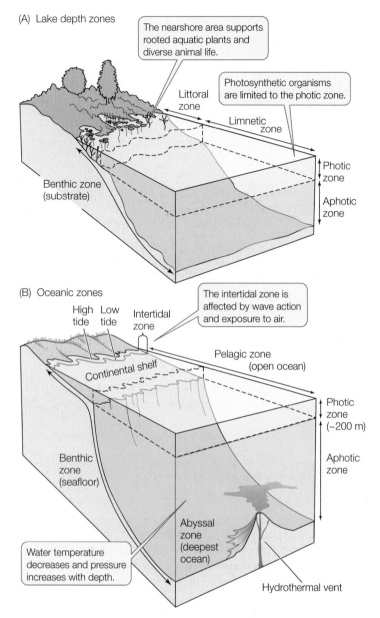

FIGURE 42.13 Water-Depth Zones Freshwater (A) and marine (B) environments can be divided into water-depth zones.

in the deepest **abyssal zone** of the oceans experience very high pressures, low oxygen levels, and (except near hydrothermal vents) cold temperatures.

Do You Understand Concept 42.3?

- What is meant by a *biome*?
- What conditions limit the distribution of photosynthetic organisms in aquatic biomes? In terrestrial biomes?
- Is it legitimate to consider the human gut a biome? Explain your answer.

We have just seen how climate, topography, and other abiotic factors influence the distributions of species. But those factors have changed over Earth's history. How have those changes affected the distributions of species?

concept 42.4 Geological History Has Shaped the Distributions of Organisms

The naturalist–explorers of the eighteenth and nineteenth centuries showed that the physical environment shapes the distributions of organisms. But it soon became clear to them that there were other factors at work as well. The observations of one explorer suggested an important role for geological history in the distributions of organisms.

Barriers to dispersal affect the distributions of species

Alfred Russel Wallace—who, along with Charles Darwin, advanced the idea that natural selection could account for the evolution of life's diversity (see Concept 15.1)—first noticed an odd pattern of species distributions during a seven-year exploration of the Malay Archipelago. He observed that dramatically different bird faunas inhabited two neighboring islands, Bali and Lombok. Wallace pointed out that these differences could not be explained by climate or by soil characteristics, because in those respects the two islands are equivalent. Instead, Wallace suggested that the Malay Archipelago was divided into two distinct halves by a line (now known as Wallace's line) that follows a deep-water channel separating Bali and Lombok (**FIGURE 42.14**). This channel is so deep that it would have remained full of water—and thus would have been a barrier to the movement of terrestrial animals—even during the glaciations of the Pleistocene era, when sea level dropped more than 100 m and Bali and the islands to its west were connected to the Asian mainland. As a consequence, the faunas on either side of Wallace's line evolved mostly in isolation over a long period.

The movement of continents accounts for biogeographic regions

Wallace's observations of animal distributions led him to divide the world into six continental-scale areas called **biogeographic regions**. Each biogeographic region encompasses multiple biomes and contains a distinct assemblage of species, many of which are phylogenetically related. Athough Wallace based his region's boundaries on animal distributions, those based on plants have similar boundaries, and the biogeographic regions as we define them today have changed little from those first proposed by Wallace (**FIGURE 42.15**). Many of the boundaries correspond to geographic barriers to movement. These include bodies of water (for terrestrial organisms); areas with extreme climates; and physical barriers such as mountain ranges. At Wallace's line, for example, the Oriental biogeographic region is separated from the Australasian biogeographic region by the deep-water channel Wallace described.

Not all biogeographic regions are bounded by obvious barriers, however, and some of them span several continents.

■ Current land surface

░ Continental shelf (exposed during the Pleistocene)

▓ Deep water (≥ 200 m below current sea level)

FIGURE 42.14 Wallace's Line Wallace's line corresponds to a deep-water channel that runs between the islands of Bali and Lombok. This channel would have blocked the movement of terrestrial organisms even during the Pleistocene glaciations, when sea level was 100 m lower than it is today.

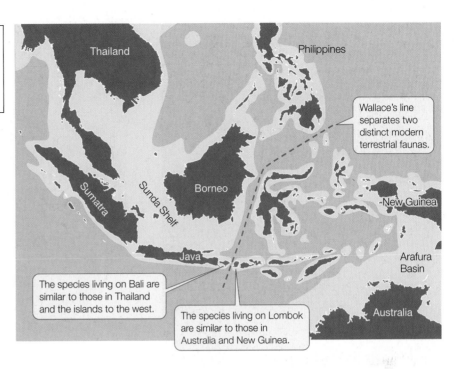

Wallace's line separates two distinct modern terrestrial faunas.

The species living on Bali are similar to those in Thailand and the islands to the west.

The species living on Lombok are similar to those in Australia and New Guinea.

FIGURE 42.15   Movement of the Continents Shaped Earth's Biogeographic Regions The biogeographic regions we recognize today are essentially the six regions Wallace proposed based on their distinctive assemblages of animals; modern biogeographers add an Antarctic region. Today we know that the boundaries delineating these regions are largely the result of continental drift. On this map, red arrows show the time (in millions of years) since land masses came together. Black arrows show the time since land masses separated.

yourBioPortal.com

Go to WEB ACTIVITY 42.1
Major Biogeographic Regions

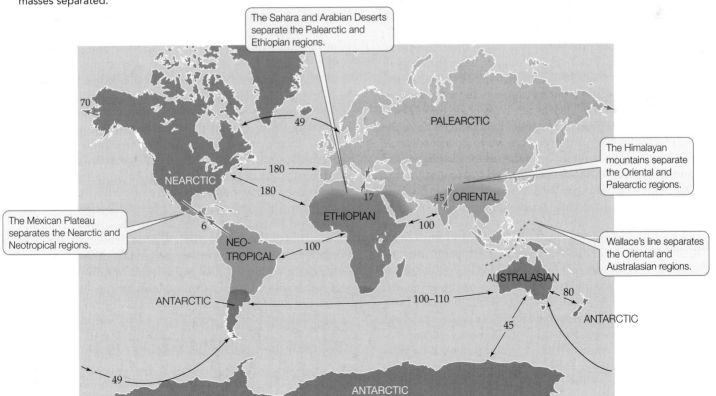

The Sahara and Arabian Deserts separate the Palearctic and Ethiopian regions.

The Himalayan mountains separate the Oriental and Palearctic regions.

The Mexican Plateau separates the Nearctic and Neotropical regions.

Wallace's line separates the Oriental and Australasian regions.

Nothofagus sp.

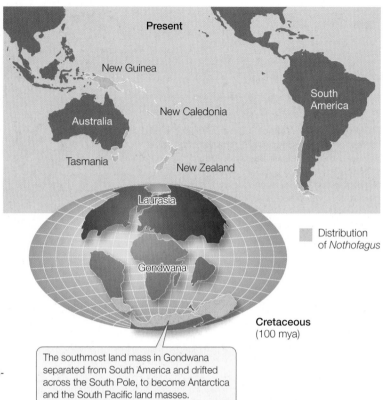

The southmost land mass in Gondwana separated from South America and drifted across the South Pole, to become Antarctica and the South Pacific land masses.

FIGURE 42.16 Distribution of *Nothofagus* The modern distribution of southern beeches is best explained by their origin on Gondwana during the Cretaceous period. The breakup of Gondwana and subsequent continental drift resulted in the modern distribution of *Nothofagus* in South America, Australia, New Zealand, and some islands of the South Pacific.

Why, for example, are some groups of organisms found at the southern tips of South America and Africa and in New Zealand? Conversely, why are there major biogeographic boundaries in Mexico, northern Africa, and Asia?

These aspects of biogeographic patterns remained a mystery until the second half of the twentieth century, when the theory of continental drift became widely accepted. Until then, biogeographers could only propose arbitrary explanations for the distributions of many organisms, such as rare cross-ocean colonization events or the transient appearances of land bridges. Once the movements of the continents over time were understood, however, it became clear that they have shaped the distributions of organisms by creating and disrupting dispersal routes.

One distribution that was puzzling to early biogeographers was that of southern beeches—trees of the genus *Nothofagus*. These trees are found in South America, New Zealand, Australia, and on certain islands of the southern Pacific Ocean (**FIGURE 42.16**). Their distribution was once thought to be the product of several separate transoceanic dispersal events. A more parsimonious explanation is that the genus originated on the southern supercontinent Gondwana during the Cretaceous period and was carried along when Gondwana broke apart and its fragments drifted to their present locations. Paleontologists have found fossilized *Nothofagus* pollen dated at over 80 mya in Australia, New Zealand, Antarctica, and South America. This fossil evidence comes from before the breakup of these portions of Gondwana, suggesting that *Nothofagus* was once widely distributed across that supercontinent.

Our understanding of continental drift makes clear that the major biogeographic regions occupy land masses that have been isolated from one another for most of the Cenozoic era—long enough to allow the organisms present at the time of isolation to undergo independent evolutionary radiations. The northern biogeographic regions (Nearctic and Palearctic) became isolated from the southern regions (Neotropical, Ethiopian, Oriental, Australasian, and Antarctic) during the Jurassic period, about 200 mya, when Pangaea broke into two great land masses, Laurasia to the north and Gondwana to the south (see Figure 18.12). The southern biogeographic regions subsequently became isolated from one another when Gondwana began to break apart during the Cretaceous period. The Neotropical, Ethiopian, Oriental, and Australasian regions were separated approximately 100 mya, and Antarctica became isolated from Australia around 45 mya. About 49 mya, the Nearctic and Palearctic regions became isolated by the opening of the Atlantic Ocean, except for a periodic connection via the Bering land bridge between Asia and North America. Thus the biotas of the seven biogeographic regions developed largely in isolation throughout the Tertiary period (about 65 to 1.8 mya), when extensive evolutionary radiations of flowering plants and vertebrates took place (see Table 18.1).

LINK The movements of the continents over geological history are depicted in Figure 18.12

Continued movements of continents have more recently eliminated some barriers to dispersal and have caused mixing of species, referred to as **biotic interchange**. The Oriental biogeographic region was formed when India collided with Asia approximately 45 mya, allowing mixing of organisms that had previously been isolated. Interchange between Oriental and Palearctic regions was prevented by formation of the Himalayas when the Indian and Asian plates collided. Exchanges of organisms between the Ethiopian and the Palearctic regions became possible when Africa collided with Eurasia about 17 mya, but the Sahara and Arabian deserts present a climatic barrier to extensive exchange. Similarly, biotic interchange between the Neotropical and the Nearctic regions began when a land bridge formed between South America and North America about 6 mya. During this Great American Interchange, many North American species displaced South American lineages and drove them to extinction. Currently the Mexican Plateau provides a barrier to the northward dispersal of tropical species.

Phylogenetic methods contribute to our understanding of biogeography

As we saw in Chapter 16, taxonomists have developed powerful methods of reconstructing phylogenetic relationships among organisms. Biogeographers use phylogenetic information, in conjunction with the fossil record and geological history, to address questions about how the modern distributions of organisms came about. They do so by superimposing on the phylogenetic tree the geographic area that is or has been occupied by a taxon. By comparing the sequence and timing of splits in the phylogenetic tree with the sequence of separation or connection of geographic areas, biogeographers can determine where a lineage originated and reconstruct the history of its diversification and dispersal. For example, if a phylogenetic split coincides in time with the appearance of a barrier to

APPLY THE CONCEPT

Geological history has shaped the distributions of organisms

Below is a phylogenetic tree of mammals based on molecular data. This tree provides estimated dates for major nodes as well as the animals' original and current continental distributions. Refer to Figures 42.15 and 42.16 for estimated times of the separation and reunion of land masses. Use these data to answer the following questions.

1. Which one node in the tree below most clearly corresponds to an event of continental drift?

2. Are there groups of mammals that appear to have dispersed to new areas following removal of a barrier to dispersal?

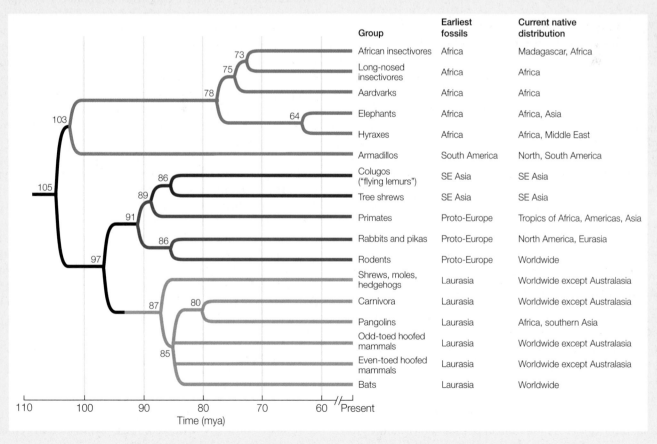

Group	Earliest fossils	Current native distribution
African insectivores	Africa	Madagascar, Africa
Long-nosed insectivores	Africa	Africa
Aardvarks	Africa	Africa
Elephants	Africa	Africa, Asia
Hyraxes	Africa	Africa, Middle East
Armadillos	South America	North, South America
Colugos ("flying lemurs")	SE Asia	SE Asia
Tree shrews	SE Asia	SE Asia
Primates	Proto-Europe	Tropics of Africa, Americas, Asia
Rabbits and pikas	Proto-Europe	North America, Eurasia
Rodents	Proto-Europe	Worldwide
Shrews, moles, hedgehogs	Laurasia	Worldwide except Australasia
Carnivora	Laurasia	Worldwide except Australasia
Pangolins	Laurasia	Africa, southern Asia
Odd-toed hoofed mammals	Laurasia	Worldwide except Australasia
Even-toed hoofed mammals	Laurasia	Worldwide except Australasia
Bats	Laurasia	Worldwide

Time (mya)

dispersal, it is reasonable to conclude that the barrier caused the split by subdividing the original range of the ancestral species. If the phylogenetic split comes before or after a barrier arises, then it is likely that a portion of the ancestral population dispersed to a new location and evolved there in isolation from its ancestors.

> **LINK** Review the effects of barriers to dispersal on speciation in Concept 16.1

Do You Understand Concept 42.4?

- What is Wallace's line?
- What kinds of barriers might limit the dispersal of terrestrial species? Of aquatic species?
- Biomes and biogeographic regions relate to two different fundamental factors that mold the distribution of organisms. Explain.

As we have seen, the makeup of ecological systems on Earth is enormously variable in time and space. Recently, however, the change has accelerated because of the activities of one species—*Homo sapiens*. How are our activities affecting the complexity and heterogeneity of ecological systems? We will address this question in the next concept.

concept 42.5 Human Activities Affect Ecological Systems on a Global Scale

Some have said that we are entering a new geological period, dubbed the "Anthropocene," or Age of Humans, because human activities are altering ecological systems on a global scale. Humans are changing many of the factors that determine the distributions of organisms. We are spreading organisms across the globe without regard to natural barriers to dispersal; we are changing vegetation and topography; and, as Concept 46.4 will describe, we are changing Earth's climates. Others say the new age should be called the "Homogecene," or Homogeneous Age, because by our activities we are making ecological systems less complex and more homogeneous.

Human-dominated ecosystems are more uniform than the natural ones they replace

Humans have converted almost half of Earth's land area into human-dominated ecosystems, such as croplands, pasturelands, and urban settlements. Agriculture is the largest human land use. Cities occupy only about 1 percent of the land, whereas an area roughly the size of South America (about 18 million km^2) is cultivated, and another roughly the size of Africa (about 35 million km^2) is managed as pasture or used as rangeland for cattle. As the human population continues to

FIGURE 42.17 Human Agricultural Practices Produce a Uniform Landscape The cultivation of crop plants for food and other uses diminishes the diversity of ecosystems. This monoculture is a recently harvested field of lavender (*Lavandula angustifolia*) in Provence, France.

grow, the land area devoted to agricultural and urban uses continues to increase as well.

Human-dominated ecosystems contain fewer interacting species than natural ecosystems and therefore are less complex. When an area is converted to agriculture, monocultures (plantings of single crops) often replace species-rich natural plant communities (**FIGURE 42.17**). Pastures or degraded rangelands often contain fewer species than natural grasslands or forests, as discussed at the start of this chapter. Furthermore, agricultural systems worldwide are dominated by relatively few plant species. Some 19 crops comprise 95 percent of total global food production. Even the highest estimate of crop diversity is a tiny fraction of the approximately 350,000 named plant species on Earth.

Agricultural systems are also more spatially and physically uniform than are natural ecological systems. Because humans provide ideal conditions for their crops by adding fertilizer or water, by planting seeds in prepared soil, and by removing enemies, cultivated species can grow in environments they would not be able to occupy on their own. Hence agricultural systems show much less spatial heterogeneity in species makeup along climatic or soil gradients than do natural ecological systems. Furthermore, because many agricultural systems are monocultures, there is little diversity in the physical structure of the habitat.

Human activities are simplifying remaining natural ecosystems

In addition to replacing natural ecological systems with human-dominated ones, human activities are reducing the complexity of the natural ecological systems that remain.

Converting land to human uses affects not only the converted land itself but also the remaining natural systems nearby. For example, when rivers are channelized or dammed to control flooding, or dewatered by irrigation diversions, species-rich natural riparian (streamside) ecosystems disappear. Similarly, land conversion reduces the area of natural systems that remain and fragments them. Human inputs, such as pollutants, alter the abiotic components of natural systems. All of these changes—habitat loss and fragmentation, pollution, water diversion or impoundment—can degrade natural ecological systems and cause species extinctions. Overexploitation of wild species for food, pharmaceuticals, or building materials can cause extinction of those species as well as others that interact with them. Even introducing new species to natural ecosystems often results in losses of native species, as we'll see in Concept 44.4.

Human-assisted dispersal of species blurs biogeographic boundaries

Humans are moving organisms around on a global scale, sometimes deliberately—as in the case of domesticated plants and animals—and sometimes inadvertently. Only about 1 percent of inadvertent introductions result in self-sustaining populations in the new locality, but the cumulative effect on geographic distributions is large nonetheless. Of the insects found both in Europe and in North America, for example, an estimated half have been transported between the two continents by humans. Similarly, more than half of the plant species on many oceanic islands are not native, and even in many continental areas the figure is 20 percent or more. The pace of introductions is astonishing: in California's San Francisco Bay, for example, an average of one new species became established every 12 weeks during the 1990s.

This human-assisted biotic interchange is homogenizing the biota of the planet, blurring the spatial heterogeneity in species composition that evolved during long periods of continental isolation. One wonders how Alfred Wallace would draw the boundaries of biogeographic regions if he were using present-day data!

Do You Understand Concept 42.5?

- Identify two human activities that alter ecosystems.
- In what ways do agricultural ecosystems differ from the natural ecosystems they replace?
- How does extinction simplify ecosystems?
- In what ways are humans making ecosystems more homogeneous?

We have seen that ecological systems are heterogeneous and complex, and that humans are altering this complexity. In order to understand and protect our planet, ecologists need new ways to make sense of ecological heterogeneity and complexity.

---

concept **42.6**

Ecological Investigation Depends on Natural History Knowledge and Modeling

Ecologists deploy the same tools of scientific inquiry that other scientists use—observation, questioning, inductive and deductive logic, comparison, and experimentation (see Concept 1.5). However, the idiosyncrasy and complexity of ecological systems present special challenges that make certain tools particularly important. These tools include natural history and mathematical modeling.

Surprising as it may seem, **natural history**—the observation of nature outside of a formal, hypothesis-testing investigation—provides important knowledge about the many components of an ecological system, their interactions, and the environmental context of those interactions. Natural history is fundamental to virtually all stages of ecological inquiry. Often natural history observations provide the source of new questions and are critical for the formulation of hypotheses as well as for the design of ecological experiments or comparisons to test them. Laboratory studies can give misleading results if conditions are too different from those in nature, so knowledge of the natural context is essential. Similarly, the design of effective field studies rests on knowing what variable factors other than the ones of interest should be controlled, or how to impose a realistic experimental treatment that achieves the desired effect.

Our lack of natural history knowledge often limits our ability to answer important questions. For example, we cannot answer the question "Has the introduction of honey bees caused declines in native bee species?" without knowing which species were present before honey bees were introduced and how abundant they were. It turns out that this information is available for only a handful of places on Earth.

FRONTIERS To overcome our limited knowledge of Earth's diverse ecological systems, scientists are devising ways to expand our ability to observe nature and assimilate the resulting natural history data. New sensor technologies that allow real-time monitoring of physical and biological variables are giving us "new eyes on the world." Citizen-science programs are giving us "more eyes on the world." And computer and internet "informatics" technologies are improving our ability to compile natural history information and make it broadly available.

Models are often needed to deduce testable predictions with complex systems

As we saw in Chapter 1, hypotheses are tested by deducing what predictions follow if the hypotheses are true, and testing those predictions. Sometimes we can use simple rules of logic to develop predictions, but often we need a computer model. What is critical is that we use knowledge about the system—natural history—to begin building the model. In the case of rangeland systems, for example, the grasses that are eaten by cattle are also affected by interactions with other plants and other herbivores, by soil fertility, by climate, and by wildfire. To predict whether removing cattle will bring back palatable grasses, we need to develop a model of the major interactions in the system that allows us to trace the various paths by which removal of cattle will affect the palatable grasses.

Now that we have explored some of the properties of ecological systems, we are ready to take a closer look, beginning with the smallest scales in the ecological hierarchy: individual organisms and populations.

QUESTION Why did the rangeland restoration method that worked in Europe fail to work in the Borderlands?

ANSWER As we have seen in this chapter, the physical environment provides a template for the biological communities that inhabit a location. We have also seen that communities are molded by their evolutionary history of geographic isolation from or connection with other locations. Land managers who understand these principles realize that grasslands around the world might look similar but actually behave very differently.

The world's grasslands differ in a number of ways. Some, including many of the pastures of Europe and eastern North America, once were forests that were cleared to make way for grasses; here "resting the range" and managing the size of grazing herds can restore much of the ecosystem (**FIGURE**

FIGURE 42.18 Harmonious Grazers In Switzerland, controlled grazing and managed herd size have led to the maintenance of species-rich grasslands.

42.18). Some, such as the temperate grasslands of midwestern North America, Eurasia, and South America and the tropical savannas of Africa, have long histories of evolution with grazing mammals. Grasslands occupy a range of climatic conditions, from relatively moist to very arid.

As yet, ecologists do not completely understand why removal of cattle has not restored grasslands in the U.S.–Mexico Borderlands. They are actively researching the interacting effects of climate, soil loss, fire, and adaptations of grasses to grazing animals.

42 SUMMARY

concept 42.1 Ecological Systems Vary in Space and over Time

- An **ecological system** consists of one or more organisms and the **biotic** and **abiotic** components of the environment with which they interact.

- Ecological systems can be studied at any scale in the biological hierarchy, from an individual organism and its immediate surroundings to populations, communities, landscapes, or the entire biosphere. **Review Figure 42.1**

- Because the components of ecological systems and their interactions are distributed unevenly in space and change over time, each system, at each time, is potentially unique.

concept 42.2 Climate and Topography Shape Earth's Physical Environments

- **Weather** is the state of atmospheric conditions in a particular place at a particular time, and **climate** is their average state and pattern of variation over longer periods.

- Latitudinal differences in solar energy input are caused by differences in the angle of the sun's incoming rays. **Seasonality** results from the tilt of Earth's axis of rotation relative to its orbit around the sun. **Review Figures 42.3 and 42.4**

- Latitudinal differences in solar energy input drive north–south patterns of atmospheric circulation. These patterns, in turn, influence latitudinal patterns of precipitation. **Review Figure 42.5**

- Prevailing surface winds result from the interaction of the north–south atmospheric circulation cells with Earth's eastward rotation. **Review Figure 42.6**

- Surface **currents** in the oceans are driven by prevailing winds and deflected by continents. **Review Figure 42.7**

- **Climate diagrams** summarize climate at a particular location in an ecologically relevant way. **Review Figure 42.8**

- Variation in the elevation of Earth's surface, known as **topography**, affects the physical environment. **Review Figure 42.9 and ANIMATED TUTORIAL 42.1**

concept 42.3 Physical Geography Provides the Template for Biogeography

- A **biome** is a distinct physical environment that is inhabited by ecologically similar organisms with similar adaptations. The species that occupy the same biome in geographically separate regions may not be closely related phylogenetically but often show convergent adaptations. **See ANIMATED TUTORIAL 42.2**

- The distribution of terrestrial biomes is broadly determined by annual patterns of temperature and precipitation. **Review Figures 42.10 and 42.11**

- Other features of the physical environment—particularly soil characteristics—interact with climate to influence the character of vegetation.

- Climate is less important in distinguishing aquatic biomes than are water depth and flow, temperature, pressure, salinity, and characteristics of the substrate. **Review Figure 42.13 and Table 42.1**

concept 42.4 Geological History Has Shaped the Distributions of Organisms

- Alfred Russel Wallace first noticed a boundary, now known as Wallace's line, between two distinct assemblages of species that was not explained by climate or soil. **Review Figure 42.14**

- The world can be divided into **biogeographic regions**, each of which contains distinct assemblages of species. The boundaries of biogeographic regions generally correspond to present or past barriers to dispersal and can be explained by continental drift. **Review Figures 42.15, 42.16, and WEB ACTIVITY 42.1**

- Biogeographers use phylogenetic information, in conjunction with the fossil record and geological history, to determine how the modern distributions of organisms came about.

concept 42.5 Human Activities Affect Ecological Systems on a Global Scale

- Humans have converted almost half of Earth's land area into human-dominated ecosystems, which are much more homogenous and less complex than natural ecosystems, and human activities are reducing the complexity of the natural ecological systems that remain.

- Humans move species around the globe without regard to natural barriers to dispersal. These movements are homogenizing the biota of the planet.

concept 42.6 Ecological Investigation Depends on Natural History Knowledge and Modeling

- Accurate and reliable knowledge of **natural history**—that is, information about the components of ecological systems, their interactions, and the environmental context of those interactions—is crucial to the study of ecology.

- Developing computer models of the major components of an ecological system and their interactions is often needed to generate testable predictions about these complex systems.

Populations

Ecologist Rick Ostfeld was studying white-footed mice (*Peromyscus leucopus*) and eastern chipmunks (*Tamias striatus*) at the Cary Institute of Ecosystem Studies in Millbrook, New York, when an unexpected event took his work in a new direction. Ostfeld and his colleague Clive Jones were working on these rodents in order to understand what triggers the dramatic once-in-a-decade outbreaks of the introduced gypsy moth (*Lymantria dispar*) that defoliate oak forests in North America. They suspected that outbreaks start when the moths escape control by mice and chipmunks, which eat moth pupae.

Soon after the study began in 1991, one of the field assistants discovered that half his face was paralyzed and he could not close one eyelid. (He devised a protective eye patch so he would be allowed to continue his fieldwork.) The eventual diagnosis was Lyme disease. Scientists at the time were only beginning to understand this disease, which was not described until 1975 and whose cause—infection by spirochete bacteria of the genus *Borrelia*—was not discovered until 1982. By 1991, epidemiologists had found out that Lyme disease is transmitted by black-legged ticks (*Ixodes scapularis*), which become infected with *Borrelia* when they bite infected hosts, and that Lyme disease risk increases with the abundance of infected ticks. Tick hosts include white-tailed deer, birds, lizards, and rodents as well as humans.

Millbrook turned out to be at the epicenter of the emerging Lyme disease epidemic, and Ostfeld focused his attention on black-legged ticks, recognizing that they are vectors that transmit the bacterial disease. He was already counting ticks on live-trapped mice and chipmunks. Now he began to study how the abundance of rodents and other vertebrate hosts affects

tick abundance and infection rate. He discovered that rodents are the primary reservoir for *Borrelia*. Ticks are not infected when they hatch from eggs as larvae; they become infected when they bite an infected host as a larva or nymph. It turns out that birds and lizards do not harbor *Borrelia*, and that most ticks are already infected by the time they are adults able to climb high into the vegetation, where they encounter large hosts such as deer.

From his previous work, Ostfeld already knew that rodent abundance soars when oaks produce a big acorn crop. Now he found that the number of tick larvae that survive to become nymphs, and of nymphs that survive to become adults, increases with rodent numbers.

A field worker sweeps for ticks in the forests near Millbrook, New York.

QUESTION

How does understanding the population ecology of disease vectors help us combat infectious diseases?

KEY CONCEPTS

43.1 Populations Are Patchy in Space and Dynamic over Time

43.2 Births Increase and Deaths Decrease Population Size

43.3 Life Histories Determine Population Growth Rates

43.4 Populations Grow Multiplicatively, but Not for Long

43.5 Extinction and Recolonization Affect Population Dynamics

43.6 Ecology Provides Tools for Managing Populations

concept 43.1 Populations Are Patchy in Space and Dynamic over Time

Individual organisms are the fundamental ecological units. We wish to understand their properties, but also those of the populations to which they belong. Recall that a *population* consists of the individuals of a species that interact with one another within a given area at a particular time. We focus on populations because the ecological role of a species in a community or ecosystem is determined not only by the properties of individuals of that species, but also by that species' relative abundance, which is a population-level property.

Humans have long had a practical interest in understanding what determines species abundance because we want to manage our ecological interactions with other species. We want to know how to increase the abundance of organisms that provide us with resources such as food or fiber, to conserve wild organisms that provide ecological services such as pollination or pest control, and to decrease the abundance of undesirable organisms such as crop pests, weeds, or pathogens. Aside from these practical considerations, we also want to know how to conserve species for aesthetic or ethical reasons. The study of population ecology is key to achieving these goals.

Population density and population size are two measures of abundance

The concept of abundance has several meanings. We might consider a species abundant because there are many individuals in a small area. If that same species occurs only in a small geographic region, however, we might consider it rare because the total number of individuals of that species is small. Depending on why they want to measure abundance, ecologists use one of two approaches. If they are interested in the causes or consequences of local abundance, they usually measure **population density**: the number of individuals per unit of area (for terrestrial organisms) or volume (for organisms that live in air, soil, or water). Alternatively, ecologists interested in the population as a whole want to know the total number of individuals in the population, or **population size**. Counting all the individuals in a population is practical only for very small populations in which individuals can be distinguished. For this reason, ecologists usually measure population density first. Then they multiply population density by the area occupied by the population to calculate population size.

Abundance varies in space and over time

The abundance of organisms varies over several spatial scales. At the largest scale, as we saw in Chapter 42, many species are found only in a particular region, and their densities are zero elsewhere. Within that region, which is called the species' **geographic range**, the species may be restricted to particular kinds of environments, often called **habitats**, which may themselves be patchily distributed. For example, Edith's checkerspot butterfly (*Euphydryas editha*) is found from southern British

Columbia and Alberta to Baja California, Nevada, Utah, and Colorado (**FIGURE 43.1A**). Within this geographic range, it occurs in the open habitats favored by the herbaceous plants its caterpillars eat: coastal chaparral, meadows, grasslands, and open woods. In the San Francisco Bay area, suitable food plants grow only in grassland habitat on outcrops of a chemically unique rock called serpentine (**FIGURE 43.1B**). These serpentine outcrops constitute **habitat patches** for the butterflies—that is, "islands" of suitable habitat separated by areas of unsuitable habitat. Habitat patchiness has consequences for abundance that we will explore in Concept 43.5.

In any given locality, population densities are *dynamic*—they change over time. In the oak forests near Millbrook, New York described in the opening story, for example, abundances

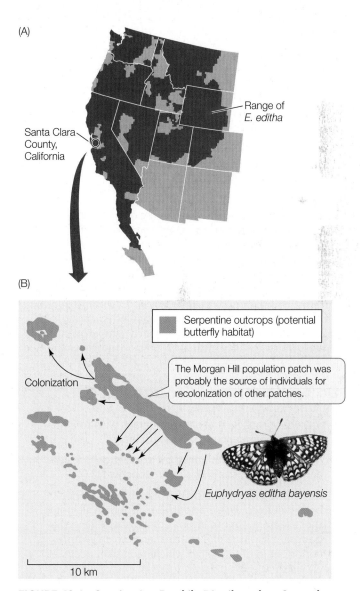

(A)

Santa Clara County, California

Range of *E. editha*

(B)

Serpentine outcrops (potential butterfly habitat)

Colonization

The Morgan Hill population patch was probably the source of individuals for recolonization of other patches.

Euphydryas editha bayensis

10 km

FIGURE 43.1 Species Are Patchily Distributed on Several Spatial Scales (A) The geographic range of Edith's checkerspot butterfly *Euphydryas editha* extends from British Columbia and Alberta to Baja California. (B) In the San Francisco Bay area, the butterflies are divided into subpopulations, each occupying a patch of suitable habitat. Arrows indicate colonization events in 1986.

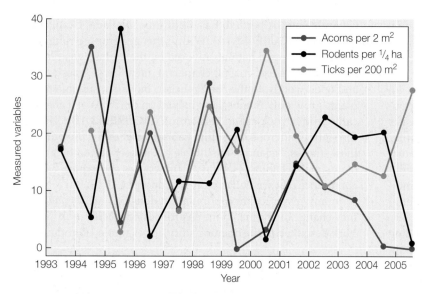

FIGURE 43.2 Population Densities Are Dynamic and Interconnected Densities of acorns, rodents, and nymphal *Ixodes scapularis* (black-legged tick) populations vary over time in the oak forests near Millbrook, New York. Note that rodent density tends to follow acorn density with a 1-year lag, and tick density tracks rodent density with the same time lag.

of rodents, black-legged ticks, and oak acorns vary greatly from year to year (**FIGURE 43.2**).

FRONTIERS Population studies are often thwarted because individual organisms cannot be marked and followed. Wild bumble bees, for example, are critical pollinators for many plants, including crops. Ecologists and crop scientists want to understand their movements in order to take best advantage of their free pollination services. But how to follow a bumble bee that weighs only as much as a few postage stamps? Ingenious new technologies such as harmonic radar and microchip transponders now allow ecologists to glue tiny tags to bumble bees and follow them as they move through real landscapes.

Do You Understand Concept 43.1?

- For what practical reasons do humans wish to influence the abundance of some species?

- What aspects of abundance do population density and population size measure, respectively?

- Barriers to dispersal and environmental tolerance both influence where a species is found. Which of these factors is more important in determining the geographic range of a species, and which is more important in determining which habitats it occupies within the geographic range?

Why do species abundances vary in time and space? To answer this question, we need to understand the processes that add individuals to or subtract them from populations. The study of these processes is known as *demography* (Greek *demos*, "population"; *-graphia*, "description").

concept 43.2 Births Increase and Deaths Decrease Population Size

The size of anything, whether a bank account or a population, changes over time through additions and subtractions. Consider the size of something in the future—for example the size of your bank account. It is equal to the size now plus what is added, minus what is taken away. What differs between bank balances and populations are the processes that add or subtract. In the case of your bank account, deposits and interest add to your balance, while withdrawals and bank fees subtract. In the case of populations, births (or the production of seeds in the case of plants) add individuals and deaths remove individuals. To spell this out,

The number of individuals in a population at some time in the future =

the number now + the number that are born – the number that die

If we turn this word equation into symbols, we have the most basic mathematical model of population growth:

$$N_{t+1} = N_t + B - D \qquad \text{Equation 43.1}$$

where N is the population size, B is the number of births in the time interval from time t to time $t + 1$, and D is the number of deaths in that same interval. Ecologists call this the "birth–death" or **BD model** of population change.

We can use Equation 43.1 to calculate population size into the future. Most often, the population size will change over time—it will either grow or shrink. How much it changes in a period of time indicates how fast it is changing. Ecologists call this rate of change the *growth rate* of the population (a somewhat confusing term until we realize that when a population shrinks, its growth rate is negative). The population growth rate can be obtained from Equation 43.1 by subtracting N_t (the present population size) from both sides of the equation:

$$N_{t+1} - N_t = \Delta N = B - D$$

The term ΔN means change in population size (using the convention that the Greek letter delta means "change in"). To make more explicit that this is a rate—a change through time—we can divide both sides by ΔT, the time interval from t to $t + 1$:

$$\frac{\Delta N}{\Delta T} = \frac{B - D}{(t+1) - t} = \frac{B - D}{1} = B - D \qquad \text{Equation 43.2}$$

This shows, logically enough, that the change in size over one time interval is the number of births minus the number of deaths during that interval.

Change in size of the total population can be measured directly only for very small, discrete populations, such as a zoo population. Ecologists can get around this difficulty by keeping track of a sample of individuals (a *cohort*) over time. From this sample, they can estimate the number of offspring that an average individual produces, called the *per capita birth rate* (*b*),

APPLY THE CONCEPT

Births increase and deaths decrease population size

Rancher Jane is thinking of shifting her operation from cattle to American bison (*Bison bison*), but she needs to know how well bison will do on her ranch. She buys 50 female bison that are already inseminated and places 10 of them, picked at random, into their own pasture. These 10 females serve as a sample from which Rancher Jane collects demographic data over one year. Use her data to answer the questions below.

1. What is the total number of births and deaths among this sample population?

2. What are the estimated average per capita birth and death rates (*b* and *d*) for the entire bison herd, based on this sample?

3. What is estimated per capita growth rate *r*?

4. Based on these estimates, what is the size of Rancher Jane's entire bison herd at the end of the year? (Hint: Use Equation 43.3.)

FEMALE #	ALIVE AT END OF YEAR?	NUMBER OF OFFSPRING
1	Yes	1
2	Yes	0
3	Yes	1
4	Yes	0
5	No	0
6	Yes	1
7	Yes	1
8	No	0
9	Yes	1
10	Yes	0

and the average individual's chance of dying, called the *per capita death rate* (*d*), in some interval of time. *Per capita* (literally "per head") means "per individual." Total birth and death rates in the population can then be estimated by multiplying per capita birth and death rates by population size, so that $B = bN_t$ and $D = dN_t$. Substituting into Equation 43.1, we have

$$N_{t+1} = N_t + bN_t - dN_t$$

or, equivalently,

$$N_{t+1} = N_t + (b - d) N_t$$

The value (*b* – *d*) is the difference between per capita birth rate and per capita death rate and represents the average individual's contribution to total population growth rate. This value is the **per capita growth rate**, which ecologists symbolize as *r*. Substituting *r* for (*b* – *d*), we have

$$N_{t+1} = N_t + rN_t \qquad \text{Equation 43.3}$$

Converting this into an equation of growth rate analogous to equation 43.2, we get

$$\frac{\Delta N}{\Delta T} = rN \qquad \text{Equation 43.4}$$

What happens if per capita birth rate is greater than per capita death rate, or death rate greater than birth rate, or they are equal? If *b* > *d*, then *r* > 0, and the population grows. If *b* < *d*, then *r* < 0, and the population shrinks. If *b* = *d*, then *r* = 0, and the population size does not change.

We are now in a better position to understand why population sizes or densities change over time: they *have to* unless the number of births exactly equals the number of deaths. We are also in a better position to understand why population densities vary in space. We saw in Chapter 42 that barriers

to dispersal can cause species to be absent from some places. Another possible reason for a species' absence from a region is a negative population growth rate. A species will not persist in a given location if its per capita growth rate, *r*, is negative there. Any population that decreases steadily in size over time eventually reaches a density of zero—that is, it goes extinct.

Do You Understand Concept 43.2?

- What is the relationship between per capita birth rate and total births in a population during a time period?

- What do we need in addition to estimates of per capita birth and death rates in order to predict population size at some future time?

- What two strategies could rancher Jane (see "Apply the Concept") use to expand her bison herd faster?

We can understand the dynamics of whole populations, and even some aspects of species' distributions, if we know about the average birth and death rates of the individuals that constitute those populations. What determines these demographic variables?

concept 43.3 Life Histories Determine Population Growth Rates

The demographic processes that influence birth, death, and population growth rates can best be understood in the context of a species' **life history**—the time course of growth and development, reproduction, and death during an average individual's life.

Life histories are quantitative descriptions of life cycles

Life histories are, in essence, quantitative descriptions of life cycles. To describe the life cycle of the black-legged tick, for example, we can begin with one of the thousands of eggs laid by an adult female in spring (**FIGURE 43.3**). The egg hatches into a larva in mid-summer. If the larva can obtain a blood meal from a rodent, bird, or lizard, it molts into a nymph and goes dormant for the winter. If it survives the winter, the nymph becomes active again the following summer and seeks another blood meal from a vertebrate host. If it is successful, it molts into the adult stage and seeks a large mammal host (often a deer; hence these are sometimes called deer ticks) in the fall. It mates on this final host and if it is a female it goes dormant for another winter, lays eggs in the spring, and then dies.

A quantitative description of a black-legged tick's life cycle would indicate at what age individuals make life-cycle transitions and how many do so successfully: the fraction of eggs that hatch successfully, of larvae that find a host and molt successfully into nymphs, of nymphs that live through the winter, find a host, and molt into adults; and the number of eggs that the average adult female lays. This information can be summarized in a **life table**.

TABLE 43.1 shows a life table for a cohort of 210 cactus ground finches (*Geospiza scandens*) on Isla Daphne in the Galápagos. Life tables contain two types of information: the fraction of individuals that survive from birth to different life stages or ages, called **survivorship**, and the average number of offspring each individual produces at those life stages or ages if they do survive, called **fecundity**. Survivorship can also be expressed as its opposite, *mortality*, which is the fraction of individuals that do not survive from birth to a given stage or age (mortality = 1 − survivorship). Life tables thus can be used to calculate per capita growth rates (*r*). Although the calculations involved are complex, in general it is not hard to see how survivorship and fecundity affect *r*. All else being equal, the higher the fecundity and the higher the survivorship, the higher *r* will be. If reproduction shifts to earlier ages, *r* will increase as well.

Life histories are diverse

Species vary considerably in how many and what types of developmental stages they go through, how old they are when they begin to reproduce, how often they reproduce, how many offspring they produce, and how long they live. Black-legged ticks spin out their lives over 2 years, whereas the life cycle of a periodic cicada takes nearly 2 decades. Other arthropods have life spans of days or weeks. Some species, like the tick, go through a discrete series of life stages of variable duration,

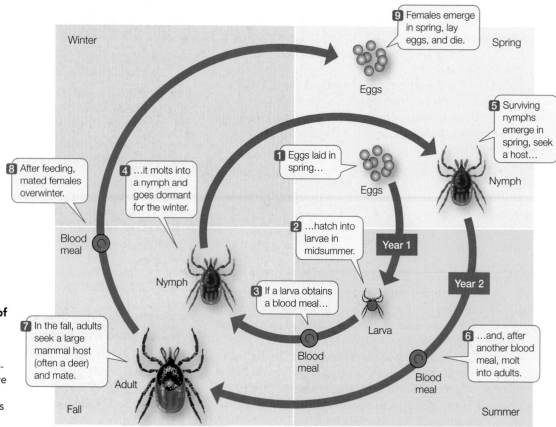

FIGURE 43.3 Life History of the Black-Legged Tick The seasonality of the ticks' environment and the difficulty in obtaining the blood meal necessary for metamorphosis have produced a 2-year life cycle that includes two long periods of winter domancy.

TABLE 43.1	Life Table for the 1978 Cohort of Cactus Ground Finch on Isla Daphne		
CALENDAR YEAR	AGE OF BIRD (YEARS)	SURVIVORSHIP[a]	FECUNDITY[b]
1978	0 (hatchlings)	1.00	0.00
1979	1	0.43	0.05
1980	2	0.37	0.67
1981	3	0.33	1.50
1982	4	0.31	0.66
1983 Increased rain	5	0.30	5.50
1984	6	0.20	0.69
1985 Drought	7	0.11	0.00
1986	8	0.07	0.00
1987	9	0.07	2.20
1988	10	0.05	0.00
1989	11	0.05	0.00

[a] Survivorship = the proportion of the original cohort (here, 210 birds) surviving from fledging to age x.

[b] Fecundity = average number of young fledged per female of age x.

each separated by a molt. Other species, such as humans, develop continuously as they age. Some plants germinate, grow, flower, produce seeds, and die all in one growing season—they are annuals (see Concept 27.2). Other plants live for centuries. Some species spend long periods in a dormant state, whereas others are continuously active. Some organisms, including the tick, reproduce only once and then die, whereas others, such as humans, can reproduce multiple times.

Life histories vary not only among species but within species as well, including among human populations. *Life expectancy* (the age to which an average person survives) among the short-statured, nomadic Aeta people of the Philippines, for example, is 16.5 years, and girls reach puberty at age 12. The taller-statured, pastoralist Turkana of East Africa experience food shortages as frequently as the Aeta but have a life expectancy of 47.5 years, and Turkana women reach sexual maturity at age 15. In well-nourished United States populations, life expectancy is almost 79 years—much greater than for either the Aeta or the Turkana—yet girls in the U.S. reach puberty at age 12 years on average.

Resources and physical conditions shape life histories

Individual organisms must acquire materials and energy to maintain homeostasis and to fuel metabolism, growth, activity, defense, and reproduction. Materials and energy—and the time available to acquire them—constitute **resources**. Organisms also need *physical conditions* that they are able to tolerate.

The primary distinction between resources and conditions is that resources can be used up (e.g., mineral nutrients, photons of light, time, space, or food items), whereas conditions are experienced but not consumed.

The rate at which an organism can acquire resources increases with the availability of resources in the environment. A plant's net photosynthetic rate, for example, increases with sunlight intensity (**FIGURE 43.4A**). Similarly, an animal's rate of food intake increases with the density of food (**FIGURE 43.4B**).

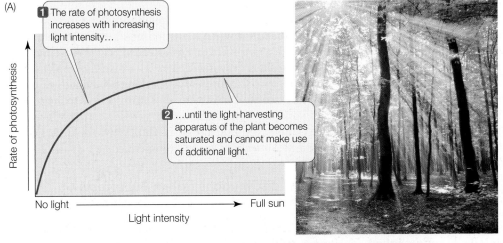

(A)

1 The rate of photosynthesis increases with increasing light intensity...

2 ...until the light-harvesting apparatus of the plant becomes saturated and cannot make use of additional light.

Rate of photosynthesis

No light ——————→ Full sun
Light intensity

(B)

Seeds harvested per second

Seed density (no. per cm²)

Dipodomys sp.

FIGURE 43.4 Resource Acquisition Increases with Resource Availability (A) Plant photosynthetic rate increases with light intensity. (B) Kangaroo rats (*Dipodomys* spp.) are able to harvest seeds faster as seed density increases. (Note the cheek pouch bulging with seeds.)

FIGURE 43.5 The Principle of Allocation Organisms use energy to acquire resources, including foraging for food. The organism then must allocate those resources among competing life functions. In general, an organism's first priority is to maintain the integrity of its body. This figure shows the same amount of energy being allocated to resource acquisition in all three cases.

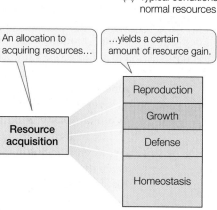

(A) Typical conditions, normal resources

Homeostatic needs must be met first. Remaining resources are divided among other activities.

(B) Typical conditions, abundant resources

When resources are abundant, more resources are gained and more are available after homeostatic needs are met.

(C) Stressful conditions, normal resources

Stressful conditions mean more resources must be expended on homoestasis and fewer are available for other purposes.

The **principle of allocation** states that once an organism has acquired a unit of some resource, it can be used for only one function at a time. This principle means that organisms have to divide the resources they are able to obtain among competing functions—maintenance, foraging, growth, defense, and reproduction (**FIGURE 43.5**). In general, an organism's first priority is to maintain the integrity of its body. Because maintaining homeostasis is more difficult under stressful physical conditions (as was illustrated for temperature in Chapter 29), investment in maintenance is greater in stressful than in benign environments. For example, the hotter their desert environment, the more time day-active lizards must spend in a cool retreat to keep body temperature below a lethal level, and thus the less time they have to look for food (see Figure 29.12). Once an organism obtains more resources than it needs for maintenance, it can allocate the excess to other functions. In general, as the average individual in a population acquires more resources, the average fecundity, survivorship, and per capita growth rate increase.

Life history variation among species and populations often reflects the principle of allocation. A species that invests heavily in growth early in life, for example, cannot simultaneously invest heavily in defense—for example, in protective structures or chemical defenses. As a consequence it reaches adult size quickly, but at the cost of lower survivorship than a species that invests more in defense. A species that starts investing in reproduction early in development grows more slowly and matures at a smaller adult size than a species that waits to reproduce. Species that invest heavily in reproduction often do so at the expense of adult survivorship. They have high fecundity, but short life spans. Such negative relationships among growth, reproduction, and survival are called *life-history tradeoffs*.

The environment shapes much of this life history variation because, together with a species' way of life, it determines the relative costs and benefits, in terms of survivorship and fecundity, of any particular allocation pattern. The Aeta people mentioned earlier, for example, live in an environment that imposes much higher mortality than the Turkana experience. High mortality means that the benefit of early reproduction (more individuals survive to reproduce at least once) outweighs the fecundity benefit of growing to a larger adult size (in general, human females produce 0.24 more offspring per additional centimeter of height). Most Aeta women reach puberty and stop growing at about 12 years of age, at an adult height of 140 cm. For the

Turkana, the lower mortality rate shifts the balance—the advantage of early reproduction is outweighed by the fecundity advantage of larger adult size. Turkana women mature at age 15 and an average height of 166 cm. In the United States, plentiful food and modern medicine reduce the cost of maturing early—girls grow fast and mature early at an adult size equivalent to the Turkana.

Similarly, the short summers in New York State make it advantageous for black-legged tick nymphs to invest in overwinter survival, even though it means they do not reproduce until their second year (see Figure 43.3). Ticks find their host by waiting: they climb up on vegetation, extend their forelegs, and climb aboard when a host bumps into them. Finding a host is such a rare event that a tick can expect at most one meal during the short growing season. As a result, nymphs do not allocate resources to further host-seeking behavior during their first season of life, but instead allocate resources to overwinter survival. The fact that ticks take 2 years to complete their life cycle is thus a consequence of the difficulty in obtaining a host and the seasonality of the environment.

Species' distributions reflect the effects of environment on per capita growth rates

Because a population cannot persist in an environment where its per capita growth rate is negative, we should be able to predict species' distributions from knowledge of how survivorship and fecundity are affected by resource availability and physical conditions. But obtaining complete knowledge of life histories is often impossible for species in the wild. Even incomplete knowledge, however, can help us understand aspects of species' distributions. **FIGURE 43.6** illustrates this point for the lizard *Sceloporus serrifer*. By combining measurements of conditions in the animal's natural environment with knowledge of its physiology and behavior, researchers were able to draw conclusions about how climate change can affect survivorship, fecundity, and distribution of these lizards.

INVESTIGATION

FIGURE 43.6 Climate Warming Stresses Spiny Lizards
Barry Sinervo and colleagues investigated whether higher daytime temperatures can reduce the number of hours that Mexican blue spiny lizards (*Sceloporus serrifer*) can remain outside their burrows without overheating. The researchers knew that climate warming was taking place in Mexico, and that the lizards cannot feed while inside their burrows.

HYPOTHESIS

Spiny lizards are able to forage for fewer hours on hotter days.

METHOD

1. Construct model "lizards" that have the same thermal properties as actual lizards.

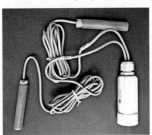

2. At 4 sites in the Yucatán, place model lizards in various sunny and shady perches known to be used by real lizards in 1975. In 2008, monitor body temperature of the models and record maximum daily air temperatures throughout the breeding season (March and April).

3. For each day, calculate the number of daylight hours when the body temperature T_b of the thermal models exceeded 31°C—the temperature at which *S. serrifer* are known to retreat to their burrows and become inactive.

4. Determine whether the number of inactive hours increased with maximum air temperature.

RESULTS

The number of hours during which lizards were inactive—and hence could not forage—increased by 0.74 hours for each °C of increase in daily maximum air temperature ($P < 0.001$).

Dashed lines represent the 95% confidence interval.

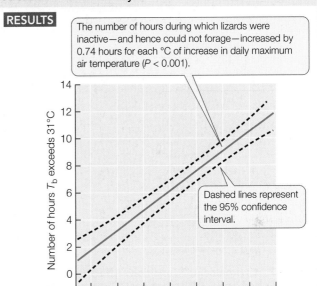

CONCLUSION

Climate warming can decrease the number of hours per day when *S. serrifer* can forage without overheating.

ANALYZE THE DATA

Between 1975 and 2008, the *S. serrifer* populations went extinct at two of the four sites surveyed. Examine the graph at right and answer the questions below.

A. How did the hours available for foraging differ between sites where the lizards had gone extinct versus where they persisted?

B. How might that difference in foraging time influence lizard fecundity, survivorship, and per capita growth rates?

C. Could global warming have caused the extinctions of the lizard populations? Explain your answer.

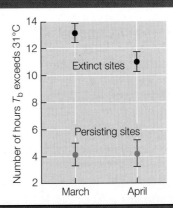

Go to **yourBioPortal.com** for original citations, discussions, and relevant links for all INVESTIGATION figures.

Sometimes it is possible to explore the links between environmental conditions, life histories, and species distributions with the help of experiments carried out in the laboratory. Charles Birch used this approach to understand why two species of beetles that infest stored grain are serious pests in some parts of Australia but not in others. In his laboratory, Birch reared populations of the rice weevil (*Sitophilus oryzae*, known in Birch's time as *Calandra oryzae*) and the lesser grain borer (*Rhyzopertha dominica*) under various conditions of temperature and humidity. He constructed life tables for each set of conditions and used the tables to calculate per capita growth rates for each species. The results, summarized in **FIGURE 43.7**, explain the more tropical distribution of *R. dominica*. In addition, both species had negative per capita growth rates in cool, dry environments, suggesting a strategy for minimizing grain losses: keep the grain cool and dry.

Sitophilus oryzae (rice weevil)

Rhyzopertha dominica (lesser grain borer)

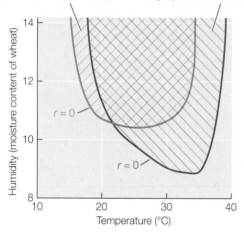

FIGURE 43.7 Environmental Conditions Affect Per Capita Growth Rates and Species Distributions Charles Birch's laboratory research revealed that *R. dominica* populations performed better under warmer conditions than did *S. oryzae* populations, thus explaining why *R. dominica* has a more tropical distribution. The lines marked r = 0 indicate the limit of the conditions under which populations of each species can grow; within the hatched areas, per capita growth rates are positive for the species.

Do You Understand Concept 43.3?

- An organism can allocate a unit of resource to which five functions?

- What is the relationship between a life history and a life table?

- Female hummingbirds abandon their young in the nest when food supply declines. What does this imply about their allocation priorities?

- Ecologists added predatory fish to some streams, leaving other streams as controls. How do you expect size and age of first reproduction of prey fish to evolve in the presence of predators? (Hint: consider the example of the Aeta and Turkana)

Knowing whether population growth rates are negative or positive in particular environments tells us why species are absent or present. But this knowledge does not help us to understand why population densities vary among locations where population growth rates are positive. To do so, we must take a closer look at the dynamics of population growth.

concept 43.4 Populations Grow Multiplicatively, but Not for Long

We saw in Concept 43.2 that a population will grow as long as the per capita growth rate, *r*, is greater than zero. Recall also that *r* refers to a specific period of time (it might be a day, a week, or a year, depending on the species) over which we have estimated per capita rates of birth and death. During this period, the population will add a number of individuals that is precisely *r* times its initial size. We actually saw this in equation 43.3, which states that $N_{t+1} = N_t + rN_t$. What may not be obvious at first is that this continues in the next period—the population will again add a multiple *r* of its numbers, since $N_{t+2} = N_{t+1} + rN_{t+1}$. Because N_{t+1} was already larger than the initial population N_t, this means that an ever-larger number of individuals is added in each successive period of time. Thus the growth is **multiplicative**. Multiplicative growth differs dramatically in its form from **additive growth**, in which a constant *number* (rather than a constant *multiple*) is added in each time period:

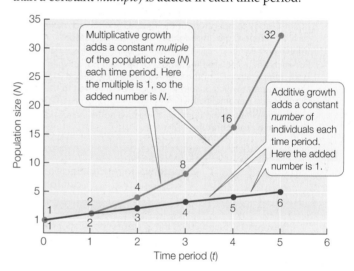

Multiplicative growth generates large numbers very quickly

We naturally tend to think in terms of additive growth, and multiplicative growth astonishes us with the phenomenal numbers it can generate in a short amount of time. For example, in 1911, L. O. Howard, then chief entomologist of the U. S. Department of Agriculture, estimated that a pair of flies reproducing for the first time on April 15 would produce a population of 5,598,720,000,000 adults by September 10 if all the offspring survived and reproduced. Other entomologists disagreed with Howard's calculation—they pegged the number much *higher*!

Charles Darwin was well aware of the power of multiplicative growth:

[E]very organic being naturally increases at so high a rate, that, if not destroyed, the earth would soon be covered by the progeny of a single pair. ...[A]s more individuals are produced than can possibly survive, there must in every case be a struggle for existence.

This ecological struggle for existence, fueled by multiplicative growth, is what drives natural selection and adaptation.

> **LINK** Darwin's theory of natural selection is described in Concepts 15.1 and 15.2.

Multiplicatively growing populations have a constant doubling time

Multiplicative growth characterizes many phenomena other than population growth that affect our everyday lives. These include interest-bearing bank accounts, fossil-fuel consumption, and radioactive decay (see Figure 18.1). Our tendency to think in additive terms impedes our ability to comprehend these phenomena and to understand their practical implications. For example, how much more money will we have in 5 years if we invest $1,000 in a bank account that earns 2% per year as opposed to 1% per year? How does the drop in per capita rate of growth of the human population, from 0.022 in 1963 to the current 0.013 per year, change our estimates of future population size? Answering these questions requires an understanding of multiplicative growth.

Multiplicative growth has a very striking property: a constant **doubling time**. Because a population growing multiplicatively adds a constant multiple of itself each time period, the amount added increases over time. At some specific time period, the amount added exactly equals the initial population, and the population has doubled. As long as r does not change, the time the population takes to double will also remain constant. The time to double can be calculated if we know r. It is easy to see the implications of multiplicative growth if we

think in terms of doubling times. By doing so, municipalities, for example, would realize that even a seemingly small growth rate of 3% per year would mean that the population would double every 23 years, and that its sewage treatment capacity and other such municipal services would need to do the same.

Density dependence prevents populations from growing indefinitely

If populations grow multiplicatively, why isn't Earth covered with flies, or any other organism? It turns out that populations do not show the J-shaped growth pattern of multiplicative growth (red line in the graph below) for very long. Instead of continuing to increase, population growth slows and usually reaches a more or less steady size. The simplest example is so-called *logistic growth*:

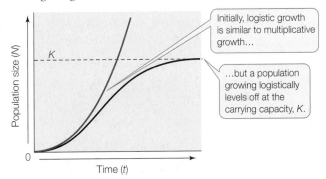

Initially, logistic growth is similar to multiplicative growth...

...but a population growing logistically levels off at the carrying capacity, *K*.

---

yourBioPortal.com

Go to WEB ACTIVITY 43.1
Population Growth

APPLY THE CONCEPT

Multiplicatively growing populations have a constant doubling time

Yellow star-thistle (*Centaurea solstitialis*) is a spiny annual plant native to the Mediterranean region. The species is a noxious weed that has invaded several regions of the United States. It is unpalatable to livestock (including American bison). Rancher Jane discovers that 1 hectare of the 128-hectare pasture into which she placed her sample cohort of 10 female bison has been invaded by star-thistle. A year later she finds the weed population has grown to cover 2 hectares.

1. Based on the information above, how many hectares do you predict the star-thistle population will cover in 1, 2, and 3 more years if the population is growing additively? How many hectares if the population is growing multiplicatively?

2. Imagine that Rancher Jane only discovers the star-thistle population after it has already covered 32 hectares of her pasture. How many years does she have until the weed completely covers the pasture if its population is growing additively? Multiplicatively?

The invasive yellow star-thistle is noxious to grazing livestock.

Why do populations stop growing? Upon examination, we see that population growth slows because r decreases as the population becomes more crowded—that is, r is **density dependent**. At low population densities, per capita growth rates are at their highest. As the population grows and becomes more crowded, birth rates tend to decrease and death rates tend to increase. As a result, r decreases as population density increases

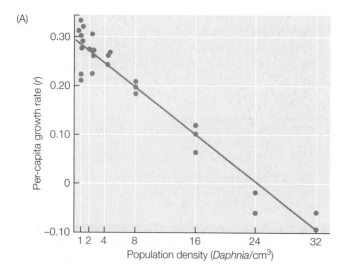

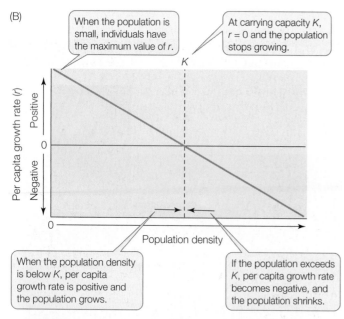

FIGURE 43.8 Per Capita Growth Rate Decreases with Population Density (A) The decline in r as population density increases for in a population of a small crustacean, *Daphnia pulex*. (B) Schematic representation of logistic growth, the simplest form of density dependence, in which r declines linearly with population density.

(FIGURE 43.8A), and eventually reaches zero. When $r = 0$, the population stops changing in size—it reaches an **equilibrium** size that is called the carrying capacity, or K (**FIGURE 43.8B**). K can be thought of as the number of individuals the environment can support indefinitely. If individuals are added to the population so that it exceeds the carrying capacity, per capita growth rate becomes negative, and the population shrinks. If the population falls below K, per capita growth rate becomes positive, and the population grows.

Why does r decrease with crowding? When an organism uses a resource, it makes it unavailable to other organisms. Therefore, as population density increases, each individual gets a smaller piece of the resource "pie" and has fewer resources to allocate to its life functions (see Figure 43.5). Birth rate decreases, death rate increases, and r decreases with population density. Once again it is useful to spell this out:

Per capita growth rate =
the maximum possible value in uncrowded conditions –
an amount that is a function of density

When population density reaches K, the carrying capacity, each individual has just the amount of resources it needs in order to produce enough offspring to exactly replace itself when it dies. When density is less than K, each individual can more than replace itself; when density is greater than K, each individual has less than it needs to replace itself.

Now we are prepared to return to the question posed at the end of Concept 43.3: What causes temporal and spatial variation in population density?

Variable environmental conditions cause the carrying capacity to change

We have learned that resource availability in the environment affects an individual's success in acquiring resources (see Figure 43.4), and that physical conditions affect the costs of maintaining homeostasis (see Figure 43.5). If these factors vary over space, then the carrying capacity will vary over space as well.

Similarly, if resources or environmental conditions are good in one year but bad in another, the population will find itself alternately above and below the current carrying capacity. As a result, populations will fluctuate around an average K over time. Figure 43.2 illustrates this dynamic for rodents and ticks in Millbrook, New York. Rodent densities increase following years of high acorn production because the rodents have enough food to produce many offspring, and they decrease following poor acorn years. Similarly, nymphal tick densities go up and down following high and low rodent years, respectively, because rodent abundance determines the success of larval ticks in obtaining a blood meal.

A look back at Table 43.1 shows how temporal changes in the environment affected the fecundity of a group (cohort) of 210 cactus ground finches born on one of the Galápagos Islands in 1978. Notice that females in the cohort produced no surviving young when they were 7 and 8 years old, and that survival dropped precipitously in those years. These were years of low food availability following a severe drought in 1985. Conversely, finch fecundity was unusually high the year they turned 5. This was 1983, an unusually wet year in the Galápagos due to an El Niño event, and the finches had abundant food.

Technology has increased Earth's carrying capacity for humans

The human population is unique among populations of large animals. Not only has it continued to grow, it has grown at an ever-faster per capita rate, as is indicated by steadily decreasing doubling times (**FIGURE 43.9A**). The current growth rate (*r*) is approximately 1.3 percent per year—lower than the high of 2.2 percent reached in 1963—because birth rates have declined faster than have death rates. Even if this trend continues, however, the human population is projected to reach 8 billion by 2025.

The human population growth rate has jumped with every technological advance that has raised carrying capacity by increasing food production or improving health. The first jump occurred with the rise of agriculture that enabled humans to build civilizations during the Neolithic (about 10,000 years ago). In the nineteenth and twentieth centuries, advances in sanitation and the development of vaccines and antibiotics decreased death rates to a level undreamed of by our ancestors (**FIGURE 43.9B**). Finally, the "Green Revolution" in agriculture more than doubled world grain production between 1950 and 1984.

Human ingenuity has repeatedly stymied attempts to predict the future of the human population. The most famous attempt was that of Thomas Robert Malthus, whose *Essay on the Principle of Population*, first published in 1798, provided Charles Darwin with a mechanism for natural selection. Malthus pointed out that the human population was growing multiplicatively, whereas its food supply was growing additively. He predicted that food shortages would limit human population growth, which he estimated would peak by the end of the nineteenth century. Malthus could not, of course, have anticipated the effects of medical discoveries and the Green Revolution.

Many believe that the human population has now overshot its carrying capacity. They give two reasons. First, most twentieth-century technological advances relied on abundant fossil fuels, and these resources are finite. Maintaining current levels of agricultural production depends heavily on fertilizers derived from natural gas, pesticides derived from oil, and petroleum-fueled engines that pump irrigation water, till the soil, and process and transport food. Similarly, the manufacture and distribution of medicines and diagnostic machinery depends on fossil fuels. World production of crude oil appears to have peaked in 2008 and is expected to decline steadily into the future. Although coal reserves are still abundant, they, too, are finite and will eventually be gone.

A second reason many believe we have exceeded our carrying capacity is the climate change and degradation of Earth's ecosystems that have been a consequence of our twentieth-century population expansion. We will consider these topics further in Chapter 46.

Population ecology tells us that any change in the size of the global human population must result from changes in birth and death rates. If the human population has indeed exceeded its carrying capacity, ultimately it will decrease. We can bring this about voluntarily if we continue to reduce per capita birth rate. Alternatively, we can allow the Four Horsemen of the Apocalypse—famine, pestilence, war, and death—to achieve a sustainable human population size by increasing mortality.

> **LINK** Voluntary reduction of birth rates can be achieved by the use of the contraceptive methods described in Table 32.1

FIGURE 43.9 Human Population Growth Advances in technology have allowed the human population to grow at an ever-faster per capita rate.

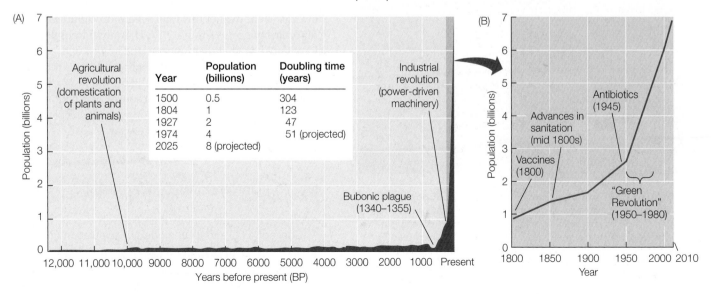

Year	Population (billions)	Doubling time (years)
1500	0.5	304
1804	1	123
1927	2	47
1974	4	51 (projected)
2025	8 (projected)	

Do You Understand Concept 43.4?

- Why don't populations grow indefinitely?
- Explain the fundamental difference between multiplicative and additive growth.
- Why do limits to population growth lead to natural selection? (Hint: Consider that individuals in a population vary in *b* and *d* due to inherited traits.)

So far we have considered effects of two demographic processes—birth and death—on populations. But there is another way—immigration—to join a population, and another way—emigration—to leave it. These among-population movements are also important for population dynamics.

concept 43.5 Extinction and Recolonization Affect Population Dynamics

Most organisms live in habitat patches that are separated from other patches by unsuitable environments (**FIGURE 43.10**; see also Figure 43.1). The population within a region consequently consists of a cluster of distinct *subpopulations* that are linked together by the dispersal of individuals among patches. This larger, regional population is known as a **metapopulation**.

Each subpopulation of a metapopulation changes as all populations do—through births and deaths—but with the added input of individuals that move into the population

FIGURE 43.10 A Metapopulation Has Many Subpopulations
The coast of Scandinavia is characterized by a many natural freshwater rockpools containing populations of small crustaceans of the genus *Daphnia* (sometimes called "water fleas"), which feed by filtering algae from the water. Each rockpool houses a subpopulation; the surrounding habitat is inhospitable to *Daphnia*, but may be crossed by enough individuals to recolonize or reinvigorate other subpopulations.

(immigrants) and loss of individuals that move away (emigrants). Adding the number of immigrants (*I*) and emigrants (*E*) to the BD growth model, we get the **BIDE model**:

The number of individuals in a population
at some time in the future =
the number now + the number that are born +
the number that immigrate –
the number that die –
the number that emigrate

The important difference between the BD and BIDE models is that the latter includes dispersal of individuals into and out of the subpopulation. In the BD model, a subpopulation that goes extinct forever stays extinct. Once the subpopulation reaches zero there are no individuals left to reproduce. With the BIDE model, however, immigration can bring an extinct population back to life.

In fact, extinction in small habitat patches is fairly common even when the environment of the patch is favorable. When subpopulations are small there is a reasonable chance that no female gives birth or that all individuals die of natural causes during a time period. Small populations are also at greater risk than large ones if there is an environmental disturbance.

Once a subpopulation goes extinct, it will remain at zero density—unless individuals disperse in and thus recolonize. Thus, if there is no dispersal among patches, the subpopulations will "wink out" one by one, and the metapopulation as a whole will eventually go extinct. The time to extinction is longer the larger each subpopulation is and the more subpopulations there are in the metapopulation, but extinction will occur nonetheless. If subpopulations are connected by dispersal, however, the metapopulation as a whole will persist much, much longer because empty patches can be repopulated and declining subpopulations can be "rescued" from extinction by immigration. The repopulated patches, in turn, can colonize others, causing them to "wink on."

We can see the recolonization process at work in the Santa Clara County, California metapopulation of Edith's checkerspot butterfly. Many subpopulations of this endangered butterfly went extinct during a severe drought that gripped California between 1975 and 1977. The only subpopulation that did not go extinct at that time was the largest one, on Morgan Hill (see Figure 43.1). The extinct subpopulations remained at zero density until 1986, when 9 suitable habitat patches were recolonized from the Morgan Hill source population. Patches closest to Morgan Hill were most likely to be recolonized because adult butterflies do not fly very far.

Do You Understand Concept 43.5?

- What is the fundamental difference between BD (see Concept 43.2) and BIDE models of population growth?
- Human-caused changes in land use often break up the original habitat of a species into spatially separated fragments. Why does this increase risk of extinction?

Extinction risk decreases as population size increases

In a study of subpopulations of *Daphnia* in Swedish rockpools (see Figure 43.10), ecologist Jan Bengtsson established a number of artificial rockpools that held either 12, 50, or 300 liters of water. He introduced *Daphnia* and then covered the pools with mesh, which prevented immigration or emigration. Over the subsequent 4 years, Bengtsson returned repeatedly to census numbers of *Daphnia*, thereby obtaining the average population size in each pool. If a population went extinct, he included only the non-zero values for population sizes before extinction in calculating the average. We can calculate the grand average of these time-based averages for pools in which the populations went extinct, and for those in which they persisted, to explore how extinction risk changes with pool and population size.

POOL SIZE (L)	NUMBER OF REPLICATE POOLS	PERCENTAGE OF POPULATIONS GOING EXTINCT	GRAND AVERAGE SIZE OF PERSISTENT POPULATIONS	GRAND AVERAGE SIZE OF EXTINCT POPULATIONS
12	27	18.5	1,500	65
50	25	16	2,705	530
300	11	9.1	9,690	1,800

1. Graph the relationship between pool size and probability of extinction, using size as the *x*-axis. What does your graph show you?

2. Within each pool size, which populations were at greatest risk of extinction?

3. How are these two results related to each other?

4. What might have happened if Bengtsson had removed the mesh from the pools after some amount of time?

In Chapter 42 we saw that ecological knowledge has many important applications in our everyday lives—in medicine, agriculture, and so on. An ability to predict the dynamics of populations and metapopulations contributes to these applications by allowing us to influence the fates of natural populations, as several examples will illustrate.

concept 43.6 Ecology Provides Tools for Managing Populations

For millennia, humans have tried to reduce populations of species they consider undesirable and maintain or increase populations of desirable or useful species. Such efforts to manage populations are most likely to be successful if they are based on knowledge of how those populations grow and what determines their densities.

Knowledge of life histories helps us to manage populations

Knowing the life history of a species helps us to identify those life stages that are most important for reproduction and survival, and hence for population growth rate. Let's look at a few examples of how this knowledge can be applied.

MANAGING FISHERIES The black rockfish (*Sebastes melanops*) is an important game fish that lives off the Pacific coast of North America. Rockfish exhibit indeterminate growth—that is, they grow continually throughout their lives. As in many other animals, the number of eggs a female produces is proportional to her size, so larger females produce more eggs than smaller females. Larger females are also better able to provision their eggs with oil droplets, which provide energy to the newly hatched larvae, allowing them to grow faster and survive better than larvae from eggs with smaller oil droplets.

These life history characteristics have important implications for the management of rockfish populations. Because fishermen prefer to catch big fish, intensive fishing off the Oregon coast from 1996 to 1999 reduced the average age of female rockfish from 9.5 to 6.5 years. Thus, in 1999, females were, on average, smaller than in 1996. This change decreased the average number of eggs produced by females in the population and reduced the average growth rate of larvae by about 50 percent, causing a rapid decline in rockfish population density. Because a relatively small number of large females can produce enough eggs to maintain the population, one strategy for maintaining productive rockfish populations without shutting fishing down completely is to set aside a few no-fishing areas where some females are protected and can grow to large sizes.

> **LINK** Strategies for maintaining the viability of a habitat for all its interacting species are discussed in Concept 45.6

REDUCING DISEASE RISK Because adult black-legged ticks feed and mate primarily on large mammals—white-tailed deer in particular—it seems logical that controlling deer densities would reduce the abundance of tick nymphs, which present the greatest risk of transmitting Lyme disease to humans. Surprisingly, experimental reductions in deer density have had very weak effects on subsequent nymph abundance. Studies of the tick's life history (see Figure 43.3) indicate that the success of larval ticks in obtaining a blood meal, not the number of them that hatch from eggs, has the greatest effect on the subsequent abundance of nymphs. As is true of rockfish, a very few adult female ticks can produce enough eggs to maintain populations. Hence, controlling the abundance of hosts for larvae (rodents) is a more effective strategy for reducing tick populations than controlling the abundance of deer.

CONSERVING ENDANGERED SPECIES The larval life stage is critical to the dynamics of Edith's checkerspot butterfly populations. Temporal and spatial variation in butterfly density are closely tied to the availability of two larval food plants: California plantain (*Plantago erecta*) and purple owl's clover (*Orthocarpus densiflorus*). Maintaining healthy populations of these plants is critical for conservation of these endangered butterflies. The two food plants are only found on serpentine soils, which have an unusual chemical makeup, but serpentine grasslands are being invaded by tall non-native grasses that suppress the low-growing *Plantago* and *Orthocarpus*. Grazing by cattle—another introduced species—can control the invasive grasses, and grazing is an important strategy for conserving *Euphydryas editha*. This shows that conservation goals often can be achieved without preventing other human uses.

Knowledge of metapopulation dynamics helps us conserve species

One goal of conservation efforts is to avoid the extinction of species. From studies of factors that affect extinction rates in individual populations, we know that risk to metapopulations is affected by the number and average size of subpopulations and rates of dispersal among them. This knowledge helps us to shape conservation strategies. Conservation plans therefore begin with an inventory of remaining areas of natural habitat and evaluation of the risks to those habitat patches. They then attempt to devise ways to protect as many habitat patches as possible.

FRONTIERS The National Aeronautics and Space Administration (NASA) and the United States Geological Survey (USGS) have teamed up to compile vast numbers of Landsat images (collected by "land-sensing" satellites in orbit around Earth) into computerized Geographic Information Systems (GIS) databases. With this information, conservation biologists can overlay maps of protected areas onto maps of endangered-species habitats, thus identifying gaps in the protection of critical habitats. This Gap Analysis Program allows them to efficiently plan for conservation of endangered species.

Priority is given to patches with the largest area because these can potentially host the largest populations. Another step is to evaluate the quality of the patches, as measured by their carrying capacity for the species, and to develop ways to restore or maintain quality. Finally, opportunities for dispersal among subpopulations must

---

INVESTIGATION

FIGURE 43.11 **Corridors Can Rescue Some Populations** Data from the experiments summarized here suggest that corridors between patches of habitat increase the chances of recolonization, and thus of subpopulation persistence.

HYPOTHESIS

Subpopulations of a fragmented metapopulation are more likely to persist if there is no barrier to recolonization.

METHOD

1. On replicate moss-covered boulders, scrape off the continuous cover of moss to create a "landscape" of moss "mainland" with patches surrounded by bare rock. A central 50 cm × 50 cm moss "mainland" (M) is surrounded by 12 circular patches of moss, each 10 cm² (subpopulations). In the "insular" treatment (I), the patches are surrounded by bare rock (which is inhospitable to moss-dwelling small arthropods, and thus a barrier to recolonization). In the "corridor" treatment (C), the patches are connected to the mainland by a 7 × 2 cm strip of live moss. In the "broken-corridor" treatment (B), the configuration is the same as the "corridor" treatment, except that the moss strip is cut by a 2-cm strip of bare rock.

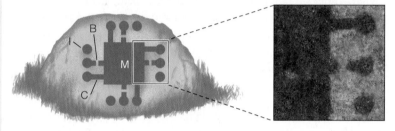

2. After 6 months, determine the number of small arthropod species present in each of the mainlands and small patches.

RESULTS

Patches connected to the mainland by corridors retained as many species as did the mainland to which they were connected. Fewer species remained in the broken-corridor and insular treatments.

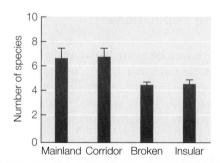

CONCLUSION

Barriers to recolonization reduce the number of subpopulations that persist in a metapopulation.

ANALYZE THE DATA

A. What percentage of the species present in the mainlands was lost, on average, in the corridor patches?

B. What percentage of the species present in the mainlands was lost, on average, in the insular and broken-corridor patches? What does this percentage tell you about the average risk of extinction for the subpopulation of each arthropod species in each patch?

C. The broken-corridor treatment controls for another factor that can affect extinction risk. What is this factor?

For more, go to Working with Data 43.1 at **yourBioPortal.com**.

Go to **yourBioPortal.com** for original citations, discussions, and relevant links for all INVESTIGATION figures.

be evaluated. For some species (such as Edith's checkerspot butterfly, which can fly through unsuitable habitat to reach a new serpentine outcrop), simple proximity of patches determines rates of recolonization. For those species, patches that are distant from one another can be connected by a series of intervening subpopulations that serve as "stepping stones." For other species, however, a continuous **corridor** of habitat through which the species can disperse is needed to connect subpopulations (**FIGURE 43.11**). Dispersal corridors can sometimes be created by maintaining vegetation along roadsides, fence lines, or streams, and by creating bridges or underpasses that allow individuals to circumvent roads or other barriers to movement (**FIGURE 43.12**).

Do You Understand Concept 43.6?

- What similarities do you see between the rockfish and black-legged tick examples summarized above?
- Name three elements of a conservation plan for a species that has a metapopulation structure.
- Conservation plans for marine species such as rockfish often lack a corridor component. Why?

You may have noticed that although this chapter has focused on single-species populations, it has also mentioned many interactions among species—rodents interacting with acorns and ticks, ticks with hosts and bacteria, butterflies with food plants and invasive plants. These ubiquitous between-species interactions, which play important roles in the population dynamics of all the interacting species, are the subject of the next two chapters.

QUESTION   How does understanding the population ecology of disease vectors help us combat infectious diseases?

ANSWER   How frequently people come into contact with a dangerous pathogen depends on its abundance and distribution and those of any vectors that may transmit it to humans. Population ecology helps us understand what determines the abundance and distribution of pathogens and their vectors, and hence allows us to devise ways of controlling their abundance or avoiding contact with them.

By studying the survivorship and fecundity of the black-legged tick through its life history (Concept 43.3), Rick Ostfeld learned that it is the abundance of hosts for larvae, rather than weather or the abundance of deer, that determines tick abundance. Furthermore, Ostfeld ascertained that ticks become infected with *Borrelia* when they feed on infected hosts, and that rodents are the most effective reservoirs for *Borrelia*. Long-term studies of rodent population dynamics had shown that their abundance is determined by acorn availability (Concept 43.4).

This knowledge suggests an immediate way to reduce the probability of contracting Lyme disease (Concept 43.6). Acorn production can be used to predict areas that are likely to become infested with disease-carrying ticks, and various measures can be taken to reduce the risk of human contact with those ticks. These measures include boosting public education efforts for several years after a big acorn crop, posting warning signs in high-risk areas, and alerting health-care specialists whose patients live or work in high-risk areas. A longer-term possibility that has yet to be assessed is to reduce the fraction of ticks that carry *Borrelia* by augmenting populations of lizards and birds—which do not harbor *Borrelia*—in the host community.

FIGURE 43.12   A Corridor for Large Mammals An overpass above the Trans Canada Highway in Banff National Park, Alberta. Bears, elk, deer and other large mammals use such overpasses, limiting the number of encounters with speeding vehicles that can have devastating effects for all involved.

43 SUMMARY

concept 43.1 Populations Are Patchy in Space and Dynamic over Time

- **Population density** is the number of individuals in a population per unit of area (or volume). **Population size** is the total number of individuals in a population.

- The region within which a species occurs is called the species' **geographic range**. Within that range, the species may be restricted to particular suitable **habitats**; these may occur in **habitat patches** separated by areas of unsuitable habitat. **Review Figure 43.1**

- Population densities vary over time as well as over space. **Review Figure 43.2**

concept 43.2 Births Increase and Deaths Decrease Population Size

- According to the **BD model**, the population size at some future time is the current size plus the total number of births minus the number of deaths that occur until that time. Equations based on the BD model can be used to calculate a population's growth rate.

- Ecologists can estimate the **per capita growth rate** (*r*) of a population by tracking births and deaths in a sample of individuals.

- If a population's per capita growth rate does not equal zero, then its size must be changing. If *r* is positive, the population is growing. If *r* is negative, the population is shrinking.

concept 43.3 Life Histories Determine Population Growth Rates

- A species' **life history**, which is a quantitative description of its life cycle, can be summarized in a **life table**. **Review Figure 43.3 and Table 43.1**

- The fraction of individuals that survive to particular life stages or ages is **survivorship**, and the average number of offspring they produce at those life stages or ages is **fecundity**. Survivorship and fecundity determine the per capita growth rate, *r*. **Review Table 43.1**

- To survive and reproduce, organisms need materials and energy and the time to acquire them—all of which constitute **resources**. Organisms also need physical **conditions** they are able to tolerate.

- The rate at which an organism can acquire resources increases with the availability of resources in its environment. **Review Figure 43.4**

- The **principle of allocation** states that once an organism has acquired a unit of some resource, it can be used for only one function at a time. **Review Figure 43.5**

- The environment shapes life histories because it determines the relative costs and benefits of any particular pattern of allocation to different functions.

- Species can persist only in environments in which their per capita growth rate is positive. **Review Figures 43.6 and 43.7**

concept 43.4 Populations Grow Multiplicatively, but Not for Long

- Populations grow **multiplicatively**, increasing or decreasing by a constant multiple of their current size in each interval of time. As a result, populations that are increasing in size have a constant **doubling time**. **Review WEB ACTIVITY 43.1 and ANIMATED TUTORIALS 43.1 and 43.2**

- As a population becomes denser and resources become scarcer, birth rates decline and death rates increase. Thus the per capita growth rate is said to be **density dependent**.

- When *r* = 0, the population reaches an **equilibrium** size. That size, called the **carrying capacity**, or *K*, is the number of individuals the environment can support indefinitely. **Review Figure 43.8**

- Changes in resource availability or physical conditions cause the carrying capacity to change.

- Technology has increased Earth's carrying capacity for humans, and the human population has responded by growing rapidly. **Review Figure 43.9**

concept 43.5 Extinction and Recolonization Affect Population Dynamics

- The regional distribution of a species often takes the form of a **metapopulation**, a cluster of distinct subpopulations in separate habitat patches linked together by the dispersal of individuals among them.

- The **BIDE model** adds immigration and emigration to the BD model. Immigration can repopulate an area where a subpopulation has gone extinct.

concept 43.6 Ecology Provides Tools for Managing Populations

- Knowing the life history of a species makes it possible to identify those life stages that are most important for its population growth rate. That knowledge can be applied by managers to reduce populations of species considered undesirable and maintain or increase populations of desirable or useful species.

- Conservation efforts give priority to protecting the largest remaining habitat patches because these patches can host the largest populations. Providing continuous **corridors** of habitat through which individuals can disperse among patches reduces extinction risk. **Review Figure 43.11, WORKING WITH DATA 43.1, and ANIMATED TUTORIAL 43.3**

Ecological and Evolutionary Consequences of Species Interactions

On a morning walk in the Arizona desert, it's hard to resist the temptation to stop and watch the parades of colorful flower and leaf fragments that march across your path. The fragments aren't moving on their own, of course—they're being carried by *Acromyrmex versicolor*, one species of leaf-cutter ant. Leaf-cutter ants are tropical insects whose geographic range barely extends into North America. If you travel south through Mexico and into the New World tropics, however, you will see more and more ant highways and more and more different leaf-cutter species. Leaf-cutter ants originated in South America about 10 million years ago, crossing into North America only after the Panamanian land bridge connected the two continents.

Leaf-cutter ants are major herbivores in the Neotropics. A single colony can contain 8 million workers, occupy 20 cubic meters of soil, and consume more than 2 kilograms of plant material per day—enough to strip a large area of vegetation.

Leaf-cutters are no ordinary herbivores, however: they're fungus farmers! They use the leaves they harvest to grow a fungus that occurs in their nests and nowhere else on Earth. The interaction between fungus and ant is an obligate one—neither species can live without the other—that involves a remarkable array of specialized adaptations on the part of both partners.

New queens carry a pellet of fungus with them when they leave the nest where they were born and use it to start their own fungus garden, which they fertilize with their feces. They lay eggs in the developing fungal mass, and the resulting larvae feed on the fungus. Once the larvae

develop into worker ants, they collect leaf material to feed the growing fungus garden, avoiding plant species toxic to their fungal crop. As the colony expands, its workers take on additional tasks as scouts, nest excavators, gardeners, nurses, and garbage collectors—even bodyguards of leaf harvesters.

Leaf-cutters also associate with other species that benefit their interaction with the fungus. A bacterium in the genus *Pseudonocardia*, carried in special organs in the ants' exoskeletons, manufactures antibiotics that suppress the crop-killing green mold *Escovopsis* but do not harm the cultivated fungus. And just recently, researchers discovered a nitrogen-fixing bacterium, *Klebsiella*, in the fungus gardens. Who knows how many more leaf-cutter associates remain to be discovered?

Leaf-cutter ants of the tropical species *Atta cephalotes* carry coca leaves through the rainforest. The ants riding on the leaves are thought to be "bodyguards," warding off parasitic flies that would otherwise lay their eggs on the larger worker ants.

Q QUESTION How could the intricate ecological relationship between leaf-cutter ants and fungi have evolved?

KEY CONCEPTS

44.1 Interactions between Species May Be Positive, Negative, or Neutral

44.2 Interspecific Interactions Affect Population Dynamics and Species Distributions

44.3 Interactions Affect Individual Fitness and Can Result in Evolution

44.4 Introduced Species Alter Interspecific Interactions

concept 44.1 Interactions between Species May Be Positive, Negative, or Neutral

The interactions between leaf-cutter ants and their nest associates represent but a small fraction of the ants' interactions with other species. When they harvest leaves, the ants interact with many plant species. The ants themselves are consumed by parasitic flies, beetles, and pathogenic fungi, and the ant corpses that are hauled to the colony's refuse dump are consumed by many species of microbes. The leaf-cutter example illustrates one of life's certainties: at some point between birth and death, every organism will encounter and interact with individuals of other species. These **interspecific interactions** have consequences that can affect the individual's fitness. Thus they can influence the densities of populations, alter the distributions of species, and lead to evolutionary change in one or both of the interacting species.

Interspecific interactions are classified by their effect on per capita growth rates

The details of interspecific interactions are bewilderingly diverse. Nonetheless, we can make considerable progress in understanding their ecological and evolutionary consequences by simply asking whether an interaction is beneficial or detrimental to individuals of the participating species—that is, whether an interaction increases, decreases, or does not affect each species' per capita growth rate.

If we consider all the possible pairwise combinations of negative (–), positive (+), and neutral (0) effects, we come up with five broad categories of interspecific interactions (**FIGURE 44.1A**): competition (–/–), consumer–resource (+/–), mutualism (+/+), commensalism (+/0), and amensalism (–/0).

Interspecific competition refers to –/– interactions in which members of two or more different species use the same resource. Recall from Concept 43.4 that when an organism consumes a resource, it makes that resource unavailable to other individuals of the same or different species, potentially reducing their resource intake and their per capita growth rate. We say "potentially" because not all resources are in short enough supply to limit per capita growth rate. However, at any one time there generally is one **limiting resource** that is in the shortest supply relative to demand. The limiting resource need not be food; it may be water, space, or—in the case of plants—sunlight (**FIGURE 44.1B**), inorganic nutrients, or pollinators. Competition can occur among predators that depend on the same prey species, among herbivores that depend on the same host plant, among pathogenic microbes that attack the same host, among neighboring plants, and even among plants vying for the attention of pollinating animals.

Consumer–resource interactions are those in which organisms gain their nutrition by eating other living organisms or are eaten themselves. It's obvious why these are +/– interactions: the consumer benefits while the consumed organism—the resource—loses. Consumer–resource interactions include

predation, in which an individual of one species (the **predator**) kills and consumes individuals of another species (the **prey**); **herbivory**, in which an animal consumes part or all of a plant (**FIGURE 44.1C**); and **parasitism**, in which a parasitic organism consumes part of a host individual but usually does not kill it. Parasites are generally smaller than their hosts. Some parasites are considered pathogens because they make their host ill by damaging tissue or producing toxins in the host's body.

A **mutualism** is an interaction that benefits both species (+/+). The interaction between leaf-cutter ants and fungi described at the start of this chapter is mutualistic: the ants feed, cultivate, and disperse the fungi; the fungi, in turn, convert inedible plant material into food the ants can eat. The interactions between plants and pollinating or seed-dispersing animals (**FIGURE 44.1D**), or between humans and the bifidobacteria in our guts (see Figure 42.2), are similarly beneficial to both partners. Mutualisms take many forms and involve many kinds of organisms. They also vary in how essential the interaction is to the partners. We have seen several examples of mutualism in this book, including interactions between mycorrhizal fungi and plants (see Concepts 22.2 and 25.2); between fungi, algae, and cyanobacteria in lichens (see Concept 22.2); and between corals and dinoflagellates (see Concept 20.4).

Competition, consumer–resource interactions, and mutualism all affect the fitness of both participants. The other two defined types of interactions affect only one of the participants.

Commensalism is an interaction in which one participant benefits while the other is unaffected (+/0). Most examples of commensalism (Latin, "eating together") involve one species whose feeding behavior makes food more accessible for another. For example, the brown-headed cowbird (*Molothrus ater*) owes its common name to its habit of following herds of grazing cattle, foraging on insects flushed from the vegetation by their hooves and teeth. (*M. ater* has been called the buffalo bird because it followed the American bison herds that were once abundant across North America; see Figure 44.1C). The ungulate–cowbird interaction is a commensal one because the cowbirds have no effect on the fitness of the ungulates.

In other cases, the feeding behavior of one species enhances the availability of food for another species by converting food that is unusable to a second consumer into a usable form. Cattle, for example, convert plants into dung, which dung beetles can use. Dung beetles, in turn, are on the giving end of another commensalism: dung-living organisms that cannot fly, such as mites, nematodes, and even fungi, attach themselves to the bodies of the beetles, which not only can fly but are very good at locating fresh dung. The hitchhikers have no apparent effect on the dung beetles' fitness.

Amensalism refers to interactions in which one participant is harmed while the other is unaffected (–/0). A herd of elephants moving through a forest or a herd of bison grazing the plains crush insects and plants with each step (see Figure 44.1C), but the large mammals are unaffected by this carnage. Amensal interactions tend to have a more accidental relationship to the biology of the participants than do other interactions.

(A)

Major Types of Species Interactions

TYPE OF INTERACTION	EFFECT ON SPECIES 1	EFFECT ON SPECIES 2
Competition	−	−
Consumer–resource: Predation, herbivory, parasitism	+	−
Mutualism	+	+
Commensalism	+	0
Amensalism	−	0

(C)

Consumer–resource
The American bison feeds on the grasses of the Great Plains.

Amensalism, Commensalism
The large mammal unwittingly destroys insects and their nests. The buffalo birds feed on insects disturbed by the bison's passage.

(B)

Competition
Green plants compete for light. The leaves of tall trees have reduced the light available to the plants growing on the forest floor.

(D) *Tegeticula yuccasella*

...and deposits the pollen on the stigma of another flower.

Mutualism
The female yucca moth collects and carries pollen grains in specialized mouthparts...

Female moths subsequently lay eggs in the plant's ovary. The resulting larvae eat some of the developing seeds.

Because commensalism and amensalism are not reciprocal (that is, only one species is affected by the abundance of the other), they do not show the same ecological or evolutionary dynamics as reciprocal interactions do. For this reason, we will not focus on commensalism or amensalism in the remainder of the chapter.

Many interactions have both positive and negative aspects

Although ecologists find it useful to group interspecific interactions into the five categories we have described, the boundaries separating these categories are not always clear. Most interactions have both beneficial and harmful aspects, and the net effect of one species on another can be positive or negative, depending on the relative strengths of these beneficial and harmful aspects. These relative strengths in turn may be contingent on environmental conditions.

Sea anemones, for example, sting and eat small fish, but a select few fish species (mostly anemonefish; genus *Amphiprion*) are protected by a special mucus coat and thus can live among the anemones' stinging tentacles (**FIGURE 44.2**). Although the

FIGURE 44.1 Types of Interspecific Interactions (A) Interactions between species can be grouped into categories based on whether their influence on the per capita growth rate of each species is positive (+), negative (−), or neutral (0). (B) The composition of canopy (treetop) and understory (low-growing) vegetation in a forest is largely a product of competition for light. (C) Commensalisms between large, grazing ungulates and insect-eating birds are found in grassland environments around the world. (D) The mutualisms between desert *Yucca* plants and their moth pollinators are highly specialized. In the case seen here, *Yucca filamentosa* has an obligate mutualism with its sole pollinator, the moth *Tegeticula yuccasella*.

yourBioPortal.com
Go to WEB ACTIVITY 44.1
Ecological Interactions

benefit of this association to the anemonefish is clear—they escape their own predators by hiding behind the anemones' nematocysts and can scavenge food caught by the anemones—the consequences for the anemones are less clear. Do they benefit because the fish defecate nitrogen-rich feces whose nutrients they can use? Do they suffer because the fish steal some

APPLY THE CONCEPT

Interactions between species may be positive, negative, or neutral

In many animals that face uncertain food availability, a tendency to collect and store more food than they can eat immediately has evolved. Desert seed-eating rodents such as Merriam's kangaroo rat (*Dipodomys merriami*; see Figure 43.4B), for example, will harvest as many seeds as you put in front of them and bury the seeds in shallow depots scattered around their territory. In some years seed production is low, and the rodents eat all the seeds they manage to store. In other years they store more seeds than they are able to eat, and the uneaten stored seeds are likely to germinate as grass seedlings.

In a classic long-term study done in the 1930s and 1940s, researchers at the Santa Rita Experimental Range in southern Arizona monitored the effect of kangaroo rats on native grasses. They established replicate enclosures, removed kangaroo rats from half of them, then monitored the abundance of grasses in the two types of enclosures over a period of 10 years. From 1931 to 1935 the region experienced a severe drought and grass populations declined. Rains returned in 1935. The year 1941 saw particularly high (for the desert) rainfall. Use the graph to answer the following questions.

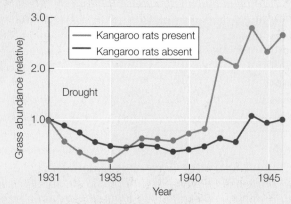

1. During the drought, did kangaroo rats have a positive, negative, or neutral effect on the per capita growth rate of grasses? Explain your reasoning.

2. Did the effect of kangaroo rats on grasses change after rains returned? Explain your reasoning.

3. What aspects of kangaroo rat behavior are detrimental to grasses? What aspects might help grasses?

4. How might year-to-year variation in rainfall influence the relative strengths of the positive and negative effects kangaroo rats have on grass growth rates?

---

of their prey? The net effect of the interaction on the anemones may well depend on the availability of nitrogen-rich food in their environment.

Amphiprion ocellaris

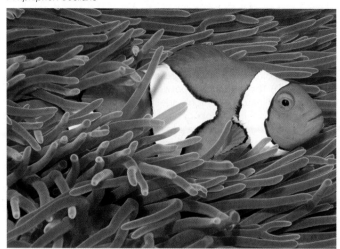

FIGURE 44.2   Interactions between Species Are Not Always Clear-Cut   Ecologists long believed that the relationship between sea anemones and anemonefish was a commensalism: that the fish, by living among the anemones' stinging tentacles, gained protection from predators. But could it also be considered a mutualism—if the fishes' feces provide the anemones with beneficial nutrients—or competition—if the fish occasionally steal the anemones' prey?

Do You Understand Concept 44.1?

- What is the criterion for classifying interspecific interactions?

- What type of interspecific interaction would you consider human cultivation and consumption of crop plants to be?

- Describe an experiment that would allow you to determine whether the effect of anemonefish on sea anemones is positive, negative, or neutral.

Because interspecific interactions affect per capita growth rates, it is clear that they affect the population dynamics of the participant species. How exactly are those dynamics affected?

concept 44.2   Interspecific Interactions Affect Population Dynamics and Species Distributions

As we saw in Concept 43.2, the growth rate of a population equals the average individual's contribution to change in population size (the per capita growth rate, r) multiplied by the number of individuals in the population. If r does not change, population size either follows a J-shaped trajectory to infinity (if r is positive) or decreases to zero (if r is negative). We also saw in Concept 43.4 that r usually decreases with population density, and that this density dependence dramatically alters the trajectory of growth: instead of growing to infinity, populations stop growing and fluctuate around a carrying capacity (K).

Interspecific interactions can modify per capita growth rates

Density-dependent population growth reflects *intraspecific*, or within-species, interactions among the individuals in a population. These intraspecific interactions are usually detrimental because per capita resource availability decreases as a population's density increases (see Concept 43.4). Density-dependent population growth therefore generally describes **intraspecific competition**.

The word equation on page 852 describes the intraspecific effects of density dependence, but can be extended to include competition between two species:

per capita growth rate (r) of species A =

{maximum possible r for species A in uncrowded conditions –

an amount that is a function of A's own population's density} –

{an amount that is a function of the population density of competing species B}

Notice that we need to write a pair of equations—one each for competing species A and B. Each equation consists of an expression that describes growth of one species in the absence of the other, minus an expression that describes the effect of the other species.

Interspecific interactions other than competition can be described in a similar way. Depending on the type of interaction, we must either subtract or add the effect of the other species (see Figure 44.1A). In consumer–resource interactions, we subtract the effect of the consumer (e.g., the predator) in the equation for the resource species (e.g., the prey), since the consumer increases the mortality of the resource species. Conversely, we add an effect of the resource in the equation for the consumer, since the consumer benefits from the presence of the resource. In mutualistic interactions, we add an effect of each mutualist to the equation for the other. In commensal interactions we add a term in the equation of the species that benefits, and in amensal interactions we subtract a term in the equation of the species that is harmed. (For commensal and amensal interactions, the equations for the partner species contain no reciprocal terms, because by definition the partner is unaffected by the interaction.)

Interspecific interactions can lead to extinction

Because interspecific interactions modify per capita growth rates, populations show different dynamics in the presence and absence of other species. The Russian ecologist G. F. Gause demonstrated the effects of interspecific interactions on the population dynamics of several species of the protist *Paramecium* in a classic laboratory study published in 1934. Gause grew these unicellular organisms (see Figure 20.6) both in single-species cultures and in cultures containing a competing species of *Paramecium*. In order to better understand the different outcomes he observed, Gause used both word models such as the one we gave earlier, and mathematical expressions of the same interactions. In the many decades since Gause's work, mathematical models have proved to be useful tools for understanding how interspecific interactions affect population dynamics. In the case of competition, all models lead to the following conclusions:

- The presence of a competitor always reduces population growth rate. In **FIGURE 44.3A**, notice that the slope of the curve of population density versus time at any given population density is shallower in the presence of a competitor, because the competitor reduces the per capita growth rate from what it would be with just intraspecific density dependence.

- In addition, when two species coexist as in Figure 44.3A, they achieve lower equilibrium population densities than either would achieve alone.

- In some cases, competition causes one species to go extinct (**FIGURE 44.3B**).

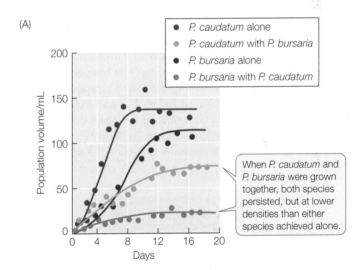

(A)

When *P. caudatum* and *P. bursaria* were grown together, both species persisted, but at lower densities than either species achieved alone.

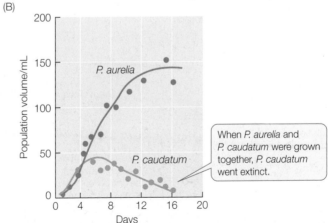

(B)

When *P. aurelia* and *P. caudatum* were grown together, *P. caudatum* went extinct.

FIGURE 44.3 Interspecific Competition Affects Population Growth (A) Populations of the unicellular protist *Paramecium* grow more slowly and reach lower equilibrium densities when a competing species is present. (B) Competition can result in extinction.

Other types of interspecific interactions have similar consequences:

- In all cases, the per capita growth rate of each interacting species is modified by the presence of the other, sometimes positively and sometimes negatively.

- In all cases, the average population densities of each species differ in the presence and absence of the other species. Population densities are increased in positive interactions and decreased in negative interactions.

- Finally, in interactions that have negative effects—that is, consumer–resource interactions, competition, and amensalism—extinction of one or both of the interacting species is possible.

Interspecific interactions can affect the distributions of species

In Chapters 42 and 43 we saw that a species may be absent from a particular location, either because barriers to dispersal have prevented the species from colonizing it or because the environment is unfavorable for the species' population growth. We can now see that even if a species reaches a site, interactions with other species may affect its distribution. A site can be unfavorable because it lacks suitable prey or because it harbors a competitor or predator; conversely, an otherwise unfavorable site may be improved by the presence of a mutualist.

FIGURE 44.4 shows a classic example of how competitive interactions can restrict the habitats in which species occur. The rock barnacle (*Semibalanus balanoides*) and Poll's stellate barnacle (*Chthamalus stellatus*) compete for space on the rocky shorelines of the North Atlantic Ocean. In these organisms, occupying a space on a rock amounts to having the opportunity

to feed because adult barnacles attach permanently to rocks, filtering food particles from the water that flows past them. The planktonic (swimming) larvae of both *S. balanoides* and *C. stellatus* settle in the intertidal zone and metamorphose into sessile adults. There is little overlap between the areas occupied by adults; the smaller stellate barnacles generally live higher in the intertidal, where they face longer periods of exposure and desiccation, than do the rock barnacles.

In a famous experiment conducted almost 50 years ago, Joseph Connell removed each of the species from its characteristic zone and observed the response of the other species. Stellate barnacle larvae settled in large numbers throughout much of the intertidal zone, including the lower levels where rock barnacles are normally found, but they thrived at those levels only when rock barnacles were not present. The rock barnacles grew so fast that they smothered, crushed, or undercut the stellate barnacle larvae. In other words, by outcompeting them for space, rock barnacles made the lower intertidal zone unsuitable for stellate barnacles. In contrast, removing stellate barnacles from higher in the intertidal zone did not lead to their replacement by rock barnacles; the rock barnacles were less tolerant of desiccation and so failed to thrive there even when stellate barnacles were not present.

Rarity advantage promotes species coexistence

Mathematical models of interspecific interactions have been useful not only in helping us understand how interactions affect population dynamics and species distributions, but also by giving us insights into the conditions that allow interacting species to coexist.

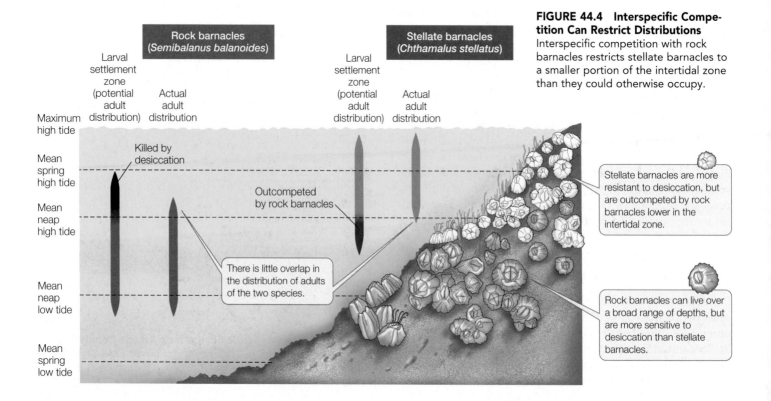

FIGURE 44.4 Interspecific Competition Can Restrict Distributions Interspecific competition with rock barnacles restricts stellate barnacles to a smaller portion of the intertidal zone than they could otherwise occupy.

Rock barnacles (*Semibalanus balanoides*)

Stellate barnacles (*Chthamalus stellatus*)

Larval settlement zone (potential adult distribution)

Actual adult distribution

Maximum high tide

Mean spring high tide

Mean neap high tide

Mean neap low tide

Mean spring low tide

Killed by desiccation

Outcompeted by rock barnacles

There is little overlap in the distribution of adults of the two species.

Stellate barnacles are more resistant to desiccation, but are outcompeted by rock barnacles lower in the intertidal zone.

Rock barnacles can live over a broad range of depths, but are more sensitive to desiccation than stellate barnacles.

Models of interspecific competition indicate that two competitors will coexist when each species suppresses its own per capita growth rate more than it suppresses the per capita growth rate of its competitor. In other words, intraspecific competition must be stronger than interspecific competition. When this is the case, a species gains a growth advantage when it is at a low density and its competitor is at a high density—that is, when the species is rare—and this *rarity advantage* prevents the species from decreasing to zero. The result is coexistence.

One mechanism that can cause intraspecific competition to be stronger than interspecific competition is **resource partitioning**: differences between competing species in resource use. For example, *Paramecium caudatum* can coexist with *P. bursaria* but not with *P. aurelia* (see Figure 44.3), because *P. bursaria* can feed on bacteria in the low-oxygen sediment layer at the bottom of culture flasks—a habitat that *P. caudatum* cannot tolerate. This partitioning of habitat is possible because *P. bursaria* harbors symbiotic algae that provide it with additional oxygen as a by-product of algal photosynthesis.

FIGURE 44.5 illustrates how, in general, differences between species in their use of resources cause individuals to have a larger effect on the resources available to another individual of their own species than to the typical individual of another species. If differences in resource use are sufficiently large, competing species can coexist, as we will discuss again in Concept 44.3.

In the case of predators and prey, a number of processes can result in a rarity advantage that protects prey from being driven extinct by their predators. For example, prey may become harder to find as they become more rare because they all can hide in the best refuges; or they can invest in better defense when they are at low densities and have more resources available per capita. Predators may also stop looking for rare prey and switch to another prey. Alternatively, some other limiting factor (such as availability of nest sites) may prevent the predators from becoming numerous enough to eat all the prey. The effect of such processes is that per capita growth rates may rebound when prey become rare, thus making prey extinction less likely.

Do You Understand Concept 44.2?

- Describe the components of an equation for the per capita growth rate of a species that interacts with another species.

- What are the possible outcomes for the final population sizes of two species involved in a predator–prey interaction?

- Explain, in terms of resource use, why the equilibrium population size of one species is reduced by the presence of a competing species.

The effects of interspecific and intraspecific interactions on individual members of a population are not fixed, but depend on the individuals' characteristics. For example, the overlap between seed-eating birds in the size of the seeds they are able to eat depends on just how different their beak sizes are. How successful a lynx is at catching a snowshoe hare depends on how

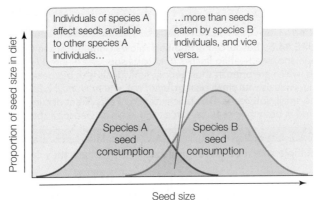

FIGURE 44.5 Resource Partitioning Can Result in Intraspecific Competition Being Greater than Interspecific Competition When two species differ in resource use, individuals will have a greater affect on resource availability to individuals of the same species than to individuals of the second species.

conspicuous the hare is and how fast or alert the lynx is. If individuals vary in attributes that determine how much a competitor affects their access to resources or their risk of being eaten by a predator, then the interaction will result in natural selection in both interacting species.

concept 44.3   Interactions Affect Individual Fitness and Can Result in Evolution

As we saw in Chapter 43, an individual's contribution to population growth summarizes its success in surviving and in reproducing. We have seen that both intraspecific and interspecific interactions affect per capita growth rates. If individuals involved in an interaction are affected by it to different extents, they will differ in *fitness*: those phenotypes that gain the most from a positive interaction or suffer the least from a negative interaction will increase in frequency in the population, and the population will evolve (see Chapter 15). The mixes of positive and negative effects that distinguish the various types of interactions cause different evolutionary dynamics, as we will now see. We will begin with intraspecific competition.

Intraspecific competition can increase carrying capacity

Equations of density-dependent growth assume that all the individuals in a population are equally affected by its density. However, individuals often vary in traits that affect how well they can use available resources. Suppose, for example, that in a particular population of seed-eating birds, individuals with small beaks can feed on both large and small seeds, whereas birds with larger beaks cannot feed efficiently on the small seeds. In this situation, more resources are available to the small-beaked individuals, their per capita growth rate—their fitness—exceeds that of the larger-beaked individuals, and the frequency of small-beaked individuals in the population will increase.

As small-beaked individuals come to dominate the population, the average beak size in the population will get smaller,

INVESTIGATION

FIGURE 44.6 Resource Partitioning Allows Competitors to Coexist In the Galápagos archipelago, seed-eating finches (*Geospiza* spp.; see Figure 17.7) use their beaks to crack open seeds, which are often in short supply. Individuals with big beaks can crack large, hard seeds that individuals with small beaks cannot eat, whereas small-beaked birds are more efficient at eating small, soft seeds. Dolph Schluter and Peter Grant documented how 15 of the islands differed in both their seed resources and in which *Geospiza* species were present.

HYPOTHESIS

Multiple species of *Geospiza* coexist only where they can partition seed resources.

METHOD

1. Determine which species of seed can be eaten efficiently by *Geospiza* with different beak sizes.
2. Measure the abundances of seeds of different sizes on 15 islands.
3. Characterize the availability of food for finches with different beak sizes, expressing availability as the population density that can be supported on each island (a measure of carrying capacity).
4. Determine the average beak sizes of finch species present on each island.

RESULTS

Islands supported between 1 and 3 finch species (shown here for 6 of the 15 islands). The number of species present (indicated by colored dots) increased with the number of seed sizes available (reflected in the number of distinct peaks in carrying capacity).

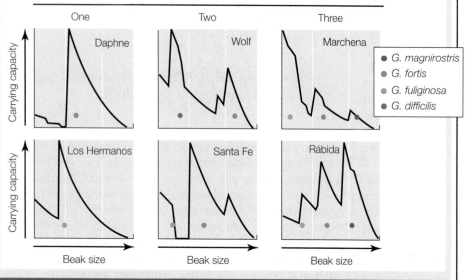

Number of finch species present on island

- G. magnirostris
- G. fortis
- G. fuliginosa
- G. difficilis

CONCLUSION

Coexisting *Geospiza* species differ in beak sizes and thus in resource use.

ANALYZE THE DATA

A. The average beak sizes of finch species present on an island (indicated by the locations of the colored dots) closely match the peaks in carrying capacity. Why might this be so?

B. Why do peaks in seed availability have no more than one associated finch species?

C. Why might some peaks in seed availability lack an associated finch species? (Hint: See Concept 45.1.)

Go to **yourBioPortal.com** for original citations, discussions, and relevant links for all INVESTIGATION figures.

LINK The three patterns of selection—directional, stabilizing, and disruptive—are described in Concept 15.4

Interspecific competition can lead to resource partitioning and coexistence

Just as individuals can vary in traits that affect their sensitivity to intraspecific competition, they can vary in traits that affect their sensitivity to interspecific competition. Studies from the Galápagos Islands show that beak sizes of a given finch species can vary from island to island, that beak sizes appear to match the sizes of available seeds, and that no island supports more than one finch species with a given beak size (**FIGURE 44.6**). These results are interpreted as another example of resource partitioning, in this case as an evolutionary response to interspecific competition. To see how this might occur, consider another example from the Galápagos, this time between a finch and an insect.

The small ground finch (*Geospiza fuliginosa*) consumes flower nectar as well as seeds. Smaller-size birds are better at obtaining nectar than are larger birds, whereas larger birds are better at cracking seeds. Carpenter bees (*Xylocopa darwinii*) compete with *G. fuliginosa* for nectar on some islands but are absent from others. Small finches are affected more strongly by competition from these bees than are larger individuals because they are less adept at eating foods other than nectar. As a consequence, the finches are larger in size and drink less nectar on islands where carpenter bees are present (**FIGURE 44.7**). This evolutionary shift to larger size and greater use of seeds means *G. fuliginosa* resource use has diverged from that of their bee competitors on islands where the two species coexist.

Consumer–resource interactions can lead to an evolutionary arms race

Predators, parasites, and herbivores benefit if they can acquire lots of nutritious food quickly and inexpensively, whereas resource species benefit if they can deter these natural enemies at little cost. The interests of consumer and resource species are obviously at odds. These opposing interests can lead to an

and the population will grow larger because the abundance of small seeds makes the carrying capacity greater for populations with a smaller average beak size. The population will stop growing when it reaches the higher carrying capacity for smaller-beaked birds. This change in average beak size, and the associated change in carrying capacity, are the results of directional selection.

Geospiza fuliginosa

Xylocopa darwinii

Nectar Use and Size of *G. fuliginosa*

ISLAND	TIME SPENT FEEDING ON FLOWER NECTAR (%)	MEAN SIZE (WINGSPAN, mm)
Bees Absent		
Pinta	10	59.8
Marchena	28	58.2
Bees Present		
Fernandina	1	64.8
Santa Cruz	14	64.0
San Salvador	0	63.8
Española	0	64.7
Isabela	7	64.5

FIGURE 44.7 Finch Morphology Evolves in Response to Competition with Carpenter Bees   On those Galápagos islands where small ground finches are the sole pollinators of small flowers, the birds are small (as measured by their short wingspans). Small size lowers their energy requirements and may increase their ability to negotiate the flowers. On islands where carpenter bees compete with these finches for nectar, the birds are larger and have stronger beaks (which allows them to include seeds in their diet).

"evolutionary arms race," in which prey continually evolve better defenses and predators continually evolve better offenses, and neither gains any lasting advantage over the other. The Red Queen summed up such a dynamic in Lewis Carroll's *Through the Looking-Glass* with the words "it takes all the running you can do, to keep in the same place."

Resource species have various defensive strategies available to them. Some mobile animals use speed, size, or weapons to thwart predators. Others hide or use camouflage to avoid being detected, or they mimic unpalatable species (**FIGURE 44.8A**). Immobile organisms have other tricks up their sleeves; for example, they may have evolved thick armor, or they may be non-nutritive or poisonous. In response to these strategies, natural selection in populations of consumers favors greater speed, size, or strength; keen senses; armor-piercing or crushing tools (**FIGURE 44.8B**); or means of detoxifying poisons.

Plants produce a great variety of chemicals that are toxic to herbivores or pathogens, as we saw in Chapter 28. Used in small quantities, some of these chemicals spice up our lives— the caffeine of coffee and tea, the mustard oils of cabbages, the capsaicins of chilis, and the piperine of black pepper are examples of plant defensive chemicals. Ways around these defenses have evolved in some herbivores. *Heliconius* butterflies store or detoxify the cyanide-containing defensive compounds of passionflower (*Passiflora*) vines they feed on as larvae, and even use these poisons as defenses against their own predators. In turn, in some *Passiflora* species, modified leaf structures have evolved that resemble butterfly eggs (**FIGURE 44.9**).

(A)

Episyrphus balteatus

Vespula vulgaris

(B) *Anodorhyncus hyacinthinus*

FIGURE 44.8 Defense Mechanisms and "Arms Races"
(A) The harmless hoverfly (above) gains protection by mimicking a dangerous stinging wasp species (below) that predators have come to avoid. (B) Brazil nuts (the seeds of *Bertholletia excelsa*) are protected by extremely hard shells that most birds cannot penetrate. The evolution of beak strength in the hyacinth macaw, however, has kept pace with the nut's hardness.

FIGURE 44.9 Using Mimicry to Avoid Being Eaten The raised yellow bumps on these passionflower leaves resemble the eggs of the plant's principal herbivores, zebra butterflies (*Heliconius* spp.). Because female butterflies will not lay their eggs on leaves that already contain eggs, these "false eggs" protect the plant from being eaten by caterpillars.

Since female butterflies will not lay eggs on plants that already contain eggs, the egg-mimicking structures reduce the plant's probability of being eaten by *Heliconius* caterpillars.

FRONTIERS Molecular genetic methods are allowing ecologists to better understand how coevolution—the reciprocal evolution of traits in interacting species—occurs at the level of genes. Scientists can alter the expression of single genes affecting traits that are important in interspecific interactions, and study the ecological consequences for the organisms involved. For example, ecologists have used molecular methods to "turn off" plant genes that control the production of defensive chemicals. This allows researchers to determine precisely how those chemicals influence insect herbivory, and thereby plant fitness. Besides improving our understanding of coevolution, such research has potential applications in agriculture.

Mutualisms can involve exploitation and cheating

Mutualisms are often misunderstood. How often do we hear that "bees visit flowers to pollinate them" or that "flowers provide food for their pollinators"? The problem with such statements is they imply an impossible evolutionary process in

---

APPLY THE CONCEPT

Interactions affect individual fitness and can result in evolution

Crabs feed on mussels, and individual mussels can respond to the presence of crabs in their environment by thickening their shells—*if* they can perceive the threat. On the east coast of North America, blue mussels (*Mytilus edulis*) face two non-native crab predators. The European green crab (*Carcinus maenas*) was introduced to the United States almost 200 years ago, and the Asian shore crab (*Hemigrapsus sanguineus*) arrived only 20 years ago. *C. maenas* has spread all along the Atlantic coast of North America, whereas *H. sanguineus* has not yet reached northern Maine. Have mussels evolved the ability to detect this new enemy and respond to it?

Researchers collected very young *M. edulis* from northern populations that have experienced *C. maenas* but not *H. sanguineus* and from southern populations that have experienced both predator species. They grew mussels from each population on floating docks under three different conditions: no crab nearby (controls); a hungry *C. maenas* caged nearby; and a hungry *H. sanguineus* caged nearby. After 3 months, researchers measured mussel shell thickness.

Use the graph at right to answer the following questions. Error bars indicate 95% confidence intervals; bars with different letters are statistically different from one another (*P* < 0.05).

1. Have mussel populations exposed to predation by non-native crabs evolved the ability to detect and respond to them by thickening their shells?

2. Which comparison is critical to your conclusion?

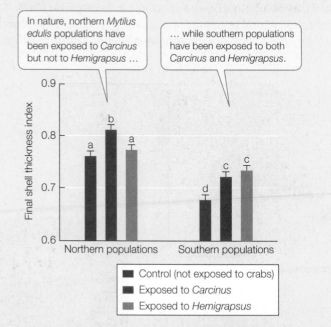

3. Do mussels in southern populations have thinner shells than in northern populations? Would a difference in average shell thickness between the two populations invalidate your conclusion?

4. What do these data suggest about how quickly evolution can take place?

which species display traits that evolved solely for the benefit of another species. As Charles Darwin said,

> If it could be proved that any part of the structure of any one species had been formed for the exclusive good of another species, it would annihilate my theory, for such could not have been produced through natural selection.

In truth, most pollinators visit flowers to get food and happen to pollinate the flowers in the process; flowers provide food (usually as little as possible) to lure the animals, not to fatten them up; birds and bees "like" any conspicuous color that signals the presence of food; and most flowers are visited by any animal that gains economically by doing so, whether it pollinates the plant or not. In the rough-and-tumble world of ecological interactions, species benefit other species not because they are altruistic, but because acting in their own self-interest happens to benefit others.

> **LINK** The coevolution of plants and pollinators is discussed in Concept 21.5

All mutualisms involve "biological barter": the exchange of resources and services. Plants and pollinators exchange food (nectar or pollen) for pollen transport. Plants and frugivores exchange food (fruit) for seed dispersal. Some mutualists (such as leaf-cutter ants and fungi) exchange food for care and dispersal. Others (such as ants and plants) exchange food or shelter for defense or parasite control. Still others (such as plants and mycorrhizae, or mammals and gut bacteria) exchange one kind of nutrient for another. In such "barter" systems, the fitness effect of the mutualism can vary depending on environmental conditions that determine the value and cost of the goods and services that are exchanged. Mycorrhizae, for example, benefit plants in nutrient-poor soils, but they can be a liability to plants in nutrient-rich soils, where the cost of feeding the mycorrhizae outweighs their value in nutrient uptake.

> **yourBioPortal.com**
> **Go to ANIMATED TUTORIAL 44.1**
> **An Ant–Plant Mutualism**

"Cheating" is common in mutualisms. Some flowers mimic the form and smell of female insects and are pollinated when males attempt to copulate with them (see Figure 17.9). Others cheat pollinators by providing no nectar. And most provide less food than their visitors would like. Pollinators cheat too when it is cost-effective to do so. Some bees "rob" flowers by extracting nectar through holes they bite in the base of the flower, consuming the nectar without pollinating the flower. Other pollinators become "secondary robbers" by reusing these holes. And some yucca moth species (see Figure 44.1D) cheat by laying eggs in *Yucca* flowers without depositing a pollen mass, or by laying so many eggs that the larvae consume all the flower's seeds.

Because mutualisms are diverse and often complex, their evolutionary consequences are difficult to predict. Some mutualisms evolve into obligate, one-on-one relationships with ever

tighter reciprocal adaptation. Others evolve into diffuse, many-species relationships with generalized adaptations. Some are unstable and dissolve into consumer–resource relationships. Understanding the conditions under which mutualisms follow different evolutionary trajectories is an active area of current research.

FRONTIERS Ecologists are collaborating with sensory physiologists and neurobiologists to learn how different animals perceive the world and how perception influences their ecological interactions. This new field of cognitive ecology is exploring how pollinating insects distinguish flowers from other objects. For example, although it has long been known that bees have color vision, one surprising new discovery is that they cannot distinguish colored objects such as flowers until those objects are close enough to simultaneously stimulate dozens of adjacent ommatidia (see Figure 35.15). The results of such studies will improve our understanding of plant–pollinator interactions.

Do You Understand Concept 44.3?

- What are the two components of fitness?
- How are these two components related to *r*? (Hint: Recall from p. 845 in Chapter 43 the ten female bison that rancher Jane followed for a year.)
- Why might individual plants in populations within the geographic range of an insect herbivore contain high levels of defensive chemicals while individuals in populations outside of the insect's range produce no defensive chemicals?

We have now discussed how interspecific interactions influence the population growth rates and the adaptations of the interacting partners. What happens when the partners are suddenly forced to interact with new species?

concept 44.4 Introduced Species Alter Interspecific Interactions

As we saw in Concept 42.5, humans are introducing many species to new places, both intentionally and unintentionally. These introductions not only blur biogeographic boundaries but have the potential to alter interactions among native species.

Introduced species can become invasive

Species that are introduced into a region where their natural enemies are absent may reach very high population densities. Such introduced species may become **invasive**—that is, reproduce rapidly and spread widely. They are likely to have negative effects on native species that lack adaptations to compete with or defend themselves against the newcomers. However, introduced species may benefit some native species, if they provide a resource that was not available previously.

Invasive species are spread in several ways. Marine organisms, for example, have been spread throughout the oceans by ballast water from ships, taken on at the port of departure and discharged at the destination port along with its content of surviving animals and plants. Similarly, many terrestrial plants and animals have "hitchhiked," unbeknownst to their carriers, as humans have moved themselves and their belongings to new locations. All such movements can be thought of as accidental. In addition, there are deliberate introductions by humans, who wish to reconstruct their familiar surroundings in a new location, or who believe that an introduction improves a native ecological system. For example, Europeans colonizing new continents deliberately brought a wide range of plants and animals to their new homes. This transport is still ongoing in the form of ornamental plants, exotic pets, and so on. Some species deliberately introduced to control other invasive species have themselves become invasive and caused further problems.

Introduced species alter ecological relationships of native species

The ways in which invasive species may harm native species are diverse. For example, introduced flowering plants can alter relationships between native plants and their pollinators. Purple loosestrife (*Lythrum salicaria*), introduced into North America in the early 1800s, has expanded its range dramatically and now dominates many wetlands in the United States (**FIGURE 44.10**). This species competes for pollination with the native species *Lythrum alatum*, which receives fewer visits from pollinators and produces fewer seeds when purple loosestrife is present.

Some invaders alter ecological interactions by causing extinction of native species. Concept 22.3 described the case of an introduced sac fungus and cause of a blight that drove the once-dominant American chestnut (*Castanea dentata*) to extinction. The chestnut has been replaced by oaks throughout the forests of the northeastern United States. Because acorn production varies greatly from year to year, whereas chestnuts produced consistent nut crops every year, the loss of chestnuts has contributed to the year-to-year fluctuations in rodents, ticks, and Lyme disease (see the opening story in Chapter 43).

Organisms that are deliberately introduced to control pests can alter interactions among native species if they do not confine their attention to the pest species. *Rhinocyllus conicus*, for example, is a Eurasian weevil whose larvae eat developing seeds. It was introduced into North America in 1968 to control the invasive musk thistle (*Carduus nutans*). When the abundance of musk thistle declined, the weevil moved over to the native Platte thistle (*Cirsium canescens*), and then to wavyleaf thistle (*C. undulatum*). By entering into consumer–resource interactions with native thistles, the introduced weevil has also become a competitor of the native insects that eat thistles.

Do You Understand Concept 44.4?

- What are two ways in which species are introduced into new areas by human activities?
- Why do some introduced species become invasive?

The effects of invasive species on interspecific interactions highlight the fact that these interactions occur within a larger web of multispecies interactions. As we have seen, these interactions determine which species can coexist in an area, and they affect the abundances of those species. The number of species present and their relative abundances determine the structure of communities—the topic of the next chapter.

FIGURE 44.10 An Invasive Species Purple loosestrife (*Lythrum salicaria*) was introduced into North America and now dominates many wetlands, especially in the northeastern United States. Thriving in the absence of its natural herbivores, the invasive plant has crowded out native species such as cattails, affecting the life cycles of wetland organisms such as waterfowl and amphibians.

QUESTION How could the intricate ecological relationship between leaf-cutter ants and fungi have evolved?

ANSWER We have seen that interactions between species can be mutually beneficial, positive for one species and negative or neutral for the other, negative for one species and neutral for the other, or mutually detrimental (Concept 44.1). These interspecific interactions affect the per capita growth rates of the respective populations, and hence the fitness of the individuals involved in the interaction (Concept 44.2). Any traits that increase the benefit of a positive interaction, or decrease the fitness loss of a negative interaction, will give individuals with those traits a fitness advantage. Such favored phenotypes will increase in frequency in the population (Concept 44.3). In the case of the mutually beneficial interaction between leaf-cutter ants and their farmed fungus, both ants and fungus gain nutrition from their interaction. In addition, the ants disperse the fungus and protect it from pathogens.

Fungus cultivation presumably originated when fungus-eating ancestors of the leaf-cutters stabilized their food supply by eating fungi growing on the refuse in their nests. Ants that provided better growing conditions for these fungi had more fungus to eat (**FIGURE 44.11**), and fungi that provided better ant food benefited from faster ant colony growth and queen production. Ants that chose the best fungus cultivars benefited from more productive crops, and fungi that provided ants with nutrient-rich and easy-to-harvest structures were more likely to be propagated by ants. Leaf-cutter ants expanded their food base by feeding fresh leaf material (which ants cannot digest) to their fungus; the fungus then converted leaf matter into food the ants could consume. The fungi benefitted by having access to food that they would not be able to use if ants did not chop it up for them. The ants' behavior contributed to the great ecological success of leaf-cutter ants and their fungus; leaf-cutter ants are not only major herbivores in the Neotropics, they have expanded into dry environments, such as the Arizona desert, that are normally hostile to fungi.

FIGURE 44.11 A Fungal Garden Cutaway view of the nest chamber of the leaf-cutter ant *Acromyrmex octospinosus*. The ants cultivate and feed on the white fungus.

44 | SUMMARY

concept 44.1 | Interactions between Species May Be Positive, Negative, or Neutral

- **Interspecific interactions** have consequences that affect the fitness of individuals and thus the growth rates and densities of populations and the distributions of species.

- Interspecific interactions can be divided into five broad categories depending on whether their effects on the interacting species are positive, negative, or neutral. **Review Figure 44.1 and WEB ACTIVITY 44.1**

- **Interspecific competition** refers to interactions in which members of two or more species use the same resource. There generally is one **limiting resource** that is in the shortest supply relative to demand.

- In **consumer–resource interactions**, the consumer gains nutrition and the resource species is killed or harmed. Consumer–resource interactions include **predation**, **herbivory**, and **parasitism**.

- A **mutualism** is an interspecific interaction that benefits both participants.

- **Commensalism**, in which one participant benefits while the other is unaffected, and **amensalism**, in which one participant is harmed while the other is unaffected, are not reciprocal and thus do not show the same ecological or evolutionary dynamics as other types of interspecific interactions.

- Many interspecific interactions have both beneficial and harmful aspects, and their effects may be contingent on environmental conditions. **Review Figure 44.2**

concept 44.2 | Interspecific Interactions Affect Population Dynamics and Species Distributions

- The equation for density-dependent population growth, which describes **intraspecific competition**, can be extended to include the effects of interspecific competitors on per capita growth rates. Consumer–resource interactions and mutualisms can be described in a similar way by equations.

- The per capita growth rate and the average population size of each interacting species are modified by the presence of the other in a way that depends on the type of interaction.

- In interactions with negative effects on one species (consumer–resource interactions, competition, and amensalism), extinction of one or both of the interacting species is possible. **Review Figure 44.3**

- The presence or absence of other species can make an environment favorable or unfavorable for a species and so affect its distribution. **Review Figure 44.4**

- Coexistence of species competing for resources, and of prey species with their predators, is possible if the populations of such species can increase even at low densities. Differences between competing species in their use of resources, known as **resource partitioning**, often are great enough to generate such a rarity advantage. **Review Figure 44.4**

concept 44.3 Interactions Affect Individual Fitness and Can Result in Evolution

- If intraspecific competition results in directional selection for traits that allow individuals to use new resources, or to use resources more efficiently, it can increase the carrying capacity of the population.

- If interspecific competition results in directional selection for traits that allow individuals of one species to use resources different from those a competitor species uses, resource partitioning and coexistence may result. **Review Figures 44.6 and 44.7**

- The opposing interests of consumer species and resource species can lead to an "evolutionary arms race" in which resource species continually evolve better defenses and consumer species continually evolve better offenses.

- Mutualisms involve mutual exploitation and the exchange of goods and services. The fitness effects of mutualisms may vary with environmental conditions that determine the value and cost of the goods and services that are exchanged. Mutualisms often involve cheating. **See ANIMATED TUTORIAL 44.1**

concept 44.4 Introduced Species Alter Interspecific Interactions

- Species introduced into a new region by humans, whether accidentally or intentionally, can alter interactions among the native species of that region.

- Species introduced into a region where their natural enemies are absent may become **invasive**, reproducing rapidly and spreading widely.

- Invasive species are likely to have negative effects on native species that lack adaptations to compete with or defend themselves against the invaders. Invasive species may even drive native species extinct.

Ecological Communities

We have *Coffea arabica* to thank for 70 percent of our morning cup of coffee. This understory shrub or small tree of tropical forests originated in the southwestern Ethiopian highlands as a natural hybrid between *C. canephora* ("robusta" coffee, from which we get the other 30 percent) and *C. eugenioides*, both of which hail from west and central Africa.

Coffea arabica has probably been cultivated in Ethiopia for 1,500 years, although the earliest written description of the plant dates to a tenth-century set of physician's notes. Molecular genetic evidence indicates that cultivation spread to Yemen about A.D. 575. Coffee was widely traded within the Muslim world, in spite of controversy there about use of a stimulant (see the opening of Chapter 5), and the first coffeehouses were in Arabia. The beverage reached Italy in the 1600s and later spread through Europe. Arabs tried to keep a corner on the coffee market by banning export of plants and fertile seeds, but the Dutch managed to smuggle some live plants out of Yemen and introduced the crop into their colonies.

By the 1700s, coffee cultivation had spread from Africa to Southeast Asia and the Americas. Coffee is now one of the most valuable export commodities of the developing world, second only to oil. As demand has grown, worldwide production has risen from about 5 million metric tons per year in the 1970s to about 7 million metric tons today, and coffee production employs some 25 million people in developing countries.

Most coffee is grown in tropical regions using a variety of cultivation methods. The least intensive, traditional method is to plant coffee in the shade of undisturbed

forests. This practice embeds coffee in natural forest communities, with minimal loss of their original plant and animal species. Medium-intensity shade cultivation replaces the forest with a low-diversity plantation of trees, under which the coffee is grown. High-intensity sun cultivation uses no shade trees; modern genetic cultivars are planted at high densities, heavily pruned, and subsidized with fertilizer, herbicides, and pesticides. Thus the more intense methods increasingly replace natural communities with human-constructed communities that contain far fewer species. High-intensity methods increase coffee yields, but they sacrifice the greater biological diversity that traditional methods maintain, cause greater pollution, and require costly chemicals.

This coffee plantation uses high-intensity sun cultivation techniques that increase crop yield at the expense of species diversity.

QUESTION Can we use principles of community ecology to improve methods of coffee cultivation?

KEY CONCEPTS

45.1 Communities Contain Species That Colonize and Persist

45.2 Communities Change over Space and Time

45.3 Trophic Interactions Determine How Energy and Materials Move through Communities

45.4 Species Diversity Affects Community Function

45.5 Diversity Patterns Provide Clues to Determinants of Diversity

45.6 Community Ecology Suggests Strategies for Conserving Community Function

concept 45.1 Communities Contain Species That Colonize and Persist

The interactions between pairs of species that we explored in the last chapter are only a part of the much broader web of interactions taking place in ecological communities. A **community** is a group of species that coexist and interact with one another within a defined geographic area. Ecologists sometimes specify the boundaries of a community in terms of discontinuities in habitat; the edge of a pond, for example, defines a community of aquatic species that interact much more with one another than with terrestrial species outside the pond. Alternatively, ecologists may define a community somewhat arbitrarily by designating the boundaries of an area they wish to study.

Because it is impossible or impractical to study all the species within a defined area, ecologists often restrict their attention to particular subsets of species. Those subsets may be defined taxonomically or by types of interactions. Hence biologists may speak of the bird community of an island, the *Daphnia* community of rock pools (see Figure 43.10), the community of beetles that live in dung, or the pollinator community of a forest.

We characterize communities in terms of their **species composition**: the particular mix of species they contain and the relative abundances of those species. The mix of species in a community is determined by the same factors that explain the distributions of species—as we saw in Chapter 43, a species can occur in a location only if it is able to colonize that location and to persist there. A community thus contains those species that have colonized minus those that have gone extinct locally. Extinction can occur for several reasons. Individuals may be unable to tolerate environmental conditions, or a resource may be lacking. Populations may be excluded by competitors, predators, or pathogens. Finally, some species may go locally extinct because their populations are so small that, by chance, all individuals die during a time interval without reproducing (see Concept 43.5). Abundances of the species that remain in the community are determined by population-level processes that were discussed in Chapters 43 and 44.

We can therefore think of communities as being *assembled* through the gain and loss of individual species. This process is well illustrated by Krakatau, a small (17 km^2) volcanic island in the Sunda Strait of Indonesia (**FIGURE 45.1A**). The volcano exploded in 1883, sterilizing what was left of the island and covering it with a thick layer of ash. Scientists quickly mounted expeditions to observe the return of life to the island and have re-surveyed it periodically ever since. By 1886, seeds of 10 plant species that grow on nearby tropical beaches had floated to the island (**FIGURE 45.1B**), and the wind had brought seeds or spores of 14 additional species of grasses and ferns. By the 1920s, the wind had carried tree seeds to the island. As the forest canopy closed in, some pioneering plant species that require high levels of light disappeared from the island's now-shady interior.

Once forests developed, fruit-eating birds and bats began to be attracted to the island, bringing new animal-dispersed seeds with them. In the many years since Krakatau exploded, the types of communities found there have stabilized. Nonetheless, the species composition of those communities continues to change as new species colonize, earlier colonists go extinct, and species abundances shift.

Do You Understand Concept 45.1?

- Describe at least two ways in which ecologists define which species to include in the study of a community.
- Describe the processes that are involved in community assembly.
- Why would abundances of the species on Krakatau shift as new species arrive? (Hint: See Concept 44.2.)

Krakatau provides a dramatic illustration of a general feature of communities: their species composition changes over time. Communities also change in space. Are there general rules that govern which species we find where and when?

FIGURE 45.1 Vegetation Recolonized Krakatau (A) Almost 125 years after a massive eruption left the island an ash-covered shell, vegetation has once again covered much of the island. (B) Beach vegetation such as *Ipomoea pes-caprae* (beach morning glory) was among the first to take hold on the denuded island.

(A)

(B)

Communities Change over Space
and Time

Ecologists have noted repeated patterns to spatial and temporal changes in species composition, indicating that the species we find in one place and time are not a random subset of those that potentially could be there. Let's take a closer look at these patterns and what they mean.

Species composition varies along environmental gradients

Even a small island like Krakatau contains gradients in physical conditions, ranging from shoreline habitat that is buffeted by wind and waves to inland habitat that varies in elevation (from 0 m at sea level to 800 m at the highest point) and from gentle ash-covered slopes to steep rocky escarpments. When we look at plant species composition, we find that it is similar in locations on the island with similar physical conditions, and different in locations with different conditions. In other words, species composition changes along environmental gradients. Sandy beaches all around the island are dominated by low-growing herbs and vines, bordered inland by a fringe of short trees and shrubs. Gentle inland slopes are dominated by taller trees with an understory of vines and ferns. The summit, which often is in the clouds, is covered in a thick growth of short, moss-covered trees (see Figure 45.1).

Such spatial variation in environmental conditions is generally associated with predictable changes in species composition. The serpentine outcrops that provide habitat for Edith's checkerspot butterfly (see Figure 43.1), for example, contain soils that are chemically distinct from those of surrounding areas. Many plants cannot tolerate the high content of heavy metals in serpentine soils, which therefore support very different plant communities than do non-serpentine soils. If we measure off a *transect*—a straight line used for ecological surveys—running from a non-serpentine to a serpentine area and identify the plants that occur along it, we will find different species in each type of soil the transect encounters. As we move along the transect, we will observe *species turnover* through space—that is, some species drop out and new ones appear (**FIGURE 45.2**).

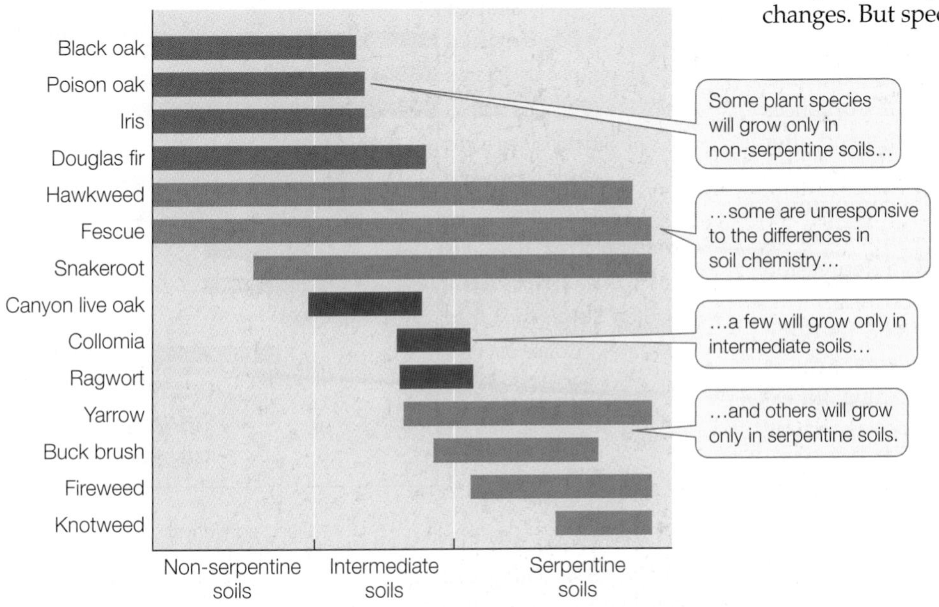

The early naturalist–explorers who documented associations of climate and vegetation—the *biomes* described in Concept 42.3—likewise documented associations of vegetation with particular animal species. In some cases, such as Edith's checkerspot butterfly, animal species are found in certain plant communities because the plants that they eat grow there. In other cases, animals are associated with particular plant communities because plants modify physical conditions such as temperature and humidity and contribute to *habitat structure*—characteristics of the surfaces and the horizontal and vertical distribution of objects. The vertical distribution of leaves and the density, size, and surface texture of plant stems differ considerably between forest and grassland communities, for example. Animals typically are associated with habitats of particular structure because structure determines the effectiveness of an animal's particular method for obtaining food or escaping predators (**FIGURE 45.3**).

Several processes cause communities to change over time

As we saw in the case of Krakatau, community composition changes not only over space, but also through time. Krakatau is not unique—all communities, not just those in the early stages of assembly, are dynamic. Over very long periods, of course, the species composition of communities reflects evolutionary changes. But species composition also changes over short periods. Three major processes contribute to this dynamism—extinction and colonization, disturbance, and climate change.

EXTINCTION AND COLONIZATION Ecological communities are unlikely to remain unchanged even in a constant environment because species may go locally extinct and be replaced by new species that arrive from time to time. As we saw in Concept 45.1, extinction has several causes. As we saw in Chapters 42 and 43, dispersal is an ongoing process that delivers a constant influx of new individuals to all but the most isolated locations. When those individuals arrive at a location that already contains a population of the same species, they add to its size; when those individuals are members of a new species, they can establish a new population. The result of ongoing extinction and colonization is steady turnover in

Black oak
Poison oak
Iris
Douglas fir
Hawkweed
Fescue
Snakeroot
Canyon live oak
Collomia
Ragwort
Yarrow
Buck brush
Fireweed
Knotweed

Some plant species will grow only in non-serpentine soils…

…some are unresponsive to the differences in soil chemistry…

…a few will grow only in intermediate soils…

…and others will grow only in serpentine soils.

Non-serpentine soils   Intermediate soils   Serpentine soils

FIGURE 45.2 Species Turnover along an Environmental Gradient As ecologists identified plant species along a transect running from non-serpentine to serpentine soils, they found that some species dropped out and new ones appeared.

(A)

(B)

FIGURE 45.3 Many Animals Associate with Habitats of a Particular Structure (A) The legs and feet of the red-breasted nuthatch (*Sitta canadensis*) allow it to cling to the trunks, branches, and twigs of trees as it gleans small arthropods and seeds. It prefers conifer forests. (B) Elk (*Cervus canadensis*) prefer open habitats that present no obstacles to escape from wolves.

the species composition of communities, as is occurring on present-day Krakatau.

DISTURBANCE Species turnover can be caused by sudden environmental change. Examples are volcanic eruptions, wildfires, hurricanes, landslides, and human activities. Such disturbances wipe out some or all of the species in communities and alter the physical environment. Sudden environmental change also can occur without the destruction of a resident community. New environments can suddenly appear, for example when an island rises out of the sea, a glacier melts away, a depression fills with rainwater—or a large mammal deposits dung.

After a sudden change in the environment, species often replace one another in a relatively predictable sequence called **succession**. A freshly deposited patch of elephant dung, for example, is colonized by a predictable sequence of species of dung beetles (Scarabidae). These animals eat dung and lay their eggs on it; some species typically arrive early and are replaced by later arrivals (**FIGURE 45.4**).

Several factors are responsible for successional sequences. One is that species vary in their colonizing ability. The first species to arrive are good dispersers. Early-arriving dung beetles tend to be strong fliers with excellent senses of smell, or "hitchhikers" that ride on dung-producers and colonize the dung as it

is produced. Similarly, the first plant species on Krakatau were those with seeds that are readily dispersed by sea or wind.

A second factor in succession is that environmental conditions in the disturbed or new site change over time, and species differ in their environmental tolerances. Early-arriving dung beetles must cope with wet conditions; pioneering species on Krakatau had to tolerate full sun and volcanic ash. Later-arriving species experience an environment that has changed because of physical processes and the presence and activities of early-arriving species. Dung dries out over time, so late-arriving beetle species must cope with a drier environment. Similarly, the seeds of plants that colonized Krakatau after the forest canopy had closed had to be able to germinate in the shade of other plants and to compete successfully for water and nutrients in soils that earlier colonists had modified.

When a disturbance destroys a pre-existing community, succession often leads eventually to reestablishment of a community that resembles the original one. On Krakatau, the tropical forests destroyed by the volcanic eruption eventually came back. After some types of disturbance, however, the original community is not reestablished. Instead, there is an **ecological transition** to a distinctly different community. The conversion of grasslands to shrublands in the U.S.–Mexico Borderlands following intensive cattle grazing (see the opening of Chapter 42) appears to be an example of such a transition.

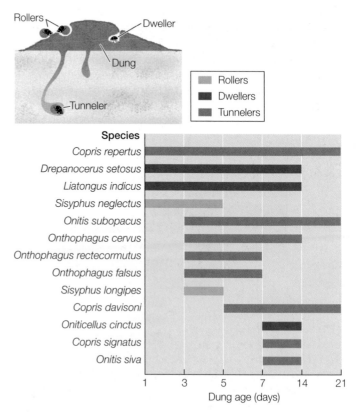

FIGURE 45.4 Dung Beetle Species Composition Changes over Time The Indian dung beetles in this study feed on and lay eggs in piles of elephant dung. Exactly which species are present in a given pile of dung varies over time in a predictable way, dictated by the adaptations of each species for finding and using dung.

FIGURE 45.5 Species Composition Changes with Climate Change (A) Fossilized plant remains extracted from ancient packrat middens indicate that 14,000 years ago, when the climate was moist, valleys in the U.S.–Mexico Borderlands held lakes whose shores were clothed in piñon–juniper woodland. (B) In today's more arid climate, the lakes have dried up and the woodland has given way to desert grassland and shrubland.

(A) 14,000 years ago

(B) Present

☐ Desert
☐ Piñon-juniper woodlands
■ Mixed-conifer forest
■ Spruce–fir forest
▨ Grassland/shrubland
▨ Ponderosa pine forest
☐ Alpine
▨ Pluvial lakes

yourBioPortal.com
Go to ANIMATED TUTORIAL 45.1
Succession on a Glacial Moraine

CLIMATE CHANGE A final cause of temporal variation in communities is climate change. As physical conditions change, the geographic ranges of species necessarily change with them. One way in which ecologists have been able to reconstruct the history of such changes is by analyzing fossilized remains of plants found in middens—refuse piles—left by packrats (*Neotoma* spp.). These rodents bring plant material into their nests, which, when preserved, provides a record of the vegetation that was present when the packrats were alive. Using carbon-14 (see Figure 18.1) to date packrat midden fossils, biologists have been able to show how plant communities of the Borderlands changed over the last 14,000 years as the climate became drier (**FIGURE 45.5**).

Do You Understand Concept 45.2?

- Why are animal species often associated with particular plant communities?

- How do successional sequences differ from species turnover resulting from ongoing colonization and extinction?

- Give an example of an environmental gradient that you have observed yourself, explaining what factors change along it and what visible changes you see in the species present.

Now that we have seen what factors can affect the species composition of a community, let's examine how interactions among those species influence the movement of energy and materials through the community.

concept **45.3**
Trophic Interactions Determine How Energy and Materials Move through Communities

Each species in a community has a unique **niche**. This central ecological concept refers to the set of environmental tolerances of a species, which define where it can live. It also refers to the way in which a species obtains energy and materials and to its pattern of interaction with other species in the community, both of which affect the dynamics of the species' population (see Chapters 43 and 44).

Interactions among the members of a community are extensive. For example, most plants are pollinated by more than one pollinator species, and pollinators rarely visit only one plant species. Furthermore, each species in a web of pollination interactions is connected to a much larger set of species that act as its competitors, hosts, parasites, consumers, and so forth. *Prosopis* species (mesquite; see the opening of Chapter 42) in southeastern Arizona are pollinated by many species of native bees and other insects, and their leaves are eaten by dozens of species of herbivorous insects. The seeds are eaten (and dispersed) by various ants, rodents, and beetles as well as by coyotes, foxes, and deer. The roots interact with an entire community of soil fungi and bacteria as well as with roots of other plants. Each of the species that interacts with *Prosopis* in turn interacts with a constellation of other species.

Consumer–resource interactions determine an important property of communities

Although the extensive network of interspecific interactions within a community may seem bewildering, one way to simplify it is to focus on how these interactions contribute to an important community-level property: the flow of energy and materials through the community. Most interspecific interactions directly or indirectly involve the exchange of energy and materials. In predator–prey and host–parasite interactions, one species (the prey or host) is a resource—a source of energy and materials—to a consumer (the predator or parasite). Most competition occurs between consumers that use some of the same sources of energy and materials—for example, sunlight, chemical nutrients, or food. Even mutualisms usually involve consumers and resources. In most plant–pollinator interactions, the service of pollination is a by-product of the consumption of a resource (nectar or pollen) by a consumer (the pollinator). Protists that live inside corals photosynthesize, producing carbohydrates that the corals consume (see Figure 20.19).

Consumer–resource, or **trophic** interactions (Greek *trophes*, "nourishment") cause energy and materials to flow through a community. Most biologically available energy enters communities via photosynthetic organisms, which use solar energy to manufacture the carbohydrates that fuel their metabolism. Because these organisms convert solar energy into a form that can be used by the rest of the community, they are called **primary producers**. They are also called **autotrophs** ("self-feeders") because they create their own "food" from inorganic sources.

Species that obtain energy by breaking apart organic compounds that have been assembled by other organisms are called **heterotrophs** ("other-feeders"). Heterotrophs that dine on primary producers are called **primary consumers**, or herbivores. Those that consume herbivores are called **secondary consumers**, or primary carnivores; those that consume primary carnivores are called **tertiary consumers**, or secondary carnivores, and so on. These feeding positions—primary producers and primary, secondary, and tertiary consumers—are called **trophic levels**. Some organisms, called **omnivores**, feed from multiple trophic levels. **Decomposers**, also called **detritivores**, feed on

waste products or dead bodies of organisms. Decomposers are largely responsible for the recycling of materials within ecosystems—they break down organic matter into inorganic components that primary producers can absorb (**TABLE 45.1**).

Trophic interactions can be described in diagrams that show linkages between interacting species as arrows that indicate the direction of the flow of energy and materials—that is, who eats whom—and to organize those linkages vertically according to trophic level. An interaction diagram organized in this way is called a **food web** (**FIGURE 45.6**).

Energy is lost as it moves through a food web

The total amount of energy that primary producers capture and convert to chemical energy during some period of time is called **gross primary productivity** (**GPP**). The energy that is contained in the tissues that primary producers have produced during that time is called **net primary productivity** (**NPP**), and it is what is available for consumption. GPP and NPP can be measured from the rate of uptake of carbon dioxide during photosynthesis and its release during respiration. Although new technologies allow us to measure such gas exchange in nature, it is often more convenient to use change in *biomass* of primary producers (the dry mass of their tissues) during a period of time as an approximation for NPP.

The flow of energy through trophic levels is shown in simplified form in **FIGURE 45.7**. Only about 10 percent of the energy

FIGURE 45.6 A Food Web in the Yellowstone Grasslands
The arrows show who eats whom. Even as highly simplified here, the web is complex. Species whose sole source of food is plants (green arrows) are primary consumers. Secondary and tertiary consumers are carnivores who kill and eat live animals (red arrows). Omnivores such as grizzly bears, coyotes, and ravens eat plant and animal tissues; ravens and grizzlies also eat carrion (dashed red arrows), so these species are also decomposers.

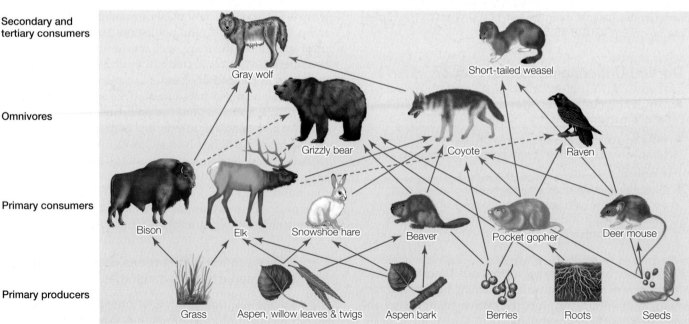

TABLE 45.1 The Major Trophic Levels

TROPHIC LEVEL	SOURCE OF ENERGY	EXAMPLES
Primary producers (photosynthesizers)	Solar energy	Green plants, photosynthetic bacteria and protists
Primary consumers (herbivores)	Tissues of primary producers	Elk, grasshoppers, gypsy moth larvae, pollinating bees, geese
Secondary consumers (carnivores)	Tissues of herbivores	Spiders, great tits, cheetahs, parasites
Tertiary consumers (carnivores)	Tissues of carnivores	Tuna, killer whales, parasites
Omnivores	Several trophic levels	Coyotes, opossums, crabs, robins, white-footed mice
Decomposers (detritivores)	Dead bodies and waste products of other organisms	Fungi, many bacteria, vultures, earthworms, dung beetles

incorporated in primary producer biomass is incorporated in primary consumer biomass. Similarly, only about 10 percent of the energy in primary consumer biomass becomes secondary consumer biomass. This **ecological efficiency** of roughly 10 percent means that, on average, the total biomass of each trophic level is about one-tenth that of the level it feeds on. This loss of available energy at successive levels limits the number of trophic levels a community can support; few communities support more than four.

There are several reasons for the incomplete transfer of energy between trophic levels. First, not all the biomass at one trophic level is ingested by the next one—for example, herbivores routinely miss hard-to-get plant parts or avoid eating plants that are well defended chemically. Second, some ingested matter is indigestible and is excreted as waste. Tree bark, for example, contains lignin and cellulose, which most herbivores cannot digest. Finally, and most important, consumers as well as primary producers use much of the energy they assimilate (the digested portion of what they ingest) to fuel their own metabolism rather than to add biomass through growth or reproduction.

Trophic interactions can change the species composition of communities

As we saw in Chapter 44, interspecific interactions affect the per capita growth rates, and densities, of the participating species—positively in the case of mutualisms, negatively in the case of competition, and positively for one species and negatively for the other in consumer–resource interactions. The per capita growth rate of a species is a function of the sum of the positive and negative contributions across all of the species with which it interacts. The extensive web of interaction among species in a community means that a change in the density of any one species (or the addition of a new species) can precipitate changes in density that ripple through the community. Succession can be driven by such interactions when, for example, later-colonizing species of dung beetles inhibit early colonizers by competing for dung nutrients, or late-arriving plants

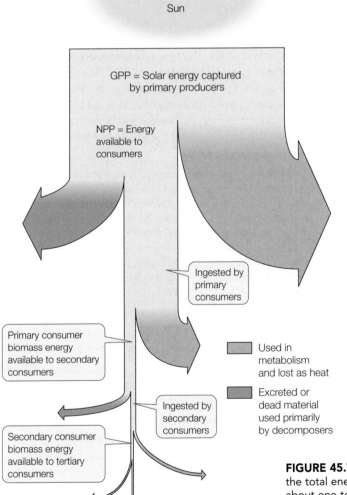

yourBioPortal.com
Go to WEB ACTIVITIES 45.1 and 45.2
The Major Trophic Levels
Energy Flow through an Ecological Community

FIGURE 45.7 Energy Flow through Ecological Communities On average, the total energy incorporated into the biomass of a trophic level per unit time is about one-tenth that of the level above it (i.e., the level that it feeds on). Of the 90 percent of energy *not* available to the next trophic level, some is waste or dead material used by decomposers, and some is used up in the cellular respiration that supports metabolism.

(A)

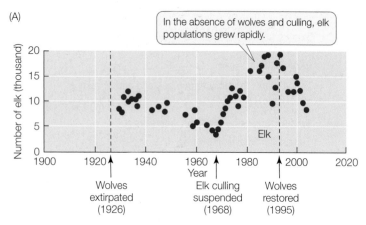

In the absence of wolves and culling, elk populations grew rapidly.

Wolves extirpated (1926)

Elk culling suspended (1968)

Wolves restored (1995)

(B)

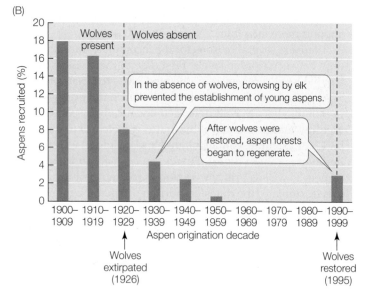

In the absence of wolves, browsing by elk prevented the establishment of young aspens.

After wolves were restored, aspen forests began to regenerate.

Wolves extirpated (1926)

Wolves restored (1995)

(C) *Populus tremuloides*

FIGURE 45.8 Removing Wolves Initiated a Trophic Cascade (A) The elk population in Yellowstone National Park increased rapidly once wolves were removed (extirpated) and the park stopped culling the elk herd. (B) Young aspens did not become established when wolves were absent and the abundant elk browsed the trees. (C) Researchers constructed an "elk exclusion" fence to demonstrate that aspen regenerated only inside the fence, where elk were absent.

shade out pioneering species. Consumer–resource interactions are so prevalent in communities that ripple effects often cross trophic levels, resulting in a pattern called a **trophic cascade**.

An example of an anthropogenic (human-caused) trophic cascade comes from Wyoming's Yellowstone National Park. Wolves in the park feed on a variety of large mammals, including elk (see Figure 45.6). Wolves exert particularly strong effects on the park community's structure and dynamics, as demonstrated by the effects of their absence during most of the twentieth century. For much of the twentieth century, there were no restrictions on wolf hunting, and by 1926 wolves had been eliminated from the park. Their removal had several effects on species at other trophic levels.

Park managers initiated annual censuses of elk in 1920. Concerned by the rapid increase in elk numbers after wolves disappeared, the park deliberately reduced elk densities (that is, culled some members of the herd) until public pressure caused them to stop in 1968. Elk densities subsequently increased (**FIGURE 45.8A**). The elk browsed aspen trees so heavily that no young trees were added to the population after 1960 (**FIGURE 45.8B**). The elk also browsed streamside willows, with the result that beavers, which depend on willows for food, nearly went extinct locally.

In 1995 park managers reintroduced wolves to Yellowstone. Not only did elk populations decrease, but elk began to avoid

aspen groves and willow thickets, where they are especially vulnerable to wolf predation (see Figure 45.3B). The consequence is that young aspen now survive, willows regrow, and species associated with them, such as beavers, have increased. Thus by eliminating and then restoring one predator species, humans caused trophic cascades that affected species at two lower trophic levels.

FRONTIERS The complexity of ecological interactions makes mathematical modeling an essential tool for studying community properties and dynamics. Ecologists are studying webs of ecological interactions with the same mathematical theory that is used to understand human social interaction networks (Facebook, for example) and communication networks (the Internet, for example). Studies of ecological interaction webs are providing insights into such issues as whether losses of species from a community will lead to further species extinctions.

Do You Understand Concept 45.3?

- What is the major source of energy for communities, and how does it enter them?
- At what trophic level would you place humans?
- What is the general rule of "10 percent ecological efficiency," what explains it, and how does it influence the number of trophic levels we see in communities and the total biomass at each level?
- What would a terrestrial landscape look like if its communities contained no decomposers?

Ecologists view trophic interactions and the resulting flow of energy and materials as manifestations of *community function.* In the next concept we will see what community function is and how it is affected by another community-level property, *species diversity.*

concept 45.4 Species Diversity Affects Community Function

Communities differ in their biological variety, often to a striking degree. Most people who visit a tropical rainforest, for example, are struck by the impression that each plant or animal they encounter differs from the last one, whereas other types of communities are more monotonous. It is these differences among communities that we wish to capture in a measure of their species diversity. Such measures are important because species diversity influences many functional properties of a community, including ones that affect humans.

The number of species and their relative abundances contribute to species diversity

Ecologists recognize two primary components of species diversity, which are illustrated in **FIGURE 45.9**. One is simply the number of species in the community, called **species richness**. A community that contains more species is more diverse, all else being equal. A less intuitive component of diversity has to do with the distribution of species' abundances, a property called **species evenness**. A community that has four equally abundant species is more diverse than one in which 75 percent of individuals are one species and the remaining 25 percent are spread among three other species. The former community has a more even distribution of species abundances.

Ecologists can use species richness as a simple measure of diversity, but they often prefer to use a mathematical diversity index that incorporates both richness and evenness. This is because both aspects of diversity affect *community function.* A species' influence in a community depends not only on its pattern of interaction, but also on its abundance. Properties

of a community with a few very abundant species are largely defined by them, rather than by the many rare ones.

Species diversity affects community processes and outputs

As we saw in Chapter 42, an ecological community can be thought of as a system, analogous in some ways to a man-made machine that has inputs, "internal workings," and outputs. Communities function by taking in, transforming, and putting out energy and materials. The function of a community can be measured in terms of these inputs and outputs. Important measures of overall function are the total flow of energy into the community, which is GPP, and the net energy that is made available for consumption by heterotrophs, which is NPP. We also can measure function for parts of the community. Wolves convert elk into new wolves, for example—a conversion of primary consumer biomass (an input) into secondary consumer biomass (an output). The output of seeds and fruits by most flowering plants is a function of an input by insect pollinators. Retention of nutrients within an ecosystem is a function of the output of the community's decomposers—the recyclers. Many of these outputs affect not only species in the community but also humans and other ecosystems—they represent "goods" and "services," as we will see in Concept 45.6.

Ecologists have discovered that community outputs vary with species diversity. Within a community type, NPP generally is greater and more stable over time as species diversity increases. Control of agricultural pests by natural predators and parasites is enhanced when multiple crops are planted together or when a diverse natural plant community borders agricultural fields. Diverse plant communities use inputs of soil nutrients more efficiently. These are just a few examples.

David Tilman and his colleagues at the University of Minnesota have undertaken a long-term study of prairie plant communities. They cleared several outdoor plots into which they seeded

FIGURE 45.9 Species Richness and Species Evenness Contribute to Diversity These hypothetical communities of fungi (mushrooms) are all the same size (12 individuals) but differ in species richness (3 versus 4 species) and the species' relative abundances, both of which affect diversity.

Community A

Community B

Community C

Community A is less diverse than community B because it contains three equally abundant species rather than four.

With four equally abundant species, community B is the most diverse.

Community C is less diverse than community B because it has an uneven distribution of the four species.

between 1 and 32 native prairie species. After the plots had grown for 3 years, they harvested the plants and measured their aboveground biomass. They found that this measure of NPP increased as the species diversity of a plot increased (**FIGURE 45.10A**).

Ecologists are still discovering why species diversity affects community function, and this is an active area of research. Two basic possibilities exist. First, the relationship may be an effect of *sampling*: communities that contain more species are more likely, by chance, to contain some that have a strong influence on total community output. These species may be particularly good at pollination or pest control, or particularly productive, or particularly efficient in their resource use. Alternatively, diversity may affect community function because of *niche complementarity*: communities that contain more species may be better able to use all available resources because they are more likely

to include species that have complementary niches. Indeed, Tilman and colleagues found that their most species-rich prairie plots also contained the largest number of plant *functional groups* (**FIGURE 45.10B**)—groups that differ in such traits as their ability to grow in warm versus cool seasons, in their associations with nitrogen-fixing bacteria, in their allocation to growth versus reproduction, and so on.

Do You Understand Concept 45.4?

- What are the two elements of species diversity?
- How are these two elements parallel to aspects of cultural and ethnic diversity in human societies?
- What outcome of Tilman and colleagues' experiment would have suggested that the effect of species diversity on community function is due to "sampling"?

We have now seen that species diversity is an important characteristic of communities because it affects how the community functions as an ecological system. But what factors influence species diversity?

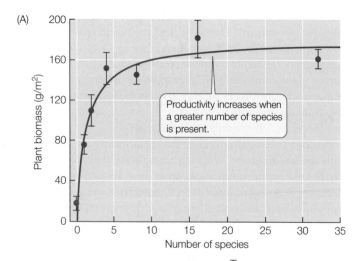

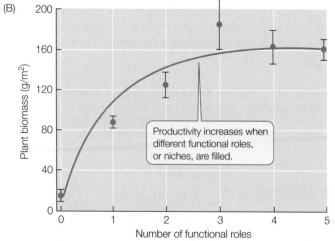

FIGURE 45.10 Species and Functional Group Diversity Affect Grassland Productivity (A) Primary productivity (measured as plant biomass produced over a 3-year period) increased with greater species richness. (B) Tilman and colleagues realized that species-rich plots also contain a greater diversity of plant functional groups (e.g., both C_3 and C_4 grasses, legumes with nitrogen-fixing bacteria, and woody shrubs) that differ in their growth responses to temperature, nutrient availability, and drought. These results suggest that productivity is enhanced by a greater variety of species with complementary functional roles.

concept 45.5 Diversity Patterns Provide Clues to Determinants of Diversity

The explorer–naturalists of the eighteenth and nineteenth centuries provided the first catalogues of how many species occurred in the various places they visited during their travels. Geographic patterns in species richness soon became apparent. These patterns indicate possible factors that affect diversity by modifying the balance between processes that add species to and remove species from communities.

Species richness varies with latitude

About 200 years ago, when the German naturalist Alexander von Humboldt traveled around Central and South America, Europe, and Asia, he remarked in the account of his voyages that "the nearer we approach the tropics, the greater the increase in variety of structure … of organic life." Humboldt's observation that species richness varies with latitude has been amply confirmed for many taxonomic groups. A tropical forest in Malaysia, for example, may contain 227 tree species within a 2-hectare plot, whereas a comparable area of temperate forest in Michigan may contain 10–15 tree species. There are 56 bird species that breed in Greenland, 105 in New York State, 469 in Guatemala, and 1,395 in Colombia. Many other taxonomic groups (although not all) show similar latitudinal patterns (**FIGURE 45.11**).

What is it about the tropics that supports such diversity? Ecologists have tackled this fascinating and still-unresolved question by looking for correlations between latitude and environmental factors that can affect diversity. They have identified many possible causes of latitudinal diversity gradients. One possibility is

FIGURE 45.11 Species Richness Increases toward the Equator Among swallowtail butterflies (Papilionidae), species richness decreases with latitude both north and south of the equator. Similar latitudinal gradients of species richness have been observed in many other taxa.

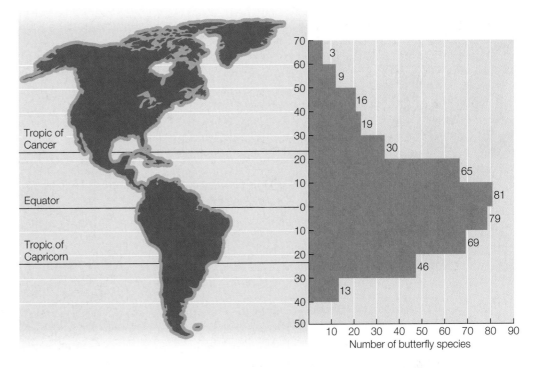

that climatic conditions have been more stable in the tropics because they were not subjected to the effects of the dramatic glacial cycles that disrupted temperate regions, causing massive shifts in geographic ranges and many extinctions. Tropical communities are certainly dynamic, and they are strongly influenced by disturbance such as the death and fall of individual trees—but the absence of disturbance at large spatial scales may have allowed these communities to retain more of their species, leaving them with higher diversity today.

We also know that climate changes with latitude. The tropics receive abundant solar energy input, which makes them warm and wet (see Concept 42.2). These conditions promote rapid growth of primary producers, so NPP, like diversity, increases toward the equator. Productivity can enhance diversity (as well as be enhanced by diversity; see Concept 45.4) if greater energy input into food webs supports larger population sizes. It can also enhance diversity if greater energy input allows species to maintain viable population sizes even as they become more specialized in their use of resources or habitats. In this way, greater energy flow through communities could facilitate coexistence of a greater number of species with narrow, specialized niches.

LINK   Some ways in which differences in resource use evolve and promote coexistence are described in Concepts 44.2 and 44.3

Spatial heterogeneity could interact with greater niche specialization to amplify tropical diversity. In general, diversity is higher in more structurally complex habitats. Bird species diversity, for example, increases with the diversity of foliage height in deciduous forests (**FIGURE 45.12**). If tropical species have more specialized habitat requirements, and if tropical vegetation is more structurally complex by virtue of greater plant species richness, then environmental heterogeneity and productivity may interact to increase tropical diversity.

Although ecologists still have much to learn about the relative contributions of these various factors to the puzzle of tropical diversity, the research stimulated by that puzzle has demonstrated that disturbance patterns, productivity, and environmental heterogeneity all affect the diversity of at least some taxa in some locations.

Diversity represents a balance between colonization and extinction

In addition to documenting latitudinal diversity gradients, the explorer–naturalists noted that oceanic islands contain fewer species than areas of comparable size on nearby mainlands. They also noted that small islands contain fewer species than large islands, and isolated islands contain fewer species than comparable-size islands closer to a mainland. These patterns, shown for birds in **FIGURE 45.13**, have been documented for a wide variety of taxa, and for islands worldwide.

Island diversity patterns remained mysterious for a long time—they could not be explained by productivity, habitat heterogeneity, or disturbance rate because those factors do not vary consistently with island size or isolation. In 1963, Edward O. Wilson and Robert MacArthur realized that, just as population size reflects the balance between additions (births)

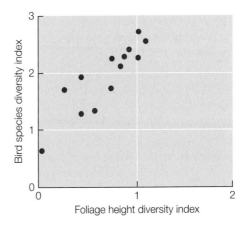

FIGURE 45.12 Structurally Complex Habitats Support Greater Diversity The greater the foliage height diversity is in a deciduous forest, the greater its bird species diversity.

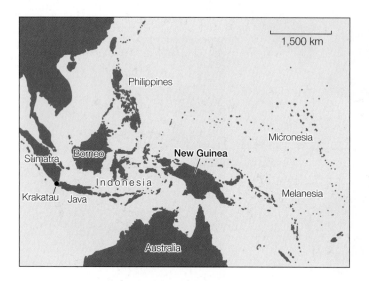

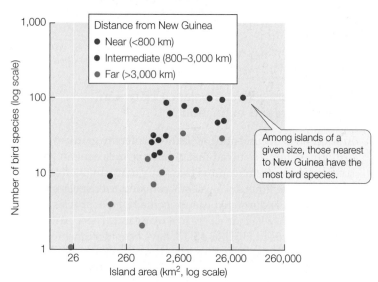

FIGURE 45.13 Area and Isolation Influence Species Richness on Islands Species richness increases with island area and decreases with distance from a source of colonists.

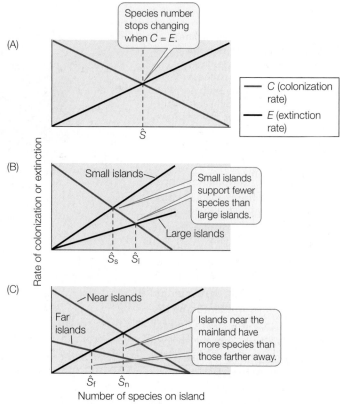

FIGURE 45.14 MacArthur and Wilson's Theory of Island Biogeography (A) Species richness reaches a stable equilibrium when the colonization rate equals the extinction rate. (B) Smaller islands have larger per species extinction rates, and hence lower equilibrium species richness. (C) Islands more distant from the mainland have lower colonization rates, and hence lower equilibrium species richness.

and losses (deaths), species richness on islands might represent a balance between the rate at which new species are added to a community and the rate at which resident species go extinct. Their theory of **island biogeography** accounts for island diversity patterns as follows:

- An offshore island gains species only if they colonize from elsewhere—usually from nearby islands or continents, whose communities make up the pool of potential colonists.

- The rate at which new species arrive on an island must decline as the island fills with species. At first, every immigrant represents a species new to the island, but as the island fills up, more and more of the arriving individuals represent species already resident there. (If all species in the mainland species pool were present, the colonization rate would be zero.)

- The overall rate at which species are lost from the island—the extinction rate—must increase as the island fills with species. In a time period, each species has some chance of going extinct from

various causes, and as the number of species increases, so does the overall number that go extinct per unit of time.

- The number of species on an island—its species richness—stops changing if the colonization rate equals the extinction rate. When we plot colonization and extinction rates on a graph, this equilibrium species richness is represented by the point at which the two curves cross, $\hat{S}$ (**FIGURE 45.14A**).

This theory predicts that an island's equilibrium species richness will depend on relative rates of colonization and extinction. And those rates, as MacArthur and Wilson pointed out, should vary both with island size and island isolation. Population sizes will decline as island size decreases, and since small populations are more at risk of extinction, the extinction rate for small islands should rise more steeply than that of large islands as a function of total species on the islands. Thus the equilibrium species richness should be lower on small islands (**FIGURE 45.14B**). Similarly, fewer wind- and waterborne seeds, and fewer dispersing animals, will encounter a distant island than one closer to the mainland species pool. As a result, the colonization rate, and thus equilibrium species richness, should be lower for more isolated islands (**FIGURE 45.14C**).

The theory of island biogeography has been tested in natural communities and has proved to be one of the most successful explanatory theories in ecology (**FIGURE 45.15**).

INVESTIGATION

FIGURE 45.15 The Theory of Island Biogeography Can Be Tested By experimentally removing all the arthropods on four small mangrove islands of equal size but different distance from the mainland, Simberloff and Wilson were able to observe the process of recolonization and compare the results with the predictions of island biogeography theory.

HYPOTHESIS

The rate at which experimentally defaunated islands accumulate species initially, and their equilibrium species number, will decrease with distance from a mainland source of colonists.

METHOD

1. Census the terrestrial arthropods on 4 small mangrove islands of equal size (11–12 m diameter) but different distance from a mainland source of colonists.
2. Erect scaffolding and tent the islands. Fumigate with methyl bromide (a chemical that kills arthropods but does not harm plants).

3. Remove tenting. Monitor recolonization for the following 2 years, periodically censusing terrestrial arthropod species.

RESULTS

Recolonization was fastest on the closer islands, slowest on the one farthest from the mainland. Two years after defaunation, each island had about the same number of species it had before the experiment.

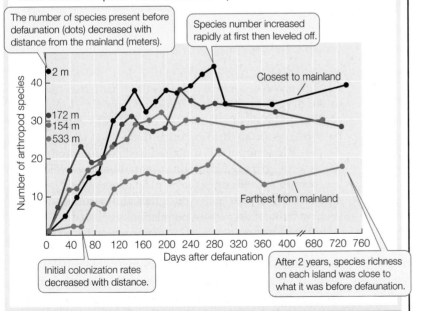

The number of species present before defaunation (dots) decreased with distance from the mainland (meters).

Species number increased rapidly at first then leveled off.

Initial colonization rates decreased with distance.

After 2 years, species richness on each island was close to what it was before defaunation.

CONCLUSION

The data support the theory that species richness on islands represents a dynamic balance between colonization and extinction rates.

For more, go to Working with Data 45.1 at **yourBioPortal.com**.

Go to **yourBioPortal.com** for original citations, discussions, and relevant links for all INVESTIGATION figures.

---

yourBioPortal.com

Go to ANIMATED TUTORIAL 45.2
Biogeography Simulation

Do You Understand Concept 45.5?

- Describe at least two factors that may contribute to the high species diversity of tropical forests.
- Tropical coral reef communities are among the most species-rich of any on Earth. Speculate on factors that might explain their diversity.
- What do Figures 45.14B and C tell you about the predicted rate of change in species composition (species turnover rate) at equilibrium on small versus large islands and on islands distant from versus close to a mainland? (Hint: think of turnover rate as the number of new species that replace previous species in a period of time.)

Quantitative studies of species richness and island biogeography have contributed greatly to our knowledge of the structure and function of ecological communities. We will next turn to some ways in which this knowledge can be applied to conserving these communities and their valuable services.

---

concept **45.6**
Community Ecology Suggests Strategies for Conserving Community Function

Ecological communities provide humans with critical goods and services, and a community's ability to deliver them often depends on its diversity. We have discussed several factors that influence the diversity of communities, including the balance between colonization and extinction and how complex the habitat is in terms of its physical structure. How can we apply this knowledge to maintain the diversity, and therefore the function, of ecological communities?

Ecological communities provide humans with goods and services

Humans depend on the ecological interactions that occur in communities and the interactions between communities and their physical environment. Although this seems obvious today, explicit recognition of the value of ecological systems is recent. Environmental writers introduced the idea of "natural

Diversity patterns provide clues to determinants of diversity

An alternative to MacArthur and Wilson's explanation for patterns of species richness has been proposed: that any island has a particular mix of "niches" that support an island-specific community. For Simberloff and Wilson's study (see Figure 45.15), this would mean that each island gradually reassembled its original arthropod community, and the number of species stopped changing when the full complement of species was restored. Review the two additional results from that study shown below and answer the associated questions.

1. Figure A is a sample of census results for individual insect species on one island. Do these census results support or refute the alternative hypothesis? Explain your answer.

2. Figure B shows the accumulation of insect species in three functional groups—herbivores, carnivores, and parasites—for the same island. Did the ratio of herbivores : carnivores : parasites change over time? Does your answer suggest that niches may play a role in the colonization sequence? Explain your reasoning.

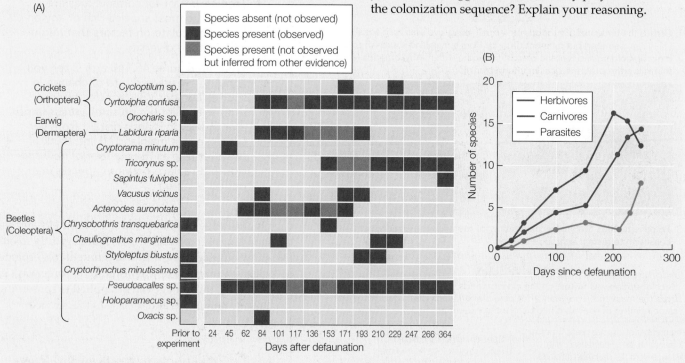

capital" in the 1940s; it was in 1970 that ecological systems were first said to provide people with "goods and services." The goods include food, clean water, clean air, fiber, building materials, and fuel; services include flood control, soil stabilization, pollination of plants, and climate regulation.

Because many of these goods and services involve exchanges between communities and the physical environment, they are called **ecosystem services**; a short list of them is given in **TABLE 45.2**. We will consider those that involve physical and geological processes in more detail in Chapter 46.

Goods that are of direct biological origin are obvious to us all—everybody knows that we get food from plants and animals, wood from trees, wool from sheep, and aesthetic enjoyment from flowers and birds. The roles of properly functioning communities in providing ecosystem services are less obvious, however, and we often take them for granted—at least until something goes wrong.

European settlers of Australia learned the hard way about the importance of the services provided by decomposers. When the first settlers arrived in 1788, they brought cattle with them. From the beginning, however, the Australian cattle industry

ran into problems: native dung beetles, adapted as they were to the dry, fibrous dung of forest marsupials, shunned the copious amounts of wet manure produced by cattle in open pastures. As a consequence, cattle dung piled up higher and deeper. The accumulated dung reduced pasture productivity (because the nitrogen-rich dung was not being recycled, and because cattle avoid eating fouled plant material) and provided a perfect nursery for native bush flies and introduced blood-sucking buffalo flies, whose populations exploded.

In 1964 Australia's national scientific research organization, CSIRO, finally tackled the problem. Insect biologists traveled the world to find dung beetle species that could process cattle dung in Australian environments without disrupting native beetle communities—an important precaution since many well-intended introductions have ended up having catastrophic impacts (see Concept 44.4). CSIRO introduced the first beetle species in 1968; by 1984, over 50 species had been introduced. The Australian Dung Beetle Project was successful: soil nitrogen content and water retention improved, pasture productivity increased, fly numbers plummeted, cattle became healthier, and native dung beetles were unharmed.

TABLE 45.2	Some Major Ecosystem Goods and Services	
GOOD OR SERVICE	**COMMUNITY PROCESS**	**EXAMPLES**
Food production	Trophic interactions	Production of wild food, production of crops and livestock
Materials production	Trophic interactions	Production of lumber, fuel, fiber
Pollination and seed dispersal	Plant–pollinator and plant–disperser interactions	Plant reproduction and dispersal
Maintenance of fertile soil	Decomposition; composition of vegetation; plant- and animal-microbe mutualisms	Nitrogen fixation, nutrient recycling, erosion control
Waste treatment	Decomposition	Breakdown of toxins and wastes
Pest control	Predator–prey, host–parasite, and competitive interactions	Removal of pests and pest breeding sites
Water supply	Regulation of water infiltration and runoff by vegetation structure and animal activity	Retention of water in watersheds, reservoirs, and aquifers
Climate regulation	Metabolic gas exchange	Regulation of greenhouse gases and cloud formation
Disturbance control	Composition and biomass of vegetation	Damping of storm winds and wave surges; flood control
Recreational and cultural opportunities	Ecological processes; species diversity	Ecotourism, outdoor recreation; aesthetic, educational, spiritual, scientific values

Ecosystem services have economic value

Ecosystem services have economic value, as the Australian Dung Beetle Project shows. Similarly, in the United States wild native pollinating insects (not including domesticated honey bees) contribute $3 billion annually to crop production.

Services related to food production are relatively straightforward to value because they are included in commercial markets. Many others—such as greenhouse gas regulation—are more difficult to value, but that does not mean they lack value. Consider the case of New York City's water, which comes from reservoirs in the Catskill Mountains. For many years, natural ecological systems in the Catskills provided pure water that met U.S. Environmental Protection Agency standards. However, changes in land use began to convert forests to agricultural fields and housing developments. The loss of forest cover decreased water infiltration into the soil, thus increasing discharges of sewage, fertilizers, pesticides, and sediment into streams. Water quality deteriorated to the point that New York was faced with the prospect of a new water treatment facility that would cost $6–$8 billion to build and $300 million annually to run. When the city considered what it would cost to restore the Catskills' natural water-purifying function, it realized that it could meet EPA water-quality standards by investing $1.5 billion in land protection and better sewage treatment in the Catskills. This is a lot of money, to be sure, but much less than the technological alternative. New York City's Long-Term Watershed Protection Plan, completed in 2006, has been a complete success.

The New York City example is not an isolated one. Many types of ecosystems can provide more goods and services, at lower cost, than artificial substitutes. But as we have seen, the community functions that deliver those goods and services depend on community diversity. Therefore, biological communities can best meet our needs if we do not manage them so intensively, or disturb them so severely, that their ability to provide natural goods and services is compromised.

Island biogeography suggests strategies for maintaining community diversity

As we saw in Chapter 42, humans are rapidly converting natural communities into less diverse, human-managed communities such as croplands, pastures, and urban settlements. The effect of this land conversion is that once-continuous large areas of habitat are reduced to scattered small fragments—habitat "islands" surrounded by "seas" of human-modified habitat (**FIGURE 45.16**). Habitat fragmentation can be expected to cause losses of species from communities for several reasons—the total amount of habitat decreases, the average patch size decreases, and patches become more isolated from one another. Given what you have learned about the theory of island biogeography, you now can predict the results. Populations become smaller as habitat area shrinks, and are therefore more prone to extinction. The matrix of human-dominated habitat surrounding the patches may serve as a barrier to dispersal, decreasing colonization rates. The net consequence of fragmentation is lower species richness in the fragments than in the original habitat.

MacArthur and Wilson's insight that species richness reflects a balance between colonization and extinction suggests that the detrimental effects of fragmentation can be minimized by enhancing colonization and reducing extinction. Colonization can be increased either by clustering habitat fragments close to one another or to a large "mainland" patch, or by connecting fragments with dispersal corridors, as described in Concept 43.6. Species extinction can similarly be minimized

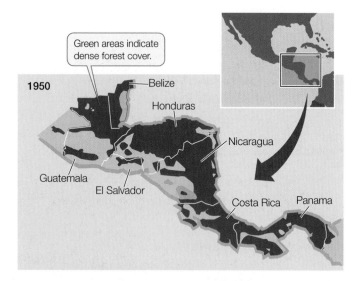

Green areas indicate dense forest cover.

1950

Belize
Honduras
Nicaragua
Guatemala
El Salvador
Costa Rica   Panama

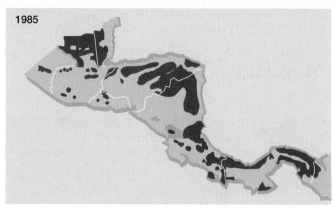

1985

FIGURE 45.16 Habitat Fragmentation in Tropical Forests
Less than half of the tropical forest that existed in Central America in 1950 remained by 1985, most of it in small patches. Although the rate of deforestation slowed somewhat after 1985, an additional 19 percent of the forest had disappeared by 2005.

by retaining large patches of the original habitat and by preserving the capacity of remaining patches to support healthy populations.

Several of these predictions have been tested in a remarkable large-scale experiment. Land owners in Brazil agreed to preserve tropical forest patches of certain sizes and configurations laid out by biologists (**FIGURE 45.17**), and they allowed the biologists to survey the parcels before and after intervening forest was cleared for pasture. Species soon began to disappear from the patches. The first species lost were monkeys that travel over large areas of forest. Army ants and the birds that follow their columns also disappeared rapidly. Species were lost more rapidly in isolated fragments than in ones of equal area that were connected to nearby unfragmented forest, and small isolated fragments lost species more rapidly than larger isolates.

Results of studies such as these now guide land use planning worldwide. They suggest that dispersal corridors and large habitat patches are necessary to maintain healthy, diverse communities.

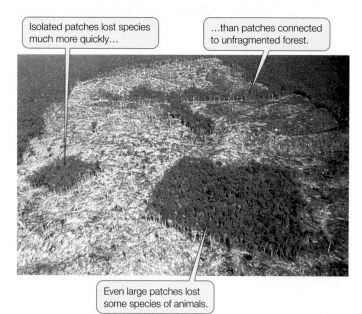

Isolated patches lost species much more quickly...

...than patches connected to unfragmented forest.

Even large patches lost some species of animals.

FIGURE 45.17 A Large-Scale Study of Habitat Fragmentation
Biologists studied patches of tropical evergreen forest near Manaus, Brazil, before and after the parcels were isolated by forest clearing. The results of the study demonstrated that small, isolated habitat fragments support fewer species than larger or less isolated patches of the same habitat.

yourBioPortal.com
Go to ANIMATED TUTORIAL 45.3
Edge Effects

Trophic cascades suggest the importance of conserving critical components of food webs

It is all well and good to plan for dispersal corridors and large patches, but species differ considerably in their habitat area requirements and in the kinds of dispersal corridors they will use. How big should natural areas be? What constitutes an effective corridor? These decisions can often be facilitated if they target species that play particularly important roles in communities.

The wolves of Yellowstone National Park provide an example of a species whose presence has a major effect on community structure and function. Wolves are crucial for maintaining healthy aspen forests and watersheds via trophic cascades that also involve elk, aspen and willow, and beavers. Conservation planning is now targeting such critical species. The Yellowstone to Yukon Conservation Initiative, for example, is a program that has as its goal the maintenance of a continuous corridor of wolf habitat between Yellowstone National Park and areas of similar habitat to the north. It is hoped that the corridors will maintain the Yellowstone wolf population, and hence a healthy Yellowstone ecosystem.

The relationship of diversity to community function suggests strategies for restoring degraded habitats

We have seen that even highly disturbed communities, such as those on Krakatau or on experimentally defaunated mangrove islands, can recover. This insight from community ecology suggests that degraded ecological systems can often be

INVESTIGATION

FIGURE 45.18 Species Richness Can Enhance Wetland Restoration In one large-scale field experiment, ecologists compared different methods for restoring denuded areas of the Tijuana Estuary, a wetlands environment near San Diego, California. They found that several measures of community function improved more rapidly in species-rich than in species-poor plantings.

HYPOTHESIS

Faster progress toward restoring the community's original condition will be made by planting mixtures of species than by planting single species alone.

METHOD

1. In an area of wetland denuded of vegetation, mark off replicate small experimental plots, all of the same size.

2. Choose 8 native species typical of the region. Plant some plots with each of the 8 species by itself, others with different 3-species subsets, and others with different 6-species subsets. Plant the same total number of seedlings per plot. Leave some plots unplanted as controls.

3. Return over the next 18 months to measure the vegetation and soil nitrogen levels.

RESULTS

Vegetation covered the bare ground more quickly in those plots with higher species richness. Those same plots developed complex vertical structure more quickly and accumulated more nitrogen in plant roots per m^2 of area.

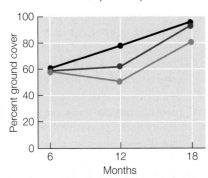
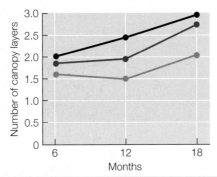
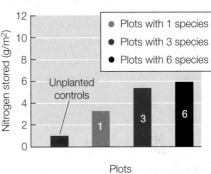

CONCLUSION

Planting a mixture of species leads to more rapid restoration of wetlands

ANALYZE THE DATA

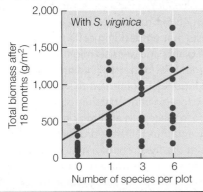
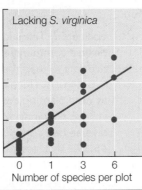

A. The unplanted control plots (above) showed non-zero values of stored nitrogen after 18 months. Is this a surprise? Why or why not?

B. One of the 8 species in the study, *Salicornia virginica*, contributed a disproportionate share of the biomass, cover, and nitrogen in mixtures in which it was included. In the two graphs at left, each circle represents the performance of a particular plot; the lines graph the average values. Based on these two graphs, what do you conclude about the mechanism for the higher success of species-rich plantings? Is the dominance of *S. virginica* the likely explanation?

Go to **yourBioPortal.com** for original citations, discussions, and relevant links for all INVESTIGATION figures.

---

restored—that their function can be improved. Devising ways to do so is the goal of the field of *restoration ecology*. The relationship between diversity and community function suggests that one of the best ways to improve the functioning of a damaged system is to restore the original species diversity. To do so, restoration ecologists draw on their knowledge of the factors that shape diversity (**FIGURE 45.18**).

As we have also seen, however, disturbance sometimes results in an ecological transition to a community that is very different from the original community. In such cases it may be difficult to find a way to reverse the transition and restore the function of the original community (recall the opening question in Chapter 42). This is a sobering reminder that restoration can be challenging and requires deep understanding of individual ecological systems.

Do You Understand Concept 45.6?

- Give two examples each of goods and services provided to humans by natural ecological systems, and by ecological systems constructed and managed by humans.

- If you were working as a restoration ecologist to restore a specific system, what measures might you use to assess whether your efforts were successful?

FIGURE 45.19 Traditional Coffee Cultivation and Community Diversity Low-intensity, high-diversity coffee cultivation (often marketed as "shade-grown organic coffee") results in lower yields but can net a higher profit by avoiding the costs of fertilizers and other chemicals—with the added benefit of maintaining the services provided by natural forests.

In this chapter we have discussed factors that influence the species composition of communities and the interactions that shape the flow of energy and materials through them. We have seen how species composition affects a community's function, including its outputs of goods and services. But outputs are limited by inputs, and to understand what determines inputs to communities, we need to consider their abiotic environments. The next chapter applies this ecosystem perspective, and considers how humans are influencing the global ecosystem.

 Can we use principles of community ecology to improve methods of coffee cultivation?

ANSWER Croplands are ecological communities that are governed by the same factors that govern natural communities. We can therefore use our understanding of how natural communities function to improve the function of agricultural communities. For example, we know from studying natural communities that species diversity often enhances community function (Concept 45.4)—whether we are talking about pollination, pest control, or primary productivity. We also know that species diversity is increased when there are opportunities for species with diverse niches to coexist and when colonization rates are high and extinction rates are low (Concept 45.5). We can use these ecological principles in many situations to produce more food at less cost and with fewer negative effects on the environment. For example, we can enhance

crop production by growing crop plants in complementary functional groups together.

The desired output of coffee plantations is, of course, seeds (coffee "beans"). Although *Coffea arabica* flowers can pollinate themselves, seed production is enhanced when bees move pollen between individual plants. From our knowledge of community ecology, we might predict that diverse bee communities provide the best pollination overall, because different species may pollinate in complementary ways. Recent research in Indonesia confirms this prediction: seed production was 50 percent higher, and varied less from year to year, in plantations that hosted 20 wild bee species than in those that hosted 3 bee species.

We also know that pollination of flowers is a free ecosystem service (Concept 45.6) provided by wild species of pollinators. This suggests that embedding coffee plants in diverse communities of plants and animals might be an effective way to reap the benefits of better pollination. Research in both Costa Rica and Indonesia confirmed this prediction as well: planting coffee close to intact patches of forest and leaving flowering weeds in place was all it took to attract a diverse bee community and thus increase production (**FIGURE 45.19**). This research is at the forefront of a new movement to enhance pollination and pest-control services for crops by bordering croplands with species-rich and structurally diverse communities.

Traditional low-intensity cultivation methods therefore seem a cost-effective way to increase coffee production by maintaining high diversity of native bees. Although traditional methods may not provide the highest yields per acre, they may nevertheless be very profitable. A study of nine coffee plantations in Mexico reported that high-intensity sun cultivation produced more coffee per hectare, but that low-intensity plantations were the most profitable because they avoided costs of chemicals and labor. That low-intensity cultivation also avoids pollution and helps maintain natural communities and their species are added benefits of taking advantage of natural ecosystem services.

45 SUMMARY

concept 45.1 Communities Contain Species That Colonize and Persist

- A **community** is a group of species that coexist and interact with one another within a defined geographic area.

- A community is characterized by its **species composition**: the particular mix of species it contains and the abundances of those species.

- A community contains those species that are able to colonize an area and persist there.

concept 45.2 Communities Change over Space and Time

- Variation in space in environmental conditions and habitat structure is associated with predictable turnover in species composition. **Review Figure 45.2**

- Even in a constant environment, ongoing extinction and colonization result in a steady turnover in the species composition of communities through time.

- After a sudden disturbance, species sometimes replace one another in a more or less predictable sequence called **succession**. Succession following a disturbance that removes the original community often results in a community that resembles the original one. **Review Figure 45.4 and ANIMATED TUTORIAL 45.1**

- Some types of disturbance lead to an **ecological transition**, in which an original community is replaced by a different type of community.

- Climate change can cause the species composition of a community to change over time. **Review Figure 45.5**

concept 45.3 Trophic Interactions Determine How Energy and Materials Move through Communities

- Each species has a unique **niche**—a set of environmental tolerances and pattern of interactions with other species.

- The **trophic interactions** by which a species obtains energy are a critical aspect of its niche. These interactions determine how energy and materials flow through communities.

- Energy enters communities through **primary producers**, which use photosynthesis to synthesize organic compounds that can be used by the rest of the community.

- Species that feed on primary producers are called **primary consumers**, or herbivores. Those that consume primary consumers are **secondary consumers** (or primary carnivores), and so on. These feeding positions are called **trophic levels**. **Review Table 45.1 and WEB ACTIVITY 45.1**

- Trophic interactions can be described in a diagram called a **food web**. **Review Figure 45.6**

- The total amount of energy that primary producers capture and convert into chemical energy during some period of time is called **gross primary productivity** (**GPP**). The portion of GPP that becomes available to consumers during that time is called **net primary productivity** (**NPP**). **Review WEB ACTIVITY 45.2**

- The **ecological efficiency** of energy transfer from one trophic level to the next is only about 10 percent. **Review Figure 45.7**

- Changes in species composition or species densities at any trophic level may cause a **trophic cascade** of changes at other trophic levels. **Review Figure 45.8**

concept 45.4 Species Diversity Affects Community Function

- Species diversity has two primary components: the number of species in the community, called **species richness**, and the distribution of those species' abundances, called **species evenness**. **Review Figure 45.9**

- Ecologists often quantify diversity by using a mathematical diversity index that incorporates both species richness and species evenness.

- The outputs of the interactions between species in a community and between species and their physical environment are measures of community function. Community outputs increase with increasing species diversity of the community. For example, NPP is greater, and more stable over time, in diverse communities. **Review Figure 45.10**

concept 45.5 Diversity Patterns Provide Clues to Determinants of Diversity

- Species richness is greatest in the tropics and decreases with increasing latitude. Possible causes of tropical species diversity include greater tropical climate stability, greater energy input, and greater habitat complexity. **Review Figures 45.11 and 45.12**

- Species richness is greater on large islands than on small islands and greater on islands closer to a source of colonists than on more distant islands. **Review Figure 45.13**

- The theory of **island biogeography** proposes that community diversity on islands represents a balance between the rate at which new species colonize and the rate at which resident species go extinct. **Review Figures 45.14, 45.15, ANIMATED TUTORIAL 45.2, and WORKING WITH DATA 45.1**

concept 45.6 Community Ecology Suggests Strategies for Conserving Community Function

- Ecological systems provide humans with a variety goods and services. These **ecosystem services** have economic value. **Review Table 45.2**

- Production of ecosystem services is threatened by human activities, including the fragmentation of natural habitats. The detrimental effects of habitat fragmentation can be mitigated by land-use management that minimizes extinction in each habitat "island" and maximizes dispersal among them. **Review Figure 45.17 and ANIMATED TUTORIAL 45.3**

- Ecosystem function often can be preserved or restored by focusing on particular species that play especially important roles in the community, and by maintaining or restoring overall species diversity. **Review Figure 45.18**

The Global Ecosystem — 46

The 1800s were a golden age for Earth science. Geologists deduced the astonishing age of the planet and discovered that past climates differed from modern ones—for example, that vast sheets of ice covered parts of the continents only 15,000 years ago.

What could cause such dramatic climate variation? One possibility, proposed in 1861, pointed to atmospheric gases such as CO_2 and water vapor, which had been found to absorb infrared (heat) radiation. Scientists hypothesized that these gases might influence Earth's climate by trapping heat in a manner analogous to the glass of a greenhouse. By the end of the nineteenth century, they had outlined a "carbon dioxide theory" for how fluctuating atmospheric CO_2 could cause fluctuations in this *greenhouse effect* and explain past ice ages.

Scientists pursuing the carbon dioxide theory had an additional insight. They realized that coal and petroleum contained fossilized carbon from dead plants and animals. Since the Industrial Revolution, humans had been burning these *fossil fuels*, releasing carbon as CO_2. Might this be increasing CO_2 in the atmosphere, and thus the greenhouse effect? Some skeptics felt that any effect of CO_2 must be insignificant because it comprises only about 300 parts per million (ppm) of the molecules in the atmosphere. In the 1930s, however, physicists and chemists demonstrated that changes in atmospheric CO_2 could affect climate. In the 1950s the American scientist Gilbert Plass estimated the rate at which atmospheric CO_2 was increasing and used one of the first electronic computers to forecast that increased CO_2 could significantly raise the average global temperature.

Mauna Loa observatory on the "Big Island" of Hawaii provides scientists with climatological data.

A big problem remained, however: nobody had yet been able to measure atmospheric CO_2 with precision. The challenge was met by a young American chemist, Dave Keeling, who developed sensitive new instruments and set them up atop Mauna Loa in Hawaii, 4,000 meters above sea level and far from most sources of human-generated pollution. Keeling recognized that the challenge was to see if the estimated rise in atmospheric CO_2 could be detected, and he pledged himself to a long-term study. He began in 1957, the International Geophysical Year (IGY), one of the first international scientific collaborations supported by governments. But the IGY money soon dried up. Undaunted by periods of insufficient funding, Keeling continued his research on atmospheric CO_2 and the carbon cycle over the next four decades.

Q QUESTION How did Keeling's research contribute to our understanding of the global ecosystem?

KEY CONCEPTS

46.1 Climate and Nutrients Affect Ecosystem Function

46.2 Biological, Geological, and Chemical Processes Move Materials through Ecosystems

46.3 Certain Biogeochemical Cycles Are Especially Critical for Ecosystems

46.4 Biogeochemical Cycles Affect Global Climate

46.5 Rapid Climate Change Affects Species and Communities

46.6 Ecological Challenges Can Be Addressed through Science and International Cooperation

concept 46.1 Climate and Nutrients Affect Ecosystem Function

As we saw in Concept 42.1, the term "ecosystem" is generally applied to an ecological community plus the abiotic environment with which it exchanges energy and materials. Most of the interactions among components of ecosystems happen at a local scale. A plant, for example, can only absorb nitrate molecules that are in physical contact with root hairs; an elk in Manitoba cannot fall prey to a wolf living in Wyoming's Yellowstone National Park; and a raindrop interacts directly only with soil particles it contacts. For this reason, most studies of ecosystems involve relatively small spatial scales, such as a single river drainage, lake, or patch of forest.

Ecosystems are linked, however, by slower exchanges that occur on spatial scales that range from tens of kilometers within regional landscapes to thousands of kilometers within the entire biosphere. These long-distance exchanges occur when organisms disperse or migrate among communities and when materials are moved around the planet by the physical circulation of the atmosphere and the oceans, the movement of flowing water, and by geological processes within Earth's crust. These large-scale linkages ultimately influence patterns of input or loss of materials to or from local ecosystems, making it impossible to understand a local ecosystem completely without considering it in the context of the larger systems of which it is a part. The present chapter therefore takes both local and global perspectives.

NPP is a measure of ecosystem function

Ecosystems, like other ecological systems, can be characterized by their components and by patterns of interaction among the components. The biological community as a whole is often treated as one component, and distinct portions of the abiotic environment, such as soil, atmosphere, or water, as other components. The dynamic processes by which these components exchange and transform energy and materials are aspects of *ecosystem function*. Ultimately, movements of materials between the abiotic and biotic components of ecosystems are tied to carbon, life's energy currency. Primary producers convert inorganic carbon into carbohydrates, and energy stored in the chemical bonds of those molecules fuels the synthesis of additional biomass, which ultimately contains all the atomic constituents of the bodies of organisms. Calcium in the bones of a wolf, for example, came from the elk it ingested; the elk obtained calcium from the plants it ate; and the plants absorbed calcium from the soil in a metabolic process fueled by energy released by breakdown of the carbohydrates produced during photosynthesis. The rate per unit of area at which an ecosystem produces primary-producer biomass—its *net primary productivity*, or *NPP* (see Concept 45.3)—is therefore a measure of the influx of materials, as well as energy, into that community. This measure of ecosystem function does not incorporate all exchanges of materials between organisms and the abiotic environment—organisms take in or excrete inorganic water

or salts to maintain homeostasis, for example. Rates of such exchanges, however, are indirectly coupled with primary productivity because they are fueled by the energy that enters the community via primary producers.

Scientists are now able to estimate NPP with instruments, carried by orbiting satellites, that measure the amount of different wavelengths of light reflected by Earth's surface. This "view from space" allows them to calculate how much sunlight is absorbed by chlorophyll and to map the distribution of photosynthetic biomass—which is correlated with NPP—across the planet. These maps show that NPP is not the same everywhere. As we will see next, patterns of variation in NPP suggest factors that limit ecosystem function.

NPP varies predictably with climate and nutrients

Net primary productivity varies considerably among ecosystem types. Among terrestrial ecosystems, tropical forests are the most productive per unit of area, followed by temperate forests; tundra and deserts are the least productive (**FIGURE 46.1**). Among aquatic ecosystems, swamps and marshes and near-shore algal beds and coral reefs are the most productive, followed by estuarine habitats; open ocean and the benthic zone of the deep ocean are the least productive (although the latter zone is astonishingly high in species richness). You will notice that, despite our best efforts, cultivated land is less productive than many natural ecosystems. Much of this pattern derives from variation in climate and nutrient availability.

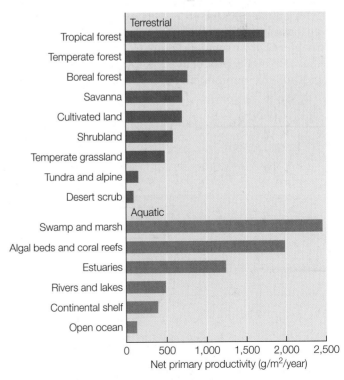

FIGURE 46.1 NPP Varies among Ecosystem Types Net primary productivity is expressed here as grams of biomass produced per square meter of area per year.

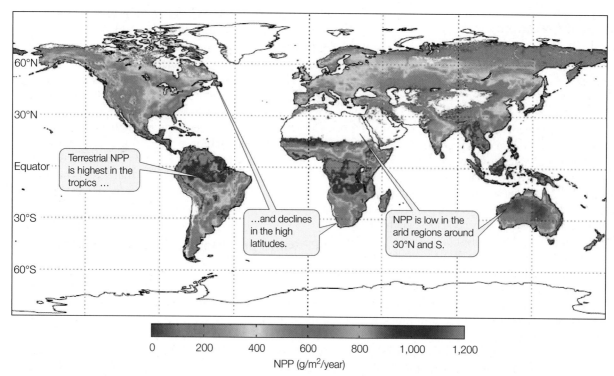

FIGURE 46.2 Terrestrial NPP Corresponds to Climate The map of estimated terrestrial NPP is based on satellite sensor data accumulated over the period 2000–2005. The white spaces on land represent unvegetated areas, including deserts and ice caps. Note the high productivity of the equatorial regions.

Satellite images of terrestrial ecosystems show that NPP varies with latitude (**FIGURE 46.2**). This is what we would expect, given that solar input and climate vary with latitude, as we saw in Chapter 42. If we compare Figure 46.2 with Figure 42.11, we can see that highly productive areas are concentrated in the tropics and that unproductive areas are concentrated at high latitudes and in the dry subtropics. These trends suggest that terrestrial NPP is affected both by temperature and moisture. Indeed, terrestrial NPP generally increases with temperature (**FIGURE 46.3A**), which makes sense when we consider that the activity of photosynthetic enzymes, like that of other enzymes, increases with temperature up to the critical temperature at which enzymes denature (see Concept 3.4). The critical temperature varies among plant species that are adapted to different

FIGURE 46.3 Terrestrial NPP Varies with Temperature and Precipitation NPP is given in megagrams (1 Mg equals 10^6 g or 1 metric ton per ha per year).

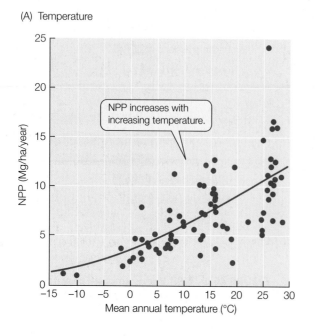

(A) Temperature

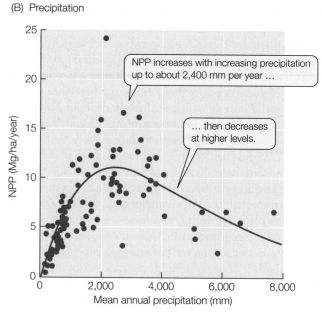

(B) Precipitation

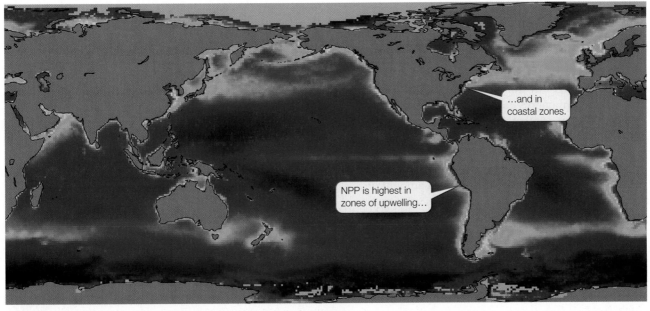

NPP (g/m²/year)

0   200   400   600   800

FIGURE 46.4 Marine NPP Is Highest around Coastlines
Photosynthesis is possible only in the top layers of oceans, where sunlight penetrates. The availability of nutrients determines how much NPP occurs in any given part of this photic zone. Highest NPP occurs where runoff from land brings nutrients into shallow coastal waters, and where upwelling of deep waters brings nutrients from the benthic zone closer to the surface.

environments but is usually about 40°C–50°C. Terrestrial NPP also increases with precipitation up to about 2,400 millimeters a year, then decreases at higher levels (**FIGURE 46.3B**), which also makes sense. Moist soil conditions facilitate water and nutrient uptake by roots and allow plants to maintain turgor pressure without closing their stomata, thus facilitating CO_2 uptake (see Concept 25.3). The decrease in NPP in extremely moist climates may result from increased cloud cover and lower solar input or from a lack of oxygen in saturated soils. Finally, we also know that elements such as nitrogen, phosphorus, and potassium are critical for plant growth and can be in short enough supply to limit NPP. As we saw in Concept 42.3, for example, soils in southwestern Australia are so nutrient-poor that plants cannot afford to waste nitrogen on defensive compounds; instead they allocate scarce nitrogen to proteins that are essential for metabolism. Southwestern Australia has been inactive geologically for hundreds of millions of years; over this long span of time most nutrients have weathered out of the bedrock and have been leached from soils by the movement of water.

Satellite images of oceanic ecosystems, in turn, suggest that aquatic NPP is more strongly a function of nutrient availability than of temperature (**FIGURE 46.4**). NPP occurs only in the top layers of aquatic systems, where light can penetrate and photosynthesis can occur (the *photic zone*; see Concept 42.3). Nutrients are very limited in this zone. They are most abundant in coastal marine areas and near-shore parts of lakes and in shallow wetlands, where rivers and streams discharge nutrients leached from terrestrial ecosystems, and in areas where water from the bottom of oceans or lakes reaches the surface, bringing with it nutrients from benthic sediments. Shallow

waters and areas of upwelling support high rates of primary productivity by aquatic plants and phytoplankton, which in turn support dense consumer populations. The only productive deep-water areas are rare hydrothermal vents, where chemolithotrophs (see Concept 19.3) generate new biomass by using chemical energy rather than sunlight and by using nutrients in the benthos and in water from the vents.

Do You Understand Concept 46.1?

- In Figure 46.1 we see that NPP is generally higher in freshwater wetlands (swamps and marshes) than in shallow marine waters (algal beds, reefs, and estuaries). Why might this be so?

- Outline an experiment that would allow you to determine what specific nutrient limits NPP in a given terrestrial ecosystem.

We now have an overview of global patterns of ecosystem function, as measured by NPP, and of some of the factors that determine these patterns. We have seen that nutrient availability plays an important role in NPP. This makes sense because nutrients are chemical elements that organisms require for their metabolism and as materials to build their bodies. What are the processes that affect the availability of nutrients?

concept

concept 46.2 Biological, Geological, and Chemical Processes Move Materials through Ecosystems

As we have seen throughout this book, the sun provides a steady input of radiant energy to Earth, some of which is captured by photosynthetic organisms. In chemical form, this energy then fuels the metabolism of all organisms in the community and ultimately is lost as waste heat. In contrast, the chemical elements that make up organisms come from within the Earth system itself. There is essentially a fixed amount of each element because little matter escapes Earth's gravity or enters from space. Whereas Earth is an *open system* with respect to energy, it is a *closed system* with respect to matter. This does not mean the distribution of matter is static, however—energy from the sun and heat from Earth's interior drive biological, geological, and chemical processes that transform matter and move it around the planet. But matter must be in the right place and in the right chemical form for organisms to be able to use it.

The form and location of elements determine their accessibility to organisms

Imagine an atom. At any time, this atom occurs in a particular molecular form and occupies a particular location in Earth's closed system. It can be part of a living organism or of dead organic matter. It can be part of a gas molecule in the atmosphere or dissolved in a body of fresh or salt water, or a component of soil, sediment, or rock. If in rock, it can be part of the surface crust of Earth or more deeply buried. These alternative forms and locations of matter can be thought of as *compartments*. Atoms cycle among compartments when they are transformed or physically moved (**FIGURE 46.5**). The rocks that make up continents, for example, have gone through repeated cycles of being buried, compressed, heated, unburied, and eroded (see Figure 18.2).

The chemical form in which an atom exists determines whether it is accessible to life. With rare exceptions, autotrophs take up carbon in the form of CO_2, and they absorb nutrients such as nitrogen, phosphorus, or potassium in the form of ions dissolved in water (see Concept 25.1). Heterotrophs extract most of the materials they need from living or dead organic matter. Microbes have the most diverse nutritional capabilities of all, as we saw

in Concept 19.3. But even when an atom is in the right chemical form, organisms cannot use it if it is in the wrong place. Matter that is buried too deeply, or that occurs in places that are too hot, too cold, or otherwise too hostile, is inaccessible to life.

The processes that supply matter to organisms occur in the *biosphere*, the thin skin at Earth's surface where atmosphere, land, and water are in contact with one another and where organisms live. The biosphere is only about 23 kilometers thick—a mere 1/277 of Earth's radius—extending up to the stratosphere and down to the abyssal zone of the oceans.

Fluxes of matter are driven by biogeochemical processes

Movement of matter among Earth's compartments is caused by processes that are biological, geological, and chemical; thus these movements are called **biogeochemical cycles**. Biological and abiotic chemical processes combine elements with other elements and compounds, converting them between inorganic and organic forms, or between oxidized and reduced states. These reactions often involve the abundant elements oxygen and hydrogen. Chemical processes are also involved in the weathering of rock during soil formation. Geological processes—the circulation of air and water and convective flows within Earth's mantle—move matter around, alternately exposing it to air, water, and solar radiation or to heat and pressure.

LINK Oxidation–reduction ("redox") reactions are described in Concept 6.1

The total amount, or **pool**, of an element or molecule in a given compartment depends on its flows, or **fluxes**, into and out of the compartment. Unless the fluxes in both directions are identical, the pool will grow or shrink. Oxidation of organic matter by respiration, fire, or fuel combustion, for example, will increase the atmospheric pool of carbon unless photosynthesis

FIGURE 46.5 Chemical Elements Cycle among Compartments of the Biosphere The different chemical forms and locations of the elements determine whether or not they are accessible to living organisms. These different forms of matter represent compartments, as in the diagram shown here. Biological, geological, and chemical processes cycle matter among the compartments.

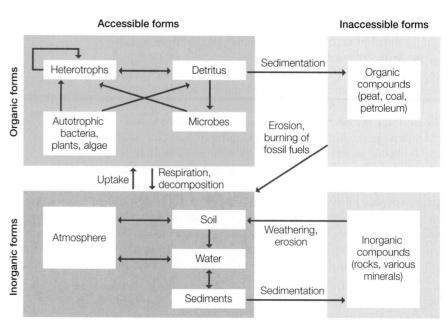

and other biogeochemical processes remove carbon at an identical or greater rate.

All of the materials contained in the bodies of living organisms are ultimately derived from abiotic sources, either atmospheric gases or material that is dissolved in water. Primary producers (see Concept 45.3) take up elements from inorganic pools, accumulating them as biomass. Trophic interactions pass those elements on to heterotrophs, and eventually they enter a pool of dead and waste organic matter. Decomposers break down this dead and waste matter into its constituents, which renders it available again for uptake by primary producers. Much of this *recycling* of materials occurs locally, within the original local ecosystem, but some occurs on a larger scale if the materials diffuse away in gaseous form or are carried out of the local system by wind or water.

Do You Understand Concept 46.2?

- What is the distinction between the concepts of *pool* and *flux* with regard to a given chemical element?
- Explain the statement that energy flows through ecosystems whereas materials cycle within ecosystems.

Now that the basics of biogeochemical cycles have been explained, let's next consider three of these cycles that are especially critical for ecosystem function.

Certain Biogeochemical Cycles Are Especially Critical for Ecosystems

In Concept 46.2 we saw that the materials that make up living tissue—chemical elements and compounds—cycle through ecosystems. The relevant spatial scale of the cycle and the processes that drive it vary with the material, but all cycles have features in common. As Figure 46.5 suggests, we can describe cycles in terms of the compartments involved, the amounts of the material contained in each compartment at a time (pools), and the rates at which the material enters and leaves those compartments (fluxes). In this concept we will describe cycles for three materials—water, nitrogen, and carbon—that are essential for life. We will point out how each of the three cycles affects physical properties of the Earth system—in particular climate—and how human activities are altering each of the cycles.

Water transports materials among compartments

Water is a most remarkable molecule, essential for life. It makes up more than 70 percent of living biomass. It is the medium for metabolism and the solvent in which biologically accessible forms of many nutrients are dissolved. Precipitation of water as rain or snow transports materials from the atmosphere to Earth's surface, and flowing water is the agent of erosion and sediment transport, responsible for much of the physical movement of materials around the planet. By virtue of its high heat capacity and its ability to change between solid, liquid, and gaseous states at normal Earth temperatures, water redistributes heat around the planet as it circulates through the oceans and atmosphere. And, as we shall soon see, water in the atmosphere is an important player in the global radiation balance.

Earth, the blue planet, contains an enormous amount of water—over 1.4 billion cubic kilometers, which make up 0.02% of Earth's total mass. About 96.5% of Earth's water is in the oceans (**FIGURE 46.6**). Other compartments include ice and snow (1.76%),

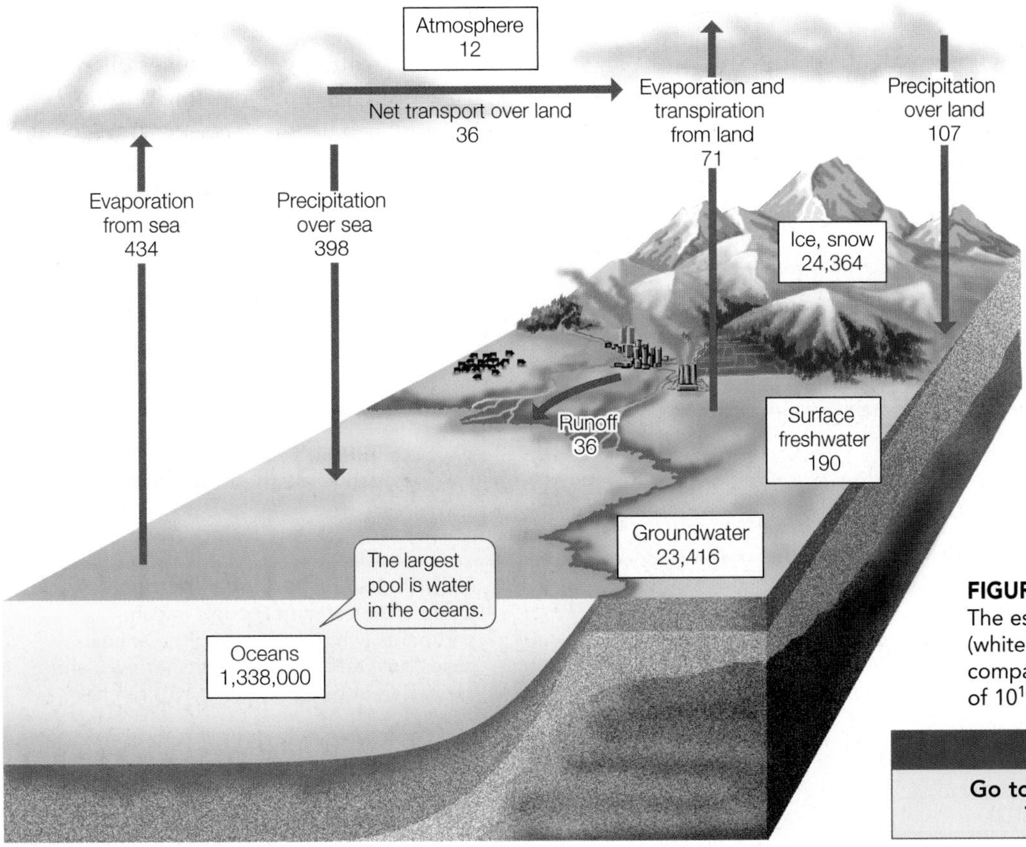

FIGURE 46.6 The Global Water Cycle
The estimated pools in major compartments (white boxes) and the annual fluxes between compartments (arrows) are expressed in units of 10^{18} grams.

Atmosphere 12
Net transport over land 36
Evaporation and transpiration from land 71
Precipitation over land 107
Evaporation from sea 434
Precipitation over sea 398
Ice, snow 24,364
Runoff 36
Surface freshwater 190
Groundwater 23,416
The largest pool is water in the oceans.
Oceans 1,338,000

yourBioPortal.com
Go to ANIMATED TUTORIAL 46.1
The Global Water Cycle

groundwater (1.70%), the atmosphere (0.001%), and fresh water in lakes and rivers (surface fresh water; 0.013%). Even though most of biomass is water, only about 0.0001% of Earth's water is found in living organisms.

> **LINK** The molecular nature of water and its crucial role in the evolution of life as we know it is covered in Chapter 2

Some of the movement among these compartments involves gravity-driven flows of liquid water from the atmosphere to Earth's surface (precipitation) and from land to the oceans (in runoff). Other among-compartment fluxes of water, however, are associated with changes in its physical state—from liquid to gas (evaporation and transpiration), from gas to liquid (condensation), between liquid and solid (freezing and thawing), and between solid and gas.

The driving force of the water, or *hydrological*, cycle is solar-powered evaporation that moves water from ocean and land surfaces into the atmosphere. It takes considerable energy—2.24 kJ (about the energy stored in one AA battery)—to evaporate 1 gram of water. On a global scale, the rate of evaporation is enormous. It uses about one-third of the total input of solar energy to Earth's surface. This energy is released again as heat when the water vapor condenses. Fluxes from Earth's surface to the atmosphere and from the atmosphere to the surface are approximately equal on a global basis, although

precipitation exceeds evaporation over land, and evaporation exceeds precipitation over the oceans (see Figure 46.6).

Humans affect the global water cycle when they change how land is used. Reduction in vegetation cover reduces the amount of precipitation that is retained by soil and recycled through the local ecosystem and increases the amount that leaves the local system as surface runoff. As a result, deforestation, overgrazing, and cultivation of crops all tend to dry out local ecosystems. Groundwater pumping depletes aquifers, moving groundwater to the surface, where it either evaporates or runs off into streams and, eventually, into oceans. The water cycle also affects and is affected by the global climate. Warming of the climate is melting polar ice caps and glaciers, increasing the total amount of water in the oceans and causing sea levels to rise. With more liquid water there is more evaporation, and thus more water entering in the atmosphere and more precipitation globally. Water vapor is a *greenhouse gas*, (it absorbs infrared radiation), but it also forms clouds that reflect incoming sunlight. The net effects on climate of such complex changes in the water cycle are difficult to predict.

Nitrogen is often a limiting nutrient

Nitrogen is abundant on earth, yet it is commonly in short supply in biological communities. Unlike water—whose availability to organisms depends only on where it occurs and its physical state—nitrogen is an element whose accessibility to life depends not only on where it occurs, but also on what other elements it is attached to. As a result, the nitrogen cycle, unlike the water cycle, involves chemical transformations (**FIGURE 46.7**).

The gas N_2 constitutes about 78 percent of the molecules in Earth's atmosphere, but most organisms cannot use nitrogen in this form because they cannot break the strong triple bond between the two nitrogen atoms. The exceptions are microbes in terrestrial and aquatic ecosystems that reduce N_2 to ammonium (NH_4^+) through a metabolic process called **nitrogen fixation**.

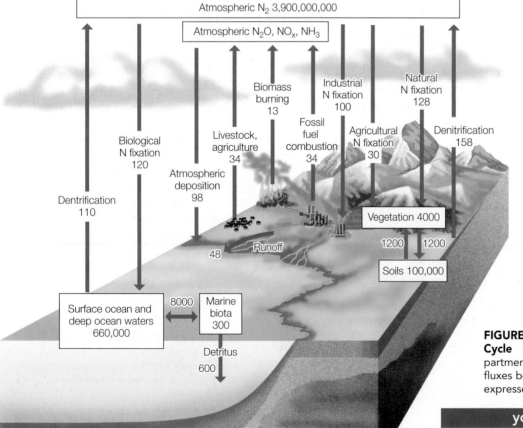

FIGURE 46.7 The Global Nitrogen Cycle The estimated pools in compartments (white boxes) and the annual fluxes between compartments (arrows) are expressed in teragrams (Tg; 10^{12} grams).

> **yourBioPortal.com**
>
> Go to **ANIMATED TUTORIAL 46.2**
> **The Global Nitrogen Cycle**

Atmospheric N_2 3,900,000,000

Atmospheric N_2O, NO_x, NH_3

Biological N fixation 120

Dentrification 110

Biomass burning 13

Livestock, agriculture 34

Atmospheric deposition 98

Industrial N fixation 100

Fossil fuel combustion 34

Runoff 48

Agricultural N fixation 30

Natural N fixation 128

Denitrification 158

Vegetation 4000

1200   1200

Soils 100,000

Surface ocean and deep ocean waters 660,000

8000

Marine biota 300

Detritus 600

Benthic sediments and rocks 400,000,000

INVESTIGATION

FIGURE 46.8 Where Does the Extra Nitrogen Come From? Leaf-cutter ants require more nitrogen than is contained in the fresh leaves they feed to their fungal garden. Perhaps the fungus concentrates nitrogen from the leaves into the fungal structures that the ants eat. If so, spent leaf material in the refuse dump outside the ant nest should be lower in nitrogen than fresh leaves—but the opposite turns out to be true. One possible explanation for this nitrogen enrichment is that ants eat protein-rich insects as well as fungus, and fertilize the fungal garden with their feces. Another is that the fungus absorbs additional nitrogen from the soil. A third possibility is that ant nests harbor nitrogen-fixing organisms.

HYPOTHESIS

Nitrogen-fixing organisms in ant nests supply nitrogen to leaf-cutter ants

METHOD

1. Bring ant colonies into the laboratory and allow them to function in an environment with no insects and no soil (i.e., where the only non-atmospheric source of N is fresh leaves).
2. Measure the nitrogen content of leaves, fungus, ants, and leaf refuse.

RESULTS

The cultivated fungus, the bodies of worker ants, the contents of the refuse dump, and fresh leaves all differed significantly from each other in nitrogen content.

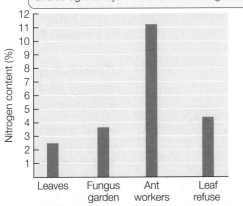

CONCLUSION

Nitrogen is being fixed within the ant nest.

ANALYZE THE DATA

Scientists measured activity of nitrogenase, an enzyme in the bacterial metabolic pathway that fixes atmospheric nitrogen. The results are shown in the figure below. Different letters indicate significant differences in nitrogen content.

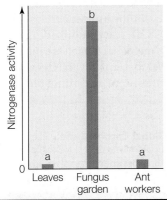

A. Where in the nest is nitrogen fixation occurring?

B. Taken together, the two graphs in this figure support the hypothesis that the fungus concentrates nitrogen in the fungal structures eaten by the ants. Explain. (Hint: Consider that the four values in the Results graph are all significantly different from each other.)

C. Can the researchers now conclude that the extra nitrogen comes from nitrogen-fixing bacteria?

Go to **yourBioPortal.com** for original citations, discussions, and relevant links for all INVESTIGATION figures.

In terrestrial ecosystems, nitrogen fixation is carried out mostly by free-living bacteria in the soil and by symbiotic bacteria associated with plant roots, although it also occurs in other places such as the fungal gardens of leaf-cutter ants (**FIGURE 46.8**). Different microbial species convert ammonium into nitrate (NO_3^-) and other oxides of nitrogen (NO_x) in terrestrial ecosystems. Analogous biochemical processes in aquatic ecosystems are carried out by free-living microorganisms and in some cases by those living in the bodies of phytoplankton. All of these nitrogen-fixing reactions are reversed by another group of microbes in a process called *denitrification*, which ultimately completes the cycle by returning N_2 gas to the atmosphere (see Concept 19.3).

The microbes that contribute to the nitrogen cycle make the rest of life possible because they supply forms of nitrogen that are accessible to primary producers—most notably ammonium and nitrate. Once these soluble chemical forms are taken up by plant roots in terrestrial ecosystems, much of the nitrogen is recycled locally through decomposition of dead organic matter, which again releases ammonium and nitrate. Some nitrogen, however, is leached from soils and carried away by the movement of water, and some of this accessible nitrogen ultimately enters lakes and oceans. In aquatic systems, primary productivity occurs in surface waters, and some nitrogen is recycled locally in these waters. However, much is recycled at greater depths in the water column, as organisms in deep waters intercept and consume sinking detritus. Some of the nitrogen that sinks to the bottom and accumulates in sediments is returned to the surface by upwelling (see Concept 46.1).

LINK The metabolic processes by which microbes transform nitrogen are described in Concepts 19.3 and 25.2

Human activities are affecting nitrogen fluxes and pools. **Fossil fuels** such as coal and petroleum contain nitrogen, so burning them releases oxides of nitrogen, which contribute to atmospheric smog and acid rain. Rice cultivation and raising of livestock also release NO_x, nitrous oxide (N_2O), and ammonia (NH_3) into the atmosphere. This atmospheric nitrogen can be deposited locally or far from the source, in some areas adding as much nitrogen as farmers place on their crops. Humans also fix nitrogen by an industrial process in order to manufacture fertilizer and explosives. The total of industrial nitrogen fixation (see Concept 25.1) and fixation by legume crops such as soybeans now rivals the rate of natural terrestrial fixation (see Figure 46.7). Losses of topsoil and dissolved nitrates from fertilized croplands and deforested areas by wind

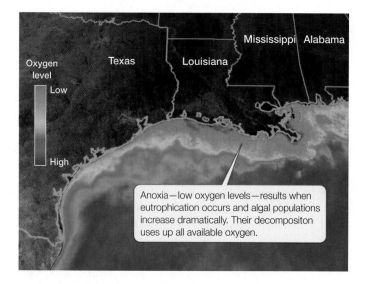

FIGURE 46.9 High Nutrient Input Creates Dead Zones
Runoff from agricultural lands in the midwestern United States carries nitrogen and phosphorus down the Mississippi River and into the Gulf of Mexico. The resulting eutrophication creates a "dead zone" in which few aquatic organisms can survive.

eutrophication—can result in such rapid growth of phytoplankton that consumers cannot eat them all. Respiration by phytoplankton and by decomposers that process their dead bodies depletes oxygen levels below the tolerances of many aquatic organisms, including fish and crustaceans. This can lead to **dead zones** devoid of aquatic life, for example where the Mississippi River discharges nutrient-rich agricultural runoff into the Gulf of Mexico (**FIGURE 46.9**). In nutrient-poor terrestrial ecosystems, excess nitrogen can change the species composition of plant communities. Species that are adapted to low nutrient levels grow slowly, even when fertilized, and can be easily displaced by faster-growing species that can take advantage of the additional nutrient supply. In the Netherlands, nitrogen deposition from upwind industrial and agricultural regions has caused species-rich plant communities to be replaced by species-poor communities. This terrestrial eutrophication is estimated to have caused 13 percent of the recent loss of plant species diversity in the Netherlands.

or water runoff are increasingly exporting nitrogen from terrestrial ecosystems and depositing it in aquatic systems.

Although increased nitrogen inputs can increase primary production, too much can be a bad thing. In aquatic systems, nutrient-stimulated increases in primary productivity—called

Carbon flux is linked to energy flow through ecosystems

All of the macromolecules that make up living organisms contain carbon, and much of the energy that organisms need to fuel their metabolic activities comes from the oxidation of organic carbon compounds. The flux of carbon into, through, and out of communities is therefore intimately linked to energy flow through ecosystems, and biomass is an important pool of the carbon cycle (**FIGURE 46.10**). Carbon occurs in the atmosphere as the gases CO_2 (carbon dioxide) and CH_4 (methane), which are mixed on a global scale by the dynamic circulation of the atmosphere.

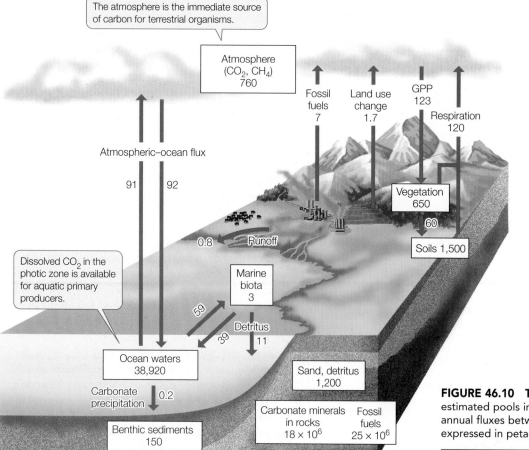

FIGURE 46.10 The Global Carbon Cycle The estimated pools in compartments (white boxes) and annual fluxes between compartments (arrows) are expressed in petagrams (Pg; 10^{15} grams).

yourBioPortal.com

**Go to ANIMATED TUTORIAL 46.3
The Global Carbon Cycle**

APPLY THE CONCEPT

Certain biogeochemical cycles are especially critical for ecosystems

Figure 46.10 depicts the global carbon cycle, showing organic and inorganic pools in various physical compartments as well as fluxes. Refer to that figure to answer the following questions.

1. What percentage of the flux of carbon from terrestrial ecosystems into the atmosphere is due to human activities?

2. What other flux of carbon from terrestrial ecosystems is being influenced by humans, and how?

3. Given the biogeochemical cycles we have discussed so far, which additional two cycles of materials would you wish to explore next in order to understand ecosystem function, and why? (Hint: consider the major chemical constituents of living tissue, discussed in Concept 2.1.)

These same forms are found dissolved in the oceans and fresh water, and aquatic systems also harbor carbonate (CO_3^{2-}). By far the largest pools of carbon occur in fossil fuels and carbonate rocks.

Fluxes of carbon are driven by biological, chemical, and physical processes. The biochemical process of photosynthesis moves carbon from inorganic compartments in the atmosphere and water into the organic compartment; respiration reverses this flux. CO_2 and CH_4 move from their atmospheric pools into the much larger pools in oceans when they dissolve and return to the atmosphere when they outgas; both of these are purely physical processes. The rate at which CO_2 dissolves slightly exceeds the rate at which it outgasses, for two reasons. First, some of the dissolved CO_2 is converted into organic compounds by primary production; most of this carbon is then recycled in surface and deeper waters through the trophic interactions of aquatic organisms, but gravity moves a steady rain of organic detritus into the benthic zone. Second, some of the CO_2 is transformed through chemical reactions with water and other dissolved elements into relatively insoluble carbonate compounds, especially calcium carbonate, which ultimately also sink to the bottom. Some of the organic detritus that accumulates in sediments is transformed through time and under pressure into fossil fuels, whereas some of the carbonates are transformed into carbonate rocks such as limestone.

Human activities influence the carbon cycle in a number of ways. Anything that changes primary productivity, such as nitrogen deposition or altered land use, potentially alters carbon flux between inorganic and organic compartments. Any activity that affects water runoff, such as deforestation or impoundment or alteration of river flows, affects the movement of carbon between the terrestrial and aquatic compartments. Deforestation and burning of fossil fuels increase the atmospheric pool of CO_2. The atmospheric pool of CH_4, in turn, is increased through livestock production, rice cultivation, and water storage in reservoirs, because microbes in the guts of cattle and in waterlogged sediments break down organic compounds anaerobically to produce this gas. Although the atmospheric pool of CH_4 is far smaller than that of CO_2, both are potent greenhouse gases and affect Earth's radiation balance (see Figure 46.11).

Biogeochemical cycles interact

The fluxes of many materials through ecosystems are positively related. If the rate at which primary producers take up carbon increases, for example, fluxes of nitrogen, phosphorus, and other chemical elements into organic compartments also increase, because all are components of the molecules that make up living tissue. When the rate of decomposition increases, all of these elements move back into inorganic compartments. Similarly, the rates of loss of multiple water-soluble nutrients from the soil all increase or decrease with variation in runoff.

However, materials that are critical for organisms differ in their pools as well as in the rates of chemical and physical transport and transformation among compartments. They also vary in their availability relative to the needs of organisms. What this means is that any number of nutrients can limit biological functions: generally, the limiting one is the one that is in lowest supply relative to demand. It also means that biogeochemical cycles can interact in hard-to-predict ways. Increased atmospheric concentrations of CO_2, for example, can increase the water-use efficiency of terrestrial plants. These plants obtain the CO_2 they need for photosynthesis by opening their stomata, which also leads to loss of water vapor from their tissues. In a high-CO_2 environment they leave their stomata open for a shorter time and therefore transpire less water, which slows the rate of water movement from soil to the atmosphere.

Do You Understand Concept 46.3?

- Explain why rates of outgassing of CO_2 from oceans and fresh waters are lower than rates of its dissolution in those waters.

- Why is nitrogen often a limiting nutrient for plant growth, even though it is the most abundant element in Earth's atmosphere?

- How is the cycling of water important to the cycling of other materials that are essential for terrestrial and aquatic ecosystems?

concept 46.4 Biogeochemical Cycles Affect Global Climate

Concept 42.2 discussed how variation in the input of solar radiation influences Earth's climate at different latitudes. We now will return to this theme, taking a slightly different perspective. In Concept 46.3 we explained how solar input drives a critical biogeochemical cycle, the hydrological cycle, by providing the energy to transform water from a liquid to a gaseous state. Water vapor in the atmosphere—gaseous water—has an important impact on the climate as a greenhouse gas. You also may recall from Concept 46.3 that atmospheric CO_2, and CH_4, are greenhouse gases. What is the meaning of this term, and the implications for climate? Because incoming and outgoing radiation passes through Earth's atmosphere, any interactions between photons and atmospheric gas molecules affect the radiation balance of the planet, as we will now see.

Earth's surface is warm because of the atmosphere

All objects that are warmer than absolute zero—our sun included—emit electromagnetic radiation in wavelengths that range from short (ultraviolet), through wavelengths visible to the human eye, to long wavelengths that we perceive as heat (infrared). The distribution of the wavelengths emitted by an object depends on its temperature. Because the surface of the sun is so hot—about 6000°C—its distribution peaks in the visible wavelengths. What happens to incoming solar radiation? Approximately 23% of photons hit aerosols (tiny airborne particles and liquid droplets) and gas molecules in the atmosphere and are reflected back into space. Another 8% of photons are reflected back into space when they reach Earth's surface. About 20% are absorbed rather than reflected by gas molecules in the atmosphere. Finally, about 49% of incoming solar energy (167 watts per m^2, averaged across day and night, all seasons, and all locations) is absorbed by Earth's surface (**FIGURE 46.11**).

Earth's surface, in turn, re-emits photons, but in much longer, less energetic infrared wavelengths. Some of this infrared radiation is absorbed by gas molecules in the atmosphere rather than escaping immediately into space. These molecules are warmed and radiate photons back to Earth's surface, keeping the energy within the Earth system as heat. The warming of Earth that results from retention of heat in its atmosphere is called the **greenhouse effect**. Logically enough, the gases involved are called **greenhouse gases**, and they are those components of the atmosphere that absorb strongly in infrared wavelengths.

We have already mentioned that water vapor, carbon dioxide (CO_2), and methane (CH_4) are greenhouse gases, and it turns out

---

yourBioPortal.com
Go to ANIMATED TUTORIAL 46.4
Earth's Radiation Balance

FIGURE 46.11 Earth's Radiation Balance The average annual energy gain from incoming solar radiation balances the average annual loss from infrared radiation that leaves the Earth system. Gains and losses are given as percentages of the average solar radiant energy striking the top of the atmosphere (342 watts/m²).

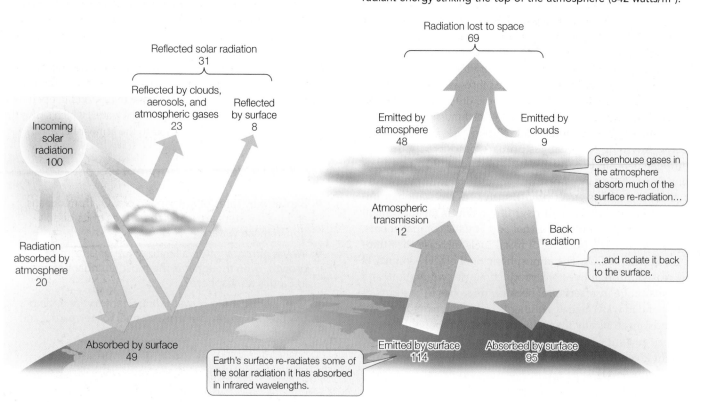

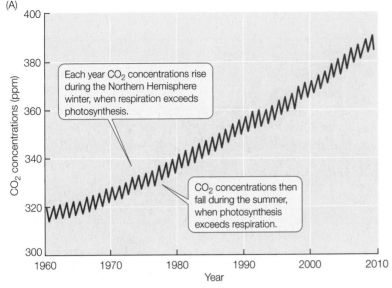

(A)

Each year CO₂ concentrations rise during the Northern Hemisphere winter, when respiration exceeds photosynthesis.

CO₂ concentrations then fall during the summer, when photosynthesis exceeds respiration.

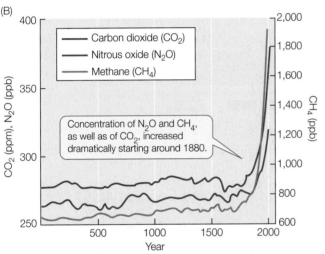

(B)

Concentration of N_2O and CH_4, as well as of CO_2, increased dramatically starting around 1880.

FIGURE 46.12 Atmospheric Greenhouse Gas Concentrations Are Increasing (A) The Keeling curve is based on measurements of atmospheric CO_2 concentrations atop of Mauna Loa, Hawaii, expressed as parts per million by volume of dry air. (B) Measurements of air trapped in glacial ice have allowed researchers to extend our record of greenhouse gas concentrations back 2,000 years.

Recent increases in greenhouse gases are warming Earth's surface

Improved methods for measuring the content of air samples accurately—such as Dave Keeling's precise methods for measuring CO_2—show us that the composition of the atmosphere is changing. Keeling's measurements reveal a steady increase in atmospheric CO_2 over the last half century (**FIGURE 46.12A**), and analyses of air trapped in glacial ice demonstrate that CO_2 and other greenhouse gases began increasing after about 1880 (**FIGURE 46.12B**). Average annual global temperatures have followed suit (**FIGURE 46.13**)—as they must, unless greenhouse gas increases are accompanied by compensating increases in clouds and aerosols that reflect more incoming solar radiation away from Earth.

These higher global temperatures are affecting climate worldwide. A warmer Earth means not only hotter air temperatures, but also a more intense water cycle, with greater overall evaporation and precipitation. Hadley cells (see Concept 42.2) are expected to expand poleward because the warmer tropical air will rise higher in the atmosphere and expand farther toward the poles before sinking. This should cause precipitation to increase near the equator and at high latitudes and to decrease at mid-latitudes. Because warming is uneven in space (the tropical Pacific and western Indian Oceans are warming more than other waters, for example), changes in precipitation will be season- and region-specific. But in general, wet regions are expected to get wetter and dry regions drier. Precipitation trends in the twentieth century appear to support these expectations (**FIGURE 46.14**). Warming is also expected to increase storm intensity. Recent analyses of the intensities of tropical storms indicate that although the overall hurricane number has not increased, strong hurricanes (category 4 and 5) have become more frequent since the 1970s.

Human activities are contributing to changes in Earth's radiation balance

As we saw in Concept 46.3, human activities are affecting Earth's biogeochemical cycles. Burning of fossil fuels and clearing of forests are adding CO_2 to the atmosphere, and expansion of livestock and wetland crop production is adding CH_4 and N_2O. Other human activities are also influencing Earth's radiation balance. Deposition of dust and dark-colored soot particles ("black carbon") from fossil fuel burning increases the amount of incoming solar energy absorbed by snow and ice, thereby increasing the melting of persistent ice fields such as

that nitrous oxide (N_2O) is as well. Fluxes of these molecules into and out of the atmosphere influence Earth's radiation balance. Without the atmosphere, Earth's average surface temperature would be –18°C—about 34°C colder than it is at present. The greenhouse effect influences any planet with an atmosphere; its magnitude depends on the amount and type of material in the atmosphere. Mars's thin atmosphere warms its surface by only 3°C, whereas the thick atmosphere of Venus warms that planet's surface by a whopping 468°C.

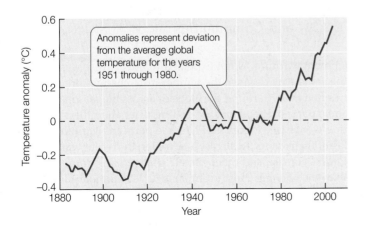

Anomalies represent deviation from the average global temperature for the years 1951 through 1980.

FIGURE 46.13 Global Temperatures Are Increasing A global increase in average annual temperature has been occurring in parallel with the increase atmospheric greenhouse gas concentrations.

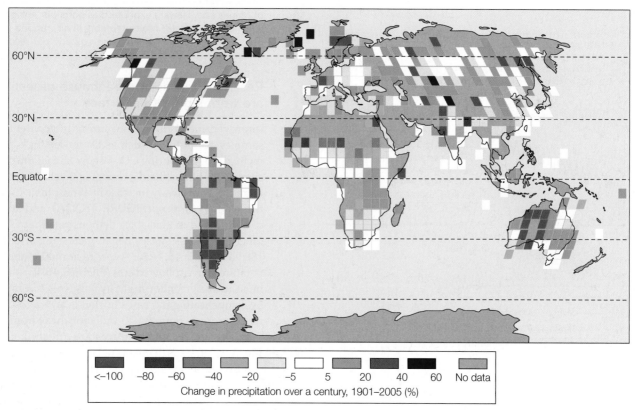

<-100   -80   -60   -40   -20   -5   5   20   40   60   No data

Change in precipitation over a century, 1901–2005 (%)

FIGURE 46.14 Global Precipitation Patterns Have Changed Over the past century, precipitation has increased at high latitudes and in some tropical regions, while it has decreased in some already-dry mid-latitudes. Note, however, that much regional variation is apparent.

the Greenland ice sheet and Arctic sea ice, as well as advancing the timing of spring snowmelt in some temperate regions.

It is possible that reflectance of solar radiation is increased sufficiently by other human activities to offset the increase in greenhouse gases and the deposition of soot and dust on snow and ice. Reflectance rises with increasing amounts of aerosols in the atmosphere and with clearing of land. However, when climate scientists incorporate all of these human effects into quantitative computer models of Earth's climate, they conclude that overall, human activities have contributed significantly to recent climate warming (see Apply the Concept, p. 905).

FRONTIERS Our ability to forecast future climates is enhanced by computing power. Global climate models are computer simulations consisting of equations that describe incoming and outgoing radiation, circulation of the atmosphere and oceans, and interactions with Earth's surface. These models divide Earth into grid cells and solve the equations in time steps. Researchers use global climate models to predict how changes in vegetation and greenhouse gases will alter the climate at any location. Millions of calculations for each time step are needed to achieve the desired 1 km² resolution, so it is no surprise that accurate predictions require supercomputers.

Do You Understand Concept 46.4?

- What is the greenhouse effect?
- Is the greenhouse effect something that occurs without humans, or is it caused by humans?
- What factors caused the increases in methane and carbon dioxide in the atmosphere beginning in the 1800s?

Now that we have seen how human activities are changing Earth's biogeochemical cycles, and how those changes are altering Earth's climate, we will consider how ecological systems are likely to respond to those changes.

concept 46.5 Rapid Climate Change Affects Species and Communities

Climate warming is not expected to affect all ecosystems in the same way because the temperature changes are not distributed evenly around the globe or across the seasons. Just as some parts of the planet are getting drier and others are getting wetter (see Figure 46.14), some parts are warming much faster than others. In all cases, however, the recent warming of Earth's climate, and other climate changes, are far more rapid than the changes most organisms have experienced in their evolutionary histories.

APPLY THE CONCEPT

Biogeochemical cycles affect global climate

Shown below are results of computer simulations of Earth's average temperature between 1850 and 2000 (blue-shaded areas), with the actual observed temperatures superimposed (red lines). Shaded regions indicate the range of outcomes from multiple simulations. The simulations were run using three sets of climate-affecting variables: (A) only natural factors such as solar input, aerosols from volcanic eruptions, and oscillations in ocean circulation; (B) only anthropogenic (human-generated) factors such as land clearing, soot deposition, and emission of aerosols and greenhouse gases; and (C) both natural and anthropogenic factors. Temperatures are expressed as deviation (anomaly) from the average temperature of the last part of the nineteenth century. Using these graphs, answer the following questions and explain your reasoning.

1. What aspects of observed trends in global temperature are not accounted for by natural factors?

2. What aspects of observed trends in global temperature are not accounted for by human-related factors?

3. Which set of factors best accounts for observed global temperature trends?

4. The arrows in panel (C) indicate five major volcanic eruptions. What effect did these natural events have on global temperatures?

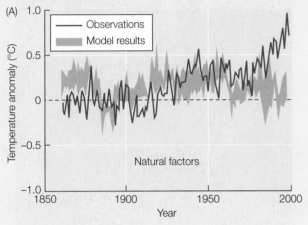

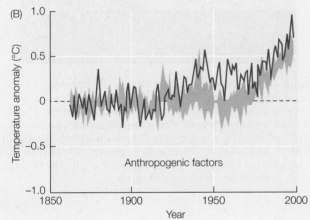

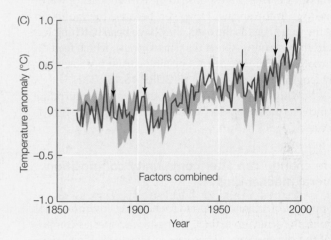

Rapid climate change can leave species behind

The life cycles of organisms that live in seasonal environments evolve so that crucial events of the cycle occur at favorable times of the year. Alpine plants, for example, start growing when the snow melts; seeds of desert plants germinate in response to certain combinations of soil temperature and moisture; changes in day length often trigger plant flowering, bird migration, and onset or cessation of insect dormancy. But climate change is altering the timing of some environmental cues. Although species' responses can change over time, evolving to match new cues, the rate of such evolution may be too slow to keep up with an environment that changes too rapidly or that continues to change directionally (consistently warming, for example) for too long.

Over the short term, there is evidence that many species are continuing to track climatic cues so that events in their life histories remain within favorable periods. For example, biological events that mark the end of the winter season in temperate ecosystems—such as the leafing out of trees—have been taking place at earlier dates as the climate has warmed. In Britain, the first flowers of a sample of 385 plant species now appear, on average, 4.5 days earlier than they did only 25 years ago, during which time the average temperature has increased 1°C at the latitude of the study. However, this is an *average* value for advances in flowering, and not all the species responded in the same way—some did not shift their time of first flowering, and some now flower later in the spring rather than earlier. Thus some species may not respond adaptively to climate change at all. Even for those that do, it is unclear that the adaptive tracking of change can be sustained over the long term.

Changes in seasonal timing can disrupt interspecific interactions

Not all environmental cues respond to the same degree, or in the same direction, to climate change (indeed, some, such as day length, are independent of climate), so temporal relationships among cues are shifting. Because different species make use of different cues, they respond differently to climate change, as we saw with British plant studies. These differences raise the possibility of timing mismatches among species in a community, which can disrupt interactions between consumers and their resources or between mutualists such as plants and pollinators.

A well-documented example of such a mismatch involves oaks, moths, and insect-eating birds. In the Netherlands, the eggs of the winter moth (*Operophtera brumata*) hatch when temperatures warm after a fixed number of frosts, whereas the leaves of the oak trees on which the moths feed (*Quercus robur*) emerge when warm days follow a certain number of cool days. The moths have been hatching earlier, and oak leaves have been emerging earlier, in response to the earlier onset of warm temperatures in the spring. However, because the moths do not use the same cues as the oaks, they have been hatching too early and then starving. Great tits (*Parus major*), which feed on winter moth caterpillars, are not nesting earlier because they use day length to cue egg laying (**FIGURE 46.15**). These birds are now feeding their nestlings after the peak in caterpillar abundance has passed, and as a result are having lower breeding success.

Climate change can alter community composition by several mechanisms

If local populations cannot respond to changing environmental conditions, they may go extinct, changing the species composition of the community. As we saw in Figure 43.6, researchers found that hotter temperatures during the spring breeding season in Mexico's Yucatán Peninsula are restricting the hours during which *Sceloporus* lizards can forage. In fact, the lizards have gone locally extinct at some sites, and lack of sufficient feeding time to support breeding is implicated as the cause. Extending their analysis to lizards worldwide, the researchers predict that a continued increase in global temperatures will place most lizard populations at risk of extinction by 2080. Losses of lizards will affect populations of their insect prey, of the plants those insects feed on, and of other animals that also eat the insects. As we saw in Concept 45.4, the loss of diversity in communities leads to overall loss of community function.

Altered relationships between environmental cues can also change communities even if species do not go extinct locally. In southwestern North America, the winter rainy season now arrives later in the fall, after soil temperatures have cooled. This shift favors plant species whose seeds use cool-season moisture as the cue to germinate, at the expense of species whose

FIGURE 46.15   Climate Change Affects Life Histories A long-term study done in the Netherlands shows that warming temperatures have affected the seasonal cycle of the winter moth *Operophtera brumata*, whose caterpillars are now emerging before the leaves on which they feed have expanded. The caterpillars starve, and great tits (*Parus major*), whose nesting habits have not yet adapted to warmer temperatures, have difficulty finding enough caterpillars to feed their nestlings.

germination is cued by warm-season moisture. Thus the relative abundance of the cool-season species is increasing in these communities.

Climate-caused shifts in the geographic distributions of species can also lead to the assembly of novel communities, as was noted in Concept 45.2. Species have moved up mountains as their low-elevation habitats become too warm and higher-elevation habitats become more favorable. Others have moved toward higher latitudes for similar reasons; such poleward shifts are now documented for numerous taxonomic groups. Different species shift to different degrees or not at all (as was true of shifts in timing of life-cycle events), so the mixture of species at any place changes. Distributional shifts also influence species interactions. Warmer temperatures in western North America, for example, are speeding up the life cycle of the mountain bark beetle (*Dendroctonus ponderosae*) and decreasing its winter mortality rate. As a result, beetle populations have increased and expanded, causing widespread tree mortality in pine and spruce forests from New Mexico to Alaska. Loss of the conifers has cascading effects on the consumers that depend on them, and on water runoff, wildfire frequency, and NPP.

Extreme climate events also have an impact

One of the predictions from climate modeling is that the frequency of extreme climate events will increase. Severe storms, flooding, drought, heat waves, and other such events can elicit sudden shifts in species distributions and in community composition. For example, a brief but intense drought in the 1950s caused a ponderosa pine forest in northern New Mexico to shrink abruptly, and drought-adapted piñon–juniper woodland to expand, by more than 2 kilometers in less than five years. The new community boundary remained stable even after the drought ended. This is the most rapid shift in the boundary between two terrestrial plant communities ever documented at the scale of an entire landscape.

Such a response to an out-of-the-ordinary event might suggest that species are always able to shift their geographic distributions rapidly in response to climate change. But as was true with the response to temporal shifts in cues, it remains virtually certain that many species will be left behind by continued climate change, and that the number of species, their relative abundances, and their interactions will shift.

Do You Understand Concept 46.5?

- Extrapolating from the British data, what change might you anticipate in average first flowering date by the middle of this century, when atmospheric CO_2 is predicted to reach double its preindustrial concentration and temperature in Britain is expected to rise by an additional 3°C?

- How might changes in the seasonal timing of some events (such as the onset of warm weather at the end of winter) affect mutualistic interactions between plants and insect pollinators?

The changes human activities are causing in Earth's biogeochemical cycles and climate have potentially serious ecological effects. What can we do to address these effects?

concept 46.6 Ecological Challenges Can Be Addressed through Science and International Cooperation

The current period of climate change is not unprecedented. Throughout the history of life on Earth, wobble in the planet's orbit around the sun, continental drift, volcanic activity, sunspots, and even asteroid impacts have caused Earth's temperatures to change, precipitating five major episodes of mass extinction (see Figure 18.12). There is even precedent for changes in Earth's atmosphere induced by organisms. The first photosynthetic microbes were responsible for increasing atmospheric oxygen concentrations to a level that was toxic to the anaerobic prokaryotes that inhabited Earth at the time, and the first land plants caused another rise in oxygen concentrations 250 million years ago. What is unprecedented about present climate change is that it has been precipitated by diverse activities of a single component of the biosphere: *Homo sapiens*.

It is sobering to realize that one species can wield such power. On the other hand, it is encouraging to realize that, of all species on Earth, humans are best able to address the problems they have caused—not only because science equips us to understand the natural world and to devise solutions to problems, but also because *Homo sapiens* has a remarkable capacity for cooperative action. Cooperative interactions are central features of all human societies. While other group-living animals cooperate as well, cooperation with unrelated individuals is especially highly developed in humans. We routinely support cooperative societal ventures, we police bad behavior, we even support scientific discovery—and we are unique, as far as we now know, in caring about other species with which we share Earth.

FRONTIERS Are humans truly cooperative? Biologists are fascinated by social behavior, which is found scattered throughout the tree of life. Recent advances in *game theory* reveal a surprisingly broad range of conditions under which social cooperation, even with strangers, is more adaptive than selfishness. Complementing these theoretical advances, recent studies of human infants show that they help other infants unknown to them, even when they receive no reward. Infant chimpanzees show similar, but less well developed, helping behavior. Such emerging results suggest that *Homo sapiens* has evolved an unusual tendency toward cooperation.

Governments of separate nations have cooperated in several global-scale initiatives to tackle complex environmental issues, from the IGY, which supported the first years of Dave Keeling's work, to the United Nations' Intergovernmental Panel on Climate Change and World Meteorological Organization, which are operating today. Earth's nations have also negotiated international agreements to achieve environmental goals, including the Montreal Protocol to prevent depletion of UV-absorbing atmospheric ozone, the Kyoto Protocol to reduce emissions of greenhouse gases, and the Convention on International Trade in Endangered Species of Wild Fauna and Flora (CITES), which seeks to conserve species by eliminating the economic benefits of exploiting them.

yourBioPortal.com

Go to WEB ACTIVITY 46.1
The Benefits of Cooperation

Nonetheless, humans face huge challenges to achieving effective cooperation on a global scale. A major challenge is that the economic policies of virtually every nation are structured to achieve continual economic growth—ever-increasing production and consumption of goods and services—despite the fact that Earth has a finite capacity to provide those goods and services. We will need to make the transition to sustainable, steady-state economies, and that will require an overhaul of economic models and institutions. Another, related, challenge is the continued multiplicative growth of the human population (see Concept 43.4). On a crowded planet, with competition among societies for limited resources, cooperation inevitably becomes more difficult. Addressing both challenges will require that we devise international systems for establishing—and enforcing—rules of acceptable behavior among groups and nations.

 QA **QUESTION** How did Keeling's research contribute to our understanding of the global ecosystem?

ANSWER Dave Keeling was successful in obtaining long-term measurements of CO_2. His dedication left an important legacy, the Keeling curve (see Figure 46.12A), which documents how atmospheric CO_2 is increasing (Concept 46.4). Almost immediately, Keeling discovered that CO_2 concentrations do not vary erratically, as earlier crude methods of measurement suggested, but instead change seasonally as summer warmth, and the primary productivity that goes with it, shifts from the Northern Hemisphere (where most of Earth's land mass lies) to the Southern Hemisphere (notice the saw-teeth of the Keeling curve). Keeling's measurements quickly contributed to a better understanding of the pools and fluxes of the global carbon cycle (Concept 46.3), including the influence of fossil-fuel burning (currently pegged to add over 8 billion metric tons of carbon to the atmosphere annually). Better understanding of the carbon cycle contributed to greatly improved global climate models. The measurements on Mauna Loa and elsewhere continue to the present day, and they show that atmospheric CO_2 increased from 315 ppm in 1958 to 390 ppm in 2010.

Keeling's results were noticed almost immediately. In 1965 the President's Science Advisory Committee warned of the increased greenhouse effect, and in 1966 the National Academy of Sciences issued a scholarly report on the topic. The first World Climate Conference was held in Geneva, Switzerland, in 1979, at a time when there was growing scientific consensus that climate change poses a critical environmental challenge to humans. The number of scientists studying climate change, and the sophistication of their climate experiments and studies, grew rapidly. Measurements by the end of the twentieth century showed that the average temperature of the planet had increased by 0.7°C since 1900—an increase very close to the predictions of global climate models.

The Keeling curve forms a crucial part of our understanding of the Earth system. It is an example of the carefully documented scientific information that is used by the Intergovernmental Panel on Climate Change (IPCC). The IPCC was formed in 1988 by another scientific collaboration of governments (Concept 46.6)—a sort of "grandchild" of IGY—and its many scientists continue to summarize and report on the latest evidence for natural and human-caused climate change. The 2009 IPCC report predicts a further human-caused increase in global average temperature between 1.8°C and 4.0°C by the end of this century.

We have a civilization based on science and technology, and we've cleverly arranged things so that almost nobody understands science and technology. That is as clear a prescription for disaster as you can imagine. While we might get away with this combustible mixture of ignorance and power for a while, sooner or later it's going to blow up in our faces. The powers of modern technology are so formidable that it's insufficient just to say, 'Well, those in charge, I'm sure, are doing a good job.' This is a democracy, and for us to make sure that the powers of science and technology are used properly and prudently, we ourselves must understand science and technology. We must be involved in the decision-making process.

Carl Sagan

concept 46.1 Climate and Nutrients Affect Ecosystem Function

- An ecological community in its abiotic context is an ecosystem.

- A community's rate of biomass production per unit of area—its net primary productivity (NPP)—is related to its exchange of energy and materials with its abiotic surroundings, and thus is a measure of ecosystem function.

- NPP varies considerably among ecosystem types. **Review Figure 46.1**

- Variation of terrestrial NPP shows that it is influenced by both temperature and precipitation. Soil nutrients also play a role in terrestrial NPP. **Review Figures 46.2 and 46.3**

- Aquatic NPP varies most strongly with availability of light and nutrients. **Review Figures 46.4**

concept 46.2 Biological, Geological, and Chemical Processes Move Materials through Ecosystems

- The molecular form and physical state in which an element exists and the physical compartment of the biosphere in which it is located determine whether the element is accessible to life.

- The transformations of materials and their movements among compartments are called **biogeochemical cycles**. **Review Figure 46.5**

- The **pool**, or amount, of an element or molecule in a compartment depends on its flows, or **fluxes**, into and out of that compartment.

concept 46.3 Certain Biogeochemical Cycles Are Especially Critical for Ecosystems

- Evaporation drives the water cycle, moving water into the atmosphere in gaseous form; precipitation returns liquid water to Earth's surface. These transitions between gaseous and liquid states involve exchanges of heat energy between water molecules and their surroundings. Gravity-driven flows move water from land to the oceans, transporting many other materials as well. **Review Figure 46.6 and ANIMATED TUTORIAL 46.1**

- The availability of nitrogen to living organisms depends on **nitrogen fixation** and other biochemical processes carried out by microbes. These processes change gaseous N_2 into forms that primary producers can absorb and that are recycled locally, and return N_2 to the atmosphere. **Review Figure 46.7 and ANIMATED TUTORIAL 46.2**

- Human activities add various chemical forms of nitrogen to the atmosphere and to aquatic systems, contributing to smog and causing **eutrophication** and oxygen-poor **dead zones** devoid of life. **Review Figure 46.9**

- Photosynthesis and respiration move carbon between the inorganic and organic compartments of the biosphere. As a result, energy flow and the flux of carbon through biological communities are intimately linked.

- Some carbon absorbed from the atmosphere by surface waters is taken up through photosynthesis, some precipitates as calcium carbonate, and some falls to the benthic zone as organic detritus. Calcium carbonate and detritus that accumulate in sediments and soils are transformed into carbonate rocks and **fossil fuels**.

Use of fossil fuels by humans increases the carbon content of the atmosphere. **Review Figure 46.10 and ANIMATED TUTORIAL 46.3**

- The biogeochemical cycles of different materials can interact in ways that are hard to predict.

concept 46.4 Biogeochemical Cycles Affect Global Climate

- Earth absorbs incoming solar energy and re-emits it in infrared wavelengths. Some of this infrared radiation is absorbed by atmospheric gases and is re-radiated back to Earth's surface. The resulting retention of heat within the Earth system is called the **greenhouse effect**. **Review Figure 46.11 and ANIMATED TUTORIAL 46.4**

- Gases such as water vapor, carbon dioxide , methane, and nitrous oxide that absorb strongly in infrared wavelengths are called **greenhouse gases**. Fluxes of these molecules into and out of the atmosphere influence Earth's radiation balance.

- Atmospheric concentrations of CO_2 and other greenhouse gases have been increasing since about 1880, and average annual global temperatures have followed suit. **Review Figures 46.12 and 46.13**

- Global warming is changing Earth's climates. High latitudes are warming more than low latitudes, precipitation patterns are changing, and storm intensities are increasing. **Review Figure 46.14**

- Computer models of the Earth system show that human activities have contributed significantly to the recent warming of the climate.

concept 46.5 Rapid Climate Change Affects Species and Communities

- Climate change is altering the timing of some seasonal environmental cues but not others. Although species can evolve new responses to such cues, the rate of evolution will not keep up with an environment that changes too rapidly.

- Because different species respond to different seasonal cues, altered cues result in timing mismatches among species in a community and thus disrupt their interactions.

- Climate change is altering the distribution and abundance of species, resulting in the assembly of novel communities.

- Climate change is increasing the frequency of extreme weather events, which can cause sudden shifts in species distributions and in community composition.

concept 46.6 Ecological Challenges Can Be Addressed through Science and International Cooperation

- Humans are causing major changes in the biosphere and in other aspects of the Earth system. However, we are also uniquely equipped to address these changes, not only because science enables us to understand the natural world and to devise solutions to problems, but also because of our capacity for cooperative action. **Review WEB ACTIVITY 46.1**

See WEB ACTIVITY 46.2 for a concept review of this chapter.

Appendix A: The Tree of Life

Phylogeny is the organizing principle of modern biological taxonomy. A guiding principle of modern phylogeny is monophyly. A monophyletic group is considered to be one that contains an ancestral lineage and all of its descendants. Any such group can be extracted from a phylogenetic tree with a single cut.

The tree shown here provides a guide to the relationships among the major groups of extant (living) organisms in the tree of life as we have presented them throughout this book. The position of the branching "splits" indicates the relative branching order of the lineages of life, but the time scale is not meant to be uniform. In addition, the groups appearing at the branch tips do not necessarily carry equal phylogenetic "weight." For example, the ginkgo [75] is indeed at the apex of its lineage; this gymnosperm group consists of a single living species. In contrast, a phylogeny of the eudicots [83] could continue on from this point to fill many more trees the size of this one.

The glossary entries that follow are informal descriptions of some major features of the organisms described in Part Seven of this book. Each entry gives the group's common name, followed by the formal scientific name of the group (in parentheses). Numbers in square brackets reference the location of the respective groups on the tree.

It is sometimes convenient to use an informal name to refer to a collection of organisms that are not monophyletic but nonetheless all share (or all lack) some common attribute. We call these "convenience terms"; such groups are indicated in these entries by quotation marks, and we do not give them formal scientific names. Examples include "prokaryotes," "protists," and "algae." Note that these groups cannot be removed with a single cut; they represent a collection of distantly related groups that appear in different parts of the tree. We also use quotation marks here to designate two groups of fungi that are not believed to be monophyletic.

An interactive version of this tree, with links to much greater detail (such as photos, distribution maps, species lists, and identification keys), can be found at yourBioPortal.com; see also http://tolweb.org/tree.

– A –

acorn worms (*Enteropneusta*) Benthic marine hemichordates [119] with an acorn-shaped proboscis, a short collar (neck), and a long trunk.

"algae" Convenience term encompassing various distantly related groups of aquatic, photosynthetic eukaryotes [4].

alveolates (*Alveolata*) [5] Unicellular eukaryotes with a layer of flattened vesicles (alveoli) supporting the plasma membrane. Major alveolate groups include the dinoflagellates [51], apicomplexans [50], and ciliates [49].

amborella (*Amborella*) [78] An understory shrub or small tree found on the South Pacific island of New Caledonia. Thought to be the sister group of the remaining living angiosperms [15].

ambulacrarians (*Ambulacraria*) [29] The echinoderms [118] and hemichordates [119].

amniotes (*Amniota*) [36] Mammals, reptiles, and their extinct close relatives. Characterized by many adaptations to terrestrial life, including an amniotic egg (with a unique set of membranes—the amnion, chorion, and allantois), a water-repellant epidermis (with epidermal scales, hair, or feathers), and, in males, a penis that allows internal fertilization.

amoebozoans (*Amoebozoa*) [84] A group of eukaryotes [4] that use lobe-shaped pseudopods for locomotion and to engulf food. Major amoebozoan groups include the loboseans, plasmodial slime molds, and cellular slime molds.

amphibians (*Amphibia*) [128] Tetrapods [35] with glandular skin that lacks epidermal scales, feathers, or hair. Many amphibian species undergo a complete metamorphosis from an aquatic larval form to a terrestrial adult form, although direct development is also common. Major amphibian groups include frogs and toads (anurans), salamanders, and caecilians.

amphipods (*Amphipoda*) Small crustaceans [116] that are abundant in many marine and freshwater habitats. They are important herbivores, scavengers, and micropredators, and are an important food source for many aquatic organisms.

angiosperms (*Anthophyta* or *Magnoliophyta*) [15] The flowering plants. Major angiosperm groups include the monocots [82], eudicots [83], and magnoliids [81].

animals (*Animalia* or *Metazoa*) [19] Multicellular heterotrophic eukaryotes. The majority of animals are bilaterians [22]. Other groups of animals include the cnidarians [97], ctenophores [96], placozoans [95], and sponges [20]. The closest living relatives of the animals are the choanoflagellates [91].

annelids (*Annelida*) [105] Segmented worms, including earthworms, leeches, and polychaetes. One of the major groups of lophotrochozoans [24].

anthozoans (*Anthozoa*) One of the major groups of cnidarians [97]. Includes the sea anemones, sea pens, and corals.

anurans (*Anura*) Comprising the frogs and toads, this is the largest group of living amphibians [128]. They are tail-less, with a shortened vertebral column and elongate hind legs modified for jumping. Many species have an aquatic larval form known as a tadpole.

apicomplexans (*Apicomplexa*) [50] Parasitic alveolates [5] characterized by the possession of an apical complex at some stage in the life cycle.

arachnids (*Arachnida*) Chelicerates [114] with a body divided into two parts: a cephalothorax that bears six pairs of appendages (four pairs of which are usually used as legs) and an abdomen that bears the genital opening. Familiar arachnids include spiders, scorpions, mites and ticks, and harvestmen.

arbuscular mycorrhizal fungi (*Glomeromycota*) [88] A group of fungi [17] that associate with plant roots in a close symbiotic relationship.

archaeans (*Archaea*) [3] Unicellular organisms lacking a nucleus and lacking peptidoglycan in the cell wall. Once grouped with the bacteria, archaeans possess distinctive membrane lipids.

archosaurs (*Archosauria*) [38] A group of reptiles [37] that includes dinosaurs and crocodilians [133]. Most dinosaur groups became extinct at the end of the Cretaceous; birds [132] are the only surviving dinosaurs.

arrow worms (*Chaetognatha*) [98] Small planktonic or benthic predatory marine worms with fins and a pair of hooked, prey-grasping spines on each side of the head.

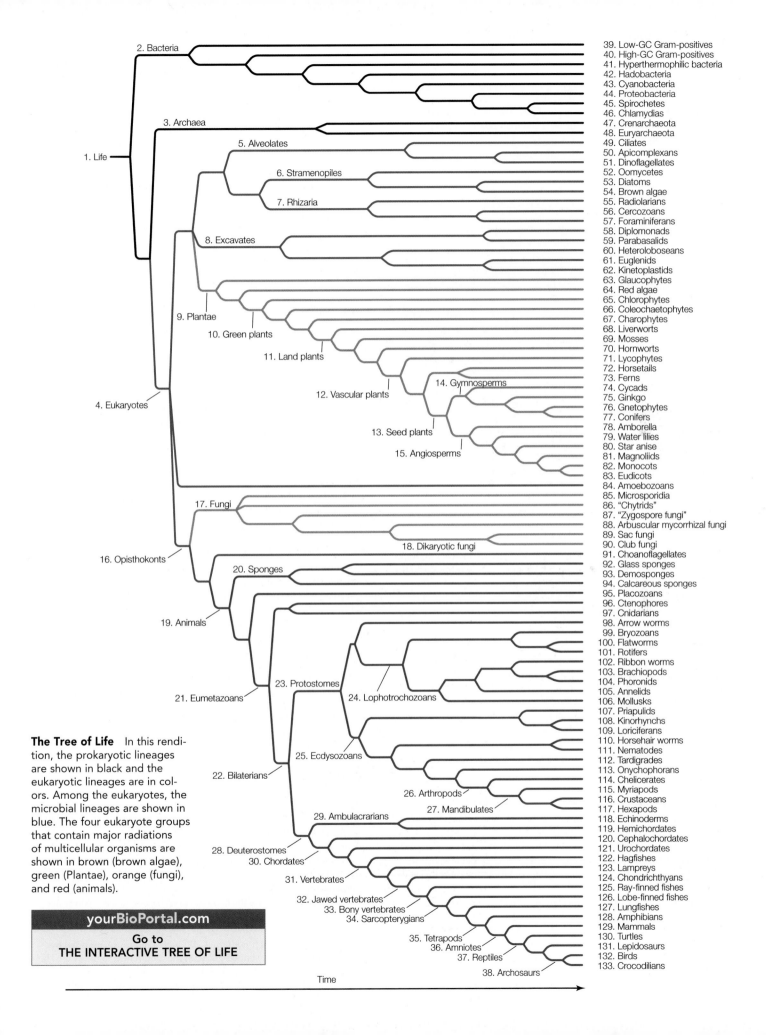

The Tree of Life In this rendition, the prokaryotic lineages are shown in black and the eukaryotic lineages are in colors. Among the eukaryotes, the microbial lineages are shown in blue. The four eukaryote groups that contain major radiations of multicellular organisms are shown in brown (brown algae), green (Plantae), orange (fungi), and red (animals).

yourBioPortal.com

Go to
THE INTERACTIVE TREE OF LIFE

Time

2. Bacteria
3. Archaea
1. Life
5. Alveolates
6. Stramenopiles
7. Rhizaria
8. Excavates
4. Eukaryotes
9. Plantae
10. Green plants
11. Land plants
14. Gymnosperms
12. Vascular plants
13. Seed plants
15. Angiosperms
17. Fungi
18. Dikaryotic fungi
16. Opisthokonts
20. Sponges
19. Animals
21. Eumetazoans
23. Protostomes
24. Lophotrochozoans
25. Ecdysozoans
22. Bilaterians
26. Arthropods
27. Mandibulates
29. Ambulacrarians
28. Deuterostomes
30. Chordates
31. Vertebrates
32. Jawed vertebrates
33. Bony vertebrates
34. Sarcopterygians
35. Tetrapods
36. Amniotes
37. Reptiles
38. Archosaurs

39. Low-GC Gram-positives
40. High-GC Gram-positives
41. Hyperthermophilic bacteria
42. Hadobacteria
43. Cyanobacteria
44. Proteobacteria
45. Spirochetes
46. Chlamydias
47. Crenarchaeota
48. Euryarchaeota
49. Ciliates
50. Apicomplexans
51. Dinoflagellates
52. Oomycetes
53. Diatoms
54. Brown algae
55. Radiolarians
56. Cercozoans
57. Foraminiferans
58. Diplomonads
59. Parabasalids
60. Heteroloboseans
61. Euglenids
62. Kinetoplastids
63. Glaucophytes
64. Red algae
65. Chlorophytes
66. Coleochaetophytes
67. Charophytes
68. Liverworts
69. Mosses
70. Hornworts
71. Lycophytes
72. Horsetails
73. Ferns
74. Cycads
75. Ginkgo
76. Gnetophytes
77. Conifers
78. Amborella
79. Water lilies
80. Star anise
81. Magnoliids
82. Monocots
83. Eudicots
84. Amoebozoans
85. Microsporidia
86. "Chytrids"
87. "Zygospore fungi"
88. Arbuscular mycorrhizal fungi
89. Sac fungi
90. Club fungi
91. Choanoflagellates
92. Glass sponges
93. Demosponges
94. Calcareous sponges
95. Placozoans
96. Ctenophores
97. Cnidarians
98. Arrow worms
99. Bryozoans
100. Flatworms
101. Rotifers
102. Ribbon worms
103. Brachiopods
104. Phoronids
105. Annelids
106. Mollusks
107. Priapulids
108. Kinorhynchs
109. Loriciferans
110. Horsehair worms
111. Nematodes
112. Tardigrades
113. Onychophorans
114. Chelicerates
115. Myriapods
116. Crustaceans
117. Hexapods
118. Echinoderms
119. Hemichordates
120. Cephalochordates
121. Urochordates
122. Hagfishes
123. Lampreys
124. Chondrichthyans
125. Ray-finned fishes
126. Lobe-finned fishes
127. Lungfishes
128. Amphibians
129. Mammals
130. Turtles
131. Lepidosaurs
132. Birds
133. Crocodilians

arthropods (*Arthropoda*)   The largest group of ecdysozoans [25]. Arthropods are characterized by a stiff exoskeleton, segmented bodies, and jointed appendages. Includes the chelicerates [114], myriapods [115], crustaceans [116], and hexapods (insects and their relatives) [117].

ascidians (*Ascidiacea*)   "Sea squirts"; the largest group of urochordates [121]. Also known as tunicates, they are sessile (as adults), marine, saclike filter feeders.

– B –

bacteria (*Eubacteria*) [2]   Unicellular organisms lacking a nucleus, possessing distinctive ribosomes and initiator tRNA, and generally containing peptidoglycan in the cell wall. Different bacterial groups are distinguished primarily on nucleotide sequence data.

barnacles (*Cirripedia*)   Crustaceans [116] that undergo two metamorphoses—first from a feeding planktonic larva to a nonfeeding swimming larva, and then to a sessile adult that forms a "shell" composed of four to eight plates cemented to a hard substrate.

bilaterians (*Bilateria*) [22]   Those animal groups characterized by bilateral symmetry and three distinct tissue types (endoderm, ectoderm, and mesoderm). Includes the protostomes [23] and deuterostomes [28].

birds (*Aves*) [132]   Feathered, flying (or secondarily flightless) tetrapods [35].

bivalves (*Bivalvia*)   Major mollusk [106] group; clams and mussels. Bivalves typically have two similar hinged shells that are each asymmetrical across the midline.

bony vertebrates (*Osteichthyes*) [33]   Vertebrates [31] in which the skeleton is usually ossified to form bone. Includes the ray-finned fishes [125], coelacanths [126], lungfishes [127], and tetrapods [35].

brachiopods (*Brachiopoda*) [103]   Lophotrochozoans [24] with two similar hinged shells that are each symmetrical across the midline. Superficially resemble bivalve mollusks, except for the shell symmetry.

brittle stars (*Ophiuroidea*)   Echinoderms [118] with five long, whip-like arms radiating from a distinct central disk that contains the reproductive and digestive organs.

brown algae (*Phaeophyta*) [54]   Multicellular, almost exclusively marine stramenopiles [6] generally containing the pigment fucoxanthin as well as chlorophylls *a* and *c* in their chloroplasts.

bryozoans (*Ectoprocta* or *Bryozoa*) [99]   A group of marine and freshwater lophotrochozoans [24] that live in colonies attached to substrata; also known as ectoprocts or moss animals.

– C –

caecilians (*Gymnophiona*)   A group of burrowing or aquatic amphibians [128]. They are elongate, legless, with a short tail (or none at all), reduced eyes covered with skin or bone, and a pair of sensory tentacles on the head.

calcareous sponges (*Calcarea*) [94]   Filter-feeding marine sponges with spicules composed of calcium carbonate.

cellular slime molds (*Dictyostelida*)   Amoebozoans [84] in which individual amoebas aggregate under stress to form a multicellular pseudoplasmodium.

cephalochordates (*Cephalochordata*) [120]   A group of weakly swimming, eel-like benthic marine chordates [30]; also called lancelets.

cephalopods (*Cephalopoda*)   Active, predatory mollusks [106] in which the molluscan foot has been modified into muscular hydrostatic arms or tentacles. Includes octopuses, squids, and nautiluses.

cercozoans (*Cercozoa*) [56]   Unicellular eukaryotes [4] that feed by means of threadlike pseudopods. Group together with foraminiferans [57] and radiolarians [55] to comprise the rhizaria [7].

charophytes (*Charales*) [67]   Multicellular green algae with branching, apical growth and plasmodesmata between adjacent cells. The closest living relatives of the land plants [11], they retain the egg in the parent organism.

chelicerates (*Chelicerata*) [114]   A major group of arthropods [26] with pointed appendages (chelicerae) used to grasp food (as opposed to the chewing mandibles of most other arthropods). Includes the arachnids, horseshoe crabs, pycnogonids, and extinct sea scorpions.

chimaeras (*Holocephali*)   A group of bottom-dwelling, marine, scaleless chondrichthyan fishes [124] with large, permanent, grinding tooth plates (rather than the replaceable teeth found in other chondrichthyans).

chitons (*Polyplacophora*)   Flattened, slow-moving mollusks [106] with a dorsal protective calcareous covering made up of eight articulating plates.

chlamydias (*Chlamydiae*) [46]   A group of very small Gram-negative bacteria; they live as intracellular parasites of other organisms.

chlorophytes (*Chlorophyta*) [65]   The most abundant and diverse group of green algae, including freshwater, marine, and terrestrial forms; some are unicellular, others colonial, and still others multicellular. Chlorophytes use chlorophylls *a* and *c* in their photosynthesis.

choanoflagellates (*Choanozoa*) [91]   Unicellular eukaryotes [4] with a single flagellum surrounded by a collar. Most are sessile, some are colonial. The closest living relatives of the animals [19].

chondrichthyans (*Chondrichthyes*) [124]   One of the two main groups of jawed vertebrates [32]; includes sharks, rays, and chimaeras. They have cartilaginous skeletons and paired fins.

chordates (*Chordata*) [30]   One of the two major groups of deuterostomes [28], characterized by the presence (at some point in development) of a notochord, a hollow dorsal nerve cord, and a post-anal tail. Includes the cephalochordates [120], urochordates [121], and vertebrates [31].

"chytrids" [90]   Convenience term used for a paraphyletic group of mostly aquatic, microscopic fungi [17] with flagellated gametes. Some exhibit alternation of generations.

ciliates (*Ciliophora*) [49]   Alveolates [5] with numerous cilia and two types of nuclei (micronuclei and macronuclei).

clitellates (*Clitellata*)   Annelids [105] with gonads contained in a swelling (called a clitellum) toward the head of the animal. Includes earthworms (oligochaetes) and leeches.

club fungi (*Basidiomycota*) [90]   Fungi [17] that, if multicellular, bear the products of meiosis on club-shaped basidia and possess a long-lasting dikaryotic stage. Some are unicellular.

club mosses (*Lycopodiophyta*) [71]   Vascular plants [12] characterized by microphylls. See lycophytes.

cnidarians (*Cnidaria*) [97]   Aquatic, mostly marine eumetazoans [21] with specialized stinging organelles (nematocysts) used for prey capture and defense, and a blind gastrovascular cavity. The closest living relatives of the ctenophores [96].

coelacanths (*Actinista*) [126]   A group of marine sarcopterygians [34] that was diverse from the Middle Devonian to the Cretaceous, but is now known from just two living species. The pectoral and anal fins are on fleshy stalks supported by skeletal elements, so they are also called lobe-finned fishes.

coleochaetophytes (*Coleochaetales*) [66]   Multicellular green algae characterized by flattened growth form composed of thin-walled cells. Thought to be the sister-group to the charophytes [67] plus land plants [11].

conifers (*Pinophyta* or *Coniferophyta*) [77]   Cone-bearing, woody seed plants [13].

copepods (*Copepoda*)   Small, abundant crustaceans [116] found in marine, freshwater, or wet terrestrial habitats. They have a single eye, long antennae, and a body shaped like a teardrop.

craniates (*Craniata*)   Some biologist exclude the hagfishes [122] from the vertebrates [31], and use the term craniates to refer to the two groups combined.

crenarchaeotes (*Crenarchaeota*) [47]   A major and diverse group of archaeans [3], defined on the basis of rRNA base sequences. Many are extremophiles (inhabit extreme environments), but the group may also be the most abundant archaeans in the marine environment.

crinoids (*Crinoidea*)   Echinoderms [118] with a mouth surrounded by feeding arms, and a U-shaped gut with the mouth next to the anus. They attach to the substratum by a stalk or are free-swimming. Crinoids were abundant in the middle and late Paleozoic, but only a few hundred species have survived to the present. Includes the sea lilies and feather stars.

crocodilians (*Crocodylia*) [133]   A group of large, predatory, aquatic archosaurs [38]. The closest living relatives of birds [132]. Includes alligators, caimans, crocodiles, and gharials.

crustaceans (*Crustacea*) [116]   Major group of marine, freshwater, and terrestrial arthropods [26] with a head, thorax, and abdomen (although the head and thorax may be fused), covered with a thick exoskeleton, and with two-part appendages. Crustaceans undergo metamorphosis from a nauplius larva. Includes decapods, isopods, krill, barnacles, amphipods, copepods, and ostracods.

ctenophores (*Ctenophora*) [96]   Radially symmetrical, diploblastic marine animals [19], with a complete gut and eight rows of fused plates of cilia (called ctenes).

cyanobacteria (*Cyanobacteria*) [43]   A group of unicellular, colonial, or filamentous bacteria that conduct photosynthesis using chlorophyll *a*.

cycads (*Cycadophyta*) [74]   Palmlike gymnosperms with large, compound leaves.

cyclostomes (*Cyclostomata*)   This term refers to the possibly monophyletic group of lampreys [123] and hagfishes [122]. Molecular data support this group, but morphological data suggest that lampreys are more closely related to jawed vertebrates [32] than to hagfishes.

– D –

decapods (*Decapoda*)   A group of marine, freshwater, and semiterrestrial crustaceans [116] in which five of the eight pairs of thoracic appendages function as legs (the other three pairs, called maxillipeds, function as mouthparts). Includes crabs, lobsters, crayfishes, and shrimps.

demosponges (*Demospongiae*) [93]   The largest of the three groups of sponges [20], accounting for 90 percent of all sponge species. Demosponges have spicules made of silica, spongin fiber (a protein), or both.

deuterostomes (*Deuterostomia*) [28]   One of the two major groups of bilaterians [22], in which the mouth forms at the opposite end of the embryo from the blastopore in early development (contrast with protostomes). Includes the ambulacrarians [29] and chordates [30].

diatoms (*Bacillariophyta*) [53]   Unicellular, photosynthetic stramenopiles [6] with glassy cell walls in two parts.

dikaryotic fungi (*Dikarya*) [18]   A group of fungi [17] in which two genetically different haploid nuclei coexist and divide within the same hypha; includes club fungi [90] and sac fungi [89].

dinoflagellates (*Dinoflagellata*) [51]   A group of alveolates [5] usually possessing two flagella, one in an equatorial groove and the other in a longitudinal groove; many are photosynthetic.

diplomonads (*Diplomonadida*) [58]   A group of eukaryotes [4] lacking mitochondria; most have two nuclei, each with four associated flagella.

– E –

ecdysozoans (*Ecdysozoa*) [25]   One of the two major groups of protostomes [23], characterized by periodic molting of their exoskeletons. Nematodes [111] and arthropods [26] are the largest ecdysozoan groups.

echinoderms (*Echinodermata*) [118]   A major group of marine deuterostomes [28] with fivefold radial symmetry (at some stage of life) and an endoskeleton made of calcified plates and spines. Includes sea stars, crinoids, sea urchins, sea cucumbers, and brittle stars.

elasmobranchs (*Elasmobranchii*)   The largest group of chondrichthyan fishes [124]. Includes sharks, skates, and rays. In contrast to the other group of living chondrichthyans (the chimaeras), they have replaceable teeth.

embryophytes   *See* land plants [11].

eudicots (*Eudicotyledones*)[83]   A group of angiosperms [15] with pollen grains possessing three openings. Typically with two cotyledons, net-veined leaves, taproots, and floral organs typically in multiples of four or five.

euglenids (*Euglenida*) [61]   Flagellate excavates characterized by a pellicle composed of spiraling strips of protein under the plasma membrane; the mitochondria have disk-shaped cristae. Some are photosynthetic.

eukaryotes (*Eukarya*) [4]   Organisms made up of one or more complex cells in which the genetic material is contained in nuclei. Contrast with archaeans [3] and bacteria [2].

eumetazoans (*Eumetazoa*) [21]   Those animals [19] characterized by body symmetry, a gut, a nervous system, specialized types of cell junctions, and well-organized tissues in distinct cell layers (although there have been secondary losses of some of these characteristics in some eumetazoans).

euphyllophytes (*Euphyllophyta*)   The group of vascular plants [12] that is sister to the lycophytes [71] and which includes all plants with megaphylls.

euryarchaeotes (*Euryachaeota*) [48]   A major group of archaeans [3], diagnosed on the basis of rRNA sequences. Includes many methanogens, extreme halophiles, and thermophiles.

eutherians (*Eutheria*)   A group of viviparous mammals [129], eutherians are well developed at birth (contrast to prototherians and marsupials, the other two groups of mammals). Most familiar mammals outside the Australian and South American regions are eutherians (see Table 23.3).

excavates (*Excavata*) [8]   Diverse group of unicellular, flagellate eukaryotes, many of which possess a feeding groove; some lack mitochondria.

– F –

"ferns"   Vascular plants [12] usually possessing large, frondlike leaves that unfold from a "fiddlehead." Not a monophyletic group, although most fern species are encompassed in a monophyletic clade, the leptosporangiate ferns [73].

flatworms (*Platyhelminthes*) [100]   A group of dorsoventrally flattened and generally elongate soft-bodied lophotrochozoans [24]. May be free-living or parasitic, found in marine, freshwater, or damp terrestrial environments. Major flatworm groups include the tapeworms, flukes, monogeneans, and turbellarians.

flowering plants   *See* angiosperms [15].

flukes (*Trematoda*)   A group of wormlike parasitic flatworms [100] with complex life cycles that involve several different host species. May be paraphyletic with respect to tapeworms.

foraminiferans (*Foraminifera*) [57]   Amoeboid organisms with fine, branched pseudopods that form a food-trapping net. Most produce external shells of calcium carbonate.

fungi (*Fungi*) [17]   Eukaryotic heterotrophs with absorptive nutrition based on extracellular digestion; cell walls contain chitin. Major fungal groups include the microsporidia [85], "chytrids" [86], "zygospore fungi" [87], arbuscular mycorrhizal fungi [88], sac fungi [89], and club fungi [90].

– G –

gastropods (*Gastropoda*)   The largest group of mollusks [106]. Gastropods possess a well-defined head with two or four sensory tentacles (often terminating in eyes) and a ventral foot. Most species have a single coiled or spiraled shell. Common in marine, freshwater, and terrestrial environments.

ginkgo (*Ginkgophyta*) [75]   A gymnosperm [14] group with only one living species. The ginkgo seed is surrounded by a fleshy tissue not derived from an ovary wall and hence not a fruit.

glass sponges (*Hexactinellida*) [92]   Sponges [20] with a skeleton composed of four- and/or six-pointed spicules made of silica.

glaucophytes (*Glaucophyta*) [63]   Unicellular freshwater algae with chloroplasts containing traces of peptidoglycan, the characteristic cell wall material of bacteria.

gnathostomes (*Gnathostomata*)   *See* jawed vertebrates [32].

gnetophytes (*Gnetophyta*) [76]   A gymnosperm [14] group with three very different lineages; all have wood with vessels, unlike other gymnosperms.

green plants (*Viridiplantae*) [10]   Organisms with chlorophylls *a* and *b*, cellulose-containing cell walls, starch as a carbohydrate storage product, and chloroplasts surrounded by two membranes.

gymnosperms (*Gymnospermae*) [14]   Seed plants [13] with seeds "naked" (i.e., not enclosed in carpels). Probably monophyletic, but status still in doubt. Includes the conifers [77], gnetophytes [76], ginkgo [75], and cycads [74].

– H –

hadobacteria (*Hadobacteria*)[42]   A group of extremophilic bacteria [2] that includes the genera *Deinococcus* and *Thermus*.

hagfishes (*Myxini*) [122]   Elongate, slimy-skinned vertebrates [31] with three small accessory hearts, a partial cranium, and no stomach or paired fins. *See also* craniata; cyclostomes.

hemichordates (*Hemichordata*) [119]   One of the two primary groups of ambulacrarians [29]; marine wormlike organisms with a three-part body plan.

heteroloboseans (*Heterolobosea*) [60]   Colorless excavates [8] that can transform among amoeboid, flagellate, and encysted stages.

hexapods (*Hexapoda*) [117]   Major group of arthropods [26] characterized by a reduction (from the ancestral arthropod condition) to six walking appendages, and the consolidation of three body segments to form a thorax. Includes insects and their relatives (see Table 23.2).

high-GC Gram-positives (*Actinobacteria*) [40]   Gram-positive bacteria with a relatively high (G+C)/(A+T) ratio of their DNA, with a filamentous growth habit.

hornworts (*Anthocerophyta*) [70]   Nonvascular plants with sporophytes that grow from the base. Cells contain a single large, platelike chloroplast.

horsehair worms (*Nematomorpha*) [110]   A group of very thin, elongate, wormlike freshwater ecdysozoans [25]. Largely nonfeeding as adults, they are parasites of insects and crayfish as larvae.

horseshoe crabs (*Xiphosura*)   Marine chelicerates [114] with a large outer shell in three parts: a carapace, an abdomen, and a tail-like telson. There are only five living species, but many additional species are known from fossils.

horsetails (*Sphenophyta* or *Equisetophyta*) [72]   Vascular plants [12] with reduced megaphylls in whorls.

hydrozoans (*Hydrozoa*)   A group of cnidarians [97]. Most species go through both polyp and mesuda stages, although one stage or the other is eliminated in some species.

hyperthermophilic bacteria [41]   A group of thermophilic bacteria [2] that live in volcanic vents, hot springs, and in underground oil reservoirs; includes the genera *Aquifex* and *Thermotoga*.

– I –

insects (*Insecta*)   The largest group within the hexapods [117]. Insects are characterized by exposed mouthparts and one pair of antennae containing a sensory receptor called a Johnston's organ. Most have two pairs of wings as adults. There are more described species of insects than all other groups of life [1] combined, and many species remain to be discovered. The major insect groups are described in Table 23.2.

"invertebrates"   Convenience term encompassing any animal [19] that is not a vertebrate [31].

isopods (*Isopoda*)   Crustaceans [116] characterized by a compact head, unstalked compound eyes, and mouthparts consisting of four pairs of appendages. Isopods are abundant and widespread in salt, fresh, and brackish water, although some species (the sow bugs) are terrestrial.

– J –

jawed vertebrates (*Gnathostomata*) [32]   A major group of vertebrates [31] with jawed mouths. Includes chondrichthyans [124], ray-finned fishes [125], and sarcopterygians [34].

– K –

kinetoplastids (*Kinetoplastida*) [62]   Unicellular, flagellate organisms characterized by the presence in their single mitochondrion of a kinetoplast (a structure containing multiple, circular DNA molecules).

kinorhynchs (*Kinorhyncha*) [108]   Small (<1 mm) marine ecdysozoans [25] with bodies in 13 segments and a retractable proboscis.

korarchaeotes (*Korarchaeota*)   A group of archaeans [3] known only by evidence from nucleic acids derived from hot springs. Its phylogenetic relationships within the Archaea are unknown.

krill (*Euphausiacea*)   A group of shrimplike marine crustaceans [116] that are important components of the zooplankton.

– L –

lampreys (*Petromyzontiformes*) [123]   Elongate, eel-like vertebrates [31] that often have rasping and sucking disks for mouths.

lancelets (*Cephalochordata*)   *See* cephalochordates [120].

land plants (*Embryophyta*) [11]   Plants with embryos that develop within protective structures; also called embryophytes. Sporophytes and gametophytes are multicellular. Land plants possess a cuticle. Major groups are the liverworts [68], mosses [69], hornworts [70], and vascular plants [12].

larvaceans (*Larvacea*)   Solitary, planktonic urochordates [121] that retain both notochords and nerve cords throughout their lives.

lepidosaurs (*Lepidosauria*) [131]   Reptiles [37] with overlapping scales. Includes tuataras and squamates (lizards, snakes, and amphisbaenians).

leptosporangiate ferns (*Pteridopsida* or *Polypodiopsida*) [73]   Vascular plants [12] usually possessing large, frondlike leaves that unfold from a "fiddlehead," and possessing thin-walled sporangia.

life (*Life*) [1]   The monophyletic group that includes all known living organisms. Characterized by a nucleic-acid based genetic system (DNA or RNA), metabolism, and cellular structure. Some parasitic forms, such as viruses, have secondarily lost some of these features and rely on the cellular environment of their host.

liverworts (*Hepatophyta*) [68]   Nonvascular plants lacking stomata; stalk of sporophyte elongates along its entire length.

loboseans (*Lobosea*)   A group of unicellular amoebozoans [84]; includes the most familiar amoebas (e.g., *Amoeba proteus*).

"lophophorates"   Convenience term used to describe several groups of lophotrochozoans [24] that have a feeding structure called a lophophore (a circular or U-shaped ridge around the mouth that bears one or two rows of ciliated, hollow tentacles). Not a monophyletic group.

lophotrochozoans (*Lophotrochozoa*) [24]   One of the two main groups of protostomes [23]. This group is morphologically diverse, and is supported primarily on information from gene sequences. Includes bryozoans [99], flatworms [100], rotifers [101], ribbon worms [102], brachiopods [103], phoronids [104], annelids [105], and mollusks [106].

loriciferans (*Loricifera*) [109]   Small (< 1 mm) ecdysozoans [25] with bodies in four parts, covered with six plates.

low-GC Gram-positives (*Firmicutes*) [39]   A diverse group of bacteria [2] with a relatively low (G+C)/(A+T) ratio of their DNA, often but not always Gram-positive, some producing endospores.

lungfishes (*Dipnoi*) [127]   A group of aquatic sarcopterygians [34] that are the closest living relatives of the tetrapods [35]. They have a modified swim bladder used to absorb oxygen from air, so some species can survive the temporary drying of their habitat.

lycophytes (*Lycopodiophyta*) [71]   Vascular plants [12] characterized by microphylls; includes club mosses, spike mosses, and quillworts.

– M –

magnoliids (*Magnoliidae*) [81]   A major group of angiosperms [15] possessing two cotyledons and pollen grains with a single opening. The group is defined primarily by nucleotide sequence data; it is more closely related to the eudicots and monocots than to three other small angiosperm groups.

mammals (*Mammalia*) [129]   A group of tetrapods [35] with hair covering all or part of their skin; females produce milk to feed their developing young. Includes the prototherians, marsupials, and eutherians.

mandibulates (*Mandibulata*) [27]   Arthropods [26] that include mandibles as mouth parts. Includes myriapods [115], crustaceans [116], and hexapods [117].

marsupials (*Marsupialia*)   Mammals [129] in which the female typically has a marsupium (a pouch for rearing young, which are born at an extremely early stage in development). Includes such familiar mammals as opossums, koalas, and kangaroos.

metazoans (*Metazoa*)   *See* animals [19].

microbial eukaryotes   *See* "protists."

microsporidia (*Microsporidia*) [85]   A group of parasitic unicellular fungi [17] that lack mitochondria and have walls that contain chitin.

mollusks (*Mollusca*) [106]   One of the major groups of lophotrochozoans [24], mollusks have bodies composed of a foot, a mantle (which often secretes a hard, calcareous shell), and a visceral mass. Includes monoplacophorans, chitons, bivalves, gastropods, and cephalopods.

monilophytes (*Monilophyta*)   A group of vascular plants [12], sister to the seed plants [13], characterized by overtopping and possession of megaphylls; includes the horsetails [72] and ferns [73].

monocots (*Monocotyledones*) [82]   Angiosperms [15] characterized by possession of a single cotyledon, usually parallel leaf veins, a fibrous root system, pollen grains with a single opening, and floral organs usually in multiples of three.

monogeneans (*Monogenea*)   A group of ectoparasitic flatworms [100].

monoplacophorans (*Monoplacophora*)   Mollusks [106] with segmented body parts and a single, thin, flat, rounded, bilateral shell.

mosses (*Bryophyta*) [69]   Nonvascular plants with true stomata and erect, "leafy" gametophytes; sporophytes elongate by apical cell division.

moss animals   See bryozoans [99].

myriapods (*Myriapoda*) [115]   Arthropods [26] characterized by an elongate, segmented trunk with many legs. Includes centipedes and millipedes.

– **N** –

nanoarchaeotes (*Nanoarchaeota*)   A group of extremely small, thermophilic archaeans [3] with a much-reduced genome. The only described example can survive only when attached to a host organism.

nematodes (*Nematoda*) [111]   A very large group of elongate, unsegmented ecdysozoans [25] with thick, multilayer cuticles. They are among the most abundant and diverse animals, although most species have not yet been described. Include free-living predators and scavengers, as well as parasites of most species of land plants [11] and animals [19].

neognaths (*Neognathae*)   The main group of birds [132], including all living species except the ratites (ostrich, emu, rheas, kiwis, cassowaries) and tinamous. *See* palaeognaths.

– **O** –

oligochaetes (*Oligochaeta*)   Annelid [105] group whose members lack parapodia, eyes, and anterior tentacles, and have few setae. Earthworms are the most familiar oligochaetes.

onychophorans (*Onychophora*) [113]   Elongate, segmented ecdysozoans [25] with many pairs of soft, unjointed, claw-bearing legs. Also known as velvet worms.

oomycetes (*Oomycota*) [52]   Water molds and relatives; absorptive heterotrophs with nutrient-absorbing, filamentous hyphae.

opisthokonts (*Opisthokonta*) [16]   A group of eukaryotes [4] in which the flagellum on motile cells, if present, is posterior. The opisthokonts include the fungi [17], animals [19], and choanoflagellates [91].

ostracods (*Ostracoda*)   Marine and freshwater crustaceans [116] that are laterally compressed and protected by two clamlike calcareous or chitinous shells.

– **P** –

palaeognaths (*Palaeognathae*)   A group of secondarily flightless or weakly flying birds [132]. Includes the flightless ratites (ostrich, emu, rheas, kiwis, cassowaries) and the weakly flying tinamous.

parabasalids (*Parabasalia*) [59]   A group of unicellular eukaryotes [4] that lack mitochondria; they possess flagella in clusters near the anterior of the cell.

phoronids (*Phoronida*) [104]   A small group of sessile, wormlike marine lophotrochozoans [24] that secrete chitinous tubes and feed using a lophophore.

placoderms (*Placodermi*)   An extinct group of jawed vertebrates [32] that lacked teeth. Placoderms were the dominant predators in Devonian oceans.

placozoans (*Placozoa*) [95]   A poorly known group of structurally simple, asymmetrical, flattened, transparent animals found in coastal marine tropical and subtropical seas. Most evidence suggests that placozoans are the sister-group of eumetazoans [21].

Plantae [9]   The most broadly defined plant group. In most parts of this book, we use the word "plant" as synonymous with "land plant" [11], a more restrictive definition.

plasmodial slime molds (*Myxogastrida*)   Amoebozoans [84] that in their feeding stage consist of a coenocyte called a plasmodium.

pogonophorans (*Pogonophora*)   Deep-sea annelids [105] that lack a mouth or digestive tract; they feed by taking up dissolved organic matter, facilitated by endosymbiotic bacteria in a specialized organ (the trophosome).

polychaetes (*Polychaeta*)   A group of mostly marine annelids [105] with one or more pairs of eyes and one or more pairs of feeding tentacles; parapodia and setae extend from most body segments. May be paraphyletic with respect to the clitellates.

priapulids (*Priapulida*) [107]   A small group of cylindrical, unsegmented, wormlike marine ecdysozoans [25] that takes its name from its phallic appearance.

"prokaryotes"   Not a monophyletic group; as commonly used, includes the bacteria [2] and archaeans [3]. A term of convenience encompassing all cellular organisms that are not eukaryotes.

proteobacteria (*Proteobacteria*) [44]   A large and extremely diverse group of Gram-negative bacteria that includes many pathogens, nitrogen fixers, and photosynthesizers. Includes the alpha, beta, gamma, delta, and epsilon proteobacteria.

"protists"   This term of convenience is used to encompass a large number of distinct and distantly related groups of eukaryotes, many but far from all of which are microbial and unicellular. Essentially a "catch-all" term for any eukaryote group not contained within the land plants [11], fungi [17], or animals [19].

protostomes (*Protostomia*) [23]   One of the two major groups of bilaterians [22]. In protostomes, the mouth typically forms from the blastopore (if present) in early development (contrast with deuterostomes). The major protostome groups are the lophotrochozoans [24] and ecdysozoans [25].

protertherians (*Prototheria*)   A mostly extinct group of mammals [129], common during the Cretaceous and early Cenozoic. The five living species—four echidnas and the duck-billed platypus—are the only extant egg-laying mammals.

pterobranchs (*Pterobranchia*)   A small group of sedentary marine hemichordates [119] that live in tubes secreted by the proboscis. They have one to nine pairs of arms, each bearing long tentacles that capture prey and function in gas exchange.

pycnogonids (*Pycnogonida*)   Treated in this book as a group of chelicerates [114], but sometimes considered an independent group of arthropods [26]. Pycnogonids have reduced bodies and very long, slender legs. Also called sea spiders.

– **R** –

radiolarians (*Radiolaria*) [55]   Amoeboid organisms with needlelike pseudopods supported by microtubules. Most have glassy internal skeletons.

ray-finned fishes (*Actinopterygii*) [125]   A highly diverse group of freshwater and marine bony vertebrates [33]. They have reduced swim bladders that often function as hydrostatic organs and fins supported by soft rays (lepidotrichia). Includes most familiar fishes.

red algae (*Rhodophyta*) [64]   Mostly multicellular, marine and freshwater algae characterized by the presence of phycoerythrin in their chloroplasts.

reptiles (*Reptilia*) [37]   One of the two major groups of extant amniotes [36], supported on the basis of similar skull structure and gene sequences. The term "reptiles" traditionally excluded the birds [132], but the resulting group is then clearly paraphyletic. As used in this book, the reptiles include turtles [130], lepidosaurs [131], birds [132], and crocodilians [133].

rhizaria (*Rhizaria*) [7]   Mostly amoeboid unicellular eukaryotes with pseudopods, many with external or internal shells. Includes the foraminiferans [57], cercozoans [56], and radiolarians [55].

rhyniophytes (*Rhyniophyta*)   A group of early vascular plants [12] that appeared in the Silurian and became extinct in the Devonian. Possessed dichotomously branching stems with terminal sporangia but no true leaves or roots.

ribbon worms (*Nemertea*) [102]   A group of unsegmented lophotrochozoans [24] with an eversible proboscis used to capture prey. Mostly marine, but some species live in fresh water or on land.

rotifers (*Rotifera*) [101]   Tiny (< 0.5 mm) lophotrochozoans [24] with a pseudocoelomic body cavity that functions as a hydrostatic organ and a ciliated feeding organ called the corona that surrounds the head. They live in freshwater and wet terrestrial habitats.

roundworms (*Nematoda*) [111]   See nematodes.

– S –

sac fungi (*Ascomycota*) [89] Fungi that bear the products of meiosis within sacs (asci) if the organism is multicellular. Some are unicellular.

salamanders (*Caudata*) A group of amphibians [128] with distinct tails in both larvae and adults and limbs set at right angles to the body.

salps *See* thaliaceans.

sarcopterygians (*Sarcopterygii*) [34] One of the two major groups of bony vertebrates [33], characterized by jointed appendages (paired fins or limbs).

scyphozoans (*Scyphozoa*) Marine cnidarians [97] in which the medusa stage dominates the life cycle. Commonly known as jellyfish.

sea cucumbers (*Holothuroidea*) Echinoderms [118] with an elongate, cucumber-shaped body and leathery skin. They are scavengers on the ocean floor.

sea spiders *See* pycnogonids.

sea squirts *See* ascidians.

sea stars (*Asteroidea*) Echinoderms [118] with five (or more) fleshy "arms" radiating from an indistinct central disk. Also called starfishes.

sea urchins (*Echinoidea*) Echinoderms [118] with a test (shell) that is covered in spines. Most are globular in shape, although some groups (such as the sand dollars) are flattened.

"seed ferns" A paraphyletic group of loosely related, extinct seed plants that flourished in the Devonian and Carboniferous. Characterized by large, frondlike leaves that bore seeds.

seed plants (*Spermatophyta*) [13] Heterosporous vascular plants [12] that produce seeds; most produce wood; branching is axillary (not dichotomous). The major seed plant groups are gymnosperms [14] and angiosperms [15].

sow bugs *See* isopods.

spirochetes (*Spirochaetes*) [45] Motile, Gram-negative bacteria with a helically coiled structure and characterized by axial filaments.

sponges (*Porifera*) [20] A group of relatively asymmetric, filter-feeding animals that lack a gut or nervous system and generally lack differentiated tissues. Includes glass sponges [92], demosponges [93], and calcareous sponges [94].

springtails (*Collembola*) Wingless hexapods [117] with springing structures on the third and fourth segments of their bodies. Springtails are extremely abundant in some environments (especially in soil, leaf litter, and vegetation).

squamates (*Squamata*) The major group of lepidosaurs [131], characterized by the possession of movable quadrate bones (which allow

the upper jaw to move independently of the rest of the skull) and hemipenes (a paired set of eversible penises, or penes) in males. Includes the lizards (a paraphyletic group), snakes, and amphisbaenians.

star anise (*Austrobaileyales*) [80] A group of woody angiosperms [15] thought to be the sister-group of the clade of flowering plants that includes eudicots [83], monocots [82], and magnoliids [81].

starfish (*Asteroidea*) *See* sea stars.

stramenopiles (*Heterokonta* or *Stramenopila*) [6] Organisms having, at some stage in their life cycle, two unequal flagella, the longer possessing rows of tubular hairs. Chloroplasts, when present, surrounded by four membranes. Major stramenopile groups include the brown algae [54], diatoms [53], and oomycetes [52].

– T –

tapeworms (*Cestoda*) Parasitic flatworms [100] that live in the digestive tracts of vertebrates as adults, and usually in various other species of animals as juveniles.

tardigrades (*Tardigrada*) [112] Small (< 0.5 mm) ecdysozoans [25] with fleshy, unjointed legs and no circulatory or gas exchange organs. They live in marine sands, in temporary freshwater pools, and on the water films of plants. Also called water bears.

tetrapods (*Tetrapoda*) [35] The major group of sarcopterygians [34]; includes the amphibians [128] and the amniotes [36]. Named for the presence of four jointed limbs (although limbs have been secondarily reduced or lost completely in several tetrapod groups).

thaliaceans (*Thaliacea*) A group of solitary or colonial planktonic marine urochordates [121]. Also called salps.

therians (*Theria*) Mammals [129] characterized by viviparity (live birth). Includes eutherians and marsupials.

theropods (*Theropoda*) Archosaurs [38] with bipedal stance, hollow bones, a furcula ("wishbone"), elongated metatarsals with three-fingered feet, and a pelvis that points backwards. Includes many well-known extinct dinosaurs (such as *Tyrannosaurus rex*), as well as the living birds [132].

tracheophytes *See* vascular plants [12].

trilobites (*Trilobita*) An extinct group of arthropods [26] related to the chelicerates [114]. Trilobites flourished from the Cambrian through the Permian.

tuataras (*Rhyncocephalia*) A group of lepidosaurs [131] known mostly from fossils; there are just two living tuatara species. The quadrate bone of the upper jaw is fixed firmly to the skull. Sister group of the squamates.

tunicates *See* ascidians.

turbellarians (*Turbellaria*) A group of free-living, generally carnivorous flatworms [100]. Their monophyly is questionable.

turtles (*Testudines*) [130] A group of reptiles [37] with a bony carapace (upper shell) and plastron (lower shell) that encase the body in a fashion unique among the vertebrates.

– U –

urochordates (*Urochordata*) [121] A group of chordates [30] that are mostly saclike filter feeders as adults, with motile larval stages that resemble tadpoles.

– V –

vascular plants (*Tracheophyta*) [12] Plants with xylem and phloem. Major groups include the lycophytes [71] and euphyllophytes.

vertebrates (*Vertebrata*) [31] The largest group of chordates [30], characterized by a rigid endoskeleton supported by the vertebral column and an anterior skull encasing a brain. Includes hagfishes [122], lampreys [123], and the jawed vertebrates [32], although some biologists exclude the hagfishes from this group. *See also* craniates.

– W –

water bears *See* tardigrades.

water lilies (*Nymphaeaceae*) [79] A group of aquatic, freshwater angiosperms [15] that are rooted in soil in shallow water, with round floating leaves and flowers that extend above the water's surface. They are the sister-group to most of the remaining flowering plants, with the exception of the genus *Amborella* [78].

– Y –

"yeasts" Convenience term for several distantly related groups of unicellular fungi [17].

– Z –

"zygospore fungi" (*Zygomycota*, if monophyletic) [87] A convenience term for a probably paraphyletic group of fungi [17] in which hyphae of differing mating types conjugate to form a zygosporangium.

Appendix B: Statistics Primer

This appendix is designed to help you conduct simple statistical analyses and understand their application and importance. This introduction will help you complete the Apply the Concept and Analyze the Data problems throughout this book. The formulas for a number of statistical tests are presented here, but the presentation is designed primarily to help you understand the purpose and reasoning of the various tests. Once you understand the basis of the analysis, you may wish to use one of many free, online web sites for conducting the tests and calculating relevant test statistics (such as http://faculty.vassar.edu/lowry/VassarStats.html).

Why Do We Do Statistics?

ALMOST EVERYTHING VARIES We live in a variable world, but within the variation we see among biological organisms there are predictable patterns. We use statistics to find and analyze these patterns. Consider any group of common things in nature—all women aged 22, all the cells in your liver, or all the blades of grass in your yard. Although they will have many similar characteristics, they will also have important differences. Men aged 22 tend to be taller than women aged 22, but, of course, not every man will be taller than every woman in this age group.

Natural variation can make it difficult to find general patterns. For example, scientists have determined that smoking increases the risk of getting lung cancer. But we know that not all smokers will develop lung cancer and not all nonsmokers will remain cancer-free. If we compare just one smoker to just one nonsmoker, we may end up drawing the wrong conclusion. So how did scientists discover this general pattern? How many smokers and nonsmokers did they examine before they felt confident about the risk of smoking?

Statistics helps us to find general patterns, even when nature does not always follow those patterns.

AVOIDING FALSE POSITIVES AND FALSE NEGATIVES When a woman takes a pregnancy test, there is some chance that it will be positive even if she is not pregnant, and there is some chance that it will be negative even if she is pregnant. We call these kinds of mistakes *false positives* and *false negatives*.

Doing science is a bit like taking a medical test. We observe patterns in the world, and we try to draw conclusions about how the world works from those observations. Sometimes our observations lead us to draw the wrong conclusions. We might conclude that a phenomenon occurs, when it actually does not; or we might conclude that a phenomenon does not occur, when it actually does.

For example, the planet Earth has been warming over the past century (see Concept 46.4). Ecologists are interested in whether plant and animal populations have been affected by global warming. If we have long-term information about the locations of species and temperatures in certain areas, we can determine whether species movements coincide with temperature changes. Such information can, however, be very complicated. Without proper statistical methods, one may not be able to detect the true impact of temperature or, instead, may think a pattern exists when it does not.

Statistics helps us to avoid drawing the wrong conclusions.

How Does Statistics Help Us Understand the Natural World?

Statistics is essential to scientific discovery. Most biological studies involve five basic steps, each of which requires statistics:

- **Step 1: Experimental Design**
  Clearly define the scientific question and the methods necessary to tackle the question.

- **Step 2: Data Collection**
  Gather information about the natural world through experiments and field studies.

- **Step 3: Organize and Visualize the Data**
  Use tables, graphs, and other useful representations to gain intuition about the data.

- **Step 4: Summarize the Data**
  Summarize the data with a few key statistical calculations.

- **Step 5: Inferential Statistics**
  Use statistical methods to draw general conclusions from the data about the way the world works.

Step 1: Experimental Design

We conduct experiments to gain knowledge about the world. Scientists come up with scientific ideas based on prior research and their own observations. These ideas may take the form of a question like "Does smoking cause cancer?," a hypothesis like "Smoking increases the risk of cancer," or a prediction like "If a person smokes, he/she will increase his/her chances of developing cancer." Experiments allow us to test such scientific ideas, but designing a good experiment can be quite challenging.

We use statistics to guide us in designing experiments so that we end up with the right kinds of data. Before embarking on an experiment, we use statistics to determine how much data will be required to test our idea, and to prevent extraneous factors from misleading us. For example, suppose we want to conduct an experiment on fertilizers to test the hypothesis that nitrogen increases plant growth. If we include too few plants, we will not be able to determine whether or not nitrogen has an effect on growth, and the experiment will be for naught. If we include too many plants, we will waste valuable time and resources. Furthermore, we should design the experiment so that we can detect differences that are actually caused by nitrogen fertilization rather than by variation, for example, in sunlight or precipitation experienced by the plants.

Step 2: Data Collection

TAKING SAMPLES When biologists gather information about the natural world, they typically collect a few representative pieces of information. For example, when evaluating the efficacy of a candidate drug for medulloblastoma brain cancer, scientists may test the drug on tens or hundreds of patients, and then draw conclusions about its efficacy for all patients with these tumors. Similarly, scientists studying the relationship between body weight and clutch size (number of eggs) for female spiders of a particular species may examine tens to hundreds of spiders to make their conclusions.

We use the expression "sampling from a population" to describe this general method of taking representative pieces of information from the system under investigation (**FIGURE B1**). The pieces of information in a **sample** are called **observations**. In the cancer therapy example, each observation was the change in a patient's tumor size six months after initiating treatment, and the population of interest was all individuals with medulloblastoma tumors. In the spider example, each observation was a pair of measurements—body size and clutch size—for a single female spider, and the population of interest was all female spiders of this species.

Sampling is a matter of necessity, not laziness. We cannot hope (and would not want) to collect *all* of the female spiders of the species of interest on Earth! Instead, we use statistics to determine how many spiders we must collect in order to

TABLE B1	Poinsettia Colors	
COLOR	FREQUENCY	PROPORTION
Red	108	0.59
Pink	34	0.19
White	40	0.22
Total	**182**	**1.0**

confidently infer something about the general population and then use statistics again to make such inferences.

DATA COME IN ALL SHAPES AND SIZES In statistics, we use the word *variable* to mean a measurable characteristic of an individual or a system. Some variables are on a numerical scale, like the daily high temperature (a numerical value constrained by the precision of our thermometer), or the clutch size of a spider (a whole number: 0, 1, 2, 3,...). We call these **quantitative variables**. Quantitative variables that only take on whole number values are called **discrete variables**, whereas variables that can also take on any fractional value are called **continuous variables**.

Other variables take categories as values, like a human blood type (A, B, AB, or O) or an ant caste (queen, worker, or male). We call these **categorical variables**. Categorical variables with a natural ordering, like a final grade in Introductory Biology (A, B, C, D, or F), are called **ordinal variables**.

Each class of variables comes with its own set of statistical methods. We will introduce a few common methods in this Appendix that will help you work on the problems presented in this book, but you should consult a biostatistics textbook for more advanced tests and analyses for other data sets and problems.

Step 3: Organize and Visualize the Data

Tables and graphs can help you gain intuition about your data, design appropriate statistical tests, and anticipate the outcome of your analysis. A **frequency distribution** lists all possible

TABLE B2	Fish Weights of *Abramis brama* from Lake Laengelmavesi	
WEIGHT (GRAMS)	FREQUENCY	RELATIVE FREQUENCY
201–300	2	0.06
301–400	3	0.09
401–500	8	0.24
501–600	3	0.09
601–700	8	0.24
701–800	3	0.09
801–900	1	0.03
901–1000	6	0.18
Total	**34**	**1.0**

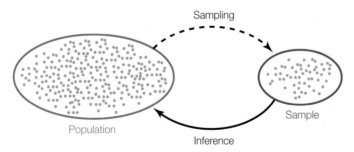

FIGURE B1 Sampling From a Population Biologists take representative samples from a population, use descriptive statistics to characterize their samples, and then use inferential statistics to draw conclusions about the original population.

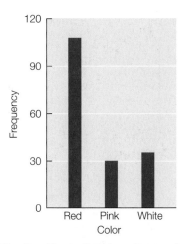

FIGURE B2   Bar Charts Compare Categorical Data   This bar chart shows the frequency of three poinsettia colors that result from an experimental cross.

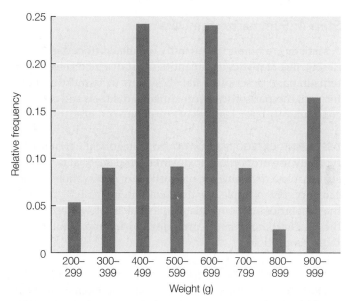

FIGURE B4   Histograms Depict Frequency Distributions of Quantitative Data   This histogram shows the relative frequency of different weight-classes of fish (*Abramis brama*).

values and the number of occurrences of each value in the sample.

TABLE B1 shows a frequency distribution of the colors of 182 poinsettia plants (red, pink, or white) resulting from an experimental cross between two parent plants. For categorical data like this, we can visualize the frequency distribution by constructing a **bar chart**. The heights of the bars indicate the number of observations in each category (**FIGURE B2**). Another way to display the same data is in a **pie chart**, which shows the proportion of each category represented like pieces of a pie (**FIGURE B3**).

For quantitative data, it is often useful to condense your data by grouping (or binning) it into **classes**. In **TABLE B2**, we see a grouped frequency distribution of fish weights for a sample of 34 fish (*Abramis brama*) caught in Lake Laengelmavesi in Finland. The second column (*Frequency*) gives the number of observations in each class and the third column (*Relative Frequency*) gives the overall proportion of observations falling into each class.

Histograms depict frequency distributions for quantitative data. The histogram in **FIGURE B4** shows the relative frequencies of each weight class in this study. When grouping quantitative data, it is necessary to decide how many classes to include.

It is often useful to look at multiple histograms before deciding which grouping offers the best representation of the data.

Sometimes we wish to compare two quantitative variables. For example, the researchers at Lake Laengelmavesi investigated the relationship between fish weight and length and thus also measured the length of each fish. We can visualize this relationship using a **scatter plot** in which the weight and length of each fish is represented as a single point (**FIGURE B5**). We say that these two variables have a **linear relationship** since the points in their scatter plot fall roughly on a straight line.

Tables and graphs are critical to interpreting and communicating data, and thus should be as self-contained and comprehensible as possible. Their content should be easily understood simply by looking at them. Axes, captions, and units should be clearly labeled, statistical terms should be defined, and appropriate groupings should be used when tabulating or graphing quantitative data.

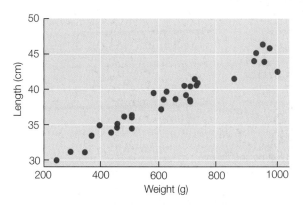

FIGURE B5   Scatter Plots Contrast Two Variables   Scatter plot of *Abramis brama* weights and lengths (measured from nose to end of tail). These two variables have a linear relationship since the data points lie close to a straight line.

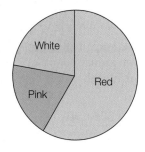

FIGURE B3   Pie Charts Show Proportions of Categories   This pie chart shows the proportions of the three poinsettia colors presented in Table B1.

Step 4: Summarize the Data

A **statistic** is a numerical quantity calculated from data, while **descriptive statistics** are quantities that describe general patterns in data. Descriptive statistics allow us to make straightforward comparisons between different data sets and concisely communicate basic features of our data.

DESCRIBING CATEGORICAL DATA   For categorical variables, we typically use proportions to describe our data. That is, we construct tables containing the proportions of observations in each category. For example, the third column in Table B1 provides the proportions of poinsettia plants in each color category, and the pie chart in Figure B3 provides a visual representation of those proportions.

DESCRIBING QUANTITATIVE DATA   For quantitative data, we often start by calculating the average value or **mean** of our sample. This familiar quantity is simply the sum of all the values in the sample divided by the number of observations in our sample (**FIGURE B6**). The mean is only one of several quantities that roughly tell us where the *center* of our data lies. We call these quantities **measures of center**. Other commonly used

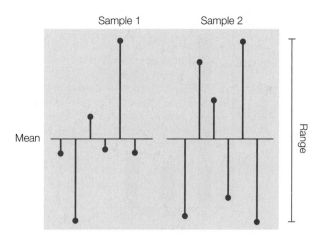

Sample 1   Sample 2

Mean

Range

FIGURE B7   Measures of Dispersion   Two samples with the same mean (black horizontal lines) and range (blue vertical line). Red lines show the deviations of each observation from the mean. Samples with large deviations have large standard deviations. The left sample has a smaller standard deviation than the right sample.

measures of center are the **median**—the value that literally lies in the middle of the sample—and the **mode**—the most frequent value in the sample.

It is often just as important to quantify the variation in the data as it is to calculate its center. There are several statistics that tell us how much the values differ from one another. We call these **measures of dispersion**. The easiest to one understand and calculate is the **range**, which is simply the largest value in the sample minus the smallest value. The most commonly used measure of dispersion is the **standard deviation**, which calculates the extent to which the data are spread out from the mean. A deviation is the difference between an observation and the mean of the sample, and the standard deviation is a number that summarizes all of the deviations. Two samples can have the same range, but very different standard deviations if one is clustered closer to the mean than the other. In **FIGURE B7**, for example, the left sample has a lower standard deviation ($s = 2.6$) than the right sample ($s = 3.6$), even though the two samples have the same means and ranges.

To demonstrate these descriptive statistics, we return to the Lake Laengelmavesi study. The researchers also caught and recorded the weights of six fish in the species *Leusiscus idus*: 270, 270, 306, 540, 800, and 1,000 grams. The mean weight in this sample (equation 1 in Figure B6) is:

$$\bar{x}\frac{\Delta N}{\Delta T} = \frac{(270 + 270 + 306 + 540 + 800 + 1000)}{6} = 531$$

Since there is an even number of observations in the sample, then the median weight is the value halfway between the two middle values:

$$\frac{306 + 540}{2} = 423$$

The mode of the sample is 270, the only value that appears more than once. The standard deviation (equation 2 in Figure B6) is:

RESEARCH TOOLS

FIGURE B6   Descriptive Statistics for Quantitative Data

Below are the equations used to calculate the descriptive statistics we discuss in this appendix. You can calculate these statistics yourself, or use free internet resources to help you make your calculations.

Notation:
$x_1, x_2, x_3, \ldots x_n$ are the n observations of variable X in the sample.

$$\sum_{i=1}^{n} x_i = x_1 + x_2 + x_3, \ldots + x_n$$ is the sum of all of the observations. (The Greek letter sigma, Σ, is used to denote "sum of.")

In regression, the independent variable is X, and the dependent variable is Y. b_0 is the vertical intercept of a regression line. b_1 is the slope of a regression line.

Equations

1. Mean: $$\bar{x} = \frac{\sum_{i=1}^{n} x_i}{n}$$

2. Standard deviation: $$s = \sqrt{\frac{\sum (x_i - \bar{x})^2}{n-1}}$$

3. Correlation coefficient: $$r = \frac{\sum (x_i - \bar{x})(y_i - \bar{y})}{\sqrt{\sum (x_i - \bar{x})^2 (y_i - \bar{y})^2}}$$

4. Least-squares regression line: $Y = b_0 + b_1 X$
   where $$b_1 = \frac{\sum (x_i - \bar{x})(y_i - \bar{y})}{\sum (x_i - \bar{x})^2}$$ and $b_0 = \bar{y} - b_1 \bar{x}$

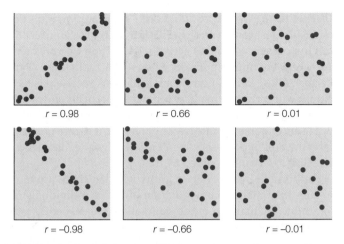

r = 0.98   r = 0.66   r = 0.01

r = –0.98   r = –0.66   r = –0.01

FIGURE B8 Correlation Coefficients The correlation coefficient (*r*) indicates both the strength and the direction of the relationship.

$$s = \sqrt{\frac{(270-531)^2 + (270-531)^2 + (306-531)^2 + (540-531)^2 + (800-531)^2 + (1000-531)^2}{5}} = 309.6$$

and the range is 1000 – 270 = 730.

DESCRIBING THE RELATIONSHIP BETWEEN TWO QUANTITATIVE VARI-ABLES Biologists are often interested in understanding the relationship between two different quantitative variables: How does the height of an organism relate to its weight? How does air pollution relate to the prevalence of asthma? How does lichen abundance relate to levels of air pollution? Recall that scatter plots visually represent such relationships.

We can quantify the strength of the relationship between two quantitative variables using a single value called the Pearson product–moment **correlation coefficient** (equation 3 in Figure B6). This statistic ranges between –1 and 1, and tells us how closely the points in a scatter plot conform to a straight line. A negative correlation coefficient indicates that one variable decreases as the other increases; a positive correlation coefficient indicates that the two variables increase together, and a correlation coefficient of zero indicates that there is no linear relationship between the two variables (**FIGURE B8**).

One must always keep in mind that *correlation does not mean causation*. Two variables can be closely related without one causing the other. For example, the number of cavities in a child's mouth correlates positively with the size of their feet. Clearly cavities do not enhance foot growth; nor does foot growth cause tooth decay. Instead the correlation exists because both quantities tend to increase with age.

Intuitively, the straight line that tracks the cluster of points on a scatter plot tells us something about the *typical* relationship between the two variables. Statisticians do not, however, simply eyeball the data and draw a line by hand. They often use a method called least-squares **linear regression** to fit a straight line to the data (equation 4 in Figure B6). This method calculates the line that minimizes the overall vertical distances between the points in the scatter plot and the line itself. These distances are called **residuals** (**FIGURE B9**). Two parameters describe the regression line: b_0 (the vertical intercept of the line,

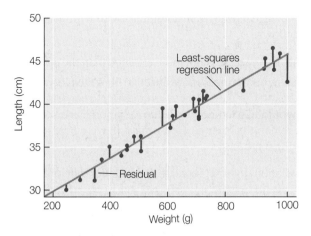

FIGURE B9 Linear Regression Estimates the Typical Relationship Between Two Variables Linear-least squares regression line for *Abramis brama* weights and lengths (measured from nose to end of tail). The regression line (blue line) is given by the equation $Y = 26.1 + 0.02X$. It is the line that minimizes the sum of the squares of the residuals (red lines).

or the expected value of variable Y when $X = 0$), and b_1 (the slope of the line, or how much values of Y are expected to change with changes in values of X).

Step 5: Inferential Statistics

Data analysis often culminates with statistical inference—an attempt to draw general conclusions about the system under investigation. As depicted in Figure B1, the primary reason we collect data is to gain insight into the larger system from which the data are collected. When we test a new medulloblastoma brain cancer drug on ten patients, we do not simply want to know the fate of those ten individuals; rather, we hope to predict its efficacy on the much larger group of all medulloblastoma patients.

STATISTICAL HYPOTHESES When it comes to inferring something about the real world from our data, we often have a *"Whether or not"* question in mind. For example, we would like to know whether or not global warming impacts biodiversity; whether or not the clutch size of a spider increases with body size; or whether or not soil nitrogen increases the growth of a particular plant species.

Before making statistical inferences from data, we must formalize our *"Whether or not"* question into a pair of opposing hypotheses—a **null hypothesis** (denoted H_0) and an **alternative hypothesis** (denoted H_A). The alternative hypothesis is the *"Whether"*—it is formulated to describe the effect that we expect our data to support; the null hypothesis is the *"or not"*—it is formulated to represent the absence of the effect. In other words, we typically conduct our experiment seeking to demonstrate something new (the alternative hypothesis) and thereby reject the idea that it does not occur (the null hypothesis).

Suppose, for example, we would like to know *whether or not* a new vaccine is more effective than an existing vaccine at immunizing children against influenza. Our hypotheses would be as follows:

H_0: The new vaccine is not more effective than the old vaccine.

H_A: The new vaccine is more effective than the old vaccine.

If we would like to know whether radiation increases the mutation rate in the bacteria *Escherichia coli*, we would set up the following hypotheses:

H_0: Radiation does not increase the mutation rate of *E. coli*.

H_A: Radiation does increase the mutation rate of *E. coli*.

STATISTICAL BURDEN OF PROOF In the U.S. justice system, people are innocent until proven guilty. In statistics, the world is *null until proven alternative*. Statistics requires overwhelming proof in favor of the alternative hypothesis before rejecting the null hypothesis. In other words, scientists favor existing ideas and resist adopting new ideas until compelling evidence suggests otherwise. This is based on a philosophy that it is worse to accept new claims when they are false than to miss out on discovering some true facts about world.

When testing a new influenza vaccine, the burden of proof is on the new vaccine. Suppose we were to vaccinate three children with the new vaccine (Group A), three with the old vaccine (Group B) and leave three children unvaccinated (Group C). If no children from Group A, one child from Group B, and one child from Group C became infected, would we have enough evidence to conclude that the new vaccine is superior to the old vaccine? No, we would not. If the study were enlarged, and two out of 100 children in group A, seven out of 100 children in group B, and 22 out of 100 children in group C become infected, would we then have sufficient evidence to choose the new vaccine? Perhaps, but we need to use statistics to be sure.

This is the traditional burden of proof in biology and science in general. As a consequence, scientists are more likely to miss out on discovering something new (and true) about the world than they are to make a false discovery. In recent years, scientists have begun to question this approach and develop an alternative statistical approach, called **Bayesian inference**, which makes it easier to favor new hypotheses. In this primer, we discuss only traditional statistical methods, often called **frequentist statistics**.

The real world

	Null hypothesis true *(not more females)*	Null hypothesis false *(more females)*
Null hypothesis true *(not more females)*	✓	Type 2 error *(false negative)*
Null hypothesis false *(more females)*	Type 1 error *(false positive)*	✓

(left axis label: Our conclusion)

FIGURE B10 Two Types of Error Possible outcomes of a statistical test. Statistical inference can result in correct and incorrect conclusions about the population of interest.

JUMPING TO THE WRONG CONCLUSIONS There are two ways that a statistical test can go wrong (**FIGURE B10**). We can reject the null hypothesis when it is actually true (**Type I error**) or we can accept the null hypothesis when it is actually false (**Type II error**). These kinds of errors are analogous to false positives and false negatives in medical testing, respectively. If we mistakenly reject the null hypothesis when it is actually true, then we falsely endorse the incorrect hypothesis. If we are unable to reject the null hypothesis when it is actually false, then we fail to realize a yet undiscovered truth.

Suppose we would like to know whether there are more females than males in a population of 10,000 individuals. To determine the makeup of the population, we choose 20 individuals randomly and record their sex. Our null hypothesis is that there are *not* more females than males; and our alternative hypothesis is that there are. The following scenarios illustrate the possible mistakes we might make:

- *Scenario 1*: The population actually has 40% females and 60% males. Although our random sample of 20 people is likely to be dominated by males, it is certainly possible that, by chance, we will end up choosing more females than males. If this occurs, and we mistakenly reject the null hypothesis (that there are *not* more females than males), then we make a Type I error.

- *Scenario 2*: The population actually has 60% females and 40% males. If, by chance, we end up with a majority of males in our sample and thus fail reject the null hypothesis, then we make a Type II error.

Fortunately, statistics has been developed precisely to avoid these kinds of errors and inform us about the reliability of our conclusions. The methods are based on calculating the **probabilities** of different possible outcomes. Although you may have heard or even used the word "probability" on multiple occasions, it is important that you understand its mathematical meaning. A probability is a numerical quantity that expresses the likelihood of some event. It ranges between zero and one; zero means that there is no chance the event will occur and one means that the event is guaranteed to occur. This only makes sense if there is an element of chance, that is, if it is possible the event will occur and possible that it will not occur. For example, when we flip a fair coin, it will land on heads with probability 0.5 and land on tails with probability 0.5. When we select individuals randomly from a population with 60% females and 40% males, we will encounter a female with probability 0.6 and a male with probability 0.4.

Probability plays a very important role in statistics. To draw conclusions about the real world (the population) from our sample, we first calculate the probability of obtaining our sample if the null hypothesis is true. Specifically, statistical inference is based on answering the following question:

Suppose the null hypothesis is true. What is the probability that a random sample would, by chance, differ from the null hypothesis as much as our sample differs from the null hypothesis?

If our sample is highly improbable under the null hypothesis, then we rule it out in favor of our alternative hypothesis. If, instead, our sample has a reasonable probability of occurring under the null hypothesis, then we conclude that our data are consistent with the null hypothesis and we do not reject it.

Returning to the sex ratio example, we consider two new scenarios:

- *Scenario 3*: Suppose we want to infer whether or not females constitute the majority of the population (our alternative hypothesis) based on a random sample containing 12 females and eight males. We would calculate the probability that a random sample of 20 people includes at least 12 females assuming that the population, in fact, has a 50:50 sex ratio (our null hypothesis). This probability is 0.13, which is too high to rule out the null hypothesis.

- *Scenario 4*: Suppose now that our sample contains 17 females and three males. If our population is truly evenly divided, then this sample is much less likely than the sample in scenario 3. The probability of such an extreme sample is 0.0002, and would lead us to rule out the null hypothesis and conclude that there are more females than males.

This agrees with our intuition. When choosing 20 people randomly from an evenly divided population, we would be surprised if almost all of them were female, but would not be surprised at all if we ended up with a few more females than males (or a few more males than females). Exactly how many females do we need in our sample before we can confidently infer that they make up the majority of the population? And how confident are we when we reach that conclusion? Statistics allows us to answer these questions precisely.

STATISTICAL SIGNIFICANCE: AVOIDING FALSE POSITIVES Whenever we test hypotheses, we calculate the probability just discussed, and refer to this value as the ***P*-value** of our test. Specifically, the P-value is the probability of getting data as extreme as our data (just by chance) if the null hypothesis is, in fact, true. In other words, it is the likelihood that chance alone would produce data that differ from the null hypothesis as much as our data differ from the null hypothesis. How we measure the difference between our data and the null hypothesis depends on the kind of data in our sample (categorical or quantitative) and the nature of the null hypothesis (assertions about proportions, single variables, multiple variables, differences between variables, correlations between variables, etc.).

For many statistical tests, P-values can be calculated mathematically. One option is to quantify the extent to which the data depart from the null hypothesis and then use look-up tables (available in most statistics textbooks, or on the internet) to find the probability that chance alone would produce a difference of that magnitude. Most scientists, however, find P-values primarily by using statistical software rather than hand calculations combined with look-up tables. Regardless of the technology, the most important steps of the statistical analysis are still left to the researcher: constructing appropriate null and alternative hypotheses, choosing the correct statistical test, and drawing correct conclusions.

After we calculate a P-value from our data, we have to decide whether it is small enough to conclude that our data are inconsistent with the null hypothesis. This is decided by comparing the P-value to a threshold called the **significance level**, which is often chosen even before making any calculations. We reject the null hypothesis only when the P-value is less than or equal to the significance level, denoted α. This ensures that, if the null hypothesis is true, we have at most a probability α of accidentally rejecting it. Therefore, the lower the value of α, the less likely you are to make a Type I error (lower left cell of Figure B10). The most commonly used significance level is $\alpha = 0.05$, which limits the probability of a Type I error to 5%.

If our statistical test yields a P-value that is less than our significance level α, then we conclude that the effect described by our alternative hypothesis is statistically significant at the level α and we reject the null hypothesis. If our P-value is greater than α, then we conclude that we are unable to reject the null hypothesis. In this case, we do not actually reject the alternative hypothesis, rather we conclude that we do not yet have enough evidence to support it.

POWER: AVOIDING FALSE NEGATIVES The **power** of a statistical test is the probability that we will correctly reject the null hypothesis when it is false (lower right cell of Figure B10). Therefore, the higher the power of the test, the less likely we are to make a Type II error (upper right cell of Figure B10). The power of a test can be calculated, and such calculations can be used to improve your methodology. Generally, there are several steps that can be taken to increase power and thereby avoid false negatives:

- **Decrease the significance level, α.** The higher the value of α, the harder it is to reject the null hypothesis, even if it is actually false.

- **Increase the sample size.** The more data one has, the more likely one is to find evidence against the null hypothesis, if it is actually false.

- **Decrease variability in the sample.** The more variation there is in the sample, the harder it is to discern a clear effect (the alternative hypothesis) when it actually exists.

It is always a good idea to design your experiment to reduce any variability that may obscure the pattern you seek to detect. For example, it is possible that the chance of a child contracting influenza varies depending on whether he or she lives in a crowded (e.g., urban) environment or one that is less so (e.g., rural). To reduce variability, a scientist might choose to test a new influenza vaccine only on children from one environment or the other. After you have minimized such extraneous variation, you can use power calculations to choose the right combination of α and sample size to reduce the risks of Type I and Type II errors to desirable levels.

There is a trade-off between Type I and Type II errors: As α increases, the risk of a Type I decreases but the risk of a Type II error increases. As discussed above, scientists tend to be more concerned about Type I errors than Type II errors. That is, they believe that it is worse to mistakenly believe a false hypothesis than it is to fail to make a new discovery. Thus, they prefer to use low values of α. However, there are many real-world scenarios in which it would be worse to make a Type II error than a Type I error. For example, suppose a new cold medication is being tested for dangerous (life-threatening) side effects. The null hypothesis is that there are no such side effects. A Type II error might lead regulatory agencies to approve a harmful medication that could cost human lives. In contrast, a Type I error would simply mean one less cold medication among the many that already line pharmacy shelves. In such cases, policymakers take steps to avoid a Type II error, even if, in doing so, they increase the risk of a Type I error.

STATISTICAL INFERENCE WITH QUANTITATIVE DATA There are many forms of statistical inference for quantitative data. When measuring a single quantitative variable, like birth weight in lambs, calcium concentration in the blood of pregnant women, or migration rate of birds, we often wish to infer the mean value of the population from which we drew the sample. However, the mean of a randomly chosen sample will not necessarily be the same or even close to the population mean. Suppose we wanted to know the average weight of newborn lambs on a particular farm. By chance, we may end up with a random sample that includes an excess of lightweight lambs and therefore a sample mean that is less than the overall mean in the population.

To infer the population mean from the sample data, we can calculate a **confidence interval for the mean**. This is a statistically derived range of values that is centered on the sample mean and is likely to include the population mean. For example, based on the sample of 34 *Abramis brama* weights from Lake Laengelmavesi (see Table B2; Figure B4), the 95% confidence interval for the mean weight ranges from 554 grams to 698 grams. The true average weight for this species of fish is likely, but not guaranteed, to fall within this range.

Biologists frequently wish to compare the mean values in two or more groups; for example, newborn lamb weights on several different farms, calcium concentration in women in early and late stages of pregnancy, or migration rates in birds of different species. Based on the means and standard deviations calculated for each of the samples, they infer whether or not the means in the different populations are statistically different from one another. There are several statistical methods for this, and the correct method depends on the number of groups, the experimental design, and the nature of the data.

FIGURE B11 describes the steps of a *t*-test, a simple method for comparing the means in two different groups. To illustrate, we can apply a *t*-test to the Lake Laengelmavesi data to assess whether the two fish species *Abramis brama* and *Leusiscus idus* have significantly different mean weights. We begin by stating our hypotheses and choosing a significance level:

H_0: *Abramis brama* and *Leusiscus idus* have the same mean weight.

H_A: *Abramis brama* and *Leusiscus idus* have different mean weights.

$\alpha = 0.05$

The test statistic is calculated using the means, standard deviations, and sizes of the two samples:

$$t_s = \frac{626 - 531}{\sqrt{\dfrac{207^2}{34} + \sqrt{\dfrac{310^2}{6}}}} = 0.724$$

We can use statistical software or one of the free statistical sites on the internet to find the *P*-value for this result to be $P = 0.497$. Since *P* is considerably greater than α, we fail to reject the null hypothesis and conclude that our study does not provide evidence that the two species have different mean weights.

You may want to consult an introductory statistics textbook to learn more about confidence intervals, *t*-tests, and other basic statistical tests.

STATISTICAL INFERENCE WITH CATEGORICAL DATA With categorical data, we often wish to infer the distribution of the different categories within the populations from which our samples are drawn. In the simplest case, we have a single categorical variable with two or more categories. If there are just two categories, we can construct a **confidence interval for the proportion** of the population that belongs to one of the two categories. This is a statistically derived range of values that is centered on the sample proportion and is likely to include the population proportion. If there are three or more categories, we can use a **chi-square goodness-of-fit** test to determine whether the distribution of the different categories in the population is consistent with a specific distribution.

FIGURE B12 outlines the steps of a chi-square goodness-of-fit-test. As an example, consider the data described in Table B1. Many plant species have simple Mendelian genetic systems in which parent plants produce progeny with three different colors of flowers in a ratio of 2:1:1. However, a botanist believes that these particular poinsettia plants have a different genetic system that does not produce a 2:1:1 ratio of red, pink, and

---

RESEARCH TOOLS

FIGURE B11   The *t*-test

What is the *t*-test? It is a standard method for assessing whether the means of two groups are statistically different from each another.

Step 1: State the null and alternative *hypotheses*:
  H_0: The two populations have the same mean.
  H_A: The two populations have different means.

Step 2: Choose a significance level, α, to limit the risk of a Type 1 error.

Step 3: Calculate the *test statistic*: $t_s = \dfrac{\bar{y}_1 - \bar{y}_2}{\sqrt{\dfrac{s_1^2}{n_1} + \dfrac{s_2^2}{n_2}}}$

  Notation: $\bar{y}_1$ and $\bar{y}_2$ are the sample means; s_1 and s_2 are the sample standard deviations; and n_1 and n_2 are the sample sizes.

Step 4: Use the test statistic to assess whether the data are consistent with the null hypothesis:

  Calculate the *P-value* (*P*) using statistical software or by hand using statistical tables.

Step 5: Draw conclusions from the test:

  If $P \le \alpha$, then reject H_0, and conclude that the population distribution is significantly different.

  If $P \le \alpha$, then we do not have sufficient evidence to conclude that the means differ.

RESEARCH TOOLS

FIGURE B12   The Chi-Square Goodness-of-Fit Test

What is the chi-square goodness-of-fit test? It is a standard method for assessing whether a sample came from a population with a specific distribution.

Step 1: State the null and alternative *hypotheses*:

H_0: The population has the specified distribution.

H_A: The population does not have the specified distribution.

Step 2: Choose a significance level, α, to limit the risk of a Type 1 error.

Step 3: Determine the *observed frequency* and *expected frequency* for each category:

The observed frequency of a category is simply the number of observations in the sample of that type.

The expected frequency of a category is the probability of the category specified in H_0 multiplied by the overall sample size.

Step 4: Calculate the *test statistic*: $\chi_s^2 = \sum_{i=1}^{c} \frac{(O_i - E_i)^2}{E_i}$

Notation: C is the total number of categories, O_i is the observed frequency of category i, and E_i is the expected frequency of category i.

Step 5: Use the test statistic to assess whether the data are consistent with the null hypothesis:

Calculate the *P-value* (P) using statistical software or by hand using statistical tables.

Step 6: Draw conclusions from the test:

If $P \leq \alpha$, then reject H_0, and conclude that the population distribution is significantly different than the distribution specified by H_0.

If $P \leq \alpha$, then we do not have sufficient evidence to conclude that population has a different distribution.

We find the *P*-value for this result to be $P = 0.0343$ using statistical software. Since P is less than α, we reject the null hypothesis and conclude that the botanist is correct: The plant color patterns cannot be explained by the simple Mendelian genetic model under consideration.

This introduction is only meant to provide a brief introduction to the concepts of statistical analysis, with a few example tests. **FIGURE B13** provides a summary of some of the commonly used statistical tests that you may encounter in biological studies.

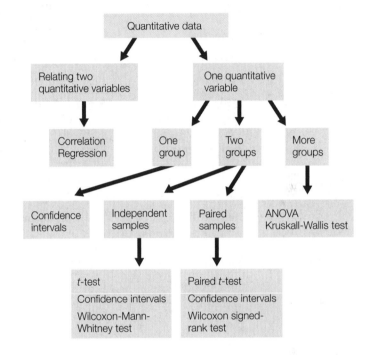

white plants. A chi-square goodness-of-fit can be used to assess whether or not the data are consistent with this ratio, and thus whether or not this simple genetic explanation is valid. We start by stating our hypotheses and significance level:

H_0: The progeny of this type of cross have the following probabilities of each flower color:

$Pr\{Red\} = .50, Pr\{Pink\} = .25, Pr\{White\} = .25$

H_A: At least one of the probabilities of H_0 is incorrect.

$\alpha = 0.05$

We next use the probabilities in H_0 and the sample size to calculate the expected frequencies:

	Red	Pink	White
Observed	108	34	40
Expected	(.50)(182) = 91	(.25)(182) = 45.5	(.25)(182) = 45.5

Based on these quantities, we calculate the chi-square test statistic:

$$\chi_s^2 = \sum_{i=1}^{C} \frac{(O_i - E_i)^2}{E_i} = \frac{(108 - 91)^2}{91} + \frac{(34 - 45.5)^2}{45.5} + \frac{(40 - 45.5)^2}{45.5} = 6.747$$

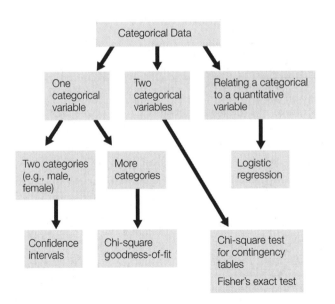

FIGURE B13   Some Common Methods of Statistical Inference
This flow-chart shows some of the commonly used methods of statistical inference for different combinations of data. Detailed descriptions of these methods can be found in most introductory biostatistics textbooks.

Appendix C: Some Measurements Used in Biology

MEASURES OF	UNIT	EQUIVALENTS	METRIC → ENGLISH CONVERSION
Length	meter (m)	base unit	1 m = 39.37 inches = 3.28 feet = 1.196 yards
	kilometer (km)	1 km = 1000 (10^3) m	1 km = 0.62 miles
	centimeter (cm)	1 cm = 0.01 (10^{-2}) m	1 cm = 0.39 inches
	millimeter (mm)	1 mm = 0.1 cm = 10^{-3} m	1 mm = 0.039 inches
	micrometer (μm)	1 μm = 0.001 mm = 10^{-6} m	
	nanometer (nm)	1 nm = 0.001 μm = 10^{-9} m	
Area	square meter (m^2)	base unit	1 m^2 = 1.196 square yards
	hectare (ha)	1 ha = 10,000 m^2	1 ha = 2.47 acres
Volume	liter (L)	base unit	1 L = 1.06 quarts
	milliliter (mL)	1 mL = 0.001 L = 10^{-3} L	1 mL = 0.034 fluid ounces
	microliter (μL)	1 μL = 0.001 mL = 10^{-6} L	
Mass	gram (g)	base unit	1 g = 0.035 ounces
	kilogram (kg)	1 kg = 1000 g	1 kg = 2.20 pounds
	metric ton (mt)	1 mt = 1000 kg	1 mt = 2,200 pounds = 1.10 ton
	milligram (mg)	1 mg = 0.001 g = 10^{-3} g	
	microgram (μg)	1 μg = 0.001 mg = 10^{-6} g	
Temperature	degree Celsius (°C)	base unit	°C = (°F − 32)/1.8
			0°C = 32°F (water freezes)
			100°C = 212°F (water boils)
			20°C = 68°F ("room temperature")
			37°C = 98.6°F (human internal body temperature)
	Kelvin (K)*	K = °C − 273	0 K = −460°F
Energy	joule (J)		1 J ≈ 0.24 calorie = 0.00024 kilocalorie[†]

*0 K (−273°C) is "absolute zero," a temperature at which molecular oscillations approach 0—that is, the point at which motion all but stops.

[†]A *calorie* is the amount of heat necessary to raise the temperature of 1 gram of water 1°C. The *kilocalorie,* or nutritionist's calorie, is what we commonly think of as a calorie in terms of food.

Glossary

A horizon *See* top soil.

abiotic (a' bye ah tick) [Gk. *a*: not + *bios*: life] Nonliving. (Contrast with biotic.)

abomasum The true stomach of a ruminant.

abortion Any termination of pregnancy that occurs after a fertilized egg is successfully implanted in the uterus.

abscisic acid (ABA) (ab sighs' ik) A plant growth substance with growth-inhibiting action. Causes stomata to close; involved in a plant's response to salt and drought stress.

abscission (ab sizh' un) [L. *abscissio*: break off] The process by which leaves, petals, and fruits separate from a plant.

absorption (1) Of light: complete retention, without reflection or transmission. (2) Of water or other molecules: soaking up (taking in through pores or by diffusion).

absorption spectrum A graph of light absorption versus wavelength of light; shows how much light is absorbed at each wavelength.

absorptive heterotroph An organism (usually a fungus) that obtains its food by secreting digestive enzymes into the environment to break down large food molecules, then absorbing the breakdown products.

absorptive state State in which food is in the gut and nutrients are being absorbed. (Contrast with postabsorptive state.)

abyssal zone (uh biss' ul) [Gk. *abyssos*: bottomless] The deepest parts of the ocean.

accessory pigments Pigments that absorb light and transfer energy to chlorophylls for photosynthesis.

accessory sex organs Anatomical structures that allow transfer of sperm from male to female for internal fertilization.

acetyl coenzyme A (acetyl CoA) A compound that reacts with oxaloacetate to produce citrate at the beginning of the citric acid cycle; a key metabolic intermediate in the formation of many compounds.

acetylcholine (ACh) A neurotransmitter that carries information across vertebrate neuromuscular junctions and some other synapses. It is then broken down by the enzyme acetylcholinesterase (AChE).

acid [L. *acidus*: sharp, sour] A substance that can release a proton in solution. (Contrast with base.)

acid growth hypothesis The hypothesis that auxin increases proton pumping, thereby lowering the pH of the cell wall and activating enzymes that loosen polysaccharides. Proposed to explain auxin-induced cell expansion in plants.

acid rain Precipitation that has a lower pH than normal as a result of acid-forming precursor molecules introduced into the atmosphere by human activities.

acidic Having a pH of less than 7.0 (a hydrogen ion concentration greater than 10^{-7} molar). (Contrast with basic.)

acoelomate An animal that does not have a coelom.

acrosome (a' krow soam) [Gk. *akros*: highest + *soma*: body] The structure at the forward tip of an animal sperm which is the first to fuse with the egg membrane and enter the egg cell.

ACTH *See* corticotropin.

actin [Gk. *aktis*: ray] A protein that makes up the cytoskeletal microfilaments in eukaryotic cells and is one of the two contractile proteins in muscle.

action potential An impulse in a neuron taking the form of a wave of depolarization or hyperpolarization.

action spectrum A graph of a biological process versus light wavelength; shows which wavelengths are involved in the process.

activation energy (E_a) The energy barrier that blocks the tendency for a chemical reaction to occur.

activator A transcription factor that stimulates transcription when it binds to a gene's promoter. (Contrast with repressor.)

active site The region on the surface of an enzyme or ribozyme where the substrate binds, and where catalysis occurs.

active transport The energy-dependent transport of a substance across a biological membrane against a concentration gradient—that is, from a region of low concentration (of that substance) to one of high concentration. (*See also* primary active transport, secondary active transport; contrast with facilitated diffusion, passive transport.)

adaptation (a dap tay' shun) (1) In evolutionary biology, a particular structure, physiological process, or behavior that makes an organism better able to survive and reproduce. Also, the evolutionary process that leads to the development or persistence of such a trait. (2) In sensory neurophysiology, a sensory cell's loss of sensitivity as a result of repeated stimulation.

adaptive immunity In animals, one of two general types of defenses against pathogens. Involves antibody proteins and other proteins that recognize, bind to, and aid in the destruction of specific viruses and bacteria. Present only in vertebrate animals. (Contrast with innate immunity.)

additive growth Population growth in which a constant number of individuals is added to the population during successive time intervals. (Contrast with multiplicative growth.)

adenine (A) (a' den een) A nitrogen-containing base found in nucleic acids, ATP, NAD, and other compounds.

adenosine triphosphate *See* ATP.

adiabatic cooling A decrease in air temperature that occurs when air rises and then expands as pressure decreases at higher altitudes.

adrenal gland (a dree' nal) [L. *ad*: toward + *renes*: kidneys] An endocrine gland located near the kidneys of vertebrates, consisting of two glandular parts, the cortex and medulla.

adrenaline *See* epinephrine.

adrenergic receptors G protein–linked receptor proteins that bind to the hormones epinephrine and norepinephrine, triggering specific responses in the target cells.

adrenocorticotropic hormone *See* corticotropin.

adsorption Binding of a gas or a solute to the surface of a solid.

adventitious roots (ad ven ti' shus) [L. *adventitius*: arriving from outside] Roots originating from the stem at ground level or below; typical of the fibrous root system of monocots.

aerenchyma In plants, parenchymal tissue containing air spaces.

aerobic (air oh' bic) [Gk. *aer*: air + *bios*: life] In the presence of oxygen; requiring oxygen. (Contrast with anaerobic.)

afferent (af' ur unt) [L. *ad*: toward + *ferre*: to carry] Carrying to, as in a neuron that carries impulses to the central nervous system (afferent neuron), or a blood vessel that carries blood to a structure. (Contrast with efferent.)

AIDS Acquired immune deficiency syndrome, a condition caused by human immunodeficiency virus (HIV) in which the body's T-helper cells are reduced, leaving the victim subject to opportunistic diseases.

air sacs Structures in the respiratory system of birds that receive inhaled air; they keep fresh air flowing unidirectionally through the lungs, but are not themselves gas exchange surfaces.

aldosterone (al dohs' ter own) A steroid hormone produced in the adrenal cortex of mammals. Promotes secretion of potassium and reabsorption of sodium in the kidney.

aleurone layer In some seeds, a tissue that lies beneath the seed coat and surrounds the endosperm. Secretes digestive enzymes that break down macromolecules stored in the endosperm.

allantoic membrane In animal development, an outgrowth of extraembryonic endoderm plus adjacent mesoderm that forms the allantois, a saclike structure that stores metabolic wastes produced by the embryo.

allantois (al' lun toh is) [Gk. *allant*: sausage] An extraembryonic membrane enclosing a sausage-shaped sac that stores the embryo's nitrogenous wastes.

allele (a leel') [Gk. *allos*: other] The alternate form of a genetic character found at a given locus on a chromosome.

allele frequency The relative proportion of a particular allele in a specific population.

allergic reaction [Ger. *allergie*: altered] An overreaction of the immune system to amounts of an antigen that do not affect most people; often involves IgE antibodies.

allopatric speciation (al' lo pat' rick) [Gk. *allos*: other + *patria*: homeland] The formation of two species from one when reproductive isolation occurs because of the interposition of (or crossing of) a physical geographic barrier such as a river. Also called geographic speciation. (Contrast with sympatric speciation.)

allopolyploidy The possession of more than two chromosome sets that are derived from more than one species.

allosteric regulation (al lo steer' ik) [Gk. *allos*: other + *stereos*: structure] Regulation of the activity of a protein (usually an enzyme) by the binding of an effector molecule to a site other than the active site.

α (alpha) helix A prevalent type of secondary protein structure; a right-handed spiral.

alternation of generations The succession of multicellular haploid and diploid phases in some sexually reproducing organisms, notably plants.

alternative splicing A process for generating different mature mRNAs from a single gene by splicing together different sets of exons during RNA processing.

altruism Pertaining to behavior that benefits other individuals at a cost to the individual who performs it.

alveolus (al ve' o lus) (plural: alveoli) [L. *alveus*: cavity] A small, baglike cavity, especially the blind sacs of the lung.

amensalism (a men' sul ism) Interaction in which one animal is harmed and the other is unaffected. (Contrast with commensalism, mutualism.)

amine An organic compound containing an amino group (NH_2).

amine hormones Small hormone molecules synthesized from single amino acids (e.g., thyroxine and epinephrine).

amino acid An organic compound containing both NH_2 and COOH groups. Proteins are polymers of amino acids.

amino acid replacement A change in the nucleotide sequence that results in one amino acid being replaced by another.

ammonia NH_3, the most common nitrogenous waste.

ammonotelic (am moan' o teel' ic) [Gk. *telos*: end] Pertaining to an organism in which the final product of breakdown of nitrogen-containing compounds (primarily proteins) is *ammonia*. (Contrast with ureotelic, uricotelic.)

amnion (am' nee on) The fluid-filled sac within which the embryos of reptiles (including birds) and mammals develop.

amniote egg A shelled egg surrounding four extraembryonic membranes and embryo-nourishing yolk. This evolutionary adaptation permitted mammals and reptiles to live and reproduce in drier environments than can most amphibians.

amphipathic (am' fi path' ic) [Gk. *amphi*: both + *pathos*: emotion] Of a molecule, having both hydrophilic and hydrophobic regions.

amplitude The magnitude of change over the course of a regular cycle.

amygdala A component of the limbic system that is involved in fear and fear memory.

amylase (am' ill ase) An enzyme that catalyzes the hydrolysis of starch, usually to maltose or glucose.

anabolic reaction (an uh bah' lik) [Gk. *ana*: upward + *ballein*: to throw] A synthetic reaction in which simple molecules are linked to form more complex ones; requires an input of energy and captures it in the chemical bonds that are formed. (Contrast with catabolic reaction.)

anaerobic (an ur row' bic) [Gk. *an*: not + *aer*: air + *bios*: life] Occurring without the use of molecular oxygen, O_2. (Contrast with aerobic.)

anaphase (an' a phase) [Gk. *ana*: upward] The stage in cell nuclear division at which the first separation of sister chromatids (or, in the first meiotic division, of paired homologs) occurs.

ancestral trait The trait originally present in the ancestor of a given group; may be retained or changed in the descendants of that ancestor.

androgen (an' dro jen) Any of the several male sex steroids (most notably testosterone).

aneuploidy (an' you ploy dee) A condition in which one or more chromosomes or pieces of chromosomes are either lacking or present in excess.

angiosperms Flowering plants; one of the two major groups of living seed plants. (*See also* gymnosperms.)

angiotensin (an' jee oh ten' sin) A peptide hormone that raises blood pressure by causing peripheral vessels to constrict. Also maintains glomerular filtration by constricting efferent vessels and stimulates thirst and the release of aldosterone.

animal hemisphere The metabolically active upper portion of some animal eggs, zygotes, and embryos; does not contain the dense nutrient yolk. (Contrast with vegetal hemisphere.)

anion (an' eye on) [Gk. *ana*: upward progress] A negatively charged ion. (Contrast with cation.)

annual A plant whose life cycle is completed in one growing season. (Contrast with biennial, perennial.)

antenna system *See* light-harvesting complex.

anterior pituitary The portion of the vertebrate pituitary gland that derives from gut epithelium and produces tropic hormones.

anterior Toward or pertaining to the tip or headward region of the body axis. (Contrast with posterior.)

anther (an' thur) [Gk. *anthos*: flower] A pollen-bearing portion of the stamen of a flower.

antheridium (an' thur id' ee um) [Gk. *antheros*: blooming] The multicellular structure that produces the sperm in nonvascular land plants and ferns.

antibody One of the myriad proteins produced by the immune system that specifically binds to a foreign substance in blood or other tissue fluids and initiates its removal from the body.

anticodon The three nucleotides in transfer RNA that pair with a complementary triplet (a codon) in messenger RNA.

antidiuretic hormone (ADH) A hormone that promotes water reabsorption by the kidney. ADH is produced by neurons in the hypothalamus and released from nerve terminals in the posterior pituitary. Also called vasopressin.

antigen (an' ti jun) Any substance that stimulates the production of an antibody or antibodies in the body of a vertebrate.

antigen-presenting cell In cellular immunity, a cell that ingests and digests an antigen, and then exposes fragments of that antigen to the outside of the cell, bound to proteins in the cell's plasma membrane.

antigenic determinant The specific region of an antigen that is recognized and bound by a specific antibody. Also called an epitope.

antiparallel Pertaining to molecular orientation in which a molecule or parts of a molecule have opposing directions.

antiporter A membrane transport protein that moves one substance in one direction and another in the opposite direction. (Contrast with symporter, uniporter.)

antisense RNA A single-stranded RNA molecule complementary to, and thus targeted against, an mRNA of interest to block its translation.

anus (a' nus) An opening through which solid digestive wastes are expelled, located at the posterior end of a tubular gut.

aorta (a or' tah) [Gk. *aorte*: aorta] The main trunk of the arteries leading to the systemic (as opposed to the pulmonary) circulation.

aortic body A chemosensor in the aorta that senses a decrease in blood supply or a dramatic decrease in partial pressure of oxygen in the blood.

aortic valve A one-way valve between the left ventricle of the heart and the aorta that prevents backflow of blood into the ventricle when it relaxes.

apex (a' pecks) The tip or highest point of a structure, as of a growing stem or root.

aphasia a deficit in the ability to use or understand words.

aphotic zone In bodies of water (lakes and oceans), the region below the reach of light.

apical (a' pi kul) Pertaining to the *apex*, or tip, usually in reference to plants.

apical dominance In plants, inhibition by the apical bud of the growth of axillary buds.

apical hook A form taken by the stems of many eudicot seedlings that protects the delicate shoot apex while the stem grows through the soil.

apical meristem The meristem at the tip of a shoot or root; responsible for a plant's primary growth.

apomixis (ap oh mix' is) [Gk. *apo*: away from + *mixis*: sexual intercourse] The asexual production of seeds.

apoplast (ap' oh plast) In plants, the continuous meshwork of cell walls and extracellular spaces through which material can pass without crossing a plasma membrane. (Contrast with symplast.)

apoptosis (ap uh toh' sis) A series of genetically programmed events leading to cell death.

appendix In the human digestive system, the vestigial equivalent of the cecum, which serves no digestive function.

aquaporin A transport protein in plant and animal cell membranes through which water passes in osmosis.

aquatic (a kwa' tic) [L. *aqua*: water] Pertaining to or living in water. (Contrast with marine, terrestrial.)

aqueous (a' kwee us) Pertaining to water or a watery solution.

aquifer A large pool of groundwater.

archegonium (ar' ke go' nee um) The multicellular structure that produces eggs in nonvascular land plants, ferns, and gymnosperms.

archenteron (ark en' ter on) [Gk. *archos*: first + *enteron*: bowel] The earliest primordial animal digestive tract.

arms race A series of reciprocal adaptations between species involved in antagonistic interactions, in which adaptations that increase the fitness of a consumer species exert selection pressure on its resource species to counter the consumer's adaptation, and vice versa.

arteriole A small blood vessel arising from an artery that feeds blood into a capillary bed.

artery A muscular blood vessel carrying oxygenated blood away from the heart to other parts of the body. (Contrast with vein.)

artificial insemination An infertility treatment that involves the artificial introduction of sperm into the woman's reproductive tract.

artificial selection The selection by plant and animal breeders of individuals with certain desirable traits.

ascoma The fruiting structure of sexually reproducing sac fungi; often a cup-shaped structure.

ascus (ass' cus) (plural: asci) [Gk. *askos*: bladder] In sac fungi, the club-shaped sporangium within which spores (ascospores) are produced by meiosis.

asexual reproduction Reproduction without sex.

assisted reproductive technologies (ARTs) Any of several procedures that remove unfertilized eggs from the ovary, combine them with sperm outside the body, and then place fertilized eggs or egg–sperm mixtures in the appropriate location in a female's reproductive tract for development.

association cortex In the vertebrate brain, the portion of the cortex involved in higher-order information processing, so named because it integrates, or associates, information from different sensory modalities and from memory.

associative learning A form of learning in which two unrelated stimuli become linked to the same response.

astrocyte [Gk. *astron*: star] A type of glial cell that contributes to the blood–brain barrier by surrounding the smallest, most permeable blood vessels in the brain.

atom [Gk. *atomos*: indivisible] The smallest unit of a chemical element. Consists of a nucleus and one or more electrons.

atomic mass *See* atomic weight.

atomic number The number of protons in the nucleus of an atom; also equals the number of electrons around the neutral atom. Determines the chemical properties of the atom.

atomic weight The average of the mass numbers of a representative sample of atoms of an element, with all the isotopes in their normally occurring proportions. Also called atomic mass.

ATP (adenosine triphosphate) An energy-storage compound containing adenine, ribose, and three phosphate groups. When it is formed from ADP, useful energy is stored; when it is broken down (to ADP or AMP), energy is released to drive endergonic reactions.

ATP synthase An integral membrane protein that couples the transport of protons with the formation of ATP.

atrial natriuretic peptide A hormone released by the atrial muscle fibers of the heart when they are overly stretched, which decreases reabsorption of sodium by the kidney and thus blood volume.

atrioventricular node A modified node of cardiac muscle that organizes the action potentials that control contraction of the ventricles.

atrium (a' tree um) [L. *atrium*: central hall] An internal chamber. In the hearts of vertebrates, the thin-walled chamber(s) entered by blood on its way to the ventricle(s). Also, the outer ear.

auditory system A sensory system that uses mechanoreceptors to convert pressure waves into receptor potentials; includes structures that gather sound waves, direct them to a sensory organ, and amplify their effect on the mechanoreceptors.

autocatalysis [Gk. *autos*: self + *kata*: to break down] A positive feedback process in which an activated enzyme acts on other inactive molecules of the same enzyme to activate them.

autocrine A chemical signal that binds to and affects the cell that makes it. (Contrast with paracrine.)

autoimmune diseases Diseases (e.g., rheumatoid arthritis) that result from failure of the immune system to distinguish between self and nonself, causing it to attack tissues in the organism's own body.

autoimmunity An immune response by an organism to its own molecules or cells.

autonomic nervous system (ANS) The portion of the peripheral nervous system that controls such involuntary functions as those of guts and glands. Also called the involuntary nervous system.

autopolyploidy The possession of more than two entire chromosomes sets that are derived from a single species.

autoregulatory mechanisms In mammalian circulatory systems, local control of blood flow through capillary beds by constriction or dilation of incoming arterioles in response to local metabolite concentrations.

autosome Any chromosome (in a eukaryote) other than a sex chromosome.

autotroph (au' tow trowf') [Gk. *autos*: self + *trophe*: food] An organism that is capable of living exclusively on inorganic materials, water, and some energy source such as sunlight (photoautotrophs) or chemically reduced matter (see chemolithotrophs). (Contrast with heterotroph.)

auxin (awk' sin) [Gk. *auxein*: to grow] In plants, a substance (the most common being indoleacetic acid) that regulates growth and various aspects of development.

avirulence (Avr) genes Genes in a pathogen that may trigger defenses in plants. *See* gene-for-gene resistance.

Avogadro's number The number of atoms or molecules in a mole (weighed out in grams) of a substance, calculated to be 6.022 $\times 10^{23}$.

axillary bud A bud that forms in the angle (axil) where a leaf meets a stem.

axon [Gk. axle] The part of a neuron that conducts action potentials away from the cell body.

axon hillock The junction between an axon and its cell body, where action potentials are generated.

axon terminals The endings of an axon; they form synapses and release neurotransmitter.

– B –

B cell A type of lymphocyte involved in the humoral immune response of vertebrates. Upon recognizing an antigenic determinant, a B cell develops into a plasma cell, which secretes an antibody. (Contrast with T cell.)

B horizon *See* subsoil.

bacillus (bah sil' us) [L: little rod] Any of various rod-shaped bacteria.

bacterial conjugation *See* conjugation.

bacteriophage (bak teer' ee o fayj) [Gk. *bakterion*: little rod + *phagein*: to eat] Any of a group of viruses that infect bacteria. Also called phage.

bacteroids Nitrogen-fixing organelles that develop from endosymbiotic bacteria.

bark All tissues external to the vascular cambium of a plant.

barometric pressure Atmospheric pressure; the total pressure of the gas mixture in air.

baroreceptor [Gk. *baros*: weight] A pressure-sensing cell or organ. Sometimes called a stress receptor.

basal metabolic rate (BMR) The minimum rate of energy turnover in an awake (but resting) bird or mammal that is not expending energy for thermoregulation.

base (1) A substance that can accept a hydrogen ion in solution. (Contrast with acid.) (2) In nucleic acids, the purine or pyrimidine that is attached to each sugar in the sugar–phosphate backbone.

base pair (bp) In double-stranded DNA, a pair of nucleotides formed by the *complementary base pairing* of a purine on one strand and a pyrimidine on the other.

basic Having a pH greater than 7.0 (i.e., having a hydrogen ion concentration lower than 10^{-7} molar). (Contrast with acidic.)

basidioma (plural: basiomata) A fruiting structure produced by club fungi.

basidium (bass id' ee yum) In club fungi, the characteristic sporangium in which four spores are formed by meiosis and then borne externally before being shed.

basilar membrane A membrane in the human inner ear whose flexion in response to sound waves activates hair cells; flexes at different locations in response to different pitches of sound.

basophil A type of phagocytic white blood cell that releases histamine and may promote T cell development.

BD model The "birth–death model": $N_{t+1} = N_t + B - D$. The most basic mathematical model of population growth.

behaviorism One of two classical approaches to the study of proximate causes of animal behavior, derived from the discoveries of Ivan Pavlov and focused on laboratory studies. (Compare with ethology.)

benefit An improvement in survival and reproductive success resulting from performing a behavior or having a trait.

benthic zone [Gk. *benthos*: bottom] The bottom of the ocean.

β (beta) pleated sheet A type of protein secondary structure; results from hydrogen bonding between polypeptide regions running antiparallel to each other.

bicarbonate ion Ion (HCO_3^-) resulting from dissociation of carbonic acid in water; important in pH regulation and carbon dioxide (CO_2) transport.

BIDE model A model of population growth that takes into account immigration and emigration, in addition to births and deaths.

biennial A plant whose life cycle includes vegetative growth in the first year and flowering and senescence in the second year. (Contrast with annual, perennial.)

bilateral symmetry The condition in which only the right and left sides of an organism, divided by a single plane through the midline, are mirror images of each other.

bilayer A structure that is two layers in thickness. In biology, most often refers to the phospholipid bilayer of membranes. (*See* phospholipid bilayer.)

bile A secretion of the liver made up of bile salts synthesized from cholesterol, various phospholipids, and bilirubin (the breakdown product of hemoglobin). Emulsifies fats in the small intestine.

binary fission Reproduction of a prokaryote by division of a cell into two comparable progeny cells.

binocular vision Overlapping visual fields of an animal's two eyes; allows the animal to see in three dimensions.

binomial nomenclature A taxonomic naming system in which each species is given two names (a genus name followed by a species name).

biofilm A community of microorganisms embedded in a polysaccharide matrix, forming a highly resistant coating on almost any moist surface.

biogeochemical cycle Movement of inorganic elements such as nitrogen, phosphorus, and carbon through living organisms and the physical environment.

biogeographic region One of several defined, continental-scale regions of Earth, each of which has a biota distinct from that of the others. (Contrast with biome.)

biogeography The scientific study of the patterns of distribution of populations, species, and ecological communities across Earth.

bioinformatics The use of computers and/or mathematics to analyze complex biological information, such as DNA sequences.

biological species concept The definition of a species as a group of actually or potentially interbreeding natural populations that are reproductively isolated from other such groups. (Contrast with lineage species concept; morphological species concept.)

biology [Gk. *bios*: life + *logos*: study] The scientific study of living things.

bioluminescence The production of light by biochemical processes in an organism.

biomass The total weight of all the organisms, or some designated group of organisms, in a given area.

biome (bye' ome) A major division of the ecological communities of Earth, characterized primarily by distinctive vegetation. A given biogeographic region contains many different biomes.

bioremediation The use by humans of other organisms to remove contaminants from the environment.

biosphere (bye' oh sphere) All regions of Earth (terrestrial and aquatic) and Earth's atmosphere in which organisms can live.

biota (bye oh' tah) All of the organisms—animals, plants, fungi, and microorganisms—found in a given area. (Contrast with flora, fauna.)

biotechnology The use of cells or living organisms to produce materials useful to humans.

biotic (bye ah' tick) [Gk. *bios*: life] Alive. (Contrast with abiotic.)

biotic interchange The mixing of biotas previously separated by physical, climatic, or other barriers, for example when two formerly separated land masses fuse.

blade The thin, flat portion of a leaf.

blastocoel (blass' toe seal) [Gk. *blastos*: sprout + *koilos*: hollow] The central, hollow cavity of a blastula.

blastocyst (blass' toe cist) An early embryo formed by the first divisions of the fertilized egg (zygote). In mammals, a hollow ball of cells.

blastodisc (blass' toe disk) An embryo that forms as a disk of cells on the surface of a large yolk mass; comparable to a blastula, but occurring in animals such as birds and reptiles, in which the massive yolk restricts complete cleavage.

blastomere Any of the cells produced by the early divisions of a fertilized animal egg.

blastopore The opening created by the invagination of the vegetal pole during gastrulation of animal embryos.

blastula (blass' chu luh) An early stage of the animal embryo; in many species, a hollow sphere of cells surrounding a central cavity, the blastocoel. (Contrast with blastodisc.)

block to polyspermy Any of several responses to entry of a sperm into an egg that prevent more than one sperm from entering the egg.

blood A fluid tissue that is pumped around the body; a component of the circulatory system.

blood clotting A cascade of events involving platelets and circulating proteins (clotting factors) that seals damaged blood vessels.

blood–brain barrier A property of blood vessels in the brain that prevents most chemicals from diffusing from the blood into the brain.

blue-light receptors Pigments in plants that absorb blue light (400–500 nm). These pigments mediate many plant responses including photo-tropism, stomatal movements, and expression of some genes.

body plan The general structure of an animal, the arrangement of its organ systems, and the integrated functioning of its parts.

Bohr effect A shift in the O_2 binding curve of hemoglobin in response to excess H^+ ions such that the hemoglobin releases more O_2 in tissues where pH is low.

Bohr model A model for atomic structure that depicts the atom as largely empty space, with a central nucleus surrounded by electrons in orbits, or electron shells, at various distances from the nucleus.

bond *See* chemical bond.

bone A rigid component of vertebrate skeletal systems that contains an extracellular matrix of insoluble calcium phosphate crystals as well as collagen fibers.

bottleneck *See* population bottleneck.

Bowman's capsule An elaboration of the renal tubule, composed of podocytes, that surrounds and collects the filtrate from the glomerulus.

brain The centralized integrative center of a nervous system.

brainstem The portion of the vertebrate brain between the spinal cord and the forebrain, made up of the medulla, pons, and midbrain.

brassinosteroids Plant steroid hormones that mediate light effects promoting the elongation of stems and pollen tubes.

Broca's area A portion of the human brain essential for speech. Located in the frontal lobe just in front of the primary motor cortex.

bronchioles The smallest airways in a vertebrate lung, branching off the bronchi.

bronchus (plural: bronchi) The major airway(s) branching off the trachea into the vertebrate lung.

brown fat In mammals, fat tissue that is specialized to produce heat. It has many mitochondria and capillaries, and a protein that uncouples oxidative phosphorylation.

budding Asexual reproduction in which a more or less complete new organism grows from the body of the parent organism, eventually detaching itself.

buffer A substance that can transiently accept or release hydrogen ions and thereby resist changes in pH.

bulbourethral glands Secretory structures of the human male reproductive system that produce a small volume of an alkaline, mucoid secretion that helps neutralize acidity in the urethra and lubricate it to facilitate the passage of semen.

bulk flow The movement of a solution from a region of higher pressure potential to a region of lower pressure potential.

bundle of His Fibers of modified cardiac muscle that conduct action potentials from the atria to the ventricular muscle mass.

bundle sheath cell Part of a tissue that surrounds the veins of plants.

– C –

C horizon *See* parent rock.

calcitonin Hormone produced by the thyroid gland; lowers blood calcium and promotes bone formation. (Contrast with parathyroid hormone.)

calcitriol A hormone derived from vitamin D whose actions include stimulating the cells of the digestive tract to absorb calcium from ingested food.

calorie [L. *calor*: heat] The amount of heat required to raise the temperature of 1 gram of water by 1°C. Physiologists commonly use the kilocalorie (kcal) as a unit of measure (1 kcal = 1,000 calories). Nutritionists also use the kilocalorie, but refer to it as the *Calorie* (capital C).

Calvin cycle The stage of photosynthesis in which CO_2 reacts with RuBP to form 3PG, 3PG is reduced to a sugar, and RuBP is regenerated, while other products are released to the rest of the plant. Also known as the Calvin–Benson cycle.

calyx (kay' licks) [Gk. *kalyx*: cup] All of the sepals of a flower, collectively.

Cambrian explosion The rapid diversification of multicellular life that took place during the Cambrian period.

cAMP (cyclic AMP) A compound formed from ATP that acts as a second messenger.

cancellous bone A type of bone with numerous internal cavities that make it appear spongy, although it is rigid. (Contrast with compact bone.)

canopy The leaf-bearing part of a tree. Collectively, the aggregate of the leaves and branches of the larger woody plants of an ecological community.

5′ cap *See* G cap.

capillaries [L. *capillaris*: hair] Very small tubes, especially the smallest blood-carrying vessels of animals between the termination of the arteries and the beginnings of the veins. Capillaries are the site of exchange of materials between the blood and the interstitial fluid.

capillary beds Networks of capillaries where materials are exchanged between the blood and the interstitial fluid.

capsid The outer shell of a virus that encloses its nucleic acid.

carbohydrates Organic compounds containing carbon, hydrogen, and oxygen in the ratio 1:2:1 (i.e., with the general formula $C_nH_{2n}O_n$). Common examples are sugars, starch, and cellulose.

carbon skeleton The chains or rings of carbon atoms that form the structural basis of organic molecules. Other atoms or functional groups are attached to the carbon atoms.

carbon-fixation reactions The phase of photosynthesis in which chemical energy captured in the light reactions is used to drive the reduction of CO_2 to form carbohydrates.

carboxylase An enzyme that catalyzes the addition of carboxyl groups to a substrate.

cardiac (kar' dee ak) [Gk. *kardia*: heart] Pertaining to the heart and its functions.

cardiac cycle Contraction of the two atria of the heart, followed by contraction of the two ventricles and then relaxation.

cardiac muscle A type of muscle tissue that makes up, and is responsible for the beating of, the heart. Characterized by branching cells with single nuclei and a striated (striped) appearance. (Contrast with smooth muscle, skeletal muscle.)

cardiovascular system [Gk. *kardia*: heart + L. *vasculum*: small vessel] The heart, blood, and vessels are of a circulatory system.

carnivore [L. *carn*: flesh + *vorare*: to devour] An organism that eats animal tissues. (Contrast with detritivore, frugivore, herbivore, omnivore.)

carotenoid (ka rah′ tuh noid) A yellow, orange, or red lipid pigment commonly found as an accessory pigment in photosynthesis; also found in fungi.

carotid body A chemosensor in the carotid artery that senses a decrease in blood supply or a dramatic decrease in partial pressure of oxygen in the blood.

carpel (kar′ pel) [Gk. *karpos*: fruit] The organ of the flower that contains one or more ovules.

carrier (1) In facilitated diffusion, a membrane protein that binds a specific molecule and transports it through the membrane. (2) In respiratory and photosynthetic electron transport, a participating substance such as NAD that exists in both oxidized and reduced forms. (3) In genetics, a person heterozygous for a recessive trait.

carrying capacity (K) The number of individuals in a population that the resources of its environment can support.

cartilage bone A type of bone that begins its development as a cartilaginous structure resembling the future mature bone, then gradually hardens into mature bone. (Contrast with membranous bone.)

cartilage In vertebrates, a tough connective tissue found in joints, the outer ear, and elsewhere. Forms the entire skeleton in some animal groups.

Casparian strip A band of cell wall containing suberin and lignin, found in the endodermis. Restricts the movement of water across the endodermis.

caspase One of a group of proteases that catalyze cleavage of target proteins and are active in apoptosis.

catabolic reaction (kat uh bah′ lik) [Gk. *kata*: to break down + *ballein*: to throw] A synthetic reaction in which complex molecules are broken down into simpler ones and energy is released. (Contrast with anabolic reaction.)

catalyst (kat′ a list) [Gk. *kata*: to break down] A chemical substance that accelerates a reaction without itself being consumed in the overall course of the reaction. Catalysts lower the activation energy of a reaction. Enzymes are biological catalysts.

cation (cat′ eye on) An ion with one or more positive charges. (Contrast with anion.)

caudal [L. *cauda*: tail] Pertaining to the tail, or to the posterior part of the body.

cDNA library A collection of complementary DNAs derived from mRNAs of a particular tissue at a particular time in the life cycle of an organism.

cDNA *See* complementary DNA.

cecum (see′ cum) [L. blind] A blind branch off the large intestine. In many nonruminant mammals, the cecum contains a colony of microorganisms that contribute to the digestion of food.

cell adhesion molecules Molecules on animal cell surfaces that affect the selective association of cells into tissues during development of the embryo. Also a component of desmosomes.

cell cycle checkpoints Points of transition between different phases of the cell cycle, which are regulated by cyclins and cyclin-dependent kinases (Cdk's).

cell cycle The stages through which a cell passes between one division and the next. Includes all stages of interphase and mitosis.

cell division The reproduction of a cell to produce two new cells. In eukaryotes, this process involves nuclear division (mitosis) and cytoplasmic division (cytokinesis).

cell fate The type of cell that an undifferentiated cell in an embryo will become in the adult.

cell junctions Specialized structures associated with the plasma membranes of epithelial cells. Some contribute to cell adhesion, others to intercellular communication.

cell recognition Binding of cells to one another mediated by membrane proteins or carbohydrates.

cell The simplest structural unit of a living organism. In multicellular organisms, the building blocks of tissues and organs.

cell theory States that cells are the basic structural and physiological units of all living organisms, and that all cells come from preexisting cells.

cell wall A relatively rigid structure that encloses cells of plants, fungi, many protists, and most prokaryotes, and which gives these cells their shape and limits their expansion in hypotonic media.

cellular immune response Immune system response mediated by T cells and directed against parasites, fungi, intracellular viruses, and foreign tissues (grafts). (Contrast with humoral immune response.)

cellular respiration The catabolic pathways by which electrons are removed from various molecules and passed through intermediate electron carriers to O_2, generating H_2O and releasing energy.

cellular specialization In multicellular organisms, the division of labor such that different cell types become responsible for different functions (e.g., reproduction or digestion) within the organism.

cellulose (sell′ you lowss) A straight-chain polymer of glucose molecules, used by plants as a structural supporting material.

central nervous system (CNS) That portion of the nervous system that is the site of most information processing, storage, and retrieval; in vertebrates, the brain and spinal cord. (Contrast with peripheral nervous system.)

central vacuole In plant cells, a large organelle that stores the waste products of metabolism and maintains turgor.

centrifuge [L. *centrum*: center + *fugere*: to flee] A laboratory device in which a sample is spun around a central axis at high speed. Used to separate suspended materials of different densities.

centriole (sen′ tree ole) A paired organelle that helps organize the microtubules in animal and protist cells during nuclear division.

centromere (sen′ tro meer) [Gk. *centron*: center + *meros*: part] The region where sister chromatids join.

centrosome (sen′ tro soam) The major microtubule organizing center of an animal cell.

cephalization (sef ah luh zay′ shun) [Gk. *kephale*: head] The evolutionary trend toward increasing concentration of brain and sensory organs at the anterior end of the animal.

cerebellum (sair uh bell′ um) [L. diminutive of *cerebrum*, brain] The brain region that controls muscular coordination; located at the anterior end of the hindbrain.

cerebral cortex The thin layer of gray matter (neuronal cell bodies) that overlies the cerebrum.

cerebral hemispheres The major information-processing areas of the vertebrate brain. *See also* cerebrum.

cerebrum (su ree′ brum) [L. brain] The dorsal anterior portion of the forebrain, making up the largest part of the brain of mammals; the chief coordination center of the nervous system; consists of two *cerebral hemispheres*.

cervix (sir′ vix) [L. neck] The opening of the uterus into the vagina.

cGMP (cyclic guanosine monophosphate) An intracellular messenger that is part of signal transmission pathways involving G proteins. (*See* G protein.)

channel protein An integral membrane protein that forms an aqueous passageway across the membrane in which it is inserted and through which specific solutes may pass.

character In genetics, an observable feature, such as eye color. (Contrast with trait.)

chemical bond An attractive force stably linking two atoms.

chemical reaction The change in the composition or distribution of atoms of a substance with consequent alterations in properties.

chemical synapse Neural junction at which neurotransmitter molecules released from a presynaptic cell induce changes in a postsynaptic cell. (Contrast with electrical synapse.)

chemically gated channel A type of gated channel that opens or closes depending on the presence or absence of a specific molecule, which binds to the channel protein or to a separate receptor that in turn alters the three-dimensional shape of channel protein.

chemiosmosis Formation of ATP in mitochondria and chloroplasts, resulting from a pumping of protons across a membrane (against a gradient of electrical charge and of pH), followed by the return of the protons through a protein channel with ATP synthase activity.

chemoautotroph *See* chemolithotroph.

chemoheterotroph An organism that must obtain both carbon and energy from organic substances. (Contrast with chemolithotroph, photoautotroph, photoheterotroph.)

chemolithotroph [Gk. *lithos*: stone, rock] An organism that uses carbon dioxide as a carbon source and obtains energy by oxidizing inorganic substances from its environment; also called chemoautotroph. (Contrast with chemoheterotroph, photoautotroph, photoheterotroph.)

chemoreceptor A sensory receptor cell that senses specific molecules (such as odorant molecules or pheromones) in the environment.

chiasma (kie az′ muh) (plural: chiasmata) [Gk. cross] An X-shaped connection between paired homologous chromosomes in prophase I of meiosis. A chiasma is the visible manifestation of crossing over between homologous chromosomes.

chitin (kye′ tin) [Gk. *kiton*: tunic] The characteristic tough but flexible organic component of the exoskeleton of arthropods, consisting of a complex, nitrogen-containing polysaccharide. Also found in cell walls of fungi.

chlorophyll (klor′ o fill) [Gk. *kloros*: green + *phyllon*: leaf] Any of several green pigments associated with chloroplasts or with certain bacterial membranes; responsible for trapping light energy for photosynthesis.

chloroplast [Gk. *kloros*: green + *plast*: a particle] An organelle bounded by a double membrane containing the enzymes and pigments that perform photosynthesis. Chloroplasts occur only in eukaryotes.

choanocyte (ko′ an uh site) The collared, flagellated feeding cells of sponges.

cholecystokinin (ko′ luh sis tuh kai′ nin) A hormone produced and released by the lining of the duodenum when it is stimulated by undigested fats and proteins. It stimulates the gallbladder to release bile and slows stomach activity.

chorion (kor′ ee on) [Gk. *khorion*: afterbirth] The outermost of the membranes protecting mammal, bird, and reptile embryos; in mammals it forms part of the placenta.

chromatid (kro′ ma tid) A newly replicated chromosome, from the time molecular duplication occurs until the time the centromeres separate (during anaphase of mitosis or of meiosis II).

chromatin remodeling A mechanism for epigenetic gene regulation by the alteration of chromatin structure.

chromatin The nucleic acid–protein complex that makes up eukaryotic chromosomes.

chromosomal mutation Loss of or changes in position/direction of a DNA segment on a chromosome.

chromosome (krome′ o sowm) [Gk. *kroma*: color + *soma*: body] In bacteria and viruses, the DNA molecule that contains most or all of the genetic information of the cell or virus. In eukaryotes, a structure composed of DNA and proteins that bears part of the genetic information of the cell.

chylomicron (ky low my′ cron) Particles of lipid coated with protein, produced in the gut from dietary fats and secreted into the extracellular fluids.

chyme (kime) [Gk. *kymus*: juice] Created in the stomach; a mixture of ingested food with the digestive juices secreted by the salivary glands and the stomach lining.

cilium (sil′ ee um) (plural: cilia) [L. eyelash] Hairlike organelle used for locomotion by many unicellular organisms and for moving water and mucus by many multicellular organisms. Generally shorter than a flagellum.

circadian rhythm (sir kade′ ee an) [L. *circa*: approximately + *dies*: day] A rhythm of growth or activity that recurs about every 24 hours.

circannual rhythm [L. *circa*: + *annus*: year] A rhythm of growth or activity that recurs on a yearly basis.

circulatory system A system consisting of a muscular pump (heart), a fluid (blood or hemolymph), and a series of conduits (blood vessels) that transports materials around the body.

citric acid cycle In cellular respiration, a set of chemical reactions whereby acetyl CoA is oxidized to carbon dioxide and hydrogen atoms are stored as NADH and $FADH_2$. Also called the Krebs cycle.

clade [Gk. *klados*: branch] A monophyletic group made up of an ancestor and all of its descendants.

class I MHC molecules Cell surface proteins that participate in the cellular immune response directed against virus-infected cells.

class II MHC molecules Cell surface proteins that participate in the cell–cell interactions (of T-helper cells, macrophages, and B cells) of the humoral immune response.

cleavage The first few cell divisions of an animal zygote. *See also* complete cleavage, incomplete cleavage.

climate The long-term average atmospheric conditions (temperature, precipitation, humidity, wind direction and velocity) found in a region. (Contrast with weather.)

climate diagram A way of graphically summarizing the climate in a given location by superimposing graphs of average monthly temperature and average precipitation through a year.

cloaca The opening through which both urinary wastes and digestive wastes are expelled in most amphibians and in reptiles (including birds).

clonal deletion Inactivation or destruction of lymphocyte clones that would produce immune reactions against the animal's own body.

clonal lineages Asexually reproduced groups of nearly identical organisms.

clonal selection Mechanism by which exposure to antigen results in the activation of selected T- or B-cell clones, resulting in an immune response.

clone [Gk. *klon*: twig, shoot] (1) Genetically identical cells or organisms produced from a common ancestor by asexual means. (2) To produce many identical copies of a DNA sequence by its introduction into, and subsequent asexual reproduction of, a cell or organism.

closed circulatory system Circulatory system in which the circulating fluid is contained within a continuous system of vessels. (Contrast with open circulatory system.)

CO (CONSTANS) Gene coding for a transcription factor that activates the synthesis of florigen (FT); involved in the induction of flowering.

co-repressor In the regulation of bacterial operons, a molecule that binds to the repressor, causing it to change shape and bind to the operator, thereby inhibiting transcription.

coccus (kock′ us) (plural: cocci) [Gk. *kokkos*: berry, pit] Any of various spherical or spheroidal bacteria.

cochlea (kock′ lee uh) [Gk. *kokhlos*: snail] A spiral tube in the inner ear of vertebrates; it contains the sensory cells involved in hearing.

codominance A condition in which two alleles at a locus produce different phenotypic effects and both effects appear in heterozygotes.

codon Three nucleotides in messenger RNA that direct the placement of a particular amino acid into a polypeptide chain. (Contrast with anticodon.)

coelom (see′ loam) [Gk. *koiloma*: cavity] An animal body cavity, enclosed by muscular mesoderm and lined with a meso-dermal layer called peritoneum that also surrounds the internal organs.

coenocytic   (seen' a sit ik) [Gk. *koinos*: common + *kytos*: container] Referring to the condition, found in some fungal hyphae, of "cells" containing many nuclei but enclosed by a single plasma membrane. Results from nuclear division without cytokinesis.

coenzyme A (CoA)   A coenzyme used in various biochemical reactions as a carrier of acyl groups.

coenzyme   A nonprotein organic molecule that plays a role in catalysis by an enzyme.

coevolution   *See* reciprocal adaptation.

cofactor   An inorganic ion that is weakly bound to an enzyme and required for its activity.

cohesin   A protein involved in binding chromatids together.

cohesion   The tendency of molecules (or any substances) to stick together.

cohort   (co' hort) [L. *cohors*: company of soldiers] A group of similar-aged organisms.

cold-hardening   A process by which plants can acclimate to cooler temperatures; requires repeated exposure to cool temperatures over many days.

coleoptile   A sheath that surrounds and protects the shoot apical meristem and young primary leaves of a grass seedling as they move through the soil.

collagen   [Gk. *kolla*: glue] A fibrous protein found extensively in bone and connective tissue.

collecting duct   In vertebrates, a tubule that receives urine produced in the neph-rons of the kidney and delivers that fluid to the ureter for excretion.

collenchyma   (cull eng' kyma) [Gk. *kolla*: glue + *enchyma*: infusion] A type of plant cell, living at functional maturity, which lends flexible support by virtue of primary cell walls thickened at the corners. (Contrast with parenchyma, sclerenchyma.)

colon   [Gk. *kolon*] The portion of the gut between the small intestine and the anus. Also called the large intestine.

commensalism   [L. *com*: together + *mensa*: table] A type of interaction between species in which one participant benefits while the other is unaffected.

communication   A signal from one organ-ism (or cell) that alters the functioning or behavior of another organism (or cell).

community   Any ecologically integrated group of species of microorganisms, plants, and animals inhabiting a given area.

compact bone   A type of bone with a solid, hard structure. (Contrast with cancellous bone.)

companion cell   In angiosperms, a spe-cialized cell found adjacent to a sieve tube element.

comparative experiment   Experimental design in which data from various

unmanipulated samples or populations are compared, but in which variables are not controlled or even necessarily identified. (Contrast with controlled experiment.)

comparative genomics   Computer-aided comparison of DNA sequences between different organisms to reveal genes with related functions.

competition   In ecology, use of the same resource by two or more species when the resource is present in insufficient supply for the combined needs of the species.

competitive inhibitor   A nonsubstrate that binds to the active site of an enzyme and thereby inhibits binding of its substrate. (Contrast with noncompetitive inhibitor.)

complement system   A group of eleven proteins that play a role in some reactions of the immune system. The complement pro-teins are not immunoglobulins.

complementary base pairing   The AT (or AU), TA (or UA), CG, and GC pairing of bases in double-stranded DNA, in transcrip-tion, and between tRNA and mRNA.

complementary DNA (cDNA)   DNA formed by reverse transcriptase acting with an RNA template; essential intermediate in the reproduction of retroviruses; used as a tool in recombinant DNA technology; lacks introns.

complete cleavage   Pattern of cleavage that occurs in eggs that have little yolk. Early cleavage furrows divide the egg com-pletely and the blastomeres are of similar size. (Contrast with incomplete cleavage.)

complete metamorphosis   A change of state during the life cycle of an organism in which the body is almost completely rebuilt to produce an individual with a very differ-ent body form. Characteristic of insects such as butterflies, moths, beetles, ants, wasps, and flies.

compound   (1) A substance made up of atoms of more than one element. (2) Made up of many units, as in the *compound eyes* of arthropods.

condensation reaction   A chemical reac-tion in which two molecules become con-nected by a covalent bond and a molecule of water is released (AH + BOH → AB + H$_2$O.) (Contrast with hydrolysis reaction.)

conditional mutation   A mutation that results in a characteristic phenotype only under certain environmental conditions.

conditioned reflex   A form of associative learning first described by Ivan Pavlov, in which a natural response (such as salivation in response to food) becomes associated with a normally unrelated stimulus (such as a ringing bell).

conduction   The transfer of heat from one object to another through direct contact.

cone   (1) In conifers, a reproductive struc-ture consisting of spore-bearing scales extending from a central axis. (Contrast with strobilus.) (2) In the vertebrate retina, a type

of photoreceptor cell responsible for color vision.

conidium   (ko nid' ee um) (plural: conidia) [Gk. *konis*: dust] A type of haploid fun-gal spore borne at the tips of hyphae, not enclosed in sporangia.

conjugation   (kon ju gay' shun) [L. *conju-gare*: yoke together] (1) A process by which DNA is passed from one cell to another through a *conjugation tube*, as in bacteria. (2) A nonreproductive sexual process by which *Paramecium* and other ciliates exchange genetic material.

connective tissue   A type of tissue that connects or surrounds other tissues; its cells are embedded in a collagen-containing matrix. One of the four major tissue types in multicellular animals.

consensus sequences   Short stretches of DNA that appear, with little variation, in many different genes.

conservation biology   An applied science that carries out investigations with the aim of maintaining the diversity of life on Earth.

conserved   Pertaining to a gene or trait that has evolved very slowly and is similar or even identical in individuals of highly divergent groups.

conspecifics   Individuals of the same species.

constant region   The portion of an immu-noglobulin molecule whose amino acid composition determines its class and does not vary among immunoglobulins in that class. (Contrast with variable region.)

constitutive   Always present; produced continually at a constant rate. (Contrast with inducible.)

constitutive genes   Genes that are expressed all the time. (Contrast with induc-ible genes.)

consumer   An organism that eats the tis-sues of some other organism.

consumer–resource interactions   Inter-actions in which organisms gain their nutri-tion by eating other living organisms or are eaten themselves.

continental drift   The gradual movements of the world's continents that have occurred over billions of years.

contraception   Birth control methods that prevent fertilization or implantation (conception).

contractile vacuole   (kon trak' tul) A spe-cialized vacuole that collects excess water taken in by osmosis, then contracts to expel the water from the cell.

controlled experiment   An experiment in which a sample is divided into groups whereby experimental groups are exposed to manipulations of an independent variable while one group serves as an untreated con-trol. The data from the various groups are then compared to see if there are changes in a dependent variable as a result of the experimental manipulation. (Contrast with comparative experiment.)

controlled system A set of components in a physiological system that is controlled by commands from a regulatory system. (Contrast with regulatory system.)

convection The transfer of heat to or from a surface via a moving stream of air or fluid.

convergent evolution Independent evolution of similar features from different ancestral traits.

copulation Reproductive behavior that results in a male depositing sperm in the reproductive tract of a female.

cork cambium [L. *cambiare*: to exchange] In plants, a lateral meristem that produces secondary growth, mainly in the form of waxy-walled protective cells, including some of the cells that become bark.

cork In plants, a protective outermost tissue layer composed of cells with thick walls waterproofed with suberin.

cornea The clear, transparent tissue that covers the eye and allows light to pass through to the retina.

corolla (ko role' lah) [L. *corolla*: a small crown] All of the petals of a flower, collectively.

coronary artery (kor' oh nair ee) An artery that supplies blood to the heart muscle.

corpus luteum (kor' pus loo' tee um) (plural: corpora lutea) [L. yellow body] A structure formed from a follicle after ovulation; produces hormones important to the maintenance of pregnancy.

corridor A connection between habitat patches through which organisms can disperse; plays a critical role in maintaining subpopulations.

cortex [L. *cortex*: covering, rind] (1) In plants, the tissue between the epidermis and the vascular tissue of a stem or root. (2) In animals, the outer tissue of certain organs, such as the adrenal gland (adrenal cortex) and the brain (cerebral cortex).

corticosteroids Steroid hormones produced and released by the cortex of the adrenal gland.

corticotropin A tropic hormone produced by the anterior pituitary hormone that stimulates cortisol release from the adrenal cortex. Also called adrenocorticotropic hormone (ACTH).

corticotropin-releasing hormone A releasing hormone produced by the hypothalamus that controls the release of cortisol from the anterior pituitary.

cortisol A corticosteroid that mediates stress responses.

cost A decrease in fitness resulting from performing a behavior or having a trait.

cost–benefit approach An approach to evolutionary studies that assumes an animal has a limited amount of time and energy to devote to each of its activities, and that each activity has fitness costs as well as benefits. (*See also* trade-off.)

cotyledon (kot' ul lee' dun) [Gk. *kotyledon*: hollow space] A "seed leaf." An embryonic organ that stores and digests reserve materials; may expand when seed germinates.

countercurrent exchange *See* countercurrent flow.

countercurrent flow An arrangement that promotes the maximum exchange of heat, or of a diffusible substance, between two fluids by having the fluids flow in opposite directions through parallel vessels close together.

countercurrent heat exchanger In "hot" fish, an adaptation of the circulatory system such that arterial blood flowing into the muscles is warmed by venous blood flowing out of the muscles, thereby conserving body heat by countercurrent exchange.

countercurrent multiplier The mechanism that increases the concentration of the interstitial fluid in the mammalian kidney through countercurrent flow in the loops of Henle and selective permeability and active transport of ions by segments of the loops of Henle.

covalent bond Chemical bond based on the sharing of electrons between two atoms.

CpG islands DNA regions rich in C residues adjacent to G residues. Especially abundant in promoters, these regions are where methylation of cytosine usually occurs.

critical night length In the photoperiodic flowering response of short-day plants, the length of night above which flowering occurs and below which the plant remains vegetative. (The reverse applies in the case of long-day plants.)

critical period *See* sensitive period.

crop A simple food storage sac, the first of two stomachlike organs in many animals (including reptiles, earthworms, and various insects). (*See also* gizzard.)

cross section A section taken perpendicular to the longest axis of a structure. Also called a transverse section.

crossing over The mechanism by which linked genes undergo recombination. In general, the term refers to the reciprocal exchange of corresponding segments between two homologous chromatids.

cryptochromes [Gk. *kryptos*: hidden + *kroma*: color] Photoreceptors mediating some blue-light effects in plants and animals.

ctene (teen) [Gk. *cteis*: comb] In ctenophores, a comblike row of cilia-bearing plates. Ctenophores move by beating the cilia on their eight ctenes.

culture (1) A laboratory association of organisms under controlled conditions. (2) The collection of knowledge, tools, values, and rules that characterize a human society.

cumulus A thick gelatinous layer that protects a mammalian ovum.

currents Circulation patterns in the surface waters of oceans driven by the prevailing winds.

cuticle (1) In plants, a waxy layer on the outer body surface that retards water loss. (2) In ecdysozoans, an outer body covering that provides protection and support and is periodically molted.

cyclic AMP *See* cAMP.

cyclic electron transport In photosynthetic light reactions, the flow of electrons that produces ATP but no NADPH or O_2.

cyclin A protein that activates a cyclin-dependent kinase, bringing about transitions in the cell cycle.

cyclin-dependent kinase (Cdk) A protein kinase whose target proteins are involved in transitions in the cell cycle and which is active only when complexed with additional protein subunits, called cyclins.

cytokine A regulatory protein made by immune system cells that affects other target cells in the immune system.

cytokinesis (sy' toe kine ee' sis) [Gk. *kytos*: container + *kinein*: to move] The division of the cytoplasm of a dividing cell. (Contrast with mitosis.)

cytokinin (sy' toe kine' in) A member of a class of plant growth substances that plays roles in senescence, cell division, and other phenomena.

cytoplasm The contents of the cell, excluding the nucleus.

cytoplasmic determinants In animal development, gene products whose spatial distribution may determine such things as embryonic axes.

cytoplasmic segregation The asymmetrical distribution of cytoplasmic determinants in a developing animal embryo.

cytosine (C) (site' oh seen) A nitrogen-containing base found in DNA and RNA.

cytoskeleton The network of microtubules and microfilaments that gives a eukaryotic cell its shape and its capacity to arrange its organelles and to move.

cytosol The fluid portion of the cytoplasm, excluding organelles and other solids.

cytotoxic T cells (T_C) Cells of the cellular immune system that recognize and directly eliminate virus-infected cells. (Contrast with T-helper cells.)

– D –

DAG *See* diacylglycerol.

data Quantified observations about a system under study.

daughter chromosomes During mitosis, the separated chromatids from the beginning of anaphase onward.

dead space The lung volume that fails to be ventilated with fresh air (because the lungs are never completely emptied during exhalation).

dead zones Regions in aquatic ecosystems that are devoid of aquatic life because eutrophication has resulted in severe oxygen depletion.

deciduous [L. *deciduus*: falling off] Pertaining to a woody plant that sheds its leaves but does not die.

declarative memory Memory of people, places, events, and things that can be consciously recalled and described. (Contrast with procedural memory.)

decomposer An organism that metabolizes organic compounds in debris and dead organisms, releasing inorganic material; found among the bacteria, protists, and fungi. *See also* detritivore, saprobe.

defensin A type of protein made by phagocytes that kills bacteria and enveloped viruses by insertion into their plasma membranes.

deficiency disease A condition (e.g., scurvy and beriberi) caused by chronic lack of any essential nutrient.

degeneracy The situation in which a single amino acid may be represented by any of two or more different codons in messenger RNA. Most of the amino acids can be represented by more than one codon.

deletion A mutation resulting from the loss of a continuous segment of a gene or chromosome. Such mutations almost never revert to wild type. (Contrast with duplication, point mutation.)

demethylase An enzyme that catalyzes the removal of the methyl group from cytosine, reversing DNA methylation.

demography The study of population structure and of the processes by which it changes.

denaturation Loss of activity of an enzyme or nucleic acid molecule as a result of structural changes induced by heat or other means.

dendrite [Gk. *dendron*: tree] A fiber of a neuron which often cannot carry action potentials. Usually much branched and relatively short compared with the axon, and commonly carries information to the cell body of the neuron.

denitrification Metabolic activity by which nitrate and nitrite ions are reduced to form nitrogen gas; carried out by certain soil bacteria.

denitrifiers Bacteria that release nitrogen to the atmosphere as nitrogen gas (N_2).

density-dependent Pertaining to a factor with an effect on population size that increases in proportion to population density.

density-independent Pertaining to a factor with an effect on population size that acts independently of population density.

deoxyribonucleic acid *See* DNA.

deoxyribonucleoside triphosphates (dNTPs) The raw materials for DNA synthesis: deoxyadenosine triphosphate (dATP), deoxythymidine triphosphate (dTTP), deoxycytidine triphosphate (dCTP), and deoxyguanosine triphosphate (dGTP). Also called deoxyribonucleotides.

deoxyribose A five-carbon sugar found in nucleotides and DNA.

depolarization A change in the resting potential across a membrane so that the inside of the cell becomes less negative, or even positive, compared with the outside of the cell. (Contrast with hyperpolarization.)

derived trait A trait that differs from the ancestral trait. (Contrast with synapomorphy.)

dermal tissue system The outer covering of a plant, consisting of epidermis in the young plant and periderm in a plant with extensive secondary growth. (Contrast with ground tissue system and vascular tissue system.)

desmosome (dez′ mo sowm) [Gk. *desmos*: bond + *soma*: body] An adhering junction between animal cells.

determinate growth A growth pattern in which the growth of an organism or organ ceases when an adult state is reached; characteristic of most animals and some plant organs. (Contrast with indeterminate growth.)

determination In development, the process whereby the fate of an embryonic cell or group of cells (e.g., to become epidermal cells or neurons) is set.

detritivore (di try′ ti vore) [L. *detritus*: worn away + *vorare*: to devour] An organism that obtains its energy from the dead bodies or waste products of other organisms.

development The process by which a multicellular organism, beginning with a single cell, goes through a series of changes, taking on the successive forms that characterize its life cycle.

developmental module A functional entity in the embryo encompassing genes and signaling pathways that determine a physical structure independently of other such modules.

diabetes mellitus A condition caused by lack of insulin or inability to respond to insulin, characterized by an inability of cells to take up glucose from the blood.

diacylglycerol (DAG) In hormone action, the second messenger produced by hydrolytic removal of the head group of certain phospholipids.

diaphragm (dye′ uh fram) [Gk. *diaphrassein*: barricade] (1) A sheet of muscle that separates the thoracic and abdominal cavities in mammals; responsible for breathing. (2) A method of birth control in which a sheet of rubber is fitted over the woman′s cervix, blocking the entry of sperm.

diastole (dye ass′ toll ee) [Gk. dilation] The portion of the cardiac cycle when the heart muscle relaxes. (Contrast with systole.)

diencephalon The portion of the vertebrate forebrain that develops into the thalamus and hypothalamus.

differential gene expression The hypothesis that, given that all cells contain all genes, what makes one cell type different from another is the difference in transcription and translation of those genes.

differentiation The process whereby originally similar cells follow different developmental pathways; the actual expression of determination.

diffusion Random movement of molecules or other particles, resulting in even distribution of the particles when no barriers are present.

digestive vacuole In protists, an organelle specialized for digesting food ingested by endocytosis.

dihybrid cross A mating in which the parents differ with respect to the alleles of two loci of interest.

dikaryon (di care′ ee ahn) [Gk. *di*: two + *karyon*: kernel] A cell or organism carrying two genetically distinguishable nuclei. Common in fungi.

dioecious (die eesh′ us) [Gk. *di*: two + *oikos*: house] Pertaining to organisms in which the two sexes are "housed" in two different individuals, so that eggs and sperm are not produced in the same individuals. Examples: humans, fruit flies, date palms. (Contrast with monoecious.)

diploblastic Having two cell layers. (Contrast with triploblastic.)

diploid (dip′ loid) [Gk. *diplos*: double] Having a chromosome complement consisting of two copies (homologs) of each chromosome. Designated $2n$. (Contrast with haploid.)

directional selection Selection in which phenotypes at one extreme of the population distribution are favored. (Contrast with disruptive selection, stabilizing selection.)

disaccharide A carbohydrate made up of two monosaccharides (simple sugars).

discoidal cleavage In animal development, a type of incomplete cleavage that is common in fishes, reptiles, and birds, the eggs of which contain a dense yolk mass.

dispersal Movement of organisms away from a parent organism or from an existing population.

disruptive selection Selection in which phenotypes at both extremes of the population distribution are favored. (Contrast with directional selection; stabilizing selection.)

distal Away from the point of attachment or other reference point. (Contrast with proximal.)

distal convoluted tubule The portion of a renal tubule from where it reaches the renal cortex, just past the loop of Henle to where it joins a collecting duct. (Compare with proximal convoluted tubule.)

disturbance A short-term event that disrupts populations, communities, or ecosystems by changing the environment.

disulfide bridge The covalent bond between two sulfur atoms (–S—S–) linking two molecules or remote parts of the same molecule.

DNA (deoxyribonucleic acid)   The fundamental hereditary material of all living organisms. In eukaryotes, stored primarily in the cell nucleus. A nucleic acid using deoxyribose rather than ribose.

DNA fingerprint   An individual's unique pattern of allele sequences, commonly short tandem repeats and single nucleotide polymorphisms.

DNA ligase   Enzyme that unites broken DNA strands during replication and recombination.

DNA methylation   The addition of methyl groups to bases in DNA, usually cytosine or guanine.

DNA methyltransferase   An enzyme that catalyzes the methylation of DNA.

DNA microarray   A small glass or plastic square onto which thousands of single-stranded DNA sequences are fixed so that hybridization of cell-derived RNA or DNA to the target sequences can be performed.

DNA polymerase   Any of a group of enzymes that catalyze the formation of DNA strands from a DNA template.

DNA transposons   Mobile genetic elements that move without making an RNA intermediate. (Contrast with retrotransposons.)

domain   (1) An independent structural element within a protein. Encoded by recognizable nucleotide sequences, a domain often folds separately from the rest of the protein. Similar domains can appear in a variety of different proteins across phylogenetic groups (e.g., "homeobox domain"; "calcium-binding domain"). (2) In phylogenetics, the three monophyletic branches of life (Bacteria, Archaea, and Eukarya).

dominance   In genetics, the ability of one allelic form of a gene to determine the phenotype of a heterozygous individual in which the homologous chromosomes carry both it and a different (recessive) allele. (Contrast with recessive.)

dormancy   A condition in which normal activity is suspended, as in some spores, seeds, and buds.

dorsal   [L. *dorsum*: back] Toward or pertaining to the back or upper surface. (Contrast with ventral.)

dorsal lip   In amphibian embryos, the dorsal segment of the blastopore. Also called the "organizer," this region directs the development of nearby embryonic regions.

double fertilization   In angiosperms, a process in which the nuclei of two sperm fertilize one egg. One sperm's nucleus combines with the egg nucleus to produce a zygote, while the other combines with the same egg's two polar nuclei to produce the first cell of the triploid endosperm (the tissue that will nourish the growing plant embryo).

double helix   Refers to DNA and the (usually right-handed) coil configuration of two complementary, antiparallel strands.

downregulation   A negative feedback process in which continuous high concentrations of a hormone can decrease the number of its receptors. (Contrast with upregulation.)

duodenum   (do' uh dee' num) The beginning portion of the vertebrate small intestine. (Contrast with ileum, jejunum.)

duplication   A mutation in which a segment of a chromosome is duplicated, often by the attachment of a segment lost from its homolog. (Contrast with deletion.)

dynamic instability   A property of actin filaments in the cytoskeleton, characterized by rapid shortening or lengthening of individual filaments.

– E –

ecdysone   (eck die' sone) [Gk. *ek*: out of + *dyo*: to clothe] In insects, a hormone that induces molting.

ecological efficiency   The overall transfer of energy from one trophic level to the next, expressed as the ratio of consumer production to producer production.

ecological system   One or more organisms plus the external environment with which they interact.

ecological transition   A type of succession in which a community distinctly different from the original community is established after a disturbance.

ecology   [Gk. *oikos*: house] The scientific study of the interaction of organisms with their living (biotic) and nonliving (abiotic) environments.

ecosystem   (eek' oh sis tum) The organisms of a particular habitat, such as a pond or forest, together with the physical environment in which they live.

ecosystem services   Processes by which ecosystems maintain resources that benefit human society.

ectoderm   [Gk. *ektos*: outside + *derma*: skin] The outermost of the three embryonic germ layers first delineated during gastrulation. Gives rise to the skin, sense organs, and nervous system.

ectotherm   [Gk. *ektos*: outside + *thermos*: heat] An animal that is dependent on external heat sources for regulating its body temperature (Contrast with endotherm.)

edema   (i dee' mah) [Gk. *oidema*: swelling] Tissue swelling caused by the accumulation of fluid.

effector cells   In cellular immunity, B cells and T cells that attack an antigen, either by secreting antibodies that bind to the antigen or by releasing molecules that destroy any cell bearing the antigen.

efferent   (ef' ur unt) [L. *ex*: out + *ferre*: to bear] Carrying outward or away from, as in a neuron that carries impulses outward from the central nervous system (efferent neuron), or a blood vessel that carries blood away from a structure. (Contrast with afferent.)

efferent neurons   In a neural network, nerve cells that carry commands to physiological and behavioral effectors such as muscles and glands.

egg   In all sexually reproducing organisms, the female gamete; in birds, reptiles, and some other vertebrates, a structure within which early embryonic development occurs. See also amniote egg, ovum.

electrical synapse   A type of synapse at which action potentials spread directly from presynaptic cell to postsynaptic cell. (Contrast with chemical synapse.)

electrocardiogram (ECG or EKG)   A graphic recording of electrical potentials from the heart.

electrochemical gradient   The concentration gradient of an ion across a membrane plus the voltage difference across that membrane.

electroencephalogram (EEG)   A graphic recording of electrical potentials from the brain.

electromagnetic radiation   A self-propagating wave that travels though space and has both electrical and magnetic properties.

electron   A subatomic particle outside the nucleus carrying a negative charge and very little mass.

electron shell   The region surrounding the atomic nucleus at a fixed energy level in which electrons orbit.

electron transport   The passage of electrons through a series of proteins with a release of energy which may be captured in a concentration gradient or in chemical form such as NADH or ATP.

electronegativity   The tendency of an atom to attract electrons when it occurs as part of a compound.

electrophoresis   See gel electrophoresis.

element   A substance that cannot be converted to simpler substances by ordinary chemical means.

elongation   (1) In molecular biology, the addition of monomers to make a longer RNA or protein during transcription or translation. (2) Growth of a plant axis or cell primarily in the longitudinal direction.

embryo   [Gk. *en*: within + *bryein*: to grow] A young animal, or young plant sporophyte, while it is still contained within a protective structure such as a seed, egg, or uterus.

embryo sac   In angiosperms, the female gametophyte. Found within the ovule, it consists of eight or fewer cells, membrane bounded, but without cellulose walls between them.

embryonic stem cell (ESC)   A pluripotent cell in the blastocyst.

emigration   The deliberate and usually oriented departure of an organism from the habitat in which it has been living.

3' end   (3 prime) The end of a DNA or RNA strand that has a free hydroxyl group at the 3′ carbon of the sugar (deoxyribose or ribose).

5′ end   (5 prime) The end of a DNA or RNA strand that has a free phosphate group at the 5′ carbon of the sugar (deoxyribose or ribose).

endemic (en dem′ ik) [Gk. *endemos*: native] Confined to a particular region, thus often having a comparatively restricted distribution.

endergonic   A chemical reaction in which the products have higher free energy than the reactants, thereby requiring free energy input to occur. (Contrast with exergonic.)

endocrine cells   Cells that secrete substances into the extracellular fluid. (*See also* endocrine gland.)

endocrine gland   (en′ doh krin) [Gk. *endo*: within + *krinein*: to separate] An aggregation of secretory cells that secretes hormones into the blood. The *endocrine system* consists of all *endocrine cells* and endocrine glands in the body that produce and release hormones. (Contrast with exocrine gland.)

endocytosis   A process by which liquids or solid particles are taken up by a cell through invagination of the plasma membrane. (Contrast with exocytosis.)

endoderm [Gk. *endo*: within + *derma*: skin] The innermost of the three embryonic germ layers delineated during gastrulation. Gives rise to the digestive and respiratory tracts and structures associated with them.

endodermis   In plants, a specialized cell layer marking the inside of the cortex in roots and some stems. Frequently a barrier to free diffusion of solutes.

endomembrane system   A system of intracellular membranes that exchange material with one another, consisting of the Golgi apparatus, endoplasmic reticulum, and lysosomes when present.

endometrium   The epithelial lining of the uterus.

endoplasmic reticulum (ER) [Gk. *endo*: within + L. *reticulum*: net] A system of membranous tubes and flattened sacs found in the cytoplasm of eukaryotes. Exists in two forms: rough ER, studded with ribosomes; and smooth ER, lacking ribosomes.

endorphins   Molecules in the mammalian brain that act as neurotransmitters in pathways that control pain.

endoskeleton [Gk. *endo*: within + *skleros*: hard] An internal skeleton covered by other, soft body tissues. (Contrast with exoskeleton.)

endosperm [Gk. *endo*: within + *sperma*: seed] A specialized triploid seed tissue found only in angiosperms; contains stored nutrients for the developing embryo.

endospore [Gk. *endo*: within + *spora*: to sow] In some bacteria, a resting structure that can survive harsh environmental conditions.

endosymbiosis theory [Gk. *endo*: within + *sym*: together + *bios*: life] The theory that the eukaryotic cell evolved via the engulfing of one prokaryotic cell by another.

endothelium   The single layer of epithelial cells lining the interior of a blood vessel.

endotherm [Gk. *endo*: within + *thermos*: heat] An animal that can control its body temperature by the expenditure of its own metabolic energy. (Contrast with ectotherm.)

endotoxin   A lipopolysaccharide that forms part of the outer membrane of certain Gram-negative bacteria that is released when the bacteria grow or lyse. (Contrast with exotoxin.)

energetic cost   The difference between the energy an animal expends in performing a behavior and the energy it would have expended had it rested.

energy   The capacity to do work or move matter against an opposing force. The capacity to accomplish change in physical and chemical systems.

enkephalins   Molecules in the mammalian brain that act as neurotransmitters in pathways that control pain.

enteric nervous system   The nerve nets in the submucosa and between the smooth muscle layers of the vertebrate gut.

enthalpy (H)   The total energy of a system.

entrain   To phase-advance or phase-delay an organism's circadian clock each day so that it is in phase with the light-dark cycle of the organism's environment.

entropy (S)   (en′ tro pee) [Gk. *tropein*: to change] A measure of the degree of disorder in any system. Spontaneous reactions in a closed system are always accompanied by an increase in entropy.

environment   Whatever surrounds and interacts with or otherwise affects a population, organism, or cell. May be external or internal.

enzyme   (en′ zime) [Gk. *zyme*: to leaven (as in yeast bread)] A catalytic protein that speeds up a biochemical reaction.

enzyme–substrate complex (ES)   An intermediate in an enzyme-catalyzed reaction; consists of the enzyme bound to its substrate(s).

epi- [Gk. upon, over] A prefix used to designate a structure located on top of another; for example, epidermis, epiphyte.

epiblast   The upper or overlying portion of the avian blastula which is joined to the hypoblast at the margins of the blastodisc.

epiboly   The movement of cells over the surface of the blastula toward the forming blastopore.

epidermis [Gk. *epi*: over + *derma*: skin] In plants and animals, the outermost cell layers. (Only one cell layer thick in plants.)

epididymis   (epuh did′ uh mus) [Gk. *epi*: over + *didymos*: testicle] Coiled tubules in the testes that store sperm and conduct sperm from the seminiferous tubules to the vas deferens.

epigenetics   The scientific study of changes in the expression of a gene or set of genes that occur without change in the DNA sequence.

epinephrine   (ep i nef′ rin) [Gk. *epi*: over + *nephros*: kidney] The "fight or flight" hormone produced by the medulla of the adrenal gland; it also functions as a neurotransmitter. (Also known as adrenaline.)

epistasis   Interaction between genes in which the presence of a particular allele of one gene determines whether another gene will be expressed.

epithelium   A type of animal tissue made up of sheets of cells that lines or covers organs, makes up tubules, and covers the surface of the body; one of the four major tissue types in multicellular animals.

epitope   *See* antigenic determinant.

equilibrium   Any state of balanced opposing forces and no net change.

equilibrium potential   The membrane potential at which an ion is at electrochemical equilibrium, i.e., there is no net flux of the ion across the membrane.

ER   *See* endoplasmic reticulum.

erection   The process by which a penis becomes enlarged and engorged with blood.

error signal   In regulatory systems, any difference between the set point of the system and its current condition.

erythrocyte   (ur rith′ row site) [Gk. *erythros*: red + *kytos*: container] A red blood cell.

erythropoietin   A hormone produced by the kidney in response to lack of oxygen that stimulates the production of red blood cells.

esophagus   (i soff′ i gus) [Gk. *oisophagos*: gullet] That part of the gut between the pharynx and the stomach.

essential amino acids   Amino acids that an animal cannot synthesize for itself and must obtain from its food.

essential element   A mineral nutrient required for normal growth and reproduction in plants and animals.

essential fatty acids   Fatty acids that an animal cannot synthesize for itself and must obtain from its food.

essential mineral   *See* essential element.

ester linkage   A condensation (water-releasing) reaction in which the carboxyl group of a fatty acid reacts with the hydroxyl group of an alcohol. Lipids are formed in this way.

estivation   (ess tuh vay′ shun) [L. *aestivalis*: summer] A state of dormancy and hypometabolism that occurs during the summer; usually a means of surviving drought and/or intense heat. (Contrast with hibernation.)

estrogen   Any of several steroid sex hormones; produced chiefly by the ovaries in mammals.

estrus   (es′ trus) [L. *oestrus*: frenzy] The period of heat, or maximum sexual receptivity, in some female mammals. Ordinarily, the estrus is also the time of release of eggs in the female.

ethology [Gk. *ethos*: character + *logos*: study] An approach to the study of animal behavior that focuses on studying many species in natural environments and addresses questions about the evolution of behavior. (Compare with behaviorism.)

ethylene One of the plant growth hormones, the gas $H_2C=CH_2$. Involved in fruit ripening and other growth and developmental responses.

eudicots Angiosperms with two embryonic cotyledons; one of the two largest clades of angiosperms. (See also monocots.)

Eukarya One of the three domains of life; organisms made up of one or more eukaryotic cells. (See also eukaryotes.)

eukaryotes (yew car′ ree oats) [Gk. *eu*: true + *karyon*: kernel or nucleus] Organisms whose cells contain their genetic material inside a nucleus. Includes all life other than the viruses, archaea, and bacteria. (Contrast with prokaryotes.)

eusocial Pertaining to a social group that includes nonreproductive individuals, as in honey bees.

eustachian tube A connection between the middle ear and the throat that allows air pressure to equilibrate between the middle ear and the outside world.

eutrophication (yoo trofe′ ik ay′ shun) [Gk. *eu*: truly + *trephein*: to flourish] The addition of nutrient materials to a body of water, resulting in changes in ecological processes and species composition therein.

evaporation The transition of water from the liquid to the gaseous phase.

evolution Any gradual change. Most often refers to organic or Darwinian evolution, which is the genetic and resulting phenotypic change in populations of organisms from generation to generation. (See macroevolution, microevolution; contrast with speciation.)

evolutionary developmental biology (evo-devo) The study of the interplay between evolutionary and developmental processes, with a focus on the genetic changes that give rise to novel morphology. Key concepts of evo-devo include modularity, genetic toolkits, genetic switches, and heterochrony.

evolutionary radiation The proliferation of many species within a single evolutionary lineage.

evolutionary reversal The reappearance of an ancestral trait in a group that had previously acquired a derived trait.

evolutionary theory The understanding and application of the mechanisms of evolutionary change to biological problems.

excitatory Input from a neuron that causes depolarization of the recipient cell.

excited state The state of an atom or molecule when, after absorbing energy, it has more energy than in its normal, ground state.

excretion Release of metabolic wastes by an organism.

excretory systems In animals, organs that maintain the volume, solute concentration, and composition of the extracellular fluid by excreting water, solutes, and nitrogenous wastes in the form of urine.

exergonic A chemical reaction in which the products of the reaction have lower free energy than the reactants, resulting in a release of free energy. (Contrast with endergonic.)

exocrine gland (eks′ oh krin) [Gk. *exo*: outside + *krinein*: to separate] Any gland, such as a salivary gland, that secretes to the outside of the body or into the gut. (Contrast with endocrine gland.)

exocytosis A process by which a vesicle within a cell fuses with the plasma membrane and releases its contents to the outside. (Contrast with endocytosis.)

exon A portion of a DNA molecule, in eukaryotes, that codes for part of a polypeptide. (Contrast with intron.)

exoskeleton (eks′ oh skel′ e ton) [Gk. *exos*: outside + *skleros*: hard] A hard covering on the outside of the body to which muscles are attached. (Contrast with endoskeleton.)

exotoxin A highly toxic, usually soluble protein released by living, multiplying bacteria. (Contrast with endotoxin.)

experiment A testing process to support or disprove hypotheses and to answer questions. The basis of the scientific method. *See* comparative experiment, controlled experiment.

expiratory reserve volume The amount of air that can be forcefully exhaled beyond the normal tidal expiration. (Contrast with inspiratory reserve volume, tidal volume, vital capacity.)

exponential growth *See* multiplicative growth.

expression vector A DNA vector, such as a plasmid, that carries a DNA sequence for the expression of an inserted gene into mRNA and protein in a host cell.

expressivity The degree to which a genotype is expressed in the phenotype; may be affected by the environment.

extensor A muscle that extends an appendage.

external fertilization The release of gametes into the environment; typical of aquatic animals. Also called spawning. (Contrast with internal fertilization.)

external gills Highly branched and folded extensions of the body surface that provide a large surface area for gas exchange with water; typical of larval amphibians and many larval insects.

extinction The termination of a lineage of organisms.

extracellular matrix A material of heterogeneous composition surrounding cells and performing many functions including adhesion of cells.

extraembryonic membranes Four membranes that support but are not part of the developing embryos of reptiles, birds, and mammals, defining these groups phylogenetically as amniotes. (See amnion, allantois, chorion, and yolk sac.)

extreme halophiles A group of euryarchaeotes that live exclusively in very salty environments.

extremophiles Archaea and bacteria that live and thrive under conditions (e.g., extremely high temperatures) that would kill most organisms.

eye cups Photosensory organs in flatworms; components of one of the simplest visual systems in animals.

– F –

F-box A protein domain that facilitates protein–protein interactions necessary for protein degradation. Often involved in the regulation of gene expression (through the breakdown of regulatory proteins such as repressors), especially in plants.

F_1 The first filial generation; the immediate progeny of a parental (P) mating.

F_2 The second filial generation; the immediate progeny of a mating between members of the F_1 generation.

facilitated diffusion Passive movement through a membrane involving a specific carrier protein; does not proceed against a concentration gradient. (Contrast with active transport, diffusion.)

facultative anaerobe A prokaryote that can shift its metabolism between anaerobic and aerobic modes depending on the presence or absence of O_2. (Alternatively, facultative aerobe.)

fast-twitch fibers Skeletal muscle fibers that can generate high tension rapidly, but fatigue rapidly ("sprinter" fibers). Characterized by an abundance of enzymes of glycolysis. (Compare to slow-twitch fibers.)

fat A triglyceride that is solid at room temperature. (Contrast with oil.)

fate map A diagram of the blastula showing which cells (blastomeres) are "fated " to contribute to specific tissues and organs in the mature body.

fatty acid A molecule made up of a long nonpolar hydrocarbon chain and a polar carboxyl group. Found in many lipids.

fauna (faw′ nah) All the animals found in a given area. (Contrast with flora.)

FD (FLOWERING LOCUS D) Gene coding for a transcription factor in the shoot apical meristem that binds to florigen (FT); involved in the induction of flowering.

feces [L. *faeces*: dregs] Waste excreted from the digestive system.

fecundity The average number of offspring produced by each female.

feedback information In regulatory systems, information about the relationship between the set point of the system and its current state.

feedback inhibition   A mechanism for regulating a metabolic pathway in which the end product of the pathway can bind to and inhibit the enzyme that catalyzes the first committed step in the pathway. Also called end-product inhibition.

feedforward information   In regulatory systems, information that changes the set point of the system.

fermentation   (fur men tay' shun) [L. *fermentum*: yeast] The anaerobic degradation of a substance such as glucose to smaller molecules such as lactic acid or alcohol with the extraction of energy.

fertilization   Union of gametes. Also known as syngamy.

fertilizer   Any of a number of substances added to soil to improve the soil's capacity to support plant growth. May be organic or inorganic.

fetus   Medical and legal term for the stages of a developing human embryo from about the eighth week of pregnancy (the point at which all major organ systems have formed) to the moment of birth.

fiber   In angiosperms, an elongated, tapering sclerenchyma cell, usually with a thick cell wall, that serves as a support function in xylem. (*See also* muscle fiber.)

fibrin   A protein that polymerizes to form long threads that provide structure to a blood clot.

fibrinogen   A circulating protein that can be stimulated to fall out of solution and provide the structure for a blood clot.

fibrous root system   A root system typical of monocots composed of numerous thin adventitious roots that are all roughly equal in diameter. (Contrast with taproot system.)

Fick's law of diffusion   An equation that describes the factors that determine the rate of diffusion of a molecule from an area of higher concentration to an area of lower concentration.

fight-or-flight response   A rapid physiological response to a sudden threat mediated by the hormone epinephrine.

filament   In flowers, the part of a stamen that supports the anther.

filter feeder   An organism that feeds on organisms much smaller than itself that are suspended in water or air by means of a straining device.

first filial generation   *See* F₁.

first law of thermodynamics   The principle that energy can be neither created nor destroyed.

first polar body   In oogenesis, the daughter cell of the first meiotic division that receives almost no cytoplasm.

fission   *See* binary fission.

fitness   The contribution of a genotype or phenotype to the genetic composition of subsequent generations, relative to the

contribution of other genotypes or phenotypes. (*See* also inclusive fitness.)

fixed action pattern   In ethology, a genetically determined behavior that is performed without learning, stereotypic (performed the same way each time), and not modifiable by learning.

flagellum   (fla jell' um) (plural: flagella) [L. *flagellum*: whip] Long, whiplike appendage that propels cells. Prokaryotic flagella differ sharply from those found in eukaryotes.

flexor   A muscle that flexes an appendage.

flora   (flore' ah) All of the plants found in a given area. (Contrast with fauna.)

floral meristem   In angiosperms, a meristem that forms the floral organs (sepals, petals, stamens, and carpels).

floral organ identity genes   In angiosperms, genes that determine the fates of floral meristem cells; their expression is triggered by the products of meristem identity genes.

florigen   A plant hormone involved in the conversion of a vegetative shoot apex to a flower.

flower   The sexual structure of an angiosperm.

fluid feeder   An animal that feeds on fluids it extracts from the bodies of other organisms; examples include nectar-feeding birds and blood-sucking insects.

fluid mosaic model   A molecular model for the structure of biological membranes consisting of a fluid phospholipid bilayer in which suspended proteins are free to move in the plane of the bilayer.

flux   In ecology, the flow of an element into or out of a compartment of the biosphere.

follicle   [L. *folliculus*: little bag] In female mammals, an immature egg surrounded by nutritive cells.

follicle-stimulating hormone (FSH)   A gonadotropin produced by the anterior pituitary.

food vacuole   Membrane-enclosed structure formed by phagocytosis in which engulfed food particles are digested by the action of lysosomal enzymes.

food web   The complete set of food links between species in a community; a diagram indicating which ones are the eaters and which are eaten.

forebrain   The region of the vertebrate brain that comprises the cerebrum, thalamus, and hypothalamus.

fossil   Any recognizable structure originating from an organism, or any impression from such a structure, that has been preserved over geological time.

fossil fuels   Fuels, including oil, natural gas, coal, and peat, formed over geologic time from organic material buried in anaerobic sediments.

founder effect   Random changes in allele frequencies resulting from establishment

of a population by a very small number of individuals.

fovea   [L. *fovea*: a small pit] In the vertebrate retina, the area of most distinct vision.

frame-shift mutation   The addition or deletion of a single or two adjacent nucleotides in a gene's sequence. Results in the misreading of mRNA during translation and the production of a nonfunctional protein. (Contrast with missense mutation, nonsense mutation, silent mutation.)

Frank–Starling law   The stroke volume of the heart increases with increased return of blood to the heart.

free energy (G)   Energy that is available for useful work, after allowance is made for the increase or decrease of disorder.

frontal lobe   The largest of the brain lobes in humans; involved with feeling and planning functions; includes the primary motor cortex.

frugivore   [L. *frugis*: fruit + *vorare*: to devour] An animal that eats fruits.

fruit   In angiosperms, a ripened and mature ovary (or group of ovaries) containing the seeds. Sometimes applied to reproductive structures of other groups of plants.

FT (FLOWERING LOCUS T)   Gene that codes for florigen, a small, diffusible protein involved in the induction of flowering.

functional genomics   The assignment of functions to the proteins encoded by genes identified by sequencing entire genomes.

functional group   A characteristic combination of atoms that contribute specific properties when attached to larger molecules.

– G –

G cap   A chemically modified GTP added to the 5′ end of mRNA; facilitates binding of mRNA to ribosome and prevents mRNA breakdown.

G protein   A membrane protein involved in signal transduction; characterized by binding GDP or GTP.

G protein–linked receptors   A class of receptors that change configuration upon ligand binding such that a G protein binding site is exposed on the cytoplasmic domain of the receptor, initiating a signal transduction pathway.

G1   In the cell cycle, the gap between the end of mitosis and the onset of the S phase.

G1–S transition   In the cell cycle, the point at which G1 ends and the S phase begins.

G2   In the cell cycle, the gap between the S (synthesis) phase and the onset of mitosis.

gain of function mutation   A mutation that results in a protein with a new function. (Contrast with loss of function mutation.)

gallbladder   In the human digestive system, an organ in which bile is stored.

gametangium   (gam uh tan' gee um) (plural: gametangia) [Gk. *gamos*: marriage + *angeion*: vessel] Any plant or fungal structure within which a gamete is formed.

gamete   (gam' eet) [Gk. *gamete/gametes*: wife, husband] The mature sexual reproductive cell: the egg or the sperm.

gametogenesis   (ga meet' oh jen' e sis) The specialized series of cellular divisions that leads to the production of gametes. (*See also* oogenesis, spermatogenesis.)

gametophyte   (ga meet' oh fyte) In plants and photosynthetic protists with alternation of generations, the multicellular haploid phase that produces the gametes. (Contrast with sporophyte.)

ganglion   (gang' glee un) (plural: ganglia) [Gk. tumor] A cluster of neurons that have similar characteristics or function.

ganglion cells   Cells at the front of the human retina that transmit information from the bipolar cells to the brain.

gap genes   In *Drosophila* (fruit fly) development, segmentation genes that define broad areas along the anterior–posterior axis of the early embryo. Part of a developmental cascade that includes maternal effect genes, pair rule genes, segment polarity genes, and Hox genes.

gap junction   A 2.7-nanometer gap between plasma membranes of two animal cells, spanned by protein channels. Gap junctions allow chemical substances or electrical signals to pass from cell to cell.

gastric pits   Deep infoldings in the walls of the stomach lined with secretory cells.

gastrin   A hormone secreted by cells in the lower region of the stomach that stimulates the secretion of digestive juices as well as movements of the stomach.

gastrovascular cavity   Serving for both digestion (gastro) and circulation (vascular); in particular, the central cavity of the body of jellyfish and other cnidarians.

gastrulation   Development of a blastula into a gastrula. In embryonic development, the process by which a blastula is transformed by massive movements of cells into a *gastrula*, an embryo with three germ layers and distinct body axes.

gated channel   A membrane protein that changes its three-dimensional shape, and therefore its ion conductance, in response to a stimulus. When open, it allows specific ions to move across the membrane.

gel electrophoresis   (e lek' tro fo ree' sis) [L. *electrum*: amber + Gk. *phorein*: to bear] A technique for separating molecules (such as DNA fragments) from one another on the basis of their electric charges and molecular weights by applying an electric field to a gel.

gene   [Gk. *genes*: to produce] A unit of heredity. Used here as the unit of genetic function which carries the information for a polypeptide or RNA.

gene duplication   The generation of extra copies of a gene in a genome over evolutionary time. A mechanism by which genomes can acquire new functions.

gene family   A set of similar genes derived from a single parent gene; need not be on the same chromosomes. The vertebrate globin genes constitute a classic example of a gene family.

gene flow   Exchange of genes between populations through migration of individuals or movements of gametes.

gene pool   All of the different alleles of all of the genes existing in all individuals of a population.

gene tree   A graphic representation of the evolutionary relationships of a single gene in different species or of the members of a gene family.

gene-for-gene resistance   In plants, a mechanism of resistance to pathogens in which resistance is triggered by the specific interaction of the products of a pathogen's *Avr* genes and a plant's *R* genes.

general transcription factors   In eukaryotes, transcription factors that bind to the promoters of most protein-coding genes and are required for their expression. Distinct from transcription factors that have specific regulatory effects only at certain promoters or classes of promoters.

genetic code   The set of instructions, in the form of nucleotide triplets, that translate a linear sequence of nucleotides in mRNA into a linear sequence of amino acids in a protein.

genetic drift   Changes in gene frequencies from generation to generation as a result of random (chance) processes.

genetic linkage   Association between genes on the same chromosome such that they do not show random assortment and seldom recombine; the closer the genes, the lower the frequency of recombination.

genetic map   The positions of genes along a chromosome as revealed by recombination frequencies.

genetic marker   (1) In gene cloning, a gene of identifiable phenotype that indicates the presence of another gene, DNA segment, or chromosome fragment. (2) In general, a DNA sequence such as a single nucleotide polymorphism whose presence is correlated with the presence of other linked genes on that chromosome.

genetic screen   A technique for identifying genes involved in a biological process of interest. Involves creating a large collection of randomly mutated organisms and identifying those individuals that are likely to have a defect in the pathway of interest. The mutated gene(s) in those individuals can then be isolated for further study.

genetic structure   The frequencies of different alleles at each locus and the frequencies of different genotypes in a Mendelian population.

genetic switches   Mechanisms that control how the genetic toolkit is used, such as promoters and the transcription factors that bind them. The signal cascades that converge on and operate these switches

determine when and where genes will be turned on and off.

genetic toolkit   A set of developmental genes and proteins that is common to most animals and is hypothesized to be responsible for the evolution of their differing developmental pathways.

genetics   The scientific study of the structure, functioning, and inheritance of genes, the units of hereditary information.

genome   (jee' nome) The complete DNA sequence for a particular organism or individual.

genomic equivalence   The principle that no information is lost from the nuclei of cells as they pass through the early stages of embryonic development.

genomic library   All of the cloned DNA fragments generated by the breakdown of genomic DNA into smaller segments.

genomics   The scientific study of entire sets of genes and their interactions.

genotype   (jean' oh type) [Gk. *gen*: to produce + *typos*: impression] An exact description of the genetic constitution of an individual, either with respect to a single trait or with respect to a larger set of traits. (Contrast with phenotype.)

genotype frequency   The proportion of a genotype among individuals in a population.

genus   (jean' us) (plural: genera) [Gk. *genos*: stock, kind] A group of related, similar species recognized by taxonomists with a distinct name used in binomial nomenclature.

geographic range   The region within which a species occurs.

germ cell   [L. *germen*: to beget] A reproductive cell or gamete of a multicellular organism. (Contrast with somatic cell.)

germ layers   The three embryonic layers formed during gastrulation (ectoderm, mesoderm, and endoderm). Also called cell layers or tissue layers.

germ line mutation   Mutation in a cell that produces gametes (i.e., a germ line cell). (Contrast with somatic mutation.)

germination   Sprouting of a seed or spore.

gestation   (jes tay' shun) [L. *gestare*: to bear] The period during which the embryo of a mammal develops within the uterus. Also known as pregnancy.

ghrelin   A hormone produced and secreted by cells in the stomach that stimulates appetite.

gibberellin   (jib er el' lin) A class of plant growth hormones playing roles in stem elongation, seed germination, flowering of certain plants, etc.

gill   An organ specialized for gas exchange with water.

gizzard   (giz' erd) [L. *gigeria*: cooked chicken parts] The second of two stomachlike organs in birds, other reptiles, earthworms, and various insects, that grinds up food, sometimes with the aid of fragments of stone. (*See also* crop.)

glia   (glee' uh) [Gk. *glia*: glue] Cells of the nervous system that do not conduct action potentials.

glomerular filtration rate (GFR)   The rate at which the blood is filtered in the glomeruli of the kidney.

glomerulus   (glo mare' yew lus) [L. *glomus*: ball] Sites in the kidney where blood filtration takes place. Each glomerulus consists of a knot of capillaries served by afferent and efferent arterioles.

glucagon   Hormone produced by alpha cells of the pancreatic islets of Langerhans. Glucagon stimulates the liver to break down glycogen and release glucose into the circulation.

glucocorticoids   One of two main types of corticosteroid hormones. Produced and released by the cortex of the adrenal gland and involved in regulating blood glucose and fat, protein, and carbohydrate metabolism. (*See also* mineralocorticoids.)

gluconeogenesis   The biochemical synthesis of glucose from other substances, such as amino acids, lactate, and glycerol.

glucose   [Gk. *gleukos*: sugar, sweet] The most common monosaccharide; the monomer of the polysaccharides starch, glycogen, and cellulose.

glyceraldehyde 3-phosphate (G3P)   A phosphorylated three-carbon sugar; an intermediate in glycolysis and photosynthetic carbon fixation.

glycerol   (gliss' er ole) A three-carbon alcohol with three hydroxyl groups; a component of phospholipids and triglycerides.

glycogen   (gly' ko jen) An energy storage polysaccharide found in animals and fungi; a branched-chain polymer of glucose, similar to starch.

glycolipid   A lipid to which sugars are attached.

glycolysis   (gly kol' li sis) [Gk. *gleukos*: sugar + *lysis*: break apart] The enzymatic breakdown of glucose to pyruvic acid.

glycoprotein   A protein to which sugars are attached.

glycosidic linkage   Bond between carbohydrate (sugar) molecules through an intervening oxygen atom (–O–).

glycosylation   The addition of carbohydrates to another type of molecule, such as a protein.

glyoxysome   (gly ox' ee soam) An organelle found in plants, in which stored lipids are converted to carbohydrates.

Golgi apparatus   (goal' jee) A system of concentrically folded membranes found in the cytoplasm of eukaryotic cells; functions in secretion from the cell by exocytosis.

gonad   (go' nad) [Gk. *gone*: seed] An organ that produces gametes in animals: either an ovary (female gonad) or testis (male gonad).

gonadotropin   A type of trophic hormone that stimulates the gonads.

gonadotropin-releasing hormone (GnRH)   Hormone produced by the hypothalamus that stimulates the anterior pituitary to secrete ("release") gonadotropins.

Gondwana   The large southern land mass that existed from the Cambrian (540 mya) to the Jurassic (138 mya). Present-day remnants are South America, Africa, India, Australia, and Antarctica.

graded membrane potentials   Small local changes in membrane potential caused by opening or closing of ion channels.

grafting   Artificial transplantation of tissue from one organism to another. In horticulture, the transfer of a bud or stem segment from one plant onto the root of another as a form of asexual reproduction.

Gram stain   A differential purple stain useful in characterizing bacteria. The peptidoglycan-rich cell walls of Gram-positive bacteria stain purple; cell walls of Gram-negative bacteria generally stain orange.

gravitropism   [Gk. *tropos*: to turn] A directed plant growth response to gravity.

gray crescent   In frog development, a band of diffusely pigmented cytoplasm on the side of the egg opposite the site of sperm entry. Arises as a result of cytoplasmic rearrangements that establish the anterior–posterior axis of the zygote.

gray matter   In the nervous system, tissue that is rich in neuronal cell bodies. (Contrast with white matter.)

greenhouse effect   The warming of Earth that results from retention of heat in its atmosphere; caused by the presence of greenhouse gases in the atmosphere.

greenhouse gases   Gases in the atmosphere, such as carbon dioxide and methane, that are transparent to sunlight, but trap heat radiating from Earth's surface, causing heat to build up at Earth's surface.

gross primary productivity (GPP)   The rate at which the primary producers in a community turn solar energy into stored chemical energy via photosynthesis.

ground meristem   That part of an apical meristem that gives rise to the ground tissue system of the primary plant body.

ground tissue system   Those parts of the plant body not included in the dermal or vascular tissue systems. Ground tissues function in storage, photosynthesis, and support.

growth   An increase in the size of the body and its organs by cell division and cell expansion.

growth factor   A chemical signal that stimulates cells to divide.

growth hormone   A peptide hormone released by the anterior pituitary that stimulates many anabolic processes.

guanine (G)   (gwan' een) A nitrogen-containing base found in DNA, RNA, and GTP.

guard cells   In plants, specialized, paired epidermal cells that surround and control the opening of a stoma (pore). *See* stoma.

gustation   The sense of taste.

gut   An animal's digestive tract.

gymnosperms   Seed plants that do not produce flowers or fruits; one of the two major groups of living seed plants. (*See also* angiosperms.)

– H –

habitat patches   Areas of suitable habitat for a species that are separated by areas of unsuitable habitat.

habitat   The particular environment in which an organism lives. A *habitat patch* is an area of a particular habitat surrounded by other habitat types that may be less suitable for that organism.

Hadley cells   Patterns of vertical atmospheric circulation that influence surface winds and precipitation patterns according to latitude.

hair cell   A type of mechanoreceptor in animals. Detects sound waves and other forms of motion in air or water.

half-life   The time required for half of a sample of a radioactive isotope to decay to its stable, nonradioactive form, or for a drug or other substance to reach half its initial dosage.

halophyte   (hal' oh fyte) [Gk. *halos*: salt + *phyton*: plant] A plant that grows in a saline (salty) environment.

Hamilton's rule   The principle that, for an apparent altruistic behavior to be adaptive, the fitness benefit of that act to the recipient times the degree of relatedness of the performer and the recipient must be greater than the cost to the performer.

haplodiploidy   A sex determination mechanism in which diploid individuals (which develop from fertilized eggs) are female and haploid individuals (which develop from unfertilized eggs) are male; typical of hymenopterans.

haploid   (hap' loid) [Gk. *haploeides*: single] Having a chromosome complement consisting of just one copy of each chromosome; designated 1*n* or *n*. (Contrast with diploid.)

haplotype   Linked nucleotide sequences that are usually inherited as a unit (as a "sentence" rather than as individual "words").

Hardy–Weinberg equililbrium   In a sexually reproducing population, the allele frequency at a given locus that is not being acted on by agents of evolution; the conditions that would result in no evolution in a population.

haustorium   (haw stor' ee um) (plural: haustoria)[L. *haustus*: draw up] A specialized hypha or other structure by which fungi and some parasitic plants draw nutrients from a host plant.

heart   In circulatory systems, a muscular pump that moves extracellular fluid around the body.

heat budget equation   An expression of the balance between thermal energy flowing into the body of an animal from the environment and from metabolism, and the thermal energy leaving the body.

heat of vaporization   The energy that must be supplied to convert a molecule from a liquid to a gas at its boiling point.

heat shock proteins   Chaperone proteins expressed in cells exposed to high or low temperatures or other forms of environmental stress.

helical   Shaped like a screw or spring (helix); this shape occurs in DNA and proteins.

helper T cells   *See* T-helper cells.

hemiparasite   A parasitic plant that can photosynthesize, but derives water and mineral nutrients from the living body of another plant. (Contrast with holoparasite.)

hemizygous   (hem' ee zie' gus) [Gk. *hemi*: half + *zygotos*: joined] In a diploid organism, having only one allele for a given trait, typically the case for X-linked genes in male mammals and Z-linked genes in female birds. (Contrast with homozygous, heterozygous.)

hemoglobin   (hee' mo glow bin) [Gk. *heaema*: blood + L. *globus*: globe] Oxygen-transporting protein found in the red blood cells of vertebrates (and found in some invertebrates).

Hensen's node   In avian embryos, a structure at the anterior end of the primitive groove; determines the fates of cells passing over it during gastrulation.

hepatic   (heh pat' ik) [Gk. *hepar*: liver] Pertaining to the liver.

herbivore   (ur' bi vore) [L. *herba*: plant + *vorare*: to devour] An animal that eats plant tissues. (Contrast with carnivore, detritivore, omnivore.)

heritable trait   A trait that is at least partly determined by genes.

hermaphroditism   (her maf' row dite ism) The coexistence of both female and male sex organs in the same organism.

hetero-   [Gk.: *heteros*: other, different] A prefix indicating two or more different conditions, structures, or processes. (Contrast with homo-.)

heterochrony   Alteration in the timing of developmental events, leading to different results in the adult organism.

heterocyst   A large, thick-walled cell type in the filaments of certain cyanobacteria that performs nitrogen fixation.

heteromorphic   (het' er oh more' fik) [Gk. *heteros*: different + *morphe*: form] Having a different form or appearance, as two heteromorphic life stages of a plant. (Contrast with isomorphic.)

heterosis   The superior fitness of heterozygous offspring as compared with that of their dissimilar homozygous parents. Also called hybrid vigor.

heterosporous   (het' er os' por us) Producing two types of spores, one of which gives rise to a female megaspore and the other to a male microspore. (Contrast with homosporous.)

heterotroph   (het' er oh trof) [Gk. *heteros*: different + *trophe*: feed] An organism that requires preformed organic molecules as food. (Contrast with autotroph.)

heterotypic   Pertaining to adhesion of cells of different types.

heterozygous   (het' er oh zie' gus) [Gk. *heteros*: different + *zygotos*: joined] In diploid organisms, having different alleles of a given gene on the pair of homologs carrying that gene. (Contrast with homozygous.)

heterozygous carrier   An individual that carries a recessive allele for a phenotype of interest (e.g., a genetic disease); the individual does not show the phenotype, but may have progeny with the phenotype if the other parent also carries the recessive allele.

hexose   [Gk. *hex*: six] A sugar containing six carbon atoms.

hibernation   [L. *hibernum*: winter] The state of inactivity of some animals during winter; marked by a drop in body temperature and metabolic rate. (Contrast with estivation)

high-density lipoproteins (HDLs)   Lipoproteins that remove cholesterol from tissues and carry it to the liver; HDLs are the "good" lipoproteins associated with good cardiovascular health.

highly repetitive sequences   Short (less than 100 bp), nontranscribed DNA sequences, repeated thousands of times in tandem arrangements.

hindbrain   The region of the developing vertebrate brain that gives rise to the medulla, pons, and cerebellum.

hippocampus   [Gk. sea horse] A part of the forebrain that takes part in long-term memory formation.

histamine   (hiss' tah meen) A substance released by damaged tissue, or by mast cells in response to allergens. Histamine increases vascular permeability, leading to edema (swelling). (Contrast with histone deacetylase.)

histone acetyltransferases   Enzymes involved in chromatin remodeling. Add acetyl groups to the tail regions of histone proteins.

histone   Any one of a group of proteins forming the core of a nucleosome, the structural unit of a eukaryotic chromosome.

histone deacetylase   In chromatin remodeling, an enzyme that removes acetyl groups from the tails of histone proteins. (Contrast with histone acetyltransferases.)

HIV   Human immunodeficiency virus, the retrovirus that causes acquired immune deficiency syndrome (AIDS).

holoparasite   A fully parasitic plant (i.e., one that does not perform photosynthesis).

homeobox   180-base-pair segment of DNA found in certain homeotic genes; regulates the expression of other genes and thus controls large-scale developmental processes.

homeostasis   (home' ee o sta' sis) [Gk. *homos*: same + *stasis*: position] The maintenance of a steady state, such as a constant temperature, by means of physiological or behavioral feedback responses.

homeotic genes   Genes that act during development to determine the formation of an organ from a region of the embryo.

homeotic mutation   Mutation in a homeotic gene that results in the formation of a different organ than that normally made by a region of the embryo.

homing   In animal navigation, the ability to return to a nest site, burrow, or other specific location.

homo-   [Gk. *homos*: same] A prefix indicating two or more similar conditions, structures, or processes. (Contrast with hetero-.)

homolog   (1) In cytogenetics, one of a pair (or larger set) of chromosomes having the same overall genetic composition and sequence. In diploid organisms, each chromosome inherited from one parent is matched by an identical (except for mutational changes) chromosome—its homolog—from the other parent. (2) In evolutionary biology, one of two or more features in different species that are similar by reason of descent from a common ancestor.

homologous pair   A pair of matching chromosomes made up of a chromosome from each of the two sets of chromosomes in a diploid organism.

homologous recombination   Exchange of segments between two DNA molecules based on sequence similarity between the two molecules. The similar sequences align and crossover. Used to create knockout mutants in mice and other organisms.

homology   (ho mol' o jee) [Gk. *homologia*: of one mind; agreement] A similarity between two or more features that is due to inheritance from a common ancestor. The structures are said to be *homologous*, and each is a *homolog* of the others.

homoplasy   (home' uh play zee) [Gk. *homos*: same + *plastikos*: shape, mold] The presence in multiple groups of a trait that is not inherited from the common ancestor of those groups. Can result from convergent evolution, evolutionary reversal, or parallel evolution.

homosporous   Producing a single type of spore that gives rise to a single type of gametophyte, bearing both female and male reproductive organs. (Contrast with heterosporous.)

homozygous   (home' oh zie' gus) [Gk. *homos*: same + *zygotos*: joined] In diploid organisms, having identical alleles of a given gene on both homologous chromosomes. An individual may be a homozygote with respect to one gene and a heterozygote with respect to another. (Contrast with heterozygous.)

horizons   The horizontal layers of a soil profile, including the topsoil (A horizon), subsoil (B horizon) and parent rock or bedrock (C horizon).

hormone   (hore' mone) [Gk. *hormon*: to excite, stimulate] A chemical signal produced in minute amounts at one site in a multicellular organism and transported to another site where it acts on target cells.

host   An organism that harbors a parasite or symbiont and provides it with nourishment.

Hox genes   Conserved homeotic genes found in vertebrates, *Drosophila*, and other animal groups. Hox genes contain the homeobox domain and specify pattern and axis formation in these animals.

human chorionic gonadotropin (hCG)   A hormone secreted by the placenta which sustains the corpus luteum and helps maintain pregnancy.

Human Genome Project   A publicly and privately funded research effort, successfully completed in 2003, to produce a complete DNA sequence for the entire human genome.

humoral immune response   The response of the immune system mediated by B cells that produces circulating antibodies active against extracellular bacterial and viral infections. (Contrast with cellular immune response.)

humus   (hew' mus) The partly decomposed remains of plants and animals on the surface of a soil.

hybrid   (high' brid) [L. *hybrida*: mongrel] (1) The offspring of genetically dissimilar parents. (2) In molecular biology, a double helix formed of nucleic acids from different sources.

hybrid vigor   *See* heterosis.

hybrid zone   A region of overlap in the ranges of two closely related species where the species may hybridize.

hybridize   (1) In genetics, to combine the genetic material of two distinct species or of two distinguishable populations within a species. (2) In molecular biology, to form a double-stranded nucleic acid in which the two strands originate from different sources.

hydrocarbon   A compound containing only carbon and hydrogen atoms.

hydrogen bond   A weak electrostatic bond which arises from the attraction between the slight positive charge on a hydrogen atom and a slight negative charge on a nearby oxygen or nitrogen atom.

hydrological cycle   The movement of water from the oceans to the atmosphere, to the soil, and back to the oceans.

hydrolysis reaction   (high drol' uh sis) [Gk. *hydro*: water + *lysis*: break apart] A chemical reaction that breaks a bond by inserting the components of water (AB +

$H_2O \rightarrow AH + BOH$). (Contrast with condensation reaction.)

hydrophilic   (high dro fill' ik) [Gk. *hydro*: water + *philia*: love] Having an affinity for water. (Contrast with hydrophobic.)

hydrophobic   (high dro foe' bik) [Gk. *hydro*: water + *phobia*: fear] Having no affinity for water. Uncharged and nonpolar groups of atoms are hydrophobic. (Contrast with hydrophilic.)

hydroponic   Pertaining to a method of growing plants with their roots suspended in nutrient solutions instead of soil.

hydrostatic pressure   Pressure generated by compression of liquid in a confined space. Generated in plants, fungi, and some protists with cell walls by the osmotic uptake of water. Generated in animals with closed circulatory systems by the beating of a heart.

hydrostatic skeleton   A fluid-filled body cavity that transfers forces from one part of the body to another when acted on by surrounding muscles.

hydroxyl group   The —OH group found on alcohols and sugars.

hyper-   [Gk. *hyper*: above, over] Prefix indicating above, higher, more. (Contrast with hypo-.)

hyperaccumulators   Species of plants that store large quantities of heavy metals such as arsenic, cadmium, nickel, aluminum, and zinc.

hyperpolarization   A change in the resting potential across a membrane so that the inside of a cell becomes more negative compared with the outside of the cell. (Contrast with depolarization.)

hypersensitive response   A defensive response of plants to microbial infection in which phytoalexins and pathogenesis-related proteins are produced and the infected tissue undergoes apoptosis to isolate the pathogen from the rest of the plant.

hypertonic   Having a greater solute concentration. Said of one solution compared with another. (Contrast with hypotonic, isotonic.)

hypha   (high' fuh) (plural: hyphae) [Gk. *hyphe*: web] In the fungi and oomycetes, any single filament.

hypo-   [Gk. *hypo*: beneath, under] Prefix indicating underneath, below, less. (Contrast with hyper-.)

hypoblast   The lower tissue portion of the avian blastula which is joined to the epiblast at the margins of the blastodisc.

hypothalamus   The part of the brain lying below the thalamus; it coordinates water balance, reproduction, temperature regulation, and metabolism.

hypothesis   A tentative answer to a question, from which testable predictions can be generated. (Contrast with theory.)

hypotonic   Having a lesser solute concentration. Said of one solution in comparing

it to another. (Contrast with hypertonic, isotonic.)

hypoxia   A deficiency of oxygen.

– I –

ileum   The final segment of the small intestine. (*See also* duodenum, jejunum.)

imbibition   Water uptake by a seed; first step in germination.

immediate memory   A form of memory for events happening in the present that is almost perfectly photographic, but lasts only seconds.

immune system   [L. *immunis*: exempt from] A system in an animal that recognizes and attempts to eliminate or neutralize foreign substances such as bacteria, viruses, and pollutants.

immunity   In animals, the ability to avoid disease when invaded by a pathogen by deploying various defense mechanisms.

immunoassay   The use of antibodies to measure the concentration of an antigen in a sample.

immunoglobulins   A class of proteins containing a tetramer consisting of four polypeptide chains—two identical light chains and two identical heavy chains—held together by disulfide bonds; active as receptors and effectors in the immune system.

immunological memory   The capacity to more rapidly and massively respond to a second exposure to an antigen than occurred on first exposure.

imperfect flower   A flower lacking either functional stamens or functional carpels. (Contrast with perfect flower.)

implantation   The process by which the early mammalian embryo becomes attached to and embedded in the lining of the uterus.

imprinting   In animal behavior, a rapid form of learning in which an animal learns, during a brief critical period, to make a particular response, which is maintained for life, to some object or other organism.

in vitro   [L. in glass] A biological process occurring outside of the organism, in the laboratory. (Contrast with in vivo.)

in vitro evolution   A method based on natural molecular evolution that uses artificial selection in the laboratory to rapidly produce molecules with novel enzymatic and binding functions.

in vivo   [L. alive] A biological process occurring within a living organism or cell. (Contrast with in vitro.)

inbreeding   Breeding among close relatives.

inclusive fitness   The sum of an individual's genetic contribution to subsequent generations both via production of its own offspring and via its influence on the survival of relatives who are not direct descendants.

incomplete cleavage   A pattern of cleavage that occurs in many eggs that have a lot of yolk, in which the cleavage furrows do not penetrate all of it. (*See also* discoidal cleavage, superficial cleavage; contrast with complete cleavage.)

incomplete dominance Condition in which the heterozygous phenotype is intermediate between the two homozygous phenotypes.

incomplete metamorphosis Insect development in which changes between instars are gradual.

independent assortment During meiosis, the random separation of genes carried on nonhomologous chromosomes into gametes so that inheritance of these genes is random. This principle was articulated by Mendel as his second law.

indeterminate growth A open-ended growth pattern in which an organism or organ continues to grow as long as it lives; characteristic of some animals and of plant shoots and roots. (Contrast with determinate growth.)

individual fitness That component of inclusive fitness resulting from an organism producing its own offspring. (Contrast with kin selection.)

induced fit A change in the shape of an enzyme caused by binding to its substrate that exposes the active site of the enzyme.

induced mutation A mutation resulting from exposure to a mutagen from outside the cell. (Contrast with spontaneous mutation.)

induced pluripotent stem cells (iPS cells) Multipotent or pluripotent animal stem cells produced from differentiated cells in vitro by the addition of several genes that are expressed.

induced responses Defensive responses that a plant produces only in the presence of a pathogen, in contrast to constitutive defenses, which are always present.

inducer (1) A compound that stimulates the synthesis of a protein. (2) In embryonic development, a substance that causes a group of target cells to differentiate in a particular way.

inducible genes Genes that are expressed only when their products are needed. (Contrast with constitutive genes.)

inducible Produced only in the presence of a particular compound or under particular circumstances. (Contrast with constitutive.)

induction In embryonic development, the process by which a factor produced and secreted by certain cells determines the fates other cells.

inflammation A nonspecific defense against pathogens; characterized by redness, swelling, pain, and increased temperature.

inflorescence A structure composed of several to many flowers.

inflorescence meristem A meristem that produces floral meristems as well as other small leafy structures (bracts).

ingroup In a phylogenetic study, the group of organisms of primary interest. (Contrast with outgroup.)

inhibitor A substance that blocks a biological process.

inhibitory Input from a neuron that causes hyperpolarization of the recipient cell.

initiation complex In protein translation, a combination of a small ribosomal subunit, an mRNA molecule, and the tRNA charged with the first amino acid coded for by the mRNA; formed at the onset of translation.

initiation In molecular biology, the beginning of transcription or translation.

initiation site The place within a promoter where transcription begins.

innate immunity In animals, one of two general types of defenses against pathogens. Nonspecific and present in most animals. (Contrast with adaptive immunity.)

inner cell mass Derived from the mammalian blastula (bastocyst), the inner cell mass that will give rise to the yolk sac (via hypoblast) and embryo (via epiblast).

inorganic fertilizer A chemical or combination of chemicals applied to soil or plants to make up for a plant nutrient deficiency. Often contains the macronutrients nitrogen, phosphorus, and potassium (N-P-K).

inspiratory reserve volume The amount of air that can be inhaled above the normal tidal inspiration. (Contrast with expiratory reserve volume, tidal volume, vital capacity.)

instar (in′ star) An immature stage of an insect between molts.

insulin (in′ su lin) [L. *insula*: island] A hormone synthesized in islet cells of the pancreas that promotes the conversion of glucose into the storage material, glycogen.

integral membrane proteins Proteins that are at least partially embedded in the plasma membrane. (Contrast with peripheral membrane proteins.)

integrin In animals, a transmembrane protein that mediates the attachment of epithelial cells to the extracellular matrix.

integument [L. *integumentum*: covering] A protective surface structure. In gymnosperms and angiosperms, a layer of tissue around the ovule which will become the seed coat.

intercostal muscles Muscles between the ribs that can augment breathing movements by elevating and suppressing the rib cage.

interference RNA (RNAi) *See* RNA interference.

interferon A glycoprotein produced by virus-infected animal cells; increases the resistance of neighboring cells to the virus.

intermediate filaments Components of the cytoskeleton whose diameters fall between those of the larger microtubules and those of the smaller microfilaments.

internal environment In multicelluar organisms, the extracellular fluid surrounding the cells.

internal fertilization The release of sperm into the female reproductive tract; typical of most terrestrial animals. (Contrast with external fertilization.)

internal gills Gills enclosed in protective body cavities; typical of mollusks, arthropods, and fishes.

interneuron A neuron that communicates information between two other neurons.

internode The region between two nodes of a plant stem.

interphase In the cell cycle, the period between successive nuclear divisions during which the chromosomes are diffuse and the nuclear envelope is intact. During interphase the cell is most active in transcribing and translating genetic information.

interspecific competition Competition between members of two or more species. (Contrast with intraspecific competition.)

interspecific interactions Interactions between members of different species.

interstitial fluid Extracellular fluid that is not contained in the vessels of a circulatory system.

intertidal zone A nearshore region of oceans that is periodically exposed to the air as the tides rise and fall.

intestine The portion of the gut following the stomach, in which most digestion and absorption occurs.

intraspecific competition Competition among members of the same species. (Contrast with interspecific competition.)

intron Portion of a of a gene within the coding region that is transcribed into pre-mRNA but is spliced out prior to translation. (Contrast with exon.)

invasive species An exotic species that reproduces rapidly, spreads widely, and has negative effects on the native species of the region to which it has been introduced.

invasiveness The ability of a pathogen to multiply in a host's body. (Contrast with toxigenicity).

inversion A rare 180° reversal of the order of genes within a segment of a chromosome.

involution Cell movements that occur during gastrulation of frog embryos, giving rise to the archenteron.

ion (eye′ on) [Gk. *ion*: wanderer] An electrically charged particle that forms when an atom gains or loses one or more electrons.

ion channel An integral membrane protein that allows ions to diffuse across the membrane in which it is embedded.

ion exchange A process by which protons produced by a plant root displace mineral cations from clay particles in the surrounding soil.

ionic bond An electrostatic attraction between positively and negatively charged ions.

ionotropic receptors   A receptor that directly alters membrane permeability to a type of ion when it combines with its ligand.

iris   (eye' ris) [Gk. *iris*: rainbow] The round, pigmented membrane that surrounds the pupil of the eye and adjusts its aperture to regulate the amount of light entering the eye.

island biogeography   A theory proposing that the number of species on an island (or in another geographically defined and isolated area) represents a balance, or equilibrium, between the rate at which species immigrate to the island and the rate at which resident species go extinct.

islets of Langerhans   Clusters of hormone-producing cells in the pancreas.

iso-   [Gk. *iso*: equal] Prefix used for two separate entities that share some element of identity.

isomers   Molecules consisting of the same numbers and kinds of atoms, but differing in the bonding patterns by which the atoms are held together.

isomorphic   (eye so more' fik) [Gk. *isos*: equal + *morphe*: form] Having the same form or appearance, as when the haploid and diploid life stages of an organism appear identical. (Contrast with heteromorphic.)

isotonic   Having the same solute concentration; said of two solutions. (Contrast with hypertonic, hypotonic.)

isotope   (eye' so tope) [Gk. *isos*: equal + *topos*: place] Isotopes of a given chemical element have the same number of protons in their nuclei (and thus are in the same position on the periodic table), but differ in the number of neutrons.

isozymes   Enzymes of an organism that have somewhat different amino acid sequences but catalyze the same reaction.

– J –

jasmonate   Also called jasmonic acid, a plant hormone involved in triggering responses to pathogen attack as well as other processes.

jejunum   (jih jew' num) The middle division of the small intestine, where most absorption of nutrients occurs. (*See also* duodenum, ileum.)

jelly coat   The outer protective layer of a sea urchin egg, which triggers an acrosomal reaction in sperm.

joint   In skeletal systems, a junction between two or more bones.

juvenile hormone   In insects, a hormone maintaining larval growth and preventing maturation or pupation.

– K –

karyogamy   The fusion of nuclei of two cells. (Contrast with plasmogamy.)

karyotype   The number, forms, and types of chromosomes in a cell.

kidneys   A pair of excretory organs in vertebrates.

kilocalorie (kcal)   *See* calorie.

kin selection   That component of inclusive fitness resulting from helping the survival of relatives containing the same alleles by descent from a common ancestor. (Contrast with individual fitness.)

kinase   *See* protein kinase.

kinetic energy   (kuh-net' ik) [Gk. *kinetos*: moving] The energy associated with movement. (Contrast with potential energy.)

kinetochore   (kuh net' oh core) Specialized structure on a centromere to which microtubules attach.

knockout   A molecular genetic method in which a single gene of an organism is permanently inactivated.

Koch's postulates   A set of rules for establishing that a particular microorganism causes a particular disease.

Krebs cycle   *See* citric acid cycle.

– L –

lagging strand   In DNA replication, the daughter strand that is synthesized in discontinuous stretches. (*See* Okazaki fragments.)

large intestine   *See* colon.

larva   (plural: larvae) [L. *lares*: guiding spirits] An immature stage of any animal that differs dramatically in appearance from the adult.

lateral   [L. *latus*: side] Pertaining to the side.

lateral gene transfer   The transfer of genes from one species to another, common among bacteria and archaea.

lateral line   A sensory system in fishes consisting of a canal filled with water and hair cells running down each side under the surface of the skin, which senses disturbances in the surrounding water.

lateral meristem   Either of the two meristems, the vascular cambium and the cork cambium, that give rise to a plant's secondary growth.

lateral root   A root extending outward from the taproot in a taproot system; typical of eudicots.

lateralization   A phenomenon in humans in which language functions come to reside in one cerebral hemisphere, usually the left.

laticifers   (luh tiss' uh furs) In some plants, elongated cells containing secondary plant products such as latex.

Laurasia   The northernmost of the two large continents produced by the breakup of Pangaea.

law of independent assortment   *See* independent assortment.

law of segregation   *See* segregation.

laws of thermodynamics   [Gk. *thermos*: heat + *dynamis*: power] Laws derived from studies of the physical properties of energy and the ways energy interacts with matter. (*See also* first law of thermodynamics, second law of thermodynamics.)

leaching   In soils, a process by which mineral nutrients in upper soil horizons are dissolved in water and carried to deeper horizons, where they are unavailable to plant roots.

leading strand   In DNA replication, the daughter strand that is synthesized continuously. (Contrast with lagging strand.)

leaf   (plural: leaves) In plants, the chief organ of photosynthesis.

leaf primordium   (plural: primordia) An outgrowth on the side of the shoot apical meristem that will eventually develop into a leaf.

leghemoglobin   In nitrogen-fixing plants, an oxygen-carrying protein in the cytoplasm of nodule cells that transports enough oxygen to the nitrogen-fixing bacteria to support their respiration, while keeping free oxygen concentrations low enough to protect nitrogenase.

lek   A display ground within which male animals compete for and defend small display areas as a means of demonstrating their territorial prowess and winning opportunities to mate.

lens   In the vertebrate eye, a crystalline protein structure that makes fine adjustments in the focus of images falling on the retina.

lenticel   (len' ti sill) In plants, a spongy region in the periderm that allows gas exchange.

leptin   A hormone produced by fat cells that is believed to provide feedback information to the brain about the status of the body's fat reserves.

leukocyte   *See* white blood cell.

lichen   (lie' kun) An organism resulting from the symbiotic association of a fungus and either a cyanobacterium or a unicellular alga.

life cycle   The entire span of the life of an organism from the moment of fertilization (or asexual generation) to the time it reproduces in turn.

life history strategy   The way in which an organism partitions its time and energy among growth, maintenance, and reproduction.

life history   The time course of growth and development, reproduction, and death during an average individual organism's life.

life table   A summary of information about the progression of individuals in a population through the various stages of their life cycles.

ligament   A band of connective tissue linking two bones in a joint.

ligand   (lig' and) Any molecule that binds to a receptor site of another (usually larger) molecule.

light reactions   The initial phase of photosynthesis, in which light energy is converted into chemical energy.

light-harvesting complex   In photosynthesis, a group of different molecules that cooperate to absorb light energy and transfer it to a reaction center. Also called *antenna system*.

lignin   A complex, hydrophobic polyphenolic polymer in plant cell walls that cross-links other wall polymers, strengthening the walls, especially in wood.

limbic system   A group of evolutionarily primitive structures in the vertebrate telencephalon that are involved in emotions, drives, instinctive behaviors, learning, and memory.

limnetic zone   The open-water region of a lake.

lineage   A series of populations, species, or genes descended from a single ancestor over evolutionary time.

lineage species concept   The definition of a species as a branch on the tree of life, which has a history that starts at a speciation event and ends either at extinction or at another speciation event. (Contrast with biological species concept; morphological species concept.)

linkage   *See* genetic linkage.

lipase   (lip′ ase; lye′ pase) An enzyme that digests fats.

lipid   (lip′ id) [Gk. *lipos*: fat] Nonpolar, hydrophobic molecules that include fats, oils, waxes, steroids, and the phospholipids that make up biological membranes.

lipid bilayer   *See* phospholipid bilayer.

lipoproteins   Lipids packaged inside a covering of protein so that they can be circulated in the blood.

littoral zone   The nearshore region of a lake that is shallow and is affected by wave action and fluctuations in water level.

liver   A large digestive gland. In vertebrates, it secretes bile and is involved in the formation of blood.

loam   A type of soil consisting of a mixture of sand, silt, clay, and organic matter. One of the best soil types for agriculture.

locus   (low′ kus) (plural: loci, low′ sigh) In genetics, a specific location on a chromosome. May be considered synonymous with *gene*.

logistic growth   Growth, especially in the size of an organism or in the number of organisms in a population, that slows steadily as the entity approaches its maximum size. (Contrast with multiplicative growth.)

long-day plant (LDP)   A plant that requires long days (actually, short nights) in order to flower. (Compare to short-day plant.)

long-term potentiation (LTP)   A long-lasting increase in the responsiveness of a neuron resulting from a period of intense stimulation.

loop of Henle   (hen′ lee) Long, hairpin loop of the mammalian renal tubule that runs from the cortex down into the medulla and back to the cortex; creates a concentration gradient in the interstitial fluids in the medulla.

lophophore   A U-shaped fold of the body wall with hollow, ciliated tentacles that encircles the mouth of animals in several different groups. Used for filtering prey from the surrounding water.

loss of function mutation   A mutation that results in the loss of a functional protein. (Contrast with gain of function mutation.)

low-density lipoproteins (LDLs)   Lipoproteins that transport cholesterol around the body for use in biosynthesis and for storage; LDLs are the "bad" lipoproteins associated with a high risk of cardiovascular disease.

lumen   (loo′ men) [L. *lumen*: light] The open cavity inside any tubular organ or structure, such as the gut or a renal tubule.

lung   An internal organ specialized for respiratory gas exchange with air.

luteinizing hormone (LH)   A gonadotropin produced by the anterior pituitary that stimulates the gonads to produce sex hormones.

lymph   [L. *lympha*: liquid] A fluid derived from blood and other tissues that accumulates in intercellular spaces throughout the body and is returned to the blood by the lymphatic system.

lymph node   A specialized structure in the vessels of the lymphatic system. Lymph nodes contain lymphocytes, which encounter and respond to foreign cells and molecules in the lymph as it passes through the vessels.

lymphatic system   A system of vessels that returns interstitial fluid to the blood.

lymphocyte   One of the two major classes of white blood cells; includes T cells, B cells, and other cell types important in the immune system.

lymphoid tissues   Tissues of the immune system that are dispersed throughout the body, consisting of the thymus, spleen, bone marrow, and lymph nodes.

lysis   (lie′ sis) [Gk. *lysis*: break apart] Bursting of a cell.

lysogeny   A form of viral replication in which the virus becomes incorporated into the host chromosome and remains inactive. Also called a lysogenic cycle. (Contrast with lytic cycle.)

lysosome   (lie′ so soam) [Gk. *lysis*: break away + *soma*: body] A membrane-enclosed organelle originating from the Golgi apparatus and containing hydrolytic enzymes. (Contrast with secondary lysosome.)

lysozyme   (lie′ so zyme) An enzyme in saliva, tears, and nasal secretions that hydrolyzes bacterial cell walls.

lytic cycle   A viral reproductive cycle in which the virus takes over a host cell's synthetic machinery to replicate itself, then bursts (lyses) the host cell, releasing the new viruses. (Contrast with lysogeny.)

– M –

M phase   The portion of the cell cycle in which mitosis takes place.

macroevolution   [Gk. *makros*: large] Evolutionary changes occurring over long time spans and usually involving changes in many traits. (Contrast with microevolution.)

macromolecule   A giant (molecular weight > 1,000) polymeric molecule. The macromolecules are the proteins, polysaccharides, and nucleic acids.

macronutrient   In plants, a mineral element required in concentrations of at least 1 milligram per gram of plant dry matter; in animals, a mineral element required in large amounts. (Contrast with micronutrient.)

macrophage   (mac′ roh faj) Phagocyte that engulfs pathogens by endocytosis.

maintenance methylase   An enzyme that catalyzes the methylation of the new DNA strand when DNA is replicated.

major histocompatibility complex (MHC)   A complex of linked genes, with multiple alleles, that control a number of cell surface antigens that identify self and can lead to graft rejection.

malnutrition   A condition caused by lack of any essential nutrient.

Malpighian tubule   (mal pee′ gy un) A type of protonephridium found in insects.

map unit   The distance between two genes as calculated from genetic crosses; a recombination frequency.

marine   [L. *mare*: sea, ocean] Pertaining to or living in the ocean. (Contrast with aquatic, terrestrial.)

mass extinction   A period of evolutionary history during which rates of extinction are much higher than during intervening times.

mass number   The sum of the number of protons and neutrons in an atom's nucleus.

mast cells   Cells, typically found in connective tissue, that release histamine in response to tissue damage.

maternal effect genes   Genes coding for morphogens that determine the polarity of the egg and larva in fruit flies. Part of a developmental cascade that includes gap genes, pair rule genes, segment polarity genes, and Hox genes.

mating type   A particular strain of a species that is incapable of sexual reproduction with another member of the same strain but capable of sexual reproduction with members of other strains of the same species.

maximum likelihood   A statistical method of determining which of two or more hypotheses (such as phylogenetic trees) best fit the observed data, given an explicit model of how the data were generated.

mechanically gated channel   A molecular channel that opens or closes in response to mechanical force applied to the plasma membrane in which it is inserted.

mechanoreceptor   A cell that is sensitive to physical movement and generates action potentials in response.

medulla   (meh dull' luh) (1) The inner, core region of an organ, as in the adrenal medulla (adrenal gland) or the renal medulla (kidneys). (2) The portion of the brainstem that connects to the spinal cord.

medusa   (plural: medusae) In cnidarians, a free-swimming, sexual life cycle stage shaped like a bell or an umbrella.

megagametophyte   In heterosporous plants, the female gametophyte; produces eggs. (Contrast with microgametophyte.)

megaphyll   The generally large leaf of a fern, horsetail, or seed plant, with several to many veins. (Contrast with microphyll.)

megaspore   [Gk. *megas*: large + *spora*: to sow] In plants, a haploid spore that produces a female gametophyte.

megastrobilus   In conifers, the female (seed-bearing) cone. (Contrast with microstrobilus.)

meiosis   (my oh' sis) [Gk. *meiosis*: diminution] Division of a diploid nucleus to produce four haploid daughter cells. The process consists of two successive nuclear divisions with only one cycle of chromosome replication. In *meiosis I*, homologous chromosomes separate but retain their chromatids. The second division *meiosis II*, is similar to mitosis, in which chromatids separate.

melatonin   A hormone released by the pineal gland. Involved in photoperiodicity and circadian rhythms.

membrane potential   The difference in electrical charge between the inside and the outside of a cell, caused by a difference in the distribution of ions.

membranous bone   A type of bone that develops by forming on a scaffold of connective tissue. (Contrast with cartilage bone.)

memory cells   Long-lived lymphocytes produced after exposure to antigen. They persist in the body and are able to mount a rapid response to subsequent exposures to the antigen.

Mendel's laws   *See* independent assortment; segregation.

menopause   In human females, the end of fertility and menstrual cycling.

menstruation   The process by which the endometrium breaks down, and the sloughed-off tissue, including blood, flows from the body.

meristem   [Gk. *meristos*: divided] Plant tissue made up of undifferentiated actively dividing cells.

meristem culture   A method for the asexual propagation of plants, in which pieces of shoot apical meristem are cultured to produce plantlets.

meristem identity genes   In angiosperms, a group of genes whose expression initiates flower formation, probably by switching meristem cells from a vegetative to a reproductive fate.

mesenchyme   (mez' en kyme) [Gk. *mesos*: middle + *enchyma*: infusion] Embryonic or unspecialized cells derived from the mesoderm.

mesoderm   [Gk. *mesos*: middle + *derma*: skin] The middle of the three embryonic germ layers first delineated during gastrulation. Gives rise to the skeleton, circulatory system, muscles, excretory system, and most of the reproductive system.

mesoglea   (mez' uh glee uh) [Gk. *mesos*: middle + *gloia*, glue] A thick, gelatinous noncellular layer that separates the two cellular tissue layers of ctenophores, cnidarians, and scyphozoans.

mesophyll   (mez' uh fill) [Gk. *mesos*: middle + *phyllon*: leaf] Chloroplast-containing, photosynthetic cells in the interior of leaves.

messenger RNA (mRNA)   Transcript of a region of one of the strands of DNA; carries information (as a sequence of codons) for the synthesis of one or more proteins.

meta-   [Gk.: between, along with, beyond] Prefix denoting a change or a shift to a new form or level; for example, as used in metamorphosis.

metabolic pathway   A series of enzyme-catalyzed reactions so arranged that the product of one reaction is the substrate of the next.

metabolism   (meh tab' a lizm) [Gk. *metabole*: change] The sum total of the chemical reactions that occur in an organism, or some subset of that total (as in respiratory metabolism).

metabolome   The quantitative description of all the small molecules in a cell or organism.

metabolomics   The study of the metabolome as it relates to the physiological state of a cell or organism.

metabotropic receptor   A receptor that that indirectly alters membrane permeability to a type of ion when it combines with its ligand.

metagenomics   The practice of analyzing DNA from environmental samples without isolating intact organisms.

metamorphosis   (met' a mor' fo sis) [Gk. *meta*: between + *morphe*: form, shape] A change occurring between one developmental stage and another, as for example from a tadpole to a frog. (*See* complete metamorphosis, incomplete metamorphosis.)

metanephridia   The paired excretory organs of annelids.

metaphase   (met' a phase) The stage in nuclear division at which the centromeres of the highly supercoiled chromosomes are all lying on a plane (the metaphase plane or plate) perpendicular to a line connecting the division poles.

metapopulation   A population divided into subpopulations, among which there are occasional exchanges of individuals.

methylation   The addition of a methyl group (—CH$_3$) to a molecule.

MHC   *See* major histocompatibility complex.

micelle   A particle of lipid covered with bile salts that is produced in the duodenum and facilitates digestion and absorption of lipids.

microevolution   Evolutionary changes below the species level, affecting allele frequencies. (Contrast with macroevolution.)

microfibril   Crosslinked cellulose polymers, forming strong aggregates in the plant cell wall.

microfilament   In eukaryotic cells, a fibrous structure made up of actin monomers. Microfilaments play roles in the cytoskeleton, in cell movement, and in muscle contraction.

microgametophyte   In heterosporous plants, the male gametophyte; produces sperm. (Contrast with megagametophyte.)

microglia   Glial cells that act as macrophages and mediators of inflammatory responses in the central nervous system.

micronutrient   In plants, a mineral element required in concentrations of less than 100 micrograms per gram of plant dry matter; in animals, a mineral element required in concentrations of less than 100 micrograms per day. (Contrast with macronutrient.)

microphyll   A small leaf with a single vein, found in club mosses and their relatives. (Contrast with megaphyll.)

micropyle   (mike' roh pile) [Gk. *mikros*: small + *pylon*: gate] Opening in the integument(s) of a seed plant ovule through which pollen grows to reach the female gametophyte within.

microRNA   A small, noncoding RNA molecule, typically about 21 bases long, that binds to mRNA to inhibit its translation.

microspore   [Gk. *mikros*: small + *spora*: to sow] In plants, a haploid spore that produces a male gametophyte.

microstrobilus   In conifers, male pollen-bearing cone. (Contrast with megastrobilus.)

microtubules   Tubular structures found in centrioles, spindle apparatus, cilia, flagella, and cytoskeleton of eukaryotic cells. These tubules play roles in the motion and maintenance of shape of eukaryotic cells.

microvilli   (sing.: microvillus) Projections of epithelial cells, such as the cells lining the small intestine, that increase their surface area.

midbrain   One of the three regions of the vertebrate brain. Part of the brainstem, it serves as a relay station for sensory signals sent to the cerebral hemispheres.

middle lamella   (la mell' ah) [L. *lamina*: thin sheet] A layer of polysaccharides that separates plant cells; a shared middle lamella lies outside the primary walls of the two cells.

mineral nutrients Inorganic ions required by organisms for normal growth and reproduction.

mineralocorticoids One of two main types of corticosteroid hormones. Produced and released by the cortex of the adrenal gland and involved in regulating salt and water balance of the extracellular fluid. (See also glucocorticoids.)

mismatch repair A mechanism that scans DNA after it has been replicated and corrects any base-pairing mismatches.

missense mutation A change in a gene's sequence that changes the amino acid at that site in the encoded protein. (Contrast with frame-shift mutation, nonsense mutation, silent mutation.)

mitochondrial matrix The fluid interior of the mitochondrion, enclosed by the inner mitochondrial membrane.

mitochondrion (my' toe kon' dree un) (plural: mitochondria) [Gk. *mitos*: thread + *chondros*: grain] An organelle in eukaryotic cells that contains the enzymes of the citric acid cycle, the respiratory chain, and oxidative phosphorylation.

mitosis (my toe' sis) [Gk. *mitos*: thread] Nuclear division in eukaryotes leading to the formation of two daughter nuclei, each with a chromosome complement identical to that of the original nucleus.

model systems Also known as model organisms, these include the small group of species that are the subject of extensive research. They are organisms that adapt well to laboratory situations and findings from experiments on them can apply across a broad range of species. Classic examples include white rats and the fruit fly *Drosophila*.

moderately repetitive sequences DNA sequences repeated 10–1,000 times in the eukaryotic genome. They include the genes that code for rRNAs and tRNAs, as well as the DNA in telomeres.

Modern Synthesis An understanding of evolutionary biology that emerged in the early twentieth century as the principles of evolution were integrated with the principles of modern genetics.

modularity In evolutionary developmental biology, the principle that the molecular pathways that determine different developmental processes operate independently from one another. *See also* developmental module.

mole A quantity of a compound whose weight in grams is numerically equal to its molecular weight expressed in atomic mass units. Avogadro's number of molecules: 6.023×10^{23} molecules.

molecular clock The approximately constant rate of divergence of macromolecules from one another over evolutionary time; used to date past events in evolutionary history.

molecular evolution The scientific study of the mechanisms and consequences of the evolution of macromolecules.

molecular toolkit *See* genetic toolkit.

molecular weight The sum of the atomic weights of the atoms in a molecule.

molecule A chemical substance made up of two or more atoms joined by covalent bonds or ionic attractions.

molting The process of shedding part or all of an outer covering, as the shedding of feathers by birds or of the entire exoskeleton by arthropods.

monocots Angiosperms with a single embryonic cotyledon; one of the two largest clades of angiosperms. (*See also* eudicots.)

monoculture In agriculture, a large-scale planting of a single species of domesticated crop plant.

monoecious (mo nee' shus) [Gk. *mono*: one + *oikos*: house] Pertaining to organisms in which both sexes are "housed" in a single individual that produces both eggs and sperm. (In some plants, these are found in different flowers within the same plant.) Examples include corn, peas, earthworms, hydras. (Contrast with dioecious.)

monohybrid cross A mating in which the parents differ with respect to the alleles of only one locus of interest.

monomer [Gk. *mono*: one + *meros*: unit] A small molecule, two or more of which can be combined to form oligomers (consisting of a few monomers) or polymers (consisting of many monomers).

monophyletic (mon' oh fih leht' ik) [Gk. *mono*: one + *phylon*: tribe] Pertaining to a group that consists of an ancestor and all of its descendants. (Contrast with paraphyletic, polyphyletic.)

monosaccharide A simple sugar. Oligosaccharides and polysaccharides are made up of monosaccharides.

monosomic Pertaining to an organism with one less than the normal diploid number of chromosomes.

monosynaptic reflex A neural reflex that begins in a sensory neuron and makes a single synapse before activating a motor neuron.

morphogen A diffusible substance whose concentration gradient determines a developmental pattern in animals and plants.

morphogenesis (more' fo jen' e sis) [Gk. *morphe*: form + *genesis*: origin] The development of form; the overall consequence of determination, differentiation, and growth.

morphological species concept The definition of a species as a group of individuals that look alike. (Contrast with biological species concept; lineage species concept.)

morphology (more fol' o jee) [Gk. *morphe*: form + *logos*: study, discourse] The scientific study of organic form, including both its development and function.

mosaic development Pattern of animal embryonic development in which each blastomere contributes a specific part of the adult body. (Contrast with regulative development.)

motif *See* structural motif.

motile (mo' tul) Able to move from one place to another. (Contrast with sessile.)

motor cortex The region of the cerebral cortex that contains motor neurons that directly stimulate specific muscle fibers to contract.

motor end plate The depression in the postsynaptic membrane of the neuromuscular junction where the terminals of the motor neuron sit.

motor neuron A neuron carrying information from the central nervous system to a cell that produces movement.

motor proteins Specialized proteins that use energy to change shape and move cells or structures within cells.

motor unit A motor neuron and the muscle fibers it controls.

mouth An opening through which food is taken in, located at the anterior end of a tubular gut.

mRNA *See* messenger RNA.

mucosa The innermost layer of the vertebrate gut. (*See also* mucosal epithelium.)

mucosal epithelium An epithelial cell layer containing cells that secrete mucus; found in the digestive and respiratory tracts. Also called mucosa.

mucus A slippery substance secreted by mucous membranes (e.g., mucosal epithelium). A barrier defense against pathogens in innate immunity in animals.

multiplicative growth Population growth in which a constant multiple of the population size is added to the population during successive time intervals. Also known as exponential growth. (Contrast with additive growth.)

multipotent Having the ability to differentiate into a limited number of cell types. (Contrast with pluripotent, totipotent.)

muscle fiber A single muscle cell. In the case of skeletal muscle, a syncitial, multinucleate cell.

muscle tissue Excitable tissue that can contract through the interactions of actin and myosin; one of the four major tissue types in multicellular animals. There are three types of muscle tissue: skeletal, smooth, and cardiac.

mutagen (mute' ah jen) [L. *mutare*: change + Gk. *genesis*: source] Any agent (e.g., a chemical, radiation) that increases the mutation rate.

mutation A change in the genetic material not caused by recombination.

mutualism A type of interaction between species that benefits both species.

mycelium (my seel' ee yum) [Gk. *mykes*: fungus] In the fungi, a mass of hyphae.

mycorrhiza (my′ ko rye′ za) (plural: mycorrhizae) [Gk. *mykes*: fungus + *rhiza*: root] An association of the root of a plant with the mycelium of a fungus.

myelin (my′ a lin) Concentric layers of plasma membrane that form a sheath around some axons; myelin provides the axon with electrical insulation and increases the rate of transmission of action potentials.

MyoD The protein encoded by the *myo-blast determing* gene. A transcription factor involved in the differentiation of myoblasts (muscle precursor cells).

myofibril (my′ oh fy′ bril) [Gk. *mys*: muscle + L. *fibrilla*: small fiber] A polymeric unit of actin or myosin in a muscle.

myoglobin (my′ oh globe′ in) [Gk. *mys*: muscle + L. *globus*: sphere] An oxygen-binding molecule found in muscle. Consists of a heme unit and a single globin chain; carries less oxygen than hemoglobin.

myosin One of the two contractile proteins of muscle.

– N –

natural history The characteristics of a group of organisms, such as how the organisms get their food, reproduce, behave, regulate their internal environments (their cells, tissues, and organs), and interact with other organisms.

natural killer cell A type of lymphocyte that attacks virus-infected cells and some tumor cells as well as antibody-labeled target cells.

natural selection The differential contribution of offspring to the next generation by various genetic types belonging to the same population. The mechanism of evolution proposed by Charles Darwin.

necrosis (nec roh′ sis) [Gk. *nekros*: death] Premature cell death caused by external agents such as toxins.

negative feedback In regulatory systems, information that decreases a regulatory response, returning the system to the set point. (Contrast with positive feedback.)

negative regulation A type of gene regulation in which a gene is normally transcribed, and the binding of a repressor protein to the promoter prevents transcription. (Contrast with positive regulation.)

nematocyst (ne mat′ o sist) [Gk. *nema*: thread + *kystis*: cell] An elaborate, threadlike structure produced by cells of jellyfishes and other cnidarians, used chiefly to paralyze and capture prey.

nephron (nef′ ron) [Gk. *nephros*: kidney] The functional unit of the kidney, consisting of a structure for receiving a filtrate of blood and a tubule that reabsorbs selected parts of the filtrate.

Nernst equation A mathematical statement that calculates the potential across a membrane permeable to a single type of ion that differs in concentration on the two sides of the membrane.

nerve A structure consisting of many neuronal axons and connective tissue.

nervous tissue Tissue specialized for processing and communicating information; one of the four major tissue types in multicellular animals.

net primary productivity (NPP) The rate at which energy captured by photosynthesis is incorporated into the bodies of primary producers through growth and reproduction.

neural crest cells During vertebrate neurulation, cells that migrate outward from the neural plate and give rise to connections between the central nervous system and the rest of the body.

neural network An organized group of neurons that contains three functional categories of neurons—afferent neurons, interneurons, and efferent neurons—and is capable of processing information.

neural tube An early stage in the development of the vertebrate nervous system consisting of a hollow tube created by two opposing folds of the dorsal ectoderm along the anterior–posterior body axis.

neurohormone A chemical signal produced and released by neurons that subsequently acts as a hormone.

neuromuscular junction Synapse (point of contact) where a motor neuron axon stimulates a muscle fiber cell.

neuron (noor′ on) [Gk. *neuron*: nerve] A nervous system cell that can generate and conduct action potentials along an axon to a synapse with another cell.

neurotransmitter A substance produced in and released by a neuron (the presynaptic cell) that diffuses across a synapse and excites or inhibits another cell (the postsynaptic cell).

neurulation Stage in vertebrate development during which the nervous system begins to form.

neutral allele An allele that does not alter the functioning of the proteins for which it codes.

neutral theory A view of molecular evolution that postulates that most mutations do not affect the amino acid being coded for, and that such mutations accumulate in a population at rates driven by genetic drift and mutation rates.

neutron (new′ tron) One of the three fundamental particles of matter (along with protons and electrons), with mass slightly larger than that of a proton and no electrical charge.

next-generation DNA sequencing A rapid, relatively low cost method for sequencing DNA that involves attaching short, single-stranded DNA fragments to a solid surface. A primer and DNA polymerase are added, and tagged nucleotides are detected by a laser as they are added to the complementary DNA strand. Enables

the analysis of many sequences in parallel, which are then assembled by a computer.

niche (nitch) [L. *nidus*: nest] The set of physical and biological conditions a species requires to survive, grow, and reproduce.

nitrate reduction The process by which nitrate (NO_3^-) is reduced to ammonia (NH_3).

nitric oxide (NO) An unstable molecule (a gas) that serves as a second messenger causing smooth muscle to relax. In the nervous system it operates as a neurotransmitter.

nitrifiers Chemolithotrophic bacteria that oxidize ammonia to nitrate in soil and in seawater.

nitrogen fixation Conversion of atmospheric nitrogen gas (N_2) into a more reactive and biologically useful form (ammonia), which makes nitrogen available to living things. Carried out by nitrogen-fixing bacteria, some of them free-living and others living within plant roots.

nitrogenase An enzyme complex found in nitrogen-fixing bacteria that mediates the stepwise reduction of atmospheric N_2 to ammonia and which is strongly inhibited by oxygen.

nitrogenous wastes The potentially toxic nitrogen-containing end products of protein and nucleic acid catabolism. Eliminated from the body by excretion.

node [L. *nodus*: knob, knot] In plants, a (sometimes enlarged) point on a stem where a leaf is or was attached.

node of Ranvier A gap in the myelin sheath covering an axon; the point where the axonal membrane can fire action potentials.

nodule A specialized structure in the roots of nitrogen-fixing plants that houses nitrogen-fixing bacteria, in which oxygen is maintained at a low level by leghemoglobin.

non-REM sleep A state of deep, restorative sleep characterized by high-amplitude slow waves in the EEG. (Contrast with REM sleep.)

noncompetitive inhibitor A nonsubstrate that inhibits the activity of an enzyme by binding to a site other than its active site. (Contrast with competitive inhibitor.)

noncyclic electron transport In photosynthesis, the flow of electrons that forms ATP, NADPH, and O_2.

nondisjunction Failure of sister chromatids to separate in meiosis II or mitosis, or failure of homologous chromosomes to separate in meiosis I. Results in aneuploidy.

nonpolar Having electric charges that are evenly balanced from one end to the other. (Contrast with polar.)

nonrandom mating Selection of mates on the basis of a particular trait or group of traits.

nonsense mutation Change in a gene's sequence that prematurely terminates translation by changing one of its codons to a stop codon.

nonsynonymous substitution A change in a gene from one nucleotide to another that changes the amino acid specified by the corresponding codon (i.e., AGC → AGA, or serine → arginine). (Contrast with synonymous substitution.)

norepinephrine A neurotransmitter found in the central nervous system and also at the postganglionic nerve endings of the sympathetic nervous system. Also called noradrenaline.

normal flora Microorganisms that normally live and reproduce on or in the body without causing disease, and which form a nonspecific defense against pathogens by competing with them for space and nutrients.

notochord (no' tow kord) [Gk. *notos*: back + *chorde*: string] A flexible rod of gelatinous material serving as a support in the embryos of all chordates and in the adults of tunicates and lancelets.

nucleic acid (new klay' ik) A polymer made up of nucleotides, specialized for the storage, transmission, and expression of genetic information. DNA and RNA are nucleic acids.

nucleic acid hybridization A technique in which a single-stranded nucleic acid probe is made that is complementary to, and binds to, a target sequence, either DNA or RNA. The resulting double-stranded molecule is a hybrid.

nucleoid (new' klee oid) The region that harbors the chromosomes of a prokaryotic cell. Unlike the eukaryotic nucleus, it is not bounded by a membrane.

nucleolus (new klee' oh lus) A small, generally spherical body found within the nucleus of eukaryotic cells. The site of synthesis of ribosomal RNA.

nucleoside A nucleotide without the phosphate group; a nitrogenous base attached to a sugar.

nucleosome A portion of a eukaryotic chromosome, consisting of part of the DNA molecule wrapped around a group of histone molecules, and held together by another type of histone molecule. The chromosome is made up of many nucleosomes.

nucleotide substitution A change of one base pair to another in a DNA sequence.

nucleotide The basic chemical unit in nucleic acids, consisting of a pentose sugar, a phosphate group, and a nitrogen-containing base.

nucleus (new' klee us) [L. *nux*: kernel or nut] (1) In cells, the centrally located compartment of eukaryotic cells that is bounded by a double membrane and contains the chromosomes. (2) In the brain, an identifiable group of neurons that share common characteristics or functions.

null hypothesis In statistics, the premise that any differences observed in an experiment are simply the result of random differences that arise from drawing two finite samples from the same population.

nutrient A food substance; or, in the case of mineral nutrients, an inorganic element required for completion of the life cycle of an organism.

– O –

obligate anaerobe An anaerobic prokaryote that cannot survive exposure to O_2.

occipital lobe One of the four lobes of the brain's cerebral hemisphere; processes visual information.

odorant A molecule that can bind to an olfactory receptor.

oil A triglyceride that is liquid at room temperature. (Contrast with fat.)

Okazaki fragments Newly formed DNA making up the lagging strand in DNA replication. DNA ligase links Okazaki fragments together to give a continuous strand.

olfactory [L. *olfacere*: to smell] Pertaining to the sense of smell (*olfaction*).

oligodendrocyte A type of glial cell that myelinates axons in the central nervous system.

oligosaccharide A polymer containing a small number of monosaccharides.

omasum One of the four chambers of the stomach in ruminants; concentrates food by water absorption before it enters the true stomach (abomasum).

ommatidia [Gk. *omma*: eye] The units that make up the compound eye of some arthropods.

omnivore [L. *omnis*: everything + *vorare*: to devour] An organism that eats both animal and plant material. (Contrast with carnivore, detritivore, herbivore.)

oncogene [Gk. *onkos*: mass, tumor + *genes*: born] A gene that codes for a protein product that stimulates cell proliferation. Mutations in oncogenes that result in excessive cell proliferation can give rise to cancer.

one gene–one polypeptide The idea, since shown to be an oversimplification, that each gene in the genome encodes only a single polypeptide—that there is a one-to-one correspondence between genes and polypeptides.

oocyte *See* primary oocyte, secondary oocyte.

oogenesis (oh' eh jen e sis) [Gk. *oon*: egg + *genesis*: source] Gametogenesis leading to production of an ovum.

oogonium (oh' eh go' nee um) (plural: oogonia) (1) In some algae and fungi, a cell in which an egg is produced. (2) In animals, the diploid progeny of a germ cell in females.

ootid In oogenesis, the daughter cell of the second meiotic division that differentiates into the mature ovum.

open circulatory system Circulatory system in which extracellular fluid leaves the vessels of the circulatory system, percolates between cells and through tissues, and then flows back into the circulatory system to be pumped out again. (Contrast with closed circulatory system.)

operator The region of an operon that acts as the binding site for the repressor.

operon A genetic unit of transcription, typically consisting of several structural genes that are transcribed together; the operon contains at least two control regions: the promoter and the operator.

opportunity cost The sum of the benefits an animal forfeits by not being able to perform some other behavior during the time when it is performing a given behavior.

opsin (op' sin) [Gk. *opsis*: sight] The protein portion of the visual pigment rhodopsin; associated with the pigment molecule 11-*cis*-retinal. (*See* rhodopsin.)

orbital A region in space surrounding the atomic nucleus in which an electron is most likely to be found.

organ [Gk. *organon*: tool] A body part, such as the heart, liver, brain, root, or leaf. Organs are composed of different tissues integrated to perform a distinct function. Organs, in turn, are integrated into organ systems.

organ identity genes In angiosperms, genes that specify the different organs of the flower. (Compare with homeotic genes.)

organ of Corti Structure in the inner ear that transforms mechanical forces produced from pressure waves ("sound waves") into action potentials that are sensed as sound.

organ system An interrelated and integrated group of tissues and organs that work together in a physiological function.

organelle (or gan el') Any of the membrane-enclosed structures within a eukaryotic cell. Examples include the nucleus, endoplasmic reticulum, and mitochondria.

organic (1) Pertaining to any chemical compound that contains carbon. (2) Pertaining to any aspect of living matter, e.g., to its evolution, structure, or chemistry.

organic fertilizers Substances added to soil to improve the soil's fertility; derived from partially decomposed plant material (compost) or animal waste (manure).

organism Any living entity.

organizer Region of the early amphibian embryo that directs early embryonic development. Also known as the primary embryonic organizer.

organogenesis The formation of organs and organ systems during development.

origin of replication (*ori*) DNA sequence at which helicase unwinds the DNA double helix and DNA polymerase binds to initiate DNA replication.

orthologs Genes that are related by orthology.

orthology (or thol' o jee) Type of homology in which the divergence of homologous genes can be traced to speciation events.

osmoconformer An aquatic animal that equilibrates the osmolarity of its extracellular fluid to be the same as that of the external environment. (Contrast with osmoregulator.)

osmolarity   The concentration of osmotically active particles in a solution.

osmoregulation   Regulation of the chemical composition of the body fluids of an organism.

osmoregulator   An aquatic animal that actively regulates the osmolarity of its extracellular fluid. (Contrast with osmoconformer.)

osmosis   (oz mo' sis) [Gk. *osmos*: to push] Movement of water across a differentially permeable membrane, from one region to another region where the water potential is more negative.

ossicle   (oss' ick ul) [L. *os*: bone] The calcified construction unit of echinoderm skeletons.

osteoblast   (oss' tee oh blast) [Gk. *osteon*: bone + *blastos*: sprout] A cell that lays down the protein matrix of bone.

osteoclast   (oss' tee oh clast) [Gk. *osteon*: bone + *klastos*: broken] A cell that dissolves bone.

osteocyte   An osteoblast that has become enclosed in lacunae within the bone it has built.

outgroup   In phylogenetics, a group of organisms used as a point of reference for comparison with the groups of primary interest (the ingroup).

oval window   The flexible membrane that, when moved by the bones of the middle ear, produces pressure waves in the inner ear.

ovarian cycle   In human females, the monthly cycle of events by which eggs and hormones are produced. (Contrast with uterine cycle).

ovary   (oh' var ee) [L. *ovum*: egg] Any female organ, in plants or animals, that produces an egg.

overtopping   Plant growth pattern in which one branch differentiates from and grows beyond the others.

oviduct   In mammals, the tube serving to transport eggs to the uterus or to the outside of the body.

oviparity   Reproduction in which eggs are released by the female and development is external to the mother's body. (Contrast with viviparity.)

ovulation   Release of an egg from an ovary.

ovule   (oh' vule) In plants, a structure comprising the megasporangium and the integument, which develops into a seed after fertilization.

ovum   (oh' vum) (plural: ova) [L. egg] The female gamete.

oxidation   (ox i day' shun) Relative loss of electrons in a chemical reaction; either outright removal to form an ion, or the sharing of electrons with substances having a greater affinity for them, such as oxygen.

Most oxidations, including biological ones, are associated with the liberation of energy. (Contrast with reduction.)

oxidative phosphorylation   ATP formation in the mitochondrion, associated with flow of electrons through the respiratory chain.

oxygenase   An enzyme that catalyzes the addition of oxygen to a substrate from O_2.

oxytocin   A hormone released by the posterior pituitary that promotes social bonding.

– P –

pacemaker cells   Cardiac cells that can initiate action potentials without stimulation from the nervous system, allowing the heart to initiate its own contractions.

pair rule genes   In *Drosophila* (fruit fly) development, segmentation genes that divide the early embryo into units of two segments each. Part of a developmental cascade that includes maternal effect genes, gap genes, segment polarity genes, and Hox genes.

paleomagnetic dating   A method for determining the age of rocks based on properties relating to changes in the patterns of Earth's magnetism over time.

pancreas   (pan' cree us) A gland located near the stomach of vertebrates that secretes digestive enzymes into the small intestine and releases insulin into the bloodstream.

Pangaea   (pan jee' uh) [Gk. *pan*: all, every] The single land mass formed when all the continents came together in the Permian period.

para-   [Gk. *para*: akin to, beside] Prefix indicating association in being along side or accessory to.

parabronchi   Passages in the lungs of birds through which air flows.

paracrine   [Gk. *para*: near] Pertaining to a chemical signal, such as a hormone, that acts locally, near the site of its secretion. (Contrast with autocrine.)

parallel phenotypic evolution   The repeated evolution of similar traits, especially among closely related species; facilitated by conserved developmental genes.

paraphyletic   (par' a fih leht' ik) [Gk. *para*: beside + *phylon*: tribe] Pertaining to a group that consists of an ancestor and some, but not all, of its descendants. (Contrast with monophyletic, polyphyletic.)

parasite   An organism that consumes parts of an organism much larger than itself (known as its host). Parasites sometimes, but not always, kill their host.

parasympathetic nervous system   The division of the autonomic nervous system that works in opposition to the sympathetic nervous system. (Contrast with sympathetic nervous system.)

parathyroid glands   Four glands on the posterior surface of the thyroid gland that produce and release parathyroid hormone.

parathyroid hormone (PTH)   A hormone secreted by the parathyroid glands that stimulates osteoclast activity and raises blood calcium levels. Also called parathormone.

parenchyma   (pair eng' kyma) A plant tissue composed of relatively unspecialized cells without secondary walls.

parent rock   The soil horizon consisting of the rock that is breaking down to form the soil. Also called bedrock, or the C horizon.

parental (P) generation   The individuals that mate in a genetic cross. Their offspring are the first filial (F_1) generation.

parietal lobe   One of four lobes of the cerebral hemisphere; processes complex stimuli and includes the primary somatosensory cortex.

parsimony   Preferring the simplest among a set of plausible explanations of any phenomenon.

parthenocarpy   Formation of fruit from a flower without fertilization.

parthenogenesis   [Gk. *parthenos*: virgin] Production of an organism from an unfertilized egg.

partial pressure of oxygen   The contribution of O_2 to total air pressure; about 159 mm Hg at sea level.

particulate theory   In genetics, the theory that genes are physical entities that retain their identities after fertilization.

passive transport   Diffusion across a membrane; may or may not require a channel or carrier protein. (Contrast with active transport.)

pathogen   (path' o jen) [Gk. *pathos*: suffering + *genesis*: source] An organism that causes disease.

pathogenesis-related (PR) proteins   A component of the hypersensitive response of plants to pathogens. Some PR proteins function as direct defenses against pathogens, others initiate other defensive responses.

pattern formation   In animal embryonic development, the organization of differentiated tissues into specific structures such as wings.

pedigree   The pattern of transmission of a genetic trait within a family.

pelagic zone   [Gk. *pelagos*: sea] The open ocean.

penetrance   The proportion of individuals with a particular genotype that show the expected phenotype.

penis   An accessory sex organ of male animals that enables the male to deposit sperm in the female's reproductive tract.

pentaradial symmetry   Symmetry in five or multiples of five; a feature of adult echinoderms.

pentose   [Gk. *penta*: five] A sugar containing five carbon atoms.

pepsin   [Gk. *pepsis*: digestion] An enzyme in gastric juice that digests protein.

pepsinogen Inactive secretory product that is converted into pepsin by low pH or by enzymatic action.

peptide hormones Relatively large hormone molecules made up of amino acids; encoded by genes and produced by translation.

peptide linkage The bond between amino acids in a protein; formed between a carboxyl group and amino group (—CO—NH—) with the loss of water molecules.

peptidoglycan The cell wall material of many bacteria, consisting of a single enormous molecule that surrounds the entire cell.

peptidyl transferase A catalytic function of the large ribosomal subunit that consists of two reactions: breaking the bond between an amino acid and its tRNA in the P site, and forming a peptide bond between that amino acid and the amino acid attached to the tRNA in the A site.

per capita birth rate (b) In population growth models, the number of offspring that an average individual produces in some time interval.

per capita death rate (d) In population growth models, the average individual's chance of dying in some time interval.

per capita growth rate (r) In population models, the average individual's contribution to total population growth rate.

perennial (per ren' ee al) [L. *per*: throughout + *annus*: year] A plant that survives from year to year. (Contrast with annual, biennial.)

perfect flower A flower with both stamens and carpels; a hermaphroditic flower. (Contrast with imperfect flower.)

pericycle [Gk. *peri*: around + *kyklos*: ring or circle] In plant roots, tissue just within the endodermis, but outside of the root vascular tissue. Meristematic activity of pericycle cells produces lateral root primordia.

periderm The outer tissue of the secondary plant body, consisting primarily of cork.

period (1) A category in the geological time scale. (2) The duration of a single cycle in a cyclical event, such as a circadian rhythm.

peripheral membrane proteins Proteins associated with but not embedded within the plasma membrane. (Contrast with integral membrane proteins.)

peripheral nervous system (PNS) The portion of the nervous system that transmits information to and from the central nervous system, consisting of neurons that extend or reside outside the brain or spinal cord and their supporting cells. (Contrast with central nervous system.)

peristalsis (pair' i stall' sis) Wavelike muscular contractions proceeding along a tubular organ, propelling the contents along the tube.

peritoneum The mesodermal lining of the body cavity in coelomate animals.

peroxisome An organelle that houses reactions in which toxic peroxides are formed and then converted to water.

petal [Gk. *petalon*: spread out] In an angiosperm flower, a sterile modified leaf, nonphotosynthetic, frequently brightly colored, and often serving to attract pollinating insects.

petiole (pet' ee ole) [L. *petiolus*: small foot] The stalk of a leaf.

P$_{fr}$ *See* phytochrome.

pH The negative logarithm of the hydrogen ion concentration; a measure of the acidity of a solution. A solution with pH = 7 is said to be neutral; pH values higher than 7 characterize basic solutions, while acidic solutions have pH values less than 7.

phage (fayj) *See* bacteriophage.

phagocyte [Gk. *phagein*: to eat + *kystos*: sac] One of two major classes of white blood cells; one of the nonspecific defenses of animals; ingests invading microorganisms by phagocytosis.

phagocytosis Endocytosis by a cell of another cell or large particle.

pharmacogenomics The study of how an individual's genetic makeup affects his or her response to drugs or other agents, with the goal of predicting the effectiveness of different treatment options.

pharming The use of genetically modified animals to produce medically useful products in their milk.

pharynx [Gk. throat] The part of the gut between the mouth and the esophagus.

phenotype (fee' no type) [Gk. *phanein*: to show] The observable properties of an individual resulting from both genetic and environmental factors. (Contrast with genotype.)

pheromone (feer' o mone) [Gk. *pheros*: carry + *hormon*: excite, arouse] A chemical substance used in communication between organisms of the same species.

phloem (flo' um) [Gk. *phloos*: bark] In vascular plants, the vascular tissue that transports sugars and other solutes from sources to sinks.

phosphate group The functional group —OPO$_3$H$_2$.

phosphodiester linkage The connection in a nucleic acid strand, formed by linking two nucleotides.

phospholipid A lipid containing a phosphate group; an important constituent of cellular membranes. (*See* lipid.)

phospholipid bilayer The basic structural unit of biological membranes; a sheet of phospholipids two molecules thick in which the phospholipids are lined up with their hydrophobic "tails" packed tightly together and their hydrophilic, phosphate-containing "heads" facing outward. Also called lipid bilayer.

phosphorylation Addition of a phosphate group.

photic zone The region of lakes and oceans that is penetrated by light and therefore supports photosynthetic organisms.

photoautotroph An organism that obtains energy from light and carbon from carbon dioxide. (Contrast with chemolithotroph, chemoheterotroph, photoheterotroph.)

photoheterotroph An organism that obtains energy from light but must obtain its carbon from organic compounds. (Contrast with chemolithotroph, chemoheterotroph, photoautotroph.)

photomorphogenesis In plants, a process by which physiological and developmental events are controlled by light.

photon (foe' ton) [Gk. *photos*: light] A quantum of visible radiation; a "packet" of light energy.

photoperiodicity Control of an organism's physiological or behavioral responses by the length of the day or night (photoperiod).

photoreceptor (1) In plants, a pigment that triggers a physiological response when it absorbs a photon. (2) In animals, a sensory receptor cell that senses and responds to light energy.

photosensitivity The ability to detect light.

photosynthesis (foe tow sin' the sis) [literally, "synthesis from light"] Metabolic processes carried out by green plants and cyanobacteria, by which visible light is trapped and the energy used to convert CO$_2$ into organic compounds.

photosystem [Gk. *phos*: light + *systema*: assembly] A light-harvesting complex in the chloroplast thylakoid composed of pigments and proteins.

photosystem I In photosynthesis, the complex that absorbs light at 700 nm, passing electrons to ferrodoxin and thence to NADPH.

photosystem II In photosynthesis, the complex that absorbs light at 680 nm, passing electrons to the electron transport chain in the chloroplast.

phototropins A class of blue light receptors that mediate phototropism and other plant responses.

phototropism [Gk. *photos*: light + *trope*: turning] A directed plant growth response to light.

phycobilin Photosynthetic pigment that absorbs red, yellow, orange, and green light and is found in cyanobacteria and some red algae.

phylogenetic tree A graphic representation of lines of descent among organisms or their genes.

phylogeny (fy loj' e nee) [Gk. *phylon*: tribe, race + *genesis*: source] The evolutionary history of a particular group of organisms or their genes.

physical geography The study of the distributions of Earth's climates and surface features.

physiology (fiz' ee ol' o jee) [Gk. *physis*: natural form] The scientific study of the functions of living organisms and the individual organs, tissues, and cells of which they are composed.

phytoalexins Substances toxic to pathogens, produced by plants in response to fungal or bacterial infection.

phytochrome (fy' tow krome) [Gk. *phyton*: plant + *chroma*: color] A plant pigment regulating a large number of developmental and other phenomena in plants. It has two isomers: P_r, which absorbs red light, and P_{fr}, which absorbs far red light. P_{fr} is the active form.

phytomers In plants, the repeating modules that compose a shoot, each consisting of one or more leaves, attached to the stem at a node; an internode; and one or more axillary buds.

phytoplankton Photosynthetic plankton.

phytoremediation A form of bioremediation that uses plants to clean up environmental pollution.

pigment A substance that absorbs visible light.

piloting A form of navigation in which an animal finds its way by knowing and remembering the structure of its environment.

pineal gland Gland located between the cerebral hemispheres that secretes melatonin.

pinocytosis Endocytosis by a cell of liquid containing dissolved substances.

pistil [L. *pistillum*: pestle] The structure of an angiosperm flower within which the ovules are borne. May consist of a single carpel, or of several carpels fused into a single structure. Usually differentiated into ovary, style, and stigma.

pith In plants, relatively unspecialized tissue found within a cylinder of vascular tissue.

pituitary gland A small gland attached to the base of the brain in vertebrates. Its hormones control the activities of other glands. Also known as the hypophysis.

placenta (pla sen' ta) The organ in female mammals that provides for the nourishment of the fetus and elimination of the fetal waste products.

plankton Free-floating small aquatic organisms. Photosynthetic members of the plankton are referred to as phytoplankton.

planula (plan' yew la) [L. *planum*: flat] A free-swimming, ciliated larval form typical of the cnidarians.

plaque (plack) [Fr.: a metal plate or coin] (1) A circular clearing in a layer (lawn) of bacteria growing on the surface of a nutrient agar gel. (2) An accumulation of prokaryotic organisms on tooth enamel. Acids produced by these microorganisms cause tooth decay. (3) A region of arterial wall invaded by fibroblasts and fatty deposits.

plasma (plaz' muh) The liquid portion of blood, in which blood cells and other particulates are suspended.

plasma cell An antibody-secreting cell that develops from a B cell; the effector cell of the humoral immune system.

plasma membrane The membrane that surrounds the cell, regulating the entry and exit of molecules and ions. Every cell has a plasma membrane.

plasmid A DNA molecule distinct from the chromosome(s); that is, an extrachromosomal element; found in many bacteria. May replicate independently of the chromosome.

plasmodesma (plural: plasmodesmata) [Gk. *plassein*: to mold + *desmos*: band] A cytoplasmic strand connecting two adjacent plant cells.

plasmogamy The fusion of the cytoplasm of two cells. (Contrast with karyogamy.)

plastid A class of plant cell organelles that includes the chloroplast, which houses biochemical pathways for photosynthesis.

plate tectonics [Gk. *tekton*: builder] The scientific study of the structure and movements of Earth's lithospheric plates, which are the cause of continental drift.

platelet A membrane-bounded body without a nucleus, arising as a fragment of a cell in the bone marrow of mammals. Important to blood-clotting action.

pleural membrane [Gk. *pleuras*: rib, side] The membrane lining the outside of the lungs and the walls of the thoracic cavity. Inflammation of these membranes is a condition known as pleurisy.

pluripotent [L. *pluri*: many + *potens*: powerful] Having the ability to form all of the cells in the body. (Contrast with multipotent, totipotent.)

podocytes Cells of Bowman's capsule of the nephron that cover the capillaries of the glomerulus, forming filtration slits.

point mutation A mutation that results from the gain, loss, or substitution of a single nucleotide.

polar body A nonfunctional nucleus produced by meiosis during oogenesis.

polar covalent bond A covalent bond in which the electrons are drawn to one nucleus more than the other, resulting in an unequal distribution of charge.

polar Having separate and opposite electric charges at two ends, or poles. (Contrast with nonpolar.)

polar nuclei In angiosperms, the two nuclei in the central cell of the megagametophyte; following fertilization they give rise to the endosperm.

polarity (1) In chemistry, the property of unequal electron sharing in a covalent bond that defines a polar molecule. (2) In development, the difference between one end of an organism or structure and the other.

pollen [L. *pollin*: fine flour] In seed plants, microscopic grains that contain the male gametophyte (microgametophyte) and gamete (microspore).

pollen tube A structure that develops from a pollen grain through which sperm are released into the megagametophyte.

pollination The process of transferring pollen from an anther to the stigma of a pistil in an angiosperm or from a strobilus to an ovule in a gymnosperm.

poly A tail A long sequence of adenine nucleotides (50–250) added after transcription to the 3' end of most eukaryotic mRNAs.

poly- [Gk. *poly*: many] A prefix denoting multiple entities.

polyandry Mating system in which one female mates with multiple males.

polygyny Mating system in which one male mates with multiple females.

polymer [Gk. *poly*: many + *meros*: unit] A large molecule made up of similar or identical subunits called monomers. (Contrast with monomer.)

polymerase chain reaction (PCR) An enzymatic technique for the rapid production of millions of copies of a particular stretch of DNA where only a small amount of the parent molecule is available.

polymorphic (pol' lee mor' fik) [Gk. *poly*: many + *morphe*: form, shape] Coexistence in a population of two or more distinct traits.

polyp (pah' lip) [Gk. *poly*: many + *pous*: foot] In cnidarians, a sessile, asexual life cycle stage.

polypeptide A large molecule made up of many amino acids joined by peptide linkages. Large polypeptides are called proteins.

polyphyletic (pol' lee fih leht' ik) [Gk. *poly*: many + *phylon*: tribe] Pertaining to a group that consists of multiple distantly related organisms, and does not include the common ancestor of the group. (Contrast with monophyletic, paraphyletic.)

polyploidy (pol' lee ploid ee) The possession of more than two entire sets of chromosomes.

polyribosome (polysome) A complex consisting of a threadlike molecule of messenger RNA and several (or many) ribosomes. The ribosomes move along the mRNA, synthesizing polypeptide chains as they proceed.

polysaccharide A macromolecule composed of many monosaccharides (simple sugars). Common examples are cellulose and starch.

pons [L. *pons*: bridge] Region of the brainstem anterior to the medulla.

pool The total amount of an element in a given compartment of the biosphere.

population Any group of organisms coexisting at the same time and in the same place and capable of interbreeding with one another.

population bottleneck A period during which only a few individuals of a normally large population survive.

population density The number of individuals in a population per unit of area or volume.

population dynamics The patterns and processes of change in populations.

population genetics The study of genetic variation and its causes within populations.

population size The total number of individuals in a population.

positional information In development, the basis of the spatial sense that induces cells to differentiate as appropriate for their location within the developing organism; often comes in the form of a morphogen gradient.

positive feedback In regulatory systems, information that amplifies a regulatory response, increasing the deviation of the system from the set point. (Contrast with negative feedback.)

positive regulation A form of gene regulation in which a regulatory macromolecule is needed to turn on the transcription of a structural gene; in its absence, transcription will not occur. (Contrast with negative regulation.)

post- [L. *postere*: behind, following after] Prefix denoting something that comes after.

postabsorptive state State in which no food remains in the gut and thus no nutrients are being absorbed. (Contrast with absorptive state.)

posterior pituitary A portion of the pituitary gland that is derived from neural tissue and is involved in the storage and release of antidiuretic hormone and oxytocin. (*See also* pituitary gland and anterior pituitary.)

posterior Toward or pertaining to the rear. (Contrast with anterior.)

postsynaptic cell *See* postsynaptic neuron.

postsynaptic neuron The cell that receives information from a neuron at a synapse. (Contrast with presynaptic neuron.)

postzygotic isolating mechanisms Barriers to the reproductive process that occur after the union of the nuclei of two gametes. (Contrast with prezygotic isolating mechanisms.)

potential energy Energy not doing work, such as the energy stored in chemical bonds. (Contrast with kinetic energy.)

P_r P_r *See* phytochrome.

pre-mRNA (precursor mRNA) Initial gene transcript before it is modified to produce functional mRNA. Also known as the primary transcript.

Precambrian The first and longest period of geological time, during which life originated.

precapillary sphincter A cuff of smooth muscle that can shut off the blood flow to a capillary bed.

predator An organism that kills and eats other organisms.

pre-replication complex In eukaryotes, a complex of proteins that binds to DNA at the initiation of DNA replication.

pressure flow model An effective model for phloem transport in angiosperms. It holds that sieve element transport is driven by an osmotically generated pressure gradient between source and sink.

pressure potential (Ψ_p) The hydrostatic pressure of an enclosed solution in excess of the surrounding atmospheric pressure. (Contrast with solute potential, water potential.)

presynaptic neuron The neuron that transmits information to another cell at a synapse. (Contrast with postsynaptic cell.)

prey [L. *praeda*: booty] An organism consumed by a predator as an energy source.

prezygotic isolating mechanisms Barriers to the reproductive process that occur before the union of the nuclei of two gametes (Contrast with postzygotic isolating mechanisms.)

primary active transport Active transport in which ATP is hydrolyzed, yielding the energy required to transport an ion or molecule against its concentration gradient. (Contrast with secondary active transport.)

primary cell wall In plant cells, a structure that forms at the middle lamella after cytokinesis, made up of cellulose microfibrils, hemicelluloses, and pectins. (Contrast with secondary cell wall.)

primary consumer An organism (herbivore) that eats plant tissues.

primary endosymbiosis The engulfment of a cyanobacterium by a larger eukaryotic cell that gave rise to the first photosynthetic eukaryotes with chloroplasts.

primary growth In plants, growth that is characterized by the lengthening of roots and shoots and by the proliferation of new roots and shoots through branching. (Contrast with secondary growth.)

primary immune response The first response of the immune system to an antigen, involving recognition by lymphocytes and the production of effector cells and memory cells. (Contrast with secondary immune response.)

primary lysosome *See* lysosome.

primary meristem Meristem that produces the tissues of the primary plant body.

primary motor cortex An area of the frontal lobe that controls muscles; neurons in this area are arranged according to the parts of the body with which they communicate.

primary oocyte (oh' eh site) [Gk. *oon*: egg + *kytos*: container] The diploid progeny of an oogonium. In many species, a primary oocyte enters prophase of the first meiotic division, then remains in developmental arrest for a long time before resuming meio-

sis to form a secondary oocyte and a polar body.

primary plant body That part of a plant produced by primary growth. Consists of all the *nonwoody* parts of a plant; many herbaceous plants consist entirely of a primary plant body. (Contrast with secondary plant body.)

primary producer A photosynthetic or chemosynthetic organism that synthesizes complex organic molecules from simple inorganic ones.

primary somatosensory cortex An area of the parietal lobe that receives touch and pressure information from the body; neurons in this area are arranged according to the parts of the body with which they communicate.

primary spermatocyte The diploid progeny of a spermatogonium; undergoes the first meiotic division to form secondary spermatocytes.

primary structure The specific sequence of amino acids in a protein.

primase An enzyme that catalyzes the synthesis of a primer for DNA replication.

primer Strand of nucleic acid, usually RNA, that is the necessary starting material for the synthesis of a new DNA strand, which is synthesized from the 3′ end of the primer.

primordium (plural: primordia) [L. origin] The most rudimentary stage of an organ or other part.

principle of allocation The idea that a unit of some resource acquired by an organism can be used for only one function at a time, meaning that resources must be divided among competing functions.

pro- [L.: first, before, favoring] A prefix often used in biology to denote a developmental stage that comes first or an evolutionary form that appeared earlier than another. For example, prokaryote, prophase.

probe A segment of single-stranded nucleic acid used to identify DNA molecules containing the complementary sequence.

procambium Primary meristem that produces the vascular tissue.

procedural memory Memory of motor tasks. Cannot be consciously recalled and described. (Contrast with declarative memory.)

processive Pertaining to an enzyme that catalyzes many reactions each time it binds to a substrate, as DNA polymerase does during DNA replication.

products The molecules that result from the completion of a chemical reation.

progesterone [L. *pro*: favoring + *gestare*: to bear] A female sex hormone that maintains pregnancy.

prokaryotes Unicellular organisms that do not have nuclei. (Contrast with eukaryotes.)

prolactin   A hormone released by the anterior pituitary, one of whose functions is the stimulation of milk production in female mammals.

prometaphase   The phase of nuclear division that begins with the disintegration of the nuclear envelope.

promoter   A DNA sequence to which RNA polymerase binds to initiate transcription.

prop roots   Adventitious roots in some monocots that function as supports for the shoot.

prophage   (pro' fayj) The noninfectious units that are linked with the chromosomes of the host bacteria and multiply with them but do not cause dissolution of the cell. Prophage can later enter into the lytic phase to complete the virus life cycle.

prophase   (pro' phase) The first stage of nuclear division, during which chromosomes condense from diffuse, threadlike material to discrete, compact bodies.

prostaglandin   Any one of a group of specialized lipids with hormone-like functions. It is not clear that they act at any considerable distance from the site of their production.

prostate gland   In male humans, surrounds the urethra at its junction with the vas deferens; supplies an acid-neutralizing fluid to the semen.

prosthetic group   Any nonprotein portion of an enzyme.

proteasome   In the eukaryotic cytoplasm, a huge protein structure that binds to and digests cellular proteins that have been tagged by ubiquitin.

protein   (pro' teen) [Gk. *protos*: first] Long-chain polymer of amino acids with twenty different common side chains. Occurs with its polymer chain extended in fibrous proteins, or coiled into a compact macromolecule in enzymes and other globular proteins.

protein kinase   (kye' nase) An enzyme that catalyzes the addition of a phosphate group from ATP to a target protein.

protein kinase cascade   A series of reactions in response to a molecular signal, in which a series of protein kinases activate one another in sequence, amplifying the signal at each step.

proteoglycan   A glycoprotein containing a protein core with attached long, linear carbohydrate chains.

proteolysis   [protein + Gk. *lysis*: break apart] An enzymatic digestion of a protein or polypeptide.

proteome   The set of proteins that can be made by an organism. Because of alternative splicing of pre-mRNA, the number of proteins that can be made is usually much larger than the number of protein-coding genes present in the organism's genome.

proteomics   The study of the proteome—the complete complement of proteins produced by an organism.

prothrombin   The inactive form of thrombin, an enzyme involved in blood clotting.

protoderm   Primary meristem that gives rise to the plant epidermis.

proton   (pro' ton) [Gk. *protos*: first, before] (1) A subatomic particle with a single positive charge. The number of protons in the nucleus of an atom determine its element. (2) A hydrogen ion, H^+.

proton pump   An active transport system that uses ATP energy to move hydrogen ions across a membrane, generating an electric potential.

proton-motive force   Force generated across a membrane having two components: a chemical potential (difference in proton concentration) plus an electrical potential due to the electrostatic charge on the proton.

protoplast   The living contents of a plant cell; the plasma membrane and everything contained within it.

provirus   Double-stranded DNA made by a virus that is integrated into the host's chromosome and contains promoters that are recognized by the host cell's transcription apparatus.

proximal convoluted tubule   The initial segment of a renal tubule, closest to the glomerulus. (Compare with distal convoluted tubule.)

proximal   Near the point of attachment or other reference point. (Contrast with distal.)

proximate cause   The immediate genetic, physiological, neurological, and developmental mechanisms responsible for a behavior or morphology. (Contrast with ultimate cause.)

pseudocoelomate   (soo' do see' low mate) [Gk. *pseudes*: false + *koiloma*: cavity] Having a body cavity, called a pseudocoel, consisting of a fluid-filled space in which many of the internal organs are suspended, but which is enclosed by mesoderm only on its outside.

pseudogene   [Gk. *pseudes*: false] A DNA segment that is homologous to a functional gene but is not expressed because of changes to its sequence or changes to its location in the genome.

pseudopod   (soo' do pod) [Gk. *pseudes*: false + *podos*: foot] A temporary, soft extension of the cell body that is used in location, attachment to surfaces, or engulfing particles.

puberty   The process of sexual maturation in humans.

pulmonary   [L. *pulmo*: lung] Pertaining to the lungs.

pulmonary circuit   The portion of the circulatory system by which blood is pumped from the heart to the lungs or gills for oxygenation and back to the heart for distribution. (Contrast with systemic circuit.)

pulmonary valve   A one-way valve between the right ventricle of the heart and the pulmonary artery that prevents backflow of blood into the ventricle when it relaxes.

Punnett square   Method of predicting the results of a genetic cross by arranging the gametes of each parent at the edges of a square.

pupa   (pew' pa) [L. *pupa*: doll, puppet] In certain insects (the Holometabola), the encased developmental stage between the larva and the adult.

pupil   The opening in the vertebrate eye through which light passes.

purine   (pure' een) One of the two types of nitrogenous bases in nucleic acids. Each of the purines—adenine and guanine—pairs with a specific pyrimidine.

Purkinje fibers   Specialized heart muscle cells that conduct excitation throughout the ventricular muscle.

pyrimidine   (per im' a deen) One of the two types of nitrogenous bases in nucleic acids. Each of the pyrimidines—cytosine, thymine, and uracil—pairs with a specific purine.

pyruvate   The ionized form of pyruvic acid, a three-carbon acid; the end product of glycolysis and the raw material for the citric acid cycle.

pyruvate oxidation   Conversion of pyruvate to acetyl CoA and CO_2 that occurs in the mitochondrial matrix in the presence of O_2.

– Q –

Q_{10}   A value that compares the rate of a biochemical process or reaction over 10°C temperature ranges. A process that is not temperature-sensitive has a Q_{10} of 1; values of 2 or 3 mean the reaction speeds up as temperature increases.

quantify   To assign numerical values to observations through measurement.

quantitative trait loci   A set of genes that determines a complex character that exhibits quantitative variation (*quantitative trait*).

quaternary structure   The specific three-dimensional arrangement of protein subunits.

– R –

R group   The distinguishing group of atoms of a particular amino acid; also known as a side chain.

radial symmetry   The condition in which any two halves of a body are mirror images of each other, providing the cut passes through the center; a cylinder cut lengthwise down its center displays this form of symmetry.

radiation   The transfer of heat from warmer objects to cooler ones via the exchange of infrared radiation. *See also* electromagnetic radiation; evolutionary radiation.

radicle   An embryonic root.

radioisotope   A radioactive isotope of an element. Examples are carbon-14 (^{14}C) and hydrogen-3, or tritium (^3H).

radiometric dating A method for determining the age of objects such as fossils and rocks based on the decay rates of radioactive isotopes.

rapid eye movement sleep *See* REM sleep.

reactant A chemical substance that enters into a chemical reaction with another substance.

reaction center A group of electron transfer proteins that receive energy from light-absorbing pigments and convert it to chemical energy by redox reactions.

receptive field The area of visual space that activates a particular cell in the visual system.

receptor potential The change in the resting potential of a sensory cell when it is stimulated.

receptor protein A protein that can bind to a specific molecule, or detect a specific stimulus, within the cell or in the cell's external environment.

receptor *See* receptor protein, sensory receptor cell.

receptor-mediated endocytosis Endocytosis initiated by macromolecular binding to a specific membrane receptor.

recessive In genetics, an allele that does not determine phenotype in the presence of a dominant allele. (Contrast with dominance.)

reciprocal adaptation Evolutionary processes in which an adaptation in one species leads to the evolution of an adaptation in a species with which it interacts; also known as coevolution.

reciprocal crosses A pair of matings in one of which a female of genotype A mates with a male of genotype B and in the other of which a female of genotype B mates with a male of genotype A.

recognition sequence *See* restriction site.

recombinant DNA A DNA molecule made in the laboratory that is derived from two or more genetic sources.

recombinant Pertaining to an individual, meiotic product, or chromosome in which genetic materials originally present in two individuals end up in the same haploid complement of genes.

recombination frequency The proportion of offspring of a genetic cross that have phenotypes different from the parental phenotypes due to crossing over between linked genes during gamete formation.

rectum The terminal portion of the gut, ending at the anus.

redox reaction A chemical reaction in which one reactant becomes oxidized and the other becomes reduced. Short for reduction–oxidation reaction.

reduction Gain of electrons by a chemical reactant; any reduction is accompanied by an oxidation. (Contrast with oxidation.)

refractory period The time interval after an action potential during which another action potential cannot be elicited from an excitable membrane.

regeneration The development of a complete individual from a fragment of an organism.

regulative development A pattern of animal embryonic development in which the fates of the first blastomeres are not absolutely fixed. (Contrast with mosaic development.)

regulatory gene A gene that codes for a protein (or RNA) that in turn controls the expression of another gene.

regulatory sequence A DNA sequence to which the protein product of a regulatory gene binds.

regulatory system A system that uses feedback information to maintain a physiological function or parameter at an optimal level. (Contrast with controlled system.)

regulatory T cells (T$_{reg}$) The class of T cells that mediates tolerance to self antigens.

reinforcement The evolution of enhanced reproductive isolation between populations due to natural selection for greater isolation.

Reissner's membrane One of two membranes (the other is the basilar membrane) that extend along the length of the cochlea in the human ear.

releaser Sensory stimulus that triggers performance of a stereotyped behavior pattern.

REM (rapid-eye-movement) sleep A sleep state characterized by vivid dreams, skeletal muscle relaxation, and rapid eye movements. (Contrast with non-REM sleep.)

renal [L. *renes*: kidneys] Relating to the kidneys.

renal tubule A structural unit of the kidney that collects filtrate from the blood, reabsorbs specific ions, nutrients, and water and returns them to the blood, and concentrates excess ions and waste products such as urea for excretion from the body.

renin An enzyme released from the kidneys in response to a drop in the glomerular filtration rate. Together with angiotensin converting enzyme, converts an inactive protein in the blood into angiotensin.

replication complex The close association of several proteins operating in the replication of DNA.

replication fork A point at which a DNA molecule is replicating. The fork forms by the unwinding of the parent molecule.

replication The duplication of genetic material.

replicon A region of DNA replicated from a single origin of replication.

reporter gene A genetic marker included in recombinant DNA to indicate the presence of the recombinant DNA in a host cell.

repressor A protein encoded by a regulatory gene that can bind to a promoter and prevent transcription of the associated gene. (Contrast with activator.)

reproductive isolation Condition in which two divergent populations are no longer exchanging genes. Can lead to speciation.

residual volume (RV) In tidal ventilation, the dead space that remains in the lungs at the end of exhalation.

resistance (R) genes Plant genes that confer resistance to specific strains of pathogens.

resource partitioning A situation in which selection pressures resulting from interspecific competition cause changes in the ways in which the competing species use the limiting resource, thereby allowing them to coexist.

resource Something in the environment required by an organism for its maintenance and growth that is consumed in the process of being used.

respiration (res pi ra' shun) [L. *spirare*: to breathe] (1) Cellular respiration. (2) Breathing.

respiratory chain The terminal reactions of cellular respiration, in which electrons are passed from NAD or FAD, through a series of intermediate carriers, to molecular oxygen, with the concomitant production of ATP.

respiratory gases Oxygen (O_2) and carbon dioxide (CO_2); the gases that an animal must exchange between its internal body fluids and the outside medium (air or water).

resting potential The membrane potential of a living cell at rest. In cells at rest, the interior is negative to the exterior. (Contrast with action potential.)

restoration ecology The science and practice of restoring damaged or degraded ecosystems.

restriction enzyme Any of a type of enzyme that cleaves double-stranded DNA at specific sites; extensively used in recombinant DNA technology. Also called a restriction endonuclease.

restriction fragment length polymorphism *See* RFLP.

restriction point (R) The specific time during G1 of the cell cycle at which the cell becomes committed to undergo the rest of the cell cycle.

restriction site A specific DNA base sequence that is recognized and acted on by a restriction endonuclease.

reticular system A central region of the vertebrate brainstem that includes complex fiber tracts conveying neural signals between the forebrain and the spinal cord, with collateral fibers to a variety of nuclei that are involved in autonomic functions, including arousal from sleep.

reticulum One of the four chambers of the ruminant stomach. Along with the rumen, where food is partially digested with the assistance of gut bacteria.

retina (rett' in uh) [L. *rete*: net] The light-sensitive layer of cells in the vertebrate or cephalopod eye.

11-*cis*-retinal The nonprotein, light-absorbing component of the visual system pigment rhodopsin; associated with the protein opsin. (*See* rhodopsin.)

retinoblastoma protein A protein that inhibits an animal cell from passing through the restriction point; inactivation of this protein is necessary for the cell cycle to proceed.

retrotransposons Mobile genetic elements that are reverse transcribed into RNA as part of their transfer mechanism. (Contrast with DNA transposons.)

retrovirus An RNA virus that contains reverse transcriptase. Its RNA serves as a template for cDNA production, and the cDNA is integrated into a chromosome of the host cell.

reverse transcriptase An enzyme that catalyzes the production of DNA (cDNA), using RNA as a template; essential to the reproduction of retroviruses.

RFLP Restriction fragment length polymorphism, the coexistence of two or more patterns of restriction fragments resulting from underlying differences in DNA sequence.

rhizoids (rye' zoids) [Gk. root] Hairlike extensions of cells in mosses, liverworts, and a few vascular plants that serve the same function as roots and root hairs in vascular plants. The term is also applied to branched, rootlike extensions of some fungi and algae.

rhizome (rye' zome) An underground stem (as opposed to a root) that runs horizontally beneath the ground.

rhodopsin A photopigment used in the visual process of transducing photons of light into changes in the membrane potential of photoreceptor cells.

ribonucleic acid *See* RNA.

ribose A five-carbon sugar in nucleotides and RNA.

ribosomal RNA (rRNA) Several species of RNA that are incorporated into the ribosome. Involved in peptide bond formation.

ribosome A small particle in the cell that is the site of protein synthesis.

ribozyme An RNA molecule with catalytic activity.

ribulose bisphosphate carboxylase/oxygenase *See* rubisco.

risk cost The increased chance of being injured or killed as a result of performing a behavior, compared to resting.

RNA (ribonucleic acid) An often single-stranded nucleic acid whose nucleotides use ribose rather than deoxyribose and in which the base uracil replaces thymine found in DNA. Serves as genome from some viruses. (*See* ribosomal RNA, transfer RNA, messenger RNA, and ribozyme.)

RNA interference (RNAi) A mechanism for reducing mRNA translation whereby a double-stranded RNA, made by the cell or synthetically, is processed into a small, single-stranded RNA, whose binding to a target mRNA results in the latter's breakdown.

RNA polymerase An enzyme that catalyzes the formation of RNA from a DNA template.

RNA splicing The last stage of RNA processing in eukaryotes, in which the transcripts of introns are excised through the action of small nuclear ribonucleoprotein particles (snRNP).

rod cells Light-sensitive cells in the vertebrate retina; these sensory receptor cells are sensitive in extremely dim light and are responsible for dim light, black and white vision.

root apical meristem Undifferentiated tissue at the apex of the root that gives rise to the organs of the root.

root cap A thimble-shaped mass of cells, produced by the root apical meristem, that protects the meristem; the organ that perceives the gravitational stimulus in root gravitropism.

root hair A long, thin process from a root epidermal cell that absorbs water and minerals from the soil solution.

root system The organ system that anchors a plant in place, absorbs water and dissolved minerals, and may store products of photosynthesis from the shoot system.

root The organ responsible for anchoring the plant in the soil, absorbing water and minerals, and producing certain hormones. Some roots are storage organs.

rough endoplasmic reticulum (RER) The portion of the endoplasmic reticulum whose outer surface has attached ribosomes. (Contrast with smooth endoplasmic reticulum.)

round window A flexible membrane at the end of the lower canal of the cochlea in the human ear. (*See also* oval window.)

rRNA *See* ribosomal RNA.

rubisco Contraction of ribulose bisphosphate carboxylase/oxygenase, the enzyme that combines carbon dioxide or oxygen with ribulose bisphosphate to catalyze the first step of photosynthetic carbon fixation or photorespiration, respectively.

rumen One of the four chambers of the ruminant stomach. Along with the reticulum, where food is partially digested with the assistance of gut bacteria.

ruminant Herbivorous, cud-chewing mammals such as cows or sheep, characterized by a stomach that consists of four compartments: the rumen, reticulum, omasum, and abomasum.

– S –

S phase In the cell cycle, the stage of interphase during which DNA is replicated. (Contrast with G1 phase, G2 phase, M phase.)

salt glands Glands on the leaves of some halophytic plants that secrete salt, thereby ridding the plants of excess salt.

saltatory conduction [L. *saltare*: to jump] The rapid conduction of action potentials in myelinated axons; so called because action potentials appear to "jump" between nodes of Ranvier along the axon.

saprobe [Gk. *sapros*: rotten] An organism (usually a bacterium or fungus) that obtains its carbon and energy by absorbing nutrients from dead organic matter.

sarcomere (sark' o meer) [Gk. *sark*: flesh + *meros*: unit] The contractile unit of a skeletal muscle.

sarcoplasm The cytoplasm of a muscle cell.

sarcoplasmic reticulum The endoplasmic reticulum of a muscle cell.

saturated fatty acid A fatty acid in which all the bonds between carbon atoms in the hydrocarbon chain are single bonds—that is, all the bonds are saturated with hydrogen atoms. (Contrast with unsaturated fatty acid.)

Schwann cell A type of glial cell that myelinates axons in the peripheral nervous system.

scientific method A means of gaining knowledge about the natural world by making observations, posing hypotheses, and conducting experiments to test those hypotheses.

scion In horticulture, the bud or stem from one plant that is grafted to a root or root-bearing stem of another plant (the stock).

sclereid One of the principle types of cells in sclerenchyma.

sclerenchyma (skler eng' kyma) [Gk. *skleros*: hard + *kymus*: juice] A plant tissue composed of cells with heavily thickened cell walls. The cells are dead at functional maturity. The principal types of sclerenchyma cells are fibers and sclereids.

scrotum In most mammals, a pouch outside the body cavity that contains the testes.

seasonality A aspect of climate characterized by fluctuations in temperature over the course of a year.

second filial generation See F_2.

second law of thermodynamics The principle that when energy is converted from one form to another, some of that energy becomes unavailable for doing work.

second messenger A compound, such as cAMP, that is released within a target cell after a hormone (the first messenger) has bound to a surface receptor on a cell; the second messenger triggers further reactions within the cell.

second polar body In oogenesis, the daughter cell of the second meiotic division that subsequently degenerates. (*See also* ootid.)

secondary active transport A form of active transport that does not use ATP as an energy source; rather, transport is coupled to ion diffusion down a concentration gradient established by primary active transport.

secondary cell wall   A thick, cellulosic structure internal to the primary cell wall formed in some plant cells after cell expansion stops (Contrast with primary cell wall.)

secondary consumer   An organism that eats primary consumers.

secondary endosymbiosis   The engulfment of a photosynthetic eukaryote by another eukaryotic cell that gave rise to certain groups of photosynthetic eukaryotes (e.g., euglenids).

secondary growth   In plants, growth that contributes to an increase in girth. (Contrast with primary growth.)

secondary immune response   A rapid and intense response to a second or subsequent exposure to an antigen, initiated by memory cells. (Contrast with primary immune response.)

secondary lysosome   Membrane-enclosed organelle formed by the fusion of a primary lysosome with a phagosome, in which macromolecules taken up by phagocytosis are hydrolyzed into their monomers. (Contrast with lysosome.)

secondary metabolite   A compound synthesized by a plant that is not needed for basic cellular metabolism. Typically has an antiherbivore or antiparasite function.

secondary oocyte   In oogenesis, the daughter cell of the first meiotic division that receives almost all the cytoplasm. (*See also* first polar body.)

secondary plant body   That part of a plant produced by secondary growth; consists of woody tissues. (Contrast with primary plant body.)

secondary spermatocyte   One of the products of the first meiotic division of a primary spermatocyte.

secondary structure   Of a protein, localized regularities of structure, such as the α helix and the β pleated sheet.

secretin   (si kreet' in) A peptide hormone secreted by the upper region of the small intestine when acidic chyme is present. Stimulates the pancreatic duct to secrete bicarbonate ions.

sedimentary rock   Rock formed by the accumulation of sediment grains on the bottom of a body of water.

seed   A fertilized, ripened ovule of a gymnosperm or angiosperm. Consists of the embryo, nutritive tissue, and a seed coat.

seedling   A plant that has just completed the process of germination.

segment polarity genes   In *Drosophila* (fruit fly) development, segmentation genes that determine the boundaries and anterior–posterior organization of individual segments. Part of a developmental cascade that includes maternal effect genes, gap genes, pair rule genes, and Hox genes.

segmentation   Division of an animal body into segments.

segmentation genes   Genes that determine the number and polarity of body segments.

segregation   In genetics, the separation of alleles, or of homologous chromosomes, from each other during meiosis so that each of the haploid daughter nuclei produced contains one or the other member of the pair found in the diploid parent cell, but never both. This principle was articulated by Mendel as his first law.

selectable marker   A gene, such as one encoding resistance to an antibiotic, that can be used to identify (select) cells that contain recombinant DNA from among a large population of untransformed cells.

selective permeability   Allowing certain substances to pass through while other substances are excluded; a characteristic of membranes.

self-incompatability   In plants, the possession of mechanisms that prevent self-fertilization.

semen   (see' men) [L. *semin*: seed] The thick, whitish liquid produced by the male reproductive system in mammals, containing the sperm.

semicircular canals   Three canals in the human inner ear that form part of the vestibular system. (*See* vestibular system.)

semiconservative replication   The way in which DNA is synthesized. Each of the two partner strands in a double helix acts as a template for a new partner strand. Hence, after replication, each double helix consists of one old and one new strand.

seminiferous tubules   The tubules within the testes within which sperm production occurs.

senescence   [L. *senescere*: to grow old] Aging; deteriorative changes with aging; the increased probability of dying with increasing age.

sensitive period   The life stage during which some particular type of learning must take place, or during which it occurs much more easily than at other times. Typical of song learning among birds.

sensor   *See* sensory receptor cell.

sensory neuron   A specialized neuron that transduces a particular type of sensory stimulus into action potentials.

sensory receptor cell   Cell that is responsive to a particular type of physical or chemical stimulation.

sensory system   A set of organs and tissues for detecting a stimulus; consists of sensory cells, the associated structures, and the neural networks that process the information.

sensory transduction   The transformation of environmental stimuli or information into neural signals.

sepal   (see' pul) [L. *sepalum*: covering] One of the outermost structures of the flower, usually protective in function and enclosing the rest of the flower in the bud stage.

sepsis   Generalized inflammation caused by bacterial infection. Can cause a dangerous drop in blood pressure.

septum   (plural: septa) [L. wall] (1) A partition or cross-wall appearing in the hyphae of some fungi. (2) The bony structure dividing the nasal passages.

Sertoli cells   Cells in the seminiferous tubules that nurture the developing sperm.

sessile   (sess' ul) [L. *sedere*: to sit] Permanently attached; not able to move from one place to another. (Contrast with motile.)

set point   In a regulatory system, the threshold sensitivity to the feedback stimulus.

sex chromosome   In organisms with a chromosomal mechanism of sex determination, one of the chromosomes involved in sex determination.

sex linkage   The pattern of inheritance characteristic of genes located on the sex chromosomes of organisms having a chromosomal mechanism for sex determination. Also called *sex-linked inheritance*.

sex pilus   A thin connection between two bacteria through which genetic material passes during conjugation.

sex-linked inheritance   Inheritance of a gene that is carried on a sex chromosome. Also called *sex linkage*.

sexual reproduction   Reproduction involving the union of gametes.

sexual selection   Selection by one sex of characteristics in individuals of the opposite sex. Also, the favoring of characteristics in one sex as a result of competition among individuals of that sex for mates.

shared derived trait   *See* synapomorphy.

shoot apical meristem   Undifferentiated tissue at the apex of the shoot that gives rise to the organs of the shoot.

shoot system   In plants, the organ system consisting of the leaves, stem(s), and flowers.

short-day plant (SDP)   A plant that flowers when nights are longer than a critical length specific for that plant's species. (Compare to long-day plant.)

short tandem repeat (STR)   A short (1–5 base pairs), moderately repetitive sequence of DNA. The number of copies of an STR at a particular location varies between individuals and is inherited.

side chain   *See* R group.

sieve tube element   The characteristic cell of the phloem in angiosperms, which contains cytoplasm but relatively few organelles, and whose end walls (*sieve plates*) contain pores that form connections with neighboring cells.

sigma factor   In prokaryotes, a protein that binds to RNA polymerase, allowing the complex to bind to and stimulate the transcription of a specific class of genes (e.g., those involved in sporulation).

signal sequence The sequence within a protein that directs the protein to a particular organelle.

signal transduction pathway The series of biochemical steps whereby a stimulus to a cell (such as a hormone or neurotransmitter binding to a receptor) is translated into a response of the cell.

silent mutation A change in a gene's sequence that has no effect on the amino acid sequence of a protein because it occurs in noncoding DNA or because it does not change the amino acid specified by the corresponding codon. (Contrast with frameshift mutation, missense mutation, nonsense mutation.)

simple diffusion Diffusion that doesn't involve a direct input of energy or assistance by carrier proteins.

single nucleotide polymorphisms (SNPs) Inherited variations in a single nucleotide base in DNA that differ between individuals.

sink In plants, any organ that imports the products of photosynthesis, such as roots, developing fruits, and immature leaves. (Contrast with source.)

sinoatrial node (sigh′ no ay′ tree al) [L. *sinus*: curve + *atrium*: chamber] The pacemaker of the mammalian heart.

siRNAs (small interfering RNAs) Short, double-stranded RNA molecules used in RNA interference.

sister chromatid Each of a pair of newly replicated chromatids.

sister clades Two phylogenetic groups that are each other's closest relatives.

sister species Two species that are each other's closest relatives.

skeletal muscle A type of muscle tissue characterized by multinucleated cells containing highly ordered arrangements of actin and myosin microfilaments. Also called striated muscle. (Contrast with cardiac muscle, smooth muscle.)

skeletal systems Organ systems that provide rigid supports against which muscles can pull to create directed movements.

sliding filament theory Mechanism of muscle contraction based on the formation and breaking of crossbridges between actin and myosin filaments, causing the filaments to slide together.

slow-twitch fibers Skeletal muscle fibers specialized for sustained aerobic work; contain myoglobin and abundant mitochondria, and are well-supplied with blood vessels. Also called oxidative or red muscle fibers. (Compare to fast-twitch fibers.)

slow-wave sleep See non-REM sleep.

small intestine The portion of the gut between the stomach and the colon; consists of the duodenum, the jejunum, and the ileum.

small nuclear ribonucleoprotein particle (snRNP) A complex of an enzyme and a small nuclear RNA molecule, functioning in RNA splicing.

smooth endoplasmic reticulum (SER) Portion of the endoplasmic reticulum that lacks ribosomes and has a tubular appearance. (Contrast with rough endoplasmic reticulum.)

smooth muscle Muscle tissue consisting of sheets of mononucleated cells innervated by the autonomic nervous system. (Contrast with cardiac muscle, skeletal muscle.)

sodium–potassium (Na$^+$–K$^+$) pump Anti-porter responsible for primary active transport; it pumps sodium ions out of the cell and potassium ions into the cell, both against their concentration gradients. Also called a sodium–potassium ATPase.

soil horizon See horizons.

soil solution The aqueous portion of soil, from which plants take up dissolved mineral nutrients.

solute A substance that is dissolved in a liquid (solvent) to form a solution.

solute potential (Ψ_s) A property of any solution, resulting from its solute contents; it may be zero or have a negative value. The more negative the solute potential, the greater the tendency of the solution to take up water through a differentially permeable membrane. (Contrast with pressure potential, water potential.)

solution A liquid (the solvent) and its dissolved solutes.

solvent Liquid in which a substance (solute) is dissolved to form a solution.

somatic cell [Gk. *soma*: body] All the cells of the body that are not specialized for reproduction. (Contrast with germ cell.)

somatic mutation Permanent genetic change in a somatic cell. These mutations affect the individual only; they are not passed on to offspring. (Contrast with germ line mutation.)

somatosensory cortex The region of the cerebral cortex that receives input from mechanosensors distributed throughout the body.

somatostatin Peptide hormone made in the hypothalamus that inhibits the release of other hormones from the pituitary and intestine.

somite (so′ might) One of the segments into which an embryo becomes divided longitudinally, leading to the eventual segmentation of the animal as illustrated by the spinal column, ribs, and associated muscles.

source In plants, any organ that exports the products of photosynthesis in excess of its own needs, such as a mature leaf or storage organ. (Contrast with sink.)

spatial summation In the production or inhibition of action potentials in a postsynaptic cell, the interaction of depolarizations and hyperpolarizations produced at differ-

ent sites on the postsynaptic cell. (Contrast with temporal summation.)

spawning See external fertilization.

speciation (spee′ see ay′ shun) The process of splitting one population into two populations that are reproductively isolated from one another.

species (spee′ sees) [L. kind] The base unit of taxonomic classification, consisting of an ancestor–descendant group of populations of evolutionarily closely related, similar organisms. The more narrowly defined "biological species" consists of individuals capable of interbreeding with each other but not with members of other species.

species composition The particular mix of species a community contains and the abundances of those species.

species evenness A measure of species diversity that reflects the distribution of the species' abundances in a community.

species richness The total number of species living in a region.

specific heat The amount of energy that must be absorbed by a gram of a substance to raise its temperature by one degree centigrade. By convention, water is assigned a specific heat of one.

sperm [Gk. *sperma*: seed] The male gamete.

spermatid One of the products of the second meiotic division of a primary spermatocyte; four haploid spermatids, which remain connected by cytoplasmic bridges, are produced for each primary spermatocyte that enters meiosis.

spermatogenesis (spur mat′ oh jen′ e sis) [Gk. *sperma*: seed + *genesis*: source] Gametogenesis leading to the production of sperm.

spermatogonia In animals, the diploid progeny of a germ cell in males.

spherical symmetry The simplest form of symmetry, in which body parts radiate out from a central point such that an infinite number of planes passing through that central point can divide the organism into similar halves.

sphincter (sfink′ ter) [Gk. *sphinkter*: something that binds tightly] A ring of muscle that can close an orifice, for example, at the anus.

spicule [L. arrowhead] A hard, calcareous skeletal element typical of sponges.

spinal cord Along with the brain, part of the central nervous system; transmits information between the body and the brain and mediates simple reflexes.

spinal reflex The conversion of afferent to efferent information in the spinal cord without participation of the brain.

spindle Array of microtubules emanating from both poles of a dividing cell during mitosis and playing a role in the movement of chromosomes at nuclear division. Named for its shape.

spleen   Organ that serves as a reservoir for venous blood and eliminates old, damaged red blood cells from the circulation.

spliceosome   RNA–protein complex that splices out introns from eukaryotic pre-mRNAs.

splicing   *See* RNA splicing.

spontaneous mutation   A genetic change caused by internal cellular mechanisms, such as an error in DNA replication. (Contrast with induced mutation.)

sporangiophore   A stalked reproductive structure produced by zygospore fungi that extends from a hypha and bears one or many sporangia.

sporangium   (spor an' gee um) (plural: sporangia) [Gk. *spora*: seed + *angeion*: vessel or reservoir] In plants and fungi, any specialized stucture within which one or more spores are formed.

spore   [Gk. *spora*: seed] (1) Any asexual reproductive cell capable of developing into an adult organism without gametic fusion. In plants, haploid spores develop into gametophytes, diploid spores into sporophytes. (2) In prokaryotes, a resistant cell capable of surviving unfavorable periods.

sporocyte   Specialized cells of the diploid sporophyte that will divide by meiosis to produce four haploid spores. Germination of these spores produces the haploid gametophyte.

sporophyte   (spor' o fyte) [Gk. *spora*: seed + *phyton*: plant] In plants and protists with alternation of generations, the diploid phase that produces the spores. (Contrast with gametophyte.)

stabilizing selection   Selection against the extreme phenotypes in a population, so that the intermediate types are favored. (Contrast with disruptive selection.)

stamen   (stay' men) [L. *stamen*: thread] A male (pollen-producing) unit of a flower, usually composed of an anther, which bears the pollen, and a filament, which is a stalk supporting the anther.

starch   [O.E. *stearc*: stiff] A polymer of glucose; used by plants to store energy.

Starling's forces   The two opposing forces responsible for water movement across capillary walls: blood pressure, which squeezes water and small solutes out of the capillaries, and osmotic pressure, which pulls water back into the capillaries.

start codon   The mRNA triplet (AUG) that acts as a signal for the beginning of translation at the ribosome. (Contrast with stop codon.)

stele   (steel) [Gk. *stylos*: pillar] The central cylinder of vascular tissue in a plant stem.

stem   In plants, the organ that holds leaves and/or flowers and transports and distributes materials among the other organs of the plant.

stem cell   In animals, an undifferentiated cell that is capable of continuous proliferation. A stem cell generates more stem cells and a large clone of differentiated progeny cells. (*See also* embryonic stem cell.)

stereocilia   Fingerlike extensions of hair cell membranes whose bending initiates sound perception. (*See* hair cell.)

steroid   Any of a family of lipids whose multiple rings share carbons. The steroid cholesterol is an important constituent of membranes; other steroids function as hormones.

sticky ends   On a piece of two-stranded DNA, short, complementary, one-stranded regions produced by the action of a restriction endonuclease. Sticky ends facilitate the joining of segments of DNA from different sources.

stigma   [L. *stigma*: mark, brand] The part of the pistil at the apex of the style that is receptive to pollen, and on which pollen germinates.

stimulus   [L. *stimulare*: to goad] Something causing a response; something in the environment detected by a receptor.

stock   In horticulture, the root or root-bearing stem to which a bud or piece of stem from another plant (the scion) is grafted.

stoma   (plural: stomata) [Gk. *stoma*: mouth, opening] Small opening in the plant epidermis that permits gas exchange; bounded by a pair of guard cells whose osmotic status regulates the size of the opening.

stomach   An organ that physically (and sometimes enzymatically) breaks down food, preparing it for digestion in the midgut.

stomatal crypt   In plants, a sunken cavity below the leaf surface in which a stoma is sheltered from the drying effects of air currents.

stop codon   Any of the three mRNA codons that signal the end of protein translation at the ribosome: UAG, UGA, UAA.

stratosphere   The upper part of Earth's atmosphere, above the troposphere; extends from approximately 18 kilometers upward to approximately 50 kilometers above Earth's surface.

stratum   (plural strata) [L. *stratos*: layer] A layer of sedimentary rock laid down at a particular time in the past.

stretch receptor   A modified muscle cell embedded in the connective tissue of a muscle that acts as a mechanoreceptor in response to stretching of that muscle.

striated muscle   *See* skeletal muscle.

strigolactones   Signaling molecules produced by plant roots that attract the hyphae of mycorrhizal fungi.

strobilus   (plural: strobili) One of several conelike structures in various groups of plants (including club mosses, horsetails, and conifers) associated with the production and dispersal of reproductive products.

stroma   The fluid contents of an organelle such as a chloroplast or mitochondrion.

structural gene   A gene that encodes the primary structure of a protein not involved in the regulation of gene expression.

structural motif   A three-dimensional structural element that is part of a larger molecule. For example, there are four common motifs in DNA-binding proteins: helix-turn-helix, zinc finger, leucine zipper, and helix-loop-helix.

style   [Gk. *stylos*: pillar or column] In the angiosperm flower, a column of tissue extending from the tip of the ovary, and bearing the stigma or receptive surface for pollen at its apex.

sub-   [L. under] A prefix used to designate a structure that lies beneath another or is less than another. For example, subcutaneous (beneath the skin); subspecies.

suberin   A waxlike lipid that is a barrier to water and solute movement across the Casparian strip of the endodermis.

submucosa   (sub mew koe' sah) The tissue layer just under the epithelial lining of the lumen of the digestive tract.

subsoil   The soil horizon lying below the topsoil and above the parent rock (bedrock); the zone of infiltration and accumulation of materials leached from the topsoil. Also called the B horizon.

substrate   (sub' strayte) (1) The molecule or molecules on which an enzyme exerts catalytic action. (2) The base material on which a sessile organism lives.

succession   The gradual, sequential series of changes in the species composition of a community following a disturbance.

succulence   In plants, possession of fleshy, water-storing leaves or stems; an adaptation to dry environments.

superficial cleavage   A variation of incomplete cleavage in which cycles of mitosis occur without cell division, producing a syncytium (a single cell with many nuclei).

suprachiasmatic nuclei (SCN)   In mammals, two clusters of neurons just above the optic chiasm that act as the master circadian clock.

surface area-to-volume ratio   For any cell, organism, or geometrical solid, the ratio of surface area to volume; this is an important factor in setting an upper limit on the size a cell or organism can attain.

surface tension   The attractive intermolecular forces at the surface of liquid; an especially important property of water.

surfactant   A substance that decreases the surface tension of a liquid. Lung surfactant, secreted by cells of the alveoli, is mostly phospholipid and decreases the amount of work necessary to inflate the lungs.

survivorship   The fraction of individuals that survive from birth to a given life stage or age.

suspensor   In the embryos of seed plants, the stalk of cells that pushes the embryo into the endosperm and is a source of nutrient transport to the embryo.

symbiosis (sim' bee oh' sis) [Gk. *sym*: together + *bios*: living] The living together of two or more species in a prolonged and intimate relationship.

symmetry Pertaining to an attribute of an animal body in which at least one plane can divide the body into similar, mirror-image halves. (*See* bilateral symmetry, radial symmetry.)

sympathetic nervous system The division of the autonomic nervous system that works in opposition to the parasympathetic nervous system. (Contrast with parasympathetic nervous system.)

sympatric speciation (sim pat' rik) [Gk. *sym*: same + *patria*: homeland] Speciation due to reproductive isolation without any physical separation of the subpopulation. (Contrast with allopatric speciation.)

symplast The continuous meshwork of the interiors of living cells in the plant body, resulting from the presence of plasmodesmata. (Contrast with apoplast.)

symporter A membrane transport protein that carries two substances in the same direction. (Contrast with antiporter, uniporter.)

synapomorphy A trait that arose in the ancestor of a phylogenetic group and is present (sometimes in modified form) in all of its members, thus helping to delimit and identify that group. Also called a shared derived trait.

synapse (sin' aps) [Gk. *syn*: together + *haptein*: to fasten] A specialized type of junction where a neuron meets its target cell (which can be another neuron or some other type of cell) and information in the form of neurotransmitter molecules is exchanged across a synaptic cleft.

synapsis (sin ap' sis) The highly specific parallel alignment (pairing) of homologous chromosomes during the first division of meiosis.

synaptic cleft The space between the presynaptic cell and the postsynaptic cell in a chemical synapse.

synergids [Gk. *syn*: together + *ergos*: work] In angiosperms, the two cells accompanying the egg cell at one end of the megagametophyte.

syngamy *See* fertilization.

synonymous (silent) substitution A change of one nucleotide in a sequence to another when that change does not affect the amino acid specified (i.e., UUA → UUG, both specifying leucine). (Contrast with nonsynonymous substitution, missense mutation, nonsense mutation.)

systematics The scientific study of the diversity and relationships among organisms.

systemic acquired resistance A general resistance to many plant pathogens following infection by a single agent.

systemic circuit Portion of the circulatory system by which oxygenated blood from the lungs or gills is distributed throughout the rest of the body and returned to the heart. (Contrast with pulmonary circuit.)

systems biology The scientific study of an organism as an integrated and interacting system of genes, proteins, and biochemical reactions.

systole (sis' tuh lee) [Gk. *systole*: contraction] Contraction of a chamber of the heart, driving blood forward in the circulatory system. (Contrast with diastole.)

– T –

T cell A type of lymphocyte involved in the cellular immune response. The final stages of its development occur in the thymus gland. (Contrast with B cell; *see also* cytotoxic T cell, T-helper cell.)

T cell receptor A protein on the surface of a T cell that recognizes the antigenic determinant for which the cell is specific.

T tubules A system of tubules that runs throughout the cytoplasm of a muscle fiber, through which action potentials spread.

T-helper (T_H) cell Type of T cell that stimulates events in both the cellular and humoral immune responses by binding to the antigen on an antigen-presenting cell; target of the HIV-I virus, the agent of AIDS. (Contrast with cytotoxic T cells.)

taproot system A root system typical of eudicots consisting of a primary root (*taproot*) that extends downward by tip growth and outward by initiating lateral roots. (Contrast with fibrous root system.)

target cell A cell with the appropriate receptors to bind and respond to a particular hormone or other chemical mediator.

taste bud A structure in the epithelium of the tongue that includes a cluster of chemoreceptors innervated by sensory neurons.

TATA box An eight-base-pair sequence, found about 25 base pairs before the starting point for transcription in many eukaryotic promoters, that binds a transcription factor and thus helps initiate transcription.

taxon (plural: taxa) [Gk. *taxis*: arrange, put in order] A biological group (typically a species or a clade) that is given a name.

teeth Parts of the jaw in mammals that are variously adapted for grinding, stabbing, or cutting food.

telencephalon The outer, surrounding structure of the embryonic vertebrate forebrain, which develops into the cerebrum.

telomerase An enzyme that catalyzes the addition of telomeric sequences lost from chromosomes during DNA replication.

telomeres (tee' lo merz) [Gk. *telos*: end + *meros*: units, segments] Repeated DNA sequences at the ends of eukaryotic chromosomes.

telophase (tee' lo phase) [Gk. *telos*: end] The final phase of mitosis or meiosis during which chromosomes become diffuse, nuclear envelopes re-form, and nucleoli begin to reappear in the daughter nuclei.

template A molecule or surface on which another molecule is synthesized in complementary fashion, as in the replication of DNA.

template strand In double-stranded DNA, the strand that is transcribed to create an RNA transcript that will be processed into a protein. Also refers to a strand of RNA that is used to create a complementary RNA.

temporal lobe One of the four lobes of the cerebral hemisphere; receives and processes auditory and visual information; involved in recognizing, identifying, and naming objects.

temporal summation In the production or inhibition of action potentials in a postsynaptic cell, the interaction of depolarizations or hyperpolarizations produced by rapidly repeated stimulation of a single point on the postsynaptic cell. (Contrast with spatial summation.)

tendon A collagen-containing band of tissue that connects a muscle with a bone.

tepal A sterile, modified, nonphotosynthetic leaf of an angiosperm flower that cannot be distinguished as a petal or a sepal.

termination In molecular biology, the end of transcription or translation.

terminator A sequence at the 3' end of mRNA that causes the RNA strand to be released from the transcription complex.

terrestrial (ter res' tree al) [L. *terra*: earth] Pertaining to or living on land. (Contrast with aquatic, marine.)

territorial behavior Aggressive actions engaged in to defend a habitat or resource such that other animals are denied access.

tertiary consumers Carnivores that consume primary carnivores (secondary consumers).

tertiary endosymbiosis The mechanism by which some eukaryotes acquired the capacity for photosynthesis; for example, a dinoflagellate that apparently lost its chloroplast became photosynthetic by engulfing another protist that had acquired a chloroplast through secondary endosymbiosis.

tertiary structure In reference to a protein, the relative locations in three-dimensional space of all the atoms in the molecule. The overall shape of a protein. (Contrast with primary, secondary, and quaternary structures.)

test cross Mating of a dominant-phenotype individual (who may be either heterozygous or homozygous) with a homozygous-recessive individual.

testis (tes' tis) (plural: testes) [L. *testis*: witness] The male gonad; the organ that produces the male gametes.

tetanus [Gk. *tetanos*: stretched] (1) A state of sustained maximal muscular contraction caused by rapidly repeated stimulation. (2) In medicine, an often fatal disease ("lockjaw") caused by the bacterium *Clostridium tetani*.

tetrad [Gk. *tettares*: four] During prophase I of meiosis, the association of a pair of homologous chromosomes or four chromatids.

thalamus [Gk. *thalamos*: chamber] A region of the vertebrate forebrain; involved in integration of sensory input.

theory [Gk. *theoria*: analysis of facts] A far-reaching explanation of observed facts that is supported by such a wide body of evidence, with no significant contradictory evidence, that it is scientifically accepted as a factual framework. Examples are Newton's theory of gravity and Darwin's theory of evolution. (Contrast with hypothesis.)

thermoneutral zone (TNZ) [Gk. *thermos*: temperature] The range of temperatures over which an endotherm does not have to expend extra energy to thermoregulate.

thermophile (ther' muh fyle)[Gk. *thermos*: temperature + *philos*: loving] An organism that lives exclusively in hot environments.

thoracic cavity [Gk. *thorax*: breastplate] The portion of the mammalian body cavity bounded by the ribs, shoulders, and diaphragm. Contains the heart and the lungs.

thoracic duct The connection between the lymphatic system and the circulatory system.

threshold The level of depolarization that causes an electrically excitable membrane to fire an action potential.

thrombin An enzyme involved in blood clotting; cleaves fibrinogen to form fibrin.

thrombus (throm' bus) [Gk. *thrombos*: clot] A blood clot that forms within a blood vessel and remains attached to the wall of the vessel.

thylakoid (thigh la koid) [Gk. *thylakos*: sack or pouch] A flattened sac within a chloroplast. Thylakoid membranes contain all of the chlorophyll in a plant, in addition to the electron carriers of photophosphorylation. Thylakoids stack to form grana.

thymine (T) Nitrogen-containing base found in DNA.

thymus [Gk. *thymos*: warty] A ductless, glandular lymphoid tissue, involved in development of the immune system of vertebrates. In humans, the thymus degenerates during puberty.

thyroid gland [Gk. *thyreos*: door-shaped] A two-lobed gland in vertebrates. Produces the hormone thyroxine.

thyrotropin Hormone produced by the anterior pituitary that stimulates the thyroid gland to produce and release thyroxine. Also called thyroid-stimulating hormone (TSH).

thyrotropin-releasing hormone (TRH) Hormone produced by the hypothalamus that stimulates the anterior pituitary to release thyrotropin.

thyroxine Hormone produced by the thyroid gland; controls many metabolic processes.

tidal The bidirectional form of ventilation used by all vertebrates except birds; air enters and leaves the lungs by the same route.

tidal volume (TV) The amount of air that is exchanged during each breath when a person is at rest.

tight junction A junction between epithelial cells in which there is no gap between adjacent cells.

tissue A group of similar cells organized into a functional unit; usually integrated with other tissues to form part of an organ.

tissue system In plants, any of three organized groups of tissues—dermal tissue, vascular tissue, and ground tissue—that are established during embryogenesis and have distinct functions.

titin A protein that holds bundles of myosin filaments in a centered position within the sarcomeres of muscle cells. The largest protein in the human body.

TNZ See thermoneutral zone.

tonoplast The membrane of the plant central vacuole.

topography The variations in the elevation of Earth's surface that form, for example, mountains and valleys.

topsoil The uppermost soil horizon; contains most of the organic matter of soil, but may be depleted of most mineral nutrients by leaching. Also called the A horizon.

totipotent [L. *toto*: whole, entire + *potens*: powerful] Possessing all the genetic information and other capacities necessary to form an entire individual. (Contrast with multipotent, pluripotent.)

toxigenicity The ability of some pathogenic bacteria to produce chemical substances that harm the host.

trachea (tray' kee ah) [Gk. *trakhoia*: tube] A tube that carries air to the bronchi of the lungs of vertebrates. When plural (*tracheae*), refers to the major airways of insects.

tracheary element Either of two types of xylem cells—tracheids and vessel elements—that undergo apoptosis before assuming their transport function.

tracheid (tray' kee id) A type of tracheary element found in the xylem of nearly all vascular plants, characterized by tapering ends and walls that are pitted but not perforated. (Contrast with vessel element.)

trade-off The relationship between the fitness benefits conferred by an adaptation and the fitness costs it imposes. For an adaptation to be favored by natural selection, the benefits must exceed the costs.

trait In genetics, a specific form of a character: eye color is a character; brown eyes and blue eyes are traits. (Contrast with character.)

transcription factors Proteins that assemble on a eukaryotic chromosome, allowing RNA polymerase II to perform transcription.

transcription initiation site The part of a gene's promoter where synthesis of the gene's RNA transcript begins.

transcription The synthesis of RNA using one strand of DNA as a template.

transduction (1) Transfer of genes from one bacterium to another by a bacteriophage. (2) In sensory cells, the transformation of a stimulus (e.g., light energy, sound pressure waves, chemical or electrical stimulants) into action potentials.

transfection Insertion of recombinant DNA into animal cells.

transfer RNA (tRNA) A family of folded RNA molecules. Each tRNA carries a specific amino acid and anticodon that will pair with the complementary codon in mRNA during translation.

transformation (1) A mechanism for transfer of genetic information in bacteria in which pure DNA from a bacterium of one genotype is taken in through the cell surface of a bacterium of a different genotype and incorporated into the chromosome of the recipient cell. (2) Insertion of recombinant DNA into a host cell.

transgenic Containing recombinant DNA incorporated into the genetic material.

transition state In an enzyme-catalyzed reaction, the reactive condition of the substrate after there has been sufficient input of energy (activation energy) to initiate the reaction.

translation The synthesis of a protein (polypeptide). Takes place on ribosomes, using the information encoded in messenger RNA.

translational repressor A protein that blocks translation by binding to mRNAs and preventing their attachment to the ribosome. In mammals, the production of ferritin protein is regulated by a translational repressor.

translocation (1) In genetics, a rare mutational event that moves a portion of a chromosome to a new location, generally on a nonhomologous chromosome. (2) In vascular plants, movement of solutes in the phloem.

transmembrane protein An integral membrane protein that spans the phospholipid bilayer.

transpiration [L. *spirare*: to breathe] The evaporation of water from plant leaves and stem, driven by heat from the sun, and providing the motive force to raise water (plus mineral nutrients) from the roots.

transpiration–cohesion–tension mechanism Theoretical basis for water movement in plants: evaporation of water from cells within leaves (transpiration) causes an increase in surface tension, pulling water up through the xylem. Cohesion of water occurs because of hydrogen bonding.

transposable element   A segment of DNA that can move to, or give rise to copies at, another locus on the same or a different chromosome.

transposon   Mobile DNA segment that can insert into a chromosome and cause genetic change.

triglyceride   A simple lipid in which three fatty acids are combined with one molecule of glycerol.

triploblastic   Having three cell layers.

trisomic   Containing three rather than two members of a chromosome pair.

tRNA   *See* transfer RNA.

trochophore   (troke' o fore) [Gk. *trochos*: wheel + *phoreus*: bearer] A radially symmetrical larval form typical of annelids and mollusks, distinguished by a wheel-like band of cilia around the middle.

trophic cascade   The progression over successively lower trophic levels of the indirect effects of a predator.

trophic interactions   The consumer–resource relationships among species in a community.

trophic level   [Gk *trophes*: nourishment] A group of organisms united by obtaining their energy from the same part of the food web of a biological community.

trophoblast   [Gk *trophes*: nourishment + *blastos*: sprout] At the 32-cell stage of mammalian development, the outer group of cells that will become part of the placenta and thus nourish the growing embryo. (Contrast with inner cell mass.)

tropic hormones   Hormones produced by the anterior pituitary that control the secretion of hormones by other endocrine glands.

tropomyosin   [troe poe my' oh sin] One of the three protein components of an actin filament; controls the interactions of actin and myosin necessary for muscle contraction.

troponin   One of the three components of an actin filament; binds to actin, tropomyosin, and Ca^{2+}.

true-breeding   A genetic cross in which the same result occurs every time with respect to the trait(s) under consideration, due to homozygous parents.

trypsin   A protein-digesting enzyme. Secreted by the pancreas in its inactive form (trypsinogen), it becomes active in the duodenum of the small intestine.

tube feet   A unique feature of echinoderms; extensions of the water vascular system, which functions in gas exchange, locomotion, and feeding.

tubulin   A protein that polymerizes to form microtubules.

tumor   [L. *tumor*: a swollen mass] A disorganized mass of cells. Malignant tumors spread to other parts of the body.

tumor necrosis factor   A family of cytokines (growth factors) that causes cell death and is involved in inflammation.

tumor suppressor   A gene that codes for a protein product that inhibits cell proliferation; inactive in cancer cells. (Contrast with oncogene.)

turgor pressure   [L. *turgidus*: swollen] *See* pressure potential.

twitch   A muscle fiber's minimum unit of contraction, stimulated by a single action potential.

tympanic membrane   [Gk. *tympanum*: drum] The eardrum.

– U –

ubiquitin   A small protein that is covalently linked to other cellular proteins identified for breakdown by the proteosome.

ultimate cause   In ethology, the evolutionary processes that produced an animal's capacity and tendency to behave in particular ways. (Contrast with proximate cause.)

uniporter   [L. *unus*: one + *portal*: doorway] A membrane transport protein that carries a single substance in one direction. (Contrast with antiporter, symporter.)

unsaturated fatty acid   A fatty acid whose hydrocarbon chain contains one or more double bonds. (Contrast with saturated fatty acid.)

upregulation   A process by which the abundance of receptors for a hormone increases when hormone secretion is suppressed. (Contrast with downregulation.)

upwelling zones   Areas of the ocean where cool, nutrient-rich water from deeper layers rises to the surface.

uracil (U)   A pyrimidine base found in nucleotides of RNA.

urea   A compound that is the main form of nitrogen excreted by many animals, including mammals.

ureotelic   Pertaining to an organism in which the final product of the breakdown of nitrogen-containing compounds (primarily proteins) is urea. (Contrast with ammonotelic, uricotelic.)

ureter   (your' uh tur) Long duct leading from the vertebrate kidney to the urinary bladder or the cloaca.

urethra   (you ree' thra) In most mammals, the canal through which urine is discharged from the bladder and which serves as the genital duct in males.

uric acid   A compound that serves as the main excreted form of nitrogen in some animals, particularly those which must conserve water, such as birds, insects, and reptiles.

uricotelic   Pertaining to an organism in which the final product of the breakdown of nitrogen-containing compounds (primarily proteins) is uric acid. (Contrast with ammonotelic, ureotelic.)

urinary bladder   A structure in which urine is stored until it can be excreted to the outside of the body.

urine   (you' rin) In vertebrates, the fluid waste product containing the toxic nitrogenous by-products of protein and nucleic acid metabolism.

uterine cycle   In human females, the monthly cycle of events by which the endometrium is prepared for the arrival of a blastocyst. (Contrast with ovarian cycle).

uterus   (yoo' ter us) [L. *utero*: womb] A specialized portion of the female reproductive tract in mammals that receives the fertilized egg and nurtures the embryo in its early development. Also called the womb.

– V –

vaccination   Injection of virus or bacteria or their proteins into the body, to induce immunity. The injected material is usually attenuated (weakened) before injection and is called a *vaccine*.

vacuole   (vac' yew ole) Membrane-enclosed organelle in plant cells that can function for storage, water concentration for turgor, or hydrolysis of stored macromolecules.

vagina   (vuh jine' uh) [L. sheath] In female animals, the entry to the reproductive tract.

van der Waals forces   Weak attractions between atoms resulting from the interaction of the electrons of one atom with the nucleus of another. This type of attraction is about one-fourth as strong as a hydrogen bond.

variable   In a controlled experiment, a factor that is manipulated to test its effect on a phenomenon.

variable region   The portion of an immunoglobulin molecule or T cell receptor that includes the antigen-binding site and is responsible for its specificity. (Contrast with constant region.)

vas deferens   (plural: vasa deferentia) Duct that transfers sperm from the epididymis to the urethra.

vasa recta   Blood vessels that parallel the loops of Henle and the collecting ducts in the renal medulla of the kidney.

vascular   (vas' kew lar) [L. *vasculum*: a small vessel] Pertaining to organs and tissues that conduct fluid, such as blood vessels in animals and xylem and phloem in plants.

vascular bundle   In vascular plants, a strand of vascular tissue, including xylem and phloem as well as thick-walled fibers.

vascular cambium   (kam' bee um) [L. *cambiare*: to exchange] In plants, a lateral meristem that gives rise to secondary xylem and phloem.

vascular tissue system   The transport system of a vascular plant, consisting primarily of xylem and phloem.

vasopressin   *See* antidiuretic hormone.

vector (1) An agent, such as an insect, that carries a pathogen affecting another species. (2) A plasmid or virus that carries an inserted piece of DNA into a bacterium for cloning purposes in recombinant DNA technology.

vegetal hemisphere The lower portion of some animal eggs, zygotes, and embryos, in which the dense nutrient yolk settles. The *vegetal pole* is to the very bottom of the egg or embryo. (Contrast with animal hemisphere.)

vegetative cells In filamentous colonies of cyanobacteria the cells that perform photosynthesis. (Contrast with spores and heterocysts.)

vegetative meristem An apical meristem that produces leaves.

vegetative Nonreproductive, nonflowering, or asexual.

vegetative reproduction Asexual reproduction through the modification of stems, leaves, or roots.

vein [L. *vena*: channel] A blood vessel that returns blood to the heart. (Contrast with artery.)

vena cavae In the circulatory systems of crocodilians, birds, and mammals, large veins that empty into the right atrium of the heart.

ventral [L. *venter*: belly, womb] Toward or pertaining to the belly or lower side. (Contrast with dorsal.)

ventricle A muscular heart chamber that pumps blood through the lungs or through the body.

venule A small blood vessel draining a capillary bed that joins others of its kind to form a vein. (Contrast with arteriole.)

vernalization [L. *vernalis*: spring] Events occurring during a required chilling period, leading eventually to flowering.

vertebral column [L. *vertere*: to turn] The jointed, dorsal column that is the primary support structure of vertebrates.

very low-density lipoproteins (VLDLs) Lipoproteins that consist mainly of triglyceride fats, which they transport to fat cells in adipose tissues throughout the body; associated with excessive fat deposition and high risk for cardiovascular disease.

vesicle Within the cytoplasm, a membrane-enclosed compartment that is associated with other organelles; the Golgi complex is one example.

vessel element A type of tracheary element with perforated end walls; found only in angiosperms. (Contrast with tracheid.)

vestibular system (ves tib' yew lar) [L. *vestibulum*: an enclosed passage] Structures within the inner ear that sense changes in position or momentum of the head, affecting balance and motor skills.

villus (vil' lus) (plural: villi) [L. *villus*: shaggy hair or beard] A hairlike projection from a membrane; for example, from many gut walls.

virion (veer' e on) The virus particle, the minimum unit capable of infecting a cell.

virulence [L. *virus*: poison, slimy liquid] The ability of a pathogen to cause disease and death.

virus Any of a group of ultramicroscopic particles constructed of nucleic acid and protein (and, sometimes, lipid) that require living cells in order to reproduce. Viruses evolved multiple times from different cellular species.

vital capacity (VC) The maximum capacity for air exchange in one breath; the sum of the tidal volume and the inspiratory and expiratory reserve volumes.

vitamin [L. *vita*: life] An organic compound that an organism cannot synthesize, but nevertheless requires in small quantities for normal growth and metabolism.

vitelline envelope The inner, proteinaceous protective layer of a sea urchin egg.

viviparity (vye vi par' uh tee) Reproduction in which fertilization of the egg and development of the embryo occur inside the mother's body. (Contrast with oviparity.)

vivipary Premature germination in plants.

voltage A measure of the difference in electrical charge between two points.

voltage-gated channel A type of gated channel that opens or closes when a certain voltage exists across the membrane in which it is inserted.

vomeronasal organ (VNO) Chemosensory structure embedded in the nasal epithelium of amphibians, reptiles, and many mammals. Often specialized for detecting pheromones.

– W –

water potential (psi, Ψ) In osmosis, the tendency for a system (a cell or solution) to take up water from pure water through a differentially permeable membrane. Water flows toward the system with a more negative water potential. (Contrast with solute potential, pressure potential.)

water vascular system In echinoderms, a network of water-filled canals that functions in gas exchange, locomotion, and feeding.

wavelength The distance between successive peaks of a wave train, such as electromagnetic radiation.

weather The state of atmospheric conditions in a particular place at a particular time. (Contrast with climate.)

weathering The mechanical and chemical processes by which rocks are broken down into soil particles.

Wernicke's area A region in the temporal lobe of the human brain that is involved with the sensory aspects of language.

white blood cells Cells in the blood plasma that play defensive roles in the immune system. Also called leukocytes.

white matter In the central nervous system, tissue that is rich in axons. (Contrast with gray matter.)

wild type Geneticists' term for standard or reference type. Deviants from this standard, even if the deviants are found in the wild, are usually referred to as mutant. (Note that this terminology is not usually applied to human genes.)

wood Secondary xylem tissue.

– X –

xerophyte (zee' row fyte) [Gk. *xerox*: dry + *phyton*: plant] A plant adapted to an environment with limited water supply.

xylem (zy' lum) [Gk. *xylon*: wood] In vascular plants, the tissue that conducts water and minerals; xylem consists, in various plants, of tracheids, vessel elements, fibers, and other highly specialized cells.

– Y –

yolk [M.E. *yolke*: yellow] The stored food material in animal eggs, rich in protein and lipids.

yolk sac In reptiles, birds, and mammals, the extraembryonic membrane that forms from the endoderm of the hypoblast; it encloses and digests the yolk.

– Z –

zeaxanthin A blue-light receptor involved in the opening of plant stomata.

zona pellucida A jellylike substance that surrounds the mammalian ovum when it is released from the ovary.

zone of cell division The apical and primary meristems of a plant root; the source of all cells of the root's primary tissues.

zone of cell elongation The part of a plant root, generally above the zone of cell division, where cells are expanding (growing), primarily in the longitudinal direction.

zone of cell maturation The part of a plant root, generally above the zone of cell elongation, where cells are differentiating.

zoospore (zoe' o spore) [Gk. *zoon*: animal + *spora*: seed] In algae and fungi, any swimming spore. May be diploid or haploid.

zygospore Multinucleate, diploid cell that is a resting stage in the life cycle of zygospore fungi.

zygote (zye' gote) [Gk. *zygotos*: yoked] The cell created by the union of two gametes, in which the gamete nuclei are also fused. The earliest stage of the diploid generation.

zymogen The inactive precursor of a digestive enzyme; secreted into the lumen of the gut, where a protease cleaves it to form the active enzyme.

Illustration Credits

Table of Contents Page xix: © SPL/Photo Researchers, Inc. Page xxi: © Andrew Syred/ Photo Researchers, Inc. Page xxii: © Picture Partners/Alamy. Page xxiii: Courtesy of Scott Bauer/USDA. Page xxiv: © Colin Carson/ AGE Fotostock. Page xxv: © Tom Stack/ Waterframe/AGE Fotostock. Page xxvi: © Eye of Science/Photo Researchers, Inc. Page xxvii: © Paul van Gaalen/ANP Photo/AGE Fotostock. Page xxix: © Mark Bolton/Garden Picture Library/Photolibrary.com. Page xxxi: © OSF/Photolibrary.com. Page xxxii: © Rolf Nussbaumer Photography/Alamy. Page xxxiii: © SPL/Photo Researchers, Inc. Page xxxiv: © Juan Carlos Muñoz/AGE Fotostock. Page xxxv: © Nigel Pavitt/John Warburton-Lee Photography/Photolibrary.com. Page xxxvii: Reto Stöckli/NASA.

Part 1 Opener (page 15): © Biophoto Associates/Photo Researchers, Inc.
Part 2 Opener (page 123): © Conly Rieder and Alexey Khodjakov/Visuals Unlimited, Inc.
Part 3 Opener (page 287): © José Antonio Moreno/AGE Fotostock.
Part 4 Opener (page 365): © Ted Mead/ Photolibrary.com.
Part 5 Opener (page 505): David McIntyre.
Part 6 Opener (page 587): © Mike Anich/ AGE Fotostock.
Part 7 Opener (page 821): © David Nunuk/ All Canada Photos/Photolibrary.com.

Chapter 1 *Opener*: © Lee Grismer. 1.2: © Eye of Science/Photo Researchers, Inc. 1.3A: © Francois Gohier/Photo Researchers, Inc. 1.3B: © Doug Perrine/Peter Arnold Images/ Photolibrary.com. 1.6 *Biosphere*: NASA images by Reto Stöckli, based on data from NASA and NOAA. 1.7A: © Arco Images GmbH/Alamy. 1.7B: © Heather Angel/Natural Visions/ Alamy. 1.7C: © Juniors Bildarchiv/Alamy. 1.7D: © Stephen Dalton/Photoshot Holdings Ltd/Alamy. 1.9: From T. Hayes et al., 2003. *Environ. Health Perspect.* 111: 568.

Chapter 2 *Opener*: Courtesy of NASA/ JPL-Caltech/MSSS. 2.10C *left*: © Biophoto Associates/Photo Researchers, Inc. 2.10C *center*: © Dennis Kunkel Microscopy, Inc. 2.10C *right*: © Barry King/Biological Photo Service.

Chapter 3 *Opener*: © blickwinkel/Alamy. 3.9: Data from PDB 3LPO. L. Sim et al., 2010. *J. Biol. Chem.* 285: 17763. 3.14: Data from PDB 1IG8 (P. R. Kuser et al., 2000. *J. Biol. Chem.* 275: 20814) and 1BDG (A. M. Mulichak et al., 1998 *Nat. Struct. Biol.* 5: 555). 3.22: Data from PDB 1PTH. P. J. Loll et al., 1995. *Nat. Struct. Biol.* 2: 637.

Chapter 4 *Opener*: Courtesy of Tom Deerinck and Mark Ellisman of the National Center for Microscopy and Imaging Research at the University of California at San Diego. 4.1: After N. Campbell, 1990. *Biology*, 2nd Ed., Benjamin Cummings. 4.1 *Protein*: Data from PDB 1IVM. T. Obita, T. Ueda, & T. Imoto, 2003. *Cell. Mol. Life Sci.* 60: 176. 4.1 *T4*: © Dept. of Microbiology, Biozentrum/ SPL/Photo Researchers, Inc. 4.1 *Organelle*: © E.H. Newcomb & W.P. Wergin/Biological Photo Service. 4.1 *Bacterium*: © Jim Biddle/ Centers for Disease Control. 4.1 *Plant cells*: © Michael Eichelberger/Visuals Unlimited, Inc. 4.1 *Butterfly egg*: David McIntyre. 4.1 *Bird*: © Johan Pienaar/Shutterstock. 4.1 *Human*: © Jose Manuel Gelpi Diaz/istock. 4.1 *Redwood*: © urosr/Shutterstock. 4.3A *Light microscope*: © Olaf Doering/Alamy. 4.3A *Light micrograph*: © Dr. David J. Patterson/ Photo Researchers, Inc. 4.3A *Electron microscope*: © Enzo/AGE Fotostock. 4.3A *Electron micrograph*: © Aaron J. Bell/Photo Researchers, Inc. 4.5: © J. J. Cardamone Jr. & B. K. Pugashetti/Biological Photo Service. 4.6A: © Dennis Kunkel Microscopy, Inc. 4.6B: Courtesy of David DeRosier, Brandeis U. 4.7 *Mitochondrion*: © K. Porter, D. Fawcett/ Visuals Unlimited, Inc. 4.7 *Cytoskeleton*: © Don Fawcett, John Heuser/Photo Researchers, Inc. 4.7 *Nucleolus*: © Richard Rodewald/ Biological Photo Service. 4.7 *Peroxisome*: © E. H. Newcomb & S. E. Frederick/Biological Photo Service. 4.7 *Cell wall*: © Biophoto Associates/Photo Researchers, Inc. 4.7 *Ribosome*: From M. Boublik et al., 1990. *The Ribosome*, p. 177. Courtesy of American Society for Microbiology. 4.7 *Centrioles*: © Conly L. Rieder/Biological Photo Service. 4.7 *Plasma membrane*: Courtesy of J. David Robertson, Duke U. Medical Center. 4.7 *Rough ER*: © Don Fawcett/Science Source/Photo Researchers, Inc. 4.7 *Smooth ER*: © Don Fawcett, D. Friend/ Science Source/Photo Researchers, Inc. 4.7 *Chloroplast*: © E.H. Newcomb & W.P. Wergin/ Biological Photo Service. 4.7 *Golgi apparatus*: Courtesy of L. Andrew Staehelin, U. Colorado. 4.9: © Sanders/Biological Photo Service. 4.10: Courtesy of Vic Small, Austrian Academy of Sciences, Salzburg, Austria. 4.11A *upper*: © Dennis Kunkel Microscopy, Inc. 4.11A *lower*, 4.11B: © W. L. Dentler/Biological Photo Service. 4.13: From N. Pollock et al., 1999. *J. Cell Biol.* 147: 493. Courtesy of R. D. Vale. 4.14: © Roland Birke/Peter Arnold Images/Photolibrary.com. 4.15: © Biophoto Associates/Photo Researchers, Inc. 4.16 *left*: Courtesy of David Sadava. 4.16 *upper right*: From J. A. Buckwalter & L. Rosenberg, 1983. *Coll. Rel. Res.* 3: 489. Courtesy of L. Rosenberg. 4.16 *lower right*: © J. Gross, Biozentrum/

SPL/Photo Researchers, Inc. 4.18A: © Barry F. King/Biological Photo Service. 4.18B: Courtesy of Darcy E. Kelly, U. Washington. 4.18C: Courtesy of C. Peracchia. 4.19: Courtesy of Janet Iwasa, Szostak group, MGH/ Harvard. Page 61: From Y.-L. Shih, T. Le, and L. Rothfield. 2003. *PNAS USA* 100: 7865. Courtesy of L. Rothfield.

Chapter 5 *Opener*: David McIntyre and Chris Small. 5.3A *left*: © Stanley Flegler/Visuals Unlimited, Inc. 5.3A *center, right*: © David M. Phillips/Photo Researchers, Inc. 5.3B: © Ed Reschke/Peter Arnold Images/Photolibrary. com. 5.5: From G. M. Preston et al., 1992. *Science* 256: 385. 5.9: From M. M. Perry, 1979. *J. Cell Sci.* 39: 26. 5.19: Data from PDB 3EML. V. P. Jaakola et al., 2008. *Science* 322: 1211.

Chapter 6 *Opener*: © Raymond Forbes/AGE Fotostock. 6.17: © Nigel Cattlin/Alamy. 6.23: David McIntyre.

Chapter 7 *Opener*: Dr. Gopal Murti/Photo Researchers, Inc. 7.1A: © Stanley C. Holt/ Biological Photo Service. 7.1B: © Ed Reschke/ Peter Arnold Images/Photolibrary.com. 7.1C: © Fabio Liverani/Naturepl.com. 7.2: © Robert E. Ford/Biological Photo Service. 7.3 *left*: © Dr. Jeremy Burgess/SPL/Photo Researchers, Inc. 7.3 *center*: David McIntyre. 7.3 *right*: Courtesy of Andrew D. Sinauer. 7.5 *Chromosome*: © Biophoto Associates/Photo Researchers, Inc. 7.5 *Nucleus*: © Richard Rodewald/Biological Photo Service. 7.6: © Conly L. Rieder/ Biological Photo Service. 7.7A: © Robert Brons/Biological Photo Service. 7.7B: © B. A. Palevitz, E. H. Newcomb/Biological Photo Service. 7.12: © C. A. Hasenkampf/Biological Photo Service. 7.14A: © Gopal Murti/ Phototake, Inc./Alamy.

Chapter 8 *Opener*: Courtesy of Chesdovi/ Wikipedia. 8.9 *Dark*: © Marina Golskaya/ istock. 8.9 *Chinchilla*: © purelook/istock. 8.9 *Point*: © Carolyn A. McKeone/Photo Researchers, Inc. 8.9 *Albino*: © ZTS/ Shutterstock. 8.12 *Black*: © Debbi Smirnoff/ istock. 8.12 *Chocolate*: © Erik Lam/istock. 8.12 *Yellow*: © claire norman/istock. 8.19A: © David Scharf/Peter Arnold Images/Photolibrary. com. Page 154: Courtesy of the Plant and Soil Sciences eLibrary (plantandsoil.unl.edu); used with permission from the Institute of Agriculture and Natural Resources at the University of Nebraska.

Chapter 9 *Opener*: © De Agostini Editore/ Photolibrary.com. 9.1A: © Tomasz Szul/ Phototake/Alamy. 9.2: © Lee D. Simon/ Photo Researchers, Inc. 9.3 *X-ray crystallograph*:

Courtesy of Prof. M. H. F. Wilkins, Dept. of Biophysics, King's College, U. London. 9.4B: © Jewish Chronical/Photolibrary.com. 9.5A: © A. Barrington Brown/Photo Researchers, Inc. 9.5B: Data from S. Arnott & D. W. Hukins, 1972. *Biochem. Biophys. Res. Commun.* 47(6): 1504. 9.10A: Data from PDB 1SKW. Y. Li et al., 2001. *Nat. Struct. Mol. Biol.* 11: 784. 9.13C: © Peter Lansdorp/Visuals Unlimited, Inc. 9.20: Courtesy of Christoph P. E. Zollikofer, Marcia S. Ponce de León, and Elisabeth Daynès. Page 166 *Chromosomes*: © Ed Reschke/Peter Arnold Images/Photolibrary.com. Page 180: © Krissi Lundgren/istock. Page 181: © Stanley Flegler/Visuals Unlimited, Inc.

Chapter 10 *Opener*: © Dennis Kunkel Microscopy, Inc. 10.4: Data from PDB 1MSW. Y. W. Yin & T. A. Steitz, 2002. *Science* 298: 1387. 10.8: From D. C. Tiemeier et al., 1978. *Cell* 14: 237. 10.13: Data from PDB 1EHZ. H. Shi & P. B. Moore, 2000. *RNA* 6: 1091. 10.14: Data from PDB 1GIX and 1G1Y. M. M. Yusupov et al., 2001. *Science* 292: 883. 10.18B: Courtesy of J. E. Edström and *EMBO J.* 10.22: Data from PDB 1HNW. D. E. Brodersen et al. 2000. *Cell* 103: 1143.

Chapter 11 *Opener*: © Lenartowski/AGE Fotostock. 11.11: Data from PDB 1A66. P. Zhou et al., 1998. *Cell* 92: 687. 11.14: Courtesy of the Centers for Disease Control. 11.20: Data from PDB 1T2K. D. Panne et al., 2004. *EMBO J.* 23: 4384.

Chapter 12 *Opener*: © Phil Banko/Flirt Collection/Photolibrary.com. 12.4: From P. H. O'Farrell, 1975. High resolution two-dimensional electrophoresis of proteins. *J. Biol. Chem.* 250: 4007. Courtesy of Patrick H. O'Farrell. 12.15B: Bettmann/Corbis. 12.16A *left*: © Bruce Stotesbury/PostMedia News/Zuma Press. 12.16A *right*: © kostudio/Shutterstock. 12.16B: © Yann Arthus-Bertrand/Corbis.

Chapter 13 *Opener*: © Ted Spiegel/Corbis. 13.2: © P. Potter & S. McInnes/CSIRO/Photolibrary.com. 13.6: © Martin Shields/Alamy. 13.12: David McIntyre. 13.15: Courtesy of the Golden Rice Humanitarian Board, www.goldenrice.org. 13.16: Courtesy of Eduardo Blumwald. 13.17: McKinnon Films Ltd./OSF/Photolibrary.com. Page 259: © Dr. George Chapman/Visuals Unlimited, Inc.

Chapter 14 *Opener*: © Benoit Photo. 14.4: © Roddy Field, the Roslin Institute. 14.8B: © Cold Spring Harbor Press; courtesy of Dustin Updike and Susan Strome. 14.9: From J. E. Sulston & H. R. Horvitz, 1977. *Dev. Bio.* 56: 100. 14.11C: David McIntyre. 14.13 *upper*: Courtesy of C. Rushlow and M. Levine. 14.13 *center*: Courtesy of T. Karr. 14.13 *lower*: Courtesy of S. Carroll and S. Paddock. 14.14A: © Eye of Science/Photo Researchers, Inc. 14.14B: © Science VU/Dr. F. Rudolph Turner/Visuals Unlimited, Inc. 14.16: David McIntyre. 14.17: © Bone Clones, www.boneclones.com. 14.18: Courtesy of J. Hurle and E. Laufer. 14.19 *Insect*: © flavijus piliponis/istock. 14.19 *Centipede*: © jeridu/istock. 14.21: Courtesy of Mike Shapiro and David Kingsley. 14.22 *Stem cells*: © 2007, photo used with permission of Anant Kamath, Cellular Engineering Technologies. 14.22 *Bone*: © Ed Reschke/Peter Arnold Images/Photolibrary.com. 14.22 *Cartilage*: ©

franco valoti/Marka/AGE Fotostock. 14.22 *Fat*: Courtesy of Thomas Eisner. 14.22 *Muscle*: © Carolina Biological Supply Company/Phototake Inc./Alamy. 14.22 *Heart, Blood*: © Ed Reschke/Peter Arnold Images/Photolibrary.com. 14.22 *Nerve*: © Volker Steger/Peter Arnold Images/Photolibrary.com. Page 276: Courtesy of D. Daily and W. Sullivan.

Chapter 15 *Opener*: Courtesy of the National Museum of Health and Medicine, Armed Forces Institute of Pathology, Washington, D.C. 15.5A: © Luis César Tejo/Shutterstock. 15.5B: © Duncan Usher/Alamy. 15.5C: © PetStockBoys/Alamy. 15.5D: © Arco Images GmbH/Alamy. 15.8: © Stan Osolinski/OSF/Photolibrary.com. 15.14: Courtesy of David Hillis. 15.18A *Langur*: © Cyril Ruoso/Photolibrary.com. 15.18A *Longhorn*: Courtesy of David Hillis. 15.18B: © Marshall Bruce/istock. 15.24: © Pasieka/Photo Researchers, Inc. Page 289 *Darwin*: © The Art Gallery Collection/Alamy. Page 289 *Beagle*: Painting by Ronald Dean, reproduced by permission of the artist and Richard Johnson, Esquire.

Chapter 16 *Opener*: Courtesy of Misha Matz. 16.4 *Tunicate larva*: Courtesy of William Jeffery. 16.4 *Tunicate adult*: © WaterFrame/Alamy. 16.4 *Frog larva*: © Michael & Patricia Fogden/Minden Pictures. 16.4 *Frog adult*: © Mark Kostich/istock. 16.6 *L. bicolor*: Courtesy of Steve Matson. 16.6 *L. liniflorus*: Courtesy of Anthony Valois/National Park Service. 16.8: © Alexandra Basolo. 16.12A: © Mark Herreid/Shutterstock. 16.12B: Courtesy of Dr. Thomas Barnes/U.S. Fish and Wildlife Service. 16.12C: Courtesy of Steve Hillebrand/U.S. Fish and Wildlife Service. 16.14: Courtesy of Misha Matz.

Chapter 17 *Opener*: David McIntyre, courtesy of Exotic Fish and Pet World, Southampton, MA. 17.1A *left*: © Roger K. Burnard/Biological Photo Service. 17.1A *right*: © Richard Codington/Alamy. 17.1B: © Stubblefield Photography/Shutterstock. 17.2A: David McIntyre. 17.2B: © Gerry Bishop/Visuals Unlimited, Inc. 17.4: © Barry Mansell/Naturepl.com. 17.8: © OSF/Photolibrary.com. 17.10 *G. olivacea*: © Phil A. Dotson/Photo Researchers, Inc. 17.10 *G. carolinensis*: © Suzanne L. Collins/Photo Researchers, Inc. 17.11A: © Gustav Verderber/Visuals Unlimited, Inc. 17.11B: © Nathan Derieg. 17.11C: © J. S. Sira/Photolibrary.com. 17.11D: © Daniel L. Geiger/SNAP/Alamy. 17.12: Courtesy of Donald A. Levin. 17.13 *upper*: © Gerhard Schulz/AGE Fotostock. 17.13 *lower*: © Christophe Courteau/Naturepl.com. 17.14: Courtesy of William R. Rice.

Chapter 18 *Opener*: © Graham Cripps/NHMPL. *Opener inset*: © Natasha Litova/istock. 18.4: © UNEP/Photolibrary.com. 18.5: © Martin Bond/SPL/Photo Researchers, Inc. 18.6A: © Ted Kinsman/Photo Researchers, Inc. 18.6B: © Georgette Douwma/Photo Researchers, Inc. 18.9: David McIntyre. 18.11 *left*: © Sinclair Stammers/Photo Researchers, Inc. 18.11 *center*: © Sinclair Stammers/Photo Researchers, Inc. 18.11 *right*: Courtesy of Martin Smith. 18.12 *Cambrian*: © John Sibbick/NHMPL. 18.12 *Marella*: Courtesy of the Amherst College Museum of Natural History, The Trustees of Amherst College. 18.12

Ottoia: © Alan Sirulnikoff/Photo Researchers, Inc. 18.12 *Anomalocaris*: © Kevin Schafer/Alamy. 18.12 *Devonian*: © The Field Museum, #GEO86500_125d. 18.12 *Archaeopteris*: © John Cancalosi/Peter Arnold Images/Photolibrary.com. 18.12 *Eusthenopteron*: © Wolfgang Kaehler/Alamy. 18.12 *Permian*: © Karen Carr Studio Inc. 18.12 *Dragonfly*: Image by Roy J. Beckemeyer. 18.12 *Walchia*: © The Natural History Museum, London. 18.12 *Triassic*: © OSF/Photolibrary.com. 18.12 *Ferns*: © Ken Lucas/Visuals Unlimited, Inc. 18.12 *Coelophysis*: © Ken Lucas/Visuals Unlimited, Inc. 18.12 *Cretaceous*: © Anness Publishing/NHMPL. 18.12 *Chasmosaurus*: © Oleksiy Maksymenko/Alamy. 18.12 *Sapindopsis*: © Barbara J. Miller/Biological Photo Service. 18.12 *Tertiary*: © Publiphoto/Photo Researchers, Inc. 18.12 *Hyracotherium*: Courtesy of the Amherst College Museum of Natural History, The Trustees of Amherst College. 18.12 *Plesiadapis*: © The Natural History Museum, London. 18.13: Courtesy of Conrad C. Labandeira, Department of Paleobiology, National Museum of Natural History, Smithsonian Institution.

Chapter 19 *Opener*: "Milky seas" data prepared by Steven Miller of the Naval Research Laboratory, Monterey, CA. Background data (*Earth at Night* image) courtesy Marc Imhoff (NASA/GSFC) and Christopher Elvidge (NOAA/NGDC); image by Craig Mayhew (NASA/GSFC) and Robert Simmon (NASA/GSFC). 19.2A: © Manfred Kage/Peter Arnold Images/Photolibrary.com. 19.2B: Courtesy of the Centers for Disease Control. 19.3: © Dennis Kunkel Microscopy, Inc. 19.5: © Dr. George Chapman/Visuals Unlimited, Inc. 19.6: © David Scharf/Peter Arnold Images/Photolibrary.com. 19.7: © Don W. Fawcett/Photo Researchers, Inc. 19.8: © David Phillips/Visuals Unlimited, Inc. 19.9A: © Paul W. Johnson/Biological Photo Service. 19.9B: © Dr. Terry Beveridge/Visuals Unlimited, Inc. 19.9C: © D. Harms/Wildlife/Photolibrary.com. 19.10A: © J. A. Breznak & H. S. Pankratz/Biological Photo Service. 19.10B: © James Cavallini/Photo Researchers, Inc. 19.11: Courtesy of Randall C. Cutlip. 19.12: © Kwangshin Kim/Photo Researchers, Inc. 19.13: © Dean A. Glawe/Biological Photo Service. 19.14: From K. Kashefi & D. R. Lovley, 2003. *Science* 301: 934. Courtesy of Kazem Kashefi. 19.16: © eye35.com/Alamy. 19.17: © Nancy Nehring/istock. 19.18: From H. Huber et al., 2002. *Nature* 417: 63. © Macmillan Publishers Ltd. Courtesy of Karl O. Stetter. 19.19B: © Science Photo Library RF/Photolibrary.com. 19.21: © Juergen Berger/Photo Researchers, Inc. 19.22A: © Pasieka/Photo Researchers, Inc. 19.22B, C: © Russell Kightley/Photo Researchers, Inc. 19.22D: © Science Photo Library RF/Photolibrary.com. 19.22E: © animate4.com ltd./Photo Researchers, Inc. 19.22F: © Russell Kightley/Photo Researchers, Inc. 19.23: © Nigel Cattlin/Alamy.

Chapter 20 *Opener*: © Bill Bachman/Photo Researchers, Inc. 20.4: © Paul W. Johnson/Biological Photo Service. 20.5A: © SPL/Photo Researchers, Inc. 20.5B: © Aaron Bell/Visuals Unlimited, Inc. 20.5C: © Steve Gschmeissner/Photo Researchers, Inc. 20.7A: Courtesy of Dr. Stan Erlandsen/CDC. 20.7B: © Dennis Kunkel Microscopy, Inc. 20.9: © Scenics &

Science/Alamy. 20.10A: © Marevision/AGE Fotostock. 20.10B: © Jan Hodder & Michael Graybill/Biological Photo Service. 20.11: © James W. Richardson/Visuals Unlimited, Inc. 20.12: © Andrew Syred/Photo Researchers, Inc. 20.13A: © Robert Brons/Biological Photo Service. 20.13B: © Manfred Kage/Peter Arnold Images/Photolibrary.com. 20.14: © Michael Abbey/Photo Researchers, Inc. 20.15: © Wim van Egmond/Visuals Unlimited, Inc. 20.16A: © Matt Meadows/Peter Arnold Inc./Photolibrary.com. 20.16B: © Ed Reschke/Peter Arnold Inc./Photolibrary.com. 20.17: Courtesy of R. Blanton and M. Grimson. 20.19: Courtesy of M.A. Coffroth and Cindy Lewis, University at Buffalo. 20.20: © Per-Andre Hoffmann/Picture Press/Photolibrary.com.

Chapter 21 *Opener:* © Mitsuhiko Imamori/Minden. 21.2: © Philippe Clement/Naturepl.com. 21.3A: © Marevision/AGE Fotostock. 21.3B: © Larry Mellichamp/Visuals Unlimited, Inc. 21.3C: © Andre Seale/Alamy. 21.5A: © Ed Reschke/Peter Arnold Inc./Photolibrary.com. 21.5B: © Ed Reschke/Peter Arnold Inc./Photolibrary.com. 21.5C: © Daniel Vega/AGE Fotostock. 21.6 *upper:* © Biodisc/Visuals Unlimited/Alamy. 21.6 *lower:* © Ed Reschke/Peter Arnold Inc./Photolibrary.com. 21.7: © Publiphoto/Photo Researchers, Inc. 21.8A: © Ed Reschke/Peter Arnold Inc./Photolibrary.com. 21.8B: © Stanislav Sokolov/istock. 21.8C: © Ted Mead/Ticket/Photolibrary.com. 21.8D: Courtesy of the Talcott Greenhouse, Mount Holyoke College. 21.9 *inset:* © John N. A. Lott/Biological Photo Service. 21.10A: Courtesy of the Biology Department Greenhouses, U. Massachusetts, Amherst. 21.10B: David McIntyre. 21.13: © John Robinson/Still Pictures/Photolibrary.com. 21.14A: © Susumu Nishinaga/Photo Researchers, Inc. 21.15A: © Walter H. Hodge/Peter Arnold Inc./Photolibrary.com. 21.15B: David McIntyre. 21.15C: © Dieter Herrmann/AGE Fotostock. 21.15D: © Mason Vranish/Alamy. 21.16A *left:* © Fritz Pölking/Peter Arnold Images/Photolibrary.com. 21.16A *right:* © Stan W. Elems/Visuals Unlimited, Inc. 21.16B *left:* © Gunter Marx/Alamy. 21.16B *right:* © Dr. John D. Cunningham/Visuals Unlimited, Inc. 21.18A, B: David McIntyre. 21.18C: © Nigel Cattlin/Alamy. 21.19A: © Phiseksit/Shutterstock. 21.19B: David McIntyre. 21.23: © Ted Kinsman/Photo Researchers, Inc. 21.24A: Courtesy of Keith Weller/USDA ARS. 21.24B: © blickwinkel/Alamy. 21.24C: © Brian A Jackson/Shutterstock. 21.24D: © Rob Walls/Alamy. 21.24E: © Arco Images GmbH/Alamy. 21.24F: Courtesy of Keith Weller/USDA ARS. 21.27A: Photo by David McIntyre, courtesy of the U. Massachusetts Biology Department Greenhouses. 21.27B: © Cerealphoto/AGE Fotostock. 21.27C: © Holmes Garden Photos/Alamy. 21.27D: © dora modly-paris/Shutterstock. 21.27E: © Florapix/Alamy. 21.28A: © Floris Slooff/istock. 21.28B: © George Clerk/istock. 21.28C: © Jose B. Ruiz/Naturepl.com. 21.28D: © rotofrank/istock. 21.29A: © Garden Picture Library/Photolibrary.com. 21.29B: David McIntyre. 21.29C: © A & J Visage/Alamy. 21.29D: Courtesy of David Hillis.

Chapter 22 *Opener:* © Biophoto Associates/Photo Researchers, Inc. 22.2: © Manfred Kage/OSF/Photolibrary.com. 22.3B: © Dr. Jeremy Burgess/Photo Researchers, Inc. 22.4: © Arco Images GmbH/Alamy. 22.5A: © Biophoto Associates/Photo Researchers, Inc. 22.6: © N. Allin & G. L. Barron/Biological Photo Service. 22.7A: © Rafael Campillo/AGE Fotostock. 22.7B: David McIntyre. 22.9A: © R. L. Peterson/Biological Photo Service. 22.9B: © M. F. Brown/Biological Photo Service. 22.12: © Eye of Science/Photo Researchers, Inc. 22.13A: © J. Robert Waaland/Biological Photo Service. 22.13B: © Dr. Jeremy Burgess; Photo Researchers, Inc. 22.14: Photo by David McIntyre; manure courtesy of Myrtle Jackson. 22.15A: © Dr. Cecil H. Fox/Photo Researchers, Inc. 22.15B: © Biophoto Associates/Photo Researchers, Inc. 22.16A: © blickwinkel/Alamy. 22.16B: © Matt Meadows/Peter Arnold Inc./Photolibrary.com. 22.17: © Dennis Kunkel Microscopy, Inc. 22.18A: David McIntyre. 22.18B: © Roger Eritja/AGE Fotostock. 22.19: Courtesy of David Hillis. 22.20: © Biophoto Associates/Photo Researchers, Inc.

Chapter 23 *Opener:* © Philippe Psaila/Photo Researchers, Inc. 23.3A: © Ed Robinson/Photolibrary.com. 23.3B: © Steve Gschmeissner/Photo Researchers, Inc. 23.3C: © DEA/Christian Ricci/Photolibrary.com. 23.4A: © Marevision/AGE Fotostock. 23.4B: © John Bell/AGE Fotostock. 23.5A: © Borut Furlan/WaterFrame/Alamy. 23.5B: Courtesy of NOAA/Monterey Bay Aquarium Research Institute. 23.5C: © Robert Brons/Biological Photo Service. 23.6A: Courtesy of Wim van Egmond. 23.7B: © Larry Jon Friesen. 23.8: Adapted from F. M. Bayerand & H. B. Owre, 1968. *The Free-Living Lower Invertebrates*, Macmillan Publishing Co. 23.9A: © Charles Wyttenbach/Biological Photo Service. 23.9B: © Larry Jon Friesen. 23.9C: © Mark Conlin/OSF/Photolibrary.com. 23.10A: © Robert Brons/Biological Photo Service. 23.10B: © WaterFrame/Alamy. 23.11A: © David Fleetham/Alamy. 23.12B: © Roland Birke/Peter Arnold Images/Photolibrary.com. 23.13B: © Paul Kay/OSF/Photolibrary.com. 23.14A: © Lawrence Naylor/Photo Researchers, Inc. 23.15: © Andrew J. Martinez/Photo Researchers, Inc. 23.16: © jerrysa/Shutterstock. 23.17A: © WaterFrame/Alamy. 23.17B: © Buena Vista Pictures/Zuma Press. 23.17C: © Larry Jon Friesen. 23.18B: © Marevision/AGE Fotostock. 23.18C: © Francesco Tomasinelli/Photo Researchers, Inc. 23.18D: © Marevision/AGE Fotostock. 23.18E: © moodboard/Photolibrary.com. 23.19A: © Larry Jon Friesen. 23.19B: © Jeff Rotman/Naturepl.com. 23.20A: From D. C. García-Bellido & D. H. Collins, 2004. *Nature* 429: 40. Courtesy of Diego García-Bellido Capdevila. 23.20B: © Nature's Images/Photo Researchers, Inc. 23.21A: Courtesy of Dr. Mae Melvin/CDC. 23.21B: © Walter Dawn/Photo Researchers, Inc. 23.22: © Pascal Goetgheluck/Photo Researchers, Inc. 23.23A: Courtesy of Jen Grenier and Sean Carroll, U.Wisconsin. 23.23B: © David Scharf/Peter Arnold Images/Photolibrary.com. 23.23C: Courtesy of Reinhardt Møbjerg Kristensen. 23.24A: © Morley Read/Naturepl.com. 23.24B: © Manfred Kage/Peter Arnold Images/Photolibrary.com. 23.25A: © WaterFrame/Alamy. 23.25B: © Jan Hodder & Michael Graybill/Biological Photo Service. 23.26A: © Cathy Keifer/Shutterstock. 23.26B: © EcoPrint/Shutterstock. 23.26C: © Clive Bromhall/OSF/Photolibrary.com. 23.26D: Photo by Eric Erbe, digital colorization by Chris Pooley, courtesy of the USDA ARS. 23.27A: © Emanuele Biggi/OSF/Photolibrary.com. 23.27B: © Michael P. Gadomski/Photo Researchers, Inc. 23.28A: © Michael Lustbader/Photo Researchers, Inc. 23.28B: David McIntyre. 23.28C: © blickwinkel/Alamy. 23.28D: © Kathie Atkinson/OSF/Photolibrary.com. 23.28E: © Larry Jon Friesen. 23.30A, B: © Peter J. Bryant/Biological Photo Service. 23.30C: © Gregory G. Dimijian, M.D./Photo Researchers, Inc. 23.30D: David McIntyre. 23.30E: © orionmystery@flickr/Shutterstock. 23.30F: © Miles Boyer/Shutterstock. 23.30G: David McIntyre. 23.30H: © blickwinkel/Alamy. 23.32A: © WaterFrame/Alamy. 23.32B: © WaterFrame/Alamy. 23.32C: © Peter Scoones/Photo Researchers, Inc. 23.32D: © Marevision/AGE Fotostock. 23.32E: © Robert L. Dunne/Photo Researchers, Inc. 23.33A: © C. R. Wyttenbach/Biological Photo Service. 23.34A: © Stan Elems/Visuals Unlimited, Inc. 23.34B: © Larry Jon Friesen. 23.35: © imagebroker/Alamy. 23.37A: © Ken Lucas/Biological Photo Service. 23.37B *left:* © Marevision/AGE Fotostock. 23.37B *right:* © anne de Haas/istock. 23.40A: © Larry Jon Friesen. 23.40B: © blickwinkel/Alamy. 23.40C: © WaterFrame/Alamy. 23.41A: © Images & Stories/Alamy. 23.41B: © David Fleetham/Alamy. 23.42A: © Hoberman Collection UK/Photolibrary.com. 23.42B: © Tom McHugh/Photo Researchers, Inc. 23.44A: © Michael Fogden/OSF/Photolibrary.com. 23.44B: © Michael Fogden/OSF/Photolibrary.com. 23.44C: © Edmund D. Brodie, Jr./Biological Photo Service. 23.44D: Courtesy of David Hillis. 23.46A: © Adam Jones/Photo Researchers, Inc. 23.46B: © John Cancalosi/AGE Fotostock. 23.46C: © Larry Jon Friesen. 23.46D: © Chris Mattison/Alamy. 23.47A: Courtesy of Sarah McCans. 23.47B: © Naturepix/Alamy. 23.48A: From X. Xu et al., 2003. *Nature* 421: 335. © Macmillan Publishers Ltd. 23.48B: © James L. Amos/Photo Researchers, Inc. 23.49A: © Larry Jon Friesen. 23.49B: © mlorenz/Shutterstock. 23.49C: © PetrP/Shutterstock. 23.50A: © Dave Watts/Alamy. 23.50B: © R. Wittek/Arco Images/AGE Fotostock. 23.50C: © Rolf Nussbaumer Photography/Alamy. 23.50D: © ANT Photo Library/Photo Researchers, Inc. 23.50E: Courtesy of Andrew D. Sinauer. 23.50F: © Yann Hubert/Bios/Photolibrary.com. Page 459: Courtesy of J. B. Morrill.

Chapter 24 *Opener:* © paolo negri/Alamy. 24.6 *upper:* © Biodisc/Visuals Unlimited/Alamy. 24.6 *lower:* © M. I. Walker/Photo Researchers, Inc. 24.7B: © John N. A. Lott/Biological Photo Service. 24.8A: © James Solliday/Biological Photo Service. 24.8B: © John N. A. Lott/Biological Photo Service. 24.9A: © modesigns58/istock. 24.9B: © Adrian Sherratt/Alamy. 24.9C: © Danita Delimont/Alamy. 24.10A *left:* David McIntyre. 24.10A *right:* © Andrew Syred/Photo Researchers, Inc. 24.10B *left:* © Ed Reschke/Peter Arnold Images/Photolibrary.com. 24.10B *right:* © Steve Gschmeissner/Photo Researchers, Inc. 24.11A, B: David McIntyre. 24.11C: © Carl May/Biological Photo Service. 24.14: © David M. Dennis/OSF/Photolibrary.com. 24.16: Courtesy of Scott Bauer/USDA. Page 510 *Parenchyma:* © Dr. Ken Wagner/Visuals

Unlimited, Inc. Page 510 *Collenchyma*: © Phil Gates/Biological Photo Service. Page 510 *Sclerenchyma*: © Biophoto Associates/Photo Researchers, Inc. Page 510 *Sclereids*: © Dr. Jack Bostrack/Visuals Unlimited, Inc. Page 510 *Tracheids*: © Dr. John D. Cunningham/Visuals Unlimited, Inc. Page 510 *Vessel elements*: © J. Robert Waaland/Biological Photo Service. Page 511: © Herve Conge/ISM/Phototake.

Chapter 25 *Opener*: Courtesy of the NASA/Goddard Space Flight Center Scientific Visualization Studio. 25.1: David McIntyre. 25.3: © G. R. 'Dick' Roberts/NSIL/Visuals Unlimited, Inc. 25.7A: © blickwinkel/Alamy. 25.7B: © Gilles Nicolet/BIOS/Photolibrary. com. 25.9: © David Cook/blueshiftstudios/Alamy. 25.13A: © Susumu Nishinaga/Photo Researchers, Inc. 25.15: Courtesy of the USDA ARS. Page 535: © M. H. Zimmermann.

Chapter 26 *Opener*: © Micheline Pelletier/Sygma/Corbis. 26.2: From J. M. Alonso and J. R. Ecker, 2006. *Nat. Rev. Genet.* 7: 524. 26.3A: Courtesy of J. A. D. Zeevaart, Michigan State U. 26.3B: From W. M. Gray, 2004. *PLoS Biol* 2(9): e311. 26.11: David McIntyre. 26.13: Courtesy of Drs. Matsuoka and Ashikari. Page 544: © Sylvan Wittwer/Visuals Unlimited, Inc. Page 545: © Ed Reschke/Peter Arnold Images/Photolibrary.com. Page 549: Courtesy of Adel A. Kader. Page 550: David McIntyre. Page 551: Courtesy of Eugenia Russinova, VIB Department of Plant Systems Biology, Ghent University, Belgium.

Chapter 27 *Opener*: © Rich Iwasaki/Tips Italia/Photolibrary.com. 27.1A: © kukuruxa/Shutterstock. 27.1B *Male*: © Tish1/Shutterstock. 27.1B *Female*: © Pierre BRYE/Alamy. 27.1C: © Bill Beatty/Visuals Unlimited, Inc. 27.5A: David McIntyre. 27.5B: © Michael Moreno/istock. 27.5C: © Scenics & Science/Alamy. 27.11: Courtesy of Richard Amasino and Colleen Bizzell. 27.12A: © ooyoo/istock. 27.12B: © Nigel Cattlin/Alamy. 27.12C: © Ed Reschke/Peter Arnold Images/Photolibrary. com. 27.14: © AGStockUSA/Alamy. 27.15: © Jaroslaw Pyrih/Alamy. Page 561 *Thistle*: © John N. A. Lott/Biological Photo Service. Page 561 *Burrs*: © Scott Camazine/Alamy.

Chapter 28 *Opener*: Courtesy of Robert Bowden/Kansas State University. 28.3: © Nigel Cattlin/Alamy. 28.4A: © Arco Images/J. Meul/AGE Fotostock. 28.4B: © Lee & Marleigh Freyenhagen/Shutterstock. 28.5: After A. Steppuhn et al., 2004. *PLoS Biology* 2: 1074. 28.7: © Jon Mark Stewart/Biological Photo Service. 28.8: © John N. A. Lott/Biological Photo Service. 28.9: © TH Foto/Alamy. 28.10: © blickwinkel/Alamy. 28.11: © John N. A. Lott/Biological Photo Service. 28.13: Courtesy of Scott Bauer/USDA. 28.14: © Jurgen Freund/Naturepl.com. 28.15: Courtesy of Ryan Somma. 28.16: From M. Ayliffe, R. Singh, and E. Lagudah. 2008. *Curr. Opin. Plant Biol.* 11: 187. Page 580 *Beetle*: Courtesy of Thomas Eisner, Cornell U.

Chapter 29 *Opener*: © Konrad Wothe/imagebroker RF/Photolibrary.com. 29.6: © Greg Epperson/istock. 29.7: From M. H. Ross, W. Pawlina, and T. A. Barnash, 2009. *Atlas of Descriptive Histology*. Sinauer Associates:

Sunderland, MA. 29.9A: © Rick & Nora Bowers/Alamy. 29.9B: © Shutterstock. 29.11: Courtesy of Anton Stabentheiner. 29.12: © WoodyStock/Alamy.

Chapter 30 *Opener*: © David Reed/Alamy. 30.1A *Insulin*: Data from PDB 2HIU. Q. X. Hua et al., 1995. *Nat. Struct. Biol.* 2: 129. 30.1A *HGH*: Data from PDB 1HGU. L. Chantalat et al., 1995. *Protein Pept. Lett.* 2: 333. 30.10: © Mike Goldwater/Alamy. 30.14: Courtesy of Kimberly Saviano/AISSG-USA.

Chapter 31 *Opener*: © The Bridgeman Art Library/Getty Images. Page 623: © Science Photo Library/Photo Researchers, Inc.

Chapter 32 *Opener*: © Bildagentur RM/Tips Italia/Photolibrary.com. 32.1A: © Oxford Scientific Films/Photolibrary.com. 32.1B: © Georgie Holland/AGE Fotostock. 32.1C: © Patricia J. Wynne. 32.4: © david gregs/Alamy. 32.6B, 32.9: © Ed Reschke/Peter Arnold Images/Photolibrary.com.

Chapter 33 *Opener*: © Nathan Maxfield/istock. 33.1: Courtesy of Richard Elinson, U. Toronto. 33.3A *left*: From H. W. Beams and R. G. Kessel, 1976. *Am. Sci.* 64: 279. 33.3A *center, right*: © Dr. Lloyd M. Beidler/Photo Researchers, Inc. 33.3B: From H. W. Beams and R. G. Kessel, 1976. *Am. Sci.* 64: 279. 33.3C: Courtesy of D. Daily and W. Sullivan. 33.4A: From J. G. Mulnard, 1967. *Arch. Biol.* (Liege) 78: 107. Courtesy of J. G. Mulnard. 33.12D: Courtesy of K. W. Tosney and G. Schoenwolf. 33.14B: Courtesy of K. W. Tosney. 33.17: Courtesy of Dr. Lynn James, USDA Poisonous Plant Research Laboratory.

Chapter 34 *Opener*: © Randy Lincks/Flirt Collection/Photolibrary.com. 34.2B: © C. Raines/Visuals Unlimited, Inc. 34.18: © Wellcome Dept. of Cognitive Neurology/SPL/Photo Researchers, Inc. 34.20: Courtesy of Dr. Kevin LaBar, Duke U.

Chapter 35 *Opener*: © Rolf Nussbaumer Photography/Alamy. 35.4A: © Hans Pfletschinger/Peter Arnold Images/Photolibrary.com. 35.4B: © Oxford Scientific Films/Photolibrary.com. 35.8A: © Dr. Fred Hossler/Visuals Unlimited, Inc. 35.15A: © Cheryl Power/Photo Researchers, Inc. 35.17A: © Omikron/Photo Researchers, Inc.

Chapter 36 *Opener*: © AFP/Getty Images. 36.1 *Micrograph*: © Don W. Fawcett/Photo Researchers, Inc. 36.2: © James Dennis/Phototake. 36.4: © Kent Wood/Peter Arnold Images/Photolibrary.com. 36.10: Courtesy of Jesper L. Andersen. 36.19: © Oxford Scientific Films/Photolibrary.com. Page 714 *Skeletal muscle*: From M. H. Ross, W. Pawlina, and T. A. Barnash, 2009. *Atlas of Descriptive Histology*. Sinauer Associates: Sunderland, MA. Page 718 *Cardiac muscle*: © Manfred Kage/Peter Arnold Images/Photolibrary.com. Page 718 *Smooth muscle*: © Biophoto Associates/Photo Researchers, Inc.

Chapter 37 *Opener Geese*: © John Downer/OSF/Photolibrary.com. *Mountains*: © Stefan Auth/Imagebroker RF/Photolibrary.com. 37.2B: © Creatas/Photolibrary.com. 37.2C:

Courtesy of Thomas Eisner, Cornell U. 37.6 *Bronchi*: © Ralph Hutchings/Visuals Unlimited, Inc. 37.6 *Alveoli*: © Dr. David Phillips/Visuals Unlimited, Inc. 37.9: After C. R. Bainton, 1972. *J. Appl. Physiol.* 33: 775. Page 731 *Flat worm*: © Leslie Newman & Andrew Flowers/Photo Researchers, Inc. Page 731 *Sponge*: © Andrew J. Martinez/Photo Researchers, Inc.

Chapter 38 *Opener*: © Brand X Pictures/Photolibrary.com. 38.7: After N. Campbell, 1990. *Biology*, 2nd Ed., Benjamin Cummings. 38.8B: © Steve Gschmeissner/Science Photo Library/Alamy. 38.10: © Ed Reschke/Peter Arnold Images/Photolibrary.com.

Chapter 39 *Opener*: © Marilyn "Angel" Wynn/Nativestock.com. 39.1A: © John Cancalosi/AGE Fotostock. 39.1B: © blickwinkel/Alamy. 39.4C *Microvilli*: © Dennis Kunkel Microscopy, Inc. 39.14: © Science VU/Jackson/Visuals Unlimited, Inc.

Chapter 40 *Opener*: © Barry Mansell/Naturepl.com. 40.1 *inset*: © Kim Taylor/Naturepl.com. 40.6A: © Steve Gschmeissner/Science Photo Library/Alamy. 40.6B: © Steve Gschmeissner/Photo Researchers, Inc. 40.11: © A. Held/blickwinkel/AGE Fotostock.

Chapter 41 *Opener*: © Alexander Gabrysch/Picture Press/Photolibrary.com. 41.2A: © blickwinkel/Alamy. 41.10B: © Don Paulson/Superstock/Photolibrary.com. 41.13: © Tui De Roy/Minden Pictures. 41.15: © J. Jarvis/Visuals Unlimited, Inc. 41.16B: © Richard R. Hansen/Photo Researchers, Inc.

Chapter 42 *Opener*: © Bradley Sauter/Alamy. 42.1 *Biosphere*: Courtesy of NASA. 42.12A: © Bill Bachman/Alamy. 42.12B: © Carl W. May/Biological Photo Service. 42.12C: © danilo donadoni/AGE Fotostock. 42.12D: © Oxford Scientific Films/Photolibrary. com. 42.16: © Nico Stengert/Imagebroker/Photolibrary.com. 42.17: © blickwinkel/Alamy. 42.18: © Alberto Nardi/Tips Italia/Photolibrary.com.

Chapter 43 *Opener*: Courtesy of Richard S. Ostfeld Laboratory. 43.1 *inset*: © T. W. Davies/California Academy of Sciences. 43.4A: © irin-k/Shutterstock. 43.4B: © Mary McDonald/Naturepl.com. 43.6: Courtesy of Barry Sinervo. 43.7 *left*: Courtesy of Joseph Berger, Bugwood.org. 43.7 *right*: Courtesy of Clemson University/USDA Cooperative Extension Slide Series/Bugwood.org. 43.10: © Mikko Mattila/Alamy. 43.10 *inset*: © Nancy Nehring/istock. 43.11: David McIntyre. 43.12: © Joel Sartore/National Geographic Stock. Page 851 *Field*: Courtesy of Steve Dewey, Utah State University, Bugwood.org. Page 851 *inset*: Courtesy of Peggy Greb, USDA Agricultural Research Service, Bugwood.org.

Chapter 44 *Opener*: © Kim Taylor/Naturepl. com. 44.1B: © Inge Johnsson/Alamy. 44.1C: © Werner Bollmann/AGE Fotostock. 44.1D: Courtesy of Olle Pellmyr. 44.2: © cbpix/Shutterstock. 44.7 *left*: © Krystyna Szulecka/Alamy. 44.7 *right*: © Photoshot/Alamy. 44.8A *upper*: © Juniors Bildarchiv/Alamy. 44.8A *lower*: © ranplett/istock. 44.8B: © Roy Toft/

Index

Numbers in *italic* indicate that the information will be found in an illustration or table.

BIO PORTAL Resources—available at yourBioPortal.com

Animated Tutorials **AT**, Web Activities **WA**, Interactive Tutorials **IT**, and Working with Data Exercises **WD**

BIO PORTAL Resources— available at yourBioPortal.com

Animated Tutorials **AT**, Web Activities **WA**, Interactive Tutorials **IT**, and Working with Data Exercises **WD**